2009 FUNDAMENTALS

W9-AXH-774

2008 HVAC SYSTEMS AND EQUIPMENT

The CD-ROM in the sleeve on the opposite page contains the 2011 *ASHRAE Handbook—HVAC Applications* in both I-P and SI editions, with PDFs of chapters easily viewable using Adobe® Reader®.

Minimum System Requirements:

Windows®: CD-ROM Drive · Intel® Pentium® III Processor · Microsoft® Windows XP® Professional, Home; or Windows Vista® Home Basic, Home Premium, Ultimate, Business, or Enterprise with Service Pack 1 or 2 (32-bit of 64-bit editions); or Microsoft Windows 7 Starter, Home Premium, Professional, Ultimate, or Enterprise (32-bit and 64-bit) · 256 MB of RAM · PDF viewing software · 125 MB hard-disk space · Adobe Reader 8 or higher (Reader 9.3 is on the CD) · Internet Explorer® 7.0 or higher

Macintosh®: Mac OS® X v. 10.4.11 or higher · 128 MB of RAM (256 MB recommended) · 125 MB hard-disk space · Firefox® or Safari®

Windows Installation: This CD-ROM has a setup program for installing, and a CD license key attached to the sleeve. The setup program requires an Internet connection for product installation only (one-time). It will allow installation of the product on two computers using the provided license key. Two additional activations are permitted as needed. Upon successful authentication of the license key, the setup program will install the product content files on the computer. The setup program will also stamp all PDF files with user name, license date, and license key on the header and footer of each page. **Please read the installation file on the CD for more details.**

Macintosh Installation: Go to http://software.ashrae.org for instructions.

Customer Support: cdsupport@ashrae.org

Important: Please keep the CD-ROM product key attached to the CD sleeve for future reference. This number is required to install the Handbook.

© 2011 American Society of Heating, Refrigerating and Air-Conditioning Engineers, Inc.
1791 Tullie Circle NE, Atlanta, GA 30329-2305
U.S. & Canada toll free—800/527-4723 or Worldwide 404/636-8400
Customer comments/help: cdsupport@ashrae.org

2011 ASHRAE® HANDBOOK

Heating, Ventilating, and Air-Conditioning APPLICATIONS

Inch-Pound Edition

American Society of Heating, Refrigerating and Air-Conditioning Engineers, Inc.

1791 Tullie Circle, N.E., Atlanta, GA 30329

(404) 636-8400

http://www.ashrae.org

DEDICATED TO THE ADVANCEMENT OF

THE PROFESSION AND ITS ALLIED INDUSTRIES

Volunteer members of ASHRAE Technical Committees and others compiled the information in this handbook, and it is generally reviewed and updated every four years. Comments, criticisms, and suggestions regarding the subject matter are invited. Any errors or omissions in the data should be brought to the attention of the Editor. Additions and corrections to Handbook volumes in print will be published in the Handbook published the year following their verification and, as soon as verified, on the ASHRAE Internet Web site.

DISCLAIMER

ISBN 978-1-936504-06-0
ISSN 1078-6074

The paper for this book was manufactured in an acid- and
elemental-chlorine-free process with pulp obtained from sources
using sustainable forestry practices. The printing used soy-based inks.

CONTENTS

Contributors

ASHRAE Technical Committees, Task Groups, and Technical Resource Groups

ASHRAE Research: Improving the Quality of Life

Preface

CONTRIBUTORS

In addition to the Technical Committees, the following individuals contributed significantly
to this volume. The appropriate chapter numbers follow each contributor's name.

Eric Berg (1)
Lennox Industries, Inc.

Lorenzo Cremaschi (1)
Oklahoma State University

John E. Wolfert (2)

Itzhak Maor (3, 7)
Johnson Controls

Harvey Brickman (4)
Tishman Realty & Const.

Mark Fly (4)
AAON, Inc.

Tom Kroschel (4)

Peter Simmonds (4)
IBE Consulting

William Webb (4)

Lynn Werman (4)
Ferris Engineering

Ralph Kittler (5)
Seresco, Inc.

Reinhold Kittler (5)
Hudson Industrial Consulting, Inc.

Mark Scott (5)
Wiss. Janney, Elstner Associates, Inc.

Frank Mills (5, 6)
Sinclair Knight Merz

Peter Langowski (8)
BSA LifeStructures, Inc.

Kenneth R. Mead (8, 31, 32)
Centers for Disease Control and
Prevention/National Institute for Occupational
Safety and Health

Snehal R. Desai (9)
Federal Bureau of Prisons

Edward D. Fitts (9, 55)
Fitts HVAC Consulting

Gursaran D. Mathur (10, 52)
Calsonic Kansei North America

Hugh Ferdows (11)
Sutrak Corporation

Robert L. May (11)
LTK Engineering Services

Gary Prusak (11)
Bombardier Transportation

James J. Bushnell (11, 13)
HVAC Consulting Services

Raymond H. Horstman (12)
Boeing Commercial Airplane Group

Augusto San Cristobal (13)
Bronswerk Marine, Inc.

Adam Smith (13)
Bronswerk Marine, Inc.

Douglass S. Abramson (14)
Handy Hubby

Richard A. Evans (14)
Evans Associates

Ravisankar Ganta (14, 22, 28, 29)
Shaw Group

Vernon Peppers (14, 26)
Peppers Engineering

Jarrod Alston (15)
ARUP

Arthur Bendelius (15)
A&G Consultants, Inc.

Craig Quaglini (15)
ARUP

Mohammad Tabarra (15)
ARUP

Jeffrey Tubbs (15, 53)
ARUP

Louis Hartman (16)
Harley Ellis Deveraux

Al Woody (17)
Ventilation/Energy Applications, PLLC

Art Giesler (18)
PermAlert ESP

Larry J. Hughes (18)
Alpha Engineering, Inc.

Gary Shamshoian (18)
Genetech

Michael Shelton (18)
Bahnson Environmental Specialties
Environmental Chambers Mfg.

Wei Sun (18)
Engsysco, Inc.

Charles Chun-Lun Shieh (18, 26, 27, 43, 58)
Fluor Corporation

Craig A. Crader (19)
Bick Group

Edward L. Gutowski (19)
Facilities Engineering Associates, Inc.

Magnus K. Herrlin (19)
ANCIS Incorporated

Douglas K. McLellan (19)
Hewlett-Packard Co.

John Peterson (19)
Hewlett-Packard Co.

David Quirk (19)
Verizon Wireless

Jeff Trower (19)
Data Aire, Inc.

Joseph Marino (20)
Newsday

Norm Maxwell (20)
Environmental Air Quality

Michael C. Connor (21, 30)
Connor Engineering Solutions

James W. Carty (22)
Kodak Project Management Division

Cecily Grzywacz (23)
National Gallery of Art

Phil Maybee (23)
The Filter Man, Ltd.

Jean Tétreault (23)
Canadian Heritage

Thomas Axley (27)
Tennessee Valley Authority

Erich Binder (27)
Erich Binder Consulting, Ltd.

Deep Ghosh (27, 28)
Southern Company

Matt Hargan (28)
Hargan Engineering

John McKernan (32)
U.S. Environmental Protection Agency

Steve Brown (33)
LC Systems, Inc.

Frank Kohout (33)
McDonald's Corp.

Jay Parikh (33)
Compliance Solutions International

Derek Schrock (33)
Halton Company

Scott Hackel (34)
Energy Center of Wisconsin

Steve Kavanaugh (34)
University of Alabama

Kevin Rafferty (34)
Wapiti Engineering

Mark Hertel (35)
SunEarth, Inc.

Dieter Bartel (36)
Manitoba Hydro

Janice Peterson (36)
NW Energy Efficiency Alliance

Klas C. Haglid (37)
Haglid Engineering & Associates

Michael Brambley (39)
Pacific Northwest National Laboratory

Richard Dames (39)
Boone County Schools

Richard Danks (39)
NASA Glenn Research Center

Robyn Ellis (39)
St. Michael's Hospital

John M. House (39)
Johnson Controls

Michael Khaw (39)
Isotherm Engineering

Angela Lewis (39)
The University of Reading

Haorong Li (39)
University of Nebraska-Lincoln

William McCartney (39)
Isotherm Engineering

ASHRAE TECHNICAL COMMITTEES, TASK GROUPS, AND TECHNICAL RESOURCE GROUPS

SECTION 1.0—FUNDAMENTALS AND GENERAL

- 1.1 Thermodynamics and Psychrometrics
- 1.2 Instruments and Measurements
- 1.3 Heat Transfer and Fluid Flow
- 1.4 Control Theory and Application
- 1.5 Computer Applications
- 1.6 Terminology
- 1.7 Business, Management, and General Legal Education
- 1.8 Mechanical Systems Insulation
- 1.9 Electrical Systems
- 1.10 Cogeneration Systems
- 1.11 Electric Motors and Motor Control
- 1.12 Moisture Management in Buildings
- TG1 Optimization (OPT)

SECTION 2.0—ENVIRONMENTAL QUALITY

- 2.1 Physiology and Human Environment
- 2.2 Plant and Animal Environment
- 2.3 Gaseous Air Contaminants and Gas Contaminant Removal Equipment
- 2.4 Particulate Air Contaminants and Particulate Contaminant Removal Equipment
- 2.5 Global Climate Change
- 2.6 Sound and Vibration Control
- 2.7 Seismic and Wind Restraint Design
- 2.8 Building Environmental Impacts and Sustainability
- 2.9 Ultraviolet Air and Surface Treatment
- TG2 Heating, Ventilation, and Air-Conditioning Security (HVAC)

SECTION 3.0—MATERIALS AND PROCESSES

- 3.1 Refrigerants and Secondary Coolants
- 3.2 Refrigerant System Chemistry
- 3.3 Refrigerant Contaminant Control
- 3.4 Lubrication
- 3.6 Water Treatment
- 3.8 Refrigerant Containment
- TG3 HVAC&R Contractors and Design-Build Firms (CDBF)

SECTION 4.0—LOAD CALCULATIONS AND ENERGY REQUIREMENTS

- 4.1 Load Calculation Data and Procedures
- 4.2 Climatic Information
- 4.3 Ventilation Requirements and Infiltration
- 4.4 Building Materials and Building Envelope Performance
- 4.5 Fenestration
- 4.7 Energy Calculations
- 4.10 Indoor Environmental Modeling
- TRG4 Indoor Air Quality Procedure Development (IAQP)

SECTION 5.0—VENTILATION AND AIR DISTRIBUTION

- 5.1 Fans
- 5.2 Duct Design
- 5.3 Room Air Distribution
- 5.4 Industrial Process Air Cleaning (Air Pollution Control)
- 5.5 Air-to-Air Energy Recovery
- 5.6 Control of Fire and Smoke
- 5.7 Evaporative Cooling
- 5.8 Industrial Ventilation Systems
- 5.9 Enclosed Vehicular Facilities
- 5.10 Kitchen Ventilation
- 5.11 Humidifying Equipment

SECTION 6.0—HEATING EQUIPMENT, HEATING AND COOLING SYSTEMS AND APPLICATIONS

- 6.1 Hydronic and Steam Equipment and Systems
- 6.2 District Energy
- 6.3 Central Forced-Air Heating and Cooling Systems
- 6.5 Radiant Heating and Cooling
- 6.6 Service Water Heating Systems
- 6.7 Solar Energy Utilization
- 6.8 Geothermal Energy Utilization
- 6.9 Thermal Storage
- 6.10 Fuels and Combustion

SECTION 7.0—BUILDING PERFORMANCE

- 7.1 Integrated Building Design
- 7.2 HVAC&R Construction and Design Build Technologies
- 7.3 Operation and Maintenance Management
- 7.4 Exergy Analysis for Sustainable Buildings (EXER)
- 7.5 Smart Building Systems
- 7.6 Systems Energy Utilization
- 7.7 Testing and Balancing
- 7.8 Owning and Operating Costs
- 7.9 Building Commissioning
- TRG7 Underfloor Air Distribution (UFAD)

SECTION 8.0—AIR-CONDITIONING AND REFRIGERATION SYSTEM COMPONENTS

- 8.1 Positive-Displacement Compressors
- 8.2 Centrifugal Machines
- 8.3 Absorption and Heat-Operated Machines
- 8.4 Air-to-Refrigerant Heat Transfer Equipment
- 8.5 Liquid-to-Refrigerant Heat Exchangers
- 8.6 Cooling Towers and Evaporative Condensers
- 8.7 Variable Refrigerant Flow
- 8.8 Refrigerant System Controls and Accessories
- 8.9 Residential Refrigerators and Food Freezers
- 8.10 Mechanical Dehumidification Equipment and Heat Pipes
- 8.11 Unitary and Room Air Conditioners and Heat Pumps
- 8.12 Desiccant Dehumidification Equipment and Components

SECTION 9.0—BUILDING APPLICATIONS

- 9.1 Large-Building Air-Conditioning Systems
- 9.2 Industrial Air Conditioning
- 9.3 Transportation Air Conditioning
- 9.5 Residential and Small-Building Applications
- 9.6 Health-Care Facilities
- 9.7 Educational Facilities
- 9.8 Large-Building Air-Conditioning Applications
- 9.9 Mission-Critical Facilities, Technology Spaces, and Electronic Equipment
- 9.10 Laboratory Systems
- 9.11 Clean Spaces
- 9.12 Tall Buildings
- TG9 Justice Facilities (JF)

SECTION 10.0—REFRIGERATION SYSTEMS

- 10.1 Custom-Engineered Refrigeration Systems
- 10.2 Automatic Icemaking Plants and Skating Rinks
- 10.3 Refrigerant Piping
- 10.4 Ultralow-Temperature Systems and Cryogenics
- 10.5 Refrigerated Distribution and Storage Facilities
- 10.6 Transport Refrigeration
- 10.7 Commercial Food and Beverage Cooling, Display, and Storage
- 10.8 Refrigeration Load Calculations
- 10.9 Refrigeration Application for Foods and Beverages
- 10.10 Management of Lubricant in Circulation

ASHRAE Research: Improving the Quality of Life

The American Society of Heating, Refrigerating and Air-Conditioning Engineers is the world's foremost technical society in the fields of heating, ventilation, air conditioning, and refrigeration. Its members worldwide are individuals who share ideas, identify needs, support research, and write the industry's standards for testing and practice. The result is that engineers are better able to keep indoor environments safe and productive while protecting and preserving the outdoors for generations to come.

One of the ways that ASHRAE supports its members' and industry's need for information is through ASHRAE Research. Thousands of individuals and companies support ASHRAE Research annually, enabling ASHRAE to report new data about material properties and building physics and to promote the application of innovative technologies.

Chapters in the ASHRAE Handbook are updated through the experience of members of ASHRAE Technical Committees and through results of ASHRAE Research reported at ASHRAE conferences and published in ASHRAE special publications and in *ASHRAE Transactions*.

For information about ASHRAE Research or to become a member, contact ASHRAE, 1791 Tullie Circle, Atlanta, GA 30329; telephone: 404-636-8400; www.ashrae.org.

Preface

The 2011 *ASHRAE Handbook—HVAC Applications* comprises over 60 chapters covering a broad range of facilities and topics, and is written to help engineers design and use equipment and systems described in other Handbook volumes. ASHRAE Technical Committees have revised nearly every chapter to cover current requirements, technology, and design practice. An accompanying CD-ROM contains all the volume's chapters in both I-P and SI units.

This edition includes *two* new chapters:

- Chapter 4, Tall Buildings, focuses on HVAC issues unique to tall buildings, including stack effect, system selection, mechanical room location, water distribution, vertical transportation, and life safety.
- Chapter 60, Ultraviolet Air and Surface Treatment, covers ultraviolet germicidal irradiation (UVGI) systems and relevant guidelines, standards, and practices, as well as energy use and economic considerations.

Here are selected highlights of the other revisions and additions:

- Chapter 3, Commercial and Public Buildings, now covers office buildings, transportation centers, and warehouses and distribution centers, with new sections on commissioning, sustainability, energy efficiency, energy benchmarking, renewable energy, value engineering, and life-cycle cost analysis.
- Chapter 7, Educational Facilities, has added content on higher education facilities, commissioning, dedicated outdoor air systems (DOAS), combined heat and power (CHP), and sustainability and energy efficiency.
- Chapter 8, Health-Care Facilities, has been updated to reflect ASHRAE *Standard* 170-2008 and has revised discussion on design criteria for pharmacies.
- Chapter 18, Clean Spaces, has updated content on standards, filters, barrier technology, and sustainability plus a new section on installation and test procedures.
- Chapter 19, Data Processing and Telecommunication Facilities, has a new title and revised and/or new content on design temperatures, change rate, humidity, power usage effectiveness (PUE), aisle containment, economizer cycles, and computer room air-handling (CRAH) units.
- Chapter 33, Kitchen Ventilation, largely rewritten, covers key sustainability impacts and recent research results.
- Chapter 34, Geothermal Energy, has updated tables and graphs, with new, step-by-step design guidance on vertical systems, and expanded content on hybrid systems, ISO rating, and system efficiency.
- Chapter 36, Energy Use and Management, has updates on ASHRAE's Building Energy Quotient (eQ) labeling program.

- Chapter 40, Computer Applications, updated throughout, has new content on building information modeling (BIM) and wireless applications.
- Chapter 41, Building Energy Monitoring, has a new section on simplifying methodology for small projects.
- Chapter 42, Supervisory Control Strategies and Optimization, has been reorganized, with new content on thermal storage and thermally active building systems (TABS), hybrid cooling plants, and predictive control.
- Chapter 43, HVAC Commissioning, has been updated throughout to reflect ASHRAE *Guideline* 1.1-2007.
- Chapter 44, Building Envelopes, has reorganized and expanded content on nonresidential and existing buildings, durability, and common building envelope assemblies.
- Chapter 48, Noise and Vibration Control, has a new title plus reorganized and new content on noise criteria, chiller noise, and vibration measurement.
- Chapter 50, Service Water Heating, has expanded content on sizing tankless water heaters plus new data on piping heat loss.
- Chapter 55, Seismic- and Wind-Resistant Design, has a new title and reflects changes to building codes, standards for anchor bolt design, and other new requirements.
- Chapter 57, Room Air Distribution, has extensive new application guidelines plus new content on indoor air quality (IAQ), sustainability, and chilled beams.
- Chapter 59, HVAC Security, has a new title, with updates from ASHRAE *Guideline* 29-2009 and new sections on risk evaluation, requirements analysis, and system design.

This volume is published, both as a bound print volume and in electronic format on a CD-ROM, in two editions: one using inch-pound (I-P) units of measurement, the other using the International System of Units (SI).

Corrections to the 2008, 2009, and 2010 Handbook volumes can be found on the ASHRAE Web site at http://www.ashrae.org and in the Additions and Corrections section of this volume. Corrections for this volume will be listed in subsequent volumes and on the ASHRAE Web site.

Reader comments are enthusiastically invited. To suggest improvements for a chapter, **please comment using the form on the ASHRAE Web site** or, using the cutout pages at the end of this volume's index, write to Handbook Editor, ASHRAE, 1791 Tullie Circle, Atlanta, GA 30329, or fax 678-539-2187, or e-mail mowen@ashrae.org.

Mark S. Owen
Editor

CHAPTER 1

RESIDENCES

SPACE-CONDITIONING systems for residential use vary with both local and application factors. Local factors include energy source availability (present and projected) and price; climate; socioeconomic circumstances; and availability of installation and maintenance skills. Application factors include housing type, construction characteristics, and building codes. As a result, many different systems are selected to provide combinations of heating, cooling, humidification, dehumidification, ventilation, and air filtering. This chapter emphasizes the more common systems for space conditioning of both single-family (i.e., traditional site-built and modular or manufactured homes) and multifamily residences. Low-rise multifamily buildings generally follow single-family practice because constraints favor compact designs; HVAC systems in high-rise apartment, condominium, and dormitory buildings are often of commercial types similar to those used in hotels. Retrofit and remodeling construction also adopt the same systems as those for new construction, but site-specific circumstances may call for unique designs.

SYSTEMS

Common residential systems are listed in Table 1. Three generally recognized groups are central forced air, central hydronic, and zoned systems. System selection and design involve such key decisions as (1) source(s) of energy, (2) means of distribution and delivery, and (3) terminal device(s).

Climate determines the services needed. Heating and cooling are generally required. Air cleaning, by filtration or electrostatic devices, is present in most systems. Humidification, which is commonly added to all but the most basic systems, is provided in heating systems for thermal comfort (as defined in ASHRAE *Standard* 55), health, and reduction of static electricity discharges. Cooling

systems usually dehumidify air as well as lowering its temperature. Typical forced-air residential installations are shown in Figures 1 and 2.

Figure 1 shows a gas furnace, split-system air conditioner, humidifier, and air filter. Air from the space enters the equipment through a return air duct. It passes initially through the air filter. The circulating blower is an integral part of the furnace, which supplies heat during winter. An optional humidifier adds moisture to the heated air, which is distributed throughout the home via the supply duct. When cooling is required, heat and moisture are removed from the circulating air as it passes across the evaporator coil. Refrigerant lines connect the evaporator coil to a remote condensing unit located outdoors. Condensate from the evaporator is removed through a drainline with a trap.

Figure 2 shows a split-system heat pump, supplemental electric resistance heaters, humidifier, and air filter. The system functions as follows: Air from the space enters the equipment through the return air duct, and passes through a filter. The circulating blower is an integral part of the indoor air-handling portion of the heat pump system, which supplies heat through the indoor coil during the heating season. Optional electric heaters supplement heat from the heat pump during periods of low outdoor temperature and counteract indoor airstream cooling during periodic defrost cycles. An optional

Table 1   Residential Heating and Cooling Systems

	Central Forced Air	Central Hydronic	Zoned
Most common energy sources	Gas Oil Electricity	Gas Oil Electricity	Gas Electricity
Distribution medium	Air	Water Steam	Air Water Refrigerant
Distribution system	Ducting	Piping	Ducting Piping or Free delivery
Terminal devices	Diffusers Registers Grilles	Radiators Radiant panels Fan-coil units	Included with product or same as forced-air or hydronic systems

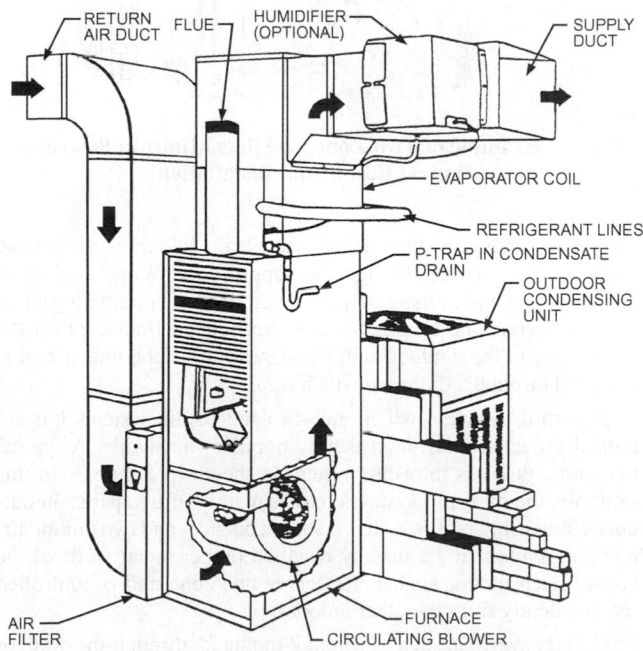

Fig. 1   Typical Residential Installation of Heating, Cooling, Humidifying, and Air Filtering System

---

The preparation of this chapter is assigned to TC 8.11, Unitary and Room Air Conditioners and Heat Pumps.

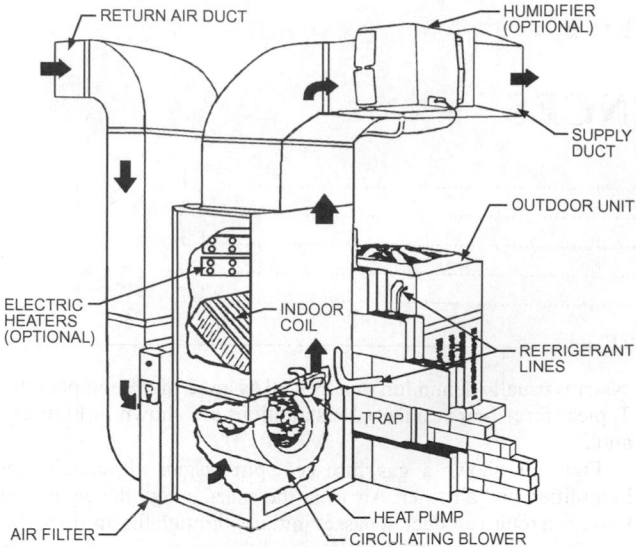

Fig. 2 Typical Residential Installation of Air-Coupled Heat Pump

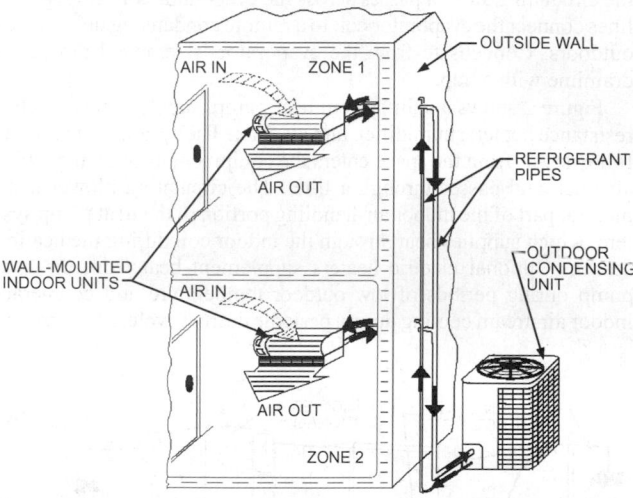

Fig. 3 Example of Two-Zone, Ductless Minisplit System in Typical Residential Installation

humidifier adds moisture to the heated air, which is distributed throughout the home through the supply duct. When cooling is required, heat and moisture are removed from the circulating air as it passes across the evaporator coil. Refrigerant lines connect the indoor coil to the outdoor unit. Condensate from the indoor coil is removed through a drainline with a trap.

Minisplit systems, which are similar to split systems but are typically ductless, are increasingly popular worldwide. A typical two-zone, ductless minisplit system is shown in Figure 3. In this example, the minisplit system consists mainly of two parts: an outdoor condensing unit, which is installed outside, and two indoor air-handling units that are usually installed on perimeter walls of the house. Each indoor air handler serves one zone and is controlled independently from the other indoor unit.

Unitary systems, such as window-mounted, through-the-wall, or rooftop units where all equipment is contained in one cabinet, are also popular. Ducted versions are used extensively in regions where residences have duct systems in crawlspaces beneath the main floor

and in areas such as the southwestern United States, where rooftop-mounted packages connect to attic duct systems.

Central hydronic heating systems are popular both in Europe and in parts of North America where central cooling has not normally been provided. New construction, especially in multistory homes, now typically includes forced-air cooling.

Zoned systems are designed to condition only part of a home at any one time. They may consist of individual room units or central systems with zoned distribution networks. Multiple central systems that serve individual floors or the sleeping and common portions of a home separately are sometimes used in large single-family residences.

The energy source is a major consideration in system selection. For heating, natural gas and electricity are most widely used in North America, followed by fuel oil, propane, wood, corn, solar energy, geothermal energy, waste heat, coal, district thermal energy, and others. Relative prices, safety, and environmental concerns (both indoor and outdoor) are further factors in heating energy source selection. Where various sources are available, economics strongly influence the selection. Electricity is the dominant energy source for cooling.

EQUIPMENT SIZING

The heat loss and gain of each conditioned room and of ductwork or piping run through unconditioned spaces in the structure must be accurately calculated to select equipment with the proper heating and cooling capacity. To determine heat loss and gain accurately, the floor plan and construction details, including information on wall, ceiling, and floor construction as well as the type and thickness of insulation, must be known. Window design and exterior door details are also needed. With this information, heat loss and gain can be calculated using the Air-Conditioning Contractors of America (ACCA) *Manual* J® or similar calculation procedures. To conserve energy, many jurisdictions require that the building be designed to meet or exceed the requirements of ASHRAE *Standard* 90.2 or similar requirements.

Proper matching of equipment capacity to the building heat loss and gain is essential. The heating capacity of air-source heat pumps is usually supplemented by auxiliary heaters, most often of the electric resistance type; in some cases, however, fossil fuel furnaces or solar systems are used.

Undersized equipment will be unable to maintain the intended indoor temperature under conditions of extreme outdoor temperatures. Some oversizing may be desirable to enable recovery from setback and to maintain indoor comfort during outdoor conditions that are more extreme than the nominal design conditions. Grossly oversized equipment can cause discomfort because of short on-times, wide indoor temperature swings, and inadequate dehumidification when cooling. Gross oversizing may also contribute to higher energy use by increasing cyclic thermal losses and off-cycle losses. Variable-capacity equipment (heat pumps, air conditioners, and furnaces) can more closely match building loads over broad ambient temperature ranges, usually reducing these losses and improving comfort levels; in the case of heat pumps, supplemental heat needs may also be reduced.

Residences of tight construction may have high indoor humidity and a build-up of indoor air contaminants at times. Air-to-air heat recovery equipment may be used to provide tempered ventilation air to tightly constructed houses. Outdoor air intakes connected to the return duct of central systems may also be used when reducing installed costs is the most important task. Simple exhaust systems with or without passive air intakes are also popular. Natural ventilation by operable windows is also popular in some climates. Excessive accumulation of radon is of concern in all buildings; lower-level spaces should not be depressurized, which causes increased migration of soil gases into buildings. All ventilation schemes increase heating and cooling loads and thus the required system capacity, thereby resulting in greater energy consumption. In all cases, minimum ventilation rates, as described in ASHRAE *Standards* 62.1 and 62.2, should be maintained.

SINGLE-FAMILY RESIDENCES

Heat Pumps

Heat pumps for single-family houses are normally unitary or split systems, as illustrated in Figures 2 and 3.

Most commercially available heat pumps, particularly in North America, are electrically powered, air-source systems. Supplemental heat is generally required at low outdoor temperatures or during defrost. In most cases, supplemental or back-up heat is provided by electric resistance heating elements.

Heat pumps may be classified by thermal source and distribution medium in the heating mode as well as the type of fuel used. The most commonly used classes of heat pump equipment are air-to-air and water-to-air. Air-to-water and water-to-water types are also used.

Heat pump systems are generally described as air-source or ground-source. The thermal sink for cooling is generally assumed to be the same as the thermal source for heating. Air-source systems using ambient air as the heat source/sink are generally the least costly to install and thus the most commonly used. Ground-source systems usually use water-to-air heat pumps to extract heat from the ground using groundwater or a buried heat exchanger.

Ground-Source (Geothermal) Systems. As a heat source/sink, groundwater (from individual wells or supplied as a utility from community wells) offers the following advantages over ambient air: (1) heat pump capacity is independent of ambient air temperature, reducing supplementary heating requirements; (2) no defrost cycle is required; (3) although operating conditions for establishing rated efficiency are not the same as for air-source systems, seasonal efficiency is usually higher for heating and for cooling; and (4) peak heating energy consumption is usually lower. Two other system types are ground-coupled and surface-water-coupled systems. Ground-coupled systems offer the same advantages, but because surface water temperatures track fluctuations in air temperature, surface-water-coupled systems may not offer the same benefits as other ground-source systems. Both system types circulate brine or water in a buried or submerged heat exchanger to transfer heat from the ground or water. Direct-expansion, ground-source systems, with evaporators buried in the ground, also are available but are seldom used. Water-source systems that extract heat from surface water (e.g., lakes or rivers) or city (tap) water are sometimes used where local conditions allow. Further information may be found in Chapter 48 of the 2008 *ASHRAE Handbook—HVAC Systems and Equipment.*

Water supply, quality, and disposal must be considered for groundwater systems. Caneta Research (1995) and Kavanaugh and Rafferty (1997) provide detailed information on these subjects. Secondary coolants for ground-coupled systems are discussed in Caneta Research (1995) and in Chapter 31 of the 2009 *ASHRAE Handbook—Fundamentals.* Buried heat exchanger configurations may be horizontal or vertical, with the vertical including both multiple-shallow- and single-deep-well configurations. Ground-coupled systems avoid water quality, quantity, and disposal concerns but are sometimes more expensive than groundwater systems. However, ground-coupled systems are usually more efficient, especially when pumping power for the groundwater system is considered. Proper installation of the ground coil(s) is critical to success.

Add-On Heat Pumps. In add-on systems, a heat pump is added (often as a retrofit) to an existing furnace or boiler/fan-coil system. The heat pump and combustion device are operated in one of two ways: (1) alternately, depending on which is most cost-effective, or (2) in parallel. In unitary bivalent heat pumps, the heat pump and combustion device are grouped in a common chassis and cabinets to provide similar benefits at lower installation costs.

Fuel-Fired Heat Pumps. Extensive research and development has been conducted to develop fuel-fired heat pumps. They have been marketed in North America. More information may be found in Chapter 48 of the 2008 *ASHRAE Handbook—HVAC Systems and Equipment.*

Water-Heating Options. Heat pumps may be equipped with desuperheaters (either integral or field-installed) to reclaim heat for domestic water heating when operated in cooling mode. Integrated space-conditioning and water-heating heat pumps with an additional full-size condenser for water heating are also available.

Furnaces

Furnaces are fueled by gas (natural or propane), electricity, oil, wood, or other combustibles. Gas, oil, and wood furnaces may draw combustion air from the house or from outdoors. If the furnace space is located such that combustion air is drawn from the outdoors, the arrangement is called an isolated combustion system (ICS). Furnaces are generally rated on an ICS basis. Outdoor air is ducted to the combustion chamber (a direct-vent system) for manufactured home applications and some mid- and high-efficiency equipment designs. Using outside air for combustion eliminates both infiltration losses associated with using indoor air for combustion and stack losses associated with atmospherically induced draft-hood-equipped furnaces.

Two available types of high-efficiency gas furnaces are noncondensing and condensing. Both increase efficiency by adding or improving heat exchanger surface area and reducing heat loss during furnace off-times. The higher-efficiency condensing type also recovers more energy by condensing water vapor from combustion products. Condensate is formed in a corrosion resistant heat exchanger and is disposed of through a drain line. Care must be taken to prevent freezing the condensate when the furnace is installed in an unheated space such as an attic. Condensing furnaces generally use PVC for vent pipes and condensate drains.

Wood-, corn-, and coal-fueled furnaces are used in some areas as either the primary or supplemental heating unit. These furnaces may have catalytic converters to enhance the combustion process, increasing furnace efficiency and producing cleaner exhaust.

Chapters 30 and 32 of the 2008 *ASHRAE Handbook—HVAC Systems and Equipment* include more detailed information on furnaces and furnace efficiency.

Hydronic Heating Systems

With the growth of demand for central cooling systems, hydronic systems have declined in popularity in new construction, but still account for a significant portion of existing systems in colder climates. The fluid is heated in a central boiler and distributed by piping to terminal units in each room. Terminal units are typically either radiators or baseboard convectors. Other terminal units include fan-coils and radiant panels. Most recently installed residential systems use a forced-circulation, multiple-zone hot-water system with a series-loop piping arrangement. Chapters 12 and 35 of the 2008 *ASHRAE Handbook—HVAC Systems and Equipment* have more information on hydronics.

Design water temperature is based on economic and comfort considerations. Generally, higher temperatures result in lower first costs because smaller terminal units are needed. However, losses tend to be greater, resulting in higher operating costs and reduced comfort because of the concentrated heat source. Typical design temperatures range from 180 to 200°F. For radiant panel systems, design temperatures range from 110 to 170°F. The preferred control method allows the water temperature to decrease as outdoor temperatures rise. Provisions for expansion and contraction of piping and heat distributing units and for eliminating air from the hydronic system are essential for quiet, leak-tight operation.

Fossil fuel systems that condense water vapor from the flue gases must be designed for return water temperatures in the range of 120 to 130°F for most of the heating season. Noncondensing systems must maintain high enough water temperatures in the boiler to prevent this condensation. If rapid heating is required, both terminal unit and boiler size must be increased, although gross oversizing should be avoided.

Another concept for multi- or single-family dwellings is a combined water-heating/space-heating system that uses water from the domestic hot-water storage tank to provide space heating. Water circulates from the storage tank to a hydronic coil in the system air handler. Space heating is provided by circulating indoor air across the coil. A split-system central air conditioner with the evaporator located in the system air handler can be included to provide space cooling.

Zoned Heating Systems

Most moderate-cost residences in North America have single-thermal-zone HVAC systems with one thermostat. Multizoned systems, however, offer the potential for improved thermal comfort. Lower operating costs are possible with zoned systems because unoccupied areas (e.g., common areas at night, sleeping areas during the day) can be kept at lower temperatures in the winter.

One form of this system consists of individual heaters located in each room. These heaters are usually electric or gas-fired. Electric heaters are available in the following types: baseboard free-convection, wall insert (free-convection or forced-fan), radiant panels for walls and ceilings, and radiant cables for walls, ceilings, and floors. Matching equipment capacity to heating requirements is critical for individual room systems. Heating delivery cannot be adjusted by adjusting air or water flow, so greater precision in room-by-room sizing is needed. Most individual heaters have integral thermostats that limit the ability to optimize unit control without continuous fan operation.

Individual heat pumps for each room or group of rooms (zone) are another form of zoned electric heating. For example, two or more small unitary heat pumps can be installed in two-story or large one-story homes.

The multisplit heat pump consists of a central compressor and an outdoor heat exchanger to service multiple indoor zones. Each zone uses one or more fan-coils, with separate thermostatic controls for each zone. Such systems are used in both new and retrofit construction.

A method for zoned heating in central ducted systems is the zone-damper system. This consists of individual zone dampers and thermostats combined with a zone control system. Both variable-air-volume (damper position proportional to zone demand) and on/off (damper fully open or fully closed in response to thermostat) types are available. These systems sometimes include a provision to modulate to lower capacities when only a few zones require heating.

Solar Heating

Both active and passive solar thermal energy systems are sometimes used to heat residences. In typical active systems, flat-plate collectors heat air or water. Air systems distribute heated air either to the living space for immediate use or to a thermal storage medium (e.g., a rock pile). Water systems pass heated water from the collectors through a heat exchanger and store heat in a water tank. Because of low delivered-water temperatures, radiant floor panels requiring moderate temperatures are often used. A water-source heat pump between the water storage tank and the load can be used to increase temperature differentials.

Trombe walls, direct-gain, and greenhouse-like sunspaces are common passive solar thermal systems. Glazing facing south (in the northern hemisphere), with overhangs to reduce solar gains in the summer, and movable night insulation panels reduce heating requirements.

Some form of back-up heating is generally needed with solar thermal energy systems. Solar electric systems are not normally used for space heating because of the high energy densities required and the economics of photovoltaics. However, hybrid collectors, which combine electric and thermal capabilities, are available. Chapter 35 has information on sizing solar heating equipment.

Unitary Air Conditioners

In forced-air systems, the same air distribution duct system can be used for both heating and cooling. Split-system central cooling, as illustrated in Figure 1, is the most widely used forced-air system. Upflow, downflow, and horizontal-airflow indoor units are available. Condensing units are installed on a noncombustible pad outside and contain a motor- or engine-driven compressor, condenser, condenser fan and fan motor, and controls. The condensing unit and evaporator coil are connected by refrigerant tubing that is normally field-supplied. However, precharged, factory-supplied tubing with quick-connect couplings is also common where the distance between components is not excessive.

A distinct advantage of split-system central cooling is that it can readily be added to existing forced-air heating systems. Airflow rates are generally set by the cooling requirements to achieve good performance, but most existing heating duct systems are adaptable to cooling. Airflow rates of 350 to 450 cfm per nominal ton of refrigeration are normally recommended for good cooling performance. As with heat pumps, these systems may be fitted with desuperheaters for domestic water heating.

Some cooling equipment includes forced-air heating as an integral part of the product. Year-round heating and cooling packages with a gas, oil, or electric furnace for heating and a vapor-compression system for cooling are available. Air-to-air and water-source heat pumps provide cooling and heating by reversing the flow of refrigerant.

Distribution. Duct systems for cooling (and heating) should be designed and installed in accordance with accepted practice. Useful information is found in ACCA *Manuals* D® and S®. Chapter 9 of the 2008 *ASHRAE Handbook—HVAC Systems and Equipment* also discusses air distribution design for small heating and cooling systems.

Because weather is the primary influence on the load, the cooling and heating load in each room changes from hour to hour. Therefore, the owner or occupant should be able to make seasonal or more frequent adjustments to the air distribution system to improve comfort. Adjustments may involve opening additional outlets in second-floor rooms during summer and throttling or closing heating outlets in some rooms during winter. Manually adjustable balancing dampers may be provided to facilitate these adjustments. Other possible refinements are installing a heating and cooling system sized to meet heating requirements, with additional self-contained cooling units serving rooms with high summer loads, or separate central systems for the upper and lower floors of a house. On deluxe applications, zone-damper systems can be used. Another way of balancing cooling and heating loads is to use variable-capacity compressors in heat pump systems.

Operating characteristics of both heating and cooling equipment must be considered when zoning is used. For example, a reduction in air quantity to one or more rooms may reduce airflow across the evaporator to such a degree that frost forms on the fins. Reduced airflow on heat pumps during the heating season can cause overloading if airflow across the indoor coil is not maintained above 350 cfm per ton. Reduced air volume to a given room reduces the air velocity from the supply outlet and might cause unsatisfactory air distribution in the room. Manufacturers of zoned systems normally provide guidelines for avoiding such situations.

Special Considerations. In residences with more than one story, cooling and heating are complicated by air buoyancy, also known as the stack effect. In many such houses, especially with single-zone systems, the upper level tends to overheat in winter and undercool in summer. Multiple air outlets, some near the floor and others near the ceiling, have been used with some success on all levels. To control airflow, the homeowner opens some outlets and closes others from season to season. Free air circulation between floors can be reduced by locating returns high in each room and keeping doors closed.

In existing homes, the cooling that can be added is limited by the air-handling capacity of the existing duct system. Although the existing duct system is usually satisfactory for normal occupancy, it may be inadequate during large gatherings. In all cases where new cooling (or heating) equipment is installed in existing homes, supply air ducts and outlets must be checked for acceptable air-handling capacity and air distribution. Maintaining upward airflow at an effective velocity is important when converting existing heating systems with floor or baseboard outlets to both heat and cool. It is not necessary to change the deflection from summer to winter for registers located at the perimeter of a residence. Registers located near the floor on the inside walls of rooms may operate unsatisfactorily if the deflection is not changed from summer to winter.

Occupants of air-conditioned spaces usually prefer minimum perceptible air motion. Perimeter baseboard outlets with multiple slots or orifices directing air upwards effectively meet this requirement. Ceiling outlets with multidirectional vanes are also satisfactory.

A residence without a forced-air heating system may be cooled by one or more central systems with separate duct systems, by individual room air conditioners (window-mounted or through-the-wall), or by minisplit room air conditioners.

Cooling equipment must be located carefully. Because cooling systems require higher indoor airflow rates than most heating systems, sound levels generated indoors are usually higher. Thus, indoor air-handling units located near sleeping areas may require sound attenuation. Outdoor noise levels should also be considered when locating the equipment. Many communities have ordinances regulating the sound level of mechanical devices, including cooling equipment. Manufacturers of unitary air conditioners often rate the sound level of their products according to an industry standard (AHRI *Standard* 270). AHRI *Standard* 275 gives information on how to predict the dBA sound level when the AHRI sound rating number, the equipment location relative to reflective surfaces, and the distance to the property line are known.

An effective and inexpensive way to reduce noise is to put distance and natural barriers between sound source and listener. However, airflow to and from air-cooled condensing units must not be obstructed; for example, plantings and screens must be porous and placed away from units so as not to restrict intake or discharge of air. Most manufacturers provide recommendations regarding acceptable distances between condensing units and natural barriers. Outdoor units should be placed as far as is practical from porches and patios, which may be used while the house is being cooled. Locations near bedroom windows and neighboring homes should also be avoided. In high-crime areas, consider placing units on roofs or other semisecure areas.

Evaporative Coolers

In climates that are dry throughout the entire cooling season, evaporative coolers can be used to cool residences. They must be installed and maintained carefully to reduce the potential for water and thus air quality problems. Further details on evaporative coolers can be found in Chapter 40 of the 2008 *ASHRAE Handbook—HVAC Systems and Equipment* and in Chapter 52 of this volume.

Humidifiers

For improved winter comfort, equipment that increases indoor relative humidity may be needed. In a ducted heating system, a central whole-house humidifier can be attached to or installed within a supply plenum or main supply duct, or installed between the supply and return duct systems. When applying supply-to-return duct humidifiers on heat pump systems, care should be taken to maintain proper airflow across the indoor coil. Self-contained portable or tabletop humidifiers can be used in any residence. Even though this type of humidifier introduces all the moisture to one area of the home, moisture migrates and raises humidity levels in other rooms.

Overhumidification should be avoided: it can cause condensate to form on the coldest surfaces in the living space (usually windows). Also, because moisture migrates through all structural materials, vapor retarders should be installed near the warmer inside surface of insulated walls, ceilings, and floors in most temperature climates. Lack of attention to this construction detail allows moisture to migrate from inside to outside, causing damp insulation, mold, possible structural damage, and exterior paint blistering.

Central humidifiers may be rated in accordance with AHRI *Standard* 610. This rating is expressed in the number of gallons per day evaporated by 140°F entering air. Some manufacturers certify the performance of their product to the AHRI standard. Selecting the proper size humidifier is important and is outlined in AHRI *Guideline* F.

Humidifier cleaning and maintenance schedules must be followed to maintain efficient operation and prevent bacteria build-up.

Chapter 21 of the 2008 *ASHRAE Handbook—HVAC Systems and Equipment* contains more information on residential humidifiers.

Dehumidifiers

Many homes also use dehumidifiers to remove moisture and control indoor humidity levels. In cold climates, dehumidification is sometimes required during the summer in basement areas to control mold and mildew growth and to reduce zone humidity levels. Traditionally, portable dehumidifiers have been used to control humidity in this application. Although these portable units are not always as efficient as central systems, their low first cost and ability to serve a single zone make them appropriate in many circumstances.

In hot, humid climates, providing sufficient dehumidification with sensible cooling is important. Although conventional air-conditioning units provide some dehumidification as a consequence of sensible cooling, in some cases space humidity levels can still exceed comfortable levels.

Several dehumidification enhancements to conventional air-conditioning systems are possible to improve moisture removal characteristics and lower the space humidity level. Some simple improvements include lowering the supply airflow rate and eliminating off-cycle fan operation. Additional equipment options such as condenser/reheat coils, sensible-heat-exchanger-assisted evaporators (e.g., heat pipes), and subcooling/reheat coils can further improve dehumidification performance. Desiccants, applied as either thermally activated units or heat recovery systems (e.g., enthalpy wheels), can also increase dehumidification capacity and lower the indoor humidity level. Some dehumidification options add heat to the conditioned zone that, in some cases, increases the load on the sensible cooling equipment.

Air Filters

Most comfort conditioning systems that circulate air incorporate some form of air filter. Usually, they are disposable or cleanable filters that have relatively low air-cleaning efficiency. Higher-efficiency alternatives include pleated media filters and electronic air filters. These high-efficiency filters may have high static pressure drops. The air distribution system should be carefully evaluated before installing such filters so that airflow rates are not overly reduced with their use. Airflow must be evaluated both when the filter is new and when it is in need of replacement or cleaning.

Air filters are mounted in the return air duct or plenum and operate whenever air circulates through the duct system. Air filters are rated in accordance with AHRI *Standard* 680, which was based on ASHRAE *Standard* 52.1. Atmospheric dust spot efficiency levels are generally less than 20% for disposable filters and vary from 60 to 90% for electronic air filters. However, increasingly, the minimum efficiency rating value (MERV) from ASHRAE *Standard* 52.2 is given instead; a higher MERV implies greater particulate removal, but also typically increased air pressure drop across the filter.

To maintain optimum performance, the collector cells of electronic air filters must be cleaned periodically. Automatic indicators are often used to signal the need for cleaning. Electronic air filters have higher initial costs than disposable or pleated filters, but generally last the life of the air-conditioning system. Also available are gas-phase filters such as those that use activated carbon. Chapter 28 of the 2008 *ASHRAE Handbook—HVAC Systems and Equipment* covers the design of residential air filters in more detail.

Ultraviolet (UV) germicidal light as an air filtration system for residential applications has become popular recently. UV light has been successfully used in health care facilities, food-processing plants, schools, and laboratories. It can break organic molecular bonds, which translates into cellular or genetic damages for microorganisms. Single or multiple UV lamps are usually installed in the return duct or downstream of indoor coils in the supply duct. Direct exposure of occupants to UV light is avoided because UV light does not pass through metal, glass, or plastic. This air purification method effectively reduces the transmission of airborne germs, bacteria, molds, viruses, and fungi in the air streams without increasing duct pressure losses. The power required by each UV lamp might range between 30 and 100 W, depending on the intensity and exposure time required to kill the various microorganisms. Chapter 16 of the 2008 *ASHRAE Handbook—HVAC Systems and Equipment* and Chapter 60 of this volume cover the design and application of UV lamp systems in more detail.

Controls

Historically, residential heating and cooling equipment has been controlled by a wall thermostat. Today, simple wall thermostats with bimetallic strips are often replaced by programmable microelectronic models that can set heating and cooling equipment at different temperature levels, depending on the time of day or week. This has led to night setback, workday, and vacation control to reduce energy demand and operating costs. For heat pump equipment, electronic thermostats can incorporate night setback with an appropriate scheme to limit use of resistance heat during recovery. Chapter 47 contains more details about automatic control systems.

MULTIFAMILY RESIDENCES

Attached homes and low-rise multifamily apartments generally use heating and cooling equipment comparable to that used in single-family dwellings. Separate systems for each unit allow individual control to suit the occupant and facilitate individual metering of energy use; separate metering and direct billing of occupants encourages energy conservation.

Forced-Air Systems

High-rise multifamily structures may also use unitary or minisplit heating and cooling equipment comparable to that used in single-family dwellings. Equipment may be installed in a separate mechanical equipment room in the apartment, in a soffit or above a dropped ceiling over a hallway or closet, or wall-mounted. Split systems' condensing or heat pump units are often placed on roofs, balconies, or the ground.

Small residential warm-air furnaces may also be used, but a means of providing combustion air and venting combustion products from gas- or oil-fired furnaces is required. It may be necessary to use a multiple-vent chimney or a manifold-type vent system. Local codes must be consulted. Direct-vent furnaces that are placed near or on an outside wall are also available for apartments.

Hydronic Systems

Individual heating and cooling units are not always possible or practical in high-rise structures. In this case, applied central systems are used. Two- or four-pipe hydronic central systems are widely used in high-rise apartments. Each dwelling unit has either individual room units or ducted fan-coil units.

The most flexible hydronic system with usually the lowest operating costs is the four-pipe type, which provides heating or cooling for each apartment dweller. The two-pipe system is less flexible because it cannot provide heating and cooling simultaneously. This limitation causes problems during the spring and fall when some apartments in a complex require heating while others require cooling because of solar or internal loads. This spring/fall problem may be overcome by operating the two-pipe system in a cooling mode and providing the relatively low amount of heating that may be required by means of individual electric resistance heaters.

See the section on Hydronic Heating Systems for description of a combined water-heating/space-heating system for multi- or single-family dwellings. Chapter 12 of the 2008 *ASHRAE Handbook—HVAC Systems and Equipment* discusses hydronic design in more detail.

Through-the-Wall Units

Through-the-wall room air conditioners, packaged terminal air conditioners (PTACs), and packaged terminal heat pumps (PTHPs) can be used for conditioning single rooms. Each room with an outside wall may have such a unit. These units are used extensively in renovation of old buildings because they are self-contained and typically do not require complex piping or ductwork renovation.

Room air conditioners have integral controls and may include resistance or heat pump heating. PTACs and PTHPs have special indoor and outdoor appearance treatments, making them adaptable to a wider range of architectural needs. PTACs can include gas, electric resistance, hot water, or steam heat. Integral or remote wall-mounted controls are used for both PTACs and PTHPs. Further information may be found in Chapter 49 of the 2008 *ASHRAE Handbook—HVAC Systems and Equipment* and in AHRI *Standard* 310/380.

Water-Loop Heat Pumps

Any mid- or high-rise structure having interior zones with high internal heat gains that require year-round cooling can efficiently use a water-loop heat pump. Such systems have the flexibility and control of a four-pipe system but use only two pipes. Water-source heat pumps allow individual metering of each apartment. The building owner pays only the utility cost for the circulating pump, cooling tower, and supplemental boiler heat. Existing buildings can be retrofitted with heat flow meters and timers on fan motors for individual metering. Economics permitting, solar or ground heat energy can provide the supplementary heat in lieu of a boiler. The ground can also provide a heat sink, which in some cases can eliminate the cooling tower. In areas where the water table is continuously high and the soil is porous, groundwater from wells can be used.

Special Concerns for Apartment Buildings

Many ventilation systems are used in apartment buildings. Local building codes generally govern outdoor air quantities. ASHRAE *Standard* 62.1-2004 requires minimum outdoor air quantities of 50 cfm intermittent or 20 cfm continuous or operable windows for baths and toilets, and 100 cfm intermittent or 25 cfm continuous or operable windows for kitchens.

In some buildings with centrally controlled exhaust and supply systems, the systems are operated on time clocks for certain periods of the day. In other cases, the outside air is reduced or shut off during extremely cold periods. If known, these factors should be considered when estimating heating load.

Another important load, frequently overlooked, is heat gain from piping for hot-water services.

Buildings using exhaust and supply air systems 24 h/day may benefit from air-to-air heat recovery devices (see Chapter 25 of the 2008 *ASHRAE Handbook—HVAC Systems and Equipment*). Such recovery devices can reduce energy consumption by transferring 40 to 80% of the sensible and latent heat between the exhaust air and supply air streams.

Infiltration loads in high-rise buildings without ventilation openings for perimeter units are not controllable year-round by general building pressurization. When outer walls are penetrated to supply outdoor air to unitary or fan-coil equipment, combined wind and thermal stack effects create other infiltration problems.

Interior public corridors in apartment buildings need conditioning and smoke management to meet their ventilation and thermal needs, and to meet the requirements of fire and life safety codes. Stair towers, however, are normally kept separate from hallways to maintain fire-safe egress routes and, if needed, to serve as safe havens until rescue. Therefore, great care is needed when designing buildings with interior hallways and stair towers. Chapter 53 provides further information.

Air-conditioning equipment must be isolated to reduce noise generation or transmission. The design and location of cooling towers must be chosen to avoid disturbing occupants within the building and neighbors in adjacent buildings. Also, for cooling towers, prevention of *Legionella* is a serious concern. Further information on cooling towers is in Chapter 39 of the 2008 *ASHRAE Handbook—HVAC Systems and Equipment*.

In large apartment houses, a central building energy management system may allow individual apartment air-conditioning systems or units to be monitored for maintenance and operating purposes.

MANUFACTURED HOMES

Manufactured homes are constructed in factories rather than site-built, and in 2001 constituted over 6.4% of all housing units and about 11% of all new single-family homes sold in the United States (DOE 2005). Heating and cooling systems in manufactured homes, as well as other facets of construction such as insulation levels, are regulated in the United States by HUD Manufactured Home Construction and Safety Standards. Each complete home or home section is assembled on a transportation frame (a chassis with wheels and axles) for transport. Manufactured homes vary in size from small, single-floor section units starting at 400 ft^2 to large, multiple sections, which when joined together can provide over 2500 ft^2 and have an appearance similar to site-constructed homes.

Heating systems are factory-installed and are primarily forced-air downflow units feeding main supply ducts built into the subfloor, with floor registers located throughout the home. A small percentage of homes in the far southern and southwestern United States use upflow units feeding overhead ducts in the attic space. Typically, there is no return duct system. Air returns to the air handler from each room through door undercuts, hallways, and a grilled door or louvered panel. The complete heating system is a reduced-clearance type with the air-handling unit installed in a small closet or alcove, usually in a hallway. Sound control measures may be required if large forced-air systems are installed close to sleeping areas. Gas, oil, and electric furnaces or heat pumps may be installed by the home manufacturer to satisfy market requirements.

Gas and oil furnaces are compact direct-vent types approved for installation in a manufactured home. The special venting arrangement used is a vertical through-the-roof concentric pipe-in-pipe system that draws all air for combustion directly from the outdoors and discharges combustion products through a windproof vent terminal. Gas furnaces must be easily convertible from liquefied petroleum to natural gas and back as required at the final site.

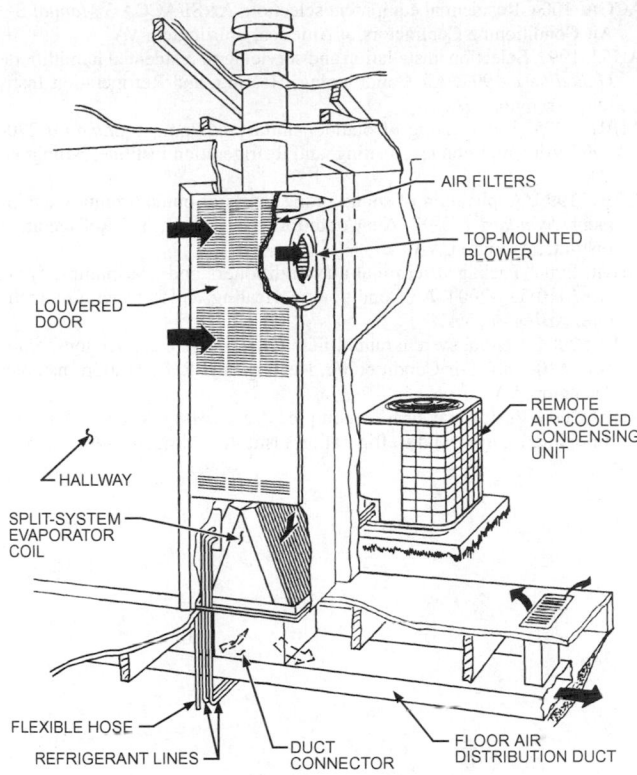

Fig. 4 Typical Installation of Heating and Cooling Equipment for Manufactured Home

Manufactured homes may be cooled with add-on split or single-package air-conditioning systems when supply ducts are adequately sized and rated for that purpose according to HUD requirements. The split-system evaporator coil may be installed in the integral coil cavity provided with the furnace. A high-static-pressure blower is used to overcome resistance through the furnace, evaporator coil, and compact air distribution system. Single-package air conditioners are connected with flexible air ducts to feed existing factory in-floor or overhead ducts. Dampers or other means are required to prevent the cooled, conditioned air from backflowing through a furnace cabinet.

A typical installation of a downflow gas or oil furnace with a split-system air conditioner is illustrated in Figure 4. Air enters the furnace from the hallway, passing through a louvered door on the front of the furnace. The air then passes through air filters and is drawn into the top-mounted blower, which during winter forces air down over the heat exchanger, where it picks up heat. For summer cooling, the blower forces air through the furnace heat exchanger and then through the split-system evaporator coil, which removes heat and moisture from the passing air. During heating and cooling, conditioned air then passes through a combustible floor base via a duct connector before flowing into the floor air distribution duct. The evaporator coil is connected with quick-connect refrigerant lines to a remote air-cooled condensing unit. The condensate collected at the evaporator is drained by a flexible hose, routed to the exterior through the floor construction, and connected to a suitable drain.

REFERENCES

ACCA. 2009. Residential duct systems, 3rd ed. ANSI/ACCA 1 *Manual* D®. Air Conditioning Contractors of America, Shirlington, VA

ACCA. 2006. Residential load calculation, 8th ed., v. 2. ANSI/ACCA 2 *Manual* F®. Air Conditioning Contractors of America, Shirlington, VA

ACCA. 2004. Residential equipment selection. ANSI/ACCA 3 *Manual* S®. Air Conditioning Contractors of America, Shirlington, VA

AHRI. 1997. Selection, installation and servicing of residential humidifiers. *Guideline* F-1997. Air Conditioning, Heating, and Refrigeration Institute, Arlington, VA.

AHRI. 1995. Sound rating of outdoor unitary equipment. *Standard* 270-2008. Air Conditioning, Heating, and Refrigeration Institute, Arlington, VA.

AHRI. 1997. Application of sound rating levels of outdoor unitary equipment. *Standard* 275-97. Air Conditioning, Heating, and Refrigeration Institute, Arlington, VA.

AHRI. 2004. Packaged terminal air-conditioners and heat pumps. *Standard* 310/380-2004. Air Conditioning, Heating, and Refrigeration Institute, Arlington, VA.

AHRI. 2004. Central system humidifiers for residential applications. *Standard* 610-2004. Air Conditioning, Heating, and Refrigeration Institute, Arlington, VA.

AHRI. 2004. Residential air filter equipment. *Standard* 680-2004. Air Conditioning, Heating, and Refrigeration Institute, Arlington, VA.

ASHRAE. 1992. Gravimetric and dust spot procedures for testing air-cleaning devices used in general ventilation for removing particulate matter. *Standard* 52.1-1992.

ASHRAE. 2010. Thermal environmental conditions for human occupancy. ANSI/ASHRAE *Standard* 55-2010.

ASHRAE. 2010. Ventilation for acceptable indoor air quality. ANSI/ASHRAE *Standard* 62.1-2010.

ASHRAE. 2010. Ventilation and acceptable indoor air quality in low-rise residential buildings. *Standard* 62.2-2010.

ASHRAE. 2007. Energy-efficient design of low-rise residential buildings. ANSI/ASHRAE *Standard* 90.2-2007.

Caneta Research. 1995. *Commercial/institutional ground-source heat pump engineering manual.* ASHRAE.

DOE. 2005. *Buildings energy databook.* U.S. Department of Energy, Office of Energy Efficiency and Renewable Energy, Washington, D.C.

Kavanaugh, S.P. and K. Rafferty. 1997. *Ground source heat pumps—Design of geothermal systems for commercial and institutional buildings.* ASHRAE.

RETAIL FACILITIES

THIS chapter covers design and application of air-conditioning and heating systems for various retail merchandising facilities. Load calculations, systems, and equipment are covered elsewhere in the Handbook series.

GENERAL CRITERIA

To apply equipment properly, the construction of the space to be conditioned, its use and occupancy, the time of day in which greatest occupancy occurs, physical building characteristics, and lighting layout must be known.

The following must also be considered:

- Electric power—size of service
- Heating—availability of steam, hot water, gas, oil, or electricity
- Cooling—availability of chilled water, well water, city water, and water conservation equipment
- Internal heat gains
- Equipment locations
- Structural considerations
- Rigging and delivery of equipment
- Obstructions
- Ventilation—opening through roof or wall for outdoor air duct
- Exposures and number of doors
- Orientation of store
- Code requirements
- Utility rates and regulations
- Building standards

Specific design requirements, such as the increase in outdoor air required to make up for kitchen exhaust, must be considered. Ventilation requirements of ASHRAE Standard 62.1 must be followed. Objectionable odors may necessitate special filtering, exhaust, and additional outdoor air intake.

Security requirements must be considered and included in the overall design and application. Minimum considerations require secure equipment rooms, secure air-handling systems, and outdoor air intakes located on the top of facilities. More extensive security measures should be developed based on overall facility design, owner requirements, and local authorities.

Load calculations should be made using the procedures outlined in the ASHRAE Handbook—Fundamentals.

Almost all localities have some form of energy code in effect that establishes strict requirements for insulation, equipment efficiencies, system designs, etc., and places strict limits on fenestration and lighting. The requirements of ASHRAE Standard 90.1 must be met as a minimum guideline for retail facilities. The Advanced Energy Design Guide for Small Retail Buildings (ASHRAE 2006) provides additional energy savings suggestions.

Retail facilities often have a high internal sensible heat gain relative to the total heat gain. However, the quantity of outdoor air required by ventilation codes and standards may result in a high latent heat removal demand at the equipment. The high latent heat removal requirement may also occur at outdoor dry-bulb temperatures below design. Unitary HVAC equipment and HVAC systems should be designed and selected to provide the necessary sensible and latent heat removal. The equipment, systems, and controls should be designed to provide the necessary temperature, ventilation, filtration, and humidity conditions.

HVAC system selection and design for retail facilities are normally determined by economics. First cost is usually the determining factor for small stores. For large retail facilities, owning, operating, and maintenance costs are also considered. Decisions about mechanical systems for retail facilities are typically based on a cash flow analysis rather than on a full life-cycle analysis.

SMALL STORES

Small stores are typically located in convenience centers and may have at least the store front exposed to outdoor weather, although some are free standing. Large glass areas found at the front of many small stores may cause high peak solar heat gain unless they have northern exposures or large overhanging canopies. High heat loss may be experienced on cold, cloudy days in the front of these stores. The HVAC system for this portion of the small store should be designed to offset the greater cooling and heating requirements. Entrance vestibules, entry heaters, and/or air curtains may be needed in some climates.

Design Considerations

System Design. Single-zone unitary rooftop equipment is common in store air conditioning. Using multiple units to condition the store involves less ductwork and can maintain comfort in the event of partial equipment failure. Prefabricated and matching curbs simplify installation and ensure compatibility with roof materials.

Air to air heat pumps, offered as packaged equipment, are readily adaptable to small-store applications. Ground-source and other closed-loop heat pump systems have been provided for small stores where the requirements of several users may be combined. Winter design conditions, utility rates, maintenance costs, and operating costs should be compared to those of conventional heating HVAC systems before this type of system is chosen.

Water-cooled unitary equipment is available for small-store air conditioning. However, many communities restrict the use of city water and groundwater for condensing purposes and may require installation of a cooling tower. Water-cooled equipment generally operates efficiently and economically.

Air Distribution. External static pressures available in small-store air-conditioning units are limited, and air distribution should be designed to keep duct resistances low. Duct velocities should not exceed 1200 fpm, and pressure drop should not exceed 0.10 in. of water per 100 ft. Average air quantities, typically range from 350 to 450 cfm per ton of cooling in accordance with the calculated internal sensible heat load.

Attention should be paid to suspended obstacles (e.g., lights, soffits, ceiling recesses, and displays) that interfere with proper air distribution.

The preparation of this chapter is assigned to TC 9.8, Large-Building Air-Conditioning Applications.

The duct system should contain enough dampers for air balancing. Volume dampers should be installed in takeoffs from the main supply duct to balance air to the branch ducts. Dampers should be installed in the return and outdoor air ducts for proper outdoor air/ return air balance and for economizer operation.

Control. Controls for small stores should be kept as simple as possible while still providing the required functions. Unitary equipment is typically available with manufacturer-supplied controls for easy installation and operation.

Automatic dampers should be placed in outdoor air inlets and in exhausts to prevent air entering when the fan is turned off.

Heating controls vary with the nature of the heating medium. Duct heaters are generally furnished with manufacturer-installed safety controls. Steam or hot-water heating coils require a motorized valve for heating control.

Time clock control can limit unnecessary HVAC operation. Unoccupied reset controls should be provided in conjunction with timed control.

Maintenance. To protect the initial investment and ensure maximum efficiency, maintenance of air-conditioning units in small stores should be provided by a reliable service company on a yearly basis. The maintenance agreement should clearly specify responsibility for filter replacements, lubrication, belts, coil cleaning, adjustment of controls, refrigeration cycle maintenance, replacement of refrigerant, pump repairs, electrical maintenance, winterizing, system start-up, and extra labor required for repairs.

Improving Operating Cost. Outdoor air economizers can reduce the operating cost of cooling in most climates. They are generally available as factory options or accessories with roof-mounted units. Increased exterior insulation generally reduces operating energy requirements and may in some cases allow the size of installed equipment to be reduced. Most codes now include minimum requirements for insulation and fenestration materials. The *Advanced Energy Design Guide for Small Retail Buildings* (ASHRAE 2006) provides additional energy savings suggestions.

DISCOUNT, BIG-BOX, AND SUPERCENTER STORES

Large discount, big-box, and supercenter stores attract customers with discount prices. These stores typically have high-bay fixture displays and usually store merchandise in the sales area. They feature a wide range of merchandise and may include such diverse areas as a food service area, auto service area, supermarket, pharmacy, bank, and garden shop. Some stores sell pets, including fish and birds. This variety of activity must be considered in designing the HVAC systems. The design and application suggestions for small stores also apply to discount stores.

Each specific area is typically treated as a traditional stand-alone facility would be. Conditioning outdoor air for all areas must be considered to limit the introduction of excess moisture that will migrate to the freezer aisles of a grocery area.

Hardware, lumber, furniture, etc., is also sold in big-box facilities. A particular concern in this type of facility is ventilation for merchandise and material-handling equipment, such as forklift trucks.

In addition, areas such as stockrooms, rest rooms, break rooms, offices, and special storage rooms for perishable merchandise may require separate HVAC systems or refrigeration.

Load Determination

Operating economics and the spaces served often dictate inside design conditions. Some stores may base summer load calculations on a higher inside temperature (e.g., 80°F db) but then set the thermostats to control at 72 to 75°F db. This reduces the installed equipment size while providing the desired inside temperature most of the time.

Heat gain from lighting is not uniform throughout the entire area. For example, jewelry and other specialty displays typically have lighting heat gains of 6 to 8 W per square foot of floor area, whereas the typical sales area has an average value of 2 to 4 W/ft^2. For stockrooms and receiving, marking, toilet, and rest room areas, a value of 2 W/ft^2 may be used. When available, actual lighting layouts rather than average values should be used for load computation.

ASHRAE *Standards* 62.1 and 90.1 provide data and population density information to be used for load determination. Chapter 33 of this volume has specific information on ventilation systems for kitchens and food service areas. Ventilation and outdoor air must be provided as required in ASHRAE *Standard* 62.1 and local codes.

Data on the heat released by special merchandising equipment, such as amusement rides for children or equipment used for preparing speciality food items (e.g., popcorn, pizza, frankfurters, hamburgers, doughnuts, roasted chickens, cooked nuts, etc.), should be obtained from the equipment manufacturers.

Design Considerations

Heat released by installed lighting is often sufficient to offset the design roof heat loss. Therefore, interior areas of these stores need cooling during business hours throughout the year. Perimeter areas, especially the storefront and entrance areas, may have highly variable heating and cooling requirements. Proper zone control and HVAC design are essential. Location of checkout lanes in the storefront or entrance areas makes proper environmental zone control even more important.

System Design. The important factors in selecting discount, big-box, and supercenter store air-conditioning systems are (1) installation costs, (2) floor space required for equipment, (3) maintenance requirements, (4) equipment reliability, and (5) simplicity of control. Roof-mounted units are most commonly used.

Air Distribution. The air supply for large interior sales areas should generally be designed to satisfy the primary cooling requirement. For perimeter areas, the variable heating and cooling requirements must be considered.

Because these stores require high, clear areas for display and restocking, air is generally distributed from heights of 14 ft and greater. Air distribution at these heights requires high discharge velocities in the heating season to overcome the buoyancy of hot air. This discharge air velocity creates turbulence in the space and induces airflow from the ceiling area to promote complete mixing. Space-mounted fans, and radiant heating at the perimeter, entrance heaters, and air curtains may be required.

Control. Because the controls are usually operated by personnel who have little knowledge of air conditioning, systems should be kept as simple as possible while still providing the required functions. Unitary equipment is typically available with manufacturer-supplied controls for easy installation and operation.

Automatic dampers should be placed in outdoor air inlets and in exhausts to prevent air entering when the fan is turned off.

Heating controls vary with the nature of the heating medium. Duct heaters are generally furnished with manufacturer-installed safety controls. Steam or hot-water heating coils require a motorized valve for heating control.

Time clock control can limit unnecessary HVAC operation. Unoccupied reset controls should be provided in conjunction with timed control.

Maintenance. Most stores do not employ trained HVAC maintenance personnel; they rely instead on service contracts with either the installer or a local service company. (See the section on Small Stores).

Improving Operating Cost. See the section on Small Stores.

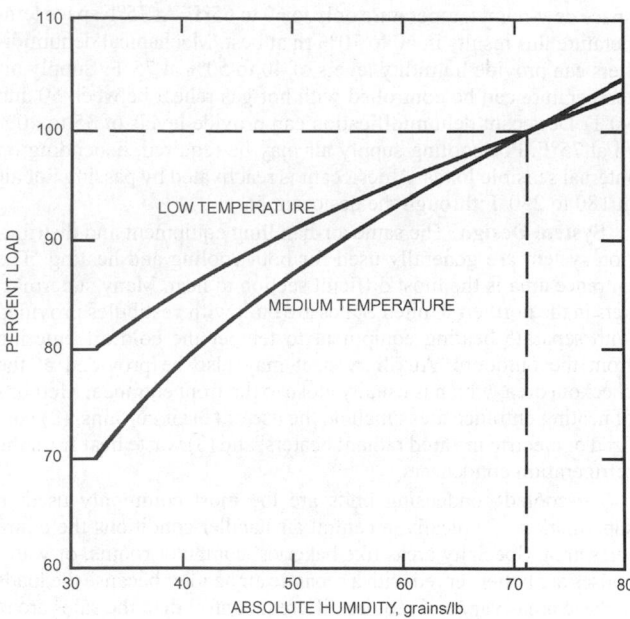

Fig. 1 Refrigerated Case Load Variation with Store Air Humidity

Table 1 Refrigerating Effect (RE) Produced by Open Refrigerated Display Fixtures

Display Fixture Types	RE on Building Per Unit Length of Fixture*			
	Latent Heat, Btu/h·ft	% Latent to Total RE	Sensible Heat, Btu/h·ft	Total RE, Btu/h·ft
Low-temperature (frozen food)				
Single-deck	38	15	207	245
Single-deck/double-island	70	15	400	470
2-deck	144	20	576	720
3-deck	322	20	1288	1610
4- or 5-deck	400	20	1600	2000
Ice cream				
Single-deck	64	15	366	430
Single-deck/double-island	70	15	400	470
Standard-temperature				
Meats				
Single-deck	52	15	298	350
Multideck	219	20	876	1095
Dairy, multideck	196	20	784	980
Produce				
Single-deck	36	15	204	240
Multideck	192	20	768	960

*These figures are general magnitudes for fixtures adjusted for average desired product temperatures and apply to store ambients in front of display cases of 72 to 74°F with 50 to 55% rh. Raising the dry bulb only 3 to 5°F and the humidity to 5 to 10% can increase loads (heat removal) 25% or more. Lower temperatures and humidities, as in winter, have an equally marked effect on lowering loads and heat removal from the space. Consult display case manufacturer's data for the particular equipment to be used.

SUPERMARKETS

Load Determination

Heating and cooling loads should be calculated using the methods outlined in Chapter 18 of the 2009 *ASHRAE Handbook—Fundamentals*. In supermarkets, space conditioning is required both for human comfort and for proper operation of refrigerated display cases. The air-conditioning unit should introduce a minimum quantity of outdoor air, either the volume required for ventilation based on ASHRAE *Standard* 62.1 or the volume required to maintain slightly positive pressure in the space, whichever is larger.

Many supermarkets are units of a large chain owned or operated by a single company. The standardized construction, layout, and equipment used in designing many similar stores simplify load calculations.

It is important that the final air-conditioning load be correctly determined. Refer to manufacturers' data for information on total heat extraction, sensible heat, latent heat, and percentage of latent to total load for display cases. Engineers report considerable fixture heat removal (case load) variation as the relative humidity and temperature vary in comparatively small increments. Relative humidity above 55% substantially increases the load; reduced absolute humidity substantially decreases the load, as shown in Figure 1. Trends in store design, which include more food refrigeration and more efficient lighting, reduce the sensible component of the load even further.

To calculate the total load and percentage of latent and sensible heat that the air conditioning must handle, the refrigerating effect imposed by the display fixtures must be subtracted from the building's gross air-conditioning requirements (Table 1).

Modern supermarket designs have a high percentage of closed refrigerated display fixtures. These vertical cases have large glass display doors and greatly reduce the problem of latent and sensible heat removal from the occupied space. The doors do, however, require heaters to minimize condensation and fogging. These heaters should cycle by automatic control.

For more information on supermarkets, see Chapter 15 in the 2010 *ASHRAE Handbook—Refrigeration*.

Design Considerations

Store owners and operators frequently complain about cold aisles, heaters that operate even when the outdoor temperature is above 70°F, and air conditioners that operate infrequently. These problems are usually attributed to spillover of cold air from open refrigerated display equipment.

Although refrigerated display equipment may cause cold stores, the problem is not excessive spillover or improperly operating equipment. Heating and air-conditioning systems must compensate for the effects of open refrigerated display equipment. Design considerations include the following:

- Increased heating requirement because of removal of large quantities of heat, even in summer.
- Net air-conditioning load after deducting the latent and sensible refrigeration effect. The load reduction and change in sensible-latent load ratio have a major effect on equipment selection.
- Need for special air circulation and distribution to offset the heat removed by open refrigerating equipment.
- Need for independent temperature and humidity control.

Each of these problems is present to some degree in every supermarket, although situations vary with climate and store layout. Methods of overcoming these problems are discussed in the following sections. Energy costs may be extremely high if the year-round air-conditioning system has not been designed to compensate for the effects of refrigerated display equipment.

Heat Removed by Refrigerated Displays. The display refrigerator not only cools a displayed product but also envelops it in a blanket of cold air that absorbs heat from the room air in contact with it. Approximately 80 to 90% of the heat removed from the room by vertical refrigerators is absorbed through the display opening. Thus, the open refrigerator acts as a large air cooler, absorbing heat from the room and rejecting it via the condensers outside the building. Occasionally, this conditioning effect can be greater than the design air-conditioning capacity of the store. The heat removed by the refrigeration equipment *must* be considered in the design of the air-conditioning and heating systems because this heat is being removed constantly, day and night, summer and winter, regardless of the store temperature.

Display cases increase the building heating requirement such that heat is often required at unexpected times. The following example illustrates the extent of this cooling effect. The desired store temperature is 75°F. Store heat loss or gain is assumed to be 15,000 Btu/h per °F of temperature difference between outdoor and store temperature. (This value varies with store size, location, and exposure.) The heat removed by refrigeration equipment is 190,000 Btu/h. (This value varies with the number of refrigerators.) The latent heat removed is assumed to be 19% of the total, leaving 81% or 154,000 Btu/h sensible heat removed, which cools the store 154,000/15,000 = 10°F. By constantly removing sensible heat from its environment, the refrigeration equipment in this store will cool the store 10°F below outdoor temperature in winter and in summer. Thus, in mild climates, heat must be added to the store to maintain comfort conditions.

The designer can either discard or reclaim the heat removed by refrigeration. If economics and store heat data indicate that the heat should be discarded, heat extraction from the space must be included in the heating load calculation. If this internal heat loss is not included, the heating system may not have sufficient capacity to maintain design temperature under peak conditions.

The additional sensible heat removed by the cases may change the air-conditioning latent load ratio from 32% to as much as 50% of the net heat load. Removing a 50% latent load by refrigeration alone is very difficult. Normally, it requires specially designed equipment with reheat or chemical adsorption.

Multishelf refrigerated display equipment requires 55% rh or less. In the dry-bulb temperature ranges of average stores, humidity in excess of 55% can cause heavy coil frosting, product zone frosting in low-temperature cases, fixture sweating, and substantially increased refrigeration power consumption.

A humidistat can be used during summer cooling to control humidity by transferring heat from the condenser to a heating coil in the airstream. The store thermostat maintains proper summer temperature conditions. Override controls prevent conflict between the humidistat and the thermostat.

The equivalent result can be accomplished with a conventional air-conditioning system by using three- or four-way valves and reheat condensers in the ducts. This system borrows heat from the standard condenser and is controlled by a humidistat. For higher energy efficiency, specially designed equipment should be considered. Desiccant dehumidifiers and heat pipes have also been used.

Humidity. Cooling from refrigeration equipment does not preclude the need for air conditioning. On the contrary, it increases the need for humidity control.

With increases in store humidity, heavier loads are imposed on the refrigeration equipment, operating costs rise, more defrost periods are required, and the display life of products is shortened. The dew point rises with relative humidity, and sweating can become so profuse that even nonrefrigerated items such as shelving superstructures, canned products, mirrors, and walls may sweat.

Lower humidity results in lower operating costs for refrigerated cases. There are three methods to reduce the humidity level: (1) standard air conditioning, which may overcool the space when the latent load is high and sensible load is low; (2) mechanical dehumidification, which removes moisture by lowering the air temperature to its dew point, and uses hot-gas reheat when needed to discharge at any desired temperature; and (3) desiccant dehumidification, which removes moisture independent of temperature, supplying warm air to the space unless postcooling is provided to discharge at any desired temperature.

Each method provides different dew-point temperatures at different energy consumption and capital expenditures. The designer should evaluate and consider all consequential tradeoffs. Standard air conditioning requires no additional investment but reduces the space dew-point temperature only to 60 to 65°F. At 75°F space temperature this results in 60 to 70% rh at best. Mechanical dehumidifiers can provide humidity levels of 40 to 50% at 75°F. Supply air temperature can be controlled with hot-gas reheat between 50 and 90°F. Desiccant dehumidification can provide levels of 35 to 40% rh at 75°F. Postcooling supply air may be required, depending on internal sensible loads. A desiccant is reactivated by passing hot air at 180 to 250°F through the desiccant base.

System Design. The same air-handling equipment and distribution system are generally used for both cooling and heating. The entrance area is the most difficult section to heat. Many supermarkets in the northern United States are built with vestibules provided with separate heating equipment to temper the cold air entering from the outdoors. Auxiliary heat may also be provided at the checkout area, which is usually close to the front entrance. Methods of heating entrance areas include the use of (1) air curtains, (2) gas-fired or electric infrared radiant heaters, and (3) waste heat from the refrigeration condensers.

Air-cooled condensing units are the most commonly used in supermarkets. Typically, a central air handler conditions the entire sales area. Specialty areas like bakeries, computer rooms, or warehouses are better served with a separate air handler because the loads in these areas vary and require different control than the sales area.

Most installations are made on the roof of the supermarket. If air-cooled condensers are located on the ground outside the store, they must be protected against vandalism as well as truck and customer traffic. If water-cooled condensers are used on the air-conditioning equipment and a cooling tower is required, provisions should be made to prevent freezing during winter operation.

Air Distribution. Designers overcome the concentrated load at the front of a supermarket by discharging a large portion of the total air supply into the front third of the sales area.

The air supply to the space with a standard air-conditioning system is typically 1 cfm per square foot of sales area. This value should be calculated based on the sensible and latent internal loads. The desiccant system typically requires less air supply because of its high moisture removal rate, typically 0.5 cfm per square foot. Mechanical dehumidification can fall within these parameters, depending on required dew point and suction pressure limitations.

Being denser, air cooled by the refrigerators settles to the floor and becomes increasingly colder, especially in the first 36 in. above the floor. If this cold air remains still, it causes discomfort and does not help to cool other areas of the store that need more cooling. Cold floors or areas in the store cannot be eliminated by the simple addition of heat. Reduction of air-conditioning capacity without circulation of localized cold air is analogous to installing an air conditioner without a fan. To take advantage of the cooling effect of the refrigerators and provide an even temperature in the store, the cold air must be mixed with the general store air.

To accomplish the necessary mixing, air returns should be located at floor level; they should also be strategically placed to remove the cold air near concentrations of refrigerated fixtures. Returns should be designed and located to avoid creating drafts. There are two general solutions to this problem:

- **Return Ducts in Floor.** This is the preferred method and can be accomplished in two ways. The floor area in front of the refrigerated display cases is the coolest area. Refrigerant lines are run to all of these cases, usually in tubes or trenches. If the trenches or tubes are enlarged and made to open under the cases for air return, air can be drawn in from the cold area (Figure 2). The air is returned to the air-handling unit through a tee connection to the trench before it enters the back room area. The opening through which the refrigerant lines enter the back room should be sealed.

  If refrigerant line conduits are not used, air can be returned through inexpensive underfloor ducts. If refrigerators have insufficient undercase air passage, the manufacturer should be

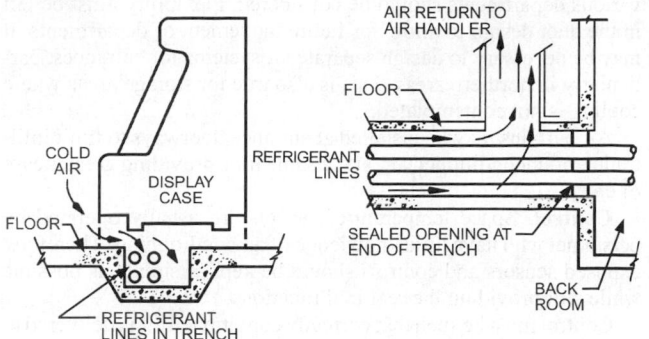

Fig. 2   Floor Return Ducts

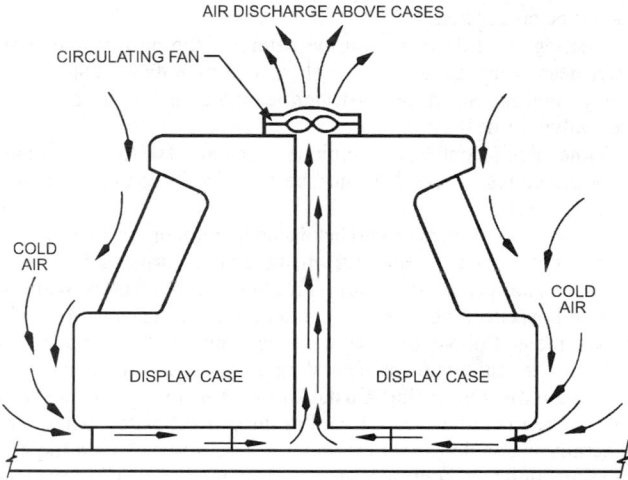

Fig. 3   Air Mixing Using Fans Behind Cases

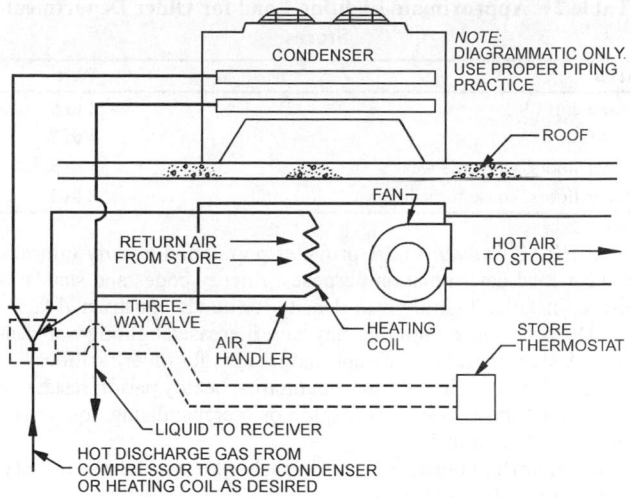

Fig. 4   Heat Reclaiming Systems

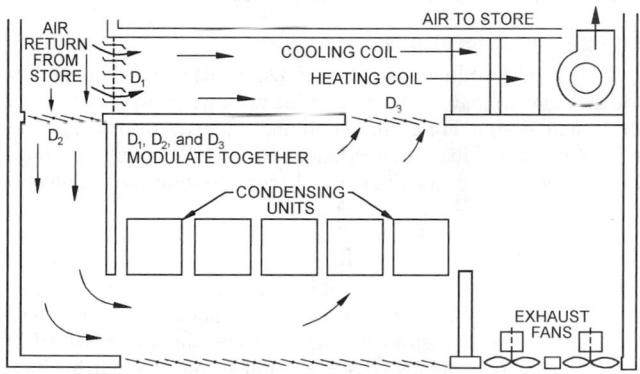

Fig. 5   Machine Room with Automatic Temperature Control Interlocked with Store Temperature Control

consulted. Often they can be raised off the floor approximately 1.5 in. Floor trenches can also be used as ducts for tubing, electrical supply, and so forth.

Floor-level return relieves the problem of localized cold areas and cold aisles and uses the cooling effect for store cooling, or increases the heating efficiency by distributing the air to areas that need it most.

• **Fans Behind Cases.** If ducts cannot be placed in the floor, circulating fans can draw air from the floor and discharge it above the cases (Figure 3). Although this approach prevents objectionable cold aisles in front of the refrigerated display cases, it does not prevent an area with a concentration of refrigerated fixtures from remaining colder than the rest of the store.

Control. Store personnel should only be required to change the position of a selector switch to start or stop the system or to change from heating to cooling or from cooling to heating. Control systems for heat recovery applications are more complex and should be coordinated with the equipment manufacturer.

Maintenance and Heat Reclamation. Most supermarkets, except large chains, do not employ trained maintenance personnel, but rather rely on service contracts with either the installer or a local service company. This relieves store management of the responsibility of keeping the air conditioning operating properly.

Heat extracted from the store and heat of compression may be reclaimed for heating cost saving. One method of reclaiming rejected heat is to use a separate condenser coil located in the air conditioner's air handler, either alternately or in conjunction with the main refrigeration condensers, to provide heat as required (Figure

4). Another system uses water-cooled condensers and delivers its rejected heat to a water coil in the air handler.

The heat rejected by conventional machines using air-cooled condensers may be reclaimed by proper duct and damper design (Figure 5). Automatic controls can either reject this heat to the outdoors or recirculate it through the store.

DEPARTMENT STORES

Department stores vary in size, type, and location, so air-conditioning design should be specific to each store. Essential features of a quality system include (1) an automatic control system properly designed to compensate for load fluctuations, (2) zoned air distribution to maintain uniform conditions under shifting loads, and (3) use of outdoor air for cooling during favorable conditions. It is also desirable to adjust inside temperature for variations in outdoor temperature. Although close control of humidity is not necessary, a properly designed system should operate to maintain relative humidity at 50% or below. This humidity limit eliminates musty odors and retards perspiration, particularly in fitting rooms.

Load Determination

Because the occupancy (except store personnel) is transient, inside conditions are commonly set not to exceed 78°F db and 50% rh at outdoor summer design conditions, and 70°F db at outdoor winter design conditions. Winter humidification is seldom used in store air conditioning.

Table 2 Approximate Lighting Load for Older Department Stores

Area	W/ft²
Basement	3 to 5
First floor	4 to 7
Upper floors, women's wear	3 to 5
Upper floors, house furnishings	2 to 3

ASHRAE *Standard* 62.1 provides population density information for load determination purposes. Energy codes and standards restrict installed lighting watt density for newly constructed facilities. However, older facilities may have increased lighting watt densities. Values in Table 2 are approximations for older facilities.

Other loads, such as those from motors, beauty parlor, restaurant equipment, and any special display or merchandising equipment, should be determined.

Minimum outdoor air requirements should be as defined in ASHRAE *Standard* 62.1 or local codes.

Paint shops, alteration rooms, rest rooms, eating places, and locker rooms should be provided with positive exhaust ventilation, and their requirements must be checked against local codes.

Design Considerations

Before performing load calculations, the designer should examine the store arrangement to determine what will affect the load and the system design. For existing buildings, actual construction, floor arrangement, and load sources can be surveyed. For new buildings, examination of the drawings and discussion with the architect or owner is required.

Larger stores may contain beauty parlors, food service areas, extensive office areas, auditoriums, warehouse space, etc. Some of these special areas may operate during hours in addition to the normal store-open hours. If present or future operation could be compromised by such a strategy, these spaces should be served by separate HVAC systems. Because of the concentrated load and exhaust requirements, beauty parlors and food service areas should be provided with separate ventilation and air distribution.

Future plans for the store must be ascertained because they can have a great effect on the type of air conditioning and refrigeration to be used.

System Design. Air conditioning systems for department stores may use unitary or central station equipment. Selection should be based on owning and operating costs as well as special considerations for the particular store, such as store hours, load variations, and size of load.

Large department stores have often used central-station systems consisting of air-handling units having chilled-water cooling coils, hot-water heating coils, fans, and filters. Some department stores now use large unitary units. Air systems must have adequate zoning for varying loads, occupancy, and usage. Wide variations in people loads may justify considering variable-volume air distribution systems. Water chilling and heating plants distribute water to the various air handlers and zones and may take advantage of some load diversity throughout the building.

Air-conditioning equipment should not be placed in the sales area; instead, it should be located in mechanical equipment room areas or on the roof whenever practicable. Ease of maintenance and operation must be considered in the design of equipment rooms and locations.

Many locations require provisions for smoke removal. This is normally accommodated through the roof and may be integrated with the HVAC system.

Air Distribution. All buildings must be studied for orientation, wind exposure, construction, and floor arrangement. These factors affect not only load calculations, but also zone arrangements and duct locations. In addition to entrances, wall areas with significant glass, roof areas, and population densities, the expected locations of various departments should be considered. Flexibility must be left in the duct design to allow for future movement of departments. It may be necessary to design separate air systems for entrances, particularly in northern areas. This is also true for storage areas where cooling is not contemplated.

Air curtains may be installed at entrance doorways to limit infiltration of unconditioned air, at the same time providing greater ease of entry.

Control. Space temperature controls are usually operated by personnel who have little knowledge of air conditioning. Therefore, exposed sensors and controls should be kept as simple as possible while still providing the required functions.

Control must be such that correctly conditioned air is delivered to each zone. Outdoor air intake should be automatically controlled to operate at minimum cost while providing required airflow. Partial or full automatic control should be provided for cooling to compensate for load fluctuations. Completely automatic refrigeration plants should be considered.

Heating controls vary with the nature of the heating medium. Duct heaters are generally furnished with manufacturer-installed safety controls. Steam or hot-water heating coils require a motorized valve for heating control.

Time clock control can limit unnecessary HVAC operation. Unoccupied reset controls should be provided in conjunction with timed control.

Automatic dampers should be placed in outdoor air inlets and in exhausts to prevent air entering when the fan is turned off.

Maintenance. Most department stores employ personnel for routine housekeeping, operation, and minor maintenance, but rely on service and preventive maintenance contracts for refrigeration cycles, chemical treatment, central plant systems, and repairs.

Improving Operating Cost. An outdoor air economizer can reduce the operating cost of cooling in most climates. These are generally available as factory options or accessories with the air-handling units or control systems. Heat recovery and desiccant dehumidification should also be analyzed.

CONVENIENCE CENTERS

Many small stores, discount stores, supermarkets, drugstores, theaters, and even department stores are located in convenience centers. The space for an individual store is usually leased. Arrangements for installing air conditioning in leased space vary. Typically, the developer builds a shell structure and provides the tenant with an allowance for usual heating and cooling and other minimum interior finish work. The tenant must then install an HVAC system. In another arrangement, developers install HVAC units in the small stores with the shell construction, often before the space is leased or the occupancy is known. Larger stores typically provide their own HVAC design and installation.

Design Considerations

The developer or owner may establish standards for typical heating and cooling that may or may not be sufficient for the tenant's specific requirements. The tenant may therefore have to install systems of different sizes and types than originally allowed for by the developer. The tenant must ascertain that power and other services will be available for the total intended requirements.

The use of party walls in convenience centers tends to reduce heating and cooling loads. However, the effect an unoccupied adjacent space has on the partition load must be considered.

REGIONAL SHOPPING CENTERS

Regional shopping centers generally incorporate an enclosed, heated and air-conditioned mall. These centers are normally owned by a developer, who may be an independent party, a financial institution, or one of the major tenants in the center.

Some regional shopping centers are designed with an open pedestrian mall between rows of stores. This open-air concept results in tenant spaces similar to those in a convenience center. Storefronts and other perimeters of the tenant spaces are exposed to exterior weather conditions.

Major department stores in shopping centers are typically considered separate buildings, although they are attached to the mall. The space for individual small stores is usually leased. Arrangements for installing air conditioning in the individually leased spaces vary, but are similar to those for small stores in convenience centers.

Table 3 presents typical data that can be used as check figures and field estimates. However, this table should not be used for final determination of load, because the values are only averages.

Design Considerations

The owner or developer provides the HVAC system for an enclosed mall. The regional shopping center may use a central plant or unitary equipment. The owner generally requires that the individual tenant stores connect to a central plant and includes charges for heating and cooling services. Where unitary systems are used, the owner generally requires that the individual tenant install a unitary system of similar design.

The owner may establish standards for typical heating and cooling systems that may or may not be sufficient for the tenant's specific requirements. Therefore, the tenant may have to install systems of different sizes than originally allowed for by the developer.

Leasing arrangements may include provisions that have a detrimental effect on conservation (such as allowing excessive lighting and outdoor air or deleting requirements for economizer systems). The designer of HVAC for tenants in a shopping center must be well aware of the lease requirements and work closely with leasing agents to guide these systems toward better energy efficiency.

Many regional shopping centers contain specialty food court areas that require special considerations for odor control, outdoor air requirements, kitchen exhaust, heat removal, and refrigeration equipment.

System Design. Regional shopping centers vary widely in physical arrangement and architectural design. Single-level and smaller centers usually use unitary systems for mall and tenant air conditioning; multilevel and larger centers usually use a central system. The owner sets the design of the mall and generally requires that similar systems be installed for tenant stores.

A typical central system may distribute chilled air to individual tenant stores and to the mall air-conditioning system and use variable-volume control and electric heating at the local use point. Some plants distribute both hot and chilled water. Some all-air systems also distribute heated air. Central plant systems typically provide improved efficiency and better overall economics of operation. Central systems may also provide the basic components required for smoke removal.

Air Distribution. Air distribution in individual stores should be designed for the particular space occupancy. Some tenant stores maintain a negative pressure relative to the public mall for odor control.

The total facility HVAC system should maintain a slight positive pressure relative to atmospheric pressure and a neutral pressure relative between most of the individual tenant stores. Exterior entrances should have vestibules.

Smoke management is required by many building codes, so air distribution should be designed to easily accommodate smoke control requirements.

Maintenance. Methods for ensuring the operation and maintenance of HVAC systems in regional shopping centers are similar to those used in department stores. Individual tenant stores may have to provide their own maintenance.

Improving Operating Cost. Methods for lowering operating costs in shopping centers are similar to those used in department stores. Some shopping centers have successfully used cooling tower heat exchanger economizers.

Central plant systems for regional shopping centers typically have lower operating costs than unitary systems. However, the initial cost of the central plant system is typically higher.

MULTIPLE-USE COMPLEXES

Multiple-use complexes are being developed in many metropolitan areas. These complexes generally combine retail facilities with other facilities such as offices, hotels, residences, or other commercial space into a single site. This consolidation of facilities into a single site or structure provides benefits such as improved land use; structural savings; more efficient parking; utility savings; and opportunities for more efficient electrical, fire protection, and mechanical systems.

Load Determination

The various occupancies may have peak HVAC demands that occur at different times of the day or year. Therefore, the HVAC loads of these occupancies should be determined independently. Where a combined central plant is considered, a block load should also be determined.

Design Considerations

Retail facilities are generally located on the lower levels of multiple-use complexes, and other commercial facilities are on upper levels. Generally, the perimeter loads of the retail portion differ from those of the other commercial spaces. Greater lighting and population densities also make HVAC demands for the retail space different from those for the other commercial space.

The differences in HVAC characteristics for various occupancies within a multiple-use complex indicate that separate air handling and distribution should be used for the separate spaces. However, combining the heating and cooling requirements of various facilities into a central plant can achieve a substantial saving. A combined central heating and cooling plant for a multiple-use complex also provides good opportunities for heat recovery, thermal storage, and other similar functions that may not be economical in a single-use facility.

Many multiple-use complexes have atriums. The stack effect created by atriums requires special design considerations for tenants and space on the main floor. Areas near entrances require special measures to prevent drafts and accommodate extra heating requirements.

System Design. Individual air-handling and distribution systems should be designed for the various occupancies. The central heating

Table 3 Typical Installed Cooling Capacity and Lighting Levels—Midwestern United States

Type of Space	Area per Unit of Installed Cooling, ft²/ton	Installed Cooling per Unit of Area, Btu·h/ft²	Lighting Density of Area, W/ft²	Annual Lighting Energy Use,[a] kWh/ft²
Dry retail[b]	367	33	4.0	16.2
Restaurant	136	88	2.0	8.1
Fast food				
food court tenant area	160	75	3.0	12.2
food court seating area	147	82	3.0	12.2
Mall common area	282	43	3.0	12.2[c]
Total	264	45	3.6	14.6

[a]Hours of operating lighting assumes 12 h/day and 6.5 days/week.
[b]Jewelry, high-end lingerie, and some other occupancy lighting levels are typically 6 to 8 W/ft² and can range to 11 W/ft². Cooling requirements for these spaces are higher.
[c]5.8 kWh/ft² for centers that shut off lighting during daylight, assuming 6 h/day and 6.2 days/week.

and cooling plant may be sized for the block load requirements, which may be less than the sum of each occupancy's demand.

Control. Multiple-use complexes typically require centralized control. It may be dictated by requirements for fire and smoke control, security, remote monitoring, billing for central facilities use, maintenance control, building operations control, and energy management.

REFERENCES

ASHRAE. 2006. *Advanced energy design guide for small retail buildings.*

ASHRAE. 2010. Ventilation for acceptable indoor air quality. ANSI/ASH-RAE *Standard* 62.1-2010.

ASHRAE. 2007. Energy standard for buildings except low-rise residential buildings. ANSI/ASHRAE/IES *Standard* 90.1-2007.

CHAPTER 3

COMMERCIAL AND PUBLIC BUILDINGS

THIS chapter contains technical, environmental, and design considerations to assist the design engineer in the proper application of HVAC systems and equipment for commercial and public buildings.

OFFICE BUILDINGS

General Design Considerations

Despite cyclical market fluctuations, office buildings are considered the most complex and competitive segments of real estate development. Survey data of 824,000 office buildings (EIA 2003) demonstrate the distribution of the U.S. office buildings by the numbers and the area, as shown in Table 1.

According to Gause (1998), an office building can be divided into the following categories:

Class. The most basic feature, class represents the building's quality by taking into account variables such as age, location, building materials, building systems, amenities, lease rates, etc. Office buildings are of three classes: A, B, and C. **Class A** is generally the most desirable building, located in the most desirable locations, and offering first-rate design, building systems, and amenities. **Class B** buildings are located in good locations, have little chance of functional obsolescence, and have reasonable management. **Class C** buildings are typically older, have not been modernized, are often functionally obsolete, and may contain asbestos. These low standards make Class C buildings potential candidates for demolition or conversion to another use.

Size and Flexibility. Office buildings are typically grouped into three categories: **high rise** (16 stories and above), **mid rise** (four to 15 stories), and **low rise** (one to three stories).

Location. An office building is typically in one of three locations: **downtown** (usually high rises), **suburban** (low- to mid-rise

buildings), or **business/industrial park** (typically one- to three-story buildings).

Floorplate (Floor Space Area). Size typically ranges from 18,000 to 30,000 ft² and averages from 20,000 to 25,000 ft².

Use and Ownership. Office buildings can be single tenant or multitenant. A single-tenant building can be owned by the tenant or leased from a landlord. From an HVAC&R systems standpoint, a single tenant/owner is more cautious considering issues such as life-cycle cost and energy conservation. In many cases, the systems are not selected based on the lowest first cost but on life-cycle cost. Sometimes, the developer may wish to select a system that allows individual tenants to pay directly for the energy they consume.

Building Features and Amenities. Examples include typically parking, telecommunications, HVAC&R, energy management, restaurants, security, retail outlets, health club, etc.

Typical areas that can be found in office buildings are:

Offices

- Offices: (private or semiprivate acoustically and/or visually).
- Conference rooms

Employee/Visitor Support Spaces

- Convenience store, kiosk, or vending machines
- Lobby: central location for building directory, schedules, and general information
- Atria or common space: informal, multipurpose recreation and social gathering space
- Cafeteria or dining hall
- Private toilets or restrooms
- Child care centers
- Physical fitness area
- Interior or surface parking areas

Administrative Support Spaces

- May be private or semiprivate acoustically and/or visually.

Operation and Maintenance Spaces

- General storage: for items such as stationery, equipment, and instructional materials.
- Food preparation area or kitchen
- Computer/information technology (IT) closets
- Maintenance closets
- Mechanical and electrical rooms

A well-designed and functioning HVAC system should provide the following:

- Comfortable and consistent temperature and humidity
- Adequate amounts of outdoor air at all time to satisfy ventilation requirements
- Remove odors and contaminates from circulated air

The major factors affecting sizing and selection of the HVAC systems are as follows:

Table 1   Data for U.S. Office Buildings

	Number of Buildings (Thousands)	Percent of Total Number of Buildings	Total Floor Space (Million ft²)	Percent of Total Floor Space
Total	824	100.0	12,208	100.0
1,001 to 5,000 ft²	503	61.0	1,382	11.32
5,001 to 10,000 ft²	127	15.4	938	7.68
10,001 to 25,000 ft²	116	14.1	1,887	15.46
25,001 to 50,000 ft²	43	5.2	1,506	12.34
50,001 to 100,000 ft²	17	2.1	1,209	9.90
100,001 to 200,000 ft²	11	1.3	1,428	11.70
200,001 to 500,00 ft²	5	0.6	1,493	12.23
>500,000 ft²	2	0.2	2,365	19.37

Source: EIA (2003).

The preparation of this chapter is assigned to TC 9.8, Large-Building Air-Conditioning Applications.

- Building size, shape and number of floors
- Amount of exterior glass
- Orientation, envelope
- Internal loads, occupants, lighting
- Thermal zoning (number of zones, private offices, open areas, etc.)

Office HVAC systems generally range from small, unitary, decentralized cooling and heating up to large systems comprising central plants (chillers, cooling towers, boilers, etc.) and large air-handling systems. Often, several types of HVAC systems are applied in one building because of special requirements such as continuous operation, supplementary cooling, etc. In office buildings, the class of the building also affects selection of the HVAC systems. For example, in a class A office building, the HVAC&R systems must meet more stringent criteria, including individual thermal control, noise, and flexibility; HVAC systems such as single-zone constant-volume, water-source heat pump, and packaged terminal air conditioners (PTACs) might be inapplicable to this class, whereas properly designed variable-air-volume (VAV) systems can meet these requirements.

Design Criteria

A typical HVAC design criteria covers parameters required for thermal comfort, indoor air quality (IAQ), and sound. Thermal comfort parameters (temperature and humidity) are discussed in ASHRAE *Standard* 55-2010 and Chapter 9 of the 2009 *ASHRAE Handbook—Fundamentals*. Ventilation and IAQ are covered by ASHRAE *Standard* 62.1-2010, the user's manual for that standard (ASHRAE 2010), and Chapter 16 of the 2009 *ASHRAE Handbook—Fundamentals*. Sound and vibration are discussed in Chapter 48 of this volume and Chapter 8 of the 2009 *ASHRAE Handbook—Fundamentals*.

Thermal comfort is affected by air temperature, humidity, air velocity, and mean radiant temperature (MRT), as well as nonenvironmental factors such as clothing, gender, age, and physical activity. These variables and how they correlate to thermal comfort can be evaluated by the *Thermal Comfort Tool CD* (ASHRAE 1997) in conjunction with ASHRAE *Standard* 55. General guidelines for temperature and humidity applicable for areas in office buildings are shown in Table 2.

All office, administration, and support areas need outdoor air for ventilation. Outdoor air is introduced to occupied areas and then exhausted by fans or exhaust openings, removing indoor air pollutants generated by occupants and any other building-related sources. ASHRAE *Standard* 62.1 is used as the basis for many building codes. To define the ventilation and exhaust design criteria, consult local applicable ventilation and exhaust standards. Table 3 provides

recommendations for ventilation design based on the ventilation rate procedure method and filtration criteria for office buildings.

Acceptable noise levels in office buildings are important for office personnel; see Table 4 and Chapter 48.

Load Characteristics

Office buildings usually include both peripheral and interior zone spaces. The peripheral zone extends 10 to 12 ft inward from the outer wall toward the interior of the building, and frequently has a large window area. These zones may be extensively subdivided. Peripheral zones have variable loads because of changing sun position and weather. These zones typically require heating in winter. During intermediate seasons, one side of the building may require

Table 3 Typical Recommended Design Criteria for Ventilation and Filtration for Office Buildings

Category	Ventilation and Exhaust[a,b]				
	Combined Outdoor Air (Default Value) cfm per Person	Occupant Density,[f] per 1000 ft²	Outdoor Air		Minimum Filtration Efficiency, MERV[c]
			cfm/ft²	cfm per Unit	
Office areas	17	5			6 to 8
Reception areas	7	30			6 to 8
Main entry lobbies	11	10			6 to 8
Telephone/data entry	6	60			6 to 8
Cafeteria	9	100			6 to 8
Kitchen[d,e]			0.7 (exhaust)		NA
Toilets				70 (exhaust)	NA
Storage[g]			0.12		1 to 4

Notes:
[a]Based on ASHRAE *Standard* 62.1-2010, Tables 6-1 and 6-4. For systems serving multiple zones, apply multiple-zone calculations procedure. If DCV is considered, see the section on Demand Control Ventilation (DCV).
[b]This table should not be used as the only source for design criteria. Governing local codes, design guidelines, ANSI/ASHRAE *Standard* 62.1-2010 and user's manual, (ASHRAE 2010) must be consulted.
[c]MERV = minimum efficiency reporting values, based on ASHRAE *Standard* 52.2-2007.
[d]See Chapter 33 for additional information on kitchen ventilation. For kitchenette use 0.3 cfm/ft²
[e]Consult local codes for kitchen exhaust requirements.
[f]Use default occupancy density when actual occupant density is not known.
[g]This recommendation for storage might not be sufficient when the materials stored have harmful emissions.

Table 4 Typical Recommended Design Guidelines for HVAC-Related Background Sound for Areas in Office Buildings

Category	Sound Criteria[a,b]	
	RC (N); QAI ≤ 5 dB	Comments
Executive and private office	25 to 35	
Conference rooms	25 to 35	
Teleconference rooms	≤25	
Open-plan office space	≤40	
	≤35	With sound masking
Corridors and lobbies	40 to 45	
Cafeteria	35 to 45	Based on service/support for hotels
Kitchen	35 to 45	Based on service/support for hotels
Storage	35 to 45	Based on service/support for hotels
Mechanical rooms	35 to 45	Based on service/support for hotels

Notes:
[a]Based on Table 1 in Chapter 48.
[b]RC (room criterion), QAI (quality assessment index) from Chapter 8 of the 2009 *ASHRAE Handbook—Fundamentals*.

Table 2 Typical Recommended Indoor Temperature and Humidity in Office Buildings

Area	Indoor Design Conditions		
	Temperature, °F/ Relative Humidity, %		
	Winter	Summer	Comments
Offices, conference rooms, common areas	70.0 to 74.0 20 to 30%	74.0 to 78.0 50 to 60%	
Cafeteria	70.0 to 73.5 20 to 30%	78.5 50%	
Kitchen	70.0 to 73.5	84.0 to 88.0	No humidity control
Toilets	72.0		Usually not conditioned
Storage	64.0		No humidity control
Mechanical rooms	61.0		Usually not conditioned

cooling, while another side requires heating. However, the interior zone spaces usually require a fairly uniform cooling rate throughout the year because their thermal loads are derived almost entirely from lights, office equipment, and people. Interior space conditioning is often by systems that have VAV control for low- or no-load conditions.

Most office buildings are occupied from approximately 8:00 AM to 6:00 PM; many are occupied by some personnel from as early as 5:30 AM to as late as 7:00 PM. Some tenants' operations may require night work schedules, usually not beyond 10:00 PM. Office buildings may contain printing facilities, information and computing centers, or broadcasting studios, which could operate 24 h per day. Therefore, for economical air-conditioning design, the intended uses of an office building must be well established before design development.

Occupancy varies considerably. In accounting or other sections where clerical work is done, the maximum density is approximately one person per 75 ft^2 of floor area. Where there are private offices, the density may be as little as one person per 200 ft^2. The most serious cases, however, are the occasional waiting rooms, conference rooms, or directors' rooms, where occupancy may be as high as one person per 20 ft^2.

The lighting load in an office building can be a significant part of the total heat load. Lighting and normal equipment electrical loads average from 1 to 5 W/ft^2 but may be considerably higher, depending on the type of lighting and amount of equipment. Buildings with computer systems and other electronic equipment can have electrical loads as high as 5 to 10 W/ft^2. The amount, size, and type of computer equipment anticipated for the life of the building should be accurately appraised to size the air-handling equipment properly and provide for future installation of air-conditioning apparatus.

Total lighting heat output from recessed fixtures can be withdrawn by exhaust or return air and thus kept out of space-conditioning supply air requirements. By connecting a duct to each fixture, the most balanced air system can be provided. However, this method is expensive, so the suspended ceiling is often used as a return air plenum with air drawn from the space to above the suspended ceiling.

Miscellaneous allowances (for fan heat, duct heat pickup, duct leakage, and safety factors) should not exceed 12% of the total load.

Building shape and orientation are often determined by the building site, but some variations in these factors can increase refrigeration load. Shape and orientation should therefore be carefully analyzed in the early design stages.

Design Concepts

The variety of functions and range of design criteria applicable to office buildings have allowed the use of almost every available air-conditioning system. Multistory structures are discussed here, but the principles and criteria are similar for all sizes and shapes of office buildings.

Attention to detail is extremely important, especially in modular buildings. Each piece of equipment, duct and pipe connections, and the like may be duplicated hundreds of times. Thus, seemingly minor design variations may substantially affect construction and operating costs. In initial design, each component must be analyzed not only as an entity, but also as part of an integrated system. This systems design approach is essential for achieving optimum results.

As discussed under General Design Considerations, there are several classes of office buildings, determined by the type of financing required and the tenants who will occupy the building. Design evaluation may vary considerably based on specific tenant requirements; it is not enough to consider typical floor patterns only. Many larger office buildings include stores, restaurants, recreational facilities, data centers, telecommunication centers, radio and television studios, and observation decks.

Built-in system flexibility is essential for office building design. Business office procedures are constantly being revised, and basic building services should be able to meet changing tenant needs.

The type of occupancy may have an important bearing on air distribution system selection. For buildings with one owner or lessee, operations may be defined clearly enough that a system can be designed without the degree of flexibility needed for a less well-defined operation. However, owner-occupied buildings may require considerable design flexibility because the owner will pay for all alterations. The speculative builder can generally charge alterations to tenants. When different tenants occupy different floors, or even parts of the same floor, the degree of design and operation complexity increases to ensure proper environmental comfort conditions to any tenant, group of tenants, or all tenants at once. This problem is more acute if tenants have seasonal and variable overtime schedules.

Certain areas may have hours of occupancy or design criteria that differ substantially from those of the office administration areas; such areas should have their own air distribution systems and, in some cases, their own heating and/or refrigeration equipment.

Main entrances and lobbies are sometimes served by a separate and self contained system because they buffer the outdoor atmosphere and the building interior. Some engineers prefer to have a lobby summer temperature 4 to 6°F above office temperature to reduce operating cost and temperature shock to people entering or leaving the building. In cases where lobbies or main entrances have longer (or constant) operation, a dedicated/self-contained HVAC system is recommended to allow turning off other building systems.

The unique temperature and humidity requirements of server rooms or computer equipment/data processing installations, and the fact that they often run 24 h per day for extended periods, generally warrant separate refrigeration and air distribution systems. Separate back-up systems may be required for data processing areas in case the main building HVAC system fails. Chapter 19 has further information.

The degree of air filtration required should be determined. Service cost and effect of air resistance on energy costs should be analyzed for various types of filters. Initial filter cost and air pollution characteristics also need to be considered. Activated charcoal filters for odor control and reduction of outdoor air requirements are another option to consider.

Providing office buildings with continuous 100% outdoor air (OA) is seldom justified, so most office buildings are designed to minimize outdoor air use, except during economizer operation. However, attention to indoor air quality may dictate higher levels of ventilation air. In addition, the minimum volume of outdoor air should be maintained in variable-volume air-handling systems. Dry-bulb- or enthalpy-controlled economizer cycles should be considered for reducing energy costs. Consult ASHRAE *Standard* 90.1-2010 for the proper air economizer system (dry-bulb or enthalpy). When an economizer cycle is used, systems should be zoned so that energy is not wasted by heating outdoor air. This is often accomplished by a separate air distribution system for the interior and each major exterior zone. A dedicated outdoor air system (DOAS) can be considered where the zones are served by in-room terminal systems (fan coils, induction unit systems, etc.) or decentralized systems [e.g., minisplit HVAC, water-source heat pump (WSHP)]. Because the outdoor air supply is relatively low in office buildings, air-to-air heat recovery is not cost effective; instead, a DOAS with enhanced cooling and dehumidification systems can be used.

These systems typically use hot-gas reheat or other means of free reheat (e.g., heat pipes, plate-frame heat exchangers). In hot, humid climates, these systems can significantly improve space conditions. By having a DOAS, the OA supply can be turned off during unoccupied hours (which can be significant in office buildings). In unoccupied mode, the in-room unit needs to maintain only the desired space conditions (for example night/weekend setback temperature).

High-rise office buildings have traditionally used perimeter fan-powered VAV terminals, induction, or fan-coil systems. Separate all-air systems have generally been used for the interior and/or the exterior for the fan-powered VAV perimeter terminals; modulated air diffusers and fan-powered perimeter unit systems have also been used. If variable-air-volume systems serve the interior, perimeters are usually served by variable-volume fan-powered terminals, typically equipped with hydronic (hot-water) or electric reheat coils. In colder climates, perimeter baseboard heaters are commonly applied. Baseboards are typically installed under windows to minimize the effect of the cold surface.

Many office buildings without an economizer cycle have a bypass multizone unit installed on each floor or several floors with a heating coil in each exterior zone duct. VAV variations of the bypass multizone and other floor-by-floor, all-air, or self-contained systems are also used. These systems are popular because of their low fan power and initial cost, and the energy savings possible from independent operating schedules between floors occupied by tenants with different operating hours.

Perimeter radiation or infrared systems with conventional, single-duct, low-velocity air conditioning that furnishes air from packaged air-conditioning units may be more economical for small office buildings. The need for a perimeter system, which is a function of exterior glass percentage, external wall thermal value, and climate severity, should be carefully analyzed.

A perimeter heating system separate from the cooling system is preferable, because air distribution devices can then be selected for a specific duty rather than as a compromise between heating and cooling performance. The higher cost of additional air-handling or fan-coil units and ductwork may lead the designer to a less expensive option, such as fan-powered terminal units with heating coils serving perimeter zones in lieu of a separate heating system. Radiant ceiling panels for perimeter zones are another option.

Interior space use usually requires that interior air-conditioning systems allow modification to handle all load situations. Variable-air-volume systems are often used. When using these systems, low-load conditions should be carefully evaluated to determine whether adequate air movement and outdoor air can be provided at the proposed supply air temperature without overcooling. Increases in supply air temperature tend to nullify energy savings in fan power, which are characteristic of VAV systems. Low-temperature air distribution for additional savings in transport energy is seeing increased use, especially when coupled with an ice storage system.

In small to medium-sized office buildings, air-source heat pumps or minisplit systems (cooling only, heat pump, or combination) such as variable refrigerant flow (VRF) may be chosen. VRF systems that can cool and heat simultaneously are available, and allow users to provide heating in perimeter zones and cooling in interior zones in a similar fashion to four-pipe fan coil (FPFC) systems. In larger buildings, water-source heat pump (WSHP) systems are feasible with most types of air-conditioning systems. Heat removed from core areas is rejected to either a cooling tower or perimeter circuits. The water-source heat pump can be supplemented by a central heating system or electrical coils on extremely cold days or over extended periods of limited occupancy. Removed excess heat may also be stored in hot-water tanks. Note that in-room systems (e.g., VRF, WSHP) might need a DOAS to provide the required outdoor air.

Many heat recovery or water-source heat pump systems exhaust air from conditioned spaces through lighting fixtures. This reduces required air quantities, and extends lamp life by providing a much cooler ambient operating environment.

Suspended-ceiling return air plenums eliminate sheet metal return air ductwork to reduce floor-to-floor height requirements. However, suspended-ceiling plenums may increase the difficulty of proper air balancing throughout the building. Problems often connected with suspended ceiling return plenums include

- Air leakage through cracks, with resulting smudges

- Tendency of return air openings nearest to a shaft opening or collector duct to pull too much air, thus creating uneven air motion and possible noise
- Noise transmission between office spaces

Air leakage can be minimized by proper workmanship. To overcome drawing too much air, return air ducts can be run in the suspended ceiling pathway from the shaft, often in a simple radial pattern. Ends of ducts can be left open or dampered. Generous sizing of return air grilles and passages lowers the percentage of circuit resistance attributable to the return air path. This bolsters effectiveness of supply-air-balancing devices and reduces the significance of air leakage and drawing too much air. Structural blockage can be solved by locating openings in beams or partitions with fire dampers, where required.

Systems and Equipment Selection

Selection of HVAC equipment and systems depends on whether the facility is new or existing, and whether it is to be totally or partially renovated. For minor renovations, existing HVAC systems are often expanded in compliance with current codes and standards with equipment that matches the existing types. For major renovations or new construction, new HVAC systems and equipment should be installed. When applicable, the remaining useful life of existing equipment and distribution systems should be considered.

HVAC systems and equipment energy use and associated life cycle costs should be evaluated. Energy analysis may justify new HVAC equipment and systems when an acceptable return on investment can be shown. The engineer must review all assumptions in the energy analysis with the owner. Other considerations for existing facilities are (1) whether the central plant is of adequate capacity to handle additional loads from new or renovated facilities; (2) age and condition of existing equipment, pipes, and controls; and (3) capital and operating costs of new equipment.

Chapter 1 of the 2008 *ASHRAE Handbook—HVAC Systems and Equipment* provides general guidelines on HVAC systems analysis and selection procedures. Although in many cases system selection is based solely on the lowest first cost, it is suggested that the engineer propose a system with the lowest life-cycle cost (LCC). LCC analysis typically requires hour-by-hour building energy simulation for annual energy cost estimation. Detailed first and maintenance cost estimates of proposed design alternatives, using sources such as R.S. Means (R.S. Means 2010a, 2010b), can also be used for the LCC analysis along with software such as BLCC 5.1 (FEMP 2003). Refer to Chapters 37 and 58 and the Value Engineering and Life-Cycle Cost Analysis section of this chapter for additional information.

System Types. HVAC systems for office buildings may be centralized, decentralized, or a combination of both. Centralized systems typically incorporate secondary systems to treat the air and distribute it. The cooling and heating medium is typically water or brine that is cooled and/or heated in a primary system and distributed to the secondary systems. Centralized systems comprise the following systems:

Secondary Systems

- Air handling and distribution (see Chapter 4 of the 2008 *ASHRAE Handbook—HVAC Systems and Equipment*)
- In-room terminal systems (see Chapter 5 of the 2008 *ASHRAE Handbook—HVAC Systems and Equipment*)
- Dedicated outdoor air systems (DOAS) with chilled water for cooling and hot water, steam, or electric heat for heating (for special areas when required)

Primary Systems

- Central cooling and heating plant (see Chapter 3 of the 2008 *ASHRAE Handbook—HVAC Systems and Equipment*)

Table 5   Applicability of Systems to Typical Office Buildings

Building Area/Stories	Cooling/Heating Systems								
	Centralized			Decentralized				Heating Only	
	SZ[a]	VAV/ Reheat	Fan Coil (Two-and Four-Pipe)	PSZ/SZ* Split/ VRF	PVAV/ Reheat	WSHP	Geothermal Heat Pump and Hybrid Geothermal Heat Pump	Perimeter Baseboard/ Radiators	Unit Heaters
<25,000 ft², one to three stories				X		X	X	X	Special areas
25,000 to 150,000 ft², one to five stories	X	X	X	X	X	X	X	X	Special areas
>150,000 ft², low rise and high rise	X	X	X			X	X	X	Special areas

*SZ = single zone	PSZ = packaged single zone	WSHP = water-source heat pump
VAV = variable-air-volume	PVAV = packaged variable-air-volume	VRF = variable refrigerant flow

More detailed information on systems selection by application can be found in Table 5.

Typical decentralized systems (dedicated systems serving a single zone, or packaged systems such as packaged variable air volume) include the following:

- Water-source heat pumps (WSHP), also known as water-loop heat pumps (WLHP)
- Geothermal heat pumps (e.g., groundwater heat pumps, ground-coupled heat pumps)
- Hybrid geothermal heat pumps (combination of groundwater heat pumps, ground-coupled heat pumps, and an additional heat rejection device) for cases with limited area for the ground-coupled heat exchanger or where it is economically justified
- Packaged single-zone and variable-volume units
- Light commercial split systems
- Minisplit and variable refrigerant flow (VRF) units

Chapters 2, 8, 48, and 49 of the 2008 *ASHRAE Handbook—HVAC Systems and Equipment* provide additional information on decentralized HVAC systems. Additional information on geothermal energy can be found in Chapter 34 of this volume.

Whereas small office buildings (<25,000 ft²) normally apply packaged unitary and split systems equipment, larger office buildings can use a combination of packaged, unitary, split, and/or centralized systems, or large packaged rooftop systems. The building class also must be considered during system selection.

Systems Selection by Application. Table 5 shows the applicability of several systems for office buildings.

Special Systems

The following is a list of systems that can be considered for special areas in office buildings. Chapter 57 of this volume, Chapter 6 of the 2008 *ASHRAE Handbook—HVAC Systems and Equipment*, and Skistad et al. (2002) provide additional information of these systems.

- Displacement ventilation
- Underfloor air distribution (UFAD)
- Active (induction) and passive chilled beams

Demand-Controlled Ventilation (DCV). Demand-controlled ventilation can reduce the operating cost of HVAC systems. Areas such as auditoriums, large conference rooms, and other spaces designed for large numbers of occupants and intermittent occupancy can use DCV. This approach is most cost effective when one dedicated air handling system serves each of these zones. Special attention is required when DCV is applied to VAV systems. In these cases, it is insufficient to use only one CO_2 sensor in the return air plenum of the central AHU, because the readings are the average of all the zones. To address properly DCV in a VAV system, a CO_2 sensor is required in every controlled zone.

Spatial Requirements

Total office building electromechanical space requirements vary tremendously based on types of systems planned; however, the average is approximately 8 to 10% of the gross area. Clear height required for fan rooms varies from approximately 10 to 18 ft, depending on the distribution system and equipment complexity. On office floors, perimeter fan-coil or induction units require approximately 1 to 3% of the floor area. Interior air shafts and pipe chases require approximately 3 to 5% of the floor area. Therefore, ducts, pipes, and equipment require approximately 4 to 8% of each floor's gross area.

Where large central units supply multiple floors, shaft space requirements depend on the number of fan rooms. In such cases, one mechanical equipment room usually furnishes air requirements for 8 to 20 floors (above and below for intermediate levels), with an average of 12 floors. The more floors served, the larger the duct shafts and equipment required. This results in higher fan room heights and greater equipment size and mass.

The fewer floors served by an equipment room, the greater the flexibility in serving changing floor or tenant requirements. Often, one mechanical equipment room per floor and complete elimination of vertical shafts requires no more total floor area than fewer larger mechanical equipment rooms, especially when there are many small rooms and they are the same height as typical floors. Equipment can also be smaller, although maintenance costs are higher. Energy costs may be reduced with more equipment rooms serving fewer areas, because equipment can be shut off in unoccupied areas, and high-pressure ductwork is not required. Equipment rooms on upper levels generally cost more to install because of rigging and transportation logistics.

In all cases, mechanical equipment rooms must be thermally and acoustically isolated from office areas.

Cooling Towers. Cooling towers can be the largest single piece of equipment required for air-conditioning systems. Cooling towers require approximately 1 ft² of floor area per 400 ft² of total building area and are 13 to 40 ft high. If towers are located on the roof, the building structure must be able to support the cooling tower and dunnage, full water load (approximately 120 to 150 lb/ft²), and seismic and wind load stresses.

Where cooling tower noise may affect neighboring buildings, tower design should include sound traps or other suitable noise baffles. This may affect tower space, mass of the units, and motor power. Slightly oversizing cooling towers can reduce noise and power consumption because of lower speeds and also the ability to reduce the condenser water temperature, which reduces cooling energy. The size increase may increase initial cost.

Cooling towers are sometimes enclosed in a decorative screen for aesthetic reasons; therefore, calculations should ascertain that the screen has sufficient free area for the tower to obtain its required air quantity and to prevent recirculation.

If the tower is placed in a rooftop well or near a wall, or split into several towers at various locations, design becomes more complicated, and initial and operating costs increase substantially. Also, towers should not be split and placed on different levels because hydraulic problems increase. Finally, the cooling tower should be built high enough above the roof so that the bottom of the tower and the roof can be maintained properly.

Special Considerations

Office building areas with special ventilation and cooling requirements include elevator machine rooms, electrical and telephone closets, electrical switchgear, plumbing rooms, refrigeration rooms, and mechanical equipment rooms. The high heat loads in some of these rooms may require air-conditioning units for spot cooling.

In larger buildings with intermediate elevator, mechanical, and electrical machine rooms, it is desirable to have these rooms on the same level or possibly on two levels. This may simplify horizontal ductwork, piping, and conduit distribution systems and allow more effective ventilation and maintenance of these equipment rooms.

An air-conditioning system cannot prevent occupants at the perimeter from feeling direct sunlight. Venetian blinds and drapes are often provided but seldom used. External shading devices (screens, overhangs, etc.) or reflective glass are preferable.

Tall buildings in cold climates experience severe stack effect. The extra amount of heat provided by the air-conditioning system in attempts to overcome this problem can be substantial. The following features help combat infiltration from stack effect:

- Revolving doors or vestibules at exterior entrances
- Pressurized lobbies or lower floors
- Tight gaskets on stairwell doors leading to the roof
- Automatic dampers on elevator shaft vents
- Tight construction of the exterior skin
- Tight closure and seals on all dampers opening to the exterior

TRANSPORTATION CENTERS

Major transportation facilities include transit facilities (rail transit, bus terminals), airports, and cruise terminals. Other areas that can be found in transportation centers are airplane hangars and freight and mail buildings, which can be treated as warehouse facilities. Bus terminals are covered partially in this chapter, but Chapter 15 provides more detail.

Airports

Airports are large, complex, and highly profitable enterprise. Most U.S. airports are public nonprofits, run directly by government entities or by government-created authorities known as airport or port authorities. There are three main types of airports:

- **International airports** serving over 20 million passengers a year.
- **National airports** serving between 2 to 20 million passengers a year.
- **Regional airport** serving up to 2 million passengers a year.

Airports typically consists the following:

- Runways and taxiing areas
- Air traffic control buildings
- Aircraft maintenance buildings and hangars
- Passenger terminals and car parking (open, partially open, or totally enclosed)
- Freight warehouses
- Lodging facilities (hotels)

In addition, support areas such as administration buildings, central utility plants, and transit facilities (rail and bus) are common in airport facilities.

Areas such as hangars, hotels, and car parking are not covered in this section. Information about hotels and parking garages can be found in Chapters 6 and 15, respectively. Warehouses are discussed in the next section of this chapter.

Most terminals can be divided into the following sections and subsections:

Departure

- Entrance concourse
- Check-in and ticketing
- Security and passports
- Shops, restaurants, banks, medical services, conference and business facilities, etc.
- Departure lounge
- Departure gates

Arrival

- Arrival lounge
- Baggage claim
- Customs, immigration, and passport control
- Exit concourse

Cruise Terminals

Cruise terminals typically have three main areas: departure/arrival concourse, ticketing, and baggage handling. These areas are open and large, and are designed to provide acceptable thermal comfort to the passenger during embarkation and debarkation.

Design Criteria

Transportation centers consist of a variety of areas, such as administration, large open areas, shops, and restaurants. Design criteria for these areas should be based on information on relevant chapters from this volume or ASHRAE *Standard* 62.1.

Load Characteristics

Airports, cruise terminals, and bus terminals operate on a 24 h basis, with a reduced schedule during late night and early morning hours. To better understand the load characteristics of these facilities, computer-based building energy modeling and simulation tools should be used; this chapter provides basic information and references for energy modeling. Given the dynamic nature of transportation facilities, well-supported assumptions of occupancy schedules should be established during the analysis process.

Airports. Terminal buildings consist of large, open circulating areas, one or more floors high, often with high ceilings, ticketing counters, and various types of stores, concessions, and convenience facilities. Lighting and equipment loads are generally average, but occupancy varies substantially. Exterior loads are, of course, a function of architectural design. The largest single problem often is thermal drafts created by large entranceways, high ceilings, and long passageways that have openings to the outdoors.

Cruise Terminals. Freight and passenger docks consist of large, high-ceilinged structures with separate areas for administration, visitors, passengers, cargo storage, and work. The floor of the dock is usually exposed to the outdoors just above the water level. Portions of the sidewalls are often open while ships are in port. In addition, the large ceiling (roof) area presents a large heating and cooling load. Load characteristics of passenger dock terminals generally require roof and floors to be well insulated. Occasional heavy occupancy loads in visitor and passenger areas must be considered.

Bus Terminals. These buildings consist of two general areas: the terminal, which contains passenger circulation, ticket booths, and stores or concessions; and the bus loading area. Waiting rooms and passenger concourse areas are subject to a highly variable occupant load: density may reach 10 ft^2 per person and, at extreme periods, 3 to 5 ft^2 per person. Chapter 15 has further information on bus terminals.

Design Concepts

Heating and cooling is generally centralized or provided for each building or group in a complex. In large, open-circulation areas of transportation centers, any all-air system with zone control can be used. Where ceilings are high, air distribution is often along the side wall to concentrate air conditioning where desired and avoid disturbing stratified air. Perimeter areas may require heating by radiation, a fan-coil system, or hot air blown up from the sill or floor grilles, particularly in colder climates. Hydronic perimeter radiant ceiling panels may be especially suited to these high-load areas.

Airports. Airports generally consist of one or more central terminal buildings connected by long passageways or trains to rotundas containing departure lounges for airplane loading. Most terminals have portable telescoping-type loading bridges connecting departure lounges to the airplanes. These passageways eliminate heating and cooling problems associated with traditional permanent passenger-loading structures.

Because of difficulties in controlling the air balance and because of the many outdoor openings, high ceilings, and long, low passageways (which often are not air conditioned), the terminal building (usually air conditioned) should be designed to maintain a substantial positive pressure. Zoning is generally required in passenger waiting areas, in departure lounges, and at ticket counters to take care of the widely variable occupancy loads.

Main entrances may have vestibules and windbreaker partitions to minimize undesirable air currents in the building.

Hangars must be heated in cold weather, and ventilation may be required to eliminate possible fumes (although fueling is seldom permitted in hangars). Gas-fired, electric, and low- and high-intensity radiant heaters are used extensively in hangars because they provide comfort for employees at relatively low operating costs.

Hangars may also be heated by large air blast heaters or floor-buried heated liquid coils. Local exhaust air systems may be used to evacuate fumes and odors that occur in smaller ducted systems. Under some conditions, exhaust systems may be portable and may include odor-absorbing devices.

Cruise Terminals. In severe climates, occupied floor areas may contain heated floor panels. The roof should be well insulated, and, in appropriate climates, evaporative spray cooling substantially reduces the summer load. Freight docks are usually heated and well ventilated but seldom cooled.

High ceilings and openings to the outdoors may present serious draft problems unless the systems are designed properly. Vestibule entrances or air curtains help minimize cross drafts. Air door blast heaters at cargo opening areas may be quite effective.

Ventilation of the dock terminal should prevent noxious fumes and odors from reaching occupied areas. Therefore, occupied areas should be under positive pressure, and cargo and storage areas exhausted to maintain negative air pressure. Occupied areas should be enclosed to simplify any local air conditioning.

In many respects, these are among the most difficult buildings to heat and cool because of their large open areas. If each function is properly enclosed, any commonly used all-air or large fan-coil system is suitable. If areas are left largely open, the best approach is to concentrate on proper building design and heating and cooling of the openings. High-intensity infrared spot heating is often advantageous (see Chapter 15 of the 2008 *ASHRAE Handbook—HVAC Systems and Equipment*). Exhaust ventilation from tow truck and cargo areas should be exhausted through the roof of the dock terminal.

Bus Terminals. Conditions are similar to those for airport terminals, except that all-air systems are more practical because ceiling heights are often lower, and perimeters are usually flanked by stores or office areas. The same systems are applicable as for airport terminals, but ceiling air distribution is generally feasible.

Properly designed radiant hydronic or electric ceiling systems may be used if high-occupancy latent loads are fully considered. This may result in smaller duct sizes than are required for all-air systems and may be advantageous where bus-loading areas are above the terminal and require structural beams. This heating and cooling system reduces the volume of the building that must be conditioned.

In areas where latent load is a concern, heating-only panels may be used at the perimeter, with a cooling-only interior system.

The terminal area air supply system should be under high positive pressure to ensure that no fumes and odors infiltrate from bus areas. Positive exhaust from bus loading areas is essential for a properly operating total system (see Chapter 15).

Systems and Equipment Selection

Given the size and magnitude of the systems in airports and cruise terminals, the selection of the HVAC equipment and systems tend to be centralized. Depending on the area served and site limitations, decentralized systems can also be considered for these specific cases.

Centralized systems typically incorporate secondary systems to treat and distribute air. The cooling and heating medium is typically water or brine that is cooled and/or heated in a primary system and distributed to the secondary systems. Centralized systems comprise the following systems:

Secondary Systems

- Air-handling and distribution (see Chapter 4 of the 2008 *ASHRAE Handbook—HVAC Systems and Equipment*)
- In-room terminal systems (see Chapter 5 of the 2008 *ASHRAE Handbook—HVAC Systems and Equipment*)
- Secondary systems such as variable-air-volume (VAV) are common in airports. Small, single-zone areas can be treated by constant-volume systems or fan coils.

Primary Systems

- Central cooling and heating plant (see Chapter 3 of the 2008 *ASHRAE Handbook—HVAC Systems and Equipment*)
- For cases where decentralized systems (dedicated systems serving a single zone or packaged systems such as packaged variable-air-volume) are:
- Water-source heat pumps (WSHP) (also known as water-loop heat pumps or WLHP)
- Packaged single-zone and variable-volume units
- Light commercial split systems
- Mini-split and variable-refrigerant-flow (VRF) units

Special Considerations

Airports. Filtering outdoor air with activated charcoal filters should be considered for areas subject to excessive noxious fumes from jet engine exhausts. However, locating outdoor air intakes as remotely as possible from airplanes is a less expensive and more positive approach.

Where ionization filtration enhancers are used, outdoor air quantities are sometimes reduced because the air is cleaner. However, care must be taken to maintain sufficient amounts of outdoor air for space pressurization.

Cruise Terminals. Ventilation design must ensure that fumes and odors from forklifts and cargo in work areas do not penetrate occupied and administrative areas.

Bus Terminals. The primary concerns with enclosed bus loading areas are health and safety problems, which must be handled by proper ventilation (see Chapter 15). Although diesel engine fumes are generally not as noxious as gasoline fumes, bus terminals often have many buses loading and unloading at the same time, and the total amount of fumes and odors may be disturbing.

In terms of health and safety, enclosed bus loading areas and automobile parking garages present the most serious problems. Three major problems are encountered, the first and most serious of which is emission of carbon monoxide (CO) by cars and oxides of nitrogen (NO_x) by buses, which can cause serious illness and possibly death. Oil and gasoline fumes, which may cause nausea and headaches and can create a fire hazard, are also of concern. The third issue is lack of air movement and the resulting stale atmosphere caused by increased CO content in the air. This condition may cause headaches or grogginess. Most codes require a minimum ventilation rate to ensure that the CO concentration does not exceed safe limits. Chapter 15 covers ventilation requirements and calculation procedures for enclosed vehicular facilities in detail.

All underground garages should have facilities for testing the CO concentration or should have the garage checked periodically. Problems such as clogged duct systems; improperly operating fans, motors, or dampers; or clogged air intake or exhaust louvers may not allow proper air circulation. Proper maintenance is required to minimize any operational defects.

WAREHOUSES AND DISTRIBUTION CENTERS

General Design Considerations

Warehouses can be defined as facilities that provide proper environment for the purpose of storing goods and materials. They are also used to store equipment and material inventory at industrial facilities. At times, warehouses may be open to the public. The buildings are generally not air conditioned, but often have sufficient heat and ventilation to provide a tolerable working environment. In many cases, associated facilities occupied by office workers, such as shipping, receiving, and inventory control offices, are air conditioned. Warehouses must be designed to accommodate the loads of materials to be stored, associated handling equipment, receiving and shipping operations and associated trucking, and needs of operating personnel. Types of warehouses include the following:

- **Heated and unheated general warehouses** provide space for bulk, rack, and bin storage, aisle space, receiving and shipping space, packing and crating space, and office and toilet space. As indicated some areas are typically equipped with small-decentralized air conditioning systems for the support personnel.
- **Conditioned general warehouses** are similar to heated and unheated general warehouses, but can provide space cooling to meet the stored goods' requirements.
- **Refrigerated warehouses** are designed to preserve the quality of perishable goods and general supply materials that require refrigeration. This includes freeze and chill spaces, processing facilities, and mechanical areas. For information on this type of warehouse, see Chapters 23 and 24 in the 2010 *ASHRAE Handbook—Refrigeration*.
- **Controlled humidity (CH) and dry-air storage warehouses** are similar to general warehouses except that they are constructed with vapor barriers and contain humidity control equipment to maintain humidity at desired levels. For additional information, see Chapter 29 of Harriman et al. (2001).
- **Specialty warehouses** includes storing facilities with special and in some instances strict requirements for temperature, humidity, cleanliness, minimum ventilation rates, etc. These facilities are typically conditioned to achieve the required space conditions. These warehouses can be found in industrial and manufacturing facilities or can be standalone buildings. Examples include
  - Pharmaceutical and life sciences facilities. Good manufacturing practices (GMP) may be required.
  - Liquid storage (fuel and nonpropellants), flammable and combustible storage, radioactive material storage, hazardous chemical storage, and ammunition storage.
  - Automated storage and retrieval systems (AS/RS), which are designed for maximum storage and minimum personnel on site. They are built for lower-temperature operation with minimal heat and light needed, but require a tall structure with extremely level floors. In some cases, specialty HVAC equipment is required for servers and other computer areas in AS/RS facility.

Features already now common in warehouse designs are higher bays, sophisticated materials-handling equipment, broadband connectivity access, and more distribution networks. A wide range of storage alternatives, picking alternatives, material-handling equipment, and software exist to meet the physical and operational requirements. Warehouse spaces must also be flexible to accommodate future operations and storage needs as well as mission changes.

Areas that can be found in warehouses and distribution centers include the following:

- Storage areas
- Office and administrative areas
- Loading docks
- Light industrial spaces
- Computer/server rooms

Other areas can be site specific.

Design Criteria

Design criteria (temperature, humidity, noise, etc.) for warehouses are space specific; the designer should refer to the relevant sections and chapters (e.g., the section on Office Buildings for office and administration areas). For conditioned storage areas, the special requirements of the product stores dictate the design conditions.

Outdoor air for ventilation of office, administration, and support areas should be based on local code requirements or ASHRAE *Standard* 62.1. For general warehouses where special ventilation or minimum ventilation rates are not specifically defined, *Standard* 62.1 can be used as the criterion for minimum outdoor air. To define the specific ventilation and exhaust design criteria, consult local applicable ventilation and exhaust standards. Table 6-1 of *Standard* 62.1 recommends 0.06 cfm/ft^2 of ventilation as a design criterion for warehouse ventilation, although this amount may be insufficient when stored materials have harmful emissions.

Load Characteristics

Given the variety of warehouses facilities, every case should be analyzed carefully. In general, internal loads from lighting, people, and miscellaneous sources are low. Most of the load is thermal transmission and infiltration. An air-conditioning load profile tends to flatten where materials stored are massive enough to cause the peak load to lag. In humid climates, special attention should be given to the sensible and latent loads' variations for cases where the warehouse or distribution center is conditioned or cooled by thermostatically controlled packaged HVAC equipment. In these climates, it is common to satisfy the space temperature (i.e., very low or no sensible cooling load), but, because of infiltration of moist air and without proper cooling (i.e., the cooling equipment is off), for space humidity to be unacceptably high.

Design Concepts

Most warehouses are only heated and ventilated. Forced-flow unit heaters may be located near entrances and work areas. Large central heating and ventilating units are also widely used. Even though comfort for warehouse workers may not be considered, it may be necessary to keep the temperature above 40°F to protect sprinkler piping or stored materials from freezing.

A building designed for adding air conditioning at a later date requires less heating and is more comfortable. For maximum summer comfort without air conditioning, excellent ventilation with noticeable air movement in work areas is necessary. Even greater comfort can be achieved in appropriate climates by adding roof-spray cooling. This can reduce the roof's surface temperature, thereby reducing ceiling radiation inside. Low- and high-intensity radiant heaters can be used to maintain the minimum ambient temperature throughout a facility above freezing. Radiant heat may also be used for occupant comfort in areas permanently or frequently open to the outdoors.

If the stored product requires specific inside conditions, an air-conditioning system must be added. Using only ventilation may help maintain lower space temperatures, but care should be taken

Table 6 Applicability of Systems to Typical Warehouse Building Areas

	Cooling/Heating Systems		Heating Only	
	Centralized	Decentralized	Heating and Ventilating Units	Local Unit Heaters
Warehouse Area	SZ	PSZ/SZ Split/VRF		
Storage areas	X	X	X	X
Office and administration areas	X	X		
Loading docks			X	X
Light industrial spaces	X	X	X	
Computer/server rooms	X (also CHW, CRAC Unit)	X (also DX, CRAC Unit)		

SZ = single zone   CHW = chilled water
PSZ = packaged single zone   CRAC = computer room air conditioning
VRF = variable refrigerant flow

not to damage the stored product with uncontrolled humidity. Direct or indirect evaporative cooling may also be an option.

Systems and Equipment Selection

Selection of HVAC equipment and systems depends on type of warehouse. As indicated previously the warehouse might need only heating, cooling in admin areas, or in some cases highly sophisticated HVAC system to address special ambient conditions required by the product stored in this warehouse. The same principles and procedures of selecting the HVAC systems described in the office building section of this chapter should be followed.

Selection by Application. Table 6 depicts typical systems applied for warehouse facilities. Centralized systems refer to warehouses where central chilled-water and/or hot-water/steam system is available. Decentralized systems are typically direct expansion (DX) systems with gas-fired heating or other available heating source.

Special systems are typically required when special ambient conditions have to be maintained: usual examples are desiccant dehumidification, mechanical dehumidification, and humidification.

In hot and humid climates, a combination of desiccant-based dehumidification equipment along with standard DX, packaged, single-zone units can be considered. This approach allows separation of sensible cooling load from latent load, thereby enhancing humidity control under most ambient conditions, reducing energy consumption, and allowing optimal equipment sizing and use.

Spatial Requirements

Total building electromechanical space requirements vary based on types of systems planned. Typically, the HVAC equipment can be roof mounted, slab, indoor, or ceiling mounted. Ductwork and air discharge plenums usually are not concealed; often, the systems are free discharge.

Special Considerations

Forklifts and trucks powered by gasoline, propane, and other fuels are often used inside warehouses. Proper ventilation is necessary to alleviate build-up of CO and other noxious fumes. Proper ventilation of battery-charging rooms for electrically powered forklifts and trucks is also required.

SUSTAINABILITY AND ENERGY EFFICIENCY

In the context of this chapter, sustainable refers to a building that minimizes the use of energy, water, and other natural resources and provides a healthy and productive indoor environment (e.g., IAQ, lighting, noise). The HVAC&R designer plays a major role in supporting the design team in designing, demonstrating, and verifying these goals, particularly in the areas of energy efficiency and indoor environmental quality (mainly IAQ).

Several tools and mechanisms are available to assist the HVAC&R designer in designing and demonstrating sustainable commercial facilities; see the References and Bibliography in this chapter, the Sustainability and Energy Efficiency section in Chapter 7, and Chapter 35 in the 2009 *ASHRAE Handbook—Fundamentals*.

Energy Considerations

Energy standards such as ANSI/ASHRAE/IESNA *Standard* 90.1-2007 and local energy codes should be followed for minimum energy conservation criteria. Note that additional aspects such as lighting, motors/drives, building envelope, and electrical services should also be considered for energy reduction. Energy procurement/supply-side opportunities should also be investigated for energy cost reduction. Table 14 in Chapter 7 depicts a list of selected energy conservation opportunities.

Energy Efficiency and Integrated Design Process for Commercial Facilities

The integrated design process (IDP) is vital for the design of high-performance commercial facilities. For background and details on integrated building design (IBD) and IDP, see Chapter 58.

Unlike the sequential design process (SDP), where the elements of the built solution are defined and developed in a systematic and sequential manner, IDP encourages holistic collaboration of the project team during the all phases of the project, resulting in cost-effective and environmentally friendly design. IDP responds to the project objectives, which typically are established by the owner before team selection. Typical IDP includes the following elements:

- Owner planning
- Predesign
- Schematic design
- Schematic design
- Design development
- Construction documents
- Procurement
- Construction
- Operation

Detailed information on each element can be found in Chapter 58.

In high-performance buildings, these objectives are typically sustainable sites, water efficiency, energy and atmosphere quality, materials and resources, and indoor environmental quality. These objectives are the main components of several rating systems. Energy use objectives are typically the following:

- Meeting minimum prescriptive compliance (mainly local energy codes, ASHRAE *Standard* 90.1, etc.)
- Improving energy performance by an owner-defined percentage beyond the applicable code benchmark
- Demonstrating minimum energy performance (or prerequisite) and enhanced energy efficiency (for credit points) for sustainable design rating [e.g., U.S. Green Building Council (USGBC) Leadership in Energy and Environmental Design (LEED®)]
- Providing a facility/building site energy density [e.g., energy utilization index (EUI)] less than an owner-defined target [e.g., U.S. Environmental Protection Agency (EPA) ENERGY STAR guidelines)
- Provide an owner-defined percentage of facility source energy from renewable energy

Building Energy Modeling

Building energy modeling has been one of the most important tools in the process of IDP and sustainable design. Building energy modeling uses sophisticated methods and tools to estimate the energy consumption and behavior of buildings and building systems. To better illustrate the concept of energy modeling, the difference between HVAC sizing and selection programs and energy modeling tools will be described.

Design, sizing selection, and equipment sizing tools are typically used for design and sizing of HVAC&R systems, normally at the **design** process. Examples include cooling/heating load calculations tools, ductwork design software, piping design programs, acoustics software, and selection programs for specific types of equipment. The results are used to specify cooling and heating capacities, airflow, water flow, equipment size, etc., during the design as defined and agreed by the client.

Energy modeling [also known as building modeling and simulation (BMS)] is used to model the building's thermal behavior and the building energy systems' performance. Unlike design tools, which are used for one design point (or for sizing), the building energy simulation analyzes the building and the building systems up to 8760 times: hour by hour, or even in smaller time intervals.

A building energy simulation tool is a computer program consisting of mathematical models of building elements and HVAC&R equipment. To run a building energy simulation, the user must define the building elements, equipment variables, energy cost, etc. The simulation engine then solves mathematical models of the building elements, equipment, and so on 8760 times (one for every hour), usually through a sequential process. Common results include annual energy consumption, annual energy cost, hourly profiles of cooling loads, and hourly energy consumption. Chapter 19 of the 2009 *ASHRAE Handbook—Fundamentals* provides detailed information on energy modeling techniques.

Typically, energy modeling tools must meet minimum requirements to be accepted by rating authorities such as USGBC or local building codes. The following is typical of minimum modeling capabilities:

- 8760 h per year
- Hourly variations in occupancy, lighting power, miscellaneous equipment power, thermostat set points, and HVAC system operation, defined separately for each day of the week and holidays
- Thermal mass effects
- Ten or more thermal zones
- Part-load performance curves for mechanical equipment
- Capacity and efficiency correction curves for mechanical heating and cooling equipment
- Air-side economizers with integrated control
- Design load calculations to determine required HVAC equipment capacities and air and water flow rates in accordance with generally accepted engineering standards and practice
- Tested according to ASHRAE *Standard* 140

Energy modeling is typically used in the following ways:

- As a decision support tool for energy systems in new construction and retrofit projects; that is, it allows analyzing several design alternatives and the selection of the optimal solution for a given criterion
- To provide vital information to the engineer about the building behavior and systems performance during design
- To demonstrates compliance with energy standards such as ASHRAE *Standard* 90.1 (energy cost budget method)
- To support USGBC LEED certification in the Energy and Atmosphere (EA) section
- To model existing buildings and systems and analyzing proposed energy conservation measures (ECMs) by performing calibrated simulation
- Demonstrate energy cost savings as part of measurements and verification (M&V) protocol (by using calibrated simulation procedures)

Energy modeling is used intensively in LEED for New Construction (USGBC 2009), Energy & Atmosphere (EA), prerequisite 2 (minimum energy performance), and for EA credit 1 (Optimize Energy Performance). An energy simulation program (with the requirements shown above) along with ASHRAE *Standard* 90.1 is used to perform whole-building energy simulation for demonstrating energy cost savings. The number of credits awarded is in correlation to the energy cost reduction.

Energy Benchmarking and Benchmarking Tools

Energy benchmarking is an important element of energy use evaluation and tracking. It involves comparing building normalized energy consumption to that of other similar buildings. The most common normalization factor is the gross floor area. Energy benchmarking is less accurate then other energy analysis methods, but can provide a good overall picture of relative energy use.

Relative energy use is commonly expressed by the energy utilization index (EUI), which is the energy use per unit area per year. Typically, EUI defined in terms of Btu/ft^2 per year. In some cases, the user is interested in energy cost benchmarking, which is known as the cost utilization index (CUI). CUI units are \$/ft^2 per year. It is important to differentiate between site EUI (actual energy used on site) and source EUI (energy used at the energy source); about two-thirds of the primary energy that goes into an electric power plant is lost in the process as waste heat.

One of the most important sources of energy benchmarking data is the Commercial Building Energy Consumption Survey (CBECS) by the U.S. Department of Energy's Energy Information Administration (DOE/EIA). Table 2 of Chapter 36 shows an example of EUI calculated based on DOE/EIA 2003 CBECS; the mean site EUI for mixed-use office space is 88 kBtu/ft^2·yr. Other EUIs for commercial facilities can be found in the same table.

Common energy benchmarking tools include the following:

- U.S. EPA ENERGY STAR Portfolio Manager (http://www.energystar.gov/benchmark)
- Lawrence Berkeley National Laboratory (LBNL) ARCH (http://poet.lbl.gov/arch/)
- CAL-ARCH for the state of California (http://poet.lbl.gov/cal-arch/)

Comprehensive information on energy benchmarking and available benchmarking tools can be found in Glazer (2006) and Chapter 36.

Combined Heat and Power in Commercial Facilities

Combined heat and power (CHP) plants and building cooling heating and power (BCHP) can be considered for large facilities such as large office buildings and campuses and airports when economically justifiable. Chapter 7 of the 2008 *ASHRAE Handbook—HVAC Systems and Equipment* and other sources such as Meckler and Hyman (2010), Orlando (1996), and Petchers (2002) provide information on CHP systems. Additional Internet-based sources for CHP include the following:

- U.S. EPA Combined Heat and Power (CHP) Partnership at http://www.epa.gov/chp/; procedures for feasibility studies and evaluations for CHP integration are available at http://www.epa.gov/chp/project-development/index.html
- U.S. Department of Energy, Energy Efficiency and Renewable Energy at http://www1.eere.energy.gov/industry/distributedenergy/
- The Midwest CHP Application Center (MAC) at http://www.chpcentermw.org
- A database of CHP installations can be found at http://www.eea-inc.com/chpdata/index.html.

Maor and Reddy (2008) show a procedure to optimize the size of the prime mover and thermally operated chiller for large office buildings by combining a building energy simulation program and CHP optimization tools.

CHP systems can be applied in large district cooling and heating facilities and infrastructure to use waste heat efficiently. The type of the prime mover is heavily dependent on the electrical and thermal loads, ability to use waste heat efficiently, and utility rates. Table 1 in Chapter 7 of the 2008 *ASHRAE Handbook—HVAC Systems and Equipment* provides information on the applicability of CHP.

Renewable Energy

Renewable energy (RE) technologies, including solar, wind, and biomass, can be considered when applicable and economically justifiable. Renewable energy use can add LEED credits (USGBC 2009) under Energy and Atmosphere (credit 2), depending on the percentage of renewable energy used.

Given the increased number and popularity of solar systems, only these systems will be discussed in this chapter. Geothermal energy is also considered to be renewable energy; these systems are discussed earlier in this chapter, and in more detail in Chapter 34.

Solar/Photovoltaic. Photovoltaic (PV) technology is the direct conversion of sunlight to electricity using semiconductor devices called solar cells. Photovoltaic are almost maintenance-free and seem to have a long lifespan. Given the longevity, no pollution, simplicity, and minimal resources, this technology is highly sustainable, and the proper financing mechanisms can make this system economically justifiable.

Airport facilities can be considered good candidates for PV technology for the following reasons:

• Large, low-rise buildings with available roof for PV collectors
• Little or no shading
• Large open area (open areas, parking lots, etc.)
• Hours and seasons of operation

The most common technology in use today is single-crystal PV, which uses wafers of silicon wired together and attached to a module substrate. Thin-film PV, such as amorphous silicon technology, uses silicon and other chemicals deposited directly on a substrate such as glass or flexible stainless steel. Thin films promise lower cost per unit area, but also have lower efficiency and produce less electricity per unit area compared to single-crystal PVs. Typical values for dc electrical power generation are around 6 W/ft^2 for thin film and up to 15 W/ft^2 for single-crystal PV.

PV panels produce direct current, not the alternating current used to power most building equipment. Direct current is easily stored in batteries; an inverter is required to transform the direct current to alternating current. The costs of an inverter and of reliable batteries to store electricity increase the overall cost of a system, which is usually $5 to $7/W (Krieth and Goswami 2007).

Another option is concentrated PV (CPV). CPV uses high-concentration lenses or mirrors to focus sunlight onto miniature solar cells. CPV systems must track the sun to keep the light focused on the PV cells. The main advantage of this system is higher efficiency than other technologies. Reliability, however, is an important technical challenge for this emerging technology: the systems generally require highly sophisticated tracking devices.

Being able to transfer excess electricity generated by a photovoltaic system back into the utility grid can be advantageous. Most utilities are required to buy excess site-generated electricity back from the customer. In many states, public utility commissions or state legislatures have mandated **netmetering**, which means that utilities pay and charge equal rates regardless of which way the electricity flows. A good source of rebates and incentives in the United States for solar systems and other renewable technologies is the Database of State Incentives for Renewable and Efficiency (DSIRE), available at http://www.dsireusa.org/ (North Carolina State University 2011). DSIRE is a comprehensive source of information on state, local, utility, and federal incentives and policies that promote renewable energy and energy efficiency, as well as state requirements for licensed solar contractors.

PV systems should be integrated during the early stages of the design. In existing facilities, a licensed contractor can be employed for a turnkey project, which includes sizing, analysis, economic analysis, design documents, specifications, permits, and documentation for incentives.

Available tools for analysis during design and installation of PV systems include the following:

• PVsyst, a PC software package for the study, sizing, simulation and data analysis of complete PV systems (University of Geneva 2010) at http://www.pvsyst.com/5.2/index.php
• Hybrid Optimization Modeling Software (HOMER 2010), a program for analyzing and optimizing renewable energy technologies (http://www.homerenergy.com/)
• RETScreen (Natural Resources Canada 2010), a free decision support tool (which supports 35 languages) developed to help evaluate energy production and savings, costs, emission reductions, financial viability, and risk for various types of renewable energy technologies, at http://www.retscreen.net/ang/home.php
• eQUEST (Quick Energy Simulation Tool), a full-scale building energy simulation program capable of performing a complete building energy evaluation, at http://www.doe2.com/

Financing PV projects in the public sector can be more complex because of tax exemptions and efficient allocation of public funds and leverage incentives. The primary mechanism for financing public-sector PV projects is a third-party ownership model, which allows the public sector take advantage of all the federal tax and other incentives without large up-front outlay of capital. The public sector does not own the solar PV, but only hosts it on its property. The cost of electrical power generated is then secured at a fixed rate, which is lower than the retail price for 15 to 25 years. Cory et al. (2008) discuss solar photovoltaic financing for the public sector in detail.

Solar/Thermal. Some commercial facilities can consider active thermal solar heating systems. Solar hot-water systems usually can reduce the energy required for service hot water. Solar heating design and installation information can be found in ASHRAE (1988, 1991). Chapter 36 of the 2008 *ASHRAE Handbook—HVAC Systems and Equipment* and Krieth and Goswami (2007) are good sources of information for design and installation of active solar systems, as are Web-based sources such as U.S. Department of Energy's Energy Efficiency and Renewable Energy page at http://www.eere.energy.gov/topics/solar.html.

Value Engineering and Life-Cycle Cost Analysis

Use of value engineering (VE) and life-cycle cost analysis (LCCA) studies is growing in all types of construction and as part of the integrated design process (IDP). VE and LCCA are logical, structured, systematic processes used as decision support tools to achieve overall cost reduction, but they are two distinct tools (Anderson et al. 2004).

Value engineering refers to a process where the project team examines the proposed design components in relation to the project objectives and requirements. The intent is to provide essential functions, while exploring cost savings opportunities through modification or elimination of nonessential design elements. Examples are alternative systems, substitute equipment, etc. VE typically includes seven steps, as shown in Figure 11 of Chapter 7.

Life-cycle cost analysis is used as part of VE to evaluate design alternatives (e.g., alternative systems, equipment substitutions) that meet the facility design criteria with reduced cost or increased value over the life of the facility or system.

The combination of VE and LCCA is suitable for public facilities, which are often government funded and intended for longer lifespans than commercial facilities. Unfortunately, these tools often are not included in the early stages of the design, which results in a last-minute effort to reduce cost and stay within the budget, compromising issues such as energy efficiency and overall value of the facility. To avoid this, VE and LCCA should be deployed in the early stages of the project.

LCCA is recommended as part of any commercial building construction for economic evaluation. Chapters 37 and 58 discuss LCCA in detail. Other methodologies such as simple payback should be avoided because of inaccuracies and the need to take into account the time value of money. Life-cycle cost is more accurate because it captures all the major initial costs associated with each item, the costs occurring during the life of the system, and the value of money for the entire life of the system.

COMMISSIONING AND RETROCOMMISSIONING

Commissioning (Cx) is a quality assurance process for buildings from predesign through design, construction, and operations. It involves achieving, verifying, and documenting the performance of each system to meet the building operational needs. Given the growing demand for enhanced indoor air quality, thermal comfort, noise, etc., in commercial facilities and the application of equipment and systems such as DOAS, EMS, and occupancy sensors, it is important to follow the commissioning process as described in Chapter 43 and ASHRAE *Guideline* 0-2005. The technical requirements for the commissioning process are described in detail in ASHRAE *Guideline* 1.1-2007. Another source is ACG (2005). Proper commissioning ensures fully functional systems that can be operated and maintained properly throughout the life of the building. Although commissioning activities should be implemented by qualified commissioning professional [commissioning authority (CA)], it is important for other professionals to understand the basic definitions and processes in commissioning, such as the following:

- Owner project requirements (OPR), which is a written document that details the functional requirements of the project and the expectations of how it will be used and operated.
- Commissioning refers to a quality-focused process for enhancing the delivery of a project. The process focuses upon verifying and documenting that the facility and all its systems and assemblies are planned, installed, tested, and maintained to meet the OPR.
- Recommissioning is an application of the commissioning process to a project that has been delivered using the commissioning process.
- Retrocommissioning is applied to an existing facility that was not previously commissioned.
- Ongoing commissioning is a continuation of the commissioning process well into the occupancy and operation phase.

Commissioning: New Construction

Table 7 shows the phases of commissioning a new building, as defined by ASHRAE *Guideline* 1.1.

ACG 2005 refers to the following HVAC commissioning processes for new construction:

- Comprehensive HVAC commissioning starts at the inception of a building project from the predesign phase till postacceptance)
- Construction HVAC commissioning occurs during construction, acceptance, and postacceptance (predesign and design phases are not included in this process)

Commissioning is an important element in LEED for new construction (USGBC 2009). As a prerequisite (Energy and Atmosphere, prerequisite 1), commissioning must verify that the project's energy-related systems are installed and calibrated, and perform according to the OPR, BOD, and the construction document. Additional credits (Energy and Atmosphere, credit 3—Enhanced

Commissioning) can be obtained by applying the entire commissioning process (or the comprehensive HVAC commissioning as described previously.

Commissioning: Existing Buildings

HVAC commissioning in existing buildings covers the following:

- Recommissioning
- Retrocommissioning (RCx)
- HVAC systems modifications

Although the methodology for both is identical, there is a difference between recommissioning and retrocommissioning. Recommissioning is initiated by the building owner and seeks to resolve ongoing problems or to ensure that systems continue to meet the facility's requirements. There are can be changes in the building's occupancy or design strategies, outdated equipment, degraded equipment efficiency, occupant discomfort, and IAQ problems that can initiate the need for recommissioning. Typical recommissioning activities are shown in Table 8.

Commissioning is also an important element in existing buildings. USGBC (2009), *LEED for Existing Buildings & Operation Maintenance* awards up to six credits for commissioning systems in existing buildings in the Energy and Atmosphere (EA) section

HVAC systems modifications can vary from minor modification to HVAC systems up to complete reconstruction of all or part of building HVAC system. The process for this type of project should follow the process described previously for new construction.

Table 7 Key Commissioning Activities for New Building

Phase	Key Commissioning Activities
Predesign	Preparatory phase in which the OPR is developed and defined.
Design	OPR is translated into construction documents, and basis of design (BOD) document is created to clearly convey assumptions and data used to develop the design solution. See informative annex k of ASHRAE *Guideline* 1.1-2007 for detailed structure and an example of a typical bod.
Construction	The commissioning team is involved to ensure that systems and assemblies installed and placed into service meet the OPR.
Occupancy and operation*	The commissioning team is involved to verify ongoing compliance with the OPR.

Source: ASHRAE *Guideline* 1.1-2007.
*Also known as acceptance and post-acceptance in ACG (2005).

Table 8 Key Commissioning Activities for Existing Building

Phase	Key Commissioning Activities
Planning	Define HVAC goals
	Select a commissioning team
	Finalize recommissioning scope
	Documentation and site reviews
	Site survey
	Preparation of recommissioning plan
Implementation	Hire testing and balancing (TAB) agency and automatic temperature control (ATC) contractor
	Document and verify tab and controls results
	Functional performance tests
	Analyze results
	Review operation and maintenance (O&M) practices
	O&M instruction and documentation
	Complete commissioning report

Source: ACG (2005).

SEISMIC AND WIND RESTRAINT CONSIDERATIONS

Seismic bracing of HVAC equipment should be considered. Wind restraint codes may also apply in areas where tornados and hurricanes necessitate additional bracing. This consideration is especially important if there is an agreement with local officials to use the facility as a disaster relief shelter. See Chapter 55 for further information.

REFERENCES

ACG. 2005. *ACG commissioning guideline*. AABC Commissioning Group. Washington, D.C. Available from http://www.commissioning.org/commissioningguideline/.

Anderson, D.R., J. Macaluso, D.J. Lewek, and B.C. Murphy. 2004. *Building and renovating schools: Design, construction management, cost control*. Reed Construction Data, Kingston, MA.

ASHRAE. 1988. *Active solar heating systems design manual*.

ASHRAE. 1991. *Active solar heating systems installation manual*.

ASHRAE. 1997. *Thermal comfort tool CD*.

ASHRAE. 2008. *Advanced energy design guide for small office buildings*. Available from http://www.ashrae.org/publications/page/1604.

ASHRAE. 2008. *Advanced energy design guide for small warehouse and self storage buildings*. Available from http://www.ashrae.org/publications/page/1604.

ASHRAE. 2010. *Standard 62.1-2010 user's manual*.

ASHRAE. 2005. The commissioning process. *Guideline* 0-2005.

ASHRAE. 2007. HVAC&R technical requirements for the commissioning process. *Guideline* 1.1-2007

ASHRAE. 2007. Method of testing general ventilation air cleaning devices for removal efficiency by particle size. *Standard* 52.2-2007.

ASHRAE. 2010. Thermal environmental conditions for human occupancy. ANSI/ASHRAE *Standard* 55-2010.

ASHRAE. 2007. Ventilation for acceptable indoor air quality. ANSI/ASHRAE *Standard* 62.1-2007.

ASHRAE. 2007. Energy standard for buildings except low-rise residential buildings. ANSI/ASHRAE/IESNA *Standard* 90.1-2007

ASHRAE. 2009. Standard for the design of high-performance green buildings except low-rise residential buildings. ANSI/ASHRAE/USGBC/IES *Standard* 189.1-2009.

Cory, K., J. Coughlin, and C. Coggeshall. 2008. Solar photovoltaic financing: Deployment on public property by state and government. NREL *Technical Report* NREL/TP-670-43115.

EIA. 2003. *2003 CBECS details tables. U.S. Energy Information Administration*, Washington, D.C. http://www.eia.doe.gov/emeu/cbecs/cbecs2003/detailed_tables_2003/detailed_tables_2003.html.

FEMP. 2003. *BLLC 5.1: Building life cycle cost*. Federal Energy Management Program, Washington, D.C. http://www.eere.energy.gov/femp.

Gause, J.A., M.J. Eppli, M.E. Hickok, and W. Ragas. 1998. *Office development handbook*, 2nd ed. Urban Land Institute, Washington, D.C.

Glazer, J. 2006. Evaluation of building performance rating protocols. ASHRAE Research Project RP-1286, *Final Report*.

Harriman, L.G., G.W. Brundrett, and R. Kittler. 2001. *Humidity control design guide for commercial and institutional buildings*. ASHRAE.

Homer. 2010. *HOMER: Energy modeling software for hybrid renewable energy systems*. HOMER ENERGY LLC, Boulder, CO. http://www.homerenergy.com/index.asp.

Kriethm F. and Y. Goswami. 2007. *Handbook of energy efficiency and renewable energy*. CRC Press, Boca Raton, FL.

Maor, I. and T.A. Reddy. 2008. Near-optimal scheduling control of combined heat and power systems for buildings, Appendix E. ASHRAE Research Project RP-1340, *Final Report*.

Meckler, M. and L. Hyman. 2010. *Sustainable on-site CHP systems: Design, construction, and operations*. McGraw-Hill.

Natural Resources Canada. 2010. *RETScreen international*. http://www.retscreen.net/ang/home.php.

North Carolina State University. 2011. *Database of state incentives for renewables and efficiency*. http://www.dsireusa.org/.

Orlando, J.A. 1996. *Cogeneration design guide*. ASHRAE.

Petchers, N. 2002. *Combined heating, cooling & power handbook: Technologies & applications*. Fairmont Press, Lilburn, GA.

R.S. Means. 2010a. *Means mechanical cost data*. R.S. Means Company, Kingston, MA.

R.S. Means. 2010b. *Means maintenance and repair cost data*. R.S. Means Company, Kingston, MA.

Skistad, H., E. Mundt, P.V. Nielsen, K. Hagstrom, and J. Railio. 2002. *Displacement ventilation in non-industrial premises*. Federation of European Heating and Air-Conditioning Associations (REHVA), Brussels.

USGBC. 2009. *LEED-2009 for new construction and major renovations*. U.S. Green Building Council, Washington, D.C.

USGBC. 2009. *LEED-2009 for existing buildings & operation maintenance*. U.S. Green Building Council, Washington, D.C.

BIBLIOGRAPHY

ASHRAE. 2004. *Advanced energy design guide for small office buildings*.

ASHRAE. 2008. *Advanced energy design guide for small warehouses and self storage buildings*.

ASHRAE. 2010. *ASHRAE greenguide*, 3rd ed.

ASHRAE. 2010. *Standard 90.1-2010 user's manual*.

ASHRAE. 2006. Weather data for building design standards. ANSI/ASHRAE *Standard* 169-2006

ASHRAE. 2009. Standard for the design of high-performance green buildings. ANSI/ASHRAE/USGBC/IES *Standard* 189-2009.

Chen, Q. and L. Glicksman. 2002. *System performance evaluation and design guidelines for displacement ventilation*. ASHRAE.

Dell'Isola, A.J. 1997. *Value engineering: Practical applications*. R.S. Means Company, Kingston, MA.

Ebbing, E. and W. Blazier, eds. 1998. *Application of manufacturers' sound data*. ASHRAE.

Edwards, B. 2005. *The modern airport terminal*, 2nd ed. Spon Press, New York.

Harriman, L.G. and J. Judge. 2002. Dehumidification equipment advances. *ASHRAE Journal* 44(8):22-27.

Kavanaugh, S.P. and K. Rafferty. 1997. *Ground-source heat pumps*. ASHRAE.

Mumma, S.A. 2001. Designing dedicated outdoor air systems. *ASHRAE Journal* 43(5):28-31.

Schaffer, M.E. 1993. *A practical guide to noise and vibration control for HVAC systems*. ASHRAE.

U.S. DOE. 2011. *ENERGY STAR*. http://www.energystar.gov.

USGBC. 2009. *Leadership in energy and environmental design (LEED®)*. U.S. Green Building Council, Washington, D.C.

Wolf, M. and J. Smith. 2009. Optimizing dedicated outdoor-air systems. *HPAC Engineering* (Dec.).

Wulfinghoff, D.R. 2000. *Energy efficiency manual*. Energy Institute, Wheaton, MD.

CHAPTER 4

TALL BUILDINGS

TALL buildings have existed for more than 100 years and have been built in cities worldwide. Tall building's only became possible after the invention of the elevator safety braking system in 1853; subsequent population and economic growth in cities made these taller buildings very popular. This chapter focuses on the specific HVAC system requirements unique to tall buildings.

ASHRAE Technical Committee (TC) 9.12, Tall Buildings, defines a tall building as one whose height is greater than 300 ft. The Council on Tall Buildings and Urban Habitat defines a tall building as one in which the height strongly influences planning, design, or use.

Traditionally, model codes in the United States were adopted on a regional basis, but recently the three leading code associations united to form the International Code Council (ICC), which publishes the unified *International Building Code®* [IBC (2009)]. Another important national code, developed by the National Fire Protection Association (NFPA), is NFPA 5000®.

The overall cost of a tall building is affected by the floor-to-floor height. A small difference in this height, when multiplied by the number of floors and the area of the perimeter length of the building, results in an increase in the area that must be added to the exterior skin of the building. The final floor-to-floor height of the office occupancy floors of any building is jointly determined by the owner, architect, and structural, HVAC, and electrical engineers.

Much of the material in this chapter derives from Ross (2004).

STACK EFFECT

Stack effect occurs in tall buildings when the outdoor temperature is lower than the temperature of the spaces inside. A tall building acts like a chimney in cold weather, with natural convection of air entering at the lower floors, flowing through the building, and exiting from the upper floors. It results from the difference in density between the cold, denser air outside the building and the warm, less dense air inside the building. The pressure differential created by stack effect is directly proportional to building height as well as to the difference between the warm inside and cold outdoor temperatures.

When the temperature outside the building is warmer than the temperature inside the building, the stack effect phenomenon is reversed. This means that, in very warm climates, air enters the building at the upper floors, flows through the building, and exits at the lower floors. The cause of **reverse stack effect** is the same in that it is caused by the differences in density between the air in the building and the air outside the building, but in this case the heavier, denser air is inside the building.

Reverse stack effect is not as significant a problem in tall buildings in warm climates because the difference in temperature between inside and outside the building is significantly less than the temperatures difference in very cold climates. Accordingly, this section focuses on the problems caused by stack effect in cold climates.

Theory

For a theoretical discussion of stack effect, see Chapter 16 in the 2009 *ASHRAE Handbook—Fundamentals*. That chapter describes calculation of the theoretical total stack effect for alternative temperature differences between the inside and outside of the building. It also points out that every building has a neutral pressure level (NPL): the point at which interior and exterior pressures are equal at a given temperature differential. The location of the NPL is governed by the actual building, the permeability of its exterior wall, the internal partitions, and the construction and permeability of stairs and shafts, including the elevator shafts and shafts for ducts and pipes. Other factors include the air-conditioning systems; exhaust systems tend to raise the NPL, thereby increasing the total pressure differential experienced at the base of the building. This also increases infiltration of outside air, which tends to lower the NPL, thus decreasing the total pressure differential experienced at the base of the building. Finally, wind pressure, which typically increases with elevations and is stronger at the upper floors of a building, also can shift the neutral plane, and should be considered as an additional pressure to stack effect when locating the neutral plane.

Figure 1 diagrammatically depicts airflow into and out of a building when the outside temperature is cold (stack effect) and hot (reverse stack effect). Not shown is the movement of air up or down in the building as a function of stack effect. Assuming there are no openings in the building, the NPL is the point in the building elevation where air neither enters nor leaves the building. Vertical movement of air in the building occurs at the paths of least resistance, including but not limited to shafts and stairs in the building as well as any other openings at the slab edge or in vertical piping sleeves that are less than totally sealed. Figure 1 also indicates that air movement into and out of the building increases as the distance from the NPL increases. The total theoretical pressure differential can be calculated for a building of a given height and at various differences in temperature between indoor and outdoor air.

The theoretical stack effect pressure gradient for alternative temperature differences and building heights is shown in Figure 2. The diagram illustrates the potential maximum differentials that can occur (which are significant), but these plotted values are based on a building with no internal subdivisions in the form of slabs and partitions. The plot, therefore, includes no provisions for resistance to airflow in the building. Further, the outside wall's permeability influences the values on the diagram and, as noted previously, the wind effect and operation of the building air-handling systems and fans

The preparation of this chapter is assigned to TC 9.12, Tall Buildings.

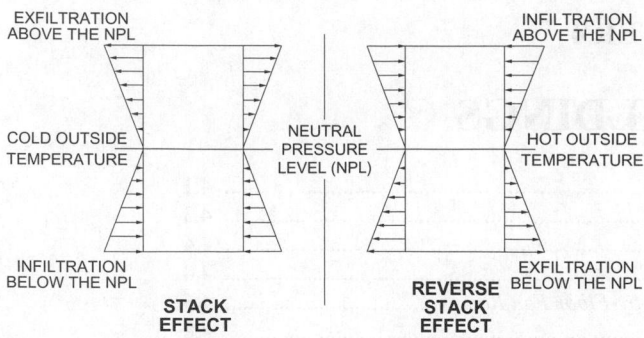

**Fig. 1   Airflow due to Stack Effect and Reverse
Stack Effect**
(Ross 2004)

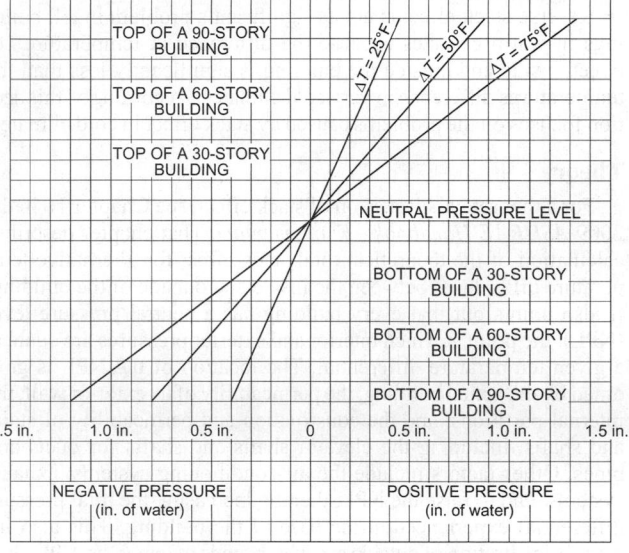

NOTES:
1. Δ*T* equals differences between condition inside and outside building.
2. Floor-to-floor height for alternative buildings is assumed to be 13 ft 0 in.

**Fig. 2   Theoretical Stack Effect Pressure Gradient
for Various Building Heights at Alternative
Temperature Differences**
(Ross 2004)

also affect this theoretical value. Thus, the diagram should be considered an illustration of the possible magnitude of stack effect, not as an actual set of values for any building. The actual stack effect and location of the NPL in any building are difficult (if not in a practical sense impossible) to determine. Nevertheless, stack effect can be troublesome, and its possible effects must be recognized in the design documentation for a project.

Practical Considerations of Stack Effect

Stack effect in tall buildings often presents major problems:

- **Elevator doors** may fail to close properly because of the pressure differential across the doors, which causes the door to bind in its guideway enough that the closing mechanism does not generate sufficient force to overcome it.
- **Manual doors** may be difficult to open and close because of strong pressure created by stack effect.
- **Smoke and odor propagation** through the air path of stack effect can also occur.

- **Heating problems** can occur in lower areas of the building may be difficult to heat because of a substantial influx of cold air through entrances and across the building's outside wall (caused by higher-than-anticipated wall permeability). Heating problems can be so severe as to freeze water in sprinkler system piping, cooling coils, and other water systems on lower floors. The National Association of Architectural Metal Manufacturers (NAAMM) specifies a maximum leakage per unit of exterior wall area of 0.06 cfm/ft² at a pressure difference of 0.30 in. of water exclusive of leakage through operable windows. In reality, tall buildings in cold climates can exceed this pressure difference through a combination of stack, wind, and HVAC system pressure. Even when leakage similar to the NAAMM criterion is included in project specification, it is not always met in actual construction, thereby causing potential operational problems.

Two actual examples, although extreme, illustrate the degree to which stack effect can cause major problems in building in cold climates.

A very tall commercial building in Chicago was partially occupied in September: the lower 30% of the building was occupied, and the top of the building was still under construction and open to atmosphere. There were few operating problems as the construction of the top portion of the building continued into the fall. Major problems only occurred when winter hit the area and temperatures 20°F and below were experienced. At this time, because of the building's open top, its neutral pressure level was raised substantially above the midpoint. (In a practical sense, the neutral pressure level was at the roof and the entire theoretical pressure differential was experienced at the entrance level.) The result was the collapse of revolving doors, an inability to close elevator doors, and inability to adequately heat the entrance levels of the building. Additional heated outdoor air was introduced at the entrance level, stairs at the point where occupancy stopped were sealed, and construction at the top of the building was expedited to close that portion of the building. By midwinter, these efforts minimized the problems and allowed more conventional use of the occupied lower floors.

A second problem developed in a 64-story building in New York City that was built, in part, over a major transportation hub with a direct open connection from the building to the hub itself. The transportation center, with train tunnels entering and leaving the hub and multiple doors that open and close as passengers enter the hub, is effectively open to atmosphere. With large volumes of outdoor air entering the train hub and able to pass directly to the connected office building, the result on cold winter days was such that the elevator doors would not close and comfort conditions could not be maintained in the lobby areas of the office building.

This problem was solved by providing a glass enclosure with revolving doors between the office building lobby and the escalators that allowed individuals to enter the train station. The practical closure of the openings to the train station solved the elevator door and heating problems, and the glass enclosure maintained the desired sense of openness.

Minimizing Stack Effect

During design, the architect and HVAC design engineer should take steps to minimize air leakage into or out of (and vertically within) the building. Although it is not possible to completely seal any building, this approach can help mitigate potential problems that could be caused by stack effect.

Outdoor air infiltration points include building entry doors, doors that open to truck docks, outside air intake or exhaust louvers, construction overhangs with light fixtures that are located immediately above the ground level and are not properly sealed against leakage or provided with heat, and any small fissures in the exterior wall itself. Internally, the building allows air passage through fire stairs, elevator shafts, mechanical shafts for ducts and piping, and any other vertical penetrations for piping or conduit or at the edge of the floor slab at the exterior wall. All these are candidates for careful review to ensure, as much as possible, that the exterior wall is tight, all shafts are closed, and all penetrations sealed. Vestibules or airlocks can be provided for loading docks with good door seals on the doors to and from the loading dock.

Entrances for tall buildings in cold climates should be revolving doors. Doors of this type are balanced, with equal pressure in opposite directions on the panels on either side of the central pivot, making operation relatively simple and requiring no special effort to turn. Their gasketing also provides closure at all times.

Two-door vestibules are acceptable for the loading dock, assuming the doors are properly spaced to allow them to be operated independently and with one door to the vestibule always closed, and sufficient heat is provided in the space between the doors. If properly spaced, the simultaneous opening of both doors on either side of the vestibule can be controlled. However, two-door vestibules in cold climates are inadequate for personnel entry because, with large numbers of people entering the building at various times, both doors will be open simultaneously and major quantities of air can enter the building. In cold climates, it is strongly recommended that revolving doors be used at all points of personnel entry.

To control airflow into the elevator shaft, consider adding doors at the entry to the elevator banks. This creates an elevator vestibule on each floor that minimizes flow through open elevator doors. Elevator shafts are also a problem because an air opening may be required at the top of the shaft. All shafts, however, can be sealed in their vertical faces to minimize inflow that would travel vertically in the shaft to the openings at its top.

It can be helpful to interrupt stairs with well-sealed doors to minimize vertical airflow through buildings. This is particularly useful for fire stairs that run the height of the building. Entrances to fire stairs should be provided with good door and sill gaskets.

The last key item is to ensure a tight exterior wall through specification, proper testing, and hiring a contractor to erect the wall.

The preceding precautions involve the architect and allied trades. The HVAC designer primarily must ensure that mechanical air-conditioning and ventilation systems supply more outdoor air than they exhaust, to pressurize the building above atmospheric pressure. This is true of all systems where a full air balance should be used for the entire building, with a minimum of 5% more outdoor air than the combination of spill and exhaust air provided at all operating conditions, to ensure pressurization. In addition, it is good design, and often required by code for smoke control, to have a separate system for the entrance lobby. Although not always required, this system can be designed to operate in extreme winter outside air conditions with 100% outdoor air. This air is used to pressurize the building lobby, which is a point of extreme vulnerability in minimizing stack effect.

TYPICAL HVAC DESIGN PROCESS

Design of an HVAC system typically proceeds through the following phases. The process is subject to economic and time constraints, to availability of or preferences for certain equipment, and to modifications in design criteria as the project progresses.

Program Phase

The functional use(s) of the building, owner and occupant requirements, architectural concepts, and budget limitations are typically determined during this phase. These requirements should be documented in a statement of design intent for the building systems, which is refined and updated as the design progresses. If cold-air distribution is to be considered for the project, it is best to advance the option at this point, so the entire design can be developed to best realize its benefits. For example, the reduced space required for ductwork may allow revisions to the structural design

and substantial cost savings. Reduced duct space may also be a solution for architectural design concepts such as vaulted ceilings. Similarly, if underfloor air distribution (UFAD) is to be considered, this is the time to evaluate that option, because it affects building height and HVAC system design.

The program phase, sometimes called preliminary design or concept phase, begins when the owner identifies the need for the project and the program is developed. The two end products of the program phase for the mechanical engineer are

- A detailed design intent statement describing the selected HVAC system
- A commissioning or implementation plan outlining the process by which the system will be designed, installed, and commissioned

The design team performs a feasibility study to evaluate system options and establish a budget. At the conclusion of the program phase, the owner selects the desired system, and the design team produces the design intent statement.

The owner's program includes information to develop the following design criteria:

- Building area, height, and number of stories
- Geographic site location
- Building construction materials, area and type of glazing, and insulation levels
- Conceptual architectural drawings
- Functions of each area in the building
- Occupancy and operating schedules
- Environmental requirements, such as temperature and humidity set points
- Process loads, temperature, and flow requirements
- Possibilities for future growth or expansion
- Acoustic requirements
- Applicable standards and codes
- Space available for equipment
- Reliability requirements
- Budgets for capital costs and operating costs
- Sustainability requirements

Project Team. In some cases, the project team is selected at the beginning of the program phase so that the facility concepts can be firmly established with these experts' input early in project development. In other cases, the full project team may not be selected until the completion of the program phase, after the owner has chosen the preferred type of system and the design intent statement has been completed.

Team members may include the owner's team, typical design team members (e.g., architect engineers, landscape and geotechnical experts), construction manager, and/or contractor(s) and may also include specialists in fields such as vibration and acoustics, lighting, security, facades, and computational fluid dynamics (CFD).

Schematic Design

In this phase, the designer selects and compares appropriate candidate systems to meet the building's functional needs and develops recommendations for the owner. This phase typically involves evaluation of the benefits and tradeoffs of different design approaches. Annual energy consumption, demand, and costs, as well as system-installed costs, are estimated. The level of analysis may range from simple rule-of-thumb estimates to hour-by-hour simulations with detailed cost comparisons.

Preliminary Design or Design Development

During this phase, the mechanical designer coordinates the HVAC system design with the architectural, structural, and electrical systems to resolve potential conflicts. Heat loss, heat gain, and ventilation calculations are refined, equipment sizes and capacities are selected, and system layouts are developed.

Final Design and Preparation of Construction Documents

In this stage, the preliminary layouts and equipment selections are fully developed and the final drawings and specifications are completed. Some projects involve preparing alternative designs, with bids solicited for all options. In such cases, the primary option generally receives the most effort during the design.

Construction Phase

The design engineer's responsibilities during this phase typically include reviewing shop drawings, observing construction, and conducting or witnessing system performance tests. Contractors may require guidance to ensure that they follow the project specifications rather than their standard methods.

Acceptance or Commissioning Phase

The term "commissioning" is often used to refer to the start-up and functional performance testing of mechanical systems. However, ASHRAE *Guidelines* 0 and 1 outline a much more comprehensive process for ensuring that an HVAC system is designed, installed, tested, and operated in accordance with the design intent. Acceptance-phase procedures are an important part of this process, but other important elements must take place during each phase of a project. Experience has shown that a comprehensive commissioning procedure is essential to the success of a project. For details, see Chapter 43.

Postoccupancy Services

Designers may be called upon to assist with training operating personnel or optimizing system operation. These services are often included in comprehensive commissioning.

Evolution of Design Intent Statement

During the design phase, the statement of design intent that was developed in the program phase is expanded. The following items are added to the previously defined criteria:

• Narrative description of the system
• Energy performance goals
• Hourly operating profile
• Schematic diagrams
• Control sequence description

Each of these items begins in the initial design phase as a general description of the intended final result. As the design develops, they are refined and expanded, so that the final statement of design intent provides a detailed description of the intended configuration, operation, and control of the system.

Safety Factors

System designers typically apply safety factors at various points in the design process to avoid undersizing equipment. Judicious use of safety factors is good engineering practice. However, safety factors are too often misapplied as a substitute for engineering design, and this practice typically results in grossly oversized equipment. Therefore, care is necessary in applying safety factors.

SYSTEMS

Systems used in tall buildings have evolved to address owners' goals, occupants' needs, energy costs, and environmental concerns (including indoor air quality).

Chapter 37 discusses mechanical maintenance and life-cycle costing, which may be useful in the evaluation process with regard to alternative systems. Chapter 1 of the 2008 *ASHRAE Handbook—HVAC Systems and Equipment* provides guidelines to allow a quantitative evaluation of alternative systems that should be considered in the system selection process. Chapter 19 of the 2009 *ASHRAE Handbook—Fundamentals* provides means for estimating annual energy costs. Ross (2004) provides a more detailed discussion of systems to be considered.

SYSTEM SELECTION CONSIDERATIONS

In a fully developed building (including the core and shell as well as space developed for occupancy), the cost of mechanical and electrical trades (i.e., HVAC, electrical, plumbing, and fire protection) is typically 30 to 35%, and for a high-rise commercial building is usually over 25%, of the overall cost (exclusive of land). In addition, the mechanical and electrical equipment and associated shafts can consume 7 to 10% of the gross building area. The architectural design of the building's exterior and the building core is fundamentally affected by the system chosen. Consequently, HVAC system selection for any tall building should involve the entire building design team (i.e., owner, architect, engineers, and contractors), because the entire team is affected by this decision.

The points of concern and analysis methods do not differ in any way from the process that would be followed for a low-rise building. Possible alternative systems also are very similar, but the choices for high-rise buildings are typically more limited.

Air-Conditioning System Alternatives

Several alternative systems are used in tall buildings. Although the precise system configurations are subject to the experience and imagination of the design HVAC engineer, the most common ones are variations of generic all-air and air/water systems.

Unitary, refrigerant-based systems, such as through-the-wall units, are used in conjunction with all-air systems providing conditioned ventilation air from the interior zone, but this combined solution has been limited to retrofits of older buildings that were not previously air-conditioned and smaller low-rise projects. They are seldom used in first-class tall commercial buildings.

Another option is panel-cooling-type systems, including chilled-ceiling and chilled-beam systems. Though not common in the United States, these systems are used in Europe as a retrofit alternative in existing buildings that were not previously air-conditioned, because these systems can be installed with minimal effect on existing floor-to-ceiling dimension.

All-Air Variable-Air-Volume Systems. All-air variable-air-volume (VAV) systems in various configurations are one of the most common solutions in tall buildings. Conditioned air for VAV systems can be provided from a central fan room or from local floor-by-floor air conditioning units. These alternative means of delivering conditioned air are discussed in the section on Central Mechanical Equipment Room Versus Floor-by-Floor Fan Rooms. The current section is primarily concerned with system functioning, configurations in use, and possible variations in system design.

VAV systems control space temperature by directly varying the quantity of cold supply air in response to the cooling load requirements. VAV terminals or boxes are available in many configurations; pressure-independent terminal units are recommended. Interior spaces that have a year-round cooling load regardless of outside air temperature can use any of the alternative types of VAV boxes:

• **A pinch-off box** reduces supply air volume directly with a reduction of the cooling load. This is a very common terminal in commercial projects, and has the smallest height of any terminal used in office buildings. Usually a stop is used to maintain minimum airflow, for proper ventilation.
• **A series-flow fan-powered VAV terminal** maintains constant airflow into a space by mixing the required amount of cold supply air with return air from the space. The VAV terminal contains a small fan to deliver constant airflow to the space. The fan operates any time the building is occupied. The primary advantage of the fan-powered box is that airflow in the space it supplies is constant

at all conditions of load. This is of particular import if low-temperature air is used to reduce the distributed air quantity and the energy necessary to distribute the system air.

- **A parallel-flow fan-powered VAV terminal** maintains variable airflow into a space and mixes the required amount of cold supply air at minimum flow requirements with return air from the space. The VAV terminal contains a small fan that starts only in heating mode to deliver mixed primary and return airflow to the space. The fan operates only when heating is required to deliver warm return air, mixed with cool primary air when the building is occupied. Unlike the series-flow box, this option delivers increased airflow to the space during heating but can also shut off primary air and operate only the fan to deliver return air during unoccupied periods. A box-mounted heating coil (hot-water or electric) supplements the heat provided by return air when heating requirements increase. The parallel approach does not ensure constant air volume to the space, as can be obtained with the series approach, but it does provide a minimum airflow at significantly lower operating cost.

- **An induction box that reduces supply air volume and induces room air to mix with supply air, thus maintaining a constant supply airflow to the space.** These units require higher inlet static pressure to achieve velocities necessary for induction, with a concomitant increase in supply fan energy requirements. Moreover, operational problems have been experienced, especially at reduced primary airflow quantities. Thus, these boxes are now seldom used in commercial projects.

The exterior zone can use any VAV box type, but in geographical locations requiring heat, the system must be designed with an auxiliary means of providing the necessary heat. This can be done by installing hot-water baseboard, controlled either directly by thermostat or by resetting the hot-water temperature inversely with the outdoor air temperature. Other alternatives are thermostatically controlled electric baseboard on the exterior wall, or either electric or hot-water heating coils in the perimeter VAV boxes.

Low-Temperature Air VAV Systems. All of the preceding variations can be designed using conventional temperature differentials (16 and 18°F) between the supply air and room temperature. Buildings have been successfully designed, installed, and operated for decades with low-temperature supply air between 48 and 50°F. This increases the temperature supply differential to approximately 28°F, thus dramatically reducing primary air quantities.

This lower-temperature air can be obtained by operating the refrigeration machines with chilled water leaving at 40°F or by using ice storage. If the chiller supplies 40°F chilled water, operating costs of the refrigeration plant increase and the chiller must operate for a longer time before an economizer cycle can occur. Moreover, the use of absorption refrigeration machines may not be possible, because they are usually not capable of providing chilled water as cold as 40°F.

However, the reduced quantity of air distributed also reduces fan horsepower, which more than offsets the additional energy used by the chiller. This lower-temperature air requires fan-powered variable-air-volume terminals or induction-type air supply terminals to avert the problem of reduced airflow at less than design loads, particularly in the interior zone. The function of the air delivery terminals is to mix room and cold supply air to deliver warmer air to the space to offset heat gain.

Using low-temperature supply air requires elimination of air leaks and the proper installation of the correct thickness of duct insulation to prevent moisture condensation. Note that the decrease in supply duct size when cold air is used can make lower floor-to-floor heights more practical.

Air/Water Systems. Air/water systems historically included induction systems, but modern systems quite often use fan-coil units outside the building, with interior spaces typically supplied by an all-air variable-air-volume system. Exterior zones are typically provided with a constant volume of air from either (1) the interior VAV system in sufficient quantities to meet requirements of ASHRAE *Standard* 62.1's multiple-spaces equation or (2) a separate dedicated outside air system providing exterior-zone outdoor air ventilation. Fan-coil units in a tall building that requires winter heat are usually designed with a four-pipe secondary water system to provide simultaneous building heating and cooling.

An advantage of the air/water system is it reduces the required capacity of the central supply and return air systems and the size of distribution air ducts, compared to those required with an all-air system (including low-temperature all-air). At the same time, it reduces the air-conditioning supply system's mechanical equipment room space needs. However, air/water systems require space for heat exchangers and pumps to obtain the hot and cold secondary water needed by the fan-coil unit system.

Underfloor Air Distribution (UFAD) Systems. In underfloor air distribution (UFAD) systems, the space beneath a raised floor is used as a distribution plenum. Most installations use manually adjustable supply diffusers or automatically controlled terminal units beneath the floor to control air delivered to the space above. (In contrast, for more traditional systems, terminal units are installed above the ceiling and supply air is delivered from above.) When properly designed, either underfloor or ceiling-mounted air distribution systems can meet occupants' comfort requirements. UFAD systems typically have a higher first cost because of the raised floor, but operating costs are usually lower because less fan horsepower is required. However, if a raised floor is a design requirement for electrical distribution and information technology cabling, UFAD may offer savings in overall first and operating costs.

The UFAD system can use central fan rooms or floor-by-floor fan units. Conditioned air is typically provided at 60 to 64°F in the raised-floor plenum (between the structural slab and the raised floor), but in locations requiring dehumidification, the air must first be cooled to approximately 55°F to remove moisture and then blended with return air (often using an underfloor-mounted series fan powered box or similar arrangement) to achieve supply air temperatures of 60 to 64°F. The suspended ceiling acts as a return plenum but can be reduced in depth because of the absence of supply ductwork.

A major concern with UFAD in tall buildings is the perimeter zone, which has widely varying loads between summer and winter conditions, especially in buildings with large glass exterior elements. Thermostatically controlled fan-coils beneath the floor can be a cost-effective solution. Additionally, extreme caution is needed in sealing all structural floor penetrations to prevent short-circuiting of supply air.

Underfloor air conditioning for a tall building must be selected early in the design process, because it affects architectural (e.g., floor-to-floor heights, exterior facade treatment, stairs, elevators), structural (e.g., depressed structural slabs), and electrical (e.g., plenum-rated cabling) design considerations. All design disciplines must be involved in this decision process.

The combination of system components and the resultant system configuration for a specific building are limited only by the designer's imagination. The chosen alternative is of interest and concern to the owner, architect, and other engineering consultants, and therefore should be subjected to scrutiny and review by the entire design team before final selection is made.

CENTRAL MECHANICAL EQUIPMENT ROOM VERSUS FLOOR-BY-FLOOR FAN ROOMS

Project needs for conditioned air can be met by one or more central mechanical equipment room(s) serving multiple floors, or by systems installed in separate, local fan rooms on each floor, supplying air only to the floor on which the system is installed. Further, the decision to use chilled-water cooling or self-contained air-conditioning units in the floor-by-floor scheme must also be made. The choice of any of the three alternative schemes is one of the most fundamental decisions made during the conceptual design phase. This issue concerns the owner, each member of the design team, and the constructing contractors, because it affects space requirements, space distribution, standard versus custom HVAC equipment, and piping and electrical distribution costs.

Central Fan Room (Alternative 1)

In central fan rooms, the supply of conditioned air for each office floor originates from multiple air-handling systems located in one or more central fan room(s), which are frequently identified as central mechanical equipment rooms (MERs). Each air-handling system can be provided with an outdoor air economizer through minimum and variable outside air dampers, as dictated by the annual ambient temperature and humidity conditions and building code requirements. Multiple systems in a fan room can be interconnected by delivering supply air into a common discharge plenum from all supply systems on that floor.

Air from the central fan room(s) is distributed to each floor by means of vertical duct risers in fire-rated shafts (typically 2 h rated) within the core of the building. At each floor, horizontal duct taps are made into each riser. This horizontal duct tap contains a fire damper or a fire/smoke damper, as required by the local building code, that must be installed where the supply air duct exits the rated shaft enclosure. In many situations, an automatic, remotely controlled two-position damper, which can be rated as a smoke damper, provides individual-floor overtime operation and smoke control. The position (open or closed) is typically controlled by the building management system in either an occupancy schedule or by occupancy sensor or manual reset switch.

Return air from each floor's ceiling plenum also enters the vertical shaft though a return air fire damper at each floor.

Return air is often not ducted within the shaft, so the air is carried back to the central fan room in the 2 h rated drywall shaft. In each central fan room, multiple return air fans draw return air from the return air shafts and deliver it to a headered return air duct system in the central room and then to each air-handling unit.

Where an outdoor air economizer is used, return air is either returned to the supply air system or exhausted to atmosphere, as determined by the relative dry-bulb temperature (or enthalpy) of the return air and the outdoor air being provided to the building. The quantities of outdoor and return air depend on the season and the resultant outdoor temperature and humidity. In warmer geographic areas where the systems operate on minimum outdoor air at all times, return air is always returned to the supply air system except during morning start-up or where the fans are operating under a smoke control mode.

A typical central fan room and supply and return air shaft arrangements are shown in Figure 3.

Floor-by-Floor Fan Rooms with Chilled-Water Units (Alternative 2)

The air supply for each office floor under this alternative originates from a local floor fan room, typically located in the building core. This room contains a chilled-water air-handling unit with a cooling coil, filters, and fan(s). Morning heating at start-up in cold climates can be provided by a heating coil in the air-handling unit, a unit heater installed in the local fan room, or heating coils in the VAV or fan-powered VAV (FPVAV) boxes. The unit on a given floor usually only supplies the floor on which the unit is installed. Typically, one unit is installed on each floor, but multiple units may be used with interconnected air systems on large floors. Chilled water for the cooling coil is provided by a central chilled-water plant in the building, sized to meet the combined capacity requirements of all of the cooling and heating needs. The supply air fan in

the air-conditioning system serves to both supply air and return it from the zone served. The return air is typically directed to the fan room through the ceiling plenum, but the return may be either ducted or unducted in the fan room. In most cases, however, the fan room acts as a return air plenum.

This system typically operates on minimum outdoor air during all periods of occupancy. Outdoor air for the system is provided by an air-handling unit serving as a dedicated outdoor air system (DOAS), located on the roof or in a central mechanical equipment room. This unit provides conditioned outdoor air to the unit on each floor by a vertical air riser routed to each air-handling unit. The outdoor air unit may include preheat and cooling coils to treat incoming outdoor air, and should contain filtration to clean this air. Alternatively, this unit can contain heat recovery to precondition the outdoor air by recovering heat or cool from exhaust air.

Although chilled water is typically provided by a central refrigeration plant, economizer requirements can be provided by cooling the chilled water in mild weather by the condenser water from the cooling tower. During periods of low wet-bulb temperature, the condenser water cools the chilled water through a heat exchanger in the central chilled-water plant or by refrigerant migration through the refrigeration unit.

A typical local fan room supply, return, and outdoor air arrangement is shown in Figure 4. The unit heater shown provides morning heat. It can use electric energy or hot water as its heat source.

As shown in Figure 4, the walls around the local floor fan room are not fire-rated because the duct penetration serves only this floor. The vertical shaft that contains the outdoor air duct from the central fan room, and perhaps the smoke exhaust ducts, constitute a fire-rated shaft. Accordingly, fire dampers are only provided at the point where ducts penetrate the shaft wall, not as they leave or enter the local floor fan room itself. Although fire dampers are shown in the smoke exhaust ducts, many codes prohibit their use in an engineered smoke control system to avoid the possibility of having a closed damper when smoke removal is required.

Floor-by-Floor Fan Rooms with Direct Expansion Units (Alternative 3)

A second variation of the floor-by-floor alternative consists of a floor-by-floor air-conditioning supply system that is virtually identical to that in the chilled-water alternative. In this alternative, a packaged, self-contained, water-cooled direct-expansion (DX) unit, complete with one or more refrigeration compressors and water-cooled condensers, is used to produce the cooling. The heat of rejection from the compressor is handled by a circulating condenser water system and cooling tower. If geographic location dictates an economizer, it can be met by a free-cooling coil that is installed in the packaged unit that will only operate when condenser water delivered to the unit is cold enough to provide effective cooling. The only central cooling equipment is a cooling tower, condenser water pumps, and the central outdoor air supply unit. If an open tower system is used, consider providing a way to remove particulates from the circulating condenser water. Depending on the size of anticipated particles, typical options include sand filtration, media filtration, and centrifugal separators. For an open system, condensers should be cleanable. Bear in mind that significant water will end up on the floor during condenser cleaning, so it is important to ensure that the room has a recessed floor drain and that the floor is moisture sealed.

The physical arrangement of the supply air unit does not differ from that shown in Figure 4, except that the chilled-water risers are replaced by condenser water piping.

Floor-by-Floor Units Located on an Outside Wall

A popular variant location for a packaged floor-by-floor unit is on an outside wall. This location obviates the need for a separate outdoor air unit in a central fan room. Outdoor air can be directly

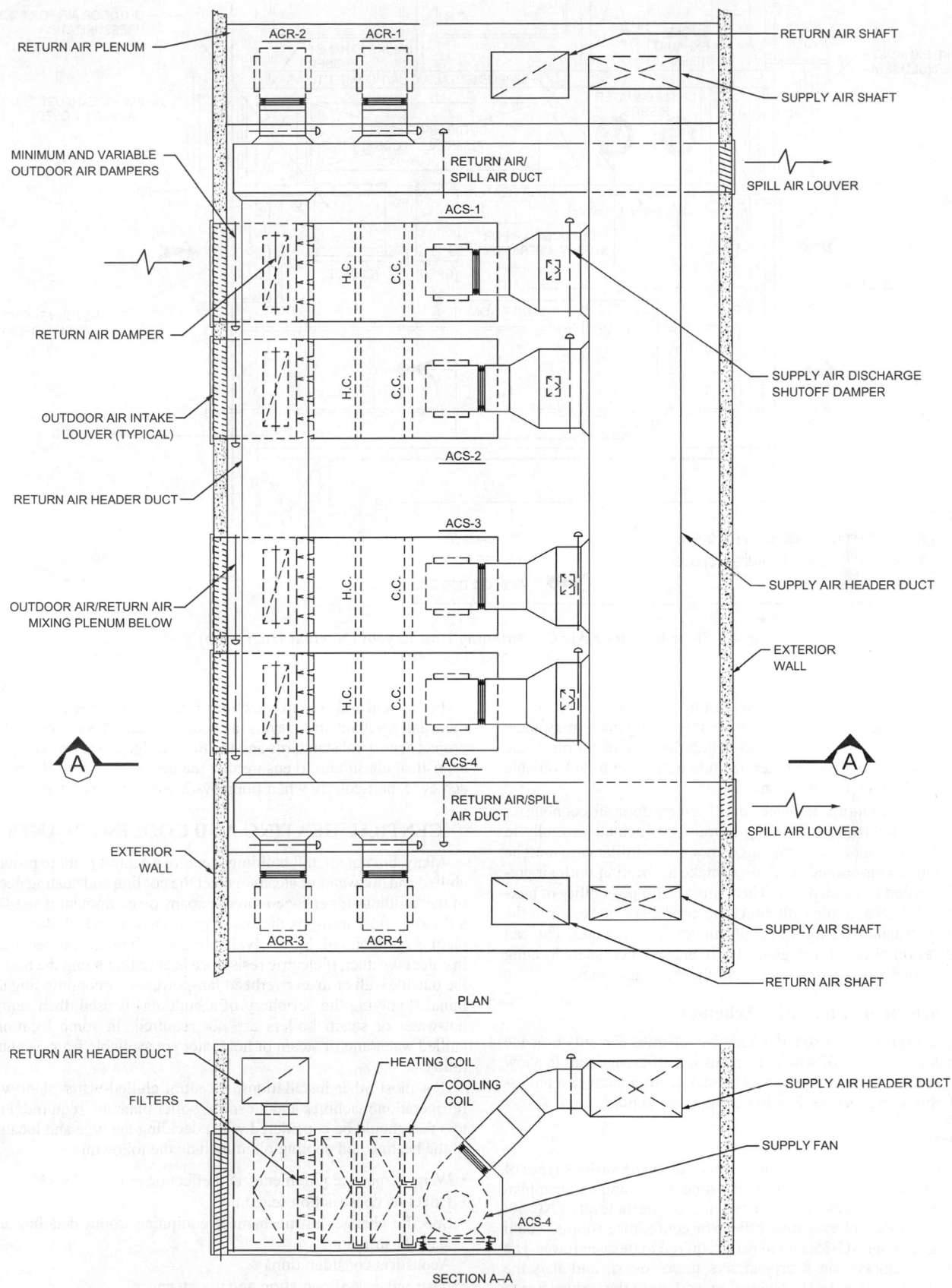

RETURN AIR PLENUM

ACR-2   ACR-1

RETURN AIR SHAFT

SUPPLY AIR SHAFT

MINIMUM AND VARIABLE
OUTDOOR AIR DAMPERS

RETURN AIR/
SPILL AIR DUCT

SPILL AIR LOUVER

ACS-1

H.C.   C.C.

RETURN AIR DAMPER

SUPPLY AIR DISCHARGE
SHUTOFF DAMPER

H.C.   C.C.

OUTDOOR AIR INTAKE
LOUVER (TYPICAL)

ACS-2

RETURN AIR HEADER DUCT

ACS-3

H.C.   C.C.

SUPPLY AIR HEADER DUCT

OUTDOOR AIR/RETURN AIR
MIXING PLENUM BELOW

H.C.   C.C.

EXTERIOR
WALL

A

A

ACS-4

RETURN AIR/SPILL
AIR DUCT

EXTERIOR
WALL

SPILL AIR LOUVER

ACR-3   ACR-4

SUPPLY AIR SHAFT

RETURN AIR SHAFT

PLAN

RETURN AIR HEADER DUCT

HEATING COIL

COOLING
COIL

FILTERS

SUPPLY AIR HEADER DUCT

SUPPLY FAN

ACS-4

SECTION A–A

Fig. 3   Central Fan Room Arrangement
(Adapted from Ross 2004)

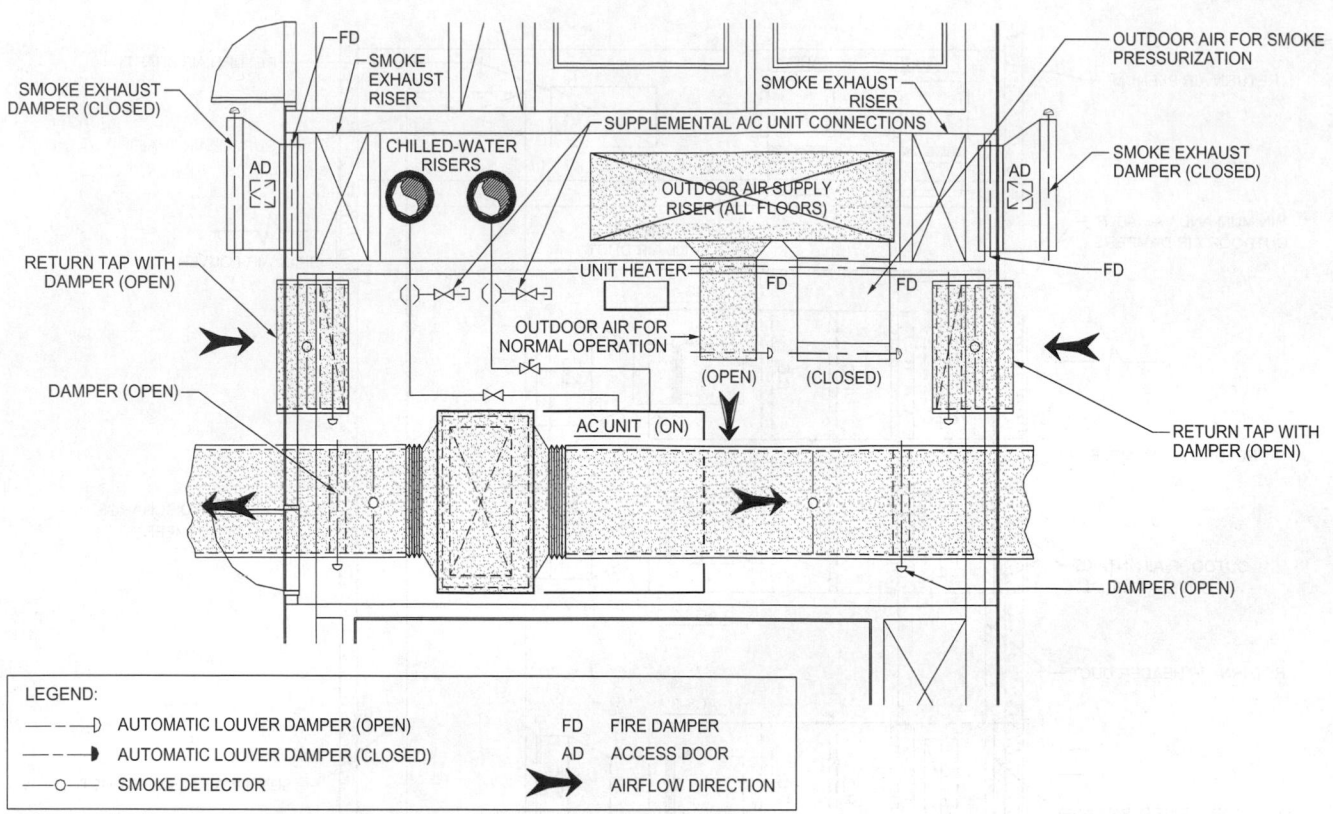

Fig. 4 Floor-By-Floor Air Conditioning Unit Layout (Normal Operation)
(Adapted from Ross 2004)

introduced to the floor-by-floor unit through a louver and automatic louver damper for each unit. Moreover, this arrangement may allow use of an air-cooled condenser to handle heat of rejection. If the location requires an economizer, include a minimum and variable air damper behind the outside air louver.

Several precautions are necessary. If an outdoor air economizer is used, the return air spill damper must be located carefully to ensure that outdoor air and spill air do not mix. Similar care must be taken to avoid air-cooled condenser intake air mixing with air previously spilled to atmosphere. There must be no possibility of mixing heated discharge air with either the condenser intake air or the outdoor ventilation air for the supply air conditioning unit. This can become a complicated arrangement, which may necessitate locating the air-cooled condenser remote from the local fan room.

Comparison of Alternative Schemes

An accurate comparison of alternative schemes can only be made with a developed set of schematic plans in sufficient detail to allow a cost estimate to be completed by the contracting team or a professional estimating service. For an example, see Table 1.

Acoustics

Acoustical criteria should be established for the various types of occupancy that are expected in the building. For example, open-plan office space can be designed to meet a noise criteria level of NC-40, whereas private and executive offices or conference rooms should be no higher than NC-35, and may be required to be even lower. The acoustical engineer on a project sets these levels, and it is the responsibility of the HVAC designer to work with the acoustician to see that the criteria established are achieved in the final installation. (For details on sound levels, see Chapter 48 in this volume and Chapter 8 in the 2009 *ASHRAE Handbook—Fundamentals*).

Equipment and system selection affects the required sound treatment and resultant noise levels in occupied areas. It is important that project acoustical standards and the final design are reviewed by the acoustical consultant to ensure that the desired noise levels can be achieved, particularly when floor-by-floor fan rooms are used.

CENTRAL HEATING AND COOLING PLANTS

Many, but not all, tall buildings require a central plant to provide chilled and hot water or steam to meet the cooling and heating needs of the building. If packaged direct-expansion equipment is used on a floor-by-floor basis, as discussed previously, then a chilled-water plant is not required. Similarly, in climates where heat is necessary in colder weather, if electric resistance heat (either along the base of the outside wall or in an overhead fan-powered air conditioning terminal supplying the periphery of a building) is used, then central hot-water or steam boilers are not required. In some locations, chilled water and/or steam or hot water are available from a central utility.

For most other installations, a central chilled-water plant with refrigeration machines and a central boiler plant are required. Factors that should be considered when deciding the type and location of the heating and cooling plant include the following:

- Weight, space requirements, and effect on structural system
- Effect on construction schedule
- Specific changes in mechanical equipment room detailing and slab construction
- Acoustical considerations
- Ease and cost of operation and maintenance
- Available energy sources
- Annual operating costs and possibly life-cycle costs of each alternative

Table 1 Comparison of Construction Alternatives

Alternative 1	Alternative 2	Alternative 3
Central Fan Systems Central Chilled Water	Floor-by-Floor Fan Systems Central Chilled Water	Floor-by-Floor DX Systems Central Cooling Tower

First-Cost Considerations

HVAC

Fewer units, field erected. More complex and expensive duct systems. More complex field-installed controls. Central chilled-water plant.	More units, factory-fabricated and assembled. Simpler ductwork. Field-installed control system. Central chilled-water plant.	More units, factory-fabricated and assembled. Simpler ductwork. Factory-installed control system. No central chilled-water plant; cooling tower only.

Building Management System

Complex controls and interface with building management system (BMS) and smoke control system.	Controls are relatively simple but field installed. Interface with BMS and smoke control system less complex.	Unit controls provided by manufacturer. Interface with BMS and smoke control system simple.

Electrical

Electrical loads concentrated in central location. Probably lowest electrical cost.	Minor cost premium for distributed fan motors. Probably higher electrical cost than alternative 1.	Additional cost for electrical distribution to local DX units. Highest electrical cost.

General Construction

Additional gross floor space needed. No separate outside air or smoke exhaust shaft.	Additional cost of sound treatment of local floor-by-floor fan room. Need separate outdoor air and smoke exhaust shaft.	Additional cost of sound treatment of local floor-by-floor fan room. Need separate outdoor air and smoke exhaust shaft.

Construction Schedule

General Complexity of Installation

Central mechanical equipment room space and complex construction technology for both chiller plant and fan systems locations. Requires piping of a major chiller plant. Chiller plant location critical to construction schedule. Heavier slab construction at central mechanical equipment room. Extensive complex ductwork in central mechanical equipment room.	Chiller plant space is required, with need for more complex construction technology. Requires piping a major chiller plant. Chiller plant location critical to construction schedule. Heavier slab construction for chiller plant only. Limited ductwork, repetitive fan room arrangement on each floor.	Areas that contain complex construction technology are limited. No major chiller plant. Cooling tower only. Chiller plant is not required. Very limited special slab construction. Limited ductwork, repetitive fan room arrangement on each floor.

Owner Issues

Marketing/Electric Metering

Tenant lights and small power can be metered directly. Fan energy and chiller plant energy, as well as heating energy, operating costs are allocated unless heating is by electric resistance heat. Other common building operating costs are allocated.	Tenant lights, small power, and fan energy can be metered directly for any floor with a single tenant. Multitenanted floors require allocation of fan energy only. Chiller plant energy, as well as heating energy, operating costs are allocated unless heating is by electric resistance heat. Other common building operating costs are allocated.	Tenant lights, small power, fan, and cooling energy can all be metered for any floor with a single tenant. Multitenanted floors require allocation of fan energy and cooling energy only. Heating energy operating cost must be allocated unless heating is by electric resistance heat. Other common building operating costs are allocated.

Operating Costs

For normal operating day, operating costs for all floors occupied are lower than alternative 3. Approximately equal to alternative 2. Overtime operation requires the chiller plant to operate in the summer. With variable-speed fan control and headered supply and return fans, energy costs equal to alternative 2. Operation more cumbersome. Fan and chiller plant costs must be allocated.	For summer operating day, operating costs for all floors occupied are lower because of lower energy consumption than alternative 3. Approximately equal to alternative 1. Overtime operation requires chiller plant to operate in summer but otherwise is simple. Chiller plant cost must be allocated.	For the summer operating day, operating costs for all floors occupied are higher due to higher energy consumption than alternatives 1 or 2 because of less efficient DX compressors. Overtime operation simplest but probably higher in cost than alternatives 1 or 2. Single-floor tenant cost for cooling tower only must be allocated.

Equipment Issues

Equipment Maintenance

All equipment is installed in central mechanical equipment room with centralized maintenance.	Requires more maintenance than alternative 1 but less than alternative 3, because of larger number of units with filters, motors, fan drives, bearings, etc. Chiller is in central mechanical equipment room, allowing centralized maintenance.	Requires more maintenance than alternatives 1 or 2 because of larger number of units with filters, motors, fan drives, bearings, etc., plus compressor equipment on each floor.

Table 1 Comparison of Construction Alternatives (*Continued*)

Alternative 1	Alternative 2	Alternative 3
Equipment Redundancy and Flexibility		
Can operate in reduced mode in case of limited failure due to headered fan arrangement. Can handle changing cooling loads and/or uneven cooling loads on a floor-by-floor basis within limits. Can usually turn down system operation to supply air to a single floor.	If unit fails, floor is without air conditioning. Cannot handle changing cooling loads or uneven cooling loads on a floor-to-floor basis without building in additional system capacity at design.	If unit fails, floor is without air conditioning. Cannot handle changing cooling loads or uneven cooling loads on a floor-to-floor basis without building in additional system capacity at design.
Equipment Life Expectancy		
Life expectancy of equipment is in excess of 25 years.	Life expectancy of equipment is in excess of 25 years.	Compressor life expectancy is probably approximately 10 years. Remainder of installation life expectancy is in excess of 25 years.
Architectural Issues		
Building Massing		
Central fan rooms usually require two-story MER. Chiller plant room usually requires two-story MER.	Local fan room fits within floor-to-floor height of the office floor. Chiller plant room usually requires two-story MER.	Local fan room fits within floor-to-floor height of the office floor. No central chiller plant room required.
Usable Area		
Takes the least area per office floor. Maximum usable area per office floor.	Takes a greater area per floor. Less usable area per office floor than alternative 1.	Takes a greater area per floor. Less usable area per office floor than alternative 1.
Gross Area		
Takes more gross building area than alternatives 2 or 3.	Takes more gross building area than alternative 3 but less than alternative 1.	Takes less gross building area than alternatives 1 or 2.

Calculation of owning and operating costs is discussed in Chapter 37. Alternative refrigeration technologies are detailed in Chapters 1 to 3 of the 2010 *ASHRAE Handbook—Refrigeration*, and boilers are covered in Chapter 31 of the 2008 *ASHRAE Handbook—HVAC Systems and Equipment*. Useful reference information is also contained in ASME (2010).

Plant Economic Considerations

Detailed analysis is needed to determine the cooling method that should be installed in a project. The choices are usually limited to either centrifugal refrigeration or absorption chilled-water machines, although recent developments have made screw chillers more relevant for use in tall buildings. Centrifugal machines can be electric drive or steam drive; screw machines are available only with electric motor drives, and both are almost always water-cooled. The absorption machines can be single- or double-effect, but the latter require high-pressure steam to achieve their lower energy costs. High-pressure steam is rare in today's commercial projects unless the steam is available from a central utility.

Air-cooled refrigeration machines have been installed in tall buildings, but infrequently: commercially available sizes of air-cooled refrigeration equipment are limited, and space requirements are comparatively excessive. The largest air-cooled refrigeration machine that currently can be purchased this time is approximately 400 tons. Tall buildings, by nature, are typically large, and the number of air-cooled refrigeration machines and relatively large equipment space that would be required usually make air cooling not viable. In addition, air-cooled equipment's operating costs may be higher because of higher condensing temperatures developed by the refrigeration equipment caused by outdoor dry-bulb temperatures that are higher than the coincident wet-bulb temperature. Water-cooled equipment's refrigerant condensing temperature, on the other hand, is driven by the lower outdoor air wet-bulb temperature. This operating cost difference exists even though there is no cooling tower fan or condenser water pump.

Air-cooled equipment may, however, find application in tall buildings where water for cooling tower makeup either is not available or is prohibitively expensive.

For tall buildings that do not use electric resistance heat, the fuel-fired heating plant includes boilers fired by oil or gas, by both fuels (with oil as a standby fuel), or by electricity. These boilers provide hydronic heat and low-pressure steam for distribution to spaces in the building, or act as supplements to heat pumps or heat recovery systems. Choosing the correct solution for a building is subject to an economic analysis that considers space requirements, first cost, and operating expense.

Central Plant Location

Further complicating the energy transfer source decision is the location of the equipment within the building. This affects structural costs, architectural design, construction time, and availability of cooling or heating relative to the initial occupancy schedule. A below-grade location could potentially provide early heating availability, but also could complicate the design process and result in higher overall project costs. Locating cooling and heating plants on floors above grade, up to and including space immediately below the roof, is common and may be desirable for simplicity of construction and ease of providing the necessary ventilation air and other services to the equipment. Moreover, the two types of plants need not be installed at the same level in the building, because there is usually no direct interconnection of the two plants.

Virtually any location in a tall building can be used for the heating and cooling equipment. When choosing the location, consider the following:

- If a boiler is installed above grade, fuel (i.e., oil, gas, electricity) must be brought to the boiler and a flue and combustion air, in the case of a fuel-fired boiler, must be taken from the boiler to atmosphere.
- Boiler plant location should be determined by analysis following previously outlined parameters.

- Regardless of where it is installed, the design must include appropriate acoustical design considerations and vibration isolation.

Considerations for the refrigeration plant location are more complex. Not only must electricity, gas, oil, or steam be brought to the machine to provide energy to operate the equipment, but chilled and condenser water also must be pumped from the refrigeration plant to the air-conditioning supply equipment. In addition, the cooling tower and the working pressure of the refrigeration machines, piping, fittings, and valves must be reviewed based on the static height of liquid above this equipment, as discussed in Ross (2004) and Chapter 39 of the 2008 *ASHRAE Handbook—HVAC Systems and Equipment*.

Acoustical Considerations of Central Plant Locations

Acoustics and vibration also are key considerations during architectural, structural, and mechanical design. The HVAC designer and project acoustician should place mechanical equipment to achieve the desired acoustical levels in spaces above, below, or adjacent to the central plant. Achieving the proper solution involves understanding the characteristics of sound generated by the equipment and the various paths (e.g., through floors, ceilings, walls, building structure) for transmission of that noise and vibration to occupied areas of the building.

Regardless of the type of equipment being installed on a project, it is prudent to specify a maximum permissible sound level for equipment. Sound and vibration generation, transmission, and correction are discussed in Chapter 48 in this volume and in Chapter 8 of the 2009 *ASHRAE Handbook—Fundamentals*.

Effect of Central Plant Location on Construction Schedule

The locations of the boiler and chiller plant also affect the construction schedule. This concern is especially critical for the refrigeration plant, which is a complex installation that involves a significant amount of labor because of the need to complete the chilled-water, condenser water, and possible steam piping as well as provide for the electrical capacity requirements of the machines. The heaviest piping and most difficult installation process for piping in the building occur at the refrigeration plant. As a result, if the refrigeration plant is on the uppermost level of the building, installation of the machines and their associated piping can delay the overall schedule. Accordingly, if the refrigeration equipment cannot be installed in the below-grade level because that space has other priorities (e.g., parking, storage), the refrigeration plant may be best located above the lobby level and below the uppermost levels of the building.

WATER DISTRIBUTION SYSTEMS

Water distribution systems for a tall building require special consideration, primarily because the building height creates high static pressure on the piping system. This pressure can affect the design of the piping systems, including domestic water and sprinkler piping systems. This section addresses chilled-, hot-, and condenser water systems.

The chilled- and hot-water systems are always closed systems (i.e., pumped fluid is not exposed to the atmosphere), whereas the condenser water system is usually open. Closed systems contain an expansion tank, which can be either open or closed. An open expansion tank is located at the highest point of the piping system and is open to atmosphere; the exposed surface area of the water in the open tank is insignificant and the system is still considered closed.

In an open system, the pumped fluid is exposed to atmospheric pressure at one or more points in the piping system. The condenser water piping distribution system is typically considered open because the water is exposed to atmosphere by the clean break in the piping at the open cooling tower.

As stated in Chapter 12 of the 2008 *ASHRAE Handbook—HVAC Systems and Equipment*, the "major difference in hydraulics between open and closed systems is that certain hydraulic characteristics of open systems cannot occur in closed systems. For example, in contrast to the hydraulics of an open system, in a closed system (1) flow cannot be motivated by static head differences, (2) pumps do not provide static lift, and (3) the entire piping system is always filled with water."

If an evaporative cooler or dry cooler (commonly called an industrial fluid cooler) were used for the condenser water rather than a cooling tower, the piping system would be closed rather than open. Using evaporative or dry coolers for an entire large commercial office building is extremely rare. However, they are used in portions of tall buildings to handle the heat of rejection from supplemental cooling systems that may be required for spaces or equipment that require additional cooling capacity.

Hydrostatic Considerations

A major consideration in piping system design for a tall building is the hydrostatic pressure created by the height of the building. This hydrostatic pressure affects not only the piping and its associated valves and fittings, but also equipment in the building; in the chilled-water system, this includes refrigeration machines, casings for chilled-water pumps, cooling coils in air-conditioning systems, heat exchangers, and any fan-coil units at the exterior wall of the building. A similar list of devices beyond piping, valves, and fittings can be developed for other pumped systems such as the condenser water or any hot-water system.

Dynamic pressures created by the pumps also must be added to the static pressure to determine the working pressure on any element in the piping system. This dynamic pressure is the total of the following elements:

- Friction loss through piping, valves, and fittings
- Residual pressure required at the most remote piece of heat transfer equipment for its proper operation (includes pressure loss through the equipment's control valve as well as drop through the equipment itself)
- Any excess pressure caused by pumps operating at reduced flow close to their shutoff pressure

The working pressure of the piping and connected equipment at various elevations in the building must be known. This is found by adding the hydrostatic pressure at the specific location to the dynamic pressure that can be developed by the pumps at that location. The dynamic pressure at any point should include the pump pressure at or close to pump shutoff at full speed, even if variable-speed pumps are used, because it is possible for the pumps to operate at this shutoff point in the event of a VFD failure. This working pressure on piping and equipment invariably lessens as the static pressure at a specific location is reduced.

Effect of Refrigeration Machine Location

The level on which the refrigeration machines and the supporting chilled- and condenser water pumps are located in a building can affect the cost of refrigeration equipment, the pumps, the piping, and the fittings and valves associated with the piping. There is economic impact because of the working pressure to which the equipment, piping, fittings and valves will be subjected by the height of the system above.

Figure 5 shows the effects of three alternative chiller locations in a 70-story, 900 ft tall building: at basement level, a mid-level mechanical equipment room, and a mechanical equipment room on the roof. There is an open expansion tank at the top of the building (the highest point in the system) in all three alternatives. If a closed expansion tank is used, the maximum pressure must be established and considered in the determination of the system's working pressure.

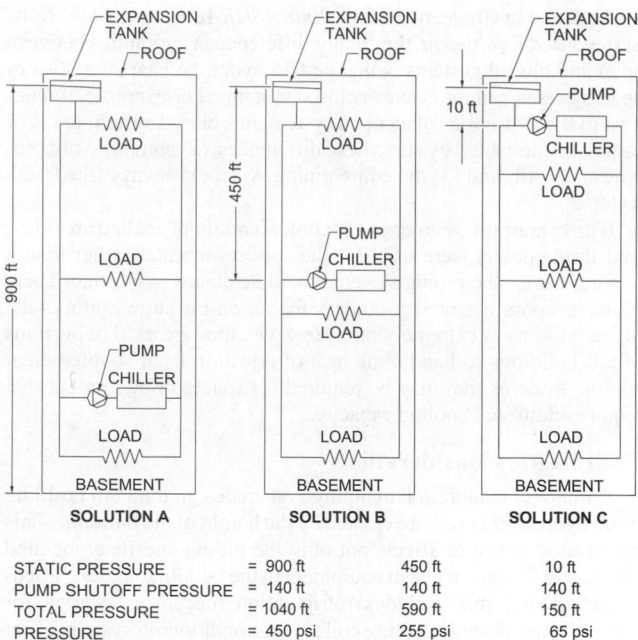

	SOLUTION A	SOLUTION B	SOLUTION C
STATIC PRESSURE	= 900 ft	450 ft	10 ft
PUMP SHUTOFF PRESSURE	= 140 ft	140 ft	140 ft
TOTAL PRESSURE	= 1040 ft	590 ft	150 ft
PRESSURE	= 450 psi	255 psi	65 psi

Fig. 5   Chiller Working Pressure Considerations in 70-Story, 900 ft Building
(Adapted from Ross 2004)

The working pressure on equipment or piping, valves, and fittings at any location in a building is the sum of the hydrostatic height of the water in the piping above the point being considered plus the dynamic pressure created by the pump at the point being analyzed. The hydrostatic and dynamic pressures are determined in feet of water. Their sum, when added together, is the total pressure or working pressure in feet at the referenced point. To determine the working pressure in psig, divide the total pressure in feet by 2.31.

For example, in solution A shown in Figure 5, the vertical height of the column of water above the refrigeration machine is 900 ft. The pump delivering water through the machines has a maximum shutoff pressure of 140 ft. The total pressure is therefore the sum of these two pressures, or 1040 ft or 450 psi.

Calculations for alternative refrigeration plant locations (at midlevel and top of the building) are also shown in Figure 5. Working pressure on the refrigeration equipment at the midlevel of the building is 255 psig, and at the top of the building is 65 psig.

The standard working pressure for the coolers and condensers on large refrigeration machines from all of the major manufacturers in the United States is 150 psig. These machines can be manufactured for any working pressure above 150 psig for additional cost. The incremental increase in the cost of a given vessel becomes larger with each unit of increase in the working pressure. Accordingly, it is necessary for the HVAC design engineer to accurately determine and separately specify the working pressure on both the cooler and the condenser of the refrigeration machines.

Working pressure on the refrigeration machine can be reduced by locating the chilled-water pump on the discharge side rather than the suction side. If this is done, the residual pump pressure on the refrigeration machine water boxes is reduced to the sum of the hydrostatic pressure and this nominal value of dynamic pressure from the pumps. This can reduce the cost of the refrigeration machines, but does not alter the pressure on the pump casing and flanges, which must still be the sum of the static and dynamic pressures.

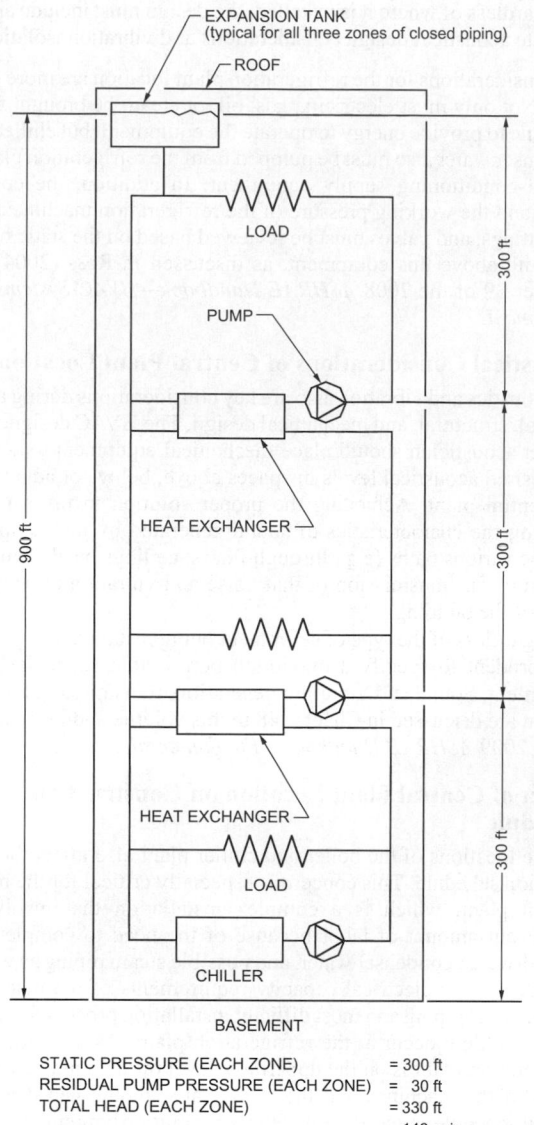

STATIC PRESSURE (EACH ZONE)	= 300 ft
RESIDUAL PUMP PRESSURE (EACH ZONE)	= 30 ft
TOTAL HEAD (EACH ZONE)	= 330 ft
	= 143 psig

Fig. 6   Zoned Chilled Water for 70-Story, 900 ft Building
(Adapted from Ross 2004)

Chilled-Water Pressure Reduction

Pressure on (and cost of) refrigeration equipment can be reduced by locating it above the basement; this, however, will not alter the maximum pressure experienced by the pipe, fittings, and valves at any location that is used. It is possible, however, to reduce the chilled-water working pressure on both the machines and piping by using plate-and-frame heat exchangers, which segregate groups of floors into separate static pressure zones.

In the 900 ft tall example building with the refrigeration machine in the basement, it is possible to break the chilled-water system into three separate zones (Figure 6).

Each zone has static pressure of one-third of the total building height, or 300 ft. All of the pumps are located on the discharge side of the refrigeration machines or the secondary zone heat exchangers. The result is that the maximum head of each zone is 143 psig, which is below the threshold design pressure of 150 psig, or the point at which an increased pressure rating for the chiller and other heat transfer equipment must be considered.

The working pressure of the primary chilled-water pump in the basement will not change substantively from that required where no secondary systems were included, because the primary chilled-water pump must now overcome the loss through the flat-plate heat exchanger. In addition, motor-driven pumps are added at each secondary water heat exchanger. Finally, with the addition of the two additional zones and the resultant chilled-water temperature increase, there is a requisite increase in the volume of water flowing through the systems on the upper floors. Accordingly, although there are benefits in the reduction in pressure, there are partially offsetting considerations that must be analyzed to determine the overall cost effectiveness of using flat-plate heat exchangers to reduce the operating pressure on the equipment, pipe, valves, and fittings at a given level.

Use of flat-plate heat exchangers and their location in a chilled-water piping system is subject to an economic analysis by the design HVAC engineer to determine the first cost of alternative arrangements as well as the operating cost differentials, if any, for any scheme.

Using a flat-plate heat exchanger to reduce the working pressure on the condenser, although feasible, is not often considered, because the condenser water piping is usually in a single shaft with minimal (if any) offsets and a resultant small number of fittings. Valves are also only installed at the machines and are few in number. This limited number of fittings and valves may not be sufficient to offset the cost of the flat-plate heat exchanger and its valving as well as the added pump on the secondary side of the heat exchanger. Beyond that, there is an increase in the temperature of the condenser water, which increases the cost of operating the refrigeration machines.

Piping, Valves, and Fittings

The working pressure on the piping, valves, and fittings at various levels in a building must be determined so that proper piping material can be specified. In the United States, with steel pipe, Schedule 40 pipe is the standard wall thickness for pipes up to 10 in. diameter. For pipes 12 in. and larger, the pipe standard that is used has a wall thickness of 0.375 in. Either of these standards would accommodate the working pressures experienced in any expected pipe diameter in any tall building. The allowable pressures for various pipe diameters can be found in ASME *Standard* A17.1 and the *Boiler and Pressure Vessel Code* (ASME 2010) and in the publications of various pipe manufacturers. The valves used should be reviewed in the valve manufacturers' literature to ensure their ability to meet the project's requirements.

For steam condensate piping or for condenser water piping, where corrosion is a possible concern, pipe with a heavier wall thickness should be considered, although not because of the working pressure on either system.

Piping materials other than steel are often used. For pipe sizes below about 4 in., in the cases of runouts or in open condenser water piping where corrosion is a concern, copper is the usual choice. Copper pipe is rare, but copper tubing is common. The limiting factor in the use of copper tubing is usually at the joints, where the ability to handle higher working pressure is restricted.

Piping Design Considerations

The design of piping must also consider other factors, including

- Expansion and contraction in the piping and its static and dynamic loads, because they are reflected in the structural steel framing system of the building
- Access to expansion joints and the anchors and guides for the piping, which should be inspected periodically after the building is constructed
- Firestopping between the pipe and the sleeve located at all penetrations of rated slabs, walls, and partitions
- Seismic restraints (if required) on the piping systems and pumps

In addition to expansion and contraction of the piping caused by changes in the ambient temperature or of the pumped fluid in the piping, frame shortening can be a problem in concrete buildings. Concrete shrinks as it cures: over time, this shortening can be in the range of 1/8 in. per floor. Although this movement is relatively small, it amounts to about 9 in. for a 70-story building. This condition requires that pipes above, below, and between anchor points be flexible enough to allow for pipe movement with respect to the structure. To properly design for this condition, the HVAC designer should obtain from the structural engineer the exact amount of movement that the piping system can experience.

Economics of Temperature Differentials

Traditionally, rules of thumb for selecting refrigeration machines in the United States have used a 10 or 12°F temperature differential between entering and leaving water in the chiller and a 10°F differential or 3 gpm per ton of capacity for the condenser. These guidelines are appropriate for small buildings, because they have little effect on project cost, but may be less ideal for large buildings, particularly tall buildings. In projects of this type, the capital costs of piping, valves, and fittings can be substantially reduced, with a possible penalty in refrigeration machine operating cost, by using larger temperature differentials with lower water flow and a consequent reduction in piping diameter.

For a large project with a total cooling capacity requirement of 4000 tons and chilled-water flow at a 10°F temperature differential, 9600 gpm is circulated through 20 in. piping at approximately 10 fps. If a 16°F temperature differential is used, total flow from the refrigeration plant is 6000 gpm and the piping is 16 in. Cost savings on the piping using the greater temperature differential would be significant. Also, although the kilowatts per unit of cooling under both conditions should be studied, with the same discharge temperature, the operating energy consumption probably is unchanged.

For the 4000 ton refrigeration plant with a 10°F temperature differential, the condenser water flow is 12,000 gpm and 24 in. piping is required. If this temperature differential were increased to 15°F, condenser water would be reduced to 8000 gpm, and the piping to 20 in. Again, this change results in a significant first-cost savings, depending on the distance between the refrigeration machines and the cooling towers.

Energy consumption for the refrigeration machines might marginally increase, because the condensing temperature of the refrigerant and the resultant energy usage is largely (but not solely) a function of the leaving condenser water temperature.

VERTICAL TRANSPORTATION

The HVAC designer's main involvement with elevators in a tall building is to provide cooling in the elevator machine room to ensure reliable operation. Many codes now require that this machine room be conditioned by a separate HVAC system that is independent of other building systems. This section addresses the possible code requirement of elevator shaft and machine room ventilation to atmosphere.

Elevator Machine Room Cooling

The elevator machine room's cooling loads consist not only of the electric motor that drives the hoisting mechanism but also of extensive heat-generating electronic elevator controls. The electronic components that are part of the system require that the elevator machine room be maintained at a temperature between 80 and 60°F. This can be accomplished by means of a packaged DX condenser water-cooled unit in the elevator machine room; however, because of possible significant operational availability restrictions on the use of water in the machine room, the HVAC designer should review this alternative with the building developer and possibly code officials. Using a packaged DX condenser water unit may be necessary for a low- or mid-rise elevator bank with its machine room in the middle of the building, without easy access to outdoor air unless the remainder of the floor is used as a mechanical equipment room. At the top of the building, the cooling equipment can be air-cooled.

The ultimate size of DX units is determined by information provided by the elevator manufacturer. The elevator consultant can provide the necessary general information to allow the design to proceed through bidding. The amount of cooling for this equipment can be significant: as much as 10 to 15 tons for a single elevator equipment room.

Elevator Hoistway and Machine Room Venting

All elevators installed in the United States must conform to ASME *Standard* A17.1, as modified by local authority and applicable building code. One requirement of many codes is to include a vent opening at the top of each elevator shaft that is 3.5% of the plan area of the hoistway or 3 ft^2 per elevator, whichever is greater. The purpose of this requirement is to allow venting of smoke during a building fire. To accomplish this, a duct must be provided from the vent to atmosphere. This is simple at the top of the building, but for low- and mid-rise elevators, where the elevator equipment room is not located in a mechanical room with perimeter access, extending the connecting duct to atmosphere may be difficult.

Under many codes, including the model *International Building Code®* [IBC (ICC 2009)], for a building that is fully sprinklered, the need for the vent and its extension to atmosphere may be waived for passenger elevators. The vent is typically still required for a dedicated service elevator car.

In addition, under the IBC, the vent may be closed under normal building operating conditions by including an automatic damper in the atmospheric vent or, under some code jurisdictions, by installing a piece of glass that will break in a fire. This damper must open upon the detection of smoke by any of the elevator lobby smoke detectors. Dampers have a distinct advantage in that they are manually and remotely resettable.

Where elevator speeds are greater than 1400 fpm, vents at the bottom of the shafts may be required by code to allow rapid escape of air when the high-speed car is descending.

LIFE SAFETY IN TALL BUILDINGS

Life safety challenges for tall buildings are similar to those of shorter high-rise buildings. It is impractical to rely on stairs as the means of egress to grade. Elevators should play a major role in safe evacuation of occupants and response of emergency forces. Areas or floors of refuge are needed to provide staging points for occupants evacuating and emergency forces responding. Codes have developed means to confront this challenge. This following provides a brief review of those life safety measures.

Codes and Standards

In the United States, the *International Building Code®* (IBC) is the predominant building code; in Canada, it is the *National Building Code of Canada* (NRC 2005). The National Fire Protection Association's (NFPA) *Standard* 5000 generally incorporates NFPA *Standard* 101. These codes do not define a "tall building," but have additional requirements for a high-rise building greater than or equal to 420 ft in height.

Components of Life Safety Systems for Tall Buildings

Tall buildings share many of the code requirements of other high-rise buildings. The IBC defines a high-rise building as, "a building with an occupied floor located more than 75 ft above the lowest floor of fire department vehicle access." Additional requirements are imposed for buildings 120 and 420 ft above grade. No specific definition of "tall building" is contained in the codes.

Key fire safety provisions for tall buildings should include the following:

- Smoke detection for elevator lobbies, elevator machine rooms, and HVAC systems
- Complete automatic sprinkler protection
- Fire standpipe system
- Smoke management system for enclosed exits, stairs, elevators, and areas or floors of refuge
- Emergency power for life safety systems
- Fire department or first-responder elevator
- Redundant exit stair or elevator emergency evacuation provisions
- Area or floor of refuge
- Fire command center

Detection

Automatic smoke detection should be provided in elevator lobbies, elevator machine rooms, mechanical and electrical equipment rooms, and any other spaces not provided with automatic sprinklers. The detection system should be connected to the automatic fire alarm system. Duct smoke detectors should be provided in the main return air and exhaust air plenum of each air-conditioning system with a capacity greater than 2000 cfm. Duct smoke detectors are also needed at each connection to a vertical duct or riser serving two or more floors from a return air duct or plenum.

The smoke detection system should be designed in accordance with NFPA *Standard* 72.

Residential buildings should have smoke alarms in each room used for sleeping purposes and on the ceiling or wall outside of each separate sleeping area. The smoke alarms should be interconnected so that activation of any smoke alarm in the dwelling unit activates all of the smoke alarms in that unit. This does not require activating smoke alarms in other apartments in the building.

Automatic Sprinkler Protection

Complete automatic sprinkler protection should be provided in accordance with NFPA *Standard* 13.

Standpipe System

Standpipe systems should be provided in accordance with NFPA *Standard* 14.

Smoke Management

The essential features of smoke management design are described in Chapter 53. Additional information is contained in NFPA *Standards* 92A and 92B.

The IBC requires exit stairs to be smoke protected. One way to achieve this is with a smokeproof tower of pressurized stairs. To enhance egress for buildings 420 ft high or more, the codes require either an additional exit stairway beyond those required by the typical exit calculations, or pressurization of the elevator shafts. To prevent smoke spread through the elevator without elevator shaft pressurization, elevator vestibules with a minimum 1 h fire resistance rating are required.

Codes also require an elevator for use by emergency responders, with access from a vestibule directly connected to an egress stair.

Elevators to be used for occupants in an emergency require special protection, including pressurized elevator shafts, an emergency voice/alarm communication system, elevator lobbies with direct access to a exit enclosure, and a means to protect the elevator from automatic sprinkler system water infiltrating the hoistway enclosure. Automatic sprinklers are prohibited from the elevator machine room, and shunt trips for elevators shutdown should not be provided.

Emergency Power

All life safety systems are required to have standby power designed and installed in accordance with NFPA *Standards* 110 and 111, as appropriate.

Fire Command Center

A fire command center is required in a protected location at or near grade to monitor all fire safety and emergency systems. It should also have controls for the smoke management system and emergency power system.

REFERENCES

ASME. 2010. *Boiler and pressure vessel code.* American Society of Mechanical Engineers, New York.

ASME. 2007. Safety code for elevators and escalators. *Standard* A17.1/CSA 844-2007. American Society of Mechanical Engineers, New York.

Harris, D.A. (ed.) 1991. *Noise control manual.* Van Nostrand Reinhold, New York.

ICC. 2009. *International building code®.* International Code Council, Washington, D.C.

Jordan, C. 1989. Central vs. local HVAC fan systems for high rise office buildings. *ASHRAE Journal* (Sept.):48-46.

Lovatt, J.E. and A.G. Wilson. 1994. Stack effect in tall buildings. *ASHRAE Transactions* 100(2):420-431.

Linford, R.G. and S.T. Taylor. 1989. HVAC systems: Central vs. floor-by-floor. *Heating/Piping/Air Conditioning* (July):43-49, 56-57, 84.

NFPA. 2010. Installation of sprinkler systems. *Standard* 13. National Fire Protection Association, Quincy, MA.

NFPA. 2010. Installation of standpipe and hose systems. *Standard* 14. National Fire Protection Association, Quincy, MA.

NFPA. 2010. National fire alarm and signaling code. *Standard* 72. National Fire Protection Association, Quincy, MA.

NFPA. 2009. Smoke-control systems utilizing barriers and pressure differences. *Standard* 92A. National Fire Protection Association, Quincy, MA.

NFPA. 2009. Smoke management systems in malls, atria, and large spaces. *Standard* 92B. National Fire Protection Association, Quincy, MA.

NFPA. 2010. Emergency and standby power systems. *Standard* 110. National Fire Protection Association, Quincy, MA.

NFPA. 2010. Stored electrical energy emergency and standby power systems. *Standard* 111. National Fire Protection Association, Quincy, MA.

NFPA. 2009. Building construction and safety code®. *Standard* 5000. National Fire Protection Association, Quincy, MA.

NRC. 2005. *National building code of Canada.* National Research Council Canada, Ottawa, ON.

Ross, D.E. 1996. Bank of China—An integration of architecture and engineering. Total Building Design Seminar, Chicago, Illinois.

Ross, D. 2004. *An HVAC design guide for tall commercial buildings.* ASHRAE.

Stewart, W.E., Jr. 1998. Effect of air pressure differential on vapor flow through sample building walls. *ASHRAE Transactions* 104(2):17-24.

Tamblyn, R.T. 1991. Coping with air pressure problems in tall buildings. *ASHRAE Transactions* 97(1):824-827.

Tamblyn, R.T. 1993. HVAC system effects for tall buildings. *ASHRAE Transactions* 99(2):789-792.

CHAPTER 5

PLACES OF ASSEMBLY

ASSEMBLY rooms are generally large, have relatively high ceilings, and are few in number for any given facility. They usually have a periodically high density of occupancy per unit floor area, as compared to other buildings, and thus have a relatively low design sensible heat ratio.

This chapter summarizes some of the design concerns for enclosed assembly buildings. (Chapter 3, which covers general criteria for commercial and public buildings, also includes information that applies to public assembly buildings.)

GENERAL CRITERIA

Energy conservation codes and standards must be considered because they have a major impact on design and performance.

Assembly buildings may have relatively few hours of use per week and may not be in full use when maximum outdoor temperatures or solar loading occur. Often they are fully occupied for as little as 1 to 2 h, and the load may be materially reduced by precooling. The designer needs to obtain as much information as possible about the anticipated hours of use, particularly times of full seating, so that simultaneous loads may be considered to optimize performance and operating economy. Dehumidification requirements and part-load dehumidification requirements should be considered before determining equipment size. The intermittent or infrequent nature of the cooling loads may allow these buildings to benefit from thermal storage systems.

Occupants usually generate the major room cooling and ventilation load. The number of occupants is best determined from the seat count, but when this is not available, it can be estimated at 7.5 to 10 ft² per person for the entire seating area, including exit aisles but not the stage, performance areas, or entrance lobbies.

Safety and Security

Assembly buildings may need new safety and security considerations regarding extraordinary incidents. Designers should follow the recommendations outlined in Chapter 59.

Outdoor Air

Outdoor air ventilation rates as prescribed by ASHRAE *Standard* 62.1 can be a major portion of the total load. The latent load (dehumidification and humidification) and energy used to maintain relative humidity within prescribed limits are also concerns. Humidity must be maintained at proper levels to prevent mold and mildew growth and for acceptable indoor air quality and comfort.

Lighting Loads

Lighting loads are one of the few major loads that vary from one type of assembly building to another. Levels can vary from 150 footcandles in convention halls where television cameras are expected to be used, to virtually nothing, as in a movie theater. In many assembly buildings, lights are controlled by dimmers or other

means to present a suitably low level of light during performances, with much higher lighting levels during cleanup, when the house is nearly empty. The designer should ascertain the light levels associated with maximum occupancies, not only for economy but also to determine the proper room sensible heat ratio.

Indoor Air Conditions

Indoor air temperature and humidity should follow ASHRAE comfort recommendations in Chapter 9 of the 2009 *ASHRAE Handbook—Fundamentals* and ASHRAE *Standard* 55. In addition, the following should be considered:

- In arenas, stadiums, gymnasiums, and movie theaters, people generally dress informally. Summer indoor conditions may favor the warmer end of the thermal comfort scale, and the winter indoor temperature may favor the cooler end.
- In churches, concert halls, and theaters, most men wear jackets and ties and women often wear suits. The temperature should favor the middle range of design, and there should be little summer-to-winter variation.
- In convention and exhibition centers, the public is continually walking. The indoor temperature should favor the lower range of comfort conditions both in summer and in winter.
- In spaces with a high population density or with a sensible heat factor of 0.75 or less, reheat should be considered.
- Energy conservation codes must be considered in both the design and during operation.

Assembly areas generally require some reheat to maintain the relative humidity at a suitably low level during periods of maximum occupancy. Refrigerant hot gas or condenser water is well suited for this purpose. Face-and-bypass control of low-temperature cooling coils is also effective. In colder climates, it may also be desirable to provide humidification. High rates of internal gain may make evaporative humidification attractive during economizer cooling.

Filtration

Most places of assembly are minimally filtered with filters rated at 30 to 35% efficiency, as tested in accordance with ASHRAE *Standard* 52.1. Where smoking is permitted, however, filters with a minimum rating of 80% are required to remove tobacco smoke effectively. Filters with 80% or higher efficiency are also recommended for facilities having particularly expensive interior decor. Because of the few operating hours of these facilities, the added expense of higher-efficiency filters can be justified by their longer life. Low-efficiency prefilters are generally used with high-efficiency filters to extend their useful life. Ionization and chemically reactive filters should be considered where high concentrations of smoke or odors are present.

Noise and Vibration Control

The desired noise criteria (NC) vary with the type and quality of the facility. The need for noise control may be minimal in a gymnasium or natatorium, but it is important in a concert hall. Multipurpose facilities require noise control evaluation over the entire spectrum of use.

The preparation of this chapter is assigned to TC 9.8, Large-Building Air-Conditioning Applications.

In most cases, sound and vibration control is required for both equipment and duct systems, as well as in diffuser and grille selection. When designing a performance theater or concert hall, an experienced acoustics engineer should be consulted, because the quantity and quality or characteristic of the noise is very important.

Transmission of vibration and noise can be decreased by mounting pipes, ducts, and equipment on a separate structure independent of the music hall. If the mechanical equipment space is close to the music hall, the entire mechanical equipment room may need to be floated on isolators, including the floor slab, structural floor members, and other structural elements such as supporting pipes or similar materials that can carry vibrations. Properly designed inertia pads are often used under each piece of equipment. The equipment is then mounted on vibration isolators.

Manufacturers of vibration isolating equipment have devised methods to float large rooms and entire buildings on isolators. Where subway and street noise may be carried into the structure of a music hall, it is necessary to float the entire music hall on isolators. If the music hall is isolated from outdoor noise and vibration, it also must be isolated from mechanical equipment and other internal noise and vibrations.

External noise from mechanical equipment such as cooling towers should not enter the building. Avoid designs that allow noises to enter the space through air intakes or reliefs and carelessly designed duct systems.

For more details on noise and vibration control, see Chapter 48 of this volume and Chapter 8 in the 2009 *ASHRAE Handbook—Fundamentals*.

Ancillary Facilities

Ancillary facilities are generally a part of any assembly building; almost all have some office space. Convention centers and many auditoriums, arenas, and stadiums have restaurants and cocktail lounges. Churches may have apartments for clergy or a school. Many facilities have parking structures. These varied ancillary facilities are discussed in other chapters of this volume. However, for reasonable operating economy, these facilities should be served by separate systems when their hours of use differ from those of the main assembly areas.

Air Conditioning

Because of their characteristic large size and need for considerable ventilation air, assembly buildings are frequently served by single-zone or variable-volume systems providing 100% outdoor air. Separate air-handling units usually serve each zone, although multizone, dual-duct, or reheat types can also be applied with lower operating efficiency. In larger facilities, separate zones are generally provided for entrance lobbies and arterial corridors that surround the seating space. Low-intensity radiant heating is often an efficient alternative. In some assembly rooms, folding or rolling partitions divide the space for different functions, so a separate zone of control for each resultant space is best. In extremely large facilities, several air-handling systems may serve a single space, because of the limits of equipment size and also for energy and demand considerations.

Peak Load Reduction

There are several techniques currently in use to help address peak loads. **Thermal storage** is discussed in Chapter 50 of the 2008 *ASHRAE Handbook—HVAC Systems and Equipment*. Another popular technique, **precooling**, can be managed by the building operator. Precooling the building mass several degrees below the desired indoor temperature several hours before it is occupied allows it to absorb a part of the peak heat load. This cooling reduces the equipment size needed to meet short-term loads. The effect can be used if cooling time of at least 1 h is available prior to occupancy, and then only when the period of peak load is relatively short (2 h or less).

The designer must advise the owner that the space temperature will be cold to most people as occupancy begins, but will warm up as the performance progresses; this should be understood by all concerned before proceeding with precooling. Precooling works best when the space is used only occasionally during the hotter part of the day and when provision of full capacity for an occasional purpose is not economically justifiable.

Stratification

Because most assembly buildings have relatively high ceilings, some heat may be allowed to stratify above the occupied zone, thereby reducing load on the equipment. Heat from lights can be stratified, except for the radiant portion (about 50% for fluorescent and 65% for incandescent or mercury-vapor fixtures). Similarly, only the radiant effect of the upper wall and roof load (about 33%) reaches the occupied space. Stratification only occurs when air is admitted and returned at a sufficiently low elevation so that it does not mix with the upper air. Conversely, stratification may increase heating loads during periods of minimal occupancy in winter. In these cases, ceiling fans, air-handling systems, or high/low air distribution may be desirable to reduce stratification. Balconies may also be affected by stratification and should be well ventilated.

Air Distribution

In assembly buildings with seating, people generally remain in one place throughout a performance, so they cannot move away from drafts. Therefore, good air distribution is essential. Airflow-modeling software could prove helpful in predicting potential problem areas.

Heating is seldom a major problem, except at entrances or during warm-up before occupancy. Generally, the seating area is isolated from the exterior by lobbies, corridors, and other ancillary spaces. For cooling, air can be supplied from the overhead space, where it mixes with heat from the lights and occupants. Return air openings can also aid air distribution. Air returns located below seating or at a low level around the seating can effectively distribute air with minimum drafts; however, register velocities over 275 fpm may cause objectionable drafts and noise.

Because of the configuration of these spaces, supply jet nozzles with long throws of 50 to 150 ft may need to be installed on sidewalls. For ceiling distribution, downward throw is not critical if returns are low. This approach has been successful in applications that are not particularly noise-sensitive, but the designer needs to select air distribution nozzles carefully.

The air-conditioning systems must be quiet. This is difficult to achieve if supply air is expected to travel 30 ft or more from sidewall outlets to condition the center of the seating area. Because most houses of worship, theaters, and halls are large, high air discharge velocities from the wall outlets are required. These high velocities can produce objectionable noise levels for people sitting near the outlets. This can be avoided if the return air system does some of the work. The supply air must be discharged from the air outlet (preferably at the ceiling) at the highest velocity consistent with an acceptable noise level. Although this velocity does not allow the conditioned air to reach all seats, the return air registers, which are located near seats not reached by the conditioned air, pull the air to cool or heat the audience, as required. In this way, supply air blankets the seating area and is pulled down uniformly by return air registers under or beside the seats.

A certain amount of exhaust air should be taken from the ceiling of the seating area, preferably over the balcony (if there is one) to prevent pockets of hot air, which can produce a radiant effect and cause discomfort, as well as increase the cost of air conditioning. Where the ceiling is close to the audience (e.g., below balconies and mezzanines), specially designed plaques or air-distributing ceilings should be provided to absorb noise.

Regular ceiling diffusers placed more than 30 ft apart normally give acceptable results if careful engineering is applied in selecting the diffusers. Because large air quantities are generally involved and because the building is large, fairly large capacity diffusers, which tend to be noisy, are frequently selected. Linear diffusers are more acceptable architecturally and perform well if selected properly. Integral dampers in diffusers should not be used as the only means of balancing because they generate intolerable amounts of noise, particularly in larger diffusers.

Mechanical Equipment Rooms

The location of mechanical and electrical equipment rooms affects the degree of sound attenuation treatment required. Those located near the seating area are more critical because of the normal attenuation of sound through space. Those near the stage area are critical because the stage is designed to project sound to the audience. If possible, mechanical equipment rooms should be in an area separated from the main seating or stage area by buffers such as lobbies or service areas. The economies of the structure, attenuation, equipment logistics, and site must be considered in selecting locations for mechanical equipment rooms.

At least one mechanical equipment room is placed near the roof to house the toilet exhaust, general exhaust, cooling tower, kitchen, and emergency stage exhaust fans, if any. Individual roof-mounted exhaust fans may be used, thus eliminating the need for a mechanical equipment room. However, to reduce sound problems, mechanical equipment should not be located on the roof over the music hall or stage but rather over offices, storerooms, or auxiliary areas.

HOUSES OF WORSHIP

Houses of worship seldom have full or near-full occupancy more than once a week, but they have considerable use for smaller functions (meetings, weddings, funerals, christenings, or daycare) throughout the rest of the week. It is important to determine how and when the building will be used. When thermal storage is used, longer operation of equipment before occupancy may be required because of the structure's high thermal mass. Seating capacity is usually well defined. Some houses of worship have a movable partition to form a single large auditorium for special holiday services. It is important to know how often this maximum use is expected.

Houses of worship test a designer's ingenuity in locating equipment and air diffusion devices in architecturally acceptable places. Because occupants are often seated, drafts and cold floors should be avoided. Many houses of worship have high, vaulted ceilings, which create thermal stratification. Where stained glass is used, a shade coefficient equal to solar glass (SC = 0.70) is assumed.

Houses of worship may also have auxiliary rooms that should be air conditioned. To ensure privacy, sound transmission between adjacent areas should be considered in the air distribution scheme. Diversity in the total air-conditioning load requirements should be evaluated to take full advantage of the characteristics of each area.

It is desirable to provide some degree of individual control for the platform, sacristy, and bema or choir area.

AUDITORIUMS

The types of auditoriums considered are movie theaters, playhouses, and concert halls. Auditoriums in schools and the large auditoriums in some convention centers may follow the same principles, with varying degrees of complexity.

Movie Theaters

Movie theaters are the simplest of the auditorium structures discussed here. They run continuously for periods of 8 h or more and, thus, are not a good choice for precooling techniques, except for the first matinee peak. They operate frequently at low occupancy levels, and low-load performance must be considered. Additionally, they tend to have lower sensible heat factors; special care must be taken to ensure proper relative humidity levels can be maintained without overcooling the space.

Motion picture studios often require that movie theaters meet specific noise criteria. Consequently, sound systems and noise control are as critical in these applications as they are in other kinds of theaters. The lobby and exit passageways in a motion picture theater are seldom densely occupied, although some light to moderate congestion can be expected for short times in the lobby area. A reasonable design for the lobby space is one person per 20 to 30 ft^2.

Lights are usually dimmed when the house is occupied; full lighting intensity is used only during cleaning. A reasonable value for lamps above the seating area during a performance is 5 to 10% of the installed wattage. Designated smoking areas should be handled with separate exhaust or air-handling systems to avoid contamination of the entire facility.

Projection Booths. The projection booth represents a larger challenge in movie theater design. For large theaters using high-intensity lamps, projection room design must follow applicable building codes. If no building code applies, the projection equipment manufacturer usually has specific requirements. The projection room may be air conditioned, but it is normally exhausted or operated at negative pressure. Exhaust is normally taken through the housing of the projectors. Additional exhaust is required for the projectionist's sanitary facilities. Other heat sources include sound and dimming equipment, which require a continuously controlled environment and necessitate a separate system.

Smaller theaters have fewer requirements for projection booths. It is a good idea to condition the projection room with filtered supply air to avoid soiling lenses. In addition to the projector light, heat sources in the projection room include the sound equipment, as well as the dimming equipment.

Performance Theaters

Performance theaters differ from motion picture theaters in the following ways:

- Performances are seldom continuous. Where more than one performance occurs in a day, performances are usually separated by 2 to 4 h. Accordingly, precooling techniques are applicable, particularly for afternoon performances.
- Performance theaters generally play to a full or near-full house.
- Performance theaters usually have intermissions, and the lobby areas are used for drinking and socializing. The intermissions are usually relatively short, seldom exceeding 15 to 20 min; however, the load may be as dense as one person per 5 ft^2.
- Because sound amplification is less used than in motion picture theaters, background noise control is more important.
- Stage lighting contributes considerably to the total cooling load in performance theaters. Lighting loads can vary from performance to performance.

Stages. The stage presents the most complex problem. It consists of the following loads:

- A heavy, mobile lighting load
- Intricate or delicate stage scenery, which varies from scene to scene and presents difficult air distribution requirements
- Actors, who may perform tasks that require exertion

Approximately 40 to 60% of the lighting load can be eliminated by exhausting air around the lights. This procedure works for lights around the proscenium. However, it is more difficult to place exhaust air ducts directly above lights over the stage because of the scenery and light drops. Careful coordination is required to achieve an effective and flexible layout.

Conditioned air should be introduced from the low side and back stages and returned or exhausted around the lights. Some exhaust air must be taken from the top of the tower directly over the stage containing lights and equipment (i.e., the fly). Air distribution design is further complicated because pieces of scenery may consist of light materials that flutter in the slightest air current. Even the vertical stack effect created by the heat from lights may cause this motion. Therefore, low air velocities are essential and air must be distributed over a wide area with numerous supply and return registers.

With multiple scenery changes, low supply or return registers from the floor of the stage are almost impossible to provide. However, some return air at the footlights and for the prompter should be considered. Air conditioning should also be provided for the stage manager and control board areas.

In many theaters with overhead flies, the stage curtain billows when it is down. This is primarily caused by the stack effect created by the height of the main stage tower, heat from lights, and the temperature difference between the stage and seating areas. Proper air distribution and balancing can minimize this phenomenon. Bypass damper arrangements with suitable fire protection devices may be feasible.

Loading docks adjacent to stages located in cold climates should be heated. Doors to these areas may be open for long periods, for example, while scenery is being loaded or unloaded for a performance.

On the stage, local code requirements must be followed for emergency exhaust ductwork or skylight (or blow-out hatch) requirements. These openings are often sizable and should be incorporated in the early design concepts.

Concert Halls

Concert halls and music halls are similar to performance theaters. They normally have a full stage, complete with fly gallery, and dressing areas for performers. Generally, the only differences between the two are in size and decor, with the concert hall usually being larger and more elaborately decorated.

Air-conditioning design must consider that the concert hall is used frequently for special charity and civic events, which may be preceded or followed by parties (and may include dancing) in the lobby area. Concert halls often have cocktail lounge areas that become very crowded, with heavy smoking during intermissions. These areas should be equipped with flexible exhaust-recirculation systems. Concert halls may also have full restaurant facilities.

As in theatres, noise control is important. Design must avoid characterized or narrow-band noises in the level of audibility. Much of this noise is structure-borne, resulting from inadequate equipment and piping vibration isolation. An experienced acoustical engineer is essential for help in the design of these applications.

ARENAS AND STADIUMS

Functions at arenas and stadiums may be quite varied, so the air-conditioning loads will vary. Arenas and stadiums are not only used for sporting events such as basketball, ice hockey, boxing, and track meets but may also house circuses; rodeos; convocations; social affairs; meetings; rock concerts; car, cycle, and truck events; and special exhibitions such as home, industrial, animal, or sports shows. For multipurpose operations, the designer must provide highly flexible systems. High-volume ventilation may be satisfactory in many instances, depending on load characteristics and outdoor air conditions.

Load Characteristics

Depending on the range of use, the load may vary from a very low sensible heat ratio for events such as boxing to a relatively high sensible heat ratio for industrial exhibitions. Multispeed fans often improve performance at these two extremes and can aid in sound control for special events such as concerts or convocations. When

using multispeed fans, the designer should consider the performance of the air distribution devices and cooling coils when the fan is operating at lower speeds.

Because total comfort cannot be ensured in an all-purpose facility, the designer must determine the level of discomfort that can be tolerated, or at least the type of performances for which the facility is primarily intended.

As with other assembly buildings, seating and lighting combinations are the most important load considerations. Boxing events, for example, may have the most seating, because the boxing ring area is very small. For the same reason, however, the area that needs to be intensely illuminated is also small. Thus, boxing matches may represent the largest latent load situation. Other events that present large latent loads are rock concerts and large-scale dinner dances, although the audience at a rock concert is generally less concerned with thermal comfort. Ventilation is also essential in removing smoke or fumes at car, cycle, and truck events. Circuses, basketball, and hockey have a much larger arena area and less seating. The sensible load from lighting the arena area improves the sensible heat ratio. The large expanse of ice in hockey games considerably reduces both latent and sensible loads. High latent loads caused by occupancy or ventilation can create severe problems in ice arenas such as condensation on interior surfaces and fog. Special attention should be paid to the ventilation system, air distribution, humidity control, and construction materials.

Enclosed Stadiums

An enclosed stadium may have either a retractable or a fixed roof. When the roof is closed, ventilation is needed, so ductwork must be run in the permanent sections of the stadium. The large air volumes and long air throws required make proper air distribution difficult to achieve; thus, the distribution system must be very flexible and adjustable.

Some open stadiums have radiant heating coils in the floor slabs of the seating areas. Gas-fired or electric high- or low-intensity radiant heating located above the occupants is also used.

Open racetrack stadiums may present a ventilation problem if the grandstand is enclosed. The grandstand area may have multiple levels and be in the range of 1300 ft long and 200 ft deep. The interior (ancillary) areas must be ventilated to control odors from toilet facilities, concessions, and the high population density. General practice provides about four air changes per hour for the stand seating area and exhausts air through the rear of the service areas. More efficient ventilation systems may be selected if architectural considerations allow. Window fogging is a winter concern with glass-enclosed grandstands. This can be minimized by double glazing, humidity control, moving dry air across the glass, or a radiant heating system for perimeter glass areas.

Air-supported structures require continuous fan operation to maintain a properly inflated condition. The possibility of condensation on the underside of the air bubble should be considered. The U-factor of the roof should be sufficient to prevent condensation at the lowest expected ambient temperature. Heating and air-conditioning functions can be either incorporated into the inflating system or furnished separately. Solar and radiation control is also possible through the structure's skin. Applications, though increasing rapidly, still require working closely with the enclosure manufacturer to achieve proper and integrated results.

Ancillary Spaces

The concourse areas of arenas and stadiums are heavily populated during entrance, exit, and intermission periods. Considerable odor is generated in these areas by food, drink, and smoke, requiring considerable ventilation. If energy conservation is an important factor, carbon filters and controllable recirculation rates should be considered. Concourse area air systems should be considered for their flexibility of returning or exhausting air. The economics of this type of flexibility should be evaluated with regard to the associated problem of air balance and freeze-up in cold climates.

Ticket offices, restaurants, and similar facilities are often expected to be open during hours that the main arena is closed; therefore, separate systems should be considered for these areas.

Locker rooms require little treatment other than excellent ventilation, usually not less than 2 or 3 cfm per square foot. To reduce the outdoor air load, excess air from the main arena or stadium may be transferred into the locker rooms. However, reheat or recooling by water or primary air should be considered to maintain the locker room temperature. To maintain proper air balance under all conditions, locker rooms should have separate supply and exhaust systems.

Ice Rinks

Refer to Chapter 44 of the 2010 *ASHRAE Handbook—Refrigeration* for ice sheet design information. When an ice rink is designed into the facility, the concerns of groundwater conditions, site drainage, structural foundations, insulation, and waterproofing become even more important, with the potential of freezing soil or fill under the floor and subsequent expansion. The rink floor may have to be strong enough to support heavy trucks. The floor insulation also must be strong enough to take this load. Ice-melting pits of sufficient size with steam pipes may have to be furnished. If the arena is to be air conditioned, the possibility of combining the air-conditioning system with the ice rink system could be considered. The designer should be aware, however, that both systems operate at vastly different temperatures and have considerably different operation profiles. The radiant effects of the ice on the people and of heat from the roof and lights on the ice must be considered in the design and operation of the system. Low air velocities at the ice sheet level help minimize the refrigeration load. Conversely, high air velocities cause the ice to melt or sublimate.

Fog forms when moisture-laden air cools below its dew point. This is most likely to occur close to the ice surface within the boarded area (playing area). Fog can be controlled by reducing the indoor dew point with a dehumidification system or high-latent-capacity air-conditioning system and by delivering appropriate air velocities to bring the air in contact with the ice. Air-conditioning systems have had limited success in reducing the dew-point temperature sufficiently to prevent fog. The section on Ice Rink Dehumidifiers in Chapter 24 of the 2008 *ASHRAE Handbook—HVAC Systems and Equipment* has more information on fog control.

The type of lighting used over ice rinks must be carefully considered when precooling is used before hockey games and between periods. Main lights should be able to be turned off, if feasible. Incandescent lights require no warm-up time and are more applicable than types requiring warm-up. Low-emissivity ceilings with reflective characteristics successfully reduce condensation on roof structures; they also reduce lighting and, consequently, the cooling requirements.

Gymnasiums

Smaller gymnasiums, such as those in schools, are miniature versions of arenas and often have multipurpose features. For further information, see Chapter 7, Educational Facilities.

Many school gymnasiums are not air conditioned. Low-intensity perimeter radiant heaters with central ventilation supplying four to six air changes per hour are effective and energy efficient. Unit heaters on the ceiling are also effective. Ventilation must be provided because of high activity levels and resulting odors.

Most gymnasiums are located in schools. However, public and private organizations and health centers may also have gymnasiums. During the day, gymnasiums are usually used for physical activities, but in the evening and on weekends, they may be used for sports events, social affairs, or meetings. Thus, their activities fall within the scope of those of a civic center. More gymnasiums are being considered for air conditioning to make them more suitable for civic center activities. Design criteria are similar to arenas and civic centers when used for such activities. However, for schooltime use, space temperatures are often kept between 65 and 68°F during the heating season. Occupancy and the degree of activity during daytime use does not usually require high quantities of outdoor air, but if used for other functions, system flexibility is required.

CONVENTION AND EXHIBITION CENTERS

Convention and exhibition centers schedule diverse functions similar to those at arenas and stadiums and present a unique challenge to the designer. The center generally is a high-bay, long-span space, and can change weekly, for example, from an enormous computer room into a gigantic kitchen, large machine shop, department store, automobile showroom, or miniature zoo. They can also be the site of gala banquets or used as major convention meeting rooms.

Income earned by these facilities is directly affected by the time it takes to change from one activity to the next, so highly flexible utility distribution and air-conditioning equipment are needed.

Ancillary facilities include restaurants, bars, concession stands, parking garages, offices, television broadcasting rooms, and multiple meeting rooms varying in capacity from small (10 to 20 people) to large (hundreds or thousands of people). Often, an appropriately sized full-scale auditorium or arena is also incorporated.

By their nature, these facilities are much too large and diverse in their use to be served by a single air-handling system. Multiple air handlers with several chillers can be economical.

Load Characteristics

The main exhibition room is subject to a variety of loads, depending on the type of activity in progress. Industrial shows provide the highest sensible loads, which may have a connected capacity of 20 W/ft^2 along with one person per 40 to 50 ft^2. Loads of this magnitude are seldom considered because large power-consuming equipment is seldom in continuous operation at full load. An adequate design accommodates (in addition to lighting load) about 10 W/ft^2 and one person per 40 to 50 ft^2 as a maximum continuous load.

Alternative loads of very different character may be encountered. When the main hall is used as a meeting room, the load will be much more latent. Thus, multispeed fans or variable-volume systems may provide a better balance of load during these high-latent, low-sensible periods of use. Accurate occupancy and usage information is critical in any plan to design and operate such a facility efficiently and effectively.

System Applicability

The main exhibition hall is normally handled by one or more all-air systems. This equipment should be able to operate on all outdoor air, because during set-up, the hall may contain highway-size trucks bringing in or removing exhibit materials. There are also occasions when the space is used for equipment that produces an unusual amount of fumes or odors, such as restaurant or printing industry displays. It is helpful to build some flues into the structure to duct fumes directly to the outdoors. Perimeter radiant ceiling heaters have been successfully applied to exhibition halls with large expanses of glass.

Smaller meeting rooms are best conditioned either with individual room air handlers, or with variable-volume central systems, because these rooms have high individual peak loads but are not used frequently. Constant-volume systems of the dual- or single-duct reheat type waste considerable energy when serving empty rooms, unless special design features are incorporated.

Offices and restaurants often operate for many more hours than the meeting areas or exhibition areas and should be served separately.

Storage areas can generally be conditioned by exhausting excess air from the main exhibit hall through these spaces.

NATATORIUMS

Environmental Control

A natatorium requires year-round humidity levels between 40 and 60% for comfort, reasonable energy consumption, and building envelope protection. The designer must address the following concerns: humidity control, room pressure control, ventilation requirements for air quality (outdoor and exhaust air), air distribution, duct design, pool water chemistry, and evaporation rates. A humidity control system alone will not provide satisfactory results if any of these items are overlooked. See Chapter 24 of the 2008 *ASHRAE Handbook—HVAC Systems and Equipment* for additional dehumidifier application and design information.

Humidity Control

People who are wet are very sensitive to relative humidity and the resultant evaporation that occurs. Fluctuations in relative humidity outside the 50 to 60% range are not recommended. Sustained levels above 60% can promote factors that reduce indoor air quality. Relative humidity levels below 50% significantly increase the facility's energy consumption. For swimmers, 50 to 60% rh limits evaporation and corresponding heat loss from the body and is comfortable without being extreme. Higher relative humidity levels can be destructive to building components. Mold and mildew can attack wall, floor, and ceiling coverings, and condensation can degrade many building materials. In the worst case, the roof structure could fail because of corrosion from water condensing on the structure.

Load Estimation

Loads for a natatorium include heat gains and losses from outdoor air, lighting, walls, roof, and glass. Internal latent loads are generally from people and evaporation. Evaporation loads in pools and spas are significant relative to other load elements and may vary widely depending on pool features, areas of water and wet deck, water temperature, and activity level in the pool.

Evaporation. The rate of evaporation can be estimated from empirical Equation (1). This equation is valid for pools at normal activity levels, allowing for splashing and a limited area of wetted deck. Other pool uses may have more or less evaporation (Smith et al. 1993).

$$w_p = \frac{A}{Y}(p_w - p_a)(95 + 0.425\,V) \tag{1}$$

where

w_p = evaporation of water, lb/h
A = area of pool surface, ft^2
Y = latent heat required to change water to vapor at surface water temperature, Btu/lb
p_w = saturation vapor pressure taken at surface water temperature, in. Hg
p_a = saturation pressure at room air dew point, in. Hg
V = air velocity over water surface, fpm

Table 1 Typical Natatorium Design Conditions

Type of Pool	Air Temperature, °F	Water Temperature, °F	Relative Humidity, %
Recreational	75 to 85	75 to 85	50 to 60
Therapeutic	80 to 85	85 to 95	50 to 60
Competition	78 to 85	76 to 82	50 to 60
Diving	80 to 85	80 to 90	50 to 60
Elderly swimmers	84 to 90	85 to 90	50 to 60
Hotel	82 to 85	82 to 86	50 to 60
Whirlpool/spa	80 to 85	97 to 104	50 to 60

Units for the constant 95 are Btu/(h·ft^2·in. Hg). Units for the constant 0.425 are Btu·min/(h·ft^3·in. Hg).

Equation (1) may be modified by multiplying it by an activity factor F_a to alter the estimate of evaporation rate based on the level of activity supported. For Y values of about 1000 Btu/lb and V values ranging from 10 to 30 fpm, Equation (1) can be reduced to

$$w_p = 0.1A(p_w - p_a)F_a \tag{2}$$

The following activity factors should be applied to the areas of specific features, and not to the entire wetted area:

Type of Pool	Typical Activity Factor (F_a)
Baseline (pool unoccupied)	0.5
Residential pool	0.5
Condominium	0.65
Therapy	0.65
Hotel	0.8
Public, schools	1.0
Whirlpools, spas	1.0
Wavepools, water slides	1.5 (minimum)

The effectiveness of controlling the natatorium environment depends on correct estimation of water evaporation rates. Applying the correct activity factors is extremely important in determining water evaporation rates. The difference in peak evaporation rates between private pools and active public pools of comparable size may be more than 100%.

Actual operating temperatures and relative humidity conditions should be established before design. How the area will be used usually dictates design (Table 1).

Air temperatures in public and institutional pools are recommended to be maintained 2 to 4°F above the water temperature (but not above the comfort threshold of 86°F) for energy conservation through reduced evaporation and to avoid chill effects on swimmers.

Competition pools that host swim meets have two distinct operating profiles: (1) swim meets and (2) normal occupancy. It is recommended that both be fully modeled to evaluate the facility's needs. Although swim meets tend to be infrequent, the loads during meets are often considerably higher than during normal operations. To model the swim meet load accurately, it is recommended that the designer know the number of spectators, number of swimmers on the deck, and operating conditions required during the meets. The operator may request a peak relative humidity of 55%, which has a significant impact on total loads. A system designed for swim meet loads should also be designed to operate for considerable portions of the year at part loads.

Water parks and water feature (slides, spray cannons, arches, etc.) loads are not fully covered by this chapter. It is recommended that the dehumidification load generated by each water feature be calculated individually. The water toys' manufacturers should be contacted to provide specifications to allow for proper load determination. Due to the concentrated nature of the loads in these facilities, it is recommended that more supply air and outdoor air be used in these facilities compared to what is recommended for traditional pools.

Ventilation Requirements

Air Quality. Outdoor air ventilation rates prescribed by ASHRAE *Standard* 62.1 are intended to provide acceptable air quality conditions for the average pool using chlorine for primary disinfection. The ventilation requirement may be excessive for private pools and installations with low use, and may also prove inadequate for high-occupancy public or water park installations.

Air quality problems in pools and spas are often caused by water quality problems, so simply increasing ventilation rates may prove both expensive and ineffective. Water quality conditions are a direct function of pool use and the type and effectiveness of water disinfection used.

Because indoor pools usually have high ceilings, temperature stratification and stack effect (see Chapter 16 of the 2009 *ASHRAE Handbook—Fundamentals*) can have a detrimental effect on indoor air quality. Careful duct layout must ensure that the space receives proper air changes and homogeneous air quality throughout. Some air movement at the deck and pool water level is essential to ensure acceptable air quality. Complaints from swimmers indicate that the greatest chloramine (see the section on Pool Water Chemistry) concentrations occur at the water surface. Children are especially vulnerable to the ill effects of chloramine inhalation.

Exhaust air from pools is rich in moisture and may contain high levels of corrosive chloramine compounds. Although most codes allow pool air to be used as makeup for showers, toilets, and locker rooms, these spaces should be provided with separate ventilation and maintained at a positive pressure with respect to the pool.

Pool and spa areas should be maintained at a negative pressure of 0.05 to 0.15 in. of water relative to the outdoors and adjacent areas of the building to prevent chloramine odor migration. Active methods of pressure control may prove more effective than static balancing and may be necessary where outdoor air is used as a part of an active humidity control strategy. Openings from the pool to other areas should be minimized and controlled. Passageways should be equipped with doors with automatic closers and sweeps to inhibit migration of moisture and air.

Exhaust air intake grilles should be located as close as possible to the warmest body of water in the facility. Warmer waters and those with high agitation levels off gas chemicals at higher rates compared to traditional pools. This also allows body oils to become airborne. Ideally these pollutants should be removed from close to the source before they have a chance to diffuse and negatively impact the air quality. Installations with intakes directly above whirlpools have resulted in the best air quality.

Air Delivery Rates. Most codes require a minimum of six air changes per hour, except where mechanical cooling is used. This rate may prove inadequate for some occupancy and use.

Where mechanical dehumidification is provided, air delivery rates should be established to maintain appropriate conditions of temperature and humidity. The following rates are typically desired:

Pools with no spectator areas	4 to 6 air changes per hour
Spectator areas	6 to 8 air changes per hour
Therapeutic pools	4 to 6 air changes per hour

Outdoor air delivery rates may be constant or variable, depending on design. Minimum rates, however, must provide adequate dilution of contaminants generated by pool water and must maintain acceptable ventilation for occupancy.

Where a minimum outdoor air ventilation rate is established to protect against condensation in a building's structural elements, the rates are typically used for 100% outdoor air systems. These rates usually result in excessive humidity levels under most operating conditions and are generally not adequate to produce acceptable indoor air quality, especially in public facilities subject to heavy use.

Duct Design

Proper duct design and installation in a natatorium is critical. Failure to effectively deliver air where needed can result in air quality problems, condensation, stratification, and poor equipment performance. Ductwork that fails to deliver airflow at the pool deck and water surface, for example, can lead to air quality problems in those areas. The following duct construction practices apply to natatoriums:

- Duct materials and hardware must be resistant to chemical corrosion from the pool atmosphere. Stainless steels, even the 316 series,

are readily attacked by chlorides and are prone to pitting. They require treatment to adequately perform in a natatorium environment. Galvanized steel and aluminum sheet metal may be used for exposed duct systems. If galvanized duct is used, steps should be taken to adequately protect the metal from corrosion. It is recommended that, at a minimum, the galvanized ducts be properly prepared and painted with epoxy-based or other durable paint suitable to protect metal surfaces in a natatorium environment. Note that galvannealed ductwork is easier to weld and paint than hot-dip galvanized, but galvannealed is more susceptible to corrosion if left bare. Certain types of fabric duct (airtight) with appropriate grilles sewn in are also a good choice. Buried ductwork should be constructed from nonmetallic fiberglass-reinforced or PVC materials because of the more demanding environment.

- Grilles, registers, and diffusers should be constructed from aluminum. They should be selected for low static pressure loss and for appropriate throws for proper air distribution.
- Supply air should be directed against envelope surfaces prone to condensation (glass and doors). Some supply air should be directed over the water surface to move contaminated air toward an exhaust point and control chloramines released at the water surface. However, air movement over the pool water surface must not exceed 30 fpm [as per the evaporation rate w_p in Equation (1)].
- Return air inlets should be located to recover warm, humid air and return it to the ventilation system for treatment, to prevent supply air from short-circuiting and to minimize recirculation of chloramines.
- Exhaust air inlets should be located to maximize capture effectiveness and minimize recirculation of chloramines. Exhausting from directly above whirlpools is also desirable. Exhaust air should be taken directly to the outdoors, through heat recovery devices where provided.
- Filtration should be selected to provide 45 to 65% efficiencies (as defined in ASHRAE *Standard* 52.1) and be installed in locations selected to prevent condensation in the filter bank. Filter media and support materials should be resistant to moisture degradation.
- Fiberglass duct liner should not be used. Where condensation may occur, the insulation must be applied to the duct exterior.
- Air systems should be designed for noise levels listed in Table 42 of Chapter 48 (NC 45 to 50); however the room wall, floor, and ceiling surfaces should be evaluated for their reverberation times and speech intelligibility.

Envelope Design

Glazing in exterior walls becomes susceptible to condensation when the outdoor temperature drops below the pool room dew point. The design goal is to maintain the surface temperature of the glass and the window frames a minimum of 5°F above the pool room dew point. Windows must allow unobstructed air movement on inside surfaces, and thermal break frames should be used to raise the indoor temperature of the frame. Avoid recessed windows and protruding window frames. Skylights are especially vulnerable, and require attention to control condensation. Wall and roof vapor retarder designs should be carefully reviewed, especially at wall-to-wall and wall-to-roof junctures and at window, door, skylight, and duct penetrations. The pool enclosure must be suitable for year-round operation at 50 to 60% relative humidity. A vapor retarder analysis (as in Figure 12 in Chapter 27 of the 2009 *ASHRAE Handbook—Fundamentals*) should be prepared. Failure to install an effective vapor retarder will result in condensation forming in the structure, and potentially serious envelope damage.

Pool Water Chemistry

Failure to maintain proper chemistry in the pool water causes serious air quality problems and deterioration of mechanical systems and building components. Water treatment equipment and chemicals should be located in a separate, dedicated, well-ventilated space that is under negative pressure. Pool water treatment consists of primary disinfection, pH control, water filtration and purging, and water heating. For further information, refer to Kowalsky (1990).

Air quality problems are usually caused by the reaction of chlorine with biological wastes, and particularly with ammonia, which is a by-product of the breakdown of urine and perspiration. Chlorine reacts with these wastes, creating chloramines (monochloramine, dichloramine, and nitrogen trichloride) that are commonly measured as combined chlorine. Adding chemicals to pool water increases total contaminant levels. In high-occupancy pools, water contaminant levels can double in a single day of operation.

Chlorine's efficiency at reducing ammonia is affected by several factors, including water temperature, water pH, total chlorine concentration, and level of dissolved solids in the water. Because of their higher operating temperature and higher ratio of occupancy per unit water volume, spas produce greater quantities of air contaminants than pools.

The following measures have demonstrated a potential to reduce chloramine concentrations in the air and water:

- **Ozonation.** In low concentrations, ozone has substantially reduced the concentration of combined chlorine in the water. In high concentrations, ozone can replace chlorine as the primary disinfection process; however, ozone is unable to maintain sufficient residual levels in the water to maintain a latent biocidal effect. This necessitates maintenance of chlorine as a residual process at concentrations of 0.5 to 1.5 ppm.
- **Water Exchange Rates.** High concentrations of dissolved solids in water have been shown to directly contribute to high combined chlorine (chloramine) levels. Adequate water exchange rates are necessary to prevent the buildup of biological wastes and their oxidized components in pool and spa water. Conductivity measurement is an effective method to control the exchange rate of water in pools and spas to effectively maintain water quality and minimize water use. In high-occupancy pools, heat recovery may prove useful in reducing water heating energy requirements.

Energy Considerations

Natatoriums can be a major energy burden on facilities, so they represent a significant opportunity for energy conservation and recovery. ASHRAE *Standard* 90.1 offers some recommendations. Several design solutions are possible using both dehumidification and ventilation strategies. When evaluating a system, the seasonal space conditions and energy consumed by all elements should be considered, including primary heating and cooling systems, fan motors, water heaters, and pumps.

Operating conditions factor significantly in the total energy requirements of a natatorium. Although occupant comfort is a primary concern, the impact of low space temperatures and relative humidity levels below 50% (especially in winter) should be discussed with the owner/operator. Reductions in either room air temperature or relative humidity increase evaporation from the pools, thus increasing the dehumidification requirements and increasing pool water heating costs.

Natatoriums with fixed outdoor air ventilation rates without dehumidification generally have seasonally fluctuating space temperature and humidity levels. Systems designed to provide minimum ventilation rates without dehumidification are unable to maintain relative humidity conditions within prescribed limits. These systems may facilitate mold and mildew growth and may be unable to provide acceptable indoor air quality. Peak dehumidification loads vary with activity levels and during the cooling season when ventilation air becomes an additional dehumidification load to the space.

FAIRS AND OTHER TEMPORARY EXHIBITS

Occasionally, large-scale exhibits are constructed to stimulate business, present new ideas, and provide cultural exchanges. Fairs of this type take years to construct, are open from several months to several years, and are sometimes designed considering future use of some buildings. Fairs, carnivals, or exhibits, which may consist of prefabricated shelters and tents that are moved from place to place and remain in a given location for only a few days or weeks, are not covered here because they seldom require the involvement of architects and engineers.

Design Concepts

One consultant or agency should be responsible for setting uniform utility service regulations and practices to ensure proper organization and operation of all exhibits. Exhibits that are open only during spring or fall months require a much smaller heating or cooling plant than those open during peak summer or winter months. This information is required in the earliest planning stages so that system and space requirements can be properly analyzed.

Occupancy

Fair buildings have heavy occupancy during visiting hours, but patrons seldom stay in any one building for a long period. The length of time that patrons stay in a building determines the air-conditioning design. The shorter the anticipated stay, the greater the leeway in designing for less-than-optimum comfort, equipment, and duct layout. Also, whether patrons wear coats and jackets while in the building influences operating design conditions.

Equipment and Maintenance

Heating and cooling equipment used solely for maintaining comfort and not for exhibit purposes may be secondhand or leased, if available and of the proper capacity. Another possibility is to rent the air-conditioning equipment to reduce the capital investment and eliminate disposal problems when the fair is over.

Depending on the size of the fair, length of operation time, types of exhibitors, and fair sponsors' policies, it may be desirable to analyze the potential for a centralized heating and cooling plant versus individual plants for each exhibit. The proportionate cost of a central plant to each exhibitor, including utility and maintenance costs, may be considerably less than having to furnish space and plant utility and maintenance costs. The larger the fair, the more savings may result. It may be practical to make the plant a showcase, suitable for exhibit and possibly added revenue. A central plant may also form the nucleus for commercial or industrial development of the area after the fair is over.

If exhibitors furnish their own air-conditioning plants, it is advisable to analyze shortcuts that may be taken to reduce equipment space and maintenance aids. For a 6-month to 2-year maximum operating period, for example, tube pull or equipment removal space is not needed or may be drastically reduced. Higher fan and pump motor power and smaller equipment are permissible to save on initial costs. Ductwork and piping costs should be kept as low as possible because these are usually the most difficult items to salvage; cheaper materials may be substituted wherever possible. The job must be thoroughly analyzed to eliminate all unnecessary items and reduce all others to bare essentials.

The central plant may be designed for short-term use as well. However, if it is to be used after the fair closes, the central plant should be designed in accordance with the best practice for long-life plants. It is difficult to determine how much of the piping distribution system can be used effectively for permanent installations. For that reason, piping should be simply designed initially, preferably in a grid, loop, or modular layout, so that future additions can be made easily and economically.

Air Cleanliness

The efficiency of filters needed for each exhibit is determined by the nature of the area served. Because the life of an exhibit is very short, it is desirable to furnish the least expensive filtering system. If possible, one set of filters should be selected to last for the life of the exhibit. In general, filtering efficiencies do not have to exceed 30% (see ASHRAE *Standard* 52.1).

System Applicability

If a central air-conditioning plant is not built, equipment installed in each building should be the least costly to install and operate for the life of the exhibit. These units and systems should be designed and installed to occupy the minimum usable space.

Whenever feasible, heating and cooling should be performed by one medium, preferably air, to avoid running a separate piping and radiation system for heating and a duct system for cooling. Air curtains used on an extensive scale may, on analysis, simplify building structure and lower total costs.

Another possibility when both heating and cooling are required is a heat pump system, which may be less costly than separate heating and cooling plants. Economical operation may be possible, depending on the building characteristics, lighting load, and occupant load. If well or other water is available, it may produce a more economical installation than an air-source heat pump.

ATRIUMS

Atriums have diverse functions and occupancies. An atrium may (1) connect buildings; (2) serve as an architectural feature, leisure space, greenhouse, and/or smoke reservoir; and (3) afford energy and lighting conservation. The temperature, humidity, and hours of usage of an atrium are directly related to those of the adjacent buildings. Glass window walls and skylights are common. Atriums are generally large in volume with relatively small floor areas. The temperature and humidity conditions, air distribution, impact from adjacent buildings, and fenestration loads to the space must be considered in the design of an atrium.

Perimeter radiant heating (e.g., overhead, wall finned-tube, floor, or combinations thereof) is commonly used for expansive glass windows and skylights. Air-conditioning systems can heat, cool, and control smoke. Distribution of air across windows and skylights can also control heat transfer and condensation. Low supply and high return air distribution can control heat stratification, as well as wind and stack effects. Some atrium designs include a combination of high/low supply and high/low return air distribution to control heat transfer, condensation, stratification, and wind/stack effects.

The energy use of an atrium can be reduced by installing double- and triple-panel glass and mullions with thermal breaks, as well as shading devices such as external, internal, and interior screens, shades, and louvers.

Extensive landscaping is common in atriums. Humidity levels are generally maintained between 10 and 35%. Hot and cold air should not be distributed directly onto plants and trees.

REFERENCES

ASHRAE. 1992. Gravimetric and dust-spot procedures for testing air-cleaning devices used in general ventilation for removing particulate matter. ANSI/ASHRAE *Standard* 52.1-1992.

ASHRAE. 2004. Thermal environmental conditions for human occupancy. ANSI/ASHRAE *Standard* 55-2004.

ASHRAE. 2004. Ventilation for acceptable indoor air quality. ANSI/ASHRAE *Standard* 62.1-2004.

Kowalsky, L., ed. 1990. *Pool/spa operators handbook*. National Swimming Pool Foundation, Merrick, NY.

Smith, C.C., R.W. Jones, and G.O.G. Löf. 1993. Energy requirements and potential savings for heated indoor swimming pools. *ASHRAE Transactions* 99(2):864-874.

BIBLIOGRAPHY

CDC. 2003. Surveillance data from swimming pool inspections—Selected states and counties, United States, May–September 2002. *Morbidity and Mortality Weekly Report* (22):52:513-516.

Kittler, R. 1989. Indoor natatorium design and energy recycling. *ASHRAE Transactions* 95(1):521-526.

CHAPTER 6

HOTELS, MOTELS, AND DORMITORIES

HOTELS, motels, and dormitories may be single-room or multiroom, long- or short-term dwelling (or residence) units; they may be stacked sideways and/or vertically. Information in the first three sections of the chapter applies generally; the last three sections are devoted to the individual types of facilities. High energy costs and consequent environmental damage require that these type of facilities be energy efficient and sustainable. Occupants need assurance that they can afford the fuel bills and that their lifestyle is not damaging to the planet. This chapter provides advice on sustainable practices to achieve these aims.

LOAD CHARACTERISTICS

- Ideally, each room served by an HVAC unit should be able to be ventilated, cooled, heated, or dehumidified independently of any other room. If not, air conditioning for each room will be compromised, and personal comfort will not be possible.
- Typically, the space is not occupied at all times. For adequate flexibility, each unit's ventilation and cooling should be able to be shut off (except when humidity control is required), and its heating to be shut off or turned down. This can be achieved by occupant detection, use of door key fobs, or simple-to-use manual controls such as thermostatic radiator valves (TRVs) on radiators.
- Concentrations of lighting and occupancy are variable, ranging from low for those who work during the day to high and continuous for family homes and residential elderly accommodation; activity is generally sedentary or light.
- Kitchens have the potential for high appliance loads and odor and steam generation, and have large exhaust requirements, with control from low to high, to boost air extraction to suit cooking.
- Rooms generally have an exterior exposure with good daylight levels and a view to green features; however, kitchens, toilets, and dressing rooms are normally internal and require extract ventilation. The building as a whole usually has multiple exposures, as may many individual dwelling units. Design must optimize passive solar gains while avoiding overheating and glare.
- Toilet, washing, and bathing facilities are almost always incorporated in the dwelling units, and the modern trend is to provide en-suite bathrooms in every bedroom. Exhaust air is usually incorporated in each toilet and bathroom area.
- The building has a relatively high hot-water demand, generally for periods of an hour or two, several times a day. This demand can vary from a fairly moderate and consistent daily load profile in a senior citizens building to sharp, unusually high peaks at about 6:00 PM in dormitories. Chapter 50 includes details on service water heating.
- Load characteristics of rooms, dwelling units, and buildings can be well defined with little need to anticipate future changes to design loads, other than adding a service such as cooling that may not have been incorporated originally.
- The prevalence of shifting, transient interior loads and exterior exposures with glass results in high diversity factors; the long hours of use results in fairly high load factors.

DESIGN CONCEPTS AND CRITERIA

Wide load swings and diversity within and between rooms require a flexible system design for 24 h comfort. Besides opening windows, the only way to provide flexible temperature control is having individual room components under individual room control that can cool, heat, and ventilate independently of equipment in other rooms.

In some climates, summer humidity becomes objectionable because of the low internal sensible loads that result when cooling is on/off controlled. Modulated cooling and/or reheat may be required to achieve comfort. Reheat should be avoided unless some sort of heat recovery is involved.

Dehumidification can be achieved by lowering cooling coil temperatures and reducing airflow or by using desiccant dehumidifiers.

Some people have a noise threshold low enough that certain types of equipment disturb their sleep. Higher noise levels may be acceptable in areas where there is little need for air conditioning. Medium- and better-quality equipment is available with noise criteria (NC) 35 levels at 10 to 14 ft in medium to soft rooms and little sound change when the compressor cycles.

Perimeter fan coils are usually quieter than unitary systems, but unitary systems provide more redundancy in case of failure.

SYSTEMS

Energy-Efficient Systems

There is increased impetus to select energy-efficient systems for dwellings to limit potential climate impact, conserve fossil fuel reserves, and avoid fuel poverty. In Europe, the Energy Performance Directive sets out a strategy for each European country to achieve targets toward this objective; in the United Kingdom, for example, all new dwellings should be zero-carbon by 2016, which means a sliding scale from the current allowable values to zero between 2011 to 2016. Other countries have similar schemes. In North America, ASHRAE *Standard* 90.1 is setting progressive reductions also aimed zero net energy.

Where natural gas is available, gas-fired condensing boilers are used, with modulating controls linked to load monitoring such as an outside temperature detector.

Heating and cooling applications generally include water-source and air-source heat pumps. In areas with ample solar radiation, water-source heat pumps may be solar assisted, and/or solar thermal collectors can be used. Energy-efficient equipment generally has the lowest operating cost and should be kept simple, an important factor where skilled operating personnel are unlikely to be available. Most systems allow individual operation and thermostatic control. The

The preparation of this chapter is assigned to TC 9.8, Large-Building Air-Conditioning Applications.

typical system allows individual metering so that most, if not all, of the cooling and heating costs can be metered directly to the occupant (McClelland 1983). Existing buildings can be retrofitted with heat flow meters and timers on fan motors for individual metering, and there is a drive toward provision of better real-time energy use to allow occupants to make changes that reduce their costs at the right time.

The water-loop heat pump has a lower operating cost than air-cooled unitary equipment, and allows a degree of heat recovery because the condenser water loop acts to balance energy use when possible. The lower installed cost encourages its use in mid- and high-rise buildings where individual dwelling units have floor areas of 800 ft^2 or larger. Some systems incorporate sprinkler piping as the water loop.

The system has a central plant consisting of circulating pumps, heat rejection when there is surplus heat capacity in the building, and supplementary gas-fired boiler heat input when there is an overall deficit of heat. The water-loop heat pump is predominantly decentralized; individual metering allows most of the operating cost to be paid by the occupant. Its life should be longer than for other unitary systems because most of the mechanical equipment is in the building and not exposed to outdoor conditions. Also, load on the refrigeration circuit is not as severe because water temperature is controlled for optimum operation. Operating costs are low because of the system's inherent energy conservation. Excess heat may be stored during the day for the following night, and heat may be transferred from one part of the building to another.

Although heating is required in many areas during cool weather, cooling could be needed in rooms having high solar loads. This should be avoided by effective solar shading design. On a mild day, surplus heat throughout the building is frequently transferred into the hot-water loop by water-cooled condensers on cooling cycle, so that water temperature rises. The heat remains stored in the water and can be extracted at night; a water heater is therefore avoided. This heat storage is improved by the presence of a greater mass of water in the pipe loop; some systems include a storage tank for this reason, or water tank with phase-change material (PCM) thermal storage. Because the system is designed to operate during the heating season with water supplied at a temperature as low as 60°F, the water-loop heat pump lends itself to solar assist; relatively high solar collector efficiencies result from the low water temperature.

The installed cost of the water-loop heat pump is higher in very small buildings. In severe cold climates with prolonged heating seasons, even where natural gas or fossil fuels are available at reasonable cost, the operating cost advantages of this system may diminish unless heat can be recovered from some another source, such as solar collectors, geothermal, or internal heat from a commercial area served by the same system.

Energy-Neutral Systems

To qualify as energy-neutral, a system must have controls that prevent simultaneous operation of the cooling and heating cycles. Some examples are (1) packaged terminal air conditioners (PTACs) (through-the-wall units), (2) window units or radiant ceiling panels for cooling combined with finned or baseboard radiation for heating, (3) unitary air conditioners with an integrated heating system, (4) fan coils with remote condensing units, and (5) variable-air-volume (VAV) systems with either perimeter radiant panel heating or baseboard heating. For unitary equipment, control may be as simple as a heat/cool switch. For other types, dead-band thermostatic control may be required.

PTACs are frequently installed to serve one or two rooms in buildings with mostly small, individual units. In a common two-room arrangement, a supply plenum diverts some of the conditioned air serving one room into the second, usually smaller, room. Multiple PTAC units allow additional zoning in dwellings with more

rooms. Additional radiation heat is sometimes needed around the perimeter in cold climates.

Heat for a PTAC may be supplied either by electric resistance heaters or by hot-water or steam heating coils. Initial costs are lower for a decentralized system using electric resistance heat. Operating costs are lower for coils heated by combustion fuels. Despite its relatively inefficient refrigeration circuits, a PTAC's operating cost is quite reasonable, mostly because of individual thermostatic control over each machine, which eliminates the use of reheat while preventing the space from being overheated or overcooled. Also, because equipment is located in the space being served, little power is devoted to circulating the room air. Servicing is simple: a defective machine is replaced by a spare chassis and forwarded to a service organization for repair. Thus, building maintenance can be done by relatively unskilled personnel.

Noise levels are generally no higher than NC 40, but some units are noisier than others. Installations near a seacoast should be specially constructed (usually with stainless steel or special coatings) to avoid accelerated corrosion of aluminum and steel components caused by salt. In high-rise buildings of more than 12 stories, special care is required, both in design and construction of outside partitions and in installation of air conditioners, to avoid operating problems associated with leakage (caused by stack effect) around and through the machines.

Frequently, the least expensive installation is finned or baseboard radiation for heating and window-type room air conditioners for cooling. The window units are often purchased individually by the building occupants. This choice offers a reasonable operating cost and is relatively simple to maintain. However, window units have the shortest equipment life, highest operating noise level, and poorest distribution of conditioned air of any systems discussed in this section.

Fan-coils with remote condensing units are used in smaller buildings. Fan-coil units are located in closets, and the ductwork distributes air to the rooms in the dwelling. Condensing units may be located on roofs, at ground level, or on balconies.

Low-capacity residential warm-air furnaces may be used for heating, but with gas- or oil-fired units, combustion products must be vented. In a one- or two-story structure, it is possible to use individual chimneys or flue pipes, but in a high-rise structure requires a multiple-vent chimney or a manifold vent. Local codes should be consulted.

Sealed combustion furnaces draw all combustion air from outside and discharge flue products through a windproof vent to the outdoors. The unit must be located near an outside wall, and exhaust gases must be directed away from windows and intakes. In one- or two-story structures, outdoor units mounted on the roof or on a pad at ground level may also be used. All of these heating units can be obtained with cooling coils, either built-in or add-on. Evaporative-type cooling units are popular in motels, low-rise apartments, and residences in mild climates.

Desiccant dehumidification should be considered when independent control of temperature and humidity is required to avoid reheat.

Energy-Inefficient Systems

Energy-inefficient systems allow simultaneous cooling and heating. Examples include two-, three-, and four-pipe fan coil units, terminal reheat systems, and induction systems. Some units, such as the four-pipe fan coil, can be controlled so that they are energy-neutral by ensuring that the two circuits do not simultaneously serve the PTAC. They are primarily used for humidity control.

Four-pipe systems and two-pipe systems with electric heaters can be designed for complete temperature and humidity flexibility during summer and intermediate season weather, although none provides winter humidity control. Both systems provide full dehumidification and cooling with chilled water, reserving the other two pipes or an electric coil for space heating or reheat. The equipment and necessary controls are expensive, and only the four-pipe system, if equipped with an internal-source heat-recovery design for the warm coil energy, can operate at low cost. When year-round comfort is essential, four-pipe systems or two-pipe systems with electric heat should be considered.

Total Energy Systems

A total energy system is an option for any multiple or large housing facility with large year-round service water heating requirements. Total energy systems are a form of cogeneration in which all or most electrical and thermal energy needs are met by on-site systems as described in Chapter 7 of the 2008 *ASHRAE Handbook—HVAC Systems and Equipment*. A detailed load profile must be analyzed to determine the merits of using a total energy system. The reliability and safety of the heat-recovery system must also be considered.

Any of the previously described systems can perform the HVAC function of a total energy system. The major considerations as they apply to total energy in choosing a HVAC system are as follows:

- Optimum use must be made of thermal energy recoverable from the prime mover during all or most operating modes, not just during conditions of peak HVAC demand.
- Heat recoverable through the heat pump may become less useful because the heat required during many of its potential operating hours will be recovered from the prime mover. The additional investment for heat pump or heat recovery cycles may be more difficult to justify because operating savings are lower.
- The best application for recovered waste heat is for those services that use only heat (i.e., service hot water, laundry facilities, and space heating).

Special Considerations

Local building codes govern ventilation air quantities for most buildings. Where they do not, ASHRAE *Standard* 62.1 should be followed. The quantity of outdoor air introduced into rooms or corridors is usually slightly in excess of the exhaust quantities to pressurize the building. To avoid adding load to individual systems, outdoor air should be treated to conform to indoor air temperature and humidity conditions. In humid climates, special attention must be given to controlling humidity from outdoor air. Otherwise, the outdoor air may reach corridor temperature while still retaining a significant amount of moisture.

In buildings having a centrally controlled exhaust and supply, the system is regulated by a time clock or a central management system for certain periods of the day. In other cases, the outside air may be reduced or shut off during extremely cold periods, although this practice is not recommended and may be prohibited by local codes. These factors should be considered when estimating heating load.

For buildings using exhaust and supply air on a 24-hour basis, air-to-air heat recovery devices may be merited (see Chapter 25 of the 2008 *ASHRAE Handbook—HVAC Systems and Equipment*). These devices can reduce energy consumption by capturing 60 to 80% of the sensible and latent heat extracted from the air source.

Infiltration loads in high-rise buildings without ventilation openings for perimeter units are not controllable year-round by general building pressurization. When outer walls are pierced to supply outdoor air to unitary or fan-coil equipment, combined wind and thermal stack-effect forces create equipment operating problems. These factors must be considered for high-rise buildings (see Chapter 16 of the 2009 *ASHRAE Handbook—Fundamentals*).

Interior public corridors should have tempered supply air with transfer into individual area units, if necessary, to provide kitchen and toilet makeup air requirements. Transfer louvers need to be acoustically lined. Corridors, stairwells, and elevators should be pressurized for fire and smoke control (see Chapter 53).

Kitchen air can be recirculated through hoods with activated charcoal filters rather than exhausted. Toilet exhaust can be VAV with a damper operated by the light switch. A controlled source of supplementary heat in each bathroom is recommended to ensure comfort while bathing.

Air-conditioning equipment must be isolated to reduce noise generation or transmission. The cooling tower or condensing unit must be designed and located to avoid disturbing occupants of the building or of adjacent buildings.

An important but frequently overlooked load is the heat gain from piping for hot-water services. Insulation thickness should conform to the latest local energy codes and standards at a minimum. In large, luxury-type buildings, a central energy or building management system allows supervision of individual air-conditioning units for operation and maintenance.

Some facilities conserve energy by reducing indoor temperature during the heating season. Such a strategy should be pursued with caution because it could affect occupant comfort, and, for example, the competitiveness of a hotel/motel.

HOTELS AND MOTELS

Hotel and motel accommodations are usually single guest rooms with a toilet and bath adjacent to a corridor, and flanked on both sides by other guest rooms. The building may be single-story, low-rise, or high-rise. Multipurpose subsidiary facilities range from stores and offices to ballrooms, dining rooms, kitchens, lounges, auditoriums, and meeting halls. Luxury motels may be built with similar facilities. Occasional variations are seen, such as kitchenettes, multiroom suites, and outside doors to patios and balconies. Hotel classes range from the deluxe hotel to the economy hotel/motel as outlined in Table 1.

Table 1   Hotel Classes

Type of Facility	Typical Occupancy, Persons per Room	Characteristics
Deluxe hotel	1.2	Large rooms, suites, specialty restaurants
Luxury/first class, full-service hotel	1.2 to 1.3	Large rooms, large public areas, business center, pool and health club, several restaurants
Mid-scale, full-service hotel	1.2 to 1.3	Large public areas, business center, several restaurants
Convention hotel	1.4 to 1.6	Large number of rooms, very large public areas, extensive special areas, rapid shifting of peak loads
Limited-service hotel	1.1	Limited public areas, few restaurants, may have no laundry
Upscale, all-suites hotel	2.0	Rooms are two construction bays, in-room pantries, limited public areas, few restaurants
Economy, all-suites hotel	2.0 to 2.2	Smaller suites, limited public areas and restaurants
Resort hotel	1.9 to 2.4	Extensive public areas, numerous special and sport areas, several restaurants
Conference center	1.3 to 1.4	Numerous special meeting spaces, limited dining options
Casino hotel	1.5 to 1.6	Larger rooms, large gaming spaces, extensive entertainment facilities, numerous restaurants
Economy hotel/motel	1.6 to 1.8	No public areas, little or no dining, usually no laundry

Table 2 Hotel Design Criteria[a,b]

| Category | Inside Design Conditions | | | | Ventilation[d] | Exhaust[e] | Filter Efficiency[f] | Noise, RC Level |
| | Winter | | Summer | | | | | |
	Temperature	Relative Humidity[c]	Temperature	Relative Humidity				
Guest rooms	74 to 76°F	30 to 35%	74 to 78°F	50 to 60%	30 to 60 cfm per room	20 to 50 cfm per room	6 to 8 MERV	25 to 35
Lobbies	68 to 74°F	30 to 35%	74 to 78°F	40 to 60%	15 cfm per person	—	8 MERV or better	35 to 45
Conference/ meeting rooms	68 to 74°F	30 to 35%	74 to 78°F	40 to 60%	20 cfm per person	—	8 MERV or better	25 to 35
Assembly rooms	68 to 74°F	30 to 35%	74 to 78°F	40 to 60%	15 cfm per person	—	8 MERV or better	25 to 35

[a] This table should not be the only source for design criteria. Data contained here can be determined from volumes of the *ASHRAE Handbook*, standards, and governing local codes.
[b] Design criteria for stores, restaurants, and swimming pools are in Chapters 2, 3, and 5, respectively.
[c] Minimum recommended humidity.
[d] Per ASHRAE *Standard* 62.1.
[e] Air exhaust from bath and toilet area.
[f] Per ASHRAE *Standard* 52.2 (MERV = minimum efficiency reporting values).

A hotel can be divided into three main areas:

1. Guest rooms
2. Public areas
   - Lobby, atrium, and lounges
   - Ballrooms
   - Meeting rooms
   - Restaurants and dining rooms
   - Stores
   - Swimming pools
   - Health clubs
3. Back-of-the-house (BOTH) areas
   - Kitchens
   - Storage areas
   - Laundry
   - Offices
   - Service areas and equipment rooms

The two main areas of use are the guest rooms and the public areas. Maximum comfort in these areas is critical to success of any hotel. Normally the BOTH spaces are less critical than the remainder of the hotel with the exception of a few spaces where a controlled environment is required or recommended.

Guest Rooms

Air conditioning in hotel rooms should be quiet, easily adjustable, and draft free. It must also provide ample outside air. Because the hotel business is so competitive and space is at a premium, systems that require little space and have low total owning and operating costs should be selected.

Design Concepts and Criteria. Table 2 lists design criteria for hotel guest rooms. In addition, the design criteria for hotel room HVAC services must consider the following factors:

- Individual and quickly responding temperature control
- Draft-free air distribution
- Toilet room exhaust
- Ventilation (makeup) air supply
- Humidity control
- Acceptable noise level
- Simple controls
- Reliability
- Ease of maintenance
- Operating efficiency
- Use of space

Load Characteristics. The great diversity in the design, purpose, and use of hotels and motels makes analysis and load studies very important. Load diversification is possible because of guest rooms' transient occupancy and the diversity associated with support facility operation.

The envelope cooling and heating load is dominant because the guest rooms normally have exterior exposures. Other load sources such as people, lights, appliances, etc. are a relatively small part of

the space sensible and latent loads. The ventilation load can represent up to 15% of the total cooling load.

Because of the nature of the changing envelope sensible load and the transient occupancy of the guest room, large fluctuations in the space sensible load in a one-day cycle are common. The ventilation sensible cooling load can vary from 0 to 100% in a single day, whereas the ventilation latent load can remain almost constant for the entire day. A low sensible heat ratio is common in moderate to very humid climates. Usually, the HVAC equipment must only handle part or low loads and peak loads rarely occur. For example, in humid climates, introducing untreated outside air directly into the guest room or into the return air plenum of the HVAC unit operating at part or low load creates a severe high-humidity problem, which is one of the causes of mold and mildew. The situation is further aggravated when the HVAC unit operates in on/off cycle during part- or low-load conditions.

Applicable Systems. Most hotels use all-water or unitary refrigerant-based equipment for guest rooms. All-water systems include

- Two-pipe fan-coils
- Two-pipe fan-coil with electric heat
- Four-pipe fan-coils

Unitary refrigerant-based systems include

- Packaged terminal air conditioner or packaged terminal heat pump (with electric heat)
- Air-to-air heat pump (ductless, split)
- Water-source heat pump

Except for the two-pipe fan-coil, all these systems cool, heat, or dehumidify independently of any other room and regardless of the season. A two-pipe fan-coil system should be selected only when economics and design objectives dictate a compromise to performance. Selection of a particular system should be based on

- First cost
- Economical operation, especially at part load
- Maintainability

Compared to unitary refrigerant-based units, all-water systems offer the following advantages:

- Reduced total installed cooling capacity due to load diversity
- Lower operating cost due to a more efficient central cooling plant
- Lower noise level (compared to PTAC and water-source heat pump)
- Longer service life
- Less equipment to be maintained in the occupied space
- Less water in circulation (compared to water-source heat pump)
- Smaller pipes and pumps (compared to water-source heat pump)

Unitary refrigerant-based systems offer the following advantages:

- Lower first cost
- Immediate all year availability of heating and cooling

- No seasonal changeover required
- Cooling available without operating a central refrigeration plant
- Can transfer energy from spaces being cooled to spaces being heated (with water-source heat pump)
- Range of circulated water temperature requires no pipe insulation (for water-source heat pump)
- Less dependence on a central plant for heating and cooling
- Simplicity, which results in lower operating and maintenance staff costs

The type of facility, sophistication, and quality desired by the owner/operator, as well as possible code requirements typically influence the selection. An economic analysis (life-cycle cost) is particularly important when selecting the most cost-effective system. Chapter 37 has further information on economic analysis techniques. Computer software like the NIST Building Life-Cycle Cost Program (BLCC) performs life-cycle cost analyses quickly and accurately (NIST 2006).

Chapters 2 and 5 of the 2008 *ASHRAE Handbook—HVAC Systems and Equipment* provide additional information about all-water systems and unitary refrigerant-based systems.

Room fan-coils and room unitary refrigerant-based units are available in many configurations, including horizontal, vertical, exposed, and concealed. The unit should be located in the guest room so that it provides excellent air diffusion without creating unpleasant drafts. Air should not discharge directly over the head of the bed, to keep cold air away from a sleeping guest. The fan-coil/heat pump unit is most commonly located

- Above the ceiling in the guest room entry corridor or above the bathroom ceiling (horizontal air discharge),
- On the room's perimeter wall (vertical air discharge), or
- In a floor-to-ceiling enclosed chase (horizontal air discharge).

Locating the unit above the entry corridor is preferred because air can flow directly along the ceiling and the unit is relatively accessible for maintenance (see Figures 1 and 2).

Most units are designed for free-air discharge. The supply air grille should be selected according to the manufacturer's recommendations for noise and air diffusion. Also, airflow should not interfere with the room drapes or other wall treatment.

Other factors that should be considered include

- Sound levels at all operating modes, particularly with units that cycle on and off
- Adequately sized return air grille
- Access for maintenance, repair, and filter replacement

Ventilation (makeup) supply and exhaust rates must meet local code requirements. Ventilation rates vary and the load imposed by ventilation must be considered.

Providing conditioned ventilation air directly to the guest room is the preferred approach. Normally, outside air is conditioned in a primary makeup air unit and distributed by a primary air duct to every guest room. This approach controls the supply air conditions, ensures satisfactory room conditions and room air balance (room pressurization) even during part- or no-load conditions, and controls mold and mildew.

Other ventilation techniques are to

- Transfer conditioned ventilation air from the corridor to each guest room. This approach controls ventilation air conditions better; however, the air balance (makeup versus exhaust) in the guest room may be compromised. This approach is prohibited under many code jurisdictions.
- Introduce unconditioned outside air directly to the air-conditioning unit's return air plenum (perimeter wall installations). This approach can cause mold and mildew and should be avoided. During periods of part or low load, which occur during most of the cooling season, the thermostatically controlled air conditioner does not adequately condition the constant flow of outside air because the cooling coil valve closes and/or the compressor

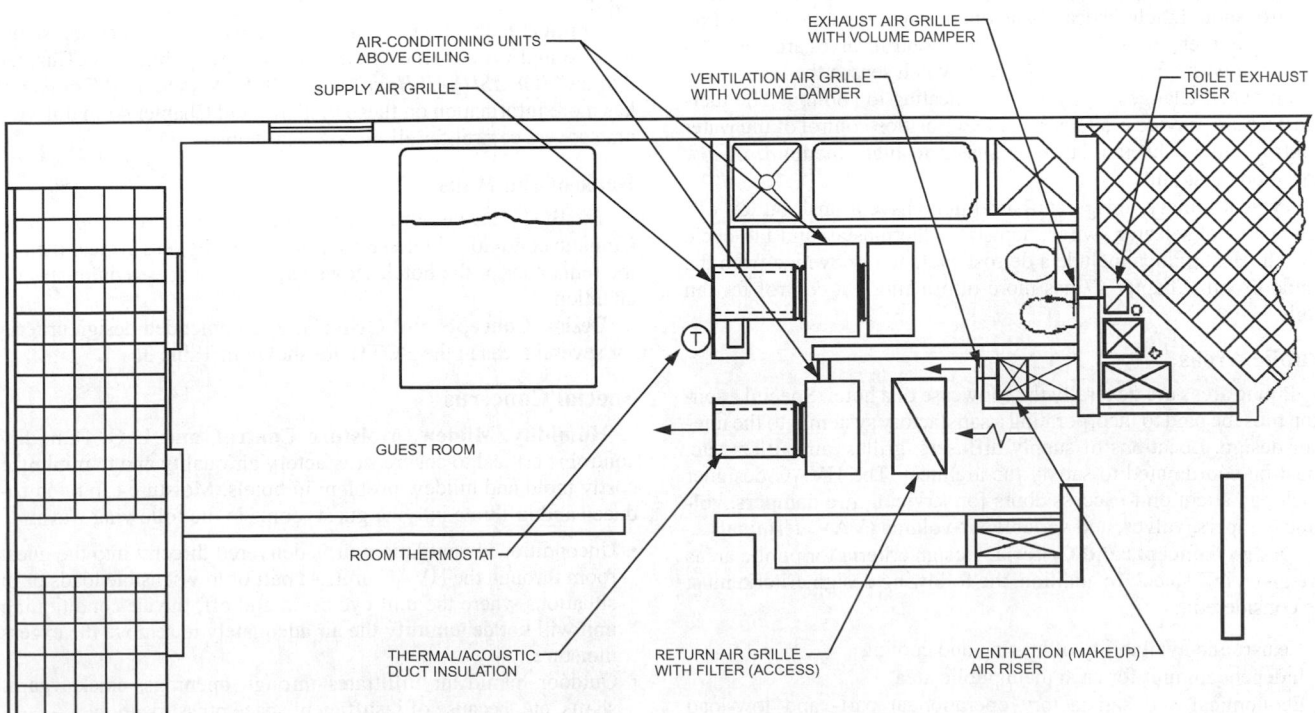

Fig. 1   Alternative Location for Hotel Guest Room Air-Conditioning Unit above Hung Ceiling

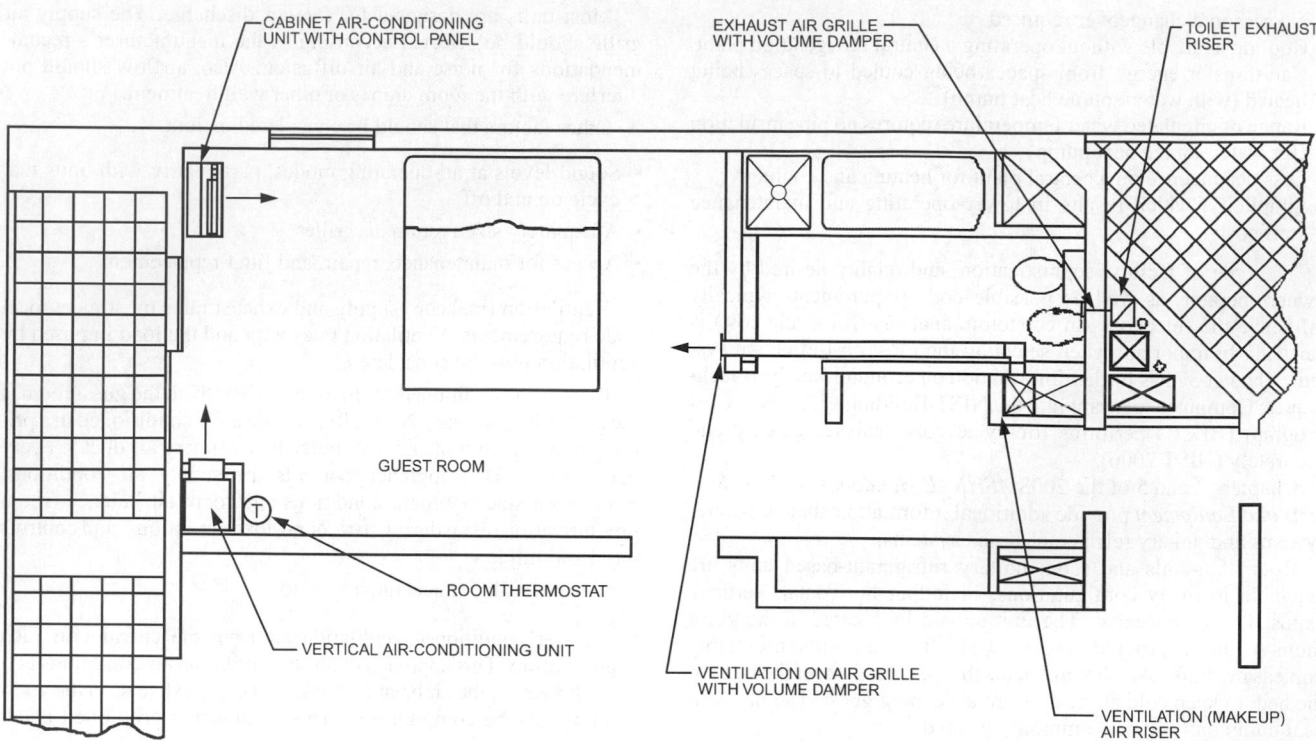

Fig. 2   Alternative Location for Hotel Guest Room Air-Conditioning Unit on Room Perimeter and Chase-Enclosed

cycles off. As a result, humidity in the room increases. Also, when the air conditioner's fan is off, outside air infiltrates through the ventilation opening and again elevates the room's humidity level.

Guest-room HVAC units are normally controlled by a room thermostat. Thermostats for fan-coils normally control valves in two-pipe, four-pipe, and two-pipe chilled-water/electric heat systems. Control should include dead-band operation to separate the heating and cooling set points. Two-pipe system control valves are normally equipped with automatic changeover, which senses the water temperature and changes operation from heating to cooling. The thermostat may provide modulation or two-position control of the water control valve. The fan can be adjusted to high, medium, or low speed on most units.

Typical unitary refrigerant-based units have a push button off/fan/heat/cool selector switch, adjustable thermostat, and fan cycle switch. Heat pumps include a defrost cycle to remove ice from the outdoor coil. Chapter 47 has more information on control for fan coils.

Public Areas

Public areas are generally the showcase of a hotel. Special attention must be paid to incorporating a satisfactory system into the interior design. Locations of supply diffusers, grilles. air outlets. etc. must be coordinated to satisfy the architect. The HVAC designer must pay attention to access doors for servicing fire dampers, volume dampers, valves, and variable-air-volume (VAV) terminals.

Design Concepts and Criteria. Design criteria for public areas are given in Table 2. In addition, the following design criteria must be considered:

- Year-round availability of heating and cooling
- Independent unit for each main public area
- Economical and satisfactory operation at part- and low-load conditions
- Coordination with adjacent back-of-the-house (BOTH) areas to ensure proper air pressurization (e.g., restaurants, kitchens)

Load Characteristics. The hours each public area is used vary widely. In many cases, the load is from internal sources from people, lights, and equipment. The main lobby normally is operational 24 hours per day. Areas like restaurants, meeting rooms, and retail areas have intermittent use, so the load changes frequently. HVAC systems that respond effectively and economically must be selected for these areas.

Applicable Systems. All-air systems, single-duct constant-volume, and VAV are most frequently used for public areas. Chapter 4 of the 2008 *ASHRAE Handbook—HVAC Systems and Equipment* has more information on these systems, and Chapter 47 in this volume covers control for all-air VAV systems.

Back-of-the-House (BOTH) Areas

The BOTH area is normally considered a service or support area. Climatic conditions in these areas are typically less critical than in the remainder of the hotel. However, a few spaces require special attention.

Design Concepts and Criteria. Recommended design criteria for several areas in the BOTH are shown in Table 3.

Special Concerns

Humidity, Mildew, Moisture Control, and IAQ. Humidity control is critical to ensure satisfactory air quality and to minimize costly mold and mildew problem in hotels. Moisture can be introduced and infiltrate into the guest rooms in the following ways:

- Unconditioned ventilation air is delivered directly into the guest room through the HVAC unit. At part or low sensible loads or in situations where the unit cycles on and off, the air-conditioning unit will not dehumidify the air adequately to remove the excess moisture.
- Outdoor humid air infiltrates through openings, cracks, gaps. shafts, etc. because of insufficient space pressurization.
- Moisture migrates through external walls and building elements because of a vapor pressure differential.
- An internal latent load or moisture is generated.

Table 3   Design Criteria for Hotel Back-of-the-House Areas[a]

Category	Inside Design Conditions	Comments
Kitchen, general[b]	82°F	Provide spot cooling
pastry[b]	76°F	
chef's office[b]	74 to 78°F 50 to 60% rh (summer) 30 to 35% rh (winter)	Fully air conditioned
Housekeeper's office	74 to 78°F 50 to 60% rh (summer) 30 to 35% rh (winter)	Fully air conditioned
Telephone equipment room	Per equipment criteria	Stand-alone air conditioner; air conditioned all year
Wine storage	Per food and beverage manager criteria	Air conditioned all year
Laundry		Spot cooling as required at workstations

[a] Governing local codes must be followed for design of the HVAC.
[b] Consult Chapter 33 for details on kitchen ventilation.

Removing water vapor from the air is the most feasible way to control mold and mildew, particularly when the problem spreads to walls and carpeting. Good moisture control can be achieved by applying the following techniques:

- Introduce adequately dried ventilation (makeup) air [i.e., with a dew point of 53°F (60 grains/lb of dry air) or less] directly to the guest room.
- Maintain slightly positive pressure in the guest room to minimize infiltration of hot and humid air into the room. Before a new HVAC system is accepted by the owner, a certified air balance contractor should be engaged to demonstrate that the volume of dry makeup air exceeds the volume of exhaust air. As the building ages, it is important to maintain this slight positive pressure; otherwise, humid air that infiltrates into the building cavities will be absorbed regardless of how dry the room is maintained (Banks 1992).
- Provide additional dehumidification capability to the ventilation (makeup air) by dehumidifying the air to a lower level than the desired space humidity ratio. For example, introducing 60 cfm of makeup air at 55 gr/lb can provide approximately 400 Btu/h of internal latent cooling (assuming 65 gr/lb is a desirable space humidity ratio).
- Allow air conditioning to operate in unoccupied rooms instead of turning the units off, especially in humid areas.
- Improve the room envelope by increasing its vapor and infiltration resistance.

The third method allows ventilation air to handle part of the internal latent load (people, internal moisture generation, and moisture migration from external walls and building elements). In addition, this method can separate the internal sensible cooling, internal latent cooling, and ventilation loads. Independent ventilation/dehumidification allows room pressurization and space humidity control regardless of the mode of operation or magnitude of the air-conditioning load. Desiccant dehumidifiers can be retrofitted to solve existing moisture problems.

Dedicated Outdoor Air Systems (DOAS). DOAS air units are designed to condition ventilation air introduced into a space and to replace air exhausted from the building. The geographic location and class of the hotel dictate the functions of the makeup air units, which may filter, heat, cool, humidify, and/or dehumidify the ventilation air. Makeup air may be treated directly or by air-to-air heat recovery (sensible or combined sensible and latent) and other heat recovery techniques. Equipment to condition the air by air-to-air

Table 4   Design Criteria for Hotel Guest Room DOAS

Winter Temperature	Winter Relative Humidity	Summer Temperature	Summer Relative Humidity	Filter Efficiency (ASHRAE *Standard* 52.2)
68 to 76°F	30 to 45%	74 to 78°F	40 to 50%	6 to 8 MERV

Notes:
1. Follow local codes when applicable.
2. Building location may dictate optimum supply condition in recommended range.
3. MERV = minimum efficiency rating values.

heat recovery and final heating, cooling, humidification, and/or dehumidification is also available.

Chapter 14 of the 2009 *ASHRAE Handbook—Fundamentals* provides design weather data for ventilation. Analyzing and selecting the proper makeup unit for the full range of entering conditions are critical for efficient and sufficient all-year operation. Air-to-air heat recovery helps stabilize entering conditions, which helps provide efficient and stable operation. However, heat recovery may not always be feasible. Often, exhaust air comes from many individual stacks. In this case, the cost of combining many exhausts for heat recovery may not be warranted.

Typical design criteria for ventilation (makeup) air units are listed in Table 4.

Makeup air units can be stand alone packaged (unitary) or integrated in an air handler. A typical makeup air unit usually has the following features:

- Heating, cooling, and dehumidification
  - Chilled/hot water or steam coils in the air handling unit
  - Unitary refrigerant-based unit (direct-expansion cooling and gas furnace or electric heat)
  - Air-to-air energy recovery combined with mechanical cooling (DX or chilled water) and heating
  - Desiccant-based dehumidifier combined with air-to-air energy recovery, indirect/direct evaporative cooling and supplementary mechanical cooling and heating
- Heating only
  - Hot water or steam coils in the air handling unit
  - Stand alone gas-fired or electric makeup units
  - Air-to-air energy recovery with supplement heat

Humidification should be considered for all cold climates. The HVAC designer must also consider avoiding coil freeze up in water based systems. Chapters 27 and 37 of the 2008 *ASHRAE Handbook—HVAC Systems and Equipment* provide information about makeup air units and air-to-air energy recovery, respectively.

Hotel location, environmental quality desired by the owner, and design sophistication determines the system selected. For example, in locations with cool summers, dehumidification with mechanical cooling only is satisfactory. For humid locations or where enhanced dehumidification is required, a desiccant-based unit can provide lower supply air humidity, to help prevent mold and mildew and provide internal latent cooling.

Central Mechanical Plant. Designing a reliable and energy-efficient mechanical plant is essential to ensuring a profitable hotel. The chiller plant must operate efficiently at part-load conditions. Some redundancy should be considered in case of equipment failure. Designs often include spare critical equipment where spare parts and qualified service are not readily available. Chillers with multistage compressors should be considered because they provide partial cooling during failures and enhance part-load operation. When using two chillers, each should provide at least 60% of the total load. Combinations of three chillers providing 40% each or four chillers providing 30% each are better for tracking part-load conditions. Cooling towers, pumps, etc., can be sized in a similar manner.

The heating plant should be designed to accommodate the winter heating load and could provide domestic hot water, swimming pool heating, and service to kitchens and laundries as well. The type of fuel used depends on location, availability, use, and cost.

Multipurpose boiler design for the kitchen and laundry should offer redundancy, effective part-load handling, and efficient operation during summer, when the HVAC heating load does not exist.

In areas with mild winters, a two-pipe system or an air-to-water heat pump chiller/heater can be considered. In any event, the HVAC designer must understand the need for all-year cooling and heating availability in the public areas. In this case, a combination of air-to-water heat pump, chiller/heater for the guest rooms, and independent heat pumps for public areas can be installed.

Acoustics and Noise Control. The sound level in guest room and public areas is a major design element. Both the level and constancy of noise generated by the HVAC unit are of concern. Normally, packaged terminal air conditioners/heat pumps and water-source heat pumps are noisier because of the compressor. Some equipment, however, has extra sound insulation, which reduces the noise significantly.

Lowering fan speed, which is usually acceptable, can reduce fan noise levels. On/off cycling of the fan and compressor can be objectionable, even if the generated noise is low. Temperature control by cycling the fan only (no flow control valve) should not be used.

Another source of noise is sound that transfers between guest rooms through the toilet exhaust duct. Internal duct lining and sound attenuators are commonly used to minimize this problem.

Noise from equipment located on the roof or in a mechanical room located next to a guest room should be avoided. Proper selection of vibration isolators should prevent vibration transmission. In critical cases, an acoustician must be consulted.

New Technology in Hotels. Modern hotels are implementing techniques to enhance comfort and convenience. For example, the telephone, radio, TV, communications, lighting, and air-conditioning unit can be integrated into one control system. Occupancy sensors conserve energy by resetting the temperature control when the room is occupied or when guests leave. As soon as a new guest checks in at the front desk, the room temperature is automatically reset. But even with this improved technology, it is important to remember that temperature reset may create humidity problems.

DORMITORIES

Dormitory buildings frequently have large commercial dining and kitchen facilities, laundering facilities, and common areas for indoor recreation and bathing. These ancillary loads may make heat pump or total energy systems appropriate, economical alternatives, especially on campuses with year-round activity.

When dormitories are shut down during cold weather, the heating system must supply enough heat to prevent freezing. If the dormitory contains non-dwelling areas such as administrative offices or eating facilities, these facilities should be designed as a separate zone or with a separate system for flexibility, economy, and odor control.

Subsidiary facilities should be controlled separately for flexibility and shutoff capability, but they may share common refrig-

eration and heating plants. With internal-source heat pumps, this interdependence of unitary systems allows reclamation of all internal heat usable for building heating, domestic water preheating, and snow melting. It is easier and less expensive to place heat reclaim coils in the building's exhaust than to use air-to-air heat recovery devices. Heat reclaim can easily be sequence controlled to add heat to the building's chilled-water system when required.

MULTIPLE-USE COMPLEXES

Multiple-use complexes combine retail, office, hotel, residential, and/or other commercial spaces into a single site. Peak HVAC demands of the various facilities may occur at different times of the day and year. Loads should be determined independently for each occupancy. Where a central plant is considered, a block load should also be determined.

Separate air handling and distribution should serve separate facilities. However, heating and cooling units can be combined economically into a central plant. A central plant provides good opportunities for heat recovery, thermal storage, and other techniques that may not be economical in a single-use facility. A multiple-use complex is a good candidate for central fire and smoke control, security, remote monitoring, billing for central facility use, maintenance control, building operations control, and energy management.

REFERENCES

ASHRAE. 2007. Method of testing general ventilation air cleaning devices for removal efficiency by particle size. ANSI/ASHRAE *Standard* 52.2-2007.

ASHRAE. 2010. Ventilation for acceptable indoor air quality. ANSI/ASHRAE *Standard* 62.1-2010.

Banks, N.J. 1992. Field test of a desiccant-based HVAC system for hotels. *ASHRAE Transactions* 98(1):1303-1310.

McClelland, L. 1983. Tenant paid energy costs in multi-family rental housing. DOE, University of Colorado, Boulder.

NIST. 2006. Building life-cycle cost (BLCC) program, v. 5.3-06. National Institute of Standards and Technology, Gaithersburg, MD.

BIBLIOGRAPHY

Haines, R.W. and D.C. Hittle. 2006. *Control systems for heating, ventilation and air conditioning*, 6th ed. Springer, New York.

Harriman, L.G., D. Plager, and D. Kosar. 1997. Dehumidification and cooling loads from ventilation air. *ASHRAE Journal* 39(11):37-45.

Kimbrough, J. 1990. The essential requirements for mold and mildew. Plant Pathology Department, University of Florida, Gainesville.

Kokayko, M.J. 1997. Dormitory renovation project reduces energy use by 69%. *ASHRAE Journal* 39(6):33-36.

Lehr, V.A. 1995. Current trends in hotel HVAC design. *Heating/Piping/Air Conditioning*, February.

Lorsch, H. 1993. *Air-conditioning system design manual*. ASHRAE.

Peart, V. 1989. *Mildew and moisture problems in hotels and motels in Florida*. Institute of Food and Agricultural Sciences, University of Florida, Gainesville.

Wong, S.P. and S.K. Wang. 1990. Fundamentals of simultaneous heat and moisture transfer between the building envelope and the conditioned space air. *ASHRAE Transactions* 96(2):73-83.

CHAPTER 7

EDUCATIONAL FACILITIES

THIS chapter contains technical, environmental, and design considerations to assist the design engineer in the proper application of heating, ventilation, and air-conditioning systems and equipment for educational facilities.

PRESCHOOLS

General Design Considerations

Commercially operated preschools are generally provided with standard architectural layouts based on owner-furnished designs. A typical preschool facility provides programs for infants (1 to 2 years old), toddlers (2 years old), and preschoolers (3 to 4 years old). Larger facilities also offer programs for older children, such as kindergarten programs (5 years old). Areas such as lobbies, libraries, and kitchens are also included to support the variety of programs. Given this range of age, special attention for the design of the HVAC systems is required to meet the needs of every age group.

All preschool facilities require quiet and economical systems. The equipment should be easy to operate and maintain, and the design should provide warm floors and no drafts. These facilities have two distinct occupant zones: (1) the floor level, where younger children play, and (2) normal adult height, for the teachers. The teacher also requires a place for a desk; consider treating this area as a separate zone.

Preschool facilities generally operate on weekdays from early in the morning to 6:00 or 7:00 PM. This schedule usually coincides with the normal working hours of the children's parents plus one hour for drop-off and pick-up. The HVAC systems therefore operate 12 to 14 h per workday, and may be off or on at night and weekends, depending on whether setback is applied.

Supply air outlets should be positioned so that the floor area is maintained at about 75°F without the introduction of drafts. Both supply and return air outlets should be placed where they will not be blocked by furniture positioned along the walls or where children can reach them. Coordination with the architect about locating these outlets is essential. Proper ventilation is crucial for controlling odors and helping prevent the spread of diseases among the children.

Floor-mounted heating equipment, such as electric baseboards heaters, should be avoided because children must be prevented from coming in contact with hot surfaces or electrical devices. However, radiant-floor systems can be used safely and effectively.

Design Criteria

Table 1 provides typical indoor design conditions for preschools. Table 2 provides typical ventilation and exhaust design criteria using the ventilation rate procedure of ANSI/ASHRAE *Standard* 62.1-2007; the user's manual (ASHRAE 2007) is also strongly recommended for additional information on applying this standard. Table 3 lists design criteria for acceptable noise in preschool facilities.

The preparation of this chapter is assigned to TC 9.7, Educational Facilities.

Table 1 Recommended Temperature and Humidity Design Criteria for Various Spaces in Preschools

Category/Humidity Criteria	Indoor Design Conditions, °F	
	Winter	Summer
Infant, Toddler, and Preschooler Classrooms[a]		
30% rh	68.5 to 75.5	74.0 to 80.0
40% rh	68.0 to 75.0	73.5 to 80.0
50% rh	68.5 to 74.5	73.0 to 79.0
60% rh	67.5 to 74.0	73.0 to 78.5
Administrative, Offices, Lobby, Kitchen		
30 to 60% rh	68.5 to 74.0	74.0 to 78.5
Storage		
No humidity control	64.0	
Mechanical Rooms[b]		
No humidity control	61.0	

Notes:
[a]Based on EPA (2000) and ASHRAE *Standard* 55-2007 for people wearing typical summer and winter clothing, at mainly sedentary activity.
[b]Usually not conditioned.

Table 2 Typical Recommended Design Criteria for Ventilation and Filtration for Preschools

Category	Ventilation and Exhaust [a, g, j]				Minimum Filtration Efficiency, MERV[h]
	Outdoor Air, cfm/ Person	Occupant Density[k] per 1000 ft²	Outdoor Air cfm/ft²	cfm/Unit	
Infant, Toddler, and Preschooler Classrooms[b]	17	25			6 to 8
Administrative and Office Space[c]	17	5			6 to 8
Kitchen[d]			0.3 (exhaust)		[i]
Toilets[e]				50 (exhaust)	NA
Storage[f]			0.12		1 to 4

Notes:
[a]Based on ANSI/ASHRAE *Standard* 62.1-2007, Table 6-1, default values for ventilation, and Table 6-4 for exhaust rates.
[b]Based on ASHRAE *Standard* 62.1-2007, Table 6-1, default values for educational facilities-daycare.
[c]Based on ASHRAE *Standard* 62.1-2007, Table 6-1, default values for office buildings/office spaces.
[d]Based on ASHRAE *Standard* 62.1-2007, Table 6-4, for kitchenettes.
[e]Based on ASHRAE *Standard* 62.1-2007, Table 6-4, for private toilets (rate is for toilet room intended to be occupied by one person).
[f]Based on ASHRAE *Standard* 62.1-2007, Table 6-1, for storage rooms.
[g]This table should not be used as the only source for design criteria. Governing local codes, design guidelines, and ASHRAE *Standard* 62.1-2007 with current addenda *must* be consulted.
[h]MERV = minimum efficiency reporting values, based on ASHRAE *Standard* 52.2-2007.
[i]See Chapter 33 for additional information on kitchen ventilation.
[j]Consult local codes for exhaust requirements.
[k]Use default occupancy density when actual occupant density is not known.

Table 3   Typical Recommended Design Guidelines for HVAC-Related Background Sound for Preschool Facilities

Category	Sound Criteria[a, b] RC (N); QAI < 5 dB	Comments
Infant, Toddler, and Preschooler Classrooms	25 to 30	
Administrative/Office Areas	≤40	For open-plan office
Service/Support Areas	35 to 45	

Notes:
[a]Based on Chapter 48.
[b]RC (Room Criterion), QAI (Quality Assessment Index) from Chapter 8 of 2009 *ASHRAE Handbook—Fundamentals*.

Load Characteristics

Preschool cooling and heating loads depend heavily on ambient conditions, because the rooms typically have exterior exposures (walls, windows, and roofs) and also relatively higher needs for ventilation. Although preschool facilities are relatively small, the design engineer must pay special attention to properly calculate the cooling, heating, dehumidification, and humidification loads. Sizing and applying the HVAC equipment is critical for handling the loads and the large amounts of outdoor air from a capacity and occurrence standpoint (peak sensible and latent loads do not always coincide).

Humidity Control

Preschool classrooms require humidity control to provide human comfort and prevent health problems. Maintaining humidity levels between 30 and 60°F dew point satisfies nearly all people nearly all the time. However, the designer should discuss comfort expectations with the owner, to avoid misunderstandings.

In hot and humid climates, it is recommended that air conditioning and/or dehumidification be operated year-round to prevent growth of mold and mildew. Dehumidification can be improved by adding optional condenser heat/reheat coils, heat pipes, or air-to-air heat exchangers in conjunction with humidity sensors in the conditioned space or return air. Additional information on humidity control is in the section on K-12 Schools.

Systems and Equipment Selection

HVAC systems for preschools are typically decentralized, using either self-contained or split air-conditioners or heat pumps (typically air- or water-source). When the preschool is part of a larger facility, utilities such as chilled water, hot water, or steam from a central plant can be used. When natural gas is available, the heating system can be a gas-fired furnace, or, when economically justifiable, electric heat can be used.

The type of HVAC equipment selected also depends on the climate and the months of operation. In hot and dry climates, for instance, evaporative cooling may be the primary type of cooling. In colder climates, heating can also be provided by a hot-water hydronic system originating from a boiler plant in conjunction with radiant floor or hot-water coils. For small, decentralized systems without central building control, a zone-level programmable temperature control is recommended (if not required by local code).

Decentralized systems are dedicated systems serving a single zone, and typically include the following:

- Direct-expansion (DX) split systems
- Rooftop packaged air conditioners or heat pumps with or without optional enhanced dehumidification (condenser reheat coil)
- Rooftop packaged air conditioners or heat pumps integrated with an energy recovery module, with optional enhanced dehumidification (condenser reheat coil; see Figure 5). ANSI/ASHRAE/IESNA *Standard* 90.1-2007, Addendum e, should be consulted for cases with a high percentage of outdoor air.

Table 4   Applicability of Systems to Typical Areas[d]

Typical Area	Decentralized Cooling/Heating Systems[c] PSZ/SZ Split	PSZ with Energy Recovery and Dehumidification	WSHP	Geothermal Heat Pump	Heating Only Radiant Floor[b]
Classrooms	X[a]	X[a]	X	X	X
Administrative Areas, Lobby	X		X	X	
Kitchen	X		X	X	
Ventilation (Outdoor Air)	DOAS		DOAS	DOAS	DOAS

SZ = single zone   PSZ = packaged single zone
WSHP = water-source heat pump   DOAS = dedicated outdoor air system
Notes:
[a]PSZ for classrooms requires individual thermostatic control.
[b]Typically with cooling system such as PSZ/SZ split.
[c]Heating system for PSZ/SZ split can be gas furnace, hot-water coil, or electric.
[d]See Table 10 for additional systems if preschool is not a stand-alone facility.

- Water-source heat pumps (with cooling tower and supplementary boiler)
- Geothermal heat pumps (ground-coupled, ground-water-source, surface-water-source)
- Packaged dedicated outdoor air systems with DX system for cooling and gas-fired furnace, electric heating, or part of water-source and geothermal heat pump system

Information about decentralized systems can be found in Chapters 2 and 48 of the 2008 *ASHRAE Handbook—HVAC Systems and Equipment*. Additional information on geothermal heat pumps can be found in Kavanaugh and Rafferty (1997) and Chapter 34 of this volume. Chapter 8 of the 2008 *ASHRAE Handbook—HVAC Systems and Equipment* provides information on radiant heating.

- Geothermal heat pumps (ground-coupled, ground-water-source, surface-water-source)
- Packaged dedicated outdoor air systems (DOASs) with DX system for cooling and gas-fired furnace, electric heating, or part of water-source and geothermal heat pump system

Information about decentralized systems can be found in Chapters 5 and 45 of the 2008 *ASHRAE Handbook—HVAC Systems and Equipment*. Additional information on geothermal heat pumps can be found in Kavanaugh and Rafferty (1997) and Chapter 32 of this volume. Chapter 6 of the 2008 *ASHRAE Handbook—HVAC Systems and Equipment* provides information on radiant heating.

Note that some decentralized systems may need additional acoustical modifications to meet the design criteria in Table 3. Therefore, it is strongly recommended to carefully check the acoustical implications of applying these systems.

Dedicated Outdoor Air Systems. Specialized DOASs should be used to treat outdoor air before it is introduced into classrooms or other areas. DOAS units can bring 100% outdoor air to at least space conditions, which allows the individual space units to handle only the space cooling and heating loads. A detailed description of DOAS is shown in the K-12 Schools section of this chapter. Additional information can be found in Chapter 24 of the 2008 *ASHRAE Handbook—HVAC Systems and Equipment*.

Systems Selection by Application. Table 4 shows the applicability of systems to areas in preschool facilities.

K-12 SCHOOLS

General and Design Considerations

K (kindergarten)-12 schools typically include elementary, middle (or junior high), and high schools. These facilities are typically one- to three-story buildings.

Elementary schools are generally comprised of 10 to 15 classrooms plus cafeteria, administration, gymnasium, and library areas. Elementary schools are typically used during the school season (late August to June); during summer, they are typically closed or have minimal activity. Current trends include science classrooms and a preschool facility. Typical elementary schools operate between 7:00 AM and 4:00 PM.

Middle schools are larger than elementary schools and include additional computer classrooms and locker rooms. Their hours of operation are longer because of extracurricular activities. A recent trend toward eliminating middle schools (retaining traditional K-8 elementary and 9-12 high schools) (Wright 2003) may require that elementary school designs incorporate some middle school features.

High schools also include a cafeteria and auditorium, and may include a natatorium, ice-skating rink, etc. High schools operate longer hours and are often open during the summer, either as a summer school or to use special facilities such as gymnasiums, natatoriums, etc.

Typical areas found in K-12 schools are shown in Table 5.

K-12 schools require an efficiently controlled atmosphere for a proper learning environment. This involves the selection of HVAC systems, equipment, and controls to provide adequate ventilation and indoor air quality (IAQ), comfort, and a quiet atmosphere. The system must also be easily maintained by the facility's maintenance staff.

The following are general design considerations for each of the areas typically found in K-12 schools:

Classrooms. Classrooms typically range between 900 and 1000 ft^2, and are typically designed for 20 to 30 students. Each classroom should be, at a minimum, heated and ventilated. Air conditioning should be seriously considered for school districts that have year-round classes in warm, humid climates. In humid climates, seriously consider providing dehumidification during summer, even if the school is unoccupied, to prevent mold and mildew.

Science Classrooms. Science rooms are now being provided for elementary schools. Although the children do not usually perform experiments, odors may be generated if the teacher demonstrates an experiment or if animals are kept in the classroom. Under these conditions, adequate ventilation is essential along with an exhaust fan with a local, timer-based (e.g., 0 to 60 min) on/off switch for occasional removal of excessive odors.

Computer Classrooms. These rooms have a high sensible heat load because of the computer equipment. They may require

additional cooling equipment such as small spot-cooling units to offset the additional load. Humidification may also be required. See Chapter 19 for additional information.

Educational Laboratories. Middle and high school laboratories and science facilities may require fume hoods with special exhaust systems. A makeup air system may be required if there are several fume hoods within a room. If there are no fume hoods, a room exhaust system is recommended for odor removal, depending on the type of experiments conducted in the room and whether animals are kept there; when applicable, a local exhaust with on/off switch and a timer can be considered. Associated storage and preparation rooms are generally exhausted continuously to remove odors and vapors emanating from stored materials. The amount of exhaust and location of exhaust grilles may be dictated by local codes or National Fire Protection Association (NFPA) standards. See Chapter 16 for further information. Additional information on laboratories can be found in McIntosh et al. (2001) and ANSI/AIHA *Standard* Z9.5-2010.

Administrative Areas. The office area should be set up for individual control because it is usually occupied during and after school hours. Because offices are also occupied before school starts in the fall, air conditioning for the area should be considered or provisions should be allowed for future upgrades.

Gymnasiums. Gyms may be used after regular school hours for evening classes, meetings, and other functions. The gym may also be used on weekends for group activities. The loads for these occasional uses should be considered when selecting and sizing the systems and equipment. Independent gymnasium HVAC systems with control capability allow for flexibility with smaller part-load conditions. If a wooden floor is installed, humidity control should be considered to avoid costly damage.

Libraries. Libraries should be air-conditioned to preserve the books and materials stored in them. See Chapters 3 and 23 for additional information.

Auditoriums. These facilities require a quiet atmosphere as well as heating, ventilation, and, in some cases, air conditioning. Auditoriums are not often used, except for assemblies, practice for programs, and special events. For other considerations, see Chapter 5.

Home Economics Rooms. These rooms usually have a high sensible heat load from appliances such as washing machines, dryers, stoves, ovens, and sewing machines. Different options should be considered for exhaust of stoves and dryers. If local codes allow, residential-style range hoods may be installed over the stoves. A central exhaust system could be applied to the dryers as well as to the stoves. If enough appliances are located within the room, a makeup air system may be required. These areas should be maintained at negative pressure in relation to adjacent classrooms and administrative areas. See Chapter 33 for more information.

Cafeteria and Kitchen. Typical schools require space for preparation and serving of meals. A well-designed school cafeteria includes the following areas: loading/receiving, storage, kitchen, serving area, dining area, dishwashing, office, and staff facilities (lockers, lavatories, and toilets). Chapter 33 provides detailed information on design criteria, load characteristics, and design concepts for these facilities.

Auto Repair Shops. These facilities require outdoor air ventilation to remove odors and fumes and to provide makeup air for exhaust systems. The shop is usually heated and ventilated but not air-conditioned. To contain odors and fumes, return air should not be supplied to other spaces, and the shop should be kept at a negative pressure relative to the surrounding spaces. Special exhaust systems such as welding exhaust or direct-connected carbon monoxide exhaust systems may be required. See Chapter 32 for more information.

Industrial Shops. These facilities are similar to auto repair shops and have special exhaust requirements for welding, soldering,

Table 5   Typical Spaces in K-12 Schools

Typical Area	School		
	Elementary (K to 5)[a]	Middle (6 to 8)[a]	High (9 to 12)[a]
Classrooms	X	X	X
Science	X	X	X
Computer	X	X	X
Laboratories and Science Facilities		X	X
Administrative Areas	X	X	X
Gymnasium	X	X	X
Libraries	X	X	X
Auditorium			X
Home Economics Room			X
Cafeteria	X	X	X
Kitchen	X	X	X
Auto Repair Shop[b]			X
Industrial Shop			X
Locker Rooms		X	X
Ice Rink[b]			X
Natatorium[b]			X
School Store[b]			X

Notes: [a]School grades can vary. [b]These zones are not typical.

and paint booths. In addition, a dust collection system is sometimes provided and the collected air is returned to the space. Industrial shops have a high sensible load from operation of the shop equipment. When calculating loads, the design engineer should consult the teacher about shop operation, and, where possible, diversity factors should be applied. See Chapter 32 for more information.

Locker Rooms. Building codes in the United States require that these facilities be exhausted directly to the outside when they contain toilets and/or showers. They are usually heated and ventilated only. These areas typically require makeup air and exhaust systems that should operate only when required. Where applicable, energy recovery systems can be considered.

Ice Rinks. These facilities require special HVAC and dehumidification systems to keep spectators comfortable, and to prevent roof condensation and fog formation at the surface. Where applicable, energy recovery systems can be considered. See Chapter 5 of this volume, Chapter 44 of the 2010 *ASHRAE Handbook—Refrigeration*, and Harriman et al. (2001) for more on these systems.

Natatoriums. These facilities, like ice rinks, require special humidity control systems. In addition, special construction materials are required. Where applicable, energy recovery systems can be considered. See Chapter 5 and Harriman et al. (2001) for more on these systems.

School Stores. These facilities contain school supplies and paraphernalia and are usually open for short periods. The heating and air-conditioning systems serving these areas should be able to be shut off when the store is closed to save energy.

Design Criteria

A typical HVAC design criteria covers parameters required for thermal comfort, indoor air quality (IAQ), and sound. Thermal comfort parameters (temperature and humidity) are well covered by ANSI/ASHRAE *Standard* 55-2004 and Chapter 9 of the 2009 *ASHRAE Handbook—Fundamentals*. Ventilation and IAQ are covered by ANSI/ASHRAE *Standard* 62.1-2007, the user manual for that standard (ASHRAE 2007), and Chapter 16 of the 2009 *ASHRAE Handbook—Fundamentals*. Sound and vibration are discussed in Chapter 48 of this volume and Chapter 8 of the 2009 *ASHRAE Handbook—Fundamentals*.

Thermal comfort is affected by air temperature, humidity, air velocity, and mean radiant temperature (MRT). In addition, nonenvironmental factors (clothing, gender, age, and physical activity) affect thermal comfort. These variables and their correlation with thermal comfort can be evaluated by the *Thermal Comfort Tool CD* (ASHRAE 1997) in conjunction with ANSI/ASHRAE *Standard* 55-2004. It is important to indicate that, in addition to thermal comfort criteria, several zones in schools (libraries, gymnasiums, locker rooms, natatoriums, ice rinks, etc.) require additional considerations to cover issues such as mold prevention, condensation, corrosion, etc., as discussed in more detail in the section on Humidity Control. General guidelines for temperature and humidity applicable for K-12 schools are shown in Table 6.

All schools need outdoor air for ventilation. Outdoor air is introduced to occupied areas and then exhausted by fans or exhaust openings, removing indoor air pollutants generated by occupants and any other building-related sources. ANSI/ASHRAE *Standard* 62.1-2007 is used as the basis for many building codes. To define the

Table 6 Typical Recommended Temperature and Humidity Ranges for K-12 Schools

Category/Humidity Criteria	Indoor Design Conditions		
	Temperature, °F		
	Winter	Summer	Comments
Classrooms, Laboratories, Libraries, Auditoriums, Offices [a, e]			
30% rh	68.5 to 75.5	74.0 to 80.0	
40% rh	68.0 to 75.0	73.5 to 80.0	
50% rh	68.5 to 74.5	73.0 to 79.0	
60% rh	67.5 to 74.0	73.0 to 78.5	
Gymnasiums			
30 to 60% rh	68.5 to 74.0	74.0 to 78.5	For gym with wooden floor, 35 to 50% humidity recommended at all times
Shops			
20 to 60% rh	68.5 to 74.0	74.0 to 78.5	
Cafeteria [b]			
20 to 30% (winter), 50% (summer) rh	70.0 to 73.5	78.5	
Kitchen [b]			
No humidity control	70.0 to 73.5	84.0 to 88.0	
Locker/Shower Rooms			
No humidity control	75.0		Usually not conditioned
Toilets			
No humidity control	72.0		Usually not conditioned
Storage			
No humidity control	64.0		
Mechanical Rooms			
No humidity control	61.0		Usually not conditioned
Corridors			
No humidity control	68.0		Frequently not conditioned
Natatorium [c]			
50 to 60% rh	75.0 to 84.0	75.0 to 84.0	Based on recreational pool
Ice Rink [d]			
35 to 45°F dp (maximum)	50.0 (minimum)	65.0 (maximum)	Minimum 10°F temperature difference between dew point and dry bulb to prevent fog and condensation

Notes:
[a]Based on EPA (2000) for people wearing typical summer and winter clothing, at mainly sedentary activity.
[b]Based on Chapter 3.

[c]Based on Chapter 5.
[d]Based on Harriman et al. (2001).
[e]For libraries, keep minimum humidity of 30°F dp and maximum of 55% rh.

Table 7 Typical Recommended Design Criteria for Ventilation and Filtration for K-12 Schools

Category	Ventilation and Exhaust[a]				
	Combined Outdoor Air, cfm/ Person	Occupant Density,[i] per 1000 ft²	Outdoor Air		Minimum Filtration Efficiency, MERV[c]
			cfm/ft²	cfm/Unit	
Classrooms, Ages 5 to 8	15	25			6 to 8
Ages 9 and over	13	35			6 to 8
Lecture	8	65			6 to 8
Art	19	20			6 to 8
Lecture Halls (fixed seats)	8	150			6 to 8
Science Laboratories[f]	17	25			6 to 8
Computer Lab	15	25			6 to 8
Media Center	15	25			6 to 8
Music/Theatre/ Dance	12	35			6 to 8
Multiuse Assembly	8	100			6 to 8
Libraries	17	10			6 to 8
Auditorium	5	150			9 to 10[g]
Administrative/ Office Areas	17	5			6 to 8
Gymnasium (playing floors)			0.3		6 to 8
Wood/Metal Shops	19	20			6 to 8
Locker Rooms			0.5 (exhaust)		1 to 4
Cafeteria	9	100			6 to 8
Kitchen[d, e]			0.7 (exhaust)		NA
Toilets				70 (exhaust)	NA
Storage			0.12		1 to 4
Corridors			0.06		6 to 8
Natatoriums (pool and deck)			0.48		6 to 8
Ice Rinks (spectator areas)[h]	8	150			6 to 8

Notes:

[a]Based on ANSI/ASHRAE *Standard* 62.1-2007, Tables 6-1 (i.e., default values) and 6-4. For systems serving multiple zones, apply multiple-zone calculations procedure. See the section on Demand Control Ventilation (DCV) when DCV is considered.

[b]This table should not be used as the only source for design criteria. Governing local codes, design guidelines, ANSI/ASHRAE *Standard* 62.1-2007, and user's manual (ASHRAE 2007) *must* be consulted.

[c]MERV = minimum efficiency reporting values, based on ASHRAE *Standard* 52.2-2007.

[d]See Chapter 33 for additional information on kitchen ventilation.

[e]Consult local codes for kitchen exhaust requirements.

[f]This table should not be used as the only source for laboratory design criteria. Governing local codes and design guidelines such as ANSI/AIHA *Standard* Z9.5-2010 and Chapter 16 of this volume *must* be consulted.

[g]When higher filtration efficiency specified, prefiltration is recommended.

[h]Based on ANSI/ASHRAE *Standard* 62.1-2007 values for sports and entertainment; for rink playing area, use gymnasium (playing floors) design criteria. Special attention should be given to internal-combustion ice-surfacing equipment for carbon monoxide control. Consult local code for ice rink design.

[i]Use default occupancy density when actual occupant density is not known.

Table 8 Typical Recommended Design Guidelines for HVAC-Related Background Sound for K-12 Schools

Category	Sound Criteria[a, b]	
	RC (N); QAI < 5 dB	Comments
Classrooms	25 to 30	
Large Lecture Rooms	25 to 30	
Without speech amplification	25	
Science Laboratories	35 to 45	
Libraries	30 to 40	See Table 42 of Chapter 48
Auditorium	30 to 35	Use as guide only; consult acoustician
Administrative	30 to 40	For open-office space
Gymnasium	40 to 50	
Shops	35 to 45	Use as guide only; consult acoustician
Cafeteria	35 to 45	Based on service/ support for hotels
Kitchen	35 to 45	Based on service/ support for hotels
Storage	35 to 45	
Mechanical Rooms	35 to 45	
Corridors	35 to 45	
Natatoriums	40 to 50	
Ice Rinks	40 to 50	Based on values for gymnasiums and natatoriums

Notes:

[a]Based on Chapter 48, Table 42.

[b]RC (Room Criterion), QAI (Quality Assessment Index) from Chapter 7 of the 2009 *ASHRAE Handbook—Fundamentals.*

ventilation and exhaust design criteria, consult local applicable ventilation and exhaust standards. Table 7 provides recommendations for ventilation design based on the ventilation rate procedure method and filtration criteria for K-12 educational facilities.

Additional information on IAQ for educational facilities can be found in EPA (2000).

Acceptable noise levels in classrooms are critical for a proper learning environment. High noise levels reduce speech intelligibility and student's learning capability. Although Chapter 48 provides information on design noise criteria, additional sources, such as local codes and ANSI *Standard* S12.60-2002, should be consulted for adequate design criteria. Table 8 summarizes applicable noise criteria for K-12 schools.

Load Characteristics

Proper cooling, heating, dehumidification, and humidification load calculations and properly sized equipment are critical to both energy efficiency and cost effectiveness. Many computer programs and calculation methodologies, as described in Chapter 18 of the 2009 *ASHRAE Handbook—Fundamentals*, can be used for these tasks. Assumptions and data used about infiltration, lighting, equipment loads, occupancy, etc., are critical for proper load calculations. Although equipment is sized by peak cooling and heating, it is extremely important to analyze the occurrences of the peak sensible and latent cooling loads. In many instances, peak sensible cooling load does not coincide with peak latent cooling load. Ignoring this phenomenon can result in unacceptable indoor humidity. By carefully analyzing and understanding the peak loads and the load profiles, the designer can properly apply and size the most suitable equipment to meet the sensible and the latent cooling loads efficiently. Elementary schools are generally occupied from about 7:00 AM to about 3:00 PM; occupation is longer for middle and high schools. Peak cooling loads usually occur at the end of the school

Table 9 Typical Classroom Summer Latent (Moisture) Loads

Category	Moisture Loads, lb/h	Moisture Loads, %
People	7.3	22.5
Permeance	0.2	0.6
Ventilation	20.3	62.5
Infiltration	4.7	14.4
Doors	0	0
Wet Surfaces	0	0
Humid Materials	0	0
Domestic Loads	0	0

Note: Based on Harriman et al. (2001), Chapter 18, Figure 18.2.

day. Peak heating usually occurs early in the day, when classrooms begin to be occupied and outdoor air is introduced into the facility. Although K-12 schools are dominated by perimeter zones (and zones exposed to the roof), careful attention should be given to components of the loads. Typical breakdowns of moisture loads are shown in Table 9.

Typically, the dominant cooling loads in classrooms are occupants and ventilation, and ventilation and roof for heating. Given the dominance of ventilation loads, special effort should be made to effectively treat outdoor air before its introduction to the space, as discussed in more detail in the section on Systems and Equipment Selection.

Humidity Control

School buildings host many activities that require special humidity control. Harriman et al. (2001) provide detailed information on the basics of design and equipment selection for proper humidity control for several applications; Chapter 18 is dedicated to schools.

Classrooms require humidity control to provide comfort and prevent humidity-related problems (e.g., growth of dust mites and fungus, which produce allergens and even toxic by-products). Low humidity, on the other hand, favors longevity of infectious viruses, and therefore their transmission between occupants. Maintaining dew point levels between 30 and 60°F satisfies nearly all people nearly all the time. However, the designer should discuss comfort expectations with the owner, to avoid misunderstandings.

Libraries require humidity control to provide human comfort to the occupants and also to protect books and electronic records. Maintaining dew point levels between 30 and 60°F provides a comfortable environment for the library occupants. However, controlling humidity at this range does not prevent books from absorbing excess moisture. Typically, books take up moisture quickly but lose it slowly. To avoid growth of mold and mildew, a dew point above 30°F and maximum of 55% rh are recommended. As for classrooms, the principal moisture loads for the library are ventilation (the major load) and infiltration.

Gymnasiums with wooden floors require special attention; failure to control humidity in gyms with wooden floors may have costly consequences. The Maple Flooring Manufacturers Association (MFMA) specifies a floor-level humidity between 35 and 50% rh.

Showers and locker rooms require humidity control to prevent corrosion and growth of bacteria and fungus. Therefore, special attention is required to exhaust air quantities and placement of supply and exhaust air registers.

Natatoriums and ice rinks are typically isolated areas with more specialized HVAC equipment specifically designed to address ventilation and humidity control. Chapters 27 and 28 of Harriman et al. (2001) provide detailed information on humidity control for natatoriums and ice rinks, respectively.

Systems and Equipment Selection

Selection of HVAC equipment and systems depends on whether the facility is new or existing, and whether it is to be totally or partially renovated. For minor renovations, existing HVAC systems are often expanded in compliance with current codes and standards with equipment that matches the existing types. For major renovations or new construction, new HVAC systems and equipment should be installed. When applicable, the remaining useful life of existing equipment and distribution systems should be considered.

HVAC systems and equipment energy use and associated life-cycle costs should be evaluated. Energy analysis may justify new HVAC equipment and systems when an acceptable return on investment can be shown. The engineer must review all the assumptions in the energy analysis with the school administration. Assumptions, especially about hard-to-measure items such as infiltration and part-load factors, can significantly affect the energy use calculated.

Other considerations for existing facilities are (1) whether the central plant is of adequate capacity to handle additional loads from new or renovated facilities; (2) the age and condition of the existing equipment, pipes, and controls; and (3) the capital and operating costs of new equipment. Schools usually have very limited budgets. Any savings in capital expenditures and energy costs may be available for the maintenance and upkeep of the HVAC systems and equipment and for other facility needs.

The type of HVAC equipment selected also depends on the climate and months of operations. In hot, dry climates, for instance, evaporative cooling may be the primary approach. Some school districts may choose not to provide air conditioning. However, in hot, humid climates, it is recommended that air conditioning or dehumidification be operated year-round to prevent growth of mold or mildew.

Chapter 1 of the 2008 *ASHRAE Handbook—HVAC Systems and Equipment* provides general guidelines on HVAC systems analysis and selection procedures. Although in many cases system selection is based solely on the lowest first cost, it is suggested that the engineer propose a system with the lowest life-cycle cost (LCC). LCC analysis typically requires hour-by-hour building energy simulation for annual energy cost estimation. Detailed first and maintenance cost estimates of proposed design alternatives, using sources such as R.S. Means (2010a, 2010b), can also be used for the LCC analysis along with software such as BLCC 5.1 (FEMP 2010). Refer to Chapters 37 and 58, and the Value Engineering (VE) and Life-Cycle Cost Analysis (LCCA) section of this chapter for additional information.

System Types. HVAC systems for K-12 schools may be centralized, decentralized, or a combination of both. Centralized systems typically incorporate secondary systems to treat the air and distribute it. The cooling and heating medium is typically water or brine that is cooled and/or heated in a primary system and distributed to the secondary systems. Centralized systems comprise the following systems:

Secondary Systems

- Air handling and distribution (see Chapter 4 of the 2008 *ASHRAE Handbook—HVAC Systems and Equipment*)
- In-room terminal systems (see Chapter 5 of the 2008 *ASHRAE Handbook—HVAC Systems and Equipment*)
- DOAS with chilled water for cooling and hot water, steam, or electric heat for heating

Primary Systems

- Central cooling and heating plant (see Chapter 3 of the 2008 *ASHRAE Handbook—HVAC Systems and Equipment*)

Typical decentralized systems (dedicated systems serving a single zone, or packaged systems such as packaged variable-air-volume) are

- Water-source heat pumps (WSHPs), also known as water-loop heat pumps (WLHPs)
- Geothermal heat pumps (groundwater heat pumps, ground-coupled heat pumps)

- Hybrid geothermal heat pumps (combination of groundwater heat pumps, ground-coupled heat pumps, and an additional heat rejection device), for cases with limited area for the ground-coupled heat exchanger or where it is economically justified
- Packaged single-zone and variable-volume units
- Light commercial split systems
- Minisplit and variable-refrigerant-flow (VRF) units

Chapters 2, 8, 48, and 49 of the 2008 *ASHRAE Handbook—HVAC Systems and Equipment* provide additional information on decentralized HVAC systems. Additional information on geothermal energy can be found in Chapter 34 of this volume.

It is important to note that, to meet the acoustical design criteria in Table 8, designers should avoid locating HVAC equipment in classrooms, and that some centralized and decentralized systems located close to classrooms might need additional sound-attenuating features. Coordination between the HVAC designer, architect, and acoustical consultant is critical for meeting the desired noise criteria. Siebein and Likendey (2004) provide information on the applicability of systems to classrooms with regard to acoustical criteria. Additional information on how HVAC&R manufacturers' acoustical data and application information can be best used can be found in Ebbing and Blazier (1998). Schaffer (1993) provides a practical guide to noise and vibration control for HVAC systems. Commercial acoustics analysis software can also be helpful.

Dedicated Outdoor Air Systems. Although most centralized and decentralized systems are very effective at handling the space sensible cooling and heating loads, they are less effective (or ineffective) at handling ventilation air and the latent loads. As a result, a DOAS should be used. DOAS units bring 100% outdoor air to at least space conditions, which allows individual space units to handle only the space loads. It is preferable, however, to introduce the outdoor air at a lower humidity ratio than the desired space humidity ratio, to allow the zone HVAC unit to handle only the space sensible cooling load. This approach can be easily implemented in a classroom where a significant amount of outdoor air is required for ventilation.

Example. In a typical classroom with 30 students, the ventilation requirements are 450 cfm. If the outdoor air can be introduced at a humidity ratio of 48 gr/lb and the space is designed to be maintained at 70 gr/lb, the space dehumidification capability of the pre-dehumidified outdoor air is the following:

$$\text{Space dehumidification capability, Btu/h} = 0.68 \times \text{cfm} \times \left(\begin{array}{c} \text{Space humidity ratio} - \\ \text{Supply humidity ratio} \end{array} \right)$$

Then,

$$\text{Dehumidification capability, Btu/h} = 0.68 \times 450 \times (70 - 48) = 6732 \text{ Btu/h}$$

where

$$0.68 = (60/13.5)(1076/7000)$$

60 = min/h
13.5 = specific volume of moist air at 70°F and 50% rh, ft³/lb
1076 = average heat removal required to condense 1 lb of water vapor from room air, Btu/lb
7000 = grains per pound

The 6732 Btu/h of space latent load is equivalent to the latent load of 30 occupants (seated, very light work, 155 Btu/h per occupant) and the additional space latent load (e.g., infiltration latent load).

Occupant latent load = 30 Occupants × 155 Btu/h per occupant
= 4650 Btu/h

Remainder of total
dehumidification = 6732 − 4650 = 2082 Btu/h
capability

This additional dehumidification capability can help in handling infiltration latent load and others.

This simple example demonstrates the ability of pre-dehumidified outdoor air to handle the space latent load, resulting in almost full separation of the space latent cooling load treatment from the space sensible cooling load. This approach allows only thermostatic control without losing humidity control in conditioned classrooms.

Typical DOAS units are air-handling units that cool, dehumidify, heat, humidify, and filter the outdoor air before it is introduced to the conditioned space. Typical DOASs include the following major components:

- Mechanical cooling/dehumidification
  - DX coil
  - Chilled-water coil
- Desiccant-based cooling/dehumidification
  - Desiccant (dehumidification) and direct-expansion (DX) coil (post sensible cooling)
  - Desiccant (dehumidification) and chilled-water coil (post sensible cooling)
- Heating
  - Coils (hot-water, steam, electric, heat pump)
  - Gas-fired furnace
- Humidification
  - Passive (in conjunction with enthalpy wheel heat recovery)
  - Active (steam, electric-to-steam, gas-to-steam)
- Exhaust air recovery: air-to-air heat recovery
  - Rotary (enthalpy wheel, sensible wheel)
  - Fixed (heat pipe, plate heat exchanger, runaround coils)
- Dehumidification enhancements for air-to-air heat recovery
  - Heat pipe based (wraparound coil)
  - Mini plate heat exchanger based

Which DOAS configuration is most cost-effective depends on variables such as availability of utilities (chilled water, gas, steam),

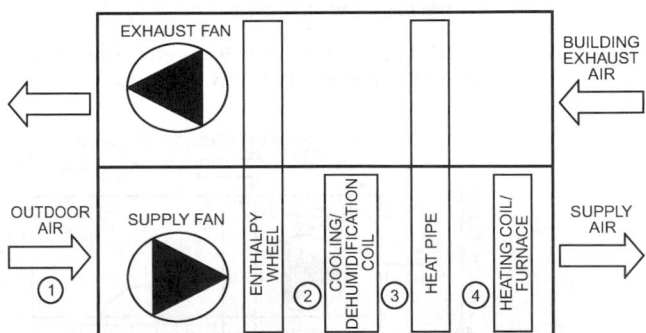

Fig. 1   Typical Configuration of DOAS Air-Handling Unit: Enthalpy Wheel with Heat Pipe for Reheat

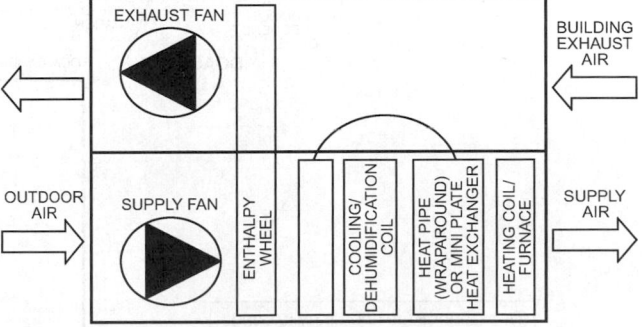

Fig. 2   Typical Configuration of DOAS Air-Handling Unit: Enthalpy Wheel with Wraparound Heat Pipe for Reheat

space constraints, climatic data, utility cost, and budget. DOAS can be configured easily by using modular components that meet the design criteria. Selection and analysis software of these systems is readily available from DOAS manufacturers, which simplifies configuration and analysis of the most cost-effective system. Typical configurations of DOAS are shown in Figures 1 and 2. A cooling/dehumidification psychrometrics process of DOAS shown in Figure 3.

Air-to-air energy recovery is an important element in a DOAS. In addition to recovering energy from the exhaust air, a well-designed energy recovery module, such as an enthalpy wheel, can enhance and stabilize operation of the cooling and heating elements in the DOAS unit. As shown in Figure 3, the process of bringing outside air from point 1 to point 2 can be defined as "compressing" the outdoor air conditions to almost return air conditions.

Given the need for more stringent and complex control schemes for outdoor air preconditioning, DOAS typically incorporate, direct

digital control (DDC) systems, either stand-alone microprocessor-based or with the ability to communicate with central energy management system. The control system can be purchased as an option or installed in the field by the controls vendor. Typical supply air conditions for a DOAS air-handling unit are shown in Table 10.

Typical arrangements of DOAS integrated with local cooling and heating systems are shown in Figure 4.

Systems with High Percentage of Outdoor Air. Air-handling systems with a high percentage of outdoor air (above 30%) can be found in several areas in educational facilities. To prevent indoor air quality problems and conserve energy, an energy recovery module can be added to pretreat the outdoor air before it is mixed with return air. Figure 5 shows a typical rooftop packaged AC unit with energy recovery module. See Chapter 25 of the 2008 *ASHRAE Handbook—HVAC Systems and Equipment* for more information on energy recovery equipment and systems.

The addition of an energy recovery module is dependent on the percentage of outdoor air and the geographic location. See Addendum e of ANSI/ASHRAE/IES *Standard* 90.1-2007 for the correlation between geographic location and percentage of outdoor air (OA). It is strongly recommended to refer to Addendum e's exceptions with regard to the applicability of energy recovery.

Systems Selection by Application. Table 11 shows the applicability of systems to areas in K-12 school facilities.

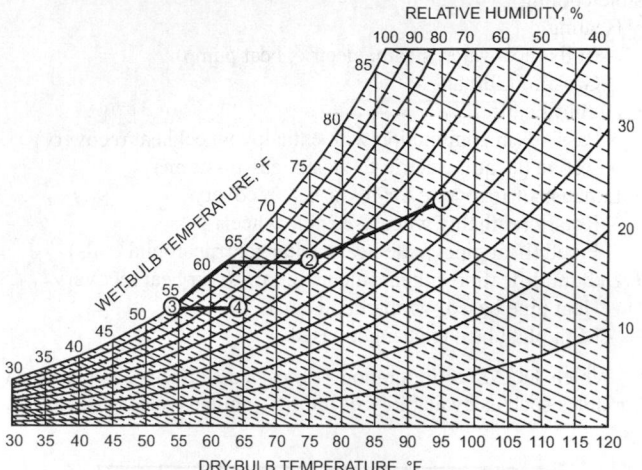

Fig. 3   Cooling/Dehumidification Psychrometric Process of Typical DOAS Air-Handling Unit in Figure 1

Table 10   Typical Design Criteria for DOAS Air-Handling Unit

	Supply Air Conditions[a]		Minimum Air Filtration Efficiency, MERV[b]
	Temperature, °F	Humidity Ratio, gr/lb	
Winter	65 to 68	30 to 40	6 to 8
Summer	60 to 65	40 to 60	6 to 8

Notes:
[a]Building location may dictate optimum supply condition in recommended range.
[b]Filter efficiency definition per ASHRAE *Standard* 52.2-2007.
MERV = minimum efficiency reporting values

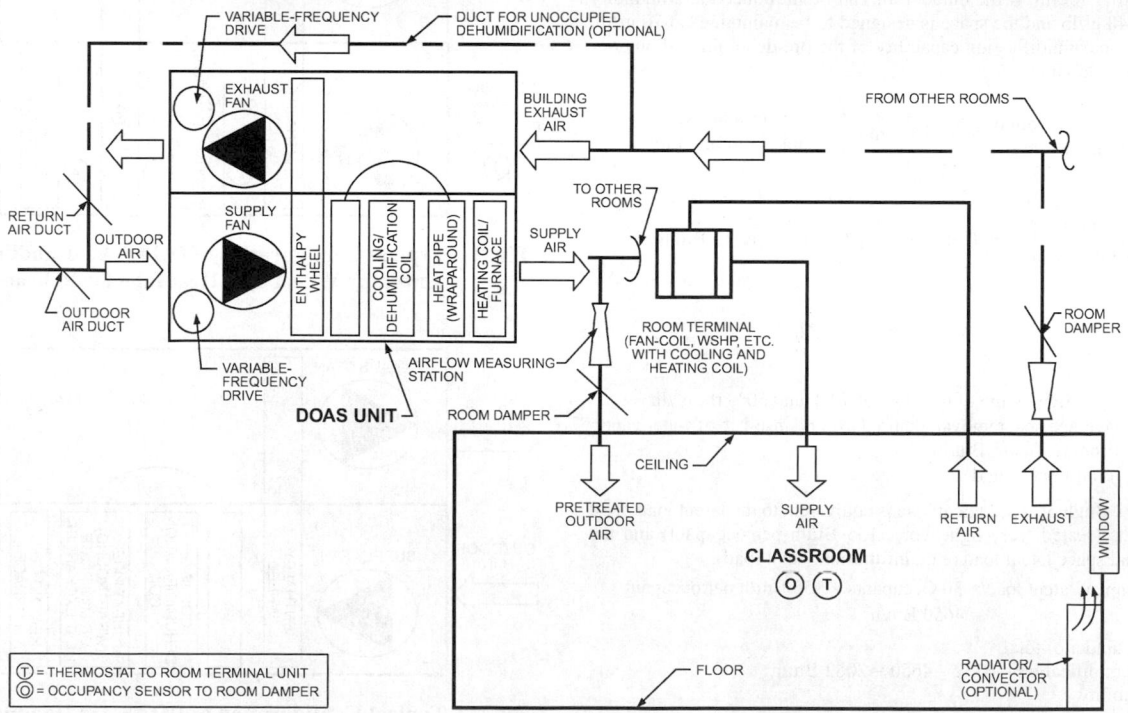

Fig. 4   Typical Schematic of DOAS with Local Classroom Cooling/Heating Terminal

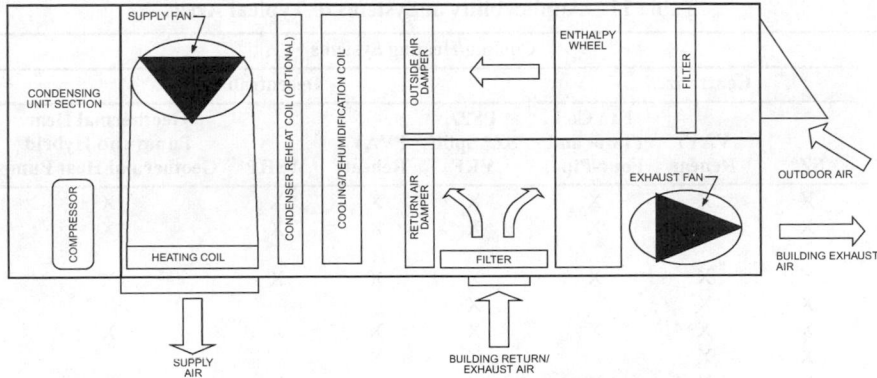

Fig. 5 Typical Configuration of Rooftop Packaged Air Conditioners with Energy Recovery Module and Enhanced Dehumidification (Condenser Reheat Coil)

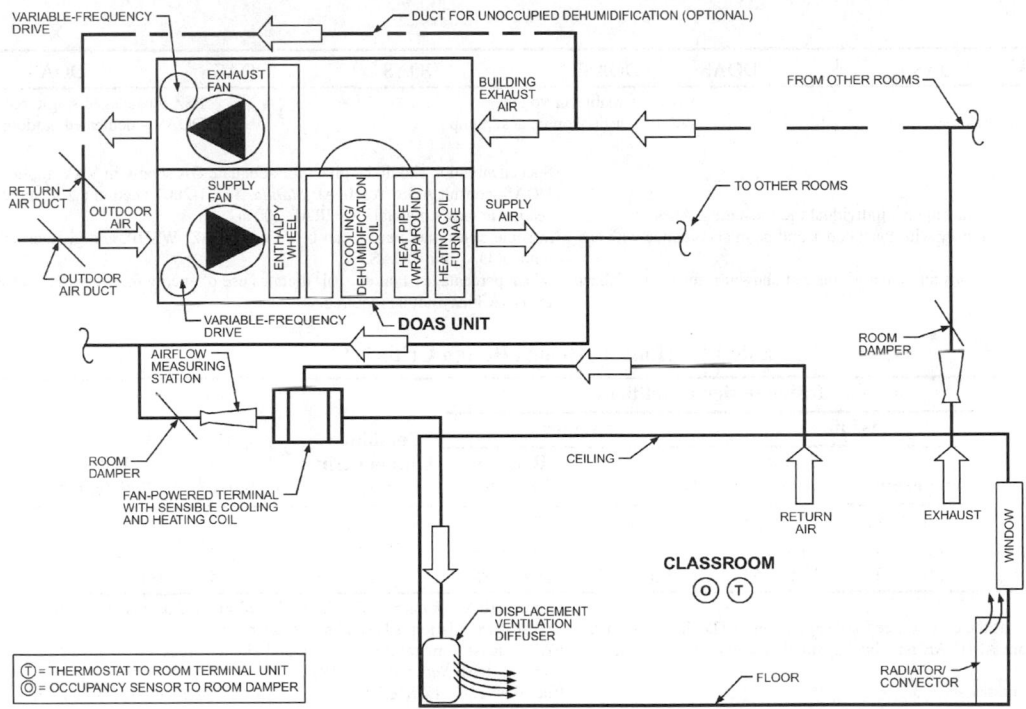

Fig. 6 Typical Displacement Ventilation System Layout

Displacement Ventilation and Active/Induction Chilled Beams

Displacement Ventilation. The use of displacement ventilation (as opposed to the more traditional mixing ventilation) for classrooms has been extended for enhanced IAQ and thermal comfort. In displacement ventilation, fresh air at colder temperature than the room air is discharged close to the floor level, and warm air is exhausted at or close to the ceiling. After being discharged at a low level, the colder supply air rises as it is heated by heat sources (e.g., people, computers), also allowing effective removal of containments generated in the room.

Guidelines and procedures for designing displacement ventilation systems can be found in Chen and Glicksman (2003), Skistad et al. (2002), California Energy Commission (2006), Chapter 19 of the 2008 *ASHRAE Handbook—HVAC Systems and Equipment,* and Chapter 57 of this volume, which provides a classroom-based example.

Typical displacement ventilation systems for classrooms include the following main subsystems (Figure 6):

- DOAS air-handling unit that can cool and dehumidify outdoor air to 60 to 62°F and 40 to 50 gr/lb for summer, and heat air to 65 to 68°F for winter
- Zone fan-powered terminal with sensible cooling capability (located outside the conditioned zone)
- Special displacement ventilation diffusers
- Heating radiators or convectors placed below windows in perimeter zones
- Control systems (thermostats and occupancy sensors)

In addition to the traditional displacement ventilation system described previously, displacement ventilation with induction can also be considered for classrooms. A displacement ventilation system with induction uses special terminals to provide additional cooling and heating with the displacement ventilation effect. These terminals are not equipped with fans, resulting in lower noise levels as required by more stringent noise criteria.

A displacement ventilation system with induction includes the following main subsystems:

Table 11   Applicability of Systems to Typical Areas

| | Cooling/Heating Systems | | | | | | | Heating Only | |
| | Centralized | | | Decentralized | | | | | |
Typical Area[c]	SZ[a]	VAV/ Reheat	Fan Coil (Two- and Four-Pipe)	PSZ/ SZ[a] Split/ VRF	PVAV/ Reheat	WSHP	Geothermal Heat Pump and Hybrid Geothermal Heat Pump	Baseboard/ Radiators	Unit Heaters
Classrooms	X	X	X	X	X	X	X	X	
Laboratories and Science Facilities[b]	X	X	X	X	X	X	X	X	
Administrative Areas	X	X	X	X	X	X	X	X	
Gymnasium[e]	X	X		X					X
Libraries	X	X	X	X	X	X	X	X	
Auditorium[e]	X	X		X	X				
Home Economics Room	X	X	X	X	X	X	X	X	
Cafeteria[e]	X			X					
Kitchen[e]	X			X					X
Auto Repair Shop									X
Industrial Shop									X
Locker Rooms								X	X
Ventilation (Outdoor Air)	DOAS	[d]	DOAS	DOAS[f]	[d]	DOAS	DOAS	DOAS	DOAS

SZ = single zone
PVAV = packaged variable air volume
VRF = variable refrigerant flow

VAV = variable air volume
WSHP = water-source heat pump

PSZ = packaged single zone
DOAS = dedicated outdoor air system

Notes:

[a]SZ and PSZ/SZ split for classrooms requires individual thermostatic control.

[b]Systems for laboratories must comply with local codes and be in accordance with current practices for laboratories.

[c]Systems and equipment for ice rinks and natatoriums not shown; refer to specialized equipment section.

[d]Special attention should be given for adequate OA supply in VAV applications without DOAS; consult ANSI/ASHRAE *Standard* 62.1-2007 Section 6.2.5, and corresponding section in user's manual (ASHRAE 2007).

[e]In some cases, these areas can be served by SZ, WSHP, and geothermal HP systems without OA from DOAS.

[f] When percentage of outdoor air dictates use of energy recovery in SZ or PSZ unit, OA for DOAS may not be required.

Table 12   Housing Rooms Design Criteria[a]

| | Inside Design Conditions | | | | | | | |
| | Winter | | Summer | | Combined Outdoor Air Rate[c] | Exhaust[d] | Filter Efficiency[e] | Noise, RC (N);QAI < 5 dB Level[f] |
Category	Temperature	Relative Humidity[b]	Temperature	Relative Humidity				
Dorm, suite rooms	70 to 72°F	30 to 35%	74 to 78°F	50 to 60%	22 cfm	NR	6 to 8 MERV	25 to 35
Apartments and studio rooms	70 to 72°F	30 to 35%	74 to 78°F	50 to 60%	85 cfm	75 cfm	6 to 8 MERV	25 to 35
Couple and faculty housing	70 to 72°F	30 to 35%	74 to 78°F	50 to 60%	85 cfm	75 cfm	6 to 8 MERV	25 to 35

NR = not required.

[a]This table should not be used as the only source for design criteria. The data contained here can be determined from ASHRAE handbooks, standards, and governing local codes.

[b]Minimum recommended humidity.

[c]Per ASHRAE *Standard* 62.1-2007, based on two occupants for room. For areas with exhaust, ventilation is based on exhaust requirements.

[d] Air exhaust from bathroom, toilet, and kitchen areas.

[e]Per ASHRAE *Standard* 52.2-1999

[f]Based on 2007 *ASHRAE Handbook—HVAC Applications*, Chapter 47.

- DOAS air-handling unit that can cool and dehumidify outdoor air to 54 to 57°F and 40 to 50 gr/lb for summer, and heat air to 65 to 68°F for winter
- Zone displacement ventilation with induction terminal, equipped with two- or four-pipe cooling and heating coil mounted along perimeter walls and windows
- Control systems (thermostats and occupancy sensors)

Active (Induction) Chilled Beams. Recently, the use of active/induction chilled beams for classrooms and other areas in educational facilities has been extended for enhanced IAQ, thermal comfort, and energy conservation. As with displacement ventilation with induction, an active/induction chilled beam terminal includes special small air jets that induce room air to flow through cooling or heating coils, depending on the system (two- or four-pipe). The primary air is outdoor air pretreated in a DOAS unit, as described previously. Figure 7 shows the principle of active/induction chilled beam terminals.

Although more room space is required for chilled-beam induction, these systems allow significant size and capacity reductions in air-handling systems, and decouple sensible cooling and heating from ventilation and humidity control. Temperatures of chilled

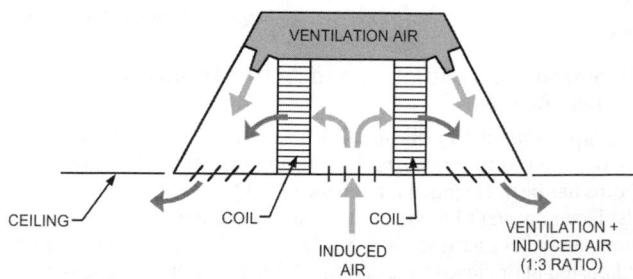

Fig. 7   Typical Active/Induction Chilled-Beam Terminal
(Rumsey and Weale 2006)

water distributed to the chilled-beam terminals are typically elevated to around 55°F, which can reduce energy consumption. Hot water can be provided from a standard hot-water boiler at 150 to 180°F, or lower if condensing boilers applied.

An active/induction chilled-beam system typically includes the following main subsystems:

- DOAS unit that can cool and dehumidify outdoor air to 54 to 57°F and 40 to 50 gr/lb for summer, and heat air to 65 to 68°F for winter
- Zone active/induction chilled-beam terminal, equipped with two- or four-pipe cooling and heating
- Control systems (thermostats and occupancy sensors)

Systems Selection by Application. Table 12 shows the applicability of systems to areas in K-12 school facilities.

Specialized Equipment. Areas such as natatoriums and ice rinks need specialized equipment to address the unique design requirements and the cooling, dehumidification, and heating characteristics of these areas. Natatoriums typically use special units that can introduce large quantities of outdoor air and allow active humidity control (mainly dehumidification). This equipment is similar to DOAS, and typically uses chilled water or a DX system for dehumidification. For systems with air-cooled condensers, condenser heat can be recovered to heat the swimming pool. See Chapter 5 of this volume for more information on natatoriums. Similarly, an ice rink requires special equipment; selection depends heavily on the school's location and seasonal use. Ice rink HVAC and dehumidification equipment can be desiccant-based or self-contained mechanical refrigeration. See Chapter 5 of this volume and Chapter 44 of the 2010 *ASHRAE Handbook—Refrigeration* for more information on ice rinks.

Chapters 27 and 28 of Harriman et al. (2001) also provide detailed information on humidity control for natatoriums and ice rinks, respectively.

Demand Control Ventilation (DCV). Demand control ventilation can reduce the cost of operating the HVAC systems. To ensure proper IAQ and comply with ANSI/ASHRAE *Standard* 62.1-2007 and local codes that permit DCV, the designer must carefully follow section 6.2.7 (Dynamic Reset) of ANSI/ASHRAE *Standard* 62.1 and corresponding sections of the user's manual (ASHRAE 2007). ANSI/ASHRAE *Standard* 62.1-2007 explicitly allows use of CO_2 levels or occupancy to reset intake airflow in response to space occupancy levels. Special attention should be given to the area served by the HVAC system and the system type. Areas such as gymnasiums and auditoriums can benefit from CO_2-based DCV, commonly used in single-zone systems without DOAS, serving one space with varying occupancy. In these cases, DCV control is simple, reliable, and cost-effective. Systems such as multizone VAV with recirculated air without DOAS require special attention to ensure adequate OA supply to multiple zones under varying loads (such as classrooms). This problem complicates the design, operation, and maintenance of DCV control systems and also adds the cost of additional sensors.

A simpler approach for DCV is in systems that use DOAS: the OA supply to each individual space can be controlled independently by occupancy sensors that can reduce the OA to a preset value (and also turn off the lights), or by CO_2 sensors (see Figure 4).

COLLEGES AND UNIVERSITIES

General and Design Considerations

College and university facilities can be a campus, cluster of buildings, or a single isolated building. Some colleges and universities have satellite campuses scattered throughout a city or a state. The design criterion for each building is established by the requirements of its users. The following are major facilities commonly found on college and university campuses.

Libraries/Learning Centers. Libraries and learning centers are central to the purpose of modern college and university. A library can be a collection of printed and electronic material and/or a place where individuals or groups of students gather for study or other academic activities. A typical library includes the following areas:

- Collection/stacks
- Library staff and services

- Main reading room
- Specialty areas (special collections, music and audiovisual resources, computer areas, etc.)
- Support areas

Temperature and humidity control is needed for maintaining the printed materials and the collections. Proper air distribution can be challenging because of different ceiling heights, stacks, mezzanines, etc. Reading rooms require air supply without draft, and special collections or rare books areas need a dedicated air handling system. Noise also is critical in libraries; an acoustic consultant must review or be part of the mechanical design. See Chapter 23 for specifics on HVAC design for libraries.

Academic Buildings and Professional Schools. These buildings accommodate classrooms, which are the core of the university teaching and learning experience. There are two main categories of classrooms, with several subcategories (Neumann 2003):

Flat-floor classrooms are typically rectangular, basic, and easily reconfigurable for different teaching needs. In most cases, the number of students is relatively low. Sometimes, a larger flat-floor room can be subdivided to smaller rooms by folding or sliding partitions.

Sloped-floor classrooms are used when the class size exceeds the point where all the students can see each other clearly in a flat-floor classroom. Sloped-floor classrooms typically have more than 40 students. Those with a capacity of 250 students or more are generally referred as auditoriums, which require theater design consideration.

Academic buildings also have faculty offices and auxiliary areas to support the teaching activities. Professional schools are typically allocated to a specific academic discipline. Each of these schools has specific needs, depending on the academic requirements. The HVAC design and systems for classrooms and other administrative areas are similar to classrooms in high schools (see Table 11).

Science Teaching and Research Facilities. College and universities science facilities accommodate highly specialized areas for teaching and research in several disciplines (e.g., chemistry, biology, physics). Teaching facilities are designed mainly for group instruction, typically with one or more instructors and 12 to 32 students; an average-sized teaching lab can accommodate 24 students. The laboratory should be designed to support a range of activities for various courses: for example, a chemistry lab should be able to handle introductory chemistry, organic chemistry, etc.

Research facilities can be part of a science teaching building or grouped in a stand-alone research facility. Research facilities are customized and designed for graduate and postgraduate students, typically under the direction and supervision of several principal investigators (PIs). Unlike teaching labs, which are designed for large group instruction, research labs should be designed to accommodate the activities of individuals or small groups. Given potentially hazardous activities in teaching and research labs, the most critical factor in designing systems for labs is safety; this concern has major implications on the design of HVAC and mechanical systems.

Teaching and research labs may contain fume hoods, machinery, lasers, vivariums, areas with controlled environments, and departmental offices. The HVAC systems and controls must be able to accommodate diverse functions of the facility, which may have 24 h, year-round operation, and yet be easy to service and quick to repair. Variable-air-volume (VAV) systems can be used. Proper control systems should be applied to introduce and extract the required quantities of supply and exhaust air. Maintaining the required space pressure differential to adjacent spaces and the minimum airflow under all circumstances is extremely critical for safe laboratory operation. Energy can be saved by recovering energy from exhaust air and tempering outdoor makeup air. Special attention should be given to containment in the exhaust air stream. Potential carryover of air from exhaust to supply, and interaction with the energy recovery device adsorbent for cases with total (sensible and latent) energy recovery, should be examined. In general, air exhausted from fume hoods should not be used for energy recovery. Where heat recovery from fume hoods exhaust is considered, careful coordination with the site health and safety (H&S) officer is required. Other energy-saving systems used for laboratory buildings include (1) active chilled beams (Rumsey and Weale 2006), (2) ice storage, (3) heat reclaim chillers to produce hot water for domestic use or for booster coils in the summer, and (4) cooling tower free cooling.

The design engineer should discuss expected contaminants and concentrations with the owner to determine construction materials for fume hoods and fume exhaust systems. Close coordination with H&S personnel is vital for safe laboratory building operation. Back-up or standby systems for emergency use should be considered, such as alarms on critical systems. Maintenance staff should be thoroughly trained in upkeep and repair of all systems, components, and controls. For design criteria and other design information on laboratories and vivariums, see Chapter 16, ANSI/AIHA *Standard* Z9.5-2010, DiBerardinis et al. (2001), and McIntosh et al. (2001). Additional information on energy conservation in labs can be found on the Labs 21 Web site (http://labs21bench marking.lbl.gov/).

Some research facilities include vivariums (animal facilities). These spaces are commonly associated with laboratories, but usually have their own separate areas. Additional areas that can found in vivariums are necropsy rooms, surgery suites, and other specialty areas. Animal facilities need close temperature control and require a significant amount of outdoor ventilation to control odors and prevent the spread of diseases among the animals. Animal facilities are discussed in Chapters 16 and 24, and by the National Research Council (NRC 1996).

Housing

Student Housing. Housing is an integral part of student's academic and social life. Student housing traditionally had few amenities, and the emphasis for years was economy and reduced construction cost. Today, more housing administrators are changing this philosophy by providing an enhanced, rich on-campus residential life. Student and staff housing facilities include the following:

• Dormitories (residence halls)
• Suites
• Apartments and studios
• Couples housing

Dormitories (residence halls) are typically for freshman students. Student living units are generally single- or double-occupancy rooms that open directly to a corridor. The building can be a high rise or low rise, depending on the setting or the location of the campus. Typically, there are two students per room, with one single-occupancy room reserved for the resident assistant. On the ground floor are public facilities, which may include a living room, reception desk, kitchen/lounge, and cafeteria. Dorm rooms typically do not have individual kitchens or bathrooms; communal bathrooms usually serve one floor.

Suites are typically occupied by older undergraduate students. The suite plan typically connects four to six double-occupancy sleeping room rooms with a shared bathroom and living room.

Apartments and **studios** are typically occupied by upper-division and graduate students, and arc basically suites with kitchens and private bathrooms. Apartments and studios are the most desirable housing and are the most expensive because of their additional plumbing and electrical systems.

Couples housing generally consists of one-, two-, or three-bedroom apartments in separated complexes. A couples housing facility may have a section for married couples, who often have young children whose safety and security needs must be considered. These facilities may have outdoor play areas and child care facilities.

Faculty Housing. Faculty members typically find housing outside the campus, but the high cost of local living has convinced many universities that offering on-campus housing will attract the best candidates to their academic institution. This type of housing is similar to typical residential housing and can include duplexes, apartments, townhouses, and single-family homes.

Air conditioning in campus housing for students and faculty should be quiet, easily adjustable, and draft free. Systems that require little space and have low total owning and operating costs should be selected. Table 12 lists design criteria for housing facilities.

Typically, decentralized systems with DOAS or air-to-air energy recovery should be used for these applications:

• Water-source heat pumps (WSHPs), also known as water-loop heat pumps (WLHPs)
• Geothermal heat pumps (groundwater heat pumps, ground-coupled heat pumps)
• Hybrid geothermal heat pumps (combination of groundwater heat pumps, ground-coupled heat pumps, and an additional heat rejection device), where there is limited area for the ground-coupled heat exchanger or where it is economically justified
• Light commercial split systems
• Minisplit and variable-refrigerant-flow (VRF) units
• Fan-coil units

When dormitories are closed during winter breaks, the heating system must supply sufficient heat to prevent freeze-up. If the dormitory contains nondwelling areas, such as administrative offices or eating facilities, these facilities should be designed as a separate zone or with a separate system for flexibility, economy, and odor control. Solar energy can be considered for domestic hot water (DHW).

Athletics and Recreational Facilities

College and university sports facilities ranging from large arenas for ice hockey, basketball, and other spectator sports, to small gymnasiums and fitness centers. College sports activities are heavily influenced by intercollegiate sports, which are governed by extensive standards and regulations of the National Collegiate Athletic Association (NCAA). A university's participation in intercollegiate sports is well known to be an important revenue source and is often critical in prospective students' decision-making processes. Typical sports facilities that can be found in universities campuses are

• Collegiate arenas (indoor sport arenas dedicated to a particular sport, or multipurpose)
• Gymnasiums (for activities such as physical education)
• Field houses (for outdoor activities to be played indoors during bad weather)
• Natatoriums
• Recreation centers (multipurpose activity courts, fitness/weight room)

Chapter 5 of this volume covers design practices for several of these facilities. For ice rinks and arenas, consult Chapter 44 of the 2010 *ASHRAE Handbook—Refrigeration* and Chapters 27 of Harriman et al. (2001). For natatoriums, see also Chapter 27 of Harriman et al. (2001).

Social and Support Facilities

Social and support facilities and campus centers include common areas designed to improve and expand student services: for example, auditoriums, lounges, lobbies, dining and food services, offices and administration, libraries, cafés and snack bars, classrooms, meeting rooms, bookstores and other retail areas, banks, printing shops, etc. Given this variety of applications, the reader should refer to Chapters 2, 3, 5, 23, and 33 of this volume, and other application-specific sources for the design of HVAC&R systems for these areas.

Cultural Centers

Universities and colleges with cultural facilities and academic programs such as music, theater, dance, and visual arts enhance the cultural and artistic lives of students. The two main cultural facilities are performing arts and visual arts centers. Several areas are common for both these areas are

- **Public support areas,** which include lobby, student common, café, gift shop, box office, coat room, and restroom facilities
- **Administration/faculty areas,** including offices, administration areas, and conference rooms
- **Back of the house,** such as loading docks, shipping and receiving, maintenance and building operation, mechanical rooms, and control rooms

Unique areas for **performing arts** are

- **Performance spaces,** including seating areas, stage, orchestra pit, dimmer room, audio rack room, and lighting and sound control
- **Backstage/performer support,** such as the green room, dressing rooms, wardrobe, laundry, and storage
- **Theatre, music, and dance instruction areas,** which include rehearsal rooms, dance studios, instrumental rehearsal rooms, listening labs, and music and instrument storage

Unique areas for **visual arts** are

- **Museums,** which include art galleries, workrooms, art storage, and conservation areas
- **Fine arts instruction rooms,** comprising design, drawing, painting, print making studios, photographic darkrooms, and library
- **General arts instruction,** such as lecture halls, classrooms, seminar rooms, and computer labs

Cultural centers encompass a large number of specialty areas, and careful attention required when designing, constructing, and maintaining the HVAC&R systems. Chapters 2, 3, 5, and 23 of this volume should be consulted.

Central Utility Plants

Universities and college campuses typically have large central utility plants or smaller mechanical rooms serving an individual building or cluster of buildings. The central utility plants can supply chilled water, steam, and electrical power or only steam or chilled water. In these cases, chilled water, steam, or hot water is generated at a building level or in one smaller utility plant serving a cluster of buildings. The setup depends heavily on site constraints, including geographic location. The central utility plant comprises chillers, boilers, steam specialties, primary and secondary pumps, cooling towers, heat exchangers, combined heat and power (CHP) prime movers, and CHP auxiliary equipment, electrical power transformers, switchgears, control systems, etc. In the 2008 *ASHRAE Handbook—HVAC Systems and Equipment,* see Chapter 3 for design of central heating and cooling plants, Chapter 7 for CHP, Chapter 10 for steam systems, and Chapter 11 for district heating and cooling.

In addition to accommodating the mechanical and electrical equipment, central utility plants also house engineering, operation, and maintenance personnel. Central plants are not conditioned but generally are heated and ventilated; storage areas, shops, and other support areas are heated, ventilated, or cooled, depending on the use. Offices, administration areas, and control rooms are typically fully conditioned.

Where economically justifiable, chilled water and steam can be purchased from an independent operator

SUSTAINABILITY AND ENERGY EFFICIENCY

A trend in the educational community to embrace the principles of sustainable design has increased in the last several years. Begun as a means to educate the students in conserving earth resources, this approach also provides benefits such as enhanced IAQ and lower operating costs.

There are several definitions of sustainability, green buildings, and high-performance buildings. In the context of this chapter, these terms refer to a building that minimizes the use of energy, water, and other natural resources and provides a healthy and productive indoor environment (e.g., IAQ, lighting, noise). The HVAC&R designer plays a major role in supporting the design team in designing, demonstrating, and verifying these goals, particularly in the areas of energy efficiency and indoor environmental quality. Because energy efficiency is the area of expertise of the HVAC&R designer, this section covers these topics in more detail.

Several tools and mechanisms are available to assist the HVAC&R designer in designing and demonstrating sustainable educational facilities; the following are the most common tools:

Advanced Energy Design Guide (AEDG) for K-12 Schools

The *Advanced Energy Design Guide for K-12 Schools* (ASHRAE 2008) was developed to help designers of K-12 facilities achieve energy savings of at least 30% compared to ANSI/ASHRAE/IESNA *Standard* 90.1-1999. The guide provides recommendations for energy-efficient design based on geographic location, covering issues such as envelope, lighting, HVAC, and service water heating (SWH).

The guide can be downloaded from ASHRAE Web site at http://www.ashrae.org/publications/page/1604.

ASHRAE/USGBC/IES *Standard* 189.1-2009

The purpose of this standard is to provide minimum requirements for the siting, design, construction, and plan for operation of high performance, green buildings to

- Balance environmental responsibility, resource efficiency, occupant comfort and well-being, and community sensitivity
- Support the goal of development that meets the needs of the present without compromising the ability of future generations to meet their own needs.

This standard provides minimum criteria that apply to the following elements of building projects:

- New buildings and their systems
- New portions of buildings and their systems
- New systems and equipment in existing buildings

The standard addresses site sustainability, water use efficiency, energy efficiency, indoor environmental quality (IEQ), and the building's impact on the atmosphere, materials, and resources.

Leadership in Energy and Environmental Design (LEED)

Many schools are seeking LEED certification from the U.S. Green Building Council (USGBC). The LEED for Schools (USGBC 2009a) rating system is unique to the design and construction of K-12 schools.

The system awards credits in seven categories:

1. Sustainable sites (SS)
2. Water efficiency (WE)
3. Energy and atmosphere (EA)
4. Materials and resources (MR)
5. Indoor environmental quality (IEQ)
6. Innovation and design process (ID)
7. Regional priority (RP)

Categories 1 through 5 include prerequisites, which are mandatory for certification, and credits. The last two categories are credits only.

Typically, the HVAC&R designer is heavily involved in the (1) energy and atmosphere (EA) and (2) indoor environmental quality (IEQ) categories. In the EA category, the HVAC&R designer, along with the architect, electrical engineers, and plumbing engineers, demonstrate compliance with prerequisite EA 2 by using the following procedures:

- Option 1: Whole-building energy simulation, by demonstrating 10% improvement over ANSI/ASHRAE/IESNA *Standard* 90.1-2007
- Option 2: Prescriptive compliance path (ASHRAE 2008), for less than 200,000 ft^2
- Option 3: Prescriptive compliance path (New Buildings Institute 2007), for less than 100,000 ft^2

Additional EA credits can be obtained by demonstrating additional energy cost savings compared to the ANSI/ASHRAE/IESNA *Standard* 90.1-2007's Appendix G and from other sections of the EA group, such as on-site renewable energy, enhanced commissioning, measurement and verification, and green power. In addition, the HVAC&R designer is involved in issues of indoor environmental quality; these issues are typically associated with minimum and enhanced ventilation, acoustics, thermal comfort, controls, daylighting, mold prevention, etc.

Details and additional information on new construction and major renovations of K-12 facilities or previous editions of LEED for Schools can be found on the USGBC Web site at http://www.usgbc. org and http://www.greenschoolbuildings.org.

For existing schools, the LEED rating system for existing buildings can be applied (see USGBC Web site).

ENERGY STAR for K-12 Facilities

Similarly to appliances, a building or manufacturing plant can earn the ENERGY STAR label. An ENERGY STAR-qualified facility meets strict energy performance standards set by the U.S. EPA and uses less energy, is less expensive to operate, and causes fewer greenhouse gas emissions than its peers. To qualify, a building must score in the top 25% based on the EPA's National Energy Performance Rating System, which considers energy use among other, similar types of facilities (including K-12 educational facilities) on a scale of 1 to 100. This rating system accounts for differences in operating conditions, regional weather data, and other important considerations.

To determine eligibility for the ENERGY STAR label, as well as LEED-EB certification, the EPA's free online tool Portfolio Manager can be used (http://www.energystar.gov/benchmark). If the school facility scores 75 or higher (of a maximum of 100) using Portfolio Manager, a professional engineer (PE) will verify and approve the analysis. Detailed procedures for earning the ENERGY STAR labels can be found at http://www.energystar.gov, including case studies, useful in-formation for educational facilities, and a list of professional engineers who provide free verification services.

Collaborative for High Performance Schools (CHPS)

CHPS (http://www.chps.net) is leading a national movement to improve student performance and the entire educational experience by building the best possible schools. CHPS provides useful information for designing and maintaining high-performance schools. The following is a list of best practices and information available from CHPS:

- Planning for high-performance schools
- Design for high-performance schools
- Maintenance and operations of high-performance schools
- Commissioning of high-performance schools
- High-performance relocatable classrooms

In addition, lists of CHPS criteria for several states are available.

Laboratories for the 21st Century (Labs21)

Laboratories for the 21st Century [Labs21; EPA (2010)] is designed to meet the needs of facility designers, engineers, owners, and facility managers of laboratory and similar high-performance facilities. Cosponsored by the EPA and DOE, Labs21 offers the opportunity for worldwide information exchange and education.

The primary guiding principle of the Labs21 approach is that improving a facility's energy efficiency and environmental performance requires examining the entire facility from a whole-building perspective. This perspective allows owners to improve the efficiency of the entire facility, rather than focusing on specific building components. The Labs21 program provides excellent information for laboratory design, energy conservation, best practices, and tools, such as the following:

- Introduction to low-energy design
- Design guide for energy-efficient research labs
- Best practice guides
- Case studies
- Energy benchmarking
- Laboratory equipment efficiency wiki
- Environmental performance criteria
- Design intent tool
- Labs21 design process manual

Additional information can be found in http://www.labs21 century.gov/index.htm.

EnergySmart Schools

The EnergySmart Schools (DOE 2009) program provides energy efficiency information on planning, financing, design build and operation and maintenance of schools at http://www1.eere.energy. gov/buildings/energysmartschools/index.html.

Other Domestic and International Rating Systems

Additional domestic and international shown in Table 13; additional information on these systems can be found in Air Quality Sciences (2009).

ENERGY CONSIDERATIONS

Energy standards such as ANSI/ASHRAE/IESNA *Standard* 90.1-2007 and local energy codes should be followed for minimum energy conservation criteria. Because the HVAC&R designer deals mostly with the mechanical systems, Table 14 presents a list of selected energy conservation measures. Noted that additional measures such as modifications to lighting, motors/drives, building envelope, and electrical services should be considered for energy reduction. Energy procurement or supply-side opportunities should also be investigated for energy cost reduction.

Table 13   Summary of Domestic and International Rating Systems

Rating System	Country
BRE Environmental Assessment Method (BREEAM)	U.K.
Comprehensive Assessment System for Building Environmental Efficiency (CASBEE)	Japan
Germany Sustainable Building Certificate (DGNB)	Germany
Green Building Evaluation Standard (Three-Star System)	China
Green Globes System	Canada
Green Star	Australia
Hong Kong Building Environmental Assessment Method (HK-BEAM)	China (Hong Kong only)
National Green Building Standard	United States

Table 14   Selected Potential Energy Conservation Measures

Category	Description	Category	Description
HVAC Air Side	DDC systems upgrade Variable-speed drives on fan motors Conversion from constant volume (CV) to variable air volume (VAV) Air-side economizer Temperature set point adjustments Exhaust fume hood controls modifications Reheat minimization DOAS and air-to-air energy recovery Destratification fans Airflow reduction and air-side retrocommissioning in laboratories Active chilled beams (classrooms, laboratories, etc.) Natural ventilation (where applicable) Evaporative cooling (where applicable)	Steam and Chilled-Water Distribution	Steam distribution pressure control Steam trap repair/replacement/program Insulation repairs/upgrade Piping balancing Variable-speed pumping Primary/secondary piping Conversion from constant flow to variable flow
Chiller Plants	Chiller plant operation optimization (hydronic system) Chiller(s) replacement Chiller energy source switching Heat recovery (from CHP) driven chiller Cooling tower repair, optimization, replacement Cooling tower water treatment optimization Cooling tower fans conversion to variable speed Water-side free cooling Conversion of DX system to chilled water Offline chiller isolation Chilled/condenser water temperature reset Thermal storage	Energy Management and Control Systems	LAN systems/network interfacing Equipment sequencing Conversion to DDC system Space temperature setback and setup Demand control ventilation (DCV) Chiller plant efficiency monitoring (see ASHRAE *Guideline* 22-2008) Boiler plant efficiency monitoring (steam flow and gas flow) Duty cycling Chiller plant control optimization Boilers sequencing optimization Load shedding Remote communications Equipment performance and energy use monitoring Preventive/predictive maintenance Automated/Web-based fault detection and diagnostics (FDD) Airflow and water flow measurements Energy metering and submetering Emissions and/or CO_2 tracking
Boiler Plants	Boiler optimization/replacement Burner optimization/replacements Oxygen and excess air trim controls Conversion of linkage-based burner control to parallel positioning (servo motors) Dual-fuel switching/capability Boiler heat recovery (stack economizer) Condensing boilers Boiler temperature reset Offline boiler isolation Automatic blowdown control Blowdown heat recovery Condensate systems upgrade and optimization Feed water delivery improvements Water treatment optimization	Central Plant Supply Side and Renewable Energy	Combined heat and power (CHP) Solar energy (thermal) Photovoltaic applications Wind energy Geothermal energy and hybrid geothermal systems
		Domestic Hot Water	Condensing water heaters Demand (tankless or instantaneous) water heaters Heat pump water heaters Solar domestic water heater and pool water heating

Source: Adapted from Petchers (2002).

Energy Efficiency and Integrated Design Process (IDP)

An integrated design process (IDP) is vital for the design of high-performance educational facilities. Chapter 58 covers the concept of integrated building design (IBD) and IDP in detail, and additional information can be found on the Northwest Energy Efficiency Alliance's BetterBricks Web site (http://www.betterbricks.com).

Unlike the sequential design process (SDP), in which the elements of the built solution are defined and developed in a systematic and sequential manner, the integrated design process (IDP) encourages holistic collaboration of the project team during all phases of the project, resulting in cost-effective and environmentally friendly design. IDP is accomplished by responding to the project objectives, which typically are established by the owner before team selection. A typical IDP approach includes the following elements:

- Owner planning
- Predesign
- Schematic design
- Schematic design
- Design development
- Construction documents
- Procurement
- Construction
- Operation

Detailed information on each element can be found in Chapter 58.

In high-performance buildings, the objectives are typically related to site sustainability, water efficiency, energy and atmosphere, materials and resources, and indoor environmental quality. These objectives are in fact the main components of several rating systems. As indicated previously, the HVAC&R designer is heavily involved in meeting energy efficiency objectives. Energy use objectives are typically the following:

- Meeting minimum prescriptive compliance (mainly local energy codes, ANSI/ASHRAE/IESNA *Standard* 90.1, etc.)
- Improving energy performance by an owner-defined percentage beyond the applicable code benchmark

- Demonstrating minimum energy performance (or prerequisite) and enhanced energy efficiency (for credit points) for sustainable design rating (e.g., USGBC; LEED; energy and atmosphere using ANSI/ASHRAE/IESNA *Standard* 90.1-2007, Appendix G)
- Providing a facility/building site energy density [e.g., energy utilization index (EUI)] less than an owner-defined target (e.g., EPA, ENERGY STAR's Portfolio Manager)
- Providing a facility/building source energy density less than an owner-defined target
- Deriving an owner-defined percentage of facility source energy from renewable energy

Building Energy Modeling

Building energy modeling has been one of the most important tools in the process of IDP and sustainable design. Building energy modeling uses sophisticated methods and tools to estimate the energy consumption and behavior of buildings and building systems. To better clarify the concept of energy modeling, the difference between HVAC sizing and selection programs and energy modeling tools will be described.

Design, sizing selection, and equipment sizing tools are typically used for design and sizing of HVAC&R systems normally at the design point. Examples include the following:

- Cooling/heating loads calculations tools
- Ductwork design
- Piping design
- Acoustics
- Equipment selection programs for air-handling units, packaged rooftop units, fans, chillers, pumps, diffusers, etc.

These tools are used to specify cooling and heating capacities, airflow, water flow, equipment size, etc., at a design point as defined and agreed by the client.

Energy modeling (or building modeling and simulation) is used to model the building's thermal behavior and the performance of building energy systems. Unlike design tools, which are used for one design point or for sizing, the building energy simulation analyzes the building and its systems up to 8760 times (or hour-by-hour, or in some cases in smaller time intervals).

A building energy simulation tool is a computer program consisting of mathematical models of building elements and HVAC&R equipment. To run a building energy simulation, the user must define the building elements, equipment variables, energy cost, and so on. After these variables are defined, the simulation engine solves mathematical models of the building elements, equipment, etc., typically through a sequential process, 8760 times (one for every hour). Results include annual energy consumption, annual energy cost, hourly profiles of cooling loads, and hourly energy consumption. Chapter 19 of the 2009 *ASHRAE Handbook—Fundamentals* provides detailed information on energy modeling techniques.

Typically, energy modeling tools (or building energy simulation programs) have to meet minimum requirements to be accepted by rating authorities such as the USGBC and local building codes. The following is a typical minimum modeling capabilities for building energy simulation program:

- 8760 h per year
- Hourly variations in occupancy, lighting power, miscellaneous equipment power, thermostat set points, and HVAC system operation are defined separately for each day of the week and holidays
- Thermal mass effects
- Ten or more thermal zones
- Part-load performance curves for mechanical equipment
- Capacity and efficiency correction curves for mechanical heating and cooling equipment
- Air-side economizers with integrated control

- Capable of performing design load calculations to determine required HVAC equipment capacities and air and water flow rates in accordance with generally accepted engineering standards and handbooks (e.g., *ASHRAE Handbook—Fundamentals*)
- Testing according to ASHRAE *Standard* 140

Energy modeling is typically used for the following applications:

- As a decision support tool to analyze several design alternatives and select the optimal solution for a given set of criteria for energy systems in new construction and retrofit projects.
- To provide vital information to the engineer about the building behavior and systems performance during the design stage
- To demonstrate compliance with energy standards such as ASHRAE *Standard* 90.1-2007, section 11 (energy cost budget method)
- To support LEED certification in the energy and atmosphere (EA) section
- To model existing buildings and systems and analyze proposed energy conservation measures (ECMs) by performing calibrated simulation
- To demonstrate energy cost savings as part of measurements and verification (M&V) protocol by using calibrated simulation procedures

Energy modeling is used intensively in LEED for Schools (USGBC 2009a), energy and atmosphere (EA), prerequisite 2 (minimum energy performance), and for EA credit 1 (optimize energy performance). An energy simulation program meeting the preceding requirements and those of ASHRAE *Standard* 90.1-2007, Appendix G, is used to perform whole-building energy simulation to demonstrate energy cost savings. The number of credits awarded is in correlation to the energy cost reduction.

Energy Benchmarking and Benchmarking Tools

Energy benchmarking is an important element of energy use evaluation and tracking, comparing a building's normalized energy consumption to that of other similar buildings. The most common normalization factor is gross floor area. Energy benchmarking is less accurate than other energy analysis methods, but can provide a good overall picture of relative energy use.

Relative energy use is commonly expressed by an energy utilization index (EUI), which is the energy use per unit area per year. Typically EUI is defined in terms of Btu/ft^2 per year. In some cases, the user is interested in energy cost benchmarking, which is known as the cost utilization index (CUI), with units of $\$/ft^2$ per year. It is important to differentiate between *site* EUI and *source* EUI. Building energy use can be reported as the actual energy used on site (i.e., site EUI), or as energy used at the energy source (i.e., source EUI). About two-thirds of the primary energy that goes into an electric power plant is lost in the process as waste heat.

One of the most important sources of energy benchmarking data is the U.S. DOE Energy Information Administration's (DOE/EIA) Commercial Building Energy Consumption Survey (CBECS). Table 2 of Chapter 36 shows an example of EUI calculated based on DOE/EIA 2003 CBECS. As shown in that table, the mean site EUI for high schools is 75 kBtu/yr per gross square foot.

The following is a list of common energy benchmarking tools:

- U.S. EPA ENERGY STAR Portfolio Manager (http://www.energystar.gov/benchmark)
- Lawrence Berkeley National Laboratory's (LBNL) Arch building comparison tool (http://poet.lbl.gov/arch//; for California, http://poet.lbl.gov/cal-arch/compare.html)
- Labs 21 for laboratory energy benchmarking (http://labs21benchmarking.lbl.gov)

An example of laboratory energy benchmarking is shown in Figure 8.

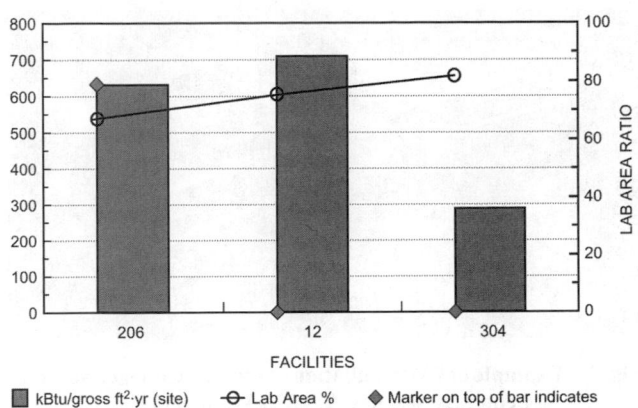

METRIC	MINIMUM	AVERAGE	MAXIMUM	COUNT
TOTAL BUILDING kBtu/gross ft²·yr (SITE)	285.78	542.37	708.61	3

FACILITY	LAB TYPE	YEAR	kBtu/gross ft²·yr (site)	LAB AREA RATIO	OCCUPANCY HOURS PER WEEK	CLIMATE
206	Biological	2007	632.73	67%	108	5A
12	Biological	2001	708.61	75%	144	5A
304	Biological	2008	285.78	82%	100	5A

Fig. 8 Example of Laboratory Building Energy Benchmarking (Labs 21)

Comprehensive information on energy benchmarking and available benchmarking tools can be found in Glazer (2006) and Chapter 36.

Combined Heat and Power in Educational Facilities

Combined heat and power (CHP) plants and building cooling, heating, and power (BCHP) can be considered for large facilities when economically justifiable. Chapter 7 of the 2008 *ASHRAE Handbook—HVAC Systems and Equipment* and other sources such as Meckler and Hyman (2010). Orlando (1996) and Petchers (2002) provide information on CHP systems. Additional Internet-based sources for CHP include the following:

- U.S. EPA Combined Heat and Power (CHP) Partnership, at http://www.epa.gov/chp/
- U.S. Department of Energy, Energy Efficiency and Renewable Energy, at http://www1.eere.energy.gov/industry/distributedenergy/
- The Midwest CHP Application Center (MAC), at http://www.chpcentermw.org.

A market analysis report by Ryan (2004) clearly suggests that secondary schools (9-12) are more suitable for BCHP than primary schools, because secondary schools

- Are more likely to operate 12 months a year
- Are more likely to contain an indoor swimming pool facility
- Are more likely to operate into the evenings and weekends, allowing longer period of BCHP operation
- Typically contain gymnasiums with shower facilities

The EPA's Combined Heat and Power (CHP) Partnership (http://www.epa.gov/chp/project-development/index.html) Web site can be consulted for procedures of conducting feasibility studies and evaluations for CHP integration.

Maor and Reddy (2008) describe a procedure to optimally size the prime mover and thermally operated chiller for a large school by combining a building energy simulation program and the CHP optimization tool (Hudson 2005).

A database of CHP installations is available at http://www.eeainc.com/chpdata/index. html.

CHP is more common for large colleges and universities than for primary or secondary schools, given their larger scale and ability to use waste heat efficiently. Because many large colleges and universities are equipped with large district cooling and heating facilities, the integration of CHP can be very cost effective.

The type of prime mover depends heavily on the electrical and thermal loads, ability to use the waste heat efficiently, and utility rates. Typically, schools are good candidates for gas-fired reciprocating engine prime movers or microturbine-based systems. Large universities can use reciprocating engine prime movers or gas-fired combustion turbines. Table 1 in Chapter 7 of the 2008 *ASHRAE Handbook—HVAC Systems and Equipment* provides information on the applicability of CHP.

Renewable Energy

The U.S. Department of Energy's EnergySmart Schools program (DOE 2009; http://www1.eere.energy.gov) discusses several renewable energy (RE) options for schools, including solar, wind, and biomass.

Renewable energy utilization can add credits for USGBC LEED for Schools (USGBC 2009a), energy and atmosphere (credit 2) by awarding credits depending o n the percentage of renewable energy used.

Given the increased number and popularity of solar systems in educational facilities, only these systems will be discussed in this chapter. Several examples of wind and biomass can be found on the EnergySmart School program Web site. Geothermal energy is also considered renewable; these systems are discussed earlier in this chapter and in Chapter 34.

Solar: Photovoltaic. Photovoltaic (PV) technology is the direct conversion of sunlight to electricity using semiconductor devices called solar cells. Photovoltaics are almost maintenance-free and seem to have a long lifespan. Their longevity, lack of pollution, simplicity, and minimal resource requirements make this technology highly sustainable, and, along with the proper financing mechanisms (as explained later), these systems can be economically justifiable.

Educational facilities are excellent candidate for PV technology due to the following reasons:

- Availability of large roof area
- Hours and seasons of operation
- Educational as a showcase of renewable energy technologies

The most common technology in use today is single-crystal PV, which uses silicon wafers wired together and attached to a module substrate. Thin-film PV, such as amorphous silicon technology, is based on depositing silicon and other chemicals directly on a substrate (e.g., glass or flexible stainless steel). Thin films promise lower cost per unit area, but also have lower efficiency and produce less electricity per unit area compared to single-crystal PVs. Typical values for DC electrical power generation are around 6 W/ft² for thin films and up to 15 W/ft² for single-crystal PV.

PV panels produce direct current, not the alternating current used to power most building equipment. Direct current is easily stored in batteries; an inverter is required to transform the direct current to alternating current. The costs of reliable storage batteries and an inverter increase the overall system cost; typically, an installed PV system costs $5 to $7/W (Krieth and Goswami 2007).

The ability to transfer excess electricity generated by a photovoltaic system back into the utility grid can be advantageous for schools. Most utilities are required to buy excess site-generated electricity back from the customer. In many states, public utility commissions or state legislatures have mandated net metering: utilities pay and charge equal rates regardless of which way the electricity flows. Schools districts in these states will find PV more economically attractive. A good source of information on rebates and incentives for solar systems and other renewable technologies is the Database of State Incentives for Renewables & Efficiency [DSIRE (NCSU 2010), http://www.dsireusa.org], which is a comprehensive source of information on state, local, utility, and federal incentives and policies that promote renewable energy and energy efficiency.

PV systems should be integrated during the early stages of the design. In existing facilities, a licensed contractor can be employed for a turnkey project, which should include sizing, analysis, economic analysis, design documents, specifications, permits, documentation for incentives, etc. The DSIRE database also provides state requirements for licensed solar contractors.

RETScreen® (Renewable Energy and Energy-Efficient Technologies) is a free decision support tool at http://www.retscreen.net/ang/home.php, developed to assist in evaluation of energy production and savings, costs, emission reductions, financial viability, and risk for various types of renewable energy technologies (RETScreen 2010). The program is available in 35 languages. In addition, several commercial tools are available for analysis of PV systems.

Financing PV projects in the educational sector can be more complex because of tax exemptions and questions of how to most efficiently allocate public funds and leverage incentives; detailed information can be found in Bolinger (2009) and Cory et al. (2008). The primary mechanism that has emerged to finance public-sector PV projects is a third-party ownership model. This model allows the public sector take advantage of all the federal tax and other incentives without large up-front outlay of capital. The public sector does not own the solar PV, but only hosts it in its property. The cost of the electrical power generated is then secured at a fixed rate, which is lower than the retail price for 15 to 25 years.

Figures 9 and 10 show examples of educational facilities' PV projects.

Solar: Thermal. Educational facilities can be good candidate for active thermal solar heating systems. In most cases, a solar hot-water system can reduce the energy required for service hot water and pool heating. Solar heating design and installation information can be found in ASHRAE (1988, 1991). Chapter 36 of the 2008 *ASHRAE Handbook—HVAC Systems and Equipment* and Krieth and Goswami (2007) are good sources of information for design and installation of active solar systems. Web-based sources include the U.S. Department of Energy, Energy Efficiency and Renewable Energy's site at http://www.eere.energy.gov.

Value Engineering (VE) and Life-Cycle Cost Analysis (LCCA)

The use of value engineering (VE) and life-cycle cost analysis (LCCA) is growing in all types of construction and as part of the integrated design process (IDP) concept. In some cases, public facilities such as schools are required to use these procedures. Both VE and LCCA are logical, structured, systematic processes used with decision support tools to achieve overall cost reduction, but there are some distinctions between them (Anderson et al. 2004).

In value engineering, the project team examines the proposed design components in relation to the project objectives and requirements. The intent is to provide essential functions while exploring cost savings opportunities by modifying or eliminating nonessential design elements. Examples are using alternative systems or substituting equipment.

Life-cycle cost analysis is used to evaluate design alternatives (or alternative systems, equipment substitutions, etc., as part of VE) that meet the facility's design criteria with reduced cost or increased value over the life of the facility or system.

The combination of VE and LCCA is suitable for schools, because they are often government funded and intended for longer lifespans than commercial facilities. Unfortunately, VE and LCCA

Fig. 9   Example of PV Installation at Ohlone College, Newark Center, Newark, CA: 450 kW, 38,000 ft²
(Esberg 2010)

Fig. 10   Example of PV Installation at Twenhofel Middle School, Independence, KY: 22 kW
(Seibert 2010)

Table 15   Key Commissioning Activities for New Building

Phase	Key Commissioning Activities
Predesign	Preparatory phase in which OPR is developed and defined.
Design	OPR is translated into construction documents, and basis of design (BOD) document is created to clearly convey assumptions and data used to develop the design solution. See Informative Annex K of ASHRAE *Guideline* 1.1-2007 for detailed structure and an example of a typical BOD.
Construction	The commissioning team is involved to ensure that systems and assemblies installed and placed into service meet the OPR.
Occupancy and operation*	The commissioning team is involved to verify ongoing compliance with the OPR.

*Also known as acceptance and post-acceptance in ACG (2005).

often are not included in the early design stages, which results in a last-minute effort to reduce cost and stay within the budget, compromising issues such as energy efficiency and overall value of the facility. Therefore, VE and LCCA should be deployed in the early stages of the project. VE and LCCA programs for large schools can add 0.1 to 0.5% in initial cost, but can save 5 to 10% of initial costs and 0.5 to 10% of operation and maintenance costs (Dell'Isola 1997).

LCCA is recommended for economic evaluation as part of any school construction. Chapters 37 and 58 discuss LCCA in detail. Other methodologies such as simple payback should be avoided because of inaccuracies and the need to take in account the time value of money. LCCA is more accurate: it captures all the major initial costs associated with each item, the costs occurring during

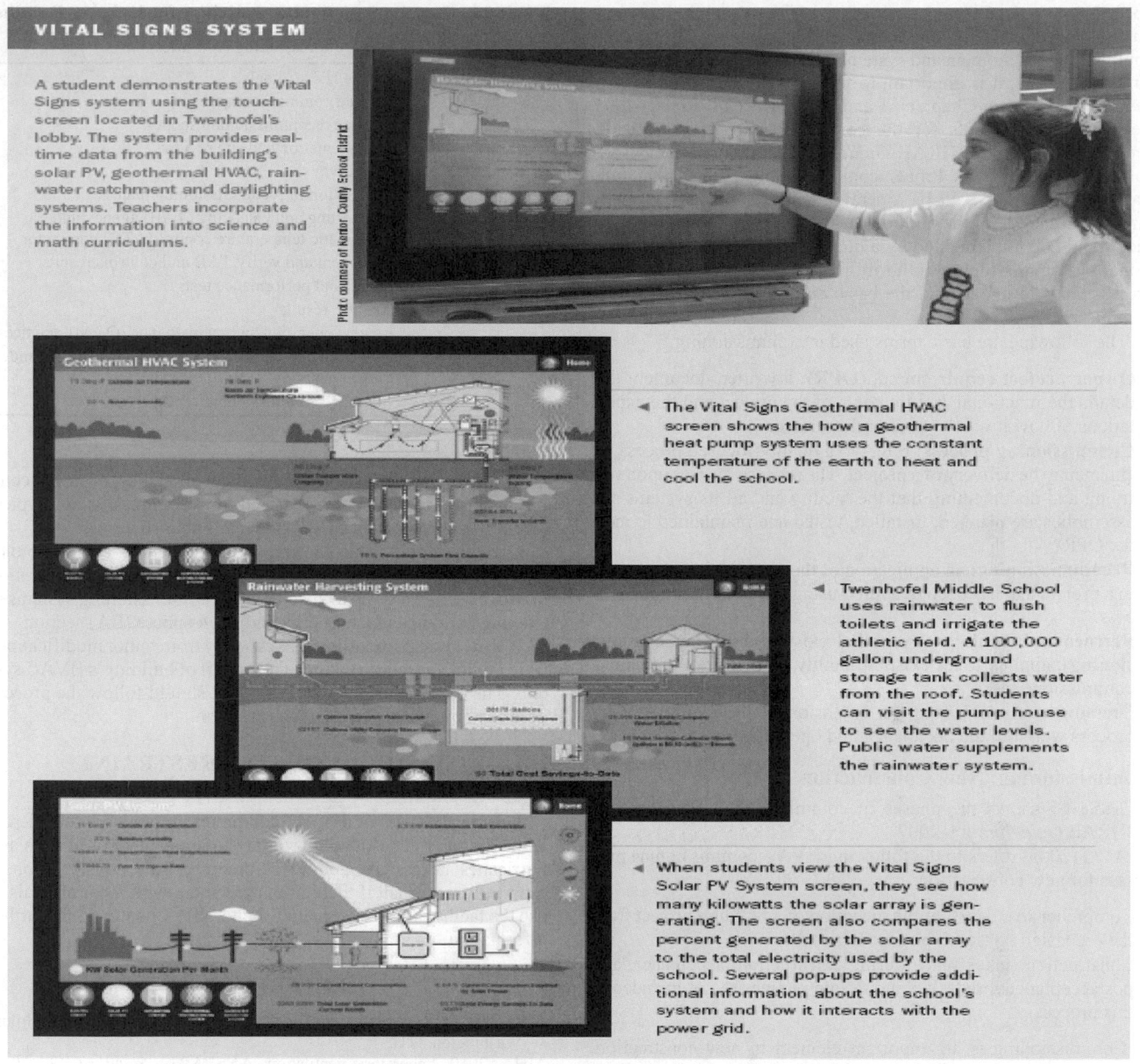

Fig. 11 Integration of Sustainability Features for Educational Purposes, Twenhofel Middle School, Independence, KY
(http://www.twhvac.kenton.kyschools.us/)
(Seibert 2010)

the life of the system, and the value of money for the entire life of the system.

Chapter 37 provides details, tools, and examples of LCCA (refer to Table 7 in that chapter). Anderson et al. (2004) provides detailed information on all the aspects of design, construction management, cost control, and other resources for building and renovating schools.

The School as a Learning Tool for Sustainability

Schools are excellent for enhancing students' interest in energy efficiency and sustainable design from a young age. USGBC (2009a)'s LEED for Schools awards one point for integrating high-performance features in the school curriculum (ID section, credit 3). Sources for this integration include the following:

• National Energy Education Development (NEED) project (http://www.need.org)

• Alliance to Save Energy's Green Schools Program (http://ase.org/programs/green-schools-program)

• National Energy Foundation educational resources (http://www.nef1.org/)

• Energy Information Administration's Web site (http://tonto.eia.doe.gov/kids/energy.cfm?page=kiddie_resources)

In addition, real-time feedback on how systems such as photovoltaic electrical generation, geothermal heat pumps, and water conservation save energy and operating costs is recommended. Seibert (2010) shows these features, as illustrated in Figure 11.

COMMISSIONING AND RETROCOMMISSIONING

Commissioning (Cx) is a quality assurance process for buildings from predesign through design, construction, and operations. The commissioning process involves achieving, verifying, and documenting the performance of each system to meet the building's operational needs. Given the criticality of issues such as indoor air quality, thermal comfort, noise, etc., in educational facilities and the application of equipment and systems such as DOAS, EMS, and occupancy sensors, it is important to follow the commissioning process as described in Chapter 43 and ASHRAE *Guideline* 0-2005. Technical requirements for the commissioning process are described in detail in ASHRAE *Guideline* 1.1-2007; another useful source is ACG (2005). Proper commissioning ensures that fully functional systems can be operated and maintained properly throughout the life of the building. Although commissioning activities should be implemented by a qualified commissioning professional or commissioning authority (CA), it is important for other professionals to understand the basic definitions and processes in commissioning.

The following are basic terms used in commissioning:

- **Owner project requirements (OPR):** a written document that details the functional requirements of the project and the expectations of how it will be used and operated.
- **Commissioning process:** refers to a quality-focused process for enhancing the delivery of a project. The process focuses upon verifying and documenting that the facility and all its systems and assemblies are planned, installed, tested and maintained to meet the OPR
- **Recommissioning:** an application of the commissioning process to a project that has been delivered using the commissioning process.
- **Retrocommissioning** (also called **existing building commissioning**): applied to an existing facility that was not previously commissioned.
- **Ongoing commissioning:** an extension of the commissioning process well into the occupancy and operation phase.

Commissioning: New Construction

Table 15 shows the phases of commissioning, as defined in ASHRAE *Guideline* 1.1-2007.

ACG (2005) refers to the following HVAC commissioning processes for new construction:

- Comprehensive (starts at the inception of a building project from the predesign phase till postacceptance)
- Construction (takes place during construction, acceptance, and postacceptance; predesign and design phases are not included in this process)

Commissioning is an important element in new construction. LEED for Schools (USGBC 2009a) requires as a prerequisite (Energy and Atmosphere, prerequisite 1) verification that the project's energy-related systems are installed and calibrated and perform according to the OPR, BOD, and construction document. Additional credits (Energy & Atmosphere, credit 3—Enhanced Commissioning) can be obtained by applying the entire commissioning process (or comprehensive HVAC commissioning), as described previously.

Commissioning Existing Buildings

HVAC commissioning in existing buildings covers the following:

- Recommissioning
- Retrocommissioning (RCx)
- HVAC systems modifications

Although there is a difference between recommissioning and retrocommissioning, the methodology for both is identical. Retrocommissioning applies to buildings that were not previously commissioned. Recommissioning is initiated by the owner of a previously commissioned building, and seeks to resolve ongoing problems or to ensure that the systems continue to meet the facility's requirements. There also could have been changes in the building's occupancy,

Table 16 Key Commissioning Activities for Existing Building

Phase	Key Commissioning Activities
Planning	Define HVAC goals
	Select a commissioning team
	Finalize recommissioning scope
	Documentation and site reviews
	Site survey
	Preparation of recommissioning plan
Implementation	Hire testing and balancing (TAB) agency and automatic temperature control (ATC) contractor
	Document and verify TAB and controls results
	Functional performance tests
	Analyze results
	Review operation and maintenance (O&M) practices
	Operation and maintenance (O&M) instruction and documentation
	Complete commissioning report

Source: ACG (2005).

design strategies, equipment or equipment efficiency, occupant comfort, or IAQ that can initiate the need for recommissioning. Typical recommissioning activities are shown in Table 16.

Commissioning is also an important element in existing buildings. USGBC (2009b) *LEED for Existing Buildings & Operation Maintenance* awards up to six credits for commissioning systems in existing buildings in the Energy and Atmosphere (EA) section.

HVAC systems modifications can vary from minor modifications up to complete reconstruction of all or part of building's HVAC system. The process for this type of project should follow the process described previously for new construction.

SEISMIC- AND WIND-RESTRAINT CONSIDERATIONS

Seismic bracing of HVAC equipment should be considered. Wind restraint codes may also apply in areas where tornados and hurricanes necessitate additional bracing. This consideration is especially important if there is an agreement with local officials to use the facility as a disaster relief shelter. See Chapter 55 for further information.

REFERENCES

ACG. 2005. *ACG commissioning guideline*. AABC Commissioning Group, Washington, D.C.

AIHA. 2010. Laboratory ventilation. ANSI/AIHA *Standard* Z9.5-2010. American Industrial Hygiene Association, Fairfax, VA.

Air Quality Sciences. 2009. *Building rating systems (certification programs): A comparison of key programs*. Air Quality Sciences, Marietta, GA. Available at http://www.aerias.org/uploads/2009.12.09_Green_Building_Programs_Comparison_PUBLISHED.pdf.

Anderson, D.R., J. Macaluso, D.J. Lewek, and B.C. Murphy. 2004. *Building and renovating schools: Design, construction management, cost control*. R.S. Means, Kingston, MA, and John Wiley & Sons, New York.

ANSI. 2002. Acoustical performance criteria, design requirements and guidelines for schools. *Standard* 12.60-2002. American National Standards Institute, Washington, D.C.

ASHRAE. 1988. *Active solar heating systems design manual*.

ASHRAE. 1991. *Active solar heating systems installation manual*.

ASHRAE. 1997. *Thermal comfort tool*.

ASHRAE. 2007. *Standard 62.1-2007 user's manual*.

ASHRAE. 2008. *Advanced energy design guide for K-12 school buildings*. Available from http://www.ashrae.org/publications/page/1604.

ASHRAE. 2005. The commissioning process. *Guideline* 0-2005.

ASHRAE. 2007. HVAC&R technical requirements for the commissioning process. *Guideline* 1.1-2007.

ASHRAE. 2008. Instrumentation for monitoring central chilled-water plant efficiency. *Guideline* 22-2008.

ASHRAE. 2007. Method of testing general ventilation air cleaning devices for removal efficiency by particle size. *Standard* 52.2-2007.

ASHRAE. 2004. Thermal environmental conditions for human occupancy. ANSI/ASHRAE *Standard* 55-2004.

ASHRAE. 2007. Ventilation for acceptable indoor air quality. ANSI/ASHRAE *Standard* 62.1-2007.

ASHRAE. 2007. Energy standard for buildings except low-rise residential buildings. ANSI/ASHRAE/IESNA *Standard* 90.1-2007

ASHRAE. 2007. Standard method of test for the evaluation of building energy analysis computer programs. ANSI/ASHRAE *Standard* 140-2007.

ASHRAE. 2006. Weather data for building design standards. ANSI/ASHRAE *Standard* 169-2006.

ASHRAE. 2009. Standard for the design of high-performance green buildings except low-rise residential buildings. ANSI/ASHRAE/USGBC/IES *Standard* 189.1-2009.

BetterBricks. 2009. *Bottom-line thinking on energy in commercial buildings: Schools.* BetterBricks, Northwest Energy Efficiency Alliance, Portland, OR. http://www.betterbricks.com/subHomePage.aspx?ID=4.

Bolinger, M. 2009. Financing non-residential photovoltaic projects; options and implications. Lawrence Berkeley National Laboratory *Report* LBNL-1410 E. Available from http://eetd.lbl.gov/ea/emp/re-pubs.html.

California Energy Commission. 2006. *Displacement ventilation design guide: K-12 schools.*

Chen, Q. and L. Glicksman. 2003. *System performance evaluation and design guidelines for displacement ventilation.* ASHRAE.

Cory, K., J. Coughlin, and C. Coggeshall. 2008. Solar photovoltaic financing: Deployment on public property by state and government. NREL Technical *Report* NREL/TP-670-43115, May 2009. http://apps1.eere.energy.gov/wip/pdfs/43115.pdf.

Dell'Isola, A.J. 1997. *Value engineering: Practical applications.* R.S. Means Company, Kingston, MA.

DiBerardinis, L.J. et al. 2001. *Guidelines for laboratory design—Health and safety considerations,* 3rd ed. John Wiley & Sons, New York.

Ebbing, E. and W. Blazier, eds. 1998. *Application of manufacturers' sound data.* ASHRAE.

EPA. 2000. *Indoor air quality—Tools for schools,* 2nd ed. U.S. Environmental Protection Agency, Washington, D.C. http://www.epa.gov/iaq.

EPA. 2010. *Laboratories for the 21st century (Labs21).* U.S. Environmental Protection Agency, Washington, D.C. http://www.epa.gov/lab21gov.

Esberg, G. 2010. Dispelling the cost myth. *High Performing Buildings* (Winter):30-42.

FEMP. 2010. *BLCC 5.3: Building life cycle cost program.* Federal Energy Management Program, Washington, D.C. http://www1.eere.energy.gov/femp/information/download_blcc.html.

Glazer, J. 2006. Evaluation of building performance rating protocols. ASHRAE Research Project RP-1286, *Final Report.*

Harriman, L.G., G.W. Brundrett, and R. Kittler. 2001. *Humidity control design guide for commercial and institutional buildings.* ASHRAE.

Hudson, R. 2005. *ORNL CHP capacity optimizer.* Oak Ridge National Laboratory, Oak Ridge TN. http://www1.eere.energy.gov/industry/best practices/software.html.

Kavanaugh, S.P. and K. Rafferty. 1997. *Ground-source heat pumps.* ASHRAE.

Krieth, F. and Y. Goswami. 2007. *Handbook of energy efficiency and renewable energy.* CRC Press, Boca Raton, FL.

Maor, I. and T.A. Reddy. 2008. Near-optimal scheduling control of combined heat and power systems for buildings, Appendix E. ASHRAE Research Project RP-1340, *Final Report.*

McIntosh, I.B.D., C.B. Dorgan, and C.E. Dorgan. 2001. *ASHRAE laboratory design guide.* ASHRAE.

Meckler, M. and L. Hyman. 2010. *Sustainable on-site CHP systems: Design, construction, and operations.* McGraw-Hill, Columbus, OH.

NCSU. 2010. *DSIRE database of state incentives for renewables & efficiency.* North Carolina State University, under National Renewable Energy Laboratory Subcontract XEU-0-99515-01. http://www.dsireusa.org/.

NRC. 1996. *Guide for the care and use of laboratory animals.* National Research Council, National Academy Press, Washington, D.C.

Neuman, D.J. 2003. *Building type basics for college and university facilities.* John Wiley & Sons, Hoboken, NJ.

New Buildings Institute. 2007. *Core performance guide: A prescriptive program to achieve significant, predictable energy savings in new commercial buildings.* New Buildings Institute, White Salmon, WA.

Orlando, J.A. 1996. *Cogeneration design guide.* ASHRAE.

Petchers, N. 2002. *Combined heating, cooling & power handbook: Technologies & applications.* Fairmont Press, Inc. Lilburn, GA

RETScreen. 2010. *RETScreen® international: Empowering cleaner energy decisions.* Natural Resources Canada. http://www.retscreen.net/ang/home.php.

R.S. Means. 2010a. *Means mechanical cost data.* R.S. Means Company, Inc., Kingston, MA.

R.S. Means. 2010b. *Means maintenance and repair cost data.* R.S. Means Company, Inc., Kingston, MA

Rumsey, P. and J. Weale. 2006. Chilled beams in labs. *ASHRAE Journal* 49(1):18-25.

Ryan, W. 2004. *Targeted CHP outreach in selected sectors of the commercial market.* Report prepared by the University of Illinois at Chicago Energy Resource Center for the U.S. Department of Energy, Energy Efficiency and Renewable Energy Program.

Schaffer, M.E. 1993. *A practical guide to noise and vibration control for HVAC systems.* ASHRAE.

Siebert, K.L. 2010. An energy education. *High Performing Buildings* (Winter):44-55.

Siebein, G.W. and R.M. Likendey. 2004. Acoustical case studies of HVAC systems in schools. *ASHRAE Journal* 46(5):35-47.

Skistad, H., E. Mundt, P.V. Nielsen, K. Hagstrom, and J. Railio. 2002. *Displacement ventilation in non-industrial premises.* Federation of European Heating and Air-Conditioning Associations (REHVA), Brussels.

USGBC. 2009a. *LEED 2009 for schools—New construction and major renovations.* U.S. Green Building Council, Washington, D.C. http://www.usgbc.org/DisplayPage.aspx?CMSPageID=1586.

USGBC 2009b. *LEED 2009 for existing buildings and operation maintenance.* U.S. Green Building Council, Washington, D.C. http://www.usgbc.org/DisplayPage.aspx?CMSPageID=221.

U.S. DOE. 2009. *EnergySmart schools.* Energy Efficiency & Renewable Energy.http://www1.eere.energy.gov/buildings/energysmartschools/index.html.

Wolf, M. and J. Smith. 2009. Optimizing dedicated outdoor-air systems. *HPAC Engineering* (December).

Wright, G. 2003. The ABC's of K-12. *Building Design & Construction,* June.

BIBLIOGRAPHY

ASHRAE. 2007. *Standard 90.1-2007 user's manual.*

ASHRAE. 2010. *ASHRAE greenguide: The design, construction and operation of sustainable buildings,* 3rd ed. J.M. Swift and T. Lawrence, eds.

CHPS. 2010. *Best practices manual.* Collaborative for High Performance Schools, CA. http://www.chps.net.

Darbeau, M. 2003. ARI's views on ANSI S-12.60-2002. *ASHRAE Journal* 45(2):27.

DOE. 2002. *National best practices manual for building high performance schools.* U.S. Department of Energy, Washington, D.C. Available from http://www.epa.gov/iaq/schools/high_performance.html.

Harriman, L.G. and J. Judge. 2002. Dehumidification equipment advances. *ASHRAE Journal* 44(8):22-27.

Lilly, J.G. 2000. Understanding the problem: Noise in the classroom. *ASHRAE Journal* 42(2):21-26.

Megerson, J.E. and C.R. Lawson. 2008. Underfloor for schools. *ASHRAE Journal* 50(5):28-30, 32.

Moxley, R.W. 2003. Prioritizing for preschoolers. *American School & University* (November).

Mumma, S.A. 2001. Designing dedicated outdoor air systems. *ASHRAE Journal* 43(5):28-31.

Nasis, R.W. and R. Tola. 2002. Environmental impact. *American School & University* (November).

Nelson, P.B. 2003. Sound in the classroom—Why children need quiet. *ASHRAE Journal* 45(2):22-25.

Perkins, B. 2001. *Building type basics for elementary and secondary schools.* John Wiley & Sons, New York.

Schaffer, M.E. 2003. ANSI standard: Complying with background noise limits. *ASHRAE Journal* 45(2):26.

Watch, D. 2001. *Building type basics for research laboratories.* John Wiley & Sons, New York.

Wulfinghoff, D.R. 2000. *Energy efficiency manual.* Energy Institute, Wheaton, MD.

CHAPTER 8

HEALTH-CARE FACILITIES

CONTINUAL advances in medicine and technology necessitate constant reevaluation of the air-conditioning needs of hospitals and medical facilities. Medical evidence has shown that proper air conditioning is helpful in preventing and treating many conditions, and ventilation requirements exist to protect against harmful occupational exposures. Although the need for clean and conditioned air in healthcare facilities is high, the relatively high cost of air conditioning demands efficient design and operation to ensure economical energy management (Demling and Maly 1989; Fitzgerald 1989; Murray et al. 1988; Woods et al. 1986).

Health care occupancy classification, based on the latest occupancy guidelines from the National Fire Protection Association (NFPA) Life Safety Code®, should be considered early in project design. Health care occupancy is important for fire protection (smoke zones, smoke control) and for future adaptability of the HVAC system for a more restrictive occupancy.

Health care facilities are increasingly diversifying in response to a trend toward outpatient services. The term **clinic** may refer to any building from the ubiquitous residential doctor's office to a specialized cancer treatment center. Prepaid health maintenance provided by integrated regional health care organizations is becoming the model for medical care delivery. These organizations, as well as long-established hospitals, are constructing buildings that look less like hospitals and more like luxury hotels and office buildings.

For the purpose of this chapter, health care facilities are divided into the following categories:

- Hospital facilities
- Outpatient health care facilities
- Nursing facilities
- Dental care facilities

The specific environmental conditions required by a particular medical facility may vary from those in this chapter, depending on the agency responsible for the environmental standard. Among the agencies that may have standards and guidelines for medical facilities are state and local health agencies, the U.S. Department of Health and Human Services, Indian Health Service, Public Health Service, Medicare/Medicaid, U.S. Department of Defense, U.S. Department of Veterans Affairs, and The Joint Commission's Hospital Accreditation Program. The Facility Guidelines Institute (FGI 2010) requires the owner to provide an infection control risk assessment (ICRA) and prepare infection control risk mitigation recommendations (ICRMR) that are intended to pre-identify and control infection risks arising from facility construction activities. The ICRMR and ICRA are then to be incorporated in the contract documents by the design professional. Therefore, it is advisable to discuss infection control objectives with the hospital's infection control committee.

The general hospital was selected as the basis for the fundamentals outlined in the first section, Hospital Facilities, because of the variety of services it provides. Environmental conditions and design criteria apply to comparable areas in other health facilities.

The general acute care hospital has a core of critical-care spaces, including operating rooms, emergency rooms, delivery rooms, and a nursery. Usually the functions of radiology, laboratory, central sterile, and pharmacy are located close to the critical care space. Inpatient nursing, including intensive care nursing, is in the complex. The facility also incorporates a kitchen, dining and food service, morgue, and central housekeeping support.

Criteria for outpatient facilities are given in the second section. Outpatient surgery is performed with the anticipation that the patient will not stay overnight. An outpatient facility may be part of an acute care facility, a freestanding unit, or part of another medical facility such as a medical office building.

Nursing facilities are addressed separately in the third section, because their fundamental requirements differ greatly from those of other medical facilities.

Dental facilities are briefly discussed in the fourth section. Requirements for these facilities differ from those of other health care facilities because many procedures generate aerosols, dusts, and particulates.

AIR CONDITIONING IN DISEASE PREVENTION AND TREATMENT

Hospital air conditioning plays a more important role than just the promotion of comfort. In many cases, proper air conditioning is a factor in patient therapy; in some instances, it is the major treatment.

Studies show that patients in controlled environments generally have more rapid physical improvement than do those in uncontrolled environments. Patients with thyrotoxicosis do not tolerate hot, humid conditions or heat waves very well. A cool, dry environment favors the loss of heat by radiation and evaporation from the skin and may save the patient's life.

Cardiac patients may be unable to maintain the circulation necessary to ensure normal heat loss. Therefore, air conditioning cardiac wards and rooms of cardiac patients, particularly those with congestive heart failure, is necessary and considered therapeutic (Burch and Pasquale 1962). Individuals with head injuries, those subjected to brain operations, and those with barbiturate poisoning may have hyperthermia, especially in a hot environment, due to a disturbance in the heat regulatory center of the brain. An important factor in recovery is an environment in which the patient can lose heat by radiation and evaporation: namely, a cool room with dehumidified air.

A hot, dry environment of 90°F db and 35% rh has been successfully used in treating patients with rheumatoid arthritis.

Dry conditions may be hazardous to the ill and debilitated by contributing to secondary infection or infection totally unrelated to

---

The preparation of this chapter is assigned to TC 9.6, Health-Care Facilities.

the clinical condition causing hospitalization. Clinical areas devoted to upper respiratory disease treatment and acute care, as well as the general clinical areas of the entire hospital, should be maintained at 30 to 60% rh.

Patients with chronic pulmonary disease often have viscous respiratory tract secretions. As these secretions accumulate and increase in viscosity, the patient's exchange of heat and water dwindles. Under these circumstances, the inspiration of warm, humidified air is essential to prevent dehydration (Walker and Wells 1961). Patients needing oxygen therapy and those with tracheotomies require special attention to ensure warm, humid supplies of inspired air. Cold, dry oxygen or bypassing the nasopharyngeal mucosa presents an extreme situation. Rebreathing techniques for anesthesia and enclosure in an incubator are special means of addressing impaired heat loss in therapeutic environments.

Burn patients need a hot environment and high relative humidity. A ward for severe burn victims should have temperature controls (and compatible architectural design and construction) that permit adjusting the room temperature up to 90°F db and relative humidity up to 95%.

HOSPITAL FACILITIES

Although proper air conditioning is helpful in preventing and treating disease, application of air conditioning to health facilities presents many problems not encountered in usual comfort conditioning design.

The basic differences between air conditioning for hospitals (and related health facilities) and that for other building types stem from the (1) need to restrict air movement in and between the various departments; (2) specific requirements for ventilation and filtration to dilute and remove contamination (odor, airborne microorganisms and viruses, and hazardous chemical and radioactive substances); (3) different temperature and humidity requirements for various areas; and (4) design sophistication needed to permit accurate control of environmental conditions.

Infection Sources and Control Measures

Bacterial Infection. Examples of bacteria that are highly infectious and transported within air or air and water mixtures are *Mycobacterium tuberculosis* and *Legionella pneumophila* (Legionnaires' disease). Wells (1934) showed that droplets or infectious agents of 5 μm or less in size can remain airborne indefinitely. Isoard et al. (1980) and Luciano (1984) have shown that 99.9% of all bacteria present in a hospital are removed by minimum efficiency reporting value (MERV) 14 filters (ASHRAE *Standard* 52.2), because bacteria are typically present in colony-forming units that are larger than 1 μm. Some authorities recommend the use of high-efficiency particulate air (HEPA) filters having dioctyl phthalate (DOP) test filtering efficiencies of 99.97% in certain areas.

Viral Infection. Examples of viruses that are transported by and virulent within air are *Varicella* (chicken pox/shingles), *Rubella* (German measles), and *Rubeola* (regular measles). Epidemiological evidence and other studies indicate that many of the airborne viruses that transmit infection are submicron in size. Although there is no known method to effectively eliminate 100% of the viable particles, HEPA and/or ultralow-penetration (ULPA) filters provide the greatest efficiency currently available. Research knowledge to deactivate viruses with ultraviolet light and chemical sprays has advanced, but design guidance and operational requirements have still not fully developed, so these methods are not recommended by most codes as a primary infection control measure. Therefore, isolation rooms and isolation anterooms with appropriate ventilation-pressure relationships are the primary means used to prevent the spread of airborne viruses in the health care environment.

Molds. Evidence indicates that some molds such as *Aspergillis* can be fatal to advanced leukemia, bone marrow transplant, and other immunocompromised patients.

Outdoor Air Ventilation. If outdoor air intakes are properly located, and areas adjacent to the intakes are properly maintained, outdoor air, in comparison to room air, is virtually free of infectious bacteria and viruses. Infection control problems frequently involve a bacterial or viral source within the hospital. Ventilation air dilutes viral and bacterial contamination within a hospital. If ventilation systems are properly designed, constructed, and maintained to preserve correct pressure relations between functional areas, they control the spread of airborne infectious agents and enable their proper containment and removal from the hospital environment.

Temperature and Humidity. These conditions can inhibit or promote the growth of bacteria and activate or deactivate viruses. Some bacteria such as *Legionella pneumophila* are basically waterborne and survive more readily in a humid environment. Codes and guidelines specify temperature and humidity range criteria in some hospital areas as a measure for infection control as well as comfort.

AIR QUALITY

Systems must also provide air virtually free of dust, dirt, odor, and chemical and radioactive pollutants. In some cases, untreated outdoor air is hazardous to patients suffering from cardiopulmonary, respiratory, or pulmonary conditions. In such instances, treatment of outdoor air as discussed in ASHRAE *Standard* 62.1 should be considered.

Outdoor Air Intakes. These intakes should be located as far as practical (on directionally different exposures whenever possible), but not less than 25 ft, from combustion equipment stack exhaust outlets, ventilation exhaust outlets from the hospital or adjoining buildings, medical-surgical vacuum systems, cooling towers, plumbing vent stacks, smoke control exhaust outlets, and areas that may collect vehicular exhaust and other noxious fumes. The bottom of outdoor air intakes serving central systems should be located as high as practical (minimum of 12 ft recommended) but not less than 6 ft above ground level or, if installed above the roof, 3 ft above the roof level.

Exhaust Air Outlets. These exhausts should be located a minimum of 10 ft above ground level and away from doors, occupied areas, and operable windows. Preferred location for exhaust outlets is at roof level projecting upward or horizontally away from outside intakes. Care must be taken in locating highly contaminated exhausts (e.g., from engines, fume hoods, biological safety cabinets, kitchen hoods, and paint booths). Prevailing winds, adjacent buildings, and discharge velocities must be taken into account (see Chapter 24 of the 2009 *ASHRAE Handbook—Fundamentals*). In critical or complicated applications, wind tunnel studies or computer modeling may be appropriate.

Air Filters. A number of methods are available for determining the efficiency of filters in removing particulates from an airstream (see Chapter 28 of the 2008 *ASHRAE Handbook—HVAC Systems and Equipment*). All central ventilation or air-conditioning systems should be equipped with filters having efficiencies no lower than those indicated in Table 1. Appropriate precautions should be observed to prevent wetting the filter media by free moisture from humidifiers. Application of filter beds should follow ASHRAE *Standard* 170. All filter efficiencies in Table 1 are based on ASHRAE *Standard* 52.2.

The following are guidelines for filter installations:

- HEPA filters should be used on air supplies serving protective-environment rooms for clinical treatment of patients with a high susceptibility to infection due to leukemia, burns, bone marrow transplant, organ transplant, or human immunodeficiency virus (HIV). HEPA filters should also be used on discharge air from fume hoods or biological safety cabinets in which infectious or highly toxic or radioactive materials are processed. Some health care facilities may also choose (or be required) to apply HEPA filters to the exhaust originating from airborne infectious isolation

Table 1 Filter Efficiencies for Central Ventilation and Air-Conditioning Systems in General Hospitals[c]

Minimum Number of Filter Beds	Area Designation	Filter Efficiencies, MERV[a] Filter Bed No. 1	Filter Bed No. 2
2	Orthopedic operating room Bone marrow transplant operating room Organ transplant operating room	7	HEPA[b]
2	General procedure operating rooms Delivery rooms Nurseries Intensive care units Patient care rooms Treatment rooms Diagnostic and related areas	7	14
1	Laboratories Sterile storage	13	
1	Food preparation areas Laundries Administrative areas Bulk storage Soiled holding areas	7	

[a]MERV = minimum efficiency reporting value based on ASHRAE *Standard* 52.2-2007.
[b]HEPA filters at air outlets.
[c]For guidance on selection and placement of filters, see ASHRAE *Standard* 170.

Table 2 Influence of Bedmaking on Airborne Bacterial Count in Hospitals

Item	Count per Cubic Foot Inside Patient Room	Hallway near Patient Room
Background	34	30
During bedmaking	140	64
10 min after	60	40
30 min after	36	27
Background	16	
Normal bedmaking	100	
Vigorous bedmaking	172	

Source: Greene et al. (1960).

rooms. When used, the filter system should be designed and equipped to permit safe removal, disposal, and replacement of contaminated filters.

• Filters should be installed to prevent leakage between filter segments and between the filter bed and its supporting frame. A small leak that permits any contaminated air to escape through the filter can destroy the usefulness of the best air cleaner.

• A manometer is recommended to be installed in the filter system to measure pressure drop across each filter bank. Visual observation is not accurate for determining filter loading.

• High-efficiency filters should be installed in the system, with adequate facilities provided for maintenance and in situ filtration performance testing without introducing contamination into the delivery system or the area served.

• Because high-efficiency filters are expensive, the hospital should project the filter bed life and replacement costs and incorporate these into the operating budget. Control sequences to monitor and alarm, including ability to normalize or benchmark pressure drops and associated airflows, can enhance indication of filter loading when air handlers operate at less than design flow.

• During construction, openings in ductwork and diffusers should be sealed to prevent intrusion of dust, dirt, and hazardous materials. Such contamination is often permanent and provides a medium for growth of infectious agents. Existing or new filters as well as coils may rapidly become contaminated by construction dust.

Air Movement

The data given in Table 2 illustrate the degree to which contamination can be dispersed into the air of the hospital environment by one of the many routine activities for normal patient care. The bacterial counts in the hallway clearly indicate the spread of this contamination.

Because of the dispersal of bacteria resulting from such necessary activities, air-handling systems should provide air movement patterns that minimize the spread of contamination. Undesirable airflow between rooms and floors is often difficult to control because of open doors, movement of staff and patients, temperature differentials, and stack effect, which is accentuated by vertical openings such as chutes, elevator shafts, stairwells, and mechanical shafts common to hospitals. Although some of these factors are beyond practical control, the effect of others may be minimized by terminating shaft openings in enclosed rooms and by designing and balancing air systems to create positive or negative air pressure within certain rooms and areas.

Systems serving highly contaminated areas, such as autopsy and airborne infectious isolation rooms, must maintain a negative air pressure in these rooms relative to adjoining rooms or the corridor (Murray et al. 1988). The negative pressure difference is obtained by supplying less air to the area than is exhausted from it (CDC 1994). Protective-environment rooms exemplify positive or negative pressure conditions. Exceptions to normally established negative and positive pressure conditions include operating rooms where highly infectious patients may be treated (e.g., operating rooms in which bronchoscopy or lung surgery is performed) and infectious isolation rooms that house immunosuppressed patients with airborne infectious diseases such as tuberculosis (TB). These areas should include an anteroom between the operating or protective-environment room and the corridor or other contiguous space. The three common approaches to anteroom relative pressurization are (1) anteroom positive to both the room and contiguous space, (2) anteroom negative to both the room and contiguous space, and (3) anteroom positive to room, negative to contiguous space. Any of these techniques minimizes cross-contamination between the patient area and surrounding areas, and may be used depending on local fire smoke management regulations.

Pressure differential causes air to flow in or out of a room through various leakage areas (e.g., perimeter of doors and windows, utility/fixture penetrations, cracks, etc.). A level of differential air pressure (0.01 in. of water) can be efficiently maintained only in a tightly sealed room. Therefore, it is important to obtain a reasonably close fit of all doors and seal all walls and floors, including penetrations between pressurized areas. Opening a door between two areas immediately reduces any existing pressure differential between them to such a degree that its effectiveness is nullified. When such openings occur, a natural interchange of air takes place between the two rooms because of turbulence created by the door opening and closing combined with personnel ingress/egress.

For critical areas requiring both the maintenance of pressure differentials to adjacent spaces and personnel movement between the critical area and adjacent spaces, the use of appropriate anterooms must be considered.

Figure 1 shows the bacterial count in a surgery room and its adjoining rooms during a normal surgical procedure. These bacterial counts were taken simultaneously. The relatively low bacterial counts in the surgery room, compared with those of the adjoining rooms, are attributable to the lower level of activity and higher air pressure in operating rooms.

In general, outlets supplying air to sensitive ultraclean areas should be located on the ceiling, and perimeter or several exhaust

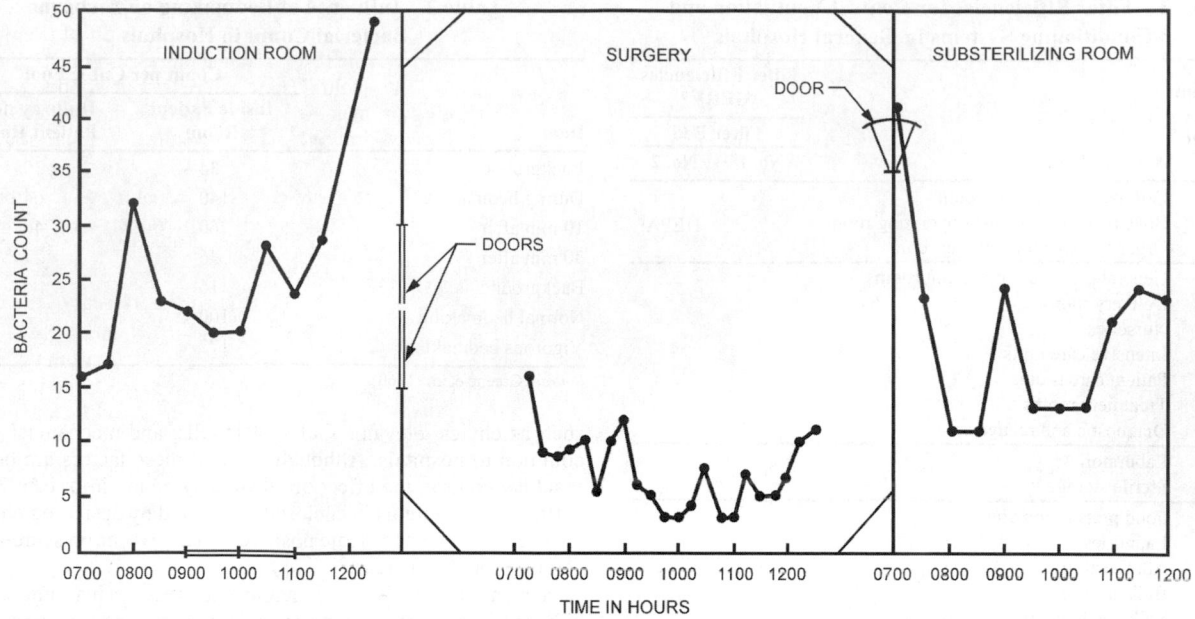

Fig. 1   Typical Airborne Contamination in Surgery and Adjacent Areas

outlets should be near the floor. This arrangement provides downward movement of clean air through the breathing and working zones to the floor area for exhaust. There are two recognized approaches to infectious isolation room air distribution. One approach locates the supply air above and near the doorway, and exhaust air from near the floor behind the patient's bed. This arrangement has the intent of controlling the flow of clean air first to parts of the room where workers or visitors are likely to be, and across the infected source into the exhaust. The limited ability of this arrangement to achieve this directional airflow movement, in view of the relatively low air exchange rates involved and the minimal influence of the exhaust outlet, led others to advocate a second arrangement in which the supply diffuser and exhaust outlet are located to maximize room air mixing, and therefore contaminant dilution and removal, typically with ceiling-mounted supply outlets paired with exhaust outlets over the patient bed or on the wall behind the bed. With this arrangement, the supply diffusers must be carefully selected and located such that primary air throw does not induce bedroom air to enter the anteroom.

The laminar airflow concept developed for industrial clean room use has attracted interest from some medical authorities. There are advocates of both vertical and horizontal laminar airflow systems, with and without fixed or movable walls around the surgical team (Pfost 1981). Some medical authorities do not advocate laminar airflow for surgeries but encourage air systems similar to those described in this chapter.

Laminar airflow in surgical operating rooms is predominantly unidirectional when not obstructed. Laminar airflow has shown promise in rooms used for treating patients who are highly susceptible to infection (Michaelson et al. 1966).

Temperature and Humidity

Specific recommendations for design temperatures and humidities are given in ASHRAE *Standard* 170, a portion of which is listed in Table 3.

Pressure Relationships and Ventilation

Table 3 covers ventilation recommendations for comfort, asepsis, and odor control in areas of acute care hospitals that directly

affect patient care. If specific organizational criteria must be met, refer to that organization's literature. Ventilation should be in accordance with ASHRAE *Standard* 170. Ventilation in accordance with ASHRAE *Standard* 62.1 should be used for areas where specific standards are not given. Where a higher outdoor air requirement is called for in ASHRAE *Standard* 62.1 than in Table 3, the higher value should be used. Specialized patient care areas, including organ transplant and burn units, should have additional ventilation provisions for air quality control as may be appropriate.

Ventilation system design must as much as possible provide air movement from clean to less clean areas. In critical-care areas, constant-volume systems should be used to ensure proper pressure relationships and ventilation. In noncritical patient care areas and staff rooms, variable air volume (VAV) systems may be considered for energy conservation. When using VAV systems in the hospital, special care should be taken to ensure that minimum ventilation rates (as required by codes) are maintained and that pressure relationships between various spaces are maintained. With VAV systems, a method such as air volume tracking between supply, return, and exhaust could be used to control pressure relationships (Lewis 1988).

The number of air changes may be reduced to 25% of the indicated value when the room is unoccupied, if provisions are made to ensure that (1) the number of air changes indicated is reestablished whenever the space is occupied, and (2) the pressure relationship with the surrounding rooms is maintained when the air changes are reduced.

In areas requiring no continuous directional control (±), ventilation systems may be shut down when the space is unoccupied and ventilation is not otherwise needed.

Because of the cleaning difficulty and potential for buildup of contamination, recirculating room heating and/or cooling units must not be used in areas marked "No." Note that the standard recirculating room unit may also be impractical for primary control where exhaust to the outside is required.

In rooms with hoods, extra air must be supplied for hood exhaust so that the designated pressure relationship is maintained. Refer to Chapter 16 for further discussion of laboratory ventilation.

For maximum energy conservation, using recirculated air is preferred as long as health and safety issues are adequately addressed. If all-outdoor air is used, an efficient heat recovery method should be considered.

Smoke Control

As the ventilation design is developed, a proper smoke control strategy must be considered. Passive systems rely on fan shutdown and smoke and fire barriers. Proper treatment of duct penetrations must be observed.

Active smoke control systems use the ventilation system to create areas of positive and negative pressures that, along with fire and smoke partitions, limit the spread of smoke. The ventilation system may be used in a smoke removal mode in which combustion products are exhausted by mechanical means. As design of active smoke control systems evolves, the engineer and code authority should carefully plan system operation and configuration. Refer to Chapter 53 and NFPA *Standards* 90A, 92A, and 101.

SPECIFIC DESIGN CRITERIA

There are seven principal divisions of an acute care general hospital: (1) surgery and critical care, (2) nursing, (3) ancillary, (4) administration, (5) diagnostic and treatment, (6) sterilizing and supply, and (7) service. Environmental requirements of each department/space in these divisions differ according to their function and procedures carried out in them. This section describes the functions of these departments/spaces and covers details of design requirements. Close coordination with health care planners and medical equipment specialists in mechanical design and construction of health facilities is essential to achieve the desired conditions.

Surgery and Critical Care

No area of the hospital requires more careful control of the aseptic condition of the environment than the surgical suite. Systems serving operating rooms, including cystoscopic and fracture rooms, require careful design to minimize the concentration of airborne organisms.

The greatest amount of bacteria found in the operating room comes from the surgical team and is a result of their activities during surgery. During an operation, most members of the surgical team are near the operating table, creating the undesirable situation of concentrating contamination in this highly sensitive area. There are three distinct classifications for surgery care areas class A, B, and C (ACS 2000) that generally support invasive procedures as defined in FGI (2010):

- **Class A surgery** involves minor surgical procedures performed under topical, local, or regional anesthesia without preoperative sedation. Excluded are intravenous, spinal, and epidural procedures, which are Class B or C surgeries.
- **Class B surgery** provides minor or major surgical procedures performed in conjunction with oral, parenteral, or intravenous sedation or performed with the patient under analgesic or dissociative drugs.
- **Class C surgery** involves major surgical procedures that require general or regional block anesthesia and/or support of vital bodily functions.

Operating Rooms. Past studies of operating-room air distribution devices and observation of installations in industrial clean rooms indicate that delivering air from the ceiling, with a downward movement to several exhaust/return openings located low on opposite walls, is probably the most effective air movement pattern for maintaining the contamination concentration at an acceptable level. Completely perforated ceilings, partially perforated ceilings, and ceiling-mounted diffusers have been applied successfully (Pfost 1981). A mixture of low and high exhaust opening locations may work better than either all low or all high locations, with supply air at face velocities of around 25 to 35 fpm from a unidirectional laminar-flow ceiling array.

Operating room suites are typically in use no more than 8 to 12 h per day (except trauma centers and emergency departments).

For energy conservation, the air-conditioning system should allow a reduction in the air supplied to some or all of the operating rooms when possible. Positive space pressure must be maintained at reduced air volumes to ensure sterile conditions. The time required for an inactive room to become usable again must be considered. Consultation with the hospital surgical staff will determine the feasibility of this feature.

A separate air exhaust system or special vacuum system should be provided for the removal of anesthetic trace gases. Medical vacuum systems have been used to remove nonflammable anesthetic gases (NFPA *Standard* 99). One or more outlets may be located in each operating room to connect the anesthetic machine scavenger hose.

Although good results have been reported from air disinfection of operating rooms by irradiation, this method is seldom used. The reluctance to use irradiation may be attributed to the need for special designs for installation, protective measures for patients and personnel, constant monitoring of lamp efficiency, and maintenance.

The following conditions are recommended for operating, catheterization, cystoscopic, and fracture rooms:

- The temperature and relative humidity set points should be adjustable by surgical staff. Systems should be capable of maintaining the space temperature in the lower end of the range (62°F) for specialized procedures such as cardiac surgery. Table 3 lists tolerable temperature ranges; however, these are not intended to be dynamic control ranges. Special or supplemental cooling equipment should be considered if this lower temperature negatively affects energy use for surrounding areas.
- Air pressure should be kept positive with respect to any adjoining rooms by supplying excess air.
- A differential-pressure-indicating device should be installed to permit air pressure readings in the rooms. Thorough sealing of all wall, ceiling, and floor penetrations, and tight-fitting doors are essential to maintaining readable pressure.
- Humidity and temperature indicators should be located for easy observation.
- Filter efficiencies should be in accordance with Table 1.
- The entire installation should conform to requirements of NFPA *Standard* 99.
- Air should be supplied at the ceiling with exhaust/return from at least two locations near the floor, with consideration given to having at least two high exhaust/return openings opposite from the low outlets. Endoscopic, laparoscopic, or thoracoscopic surgery procedures aided by camera, and robotic or robot-assisted surgery procedures, require heat-producing equipment in the operating room. Exhaust/return openings located above this equipment can capture the more buoyant heated air and prevent it from being reentrained in the ceiling supply airstream. The bottom of low openings should be at least 3 in. above the floor. Supply diffusers should be unidirectional (laminar-flow). High-induction ceiling or sidewall diffusers should be avoided.
- Total air exchange rates should address lights and equipment (e.g., blanket and blood warmers, fiber-optic equipment, robotic consoles). Suites with a large amount of electronic equipment have been reported to require 30 to 35 air changes per hour using conventional air-conditioning systems with 18 to 20°F supply temperature differentials.
- Acoustical materials should not be used as duct linings unless terminal filters of at least MERV 14 efficiency are installed downstream of the linings. Internal insulation of terminal units may be encapsulated with approved materials. Duct-mounted sound traps should be of the packless type or have polyester film linings over acoustical fill.
- Any spray-applied insulation and fireproofing should be treated with fungi growth inhibitor.

Table 3 Design Parameters for Areas Affecting Patient Care in Hospitals and Outpatient Facilities

Space Function	Pressure Relationship to Adjacent Areas[n]	Minimum Outdoor ach	Minimum Total ach	All Room Air Exhausted Directly to Outdoors[j]	Air Recirculated by Means of Room Units[a]	Relative Humidity[k] %	Design Temperature,[l] °F
Surgery and Critical Care							
Classes B and C operating rooms[m,n,o]	Positive	4	20	N/R	No	30 to 60	68 to 75
Operating/surgical cystoscopic rooms[m,n,o]	Positive	4	20	N/R	No	30 to 60	68 to 75
Delivery room (Caesarean)[m,n,o]	Positive	4	20	N/R	No	30 to 60	68 to 75
Substerile service area	N/R	2	6	N/R	No	N/R	N/R
Recovery room	N/R	2	6	N/R	No	30 to 60	70 to 75
Critical and intensive care	Positive	2	6	N/R	No	30 to 60	70 to 75
Wound intensive care (burn unit)	Positive	2	6	N/R	No	40 to 60	70 to 75
Newborn intensive care	Positive	2	6	N/R	No	30 to 60	70 to 75
Treatment room[p]	N/R	2	6	N/R	N/R	30 to 60	70 to 75
Trauma room (crisis or shock)[c]	Positive	3	15	N/R	No	30 to 60	70 to 75
Medical/anesthesia gas storage[r]	Negative	N/R	8	Yes	N/R	N/R	N/R
Laser eye room	Positive	3	15	N/R	No	30 to 60	70 to 75
ER waiting rooms[q]	Negative	2	12	Yes	N/R	max 65	70 to 75
Triage	Negative	2	12	Yes	N/R	max 60	70 to 75
ER decontamination	Negative	2	12	Yes	No	N/R	N/R
Radiology waiting rooms[q]	Negative	2	12	Yes	N/R	max 60	70 to 75
Class A Operating/Procedure room[o,d]	Positive	3	15	N/R	No	30 to 60	70 to 75
Inpatient Nursing							
Patient room (s)	N/R	2	6	N/R	N/R	max 60	70 to 75
Toilet room	Negative	N/R	10	Yes	No	N/R	N/R
Newborn nursery suite	N/R	2	6	N/R	No	30 to 60	72 to 78
Protective environment room[f,n,t]	Positive	2	12	N/R	No	max 60	70 to 75
AII room[e,n,u]	Negative	2	12	Yes	No	max 60	70 to 75
AII isolation anteroom[t,u]	N/R	N/R	10	Yes	No	N/R	N/R
Labor/delivery/recovery/postpartum (LDRP)[s]	N/R	2	6	N/R	N/R	max 60	70 to 75
Labor/delivery/recovery (LDR)[s]	N/R	2	6	N/R	N/R	max 60	70 to 75
Corridor	N/R	N/R	2	N/R	N/R	N/R	N/R
Skilled Nursing Facility							
Resident room	N/R	2	2	N/R	N/R	N/R	70 to 75
Resident gathering/activity/dining	N/R	4	4	N/R	N/R	N/R	70 to 75
Physical therapy	Negative	2	6	N/R	N/R	N/R	70 to 75
Occupational therapy	N/R	2	6	N/R	N/R	N/R	70 to 75
Bathing room	Negative	N/R	10	Yes	N/R	N/R	70 to 75
Radiology[v]							
X-ray (diagnostic and treatment)	N/R	2	6	N/R	N/R	max 60	72 to 78
X-ray (surgery/critical care and catheterization)	Positive	3	15	N/R	No	max 60	70 to 75
Darkroom[g]	Negative	2	10	Yes	No	N/R	N/R
Diagnostic and Treatment							
Bronchoscopy, sputum collection, and pentamidine administration[n]	Negative	2	12	Yes	No	N/R	68 to 73
Laboratory, general[v]	Negative	2	6	N/R	No	N/R	70 to 75
Laboratory, bacteriology[v]	Negative	2	6	Yes	No	N/R	70 to 75
Laboratory, biochemistry[v]	Negative	2	6	Yes	No	N/R	70 to 75
Laboratory, cytology[v]	Negative	2	6	Yes	No	N/R	70 to 75
Laboratory, glasswashing	Negative	2	10	Yes	No	N/R	N/R
Laboratory, histology[v]	Negative	2	6	Yes	No	N/R	70 to 75
Laboratory, microbiology[v]	Negative	2	6	Yes	No	N/R	70 to 75
Laboratory, nuclear medicine[v]	Negative	2	6	Yes	No	N/R	70 to 75
Laboratory, pathology[v]	Negative	2	6	Yes	No	N/R	70 to 75
Laboratory, serology[v]	Negative	2	6	Yes	No	N/R	70 to 75
Laboratory, sterilizing	Negative	2	10	Yes	No	N/R	70 to 75
Laboratory, media transfer[v]	Positive	2	4	N/R	No	N/R	70 to 75
Autopsy room[n]	Negative	2	12	Yes	No	N/R	68 to 75
Nonrefrigerated body-holding room[h]	Negative	N/R	10	Yes	No	N/R	70 to 75
Pharmacy[b]	Positive	2	4	N/R	N/R	N/R	N/R
Examination room	N/R	2	6	N/R	N/R	max 60	70 to 75
Medication room	Positive	2	4	N/R	N/R	max 60	70 to 75
Endoscopy	Positive	2	15	N/R	No	30-60	68 to 73
Endoscope cleaning	Negative	2	10	Yes	No	N/R	N/R
Treatment room	N/R	2	6	N/R	N/R	max 60	70 to 75
Hydrotherapy	Negative	2	6	N/R	N/R	N/R	72 to 80
Physical therapy	Negative	2	6	N/R	N/R	Max 65	72 to 80

Table 3 Design Parameters for Areas Affecting Patient Care in Hospitals and Outpatient Facilities

Space Function	Pressure Relationship to Adjacent Areas[n]	Minimum Outdoor ach	Minimum Total ach	All Room Air Exhausted Directly to Outdoors[j]	Air Recirculated by Means of Room Units[a]	Relative Humidity[k] %	Design Temperature,[l] °F
Sterilizing							
Sterilizer equipment room	Negative	N/R	10	Yes	No	N/R	N/R
Central Medical and Surgical Supply							
Soiled or decontamination room	Negative	2	6	Yes	No	N/R	72 to 78
Clean workroom	Positive	2	4	N/R	No	max 60	72 to 78
Sterile storage	Positive	2	4	N/R	N/R	max 60	72 to 78
Service							
Food preparation center[i]	N/R	2	10	N/R	No	N/R	72 to 78
Warewashing	Negative	N/R	10	Yes	No	N/R	N/R
Dietary storage	N/R	N/R	2	N/R	No	N/R	72 to 78
Laundry, general	Negative	2	10	Yes	No	N/R	N/R
Soiled linen sorting and storage	Negative	N/R	10	Yes	No	N/R	N/R
Clean linen storage	Positive	N/R	2	N/R	N/R	N/R	72 to 78
Linen and trash chute room	Negative	N/R	10	Yes	No	N/R	N/R
Bedpan room	Negative	N/R	10	Yes	No	N/R	N/R
Bathroom	Negative	N/R	10	Yes	No	N/R	72 to 78
Janitor's closet	Negative	N/R	10	Yes	No	N/R	N/R
Support Space							
Soiled workroom or soiled holding	Negative	2	10	Yes	No	N/R	N/R
Clean workroom or clean holding	Positive	2	4	N/R	N/R	N/R	N/R
Hazardous material storage	Negative	2	10	Yes	No	N/R	N/R

Note: N/R = no requirement

Notes:

a. Recirculating room HVAC units (with heating or cooling coils) are acceptable to achieve required air change rates. Because of difficulty cleaning and potential for build-up of contamination, recirculating room units should not be used in areas marked "No." Isolation and intensive care unit rooms may be ventilated by reheat induction units in which only primary air supplied from a central system passes through the reheat unit. Gravity-type heating or cooling units (e.g., radiators or convectors) should not be used in operating rooms and other special care areas.

b. Pharmacy compounding areas may have additional air change and filtering requirements beyond the minimum of this table, depending on type of pharmacy, regulatory requirements (which may include adoption of USP 797), associated level of risk of the work (see USP 797), and equipment used in the spaces.

c. The term *trauma room* here means a first-aid room and/or emergency room used for general initial treatment of accident victims. The operating room in the trauma center that is routinely used for emergency surgery is considered to be an operating room by ASHRAE *Standard* 170.

d. Pressure relationships need not be maintained when the room is unoccupied.

e. Some isolation rooms may be provided with a separate anteroom, but an anteroom is not required by ASHRAE *Standard* 170.

f. Protective environment rooms are used for high-risk immunocompromised patients. Such rooms are positively pressurized relative to all adjoining spaces to protect the patient.

g. Exception: All air need not be exhausted if darkroom equipment has a scavenging exhaust duct attached and meets ventilation standards regarding NIOSH, OSHA, and local employee exposure limits.

h. A nonrefrigerated body-holding room is applicable only to facilities that do not perform autopsies on site and use the space for short periods while waiting for the body to be transferred.

i. Minimum total air changes per hour (ach) is that required to provide proper makeup air to kitchen exhaust systems as specified in ANSI/ASHRAE *Standard* 154. In some cases, excess exfiltration or infiltration to or from exit corridors compromises the exit corridor restrictions of NFPA *Standard* 90A, the pressure requirements of NFPA *Standard* 96, or the maximum defined in the table. During operation, a reduction to the number of air changes to any extent required for odor control is permitted when the space is not in use. [See AIA (2006) in Informative Annex B: Bibliography of ASHRAE *Standard* 170.)]

j. In some areas with potential contamination and/or odor problems, exhaust air should be discharged directly to the outdoors and not recirculated to other areas. Individual circumstances may require special consideration for air exhausted to the outdoors (e.g., intensive care units in which patients with pulmonary infection are treated, rooms for burn patients). To satisfy exhaust needs, constant replacement air from the outdoors is necessary when the system is in operation.

k. The rh ranges listed are the minimum and maximum limits where control is specifically needed.

l. Systems should be able to maintain rooms within the range during normal operation. Lower or higher temperature are allowed when patients' comfort and/or medical conditions require those conditions.

m. NIOSH criteria documents regarding occupational exposure to waste anesthetic gases and vapors, and control of occupational exposure to nitrous oxide indicate a need for both local

exhaust (scavenging) systems and general ventilation of the areas in which gases are used. Refer to NFPA *Standard* 99 for other requirements.

n. If monitoring device alarms are installed, allowances should be made to prevent nuisance alarms. Short-term excursions from required pressure relationships are allowed while doors are moving or temporarily open. Simple visual methods such as smoke trail, ball-in-tube, or flutterstrip areallowed for verification of airflow direction. Recirculating devices with HEPA filters areallowed in existing facilities as interim, supplemental environmental controls to meet requirements for control of airborne infectious agents. Design of portable or fixed systems should prevent stagnation and short-circuiting of airflow. System design should also allow easy access for scheduled preventative maintenance and cleaning.

o. Surgeons or surgical procedures may require room temperatures, ventilation rates, humidity ranges, and/or air distribution methods that exceed the minimum indicated ranges.

p. Treatment rooms used for bronchoscopy should be treated as bronchoscopy rooms. Treatment rooms used for procedures with nitrous oxide shouldhave provisions for exhausting anesthetic waste gases.

q. In a recirculating ventilation system, HEPA filters areallowed instead of exhausting air from these spaces to the outdoors, provided the return air passes through HEPA filters before it is introduced into any other spaces. This requirement applies only to waiting rooms to hold patients awaiting chest x-rays for diagnosis of respiratory disease.

r. See NFPA *Standard* 99 for further requirements.

s. For patient rooms, labor/delivery/recovery rooms, and labor/delivery/recovery/postpartum rooms, 4 ach total isallowed when supplemental heating and/or cooling systems (radiant heating and cooling, baseboard heating, etc.) are used.

t. The protective environment airflow design specifications protect the patient from common environmental airborne infectious microbes (i.e., *Aspergillus* spores). Recirculation HEPA filters areallowed to increase the equivalent room air exchanges; however, the outdoor air changes are still required. Constant-volume airflow is required for consistent ventilation for the protected environment. If the design criteria indicate that AII is necessary for protective-environment patients, an anteroom should be provided. Rooms with reversible airflow provisions for the purpose of switching between protective environment and AII functions arenot allowed.

u. The AII room described in ASHRAE *Standard* 170 shouldbe used to isolate the airborne spread of infectious diseases, such as measles, varicella, or tuberculosis. Design of AII rooms should include the provision for normal patient care during periods not requiring isolation precautions. Supplemental recirculating devices using HEPA filters are allowed in the patient room to increase the equivalent room air exchanges; however, the outdoor air changes are still required. AII rooms that are retrofitted from standard patient rooms from which it is impractical to exhaust directly outside may be recirculated with air from the AII room, provided that the air first passes through a HEPA filter. HEPA-filtered exhaust air from AII rooms may mix with exhaust air that serves non-AII spaces before being discharged directly outdoors. Rooms with reversible airflow provisions for the purpose of switching between protective environment and AII functions are not allowed. See the guidelines in Informative Annex B: Bibliography of ASHRAE *Standard* 170 for more information.

v. When required, appropriate hoods and exhaust devices for the removal of noxious gases or chemical vapors should be provided in accordance with NFPA Standard 99.

- Sufficient lengths of watertight, drained stainless steel duct should be installed downstream of humidification equipment to ensure complete evaporation of water vapor before air is discharged into the room.

Control centers that monitor and allow adjustment of temperature, humidity, and air pressure may be located at the surgical supervisor's desk.

Obstetrical Areas. The pressure in the obstetrical department should be positive or equal to that in other areas.

Delivery Rooms. The delivery room design should conform to the requirements of operating rooms.

Recovery Rooms. Postoperative recovery rooms used in conjunction with operating rooms should be maintained at a relative humidity of 45 to 55%. Because the smell of residual anesthesia sometimes creates odor problems in recovery rooms, ventilation is important, and a balanced air pressure relative to that of adjoining areas should be provided.

Intensive Care Units. These units serve seriously ill patients, from postoperative to coronary patients. A variable-range temperature capability of 70 to 75°F, relative humidity of 30% minimum and 60% maximum, and positive air pressure are recommended.

Nursery Suites. Air conditioning in nurseries provides the constant temperature and humidity conditions essential to care of the newborn in a hospital environment. Air movement patterns in nurseries should be carefully designed to reduce the possibility of drafts.

Some codes or jurisdictions require that air be removed near floor level, with the bottoms of exhaust openings at least 3 in. above the floor; the relative efficacy of this exhaust arrangement has been questioned by some experts, because exhaust air outlets have a minimal effect on room air movement at the relatively low air exchange rates involved. Air system filter efficiencies should conform to Table 1. Finned tube radiation and other forms of convection heating should not be used in nurseries.

Full-Term Nurseries. A relative humidity of 30 to 60% is recommended for full-term nurseries, examination rooms, and work spaces. The maternity nursing section should be controlled similarly to protect the infant during visits with the mother. The nursery should have a positive air pressure relative to the work space and examination room, and any rooms located between the nurseries and the corridor should be similarly pressurized relative to the corridor. This prevents infiltration of contaminated air from outside areas.

Special-Care Nurseries. These nurseries require a variable range temperature capability of 75 to 80°F and a relative humidity of 30 to 60%. This type of nursery is usually equipped with individual incubators to regulate temperature and humidity. It is desirable to maintain these same conditions within the nursery proper to accommodate both infants removed from the incubators and those not placed in incubators. Pressurization of special-care nurseries should correspond to that of full-term nurseries.

Observation Nurseries. Temperature and humidity requirements for observation nurseries are similar to those for full-term nurseries. Because infants in these nurseries have unusual clinical symptoms, the air from this area should not enter other nurseries. A negative air pressure relative to that of the workroom should be maintained in the nursery. The workroom, usually located between the nursery and the corridor, should be pressurized relative to the corridor.

Emergency Rooms. Emergency rooms are typically the most highly contaminated areas in the hospital because of the soiled condition of many arriving patients and the relatively large number of persons accompanying them. Temperatures and humidities of offices and waiting spaces should be within the normal comfort range. Clean-to-dirty directional airflow and zone pressurization techniques should be considered, to reduce the airborne exposure potential for health-care personnel assigned to the emergency room reception stations.

Trauma Rooms. Trauma rooms should be ventilated in accordance with requirements in Table 3. Emergency operating rooms located near the emergency department should have the same temperature, humidity, and ventilation requirements as those of operating rooms.

Anesthesia Storage Rooms. Anesthesia storage rooms must be ventilated in conformance with NFPA *Standard* 99. However, mechanical ventilation only is recommended.

Nursing

Patient Rooms. When central systems are used to condition patient rooms, recommendations in Tables 1 and 3 should be followed to reduce cross-infection and control odor. Each patient room should have individual temperature control. Air pressure in patient suites should be neutral in relation to other areas.

Most governmental design criteria and codes require that all air from toilet rooms be exhausted directly outside. The requirement appears to be based on odor control.

Where room unit systems are used, it is common practice to exhaust through the adjoining toilet room an amount of air equal to the amount of outdoor air brought in for ventilation. Ventilation of toilets, bedpan closets, bathrooms, and all interior rooms should conform to applicable codes.

Protective Isolation Units. Immunosuppressed patients (including bone marrow or organ transplant, leukemia, burn, and AIDS patients) are highly susceptible to diseases. Some physicians prefer an isolated laminar airflow unit to protect the patient; others are of the opinion that the conditions of the laminar cell have a psychologically harmful effect on the patient and prefer flushing out the room and reducing spores in the air. An air distribution of 15 air changes per hour supplied through a nonaspirating diffuser is often recommended. With this arrangement, the sterile air is drawn across the patient and returned near the floor, at or near the door to the room.

In cases where the patient is immunosuppressed but not contagious, positive pressure should be maintained between the patient room and adjacent area. Some jurisdictions may require an anteroom, which maintains a negative pressure relative to the adjacent isolation room and an equal pressure to the corridor, nurses' station, or common area. Exam and treatment rooms should be controlled in the same manner. Positive pressure should also be maintained between the entire unit and adjacent areas to preserve sterile conditions.

When a patient is both immunosuppressed and potentially contagious, combination airborne infectious isolation/protective environment (AII/PE) rooms are provided. These AII/PE rooms require an anteroom, which should be either positive or negative to both the AII/PE room and the corridor or common space. Pressure controls in the adjacent area or anteroom must maintain the correct pressure relationship relative to the other adjacent room(s) and areas. A separate, dedicated air-handling system to serve the protective isolation unit simplifies pressure control and quality (Murray et al. 1988).

Infectious Isolation Unit. The infectious isolation room protects the rest of the hospital from patients' infectious diseases. Recent multidrug-resistant strains of tuberculosis have increased the importance of pressurization, air change rates, filtration, and air distribution design in these rooms (Rousseau and Rhodes 1993). Temperatures and humidities should correspond to those specified for patient rooms.

The designer should work closely with health care planners and the code authority to determine the appropriate isolation room design. It may be desirable to provide more complete control, with a separate anteroom used as an air lock to minimize the potential that airborne particles from the patients' area reach adjacent areas. Design approaches to airborne infectious isolation may also be found in CDC (2005).

Switchable isolation rooms (rooms that can be set to function with either positive or negative pressure) have been installed in many facilities. CDC (2005) and FGI (2010) have, respectively, recommend against and prohibited this approach. The two difficulties of this approach are (1) maintaining the mechanical dampers and controls required to accurately provide the required pressures, and (2) that it provides a false sense of security to staff who think that this provision is all that is required to change a room between protective isolation and infectious isolation, to the exclusion of other sanitizing procedures.

Floor Pantry. Ventilation requirements for this area depend on the type of food service adopted by the hospital. Where bulk food is dispensed and dishwashing facilities are provided in the pantry, using hoods above equipment, with exhaust to the outside, is recommended. Small pantries used for between-meal feedings require no special ventilation. The air pressure of the pantry should be in balance with that of adjoining areas to reduce the movement of air into or out of it.

Labor/Delivery/Recovery/Postpartum (LDRP). The procedures for normal childbirth are considered noninvasive, and rooms are controlled similarly to patient rooms. Some jurisdictions may require higher air change rates than in a typical patient room. It is expected that invasive procedures such as cesarean section are performed in a nearby delivery or operating room.

Ancillary

Radiology Department. Among the factors affecting ventilation system design in these areas are odors from certain clinical treatments and the special construction designed to prevent radiation leakage. Fluoroscopic, radiographic, therapy, and darkroom areas require special attention.

Fluoroscopic, Radiographic, and Deep Therapy Rooms. These rooms require a temperature from 78 to 80°F and a relative humidity from 40 to 50%. This relative humidity range control often requires dedicated room equipment and control. Depending on the location of air supply outlets and exhaust intakes, lead lining may be required in supply and return ducts at points of entry to various clinical areas to prevent radiation leakage to other occupied areas.

Darkroom. The darkroom is normally in use for longer periods than x-ray rooms, and should have an independent system to exhaust air to the outside. Exhaust from the film processor may be connected into the darkroom exhaust.

Laboratories. Air conditioning is necessary in laboratories for the comfort and safety of the technicians (Degenhardt and Pfost 1983). Chemical fumes, odors, vapors, heat from equipment, and the undesirability of open windows all contribute to this need.

Particular attention should be given to the size and type of equipment used in the various laboratories, as equipment heat gain usually constitutes a major portion of the cooling load.

The general air distribution and exhaust systems should be constructed of conventional materials following standard designs for the type of systems used. Exhaust systems serving hoods in which radioactive materials, volatile solvents, and strong oxidizing agents such as perchloric acid are used should be made of stainless steel. Washdown facilities and dedicated exhaust fans should be provided for hoods and ducts handling perchloric acid.

Hood use may dictate other duct materials. Hoods in which radioactive or infectious materials are to be used must be equipped with high-efficiency (HEPA) filters for the exhaust and have a procedure and equipment for safe removal and replacement of contaminated filters. Exhaust duct routing should be as short as possible with minimal horizontal offsets and, when possible, duct portions with contaminated air should be maintained under negative pressure (e.g., locate fan on clean side of filter). This applies especially to perchloric acid hoods because of the extremely hazardous, explosive nature of this material.

Determining the most effective, economical, and safe system of laboratory ventilation requires considerable study. Where laboratory space ventilation air quantities approximate the air quantities required for ventilating the hoods, the hood exhaust system may be used to exhaust all ventilation air from the laboratory areas.

The hood exhaust system should not shut off if the air system fails. Chemical storage rooms must have a constantly operating exhaust air system with a terminal fan.

Hood exhaust fans should be located at the discharge end of the duct system to prevent exhaust products entering the building. For further information on laboratory air conditioning and hood exhaust systems, see ANSI/AIHA *Standard* Z9.5, NFPA *Standard* 45, Hagopian and Hoyle (1984), and Chapter 16 of this volume.

Exhaust air from hoods in biochemistry, histology, cytology, pathology, glass washing/sterilizing, and serology-bacteriology units should be discharged to the outside with no recirculation. Use care in designing the exhaust outlet locations and arrangements: exhaust should not be reentrained in the building via outdoor air intakes or other building openings. Separation from outdoor air intake sources, wind direction and velocity, building geometry, and exhaust outlet height and velocity are important. In some laboratory exhaust systems, exhaust fans discharge vertically at a minimum of 7 ft above the roof at velocities up to 4000 fpm. The serology-bacteriology unit should be positively pressured relative to adjoining areas to reduce the possibility of infiltration of aerosols that could contaminate the specimens being processed. The entire laboratory area should be under slight negative pressure to reduce the spread of odors or contamination to other hospital areas. Temperatures and humidities should be within the comfort range.

Bacteriology Laboratories. These units should not have undue air movement, so care should be exercised to limit air velocities to a minimum. The sterile transfer room, which may be within or adjoining the bacteriology laboratory, is where sterile media are distributed and where specimens are transferred to culture media. To maintain a sterile environment, an ultrahigh-efficiency HEPA filter should be installed in the supply air duct near the point of entry to the room. The media room should be ventilated to remove odors and steam.

Infectious Disease and Virus Laboratories. These laboratories, found only in large hospitals, require special treatment. A minimum ventilation rate of 6 air changes per hour or makeup equal to hood exhaust volume is recommended for these laboratories, which should have a negative air pressure relative to any other area nearby to prevent exfiltration of any airborne contaminants. Exhaust air from fume hoods or safety cabinets must be sterilized before being exhausted to the outside. This may be accomplished by using electric or gas-fired heaters placed in series in the exhaust systems and designed to heat the exhaust air to 600°F. A more common and less expensive method of sterilizing the exhaust is to use HEPA filters in the system.

Nuclear Medicine Laboratories. Such laboratories administer radioisotopes to patients orally, intravenously, or by inhalation to facilitate diagnosis and treatment of disease. There is little opportunity in most cases for airborne contamination of the internal environment, but exceptions warrant special consideration.

One important exception involves the use of iodine-131 solution in capsules or vials to diagnose thyroid disorders. Another involves use of xenon-133 gas via inhalation to study patients with reduced lung function.

Capsules of iodine-131 occasionally leak part of their contents prior to use. Vials emit airborne contaminants when opened for preparation of a dose. It is common practice for vials to be opened and handled in a standard laboratory fume hood. A minimum face velocity of 100 fpm should be adequate for this purpose. This recommendation applies only where small quantities are handled in simple operations. Other circumstances may warrant provision of a glove box or similar confinement.

Use of xenon-133 for patient study involves a special instrument that permits the patient to inhale the gas and to exhale back into the instrument. The exhaled gas is passed through a charcoal trap mounted in lead and is often vented outside. The process suggests some potential for escape of the gas into the internal environment.

Because of the uniqueness of this operation and the specialized equipment involved, it is recommended that system designers determine the specific instrument to be used and contact the manufacturer for guidance. Other guidance is available in U.S. Nuclear Regulatory Commission *Regulatory Guide* 10.8 (NRC 1980). In particular, emergency procedures in case of accidental release of xenon-133 should include temporary evacuation of the area and/or increasing the ventilation rate of the area.

Recommendations for pressure relationships, supply air filtration, supply air volume, airborne particle counts, recirculation, and other attributes of supply and discharge systems for histology, pathology, pharmacy, and cytology laboratories are also relevant to nuclear medicine laboratories. There are, however, some special ventilation system requirements imposed by the NRC where radioactive materials are used. For example, NRC (1980) provides a computational procedure to estimate the airflow necessary to maintain xenon-133 gas concentration at or below specified levels. It also contains specific requirements as to the amount of radioactivity that may be vented to the atmosphere; the disposal method of choice is adsorption onto charcoal traps.

Autopsy Rooms. Susceptible to heavy bacterial contamination (e.g., tuberculosis) and odor, autopsy rooms, which are part of the hospital's pathology department, require special attention. Exhaust intakes should be located both at the ceiling and in the low sidewall. The exhaust system should discharge the air above the roof of the hospital, away from points of potential reentrainment, such as outdoor air intakes and building openings. The autopsy room should be negatively pressured relative to adjoining areas to prevent the spread of contamination. Where large quantities of formaldehyde are used, special exhaust systems can effectively control concentrations below legal exposure limits. A combination of localized exhaust and ventilation systems with downdraft or side-draft tables has been shown to effectively control concentrations while using smaller exhaust volumes than those required by dilution ventilation (Gressel and Hughes 1992).

In smaller hospitals where the autopsy room is used infrequently, local control of the ventilation system and an odor control system with either activated charcoal or potassium permanganate-impregnated activated alumina may be desirable.

Animal Quarters. Principally because of odor, animal quarters (found only in larger hospitals) require a mechanical exhaust system that discharges contaminated air above the hospital roof. To prevent the spread of odor or other contaminants from the animal quarters to other areas, a negative air pressure of at least 0.1 in. of water relative to adjoining areas must be maintained. Chapter 16 has further information on animal room air conditioning.

Pharmacies. Design and ventilation requirements for pharmacies can vary greatly according to the type of compounding done there. Pharmacies handling hazardous drugs and/or involved in sterile compounding activities have special ventilation requirements such as, for example, horizontal or vertical laminar-airflow workbenches (LAFW), biological safety cabinets (BSC), and compounding (barrier) isolators. Room air distribution and filtration must be coordinated with any laminar airflow benches, cabinets, and isolators that may be needed. See Chapters 16 and 18 for more information.

Sterile Compounding. Sterile pharmaceutical compounding requirements are prescribed by the U.S. Pharmacopeia (USP 2008). This chapter (797) is enforceable under the U.S. Food and Drug Administration, is adopted in whole or in part by many state boards of pharmacy, and may be incorporated into the inspection programs of health care accreditation organizations. The Joint Commission recognized USP 797 as a consensus-based safe practice guideline for sterile compounding; however, they do not require its implementation as a condition of accreditation. End users, owners, architects, and engineers should consult the most recent release of USP 797,

which is under continuous maintenance, as well as design guidance adopted by their state boards of pharmacy.

USP 797 prescribes that all sterile pharmaceutical preparations to be administered more than 1 h after preparation must be compounded entirely within a critical work zone protected by a unidirectional, HEPA-filtered airflow of ISO class 5 (former class 100 under *Federal Standard* 209E) or better air quality. This ISO class 5 environment is generally provided using a primary engineering control (PEC) such as a LAFW, BSC, or compounding isolator. USP 797 also requires that the ISO class 5 critical work zone be placed within a buffer area (also called a buffer room or cleanroom) whose air quality must meet a minimum of ISO class 7 and contain air-conditioning and humidity controls. Adjacent to the buffer area, the sterile compounding pharmacy design must incorporate an ante area for storage, hand washing, nonsterile preparation activities, donning and doffing of protective overgarments, etc. The air cleanliness in the ante area must be a minimum of ISO class 8. The ante area and buffer area constitute secondary engineering controls. Low-risk preparations that are nonhazardous and destined for administration within 12 h of compounding are granted an exemption from these secondary engineering controls if they are prepared within an ISO class 5 PEC and the compounding area is segregated from noncompounding areas. Pharmacy designers should note that the ISO class 5, 7, and 8 air cleanliness requirements are specified for dynamic conditions (USP 2008). Although ASHRAE *Standard* 170 does not prescribe a design temperature for healthcare pharmacies, USP 797 recommends a maximum temperature of 68°F because of the increased thermal insulation that results from wearing protective clothing and the adverse sterility conditions that could arise from uncomfortably warm and/or sweaty employees.

Beyond air quality requirements, the physical design features separating the buffer area from the ante area are based on the pharmacy's compounded sterile preparation (CSP) risk level (low, medium, or high) for microbial, chemical, and physical contamination. USP 797 instructs pharmacy professionals on determination of their pharmacy's CSP risk level based on purity and packaging of source materials, quantity and type of pharmaceutical, time until its administration, and various other factors. The desired CSP risk level capability should be identified before initiating the pharmacy design layout. Pharmacies intended for compounding high-risk-level CSPs require a physical barrier with a door to separate the buffer room from the anteroom and the buffer room is to be maintained at a minimum positive pressure differential of 0.02 in. of water. For medium- and low-risk level CSPs, the buffer area and ante area can be in the same room, with an obvious line of demarcation separating the two areas and with the demonstrable use of displacement airflow, flowing from the buffer area towards the ante area. Depending on the affected cross-sectional area and the moderately high velocity required to maintain the displacement uniformity (typically 40 fpm or greater), designers may find the physical barrier design to be a more energy-friendly approach. USP further prescribes areas to receive a minimum of 30 air changes per hour (with up to 15 of these provided by the PEC) if the area is designated to be ISO class 7. There is no minimum ventilation requirement prescribed for ISO class 8 ante areas (USP 2008).

Selecting pharmacy PECs can be a delicate task. Class II BSCs are currently certified following the construction and performance guidelines developed by the National Sanitation Foundation (NSF) and adopted by the American National Standards Institute (NSF/ANSI *Standard* 49-2009). However, no such national certification program exists for compounding isolators. USP 797 addresses this shortcoming by referencing isolator testing and performance guidelines developed by the Controlled Environment Testing Association (CETA 2006).

Hazardous Drugs. Compounding of hazardous drugs is another pharmaceutical operation that requires special design considerations. A NIOSH (2004) Alert warned of the dangers of occupa-

Table 4 Minimum Environmental Control Guidance for Pharmacies

Compounding Scenario	Hazardous Drug (HD) (Requires separate area)	Nonhazardous Drug
Sterile compounding to be administered within 12 h	ISO 5 CACI or BSC within negative-pressure ISO 7 buffer + ISO 7 ante areas	If immediate use and low risk: no environmental requirements if administered <12 h + ISO 5 PEC within segregated compounding area
Sterile compounding to be administered after 12 h or more	ISO 5 CACI or BSC within negative-pressure ISO 7 buffer + ISO 7 ante areas	ISO 5 PEC + ISO 7 buffer + ISO 8 ante areas - High-risk compounding requires physical barrier with min. positive pressure (0.02 in. of water) in buffer room relative to anteroom - Medium- and low-risk compounding may use physical barrier (as per high risk) or a clearly identified line of demarcation between buffer and ante areas with uniform displacement airflow (min. of 40 fpm recommended) in direction of buffer to ante areas
Nonsterile compounding	Needs compounding containment isolator or BSC	No sterility or occupational exposure controls required

*For facilities that prepare a low volume of hazardous drugs and use two tiers of containment (e.g., CSTD within CACI or BSC), a negative-pressure buffer area is not required.

tional exposures to hazardous drugs, over 130 of which were defined and identified; roughly 90 of these drugs were antineoplastic agents primarily used during cancer treatments. In addition, the NIOSH Alert provides protective recommendations, several of which can affect the ventilation design and physical layout of the pharmacy. These recommendations include the following:

- Prepare hazardous drugs in an area devoted to that purpose alone and restricted to authorized personnel.
- Prepare hazardous drugs inside a ventilated cabinet designed to prevent hazardous drugs from being released into the work environment.
- Use a high-efficiency particulate air (HEPA) filter for exhaust from ventilated cabinets and, where feasible, exhaust 100% of the filtered air to the outdoors, away from outdoor air intakes or other points of entry.
- Place fans downstream of HEPA filters so that contaminated ducts and plenums are maintained under negative pressure.
- Do not use ventilated cabinets [BSCs or compounding aseptic containment isolators (CACIs)] that recirculate air inside the cabinet or that exhaust air back into the pharmacy unless the hazardous drug(s) in use will not volatilize (evaporate or sublimate) while they are being handled or after they are captured by the HEPA filter. [*Note*: This recommendation is a shift from traditional pharmacy design practice and involves knowledge of the physical properties of drugs within the current drug formulary as well as future new drugs that might be compounded within the cabinet. Within-cabinet recirculation (e.g., BSC class II Type A2 or B1) is allowed when airstream has zero or only minute vapor drug contaminant.]
- Store hazardous drugs separately from other drugs, in an area with sufficient general exhaust ventilation to dilute and remove any airborne contaminants. Depending on the physical nature and quantity of the stored drugs, consider installing a separate, high-volume, emergency exhaust fan capable of quickly purging airborne contaminants from the storage room in the event of a spill, to prevent airborne migration into adjacent areas.

The American Society of Health Systems Pharmacists (ASHP 2006) *Guidelines on Handling Hazardous Drugs* adopted the protective equipment recommendations presented in the NIOSH Alert. In addition, hazardous drug compounding be done in a contained, negative-pressure environment or one that is protected by an airlock or anteroom (ASHP 2006).

Often, the hazardous drugs previously discussed also require sterile compounding. If so, pharmacies must have an environment suitable for both product sterility and worker protection. ASHP (2006), NIOSH (2004), and USP (2008) all address these dual objectives by recommending the use of BSCs or compounding aseptic containment isolators. The precautionary recommendations regarding in-cabinet recirculation and cabinet-to-room recirculation of air potentially contaminated with hazardous drugs still apply. In addition, USP 797 requires hazardous drug sterile compounding to be conducted within a negative-pressure compounding area and to be stored in dedicated storage areas with a minimum of 12 air changes per hour of general exhaust. When CACIs are used outside of an ISO 7 buffer area, the compounding area must maintain a negative pressure of 0.01 in. of water and also have a minimum of 12 air changes per hour.

Table 4 provides a matrix of design and equipment decision logic based on USP 797 and the NIOSH Alert on Hazardous Drugs.

Administration

This department includes the main lobby and admitting, medical records, and business offices. Admissions and waiting rooms harbor potential risks of transmitting of undiagnosed airborne infectious diseases. Local exhaust systems that move air toward the admitting patient should be considered. A separate air-handling system is considered desirable to segregate this area from the hospital proper because it is usually unoccupied at night. When interior architectural open-water features are proposed, water treatment to protect occupants from infectious or irritating aerosols should be provided.

Diagnostic and Treatment

Bronchoscopy, Sputum Collection, and Pentamidine Administration Areas. These spaces are remarkable due to the high potential for large discharges of possibly infectious water droplet nuclei into the room air. Although the procedures performed may indicate the use of a patient hood, the general room ventilation should be increased and preferably maintained under negative pressure under the assumption that higher than normal levels of airborne infectious contaminants will be generated.

Magnetic Resonance Imaging (MRI) Rooms. These rooms should be treated as exam rooms in terms of temperature, humidity, and ventilation. However, special attention is required in the control room because of the high heat release of computer equipment; in the exam room, because of the cryogens used to cool the magnet.

Treatment Rooms. Patients are brought to these rooms for special treatments that cannot be conveniently administered in the patients' rooms. To accommodate the patient, who may be brought from bed, the rooms should have individual temperature and humidity control. Temperatures and humidities should correspond to those specified for patients' rooms.

Physical Therapy Department. The cooling load of the electrotherapy section is affected by the shortwave diathermy, infrared, and ultraviolet equipment used in this area.

Hydrotherapy Section. This section, with its various water treatment baths, is generally maintained at temperatures up to 80°F. The potential latent heat buildup in this area should not be overlooked. The exercise section requires no special treatment; temperatures and humidities should be within the comfort zone. The air may be recirculated within the areas, and an odor control system is suggested.

Occupational Therapy Department. In this department, spaces for activities such as weaving, braiding, artwork, and sewing require no special ventilation treatment. Air recirculation in these areas using medium-grade filters in the system is permissible.

Larger hospitals and those specializing in rehabilitation may offer patients a greater diversity of skills to learn and craft activities, including carpentry, metalwork, plastics, photography, ceramics, and painting. The air-conditioning and ventilation requirements of the various sections should conform to normal practice for such areas and to the codes relating to them. Temperatures and humidities should be maintained in the comfort zone.

Inhalation Therapy Department. This department treats pulmonary and other respiratory disorders. The air must be very clean, and the area should have a positive pressure relative to adjacent areas.

Workrooms. Clean workrooms serve as storage and distribution centers for clean supplies and should be maintained at a positive pressure relative to the corridor.

Soiled workrooms serve primarily as collection points for soiled utensils and materials. They are considered contaminated rooms and should have a negative air pressure relative to adjoining areas. Temperatures and humidities should be in the comfort range.

Sterilizing and Supply

Used and contaminated utensils, instruments, and equipment are brought to this unit for cleaning and sterilization before reuse. The unit usually consists of a cleaning area, a sterilizing area, and a storage area where supplies are kept until requisitioned. If these areas are in one large room, air should flow from the clean storage and sterilizing areas toward the contaminated cleaning area. The air pressure relationships should conform to those indicated in Table 3. Temperature and humidity should be within the comfort range.

The following guidelines are important in the central sterilizing and supply unit:

- Insulate sterilizers to reduce heat load.
- Amply ventilate sterilizer equipment closets to remove excess heat.
- Where ethylene oxide (ETO) gas sterilizers are used, provide a separate exhaust system with terminal fan (Samuals and Eastin 1980). Provide adequate exhaust capture velocity in the vicinity of sources of ETO leakage. Install an exhaust at sterilizer doors and over the sterilizer drain. Exhaust aerator and service rooms. ETO concentration sensors, exhaust flow sensors, and alarms should also be provided. ETO sterilizers should be located in dedicated unoccupied rooms that have a highly negative pressure relative to adjacent spaces and 10 air changes per hour. Many jurisdictions require that ETO exhaust systems have equipment to remove ETO from exhaust air. See OSHA 29 CFR, Part 1910.
- Maintain storage areas for sterile supplies at a relative humidity of no more than 50%.

Service

Service areas include dietary, housekeeping, mechanical, and employee facilities. Whether these areas are conditioned or not, adequate ventilation is important to provide sanitation and a wholesome environment. Ventilation of these areas cannot be limited to exhaust systems only; provision for supply air must be incorporated into the design. Such air must be filtered and delivered at controlled temperatures. The best-designed exhaust system may prove ineffective without an adequate air supply. Experience has shown that reliance on open windows results only in dissatisfaction, particularly during the heating season. Air-to-air heat exchangers in the general ventilation system offer possibilities for sustainable operation in these areas.

Supply connections for water-using equipment (e.g., ice machines) should be copper rather than plastic tubing and should be provided with floor drains or floor sinks. Avoid designs that may include dead-end risers or branches without fixtures; in renovation projects, dead-end piping should be removed. Empty or upsized risers or branches are viable options.

Dietary Facilities. These areas usually include the main kitchen, bakery, dietitian's office, dishwashing room, and dining space. Because of the various conditions encountered (i.e., high heat and moisture production and cooking odors), special attention in design is needed to provide an acceptable environment. Refer to Chapter 33 for information on kitchen facilities.

The dietitian's office is often located within the main kitchen or immediately adjacent to it. It is usually completely enclosed for privacy and noise reduction. Air conditioning is recommended for maintaining normal comfort conditions.

The dishwashing room should be enclosed and minimally ventilated to equal the dishwasher hood exhaust. It is not uncommon for the dishwashing area to be divided into a soiled area and a clean area. In such cases, the soiled area should be kept at a negative pressure relative to the clean area.

Ventilation of the dining space should conform to local codes. The reuse of dining space air for ventilation and cooling of food preparation areas in the hospital is suggested, provided the reused air is passed through filters with a filtration efficiency of MERV 13 or better. Where cafeteria service is provided, serving areas and steam tables are usually hooded. The air-handling capacities of these hoods should be sized to accommodate exhaust flow rates (see Table 2 in Chapter 33). Ventilation systems for food preparation and adjacent areas should include an interface with hood exhaust controls to assist in maintaining pressure relationships.

Kitchen Compressor/Condenser Spaces. Ventilation of these spaces should conform to all codes, with the following additional considerations: (1) 350 cfm of ventilating air per compressor horsepower should be used for units located within the kitchen; (2) condensing units should operate optimally at 90°F maximum ambient temperature; and (3) where air temperature or air circulation is marginal, combination air- and water-cooled condensing units should be specified. It is often worthwhile to use condenser water coolers or remote condensers. Heat recovery from water-cooled condensers should be considered.

Laundry and Linen Facilities. Of these facilities, only the soiled linen storage room, the soiled linen sorting room, the soiled utility room, and the laundry processing area require special attention.

The room for storing soiled linen before pickup by commercial laundry is odorous and contaminated, and should be well ventilated and maintained at a negative air pressure.

The soiled utility room is provided for inpatient services and is normally contaminated with noxious odors. This room should be exhausted directly outside by mechanical means.

In the laundry processing area, equipment such as washers, flatwork ironers, and tumblers should have direct overhead exhaust to reduce humidity. Such equipment should be insulated or shielded whenever possible to reduce the high radiant heat effects. A canopy over the flatwork ironer and exhaust air outlets near other heat-producing equipment capture and remove heat best. Air supply inlets should be located to move air through the processing area toward the heat-producing equipment. The exhaust system from flatwork ironers and tumblers should be independent of the general exhaust system and equipped with lint filters. Air should exhaust above the roof or where it will not be obnoxious to occupants of other areas. Heat reclamation from the laundry exhaust air may be desirable and practicable.

Where air conditioning is contemplated, a separate supplementary air supply, similar to that recommended for kitchen hoods, may be located near the exhaust canopy over the ironer. Alternatively, spot cooling may be considered for personnel confined to specific areas.

Mechanical Facilities. The air supply to boiler rooms should provide both comfortable working conditions and the air quantities required for maximum combustion of the particular fuel used. Boiler and burner ratings establish maximum combustion rates, so the air quantities can be computed according to the type of fuel. Sufficient air must be supplied to the boiler room to supply the exhaust fans as well as the boilers.

At workstations, the ventilation system should limit temperatures to 90°F effective temperature. When ambient outdoor air temperature is higher, indoor temperature may be that of the outdoor air up to a maximum of 97°F to protect motors from excessive heat.

Maintenance Shops. Carpentry, machine, electrical, and plumbing shops present no unusual ventilation requirements. Proper ventilation of paint shops and paint storage areas is important because of fire hazard and should conform to all applicable codes. Maintenance shops where welding occurs should have exhaust ventilation.

CONTINUITY OF SERVICE AND ENERGY CONCEPTS

Zoning

Zoning (using separate air systems for different departments) may be indicated to (1) compensate for exposures because of orientation or for other conditions imposed by a particular building configuration, (2) minimize recirculation between departments, (3) provide flexibility of operation, (4) simplify provisions for operation on emergency power, and (5) conserve energy.

By ducting the air supply from several air-handling units into a manifold, central systems can achieve a measure of standby capacity. When one unit is shut down, air is diverted from noncritical or intermittently operated areas to accommodate critical areas, which must operate continuously. This or other means of standby protection is essential if the air supply is not to be interrupted by routine maintenance or component failure.

Separating supply, return, and exhaust systems by department is often desirable, particularly for surgical, obstetrical, pathological, and laboratory departments. The desired relative balance in critical areas should be maintained by interlocking supply and exhaust fans. Thus, exhaust should cease when supply airflow is stopped in areas otherwise maintained at positive or neutral pressure relative to adjacent spaces. Likewise, supply air should be deactivated when exhaust airflow is stopped in spaces maintained at a negative pressure.

Heating and Hot-Water Standby Service

The number and arrangement of boilers should be such that when one boiler breaks down or is temporarily taken out of service for routine maintenance, the capacity of the remaining boilers is sufficient to provide hot-water service for clinical, dietary, and patient use; steam for sterilization and dietary purposes; and heating for operating, delivery, birthing, labor, recovery, intensive care, nursery, and general patient rooms. However, reserve capacity may not be required in warmer climates, depending on individual facility building systems characteristics and operational requirements. Some codes or authorities do not require reserve capacity in climates where a design dry-bulb temperature of 25°F is equaled or exceeded for 99.6% of the total hours in any one heating period as noted in the tables in Chapter 27 of the 2009 *ASHRAE Handbook—Fundamentals*.

Boiler feed, heat circulation, condensate return, and fuel oil pumps should be connected and installed to provide both normal and standby service. Supply and return mains and risers for cooling, heating, and process steam systems should be valved to isolate the various sections. Each piece of equipment should be valved at the supply and return ends.

Some supply and exhaust systems for delivery and operating room suites should be designed to be independent of other fan systems and to operate from the hospital emergency power system in the event of power failure. Operating and delivery room suites should be ventilated such that the hospital retains some surgical and delivery capability in cases of ventilating system failure.

Boiler steam is often treated with chemicals that cannot be released in the air-handling units serving critical areas. In this case, a clean steam system should be considered for humidification.

Mechanical Cooling

The source of mechanical cooling for clinical and patient areas in a hospital should be carefully considered. The preferred method is to use an indirect refrigerating system using chilled water or antifreeze solutions. When using direct refrigerating systems, consult codes for specific limitations and prohibitions. Refer to ASHRAE *Standard* 15.

Insulation

All exposed hot piping, ducts, and equipment should be insulated to maintain energy efficiency of all systems and protect building occupants. To prevent condensation, ducts, casings, piping, and equipment with outside surface temperature below ambient dew point should be covered with insulation having an external vapor barrier. Insulation, including finishes and adhesives on the exterior surfaces of ducts, pipes, and equipment, should have a flame spread rating of 25 or less and a smoke-developed rating of 50 or less, as determined by an independent testing laboratory in accordance with NFPA *Standard* 255, as required by NFPA *Standard* 90A. The smoke-developed rating for pipe insulation should not exceed 150 (DHHS 1984).

Linings in air ducts and equipment should meet the erosion test method described in Underwriters Laboratories *Standard* 181. These linings, including coatings, adhesives, and insulation on exterior surfaces of pipes and ducts in building spaces used as air supply plenums, should have a flame spread rating of 25 or less and a smoke developed rating of 50 or less, as determined by an independent testing laboratory per ASTM *Standard* E84.

Duct linings should not be used in systems supplying operating rooms, delivery rooms, recovery rooms, nurseries, burn care units, or intensive care units, unless terminal filters of at least MERV 14 efficiency are installed downstream of linings. Duct lining should be used only for acoustical improvement; for thermal purposes, external insulation should be used.

When existing systems are modified, asbestos materials should be handled and disposed of per applicable regulations.

Energy

Health care is an energy-intensive, energy-dependent enterprise. Hospital facilities are different from other structures in that they operate 24 h/day year-round, require sophisticated back-up systems in case of utility shutdowns, use large quantities of outdoor air to combat odors and to dilute microorganisms, and must deal with problems of infection and solid waste disposal. Similarly, large quantities of energy are required to power diagnostic, therapeutic, and monitoring equipment; and support services such as food storage, preparation, and service and laundry facilities. Control strategies such as supply air temperature reset on variable-air-volume systems and hydronic reheat supply water temperature reset on variable pumping systems can often be applied with good results. Resources to help ensure efficient, economical energy management as well as reducing energy consumption in hospital facilities include ASHRAE *Standard* 90.1 and the *Advanced Energy Design for Small Hospital and Healthcare Facilities* (ASHRAE 2009). ASHRAE Standard Project Committee 189.2 is currently developing standards for design, construction, and operation of high-performance, green healthcare facilities.

Hospitals conserve energy in various ways, such as by using larger energy storage tanks and by using energy conversion devices that transfer energy from hot or cold building exhaust air to heat or cool incoming air. Heat pipes, runaround loops, and other forms of heat recovery are receiving increased attention. Solid waste incinerators, which generate exhaust heat to develop steam for laundries and hot water for patient care, are becoming increasingly common. Large health care campuses use central plant systems, which may include thermal storage, hydronic economizers, primary/secondary pumping, cogeneration, heat recovery boilers, and heat recovery incinerators.

The construction design of new facilities, including alterations of and additions to existing buildings, strongly influences the amount of energy required to provide services such as heating, cooling, and lighting. Selecting building and system components for effective energy use requires careful planning and design. Integrating building waste heat into systems and using renewable energy sources (e.g., solar under some climatic conditions) will provide substantial savings (Setty 1976).

Testing, Adjusting, and Balancing (TAB)

For existing systems, testing before the start of construction, preferably during design, can be a good investment. This early effort provides the designer with information on actual system performance and whether components are suitable for intended modifications, as well as disclosing additional modifications.

The importance of TAB for modified and new systems before patient occupancy cannot be overemphasized. Health care facilities require validation and documentation of system performance characteristics. Often, a combination of TAB with commissioning satisfies this requirement. See Chapters 38 and 43 for information on TAB and commissioning.

OUTPATIENT HEALTH CARE FACILITIES

An outpatient health care facility may be a free-standing unit, part of an acute care facility, or part of a medical facility such as a medical office building (clinic). Any surgery is performed without anticipation of overnight stay by patients (i.e., the facility operates 8 to 10 h per day).

If physically connected to a hospital and served by the hospital's HVAC systems, spaces within the outpatient health care facility should conform to requirements in the section on Hospital Facilities. Outpatient health care facilities that are totally detached and have their own HVAC systems may be categorized as diagnostic clinics, treatment clinics, or both.

DIAGNOSTIC CLINICS

A diagnostic clinic is a facility where patients are regularly seen on an ambulatory basis for diagnostic services or minor treatment, but where major treatment requiring general anesthesia or surgery is not performed. Diagnostic clinic facilities should be designed according to criteria shown in Tables 4 and 5 (see the section on Nursing Home Facilities).

TREATMENT CLINICS

A treatment clinic is a facility where major or minor procedures are performed on an outpatient basis. These procedures may render patients incapable of taking action for self-preservation under emergency conditions without assistance from others (NFPA *Standard* 101).

Design Criteria

The system designer should refer to the following paragraphs from the section on Hospital Facilities:

- Infection Sources and Control Measures
- Air Quality
- Air Movement
- Temperature and Humidity
- Pressure Relationships and Ventilation

- Smoke Control

Air-cleaning requirements correspond to those in Table 1 for operating rooms. A recovery area need not be considered a sensitive area. Infection control concerns are the same as in an acute care hospital. Minimum ventilation rates, desired pressure relationships and relative humidity, and design temperature ranges are similar to the requirements for hospitals shown in Table 3.

The following departments in a treatment clinic have design criteria similar to those in hospitals:

- Surgical: operating, recovery, and anesthesia storage rooms
- Ancillary
- Diagnostic and Treatment
- Sterilizing and Supply
- Service: soiled workrooms, mechanical facilities, and locker rooms

Continuity of Service and Energy Concepts

Some owners may desire standby or emergency service capability for the heating, air-conditioning, and service hot water systems and that these systems be able to function after a natural disaster.

To reduce utility costs, use energy-conserving measures such as recovery devices, variable air volume, load shedding, or devices to shut down or reduce ventilation of certain areas when unoccupied. Mechanical ventilation should take advantage of outdoor air by using an economizer cycle, when appropriate, to reduce heating and cooling loads.

The subsection on Continuity of Service and Energy Concepts in the section on Hospital Facilities includes information on zoning and insulation that applies to outpatient facilities as well.

NURSING FACILITIES

Nursing facilities may be classified as follows:

Extended care facilities are for recuperation of hospital patients who no longer require hospital facilities but do require the therapeutic and rehabilitative services of skilled nurses. This type of facility is either a direct hospital adjunct or a separate facility having close ties with the hospital. Clientele may be of any age, usually stay from 35 to 40 days, and usually have only one diagnostic problem.

Skilled nursing homes care for people who require assistance in daily activities; many of them are incontinent and nonambulatory, and some are disoriented. Residents may come directly from their homes or from residential care homes, are generally elderly (with an average age of 80), stay an average of 47 months, and frequently have multiple diagnostic problems.

Residential care homes are generally for elderly people who are unable to cope with regular housekeeping chores but have no acute ailments and are able to care for all their personal needs, lead normal lives, and move freely in and out of the home and the community. These homes may or may not offer skilled nursing care. The average length of stay is four years or more.

Functionally, these buildings have five types of areas that are of concern to the HVAC designer: (1) administrative and support areas inhabited by the staff, (2) patient areas that provide direct normal daily services, (3) treatment areas that provide special medical services, (4) clean workrooms for storing and distributing clean supplies, and (5) soiled workrooms for collecting soiled and contaminated supplies and for sanitizing nonlaundry items.

DESIGN CONCEPTS AND CRITERIA

Controlling bacteria levels in nursing homes is not as critical as it is in acute care hospitals. Nevertheless, the designer should be aware of the necessity for odor control, filtration, and airflow control between certain areas.

Table 5 lists recommended filter efficiencies for air systems serving specific nursing home areas. Table 6 lists recommended minimum ventilation rates and desired pressure relationships for certain areas in nursing homes.

Recommended interior winter design temperature is 75°F for areas occupied by patients and 70°F for nonpatient areas. Provisions for maintenance of minimum humidity levels in winter depend on the severity of the climate and are best left to the judgment of the designer. Where air conditioning is provided, the recommended interior summer design temperature and humidity is 75°F and 50% rh.

The general design criteria in the sections on Heating and Hot-Water Standby Service, Insulation, and Energy for hospital facilities apply to nursing home facilities as well.

APPLICABILITY OF SYSTEMS

Nursing homes occupants are usually frail, and many are incontinent. Though some occupants are ambulatory, others are bedridden, suffering from the advanced stages of illnesses. The selected HVAC system must dilute and control odors and should not cause drafts. Local climatic conditions, costs, and designer judgment determine the extent and degree of air conditioning and humidification. Odor may be controlled with large volumes of outdoor air and some form of heat recovery. To conserve energy, odor may be controlled with activated carbon or potassium permanganate-impregnated activated alumina filters instead.

Temperature control should be on an individual-room basis. In geographical areas with severe climates, patients' rooms should have supplementary heat along exposed walls. In moderate climates (i.e., where outside winter design conditions are 30°F or above), heating from overhead may be used.

DENTAL CARE FACILITIES

Institutional dental facilities include reception and waiting areas, treatment rooms (called operatories), and workrooms where supplies are stored and instruments are cleaned and sterilized; they may include laboratories where restorations are fabricated or repaired.

Many common dental procedures generate aerosols, dusts, and particulates (Ninomura and Byrns 1998). The aerosols/dusts may contain microorganisms (both pathogenic and benign), metals (such as mercury fumes), and other substances (e.g., silicone dusts, latex allergens, etc.). Some measurements indicate that levels of bioaerosols during and immediately following a procedure can be extremely high (Earnest and Loesche 1991). Lab procedures have been shown to generate dusts and aerosols containing metals. At this time, only limited information and research are available on the level, nature, or persistence of bioaerosol and particulate contamination in dental facilities.

Nitrous oxide is used as an analgesic/anesthetic gas in many facilities. The design for controlling nitrous oxide should consider that nitrous oxide (1) is heavier than air and may accumulate near the floor if air mixing is inefficient, and (2) should be exhausted directly outside. NIOSH (1996) includes recommendations for the ventilation/exhaust system.

Table 5   Filter Efficiencies for Central Ventilation and Air-Conditioning Systems in Nursing Facilities

Area Designation	Minimum Number of Filter Beds	Filter Efficiency of Main Filter Bed, MERV*
Resident care, treatment, diagnostic, and related areas	1	14
Food preparation areas and laundries	1	7
Administrative, bulk storage, and soiled holding areas	1	6

*Ratings based on ASHRAE *Standard* 52.2; MERV = minimum efficiency reporting value

Table 6   Pressure Relationships and Ventilation of Certain Areas of Nursing Facilities

Function Area	Pressure Relationship to Adjacent Areas	Minimum Air Changes of Outdoor Air per Hour Supplied to Room	Minimum Total Air Changes per Hour Supplied to Room	All Air Exhausted Directly to Outdoors	Air Recirculated Within Room Units
Resident Care					
Resident room (holding room)	*	2	4	Optional	Optional
Resident corridor	*	Optional	2	Optional	Optional
Toilet room	Negative	Optional	10	Yes	No
Resident gathering (dining, activity)	*	2	4	Optional	Optional
Diagnostic and Treatment					
Examination room	*	2	6	Optional	Optional
Physical therapy	Negative	2	6	Optional	Optional
Occupational therapy	Negative	2	6	Optional	Optional
Soiled workroom or soiled holding	Negative	2	10	Yes	No
Clean workroom or clean holding	Positive	2	4	Optional	Optional
Sterilizing and Supply					
Sterilizer exhaust room	Negative	Optional	10	Yes	No
Linen and trash chute room	Negative	Optional	10	Yes	No
Laundry, general	*	2	10	Yes	No
Soiled linen sorting and storage	Negative	Optional	10	Yes	No
Clean linen storage	Positive	Optional	2	Yes	No
Service					
Food preparation center	*	2	10	Yes	Yes
Warewashing room	Negative	Optional	10	Yes	Yes
Dietary day storage	*	Optional	2	Yes	No
Janitor closet	Negative	Optional	10	Yes	No
Bathroom	Negative	Optional	10	Yes	No
Personal services (barber/salon)	Negative	2	10	Yes	No

*Continuous directional control not required

REFERENCES

AIA. 2006. *Guidelines for design and construction of hospital and health care facilities.* The American Institute of Architects, Washington, D.C.

AIHA. 2003. Laboratory ventilation. ANSI/AIHA *Standard* Z9.5. American Industrial Hygiene Association, Fairfax, VA.

ASHP. 2006. ASHP guidelines on handling hazardous drugs. *American Journal of Health-System Pharmacy* 63:1172-1193.

ASHRAE. 2003. *HVAC design manual for hospitals and clinics.* ASHRAE Special Project 91

ASHRAE. 2009. *Advanced energy design guide for small hospitals and healthcare facilities.*

ASHRAE. 2010. Safety code for mechanical refrigeration. ANSI/ASHRAE *Standard* 15-2010.

ASHRAE. 2007. Method of testing general ventilation air-cleaning devices for removal efficiency by particle size. ANSI/ASHRAE *Standard* 52.2-2007.

ASHRAE. 2010. Ventilation for acceptable indoor air quality. ANSI/ASHRAE *Standard* 62.1-2010.

ASHRAE. 2010. Energy standard for buildings except low-rise residential buildings. ANSI/ASHRAE/IES *Standard* 90.1-2010.

ASHRAE. 2008. Ventilation of health care facilities. ANSI/ASHRAE/ASHE *Standard* 170-2008.

ASTM. 2001. Standard test method for surface burning characteristics of building materials. ANSI/ASTM *Standard* E84. American Society for Testing and Materials, West Conshohocken, PA.

Burch, G.E. and N.P. Pasquale. 1962. *Hot climates, man and his heart.* C.C. Thomas, Springfield, IL.

CDC. 1994. *Guidelines for preventing the transmission of* Mycobacterium tuberculosis *in health-care facilities, 1994.* U.S. Dept. of Health and Human Services, Public Health Service, Centers for Disease Control and Prevention, Atlanta.

CDC. 2005. Guidelines for preventing the transmission of *Mycobacterium tuberculosis* in health-care settings, 2005. *Morbidity and Mortality Weekly Report (MMWR)* 54(RR-17).

CETA. 2006. *Compounding isolator testing guide CAG-002-2006.* Controlled Environment Testing Association (CETA), Raleigh, NC.

Degenhardt, R.A. and J.F. Pfost. 1983. Fume hood design and application for medical facilities. *ASHRAE Transactions* 89(2B):558-570.

Demling, R.H. and J. Maly. 1989. The treatment of burn patients in a laminar flow environment. *Annals of the New York Academy of Sciences* 353:294-259.

DHHS. 1984. Guidelines for construction and equipment of hospital and medical facilities. *Publication* HRS-M-HF, 84-1. U.S. Department of Health and Human Services, Washington, D.C.

Earnest, R. and W. Loesche. 1991. Measuring harmful levels of bacteria in dental aerosols. *Journal of the American Dental Association* 122:55-57.

FGI. 2010. *Guidelines for design and construction of health care facilities.* Facilities Guidelines Institute, American Society for Healthcare Engineering of the American Hospital Association, Chicago.

Fitzgerald, R.H. 1989. Reduction of deep sepsis following total hip arthroplasty. *Annals of the New York Academy of Sciences* 353:262-269.

Greene, V.W., R.G. Bond, and M.S. Michaelsen. 1960. Air handling systems must be planned to reduce the spread of infection. *Modern Hospital* (August).

Gressel, M.G. and R.T. Hughes. 1992. Effective local exhaust ventilation for controlling formaldehyde exposures during embalming. *Applied Occupational and Environmental Hygiene* 7(12):840-845.

Hagopian, J.H. and E.R. Hoyle. 1984. Control of hazardous gases and vapors in selected hospital laboratories. *ASHRAE Transactions* 90(2A):341-353.

Isoard, P., L. Giacomoni, and M. Peyronnet. 1980. *Proceedings of the 5th International Symposium on Contamination Control*, Munich (September).

Lewis, J.R. 1988. Application of VAV, DDC, and smoke management to hospital nursing wards. *ASHRAE Transactions* 94(1):1193-1208.

Luciano, J.R. 1984. New concept in French hospital operating room HVAC systems. *ASHRAE Journal* 26(2):30-34.

Michaelson, G.S., D. Vesley, and M.M. Halbert. 1966. The laminar air flow concept for the care of low resistance hospital patients. Paper presented at the annual meeting of American Public Health Association, San Francisco (November).

Murray, W.A., A.J. Streifel, T.J. O'Dea, and F.S. Rhame. 1988. Ventilation protection of immune compromised patients. *ASHRAE Transactions* 94(1):1185-1192.

NFPA. 2011. Standard on fire protection for laboratories using chemicals. ANSI/NFPA *Standard* 45-2011. National Fire Protection Association, Quincy, MA.

NFPA. 2009. Standard for the installation of air conditioning and ventilation systems. ANSI/NFPA *Standard* 90A-2009. National Fire Protection Association, Quincy, MA.

NFPA. 2009. Recommended practice for smoke-control systems. ANSI/NFPA *Standard* 92A-2009. National Fire Protection Association, Quincy, MA.

NFPA. 2005. Standard for health care facilities. ANSI/NFPA *Standard* 99-2005. National Fire Protection Association, Quincy, MA.

NFPA. 2009. Life safety code®. ANSI/NFPA *Code* 101-2009. National Fire Protection Association, Quincy, MA.

NFPA. 2006. Standard method of test of surface burning characteristics of building materials. ANSI/NFPA *Standard* 255-2006. National Fire Protection Association, Quincy, MA.

Ninomura, P.T. and G. Byrns. 1998. Dental ventilation theory and applications. *ASHRAE Journal* 40(2):48-52.

NIOSH. 1975. Development and evaluation of methods for the elimination of waste anaesthetic gases and vapors in hospitals. NIOSH *Criteria Document* 75-137. National Institute for Occupational Safety and Health, Cincinnati, OH.

NIOSH. 1996. Controls of nitrous oxide in dental operatories. NIOSH *Criteria Document* 96-107 (January). National Institute for Occupational Safety and Health, Cincinnati, OH.

NIOSH. 2004. Preventing occupational exposure to antineoplastic and other hazardous drugs in health care settings. DHHS (NIOSH) *Publication* 2004-165. Department of Health and Human Services and National Institute for Occupational Safety and Health, Cincinnati, OH.

NRC. 1980. *Regulatory guide* 10.8. Nuclear Regulatory Commission.

NSF. 2010. Biosafety cabinetry: Design, construction, performance, and field certification. ANSI/NSF *Standard* 49-2010. National Sanitation Foundation and American National Standards Institute, Ann Arbor, MI.

OSHA. [Annual] *Occupational exposure to ethylene oxide.* OSHA 29 CFR, Part 1910. U.S. Department of Labor, Washington, D.C.

Pfost, J.F. 1981. A re-evaluation of laminar air flow in hospital operating rooms. *ASHRAE Transactions* 87(2):729-739.

Rousseau, C.P. and W.W. Rhodes. 1993. HVAC system provisions to minimize the spread of tuberculosis bacteria. *ASHRAE Transactions* 99(2):1201-1204.

Samuals, T.M. and M. Eastin. 1980. ETO exposure can be reduced by air systems. *Hospitals* (July).

Setty, B.V.G. 1976. Solar heat pump integrated heat recovery. *Heating, Piping and Air Conditioning* (July).

UL. 2005. Factory-made air ducts and air connectors, 10th ed. ANSI/UL *Standard* 181. Underwriters Laboratories, Northbrook, IL.

USP. 2008. Pharmaceutical compounding sterile preparations, 31st ed., Ch. 797. United States Pharmacopeial Convention, Rockville, MD.

Walker, J.E.C. and R.E. Wells. 1961. Heat and water exchange in the respiratory tract. *American Journal of Medicine* (February):259.

Wells, W.F. 1934. On airborne infection. Study II: Droplets and droplet nuclei. *American Journal of Hygiene* 20:611.

Woods, J.E., D.T. Braymen, R.W. Rasussen, G.L. Reynolds, and G.M. Montag. 1986. Ventilation requirement in hospital operating rooms—Part I: Control of airborne particles. *ASHRAE Transactions* 92(2A):396-426.

BIBLIOGRAPHY

DHHS. 1984. Energy considerations for hospital construction and equipment. *Publication* HRS-M-HF, 84-1A. U.S Department of Health and Human Services, Washington, D.C.

Gustofson, T.L. et al. 1982. An outbreak of airborne nosocomial *Varicella. Pediatrics* 70(4):550-556.

Rhodes, W.W. 1988. Control of microbioaerosol contamination in critical areas in the hospital environment. *ASHRAE Transactions* 94(1):1171-1184.

compressor speed would be independent of engine speed. These systems could be used for hybrid, electric, or fuel-cell vehicles.

REFERENCES

ACGIH. 2010. *2010 TLVs® and BEIs®*. American Council of Governmental Industrial Hygienists, Cincinnati.

ASHRAE. 2010. Thermal environmental conditions for human occupancy. ANSI/ASHRAE *Standard* 55-2010.

ASHRAE. 2010. Ventilation for acceptable indoor air quality. ANSI/ASHRAE *Standard* 62.1-2010.

Atkinson, W. 2000. Designing mobile air conditioning systems to provide occupant comfort. SAE *Paper* 2000-01-1273. Society of Automotive Engineers, Warrendale, PA.

Atkinson, W. 2008. Interior climate control committee activities. Mobile AC Climate Protection Partnership Meeting, U.S. Environmental Protection Agency, Washington, D.C.

Bang, S. 2008. Flammability evaluation: R-134a, HFO-1234yf and CO^2. Mobile AC Climate Protection Partnership Meeting, U.S. Environmental Protection Agency, Washington, D.C. Available from http://www.epa.gov/cpd/mac/4 Bang.pdf.

Chanfreau, M. and A. Farkh. 2003. The need for an electrical water valve in a thermal management intelligent system (ThemisSt). SAE *Paper* 2003-01-0274. Society of Automotive Engineers, Warrendale, PA.

DOT. 1994. Windshield defrosting and defogging systems—Passenger cars, multipurpose vehicles, trucks, and buses. Federal Motor Vehicle Safety *Standard* (FMVSS) 103. U.S. Department of Transportation, National Highway Traffic Safety Administration, Washington, D.C.

Guan, Y., M.H. Hosni, B.W. Jones, and T.P. Gielda. 2003a. Investigation of human thermal comfort under highly transient conditions for automobile applications, part 1: Experimental design and human subject testing implementation. *ASHRAE Transactions* 109(2):885-897.

Guan, Y., M.H. Hosni, B.W. Jones, and T.P. Gielda. 2003b. Investigation of human thermal comfort under highly transient conditions for automobile applications, part 2: Thermal sensation modeling. *ASHRAE Transactions* 109(2):898-907.

Guan, Y., M.H. Hosni, B.W. Jones, and T.P. Gielda. 2003c. Literature review of the advances in thermal comfort modeling. *ASHRAE Transactions* 109(2):908-916.

Hauck, A. 2003. PTC air heater with electronic control units—Innovative compact solutions. SAE *Paper* C599/058/2003. Society of Automotive Engineers, Warrendale, PA.

Hill, W. 2008. Industry evaluation of low global warming potential refrigerant HFO-1234yf. Mobile AC Climate Protection Partnership Meeting, U.S. Environmental Protection Agency, Washington, D.C. Available from http://www.epa.gov/cpd/mac/3 Hill.pdf.

Jokar, A., S.J. Eckels, and M.H. Hosni. 2004. Evaluation of heat transfer and pressure drop for the heater-core in an automotive system. *Proceedings of the ASME International Mechanical Engineering Congress*, Anaheim, CA.

Jones, B.W. 2002a. The quality of air in the passenger cabin. *Proceedings of Cabin Health 2002*, International Air Transport Association, Geneva.

Jones, B.W. 2002b. Capabilities and limitations of thermal models for use in thermal comfort standards. *Energy and Buildings* 34(6):653-659.

Koban, M. 2009. HFO-1234yf low GWP refrigerant LCCP analysis. SAE *Paper* 2009-01-0179. Society of Automotive Engineers, Warrendale, PA.

Kontomaris, K. and T.J. Leck. 2009. Low GWP refrigerants for centrifugal chillers. ASHRAE Annual Conference, Louisville, KY, June 20-24.

Kurata, S., T. Suzuki, and K. Ogura. 2007. Double-pipe internal heat exchanger for efficiency improvement in front automotive air conditioning system. SAE *Paper* 2007-01-1523. In Thermal systems & management systems, *Special Publication* SP-2132. Society of Automotive Engineers, Warrendale, PA.

Lustbader, J.A. 2005. Evaluation of advanced automotive seats to improve thermal comfort and fuel economy. *Report* NREL/CP-540-37693. National Renewable Energy Laboratory, Golden, CO. Available from http://www.nrel.gov/vehiclesandfuels/ancillary_loads/pdfs/37693.pdf.

Mackenzie, P.T., P.A. Lebbin, S.J. Eckels, and M.H. Hosni. 2004. The effects of oil in circulation on the performance of an automotive air conditioning system. *Proceedings of the ASME Heat Transfer/Fluids Engineering Summer Conference (HTFED '04)*, Charlotte, NC.

Mathur, G.D. 1998. Performance of serpentine heat exchangers. SAE *Paper* 980057. Society of Automotive Engineers, Warrendale, PA.

Mathur, G.D. 1999a. Investigation of water carryover from evaporator coils. SAE *Paper* 1999-01-1194. Society of Automotive Engineers, Warrendale, PA.

Mathur, G.D. 1999b. Predicting and optimizing thermal and hydrodynamic performance of parallel flow condensers. SAE *Paper* 1999-01-0236. Society of Automotive Engineers, Warrendale, PA.

Mathur, G.D. 2000a. Simulation of thermal and hydrodynamic performance of laminate evaporators. SAE *Paper* 2000-01-0573. Society of Automotive Engineers, Warrendale, PA.

Mathur, G.D. 2000b. Water carryover characteristics from evaporator coils during transitional airflows. SAE *Paper* 2000-01-1268. Society of Automotive Engineers, Warrendale, PA.

Mathur, G.D. 2001. Performance prediction of a laminate evaporator with hydrocarbons as the working fluids. SAE *Paper* 2001-01-1251. Society of Automotive Engineers, Warrendale, PA.

Mathur, G.D. 2002. Experimental investigation to determine the effect of laminated evaporators' tank position on heat transfer and pressure drop. SAE *Paper* 2002-01-1029. Society of Automotive Engineers, Warrendale, PA.

Mathur, G.D. 2003. Psychrometric analysis of the effect of laminate evaporator's tank position. SAE *Paper* 2003-01-0528. Society of Automotive Engineers, Warrendale, PA.

Mathur, G.D. 2004a. Experimental investigation to determine accumulation of lubricating oil in a single tank evaporator with tank at the top at different compressor operating speeds. SAE *Paper* 2004-01-0213. Society of Automotive Engineers, Warrendale, PA.

Mathur, G.D. 2004b. *Vehicle thermal management: Heat exchangers & climate control*. Society of Automotive Engineers, Warrendale, PA.

Mathur, G.D. 2005a. Influence of cowl surface temperature on air conditioning load. SAE *Paper* 2005-01-2058. Society of Automotive Engineers, Warrendale, PA.

Mathur, G.D. 2005b. Performance enhancement of mobile air conditioning system with improved air management for front end. SAE *Paper* 2005-01-1512. Society of Automotive Engineers, Warrendale, PA.

Mathur, G.D. 2006. Experimental investigation to monitor vehicle cabin indoor air quality (IAQ) in the Detroit metropolitan area. SAE *Paper* 2006-01-0269. Society of Automotive Engineers, Warrendale, PA.

Mathur, G.D. 2007a. Experimental investigation to monitor tailpipe emissions entering into vehicle cabin to improve indoor air quality (IAQ). SAE *Paper* 2007-01-0539. In *SAE Transactions*, vol. 116-6.

Mathur, G.D. 2007b. Monitoring build-up of carbon dioxide in automobile cabin to improve indoor air quality (IAQ) and safety. Vehicle Thermal Management Systems, Nottingham, UK, *Paper* 051.

Mathur, G.D. 2008. Field tests to monitor build-up of carbon dioxide in vehicle cabin with AC system operating in recirculation mode for IAQ and safety. SAE *Paper* 2008-01-0829. Society of Automotive Engineers, Warrendale, PA.

Mathur, G.D. 2009a. Measurement of carbon dioxide in vehicle cabin to monitor IAQ during winter season with HVAC operating in OSA mode. SAE *Paper* 2009-01-0542. Society of Automotive Engineers, Warrendale, PA.

Mathur, G.D. 2009b. Field monitoring of carbon dioxide in vehicle cabin to monitor indoor air quality and safety in foot and defrost modes. *Vehicle Thermal Management Systems—VTMS-8*, SAE *Paper* 2009-01-3080. Society of Automotive Engineers, Warrendale, PA.

Mathur, G.D. 2009c. Experimental investigation with cross fluted double pipe suction line heat exchanger to enhance AC system performance. SAE *Paper* 2009-01-0970. *SAE Transactions* 1118-6. Society of Automotive Engineers, Warrendale, PA.

Mathur, G.D. 2010a. Experimental investigation of AC system performance with HFO-1234yf as the working fluid. SAE *Paper* 2010-01-0041. Society of Automotive Engineers, Warrendale, PA.

Mathur, G.D. 2010b. Experimental performance of a parallel flow condenser with HFO-1234yf as the working fluid. SAE *Paper* 2010-01-0047. Society of Automotive Engineers, Warrendale, PA.

Meyer, J.J. 2008. R-1234yf system enhancements and comparison to R-134a. SAE Alternative Refrigerant Symposium, Society of Automotive Engineers, Warrendale, PA.

Meyer, J.J. 2009. Production solutions for utilization of both R-1234yf and R-134a in a single global platform. SAE *Paper* 2009-01-0172. Society of Automotive Engineers, Warrendale, PA.

Minor. B. 2008. HFO-1234yf low GWP refrigerant for MAC applications. Mobile AC Climate Protection Partnership Meeting, U.S. Environmental Protection Agency, Washington, D.C.

CHAPTER 9

JUSTICE FACILITIES

TECHNICAL and environmental factors and considerations for engineers designing HVAC systems that serve justice facilities are presented in this chapter. Most of the information presented is for facilities in the United States; regulations in other parts of the world differ significantly, and the authorities governing these facilities should be consulted directly. Refer to the 2008 *ASHRAE Handbook—HVAC Systems and Equipment* for further information on HVAC systems and equipment mentioned herein, and to other chapters of this volume for various space applications and design considerations.

TERMINOLOGY

The following terms are used throughout this chapter:

Justice Facility. Any building designated for purposes of detention, law enforcement, or rendering a legal judgment.

Cell. A room for confining one or more persons; it may contain a bed for each occupant and a toilet and wash basin.

Holding Cell. A room designed to confine a person for a short period of time; it may or may not contain a bed.

Small Jail. A facility consisting of up to 100 rooms and ancillary areas, designed for confining people.

Large Jail. A facility consisting of more than 100 rooms and ancillary areas, designed for confining people.

Prison. A facility consisting of one or several buildings and ancillary areas surrounded by high walls and/or fences, designed to confine a minimum of 500 people.

Minimum Security. A facility or area within a jail or prison that allows confined people to mix together with little supervision for periods of time during the day.

Medium Security. A facility or area within a jail or prison that allows confined people to mix together with some or total supervision for periods of time during the day.

Maximum Security. A facility or an area within a jail or prison that confines people to their cells with total supervision.

Work Release. A program that allows minimum-security occupants freedom during the day to work outside the facility, but requires them to return for the night.

Courthouse. A facility consisting of courtrooms, judges' chambers/offices, jury rooms, jury assembly rooms, attorney interview rooms, libraries, holding cells, and other support areas.

Police Stations. Facilities housing the various functions of local police departments. They may contain holding cells, evidence storage rooms, weapons storage, locker rooms, offices, conference rooms, interview rooms, and parking garages.

Juvenile Facilities. Also known as **family court** facilities, these facilities are for young offenders. Usually kept separate from adult facilities, they house their own court or hearing rooms, judges' chambers, offices for social workers and parole officers, conference

rooms, waiting areas, classrooms, sleeping rooms, intake areas, libraries, exercise rooms/areas, kitchens, dining areas, and laundry.

Inmate. A person confined to a cell, jail, prison, or juvenile facility.

Correctional Officer. A trained law officer who supervises inmates.

Correctional Officer Facilities. Areas designated for use only by correctional officers, including control rooms, break rooms, locker rooms, and storage rooms.

Inmate Areas. Areas that inmates have access to, with or without supervision, including cells, day rooms, exercise areas, outside areas, and certain ancillary areas.

Day Rooms. A room where confined people can congregate for periods of time outside of their cells during the day under supervision. The room usually contains chairs, tables, TVs, and reading and game materials.

Exercise Areas. Gymnasiums or rooms used for exercise by staff members, and areas designated for use by inmates where they can mix and exercise for short time periods during the day. This inmate area is usually outdoors or has at least one wall or the roof exposed to the outdoors.

Ancillary Areas. Support areas, including offices, kitchens, laundry, mechanical rooms/plants, electrical rooms/plants, libraries, classrooms, and rooms for exercise, health care, visitation, interviews, records, evidence, storage, fingerprinting, lineups, inmate intake, etc.

Control Room. A room that allows viewing or monitoring of various areas of the facility and/or houses electronic or pneumatic controls for door locks, lights, and other functions.

Sally Port. A room or space that encloses occupants or vehicles and allows only one door at a time to open.

Forensic Lab. Laboratory where human remains and physical evidence are examined and tested to determine whether a crime has been committed, and to identify bodies and people.

GENERAL SYSTEM REQUIREMENTS

Outside Air. All areas require outside air for ventilation to provide good air quality and makeup air for exhaust systems, and to control pressures within facilities. Minimum outside air requirements for various areas in justice (correctional) facilities can be found in publications of the American Correctional Association (ACA) and in ASHRAE *Standard* 62.1.

Equipment Locations. Access to mechanical equipment and controls must be kept secure from inmates at all times. Equipment rooms should also be located where inmates do not have access to them. Where inmates do have access, security ceilings with lockable access panels must be used when mechanical equipment and components must be located in ceiling plenums. Equipment serving areas not accessible to inmates can be located as in other facilities, unless the owner has other specific requirements. Equipment near noise-sensitive areas (e.g., courtrooms, jury rooms, attorney interview rooms) should be isolated with vibration isolators and have sound attenuation devices on supply, return, and exhaust ducts;

The preparation of this chapter is assigned to TG 9.JF, Justice Facilities.

penetrations for ducts and pipes to those areas and out of mechanical areas should be sealed for sound as well as fire protection.

Security Barriers. Where ducts or openings pass into or out of secure areas, and at exterior intakes and exhausts, security barrier bars are usually installed in ducts or openings that are at least 4 in. high and 6 in. wide. Barrier bars are usually solid steel bars or heavy-gage tubes mounted in a heavy-gage steel frame to match the duct or opening size. Space between bars or tubing must not exceed 5 in. They must be installed as an assembly in a structural wall compartment whenever possible, much like a fire damper. Barrier locations should be coordinated with the facility's owner. Include the bars in static pressure calculations for airflow systems.

Air Devices. Grilles and registers are usually security-type devices constructed of heavy-gage steel and welded or built in place in the walls or ceilings of secure areas accessible to inmates, and are designed to reduce entry of obstacles into the grilles. Locations of these devices in secure areas should be coordinated with the facility's owner. Air devices serving areas not accessible to inmates may be standard grilles, registers, and diffusers. Standard diffusers may also be installed in secure areas with ceilings over 15 ft above the floor.

Outside Air Intakes and Exhausts. Louvers and grilles associated with intake and exhaust air should be located at or above the roof level, and/or (1) where inmates do not have access to them and (2) where substances cannot be discharged into them to harm or disrupt services and personnel in the facility. Barrier bars are usually installed at these devices.

Filtration and Ultraviolet (UV) Lights. Most areas in justice facilities use pleated throwaway filters with a minimum efficiency reporting value (MERV; see ASHRAE *Standard* 52.2) of at least 8. Higher-efficiency filters, such as HEPA or MERV 14 filters, may be required for clinic areas and isolation cells, and UV lights may also be installed to reduce bacteria and the spread of disease. Grease filters must be installed in kitchen exhaust hoods over cooking surfaces. In lieu of bringing large amounts of outdoor air into the facility, normal outdoor air quantities may be saved by installing gasphase or carbon filters in recirculated air streams. Discuss filter applications with the owner and authorities having jurisdiction (AHJ). For more information on filters, see Chapter 28 of the 2008 *ASHRAE Handbook—HVAC Systems and Equipment*.

Energy Considerations

Some areas of justice facilities (e.g., cells, day rooms) are occupied 24 h/day year-round and require a large amount of outside air that is subsequently exhausted. Methods to recover exhausted tempered air and reduce the energy needed to cool and heat the outside intake air include the following:

- Enthalpy wheels for sensible and latent heat recovery or heat exchangers may be used in air-handling systems with high ventilation loads, or as required by ASHRAE *Standard* 90.1.
- Runaround heat recovery coil loops may be used when exhaust and supply airstreams are separated.
- Thermal storage is available for heating and cooling.
- Variable-speed drives may be used on cooling towers, fans, pumps, supply and exhaust fans, and chillers.
- Variable-air-volume systems may be used in office spaces and other areas not requiring constant airflow.
- Supply temperature reset based on outside air temperatures may be used on heating and cooling systems.
- Economizer cycles may be used when outside air temperature and humidity meet indoor condition requirements.
- Heat captured from boiler stacks can preheat combustion air or makeup water.
- Free-cooling heat exchangers provide cooling water by using cooling towers in lieu of the chiller when outside air conditions allow.

- Where reheat is required, water rejected from mechanical cooling or recaptured heat sources (e.g., from laundries) may offer economical paybacks.
- Smaller local systems may be installed to serve areas that are occupied at all times or operate seasonally, so that larger equipment may be shut off at certain times. Modular systems allow various modules to be staged on and off as needed to serve the same purpose.
- Night and holiday setback temperatures at least 5°F above or below the normal occupied settings, with morning warm-up or cooldown, should be used wherever possible for areas that are not always occupied or have varying occupancies.
- Evaporative cooling systems may be used in arid climates to replace water chillers and/or cooling towers. They may also be used in other regions to provide makeup air for some facilities, such as kitchens and laundries.
- Heat pumps may be used wherever possible. See Chapter 8 in the 2008 *ASHRAE Handbook—HVAC Systems and Equipment* for a discussion of these systems.
- Combined heat and power (CHP) systems may be used in larger facilities. See Chapter 7 in the 2008 *ASHRAE Handbook—HVAC Systems and Equipment* for a discussion of these systems.
- Laundry water recycling system able to reduce water consumption by 50% and save energy by reusing laundry hot water.
- Geothermal loop for remote buildings on prison campus.
- Heat recovery chillers able to capture heat from chiller for reheating or boiler water preheating.
- An intelligent hood exhaust control system for kitchen makeup air units and exhaust fans.

Whatever form of energy recovery is used, all systems should be examined for the rates of return on the cost of implementing and operating the systems.

Heating and Cooling Plants and Mechanical Rooms

Most larger justice facilities have central heating and cooling plants, with mechanical rooms located throughout the facility. Smaller facilities generally use local systems, with mechanical rooms located throughout the facility or with a combination of rooftop units or split systems. For larger facilities, central plants with water chillers, cooling towers, and fuel- or dual-fuel-fired steam or hot-water boilers are normally used to serve air-handling units, fan-coil units, reheat coils, and other equipment in mechanical rooms throughout the complex. Primary/secondary or variable-speed pumping of hydronic systems should also be considered.

The heating and cooling requirements are for continuous operation while there are occupants. Essential equipment should be backed up with standby units for use during maintenance or equipment failure. In addition, major components may need to be braced for seismic and/or wind restraint to ensure continuous service. For seismic design, HVAC systems and components need to be braced in accordance with local codes and the AHJ; see Chapter 55 and ASHRAE (2000) for details. Zoning of various areas for occupancy times and seasonal changes should be factored into system arrangements and types.

Plants and mechanical rooms should be located in areas not accessible to inmates, unless supervised maintenance and/or operation is performed by inmates. For central plants serving very large facilities, the plant may be located away from the complex (outside the fences or walls). Some of these plants use distribution tunnels from the plant to the various buildings in lieu of direct burial of the piping. Access to these tunnels must be kept secure from inmates. Vertical duct and pipe chases in facilities are usually located adjacent to cell areas, incorporated within plumbing chases, and stacked to connect to the heating, cooling, and ventilating or exhaust equipment. Service to these chases must be from outside the cell areas.

The capabilities of maintenance personnel and the training to be provided them should be considered in selecting the types of system and equipment to be used in the design. The owner and/or maintenance personnel should be consulted to determine the best combination of components, systems, and location of the plants and mechanical rooms for the facility.

Mechanical equipment in central plants and mechanical rooms must have the proper vibration isolation, flexible pipe and duct connections, and duct-mounted sound attenuators (where needed) to prevent transmission of vibration and noise to sensitive spaces. Mechanical rooms may have to be sound-treated with acoustical materials to prevent room noise from being transmitted to adjacent spaces. Equipment types may also have to be considered for noise transmission, such as fan types. See Chapter 48 for vibration and noise applications.

Controls

Controls serving HVAC systems for small facilities can be local and consist of electric, electronic, pneumatic, or a combination of all of these, and may need to be located in lockable control boxes. Controls for larger facilities are usually electronic or a combination of electronic/electric and pneumatic, and are connected to a central, computerized system or building automation system (BAS) so that operators can remotely manage and monitor systems more efficiently. Thermostats and other sensors in or near inmate areas should be inaccessible to inmates (e.g., located in return or exhaust ducts). Control panels should be locked and located within secure areas. All interconnecting wiring and pneumatic tubing should be concealed from inmates and kept secure.

Fire/Smoke Management

All confined occupants of justice facilities need to be kept safe from fire and smoke. Early detection of fires should be considered in all facilities. Installation of fire and smoke detectors should be discussed with the owner. These detectors need to be installed in secure areas or in the units and not be accessible to inmates.

Smoke control systems should also be considered to facilitate evacuation of inmates to safe areas during an emergency, especially if the facility has no other means to evacuate the inmates to secure areas outside the buildings. The owner should be aware of the costs and complexity of smoke management before implementation. See Chapter 53 for information on fire and smoke management.

Tear Gas and Pepper Spray Storage and Exhaust

Tear gas and pepper spray are used to control people during riots and other uprisings by discharging an incapacitating gas that causes their tear ducts to generate tears, blinding them. Tear gas is usually in grenade form or in canisters fired from shotguns, exploding on impact. Pepper spray can be contained in the same forms as tear gas, or in spray containers for use in close quarters. Once discharged, the gas must be evacuated from any enclosed space. Unlike smoke, both tear gas and pepper spray are heavier than air; therefore, exhaust for them usually is not designed as part of the smoke evacuation system, if incorporated into the facility design. Instead, separate, portable exhaust systems are used, such as exhaust fans blowing air toward exit doors and open windows with fans in the windows.

Storage of tear gas and pepper spray containers must follow HAZMAT requirements: the chemicals must be stored for rotation from date of purchase and removal after about three years. Shelving should be ventilated and away from walls. All persons dealing with the chemicals should have immediate access to protective masks. The storage room should be secured and located away from occupied buildings, and exhaust a minimum of 12 air changes per hour from the floor. The room should be kept at negative pressure, and at about 70°F and 50% rh year-round. Supply air should be from the ceiling near the center of the room.

Health Issues

Large prison health facilities, health care areas in large facilities, and some cells used for isolation in small facilities should be designed for negative pressurization to provide isolation from other spaces for inmates with communicable diseases such as tuberculosis (TB). These spaces should have separate, dedicated exhaust systems, alarms, and controls. Application and component requirements should be discussed with the owner. See Chapter 8 for discussions of health care systems and applications.

JAILS, PRISONS, AND FAMILY COURTS

Jails may be a stand-alone structure or part of a larger facility that confines inmates. Some are totally self-supported and have their own kitchen, laundry, intake room, fingerprinting, storage for personal belongings, sally ports, parking garage, central plants, and other support areas. Security may be anything from minimum to maximum, and may include a work release area, as well. Jails may be located within the city limits or outside of the city.

Prisons are large facilities that confine inmates for longer periods of time than jails, and may have all levels of security and fences or walls with guard towers. Prisons are usually totally self-supported and have every facility required to serve its needs in one large or several small buildings, including laundries, kitchens, dining halls, library, gyms, auditoriums, cell blocks, health clinics, offices, interview rooms, visiting areas, storage rooms for personal belongings, sally ports, intake and release areas, isolation cells or areas, central heating and cooling plants, and correctional officer facilities. Prisons are generally located outside of cities and towns.

Family courts or **juvenile detention centers** are similar to jails but house young offenders up to the age of 18. These facilities include courtrooms, judges' chambers and offices, interview rooms, exercise areas, lockable sleeping areas, classrooms, offices for social workers, kitchens, laundries, and other support facilities. Generally, offices, courtrooms, judges' chambers, interview rooms, exercise areas, classrooms, kitchens, and laundry are unoccupied after working hours, so the mechanical systems for these facilities must be able to respond to various occupied hours of operation.

HVAC Design Criteria

- Use outdoor **summer temperature conditions** equal to ASHRAE 1% design dry bulb and mean coincident wet bulb. For outdoor **winter temperature conditions**, use ASHRAE 99% design dry bulb.
- **Indoor air** should be at 74 to 78°F and maximum 50% rh for summer conditions and occupancy, and 72 ± 4°F and 20 to 35% rh for winter, unless otherwise noted.
- For **cells** or **sleeping rooms**, noise levels, use a maximum of 70 dBA (day) and 45 dBA (night), with minimum constant-volume airflows in accordance with ASHRAE *Standard* 62.1. Maintain negative room pressure, especially when the room or cell contains a toilet.
- For **classrooms**, use noise levels between 25 and 30 RC, and airflows in accordance with ASHRAE *Standard* 62.1. See Chapter 7 for discussion of educational facilities.
- **Interview rooms** should have the same noise levels as for classrooms, with minimum airflows of 6 air changes per hour (ach) of supply air with a minimum of 0.06 cfm/ft² of outdoor air through low-noise diffusers and grilles.
- When **guard stations** are separate rooms for observing inmates, minimum airflow must be in accordance with ASHRAE *Standard* 62.1. Airflows should be constant volume, and noise levels should be 35 to 45 RC. This room may be occupied 24 h a day.
- **Control rooms** should be treated as guard stations. Room loads include computer equipment and video monitors, where required. This room is occupied 24 h a day. Provide a back-up

direct-expansion (DX) system to maintain constant required temperature and humidity.

- **Laundries** are usually heated in winter when not used, and well ventilated to remove generated heat but not air conditioned when in use. All air supplied should be exhausted and the room kept at a negative pressure during operating hours. Maximum noise levels are about 45 to 50 RC. Energy conservation measures may include evaporative cooling if tempered air is supplied to the space; discharged warm laundry water may be recovered for preheating makeup water. See ASHRAE *Standard* 62.1 for minimum airflow requirements.
- **Libraries** require close space temperature and humidity control, with 75°F db year-round and 40% (winter) to 50% (summer) humidity. Constant-volume airflow meeting ASHRAE *Standard* 62.1 is required. See Chapter 23 for details on libraries.

System Requirements

HVAC equipment is generally either a triple-deck multizone, constant-volume, or variable-air-volume (VAV) system. Constant-volume and triple-deck multizone systems are usually used in dayrooms, cells or sleeping quarters, and storage areas. VAV systems may be used in administrative offices, interview rooms, and visitor areas. Cells and sleeping areas are usually exhausted to help control odors. Intake rooms are also exhausted, with supply and exhaust air outlets located high and low to sweep the room and help control odors. Systems must be able to adjust to variable loads and occupancy times, and be zoned to meet the requirements of ASHRAE *Standard* 62.1. Back-up equipment should be provided that serve dayrooms, cells, and critical spaces.

Mechanical equipment must be located in perimeter mechanical rooms outside of areas occupied by inmates, but must be accessible for maintenance. Cells, sleeping quarters, and dayrooms are usually provided with maximum-security-type grilles for supply, return, or exhaust.

Jails, family courts, and prisons located within city limits or near neighborhoods need to be concerned about noise from mechanical equipment and inmates transmitted to the outdoors. Louvers for equipment, indoor exercise rooms, or windows where inmates congregate need noise abatement. Equipment needs to be examined not only for noise transmitting into the facility but transmitted to the outdoors. See Chapter 48 for information on noise and vibration control.

Dining Halls

Dining halls are usually located in large facilities. Space loads vary, depending on occupancy schedules for food preparation and eating. Food-warming station loads also need to be included in the space loads. Smaller facilities use a central kitchen to prepare the food, which is then delivered to inmates on trays in warming carts and may then be reheated in ovens just before serving. If ovens are used, they must be included in the local space loads. Latent loads for eating must also be allowed for.

Kitchens

Kitchens are either centrally located and the prepared food is then transferred to the inmates, or they are associated with the dining halls. Kitchens are full service and include pantry, freezers, coolers, ovens, stoves, kettles, fryers, grilles, dishwashers, and exhaust hoods. In many justice facilities, inmates prepare food, so the kitchen should be designed as a secure area. See Chapter 33 and ASHRAE *Standard* 154 for information and requirements on kitchen ventilation.

Guard Stations

Guard stations either are located within the cell and dayroom area, where guards mingle with inmates, or they are remote and enclosed, where guards can observe cells and dayrooms through secure glass windows. Guard stations are staffed while inmates are awake and not in their cells, and may also be staffed during the night if the owner requires it.

Control Rooms

Control rooms use cameras to remotely monitor inmates, and control doors into and around inmate and other secure areas. These rooms are occupied at all times; room loads should include the control panels and video monitors.

Laundries

Laundries are usually located in their own building or separate rooms and contain washing machines, dryers, and pressing machines. The laundries are very warm places and may be tempered with evaporative cooling and outdoor air economizer cycles with full room exhaust. Spot cooling is recommended for personnel at work stations. Warm water discharged from the washing machines may be reused through laundry water recycling systems to reduce water and energy. Laundries usually have steam supplied from the central plant or their own boilers for heating the wash and rinse water and the pressing machines. If inmates work in laundry service, the space should be designed as a secure area.

COURTHOUSES

This section covers courtrooms in civil, bankruptcy, and criminal courthouses, as well as support divisions for judges' chambers, clerk of court, jury rooms, library, fitness center, marshal areas, jail cells, and administrative areas. Courtrooms generally do not have a clear schedule of operation; however, they generally operate between 9:00 AM until approximately noon. Support staff generally work between 8:00 AM and 5:00 PM, except in constant-occupancy spaces such as marshal areas, jail cells, and other administration areas. Jury areas are generally 9:00 AM to 5:00 PM, but may be occupied much longer, depending on the type of trial.

Courthouses (state, federal, and county) should be designed to suit operational hours and fluctuating visitors and staff occupancies for maximum energy conservation and optimum controls. Architectural features in courtrooms are generally above standard conventional design, and often include wood and ornate ceilings, which require both temperature and humidity control.

HVAC Design Criteria

- Use outdoor **summer temperature conditions** equal to ASHRAE 1% design dry bulb and mean coincident wet bulb. For outdoor **winter temperature conditions**, use ASHRAE 99% design dry bulb and mean coincident wet bulb.
- **Indoor air** should be at 74°F and 50% rh for summer conditions and occupancy, and 72°F and 20 to 35% rh for winter.
- If provided, the **smoke purge system** in the courtroom should be activated manually as well as automatically.
- All **openings** carrying piping through the slab or through partitions must be sealed with appropriate fire/smoke-resistive material. All air ducts leading to and from sensitive spaces must be acoustically treated with 2 in. thick, 3 lb/ft^3 density duct lining for at least 12 ft from the supply diffusers or return air intake.
- Design HVAC systems for **optimum flexibility in scheduling** use of courtrooms, chambers, and jury areas.
- All **fresh and exhaust air locations** should be at least 40 ft above grade, or as high as possible to protect against terrorist attack (see Chapter 59 for more information).

System Requirements

HVAC equipment is generally either constant- or variable-volume air systems. The same independent system should be used for courtrooms, judges' chambers, and jury suites. Every courtroom should have an independent system that can maintain the required

temperature and humidity set points in the space, using a thermostat and a humidistat designed to precool before scheduled occupancy. Controls for jury deliberation rooms and judges' chambers should be placed in the conditioned space and be adjustable for variable occupancy.

HVAC systems for courthouses should be zoned to meet fresh-air requirements of ASHRAE *Standard* 62.1. Central plant systems (including chillers, boilers, and air-handling units) should be designed for 24 h operation, intermittent occupancy, and after-hours activity. All VAV terminal units and reheat coils should be located outside courtrooms and deliberation rooms, and should be accessible for maintenance.

Courtrooms/Chambers

The HVAC system serving judges' chambers, courtrooms, and trial jury suites should provide an average occupied temperature of 74°F. The courtroom system zone should allow temperature sensors to be reset from the building automation system to precool to 70°F before scheduled occupancy. Humidity sensors should maintain minimum relative humidity at 20% (winter) to 50% rh (summer).

Provide a minimum of 6 ach for rooms with ceiling heights up to 15 ft, and 8 ach for rooms with higher ceilings. Systems should be designed to meet these requirements when spaces are fully occupied. These airflows should be reduced during long unoccupied hours, at night, and on weekends and holidays.

Each courtroom should be served by a dedicated fan system, and return air from each courtroom and associated areas (jury rooms, judge's chambers, etc.) must be ducted directly back to the unit or system.

Jury Facilities

Trial jury suites should be served from the same system as the associated courtrooms. (A separate temperature and humidity control for each trial jury room is desirable.)

Air distribution systems must provide separate temperature control and a high degree of acoustical isolation, particularly in grand jury and trial jury rooms. Return air must be ducted directly back to the unit or exhaust air riser. Ductwork must be treated to meet the acoustical deliberation room design criterion of a maximum of 25 to 35 RC. Before recommending underfloor air distribution, filtration, temperature, distribution, air balancing, and commissioning method should be considered.

In the jury assembly room, deliberation room, and associated toilet rooms, the system must provide 10 ach with 80 to 85% return and exhaust.

Libraries

See the discussion of libraries in the section on Jails, Prisons, and Family Courts.

Jail Cells and U.S. Marshal Spaces (24-h Spaces)

A separate air-handling system tied to the main HVAC system should be able to operate independently after hours. A separate 100% fresh and exhaust air system should be provided to jail cells; it should have security grilles and barrier bars, and negative pressure should be maintained.

Marshal spaces should be treated as normal office areas, except for cell areas and exercise rooms. Contents usually include computer and radio equipment, cells, exercise rooms, gun vaults, and perhaps sleeping rooms, and may be occupied 24 h a day. HVAC systems serving these areas should be separate from other systems.

Fitness Facilities

These facilities should be tied to the 24 h system and have a separate 100% fresh air unit able to dehumidify air to 50°F dp. Exhaust air and heat recovery systems should be provided, and the space should be maintained under negative pressure. See Chapter 5 for more information on Gymnasiums.

Acoustic Performance

Acoustic performance should be a major consideration when selecting HVAC equipment. Systems serving courtrooms and auxiliary spaces should be designed with sound attenuation to provide consistent and acceptable sound levels (25 to 40 RC). This is particularly critical in court facilities that require extensive use of sound and audio/visual (A/V) equipment for recording and presentations. Vibration and acoustic performance should be in accordance with guidance in Chapter 48.

To control noise during all modes of operation and for all load conditions, HVAC systems should be provided with one or more of the following:

- Sound traps and acoustic lining in supply and return or exhaust ductwork
- Low-velocity, low-static-pressure fan systems (pay special attention to fan types for noise levels)
- Special low-noise diffusers

Return air should be ducted, especially in courtrooms and jury rooms. Special attention should be given to location of any partitions extending to the floor above and the acoustical treatment around the penetrations of these partitions.

HVAC equipment, including air-handling units (AHUs) and VAV boxes, should not be located close to courtrooms, jury rooms, and chambers. The minimum distance should be 25 ft between the space and these units. General system design needs to provide appropriate treatment of mechanical supply/return ducts to minimize sound and voice transmission to surrounding areas.

Room criterion (RC) defines the limits that the octave-band spectrum of noise sources must not exceed; for court and jury facilities, it should range from 25 to 40. For sound level maintenance, the courtroom should be served by constant-volume air supply. The system must also support variable outside air requirements and variable cooling loads. Air ducts serving trial and grand jury suites must be lined with 2 in. thick, 3 lb/ft^3 density acoustical absorption material for at least 12 ft from the diffuser or return air intake.

FORENSIC LABS

In forensic labs, physical evidence is examined, autopsies may be performed, human remains are tested and identified, firearms are tested, evidence is stored, all aspects of suspected criminal activity are reviewed to determine whether a crime has been committed, and people are identified from evidence taken at crime scenes.

The labs contain many chemicals, fume hoods, ovens, centrifuges, microscopes, x-ray units, and other laboratory equipment that need to be considered in space loads and ventilation requirements. Some lab equipment is sensitive to changes in temperature and humidity. See Chapter 16 for more information on laboratories.

Forensic labs may be stand-alone facilities or part of other facilities, or specific departments may be separated and located in other facilities. Components may include offices, data rooms, storage rooms, laboratories, autopsy rooms, interview rooms, inspectors offices, mechanical and electrical rooms or central plants, firearms rooms, x-ray rooms, photo developing rooms, and body or evidence drying rooms.

HVAC Design Criteria

- Use outdoor **summer temperature conditions** equal to ASHRAE 1% design db and mean coincident wb. For outdoor **winter temperature conditions**, use ASHRAE 99% design dry bulb and mean coincident wet bulb.

- For **indoor air**, 74°F and 50% rh for summer conditions and occupancy, and 72°F and 20 to 35% rh for winter, depending on user requirements.
- **Autopsy rooms** may require room temperatures as low as 60°F and airflows as high as 15 ach when the medical examiners are suited in heavy or rubber garments during an autopsy, or when odors are especially noticeable. Autopsy rooms must be kept under negative pressure, and air should be exhausted high and low in the space. Specific autopsy sinks may include their own exhaust grilles and need to be exhausted when they are in use. All outdoor air may be required for odor control. Noise levels of 20 to 35 RC may be required, because of recordings made during autopsies.
- **Laboratories** should be kept at 70 to 72°F and at least 20 to 30% rh in winter, and 70 to 74°F and 50% rh in summer. These rooms often contain fume hoods and the supply, room exhaust, and hood exhaust airflows must all be controlled together to keep the space under a negative pressure, although some laboratory operations may require positive pressure. All room air must be exhausted and not recirculated. These spaces contain large equipment loads that produce high sensible and latent heat gains at various times. Systems serving these rooms must be flexible enough to react to these load changes and maintain their room sensor set points. Most labs require year-round cooling and dehumidification with reheat to prevent build-up of excessive humidity under some weather conditions. Clean steam may be required for the fume hoods and humidifiers serving the labs; fume hoods and lab benches may also require specially treated water [e.g., deionized (DI) water], inert gases, and natural gas supplies.
- **Microscope tables** need to be isolated from vibrations from mechanical equipment and from building vibrations; information on vibration isolation can be found in Chapter 48.
- Usually, **forensic labs** are occupied 24 h a day, with a small number of rooms that may only be occupied during normal business hours; equipment should be selected, zoned, and controlled to allow for these various occupancies.

System Requirements

Offices and general storage in forensic laboratories are served by normal air-handling systems and should maintain normal room airflows and temperature and humidity set points. Return air may be recirculated. Offices should be kept at positive pressure to keep odors out and to allow use as makeup air for negative-pressure spaces (e.g., storage areas for formaldehyde, which should be exhausted to the outdoors). Systems serving the offices may be required to have their own independent heating and cooling units or terminal units, because their occupancy schedules may differ from those of the labs and other areas.

HVAC systems in cold climates that use 100% outdoor air may experience maintenance problems with frozen hydronic coils unless glycol is used in the water system or the cooling coils are drained in the winter. If internal face-and-bypass dampers are used, cold air can become stratified and freeze portions of the coils. External or integral coil face-and-bypass dampers often are a better application, because they allow cold air to move around the coil or pass through a tempered coil, which reduces the freezing potential.

Ducted supply and exhaust air systems should be used for labs and autopsy rooms. The amount of negative pressure should be maintained carefully in these rooms, so that doors open without excessive door-opening force, as required for smoke control systems (see Chapter 53).

Variable-volume control of supply and exhaust fans may be provided with variable-frequency drives (VFDs). Control of the VFDs should allow for offset of supply and exhaust airflows and to adjust for variances within the systems, such as for dirty filters.

Fume hood exhaust fans should be located on the roof at least 25 ft from any outdoor air intakes, so they will discharge a vertical plume above and away from the roof. Exhaust fans should not be located inside the occupied space; this can produce a positive air pressure in the exhaust duct downstream of the fan, causing leakage back into the space. Consider using redundant exhaust fans or an $N + 1$ exhaust fan system for critical areas.

Exhaust duct and fan materials need to be checked for corrosion resistance against chemicals from fume hoods and laboratories. Coordinate with laboratory personnel about these fumes. Fiberglass, plastic, stainless steel, or coated galvanized duct materials should all be considered. Internal duct lining should never be used in exhaust ducts. Also, consider requiring the exhaust fan to be explosionproof and spark-resistant if materials exhausted may be explosive.

Care must be taken in locating supply air diffusers in relation to the fume hoods, so that exhaust air flowing into the fume hood is not disturbed and does not create turbulence in front of the hood. Supply air should be introduced slowly in a semilaminar flow pattern away from the hoods.

Intake Air Quality

The quality of outdoor air brought into forensic laboratories should be carefully controlled. Usually, these labs are in urban environments close to traffic, parking garages, industrial areas, emergency generator exhausts, restaurant exhausts, and other contaminants. Also, risks from bioterrorism should be addressed by locating outdoor air intakes where they are inaccessible to the public (see Chapter 59). MERV 13 or 14 final filters may be required for critical lab areas, such as DNA extraction labs, autopsy rooms, and toxicology labs, to prevent cross contamination from other processes in the facility or from other outside air influences.

Firearms Laboratories

Firearms testing labs often contain microscope rooms, firearms and ammunition storage rooms, workbench tool rooms, catalog reference rooms, and researcher offices.

Ballistic shooting ranges are usually kept under negative pressures because of smoke emissions. These rooms should be treated for noise attenuation to prevent noise being transmitted to other spaces and reverberating within the room. Air should be supplied near the shooter's breathing zone and exhausted at or near the bullet trap and downstream of the muzzle of the firearm. Using two-speed supply and exhaust fans or fans with VFDs is recommended, so that fan speed is lower when no shooting is occurring. All exhaust systems should be ducted to roof-mounted fans.

Catalog reference rooms and all offices may operate at different hours than the labs, and may be served by their own HVAC systems or systems with equivalent occupancy schedules. Room temperature and pressures should be the same as for general offices, and return air can be recirculated back to the air-handling system.

Acoustic Performance

Acoustic performance should be a major consideration in selecting HVAC equipment. Systems serving laboratory and autopsy spaces should be designed with sound attenuation to provide consistent and acceptable sound levels (25 to 40 RC). This is particularly critical for autopsy rooms that require extensive use of sound and A/V equipment for recordings. Vibration and acoustic performance should be in accordance with Chapter 48.

To control noise during all modes of operation and for all load conditions, the HVAC systems should be provided with one or more of the following:

- Sound traps and acoustic lining in supply and return ductwork
- Sound trap in exhaust ductwork
- Low-velocity, low-static-pressure fan systems (pay special attention to fan types for noise levels)
- Special low-noise diffusers

Pay special attention to location of any partitions extending to the floor structure above and the acoustical treatment at penetrations of

these partitions to provide sound attenuation around the perimeter of ducts and pipes to prevent noise transmissions.

Critical Spaces

Rooms containing freezers or coolers and critical computer rooms should be served by their own independent cooling systems with emergency power back-up, to ensure operation in the event of an extended power outage.

Room pressure controls and monitors should be provided in critical laboratory areas, autopsy rooms, firearms testing rooms, storage rooms that contain hazardous materials, DNA rooms, evidence vaults, trace evidence rooms, drying rooms, photo developing rooms (darkrooms), and other areas deemed necessary by the owner. Room pressures should be continually maintained by measuring supply and exhaust airflows to the room and varying the supply air rate to maintain a differential from the exhaust airflow rate.

Evidence Vaults. Humidity control and cross-contamination prevention are critical, as is exhausting odors from drugs stored in the vault. The room must be kept under negative pressure at 72 to 74°F and 30% rh in the winter and no more than 50% rh in the summer. Barrier bars must be installed on any duct penetrations for these rooms. Firestopping and combination smoke/fire dampers should be installed at duct penetrations.

Photo Developing. Photographic darkrooms must be kept under negative pressure and exhausted to the outdoors, because of the chemicals stored and used within the room. Exhaust should be located behind developing sinks and counters. Outdoor air should be supplied at a minimum of 0.5 cfm/ft^2, from low-velocity diffusers behind lab personnel. Exhaust ducts and fans must be corrosion-resistant and contain combination fire/smoke dampers and firestopping at the penetrations. Exhaust fans should also be explosionproof and spark-resistant. See Chapter 22 for information on temperatures and humidity levels.

Photo Studios. Photo studios have high heat loads because of their excessive lighting requirements. Systems serving these spaces should be designed to minimize noise and air motion and handle variable loads, because of the occupancy schedule and lighting levels.

Trace Rooms. Trace rooms are laboratories where very small amounts of evidence are examined and tested. Consequently, supply airflows need to be low, laminar flow away from the work surfaces to prevent any disturbance of materials. These rooms should be treated as any other lab for temperature, humidity, pressurization, and airflows.

Drying Rooms. Some forensic labs have rooms where evidence must be dried very slowly, to preserve it. HVAC systems serving these rooms should be separate from other systems to prevent cross contamination; temperatures should be maintained between 75 and 80°F, pressure should be negative, and all air exhausted to the outdoors. Laminar air supply should be introduced into the room and high and low exhaust inlets should be installed. Supply air should receive final HEPA filtration.

Laboratory Information Management Systems (LIMS)

Many labs use a separate laboratory information management system to document temperature, pressure, and humidity levels in critical laboratory spaces for validation and certification purposes. System requirements need to be coordinated with the owner.

Historical data storage and retrieval of selected processes and system events, system documentation, and data should be required. This function should allow report formatting and generation from archived historical data. Typical reports consist of alarm summaries, limit summaries, report time reports, all-points logs, trend listing, time of day start/stop schedules, message summaries, energy logs, and maintenance reports.

An independent commissioning agent should be retained at the beginning of system design and should perform complete, detailed commissioning services, including system start-up services, operation and maintenance training and documentation, control of shop drawings, and operation and maintenance manuals. A validation procedure may also be required to ensure the system's operational effectiveness meets both the design intent and operator's requirements.

BIBLIOGRAPHY

ACA. 2004. *Performance-based standards for adult local detention facilities*, 4th ed. American Correctional Association, Alexandria, VA.

ACA. 1991. *Standards for juvenile detention facilities*, 3rd ed. American Correctional Association, Alexandria, VA.

ACA. 1991. *Standards for small juvenile detention facilities*. American Correctional Association, Alexandria, VA.

ACA. 1991. *Standards for juvenile training schools*, 3rd ed. American Correctional Association, Alexandria, VA.

ASHRAE. 2000. *A practical guide to seismic restraint*.

ASHRAE. 2007. Method of testing general ventilation air-cleaning devices for removal efficiency by particle size. ANSI/ASHRAE *Standard* 52.2-2007.

ASHRAE. 2010. Ventilation for acceptable indoor air quality. ANSI/ASHRAE *Standard* 62.1-2010.

ASHRAE. 2003. Ventilation for commercial cooking operations. ANSI/ASHRAE *Standard* 154-2003.

Linde, J.L. and B.C. Davenport. 1995. HVAC design for minimum-, low-, and medium-security federal correctional facilities. *ASHRAE Transactions* 101(1):919-927.

Tseng, P.C., R. Krout, and D. Stanton-Hoyle. 1995. Energy program of requirements for a new detention center—Energy design criteria for prisons. *ASHRAE Transactions* 101(1):928-943.

CHAPTER 10

AUTOMOBILES

THERMAL systems in automobiles (HVAC, engine cooling, transmission, power steering) have significant energy requirements that can adversely affect vehicle performance. New and innovative approaches are required to provide the customer the desired comfort in an energy-efficient way. In recent years, efficiency of the thermal systems has increased significantly compared to systems used in the early to mid-1990s. Providing thermal comfort in an energy-efficient way has challenged the automotive industry to search for innovative approaches to thermal management. Hence, managing flows of heat, refrigerant, coolant, oil, and air is extremely important because it directly affects system performance under the full range of operating conditions. This creates significant engineering challenges in cabin and underhood thermal management. Optimization of the components and the system is required to fully understand the components' effects on the system. Thus, modeling the components and the system is essential for performance predictions. Simulation of thermal systems is becoming an essential tool in development phase of projects. Durability and reliability are also important factors in design of these systems.

Environmental control in modern automobiles usually consists of one (or two for large cars, trucks, and sport utility vehicles) in-cabin air-handling unit that performs the following functions: (1) heating, (2) defrosting, (3) ventilation, and (4) cooling and dehumidifying (air conditioning). This unit is accompanied by an underhood vapor cycle compressor, condenser, and expansion device. The basic system can be divided into three subsystems: air handling, heating, and refrigeration (cooling). All passenger cars sold in the United States must meet defroster requirements of the U.S. Department of Transportation (DOT) Federal Motor Vehicle Safety *Standard* 103 (FMVSS), so ventilation systems and heaters are included in the basic vehicle design. The most common system today integrates the defroster, heater, and ventilation system. In the United States, the vast majority of vehicles sold today are equipped with air conditioning as original equipment.

DESIGN FACTORS

General considerations for design include cabin indoor air quality (IAQ) and thermal comfort, ambient temperatures and humidity, operational environment of components, airborne contaminants, vehicle and engine concessions, physical parameters, durability, electrical power consumption, cooling capacity, occupants, infiltration, insulation, solar effect, vehicle usage profile, noise, and vibration, as described in the following sections.

Thermal Comfort and Indoor Air Quality (IAQ)

ASHRAE *Standard* 55 provides information on the airflow velocities and relative humidity required to provide thermal comfort. Effective comfort cooling system design in cars must create air movement in the vehicle, to remove heat and occupants' body

effluents and to control moisture build-up. Assuming an effective temperature of 71°F with no solar load at 75°F, 98% of people are comfortable with zero air velocity over their body. If the temperature increases to 81°F, the same number of people are comfortable with an air velocity of 500 fpm. If panel vent outlets can deliver sufficient air velocity to the occupants, comfort can be reached at a higher in-vehicle temperature than with low airflow (Figure 1).

Several modeling manikins for predicting human physiological behavior are described in Guan et al. (2003a, 2003b, 2003c), Jones (2002a, 2002b), and Rough et al. (2005).

During the increasingly common gridlock or stop-and-go conditions, **tailpipe emissions** can make outside air (OA) extremely polluted, and it is important to ensure that passengers' exposures to these gases do not exceed American Conference of Governmental Industrial Hygienists (ACGIH 2010) short- or long-term exposure limits.

Tailpipe emissions include

- Nitrogen oxides (NO_x), which include both nitric oxide (NO) and nitrogen dioxide (NO_2), which always occur together (Pearson 2001)
- Carbon monoxide (CO), which forms in the combustion chamber when oxygen supply is insufficient
- Hydrocarbons (HCs)
- Volatile organic compounds (VOCs)

Diesel engines emit mainly NO_x and HC, and gasoline engines emit mainly CO and HC. Worldwide, road transportation accounts for approximately 50% of NO_x emissions, and gasoline-powered vehicles alone account for 32% of HC emissions in the United States (Pearson 2001).

To limit passengers' exposure to tailpipe emissions, the blower unit's air intake door can be switched from outside air mode to recirculation mode during times of traffic congestion and likely poor OA quality (Mathur 2006). Once the vehicle is out of the traffic jam,

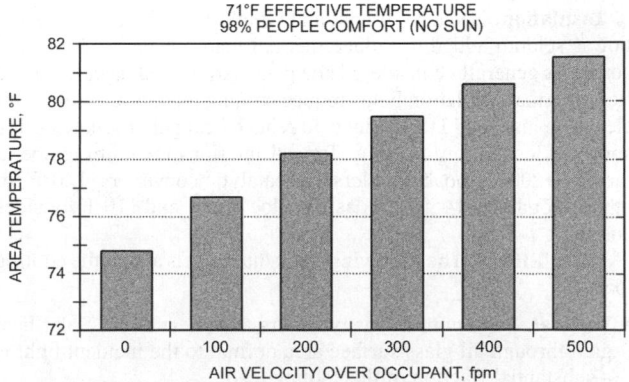

Fig. 1 Comfort as Function of Air Velocity
(Atkinson 2000)

The preparation of this chapter is assigned to TC 9.3, Transportation Air Conditioning.

the mode door can be switched back to outside air mode (Mathur 2007a).

Carbon dioxide (CO_2) from passengers' exhalations can also build up in the cabin, especially in low-body-leakage vehicles, so the vehicle's AC system should not be operated in recirculation mode for extended periods. This issue becomes critical when several occupants are in a vehicle that has 100% return air in recirculation mode. A timed strategy is recommended for recirculation; after the set time (e.g., 30 minutes) elapses, the mode automatically changes to outside to reduce CO_2 levels in the cabin. A CO_2 sensor can be installed to monitor levels in the cabin, and automatically switch to OA mode when set levels are exceeded (Mathur 2007b, 2008, 2009a, 2009b).

Relative humidity also affects cabin IAQ. Too high a level affects occupant comfort and can lead to condensation and fogging on windows. A relative humidity sensor can detect excessive humidity and intervene.

See the section on Controls under Air-Handling Subsystem for more information on cabin IAQ.

Cooling Load Factors

Occupancy. Occupancy per unit volume is high in automotive applications. The air conditioner (and auxiliary evaporators and systems) must be matched to the intended vehicle occupancy.

Infiltration. Like buildings, automobiles are not completely sealed: wiring harnesses, fasteners, and many other items must penetrate the cabin. Infiltration varies with relative wind/vehicle velocity. Unlike buildings, automobiles are intended to create a relative wind speed, and engines may emit gases other than air. Body sealing and body relief vents (also known as the drafter) are part of air-conditioning design for automobiles. Occasionally, sealing beyond that required for dust, noise, and draft control is required.

By design, vehicles are allowed to have controlled body leakage that allows air movement in the vehicle to provide comfort to the passengers. This also helps control moisture build-up and the occupants' perceived comfort level. However, excessive body leakage results in loss of heating and cooling performance. Vehicle body leakage characteristics typically are significantly different in dynamic conditions compared to static conditions. Air can leak from the vehicle's doors, windows, door handles, and trunk seals (uncontrolled exit points); drafters allow a controlled exit for air from the cabin, and should be self-closing to prevent inflow when the body pressure is negative with respect to the exterior pressure. According to the Society of Automotive Engineers (SAE) *Standard* J638, infiltration of untreated air into the passenger compartment through all controlled and uncontrolled exit points should not exceed 350 cfm at a cabin pressure of 1 in. of water (Atkinson 2000). However, each vehicle has different body leakage characteristics. Some vehicles have two drafters inside the trunk on either side, and some have only one.

Insulation. Because of cost and weight considerations, insulation is seldom added to reduce thermal load; insulation for sound control is generally considered adequate. Additional dashboard and floor thermal insulation helps reduce cooling load. Some new vehicles have insulated HVAC ducts to reduce heat gain during cooling and heat loss during heating. Typical interior maximum temperatures are 200°F above mufflers and catalytic converters, 120°F for other floor areas, 145°F for dash and toe board, and 110°F for sides and top.

Solar Effects. The following four solar effects add to the cooling load:

- **Vertical.** Maximum intensity occurs at or near noon. Solar heat gain through all glass surface area normal to the incident light is a substantial fraction of the cooling load.
- **Horizontal and reflected radiation.** Intensity is significantly less, but the glass area is large enough to merit consideration.

- **Surface heating.** Surface temperature is a function of the solar energy absorbed, the vehicle's interior and exterior colors, interior and ambient temperatures, and the automobile's velocity.
- **Vehicle colors and glazing.** The vehicle's interior and exterior colors, along with the window glazing surfaces (clear or tinted), strongly affect vehicle soak temperature. Breathing-level temperatures after a 1 h soak can be 40 to 60°F higher than ambient, with internal surfaces being 50 to 100°F above ambient (Atkinson 2000).

Ambient Temperatures and Humidity. Several ambient temperatures need to be considered. Heaters are evaluated for performance at temperatures from –40 to 70°F. Air-conditioning systems are evaluated from 40 to 110°F, although ambient temperatures above 125°F are occasionally encountered. The load on the air-conditioning system is also a function of ambient humidity (at most test conditions, this latent load is around 30% of the total). Typical design points follow the combinations of ambient temperature and humidities of higher probability, starting at around 90% rh at 90°F and with decreasing humidity as temperature increases.

Because the system is an integral part of the vehicle, the effects of vehicle-generated local heating must be considered. For interior components, the design high temperature is usually encountered during unoccupied times when the vehicle is soaked in the sun. Interior temperatures as high as 190°F are regularly recorded after soaks in the desert southwestern United States. Achieving a comfortable interior temperature after a hot soak is usually one of the design conditions for most vehicle manufacturers.

Operational Environment of Components

Underhood components may be exposed to very severe environments. Typical maximum temperatures can reach 250°F. The drive to achieve more fuel-efficient automobiles has reduced available space under the vehicle hood to a minimum. This crowding exposes many components to temperatures approaching that of exhaust system components. Heat from the vehicle also adds to the cooling loads that the air-conditioning system must handle. During idle, heat convected off the hood can raise the temperature of air entering the air inlet plenum by as much as 10 to 25°F (Mathur 2005a). A similar effect is found during idle when air from the engine compartment is reentrained into the air flowing through the condenser (Mathur 2005b). Air temperatures as high as 160°F have been encountered on parts of a vehicle's condenser during operation with a tailwind in ambient temperatures as low as 100°F. Typically, front air management is improved by using air guides and seals to prevent air bypassing either the condenser or radiator at idle. Significant improvements in vent outlet temperatures (a maximum of 7°F and cabin temperatures of 2 to 6°F) and a reduction in head pressures (30 to 77 psi) have been obtained. Recirculation of hot engine compartment air was reduced from 52.2°F over ambient (base case) to approximately 27°F over ambient. Further details are provided in the section on Vehicle Front-End Design.

Airborne Contaminants and Ventilation

Normal airborne contaminants include bacteria, pollutants, vapors from vehicle fluids, and corrosive agents (Mathur 2006). Exposure to these must also be considered when selecting materials for seals and heat exchangers. Incorporating particulate and/or carbon filters to enhance interior air quality (IAQ) is becoming common. Air-handling systems in virtually all vehicles can exceed the ventilation recommendations for buildings and public transportation in ASHRAE *Standard* 62.1. However, the driver has complete control of the HVAC system in the vehicle, and can reduce cabin airflow to virtually zero when desired (e.g., before warm-up on cold days).

Power Consumption and Availability

Many aspects of vehicle performance have a significant effect on vehicular HVAC systems. Modern vehicles have a huge variety of electric-powered systems. The need to power these systems while maintaining fuel efficiency leads manufacturers to demand a high level of efficiency in electrical power usage. On some vehicles, electrical power use is monitored and reduced during times of minimal availability. The mass of the HVAC system is also closely controlled to maintain fuel efficiency and for ride or handling characteristics. The power source for the compressor is the vehicle's engine. At engagement, the need to accelerate the rotational mass as well as pump the refrigerant can double the engine torque. This sudden surge must not be perceptible to the driver, and is controlled through careful calibration of the engine controls. Automotive compressors must provide the required cooling while compressor speed varies with the vehicle condition rather than the load requirements. Vehicle engine speeds can vary from 500 to 8000 rpm.

Physical Parameters, Access, and Durability

Durability of vehicle systems is extremely important. Hours of operation are short compared to commercial systems (160,000 miles at 40 mph = 4000 h), but the shock, vibration, corrosion, and other extreme conditions the vehicle receives or produces must not cause a malfunction or failure. Automotive systems have some unique physical parameters, such as engine motion, proximity to components causing adverse environments, and durability requirements, that are different from stationary systems. Relative to the rest of the vehicle, the engine moves both fore and aft because of inertia, and in rotation because of torque; this action is referred to as **engine rock**. Fore and aft movement may be as much as 0.5 in.; rotational movements at the compressor may be more than 0.75 in. from acceleration and 0.5 in. from deceleration when the length to center of rotation is considered. Additionally, the need for components to survive bumper impacts of up to 5 mph leads to additional clearance and strength requirements. Vehicle components may also be exposed to many different types of chemicals, such as road salt, oil, hydraulic fluid (brakes and power steering), and engine coolant.

Automobiles also increasingly incorporate electrical and electronic components and functionality. This requires manufacturers to both limit the emissions of electrical signals from components and ensure that all components work when subjected to these same types of emissions. Manufacturers' requirements for electromagnetic compatibility are increasingly stringent regarding the frequencies of radio and communication devices.

Wiring, refrigerant lines, hoses, vacuum lines, and so forth must be protected from exhaust manifold heat and sharp edges of sheet metal. Normal service items such as oil filler caps, power steering filler caps, and transmission dipsticks must be accessible. Air-conditioning components should not have to be removed to access other components.

Noise and Vibration

The temperature control system should not produce objectionable sounds. During maximum heating or cooling operation, a slightly higher noise level is acceptable. Thereafter, it should be possible to maintain comfort at a lower blower speed with an acceptable noise level. Compressor-induced vibrations, gas pulsations, blower motor vibration, and noise must be kept to a minimum. Suction and discharge mufflers are often used to reduce noise. Belt-induced noises, engine torsional vibration, and compressor mounting all require particular attention. Manufacturers have different requirements and test methods. Although it is almost impossible to predict vehicle sound level from component testing, a decrease in the sound and vibration energy at the source of noise always decreases the noise level in vehicle (assuming there is not a shift in frequency), so most automobile manufacturers require continuous improvement in overall component sound level.

Vehicle Front-End Design

Front-end design affects performance of the climate control and engine cooling systems, especially at low speeds and at idle. The design should ensure that air flowing into the front end through the bumper and/or grille does not bypass either the condenser or radiator from the sides, top, or bottom. Air takes the path of least resistance, and if not forced over the heat exchangers, it usually bypasses them. In a good design, the condenser and radiator are the same size, and should not have space between (Mathur 2005b). This eliminates the use of seals between the condenser and radiator. Typically, the front-end module has components in the following sequence: condenser, radiator, and fans (CRF); these systems are known as condenser-radiator-fan modules (CRFM). A good front-end design provides optimum performance for both air-conditioning and engine-cooling systems. Airflow over the front end couples these two systems; thus, performance of one system (e.g., air conditioning) influences the other system (engine cooling). This is most evident at idle.

In a typical design, sheet metal covers the entire area on the sides of the condenser. This prevents air from bypassing from either side of the condenser and radiator. To prevent recirculation of hot engine compartment air at idle, the front bottom part of the front end is usually covered by sheet metal or plastic sheet. To limit recirculation on the top, a seal that sits on the cross frame when the hood is closed is usually added. This prevents recirculation of the hot engine compartment air to approximately the top and bottom thirds of the condenser. Without this, condenser head pressure may increase greatly, further degrading system performance.

Enhanced R-134a systems require substantial changes to hardware and controls to achieve performance and energy requirement targets, although some of the strategies described here can be used to approach the targets.

AIR-HANDLING SUBSYSTEM

The in-cabin air-handling unit, commonly called an **air-conditioning module (ACM)**, provides air to the passenger cabin. It incorporates the following basic components: heater core, evaporator core, blower motor, air-distribution control, ram air control, body vents, and air temperature controls. In addition to the ACM, an air inlet plenum, distribution ducting, outlets, and body relief vents or drafters make up the complete air-handling subsystem. The evaporator core is a part of both the refrigeration and air-handling subsystems and links the two. The heater core is similarly the link with the heating subsystem.

The basic function of the air-handling system is as follows. The air intake valve allows air from either the exterior [taken directly from the air intake plenum or outside air (OA)] or the cabin to be recirculated to the fan. The fan then pumps air through the evaporator and into the temperature control door, which forces the air to either flow through or bypass the heater core to obtain the desired temperature. The air then moves to the distribution area of the module, where it is directed to one or more of the heater, ventilation, or defrost outlets. Air in the cabin then either is recirculated or exits the vehicle through body vents or drafter(s).

There are many variations on the basic ACM system. Common ones include regulating the air discharge temperature using coolant flow control and separating the ACM into two or more subcomponents to better fit the system in the vehicle.

Air Delivery Modes

There are three basic modes in most vehicles: heater, defroster, and air conditioning (or vent for vehicles without air conditioning). Typical mixed modes include bilevel, blend, and ambient.

Heater Mode. Heater mode is designed to provide comfort heating to vehicle occupants. Typical maximum heater airflow is 125 to 200 cfm for a midsized automobile. Heater air is generally distributed into the lower forward (foot) compartment, under the front

seat, and up into the rear compartment. Air distribution near the floor also makes the vehicle more comfortable by providing slightly cooler air at breathing level. Because the supply air temperature is relatively high, direct impingement on the occupant is not desirable. Heater air exhausts through body leakage points.

Heater mode warms air in the vehicle above the dew points of the surrounding air and of the vehicle's glass. To prevent condensation from occupant respiration or from rain or snow tracked in, most vehicles sold in North America draw only OA when in heater mode and do not allow recirculation. However, some vehicle designs do allow recirculation, avoiding the higher cost of including the electric or vacuum actuation system necessary to prevent it.

Most vehicles also provide a small bleed of air (typically 15 to 25% of total airflow) in heater (foot) mode to the windshield to isolate it from the car's interior. Properly designed, this prevents loss of visibility by window fogging under most conditions.

Defrost Mode. Defrost mode is provided to clear the windshield from frost and fog, both internally and externally. Typical maximum airflow for defrost systems is 150 to 200 cfm for a midsized automobile. Defrost mode requirements are given in the DOT's Federal Motor Vehicle Safety *Standard* (FMVSS) 103, which defines areas on the windshield for driver vision and a time frame in which they must be able to be cleared under extreme vehicle operating conditions. Most vehicles are also equipped with side window demisters that direct a small amount of heated air and/or air with lowered dew point to the front side windows. Rear windows are typically defrosted by heating wires embedded in the glass.

As in heater mode, to prevent windshield fogging, most vehicles built in North America prevent air from being recirculated in defrost mode. In addition, many vehicles automatically operate the air-conditioning system in defrost if the ambient temperature is above a threshold (usually around 40°F). This provides an extra assist and safety factor by lowering the dew point of air exiting the ACM to below ambient temperature.

Air-Conditioning (or Panel) Mode. The air-conditioning mode is provided for occupant comfort cooling and to ventilate the vehicle. Typical airflow for panel mode is 200 to 300 cfm in a midsized car. Because of the lower temperature differentials in this mode, airflow is provided in such a way that direct impingement on the occupants can be achieved if desired. A minimum air velocity of 2000 fpm at the outlet is desired, to provide adequate comfort to occupants in the front and rear of the cabin (Atkinson 2000). As discussed in the Design Factors section, the higher heat fluxes and higher initial temperature at vehicle start-up frequently require that the system be able to **spot cool**, providing the cooling airflow directly on the occupants, before lowering the overall cabin temperature. For these reasons, directability of the supply outlet on the occupants is very important. The air-conditioning system is designed to have sufficient capacity to bring the interior temperature down rapidly; panel outlets must also be positionable, to move the airflow off the occupants after a few minutes of operation.

Bilevel Mode. The most common mixed mode, bilevel mode is designed for moderate-temperature operation with high solar loading. The system provides air to both the lower outlets and the panel outlets. Typically, air from the panel outlets is 5 to 25°F cooler than the air from the lower outlets. This is to provide cooling to areas of the interior that have direct solar loading and to provide warm air to those that do not.

Blend Mode. The next most common mixed mode is blend mode, designed to provide a step between heater and defroster for times when extra heat is needed to keep the windshield clear but full defrost is not desired. A typical situation where blend mode is used is in city traffic during snowfall. The extra airflow to the windshield helps maintain a clear field of vision and still maintains adequate flow to lower outlets to keep occupants warm.

Outside Mode. This mode is also designed for mild ambients. It is intended to provide a relatively high total airflow through the

cabin but without the high local air velocities of the other modes. Typically, vehicles with outside mode are also equipped with additional panel outlets not directed toward the occupants. The most common configuration provides air toward the ceiling from outlets in the middle of the dashboard.

Controls

The HVAC control head (i.e., controls for the ACM and refrigeration system) is located within easy reach of the driver and occupants. These controls must be easy to use and not distract the driver from the road. There are many variations, from the cable-controlled manual system to fully automatic systems that control the cockpit environment. The two main classifications are manual and automatic.

Manual control is typically the base system that provides control for mode, temperature valve position, air source, and air flow rate (blower speed). In addition to air-handling controls, the control head usually also has a button to engage the compressor (i.e., to turn on the AC system). Additional functions, such as rear defrost and seat heating controls, are frequently added to the control head. Although manual control provides a temperature mix door control, this is not a temperature control; it only controls the opening of the temperature valve and fixes the amount of air that bypasses the heater core. Therefore, if there is significant variation in ambient temperature or vehicle coolant temperature, the manual system must be adjusted. Manual systems typically have four or five blower speeds.

Automatic control uses a control unit and vehicle sensors to establish a comfortable thermodynamic environment for vehicle occupants. Sensors measure air inlet temperature, vehicle cabin temperature, and ACM discharge air temperature. The automatic control then varies the mix door position, airflow rate, ACM mode, and air-conditioning compressor engagement. Some advanced systems measure cabin humidity for comfort control. Automatic systems usually have from 8 to 20 blower speeds.

Air quality control is also available in many vehicles. Most of these systems assume that a vehicle quickly passes through areas where the contamination source is prevalent. A sensor measures a **surrogate gas** (a gas that is not necessarily toxic but accompanies toxic gases that are more difficult to measure). When the surrogate gas is detected, the vehicle's air inlet door is positioned for recirculation to separate the occupants from the contamination source.

Air-Handling Subsystem Components

Air Inlet Plenum. The air inlet plenum (also called a **cowl**) is usually an integral part of the vehicle structure. There are two primary design considerations and several secondary design considerations for the air inlet plenum:

Primary

- Air that flows into the plenum should not be influenced by uncontrolled emissions from the vehicle systems (i.e., the plenum should be a source of clean air).
- The plenum should be located so that the aerodynamic effects of air movement over the vehicle increase pressure in the plenum, so when the vehicle operates with external air selected, air flows through the air-handling unit into the vehicle. This allows fresh air to flow through the vehicle and helps reduce the amount of external air that infiltrates into the vehicle from uncontrolled sources.

Secondary

- The pressure drop of the plenum should also be considered. Higher airflow pressure requires more power for the ACM blower and fan to provide adequate airflow.
- Airflow at the entrance to the ACM's blower should be uniform. In many vehicle applications, a significant loss in efficiency is caused by unbalanced airflow into the fan.

- The air inlet plenum also serves several other functions, such as water separation, protection from snow ingestion, and gross filtration (usually through a screen).
- The air inlet plenum also be located such that when the vehicle is covered by snow, the plenum still can furnish sufficient air to clear the windshield and provide fresh air to the occupants.

The air inlet plenum is usually located at base of the windshield. If properly sealed from underhood areas of the vehicle, this provides a relatively high-pressure and clean source of air. Major plenum design considerations include the following (Mathur 2005a).

Separation of Water Droplets from Airstream. It is important that openings in the plenum cover be sized carefully. Openings that are too small result in a higher pressure, which reduces airflow and increases noise. Reduced airflow increases window fogging and significantly decreases occupants' perceived comfort. Surface tension can also cause rainwater to plug small openings and get sucked into the plenum when the blower is turned on in OA mode. On the other hand, very large openings can allow snow or sleet inside, where it can accumulate and block the path of airflow. Plenum cover opening sizes should be optimized to address both these issues.

Water droplets follow the air trajectory inside the plenum. Removing the droplets requires changing the airflow direction: because their momentum is greater, the droplets do not change direction but instead hit the sheet metal wall and then drain to the bottom of the plenum channel. Otherwise, filters may become saturated with water. Adding baffles inside the plenum channel can change airflow direction, but also increases air pressure drop, which affects both airflow rate and noise levels. Angling baffles in the flow direction helps alleviate this pressure drop increase.

Expanding the plenum's cross-sectional area is another way of removing water droplets from the airstream, but is not always possible because of space limitations. This is a good approach, though, around the wiper motor and linkages, which are housed inside the plenum channel and significantly reduce airflow area.

Snow Separation. As discussed previously, plenum cover opening size is crucial in keeping precipitation out of the plenum. Even with an optimum cover design, though, accumulated snow must be removed before the blower unit is turned on in OA mode. Otherwise, dry, powderlike snow could enter the plenum and end up on the filter, saturating it and causing fogging issues.

Hard snow over the plenum cover is difficult to remove and significantly reduces airflow when the blower is turned on. As air flows over the openings, some of the ice is directly evaporated into the airstream by sublimation, increasing window fogging. To address this situation, some plenum cover openings in newer cars are under the hood, allowing some airflow into the cabin in this situation. Note, however, that this approach could be lethal in old cars that leak exhaust gases from faulty gaskets under the hood.

Distribution Ducting. Air from the air-handling unit is distributed to various areas of the vehicle through ducting. Typically, the main trunk duct exits the ACM near the middle of the dashboard. Ducting carries air from this central location to the extremes of the instrument panel, the floor, and even the rear seat (if so equipped). The design goal is to distribute air throughout the vehicle with as little pressure drop as possible, to provide sufficient airflow to the various outlets for occupant comfort. This goal is frequently compromised by the tight packaging constraints in modern vehicles. Ducts should be designed with no sharp edges inside the airflow stream, which could increase airflow rush noise.

Outlets. There are typically defrost, heater, side window, and panel air outlets in a vehicle. The defrost air outlet is located on top the instrument panel to distribute air to clear the windshield of frost and fog as quickly and efficiently as possible. Heater outlets are located on the bottom of the instrument panel to spread warm air over the floor of the vehicle. Panel outlets are designed to provide cool air to the occupants. The importance of panel outlets should not be underestimated. The ability to achieve direct air impingement on occupants with little diffusion is very important to comfort after a vehicle has been inoperative during extremely hot summer conditions. Likewise, it is important to be able to direct cool air away from occupants after the interior begins to cool down. The air pressure drop in the vent outlet changes as the direction of the vane or blade is changed, and can result in reduced airflow. This is necessary to direct the airflow over the desired area of the passenger. Being able to direct the jet air and reach the occupants under all conditions can result in satisfied consumers; the lack of this ability has led to dissatisfied consumers even in vehicles with exceptional airflow and capacity.

Body Relief Vents or Drafters. Body relief vents or drafters are designed to ensure airflow through the vehicle from front to rear. The drafters are located inside the trunk, under the carpet, on the sides near the wheel wells. Air flows from the cabin into the trunk through parcel shelf openings (holes that facilitate airflow from cabin to trunk), and then between the sheet metal and carpet to the drafters.

Typically, they are effectively low-pressure check valves, designed to allow airflow out of the vehicle when cabin pressure is above the local exterior pressure and to prevent air infiltration when the local exterior pressure is above that of the interior (i.e., when the vehicle is using recirculated air as the air source). Relief vents should be located where they will cause airflow inside the body to cover all occupant locations inside the vehicle.

A small number of openings in the vehicle body are required for wires, cables, and various attachment features; therefore, the body relief vent does not typically need to be sized large enough to exhaust the total airflow through the vehicle.

Heater Core. The heat transfer surface in an automotive heater is generally either copper/brass cellular, aluminum tube and fin, or aluminum-brazed tube and center. Each of these designs is in production in straight-through, U-flow, or W-flow configurations. The basics of each of the designs are outlined as follows:

- The **copper/brass cellular** design is not used frequently in new vehicles. It uses brass tube assemblies (0.006 to 0.016 in. wall thickness) as the water course, and convoluted copper fins (0.003 to 0.008 in. thick) held together with a lead/tin solder. The tanks and connecting pipes are usually brass (0.026 to 0.034 in. wall thickness) and are attached to the core by a lead/tin solder.
- The **aluminum tube-and-fin** design generally uses round copper or aluminum tubes, mechanically joined to aluminum fins. U tubes can take the place of a conventional return tank. The inlet/outlet tank and connecting pipes are generally plastic and attached to the core with a rubber gasket.
- The **aluminum-brazed tube-and-center** design uses flat aluminum tubes and convoluted fins or centers as the heat transfer surface. Tanks are either plastic and clinched onto the core or aluminum and brazed to the core. Connecting pipes are constructed of various materials and attached to the tanks various ways, including brazing, clinching with an O ring, fastening with a gasket, and so forth. Almost all original equipment manufacturers (OEMs) currently use brazed-aluminum heater cores (Jokar et al. 2004).

Air-side design characteristics include pressure drop and heat transfer. The pressure drop of the heater core is a function of the fin/louver geometry, fin density, and tube density. Capacity is adjusted by varying the face area of the core to increase or decrease the heat transfer surface area, adding coolant-side turbulators, or varying air-side surface geometry for turbulence.

Evaporator. Automotive evaporator materials and construction include (1) copper or aluminum tube and fin; (2) brazed-aluminum plate and fin, also known as a laminate evaporator; and (3) brazed serpentine tube and fin. This section addresses the air-side design of

the evaporator. Air-side design parameters include air pressure drop, capacity, and condensate control.

A laminate evaporator consists of a number of stamped plates and louvered fins. The plates have clad material on both sides. The plates and fins are stacked and then either vacuum-brazed or controlled-atmospheric-brazed (CAB). The advantage of using CAB is that it is a continuous process, whereas vacuum brazing is a batch process. When brazed, the plate forms internal flow passages for refrigerant. The plates have diagonal ribs (or multiple dimples) to augment heat transfer and provide strength, and central partitioning ribs that facilitate reversal of refrigerant flow. These evaporators may have tanks on both ends or on one end only. For the same airflow area, a single-tank evaporator has better performance than a double-tank evaporator, because the available heat transfer area is greater (i.e., the ratio of total heat exchange area to total volume of the core is higher for evaporators with single tanks). Laminate evaporators typically have four to six refrigerant passes. Two-phase refrigerant enters the evaporator through the inlet pipe, and vapor exits the evaporator through the outlet pipe. Two-phase refrigerant enters the evaporator through the tank and moves downward in multiflow channels (or plates) in pass 1 and then flows upward in pass 2. The refrigerant reaches the tank section at the top and then flows downward in pass 3, flows upward in pass 4, and exits the evaporator as vapor (Mathur 2000a, 2001, 2002, 2003).

Typically, an ACM is designed to provide the airflow required for cooling for the vehicle. The combination of airflow, maximum allowable current draw for the blower motor, size constraints on the ACM, and ductwork act together to establish a required evaporator air-side pressure-drop characteristic. The air-side pressure drop of the core is typically a function of fin spacing, louver design, core depth, and face area. This characteristic varies with accumulation of condensate on the core, so adequate leeway must be allowed to achieve target airflow in humid conditions.

Conditions affecting evaporator capacity are different from those in residential and commercial installations in that the average operating time, from a hot-soaked condition, is less than 20 min. Inlet air temperature at the start of operation can be as high as 160°F, but decreases as the vehicle duct system is ventilated. Capacity requirements under multiple conditions must be considered when sizing an automotive evaporator, including steady-state operation at high or low speeds, and a point in a cooldown after an initial vehicle hot soak. Some of these requirements may also be set in recirculating conditions where the temperature and humidity of inlet air decrease as the car interior temperature decreases.

The evaporator load also has a slightly higher sensible heat portion than indicated by ambient temperature. Heat gain from the vehicle and temperature rise across the blower motor must be considered when sizing the evaporator.

During longer periods of operation, the system is expected to cool the entire vehicle interior rather than just produce a flow of cool air. During sustained operation, vehicle occupants want less air noise and velocity, so the air quantity must be reduced; however, sufficient capacity must be preserved to maintain satisfactory interior temperatures.

Condensate management is very important within a motor vehicle. In the process of cooling and dehumidifying the air, the evaporator extracts moisture from the air. It is imperative that liquid condensate be prevented from entering the vehicle interior, because this will damage the vehicle. This moisture should be carried out of the vehicle and not allowed to collect inside the ACM (Mathur 2000b). A distinct odor can be identified in many cars that have plugged condensate drain holes. This odor is given off by common organisms, present almost everywhere, that grow in warm, moist environments.

Condensate management includes the following design objectives (Mathur 1999a):

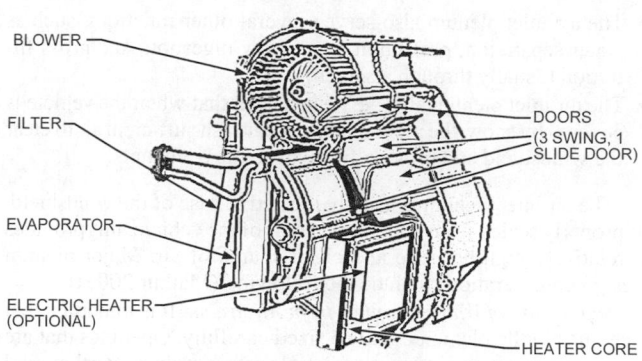

Fig. 2   Integrated HVAC Unit

- Ensure that moisture coming off the evaporator is in large enough droplets that it is not carried by the airstream (a combination of low velocity at the exit of the core and adequate ability of the core to allow surface tension to gather the water)
- Allow sufficient fin spacing for adequate condensate drainage
- Allow a large enough sump so that all water coming off the core can be collected for a short period of time when vehicle-maneuvering forces push water away from the drain
- Provide sufficient slope to the drain area so that water flows to the drain rather than collecting in the case
- Have a sufficient cross section in the drain so that water does not back up into the module, taking into account the fact that ACM interior pressure is usually 1 to 2 in. of water above the exterior pressure

Vehicle attitude (slope of the road and inclines), acceleration, and deceleration should also be considered, because these factors can significantly affect the drain system. Drains can become plugged not only by contaminants but also by road splash.

Location of HVAC Unit. The HVAC unit consists of the blower intake unit, cooling unit, and a heater unit. The system also has several ducts that feed air to different circuits. These units are mounted inside the **cockpit module (CPM)**. U.S.-made vehicles use **modular design**, in which the blower, evaporator, and heater are individual units. **Integrated units** (Figure 2) that combine all three functions into one component have been developed, but their design is complex.

Blower Motor and Fans. Airflow in an automobile must provide sufficient cooling air to passengers in both the front and rear seats. Designs for the blower motor and fan, which must fit in a relatively small space, are frequently a compromise between packaging, mass, airflow, and efficiency. Virtually all fans used in automotive ACMs are centrifugal, with fan diameters from 5.5 to 8 in. Typical motor current draws vary from 14 to 25 A, depending on factors affecting optimization of the particular application. Both forward- and backward-inclined fan blades have been used.

Reducing the noise, vibration, and harshness (NVH) characteristics of the interior has significantly improved comfort levels. During integration of the blower motor and fan into the module, careful attention must be paid to any type of vibrational excitation the fan may impart to the ACM and other underdash components.

Electronic Blower Speed Control. Historically, the blower motor speed control was simply a selector switch that selected either from direct battery voltage at the blower motor or from one of two or more resistors in series with the blower motor to reduce voltage. In general, the lowest airflow selectable is usually driven by the need to provide adequate air pressure to cool the blower motor. Modern blower motor speed controls incorporate essentially infinite speed control using devices such as **pulse-width modulating (PWM)** controllers. The newer devices are usually found on upscale vehicles using automatic climate control systems, which

reduce energy consumption and increase fuel efficiency. When incorporating devices such as PWMs into a vehicle system, careful attention needs to be paid to radio frequency interference because of the necessary proximity of all the electronics in a vehicle.

Valves. The typical ACM has valves for the air inlet source, temperature control (heater air bypass), and mode control. Some vehicles also have a ram air door, used to reduce ram effect at high speeds and provide consistent airflow. Door/valve designs are integral to the ACM design. Door types include flag style, rotary, guillotine, slider, and film valves. The optimal door type is almost always a function of the space in which a module must fit.

Actuators. Actuators on ACMs are usually cable, electric, and/ or vacuum. **Cable-based actuation** is usually the least expensive and is most frequently found in entry-level vehicles. The valving system must be designed to retain a position with minimal restraining torque, have smooth operation with essentially constant torque level, and have minimal torque required to move the valve(s). There must be a suitable cable path from the HVAC controls to the module. Cable actuation does not allow electronic control of the air-conditioning system or an interlock to ensure outside air is selected in defrost mode.

The norm for U.S. automobiles 20 years ago, **vacuum actuation**, has been replaced by electronic actuators. Vacuum actuators can provide only three position controls per actuator and require a cross-sectional area for the diaphragm proportional to the load on the door. The vacuum source is the vehicle's engine intake air manifold. Although this provides powerful control at engine idle, a great deal of variation exists in the working pressure differential, and must be taken into account in system design.

Electric actuators can control the ACM electronically and are available in variety of shapes. The possibility of linear positions allows for multiple modes, with one actuator on several doors, using a cam system. They also isolate the operator from torque variations, allowing the ACM to be optimized for other performance criteria.

Air Inlet. The air inlet interfaces the ACM with the vehicle body. If not accomplished upstream, it is necessary for the air inlet to separate out water from rain, carwashes, etc. It also provides the selection of either outside air or air recirculated from the passenger compartment. On upscale performance vehicles, the ram air door is also located here. A primary design criterion for the air inlet is to provide proper flow patterns at the inlet of the blower motor. In many applications this is compromised to fit the ACM into the vehicle. The result is either turbulence or misdistribution of air into the fan, causing noise and lower efficiency.

Mode Control. Air is usually distributed at the ACM by one or more valves directing air to the desired vehicle outlets. This system may provide several discrete modes or a continuous variation from one mode to the next. ACM valving must be designed to work with the distribution ductwork and provide the desired air distribution to occupants.

Air Distribution. Air must be distributed in a way that minimizes pressure loss, thermal lag, and heat gain. Ductwork is usually designed around other underdash components, and frequently must follow a difficult path. Air for all outlets starts at one basic plenum pressure, and variations in pressure drop versus flow rate from side to side in the vehicle must be minimized to provide even airflow to both driver and passengers. Because of the instrument cluster in front of the driver and devices such as airbags, ductwork is almost never laterally symmetrical. Computational fluid dynamics is used to ensure proper air distribution design.

Air Filter. Air filters are increasingly common, typically located in either the air inlet plenum or the ACM. Filters may be particulate, charcoal, or both; they require regular service to prevent clogging and ensure proper system function. Removal of contaminants (e.g., pollen) may also be aided by condensate on evaporator surfaces. The concentration of the volatile organic compounds (VOCs) from the vehicle's interior (plastic parts, carpet, adhesive, etc.) along with tailpipe emissions (NO_x, CO, hydrocarbons) from automobiles, buses, and trucks can be reduced by using carbon filters. These filters should be replaced regularly, based on driving conditions, because a dirty filter can be the largest source for polluting cabin air.

HEATING SUBSYSTEM

The primary heat source is the vehicle's engine. Coolant from the engine cooling system circulates through the heater core. Modern efficiency and emissions improvements have led to many types of supplemental heating, including fuel-fired heaters, refrigerant heat pumps, electrical heaters, and heat storage systems.

The heater core must be designed to work within the design of the engine cooling system. Engine coolant pressure at the heater core inlet ranges up to 40 psig in cars and 55 psig on trucks.

Modern antifreeze coolant solutions have specific heats from 0.65 to 1.0 Btu/lb·°F and boiling points from 250 to 272°F (depending on concentration) when a 15 psi radiator pressure cap is used.

Controls

Engine coolant temperature is controlled by a thermostatically operated valve that remains closed until coolant temperature reaches 160 to 205°F. Coolant flow is a function of pressure differential and system restriction, but typically ranges from 0.6 gpm at idle to 10 gpm at higher engine speed. Coolant temperature below 160°F is not desirable, because it cannot meet occupants' comfort requirements. The mechanical pump should be able to deliver sufficient coolant flow, even at idle.

Components

The minimal components of the heating subsystem are the coolant flow circuit (water pump) and temperature control, both provided by the vehicle's engine; the heater core (part of the ACM); and coolant hoses.

REFRIGERATION SUBSYSTEM

Cooling is almost universally provided by a vapor cycle system. The thermodynamics of a vapor cycle system are described in Chapter 2 of the 2009 *ASHRAE Handbook—Fundamentals*. The automotive system is unique in several ways.

Refrigeration capacity must be adequate to bring the vehicle interior to a comfortable temperature and humidity quickly and then maintain it during all operating conditions and environments. A design may be established by mathematical modeling or empirical evaluation of known and predicted factors. A design tradeoff in capacity is sought relative to criteria for vehicle weight, component size, and fuel economy. Automotive system components must meet internal and external corrosion, pressure cycle, burst, and vibration requirements.

Refrigerant-based system equipment is designed to meet the recommendations of SAE *Standard* J639, which includes several requirements for refrigerant systems. To be compliant, a system must have

- A high-pressure relief device
- Burst strength (of components subjected to high-side refrigerant pressure) at least 2.5 times the venting pressure of the relief device
- Electrical cutout of the clutch coil before pressure relief to prevent unnecessary refrigerant discharge
- Low-pressure-side components with burst strengths in excess of 300 psi

The relief device should be located as close as possible to the discharge gas side of the compressor, preferably in the compressor itself.

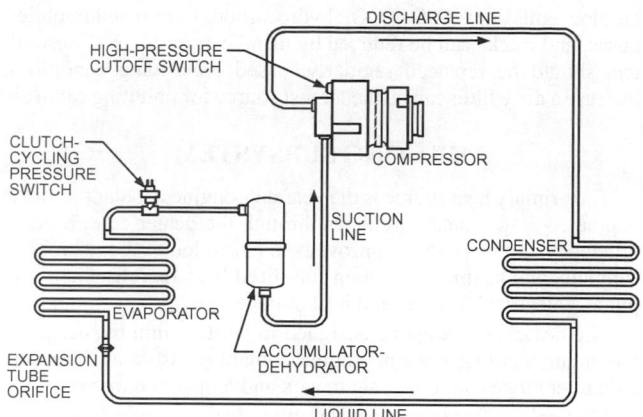

Fig. 3   **Clutch-Cycling System with Orifice Tube Expansion Device**

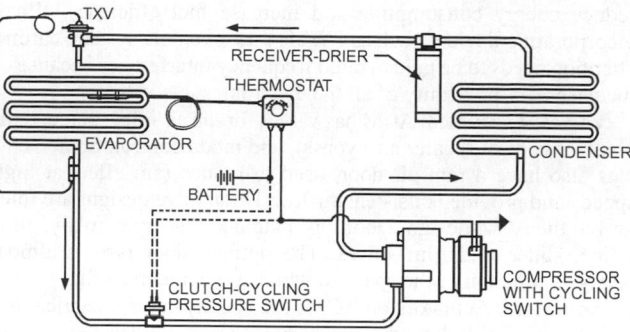

Fig. 4   **Clutch-Cycling System with Thermostatic Expansion Valve (TXV)**

Controls

Refrigerant Flow Control. Cycling-clutch designs are the most common mechanisms for controlling refrigerant flow; schematics for the two most common versions are shown in Figures 3 and 4. The clutch is cycled by either a thermostat that senses evaporator temperature or a pressure switch that senses evaporator pressure. This thermostat or pressure switch serves two functions: it prevents evaporator icing, and maintains a minimum refrigerant density at the compressor's inlet, preventing overheating. Discharge air temperature is then increased, if necessary, by passing some (or all) of the evaporator outlet air through the heater core.

The clutch-cycling switch disengages at about 25 psig and engages at about 45 psig. Thus, the evaporator defrosts on each off-cycle. The flooded evaporator has enough thermal inertia to prevent rapid clutch cycling. It is desirable to limit clutch cycling to a maximum of six cycles per minute because a large amount of heat is generated by the clutch at engagement. The pressure switch can be used with a thermostatic expansion valve in a dry evaporator if the pressure switch is damped to prevent rapid cycling of the clutch.

Cycling the clutch sometimes causes noticeable surges as the engine is loaded and unloaded by the compressor. This is more evident in cars with smaller engines. This system cools more quickly and at lower cost than a continuously running system.

For vehicles where clutch cycling is unwanted because of engine surge, or for high-end vehicles where no perceptible temperature swing is allowable, variable-displacement compressors are available, controlled either electronically or pneumatically.

In the **pneumatically controlled compressor**, a sensor (usually located in the compressor body) varies the compressor displacement so that a constant pressure is maintained at the compressor inlet. This provides a nearly uniform evaporator temperature under varied loading conditions. This type of system causes no perceptible engine surge with air-conditioning system operation.

The **electronically controlled variable-displacement compressor** opens up many possibilities for systems optimization. This type of compressor allows reduced reheat control, and evaporator temperature is maintained at such a level that comfort is achieved with less fuel consumption. A wide range of control schemes using electronically controlled variable-displacement compressors are being developed.

Other Controls. A cycling switch may be included to start an electric fan when insufficient ram air flows over the condenser. Also, output from a pressure switch or transducer may be used to put the ACM in recirculation mode, which reduces head pressure by reducing the load on the evaporator. Other possibilities include a charge loss/low-ambient switch, transducer evaporator pressure control, and thermistor control.

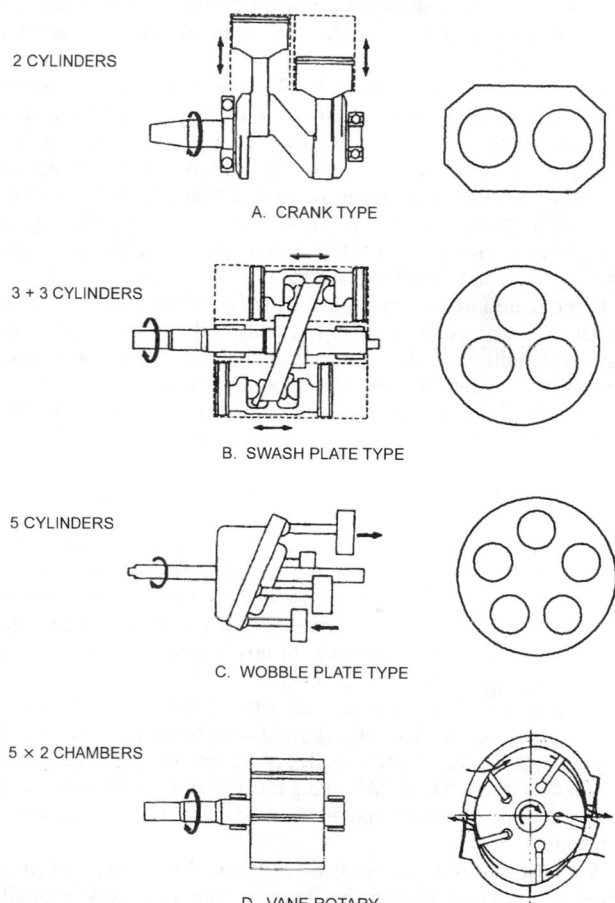

Fig. 5   **Basic Compressor Designs for Automotive Application**

Components

Compressor. Piston compressors dominate the automotive market, although scroll and rotary vane types are also significant. For detailed information on compressor design, see Chapter 37 of the 2008 *ASHRAE Handbook—HVAC Systems and Equipment*. Figure 5 illustrates basic automotive compressor types. The typical automotive compressor has the following characteristics:

- **Displacement.** Fixed-displacement compressors have displacements of 6.1 to 12.6 in³/rev. Variable-displacement piston compressors typically have a minimum displacement of about 6% of

their maximum displacement. A typical variable-capacity scroll compressor has a maximum displacement of 7.3 in³/rev and a minimum displacement of 10% of the maximum.

- **Physical size.** Fuel economy, lower hood lines, and more engine accessories all decrease compressor installation space. These features, along with the fact that smaller engines have less accessory power available, promote the use of smaller compressors.
- **Speed range.** Most compressors are belt-driven directly from the engine; they must withstand speeds of over 8000 rpm and remain smooth and quiet down to 500 rpm. The drive ratio from the vehicle engine to the compressor typically varies from 1:1 to 2:1. In the absence of a variable drive ratio, the maximum compressor speed may need to be higher to achieve sufficient pumping capacity at idle.
- **Torque requirements.** Because torque pulsations cause or aggravate vibration problems, it is best to minimize them. Minimizing peak torque benefits the compressor drive and mount systems. Multicylinder reciprocating and rotary compressors aid in reducing vibration. An economical single-cylinder compressor reduces cost; however, any design must reduce peak torques and belt loads, which are normally at a maximum in a single-cylinder design.
- **Compressor drives.** A magnetic clutch, energized by power from the vehicle engine electrical system, drives the compressor. The clutch is always disengaged when air conditioning is not required. The clutch can also be used to control evaporator temperature (see the section on Controls).
- **Variable-displacement compressors.** Both axial and wobble-plate variable-displacement compressors are available for automobile air conditioning. The angle of the plate changes in response to the suction and discharge pressure to achieve a constant suction pressure just above freezing, regardless of load. A bellows valve or electronic sensor-controlled valve routes internal gas flow to control the plate's angle. A variable-displacement compressor reduces compressor power consumption, improving fuel efficiency. These compressors improve dehumidification and comfort, have low noise and vibration, and have high reliability and efficiencies.
- **Noise, vibration, and harshness (NVH).** With decreasing mass and increasing environmental quality in automobiles, compressor design is increasingly driven by NVH concerns. Vibrational input to the structure, suction and discharge line gas pulsations, and airborne noise must all be minimized. NVH minimization is now the main impetus behind most continuous improvement efforts in the automotive compressor industry.
- **Mounting.** Compressor mounts are an important part of a successful integration of a compressor into a vehicle system. Proper mounting of the compressor minimizes structural resonances and improves the NVH characteristics of any compressor.

Compressor Oil Return. It is important that there are no areas where the lubrication oil can accumulate (Mathur 2004a). At part-load conditions, refrigerant velocities should be high enough to ensure oil return to the compressor. The presence of oil in the system affects heat exchanger performance (Mackenzie et al. 2004). Some new compressors have a built-in oil separator.

Condenser. Automotive condensers are generally of the following designs: (1) tube-and-fin with mechanically bonded fins; (2) serpentine tube with brazed, multilouvered fins; or (3) header extruded tube brazed to multilouvered fins, also known as parallel-flow (PRF) condensers, which are primarily used in automotive applications (Figure 6). To prevent air bypass, condensers generally cover the entire radiator surface. Aluminum is popular for its low cost and weight.

Operation of Parallel-Flow Condenser. A PRF condenser consists of flat tubes that have multiple flow channels. Refrigerant is supplied directly to the tubes through the header. Louvered fins are currently used in automotive heat exchangers. A typical refrigerant

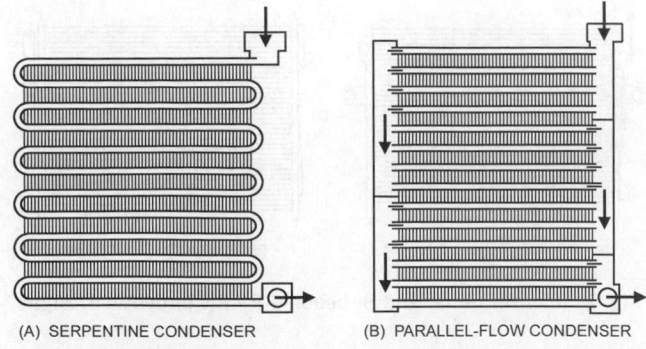

(A) SERPENTINE CONDENSER   (B) PARALLEL-FLOW CONDENSER

Fig. 6   Basic Automotive Condensers
(Mathur 1999b)

tube has 0.08 in. thick wall with tube widths ranging from 0.71 to 0.87 in., with 6 to 12 flow channels; smaller tubes are also available. Flat tubes have less projected frontal area to the airstream, which results in lower air-side pressure drop. Performance of a parallel-flow condenser is superior to that of a serpentine condenser (Mathur 1998), because the refrigerant is distributed in multiple tubes. For the same reason, refrigerant pressure drop in a PRF condenser is also much smaller.

Typically, in a PRF condenser, the first pass (see Figure 6B) has the largest number of refrigerant tubes, with fewer tubes in each successive pass. This is because the specific volume of superheated vapor coming out from the compressor is very large, and the density of refrigerant vapor is very small. This results in very high vapor velocities ($m = \rho AV$) in the tubes. At this condition, the refrigerant void fraction is unity, which results in a very high pressure drop. Therefore, this high volumetric flow must be subdivided into a large number of tubes to lower refrigerant velocities, and thus pressure drop. At some point along the condenser, refrigerant vapor temperature equals saturated temperature, and wall temperature falls below saturation temperature. At this time, condensation starts and the average density of the two-phase refrigerant mixture starts to increase. With the increase of the two-phase mixture density, the average refrigerant velocities start to decrease. This affects the condensation heat transfer coefficient and frictional pressure drop. When all vapors are condensed, the refrigerant flow becomes single-phase liquid. At this condition, the refrigerant flow velocity is lowest, which yields lower pressure drop. Thus, the last pass has the fewest tubes.

Condenser Design. Condensers must be properly sized. An undersized condenser results in high discharge pressures that reduce compressor capacity, increase compressor power requirements, and result in poorer discharge air temperatures. When the condenser is in series with the radiator, the air restriction must be compatible with the engine cooling fan and engine cooling requirements. Generally, the most critical condition occurs at engine idle under high-load conditions. An undersized condenser can raise head pressures sufficiently to stall small-displacement engines.

An oversized condenser may produce condensing temperatures significantly below the engine compartment temperature. This can result in evaporation of refrigerant in the liquid line where the liquid line passes through the engine compartment (the condenser is ahead of the engine and the evaporator is behind it). Engine compartment air has been heated not only by the condenser but also by the engine and radiator. Typically, this establishes a minimum condensing temperature between 10 and 30°F above ambient. Liquid flashing occurs more often at reduced load, when the liquid-line velocity decreases, allowing the liquid to be heated above saturation temperature before reaching the expansion valve. This is more apparent on cycling systems than on systems that have a continuous liquid flow.

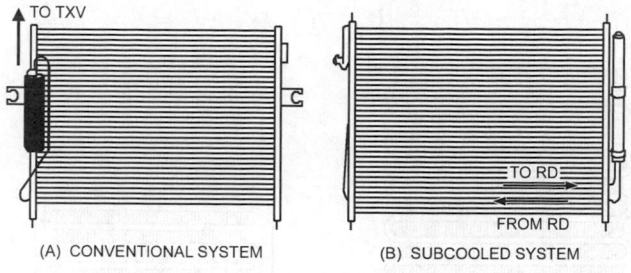

Fig. 7 Conventional and Subcooled PRF Condenser Designs

Liquid flashing is audibly detected as gas enters the expansion valve. This problem can be reduced by adding a subcooler or additional fan power to the condenser.

Internal pressure drop should be minimized to reduce compressor power requirements. Condenser-to-radiator clearances as low as 0.25 in. have been used, but 0.5 in. is preferable. Primary-to-secondary surface area ratios vary from 8:1 to 16:1. Condensers are normally painted black so they are not visible through the vehicle's grille.

Placing the condenser ahead of the engine-cooling radiator not only restricts air but also heats the air entering the radiator. Air conditioning increases requirements on the engine-cooling system, which requires an increase in radiator capacity, engine-cooling airflow, or both. Radiator capacity can be increased by adding fins, depth, or face area or by raising pump speed to increase coolant flow. Coolant velocity is not normally increased because it may cause excessive tube erosion or cavitation at the coolant pump inlet. With this configuration, engine-cooling airflow requirements increase; they are met by increasing fan size, number of blades, blade width, or blade pitch; by adding a fan shroud; or by a combination of these items. Increases in fan speed, diameter, and pitch raise the noise level and power consumption. For engine-driven fans (primarily used on trucks), temperature- and torque-sensitive drives (viscous drives or couplings) or flexible-blade fans reduce the increases in noise that come with the higher power. Virtually all automobiles rely on airflow produced by the car's forward motion to reduce the amount of air the engine-cooling fan must move to maintain adequate coolant temperatures. As vehicle speed increases, fan requirements drop, and electric fans are deenergized or engine-driven fans are decoupled by the action of the viscous drive.

Some vehicles have a side-by-side condenser and radiator, each with its own motor-driven fan. This eliminates the effect of the condenser on the engine cooling air inlet temperature, but causes other issues with fan control and potential engine bay recirculation when one system is energized and the other is not.

Subcooled Condensers. There is a trend of using subcooled condensers to improve overall air-conditioning system performance. Thermodynamically, by increasing subcooling at the end of the condenser (on a *p-h* diagram), the overall system performance is increased (see Chapter 2 of the 2009 *ASHRAE Handbook—Fundamentals*) because the overall evaporator enthalpy difference (i.e., the difference in enthalpies between evaporator outlet to inlet) increases. Figure 7 shows a conventional PRF condenser in which refrigerant flows out from the condenser to the receiver-drier. In the subcooled PRF condenser, refrigerant from the second-last pass flows to the receiver-drier and then back to the condenser in the last path to subcool the refrigerant. In a subcooled PRF condenser, the size of the receiver-drier can be reduced because the condenser has more liquid refrigerant.

Hoses. Rubber hose assemblies are installed where flexible refrigerant transmission connections are needed because of relative motion between components (usually caused by engine rock) or where stiffer connections cause installation difficulties and noise

transmission. Refrigerant permeation through the hose wall is a design concern. Permeation occurs at a reasonably slow and predictable rate that increases as pressure and temperature increase. Hose with a nylon core (**barrier hose**) is less flexible, has a smaller OD, is generally cleaner, and allows practically no permeation. However, because it is less flexible, it does not provide damping of gas pulsations as does other hose material. It is recommended for R-134a.

Reducing Noise and Vibration. Typically, refrigerant lines connected to the compressor (both suction and discharge sides) require hose that is a composite of rubber, nylon, and aluminum tube. This is necessary to eliminate or reduce transmission of clutch engagement noise to the cabin by metallic tubes. In some cases, mufflers are also used to reduce noise and vibrations from refrigerant flow.

Suction and discharge hoses and high-pressure liquid lines have connections for charging ports, sensor, and for service. Brackets and clips are also attached to the hoses and refrigerant lines to position and support the AC lines.

Expansion Devices. Virtually all modern automobiles use either a thermostatic expansion valve (TXV) or an orifice tube (or both, for dual-evaporator systems) as the expansion device (see Chapter 11 of the 2010 *ASHRAE Handbook—Refrigeration* for more on these devices). Schematics of systems that use these devices are provided in the Controls section.

Automotive TXVs operate in the same manner as those for commercial HVAC systems. Both liquid- and gas-charged power elements are common. Internally and externally equalized valves are used as dictated by system design. Externally equalized valves are necessary where high evaporator pressure drops exist. A bulbless expansion valve, usually block-style, that senses evaporator outlet pressure without the need for an external equalizer is now widely used. TXV systems use a receiver-drier-filter assembly for refrigerant and desiccant storage.

Because of their low cost and high reliability, orifice tubes have become increasingly popular with automotive manufacturers. Developing an orifice tube system requires that components be matched to obtain proper performance. The orifice tube is designed to operate at 90 to 95% quality at the evaporator outlet, which requires a suction-line accumulator to protect the compressor from floodback and to maintain oil circulation. Because the orifice tube does not fully use the latent heat in the refrigerant systems, orifice-tube systems generally require higher refrigerant flow than TXV systems to achieve the same performance. However, an orifice tube ensures that the compressor receives a continuous flow of cool refrigerant from the accumulator, offering benefits in compressor durability over a TXV system. Orifice-tube systems use an accumulator-drier-filter for refrigerant and desiccant storage.

Receiver-Drier-Filter Assembly. A receiver-driver is installed in the AC loop on the high-pressure side downstream of the condenser. Several types of desiccant are used, the most common of which is spherical molecular sieves; silica gel is occasionally used. The unit typically has desiccant either in a bag or cartridge, or sandwiched between two plates. The receiver-drier (1) serves as a reservoir for refrigerant from part- to full-load operating conditions, (2) removes moisture from the system, (3) filters out debris headed for the TXV, and (4) only allows liquid refrigerant to enter the TXV (liquid is removed from the top of the unit, and comes from the bottom via a tube connected to the top fitting).

The receiver-drier assembly accommodates charge fluctuations from changes in system load. It accommodates an overcharge of refrigerant to compensate for system leaks and hose permeation. The assembly houses the high-side filter and desiccant. Mechanical integrity (freedom from powdering) is important because of the vibration to which the assembly is exposed. For this reason, molded desiccants have not obtained wide acceptance. Moisture retention at elevated temperatures is also important. The rate of release with temperature increase and the reaction while accumulating high

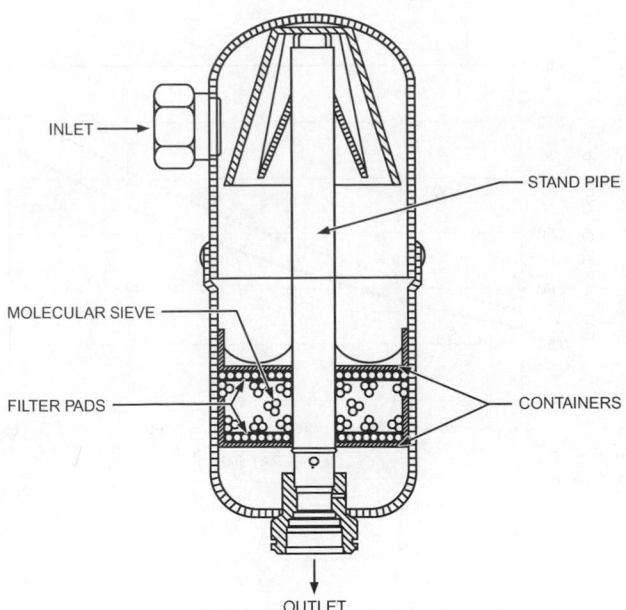

INLET

STAND PIPE

MOLECULAR SIEVE

FILTER PADS

CONTAINERS

OUTLET

Fig. 8 Schematic of Typical Accumulator-Dehydrator

concentration should be considered. Design temperatures should be at least 140°F.

Receivers are usually (though not always) mounted on or near the condenser. They should be located so that they are ventilated by ambient air. Pressure drop should be minimal. Typically, a receiver-drier has a pressure switch or a pressure transducer installed that controls AC system operation at high pressure.

Suction-Line Accumulators. A suction-line accumulator is required with an orifice tube to ensure uniform return of refrigerant and oil to the compressor, to prevent slugging, and to cool the compressor. It also stores excess refrigerant. A typical suction-line accumulator is shown in Figure 8. A bleed hole at the bottom of the standpipe meters oil and liquid refrigerant back to the compressor. The filter and desiccant are contained in the accumulator because no receiver-drier is used with this system.

Evaporator. The evaporator connects the air side of the air-conditioning system to the refrigerant side. Design aspects for the air side are discussed in the Air-Handling Subsystem section. The primary design consideration for the refrigerant side of the evaporator is low pressure drop. Because the evaporator operates at saturation, higher-pressure-drop evaporators cause nonuniform discharge temperatures unless they are designed with careful attention to pass arrangement. The space available in an automotive system does not allow for distribution manifolds and capillary tube systems outside of the evaporator envelope; this must be done within the evaporator itself.

Automotive evaporators must also mask the variation in compressor capacity that occurs with accelerating and decelerating. To avoid undesirable temperature splits, sufficient liquid refrigerant should be retained at the last pass to ensure continued cooling during acceleration.

High refrigerant pressure loss in the evaporator requires externally equalized expansion valves. A bulbless expansion valve, called a block valve, provides external pressure equalization without the added expense of an external equalizer. The evaporator must provide stable refrigerant flow under all operating conditions and have sufficient capacity to ensure rapid cooldown of the vehicle after it has been standing in the sun.

Auxiliary Evaporators. Many sport-utility vehicles, vans, and limousines are equipped with auxiliary or secondary air-conditioning modules located to cool rear-seat passengers. These system

extensions provide some unique challenges. Most of these systems operate only when there are passengers in the rear space. Consequently, sometimes there is refrigerant flow through the primary ACM and none through the secondary. Careful attention should be paid to refrigerant plumbing to avoid refrigerant and oil traps in the suction line. The auxiliary suction line must never allow liquid oil to run downhill from the front system when there is no flow to carry it back to the accumulator-dehydrator. Designing highly efficient oil separators into the line set results in frequent compressor failure.

Refrigerants and Lubricants. The 1997 Kyoto Protocol identified the almost universally used R-134a as a global warming gas, sparking a search for alternatives among vehicle manufacturers and their suppliers. The leading contender today is the hydrofluoroolefin HFO-1234yf, a low-global-warming-potential (GWP = 4) refrigerant jointly developed by Honeywell and DuPont (Koban 2009; Spatz and Minor 2008, 2009). This refrigerant has excellent environmental properties, low toxicity (similar to R-134a), and system performance similar to that of R-134a. It is being considered as a drop-in refrigerant for current MACS. Testing shows that both poly-alkylene glycol (PAG) and polyol ester (POE) lubricants are compatible with HFO-1234yf in AC systems with different types of compressors (Spatz 2009).

Figure 9 compares R-134a and HFO-1234yf AC cycles on the *p-h* diagram along with vapor density at suction temperatures (Spatz and Minor 2008). For HFO-1234yf, the latent heat of vaporization is lower and the vapor density at suction temperature is greater, compared to an R-134a system. Thus, for the same cooling capacity, the refrigerant mass flow rate for HFO-1234yf should be higher than in an R-134a system. Figure 10 compares the vapor pressures of the two fluids (Kontomaris and Leck 2009). Typical evaporating saturation pressures for HFO-1234yf are higher than for R-134a; HFO-1234yf pressure equals R-134a pressure at 100°F; and HFO-1234yf pressure is lower than R-134a above 100°F. Hence, in comparison to R-134a, a slightly higher evaporating pressure and slightly lower condensing pressure for HFO-1234yf reduces the pressure ratio, thereby improving system coefficient of performance (COP). Several OEMs and suppliers (Bang 2008; Mathur 2010a, 2010b; Meyer 2008, 2009; Minor 2008) also conducted independent tests with HFO-1234yf. SAE also conducted tests with alternative refrigerants (ARSS 2008), including HFO-1234yf (Atkinson 2008), through cooperative research projects (Hill 2008).

European Regulation of Mobile Air Conditioning. The E.U. directive (Vainio 2006) scheduled phaseout of HFC-134a to start on January 1, 2011, for all new models introduced that year. Retrofitting vehicles with HFC-134a is not allowed after January 1, 2011. The phaseout of HFC-134a does not affect air conditioning on vehicles that have not been significantly redesigned for 2011. From January 1, 2017, no new vehicles sold in the E.U. can use HFC-134a or any fluorinated gas with a global warming potential higher than 150; all models must be redesigned by the 2017 deadline. Acceptable refrigerants (with a GWP < 150) include HFC-152a, CO₂, and HFO-1234yf. Refrigerant leakage rates for single- and dual-evaporator AC systems have been defined as less than 40 and 60 g (1.4 to 2.1 oz), respectively. Using nonrefillable containers will no longer be allowed. All fluorinated gases covered by the Kyoto protocol will be recovered.

Enhanced R-134a Systems. SAE initiated a program to improve the performance of existing R-134a systems. The goals are to (1) identify technologies to reduce mobile air-conditioning system R-134a refrigerant leakage by 50%, (2) improve R-134a mobile air-conditioning system COP by 30%, (3) reduce vehicle soak and driving heat loads by 30% over current vehicles to reduce cooling requirements, and (4) reduce refrigerant loss during service and at end of life by 50%.

Suction Line Heat Exchanger. AC system performance can be improved by adding a suction line heat exchanger into the system. This directly influences the thermal comfort for the occupant along

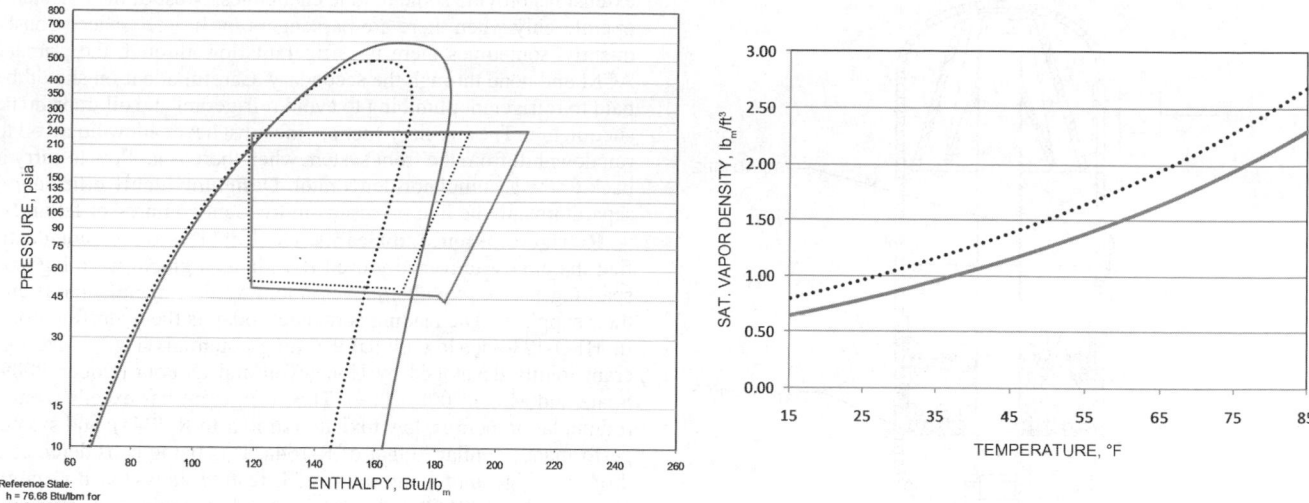

Fig. 9   Comparison of Thermodynamic Cycle Between Base Case (R-134a) and HFO-1234yf
(Spatz and Minor 2008)

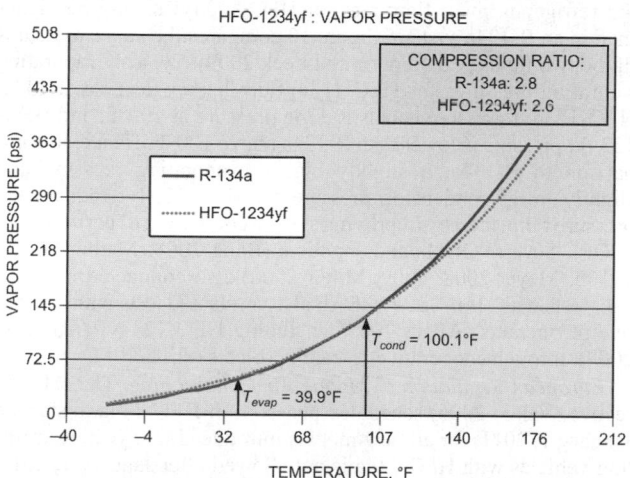

Fig. 10   Comparison of Vapor Pressure Between Base Case (R-134a) and HFO-1234yf
(Kontomaris and Leck 2009)

with fuel economy and exhaust emissions. A suction line heat exchanger in an AC loop (1) increases system performance, (2) subcools liquid refrigerant to prevent flash gas formation at inlets to the expansion valve, and (3) fully evaporates any residual liquid that may remain in the suction line before reaching the compressor. Performance of mobile air conditioning systems can be enhanced from 6 to 12% (Kurata et al. 2007; Mathur 2009c).

Advanced Technologies

HVAC suppliers are aggressively working on advancements for mobile HVAC to reduce energy consumption and improve thermal comfort for occupants. For instance, researchers are investigating using ventilated car seats to reduce air conditioning use and improve fuel efficiency without compromising thermal comfort (Lustbader 2005). Some other important technologies are as follows; Mathur (2004b) discusses them in greater detail.

Brushless Motors. These motors are simpler than standard motors and are more reliable. Advantages include the following: (1) motor efficiency is higher, (2) commutation is accomplished electronically, (3) very high speeds and torque are possible without arcing, (4) thermal resistance is lower and the operating temperature range is thus wider, and (5) the absence of brushes reduces maintenance requirements and eliminates brush residue contamination of bearings or the environment. Because there is no brush arcing or commutation, brushless motors are much quieter, both electrically and audibly.

Positive-Temperature-Coefficient (PTC) Heaters. PTC heaters are small, ceramic-based heaters that use less energy and less time to heat more quickly than conventional units. They self-regulate at a preset temperature by regulating resistance to vary their wattage. Thus, their greater thermal dissipation results in higher efficiency. These systems are maintenance-free and very reliable. PTC heaters could be used in the HVAC system to hasten cabin heating during cold start-ups by providing heat to occupants until the engine is warm (Hauck 2003).

Smart Engine Cooling Systems with Electric Water Pumps (EWPs). These cooling systems use both an electric and mechanical water pump, or replace the mechanical water pump with a EWP (Wagner et al. 2003). Typically, an EWP system includes a 100 to 600 W electric pump, four-way water valve (Chanfreau and Farkh 2003), sensor, engine control management system, software, and a variable-speed radiator fan. At cold start-ups, allowing little or no flow to the radiator hastens engine warm-up, thus reducing emissions and improving fuel economy. Because the water (or coolant) temperature is precisely maintained, thermal stresses in the engine are less. Once the engine coolant is heated, this system can provide thermal comfort, even at idle or with the engine off, by pumping coolant through the heater core and running the blower.

42 V Systems. Energy requirements of modern vehicles have increased significantly as the needs of motors, actuators, and other electrical equipment have increased. Auto manufacturers are investigating using 42 V for high-load equipment (e.g., compressors, blower, condenser fans, PTC heaters, controls), and reserving the existing 12 V grid for lighting and other smaller-load accessories. This would improve air-conditioning system performance, because

Pearson, J.K. 2001. *Improving air quality—Progress and challenges for the auto industry.* Society of Automotive Engineers, Warrendale, PA.

Rough, J., D. Bharatan, and L. Chaney. 2005. Predicting human thermal comfort in automobiles. Advanced Simulation Technologies Conference, Graz, Austria.

SAE. 1998. Motor vehicle heater test procedure. *Standard* J638. Society of Automotive Engineers, Warrendale, PA.

SAE. 2005. Safety standards for motor vehicle refrigerant vapor compressions systems. *Standard* J639. Society of Automotive Engineers, Warrendale, PA.

Spatz, M.W. 2009. HFO-1234yf technology update—Part II. VDA Winter Meeting, Austria. Verband der Automobilindustrie, Frankfurt. Available from http://www.vda-wintermeeting.de/fileadmin/downloads/presentations/HONEYWELL_M. Spatz_VDA Winter Meeeting 2009.pdf.

Spatz, M. and B. Minor. 2008. HFO-1234yf low GWP refrigerant update. Honeywell and DuPont joint collaboration. International Refrigeration and Air Conditioning Conference, Purdue University, West Lafayette, IN.

Spatz, M. and B. Minor. 2009. Low GWP Refrigerant update: Honeywell/DuPont joint collaboration. International Refrigeration and Air Conditioning Conference, Purdue University, West Lafayette, IN.

Vainio, M. 2006. European regulation of mobile air conditioning and global implications. VDA Winter Meeting, Saalfelden, Austria. Verband der Automobilindustrie, Frankfurt. Available from http://vda-wintermeeting.de/fileadmin/downloads2006/Matti_Vainio.pdf.

Wagner, J.R., V. Srinivasan, D.M. Dawson, and E. Marotta. 2003. Smart thermostat and coolant pump control for engine thermal management systems. SAE *Paper* 2003-01-0272. Society of Automotive Engineers, Warrendale, PA.

BIBLIOGRAPHY

Bhatti, M.S. 1997. A critical look at R-744 and R-134a for mobile air conditioning systems. SAE *Paper* 970527. Society of Automotive Engineers, Warrendale, PA.

Bhatti, M.S. 1999. Evolution of automotive heating—Riding in comfort: Part I. *ASHRAE Journal* 41(8):51-57.

Bhatti, M.S. 1999. Evolution of automotive air conditioning—Riding in comfort: Part II. *ASHRAE Journal* 41(9):44-50.

DOT. 1972. Flammability of interior materials—Passenger cars, multipurpose passenger vehicles, trucks, and buses. Federal Motor Vehicle Safety *Standard* (FMVSS) 302. U.S. Department of Transportation, National Highway Traffic Safety Administration, Washington, D.C.

Giles, G.R., R.G. Hunt, and G.F. Stevenson. 1997. Air as a refrigerant for the 21st century. *Proceedings of ASHRAE/NIST Refrigerants Conference: Refrigerants for the 21st Century.*

Jones, B.W. and Q. He. 1993. *User manual: Transient human heat transfer model* (includes application TRANMOD). Institute of Environmental Research, Kansas State University, Manhattan.

Jones, B.W., Q. He, J.M. Sipes, and E.A. McCullough. 1994. The transient nature of thermal loads generated by people. *ASHRAE Transactions* 100(2):432-438.

Mathur, G.D. 1998. Heat transfer coefficients for propane (R-290), isobutane (R-600a), and 50/50 mixture of propane and isobutane. *ASHRAE Transactions* 104(2):1159-1172.

Mathur, G.D. 2000. Carbon dioxide as an alternate refrigerant for automotive air conditioning systems. *Paper* AIAA-200-2858. American Institute of Aeronautics and Astronautics, Reston, VA.

Mathur, G.D. 2000d. Hydrodynamic characteristics of propane (R-290), isobutane (R-600a), and 50/50 mixture of propane and isobutane. *ASHRAE Transactions* 106(2):571-582.

Mathur, G.D. 2001. Simulating performance of a parallel flow condenser using hydrocarbons as the working fluids. SAE *Paper* 2001-01-1744. Society of Automotive Engineers, Warrendale, PA.

Mathur, G.D. 2003b. Heat transfer coefficients and pressure gradients for refrigerant R-152a. Alternative Refrigerant Systems Symposium, Scottsdale, AZ. Society of Automotive Engineers, Warrendale, PA.

Mathur, G.D. and S. Furuya. 1999c. A CO_2 refrigerant system for vehicle air conditioning. Alternative Refrigerant Systems Symposium, Scottsdale, AZ. Society of Automotive Engineers, Warrendale, PA.

Spatz, M.W. 2006. Ultra-low GWP refrigerant for mobile air conditioning applications. JSAE Automotive Air-Conditioning Conference, Tokyo. Society of Automotive Engineers of Japan, Tokyo.

MASS TRANSIT

THIS chapter describes air-conditioning and heating systems for buses, rail cars, and fixed-guideway vehicles that transport large numbers of people, often in crowded conditions. Air-conditioning systems for these vehicles generally use commercial components, but are packaged specifically for each application, often integral with the styling. Weight, envelope, power consumption, maintainability, and reliability are important factors. Power sources may be electrical (ac or dc), engine crankshaft, compressed air, or hydraulic. These sources are often limited, variable, and interruptible. Characteristics specific to each application are discussed in the following sections. Design aspects common to all mass-transit HVAC systems include passenger comfort (ventilation, thermal comfort, air quality, expectation) and thermal load analysis (passenger dynamic metabolic rate, solar loading, infiltration, multiple climates, vehicle velocity, and, in urban applications, rapid interior load change).

VENTILATION AND THERMAL COMFORT

The requirements of ASHRAE *Standards* 55 and 62.1 apply for transportation applications, with special considerations, because passengers in transit have different perceptions and expectations than typical building occupants. These considerations involve length of occupancy, occupancy turnover, infiltration, outdoor air quality, frequency and duration of door openings, personal preference, interior contamination sources such as smoking, and exterior contamination sources such as engine exhaust.

Historically, in nonsmoking air-conditioning and heating applications, outdoor air has been supplied to the vehicle interior by fans at 5 to 10 cfm per passenger at a predetermined nominal passenger loading. Nominal passenger load is based on the number of seats and may include a number of standees, up to the maximum number of standees possible if this type of loading is frequent. There are a few examples of no outdoor air being supplied by fans, but they are on short-duration trips such as people movers or urban buses with frequent door openings. Besides providing for survival, ventilation provides odor and contamination control. The amount needed for survival is less than the latter. Contamination control from interior sources is a factor in building design, but is less of a factor in vehicle design because of the ratio of people to furnishings and the lack of interior processes such as copy machines. Exterior contamination, such as from tunnel fumes, can be a problem, however. Door openings, if frequent enough, provide some additional intermittent ventilation, although this infiltration should be minimized for thermal comfort. Ventilation from doors may not be effective in controlling odors away from the doors. Fan-supplied outdoor air must be distributed equally in the vehicle for effective ventilation. Symptoms of inadequate ventilation are odors noticeable to passengers initially entering an occupied vehicle or when moving from section to section. Passengers on board who are exposed to slowly increasing odor levels may not be aware of them.

Based on ASHRAE research, ASHRAE *Standard* 161 established a ventilation rate for aircraft passengers at 7.5 cfm per passenger. This rate was based in part on the consideration that not all spaces in the enclosed area achieve 100% ventilation effectiveness. The minimum effective ventilation rate for several crowded but larger-volume spaces, as defined in ASHRAE *Standard* 62.1, is 5 cfm per person. It is recommended that ground mass transit applications use 7.5 cfm of outdoor air per passenger for most transit applications.

Emergency ventilation, such as windows or exits that can be opened or battery-powered ventilators, should be provided in case other systems fail. For example, a power interruption or a propulsion system failure may strand passengers in a situation where exit is not possible. Emergency situations include overtemperature, oxygen depletion, smoke, or toxic fumes. Operator-controlled dampers are now provided on some vehicles to close off fresh air when smoke or toxic fumes are encountered in tunnels. The duration that the dampers remain closed must be limited to avoid oxygen depletion, even though the air-conditioning system remains in operation. Fresh-air supply alone or battery-powered ventilators will not prevent overtemperatures when a full passenger load is present and/or a solar load exists in combination with high ambient temperature. Each emergency situation requires an independent solution.

The nature of the transit service may be roughly categorized by average journey time per passenger and interval between station stops, and this service type affects the necessary interior conditions in the vehicle. For example, a commuter rail or intercity bus passenger may have a journey time of an hour or more, with few stops; passengers may remove heavy outer clothing before being seated. In contrast, a subway or transit urban bus rider typically does not remove heavy clothing during a 10 min ride. Clothing and the environment from which passengers come, including how long they were exposed to those conditions and what they were doing (e.g., waiting for the train outdoors in winter), are important factors in transit comfort. At the opposite extreme, many subway stations are not climate controlled, and often reach dry-bulb temperatures over 100°F in the summer. Thus, when boarding a climate-controlled vehicle, these passengers immediately perceive a significant increase in comfort. However, a passenger adjusts to a new environment in about 10 to 20 min; after that, the traditional comfort indices begin to apply, and the same interior conditions that were perceived as comfortable may now be perceived as less than comfortable. Before stabilization, a passenger may prefer higher-velocity air or cooler or warmer temperatures, depending to some extent on clothing. At the same time, other passengers may already have stabilized and have completely different comfort control desires. Therefore, the transit system designer is presented with a number of unusual requirements in providing comfort for all.

Jones et al. (1994) evaluated the heat load imposed by people under transient weather and activity conditions as opposed to traditional steady-state metabolic rates. An application program, TRANMOD, was developed that allows a designer to predict the thermal loads imposed by passengers (Jones and He 1993).

The preparation of this chapter is assigned to TC 9.3, Transportation Air Conditioning.

Variables are activity, clothing, wet- and dry-bulb temperatures, and precipitation.

European Committee for Standardization (CEN) *Standard* EN 13129-1 provides guidance in the area of railroad passenger comfort. Although this standard does not apply to countries outside the CEN, the information is valuable and may not be readily available elsewhere.

THERMAL LOAD ANALYSIS

Cooling Design Considerations

Thermal load analysis for transit applications differs from stationary, building-based systems because vehicle orientation and occupant density change regularly on street-level and subway vehicles and, to a lesser degree, on commuter and long-distance transportation. Summer operation is particularly affected because cooling load is affected more by solar and passenger heat gain than by outdoor air conditions. ASHRAE *Standard* 55 design parameters for occupant comfort may not always apply. Vehicle construction does not allow the low thermal conductivity levels of buildings, and fenestration material must have safety features not necessary in other applications. For these reasons, thermal loads must be calculated differently. Because main line passenger rail cars and buses must operate in various parts of the country, the air conditioning must be designed to handle the national seasonal extreme design days. Commuter and local transit vehicles operate in a small geographical area, so only local design ambient conditions need be considered.

The following cooling load components should be considered:

- Ambient air conditions for locations in North America and worldwide are given in Chapter 14 of the 2009 *ASHRAE Handbook—Fundamentals*. For vehicles operating in an urban area, the heat island effect should be considered if the Handbook design values are derived from remote reporting stations. For subway car operation, tunnel temperatures should be considered. In humid regions, consider the wet-bulb temperature coincident with dry-bulb temperature relative to fresh-air loads.
- For vehicle interior comfort conditions, consult Figure 5 in Chapter 9 of the 2009 *ASHRAE Handbook—Fundamentals*. Total heat gain from passengers depends on passenger activity before boarding the vehicle, waiting time, journey time, and whether they are standing or seated during the journey. Representative values are given in Table 1 in Chapter 18 of the 2009 *ASHRAE Handbook—Fundamentals*.
- Ventilation air loads should be calculated using the method in Chapter 18 of the 2009 *ASHRAE Handbook—Fundamentals*, in the section on Infiltration and Moisture Migration Heat Gains. Air leakage and air entering during door dwell time should be taken into account.
- Interior heat includes that produced by the evaporator fan motor, inside lighting, and electrical controls.
- The vehicle's conductivity, in Btu/h·°F, should be provided by the vehicle designers. For outside skin temperature guidance, use the values in Table 1 in Chapter 29 of the 1997 *ASHRAE Handbook—Fundamentals*; however, consider that air over a vehicle in motion reduces these temperatures The car design dry bulb should be used as the interior temperature.
- The instantaneous solar gain through the glazing should be calculated using summer midafternoon data listed in Chapter 29 of the 1997 *ASHRAE Handbook—Fundamentals*, and the glass shading coefficient. The glass shading coefficient must be obtained from the window supplier. Adjustments for frequent change in vehicle direction or intermittent solar exposure may be justified. Additional information is shown in Chapter 15 of the 2009 *ASHRAE Handbook—Fundamentals*

The summer cooling analysis should be completed for different times of the day and different passenger densities to verify a reliable result. Cooling equipment capacity should consider fouling and eventual deterioration of heat transfer surfaces.

Heating Design Considerations

Winter outdoor design conditions can be taken from Chapter 14 of the 2009 *ASHRAE Handbook—Fundamentals*. Interior temperatures can be taken from Figure 5 in Chapter 9 of the 2009 *ASHRAE Handbook—Fundamentals*. During winter, conductivity is the major heat loss. The heat required to temper ventilation air and to counteract infiltration through the body and during door openings must also be considered.

Other Considerations

Harsh environments and the incursion of dirt and dust inhibit the efficiency of HVAC units. Specifications should include precise maintenance instructions to avoid capacity loss and compromised passenger comfort.

BUS AIR CONDITIONING

In general, bus air-conditioning systems can be classified as interurban, urban, or small/shuttle bus systems. Bus air-conditioning design differs from other air-conditioning applications because of climatic conditions in which the bus operates, equipment size limitations, vehicle engine, electrical generator, and compressor rpm. Providing a comfortable climate inside a bus passenger compartment is challenging because the occupancy rate per unit of surface and air recirculation volume is high, glazed area is very large, and outdoor conditions are highly variable. Factors such as high ambient temperatures, dust, rain, snow, road shocks, hail, and sleet should be considered in the design. Units should operate satisfactorily in ambient conditions from –22 to 122°F.

Ambient air quality must also be considered. Air intakes are usually subjected to thermal contamination from road surfaces, condenser air recirculation, or vehicle engine radiator air discharge. Vehicle motion also introduces pressure variables that affect condenser fan performance. In addition, engine speed governs compressor speed, which affects compressor capacity. R-134a is the current refrigerant of choice, but some units operate with refrigerants such as R-22 (pre-2010 production) and R-407C.

Bus air conditioners are initially performance-tested as units in a climate-controlled test cell. Performance tests encompass unit operation at different compressor speeds to make sure the compressor performance parameters [e.g., unit operation at maximum and minimum ambient conditions, thermostatic expansion valve (TXV) sizing, oil return, and vibration/shock] are within boundaries. In addition, individual components should be qualified before use. Larger test cells that can hold a bus are commonly used to verify installed unit performance. These tests are to measure the amount of time required to reduce the vehicle's interior temperature to a specified value, and they vary in performance and time requirements. Some commonly accepted tests include the Houston pulldown (extreme heat or performance when using higher-pressure refrigerant gas such as R-407C), modified pulldown (mild to hot climates with R-134a or equivalent), white book pulldown (mild to hot climates), and the profile test (mild to hot climates, 95 and 115°F ambient). All these tests are described in American Public Transportation Association (APTA) standard bus procurement and recommended practices for transit bus HVAC system instrumentation and performance testing.

Reliability and ease of maintenance are also important design considerations. All parts requiring service or regular maintenance should be readily accessible, and repairs should be achievable without removing any additional components and within a minimum time.

Heat Load

The main parameters that must be considered in bus air-conditioning system design include

- Occupancy data (number of passengers, distance traveled, distance traveled between stops, typical permanence time)
- Dimensions and optical properties of glass
- Outdoor weather conditions (temperature, relative humidity, solar radiation)
- Dimensions and thermal properties of materials in bus body
- Indoor design conditions (temperature, humidity, air velocity)
- Power and torque limitations of bus engine

The heating or cooling load in a passenger bus may be estimated by summing the heat flux from the following loads:

- Solid walls (side panels, roof, floor)
- Glass (side, front, and rear windows)
- Passengers
- Engine and ventilation (difference in enthalpy between outdoor and indoor air)
- Evaporator fan motor

Extreme loads for both summer and winter should be calculated. The cooling load is the most difficult load to handle; the heating load is normally handled by heat recovered from the engine, external heater, or electrical heat elements. An exception is that an idling engine provides marginal heat in very cold climates. Andre et al. (1994) and Jones and He (1993) describe computational models for calculating the heat load in vehicles, as well as for simulating the thermal behavior of the passenger compartment.

The following conditions can be assumed for calculating the summer heat load in an interurban vehicle similar to that shown in Figure 1:

- Capacity of 50 passengers
- Insulation thickness of 1 to 1.5 in.
- Double-pane tinted windows
- Outdoor air intake of 400 cfm
- Road speed of 65 mph
- Inside design temperatures of 60 to 80°F and 50% rh
- Ambient temperatures for location as listed in Chapter 14 of the 2009 *ASHRAE Handbook—Fundamentals*

Loads from 3.5 to 10 tons are calculated, depending on outdoor weather conditions and geographic location. The typical distribution of the different heat loads during a summer day at 40° north latitude is shown in Figure 2.

Air Distribution

Air-conditioning units are configured to deliver air through ducts to outlets above the windows and to the middle aisle or to act as free-blow units. In the case of free-blow units, louvers guide the air distribution inside the bus.

Interurban Buses

These buses are designed to accommodate up to 56 passengers. The air-conditioning system is usually designed to handle extreme conditions. Interurban buses produced in North America are likely to have the evaporator and heater located under the passenger compartment floor. A four- or six-cylinder reciprocating compressor, in which some cylinders are equipped with unloaders, is popular. Some interurban buses have a separate engine-driven compressor, preferably scroll, to give more constant system performance. Figure 3 shows a typical air-conditioning arrangement for an interurban bus.

Urban Buses

Urban bus heating and cooling loads are greater than those of the interurban bus. A city bus may seat up to 50 passengers and carry a "crush load" of standing passengers. The fresh-air load is greater

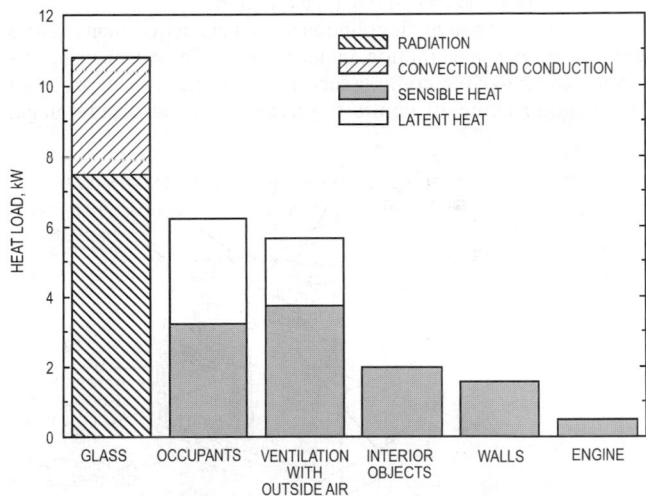

Fig. 2   Typical Main Heat Fluxes in Bus

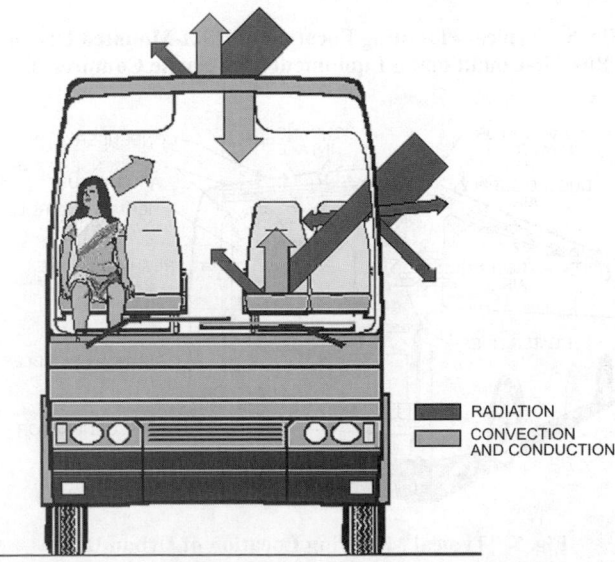

Fig. 1   Distribution of Heat Load (Summer)

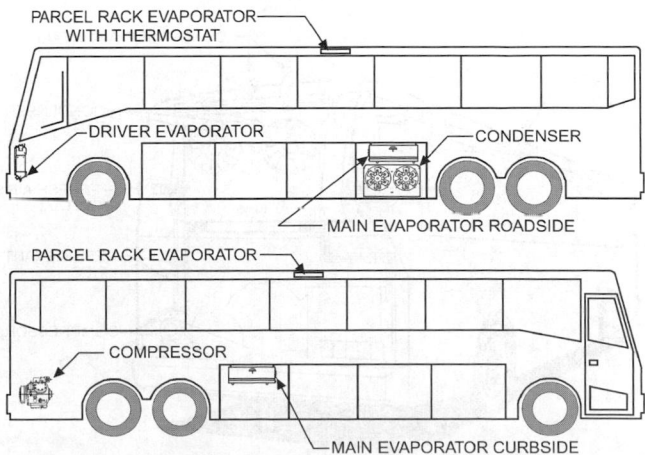

Fig. 3   Typical Arrangement of Air-Conditioning in Interurban Bus

because of the number of door openings and the infiltration around doors. Cooling capacity required for a typical 50-seat urban bus is from 6 to 10 tons. The buses are usually equipped with a roof- or rear-mounted unit, as shown in Figure 4. One or two compressors are usually belt- or shaft-driven from the propulsion engine. Capacity control is very important, because the compressor may turn more quickly than necessary at high engine speeds. Therefore, capacity control must compensate for not only the thermal load but also the engine-induced load. Cylinder unloaders are the primary means of capacity control, although evaporator pressure regulators have been used with non-unloading compressors, as shown in Figure 4. This configuration was used on buses produced between 1975 and 1995.

The heater is located just downstream of the evaporator. Hot coolant from the engine-cooling system provides sufficient heat for most operations; however, additional sources may be required in colder climates for longer idling durations. Additional floor heaters may also be required to reduce the effects of stratification. Conditioned air is delivered through overhead combination light fixture/diffuser ducts (see Figure 5).

Low-profile, self-contained, rooftop-mounted units are used for urban and interurban buses. These units contain the entire air-conditioning system except for the compressor, which is shaft- or belt-driven from the bus engine (see Figure 6).

Because of increased air pollution and other environmental issues (e.g., noise, fuel consumption, unnecessary engine wear), using traditional engine-driven compressors for interurban, urban, or school bus or motor home air comfort systems is a great disadvantage,

especially for parked vehicles. In response to these issues, most modern and efficient buses use unitized electric packaged air-conditioning (UEPAC) units, as shown in Figures 7 and 8. UEPACs have a self-contained, lightweight, integrated, modular design incorporating evaporators, condensers, valves, liquid receiver, filter-drier, electric heater elements, automatic climate controls, and scroll compressors. Electric power is supplied to the UEPAC system from onboard sources for hybrid electric and fuel-cell buses, or by a main-engine-driven generator on more traditional fuel or hybrid applications without an accessory power option. These systems enable the use of shore (wayside) power while parked, eliminating idling where power is available.

Small or Shuttle Buses

For small or shuttle buses such as those typically operating around airports or for schools, the evaporator is usually mounted in the rear and the condenser on the side or the roof of the bus. The evaporator unit is typically a free-blow unit.

Refrigerant Piping

Refer to Chapters 1 and 8 of the 2010 *ASHRAE Handbook—Refrigeration* for standard refrigerant piping practices. All components in the bus air-conditioning system are interconnected by

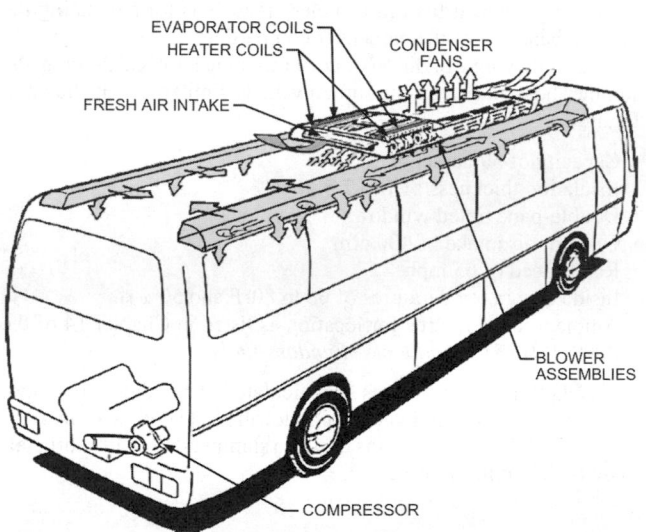

Fig. 6 Typical Mounting Location of Roof-Mounted Urban Bus Air-Conditioning Equipment with Single Compressor

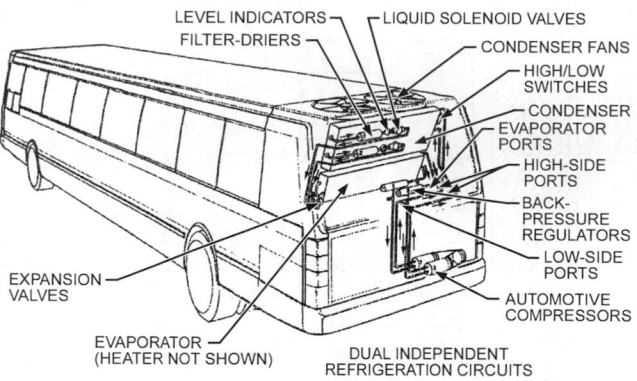

Fig. 4 Typical Mounting Location of Urban Bus Air-Conditioning Equipment

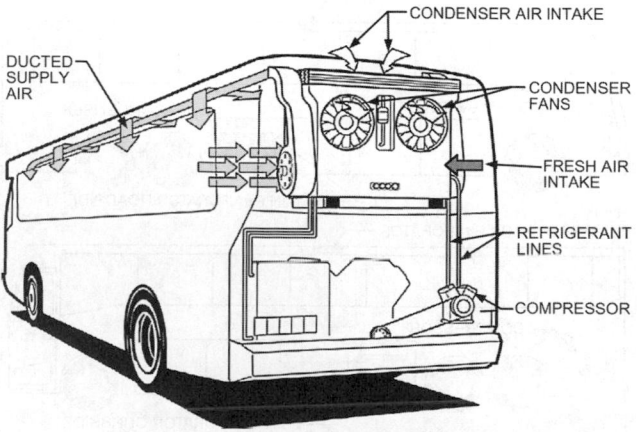

Fig. 5 Typical Mounting Location of Urban Bus Air-Conditioning Equipment with Single Compressor

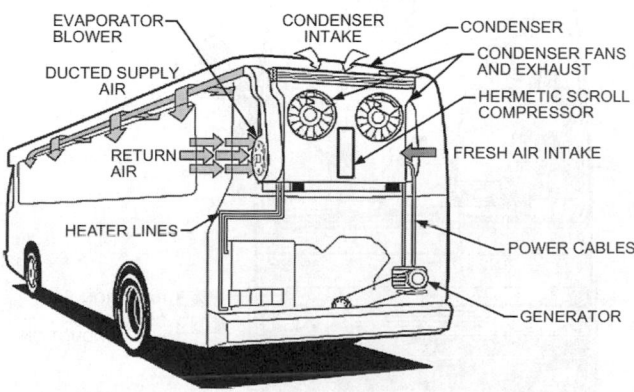

Fig. 7 Typical Mounting Location of Urban Bus Fully Electric Rear-Mounted Air-Conditioning Equipment with ac Generator

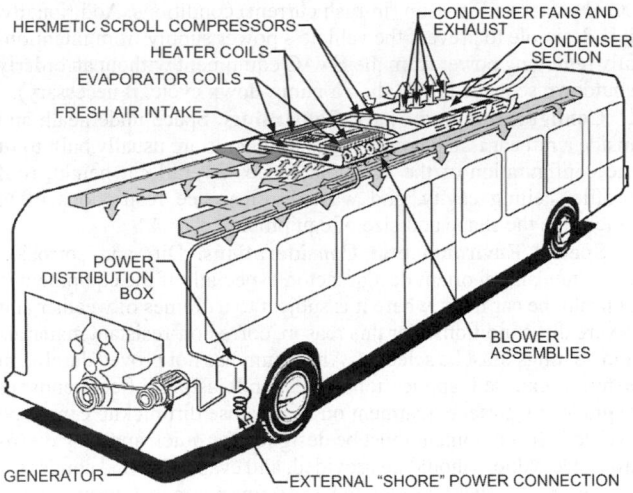

**Fig. 8   Typical Mounting Location of Urban Bus
Fully Electric Roof-Mounted Air Conditioning
Equipment with ac Generator**

copper tubing or refrigerant hose. When using copper tubing, care should be taken to analyze the effect of vibration on the tubing. Vibrational effects can be minimized by using vibration absorbers or other shock-cushioning devices. When using refrigerant hose, properties such as moisture ingression, effusion, maximum operating temperature, and burst pressure need to be taken into account. The refrigerant hose chosen should have the minimum amount of wax extractables on interaction with oil and the refrigerant.

Shock and Vibration

Most transport air-conditioning manufacturers design components for shock loading and vibrational inputs. Vibration eliminators, flexible lines, and other shock-cushioning devices interconnect the various air-conditioning components. The vibration characteristics of each component are different; in addition, the evaporator and the condenser must undergo individual vibration and shake tests. The input levels for the shake test can be based on the worst road conditions that the bus will encounter. This input level will vary because of the weight of the unit and its mounting.

System Safety

Per the U.S. Department of Transportation, all buses with air-conditioning systems operating in North America should conform to Federal Motor Vehicle Safety *Standard* (FMVSS) 302 for flammability standards. In addition, all evaporator units inside the vehicle should be mounted away from the head impact zone, as specified by FMVSS 222.

Controls

Most buses have a simple driver control to select air conditioning, heating, or automatic operation (air conditioning, heating, and reheat). In both modes, a thermal sensing element controls these systems with on/off circuitry and actuators. Many systems use solid-state control modules to interpret the bus interior and outdoor ambient temperatures and to generate signals to operate full or partial cooling, reheat, or heating functions. These systems use thermistor temperature sensors, which are usually more stable and reliable than electromechanical controls. Control systems for urban buses can also include an outdoor-air ventilation cycle. The percentage of fresh-air intake during the ventilation cycle can vary based on individual requirements.

RAIL CAR AIR CONDITIONING

Passenger rail car air-conditioning systems are generally electromechanical, direct-expansion units. R-22, a hydrochlorofluorocarbon (HCFC), has been the refrigerant most commonly used since the phase-out for R-12. R-134a, a medium-pressure refrigerant, has been used as a retrofit refrigerant in North America on systems originally designed to operate with R-12, and is commonly used in Europe for new equipment, mainly variable-speed screw compressors that are competitive in weight to R-22 reciprocating compressors. Most equipment placed in service before the January 1, 2010, ban on manufacturing new R-22 equipment has used R-407C as the refrigerant. R-410A has been used in some equipment; however, it can only be used in relatively mild climates because the condensing temperatures found in transit applications may approach the refrigerant's critical point. In 2009, the U.S. Environmental Protection Agency (EPA) added R-438A to its significant new alternatives policy (SNAP) list of approved refrigerants for motor vehicle air conditioning use.

Electronic, automatic controls are common, with a trend toward microprocessor control with increasing capability for fault monitoring and logging. Electric heating elements in the air-conditioning unit or supply duct temper outdoor air brought in for ventilation and are also used to control humidity by reheating the conditioned supply air during cooling partial-load conditions.

Air-cycle technology has been tested for passenger rail car air conditioning in Germany (Giles et al. 1997); however, issues of greater weight, higher cost, and low efficiency need to be addressed before it is widely accepted.

Vehicle Types

Main-line intercity passenger rail service generally operates single and multilevel cars hauled by a locomotive. Locomotive-driven alternators or solid-state inverters distribute power via an intercar cable power bus to air-conditioning equipment in each car. A typical rail car has a control package and two air-conditioning systems. The units are usually either split, with the compressor/condenser units located in the car undercarriage area and the evaporator-blower portion mounted in the ceiling area, or self-contained packages mounted in interior equipment rooms. Underfloor and roof-mounted package units are less common in intercity cars.

Commuter cars used to provide passenger service from the suburbs into and around large cities are similar in size to main-line cars. Air-conditioning equipment generally consists of two evaporator-heater fan units mounted above the ceiling with a common or two separate underfloor-mounted compressor-condenser unit(s) and a control package, or self-contained packaged units mounted on the roof. These cars may be locomotive hauled, with air-conditioning arrangements similar to main-line intercity cars, but they are often self propelled by high-voltage direct-current (dc) or alternating-current (ac) power supplied from an overhead catenary or from a dc-supplied third rail system. On such cars, the air conditioning may operate on ac or dc power. Self-propelled diesel-driven vehicles that use onboard-generated power for the air-conditioning systems still operate in a few areas.

Subway and **elevated rapid-transit cars** usually operate on a third-rail dc power supply. In the past, the air-conditioning system motors were commonly powered directly from the third-rail dc supply voltage. Most new equipment operates from three-phase ac power provided by a solid-state inverter. The inverter may be either an independent system or a component of the HVAC system. Split air-conditioning systems are common, with evaporators in the interior ceiling area and underfloor-mounted condensing sections, although unitary package units mounted on the roof or under the floor are increasingly common.

Streetcars and **light-rail vehicles** usually run on ac or dc power transmitted via an overhead catenary wire, and have air-conditioning

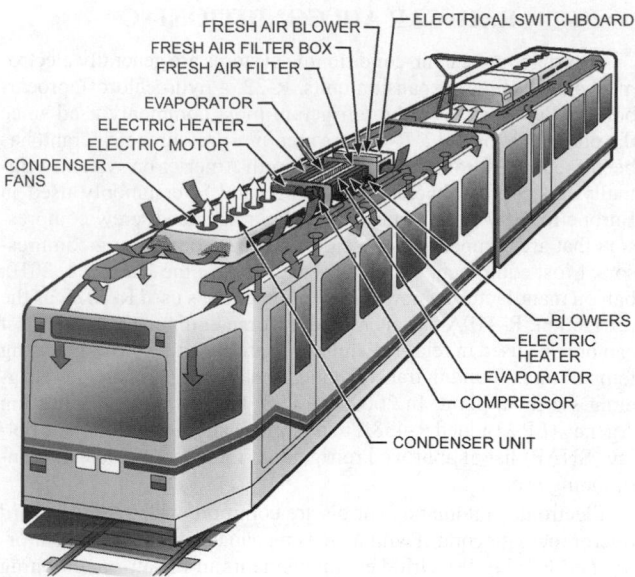

Fig. 9 Typical Light Rail Vehicle with Roof-Mounted HVAC System

equipment similar to rapid-transit cars. Roof-mounted packages are used more often than undercar or split systems. This is largely because of the lack of undercar space. Figure 9 shows a typical configuration for these vehicles.

Equipment Design Considerations

Design considerations unique to transit HVAC equipment include the characteristics of the available power supply, weight limits, type of vehicle, and vehicle service parameters. Thus, ac-powered, semihermetic or hermetic compressors, which are lighter than open machines with dc motor drives, are a common choice. However, each car design must be examined in this respect because dc/ac inverters may increase not only the total weight, but also the total power draw, because of conversion losses.

Other concerns in equipment selection include the space required, location, accessibility, reliability, and maintainability. Interior and exterior equipment noise levels must be considered both during the early stages of design and later, when the equipment is coordinated with the car builder's ductwork and grilles.

Compressors. Reciprocating and vane compressors are commonly used, although scroll compressors are becoming increasingly common. The scroll compressor is inherently more tolerant of flooded starts and liquid slugging common in the rail application than any other type of positive-displacement compressor. The low clearance volume of the scroll compressor allows it to operate at high discharge pressure more effectively than reciprocating compressors. Lower weight and less vibration and noise are benefits, as well.

Power Supply Characteristics. Vehicles that draw their power from a stationary supply, such as a third rail or overhead catenary wire, are subject to frequent power interruptions as the train passes through gaps in the third rail or phase breaks in the overhead. These interruptions cause the HVAC equipment to shut down independently of the control system, and the design must take into account these losses of power and the subsequent need to restart the equipment. Vehicles that generate electrical power from an onboard source are less affected by power interruptions, although their capacity is limited. In either case, HVAC system control design must be coordinated with the vehicle's power supply and distribution system to avoid overloading vehicle systems during both

steady-state and start-up (in-rush current) conditions. Additionally, it is desirable to prevent the vehicle's power supply from intentionally removing power from the HVAC equipment without an orderly shutdown sequence (including a pump-down cycle, if necessary).

Configuration and Space Constraints. Space underneath and inside a rail car is at a premium. Components are usually built to fit the configuration of the available space. Overall car height, roof profile, ceiling cavity, and wayside clearance restrictions often determine the shape and size of equipment.

Special Environmental Considerations. Dirt and corrosion constitute an important design factor, especially if the equipment is beneath the car floor, where it is subject to extremes of weather and severe dirt conditions. For this reason, corrosion-resistant materials and coatings must be selected. Aluminum has not proved durable in exterior exposed applications; the sandblasting effect tends to degrade any surface treatment on it. Because dirt pickup cannot be avoided, the equipment must be designed for quick and easy cleaning; access doors should be provided, and evaporator and condenser fin spacing is usually limited to 8 to 10 fins per inch. Closer spacing causes more rapid dirt build-up and higher cleaning costs. Dirt and severe environmental conditions must also be considered in selecting motors and controls.

Maintenance Provisions. Railroad HVAC equipment is subjected to mechanical shock and vibration during operation, is frequently required to operate under conditions of elevated condensing temperature and pressure, and is subjected to frequent on/off cycling because of power supply interruptions and other conditions that are not typical for a stationary application. As a consequence, the rail HVAC system's components are more highly stressed than equivalent components in a stationary system, and thus require more frequent maintenance and servicing. Because a passenger rail car with sealed windows and a well-insulated structure becomes almost unusable if the air conditioning fails, high reliability is important. Equipment design needs to consider the ease of routine service and time needed to diagnose and repair the system. The control equipment thus often incorporates monitoring and diagnostic capabilities to allow quick diagnosis and correction of a failure. However, many trains are designed with several individual vehicles permanently coupled together, in which case the failure of a single HVAC unit causes multiple cars to become unavailable for service while the HVAC system is diagnosed and repaired. The time to diagnose and repair a system varies. Railroads, by their nature, are schedule driven, and varying, unknown repair time is incompatible with the need to provide scheduled service. Therefore, many users are moving away from fully on-car-serviceable air conditioners, and toward modular, self-contained units with hermetically sealed refrigerant systems. These units are designed for rapid removal and replacement to allow the vehicle to return to service in a short, predictable time. The faulty HVAC equipment is diagnosed and repaired off-car in a dedicated air-conditioning service area.

Safety. Security of the air-conditioning equipment attachment to the vehicle must be considered, especially on equipment located beneath the car. Vibration isolators and supports should be designed to safely retain the equipment on the vehicle, even if the vibration isolators or fasteners fail completely. A piece of equipment that dangles or drops off could cause a train derailment. All belt drives and other rotating equipment must be safety guarded. High-voltage controls and equipment must be labeled by approved warning signs. Pressure vessels and coils must meet ASME test specifications for protection of passengers and maintenance personnel. Materials selection criteria include low flammability, low toxicity, and low smoke emission.

Special Design Considerations. The design, location, and installation of air-cooled condenser sections must allow for the possibility of hot condenser discharge air recirculation into the condenser inlet (in the case of split systems), or into the outdoor air intakes (in the case of roof-mounted unitary systems), as well as the hot condenser discharge from trains on adjacent tracks that may occur at passenger loading platforms or in tunnels. To prevent a total system shutdown because of high discharge pressure, a capacity reduction control device is typically used to reduce the cooling capacity before system pressure reaches the high-pressure safety switch setting, thus temporarily reducing discharge pressure.

Even with coordination between the HVAC controls and the vehicle's power supply or distribution system, abrupt shutdown of the refrigeration system caused by power loss is common. The typical split-system arrangement places the compressor at or near the low point in the system. The combination of these factors results in undesired migration of refrigerant to the compressor during the off cycle. To reduce the likelihood of flooded compressor starts, using a suction line accumulator and crankcase heater is recommended.

Other Requirements

Most cars are equipped with both overhead and floor heat, typically provided by electric resistance elements. The control design commonly uses overhead heat to raise the temperature of the recirculated and ventilation air mixture to slightly above the car design temperature, while floor heat offsets heat loss through the car body. This arrangement is intended to limit stratification in the passenger compartment by promoting buoyant, convective air circulation. Times of maximum occupancy, outdoor ambient, and solar gain must be ascertained. The peak cooling load on urban transit cars usually coincides with the evening rush hour, and the peak load on intercity rail cars occurs in the midafternoon.

Heating capacity for the car depends on body construction, car size, and the design area-averaged relative wind-vehicle velocity. In some instances, minimum car warm-up time may be the governing factor. On long-distance trains, the toilets, galley, and lounges often have exhaust fans. Ventilation airflow must exceed forced exhaust air rates sufficiently to maintain positive car pressure. Ventilation air pressurizes the car and reduces infiltration.

Air Distribution and Ventilation

The most common air distribution system is a centerline supply duct running the length of the car between the ceiling and the roof. Air outlets are usually ceiling-mounted linear slot air diffusers. Louvered or egg crate recirculation grilles are positioned in the ceiling beneath the evaporator units. The main supply duct must be insulated from the ceiling cavity to prevent thermal gain/loss and condensation. Taking ventilation air from both sides of the roof line helps overcome the effect of wind. Adequate snow and rain louvers and, in some cases, internal baffles, must be installed on the outdoor air intakes. Separate outdoor air filters are usually combined with either a return or mixed-air filter. Disposable media or permanent, cleanable air filters are used and are usually serviced every month. Some long-haul cars, such as sleeper cars, require a network of delivered-air and return ducts. Duct design should consider noise and static pressure losses.

Piping Design

Standard refrigerant piping practice is followed. Pipe joints should be accessible for inspection and, on split systems, not concealed in car walls. Evacuation, leak testing, and dehydration must be completed successfully after installation and before charging. Piping should be supported adequately and installed without traps that could retard the flow of lubricant back to the compressor. Pipe sizing and arrangement should be in accordance with Chapter 1 of the 2010 *ASHRAE Handbook—Refrigeration*. Evacuation, dehydration, and charging should be performed as described in Chapter 8 of that volume. Piping on packaged units should also conform to these recommendations.

Control Requirements

Rail HVAC control systems typically automatically transition between cooling and heating operation, based on interior and exterior dry-bulb temperature. The cooling and heating set points are generally different. This difference provides a control dead band to prevent the system from cycling directly between cooling and heating, and accommodates passengers' seasonal clothing. System capacity is matched to part-load conditions with some combination of evaporator coil staging, evaporator fan speed control, compressor cylinder unloading, or variable-speed compressor control in cooling mode, and staging or duty cycling of heat in heating mode. The control system typically does not consider latent heat information in the control algorithm, although reheat is commonly used to increase the apparent interior sensible load as the interior dry-bulb temperature falls below the desired cooling set point, to maintain humidity removal. Unitary systems may use hot-gas bypass for this purpose rather than electric reheat. If the interior dry-bulb temperature falls below the desired cooling set point, even with capacity reduction and reheat, the refrigeration system will shut down and the HVAC system will provide ventilation only. If the interior temperature drops to the heating set point, the system transitions to heating mode. Before the development of analog electronic or microprocessor control systems, this dry-bulb based control algorithm was implemented by banks of thermostats. This arrangement resulted in multiple, load-dependent interior set points as the system established quasi-equilibrium conditions within the dead band of each individual thermostat. When analog electronic controls were introduced in the early 1980s, they emulated this thermostat-based control algorithm, which is still often followed today in North America. Recently, several European and Asian HVAC manufacturers have introduced proportional-integral-derivative (PID) control systems, common in those markets for several years, to the North American market. Higher energy costs and greater environmental concern in Europe and Asia have led some manufacturers to include energy conservation algorithms in controls intended for use in those markets.

The availability of robust, low-cost humidity sensors may lead to the use of latent heat information in control algorithms.

A pumpdown cycle and low-ambient lockout are recommended on split systems to protect the compressor from damage caused by liquid flooding the compressor and subsequent flooded starts. In addition, the compressor may be fitted with a crankcase heater that is energized during the compressor off cycle.

FIXED-GUIDEWAY VEHICLE AIR CONDITIONING

Fixed-guideway (FGW) systems, commonly called people movers, can be monorails or rubber-tired cars running on an elevated or grade-level guideway, as seen at airports and in urban areas. The guideway directs and steers the vehicle and provides electrical power to operate the car's traction motors (in some cases, the vehicle is propelled by a metal cable, driven by a motor mounted at the end of the guideway), lighting, electronics, air conditioner, and heater. People movers are usually unstaffed and computer-controlled from a central point. Operations control determines vehicle speed, headway, and the length of time doors stay open, based on telemetry from individual cars or trains. Therefore, reliable and effective environmental control is essential.

People movers are usually smaller than most other mass-transit vehicles, generally having spaces for 8 to 40 seated passengers and generous floor space for standing passengers. Under some conditions of passenger loading, a 40 ft car can accommodate 100 passengers. The wide range of passenger loading and solar exposure make it essential that the car's air conditioner be especially responsive to the amount of cooling required at a given moment.

System Types

The HVAC for a people mover is usually one of three types:

- Conventional undercar condensing unit and compressor unit (which includes control box) connected with refrigerant piping to an evaporator/blower unit mounted above the car ceiling
- Packaged, roof-mounted unit having all components in one enclosure and mated to an air distribution system built into the car ceiling
- Packaged, undercar-mounted unit mated to supply and return air ducts built into the car body

Some vehicles are equipped with two systems, one at each end; each system provides one-half of the maximum cooling requirement. U.S. systems usually operate on the guideway's power supply of 460 to 600 V (ac), 60 Hz. Some newer systems with dc track power operate on 240 V (ac), 60 Hz from an inverter. Figures 10 and 11 show some arrangements used with fixed-guideway people mover vehicles, although similar arrangements could also apply to rail.

Refrigeration Components

Because commercial electrical power is available, standard semihermetic reciprocating compressors and commercially available fan motors and other components can be used. Compressors generally have one or two stages of unloaders, and/or hot-gas bypass is used to maintain cooling at low loads. Newer systems use scroll compressors with speed control, displacement control, or hot-gas bypass to control capacity. Condenser and evaporator coils

are copper tube with copper or aluminum fins. Generally, flat fins are preferred for undercar condensers to make it simpler to clean the coils. Evaporator/blower sections must often be designed for the specific vehicle and fitted to its ceiling contours. Condensing units must also be arranged to fit in the limited space available and still ensure good airflow across the condenser coil. Because of the phaseout of R-22, R-407c is commonly used to meet environmental standards (zero ozone depletion potential). Some existing R-22 systems are being retrofitted with R-407c and R-422d.

Heating

Where heating must be provided, electric resistance heaters that operate on the guideway power supply are installed at the evaporator unit discharge. One or two stages of heat control are used, depending on the size of the heaters.

Controls

A solid-state control is usually used to maintain interior conditions, although newer systems use programmable logic controller (PLC) microprocessor-based controllers. The cooling set point is typically between 74 and 76°F. For heating, the set point is 60 to 68°F. Some controls provide humidity control by using electric heat. Between the cooling and heating set points, blowers continue to operate on a ventilation cycle. On rare occasions, two-speed blower motors are used, switching to low speed for the heating cycle. Some controls have internal diagnostic capability and can signal the operations center when a cooling or heating malfunction occurs.

Ventilation

With overhead air-handling equipment, outdoor air is introduced into the return airstream at the evaporator entrance. Outdoor air is usually taken from a grilled or louvered opening in the end or side of the car. Depending on the configuration of components, fresh air is filtered separately or directed so that the return air filter can handle both airstreams. For undercar systems, a similar procedure is used, except air is introduced into the system through an intake in the undercar enclosure. In some cases, a separate fan is used to induce outdoor air into the system.

The amount of mechanical outdoor air ventilation is usually expressed as cubic feet per minute per passenger on a full-load continuous basis. Passenger loading is not continuous at full load in this application, with the net result that more outdoor air is provided than indicated. The passengers may load and unload in groups, which causes additional air exchange with the outside. Frequent door openings, sometimes on both sides at once, allows additional natural ventilation. The effective outdoor air ventilation per passenger is a summation of all these factors. The amount of outdoor air

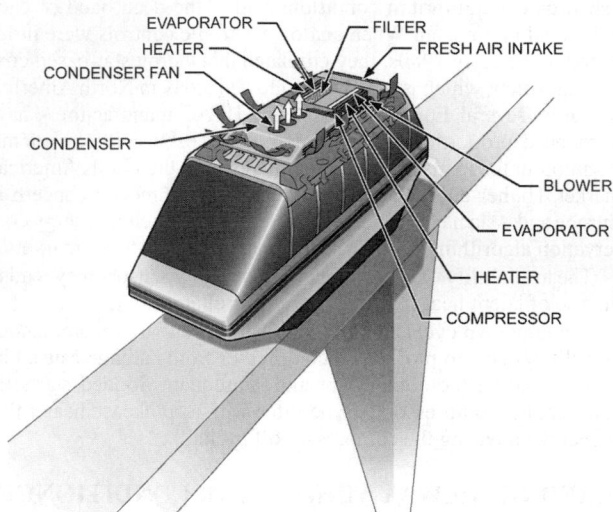

Fig. 10   Typical Small Fixed-Guideway Vehicle with Roof-Mounted HVAC System

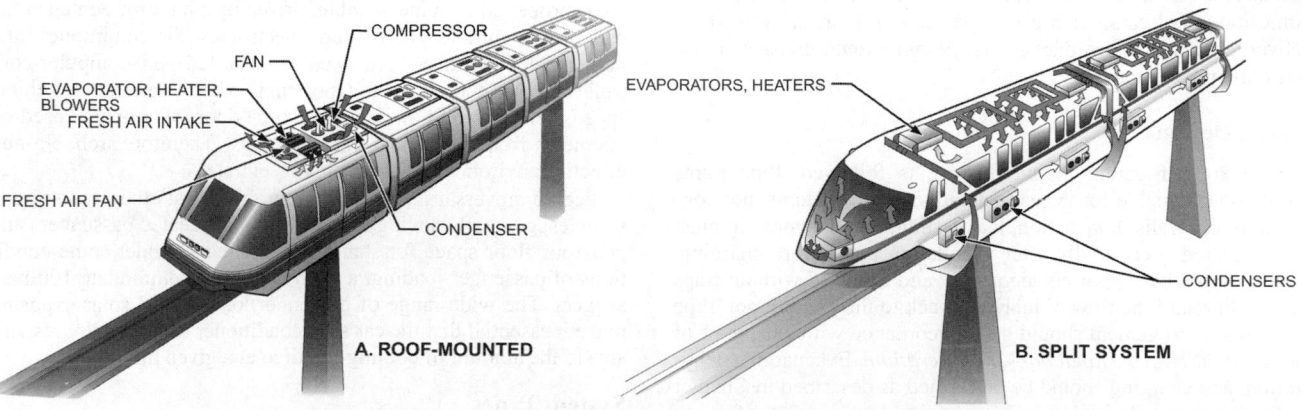

A. ROOF-MOUNTED   **B. SPLIT SYSTEM**

Fig. 11   Example Monorail HVAC System Configurations

introduced through the HVAC system varies. Some new vehicles have no mechanical outdoor air supply, whereas others provide up to 9 cfm per passenger. Lower values of mechanical ventilation, typically 3 to 5 cfm or less per passenger, are associated with travel times of less than 2 min and large passenger turnover. Longer rides justify higher rates of mechanical ventilation.

Green initiatives have caused designers to take a closer look at all aspects of energy savings. Some systems are now designed with variable outdoor air rates, which are automatically lowered under low passenger load conditions or extreme temperature loads. This approach yields lower system cooling capacities and saves energy.

Air Distribution

With overhead equipment, air is distributed through linear ceiling diffusers that are often constructed as a part of the overhead lighting fixtures. Undercar equipment usually makes use of the void spaces in the sidewalls and below fixed seating. In all cases, the spaces used for air supply must be adequately insulated to prevent condensation on surfaces and, in the case of voids below seating, to avoid cold seating surfaces. The supply air discharge from undercar systems can be from overhead diffusers through sidewall duct or a windowsill diffuser. Recirculation air from overhead equipment flows through ceiling-mounted grilles. For undercar systems, return air grilles are usually found in the door wells or beneath seats.

Because of the vehicle's typical small size and low ceilings, care must be taken to design the air supply so that it does not blow directly on passengers' heads or shoulders. Because high flow rates are necessary to achieve capacities, diffuser design and placement are important. Some systems are designed so the air supply discharge hugs the vehicle's ceiling and walls to avoid drafts on passengers. Total air quantity and discharge temperature must be carefully calculated to provide passenger comfort. Interior noise levels are typically 72 to 74 dBA for a stationary vehicle with doors shut.

REFERENCES

Andre, J.C.S., E.Z.E. Conceição, M.C.G. Silva, and D.X. Viegas. 1994. Integral simulation of air conditioning in passenger buses. Fourth International Conference on Air Distribution in Rooms (ROOMVENT '94).

ASHRAE. 2010. Thermal environmental conditions for human occupancy. ANSI/ASHRAE *Standard* 55-2010.

ASHRAE. 2010. Ventilation for acceptable indoor air quality. ANSI/ASHRAE *Standard* 62.1-2010.

ASHRAE. 2007. Air quality within commercial aircraft. ANSI/ASHRAE *Standard* 161-2007.

CEN. 2002. Railway applications—Air conditioning for mail line rolling stock—Part 1: Comfort parameters. *Standard* EN 13129.1-2002. European Committee for Standardization, Brussels.

DOT. 1977. School bus passenger seating and crash protection. Federal Motor Vehicle Safety *Standard* (FMVSS) 222. U.S. Department of Transportation, National Highway Traffic Safety Administration, Washington, D.C.

DOT. 1972. Flammability of interior materials—Passenger cars, multipurpose passenger vehicles, trucks, and buses. Federal Motor Vehicle Safety *Standard* (FMVSS) 302. U.S. Department of Transportation, National Highway Traffic Safety Administration, Washington, D.C.

Giles, G.R., R.G. Hunt, and G.F. Stevenson. 1997. Air as a refrigerant for the 21st century. *Proceedings ASHRAE/NIST Refrigerants Conference: Refrigerants for the 21st Century.*

Jones, B.W. and Q. He. 1993. *User manual: Transient human heat transfer model* (includes application TRANMOD). Institute of Environmental Research, Kansas State University, Manhattan.

Jones, B.W., Q. He, J.M. Sipes, and E.A. McCullough. 1994. The transient nature of thermal loads generated by people. *ASHRAE Transactions* 100 (2):432-438.

U.S. EPA. 2010. *SNAP list for motor vehicles.* U.S. Environmental Protection Agency, Washington, D.C. Available at http://www.epa.gov/ozone/snap/refrigerants/lists/mvacs.html.

BIBLIOGRAPHY

APTA. 1999. *Standard bus procurement guidelines for high floor diesel.* American Public Transportation Association, Washington, D.C.

APTA. 2000. *Standard bus procurement guidelines for low floor diesel.* American Public Transportation Association, Washington, D.C.

Conceição, E.Z.E., M.C.G. Silva, and D.X. Viegas. 1997. Airflow around a passenger seated in a bus. *International Journal of HVAC&R Research* (now *HVAC&R Research*) 3(4):311-323.

Conceição, E.Z.E., M.C.G. Silva, and D.X. Viegas. 1997. Air quality inside the passenger compartment of a bus. *Journal of Exposure Analysis & Environmental Epidemiology* 7:521-534.

DOT. 1994. Windshield defrosting and defogging systems—Passenger cars, multipurpose vehicles, trucks, and buses. Federal Motor Vehicle Safety *Standard* (FMVSS) 103. U.S. Department of Transportation, National Highway Traffic Safety Administration, Washington, D.C.

Guan, Y., M.H. Hosni, B.W. Jones, and T.P. Gielda. 2003. Literature review of the advances in thermal comfort modeling. *ASHRAE Transactions* 109(2):908-916.

Jones, B.W. 2002. The quality of air in the passenger cabin. *Proceedings of Cabin Health 2002*, International Air Transport Association, Geneva.

Jones, B.W. 2002. Capabilities and limitations of thermal models for use in thermal comfort standards. *Energy and Buildings* 34(6):653-659.

Silva, M.C.G. and D.X. Viegas. 1994. External flow field around an intercity bus. Second International Conference on Experimental Fluid Mechanics.

CHAPTER 12

AIRCRAFT

ENVIRONMENTAL control system (ECS) is a generic term used in the aircraft industry for the systems and equipment associated with ventilation, heating, cooling, humidity/contamination control, and pressurization in the occupied compartments, cargo compartments, and electronic equipment bays. The term ECS often encompasses other functions such as windshield defog, airfoil anti-ice, oxygen systems, and other pneumatic demands. The regulatory or design requirements of these related functions are not covered in this chapter.

DESIGN CONDITIONS

Design conditions for aircraft applications differ in several ways from other HVAC applications. Commercial transport aircraft often operate in a physical environment that is not survivable by the unprotected. In flight, the ambient air may be extremely cold and dry, and can contain high levels of ozone. On the ground, the ambient air may be hot, humid, and contain many pollutants such as particulate matter, aerosols, and hydrocarbons. These conditions change quickly from ground operations to flight. A hot-day, high-humidity ground condition usually dictates the thermal capacity of the air-conditioning equipment, and flight conditions determine the supply air compressor's capacity. Maximum heating requirements can be determined by either cold-day ground or flight operations.

In addition to essential safety requirements, the ECS should provide a comfortable cabin environment for the passengers and crew. This presents a unique challenge because of the high-density seating of the passengers. Furthermore, aircraft systems must be lightweight, accessible for quick inspection and servicing, highly reliable, able to withstand aircraft vibratory and maneuver loads, and able to compensate for various possible system failures.

Ambient Temperature, Humidity, and Pressure

Figure 1 shows typical design ambient temperature profiles for hot, standard, and cold days. The ambient temperatures used for the design of a particular aircraft may be higher or lower than those shown in Figure 1, depending on the regions in which the aircraft is to be operated. The design ambient moisture content at various altitudes that is recommended for commercial aircraft is shown in Figure 2. However, operation at moisture levels exceeding 200 gr/lb of dry air is possible in some regions. The variation in ambient pressure with altitude is shown in Figure 3. Refer to the psychrometric chart for higher altitudes for cabin humidity calculations. Figure 4 shows a psychrometric chart for 8000 ft altitude.

Heating/Air Conditioning Load Determination

The cooling and heating loads for a particular aircraft model are determined by a heat transfer study of the several elements that comprise the air-conditioning load. Heat transfer involves the following factors:

- Convection between the boundary layer and the outer aircraft skin
- Radiation between the outer aircraft skin and the external environment

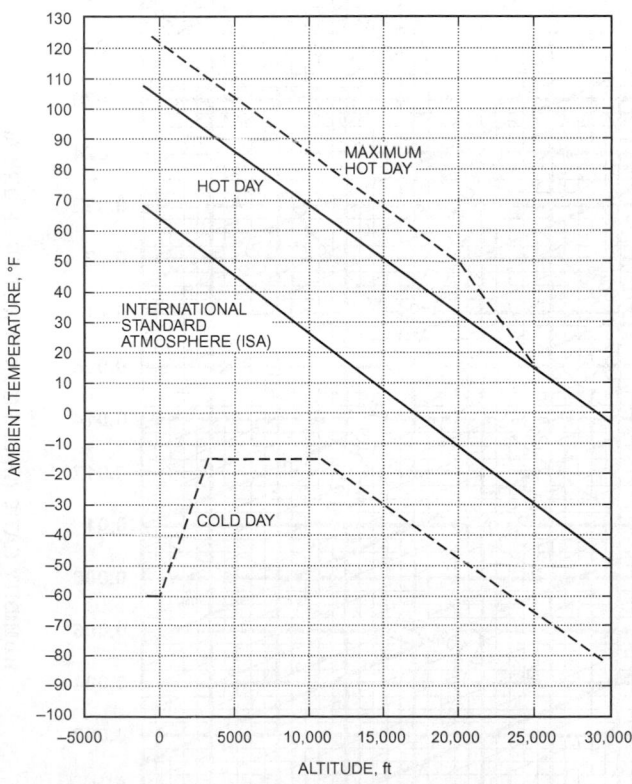

Fig. 1   Ambient Temperature Profiles

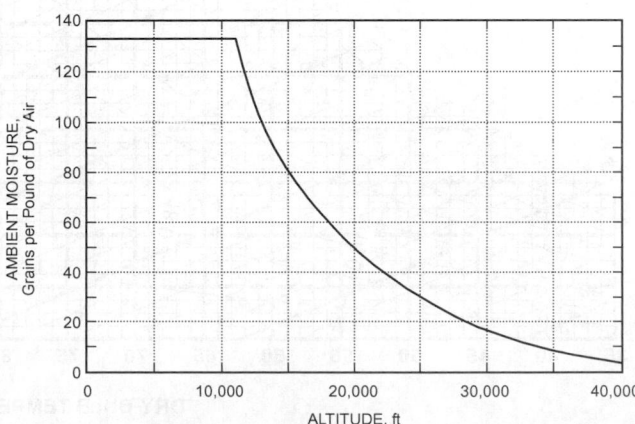

Fig. 2   Design Humidity Ratio

The preparation of this chapter is assigned to TC 9.3, Transportation Air Conditioning.

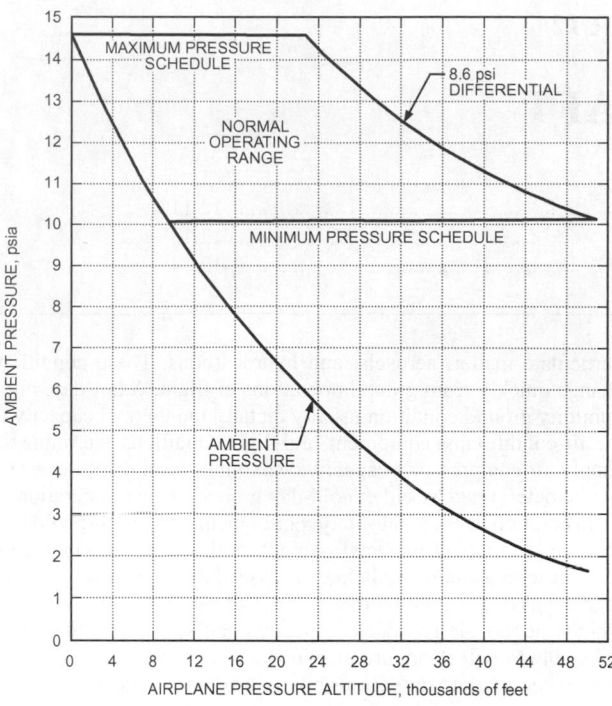

Fig. 3 Cabin Pressure Versus Altitude

- Solar radiation through windows, on the fuselage and reflected from the ground.
- Conduction through cabin walls and the aircraft structure
- Convection between the interior cabin surface and the cabin air
- Convection and radiation between the cabin and occupants
- Convection and radiation from internal sources of heat (e.g., electrical equipment)
- Latent heat from vapor cycle systems

Ambient Air Temperature in Flight

During flight, very cold ambient air adjacent to the outer surface of the aircraft increases in temperature through ram effects, and may be calculated from the following equations:

$$T_{AW} = T_\infty + r(T_T - T_\infty)$$

$$T_T = T_\infty \left(1 + \frac{k-1}{2}M^2\right)$$

or

$$T_{AW} = T_\infty \left(1 + r\frac{k-1}{2}M^2\right)$$

$$r = Pr^{1/3}$$

where

Pr = Prandtl number for air (e.g., Pr = 0.73 at 432°R
T_∞ = ambient static temperature, °R
T_T = ambient total temperature, °R
k = ratio of specific heat; for air, $k = 1.4$

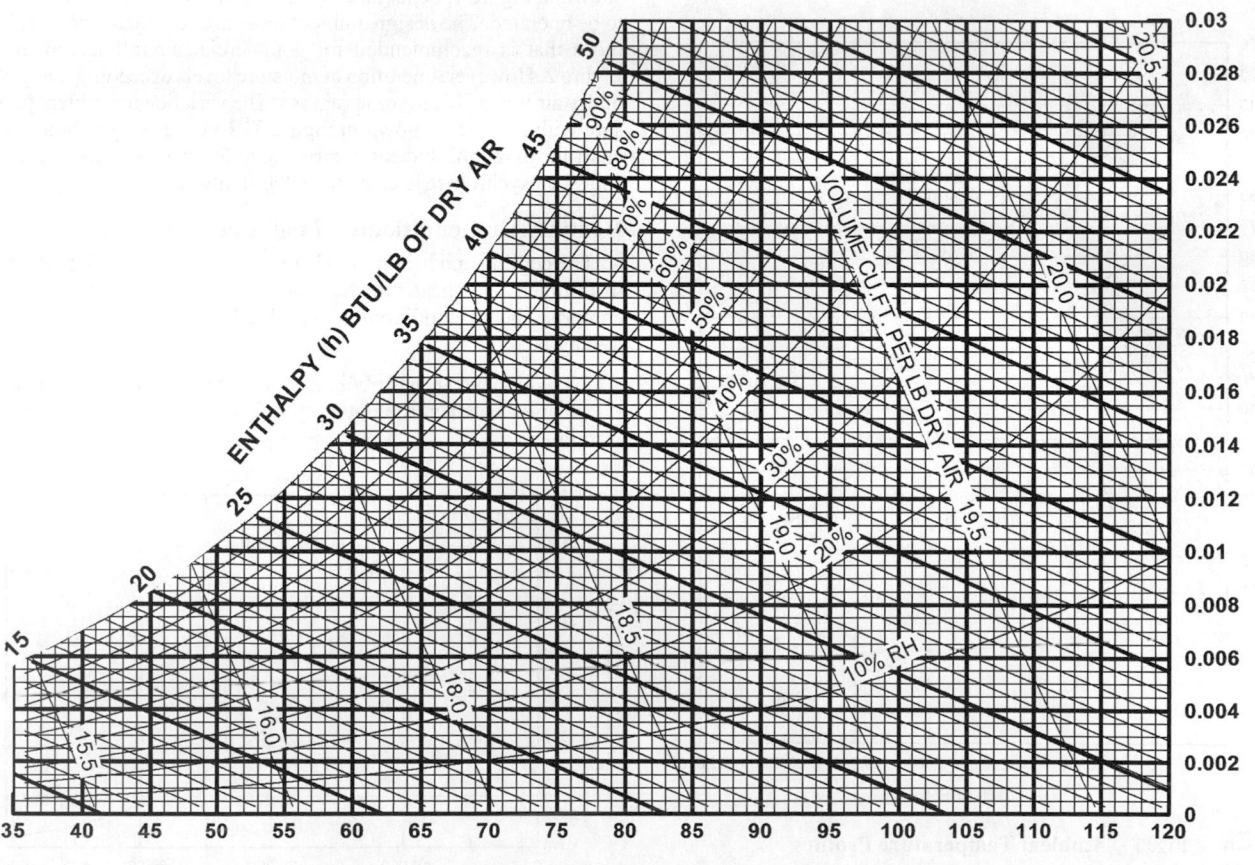

Fig. 4 Psychrometric Chart for Cabin Altitude of 8000 ft

M = airplane Mach number

r = recovery factor for turbulent boundary layer (i.e., fraction of total temperature recovered in boundary layer as air molecules rest on the surface)

T_{AW} = recovery temperature (or adiabatic wall temperature), °R

Example 1. The International Civil Aviation Organization (ICAO) cold day at 30,000 to 40,000 ft altitude has a static temperature of –85°F (375°R) and a Prandtl number of 0.739. If an airplane is traveling at 0.8 Mach, what would the external temperature be at the airplane's skin?

Solution: Iteration is usually required. First guess for $r \approx 0.9$:

$$\text{Pr} = 0.728 \text{ at } 0.9(423 - 375) + 375 = 418°R$$
$$r = \text{Pr}^{1/3} = (0.728)^{1/3} = 0.8996$$

$$T_T = T_\infty\left(1 + \frac{k-1}{2}M^2\right) = 375\left(1 + \frac{1.4-1}{2}[0.8]^2\right) = 423°R$$

$$T_{AW} = T_\infty + r(T_T - T_\infty) = 375 + 0.8996(423 - 375)$$
$$= 418.2°R \ (-41.5°F)$$

Air Speed and Mach Number

The airplane airspeed is related to the airplane Mach number by the local speed of sound:

$$u_\infty = M\sqrt{kRT_\infty}$$

where

k = ratio of specific heats; 1.4 for air
R = gas constant; 1716 ft²/s²·°R
M = airplane Mach number
u_∞ = airplane airspeed, fps

Ambient Pressure in Flight

The static pressure over most of the fuselage (the structure around the cabin) is essentially equal to the ambient pressure at the appropriate altitude.

$$P_s = P_{inf} + C_p\frac{1}{2}\rho_\infty u_\infty^2$$

where

P_s = pressure surrounding the fuselage, lb/ft²
C_P = pressure coefficient, dimensionless; approximately zero for passenger section of fuselage
ρ_∞ = free-stream or ambient air density, slug/ft³

External Heat Transfer Coefficient in Flight

The fact that the fuselage is essentially at free-stream static pressure implies that a flat-plate analogy can be used to determine the external heat transfer coefficient at any point on the fuselage:

$$h = \rho_w c_p u_{inf} 0.185(\log_{10}\text{Re}_x)^{-2.584}\text{Pr}^{-2/3}$$

$$\text{(note: } 10^7 < \text{Re}_x < 10^9)$$

$$\text{Re}_x = \frac{\rho_w u_\infty x}{\mu}$$

$\rho_w, c_p, \mu, \text{Pr}$

$$\text{evaluated at } T^* = \frac{T_{AW} + T_\infty}{2} + 0.22(T_{AW} - T_\infty)$$

$$q = hA(T - T_{AW})$$

where

h = external heat transfer coefficient, Btu/s·ft²·°F

Re_x = local Reynolds number, dimensionless
x = distance along the fuselage from nose to point of interest, 10.21 × 10⁻¹⁰(T^*)^{3/2}
c_p = constant-pressure specific heat; for air, 0.24 Btu/lb·°F
ρ_w = ambient air (weight) density at film temperature T^*, lb/ft³
μ = absolute viscosity of air at T^*; $1.021 \times 10^{-9}(T^*)^{3/2}[734.7/(T^* + 216)]$ lb_m/ft·s
A = outside surface area, ft²
T = outer skin temperature, °R
q = convective heat loss from outer skin, Btu/h
u_{inf} = airplane airspeed, ft/h

External Heat Transfer Coefficient on Ground

The dominant means of convective heat transfer depends on wind speed, fuselage temperature, and other factors. The (free convection) heat transfer coefficient for a large, horizontal cylinder in still air is entirely buoyancy-driven and is represented as follows:

$$\text{Gr} = \frac{g(\beta)(\Delta T)d^3}{v^2}$$

for $10^9 \leq \text{GrPr} \leq 10^{12}$:

$$h_{free} = \frac{0.13k(\text{GrPr})^{1/3}}{d}$$

where

g = gravitational acceleration, 32.2 ft/s²
k = thermal conductivity of air, Btu·ft/h·ft²·°R
v = kinematic viscosity, ft²/s
d = fuselage diameter, ft
h_{free} = free-convection heat transfer coefficient, Btu/h·ft²·°R
β = expansion coefficient of air = $1/T_f$, where $T_f = (T_{skin} + T_{inf})/2$, °R
$\Delta T = T_{skin} - T_{inf}$
T_{skin} = skin temperature, °R
T_{inf} = ambient temperature, °R
Gr = Grashof number
Pr = Prandtl number

A relatively light breeze introduces a significant amount of heat loss from the same horizontal cylinder. The forced-convection heat transfer coefficient for a cylinder may be extrapolated from the following:

$$\text{Re} = \frac{Vd}{v}$$

for $4 \times 10^4 \leq \text{Re} \leq 4 \times 10^5$

$$h_{forced} = \frac{0.0266k(\text{Re})^{0.805}\text{Pr}^{1/3}}{d}$$

where V is wind speed in ft/s, and v is evaluated at $T_f = (T_{skin} + T_{inf})/2$.

Example 2. One approximation of the fuselage is a cylinder in cross-flow. The fuselage is 12 ft in diameter and 120 ft long, in a 14 fps crosswind and a film temperature of 575°R. The surface temperature varies with the paint color and the degree of solar heating. For instance, a typical white paint could be 30°F higher than the ambient air temperature, so the heat transfer from the fuselage would be

Free convection:

$$\text{Gr} = \frac{g(\beta)(\Delta T)d^3}{v^2} = \frac{32.2(0.00174)(30)(12)^3}{(1.9 \times 10^{-4})^2} = 8.05 \times 10^{10}$$

for $10^9 \leq \text{GrPr} \leq 10^{12}$.

$$h_{free} = \frac{0.13k(GrPr)^{1/3}}{d} = \frac{0.13(0.016)[8.05 \times 10^{10}(0.704)]^{1/3}}{12}$$

$$= 0.67 \ \text{Btu/h} \cdot \text{ft}^2 \cdot °F$$

Forced convection:

$$Re = \frac{Vd}{\nu} = \frac{14(12)}{1.9 \times 10^{-4}} = 8.84 \times 10^5$$

for $4 \times 10^4 \le Re \le 4 \times 10^5$. Note that, although this Reynolds number is beyond the recommended range, the extrapolation has about a 10% error (underprediction) when compared to other more complicated methods.

$$h_{forced} = \frac{0.0266k(Re)^{0.805}Pr^{1/3}}{d}$$

$$= \frac{0.0266(0.016)(8.84 \times 10^5)^{0.805}(0.704)^{1/3}}{12}$$

$$= 1.93 \ \text{Btu/h} \cdot \text{ft}^2 \cdot °F$$

Comparison of heat transfer coefficients shows that, in this situation, heat transfer is dominated by forced convection, so the free-convection aspect can be ignored.

External Radiation

The section of airplane fuselage that surrounds the cabin radiates primarily to the sky. At sea level, the sky temperature is about 30°F cooler than the surrounding air temperature (depending on humidity and other factors). As the airplane climbs, there is a decreasing amount of air above to radiate to, so the difference between air temperature and sky temperature increases. For example, at a cruising altitude of 30,000 to 35,000 ft, the sky temperature is about 100°F cooler than the air temperature (free-stream static). The limiting condition, of course, is outer space, where the sky temperature is the cosmic background radiation (CBR). The sky temperature in this case is only about 6°R. The heat loss to the sky by radiation is

$$qR = A\sigma(\varepsilon)\left(T^4 - T_{sky}^4\right)$$

where

q_R = radiation heat loss from outer skin, Btu/h
A = outside surface area, ft²
T = outer skin temperature, °R
σ = Stephan-Boltzmann constant, 1.712×10^{-9} Btu/h·ft²·R⁴
ε = emissivity of surface, paint, etc.
T_{sky} = sky temperature, °R

Solar radiation on the ground is covered in detail elsewhere (e.g., Chapter 35); however, during cruising, the incident solar radiation should be adjusted for altitude. The column of air between the sun and the airplane varies with time of day (angle) and altitude. Standard sea-level solar flux, for a given latitude and time of day, can be adjusted for altitude using Beer's law:

$$I_{SL} = I_o e^{-na_{ms}}$$

$$I_y = I_o e^{-na_{ms}m}$$

$$C_y = I_y/I_{SL}$$

$$q_s = A\alpha(C_y)I$$

where

I = solar radiation to a surface at sea level after accounting for latitude and time of day, Btu/h·ft²
I_o = solar constant, 429.5 Btu/h·ft²
I_y = normal solar flux at altitude, Btu/h·ft²
I_{SL} = normal solar flux at sea level, Btu/h·ft²

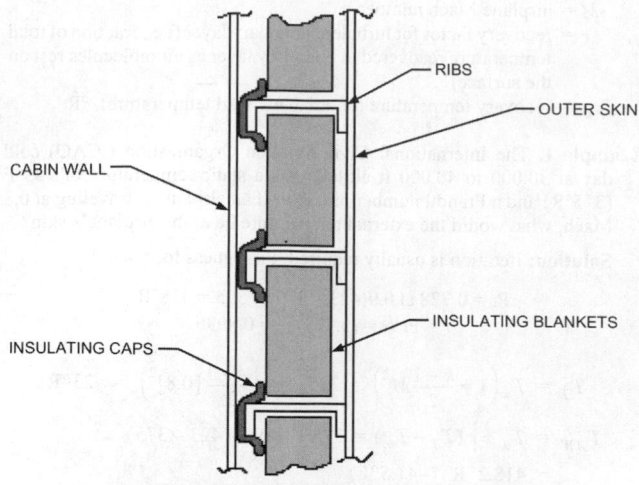

Fig. 5 Example of Aircraft Insulation Arrangement

C_y = correction factor for altitude
n = turbidity factor: 2.0 for clear air, 4 to 5 for smog
m = relative thickness of air mass, P_y/P_{SL} = (altitude pressure)/(sea level barometric pressure)
a_{ms} = molecular scattering coefficient = $0.128 - 0.054 \log_{10}(m)$
α = solar absorptivity of surface, window, paint, etc.
A = outside surface area, ft²

Conduction

The conductive path from the air in the cabin to the surrounding environment is generally described as several heat transfer elements in series and in parallel with each other. The structure is typically quite conductive (e.g., aluminum), and must be insulated to avoid a direct heat path from inside to outside. The structure typically has an outer skin supported by circumferential and longitudinal ribs. The members require a structurally efficient attachment, which often is also thermally efficient, so that the entire structure is essentially at the same temperature. As a result, the effective "fin" area may be much larger than the simple outside surface area of the fuselage. Figure 5 shows an example of an aircraft insulation arrangement.

For occupant comfort, the cabin wall temperature should not be drastically different from the air temperature within the cabin, because the passengers frequently are in contact with it and other interior surfaces. The insulation accommodates this requirement as well as noise reduction, which on occasion is the dominant requirement.

Stack Pressure across Cabin Wall

The cold outer skin during flight generates buoyancy-driven flow between the cabin and the cavity formed by the cabin wall and the outer skin. Because this cavity is normally filled with insulation blankets, it may be relatively porous to airflow. The outer skin is frequently below the cabin dew point (and below freezing), so water condenses on the structure and ice may build up with time. The amount of flow in and out of the cavity depends on the leakage area of the cabin wall. Leakage commonly occurs through panel joints and gaps surrounding penetrations, as well as around doors, where additional structure and mechanisms may provide addition thermal conductivity and air passages to the outer skin. A certain amount of flow in and out of the cavity is unavoidable because of normal pressurization and depressurization of the cabin during descent and climb. The driving pressure, or stack pressure, is simply the density difference between the connected volumes:

$$\Delta P_{stack} = (\rho_{cavity1} - \rho_{cavity2})gh$$

$$= \frac{P_{cabin}\left(\dfrac{1}{T_{cavity1}} - \dfrac{1}{T_{cavity2}}\right)gy}{R}$$

where

y = cavity height, ft
T = temperature, °R
ρ = air density, slug/ft³
g = gravitational constant = 32.3 ft/s²
R = gas constant, 1716 ft²/s²·°R
ΔP = stack pressure, lb/ft²
P_{cabin} = absolute cabin pressure, lb/ft²

Metabolic Heat from Occupants

A thorough treatment of metabolic heat from humans is covered in Chapter 9 of the 2009 *ASHRAE Handbook—Fundamentals*. Because an airplane cabin is frequently at higher altitudes, the balance between sensible and latent heat changes slightly from that given in that chapter. To correct for altitude, the following approach is recommended. First, examining the heat transfer coefficient:

For low air velocity ($V < 40$ fpm), flow is dominated by natural convection:

$$h = \frac{k}{d}C(Gr\,Pr)^m = \frac{k}{d}C\left(\frac{\rho^2 g\beta(\Delta T)d^3\,Pr}{\mu^2}\right)^m \rightarrow h \propto \rho^{2m}$$

For a cylindrical approximation of an adult at rest, $Gr = 10^7$, so $C = 0.59$ and $m = 1/4$, which leads to

$$h_{alt} = h_{SL}\left(\frac{\rho_{alt}}{\rho_{S.L.}}\right)^{2(1/4)} = h_{SL}\left(\frac{P_{alt}}{P_{S.L.}}\right)^{0.5}$$

For higher air velocity ($40 < V < 800$ fpm), flow is dominated by forced convection:

$$h = \frac{k}{d}C(Re)^n Pr^{1/3} = \frac{k}{d}C\left(\frac{\rho Vd}{\mu}\right)^n Pr^{1/3} \rightarrow h \propto \rho^n$$

For a cylindrical approximation of an adult at rest, $Re > 4000$, so $C = 0.193$ and $n = 0.618$, which leads to

$$h_{alt} = h_{SL}\left(\frac{\rho_{alt}}{\rho_{S.L.}}\right)^{0.618} = h_{SL}\left(\frac{P_{alt}}{P_{SL}}\right)^{0.618}$$

These two correction factors have been combined [see Equation (37) in Chapter 9 of the 2009 *ASHRAE Handbook—Fundamentals*] to produce a simpler relationship that applies to the full velocity range ($0 < V < 800$ fpm):

$$h_{alt} \approx h_{SL}\left(\frac{P_{alt}}{P_{SL}}\right)^{0.55}$$

Next, examining the evaporation or mass transfer from the occupants, the evaporative heat transfer coefficient varies inversely with the ambient pressure [see Equation (38) in Chapter 9 of the 2009 *ASHRAE Handbook—Fundamentals*]:

$$h_e = (LR)(h)$$

Table 1 Heat and Mass Transfer Coefficients for Human Body Versus Altitude

Altitude, ft	Pressure, lb/ft²	Convection h, Btu/h·ft²·°F		Evaporation h_e, Btu/h·ft²·°F	
		$V < 40$	$40 < V < 800$	$V < 40$	$40 < V < 800$
0	2116	0.55	$0.061V^{0.6}$	113	$12.5V^{0.6}$
1000	2041	0.54	$0.060V^{0.6}$	115	$12.7V^{0.6}$
2000	1968	0.53	$0.059V^{0.6}$	117	$12.9V^{0.6}$
3000	1897	0.52	$0.057V^{0.6}$	119	$13.1V^{0.6}$
4000	1828	0.51	$0.056V^{0.6}$	121	$13.4V^{0.6}$
5000	1761	0.50	$0.055V^{0.6}$	123	$13.6V^{0.6}$
6000	1696	0.49	$0.054V^{0.6}$	125	$13.8V^{0.6}$
7000	1633	0.48	$0.053V^{0.6}$	127	$14.0V^{0.6}$
8000	1572	0.47	$0.052V^{0.6}$	129	$14.3V^{0.6}$

$$h_{e,alt} = LR_{alt}h_{SL}\left(\frac{P_{alt}}{P_{SL}}\right)^{0.55} \quad \text{and} \quad h_{SL} = \frac{h_{e,SL}}{LR_{SL}}$$

$$h_{e,alt} = LR_{alt}h_{SL}\left(\frac{h_{e,SL}}{LR_{SL}}\right)\left(\frac{P_{alt}}{P_{SL}}\right)^{0.55}$$

Substitute

$$LR_{alt} = \frac{R_{air}h_{fg}(144)\left(\dfrac{D_v}{\alpha}\right)^{2/3}}{P_{alt}c_{p,air}R_w}$$

$$LR_{SL} = \frac{R_{air}h_{fg}(144)\left(\dfrac{D_v}{\alpha}\right)^{2/3}}{P_{SL}c_{p,air}R_w}$$

(a conversion of pressure by 144 is shown here because LR is customarily in units of °F/psi)

$$h_{e,alt} \approx h_{e,SL}\left(\frac{P_{S.L.}}{P_{alt}}\right)^{0.45}$$

where

LR = Lewis relation, °F/psi
h = heat transfer coefficient, Btu/h·ft²·°F
h_{SL} = heat transfer coefficient at sea level, Btu/h·ft²·°F
h_e = evaporative heat transfer coefficient, Btu/h·ft²·psi
$h_{e,alt}$ = evaporative heat transfer coefficient at altitude, Btu/h·ft²·psi
$h_{e,SL}$ = evaporative heat transfer coefficient at sea level, Btu/h·ft²·psi
P_{alt} = cabin pressure at altitude, lb/ft²
P_{SL} = pressure at sea level; 2116 lb/ft²²
R_{air} = gas constant for air; 1716 ft²/s²·°R
h_{fg} = evaporation enthalpy at human skin temperature, 1037 Btu/lb
D_v = mass diffusivity of water vapor in air; 0.99 ft²/h
α = diffusivity; 0.8366 ft²/h
$c_{p,air}$ = specific heat of air; 0.24 Btu/lb·°F
R_w = gas constant for water vapor; approximately 2758 ft²/s²·°R

About 70% of the metabolic heat is lost through convection/radiation (245 Btu/h) sensible or 98 Btu/h convection, $h = 0.55$, plus 147 Btu/h radiation, $h_r = 0.83$) and 30% through evaporation (105 Btu/h latent) while seated at rest at sea level (see Table 1). At 8000 ft cabin altitude, in still air, the sensible heat would drop to $(0.47/0.55)98 = 84$ Btu/h convection, the radiation would remain at 147 Btu/h, and the latent heat would rise to $(129/113)105 = 120$ Btu/h for a total of 351 Btu/h. This would indicate a slightly higher temperature for comfort, or a net effect of a slightly cooler sensation at altitude, compared to the 350 Btu/h total required.

Internal Heat Sources

When considering heat sources in the cabin, there are several parallels to commercial and residential HVAC. Many heat sources such as appliances (refrigerators, conventional ovens, microwave ovens), lighting, and entertainment (TV, stereo), may be in the cabin. In addition, the electronics and equipment associated with the operation of a commercial aircraft put demands on the airplane's environmental control system.

Cooling Requirements. The sizing criteria for air conditioning are usually ground operation on a hot, humid day with the aircraft fully loaded and the doors closed. A second consideration is cooldown of an empty, heat-soaked aircraft before passenger loading; a cooldown time of less than 30 min is usually desired. A cabin temperature of between 75 and 80°F is typically specified for these hotday ground design conditions. During cruise, the system should maintain a cabin temperature of 75°F with a full passenger load. The cooling load is entirely sensible in most cases when air-cycle machines are used. When a vapor-cycle recirculation system is used, latent heat is added.

Heating Requirements. Heating requirements are based on a partially loaded aircraft on a very cold day. Cabin temperature warm-up within 30 min for a cold-soaked aircraft is also desired. A cabin temperature of 70°F is typically specified for cold-day ground-operating conditions. During cruise, the system should able to maintain a cabin temperature of 75°F with a 20% passenger load, a cargo compartment temperature above 40°F, and cargo floor temperatures above 32°F to prevent freezing of cargo.

Temperature Control

Whenever a section of the cabin or flight deck has capability for independent supply temperature control, it is termed a **zone**. Commercial aircraft (over 19 passengers) can have as few as two zones (cockpit and cabin) and as many as seven. These crew and passenger zones are individually temperature-controlled to a crew-selected temperature for each zone, ranging from 65 to 85°F. Some systems have limited temperature control in the passenger zones that can be adjusted by the flight attendants. The selected zone temperature is controlled to within 2°F of the sensed temperature, and temperature uniformity in the zone should be within 5°F. Separate temperature controls can be provided for cargo compartments.

Temperature control may also be the predominant driver of ventilation requirements. The interior of the fuselage has several electronic/electrical heat sources that are required for the aircraft's operation, as well as heat loads from ambient and from occupants and their activities. These increasing heat loads are accommodated by reducing supply temperatures:

$$T_{supply} = T_{cabin} - \frac{q_{sources}}{c_p \dot{w}}$$

where

$q_{sources}$ = all heat into cabin, Btu/h
$\dot{w}$ = air weight flow, lb/h
c_p = specific heat; 0.24 Btu/lb·°F for air
T = temperature, °F

Supply temperatures in each of the zones have practical limits, such as the freezing temperature of water (humidity), when either the heat loads are too large or the mass flow is too low.

Air Velocity

The passenger cabin is most similar to buildings with very high occupant densities, such as theaters or lecture halls. In these situations, the air-conditioning system is typically in cooling mode (i.e.,

the supply diffuser temperature is cooler than the room temperature). The ducting and diffuser networks are best described as coldair systems, in which the duct velocities are higher, duct temperatures lower, and the fraction of recirculated air smaller (about 50% of the mixture) than in buildings (which use up to 95%). The coldair diffuser is also in much closer proximity to the occupants in an aircraft cabin. The design challenge is to deliver cool air to the passengers without uncomfortable drafts.

The velocity characteristics of an airplane cabin are uniquely affected by transitional flow behavior. The supply diffuser Reynolds number is typically between 3000 and 5000. Turbulence induced by the diffuser affects the perceived draftiness. Figure 6 shows the unsteady velocity variations measured in an aircraft cabin.

At any instant in time, the velocity field in the cabin will change, but an overall pattern develops for the time-averaged velocity. An example of this comes from computational fluid dynamics (CFD) modeling of a passenger cabin. Ventilation air enters the cabin near the center and blows outward in two directions. Air leaves near the floor on both sides, as shown in Figure 7.

Several comfort indices are used to evaluate air velocity, such as predicted percent dissatisfied (PPD) or predicted mean vote (PMV). Draft-sensitive areas of the body such as the ankles or neck receive special attention during air distribution system design. The velocity requirements are described in detail in Chapter 9 of the 2009 *ASHRAE Handbook—Fundamentals* and ASHRAE *Standard* 55.

Ventilation

Air drawn from the compressor section of the jet engine is called **bleed air** (also known as outside air, fresh air, outdoor air, or ambient air). When air is provided by sources other than the engine, it is not bleed air in the strict sense, because it is no longer bled from the engines. The current FAA requirement is to provide 0.55 lb/min of fresh air per person. Because this requirement is expressed as a weight flow, the fresh-air ventilation rate as a volumetric flow varies with cabin pressure and temperature. Cabin altitude is a convenient way of expressing cabin pressure by referencing the cabin pressure to a standard atmosphere. The required 0.55 lb/min is equivalent to about 7 cfm/person at sea level and 75°F. At other cabin pressures or altitudes, flow volumes can be found with the following equation:

$$Q_{FR} = \frac{\dot{w}}{\rho} = \frac{\dot{w}}{\left(\dfrac{P_c g}{R T_c}\right)} = \frac{\dot{w} R T_c}{P_c g}$$

where

$\dot{w}$ = 0.55 lb/min
R = 1716 ft²/s²·°R 0.2865 m²/(s²·K)

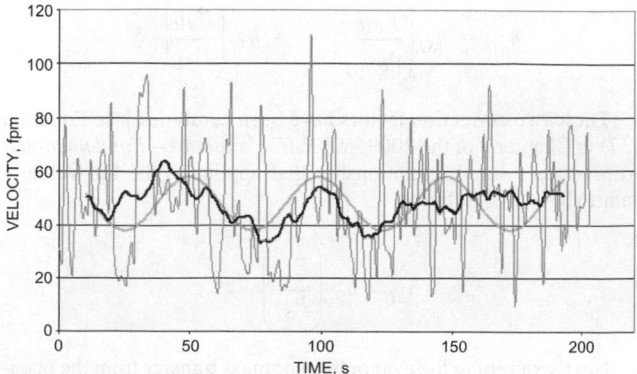

Red curve shows instantaneous anemometer data, a 5 s accumulative is shown in black, and a sinusoidal approximation is shown in blue.

Fig. 6 Transient Air Velocity Measured in Seated Area of Aircraft Cabin

$g = 32.2$ ft/s²
T_c = cabin temperature; 70°F = 529.67°R
P_c = cabin pressure from Table 2, lb/ft²

In many aircraft, the fresh/outdoor air flow is augmented by filtered or recirculated air. There are currently no regulatory requirements on the amount of recirculated air that enters the cabin, but it is customary to provide about 10 cfm per person of recirculated air in addition to the fresh air required by regulation, for a total of about 20 cfm per person at 8000 ft cabin altitude.

Ventilation Effectiveness. ASHRAE *Standard* 62.1-2010 defines ventilation effectiveness (VE) as a measure of mixing within the volume relative to a perfectly mixed system and is described with the following equations:

$$VE = \frac{c_{mixed} - c_{in}}{c_{local} - c_{in}}$$

where c = contaminant concentration, or

$$VE = \frac{Q_{local}}{Q_{cabin}}$$

where

c_{local} = local contaminant concentration by volume
c_{in} = inlet contaminant concentration
c_{mixed} = concentration if perfectly mixed
Q_{cabin} = contaminant flow to cabin or zone, cfm
Q_{local} = flow delivered to breathing zone, cfm

Contaminant concentrations in a perfectly mixed system are the same in the cabin volume as at the exit (floor grilles), and the concentration in the exit is based only on the ventilation rate and generation rate (including source and sink). Therefore, ventilation effectiveness indicates the degree of contaminant stratification with the volume. VE > 1 means that concentrations in the breathing zone are lower than in a perfectly mixed system; VE < 1 means they are higher.

There is a distinction between VE for fresh air and VE for total ventilation. For fresh air, the inlet concentration c_{in} is the concentration of gases in the supply air to the entire system (i.e., bleed air concentration). The local concentration will be larger than the inlet concentration only if the contaminant is generated within the cabin. For total ventilation, VE uses the c_{in} at the nozzle (i.e., supply mixture concentration) and includes contaminants from the recirculation system. The practical use of this VE applies to particulate levels in the cabin, because the recirculated air is equivalent to fresh air in this regard.

Contaminant concentrations in the cabin can be converted to flows delivered to the breathing zone Q_{local} using the following relationship:

Table 2   FAA-Specified Fresh Air Flow per Person

Cabin Pressure		Required Flow per Person
psia	lb/ft²	Altitude, ft	at 75°F, cfm
14.696	2116	0	7.4
14.432	2078	500	7.5
14.173	2041	1000	7.7
13.917	2004	1500	7.8
13.664	1968	2000	8.0
13.416	1932	2500	8.1
13.171	1897	3000	8.3
12.930	1862	3500	8.4
12.692	1828	4000	8.6
12.458	1794	4500	8.7
12.228	1761	5000	8.9
12.001	1728	5500	9.1
11.777	1696	6000	9.2
11.557	1664	6500	9.4
11.340	1633	7000	9.6
11.126	1602	7500	9.8
10.916	1572	8000	10.0

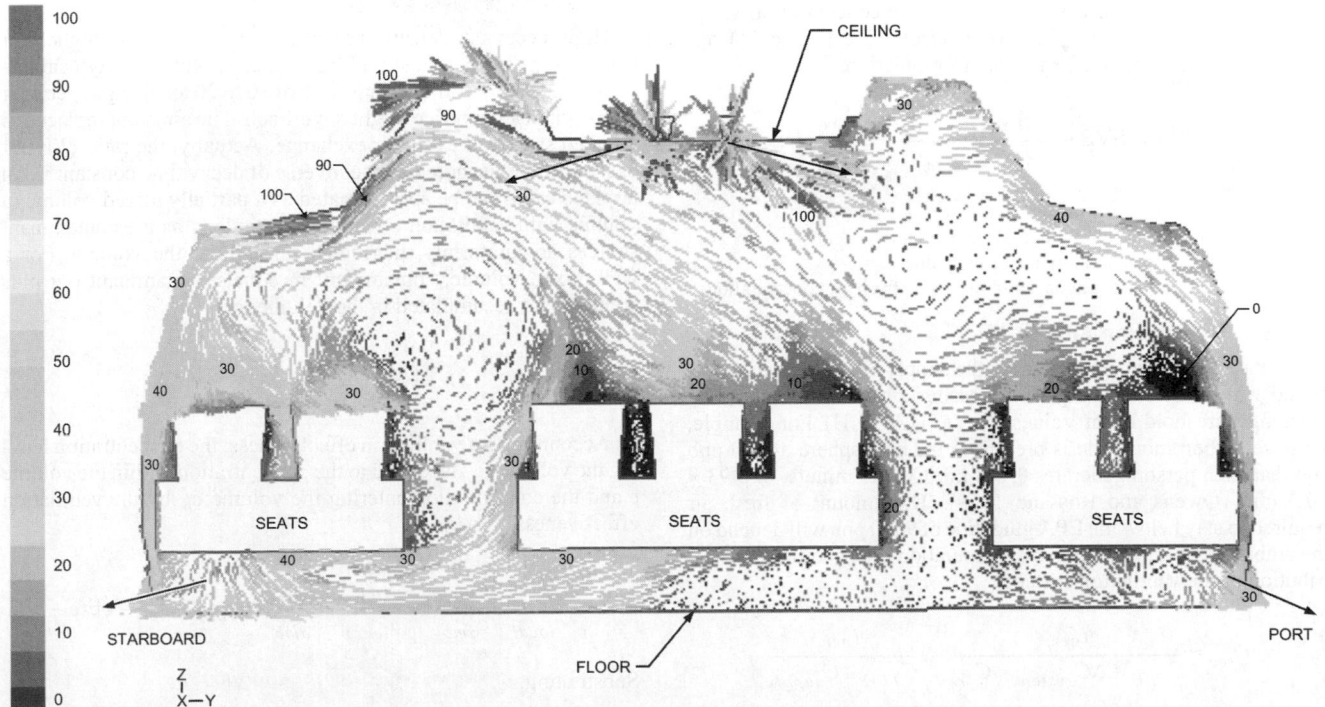

Fig. 7   Cabin Air Velocities from CFD, fpm
(Lin et al. 2005)

$$Q_{local} = \frac{q_{gen}}{c_{local} - c_{in}}$$

Substitute

$$q_{gen} = Q_{supplied}(c_{mixed} - c_{in})$$

$$Q_{local} = Q_{supplied}\frac{c_{mixed} - c_{in}}{c_{local} - c_{in}}$$

where

q_{gen} = CO_2 generation rate, 0.0105 scfm/person
c_{local} = local CO_2 concentration by volume
c_{in} = inlet CO_2 concentration
$Q_{supplied}$ = flow to cabin or zone, cfm
Q_{local} = flow delivered to breathing zone, cfm

Some consideration can be given to distribution effectiveness (DE), where flows to higher-occupant-density sections of the cabin (such as coach) are used to set minimum flows to the cabin, and lower-density sections (such as first class) may subsequently be overventilated:

$$DE = \frac{Q_{zone}/n_{zone}}{Q_{cabin}/n_{cabin}}$$

where

Q_{zone}/n_{zone} = flow per person in zone
Q_{cabin}/n_{cabin} = average flow per person for entire cabin

Distribution effectiveness accounts for a system that provides a uniform flow per length of cabin yet has varying seating densities along the length. For fresh-air distribution, this effectiveness is tempered somewhat by occupant diversity D (see ASHRAE *Standard* 62.1), because underventilated zones feed into the same recirculation flow. For total flow (fresh + recirculated) and for systems without recirculation, however, occupant diversity does not apply.

System ventilation efficiency (SVE) is a measure of how well mixed the recirculated air is with the fresh air before it enters the cabin. The SVE can be determined from the concentration variations in the ducts leaving the mix manifold (see Figure 11), for instance. The SVE is similar to VE in formulation:

$$SVE = \frac{c_{all\ zones} - c_{amb}}{c_{zone} - c_{amb}}$$

where

$c_{all\ zones}$ = average concentration of all supply ducts
c_{zone} = concentration in individual supply duct
c_{amb} = ambient reference concentration = C_{fr} (fresh air concentration)

Dilution Ventilation and TLV

Contaminants that are present in the supply air and are also generated within the cabin require increasing dilution flows to avoid reaching Threshold Limit Values (TLVs®) (ACGIH). For example, suppose carbon monoxide is present in the atmosphere at 210 ppb and that each person generates 0.168 mL CO per minute, or 5.93×10^{-6} cfm (Owens and Rossano 1969). The amount of fresh air required to stay below the EPA guideline of 9000 ppb will depend on the ambient CO levels, the human generation rate, and the CO contribution of the ventilation system:

$$Q_{req} = \frac{q_{gen}}{C_{TLV} - \Delta C_{system} - C_{fr}} = \frac{q_{gen}}{C_{TLV} - C_{supply}}$$

where

C_{TLV} = allowable concentration
ΔC_{system} = concentration rise from system

C_{supply} = concentration in supply (air entering cabin)
C_{fr} = concentration in fresh air
q_{gen} = CO generated per person

Example 3. If the ventilation system does not contribute carbon monoxide to the supply air, then the required ventilation rate to stay below the threshold is

C_{TLV} = 9000 ppb = 0.000009

q_{gen} = 5.93×10^{-6} cfm

ΔC_{system} = 0

C_{fr} = 210 ppb = 2.1×10^{-7}

$$Q_{req} = \frac{q_{gen}}{C_{TLV} - \Delta C_{system} - C_{fr}} = \frac{5.93 \times 10^{-6}}{0.000009 - 0 - 2.1 \times 10^{-7}}$$

$$= 0.67\ cfm/person$$

If, however, the ventilation system produces a 1000 ppb rise in carbon monoxide, then the required ventilation is

C_{TLV} = 9000 ppb = 0.000009

q_{gen} = 5.93×10^{-6} cfm

ΔC_{system} = 1000 ppb = 0.000001

C_{fr} = 210 ppb = 2.1×10^{-7}

$$Q_{req} = \frac{q_{gen}}{C_{TLV} - \Delta C_{system} - C_{fr}} = \frac{5.93 \times 10^{-6}}{0.000009 - 0.000001 - 2.1 \times 10^{-7}}$$

$$= 0.76\ cfm/person$$

It is important to note that, under certain circumstances, q_{gen} and ΔC_{system} may change sign as contaminant sources become contaminant sinks. This simplified approach shown here is more conservative, and could overpredict contaminant levels in real situations.

Air Exchange

High occupant density ventilation systems have higher air exchange rates than most buildings (i.e., offices). The typical airplane may have an air exchange rate of 10 to 20 air changes per hour (ach), whereas an office might have 1 ach. The air is not replaced in a mixed system at every air exchange. Actually, the ratio Q/V (air exchange rate) is more like the inverse of decay time constant τ. An airplane cabin can be approximated as a partially mixed volume (a volume with ventilation effectiveness) as long as the contaminant sources are uniformly distributed throughout the volume. For a well-mixed volume, contaminant in equals contaminant out plus contaminant accumulated in the volume, or

$$Qc_{in} = Qc_{out} + V\frac{dc}{dt}$$

Accounting for ventilation effectiveness, the concentration leaving the volume c_{out} is related to the concentration within the volume c and the concentration entering the volume c_{in} by the ventilation effectiveness VE:

$$VE = \frac{c_{mixed} - c_{amb}}{c_{local} - c_{amb}} = \frac{c_{out} - c_{amb}}{c_{local} - c_{amb}} \rightarrow c_{out} = c_{in} + VE(c - c_{in})$$

Substituting,

$$Qc_{in} = Q\left[c_m + VE(c - c_{in})\right] + V\frac{dc}{dt}$$

which leads to

$$c = c_{in} - (c_{in} - c_o)e^{-\frac{Q(VE)}{V}t}$$

Although air exchange rates are occasionally used as requirements on the ventilation system, in the case of cabin ventilation, there is no basis for setting one. Air exchange rate can be a surrogate (only for similarly sized volumes) for temperature uniformity, air quality, or smoke clearance. The flow-per-person specification is preferred, because it can be related to the predominant pollutant source more directly. Air exchange rates therefore indirectly provide valid ventilation comparisons between airplanes of similar volume and seating density. However, comparisons with buildings are misleading: occupant densities could be 30 times higher in aircraft, and bioeffluent doses (defined here as the time integral of the concentration of occupant-generated contaminants) for the same ventilation rate per person are greater in aircraft passenger cabins, depending upon occupancy times (see the section on Air Quality).

Filtration

Most airplane manufacturers have provisions for recirculated air filtration. Common practice is to install high-efficiency particulate air (HEPA) filters. The current industry standard for new build production aircraft is EU class II13 (i.e., 99.99% minimum removal efficiency by sodium flame test) (Eurovent 4/4, BS3928). This is equivalent to 99.97% minimum removal efficiency at 0.3 μm when tested according to Institute of Environmental Sciences and Technology *Recommended Practice* RP-CC001.5 (IEST 1997).

Filters are required to have sufficient particulate capacity to remain effective between normal maintenance intervals. The life of the filter is related to the recirculation system pressure drop, system operating pressure, and the recirculation fan curve. As the filter becomes loaded, pressure drop increases. When added to the system losses, the effect is a reduction in flow, as shown in Figure 8.

It is important to change the filters at least as often as recommended by the manufacturer to maintain flow capacity.

Most systems have no filtration of the engine bleed air supply as standard, although some technologies (e.g., combined VOC/ozone converters and bleed air centrifugal cleaners) are sometimes offered as optional equipment.

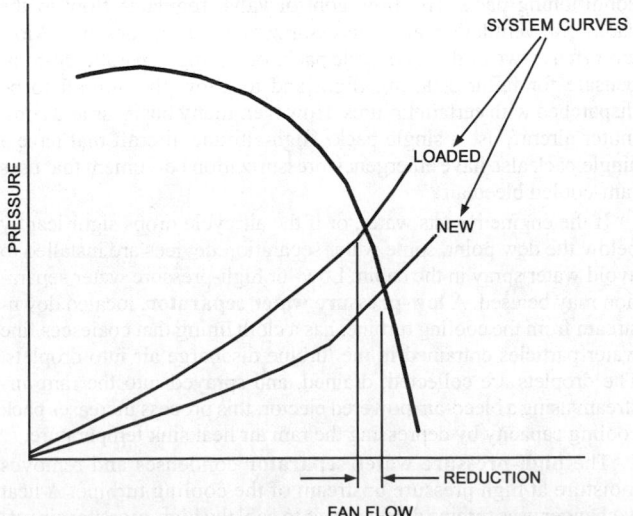

Fig. 8   Flow Reduction Caused by Filter Loading

Pressurization/Oxygen

Cabin pressurization achieves the required partial pressures of oxygen for the crew and passengers during high-altitude flight. At altitudes above 8000 ft, the occupied cabin must be pressurized to an equivalent altitude of 8000 ft or less to allow normal physiological functions without supplemental oxygen. The maximum pressure difference between the cabin and outside environment is limited by aircraft structural design limits. The differential pressure control provides a cabin pressure based on the flight altitude of the aircraft. A typical cabin altitude schedule is shown in Figure 3. Additional provisions that are separate from normal cabin pressure controls must be provided for positive- and negative-pressure relief to protect the aircraft structure.

A DOT-sponsored study concluded that current pressurization criteria and regulations are generally adequate to protect the traveling public. The study also noted that the normal maximum rates of change of cabin pressure (approximately 500 ft/min in increasing altitude and 300 ft/min in decreasing altitude) do not pose a problem for the typical passenger.

However, pressurization of the cabin to equivalent altitudes of up to 8000 ft, as well as changes in the normal rates of pressure during climb and descent, may create discomfort for some people, such as those suffering from upper respiratory or sinus infections, obstructive pulmonary diseases, anemia, or certain cardiovascular conditions. In those cases, supplemental oxygen may be recommended. Children and infants sometimes experience discomfort or pain because of pressure changes during climb and descent. Injury to the middle ear has occurred in susceptible people, but is rare.

During a sudden cabin depressurization in flight, passengers and crew are provided with overhead masks supplying supplemental oxygen. Passengers with respiratory diseases can bring portable oxygen containers on board.

Humans at rest breathe at a rate of approximately 0.32 cfm while consuming oxygen at a rate of 0.014 cfm at 8000 ft. The percent oxygen makeup of the supply air remains at approximately 21% at cruise altitude. A person receiving 10 cfm of outside air and 10 cfm of recirculation air would therefore receive approximately 4.2 cfm of oxygen. The level drops to 4.186 cfm as it leaves the cabin. Consequently, the content of oxygen in cabin air is little affected by breathing (i.e., it drops 0.33%). Although the percentage of oxygen in cabin air remains virtually unchanged (20.93%) at all normal flight altitudes, the partial pressure of oxygen decreases with increasing altitude, which decreases the amount of oxygen held by the blood's hemoglobin. The increase in cabin altitude may cause low grade hypoxia (reduced tissue oxygen levels) in some people. However, the National Academy of Sciences (NAS 1986, 2002) concluded that pressurization of the cabin to an equivalent altitude of 5000 to 8000 ft is physiologically safe for healthy individuals: no supplemental oxygen is needed to maintain sufficient arterial oxygen saturation.

System Description

The outdoor air supplied to the airplane cabin is usually provided by the compressor stages of the engine, and cooled by air-conditioning packs located under the wing center section. An air-conditioning pack uses the compressed ambient air as the refrigerant in air-cycle cooling.

Air is supplied and exhausted from the cabin on a continuous basis. As shown in Figure 9, air enters the passenger cabin from supply nozzles that run the length of the cabin. Exhaust air leaves the cabin through return air grilles located in the sidewalls near the floor, running the length of the cabin on both sides. Exhaust air is continuously extracted from below the cabin floor by recirculation fans that return part of the air to the distribution system. The remaining exhaust air passes to an outflow valve, which directs the air overboard. The cabin ventilation system is designed to deliver air uniformly along the length of the cabin.

Pneumatic System

The pneumatic system, or **engine bleed air system**, extracts a small amount of the gas turbine engine compressor air to ventilate and pressurize the aircraft compartments. A schematic of a typical system is shown in Figure 10. During climb and cruise, bleed air is usually taken from the mid-stage engine bleed port for minimum-horsepower extraction (bleed penalty). During idle descent it is taken from the high-stage engine bleed port, where maximum available pressure is required to maintain cabin pressure and ventilation. The auxiliary power unit (APU) is also capable of providing the pneumatic system with compressed air on the ground and in flight. Bleed air is pressure-controlled to meet the requirements of the system using it, and it is usually cooled to limit bleed manifold temperatures to meet fuel safety requirements. In fan jets, engine fan air is extracted for use as a heat sink for bleed air using an air-to-air heat exchanger called a precooler; for turboprop engines, ram air is used, which usually requires an ejector or fan for static operation. Other components include bleed-shutoff and modulating valves, a fan-air-modulating valve, sensors, controllers, and ozone converters. The pneumatic system is also used intermittently for airfoil and engine cowl anti-icing, engine start, and several other pneumatic functions.

Each engine has an identical bleed air system for redundancy and to equalize the compressor air bled from the engines. The equipment is sized to provide the necessary temperature and airflow for airfoil and cowl anti-icing, or cabin pressurization and air conditioning with one system or engine inoperative. The bleed air used for airfoil anti-icing is controlled by valves feeding piccolo tubes extending along the wing leading edge. Similar arrangements may be used for anti-icing the engine cowl and tail section.

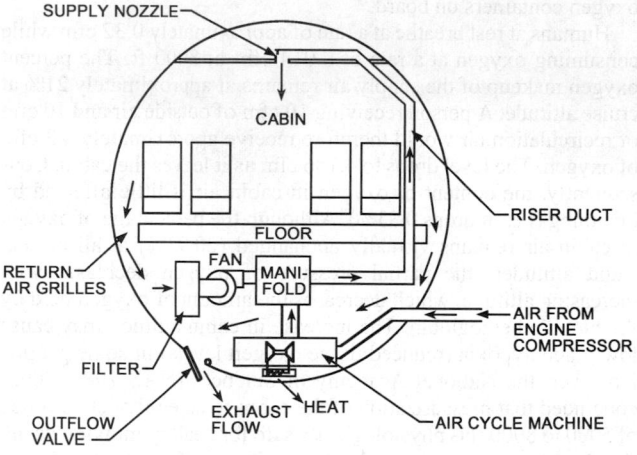

Fig. 9   Cabin Airflow Path

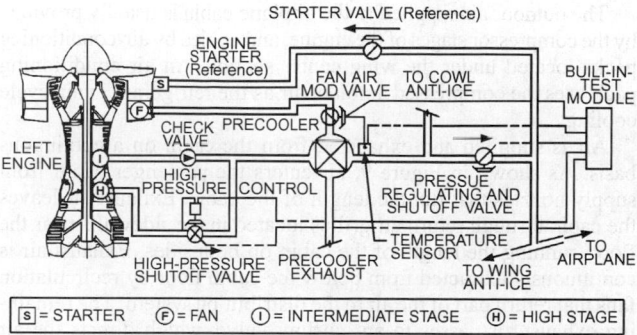

Fig. 10   Engine/APU Bleed System

Air Conditioning

Air-cycle refrigeration is the predominant means of air conditioning for commercial and military aircraft. The reverse-Brayton cycle or Brayton refrigeration cycle is used, as opposed to the Brayton power cycle that is used in gas turbine engines. The difference between the two cycles is that, in the power cycle, fuel in a combustion chamber adds heat, and in the refrigeration cycle, a ram-air heat exchanger removes heat. The familiar Rankine vapor cycle, which is used in building and automotive air conditioning and in domestic and commercial refrigeration, is used for military aircraft as well as galley cooling on larger commercial transports.

In an air cycle, compression of the ambient air by the gas turbine engine compressor provides the power input. The heat of compression is removed in a heat exchanger using ambient air as the heat sink. This cooled air is refrigerated by expansion across a turbine powered by the compressed bleed air. The turbine energy resulting from the isentropic expansion is absorbed by a second rotor, which is either a ram air fan, bleed air compressor, or both. This assembly is called an **air cycle machine (ACM)**.

The most common types of air-conditioning cycles for commercial transport aircraft are shown in Figure 11. All equipment in common use on commercial and military aircraft is open loop, although many commercial aircraft systems include various means of recirculating cabin air to minimize engine bleed air use without sacrificing cabin comfort. The basic differences between the systems are the type of air cycle machine used and its means of water separation.

The most common of these air cycle machines in use are the bootstrap ACM consisting of a turbine and compressor; the three-wheel ACM consisting of a turbine, compressor, and fan; and the four wheel ACM consisting of two turbines, a compressor, and a fan. The bootstrap ACM is most commonly used for military applications, although many older commercial aircraft models use the bootstrap cycle. The three-wheel ACM (simple bootstrap cycle) is used on most of the newer commercial aircraft, including commuter aircraft and business aircraft. The four-wheel ACM (condensing cycle) was first applied in 777 aircraft.

The compartment supply temperature may be controlled by mixing ram-cooled bleed air with the refrigerated air to satisfy the range of heating and cooling. Other more sophisticated means of temperature control are often used; these include ram air modulation, various bypass schemes in the air-conditioning pack, and downstream controls that add heat for individual zone temperature control.

The bleed airflow is controlled by a valve at the inlet of the air-conditioning pack. The flow control valve regulates flow to the cabin for ventilation and repressurization during descent. Most aircraft use two or three air cycle packs operating in parallel to compensate for failures during flight and to allow the aircraft to be dispatched with certain failures. However, many business and commuter aircraft use a single pack. High-altitude aircraft that have a single pack also have emergency pressurization equipment that uses ram-cooled bleed air.

If the engine ingests water, or if the air cycle drops significantly below the dew point, some water separation devices are installed to avoid water spray in the cabin. Low- or high-pressure water separation may be used. A **low-pressure water separator**, located downstream from the cooling turbine, has a cloth lining that coalesces fine water particles entrained in the turbine discharge air into droplets. The droplets are collected, drained, and sprayed into the ram airstream using a bleed-air-powered ejector; this process increases pack cooling capacity by depressing the ram air heat sink temperature.

The **high-pressure water separator** condenses and removes moisture at high pressure upstream of the cooling turbine. A heat exchanger uses turbine discharge air to cool the high-pressure air sufficiently to condense most of the moisture present in the bleed air supply. The moisture is collected and sprayed into the ram airstream.

In the condensing cycle one turbine removes the high-pressure water and the second turbine does the final expansion to subfreezing temperature air that is to be mixed with filtered, recirculated cabin air. Separating these functions recovers the heat of condensation, which results in a higher cycle efficiency. It also eliminates condenser freezing problems because the condensing heat exchanger is operated above freezing conditions.

The air-conditioning packs are located in unpressurized areas of the aircraft to minimize structural requirements of the ram air circuit that provides the necessary heat sink for the air-conditioning cycle. This location also provides protection against cabin depressurization in the event of a bleed or ram air duct rupture. The most common areas for the air-conditioning packs are the underwing/wheel well area and the tail cone area aft of the rear pressure bulkhead. Other areas include the areas adjacent to the nose wheel and overwing fairing. The temperature control components and recirculating fans are located throughout the distribution system in the pressurized compartments. The electronic pack and zone temperature controllers are located in the E/E bay. The air-conditioning control panel is located in the flight deck. A schematic of a typical air-conditioning system is shown in Figure 12.

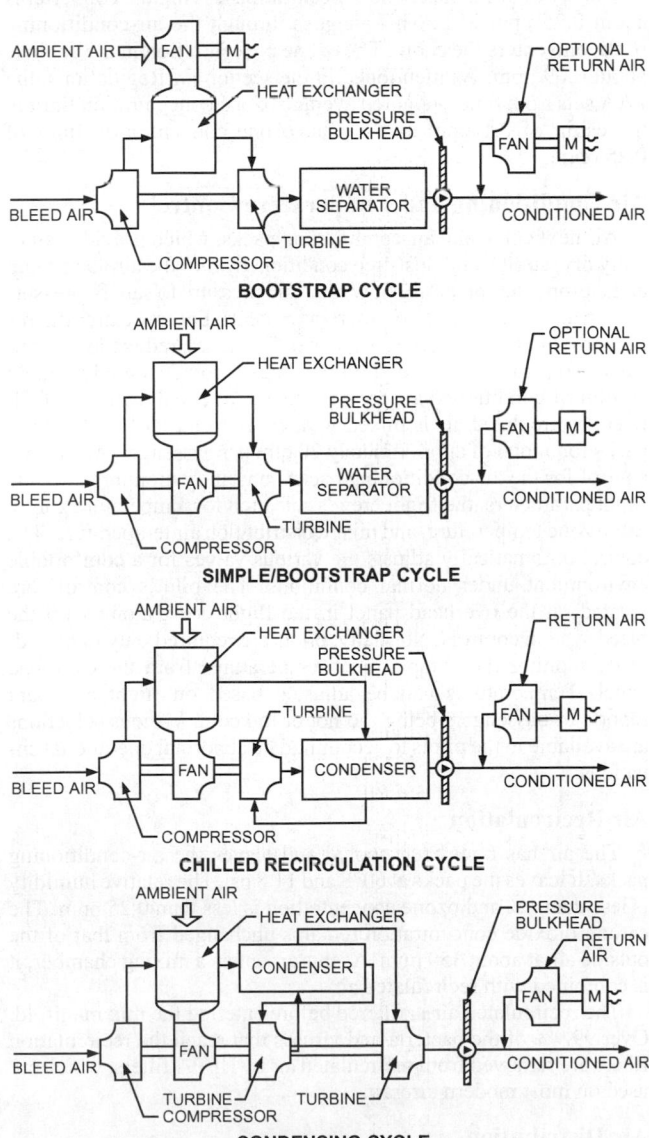

Fig. 11   Some Aircraft Refrigeration Cycles

Cabin Pressure Control

Cabin pressure is controlled by modulating airflow discharged from the pressurized cabin through one or more cabin outflow valves. The cabin pressure control includes the outflow valves, controller, selector panel, and redundant positive-pressure relief valves. Provisions for negative-pressure relief are incorporated in the relief valves and/or included in the aircraft structure (door). The system controls the cabin ascent and descent rates to acceptable comfort levels, and maintains cabin pressure altitude in accordance with cabin-to-ambient differential pressure schedules. Modern controls usually set landing field altitude, if not available from the flight management system (FMS), and monitor aircraft flight through the FMS and the air data computer (ADC) to minimize cabin pressure altitude and rate of change.

The cabin-pressure-modulating and safety valves (positive-pressure relief valves) are located either on the aircraft skin, in the case of large commercial aircraft, or on the fuselage pressure bulkhead, in the case of commuter, business, and military aircraft. Locating outflow valves on the aircraft skin precludes handling of large airflows in the unpressurized tailcone or nose areas and provides some thrust recovery; however, these double-gate valves are more complex than the butterfly or poppet-type valves used for bulkhead installations. Safety valves are poppet-type valves for either installation. Most commercial aircraft have electronic controllers located in the electrical/electronic (E/E) bay. The cabin pressure selector panel is located in the flight deck.

TYPICAL FLIGHT

A typical flight scenario from London's Heathrow Airport to Los Angeles International Airport would be as follows:

While the aircraft is at the gate and the engines have not been started yet, the ECS can be powered by compressed air supplied by the auxiliary power unit (APU), or bleed air from a ground cart. The APU or ground-cart bleed air is ducted directly to the bleed air manifold upstream of the air-conditioning packs. Once started, the engines become the compressed air source and the ground carts are disconnected.

Taxiing from the gate at Heathrow, the outside air temperature is 59°F with an atmospheric pressure of 14.7 psia. The aircraft engines are at low thrust, pushing the aircraft slowly along the taxiway.

Engine Bleed Air Control

As air from outside enters the compressor stages of the engine, it is compressed to 32 psig and a temperature of 330°F. Some of this air is then extracted from the engine core through one of two

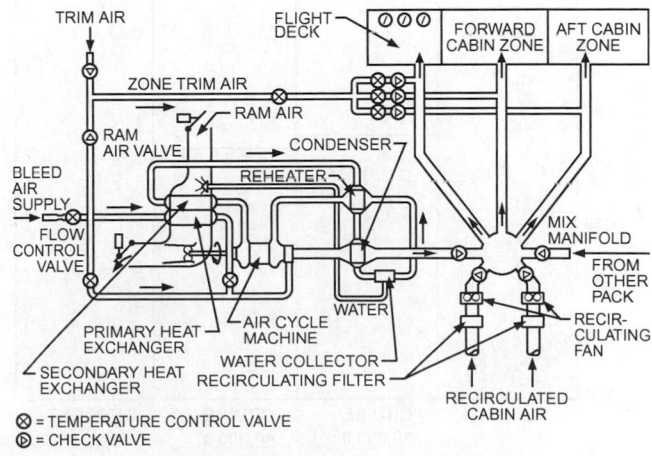

Fig. 12   Aircraft Air-Conditioning Schematic

openings (bleed ports) in the side of the engine. Which bleed port extracts the air depends on the positioning of valves that control the ports. One bleed port is at a higher engine compressor stage (e.g., fifteenth stage), commonly called high stage. The second is at a lower compressor stage (e.g., eighth stage), commonly called low stage or intermediate stage. The exact stage varies depending on engine type. At low engine power, the high stage is the only source of air at sufficient pressure to meet the needs of the bleed system. Bleed stage selection is totally automatic, except for a shutoff selection available to the pilots on the overhead panel in the flight deck.

As the aircraft turns onto the runway, the pilots advance the engine thrust to takeoff power. The engine's high stage compresses the air to 1200°F and 430 psia. This energy level exceeds the requirements for the air-conditioning packs and other pneumatic services; approximately 50% of the total energy available at the high-stage port cannot be used, so the bleed system automatically switches to the low-stage port to conserve energy.

Because the engine must cope with widely varying conditions from ground level to flight at an altitude of up to 43,100 ft, during all seasons and throughout the world, air at the high or low stage of the engine compressor seldom exactly matches the pneumatic systems' needs. Excess energy must be discarded as waste heat. The bleed system constantly monitors engine conditions and selects the least wasteful port. Even so, bleed port temperatures often exceed fuel auto-ignition temperatures. The precooler automatically discharges excess energy to the atmosphere to ensure that the temperature of the pneumatic manifold is well below that which could ignite fuel in the event of a fuel leak.

The aircraft climbs to a cruise altitude of 39,000 ft, where the outside air temperature is –70°F at an atmospheric pressure of 2.9 psia, and the partial pressure of oxygen is 0.6 psi. Until the start of descent to Los Angeles, the low-stage compressor is able to compress the low-pressure cold outdoor air to more than 30 psia and above 400°F. This conditioning of the air is all accomplished through the heat of compression: fuel is added only after the air has passed through the compressor stages of the engine core.

Figure 13 shows the temperature of the air leaving the bleed system (labeled "to airplane" in Figure 10) from the time of departure to the time of arrival at Los Angeles.

The air then passes through an ozone converter on its way to the air-conditioning packs located under the wing at the center of the aircraft.

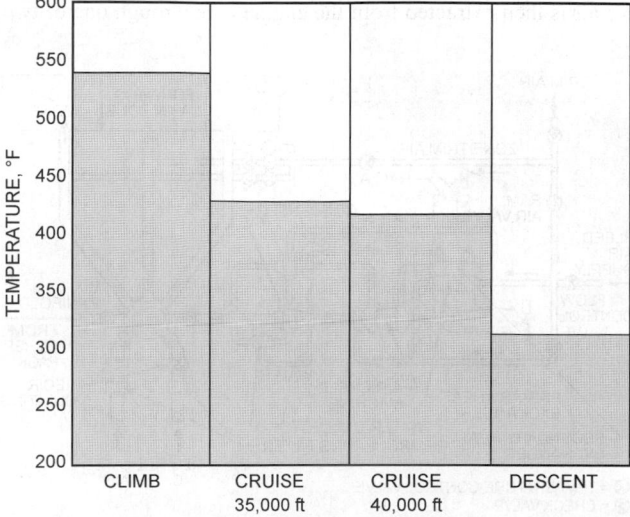

Fig. 13   Bleed Air Temperatures

Ozone Protection

While flying at 39,000 ft, several ozone plumes are encountered. Some have ozone concentrations as high as 0.8 ppm, or 0.62 ppm sea-level equivalent (SLE). This assumes a worst-case flight during the month of April, when ozone concentrations are highest. If this concentration of ozone were introduced into the cabin, passengers and crew could experience chest pain, coughing, shortness of breath, fatigue, headache, nasal congestion, and eye irritation.

Atmospheric ozone dissociation occurs when ozone goes through the compressor stages of the engine, the ozone catalytic converter (which is on aircraft with a route structure that can encounter high ozone concentrations), and the air-conditioning packs. The ozone further dissociates when contacting ducts, interior surfaces, and the recirculation system. The ozone converter dissociates ozone to oxygen molecules by using a noble catalyst such as palladium. A new converter dissociates approximately 95% of the ozone entering the converter to oxygen. It has a useful life of about 12,000 flight hours.

As the air leaves the ozone converter, it is still at 400°F and a pressure of 30 psia. Assuming a worst case when the converter is approaching the end of its useful life, with an ozone conversion efficiency of 60%, the ozone concentration leaving the converter is about 0.25 ppm SLE. This air goes through the air-conditioning packs and enters the cabin. The ozone concentration in the cabin is about 0.09 ppm. As mentioned in the section on Regulations, the FAA sets a 3 h time-weighted average ozone concentration limit in the cabin of 0.1 ppm and a peak ozone concentration limit of 0.25 ppm.

Air Conditioning and Temperature Control

Air next enters the air-conditioning packs, which provide essentially dry, sterile, and dust-free conditioned air to the airplane cabin at the proper temperature, flow rate, and pressure to satisfy pressurization and temperature control requirements. For most aircraft, this is approximately 5 cfm per passenger. To ensure redundancy, typically two (or more) air-conditioning packs provide a total of about 10 cfm of conditioned air per passenger. An equal quantity of filtered, recirculated air is mixed with air from the air-conditioning packs for a total of approximately 20 cfm per passenger. Automatic control for the air-conditioning packs constantly monitors airplane flight parameters, the flight crew's selection for temperature zones, cabin zone temperature, and mixed distribution air temperature. The control automatically adjusts the various valves for a comfortable environment under normal conditions. The pilot's controls are located on the overhead panel in the flight deck, along with the bleed system controls. Normally, pilots are required only to periodically monitor the compartment temperatures from the overhead panel. Temperatures can be adjusted based on flight attendant reports of passengers being too hot or too cold. Various selections are available to the pilots to accommodate abnormal operational situations.

Air Recirculation

The air has now been cooled and leaves the air-conditioning packs. It leaves the packs at 60°F and 11.8 psi. The relative humidity is less than 1% and ozone concentration is less than 0.25 ppm. The carbon dioxide concentration remains unchanged from that of the outside air at about 350 ppm. As this air enters a mixing chamber, it is combined with recirculated air.

The recirculated air is filtered before entering the mix manifold. Over 99.9% of the bacteria and viruses that reach the recirculation filters are removed from recirculated air by HEPA filters, which are used on most modern aircraft.

Air Distribution

The filtered and fresh-air mixture leaves the mixing chamber on its way through the air distribution system. At this time, its humidity has increased relative to fresh air by about 5 to 10% rh. The temperature of the mixture is determined by the cooling requirements of the dominant zone. Control for the remaining zones is achieved by adding hot air to the zone supply. The hot-air source is the same bleed supply as the packs, so very small amounts of air are required to adjust the temperature.

Carbon dioxide levels in the distribution system are about halfway between the levels in fresh air and in the cabin. At a 6000 ft cabin altitude, the level is about 1000 ppm in the distribution system.

The mixture leaves the air distribution system and enters the cabin through high-velocity diffusers. The diffusers run the length of the cabin. In order to minimize fore-to-aft flow and mixing between zones, flow is provided at a uniform amount per unit length of cabin. Even though the air change rates are high compared to buildings, they are low when looking at the plug flow velocity. If ventilation air were provided uniformly across the cabin, as in plug flow, the velocity would be less than 5 fpm. Momentum from the diffusers increases velocity up to comfortable levels of 15 to 65 fpm.

Once the air mixes with the air in the cabin, the humidity rises by another 5 to 10% rh to stabilize at 10 to 20% rh, and the carbon dioxide level rests at about 1700 ppm (at 6000 ft cabin altitude).

Cabin Pressure Control

The cabin pressure control system continuously monitors ground and flight modes, altitude, climb, cruise or descent modes, and the airplane's holding patterns at various altitudes. It uses this information to position the cabin pressure outflow valve to maintain cabin pressure as close to sea level as practical, without exceeding a cabin-to-outside pressure differential of 8.60 psi. At a 39,000 ft cruise altitude, the cabin pressure is equivalent to 6900 ft or a pressure of 11.5 psia. In addition, the outflow valve repositions itself to allow more or less air to escape as the airplane changes altitude. The resulting cabin altitude is consistent with airplane altitude within the constraints of keeping pressure changes comfortable for passengers. The cabin pressure control system panel is located in the pilot's overhead panel near the other air-conditioning controls. Normally, the cabin pressure control system is totally automatic, requiring no attention from the pilots.

Finally, as descent to LAX begins, the cabin pressure controller follows a prescribed schedule for repressurization. The cabin altitude eventually reaches sea level, the doors can then be opened at the gate, and passengers depart.

AIR QUALITY

Factors Affecting Perceived Air Quality

Several cabin environmental parameters, in combination with maintenance, operations, individual, and job-related factors, collectively influence cabin crew and passenger perceptions of the cabin environment, comfort, and cabin air quality. Cabin environmental quality (CEQ) must be differentiated from cabin air quality (CAQ), because many symptoms (such as eye irritation) may be caused by humidity (CEQ) as well as contaminants (CAQ), for example.

Strictly, air quality is a measure of pollutant levels. Aircraft cabin air quality is function of many variables including: the quantity of ventilation flow, ambient air quality, the design of the cabin volume, the design of the ventilation and pressurization systems, the way the systems are operated and maintained, the presence of sources of contaminants, and the strength of such sources.

Figure 14 diagrams the three groups that can influence cabin environmental quality: manufacturers, airlines, and the occupants themselves. Airplane manufacturers influence the physical environment by the design of the environmental control system integrated with the rest of the systems on the airplane. Airlines affect the environmental conditions in the cabin by seating configuration, amenities offered, and procedures for maintaining and operating the aircraft. Finally,

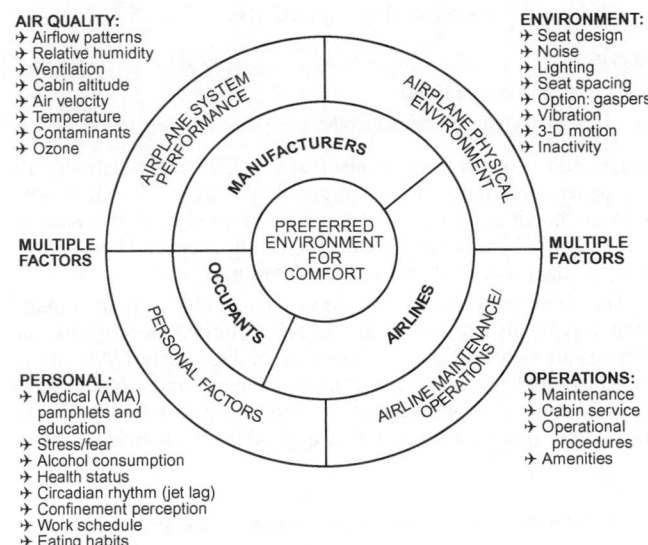

Fig. 14 Multiple Comfort Factors
(Adapted, with permission, from STP 1393—Air Quality and Comfort in Airliner Cabins, copyright ASTM International, 100 Barr Harbor Drive, West Conshohocken, PA 19428)

cabin environmental comfort is influenced by the individual and job-related activities of the cabin crew and passengers.

Airflow

The airflow per unit length of the airplane is typically the same for all sections. However, economy class has a lower airflow per passenger because of its greater seating density compared to first class and business class.

The flight deck is provided with a higher airflow per person than the cabin in order to (1) maintain a positive pressure in the cockpit to prevent smoke ingress from adjacent areas (abnormal condition), (2) provide cooling for electrical equipment, and (3) account for increased solar loads and night heat loss through the airplane skin and windows.

The fresh-air quantity supplied on some aircraft models can be lowered by shutting off one air-conditioning pack. The flight crew has control of these packs to provide flexibility in case of a system failure or for special use of the aircraft. However, packs should be in full operation whenever passengers are on board.

Air Changes

Confusion abounds over the use of air exchange rate (also called "air change rate") when making comparisons between dissimilar systems. Further, there is no air quality equivalence in the comparison of systems unless the occupied volumes are equal. This is because high air exchange rates can be achieved in two ways. As airflow increases, the air change rate increases; however, as volume decreases, the air change rate also increases but without a proportionate increase in air quality. Air exchange rate is the ratio of ventilation flow to volume:

$$\text{ACR} = \frac{Q}{V}$$

where
   ACR = air change rate, h^{-1}
   Q = flow, ft^3/h
   V = volume, ft^3

A close inspection of the definition reveals the subtle relationship between air quality $c = q/Q$ (steady state) and air exchange rate:

$$c = \text{air quality} = q/(\text{ACR})V$$

where

c = contaminant levels
q = contaminant generation rate

It would be incorrect to assume that a smaller single-aisle aircraft has better air quality than a larger double-aisle aircraft simply because the air exchange rate is higher. Similarly, comparisons to buildings would be in error. Remember, air quality is related to flow, which is the product of air change and volume.

The air exchange rate (fresh air to volume ratio) for an airplane cabin is typically between 11 and 15 ach. Dilution rates for these air changes are between 25 and 18 min for replacement of 99% of the air. The particulate equivalent of an air exchange rate (total ventilation to volume ratio, where total ventilation = fresh + HEPA filtered recirculation) is between 20 and 30 equivalent air changes per hour.

Ozone

Ozone is present in the atmosphere as a consequence of the photochemical conversion of oxygen by solar ultraviolet radiation. Ozone levels vary with season, altitude, latitude, and weather systems. A marked and progressive increase in ozone concentration occurs in the flight altitude of commercial aircraft. The mean ambient ozone concentration increases with increasing latitude, is maximal during the spring (fall season for southern latitudes), and often varies when weather causes high ozone plumes to descend.

Residual cabin ozone concentration is a function of the ambient concentration; design, operation, and maintenance of the air distribution system; and whether catalytic ozone converters are installed. Cabin ozone limits are set by FAR *Standards* 121.578 and 25.832. Catalytic ozone converters are generally required on airplanes flying mission profiles where the cabin ozone levels are predicted to exceed these limits (refer to the FAA Code of Federal Regulations for other compliance methods).

Microbial Aerosols

Biologically derived particles that become airborne include viruses, bacteria, actinomycetes, fungal spores and hyphae, arthropod fragments and droppings, and animal and human dander. One study has documented the occurrence of an outbreak of infectious disease related to airplane use. In 1977, because of an engine malfunction, an airliner with 54 persons onboard was delayed on the ground for 3 h, during which the airplane ventilation system was reportedly turned off. Within 3 days of the incident, 72% of the passengers became ill with influenza. One passenger (the index case) was ill while the airplane was delayed. With the ventilation system shut off, no fresh air was introduced into the cabin to dilute microbial aerosols and CO_2 or to control cabin temperatures.

The airplane ventilation system should never be shut off when passengers are on board, although the air packs (but not recirculation fans) may be shut off for a short time during takeoff only.

To remove particulates and biological particles from the recirculated air, use filter assemblies that contain a HEPA filter with a minimum efficiency of 99.97% on a dioctyl phthalate (DOP) test, as measured by MIL-STD-282. A HEPA filter is rated using 0.3 μm size particles. A filter's efficiency increases over time as particulates become trapped by the filter. However, system performance degrades because of increased pressure drop. Overlapping capture mechanisms in a filter also increase efficiency for particles smaller and larger than the most penetrating particle size (MPPS). For an airplane filter, the MPPS is about 0.1 to 0.2 μm.

Viruses typically range from about 0.01 to 0.2 μm, and are effectively removed by the air filtration mechanism of diffusional interception. Bacteria are typically about 0.5 to 1.5 μm, and are effectively removed by inertial impaction.

Activity Levels

Respiratory rates and hence air contaminant dose vary with activity level. Elevated activity levels increase respiration rate, and thereby may increase the dose of some airborne contaminants. Breathing rates range from approximately 0.3 cfm for a seated passenger to 0.6 cfm for a working flight attendant.

Volatile Organic Compounds

Volatile organic compounds (VOCs) can be emitted by material used in furnishings, pesticides, disinfectants, cleaning fluids, and food and beverages.

Carbon Dioxide

Carbon dioxide is the product of normal human metabolism, which is the predominant source in aircraft cabins. Concentration in the cabin varies with fresh-air flow rate; number of people present; and their individual rates of CO_2 production, which vary with activity and (to a smaller degree) with diet and health. CO_2 has been widely used as an indicator of indoor air quality, typically serving the function of a surrogate. According to the DOT (1989), measured cabin CO_2 values of 92 randomly selected smoking and nonsmoking flights averaged 1500 ppm.

The environmental exposure limit adopted by the American Conference of Governmental Industrial Hygienists (ACGIH) is 5000 ppm as the time-weighted average (TWA) limit for CO_2; this value corresponds to a fresh air ventilation rate of about 2.3 cfm per person at sea level, if the only source of CO_2 is the occupants at rest. Other sources of CO_2 within the cabin or cargo (e.g., dry ice) would of course require more ventilation. 14CFR/CS/JAR 25.831 also limits CO_2 to 5000 ppm (0.5%).

REGULATIONS

The Federal Aviation Administration (FAA) regulates the design of transport category aircraft for operation in the United States under section 14 of the Code of Federal Regulation (CFR) Part 25 [commonly referred to as the Federal Aviation Regulations (FARs)]. ECS equipment and systems must meet these requirements, which are primarily related to safety of the occupants. Certification and operation of these aircraft in the United States is regulated by the FAA in FAR Part 121. Similar regulations are applied to European nations by the European Aviation Safety Agency (EASA), formerly the Joint Aviation Authorities (JAA), which represents the combined requirements of the airworthiness authorities of the participating nations; the current equivalent design regulation is Certification Specification (CS) 25, although many airplanes were designed and certified to the former Joint Aviation Regulations (JARs) Part 25. Operating rules based on FAA or EASA/JAA regulations are applied individually by the nation of registry. Regulatory agencies may impose special conditions on the design, and compliance is mandatory.

Several 14 CFR and CS/JAR Part 25 paragraphs apply directly to transport category aircraft ECS. Those most germane to the ECS design requirements of this chapter are as follows:

1. 14CFR/CS/JAR 25.831   Ventilation
2. 14CFR/CS 25.832   Cabin ozone concentration
3. 14CFR/CS/JAR 25.841   Pressurized cabins
4. 14CFR/CS/JAR 25.1301   Function and installation
5. 14CFR/CS/JAR 25.1309   Equipment, systems, and installations
6. 14CFR/CS/JAR 25.1438   Pressurization and pneumatic systems
7. 14CFR/CS/JAR 25.1461   Equipment containing high energy rotors

These regulatory requirements are summarized in the following sections; however, the applicable FAR, CS and JAR paragraphs, amendments and advisory material should be consulted for the latest revisions and full extent of the rules.

14 CFR/CS/JAR Paragraph 25.831: Ventilation

- Each passenger and crew compartment must be ventilated.
- Each crew member must have enough fresh air to perform their duties without undue fatigue or discomfort (minimum of 10 cfm).
- Crew and passenger compartment air must be free from hazardous concentration of gases and vapors:
  - CO limit is 1 part in 20,000 parts of air
  - CO_2 limit is 0.5% by volume, sea-level equivalent. Many airplanes were designed/certified to a carbon dioxide limit of 3% by volume (the former requirement)
  - CO and CO_2 limits must be met after reasonably probable failures
- Smoke evacuation from the cockpit must be readily accomplished without depressurization.
- Occupants of the flight deck, crew rest area, and other isolated areas must be able to control the temperature and quantity of ventilating air to their compartments independently.

14 CFR 25.831, Amendment 25-87 (specifies new requirements)

- Under normal operating conditions, the ventilation system must be designed to provide each occupant with airflow containing at least 0.55 lb of fresh air per minute (or about 10 cfm at 8000 ft).
- The maximum exposure at any given temperature is specified as a function of the temperature exposure.

FAA Advisory Circular (AC)/CS AMJ/JAR ACJ: Acceptable Means of Compliance/Advisory Circular-Joint 25.831

- In the event of loss of one source, the supply of fresh air should not be less than 0.4 lb/min per person for any period exceeding 5 min. However, reductions below this flow rate may be accepted if the compartment environment can be maintained at a level that is not hazardous to the occupant.
- Any supplemental recirculating system should be able to be stopped (JAR/CS only).

14 CFR/CS 25.832: Cabin Ozone Concentration

Specifies the cabin ozone concentration during flight must be shown not to exceed the following:

- 0.25 ppm by volume, sea-level equivalent, at any time above flight level 320 (32,000 ft)
- 0.10 ppm by volume, sea-level equivalent, time-weighted average during any 3 h interval above flight level 270 (27,000 ft)

At present, JAR 25 has no requirement for cabin ozone concentration.

14 CFR/CS/JAR 25.841: Pressurized Cabins

- Maximum cabin pressure altitude is limited to 8000 ft at the maximum aircraft operating altitude under normal operating conditions.
- For operation above 25,000 ft, a cabin pressure altitude of not more than 15,000 ft must be maintained in the event of any reasonably probable failure or malfunction in the pressurization system.
- The makeup of the cabin pressure control components, instruments, and warning indication is specified to ensure the necessary redundancy and flight crew information.

14 CFR Amendment 25-87

This revision imposes additional rules for high-altitude operation.

14 CFR/CS/JAR 25.1301: Function and Installation

Each item of installed equipment must be of a kind and design appropriate to its intended function, be properly labeled, be installed according to limitations specified for that equipment, and function properly.

14 CFR/CS/JAR 25.1309: Equipment, Systems, and Installations

- Systems and associated components must be designed such that any failure that would prevent continued safe flight and landing is extremely improbable, and any other failure that reduces the ability of the aircraft or crew to cope with adverse operating conditions is improbable.
- Warning information must be provided to alert the crew to unsafe system operating conditions so they can take corrective action.
- Analysis in compliance with these requirements must consider possible failure modes, probability of multiple failures, undetected failures, current operating condition, crew warning, and fault detection.

FAR *Advisory Circular* AC 25.1309-1A, CS AMJ 25.1309, and JAR ACJs 1 to 25.1309 define the required failure probabilities for the various failure classifications: probable, improbable, and extremely improbable for the FAR requirements; and frequent, reasonably probable, remote, and extremely remote for the CS/JAR requirements.

14 CFR/CS 25.1438: Pressurization and Pneumatic Systems

This standard specifies the proof and burst pressure factors for pressurization and pneumatic systems as follows:

- Pressurization system elements
  - Proof pressure: 1.5 times max normal pressure
  - Burst pressure: 2.0 times maximum normal pressure
- Pneumatic system elements
  - Proof pressure: 1.5 times maximum normal pressure
  - Burst pressure: 3.0 times maximum normal pressure

CS/JAR 25.1438 and AMJ/ACJ 25.1438 specify the proof and burst pressure factors for pressurization and pneumatic systems as follows:

- Proof pressure
  - 1.5 times worst normal operation
  - 1.33 times worst reasonable probable failure
  - 1.0 times worst remote failure
- Burst pressure
  - 3.0 times worst normal operation
  - 2.66 times worst reasonably probable failure
  - 2.0 times worst remote failure
  - 1.0 times worst extremely remote failure

14 CFR/CS/JAR 25.1461: Equipment Containing High-Energy Rotors

Equipment must comply with at least one of the following three requirements:

- High-energy rotors contained in equipment must be able to withstand damage caused by malfunctions, vibration, and abnormal temperatures.
  - Auxiliary rotor cases must be able to contain damage caused by high-energy rotor blades.
  - Equipment control devices must reasonably ensure that operating limitations affecting the integrity of high-energy rotors will not be exceeded in service.
- Testing must show that equipment containing high-energy rotors can contain any failure that occurs at the highest speed attainable with normal speed control devices inoperative.

- Equipment containing high-energy rotors must be located where rotor failure will neither endanger the occupants nor adversely affect continued safe flight.

Categories and Definitions

Commercial users categorize their ECS equipment in accordance with the Air Transport Association of America (ATAA) *Specification* 100. The following ATAA chapters define ECS functions and components:

- **Chapter 21**, Air Conditioning, discusses heating, cooling, moisture/contaminant control, temperature control, distribution, and cabin pressure control. Common system names are the air-conditioning system (ACS) and the cabin pressure control system (CPCS).
- **Chapter 30**, Ice and Rain Protection, covers airfoil ice protection; engine cowl ice protection; and windshield ice, frost, or rain protection.
- **Chapter 35**, Oxygen, includes components that store, regulate, and deliver oxygen to the passengers and crew.
- **Chapter 36**, Pneumatic, covers ducts and components that deliver compressed (bleed) air from a power source (main engine or auxiliary power unit) to connecting points for the using systems (which are detailed in Chapters 21, 30, and Chapter 80, Starting). The pneumatic system is also commonly called the engine bleed air system (EBAS).

REFERENCES

ACGIH. Annual. *TLVs*® and *BEIs*®. American Conference of Governmental Industrial Hygienists, Cincinnati, OH.

ATAA. (no date). Specification for manufacturers technical data. *Specification* 100. Air Transport Association of America, Washington, D.C.

ASHRAE. 2010. Thermal environmental conditions for human occupancy. ANSI/ASHRAE *Standard* 55-2010.

ASHRAE. 2010. Ventilation for acceptable indoor air quality. ANSI/ASHRAE *Standard* 62.1-2010.

CFR. (annual). Aeronautics and space: Airworthiness standards: Transport category airplanes: Ventilation. 14CFR/CS/JAR 25.831. *Code of Federal Regulations*, U.S. Government Printing Office, Washington, D.C. http://edocket.access.gpo.gov/cfr_2010/janqtr/pdf/14cfr25.831.pdf.

DOT. 1989. *Airliner cabin environment: Contaminant measurements, health risks, and mitigation options*. U.S. Department of Transportation, Washington, D.C.

FAA. (no date). Airworthiness standards: Transport category airplanes. *Federal Aviation Regulations*, Part 25.

FAA. (no date). Certification and operations: Domestic, flag and supplemental air carriers and commercial operators of large aircraft. *Federal Aviation Regulations*, Part 121.

IEST. (no date). HEPA and ULPA filters. *Recommended Practice* IEST-RP-CC001.5. Institute of Environmental Sciences and Technology, Arlington Heights, IL.

JAA. (no date). *Joint airworthiness requirements: Part 25: Large aeroplanes*. Airworthiness Authorities Steering Committee. Civil Aviation Authority, Cheltenham, U.K.

Lin, C.H., R. Horstman, M.F. Ahlers, L.M. Sedgwick, K.H. Dunn, J.L. Topmiller, J.S. Bennett, and S. Wirogo. 2005. Numerical simulation of airflow and airborne pathogen transport in aircraft cabins—Part I: Numerical simulation of the flow field. *ASHRAE Transactions* 111(1): 755-763.

NAS. 1986. *The airliner cabin environment: Air quality and safety*. National Academy of Sciences, National Academy Press, Washington, D.C.

NAS. 2002. *The airliner cabin environment and the health of passengers and crew*. National Academy of Sciences, National Academy Press, Washington, D.C.

Owens, D.F. and A.T. Rossano. 1969. Design procedures to control cigarette smoke and other air pollutants. *ASHRAE Transactions*.

BIBLIOGRAPHY

ATAA. 1994. *Airline cabin air quality study*. Air Transport Association of America, Washington, D.C.

ASHRAE. 2007. Air quality within commercial aircraft. ANSI/ASHRAE *Standard* 161-2007.

Loukusa, S.M. 2010. *EASA/JAA FAA regulations and certification*. Boeing internal communication, May.

Thibeault, C. 1997. *Special committee report on cabin air quality*. Aerospace Medical Association, Alexandria, VA.

Space, D.R., R.A. Johnson, W.L. Rankin, and N.L. Nagda. 2000. The airplane cabin environment: Past, present and future research. In *Air Quality and Comfort in Airliner Cabins*, ASTM STP 1393, N.L. Nagda, ed. American Society for Testing and Materials, West Conshohocken, PA.

Walkinshaw, D.S. 2001. Investigating the impacts of occupancy density & ventilation on IAQ in offices, classrooms and aircraft. Seminar presented at 2001 ASHRAE Annual Meeting.

Washington State Department of Health. 2007. *Indoor air quality*. http://www.doh.wa.gov/hws/doc/EH/EH_INAQ2007.pdf.

SHIPS

THIS chapter covers air conditioning for oceangoing surface vessels, including naval ships, commercial vessels, fishing boats, luxury liners, pleasure craft, and inland and coastal boats, as well as oil rigs. Although the general principles of air conditioning for land installations also apply to marine applications, factors such as weight, size, fire protection, smoke control, and corrosion resistance take on greater importance, and new factors (e.g., tolerance for pitch and roll, shipboard vibration, watertightness) come into play.

The importance of shipboard air conditioning depends on a ship's mission. On passenger vessels that focus completely on passenger comfort, such as cruise ships and casino vessels, air conditioning is vital. Aboard commercial vessels (tankers, bulkers, container ships, etc.), air conditioning provides an environment in which personnel can live and work without heat stress. Shipboard air conditioning also improves reliability of electronic and other critical equipment, as well as weapons systems aboard naval ships.

This chapter discusses merchant ships, which includes passenger and commercial vessels, and naval surface ships. In general, the details of merchant ship air conditioning also apply to warships. However, all ships are governed by their specific ship specifications, and warships are often also governed by military specifications, which ensure air-conditioning system and equipment performance in the extreme environment of warship duty.

MERCHANT SHIPS

Load Calculations

The cooling load estimate considers the following factors discussed in Chapter 18 of the 2009 *ASHRAE Handbook—Fundamentals*:

- Solar radiation
- Heat transmission through hull, decks, and bulkheads
- Heat (latent and sensible) dissipation from occupants
- Heat gain from lights
- Heat (latent and sensible) gain from ventilation air
- Heat gain from motors or other electrical equipment
- Heat gain from piping, machinery, and equipment

The heating load estimate should include the following:

- Heat losses through decks and bulkheads
- Ventilation air
- Infiltration (when specified)

In addition, the construction and transient nature of ships present some complications, as addressed in the following:

SNAME. The Society of Naval Architects and Marine Engineers (SNAME) *Technical and Research Bulletin* 4-16 can be used as a guide for shipboard load calculations.

ISO. The International Organization for Standardization's (ISO) *Standard* 7547 (2002) discusses design conditions and calculations for marine HVAC systems.

Outside Ambient Temperature and Humidity. The service and type of vessel determine the proper outside design temperature,

which should be based on temperatures prevalent in a ship's area of operation. Chapter 14 of the 2009 *ASHRAE Handbook—Fundamentals* should be used to select ambient conditions, with special attention paid to high-wet-bulb data; a ship's load is often driven by the latent load associated with the outside air. It is also common for different locations to be used for cooling and heating criteria. In general, for cooling, outside design conditions are 95°F db and 78°F wb; for semitropical runs, 95°F db and 80°F wb; and for tropical runs, 95°F db and 82°F wb. For heating, 0°F is usually the design temperature, unless the vessel will always operate in warmer climates. Design temperatures for seawater are 90°F in summer and 28°F in winter.

Solar Gain. Ships require special consideration for solar gain because (1) they do not constantly face in one direction and (2) the reflective properties of water increase solar load on outside boundaries not directly exposed to sunlight. For compartments with only one outside boundary, the temperature difference (outside dry-bulb temperature − inside dry-bulb temperature) across horizontal surfaces should be increased by 50°F and vertical surfaces by 30°F. For compartments with more than one outside boundary, the temperature difference should be increased by 35°F for horizontal surfaces and 20°F for vertical surfaces. For glass surfaces, the solar cooling load (SCL) is taken to be 160 Btu/h·ft² for spaces with one outside boundary and 120 Btu/h·ft² for spaces with more than one outside boundary.

Infiltration. Infiltration through weather doors is generally disregarded. However, specifications for merchant ships occasionally require an assumed infiltration load for heating steering gear rooms and the pilothouse.

Transmission Between Spaces. For heating loads, heat transmission through boundaries of machinery spaces in either direction is not considered. Allowances are not made for heat gain from warmer adjacent spaces. For cooling loads, the cooling effect of adjacent spaces is not considered unless temperatures are maintained with refrigeration or air-conditioning equipment.

Ventilation Requirements. Ventilation must meet the requirements of ASHRAE *Standard* 62.1-2010, unless otherwise stated in the ship's specification.

Heat Transmission Coefficients. The overall heat transmission coefficients U for the composite structures common to shipboard construction do not lend themselves to theoretical derivation; they are usually obtained from full-scale panel tests. SNAME *Bulletin* 4-7 gives a method to determine these coefficients when tested data are unavailable. ISO *Standard* 7547 also gives some guidance in this area, as well as default values if better information is not available.

Inside Air Temperature and Humidity. Thermal environmental conditions for human occupancy are given in ASHRAE *Standard* 55-2010.

People. Ships normally carry a fixed number of people. The engineer must select the location where the ship's fixed complement of people creates the greatest heat load, and then not apply the people load elsewhere. Note that occupants are only counted once when determining the chiller or condensing-unit load; however, air coils in each zone must be capable of removing the heat load associated with the maximum number of people in the zone.

Ventilation in the zone can also be reduced when occupants are not present. For the ventilation load, occupants are counted once, in the location where they create the greatest ventilation requirement. The practical way to apply this concept is by measuring CO_2 levels in a space and adjusting outside air accordingly. Although using this principle can reduce required chiller or condensing-unit capacity on all ships, it is most significant aboard passenger ships.

Equipment

In general, equipment used for ships is much more rugged than that for land use. Sections 6 through 10 of ASHRAE *Standard* 26 list HVAC equipment requirements for marine applications. When selecting marine duty air-conditioning equipment, the following should be considered:

- It should function properly under dynamic roll and pitch and static trim and heel conditions. This is especially important for compressor oil sumps, oil separators, refrigerant drainage from a condenser and receiver, accumulators, and condensate drainage from drain pans.
- Construction materials should withstand the corrosive effects of salt air and seawater. Materials such as stainless steel, nickel-copper, copper-nickel, bronze alloys, and hot-dipped galvanized steel are used extensively.
- It should be designed for uninterrupted operation during the voyage and continuous year-round operation. Because ships en route cannot be easily serviced, some standby capacity, spare parts for all essential items, and extra oil and refrigerant charge should be carried.
- It should have no objectionable noise or vibration, and must meet noise criteria required by the ship's specification.
- It should occupy minimum space, commensurate with its cost and reliability. Weight should also be minimized.
- A ship may pass through one or more complete cycles of seasons on a single voyage and may experience a change from winter to summer operation in a matter of hours. Systems should be flexible enough to compensate for climatic changes with minimal attention from the ship's crew.

The following general items should be considered when selecting specific air conditioning components:

Fans. Fans must be selected for stable performance over their full range of operation and should have adequate isolation to prevent transmitting vibration to the deck. Because fan rooms are often adjacent to or near living quarters, effective sound treatment is essential.

Cooling Coils. If more than 30% outside air is brought across a cooling coil, the use of copper tube, copper fin, epoxy-coated coils, or other special treatment must be considered. To account for the ship's movement, drain pans should have two drain connections. Because of size constraints, care must be taken to prevent moisture carryover. Face velocity limits (in fpm) for different coil materials and different fin spacing are as follows:

Fins per Inch (fpi)	Aluminum Fins	Copper or Coated Fins
8	550	500
11	550	425
14	550	375

Off-coil temperatures are another concern. Ships typically have low ceiling heights and can not tolerate low air-introduction temperatures. Typically 55°F db and 54°F wb are used as limiting off-coil temperatures.

Electric Heaters. U.S. Coast Guard (USCG) approved sheathed-element heaters are typically required. The only exception is when the electric heaters, approved by a regulatory body such as UL, are incorporated in a packaged unit.

Air Diffusers. Care must be taken with the selection of air diffusers because of the low ceilings typical of shipboard applications.

Air-Conditioning Compressors. Compressors of all types are used for marine applications. Care must be taken when using a nonpositive-displacement compressor (such as centrifugal) because low-load, high-condensing temperature is a common off-load condition.

When high discharge temperatures are a concern, seawater-cooled heads are not normally an option; other methods such as fan cooling or liquid injection must be considered for maintaining acceptable discharge temperatures.

Typical Systems

All types of systems may be considered for each marine application. The systems are the same as in land applications; the difference is the relative weighting of their advantages and disadvantages for marine use. This section does not review all the systems used aboard ships, but rather some of the more common ones.

Direct refrigerant cooling systems are often used for small, single-zone applications. Aboard ships, places like control rooms and pilot houses lend themselves to a direct refrigerant system. For larger spaces, air distribution is of more concern; direct refrigerant cooling is thus less likely to be the optimum solution.

Two-pipe and **four-pipe fan coil systems** are often used for large systems. The water piping used in these systems takes up only a fraction of the space used by an all-air ducted system. The disadvantage is fan noise in the space being cooled. In addition, limited humidity control and fresh-air requirements often need to be addressed separately.

Many types of **all-air systems** are used aboard ships. Space, cost, noise, and complexity are among the leading parameters when comparing different all-air systems. Using high-velocity air distribution for an all-air system offers many advantages; unitary (factory-assembled) central air-handling equipment and prefabricated piping, clamps, and fittings facilitate installation for both new construction and conversions. Substantial space-saving is possible compared to conventional low-velocity sheet metal ducts. Maintenance is also reduced. Noise is the one major drawback of a high-velocity system, which often leads to selection of a low-velocity system.

Terminal reheat air conditioning (described in Chapter 4 of the 2008 *ASHRAE Handbook—HVAC Systems and Equipment*) is commonly used because of its simplicity and good zone control characteristics. However, as systems become larger, the energy inefficiency of this system becomes a significant drawback.

Dual-duct systems (also described in Chapter 4 of the 2008 *ASHRAE Handbook—HVAC Systems and Equipment*) have the following advantages:

- All conditioning equipment is centrally located, simplifying maintenance and operation
- Can heat and cool adjacent spaces simultaneously without cycle changeover and with minimum automatic controls
- Because only air is distributed from fan rooms, no water or steam piping, electrical equipment, or wiring are in conditioned spaces

The major drawback is the inability to finely control temperature and humidity. This disadvantage is enough to preclude using these systems in many passenger vessel applications.

Aboard ships, **constant-volume systems** are most common. Their advantages include simplicity (for maintenance, operation, and repair) and low cost. However, for large passenger vessels, the energy efficiency and the tight control of zone temperature make **variable-volume/temperature systems** very attractive.

Air Distribution Methods

Good air distribution in staterooms and public spaces is difficult to achieve because of low ceiling heights and compact space arrangements. Design should consider room dimensions, ceiling height, volume of air handled, air temperature difference between supply and room air, location of berths, and allowable noise. For

Table 1 Minimum Thickness of Steel Ducts

All vertical exposed ducts	16 USSG	0.0598 in.
Horizontal or concealed vertical ducts		
less than 6 in.	24 USSG	0.0239 in.
6.5 to 12 in.	22 USSG	0.0299 in.
12.5 to 18 in.	20 USSG	0.0359 in.
18.5 to 30 in.	18 USSG	0.0476 in.
over 30 in.	16 USSG	0.0598 in.

major installations, mock-up tests are often used to establish exacting performance criteria.

Air usually returns from individual small spaces either by a sight-tight louver mounted in the door or by an undercut in the door leading to the passageway. An undercut door can only be used with air quantities of 75 cfm or less. Louvers are usually sized for face velocity of 400 fpm based on free area.

Ductwork on merchant ships is generally constructed of steel. Ducts, other than those requiring heavier construction because of susceptibility to damage or corrosion, are usually made with riveted seams sealed with hot solder or fire-resistant duct sealer, welded seams, or hooked seams and laps. They are made of hot-dipped, galvanized, copper-bearing sheet steel, suitably stiffened externally. The minimum thickness of material is determined by the diameter of round ducts or by the largest dimension of rectangular ducts, as listed in Table 1.

The increased use of high-velocity, high-pressure systems has resulted in greater use of prefabricated round pipe and fittings, including spiral-formed sheet metal ducts. It is important that field-fabricated ducts and fittings be airtight. Using factory-fabricated fittings, clamps, and joints effectively minimizes air leakage for these high-pressure ducts.

In addition to the space advantage, small ductwork saves weight, another important consideration for this application.

Control

The conditioning load, even on a single voyage, varies over a wide range in a short period. Not only must the refrigeration plant meet these load variations, but the controls must readily adjust the system to sudden climatic changes. Accordingly, it is general practice to equip the plant with automatic controls.

Regulatory Agencies

Merchant vessels that operate under the U.S. flag come under the jurisdiction of the U.S. Coast Guard. Accordingly, the installation and components must conform to the Marine Engineering Rules and Marine Standards of the Coast Guard covered under the *Guide to Structural Fire Protection* (USDOT 2010).

Certified pressure vessels and electric components approved by independent agencies (e.g., ASME, UL) must be used. Wherever possible, equipment used should comply with ABS rules and regulations. This is important when vessels are equipped for carrying cargo refrigeration, because air-conditioning compressors may serve as standby units in the event of a cargo compressor failure. This compliance eliminates the need for a separate, spare cargo compressor. The International Convention for the Safety of Life at Sea (SOLAS) (IMO 2009) governs the use of fire-dampers and duct wall thickness when passageways or fire boundaries are crossed.

NAVAL SURFACE SHIPS

Design Criteria

Outside Ambient Temperature. Design conditions for naval vessels have been established as a compromise, considering the large cooling plants required for internal heat loads generated by machinery, weapons, electronics, and personnel. Temperatures of 90°F db and 81°F wb are used for worldwide applications, with 85°F seawater temperatures. Heating-season temperatures are 10°F for outside air and 28°F for seawater.

Inside Temperature. Naval ships are generally designed for space temperatures of 80°F db with a maximum of 55% rh for most areas requiring air conditioning. The *Air Conditioning, Ventilation and Heating Design Criteria Manual for Surface Ships of the United States Navy* (USN 1969) gives design conditions established for specific areas. *Standard Specification for Cargo Ship Construction* (USMA 1965) gives temperatures for ventilated spaces.

Ventilation Requirements. Ventilation must meet the requirements of ASHRAE *Standard* 62.1-2010, except when ship's specification requires otherwise.

Air-Conditioned Spaces. Naval ship design requires that air-conditioning systems serving living and berthing areas on surface ships replenish air in accordance with damage control classifications, as specified in USN (1969):

- Class Z systems: 5 cfm per person
- Class W systems for troop berthing areas: 5 cfm per person
- All other Class W systems: 10 cfm per person. The flow rate is increased only to meet either a 75 cfm minimum branch requirement or to balance exhaust requirements. Outside air should be kept at a minimum to minimize the size of the air-conditioning plant.

Load Determination

The cooling load estimate consists of coefficients from *Design Data Sheet* DDS511-2 of USN *General Specifications for Building Naval Ships* or USN (1969) and has allowances for the following:

- Solar radiation
- Heat transmission through hull, decks, and bulkheads
- Heat (latent and sensible) gain of occupants
- Heat gain from lights
- Heat (latent and sensible) gain from ventilation air
- Heat gain from motors or other electrical equipment
- Heat gain from piping, machinery, and equipment

Loads should be derived from requirements indicated in USN (1969). The heating load estimate should include the following:

- Heat losses through hull, decks, and bulkheads
- Ventilation air
- Infiltration (when specified)

Some electronic spaces listed in USN (1969) require adding 15% to the calculated cooling load for future growth and using one-third of the cooling-season equipment heat dissipation (less the 15% added for growth) as heat gain in the heating season.

Heat Transmission Coefficients. The overall heat transmission coefficient U between the conditioned space and the adjacent boundary should be estimated from *Design Data Sheet* DDS511-2. Where new materials or constructions are used, new coefficients may be used from SNAME (1980) or calculated using methods found in DDS511-2 and SNAME.

Heat Gain from People. USN (1969) gives heat gain values for people in various activities and room conditions.

Heat Gain from Sources Within the Space. USN (1969) gives heat gain from lights and motors driving ventilation equipment. Heat gain and use factors for other motors and electrical and electronic equipment may be obtained from the manufacturer or from Chapter 18 of the 2009 *ASHRAE Handbook—Fundamentals*.

Equipment Selection

The equipment described for merchant ships also applies to U.S. naval vessels, except as follows:

Fans. A family of standard fans is used by the navy, including vaneaxial, tubeaxial, and centrifugal fans. Selection curves used for system design are found on NAVSEA *Standard Drawings* 810-921984, 810-925368, and 803-5001058. Manufacturers are required to furnish fans dimensionally identical to the standard plan and within 5% of the delivery. No belt-driven fans are included.

Cooling Coils. The U.S. Navy uses eight standard sizes of direct-expansion and chilled-water cooling coils. All coils have eight rows in the direction of airflow, with a range in face area of 0.6 to 10.0 ft^2.

Coils are selected for a face velocity of 500 fpm maximum; however, sizes 54 DW to 58 DW may have face velocity up to 620 fpm if the bottom of the duct on the discharge is sloped up at 15° for a distance equal to the height of the coil. Construction and materials are specified in MIL-C-2939.

Chilled-water coils are most common and are selected based on 45°F inlet water with approximately a 6.7°F rise in water temperature through the coil. This is equivalent to 3.6 gpm per ton of cooling.

Heating Coils. The standard naval steam and electric duct heaters have specifications as follows:

Steam Duct Heaters
- Maximum face velocity is 1800 fpm.
- Preheater leaving air temperature is 42 to 50°F.
- Steam heaters are served from a 50 psig steam system.

Electric Duct Heaters
- Maximum face velocity is 1400 fpm.
- Temperature rise through the heater is per MIL-H-22594A, but is in no case more than 48°F.
- Power supply for the smallest heaters is 120 V, three-phase, 60 Hz. All remaining power supplies are 440 V, three-phase, 60 Hz.
- Pressure drop through the heater must not exceed 0.35 in. of water at 1000 fpm. Use manufacturers' tested data in system design.

Filters. Characteristics of the seven standard filter sizes the U.S. Navy uses are as follows:

- Filters are available in steel or aluminum.
- Filter face velocity is between 375 and 900 fpm.
- A filter-cleaning station on board ship includes facilities to wash, oil, and drain filters.

Air Diffusers. Although it also uses standard diffusers for air conditioning, the U.S. Navy generally uses a commercial type similar to those used for merchant ships.

Air-Conditioning Compressors. In the past, the U.S. Navy primarily used reciprocating compressors up to approximately 150 tons; for larger capacities, open, direct-drive centrifugal compressors are used. On new designs, the U.S. Navy primarily uses rotary compressors (e.g., screw and centrifugal). R-134a is the U.S. Navy's primary refrigerant. Seawater is used for condenser cooling at 5 gpm per ton for reciprocal compressors and 4 gpm per ton for centrifugal compressors.

Typical Air Systems

On naval ships, zone reheat is used for most applications. Some ships with sufficient electric power use low-velocity terminal reheat systems with electric heaters in the space. Some newer ships use a fan-coil unit with fan, chilled-water cooling coil, and electric heating coil in spaces with low to medium sensible heat per unit area of space requirements. The unit is supplemented by conventional systems serving spaces with high sensible or latent loads.

Air Distribution Methods

Methods used on naval ships are similar to those discussed in the section on Merchant Ships. The minimum thickness of materials for ducts is listed in Table 2.

Control

The navy's principal air-conditioning control uses a two-position dual thermostat that controls a cooling coil and an electric or steam reheater. This thermostat can be set for summer operation and does not require resetting for winter operation.

Table 2   Minimum Thickness of Materials for Ducts

	Sheet for Fabricated Ductwork			
	Nonwatertight		Watertight	
Diameter or Longer Side	**Galvanized Steel**	**Aluminum**	**Galvanized Steel**	**Aluminum**
Up to 6	0.018	0.025	0.075	0.106
6.5 to 12	0.030	0.040	0.100	0.140
12.5 to 18	0.036	0.050)	0.118	0.160
18.5 to 30	0.048	0.060	0.118	0.160
Above 30	0.060	0.088	0.118	0.160

Welded or Seamless Tubing		
Tubing Size	**Nonwatertight Aluminum**	**Watertight Aluminum**
2 to 6	0.035	0.106
6.5 to 12	0.050	0.140

Spirally Wound Duct (Nonwatertight)		
Diameter	**Steel**	**Aluminum**
Up to 8	0.018	0.025
Over 8	0.030	0.032

Note: All dimensions in inches.

Steam preheaters use a regulating valve with (1) a weather bulb controlling approximately 25% of the valve's capacity to prevent freeze-up, and (2) a line bulb in the duct downstream of the heater to control the temperature between 42 and 50°F.

Other controls are used to suit special needs. Pneumatic/electric controls can be used when close tolerances in temperature and humidity control are required, as in operating rooms. Thyristor controls are sometimes used on electric reheaters in ventilation systems.

REFERENCES

ASHRAE. 2006. Mechanical refrigeration and air conditioning installations aboard ship. ANSI/ASHRAE *Standard* 26-1996 (RA 2006).

ASHRAE. 2010. Thermal environmental conditions for human occupancy. ANSI/ASHRAE *Standard* 55-2010.

ASHRAE. 2010. Ventilation for acceptable inside air quality. ANSI/ASHRAE *Standard* 62.1-2010.

ISO. 2002. Ships and marine technology—Air-conditioning and ventilation of accommodation spaces—Design conditions and basis of calculations. *Standard* 7547-2002. International Organization for Standardization, Geneva.

IMO. 2009. International convention for the safety of life at sea (SOLAS). International Maritime Organization, London.

SNAME. 1963. Thermal insulation report. *Technical and Research Bulletin* 4-7. Society of Naval Architects and Marine Engineers, Jersey City, NJ.

SNAME. 1980. Calculations for merchant ship heating, ventilation and air conditioning design. *Technical and Research Bulletin* 4-16. Society of Naval Architects and Marine Engineers, Jersey City, NJ.

USDOT. 2010. Guide to structural fire protection. *Publication* COMDT PUB 16700.4, NVIC 9-97, CH1. U.S. Department of Transportation, Washington, D.C.

USMA. 1965. *Standard specification for cargo ship construction.* U.S. Maritime Administration, Washington, D.C.

USN. 1969. The air conditioning, ventilation and heating design criteria manual for surface ships of the United States Navy. *Document* 0938-018-0010. Naval Sea Systems Command, Department of the Navy, Washington, D.C.

USN. NAVSEA *Drawing* 810-921984, NAVSEA *Drawing* 810-925368, and NAVSEA *Drawing* 803-5001058. Naval Sea Systems Command, Department of the Navy, Washington, D.C.

USN. *Guidance in selection of heat transfer coefficients.* DDS511-2. Naval Sea Systems Washington, D.C.

USN. *General specifications for building naval ships.* Naval Sea Systems Command, Department of the Navy, Washington, D.C.

Note: MIL specifications are available from Commanding Officer, Naval Publications and Forms Center, ATTN: NPFC 105, 5801 Tabor Ave., Philadelphia, PA 19120.

BIBLIOGRAPHY

SNAME. 1992. *Marine engineering.* R. Harrington, ed. Society of Naval Architects and Marine Engineers, Jersey City, NJ.

INDUSTRIAL AIR CONDITIONING

INDUSTRIAL facilities such as manufacturing plants, laboratories, processing plants, and nuclear power plants are designed for processes and environmental conditions that include proper temperature, humidity, air motion, air quality, and cleanliness. Generated airborne contaminants must be collected and treated before being discharged from the building or recirculated.

Many industrial buildings require large quantities of energy, both in manufacturing and maintaining building environmental conditions. Energy can be saved by proper use of insulation and ventilation, and by recovery of waste heat.

For worker efficiency, the building environment should be comfortable and healthful, and should minimize fatigue and facilitate communications. The HVAC systems should control temperature and humidity, have low noise levels, control health-threatening fumes, and provide spot cooling to prevent heat stress.

GENERAL REQUIREMENTS

Typical temperatures, relative humidities, and specific filtration requirements for storage, manufacture, and processing of various commodities are listed in Table 1. Requirements for a specific application may differ from those in the table. Improvements in processes and increased knowledge may cause further variations; thus, systems should be flexible to accommodate future requirements.

Inside temperature, humidity, cleanliness, and allowable variations should be established by agreement with the owner. A compromise between the requirements for product or process conditions and those for comfort may optimize quality and production costs.

An environment that allows a worker to perform assigned duties without fatigue from the effects of temperature and humidity results in better continuous performance. It may also improve morale and reduce absenteeism.

Special Warning: Certain industrial spaces may contain flammable, combustible, and/or toxic concentrations of vapors or dusts under either normal or abnormal conditions. In spaces such as these, there are life safety issues that this chapter may not completely address. Special precautions must be taken in accordance with requirements of recognized authorities such as the National Fire Protection Association (NFPA), the Occupational Safety and Health Administration (OSHA), and the American National Standards Institute (ANSI). In all situations, engineers, designers, and installers who encounter conflicting codes and standards must defer to the code or standard that best addresses and safeguards life safety.

PROCESS AND PRODUCT REQUIREMENTS

A product or process may require control of one or more of the following factors.

The preparation of this chapter is assigned to TC 9.2, Industrial Air Conditioning.

Rate of Chemical Reaction

Some processes require temperature and humidity control to regulate chemical reactions. In rayon manufacturing, for example, pulp sheets are conditioned, cut to size, and mercerized. The temperature directly controls the rate of reaction, and the relative humidity maintains the solution at a constant strength and rate of evaporation.

In drying varnish, oxidizing depends on temperature. Desirable temperatures vary with the type of varnish. High relative humidity retards surface oxidation and allows internal gases to escape as chemical oxidizers cure the varnish from within. Thus, a bubble-free surface is maintained with a homogeneous film throughout.

Rate of Crystallization

The cooling rate determines the size of crystals formed from a saturated solution. Both temperature and relative humidity affect the cooling rate and change the solution density by evaporation.

In coating pans for pills, a heavy sugar solution is added to the tumbling mass. As water evaporates, sugar crystals cover each pill. Moving the correct quantity of air over the pills at the correct temperature and relative humidity forms a smooth opaque coating. If cooling and drying are too slow, the coating will be rough, translucent, and have an unsatisfactory appearance. If the cooling and drying are too fast, the coating will chip through to the interior.

Rate of Biochemical Reaction

Fermentation requires both temperature and humidity control to regulate the rate of biochemical reactions. Many fermentation vessels are jacketed to maintain consistent internal temperatures. Fermentors are held at different temperatures depending on the process involved. In brewing, typical fermentor temperatures range from 45 to 52°F. Because of vessel jacketing, tight control of room temperature may not be required. Usually, space temperatures should be held as close as practical to the process temperature inside the fermentation vessel.

Designing such spaces should take into account gases and other by-products generated by fermentation. Typically, carbon dioxide is the most prevalent by-product of fermentation in brewing and presents the greatest potential hazard if a fermentor overpressurizes the seal. Adequate ventilation should be provided in case carbon dioxide escapes the process.

In biopharmaceutical processes, hazardous organisms can escape a fermentor; design of spaces using those fermentors should allow containment. Heat gains from steam-sparged vessels should also be accounted for in such spaces.

Product Accuracy and Uniformity

Air temperature and cleanliness affect quality in manufacturing precision instruments, lenses, and tools. When manufacturing tolerances are within 0.0002 in., close temperature control prevents expansion and contraction of the material; constant temperature is more important than the temperature level. Usually, conditions are

Table 1 Temperatures and Humidities for Industrial Air Conditioning

Process	Dry Bulb, °F	rh, %	Process	Dry Bulb, °F	rh, %
ABRASIVE			**FOUNDRIES***		
Manufacture	79	50	Core making	60 to 70	
CERAMICS			Mold making		
Refractory	110 to 150	50 to 90	Bench work	60 to 70	
Molding room	80	60 to 70	Floor work	55 to 65	
Clay storage	60 to 80	35 to 65	Pouring	40	
Decalcomania production	75 to 80	48	Shakeout	40 to 50	
Decorating room	75 to 80	48	Cleaning room	55 to 65	

Use high-efficiency filtration in decorating room. To minimize the danger of silicosis in other areas, a dust-collecting system or medium-efficiency particulate air filtration may be required.

DISTILLING					
General manufacturing	60 to 75	45 to 60			
Aging	65 to 72	50 to 60			

Low humidity and dust control are important where grains are ground. Use high-efficiency filtration for all areas to prevent mold spore and bacteria growth. Use ultrahigh-efficiency filtration where bulk flash pasteurization is performed.

ELECTRICAL PRODUCTS		
Electronics and x-ray		
Coil and transformer winding	72	15
Semiconductor assembly	68	40 to 50
Electrical instruments		
Manufacture and laboratory	70	50 to 55
Thermostat assembly and calibration	75	50 to 55
Humidistat assembly and calibration	75	50 to 55
Small mechanisms		
Close tolerance assembly	72*	40 to 45
Meter assembly and test	75	60 to 63
Switchgear		
Fuse and cutout assembly	73	50
Capacitor winding	73	50
Paper storage	73	50
Conductor wrapping with yarn	75	65 to 70
Lightning arrester assembly	68	20 to 40
Thermal circuit breakers assembly and test	75	30 to 60
High-voltage transformer repair	79	5
Water wheel generators		
Thrust runner lapping	70	30 to 50
Rectifiers		
Processing selenium and copper oxide plates	73	30 to 40

*Temperature to be held constant.

Dust control is essential in these processes. Minimum control requires medium-efficiency filters. Degree of filtration depends on the type of function in the area. Smaller tolerances and miniature components suggest high-efficiency particulate air (HEPA) filters.

FLOOR COVERING		
Linoleum		
Mechanical oxidizing of linseed oil*	90 to 100	
Printing	80	
Stoving process	160 to 250	

*Precise temperature control required.

Medium-efficiency particulate air filtration is recommended for the stoving process.

*Winter dressing room temperatures. Spot coolers are sometimes used in larger installations.

In mold making, provide exhaust hoods at transfer points with wet-collector dust removal system. Use 600 to 800 cfm per hood.

In shakeout room, provide exhaust hoods with wet-collector dust removal system. Exhaust 400 to 500 cfm in grate area. Room ventilators are generally not effective.

In cleaning room, provide exhaust hoods for grinders and cleaning equipment with dry cyclones or bag-type collectors. In core making, oven and adjacent cooling areas require fume exhaust hoods. Pouring rooms require two-speed powered roof ventilators. Design for minimum of 2 cfm per square foot of floor area at low speed. Shielding is required to control radiation from hot surfaces. Proper introduction of air minimizes preheat requirements.

FUR		
Drying	110	
Shock treatment	18 to 20	
Storage	40 to 50	55 to 65

Shock treatment or eradication of any insect infestations requires lowering the temperature to 18 to 20°F for 3 to 4 days, then raising it to 60 to 70°F for 2 days, then lowering it again for 2 days and raising it to the storage temperature.

Furs remain pliable, oxidation is reduced, and color and luster are preserved when stored at 40 to 50°F.

Humidity control is required to prevent mold growth (which is prevalent with humidities above 80%) and hair splitting (which is common with humidities lower than 55%).

GUM		
Manufacturing	77	33
Rolling	68	63
Stripping	72	53
Breaking	73	47
Wrapping	73	58

LEATHER		
Drying	68 to 125	75
Storage, winter room temperature	50 to 60	40 to 60

After leather is moistened in preparation for rolling and stretching, it is placed in an atmosphere of room temperature and 95% relative humidity.

Leather is usually stored in warehouses without temperature and humidity control. However, it is necessary to keep humidity sufficiently low to prevent mildew. Medium-efficiency particulate air filtration is recommended for fine finish.

LENSES (OPTICAL)		
Fusing	75	45
Grinding	80	80

Table 1 Temperatures and Humidities for Industrial Air Conditioning (*Continued*)

Process	Dry Bulb, °F	rh, %	Process	Dry Bulb, °F	rh, %
MATCHES			**PLASTICS**		
Manufacture	72 to 73	50	Manufacturing areas		
Drying	70 to 75	60	Thermosetting molding compounds	80	25 to 30
Storage	60 to 63	50	Cellophane wrapping	75 to 80	45 to 65

Water evaporates with the setting of the glue. The amount of water evaporated is 18 to 20 lb per million matches. The match machine turns out about 750,000 matches per hour.

In manufacturing areas where plastic is exposed in the liquid state or molded, high-efficiency particulate air filters may be required. Dust collection and fume control are essential.

Process	Dry Bulb, °F	rh, %	Process	Dry Bulb, °F	rh, %
PAINT APPLICATION			**PLYWOOD**		
Lacquers: Baking	300 to 360		Hot pressing (resin)	90	60
Oil paints: Paint spraying	60 to 90	80	Cold pressing	90	15 to 25

The required air filtration efficiency depends on the painting process. On fine finishes, such as car bodies, high-efficiency particulate air filters are required for the outdoor air supply. Other products may require only low- or medium-efficiency filters.

Makeup air must be preheated. Spray booths must have 100 fpm face velocity if spraying is performed by humans; lower air quantities can be used if robots perform spraying. Ovens must have air exhausted to maintain fumes below explosive concentration. Equipment must be explosion-proof. Exhaust must be cleaned by filtration and solvents reclaimed or scrubbed.

Process	Dry Bulb, °F	rh, %
RUBBER-DIPPED GOODS		
Manufacture	90	
Cementing	80	25 to 30*
Dipping surgical articles	75 to 80	25 to 30*
Storage prior to manufacture	60 to 75	40 to 50*
Testing laboratory	73	50*

*Dew point of air must be below evaporation temperature of solvent.

Solvents used in manufacturing processes are often explosive and toxic, requiring positive ventilation. Volume manufacturers usually install a solvent-recovery system for area exhaust systems.

Process	Dry Bulb, °F	rh, %
PHOTO STUDIO		
Dressing room	72 to 74	40 to 50
Studio (camera room)	72 to 74	40 to 50
Film darkroom	70 to 72	45 to 55
Print darkroom	70 to 72	45 to 55
Drying room	90 to 100	35 to 45
Finishing room	72 to 75	40 to 55
Storage room (black and white film and paper)	72 to 75	40 to 60
Storage room (color film and paper)	40 to 50	40 to 50
Motion picture studio	72	40 to 55

Process	Dry Bulb, °F	rh, %
TEA		
Packaging	65	65

Ideal moisture content is 5 to 6% for quality and mass. Low-limit moisture content for quality is 4%.

Process	Dry Bulb, °F	rh, %
TOBACCO		
Cigar and cigarette making	70 to 75	55 to 65*
Softening	90	85 to 88
Stemming and stripping	75 to 85	70 to 75
Packing and shipping	73 to 75	65
Filler tobacco casing and conditioning	75	75
Filter tobacco storage and preparation	77	70
Wrapper tobacco storage and conditioning	75	75

*Relative humidity fairly constant with range as set by cigarette machine.

Before stripping, tobacco undergoes a softening operation.

The above data pertain to average conditions. In some color processes, elevated temperatures as high as 105°F are used, and a higher room temperature is required.

Conversely, ideal storage conditions for color materials necessitate refrigerated or deep-freeze temperatures to ensure quality and color balance when long storage times are anticipated.

Heat liberated during printing, enlarging, and drying processes is removed through an independent exhaust system, which also serves the lamp houses and dryer hoods. All areas except finished film storage require a minimum of medium-efficiency particulate air filters.

selected for personnel comfort and to prevent a film of moisture on the surface. A high-efficiency particulate air (HEPA) or ultralow-penetration air (ULPA) filter may be required.

Product Formability

Manufacturing pharmaceutical tablets requires close control of humidity for optimum tablet formation.

Moisture Regain

Air temperature and relative humidity markedly influence production rate and product mass, strength, appearance, and quality in manufacturing or processing hygroscopic materials such as textiles, paper, wood, leather, and tobacco. Moisture in vegetable and animal materials (and some minerals) reaches equilibrium with the moisture in the surrounding air by **regain** (the percentage of absorbed moisture in a material compared to that material's bone-dry mass). For example, if a material sample with a mass of 5.5 lb has a mass of only 5 lb after thorough drying under standard conditions of 220 to 230°F, the mass of absorbed moisture is 0.5 lb, 10% of the sample's bone-dry mass. Therefore, the regain is 10%.

Table 2 lists typical regain values for materials at 75°F in equilibrium at various relative humidities. Temperature change affects the rate of absorption or drying, which generally varies with the thickness, density, and nature of the material. Sudden temperature changes cause slight change in regain even with fixed relative humidity, but the major change occurs as a function of relative humidity.

Hygroscopic materials deliver sensible heat to the air in an amount equal to the latent heat of the absorbed moisture. The amount of heat liberated should be added to the cooling load if it is significant, but it is usually quite small. Manufacturing economy requires regain to be maintained at a level suitable for rapid and satisfactory manipulation. Uniform relative humidity allows high-speed machinery to operate efficiently.

Some materials may be exposed to the required humidity during manufacturing or processing, others may be treated separately after conditioning and drying. Conditioning removes or adds hygroscopic moisture. Drying removes both hygroscopic moisture and

Table 2   Regain of Hygroscopic Materials*

Classification	Material	Description	Relative Humidity								
			10	20	30	40	50	60	70	80	90
Natural textile fibers	Cotton	Sea island—roving	2.5	3.7	4.6	5.5	6.6	7.9	9.5	11.5	14.1
	Cotton	American—cloth	2.6	3.7	4.4	5.2	5.9	6.8	8.1	10.0	14.3
	Cotton	Absorbent	4.8	9.0	12.5	15.7	18.5	20.8	22.8	24.3	25.8
	Wool	Australian merino—skein	4.7	7.0	8.9	10.8	12.8	14.9	17.2	19.9	23.4
	Silk	Raw chevennes—skein	3.2	5.5	6.9	8.0	8.9	10.2	11.9	14.3	18.3
	Linen	Table cloth	1.9	2.9	3.6	4.3	5.1	6.1	7.0	8.4	10.2
	Linen	Dry spun—yarn	3.6	5.4	6.5	7.3	8.1	8.9	9.8	11.2	13.8
	Jute	Average of several grades	3.1	5.2	6.9	8.5	10.2	12.2	14.4	17.1	20.2
	Hemp	Manila and sisal rope	2.7	4.7	6.0	7.2	8.5	9.9	11.6	13.6	15.7
Rayons	Viscose nitrocellulose	Average skein	4.0	5.7	6.8	7.9	9.2	10.8	12.4	14.2	16.0
	Cuprammonium cellulose acetate		0.8	1.1	1.4	1.9	2.4	3.0	3.6	4.3	5.3
Paper	M.F. newsprint	Wood pulp—24% ash	2.1	3.2	4.0	4.7	5.3	6.1	7.2	8.7	10.6
	H.M.F. writing	Wood pulp—3% ash	3.0	4.2	5.2	6.2	7.2	8.3	9.9	11.9	14.2
	White bond	Rag—1% ash	2.4	3.7	4.7	5.5	6.5	7.5	8.8	10.8	13.2
	Comm. ledger	75% rag—1% ash	3.2	4.2	5.0	5.6	6.2	6.9	8.1	10.3	13.9
	Kraft wrapping	Coniferous	3.2	4.6	5.7	6.6	7.6	8.9	10.5	12.6	14.9
Miscellaneous organic materials	Leather	Sole oak—tanned	5.0	8.5	11.2	13.6	16.0	18.3	20.6	24.0	29.2
	Catgut	Racquet strings	4.6	7.2	8.6	10.2	12.0	14.3	17.3	19.8	21.7
	Glue	Hide	3.4	4.8	5.8	6.6	7.6	9.0	10.7	11.8	12.5
	Rubber	Solid tires	0.11	0.21	0.32	0.44	0.54	0.66	0.76	0.88	0.99
	Wood	Timber (average)	3.0	4.4	5.9	7.6	9.3	11.3	14.0	17.5	22.0
	Soap	White	1.9	3.8	5.7	7.6	10.0	12.9	16.1	19.8	23.8
	Tobacco	Cigarette	5.4	8.6	11.0	13.3	16.0	19.5	25.0	33.5	50.0
Miscellaneous inorganic materials	Asbestos fiber	Finely divided	0.16	0.24	0.26	0.32	0.41	0.51	0.62	0.73	0.84
	Silica gel		5.7	9.8	12.7	15.2	17.2	18.8	20.2	21.5	22.6
	Domestic coke		0.20	0.40	0.61	0.81	1.03	1.24	1.46	1.67	1.89
	Activated charcoal	Steam activated	7.1	14.3	22.8	26.2	28.3	29.2	30.0	31.1	32.7
	Sulfuric acid		33.0	41.0	47.5	52.5	57.0	61.5	67.0	73.5	82.5

*Moisture content expressed in percent of dry mass of the substance at various relative humidities, temperature 75°F.

free moisture in excess of that in equilibrium. Drying and conditioning can be combined to remove moisture and accurately regulate the final moisture content in products such as tobacco and textiles. Conditioning or drying is frequently a continuous process in which the material is conveyed through a tunnel and subjected to controlled atmospheric conditions. For more detail, see Chapter 23 of the 2008 *ASHRAE Handbook—HVAC Systems and Equipment.*

Corrosion, Rust, and Abrasion

In manufacturing metal products, temperature and relative humidity need to be kept sufficiently low to prevent hands from sweating, thus protecting the finished article from fingerprints, tarnish, and/or etching. Salt and acid in perspiration can cause corrosion and rust in a few hours. Manufacture of polished surfaces and of steel-belted radial tires usually requires medium-efficiency to HEPA filtering to prevent surface abrasion.

Air Cleanliness

Each application must be evaluated to determine the filtration needed to counter the adverse effects on the product or process of dust particles, airborne bacteria, smoke, spores, pollen, and radioactive particles. These effects include chemically altering production material, spoiling perishable goods, and clogging small openings in precision machinery. See Chapter 28 of the 2008 *ASHRAE Handbook—HVAC Systems and Equipment* for details.

Static Electricity

Static electricity is often detrimental in processing light materials such as textile fibers and paper and extremely dangerous where potentially explosive atmospheres or materials are present. Static electric charges are generally minimized when relative humidity is

above 35%. Room relative humidity may need to be maintained at 65% or higher because machinery heat raises the machine ambient temperature well above the room temperature, creating localized areas of low relative humidity. Such areas could be sources of static electricity. The parts assembly area of an ammunition plant should have design conditions of 75°F and 40 to 60% rh. In addition, air-moving equipment (fans) should be spark resistant.

EMPLOYEE REQUIREMENTS

Space conditions required by health and safety standards to avoid excess exposure to high temperatures and airborne contaminants are often established by the American Conference of Governmental Industrial Hygienists (ACGIH). In the United States, the National Institute of Occupational Safety and Health (NIOSH) does research and recommends guidelines for workplace environments. The Occupational Safety and Health Administration (OSHA) sets standards based on these guidelines, with enforcement usually assigned to a corresponding state agency.

Standards for safe levels of contaminants in the work environment or in air exhausted from facilities do not cover everything that may be encountered. Minimum safety standards and design criteria are available from U.S. Department of Health agencies such as the National Institute of Health, National Cancer Institute, and Public Health Service. The U.S. Department of Energy and Nuclear Regulatory Commission establish standards for radioactive substances.

Thermal Control Levels

Industrial plants are usually designed for an internal temperature of 60 to 90°F and a maximum of 60% rh. Tighter controls are often dictated by the specific operations and processes located in the building. ACGIH (2007) established guidelines to evaluate high temperature and humidity levels in terms of heat stress (Dukes-Dobos and Henschel 1971). See Chapter 9 of the 2009 *ASHRAE Handbook—Fundamentals* for a more detailed analysis of work rate, air velocity, rest, and the effects of radiant heat.

Temperature control becomes tighter and more specific if personnel comfort rather than avoidance of heat stress becomes the criterion. Nearly sedentary workers prefer a winter temperature of 72°F and a summer temperature of 78°F at a maximum of 60% rh. Workers at a high rate of activity prefer 65°F; they are less sensitive to temperature changes and can be cooled by increasing the air velocity. ASHRAE *Standard* 55 provides more detailed information.

Contamination Control Levels

Toxic and/or hazardous materials are present in many industrial plants and laboratories. Gases and vapors are found near acid baths and tanks holding process chemicals. Plating operations, spraying, mixing, abrasive cleaning, and other processes generate dust, fumes, and mists. Many animal and laboratory procedures (e.g., grinding, blending, sonication, weighing) generate aerosols. Air-conditioning and ventilation systems must minimize exposure to these materials. When airborne, these materials greatly expand their range and potential for affecting more employees. Chapter 11 of the 2009 *ASHRAE Handbook—Fundamentals*, OSHA requirements, and ACGIH (2007) give guidance on the health impact of various materials.

Concentrations of gaseous flammable substances must also be kept below explosive limits. Acceptable concentrations of these substances are a maximum of 25% of the lower explosive limit. Chapter 11 of the 2009 *ASHRAE Handbook—Fundamentals* provides data on flammable limits and their means of control.

Instruments are available to measure concentrations of common gases and vapors, but specific monitoring requirements and methods must be developed for uncommon ones.

DESIGN CONSIDERATIONS

Required environmental conditions for equipment, process and personnel comfort must be known before selecting HVAC equipment. The engineer and owner jointly establish design criteria, including the space-by-space environment in the facilities, process heat loads and exhaust requirements, heat and cooling energy recovery, load factors and equipment diversity, lighting, cleanliness, etc. Consideration should be given to the method of separating dirty processes from areas that require progressively cleaner air.

Insulation should be evaluated for initial cost and operating and energy cost savings. When high levels of moisture are required within the building, the air-conditioning and structural envelope must prevent unwanted condensation and ensure a high-quality product. Condensation can be prevented by eliminating thermal short circuits, installing proper insulation, and using vapor barriers. See Chapters 25 and 27 of the 2009 *ASHRAE Handbook—Fundamentals* for further details.

Personnel engaged in some industrial processes may be subject to a wide range of activity levels for which a broad range of temperatures and humidities are desirable. Chapter 9 of the 2009 *ASHRAE Handbook—Fundamentals* addresses recommended indoor conditions for a variety of activity levels.

If layout and construction drawings are not available, a complete survey of existing premises and a checklist for proposed facilities are necessary (Table 3).

New industrial buildings are typically single-story with a flat roof and ample height to distribute air and utilities without interfering with process operations. Fluorescent fixtures are commonly

Table 3   Facilities Checklist

Construction
1. Single or multistory
2. Type and location of doors, windows, crack lengths
3. Structural design live loads
4. Floor construction
5. Exposed wall materials
6. Roof materials and color
7. Insulation type and thicknesses
8. Location of existing exhaust equipment
9. Building orientation

Use of Building
1. Product needs
2. Surface cleanliness; acceptable airborne contamination level
3. Process equipment: type, location, and exhaust requirements
4. Personnel needs, temperature levels, required activity levels, and special workplace requirements
5. Floor area occupied by machines and materials
6. Clearance above floor required for material-handling equipment, piping, lights, or air distribution systems
7. Unusual occurrences and their frequency, such as large cold or hot masses of material moved inside
8. Frequency and length of time doors open for loading or unloading
9. Lighting, location, type, and capacity
10. Acoustical levels
11. Machinery loads, such as electric motors (size, diversity), large latent loads, or radiant loads from furnaces and ovens
12. Potential for temperature stratification

Design Conditions
1. Design temperatures—indoor and outdoor dry and wet bulb
2. Altitude
3. Wind velocity
4. Makeup air required
5. Indoor temperature and allowable variance
6. Indoor relative humidity and allowable variance
7. Indoor air quality definition and allowable variance
8. Outdoor temperature occurrence frequencies
9. Operational periods: one, two, or three
10. Waste heat availability and energy conservation
11. Pressurization required
12. Mass loads from the energy release of productive materials

Code and Insurance Requirements
1. State and local code requirements for ventilation rates, etc.
2. Occupational health and safety requirements
3. Insuring agency requirements

Utilities Available and Required
1. Gas, oil, compressed air (pressure), electricity (characteristics), steam (pressure), water (pressure), wastewater, interior and site drainage
2. Rate structures for each utility
3. Potable water

mounted at heights up to 12 ft, high output fluorescent fixtures up to 20 ft, and high pressure sodium or metal halide fixtures above 20 ft. Lighting design considers light quality, diffusion, room size, mounting height, and economics. Illumination levels should conform to recommendations of the Illuminating Engineering Society of North America.

Air-conditioning systems can be located on the roof of the building. Air intakes should not be located too close to loading docks or other sources of contamination. (See the section on Air Filtration Systems.) HVAC system installation must be coordinated with other systems and equipment that compete for space at the top of the building, such as fire sprinklers, lighting, cranes, structural elements, etc.

Operations in the building must also be considered: some require close control of temperature, humidity, and/or contaminants. A schedule of operations is helpful in determining heating and cooling loads.

LOAD CALCULATIONS

Table 1 and specific product chapters of this Handbook discuss product requirements. Chapter 18 of the 2009 *ASHRAE Handbook—Fundamentals* provides appropriate heating and cooling load calculation techniques.

Solar and Transmission

The roof load is usually the largest solar load on the envelope. Solar loads on walls are often insignificant particularly because modern factory buildings tend to be windowless. Insulating building walls and roof almost always benefits HVAC cost and performance.

Internal Heat Generation

Internal heat generated by equipment and processes, as well as products, lighting, people and utilities, may satisfy heating load requirements. Understanding equipment operating schedules allows an appropriate diversity factor to be applied to the actual power consumption. Using connected loads may greatly oversize the system. Processes tend to operate continuously but may be shut down on weekends or at night. Heating to some minimal level without equipment and/or process load should be considered.

The latent load in most industrial facilities is minimal, with people and outside air being the primary contributors. Some processes and products do generate a latent load. They need to be understood because this latent load can dominate the HVAC system design. Moisture condensation on cold surfaces must be managed when the latent load becomes very large.

Stratification Effect

The cooling load may be dramatically reduced in a work space that takes advantage of temperature stratification. A stagnant blanket of warm air directly under the roof will have little effect on occupants or equipment as long as it remains undisturbed. Heat sources near the stagnant air will have little effect on the cooling load. When the ceiling or roof is high, 20 to 60% of the heat energy rises out of the cooling zone, depending on building construction and the temperature of heat sources. Switching to a return air location near the roof could be cost effective, because it takes advantage of higher temperatures at the roof.

Supply and return air ducts should be installed as low as practical to avoid mixing the warm boundary layers in cooling mode. The location of supply air diffusers generally establishes the stratified air boundary. Spaces with a low occupant-to-floor-area ratio adapt well to using low quantities of supply air with spot cooling for personnel.

Makeup Air

Makeup air provides ventilation and building pressurization. It must be filtered and conditioned to blend with return air and then distributed to the conditioned space. The quantity of makeup air must exceed that of the exhaust air to positively pressurize the building. Makeup air quantity may be varied to accommodate an exhaust system with intermittently operating elements. Heat and cooling recovery from the exhaust airstream can substantially reduce the outside air load.

Processes requiring an extensive amount of exhaust air should ideally be placed in an area of the plant provided with minimal heating and no refrigerated air conditioning. Ventilation air may be required to reduce the quantity of health-threatening fumes, airborne bacteria, or radioactive particles. Minimum ventilation rates must meet the requirements of ASHRAE *Standard* 62.1.

Economizers can take advantage of ambient conditions and possibly satisfy HVAC loads without added heating or cooling for much of the year.

Fan Heat

Heat is generated by fans used to move and pressurize the air. This heat is not felt by the occupants but does add to the cooling load. The discharge air temperature of a draw-through cooling arrangement requires cooler air to the fan to accommodate the temperature increase of air passing through the fan. The increase is more significant in systems with higher discharge air pressures.

SYSTEM AND EQUIPMENT SELECTION

Industrial air-conditioning equipment includes heating and cooling sources, air-handling and air-conditioning apparatus, filters, and an air distribution system. Components should be selected and the system designed for long life with low maintenance and operating costs to provide low life-cycle cost.

Systems may consist of the following:

- Heating-only in cool climates, where ventilation air provides comfort for workers
- Air washer systems, where high humidities are desired and where the climate requires cooling
- Heating and evaporative cooling, where the climate is dry
- Heating and mechanical cooling, where temperature and humidity control are required and other means of cooling are insufficient

All systems include air filtration appropriate to the contaminant control required.

Careful evaluation should determine zones that require control, especially in large, high-bay areas where the occupied zone is a small portion of space volume. ASHRAE *Standard* 55 defines the occupied zones as 3 to 72 in. high and more than 24 in. from the walls.

HEATING SYSTEMS

Floor Heating

Floor heating is often desirable in industrial buildings, particularly in large, high-bay buildings, garages, and assembly areas where workers must be near the floor, or where large or fluctuating outside air loads make maintaining ambient temperature difficult.

Floors may be tempered to 65 or 70°F by embedded hydronic systems, electrical resistance cables, or warm air ducts as an auxiliary to the main heating system. Heating elements may be buried deep in the floor (6 to 18 in.) to allow slab warm-up at off-peak times, thus using the floor mass as heat storage to save energy during periods of high use.

Floor heating may be the primary or sole heating means, but floor temperatures above 85°F are uncomfortable, so such use should be limited to small, well-insulated spaces.

Unit and Ducted Heaters

Gas, oil, electric, hot-water, or steam-fired unit heaters with centrifugal or propeller fans are used for spot heating areas or are arranged in multiples for heating an entire building. Temperatures can be varied by individual thermostatic control. Unit heaters should be located so that the discharge (throw) will reach the floor adjacent to and parallel with the outside wall, and spaced to produce a ring of warm air moving peripherally around the building. In industrial buildings with heat-producing processes, heat tends to stratify in high-bay areas. In large buildings, additional heaters should be placed in the interior so that their discharge reaches the floor to reduce stratification. Downward-discharge unit heaters in high bays and large areas may have a revolving discharge. Gas- and oil-fired unit heaters should not be used where corrosive vapors are present.

Ducted heaters include large direct- or indirect-fired heaters, door heaters, and heating and ventilating units. They usually have centrifugal fans. Direct-fired gas heaters, in which the gas burns in the air being supplied to the space, may be used for makeup air heating because they are self-venting, thus controlling the products of combustion.

Unit heaters and makeup air heaters commonly temper outside air that enters buildings through open doors. Mixing quickly brings the space temperature back to the desired setting after the door is closed. The makeup air heater should be applied as a door heater in buildings where the doors are large and open for extended periods, such as doors for large trucks or railroad cars. Unit heaters are also needed in buildings that have considerable leakage or a sizeable negative pressure. These units help pressurize the door area, mix the incoming cold air, temper it, and quickly bring the area back to the desired temperature after the door is closed.

Door heating units that resemble a vestibule operate with airflow down across the opening and recirculated from the bottom, which help reduce cold drafts across the floor. These units are effective on high-usage doors under 10 ft tall. Additional information on heating is given in Chapter 27 of the 2008 *ASHRAE Handbook—HVAC Systems and Equipment*.

Infrared Heaters

High-intensity gas, oil, or electric infrared heaters transfer heat directly to the occupants, equipment, and floor in the space without appreciably warming the air, though some air heating occurs by convection from objects heated by the infrared heaters. These heaters are classified as near- or far-infrared heaters, depending on how close the wavelengths they emit are to visible light. Near-infrared heaters emit a substantial amount of visible light.

Both vented and unvented gas-fired infrared heaters are available as individual radiant panels, or as a continuous radiant pipe with burners 15 to 30 ft apart and an exhaust vent fan at the end of the pipe. Unvented heaters require exhaust ventilation to remove flue products from the building and prevent moisture from collecting on the walls and ceiling. Insulation reduces the ventilation requirement.

Infrared heaters are common in the following applications:

- High-bay buildings, where heaters are usually mounted 10 to 30 ft above the floor, along outside walls, and tilted to direct maximum radiation to the floor. If the building is poorly insulated, the controlling thermostat should be shielded to avoid influence from the radiant effect of the walls and the cold sky.
- Semi-open and outside areas, where people can be comfortably heated directly and objects can be heated to avoid condensation.
- Loading docks, where snow and ice can be controlled by strategic placement of near-infrared heaters.

Additional information on both electric and gas infrared heating is given in Chapter 15 of the 2008 *ASHRAE Handbook—HVAC Systems and Equipment*.

COOLING SYSTEMS

Common cooling systems include refrigeration equipment, evaporative coolers, and high-velocity ventilation air.

For manufacturing operations, particularly in heavy industry where mechanical cooling cannot be economically justified, evaporative cooling systems often provide good working conditions. If the operation requires heavy physical work, spot cooling by ventilation, evaporative coolers, or refrigerated air can be used. To minimize summer discomfort, high outside ventilation rates may be adequate in some hot-process areas. A mechanical air supply with good distribution is needed in all these operations.

Refrigerated Cooling Systems

The most commonly used refrigerated cooling systems are roof-mounted, direct-expansion packaged units. Larger systems may use chilled water distributed to air-handling units.

Central system condenser water rejects heat through a cooling tower. Refrigerated heat recovery is particularly advantageous in buildings with simultaneous need to heat exterior spaces and cool interior spaces.

Mechanical cooling equipment should be selected in multiple units. This enables the equipment to match its response to fluctuations in the load and to allow maintenance during off-peak operation periods. Packaged refrigeration equipment commonly uses positive-displacement (reciprocating, scroll, or screw) compressors with air-cooled condensers. When equipment is on the roof, the condensing temperature may be affected by warm ambient air, often 10 to 20°F higher than design outside air temperature. ASHRAE *Standard* 15 provides rules for the type and quantity of refrigerant in direct air-to-refrigerant exchangers.

Desiccant-based systems should be considered for processes that require dew points below 50°F (e.g., pharmaceutical processing).

Evaporative Cooling Systems

Evaporative cooling systems may be direct or indirect evaporative coolers or air washers. Evaporative coolers have water sprayed directly on wet surfaces through which air passes. Any excess water is drained off. An air washer recirculates water, and the air flows through a heavily misted area. Water atomized in the airstream evaporates, cooling the air. Refrigerated water simultaneously cools and dehumidifies the air. For spaces that require an air washer and high relative humidities (e.g., tobacco and textile processing areas), heat provided to the sump should provide sufficient energy for humidification beyond that from heat recovered in the return airstream.

Evaporative cooling conserves energy, particularly in mild weather. Air washers may control both temperature and humidity using refrigerated spray water and reheat coils. Temperature and humidity of the exit airstream may be controlled by varying the temperature of the chilled water and reheat coil and by varying the quantity of air passing through the reheat coil with a dew-point thermostat.

Ensure that accumulation of dust or lint does not clog the nozzles or evaporating pads of evaporative cooling systems. It may be necessary to filter air entering the evaporative cooler. Chemical treatment of the water may be necessary to prevent mineral build-up or biological growth on the pads or in the pans.

AIR FILTRATION SYSTEMS

Air filtration systems remove contaminants from the building supply or exhaust airstream. Supply air filtration at the equipment intake removes particulate contamination that may foul heat exchange surfaces, contaminant products, or present a health hazard to people, animals, or plants. Gaseous contaminants must sometimes be removed to prevent exposing personnel to odors or health-threatening fumes. Return air with a significant potential for carrying contaminants should be recirculated only if it can be filtered enough to minimize personnel exposure. Return air should be exhausted if monitoring and contaminant control cannot be ensured.

The supply filtration system usually includes collection media or a filter, a media-retaining device or filter frame, and a filter housing or plenum. The filter medium is the most important part of the system. A mat of randomly distributed small-diameter fibers is commonly used. For more on filtration systems, see Chapter 28 of the 2008 *ASHRAE Handbook—HVAC Systems and Equipment*.

Exhaust Air Filtration Systems

Exhaust air systems are either (1) general systems that remove air from large spaces or (2) local systems that capture aerosols, heat, or gases at specific locations in a room and transport them so they can be collected, inactivated, and safely discharged to the atmosphere. Air in a general system usually requires minimal or no treatment before being discharged to the atmosphere. Air from local exhaust systems can sometimes be safely discharged to the atmosphere, but may require contaminant removal before being discharged. All emitted air must meet appropriate air quality standards. Chapters 31 and 32 of this volume and Chapter 29 of the 2008 *ASHRAE Handbook—HVAC Systems and Equipment* have more information on industrial ventilation and exhaust systems.

In exhaust air emission control, fabric-bag filters, glass-fiber filters, venturi scrubbers, and electrostatic precipitators all collect particles. Packed-bed or sieve towers can absorb toxic gases. Activated carbon columns or beds, often with oxidizing agents, are frequently used to absorb toxic or odorous organics and radioactive gases.

Outside air intakes should be carefully located to avoid recirculating contaminated exhaust air. Wind direction, building shape, and location of effluent source strongly influence concentration patterns.

Air patterns from wind flowing over buildings are discussed in Chapter 24 of the 2009 *ASHRAE Handbook—Fundamentals*. The leading edge of the roof interrupts smooth airflow, reducing air pressure at the roof and on the lee side. Exhaust air must be discharged through either a vertical stack terminating above the building turbulent air boundary or a shorter stack with a high enough discharge velocity to project the effluent through the air boundary into the undisturbed air passing over the building. The high discharge prevents fume damage to both the roof and roof-mounted equipment, and keeps fumes away from building air intakes. A high vertical stack is the safest, simplest solution to fume dispersal.

Contamination Control

In addition to maintaining thermal conditions, air-conditioning systems should control contaminant levels to provide a safe and healthy environment, good housekeeping, and quality control for the processes. Contaminants may be gases, fumes, mists, or airborne particulate matter; they may be produced by a process in the building or contained in the outside air.

Contamination can be controlled by preventing the release of aerosols or gases into the room and by diluting room air contaminants. If the process cannot be enclosed, it is best to capture aerosols and gases near their source with a local exhaust system that includes a hood or enclosure, ducts, fan, motor, and exhaust stack.

Dilution controls contamination in many applications but may not provide uniform safety for personnel. High local concentrations of contaminants can exist despite a high overall dilution rate.

EXHAUST SYSTEMS

An exhaust system draws a contaminant away from its source and removes it from the space. An exhaust hood surrounding the point of generation contains the contaminant as much as is practical. The contaminant is transported through ductwork from the space, cleaned as required, and exhausted to the atmosphere. The hood inlet air quantity is established by the velocities required to convey the airborne contaminant. Chapter 32 has more information on local exhaust systems.

Design values for average and minimum face velocities are a function of the characteristics of the most hazardous material the hood is expected to handle. Minimum values may be prescribed in codes for exhaust systems. Contaminants with greater mass may require higher face velocities for control. Design face velocities should be set carefully: too high a velocity can be as hazardous as one too low. Refer to ACGIH (2007) and ASHRAE *Standard* 110 for more information.

Properly sized ductwork keeps the contaminants flowing. This requires very high velocities for heavy materials. Selection of materials and construction of exhaust ductwork and fans depend on the nature of the contaminant, ambient temperature, lengths and arrangement of ducts, method of hood fan operation, and flame and smoke spread ratings.

Exhaust systems remove gases, vapors, or smokes from acids, alkalis, solvents, and oils. The following should be minimized:

- **Corrosion.** Commonly used reagents in laboratories include hydrochloric, sulfuric, and nitric acids (singly or in combination)

and ammonium hydroxide. Commonly used organic chemicals include acetone, benzene, ether, petroleum, chloroform, carbon tetrachloride, and acetic acid.

- **Dissolution.** Coatings and plastics are subject to dissolving, particularly by solvent and oil fumes.
- **Melting.** Certain plastics and coatings at elevated hood operating temperatures can melt.

Low temperatures that cause condensation in ferrous metal ducts increase chemical destruction. Ducts are less subject to attack when runs are short and direct to the terminal discharge point. The longer the runs, the longer the exposure to fumes and the greater the condensation. Horizontal runs allow moisture to remain longer than on vertical surfaces. Intermittent fan operation can contribute to longer periods of wetness than continuous operation. High loading of condensables in exhaust systems should be avoided by installing condensers or scrubbers as close to the source as possible.

OPERATION AND MAINTENANCE

All designs should allow ample room to clean, service, and replace any component quickly so that design conditions are affected as little as possible. Maintenance of refrigeration and heat rejection equipment is essential for proper performance without energy waste. Maintenance includes changing system filters periodically. Industrial applications are dirty, so proper selection of filters, careful installation to avoid air bypassing the filter, and prudent filter changing to prevent overloading and blowout are required. Dirt that lodges on the tips of forward-curved fan blades reduces air-handling capacity appreciably. Fan and motor bearings require lubrication, and fan belts need periodic inspection. Direct- and indirect-fired heaters should be inspected annually. Steam and hot-water heaters have fewer maintenance requirements than comparable equipment with gas or oil burners.

For system compatibility, water treatment is essential. Air washers and cooling towers should not be operated unless the water is properly treated.

HEAT RECOVERY AND ENERGY CONSERVATION

Process industry presents unique opportunities to recover heat from the exhaust airstream for use in preconditioning makeup air. Extreme care must be taken to ensure compatibility of heat exchanger components and materials with contaminants often found in exhaust streams. For example, brewery spaces are held between 35 and 50°F. Exhaust air passes over a heat recovery wheel to precondition outside makeup air, which in turn controls the level of carbon dioxide contamination. Coated aluminum heat recovery wheels can be subject to premature failure because of caustic cleaning materials conveyed in the exhaust system.

Additional consideration should be given to the assessment of risk associated with the heat recovery strategy. Frequently, downtime in large industrial facilities can exceed millions of dollars per hour. Costs associated with failure of a heat recovery device can easily overcome savings in energy costs if the result is a facility shutdown.

REFERENCES

ACGIH. 2007. *Industrial ventilation: A manual of recommended practice*, 26th ed. American Conference of Governmental Industrial Hygienists, Cincinnati, OH.

ASHRAE. 2010. Safety code for mechanical refrigeration. ANSI/ASHRAE *Standard* 15-2010.

ASHRAE. 2010. Thermal environmental conditions for human occupancy. ANSI/ASHRAE *Standard* 55-2010.

ASHRAE. 2010. Ventilation for acceptable indoor air quality. ANSI/ASHRAE *Standard* 62.1-2010.

ASHRAE. 1995. Method of testing performance of laboratory fume hoods. ANSI/ASHRAE *Standard* 110-1995.

Dukes-Dobos, F. and A. Henschel. 1971. The modification of the WNGT Index for establishing permissible heat exposure limits in occupational work. U.S. Public Health Service *Publication* TR-69.

BIBLIOGRAPHY

Azer, N.Z. 1982. Design guidelines for spot cooling systems. Parts 1 and 2. *ASHRAE Transactions* 88(2):81-95 and 88(2):97-116.

Gorton, R.L. and H.M. Bagheri. 1987a. Verification of stratified air conditioning design. *ASHRAE Transactions* 93(2):211-227.

Gorton, R.L. and H.M. Bagheri. 1987b. Performance characteristics of a system designed for stratified cooling operation during the heating season. *ASHRAE Transactions* 93(2):367-381.

West, D.L. 1977. Contamination dispersion and dilution in a ventilated space. *ASHRAE Transactions* 83(1):125-140.

Yamazaki, K. 1982. Factorial analysis on conditions affecting the sense of comfort of workers in the air conditioned work environment. *ASHRAE Transactions* 88(1):241-254.

CHAPTER 15

ENCLOSED VEHICULAR FACILITIES

ENCLOSED vehicular facilities include buildings and infrastructure through which vehicles travel, are stored, or are repaired, and can include vehicles driven by internal combustion engines or electric motors. Ventilation requirements for these facilities are provided for climate and temperature control, contaminant level control, and emergency smoke management. Design approaches for various natural and mechanical ventilation systems are covered in this chapter.

The chapter is structured to address general tunnel issues first and then address the unique aspects of rail and road tunnels, rail stations, bus garages, bus terminals, and enclosed spaces for equipment maintenance later in the chapter. Finally, information on applicable ventilation equipment is presented.

TUNNELS

Transport tunnels are unique, in that vehicles travel at normal speeds, possibly carrying cargo (which may be unknown in road tunnels), and may include the traveling public (as passengers and/or motorists) during both normal and emergency operations. A tunnel is a linear-configured facility, as opposed to most buildings, which are typically more rectangular. This concept is important when confronting the need to fight a fire within a tunnel. A tunnel cannot be compartmentalized as readily as a building, which means the fire can only be fought from within the actual fire zone. Limited access and compartmentation create difficulties with containing and suppressing a fire. This combination of circumstances requires unique design approaches to both normal and emergency operation.

Tunnel Ventilation Concepts

Tunnel ventilation must accommodate normal, congested and emergency conditions. In some cases, temporary ventilation may also be necessary.

Normal Mode. Normal ventilation is required during normal operations to control temperature, provide comfort, or control level of pollutants in the facility during normal operations and under normal operating conditions, primarily to protect the health and provide comfort for the patrons and employees.

Congested Mode. Congested ventilation is required during service periods where traffic is slow moving, leading to a reduction or elimination of piston effect. The goals are the same as for normal mode.

Emergency Mode. Emergency ventilation is required during an emergency to facilitate safe evacuation and to support firefighting and rescue operations. This is often due to a fire, but it can be any nonnormal incident that requires unusual control of the environment in the facility. This includes control of smoke and high temperature from a fire, control of exceedingly high levels of contaminants, and/or control of other abnormal environmental condition.

Temporary Mode. Temporary ventilation is needed during original construction or while maintenance-related work is carried out

in a tunnel, usually during nonoperational hours. The temporary ventilation is typically removed after construction or after the maintenance work is completed. Ventilation requirements for such temporary systems are specified by either state or local mining laws, industrial codes, or the U.S. Occupational Safety and Health Administration (OSHA) and are not addressed specifically in this chapter.

Tunnel Ventilation Systems

There are two categories of ventilation systems used in most tunnels: natural and mechanical.

Natural Ventilation. Naturally ventilated facilities rely primarily on atmospheric conditions to maintain airflow and provide a satisfactory environment in the facility. The chief factor affecting the facility environment is the pressure differential created by differences in elevation, ambient air temperature, or wind effects at the boundaries of the facility. Unfortunately, most of these factors are highly variable with time, and thus the resultant natural ventilation is often neither reliable nor consistent. If vehicles are moving through a tunnel-type facility, the piston effect created by the moving vehicles may provide additional natural airflow.

Mechanical Ventilation. A tunnel that is long, has a heavy traffic flow, or experiences frequent adverse atmospheric conditions requires fan-based mechanical ventilation. Among the alternatives available are longitudinal and transverse ventilation.

Longitudinal Ventilation. This type of ventilation introduces or removes air from the tunnel at a limited number of points, primarily creating longitudinal airflow along its length. Longitudinal ventilation can be accomplished either by injection, using central fans, using jet fans mounted in the facility, or a combination of injection and extraction at intermediate points.

Transverse Ventilation. Transverse ventilation uses both a supply duct system and an exhaust duct system to uniformly distribute supply air and collect vitiated air throughout the length of the facility. The supply and exhaust ducts are served by a series of fixed fans, usually housed in a ventilation building or structure. A variant of this type of ventilation is **semitransverse ventilation**, which uses either a supply or exhaust duct, not both. The balance of airflow is made up via the tunnel portals.

Design Approach

General Design Criteria. The air quality and corresponding ventilation system airflow requirements in enclosed vehicular spaces are determined primarily by the type and quantity of contaminants that are generated or introduced into the tunnel and the amount of ventilation needed to limit the high air temperatures or concentrations of these contaminants to acceptable levels for the specific time exposures.

Normal and Congested Modes. The maximum allowable concentrations and levels of exposure for most contaminants are determined by national governing agencies such as the U.S. Environmental Protection Agency (EPA), OSHA, the American Conference of Governmental Industrial Hygienists (ACGIH).

The preparation of this chapter is assigned to TC 5.9, Enclosed Vehicular Facilities.

The contaminant generators can be as varied as gasoline or diesel automobiles, diesel or compressed natural gas (CNG) buses and trucks, and diesel locomotives. Even heat generated by air conditioning on electric trains stopped at stations and the pressure transients generated by rapid-transit moving trains can be considered contaminants, the effects of which need to be mitigated.

Emergency Mode. Design provisions may be necessary to manage smoke and other products of combustion released during fires to allow safe evacuation, to support fire fighting and rescue operations, and to protect the tunnel structure and station infrastructure during fires (Bendelius 2008).

In designing for fires, the design fire scenario and associated fire heat release rate needs to be quantified. Depending on the level of analysis, the generation of smoke and other products of combustion may also need to be quantified. As a minimum, design for life safety during fires must conform to the specific standards or guidelines of the National Fire Protection Association (NFPA), where applicable. NFPA's ventilation requirements are for systems to maintain a "tenable environment along the pathway of egress from the fire." NFPA *Standard* 130 defines a tenable environment as "an environment that permits self-rescue of occupants for a specific period of time"; NFPA *Standard* 502 includes a similar definition.

Other NFPA codes and standards; ICC (2009a, 2009b, 2009c) building, mechanical, and fire codes; and other statutory requirements may apply. Separation and pressurization requirements between adjacent facilities should also be considered.

Temporary. A temporary mode may be necessary during construction or other special condition.

Technical Approach. The technical approach differs depending on facility type; however, there are many similarities in the initial stages of the design process.

Determining the length, gradient, and cross section for tunnels is an important first step. Establishing the facility's dynamic clearance envelope is of extreme importance, especially for a tunnel, because all appurtenances, equipment, ductwork, jet fans, etc., must be located outside the envelope, and this may eventually determine the type of ventilation system used.

Vehicle speeds, vehicle cross-sectional areas, vehicle design fire scenarios, and fuel-carrying capacity are important considerations for road tunnels, as are train speeds, train headway, and rail car combustibility and design fire scenarios for rapid transit and railroad tunnels.

Types of cargo to be allowed through the facility, and their respective design fire scenarios, should be investigated to determine the ventilation rates and the best system for the application. Similarly, for railroad tunnels, it should be determined whether passenger or freight or both types of trains will be using the facility, and if the passenger trains will be powered by diesel/electric power or by electric traction power.

The emergency ventilation approach must be fully coordinated with the overall fire protection strategy, the evacuation plan, and the emergency response plan, providing a comprehensive overall life safety program for the tunnel or station. Egress systems must provide for safe evacuation under a wide range of emergency conditions. The emergency response plan must help facilitate evacuation and allow for appropriate response to emergencies.

Rail and bus stations are large unique structures designed to allow efficient movement of large populations and to serve occupants that often arrive in large groups. Stations can be below ground, above ground or at grade. Although each type of station poses specific challenges, underground facilities tend be the most challenging. Stations can be further complicated by connections to non-transit structures (Tubbs and Meacham 2007).

Rail and road tunnels pose a different set of evacuation challenges. These facilities are long, narrow, and underground, often with limited opportunities for stairwells to grade. The linear nature limits initial evacuation, which can pose challenges to the ventilation design. Further, the trackway in rail tunnels can be a dangerous environment for untrained occupants.

The ventilation and other protection systems must support the evacuation plan. NFPA *Standards* 502 and 130 provide specific criteria for components of the life safety and evacuation systems, but are not universally adopted by authorities. Where road and rail infrastructure interface with buildings, the *International Building Code* and *International Fire Code* may apply. Several documents are available to provide additional guidance on life safety concepts, evacuation strategies, and calculation methodologies (Bendelius 2008; Colino and Rosenstein 2006; Fruin 1987; Gwynne and Rosenbaum 2008; Proulx 2008; Tubbs and Meacham 2007).

Critical Velocity. Manual calculations and resources for the emission and combustion data are given in the respective sections for each enclosed vehicular facility type. A first step in determining the order of magnitude for the ventilation rate required to control the movement of the heat and smoke layer generated by a fire in a tunnel is to apply the critical velocity criterion. This approach is described here, and can be used for all types of tunnel applications.

The simultaneous solution of Equations (1) and (2), by iteration, determines the critical velocity (Kennedy et al. 1996), which is the minimum steady-state average bulk velocity of ventilation air moving toward the fire needed to prevent backlayering:

$$V_C = K_1 K_G \left(\frac{gHq}{\rho c_p A T_F} \right)^{1/3} \tag{1}$$

$$T_F = \left(\frac{q}{\rho c_p A V_C} \right) + T \tag{2}$$

where

V_C = critical velocity, ft/s
T_F = average temperature of fire site gases, °R
K_1 = 0.606
K_G = grade factor (see Figure 1)
g = acceleration caused by gravity, ft/s²
H = height of duct or tunnel at fire site, ft
q = heat that fire adds directly to air at fire site, Btu/s
ρ = average density of approach (upstream) air, lb/ft³
c_p = specific heat of air, Btu/lb·°R
A = area perpendicular to flow, ft²
T = temperature of approach air, °R

It is usual to study several alternative ventilation schemes, each using different variants and/or combinations of ventilation system (longitudinal, transverse, etc.). Some types of systems, such as fully transverse, are almost exclusively used on road tunnels only.

When selecting ventilation equipment and the number of fans and types of drives, consideration should be given to efficiency, reliability, and noise. Most of these equipment attributes are reflected in a life-cycle cost analysis of the alternatives.

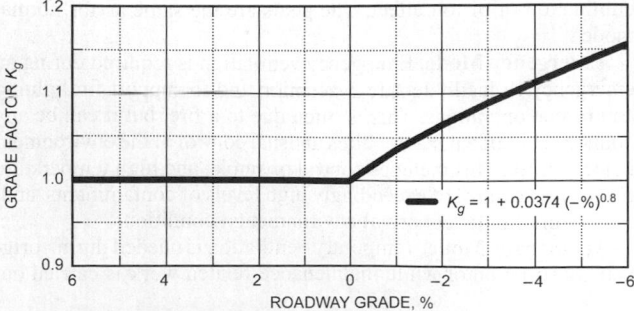

Fig. 1 Roadway Grade Factor

If the effectiveness of the system to provide for fire life-safety conditions is not evident from the manual analysis or one-dimensional computer models such as subway environment simulation (SES), the designer should investigate using a computational fluid dynamics (CFD) program to accurately determine the smoke and temperature distribution in both the steady-state and transient conditions.

Computer Modeling and Simulation. The applicable NFPA standards for road tunnels (NFPA *Standard* 502) and for railroad rapid transit tunnels (NFPA *Standard* 130) require engineering analysis for tunnels greater than a certain length, to prove that the smoke and heat layer is controlled. Often the best way to show that the requirements are met is by using a CFD program with post-processing capabilities that feed the results into another program capable of producing a still picture and/or animated graphical representation of the results. All the commonly used computer programs and their specific capabilities are discussed in the following paragraphs.

SES. The predominant worldwide tool for analyzing the aerothermodynamic environment of rapid transit rail tunnels is the Subway Environment Simulation (SES) computer program (DOT 1997a). SES is a one-dimensional network model that is used to evaluate longitudinal airflow in tunnels. The model predicts airflow rates, velocities and temperatures in the subway environment caused by train movement or fans, as well as the station cooling loads required to maintain the public areas of the station to predetermined design conditions throughout the year. This program contains a fire model that can simulate longitudinal airflow required to overcome backlayering and control smoke movement in a tunnel. Output from the SES can be applied as boundary or initial conditions for three-dimensional CFD modelling of the tunnel and station environments. The SES program is in the public domain, available from the Volpe National Transportation Systems Center in Cambridge, MA.

TUNVEN. This program solves coupled one-dimensional, steady-state tunnel aerodynamic and advection equations. It can predict quasi-steady-state longitudinal air velocities and concentrations of CO, NO_x, and total hydrocarbons along a road tunnel for a wide range of tunnel designs, traffic loads, and external ambient conditions.

The program can also be used to model all common road tunnel ventilation systems (i.e., natural, longitudinal, semitransverse, and transverse). The user must update emissions data for the calendar year of interest. The program is available from the National Technical Information Service (NTIS 1980).

Computational Fluid Dynamics (CFD). CFD software can model operating conditions in tunnels and stations and predict the resulting environment. In areas of geometrical complexity, CFD is the appropriate tool to predict three-dimensional patterns of airflow, temperature, and other flow variables, including concentration of species, which may vary with time and space. Computational fluid dynamics software is the design tool of choice to obtain an optimum design, because experimental methods are costly, complex, and yield limited information.

SOLVENT. SOLVENT is a specific CFD model developed as part of the Memorial Tunnel Fire Ventilation Test Program for simulating road tunnel fluid flow, heat transfer, and smoke transport. SOLVENT can be applied to all ventilation systems used in road tunnels, including those based on natural airflow. The program results have been validated against data from Massachusetts Highway Department and Federal Highway Authority (MHD/FHWA 1995).

Fire Dynamics Simulator (FDS). FDS is a Computational Fluid Dynamics (CFD) model of buoyancy-driven fluid flow from a fire. A separate code called Smokeview is used to visualize data output from FDS. These applications can also be configured to model pollutant levels outside the portals and around the exhaust stacks of tunnels. Both of these public domain programs are under active development and can be obtained from National Institute of Standards and Technology (NIST).

Other CFD programs, both commercially available and in the public domain, have been used to model fire scenarios in road and rapid transit tunnels and stations, the list of which is too numerous to include here. The strengths and weaknesses of each program should be investigated beforehand, and validation of results against experimental data or an equivalent program is encouraged.

Tunnel Fires

Fires occurring in tunnels are more difficult to deal with than those occurring in one of the other enclosed vehicular facilities, in a normal building, or in the open. In a tunnel, firefighting is extremely complex, because access to the tunnel is difficult in the event of a fire. The fire cannot be fought from outside the tunnel, as can be done with a building; it must be fought from within the tunnel, often in the same space where the fire is burning.

Fires occur in tunnels far less frequently than in buildings; however, because of the unique nature of a tunnel fire, they are more difficult to suppress and extinguish and usually get more attention. There is a long list of tunnel fires; the most complete history of fires in tunnels exists for road tunnels, a partial listing of which is included in Table 1. Similar information is available for rail fires (Meacham et al. 2010).

Design Fires. Design fires form the base input for emergency ventilation design analyses and are defined in terms of heat release rate, species output, and soot yields as functions of time. A design fire scenario is an input parameter that defines the ignition source, fire growth on the first item, possible spread of fire to adjacent combustibles, interaction between the fire and the enclosure and environment, and eventual fire decay and extinction.

Limited data are available regarding the magnitude and severity of vehicle design fires. In the absence of more specific data, the information available provides first-order guidance in selecting an appropriate design fire for the evaluation of an enclosed vehicular facility such as a tunnel (road or rail) or station (bus or rail).

PIARC (1999) and NFPA *Standard* 502 provide summaries of vehicle fire tests. Additional information can be found in Atkinson et al. (2001), Ingason (1994), Joyeux (1997), and Mangs and Keski-Rahkonen (1994a, 1994b).

Fire Detection. Fire detection systems are necessary to alert tunnel operators of potential unsafe condition. There are a range of methods available to detect fire and smoke within road/rail tunnels and rail stations, including linear (line-type) heat detection, CCTV video image smoke detection, flame detection, smoke and heat detectors, and spot-type detection. Fire detection systems should be selected to support the fire safety goals and objectives and the overall fire safety program, which can include notifying occupants to allow for safe evacuation, modifying tunnel ventilation or operations, and notifying emergency responders.

NFPA *Standards* 130 and 502 provide general requirements for fire detections systems in transportation tunnels. These documents reference codes such as NFPA *Standard* 72, which provide design requirements for fire detection and occupant notification. Publications developed by the Road Tunnel Operation Technical Committee of PIARC (2007b, 2008) include specific guidance on the application of these systems. There have been several research projects that can also provide additional information to assist with developing detection system concepts and designs (Liu et al. 2006, 2009; Kashef et al. 2009; Zalosh and Chantranuwat 2003). Bendelius (2008) provides information on advantages and disadvantages and selection of fire detection methods in tunnels.

Road Tunnels

A road tunnel is an enclosed vehicular facility with an operating roadway for motor vehicles passing through it. Road tunnels may be underwater (subaqueous), mountain, or urban, or may be created by air-right structures over a roadway or overbuilds of a roadway.

Table 1   List of Road Tunnel Fires

Year	Tunnel	Country	Length, ft	Fire Duration	Damage People	Damage Vehicles	Damage Structure
1949	Holland	United States	8,365	4 h	66 injured	10 trucks 13 cars	Serious
1974	Mont Blanc	France/Italy	38,053	15 min	1 injured	—	—
1976	Crossing BP	France	1,411	1 h	12 injured	1 truck	Serious
1978	Velsen	Netherlands	2,526	1 h 20 min	5 dead 5 injured	4 trucks 2 cars	Serious
1979	Nihonzaka	Japan	6,708	159 h	7 dead 1 injured	127 trucks 46 cars	Serious
1980	Kajiwara	Japan	2,427	—	1 dead	2 trucks	Serious
1982	Caldecott	United States	3,372	2 h 40 min	7 dead 2 injured	3 trucks 1 bus 4 cars	Serious
1983	Pecorila Galleria	Italy	2,172	—	9 dead 22 injured	10 cars	Limited
1986	L'Arme	France	3,625	—	3 dead 5 injured	1 truck 4 cars	Limited
1987	Gumefens	Switzerland	1,125	2 h	2 dead	2 trucks 1 van	Slight
1990	Røldal	Norway	15,274	50 min	1 injured	—	Limited
1990	Mont Blanc	France/Italy	38,053	—	2 injured	1 truck	Limited
1993	Serra Ripoli	Italy	1,450	2 h 30 min	4 dead 4 injured	5 trucks 11 cars	Limited
1993	Hovden	Norway	4,232	1 h	5 injured	1 motorcycle 2 cars	Limited
1994	Huguenot	South Africa	12,839	1 h	1 dead 28 injured	1 bus	Serious
1995	Pfander	Austria	22,041	1 h	3 dead 4 injured	1 truck 1 van 1 car	Serious
1996	Isola delle Femmine	Italy	485	—	5 dead 20 injured	1 tanker 1 bus 18 cars	Serious
1999	Mont Blanc	France/Italy	38,053	—	39 dead	23 trucks 10 cars 1 motorcycle 2 fire engines	Serious
1999	Tauern	Austria	20,998	—	12 dead 49 injured	14 trucks 26 cars	Serious
2000	Seljestad	Norway	4,173	45 min	6 injured	1 truck 4 cars 1 motorcycle	—
2001	Praponti	Italy	14,463	—	19 injured	—	Serious
2001	Gleinalm	Austria	27,293	—	5 dead 4 injured	—	—
2001	Propontin	Italy	14,463	—	14 injured	1 car	—
2001	Gleinalm	Austria	27,231	—	5 dead 4 injured	—	—
2001	Guldborgsund	Denmark	1,509	—	5 dead 6 injured	—	—
2001	St. Gotthard	Switzerland	55,512	—	11 dead	2 heavy-goods vehicle	—
2002	Ostwaldiberg	Austria	—	—	1 dead	—	—
2003	44-France	France	2,028	—	2 dead	1 car 1 motorcycle	—
2003	Baregg	Switzerland	4,560	—	2 dead 21 injured	4 trucks 3 fire engines	Serious
2004	Baregg	Switzerland	3,543	—	1 dead 1 injured	1 car 1 truck	—
2005	Frejus	France-Italy	42,306	6 h	2 dead	4 trucks 1 fire engine	—
2006	Viamala	Switzerland	3,609	—	9 dead 6 injured	—	—

Source: PIARC (2007a, 2007b)

All road tunnels require ventilation to remove contaminants produced during normal engine operation. Normal ventilation may be provided by natural means, by traffic-induced piston effects, or by mechanical equipment. The method selected should be the most economical in both construction and operating costs.

Ventilation must also provide control of smoke and heated gases from a fire in the tunnel. Smoke flow control is needed to provide an environment suitable for both evacuation and rescue in the evacuation path. Emergency ventilation can be provided by natural means, by taking advantage of the buoyancy of smoke and hot gases, or by mechanical means.

Ventilation Modes. A range of mechanical ventilation are typically considered for road tunnels: normal, congested, emergency, and temporary, as discussed in the section on Tunnel Ventilation Concepts.

Ventilation Systems. Ventilation must dilute contaminants during normal and congested tunnel operations and control smoke during emergency operations. Factors affecting ventilation system selection include tunnel length, cross section, and grade; surrounding environment; traffic volume, direction (i.e., unidirectional or bidirectional), and mix; and construction cost.

Natural and traffic-induced ventilation systems are adequate for relatively short tunnels, and for those with low traffic volume or density. Long, heavily traveled tunnels should have mechanical ventilation systems. The tunnel length at which this change takes effect is somewhere between 1200 and 2200 ft.

Natural Ventilation. Airflow through a naturally ventilated tunnel can be portal-to-portal (Figure 2A) or portal-to-shaft (Figure 2B). Portal-to-portal flow functions best with unidirectional traffic, which produces a consistent, positive airflow. In this case, air speed in the roadway area is relatively uniform, and the contaminant concentration increases to a maximum at the exit portal. Under adverse atmospheric conditions, air speed may decrease and contaminant concentration may increase, as shown by the dashed line in Figure 2A.

Introducing bidirectional traffic into such a tunnel further reduces longitudinal airflow and increases the average contaminant concentration. The maximum contaminant level in a tunnel with bidirectional traffic will not likely occur at the portal, and will not necessarily occur at the midpoint of the tunnel.

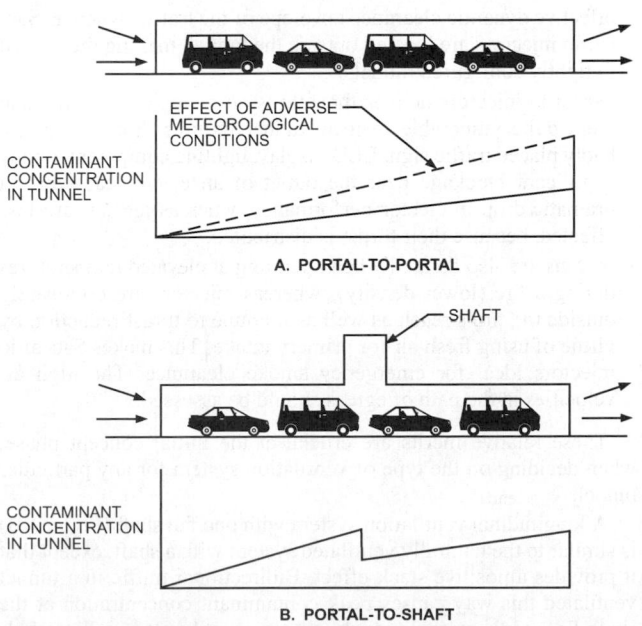

A. PORTAL-TO-PORTAL

B. PORTAL-TO-SHAFT

Fig. 2 Natural Ventilation

A naturally ventilated tunnel with an intermediate shaft (Figure 2B) is better suited for bidirectional traffic; however, airflow through the shaft is also affected by adverse atmospheric conditions. The stack effect benefit of the shaft depends on air/rock temperatures, wind, and shaft height. Adding more than one shaft to a tunnel may be more of a disadvantage than an advantage, because a pocket of contaminated air can be trapped between the shafts.

Naturally ventilated tunnels over 1000 ft long require emergency ventilation to extract smoke and hot gases generated during a fire, as recommended by NFPA *Standard* 502. This standard also further recommends that tunnels between 800 and 1000 ft long require engineering analyses to determine the need for emergency ventilation. Emergency ventilation systems may also be used to remove stagnant contaminants during adverse atmospheric conditions. Because of the uncertainties of natural ventilation, especially the effects of adverse meteorological and operating conditions, reliance on natural ventilation to maintain carbon monoxide (CO) levels for tunnels over 800 ft long should be thoroughly evaluated. This is particularly important for tunnels with anticipated heavy or congested traffic. If natural ventilation is deemed inadequate, a mechanical system should be considered for normal operations.

Smoke from a fire in a tunnel with only natural ventilation is driven primarily by the buoyant effects of hot gases, and tends to flow upgrade. The steeper the grade, the faster the smoke moves, thus restricting the ability of motorists trapped between the incident and a portal at higher elevation to evacuate the tunnel safely. As shown in Table 2, the Massachusetts Highway Department and Federal Highway Administration (MHD/FHWA) (1995) demonstrated how smoke moves in a naturally ventilated tunnel.

Mechanical Ventilation. A tunnel that is long, has a heavy traffic flow, or experiences frequent adverse atmospheric conditions, requires fan-based mechanical ventilation. Options include longitudinal ventilation, semitransverse ventilation, and full transverse ventilation.

Longitudinal ventilation introduces or removes air from the tunnel at a limited number of points, creating longitudinal airflow along the roadway. Longitudinal ventilation can be accomplished either by push-pull vent shafts, injection, jet fan operation, or a combination of injection and extraction at intermediate points in the tunnel. Injectors and jet fans are classified as impulse systems, because they impart a momentum to the tunnel flow, as the primary high-velocity jet diffuses out. At start-up, this thrust causes the air in the tunnel to accelerate until equilibrium is established between this force and the opposing drag forces due to viscous friction and the additional pressure losses at the tunnel portals, traffic, wind, and fire, etc.

Injection longitudinal ventilation, frequently used in rail tunnels, uses externally located fans to inject air into the tunnel through a high-velocity Saccardo nozzle, as shown in Figure 3A. This air injection, usually in the direction of traffic flow, induces additional longitudinal airflow. The Saccardo nozzle functions on the principle that a high-velocity air jet injected at a small angle to the tunnel axis can induce a high-volume longitudinal airflow in the tunnel. The amount of induced flow depends primarily on the nozzle area, discharge velocity and angle of the nozzle, as well as downstream air resistances. This type of ventilation is most effective with unidirectional traffic flow.

With injection longitudinal ventilation, air speed remains uniform throughout the tunnel, and the contaminant concentration increases from zero at the entrance to a maximum at the exit. Adverse atmospheric conditions can reduce system effectiveness. The contaminant level at the exit increases as airflow decreases or tunnel length increases.

Injection longitudinal ventilation, with supply at a limited number of tunnel locations, is economical because it requires the fewest fans, places the least operating burden on fans, and requires no distribution air ducts. As the length of the tunnel increases, however,

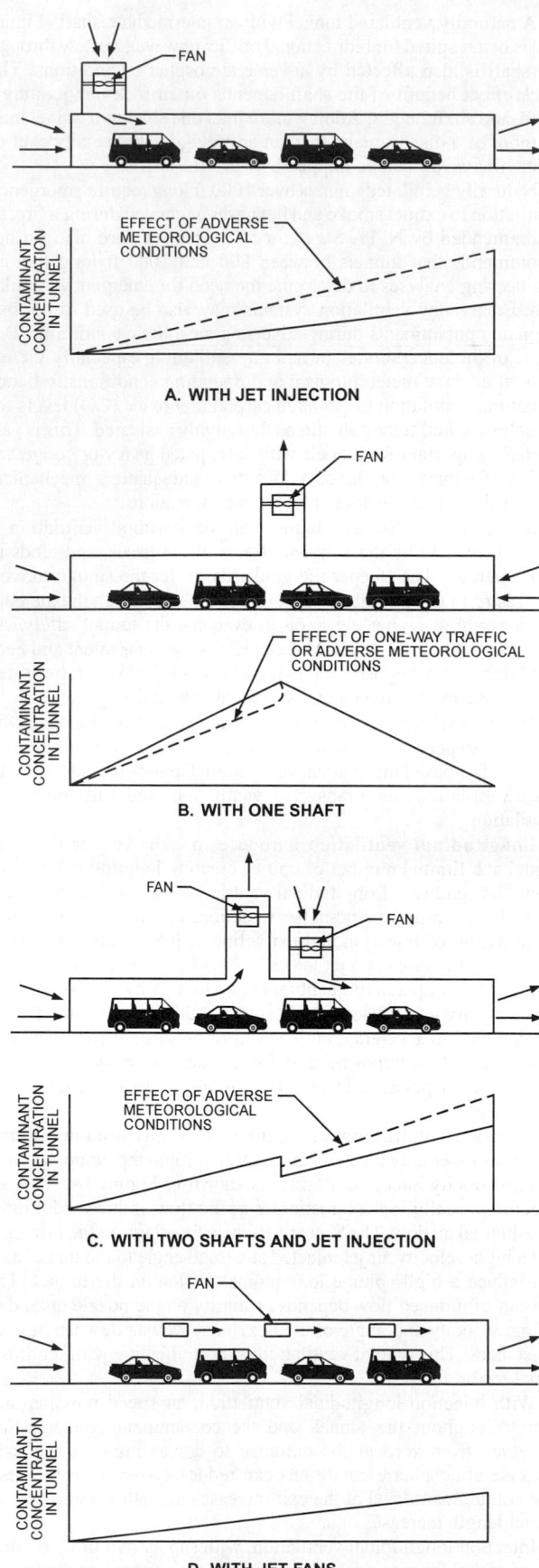

Fig. 3   Longitudinal Ventilation

Table 2   Smoke Movement During Natural Ventilation Tests
(Memorial Tunnel Fire Ventilation Test Program)

Test No.	Fire Heat Release Rate, 10^6 Btu/h		Smoke Layer Begins Descent, min	Smoke Fills Tunnel Roadway, min	Peak Smoke Velocity, fpm
	Nominal	Peak			
501	68	99	3+	5	1200
502	170	194	1+	3	1600

Note: Tunnel grade is 3.2%.

disadvantages become apparent, such as excessive air velocities in the roadway and smoke being drawn the entire length of the roadway during an emergency.

The main aerodynamic differences between the jet fan and Saccardo injectors are that the injectors impart thrust at one location in the tunnel, whereas in jet fan systems this thrust is distributed along the tunnel. Injectors use outdoor air as primary flow, whereas the primary airflow in jet fans enters the fan inlet from the tunnel.

Saccardo injectors may operate in a flow induction mode (low tunnel air resistance), or in flow rejection mode (high tunnel air resistance); both modes are acceptable. This means there may be flow reversal at the nozzle position with flow exiting the near portal, whereas jet fans always induce flow from one portal to the other. Flow under jet fans in a highly resistive tunnel may recirculate, but this is a strictly local feature.

A brief comparison of the technical and economic features of the two longitudinal impulse ventilation systems reveals the following:

- Jet fans have little or no civil engineering costs for installation, but have significant electrical cabling costs. Saccardo injectors require expensive civil engineering work to install the fans at the tunnel portal, with no cabling distribution costs.
- Routine maintenance or emergency repair work on jet fans usually requires disruption of normal tunnel service and availability; this is not the case for Saccardo injectors, which can be accessed externally.
- Saccardo injectors eliminate electrical cabling in the tunnel, providing a clear safety and cost advantage over jet fans.
- Jet fans take up headroom in the tunnel ceiling, which limits the effective dynamic clearance envelope of the traffic, whereas Saccardo injectors are located outside the tunnel, making them ideal in tightly configured tunnels.
- Saccardo injectors deliver their thrust at a single point, making them quite vulnerable to local tunnel fixtures. For example, a badly placed traffic sign, LED display, lighting equipment, or any significant blockage near the outlet of an ejector can cause a dramatic drop in ejector performance, whereas jet fans are less affected, because their thrust is distributed.
- Jet fans are also derated when operating at elevated temperatures during a fire (lower density), whereas injectors are both safely outside the fire's reach as well as immune to thrust reduction by virtue of using fresh air for primary intake. This makes Saccardo injectors ideal for emergency smoke clearance. The high air velocities in the path of egress should be assessed.

These relative merits are crucial at the initial concept phase, when deciding on the type of ventilation system for any particular tunnel.

A longitudinal ventilation system with one fan shaft (Figure 3B) is similar to the naturally ventilated system with a shaft, except that it provides a positive stack effect. Bidirectional traffic in a tunnel ventilated this way causes peak contaminant concentration at the shaft. For unidirectional tunnels, contaminant levels become unbalanced.

Another form of longitudinal system has two shafts near the center of the tunnel: one for exhaust and one for supply (Figure 3C). In this arrangement, part of the air flowing in the roadway is replaced by the interaction at the shafts, which reduces the concentration of contaminants in the second half of the tunnel. This concept is only effective for tunnels with unidirectional traffic flow. Adverse wind conditions can reduce tunnel airflow by short-circuiting the flow of air from the supply fan shaft/injection port to the exhaust fan/shaft, which causes contaminant concentrations to increase in the second half of the tunnel.

Construction costs of two-shaft tunnels can be reduced if a single shaft with a dividing wall is constructed. However, this significantly increases the potential for short-circuited airflows from supply shaft to exhaust shaft; under these circumstances, the separation between exhaust shaft and intake shaft should be maximized.

Jet fan longitudinal ventilation has been installed in a number of tunnels worldwide. With this scheme, specially designed axial fans (jet fans) are mounted at the tunnel ceiling (Figure 3D). This system eliminates the space needed to house ventilation fans in a separate structure or ventilation building, but may require greater tunnel height or width to accommodate the jet fans so that they are outside of the tunnel's dynamic clearance envelope. This envelope, formed by the vertical and horizontal planes surrounding the roadway in a tunnel, defines the maximum limits of the predicted vertical and lateral movement of vehicles traveling on the roadway at design speed. As tunnel length increases, however, disadvantages become apparent, such as excessive air speed in the roadway and smoke being drawn the entire length of the roadway during an emergency.

Longitudinal ventilation is the most effective method of smoke control in a road tunnel with unidirectional traffic. A ventilation system must generate sufficient longitudinal air velocity to prevent **backlayering** of smoke (movement of smoke and hot gases against ventilation airflow in the tunnel roadway). The air velocity necessary to prevent backlayering over stalled or blocked motor vehicles is the minimum velocity needed for smoke control in a longitudinal ventilation system and is known as the **critical velocity**.

Semitransverse ventilation can be configured for supply or exhaust. This type of ventilation involves the uniform distribution (supply) or collection (exhaust) of air throughout the length of a road tunnel. Semitransverse ventilation is normally used in tunnels up to about 7000 ft; beyond that length, tunnel air velocity near the portals becomes excessive.

Supply semitransverse ventilation in a tunnel with bidirectional traffic produces a uniform level of contaminants throughout, because air and vehicle exhaust gases enter the roadway area at the same uniform rate. With unidirectional traffic, additional airflow is generated by vehicle movement, thus reducing the contaminant level in the first half of the tunnel (Figure 4A).

Because tunnel airflow is fan-generated, this type of ventilation is not adversely affected by atmospheric conditions. Air flows the length of the tunnel in a duct with supply outlets spaced at predetermined distances. Fresh air is best introduced at vehicle exhaust pipe level to dilute exhaust gases immediately. The pressure differential between the duct and the roadway must be enough to counteract the effects of piston action and adverse atmospheric winds.

If a fire occurs in the tunnel, the supply air initially dilutes the smoke. Supply semitransverse ventilation should be operated in reverse mode for the emergency, so that fresh air enters through the portals and creates a tenable environment for both emergency egress and firefighter ingress. Therefore, a supply semitransverse ventilation system should preferably have a ceiling supply (in spite of the disadvantage during normal operations), and reversible fans, so that smoke can be drawn up to the ceiling during a tunnel fire.

Exhaust semitransverse ventilation (Figure 4B) in a tunnel with unidirectional traffic flow produces a maximum contaminant concentration at the exit portal. In a tunnel with bidirectional traffic flow, the maximum concentration of contaminants is located near

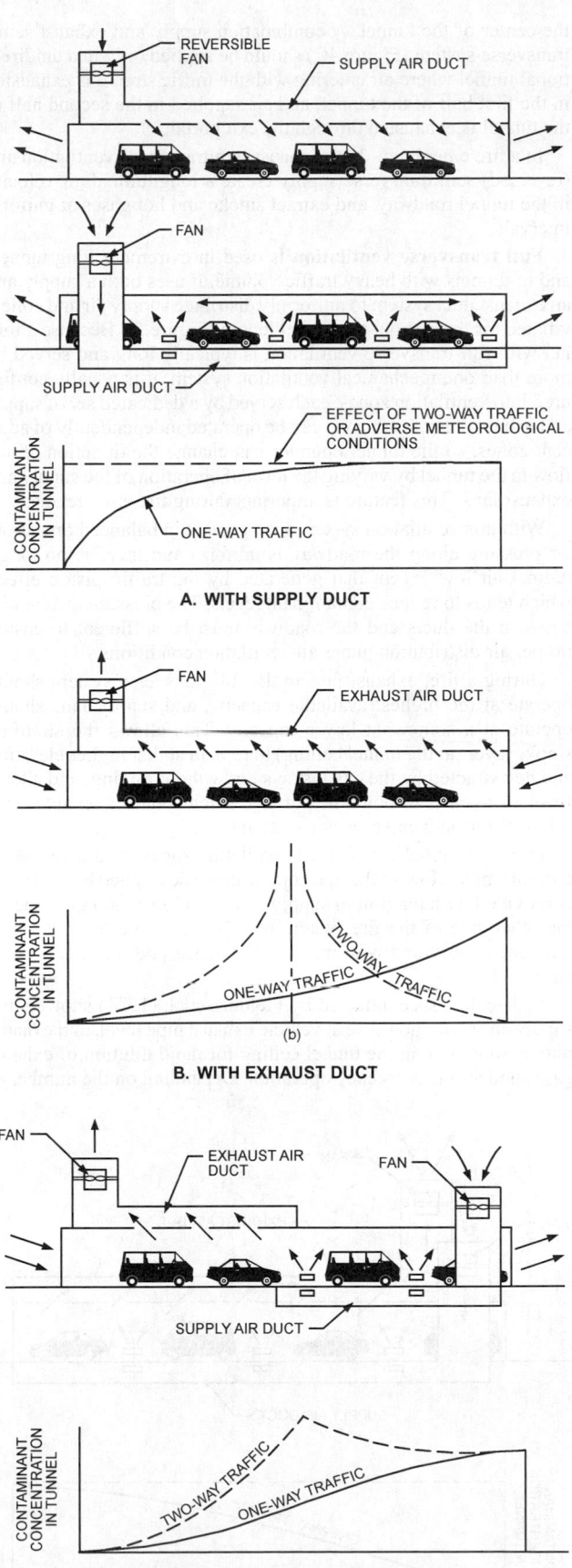

A. WITH SUPPLY DUCT

B. WITH EXHAUST DUCT

C. WITH SUPPLY AND EXHAUST DUCT

Fig. 4   Semitransverse Ventilation

the center of the tunnel. A combination supply and exhaust semi-transverse system (Figure 4C) should be applied only in a unidirectional tunnel where air entering with the traffic stream is exhausted in the first half of the tunnel, and air supplied in the second half of the tunnel is exhausted through the exit portal.

In a fire emergency, both exhaust semitransverse ventilation and (reversed) semitransverse supply create a longitudinal air velocity in the tunnel roadway, and extract smoke and hot gases at uniform intervals.

Full transverse ventilation is used in extremely long tunnels and in tunnels with heavy traffic volume. It uses both a supply and an exhaust duct system to uniformly distribute supply air and collect vitiated air throughout the tunnel length (Figure 5). Because a tunnel with full transverse ventilation is typically long and served by more than one mechanical ventilation system, it is usually configured into ventilation zones, each served by a dedicated set of supply and exhaust fans. Each zone can be operated independently of adjacent zones, so the tunnel operator can change the direction of airflow in the tunnel by varying the level of operation of the supply and exhaust fans. This feature is important during fire emergencies.

With this ventilation system arrangement in balanced operation, air pressure along the roadway is uniform and there is no longitudinal airflow except that generated by the traffic piston effect, which tends to reduce contaminant levels. The pressure differential between the ducts and the roadway must be sufficient to ensure proper air distribution under all ventilation conditions.

During a fire, exhaust fans in the full transverse system should operate at the highest available capacity, and supply fans should operate at a somewhat lower capacity. This allows the stratified smoke layer (at the tunnel ceiling) to remain at that higher elevation and be extracted by the exhaust system without mixing, and allows fresh air to enter through the portals, which creates a tenable environment for both emergency egress and firefighter ingress.

In longer tunnels, individual ventilation zones should be able to control smoke flow so that the zone with traffic trapped behind a fire is provided with maximum supply and no exhaust, and the zone on the other side of the fire (where unimpeded traffic has continued onward) is provided with maximum exhaust and minimum or no supply.

Full-scale tests conducted by Fieldner et al. (1921) showed that supply air inlets should be at vehicle exhaust pipe level, and exhaust outlets should be in the tunnel ceiling for rapid dilution of exhaust gases under nonemergency operation. Depending on the number of

traffic lanes and tunnel width, airflow can be concentrated on one side, or divided over two sides.

Other Ventilation Systems. There are many variations and combinations of the road tunnel ventilation systems described here. Most hybrid systems are configured to solve a particular problem faced in the development and planning of a specific tunnel, such as excessive air contaminants exiting at the portal(s). Figure 6 shows a hybrid system developed for a tunnel with a near-zero level of acceptable contaminant discharge at one portal. This system is essentially a semitransverse supply system, with a semitransverse exhaust system added in section 3. The exhaust system minimizes pollutant discharge at the exit portal, which is located near extremely sensitive environmental receptors.

Ventilation System Enhancements. **Single-point extraction** is an enhancement to a transverse system that adds large openings to the extraction (or exhaust) duct. These openings include devices that can be operated during a fire emergency to extract a large volume of smoke as close to the fire source as possible. Tests proved this concept effective in reducing air temperature and smoke volume in the tunnel. The size of the duct openings tested ranged from 100 to 300 ft^2 (MHD/FHWA 1995).

Oversized exhaust ports are simply expanded exhaust ports installed in the exhaust duct of a transverse or semitransverse ventilation system. Two methods are used to create this configuration. One is to install a damper with a fusible link; another uses a material that, when heated to a specific temperature, melts and opens the airway. Meltable materials showed only limited success in testing (MHD/FHWA 1995).

Normal Ventilation Air Quantities.

Contaminant Emission Rates. Because of the asphyxiate nature of the gas, CO is the exhaust gas constituent of greatest concern from spark-ignition engines. From compression-ignition (diesel) engines, the critical contaminants are nitrogen oxides (NO$_x$), such as nitric oxide (NO) and nitrogen dioxide (NO$_2$). Tests and operating experience indicate that, when CO level is properly diluted, other dangerous and objectionable exhaust by-products are also diluted to acceptable levels, although this trend needs reviewing with newer vehicle fleets. An exception is the large amount of unburned hydrocarbons from vehicles with diesel engines; when diesel-engine vehicles exceed 15% of the traffic mix, visibility in the tunnel can become a serious concern. In addition, suspended particles from tires and general road dust are gradually forming a larger percentage of particulate matter in tunnel environment, and must be considered in addition to engine emissions. The section on Bus Terminals includes further information on diesel engine contaminants and their dilution.

Vehicle emissions of CO, NO$_x$, and hydrocarbons for any given calendar year can be predicted for cars and trucks operating in the United States by using the MOBILE models, developed and maintained by the U.S. Environmental Protection Agency. The current version is MOBILE6.2 (EPA 2002). In contaminant emission rate analyses, the following practices and assumptions may be implemented:

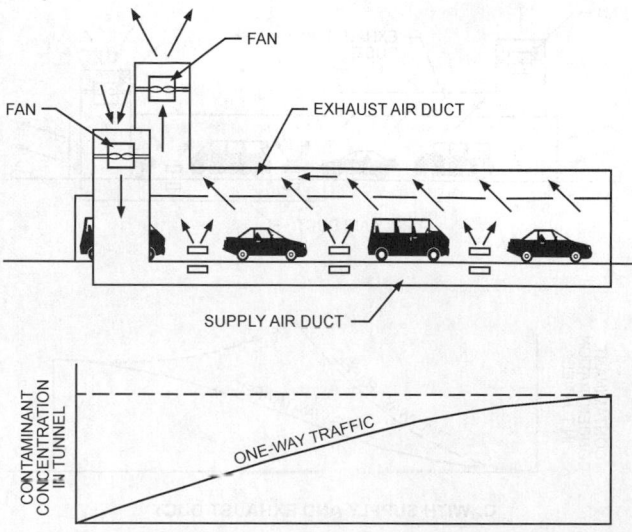

Fig. 5   **Full Transverse Ventilation**

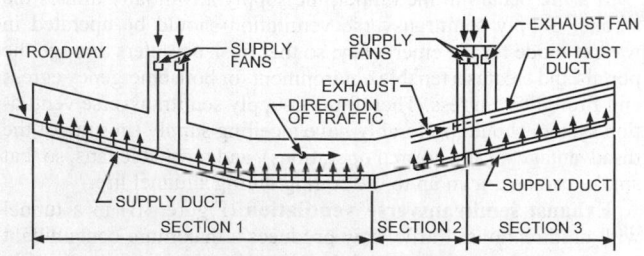

Fig. 6   **Combined Ventilation System**

Table 3 Average Dimensional Data for Automobiles Sold in the United States

Size/Class	Wheelbase, ft	Length, ft	Frontal Area, ft²
Subcompact	8.0	14.0	17.3
Compact	9.0	15.9	19.8
Midsize	9.7	18.0	21.9
Large	10.0	18.5	22.4
Average	9.18	16.60	20.35

- CO emission rates are higher during acceleration and deceleration than at constant speed; this effect may be accounted for by adding a 10% safety factor to the computations.
- The effect of positive or negative grades up to 2% is usually neglected. Engineers should use judgment, or available data, in applying correction factors for positive grades greater than 2%.
- Traffic is assumed to move as a unit, with a constant space interval between vehicles, regardless of roadway grade.
- Average passenger vehicle dimensions may be assumed where specific vehicle data are unavailable.

Table 3 presents typical physical data for automobiles for use in normal ventilation air quantity analyses.

Allowable Carbon Monoxide. EPA's (1975) supplement to its *Guidelines for Review of Environmental Impact Statements* concerns the concentration of CO in tunnels. This supplement evolved into a design approach based on keeping CO concentration at or below 125 ppm, for a maximum 1 h exposure time, for tunnels located at or below an altitude of 3280 ft. In 1989, the EPA revised its recommendations for maximum CO levels in tunnels located at or below an altitude of 5000 ft to the following:

- A maximum of 120 ppm for 15 min exposure
- A maximum of 65 ppm for 30 min exposure
- A maximum of 45 ppm for 45 min exposure
- A maximum of 35 ppm for 60 min exposure

These guidelines do not apply to tunnels in operation before the adoption date.

At higher elevations, vehicle CO emissions are greatly increased, and human tolerance to CO exposure is reduced. For tunnels above 5000 ft, the engineer should consult with medical authorities to establish a proper design value for CO concentrations. Unless otherwise specified, the material in this chapter refers to tunnels at or below an altitude of 5000 ft.

Outdoor air standards and regulations such as those from the Occupational Safety and Health Administration (OSHA) and the American Conference of Governmental Industrial Hygienists (ACGIH) are discussed in the section on Bus Terminals.

Emergency Ventilation Air Quantities. A road tunnel ventilation system must be able to protect the traveling public during the most adverse and dangerous conditions (e.g., fires), as well as during normal conditions. Establishing the requisite air volume requirements is difficult because of many uncontrollable variables, such as the possible number of vehicle combinations and traffic situations that could occur during the lifetime of the facility.

For many years, the rule of thumb has been 100 cfm per lane-foot. The Memorial Tunnel Fire Ventilation Test Program (MHD/FHWA 1995) showed that this value is, in fact, a reasonable first pass at an emergency ventilation rate for a road tunnel.

Longitudinal flow, single-point extraction, and dilution are three primary methods for controlling smoke flow in a tunnel. Both longitudinal flow and single-point extraction depend on the ability of the emergency ventilation system to generate the critical velocity necessary to prevent backlayering.

Critical Velocity. The concept of critical velocity is addressed in the section on Design Approach, under Tunnels.

Table 4 Typical Fire Size Data for Road Vehicles

Cause of Fire	Peak Fire Heat Release Rate, 10⁶ Btu/h
Passenger car	17 to 34
Multiple passenger cars (2 to 4 Vehicles)	34 to 68
Bus	68 to 102
Heavy goods truck	239 to 682
Tanker[3]	682 to 1023

Source: NFPA *Standard* 502 (2008).
Notes:
1. The designer should consider rate of fire development (peak heat release rates may be reached within 10 min), number of vehicles that could be involved in fire, and potential for fire to spread from one vehicle to another.
2. Temperatures directly above fire can be expected to be as high as 1800 to 2550°F.
3. Flammable and combustible liquids for tanker fire design should include adequate drainage to limit area of pool fire and its duration. Heat release rate may be greater than listed if more than one vehicle is involved.

Table 5 Maximum Air Temperatures at Ventilation Fans During Memorial Tunnel Fire Ventilation Test Program

Nominal FHRR, 10⁶ Btu/h	Temperature at Central Fans,[a] °F	Temperature at Jet Fans,[b] °F
68	225	450
170	255	700
340	325	1250

Source: MHD/FHWA (1995)
FHRR = Fire heat release rate
[a]Central fans located 700 ft from fire site.
[b]Jet fans located 170 ft downstream of fire site.

Design Fire Size. The design fire size selected significantly affects the magnitude of the critical velocity needed to prevent backlayering. Table 4 provides typical fire size data for a selection of road tunnel vehicles.

Temperature. A fire in a tunnel significantly increases air temperature in the tunnel roadway and exhaust duct. Thus, both the tunnel structure and ventilation equipment are exposed to the high smoke/gas temperature. The air temperatures shown in Table 5 provide guidance in selecting design exposure temperatures for ventilation equipment.

Testing. The Memorial Tunnel Fire Ventilation Test Program was a full-scale test program conducted to evaluate the effectiveness of various tunnel ventilation systems and ventilation airflow rates to control smoke from a fire (MHD/FHWA 1995). The results are useful in developing both emergency tunnel ventilation systems and emergency operational procedures.

Pressure Evaluation. Air pressure losses in tunnel ducts must be evaluated to compute the fan pressure and drive requirements. Fan selection should be based on total pressure across the fans, not on static pressure alone.

Fan total pressure FTP is defined by ASHRAE *Standard* 51/AMCA *Standard* 210 as the algebraic difference between the total pressures at fan discharge TP_2 and fan inlet TP_1, as shown in Figure 7. The fan velocity pressure FVP is defined as the pressure VP_2 corresponding to the bulk air velocity and air density at the fan discharge:

$$FVP = VP_2 \qquad (3)$$

Fan static pressure FSP is equal to the difference between fan total pressure and the fan velocity pressure:

$$FSP = FTP - FVP \qquad (4)$$

TP_2 must equal total pressure losses $\Delta TP_{2\text{-}3}$ in the discharge duct and exit pressure TP_3. Static pressure at the exit SP_3 is equal to zero.

$$TP_2 = \Delta TP_{2\text{-}3} + TP_3 = \Delta TP_{2\text{-}3} + VP_3 \qquad (5)$$

Likewise, total pressure at fan inlet TP_1 must equal the total pressure losses in the inlet duct and the inlet pressure:

$$TP_1 = TP_0 + \Delta TP_{0\text{-}1} \qquad (6)$$

Straight Ducts. Straight ducts in tunnel ventilation systems either (1) transport air, or (2) uniformly distribute (supply) or collect (exhaust) air. Several methods have been developed to predict pressure losses in a duct of constant cross-sectional area that uniformly distributes or collects air. The most widely used method was developed for the Holland Tunnel in New York (Singstad 1929). The following relationships, based on Singstad's work, give pressure losses at any point in a duct.

Total pressure for a **supply duct**

$$P_T = P_1 + \left(\frac{12\rho_a}{\rho_w g_c}\right)\left\{\frac{V_o^2}{2}\left[\frac{\alpha L Z^3}{3H} - (1-K)\frac{Z^2}{2}\right] + \frac{\beta L Z}{2H^3}\right\} \qquad (7)$$

Static pressure loss for an **exhaust duct**

$$P_S = P_1 + \left(\frac{12\rho_a}{\rho_w g_c}\right)\left\{\frac{V_o^2}{2}\left[\frac{\alpha L Z^3}{(3+c)H} + \frac{3Z^2}{(2+c)}\right] + \frac{\beta L Z}{2H^3(1+c)}\right\} \qquad (8)$$

where

P_T = total pressure loss at any point in duct, in. of water
P_S = static pressure loss at any point in duct, in. of water
P_1 = pressure at last outlet, in. of water
ρ_a = density of air, lb/ft^3
ρ_w = density of water, lb/ft^3
V_o = velocity of air entering duct, ft/s
L = total length of duct, ft
X = distance from duct entrance to any location, ft
Z = $(L-X)/L$
H = hydraulic radius, ft
K = constant accounting for turbulence = 0.615
α = constant related to coefficient of friction for concrete = 0.0035
β = constant related to coefficient of friction for concrete
  = 0.01433 ft^4/s^2
c = constant relating to turbulence of exhaust port
  = 0.20 for exhaust rates less than 200 cfm per foot
  = 0.25 for exhaust rates greater than 200 cfm per foot
g_c = gravitational constant = 32.2 lb$_m$·ft/lb$_f$·s^2

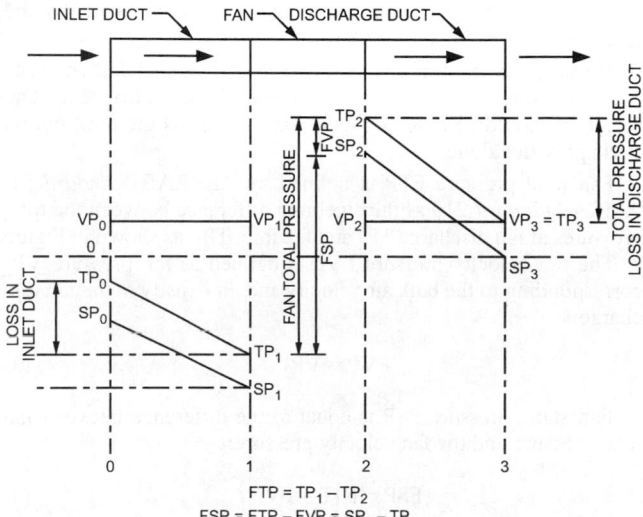

FTP = $TP_1 - TP_2$
FSP = FTP − FVP = $SP_2 - TP_1$

Fig. 7 Fan Total Pressure

The geometry of the exhaust air slot connection to the main duct is a concern in deriving the exhaust duct equation. The derivation is based on a 45° angle between the slot discharge and the main airstream axes. Variations in this angle can greatly affect the energy losses at the convergence from each exhaust slot, with total pressure losses for a 90° connection increasing by 50 to 100% over those associated with 45° angles (Haerter 1963).

For **distribution ducts** with sections that differ along their length, these equations may also be solved sequentially for each constant-area section, with transition losses considered at each change in section area. For a **transport duct** with constant cross-sectional area and constant air velocity, pressure losses are due to friction alone and can be computed using the standard expressions for losses in ducts and fittings (see Chapter 21 of the 2009 *ASHRAE Handbook—Fundamentals*).

Carbon Monoxide Analyzers and Recorders. Air quality in a tunnel should be monitored continuously at several key points. CO is the contaminant usually selected as the prime indicator of tunnel air quality, although in some of the more recent European road tunnels, NO$_x$ and visibility levels are now the main indicators driving the ventilation requirements, perhaps because of the prominence of diesel cars. CO-analyzing instruments base their measurements on one of the following three processes:

- **Catalytic oxidation (metal oxide)** analysis offers reliability and stability at a moderate initial cost. Maintenance requirements are low, plus these instruments can be calibrated and serviced by maintenance personnel after only brief instruction.

- **Infrared** analysis is sensitive and responsive, but has a high initial cost. This instrument is precise but complex, and requires a highly trained technician for maintenance and servicing.

- **Electrochemical** analysis is precise; the units are compact, lightweight, and moderately priced, but they have a limited life (usually not exceeding two years) and thus require periodic replacement.

As shown in Figures 1 to 4, the location of the peak emission concentration level in a road tunnel is a function of both traffic operation (unidirectional versus bidirectional) and type of ventilation provided (natural, longitudinal, semitransverse, or full transverse). Generally, time-averaged CO concentrations for the full length of the tunnel are needed to determine appropriate ventilation rates and/or required regulatory reporting. Time-averaged concentrations are particularly important in road tunnels where the ventilation system control is integrated with the CO monitoring system.

CO sampling locations in a road tunnel should be selected carefully to ensure meaningful results. For example, samples taken too close to an entry or exit portal do not accurately represent the overall level that can be expected throughout the tunnel. Multiple sampling locations are recommended to ensure that a reasonable average is reported. Multiple analyzers are also recommended to provide a reasonable level of redundancy in case of analyzer failure or loss of calibration. In longer road tunnels, which may have multiple, independently operated ventilation zones, the selected sampling locations should provide a representative CO concentration level for each ventilation zone. Strip chart recorders and microprocessors are commonly used to keep a permanent record of road tunnel CO levels.

CO analyzers and their probes should not be located directly in a roadway tunnel or in its exhaust plenum. Instead, an air pump should draw samples from the tunnel/exhaust duct through a sample line to the CO analyzer. This configuration eliminates the possibility of in-tunnel air velocities adversely affecting the instrument's accuracy. The length of piping between sampling point and CO analyzer should be as short as possible to maintain a reasonable air sample transport time.

Haze or smoke detectors have been used on a limited scale, but most of these instruments are optical devices and require frequent or constant cleaning with a compressed air jet. If traffic is predominantly diesel-powered, smoke haze and NO_x gases require individual monitoring in addition to that provided for CO.

Local regulations should be reviewed to determine whether ventilation exhaust monitoring is required for a particular road tunnel. If so, for tunnels using full transverse ventilation systems, CO and NO_2 pollutant sampling points should be placed carefully within the exhaust stacks/plenums. For longitudinally ventilated tunnels, sampling points should be located at least 100 ft in from the exit portal.

Controls.

Centralized Control. To expedite emergency response and to reduce the number of operating personnel for a given tunnel, all ventilating equipment should be controlled at a central location. New tunnels are typically provided with computer-based control systems, which function from operational control centers. In some older tunnel facilities, fan operation is manually controlled by an operator at a central control board. The control structure for newer road tunnel ventilation systems is typically supervisory control and data acquisition (SCADA), with programmable logic controllers (PLCs) providing direct control hardware over the associated electrical equipment. The operational control center varies from one stand-alone PC (with SCADA software providing dedicated ventilation control), to redundant client/server configurations providing an integrated control system and real-time database and alarm systems for tunnel operations (Buraczynski 1997). Communication links are required between the supervisory SCADA and PLCs.

The SCADA system operator controls the ventilation equipment through a graphical user interface, developed as part of the ventilation system design. Preprogrammed responses allow the operator to select the appropriate ventilation plan or incident response mode.

The SCADA system allows the operator to view equipment status, trend data values, log data, and use an alarm system. Whereas older tunnel facilities used chart recorders for each sampling point to demonstrate that the tunnel was sufficiently ventilated and compliant with environmental air quality standards, new tunnels use SCADA to log CO levels directly onto a nonvolatile medium, such as a CD-ROM.

Emergency response functions for road tunnel ventilation require that control system design meets life safety system standards. A high-availability system is required to respond on demand to fire incidents. High availability is obtained by using high-quality industrial components, and by adding built-in redundancy. The design must protect the system against common event failures; therefore, redundant communication links are segregated, and physically routed in separate raceways. High-integrity software for both the PLCs and the SCADA system is another major consideration.

Once a supervisory command is received, the PLC control handles equipment sequencing (e.g., fan and damper start-up sequence), least-hours-run algorithms, staggered starting of fans, and all interlocks. The PLC also receives instrumentation data from the fan and fan motor, and can directly shut down the fan if needed (e.g., because of high vibration). Conditions such as high vibration and high temperature are tolerated during emergency operation.

CO-Based Control. When input to the PLC, recorded tunnel air quality data allow fan control algorithms to be run automatically. The PLC controls fans during periods of rising and falling CO levels. Fan operations are usually based on the highest level recorded from several analyzers. Spurious high levels can occur at sampling points; the PLC control algorithm prevents the ventilation system from responding to short-lived high or low levels. PLC control also simplifies hardwired systems in older tunnel facilities, and increases flexibility through program changes.

Timed Control. This automatic fan control system is best suited for installations that experience heavy rush-hour traffic. With timed control, the fan operation schedule is programmed to increase the ventilation level, in preset increments, before the anticipated traffic increase; it can also be programmed for weekend and public holiday conditions. The timed control system is relatively simple and is easily revised to suit changing traffic patterns. Because it anticipates an increased airflow requirement, the associated ventilation system can be made to respond slowly, and thus avoid expensive demand charges from the local utility company. One variation of timed control is to schedule the minimum anticipated number of fans to run, and to start additional fans if high CO levels are experienced. As with the CO-based control system, a manual override is needed to cope with unanticipated conditions.

Traffic-Actuated Control. Several automatic fan control systems have been based on the recorded flow of traffic. Most require installation of computers and other electronic equipment needing specific maintenance expertise.

Local Fan Control. Local control panels are typically provided for back-up emergency ventilation control and for maintenance/servicing requirements. The local panels are often hardwired to the fan starters to make them independent from the normal SCADA/PLC control system. Protocols for handing over fan control from the SCADA/PLC system to the local panel must also be established, so that fans do not receive conflicting operational signals during an emergency.

Rapid Transit Tunnels and Stations

Modern high-performance, air-conditioned subway vehicles consume most of the energy required to operate rapid transit and are the greatest source of heat in the underground areas of a transit system. An environmental control system (ECS) is intended to maintain reasonable comfort during normal train operations, and help keep passengers safe during a fire emergency. Minimizing traction power consumption and vehicle combustible contents reduces ventilation requirements. The large amount of heat produced by rolling stock, if not properly controlled, can cause passenger discomfort, shorten equipment life, and increase maintenance requirements. Tropical climates present additional concerns for underground rail transit systems and make environment control more critical.

Temperature, humidity, air velocity, air pressure change, and rate of air pressure change help determine ECS performance. These conditions are affected by time of day (i.e., morning peak, evening peak, or off-peak), circumstance (i.e., normal, congested, or emergency operations), and location in the system (i.e., tunnel, station platform, entrance, or stairway). The *Subway Environmental Design Handbook* (SEDH) (DOT 1976) provides comprehensive and authoritative design aids on ECS performance; information in the SEDH is based on design experience, validated by field and model testing.

Normal operations involve trains moving through the subway system and stopping at stations according to schedule, and passengers traveling smoothly through stations to and from transit vehicles. The piston action of moving trains is the chief means of providing ventilation and maintaining an acceptable environment (i.e., air velocity and temperature) in the tunnels. Because normal operations are predominant, considerable effort should be made to optimize ECS performance during this mode.

One concern is limiting the air velocity caused by approaching trains on passengers waiting on the platform. Piston-induced platform air velocities can be reduced by providing a pressure relief shaft (also known as a blast shaft) at each end of affected platforms.

During normal train operations, platform passenger comfort is a function of the temperature and humidity of ambient and station air, platform air velocity, and duration of exposure to the station environment. For example, a person entering an 84°F station from 90°F outdoor conditions will momentarily feel more comfortable, particularly after a fast-paced walk ending with total rest, even if standing. However, in a short time, usually about 6 min, the person's metabolism adjusts to the new environment and produces a similar level of comfort as before. If a train were to arrive during this period, a relatively high station air temperature would be acceptable. Traditionally, the relative warmth index (RWI) has quantified this transient effect, allowing the designer to select an appropriate design air temperature for the station based on the transient, rather than steady-state, sensation of comfort. More recently, new transient thermal comfort models have been developed, leading to more advanced comfort indices being proposed (Gilbey 2006; Guan et al. 2009). Design temperatures based on the transient approach are typically higher (often 5 to 9°F) than those selected by the steady-state approach, and hence result in reduced cooling load and air-conditioning system requirements.

Congested operations result from delays or operational problems that prevent the normal dispatch of trains, such as missed headways or low-speed train operations. Trains may wait in stations, or stop at predetermined locations in tunnels during congested operations. Delays usually range from 30 s to 20 min, although longer delays may occasionally be experienced. Passenger evacuations or endangerment are not expected to occur. Congested ventilation analyses should focus on the potential need for forced (mechanical) ventilation, which may be required to control tunnel air temperatures in support of continued operation of train air-conditioning units. The aim of forced ventilation is to maintain onboard passenger comfort during congestion by operating the vehicle air conditioning system to prevent passengers from evacuating the train.

Emergency operations occur as a result of a fire in a subway tunnel or station. Fire emergencies include trash fires, track electrical fires, train electrical fires, and acts of arson. Some fires may involve entire train cars. Station fires are mostly trashcan fires. Statistically, most fire incidents reported in mass transit systems (up to 99%) are small, and low in smoke generation; these fires typically cause only minor injuries and operational disturbances. The most serious emergency condition is a fire on a stopped train in a tunnel; this event disrupts traffic and requires passenger evacuation. For this case, adequate tunnel ventilation is required to control smoke flow and enable safe passenger evacuation and safe ingress of emergency response personnel. Though rare, tunnel fires must be considered because of their potential life-safety ramifications.

Design Concepts. Elements of underground rail transit ventilation design may be divided into four interrelated categories: natural, mechanical, and emergency ventilation; and station air conditioning.

Natural Ventilation. Natural ventilation (e.g., ambient air infiltration and exfiltration) in subway systems primarily results from trains moving in tightly fitting tunnels, where air generally moves in the direction of train travel. The positive air pressure generated in front of a moving train expels warm air from the subway through tunnel portals, pressure relief shafts, station entrances and other openings; the negative pressure in the wake induces airflow into the subway through these same openings.

Considerable short-circuiting of airflow occurs in subways when two trains, traveling in opposite directions, pass each other; especially in stations or tunnels with porous walls (those with intermittent openings to allow air passage between trackways). Short-circuiting can also occur in stations and tunnels with nonporous walls where alternative airflow paths (e.g., open bypasses, cross-passageways, adits, crossovers) exist between the trackways. This short-circuited airflow reduces the net ventilation rate and increases air velocities on platforms and in entrances. During peak operating periods and high ambient temperatures, short-circuited airflow can cause undesirable heat build-up in the station.

To counter the negative effects of short-circuiting airflow, ventilation shafts are customarily located near interfaces between tunnels and stations. Shafts in station approach tunnels are often called blast shafts, because part of the tunnel air pushed by an approaching train is expelled through them before it affects the station environment. Shafts in station departure tunnels are known as relief shafts, because they relieve the negative air pressure created by departing trains. Relief shafts also induce outside airflow through the shaft, rather than through station entrances.

Additional shafts may be provided for natural ventilation between stations (or between portals, for underwater crossings), as dictated by tunnel length. The high cost of such ventilation structures necessitates a design that optimizes effectiveness and efficiency. Internal resistance from offsets and bends in the ventilation shaft should be kept to a minimum; shaft cross-sectional area should approximately equal the cross-sectional area of a single-track tunnel (DOT 1976).

Mechanical Ventilation. Mechanical ventilation in subways (1) supplements the natural ventilation effects of moving trains, (2) expels warm air from the system, (3) introduces fresh outside air, (4) supplies makeup air for exhaust, (5) restores the cooling potential of the tunnel heat sink by extracting heat stored during off hours or system shutdown, (6) reduces airflow between the tunnel and station, (7) provides outside air for passengers in stations or tunnels during an emergency or other unscheduled interruptions of traffic, and (8) purges smoke from the system during a fire, protecting the passengers' evacuation.

The most cost-effective design for a mechanical ventilation system serves multiple purposes. For example, a vent shaft designed for natural ventilation may also be used for emergency ventilation if a fan is installed in parallel, as part of a bypass (Figure 8). Current safety standards require emergency fans to be reversible (NFPA *Standard* 130).

Several ventilation shafts and fan plants may be required to work together to achieve many, if not all, of the eight design objectives. Depending on the shaft location, design, and local train operating characteristics, a shaft with an open bypass damper and a closed fan damper may serve as a blast or relief shaft. With the fan damper open and the bypass damper closed, air can be mechanically supplied to or exhausted from the tunnel, depending on fan rotation direction. Except for emergency ventilation, fan rotation direction is usually predetermined for various operating modes.

If a station is not air conditioned, warm air in the subway should be exchanged, at the maximum rate possible, with cooler outside air. If a station is air conditioned below the ambient temperature, inflow of warmer outside air should be limited and controlled.

Figure 9 shows a typical tunnel ventilation system between two subway stations. Here, flow of warm tunnel air into the station is minimized by either normal or mechanical ventilation effects. In Figure 9A, air pushed ahead of the train on track 2 diverts partially to the bypass ventilation shaft and partially into the wake of a train on track 1, as a result of pressure differences. Figure 9B shows an alternative operation with the same ventilation system where mid-tunnel fans operate in exhaust mode; when outdoor air conditions are favorable, makeup air is introduced through the bypass ventilation shafts. This alternative can also either provide or supplement

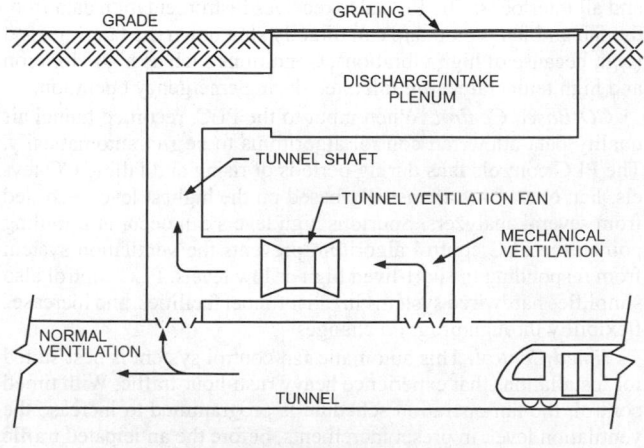

Fig. 8   Tunnel Ventilation Shaft

station ventilation. To achieve this, the bypass shafts are closed, and makeup air for the mid-tunnel exhaust fans enters through station entrances.

For forced air flow blown under car brake resistor grids, a more direct mechanical ventilation system (Figure 10) can be designed to remove station heat at its primary source, the underside of the train. Field tests have shown that trackway ventilation systems not only reduce upwelling of warm air into the platform areas, but also remove significant portions of heat generated by other undercar sources, such as dynamic-braking resistor grids and, in some cases, air-conditioning condenser units (DOT 1976), as long as consistent and steady air movement can be maintained from the heat source towards the exhaust grille. Ideally, makeup air for trackway exhaust should be introduced at track level, as in Figure 10A, to provide positive control over the direction of airflow; however, obstructions in the vehicle undercarriage area must be avoided when planning underplatform exhaust port and makeup air supply locations.

A more direct mechanical ventilation system (Figure 10) can be designed to remove station heat at its primary source, the underside of the train. Field tests have shown that trackway ventilation systems not only reduce upwelling of warm air into the platform areas, but also remove significant portions of heat generated by other undercar sources, such as dynamic-braking resistor grids and, in some cases, air-conditioning condenser units (DOT 1976). Ideally, makeup air for trackway exhaust should be introduced at track level, as in Figure 10A, to provide positive control over the direction of airflow; however, obstructions in the vehicle undercarriage area must be avoided when planning underplatform exhaust port and makeup air supply locations.

A trackway ventilation system without a dedicated makeup air supply (Figure 10B), also known as an underplatform exhaust (UPE) system, is the least effective alternative for heat removal. General design experience shows that where UPE grilles cannot be placed in close proximity to the source of undercar heat because of space constraints, or when a steady airflow cannot be established over the heat source towards the UPE grilles, heated undercar air can escape up through the gap between the car and platform edge, and the UPE effectiveness is reduced (Tabarra and Guan 2009). With a UPE system, a quantity of air equal to that withdrawn by the underplatform exhaust enters the station control volume, either from outside or from the tunnels. When the ambient, or tunnel, air temperature is higher than the station design air temperature, a UPE system reduces station heat load by removing undercar heat, but it also increases station heat load by drawing in warmer air, which

may affect platform passenger comfort. Because of these drawbacks, the effectiveness of a UPE system should be carefully considered and if possible modeled early, before the station design advances too far.

Figure 10C shows a cost-effective compromise: makeup air is introduced from the ceiling above the platform. Although heat removal effectiveness of this system may be less than that of the system with track-level makeup air, the inflow of warm tunnel air that may occur in a system without makeup air supply is negated.

Newer vehicles have air-conditioning grids above, generating heat near the ceiling during dwell time in the station. To exhaust this heat, an overtrack exhaust (OTE) system should be provided. OTE may be appropriate to remove fire smoke and heat. If analysis indicates that acceptable environmental conditions are achieved with OTE under normal and emergency conditions, the designer may consider evaluating the efficiency of the UPE system. The relative geometries of heat sources must be verified early in the design cycle, to enable the designer to make an informed decision.

Emergency Ventilation. During a subway tunnel fire, mechanical ventilation is an important part of the response and smoke control strategy. Within subway systems or other enclosed trainways, an emergency ventilation system is necessary to control the direction of smoke migration and allow safe evacuation of passengers and access by firefighters (see NFPA *Standard* 130). Depending on vehicle configuration, ventilation fan sizes, and tunnel geometry, emergency ventilation has the potential to affect fire size and smoke generation.

The most common method of ventilating a tunnel during a fire is push-pull fan operation: fans on one side of the fire operate in supply mode, while fans on the opposite side operate in exhaust mode. Emergency ventilation analyses should focus on determining the

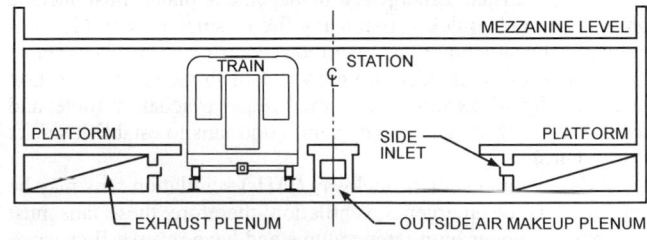

A. OUTSIDE AIR MAKEUP AT TRACK LEVEL

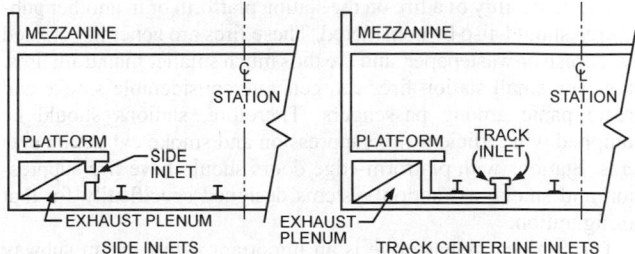

B. NO MAKEUP AIR SUPPLY

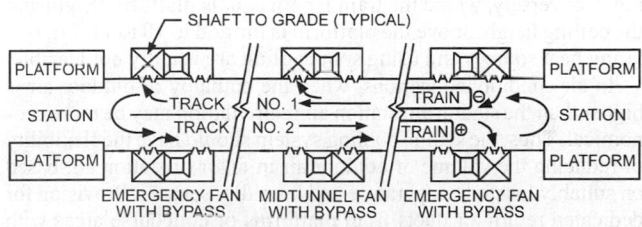

A. NORMAL VENTILATION BETWEEN STATIONS

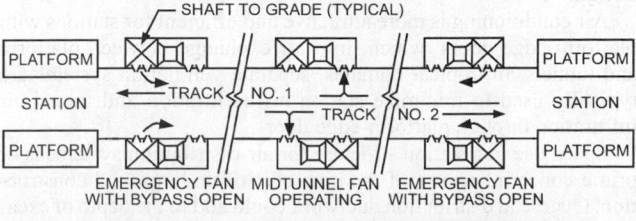

B. MECHANICAL VENTILATION BETWEEN STATIONS

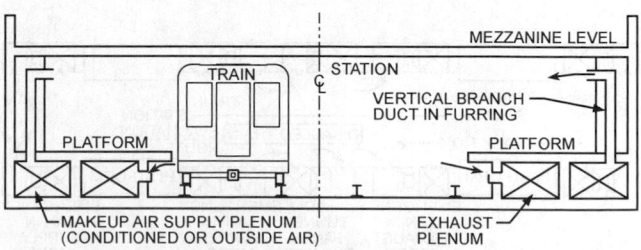

C. MAKEUP AIR SUPPLY AT CEILING LEVEL

Fig. 9   Tunnel Ventilation Concept

Fig. 10   Trackway Ventilation Concept (Cross-Sections)

airflow required to preserve tenable conditions in a single evacuation path from the train. The criterion used to design emergency ventilation for underground transit systems is critical velocity, similar to that presented in the section on Road Tunnels. The presence of nonincident trains should be considered in planning the emergency ventilation system response to specific fire incidents.

Emergency ventilation system design must allow for the unpredictable location of both the disabled train and the fire source. Therefore, emergency ventilation fans should have full reverse-flow capability, so that fans on either side of a disabled train can operate together to control airflow direction and counteract undesired smoke migration.

When a disabled train is stopped between two stations and fire or smoke is discovered, outside air is supplied by the emergency ventilation fans at the nearest station, and smoke-laden air is exhausted past the opposite end of the train by emergency ventilation fans at the next station, unless the location of the fire dictates otherwise. Passengers can then be evacuated along the tunnel walkways via the shortest possible route (Figure 11).

Emergency ventilation analysis should consider the possibility of nonincident trains stopped behind the disabled train. In this case, emergency fans should be operated so that nonincident trains are kept in the fresh airstream; if possible, they may be used to evacuate incident-train passengers. For long subway tunnels, in particular, analysis should also consider evacuating passengers to a nonincident trackway (through cross passageways), where a dedicated rescue train can move them to safety. Emergency ventilation analyses should identify passenger evacuation/firefighter ingress routes for evaluated scenarios, and fan modes to preserve tenable conditions in those routes.

When a train fire is discovered, the train should be moved if possible to the next station, to make passenger evacuation and fire suppression easier. Emergency management plans must include provisions to (1) quickly assess any fire or smoke event, (2) communicate the situation to an operations control center, (3) establish the location of the incident train, (4) establish the general location of the fire, (5) determine the best passenger evacuation route, and (6) quickly activate emergency ventilation fans to establish smoke flow control.

Midtunnel and station trackway (OTE) ventilation fans may be used to enhance emergency ventilation; therefore, these fans must also operate under high temperatures and have reverse-flow capability.

The possibility of a fire on the station platform or in another public area should also be considered. These fires are generally created by rubbish or wastepaper, and are thus much smaller than train fires. However, small station fires can generate considerable smoke and create panic among passengers. Therefore, stations should be equipped with efficient fire suppression and smoke extraction systems. Stations with platform-edge doors should have fire suppression and smoke extraction systems designed specifically for that configuration.

The fire heat release rate is an important parameter in subway emergency ventilation system design. The fire heat release rate for each vehicle type depends on initiation fire, combustibility of interior materials, size of the compartment, and ventilation (door and window openings), and thus must be established individually (see the Design Fires section for more information). Typical fire size data for single transit vehicles are as follows:

- Older transit vehicle $\approx 50 \times 10^6$ Btu/h
- New, hardened vehicle $\approx 35 \times 10^6$ Btu/h
- Light rail vehicle $\approx 30 \times 10^6$ Btu/h

Smoke obscuration is a key factor in defining a tenable environment for passenger evacuation, and visibility is often the governing criterion for station design. The smoke release rate should be calculated following acceptable procedures [e.g., Society of Fire Protection Engineers (SFPE) 2008].

Station Air Conditioning. Faster station approach speeds and closer headways, both made possible by computerized train control, have increased heat gains in subway stations. The net internal sensible heat gain for a typical two-track subway station, with 40 trains per hour per track traveling at a top speed of 50 mph, may reach 5.0×10^6 Btu/h, even after some tunnel heat is removed by the heat sink, station underplatform exhaust system, or tunnel ventilation system. To remove this heat from a station with a ventilation system using outside air and a maximum air temperature increase of 3°F, for example, would require roughly 1.4×10^6 cfm of outside air. This would be costly, and air velocities on the platforms would be objectionable to passengers.

The same amount of sensible heat gain, plus the latent heat and outside air loads (based on a station design air temperature 7°F lower than ambient), could be handled by about 630 tons of refrigeration. Even if station air conditioning is more expensive at the outset, long-term benefits include (1) reduced design airflow rates, (2) reduced ventilation shaft/duct sizing, (3) improved passenger comfort, (4) increased service life of other station equipment (e.g., escalators, elevators, fare collection), (5) reduced maintenance requirements for station equipment and structures, and (6) increased acceptance of the subway as a viable means of public transportation. Air conditioning should also be considered for other station ancillary areas, such as concourse levels and transfer levels. However, unless these walk-through areas are designed to attract patronage to concessions, the cost of air conditioning is usually not warranted.

The physical configuration of the station platform level usually determines the cooling distribution pattern. Platform areas with high ceilings, local warm spots created by trains, high-density passenger accumulation, or high-level lighting may need spot cooling. Conversely, where the train length equals platform length and the ceiling height above the platform is limited to 10 to 11.5 ft, isolating heat sources and using spot cooling are usually not feasible.

In air-conditioned stations, when the enthalpy of outdoor air is higher than the station air, station air recirculation may be more economical. Thus, the station cooling system should have the flexibility of reducing the volume of outdoor air in favor of station air, based on suitably located temperature and humidity sensors. Provision for dedicated return air ducts from platforms or concourse areas with accessible filters should be considered early in station cooling design.

Air conditioning is more attractive and efficient for stations with platform-edge doors, which limit air exchange between platform and tunnels. In tropical climates, separate ventilation systems are typically used to minimize station air exfiltration and tunnel air infiltration through platform-edge doors.

Space use in a station structure for air-distribution systems is of prime concern because of the high cost of underground construction. Overhead distribution ductwork could add to the depth of excavation during subway construction. The space beneath a subway station platform is normally an excellent area for low-cost distribution of supply, return, and/or exhaust air.

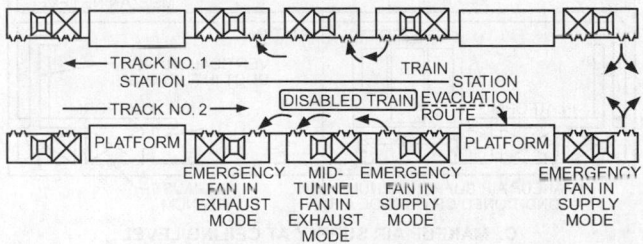

Fig. 11   Emergency Ventilation Concept

Design Method. Subways typically have two discrete sets of environmental criteria: one for normal and congested train operations and one for emergency fire/smoke operations. Criteria for normal operations include limits on tunnel air temperature (through tunnel ventilation or tunnel cooling) and humidity for various times of the year, minimum ventilation rates to dilute contaminants generated in the subway, and limits on the air velocity and rate of air pressure change to which passengers may be exposed. Some of these criteria are subjective and may vary based on demographics. Criteria for emergency operations include a minimum purge time to remove smoke from a subway, critical air velocity for smoke flow control during a tunnel fire, and minimum and maximum fan-induced tunnel air velocities.

Given a set of criteria, outdoor design conditions, and appropriate tools for estimating interior heat loads, heat sink effect, ventilation requirements, tunnel air velocity, and rate of air pressure changes, design engineers can select components for the environmental control system (ECS). ECS design should consider controls for tunnel air temperature, velocity, and quality, and the air pressure change rate. Systems selected generally combine natural and mechanical ventilation, overtrack and underplatform exhaust, and station air conditioning.

Train propulsion/braking systems and configuration of the tunnels and stations greatly affect the subway environment. Therefore, the ECS must often be considered during the early stages of subway system design. Factors affecting a subway environmental control system are discussed in this section. The *Subway Environmental Design Handbook* (SEDH) (DOT 1976) and NFPA *Standard* 130 have additional information.

Analytical Data. ECS design should be based on all the parameters affecting its operation, including ambient air conditions, train operating characteristics, applicable ventilation methods, new or existing ventilation structures, and calculated heat loads. ECS efficiency should be addressed early during transit system design. The tunnel ventilation system should be integrated with the design of other tunnel systems (including power, signaling, communications, and fire/life safety systems) and with the station ventilation system design. The ECS design must satisfy the project design criteria and comply with applicable local and national (or international) codes, standards, and regulations.

The ventilation engineer should be familiar with these requirements and apply suitable design techniques, such as computer modeling and simulations (using verified/validated engineering software).

Comfort Criteria. Because passenger exposure to the subway environment is transient, comfort criteria are not as strict as those for continuous occupancy. As a general principle, the station environment should provide a smooth transition between outside air conditions and thermal conditions in the transit vehicles. Except where platform edge doors are installed, train movement usually generates desirable air movement in stations, but air velocity should not exceed 1000 fpm in public areas during normal train operations.

Air Quality. Air quality in a subway system is influenced by many factors, some of which are not under the direct control of the HVAC engineer. Some particulates, gaseous contaminants, and odorants in the ambient air can be prevented from entering the subway system by judicious selection of ventilation shaft locations. Particulate matter, including iron and graphite dust generated by normal train operations, is best controlled by regularly cleaning stations and tunnels. However, the only viable way to control gaseous contaminants in a subway system, such as ozone (produced by electrical equipment) and CO_2 (from human respiration), is through adequate ventilation with outside air.

Subway system air quality should be analyzed either by engineering calculations or by computer modeling and simulations. The analysis should consider both the tunnel airflow induced by the piston effect of moving trains and the outside airflow required to dilute gaseous contaminants to acceptable levels. The results should comply with the *Subway Environmental Design Handbook* (DOT 1976) recommendation for at least 4 ach, as well as the recommendation of ANSI/ASHRAE *Standard* 62.1 to have a minimum of 15 cfm outside air per person. Maximum station occupancy should be used in the analysis.

Pressure Transients. Trains passing through aerodynamic discontinuities in a subway cause changes in tunnel static pressure, which can irritate passengers' ears and sinuses. Based on nuisance factor criteria, if the total change in the air pressure is greater than 2.8 in. of water, the rate of static pressure change should be kept below 1.7 in. of water per second. Pressure transients also add to the dynamic load on various equipment (e.g., fans, dampers) and appurtenances (e.g., acoustical panels). The formula and methodology of pressure transient calculations are complex; this information is presented in the SEDH (DOT 1976).

Air Velocity. During fires, emergency ventilation must be provided in the tunnels to control smoke flow and reduce air temperatures to permit both passenger evacuations and firefighting operations. The minimum air velocity in the affected tunnel should be sufficient to prevent smoke from backlayering (flowing in the upper cross section of the tunnel in the direction opposite the forced ventilation airflow). The method for ascertaining this critical air velocity is provided in the section on Design Approach, under Tunnels. The maximum tunnel air velocity experienced by evacuating passengers should not exceed 2200 fpm.

Interior Heat Loads. Heat in a subway is generated mostly by the following sources:

- Train deceleration/braking: Between 40 and 50% of heat generated in a subway arises from train deceleration/braking. Many vehicles use nonregenerative braking systems, in which the kinetic energy of the train is dissipated to the tunnel as heat, through dynamic and/or frictional brakes, rolling resistance, and aerodynamic drag. Regenerative systems dissipate less braking heat.

- Train acceleration: Heat is also generated as a train accelerates. Many vehicles use cam-controlled variable-resistance elements to regulate voltage across dc traction motors during acceleration. Electrical power is dissipated by these resistors (and the third rail) as heat into the subway. The heat released during train acceleration also comes from traction motor losses, rolling resistance, and aerodynamic drag. Heat from acceleration generally amounts to 10 to 20% of the total heat released in a subway system.

In subway systems with closely spaced stations, more heat is generated because of the frequent acceleration and deceleration.

- Vehicle air conditioning: Most new transit vehicles are fully climate-controlled. Air-conditioning equipment removes passenger and lighting heat from the cars and transfers it, along with condenser fan and compressor heat, into the subway. Vehicle air-conditioning system capacities generally range from 10 tons per vehicle for shorter rail cars (about 50 ft long), up to about 20 tons for longer rail cars (about 70 ft long). Heat from vehicle air conditioning and other accessories is generally 25 to 30% of total heat generated in a subway.

- Other sources: Tunnel heat also comes from people, lighting, induced outside air, miscellaneous equipment (e.g., fare collecting machines, escalators), and third-rail/catenary systems. These sources can generate 10 to 30% of the total heat released in a subway.

In a typical subway heat balance analysis, a control volume is defined around each station and heat sources are identified and quantified. The control volume usually includes the station and its various approach/departure tunnels. Typical values for heat emission/rejection data are given in Table 6.

Table 6 Typical Heat Source Emission Values

Source of Heat	Heat Rejection, Btu/h
Train A/C system (per vehicle)	144,000
Escalator (10 hp, 75% load factor)	19,100[a]
Fare collection machine	2730 [a]
Station lighting	10.2 per square foot [a]
People (walking, standing)	250 sensible[b]
	250 latent[b]

[a]See *Subway Environmental Design Handbook*, Part 3 (DOT 1976).
[b]See 2009 *ASHRAE Handbook—Fundamentals*, Chapter 9.

Heat Sink. The amount of heat flow from tunnel air to subway walls varies seasonally, as well as during morning and evening rush-hour operations. Short periods of abnormally high or low outside temperature may cause a temporary departure from the normal heat sink effect in unconditioned areas of the subway, changing the average tunnel air temperature. However, any change from the normal condition is diminished by the thermal inertia of the subway structure. During abnormally hot periods, heat flow from the tunnel air to subway walls increases. Similarly, during abnormally cold periods, heat flow from the subway walls to tunnel air increases.

For subway systems where daily station air temperatures are held constant by dedicated heating and cooling systems, heat flux from station walls is negligible. Depending on the amount of station air flowing into adjoining tunnels, heat flux from tunnel sections may also be reduced. Other factors affecting the heat sink component are soil type (dense rock or light, dry soil), extent of migrating groundwater or the local water table, and surface configuration of tunnel walls (ribbed or flat).

Measures to Limit Heat Loads. Various measures have been proposed to limit interior heat loads in subway systems, including regenerative braking, thyristor motor controls, track profile optimization, underplatform exhaust systems, and cooling dumping.

Electrical regenerative braking converts kinetic energy into electrical energy for use by other trains. Flywheel energy storage, an alternative form of regenerative braking, stores part of the braking energy in high-speed flywheels for use during vehicle acceleration. These methods can reduce the heat generated in train braking by approximately 25%.

Cam-controlled propulsion applies a set of resistance elements to regulate traction motor current during acceleration. Electrical energy dissipated by these resistors appears as waste heat in a subway. **Thyristor motor controls** replace the acceleration resistors with solid-state controls, which reduce acceleration-related heat losses by about 10% on high-speed subways, and by about 25% on low-speed subways.

Track profile optimization refers to a tunnel design that is lower between the stations. Less power is used for acceleration, because some of the potential energy of a standing train is converted to kinetic energy as the train accelerates toward the tunnel low point. Conversely, some of the kinetic energy of a train at maximum speed is converted to potential energy during braking, as the train approaches the next station. Track profile optimization reduces the maximum vehicle heat loss from acceleration and braking by about 10%.

An **overtrack exhaust (OTE)** and/or **underplatform exhaust (UPE)** system, described in the section on Mechanical Ventilation, uses extract grilles at regular intervals to remove heat generated by vehicle equipment located either at car roof level or under the car (e.g., resistors, compressors, air-conditioning condensers) from the station environment. For forced-blown resistor grids and cases where the airflow pattern is well controlled over the source of the undercar heat, SEDH (DOT 1976) provides a table (based on field test results in a given station platform geometry) of various UPE airflow rates versus UPE system efficiency. Care should be taken when extending these data to other platform geometries. For preliminary calculations, it may be assumed that (1) the train heat release

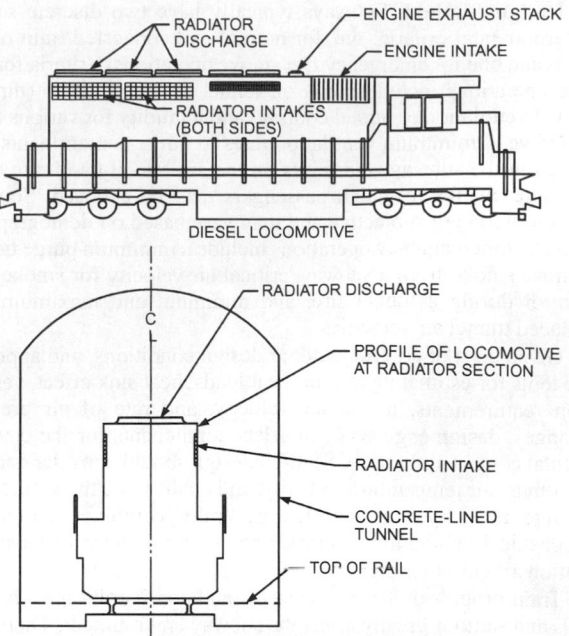

Fig. 12 Typical Diesel Locomotive Arrangement

(from braking and air conditioning) in the station box is about two-thirds of the control-volume heat load, and (2) the UPE is about 50% effective (provided the geometry and airflow pattern conditions are fulfilled). Sanchez (2003) studied the impact of OTE/UPE for air-conditioned stations.

In tropical areas, where there are only small daily differences in the ambient air temperature, tunnel walls do not cool off during the night; consequently the heat sink effect is negligible. In such cases, **cooling dumping** (releasing cooler air from the vehicle or its air-conditioning system) can be considered to limit heat accumulation in subway tunnels. However, the effect of cooling dumping on vehicle air-conditioning systems must be considered.

Railroad Tunnels

Railroad tunnels for diesel locomotives require ventilation to remove residual diesel exhaust, so that each succeeding train is exposed to a relatively clean air environment. Ventilation is also required to prevent locomotives from overheating while in the tunnel. For short tunnels, ventilation generated by the piston effect of a train, followed by natural ventilation, is usually sufficient to purge the tunnel of diesel exhaust in a reasonable time period. Mechanical ventilation for locomotive cooling is usually not required in short tunnels, because the time that a train is in the tunnel is typically less than the time it would take for a locomotive to overheat. However, under certain conditions, such as for excessively slow trains or during hot weather, locomotive overheating can still become a problem. For long tunnels, mechanical ventilation is required to purge the tunnel of diesel exhaust, and may also be required for locomotive cooling, depending on the speed of the train and the number and arrangement of locomotives used.

The diesel locomotive is essentially a fuel-driven, electrically powered vehicle. The diesel engine drives a generator, which in turn supplies electrical power to the traction motors. The power of these engines ranges from about 1000 to 6000 hp. Because the overall efficiency of the locomotive is generally under 30%, most of the energy generated by the combustion process must be dissipated as heat to the surrounding environment. Most of this heat is released above the locomotive through the engine exhaust stack and the radiator discharge (Figure 12).

In a tunnel, this heat is confined to the region surrounding the train. Most commercial trains are powered by more than one locomotive, so the last unit is subjected to heat and exhaust smoke released by preceding units. If sufficient ventilation is not provided, the air temperature entering the radiator of the last locomotive will exceed its allowable limit. Depending on the engine protection system, this locomotive will then either shut down or drop to a lower throttle position. In either event, the train will slow down. But, as discussed in the next section, a train relies on its speed to generate sufficient ventilation for cooling. As a result of the train slowing down, a domino effect takes place, which may cause the train to stall in the tunnel.

Design Concepts. Most long railroad tunnels (over 5 mi) in the western hemisphere that serve diesel operation use a ventilation concept using both a tunnel door and a system of fans and dampers, all located at one end of the tunnel. When a train moves through the tunnel, ventilation air for locomotive cooling is generated by the piston effect of the train moving toward (or away from) the closed portal door. This effect often creates a sufficient flow of air past the train for self-cooling.

Under certain conditions, when the piston effect cannot provide required airflow, fans supplement the flow and cool the tunnel. When the train exits at the portal, the tunnel is purged of residual smoke and diesel contaminants by running the fans (with the door closed) to move fresh air from one end of the tunnel to the other. Because the airflow and pressure required for cooling and purge modes may be substantially different, multiple fan systems or variable-volume fans may be required for the two operations. Also, dampers are provided to relieve the pressure across the door, which facilitates its operation while the train is in the tunnel.

Application of this basic ventilation concept varies depending on the length and grade of the tunnel, type and speed of the train, environmental and structural site constraints, and train traffic flow. One design, for a 9 mi long tunnel (Levy and Danziger 1985), extended the basic concept by including a mid-tunnel door and a partitioned shaft, which was connected to the tunnel on both sides of the mid-tunnel door. The combination of mid-tunnel door and partitioned shaft divided the tunnel into two segments, each with its own ventilation system. Thus, the ventilation requirement of each segment was satisfied independently. The need for such a system was dictated by the length of the tunnel, relatively low speed of the trains, and traffic pattern.

Locomotive Cooling Requirements. A breakdown of the heat emitted by a locomotive to the surrounding air can be determined by performing an energy balance. Starting with the fuel consumption rate (as a function of the throttle position), the heat release rates (as provided by the engine manufacturer) at the engine exhaust stack and radiator discharge, and the gross power delivered by the engine shaft (as determined from manufacturer's data), the amount of miscellaneous heat radiated by a locomotive can be determined as follows:

$$q_M = FH - q_S - q_R - P_G \qquad (9)$$

where

q_M = miscellaneous heat radiated from locomotive engine, Btu/h
F = locomotive fuel consumption, lb/h
H = heating value of fuel, Btu/lb
q_S = heat rejected at engine exhaust stack, Btu/h
q_R = heat rejected at radiator discharge, Btu/h
P_G = gross power at engine shaft, Btu/h

Because locomotive auxiliaries are driven off the engine shaft, with the remaining power used for traction power through the main engine generator, heat released by the main engine generator can be determined as follows:

$$q_G = (P_G - L_A)(1 - \varepsilon_G) \qquad (10)$$

where

q_G = main generator heat loss, Btu/h

L_A = power driving locomotive auxiliaries, Btu/h
ε_G = main generator efficiency

Heat loss from the traction motors and gear trains can be determined as follows:

$$q_{TM} = P_G - L_A - q_G - P_{TE} \qquad (11)$$

where

q_{TM} = heat loss from traction motors and gear trains, Btu/h
P_{TE} = locomotive tractive effort power, Btu/h

The total locomotive heat release rate q_T can then be determined:

$$q_T = q_S + q_R + q_m + L_A + q_G + q_{TM} \qquad (12)$$

For a train with N locomotives, the average air temperature approaching the last locomotive is determined from

$$t_{AN} = t_{AT} + \frac{q_T(N-1)}{\rho c_p Q_R} \qquad (13)$$

where

t_{AN} = average tunnel air temperature approaching Nth locomotive, °F
t_{AT} = average tunnel air temperature approaching locomotive consist, °F
ρ = density of tunnel air approaching locomotive consist, lb/ft³
c_p = specific heat of air, Btu/lb$_m$·°F
Q_R = tunnel airflow rate relative to train, ft³/h

The inlet air temperature to the locomotive radiators is used to judge the adequacy of the ventilation system. For most locomotives running at maximum throttle position, the maximum inlet air temperature recommended by manufacturers is about 115°F. Field tests in operating tunnels (Aisiks and Danziger 1969; Levy and Elpidorou 1991) showed, however, that some units can operate continuously with radiator inlet air temperatures as high as 135°F. The allowable inlet air temperature for each locomotive type should be obtained from the manufacturer when contemplating a design.

To determine the airflow rate required to prevent a locomotive from overheating, the relationship between the average tunnel air temperature approaching the last unit and the radiator inlet air temperature must be known or conservatively estimated. This relationship depends on variables such as the number of locomotives in the consist, air velocity relative to the train, tunnel cross-sectional area/configuration, type of tunnel lining, and locomotive orientation (i.e., facing forward or backward). For trains traveling under 20 mph, Levy and Elpidorou (1991) showed that a reasonable estimate is to assume the radiator inlet air temperature to be about 10°F higher than the average air temperature approaching the unit. For trains moving at 30 mph or more, a reasonable estimate is to assume that the radiator inlet air temperature equals the average air temperature approaching the unit. When the last unit of the train consist faces forward, thereby putting the exhaust stack ahead of its own radiators, the stack heat release rate must be included when evaluating the radiator inlet air temperature.

Tunnel Aerodynamics. When designing a ventilation system for a railroad tunnel, airflow and pressure distribution throughout the tunnel (as a function of train type, train speed, and ventilation system operating mode) must be determined. This information is required to determine (1) whether sufficient ventilation is provided for locomotive cooling, (2) the pressure that the fans are required to deliver, and (3) the pressure that the structural and ventilation elements of the tunnel must be designed to withstand.

The following equation, from DOT (1997a), relates the piston effect of the train, steady-state airflow from fans to the tunnel, and pressure across the tunnel door. This expression assumes that air leakage across the tunnel door is negligible. Figure 13 shows the dimensional variables on a schematic of a typical tunnel.

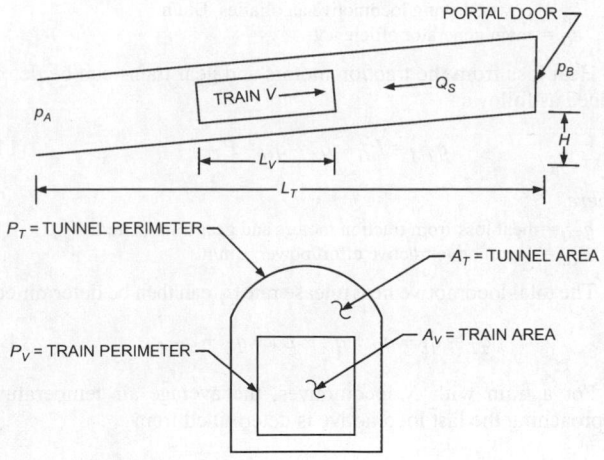

Fig. 13   Railroad Tunnel Aerodynamic Related Variables

$$\frac{\Delta p}{\rho}\left(\frac{g_C}{g}\right) = \frac{(p_A - p_B)}{\rho} \times \left(\frac{g_C}{g}\right) - H$$

$$+ \left[\frac{(A_V^2 + A_V A_T C_{DVB})}{(A_T - A_V)^2} + \frac{A_V C_{DVF}}{A_T}\right]\frac{(A_T V + Q_S)^2}{2 A_T^2 g}$$

$$+ \frac{f_T L_V P_T (A_V V + Q_S)^2}{8(A_T - A_V)^3 g} + \frac{\lambda_V L_V P_V (A_T V + Q_S)^2}{8(A_T - A_V)^3 g}$$

$$+ \frac{f_T (L_T - L_V) P_T Q_S^2}{8 A_T^3 g} + \frac{K Q_S^2}{2 A_T^2 g} \tag{14}$$

where

Δp = static pressure across tunnel door, lb_f/ft^2
ρ = density of air, lb_m/ft^3
p_A = barometric pressure at portal A, lb_f/ft^2
p_B = barometric pressure at portal B, lb_f/ft^2
H = difference in elevation between portals, ft
g = acceleration of gravity = 32.2 ft/s^2
g_C = gravitational constant = 32.2 $ft \cdot lb_m/lb_f \cdot s^2$
A_V = train cross-sectional area, ft^2
A_T = tunnel cross-sectional area, ft^2
C_{DVB} = drag coefficient at back end of train
C_{DVF} = drag coefficient at front end of train
V = velocity of train, ft/s
Q_S = airflow delivered by fan, ft^3/s
f_T = tunnel wall friction factor
L_T = tunnel length, ft
L_V = train length, ft
P_T = tunnel perimeter, ft
P_V = train perimeter, ft
λ_V = train skin friction factor
K = miscellaneous tunnel loss coefficient

The pressure across the tunnel door generated only by train piston action is evaluated by setting Q_S equal to zero. The airflow rate, relative to the train, required to evaluate locomotive cooling requirements is

$$Q_{rel} = A_T V + Q_S \tag{15}$$

where Q_{rel} is the airflow rate relative to the train, ft^3/s.

Typical values for C_{DVB} and C_{DVF} are about 0.5 and 0.8, respectively. Because trains passing through a railroad tunnel are often more than 1 mi long, the parameter that most affects the generated air pressure is the train skin friction coefficient. For dedicated coal or grain trains, which essentially use uniform cars throughout, a value of 0.09 for the skin friction coefficient results in air pressure predictions that conform closely to those observed in various railroad tunnels. For trains with nonuniform car distribution, the skin friction coefficient may be as high as 1.5 times that for a uniform car distribution.

The wall surface friction factor corresponds to the coefficient used in the Darcy-Weisbach equation for friction losses in pipe flow. Typical effective values for tunnels constructed with a formed concrete lining and having a ballasted track range from 0.015 to 0.017.

Tunnel Purge. The leading end of a locomotive must be exposed to an environment that is relatively free of smoke and diesel contaminants emitted by preceding trains. Railroad tunnels are usually purged by displacing contaminated tunnel air with fresh air by mechanical means after a train has left the tunnel. With the tunnel door closed, air is either supplied to or exhausted from the tunnel, moving fresh air from one end of the tunnel to the other. Observations at the downstream end of tunnels have found that an effective purge time is usually based on displacing 1.25 times the tunnel volume with outside air.

The time required for purging is primarily determined by operations schedule needs. A long purging time limits traffic; a short purging time may necessitate very high ventilation airflow rates, and result in high electrical energy demand and consumption. Consequently, multiple factors must be considered, including the overall ventilation concept, when establishing the purge rate.

PARKING GARAGES

Automobile parking garages can be either fully enclosed or partially open. Fully enclosed parking areas are often underground and require mechanical ventilation. Partially open parking garages are generally above-grade structural decks having open sides (except for barricades), with a complete deck above. Natural ventilation, mechanical ventilation, or a combination can be used for partially open garages.

Operating automobiles in parking garages presents two concerns. The more serious is emission of CO, with its known risks. The other concern is oil and gasoline fumes, which may cause nausea and headaches and also represent potential fire hazards. Additional concerns about NO_x and smoke haze from diesel engines may also require consideration. However, the ventilation rate required to dilute CO to acceptable levels is usually satisfactory to control the level of other contaminants as well, provided the percentage of diesel vehicles does not exceed 20%.

For many years, the various model codes, ANSI/ASHRAE *Standard* 62.1, and its predecessor standards recommended a flat exhaust rate of either 1.5 cfm/ft² or 6 ach for enclosed parking garages. But because vehicle emissions have been reduced over the years, ASHRAE sponsored a study to determine ventilation rates required to control contaminant levels in enclosed parking facilities (Krarti and Ayari 1998). The study found that, in some cases, much less ventilation than 1.5 cfm/ft² was satisfactory. The study's methodology for determining whether a reduced ventilation rate would be effective is included below. However, ANSI/ASHRAE *Standard* 62.1 and the International Code Council's *International Mechanical Code* (ICC 2009a) allow 0.75 cfm/ft² ventilation, whereas NFPA *Standard* 88A recommends a minimum of 1.0 cfm/ft², so the engineer must understand the specific codes and standards that apply. The engineer may be required to request a variation, or waiver, from authorities having jurisdiction before implementing a lesser ventilation system design.

If larger fans are installed to meet code requirements, they will not necessarily increase overall power consumption; with proper CO level monitoring and ventilation system control, fans will run for shorter time periods to maintain acceptable CO levels. With increased attention on reducing energy consumption, CO-based ventilation system control can provide substantial cost savings in the operation of parking garages.

Ventilation Requirements and Design

ASHRAE research project RP-945 (Krarti and Ayari 1998) found that the design ventilation rate required for an enclosed parking facility depends chiefly on four factors:

- Acceptable level of contaminants in the parking facility
- Number of cars in operation during peak conditions
- Length of travel and the operating time for cars in the garage
- Emission rate of a typical car under various conditions

Contaminant Level Criteria. ACGIH (1998) recommends a threshold CO limit of 25 ppm for an 8 h exposure, and the U.S. EPA (2000) determined that exposure, at or near sea level, to a CO concentration of 35 ppm for up to 1 h is acceptable. For parking garages more than 3500 ft above sea level, more stringent limits are required.

In Europe, an average concentration of 35 ppm and a maximum level of 200 ppm are usually maintained in parking garages.

Various agencies and countries differ on the acceptable level of CO in parking garages, but a reasonable solution is a ventilation rate designed to maintain a CO level of 35 ppm for 1 h exposure, with a maximum of 25 ppm for an 8 h exposure. Because the time associated with driving in and parking, or driving out of a garage, is on the order of minutes, 35 ppm is probably an acceptable level of exposure. However, Figure 14 provides nomographs for 15 and 25 ppm maximum exposures as well, to allow the designer to conform to more stringent regulations.

Number of Cars in Operation. The number of cars operating at any one time depends on the type of facility served by the parking garage. For distributed, continuous use, such as an apartment building or shopping area, the variation is generally 3 to 5% of the total vehicle capacity. The operating capacity could reach 15 to 20% in other facilities, such as sports stadiums or short-haul airports.

Length of Time of Operation. The length of time that a car remains in operation in a parking garage is a function of the size and layout of the garage, and the number of cars attempting to enter or exit at a given time. The operating time could vary from as much as 60 to 600 s, but on average usually ranges from 60 to 180 s. Table 7 lists approximate data for average vehicle entrance and exit times; these data should be adjusted to suit the specific physical configuration of the facility.

Car Emission Rate. Operating a car in a parking garage differs considerably from normal vehicle operation, including that in a road tunnel. Most car movements in and around a parking garage occur in low gear. A car entering a garage travels slowly, but the engine is usually hot. As a car exits from a garage, the engine is usually cold and operating in low gear, with a rich fuel mixture. Emissions for a cold start are considerably higher, so the distinction between hot and cold emission plays a critical role in determining the ventilation rate. Motor vehicle emission factors for hot- and cold-start operation are presented in Table 8. An accurate analysis requires correlation of CO readings with the survey data on car movements (Hama et al. 1974); the data should be adjusted to suit the specific physical configuration of the facility and the design year.

Design Method. To determine the design airflow rate to ventilate an enclosed parking garage, the following procedure can be used:

Table 7 Average Entrance and Exit Times for Vehicles

Level	Average Entrance Time, s	Average Exit Time, s
1	35	45
3*	40	50
5	70	100

Source: Stankunas et al. (1980).   *Average pass-through time = 30 s.

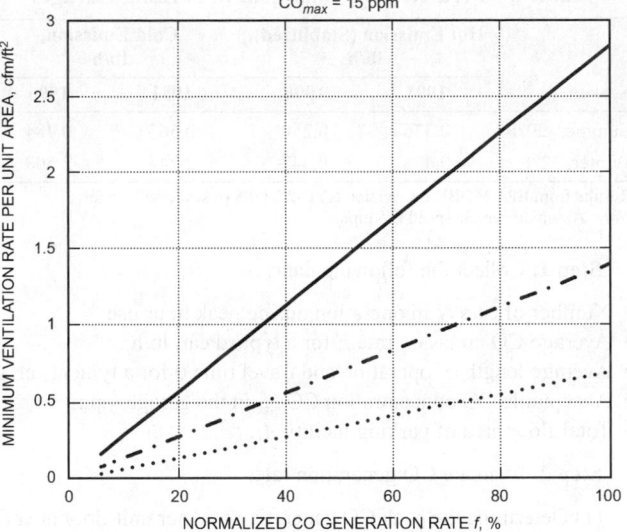

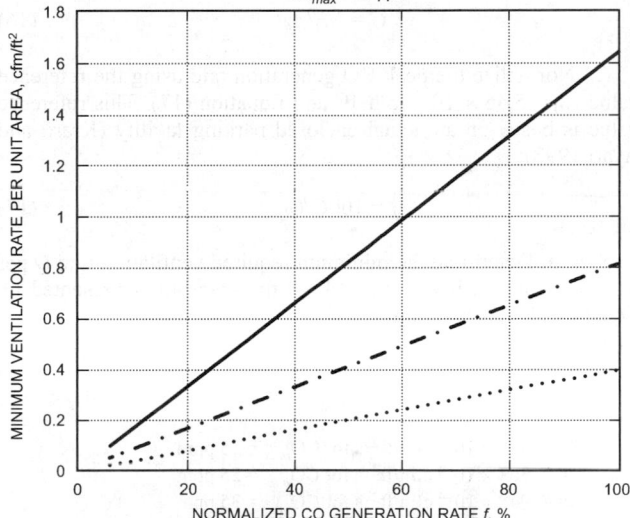

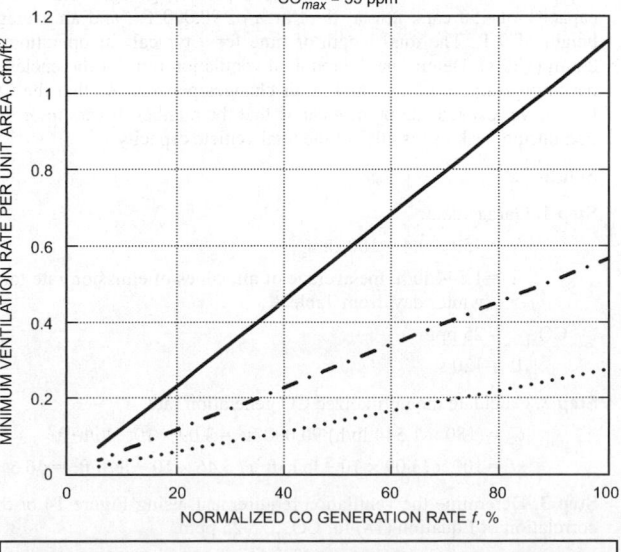

Fig. 14 Ventilation Requirement for Enclosed Parking

Table 8 Predicted CO Emissions in Parking Garages

Season	Hot Emission (Stabilized), lb/h		Cold Emission, lb/h	
	1991	1996	1991	1996
Summer, 90°F	0.336	0.250	0.565	0.484
Winter, 32°F	0.478	0.447	2.744	2.508

Results from EPA MOBILE3, version NYC-2.2 (1984); sea level location.
Note: Assumed vehicle speed is 5 mph.

Step 1. Collect the following data:

- Number of cars N in operation during peak hour use
- Average CO emission rate E for a typical car, lb/h
- Average length of operation and travel time θ for a typical car, s
- Acceptable CO concentration CO_{max} in the garage, ppm
- Total floor area of parking facility A_f, ft²

Step 2. Evaluate CO generation rate:

(1) Determine the peak CO generation rate per unit floor area G, in lb/h·ft², for the parking garage:

$$G = NE/A_f \tag{16}$$

(2) Normalize the peak CO generation rate using the reference value $G_0 = 5.46 \times 10^{-3}$ lb/h·ft² and Equation (17). This reference value is based on an actual enclosed parking facility (Krarti and Ayari 1998):

$$f = 100G/G_0 \tag{17}$$

Step 3. Determine the minimum required ventilation rate Q per unit floor area using Figure 14, or the correlation presented by Equation (18), depending on CO_{max}:

$$Q = Cf\theta \tag{18}$$

where

$C = 2.370 \times 10^{-4}$ cfm/ft²·s for CO_{max} = 15 ppm
$= 1.363 \times 10^{-4}$ cfm/ft²·s for CO_{max} = 25 ppm
$= 0.948 \times 10^{-4}$ cfm/ft²·s for CO_{max} = 35 ppm

Example 1. Consider a two-level enclosed parking garage with a total capacity of 450 cars, a total floor area of 90,000 ft², and an average height of 9 ft. The total length of time for a typical car operation is 2 min (120 s). Determine the required ventilation rate for the enclosed parking garage in cfm/ft² and in air changes per hour so that the CO level never exceeds 25 ppm. Assume that the number of cars in operation during peak use is 40% of the total vehicle capacity.

Solution:

Step 1. Garage data:

$N = 450 \times 0.4 = 180$ cars

$E = 1.544$ lb/h, the average of all values of emission rate for a winter day, from Table 8

$CO_{max} = 25$ ppm

$\theta = 120$ s

Step 2. Calculate the normalized CO generation rate:

$G = (180 \times 1.544 \text{ lb/h})/90{,}000 \text{ ft}^2 = 3.09 \times 10^{-3}$ lb/h·ft²

$f = 100 \times (3.09 \times 10^{-3} \text{ lb/h·ft}^2) / 5.46 \times 10^{-3} \text{ lb/h·ft}^2 = 56.6$

Step 3. Determine the ventilation requirement, using Figure 14 or the correlation of Equation (18) for CO_{max} = 25 ppm.

$Q = 1.363 \times 10^{-4} \text{ cfm/s·ft}^2 \times 56.6 \times 120 \text{ s} = 0.93$ cfm/ft²

Or, for air changes per hour,

$(0.93 \text{ cfm/ft}^2 \times 60 \text{ min/h})/9 \text{ ft} = 6.2$ ach

Notes:

1. If the average vehicle CO emission rate is reduced to $E = 0.873$ lb/h, because of, for instance, better emission standards or better maintained cars, the required minimum ventilation rate decreases to 0.52 cfm/ft² or 3.5 ach.

2. Once calculations are made and a decision reached to use CO demand ventilation control, increasing airflow through a safety margin does not increase operating costs; larger fans work for shorter periods to sweep the garage and maintain satisfactory conditions.

CO Demand Ventilation Control. Whether mechanical, natural, or both, a parking garage ventilation system should meet applicable codes and maintain acceptable contaminant levels. If permitted by local codes, the ventilation airflow rate should be varied according to CO levels to conserve energy. For example, the ventilation system could consist of multiple fans, with single- or two-speed motors, or variable-pitch blades. In multilevel parking garages or single-level structures of extensive area, independent fan systems with individual controls are preferred. The *International Mechanical Code* (ICC 2009a) allows ventilation system operation to be reduced from 0.75 to 0.05 cfm/ft² with the use of a CO monitoring system that restores full ventilation when CO levels of 25 ppm are detected.

Figure 15 shows the maximum CO level in a tested parking garage (Krarti and Ayari1998) for three car movement profiles (Figure 16) and the following ventilation control strategies:

- Constant-volume (CV), where the ventilation system is kept on during the entire occupancy period
- On/off control, with fans stopped and started based on input from CO sensors
- Variable-air-volume (VAV) control, using either two-speed fans or axial fans with variable-pitch blades, based on input from CO sensors

Figure 15 also shows typical fan energy savings achieved by on/off and VAV systems relative to constant-volume systems. Significant fan energy savings can be obtained using a CO-based demand ventilation control strategy to operate the ventilation system, maintaining CO levels below 25 ppm. Wear and tear and maintenance on mechanical and electrical equipment are reduced with a CO-based demand strategy.

Figure 16 is based on maintaining a 25 ppm CO level. With most systems, actual energy usage is further reduced if 35 ppm is maintained.

In cold climates, the additional cost of heating makeup air is also reduced with a CO-based demand strategy. Energy stored in the mass of the structure usually helps maintain the parking garage air temperature at an acceptable level. If only outside air openings are used to draw in ventilation air, or if infiltration is allowed, the stored energy is lost to the incoming cold air.

Ventilation System Configuration. Parking garage ventilation systems can be classified as supply-only, exhaust-only, or combined. Regardless of which system design is chosen, the following elements should be considered in planning the system configuration:

- Accounting for the contaminant level of outside air drawn in for ventilation
- Avoiding short-circuiting supply air
- Avoiding a long flow field that allows contaminants to exceed acceptable levels at the end of the flow field
- Providing short flow fields in areas of high contaminant emission, thereby limiting the extent of mixing
- Providing efficient, adequate airflow throughout the structure
- Accounting for stratification of engine exhaust gases when stationary cars are running in enclosed facilities

Other Considerations. Access tunnels or long, fully enclosed ramps should be designed in the same way as road tunnels. When natural ventilation is used, wall openings or free area should be as large as possible. Part of the free area should be at floor level.

For parking levels with large interior floor areas, a central emergency smoke exhaust system should be considered for removing smoke (in conjunction with other fire emergency systems) or vehicle fumes under normal conditions.

Noise. In general, parking garage ventilation systems move large quantities of air through large openings without extensive ductwork. These conditions, and the highly reverberant nature of the space, contribute to high noise levels, so sound attenuation should be considered in the ventilation system design. This is a pedestrian safety concern, as well, because high fan noise levels in a parking garage may mask the sound of an approaching vehicle.

Ambient Standards and Contaminant Control. Air exhausted from a parking garage should meet state and local air pollution control requirements.

AUTOMOTIVE REPAIR FACILITIES

Automotive repair activities are defined as any repair, modification, service, or restoration activity to a motor vehicle. This includes, but is not limited to, brake work, engine work, machining operations, and general degreasing of engines, motor vehicles, parts, or tools.

ANSI/ASHRAE *Standard* 62.1 recommends a ventilation rate of 1.5 cfm/ft^2 for automotive service stations; the *International Mechanical Code* (ICC 2009a) allows 0.75 cfm/ft^2. The designer

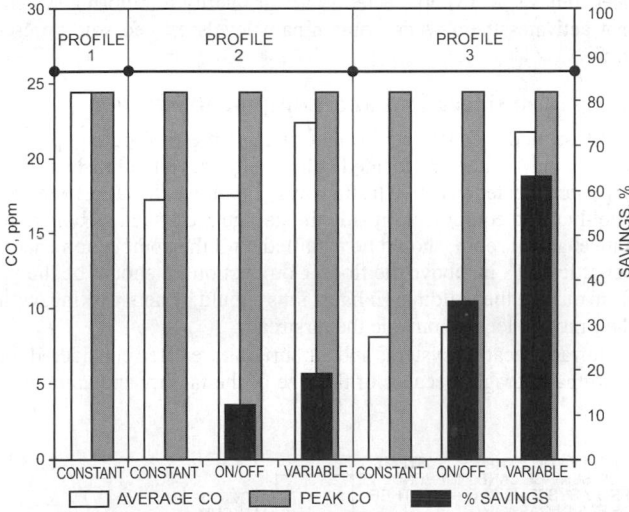

Fig. 15 Typical Energy Savings and Maximum CO Level Obtained for Demand CO-Ventilation Controls

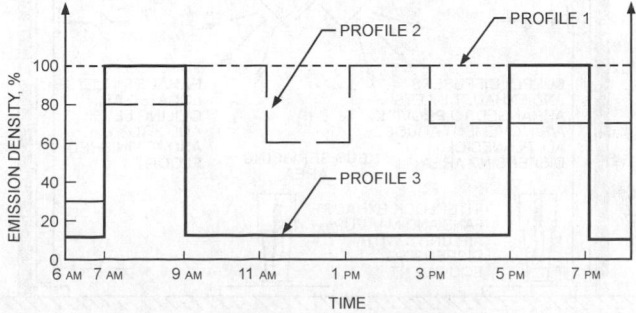

Fig. 16 Three Car Movement Profiles

must determine which code is applicable. The high ventilation rate indicates that contaminants are not related to the occupants, but are produced by the variety of tasks and materials used in the facility. Outdoor ventilation is introduced into the space and an approximately equal quantity is exhausted through a dedicated exhaust system.

As repairs or maintenance are performed on vehicles, it may be necessary to operate the vehicle inside the facility to test and validate the work. Additional mechanical ventilation is required to exhaust combustion by-products directly outdoors. An independent source capture system that connects directly to the exhaust pipe of the vehicle must be installed in the facility. These systems are available in either an above- or belowground configuration. Flow rates for individual service bays vary from 50 to 400 cfm for automobiles. A large diesel truck will require considerably more airflow per service bay than an automobile.

The above-grade system consists of an exhaust fan, associated ductwork, and flexible hoses that attach to the tailpipe of the vehicle in operation. Generally, the system is installed at a high elevation to maintain maximum clearances above floor level. The hose connections are stored in reels positioned near each service bay. The service technician pulls the hose down and attaches it to the tailpipe by a proprietary connection.

The below-grade system is similar in design to an overhead exhaust system. Care must be taken to select an appropriate corrosion-resistant material to be installed underground, because the condensing products of combustion are corrosive to traditional duct materials. The flexible tailpipe exhaust connectors are stored inside the underground duct. After sliding the flex back inside the duct, a hinged cover plate covers the opening flush to the floor.

Although there is a diversity factor in the system capacity calculations, both systems must be designed to operate at 100% capacity. A constant-volume fan is used, with all air being exhausted from the space. With a single outlet in use, some means of relief is provided to maintain constant flow through the fan. This equipment can be set up to run continuously or intermittently. Intermittent use requires the general exhaust system to vary between the maximum supply air delivered to the space when the capture system is in use and a lower exhaust flow rate reduced by the amount of air exhausted through the capture system.

BUS GARAGES

Bus garages generally include a maintenance and repair area, service lane (where buses are fueled and cleaned), storage area (where buses are parked), and support areas such as offices, stock room, lunch room, and locker rooms. The location and layout of these spaces can depend on factors such as local climate, size of the bus fleet, and type of fuel used by the buses. Bus servicing and storage areas may be located outside in a temperate region, but are often inside in colder climates. However, large bus fleets cannot always be stored indoors; for smaller fleets, maintenance areas may double as storage space. Local building and/or fire codes may also prohibit dispensing certain types of fuel indoors.

In general, bus maintenance or service areas should be ventilated using 100% outside air with no recirculation. Therefore, using heat recovery devices should be considered in colder climates.

Tailpipe emissions should be exhausted directly from buses at fixed inspection and repair stations in maintenance areas. Offices and similar support areas should be kept under positive pressure to prevent infiltration of bus emissions.

Maintenance and Repair Areas

ANSI/ASHRAE *Standard* 62.1 recommends a minimum ventilation of 1.5 cfm/ft^2 and the *International Mechanical Code* (ICC 2009a) recommends 0.75 cfm/ft^2 of floor area in vehicle repair garages, with no recirculation. The designer should determine which code is applicable. However, because the interior ceiling height may vary greatly from garage to garage, the designer should consider making a volumetric analysis of contaminant generation and air exchange rates. The section on Bus Terminals contains information on diesel engine emissions and ventilation airflow rates needed to control contaminant concentrations in areas where buses are operated.

Maintenance and repair areas often include below-grade inspection and repair pits for working underneath buses. Because vapors produced by conventional bus fuels are heavier than air, they tend to settle in these pit areas, so a separate exhaust system should be provided to prevent their accumulation. NFPA *Standard* 30A recommends a minimum of 1 cfm/ft² in pit areas and the installation of exhaust registers near the floor of the pit.

Fixed repair stations, such as inspection/repair pits or hydraulic lift areas, should include a direct exhaust system for tailpipe emissions. Such direct exhaust systems have a flexible hose and coupling attached to the bus tailpipe; emissions are discharged to the outdoors by an exhaust fan. The system may be of the overhead reel, overhead tube, or underfloor duct type, depending on the tailpipe location. For heavy diesel engines, a minimum exhaust rate of 600 cfm per station is recommended to capture emissions without creating excessive backpressure in the vehicle. Fans, ductwork, and hoses should be able to receive vehicle exhaust at temperatures exceeding 500°F without degradation.

Bus garages often include areas for battery charging, which can produce potentially explosive concentrations of corrosive, toxic gases. There are no published code requirements for ventilating battery-charging areas, but DuCharme (1991) suggested using a combination of floor and ceiling exhaust registers to remove gaseous by-products. The recommended exhaust rates are 2.25 cfm/ft² of room area at floor level to remove acid vapors, and 0.75 cfm/ft² of room area at ceiling level, to remove hydrogen gases. The associated supply air volume should be 10 to 20% less than exhaust air volume, but designed to provide a minimum terminal velocity of 100 fpm at floor level. If the battery-charging space is located in the general maintenance area rather than in a dedicated space, an exhaust hood should be provided to capture gaseous by-products. Chapter 32 contains specific information on exhaust hood design. Makeup air should be provided to replace that removed by the exhaust hood.

Garages may also contain spray booths, or rooms for painting buses. Most model codes reference NFPA *Standard* 33 for spray booth requirements; this standard should be reviewed when designing heating and ventilating systems for such areas.

Servicing Areas

For indoor service lanes, ANSI/ASHRAE *Standard* 62.1 recommends a minimum ventilation of 1.5 cfm/ft² and the *International Mechanical Code* (ICC 2009a) recommends 0.75 cfm/ft² of floor area in vehicle repair garages, with no recirculation. The designer should determine which code is applicable. However, because the interior ceiling height may vary greatly from garage to garage, the designer should consider making a volumetric analysis of contaminant generation and air exchange rates. The section on Bus Terminals contains information on diesel engine emissions and ventilation airflow rates needed to control contaminant concentrations in areas where buses are operated.

Because of the increased potential for concentrations of flammable or combustible vapor, HVAC systems for bus service lanes should not be interconnected with systems serving other parts of the bus garage. Service-lane HVAC systems should be interlocked with fuel-dispensing equipment, to prevent operation of the latter if the former is shut off or fails. Exhaust inlets should be located both at ceiling level and 3 to 12 in. above the finished floor, with supply and exhaust diffusers/registers arranged to provide air movement across all planes of the dispensing area. A typical equipment arrangement is shown in Figure 17.

Another feature in some service lanes is the cyclone cleaning system: these devices have a dynamic connection to the front door(s) of the bus, through which a large-volume fan vacuums dirt and debris from inside the bus. A large cyclone assembly then removes dirt and debris from the airstream and deposits it into a large hopper for disposal. Because of the large volume of air involved, the designer should consider the discharge and makeup air systems required to complete the cycle. Recirculation and energy recovery should be considered, especially during winter. To aid in contaminant and heat removal during summer, some systems discharge the cyclone air to the outside and provide untempered makeup air through relief hoods above the service lane.

Storage Areas

Where buses are stored inside, the minimum ventilation standard is based upon the applicable code: 0.75 cfm/ft² for the *International Mechanical Code*, or 1.5 cfm/ft² for ANSI/ASHRAE *Standard* 62.1, subject to volumetric considerations. The designer should also consider the increased contaminant levels present during peak traffic periods.

One example is morning pullout, when the majority of the fleet is dispatched for rush-hour commute. It is common practice to start and idle a large number of buses during this period to warm up the engines and check for defects. As a result, the emissions concentration in the storage area rises, and additional ventilation may be required to maintain contaminant levels in acceptable limits. Using supplemental purge fans is a common solution to this problem. These purge fans can either be (1) interlocked with a timing device to operate during peak traffic periods, (2) started manually on an as-needed basis, or (3) connected to an air quality monitoring system that activates them when contaminant levels exceed some preset limit.

Design Considerations and Equipment Selection

Most model codes require that open-flame heating equipment, such as unit heaters, be located at least 8 ft above the finished floor or, where located in active trafficways, 2 ft above the tallest vehicle. Fuel-burning equipment outside the garage area, such as boilers in a mechanical room, should be installed with the combustion chamber at least 18 in. above the floor. Combustion air should be drawn from outside the building. Exhaust fans should be nonsparking, with their motors located outside the airstream.

Infrared heating systems and air curtains are often considered for bus repair garages because of the size of the facility and amount of

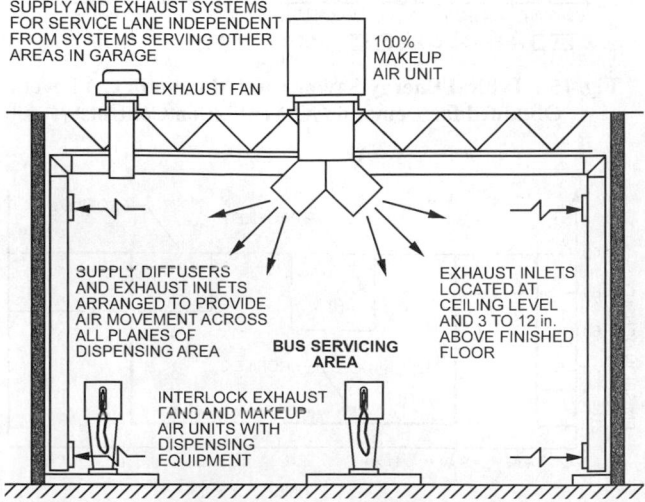

Fig. 17 Typical Equipment Arrangement for Bus Garage

infiltration through the large doors needed to move buses in and out of the garage. However, infrared heating must be used cautiously in areas where buses are parked or stored for extended periods, because the buses may absorb most of the heat, which is then lost when the buses leave the garage. This is especially true during morning pullout. Infrared heating can be applied with more success in the service lane, or at fixed repair positions. Air curtains should be considered for high-traffic doorways to limit both heat loss and infiltration of cold air.

Where air quality monitoring systems control ventilation equipment, maintainability is a key factor in determining success of the application. The high concentration of particulate matter in bus emissions can adversely affect monitoring equipment, which often has filtering media at sampling ports to protect sensors and instrumentation. The location of sampling ports, effects of emissions fouling, and calibration requirements should be considered when selecting monitoring equipment to control ventilation systems and air quality of a bus garage. NO_2 and CO exposure limits published by OSHA and the EPA should be consulted to determine contaminant levels at which exhaust fans should be activated.

Effects of Alternative Fuel Use

Because of legislation limiting contaminant concentrations in diesel bus engine emissions, the transportation industry has begun using buses that operate on alternative fuels, including methanol, ethanol, hydrogen (and fuel cells), compressed natural gas (CNG), liquefied natural gas (LNG), and liquefied petroleum gas (LPG). Flammability, emission, and vapor dispersion characteristics of these fuels differ from those of conventional fuels, for which current code requirements and design standards were developed. Thus, established ventilation requirements may not be valid for bus garage facilities used by alternative-fuel vehicles. The designer should consult current literature on HVAC system design for these facilities rather than relying on conventional practices. One source is the Alternative Fuels Data Center at the U.S. Department of Energy in Washington, D.C; their Web site (www.eere.energy.gov/afdc) includes design recommendations for various alternative fuels. The DOT (1996a, 1996b, 1996c, 1997b, 1998) Volpe Transportation Center has also issued several guidelines for alternative-fuel bus facilities, which can be consulted for additional suggestions.

CNG Vehicle Facilities. For CNG bus facilities, NFPA *Standard* 52 recommends a separate mechanical ventilation system providing at least 1 cfm per 12 ft³, or 5 ach, for indoor fueling and gas processing/storage areas. The ventilation system should operate continuously, or be activated by a continuously monitoring natural gas detector when a gas concentration of not more than 20% of the lower flammability limit (LFL) is present. The fueling or fuel-compression equipment should be interlocked to shut down if the mechanical ventilation system fails. Supply inlets should be located near floor level; exhaust outlets should be located high in the roof or exterior wall structure. The *International Mechanical Code* (ICC 2009a) has identical requirements, except that it requires activation of the ventilation system at 25% of the LFL, and the requirements apply to maintenance and repair areas as well as indoor fueling facilities.

DOT (1996a) guidelines for CNG facilities address bus storage and maintenance areas, as well as bus fueling areas. DOT recommendations include (1) minimizing potential for dead-air zones and gas pockets (which may require coordination with architectural and structural designers); (2) using a normal ventilation rate of 6 ach, with provisions to increase that rate by an additional 6 ach in the event of a gas release; (3) using nonsparking exhaust fans rated for use in Class 1, Division 2 areas (as defined by NFPA *Standard* 70); and (4) increasing the minimum ventilation rate in smaller facilities to maintain dilution levels similar to those in larger facilities. Open-flame heating equipment should not be used, and the surface temperature of heating units should not exceed 800°F. In the event of a gas release, deenergizing supply fans that discharge

near the ceiling level should be considered, to avoid spreading the gas plume.

LNG Vehicle Facilities. The 2006 edition of NFPA *Standard* 57 includes requirements for LNG bus facilities. The standard recommends a separate mechanical ventilation system providing at least 1 cfm per 12 ft³, or 5 ach, for indoor fueling areas. The ventilation system should operate continuously, or be activated by a continuously monitoring natural gas detection system when a gas concentration of not more than 20% of the LFL is present. Fueling equipment should be interlocked to shut down in case the mechanical ventilation system fails. DOT (1997b) provides further information on LNG fuel.

LPG Vehicle Facilities. NFPA *Standard* 58 and the *International Fuel Gas Code* (ICC 2009d) contain similar provisions relating specifically to LPG-fueled vehicles. Both standards prohibit indoor fueling of all LPG vehicles, allowing only an adequately ventilated weather shelter or canopy for fueling operations. However, the term "adequately ventilated" is not defined by any prescriptive rate. Vehicles are permitted to be stored and serviced indoors under NFPA *Standard* 58, provided they are not parked near sources of heat, open flames (or similar sources of ignition), or "inadequately ventilated" pits. That standard does not recommend a ventilation rate for bus repair and storage facilities, but it does recommend a minimum of 1 cfm/ft² in buildings and structures housing LPG distribution facilities. DOT (1996b) provides additional information on LPG fuel.

Hydrogen Vehicle Facilities. The 2006 edition of NFPA *Standard* 52 includes requirements for gaseous and liquid hydrogen bus facilities. The standard recommends a separate mechanical ventilation system providing at least 1 cfm/ft², but not less than 1 cfm per 12 ft³, or 5 ach, for indoor gaseous hydrogen fueling areas. The ventilation system should operate continuously, or be activated by a continuously monitoring natural gas detection system when a gas concentration of not more than 25% of the LFL is present. Fueling equipment should be interlocked to shut down in case the mechanical ventilation system fails. Liquid hydrogen fueling facilities are prohibited indoors. The *International Mechanical Code* (ICC 2009a) has the same requirements, which apply to maintenance and repair areas as well as indoor fueling facilities.

DOT (1998) provides additional information on hydrogen fuel.

BUS TERMINALS

The physical configuration of bus terminals varies considerably. Most terminals are fully enclosed spaces containing passenger waiting areas, ticket counters, and some retail areas. Buses load and unload outside the building, generally under a canopy for weather protection. In larger cities, where space is at a premium and bus service is extensive or integrated with subway service, bus terminals may have comprehensive customer services and enclosed (or semi-enclosed) multilevel structures, busway tunnels, and access ramps. Waiting rooms and consumer spaces should have controlled environments in accordance with normal HVAC system design practices for public terminal occupancies. In addition to providing the recommended ventilation air rate in accordance with ANSI/ ASHRAE *Standard* 62.1, the space should be pressurized against infiltration from the busway environment. Pressurized vestibules should be installed at each doorway to further reduce contaminant migration and to maintain acceptable air quality. Waiting rooms, passenger concourse areas, and platforms are typically subjected to a highly variable people load. The average occupant density may reach 10 ft² per person and, during periods of extreme congestion, 3 to 5 ft² per person.

The choice between natural and mechanical ventilation should be based on the physical characteristics of the bus terminal and the airflow required to maintain acceptable air quality. When natural ventilation is selected, the individual levels of the bus terminal should be open on all sides, and the slab-to-ceiling dimension should be sufficiently high, or the space contoured, to allow free air circulation. Jet fans can be used to improve natural airflow in the busway, with relatively low energy consumption. Mechanical systems that ventilate open platforms or gate positions should be configured to serve bus operating areas, as shown in Figures 18 and 19.

Platforms

Platform design and orientation should be tailored to expedite passenger loading and unloading, to minimize both passenger exposure to the busway environment and dwell time of an idling bus in an enclosed terminal. Naturally ventilated drive-through platforms may expose passengers to inclement weather and strong winds. An enclosed platform (except for an open front), with the appropriate mechanical ventilation system, should be considered. Partially enclosed platforms can trap contaminants and may require mechanical ventilation to achieve acceptable air quality.

Multilevel bus terminals have limited headroom, which restricts natural ventilation system performance. These terminals should have mechanical ventilation, and all platforms should be either partially or fully enclosed. The platform ventilation system should not induce contaminated airflow from the busway environment. Supply air velocity should also be limited to 250 fpm to avoid drafts on the platform. Partially enclosed platforms require large amounts of outside air to hinder fume penetration; experience indicates that a minimum of 17 cfm per square foot of platform area is typically required during rush hours, and about half this rate is required during other periods. Figure 18 shows a partially enclosed drive-through platform with an air distribution system.

Platform air quality should remain essentially the same as that of the ventilation air introduced. Because of the piston effect, however, some momentarily high concentrations of contaminants may occur on the platform. Separate ventilation systems with two-speed fans (for each platform) allow operational flexibility, in both fan usage frequency and supply airflow rate for any one platform. Fans should be controlled automatically to conform to bus operating schedules. In cold climates, mechanical ventilation may need to be reduced or heated during extreme winter weather conditions.

For large terminals with heavy bus traffic, fully enclosed platforms are strongly recommended. Fully enclosed platforms can be adequately pressurized and ventilated with normal heating and cooling air quantities, depending on the construction tightness and number of boarding doors and other openings. Conventional air distribution can be used; air should not be recirculated. Openings around doors and in the enclosure walls are usually adequate to relieve air pressure, unless the platform construction is extraordinarily tight. Figure 19 shows a fully enclosed waiting room with sawtooth gates.

Doors between sawtooth gates and the waiting room should remain closed, except for passenger loading and unloading. The waiting room ventilation system should provide positive pressurization to minimize infiltration of contaminants from the busway environment. Supply air from a suitable source should be provided at the passenger boarding area to dilute local contaminants to acceptable levels.

Bus Operation Areas

Ventilation for bus operation areas should be designed and evaluated to maintain engine exhaust contaminant concentrations within the limits set by federal and local regulations and guidelines. With the proliferation of alternative fuels, such as biodiesel, ethanol, methanol, compressed natural gas (CNG), and liquefied natural gas (LNG), a bus terminal ventilation system should not only be designed for maintaining acceptable air quality, but should also consider the safety risks associated with potential leakage from buses operating with alternative fuel loads. In an enclosed or semienclosed area, a comprehensive risk assessment should be performed for the specific types of buses operating in the bus terminal. The nature of the bus engines should be determined for each project.

Contaminants. Of all the different types of buses in operation, engine exhaust from diesel buses has the most harmful quantities of contaminants. Some diesel buses also have small auxiliary gasoline engines to drive the vehicle air-conditioning system. Excessive exposure to diesel exhaust can cause adverse health effects, ranging from headache and nausea to cancer and respiratory disease. Tests

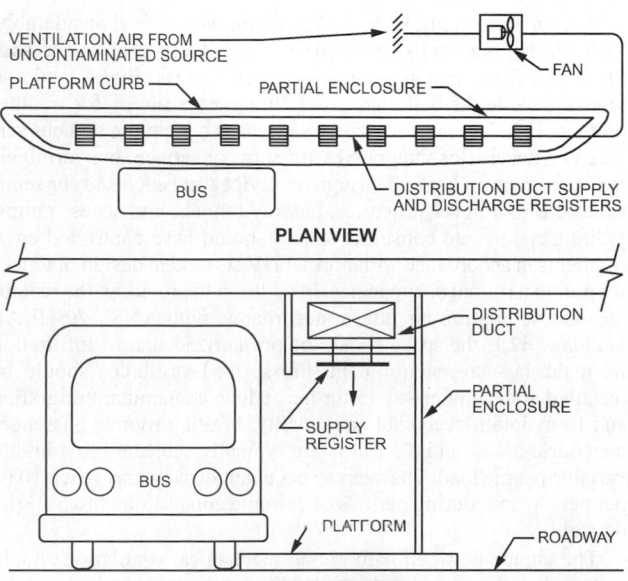

Fig. 18 Partially Enclosed Platform, Drive-Through Type

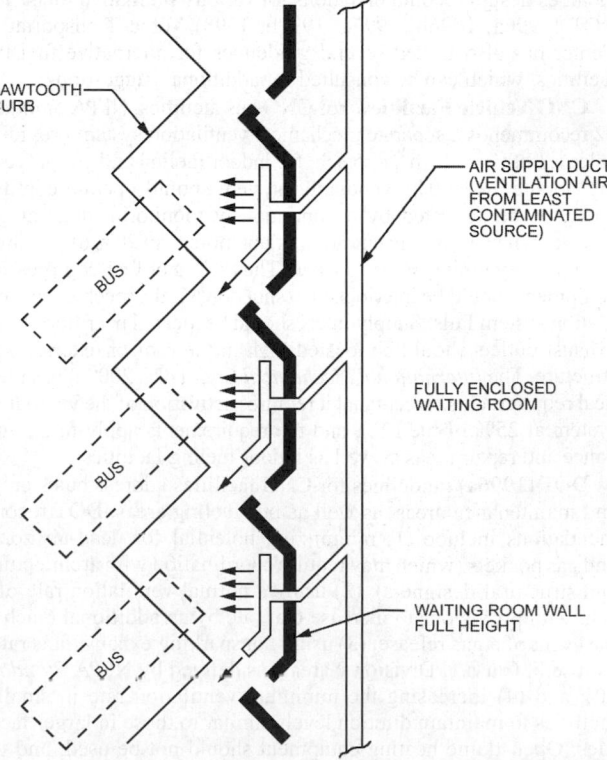

Fig. 19 Fully Enclosed Waiting Room with Sawtooth Gates

Table 9 8 h TWA Exposure Limits for Gaseous Pollutants from Diesel Engine Exhaust, ppm

Substance	OSHA PEL	ACGIH TLV
Carbon monoxide (CO)	50	25
Carbon dioxide (CO_2)	5000	5000
Nitric oxide (NO)	25	25
Nitrogen dioxide (NO_2)	5.0*	3.0
Formaldehyde (HCHO)	0.75	0.30*
Sulfur dioxide (SO_2)	5.0	2.0

*Ceiling value

Note: For data on diesel bus and truck engine emissions, see Watson et al. (1988).

Table 10 EPA Emission Standards for Urban Bus Diesel Engines

Model Year	Emissions, $lb/min \cdot bhp \times 10^{-5}$			
	Hydrocarbons (HC)	Carbon Monoxide (CO)	Oxides of Nitrogen (NO_x)	Particulate Matter (PM)
1991	4.78	57.0	18.4	0.919
1993	4.78	57.0	18.4	0.368
1994	4.78	57.0	18.4	0.257
1996	4.78	57.0	18.4	0.184*
1998 to 2003	4.78	57.0	14.7	0.184*
2004 to 2006	4.78	57.0	7.35 to 9.19	0.184*
2007 and later	0.515	57.0	0.074	0.037

*In-use PM standard 0.257×10^{-5} $lb/min \cdot bhp$

on the volume and composition of exhaust gases emitted from diesel engines during various traffic conditions indicate large variations depending on the (1) local air temperature and humidity; (2) manufacturer, size, and adjustment of the engine; and (3) type of fuel used.

Components of diesel engine exhaust gases that affect the ventilation system design are NO_x, hydrocarbons, formaldehyde, odor constituents, aldehydes, smoke particles, sulfur dioxide, and a relatively small amount of CO. Diesel engines operating in enclosed spaces also reduce visibility, and generate both odors and particulate matter.

Table 9 lists major health-threatening contaminants found in diesel engine exhaust and the exposure limits set by OSHA and ACGIH.

OSHA permissible exposure limits (PEL) are legally enforceable limits, whereas the ACGIH threshold limit values (TLV) are industrial hygiene recommendations. All the limits are time-weighted averages (TWAs) for 8 h exposure, unless noted as a ceiling value.

NO_x occurs in two basic forms: nitrogen dioxide (NO_2) and nitric oxide (NO). NO_2 is the major contaminant considered in bus terminal ventilation system design. Prolonged exposure to NO_2 concentrations of more than 5 ppm causes health problems. Furthermore, NO_2 affects light transmission and thereby reduces visibility. NO_2 is intensely colored and absorbs light over the entire visible spectrum, especially at shorter wavelengths. Odor perception of NO_2 is immediate at 0.42 ppm, but can be perceived by some at levels as low as 0.12 ppm.

Bus terminal operations also affect the quality of surrounding ambient air. The ventilation airflow rate, contaminant levels in exhaust air, and location and design of the air intakes and discharges determine the effect of the bus terminal on local ambient air quality.

State and local regulations, which require consideration of local atmospheric conditions and ambient contaminant levels in bus terminal ventilation system design, must be followed.

Calculation of Ventilation Rate

To calculate the ventilation rate, the total amount of engine exhaust gases should be determined using the bus operating schedule and amount of time that the buses are in various modes of operation (i.e., cruising, decelerating, idling, and accelerating). The designer must ascertain the grade (if any) in the terminal, and whether platforms are drive-through, drive-through with bypass lanes, or sawtooth. Bus headway, bus speed, and various platform departure patterns must also be considered. For instance, with sawtooth platforms, the departing bus must accelerate backward, brake, and then accelerate forward. The drive-through platform requires a different pattern of departure.

Certain codes prescribe a maximum idling time for bus engines, usually 3 to 5 min. Normally, 1 to 2 min of engine operation is required to build up brake air pressure. EPA emission standards for urban bus engines are summarized in Table 10 (bus emission standards in the state of California are more restrictive). The latest

version of the EPA emission factor algorithm should be used to estimate bus tailpipe emissions. MOBILE6.2 (EPA 2002) has been replaced by MOVES2010 (EPA 2009, 2010). Input parameters (e.g., local vehicular inspection and maintenance requirements) suitable for a specific facility should be obtained from the appropriate air quality regulatory agency.

Discharged contaminant quantities should be diluted by natural and/or mechanical ventilation to accepted, legally prescribed levels. To maintain odor control and visibility, exhaust gas contaminants should be diluted with outside air in the proportion 75 to 1.

Where urban-suburban bus operations are involved, the ventilation rate varies considerably throughout the day, and also between weekdays and weekends. Fan speed or blade pitch control should be used to conserve energy. The required ventilation airflow may be reduced by removing contaminant emissions as quickly as possible. This can be achieved by mounting exhaust capture hoods in the terminal ceiling, above each bus exhaust stack. Exhaust air collected by the hoods is then discharged outside of the facility through a dedicated exhaust system.

Effects of Alternative Fuel Use. As discussed in the section on Bus Garages, alternative fuels are being used more widely in lieu of conventional diesel fuel, especially for urban-suburban bus routes, as opposed to long-distance bus service.

Current codes and design standards developed for conventional fuels may not be valid for alternative-fuel buses. Comprehensive design guidelines are not yet available; there is a lack of design standards and long-term safety records for the alternative-fuel buses and their components. Special attention should be given to both risk assessment and design of HVAC and electrical systems for these facilities with regard to a fuel tank or fuel line leak. Research is continuing in this application; further information may be available from the DOT Volpe Transportation Center and NFPA *Standards* 52 and 58.

Bus terminal design should include a risk assessment to review terminal operations and identify potential hazards from alternative fuel buses. Facility managers should adopt safety principles to determine the acceptability of these hazards, based on severity and frequency of occurrence. All hazards deemed undesirable or unacceptable should be eliminated by system design, or by modifications to operations.

Natural Gas (NG) Buses. Fuel burned in LNG and CNG buses has a composition of up to 98% methane (CH_4). Methane burns in a self-sustained reaction only when the volume percentage of fuel and air is in specific limits. The lower and upper flammability, or explosive, limits (LEL and UEL) for methane are 5.3% and 15.0% by volume, respectively. At standard conditions, the fuel/air mixture burns only in this range and in the presence of an ignition source, or when the spontaneous ignition temperature of 1003°F is exceeded.

Electrical and mechanical systems in a bus terminal facility should be designed to minimize the number of ignition sources at locations where an explosive natural gas mixture can accumulate.

Although emissions from an NG bus engine include unburned methane, design of the bus terminal ventilation system must be based on maintaining facility air quality below the LEL in the event of a natural gas leak. A worst-case scenario for natural gas accumulation in a facility is a leak from the bus fuel line or fuel tank, or a sudden high-pressure release of natural gas from a CNG bus fuel tank through its pressure relief device (PRD). For instance, a typical CNG bus may have multiple fuel tanks, each holding gas at 3600 psig and 70°F. If the PRD on a single tank were to open, the tank contents would escape rapidly. After 1 min, 50% of the fuel would be released to the surroundings, after 2 min, 80% would be released, and 90% would be released after 3 min.

Because such a large quantity of fuel is released so quickly, prompt activation of a ventilation purge mode is essential. Where installed, a methane detection system should activate a ventilation purge and an alarm at 20% of the LEL. Placement of methane detectors is very important; stagnant areas, bus travel lanes, and bus loading areas must be considered. In addition, although methane is lighter than air (the relative density of CH_4 is 0.55), some research indicates that it may not rise immediately after a leak. In a natural gas release from a PRD, the rapid throttle-like flow through the small-diameter orifice of the device may actually cool the fuel, making it heavier than air. Under these conditions, the fuel may migrate toward the floor until reaching thermal equilibrium with the surrounding environment; then, natural buoyancy forces drive the fuel/air mixture to the ceiling. Thus, the designer may consider locating methane detectors at both ceiling and floor levels of the facility.

Although no specific ventilation criteria have been published for natural gas vehicles in bus terminals, NFPA *Standard* 52 recommends a blanket rate of 5 ach in fueling areas. DOT (1996a) guidelines for CNG transit facility design recommend a slightly more conservative 6 ach for normal ventilation rates in bus storage areas, with capability for 12 ach ventilation purge rate (on activation by the methane sensors). The designer can also calculate a ventilation purge rate based on the volumetric flow rate of methane released, duration of the release, and size of the facility.

The size of the bus terminal significantly affects the volume flow of ventilation air required to maintain the average concentration of methane below 10% of the LEL. The larger the facility, the lower the number of air changes required. However, a methane concentration that exceeds the LEL can be expected in the immediate area of the leak, regardless of the ventilation rate used. The size of the plume and location/duration of the unsafe methane concentration may be determined using comprehensive modeling analysis, such as computational fluid dynamics.

Source of Ventilation Air. Because dilution is the primary means of contaminant level control, the ventilation air source is extremely important. The cleanest available ambient air should be used for ventilation; in an urban area, the cleanest air is generally above roof level. Surveys of contaminant levels in ambient air should be conducted, and the most favorable source of ventilation air should be used. The possibility of short-circuiting exhaust air, because of prevailing winds and/or building airflow patterns, should also be evaluated.

If the only available ambient air has contaminant levels exceeding EPA ambient air quality standards, the air should be treated to control offending contaminants. Air-cleaning systems for removing gases, vapors, and dust should be installed to achieve necessary air quality.

Control by Contaminant Level Monitoring. Time clocks are one of the most practical means of controlling a bus terminal ventilation system. Time-clock-based ventilation control systems are typically coordinated with both bus movement schedules and installed smoke monitoring devices (i.e., obscurity meters). A bus terminal ventilation system can also be controlled by monitoring levels of individual gases, such as CO, CO_2, NO_2, methane, or other toxic or combustible gases.

Dispatcher's Booth. The bus dispatcher's booth should be kept under positive air pressure to prevent infiltration of engine exhaust fumes. Because the booth is occupied for sustained periods, both normal interior comfort conditions and minimized gas contaminant levels must be maintained during the hours of occupancy.

TOLLBOOTHS

Toll plazas for vehicular tunnels, bridges, and toll roads generally include a series of individual tollbooths. An overhead weather canopy and a utility tunnel (located below the roadway surface) are frequently provided for each toll plaza. The canopy allows installation of roadway signs, air distribution ductwork, and lighting. The utility tunnel is used to install electrical and mechanical systems; it also provides access to each tollbooth. An administration building is usually situated nearby. The current trend in toll collection facility design favors automatic toll collection methods that use magnetic tags. However, new and retrofit toll plazas still include a number of manual toll collection lanes with individual tollbooths.

Toll collectors and supervisors are exposed to adverse environmental conditions similar to those in bus terminals and underground parking garages. Automotive emission levels are considerably higher at a toll facility than on a highway because of vehicle deceleration, idling, and acceleration. Increased levels of CO, NO_x, diesel particulates, gasoline fumes, and other automotive emissions have a potentially detrimental effect on health.

Toll collectors cannot totally rely on physical barriers to isolate them from automotive emissions, because open windows are necessary for collecting tolls. Frequent opening and closing of the window makes the heating and cooling loads of each booth fluctuate independently. Heat loss or gain is extremely high, because all four sides (and frequently the ceiling) of the relatively small tollbooth are exposed to the outdoor ambient air temperature.

HVAC air distribution requirements for a toll facility should be carefully evaluated to maintain an acceptable environment inside the tollbooth and minimize the adverse ambient conditions to which toll-collecting personnel are exposed.

Air Quality Criteria

Workplace air quality standards are mandated by local, state, and federal agencies. Government health agencies differ on acceptable CO levels. ACGIH (1998) recommends a threshold limit of 25 ppm of CO for an 8 h exposure. OSHA (2001a) regulations are for 50 ppm for repeated daily 8 h exposure to CO in the ambient air. The U.S. National Institute for Occupational Safety and Health (NIOSH 1994) recommends maintaining an average of 35 ppm and a maximum level of 200 ppm. Criteria for maximum acceptable CO levels should be developed with the proper jurisdiction. As a minimum, the ventilation system should be designed to maintain CO levels below the threshold limit for an 8 h exposure. Deceleration, idling, and acceleration of vehicles, and varying traffic patterns make it difficult to estimate CO levels around specific toll-collecting facilities without using computer programs.

Longitudinal tunnel ventilation systems with jet fans or Saccardo nozzles are increasingly popular for vehicular tunnels with unidirectional traffic flow. These longitudinal ventilation systems discharge air contaminants from the tunnel through the exit portal. If toll plazas are situated near the exit portal, resultant CO levels around the facilities may be higher than for other toll facilities.

If a recirculating HVAC system were used for a toll collection facility, any contaminants entering a particular tollbooth would remain in the ventilation air. Therefore, tollbooth ventilation systems should distribute 100% outside air to each booth to prevent both intrusion and recirculation of airborne contaminants.

Design Considerations

The toll plaza ventilation system should pressurize booths to keep out contaminants emitted by traffic. Opening the window during toll collection varies depending on booth design and the habits of the individual toll collector. The amount of ventilation air required for pressurization similarly varies.

Variable-air-volume (VAV) systems that are achievable with controls now available can vary the air supply rate based on either the pressure differential between the tollbooth and the outdoor environment, or the position of the tollbooth window. A fixed (maximum/minimum) volume arrangement may also be used at toll plazas with a central VAV system.

Because the area of the window opening varies with individual toll collector habits and booth architecture, the design air supply rate may be determined based on an estimated average window open area. The minimum air supply (when the booth window is closed) should be based on the amount of air required to meet the heating/cooling requirements of the booth, and that required to prevent infiltration of contaminants through the door and window cracks. Where the minimum supply rate exceeds the exfiltration rate, provisions to relieve excess air should be made to prevent overpressurization.

The space between the booth roof and the overhead canopy may be used to install individual HVAC units, fan-coil units, or VAV boxes. Air ducts and HVAC piping may be installed on top of the plaza canopy or in the utility tunnel. The ducts or piping should be insulated as needed.

The amount of ventilation air is typically high compared to the size of the booth; the resulting rate of air change is also high. Supply air outlets should be sized and arranged to deliver air at low velocity. Air reheating should be considered where the supply air temperature is considered too low.

In summer, the ideal air supply location is the ceiling of the booth, which allows cooler air to descend through the booth. In winter, the ideal air supply location is from the bottom of the booth, or at floor level. It is not always possible to design ideal distribution for both cooling and heating. When air is supplied from the ceiling, other means for providing heat at floor level (e.g., electric forced-air heaters, electric radiant heating, heating coils in the floor) should be considered.

The supply air intake should be located so that air drawn into the system is as free as practicable of vehicle exhaust fumes. The prevailing wind should be considered when locating the intake, which should be as far from the roadway as is practicable to provide better-quality ventilation air. Particle filtration of supply air for booths should be carefully evaluated. The specific level and type of filtering should be based on the ambient level of particulate matter and the desired level of removal. See Chapter 11 of the 2009 *ASHRAE Handbook—Fundamentals* and Chapter 28 in the 2008 *ASHRAE Handbook—HVAC Systems and Equipment* for more information.

Equipment Selection

Individual HVAC units and central HVAC are commonly used for toll plazas. Individual HVAC units allow each toll collector to choose between heating, cooling, or ventilation modes. Maintenance of individual units can be performed without affecting HVAC units in other booths. In contrast, a central HVAC system should have redundancy to avoid a shutdown of the entire toll plaza system during maintenance operations.

The design emphasis on booth pressurization requires using 100% outside air; high-efficiency air filters should therefore be considered. When a VAV system is used to reduce operating cost, varying the supply rate of 100% outside air requires a complex temperature control system that is not normally available for individual HVAC units. Individual HVAC units should be considered only where the toll plaza is small, or where the tollbooths are so dispersed that a central HVAC system is not economically justifiable.

Where hot-, chilled-, or secondary water service is available from an adjacent administration building, an individual fan-coil for each tollbooth and a central air handler for supplying the total volume of ventilation air may be economical. When the operating hours for the booths and administration building are significantly different, separate heating and cooling for the toll-collecting facility should be considered. Central air distribution system selection should be based on the maximum number of open traffic lanes during peak hours and the minimum number of open traffic lanes during off-peak hours.

The HVAC system for a toll plaza is generally required to operate continuously. Minimum ventilation air may be supplied to unoccupied tollbooths to prevent infiltration of exhaust fumes. Otherwise, consideration should be given to remotely flushing the closed tollbooths with ventilation air before their scheduled occupancy.

DIESEL LOCOMOTIVE FACILITIES

Diesel locomotive facilities include shops where locomotives are maintained and repaired, enclosed servicing areas where supplies are replenished, and overbuilds where locomotives routinely operate inside an enclosed space and where railroad workers and/or train passengers may be present. In general, these areas should be kept under slightly negative air pressure to help removal of fumes and contaminants. Ventilation should use 100% outside air. However, recirculation may be used to maintain space temperature when a facility is unoccupied or when engines are not running. Heat recovery devices should be considered for facilities in colder climates, though they may require additional maintenance.

Historically, ventilation guidelines for locomotive facilities have recommended simple exhaust rates usually based on the volume of the facility. These were developed over many years of experience and were based on the assumption of nitrogen dioxide as the most critical contaminant. Because contaminant limits for constituents of diesel exhaust have been and are likely to continue changing, ASHRAE sponsored research project RP-1191 (Musser and Tan 2004), which included field measurements in several facilities and a parametric study of design options using computational fluid dynamics. The study resulted in a simplified contaminant-based design procedure that allows designers flexibility to adapt to other critical contaminants or concentrations. Both the traditional and RP-1191 approaches are discussed here.

Ventilation Guidelines and Facility Types

Maintenance and Repair Areas. ANSI/ASHRAE *Standard* 62.1 and most model codes require a minimum outdoor air ventilation rate of 1.5 cfm/ft^2 in vehicle repair garages, with no recirculation recommended. Because the ceiling is usually high in locomotive repair shops, the designer should consider making a volumetric analysis of contaminant generation and air exchange rates rather than using the 1.5 cfm/ft^2 ventilation rate as a blanket standard. The sections on Contaminant Level Criteria and Contaminant Emission Rate have more information on diesel engine exhaust emissions.

Information in the section on Bus Garages also applies to locomotive shops, especially for below-grade pits, battery charging areas, and paint spray booths. However, diesel locomotives generally have much larger engines (ranging to over 6000 hp) than buses. Ventilation is needed to reduce crew and worker exposure to exhaust gas contaminants, and to remove heat emitted from engine radiators. Where possible, diesel engines should not be operated in shops. Shop practices should restrict diesel engine activity and engine operating speeds/intervals; however, some shops require that locomotives be load-tested at high engine speeds. This should be done outdoors if possible, both to reduce indoor contaminants and to avoid problems associated with high heat (sprinkler activation, fire risk, etc.).

A dedicated area should be established for diesel engine operations; hoods should be used to capture engine exhaust in this area. If hoods are impractical because of physical obstructions, then dilution ventilation must be used.

In designing hoods, the location of each exhaust point on each type of locomotive must be identified so that each hood can be centered and located as close as possible to each exhaust point. Local and state railroad clearance regulations must be followed, along with occupational safety requirements. In some cases, high ceilings or overhead cranes may limit hood use. Some newer systems attempt to avoid this problem by using a flexible connection that attaches to the exhaust.

The hood design should not increase backpressure on locomotive exhaust; the throat velocity should be kept less than twice the exhaust discharge velocity. The associated duct design should include access doors and provisions for cleaning oily residue, which increases the risk of fire. Fans and other ventilation equipment in the airstream should be selected with regard to the elevated temperature of the exhaust air and the effects of the oily residue in the emissions.

Sometimes high ceilings or overhead cranes limit the use of hoods. The *Manual for Railway Engineering* (AREMA 2007) notes that 6 air changes per hour are usually sufficient to provide adequate dilution for both idling locomotives and short engine runs at high speed. This guideline was developed with nitrogen dioxide as the critical contaminant, with an allowable maximum concentration of 5 ppm(v). Even dilution systems can and should take advantage of thermal buoyancy by removing exhaust air at the ceiling level or a high point in the shop and introducing makeup air at floor level. If exhaust gases are allowed to cool and drop to floor level, locomotive radiator fans (if operating) can cause further mixing in the occupied zone, making removal less effective.

Shops in colder climates should be heated both for worker comfort and to prevent freezing of facility equipment and piping. The heating system may consist of a combination of perimeter convectors to offset building transmission losses, underfloor slab or infrared radiation for comfort, and makeup air units for ventilation. Where natural gas is available and local codes allow, direct-fired gas heaters can be an economical compromise to provide a high degree of worker comfort. Air curtains or door heaters are not needed in shops where doors are opened infrequently.

Enclosed Servicing Areas. Although most locomotive servicing is done outside, some railroads use enclosed servicing areas for protection from weather and extreme cold. Servicing operations include refilling fuel tanks, replenishing sand (used to aid traction), draining toilet holding tanks, checking lubrication oil and radiator coolant levels, and performing minor repairs. Generally, a locomotive spends less than 1 h in the servicing area. Ventilation is needed to reduce personnel exposure to exhaust gas contaminants and remove heat emitted from engine radiators. The designer should also consider the presence of vapors from fuel oil dispensing and silica dust from sanding. Heating may also be included in the design, depending on the need for worker comfort and the operations performed.

Ventilation for servicing areas should be similar to that for maintenance and repair areas. Where possible, hoods should be used in lieu of dilution ventilation. However, coordinating hood locations with engine exhaust points may be difficult because different types of locomotives may be coupled together in consists. Elevated sanding towers and distribution piping may also interfere. Contaminant levels might be higher in servicing areas than in the shops because of constantly idling locomotives and occasional higher-speed movements in servicing areas. For dilution ventilation, the designer should ascertain the type of operations planned for the facility and make a volumetric analysis of expected rates of contaminant generation and air exchange.

Infrared radiation should be considered for heating. As with maintenance and repair areas, direct-fired gas heaters may be economical. Door heaters or air curtains may be justified because of frequent opening of doors or a lack of doors.

Overbuilds. With increasing real estate costs, the space above trackways and station platforms is commonly built over to enclose the locomotive operation area. Ventilation is needed in overbuilds to reduce crew and passenger exposure to exhaust gases and to remove heat emitted from engine radiators and vehicle air-conditioning systems. Overbuilds are generally not heated.

Exhaust emissions from a diesel passenger locomotive operating in an overbuild are higher than from an idling locomotive because of head-end power requirements. The designer should determine the types of locomotives to be used and the operating practices in the overbuild. As with locomotive repair shops and servicing areas, hoods are recommended to capture engine exhaust. According to the *Overbuild of Amtrak Right-of-Way Design Policy* (Amtrak 2005), the air temperature at the exhaust source will be between 350 and 950°F. A typical ventilation design could have hoods approximately 18 to 23 ft above the top of the rail, with throat velocities between 30 and 36.7 fps. For dilution ventilation, the designer should perform a volumetric analysis of contaminant generation and air exchange rates.

Contaminant Level Criteria

In most locations, diesel exhaust is not regulated specifically, although concentrations of many substances found in diesel exhaust are regulated. The U.S. Occupational Safety and Health Administration (OSHA 2001a, 2001b) identifies carbon dioxide (CO_2), carbon monoxide (CO), nitrogen dioxide (NO_2), nitric oxide (NO), diesel particulate matter (DPM), and sulfur dioxide (SO_2) as major components of diesel exhaust. Thirty-one additional substances are identified as minor components, with seventeen of these being polycyclic aromatic hydrocarbons (PAH). These minor components are elements of DPM.

Federal OSHA requirements establish limits for these compounds in the United States, although a few states may set more restrictive requirements. Also, the American Council of Governmental and Industrial Hygienists (ACGIH) publishes guideline values for use in industrial hygiene that are not legally enforceable, but may evolve more quickly than OSHA requirements (ACGIH 2001). Other countries set their own contaminant limits, though these may draw heavily from the ACGIH and other U.S. publications.

When no regulations exist for DPM, nitrogen dioxide (NO_2) is present in diesel exhaust emissions at the highest levels relative to its published limits. In these circumstances, systems designed to control nitrogen dioxide will maintain other exhaust-related contaminants well below their respective limits. Table 11 shows published exposure limits in parts per million (ppm). Federal OSHA, ACGIH, and NIOSH limits are current as of at least February 2003. Other limits are taken primarily from an international database of participating countries (ILO 2003; Lu 1993). Contaminant limits are often expressed in mg/m^3, even in regions where I-P units are used.

Most authorities do not currently distinguish DPM from other particulates; however, this may change. The ACGIH recently added DPM measured as elemental carbon to its TLVs (ACGIH 2003). A $0.1 \ mg/m^3$ limit for elemental carbon in diesel environments has been established in Germany. Laws enacted by the Mine Safety and Health Administration are targeted toward limiting DPM in mining environments (MSHA 2001a, 2001b). These changes may foreshadow action by OSHA. In this changing environment, designers must check local regulations in the time and place of construction for applicable limits.

Contaminant Emission Rate

Locomotive contaminant emissions have been measured primarily for environmental reasons, and data for some models have been published in the environmental literature (Table 12). These data are classified for different duty-cycles of operation and different throttle settings, and were obtained from controlled tests conducted under steady-state operation. Engine speed, engine power, fuel rate, and engine airflow are typically reported. Emissions are usually reported for carbon monoxide (CO), oxides of nitrogen (NO_x), hydrocarbons, sulfur dioxide (SO_2), and particulates. Manufacturers can provide this information for specific engine models, and should be consulted for current and specific data for design projects.

Note that passenger locomotives consume a greater amount of power when idling with head-end power (HEP) to serve passenger-related needs. A passenger train idle at HEP can produce five times the amount of NO_x emissions as the same train idling with no HEP effects (Fritz 1994). Thus, this is an important distinction between passenger railway stations, where HEP is likely to be required, and repair facilities, where HEP is not likely to be needed.

Available emissions data have been targeted toward outdoor pollution concerns, which imposes some limitations in applying it to indoor settings. Only recent tests document exhaust temperatures, a quantity useful to design engineers concerned with sprinkler systems. Emissions data come from steady-state tests on engines whose operation has been allowed to stabilize for an hour or more, so a safety factor is suggested to allow for higher emissions related to cold start and transient operation. Also, the data include only a combined NO_x emissions value. Field measurements in locomotive facilities found that about 13% (by mass) of ambient NO_x could be attributed to NO_2 (Musser and Tan 2004). This factor can be used

estimate NO_2 source emissions from available data. The applicability of these data to design applications is supported by comparisons of CFD models based on published emissions data and field measurements taken in repair shops that showed reasonable agreement between the measured and predicted values (Musser and Tan 2004).

Table 11 Contaminant Exposure Limits for NO_2
(For information only—check updated local regulations)

Entity	NO_2, ppm(v)		
	8 h	15 min	Ceiling
OSHA: USA (PEL)			5
ACGIH: USA (TLV)	3	5	
NIOSH: USA (REL)		1	
Australia	3	5	
Belgium	3	5	
Denmark	3	5	
Finland	3	6	
France		3	
Germany			5
Japan			
Sweden	1*		
Switzerland	3	6	
United Kingdom	3	5	
China			2.6

*Limit specifically for NO_2 from exhaust fumes.

Table 12 Sample Diesel Locomotive Engine Emission Data[a]

Throttle Position (Notch)	Engine Speed, rpm	Engine Power, bhp	Engine Airflow,[b] cfm	Fuel Rate, lb/h	NO_x, lb/h	CO, lb/h	HC, lb/h	SO_2, lb/h	Particulates, lb/h
Four-Stroke Cycle, With Head End Power (HEP)									
8, Freight	1050	3268	8816	1103	81	5.9	3.9	0.51	0.96
7, HEP	900	2771	7068	929	71	13	3.0	0.43	1.1
6, HEP	900	2254	5668	762	62	11	2.6	0.35	0.92
5, HEP	900	1777	4433	609	51	10	1.6	0.28	0.64
4, HEP	900	1023	2677	369	34	3.2	1.2	0.17	0.51
3, HEP	900	713	2055	266	22	2.3	0.94	0.12	0.49
2, HEP	900	431	1656	174	14	1.6	0.88	0.08	0.54
1, HEP	900	322	1651	144	13	1.6	0.92	0.07	0.65
HEP idle	900	185	1511	81	6.3	2.0	1.1	0.04	0.91
Standby	720	512	1441	189	16	1.9	0.81	0.09	0.65
High idle	450	34	466	23	1.9	0.49	0.40	0.01	0.13
Low idle	370	22	NA[c]	17	1.1	0.70	0.35	0.01	0.10
Two-Stroke Cycle, No Head End Power (HEP)									
8	903	3210	8880	1060	86	8.1	2.0	0.49	1.5
7	821	2540	7100	833	56	4.1	1.2	0.38	0.97
6	726	1700	5310	572	38	1.5	0.85	0.26	0.64
5	647	1390	4630	480	33	1.4	0.79	0.22	0.58
4	563	1060	3950	368	28	0.61	0.64	0.17	0.42
3	489	714	3410	254	24	0.42	0.51	0.12	0.30
2	337	370	2200	142	13	0.46	0.33	0.07	0.12
1	337	207	2270	91	7.7	0.34	0.25	0.04	0.07
High idle	339	14	2390	32	2.5	0.17	0.21	0.02	0.05
Low idle	201	10	1320	14	1.3	0.08	0.08	0.01	0.02
Auxiliary Engine/Alternator for Head End Power (HEP)									
NA[c]	1800	699	1930	275	17	7.3	0.69	0.13	0.37
NA[c]	1800	566	2190	226	17	1.1	0.78	0.10	NA[c]
NA[c]	1800	438	2010	179	13	0.53	0.67	0.08	0.26
NA[c]	1800	377	1900	157	10	0.39	0.61	0.07	0.25
NA[c]	1800	305	1810	135	8.4	0.39	0.56	0.06	0.23
NA[c]	1800	238	1710	114	6.5	0.40	0.57	0.05	0.20
NA[c]	1800	173	1640	95	4.8	0.41	0.55	0.04	0.18
NA[c]	1800	31	1480	55	2.2	0.46	0.60	0.03	0.16

[a]Data from Southwest Research Institute (SwRI 1992). [b]Intake, corrected to standard air density 0.0751 lb/ft³. [c]Data not available.

Locomotive Operation

Designers need to anticipate locomotive operation during the design phase, particularly when estimating source strength based on published locomotive emissions data. Some important parameters include the number of operating locomotives and the location, duration, and throttle position at which they operate. The number of locomotives likely to be operating can be estimated based on shop or station schedules. Although it is important to remember that a locomotive could idle at any location inside a facility, there are often practical cues to identify the most common or likely locations. These include platforms, facility layout, location of equipment for servicing toilets, fuel stations, or other service equipment. In small shops, the layout may create one or two convenient positions in which locomotives are very likely to be parked.

Other operating parameters may be more difficult to estimate, particularly in shops. Field observations for ASHRAE research project RP-1191 recorded locomotive operation in several shops varying from a few minutes to an hour in duration, usually at idle and low throttle settings (Musser and Tan 2004). Operation was influenced by shop rules, practices, and conventions, which are valuable to consider during the design phase. The cooperation and involvement of shop employees in the design stage can help integrate these practices so that the design conforms to the needs of the facility, rather than the other way around.

Design Methods

General Exhaust Systems. A contaminant-based procedure using a simplified equation developed with computational fluid dynamics can be used to design general exhaust systems using the steps below. The simplified equation was developed to flexibly adapt to changes in contaminant limits. Figures 20 and 21 show schematic drawings of such a system.

Step 1: Verify that design parameters to be used in the simplified equation fall within the ranges for which the equation is valid.

- Ceiling height Z must be 20 to 45 ft.
- Fan spacing X must be 20 to 60 ft.
- Exhaust fan flow Q must provide 5 to 12 air changes per hour (ach)

Step 2: Verify that other facility characteristics show reasonable agreement with the assumptions of the parametric study:

- Fan dimensions L: Exhaust fan or duct dimensions are 5 by 5 ft.
- Fan placement: Exhaust fans or duct openings are centered above each track.
- Locomotive exhaust temperature T: 350°F.
- Locomotive exhaust flow rate F: 2000 cfm.
- Radiator fans: For many locomotive models, radiator fans do not operate when the locomotive is idling, and no radiator fan flow was modeled in this study. If radiator fans will be operating, they may alter the indoor airflow patterns.
- Operating time: The equation is based on steady state conditions, so it is not necessary to assume a maximum operating time.
- Concurrent operation: The equation allows for concurrent operation on different tracks. However, it does not include concurrent operation of more than one locomotive on the same track.
- Track-to-track spacing Y: 25 ft.
- Ambient temperature: 90°F. This was selected because warmer ambient temperatures tend to reduce the upward buoyancy of warm exhaust gases.

Step 3: Obtain emissions data for critical contaminants and determine the design indoor concentration limit for the critical contaminant.

- Emissions data for some locomotive models are published in the environmental literature, and data for specific locomotives can be obtained from the manufacturer. The emissions rate for a given

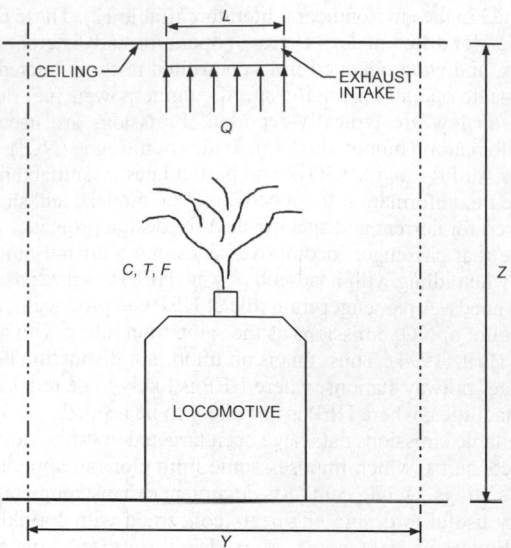

Fig. 20 Section View of Locomotive and General Exhaust System

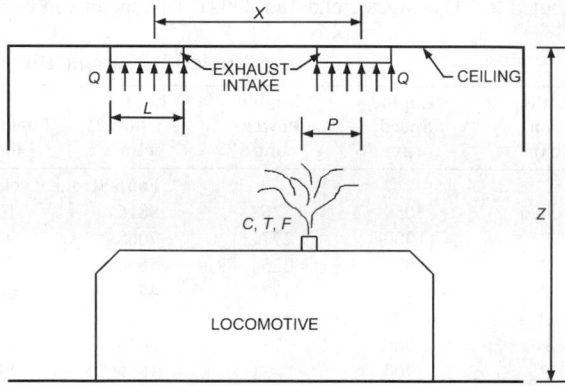

Fig. 21 Elevation View of Locomotive and General Exhaust System

locomotive model depends on throttle position and whether head-end power is used.

- Acceptable indoor concentration limits can be determined from legal requirements at the location and time of construction. The designer may also wish to consider recommended limits from organizations such as ACGIH. To allow a safety margin, a designer might choose a contaminant limit that is lower than the published legal limit.

Contaminant limits are often expressed in mg/m^3, even in regions where I-P units are used. A contaminant limit in ppm(v) can be converted to mg/m^3 for use in the simplified equation as follows (ASHRAE *Standard* 62.1):

$$ppm(v) \times \frac{\text{Molecular weight}}{24.45} = mg/m^3 \qquad (19)$$

Step 4: Select a fan flow rate and calculate the maximum concentration to which occupants would be exposed using Equation (20). Table 13 gives values for constants a to j whether occupants will be standing on the floor or a 4 ft high platform.

Table 13 Constants for Equation (20)

Constant	No Platform	With 4 ft Platform
a	20.0	22.5
b	−0.551	−0.773
c	−3.32	−2.09
d	−0.106	−0.109
e	−0.308	−0.346
f	0.0119	0.0159
g	0.235	0.236
h	0.0792	0.0407
i	0.00191	0.00190
j	−0.00505	−0.00499

$$C_{occ} = 10^{-3} C_{emissions}(a + bQ + cP + dX + eZ + \\ + fQZ + gPX + hPZ + iXZ + jPXZ) \qquad (20)$$

where

C_{occ} = maximum time-averaged concentration of critical contaminant to which occupants could be exposed, mg/m^3

$C_{emissions}$ = concentration of critical contaminant in exhaust emissions, mg/m^3

a to j = constants found in Table 13

Q = total exhaust fan flow rate required, ach; must be between 5 and 12 ach

Z = ceiling height; must be 20 to 45 ft

X = fan spacing; must be 20 to 60 ft

P = locomotive offset position, dimensionless; $P = 0$ under fan and $P = 1$ between fans. Other values for P can be calculated based on the distance of locomotive stack from the nearest exhaust fan d and fan spacing X:

$$P = 2\frac{d}{X} \qquad (21)$$

Step 5: Compare C_{occ} obtained in step 4 with the concentration limit C_{limit} determined in Step 3. If $C_{occ} < C_{limit}$, the selected flow rate is adequate. If $C_{occ} > C_{limit}$, repeat step 4 with a higher flow rate until a concentration less than the limit is obtained.

Step 6: Verify that the result is between 5 and 12 ach.

- If the flow rate obtained is between 5 and 12 ach, this is the system size.
- If the flow rate obtained is less than 5 ach, the designer could
  - Design for 5 ach.
  - Perform a more detailed analysis to verify that less than 5 ach will provide acceptable contaminant control. For rates less than the 1.5 cfm/ft^2 recommended by ASHRAE *Standard* 62.1 or in the case of unusual sources, the presence of contaminants other than those from diesel exhaust in the space (e.g., liquid fuel) should also be considered.
- If the flow rate obtained is greater than 12 ach, the designer could
  - Adjust the other parameters to attempt to reduce the air change requirement.
  - Perform a more detailed analysis to verify the necessary air flow requirement.

Example 2. Perform design calculations for a passenger locomotive repair shop.

Step 1: Verify design parameters. The planned facility ceiling height is 30 ft, which falls within the 20 to 45 ft range for which the simplified equation is valid. The planned fan spacing is 50 ft, which also falls within the required range of 20 to 60 ft.

Step 2: Verify other facility characteristics.

- Exhaust fans: Exhaust openings with an area of approximately 25 ft^2 will be used, and fans will be centered above each track.

- Locomotive: Operating locomotives are usually high idle or lower. When moving in, they will not exceed throttle position 1. Information obtained from the manufacturer of the locomotive most commonly serviced in this facility indicates an exhaust flow rate of 2300 cfm, an exhaust temperature of 375°F, and NO$_x$ generation of 7.661 lb/h. Radiator fans will not operate in the high idle position for this locomotive.
- Track-to-track spacing is 27 ft. Locomotives may operate concurrently on adjacent tracks, but concurrent operation on the same track is not planned.
- These characteristics are reasonably similar to the assumptions upon which the simplified equation is based.

Step 3: Obtain emissions data and determine the design limit.

- The critical contaminant for this design is nitrogen dioxide (NO$_2$). Emission data from the manufacturer state that the NO$_x$ generation rate is 7.661 lb/h. Field measurements conducted for ASHRAE research project RP-1191 (Musser and Tan 2004) showed that ambient NO$_2$ concentrations were about 13% of ambient NO$_x$ levels. Therefore, the NO$_2$ generation rate is estimated to be 13% of the total, or 0.996 lb/h, which converts to 452 g/h. For an exhaust flow rate of 2300 cfm, the concentration of NO$_2$ in the exhaust is 116 mg/m^3.

$$C_{emissions} = \left(\frac{452 \text{ g/h}}{2300 \text{ ft}^3/\text{min}}\right)\left(\frac{1}{60 \text{ min/h}}\right)\left(\frac{1}{0.02832 \text{ m}^3/\text{ft}^3}\right)(1000 \text{ mg/g})$$
$$= 116 \text{ mg/m}^3$$

- OSHA currently requires a 5 ppm(v) ceiling for NO$_2$, but NIOSH and other sources recommend a 1 ppm(v) 15 min short-term exposure limit (STEL). The designer decides to select the lower 1 ppm(v) limit, and to design for 0.5 ppm(v) (i.e., 0.94 mg/m^3) to allow for a safety factor for variations in emissions or operation.

$$\frac{0.5 \text{ ppm(v)} \times 46}{24.45} = 0.94 \text{ mg/m}^3$$

Step 4: Select a flow rate and solve for the contaminant concentration. Ceiling height is 30 ft, and fan spacing 50 ft. Based on the placement of services in the shop, expect that the stack of an operating locomotive will be at most 12.5 ft from the nearest exhaust fan, so $P = 0.5$. The shop does have a 4 ft high platform where workers may stand, so Equation (20) is solved using a platform. First, try fans that provide 5 ach:

$$C_{occ} = 10^{-3}(116 \text{ mg/m}^3)(22.5 - 0.773Q - 2.09P - 0.109X \\ - 0.346Z + 0.0159QZ + 0.236PX + 0.0407PZ \\ + 0.00190XZ - 0.00499PXZ) = 1.13 \text{ mg/m}^3$$

- Iterate between steps 5 and 4. With 5 ach, $C_{occ} = 1.13$ mg/m^3. This is greater than the desired limit of 0.94 mg/m^3. If the fan flow rate is increased to provide 10.5 ach, C_{occ} decreases to 0.94 mg/m^3. This meets the design criterion.

Step 5: Verify that the fan flow rate is between 5 and 12 ach. No further analysis is needed.

Exhaust Hood Design. A similar equation was also developed for design of exhaust hood systems. However, results from the parametric set of computational fluid dynamics simulations performed to develop the equation were shown to be highly specific to the situation and geometry shown in Figures 22 and 23. Therefore, these equations should not be used unless the given assumptions are exactly matched. For further information on hood design, see ACGIH (1998).

Step 1: Verify that the design parameters to be used in the simplified equation fall within the ranges for which it is valid.

- Hood mounting height H must be 3 to 8 ft.
- Hood length L must be 5 to 11 ft.
- Exhaust fan flow Q must provide 5 to 12 ach.

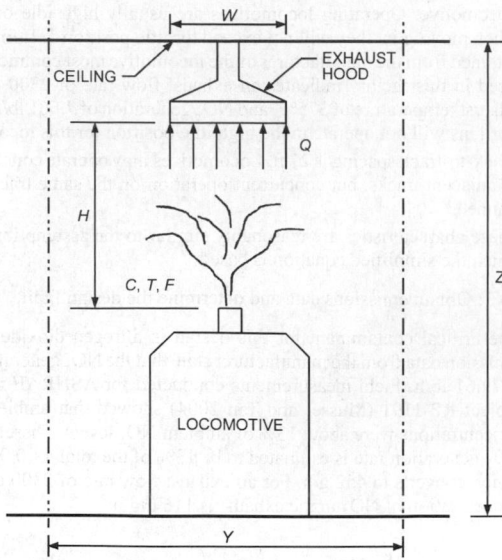

Fig. 22 Section View of Locomotive and Exhaust Hood System

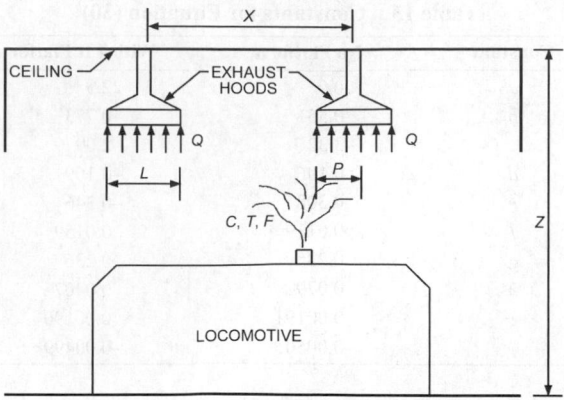

Fig. 23 Elevation View of Locomotive and Exhaust Hood System

Step 2: Verify that other facility characteristics agree with the assumptions of the parametric study. These assumptions are as follows:

- Hood width W: 5 ft.
- Hood placement: Hoods are centered above each track at 60 ft intervals.
- Hood operation: All hoods switched on together.
- Ceiling height Z: 25 ft.
- Locomotive exhaust temperature T: 350°F.
- Locomotive exhaust flow rate F: 2000 cfm.
- Radiator fans: Radiator fans do not operate.
- Operating time: The results of the study are based on steady-state conditions, so it is not necessary to assume a maximum operating time.
- Concurrent operation: The study allows for concurrent operation on different tracks, but not for concurrent operation of more than one locomotive on the same track.
- Track-to-track spacing Y: 25 ft.
- Ambient temperature: 90°F.

Step 3: Obtain emissions data for critical contaminants and determine the design indoor concentration limit for the critical contaminant. This can be done using the procedure described for general exhaust systems.

Step 4: Select a fan flow rate and calculate the maximum concentration to which occupants would be exposed using Equation (22). Table 14 gives values for constants a to l for occupants standing on the floor or on a 4 ft high platform.

$$C_{occ} = 10^{-3} C_{emissions}(a + bQ + cP + dH + eL + fQP$$
$$+ gQH + hQL + iPH + jHL + kQPH + lQHL) \quad (22)$$

where

C_{occ} = maximum time-averaged concentration of critical contaminant to which occupants could be exposed, mg/m³

$C_{emissions}$ = concentration of critical contaminant in exhaust emissions, mg/m³

a to l = constants in Table 14

Q = total exhaust fan flow rate required, ach; must be 5 to 12 ach

H = hood mounting height; must be 3 to 8 ft

L = fan spacing; must be 5 to 11 ft

Table 14 Constants for Equation 22

Constant	No Platform	With 4 ft Platform
a	0.717	2.19
b	−0.160	−0.401
c	0.900	2.18
d	−0.168	−0.283
e	−0.0508	−0.275
f	−0.129	−0.332
g	0.0381	0.0684
h	0.0245	0.0846
i	−0.174	−0.351
j	0.0294	0.0575
k	0.0290	0.0560
l	−0.00588	−0.0134

P = locomotive offset position; dimensionless; $P = 0$ centered under hood and $P = 1$ under edge of hood. Other values for P can be calculated based on distance d of locomotive stack from center of nearest exhaust hood and hood length L:

$$P = \frac{2d}{L} \quad (23)$$

Step 5: Compare C_{occ} obtained in step 4 with the concentration limit C_{limit} determined in step 3. If $C_{occ} < C_{limit}$, the selected flow rate is adequate. If $C_{occ} > C_{limit}$, repeat step 4 with a higher flow rate until a concentration less than the limit is obtained.

Step 6: Verify that the result is between 5 and 12 ach.

- If the flow rate obtained is between 5 and 12 ach, this is the system size.

- If the flow rate obtained is less than 5 ach, the designer could
  - Design for 5 ach.
  - Perform a more detailed analysis to verify that less than 5 ach will provide acceptable contaminant control. For rates less than the 1.5 cfm/ft² recommended by ASHRAE *Standard* 62.1 or in the case of unusual sources, the presence of contaminants other than those from diesel exhaust in the space should also be considered (e.g., liquid fuel).

- If the flow rate obtained is greater than 12 ach, the designer could
  - Adjust other parameters to attempt to reduce the air change requirement.
  - Perform a more detailed analysis to verify the necessary air flow requirement.

EQUIPMENT

An enclosed vehicular facility's ability to function depends mostly on the effectiveness and reliability of its ventilation system, which must operate effectively under the most adverse environmental, climatic, and vehicle traffic conditions. A tunnel ventilation system should also have more than one dependable power source, to prevent interruption of service.

Fans

Fan manufacturers should be prequalified and should be responsible under one contract for furnishing and installing the fans, bearings, drives (including any variable-speed components), motors, vibration devices, sound attenuators, discharge/inlet dampers, actuators, and limit switches. Other ventilation-related equipment, such as ductwork, may be provided under a subcontract.

The prime concerns in selecting the type, size, and number of fans include the total theoretical ventilation airflow capacity required and a reasonable comfort margin. Fan selection is also influenced by how reserve ventilation capacity is provided either when a fan is inoperative, or during maintenance or repair of either the equipment or the power supply.

Selection (i.e., number and size) of fans needed to meet normal, emergency, and reserve ventilation capacity requirements of the system is based on the principle of parallel fan operation. Actual airflow capacities can be determined by plotting fan performance and system curves on the same pressure-volume diagram.

Fans selected for parallel operation may be required to operate in a particular region of their performance curves, so that airflow capacity is not transferred back and forth between fans. This is done by selecting a fan size and speed such that the duty-point total pressure, no matter how many fans are operating, falls below the minimum total pressure characterized by the bottom of the stall dip or unstable performance range. This may require consultation with the fan manufacturer, because this information is not typically available from published fan performance data. Fans operating in parallel should be of equal size and have identical performance curves. If airflow is regulated by speed control, all fans should operate at the same speed. If airflow is regulated by dampers or by inlet vane controls, all dampers or inlet vanes should be set at the same angle. For axial-flow fans, blades on all fans should be set at the same pitch or stagger angle.

Jet fans can be used for longitudinal ventilation to provide a positive means of smoke and air temperature management in tunnels. This concept was proven as part of the Memorial Tunnel Fire Ventilation Test Program (MHD/FHWA 1995). Although jet fans deliver relatively small air quantities at high velocity, the momentum produced is transferred to the entire tunnel, inducing airflow in the desired direction. Jet fans are normally rated in terms of thrust rather than airflow and pressure, and can be either unidirectional or reversible.

Number and Size of Fans. The number and size of fans should be selected by comparing several fan arrangements based on the feasibility, efficiency, and overall economy of the arrangement, and the duty required. Factors that should be studied include (1) annual power cost for operation, (2) annual capital cost of equipment (usually capitalized over an assumed equipment life of 30 years for mass transit tunnel fans, or 50 years for highway and railroad tunnel fans), and (3) annual capital cost of the structure required to house the equipment (usually capitalized over an arbitrary structure life of 50 years).

Two views are widely held regarding the proper number and size of fans: the first advocates a few high-capacity fans, and the second prefers numerous low-capacity fans. In most cases, a compromise arrangement produces the greatest efficiency. The number and size of the fans should be selected to build sufficient flexibility into the system to meet the varying ventilation demands created by daily and

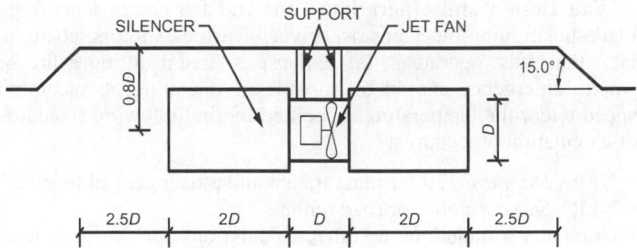

Fig. 24   Typical Jet Fan Arrangement in Niche

seasonal traffic fluctuations and emergency conditions. Consideration should be given to satisfying emergency conditions during fan outages for maintenance or unplanned downtime.

In general, when selecting the number of fans, several issues may need to be considered, ranging from redundancy and space allocation, to design issues such as determining the number of control boxes, dampers, silencers, and similar equipment. In tunnel ventilation, the required fan airflow capacity is typically very large. If one fan is installed, the fan must be large, and this design provides zero redundancy in case of failure or maintenance. However, if many fans are installed, more space is required than for a single fan. Designs need to balance space allocation with an acceptable level of redundancy.

Jet fan sizing is usually limited by space available for installation in the tunnel. Typically mounted on the tunnel ceiling (above the vehicle traffic lanes), or on the tunnel walls (outside the vehicle traffic lanes), jet fans are sometimes placed in niches to minimize the height or width of the entire tunnel boundary. However, niches must be adequately sized to avoid reducing the thrust of the fans. A typical jet fan niche arrangement is provided in Figure 24.

For longitudinal ventilation using jet fans, the required number of fans is defined (once fan size and tunnel airflow requirements have been determined) by the total thrust required to overcome the tunnel resistance (pressure loss), divided by the individual jet fan thrust, which is a function of the mean air velocity in the tunnel. Jet fans installed longitudinally should be at least 7 to 10 tunnel hydraulic diameters apart so that the jet velocity does not affect the performance of downstream fans. Jet fans installed side by side should be at least two fan diameters (centerline to centerline) apart.

Type of Fan. Normally, ventilating an enclosed vehicular facility requires a large volume of air at relatively low pressure. Some fans have low efficiencies under these conditions, so the choice of a suitable fan type is often limited to a centrifugal, vaneaxial, or jet fan.

Special Considerations. Special attention must be given to a fan installed where **airflow and pressure transients** are caused by vehicle passage. If the transient tends to increase airflow through the fan (i.e., positive flow in front of the vehicle toward an exhaust fan, or negative flow behind the vehicle toward a supply fan), blade loading must not become high enough to produce long-term fatigue failures. If the disturbance tends to decrease airflow through the fan (i.e., negative flow behind the vehicle toward an exhaust fan, or positive flow in front of the vehicle toward a supply fan), the fan performance characteristic must have adequate comfort margins to prevent an aerodynamic stall.

If the pressure pulses are large relative to the fan's total pressure capability, at either full or planned reduced-speed operation, it can result in an overblown condition. Motors, power, and mechanical systems should be designed for overblown operation if the motor needs to operate under these conditions.

The ability to **rapidly reverse** the rotation of a tunnel ventilation fan is important during an emergency. This requirement must be considered in selection and design of the fan and drive system.

Fan Design and Operation. Fans and fan components (e.g., blade-positioning mechanisms, drives, bearings, motors, controls, etc.) that must operate in the exhaust airstream during a fire or smoke emergency should be capable of operating at maximum speed under the temperatures specified by the following standards or calculation procedures:

- NFPA *Standard* 130 for mass transit and passenger rail tunnels
- NFPA *Standard* 502 for road tunnels
- Computer simulations or other calculations for the maximum expected temperatures, in railroad tunnels and other enclosed vehicular facilities

Fans and dampers that are operated infrequently or for emergency service only should be activated and tested at least once every month to ensure that all rotating elements are in good condition and properly lubricated. The period of activity should be long enough to achieve stabilized temperatures in fan bearings and motor windings.

Inlet boxes can be used to protect centrifugal fan bearings and drives from high temperatures, corrosive gases, and particulate matter in exhaust air during emergency operating mode. This arrangement requires special attention to fan shaft design, because of overhung drive loads (see the section on Fan Shafts).

Reversible axial flow fans should be able to be rapidly reversed from the maximum design speed in one direction to the maximum design speed in the opposite direction in less than 60 s. Fan design should include the effects of temperature changes associated with reversing airflow direction. All components of reversible fans should be designed for a minimum of 5000 cycles without damage.

Housings for variable-pitch axial flow fans should be furnished with instruments to measure airflow in both directions. Capped connections should be provided for measuring the pressure developed across the fan. The fan should also be protected from operating in a stall region.

To minimize blade failure in axial flow fans, the following precautions should be taken:

- Blades should be secured to the hub by positive locking devices.
- The fan inlet (and discharge, if reversible) should be protected against entry of foreign objects that could damage the rotating assembly.
- The natural frequency (static and rotating) of the blade and the maximum stress on the blade surface (for all operating points on the fan characteristic curve) should be measured during factory testing.
- For mass transit and rail systems, fans subjected to airflow and air pressure reversals caused by train passage should be designed (and tested, for verification) to withstand 4,000,000 cycles of airflow reversals.

When a fan includes a variable-frequency drive (VFD), factory testing with a production version of the VFD should be done to ensure adequate operation and compatibility with the fan system. Fans that are run by VFDs should have an installed static blade with a first bending natural frequency at least four times higher than the maximum intended running speed (e.g., an 1180 rpm fan's first bending frequency should be at least 79 Hz). If installed blades have a first bending frequency below this value, the VFD should be programmed to avoid speeds that are potentially problematic.

Jet fan blades should be strong enough to withstand the air temperatures created by a fire. Design calculations for jet fans should consider that the fire might destroy the fan(s) at the fire location, and that the jet fans downstream of the fire will operate under high temperatures and reduced thrust.

Fan Shafts. Fan shafts should be designed so that the maximum deflection of assembled fan components, including forces associated with the fan drive, does not exceed 0.005 in. per foot of shaft length between centers of the bearings. For centrifugal fans where the shaft overhangs the bearing, the maximum deflection at the centerline of the fan drive pulley should not exceed 0.005 in. per foot of shaft length between the center of the bearing and the center of the fan drive pulley.

Good practice suggests that the fundamental bending mode frequency of the assembled shaft, wheel, or rotor be more than 50% higher than the highest fan speed. The first resonant speed of all rotational components should be at least 125% above the maximum speed. The fan assembly should be designed to withstand, for at least 3 min, all stresses and loads from an overspeed test at 110% of maximum design fan speed.

Bearings. Fan and motor bearings should have a minimum equivalent L10 rated life of 10,000 h, as defined by the American Bearing Manufacturers Association (ANSI/ABMA 2000). Special attention must be given to belt-driven fans, because improper tensioning or overtensioning of belts can drastically reduce the bearing life, belt life, and possibly shaft life.

For **axial-flow fans** and **jet fans**, each fan motor bearing and fan bearing should have a monitoring system that senses individual bearing vibrations and temperatures, and provides a warning alarm if either rises above the manufacturer-specified range. Jet fan motors should have an industrial protection class (IP rating) of 55 or higher, which has bearings with washdown-rated seals.

Because of their low speed (generally less than 450 rpm), **centrifugal fans** are not always provided with bearing vibration sensors, but they do require temperature sensors with warning alarm and automatic fan shutdown. Bearing pedestals for centrifugal fans should provide rigid support for the bearings with negligible impediments to airflow. Static and dynamic loading of the shaft and the impeller, and the maximum force from tension in the belts, should be considered.

Corrosion-Resistant Materials. Choosing a particular material or coating to protect a ventilation fan from corrosive gas is a matter of economics. Selection of the material and/or coating should be based on the installation environment, fan duty, and an expected service life of 50 years.

Sound. For ventilation fan sound attenuator design, construction documents should specify the following:

- Speed and direction of airflow, and number of operating fans
- Maximum dBA rating or NC curve(s) acceptable under installed conditions, and locations of fan supply inlet and exhaust outlet where these requirements apply
- OSHA or local requirements for jet-fan-generated noise limits, which may require silencers of 1 to 2 fan diameters in length
- The dBA rating required at certain specific locations, such as intake louvers, discharge louvers, or discharge stacks, may not exceed OSHA or local requirements
- That the fan manufacturer must furnish and install the acoustical treatment needed to bring the sound level down to an acceptable value if measured sound values exceed the specified maximum values at the defined boundaries of the fan manufacturer's scope of supply
- NFPA-recommended maximum noise levels for emergency fan operations

Dampers

Dampers play a major role in overall tunnel safety and the successful operation of a tunnel ventilation system. Dampers regulate airflow into and out of the tunnel, through either natural or forced ventilation, to maintain acceptable temperatures. Dampers also relieve pressure: opening and closing dampers allows tunnel air to be driven out of ventilation shafts located in front of moving vehicles, and for fresh air to be drawn into tunnels by ventilation shafts located behind moving vehicles. Dampers are also used with fans to dilute or remove carbon monoxide (CO), flammable gases, or other toxic fumes from tunnels. However, the most important function of dampers is to direct ventilation air and smoke flow during a fire emergency. In this function, fans and dampers operate in conjunction to exhaust smoke and control its flow in the tunnel in support of passenger evacuations and firefighter ingress.

Damper Design. Tunnel ventilation damper design requires a thorough understanding of design criteria, installation methods, environmental surroundings, equipment life expectancy, maintenance requirements, and operating system. Damper construction varies, but the general construction is based on the following design criteria:

- Maximum fan operating pressure
- Normal and rogue tunnel air pressures
- Maximum air temperature
- Maximum air velocity
- Corrosion protection
- Maintainability and life expectancy of equipment
- Maximum damper module size
- Maximum air leakage

Fan Pressure. The maximum operating pressure that the damper will withstand during normal or emergency ventilation operations is typically the maximum pressure that the fan can generate at shutoff. This air pressure is generally 4 to 50 in. of water.

Normal and Rogue Tunnel Pressures. Some dampers in the track area of a train tunnel see much higher positive- and negative-pressure pulses than the maximum pressure generated by the fan. These high-pressure pulses are caused by the piston action of trains moving through the tunnel. A closed damper is subjected to positive pressures as trains approach, and to negative pressures as trains pass. This pressure reversal subjects damper blades and related components to reverse bending loads that must be considered to prevent premature fatigue failures. The magnitude of the pulsating pressure depends on factors such as maximum train speed, unidirectional or bidirectional traffic, tunnel length, blockage ratio, clearance between train and tunnel walls, and amount of air pushed through the dampers.

Pulsating pressure is part of normal tunnel operation. However, a rogue train condition (e.g., a train operating at high speed during an emergency, or a runaway train) could occur once or twice during the lifetime of a tunnel ventilation system. Dampers must be designed for both day-to-day fatigue and for maximum train-speed conditions.

Design specifications should require that the damper and its components meet reverse bending load criteria for from 1 to 6 million reverse bending cycles for normal, day-to-day train operations. This number equates to a train passing a damper once every 5 to 20 min for 30 to 50 years. The number of cycles can be adjusted for each application. In addition, the specifying engineer should indicate the pressure that could result from a (once or twice in a lifetime) rogue train condition.

Typically, actuators for tunnel dampers must be selected to operate against the maximum fan pressure. Because reversing pressures only occur briefly, and because normal train operations cease during an emergency, actuators are not expected to operate under either reverse pressure or rogue train conditions.

Temperature. The maximum temperature can vary for each tunnel project; some specifying engineers use the temperature limits recommended by NFPA. Typical equipment specifications state that dampers, actuators, and accessories should meet the operational requirements of the emergency ventilation fan system described in NFPA *Standard* 130: "Emergency ventilation fans, their motors, and all related components exposed to the exhaust airflow shall be designed to operate in an ambient atmosphere of 482°F for a minimum of 1 h with actual values to be determined by design analysis. In no case shall the operating temperatures be less than 300°F."

Some tunnel design engineers have specified higher air temperature criteria based on additional design considerations. A few road tunnels have been designed for the possibility of two tanker trucks carrying flammable liquids exploding from an accident in the tunnel, which would subject tunnel dampers to very high temperatures. Dampers for projects of this type, or others projects with special considerations, have been designed for maximum temperatures up to 800°F. The specifying engineer must evaluate design conditions for each project and determine what the maximum temperature could be.

Dampers, and especially damper actuators, must be specially constructed to operate reliably in high-temperature conditions for extended periods. It is important to verify that the proposed equipment can provide this required safety function. Because standard testing procedures have not been developed, a custom high-temperature test of a sample damper and actuator should be considered for inclusion in the equipment specifications.

Air Velocity. The maximum air velocity for a tunnel damper design is determined from the maximum airflow expected through the damper during any operating condition. Maximum airflow could be generated from more than one fan, depending on the system design. Actuators for tunnel dampers are typically selected to operate against the maximum airflow that dampers will be exposed to in a worst-case scenario. Thus, the maximum airflow must be specified. It is important that the engineer understands the effect of damper free area on expected airflow and pressure loss. Air velocity through a damper can vary significantly depending on damper construction and the installation configuration used.

A multiple-panel damper assembly usually has less free area than a single panel damper because of the additional blockage caused by its vertical and/or horizontal mullions. A multiple-panel damper assembly with 60 to 70% free area can have two to four times the pressure loss of a single-panel damper with 80% free area. Therefore, airflow through the multiple-panel damper assembly can be significantly lower than that through a comparable single-panel damper.

The configuration of the damper installation can also affect free area, airflow, and pressure loss. For example, a damper can either be mounted to the face of an opening, or in the opening itself. The damper mounted in the opening has a smaller free area because of the additional blockage of the damper frame, resulting in lower airflow and higher pressure loss. Damper performance also depends on where the damper is mounted (e.g., in a chamber, at one or the other end of a duct). AMCA *Standard* 500-D has more information on damper mounting configurations.

Corrosion Protection. Construction materials for tunnel projects vary considerably; their selection is usually determined based on one or more of the following reasons:

- Initial project cost
- Environmental conditions
- Life expectancy of the equipment
- Success or failure of previous materials used on similar projects
- Engineer's knowledge of and/or experience with the materials required to provide corrosion protection
- Design criteria (e.g., tunnel air pressure, temperature, velocity)

The corrosion resistance of a damper should be determined by the environment in which it will operate. A damper installation near a saltwater or heavy industrial area may need superior corrosion protection compared to one in a rural, nonindustrialized city. Underground or indoor dampers may need less corrosion protection. However, many underground dampers are also exposed to rain, snow, and sleet. These and other factors must be evaluated by the engineer before a proper specification can be written.

Tunnel dampers have been made from commercial-quality galvanized steel, hot-dipped galvanized steel, anodized aluminum, aluminum with a duranodic finish, carbon steel with various finishes, and stainless steel, including types 304, 304L, 316, 316L, and 317.

Maintainability and Life Expectancy of Equipment. These issues are of great concern when specifying dampers that may be difficult to access regularly for servicing, inspection, or maintenance. In addition, the equipment may be difficult to replace if it fails prematurely because it was marginally designed for the pressures, temperatures, corrosion resistance, etc., required for the application.

Thus, some specifying engineers purposely design dampers with a more robust construction. Dampers may be specified with heavier and/or more corrosion-resistant materials than may be required for the application, in hopes of reducing operational problems and maintenance costs and extending the life expectancy of the product. Typical methods used to design dampers of more robust construction include the following:

- Limiting blade, frame, and linkage deflections to a maximum of *L*/360
- Selecting actuators for 200 to 300% of the actual damper torque required
- Using large safety factors for stresses and deflections of high stress components
- Specifying heavier material sizes and gages than necessary
- Using more corrosion-resistant materials and finishes than required
- Using slower damper activation times (from full-close to full-open, and vice versa)

Many damper specifications include a quality assurance (QA) or system assurance program (SAP) to ensure that required performance levels are met. Others include an experience criterion that requires damper manufacturers to have five installations with five or more years of operating experience; a list of projects and contact names must be submitted so the current customer can communicate with past customers regarding the product performance. These requirements help ensure that reliable products are supplied.

Module Size. The maximum damper module size is one of the most important initial-cost factors. Many dampers can be made as a single-module assembly, or in several sections that can be field-assembled into a single-module damper. However, some damper openings are very large and it may not be practical to manufacture the damper in a one-piece frame construction because of shipping, handling, and/or installation problems.

Generally, initial cost is lower with fewer modules because they have fewer blades, frames, jackshafts, actuators, and mullion supports. However, other factors, such as job site access, lifting capabilities, and installation labor costs, must also be included in the initial-cost analysis. These factors vary for each project, so the specifying engineer must evaluate each application separately.

Air Leakage. The specifying engineer must consider air leakage through the damper when evaluating a design. Leakage is usually specified in terms of cubic feet per minute per square foot of damper face area, at a specific air pressure. As differential air pressure increases across the damper, so does air leakage. Leakage is, therefore, a function of air pressure and damper crack area, rather than of airflow. To reduce leakage, the number or size of leakage paths must be reduced. The most common method is adding damper blades and/or jamb seals, which can reduce leakage to an acceptable value.

Some specifications note the allowable damper air leakage as a percentage of the normal or maximum airflow. However, it is important to recognize that this is only an acceptable practice if the airflow and associated pressure are known.

Damper Applications and Types. Dampers allow or restrict airflow into a tunnel, and balance airflow in a tunnel. **Fan isolation dampers** can be installed in multiple-fan systems to (1) isolate any parallel, nonoperating fan from those operating, to prevent short-circuiting and airflow/pressure losses through the inoperative fan;

(2) prevent serious windmilling of an inoperative fan; and (3) provide a safe environment for maintenance and repair work on each fan. Single-fan installations may also have a fan isolation damper to prevent serious windmilling from natural or piston-effect drafts and facilitate fan maintenance.

Ventilation dampers control the amount of fresh air supplied to and exhausted from the tunnel and station areas. They may also serve as **smoke exhaust dampers** (SEDs), **bypass dampers** (BDs), **volume dampers** (VDs) and **fire dampers** (FDs), depending on their location and design. Two types of ventilation dampers are generally used: (1) trapdoor, which is installed in a vertical duct, such that the door lies horizontal when closed; and, (2) multiblade louver with parallel-operating blades. Both types can be driven by either an electric or pneumatic actuator; the fan controller operates the damper actuator. During normal operation, the damper usually closes when the fan is shut off and opens when the fan is turned on.

The trapdoor damper is simple and works satisfactorily where a vertical duct enters a plenum fan room through an opening in the floor. This damper is usually constructed of steel plate, with welded angle iron reinforcements; it is hinged on one side and closed by gravity against the embedded angle frame of the opening. The opening mechanism is usually a shaft sprocket-and-chain device. The drive motor and gear drive mechanism, or actuator, must develop sufficient force to open the damper door against the maximum (static) air pressure differential that the fan can develop. This pressure can be obtained from the fan performance curves. Limit switches start and stop the gear-motor drive or actuator at the proper position.

Fan isolation and ventilation dampers in places other than vertical ducts should have multiblade louvers. These dampers usually consist of a rugged channel frame, the flanges of which are bolted to the flanges of the fan, duct, wall, or floor opening. Damper blades are assembled with shafts that turn the bearings mounted on the outside of the channel frame. This arrangement requires access outside the duct for bearing and shaft lubrication, maintenance, and linkage operation space. Multiblade dampers should have blade edge and/or end seals to meet air leakage requirements for the application.

The trapdoor damper, properly fabricated, is inherently a low-leakage design because of its weight and the overlap at its edges. Multiblade dampers can also have low air leakage, but they must be carefully constructed to ensure tightness on closing. The pressure drop across a fully opened damper and the air leakage rate across a fully closed damper should be verified by the appropriate test procedure in AMCA *Standard* 500-D. A damper that leaks excessively under pressure can cause the fan to rotate counter to its power rotation, thus making restarting dangerous and possibly damaging to the fan motor drive.

Actuators and Accessory Selections. Tunnel damper specifications typically call for dampers, actuators, and accessories to meet the operational requirements of **emergency ventilation fans**, as described by NFPA *Standard* 130. Damper actuators are normally specified to be electric or pneumatic. Actuator selection is determined by the engineer or the customer and is usually decided by available power or initial and/or long-term operating cost.

Pneumatic Actuators. Pneumatic actuators are available in many sizes and designs; rack and pinion, air cylinder, and Scotch yoke are common configurations. Each can be of either double-action (i.e., air is supplied to operate the damper in both directions) or spring-return construction. A spring-return design uses air to power it in one direction, and a spring to drive it in the opposite direction; it is selected when it is desirable to have the damper fail to a set position on loss of air supply. Many manufacturers make pneumatic actuators; several manufacturers make both double-acting and spring-return designs capable of operating at 482°F for 1 h.

Electric Actuators. Electric actuators are also available in a variety of designs and sizes. They can be powered in both directions to open and close the damper; in this case the actuator usually fails in its last position on loss of power. Electric actuators that are powered in one direction and spring-driven in the opposite direction are also available. As with pneumatic actuators, spring return is selected when it is desirable to have the damper fail to a particular position on loss of power. There are fewer manufacturers of electric actuators than pneumatic actuators, and most do not make a spring-return design, especially in larger-torque models. Also, very few electric actuators are capable of operating at 482°F for 1 h, particularly for spring-return designs.

Actuator Selection. Actuators for tunnel dampers are typically sized to operate against the maximum airflow or velocity and pressure that will occur in a worst-case scenario. The maximum air velocity corresponds to the maximum airflow expected through the damper during any of its operating conditions. In addition, the maximum airflow could come from more than one fan, depending on system design. The maximum pressure on the damper during normal or emergency ventilation is typically the maximum pressure that the fan can generate at shutoff.

Actuators are sized and selected to (1) overcome the frictional resistance of blade bearings, linkage pivots, jackshafting assemblies, etc.; and (2) compress the blade and jamb seals to meet specified air leakage requirements. Therefore, the specifying engineer must determine maximum airflow (or air velocity) and pressure conditions, and maximum air leakage criteria.

Other factors in actuator selection are reliability and maintenance requirements. Although pneumatic actuators are considered more reliable than electric ones, the larger number of components in a pneumatic system and the cumulative risk of failure of any one component make the overall reliability of both systems similar.

Safety factors in actuator selection are not always addressed in tunnel damper specifications. This omission can result in operational problems if a manufacturer selects actuators too close to the required operating torque. Tunnel dampers are expected to function for many years when properly maintained. Also, damper manufacturers determine their torque requirements based on square, plumb, and true installations. These factors, plus the fact that dirt and debris build-up can increase damper torque, suggest that a minimum safety factor of at least 50% should be specified. Greater safety factors can be specified for some applications; however, larger actuators require larger drive shafts with higher initial cost.

Supply Air Intake. Supply air intakes require careful design to ensure that air drawn into the ventilation system is of the best quality available. Factors such as recirculation of exhaust air or intake of contaminants from nearby sources should be considered. Louvers or grilles are usually installed over air intakes for aesthetic, security, or safety reasons. Bird screens are also necessary if the openings between louver blades or grilles are large enough to allow birds to enter.

Because of the large volumes of air required in some ventilation systems, it may not be possible for intake louvers to have face air velocities low enough to be weatherproof. Therefore, intake plenums, ventilation shafts, fan rooms, and fan housings often need water drains. Windblown snow can also enter the fan room or plenum, but snow accumulation usually does not prevent the ventilation system from operating satisfactorily, if additional floor drains are located near the louvers.

Sound attenuation devices may be needed in fresh air intakes or exhaust outlets to keep fan-generated noise from disturbing the outside environment. If noise reduction is required, the total system (i.e., fans, housings, plenums, ventilation building, and location and size of air intakes and exhaust outlets), should be investigated. Fan selection should be based on the total system, including pressure drop from sound attenuation devices.

Exhaust Outlets. Exhaust air from ventilation systems should be discharged above street level and away from areas with human occupancy. Contaminant concentrations in exhaust air should not be

a concern if the system is working effectively. However, odors and entrained particulate matter in exhaust make discharge into occupied areas undesirable. Exhaust stack discharge velocity, usually a minimum of 2000 fpm, should be high enough to disperse contaminants into the atmosphere.

Evasé (flared) outlets have been used to regain some static pressure and thereby reduce exhaust fan energy consumption. Unless the fan discharge velocity is over 2000 fpm, the energy savings may not offset the cost of the evasé outlets.

In a vertical or near-vertical exhaust fan discharge connection to an exhaust duct or shaft, rainwater runs down the inside of the stack into the fan. This water dissolves material deposited from vehicle exhaust on the inner surface of the stack and becomes extremely corrosive. Therefore, fan housings should be corrosion-resistant or specially coated to protect the metal.

Discharge louvers and gratings should be sized and located so that their discharge is not objectionable to pedestrians or contaminating to nearby air intakes. Airflow resistance across the louver or grating should also be minimized. Discharge air velocities through sidewalk gratings are usually limited to 500 fpm. Bird screens should be provided if the exhaust airstream is not continuous (i.e., 24 h/day, 7 days/week), and the openings between louver blades are large enough to allow birds to enter.

Corrosion resistance of the louver or grating should be determined by the corrosiveness of the exhaust air and the installation environment. Pressure drop across the louvers should be verified by the design engineer using the appropriate test procedure in AMCA *Standard* 500-L.

NATIONAL AND INTERNATIONAL SAFETY STANDARDS AND GUIDELINES

National Fire Protection Association (NFPA)

NFPA developed fire protection standards for both road tunnels and for rapid transit facilities. The standard for transit systems is known as NFPA *Standard* 130, and the standard for road tunnels, bridges, and other limited-access roads is NFPA *Standard* 502.

In addition to *Standards* 130 and 502, NFPA publishes many standards and codes that are applicable to enclosed vehicular facilities, including the following:

- Standard for Portable Fire Extinguishers, NFPA 10, 2010
- Standard for the Installation of Sprinkler Systems, NFPA 13, 2010
- Standard for the Installation of Standpipe and Hose Systems, NFPA 14, 2010
- Standard for the Installation of Stationary Pumps for Fire Protection, NFPA 20, 2010
- Standard for Water Tanks for Private Fire Protection, NFPA 22, 2008
- Flammable and Combustible Liquids Code, NFPA 30, 2008
- Code for Motor Fuel Dispensing Facilities and Repair Garages, NFPA 30A, 2008
- Standard for Spray Application using Flammable or Combustible Materials, NFPA 33, 2011
- Vehicular Gaseous Fuel Systems Code, NFPA 52, 2010
- Liquefied Natural Gas (LNG) Vehicular Fuel Systems Code, NFPA 57, 2002
- Liquefied Petroleum Gas Code, NFPA 58, 2011
- National Electrical Code®, NFPA 70, 2011
- Recommended Practice for Electrical Equipment Maintenance, NFPA 70B, 2010
- National Fire Alarm and Signaling Code®, NFPA 72®, 2010
- Standard for Fire Doors and Other Opening Protectives, NFPA 80, 2010
- Standard for Parking Structures, NFPA 88A, 2011
- Standard for Repair Garages, NFPA 88B, 1997
- Life Safety Code®, NFPA 101®, 2009

- Standard for Emergency and Standby Power Systems, NFPA 110, 2010
- Standard on Stored Electrical Energy Emergency and Standby Power Systems, NFPA 111, 2010
- Standard for Safeguarding Construction, Alteration, and Demolition Operations, NFPA 241, 2009
- Standard on Emergency Services Incident Management System, NFPA 1561, 2008
- Standard for Fire Hose Connections, NFPA 1963, 2009

World Road Association (PIARC)

PIARC, or the World Road Association (formerly the Permanent International Association of Road Congresses), has for many years published technical reports on tunnels and tunnel ventilation in conjunction with their quadrennial World Road Congresses. The PIARC Technical Committee on Road Tunnel Operation (C3.3) and its working groups published several important specific documents on tunnel ventilation and fire safety:

- Classification of Tunnels, Existing Guidelines and Experiences, Recommendations, 05.03.B, 1995
- Road Tunnels: Emissions, Environment, Ventilation, 05.02.B, 1996
- Fire and Smoke Control in Road Tunnels, 05.05.B, 1999
- Pollution by Nitrogen Dioxide in Road Tunnels, 05.09.B, 2000
- Cross Section Geometry in Uni-directional Tunnels, 05.11.B, 2002
- Cross Section Design of Bidirectional Road Tunnels, 05.12.B, 2004
- Good Practice for the Operation and Maintenance of Road Tunnels, 05.13.B, 2004
- Road Tunnels: Vehicle Emissions and Air Demand for Ventilation, 05.14.B, 2004
- Traffic Incident Management Systems Used in Road Tunnels, 05.15.B, 2004
- Systems and Equipment for Fire and Smoke Control in Road Tunnels, 05.16.B, 2007
- Integrated Approach to Road Tunnel Safety, 2007R07, 2007
- Risk Analysis for Road Tunnels, 2008R02, 2008
- Management of the Operator—Emergency Teams Interface in Road Tunnels, 2008R03, 2008
- Road Tunnels: A Guide to Optimising the Air Quality Impact upon the Environment, 2008R04, 2008
- Road Tunnels: An Assessment of Fixed Fire Fighting Systems, 2008R07, 2008
- Tools for Road Tunnel Safety Management, 2009R08, 2009

Country-Specific Standards and Guidelines

Many countries publish tunnel guidelines and standards primarily for use in their country; however, many of these documents do provide an insight into numerous unique tunnel applications. A partial list of those available is as follows:

- Design Guidelines Tunnel Ventilation, RVS 9.261 & RVS 9.262, Transportation and Road Research Association, National Roads Administration, Austria, 1997
- Regulations on Technical Standards and Conditions for Design and Construction of Tunnels on Roads, Croatia, 1991
- Design of Road Tunnels, *Standard* CSN 73 7507, Czech Republic
- Road Tunnel Equipment, *Guideline* TP 98, Czech Republic
- Inter-Ministerial *Circular* 2000-63: Safety in the Tunnels of the National Highways Network, Ministry of the Establishment, Transport and Housing, France, 2000
- Guidelines for Equipment and Operation of Road Tunnels, Road and Transportation Research Association (RABT), Federal Ministry of Traffic, Germany, 2006
- Safety of Traffic in Road Tunnels with Particular Reference to Vehicles Transporting Dangerous Materials, Italy, 1999

- National Safety Standard of Emergency Facilities in Road Tunnels, Japan Road Association, Japan, 2001
- Recommendations for the Ventilation of Road Tunnels Public Works and Water Management (RWS), the Netherlands, 2005
- *Norwegian Design Guide—Road Tunnels*, Public Roads Administration, Directorate of Public Roads, Norway, 1992
- Ventilation of Road Tunnels, Sub-Committee 61, Nordisk Vejteknisk Forbund (NVF), *Report* 6, 1993
- *Manual for the Design, Construction and Operation of Tunnels*, IOS-98, Spain, 1998
- *Tunnel 2004—General Technical Specification for New Tunnels and Upgrading of Old Tunnels*, Swedish National Road Association, Sweden, 2004
- *Ventilation for Road Tunnels*, Swiss Federal Roads Authority (FEDRO), 2004
- TSI *Technical Specification for Interoperability, Safety in Railway Tunnels*, European Railway Association, 2008
- *Design of Road Tunnels*, the Highways Agency, United Kingdom, 1999
- Road Tunnel Design Guidelines, Federal Highway Administration, FHWA-IF-05-023, United States, 2004

Building and Fire Codes

Often, building and fire codes have supplementary information and requirements applicable to a specific type of facility. For example, ventilation of a vehicle parking garage is also governed by the applicable building code. Some of the commonly used codes are as follows:

- The **International Building Code (IBC)** with its own subset of mechanical codes such as the International Plumbing Code (IPC) and the International Mechanical Code (IMC), as well as the International Existing Buildings Code, International Fire Code, and International Fuel Gas Code.
- **National building codes** were the Uniform Building Code (UBC), Building Officials Code Association (BOCA), and the Southern Building Code Conference (SBCC), each of which was applicable in different parts of the country but now have been replaced by the IBC.
- Most states have their own **state building and fire codes** with specific modifications to the IBC or other as applicable for the conditions specific to the state, such as seismic requirements.
- Many cities and municipalities have their own **local building and fire codes**. The designer should be aware of the local code governing the facility. Many cities have adopted specific NFPA standards into their codes, and some amend these standards. The facility's design is required to conform to the requirements of the amended standard, unless a specific waiver is applied for and obtained.

Ancillary areas of tunnels such as electrical and mechanical equipment rooms, which are often adjacent to the tunnel they serve, are governed by the applicable building codes. For separation requirements between these ancillary spaces and the tunnel, the more stringent of the requirements between the building code and the applicable NFPA standard applies. The authority having jurisdiction should always be consulted when there is any doubt in the application of this separation requirement.

REFERENCES

ACGIH. 1998. *Industrial ventilation: A manual of recommended practice,* 23rd ed., Appendix A. American Conference of Governmental Industrial Hygienists, Cincinnati, OH.

ACGIH. 2001. *2001 TLVs and BEIs: Threshold limit values for chemical substances and physical agents & biological exposure indices.* American Conference of Governmental and Industrial Hygienists, Cincinnati, OH.

ACGIH. 2003. ACGIH Board Ratifies 2003 TLVs and BEIs. *Press Release*, Jan. 27. American Conference of Governmental and Industrial Hygienists, Cincinnati, OH.

Aisiks, E.G. and N.H. Danziger. 1969. *Ventilation research program at Cascade Tunnel, Great Northern Railway.* American Railway Engineering Association.

AMCA. 1998. Laboratory methods of testing dampers for rating. *Standard* 500-D. Air Movement and Control Association, Arlington Heights, IL.

AMCA. 1999. Laboratory methods of testing louvers for rating. *Standard* 500-L. Air Movement and Control Association, Arlington Heights, IL.

Amtrak. 2005. *Overbuild of Amtrak right-of-way design policy.* Engineering Practice EP4006 issued by the Chief Engineer, Structures, National Railroad Passenger Corporation, Philadelphia.

ANSI/ABMA. 2000. *Load and life rating for ball bearings.* American Bearing Manufacturers Association, Washington, D.C.

AREMA. 2007. Buildings and support facilities. Chapter 6, Part 4, Section 4.7 in *Manual for railway engineering.* American Railway Engineering and Maintenance-of-Way Association, Landover, MD.

ASHRAE. 1999. Laboratory methods of testing fans for rating. *Standard* 51-1999 (AMCA *Standard* 210-99).

ASHRAE. 2004. Ventilation for acceptable indoor air quality. ANSI/ASHRAE *Standard* 62.1-2004.

Atkinson, G., S. Jagger, and K. Moodie. 2001. Fire survival of rolling stock: Current standards and experience from the Ladbrook Grove crash. International Seminar: Fire in Trains, Escape and Crash Survival, Heathrow, England.

Bendelius, A.G. 2008. Road tunnels and bridges. In *Fire protection handbook*, R.E. Cote, C.C. Grant, J.R. Hall, R.E. Solomon, and P.A. Powell, eds. National Fire Protection Association, Quincy, MA.

Buraczynski, J.J. 1997. Integrated control systems at the Cumberland Gap Tunnel. Independent Technical Conferences Limited, Second International Conference: Tunnel Control and Communication, Amsterdam, The Netherlands.

Colino, M.P. and E.B. Rosenstein. 2006. Tunnel emergency egress and the mid train fire. *ASHRAE Transactions* 112(2):251-265.

DOT. 1976. *Subway environmental design handbook (SEDH).* Urban Mass Transportation Administration, U.S. Government Printing Office, Washington, D.C.

DOT. 1996a. *Design guidelines for bus transit systems using compressed natural gas as an alternative fuel.* Federal Transit Administration, U.S. Government Printing Office, Washington, D.C.

DOT. 1996b. *Design guidelines for bus transit systems using liquefied petroleum gas (LPG) as an alternative fuel.* Federal Transit Administration, U.S. Government Printing Office, Washington, D.C.

DOT. 1996c. *Design guidelines for bus transit systems using alcohol fuel (methanol and ethanol) as an alternative fuel.* Federal Transit Administration, U.S. Government Printing Office, Washington, D.C.

DOT. 1997a. Subway Environment Simulation (SES) computer program version 4: User's manual and programmer's manual. Issued as Volume II of *Subway Environmental Design Handbook.* Pub. No. FTA-MA-26-7022-97-1. US Department of Transportation, Washington, D.C. Also available from Volpe Transportation Center, Cambridge, MA.

DOT. 1997b. *Design guidelines for bus transit systems using liquefied natural gas (LNG) as an alternative fuel.* Federal Transit Administration, U.S. Government Printing Office, Washington, D.C.

DOT. 1998. *Design guidelines for bus transit systems using hydrogen as an alternative fuel.* Federal Transit Administration, U.S. Government Printing Office, Washington, D.C.

DuCharme, G.N. 1991. Ventilation for battery charging. *Heating/Piping/Air Conditioning* (February).

EPA. 1975. *Supplement to the guidelines for review of environmental impact statements.* Volume 1: Highway projects. Environmental Protection Agency, Research Triangle Park, NC.

EPA. 1984. *MOBILE3 mobile emissions factor model.* EPA 460/3-84-002. Environmental Protection Agency, Research Triangle Park, NC.

EPA. 2000. Air quality criteria for carbon monoxide. EPA/600/P-99/001F. U.S. Environmental Protection Agency, Research Triangle Park, NC.

EPA. 2002. *MOBILE6.2 mobile emissions factor model.* EPA 420-R-02-001. Environmental Protection Agency, Research Triangle Park, NC.

EPA. 2009. *Draft motor vehicle emission simulator (MOVES) 2009, software design and reference manual.* EPA-420-B-09-007. U.S. Environmental Protection Agency, Washington, D.C.

EPA. 2010. *Motor vehicle emission simulator (MOVES), user guide for MOVES2010a.* EPA-420-B-10-036. U.S. Environmental Protection Agency, Washington, D.C.

Fieldner, A.C., S.H. Katz, and S.P. Kinney. 1921. *Ventilation of vehicular tunnels.* Report of the U.S. Bureau of Mines to New York State Bridge and Tunnel Commission and New Jersey Interstate Bridge and Tunnel Commission. American Society of Heating and Ventilating Engineers (ASHVE).

Fritz, S. 1994. Exhaust emissions from two intercity passenger locomotives. *Journal of Engineering for Gas Turbines and Power* 116:774-783.

Fruin, J.J. 1987. *Pedestrian planning and design.* Elevator World, Mobile, AL.

Gilbey, M. 2006. Transient thermal comfort indices in subway. Presented at 12th International Symposium of Aerodynamics and Ventilation of Vehicle Tunnels, British Hydromechanics Research Group, Portoroz, Slovenia.

Guan, D., D. Abi-Zadeh, M. Tabarra, and H. Zhang. 2009. Transient thermal comfort model for subways. Presented at 13th International Symposium of Aerodynamics and Ventilation of Vehicle Tunnels, British Hydromechanics Research Group, New Jersey.

Gwynne, S. and E. Rosenbaum. 2008. Employing the hydraulic model in assessing emergency movement. In *SFPE handbook of fire protection engineering*, 4th ed. P.J. DiNenno, D. Drysdale, C.L. Beyler, W.D. Walton, R.L.P. Custer, J.R. Hall, and J.M. Watts, eds. National Fire Protection Association, Quincy, MA.

Haerter, A. 1963. Flow distribution and pressure change along slotted or branched ducts. *ASHVE Transactions* 69:124-137.

Hama, G.M., W.G. Frederick, and H.G. Monteith. 1974. *How to design ventilation systems for underground garages: Air engineering.* Study by the Detroit Bureau of Industrial Hygiene, Detroit (April).

ICC. 2009a. *International mechanical code.* International Code Council, Country Club Hills, IL.

ICC. 2009b. *International building code.* International Code Council, Country Club Hills, IL.

ICC. 2009c. *International fire code.* International Code Council, Country Club Hills, IL.

ICC. 2009d. *International fuel gas code.* International Code Council, Country Club Hills, IL.

ILO. 2003. *CIS chemical information database.* International Labor Organization, Occupational Safety and Health Information Centre, Geneva. Available from http://www.inchem.org/pages/about.html.

Ingason, H. 1994. Heat release rate measurements in tunnel fires. *Proceedings of the International Conference on Fires in Tunnels*, Boras, Sweden.

Joyeux, D. 1997. Natural fires in closed car parks—Car fire tests. *Report* INC-96/294d-DJ/NB, Centre Technique Industriel de la Construction Métallique, Metz, France.

Kashef, A., G.D. Lougheed, G.P. Crampton, Z. Liu, K. Yoon, G.V. Hadjisophocleous, and K.H. Almand. 2009. Findings of the international road tunnel fire detection research project. *Fire Technology* 45:221-237.

Kennedy, W.D., J.A. Gonzalez, and J.G. Sanchez. 1996. Derivation and application of the SES critical velocity equations. *ASHRAE Transactions* 102(2):40-44.

Krarti, M. and A. Ayari. 1998. Overview of existing regulations for ventilation requirements of enclosed vehicular parking facilities (RP-945). *ASHRAE Transactions* 105(2):18-26.

Levy, S.S. and N.H. Danziger. 1985. Ventilation of the Mount Macdonald Tunnel. Presented at Fifth International Symposium on Aerodynamics and Ventilation of Vehicle Tunnels, British Hydromechanics Research Group, Lille, France.

Levy, S.S. and D.P. Elpidorou. 1991. Ventilation of Mount Shaughnessy Tunnel. Presented at Seventh International Symposium on Aerodynamics and Ventilation of Vehicle Tunnels, Brighton, UK.

Liu, Z.G., A. Kashef, G.D. Lougheed, J.Z. Su, N. Bénichou, and K.H. Almand. 2006. An overview of the international road tunnel fire detection research project. Presented at 10th Fire Suppression and Detection Research Application Symposium, Orlando.

Liu, Z.G., A. Kashef, G.D. Lougheed, G.P. Crampton, Y. Ko, and G.V. Hadjisophocleous. 2009. Parameters affecting the performance of detection systems in road tunnels. Presented at 13th International Symposium on Aerodynamics and Ventilation of Vehicle Tunnels, New Brunswick, NJ.

Lu, Y. 1993. *Practical handbook of heating, ventilation, and air conditioning.* China Building Industry Press.

Mangs, J. and O. Keski-Rahkonen. 1994a. Characterisation of the fire behaviour of a burning passenger car, part I: Car fire experiments. *Fire Safety Journal* 23(1):17-35.

Mangs, J. and O. Keski-Rahkonen. 1994b. Characterization of the fire behaviour of a burning passenger car, part II: Parametrization of measured rate of heat release curves. *Fire Safety Journal* 23(1):37-49.

Meacham, B.J., N.A. Dembsey, K. Schebel, J.S. Tubbs, M.A. Johann, A. Kimball, and A. Neviackas. 2010. *Rail vehicle fire hazard guidance—Final summary report*. Worcester Polytechnic Institute/Arup, Worcester, MA, for U.S. Department of Homeland Security, Science and Technology Directorate, International Programs Division, Grant #2009-ST-108-00013).

MHD/FHWA. 1995. *Memorial Tunnel fire ventilation test program, comprehensive test report*. Massachusetts Highway Dept., Boston, and Federal Highway Administration, Washington, D.C.

MSHA. 2001a. Diesel particulate matter exposure of underground coal miners; Final Rule. 30CFR72. *Code of Federal Regulations*, U.S. Department of Labor, Mine Safety and Health Administration, Washington, D.C.

MSHA. 2001b. Diesel particulate matter exposure of underground metal and nonmetal miners; Final Rule. 30CFR57. *Code of Federal Regulations*, U.S. Department of Labor, Mine Safety and Health Administration, Washington, D.C.

Musser, A. and L. Tan. 2004. Control of diesel exhaust fumes in enclosed locomotive facilities (RP-1191). ASHRAE Research Project, *Final Report*.

NFPA. 2008. Code for motor fuel dispensing facilities and repair garages. *Standard* 30A. National Fire Protection Association, Quincy, MA.

NFPA. 2011. Standard for spray application using flammable or combustible materials. *Standard* 33. National Fire Protection Association, Quincy, MA.

NFPA. 2010. Vehicular gaseous fuel systems code. *Standard* 52. National Fire Protection Association, Quincy, MA.

NFPA. 2011. Liquefied petroleum gas code. *Standard* 58. National Fire Protection Association, Quincy, MA.

NFPA. 2011. National electrical code®. *Standard* 70. National Fire Protection Association, Quincy, MA.

NFPA. 2010. National fire alarm and signaling code. *Standard* 72. National Fire Protection Association, Quincy, MA.

NFPA. 2011. Standard for parking structures. *Standard* 88A. National Fire Protection Association, Quincy, MA.

NFPA. 1997. Standard for repair garages. *Standard* 88B. National Fire Protection Association, Quincy, MA.

NFPA. 2010. Standard for fixed guideway transit and passenger rail systems. *Standard* 130. National Fire Protection Association, Quincy, MA.

NFPA. 2011. Standard for road tunnels, bridges, and other limited access highways. *Standard* 502. National Fire Protection Association, Quincy, MA.

NIOSH. 2005. Pocket guide to chemical hazards. *Publication* 2005-149. National Institute for Occupational Safety and Health, Washington, D.C. Available from http://www.cdc.gov/niosh/npg/.

NTIS. 1980. User's guide for the TUNVEN and DUCT programs. *Publication* PB80141575. National Technical Information Service, Springfield, VA.

OSHA. 2001a. Occupational safety and health standards. 29CFR1910.1000. *Code of Federal Regulations*, U.S. Department of Labor, Occupational Safety and Health Administration, Washington, D.C.

OSHA. 2001b. *Partial list of chemicals associated with diesel exhaust*. Occupational Safety and Health Administration, U.S. Department of Labor, Washington, D.C. http://www.osha.gov/SLTC/dieselexhaust/chemical.html.

PIARC. 1995. Road tunnels. XXth World Road Congress, Montreal.

PIARC. 1999. *Fire and smoke control in road tunnels*. World Road Association (PIARC), La Défense Cedex, France.

PIARC. 2007a. *Systems and equipment for fire and smoke control in road tunnels*. World Road Association (PIARC), La Défense Cedex, France.

PIARC. 2007b. *Integrated approach to road tunnel safety*. World Road Association (PIARC), La Défense Cedex, France.

PIARC. 2008. *Management of the operator—Emergency teams interface in road tunnels*. World Road Association (PIARC), La Défense Cedex, France.

Proulx. 2008. Evacuation time. In *SFPE handbook of fire protection engineering*, 4th ed. P.J. DiNenno, D. Drysdale, C.L. Beyler, W.D. Walton, R.L.P. Custer, J.R. Hall, and J.M. Watts, eds. National Fire Protection Association, Quincy, MA.

Sanchez, J.G. 2003. Optimization of station air-conditioning systems for mass transit systems. Presented at 11th International Symposium of Aerodynamics and Ventilation of Vehicle Tunnels, British Hydromechanics Research Group, Luzern, Switzerland.

SFPE. 2008. *Handbook of fire protection engineering*, 4th ed. P.J. DiNenno, D. Drysdale, C.L. Beyler, W.D. Walton, R.L.P. Custer, J.R. Hall, and J.M. Watts, eds. National Fire Protection Association, Quincy, MA.

Singstad, O. 1929. *Ventilation of vehicular tunnels*. World Engineering Congress, Tokyo.

Stankunas, A.R., P.T. Bartlett, and K.C. Tower. 1980. Contaminant level control in parking garages. *ASHRAE Transactions* 86(2):584-605.

SwRI. 1992. *Exhaust emissions from two intercity passenger locomotives*. Report 08-4976, prepared by Steven G. Fritz for California Department of Transportation. Southwest Research Institute, San Antonio.

Tabarra, M. and D. Guan. 2009. How efficient is an under platform exhaust system? Presented at 13th International Symposium of Aerodynamics and Ventilation of Vehicle Tunnels, British Hydromechanics Research Group, New Jersey.

Tubbs, J.S. and B.J. Meacham. 2007. *Egress design solutions: A guide to evacuation and crowd management*. John Wiley & Sons, Hoboken, NJ.

Watson, A.Y., R.R. Bates, and D. Kennedy. 1988. *Air pollution, the automobile, and public health*. Sponsored by the Health Effects Institute. National Academy Press, Washington, D.C.

Zalosh, R. and P. Chantranuwat. 2003. International road fire tunnel detection research project—Phase 1. The Fire Protection Research Foundation, Quincy, MA.

BIBLIOGRAPHY

Bendelius, A.G. 1996. Tunnel ventilation. Chapter 20, *Tunnel engineering handbook*, 2nd ed., J.O. Bickel, T.R. Kuesel and E.H. King, eds. Chapman & Hall, New York.

BSI. 1999. Code of practice for fire precautions in the design and construction of railway passenger carrying trains. *British Standard* BS 6853. British Standards Institution, London.

DOE. 2002. *Alternative fuel news*. Alternative Fuels Data Center, U.S. Department of Energy, Washington, D.C.

DOT. 1995. *Summary assessment of the safety, health, environmental and system risks of alternative fuels*. Federal Transit Administration, U.S. Department of Transportation, Washington, D.C.

Klote, J.H. and J.A. Milke. 2002. *Principles of smoke management*. ASHRAE.

PIARC. 2007. *Systems and equipment for fire and smoke control in road tunnels*. World Road Association (PIARC), La Défense Cedex, France.

LABORATORIES

MODERN laboratories require regulated temperature, humidity, relative static pressure, air motion, air cleanliness, sound, and exhaust. This chapter addresses biological, chemical, animal, and physical laboratories. Within these generic categories, some laboratories have unique requirements. This chapter provides an overview of the HVAC characteristics and design criteria for laboratories, including a brief overview of architectural and utility concerns. This chapter does not cover pilot plants, which are essentially small manufacturing units.

The function of a laboratory is important in determining the appropriate HVAC system selection and design. Air-handling, hydronic, control, life safety, and heating and cooling systems must function as a unit and not as independent systems. HVAC systems must conform to applicable safety and environmental regulations.

Providing a safe environment for all personnel is a primary objective in the design of HVAC systems for laboratories. A vast amount of information is available, and HVAC engineers must study the subject thoroughly to understand all the factors that relate to proper and optimum design. This chapter serves only as an introduction to the topic of laboratory HVAC design. HVAC systems must integrate with architectural planning and design, electrical systems, structural systems, other utility systems, and the functional requirements of the laboratory. The HVAC engineer, then, is a member of a team that includes other facility designers, users, industrial hygienists, safety officers, operators, and maintenance staff. Decisions or recommendations by the HVAC engineer may significantly affect construction, operation, and maintenance costs.

Laboratories frequently use 100% outdoor air, which broadens the range of conditions to which the systems must respond. They seldom operate at maximum design conditions, so the HVAC engineer must pay particular attention to partial load operations that are continually changing due to variations in internal space loads, exhaust requirements, external conditions, and day-night variances. Most laboratories will be modified at some time. Consequently, the HVAC engineer must consider to what extent laboratory systems should be adaptable for other needs. Both economics and integration of the systems with the rest of the facility must be considered.

LABORATORY TYPES

Laboratories can be divided into the following general types:

_____
The preparation of this chapter is assigned to TC 9.10, Laboratory Systems.

- **Biological laboratories** are those that contain biologically active materials or involve the chemical manipulation of these materials. This includes laboratories that support such disciplines as biochemistry, microbiology, cell biology, biotechnology, genomics, immunology, botany, pharmacology, and toxicology. Both chemical fume hoods and biological safety cabinets are commonly installed in biological laboratories.

- **Chemical laboratories** support both organic and inorganic synthesis and analytical functions. They may also include laboratories in the material and electronic sciences. Chemical laboratories commonly contain a number of fume hoods.

- **Animal laboratories** are areas for manipulation, surgical modification, and pharmacological observation of laboratory animals. They also include animal holding rooms, which are similar to laboratories in many of the performance requirements but have an additional subset of requirements.

- **Physical laboratories** are spaces associated with physics; they commonly incorporate lasers, optics, nuclear material, high- and low-temperature material, electronics, and analytical instruments.

Laboratory Resource Materials

The following are general or specific resource materials applicable to various types of laboratories.

- ACGIH. *Industrial Ventilation: A Manual of Recommended Practice.* American Conference of Governmental Industrial Hygienists, Cincinnati, OH.

- AIA. Guidelines for Design and Construction of Hospital and Health Care Facilities. American Institute of Architects, Washington, D.C.

- AIHA. Laboratory Ventilation. ANSI/AIHA *Standard* Z9.5. American Industrial Hygiene Association, Fairfax, VA.

- CAP. *Medical Laboratory Planning and Design.* College of American Pathologists, Northfield, IL.

- DHHS. *Biosafety in Microbiological and Biomedical Laboratories.* U.S. Department of Health and Human Services (CDC).

- EEOC. *Americans with Disabilities Act Handbook.* Equal Employment Opportunity Commission.

- NFPA. *Fire Protection Guide for Hazardous Materials.* National Fire Protection Association, Quincy, MA.

- NFPA. Health Care Facilities. ANSI/NFPA *Standard* 99. National Fire Protection Association, Quincy, MA.

- NFPA. Fire Protection for Laboratories Using Chemicals. ANSI/NFPA *Standard* 45. National Fire Protection Association, Quincy, MA.
- NRC. Biosafety in the Laboratory: Prudent Practices for Handling and Disposal of Infectious Materials. National Research Council, National Academy Press, Washington, D.C.
- NRC. *Prudent Practices in the Laboratory: Handling and Disposal of Chemicals*. National Research Council, National Academy Press, Washington, D.C.
- NSF. Class II Biosafety Cabinetry. NSF/ANSI *Standard* 49.
- OSHA. *Occupational Exposure to Chemicals in Laboratories*. Appendix VII, 29 CFR 1910.1450. Available from U.S. Government Printing Office, Washington, D.C.
- SEFA. *Laboratory Fume Hoods Recommended Practices*. Scientific Equipment and Furniture Association, Hilton Head, SC.

Other regulations and guidelines may apply to laboratory design. All applicable institutional, local, state, and federal requirements should be identified before design begins.

HAZARD ASSESSMENT

Laboratory operations potentially involve some hazard; nearly all laboratories contain some type of hazardous materials. Before the laboratory is designed, the owner's designated safety officers should perform a comprehensive hazard assessment. These safety officers include, but are not limited to, the chemical hygiene officer, radiation safety officer, biological safety officer, and fire and loss prevention official. The hazard assessment should be incorporated into the chemical hygiene plan, radiation safety plan, and biological safety protocols.

Hazard study methods such as hazard and operability analysis (HAZOP) can be used to evaluate design concepts and certify that the HVAC design conforms to the applicable safety plans. The nature and quantity of the contaminant, types of operations, and degree of hazard dictate the types of containment and local exhaust devices. For functional convenience, operations posing less hazard potential are conducted in devices that use directional airflow for personnel protection (e.g., laboratory fume hoods and biological safety cabinets). However, these devices do not provide absolute containment. Operations having a significant hazard potential are conducted in devices that provide greater protection but are more restrictive (e.g., sealed glove boxes).

The design team should visit similar laboratories to assess successful design approaches and safe operating practices. Each laboratory is somewhat different. Its design must be evaluated using appropriate, current standards and practices rather than duplicating existing and possibly outmoded facilities.

DESIGN PARAMETERS

The following design parameters must be established for a laboratory space:

- Temperature and humidity, both indoor and outdoor
- Air quality from both process and safety perspectives, including the need for air filtration and special treatment (e.g., charcoal, HEPA, or other filtration of supply or exhaust air)
- Equipment and process heat gains, both sensible and latent
- Minimum ventilation rates
- Equipment and process exhaust quantities
- Exhaust and air intake locations
- Style of the exhaust device, capture velocities, and usage factors
- Need for standby equipment and emergency power
- Alarm requirements.
- Potential changes in the size and number of fume hoods
- Anticipated increases in internal loads
- Room pressurization requirements

- Biological containment provisions
- Decontamination provisions

It is important to (1) review design parameters with the safety officers and scientific staff, (2) determine limits that should not be exceeded, and (3) establish the desirable operating conditions. For areas requiring variable temperature or humidity, these parameters must be carefully reviewed with the users to establish a clear understanding of expected operating conditions and system performance.

Because laboratory HVAC systems often incorporate 100% outdoor air systems, the selection of design parameters has a substantial effect on capacity, first cost, and operating costs. The selection of proper and prudent design conditions is very important.

Internal Thermal Considerations

In addition to the heat gain from people and lighting, laboratories frequently have significant sensible and latent loads from equipment and processes. Often, data for equipment used in laboratories are unavailable or the equipment has been custom built. Information for some common laboratory equipment is listed in the appendix of the *ASHRAE Laboratory Design Guide* (Dorgan et al. 2002). Data on heat release from animals that may be housed in the space can be found in Table 2 of this chapter and in Alereza and Breen (1984).

Careful review of the equipment to be used, a detailed understanding of how the laboratory will be used, and prudent judgment are required to obtain good estimates of the heat gains in a laboratory. The convective portion of heat released from equipment located within exhaust devices can be discounted. Heat from equipment that is directly vented or heat from water-cooled equipment should not be considered part of the heat released to the room. Any unconditioned makeup air that is not directly captured by an exhaust device must be included in the load calculation for the room. In many cases, additional equipment will be obtained by the time a laboratory facility has been designed and constructed. The design should allow for this additional equipment.

Internal load as measured in watts per square foot is the average continuous internal thermal load discharged into the space. It is not a tabulation of the connected electrical load because it is rare for all equipment to operate simultaneously, and most devices operate with a duty cycle that keeps the average electrical draw below the nameplate information. When tabulating the internal sensible heat load in a laboratory, the duty cycle of the equipment should be obtained from the manufacturer. This information, combined with the nameplate data for the item, may provide a more accurate assessment of the average thermal load.

The HVAC system engineer should evaluate equipment nameplate ratings, applicable use and usage factors, and overall diversity. Much laboratory equipment includes computers, automation, sample changing, or robotics; this can result in high levels of use even during unoccupied periods. The HVAC engineer must evaluate internal heat loads under all anticipated laboratory-operating modes. Because of highly variable equipment heat gain, individual laboratories should have dedicated temperature controls.

Two cases encountered frequently are (1) building programs based on generic laboratory modules and (2) laboratory spaces that are to be highly flexible and adaptive. Both situations require the design team to establish heat gain on an area basis. The values for area-based heat gain vary substantially for different types of laboratories. Heat gains of 5 to 25 W/ft² or more are common for laboratories with high concentrations of equipment.

Architectural Considerations

Integrating utility systems into the architectural planning, design, and detailing is essential to providing successful research facilities. The architect and the HVAC system engineer must seek an early understanding of each other's requirements and develop integrated solutions. HVAC systems may fail to perform properly if the architectural requirements are not addressed correctly. Quality assurance of the installation is just as important as proper specifications. The following play key roles in the design of research facilities:

Modular Planning. Most laboratory programming and planning is based on developing a module that becomes the base building block for the floor plan. Laboratory planning modules are frequently 10 to 12 ft wide and 20 to 30 ft deep. The laboratory modules may be developed as single work areas or combined to form multiple-station work areas. Utility systems should be arranged to reflect the architectural planning module, with services provided for each module or pair of modules, as appropriate.

Development of Laboratory Units or Control Areas. National Fire Protection Association (NFPA) *Standard* 45 requires that laboratory units be designated. Similarly, the *International Building Code*® (ICC 2009) requires the development of control areas. Laboratory units or control areas should be developed, and the appropriate hazard levels should be determined early in the design process. The HVAC designer should review the requirements for maintaining separations between laboratories and note requirements for exhaust ductwork to serve only a single laboratory unit or control area.

Additionally, NFPA *Standard* 45 requires that no fire dampers be installed in laboratory exhaust ductwork. Building codes offer no relief from maintaining required floor-to-floor fire separations. These criteria and the proposed solutions should be reviewed early in the design process with the appropriate building code officials. The combination of the two requirements commonly necessitates the construction of dedicated fire-rated shafts from each occupied floor to the penthouse or building roof.

Provisions for Adaptability and Flexibility. Research objectives frequently require changes in laboratory operations and programs. Thus, laboratories must be flexible and adaptable, able to accommodate these changes without significant modifications to the infrastructure. For example, the utility system design can be flexible enough to supply ample cooling to support the addition of heat-producing equipment without requiring modifications to the HVAC system. Adaptable designs should allow programmatic research changes that require modifications to the laboratory's infrastructure within the limits of the individual laboratory area and/or interstitial and utility corridors. For example, an adaptable design would allow the addition of a fume hood without requiring work outside that laboratory space. Further, the HVAC designer should consider the impact of future programmatic changes on the sizing of main ductwork and central system components. The degree of flexibility and adaptability for which the laboratory HVAC system is designed should be determined from discussion with the researchers, laboratory programmer, and laboratory planner. The HVAC designer should have a clear understanding of these requirements and their financial impact.

Early Understanding of Utility Space Requirements. The amount and location of utility space are significantly more important in the design of research facilities than in that of most other buildings. The available ceiling space and the frequency of vertical distribution shafts are interdependent and can significantly affect the architectural planning. The HVAC designer must establish these parameters early, and the design must reflect these constraints. The designer should review alternative utility distribution schemes, weighing their advantages and disadvantages.

High-Quality Envelope Integrity. Laboratories that have stringent requirements for the control of temperature, humidity, relative static pressure, and background particle count generally require architectural features to allow the HVAC systems to perform properly. The building envelope may need to be designed to handle relatively high levels of humidification and slightly negative building pressure without moisture condensation in the winter or excessive infiltration. Some of the architectural features that the HVAC designer should evaluate include

- Vapor barriers—position, location, and kind
- Insulation—location, thermal resistance, and kind
- Window frames and glazing
- Caulking
- Internal partitions—their integrity in relation to air pressure, vapor barriers, and insulation value
- Finishes—vapor permeability and potential to release particles into the space
- Doors
- Air locks

Air Intakes and Exhaust Locations. Mechanical equipment rooms and their air intakes and exhaust stacks must be located to avoid intake of fumes into the building. As with other buildings, air intake locations must be chosen to minimize fumes from loading docks, cooling tower discharge, vehicular traffic, adjacent structures and processes, etc.

LABORATORY EXHAUST AND CONTAINMENT DEVICES

FUME HOODS

The Scientific Equipment and Furniture Association (SEFA 1996) defines a laboratory fume hood as a ventilated enclosed work space intended to capture, contain, and exhaust fumes, vapors, and particulate matter generated inside the enclosure. It consists basically of side, back and top enclosure panels, a floor or counter top, an access opening called the face, a sash(es), and an exhaust plenum equipped with a baffle system for airflow distribution. Figure 1 shows the basic elements of a general-purpose benchtop fume hood.

Fume hoods may be equipped with a variety of accessories, including internal lights, service outlets, sinks, air bypass openings,

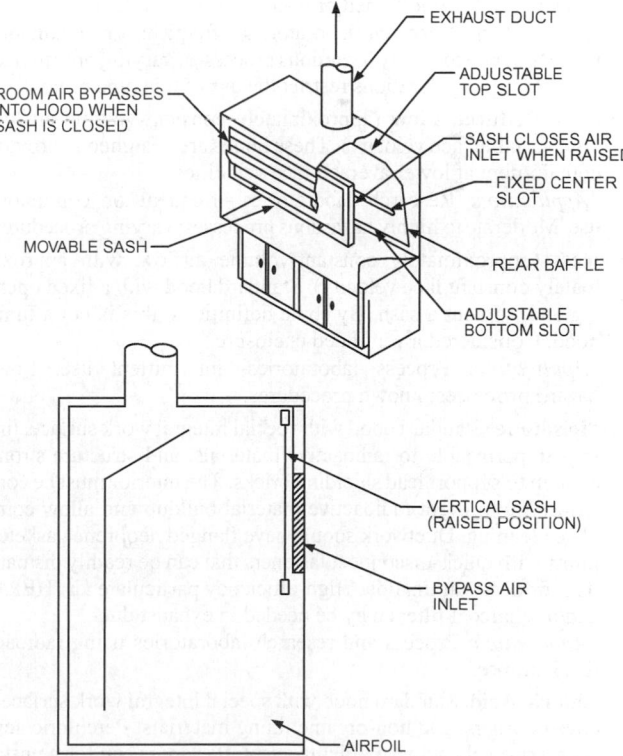

Fig. 1 Bypass Fume Hood with Vertical Sash and Bypass Air Inlet

airfoil entry devices, flow alarms, special linings, ventilated base storage units, and exhaust filters. Under counter cabinets for storage of flammable materials require special attention to ensure safe installation. NFPA *Standard* 30 does not recommend venting these cabinets; however, ventilation is often required to avoid accumulation of toxic or hazardous vapors. Ventilation of these cabinets by a separately ducted supply and exhaust that will maintain the temperature rise of the cabinet interior within the limits defined by NFPA *Standard* 30 should be considered.

Types of Fume Hoods

The following are the primary types of fume hoods and their applications:

Standard (approximately constant-volume airflow with variable face velocity). Hood that meets basic SEFA definition. Sash may be vertical, horizontal, or combination.

Application: Research laboratories—frequent or continuous use. Moderate to highly hazardous processes; varying procedures.

Bypass (approximately constant-volume airflow). Standard vertical sash hood modified with openings above and below the sash. The openings are sized to minimize the change in the face velocity, which is generally to 3 or 4 times the full-open velocity, as the sash is lowered.

Application: Research laboratories—frequent or continuous use. Moderate to highly hazardous processes; varying procedures.

Variable Volume (constant face velocity). Hood has an opening or bypass designed to provide a prescribed minimum air intake when the sash is closed and an exhaust system designed to vary airflow in accordance with sash opening. Sash may be vertical, horizontal, or a combination of both.

Application: Research laboratories—frequent or continuous use. Moderate to highly hazardous processes; varying procedures.

Auxiliary Air (approximately constant-volume airflow). A plenum above the face receives air from a secondary air supply that provides partially conditioned or unconditioned outdoor air.

Application: Research laboratories—frequent or continuous use. Moderate to highly hazardous processes; varying procedures. *Note*: Many organizations restrict the use of this type of hood.

Low or Reduced Flow (approximately constant-volume airflow with variable face velocity). These hoods are designed to provide containment at lower average face velocities.

Application: Research laboratories—frequent or continuous use. Moderate to highly hazardous processes; varying procedures.

Process (approximately constant-volume airflow with approximately constant face velocity). Standard hood with a fixed opening and without a sash. By some definitions, this is not a fume hood. Considered a ventilated enclosure.

Application: Process laboratories—intermittent use. Low-hazard processes; known procedures.

Radioisotope. Standard hood with special integral work surface, linings impermeable to radioactive materials, and structure strong enough to support lead shielding bricks. The interior must be constructed to prevent radioactive material buildup and allow complete cleaning. Ductwork should have flanged neoprene gasketed joints with quick-disconnect fasteners that can be readily dismantled for decontamination. High-efficiency particulate air (HEPA) and/or charcoal filters may be needed in exhaust duct.

Application: Process and research laboratories using radioactive isotopes.

Perchloric Acid. Standard hood with special integral work surfaces, coved corners, and non-organic lining materials. Perchloric acid is an extremely active oxidizing agent. Its vapors can form unstable deposits in the ductwork that present a potential explosion hazard. To alleviate this hazard, the exhaust system must be equipped with an internal water washdown and drainage system,

and the ductwork must be constructed of smooth, impervious, cleanable materials that are resistant to acid attack. The internal washdown system must completely flush the ductwork, exhaust fan, discharge stack, and fume hood inner surfaces. Ductwork should be kept as short as possible with minimum elbows. Perchloric acid exhaust systems with longer duct runs may need a zoned washdown system to avoid water flow rates in excess of the capacity to drain water from the hood. Because perchloric acid is an extremely active oxidizing agent, organic materials should not be used in the exhaust system in places such as joints and gaskets. Ducts should be constructed of a stainless steel material, with a chromium and nickel content not less than that of 316 stainless steel, or of a suitable nonmetallic material. Joints should be welded and ground smooth. A perchloric acid exhaust system should only be used for work involving perchloric acid.

Application: Process and research laboratories using perchloric acid. Mandatory use because of explosion hazard.

California. Special hood with sash openings on multiple sides (usually horizontal).

Application: For enclosing large and complex research apparatus that require access from two or more sides.

Floor-Mounted Hood (Walk-In). Standard hood with sash openings to the floor. Sash can be either horizontal or vertical.

Application: For enclosing large or complex research apparatus. Not designed for personnel to enter while operations are in progress.

Distillation. Standard fume hood with extra depth and 1/3- to 1/2-height benches.

Application: Research laboratory. For enclosing tall distillation apparatus.

Canopy. Open hood with an overhead capture structure.

Application: Not a true fume hood. Useful for heat or water vapor removal from some work areas. Not to be substituted for a fume hood. Not recommended when workers must bend over the source of heat or water vapor.

Fume Hood Sash Configurations

The work opening has operable glass sash(es) for observation and shielding. A sash may be vertically operable, horizontally operable, or a combination of both. A vertically operable sash can incorporate single or multiple vertical panels. A horizontally operable sash incorporates multiple panels that slide in multiple tracks, allowing the open area to be positioned across the face of the hood. The combination of a horizontally operable sash mounted within a single vertically operable sash section allows the entire hood face to be opened for setup. Then the opening area can be limited by closing the vertical panel, with only the horizontally sliding sash sections used during experimentation. Either multiple vertical sash sections or the combination sash arrangement allow the use of larger fume hoods with limited opening areas, resulting in reduced exhaust airflow requirements. Fume hoods with vertically rising sash sections should include provisions around the sash to prevent the bypass of ceiling plenum air into the fume hood.

Fume Hood Performance

Containment of hazards in a fume hood is based on the principle that a flow of air entering at the face of the fume hood, passing through the enclosure, and exiting at the exhaust port prevents the escape of airborne contaminants from the hood into the room.

The following variables affect the performance of the fume hood:

- Face velocity
- Size of face opening
- Sash position
- Shape and configuration of entrance
- Shape of any intermediate posts
- Inside dimensions and location of work area relative to face area

- Location of service fittings inside the fume hood
- Size and number of exhaust ports
- Back baffle and exhaust plenum arrangement
- Bypass arrangement, if applicable.
- Auxiliary air supply, if applicable
- Arrangement and type of replacement supply air outlets
- Air velocities near the hood
- Distance from openings to spaces outside the laboratory
- Movements of the researcher within the hood opening
- Location, size, and type of research apparatus placed in the hood
- Distance from the apparatus to the researcher's breathing zone

Air Currents. Air currents external to the fume hood can jeopardize the hood's effectiveness and expose the researcher to materials used in the hood. Detrimental air currents can be produced by

- Air supply distribution patterns in the laboratory
- Movements of the researcher
- People walking past the fume hood
- Thermal convection
- Opening of doors and windows

Caplan and Knutson (1977, 1978) conducted tests to determine the interactions between room air motion and fume hood capture velocities with respect to the spillage of contaminants into the room. Their tests indicated that the effect of room air currents is significant and of the same order of magnitude as the effect of the hood face velocity. Consequently, improper design and/or installation of the replacement air supply can lower the performance of the fume hood.

Disturbance velocities at the face of the hood should be no more than one-half and preferably one-third the face velocity of the hood. This is an especially critical factor in designs that use low face velocities. For example, a fume hood with a face velocity of 100 fpm could tolerate a maximum disturbance velocity of 50 fpm. If the design face velocity were 60 fpm, the maximum disturbance velocity would be 30 fpm.

To the extent possible, the fume hood should be located so that traffic flow past the hood is minimal. Also, the fume hood should be placed to avoid any air currents generated from the opening of windows and doors. To ensure the optimum placement of the fume hoods, the HVAC system designer must take an active role early in the design process.

Use of Auxiliary Air Fume Hoods. AIHA *Standard* Z9.5 discourages the use of auxiliary air fume hoods. These hoods incorporate an air supply at the fume hood to reduce the amount of room air exhausted. The following difficulties and installation criteria are associated with auxiliary air fume hoods:

- The auxiliary air supply must be introduced outside the fume hood to maintain appropriate velocities past the researcher.
- The flow pattern of the auxiliary air must not degrade the containment performance of the fume hood.
- The volume of auxiliary air must not be enough to degrade the fume hood's containment performance.
- Auxiliary air must be conditioned to avoid blowing cold air on the researcher; often the air must be cooled to maintain the required temperature and humidity within the hood. Auxiliary air can introduce additional heating and cooling loads in the laboratory.
- Only vertical sash should be used in the hood.
- Controls for the exhaust, auxiliary, and supply airstreams must be coordinated.
- Additional coordination of utilities during installation is required to avoid spatial conflicts caused by the additional duct system.
- Humidity control can be difficult: Unless auxiliary air is cooled to the dew point of the specified internal conditions, there is some degradation of humidity control; however, if such cooling is done, the rationale for using auxiliary air has been eliminated.

Fume Hood Performance Criteria. ASHRAE *Standard* 110 describes a quantitative method of determining the containment performance of a fume hood. The method requires the use of a tracer gas and instruments to measure the amount of tracer gas that enters the breathing zone of a mannequin; this simulates the containment capability of the fume hood as a researcher conducts operations in the hood. The following tests are commonly used to judge the performance of the fume hood: (1) face velocity test, (2) flow visualization test, (3) large-volume flow visualization, (4) tracer gas test, and (5) sash movement test. These tests should be performed under the following conditions:

- Usual amount of research equipment in the hood; the room air balance set
- Doors and windows in their normal positions
- Fume hood sash set in varying positions to simulate both static and dynamic performance

All fume hoods should be tested annually and their performance certified. The following descriptions partially summarize the test procedures. ASHRAE *Standard* 110 provides specific requirements and procedures.

Face Velocity Test

The safety officer, engineer, and the researcher should determine the desired face velocity. The velocity is a balance between safe operation of the fume hood, airflow needed for the hood operation, and energy cost. Face velocity measurements are taken on a vertical/horizontal grid, with each measurement point representing not more than 1 ft^2. The measurements should be taken with a device that is accurate in the intended operating range, and an instrument holder should be used to improve accuracy. Computerized multipoint grid measurement devices provide the greatest accuracy.

Flow Visualization

1. Swab a strip of titanium tetrachloride along both walls and the hood deck in a line parallel to the hood face and 6 in. back into the hood. *Caution*: Titanium tetrachloride forms smoke and is corrosive to the skin and extremely irritating to the eyes and respiratory system.
2. Swab an 8 in. circle on the back of the hood. Define air movement toward the face of the hood as reverse airflow and lack of movement as dead airspace.
3. Swab the work surface of the hood, being sure to swab lines around all equipment in the hood. All smoke should be carried to the back of the hood and out.
4. Test the operation of the deck airfoil bypass by running the cotton swab under the airfoil.
5. Before going to the next test, move the cotton swab around the face of the hood; if there is any outfall, the exhaust capacity test (large capacity flow visualization) should not be made.

Large-Volume Flow Visualization

Appropriate measures should be taken prior to undertaking a smoke test to avoid accidental activation of the building's smoke detection system.

1. Ignite and place a smoke generator near the center of the work surface 6 in. behind the sash. Some smoke sources generate a jet of smoke that produces an unacceptably high challenge to the hood. Care is required to ensure that the smoke generator does not disrupt the hood performance, leading to erroneous conclusions.
2. After the smoke bomb is ignited, pick it up with tongs and move it around the hood. The smoke should not be seen or smelled outside the hood.

Tracer Gas Test

1. Place the sulfur hexafluoride gas ejector in the required test locations (i.e., the center and near each side). Similarly position a mannequin with a detector in its breathing zone in the corresponding location at the hood.

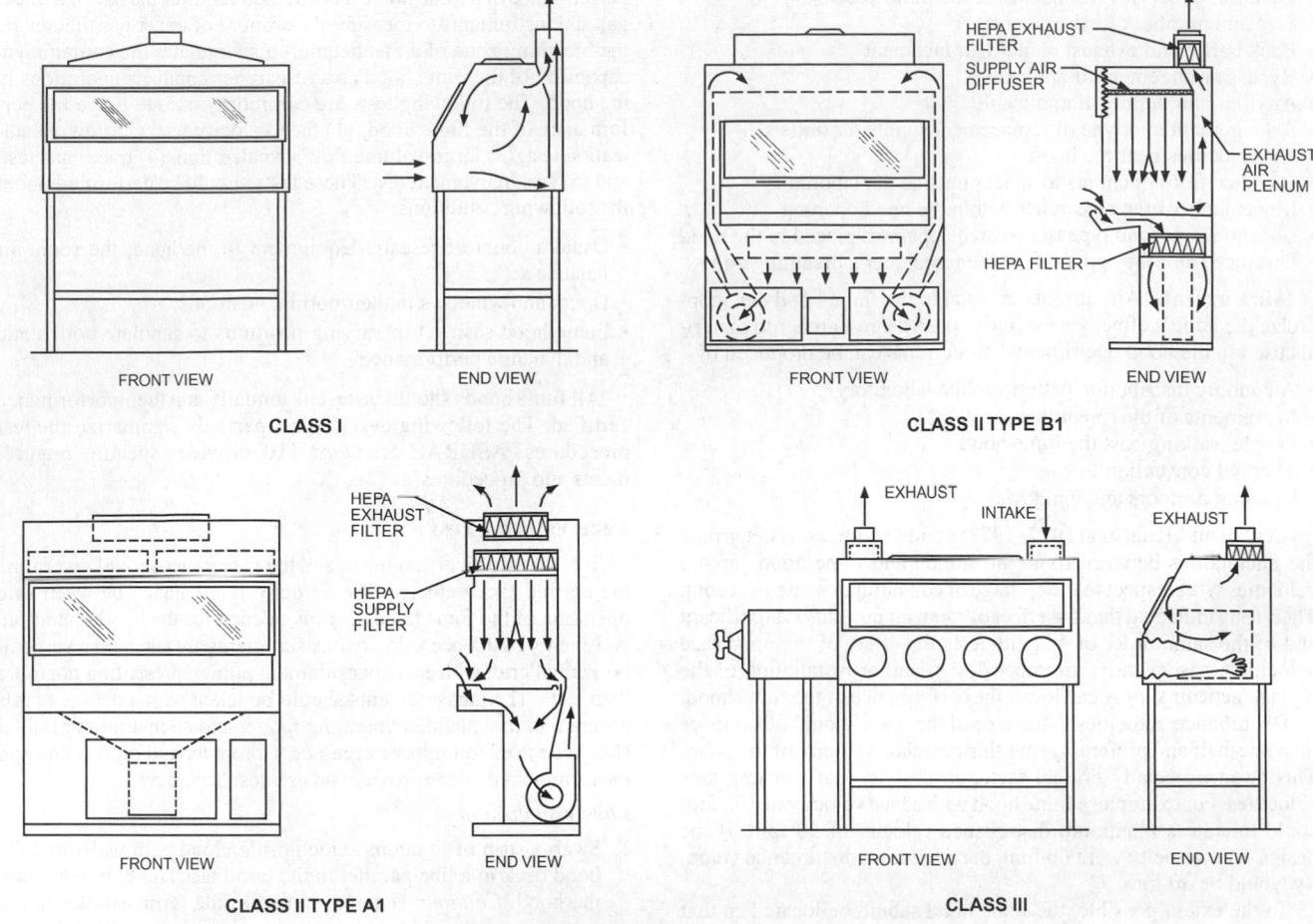

FRONT VIEW END VIEW
CLASS I

HEPA EXHAUST FILTER
SUPPLY AIR DIFFUSER
EXHAUST AIR PLENUM
HEPA FILTER
FRONT VIEW END VIEW
CLASS II TYPE B1

HEPA EXHAUST FILTER
HEPA SUPPLY FILTER
FRONT VIEW END VIEW
CLASS II TYPE A1

EXHAUST INTAKE EXHAUST
FRONT VIEW END VIEW
CLASS III

Fig. 2   Types of Biological Safety Cabinets

2. Release the tracer gas and record measurements over a 5 min time span.
3. After testing with the mannequin is complete, remove it, traverse the hood opening with the detector probe, and record the highest measurement.

Sash Movement Test

Verify containment performance of the fume during operation of the fume hood sash as described in ASHRAE *Standard* 110.

BIOLOGICAL SAFETY CABINETS

A biological safety cabinet protects the researcher and, in some configurations, the research materials as well. Biological safety cabinets are sometimes called safety cabinets, ventilated safety cabinets, laminar flow cabinets, and glove boxes. Biological safety cabinets are categorized into six groups (several are shown in Figure 2):

Class I   Similar to chemical fume hood, no research material protection, 100% exhaust through a HEPA filter

Class II

Type A1   70% recirculation within the cabinet; 30% exhaust through a HEPA filter; common plenum configuration; can be recirculated into the laboratory

Type A2   70% recirculation within the cabinet; 30% exhaust through a HEPA filter; common plenum configuration; can be recirculated to the room or exhausted to the outside

Type B1   40% recirculation within the cabinet; 60% exhaust through a HEPA filter; separate plenum configuration, must be exhausted to the outside

Type B2   100% exhaust through a HEPA filter to the outside

Class III   Special applications; 100% exhaust through a HEPA filter to the outside; researcher manipulates material within cabinet through physical barriers (gloves)

The researcher must make several key decisions before selecting a biological safety cabinet (Eagleston 1984). An important difference in biological safety cabinets is their ability to handle chemical vapors properly (Stuart et al. 1983). Of special concern to the HVAC engineer are the proper placement of the biological safety cabinet in the laboratory and the room's air distribution. Rake (1978) concluded the following:

A general rule of thumb should be that, if the cross draft or other disruptive room airflow exceeds the velocity of the air curtain at the unit's face, then problems do exist. Unfortunately, in most laboratories such disruptive room airflows are present to various extents. Drafts from open windows and doors are the most hazardous sources because they can be far in excess of 200 fpm and accompanied by substantial turbulence. Heating and air-conditioning vents perhaps pose the greatest threat to the safety cabinet because they are much less obvious and therefore seldom considered.... It is imperative then that all room airflow sources and patterns be considered before laboratory installation of a safety cabinet.

Class II biological safety cabinets should only be placed in the laboratory in compliance with NSF International *Standard* 49, Class II (Laminar Flow) Biohazard Cabinetry. Assistance in procuring, testing, and evaluating performance parameters of Class II biological safety cabinets is available from NSF as part of the standard. The cabinets should be located away from drafts, active walkways, and doors. The air distribution system should be designed to avoid air patterns that impinge on the cabinet.

The different biological safety cabinets have varying static pressure resistance requirements. Generally, Class II Type A1 cabinets have pressure drops ranging between 0.005 and 0.1 in. of water. Class II Type B1 cabinets have pressure drops in the range of 0.6 to 1.2 in. of water, and Class II Type B2 cabinets have pressure drops ranging from 1.5 to 2.3 in. of water. The manufacturer must be consulted to verify specific requirements.

Pressure requirements also vary based on filter loading and the intermittent operation of individual biological safety cabinets. Exhaust systems for biological safety cabinets must be designed with these considerations in mind. Care must be taken when manifolding biological safety cabinet exhausts to ensure that the varying pressure requirements are met.

The manufacturer of the biological safety cabinet may be able to supply the transition to the duct system. The transition should include an access port for testing and balancing and an airtight damper for decontamination. As with any containment ductwork, high-integrity duct fabrication and joining systems are necessary.

Biological safety cabinets may require periodic decontamination before service and filter replacement. During the decontamination procedure, the cabinet must be isolated or sealed from the laboratory and the exhaust system. The responsible safety officer should be consulted to determine the need for and placement of isolation dampers to facilitate decontamination operations. If provisions for decontamination are necessary, the ventilation system design should maintain laboratory airflow and pressure during the decontamination procedure.

Class I Cabinets

The Class I cabinet is a partial containment device designed for research operations with low- and moderate-risk etiologic agents. It does not provide protection for materials used in the cabinet. Room air flows through a fixed opening and prevents aerosols that may be generated in the cabinet enclosure from escaping into the room. Depending on cabinet usage, air exhausted through the cabinet may be HEPA filtered before being discharged into the exhaust system. The fixed opening through which the researcher works is usually 8 in. high. To provide adequate personnel protection, the air velocity through the fixed opening is usually at least 75 fpm.

If approved by the appropriate safety officer, it is possible to modify the Class I cabinet to contain chemical carcinogens by adding appropriate exhaust air treatment and increasing the velocity through the opening to 100 fpm. Large pieces of research equipment can be placed in the cabinet if adequate shielding is provided.

The Class I cabinet is not appropriate for containing systems that are vulnerable to airborne contamination because the air flowing into the cabinet is untreated. Also, the Class I cabinet is not recommended for use with highly infectious agents because an interruption of the inward airflow may allow aerosolized particles to escape.

Class II Cabinets

Class II cabinets provide protection to personnel, product, and the environment. The cabinets feature an open front with inward airflow and HEPA-filtered recirculated and exhaust air. Microbiological containment, product protection, and cross-contamination performance is established for certain cabinets by NSF International's *Standard* 49. Measurement techniques in NSF *Standard* 49 vary from those often used by building system testing and balancing agencies; therefore, it is important to coordinate activities of the biological safety

cabinet (BSC) certification agency and the testing and balancing agency.

The Class II Type A1 cabinet has a fixed opening with a minimum inward airflow velocity of 75 fpm. The average downward velocity is established by the manufacturer and is typically 50 to 80 fpm. The Class II Type A1 cabinet is suitable for use with agents meeting Biosafety Level 2 criteria (DHHS 1999), and, if properly certified, can meet Biosafety Level 3. However, because approximately 70% of the airflow is recirculated, the cabinet is not suitable for use with flammable, toxic, or radioactive agents.

The Class II Type A2 cabinet maintains an inward airflow velocity of 100 fpm and is similar in performance to the Class II Type A1.

The Class II Type B1 cabinet has a vertical sliding sash and maintains an inward airflow of 100 fpm at a sash opening of 8 in. The average downward velocity of the internal airflow is typically in the range of 50 to 80 fpm. The Class II Type B1 cabinet is suitable for use with agents meeting Biosafety Level 3. Approximately 70% of the internal airflow is exhausted through HEPA filters; this allows the use of biological agents treated with limited quantities of toxic chemicals and trace amounts of radionuclides, provided the work is performed in the direct exhaust area of the cabinet.

The Class II Type B2 cabinet maintains an inward airflow velocity of 100 fpm through the work opening. The cabinet is 100% exhausted through HEPA filters to the outdoors; all downward-velocity air is drawn from the laboratory or other supply source and is HEPA filtered before being introduced into the workspace. The Class II Type B2 cabinet may be used for the same level of work as the Class II Type B1, and is used when the primary consideration is protection of the material in the hood. In addition, the design permits use of small quantities of toxic chemicals and radionuclides in microbiological studies.

In Class II Type A2 cabinets, exhaust air delivered to the outlet of the cabinet by internal blowers must be handled by the laboratory exhaust system. This arrangement requires a delicate balance between the cabinet and the laboratory's exhaust system, and it may incorporate a thimble connection between the cabinet and the laboratory exhaust ductwork. Thimble (or canopy) connections incorporate an air gap between the biological safety cabinet and the exhaust duct. The purpose of the air gap is to buffer the effect of any exhaust system fluctuations on the biological safety cabinet airflow. The exhaust system must pull more air than is exhausted by the biological safety cabinet to make airflow in through the gap. The designer should confirm the amount of air to be drawn through the air gap. A minimum flow is required to provide the specified level of containment, and a maximum flow cannot be exceeded without causing an imbalance through aspiration. In the event of an exhaust system failure, the air gap allows the cabinet to maintain safe intake velocity by exhausting HEPA-filtered air through the air gap.

Class II Type B1 and Type B2 cabinets rely on the building exhaust system to pull the air from the cabinet's workspace and through the exhaust HEPA filters. The pressure resistance that must be overcome by the building exhaust system can be obtained from the cabinet manufacturer. In a fire or smoke condition, as for a fume hood, exhaust flow should continue through the cabinet; therefore, fire and smoke dampers should not be installed in the exhaust ductwork. The cabinet should be provided with a gastight damper to isolate it from the downstream ductwork to allow for decontamination. Because containment in this type of cabinet depends on the building's exhaust system, the exhaust fan(s) should have redundant back-ups and the proper controls to maintain required flow rates.

Class III Cabinets

The Class III cabinet is a gastight, negative pressure containment system that physically separates the agent from the worker. These cabinets provide the highest degree of personnel protection. Work is performed through arm-length rubber gloves attached to a sealed front panel. Room air is drawn into the cabinet through HEPA filters.

The American Glovebox Society (AGS 2007) indicates that Class III cabinets should be maintained at 0.5 in. of water below ambient pressure. Exhaust flow rate should provide a minimum of 100 fpm inward containment velocity through a glove port opening in the event of a glove being inadvertently removed. HEPA filtration or incineration before discharge to the atmosphere removes or destroys particulate material entrained in the exhaust air. A Class III system may be designed to enclose and isolate incubators, refrigerators, freezers, centrifuges, and other research equipment. Double-door autoclaves, liquid disinfectant dunk tanks, and pass boxes are used to transfer materials into and out of the cabinet.

Class III systems can contain highly infectious materials and radioactive contaminants. Although there are operational inconveniences with these cabinets, they are the equipment of choice when a high degree of personnel protection is required. Note that explosions have occurred in Class III cabinets used for research involving volatile substances.

MISCELLANEOUS EXHAUST DEVICES

Snorkels are used in laboratories to remove heat or nontoxic particles that may be generated from benchtop research equipment. Snorkels usually have funnel-shaped inlet cones connected to 3 to 6 in. diameter flexible or semi-flexible ductwork extending from the ceiling to above the benchtop level.

Typically, **canopy hoods** are used to remove heat or moisture generated by a specific piece of research apparatus (e.g., steam sterilizer) or process. Canopy hoods cannot contain hazardous fumes adequately to protect the researcher. **Benchtop slots** are used to remove nontoxic particles or fumes that may be generated by benchtop equipment.

Often, hoods are installed over weigh stations to contain and minimize disturbances from room air currents.

LAMINAR FLOW CLEAN BENCHES

Laminar flow clean benches are available in two configurations: horizontal (crossflow) and vertical (downflow). Both configurations filter the supply air and usually discharge the air out the front opening into the room. Clean benches protect the experiment or product but do not protect the researcher; therefore, they should not be used with any potentially hazardous or allergenic substances. Clean benches are not recommended for any work involving hazardous biological, chemical, or radionuclide materials.

COMPRESSED GAS STORAGE AND VENTILATION

Gas Cylinder Closets

Most laboratory buildings require storage closets for cylinders of compressed gases, which may be inert, flammable, toxic, corrosive, or poisonous. The requirements for storage and ventilation are covered in building codes and NFPA standards and codes. Water sprinklers are usually required, but other types of fire suppression may be needed based on the gases stored. Explosion containment requires a separate structural study, and closets generally require an outside wall for venting. One design used by a large chemical manufacturer to house gases with explosion potential specifies a completely welded 0.25 in. steel inner liner for the closet, heavy-duty door latches designed to hold under the force of an internal explosion, and venting out the top of the closet.

Closet temperature should not exceed 125°F per NFPA *Standard* 55. Ventilation for cylinder storage is established in NFPA *Standard* 55 at a minimum of 1 cfm/ft^2. Ventilation rates can be calculated by determining both the amount of gas that could be released by complete failure of the cylinder outlet piping connection and the time the release would take, and then finding the dilution airflow required to reduce any hazard below the maximum allowable limit.

Ventilation air is usually exhausted from the closet; makeup air comes from the surrounding space through openings in and around the door or through a transfer duct. That makeup air must be added into the building air balance. Ventilation for a closet to contain materials with explosion potential must be carefully designed, with safety considerations taken into account. NFPA *Standard* 68 is a reference on explosion venting.

Cylinder closet exhausts should be connected through a separate duct system to a dedicated exhaust fan or to a manifold system in which constant volume can be maintained under any possible manifold condition. A standby source of emergency power should be considered for the exhaust system fan(s).

Gas Cylinder Cabinets

Compressed gases that present a physical or health hazard are often placed in premanufactured gas cylinder cabinets. Gas cylinder cabinets are available for single-, dual-, or triple-cylinder configurations and are commonly equipped with valve manifolds, fire sprinklers, exhaust connections, access openings, and operational and safety controls. The engineer must fully understand safety, material, and purity requirements associated with specific compressed gases when designing and selecting cylinder cabinets and the components that make up the compressed gas handling system.

Exhaust from the gas cylinder cabinets is provided at a high rate. Air is drawn into the gas cylinder cabinet from the surrounding space through a filtered opening, usually on the lower front of the cylinder cabinet. Depending on the specific gas in the cabinet, the exhaust system may require emission control equipment and a source of emergency power.

LABORATORY VENTILATION

The total airflow rate for a laboratory is dictated by one of the following:

- Total amount of exhaust from containment and exhaust devices
- Cooling required to offset internal heat gains
- Minimum ventilation rate requirements

Fume hood exhaust requirements (including evaluation of alternate sash configurations as described in the section on Fume Hoods) must be determined in consultation with the safety officers. The HVAC engineer must determine the expected heat gains from the research equipment after consulting with the research staff (see the section on Internal Thermal Considerations).

Minimum ventilation rates should be established that provide a safe and healthy environment under normal and expected operating conditions. The dilution ventilation provided by this airflow is no substitute for the containment performance of a laboratory fume hood or other primary containment device regardless of the room ventilation rate. The appropriate ventilation rate for clearing a room of fugitive emissions or spills varies significantly based on the amount of release, the chemical's evaporation rate and hazard level, and ventilation system effectiveness.

Fixed minimum airflow rates in the range of 6 to 12 air changes per hour (ach) when the space is occupied have been used in the past. However, recent university research (Klein et al. 2009) showed a significant increase in dilution and clearing performance by increasing the air change rate from 6 to 8 ach with diminishing returns above 12 ach. Similarly, CFD research (Schuyler 2009) showed that increasing the lab's dilution ventilation rate from 4 to 8 ach reduced the background contaminant level by greater than a factor of 10. This indicates that minimum ventilation rates at the lower end of the 6 to 12 ach range may not be appropriate for all laboratories. Minimum ventilation rates should be established on a room-by-room basis considering the hazard level of materials expected to be used in the room and the operation and procedures to be performed. As the operation, materials, and hazard level of a room change, an increase or decrease in the minimum ventilation rate should be evaluated.

Active sensing of air quality in individual laboratories (Sharp 2010) is an alternative approach for dealing with the variability of appropriate ventilation rates, particularly when energy efficiency is important or when less may be known about the hazard level. With this approach, the minimum airflow rate is varied based on sensing the laboratory's actual air quality level or "air cleanliness." Sensors used to determine air quality should be evaluated for their ability to detect chemicals being used in the space. When air contaminants are sensed in the laboratory above a given threshold, the minimum air change rate is increased proportionally to an appropriate level to purge the room. When the air is "clean" and contaminants are below the previously mentioned threshold, lower minimum airflow rates may be appropriate. Extensive studies of lab room environmental conditions (Sharp 2010) have shown that the air in labs is typically "clean" over 98% of the time.

The maximum airflow rate for the laboratory should be reviewed to ensure that appropriate supply air delivery methods are chosen such that supply airflows do not impede the performance of the exhaust devices. Laboratory ventilation systems can be arranged for either constant-volume or variable-volume airflow. The specific type should be selected with the research staff, safety officers, and maintenance personnel. Special attention should be given to unique areas such as glass washing areas, hot and cold environmental rooms and labs, fermentation rooms, and cage washing rooms. Emergency power systems to operate the laboratory ventilation equipment should be considered based on hazard assessment or other specific requirements. Care should be taken to ensure that an adequate amount of makeup air is available whenever exhaust fans are operated on emergency power. Additional selection criteria are described in the sections on Hazard Assessment and Operation and Maintenance.

Usage Factor

In many laboratories, all hoods and safety cabinets are seldom needed at the same time. A system usage factor represents the maximum number of exhaust devices with sashes open or in use simultaneously. The system usage factor depends on the

- Type and size of facility
- Total number of fume hoods
- Number of fume hoods per researcher
- Airflow diversity
- Type of fume hood controls
- Fume hood sash configuration and minimum airflow required
- Type of laboratory ventilation systems
- Number of devices that must operate continuously due to chemical storage requirements or contamination prevention
- Number of current and projected research programs

Usage factors should be applied carefully when sizing equipment. For example, teaching laboratories may have a usage factor of 100% when occupied by students.

If too low a usage factor is selected, design airflow and containment performance cannot be maintained. It is usually expensive and disruptive to add capacity to an operating laboratory's supply or exhaust system. Detailed discussions with research staff are required to ascertain maximum usage factors as well as likely future requirements.

Noise

Noise level in the laboratory should be considered at the beginning of the design so that noise criterion (NC) levels suitable for scientific work can be achieved. For example, at the NIH, sound levels of NC 40 to 45 (including fume hoods) are required in regularly occupied laboratories. The requirement is relaxed to NC 55 for instrument rooms. If noise criteria are not addressed as part of the design, NC levels can be 65 or greater, which is unacceptable to

most occupants. Sound generated by the building HVAC equipment should be evaluated to ensure that excessive levels do not escape to the outdoors. Remedial correction of excessive sound levels can be difficult and expensive. See Chapter 48 for more information.

SUPPLY AIR SYSTEMS

Supply air systems for laboratories provide the following:

- Thermal comfort for occupants
- Minimum and maximum airflow rates
- Replacement for air exhausted through fume hoods, biological safety cabinets, or other exhaust devices
- Space pressurization control
- Environmental control to meet process or experimental criteria

The design parameters must be well defined for selection, sizing, and layout of the supply air system. Installation and setup should be verified as part of the commissioning process. Design parameters are covered in the section on Design Parameters, and commissioning is covered in the section on Commissioning. Laboratories in which chemicals and compressed gases are used generally require nonrecirculating or 100% outdoor air supply systems. The selection of 100% outdoor air supply systems versus return air systems should be made as part of the hazard assessment process, which is discussed in the section on Hazard Assessment. A 100% outdoor air system must have a very wide range of heating and cooling capacity, which requires special design and control.

Supply air systems for laboratories include constant-volume, high-low volume, and variable-volume systems that incorporate either single-duct reheat or dual-duct configurations, with distribution through low-, medium-, or high-pressure ductwork.

Filtration

Filtration for the air supply depends on the requirements of the laboratory. Conventional chemistry and physics laboratories commonly use 85% dust spot efficient filters (ASHRAE *Standard* 52.1). Biological and biomedical laboratories usually require 85 to 95% dust spot efficient filtration. HEPA filters should be provided for spaces where research materials or animals are particularly susceptible to contamination from external sources. HEPA filtration of the supply air is necessary for such applications as environmental studies, studies involving specific pathogen-free research animals or nude mice, dust-sensitive work, and electronic assemblies. In many instances, biological safety cabinets or laminar flow clean benches (which are HEPA filtered) may be used rather than HEPA filtration for the entire laboratory.

Air Distribution

Air supplied to a laboratory must be distributed to keep temperature gradients and air currents to minimum. Air outlets (preferably nonaspirating diffusers) must not discharge into the face of a fume hood, a biological safety cabinet, or an exhaust device. Acceptable room air velocities are covered in the sections on Fume Hoods and Biological Safety Cabinets. Special techniques and diffusers are often needed to introduce the large air quantities required for a laboratory without creating disturbances at exhaust devices.

EXHAUST SYSTEMS

Laboratory exhaust systems remove air from containment devices and from the laboratory itself. The exhaust system must be controlled and coordinated with the supply air system to maintain correct pressurization. Additional information on the control of exhaust systems is included in the section on Control. Design parameters must be well defined for selection, sizing, and layout of the exhaust air system. Installation and setup should be verified as part of the commissioning process. See the sections on Design Parameters and Commissioning. Laboratory exhaust systems should be designed for high reliability and ease of maintenance. This can be achieved by providing multiple exhaust fans and by sectionalizing equipment so that maintenance work may be performed on an individual exhaust fan while the system is operating. Another option is to use predictive maintenance procedures to detect problems prior to failure and to allow for scheduled shutdowns for maintenance. To the extent possible, components of exhaust systems should allow maintenance without exposing maintenance personnel to the exhaust airstream. Access to filters and the need for bag-in, bag-out filter housings should be considered during the design process.

Depending on the effluent of the processes being conducted, the exhaust airstream may require filtration, scrubbing, or other emission control to remove environmentally hazardous materials. Any need for emission control devices must be determined early in the design so that adequate space can be provided and cost implications can be recognized.

Types of Exhaust Systems

Laboratory exhaust systems can be constant-volume, variable-volume, or high-low volume systems with low-, medium-, or high-pressure ductwork, depending on the static pressure of the system. Each fume hood may have its own exhaust fan, or fume hoods may be manifolded and connected to one or more common central exhaust fans. Maintenance, functional requirements, and safety must be considered when selecting an exhaust system. Part of the hazard assessment analysis is to determine the appropriateness of variable-volume systems and the need for individually ducted exhaust systems. Laboratories with a high hazard potential should be analyzed carefully before variable-volume airflow is selected, because minimum air flow requirements could affect the design criteria. Airflow monitoring and pressure-independent control may be required even with constant-volume systems. In addition, fume hoods or other devices in which extremely hazardous or radioactive materials are used should receive special review to determine whether they should be connected to a manifolded exhaust system.

All exhaust devices installed in a laboratory are seldom used simultaneously at full capacity. This allows the HVAC engineer to conserve energy and, potentially, to reduce equipment capacities by installing a variable-volume system that includes an overall system usage factor. Selection of an appropriate usage factor is discussed in the section on Usage Factor.

Manifolded Exhaust Systems. These can be classified as pressure-dependent or pressure-independent. **Pressure-dependent systems** are constant-volume only and incorporate manually adjusted balancing dampers for each exhaust device. If an additional fume hood is added to a pressure-dependent exhaust system, the entire system must be rebalanced, and the speed of the exhaust fans may need to be adjusted. Because pressure-independent systems are more flexible, pressure-dependent systems are not common in current designs.

A **pressure-independent system** can be constant-volume, variable-volume, or a mix of the two. It incorporates pressure-independent volume regulators with each device. The system offers two advantages: (1) flexibility to add exhaust devices without having to rebalance the entire system and (2) variable-volume control.

The volume regulators can incorporate either direct measurement of the exhaust airflow rate or positioning of a calibrated pressure-independent air valve. The input to the volume regulator can be (1) a manual or timed switch to index the fume hood airflow from minimum to operational airflow, (2) sash position sensors, (3) fume hood cabinet pressure sensors, or (4) velocity sensors. The section on Control covers this topic in greater detail. Running many exhaust devices into the manifold of a common exhaust system offers the following potential benefits:

- Lower ductwork cost
- Fewer pieces of equipment to operate and maintain

- Fewer roof penetrations and exhaust stacks
- Opportunity for energy recovery
- Centralized locations for exhaust discharge
- Ability to take advantage of exhaust system diversity
- Ability to provide a redundant exhaust system by adding one spare fan per manifold

Individually Ducted Exhaust Systems. These comprise a separate duct, exhaust fan, and discharge stack for each exhaust device or laboratory. The exhaust fan can be single-speed, multiple-speed, or variable-speed and can be configured for constant volume, variable volume, or a combination of the two. An individually ducted exhaust system has the following potential benefits:

- Provision for installation of special exhaust filtration or treatment systems
- Customized ductwork and exhaust fan corrosion control for specific applications
- Provision for selected emergency power backup
- Simpler initial balancing
- Failure of an individual fan may affect smaller areas of the facility

Maintaining correct flow at each exhaust fan requires (1) periodic maintenance and balancing and (2) consideration of the flow rates with the fume hood sash in different positions. One problem encountered with individually ducted exhaust systems occurs when an exhaust fan is shut down. In this case, air can be drawn in reverse flow through the exhaust ductwork into the laboratory because the laboratory is maintained at a negative pressure.

A challenge in designing independently ducted exhaust systems for multistory buildings is to provide extra vertical ductwork, extra space, and other provisions for the future installation of additional exhaust devices. In multistory buildings, dedicated fire-rated shafts may be required from each floor to the penthouse or roof level. This issue should be evaluated in conjunction with the requirements of the relevant fire code. As a result, individually ducted exhaust systems (or vertically manifolded systems) consume greater floor space than horizontally manifolded systems. However, less height between floors may be required.

Ductwork Leakage

Ductwork should have low leakage rates and should be tested to confirm that the specified leakage rates have been attained. Leaks from positive pressure exhaust ductwork can contaminate the building. The design goal should be zero leakage from any positive-pressure exhaust ductwork. Designs that minimize the amount of positive-pressure ductwork are desirable. It is recommended (and required by some codes) that positive-pressure ductwork transporting potentially hazardous materials be located outside of the building. All positive-pressure ductwork should be of the highest possible integrity. The fan discharge should connect directly to the vertical discharge stack. Careful selection and proper installation of airtight flexible connectors at the exhaust fans are essential. Some feel that flexible connectors should be used on the exhaust fan inlet only. If flexible connectors are used on the discharge side of the exhaust fan, they must be of high quality and included on a preventative maintenance schedule because a connector failure could result in the leakage of hazardous fumes into the equipment room. Another viewpoint contends that the discharge side of the exhaust fan should be hard connected to the ductwork without the use of flexible connectors. The engineer should evaluate these details carefully. The potential for vibration and noise transmission must also be considered. Machine rooms that house exhaust fans should be ventilated to minimize exposure to exhaust effluent (e.g., leakage from the shaft openings of exhaust fans).

Containment Device Leakage

Leakage of the containment devices themselves must also be considered. For example, in vertical sash fume hoods, the clearance to allow sash movement creates an opening from the top of the fume hood into the ceiling space or area above. The air introduced through this leakage path also contributes to the exhaust airstream. The amount that such leakage sources contribute to the exhaust airflow depends on the fume hood design. Edge seals can be placed around sash tracks to minimize leaks. Although the volumetric flow of air exhausted through a fume hood is based on the actual face opening, appropriate allowances for air introduced through paths other than the face opening must be included.

Materials and Construction

The selection of materials and the construction of exhaust ductwork and fans depend on the following:

- Nature of the effluents
- Ambient temperature
- Ambient relative humidity
- Effluent temperature
- Length and arrangement of duct runs
- Constant or intermittent flow
- Flame spread and smoke developed ratings
- Duct velocities and pressures

Effluents may be classified generically as organic or inorganic chemical gases, vapors, fumes, or smoke; and qualitatively as acids, alkalis (bases), solvents, or oils. Exhaust system ducts, fans, dampers, flow sensors, and coatings are subject to (1) corrosion, which destroys metal by chemical or electrochemical action; (2) dissolution, which destroys materials such as coatings and plastics; and (3) melting, which can occur in certain plastics and coatings at elevated temperatures.

Common reagents used in laboratories include acids and bases. Common organic chemicals include acetone, ether, petroleum ether, chloroform, and acetic acid. The HVAC engineer should consult with the safety officer and scientists because the specific research to be conducted determines the chemicals used and therefore the necessary duct material and construction.

The ambient temperature in the space housing the ductwork and fans affects the condensation of vapors in the exhaust system. Condensation contributes to the corrosion of metals, and the chemicals used in the laboratory may further accelerate corrosion.

Ducts are less subject to corrosion when runs are short and direct, the flow is maintained at reasonable velocities, and condensation is avoided. Horizontal ductwork may be more susceptible to corrosion if condensate accumulates in the bottom of the duct. Applications with moist airstreams (cage washers, sterilizers, etc.) may require condensate drains that are connected to chemical sewers. The design should include provisions to minimize joint or seam corrosion problems.

If flow through the ductwork is intermittent, condensate may remain for longer periods because it will not be able to reevaporate into the airstream. Moisture can also condense on the outside of ductwork exhausting cold environmental rooms.

Flame spread and smoke developed ratings, which are specified by codes or insurance underwriters, must also be considered when selecting duct materials. In determining the appropriate duct material and construction, the HVAC engineer should

- Determine the types of effluents (and possibly combinations) handled by the exhaust system
- Classify effluents as either organic or inorganic, and determine whether they occur in the gaseous, vapor, or liquid state
- Classify decontamination materials
- Determine the concentration of the reagents used and the temperature of the effluents at the hood exhaust port (this may be impossible in research laboratories)

- Estimate the highest possible dew point of the effluent
- Determine the ambient temperature of the space housing the exhaust system
- Estimate the degree to which condensation may occur
- Determine whether flow will be constant or intermittent (intermittent flow conditions may be improved by adding time delays to run the exhaust system long enough to dry the duct interior prior to shutdown)
- Determine whether insulation, watertight construction, or sloped and drained ductwork are required
- Select materials and construction most suited for the application

Considerations in selecting materials include resistance to chemical attack and corrosion, reaction to condensation, flame and smoke ratings, ease of installation, ease of repair or replacement, and maintenance costs.

Appropriate materials can be selected from standard references and by consulting with manufacturers of specific materials. Materials for chemical fume exhaust systems and their characteristics include the following:

Galvanized steel. Subject to acid and alkali attack, particularly at cut edges and under wet conditions; cannot be field welded without destroying galvanization; easily formed; low in cost.

Stainless steel. Subject to acid and chloride compound attack depending on the nickel and chromium content of the alloy. Relatively high in cost. The most common stainless steel alloys used for laboratory exhaust systems are 304 and 316. Cost increases with increasing chromium and nickel content.

Asphaltum-coated steel. Resistant to acids; subject to solvent and oil attack; high flame and smoke rating; base metal vulnerable when exposed to coating imperfections and cut edges; cannot be field welded without destroying galvanization; moderate cost.

Epoxy-coated steel. Epoxy phenolic resin coatings on mild black steel can be selected for particular characteristics and applications; they have been successfully applied for both specific and general use, but no one compound is inert or resistive to all effluents. Requires sand blasting to prepare the surface for a shop-applied coating, which should be specified as pinhole-free, and field touch-up of coating imperfections or damage caused by shipment and installation; cannot be field welded without destroying coating; cost is moderate.

Polyvinyl-coated galvanized steel. Subject to corrosion at cut edges; cannot be field welded; easily formed; moderate in cost.

Fiberglass. When additional glaze coats are used, this is particularly good for acid applications, including hydrofluoric acid. May require special fire-suppression provisions. Special attention to hanger types and spacing is needed to prevent damage.

Plastic materials. Have particular resistance to specific corrosive effluents; limitations include physical strength, flame spread and smoke developed rating, heat distortion, and high cost of fabrication. Special attention to hanger types and spacing is needed to prevent damage.

Borosilicate glass. For specialized systems with high exposure to certain chemicals such as chlorine.

FIRE SAFETY FOR VENTILATION SYSTEMS

Most local authorities have laws that incorporate NFPA *Standard* 45. Laboratories located in patient care buildings require fire standards based on NFPA *Standard* 99. NFPA *Standard* 45-2004 design criteria include the following:

Air balance. "The air pressure in the laboratory work areas shall be negative with respect to adjacent corridors and non-laboratory areas." (Para. 8.3.4)

Controls. "Controls and dampers...shall be of a type that, in the event of failure, will fail in an open position to assure a continuous draft." (Para. 8.5.8)

Diffuser locations. "The location of air supply diffusion devices shall be chosen to avoid air currents that would adversely affect performance of laboratory hoods... ." (Para. 8.3.5)

Fire dampers. "Automatic fire dampers shall not be used in laboratory hood exhaust systems." (Para. 8.10.3.1)

Fire detection. "Fire detection and alarm systems shall not be interlocked to automatically shut down laboratory hood exhaust fans. . . ." (Para. 8.10.4)

Hood alarms. "A flow monitor shall be installed on each chemical fume hood." (Para. 8.8.7.1)

Hood placement. "Chemical fume hoods shall not be located adjacent to a single means of access or high traffic areas." (Para. 8.9.2)

Recirculation. "Air exhausted from laboratory hoods or other special local exhaust systems shall not be recirculated." (Para. 8.4.1) "Air exhausted from laboratory work areas shall not pass unducted through other areas." (Para. 8.4.3)

The designer should review the entire NFPA *Standard* 45 and local building codes to determine applicable requirements. Then the designer should inform the other members of the design team of their responsibilities (such as proper fume hood placement). Incorrect placement of exhaust devices is a frequent design error and a common cause of costly redesign work.

CONTROL

Laboratory controls must regulate temperature and humidity, control and monitor laboratory safety devices that protect personnel, and control and monitor secondary safety barriers used to protect the environment outside the laboratory from laboratory operations (West 1978). Reliability, redundancy, accuracy, and monitoring are important factors in controlling the lab environment. Many laboratories require precise control of temperature, humidity, and airflows. Components of the control system must provide the necessary accuracy and corrosion resistance if they are exposed to corrosive environments. Laboratory controls should provide fail-safe operation, which should be defined jointly with the safety officer. A fault tree can be developed to evaluate the impact of the failure of any control system component and to ensure that safe conditions are maintained.

Thermal Control

Temperature in laboratories with a constant-volume air supply is generally regulated with a thermostat that controls the position of a control valve on a reheat coil in the supply air. In laboratories with a variable-volume ventilation system, room exhaust device(s) are generally regulated as well. The room exhaust device(s) are modulated to handle greater airflow in the laboratory when additional cooling is needed. The exhaust device(s) may determine the total supply air quantity for the laboratory.

Most microprocessor-based laboratory control systems are able to use proportional-integral-derivative (PID) algorithms to eliminate the error between the measured temperature and the temperature set point. Anticipatory control strategies increase accuracy in temperature regulation by recognizing the increased reheat requirements associated with changes in the ventilation flow rates and adjusting the position of reheat control valves before the thermostat measures space temperature changes (Marsh 1988).

Constant Air Volume (CAV) Versus Variable Air Volume (VAV) Room Airflow Control

In the past, the only option for airflow in a laboratory setting was fixed airflow. Many laboratories used chemical fume hoods controlled by on-off switches located at the hood that significantly affected the actual air balance and airflow rate in the laboratory. Now, true CAV or VAV control can be successfully achieved. The question is which system is most appropriate for a contemporary laboratory.

Many laboratories that were considered CAV systems in the past were not truly constant. Even when the fume hoods operated

continuously and were of the bypass type, considerable variations in airflow could occur. Variations in airflow resulted from

- Static pressure changes due to filter loading
- Wet or dry cooling coils
- Wear of fan belts that change fan speed
- Position of chemical fume hood sash or sashes
- Outside wind speed and direction
- Position of doors and windows

Current controls can achieve good conformance to the requirements of a CAV system, subject to normal deviations in control performance (i.e., the dead band characteristics of the controller and the hysteresis present in the control system). The same is true for VAV systems, although they are more complex. Systems may be either uncontrolled or controlled. An uncontrolled CAV system can be designed with no automatic controls associated with airflow other than two-speed fan motors to reduce flow during unoccupied periods. These systems are balanced by means of manual dampers and adjustable drive pulleys. They provide reasonable airflow rates relating to design values but do not provide true CAV under varying conditions, maintain constant fume hood face velocity, or maintain relative static pressures in the spaces. For laboratories that are not considered hazardous and do not have stringent safety requirements, uncontrolled CAV may be satisfactory.

For laboratories housing potentially hazardous operations (i.e., involving toxic chemicals or biological hazards), a true CAV or VAV system ensures that proper airflow and room pressure relationships are maintained at all times. A true CAV system requires volume controls on the supply and exhaust systems.

The principal advantage of using a VAV system is its ability to (1) ensure that the face velocities of chemical fume hoods are maintained within a set range and (2) reduce energy use by reducing laboratory airflow. The appropriate safety officer and the users should concur with the choice of a VAV system or a CAV system with reduced airflow during unoccupied periods. Consideration should be given to providing laboratory users with the ability to reset VAV systems to full airflow volume in the event of a chemical spill. Education of the laboratory occupants in proper use of the system is essential. The engineer should recognize that the use of variable-volume exhaust systems may result in higher concentrations of contaminants in the exhaust airstream, which may increase corrosion, which influences the selection of materials.

Room Pressure Control

In most experimental work, the laboratory apparatus or the biological vector is considered to be the primary method of containment. The facility is considered the secondary level of containment.

The laboratory envelope acts as the secondary containment barrier. It is important that the walls surrounding, and door openings into, the laboratory be of appropriate construction. Because maintaining an airtight seal is rarely practical, the air pressure in the laboratory must be maintained slightly negative with respect to adjoining areas. Exceptions are sterile facilities or clean spaces that may need to be maintained at a positive pressure with respect to adjoining spaces. Positively pressurized spaces in which hazardous materials are used should have an anteroom or vestibule to maintain overall negative pressurization. See Chapter 28 for examples of secondary containment for negative pressure control.

Proper isolation is accomplished through the air balance/pressure relationship to adjacent areas. The pressure relationship is

- Negative, for hazardous isolation of hazardous or toxic operations (dirty operations), or
- Positive, for protective isolation of precious or delicate operations (clean operations).

Common methods of room pressure control include manual balancing, direct pressure, volumetric flow tracking, and cascade control. All methods manipulate airflow into or out of the space; however, each method measures a different variable. Regardless of the method of space pressure control, the goal is to maintain an inward flow of air through small gaps in the secondary barrier. In critical applications, airlocks may be required to ensure that pressure relationships are maintained as personnel enter or leave the laboratory.

Direct Pressure Control. This method measures the pressure differential across the room envelope and adjusts the amount of supply air into the laboratory to maintain the required differential pressure. Challenges encountered include (1) maintaining the pressure differential when the laboratory door is open, (2) finding suitable sensor locations, (3) maintaining a well-sealed laboratory envelope, and (4) obtaining and maintaining accurate pressure sensing devices. The direct pressure control arrangement requires tightly constructed and compartmentalized facilities and may require a vestibule on entry/exit doors. Engineering parameters pertinent to envelope integrity and associated flow rates are difficult to predict.

Because direct pressure control works to maintain the pressure differential, the control system automatically reacts to transient disturbances. Entry/exit doors may need a switch to disable the control system when they are open. Pressure controls recognize and compensate for unquantified disturbances such as stack effects, infiltration, and influences of other systems in the building. Expensive, complex controls are not required, but the controls must be sensitive and reliable. In non-corrosive environments, controls can support a combination of exhaust applications, and they are insensitive to minimum duct velocity conditions. Successful pressure control provides the desired directional airflow but cannot guarantee a specific volumetric flow differential.

Factors that favor direct pressure control include the following:

- High pressurization level (>10 Pa) and very tight construction
- Complex set of relative pressurization requirements
- Slow disturbances only (e.g., stack effect, filter loading)
- Poor conditions for airflow measurement

Volumetric Flow Tracking Control. This method measures both the exhaust and supply airflow and controls the amount of supply air to maintain the desired pressure differential. Volumetric control requires that the air at each supply and exhaust point be controlled. It does not recognize or compensate for unquantified disturbances such as stack effects, infiltration, and influences of other systems in the building. Flow tracking is essentially independent of room door operation. Engineering parameters are easy to predict, and extremely tight construction is not required. Balancing is critical and must be addressed across the full operating range. The flow offset required should be greater than the accuracy of the flow measurement and associated control error. The error in offset airflow should be evaluated to ensure that the space remains under proper offset control.

Controls may be located in corrosive and contaminated environments; however, the controls may be subject to fouling, corrosive attack, and/or loss of calibration. Flow measurement controls are sensitive to minimum duct velocity conditions. Volumetric control may not guarantee directional airflow.

Factors that favor volumetric flow tracking include the following:

- Low pressurization level (usually 2 to 10 Pa), less tight construction
- Fast disturbances (e.g., VAV fume hoods)
- Simple set of relative pressurization levels (one or two levels)

Cascade Control. This method measures the pressure differential across the room envelope to reset the flow tracking differential set point. Cascade control includes the merits and problems of both direct pressure control and flow tracking control; however, first cost is greater and the control system is more complex to operate and maintain.

Factors that favor cascade control include fast disturbances and a complex set of relative pressurization levels.

Fume Hood Control

Criteria for fume hood control differ depending on the type of hood. The exhaust volumetric flow is kept constant for standard, auxiliary air, and air-bypass fume hoods. In variable-volume fume hoods, exhaust flow is varied to maintain a constant face velocity. The fume hood control method should be selected in consultation with the safety officer. Regardless of control decisions, fume hoods must be equipped with an airflow indicator for the hood user.

Constant-volume fume hoods can further be split into pressure-dependent or pressure-independent systems. Although simple in configuration, the pressure-dependent system is unable to adjust the damper position in response to any fluctuation in system pressure across the exhaust damper.

Variable-volume fume hood control strategies can be grouped into two categories. The first either measures the air velocity entering a small sensor in the wall of the fume hood or determines face velocity by other techniques. The measured variable is used to infer the average face velocity based on an initial calibration. This calculated face velocity is then used to modulate the exhaust flow rate to maintain the desired face velocity.

The second category of variable-volume fume hood control measures the fume hood sash opening and computes the exhaust flow requirement by multiplying the sash opening by the face velocity set point. The controller then adjusts the exhaust device (e.g., by a variable-frequency drive on the exhaust fan or a damper) to maintain the desired exhaust flow rate. The control system may measure the exhaust flow for closed-loop control, or it may not measure exhaust flow in an open-loop control by using linear calibrated flow control dampers.

STACK HEIGHTS AND AIR INTAKES

Laboratory exhaust stacks should release effluent to the atmosphere without producing undesirable high concentrations at fresh air intakes, operable doors and windows, and locations on or near the building where access is uncontrolled. Three primary factors that influence the proper disposal of effluent gases are stack/intake separation, stack height, and stack height plus momentum. Chapter 24 of the 2009 *ASHRAE Handbook—Fundamentals* covers the criteria and formulas to calculate the effects of these physical relationships. For complex buildings or buildings with unique terrain or other obstacles to the airflow around the building, either scale model wind tunnel testing or computational fluid dynamics should be considered. However, standard *k-ε* computational fluid dynamics methods as applied to airflow around buildings need further development (Murakami et al. 1996; Zhou and Stathopoulos 1996). HVAC system designers that do not have the analytical skills required to undertake a dispersion analysis should consider retaining a specialized consultant.

Stack/Intake Separation

Separation of the stack discharge and air intake locations allows the atmosphere to dilute the effluent. Separation is simple to calculate with the use of short to medium-height stacks; however, to achieve adequate atmospheric dilution of the effluent, greater separation than is physically possible may be required, and the building roof near the stack will be exposed to higher concentrations of the effluent.

Stack Height

Chapter 24 of the 2009 *ASHRAE Handbook—Fundamentals* describes a geometric method to determine the stack discharge height high enough above the turbulent zone around the building that little or no effluent gas impinges on air intakes of the emitting building. The technique is conservative and generally requires tall stacks that may be visually unacceptable or fail to meet building code or zoning requirements. Also, the technique does not ensure acceptably low concentrations of effluents at air intakes (e.g., if there are large releases of hazardous materials or elevated intake locations on nearby buildings). A minimum stack height of 10 ft is required by AIHA *Standard* Z9.5 and is recommended by Appendix A of NFPA *Standard* 45.

Stack Height plus Momentum

To increase the effective height of the exhaust stacks, both the volumetric flow and the discharge velocity can be increased to increase the discharge momentum (Momentum Flow = Density × Volumetric Flow × Velocity). The momentum of the large vertical flow in the emergent jet lifts the plume a substantial distance above the stack top, thereby reducing the physical height of the stack and making it easier to screen from view. This technique is particularly suitable when (1) many small exhaust streams can be clustered together or manifolded prior to the exhaust fan to provide the large volumetric flow and (2) outdoor air can be added through automatically controlled dampers to provide constant exhaust velocity under variable load. The drawbacks to the second arrangement are the amount of energy consumed to achieve the constant high velocity and the added complexity of the controls to maintain constant flow rates. Dilution equations presented in Chapter 24 of the 2009 *ASHRAE Handbook—Fundamentals* or mathematical plume analysis (e.g., Halitsky 1989) can be used to predict the performance of this arrangement, or performance can be validated through wind tunnel testing. Current mathematical procedures tend to have a high degree of uncertainty, and the results should be judged accordingly.

Architectural Screens

Rooftop architectural screens around exhaust stacks are known to adversely affect exhaust dispersion. In general, air intakes should not be placed within the same screen enclosure as laboratory exhausts. Petersen et al. (1997) describe a method of adjusting dilution predictions of Chapter 24 of the 2009 *ASHRAE Handbook—Fundamentals* using a stack height adjustment factor, which is essentially a function of screen porosity.

Criteria for Suitable Dilution

An example criterion based on Halitsky (1988) is that the release of 15 cfm of pure gas through any stack in a moderate wind (3 to 18 mph) from any direction with a near-neutral atmospheric stability (Pasquill Gifford Class C or D) must not produce concentrations exceeding 3 ppm at any air intake. This criterion is meant to simulate an accidental release such as would occur in a spill of an evaporating liquid or after the fracture of the neck of a small lecture bottle of gas in a fume hood.

The intent of this criterion is to limit the concentration of exhausted gases at the air intake locations to levels below the odor thresholds of gases released in fume hoods, excluding highly odorous gases such as mercaptans. Laboratories that use extremely hazardous substances should conduct a chemical-specific analysis based on published health limits. A more lenient limit may be justified for laboratories with low levels of chemical usage. Project-specific requirements must be developed in consultation with the safety officer. The equations in Chapter 24 of the 2009 *ASHRAE Handbook—Fundamentals* are presented in terms of dilution, defined as the ratio of stack exit concentration to receptor concentration. The exit concentration, and therefore the dilution required to meet the criterion, varies with the total volumetric flow rate of the exhaust stack. For the above criterion with the emission of 15 cfm of a pure gas, a small stack with a total flow rate of 1000 cfm will have an exit concentration of 15/1000 or 15,000 ppm. A dilution of 1:5000 is needed to achieve an intake concentration of 3 ppm. A

larger stack with a flow rate of 10,000 cfm will have a lower exit concentration of 15/10,000 or 1500 ppm and would need a dilution of only 1:500 to achieve the 3 ppm intake concentration.

The above criterion is preferred over a simple dilution standard because a defined release scenario (15 cfm) is related to a defined intake concentration (3 ppm) based on odor thresholds or health limits. A simple dilution requirement may not yield safe intake concentrations for a stack with a low flow rate.

Adjacent Building Effects

The influence of adjacent building effects was studied under ASHRAE research project 897 (Wilson et al. 1998). Several guidelines were developed from this project:

• Designers should locate stacks near the edge of a roof.
• With the emitting building upwind, an adjacent building will always have higher dilution on a lower step-down roof than would occur on a flat roof at the emitting building's height. Ignoring the step-down in roof level will produce conservative designs.
• If the lower adjacent building is upwind of the emitting building, it will block flow approaching the emitting building, producing lower velocities and recirculation cavities on the emitting building roof and increasing dilution by factors of 2 to 10 on the emitting building.
• Designers should use increased exhaust velocity to produce jet dilution when the plume will be trapped in the recirculation cavity from a high upwind adjacent building.
• When the adjacent building is higher than the emitting building, designers should try to avoid placing air intakes on the adjacent building at heights above the roof level of the emitting building.

Also see Chapter 45 for more information.

APPLICATIONS

LABORATORY ANIMAL FACILITIES

Laboratory animals must be housed in comfortable, clean, temperature- and humidity-controlled rooms. Animal welfare must be considered in the design; the air-conditioning system must provide the macroenvironment for the animal room and the subsequent effect on the microenvironment in the animal's primary enclosure or cage specified by the facility's veterinarian (Besch 1975; ILAR 2010; Woods 1980). Early detailed discussions with the veterinarian concerning airflow patterns, cage layout, and risk assessment help ensure a successful animal room HVAC design. The elimination of research variables (fluctuating temperature and humidity, drafts, and spread of airborne diseases) is another reason for a high-quality air-conditioning system. See Chapter 24 for additional information on environments for laboratory animals.

Primary Uses of Animal Housing Facilities

Primary uses of animal facilities include the following:

• **Acute (short-term) studies:** generally less than 90 days in length, although the animal species and particular experiments involved could affect duration. Most frequently found in pharmaceutical, medical, or other life science laboratories, and includes
  • Assays and screens
  • Immune-suppressed animals
  • Pharmacology and metabolism
  • Infectious disease
• **Chronic (long-term) studies:** generally more than 90 days in length, although the species and experiment involved could affect the length. Includes
  • Toxicology
  • Teratology
  • Neurological
  • Quality control

- **Long-term holding of animals,** including
  - Production of materials used primarily in pharmaceuticals
  - Breeding
  - Laboratory animals
  - Companion animals
  - Food and fiber animals
- **Agricultural studies,** including food and fiber animals

Regulatory Environment

There are a number of regulations and guidelines that pertain to the housing of laboratory animals. Additional regulations cover the housing of animals that may be used some way in the production of pharmaceuticals, testing for agricultural products or used for quality control. Pertinent regulations are outlined below and are applied in the United States. Other countries have similar regulations that should be consulted when designing animal facilities located in that country. Regulations and guidelines include the following:

- Code of Federal Regulations (CFR) 21
  - Part 58; Good Laboratory Practices for Non-Clinical Laboratory Studies
  - Part 210; current Good Manufacturing Practice in Manufacture, Processing, Packing or Holding of Human and Veterinary Drugs
- Guide for the Care and Use of Laboratory Animals, National Research Council
- Biosafety in Microbiological and Biomedical Laboratories, Center for Disease Control (CDC).
- The Animal Welfare Act of 1966 and as subsequently amended. Regulatory authority is vested in the Secretary of the U.S. Department of Agriculture (USDA) and implemented by the USDA's Animal and Plant Health Inspection Service.
- American Association for Accreditation of Laboratory Animal Care (AAALAC), a nonprofit organization to which many institutions and corporations belong. This group provides accreditation based upon inspections and reports from member groups. Many organizations that build or maintain animal facilities adhere to AAALAC programs and HVAC engineers are expected to design to their guidelines.

Local ordinances or user organization requirements may also apply. HVAC engineers should confirm which regulations are applicable for any project.

Temperature and Humidity

Due to the nature of research programs, air-conditioning design temperature and humidity control points may be required. Research animal facilities require more precise environmental control than farm animal or production facilities because variations affect the experimental results. A totally flexible system permits control of the temperature of individual rooms to within ±2°F for any set point in a range of 64 to 85°F. This flexibility requires significant capital expenditure, which can be mitigated by designing the facility for selected species and their specific requirements.

Table 1 lists dry-bulb temperatures recommended by ILAR (2010) for several common species. In the case of animals in confined spaces, the range of daily temperature fluctuations should be kept to a minimum. Relative humidity should also be controlled. ASHRAE *Standard* 62.1 recommends that the relative humidity in habitable spaces be maintained between 30 and 60% to minimize growth of pathogenic organisms. ILAR (2010) suggests the acceptable range of relative humidity is 30 to 70%.

Ventilation

A guideline of 10 to 15 outdoor air changes per hour (ach) has been used for secondary enclosures (animal holding rooms) for many years. Although it is effective in many settings, the guideline

Table 1 Recommended Dry-Bulb Temperatures for Common Laboratory Animals

Animal	Temperature, °F
Mouse, rat, hamster, gerbil, guinea pig	64 to 79
Rabbit	61 to 72
Cat, dog, nonhuman primate	64 to 84
Farm animals and poultry	61 to 81

Source: ILAR (2010). Reprinted with permission.
Note: These ranges permit scientific personnel who will use the facility to select optimum conditions (set points). The ranges do not represent acceptable fluctuation ranges.

does not consider the range of possible heat loads; the species, size, and number of animals involved; the type of bedding or frequency of cage changing; the room dimensions; or the efficiency of air distribution from the secondary to the primary enclosure. In some situations, such a flow rate might overventilate a secondary enclosure that contains few animals and waste energy or underventilate a secondary enclosure that contains many animals and allow heat and odor to accumulate. As such, lower ventilation rates might be appropriate in the secondary enclosure or room, provided that they do not result in harmful or unacceptable concentrations of toxic gases, odors or particles. Active sensing of contaminants in the secondary enclosure and varying the air change rates based on the room environmental conditions is one approach that can be considered to meet these requirements in a more energy efficient manner.

For small-animal caging systems, recent studies suggest that room conditions have very little influence on the cage environments. ASHRAE research project RP-730 (Riskowski et al. 1995, 1996) found the following:

- No relationship between room ventilation rate and cage microenvironments for shoebox and microisolator cages exists. In fact, 5 ach provided the same cage ventilation rates for shoebox cages as did 10 and 15 ach.
- Diffuser type (perforated square versus radial) had only a small effect on shoebox cage ventilation rates. The radial diffuser provided higher wire cage ventilation rates.
- One high return provided the same cage ventilation rates as four high returns or as one low return.
- Room size had no effect on cage ventilation rates.

This research is further discussed in Chapter 24.

In certain types of animal rooms, usually those used for long-term studies involving high-value work or animals, the outdoor air change rate is maintained at the 10 to 15 per hour but the total airflow in the rooms ranges from 90 to 150 ach (mass flow spaces similar to clean rooms). The air supply is generally terminal-HEPA-filtered to reduce the potential for disease. These rooms are energy-intensive, and may not be required with the filter capability and caging systems available today.

The air-conditioning load and flow rate for an animal room should be determined by the following factors:

- Desired animal microenvironment (Besch 1975, 1980; ILAR 2010)
- Species of animal(s)
- Animal population
- Recommended ambient temperature (Table 1)
- Heat produced by motors on special animal housing units (e.g., laminar flow racks or HEPA-filtered air supply units for ventilated racks)
- Heat generated by the animals (Table 2)

Additional design factors include method of animal cage ventilation; operational use of a fume hood or a biological safety cabinet during procedures such as animal cage cleaning and animal examination; airborne contaminants (generated by animals, bedding, cage cleaning, and room cleaning); and institutional animal care

Table 2 Heat Generated by Laboratory Animals

Species	Weight, lb	Heat Generation, Btu/h per Normally Active Animal		
		Sensible	Latent	Total
Mouse	0.046	1.11	0.54	1.65
Hamster	0.260	4.02	1.98	6.00
Rat	0.62	7.77	3.83	11.6
Guinea pig	0.90	10.2	5.03	15.2
Rabbit	5.41	39.2	19.3	58.5
Cat	6.61	45.6	22.5	68.1
Nonhuman primate	12.0	71.3	35.1	106.0
Dog	22.7	105.0	56.4	161.0
Dog	50.0	231.0	124.0	355.0

standards (Besch 1980; ILAR 2010). It should be noted that the ambient conditions of the animal room might not reflect the actual conditions within a specific animal cage.

Animal Heat Production

Air-conditioning systems must remove the sensible and latent heat produced by laboratory animals. The literature concerning the metabolic heat production appears to be divergent, but new data are consistent. Current recommended values are given in Table 2. These values are based on experimental results and the following equation:

$$ATHG = 2.5M$$

where

ATHG = average total heat gain, Btu/h per animal
  M = metabolic rate of animal, Btu/h per animal = $6.6W^{0.75}$
  W = weight of animal, lb

Conditions in animal rooms must be maintained constant. This may require year-round availability of refrigeration and, in some cases, dual/standby chillers and emergency electrical power for motors and control instrumentation. Storage of critical spare parts is one alternative to installing a standby refrigeration system.

Design Considerations

If the entire animal facility or extensive portions of it are permanently planned for species with similar requirements, the range of individual adjustments may be reduced. Each animal room or group of rooms serving a common purpose should have separate temperature and humidity controls. The animal facility and human occupancy areas should be conditioned separately. The human areas may use a return air HVAC system and may be shut down on weekends for energy conservation. Separation prevents exposure of personnel to biological agents, allergens, and odors from animal rooms.

Control of air pressure in animal housing and service areas is important to ensure directional airflow. For example, quarantine, isolation, soiled equipment, and biohazard areas should be kept under negative pressure, whereas clean equipment and pathogen-free animal housing areas and research animal laboratories should be kept under positive pressure (ILAR 2010).

Supply air outlets should not cause drafts on research animals. Efficient air distribution for animal rooms is essential; this may be accomplished effectively by supplying air through ceiling outlets and exhausting air at floor level (Hessler and Moreland 1984). Supply and exhaust systems should be sized to minimize noise.

A study by Neil and Larsen (1982) showed that predesign evaluation of a full-size mock-up of the animal room and its HVAC system was a cost-effective way to select a system that distributes air to all areas of the animal-holding room. Wier (1983) describes many typical design problems and their resolutions. Room air distribution should be evaluated using ASHRAE *Standard* 113 procedures to evaluate drafts and temperature gradients.

HVAC ductwork and utility penetrations must present a minimum number of cracks in animal rooms so that all wall and ceiling surfaces can be easily cleaned. Exposed ductwork is not generally recommended; however, if constructed of 316 stainless steel in a fashion to facilitate removal for cleaning, it can provide a cost-effective alternative. Joints around diffusers, grilles, and the like should be sealed. Exhaust air grilles with 1 in. washable or disposable filters are normally used to prevent animal hair and dander from entering the ductwork. Noise from the HVAC system and sound transmission from nearby spaces should be evaluated. Sound control methods such as separate air-handling systems or sound traps should be used as required.

Multiple-cubicle animal rooms enhance the operational flexibility of the animal room (i.e., housing multiple species in the same room, quarantine, and isolation). Each cubicle should be treated as if it were a separate animal room, with air exchange/balance, temperature, and humidity control.

Caging Systems

Animal facilities use a number of different caging systems that can significantly affect the environment within the cage or the total heat load in the room. The purpose of the caging systems is to

- Protect the health and wellbeing of the animals
- Protect support staff from antigens released or shed by the animals
- Minimize exposure of animals to pheromones released by other animals in the space

To provide the appropriate design, the HVAC engineer must be aware of the type of caging system to be used. Some common caging systems include the following:

- Cage boxes made of sheet metal, plastic, or wire mesh, with the space inside the cage open to the room so the room's macroenvironment is essentially identical to the cage's microenvironment.
- Cage boxes made primarily of plastic, with the top shielded from the room by a filter material to provide some level of isolation from the room. The filter is usually not sealed to the cage, so some open space between the room and the interior of the cage remains. Exchange of air, vapors, particulates, and gases between the room and the cage interior does occur, but the rate of exchange is reduced by the filter. The microenvironment of the interior of the cage is usually different from that of the room.
- Plastic and wire cages that are part of a cage rack assembly, which provides varying degrees of isolation from the room. These usually provide filtered (generally HEPA-filtered) air directly to each individual or shelf of cage boxes. In some cases, both a fan-powered supply and an exhaust unit are used. In other cases, cage units are connected to the facility exhaust system to provide airflow. Facilities with this kind of caging system must be designed to accommodate the heat gain in the space if the exhaust is released in the room. Some heat gain may be excluded if the caging assembly is connected directly to the facility exhaust system. When the facility is used to provide the exhaust by direct connection to the caging assembly, the design must include provisions to control the airflow to ensure that the overall proper airflow and relative static pressure of the room and each cage rack assembly is maintained, especially when caging and rack connections may be changed over time. The temperature and specific humidity within each cage will be higher than the ambient conditions of the room.

ANCILLARY SPACES FOR ANIMAL LABORATORIES

In addition to animal holding rooms, a facility intended to provide for an animal colony generally requires other areas, such as

- **Cage washer:** Usually provided with some temperature control to minimize heat stress for occupants. In addition, specific exhaust hoods and separate exhaust ductwork should be considered for the space and equipment.

- **Feed storage:** Usually provided with temperature and humidity control to protect quality and shelf life of feed.
- **Diagnostic laboratory:** Usually provided with laboratory-quality air conditioning.
- **Treatment laboratory:** Usually provided with laboratory-quality air conditioning.
- **Quarantine spaces:** To separate incoming animals from the remainder of the colony until their health can be evaluated. These rooms are frequently located near the receiving location. Animal-room-quality air conditioning is provided.
- **Surgery suite:** Sterile-quality air conditioning is provided. The suites frequently have provisions to exhaust anesthetic gases.
- **Necropsy laboratory:** Usually provided with laboratory-quality air conditioning and frequently fitted with special exhaust tables or other means of protecting laboratory workers from exposure to chemical preservatives or biological contamination. For high-risk or high-hazard work, Type III biological safety cabinets may be provided.
- **Waste-holding room:** Usually only provided with heating and ventilation, but maintained at negative pressure relative to adjacent areas. When used to store carcasses, a refrigerated storage unit of appropriate size should be provided.

CONTAINMENT LABORATORIES

With the initiation of biomedical research involving recombinant DNA technology, federal guidelines on laboratory safety were published that influence design teams, researchers, and others. Containment describes safe methods for managing hazardous chemicals and infectious agents in laboratories. The three elements of containment are laboratory operational practices and procedures, safety equipment, and facility design. Thus, the HVAC design engineer helps decide two of the three containment elements during the design phase.

In the United States, the U.S. Department of Health and Human Services (DHHS), Centers for Disease Control and Prevention (CDC), and National Institutes of Health (NIH) classify biological laboratories into four levels—Biosafety Levels 1 through 4—listed in DHHS (1999). The USDA Agricultural Research Service (ARS) *Manual* 242.1 (ARS 2002) similarly classifies biological laboratories, and also identifies a BSL 3Ag containment level.

Biosafety Level 1

Biosafety Level 1 is suitable for work involving well-characterized agents not known to consistently cause disease in healthy adult humans, and of minimal potential hazard to laboratory personnel and the environment. The laboratory is not necessarily separated from the general traffic patterns in the building. Work is generally conducted on open benchtops using standard microbiological practices. Special containment equipment is neither required nor generally used. The laboratory can be cleaned easily and contains a sink for washing hands. Federal guidelines for these laboratories contain no specific HVAC requirements.

Biosafety Level 2

Biosafety Level 2 is suitable for work involving agents of moderate potential hazard to personnel and the environment. Laboratory access is limited when certain work is in progress. The laboratory can be cleaned easily and contains a sink for washing hands. Biological safety cabinets (Class I or IIA2) are used whenever

- Procedures with a high potential for creating infectious aerosols are conducted. These include centrifuging, grinding, blending, vigorous shaking or mixing, sonic disruption, opening containers of infectious materials, inoculating animals intranasally, and harvesting infected tissues or fluids from animals or eggs.

- High concentrations or large volumes of infectious agents are used. Federal guidelines for these laboratories contain minimum facility standards.

At this level of biohazard, most research institutions have a full-time safety officer (or safety committee) who establishes facility standards. The federal guidelines for Biosafety Level 2 contain no specific HVAC requirements; however, typical HVAC design criteria can include the following:

- 100% outdoor air systems
- 6 to 15 air changes per hour
- Directional airflow into the laboratory rooms
- Site-specified hood face velocity at fume hoods (many institutions specify 80 to 100 fpm)
- An assessment of research equipment heat load in a room.
- Inclusion of biological safety cabinets

Most biomedical research laboratories are designed for Biosafety Level 2. However, the laboratory director must evaluate the risks and determine the correct containment level before design begins.

Biosafety Level 3

Biosafety Level 3 applies to facilities in which work is done with indigenous or exotic agents that may cause serious or potentially lethal disease as a result of exposure by inhalation. The Biosafety Level 3 laboratory uses a physical barrier of two sets of self-closing doors to separate the laboratory work area from areas with unrestricted personnel access. This barrier enhances biological containment within the laboratory work area.

The ventilation system must be single-pass, nonrecirculating and configured to maintain the laboratory at a negative pressure relative to surrounding areas. Audible alarms and visual monitoring devices are recommended to notify personnel if the laboratory pressure relationship changes from a negative to a positive condition. The user may wish to have alarms reported to a remote constantly monitored location. Gastight dampers are required in the supply and exhaust ductwork to allow decontamination of the laboratory. The ductwork between these dampers and the laboratory must also be gastight. All penetrations of the Biosafety Level 3 laboratory envelope must be sealable for containment and to facilitate gaseous decontamination of the work area.

All procedures involving the manipulation of infectious materials are conducted inside biological safety cabinets. The engineer must ensure that the connection of the cabinets to the exhaust system does not adversely affect the performance of the biological safety cabinets or the exhaust system. Refer to the section on Biological Safety Cabinets for further discussion.

The exhaust air from biological safety cabinets and/or the laboratory work area may require HEPA filtration. The need for filtration should be reviewed with the appropriate safety officers. If required, HEPA filters should be equipped with provisions for bag-in, bag-out filter handling systems and gastight isolation dampers for biological decontamination of the filters.

The engineer should review with the safety officer the need for special exhaust or filtration of exhaust from any scientific equipment located in the Biosafety Level 3 laboratory.

Biosafety Level 4

Biosafety Level 4 is required for work with dangerous and exotic agents that pose a high risk of aerosol-transmitted laboratory infections and life-threatening disease. HVAC systems for these areas will have stringent design requirements that must be determined by the biological safety officer.

Biosafety Level 3Ag

Biosafety Level 3Ag is requires for work with certain biological agents in large animal species. Using the containment features of the standard BSL 3 facility as a starting point, BSL 3Ag facilities are specifically designed to protect the environment by including almost all of the features ordinarily used for BSL 4 facilities as enhancements. All BSL 3Ag containment spaces must be designed, constructed, and certified as primary containment barriers.

SCALE-UP LABORATORIES

Scale-up laboratories are defined differently depending on the nature and volume of work being conducted. For laboratories performing recombinant DNA research, large-scale experiments involve 10 L or more. Generally, the holding vessels do not exceed 100 L. A chemical or biological laboratory is defined as scale-up when the principal holding vessels are glass or ceramic. When the vessels are constructed primarily of metals, the laboratory is considered a pilot plant, which this chapter does not address. The amount of experimental materials present in scale-up laboratories is generally significantly greater than the amount found in the small-scale laboratory. Experimental equipment is also larger and therefore requires more space; these may include larger chemical fume hoods or reaction cubicles that may be of the walk-in type. Significantly higher laboratory airflow rates are needed to maintain the face velocity of the chemical fume hoods or reaction cubicles, although their size frequently presents problems of airflow uniformity over the entire face area. Walk-in hoods are sometimes entered during an experimental run, so provisions for breathing-quality air stations and other forms of personnel protection should be considered. Environmental containment or the ability to decontaminate the laboratory, the laboratory exhaust airstream, or other effluent may be needed in the event of an upset. Scale-up laboratories may be in operation for sustained periods.

For large walk-in hoods or reaction cubicles, the large volume of exhaust air required and the simultaneous requirement for supply air can result in temperature gradient problems in the space. Local specific ventilation capability is frequently provided within the laboratory space but outside the fume hood or reaction cubicle.

Large hoods, similar to what sometimes were called "California hoods," may also be provided in scale-up laboratories. These hoods are large in volume and height, provide access on multiple sides, and can be customized using standard components. Before beginning any custom hood design, the HVAC engineer, working with the user, should first determine how the hood will be used. Then the HVAC engineer can develop a custom hood design that considers

- What access is required for setup of experimental apparatus
- How the hood is expected to function during experimental runs
- Which doors or sashes should be open during a run
- Safety and ergonomic issues
- What features should be incorporated
- Airflow required to achieve satisfactory containment

Testing and balancing criteria should also be defined early in the design process. Mockups and factory testing of prototypes should be considered to avoid problems with installed hoods.

TEACHING LABORATORIES

Laboratories in academic settings can generally be classified as either those used for instruction or those used for research. Research laboratories vary significantly depending on the work being performed; they generally fit into one of the categories of laboratories described previously.

The design requirements for teaching laboratories also vary based on their function. The designer should become familiar with the specific teaching program, so that a suitable hazard assessment can be made. For example, the requirements for the number and size of fume hoods vary greatly between undergraduate inorganic and graduate organic chemistry teaching laboratories. Unique aspects of teaching laboratories include the need of the instructor to be in visual contact with the students at their work stations and to have ready access to the controls for the fume hood operations and any safety shutoff devices and alarms. Frequently, students have not received extensive safety instruction, so easily understood controls and labeling are necessary. Because the teaching environment depends on verbal communication, sound from the building ventilation system is an important concern.

CLINICAL LABORATORIES

Clinical laboratories are found in hospitals and as stand-alone operations. Work in these laboratories generally consists of handling human specimens (blood, urine, etc.) and using chemical reagents for analysis. Some samples may be infectious; because it is impossible to know which samples may be contaminated, good work practices require that all be handled as biohazardous materials. The primary protection of the staff at clinical laboratories depends on the techniques and laboratory equipment (e.g., biological safety cabinets) used to control aerosols, spills, or other inadvertent releases of samples and reagents. People outside the laboratory must also be protected.

The building HVAC system can provide additional protection with suitable exhaust, ventilation, and filtration. The HVAC engineer is responsible for providing an HVAC system that meets the biological and chemical safety requirements. The engineer should consult with appropriate senior staff and safety professionals to ascertain what potentially hazardous chemical or biohazardous conditions will be in the facility and then provide suitable engineering controls to minimize risks to staff and the community. Appropriate laboratory staff and the design engineer should consider using biological safety cabinets, chemical fume hoods, and other specific exhaust systems.

RADIOCHEMISTRY LABORATORIES

In the United States, laboratories located in Department of Energy (DOE) facilities are governed by DOE regulations. All other laboratories using radioactive materials are governed by the Nuclear Regulatory Commission (NRC), state, and local regulations. Other agencies may be responsible for the regulation of other toxic and carcinogenic materials present in the facility. Laboratory containment equipment for nuclear processing facilities are treated as primary, secondary, or tertiary containment zones, depending on the level of radioactivity anticipated for the area and the materials to be handled. Chapter 28 has additional information on nuclear laboratories.

OPERATION AND MAINTENANCE

During long-term research studies, laboratories may need to maintain design performance conditions with no interruptions for long periods. Even when research needs are not so demanding, systems that maintain air balance, temperature, and humidity in laboratories must be highly reliable, with a minimal amount of downtime. The designer should work with operation and maintenance personnel as well as users early in the design of systems to gain their input and agreement.

System components must be of adequate quality to achieve reliable HVAC operation, and they should be reasonably accessible for maintenance. Laboratory work surfaces should be protected from possible leakage of coils, pipes, and humidifiers. Changeout of supply and exhaust filters should require minimum downtime.

Centralized monitoring of laboratory variables (e.g., pressure differentials, face velocity of fume hoods, supply flows, and exhaust flows) is useful for predictive maintenance of equipment and for ensuring safe conditions. For their safety, laboratory users should be instructed in the proper use of laboratory fume hoods, safety cabinets, ventilated enclosures, and local ventilation devices. They should be trained to understand the operation of the devices and the indicators and alarms that show whether they are safe to operate. Users should request periodic testing of the devices to ensure that they and the connected ventilation systems are operating properly.

Personnel who know the nature of the contaminants in a particular laboratory should be responsible for decontamination of equipment and ductwork before they are turned over to maintenance personnel for work.

Maintenance personnel should be trained to keep laboratory systems in good operating order and should understand the critical safety requirements of those systems. Preventive maintenance of equipment and periodic checks of air balance should be scheduled. High-maintenance items should be placed outside the actual laboratory (in service corridors or interstitial space) to reduce disruption of laboratory operations and exposure of the maintenance staff to laboratory hazards. Maintenance personnel must be aware of and trained in procedures for maintaining good indoor air quality (IAQ) in laboratories. Many IAQ problems have been traced to poor maintenance due to poor accessibility (Woods et al. 1987).

ENERGY

Because of the nature of the functions they support, laboratory HVAC systems consume large amounts of energy (high flow rates; high static pressure filtration; critical cooling, heating, and humidification). Efforts to reduce energy use must not compromise standards established by safety officers. Typically, HVAC systems supporting laboratories and animal areas use 100% outdoor air and operate continuously. All HVAC systems serving laboratories can benefit from energy reduction techniques that are either an integral part of the original design or added later. Energy reduction techniques should be analyzed in terms of both appropriateness to the facility and economic payback.

Energy-efficient design is an iterative process that begins with establishing communication among all members of the design team. Each design discipline has an effect on the energy load. On a macro scale, building orientation, window shading devices, and high-performance envelopes offer opportunity for energy use reduction. On a micro scale, for example, the choice of a lighting system can affect sensible heat gain. Energy-efficient designs should recognize the variability of exhaust, envelope, and equipment loads and use systems that respond appropriately and perform efficiently during partial-load conditions.

The HVAC engineer must understand and respond to the scientific requirements of the facility. Research requirements typically include continuous control of temperature, humidity, relative static pressure, and air quality. Energy reduction systems must maintain required environmental conditions during both occupied and unoccupied modes.

Energy Efficiency

Energy can be used more efficiently in laboratories by reducing exhaust air requirements. One way to achieve this is to use variable-volume control of exhaust air through the fume hoods to reduce exhaust airflow when the fume hood sash is not fully open. Recent changes in NFPA *Standard* 45 and ANSI/AIHA *Standard* Z9.5 allow the use of a much lower fume hood minimum flow rate with variable-volume hoods (as low as about 100 cfm for a traditional 6 ft hood when the sash is closed), depending on the system design and aspects of laboratory operations. Any airflow control must be integrated with the laboratory control system, described in the section on Control, and its setting and operation must not jeopardize the safety and function of the laboratory.

Reducing ventilation requirements in laboratories and vivariums based on real-time sensing of contaminants in the room environment offers opportunities for energy conservation. This approach can potentially safely reduce lab air change rates to as low as 2 ach when the lab air is "clean" and the fume hood exhaust or room cooling load requirements do not require higher airflow rates. Research by Sharp (2010) showed that lab rooms are on average "clean" of contaminants in excess of about 98% of the time.

Setback controls that reduce ventilation rates when the laboratory is unoccupied can also reduce energy consumption. Timing devices, sensors, manual override, or a combination of these can be used to set back the controls at night. There should be no entry into the laboratory during unoccupied setback times and occupied ventilation rates should be engaged possibly 1 h or more in advance of occupancy to properly dilute any contaminants. If this strategy is used, the safety and function of the laboratory must be considered, and appropriate safety officers should be consulted.

Fume hood selection also impacts exhaust airflow requirements and energy consumption. Modern fume hood designs use several techniques to reduce airflow requirements, including reduced face opening sashes and specially designed components that allow operation with reduced inflow velocities. When considering these features, it is important to obtain approval of laboratory occupants and safety personnel.

Laboratory exhaust systems typically use constant-speed fans to discharge exhaust air at a constant velocity to prevent cross contamination with supply air intakes. Alternative approaches to reduce the considerable energy consumption of exhaust fans include using taller stacks, and real-time reduction of exhaust exit velocity based on sensing either wind direction and velocity or reduced contaminant levels in the exhaust fan plenum.

Room-cooling approaches, such as hydronic cooling using local fan-coil units or noncondensing, chilled radiant ceiling panels, passive chilled beams, or active chilled beams offer opportunities for energy conservation. These approaches decouple the room cooling function from the ventilation air requirements, potentially reducing outdoor air needs, overall HVAC capacity, and reheat energy. Less energy is needed to pump chilled water than to provide the equivalent amount of airflow required for a given level of cooling. Note that some form of dew-point sensing and possibly condensation monitoring is recommended (Rumsey et al. 2007) for noncondensing hydronic cooling approaches.

Energy Recovery

Energy can often be recovered economically from the exhaust airstream in laboratory buildings with large quantities of exhaust air. Many energy recovery systems are available, including rotary air-to-air energy exchangers or heat wheels, coil energy recovery loops (runaround cycle), twin tower enthalpy recovery loops, heat pipe heat exchangers, fixed-plate heat exchangers, thermosiphon heat exchangers, and direct evaporative cooling. Some of these technologies can be combined with indirect evaporative cooling for further energy recovery. See Chapters 25 and 45 of the 2008 *ASHRAE Handbook—HVAC Systems and Equipment* for more information.

Concerns about the use of energy recovery devices in laboratory HVAC systems include (1) the potential for cross-contamination of chemical and biological materials from exhaust air to the intake airstream, and (2) the potential for corrosion and fouling of devices located in the exhaust airstream.

Energy recovery is also possible for hydronic systems associated with HVAC. Rejected heat from centrifugal chillers can be used to produce low-temperature reheat water. Potential also exists in plumbing systems, where waste heat from washing operations can be recovered to heat makeup water.

Sustainable Design

Laboratories present unique challenges and opportunities for energy efficiency and sustainable design. Laboratory systems are complex, use significant energy, have health and life safety implications, need long-term flexibility and adaptablity, and handle potentially hazardous effluent with associated environmental impacts. Therefore, before implementing energy-efficiency and sustainable-design protocols, the engineer must be aware of the effects of these measures on the laboratory processes, which affect the safety of the staff, environment, and scientific procedures.

Several laboratory facilities have achieved high recognition for energy efficiency and sustainable design. Sustainable-design features specific to laboratory facilities include all aspects of design, construction, and operations. These features include (1) managing air and water effluent on the site; (2) reducing water used by laboratory processes; (3) rightsizing equipment and improving its energy efficiency; (4) hazardous material handling; (5) ventilation system enhancements, including modeling airflow patterns, fume hood testing, and additional safety alarming; and (6) laboratory-specific opportunities for innovation.

COMMISSIONING

In addition to HVAC systems, electrical systems and chemical handling and storage areas should be commissioned. Training of technicians, scientists, and maintenance personnel is a critical aspect of the commissioning process. Users should understand the systems and their operation.

It should be determined early in the design process whether any laboratory systems must comply with Food and Drug Administration (FDA) regulations because these systems have additional design, commissioning, and potential validation requirements. Commissioning is defined in Chapter 43, and the process is outlined in ASHRAE *Guidelines* 0 and 1.1. Laboratory commissioning can be more demanding than that described in ASHRAE *Guidelines* and includes systems that are not associated with other occupancies. Requirements for commissioning should be clearly understood by all participants, including the contractors and the owner's personnel. Roles and responsibilities should be defined, and responsibilities for documenting results should be established.

Laboratory commissioning starts with the intended use of the laboratory and should include development of a commissioning plan, as outlined in the ASHRAE guidelines. The start-up and prefunctional testing of individual components should come first; after individual components are successfully tested, the entire system should be functionally tested. This requires verification and documentation that the design meets applicable codes and standards and that it has been constructed in accordance with the design intent and owner's project requirements. Many facilities require integrated systems testing to verify the HVAC system is properly coordinated with other systems, such as fire alarm or emergency power systems. Before general commissioning begins, obtain the following data:

- Complete set of the laboratory utility drawings
- Definition of the use of the laboratory and an understanding of the work being performed
- Equipment requirements
- All test results
- Basis of design (BOD) that includes the intent of system operation
- Owner's project requirements (OPR)

For HVAC and associated integrated system commissioning, the following should be verified and documented:

- Manufacturer's requirements for airflow for biological safety cabinets and laminar flow clean benches have been met.
- Exhaust system configuration, damper locations, and performance characteristics, including any required emission equipment, are correct.
- Approved test and balance report.
- Control system operates as specified. Controls include fume hood alarm; miscellaneous safety alarm systems; fume hood and other exhaust airflow regulation; laboratory pressurization control system; laboratory temperature control system; and main ventilation unit controls for supply, exhaust, and heat recovery systems. Control system performance verification should include speed of response, accuracy, repeatability, turndown, and stability.
- Desired laboratory pressurization relationships are maintained throughout the laboratory, including entrances, adjoining areas,

air locks, interior rooms, and hallways. Balancing terminal devices within 10% of design requirements will not provide adequate results. Additionally, internal pressure relationships can be affected by airflow around the building. See Chapter 24 of the 2009 *ASHRAE Handbook—Fundamentals* for more information.
- Fume hood containment performance is within specification. ASHRAE *Standard* 110 provides criteria for this evaluation.
- Dynamic response of the laboratory's control system is satisfactory. One method of testing the control system is to open and shut laboratory doors during fume hood performance testing.
- System fault tree and failure modes are as specified, including life safety fan system shutdown impact on proper provisions for egress from the building within allowable limits of door-opening force requirements
- Standby electrical power systems function properly.
- Design noise criterion (NC) levels of occupied spaces have been met.

Training of facilities staff and laboratory occupants should also be considered part of the commissioning and design process. Training should address both the operation of individual system components and the overall system.

ECONOMICS

In laboratories, HVAC systems make up a significant part (often 30 to 50%) of the overall construction budget. The design criteria and system requirements must be reconciled with the budget allotment for HVAC early in the planning stages and continually throughout the design stages to ensure that the project remains within budget.

Every project must be evaluated on both its technical features and its economics. The following common economic terms are discussed in Chapter 37 and defined here as follows:

Initial cost: Costs to design, install, and test an HVAC system such that it is fully operational and suitable for use.

Operating cost: Cost to operate a system (including energy, maintenance, and component replacements) such that the total system can reach the end of its normal useful life.

Life-cycle cost: Cost related to the total cost over the life of the HVAC system, including initial capital cost, considering the time value of money.

Mechanical and electrical costs related to HVAC systems are commonly assigned a depreciation life based on current tax policies. This depreciation life may be different from the projected functional life of the equipment, which is influenced by the quality of the system components and of the maintenance they receive. Some portions of the system, such as ductwork, could last the full life of the building. Other components, such as air-handling units, may have a useful life of 15 to 30 years, depending on their original quality and ongoing maintenance efforts. Estimated service life of equipment is listed in Chapter 37.

Engineering economics can be used to evaluate life-cycle costs of configuration (utility corridor versus interstitial space), systems, and major equipment. The user or owner makes a business decision concerning the quality and reliability of the system and its ongoing operating costs. The HVAC engineer may be asked to provide an objective analysis of energy, maintenance, and construction costs, so that an appropriate life-cycle cost analysis can be made. Other considerations that may be appropriate include economic influences related to the long-term use of energy and governmental laws and regulations.

Many technical considerations and the great variety of equipment available influence the design of HVAC systems. Factors affecting design must be well understood to ensure appropriate comparisons between various systems and to determine the impact on either first or operating costs.

REFERENCES

AGS. 2007. *Guidelines for gloveboxes*, 3rd ed. American Glovebox Society, Santa Rosa, CA.

AIHA. 2003. Laboratory ventilation. ANSI/AIHA *Standard* Z9.5-03. American Industrial Hygiene Association, Fairfax, VA.

Alereza, T. and J. Breen, III. 1984. Estimates of recommended heat gains due to commercial appliances and equipment. *ASHRAE Transactions* 90(2A):25-58.

ARS. 2002. *Facilities design standards manual*. U.S. Department of Agriculture, Agricultural Research Service, Washington, D.C.

ASHRAE. 2010.Ventilation for acceptable indoor air quality. ANSI/ASHRAE *Standard* 62.1-2010.

ASHRAE. 2009. Method of testing for room air diffusion. ANSI/ASHRAE *Standard* 113-2009.

ASHRAE. 2007. Method of testing general ventilation air cleaning devices for removal efficiency by particle size. ANSI/ASHRAE *Standard* 52.2-2007.

ASHRAE. 1995. Method of testing performance of laboratory fume hoods. ANSI/ASHRAE *Standard* 110-1995.

ASHRAE. 2008. HVAC&R technical requirements for the commissioning process. *Guideline* 1.1-2008.

Besch, E. 1975. Animal cage room dry bulb and dew point temperature differentials. *ASHRAE Transactions* 81(2):459-458.

Besch, E. 1980. Environmental quality within animal facilities. *Laboratory Animal Science* 30(2II):385-406.

Caplan, K. and G. Knutson. 1977. The effect of room air challenge on the efficiency of laboratory fume hoods. *ASHRAE Transactions* 83(1):141-156.

Caplan, K. and G. Knutson. 1978. Laboratory fume hoods: Influence of room air supply. *ASHRAE Transactions* 84(1):511-537.

Code of Federal Regulations. (Latest edition). *Good laboratory practices*. CFR 21 Part 58. U.S. Government Printing Office, Washington, D.C.

DHHS. 1999. Biosafety in microbiological and biomedical laboratories, 4th ed. *Publication* No. (CDC) 93-8395. U.S. Department of Health and Human Services, NIH, Bethesda, MD.

Dorgan, C.B., C.E. Dorgan, and I.B.D. McIntosh. 2002. *ASHRAE laboratory design guide*. ASHRAE.

Eagleston, J., Jr. 1984. Aerosol contamination at work. In *The international hospital federation yearbook*. Sabrecrown Publishing, London.

Halitsky, J. 1988. Dispersion of laboratory exhaust gas by large jets. 81st Annual Meeting of the Air Pollution Control Association, June, Dallas.

Halitsky, J. 1989. A jet plume model for short stacks. *APCA Journal* 39(6).

Hessler, J. and A. Moreland. 1984. Design and management of animal facilities. In *Laboratory animal medicine*, J. Fox, B. Cohen, F. Loew, eds. Academic Press, San Diego, CA.

ICC. 2009. *International building code*®. International Code Council, Washington, D.C.

ILAR. 2010. *Guide for the care and use of laboratory animals*, 8ᵗʰ ed. Institute of Laboratory Animal Resources, National Academy of Sciences, the National Academies Press, Washington, D.C.

Klein, R., C. King, and A. Kosior. 2009. Laboratory air quality and room ventilation rates. *Journal of Chemical Health and Safety* (9/10).

Marsh, C.W. 1988. DDC systems for pressurization, fume hood face velocity and temperature control in variable air volume laboratories. *ASHRAE Transactions* 94(2):1947-1968.

Murakami, S., A. Mochida, R. Ooka, S. Kato, and S. Iizuka. 1996. Numerical prediction of flow around buildings with various turbulence models: Comparison of k-ε, EVM, ASM, DSM, and LES with wind tunnel tests. *ASHRAE Transactions* 102(1):741-753.

Neil, D. and R. Larsen. 1982. How to develop cost-effective animal room ventilation: Build a mock-up. *Laboratory Animal Science* (Jan-Feb):32-37.

NFPA. 1994. Guide for venting of deflagrations. ANSI/NFPA *Standard* 68-94.

NFPA. 1996. Flammable and combustible liquids code. ANSI/NFPA *Standard* 30-96.

NFPA. 1998. Storage, use, and handling of compressed and liquefied gases in portable cylinders. NFPA *Standard* 55-98.

NFPA. 1999. Health care facilities. ANSI/NFPA *Standard* 99-99.

NFPA. 2000. Fire protection for laboratories using chemicals. ANSI/NFPA *Standard* 45-2000

NIH. 1999a. *Research laboratory design policy and guidelines*. Office of Research Services, National Institutes of Health, Bethesda, MD.

NIH. 1999b. *Vivarium design policy and guidelines*. Office of Research Services, National Institutes of Health, Bethesda, MD.

NSF. 1992. Class II (laminar flow) biohazard cabinetry. *Standard* 49-92. NSF International, Ann Arbor, MI.

Petersen, R.L., J. Carter, and M. Ratcliff. 1997. The influence of architectural screens on exhaust dilution. ASHRAE *Research Project* RP-805. Draft Report approved by Technical Committee June 1997.

Rake, B. 1978. Influence of crossdrafts on the performance of a biological safety cabinet. *Applied and Environmental Microbiology* (August): 278-283.

Riskowski, et al. 1995. Development of ventilation rates and design information for laboratory animal facilities—Part I. ASHRAE Research Project RP-730. *ASHRAE Transactions* 101(2).

Riskowski, et al. 1996. Development of ventilation rates and design information for laboratory animal facilities—Part II. ASHRAE Research Project RP-730. *ASHRAE Transactions* 102(2).

Rumsey, P. and J. Weale. 2007. Chilled beams in labs: Eliminating reheat and saving energy on a budget. *ASHRAE Journal* 49(1):18-25.

Schuyler, G. 2009. The effect of air change rate on recovery from a spill. In Seminar 26, presented at 2009 ASHRAE Winter Conference, Chicago.

Sharp, G.P. 2010. Demand-based control of lab air change rates. *ASHRAE Journal* 52(2):30-41.

SEFA. *Laboratory fume hoods recommended practices*. SEFA 1. Scientific Equipment and Furniture Association, Hilton Head, SC.

Stuart, D., M. First, R. Rones, and J. Eagleston. 1983. Comparison of chemical vapor handling by three types of Class II biological safety cabinets. *Particulate & Microbial Control* (March/April).

West, D.L. 1978. Assessment of risk in the research laboratory: A basis for facility design. *ASHRAE Transactions* 84(1):547-557.

Wier, R.C. 1983. Toxicology and animal facilities for research and development. *ASHRAE Transactions* 89(2B):533-541.

Wilson, D.J., et al. 1998. Adjacent building effects on laboratory fume hood stack design. ASHRAE Research Project RP-897, *Final Report*.

Woods, J. 1980. The animal enclosure—A microenvironment. *Laboratory Animal Science* 30(2II):407-413.

Woods, J., J. Janssen, P. Morey, and D. Rask. 1987. Resolution of the "sick" building syndrome. *Proceedings of ASHRAE Conference: Practical Control of Indoor Air Problems*, pp. 338-348.

Zhou, Y. and T. Stathopoulos. 1996. Application of two-layer methods for the evaluation of wind effects on a cubic building. *ASHRAE Transactions* 102(1):754-764.

BIBLIOGRAPHY

Abramson, B. and T. Tucker. 1988. Recapturing lost energy. *ASHRAE Journal* 30(6):50-52.

Adams, J.B., Jr. 1989. Safety in the chemical laboratory: Synthesis—Laboratory fume hoods. *Journal of Chemical Education* 66(12).

Ahmed, O. and S.A Bradley. 1990. An approach to determining the required response time for a VAV fume hood control system. *ASHRAE Transactions* 96(2):337-342.

Ahmed, O., J.W. Mitchell, and S.A. Klein. 1993. Dynamics of laboratory pressurization. *ASHRAE Transactions* 99(2):223-229.

Albern, W., F. Darling, and L. Farmer. 1988. Laboratory fume hood operation. *ASHRAE Journal* 30(3):26-30.

Anderson, S. 1987. Control techniques for zoned pressurization. *ASHRAE Transactions* 93(2B):1123-1139.

Anderson, C.P. and K.M. Cunningham. 1988. HVAC controls in laboratories—A systems approach. *ASHRAE Transactions* 94(1):1514-1520.

Barker, K.A., O. Ahmed, and J.A. Parker. 1993. A methodology to determine laboratory energy consumption and conservation characteristics using an integrated building automation system. *ASHRAE Transactions* 99(2):1155-1167.

Baylie, C.L. and S.H. Schultz. 1994. Manage change: Planning for the validation of HVAC Systems for a clinical trials production facility. *ASHRAE Transactions* 100(1):1660-1668.

Bell, G.C., E. Mills, G. Sator, D. Avery, M. Siminovitch, and M.A. Piette. 1996. *A design guide for energy-efficient research laboratories*. LBNL-PUB-777. Lawrence Berkeley National Laboratory, Berkeley, CA.

Bertoni, M. 1987. Risk management considerations in design of laboratory exhaust stacks. *ASHRAE Transactions* 93(2B):2149-2164.

Bossert, K.A. and S.M. McGinley. 1994. Design characteristics of clinical supply laboratories relating to HVAC systems. *ASHRAE Transactions* 94(100):1655-1659.

Brow, K. 1989. AIDS research laboratories—HVAC criteria. In *Building Systems: Room Air and Air Contaminant Distribution*, L.L. Christianson, ed. ASHRAE cosponsored Symposium, pp. 223-225.

Brown, W.K. 1993. An integrated approach to laboratory energy efficiency. *ASHRAE Transactions* 99(2):1143-1154.

Carnes, L. 1984. Air-to-air heat recovery systems for research laboratories. *ASHRAE Transactions* 90(2A):327.

Coogan, J.J. 1994. Experience with commissioning VAV laboratories. *ASHRAE Transactions* 100(1):1635-1640.

Crane, J. 1994. Biological laboratory ventilation and architectural and mechanical implications of biological safety cabinet selection, location, and venting. *ASHRAE Transactions* 100(1):1257-65.

CRC. 2000. *CRC handbook of laboratory safety*, 5th ed. CRC Press, Boca Raton, FL.

Dahan, F. 1986. HVAC systems for chemical and biochemical laboratories. *Heating, Piping and Air Conditioning* (May):125-130.

Davis, S. and R. Benjamin. 1987. VAV with fume hood exhaust systems. *Heating, Piping and Air Conditioning* (August):75-78.

Degenhardt, R. and J. Pfost. 1983. Fume hood system design and application for medical facilities. *ASHRAE Transactions* 89(2B):558-570.

DHHS. 1981. *NIH guidelines for the laboratory use of chemical carcinogens*. Department of Health and Human Services, National Institutes of Health, Bethesda, MD.

DiBeradinis, L., J. Baum, M. First, G. Gatwood, E. Groden, and A. Seth. 1992. *Guidelines for laboratory design: Health and safety considerations*. John Wiley & Sons, Boston.

Doyle, D.L., R.D. Benzuly, and J.M. O'Brien. 1993. Variable volume retrofit of an industrial research laboratory. *ASHRAE Transactions* 99(2): 1168-1180.

Flanherty, R.J. and R. Gracilieri. 1994. Documentation required for the validation of HVAC systems. *ASHRAE Transactions* 100(1):1629-1634.

Ghidoni, D.A. and R.L. Jones, Jr. 1994. Methods of exhausting a BSC to an exhaust system containing a VAV component. *ASHRAE Transactions* 100(1):1275-1281.

Hayter, R.B. and R.L. Gorton. 1988. Radiant cooling in laboratory animal caging. *ASHRAE Transactions* 94(1):1834-1847.

Hitchings, D.T. and R.S. Shull 1993. Measuring and calculating laboratory exhaust diversity—Three case studies. *ASHRAE Transactions* 99(2): 1059-1071.

ILAR. 1996. Laboratory animal management—Rodents. *ILAR News* 20(3).

Kirkpatrick, A., et al. A mathematical model for the prediction of laboratory fume hood airflow: ASHRAE *Research Project* RP-848, Final Report.

Knutson, G. 1984. Effect of slot position on laboratory fume hood performance. *Heating, Piping and Air Conditioning* (February):93-96.

Knutson, G. 1987. Testing containment laboratory hoods: A field study. *ASHRAE Transactions* 93(2B):1801-1812.

Koenigsberg, J. and H. Schaal. 1987. Upgrading existing fume hood installations. *Heating, Piping and Air Conditioning* (October):77-82.

Koenigsberg, J. and E. Seipp. 1988. Laboratory fume hood—An analysis of this special exhaust system in the post "Knutson-Caplan" era. *ASHRAE Journal* 30(2):43-46.

Laboratory & Health Facilities Programming & Planning. *Planning Monograph 1.0*. Planning Collaborative Ltd., Reston, VA.

Lacey, D.R. 1994. HVAC for a low temperature biohazard facility. *ASHRAE Transactions* 100(1):1282-1286.

Lentz, M.S. and A.K. Seth. 1989. A procedure for modeling diversity in laboratory VAV systems. *ASHRAE Transactions* 95(1):114-120.

Maghirang, R.G., G.L. Riskowski, P.C. Harrison, H.W. Gonyou, L. Sebek, and J. McKee. 1994. An individually ventilated caging system for laboratory rats. *ASHRAE Transactions* 100(1):913-920.

Maust, J. and R. Rundquist. 1987. Laboratory fume hood systems—Their use and energy conservation. *ASHRAE Transactions* 93(2B):1813-1821.

McDiarmid, M.D. 1988. A quantitative evaluation of air distribution in full scale mock-ups of animal holding rooms. *ASHRAE Transactions* 94(1B): 685-693.

Mikell, W. and F. Fuller. 1988. Safety in the chemical laboratory: Good hood practices for safe hood operation. *Journal of Chemical Education* 65(2).

Moyer, R.C. 1983. Fume hood diversity for reduced energy consumption. *ASHRAE Transactions* 89(2B):552-557.

Moyer, R. and J. Dungan. 1987. Turning fume hood diversity into energy savings. *ASHRAE Transactions* 93(2B):1822-1834.

Murray, W., A. Streifel, T. O'Dea, and F. Rhame. 1988. Ventilation for protection of immune compromised patients. *ASHRAE Transactions* 94(1): 1185-1192.

NFPA. 1994. Gaseous hydrogen systems at consumer sites. ANSI/NFPA *Standard* 50A. National Fire Protection Association, Quincy, MA.

NFPA. 1996. Water spray fixed systems for fire protection. ANSI/NFPA *Standard* 15.

NFPA. 1996. Installation of air conditioning and ventilating systems. ANSI/NFPA *Standard* 90A.

NFPA. 1996. National electrical code. ANSI/NFPA *Standard* 70.

NFPA. 1998. Liquefied petroleum gas code. NFPA *Standard* 58.

NIH. 1996. *Clinical center design policy and guidelines*. Office of Research Services, National Institutes of Health, Bethesda, MD.

NIH. 1996. *Reference material for the design policy and guidelines*. Office of Research Services, National Institutes of Health, Bethesda, MD.

Neuman, V. 1989. Design considerations for laboratory HVAC system dynamics. *ASHRAE Transactions* 95(1):121-124.

Neuman, V. 1989. Disadvantages of auxiliary air fume hoods. *ASHRAE Transactions* 95(1):70-75.

Neuman, V. 1989. Health and safety in laboratory plumbing. *Plumbing Engineering* (March):21-24.

Neuman, V. and H. Guven. 1988. Laboratory building HVAC systems optimization. *ASHRAE Transactions* 94(2):432-451.

Neuman, V. and W. Rousseau. 1986. VAV for laboratory hoods—Design and costs. *ASHRAE Transactions* 92(1A):330-346.

Neuman, V., F. Sajed, and H. Guven. 1988. A comparison of cooling thermal storage and gas air conditioning for a lab building. *ASHRAE Transactions* 94(2):452-468.

Parker, J.A., O. Ahmed, and K.A. Barker. 1993. Application of building automation system (BAS) in evaluating diversity and other characteristics of a VAV laboratory. *ASHRAE Transactions* 99(2):1081-1089.

Peterson, R. 1987. Designing building exhausts to achieve acceptable concentrations of toxic effluents. *ASHRAE Transactions* 93(2):2165-2185.

Peterson, R.L., E.L. Schofer, and D.W. Martin. 1983. Laboratory air systems—Further testing. *ASHRAE Transactions* 89(2B):571-596.

Pike, R. 1976. Laboratory-associated infections: Summary and analysis of 3921 cases. *Health Laboratory Science* 13(2):105-114.

Rabiah, T.M. and J.W. Wellenbach. 1993. Determining fume hood diversity factors. *ASHRAE Transactions* 99(2):1090-1096.

Rhodes, W.W. 1988. Control of microbioaerosol contamination in critical areas in the hospital environment. *ASHRAE Transactions* 94(1):1171-1182.

Richardson, G. 1994. Commissioning of VAV laboratories and the problems encountered. *ASHRAE Transactions* 100(1):1641-1645.

Rizzo, S. 1994. Commissioning of laboratories: A case study. *ASHRAE Transactions* 100(1):1646-1652.

Sandru, E. 1996. Evaluation of the laboratory equipment component of cooling loads. *ASHRAE Transactions* 102(1):732-737.

Schuyler, G. and W. Waechter. 1987. Performance of fume hoods in simulated laboratory conditions. *Report No. 487-1605* by Rowan Williams Davies & Irwin, Inc., under contract for Health and Welfare Canada.

Schwartz, Leonard. 1994. Heating, ventilating and air conditioning considerations for pharmaceutical companies. *Pharmaceutical Engineering* 14(4).

Sessler, S. and R. Hoover. 1983. Laboratory fume hood noise. *Heating, Piping, and Air Conditioning* (September):124-137.

Simons, C.G. 1991. Specifying the correct biological safety cabinet. *ASHRAE Journal* 33(8).

Simons, C.G. and R. Davoodpour. 1994. Design considerations for laboratory facilities using molecular biology techniques. *ASHRAE Transactions* 100(1):1266-1274.

Smith, W. 1994. Validating the direct digital control (DDC) system in a clinical supply laboratory. *ASHRAE Transactions* 100(1):1669-1675.

Streets, R.A. and B.S.V. Setty. 1983. Energy conservation in institutional laboratory and fume hood systems. *ASHRAE Transactions* 89(2B): 542-551.

Stuart, D., R. Greenier, R. Rumery, and J. Eagleston. 1982. Survey, use, and performance of biological safety cabinets. *American Industrial Hygiene Association Journal* 43:265-270.

Varley, J.O. 1993. The measurement of fume hood use diversity in an industrial laboratory. *ASHRAE Transactions* 99(2):1072-1080.

Vzdemir, I.B., J.H. Whitelaw, and A.F. Bicen. 1993. Flow structures and their relevance to passive scalar transport in fume cupboards. *Proceedings of the Institution of Mechanical Engineers* 207:103-115.

Wilson, D.J. 1983. A design procedure for estimating air intake contamination from nearby exhaust vents. *ASHRAE Transactions* 89(2A):136.

Yoshida, K., H. Hachisu, J.A. Yoshida, and S. Shumiya. 1994. Evaluation of the environmental conditions in a filter-capped cage using a one-way airflow system. *ASHRAE Transactions* 100(1):901-905.

CHAPTER 17

ENGINE TEST FACILITIES

INDUSTRIAL testing of turbine and internal combustion engines is performed in enclosed test spaces to control noise and isolate the test for safety or security. These spaces are ventilated or conditioned to control the facility environment and fumes. Isolated engines are tested in test cells; engines inside automobiles are tested on chassis dynamometers. The ventilation and safety principles for test cells also apply when large open areas in the plant are used for production testing and emissions measurements.

Enclosed test cells are normally found in research or emissions test facilities. Test cells may require instruments to measure cooling system water flow and temperature; exhaust gas flow, temperature, and emission concentrations; fuel flow; power output; and combustion air volume and temperature. Changes in the temperature and humidity of the test cell affect these measurements. Accurate control of the testing environment is becoming more critical. For example, the U.S. Environmental Protection Agency requires tests to demonstrate control of automobile contaminants in both hot and cold environments.

Air conditioning and ventilation of test cells must (1) supply and exhaust proper quantities of air to remove heat and control temperature; (2) exhaust sufficient air at proper locations to prevent buildup of combustible vapors; (3) supply and modulate large quantities of air to meet changing conditions; (4) remove exhaust fumes; (5) supply combustion air; (6) prevent noise transmission through the system; (7) provide for human comfort and safety during setup, testing, and tear-down; and (8) treat the exhaust effluent. Supply and exhaust systems for test cells may be unitary, central, or a combination of the two. Mechanical exhaust is necessary in all cases.

Special Warning: Certain industrial spaces may contain flammable, combustible, and/or toxic concentrations of vapors or dusts under either normal or abnormal conditions. In spaces such as these, there are life-safety issues that this chapter may not completely address. Special precautions must be taken in accordance with requirements of recognized authorities such as the National Fire Protection Association (NFPA), the Occupational Safety and Health Administration (OSHA), and the American National Standards Institute (ANSI). In all situations, engineers, designers, and installers who encounter conflicting codes and standards must defer to the code or standard that best addresses and safeguards life safety.

ENGINE HEAT RELEASE

The special air-conditioning requirements of an engine test facility stem from burning the fuel used to run the engine. For internal combustion engines at full load, 10% of the total heat content of the fuel is radiated and convected into the room or test cell atmosphere, and 90% is fairly evenly divided between the shaft output (work), exhaust gas heating, and heating of the jacket cooling water.

Air-cooled engines create a forced convection load on the test space equal to the jacket water heat that it replaces. For turbine

engines, the exhaust gas carries double the heat of the internal combustion engine exhaust and there is no jacket water to heat. The engine manufacturer can provide a more precise analysis of heat release characteristics at various speeds and power outputs.

Test facilities use dynamometers to determine the power supplied by the engine shaft. The dynamometer converts shaft work into heat that must be accounted for by a cooling system or as heat load into the space. Often, shaft work is converted into electricity through a generator and the electric power is dissipated by a resistance load bank or sold to the local utility. Inefficiencies of the various pieces of equipment add to the load of the space in which they are located.

Heat released into the jacket water must also be removed. If a closely connected radiator is used, the heat load is added to the room load. Many test facilities include a heat exchanger and a secondary cooling circuit transfers the heat to a cooling tower. Some engines require an oil cooler separate from the jacket water. Whichever system is used, the cooling water flow, temperature, and pressure are usually monitored as part of the test operation and heat from these sources needs to be accommodated by the facility's air conditioning.

Exhaust systems present several challenges to engine test cell design. Exhaust gases can exit the engine at 1500°F or higher. Commonly, the exhaust gas is augmented by inserting the exhaust pipe into a larger-bore exhaust system (laboratory fixed system), which draws room air into the exhaust to both cool the gas and ventilate the test cell. Both the exhausted room air and combustion air must be supplied to the room from the HVAC or from the outdoor.

Radiation and convection from exhaust pipes, catalytic converter, muffler, etc., also add to the load. In most cases, the test cell's HVAC system should account for an engine that can fully load the dynamometer, and have capacity control for operation at partial and no load.

Large gas turbine engines have unique noise and airflow requirements; therefore, they usually are provided with dedicated test cells. Small gas turbines can often be tested in a regular engine test cell with minor modifications.

ENGINE EXHAUST

Engine exhaust systems remove combustible products, unburned fuel vapors, and water vapor. Flow loads and operating pressure need to be established for design of the supporting HVAC.

Flow loads are calculated based on the number of engines, the engine sizes and loads, and use factors or diversity.

Operating pressure is the engine discharge pressure at the connection to the exhaust. Systems may operate at positive pressure using available engine tail-pipe pressure to force the flow of gas, or at negative pressure with mechanically induced flow.

The simplest way to induce engine exhaust from a test cell is to size the exhaust pipe to minimize variations in pressure on the engine and to connect it directly to the outdoor (Figure 1A). Exhausts directly connected to the outdoor are subject to wind currents and air pressure, however, and can be hazardous because of positive pressure in the system.

The preparation of this chapter is assigned to TC 9.2, Industrial Air Conditioning.

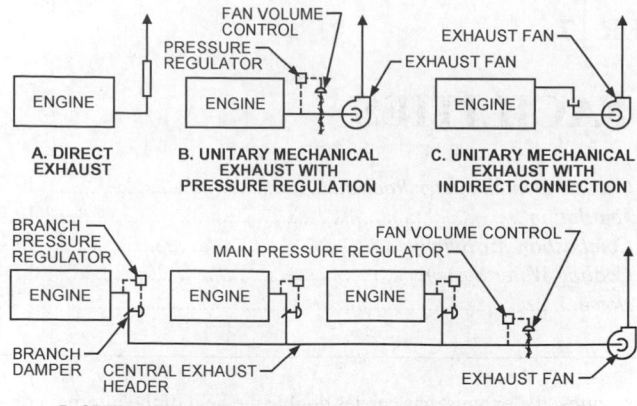

Fig. 1 Engine Exhaust Systems

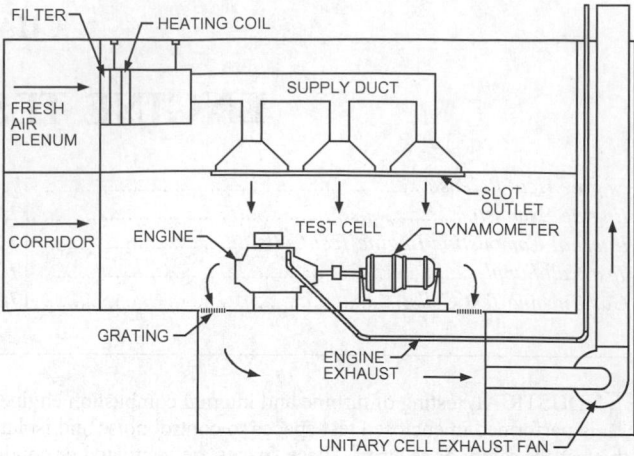

Fig. 2 Engine Test Cell Showing Direct Engine Exhaust: Unitary Ventilation System

Mechanical engine exhausts are either unitary or central. A **unitary exhaust** (Figure 1B) serves only one test cell, and can be closely regulated to match the engine's operation. A **central exhaust** (Figure 1D) serves multiple test cells with one or more exhaust fans and a duct system with branch connections to the individual test cells. Relief of a possible explosion in the ductwork should be considered.

Engine exhaust pressures fluctuate with changes in engine load and speed. Central exhausts should be designed to minimize effects of load variations in individual test cells on the system. Engine characteristics and diversity of operation determine the maximum airflow to be handled. Dampers and pressure regulators may be required to keep pressures within test tolerances.

An indirect connection between the engine exhaust pipe and mechanical exhaust gas removal (Figure 1C) eliminates variation in back pressure and augments exhaust gas flow by inducing room air into the exhaust stream. In this system the engine exhaust pipe terminates by being centered and inserted about 3 in. into the augmentation pipe, which is at least 1 in. larger in diameter. The induced room air is mixed with the exhaust gases, yielding a much cooler exhaust flow. However, the potential for increased corrosion in a cooler exhaust must be considered when selecting construction materials. The engine muffler should be located upstream of the augmentation connection to control noise. The indirect connection should be considered a potential point of ignition if the exhaust is fuel rich and the tail pipe reaches temperatures above 700°F.

Exhaust pipes and mufflers run very hot. A ventilated heat shield or a water-jacketed pipe reduces cell heat load, and some exhausts are equipped with direct water injection. Thermal expansion, stress, and pressure fluctuations must also be considered in the design of the exhaust fan and ducting. The equipment must be adequately supported and anchored to relieve the thermal expansion.

Exhaust systems for chassis dynamometer installations must capture high-velocity exhaust from the tail pipe to prevent fume buildup in the room. An exhaust flow rate of 700 cfm has been used effectively for automobiles at a simulated speed of 65 mph.

Engine exhaust should discharge through a stack extending above the roof to an elevation sufficient to allow the fumes to clear the building. Chapter 45 has further details about exhaust stacks. Codes or air emission standards may require that exhaust gases be cleaned before being discharged to atmosphere.

INTERNAL COMBUSTION ENGINE TEST CELLS

Test Cell Exhaust

Ventilation for test cells is based on exhaust requirements for (1) removal of heat generated by the engine, (2) emergency purging (removal of fumes after a fuel spill), and (3) continuous cell scavenging during nonoperating periods. Heat is transferred to the test cell by convection and radiation from all of the heated surfaces, such as the engine and exhaust system. At a standard air density of $\rho = 0.075$ lb/ft^3 and specific heat $c_p = 0.24$ Btu/lb·°F,

$$Q = \frac{q}{60\rho c_p(t_e - t_s)} = \frac{q}{1.08(t_e - t_s)}$$

where

Q = airflow, cfm
q = engine heat release, Btu/h
t_e = temperature of exhaust air, °F
t_s = temperature of supply air, °F

The constant (1.08) should be corrected for other temperatures and pressures.

Heat radiated from the engine, dynamometer, and exhaust piping warms surrounding surfaces, which release heat to the air by convection. The value for $(t_e - t_s)$ in the equation cannot be arbitrarily set when a portion of q is radiated heat. The section on Engine Heat Release discusses other factors required to determine the overall q.

Vapor Removal. The exhaust should remove vapors as quickly as possible. Emergency purging, often 10 cfm per square foot of floor area, should be controlled by a manual overriding switch for each test cell. In case of fire, provisions need to be made to shut down all equipment, close fire dampers at all openings, and shut off the fuel-flow solenoid valves.

Cell Scavenging. Exhaust air is the minimum amount of air required to keep combustible vapors from fuel leaks from accumulating. In general, the NFPA *Standard* 30 requirement of 1 cfm per square foot of floor area is sufficient. Because gasoline vapors are heavier than air, exhaust grilles should be low even when an overhead duct is used. Exhausting close to the engine minimizes the convective heat that escapes into the cell.

In some installations, all air is exhausted through a floor grating surrounding the engine bed plate and into a cubicle or duct below. In this arrangement, slots in the ceiling over the engine supply a curtain of air to remove the heat. This scheme is particularly suitable for a central exhaust (Figure 2). Water sprays in the underfloor exhaust lessen the danger of fire or explosion in case of fuel spills.

Trenches and pits should be avoided in test cells. If they exist, as in most chassis dynamometer rooms, they should be mechanically exhausted at all times. Long trenches may require multiple exhaust takeoffs. The exhaust should sweep the entire area, leaving no dead air spaces. Because of fuel spills and vapor accumulation, suspended ceilings or basements should not be located

Table 1   Exhaust Quantities for Test Cells

	Minimum Exhaust Rates per Square Foot of Floor Area	
	cfm	ach[a]
Engine testing—cell operating	10	60[b]
Cell idle	1	6
Trenches[c] and pits	10	—
Accessory testing	4	24
Control rooms and corridors	1	6

[a] Air changes per hour, based on cell height of 10 ft.
[b] For chassis dynamometer rooms, this quantity is usually set by test requirements.
[c] For large trenches, use 100 fpm across the cross-sectional area of the trench.

directly below the engine test cell. If such spaces exist, they should be ventilated continuously and have no fuel lines running through them.

Table 1 lists exhaust quantities used in current practice; the exhaust should be calculated for each test cell on the basis of heat to be removed, evaporation of possible fuel spills, and the minimum ventilation needed during downtime.

TEST CELL SUPPLY

The air supply to a test cell should be balanced to yield a slightly negative pressure. This is accomplished by having either an exhaust airflow 10% greater than the supply air or a differential pressure of the test cell at least 0.05 in. of water less than the surrounding space. Test cell air should not be recirculated. Air taken from nontest areas can be used if good ventilation practices are followed, such as using air that is free of unacceptable contaminants, is sufficient for temperature control, and can maintain the proper test cell pressure.

Ventilation air should keep heat released from the engine away from cell occupants. Slot outlets with automatic dampers to maintain a constant discharge velocity have been used with variable-volume systems.

A variation of systems C and D in Figure 3 includes a separate air supply sized for the minimum (downtime) ventilation rate and for a cooling coil with room thermostat to regulate the coil to control the temperature in the cell. This system is useful in installations where much time is devoted to the setup and preparation of tests, or where constant temperature is required for complicated or sensitive instrumentation. Except for production and endurance testing, the actual engine operating time in test cells may be surprisingly low. The average test cell is used approximately 15 to 20% of the time.

Air should be filtered to remove particulates and insects. The degree of filtration is determined by the type of tests. Facilities in relatively unpolluted areas sometimes use unfiltered outdoor air.

Heating coils are needed to temper supply air if there is danger of freezing equipment or if low temperatures adversely affect tests. For low-temperature applications, a desiccant wheel with pre- and post-cooling may be needed with appropriate environmentally friendly refrigerants.

GAS-TURBINE TEST CELLS

Large gas-turbine test cells must handle large quantities of air required by the turbine, attenuate the noise generated, and operate safely with a large flow of fuel. These cells are unitary and use the turbine to draw in untreated air and exhaust it through noise attenuators.

Small gas turbine engines can generally be tested in a conventional test cell with relatively minor modifications. The test-cell ventilation air supply and exhausts are sized for turbine-generated heat as for a conventional engine. The combustion air supply for the turbine

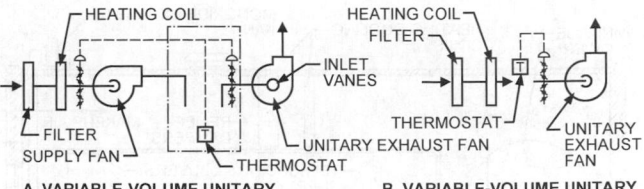

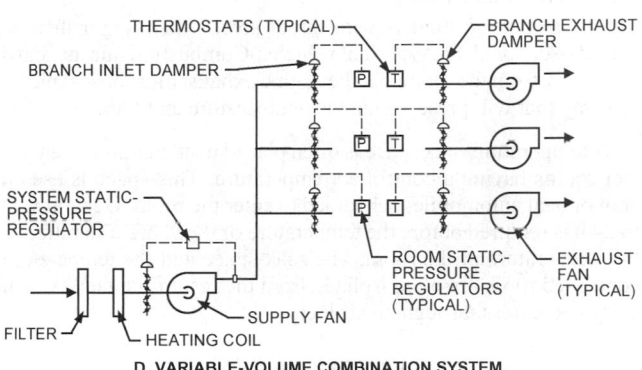

Fig. 3   Heat Removal Ventilation Systems

is considerable; it may be drawn from the cell, from outdoors, or through separate conditioning units that handle only combustion air.

Exhaust quantities are higher than from internal combustion engines and are usually ducted directly to the outdoors through muffling devices that provide little restriction to airflow. Exhaust air may be water-cooled, as temperature may exceed 1300°F.

CHASSIS DYNAMOMETER ROOMS

A chassis dynamometer (Figure 4) simulates road driving and acceleration conditions. The vehicle's drive wheels rest on a large roll, which drives the dynamometer. Air quantities, which are calibrated to correspond to air velocity at a particular road speed, flow across the front of the vehicle for radiator cooling and to approximate the effects of air speed on the body of the vehicle. Additional refinements may vary air temperature within prescribed limits from −40 to 130°F, control relative humidity, and/or add shakers to simulate road conditions. Air is usually introduced through an area approximating the frontal area of the vehicle. A duct with a return grille at the rear of the vehicle may be lowered so that air remains near the floor rather than cycling through a ceiling return air grille. Air is recirculated to air-handling equipment above the ceiling.

Chassis dynamometers are also installed in

- Cold rooms, where temperatures may be as low as −100°F
- Altitude chambers, where elevations up to 12,000 ft can be simulated
- Noise chambers for sound evaluation
- Electromagnetic cells for evaluation of electrical components

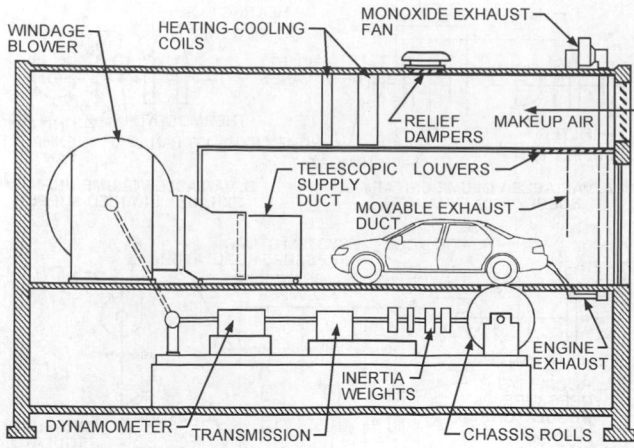

Fig. 4 Chassis Dynamometer Room

Table 2 Typical Noise Levels in Test Cells

Type and Size of Engine	Sound Level 3 ft from Engine, dBA			
	63 Hz	124 Hz	500 Hz	2000 Hz
Diesel				
Full load	105	107	98	99
Part load	70	84	56	49
Gasoline engine, 440 in³ at 5000 rpm				
Full load	107	108	104	104
Part load	75	—	—	—
Rotary engine, 100 hp				
Full load	90	90	83.5	86
Part load	79	78	75	72

- Environmental chambers
- Full-sized wind tunnels with throat areas much larger than the cross-sectional area of the vehicle. Combustion air is drawn directly from the room, but the engine exhaust must be installed in a way that will preserve the low temperature and humidity.

A temperature soak space is often placed near chassis dynamometer rooms having a controlled temperature. This space is used to cool or heat automobiles scheduled to enter the room. Generally, 18 to 24 h is required before the temperature of the vehicle stabilizes to the temperature of the room. The soak space and the temperature-controlled room are often isolated from the rest of the facility, with entry and egress through an air lock.

VENTILATION

Constant-volume systems with variable supply temperatures can be used; however, variable-volume, variable-temperature systems are usually selected. Ventilation is generally controlled on the exhaust side (see Figure 3). Unitary variable-volume systems (Figure 3A) use an individual exhaust fan and makeup air supply for each cell. Supply and exhaust fans are interlocked, and their operation is coordinated with the engine, usually by sensing the temperature of the cell. Some systems have exhaust only, with supply induced directly from outdoor (Figure 3B). The volume is varied by changing fan speed or damper position.

Ventilation with central supply fans, central exhaust fans, or both (Figure 3C) regulate air quantities by test cell temperature control of individual dampers or by two-position switches actuated by dynamometer operations. Air balance is maintained by static pressure regulation in the cell. Constant pressure in the supply duct is obtained by controlling supply fan inlet vanes, modulating dampers, or varying fan speed.

In systems with individual exhaust fans and central supply air, exhaust is controlled by cell temperature or a two-position switch actuated by dynamometer operation. The central supply system is controlled by a static pressure device in the cell to maintain room pressure (Figure 3D). Variable-volume exhaust airflow should not drop below minimum requirements. Exhaust requirements should override cell temperature requirements; thus, reheat may be needed.

Ventilation should be interlocked with fire protection to shut down the supply to and exhaust from the cell in case of fire. Exhaust fans should be nonsparking, and makeup air should be tempered.

COMBUSTION AIR SUPPLY

Combustion air is usually drawn from the test cell or introduced directly from the outdoors. Separate dedicated units can be used if combustion air must be closely regulated and conditioning of the entire test cell is impractical. These units filter, heat, and cool the supply air and regulate its humidity and pressure; they usually provide air directly to the engine air intake. Combustion air systems may be central units or portable packaged units.

COOLING WATER SYSTEMS

Dynamometers absorb and measure the useful output of an engine or its components. In a water-cooled dynamometer, engine work is converted to heat, which is absorbed by circulating water. Electrical dynamometers convert engine work to electrical energy, which can be used or dissipated as heat in resistance grids or load banks or sold to the local utility. Grids should be located outdoors or adequately ventilated.

Heat loss from electric dynamometers is approximately 8% of the measured output, plus a constant load of about 5 kW for auxiliaries in the cell. Recirculating water absorbs heat from the engine jacket water, oil coolers, and water cooled dynamometers through circulating pumps, cooling towers, or atmospheric coolers and hot- and cold-well collecting tanks.

NOISE

Noise generated by internal combustion engines and gas turbines must be considered in the design of a test cell air-handling system. Part of the engine noise is discharged through the tail pipe. If possible, internal mufflers should be installed to attenuate this noise at its source. Any ventilation ducts or pipe trenches that penetrate the cells must be insulated against sound transmission to other areas or to the outdoors. Attenuation equivalent to that provided by the cell structure should be applied to duct penetrations. Table 2 lists typical noise levels in test cells during engine operations.

BIBLIOGRAPHY

Bannasch, L.T. and G.W. Walker. 1993. Design factors for air-conditioning systems serving climatic automobile emission test facilities. *ASHRAE Transactions* 99(2):614-623.

Computer controls engine test cells. *Control Engineering* 16(75):69.

NFPA. 2008. Flammable and combustible liquids code. *Standard* 30-2008. National Fire Protection Association, Quincy, MA.

Paulsell, C.D. 1990. *Description and specification for a cold weather emissions testing facility.* U.S. Environmental Protection Agency, Washington, D.C.

Schuett, J.A. and T.J. Peckham. 1986. Advancements in test cell design. *SAE Transactions,* Paper 861215. Society of Automotive Engineers, Warrendale, PA.

CLEAN SPACES

DESIGN of clean spaces or cleanrooms covers much more than traditional temperature and humidity control. Other factors may include control of particle, microbial, electrostatic discharge (ESD), molecular, and gaseous contamination; airflow pattern control; pressurization; sound and vibration control; life safety; industrial engineering aspects; and manufacturing equipment layout. The objective of good cleanroom design is to control these variables while optimizing installation and operating costs.

TERMINOLOGY

Acceptance criteria. Upper and lower limits of a pharmaceutical critical parameter required for product or process integrity. If these limits are exceeded, the pharmaceutical product may be considered adulterated.

ach. Air changes per hour.

Air lock. A small transitional room between two other rooms of different cleanliness classification and air pressure set points.

As-built cleanroom. A cleanroom that is complete, with all services connected and functional, but not containing production equipment, materials, or personnel in the space.

Aseptic space. A space controlled such that bacterial growth is contained within acceptable limits. This is not a sterile space, in which absolutely no life exists.

At-rest cleanroom. A cleanroom that is complete with production equipment and materials and is operating, but without personnel in the room.

CFU (colony-forming unit). A measure of bacteria present in a pharmaceutical processing space, measured by sampling as part of performance qualification or routine operational testing.

Challenge. An airborne dispersion of particles of known sizes and concentration used to test filter integrity and efficiency.

Cleanroom. A specially constructed enclosed space with environmental control of particulates, temperature, humidity, air pressure, airflow patterns, air motion, vibration, noise, viable organisms, and lighting.

Clean space. A defined area in which particle concentration and environmental conditions are controlled at or below specified limits.

Contamination. Any unwanted material, substance, or energy.

Commissioning. A quality-oriented process for achieving, verifying, and documenting that the performance of facilities, systems, and assemblies meets defined objectives and criteria.

Conventional-flow cleanroom. A cleanroom with nonunidirectional or mixed airflow patterns and velocities.

Critical parameter. A space variable (e.g., temperature, humidity, air changes, room pressure, particulates, viable organisms) that, by law or per pharmaceutical product development data; affects product strength, identity, safety, purity, or quality (SISPQ).

Critical surface. The surface of the work part to be protected from particulate contamination.

Design conditions. The environmental conditions for which the clean space is designed.

DOP. Dioctyl phthalate, an aerosol formerly used for testing efficiency and integrity of HEPA filters.

ESD. Electrostatic discharge.

E.U. GMP. European Union guidelines for GMP pharmaceutical manufacturing.

Exfiltration. Air leakage from a room through material transfer openings; gaps between personnel/pass-through access doors and their respective jambs; window frame/glass interfaces; wall/ceiling and wall/floor interfaces; electrical/data outlets and other room boundary penetrations. The air leakage results from differential pressure across gaps in walls or barriers.

FDA. U.S. Food and Drug Administration.

First air. Air supplied directly from the HEPA filter before it passes over any work location.

GMP. Good manufacturing practice, as defined by *Code of Federal Regulations* (CFR) 21CFR210, 211 (also, cGMP = current GMP).

High-efficiency particulate air (HEPA) filter. A filter with a minimum efficiency of 99.97% of 0.3 μm particles.

IEST. Institute of Environmental Sciences and Technology.

Infiltration. Air leakage into a space from adjoining space(s) that are at a higher pressure.

ISPE. International Society for Pharmaceutical Engineering.

ISO. International Organization for Standardization. For comparison of ISO documents and IEST recommended practices, see Table 2.

ISO 14644-1. Specifies airborne particulate cleanliness classes in cleanrooms and clean zones. ISO (International Organization for Standardization) *Standard* 14644-1 is an international standard for cleanrooms. Table 1 and Figure 1 summarize the ISO standard classes.

Laminar flow. See Unidirectional flow.

Leakage. The movement of air into or out of a space due to its pressure relationship to surrounding space(s).

Makeup air. Outdoor air introduced to the air system for ventilation, pressurization, and replacement of exhaust air.

Minienvironment/Isolator. A barrier, enclosure, or glove box that isolates products from production personnel and other contamination sources to improve process consistency while reducing resource consumption.

Monodispersed particles. An aerosol with a narrow band of particle sizes, generally used for challenging and rating HEPA and UPLA air filters.

Nonunidirectional flow workstation. A workstation without uniform airflow patterns and velocities.

The preparation of this chapter is assigned to TC 9.11, Clean Space.

Table 1   Airborne Particle Concentration Limits from ISO *Standard* 14644-1

ISO 14644 Class	0.1 μm	0.2 μm	0.3 μm	0.5 μm	1.0 μm	5.0 μm
	Particles per m³					
1	10	2				
2	100	24	10	4		
3	1000	237	102	35	8	
4	10,000	2370	1020	352	83	
5	100,000	23,700	10,000	3520	832	29
6	1,000,000	237,000	102,000	35,200	8320	293
7				352,000	83,200	2930
8				3,520,000	832,000	29,300
9				35,200,000	8,320,000	293,000

Note: Values shown are the concentration limits for particles equal to and larger than the sizes shown.
$C_n = 10^N (0.1/D)^{2.08}$ where C_n = concentration limits in particles/m³, N = ISO class, and D = particle diameter in μm

Table 2   Filter Media Types, Efficiencies, and Applications

Filter Type	Filter Efficiency, %, at Particle Size, μm	Filter Application
A	99.97% at 0.3	Industrial, hospital, food
B	99.97% at 0.3	Nuclear
C	99.99% at 0.3	Unidirectional flow (semiconductor, pharmaceuticals)
D	99.999% at 0.3	Semiconductor, pharmaceutical
E	99.97% at 0.3	Hazardous biological
F	99.97% at 0.12	Semiconductor

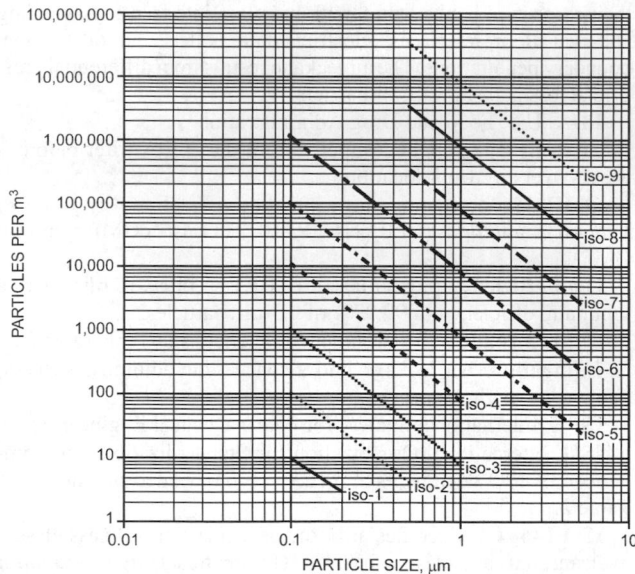

Fig. 1   Air Cleanliness Classifications ISO *Standard* 14644-1

Offset flow. The sum of all space leakage airflows; the net flow difference between supply airflow rate and exhaust and return airflow rates.

Operational cleanroom. A cleanroom in normal operation with all specified services, production equipment, materials, and personnel present and performing their normal work functions.

Oral product. A pharmaceutical product to be taken by mouth by the patient. They are usually not manufactured in aseptic spaces.

PAO. Polyalphaolefin, a substitute for DOP in testing HEPA filters.

Parenteral product. A pharmaceutical product to be injected into the patient. Parenterals are manufactured under aseptic conditions or are terminally sterilized to destroy bacteria and meet aseptic requirements.

Particle concentration. The number of individual particles per unit volume of air.

Particle size. The apparent maximum linear dimension of a particle in the plane of observation.

Polydispersed particles. An aerosol with a broad band of particle sizes, generally used to leak-test filters and filter framing systems.

Qualification. Formal, quality-driven commissioning and documenting the proper operating of a system through established installation, operational, and performance qualification procedures (with approvals).

Qualification protocol. A written description of activities necessary to qualify a specific cleanroom and its systems, with required approval signatures.

Room classification. Room air cleanliness class (Figure 1, Table 1).

SOP. Standard operating procedure.

Topical product. A pharmaceutical product to be applied to the skin or soft tissue as a liquid, cream, or ointment, which therefore does not need to be aseptic. Sterile ophthalmic products, though, are usually manufactured aseptically.

ULPA (ultralow-penetration air) filter. A filter with a minimum of 99.999% efficiency at 0.12 μm particle size.

Unidirectional flow. Formerly called laminar flow. Air flowing at constant and uniform velocity in the same direction.

Validation. A systematic, quality-driven approach for verifying and documenting that a pharmaceutical process is designed, installed, functions, and is maintained properly involving sequential executions of installation qualification, operational qualification, and performance qualification activities.

Workstation. An open or enclosed work surface with direct air supply.

CLEAN SPACES AND CLEANROOM APPLICATIONS

Use of clean space environments in manufacturing, packaging, and research continues to grow as technology advances and the need for cleaner work environments increases. The following major industries use clean spaces for their products:

Pharmaceuticals/Biotechnology. Preparations of pharmaceutical, biological, and medical products require clean spaces to control viable (living) particles that could produce undesirable bacterial growth and other contaminants.

Microelectronics/Semiconductor. Advances in semiconductor microelectronics drive cleanroom design. Semiconductor facilities are a significant percentage of all cleanrooms in operation in the United States, with most newer semiconductor cleanrooms being ISO 14644-1 Class 5 or cleaner.

Aerospace. Cleanrooms were first developed for aerospace applications to manufacture and assemble satellites, missiles, and aerospace electronics. Most applications involve large-volume spaces with cleanliness levels of ISO 14644-1 Class 8 or cleaner.

Miscellaneous Applications. Cleanrooms are also used in aseptic food processing and packaging; manufacture of artificial limbs and joints; automotive paint booths; crystal; laser/optic industries; and advanced materials research.

Hospital operating rooms may be classified as cleanrooms, but their primary function is to limit particular types of contamination rather than the quantity of particles present. Cleanrooms are used in patient isolation and surgery where risks of infection exist. For more information, see Chapter 8, Health Care Facilities.

AIRBORNE PARTICLES AND PARTICLE CONTROL

Airborne particles occur in nature as pollen, bacteria, miscellaneous living and dead organisms, and windblown dust and sea spray. Industry generates particles from combustion, chemical vapors, manipulation of material, and friction in manufacturing equipment. Personnel are a prime source of particle generation (e.g., skin flakes, hair, clothing lint, cosmetics, respiratory emissions, bacteria from perspiration). These airborne particles vary from 0.001 μm to several hundred micrometers. Particles larger than 5.0 μm tend to settle quickly by gravity, whereas particles smaller than 1.0 μm can take days to settle. In many manufacturing processes, these airborne particles are viewed as a source of contamination and can provide a pathway for biological contaminants.

Particle Sources in Clean Spaces

In general, the origins of cleanroom particles are described by two categories: external and internal.

External Sources. Externally sourced particles enter the clean space from the outside; via infiltration through doors, windows, and wall penetrations; surface contamination on personnel, material and equipment entering the space; and outdoor makeup air entering through the HVAC system.

In a typical cleanroom, external particle sources normally have little effect on overall cleanroom particle concentration because HEPA filters remove particulates from the supply air and the cleanroom is operated at a higher pressure than surrounding spaces to prevent infiltration. However, the particle concentration in clean spaces at rest relates directly to ambient particle concentrations. External sources are controlled primarily by air filtration, room pressurization, and sealing space penetrations.

Internal Sources. People, cleanroom surface shedding, process equipment, and the manufacturing process itself generate particles in the clean space. Cleanroom personnel, if not properly gowned, could be the largest source of internal particles, generating several thousand to several million particles per minute in a cleanroom. Personnel-generated particles are controlled with new cleanroom garments, proper gowning procedures, and airflow designed to continually shower personnel with clean air and direct the airflow away from critical areas and surfaces. As personnel work in the cleanroom, their movements may reentrain airborne particles from other sources. Other activities, such as writing, may also cause higher particle concentrations. Door swings or equipment challenges can produce transient differential pressure excursions, which may lead to particle infiltrations.

Though particle concentrations in the cleanroom may be used to define its cleanliness class, actual particle deposition on the product critical surface is of greater concern. The sciences of aerosols, filter theory, and fluid motions are the primary sources of understanding contamination control. Cleanroom designers may not be able to control or prevent internal particle generation completely, but they may anticipate internal sources and design control mechanisms and airflow patterns to limit their effect on the product.

Fibrous Air Filters

Proper air filtration prevents most externally generated particles from entering the cleanroom via the HVAC system. High-efficiency air filters come in two types: high-efficiency particulate air (HEPA) filters and ultralow-penetration air (ULPA) filters. HEPA and ULPA filters use glass fiber paper technology; laminates and nonglass media for special applications also have been developed. HEPA and ULPA filters are usually constructed in a minipleat form with aluminum, coated string, filter paper, or hot-melt adhesives as pleating separators. Filters pleat depths are available from 1 to 12 in. in depth; available media area increases with deeper-pleated filters and closer pleat spacing.

There are four mechanisms by which HEPA and ULPA filters capture particulate: (1) straining, (2) inertia, (3) interception, and (4) diffusion. Straining occurs in a filter when the particles enter passages between two or more fibers that have dimensions less than the particle diameter. In inertia capture, particles traveling in airstream through fiber material have too much mass to stay in the airstream as it bends through the filter fibers; it leaves the airstream and attaches to filter fibers. In interception, particles with mass small enough to stay in the airstream nevertheless touch the filter fiber and are attached. Diffusion captures very small particles that move randomly due to Brownian motion; they touch and subsequently attach to filter fibers. Theories and models verified by empirical data indicate that interception and diffusion are the dominant capture mechanisms for HEPA and ULPA filters. Fibrous filters have their lowest removal efficiency at the most penetrating particle size (MPPS), which is determined by filter fiber diameter, volume fraction or packing density, and air velocity. For most HEPA and ULPA filters, the MPPS is between 0.1 to 0.3 μm. Thus, HEPA and ULPA filters have rated efficiencies based on 0.3 and 0.12 μm particle sizes, respectively. Different types of filter media are produced to meet various cleanroom applications. See Table 2 for the different types of filter media, their filter efficiency, and their typical applications.

AIR PATTERN CONTROL

Air turbulence in the clean space is strongly influenced by air supply and return configurations, foot traffic, and process equipment layout. Optimizing airflow patterns to match operational requirements is the first step of good cleanroom design. User requirements for cleanliness level, process equipment layout, available space for installing air pattern control equipment (i.e., air handlers, clean workstations, environmental control components, etc.), and project financial considerations all influence the final air pattern design selection.

Numerous airflow pattern configurations are possible, but they fall into two general categories: nonunidirectional airflow (commonly called turbulent). and unidirectional airflow (often mistakenly called laminar flow).

Nonunidirectional Airflow

Nonunidirectional airflow has either multiple-pass circulating characteristics or nonparallel flow. Variations are based primarily on the location of supply air inlets and return/exhaust air outlets and air filter locations. Examples of unidirectional and nonunidirectional airflow of pharmaceutical cleanroom systems are shown in Figures 2 and 3. Air is typically supplied to the space through supply diffusers with HEPA filters (Figure 2) or through supply diffusers with HEPA filters in the ductwork or air handler (Figure 3). In a mixed unidirectional and nonunidirectional system, outside air is prefiltered in the supply and then HEPA-filtered at workstations in the clean space (see the left side of Figure 3).

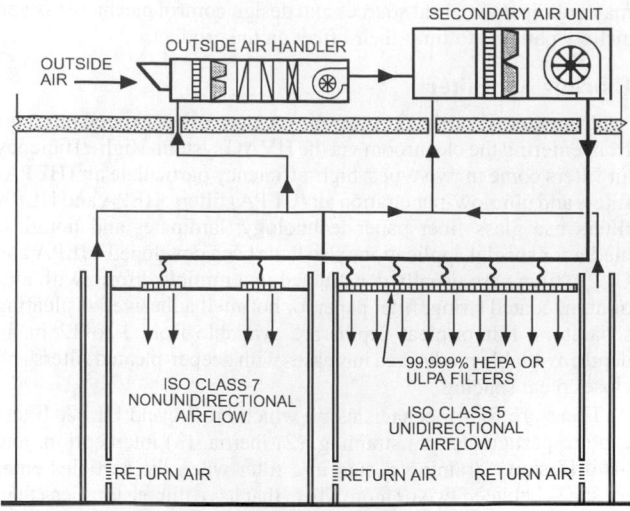

Fig. 2   ISO Class 7 Nonunidirectional Cleanroom with Ducted HEPA Filter Supply Elements and ISO Class 5 Unidirectional Cleanroom with Ducted HEPA or ULPA Filter Ceiling

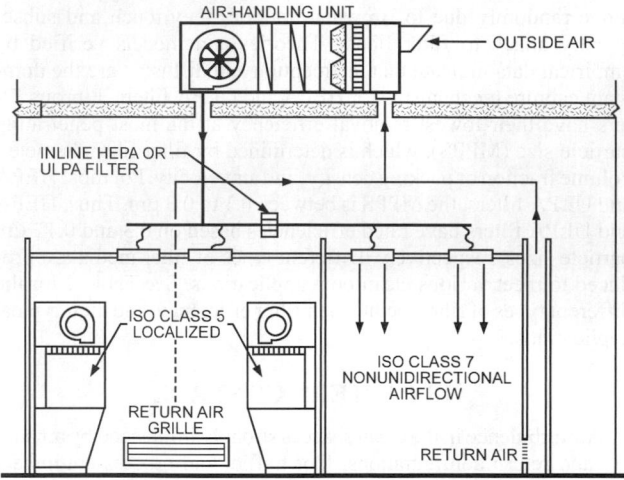

Fig. 3   ISO Class 7 Nonunidirectional Cleanroom with HEPA Filters Located in Supply Duct and ISO Class 5 Local Workstations

Nonunidirectional airflow may provide satisfactory contamination control for ISO 14644-1 Cleanliness Classes 6 through 8. Attaining desired cleanliness classes with designs similar to Figures 2 and 3 presupposes that the major space contamination is from makeup air and that contamination is removed by HEPA filters located in the airhandler or ductwork, or through HEPA filter supply devices. When internally generated particles are of primary concern, clean workstations are provided in the clean space.

Unidirectional Airflow

Unidirectional airflow, though not truly laminar, is characterized as air flowing in a single pass in a single direction through a cleanroom with generally parallel streamlines. Ideally, flow streamlines would be uninterrupted; although personnel and equipment in the airstream distort the streamlines, a state of constant velocity is approximated. Most particles that encounter an obstruction in unidirectional airflow continue around it as the airstream reestablishes itself downstream of the obstruction.

Air patterns are optimized and air turbulence is minimized in unidirectional airflow. In a **unidirectional-flow space**, air is typically introduced through ceiling HEPA or ULPA filters and returned through a raised access floor or at the base of sidewalls. Because air enters from the entire ceiling area, this configuration produces nominally parallel airflow. In a horizontal-flow cleanroom, air enters one wall and returns on the opposite wall.

A **downflow cleanroom** has a ceiling with HEPA filters. As the space cleanliness classification becomes more stringent, the space air change rate and the number of HEPA filters may increase. Typically, for an ISO Class 5 or cleaner space, the ceiling has 100% HEPA filter coverage. Ideally, a grated or perforated floor serves as the air return/exhaust. In this configuration, clean air flows downward past a contamination source, picks up the contamination particle, and removes it directly down through the floor to prevent the particle from contacting the critical surface of a product. However, this type of floor is inappropriate for pharmaceutical cleanroom applications, which typically have solid floors and low-level wall returns.

Special attention should be given to ceiling HEPA and ULPA filter design, selection, and installation to ensure a leakproof ceiling system. Properly sealed filters in the ceiling can provide the cleanest air presently available in a cleanroom.

In a **horizontal-flow cleanroom**, the supply wall consists entirely of HEPA or ULPA filters supplying air at approximately 90 fpm or less across the entire section of the space. Return/exhaust air exits through the return wall at the opposite end of the space. As with the downflow cleanroom, the horizontal-flow cleanroom removes contamination generated in the space and minimizes cross contamination perpendicular to airflow. However, a major limitation to horizontal-flow cleanrooms is that downstream air becomes contaminated. Air leaving the filter wall is the cleanest; it then becomes contaminated by the process as it flows past the first workstation. Process activities can be arranged to have the most critical operations at the clean end of the space, with progressively less critical operations located toward the return or dirty end of the space.

ISO 14644-1 does not specify velocity requirements, so the actual velocity is as specified by the owner or owner's agent. The Institute of Environmental Sciences and Technology (IEST) published recommended air change rates for various cleanliness classes, which should be reviewed by the owner; however, the basis for the ranges is not known. Acceptable cleanliness has been demonstrated at lower air change rates, suggesting that results may depend more on filter efficiency and coverage than on air changes. Careful testing should be performed to ensure that required cleanliness levels are maintained. Other reduced-air-volume designs may use a mixture of high- and low-pressure-drop HEPA filters, reduced coverage in high-traffic areas, or lower velocities in personnel corridor areas.

Unidirectional airflow systems have a predictable airflow path that airborne particles tend to follow. Without good filtration practices, unidirectional airflow only indicates a predictable path for particles. However, superior cleanroom performance may be obtained with a good understanding of unidirectional airflow, which remains parallel to below the normal work surface height of 30 to 36 in., but deteriorates when it encounters obstacles (e.g., process equipment, work benches) or over excessive distances. Personnel movement also degrades flow, resulting in a cleanroom with areas of good unidirectional airflow and areas of turbulent airflow.

Turbulent zones have countercurrents of air with high velocities, reverse flow, or no flow at all (stagnancy). Countercurrents can produce stagnant zones where small particles may cluster and settle onto surfaces or product; they may also lift particles from contaminated surfaces and deposit them on product surfaces.

Cleanroom mockups may help designers avoid turbulent airflow zones and countercurrents. Smoke, neutral-buoyancy helium-filled soap bubbles, and nitrogen vapor fogs can make air streamlines visible in the mockup.

Computational Fluid Dynamics (CFD)

CFD models of particle trajectories, transport mechanisms, and contamination propagation are commercially available. Flow analysis with computer models may compare flow fields associated with different process equipment, work benches, robots, building exterior envelope, personnel, and building structural design. Flow patterns and air streamlines are analyzed by computational fluid dynamics for laminar and turbulent flow where incompressibility and uniform thermophysical properties are assumed. Design parameters may be modified to determine the effect of airflow on particle transport and flow streamlines, thus avoiding the cost of mockups.

Major features and benefits associated with most computer flow models are

- Two- or three-dimensional modeling of cleanroom configurations, including people and equipment
- Modeling of unidirectional airflows
- Multiple air inlets and outlets of varying sizes and velocities
- Allowances for varying boundary conditions associated with walls, floors, and ceilings
- Aerodynamic effects of process equipment, workbenches, and people
- Prediction of specific airflow patterns, velocities, and temperature gradients of all or part of a cleanroom
- Simulation of space pressures by arranging supply, return, exhaust, planned exfiltration, and planned infiltration airflows
- Reduced cost associated with new cleanroom design verification
- Graphical representation of flow streamlines and velocity vectors to assist in flow analysis (Figures 4 and 5)
- Graphical representation of simulated particle trajectories and propagation (Figure 6)

Research has shown good correlation between flow modeling by computer and that done in simple mockups. However, computer flow modeling software should not be considered a panacea for cleanroom design because of the variability of individual project conditions.

Air Change Rate Determination

Cleanroom HVAC Systems can consume up to 50 times more energy than those used in commercial spaces of the same size. The airflow change rate in cleanrooms is typically higher than in general-purpose buildings. The airflow rates in cleanrooms must meet not only the heating and cooling loads, but also the dilution requirements to reduce room particle concentration. It is critical to realize that most particles in a cleanroom are not from HEPA-filtered supply air, but are generated inside the cleanroom. A very high air change rate is not typically needed for cooling, heating, or ventilation loads but mainly for dilution. Cleanroom design engineers have traditionally used conservative, simplified rule-of-thumb values published in *Federal Standard* FS-209 and IEST *Recommended Practices* RP-12.1. This existing approach solely uses the required room cleanliness class to determine an air change per hour (ach) value arbitrarily from a wide range. However, the existing method ignores many critical variables that could significantly impact the room particle concentration in terms of air change rate requirements, such as room internal particle generation rate, particle surface deposition, particle entry through filtered supply air, particle exit through return and exhaust air, and leakage air (particle loss or gain) under pressurization or depressurization. Intuitively, for example, activities that generate higher levels of dusts would need a higher air change rate to dilute particle concentration than those that generate at a lower level, but the existing table method uses an oversimplified approach that ignores such differences.

Each cleanroom facility is unique; its location, building construction, production or process activities, space configurations, HVAC

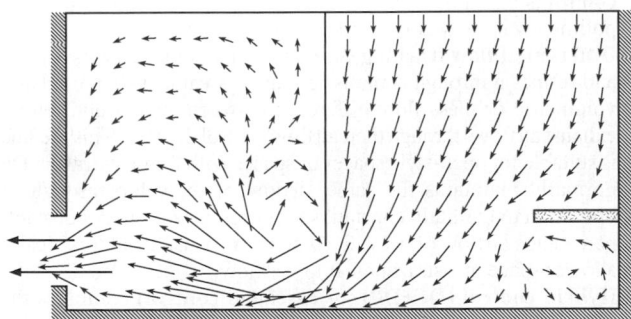

Fig. 4   Cleanroom Airflow Velocity Vectors Generated by Computer Simulation

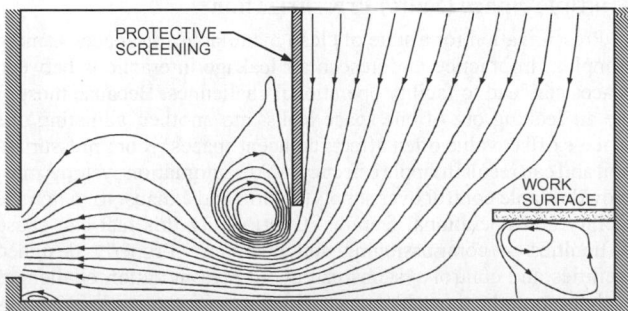

Fig. 5   Computer Modeling of Cleanroom Airflow Streamlines

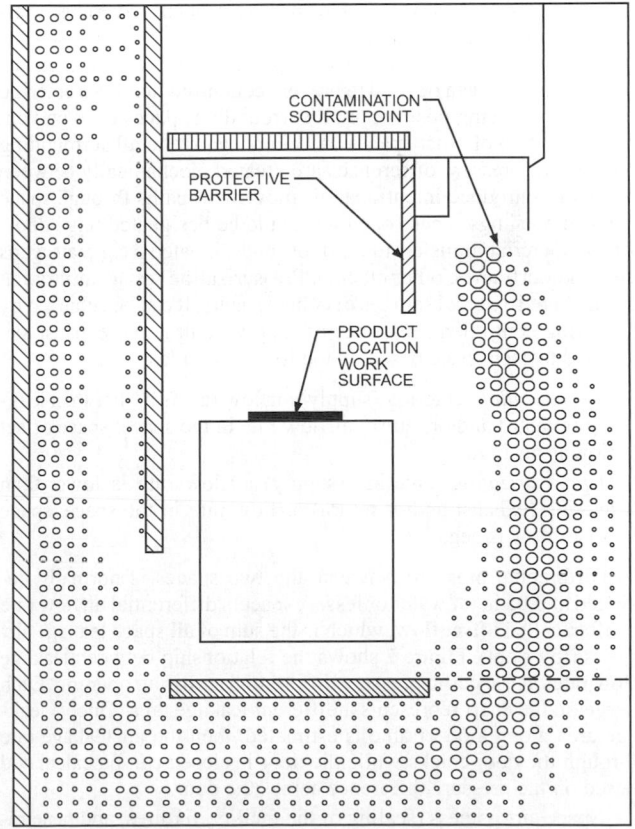

Fig. 6   Computer Simulation of Particle Propagation in Cleanroom

systems, room cleanliness requirements, etc., can impact the air change requirement for each room. Using a rough, oversimplified approach without considering all these variables could cause either significant energy waste or underdesigned HVAC systems. Many published survey reports have indicated that airflow quantities for cleanrooms often are overdesigned and cause significant energy waste. To save fan energy and precisely design related HVAC systems in cleanrooms, modeling technologies have been developed and published that provide more scientifically based, quantitative tools to calculate required airflows, rather than using previous rule-of-thumb values. As a part of this trend, IEST RP-12.2 has replaced the previous RP-12.1's "recommended" air change table with a more loosely defined "typical" air change table as an interim step. See the publications listed in the Bibliography for more detailed information.

AIRFLOW DIRECTION CONTROL BETWEEN CLEAN SPACES

Airflow direction control between clean spaces having different cleanliness classifications is complex but critical to prevent airborne cross contamination. Particulate contaminants could infiltrate a cleanroom through doors, cracks, pass-throughs, and other penetrations for pipes, ducts, conduits, etc. An effective method of contamination control is space pressurization: air moves from high pressure to low. Normally, the cleanest cleanroom(s) having the most critical operations should be designed with the highest pressure, with decreasing pressures corresponding to lower cleanliness classifications. The desired flow path should be from the area of cleanest, to less clean, to less contaminated, and then to dirty areas.

Space Pressurization

Controlling contaminants in cleanrooms requires controlling the direction of airflow between adjacent spaces that have various levels of cleanliness classification(s). This is achieved by establishing a pressure differential between the spaces. The pressurization set point for a space can be used to prevent contamination from entering the space by being positive to all surrounding spaces or to prevent contamination of other spaces by being negative to all surrounding spaces. Air pressure differences are created mechanically between spaces to introduce intentional air movement paths through space leakage openings. These openings could be designated (e.g., doorways, material transfer tunnel) or undesignated (e.g., air gaps around doorframes, other cracks). Pressurization resists infiltration of unfiltered external sources of contaminants. It can be achieved by arranging controlled flow rates of supply, return, and exhaust airstreams to each space based on the following rules:

- *Pressurization*: entering (supply) airflow rate is higher than leaving (exhaust and/or return) airflow rate in the space; space offset flow is positive.
- *Depressurization*: entering (supply) airflow rate is lower than leaving (exhaust and/or return) airflow rate in the space; space offset flow is negative.

Differential pressure between any two spaces is normally designed at 0.05 in. of water or less. A space's differential airflow rate is often called **offset flow**, which is the sum of all space leakage airflows (in or out). Figure 7 shows the relationship between leakage flow rates at a specific pressure differential across an opening. Each curve on the chart represents a different leakage area. Once a leakage area along a doorframe is estimated, then the air leakage rate through the door cracks while the door is closed can be calculated based on the pressure difference across the door.

Space airtightness (sealing of the facility, fixtures, and penetrations) is the key element in the relationship between the space's flow offset value and the resulting pressure differential, and each space's airtightness is unique and unknown unless tested. Treatment of a

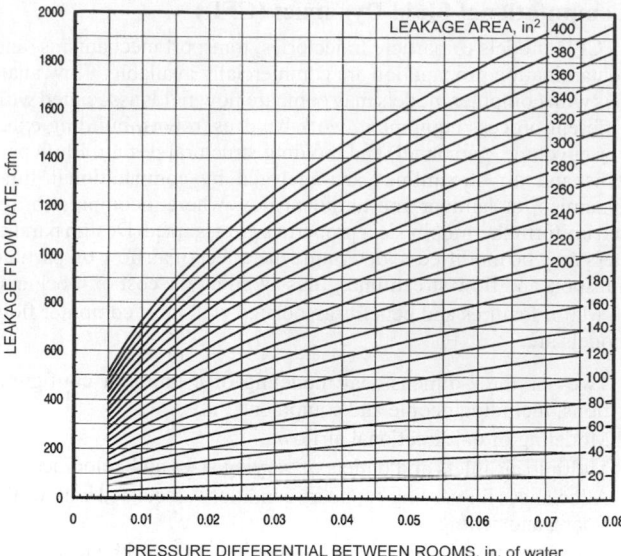

Fig. 7　Flow Rate through Leakage Area under Pressure Differential

space's offset value defines a pressurization control strategy. Typical pressurization control techniques include the following:

- **Direct pressure-differential control (DP)** uses a pressure differential sensor to measure the pressure difference between a controlled space and an adjacent space (e.g., a corridor). DP is suitable for a tightly constructed space with limited traffic. It basically ignores the specific offset value as required; instead, it directly controls the airflow control devices to achieve the required pressure differential between the controlled space and an adjacent space. A door switch is recommended to trigger a reduced pressure-differential set point if the door opens or the DP control is based on average readings over a period of time (e.g., polling every 10 seconds and averaging over a minute).

- **Differential flow tracking control (DF)** assumes an offset value and refines it through commissioning; this value is then used as a volumetric or mass flow difference between supply and return/exhaust airflows through their airflow control devices. This method is suitable for open-style spaces or spaces with frequent traffic. DF normally maintains the same airflow offset value throughout operation to maintain constant space pressurization. A constant-percentage airflow offset value is sometimes used, but this creates a lower space pressurization at lower flow.

- **Hybrid control (DF+DP)** (or **cascaded control**) combines the pressure accuracy of DP and the stability of DF. The offset value is resettable based on the pressure differential reading. The offset value reset schedule is predetermined, and the controller's parameters are adjusted or calibrated manually in the field.

Multiple-Space (Suite) Pressurization

Pressurization for a suite of clean manufacturing spaces is more complex. In practice, unforeseen air leakage interactions between spaces can lead to facility operational challenges. Because most of the air leaking out of one space leaks into another, adjusting one space's offset value often affects adjacent spaces' room pressurization and can result in ripple effects. HVAC automation systems must provide stable control over supply, return, and exhaust to match the facility and operational features. Overlooking this fact can cause difficulties in commissioning and operation. Properly designed facilities and control systems can avoid pressurization challenges such as sporadic, unstable, or unachievable pressurization requirements. For more information and procedures, consult the sources in the Bibliography.

A **room pressure and flow (P&F) diagram** for the controlled area (suite, zone, or floor) is often provided in design documents, and can be used as the basis of continuous quality control of cleanroom environmental parameters. The P&F diagram should indicate

- Airflow design settings (values) of all supply, return, and exhaust registers for each space inside the controlled area
- Desired space pressure value with an acceptable tolerance in each pressure-controlled space
- Resulting leakage flow directions (due to space pressure differentials) and their estimated leakage flow values through doors at closed-door conditions

The three traditional pressure-control methods (DP, DF, and DF+DP) require setting and field-adjusting the airflow offset value. A more robust strategy is to control all spaces' pressures together as an optimized system, instead of independently. **Adaptive DF+DP** directly accounts for leakage flows between spaces in a suite, and actively adjusts each space's airflow offset according to an online pressurization model. It uses airflow and pressure differential measurements to estimate characteristics of leakage between spaces and adjust flow offsets automatically. This adaptive approach may be more suitable for suite pressurization. For design procedures and control strategies, see the related literature in the Bibliography.

TESTING CLEAN AIR AND CLEAN SPACES

Because early cleanrooms were largely for governmental use, testing procedures were set by government standards. U.S. *Federal Standard* 209 was widely accepted, because it defined air cleanliness levels for clean spaces around the world, but it was formally withdrawn in 2001. ISO standards now govern. Standardized testing methods and practices have been published by the Institute of Environmental Sciences and Technology (IEST), American Society for Testing and Materials (ASTM), and others.

Three basic test modes are used to evaluate a facility: (1) as built, (2) at rest, and (3) operational. A cleanroom cannot be fully evaluated until it is operated in operational test mode, which includes all equipment operating and personnel present. Thus, techniques for conducting initial performance tests and operational monitoring must be similar.

As noted previously, contamination sources can be generated within the space or infiltrate into the space from an external source. The level of space contamination can be monitored using discrete particle counters, which use laser or light-scattering principles for detecting particles of 0.01 to 5 μm. For particles 5 μm and larger, microscopic counting can be used, with particles collected on a membrane filter through which a specific volume of sample air has been drawn.

HEPA filters in unidirectional flow and ISO 14644-1 Class 5 ceilings should be tested for pinhole leaks at the filter media, sealant between media and filter frame, filter frame gasket, and filter bank supporting frames. The filter frame interface with the wall or ceiling should also be tested. A filter bank pinhole leak can be extremely critical, because the leakage rate varies inversely as the square of the pressure drop across the hole. (The industry term *pinhole* used to describe the leak site is a misnomer. The size is almost never that of a hole formed by a pin, but is actually many times smaller.)

IEST testing procedures describe 12 tests for cleanrooms. The tests that are applicable to each specific cleanroom project must be determined based on the specific cleanroom's criteria.

PHARMACEUTICAL AND BIOMANUFACTURING CLEAN SPACES

Pharmaceutical product manufacturing facilities require careful assessment of many factors, including HVAC, controls, room finishes, process equipment, room operations, and utilities. Flow of equipment, personnel, and product must also be considered along with system flexibility, redundancy, and maintenance shutdown strategies. It is important to involve designers, operators, commissioning staff, quality control, maintenance, constructors, validation personnel, and the production representative during the conceptual stage of design. Critical variables for room environment and types of controls vary greatly with the clean space's intended purpose. It is particularly important to determine critical parameters with quality assurance to set limits and safety factors for temperature, humidity, room pressure, and other control requirements.

In the United States, regulatory requirements and specification documents such as 21CFR210 and 211, ISPE Design Guides, and National Fire Protection Association (NFPA) standards describe GMP requirements. The goal of GMP is to achieve a proper and repeatable method of producing therapeutic, medical, and similar products free from microbial and particle contaminants.

In the United States and other countries, the one factor that differentiates pharmaceutical processing suites from other clean spaces (e.g., for electronic and aerospace) is the requirement to meet government regulations and inspection for product licensing [e.g., U.S. Food and Drug Administration (FDA)]. It is important to include the appropriate regulatory arms, such as the FDA's Center for Biologics Evaluation and Research (CBER) or the Center for Drug Evaluation and Research (CDER), early in the concept design process.

It is important to develop a qualification plan (QP) early in the design process. Functional requirement specifications (FRS), critical parameters and acceptance criteria, installation qualification (IQ), operational qualification (OQ), and performance qualification (PQ) in the cleanroom suites are all required to ensure proper process performance and validation. IQ, OQ, and PQ protocols, in part, set the acceptance criteria and control limits for critical environmental parameters such as temperature, humidity, room pressurization, air change rates, and operating particle counts (or air classifications). These protocols must receive defined discipline approvals in compliance with the owner's quality policies. The qualification plan must also address master document updates, SOPs, preventive maintenance (PM), and operator and maintenance personnel training.

Biomanufacturing and pharmaceutical aseptic clean spaces are typically arranged in operational suites based on specific process and formulation requirements. For example, common convention positions an aseptic core (ISO 14644-1 Class 5) filling area in the innermost room, which is at the highest pressure, surrounded by areas of descending pressure and increasing particulate classes and bacterial levels (see Figure 8).

In aseptic processing facilities, the highest-cleanliness area is intentionally placed within lower-cleanliness areas and separated by space air cleanliness classification and air pressure differences by air locks. A common pressure difference is about 0.05 in. of water or less between air cleanliness classifications, with the higher-cleanliness space having the higher pressure. Lower pressure differences may be acceptable if they are proven effective. A pressure differential is generally accepted as good manufacturing practice to inhibit particles from entering a clean suite.

Where there are spaces adjoined in series that all have different cleanliness classifications, a multiple-step pressurization cascade should be implemented, which should have air flow from the cleanest spaces to the least clean spaces. Normally, three pressure steps are commonly used for biosafety level (BSL) 3 and ISO Class 6, 7, or 8 applications; four pressure steps are desirable for BSL-4, Class 5 or cleaner applications. Air locks are effective at minimizing potential particle contamination from surrounding nonclassified or less-clean areas; selection depends on the type of cleanroom (Figure 9), because some that involve fume or biological agent operations may have a containment provision. For biological agent operations, the U.S. Centers for Disease Control and Prevention (CDC) and National Institutes of Health (NIH) define four biosafety levels (BSL-1 to BSL-4), discussed in more detail in Chapter 16.

An air lock is a transitional room between adjacent rooms to prevent airborne cross contamination. Based on relative space pressure levels, air locks can be classified as follows:

- **Cascading:** Air lock pressure is between pressures in cleanroom and corridor

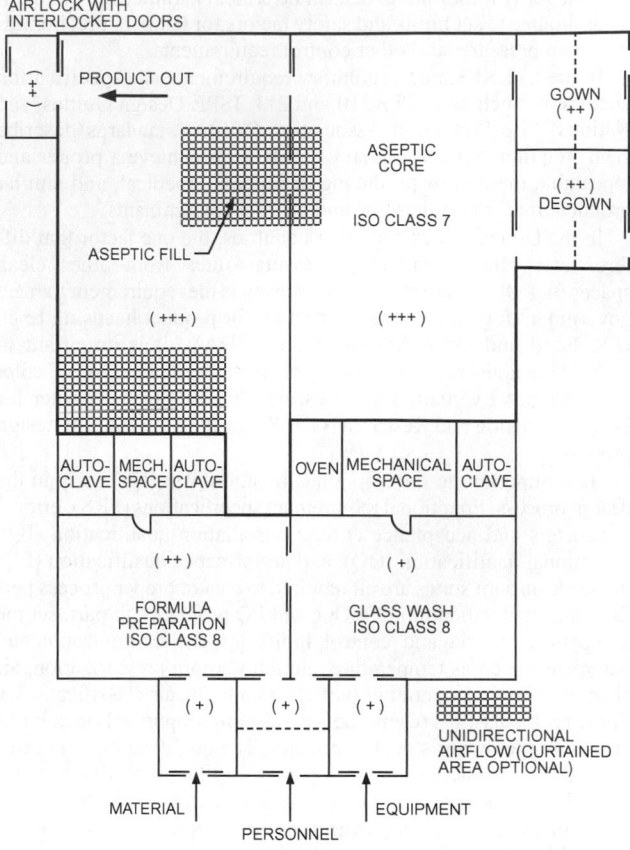

Fig. 8   Typical Aseptic Suite

- **Bubble:** Air lock pressure is above pressures in cleanroom and corridor
- **Sink:** Air lock pressure is below pressures in cleanroom and corridor
- **Dual-compartment:** A bubble and a sink air lock are connected

Double-door air locks are often used at cleanroom entrances and exits. It is important that both doors are not open at the same time, to avoid cross contamination. A **required time delay (RTD)** needs to be specified between door openings, to minimize possible contamination during door opening. The RTD should be long enough for HEPA-filtered clean supply air to partially or fully replace the entire air volume of the air lock room at least once before the second door is allowed to open. RTD operational procedures often use hard interlocks (i.e., the second door cannot be opened until after the required time delay) or soft interlocks, in which procedures are supplemented by lights or alarms.

Design Concerns for Pharmaceutical Cleanrooms

The owner and designer must define the tolerable range of variable value (**acceptance criterion**) for each critical parameter. In that range, the product's safety, identity, strength, purity, and quality are not affected. The owner should define **action alarm** points at the limits of acceptance criteria, beyond which exposed product may be adulterated. The designer should select tighter (but achievable) target design values for critical parameters (in the range of acceptance criteria) along with appropriate values for warning alerts.

Facilities manufacturing penicillin or similar antibiotics (e.g., cephalosporins) must be physically isolated from other manufacturing areas and served by a dedicated HVAC system. Other processes also require dedicated HVAC systems, including high-potency formula and formula that must have dedicated production facilities.

Facilities manufacturing aseptic/sterile products derived from chemical synthesis may have different requirements than those manufacturing biological or biotechnological products. The owner must define the inspecting agency's requirements.

The United States Pharmacopoeia (USP) limits temperatures to which finished pharmaceutical products may be exposed to 59 to 86°F. The production facility may need tighter limits than these, based on the owner's observed product data. Personnel comfort is a factor in design. Personnel perspiring in their protective overgarments can increase particulate and microbial counts, so lower

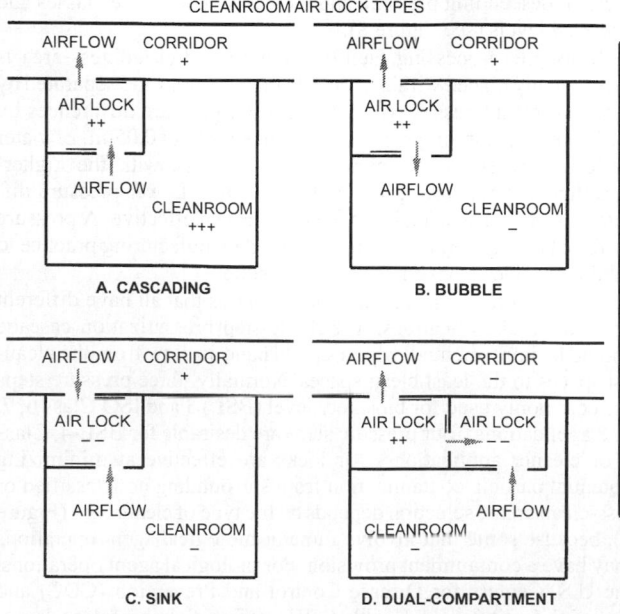

Type of Cleanroom	Air Lock Type	Purpose of Air Lock	Relative Pressure Relationship
• Positive pressure • No fume or bio agent • No containment needed	Cascading	• Prevent cleanroom being contaminated by dirty corridor air • Prevent cleanroom being contaminated from surrounding spaces through cracks	Cleanroom: +++ Air lock: ++ Corridor: +
• Negative pressure • Has fume or bio agent contamination • Containment needed	Bubble	• Prevent cleanroom being contaminated by dirty corridor air • Prevent cleanroom fume or bio agent releasing to corridor	Cleanroom: − Air lock: ++ Corridor: +
• Negative pressure • Has fume or bio agent contamination • Containment needed	Sink	• Prevent cleanroom being contaminated by dirty corridor air • Allow cleanroom fume or bio agent releasing to air lock. No personal protective equipment is needed	Cleanroom: − Air lock: − − Corridor: +
• Negative pressure • Has toxic fume or hazardous bio agent contamination, or has potent compound substances • Containment needed • Personal protection needed	Dual-Compartment	• Prevent cleanroom being contaminated by dirty corridor air • Prevent cleanroom fume or bio agent releasing to corridor • Personal protective equipment (such as pressurized suit and respirator) is required	Cleanroom: − Neg. Air lock: − − Pos. Air lock: ++ Corridor: −

Notes:
1. Excessive negative pressure in cleanroom is not recommended. If it is not surrounded by other clean spaces, untreated dirty air can infiltrate through cracks into cleanroom
2. A cleanroom service corridor often must be designed slightly positive or neutral pressure. Do not design for negative pressure unless a dual-compartment lock is used.

Fig. 9   Air Lock Types and Applications

temperatures and tighter temperature control may be advantageous.

Relative humidity may be critical to the product's integrity. Some products are processed or packaged cold and need a low room dew point to prevent condensation on equipment and vials. Some products are hygroscopic and require lower humidity than a condensing coil can provide; desiccant dehumidification should be considered. Caution must be taken in designing low-humidity (i.e., low-vapor-pressure) spaces to ensure limited moisture migration through walls and ceilings bordering an unclean space. Low-humidity spaces should be provided with air locks to reduce moisture propagation into the low-humidity cleanroom. The importance of positive pressure increases when moisture infiltration potential becomes an element of the design process. Humidification is usually needed for personnel comfort but not usually for product needs; it may also be needed where dust might present an explosion hazard or where low humidity may hinder handling of dry materials. Clean steam (free of chemicals and other additives) is preferred for humidification because it is free of bacteria, but the humidification system should be free of amines or other contaminants if space air might contact the product. Humidification control systems often require careful sensor placement in critical areas and safety shutoff monitors to prevent overhumidification.

Although airborne particles and viable organisms may be minimized by dilution (high air changes) and by supplying filtered air, the most effective control is to minimize release of these contaminants in the space. Personnel and machinery are the most common sources of contamination, and can be isolated from the product by gowning, masks, and isolation barriers. Careful study of how each space operates should reveal the most probable sources of contaminants and help the HVAC designer determine dilution air quantities and locate supply air outlets and return air inlets. Avoid duct liners and silencers in supply air ductwork where contaminants can collect and bacterial and mold spores can accumulate. Ensure special attention is paid to cleaning and degreasing of metal sheeting and air ductwork before installation. Factory-wrapped ducts and components with clean installation and inspection protocols promote cleanroom system cleanliness.

Airborne particle and microbe levels in aseptic processing areas are limited by government regulations, with lower limits for more critical spaces. European and FDA particle limits are for the space in full operational mode, and can also be used conservatively as limits for the space at rest.

Facilities complying with U.S. cGMPs for aseptic processing must meet particle levels with manufacturing under way. (An exception is aseptic powder processing, in which airborne particulate levels at powder filling heads will exceed limits.) There should be no microbial contaminants in the critical-zone airstream, where filling and other critical activities occur; this area should be ISO Class 5. The area immediately around the critical zone should be ISO Class 7. If the critical area is within an isolator, then the area outside the isolator may be ISO Class 8. Less critical support areas can be ISO Class 8. For more detail on facility design, see FDA requirements.

According to the FDA, 20 ach is usually sufficient for ISO Class 8 rooms; ISO Class 7 and 5 areas require significantly higher air change rates. Facility requirements for terminally sterilized products are not defined.

E.U. GMP also contains requirements for aseptic processing, and also addresses terminally sterilized products. Note that many facilities are constructed to meet both E.U. and U.S. GMPs.

Restricted access barrier systems (RABS) are an alternative to a conventional cleanroom or isolator. Use of RABS should be approved by the manufacturer's quality unit during design.

Once product is in containers, the need for particulate control and minimum air changes is reduced or eliminated, depending on the degree of protection provided by product packaging. The owner should determine the necessary critical environmental parameters and acceptance criteria for each space and processing step.

Return openings for space HVAC should be low on the walls, to promote downward airflow from supply to return, sweeping contaminants to the floor and away from the product. In larger spaces, internal return air columns may be necessary. Perforated floors are discouraged because of difficulty cleaning them.

Aseptic facilities usually require pinhole-scanned (integrity-tested) HEPA filters (not ULPA) on supply air. Many facilities install HEPA filters in the supply air to nonaseptic production facilities to minimize cross contamination from other manufacturing areas served by the HVAC system. To increase the life of terminal HEPA filters in aseptic facilities, and to minimize the need to rebalance the supply system because of differential loading of terminal HEPA filters, many designers install a high-capacity HEPA bank downstream of the supply air fan, with constant-volume control to compensate for primary filter pressure changes and any dehumidifier airflow. The final HEPA filter is usually in a sealed gel frame or of a one-piece lay-in design that can be caulked to the ceiling frame, maintaining the integrity of the room envelope.

Aseptic product must be protected by pressurizing the space in which it is exposed, to about 0.05 in. of water above the next lower cleanliness space classification. To keep the pressure differential from dropping to zero when a door is opened, air locks are often used between spaces of different air pressures, especially at the entrance to the aseptic fill space itself. Space pressure is a function of airflow resistance through cracks, openings, and permeable surfaces in the space shell. Consider all potential openings, slots, and door leakage that can affect the amount of air needed to pressurize the space. Because space offset airflows and space pressure are closely related, outdoor or makeup air requirements are often dictated by space pressures rather than by the number of occupants. The HVAC system should be able to handle more makeup air than needed for commissioning, because door seals can deteriorate over time.

ISO Class 5 unidirectional hoods are basically banks of HEPA filters, integrity-tested to be pinhole-free. Because it is difficult to maintain unidirectional flow for long distances or over large areas, the hood should be located as closely as possible to product critical surfaces (work surface). Hood-face velocity is usually 90 fpm or less, but the user should specify velocity and uniformity requirements. A unidirectional hood usually has clear sidewalls (curtains) to promote downward airflow and prevent entrainment of space particles into the hood's zone of protection. Curtains should extend below the product critical surface and be designed to prevent accidental disruption of airflow patterns by personnel. Many production facilities prefer rigid curtains for easier cleaning and sanitization.

Hood fan heat may become a problem, forcing the designer to overcool the space from which the hood draws its air or to provide sensible cooling air directly into the hood's circulating system.

Decontamination

Cleanrooms used for sterile operations are rarely built clean enough for their intended purpose. Before the initial use of the room or after a shutdown, the cleanroom must be decontaminated or disinfected to ensure bioburden levels are at or below acceptable limits. For some operations, such as compounding of sterile preparations, surface disinfection is considered adequate. However, larger-scale cGMP sterile manufacturing operations typically use some type of biological decontamination before final occupancy. Cleanrooms for sterile processing should be designed to accommodate decontamination or disinfection.

Having roots in small-volume spaces (i.e., sealed glove boxes), most early large-volume decontamination processes included using formaldehyde gas generated by heating paraformaldehyde in a frying pan or spraying with a mild peracetic acid and wiping all surfaces, which was very labor intensive. Today, most cleanrooms are decontaminated by using either chlorine dioxide (CD) or hydrogen peroxide. Regardless of the type of decontamination process used, the cleanroom should accommodate the process. Factors that should be considered include (1) leaktightness of the cleanroom shell, (2) compatibility of cleanroom finishes to the decontamination process, and ability to (3) remotely control the process and recirculate the gas, (4) maintain appropriate humidity levels during the decontamination process, and (5) evacuate the gas after decontamination is complete.

Sometimes it is economically feasible to integrate the gas-generating equipment with the cleanroom air ducts. This decision is dictated by the intended gassing frequency, or by the need for automated recovery preparedness following any kind of bioevent. Strategically placed, airtight dampers, gas distribution nozzles, a means to agitate the gas within the cleanroom (or suite of rooms), and exhaust equipment for evacuation are some of the components necessary for automated decontamination. As with all decontamination procedures, protocols must be developed to demonstrate efficacy.

Barrier Technology

Cleanrooms designed to meet ISO Class 5 or better require considerable equipment, space, and maintenance. Operating this equipment is expensive. Furthermore, cleanrooms typically need gowned operators inside to manipulate product and adjust machinery. Because the operator is the source of the majority of the contamination, it is better to separate the operator from the controlled environment; this allows the volume of the controlled space to be reduced to a point where only the process equipment is enclosed, which can lead to substantially reduced capital and operating costs, although this is not always the case. "Microenvironments" such as isolators (glove-boxes) and restricted access barrier system (RABs) are thus becoming increasingly popular. These systems are typically categorized under the term **barrier technology**.

Barrier technology systems must be designed to fit the specific application and can be highly customized to allow the tasks required to accomplish the process needs. Applications vary widely based on product, process equipment, and throughput volume. Barrier technology systems are typically positive-pressure envelopes around the filling equipment with multiple glove ports for operator access, constructed of polished stainless steel with clear, rigid view ports. Systems can be fully sealed or leak into the support environment via "mouse holes" used to allow passage of vials in and out of the unit. Ancillary systems designed to prevent migration of contaminants are used for passing stoppers, containers, and tools in and out of the barrier systems. These can range from simple lock chambers to highly complex alpha/beta ports fitted with features to allow sanitization of the systems or contents. Important design concerns include accessibility, ergonomics, integration with mating equipment, decontamination or sterilization/sanitization procedures, access to service equipment, filter change, filter certification, process validation, and environmental control.

Extra attention must be paid to product filling, vial, and stopper protection; access to the barrier for sterilized stoppers; interface to the vial sterilization (depyrogenation) device; sterilizing product path, including pumps and tubing; and airflow patterns inside the barrier, especially at critical points. If a vapor-forming sanitizing agent such as hydrogen peroxide is to be used as a surface sanitizer, care must be taken to ensure good circulation and adequate concentration inside the barrier, as well as removal of residual vapor in the required time frame. In addition, because many of the sanitizing agents are strong oxidizers, care must also be taken in selecting construction materials to ensure compatibility and their ability to absorb and retain or potentially outgas the sanitizing agent at a later time.

Barrier technology systems may also be designed for applications requiring operator protection from high-risk compounds (those that may have an inadvertent therapeutic effect on an operator), while maintaining a sterile internal environment. These tend to be total containment systems with totally contained product transfer ports. All internal surfaces are sealed from the external environment or potential operator exposure. Because of potential chamber leaks, its internal pressure may be kept negative compared to the ambient space via exhaust fans, posing an additional potential risk to the product that must be addressed by the owner.

Other systems, such as a nonsterile powder control booth, may incorporate more passive barrier designs. One such design incorporates a downflow sampling and weighing cubicle. This arrangement takes advantage of unidirectional airflow to wash particles down and away from the operator's breathing zone. Low-wall air returns at the rear of the cubicle capture fugitive dust. An arrangement of roughing and final filters allows air to return to the air handlers and back to the work zone through ceiling-mounted HEPA filters. Products involving noxious or solvent vapors require a once-through air design.

Barrier technology allows installation in environments that might require no special control or particulate classification. Isolators, RABs, and containment chambers are still relatively new to the pharmaceutical industry. As such, installations for sterile products should be in a controlled ambient room condition of ISO Class 8 or better.

Maintainability

A facility that considers maintainability (e.g., accessibility, frequency of maintenance, spare parts, rapid diagnostics and repair, reliability and facility uptime) in its design will be much more reliable and should have fewer operational and regulatory concerns. Many pharmaceutical facilities have been designed so that routine maintenance can be performed from outside the facility, except for unidirectional and terminal HEPA filters, which must be tested twice a year. Quality of materials is important to reliability, especially where failure can compromise a critical parameter. Consider how much exposure and risk to product and personnel are required during maintenance (e.g., how to clean the inside of a glove box contaminated by a toxic product). Beyond cleanable room surfaces that must be sanitized, consider whether and how HVAC equipment may be sanitized using the owner's procedures. Determine whether ductwork must be internally cleaned, and how. Reduced or no-shutdown HVAC system designs require energy-efficient components. Aligning HVAC system layouts with facility operational areas or suites can save significant operating costs and increase plant availability.

Controls, Monitors, and Alarms

Space pressure may be maintained by passive (statically balanced) HVAC if there are few airflow variables. For example, the HVAC system for a few pressurized spaces may be statically balanced if there is a method of maintaining supply airflow volume to compensate for filter loading to ensure minimum supply, return, and exhaust air changes. More complex designs may require dynamic pressure control. It is important to avoid multiple pressurization loops controlled from the same or interrelated parameters, because this can lead to space pressurization instabilities. Complications can result from fans in series controlling similar or related properties. Improved system stability results from controlling airflow values at the room level, and duct pressure at the branch or air handler. Pressure controls should not overreact to doors opening and closing, because it is virtually impossible to pressurize a space to 0.05 in. of water with a door standing open. A door switch is often used to send a signal to space pressure control to avoid overreaction.

If space humidity must be maintained to tolerances tighter than the broad range that normal comfort cooling can maintain, active relative humidity control should be considered. If a desiccant dehumidifier is needed, unit operation over its range of flow must not adversely affect the ability of the HVAC to deliver a constant air supply volume to the facility.

Monitor and alarm critical parameters to prove they are under control. Log alarm data and parameter values during excursions. Logging may range from a local recorder to direct digital control (DDC) data storage with controlled access. Software source code should be traceable, with changes to software under the owner's control after qualification is complete. Commercial HVAC software is usually acceptable, but should be verified with regulatory agencies before detailed design begins. Also, keep complete calibration records for sensors, alarms, and recorders of critical parameter data.

Noise Concerns

HVAC noise is a common problem resulting from an attempt to overcome the pressure drop of additional air filtration. The noise level generated must be reduced in lieu of adding duct silencers, which may harbor bacteria and are difficult to clean. Separate supply and return fans running at lower tip speeds instead of a single-fan air handler may reduce generated noise levels. HVAC noise may not be an issue if production equipment is considerably noisier.

Nonaseptic Products

Nonaseptic pharmaceutical facilities (e.g., for topical and oral products) are similar in design to those for aseptic product manufacturing, but with fewer critical components to be qualified. However, critical parameters such as space humidity may be more important, and airborne particle counts are not considered in the United States. If the product is potent, barrier isolation may still be advisable. Space differential pressures or airflow directions and air changes are usually critical (needed to control cross contamination of products), but no regulatory minimum pressure or air change values apply.

START-UP AND QUALIFICATION OF PHARMACEUTICAL CLEANROOMS

Qualification of HVAC for Aseptic Pharmaceutical Manufacturing

Qualification is a systematic, quality-based approach to ensuring and documenting that the pharmaceutical facility, systems, equipment, and processes will deliver everything required for safe and repeatable drug products, including the facility design, installation, operation, maintenance, documentation, and pharmaceutical processing, filling, capping, handling, and storage. Qualification of the pharmaceutical cleanroom HVAC is part of the overall qualification of the facility. Qualification covers equipment affecting critical parameters and their control. Other groups in the manufacturing company (e.g., safety or environmental groups) may require similar commissioning documentation for their areas of concern. The most important objectives in meeting the approving agency's requirements are to (1) state what procedures will be followed and verify that it was done, and (2) show that product is protected and space acceptance criteria are met.

Qualification Plan and Acceptance Criteria

Early in design, the owner and designer should discuss who will be responsible for as-built drawings, setting up maintenance files, and training. They should create a qualification plan for the HVAC, including (1) a functional description of what the systems do along with specific process and room requirements; (2) maps of room classification and pressurizations, airflow diagrams, and cleanliness zones served by each air handler; (3) a list of critical components to be qualified, including the automation system controlling the HVAC; (4) a list of owner's procedures that must be followed for qualification of equipment and systems that affect critical parameters; (5) a list of qualification procedures (IQ/OQ/PQ protocols) written especially for the project; and (6) a list of equipment requiring commissioning, determined through a risk-based product and process impact analysis.

The approval procedure should also be defined in the QP. It is important to measure and document critical variables of a system (e.g., space pressure), but it is also important to document and record performance requirements and results for components that affect the critical parameters (e.g., room pressure sensors, temperature sensors, airflow volume monitor) for GMP as well as business records. Documentation helps ensure that replacement parts (e.g., motors) can be specified, purchased, and installed to support critical operations.

It is important to determine all critical components and instruments that could affect critical parameters and could, through an undetected failure, lead to product adulteration. This may be accomplished by a joint effort between the HVAC engineer, owner, quality experts, and a qualified protocol writer. If performance data are in the qualification records, replacement parts of different manufacture can be installed without major change control approvals, as long as they meet performance requirements. Owner approvals for the qualification plan should be obtained during detailed design.

Qualification requires successfully completing the following activities for critical components and systems. The designer should understand the requirements for owner's approval of each protocol (usually, the owner approves the blank protocol form and the subsequently executed protocol).

The **installation qualification (IQ)** protocol documents construction inspection to verify compliance with contract documents, including completion of punch list work, for critical components. It may include material test reports, receipt verification forms, shop inspection reports, motor rotation tests, duct/equipment cleaning reports, duct leak testing, and contractor-furnished testing and balancing. It also includes calibration records for instrumentation used in commissioning and for installed instrumentation (e.g., sensors, recorders, transmitters, controllers, and actuators) traceable to National Institute of Standards and Technology (NIST) instruments.

Control software should be bench tested, and preliminary (starting) tuning parameters should be entered. Control loops should be dry-loop checked to verify that subsystem installation, addressing, operation, and graphics are correct. Equipment and instruments should be tagged and wiring labeled, then field-verified against record drawings. Commissioning documentation must attest to completion of these activities and include as-built drawings and installation/operation/maintenance (IOM) manuals from contractors and vendors.

The **operational qualification (OQ)** protocol documents start-up, including critical components. This includes individual performance testing of control loops under full operating pressure performed in a logical order (i.e., fan control before room pressure control). The commissioning agent must verify that operating parameters are within acceptance criteria.

The HVAC may be challenged under extremes of design load (where possible) to verify operation of alarms and recorders, to determine (and correct, if significant) weak points, and to verify control and door interlocks. Based on observations, informal alert values of critical parameters, which might signify abnormal operation, may be considered. Although product would not be adulterated at these parameter values, staff could assess an alarm and react to it before further deviations from normal operation occurred.

Documented smoke tests verify space pressure and airflow in critical spaces or inside containment hoods, and show airflow patterns and directions around critical parts of production equipment. Many smoke tests have been videotaped, especially when space pressure differentials are lower than acceptance criteria require and pressures cannot be corrected.

Files should include an updated description of the HVAC, describing how it operates, schematics, airflow diagrams, and space pressure maps that accompany it. Copies should be readily accessible and properly filed. Operating personnel should be familiar with the data in these records and be able to explain it to an agency inspector.

Other Documents. GMP documents should also include test reports for HEPA filters (efficiency or pinhole-scan integrity tests) at final operating velocities. If the filter installer performed the tests, the data should be part of the IQ package.

Documents should verify that instruments display, track, and store critical parameters and action alarms. (Consider recording data by exception and routine documentation of data at minimal regular frequency.)

Systems and equipment should be entered into the owner's maintenance program, including rough drafts of associated maintenance procedures (final drafts should reflect commissioning results).

Records should document the completion of these activities, including final as-built, system diagrams, facility pressurization diagrams, air change rate calculations, and air and water balance reports.

Performance qualification (PQ) is proof that the entire HVAC system performs as intended under actual production conditions. PQ is the beginning of ongoing verification (often called validation) that the system meets acceptance criteria of the product. This includes documentation of

- Maintenance record keeping and final operating and maintenance procedures in place, with recommended frequency of maintenance, and (at the owner's option) a procedure for periodic challenge of controls and alarms
- Logs of critical parameters that prove the system maintains acceptance criteria over a prescribed time
- Training records of operators and maintenance personnel
- Final loop tuning parameters

After accepting PQ, the owner's change control procedure should limit further modifications to critical components (as shown on IQ and OQ forms) that affect the product. Much of the building's HVAC equipment should not need qualification, but records for the entire facility must be kept up to date through quality change control, and problems must be corrected before they become significant. Records of corrections should also be kept.

Once the system is operational, pharmaceutical product trial lots are run in the facility (process validation) and the owner should regularly monitor levels of viable (microbial) and nonviable particles, room pressurization, and other controlled parameters in the processing areas.

SEMICONDUCTOR CLEANROOMS

Cleanroom Advances with Modern Process Technology

Since the mid-1990s, most microelectronic facilities manufacturing semiconductors have required cleanrooms providing ISO Class 3 and cleaner for wafer fabs and Class 5 to 8 for auxiliary manufacturing rooms. This state-of-the-art cleanroom technology has been driven by the decreasing size of microelectronic circuitry and larger wafer sizes. A deposited particle with a diameter of 10% of the circuit width may cause a circuit to fail. Many facilities are designed to meet as-built air cleanliness of less than one particle 0.1 μm and larger per cubic foot of air.

Currently, semiconductor manufacturing cleanroom integration is important in semiconductor facilities design. Larger wafers require larger processing equipment. Cleanroom structures are now integrated into the process and mechanical systems to reduce overall building height and construction cost, and to shorten construction duration. In addition, particle control has advanced to the level of molecular contamination control. Product contamination control also includes internal contaminations such as chemical, ionic, and static electricity control, and fire resistance performance.

Semiconductor Cleanroom Configuration

Semiconductor cleanrooms today are of two major configurations: clean tunnel or open-bay (ballroom). The **clean tunnel** is composed of narrow modular cleanrooms that may be completely isolated from each other. Fully HEPA- or ULPA-filtered pressurized plenums, ducted HEPA or ULPA filters, or individual fan modules are used. Production equipment may be located in the tunnel or installed through the wall where a lower-cleanliness (nominally ISO Class 7 or cleaner) service chase is adjacent to the clean tunnel. The service chase is used in conjunction with sidewall return or a raised floor, possibly with a basement return.

The primary advantage of the tunnel is reduced HEPA- or ULPA-filter coverage and ease of expanding additional tunnel modules into unfiltered areas. The tunnel is typically 8 to 14 ft wide. If the tunnel is narrower, production equipment cannot be placed on both sides; if wider, flow becomes too turbulent and tends to break toward the walls before it leaves the work plane. Figure 10 shows a clean tunnel.

The tunnel design has the drawback of restricting new equipment layouts. Cleanroom flexibility is valuable to semiconductor manufacturing. As processes change and new equipment is installed, the clean tunnel may restrict equipment location to the point that a new module must be added. The tunnel approach may complicate moving product from one type of equipment to another.

Fan/filter clean tunnels have a wide range of applications. However, the noise generated by fans is directly distributed to the tunnel, and the unit's efficiency is lower than in central station recirculating systems. Life-cycle cost analysis can be used (among other considerations) to determine the appropriate configuration.

Ballroom or **open-bay design** involves large (up to 100,000 ft²) open-construction cleanrooms. Interior walls may be placed wherever manufacturing logistics dictate, providing maximum equipment layout flexibility. Replacing process equipment with newer equipment is an ongoing process for most wafer fabrication facilities; support services must be designed to handle different process equipment layouts and even changes in process function. Often, a manufacturer may completely redo equipment layout if a new product is being made.

Open-bay designs can use either a pressurized plenum or ducted filter modules, although pressurized plenums are becoming more common. Pressurized plenums usually are recirculating air plenums, in which makeup air mixes with recirculating air or is ducted to the recirculating units (Figure 11). Either one large plenum with multiple supply fans or small adjacent plenums are acceptable. Small plenums allow shutting down areas of the cleanroom without disturbing other clean areas, and may also include one or more supply fans.

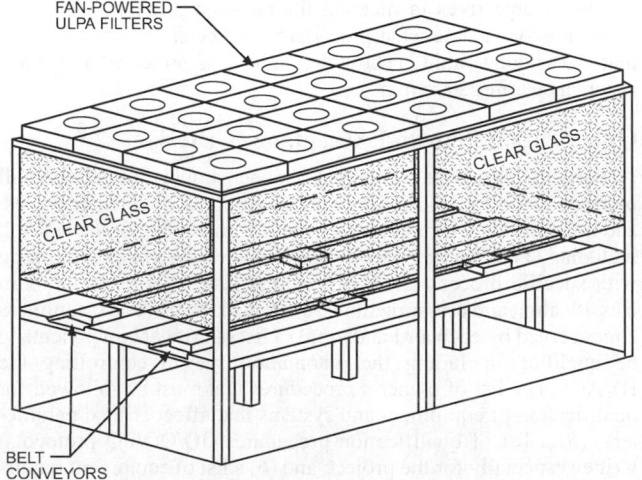

Fig. 10 Elements of a Clean Tunnel

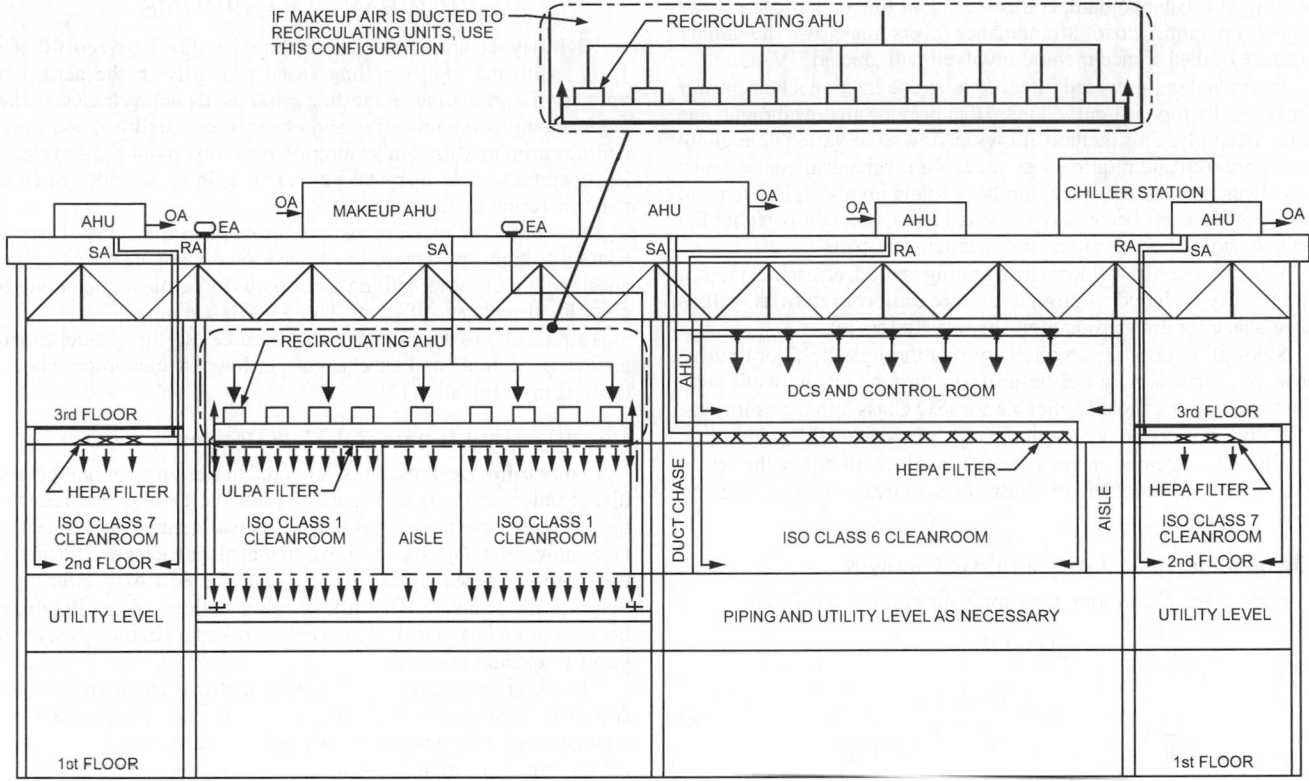

Fig. 11 Typical Semiconductor Manufacturing Plant Section View

In the past decade, silicon wafer diameter has increased three-fold. Along with the requirement for larger process equipment, the demand for larger space sizes has also become increasingly significant. Major semiconductor facilities, with total manufacturing areas of 30,000 ft² and larger, may use both open-bay and tunnel design configurations. Flexibility to allow equipment layout revisions warrants the open-bay design. Process equipment suitable for through-the-wall installation such as diffusion furnaces, may use either design. Equipment such as lithographic steppers and coaters must be located entirely in laminar flow; thus, open-bay designs are more suitable. Which method to use should be discussed among the cleanroom designer, production personnel, and contamination control specialist.

Building structure areas are sometimes used as recirculating and makeup air units, fan deck, and plenum. Some structures perform as very large cabinets of makeup air units or recirculating units. Figure 12 shows a building truss level arranged as a fan deck and air plenum. Coordination between architects, engineers (structural, process, and mechanical), semiconductor facilities personnel, and construction industries must continue to minimize building size and to satisfy ever-evolving process requirements.

Many semiconductor facilities contain separate cleanrooms for process equipment ingress into the main factory. These ingress areas are staged levels of cleanliness. For instance, the equipment receiving and vacating area may be ISO Class 8, and the preliminary equipment setup and inspection area may be ISO Class 7. The final stage, where equipment is cleaned and final installation preparations are made before the fabrication entrance, is ISO Class 6. Some staged cleanrooms must have adequate clear heights to allow forklift access for equipment subassemblies.

Airflow in Semiconductor Cleanrooms

Current semiconductor industry cleanrooms use vertical unidirectional airflow, which produces a uniform shower of clean air throughout the entire cleanroom. Particles are swept from personnel

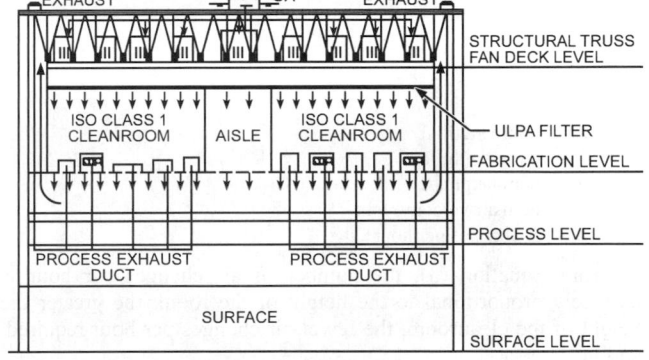

Fig. 12 Building Truss Level Arranged as Fan Deck and Air Plenum

and process equipment, with contaminated air leaving at floor level; this produces clean air for all space above the work surface.

In vertical unidirectional airflow, the cleanroom ceiling area consists of HEPA or ULPA filters set in a nominal grid size of 2 by 4 ft, T-bar-style grid with gasketed or caulked seals for many ISO Class 5 systems; Class 3 and Class 4 systems often use either low-vapor-pressure petrolatum fluid or silicone dielectric gel to seal the filters into a channel-shaped ceiling grid. Whether T-bar or channel-shaped grids are used, the HEPA or ULPA filters normally cover 85 to 95% of the ceiling area, with the rest of the ceiling area composed of grid work, lighting, and fire protection sprinkler panels.

HEPA or ULPA filters in vertical unidirectional airflow designs are installed (1) with a pressurized plenum above the filters, (2) through individually ducted filters, or (3) with individually fan-powered filter modules. A system with a plenum must provide even pressurization to maintain uniform airflow through each filter. Ducted HVAC typically has higher static pressure loss from the ducting and balance dampers, resulting in higher fan energy and higher operating cost. Maintenance costs may also be higher because of the balance method involved with ducted HVAC.

Individual fan-powered filter modules use fractional horsepower fans (usually forward-curved fans) that provide airflow through one filter assembly. This method allows airflow to be varied throughout the cleanroom and requires less space for mechanical components. Disadvantages are the large number of fans involved, low fan and motor efficiencies because of the small sizes, potentially higher fan noises, and higher operation and maintenance costs.

When through-the-floor return grating is used, a basement return is normally included to provide a more uniform return as well as floor space for dirty production support equipment.

Sidewall returns are an alternative to through-the-floor returns; however, airflow may not be uniform throughout the work area. These returns are most applicable for ISO Class 5 to 8 cleanrooms.

Prefiltration is an economical way to increase ULPA filter life. Prefilters are located in recirculation airflow, in either the return basement or air handler, to allow replacement without disrupting production.

Cleanroom Air Velocity and Air Changes

For a given cleanroom, the supply air volume Q (cfm) is

$$Q = LWv \tag{1}$$

$$ach = \frac{60Q}{LWH} \tag{2}$$

or

$$ach = \frac{60LWv}{LWH}$$

$$ach = \frac{60v}{H} \tag{3}$$

where

L = room length, ft
W = room width, ft
H = room height, ft
v = room air velocity, fpm
ach = air changes per hour

From Equation (3), the number of air changes per hour is inversely proportional to the height of the room: the greater the height of the cleanroom, the fewer air changes per hour required, and vice versa.

Air Ionization. In addition to cleanroom particle control with fiber filters, air ionization can be used to control particle attraction to product surfaces by eliminating electrostatic discharge and static charge build-up. However, the emitter tip material must be carefully selected to prevent depositing particles on the product.

HIGH-BAY CLEANROOMS

High-bay cleanrooms have ceiling heights between 40 and 160 ft, with the higher ceilings used primarily in the aerospace industry for producing and testing missiles, launch vehicles, rocket engines, and communication and observation satellites, and lower ceilings primarily used in jet aircraft assembly, painting, and cleaning operations; and in crystal-pulling areas in semiconductor chips manufacturing facilities.

Most high-bay cleanrooms are designed to meet ISO Class 7, Class 8 or higher as required by some U.S. Air Force and U.S. Navy specifications. Crystal-pulling cleanrooms for semiconductor microchips are usually specified at Class 5 to Class 6 range.

Table 3 shows approximate ranges of ceiling-height-dependent airflow per minute and air changes per hour by cleanroom classes derived from Equation (3).

Downflow and Horizontal-Flow Designs

In **downflow designs**, air is delivered in a unidirectional (or simulated unidirectional) flow pattern from the ceiling and returned through floor return openings or low sidewall returns. The objective is to shower the object from above so that all particles are flushed to the returns. The supply air terminals may be HEPA-filter or high-volume air diffusers. Downflow spaces allow space flexibility because more than one device may be worked on in the space at the same air cleanliness level.

The disadvantage is the relative difficulty of balancing airflow. High-bay cleanrooms typically have concrete floors that may include trenches to return some of the air not taken in at low sidewall returns. Special care must be taken to ensure clean air at the object because the laminar flow of the air disintegrates. At the low velocities typical of unidirectional design, pathways may be created toward the returns, causing the clean air to miss the object. Any activity in the cleanroom that generates even a small amount of heat produces updrafts in downward-flowing supply air.

Horizontal-flow designs are always unidirectional, with the cleanest air always available to wash the object in the space. Properly designed horizontal spaces are easier to balance than vertical-flow rooms because supply and return air volumes may be controlled at different horizontal levels in the space.

The main disadvantage of horizontal-flow high-bay spaces is that they provide clean air for only one object, or at best, several objects in the same plane. Once past the object, air cleanliness degrades to the extent that the process generates particles.

Downflow designs are most widely used, but certain projects such as the space telescope and space shuttle assembly room may require horizontal-airflow high-bay cleanrooms (Figure 13).

Air Handling

Because of the large volume of air in a high-bay cleanroom, central recirculating fan systems are commonly used with minimum heating and cooling capability. A separate injection air handler

Table 3 Air Changes per Hour Versus Vertical Airflow Velocities, Room Heights, and Cleanliness Classes

ISO Class	Velocity, fpm	Air Changes per Hour for Ceiling Height, ft							
		40	50	60	80	100	120	140	160
2	85 to 100	128 to 150	102 to 120	85 to 100	—	—	—	—	—
3	70 to 85	105 to 128	84 to 102	70 to 85	52 to 64	—	—	—	—
4	60 to 70	90 to 105	72 to 84	60 to 70	45 to 52	36 to 42	—	—	—
5	45 to 55	68 to 83	54 to 66	45 to 55	34 to 41	27 to 33	22 to 27	—	—
6	25 to 35	38 to 53	30 to 42	25 to 35	19 to 26	15 to 21	12 to 18	10 to 15	—
7	8 to 16	12 to 24	10 to 19	8 to 16	6 to 12	5 to 10	4 to 8	3 to 6	3 to 2
8	4 to 6	8 to 10	5 to 7	4 to 6	3 to 4	2 to 3	2 to 3	2	2
9	2 to 3	3 to 5	2 to 3	2 to 3	2	1 to 2	1 to 2	1	1

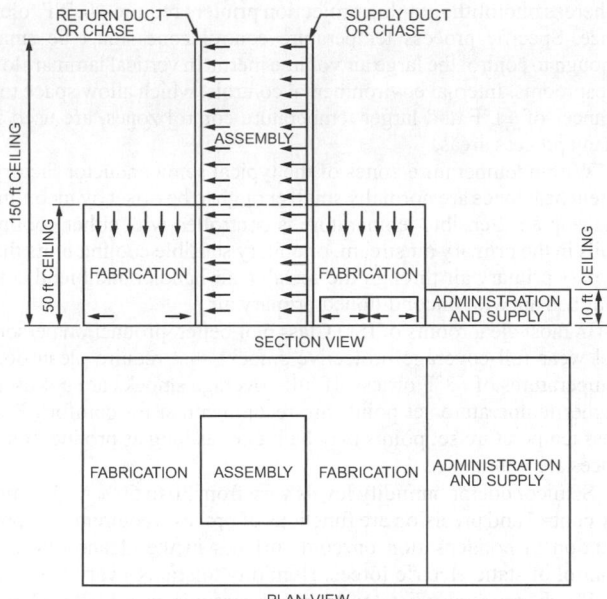

Fig. 13   High-Bay Cleanroom Scheme

provides heating, cooling, and makeup air. The injection system must include volumetric controls to ensure proper building pressure.

Equipment and Filter Access

Air-handling equipment and prefilters should be accessible from outside the cleanroom. Adequate provision must be made for changing filters if air is distributed to the cleanroom with HEPA filters at the space entry. In horizontal-flow cleanrooms, access should be from the upstream (pressure) side, and service scaffolds should be incorporated at least every 8 ft in height of the filter bank. Downflow ceiling filters in T-bar or gel-seal ceilings must be accessed from below using an approved gantry crane with full mobility across the ceiling. Prefilters in the main air supply should be placed in built-up frames with both upstream and downstream access. A HEPA filter bank remote from the space air-distribution system should be installed in a built-up bank with a gel or clamp seal. Access doors must be installed up- and downstream for certification, scanning, and qualification testing.

Prefilter Selection

In any high-bay cleanroom cleanliness classification, air will pass through a final HEPA filter before entering the space; these final filters must be protected by prefilters. HEPA filters for recirculating air should be protected with 85% rated bag or rigid media filters with as few other prefilters as required. Makeup air should include minimum 85% ASHRAE filters on the fan inlet and minimum 95% filters on the fan discharge. Tight, leakproof sealing between the filters and frame/housing improves system cleanliness and reliability.

Design Criteria and Indoor Air Quality

The indoor design temperature range for aerospace and aircraft manufacturing cleanrooms is 73 ± 5°F, with the higher temperatures commonly used in summer, and the lower ones in winter. However, the user should provide guidance on specific required space temperature requirements. In semiconductor crystal-pulling cleanroom design, space temperature is usually required at a constant level of 72 ± 0.5°F.

Another key parameter is relative humidity. For aerospace and aircraft manufacturing cleanrooms, relative humidity should not

exceed 60%; semiconductor crystal-pulling cleanrooms usually require indoor relative humidity to be 50 ± 5% as design base.

Other issues include noise and vibration from process and HVAC equipment, and dusts, fumes, smoke, odors, vapors, moistures and gaseous generated during welding, sanding, painting, washdown, fuel filling, etc. See Chapters 8 to 12 of the 2009 *ASHRAE Handbook—Fundamentals* for additional information.

ENVIRONMENTAL SYSTEMS

Cooling Loads and Cooling Methods

Two major internal heat load components in cleanroom facilities are process equipment and fans. Because most cleanrooms are located entirely within conditioned space, traditional heat sources of infiltration, fenestration, and heat conductance from adjoining spaces are typically less than 2 to 3% of the total load. Some cleanrooms have been built with windows to the outside, usually for daylight awareness, and a corridor separating the cleanroom window from the exterior window.

The major cooling sources designed to remove cleanroom heat and/or maintain environmental conditions are makeup air units, primary and secondary air units, and the process equipment cooling system. Some process heat, typically from electronic sources in computers and controllers, may be removed by process exhaust.

Fan energy is a very large heat source in ISO Class 4 or better cleanrooms. Recirculated space airflow rates of 90 fpm or less (500+ ach) are typical for these facilities.

Latent loads are primarily associated with makeup air dehumidification. A low dry-bulb leaving air temperature, associated with dehumidified makeup air, supplements sensible cooling. Supplemental cooling by makeup air may account for as much as 300 Btu/h per square foot of cleanroom.

Process cooling water (PCW) is used in process equipment heat exchangers, performing either simple heat transfer to cool internal heat sources, or process-specific heat transfer, in which the PCW contributes to the process reaction.

The diversity of manufacturing heat sources (i.e., the portion of total heat transferred to each cooling medium) should be well understood. When bulkhead or through-the-wall equipment is used, equipment heat loss to support chases versus to the production area affects the cooling design when the support chase is served by a different cooling system than the production area.

Makeup Air

Control of makeup air and cleanroom exhaust affects cleanroom pressurization, humidity, and room cleanliness. Makeup airflow requirements are dictated by the amounts required for (1) replacing process exhaust, (2) working personnel ventilation, and (3) meeting pressurization specifications. Makeup air volumes can be much greater than the total process exhaust volume to provide adequate pressurization and safe ventilation.

Makeup air is frequently introduced into the primary air path on the suction side of the primary fan(s) to enhance mixing. Makeup air volumes are adjusted with zone dampers and makeup fan controls using speed controllers, inlet vanes, etc. Opposed-blade dampers should have low leak characteristics and minimum hysteresis.

Makeup air should be filtered before injection into the cleanroom. If the makeup air is injected upstream of the cleanroom ceiling ULPA or HEPA filters, minimum 95% efficient filters (ASHRAE *Standard* 52.1) should be used to avoid high dust loading and reduced HEPA filter life.

In addition, 30% efficient prefilters followed by 85% filters may be used to prolong the life of the 95% filter. When makeup air is injected downstream of the main HEPA filter, further HEPA filtering of the makeup air should be added to the prefilters. In addition to particle filtering, many makeup air handlers require filters to remove chemical contaminants (e.g., salts and pollutants from industries and automobiles) present in outside air. If the makeup air is from an internal conditioned space (i.e., outdoor air is conditioned by the main facility HVAC system), the same filtration level may still be required to prevent the entry of volatile organic compounds (VOCs). These VOCs may be present from another active process in the facility or from building maintenance items such as cleaning agents and paints. Chemical filtration may be accomplished with absorbers such as activated carbon or potassium permanganate impregnated with activated alumina or zeolite.

Process Exhaust

Process exhausts for semiconductor facilities handle acids, solvents, toxins, pyrophoric (self-igniting) fumes, and process heat exhaust. Process exhaust should be dedicated for each fume category, by process area, or by the chemical nature of the fume and its compatibility with exhaust duct material. Typically, process exhausts are segregated into corrosive fumes, which are ducted through plastic or fiberglass-reinforced plastic (FRP) ducts, and flammable (normally from solvents) gases and heat exhaust, which are ducted in metal ducts. Care must be taken to ensure that gases cannot combine into hazardous compounds that can ignite or explode in the ductwork. Segregated heat exhausts are sometimes installed to recover heat, or hot uncontaminated air that may be exhausted into the suction side of the primary air path.

Required process exhaust volumes vary from 1 cfm per square foot of cleanroom for photolithographic process areas, to 10 cfm per square foot for wet etch, diffusion, and implant process areas. When specific process layouts are not designated before exhaust design, an average of 5 cfm per square foot is normally acceptable for fan and abatement equipment sizing. Fume exhaust ductwork should be sized at low velocities (1000 fpm) to allow for future needs.

For many airborne substances, the American Conference of Governmental Industrial Hygienists (ACGIH) established requirements to avoid excessive worker exposure. The U.S. Occupational Safety and Health Administration (OSHA) set specific standards for allowable concentrations of airborne substances. These limits are based on working experience, laboratory research, and medical data, and are subject to constant revision. See *Industrial Ventilation: A Manual of Recommended Practices* (ACGIH 2007) to determine limits.

Fire Safety for Exhaust

ICC's (2009) *International Building Code®* (IBC) designates semiconductor fabrication facilities as Group H occupancies. The Group H occupancy class should be reviewed even if the local jurisdiction does not use the IBC because it is currently the only major code in the United States specifically written for the semiconductor industry and, hence, can be considered usual practice. This review is particularly helpful if the local jurisdiction has few semiconductor facilities.

Chapter 18 in ICC's *International Fire Code* (IFC) addresses specific requirements for process exhaust relating to fire safety and minimum exhaust standards. Chapter 27, Hazardous Materials, is relevant to many semiconductor cleanroom projects because of the large quantities of hazardous materials stored in these areas. Areas covered include ventilation and exhaust standards for production and storage areas, control requirements, use of gas detectors, redundancy and emergency power, and duct fire protection.

Temperature and Humidity

Precise temperature control is required in most semiconductor cleanrooms. Specific chemical processes may change under different temperatures, or masking alignment errors may occur because of product dimensional changes as a result of the coefficient of expansion. Temperature tolerances of ±1°F are common, and precision of ±0.1 to 0.5°F is likely in wafer or mask-writing process areas. Wafer reticle writing by electron beam technology requires ±0.1°F,

whereas photolithographic projection printers require ±0.5°F tolerance. Specific process temperature control zones must be small enough to control the large air volume inertia in vertical laminar flow cleanrooms. Internal environmental controls, which allow space tolerances of ±1°F and larger temperature control zones, are used in many process areas.

Within temperature zones of the typical semiconductor factory, latent heat loads are normally small enough to be offset by incoming makeup air. Sensible temperature is controlled with either cooling coils in the primary air stream, or unitary sensible cooling units that bypass primary air through the sensible air handler and blend conditioned air with unconditioned primary air.

In most cleanrooms of ISO Class 6 or better, production personnel wear full-coverage protective smocks that require cleanroom temperatures of 68°F or less. If full-coverage smocks are not used, higher temperature set points are recommended for comfort. Process temperature set points may be higher as long as product tolerances are maintained.

Semiconductor humidity levels vary from 30 to 50% rh. Humidity control and precision are functions of process requirements, prevention of condensation on cold surfaces in the cleanroom, and control of static electric forces. Humidity tolerances vary from 0.5 to 5% rh, primarily dictated by process requirements. Photolithographic areas have more precise standards and lower set points. The exposure timing of photoresists (used in photolithography) can be affected by varying relative humidity. Negative resists typically require low (35 to 45%) relative humidity. Positive resists tend to be more stable, so the relative humidity can go up to 50% where there is less of a static electricity problem.

Independent makeup units should control the dew point in places where direct-expansion refrigeration, chilled-water/glycol cooling coils, or chemical dehumidification is used. Chemical dehumidification is rarely used in semiconductor facilities because of the high maintenance cost and potential for chemical contamination in the cleanroom. Although an operating cleanroom generally does not require reheat, systems are typically designed to provide heat to the space when new cleanrooms are being built and no production equipment has been installed.

Makeup air is humidified by steam humidifiers or atomizing equipment. Steam humidifiers are most commonly used. Take care to avoid releasing water treatment chemicals. Stainless-steel unitary packaged boilers with high-purity water and stainless-steel piping have also been used. Water sprayers in the cleanroom return use air-operated water jet sprayers. Evaporative coolers can take advantage of the sensible cooling effect in dry climates.

Pressurization

Pressurizing semiconductor cleanrooms is another method of contamination control, providing resistance to infiltration of external sources of contaminants. In nonpressurized spaces, or spaces of lesser pressure than the surrounding environment, nearby particulate contaminants enter the cleanroom by infiltration through doors, cracks, pass-throughs, and other penetrations for pipes, ducts, etc. The cleanest cleanroom should have the highest pressure, with decreasing pressure corresponding to decreasing cleanliness. A differential pressure around 0.05 in. of water is often used.

For small semiconductor cleanrooms or clean zones in ISO Classes 8 and 9, ceiling supply and low sidewall return is a typical airflow arrangement. The primary air system alone can handle the internal cooling load and the required room air change rate. Pressurization system designs are very similar to those in pharmaceutical facilities.

For cleaner (ISO Class 7 and cleaner) semiconductor cleanrooms, primary/secondary air systems are common. The secondary (makeup) HVAC unit takes care of the outside air and internal cooling loads, and the primary (recirculating) unit delivers the required room air change rate, and additional cooling if needed. A raised, perforated floor return is common for these classes. During balancing, manual or automatic balance dampers are usually set at fixed positions at air supply, return, and exhaust systems.

In vertical- and unidirectional-flow cleanrooms, single-stage constant volume for supply and return flows is common. Because internal dust generation from people and process could be lower during nonoperating or unoccupied mode than operating or occupied mode, using multiple recirculating blowers to create two- or multiple-stage supply and return flow rates is feasible as long as the room cleanliness meets the designated classification at all times, validated through continuous particle count measurement. In the nonoperating or unoccupied mode, reduced levels of supply and return airflow rates should also ensure maintaining proper room pressurization level.

Pressure level in the cleanroom is principally established by room airtightness and the **offset flow** value, which is the net flow difference between supply airflow rate and exhaust and return airflow rates. Process equipment exhaust rate is often determined by manufacturers' data, industrial hygienists, and codes. The design engineer should consult with the facility contamination control specialist to determine air change rates for each cleanroom.

One popular method of cleanroom pressurization is to fix the supply flow rate and adjust the return flow rate by volume dampers at return floor panels to create a specified positive space pressure. Return air to underfloor plenum or subfloor basement through perforated panels floor grilles or grates (usually with a 15 to 35% free area) can be balanced to ensure a fixed flow differential (offset flow) in the space. An adjustable, lockable balance damper normally is attached beneath the perforated floor panel or grate. When the damper is fully open, it normally imposes a pressure drop of 0.02 to 0.08 in. of water; higher pressure drop can be achieved when the dampers are turning toward the closed positions. Note that the balance damper could affect parallelism of the room's unidirectional flow.

Another method is to use variable-volume supply and return fans with volumetric flow tracing to ensure the required room pressure. This method could be a reasonable choice for a single, large cleanroom, but is not flexible enough to serve a suite with different room pressure requirements. For some industries, variable-volume systems may not be favorable; design engineers should consult with facility contamination control specialists before specifying variable-volume systems for cleanrooms.

Air locks typically are used between uncontrolled personnel corridors, entrance foyers, and the protective-clothing gowning area. Air locks may also be used between the gowning room and the main wafer fabrication area, and for process equipment staging areas before entering the wafer fabrication area. Install air locks only when they are really necessary, because their use along traffic paths could restrict personnel access and increase evacuation time during emergencies.

Commercial pressure differential sensors can reach accuracy at ±0.001 in. of water or better, and significant progress has been made on precision room pressure control. Many semiconductor processes affected by cleanroom pressure (e.g., glass deposition with saline gas) require process chamber pressure precision of ±0.00025 in. of water.

Pressurization calculations can be performed by using the procedures detailed in either Pedersen et al. (1998) or Spitler (2009) in the chapters on infiltration:

- Using the provided charts, calculate the building exfiltration at designated room pressurization level.
- In accordance with ASHRAE *Standard* 62.1, with the actual number of occupancy, determine the required outdoor air rate.
- Determine the total air volume of exhaust from the building.

The sum of exfiltration air volume plus exhaust rate or plus the required outdoor air, whichever is greater, is the total ventilation rate under the designated building pressurization.

To ensure the designated pressurization level, a leak test must be performed for exterior walls, interior walls, partitions, doors and windows between two adjacent areas with different pressurization levels, and for roof, exterior doors and windows, connections between wall and roof, and any building elements between two areas with different pressurization levels. All major leaks must be eliminated before start-up of HVAC systems.

Sizing and Redundancy

Environmental HVAC design must consider future requirements of the factory. Semiconductor products can become obsolete in as little as two years, and process equipment may be replaced as new product designs dictate. As new processes are added or old ones removed (e.g., wet etch versus dry etch), the function of one cleanroom may change from high-humidity requirements to low, or the heat load many increase or decrease substantially. Thus, the cleanroom designer must design for flexibility and growth. Unless specific process equipment layouts are available, maximum cooling capability should be provided in all process areas at the time of installation, along with provisions for future expansions.

Because cleanroom space relative humidity must be held to close tolerances and humidity excursions cannot be tolerated, the latent load removal capacity of the selected equipment should be based on high ambient dew points and not on the high mean coincident dry-bulb/wet-bulb data.

In addition to proper equipment sizing, redundancy is also desirable when economics dictate it. Many semiconductor wafer facilities operate 24 h per day, seven days per week, and shut down only during holidays and scheduled nonworking times. Mechanical and electrical redundancy is required if loss of equipment would shut down critical and expensive manufacturing processes. For example, process exhaust fans must operate continuously for safety reasons, and particularly hazardous exhaust should have two fans, both running. Most process equipment is computer-controlled with interlocks to provide safety for personnel and products. Electrical redundancy or uninterruptible power supplies may be necessary to prevent costly downtime during power outages. Redundancy should be based on life-cycle economics.

SUSTAINABILITY AND ENERGY CONSERVATION IN CLEANROOMS

The major operating costs associated with a cleanroom include conditioning the air, fan energy for air movement in the cleanroom, and process exhaust. The combination of environmental conditioning and control, contamination control, and process equipment electrical loads can be as much as 300 W/ft^2. Besides process equipment electrical loads, most energy is used for cooling, air movement, and process liquid transport (i.e., deionized water and process cooling water pumping). A life-cycle cost analysis should be performed to determine design choices and their total cost of ownership over time.

Energy Metrics. To evaluate design options for HVAC systems in cleanrooms, it is convenient to compare overall efficiency using standard metrics. By using a metric such as airflow per kilowatt input, it is possible to compare system efficiency for different schemes. This is a good metric for air systems because it compares the amount of energy required to move a given quantity of air, and combines equipment efficiency as well as system effects. The owner can include this metric as a design criterion.

Similarly, metrics for chilled-water system performance in terms of kilowatts per ton can be established. Chiller performance and overall chilled-water system performance issues are well documented and should be consulted to set appropriate targets.

These various energy metrics highlight the wide variation in performance and the need to establish goals for system performance.

Fan Energy. Because airflow rates in cleanrooms can range from 4 to 100 times greater than in conventional HVAC, fan systems should be closely examined for ways to conserve energy. Static pressures and total airflow requirements should be designed to reduce operating costs. The fan energy required to move recirculation air may be decreased by reducing the air volume and/or static pressure. Energy conservation operating modes should be verified during system qualification. If these modes are not part of the original design, the control procedure must be changed and the operational change validated.

Air volumes may be lowered by decreasing recirculation airflow and minimizing room volumes in high-air-change-rate suites. This could allow decreasing HEPA or ULPA filter coverage or reducing cleanroom average air velocity. Reducing air volumes can yield significant savings and often enhance room cleanliness through reduced turbulence. Based on a 90 fpm face velocity, each square foot reduction in filter coverage area in a room can save 25 to 50 W/ft^2 in fan energy and cooling load. Reducing room average velocity from 90 to 80 fpm saves 5 W/ft^2 in fan energy and in cooling energy. If the amount of air supplied to the cleanroom cannot be lowered, reducing static pressure can produce significant savings. With good fan selection and transport design, up to 15 W/ft^2 can be saved per 1 in. of water reduction in static pressure. Installing low-pressure-drop HEPA filters, pressurized plenums in lieu of ducted filters, and proper fan inlets and outlets may reduce static pressure. Many cleanrooms operate for only one shift. Air volume may be reduced during nonworking hours by using two-speed motors, variable-frequency drives, inverters, inlet vanes, and variable-pitch fans, or, in multifan systems, by using only some of the fans.

Additional energy may be saved by installing high-efficiency motors on fans instead of standard-efficiency motors. Fan selection also affects energy cost. The choice of forward-curved centrifugal fans versus backward-inclined, airfoil, or vaneaxial fans affects efficiency. The number of fans used in a pressurized plenum design influences redundancy as well as total energy use. Fan size changes impact power requirements as well. Many times, slower air through more fans provides better system efficiency and reliability; different options should be investigated to ensure optimal designs.

Makeup and Exhaust Energy. Process exhaust requirements in the typical semiconductor facility vary from 1 to 10 cfm per square foot. Makeup air requirements vary correspondingly, with an added amount for leakage and pressurization. The energy required to supply the conditioned makeup air can be quite large. Careful attention to the layout and design of the makeup air system, especially minimizing system pressure drop and specifying efficient fans and motors, is important. The type of equipment installed normally determines the quantity of exhaust in a given facility. Heat recovery has been used effectively in process exhaust; when heat recovery is used, the heat exchanger material must be selected carefully because of the potentially corrosive atmosphere; requirements for nonhazardous cleanrooms are not as significant. Also, heat recovery equipment has the potential to cross-contaminate products in pharmaceutical facilities.

Makeup air cannot normally be reduced without decreasing process exhaust, which may be difficult to do because of safety and contamination control requirements. Therefore, the costs of conditioning the makeup air should be investigated. Conventional HVAC methods such as using high-efficiency chillers, good equipment selection, and precise control design can also save energy. One energy-saving method for large facilities uses multiple-temperature chillers to bring outdoor air temperature to a desired dew point in steps.

Cleanrooms and Resource Use: Opportunities to Improve Sustainability

Because of their highly specific and complex requirements, many cleanrooms have high energy and resource demands. When possible, designers and operators should look for opportunities to reduce these demands, not only for reasons of environmental health, but also for cost savings and avoidance of risk.

Cleanrooms are more complex than, for instance, a typical office building. When developing a cleanroom-driven project, using integrated design and construction can result in major rewards in cost, schedule, and operational efficiencies. Some of the most promising areas for energy and resource use reductions include the following:

- Optimize air distribution strategies and air change rates in clean areas; reducing room volumes and air change rates saves on environmental conditioning energy, fan energy, equipment sizing, duct sizing, filter requirements (quantity, not quality), and equipment space. Implementing on-demand utility distribution, including pressure and temperature reset control strategies, provides further operational savings. See Figure 14 for information from Lawrence Berkley National Laboratory (http://hightech.lbl.gov/cleanrooms.html) on system selection efficiencies.

- Analyze and evaluate process chemistry, including cleaning materials and methods. Reducing or eliminating VOC-based solvents, heavy metals, acids, etc., in processing reduces the need for dilution air, scrubbing, treatment of effluent and other environmental and life safety issues. This step must be integrated with process developers, operators, and regulatory compliance personnel to ensure that changes do not compromise final product quality and acceptance.

- Process equipment specifications should include performance criteria for support utility needs such as process water, compressed air, exhaust air, and electrical power. More efficient equipment leads to savings in operational and capital costs. This approach may also prove attractive where process equipment is leased and will be returned to the equipment vendor, as is common in microelectronics, because of the processing technologies' rapid obsolescence. For the equipment or tool manufacturers, higher efficiency may enhance the toolset's resale value.

The effects of these broad areas of resource use reduction and energy savings on building systems should be obvious; however, there are other tangible benefits that should be considered. Reducing the resource use or environmental footprint of the cleanroom extends the site infrastructure's carrying capacity. On developed sites in developed areas, this can save significant capital and operational costs by reducing the need to increase the site's capacity or infrastructure to handle an additional building or operation. Reducing use of hazardous, toxic, or noxious materials can reduce the owner's exposure to environmental health and safety risks and the need to treat discharge air and water streams. Improving HVAC energy efficiency can reduce equipment and penthouse space requirements, capital costs, and system-generated noise and vibrations.

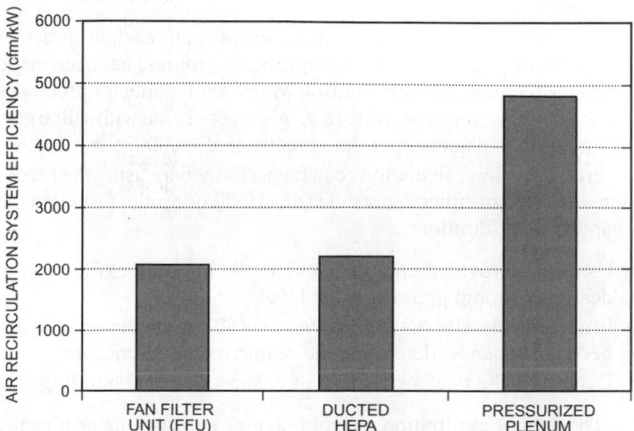

Fig. 14 Energy Efficiency of Air Recirculation Systems

NOISE AND VIBRATION CONTROL

Noise is difficult to control. Noise generated by contamination control equipment requires particular attention, although production equipment noise may be more significant than HVAC noise. Before beginning design, noise and vibration criteria should be established. Chapter 48 provides more complete information on sound control.

In normal applications of microelectronics contamination control, equipment vibration displacement levels need not be dampened below 0.5 μm in the 1 to 50 Hz range. However, electron microscopes and other ultrasensitive microelectronics cleanroom instruments may require smaller deflections in different frequency ranges. Photolithographic areas may prohibit floor deflections greater than 0.075 μm. As a general rule, displacement should not exceed one-tenth the line width.

For highly critical areas, consider using vaneaxial fans. These fans generate less noise in lower frequencies, and can be dynamically balanced to displacements of less than 4 μm, which decreases the likelihood of transmitting vibration to sensitive areas in electronics cleanrooms. Energy-efficient features of cleanroom HVAC systems, such as straight, smooth duct layouts and elimination of sound attenuators, can exacerbate noise-control issues. Instead of resorting to sound-trap additions, acoustic problems can be mitigated through proper fan selections to avoid sound generation from excessive fan-blade tip speeds.

ROOM CONSTRUCTION AND OPERATION

Control of particulate contamination from sources other than the supply air depends on the classification of the space, the type of system, and the operation involved. Typical details that may vary with the room class include the following:

Construction Finishes

- **General.** Smooth, monolithic, cleanable, and chip-resistant, with minimum seams, joints, and no crevices or moldings.
- **Floors.** Sheet vinyl, epoxy, or polyester coating with wall base carried up, or raised floor (where approved) with and without perforations using the previously mentioned materials.
- **Walls.** Plastic, epoxy-coated drywall, baked enamel, polyester, or porcelain with minimum projections.
- **Ceilings.** Gypsum wallboard or plaster, covered with plastic, epoxy, or polyester coating or with plastic-finished, clipped acoustical tiles (no tiles in pharmaceutical cleanrooms) when entire ceiling is not fully HEPA- or ULPA-filtered.
- **Lights.** Teardrop-shaped single lamp fixtures mounted between filters, sealed and installed in T-grid ceiling (gasket or gel seal) or flush-mounted and sealed.
- **Service penetrations.** All penetrations for pipes, ducts, conduit runs, etc., fully sealed or gasketed, then caulked in place. All conduits to have internal seals or pour stops to reduce infiltration/exfiltration through conduit.
- **Appurtenances.** All doors, vision panels, switches, clocks, etc., either flush-mounted or with sloped tops.
- **Windows.** All windows flush with wall; no ledges on cleanest side. Window gaskets to be close cell and windows caulked.
- **Doors.** Sliding doors perform better than swinging doors in critical cleanrooms. All door movements to be controlled for gradual, smooth motion.

Personnel and Garments

- Hands and face cleaned before entering area
- Lotions and soap containing lanolin to lessen shedding of skin particles
- No cosmetics and skin medications
- No smoking or eating
- Lint-free smocks, coveralls, gloves, head covers, and shoe covers

Materials and Equipment

- Equipment and materials are cleaned before entry.
- Nonshedding paper and ballpoint pens are used. Pencils and erasers are not permitted.
- Work parts are handled with gloved hands, finger cots, tweezers, and other methods to avoid transfer of skin oils and particles.
- Sterile pharmaceutical product containers must be handled with sterilized tools only.

Particulate Producing Operations

- Electronics grinding, welding, and soldering operations are shielded and exhausted.
- Nonshedding containers and pallets are used for transfer and storage of materials.

Entries

- Air locks and pass-throughs maintain pressure differentials and reduce contamination.

CLEANROOM INSTALLATION AND TEST PROCEDURES

Installation

Space Preparation. Building envelope construction should be completed, its insulation thoroughly installed. Insulation materials should meet cleanroom requirements. All leaks must have been eliminated, construction debris removed, and floors cleaned, washed, and blow-dried.

Cleanroom Installation. After space preparation is completed, the HVAC, plumbing, process piping, and cleanroom elements are then ready to start installation in the following sequence:

1. Install cleanroom HVAC piping, ductwork, plumbing, and process piping (prior to hookup with process equipment). All open ends of duct and piping must be temporarily sealed at end of each workday.
2. Install cleanroom ceiling, floor, and wall systems.
3. Process equipment: any process equipment package that is larger than the access doors must be moved into the cleanroom area before installation of cleanroom wall access panels. All process equipment should be protected from construction damage and remain in shipping packaging, unopened.
4. Install cleanroom access doors, pass windows, wall access panels, floor and ceiling access panels. If hard ceiling is used, do not close ceiling access before test, balance, and acceptance by the responsible HVAC engineer.
5. After completing steps 1 to 4, check the leaktightness of all access doors, pass windows, and other cleanroom openings, as well as edges between (a) ceiling and walls and (b) walls and floors. Leaks must be completely eliminated.

Cleanroom Duct and HEPA Filters.

1. Thoroughly wash and clean air-handling unit (AHU) internals, including internals of AHU fans.
2. Use compressed air to blow dry (pressure high enough to dry, but not to damage internals of the AHU).
3. Check AHU internals. If some dirt remains (especially on filter and edge areas), repeat steps 1 and 2.
4. Temporarily seal all openings on cleaned AHUs, including OA intakes, return and supply openings, water, steam connections, humidifier control box tubes, drain openings, and doors.
5. Wash clean and blow dry all internal surfaces of duct sections and immediately seal. This will prepare the installation of duct system and HEPA filters.
6. Temporarily seal all open ends in the duct system at end of each workday during installation.

7. Temporarily seal the installed duct systems to wait for the finish of architectural internal work. Leave ceiling accesses open for ceiling HEPA filter installation and HVAC system test and balance.
8. Remove all construction debris from cleanroom. Wash and dry AHU external surfaces.
9. Wash and blow dry the cleanroom floor thoroughly.
10. After step 9, installation personnel should wear cleanroom shoe covers to enter the cleaned area to continue installation work.
11. Place the originally sealed HEPA filter packets at their installation locations.
12. Unpack HEPA filters and install immediately. Do not open HEPA filter packets if not to be installed the same day.
13. Check the HVAC control system installation and pretest to ensure the control system is functioning before HEPA filter installation.
14. Check installation of fire protection, life safety, and other HVAC-related systems to ensure the systems are properly functioning.

System Start-Up, Test, and Balance.

1. Read the major equipment and controls' installation, operation, and maintenance (IOM) manual thoroughly.
2. Walk through entire system to be started up.
3. Check that all mechanical systems have been installed. Replace covers, belts, gaskets, bolts, and screws if missing or damaged.
4. Check unit base concrete slabs, roof curbs, and structural supports. All units should be firmly installed on level plane.
5. Check that all equipment, devices, and fittings are installed correctly and in operating condition.
6. Check that all dampers, louvers, and valves are set at the correct positions as shown on drawings and under the direction of test-and-balance engineer.
7. Remove all bolts and plates used for temporarily compressing internal spring isolators under AHU base during shipment.
8. Check chiller system. Ensure that the chilled-water supply and return are under operational condition.
9. If hot water is used, check the hot-water system. Check that hot-water supply and return temperature and pressure all meet HVAC system requirements.
10. If steam is used, check steam valve station. Check that the regulated steam pressure meets HVAC system required range.
11. If pneumatic control is used, check compressed air system, that the supply pressure meets control system requirement.
12. Electric wiring should be checked by the electrical engineer, who should confirm that power source voltages conform to all equipment requirements.
13. Check and correct all motors' rotation.
14. Check that the controls system has been installed, energized, and pretested by the controls contractor.
15. Check that the fire-protection system is in place, with correct links verified by the fire-protection contractor together with electrical engineer.
16. General mechanical/HVAC contractor should coordinate with all disciplines for overall status of preparation for cleanroom HVAC, control, and fire-protection systems start-up. A written report stating the completion of all of the above-listed items should be submitted to the responsible HVAC engineer at minimum two workdays before the scheduled system start-up date.
17. The responsible HVAC engineer should determine a proper day to inform the on-site commissioning authority (CA) before start-up if commissioning is required by project scope.
18. Correct all problems that may have occurred during start-up; adjust systems to meet design conditions.
19. Initial test, balance, and adjustment work should be performed by a licensed test-and-balance contractor during system start-up.

20. Check prefilters and final filters. If they have reached their pressure drop limit, change them.
21. Adjust supply, return, and exhaust fan airflows to meet design rates.
22. Verify that operational testing of all system safeties (fire alarm, high-pressure limits, etc.) is completed before releasing system for automatic operational control.
23. Keep air system operating. Set room thermostat low enough to start cooling. Check chilled-water supply and return temperatures, control valves, and condensate drain. Check room temperature. Note that the cooling performance test is under the condition without process heat. The responsible HVAC engineer should oversee if the HVAC and chiller systems are capable of satisfying the additional load with process running.
24. Keep air system operating. Set room thermostat at temperature high enough to start heating system. Check steam pressure, monitor served room temperature, and check control valves and condensate return and drain lines.
25. Keep supply air and heating system running. Set room humidistat at level high enough temporarily start humidifier. If steam humidifier is used, check steam pressure and all connections. Monitor relative humidity of served room and check control valves.
26. When all the above steps, including necessary adjustments, have been completed, the system is ready for final test and preparation for commissioning and acceptance. Clean the space for the last time to prepare for final test.
27. Cleanroom dress code enforcement begins before final test.
28. Attendees in final test include all contractors, subcontractors, the responsible HVAC engineer, the cleanroom facilities engineer, the future system lead operators, lead maintenance staff, and commissioning personnel, if appropriate.
29. All problems should be solved before the project completion. Keep complete records of all problems and solutions during start-up, testing, adjusting, and balancing.

Pressurization Test and Map

Cleanroom pressurization must be verified before commissioning and acceptance. An as-built space-to-space pressurization map should be submitted by the test-and-balance contractor to the responsible HVAC engineer for review. A retest may be performed if the HVAC engineer determines it necessary. Perform and document airflow pattern testing for final quality control verifications.

Operation Personnel Training Program

It is important that the operating and maintenance personnel responsible for systems installed under a particular project receive proper training. Usually, the training program may be offered by the control contractor under the supervision of the responsible HVAC engineer. Training should start during the functional performance testing. It is important that the operating and maintenance personnel see the systems being set up, the issues encountered, and their resolution.

Cleanliness Verification Test

Empty (as-built) cleanroom cleanliness may be verified and determined by initial testing before process equipment installation and operation. Operational cleanroom cleanliness should be tested during formal process operation to gage the influence of emissions from process materials and products, as well as the performance of process exhaust systems together with cleanroom operation rules and operating personnel activities.

For ISO Class 3 and 4 cleanrooms, the owners will most likely prefer not to have commissioning personnel walking around the clean facility during process in operation. They typically use their own professional staff to test and maintain the space cleanliness level. Therefore, as-built cleanroom cleanliness commissioning is the final step in most projects.

Commissioning

Attendees in commissioning process include the same personnel as during start-up, test, and balance, in addition to process operators, the owner's project authorities, and commissioning personnel.

Commissioning documents should include the following:

- Certificates and warrantees of system completion with complete set of as-built drawings submitted from mechanical, electrical, plumbing, controls, and fire-protection contractors
- If available, all major equipment installation, operation, and maintenance (IOM) manuals, from the equipment manufacturers
- Complete records of all problems and solutions that occurred during start-up, and tests and adjustments submitted by every individual contractor
- A certified system test and balance report with verified major equipment models and capacities, and all tested performance numbers conforming to the system criteria from the licensed test-and-balance contractor. A complete space-to-space pressurization map submitted by the test-and-balance contractor
- A control system installation, operation, and maintenance (IOM) manual submitted from the control contractor
- A certificate of test for as-built cleanroom cleanliness (tested when cleanroom facility is complete, all services are connected and functional, but without equipment and operating personnel in the cleanroom)
- If the contract scope requires, a certificate of cleanroom cleanliness at the condition of process running with operating personnel in the facility
- Commissioning protocol forms, signed and witnessed by all attendees

Process Equipment Installation (Tool Hook-up)

The process equipment installation (tool hook-up) work is covered by a separate, independent contract. It starts when the as-built cleanroom has been certified and accepted by the owner. The plant facility engineer is responsible for process equipment installation, and the project HVAC engineer monitors the cleanroom cleanliness while tool hook-up is in progress, offering consultation as needed. The following points apply to the cleanroom tool hook-up procedure:

- All cleanroom equipment installation personnel should attend a cleanroom orientation class before joining the team work.
- All installation personnel must follow the dress code entering and working in the cleanroom area for process equipment installation (tool hook-up), test, adjusting, and operation.
- Do not unpack process equipment before the cleanroom has been cleaned, tested, certified, and is ready for installation of process equipment.

- Do not unpack process equipment or open temporarily sealed pipe ends if not immediately installing or connecting to the equipment or pipe ends. Temporarily seal unfinished connection openings if not being connected immediately.
- Do not leave cleanroom doors or pass windows open anytime during installation or test operation.
- Establish a bimonthly cleanroom cleanliness retest timetable for monitoring and maintaining the cleanroom cleanliness level for the first six months. The frequency of retest can be modified according to the actual operating experience in future years.

INTEGRATION OF CLEANROOM DESIGN AND CONSTRUCTION

Integrated design and construction addresses all stages and aspects of cleanroom construction, to achieve better-quality, faster delivery; lower-cost, more optimized operation and maintenance; lower energy consumption; a cleaner environment; safer, more reliable, and more productive conditions; and longer service life. Integrated building design (IBD) is discussed in detail in Chapter 58.

A complete cleanroom project usually includes the following stages (see Figure 15): development of scope, budget, and overall project execution plan; predesign, conceptual, and schematic design; preliminary, final design, and construction documentation; and construction service.

Although the entire cleanroom building project is a large and complex operation, it may be simplified if it is considered as an integrated system with a unified overall scope of work and timeline to be achieved by an integrated design and construction team. In an integrated approach, all individual systems and their components are considered as subsystems of the overall integrated cleanroom building project, and optimizations are implemented at the component, system, and facility levels, including the following:

- **Site/utilities:** overall site plan, entrances and gates, roads and transportation, landscape, electrical substations or electrical main connection, gas or other fuel main intake pressure regulation station, water, sewer, sanitary and storm drain piping and main connections, telephone, network, security and fire protection system main connections, outdoor lighting, etc.
- **Building:** foundations, structure system, walls, roofs, ceilings, floors, elevators, electrical, gas, fuel, water, sewer, plumbing, sanitary, mechanical, HVAC, chiller, boiler, noise control, lighting, process systems, energy and process material recovery systems, exhaust air and wastewater treatment systems, hazard control systems, explosion- and corrosionproofing, instrumentation and control systems, fire protection systems, etc.
- **Cleanroom:** walls; roofs; ceilings; HEPA or ULPA filters; floors; mini-clean environment; clean tunnels; clean booths; recirculating air, makeup air, and exhaust air systems; lighting, process mechanical, chemical, electrical, and control systems; production lines; process conveyers; special gas supply systems; acoustics;

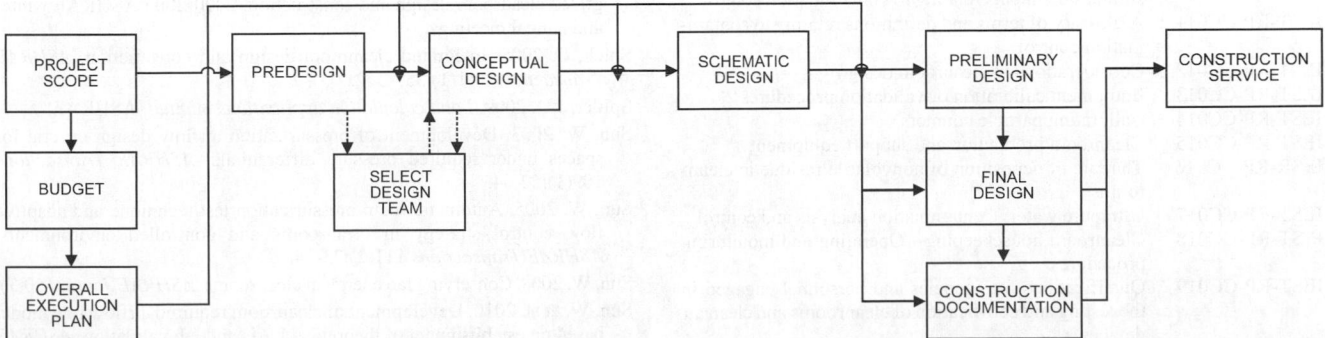

Fig. 15   General Design and Construction Procedure

operating personnel, material, and products access doors, windows, or openings; air showers; room temperature, humidity, static electricity, CO_2, pressurization, and cleanliness monitoring and control systems; fire protection and after-fire recovery systems; seismic design, emergency response facilities, etc.

- **Implementation:** design documents, submittal approvals, receiving inspections, clean construction and installation work, field inspections, system start-up, test and adjustment, balancing, commissioning, and turnover.
- **Building management:** system operation and maintenance.

Figure 12 shows an example of how an HVAC fan deck and related electrical and control systems integrated into the building structure level. In a recent semiconductor plan, the mechanical room was used as the housing for makeup air units.

BIBLIOGRAPHY

ACGIH. 1999. *Bioaerosols: Assessment and control.* American Conference of Governmental Industrial Hygienists, Cincinnati, OH.

ACGIH. 2007. *Industrial ventilation: A manual of recommended practice*, 26th ed. American Conference of Governmental Industrial Hygienists, Cincinnati, OH.

CFR. 2006. Boiler water additives. 21CFR173.310. *Code of Federal Regulations*, U.S. Government Printing Office, Washington, D.C.

E.U. 2008. *Manufacture of sterile medical products.* Revision of Annex I to the EU guide to good manufacturing practice. European Commission, Brussels.

FDA. 2008. *Current good manufacturing practice for finished pharmaceuticals.* 21CFR210, 211. Code of Federal Regulations, U.S. Government Printing Office, Washington, D.C.

FDA. 2004. *Guidance for industry: Sterile drug products produced by aseptic processing—Current good manufacturing practice.* U.S. Department of Health and Human Resources, Food and Drug Administration, Washington, D.C. http://www.fda.gov/downloads/Drugs/Guidance ComplianceRegulatoryInformation/Guidances/UCM070342.pdf.

ICC. 2009. *International building code®.* International Code Council, Washington, D.C.

ICC. 2009. *International mechanical code®.* International Code Council, Washington, D.C.

ICC. 2009. *International fire code®.* International Code Council, Washington, D.C.

The following publications are available from the Institute of Environmental Sciences and Technology, Mount Prospect, IL.

IEST-RP-CC001.3	HEPA and ULPA filters
IEST-RP-CC002	Laminar flow clean-air devices
IEST-RP-CC003.2	Garment system considerations in cleanrooms and other controlled environments
IEST-RP-CC004.2	Evaluating wiping materials used in cleanrooms and other controlled environments
IEST-RP-CC005	Gloves and finger cots used in cleanrooms and other controlled environments
IEST-RP-C-006.2	Testing cleanrooms
IEST-RP-CC007.1	Testing ULPA filters
IEST-RP CC008	Gas-phase adsorber cells
IEST-RP-CC009.2	Compendium of standards, practices, methods, and similar documents relating to contamination control
IEST-RP-CC011.2	A glossary of terms and definitions relating to contamination control
IEST-RP-CC012.2	Considerations in cleanroom design
IEST-RP-CC013	Equipment calibration or validation procedures
IEST-RP-CC014	Calibrating particle counters
IEST-RP-CC015	Cleanroom production and support equipment
IESR-RP-CC016	The rate of deposition of nonvolatile residue in cleanrooms
IEST-RP-CC017	Ultrapure water: Contamination analysis and control
IEST-RP-CC018	Cleanroom housekeeping—Operating and monitoring procedures
IEST-RP-CC019	Qualifications for agencies and personnel engaged in the testing and certification of cleanrooms and clean air devices
IEST-RP-CC020	Substrates and forms for documentation in cleanrooms
IEST-RP-CC021	Testing HEPA and ULPA filter media
IEST-RP-CC022.1	Electrostatic charge in cleanrooms and other controlled environments
IEST-RP-CC023.1	Microorganisms in cleanrooms
IEST-RP-CC024.1	Measuring and reporting vibration in microelectronics facilities
IEST-RP-CC025	Evaluation of swabs used in cleanrooms
IEST-RP-CC026.1	Cleanroom operations
IEST-RP-CC027.1	Personnel practices and procedures in cleanrooms and controlled environments
IEST-RP-CC028	Minienvironments
IEST-RP-CC029	Automotive paint spray applications
IEST-STD-CC1246D	Products cleanliness levels and contamination control program

ISO. 1999. Cleanrooms and associated controlled environments, part 1: Classification of air cleanliness. *Standard* 14644-1. International Organization for Standardization, Geneva, Switzerland.

ISO. 2000. Cleanrooms and associated controlled environments, part 2: Specifications for testing and monitoring to prove continued compliance with *Standard* 14644-1. *Standard* 14644-2. International Organization for Standardization, Geneva, Switzerland.

ISO. 2005. Cleanrooms and associated controlled environments, part 3: Test methods. *Standard* 14644-3. International Organization for Standardization, Geneva, Switzerland.

ISO. 2001. Cleanrooms and associated controlled environments, part 4: Design, construction and start-up. *Standard* 14644-4. International Organization for Standardization, Geneva, Switzerland.

ISO. 2004. Cleanrooms and associate controlled environments—Part 7: Separative devices (clean air hoods, glove boxes, isolators and minienvironments). *Standard* 14644-7. International Organization for Standardization, Geneva, Switzerland.

ISO. 2003. Cleanrooms and associated controlled environments—Biocontamination control, part 1: General principles and methods. *Standard* 14698-1. International Organization for Standardization, Geneva, Switzerland.

ISPE. 2009. *Baseline guide volume 2: Oral solid dosage forms.* International Society for Pharmaceutical Engineering, Tampa, FL.

ISPE. 1999. *Baseline guide volume 3: Sterile manufacturing facilities.* International Society for Pharmaceutical Engineering, Tampa, FL.

ISPE. 2001. *Baseline guide volume 5: Commissioning and qualification (for pharmaceutical facilities).* International Society for Pharmaceutical Engineering, Tampa, FL.

NEBB. 2009. *Procedural standards for certified testing of cleanrooms.* National Environmental Balancing Bureau, Gaithersburg, MD.

NFPA. 2009. Protection of semiconductor facilities. *Standard* 318. National Fire Protection Association, Quincy, MA.

Pedersen, C.O., D.E. Fisher, R.J. Liesen, and J.D. Spitler. 1998. *Cooling and heating load calculation principles.* ASHRAE.

Sartor, D. et al. 1999. Cleanrooms and laboratories for high-tech industries. California Energy Commission. *Final Report.* http://tinyurl.com/64qnja or http://ateam.lbl.gov/PUBS/cec/CEC_Final Report.pdf.

Shieh, C. 1990. Cleanroom HVAC design. *Proceedings of the 6th International Symposium on Heat and Mass Transfer*, Miami. International Association for Hydrogen Energy, Coral Gables, FL.

Shieh, C. 2000-2005. Abstracts. Seminar and symposium series of integrated cleanroom design and construction. 2000-2005 ASHRAE winter and annual meetings.

Shieh, C. 2005. Integrated cleanroom design and construction. *ASHRAE Transactions* 111(1):355-362.

Spitler, J.D. 2009. *Load calculation applications manual.* ASHRAE.

Sun, W. 2003. Development of pressurization airflow design criteria for spaces under required pressure differentials. *ASHRAE Transactions* 109(1):52-64.

Sun, W. 2005. Automatic room pressurization test technique and adaptive flow control strategy in cleanrooms and controlled environments. *ASHRAE Transactions* 111(2):23-34.

Sun, W. 2008. Conserving fan energy in cleanrooms. *ASHRAE Journal* 50(7).

Sun, W. et al. 2010. Development of cleanroom required airflow rate model based on establishment of theoretical basis and lab validation. *ASHRAE Transactions* 116(1):87-97.

Whyte, W. 1999. *Cleanroom design*, 2nd ed. John Wiley, New York.

CHAPTER 19

DATA PROCESSING AND TELECOMMUNICATION FACILITIES

DATACOM (data processing and telecommunications) facilities are predominantly occupied by computers, networking equipment, electronic equipment, and peripherals. The most defining HVAC characteristic of data and communications equipment centers is the potential for exceptionally high sensible heat loads (often orders of magnitude greater than a typical office building). In addition, the equipment installed in these facilities typically:

- Serves mission-critical applications (i.e., continuous operation)
- Has special environmental requirements (temperature, humidity, and cleanliness)
- Has the potential for disruptive overheating and equipment failure caused by loss of cooling

Design of any datacom facility should also address the fact that most datacom equipment will be replaced multiple times with more current technology during the life of the facility. As described in *Datacom Equipment Power Trends and Cooling Applications* (ASHRAE 2005a), typical datacom equipment product cycles are 1 to 5 years, whereas facilities and infrastructure have life cycles of 10 to 25 years. Replacement equipment has historically required more demanding power and cooling requirements.

Understanding these critical parameters is essential to datacom facility design.

DESIGN CRITERIA

Types of datacom (ASHRAE 2005a) equipment that require air conditioning to maintain proper environmental conditions include

- Computer servers (2U and greater)
- Computer servers (1U, blade, and custom)
- Communication (High-density)
- Communication (Extreme-density)
- Tape storage
- Storage servers
- Workstations (standalone)
- Other rack- and cabinet-mounted equipment

Personnel also occupy datacom facilities, but their occupancy is typically transient and environmental conditions (e.g., temperature, noise) are more typically dictated by equipment needs. However, human occupancy in smaller datacom facilities may influence the ventilation air quantity. A data center is a building or portion of a building whose primary function is to house a computer room and its support areas; data centers typically contain high-end servers and storage products with mission-critical functions. Personnel also occupy datacom facilities, but their occupancy is typically transient and environmental conditions are usually more dictated by equip-

ment needs, thereby making it more of a process cooling application rather than comfort cooling. However, human occupancy in smaller datacom facilities may influence ventilation air requirements.

Overview

Environmental requirements of datacom equipment vary depending on the type of equipment and/or manufacturer. However, a consortium of server manufacturers has agreed on a set of four standardized conditions (Classes 1 to 4), listed in *Thermal Guidelines for Data Processing Environments* (ASHRAE 2008). A fifth classification, the Network Equipment—Building Systems (NEBS) class, is typically used in telecommunications.

- **Class 1:** typically a datacom facility with tightly controlled environmental parameters (dew point, temperature, and relative humidity) and mission-critical operations; types of products typically designed for these environments are enterprise servers and storage products.
- **Class 2:** typically a datacom space or office or lab environment with some control of environmental parameters (dew point, temperature, and relative humidity); types of products typically designed for this environment are small servers, storage products, personal computers, and workstations.
- **Class 3:** typically an office, home, or transportable environment with little control of environmental parameters (temperature only); types of products typically designed for this environment are personal computers, workstations, laptops, and printers.
- **Class 4:** typically a point-of-sale or light industrial or factory environment with weather protection, sufficient winter heating, and ventilation; types of products typically designed for this environment are point-of-sale equipment, industrial controllers, or computers and handheld electronics such as PDAs.
- **NEBS:** per *Telcordia* (2001, 2006), and typically a telecommunications central office with some control of environmental parameters (dew point, temperature and relative humidity); types of products typically designed for this environment are switches, transport equipment, and routers.

Because Class 3 and 4 environments are not designed primarily for datacom equipment, they are not covered further in this chapter; refer to ASHRAE's (2008) *Thermal Guidelines for Data Center Environments* for further information.

Environmental Specifications

Table 1 lists recommended and allowable conditions for Class 1, Class 2, and NEBS environments, as defined by the footnoted sources. Figure 1A shows recommended temperature and humidity conditions for these classes on a psychrometric chart, and Figure 1B shows allowable temperature and humidity conditions. Note that dew-point temperature and relative humidity are also specified.

The preparation of this chapter is assigned to TC 9.9, Mission-Critical Facilities, Technology Spaces, and Electronic Equipment.

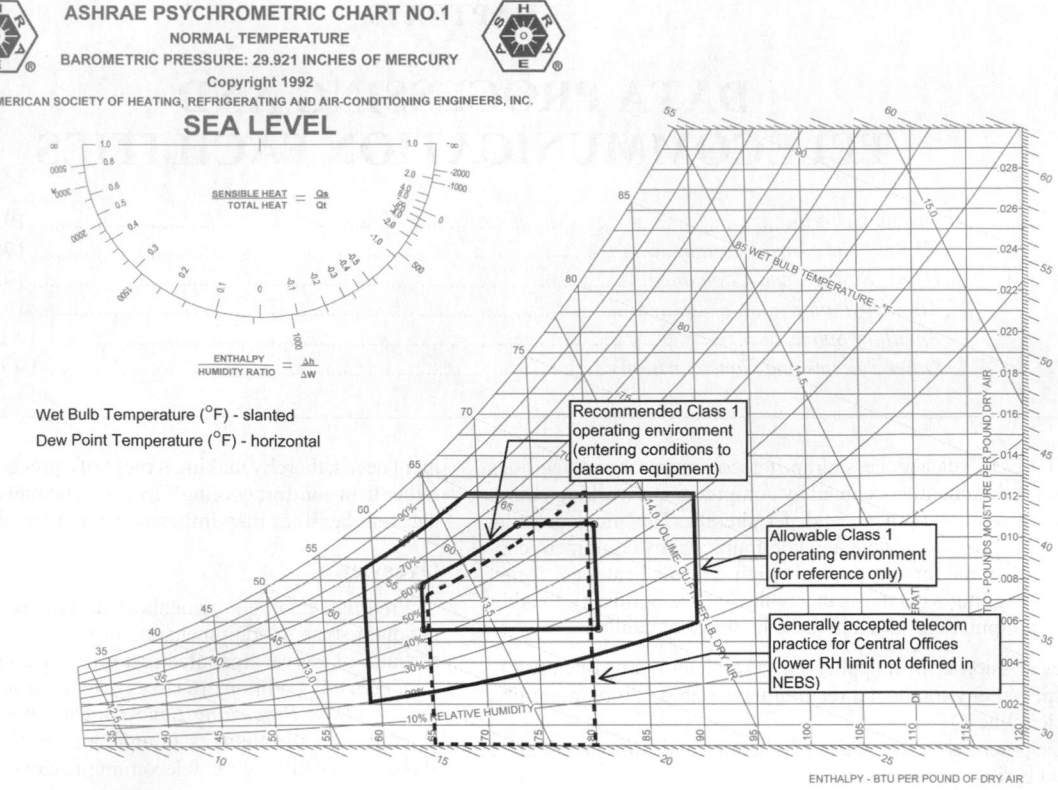

Fig. 1A   Recommended Data Center Class 1, Class 2, and NEBS Operating Conditions

Table 1   Class 1, Class 2, and Selected NEBS Design Conditions

Condition	Classes 1 and 2		NEBS	
	Allowable Level	Recommended Level	Allowable Level	Recommended Level
Temperature control range[a]	59 to 90°F[f] (Class 1) 50 to 95°F[f] (Class 2)	64.4 to 80.6°F	41 to 104°F[c,f]	65 to 80°F[d]
Maximum temperature rate of change[a]	9°F/h		(cooling) 54°F/h[c] (warming) 173°F/h[c,d]	
Relative humidity control range[a]	20 to 80%, 63°F max. dew point (Class 1) 70°F max. dew point (Class 2)	Dew point 42 to 59°F, rh less than 60%	5 to 85%, 82°F max. dew point[c]	Max 55%[e]
Filtration quality	65%, min. 30% (MERV 11, min. MERV 8)[b]		Min. 85% (Min. MERV 13)[b]	

[a]Inlet conditions recommended in ASHRAE (2008).   [c]*Telcordia* (2006).   [e]Generally accepted telecommunications practice. Telecommunications central offices are not
[b]Percentage values per ASHRAE *Standard* 52.1 dust-spot   [d]*Telcordia* (2001).   generally humidified, but personnel are often grounded to reduce electrostatic discharge (ESD).
   efficiency test. MERV values per ASHRAE *Standard* 52.2.   [f]See Figure 2 for temperature derating with altitude.

Air density also affects the ability of datacom equipment to be adequately cooled. ASHRAE's (2008) *Thermal Guidelines for Data Processing Environments* suggests that data center products be designed to operate up to 10,000 ft altitude, but recognizes that there is reduced mass flow and convective heat transfer associated with lower air density at higher elevations. To account for this effect, the guideline includes a derating chart for the maximum allowable temperature of 1°F per 550 ft altitude above 2950 ft (Classes 1 to 4). Figure 2 shows the altitude derating recommended by ASHRAE (2004) for Classes 1 and 2, and for NEBS.

The stated environmental conditions are as measured at the inlet to the data and communications equipment, and not average space or return air conditions.

Temperature

The allowable temperature range is a statement of functionality, whereas the recommended range is a statement of reliability. Thus,

equipment exposed to prolonged high temperatures (and/or to steep temperature gradients) can experience increased failure rates, reduced service life, hardware and/or software failures, and/or thermal shutdown. Exceeding the recommended limits for short periods of time should not be a problem, but running near the allowable limits for months could result in increased reliability issues. Facility designers and operators should strive for continuous operation in the recommended range. ASHRAE (2008) and Telcordia (2001) recommended range is 65 to 80°F.

Not only is air temperature into the electronics critical for reliable operation of components in the electronic box, but the air discharged from the electronics and flowing over the components (cabling, connectors, etc.) at the exit must also be addressed. The recommended ranges apply to inlets of all equipment in the data center (except where IT manufacturers specify other ranges). Attention is needed to make sure the appropriate inlet conditions are achieved for the top portion of IT equipment racks. The inlet air

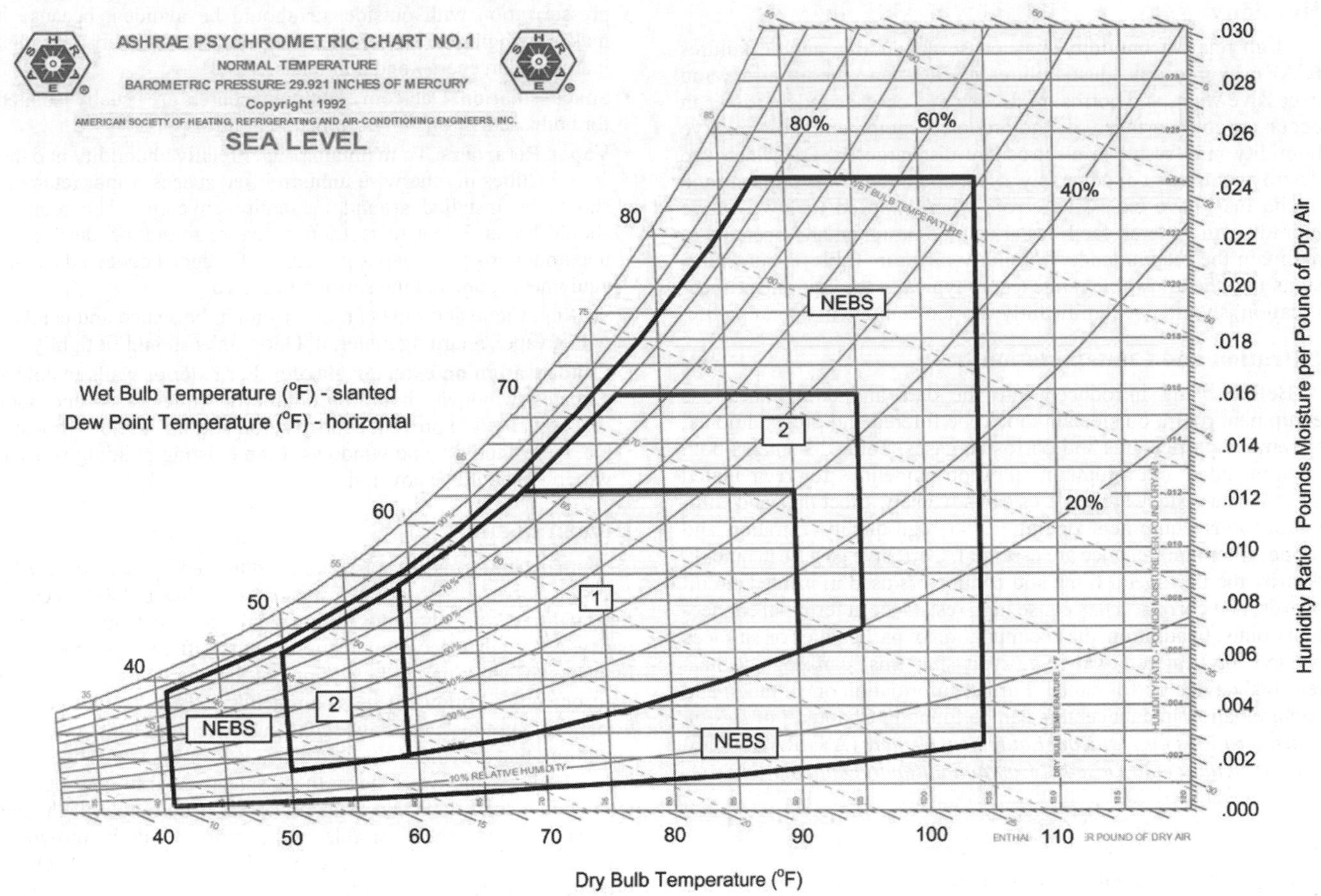

ASHRAE PSYCHROMETRIC CHART NO.1
NORMAL TEMPERATURE
BAROMETRIC PRESSURE: 29.921 INCHES OF MERCURY
Copyright 1992
AMERICAN SOCIETY OF HEATING, REFRIGERATING AND AIR-CONDITIONING ENGINEERS, INC.
SEA LEVEL

Wet Bulb Temperature (°F) - slanted
Dew Point Temperature (°F) - horizontal

Fig. 1B   Allowable Data Center Class 1, Class 2, and NEBS Operating Conditions

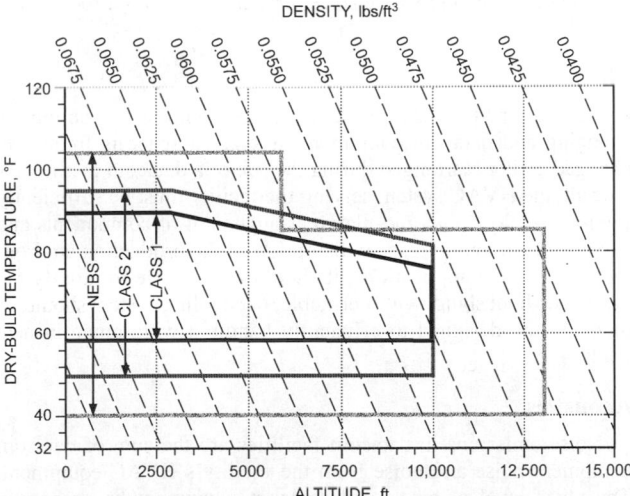

Fig. 2   Class 1, Class 2, and NEBS Allowable Temperature Range Versus Altitude

temperature in many data centers tends to be warmer near the top of racks, particularly if warm rack exhaust air does not have a direct return path to the CRACs. This warmer air also affects the relative humidity, resulting in lower values at the top of the rack. The air temperature generally follows a horizontal line on the psychometric chart where absolute humidity remains constant but relative humidity decreases.

Temperature Rate of Change

Some datacom manufacturers have established criteria for allowable rates of environmental change to prevent shock to the data and communications equipment. These criteria need to be reviewed for all installed datacom equipment. A maximum inlet temperature change of 9°F/h is recommended by ASHRAE (2008) for Classes 1 and 2. Humidity rate of change is typically most important for tape and storage products. Typical requirements for tape are a rate of change of less than 3.5°F/h and a relative humidity change of less than 5%/h (ASHRAE 2004).

In telecommunications central offices, the NEBS requirements per *Telcordia* (2006) for testing new equipment is a rate change (cooling) of 54°F/h. However, in the event of an air-conditioning failure, the rate of temperature change can easily be significantly higher. Consequently, *Telcordia* (2001, 2006) prescribes testing with a warming gradient of 173°F/h for 15 min. Manufacturers' requirements should be reviewed and fulfilled to ensure that the system functions properly during normal operation and during start-up and shutdown.

Procedures must be in place for response to an event that shuts down critical cooling systems while critical loads continue to operate, causing the space temperature to begin rising immediately. Procedures should also be in place governing how quickly elevated space temperatures can be returned to normal to avoid thermal shock damage.

Datacom equipment usually tolerates a somewhat wider range of environmental conditions when not in use [see Table 2.1 in ASHRAE (2008)]. However, it may be desirable to provide uninterruptible cooling in the room to maintain operating limits and minimize thermal shock to the equipment.

Humidity

High relative humidity may cause conductive anodic failures (CAF), hygroscopic dust failures (HDF), tape media errors and excessive wear, and corrosion. In extreme cases, condensation can occur on cold surfaces of liquid-cooled equipment. Low relative humidity may result in electrostatic discharge (ESD), which can destroy equipment or adversely affect operation. Tape products and media may have excessive errors when exposed to low relative humidity. In general, facilities should be designed and operated to maintain the recommended humidity range in Table 1, but excursions into the allowable range (more typically the equipment specification) should not significantly shorten equipment operating life.

Filtration and Contamination

Before being introduced into the data and communications equipment room, outside air should be filtered and preconditioned to remove particulates and corrosive gases. Table 1 contains both recommended and minimum filtration guidelines for recirculated air in a data center. Particulates can adversely affect data and communications equipment operation, so high-quality filtration and proper filter maintenance are essential. Corrosive gases can quickly destroy the thin metal films and conductors used in printed circuit boards, and corrosion can cause high resistance at terminal connection points. In addition, the accumulation of particulates on surfaces needed for heat removal (e.g., heat sink fins) can degrade heat removal device performance. Further information on filtration and contamination in data centers can be found in Chapter 8 of *Design Considerations for Datacom Equipment Centers* (ASHRAE 2005b) and *Particulate and Gaseous Contamination in Datacom Environments* (ASHRAE 2009).

Ventilation

Data and communications equipment room air conditioning must provide adequate outside air to achieve the following criteria:

- Maintain the room under positive pressure relative to surrounding spaces.
- Dilute indoor generated pollutants such as VOCs.
- Satisfy ASHRAE *Standard* 62.1 requirements.
- Meet local codes for ventilation for datacom facilities in all spaces, including mechanical, uninterruptible power supply (UPS), and battery rooms

The need for positive pressure to keep contaminants out of the room is usually the controlling design criterion in data and communication equipment rooms. Pressurization calculations can be performed using the procedures outlined in Chapter 16 of the 2009 *ASHRAE Handbook—Fundamentals*. Chapter 53 of this volume has calculation formulas for achieving pressurization as well as loss of pressure through cracks in walls and at windows.

Although most computer rooms have few occupants, calculations should always be performed to ensure that adequate ventilation for human occupancy is provided in accordance with ASHRAE *Standard* 62.1 and local codes. Internally generated contaminants may make the indoor air quality method the more appropriate procedure; however, maintaining positive pressure usually requires a higher outside airflow.

Envelope Considerations

In addition to meeting state, national, and local codes, there are several other parameters that should be considered in designing the envelope of datacom facilities, including pressurization, isolation, vapor retardants, sealing, and condensation.

- **Pressurization.** Datacom facilities are typically pressurized to prevent infiltration of air and pollutants through the building envelope. An air lock or mantrap is recommended for a datacom equipment room door that opens directly to the outside. Excess pressurization with outside air should be avoided, because it makes swinging doors harder to use, and wastes energy through increased fan energy and coil loads.

- **Space Isolation.** Datacom equipment centers are usually isolated for both security and environmental control.

- **Vapor Retarders.** To maintain proper relative humidity in datacom facilities in otherwise unhumidified spaces, vapor retarders should be installed around the entire envelope. The retarder should be sufficient to restrain moisture migration during the maximum projected vapor pressure difference between datacom equipment room and the surrounding areas.

- **Sealing.** Cable and pipe entrances should be sealed and caulked with a vapor-retarding material. Doorjambs should fit tightly.

- **Condensation on exterior glazing.** For exterior walls in colder climates, windows should be double or triple-glazed and door seals specified to prevent condensation and infiltration. If possible, there should be no windows. If an existing building is used, windows should be covered.

Human Comfort

Human comfort is not specifically addressed in *Thermal Guidelines for Data Processing Environments* (ASHRAE 2008) because the facilities typically have minimal and transient human occupancy. Although telecommunications central offices often have permanent staff working on the equipment, human comfort is not the main objective. Following the recommended Class 1 conditions (see Table 1) in a hot-aisle/cold-aisle configuration may result in comfort conditions that are cold in the cold aisle, and warm or even hot in the hot aisle. Personnel working in these spaces need to consider the temperature conditions that exist, and dress accordingly. If the hot aisle is excessively hot, portable spot-cooling should be provided. The National Institute for Occupational Safety and Health (NIOSH) provides detailed guidance on occupational exposure to hot environments (NIOSH 1986). Another concern is contact burns if equipment is too hot. Human tissue reaches the pain threshold at 111°F, and various levels of injury occur at levels above that (ASTM 2003). Take care that equipment surface temperatures do not represent a hazard.

Flexibility

As described in the introduction, technology is continually changing and datacom equipment in a given space is frequently changed and/or rearranged during the life of a datacom facility. As a result, the HVAC system serving the facility must be sufficiently flexible to allow plug-and-play rearrangement of components and expansion without excessive disruption of the production environment. In critical applications, it should be possible to modify the system without shutdown. If possible, the cooling system should be modular and designed to efficiently handle a wide range in heat loads.

Acoustics

Noise emissions in datacom facilities are the sum of datacom equipment noise and noise from the facility's HVAC equipment. The noise level of air-cooled datacom equipment has generally increased along with the power density and heat loads. Densely populated datacom facilities may run the risk of exceeding U.S. Occupational Safety and Health Administration (OSHA) noise limits (and thus potentially causing hearing damage without personnel protection); refer to the appropriate OSHA (1996) regulations and guidelines. European occupational noise limits are somewhat more stringent than OSHA's and are mandated in EC Directive 2003/10/EC (European Council 2003). Facility noise level calculations can be made following the methodology outlined in Chapter 48, and in Chapter 8 of the 2009 *ASHRAE Handbook—Fundamentals*. An acoustic consultant may be needed to properly predict sound levels from multiple sources and paths, as is typically the case in a datacom facility.

Manufacturers of electronic equipment typically take steps to minimize acoustic noise emissions from datacom equipment. Speed control of air-moving devices, rack- or frame-level acoustic treatments, and reduction of line-of-sight noise emissions are common techniques to reduce datacom equipment noise.

Vibration Isolation and Seismic Restraint

HVAC equipment in datacom facilities should be independently supported and isolated to prevent vibration transmission to the datacom equipment. If required, vibration isolators should be seismically rated for the specific environment into which they are installed, to comply with the appropriate codes. Consult datacom equipment manufacturers for equipment sound tolerance and specific requirements for vibration isolation. Many datacom equipment manufacturers test their equipment to the vibration and seismic requirements of *Telcordia* (2001). Additional guidance can be found in Chapter 55 and in the *International Building Code* (ICC 2009).

HVAC LOAD CONSIDERATIONS

HVAC loads in datacom facilities must be calculated in the same manner as for any other facility. Typical features of these facilities are a high internal sensible heat load from the datacom equipment itself and a correspondingly high sensible heat ratio. However, other loads exist and it is important that a composite load comprised of all sources is calculated early in the design phase, rather than relying on a generic overall "watts per square foot" estimate that neglects other potentially important loads.

Also, if the initial deployment or first-day datacom equipment load is low because of low equipment occupancy, the effect of the other loads (envelope, lighting, etc.) becomes proportionately more important in terms of part-load operation.

Datacom Equipment

The major heat source in datacom facilities is the datacom equipment itself. This heat can be highly concentrated, non-uniformly distributed, and variable. Equipment that generates large quantities of heat is normally configured with internal fans and airflow passages to transport cooling air, usually drawn from the space, through the equipment.

Information on datacom equipment heat release should be obtained from the manufacturer. Guidance on industry heat load trends can be obtained from ASHRAE's (2005a) *Datacom Equipment Power Trends and Cooling Applications*. It is important to know the approximate allocation of different types of datacom equipment when designing datacom facility environmental control, because heat loads of different types of equipment vary dramatically. Figure 3 shows projected trends of six equipment classifications through the year 2014; a sample heat load calculation based on these classifications is given in ASHRAE's (2005b) *Design Considerations for Datacom Equipment Centers*.

At the equipment level, ASHRAE (2008) includes a sample equipment thermal report that can provide heat release information in a format specifically suited for thermal design purposes. Nameplate information for data and communications equipment should not be used for thermal design, because it will yield unrealistically high design values and an oversized cooling system infrastructure.

Most current datacom equipment has variable-airflow cooling fans that depend on inlet temperature and/or load. Under typical operating conditions, the flow requirements of these fans may be low, but increase under extreme conditions such as high system inlet temperature. Consideration of this variable flow may be important to HVAC system design.

Load Considerations and Challenges

Similar to commercial loads, datacom loads often operate well below the calculated load. This can be more problematic for datacom facilities, though, because the load densities are so much greater than commercial installations. Further, the source of the load (datacom equipment) is often replaced multiple times during the life of the cooling system, requiring consideration of oversized infrastructure or phased construction to accommodate future changes. The part- and low-load conditions must be well understood and equipment selected accordingly.

It is particularly important to understand initial and future loads in detail. Otherwise, the stated initial and future loads could have compounded safety factors or, in the worst cases, guesses. *Datacom Equipment Power Trends and Cooling Applications* (ASHRAE 2005a) identifies the following topics to consider when predicting future load:

• Existing applications' floor space
• Performance growth of technology based on footprint

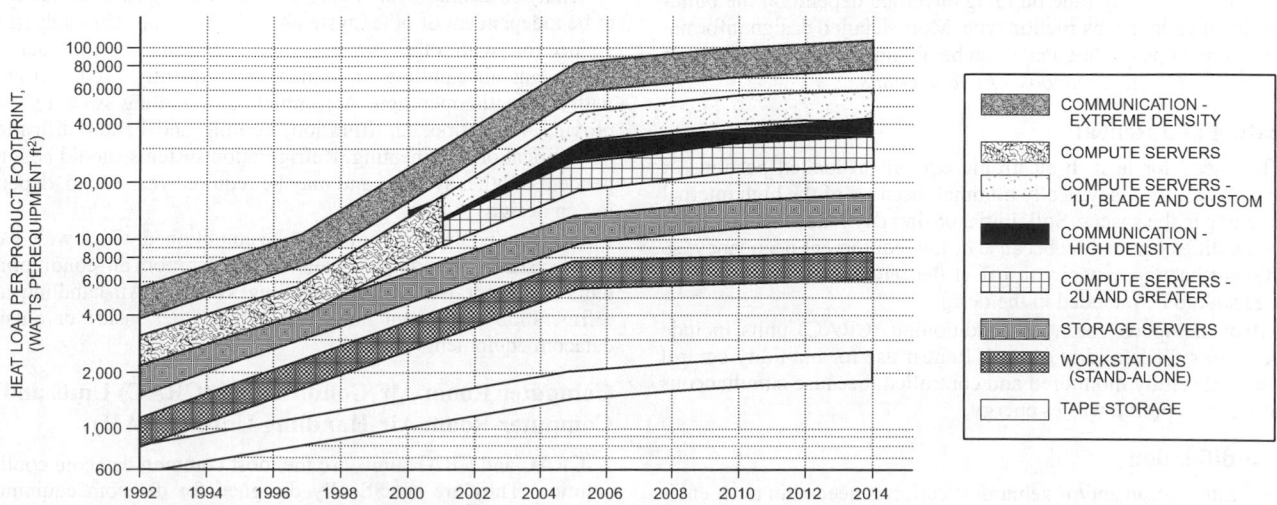

Fig. 3   Projected Power Trends of Datacom Equipment
(ASHRAE 2005a)

- Processing capability compared to storage capability
- Change in applications over time
- Asset turnover

Once initial and future loads are understood, as well as the part- and low-load conditions, equipment can be selected. This includes gaining consensus from the project stakeholders regarding acceptable amount of disruption that can occur in an operating facility for upgrades such as cooling capacity or distribution.

Ventilation and Infiltration

For load calculation protocols relating to ventilation and infiltration, refer to Chapter 16 of the 2009 *ASHRAE Handbook—Fundamentals*.

Datacom facilities' outdoor air requirements may be lower than other facilities because of the light human occupancy load. In many cases, it is advantageous to precondition this air, with the space under positive pressure, to allow for 100% sensible cooling in the space. If this approach is adopted, however, preconditioning system failure must also be addressed to avoid the potential for widespread condensation in the space.

Electrical Equipment

In some cases, power distribution units (PDUs) are located in the datacom equipment room as the final means of transforming voltage to a usable rating and distributing power to the datacom equipment. The heat dissipation from the transformers in the PDUs should be accounted for by referencing the manufacturer's equipment specifications.

Lights

High-efficiency lighting should be encouraged, as well as lighting controls, to minimize lighting heat gain. Additionally, depending on the means of fire suppression and ceiling type, unvented light fixtures should be considered.

People

Occupancy loads should be considered as light work. People often comprise the only internal latent load in a datacom facility, which may be a factor in selecting cooling coils, especially if outdoor air is supplied through a dedicated outdoor air system.

Building Envelope

Heat gains through the building envelope depend on the buildings' location and construction type. More detailed design information on envelope cooling loads can be obtained from Chapter 18 of the 2009 *ASHRAE Handbook—Fundamentals*.

Heating and Reheat

The need for heat in electronic equipment-loaded portions of datacom facilities is typically minimal, because of the high internal heat gains in the spaces. Still, initial or first day loads in many datacom facilities can be low because of low equipment occupancy, so sufficient heating capacity to offset the outdoor air and envelope losses should be included in the design.

Many computer room air-conditioning (CRAC) units include reheat coils for humidity control. Reheat use for humidity control must be carefully monitored and controlled, because simultaneous heating and cooling wastes energy.

Humidification

Humidification and/or dehumidification is needed in most environments to meet both the recommended and allowable humidity ranges specified for Class 1 and Class 2 data centers (see Table 1). In most cases, the predominant moisture load is outdoor air, but all potential loads should be considered. Where humidification is not

provided, personal grounding is typically utilized to minimize electrostatic discharge (ESD) failures.

Vapor retardant analyses should also be performed where humidity-controlled spaces contain outside walls or ceilings. Refer to Chapter 27 of the 2009 *ASHRAE Handbook—Fundamentals* for additional design information.

High-Density Loads

Heat density of some types of datacom equipment is increasing dramatically. Stand-alone server heat loads (or rack loads) can have heat loads exceeding 30 kW. These increased heat densities require a design engineer to keep abreast of latest design techniques to ensure adequate cooling, and to pay close attention to the loads of installed electronic equipment and future equipment deployments.

Ensure that local high-density loads are provided with adequate local cooling, even when the overall heat density of the general space is below the high-density threshold. Rack inlet conditions should be checked and verified as adequate to meet the manufacturer's requirements. Refer to *Thermal Guidelines for Data Processing Environments* (ASHRAE 2008) for additional information and guidance.

Increases in heat density have made it more difficult to air-cool computers, leading to increased interest in efficient liquid-cooling techniques. Liquid cooling media include water, refrigerants, high-dielectric fluorocarbons, or two-phase fluids such as dielectrics. Some computer manufacturers have already taken this approach, and there are also products available that use liquid cooling at the cabinet level. The reader is encouraged to keep abreast of research and development in this area. ASHRAE (2006) also published a book on liquid cooling, *Liquid Cooling Guidelines for Datacom Equipment Centers*.

New datacom facilities and those slated for major renovation should consider adding appropriate infrastructure (piping taps, feeders, etc.) for future use and load increases. Retrofitting these "backbones" is typically much more expensive, disruptive, and risky.

Note that local high-density areas can have power densities significantly higher than the average for the center. Ideally, high-density computing equipment should be identified during design so appropriate cooling can be provided.

HVAC SYSTEMS AND COMPONENTS

It may be desirable for HVAC systems serving datacom facilities to be independent of other systems in the building, although cross-connection with other systems may be desirable for back-up. Redundant air-handling equipment is frequently used, normally with automatic operation. A complete air-handling system should provide ventilation, air filtration, cooling and dehumidification, humidification, and heating. Refrigeration systems should be independent of other systems and may be required year-round, depending on design.

Datacom equipment rooms can be conditioned with a wide variety of systems, including packaged computer room air-conditioning units and central-station air-handling systems. Air-handling and refrigeration equipment may be located either inside or outside datacom equipment rooms.

Computer Room Air-Conditioning (CRAC) Units and Computer Room Air-Handling Units (CRAH)

CRAC and CRAH units are the most common datacom cooling solution. They are specifically designed for datacom equipment room applications and should be built and tested in accordance with the requirements of ANSI/ASHRAE *Standard* 127.

Cooling. CRAH units are special-purpose chilled-water air handlers designed for datacom applications. CRAC units are available in

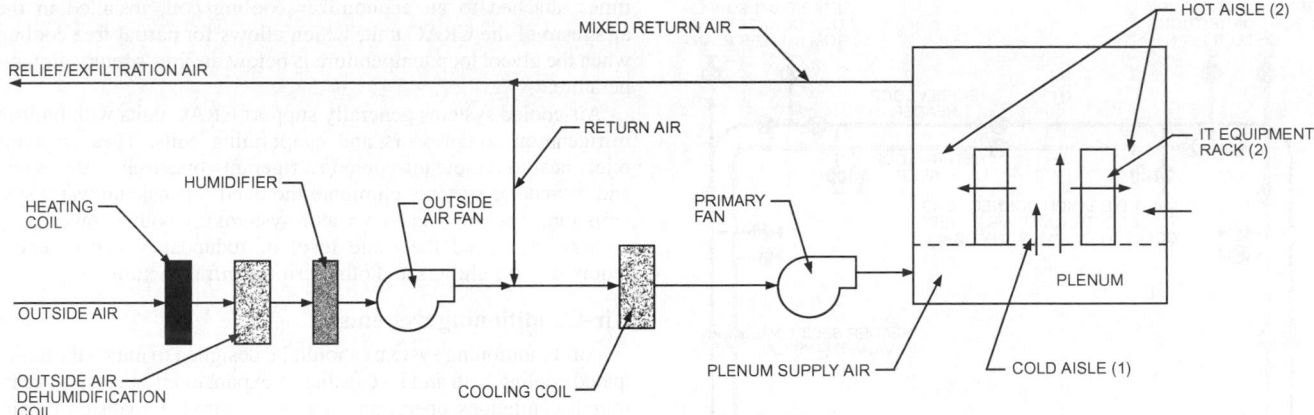

Fig. 4 Datacom Facility with Dedicated Outdoor Air Preconditioning

several types of cooling system configurations: direct expansion (DX) air-cooled, DX water-cooled, DX glycol-cooled, and dual-cooled (both chilled-water and DX). DX units typically have multiple refrigerant compressors with separate refrigeration circuits. Both CRAH and CRAC units have air filters, and integrated control systems with remote monitoring panels and interfaces. Reheat coils, variable-speed fan controls, and humidifiers are an option. Where weather conditions make this strategy economical, CRAC units may also be equipped with propylene glycol precooling coils and associated dry coolers to allow water-side economizer operation, or may also be equipped with mixing boxes to allow air-side economizer operation.

Location. CRAC/CRAH units are usually located within the datacom equipment room, but may also be remotely located and ducted to the conditioned space. With either placement, their temperature and humidity sensors should be located to properly control inlet air conditions to the datacom equipment within specified tolerances (see Table 1). Analysis of airflow patterns in the datacom equipment room [e.g., with computational fluid dynamics (CFD)] may be required to optimally locate datacom equipment, CRAC/CRAH units, and sensors, to ensure that sensors are not in a location that is not conditioned by the CRAC/CRAH unit they control, or in a nonoptimum location that forces the cooling system to expend more energy than required.

Humidity Control. Types of available humidifiers within CRAC/CRAH units may include steam, infrared, and ultrasonic. Consideration should be given to maintenance and reliability of humidifiers. It may be beneficial to relocate all humidification to a dedicated central system. Another consideration is that some humidification methods or improperly treated makeup water are more likely to carry fine particulates to the space.

Reheat is used in dehumidification mode when air is overcooled to remove moisture. On a call for reheat, sensible heat (typically from electric, hot-water, or steam coils) is introduced to supplement the actual load in the space. Using waste heat of compression (hot gas) for reheat may also be an energy-saving option. This overcooling and reheating should be tightly controlled.

Ventilation. Dedicated outdoor air systems have been installed in many datacom facilities to control space pressurization and humidity without humidifiers and reheat in either the CRAC units or other datacom central cooling systems. Figure 4 shows an independent outdoor air preconditioning system in conjunction with a sensible-only recirculation system. The humidifier in the dedicated outdoor air system often controls the humidity in the datacom equipment room based on dew point.

Central-Station Air-Handling Units

Some larger datacom facilities and most telecommunications central offices use central-station air-handling units. Some of their advantages and disadvantages are discussed here.

Coil Selection and Control. A wide range of heating and cooling coil types can be used for datacom facilities, but, ideally, any coil design or specification should include modulating control. In addition, when dehumidifying, control of the cooling coil to maintain dew point can be critical to maintaining the datacom facility within temperature and humidity set points. For more information on cooling coil design, see Chapter 22 in the 2008 *ASHRAE Handbook—HVAC Systems and Equipment*.

Humidification. Various types of central-station humidification systems can be used for datacom facility applications, with each type offering varying steam quality, level of control, and energy consumption. Available water quality and requirements for water treatment must also be considered when selecting the humidifier type.

Flexibility and Redundancy using VAV Systems. Flexibility and redundancy can be achieved by using variable-volume air distribution, oversizing, cross-connecting multiple systems, or providing standby equipment. Compared to constant-air-volume units (CAV), variable-air-volume (VAV) equipment can be sized to provide excess capacity but operate at discharge temperatures or airflow rates appropriate for optimum temperature and humidity control, reducing operational fan power requirements and the need for reheat.

Common pitfalls of VAV include shifts in underfloor pressure distribution and associated flow through tiles. Airflow should be modeled using CFD or other analytical techniques to ensure that the system can modulate without adversely affecting overall airflow and cooling capability to critical areas.

Chilled-Water Distribution Systems

Chilled-water distribution systems should be designed to the same standards of quality, reliability, and flexibility as other computer room support systems. Where growth is likely, the chilled-water system should be designed for expansion or addition of new equipment without extensive or disruptive shutdown. Figure 5 illustrates a looped chilled-water system with sectional valves and multiple valved branch connections. The branches could serve air handlers or water-cooled computer equipment.

The valve quantity and locations allow modifications or repairs without complete shutdown because chilled water can be fed from either side of the loop. This loop arrangement is a practical method of improving the reliability of a chilled-water system serving a

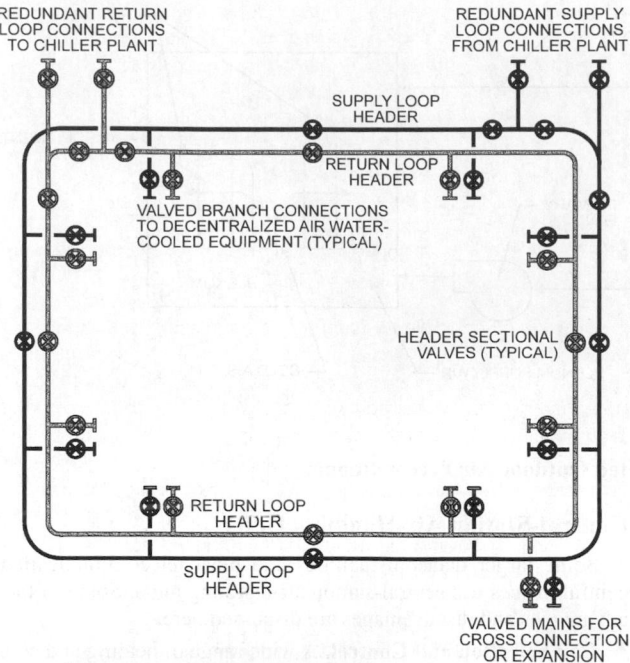

Fig. 5   Chilled-Water Loop Distribution

computer room. "Future taps" should have blind flanges with a pressure gage and drain between the flange and isolation valve to allow the valve to be exercised and checked for holding performance. Sectional valves should be suitable for bidirectional flow and tight shutoff from flow in either direction, to allow maintenance on either side of the valve. In some cases, multiple valves may be required to allow maintenance of the valves themselves.

Where chilled water serves CRAC units or other packaged equipment in the datacom equipment room, select water temperatures that satisfy the space sensible cooling loads without causing latent cooling. Because datacom equipment room loads are primarily sensible, chilled-water supply temperature can be higher than in commercial applications. Greater differentials between the supply and return chilled-water temperatures allow reduced chilled-water flow, which saves pump energy and piping installed costs.

To provide better temperature control of datacom equipment in a data center, numerous manufacturers offer products where liquid (water or refrigerant) is brought close to the datacom rack and used to remove the heat generated by the datacom equipment. Liquid-cooled heat exchangers placed in strategic locations are used to cool hot air exhausted from the datacom equipment, thereby removing either all or part of the equipment's heat load.

Chilled-water pipe insulation with a vapor barrier is required to prevent condensation, but not to prevent thermal loss in a cold plenum; therefore, minimum insulation thickness should be considered, because insulated piping can restrict underfloor air distribution.

Condenser Systems

Heat rejection in datacom facilities can be with either water-cooled or air-cooled systems. Basic information on condenser water systems can be obtained from Chapter 13 in the 2008 *ASHRAE Handbook—HVAC Systems and Equipment*. Where evaporative cooling or open-cell cooling towers are used, consider using makeup water storage as a back-up to the domestic water supply (which provides condenser makeup water).

The "dry-cooler" system incorporates a closed glycol piping loop, which transfers heat from a unit-mounted condenser to an outdoor-air-to-glycol heat exchanger. The same glycol loop is some-

times attached to an economizer cooling coil, installed in the airstream of the CRAC unit, which allows for partial free cooling when the glycol loop temperature is below the unit's return air temperature.

Air-cooled systems generally support CRAC units with built-in refrigeration compressors and evaporating coils. These systems reject heat to remote air-cooled refrigerant condensers. Air-cooled and dry-cooler systems eliminate the need for makeup water systems (and back-up makeup water systems). Cooling towers, dry coolers, etc., need the same level of redundancy and diversity required of the chillers and other critical infrastructure.

Air-Conditioning Systems

Air-conditioning systems should be designed to match the anticipated cooling load and be capable of expansion if necessary; year-round, continuous operation may be required. Expansion of air-conditioning systems while maintaining continuous operation of the data center may also be necessary. A separate system for datacom equipment room(s) may be desirable where system requirements differ from those provided for other building and process systems, or where emergency power requirements preclude combined systems.

Chillers

Because datacom facilities often use large quantities of energy, cooling systems should be designed to maximize efficiency. For many facilities, water-cooled chillers are likely the most efficient system. Basic information on chillers can be obtained from Chapter 42 of the 2008 *ASHRAE Handbook—HVAC Systems and Equipment*.

Part-load efficiency should also be considered during chiller selection, because data centers often operate at less than peak capacity. Chillers with variable-frequency drives, high evaporator temperatures, and low entering condenser water temperatures can have part-load operating efficiencies of 0.35 kW/ton or less. The relative energy efficiency of primary versus secondary pumping systems should also be analyzed to optimize energy consumption.

The recommended data center air temperature may allow higher chilled-water temperatures. Chiller and chiller plant efficiencies can be enhanced by proper selection of the chilled-water supply and return temperatures.

Heat recovery chillers may be an efficient way to recover heat from datacom equipment environments for use in other applications. The heat recovery system must provide the reliability and redundancy needed by the facility. System operation, servicing, and maintenance should not interfere with facility operation.

Pumps

Pumps and pumping system design should take into account energy efficiency, reliability and redundancy. It may be possible to design pumps with variable-speed drives, so that the redundant pump is always operational. Ramp-up to full speed occurs on a loss of an operating pump.

Basic information on pumps can be obtained from Chapter 43 of the 2008 *ASHRAE Handbook—HVAC Systems and Equipment*, and piping systems are covered in Chapter 12 of that volume.

Piping

Chilled-water and glycol piping must be pressure-tested, fully insulated, and protected with an effective vapor retardant. The test pressure should be applied in increments to all sections of pipe in the computer area during construction. In new construction, piping is often installed in trenches below the raised floor to minimize its effect on air distribution. Typically, leak detection is provided along the piping path. When installed overhead, secondary containment is often provided for all piping in datacom equipment room or critical electrical support spaces. Secondary containment systems should incorporate leak detection capability, to detect condensation and identify leaks from damaged piping, valves, fittings, etc. Leak detection should also be placed wherever water piping passes through any critical space, regardless of pipe elevation.

Piping specialty considerations should include a good-quality strainer installed at the inlet to local cooling equipment to prevent control valve and heat exchanger passages from clogging. Strategically placed drains and vents must be included locally at all equipment. Thermometers and other sensors should be installed in a serviceable manner, such as in drywells. Pressure gages should include gage cocks.

If cross connections with other systems are made, the effect of introducing dirt, scale, or other impurities on datacom equipment room systems must be addressed.

Humidifiers

Many types of humidifiers may be used to serve datacom equipment areas, including steam-generating (remote or local), pan (with immersion elements or infrared lamps) and evaporative types (wetted pad and ultrasonic). Ultrasonic devices should use deionized water to prevent formation of abrasive dusts from crystallization of dissolved solids in the water. In general, care must be taken to ensure that particulates or chemicals corrosive to datacom equipment are not used.

The humidifier must be responsive to control, maintainable, and free of moisture carryover. The humidity sensor should be located to provide control of air inlet conditions to the equipment. For additional information, see Chapter 21 of the 2008 *ASHRAE Handbook—HVAC Systems and Equipment*.

Controls and Monitoring

Controls. Control systems must be capable of reliable control of temperature, relative humidity, and, where required, pressurization within tolerance from set point. Control systems serving spaces requiring high availability must be designed so that component or communication failures do not result in failure of the controlled HVAC equipment.

There are a number of ways to accomplish this, but the general approach is to use multiple distributed control systems in a manner such that no system can cause the failure of another system. Where required, HVAC components and their power supplies should have dedicated controllers installed to ensure automatic and independent operation of redundant HVAC systems in the event of failures. In many designs, electrical power for HVAC controls must be from a UPS to maintain proper system operation during interruption of normal power.

Based on Table 1, control should be established that provides an inlet condition to data center equipment and telecommunications equipment of 65 to 80°F. Care is needed to ensure that sensors are properly located, tuned, and calibrated, especially if converting from legacy control based on return air temperature. CFD analysis as well as control system simulation may be needed for a successful retrofit.

Where multiple packaged units are provided, regular calibration of controls may also be necessary to prevent individual units from working against each other. Errors in control system calibration, differences in unit set points, and sensor drift can cause multiple-unit installations to simultaneously heat and cool, and/or humidify and dehumidify, wasting a significant amount of energy. Also consider integrated control systems that communicate from unit to unit, sharing set points and sensor data to ensure coordination, and reduce the potential for units to work against each other. Lead/lag control could also be used, if desired.

Monitoring. Datacom facilities often require extensive monitoring of the mechanical and electrical systems. Multiple interface gateways are often used to interface different monitoring and control systems to the head-end monitoring system and ensure that failure of individual system communication components does not remove access to the total system information database.

Monitoring should include control system sensors as well as independent "monitoring-only" sensors and should include datacom equipment areas, critical infrastructure equipment rooms, command/network operations centers, etc., to ensure critical parameters are maintained. Monitoring also should be sufficient to ensure that anomalies are detected early and with adequate time to allow operating staff time to mitigate and restore conditions before equipment is affected. Monitored data can facilitate trending, alarming, and troubleshooting efforts.

Examples of suitable parameters for monitoring include underfloor static air pressure, temperature, and humidity; early-warning smoke detection; ground currents; and rack inlet temperatures and humidity. Monitoring systems can be integral to or separate from control systems and can be as simple as portable data loggers or strip chart recorders or as complex as high-speed (GPS-synchronized) forensic time stamping of critical breakers and status points. New technologies allow distributed monitoring sensors to be connected to the data and communication network without separate wiring systems.

Because datacom equipment malfunctions may be caused by or attributed to improper control of the datacom room environmental conditions, it may be desirable to keep permanent records of the space temperature and humidity. Many datacom equipment manufacturers imbed temperature and humidity sensors in their equipment, which in turn can be correlated with equipment function and also provide for reduced-capacity operation or shutdown to avoid equipment damage from overheating. In the future it may be possible to connect IT sensors to building systems for monitoring and control.

As a minimum, alarms should be provided to signal when temperature or humidity limits are exceeded. Properly maintained and accurate differential pressure gages for air-handling equipment filters can help prevent loss of system airflow capacity and maintain design environmental conditions. All monitoring and alarm devices should provide local indication as well as interface to the central monitoring system.

AIR DISTRIBUTION

To provide effective cooling, air distribution should closely match load distribution. Distribution systems should be flexible enough to accommodate changes in the location and magnitude of heat gains with minimal change in the basic distribution system. Distribution system materials should ensure a clean air supply. Duct or plenum material that may erode must be avoided. Access should be maintained for cleaning or replacement as needed.

Equipment Placement and Airflow Patterns

Datacom Equipment Airflow Protocols. Datacom equipment is typically mounted in racks or cabinets arranged in rows. In a typical configuration, the "front" of cabinets, racks, or frames (i.e., the side with the air inlets) faces one aisle, and the rear, which includes cable connections, faces another aisle. The cabinets or racks in a datacom environment are usually 78 in. high, whereas telecommunications frames are generally 72 to 84 in. high. Each cabinet or rack may contain a single piece of equipment, or it may contain any number of individual items of equipment, in sizes as small as 1U, where 1U = 1.75 in. (EIA *Standard* 310).

Typically, supply air is drawn into the inlet of the datacom equipment cabinet or rack, picks up heat internal to the equipment, and is then discharged, typically from a different side of the equipment. The air then travels back to the HVAC cooling coil, where the heat is rejected.

To cool datacom equipment efficiently and effectively, there needs to be complementary directivity for airflow through the equipment and airflow through the datacom equipment room. ASHRAE's *Thermal Guidelines for Data Processing Environments* (ASHRAE 2008) and *Telcordia* (2001) define recommended airflow protocols through datacom equipment. Figure 6 shows the three communications equipment airflow protocols that are recommended for use in datacom facilities. The front-to-rear (F-R) protocol has cool air entering the front of the equipment rack (or cabinet), and exiting the rear. The F-T protocol has cool air entering the front of the equipment cabinet, and exiting the top. The front-to-top-and-rear (F-T/R) protocol has cool air entering the front of the equipment and exiting both the top and the rear. Rack-mounted equipment should follow the F-R protocol only. Cabinet-mounted systems can follow any of the three shown.

Other airflow protocols for rack-mounted datacom equipment direct airflow through the left and/or right sides of the equipment within the rack. For these installations, airflow within the rack must be managed to ensure complementary directivity of airflow between independent shelves of data and communications equipment. In addition, spacing between adjacent racks in the same line-up of equipment must be adequate to ensure the appropriate segregation of hot exhaust and cold intake air streams.

Hot Aisle/Cold Aisle Configuration. Using alternating hot and cold aisles promotes separation of the cool supply and warm return streams which generally leads to lower equipment inlet temperatures and greater energy efficiency. Figure 7 shows a schematic view of a hot aisle/cold aisle configuration.

Underfloor Plenum Supply

Datacom facilities often use an underfloor plenum to supply cooling air to the equipment. As shown in Figure 8, the CRAC units

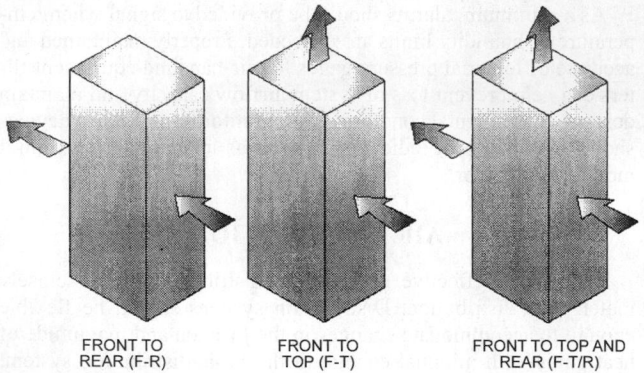

FRONT TO	FRONT TO	FRONT TO TOP AND
REAR (F-R)	TOP (F-T)	REAR (F-T/R)

Fig. 6   Recommended Equipment Airflow Directivity

push cold air into the plenum, from which it is introduced into data and communications equipment rooms via perforated floor tiles, tile cutouts, and other openings. The raised-floor design offers flexibility in placing computer equipment above the raised floor. Cool air can, in theory, be delivered to any location simply by replacing a solid floor tile by a perforated tile.

With a hot-aisle/cold aisle configuration, perforated tiles are placed in the cold aisle. Cool air delivered by the perforated tiles is drawn into the front of the racks. Warm air is exhausted from the back of the racks into the hot aisle and is ultimately returned to the CRAC units.

Often, the underfloor plenum is used for cables, electrical conduits, and pipes. These obstructions in the plenum can interfere with airflow. When determining plenum depth, below-floor obstructions must be considered.

Airflow Inlet Delivery Concerns. For good thermal management, required airflow must be supplied through the perforated tile(s) located near the inlet of each piece of datacom equipment. The heat load can vary significantly across datacom equipment rooms, and changes with addition or reconfiguration of hardware. For datacom equipment to operate reliably, the design must ensure that cool air distributes properly (i.e., the distribution of airflow rates through perforated tiles matches the cool-air needs of equipment on the raised floor).

When adequate airflow is not supplied through the perforated tiles, internal fans in the equipment racks tend to draw air through the front of the cabinet from the path of least resistance, which typically includes the space to the sides of and above the racks. Because most of this air originates in the hot aisle, its temperature is high. Thus, cooling of the sides and upper portion of the equipment racks can be seriously compromised.

Air tends to stratify, with cold supply air near the floor and hot air near the ceiling, with a temperature gradient between. High discharge air velocity through floor tile or grates is necessary to displace warm air near the highest intakes. Floor grates can be useful, because of their high mass flow discharge rates.

Pressure Variations. Distribution of cool airflow through perforated tiles is governed by the fluid mechanics of the space below the raised floor and not the large, visible, above-floor space. More specifically, the static pressure and air movement in the proportionately small underfloor space determines how much air flows through each perforated tile. Measurements from hundreds of datacom facilities confirm that flow rates from perforated tiles typically vary considerably, depending on their proximity to the CRAC unit.

Further, the pattern of airflow distribution is somewhat counterintuitive. More flow might be expected through tiles near the CRAC

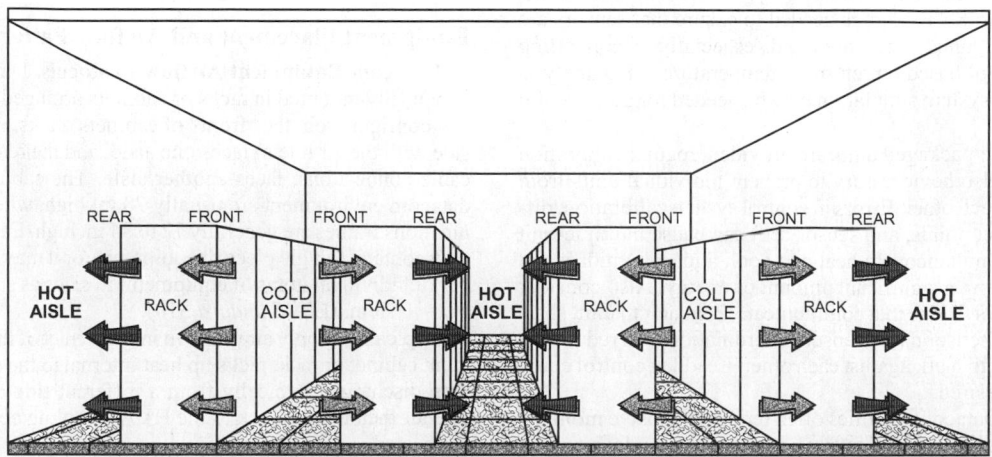

Fig. 7   Schematic of Hot-Aisle/Cold-Aisle Configuration

unit, and less away from it. In reality, there is typically very little flow near the CRAC, and greater flow through the perforated tiles located far away. Consequently, IT equipment placed near the CRAC often does not get much cool air.

The flow rate through a perforated tile depends on the pressure difference across the tile (i.e., the difference between the plenum static pressure just below the tile and the room static pressure above the raised floor). Pressure variations in data and communications equipment rooms are generally small compared to the pressure drop across the perforated tiles. The tiles are fairly restrictive (e.g., 25% or less open area). When substantial numbers of tiles with greater open area are used, airflow through the tiles may also depend on airflow dynamics above the raised floor; CFD analysis or physical measurements may be required to ensure that a design meets equipment airflow requirements.

Under some conditions, nonuniformity of airflow distribution is so severe that perforated tile airflow is directed from the room down into the floor plenum. This effect is caused when most of the CRAC unit fan's total pressure is transformed into velocity pressure by high velocity in a relatively shallow underfloor plenum. This high-velocity underfloor air can create a localized negative pressure and induce small quantities of room air into the underfloor plenum. As

distance from the supply fan increases, velocity decreases and the velocity pressure is converted to static pressure, which is required to produce airflow through perforated tiles or grates.

Other Factors Affecting Airflow Distribution. Other factors that influence distribution of airflow through perforated tiles include the following:

- Height of raised-floor plenum
- Percentage of open area of perforated tiles
- Location and size of leakage airflow paths
- Locations and redundancy of CRAC units
- Corresponding spreading of underfloor flow to various perforated tile locations
- Collision or merging of airstreams from different CRACs
- Flow disturbance caused by underfloor blockages such as pipes and cable trays

There is a common misconception that using more open tiles increases the airflow rate. Obviously, for the same static pressure in the plenum, more open tiles produce more airflow than more restrictive tiles. However, static pressure in the plenum cannot be assumed to be constant; it is a result of the tiles' flow resistance and other factors. The airflow rate is controlled by the amount of flow the CRAC

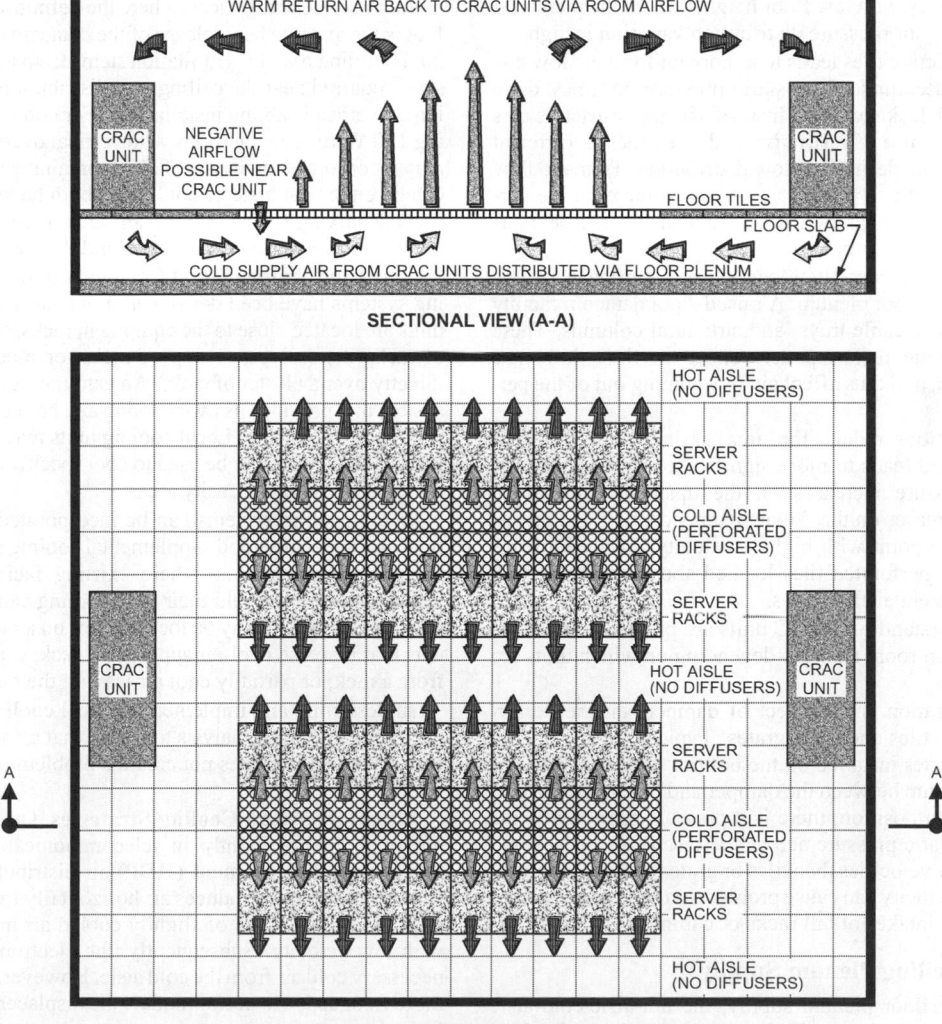

Fig. 8   Schematic of Datacom Equipment Room with Underfloor Plenum Supply Air Distribution

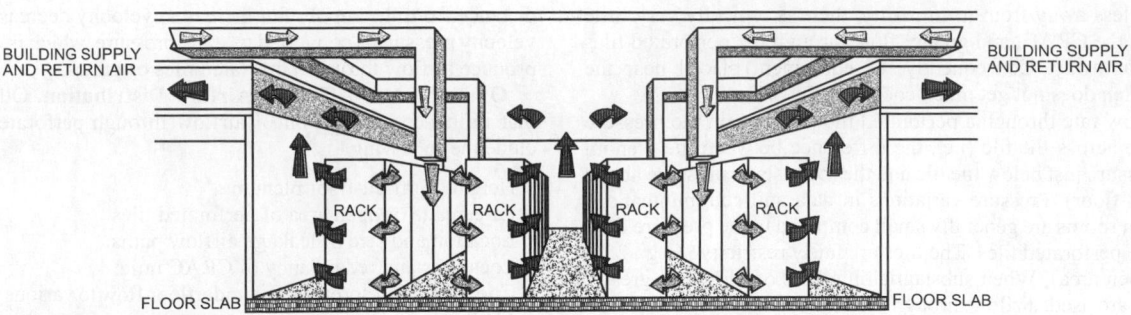

Fig. 9 Typical Ducted Ceiling Distribution Used in Datacom Facilities

unit blower is able to supply. For the blower, the controlling resistance is primarily internal to the CRAC unit, and the additional flow resistance offered by the perforated tiles is insignificant. Lowering perforated tile resistance typically does not significantly increase overall flow.

When very restrictive tiles (e.g., 5% open) are used, their flow resistance can influence the flow delivered by the CRAC unit blower. In this case, plenum pressure becomes high enough that the blower operates at a new position on the fan curve. The need for very restrictive tiles should thus be avoided, if possible, although they could be necessary with low floor heights where the plenum is restricted and underfloor pressure distribution variation is high.

Using more restrictive tiles leads to a more uniform airflow distribution, but increases underfloor static pressure and may drive more airflow through leakage paths. Instead of using restrictive tiles everywhere, selective use of restrictive and open tiles in different locations can obtain the desired airflow distribution. To make flow distribution uniform, it is typically necessary to increase the percentage open area of tiles near the CRAC unit and to decrease it further away from the CRAC unit.

Finally, underfloor obstructions can cause significant airflow variations in the underfloor plenum. A raised-floor datacom facility usually contains pipes, cable trays, and structural columns. These obstructions disturb the airflow pattern under the floor, influence pressure distribution, and thus affect airflow coming out of the perforated tiles.

Because obstructions reduce the area available for flow, air velocity increases and leads to more significant pressure changes. Usually, static pressure increases on the upstream side of an obstruction and decreases on the downstream side, with the lowest static pressure at the point with highest velocity. Because of this effect, two adjacent perforated tiles located above an obstruction may yield very different airflow rates.

When redundant (standby) CRAC units are provided, air distribution in the datacom room changes depending on which units are operating.

Another consideration is the effect of dampers on the performance of perforated tiles and floor grates. Typically, dampers are two slotted metal plates installed on the bottom surface of a tile or grate, creating a plenum between the damper and discharge surface. Even in the full open position, there is a pressure drop across the damper, reducing static pressure across the discharge surface and resulting in lower air velocities than tiles or grates without dampers. Low discharge air velocity can cause problems in supplying cooling air to the highest air intakes of tall racks or cabinets.

Overhead and Ceiling Plenum Supply

As with an underfloor plenum supply, the hot-aisle/cold-aisle configuration should also be used with an overhead supply. Currently, overhead cooling methods are more typically found in telecommunications facilities.

Ducted Supply. Overhead ducted supply, shown in Figure 9, can be used in datacom facilities without a raised-floor plenum. The **vertical overhead (VOH)** system is currently the typical and preferred configuration of the large regional phone companies, although this is changing to accommodate new technologies. This system can satisfy equipment and personnel comfort requirements, and can be fairly easily balanced to supply air to meet the distributed heat gain in the equipment room.

The vertical overhead supply system is typically limited to a cooling capacity of 130 W/ft² (*Telcordia* 2001) in mature telecommunications central offices, where the definition of watts per square foot is the average heat release of the datacom equipment in a 20 by 20 ft building bay. This limitation stems from the large physical duct sizes required near the ceiling. Air distribution is often affected by large overhead cabling installations. For cooling capacities exceeding 130 W/ft², new facilities with vertical overhead supply systems may incorporate some type of aisle containment. Aisle containment can be either hot-aisle or cold-aisle; both have physical barriers to prevent mixing of the cooling air stream from the return air path between the electronic equipment and the air handlers.

Local and Supplemental Cooling Distribution. Recently, cooling systems have been developed in which small, modular cooling units are located close to the equipment racks, either within the rows of racks themselves, on top of racks, or mounted on the ceiling directly over a cluster of racks. An extreme example is a sealed rack system that provides its own cooling and has no heating effect on air inside the data center. Local cooling units may provide all the cooling for a data center or be used to cool specific areas in the data center, typically higher-density areas.

Local cooling systems can be incorporated into new construction, and are also a good supplemental cooling approach to consider for retrofit applications where existing facilities lack sufficient infrastructure to expand their base cooling capacity. Supplemental cooling equipment may be located in or on a rack and include products that increase cool-air supply to a rack, capture hot-air exhaust from a rack, or partially cool air leaving the rack.

In designing and implementing local cooling, it is important to perform a thorough analysis to ensure that a cooling solution to one part of the facility does not create a problem in another part of the facility.

Other Overhead Cooling Strategies. Other cooling strategies are used, predominantly in telecommunications central offices. **Horizontal displacement (HDP)** air-distribution, mainly used in Europe and Asia, introduces air horizontally from one end of a cold aisle. A large volume of slightly cooled air moves along the aisle with low velocity. Subsequently, the electronic equipment draws necessary cold air from the cold aisle. However, this system requires more floor space to accommodate the displacement of the large diffusers.

Some long-distance carriers in North America use **horizontal overhead (HOH)** air distribution. This system introduces supply air horizontally above the cold aisles, and is generally used where raised floors are used for cabling.

Finally, **natural convection overhead (NOH)** air distribution, not commonly used, suspends cooling coils from the ceiling. Because the coils cool the hot air as it rises (because of buoyancy), there are no fans or ducting in this strategy.

For more information on different air-cooling strategies for telecommunications central offices, including cooling capacities, see *Telcordia* (2001).

Return Air

Return air paths discussed here can be used either with underfloor or overhead supply air systems. Most return air in datacom facilities is not ducted (i.e., heat discharged from datacom equipment enters the datacom equipment room at large and finds a pathway back to a large common return grille or to the inlet of a CRAC unit). This can be effective, but the opportunity for inefficiency, in some circumstances, is great because of potential short-circuiting and mixing of supply and return air. In addition, it is possible to draw hot equipment exhaust from one piece of equipment into the inlet of adjacent equipment. An example of how air can potentially short-circuit with an underfloor supply and unducted return configuration is highlighted in Figure 8.

Using ceiling plenums is an option for return air. Inlets should be located above hot aisles or datacom equipment with high heat dissipation to take advantage of the thermal plume created above the equipment. Ceiling plenum returns can capture part of the heat from data and communications equipment and lights directly in the return airstream. Assuming that the space is allowed to stratify, the return air being at a higher temperature than the average space temperature creates a higher temperature difference across the cooling coil, thereby allowing a reduction in airflow to the space.

Computational Fluid Dynamics Simulation

Air is the main carrier of heat and moisture in data centers. It is challenging but important to optimize the flow paths of both cold supply air and hot return to minimize mixing of these two streams as well as reduce any short-circuiting of cold air back to the air-conditioning systems.

Several factors affect airflow distribution and cooling performance of a data center. Physical measurements and field testing are not only time and labor intensive but sometimes impossible. In such a situation, computational fluid dynamics (CFD) simulations provide a feasible alternative for testing various design layouts and configurations in a relatively short time.

CFD simulations can, for example, predict air velocities, pressure, and temperature distribution in the entire data center facility; assess airflow patterns around racks and identify areas of recirculation; and provide a detailed cooling audit report for the facility, including performance evaluation of individual components such as air-conditioning units, racks, perforated tiles, and any other supplementary cooling systems in the facility.

Facilities managers, designers, and consultants can use these techniques to estimate the performance of a proposed layout (with or without performance metrics) before actually building the facility. Likewise, CFD simulations can provide appropriate insight and guidance in reconfiguring existing facilities optimize the cooling and air distribution system. For details on performing CFD, see Chapter 13 of the 2009 *ASHRAE Handbook—Fundamentals*.

ANCILLARY SPACES

Space must be allocated within a datacom facility for storing components and material, support equipment, and operating and servicing the datacom equipment. Some ancillary spaces may require environmental conditions comparable to those of the datacom equipment, whereas others may have less stringent requirements.

Component and material storage areas often require environmental conditions comparable to those of the datacom equipment. Support equipment often has substantially less stringent environmental requirements, but its continuous operation is often vital to the facility's proper functioning.

Electrical Power Distribution and Conditioning Rooms

Electrical Power Distribution Equipment. Electrical power distribution equipment can typically tolerate more variation and a wider range of temperature and humidity than datacom equipment. Equipment in this category includes incoming service/distribution switchgear, switchboard, automatic transfer switches, panel boards, and transformers. Manufacturers' data should be checked to determine the amount of heat release and design conditions for satisfactory operation. Building codes should be checked to identify when equipment must be enclosed to prevent unauthorized access or housed in a separate room.

Uninterruptible power supplies (UPSs) come in various configurations, but most often use batteries as the energy storage medium. They are usually configured to provide redundancy for the central power buses, and typically operate continuously at less than full-load capacity. They must be air-conditioned with sufficient redundancy and diversity to provide an operable system throughout an emergency or accident. The relationship between load and heat release is usually nonlinear. Verification with the equipment vendor is necessary to properly size the HVAC system.

UPS power monitoring and conditioning (rectifier and inverter) equipment is usually the primary source of heat release. This equipment usually has self-contained cooling fans that draw intake air from floor level or the equipment face and discharge heated air at the top of the equipment. Air-distribution system design should take into account the position of the UPS air intakes and discharges.

Battery Rooms

Installation of secondary battery plants as a temporary back-up power source should be in accordance with NFPA *Standard* 70; IEEE *Standard* 1187 and other applicable standards should also be referenced, in addition to a design review with the local code official. Other relevant sources of guidance are NFPA *Standards* 70E and 76.

Most codes require 1 cfm/ft^2 of forced exhaust from a battery room when hydrogen detectors are not used. The exhaust fan(s) (typically ceiling-mounted) must run continuously. When hydrogen detectors are used, the exhaust is also sized for 1 cfm/ft^2, but the fan(s) only need to run when hydrogen is detected.

Because of the potential high hazard associated with hydrogen gas build-up, battery room exhaust systems should be designed with redundancy and failure alarms. Best practice design includes both continuous fan operation and both "high" and "high-high" alarms in case one sensor goes out of calibration. If the system operates when hydrogen gas is detected, or if combustible concentrations of gas are expected, explosionproof motors and/or other provisions may be required to address an explosion hazard.

Nonsparking fan wheels may be required by code, and in any case are highly recommended. For belt-drive systems, the fans should be equipped with controls that periodically exercise them to keep the belts pliable and therefore more reliable.

Temperature in a battery area is crucial to the life expectancy and operation of the batteries. The optimum space temperature for lead-calcium batteries is 77°F. If higher temperatures are maintained, it may reduce battery life; if lower temperatures are maintained, it may reduce the batteries' ability to hold a charge (IEEE *Standard* 484).

Battery rooms should be maintained at a negative pressure to adjacent rooms and exhausted to the outside to prevent migration of fumes, gases, or electrolytic spray to other areas of the facility. It may be possible to provide makeup air from an adjacent datacom area, thereby eliminating the need for a separate HVAC system for the battery room, if temperatures are compatible. Battery rooms typically only have small heat-producing loads.

Battery rooms may require emergency eyewashes and showers. If so, these systems should include leak detection and remote alarm capabilities to alert staff of a possible leak or that an accident has occurred and emergency first aid is required.

Engine/Generator Rooms

Engine-driven generators used for primary or emergency power require large amounts of ventilation when running. This equipment is easier to start if a low ambient temperature is avoided. Low-temperature start problems are often reduced in cold climates by using engine block heaters. Design should ensure that exhaust air does not recirculate back to any building ventilation air intakes.

Spring-return motorized dampers are typically provided on air inlets and discharges and maintained normally closed when power is available to the damper actuator. Damper actuator signals are generally from the generator electrical gear as opposed to the building management system (BMS).

Where acoustical concerns exist, measures may need to be taken both inside the engine/generator room to meet the appropriate OSHA regulations and guidelines [e.g., OSHA (1996)] and on the air intake/discharge openings if the site is near an acoustically sensitive property line.

Burn-In Rooms and Test Labs

Many datacom facilities incorporate a dedicated area for the purpose of assembling, configuring and testing datacom equipment before deployment in the production environment. These areas can be used for testing equipment power supplies, dual-power capabilities, actual power draw, and cooling requirements, as well as for equipment applications testing (both software and hardware functions).

It is recommended that these areas be adjacent to production areas for convenience, yet separated with respect to power, cooling, and fire protection to prevent a power problem or fire from affecting the production environment.

Datacom Equipment Spare Parts

A spare parts room may require immediate use of parts for equipment repair. Therefore, the temperature of the space should be similar to that of the operating data center. ASHRAE's *Thermal Guidelines for Data Processing Environments* (ASHRAE 2008) provides allowable temperatures for "product power-off" conditions that include a spare parts room environment.

Storage Spaces

Storage spaces for products such as paper and tapes generally require conditions similar to those in data and communications equipment rooms, because these products absorb moisture from the air and can expand, contract, or change shape more than electronic equipment. Close-tolerance mechanical devices, such as paper feeders or tape drives, are also affected by room relative humidity.

OTHER SYSTEMS AND CONSIDERATIONS

Fire Suppression

Automatic fire extinguishing and smoke control systems afford the highest degree of protection and must be provided in accordance with the applicable national codes, local codes, and the owner's insurance underwriter. Some new code references require fire protection below the raised floor.

Exhaust Systems. Exhaust systems may be provided to ventilate datacom equipment rooms in the event of a chemical fire suppression system discharge or as required for smoke purge. Locating the

exhaust pickup point below a supply plenum floor promotes quick purging of the space.

There is no need for a purge system when datacom equipment rooms are only protected with sprinklers unless required by code. Even so, the ability to purge datacom space (including critical infrastructure rooms) can minimize the effects of combustion contaminants on datacom equipment, regardless of the type of suppression system.

For small computer rooms that use DX compressor-based cooling systems, local codes may require alarm and exhaust in the event of loss of refrigerant. The volume of refrigerant within a system should be checked against local code requirements.

Fire Smoke Dampers. When motorized fire/smoke dampers are installed to seal a clean-agent-protected room, they must be the spring-loaded type, configured to close on loss of power (or upon melting of fusible link). During start-up, it is important to verify proper operation, including making sure that dampers close fully without binding. A binding damper jeopardizes the integrity of the room seal and could prevent the clean agent from reaching the proper concentration level to extinguish a fire. To increase reliability, mechanical systems should be designed to minimize the number of motorized fire/smoke dampers.

Outdoor Air Smoke Detectors. Outside air, and any other air supply, should be equipped with smoke detection that shuts down associated fans and closes associated fire/smoke damper(s). Roof fires, nearby fires, nearby chemical spills, and even generator start-up can introduce smoke and reactive chemicals into the data center through fresh air intakes.

Additional information on fire detection and suppression systems for datacom environments can be found in *Design Considerations for Datacom Equipment Centers* (ASHRAE 2005b).

Commissioning

Commissioning of datacom facilities is critical to their proper functioning and reliable operation. The commissioning process is usually misinterpreted as focusing on systems testing only, and initiated during the construction process.

However, ASHRAE *Guideline* 0 defines commissioning as a "process focuse[d] upon verifying and documenting that the facility and all of its systems and assemblies are planned, designed, installed, tested, operated, and maintained to meet the owner's project requirements."

It is recommended that commissioning of datacom facilities begin at project inception, so that owner requirements can be better defined, addressed, and verified throughout the entire design and construction process. Five levels of commissioning are described in *Design Considerations for Datacom Equipment Centers* (ASHRAE 2005b):

- Level 1: factory acceptance tests
- Level 2: field component verification
- Level 3: system construction verification
- Level 4: site acceptance testing
- Level 5: integrated systems testing

Mission-critical facilities typically have more demanding performance requirements for responses to expected and unexpected anomalies without affecting critical operations. Systems usually include redundant components and utility feeds, and excess capacity or back up systems or equipment that, during an emergency, can be automatically or manually activated. These redundant or back-up components, systems, or groups of interrelated systems are tested during level 5 commissioning: this level is what generally sets mission-critical facility commissioning apart from typical office building commissioning.

Further information on commissioning can be found in Chapter 43 and ASHRAE *Guidelines* 0 and 1.1.

Serviceability

There are many serviceability issues to consider for HVAC equipment serving datacom facilities. Above all, the design should seek to coordinate with datacom facility operations to service and maintain equipment with the least amount of disruption to the day-to-day running of the facility. One approach is to locate all cooling equipment (e.g., CRAC or central station air-handling units) outside of the datacom equipment room in dedicated support rooms. Service and maintenance operations for this equipment is then performed in areas devoted specifically to air-conditioning equipment. System security for these spaces, however, must be addressed.

HVAC equipment serving datacom facilities can also be located on the roof, when the physical arrangement of the facility and space limitations allow.

Availability and Redundancy

It is extremely important to understand the need for uptime of the datacom facilities. Mission-critical datacom facilities, as their name implies, are often required to run 24 h, 7 days a week, all year round, and any disruption to that operation typically results in a loss of business continuity or revenue for the end user.

Availability is a percentage value representing the probability that a component or system will remain operational. Availability typically considers both component reliability and system maintenance (Beaty 2004). Values of 99.999% ("five 9s") and higher are commonly referenced in datacom facility design, but are difficult to attain. For individual components, availability is often determined through testing. For assemblies and systems, availability is often the result of a mathematical evaluation based on the availability of individual components and any redundancy or diversity that may be used.

Availability calculations for HVAC systems are seldom done and are extremely difficult, because published data on components and systems are not readily available. Further research is needed to allow for calculation of system availability as a function of component availability and level of redundancy.

System availability may be so vital that the potential cost of system failure justifies redundant systems, capacity, and/or components, as well as increased diversity. System simplicity and ease of operation should be a constant consideration; a substantial percentage of reported data center failures are related to human activity or error.

The most common method used to attain greater system availability is adding redundant parallel components to avoid a single component failure that causes a systemwide failure. HVAC system redundancy calculations commonly use the terms $N + 1$, $N + 2$, and $2N$ to indicate how many additional components are to be provided. N represents the number of pieces of equipment that it takes to satisfy the normal load. Redundant equipment is necessary to compensate for failures and allow maintenance to be performed without reducing the remaining online capacity below normal.

In the case of datacom facilities using CRAC units, if $N + 1$ redundancy is required, the number of units required to satisfy the normal N cooling load must first be determined. One additional unit would then be provided, to achieve $N + 1$ redundancy.

In theory, redundancy can be achieved with $N + 1$ (or more air-handling or CRAC units), but the dynamics of underfloor and overhead flow are such that loss of a specific unit can be critical for a specific area. CFD analysis is often performed to determine the effect of losing specific units in critical-use areas. For large spaces, consider using $N + 1$ for every X number of units, to provide one redundant unit for every set of X units required.

Take care to exercise redundant equipment frequently, to prevent conditions that enhance growth of mold and mildew in filters, insulated unit enclosures, and outside air pathways where spores and food sources for microbial growth may accumulate. Another approach is to keep redundant equipment operational at all times,

but to use variable-frequency drives (VFDs) to control fan speed. In this manner, the fans operate at a speed to match loads. If a fan fails, the other fans ramp up to maintain required airflows. No schedule is required for exercising equipment, because all equipment is operational (unless loads are so low that it is not practical to operate all equipment).

Diversity. Systems that use an alternative path for distribution are said to have diversity. In an HVAC system, diversity might be used to refer to an alternative chilled-water piping system. To be of maximum benefit, both the normal and alternative paths must each be able to support the entire normal load. One company developed a tiered classification system to rank the level of diversity and redundancy in a data center design (Turner and Brill 2003).

With dual feeds, it is often possible to perform planned datacom air-moving device infrastructure activity without shutting down critical loads, a concept called **concurrently maintainable**. **Fault-tolerant** systems do not lose power or cooling to the datacom equipment when a single component fails.

Measures to Increase Reliability. Practical ways to increase HVAC system reliability for datacom facility design may include any of the examples listed below. A fault tree analysis or other methodology should be used for each facility to examine critical failure paths and necessary design measures to increase system availability.

- Back-up utilities: power generation, second electric service, water supply, etc. Emergency power supplies probably should feed some aspects of HVAC systems as well as datacom equipment to allow for continuous equipment operation within allowable environmental conditions.
- Back-up air moving equipment: air handlers, fans, computer room units, etc.
- Back-up and/or cross-connected cooling equipment: chillers, pumps, cooling towers, dry coolers, cooling coils, makeup water supply, etc.
- Diverse piping systems: chilled water, condenser water, etc.
- Full or partial back-up of air-moving and/or cooling equipment on emergency power.
- Back-up thermal storage: chilled water, ice, makeup water, etc.

Energy Conservation

Dramatic reductions in energy use can be achieved with conservation strategies. Central-station air-conditioning systems using outside air for free cooling (where appropriate), variable-volume ventilation, and evaporative cooling/humidification strategies offer significant opportunities for reducing energy use, depending on the frequency of favorable outdoor conditions. A dew-point control strategy, often consisting of positive pressurization of the datacom facility and humidity control at the point of outdoor air intake, can eliminate the need for humidity-sensing devices and provide precise humidity control.

A significant amount of wasted energy has been identified in many existing facilities, often because of fighting between adjacent air-conditioning units attempting to maintain tight tolerances. Adopting the somewhat less stringent environmental tolerances found in ASHRAE's (2008) *Thermal Guidelines for Data Processing Environments* should minimize this historically significant problem. Control strategies such as underfloor air temperature control and additional monitoring points should be considered to identify and avoid fighting, especially where raised-floor spaces may include a mixture of heat densities in the same open area.

Baseline energy consumption of office and telecommunications equipment was estimated by Roth (2002). Case studies of energy consumption in datacom facilities have been inventoried and are available for public review (LBNL 2003). Tschudi et al. (2003) summarized existing research on energy conservation in datacom facilities and provided a roadmap for further research in this area. Areas covered include monitoring and control, electrical systems, and HVAC systems (including free cooling), and use of variable-speed compressors in CRAC units. A set of recommendations for high-performance data centers has also recently been issued (RMI 2003).

Power usage effectiveness (PUE) is a metric for characterizing and reporting overall data center infrastructure efficiency, and is defined by the following formula:

$$PUE = \left(\frac{\text{Total data center energy consumption or power}}{\text{IT energy consumption or power}} \right)$$

When calculating PUE, IT energy consumption should be measured directly at the IT load. At a minimum, it can be measured at the output of the UPS. However, the industry should progressively improve measurement capabilities over time so that measurements of only the server loads become the common practice. Once a methodology for measuring PUE is defined for a particular facility, it is important that it be calculated in the same manner over the facility's life. As such, PUE can be an effective tool in monitoring performance of the facility's infrastructure, but because IT energy for different facilities can each be measured at different points, there is little value in using it to compare different facilities. For a dedicated data center, the total energy in the PUE equation includes all energy sources at the point of utility handoff to the data center owner or operator. For a data center in a mixed-use building, the total energy is all energy required to operate the data center, similar to a dedicated data center, and should include cooling, lighting, and support infrastructure for the data center operations.

Economizer Cycles. There are several available options for economizers in datacom facilities, each with benefits and challenges specific to data centers. The following types of economizer categories are useful for discussion and evaluation:

- **Air side**
  - Direct exchange (bringing outdoor air directly into the facility)
  - Indirect exchange (heat exchangers do not introduce outdoor air into the facility)
- **Water side**
  - Direct (condenser water can mix with chilled water; cooling tower, dry, or wet coolers can be used)
  - Indirect (heat exchanger separates condenser water and chilled-water loops; cooling towers, dry, or wet coolers can be used)
  - Dry coolers (glycol cools directly with an economizer coil in the CRAC)

Application of economizers in datacom facilities requires more careful review and consideration than a typical commercial application because of potential damage to data center equipment. In many instances, economizers offer one of the greatest energy savings opportunities in datacom facilities. Careful review of the total cost of ownership (TCO) is encouraged, because it is possible that in select cases the TCO will not be attractive. Items that may affect the financial performance of an economizer include

- Ratchet clauses in the utility rate tariff
- Gas-phase filtration requirements
- Large humidification loads
- Low cost of electricity
- Large space/capital needs for louvers, ductwork, and equipment associated with the economizers

There are three major considerations when reviewing the potential number of hours of economizer operation in data centers: geographic location, return air temperatures, and indoor temperature and humidity requirements. Note that ASHRAE (2008) also allows increased economizer use in some climates, because of the higher recommended inlet air temperature range (65 to 80°F for Class 1 and 2 environments) relative to a typical office supply air

temperature of 60°F or below. In terms of potential energy savings, air-side economizers with direct exchange of outdoor air often offer greater potential energy savings than economizers with heat exchangers. Economizer processes using heat exchange have increased inefficiencies because of the required temperature difference across the heat exchanger (heat wheel, cooling towers, or dry coolers). Although DX air-side economizers may have extra energy savings benefits, they are also accompanied by some potential challenges such as gaseous and particulate contamination, as well as humidification issues. The advantage of the other three types of economizers is that they eliminate concerns over these issues, because outdoor air is not directly introduced into the building. However, beyond airborne contamination and humidity, all types of economizers introduce additional complexities to the HVAC system operation and include potential failure scenarios that must be understood before applying them to datacom facilities to ensure that the mission is not compromised. When it is determined that economizers are an appropriate fit and/or a requirement for a specific datacom application, it is still necessary to determine which type is the most appropriate. For datacom applications, this often requires a detailed evaluation with many more variables than found in a typical comfort-cooling application. When choosing between air-side economizers with indirect or direct evaporative cooling, the evaluation should include outdoor air quality as well as first cost and energy savings. See Chapter 52 for information on evaporative cooling processes.

REFERENCES

Codes, Standards, and Guidelines

ASHRAE. 2005. The commissioning process. *Guideline* 0-2005.

ASHRAE. 2007. HVAC technical requirements for the commissioning process. *Guideline* 1.1-2007.

ASHRAE. 1992. Gravimetric and dust-spot procedures for testing air-cleaning devices used in general ventilation for removing particulate matter. *Standard* 52.1-1992. (Withdrawn.)

ASHRAE. 2007. Method of testing general ventilation air-cleaning devices for removal efficiency by particle size. ANSI/ASHRAE *Standard* 52.2-2007.

ASHRAE. 2010. Ventilation for acceptable indoor air quality. ANSI/ASHRAE *Standard* 62.1-2010.

ASHRAE. 2007. Method of testing for rating computer and data processing room unitary air conditioners. ANSI/ASHRAE *Standard* 127-2007.

ASTM. 2009. Guide for heated system surface conditions that produce contact burn injuries. *Standard* C1055-03 (2009). American Society for Testing and Materials, West Conshohocken, PA.

EIA. 2005. Cabinets, racks, panels, and associated equipment. *Standard* EIA/ECA-310. Electronics Industries Alliance.

ICC. 2009. *International building code®*. International Code Council, Washington, D.C.

IEEE. 2008. Recommended practice for installation design and implementation of vented lead-acid batteries for stationary applications. IEEE *Standard* 484-2002 (R2008). Institute of Electrical and Electronics Engineers, Piscataway, NJ.

IEEE. 2002. Recommended practice for installation design and installation of valve-regulated lead-acid storage batteries for stationary applications. *Standard* 1187-2002. Institute of Electrical and Electronics Engineers, Piscataway, NJ.

NFPA. 2011. National electric code®. *Standard* 70. National Fire Protection Association, Quincy, MA.

NFPA. 2009. Standard for electrical safety in the workplace. *Standard* 70E. National Fire Protection Association, Quincy, MA.

NFPA. 2009. Recommended practice for the fire protection of telecommunications facilities. *Standard* 76. National Fire Protection Association, Quincy, MA.

OSHA. 1996. 29 CFR 1910.95: Occupational noise exposure. U.S. Department of Labor, Occupational Safety and Health Administration, Office of Information, Washington, D.C. http://www.osha.gov/pls/oshaweb/owawdisp.show_document?p_table=STANDARDS&p_id=9735.

Telcordia. 2001. Thermal management in telecommunications central offices. *Telcordia Technologies Generic Requirements* GR-3028-CORE.

Telcordia. 2006. Network equipment—Building Systems (NEBS) requirements: Physical protection. *Telcordia Technologies Generic Requirements,* Issue 3, GR-63-CORE.

Other Publications

ASHRAE. 2005a. *Datacom equipment power trends and cooling applications.*

ASHRAE. 2005b. *Design considerations for datacom equipment centers.*

ASHRAE. 2006. *Design considerations for liquid cooling in datacom and telecommunications rooms.*

ASHRAE. 2008. *Thermal guidelines for data processing environments.*

ASHRAE. 2009. *Particulate and gaseous contamination in datacom environments.*

Beaty, D.L. 2004. Reliability engineering of datacom cooling systems. Symposium, ASHRAE Winter Meeting.

European Council. 2003. *Directive 2003/10/EC of the European Parliament and of the Council of 6 February 2003 on the minimum health and safety requirements regarding the exposure of workers to the risks arising from physical agents (noise).* Available from http://osha.europa.eu/en/legislation/directives/exposure-to-physical-hazards/osh-directives/82.

Herrlin, M.K. 1996. Economic benefits of energy savings associated with: (1) Energy-efficient telecommunications equipment; and (2) Appropriate environmental control. Eighteenth International Telecommunications Energy Conference INTELEC '96, Boston.

LBNL. 2003. http://datacenters.lbl.gov/CaseStudies.html. Lawrence Berkeley National Laboratories.

NIOSH. 1986. Criteria for a recommended standard: Occupational exposure to hot environments (Revised Criteria 1986). *Report* 86-113. National Institute for Occupational Safety and Health, Washington, D.C. Available at http://www.cdc.gov/niosh/86-113.html.

RMI. 2003. *Energy efficient data centers: A Rocky Mountain Institute design charette.* Rocky Mountain Institute, Snowmass, CO.

Roth, K., F. Goldstein, and J. Kleinman. 2002. *Energy consumption by office and telecommunications equipment in commercial buildings.* vol. I: *Energy consumption baseline.* Arthur D. Little.

Tschudi, B., T. Xu, D. Sartor, and J. Stein. 2003. Roadmap for public interest research for high-performance data centers. LBNL *Report* 53483.

Turner, W.P. and K. Brill. 2003. *Industry standard tier classifications define site infrastructure performance.* The Uptime Institute, Santa Fe, NM.

BIBLIOGRAPHY

ASHRAE. 2003. *Risk management guidance for health, safety and environmental security under extraordinary incidents.*

Awbi, H.B. and G. Gan. 1994. Prediction of airflow and thermal comfort in offices. *ASHRAE Journal* 36(2):17-21.

Bash, C.E., C.O. Patel, and R.K. Sharma. 2003. Efficient thermal management of data centers—Immediate and long-term research needs. *International Journal of HVAC&R Research* (now *HVAC&R Research*) 9(2):137-152.

Beaty, D.L. 2003. Liquid cooling—Friend or foe. Symposium, ASHRAE Winter Meeting.

Brill, K., E. Orchowski, and L. Strong. 2002. *Product certification for fault-tolerance is essential for verification of high availability.* The Uptime Institute, Sante Fe, NM.

ETSI. 2009. Environmental conditions and environmental tests for telecommunications equipment; Part 1-3: Classification of environmental conditions; stationary use at weatherprotected locations. *Standard* EN 300 019-1-3 V2.3.2. European Telecommunications Standards Institute, Sophia Antipolis, France.

Herrlin, M.K. 1997. The pressurized telecommunications central office: IAQ and energy consumption. Healthy Buildings/IAQ '97, Washington D.C.

ISO. 1999. Cleanrooms and associated controlled environments—Part 1: Classification of air cleanliness. ISO/FDIS *Standard* 14644-1. International Standards Organization, Geneva.

Kang, S., R.R. Schmidt, K.M. Kelkar, A. Radmehr, and S.V. Patankar. 2001. A methodology for the design of perforated tiles in raised floor data centers using computational flow analysis, *IEEE Transactions on Components and Packaging Technologies* 24:177-183.

Karki, K.C., A. Radmehr, and S.V. Patankar. 2003. Use of computational fluid dynamics for calculating flow rates through perforated tiles in raised-floor data centers. *International Journal of HVAC&R Research* (now *HVAC&R Research*) 9(2):153-166.

Krzyzanowski, M.E. and B.T. Reagor. 1991. Measurement of potential contaminants in data processing environments. *ASHRAE Transactions* 97(1):464-476.

Lentz, M.S. 1991. Adiabatic saturation and VAV: A prescription for economy and close environmental control. *ASHRAE Transactions* 97(1):477-485.

Longberg, J.C 1991. Using a central air-handling unit system for environmental control of electronic data processing centers. *ASHRAE Transactions* 97(1):486-493.

Nakao, M., H. Hayama, and M. Nishioka. 1991. Which cooling air supply system in better for a high heat density room: Underfloor or overhead? Thirteenth International Telecommunications Energy Conference (INTELEC '91), November 1991.

Noh, H.-K., K.S. Song, and S.K. Chun. 1998. The cooling characteristics on the air supply and return flow systems in the telecommunication cabinet room. Twentieth International Telecommunications Energy Conference (INTELEC '98).

Patankar, S.V. and K. Karki. 2004. Distribution of cooling airflow in a raised-floor data center. *ASHRAE Transactions* 110(2):624-635.

Patel, C.D., R. Sharma, C.E. Bash, and A. Beitelmal. 2002. Thermal considerations in cooling large scale high compute data centers. ITHERM 2002: Eighth Intersociety Conference on Thermal and Thermomechanical Phenomena in Electronic Systems.

Patel, C.D., C.E. Bash, C. Belady, L. Stahl, and D. Sullivan. 2001. Computational fluids dynamics modeling of high compute density data centers to assure system air inlet specifications. *Paper* IPACK2001-15622, InterPack '01 Kauai.

Schmidt, R. and E. Cruz. 2002. Raised floor computer data center: effect on rack inlet temperatures of chilled air exiting both the hot and cold aisles. *IEEE 2002 Inter Society Conference on Thermal Phenomena,* pp. 580-594.

Schmidt, R. 1997. Thermal management of office data processing centers. InterPack '97, Hawaii.

Schmidt, R. 2001. Effect of data center characteristics on data processing equipment inlet temperatures. *Paper* IPACK2001-15870. InterPack '01, Kauai.

Schmidt, R.R., K.C. Karki, K.M. Kelkar, A. Radmehr, and S.V. Patankar. 2001. Measurements and predictions of the flow distribution through perforated tiles in raised-floor data centers. *Paper* IPACK2001-15728. InterPack '01, Kauai.

Stahl, L. and C. Belady. 2001. Designing an alternative to conventional room cooling, International Telecommunications and Energy Conference (INTELEC), Edinburgh.

TIA. 2003. Telecommunications infrastructure standard for data centers, draft 2.0. *Standard* PN-3-0092. Telecommunication Industry Association, Arlington, VA.

U.S. GSA. 1992. Airborne particulate cleanliness classes in clean rooms and clean zones. U.S. *Federal Standard* FS209E. General Services Administration, Washington, D.C.

Weschler, C.J. and H.C. Shields. 1991. The impact of ventilation and indoor air quality on electronic equipment. *ASHRAE Transactions* 97(1):455-463.

Yamamoto, M. and T. Abe. 1994. The new energy-saving way achieved by changing computer culture (saving energy by changing the computer room environment). *IEEE Transactions on Power Systems* 9(August).

PRINTING PLANTS

THIS chapter outlines air-conditioning requirements for key printing operations. Air conditioning of printing plants can provide controlled, uniform air moisture content and temperature in working spaces. Paper, the principal material used in printing, is hygroscopic and very sensitive to variations in the humidity of the surrounding air. Printing problems caused by paper expansion and contraction can be avoided by controlling the moisture content throughout the manufacture and printing of the paper.

DESIGN CRITERIA

The following are three basic printing methods:

- **Relief printing (letterpress).** Ink is applied to a raised surface.
- **Lithography.** Inked surface is neither in relief nor recessed.
- **Gravure (intaglio printing).** Inked areas are recessed below the surface.

Figure 1 shows the general work flow through a printing plant. The operation begins at the publisher and ends with the finished printed product and paper waste. Paper waste, which may be as much as 20% of the total paper used, affects profitability. Proper air conditioning can help reduce the amount of paper wasted.

In sheetfed printing, individual sheets are fed through a press from a stack or load of sheets and collected after printing. In webfed rotary printing, a continuous web of paper is fed through the press

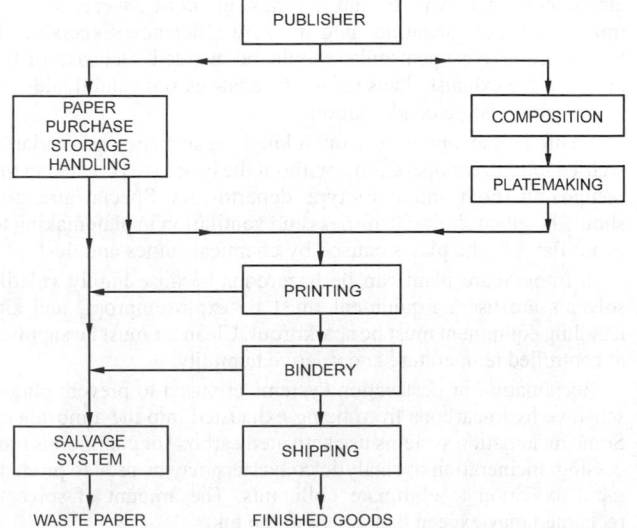

Fig. 1   Work Flow Through a Printing Plant

The preparation of this chapter is assigned to TC 9.2, Industrial Air Conditioning.

from a roll. The printed material is cut, folded, and delivered from the press as signatures, which form the sections of a book.

Sheetfed printing is a slow process in which the ink is essentially dry as the sheets are delivered from the press. **Offsetting**, the transference of an image from one sheet to another, is prevented by applying a powder or starch to separate each sheet as it is delivered from the press. Starches present a housekeeping problem: the particles (30 to 40 µm in size) tend to fly off, eventually settling on any horizontal surface.

If both temperature and relative humidity are maintained within normal human comfort limits, they have little to do with web breaks or the runnability of paper in a webfed press. At extremely low humidity, static electricity causes the paper to cling to the rollers, creating undue stress on the web, particularly with high-speed presses. Static electricity is also a hazard when flammable solvent inks are used.

Special Considerations

Special Warning: Certain industrial spaces may contain flammable, combustible, and/or toxic concentrations of vapors or dusts under either normal or abnormal conditions. In spaces such as these, there are life-safety issues that this chapter may not completely address. Special precautions must be taken in accordance with requirements of recognized authorities such as the National Fire Protection Association (NFPA), the Occupational Safety and Health Administration (OSHA), and the American National Standards Institute (ANSI). In all situations, engineers, designers, and installers who encounter conflicting codes and standards must defer to the code or standard that best addresses and safeguards life safety.

Various areas in printing plants require special attention to processing and heat loads. Engraving and platemaking departments must have very clean air: not as clean as that for industrial cleanrooms, but cleaner than that for offices. Engraving and photographic areas may also have special ventilation needs because of the chemicals used. Nitric acid fumes from powderless etching require careful duct material selection. Composing rooms, which contain computer equipment, can be treated the same as similar office areas. The excessive dust from cutting in the stitching and binding operations must be controlled. Stereotype departments have very high heat loads.

In pressrooms, air distribution must not cause the web to flutter or force contaminants or heat (which normally would be removed by roof vents) down to the occupied level. Air should be introduced immediately above the occupied zone wherever possible to minimize total flow and encourage stratification. High air exchange rates may be required where solvent- or oil-based inks are used, because of the large quantity of organic solvent vapors that may be released from nonpoint sources. Exhaust emissions from dryer systems may contain substantial concentrations of solvent vapors, which must be captured and recovered or incinerated to satisfy local air pollution

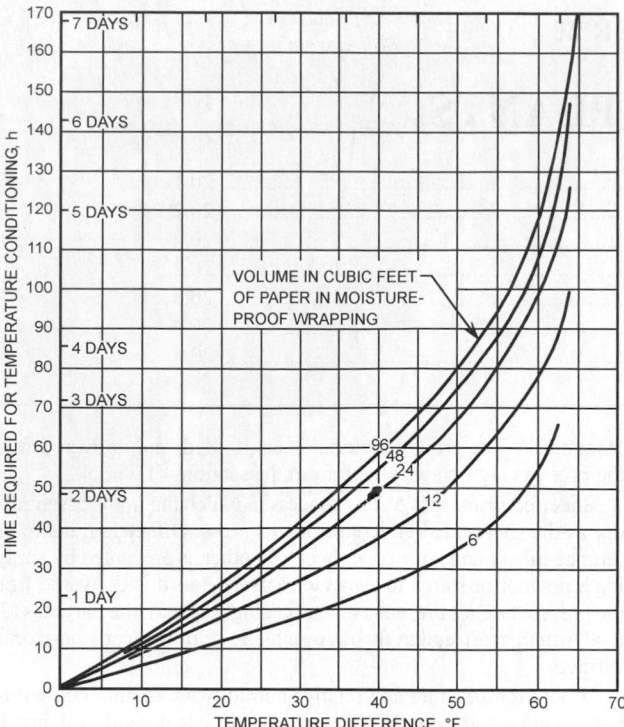

Fig. 2 Temperature-Conditioning Chart for Paper

requirements. Where these measures are required, efforts should be made to maximize point-source capture of vapors to minimize the size, cost, and energy requirements for vapor recovery/incineration equipment. These efforts also minimize the effect of these requirements on general ventilation systems.

Conventional air-conditioning and air-handling equipment, particularly rooftop equipment, may be unable to handle the high outdoor air requirements of pressroom applications effectively. Stratified ventilation may be used in high-bay installations to reduce total system airflow and air-conditioning requirements. Pressrooms using oil- or solvent-based inks should be provided with a minimum of 0.5 cfm/ft² of outdoor air to ensure adequate dilution of internally generated volatile organic compounds. Ventilation of storage areas should be about 0.5 air changes per hour (ach); bindery ventilation should be about 1 ach. Storage areas with materials piled high may need roof-mounted smoke- and heat-venting devices.

In a bindery, loads of loose signatures are stacked near equipment, which makes it difficult to supply air to occupants without scattering the signatures. One solution is to run the main ducts at the ceiling with many supply branches dropped to within 8 to 10 ft of the floor. Conventional adjustable blow diffusers, often the linear type, are used.

CONTROL OF PAPER MOISTURE CONTENT

Controlling the moisture content and temperature of paper is important in all printing, particularly multicolor lithography. Paper should be received at the printing plant in moisture-proof wrappers, which are not broken or removed until the paper is brought to the pressroom temperature. When exposed to room temperature, paper at temperatures substantially below the room temperature rapidly absorbs moisture from the air, causing distortion. Figure 2 shows the time required to temperature-condition wrapped paper. Printers usually order paper with a moisture content approximately in equilibrium with the relative humidity maintained in their pressrooms. Papermakers find it difficult to supply paper in equilibrium with a relative humidity higher than 50%.

Digital hygrometers can be used to check the hygroscopic condition of paper relative to the surrounding air. The probes contain a moisture-sensitive element that measures the electrical conductivity of the paper. Intact mill wrappings and the tightness of the roll normally protect a paper roll for about six months. If the wrapper is damaged, moisture usually penetrates no more than 0.125 in.

PLATEMAKING

Humidity and temperature control are important considerations when making lithographic and collotype plates, photoengravings, and gravure plates and cylinders. If the moisture content and temperature of the plates increase, the coatings increase in light sensitivity, which necessitates adjustments in the light intensity or the length of exposure to give uniformity.

If platemaking rooms are maintained at constant dry-bulb temperature and relative humidity, plates can be produced at known control conditions. As soon as it is dry, a bichromated colloid coating starts to age and harden at a rate that varies with the atmospheric conditions, so exposures made a few hours apart may be quite different. The rate of aging and hardening can be estimated more accurately when the space is air conditioned. Exposure can then be reduced progressively to maintain uniformity. An optimum relative humidity of 45% or less substantially increases the useful life of bichromated colloid coatings; the relative humidity control should be within 2%. A dry-bulb temperature of 75 to 80°F maintained within 2°F is good practice. The ventilation air requirements of the plate room should be investigated. A plant with a large production of deep-etch plates should consider locating this operation outside the conditioned area.

Exhausts for platemaking operations consist primarily of lateral or downdraft systems at each operation. Because of their bulkiness or weight, plates or cylinders are generally conveyed by overhead rail to the workstation, where they are lowered into the tank for plating, etching, or grinding. Exhaust ducts must be below or to one side of the working area, so lateral exhausts are generally used for open-surface tanks.

Exhaust quantities vary, depending on the nature of the solution and shape of the tank, but they should provide exhaust in accordance with the recommendations of *Industrial Ventilation* by the American Conference of Governmental Industrial Hygienists (ACGIH 2010) for a minimum control velocity of 50 fpm at the side of the tank opposite the exhaust intake. Tanks should be covered to minimize exhaust air quantities and increase efficiency. Excessive air turbulence above open tanks should be avoided. Because of the nature of the exhaust, ducts should be acid-resistant and liquidtight to prevent moisture condensation.

Webfed offset operations and related departments are similar to webfed letterpress operations, without the heat loads created in the composing room and stereotype departments. Special attention should be given to air cleanliness and ventilation in platemaking to avoid flaws in the plates caused by chemical fumes and dust.

A rotogravure plant can be hazardous because highly volatile solvents are used. Equipment must be explosionproof, and air-handling equipment must be sparkproof. Clean air must be supplied at controlled temperature and relative humidity.

Reclamation or destruction systems are used to prevent photosensitive hydrocarbons from being exhausted into the atmosphere. Some reclamation systems use activated carbon for continuous processing. Incineration or catalytic converters may be used to produce rapid oxidation to eliminate pollutants. The amount of solvents reclaimed may exceed that added to the ink.

RELIEF PRINTING

In relief printing (letterpress), rollers apply ink only to the raised surface of a printing plate. Pressure is then applied to transfer the ink from the raised surface directly to the paper. Only the raised surface touches the paper to transfer the desired image.

Air conditioning in newspaper pressrooms and other webfed letterpress printing areas minimizes problems caused by static electricity, ink mist, and expansion or contraction of the paper during printing. A wide range of operating conditions is satisfactory. The temperature should be selected for operator comfort.

At web speeds of 1000 to 2000 fpm, it is not necessary to control the relative humidity because inks are dried with heat. In some types of printing, moisture is applied to the web, and the web is passed over chill rolls to further set the ink.

Webfed letterpress ink is heat-set, made with high-boiling, slow-evaporating synthetic resins and petroleum oils dissolved or dispersed in a hydrocarbon solvent. The solvent must have a narrow boiling range with a low volatility at room temperatures and a fast evaporating rate at elevated temperatures. The solvent is vaporized in the printing press dryers at temperatures from 250 to 400°F, leaving the resins and oils on the paper. Webfed letterpress inks are dried after all colors are applied to the web.

The inks are dried by passing the web through dryers at speeds of 1000 to 2000 fpm. There are several types of dryers: open-flame gas cup, flame impingement, high-velocity hot air, and steam drum.

Exhaust quantities through a press dryer vary from about 7000 to 15,000 cfm at standard conditions, depending on the type of dryer used and the speed of the press. Exhaust temperatures range from 250 to 400°F.

Solvent-containing exhaust is heated to 1300°F in an air pollution control device to incinerate the effluent. A catalyst can be used to reduce the temperature required for combustion to 1000°F, but it requires periodic inspection and rejuvenation. Heat recovery reduces the fuel required for incineration and can be used to heat pressroom makeup air.

LITHOGRAPHY

Lithography uses a grease-treated printing image receptive to ink, on a surface that is neither raised nor depressed. Both grease and ink repel water. Water is applied to all areas of the plate, except the printing image. Ink is then applied only to the printing image and transferred to the paper in the printing process. In multicolor printing operations, the image may be printed up to four times on the same sheet of paper in different colors. Registration of images is critical to final color quality.

Offset printing transfers the image first to a rubber blanket and then to the paper. Sheetfed and web offset printing are similar to letterpress printing. The inks used are similar to those used in letterpress printing but contain water-resistant vehicles and pigment. In web offset and gravure printing, the relative humidity in the pressroom should be kept constant, and the temperature should be selected for comfort or, at least, to avoid heat stress. It is important to maintain steady conditions to ensure the dimensional stability of the paper onto which the images are printed.

The pressroom for sheet multicolor offset printing has more exacting humidity requirements than other printing processes. The paper must remain flat with constant dimensions during multicolor printing, in which the paper may make six or more passes through the press over a period of a week or more. If the paper does not have the right moisture content at the start, or if there are significant changes in atmospheric humidity during the process, the paper will not retain its dimensions and flatness, and misregistering will result. In many cases of color printing, a register accuracy of 0.005 in. is required. Figure 3 shows the close control of the air relative humidity that is necessary to achieve this register accuracy. The data shown in this figure are for composite lithographic paper.

Maintaining constant moisture content of the paper is complicated because paper picks up moisture from the moist offset blanket during printing (0.1 to 0.3% for each impression). When two or

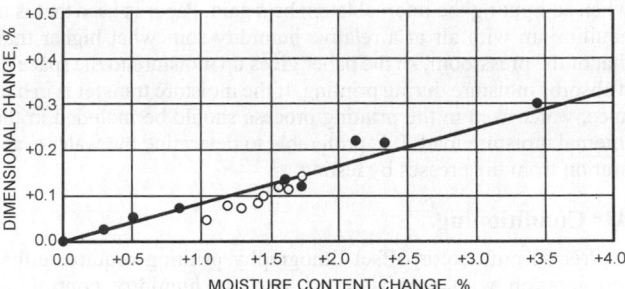

Fig. 3 Effects of Variation in Moisture Content on Dimensions of Printing Papers
(Weber and Snyder 1934)

more printings are made in close register work, the paper at the start of the printing process should have a moisture content in equilibrium with air at 5 to 8% rh above the pressroom air. At this condition, the moisture evaporated from the paper into the air nearly balances the moisture added by the press. In obtaining register, it is important to keep the sheet flat and free from wavy or tight edges. To do this, the relative humidity balance of the paper should be slightly above that of the pressroom atmosphere. This balance is not as critical in four-color roll-feed presses because the press moisture does not penetrate the paper quickly enough between colors to affect sheet dimensions or cause sheet distortion.

Recommended Environment

The Graphic Arts Technical Foundation recommends ideal conditions in a lithographic pressroom of 76 to 80°F db and 43 to 47% rh, controlled to ±2°F db and ±2% rh (Reed 1970). Comfort and economy of operation influence the choice of temperature. The effect of relative humidity variations on register can be estimated for offset paper from Figure 3. Closer relative humidity control of the pressroom air is required for multicolor printing of 76 in. sheets than for 22 in. sheets with the same register accuracy. Closer control is needed for multicolor printing, where the sheet makes two or more trips through the press, than for one-color printing.

Ink drying is affected by temperature and humidity, so uniform results are difficult to obtain without controlling the atmospheric conditions. Printing inks must dry rapidly to prevent offsetting and smearing. High relative humidity and high moisture content in paper tend to prevent ink penetration, so more ink remains on the surface than can be quickly oxidized. This affects drying time, intensity of color, and uniformity of ink on the surface. Relative humidity below 60% is favorable for drying at a comfortable temperature. Higher relative humidity may cause severe paper distortion and significant damage to the final product.

The air conditioning for the pressroom of a lithographic plant should control air temperature and relative humidity, filter the air, supply ventilation air, and distribute the air without pronounced drafts around the presses. Using antioffset sprays to set the ink creates an additional air-filtering load from the pressroom. Drafts and high airflow over the presses lead to excessive drying of the ink and water, which causes scumming or other problems.

The operating procedures of the pressroom should be analyzed to determine the heat removal load. The lighting load is high and constant throughout the day. The temperature of the paper brought into the pressroom and the length of time it is in the room should be considered to determine the sensible load from the paper. Figure 2 shows the time required for wrapped paper to reach room temperature. The press motors usually generate a large portion of the internal sensible heat gain.

Readings should be taken to obtain the running power load of the larger multicolor presses. The moisture content of the paper fed to the press and the relative humidity of the air must be considered when computing the internal latent heat gain. Paper is used that is in equilibrium with air at a relative humidity somewhat higher than that of the pressroom, so the paper gives up moisture to the space as it absorbs moisture during printing. If the moisture transfer is in balance, water used in the printing process should be included in the internal moisture load. It is preferable to determine the water evaporation from the presses by testing.

Air Conditioning

Precise multicolor offset lithography printing requires either refrigeration with provision for separate humidity control, or sorption dehumidifying equipment for independent humidity control with provision for cooling. The need for humidity control in the pressroom may be determined by calculating the dimensional change of the paper for each percent of change in relative humidity and checking this with the required register for the printing process.

Air conditioning of the photographic department is usually considered next in importance to that of the pressroom. Most of the work in offset lithography is done on film. Air conditioning controls cleanliness and comfort and maintains the size of the film for register work.

Air conditioning is important in the stripping department, both for comfort and for maintaining size and register. Curling of the film and flats, as well as shrinkage or stretch of materials, can be minimized by maintaining constant relative humidity. This is particularly important for close-register color work. The photographic area, stripping room, and platemaking area usually are maintained at the same conditions as the pressroom.

Dryers used for web offset printing are the same type as for web-fed letterpress. Drying is not as complex because less ink is applied and presses run at lower speeds (800 to 1800 fpm)).

ROTOGRAVURE

Rotogravure printing uses a cylinder with minute inkwells etched in the surface to form the printing image. Ink is applied to the cylinder, filling the wells. Excess ink is then removed from the cylinder surface by doctor blades, leaving only the ink in the wells. The image is then transferred to the paper as it passes between the printing cylinder and an impression cylinder.

In sheetfed gravure printing (as in offset printing), expansion, contraction, and distortion should be prevented to obtain correct register. The paper need not be in equilibrium with air at a relative humidity higher than that of the pressroom, because no moisture is added to the paper in the printing process. Humidity and temperature control should be exacting, like in offset printing. The relative humidity should be 45 to 50%, controlled to within ±2%, with a comfort temperature controlled to within ±2°F.

Gravure printing ink dries principally by evaporating the solvent in the ink, leaving a solid film of pigment and resin. The solvent is a low-boiling hydrocarbon, and evaporation takes place rapidly, even without the use of heat. The solvents have closed-cup flash points from 22 to 80°F and are classified as Group I or special hazard liquids by local code and insurance company standards. As a result, in areas adjacent to gravure press equipment and solvent and ink storage areas, electrical equipment must be Class I, Division 1 or 2, as described by the *National Electrical Code®* (NFPA *Standard* 70), and ventilation requirements (both supply and exhaust) are stringent. Ventilation should be designed for high reliability, with sensors to detect unsafe pollutant concentrations and then to initiate alarm or safety shutdown when necessary.

Rotogravure printing units operate in tandem, each superimposing print over that from the preceding unit. Press speeds range from 1200 to 2400 fpm. Each unit is equipped with its own dryer to prevent subsequent smearing or smudging.

A typical drying system consists of four dryers connected to an exhaust fan. Each dryer is equipped with fans to recirculate 5000 to 8000 cfm (at standard conditions) through a steam or hot water coil and then through jet nozzles. The hot air (130°F) impinges on the web and drives off the solvent-laden vapors from the ink. It is normal to exhaust half of this air. The system should be designed and adjusted to prevent solvent vapor concentration from exceeding 25% of its lower flammable limit (Marsailes 1970). If this is not possible, constant lower-flammable-limit (LFL) monitoring, concentration control, and safety shutdown capability should be included.

In exhaust design for a particular process, solvent vapor should be captured from the printing unit where paper enters and exits the dryer, from the fountain and sump area, and from the printed paper, which continues to release solvent vapor as it passes from one printing unit to another. Details of the process, such as ink and paper characteristics and rate of use, are required to determine exhaust quantities.

When dilution ventilation is used, exhaust of 1000 to 1500 cfm (at standard conditions) at the floor is often provided between each unit. The makeup air units are adjusted to supply slightly less air to the pressroom than that exhausted, to keep the pressroom negative with respect to the surrounding areas.

OTHER PLANT FUNCTIONS

Flexography

Flexography uses rubber raised printing plates and functions much like a letterpress. Flexography is used principally in the packaging industry to print labels and also to print on smooth surfaces, such as plastics and glass.

Collotype Printing

Collotype or photogelatin printing is a sheetfed printing process related to lithography. The printing surface is bichromated gelatin with varying affinity for ink and moisture, depending on the degree of light exposure received. There is no mechanical dampening as in lithography, and the necessary moisture in the gelatin printing surface is maintained by operating the press in an atmosphere of high relative humidity, usually about 85%. Because the tonal values printed are very sensitive to changes in the moisture content of the gelatin, the relative humidity should be maintained within ±2%.

Because tonal values are also very sensitive to changes in ink viscosity, temperature must be closely maintained; 80 ± 3°F is recommended. Collotype presses are usually partitioned off from the main plant, which is kept at a lower relative humidity, and the paper is exposed to high relative humidity only while it is being printed.

Salvage

Salvage systems remove paper trim and shredded paper waste from production areas, and carry airborne shavings to a cyclone or baghouse collector, where they are baled for recycling. Air quantities required are 40 to 45 ft³ per pound of paper trim, and the transport velocity in the ductwork is 4500 to 5000 fpm (Marsailes 1970). Humidification may be provided to prevent the buildup of a static charge and consequent system blockage.

Air Filtration

Ventilation and air-conditioning systems for printing plants commonly use automatic moving-curtain dry-media filters with renewable media having a weight arrestance of 80 to 90% (ASHRAE *Standard* 52.1).

In sheetfed pressrooms, a high-performance final filter is used to filter starch particles, which require about 85% ASHRAE dust-spot efficiency. In film processing areas, which require relatively dust-free conditions, high-efficiency air filters are installed, with 90 to 95% ASHRAE dust-spot efficiency.

A different type of filtration problem in printing is **ink mist** or **ink fly**, which is common in newspaper pressrooms and in heatset letterpress or offset pressrooms. Minute droplets of ink (5 to 10 µm) are dispersed by ink rollers rotating in opposite directions. The cloud of ink droplets is electrostatically charged. Suppressors, charged to repel the ink back to the ink roller, are used to control ink mist. Additional control is provided by automatic moving curtain filters.

Binding and Shipping

Some printed materials must be bound. Two methods of binding are perfect binding and stitching. In **perfect binding**, sections of a book (signatures) are gathered, ruffed, glued, and trimmed. The glued edge is flat. Large books are easily bound by this type of binding. Low-pressure compressed air and a vacuum are usually required to operate a perfect binder, and paper shavings are removed by a trimmer. The use of heated glue necessitates an exhaust system if the fumes are toxic.

In **stitching**, sections of a book are collected and stitched (stapled) together. Each signature is opened individually and laid over a moving chain. Careful handling of the paper is important. This has the same basic air requirements as perfect binding.

Mailing areas of a printing plant wrap, label, and ship the manufactured goods. Operation of the wrapper machine can be affected by low humidity. In winter, humidification of the bindery and mailing area to about 40 to 50% rh may be necessary to prevent static buildup.

REFERENCES

ACGIH. 2010. *Industrial ventilation: A manual of recommended practice*, 27th ed. American Conference of Governmental Industrial Hygienists, Cincinnati, OH.

ASHRAE. 1992. Gravimetric and dust-spot procedures for testing air-cleaning devices used in general ventilation for removing particulate matter. ANSI/ASHRAE *Standard* 52.1-1992. (Withdrawn.)

Marsailes, T.P. 1970. Ventilation, filtration and exhaust techniques applied to printing plant operation. *ASHRAE Journal* (December):27.

NFPA. 2011. National electrical code®. ANSI/NFPA *Standard* 70. National Fire Protection Association, Quincy, MA.

Reed, R.F. 1970. *What the printer should know about paper*. Graphic Arts Technical Foundation, Pittsburgh, PA.

Weber, C.G. and L.W. Snyder. 1934. Reactions of lithographic papers to variations in humidity and temperature. *Journal of Research* 12 (January). Available from National Bureau of Standards, Gaithersburg, MD.

TEXTILE PROCESSING PLANTS

THIS chapter covers (1) basic processes for making synthetic fibers, (2) fabricating synthetic fibers into yarn and fabric, (3) relevant types of HVAC and refrigerating equipment, (4) health considerations, and (5) energy conservation procedures.

Most textile manufacturing processes may be placed into one of three general classifications: synthetic fiber making, yarn making, or fabric making. Synthetic fiber manufacturing is divided into staple processing, tow-to-top conversion, and continuous fiber processing; yarn making is divided into spinning and twisting; and fabric making is divided into weaving and knitting. Although these processes vary, their descriptions reveal the principles on which air-conditioning design for these facilities is based.

TERMINOLOGY

The following is only a partial glossary of terms used in the textile industry. For more complete terminology, consult the sources in the Bibliography or the Internet search engine of your choice.

Air permeability. Porosity, or ease with which air passes through material. Air permeability affects factors such as the wind resistance of sailcloth, air resistance of parachute cloth, and efficiency of various types of air filtration media. It is also a measure of a fabric's warmness or coolness.

Bidirectional fabric. A fabric with reinforcing fibers in two directions: in the warp (machine) direction and filling (cross-machine) direction.

Calender. A machine used in finishing to impart various surface effects to fabrics. It essentially consists of two or more heavy rollers, sometimes heated, through which the fabric is passed under heavy pressure.

Denier. The weight, in grams, of 9000 m (29,528 ft) of yarn. Denier is a direct numbering system in which lower numbers represent finer sizes and higher numbers the coarser sizes. Outside the United States, the **Tex** system is used instead.

Heddle. A cord, round steel wire, or thin flat steel strip with a loop or eye near the center, through which one or more warp threads pass on the loom, so that thread movement may be controlled in weaving. Heddles are held at both ends by the harness frame. They control the weave pattern and shed as the harnesses are raised and lowered during weaving.

Lubricant. An oil or emulsion finish applied to fibers to prevent damage during textile processing, or to knitting yarns to make them more pliable.

Machine direction. The long direction within the plane of the fabric (i.e., the direction in which the fabric is being produced by the machine).

The preparation of this chapter is assigned to TC 9.2, Industrial Air Conditioning.

Pick. A single filling thread carried by one trip of the weft insertion device across the loom. Picks interface with the warp ends to form a woven fabric.

Reed. A comblike device on a loom that separates the warp yarns and also beats each succeeding filling thread against those already woven. The space between two adjacent wires of the reed is called a **dent**. The fineness of the reed is calculated by the number of dents per inch: the more dents, the finer the reed.

Selvage. The narrow edge of woven fabric that runs parallel to the warp. It is made with stronger yarns in a tighter construction than the body of the fabric, to prevent raveling. A **fast selvage** encloses all or part of the picks; a selvage is not fast when the filling threads are cut at the fabric edge after each pick.

Shuttle. A boat-shaped device usually made of wood with a metal tip that carries filling yarns through the shed in the weaving process.

Tex. The mass, in grams, of 1000 m (3281 ft) of fabric. Used primarily outside the United States. *See also* **Denier**.

Warp. The set of yarn in all woven fabrics, running lengthwise and parallel to the selvage, interwoven with the filling.

FIBER MAKING

Processes preceding fiber extrusion have diverse ventilating and air-conditioning requirements based on principles similar to those that apply to chemical plants.

Synthetic fibers are extruded from metallic spinnerets and solidified as continuous parallel filaments. This process, called **continuous spinning**, differs from the mechanical spinning of fibers or tow into yarn, which is generally referred to as **spinning**.

Synthetic fibers may be formed by melt-spinning, dry-spinning, or wet-spinning. Melt-spun fibers are solidified by cooling the molten polymer; dry-spun fibers by evaporating a solvent, leaving the polymer in fiber form; and wet-spun fibers by hardening the extruded filaments in a liquid bath. The selection of a spinning method is affected by economic and chemical considerations. Generally, nylons, polyesters, and glass fibers are melt-spun; acetates dry-spun; rayons and aramids wet-spun; and acrylics dry- or wet-spun.

For melt- and dry-spun fibers, the filaments of each spinneret are usually drawn through a long vertical tube called a **chimney** or **quench stack**, within which solidification occurs. For wet-spun fibers, the spinneret is suspended in a chemical bath where coagulation of the fibers takes place. Wet-spinning is followed by washing, applying a finish, and drying.

Synthetic continuous fibers are extruded as a heavy denier tow for cutting into short lengths called staple or somewhat longer lengths for tow-to-top conversion, or they are extruded as light denier filaments for processing as continuous fibers. Oil is then applied to lubricate, give antistatic properties, and control fiber cohesion. The extruded filaments are usually drawn (stretched) both to align the molecules along the axis of the fiber and to improve the crystalline structure of the molecules, thereby increasing the fiber's strength and resistance to stretching.

Heat applied to the fiber when drawing heavy denier or high-strength synthetics releases a troublesome oil mist. In addition, the mechanical work of drawing generates a high localized heat load. If the draw is accompanied by twist, it is called **draw-twist**; if not, it is called **draw-wind**. After draw-twisting, continuous fibers may be given additional twist or may be sent directly to warping.

When tow is cut to make staple, the short fibers are allowed to assume random orientation. The staple, alone or in a blend, is then usually processed as described in the Cotton System section. However, tow-to-top conversion, a more efficient process, has become more popular. The longer tow is broken or cut to maintain parallel orientation. Most of the steps of the cotton system are bypassed; the parallel fibers are ready for blending and mechanical spinning into yarn.

In the manufacture of glass fiber yarn, light denier multifilaments are formed by attenuating molten glass through platinum bushings at high temperatures and speeds. The filaments are then drawn together while being cooled with a water spray, and a chemical size is applied to protect the fiber. This is all accomplished in a single process prior to winding the fiber for further processing.

YARN MAKING

The fiber length determines whether spinning or twisting must be used. Spun yarns are produced by loosely gathering synthetic staple, natural fibers, or blends into rope-like form; drawing them out to increase fiber parallelism, if required; and then twisting. Twisted (continuous filament) yarns are made by twisting mile-long monofilaments or multifilaments. Ply yarns are made in a similar manner from spun or twisted yarns.

The principles of mechanical spinning are applied in three different systems: cotton, woolen, and worsted. The cotton system is used for all cotton, most synthetic staple, and many blends. Woolen and worsted systems are used to spin most wool yarns, some wool blends, and synthetic fibers such as acrylics.

Cotton System

The cotton system was originally developed for spinning cotton yarn, but now its basic machinery is used to spin all varieties of staple, including wool, polyester, and blends. Most of the steps from raw materials to fabrics, along with the ranges of frequently used humidities, are outlined in Figure 1.

Opening, Blending, and Picking. The compressed tufts are partly opened, most foreign matter and some short fibers are removed, and the mass is put in an organized form. Some blending is desired to average the irregularities between bales or to mix different kinds of fiber. Synthetic staple, which is cleaner and more uniform, usually requires less preparation. The product of the picker is pneumatically conveyed to the feed rolls of the card.

Carding. This process lengthens the lap into a thin web, which is gathered into a rope-like form called a **sliver**. Further opening and fiber separation follows, as well as partial removal of short fiber and trash. The sliver is laid in an ascending spiral in cans of various diameters.

For heavy, low-count (length per unit of mass) yarns of average or lower quality, the card sliver goes directly to drawing. For lighter, high-count yarns requiring fineness, smoothness, and strength, the card sliver must first be combed.

Lapping. In sliver lapping, several slivers are placed side by side and drafted. In ribbon lapping, the resulting ribbons are laid one on another and drafted again. The doubling and redoubling averages out sliver irregularities; drafting improves fiber parallelism. Some recent processes lap only once before combing.

Combing. After lapping, the fibers are combed with fine metal teeth to substantially remove all fibers below a predetermined length, to remove any remaining foreign matter, and to improve

fiber arrangement. The combed lap is then attenuated by drawing rolls and again condensed into a single sliver.

Drawing. Drawing follows either carding or combing and improves uniformity and fiber parallelism by doubling and drafting several individual slivers into a single composite strand. Doubling averages the thick and thin portions; drafting further attenuates the mass and improves parallelism.

Roving. Roving continues the processes of drafting and paralleling until the strand is a size suitable for spinning. A slight twist is inserted, and the strand is wound on large bobbins used for the next roving step or for spinning.

Spinning. Mechanical spinning simultaneously applies draft and twist. The packages (any form into or on which one or more ends can be wound) of roving are creeled at the top of the frame. The unwinding strand passes progressively through gear-driven drafting rolls, a yarn guide, the C-shaped traveler, and then to the bobbin. The vertical traverse of the ring causes the yarn to be placed in predetermined layers.

The difference in peripheral speed between the back and front rolls determines the draft. Twist is determined by the rate of front roll feed, spindle speed, and drag, which is related to the traveler weight.

The space between the nip or bite of the rolls is adjustable and must be slightly greater than the longest fiber. The speeds of front and back rolls are independently adjustable. Cotton spindles normally run at 8000 to 9000 rpm but may exceed 14,000 rpm. In ring

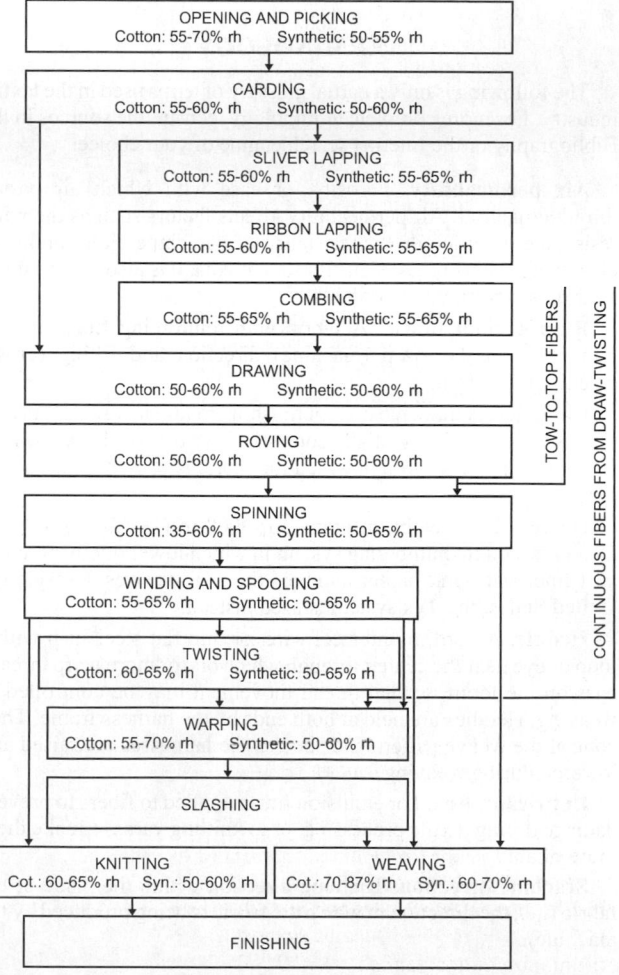

Fig. 1 Textile Process Flowchart and Ranges of Humidity

twisting, drawing rolls are omitted, and a few spindles run as high as 18,000 rpm.

Open-end or turbine spinning combines drawing, roving, lapping, and spinning. Staple fibers are fragmented as they are drawn from a sliver and fed into a small, fast-spinning centrifugal device. In this device, the fibers are oriented and discharged as yarn; twist is imparted by the rotating turbine. This system is faster, quieter, and less dusty than ring spinning.

Spinning is the final step in the cotton system; the feature that distinguishes it from twisting is the application of draft. The amount and point of draft application accounts for many of the subtle differences that require different humidities for apparently identical processes.

Atmospheric Conditions. From carding to roving, the loosely bound fibers are vulnerable to static electricity. In most instances, static can be adequately suppressed with humidity, which should not be so high as to cause other problems. In other instances, it is necessary to suppress electrostatic properties with antistatic agents. Wherever draft is applied, constant humidity is needed to maintain optimum frictional uniformity between adjacent fibers and, hence, cross-sectional uniformity.

Woolen and Worsted Systems

The woolen system generally makes coarser yarns, whereas the worsted system makes finer ones of a somewhat harder twist. Both may be used for lighter blends of wool, as well as for synthetic fibers with the characteristics of wool. The machinery used in both systems applies the same principles of draft and twist but differs greatly in detail and is more complex than that used for cotton.

Compared to cotton, wool fibers are dirtier, greasier, and more irregular. They are scoured to remove grease and are then usually reimpregnated with controlled amounts of oil to make them less hydrophilic and to provide better interfiber behavior. Wool fibers are scaly and curly, so they are more cohesive and require different treatment. Wool, in contrast to cotton and synthetic fibers, requires higher humidities in the processes prior to and including spinning than it does in the processes that follow. Approximate humidities are given in Kirk and Othmer (1993).

Twisting Filaments and Yarns

Twisting was originally applied to silk filaments; several filaments were doubled and then twisted to improve strength, uniformity, and elasticity. Essentially the same process is used today, but it is now extended to spun yarns, as well as to single or multiple filaments of synthetic fibers. Twisting is widely used in the manufacture of sewing thread, twine, tire cord, tufting yarn, rug yarn, ply yarn, and some knitting yarns.

Twisting and doubling is done on a **down-** or **ring-twister**, which draws in two or more ends from packages on an elevated creel, twists them together, and winds them into a package. Except for the omission of drafting, down-twisters are similar to conventional ring-spinning frames.

When yarns are to be twisted without doubling, an **up-twister** is used. Up-twisters are primarily used for throwing synthetic monofilaments and multifilaments to add to or vary elasticity, light reflection, and abrasion resistance. As with spinning, yarn characteristics are controlled by making the twist hard or soft, right (S) or left (Z). Quality is determined largely by the uniformity of twist, which, in turn, depends primarily on the tension and stability of the atmospheric conditions (Figure 1). Because the frame may be double- or triple-decked, twisting requires concentrations of power. The frames are otherwise similar to those used in spinning, and they present the same air distribution problems. In twisting, lint is not a serious problem.

FABRIC MAKING

Preparatory Processes

When spinning or twisting is complete, the yarn may be prepared for weaving or knitting by processes that include winding, spooling, creeling, beaming, slashing, sizing, and dyeing. These processes have two purposes: (1) to transfer the yarn from the type of package dictated by the preceding process to a type suitable for the next and (2) to impregnate some of the yarn with sizes, gums, or other chemicals that may not be left in the final product.

Filling Yarn. Filling yarn is wound on quills for use in a loom shuttle. It is sometimes predyed and must be put into a form suitable for package or skein dyeing before it is quilled. If the filling is of relatively hard twist, it may be put through a twist-setting or conditioning operation in which internal stresses are relieved by applying heat, moisture, or both.

Warp Yarn. Warp yarn is impregnated with a transient coating of size or starch that strengthens the yarn's resistance to the chafing it will receive in the loom. The yarn is first rewound onto a cone or other large package from which it will unwind speedily and smoothly. The second step is warping, which rewinds a multiplicity of ends in parallel arrangement on large spools, called **warp or section beams**. In the third step, slashing, the threads pass progressively through the sizing solution, through squeeze rolls, and then around cans, around steam-heated drying cylinders, or through an air-drying chamber. As much as several thousand pounds may be wound on a single loom beam.

Knitting Yarn. If hard-spun, knitting yarn must be twist-set to minimize kinking. Filament yarns must be sized to reduce strip-backs and improve other running qualities. Both must be put in the form of cones or other suitable packages.

Uniform tension is of great importance in maintaining uniform package density. Yarns tend to hang up when unwound from a hard package or slough off from a soft one, and both tendencies are aggravated by spottiness. The processes that require air conditioning, along with recommended relative humidities, are presented in Figure 1.

Weaving

In the simplest form of weaving, harnesses raise or depress alternate warp threads to form an opening called a **shed**. A shuttle containing a quill is kicked through the opening, trailing a thread of filling behind it. The lay and the reed then beat the thread firmly into one apex of the shed and up to the fell of the previously woven cloth. Each shuttle passage forms a pick. These actions are repeated at frequencies up to five per second.

Each warp thread usually passes through a drop-wire that is released by a thread break and automatically stops the loom. Another automatic mechanism inserts a new quill in the shuttle as the previous one is emptied, without stopping the loom. Other mechanisms are actuated by filling breaks, improper shuttle boxing, and the like, which stop the loom until it is manually restarted. Each cycle may leave a stop mark sufficient to cause an imperfection that may not be apparent until the fabric is dyed.

Beyond this basic machine and pattern are many complex variations in harness and shuttle control, which result in intricate and novel weaving effects. The most complex loom is the **jacquard**, with which individual warp threads may be separately controlled. Other variations appear in looms for such products as narrow fabrics, carpets, and pile fabrics. In the **Sulzer weaving machine**, a special filling carrier replaces the conventional shuttle. In the rapier, a flat, spring-like tape uncoils from each side and meets in the middle to transfer the grasp on the filling. In the **water jet loom**, a tiny jet of high-pressure water carries the filling through the shed of the warp. Other looms transport the filling with compressed air.

High humidity increases the abrasion resistance of the warp. Weave rooms require 80 to 85% humidity or higher for cotton and up to 70% humidity for synthetic fibers. Many looms run faster when room humidity and temperature are precisely controlled.

In the weave room, power distribution is uniform, with an average concentration somewhat lower than in spinning. The rough treatment of fibers liberates many minute particles of both fiber and size, thereby creating considerable amounts of airborne dust. In this high-humidity area, air changes average from four to eight per hour. Special provisions must be made for maintaining conditions during production shutdown periods, usually at a lower relative humidity.

Knitting

Typical knitted products are seamless articles produced on circular machines (e.g., undershirts, socks, and hosiery) and those knitted flat (e.g., full-fashioned hosiery, tricot, milanese, and warp fabrics).

Knitted fabric is generated by forming millions of interlocking loops. In its simplest form, a single end is fed to needles that are actuated in sequence. In more complex constructions, hundreds of ends may be fed to groups of elements that function more or less in parallel.

Knitting yarns may be either single strand or multifilament and must be of uniform high quality and free from neps or knots. These yarns, particularly the multifilament type, are usually treated with special sizes to provide lubrication and to keep broken filaments from stripping back.

The need for precise control of yarn tension, through controlled temperature and relative humidity, increases with the fineness of the product. For example, in finer gages of full-fashioned hosiery, a 2°F change in temperature is the limit, and a 10% change in humidity may change the length of a stocking by 3 in. For knitting, desirable room conditions are approximately 76°F db and 45 to 65% rh.

Dyeing and Finishing

Finishing, which is the final readying of a mill product for its particular market, ranges from cleaning to imparting special characteristics. The specific operations involved vary considerably, depending on the type of fiber, yarn, or fabric, and the end product usage. Operations are usually done in separate plants. These areas need not only normal heating, ventilation, and fog removal systems, but also removal of hot, dusty, and toxic fumes from continuous ovens and tenters. Packaged chilling equipment is sometimes used to control temperatures of preshrink chemicals, dyes, and coatings that are applied to textiles and yarns before finishing. Some of these processes require corrosive-resistant materials and equipment.

Inspection is the only finishing operation to which air conditioning is regularly applied, although most of the others require ventilation. Finishing operations that use wet processes usually keep their solutions at high temperatures and require special ventilation to prevent destructive condensation and fog. Spot cooling of workers may be necessary for large releases of sensible, latent, or radiant heat.

AIR-CONDITIONING DESIGN

There are many diverse and special needs of specific areas of the textile process. Generally, a meeting with the owner's representative(s), local code officials, and the owner's insurance company is helpful in satisfying the particular requirements of the process, insurance companies, and local officials. HVAC engineers designing textile projects need to have a thorough understanding of the following HVAC system elements:

- Psychrometric process in spray systems
- Humidification and dehumidification
- Draft-free air distribution
- Fog control
- Water and air filters
- Dust collectors
- Industrial ductwork
- Large built-up air handlers

- Large water chillers
- Cooling towers
- Industrial piping systems
- Pumping
- Corrosion-resistant metallurgy
- Large centrifugal air compressors
- Programmable logic controllers and supervisory control and data acquisition (SCADA) systems
- Water treatment in open sump systems

Consultation with mechanical contracting companies experienced with building and installing textile-related systems provides great insight to these attributes. Thorough understanding of the processes to be conditioned; precise calculations; familiarity with codes, regulations, and current industry standards; as well as reasonable owner/engineer/contractor relationships and adherence to the owner's budget are necessary for successful projects.

Air washers are especially important in textile manufacturing and may be either conventional low-velocity or high-velocity units in built-up systems. Unitary high-velocity equipment using rotating eliminators, although no longer common, is still found in some plants.

Contamination of air washers by airborne oils often dictates the separation of air washers and process chillers by heat exchangers, usually of the plate or frame type.

Open-Sump Chilled-Water Systems

It is common practice to use open sumps in textile processing with air washer air-handling units. Open sumps present a unique problem for the removal of lint from the basins. Many systems return the air from spinning areas, and this air carries lint and free fibers from the spinning process. These fibers are typically not completely removed by central collectors (see Figure 3). In older facilities, the central collectors may be totally ineffective or nonexistent. A rotating drum filter is commonly used to remove lint fibers from the sumps of air washers to prevent clogging of spray nozzles and fouling of spray media. The rotating drum filters are semisubmerged in the sump and are fitted with a vacuum system that traverses the part of the drum that is exposed to air, removing the lint from the drum surface and transporting it through a high-pressure blower to a bag house, where water is separated and the lint collected for future disposal.

Many textile plants have an open sump for return of chilled water from the air washers (see A in Figure 2). The chilled-water pumps draw out of these sumps through a screened inlet, C, for return of chilled water to the chillers. In designing the inlet screen, care must be taken to avoid a configuration that might lead to pump cavitation. Rotating drum filters should also be considered for these sumps to prevent fouling of chiller tubes by lint that passes the screens. These sumps must be carefully sized to receive the volume of water contained in the system when the air washers are shut off and their sumps drain down.

Integrated Systems

Many mills use a refined air washer system that combines the air-conditioning system and the collector system (see the section on Collector Systems) into an integrated unit. Air handled by the collector system fans and any air required to make up total return air are delivered back to the air-conditioning apparatus through a central duct. The quantity of air returned by individual yarn-processing machine cleaning systems must not exceed the air-conditioning supply air quantity. Air discharged by these individual suction systems is carried by return air ducts directly to the air-conditioning system. Before entering the duct, some of the cleaning system air passes over the yarn-processing machine drive motor and through a special enclosure to capture heat losses from the motor.

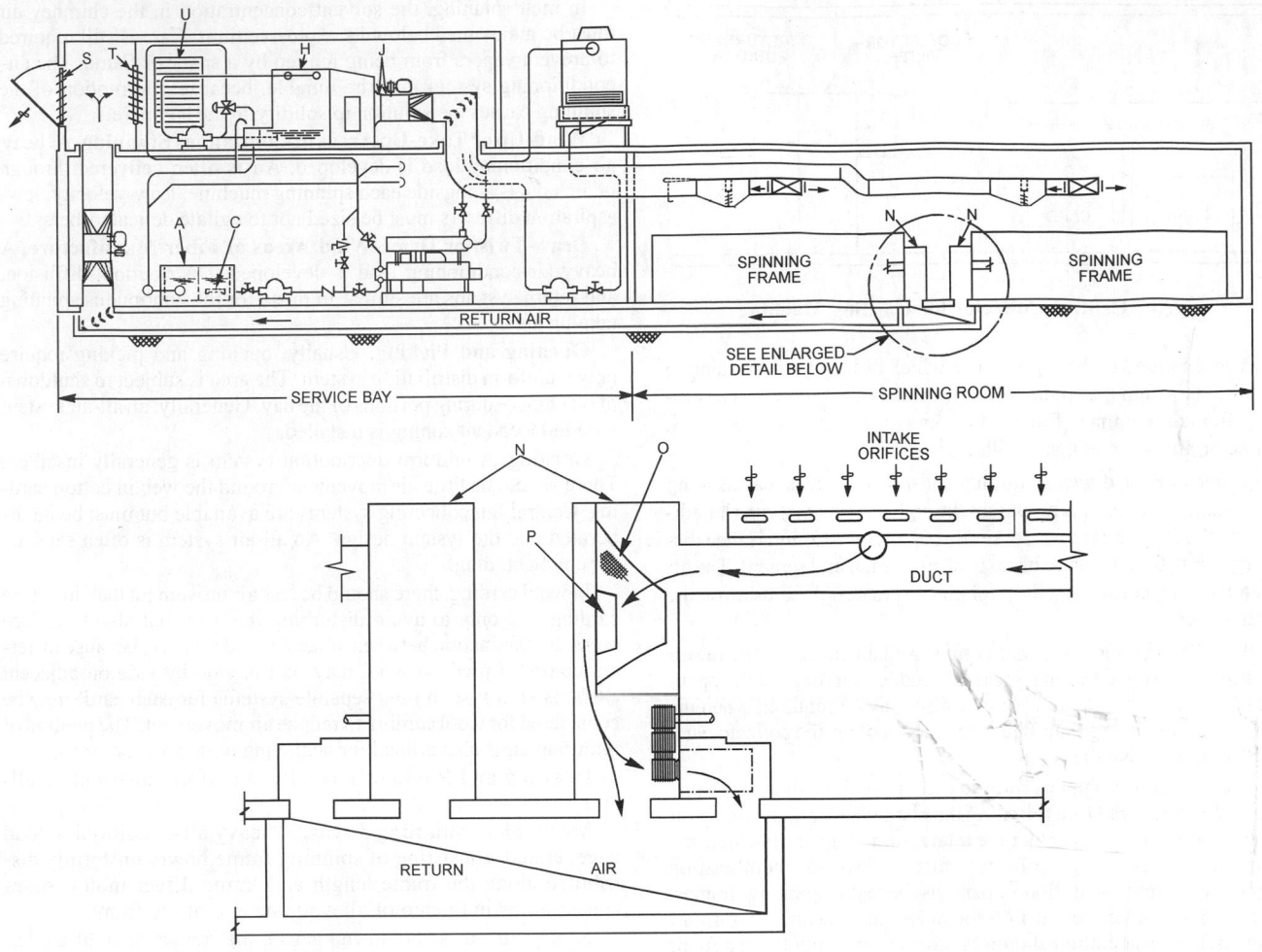

Fig. 2 Mechanical Spinning Room with Combined Air-Conditioning and Collector System

When integrated systems occasionally exceed the supply air requirements of the area served, the surplus air must be reintroduced after filtering.

Individual suction cleaning systems that can be integrated with air conditioning are available for cards, drawing frames, lap winders, combers, roving frames, spinning frames, spoolers, and warpers. The following advantages result from this integration:

- With a constant air supply, the best uniform air distribution can be maintained year-round.
- Downward airflow can be controlled; crosscurrents in the room are minimized or eliminated; drift or fly from one process to another is minimized or eliminated. Room partitioning between systems serving different types of manufacturing processes further enhances the value of this integration by controlling room air pattern year-round.
- Heat losses of the yarn-processing frame motor and any portion of the processing frame heat captured in the duct, as well as the heat of the collector system equipment, cannot affect room conditions; hot spots in motor alleys are eliminated, and although this heat goes into the refrigeration load, it does not enter the room. As a result, the supply air quantity can be reduced.
- Uniform conditions in the room improve production; conditioned air is drawn directly to the work areas on the machines, minimizing or eliminating wet or dry spots.

- Maximum cleaning use is made of the air being moved. A guide for cleaning air requirements follows:

Pickers	2500 to 4000 cfm per picker
Cards	700 to 1500 cfm per card
Spinning	4 to 8 cfm per spindle
Spooling	40 cfm per spool

Collector Systems

A collector system is a waste-capturing device that uses many orifices operating at high suction pressures. Each piece of production machinery is equipped with suction orifices at all points of major lint generation. The captured waste is generally collected in a fan and filter unit located either on each machine or centrally to accept waste from a group of machines.

A collector in the production area may discharge waste-filtered air either back into the production area or into a return duct to the air-conditioning system. It then enters the air washer or is relieved through dampers to the outdoors.

Figure 2 shows a mechanical spinning room with air-conditioning and collector systems combined into an integrated unit. In this case, the collector system returns all of its air to the air-conditioning system. If supply air from the air-conditioning system exceeds the maximum that can be handled by the collector system, additional air should be returned by other means.

Figure 2 also shows return air entering the air-conditioning system through damper T, passing through air washer H, and being

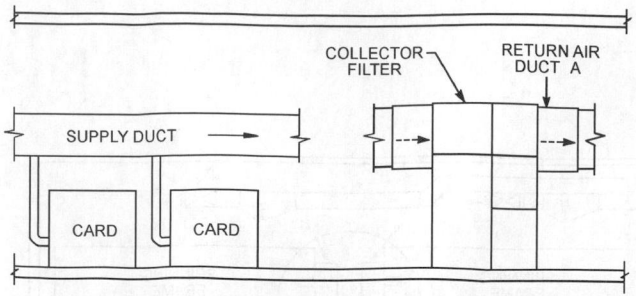

Fig. 3   Central Collector for Carding Machine

delivered by fan J to the supply duct, which distributes it to maintain conditions within the spinning room. At the other end of each spinning frame are unitary filter-collectors consisting of enclosure N, collector unit screen O, and collector unit fan P.

Collector fan P draws air through the intake orifices spaced along the spinning frame. This air passes through the duct that runs lengthwise to the spinning frame, passes through screen O, and is then discharged into the enclosure base (beneath the fan and screen). The air quantity is not constant; it drops slightly as material builds up on the filter screen.

Because the return air quantity must remain constant, and the air quantity discharged by fan P is slightly reduced at times, relief openings are necessary. Relief openings also may be required when the return air volume is greater than the amount of air the collector suction system can handle.

The discharge of fan P is split, so part of the air cools the spinning frame drive motor before rejoining the rest of the air in the return air tunnel. Regardless of whether the total return air quantity enters the return air tunnel through collector units, or through a combination of collector units and floor openings beneath spinning frames, return air fan R delivers it into the apparatus, ahead of return air damper T. Consideration should be given to filtering the return air prior to its delivery into the air-conditioning apparatus.

Mild-season operation causes more outdoor air to be introduced through damper U. This air is relieved through motorized damper S, which opens gradually as outdoor damper U opens, while return damper T closes in proportion. All other components perform as typical central station air-washer systems.

A system having the general configuration shown in Figure 2 may also be used for carding; the collector system portion of this arrangement is shown in Figure 3. A central collector filters the lint-laden air taken from multiple points on each card. This air is discharged to return air duct A and is then either returned to the air-conditioning system, exhausted outdoors, or returned directly to the room. A central collector filter may also be used with the spinning room system of Figure 2.

Air Distribution

Textile plants served by generally uniform air distribution may still require special handling for areas of load concentration.

Continuous Spinning Area. Methods of distribution are diverse and generally not critical. However, spot cooling or localized heat removal may be required. This area may be cooled by air conditioning, evaporative cooling, or ventilation.

Chimney (Quench Stack). Carefully controlled and filtered air or other gas is delivered to the chimneys; it is returned for conditioning and recovery of any valuable solvents present. Distribution of the air is of the utmost importance. Nonuniform temperature, humidity, or airflow disturbs the yarn, causing variations in fiber diameter, crystalline structure, and orientation. A fabric made of such fibers streaks when dyed.

In melt spinning, the solvent concentration in the chimney air must be maintained below its explosive limit. Care is still required to prevent vapors from being ignited by a spark or flame. The air-conditioning system must be reliable, because interruption of the spinning causes the solution to solidify in the spinnerets.

Wind-Up or Take-Up Areas of Continuous Spinning. A heavy air-conditioning load is developed. Air is often delivered through branch ducts alongside each spinning machine. Low-velocity, low-aspiration diffusers must be sized not to agitate delicate fibers.

Draw-Twist or Draw-Wind Areas of Fiber Manufacture. A heavy air-conditioning load is developed. Distribution, diffusion, and return systems are similar to those for the continuous spinning take-up area.

Opening and Picking. Usually, opening and picking require only a uniform distribution system. The area is subject to shutdown of machinery during portions of the day. Generally, an all-air system with independent zoning is installed.

Carding. A uniform distribution system is generally installed. There should be little air movement around the web in cotton carding. Central lint collecting systems are available but must be incorporated into the system design. An all-air system is often selected for cotton carding.

In wool carding, there should be less air movement than in cotton carding, not only to avoid disturbing the web, but also to reduce cross-contamination between adjacent cards. This is because different colors of predyed wool may be run side by side on adjacent cards. A split system (i.e., separate systems for each card) may be considered for wool carding to reduce air movement. The method of returning air is also critical for achieving uniform conditions.

Drawing and Roving. Generally, a uniform distribution all-air system works well.

Mechanical Spinning Areas. A heavy air-conditioning load is generated, consisting of spinning frame power uniformly distributed along the frame length and frame driver motor losses concentrated in the motor alley at one end of the frame.

Supply air ducts should run across the frames at right angles. Sidewall outlets between each of the two adjacent frames then direct the supply air down between the frames, where conditions must be maintained. Where concentrated heat loads occur, as in a double motor alley, placement of a supply air duct directly over the alley should be considered. Sidewall outlets spaced along the bottom of the duct diffuse air into the motor alley.

The collecting system, whether unitary or central, with intake points distributed along the frame length at the working level, assists in pulling supply air down to the frame, where maintenance of conditions is most important. A small percentage of the air handled by a central collecting system may be used to convey the collected lint and yarn to a central point, thus removing that air from the spinning room.

Machine design in spinning systems sometimes requires interfloor air pressure control.

Winding and Spooling. Generally, a uniform distribution, all-air system is used.

Twisting. This area has a heavy air-conditioning load. Distribution considerations are similar to those in spinning. Either all-air or split systems are installed.

Warping. This area has a very light load. Long lengths of yarn may be exposed unsupported in this area. Generally, an all-air system with uniform distribution is installed. Diffusers may be of the low-aspiration type. Return air is often near the floor.

Weaving. Generally, a uniform distribution system is necessary. Synthetic fibers are more commonly woven than natural fibers. The lower humidity requirements of synthetic fibers allow the use of an all-air system rather than the previously common split system. When lower humidity is coupled with the water jet loom, a high latent load results.

Health Considerations

For detailed information on control of industrial contaminants, see Chapter 29 of the 2008 *ASHRAE Handbook—HVAC Systems and Equipment.*

Control of Oil Mist. When textiles coated with lubricating oils are heated above 200°F in drawing operations in ovens, heated rolls, tenterframes, or dryers, an oil mist is liberated. If the oil mist is not collected at the source of emission and disposed of, a slightly odorous haze results.

Various devices have been proposed to separate oil mist from the exhaust air, such as fume incinerators, electrostatic precipitators, high-energy scrubbers, absorption devices, high-velocity filters, and condensers.

Spinning operations that generate oil mist must be provided with a high percentage (30 to 75%) of outdoor air. In high-speed spinning, 100% outdoor air is commonly used.

Operations such as drum cooling and air texturizing, which could contaminate the air with oil, require local exhausts.

Control of Monomer Fumes. Separate exhaust systems for monomers are required, with either wet- or dry-type collectors, depending on the fiber being spun. For example, caprolactam nylon spinning requires wet exhaust scrubbers.

Control of Hazardous Solvents. Provisions must be made for the containment, capture, and disposal of hazardous solvents.

Control of Cotton Dust. Byssinosis, also known as brown or white lung disease, is believed to be caused by a histamine-releasing substance in cotton, flax, and hemp dust. In the early stages of the disease, a cotton worker returning to work after a weekend experiences difficulty in breathing that is not relieved until later in the week. After 10 to 20 years, the breathing difficulty becomes continuous; even leaving the mill does not provide relief.

The U.S. Department of Labor enforces an OSHA standard of lint-free dust. The most promising means of control are improved exhaust procedures and filtration of recirculated air. Lint particles are 1 to 15 μm in diameter, so filtration equipment must be effective in this size range. Improvements in carding and picking that leave less trash in the raw cotton also help control lint.

Noise Control. The noise generated by HVAC equipment can be significant, especially if the textile equipment is modified to meet present safety criteria. For procedures to analyze and correct the noise from ventilating equipment, see Chapter 48.

Safety and Fire Protection

Special Warning: Certain industrial spaces may contain flammable, combustible, and/or toxic concentrations of vapors or dusts under either normal or abnormal conditions. In spaces such as these, there are life-safety issues that this chapter may not completely address. Special precautions must be taken in accordance with requirements of recognized authorities such as the National Fire Protection Association (NFPA), the Occupational Safety and Health Administration (OSHA), and the American National Standards Institute (ANSI). In all situations, engineers, designers, and installers who encounter conflicting codes and standards must defer to the code or standard that best addresses and safeguards life safety.

Oil mist can accumulate in ductwork and create a fire hazard. Periodic cleaning reduces the hazard, but provisions should be made to contain a fire with suppression devices such as fire-activated dampers and interior duct sprinklers.

ENERGY CONSERVATION

The following are some steps that can be taken to reduce energy consumption:

- Applying heat recovery to water and air
- Automating high-pressure dryers to save heat and compressed air
- Decreasing hot-water temperatures and increasing chilled-water temperatures for rinsing and washing in dyeing operations
- Replacing running washes with recirculating washes where practical
- Changing double-bleaching procedures to single-bleaching where practical
- Eliminating rinses and final wash in dye operations where practical
- Drying by "bump and run" process
- Modifying drying or curing oven air-circulation systems to provide counterflow
- Using energy-efficient electric motors and textile machinery
- For drying operations, using discharge air humidity measurements to control the exhaust versus recirculation rates in full economizer cycles

BIBLIOGRAPHY

Hearle, J. and R.H. Peters. 1960. *Moisture in textiles.* Textile Book Publishers, New York.

Kirk and Othmer, eds. 1993. *Kirk-Othmer encyclopedia of chemical technology,* 4th ed., vol. 9. Wiley-Interscience, New York.

Nissan, Q.H. 1959. *Textile engineering processes.* Textile Book Publishers, New York.

Press, J.J., ed. 1959. *Man made textile encyclopedia.* Textile Book Publishers, New York.

Sachs, A. 1987. Role of process zone air conditioning. *Textile Month* (October):42.

Schicht, H.H. 1987. Trends in textile air engineering. *Textile Month* (May): 41.

CHAPTER 22

PHOTOGRAPHIC MATERIAL FACILITIES

PROCESSING and storing sensitized photographic products requires temperature, humidity, and air quality control. Manufacturers of photographic products and processing equipment provide specific recommendations for facility design and equipment installation that should always be consulted. This chapter contains general information that can be used in conjunction with these recommendations. See Chapter 31 for information on general industrial ventilation.

Special Warning: Certain industrial spaces may contain flammable, combustible, and/or toxic concentrations of vapors or dusts under either normal or abnormal conditions. In spaces such as these, there are life-safety issues that this chapter may not completely address. Special precautions must be taken in accordance with requirements of recognized authorities such as the National Fire Protection Association (NFPA), the Occupational Safety and Health Administration (OSHA), and the American National Standards Institute (ANSI). In all situations, engineers, designers, and installers who encounter conflicting codes and standards must defer to the code or standard that best addresses and safeguards life safety.

STORING UNPROCESSED PHOTOGRAPHIC MATERIALS

Virtually all photosensitive materials deteriorate with age; the rate of photosensitivity deterioration depends largely on the storage conditions. Photosensitivity deterioration increases both at high temperature and at high relative humidity and usually decreases at lower temperature and humidity.

High humidity can accelerate loss of sensitivity and contrast, increase shrinkage, produce mottle (spots or blotches of different shades or colors), cause softening of the emulsion (which can lead to scratches), and promote fungal growth. Low relative humidity can increase the susceptibility of the film or paper to static markings, abrasions, brittleness, and curl.

Because different photographic products require different handling, product manufacturers should be consulted regarding proper temperature and humidity conditions for storage. Refrigerated storage may be necessary for some products in some climates.

Products not packaged in sealed vaportight containers are vulnerable to contaminants. These products must be protected from solvent, cleanser, and formaldehyde vapors (emitted by particleboard and some insulation, plastics, and glues); industrial gases; and engine exhaust. In hospitals, industrial plants, and laboratories, all photosensitive products, regardless of their packaging, must be protected from x-rays, radium, and radioactive sources. For example, films stored 25 ft away from 100 mg of radium require the protection of 3.5 in. of lead.

PROCESSING AND PRINTING PHOTOGRAPHIC MATERIALS

Ventilation with clean, fresh air maintains a comfortable working environment and prevents vapor-related complaints and health

problems. It is also necessary for high-quality processing, safe handling, and safe storage of photographic materials.

Processing produces odors, vapors, high humidity, and heat (from lamps, electric motors, dryers, mounting presses, and high-temperature processing solutions). Thus, it is important to supply plentiful clean, fresh air at the optimum temperature and relative humidity to all processing rooms. ASHRAE *Standard* 62.1 specifies 1.0 cfm/ft^2 of exhaust for darkrooms in Table 6-4.

Air Conditioning for Preparatory Operations

During receiving operations, exposed film is removed from its protective packaging for presplicing and processing. **Presplicing** combines many individual rolls of film into a long roll to be processed. At high relative humidity, photographic emulsions become soft and can be scratched. At excessively low relative humidity, the film base is prone to static, sparking, and curl deformation. The presplice work area should be maintained at 50 to 55% rh and 70 to 75°F db. Room pressures should cascade downward from areas of higher air quality to areas of lower air quality (clean to dirty).

Air Conditioning for Processing Operations

Processing exposed films or paper involves using a series of tempered chemical and wash tanks that emit heat, humidity, and vapors or gases (e.g., water vapor, acetic acid, benzyl alcohol, ammonia, sulfur dioxide). Room exhaust must be provided, along with local exhaust at noxious tanks. To conserve energy, air from pressurized presplice rooms can be used as makeup for processing room exhaust. Further supply air should maintain the processing space at a maximum of 75°F dry bulb and 50 to 55% rh.

The processed film or paper proceeds from the final wash to the dryer, which controls the moisture remaining in the product. Too little drying causes film to stick when wound, whereas too much drying causes undesirable curl. Drying can be regulated by controlling drying time, humidity, and temperature.

The volume of supply air should be sufficient to achieve the design condition. Airflow should be diffused or distributed to avoid objectionable drafts. Apart from causing personnel discomfort, drafts can cause dust problems and disturb the surface temperature uniformity of drying drums and other heated equipment. Supply and return air openings should be properly positioned (1) for good mixing and dilution of the room air, (2) to ensure efficient removal of fugitive vapors, and (3) to avoid short-circuiting of supply air into return or exhaust air openings. For automated processing equipment, tempered outdoor air should be supplied from the ceiling above the feed or head end of the machine at a minimum rate of 150 cfm per machine (Figure 1). If the machine extends through a wall into another room, both rooms need to be exhausted.

An exhaust system should be installed to remove humid or heated air and chemical vapors directly to the outdoors (process streams typically must comply with regulations pursuant to the Clean Air Act). The room air from an open machine or tank area should be exhausted to the outdoors at a rate sufficient to achieve at least the vapor dilution levels recommended by the American Conference of Governmental Industrial Hygienists (ACGIH 2010). An exhaust rate higher than the supply rate produces a negative pressure and makes

---

The preparation of this chapter is assigned to TC 9.2, Industrial Air Conditioning.

the escape of vapors or gases to adjoining rooms less likely. Depending on the process chemistry, local exhaust hoods may be needed at uncovered stabilizer tanks or at the bleach fix tanks (Figure 1).

The exhaust opening should be positioned so that the flow of exhausted air is away from the operator, as illustrated in Figure 2. This air should not be recirculated. The exhaust opening should always be as close as possible to the source of the contaminant for efficient removal [see ACGIH (2010) for more information]. For a processing tank, the exhaust hood should have a narrow opening at the back of, level with, and as wide as the top edge of the tank.

Processing tanks are often covered to reduce evaporation of heated processing chemical solutions (approximately 100°F). Covers on photographic processing equipment and chemical storage tanks can effectively minimize the amount of gases, vapors, or mists that enter the work area. If the processing tanks are enclosed and equipped with an exhaust connection, the minimum room air supply and exhaust rates may be reduced compared to an open tank (Figure 3).

A sulfide-toning sink should have a local exhaust hood to vent hydrogen sulfide. However, sulfide toners are rarely used now except for some specialized art processing and archival microfilm processing. The exhaust duct must be placed on the side opposite the operator so that vapor is not drawn toward the operator's face.

Air distribution to the drying area must provide an acceptable environment for operators as discussed in Chapter 9 of the 2009 *ASHRAE Handbook—Fundamentals* and ACGIH (2010). Exposed sides of the dryer should be insulated as much as is practical to reduce the large radiant and convected heat gain to the space. Exhaust grilles above the dryer can directly remove much of its rejected heat and moisture. Supply air should be directed to offset the remaining radiant heat gain to the space.

Using processor dryer heat to preheat cold incoming air during winter conditions can save energy. An economic evaluation is necessary to determine whether the energy savings justify the additional cost of the heat recovery equipment.

A canopy exhaust hood over the drying drum of continuous paper processors extracts heat and moisture. It is important to follow the processing equipment manufacturer's recommendations for venting the dryer section of the processor. Whenever possible, dryer vents should be exhausted to the outdoors to prevent build-up of excessive temperature and humidity in the workplace.

When drying motion picture film, exhaust should draw off vapor from the solvent and wax mixture that is normally applied for lubrication.

Air Conditioning for the Printing/Finishing Operation

In printing, where a second sensitized product is exposed through the processed original, the amount of environmental control needed depends on the size and type of operation. For small-scale printing, close control of the environment is not necessary, except to minimize dust. In photofinishing plants, printers for colored products emit substantial heat. The effect on the room can be reduced by removing the lamphouse heat directly. Computer-controlled electronic printers transport the original film and raw film or paper at high speed. Proper temperature and humidity are especially important because, in some cases, two or three images from many separate films may be superimposed in register onto one film. For best results, the printing room should be maintained at between 70 and 75°F and at 50 to 60% rh to prevent curl, deformation, and static. Curl and film deformation affect the register and sharpness of the images produced. Static charge should be eliminated because it leaves static marks and may also attract dust to the final product.

Mounting of reversal film into slides is a critical finishing operation requiring a 70 to 75°F db temperature with 50 to 55% rh.

Digital printing operations use equipment that generates significant heat. An exhaust system can be directly connected to the laser printer to remove heat at a flow rate specified by the equipment

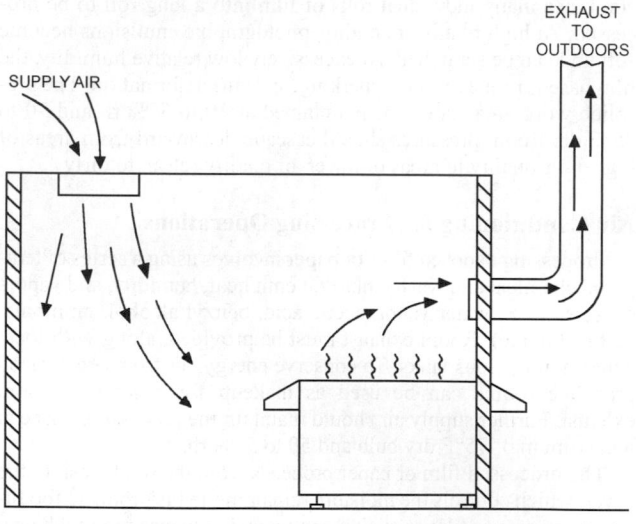

Fig. 1   Open Machine Ventilation

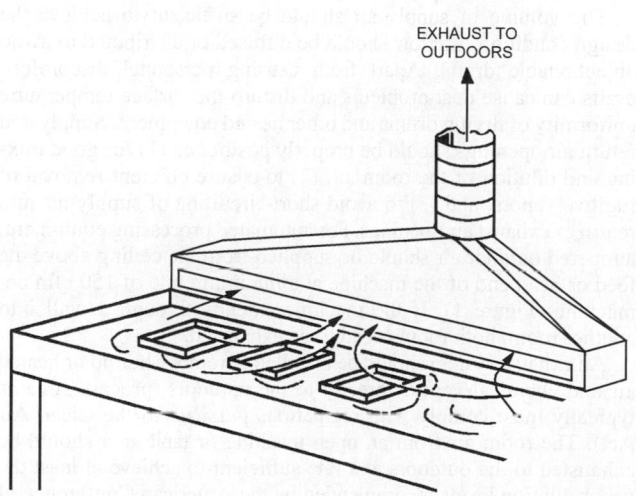

Fig. 2   Open-Tray Exhaust Ventilation from Processing Sink

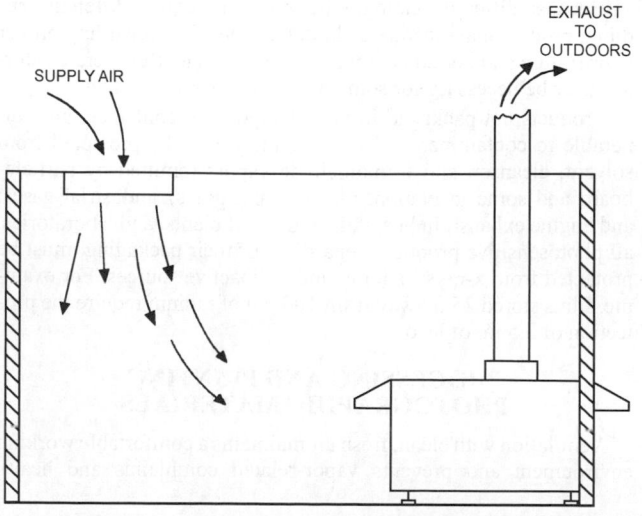

Fig. 3   Enclosed Machine Ventilation

manufacturer. Sufficient room ventilation is required so that applicable occupational exposure limits are not exceeded and a favorable operating environment is maintained.

Particulates in Air

Air conditioning for most photographic operations requires 85% efficiency disposable bag-type filters with 30% efficiency prefilters to extend the bag filter life. In critical applications (such as high-altitude aerial films) and for microminiature images, filtering of foreign matter is extremely important. These products are handled in a laminar airflow room or workbench with 95% efficiency HEPA filters plus 30% efficiency disposable prefilters.

Other Exhaust Requirements

A well-ventilated room should be provided for mixing the chemicals used in color processing and high-volume black-and-white work. The room should be furnished with movable exhaust hoods that provide a capture velocity as defined in ACGIH's (2010) *Industrial Ventilation* for the worst-case scenario. Modern photographic minilabs often use canisters of premixed processing solutions, so no chemical mixing is necessary.

If prints are lacquered regularly, a spray booth is needed. Concentrated lacquer spray is both hazardous and very objectionable to personnel; spray booth exhaust must be discharged outdoors.

Processing Temperature Control

Low processing volumes are typically handled in minilabs, which are often installed in retail locations. Minilabs are usually self-contained and equipped with temperature controls, heaters, and pumps. Typically, the owner only has to connect the minilab to water, electricity, exhaust (thimble connection), and a drain.

Higher-volume processing is handled with processors that come from the manufacturer complete with controls, heat exchangers, pumps, and control valves designed for the process that the owner has specified. Electricity, hot water, cold water, drainage, and steam may be required, depending on the manufacturer, who typically provides the specifications for these utilities.

STORING PROCESSED FILM AND PAPER

Storage of developed film and paper differs from storage of raw stock, because the developed materials are no longer photosensitive, are seldom sealed against moisture, and are generally stored for much longer periods. Required storage conditions depend on (1) the value of the records, (2) length of storage time, (3) whether the films are on nitrate or safety base, (4) whether the paper base is resin coated, and (5) type of photographic image.

Photographic materials must be protected against fire, water, mold, chemical or physical damage, high relative humidity, and high temperature. Relative humidity is much more critical than temperature. High relative humidity can cause films to stick together, (particularly roll films, but also sheet films). High humidity also damages gelatin, encourages the growth of mold, increases dimensional changes, accelerates the decomposition of nitrate support, and accelerates the deterioration of both black-and-white and color images. Low relative humidity causes a temporary increase in curl and decrease in flexibility, but when the humidity rises again, these conditions are usually reversed. An exception occurs when motion picture film is stored for a long time in loosely wound rolls at very low humidities. The curl causes the film roll to resemble a polygon rather than a circle when viewed from the side. This **spokiness** occurs because a highly curled roll of film resists being bent in the length direction when it is already bent in the width direction. When a spoky roll is stored for a long time, the film flows permanently into the spoky condition, resulting in film distortion. Very low relative humidity in storage may also cause the film or paper to crack or break if handled carelessly.

Low temperature (−10 to 50°F) is desirable for film and paper storage if (1) the relative humidity of the cold air is controlled, and (2) the material can be sufficiently warmed (for 2 to 8 h) before opening to prevent moisture condensation. High temperature can accelerate film shrinkage, which may produce physical distortions and the fading of dye images. High temperature is also detrimental to the stability of nitrate film.

Film Longevity

The American National Standards Institute (ANSI *Standard* IT9.11) defines longevities of films with a life expectancy (LE) rating. The **LE rating** is the minimum number of years that information can be retrieved if the subject film is stored under long-term storage conditions. In order to achieve the maximum LE rating, a product must be stored under long-term storage conditions. Polyester black-and-white silver gelatin films have an LE rating of 500, and acetate black-and-white silver gelatin films have an LE rating of 100. No LE ratings have been assigned to color films or black-and-white silver papers. Medium-term storage conditions have been defined for materials that are to retain their information for at least 10 years.

Medium-Term Storage

Rooms for medium-term storage of safety base film should be protected from accidental water damage by rain, flood, or pipe leaks. Air conditioning with controlled relative humidity is desirable but not always essential in moderate climates. Extremes of relative humidity are detrimental to film.

The most desirable storage relative humidity for processed film is about 50%, although 30 to 60% is satisfactory. Air conditioning is required where the relative humidity of the storage area exceeds 60% for any appreciable period. For a small room, a dehumidifier may be used if air conditioning cannot be installed. The walls should be coated with a vapor retarder, and the controlling humidistat should be set at about 40% rh. If the prevailing relative humidity is under 25% for long periods and problems from curl or brittleness are encountered, humidity should be controlled by a mechanical humidifier with a controlling humidistat set at 40%.

For medium-term storage, a room temperature between 68 and 77°F is recommended. Higher temperatures may cause shrinkage, distortion, and dye fading. Occasional peak temperatures of 95°F should not have a serious effect. Color films should be stored below 50°F to reduce dye fading. Films stored below the ambient dew point should be allowed to warm up before being opened to prevent moisture condensation.

An oxidizing or reducing atmosphere may deteriorate the film base and gradually fade the photographic image. Oxidizing agents may also cause microscopically small colored spots on fine-grain film such as microfilm (Adelstein et al. 1970). Typical gaseous contaminants include hydrogen sulfide, sulfur dioxide, peroxides, ozone, nitrogen oxides, and paint fumes. If these fumes are present in the intended storage space, they must be eliminated, or the film must be protected from contact with the atmosphere.

Long-Term Storage

For films or records that are to be preserved indefinitely, long-term storage conditions should be maintained. The recommended space relative humidity ranges from 20 to 50% rh, depending on the film type. When several film types are stored within the same area, 30% rh is a good compromise. The recommended storage temperature is below 70°F. Low temperature aids preservation, but if the storage temperature is below the dew point of the outdoor air, the records must be allowed to warm up in a closed container before they are used, to prevent moisture condensation. Temperature and humidity conditions must be maintained year-round and should be continuously monitored.

Requirements of a particular storage application can be met by any one of several air-conditioning equipment combinations. Standby equipment should be considered. Sufficient conditioned outdoor air should be provided to keep the room under a slight positive pressure for ventilation and to retard the entrance of untreated air. The air-conditioning unit should be located outside the vault for ease of maintenance, with precautions taken to prevent water leakage into the vault. The conditioner casing and all ductwork must be well insulated. Room conditions should be controlled by a dry-bulb thermostat and either a wet-bulb thermostat, humidistat, or dew-point controller.

Air-conditioning installations and fire dampers in ducts carrying air to or from the storage vault should be constructed and maintained according to National Fire Protection Association (NFPA) recommendations for air conditioning (NFPA *Standard* 90A) and for fire-resistant file rooms (NFPA *Standard* 232).

All supply air should be filtered with noncombustible HEPA filters to remove dust, which may abrade the film or react with the photographic image. As with medium-term storage, gaseous contaminants such as paint fumes, hydrogen sulfide, sulfur dioxide, peroxides, ozone, and nitrogen oxides may cause slow deterioration of the film base and gradual fading of the photographic image. When these substances cannot be avoided, an air scrubber, activated carbon adsorber, or other purification method is required.

Films should be stored in metal cabinets with adjustable shelves or drawers and with louvers or openings located to facilitate circulation of conditioned air through them. The cabinets should be arranged in the room to permit free circulation of air around them.

All films should be protected from water damage due to leaks, fire sprinkler discharge, or flooding. Drains should have sufficient capacity to keep the water from sprinkler discharge from reaching a depth of 3 in The lowest cabinet, shelf, or drawer should be at least 6 in. off the floor and constructed so that water cannot splash through the ventilating louvers onto the records.

When fire-protected storage is required, the film should be kept in either fire-resistant vaults or insulated record containers (Class 150). Fire-resistant vaults should be constructed in accordance with NFPA *Standard* 232. Although the NFPA advises against air conditioning in valuable-paper record rooms because of the possible fire hazard from outside, properly controlled air conditioning is essential for long-term preservation of archival films. The fire hazard introduced by the openings in the room for air-conditioning ducts may be reduced by fire and smoke dampers activated by smoke detectors in the supply and return ducts.

Storage of Cellulose Nitrate Base Film

Although photographic film has not been manufactured on cellulose nitrate (nitrocellulose) film base for several decades, many archives, libraries, and museums still have valuable records on this material. Preserving the cellulose nitrate film will be of considerable importance until the records have been printed on safety base.

Cellulose nitrate film base is chemically unstable and highly flammable. It decomposes slowly but continuously even under normal room conditions. The decomposition produces small amounts of nitric oxide, nitrogen dioxide, and other gases. Unless the nitrogen dioxide can escape readily, it reacts with the film base, accelerating the decomposition (Carrol and Calhoun 1955). The rate of decomposition is further accelerated by moisture and is approximately doubled with every 10°F increase in temperature.

All nitrate film must be stored in an approved vented cabinet or vault. Nitrate films should never be stored in the same vault with safety base films because any decomposition of the nitrate film will cause decomposition of the safety film. Cans in which nitrate film is stored should never be sealed, because this traps the nitrogen dioxide gas. Standards for storing nitrate film have been established (NFPA

Standard 40). The National Archives and the National Institute of Standards and Technology have also investigated the effect of a number of factors on fires in nitrate film vaults (Ryan et al. 1956).

The storage temperature should be kept as low as economically possible. The film should be kept at less than 50% rh. Temperature and humidity recommendations for the cold storage of color film in the following section also apply to nitrate film.

Storage of Color Film and Prints

All dyes fade in time. ANSI *Standard* IT9.11 does not define an LE for color films or black-and-white images on paper. However, many valuable color films and prints exist, and it is important to preserve them for as long as possible.

Light, heat, moisture, and atmospheric pollution contribute to fading of color photographic images. Storage temperature should be as low as possible to preserve dyes. For maximum permanence of images, materials should be stored in light-tight sealed containers or in moisture-proof wrapping materials at a temperature below freezing and at a relative humidity of 20 to 50%. The containers should be warmed to room temperature before opening to avoid moisture condensation on the surface. Photographic films can be brought to the recommended humidity by passing them through a conditioning cabinet with circulating air at about 20% rh for about 15 min.

An alternative is the use of a storage room or cabinet controlled at a steady (noncycled) low temperature and maintained at the recommended relative humidity. This eliminates the necessity of sealed containers, but involves an expensive installation. The dye-fading rate decreases rapidly with decreasing storage temperature.

Storage of Black-and-White Prints

The recommended storage conditions for processed black-and-white paper prints should be obtained from the manufacturer. The optimum limits for relative humidity of the ambient air are 30 to 50%, but daily cycling between these limits should be avoided.

A variation in temperature can drive relative humidity beyond the acceptable range. A temperature between 59 and 77°F is acceptable, but daily variations of more than 7°F should be avoided. Prolonged exposure to temperatures above 86°F should also be avoided. The degradative processes in black-and-white prints can be slowed considerably by low storage temperature. Exposure to airborne particles and oxidizing or reducing atmospheres should also be avoided, as mentioned for films.

Storage of Digital Images

A hard drive should only be used for temporary storage, because if that drive fails, the images could be lost forever. Digital files should be backed up on alternative media (e.g., CD-ROMs) for short-term storage. Because of rapid technological development, storage media systems in 10 to 20 years may not be compatible with current CD-ROMs. In addition, CDs are somewhat fragile and susceptible to damage and data loss if not handled properly. Digital images can be stored as photographic prints; these can last for generations when stored properly as described in the preceding sections.

REFERENCES

Adelstein, P.Z., C.L. Graham, and L.E. West. 1970. Preservation of motion picture color films having permanent value. *Journal of the Society of Motion Picture and Television Engineers* 79(November):1011.

ACGIH. 2010. *Industrial ventilation: A manual of recommended practice*, 27th ed. American Conference of Governmental Industrial Hygienists, Cincinnati, OH.

ANSI. 1998. Imaging media—Processed safety photographic films—Storage. *Standard* IT9.11-98. American National Standards Institute, New York.

ASHRAE. 2010. Ventilation for acceptable indoor air quality. ANSI/ASHRAE *Standard* 62.1-2010.

Carrol, J.F. and J.M. Calhoun. 1955. Effect of nitrogen oxide gases on processed acetate film. *Journal of the Society of Motion Picture and Television Engineers* 64(September):601.

NFPA. 2011. Storage and handling of cellulose nitrate film. ANSI/NFPA *Standard* 40-11. National Fire Protection Association, Quincy, MA.

NFPA. 2009. Installation of air-conditioning and ventilating systems. ANSI/NFPA *Standard* 90A-09. National Fire Protection Association, Quincy, MA.

NFPA. 2007. Protection of records. ANSI/NFPA *Standard* 232-07. National Fire Protection Association, Quincy, MA.

Ryan, J.V., J.W. Cummings, and A.C. Hutton. 1956. Fire effects and fire control in nitro-cellulose photographic-film storage. *Building Materials and Structures Report* 145. U.S. Department of Commerce, Washington, D.C. (April).

BIBLIOGRAPHY

ANSI. 1996. Imaging materials—Ammonia-processed diazo photographic film—Specifications for stability. ANSI/NAPM *Standard* IT9.5-96. American National Standards Institute, New York.

Carver, E.K., R.H. Talbot, and H.A. Loomis. 1943. Film distortions and their effect upon projection quality. *Journal of the Society of Motion Picture and Television Engineers* 41(July):88.

Kodak. 1997. Safe handling of photographic processing chemicals. *Publication* J-98A. Eastman Kodak, Rochester, NY.

Kodak. 2002. Indoor air quality and ventilation in photographic processing facilities. *Publication* J-314. Eastman Kodak, Rochester, NY.

Kodak. 2003. Safe handling, storage, and destruction of nitrate-based motion picture films. *Publication* H-182. Eastman Kodak, Rochester, NY.

Kodak. 2006. Health, safety, and environment. http://www.kodak.com/US/en/corp/HSE/homepage.jhtml?pq-path=2879/7196.

Kodak. 2005. Storage and care of Kodak photographic materials. *Publication* E-30. Eastman Kodak, Rochester, NY.

NFPA. 2007. Static electricity. *Standard* 77-07. National Fire Protection Association, Quincy, MA.

UL. 2001. Tests for fire resistance of record protection equipment, 15th ed. *Standard* 72-01. Underwriters Laboratories, Northbrook, IL.

CHAPTER 23

MUSEUMS, GALLERIES, ARCHIVES, AND LIBRARIES

UNDERSTANDING and appreciating humanity's diverse cultures and history dictates preserving objects including books and documents, works of art, historical artifacts, specimens of national history, examples of popular culture once-common trade goods, technological accomplishments, the products of various technologies, as well as historic buildings and sites. Museums, galleries, libraries and archives may be purpose-built buildings or existing buildings of historic significance; in some instances, the building is as (or more) important as the collection it houses. The importance of cultural heritage ranges from national to regional or even local, but all have symbolic, aesthetic, cultural, social, historical, and monetary values that are frequently impossible to estimate. Thus, their preservation is important, worthwhile, and may even be legally mandated. The loss of any one of these artifacts is a loss to all individuals.

Collections are vulnerable to many threats. Because they must be preserved indefinitely, the steps taken to protect them are sometimes extraordinary. Most threats can be addressed by properly maintained housing and professional support. The level of acceptable risk is a compromise between the theoretically ideal environment and the practical. It is possible to slow deterioration drastically, but doing so may conflict with the ultimate functions of museums, libraries, and archives: not only to preserve, but also to allow public and scholarly access. Additionally, extremely high control over all environmental parameters can help to ensure an object's survival, but at a price no cultural institution can justify or is willing to pay. Managing risk, not avoiding it altogether, is the objective.

This chapter addresses threats to collections that are mitigated by a properly designed HVAC system that provides stability for low-access storage environments and also serves high-traffic visitors' areas.

Theoretically, many systems (including passive building solutions) can successfully provide appropriate environmental control, if properly applied. From project inception, both the design objective and realistically available operation and maintenance resources must be considered.

Communication with the client is especially critical when designing systems for museums, galleries, archives, and libraries because of the uniqueness of the criteria: the inherent risk associated with environmental conditions. The design team must include not only museum administrators but also collections' managers, curators, conservators, and security. Administrators are responsible for fiscal decisions, whereas the collection managers are responsible for care of the collection. Curators build the collection and design exhibitions. Conservators are charged with preservation of the collection. Security staff is critical to safekeeping of the collection.

Many HVAC system design decisions are based on the needs of the collection and the use of the various spaces. To design the most appropriate system, all relevant parties must be part of the process. This chapter can only explain why temperature, humidity, light, and indoor air quality (IAQ) requirements are important; the team must decide the exact specifications. (*To the conservator:* Climate-induced risks should be seen in context and relation to other risks to the preservation of cultural heritage, such as natural and human-caused disasters. In some cases, it may not be the greatest risk to a collection, and available funds may be spent more effectively elsewhere. A climate-control strategy should complement mitigation strategies for other risks and should not in itself create a greater hazard (e.g., when an energy supply fails).

This chapter focuses on relative humidity, temperature, and air pollution design for HVAC systems, and describes various systems that are applicable for these spaces. The goal is to illustrate special needs of collection spaces in museums, galleries, archives, and libraries. See the References and Bibliography for additional resources.

Note that this chapter does not apply to libraries designed for public access, with collections that are not intended for archival preservation. These facilities may include collections designed for general public use or school-aged children, and may have significant quantities of electronic documents in disk format or tape. The types of controls (humidity, thermal, or particulate and molecular phase filtration) required for collections of archival preservation are not practical for these facilities (see Chapter 3 for additional information).

In nonarchival libraries, there may be no HVAC system, or the HVAC systems may be designed specifically for human comfort during occupied hours. In this application, the HVAC systems may shut off during unoccupied or low-occupancy times for energy savings.

This chapter may not apply and should be bypassed if one or more of the following conditions exist with the scope of design for a library (see Chapter 3 for additional information):

- HVAC system is cycled off during unoccupied periods
- HVAC system is turned off seasonally
- HVAC system is designed to cycle on/off with thermal satisfaction alone
- Natural ventilation is the only method of air circulation

General Factors Influencing Damage

In designing HVAC systems for collections, a good working relationship among the mechanical engineer, architect, interior designer, and owner/operator, especially client personnel responsible for preserving the collection, is critical. All expectations and limitations must be defined at the beginning of the design.

Artifacts and collections can be made of one main material (e.g., an archive of antique books), which simplifies target specifications,

---

The preparation of this chapter is assigned to TC 9.8, Large-Building Air-Conditioning Applications.

Table 1 Classification of Rooms for Museums and Libraries

		High Internal Source of Contaminants (Dirty)	Low Internal Source of Contaminants (Clean)
Collection	Non-public access	Conservation laboratories, museum workshops (VOCs, fumes, dusts) "Wet" collections (alcohol or formaldehyde evaporation from poorly sealed jars in natural history collections) Photographic collections ("vinegar syndrome" produces acetic acid vapors) Quarantine areas (potentially pest-infested objects	Most storage areas, vaults, library stacks
	Public access	Displays of conservation work in progress (unusual and temporary)	Galleries, exhibition spaces, reading rooms
Noncollection	Non-public access	Smoking offices (unusual)	Offices (nonsmoking)
	Public access	Cafeterias, rest rooms, spaces where smoking permitted	Public spaces without food preparation or smoking Some public and school library stacks and reading rooms

or combinations of materials with different levels of instability (e.g., a multimedia library that includes books, film, and paintings); in the latter case, target conditions are usually a compromise, or special localized environments may be required for some parts of a collection. For more details, see Michalski (1996a).

The building's architecture and mechanical systems must address eight types of threats to collections; mechanical engineers need to appreciate and respect these concerns even if they do not appear to relate directly to a building's mechanical systems. Respecting all the risks gives the client an increased comfort zone for threats the HVAC system is designed specifically to control. The following threats, in decreasing order of seriousness, affect all types of collections.

Light damage presents perhaps the most extensive threat to museum collections. Most materials undergo some form of undesirable, permanent photochemical or photophysical change from overexposure to light. Damage is relatively easy to control if the problem is addressed at the architectural, design, and operational levels by eliminating ultraviolet light, minimizing infrared radiation, limiting illumination intensity, and restricting total illumination duration.

Relative humidity also presents a risk. For each material, there is a level of environmental moisture content (EMC) consistent with maximum chemical, physical, or biological stability. When the EMC is significantly too low or too high the associated relative humidity becomes a risk factor. Recent literature often calls humidity-related damage "incorrect relative humidity" to emphasize the concept of ranges of acceptable moisture content rather than absolute limits. Unstable relative humidity with large variation in levels can also be damaging to certain types of objects.

Temperature ranges for materials should also be controlled. Some polymers become brittle and are more easily fractured when the temperature is too low. At temperatures that are too high, damaging chemical processes accelerate. Thermal energy not only accelerates aging, but also can magnify the effects of incorrect relative humidity. Therefore, incorrect relative humidity and temperature are often taken together when deciding ideal parameters for important classes of materials such as for paper and photography. Any temperature change also changes the relative humidity. Therefore, careful and close control of relative humidity requires temperature control in the same magnitude of importance. **Air pollution** (or **contaminants**) includes outdoor-generated gaseous and particulate contaminants that infiltrate the building and indoor-generated gaseous pollutants. Even very low levels of pollutants can adversely affect the condition of collections. Particulate filtration to control both coarse and fine particles and gaseous filtration are discussed in the System Selection and Design section.

Pest infestation primarily includes insects consuming collections for food; mold, fungi, and bacteria also qualify as pests, but they can be limited by controlling relative humidity, temperature, indoor air quality, and ventilation.

Shock and **vibration** can cause long-term damage to sensitive objects. Vibration can be transmitted to objects by service vehicles during packing and shipping. Usually, HVAC design only needs to consider this risk if vibration is transmitted through ductwork to works hung on adjacent walls or in particularly active air drafts. Additionally, excess vibration could potentially lead to objects vibrating off of exhibit and/or storage shelves.

Natural emergencies are, fortunately, rare, and most institutions have (or should have) emergency response policies.

Building and mechanical design malfunctions are usually avoidable emergencies and include water pipe failure, especially over collections and storage facilities. The infrequency of these failures leads many to forget that just one failure, however rare, could ruin a significant portion of a collection. Every effort should be made to route water lines and other utilities away from areas that house irreplaceable objects. Building systems also rely on the infrastructure to provide utilities and communications. Where the infrastructure is not reliable or of adequate capacity, provisions should be made for temporary or alternative supply.

Theft and vandalism can be addressed by limiting access to mechanical systems to improve security.

This chapter focuses on relative humidity, temperature, and air pollutant control design for HVAC systems. Many excellent books treat the subject of environmental management in museums and libraries extensively; consult the References and Bibliography for information not contained in this chapter.

ENVIRONMENTAL EFFECTS ON COLLECTIONS

DETERMINING PERFORMANCE TARGETS

Museums, archives, and libraries have two categories of indoor air requirements: general health and safety, comfort, and economy of operation as listed in ASHRAE *Standards* 55 and 62.1; and the collections' requirements, which are not yet completely understood, and often conflict across material collection types. The risk of compromising on relative humidity and temperature specifications must be assessed. The following sections summarize current information on these issues.

In terms of health and safety and the collection's requirements, building spaces can be categorized as shown in Table 1: (1) collection versus noncollection, (2) public versus nonpublic, and (3) "dirty" versus "clean." These subdivisions distinguish between areas that have very different thermal and indoor air quality requirements, outdoor ventilation rates, air supply strategies, etc. These areas often require separate HVAC systems. See Chapters 16, 31, and 32 for more information on dirty rooms. Noncollection rooms (rooms that do not contain collections) are not considered in this chapter because their HVAC requirements are similar to those in other public buildings, as discussed in Chapter 3.

The following sections provide a framework for developing appropriate climate and indoor air quality (IAQ) parameters for different types of museums, libraries, and archives. A single target is a compromise among large numbers of different, often contradictory requirements. However, many collections are uniform enough to allow useful generalizations to be made about their HVAC requirements. In this chapter, the term "effective" includes institutional value judgments as well as the science of deterioration (Michalski 1996b).

Temperature and Humidity

Current Standards. The classic reference for conservation professionals is Thomson's (1994) *The Museum Environment*. Set points of 50% rh and 68°F were listed as an example for temperate climates and unfortunately taken as ideal (in the United States, 68°F was frequently rounded up to 70°F). Misuse is a danger associated with listing standards or settings: numbers in a table can be extracted or used without understanding the associated text. This is why the parameters in this chapter are presented for different classifications of collections or building types. Environmental settings should be determined for collection types and must consider the climate zone where building is located. The history of various objects in the collection, as well as needs of especially vulnerable individual objects, may also influence design. Housing for photographic collections in a coastal, temperate climate may have different control needs than wooden furniture or porcelain in equatorial regions. The design engineer cannot be expected to know the needs of the collection, but can consult the other members of the team, especially the conservators, collection managers, and curators. Ideal set points are a compromise between comfort for museum visitors and staff and the appropriate preservation minimum temperature and relative humidity for the collection.

Davis (2006) reports that decreasing temperature from 70 to 65°F and relative humidity from 50 to 45% significantly increases expected lifetime for certain types of objects. A useful tool to measure aging rates or deterioration rates is the Preservation Calculator created by the Image Permanence Institute (IPI 2006), which can be downloaded for free from the IPI Web site at http://www.dpcalc.org. This software calculates the preservation index (PI), which expresses the preservation quality of a storage environment for organic materials. It is a useful tool to understand the effects of temperature and relative humidity on natural aging of organic collections made from organic materials. Comparing the former "ideal" set points (70°F and 50% rh) with current recommended parameters (65°F and 45% rh), the PI increases from 39 to 64 years, or 64%. IPI also developed software that includes a more sophisticated preservation calculator and takes into account collection types, light, and more.

Biological Damage. High relative humidity levels and dampness accelerate mold growth on most surfaces, corrosion of base metals, and chemical deterioration in most organic materials. Of all HVAC-controllable environmental parameters, high humidity is the most important factor.

The most comprehensive mold data are from the feed and food literature. Fortunately, this provides a conservative outer limit to dangerous conditions. Mold on museum objects occurs first on surfaces contaminated with dust, sugars, starch, oils, etc., but can also occur on objects made of grass, skin, bone, and other feed- or food-like materials. Water activity is identical to and always measured as the equilibrium relative humidity of air adjacent to the material. This provides a better measure than the EMC for mold germination and growth on a wide variety of materials (Beuchat 1987). Figure 1 shows the combined role of temperature and relative humidity. The study of the most vulnerable book materials by Groom and Panisset (1933) concurs with the general trend of culture studies from Ayerst (1968). Ohtsuki (1990) reported microscopic mold occurring on clean metal surfaces at 60% rh. The DNA helix is known to collapse

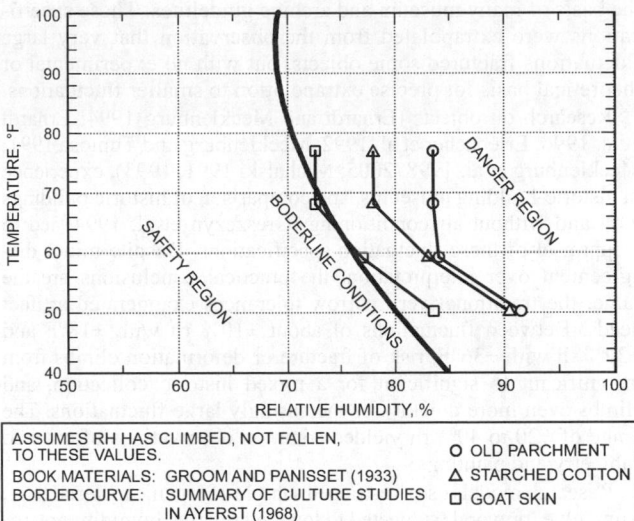

ASSUMES RH HAS CLIMBED, NOT FALLEN, TO THESE VALUES.

BOOK MATERIALS: GROOM AND PANISSET (1933)
BORDER CURVE: SUMMARY OF CULTURE STUDIES IN AYERST (1968)

○ OLD PARCHMENT
△ STARCHED COTTON
□ GOAT SKIN

Fig. 1 Temperature and Humidity for Visible Mold in 100 to 200 days

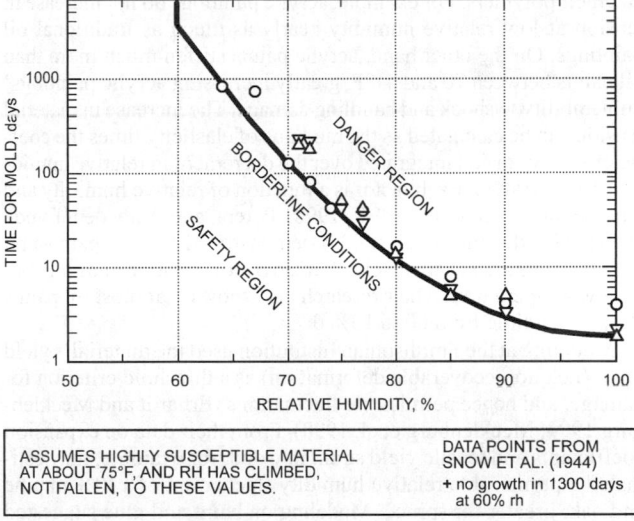

ASSUMES HIGHLY SUSCEPTIBLE MATERIAL AT ABOUT 75°F, AND RH HAS CLIMBED, NOT FALLEN, TO THESE VALUES.

DATA POINTS FROM SNOW ET AL. (1944)
+ no growth in 1300 days at 60% rh

Fig. 2 Time Required for Visible Mold Growth

near 55% rh (Beuchat 1987), so a conservative limit for no mold ever, on anything, at any temperature, is below 60% rh. Chapter 25 suggests a similar lower boundary for mold in food crops.

Snow et al. (1944) looked for visible mold growth on materials inoculated with a mixture of mold species. These are plotted in Figure 2, and follow the same trend reported by Hens (1993) for the European building industry for wall mold.

Figures 1 and 2 show practical dangers: growth in less than a summer season requires over 70% rh, and growth in less than a week requires over 85% rh. Care must be taken to avoid cold surfaces where condensation might occur, such as on windows and ductwork.

Mechanical Damage. Very low or fluctuating relative humidity or temperature can lead to mechanical damage of objects. The fundamental cause is expansion and contraction of materials, combined with some form of internal or external restraint. Very low humidity or temperature also increases the stiffness of organic materials, making them more vulnerable to fracture. Traditionally, these concerns have led to extremely narrow specifications, such as 50 ± 3% rh and 70 ± 2°F (LaFontaine 1979), which still form the basis of many museum and archive guidelines. These specifications were extrapolated from the observation that very large fluctuations fractured some objects, but with no experimental or theoretical basis for precise extrapolation to smaller fluctuations.

Research on objects (Erhardt and Mecklenburg 1994; Erhardt et al. 1996; Erlebacher et al. 1992; Mecklenburg and Tumosa 1991; Mecklenburg et al. 1998, 2005; Michalski 1991, 1993), experience in historic-building museums, and comparison of historic buildings with and without air conditioning (Oreszczyn et al. 1994) led to reappraisal of these fluctuation specifications. Despite minor disagreement over interpretation, the practical conclusions are the same: the traditional very narrow tolerances exaggerated artifact needs. Between fluctuations of about ±10% rh with ±18°F and ±20% rh with ±36°F, risk of fracture or deformation climbs from insignificant to significant for a mixed historic collection, and climbs even more quickly for increasingly large fluctuations. The range of ±20 to 40% rh yielded common observations of cracked cabinetry and paintings.

Research models use a restrained sample of organic material (e.g., paint, glue, or wood) subjected to lowered relative humidity or temperature. The material both shrinks and stiffens. Daly and Michalski (1987) and Hedley (1988) collected consistent data on the increase in tension for traditional painting materials, and Michalski (1991, 1999) found consistent results with other viscoelastic data on paints and their polymers. For example, acrylic paintings do not increase in tension at low relative humidity nearly as much as traditional oil paintings. On the other hand, acrylic paints stiffen much more than oil paints between 70 and 41°F, greatly increasing acrylic paintings' vulnerability to shock and handling damage. The increase in material tension can be calculated as the modulus of elasticity times the coefficient of expansion integrated over the decrement in relative humidity or temperature. Each factor is a function of relative humidity and temperature (Michalski 1991, 1998; Perera and Van den Eynde 1987). Mecklenburg et al. (2005) demonstrated that oil paints and acrylic paints are not adversely affected by low relative humidity but by low temperatures. Their research also shows that most oil paints shrink very little from 60 to 10% rh.

Modeling at the Smithsonian Institution used the material's yield point (i.e., nonrecoverable deformation) as a threshold criterion for damage, and hence permissible fluctuations (Erhardt and Mecklenburg 1994; Mecklenburg et al. 1998). From their data on expansion coefficients and tensile yield strain in wood, for example, they estimated a permissible relative humidity fluctuation of ±10% in pine and oak, greater for spruce. Modeling on paint and glue suggested ±15% rh as a safe range (Mecklenburg et al. 1994). More extensive data on compression yield stress across the grain are available for all useful species of wood in USDA (1999), which can be combined with elasticity data for that species and relative moisture content to obtain yield strain and yield relative humidity fluctuation. These vary widely but center near ±15% rh.

Alternative modeling at the Canadian Conservation Institute used fracture as the criterion for damage, and the general pattern of fatigue fracture in wood and polymers to extrapolate the effect of smaller multiple cycles (Michalski 1991). With a benchmark of high probability of single-cycle fracture at ±40%, known from observation of museum artifacts, fatigue threshold stress (10^7 cycles or more) can be extrapolated by an approximate factor of 0.5 in wood (±20% rh) and 0.25 in brittle polymers (±10% rh) such as old paint. This is consistent with the yield criterion: yield stress in these materials corresponds to stresses that cause very small or negligible crack growth per cycle. Researchers also noted that the coefficient of expansion in wood and other materials is minimized at moderate relative humidity because of sigmoidal adsorption isotherms, so fluctuations at lower and higher relative humidity set points tend to be even riskier.

Both models assume uniformly restrained materials. As a first approximation, many laminar objects (e.g., paintings on stretchers,

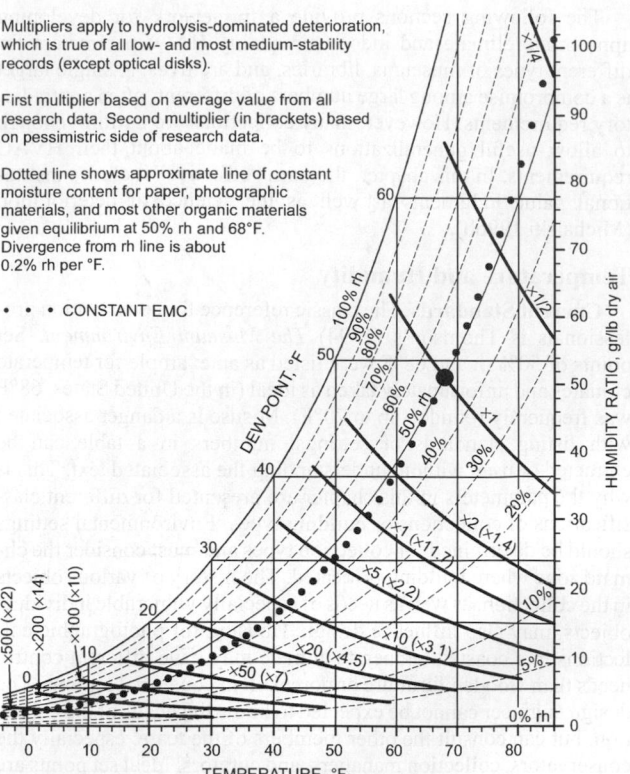

Fig. 3 Lifetime Multipliers Relative to 68°F and 50% rh

photographic records) fall in this class. Artifacts, however, tend to be complex assemblies of materials. Some are less vulnerable because of lack of restraint (e.g., floating wood panels, books or photographs with components that dilate in reasonable harmony); other assemblies contain sites of severe stress concentration that initiate early fracture. Michalski (1996b) classifies vulnerability for wooden objects as very high, high, medium (uniformly restrained components), and low. Each category differs from the next lower one by a factor of two (i.e., half the relative humidity fluctuation causes the same risk of damage).

Chemical Damage. Higher temperatures and moderate amounts of adsorbed moisture lead to rapid decay in chemically unstable artifacts, especially some paper-based and other archival materials. The most important factor for modern records is hydrolysis, which affects papers, photographic negatives, and analog magnetic media. Sebor (1995) developed a graphical format for relating these two parameters to lifetime for book papers. An improved graphical representation is shown in Figure 3. Fortunately for HVAC design purposes, dependence on relative humidity and temperature in all records is very similar. Although the precise quantification and meaning of record lifetime is debatable, all authorities agree that the most rapidly decaying records (e.g., all magnetic tapes, acidic negatives) can become unstable within a few decades at normal room conditions, and much faster in hot, humid conditions. Figure 3 shows the relative increase in record lifetime under cold, dry conditions; the range in numbers on each line reflects the spread in available data. Extension of the plots below 5% rh is uncertain; rates of chemical decay may or may not approach zero, depending on slow, nonmoisture-controlled mechanisms such as oxidation.

Lifetime improvement predictions for photographic data were devised by the Image Permanence Institute (Nishimura 1993, Reilly 1993) and implemented in a lifetime prediction wheel. These relative lifetime estimates are less optimistic about improvement with low relative humidity, but the general trend is the same as the plots in Figure 3.

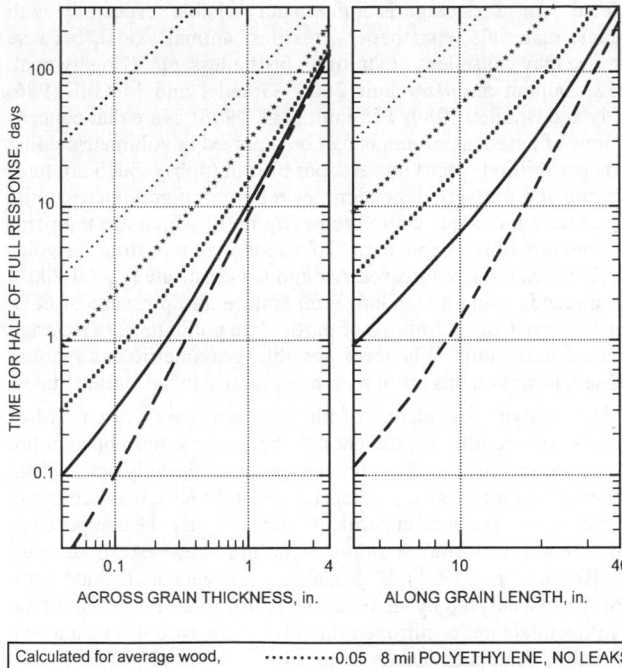

Calculated for average wood, room temperature, and 25 to 75% rh. Numbers preceding coating name on plots are permeance (perm). Use the graph that gives the fastest response.	· · · · · · · · · 0.05	8 mil POLYETHYLENE, NO LEAKS
	· · · · · · · · · 0.5	MANY COATS OF OIL PAINT
	∗ ∗ ∗ ∗ ∗ ∗ 1.8	HEAVY VARNISH OR PAINT
	· · · · · · · · 5	MEDIUM VARNISH OR PAINT
	——— 18	LIGHT VARNISH
	— — — 180	CALM AIR LAYER

Fig. 4   Calculated Humidity Response Times of Wooden Artifacts

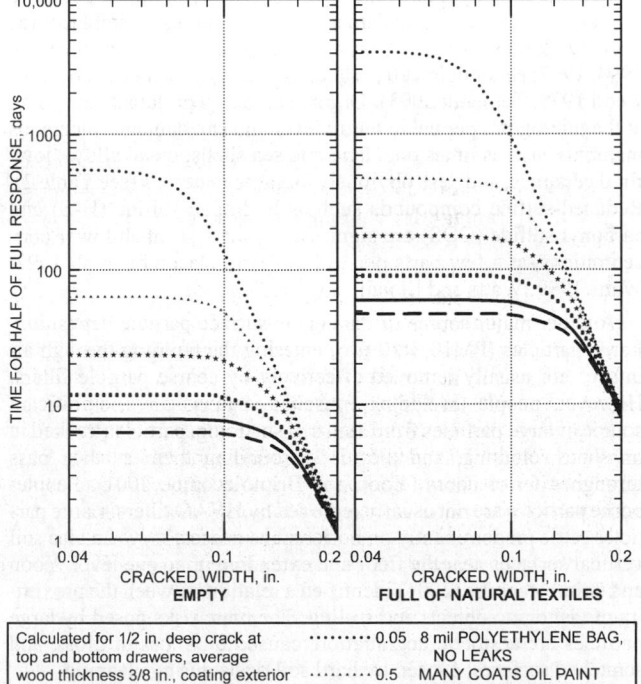

Calculated for 1/2 in. deep crack at top and bottom drawers, average wood thickness 3/8 in., coating exterior only, chest 60 in. high, 40 in. wide, 20 in. deep. Stack due to 40% rh or 2°F difference. Numbers preceding coating name on plots are permeance (perm).	· · · · · · · · 0.05	8 mil POLYETHYLENE BAG, NO LEAKS
	· · · · · · · · 0.5	MANY COATS OIL PAINT
	∗ ∗ ∗ ∗ ∗ ∗ 1.8	HEAVY VARNISH OR PAINT
	· · · · · · · · 5	MEDIUM VARNISH OR PAINT
	——— 18	LIGHT VARNISH
	— — — 180	CALM AIR LAYER

Fig. 5   Interaction of Air Leakage, Wood Coating, and Textile Buffering on Response of Wooden Chest of Drawers

Critical Relative Humidity

At some critical relative humidity, some minerals hydrate, dehydrate, or deliquesce.

When part of a salt-containing porous stone, a corroded metal, or a natural history specimen, these minerals cause disintegration of the object. Distinct critical relative humidity values are known for dozens of minerals in natural history collections (Waller 1992). Pyrites, which are contaminants of most fossils, disintegrate if held above 60% rh (Howie 1992). Bronze, one of the most important archaeological metals, has a complex chemistry of corrosion, with several critical relative humidity values (Scott 1990). This variety means there is no universal safe relative humidity; particular conditions should be achieved for specific artifacts with local cabinets or small relative-humidity-controlled packages (Waller 1992). The only generalization is that any relative humidity above 75% is dangerous.

Rapid corrosion above 75% rh occurs for two reasons: increased surface adsorption of water, and contamination by salts. Water adsorption on clean metal surfaces climbs rapidly from 3 molecules or less below 75% rh to bulk liquid layers above 75% rh (Graedel 1994). This phenomenon is aggravated by most surface contaminants, as shown in studies of the role of dust on clean steel corrosion. The most common contaminant of museum metals, sodium chloride, dissolves and liquefies (deliquesces) above 76% rh.

Response Times of Artifacts

Brief relative humidity fluctuations may not affect artifacts; very few museum objects respond significantly to fluctuations under an hour in duration. Hence, a 15 min cycle in HVAC output does not affect most artifacts, unless it is so large as to cause sudden damp conditions. Many objects take days to respond. Figure 4 shows calculated humidity response times of wooden artifacts. Figure 5 shows the interaction of air leakage, wood coatings, and textile buffering on the response of a chest of drawers; risk increases if the piece is displayed empty and open, rather than closed and full, because response time falls from months to days. However, many museum objects must be exhibited without these mitigating components.

Very long fluctuations, such as seasonal changes, are slow enough to take advantage of stress relaxation in artifact components. Data on the effective modulus of elasticity of many oil and acrylic paints as a function of time, temperature, and relative humidity (Michalski 1991) and direct stress relaxation data for both paint (Michalski 1995) and wood imply that stress caused by a given strain applied over 1 day falls to 50% or less if that stress is applied over 4 months at moderate room temperatures. Thus, a 4 month seasonal ramp of ±20% rh should cause less stress in most artifacts than a 1 week fluctuation of ±10 rh.

AIRBORNE POLLUTANTS

Research shows that gaseous outdoor pollutants can easily penetrate all types of buildings, including modern HVAC-equipped construction, when no chemical filtration exists to remove them (Cass et al. 1989; Davies et al. 1984; Druzik et al. 1990). Particulate contaminant intrusion from outside sources has also been well documented in art museums (Brimblecombe 1990; Nazaroff et al. 1993; Yoon and Brimblecombe 2001).

Collections can themselves be sources of airborne pollutants. Collections with leather, fur, and wood elements can release reduced sulfides, aldehydes, carboxylic acids, or fatty acids. These gases may instigate or accelerate deterioration of other objects. Acetic acid emitted from degrading cellulose acetate films is another good example. In general, the major risks to collections from indoor-generated pollutants are from acetic and formic acids [one precursor of which is formaldehyde (Raychaudhuri and Brimblecombe 2000), which must also be controlled]. These gases are emitted from wood and wood-based materials as well as adhesives, finishing products, etc. Reduced sulfides (e.g., hydrogen sulfide and carbonyl sulfide) can be released from wools and silks (Brimblecombe et al. 1992; Watts 1999).

During construction or renovation at the building, room, or display case levels, high amounts of suspended particles are generated. Depending on the type of HVAC system, airborne pollutant concentrations may not be reduced to acceptable levels for a few weeks up to several months after work is finished (Eremin and Tate 1999; Grzywacz 2006). Adhesives, coatings, and sealants initially release high levels of pollutants. In poorly ventilated enclosures or rooms, emission rates may be retarded because of equilibrium vapor pressures. Fortunately, levels of pollutants released by wet/aqueous products in well-ventilated rooms usually decrease rapidly, though carboxylic acid emitted by alkyd or oil-based coatings decreases at a much slower rate (Chang et al. 1998; Fortmann et al. 1998). Even after emissions level off, the level of acids released by these coatings can remain unsatisfactory for several years, even in ventilated rooms. Food preparation and service vehicles are also sources of contaminants, and require special consideration. The source of outdoor air for the HVAC system, especially intakes that deliver to collections, is important.

Sources of Airborne Pollutants

Of the hundreds of air pollutants, only a few have been identified as dangerous for collections in museums and archives. Still, there are several that are perceived to be a concern, although to date no documented damage has been attributed to them. These are included when the potential chemistry of interaction with the material substantiates the risk.

Table 2 lists major airborne pollutants, their typical sources, and collection types for which objects may be at risk. Hydrocarbons and other volatile organic compounds (VOCs), such as alcohols and ketones, may be important for other considerations such as human health and comfort, but generally are not threats to artifacts. Outdoor sources of airborne pollutants are primarily industrial and human activities. Inorganic outdoor pollutants that cause material damage are important, but as the use of alcohol-based fuels increases, organic species, especially formaldehyde and organic acids (Anderson et al. 1996; Schifter et al. 2000), may become relatively more important. Geological and biological activities release hydrogen sulfide and ammonia; the agricultural industry is also a major ammonia source (Allegrini et al. 1984; Walker et al. 2000). Inside buildings, construction materials emit organic carbonyls such as carboxylic acids and aldehydes (Gibson et al. 1997; Grzywacz and Tennent 1994, 1997; Meininghaus et al. 2000).

Materials Damage Caused by Airborne Pollutants

Pollutant-monitoring surveys of collections and laboratory studies have provided important data on the effect of airborne pollutants on materials (Table 2). Damage to materials is the sum of many parameters in the immediate environment. High levels of a single pollutant may cause serious damage even when all other conditions are ideal. However, when many pollutants act together and temperature, relative humidity, and light intensity are elevated, deterioration processes are almost always accelerated. Conservators often refer to the "inherent vice" of an object; this is when the original material, prior history of treatment, storage, or excavation enhances or aggravates chemical reactions. Tétreault's (1994, 2003) summary of materials damaged by pollutants can help determine which materials require special care. Brokerhof (1998), Craft et al. (1996), Grzywacz (1999, 2006), Hatchfield (2002), Lavédrine (2003), Ryhl-Svendsen (1999), and Tétreault (1994, 2003) summarized airborne pollution control strategies for rooms and enclosures.

Damage to artifacts is cumulative and irreversible. All efforts to minimize exposing collections to pollutants are beneficial. Both long-term exposure to low levels of pollutants and short-term exposure to high pollutant concentrations can damage susceptible objects. Not all damage is immediately visible, especially with organic materials (e.g., papers, textiles, animal skins), because damage may cause loss of strength, brittleness, etc. (Bogaty et al. 1952; Dupont and Tétreault 2000; Graedel and McGill 1986; Oddy and Bradley 1989; Tétreault et al. 1998). Note that concentrations of gaseous pollutants can be reported in volumetric units: parts per million (ppm) or parts per billion (ppb), which are temperature and pressure dependent, or reported in gravimetric units [e.g., micrograms per cubic metre ($\mu g/m^3$)], which are temperature and pressure independent. To standardize reporting for volumetric units, the Compressed Air and Gas Institute (CAGI 2002) recommends using a standard temperature and pressure of 68°F and 14.5 psi. Concentrations of particulate pollutants are reported in gravimetric units. Whenever possible, gravimetric units should be used; however, this is not common practice in the United States.

Autocatalytic degradation of film also can cause a hazard. Most cellulose nitrate film stock eventually becomes so unstable that fire and explosion are high risks. Frozen storage is the only option from a human health and safety perspective. Objects such as sculpture, decorative components on garments, etc., may also be composed of cellulose nitrate. Cellulose nitrate sculpture likewise degrades rapidly (Derrick et al. 1991). Degradation products may include nitric acid, when catalyzed by small amounts of sulfate esters remaining from manufacture, or nitrogen dioxide, in the case of uncatalyzed thermal or photochemical decomposition (Selwitz 1988).

Cellulose acetate film stock ("safety film") is also chemically unstable, but does not pose the fire or explosive hazard inherent in cellulose nitrate films. Over time, it deteriorates autocatalytically and liberates free acetic acid ("vinegar syndrome"). The amount of acetic acid liberated can be many orders of magnitude greater than any other single source in museums and archives. As such, it is a risk not only to the film itself, but also to all other acid-sensitive materials. Consequently, these films must be properly contained.

Organic carbonyl pollutants such as acetic acid, formic acid, and formaldehyde are the most damaging to collections (Brimblecombe et al. 1992; Gibson 1999; Grzywacz 2003; Grzywacz and Tennent 1994, 1997; Hatchfield 2002; Hatchfield and Carpenter 1986; Hopwood 1979; Tétreault 2003). Organic carbonyl pollutants are a risk at the microgram per cubic metre level and can damage calcareous materials such as limestone, land and sea shells, metal alloys, low-fired ceramics, and, less obviously, organic materials (see Table 2). Reduced-sulfide compounds such as hydrogen sulfide (H_2S) and carbonyl sulfide (COS) are significant threats at much lower concentrations, at a few parts per trillion (Brimblecombe et al. 1992; Watts 1999; Watts and Libedinsky 2000).

Another major source of damage is surface particle deposition. Large particles (PM10, $\geq 10\ \mu m$) entering the building through air intakes are usually removed effectively by coarse particle filters. However, people (including museum visitors) are a significant source of large particles from skin cell shedding, particles tracked in on shoes, clothing, and the air surrounding them as they pass through exterior doors (Yoon and Brimblecombe 2001). People-borne particles are not usually removed by HVAC filters. Large particles settle predominantly on horizontal surfaces; they can also soil vertical surfaces near the floor and extending up to eye level. Yoon and Brimblecombe (2001) identified a relation between the proximity of visitors to objects and soiling. The main risks posed by large particles are aesthetic degradation, caused by a loss of gloss, and scratched surfaces, caused by hard soil dusts during cleaning.

Fine particles (PM2.5) are not removed well by many coarse particle filters. Soot is the most important particle type in this size range. It does not settle out of the air easily and is influenced by energetic air molecules, natural and forced convection, and turbulent air flows. Small particles collect on all surfaces (vertical, horizontal, downward-facing, or upward-facing). Black soot particles can be a major soiling risk on large, unprotected surfaces such

Table 2 Major Gaseous Pollutants of Concern to Museums, Galleries, Archives, and Libraries: Sources and At-Risk Materials

Gaseous Pollutants[a]	Major Sources and Some Important Minor Sources	At-Risk Materials
	Important Inorganic Pollutants	
Sulfur dioxide (SO_2) Sulfur dioxide can react with water vapor in the air and form both **sulfurous acid (H_2SO_3)** and **sulfuric acid (H_2SO_4)**. Outdoors in high humidity, acidic gases coalesce into superfine, suspended droplets, or aerosols, known as acid rain. A similar reaction occurs indoors, where acidic gases are deposited on surfaces and can cause damage.	*OUTDOOR SOURCES* **Natural sources** • Marine biological activity and active volcanoes • Atmospheric reactions of hydrogen sulfide: H_2S reacts rapidly with oxygen and forms both SO_2 and H_2SO_4 **Industrial processes** • The primary source of SO_2 is the combustion of sulfur-containing fossil fuels, including coal, gasoline, and diesel fuel • Industrial processes associated with pulp and paper production, cement industry, and petroleum refineries, especially when less-expensive, higher-sulfur-content fuels are used • Fireworks (can be localized, short-term risk) *INDOOR SOURCES* • Sulfur-containing fuels, such as kerosene and coal, used to cook and heat (relevant for historic houses as well as museums in areas that predominantly use these fuels) • Firewood used to cook and heat • Vulcanized rubber • Propane- or gasoline-powered machines, equipment, and generators	• Can be absorbed onto cellulosic materials, such as **paper**, including historic wallpaper, and **textiles**, where it catalytically hydrolyzes to H_2SO_4, H_2SO_3 and H_2SO_4 can also be absorbed directly onto these materials (the acid depolymerizes the cellulosic structure, and though damage is invisible, affected materials are embrittled and weakened) • SO_2 and related acids react with animal skins, such as **leather and parchment** [this breaks down the molecular structure and weakens the material; as a result, surface becomes powdery and is easily abraded (red rot)] • With its pollutant progeny, reacts with acid-sensitive **pigments**, resulting in typically nonreversible color change (e.g., lead white is converted to black lead sulfate, resulting in darkened color); lead-tin yellow, chrome yellow, verdigris, chrome orange, emerald green, and chrome red, among others, also are darkened by sulfates • Fades **dyestuffs**, affecting color of watercolors, textiles, costumes, etc. • Most **metals**, including copper, silver, bronze alloys, and aluminum, are susceptible to corrosion by any acidic species, causing irreversible damage • Silver salts in **photographs** are attacked, darkening the image • Attacks stone such as **limestone, marble, dolomite**, and other carbonate minerals or calcareous materials such as **shells, clays,** and **tiles**; at-risk objects include sculpture, natural history collections, and low-fire ceramics • When adsorbed on soot or carbon particles on dirty objects, can be oxidized to sulfuric acid or sulfurous acid, a risk for all **acid-sensitive materials**
Nitrogen oxides (NO_x) NO_x is a collective term for **nitrogen monoxide (NO)** and **nitrogen dioxide (NO₂)**. Colorless NO is a primary vehicular pollutant. It reacts with other chemical species in the air to produce many other reactive nitrogen compounds, especially NO_2, a pungent red gas partly responsible for the color of photochemical smog. Oxidation (by ozone, UV irradiation, etc.) of nitrogen oxides generates **nitrous acid, HNO_2, and nitric acid, HNO_3.** This occurs indoors and outdoors. These acidic analogues are highly reactive.	*OUTDOOR SOURCES* **Natural sources** • Natural sources include lightning and biological processes such as soil microbes, vegetation, biomass fires, etc. • Agricultural fertilizers **Industrial processes** • Combustion of fossil fuels for industry and vehicles (high concentrations associated with urban traffic and thermal power plants) • Fireworks (temporary but relevant source) *INDOOR SOURCES* **Gas-phase reactions** • Formation of acids: reaction of NO and NO_2 on interior surfaces, including glass, and with carbon and carbonaceous aerosols generates HNO_2 and HNO_3, respectively **Indoor activities** • Stoves, heaters, fireplaces, and other sources of combustion • Tobacco smoke • Dry-process photocopiers • Degradation of cellulose nitrate objects **Construction materials/activities** • Generators and heavy equipment that use fuel-combustion engines	• Nitrous and nitric acids can damage the same **acid-sensitive materials** attacked by sulfuric or sulfurous acids • NO_2 enhances deterioration effects of SO_2 on **leather, metals, stone**, etc. • Corrosion of **copper-rich silver.** • Nitrogen pollutants fade **dyed fibers** in textiles, costumes, drapery, tapestries, etc. Reactions with **dyestuffs** alter color of textiles • Nitrogen pollutants fade certain **inks** as well as **organic pigments** in illuminated manuscripts, etc. • Nitrogen pollutants degrade **fibers** made from rayon, silk, wool, and nylon 6, causing yellowing and embrittlement • Nitrogen pollutants corrode **zinc**; synergistic effect with H_2S • Affects tarnishing of **copper** and **silver** by hydrogen sulfide
Reduced sulfur compounds **Hydrogen sulfide (H_2S)** **Carbonyl sulfide (COS)** **Carbon disulfide (CS_2)** **Hydrogen sulfide (H_2S)** smells like rotten eggs. It is responsible for the odor	*OUTDOOR SOURCES* **Natural sources** • Volcanoes, geothermal steam, geysers, sulfur wells, hot springs • Oceans, seas, marine areas • Marshes, soils, and wetlands • Biological activity, decomposition of organic material, release from vegetation, biomass burning, forest fires, tropical forests, etc.	• H_2S destroys immature **plant tissue** (relevant for natural history or botanical collections) • **Lead pigments: carbonates** (hydrocerussite or lead white) and **oxides** (e.g., red lead pigment) are susceptible to darkening by H_2S • **Metals: copper** metal exposed to H_2S develops a black copper sulfide layer, eventually replaced by a green patina of basic copper sulfate; extremely damaging to **silver** and its salts; tarnishes silver objects; reacts with silver salts in photographs; reacts with

Table 2 Major Gaseous Pollutants of Concern to Museums, Galleries, Archives, and Libraries: Sources and At-Risk Materials (Continued)

Gaseous Pollutants[a]	Major Sources and Some Important Minor Sources	At-Risk Materials
common to hot springs, as well as that noticed at wastewater treatment plants. Hydrogen sulfide is extremely toxic to people, but is detectable at very low concentrations (1 ppb). Collections are susceptible to reduced sulfur compounds at even lower levels (ppt).	**Industrial processes** • Fuel and coal combustion • Production of viscose rayon, vulcanization of rubber, etc. • Petroleum production, paper processing, and wood pulping **Atmospheric chemistry** • H_2S and COS can be formed by oxidation of carbonyl disulfide *INDOOR SOURCES* **Collections and objects** • Mineral specimens that contain pyrite (iron sulfide, FeS_2); sulfate-reducing bacteria in waterlogged objects **Display case materials** • Offgassing from sulfur-containing proteins in materials used in exhibition and display case design, especially silks, wools, and felts • Adhesives, especially those made from animal hide, such as rabbit skin glue **Construction materials/activities** • Arc welding (can be significant during renovation) **Flooring materials** • Wool carpets	silver inlays, silver gilt, etc.; reacts with **bronze**, **lead**; has synergistic effect with NO_2 to corrode **zinc** (see NO_2) • Low-fire **ceramics** • **Stone**, especially interior building stone • **Leather** (see sulfur dioxide)

Strong Oxidizing Pollutants

Gaseous Pollutants[a]	Major Sources and Some Important Minor Sources	At-Risk Materials
Ozone (O_3) is a major constituent of smog that directly affects people, plants, and property.	**Ozone-Specific Sources** *OUTDOOR SOURCES* • Smog: tropospheric (ground-level) ozone is a major secondary pollutant of vehicular and industrial emissions	• Artists' colorants: fading of **dyes** and **pigments** • Oxidation of organic compounds with double bonds (e.g., embrittles and cracks **rubber**) • **Electrical wire coatings** (of concern to industrial collections)
Peroxyacetyl Nitrate (PAN) (CH_3-COO-O-NO_2) is a principal secondary pollutant in photochemical smog	*INDOOR SOURCES* • Office/building equipment, including dry-process photocopiers and other office equipment • Electrical arcing, including electrostatic air cleaners or filter systems and electronic insect killers	• **Plants** (of concern to botanical and natural history collections) • Embrittles **fabrics, textiles,** and **cellulosic materials** • Enhances tarnishing of **silver** by reduced sulfur compounds • Causes discoloration of **photographic prints** • Attacks **paint binders**
Peroxides (—O:O—), the simplest of which is hydrogen peroxide (HO:OH), are extremely reactive because of the oxygen-oxygen bond. Strong oxidizing pollutants present a great risk to collections. Oxidants break down the structure of organic materials by attacking carbon-carbon double bonds. Oxidants can also react with other gaseous pollutants, such as NO_x, to create acidic analogs; radicals such as the hydroxyl radical, •OH; and other destructive reaction by-products [e.g., oxidation of aldehydes into acetic acid (see organic carbonyl pollutants entry)].	**PAN-Specific Sources** *OUTDOOR SOURCES* • Secondary pollutants from urban traffic emissions produced by gas-phase reactions between hydrocarbons and nitrogen oxide compounds, and reactions between organic carbonyl pollutants and hydroxy radicals (•OH) • Pollutants from ethanol-fueled vehicles • Forest fires **Peroxide-Specific Sources** *OUTDOOR SOURCES* • Secondary pollutant generated by nitrogen oxide chemical reactions with hydrocarbons, VOCs, and organic carbonyl pollutants. • By-product of atmospheric reactions of pollutants from gasohol fuels *INDOOR SOURCES* • Emission from deterioration of organic materials (e.g., rubber floor tiles) • Oil-based paints • Microorganism activities	• Affects **leather, parchment,** and **animal skins** (of concern to natural history collections) • Adsorbed onto **building products** (plasterboards, painted walls, carpet, linoleum, pinewood, and melamine-covered particleboard), where it can react with and damage surfaces or be re-released into environment

Table 2   Major Gaseous Pollutants of Concern to Museums, Galleries, Archives, and Libraries: Sources and At-Risk Materials (*Continued*)

Gaseous Pollutants[a]	Major Sources and Some Important Minor Sources	At-Risk Materials
	Organic Carbonyl Pollutants: Aldehydes and Organic Acids	
Organic carbonyl pollutants	*OUTDOOR SOURCES*	• **Metal corrosion**: non-noble metals such as **leaded**-bronzes, **copper alloys**; base metals such as **lead, copper, silver**; corrosion of **cabinetry hardware** coated with cadmium, lead, magnesium, and zinc
Aldehydes	**Atmospheric chemistry**	
• **Formaldehyde (HCHO)**	• Secondary pollutants resulting from atmospheric reactions with industrial and vehicle pollutants	• Acid hydrolysis of **cellulose** reduces degree of polymerization, which is discernible as embrittlement
• **Acetaldehyde (CH₃CHO)**	• Precipitation: gases concentrate in fog, rain, and snow	
Organic acids	**Natural sources**	• Attacks **calcareous materials**: land shells and seashells (i.e., Byne's disease), corals, limestone, calcium-rich fossils
• **Formic acid (HCOOH)**	• Biogenic emissions from vegetation	• Low-fire **ceramics**
• **Acetic acid (CH₃COOH)**	• Biomass burning, forest fires, rainforest slash and burn, etc.	• Reacts with **enamel** and **glass**, especially previously damaged and weakened glass, such as weeping glass
	• Biodeterioration of organic materials	• Stained glass: corrodes **lead joins** between glass panes
Most materials damage by organic carbonyl pollutants is attributed to acetic acid or to formaldehyde. It is suspected that damage credited to formaldehyde is really due to its oxidized form, formic acid.	*INDOOR SOURCES*	
	Gas-phase reactions	
	• Indoor reactions of outdoor pollutants that infiltrate buildings	
	• Evaporation from hot water (e.g., dishwashers, showers)	
Correspondingly, oxidation of acetaldehyde generates acetic acid. However, the risk from direct emissions of acetic acid is greater.	• Construction and building materials, especially materials used in the construction of display cases and storage cabinets; laminated materials	
	• Coatings, sealants, paints and adhesives, some polyvinyl acetate adhesives, oil-based paints	
(See also specific entries for organic acids, aldehydes, and formaldehyde.)	• New houses (in older houses off-gassing has decreased significantly or ceased): materials used in cabinet making, doors and plywood subfloors as well as floor assemblies	
	• Materials manufactured with urea formaldehyde, including foam insulation; other formaldehyde-based resins such as phenol formaldehyde wood products	
	• Wood and wood products	
	• Wood-based panel products, especially with urea formaldehyde and melamine formaldehyde binding resins (e.g., pressed wood, composite wood panels, chipboard, particleboard, medium-density fiberboard, parquet)	
	• Other wood-based building materials	
	Flooring materials	
	• Cork products	
	• Flooring, linoleum, carpets	
	Other materials	
	• Furniture and furniture coatings, varnishes	
	• Consumer and household products, such as hair spray, perfumes and cosmetics, air fresheners, cleaning agents, etc.	
	• Paper and paper products	
	• Finished fabrics	
	(See also sources specific to formic acid and acetic acid, aldehydes, and formaldehyde.)	
Organic acids	*OUTDOOR SOURCES*	(See at-risk materials under organic carbonyl pollutants.)
Acetic acid (CH₃COOH)	• Textile industry effluents and emissions	
Formic acid (HCOOH)		
(See also organic carbonyl pollutants.)	*INDOOR SOURCES*	
	Formic-Acid-Specific Sources	
	• Formaldehyde-free wood composite boards	
	• Oxidation product from reaction of formaldehyde with light	
	Acetic-Acid-Specific Sources	
	• Degradation of cellulose acetate objects	
	• Silicone sealants	

Table 2 Major Gaseous Pollutants of Concern to Museums, Galleries, Archives, and Libraries: Sources and At-Risk Materials (Continued)

Gaseous Pollutants[a]	Major Sources and Some Important Minor Sources	At-Risk Materials
Aldehydes **Formaldehyde (HCHO)** **Acetaldehyde (CH$_3$CHO)** (See also organic carbonyl pollutants and formaldehyde.)	*OUTDOOR SOURCES* **Industrial processes** • Primary pollutant from vehicles using alcohol fuels (e.g., methanol, gasohol) and ethanol/gasoline fuel blends • Automobile manufacturing, especially painting *INDOOR SOURCES* • Combustion by-products, cooking, heating, and tobacco smoke • Artists' linseed oil paints, other drying oils **Construction and building materials** • Terra cotta bricks • Ceramic manufacturing, kiln exposures • Vinyl, laminates, wallpapers, acrylic-melamine coatings • Alkyd paints • Latex and low-VOC latex paints • Secondary pollutants produced by reaction of ozone and some carpet materials	(See at-risk materials under organic carbonyl pollutants.)
Formaldehyde Formaldehyde is easily oxidized to formic acid, which is most likely the aggressive chemical.	*INDOOR SOURCES* • Natural history wet specimen collections • Consumer products, including decorative laminates, fiberglass products • Dry-process photocopiers • Textiles such as new clothes and fabrics, dry-cleaned clothes, permanent press fabrics, drapery, clothing, carpets, wall hangings, furniture coverings, unfinished fabrics, dyeing process residues, chemical finishes, etc. • Fungicide in emulsion paints and glues (e.g., wheat pastes) • PVC-backed carpeting • Floor finishes	• **Silver** tarnish or surface discoloration • Reacts with unexposed **photographic films** and **photographs**, especially black & white • Cross-links **proteins** (e.g., collagen), resulting in loss of strength in animal hides, leather objects, parchments, etc. (also attacks objects with gelatin, animal glue, or casein binders) • Reacts with **textiles, fibers** • Attacks buffered **papers** and reacts with metallic-salt inclusion in paper • Discoloration of **dyes**, fading of **organic colorants** • Chemically changes **inorganic pigments**: insoluble basic copper carbonate (main component of azurite) is converted to soluble copper acetate (See also at-risk materials in organic carbonyl pollutants.)
Other Potentially Damaging Pollutants		
Ammonia (NH$_3$) **Ammonium ion (+NH$_4$)** Most damage to museum collections is from the ammonium ion, produced when water and ammonia react.	*OUTDOOR SOURCES* **Natural sources** • Agriculture, especially fertilization, and animal wastes (animals are the largest global source of ammonia) • Biodeterioration (e.g., landfill gases), underground bacterial activity. **Industrial processes** • Fertilizer production *INDOOR SOURCES* • Household cleaning products • Museum visitors • Emulsion adhesives and paints • Alkaline silicone sealants • Concrete	• Blemishes **ebonite, natural resins** • Reacts with materials made from **cellulose nitrate**, forming ammonium salts that corrode **copper, nickel, silver,** and **zinc**

Table 2 Major Gaseous Pollutants of Concern to Museums, Galleries, Archives, and Libraries: Sources and At-Risk Materials (*Continued*)

Gaseous Pollutants[a]	Major Sources and Some Important Minor Sources	At-Risk Materials
Amines (R-NH₂) These alkali pollutants are derivatives of ammonia.	*INDOOR SOURCES* • Amine-based corrosion inhibitors [diethylaminoethanol (DEAE), cyclohexylamine (CHA), and octadecylamine (ODA)] used in humidification systems and ventilation ducts • Atmospheric reactions of nitrogen species and alkali pollutants released from new concrete • Epoxy adhesives	• Causes blemishes on **paintings**, usually when pollutant is dispersed through the ventilation system • Corrosion of **bronze, copper, and silver** • Darkens **linseed** oil and forms copper amine complexes with **copper pigments** (e.g., malachite) • Blemishes furniture **varnishes**
Fatty acids	*OUTDOOR SOURCES* • Vehicle exhaust *INDOOR SOURCES* • Animal skins, furs, taxidermy specimens, and insect collections (relevant for natural history collections and parchment) • Museum visitors • Combustion: burning candles, cooking • Adhesives • Linoleum	• Yellows **paper and photographic** documents • Corrodes **bronze, cadmium, and lead** • Blemishes **paintings**
Hydrochloric acid (HCl)	*OUTDOOR SOURCES* • Coal combustion • Oceans, sea mist spray	• HCl is an acidic gas; hence, it attacks **acid-sensitive materials**, especially under high-humidity conditions often present in coastal regions • Increases corrosion rate of metals • Affects tarnishing of **copper** and **silver** by hydrogen sulfide
Water vapor (H₂O) Water vapor (i.e., relative humidity) is a critical parameter for museum collections. Besides direct affects of humidity changes on collections, water vapor increases corrosion and decay rates, and is involved in most chemical reactions.	*OUTDOOR SOURCES* • Atmosphere (high-humidity days) • Bodies of water *INDOOR SOURCES* • Fountains • Humidifiers • People • Wet cleaning activities	• Increases hydrolysis reactions on **organic objects**, which usually leads to damage (e.g., hydrolysis of cellulose weakens paper objects) • Increases effect of nitrogen oxides on photographs • Increases hydrogen sulfide corrosion of **copper** and **silver** • Controlling factor in **bronze disease** • Increases deterioration of materials (e.g., **metal** corrosion, efflorescence of calcareous objects, and photo-oxidation of **artists' colorants**) • Increases fading of **dyestuffs** used in textiles and watercolors
Particles (fine and coarse)[b]	• *General:* atomizing humidifier; burning candles; cooking; laser printers; renovation; spray cans; shedding from clothing, carpets, packing crates, etc. (from abrasion, vibration, or wear); industrial activities; outdoor building construction; soil • *Biological and organic compounds:* microorganisms, degradation of materials and objects, visitors and animal danders, construction activities • *Soot (organic carbon):* burning candles, incense, fires, coal combustion, vehicle exhaust • *Ammonium salts (ammonium sulfate):* reaction of ammonia with SO₂ or NO₂ inside or outside or on solid surfaces	• *General:* abrasion of surfaces (critical for magnetic media); disfiguration of objects [especially critical for surfaces with interstices that entrap dust (e.g., with pores, cracks, or micro irregularities)]; may initiate or increase corrosion processes; may initiate catalysis forming reactive gases • *Soot:* disfiguration of porous surfaces (painting, frescoes, statues, books, textiles, etc.), increases rate of metals corrosion

aSource for gaseous pollutants section: Grzywacz (2006).
bSource for particle section: Tétreault (2003).

as tapestries or on small, porous, soft natural history collections and book margins.

Ligocki et al. (1990, 1993), Nazaroff and Cass (1991), Nazaroff et al. (1990, 1992, 1993), and Salmon et al. (2005) examined the origin, fate, and concentration of particles in several museums in southern California. Rates of particle deposition were established for a typical historical house and several museums with traditional air-conditioning filtration. From the deposition rate, "time for perceptible soiling" was established, to characterize the building by when soiling could be expected to be barely visible to a normal observer. Bellan et al. (2000) showed that the amount of coverage by small black particles, similar to soot, on a white surface was just detectable at a surface coverage of 2.6%, if the observer could compare a sharp edge of soiled surface against a nonsoiled surface. This value is 12 times larger than published in earlier literature (Hancock et al. 1976). Druzik and Cass (2000) determined that, if the museums studied by Nazaroff et al. (1990) are typical of most medium-sized art museums with HVAC systems, the time to onset of perceptible soiling of vertical surfaces is 24 to 86 years.

Some forms of chemical deterioration have been traced to particle deposition. Toishi and Kenjo (1975) extensively covered the risk from new concrete. However, little is known about direct chemical degradation of historic materials by these particles. It has been speculated that particle deposition may accelerate cellulose degradation and metal corrosion, based on known reaction pathways of suspended particles and other materials-science investigations, such as the corrosion of copper alloys by soluble nitrates (Hermance et al. 1971).

"Realistic" Versus "Technically Feasible" Target Specifications. Responses to environmental factors vary, and often do not appear until the materials have significantly aged. Individual objects can also react differently than their nearest relatives in a class of objects. Providing the best available technology for a reasonably large collection to guard against all contingencies is not realistic. For most objects, it is feasible to create a safe and protective environment at a reasonable cost to the client such that they can be maintained indefinitely. Expanding protection beyond certain limits for smaller artifacts and smaller collections becomes less practical at the whole-building mechanical control level. For artifacts with specific problems or risks, it is usually better to provide microenvironments.

"Best available technology" often implies protection on a diminishing cost/benefit scale. Museums, libraries, and archives are frequently nonprofit organizations on tight budgets. Insisting on extraordinary humidity control or filtering out every air pollutant that is theoretically damaging can endanger long-term viability of the effort. Insistence on best available technology may also have the unintentional effect of highlighting the lack of useful engineering solutions rather than the more important role of HVAC systems as efficient, protective, and necessary.

Risk Management Plans for Collections. Preservation of collections requires a tradeoff among many factors. There is not one golden rule; instead, risk management approaches are used (Ashley-Smith 1999; Brimblecombe 2000; Dahlin et al. 1997; Hens 1993; Michalski 1996b; Stock and Venso 1993; Tétreault 2003; Thickett et al. 1998; Waller 1999). Waller and Michalski (2005) report on a preventive conservation software tool to determine appropriate set points and ranges for temperature and relative humidity.

DESIGN PARAMETERS

PERFORMANCE TARGET SPECIFICATIONS

Temperature and Relative Humidity

No two collections are identical. Ideally, temperature and humidity targets and tolerances for each facility are developed collaboratively by a conservator with expert knowledge of the damage factors and realities of the specific collection, and a design engineer with extensive experience in designing systems that meet the needs of many different types of collections. That approach ensures specifications most likely to provide maximum life for the collection. Experienced experts are best equipped to identify areas of special risk and devise solutions, and to properly manage economic and other tradeoffs, although this level of expertise is not always easily available.

Table 3 summarizes the probable effects of various specification options, based on the best current knowledge (including all available data, research results, and judgment of conservators). Michalski (1991, 1993, 1995, 1996a, 1996b, 1999) reduced permissible fluctuations to five classes: AA, A, B, C, and D. Gradients are conservatively considered to add to short-term fluctuations because artifacts can be moved from one part of a space to another, adding a space-gradient fluctuation to the dynamic fluctuations of the HVAC.

Class AA control has the highest potential for energy consumption and affords no protection to historic buildings in cold climates. Older buildings are at most risk to damage from condensation on windows, walls, and roofs. Therefore, HVAC application in historic buildings must consider the risk to the structure and building envelope. Mecklenburg et al. (2004) discuss the concept of preserving both the collections and the buildings. Class A control has the benefit of some reduction of energy consumption and, with seasonal lowering of the relative humidity set point in the winter, affords some protection to historic buildings in cold climates. Older buildings at these lower temperature and relative humidity specifications are at less risk to damage from condensation on windows, walls, and roofs.

Class A is the optimum for most museums and galleries. Two possibilities with equivalent risks are given: a larger gradient and short-term fluctuations, or a larger seasonal swing. Stress relaxation is used to equate ±10% rh seasonal swing to a short ±5% rh. A major institution with the mandate and resources to prevent even tiny risks might move toward the narrower fluctuations of Class AA. However, design for very-long-term reliability must take precedence over narrow fluctuations.

Classes B and C are useful and feasible for many medium and small institutions, and are the best that can be done in most historic buildings. Class D recognizes that control of dampness is the only climatic issue.

Architectural Considerations. Building envelopes play an important role in controlling moisture migration into the space. Architectural design should include heavy insulation, and consider possible vapor barriers. Control of openings, window materials, and floor slab insulation are critical in the design. Rainwater runoff design should be considered in the HVAC relationship to collection storage and display areas.

Many collections are housed in existing buildings and it is not uncommon that the building is also an important part of the cultural heritage. In the case of an existing building, it is crucial to understand how the existing indoor climate is regulated by the building. If environmental improvements are necessary, it should first be investigated how these may be achieved by passive building measures. If any system is necessary, the installation should be sensitive to the historic significance of the building and all of its fabric, as well as compliance with existing historic building protection codes.

Building Envelope and Climate-Control Issues

As noted previously, climate control is critical to protect collections of cultural property. One often overlooked aspect to good climate control is the integrity of the building envelope, primarily in terms of airtightness. Persily (1999) discusses the airtightness of building envelopes, noting that commercial buildings are often quite leaky, leading to significant rates of uncontrolled infiltration entering the building. There is no reason to assume that museums, galleries, archives, and libraries are significantly tighter and have lower infiltration rates than other commercial buildings. In addition, these buildings often require mechanical engineers to provide for

Table 3 Temperature and Relative Humidity Specifications for Collections

Type	Set Point or Annual Average	Maximum Fluctuations and Gradients in Controlled Spaces			Collection Risks and Benefits
		Class of Control	Short Fluctuations plus Space Gradients	Seasonal Adjustments in System Set Point	
General Museums, Art Galleries, Libraries, and Archives	50% rh (or historic annual average for permanent collections)	**AA** Precision control, no seasonal changes, with system failure fallback	±5% rh, ±4°F	Relative humidity no change Up 9°F; down 9°F	No risk of mechanical damage to most artifacts and paintings. Some metals and minerals may degrade if 50% rh exceeds a critical relative humidity. Chemically unstable objects unusable within decades.
All reading and retrieval rooms, rooms for storing chemically stable collections, especially if mechanically medium to high vulnerability.	Temperature set between 59 and 77°F *Note*: Rooms intended for loan exhibitions must handle set point specified in loan agreement, typically 50% rh, 70°F, but sometimes 55% or 60% rh.	**A** Precision control, some gradients or seasonal changes, not both, with system failure fallback	±5% rh, ±4°F	Up 10% rh, down 10% rh Up 9°F; down 18°F	Small risk of mechanical damage to high-vulnerability artifacts; no mechanical risk to most artifacts, paintings, photographs, and books. Chemically unstable objects unusable within decades.
			±10% rh, ±4°F	RH no change Up 9°F; down 18°F	
		B Precision control, some gradients plus winter temperature setback	±10% rh, ±9°F	Up 10%, down 10% rh Up 18°F, but not above 86°F	Moderate risk of mechanical damage to high-vulnerability artifacts; tiny risk to most paintings, most photographs, some artifacts, some books; no risk to many artifacts and most books. Chemically unstable objects unusable within decades, less if routinely at 86°F, but cold winter periods double life.
		C Prevent all high-risk extremes	Within 25 to 75% rh year-round Temperature rarely over 86°F, usually below 77°F		High risk of mechanical damage to high-vulnerability artifacts; moderate risk to most paintings, most photographs, some artifacts, some books; tiny risk to many artifacts and most books. Chemically unstable objects unusable within decades, less if routinely at 86°F, but cold winter periods double life.
		D Prevent dampness	Reliably below 75% rh		High risk of sudden or cumulative mechanical damage to most artifacts and paintings because of low-humidity fracture; but avoids high-humidity delamination and deformations, especially in veneers, paintings, paper, and photographs. Mold growth and rapid corrosion avoided. Chemically unstable objects unusable within decades, less if routinely at 86°F, but cold winter periods double life.
Archives, Libraries Storing chemically unstable collections	Cold Store: −4°F, 40% rh	±10% rh, ±4°F			Chemically unstable objects usable for millennia. Relative humidity fluctuations under one month do not affect most properly packaged records at these temperatures (time out of storage becomes lifetime determinant).
	Cool Store: 50°F 30 to 50% rh	(Even if achieved only during winter setback, this is a net advantage to such collections, as long as damp is not incurred)			Chemically unstable objects usable for a century or more. Such books and papers tend to have low mechanical vulnerability to fluctuations.
Special Metal Collections	Dry room: 0 to 30% rh	Relative humidity not to exceed some critical value, typically 30% rh			

Note: Short fluctuations means any fluctuation less than the seasonal adjustment. However, as noted in the section on Response Times of Artifacts, some fluctuations are too short to affect some artifacts or enclosed artifacts.

"improved climate control" in buildings never designed for such purposes. Conrad (1995) grouped such buildings (and building parts) by their possibilities and limitations into seven categories. In an abridged version of his scheme, Table 4 lists the possible classes of fluctuation control possible each class of building. Local climate determines which possibility is most likely. For detailed guidance on air leakage and thermal and moisture performance of building envelopes, refer to Chapters 16 and 25 to 27 of the 2009 *ASHRAE Handbook—Fundamentals*.

Airborne Pollutant Targets

Assigning target concentrations for airborne pollutants is a complex task. An object's susceptibility depends on a variety of factors (including historical storage and conservation treatments, and current stability) and is linked to other environmental factors, such as temperature, relative humidity, and light levels (Gibson 1999). The conservator, collection manager, or curator considers all these factors and more.

Principal airborne pollutants for museums, galleries, archives, and libraries are listed in Table 5. Nitrogen oxides, fine particles, ozone, and sulfur dioxide are generated mainly outside. Hydrogen sulfide can be generated inside the building, depending on the finishing materials used (especially carpets and wall coverings), or infiltrate the building envelope, depending on geographic location and factors such as biomass decay, volcanoes, sulfur springs, etc. Wood, construction materials, indoor activities, and artifacts mainly

Table 4 Classification of Climate Control Potential in Buildings

Category of Control	Building Class	Typical Building Construction	Typical Type of Building	Typical Building Use	System Used	Practical Limit of Climate Control	Class of Control Possible
Uncontrolled	I	Open structure	Privy, stocks, bridge, sawmill, well	No occupancy, open to viewers all year.	No system.	None	D (if benign climate)
	II	Sheathed post and beam	Cabins, barns, sheds, silos, icehouse	No occupancy. Special event access.	Exhaust fans, open windows, supply fans, attic venting. No heat.	Ventilation	C (if benign climate) D (unless damp climate)
Partial control	III	Uninsulated masonry, framed and sided walls, single-glazed windows	Boat, train, lighthouse, rough frame house, forge	Summer tour use. Closed to public in winter. No occupancy.	Low-level heat, summer exhaust ventilation, humidistatic heating for winter control.	Heating, ventilating	C (if benign climate) D (unless hot, damp climate)
	IV	Heavy masonry or composite walls with plaster. Tight construction; storm windows	Finished house, church, meeting house, store, inn, some office buildings	Staff in isolated rooms, gift shop. Walk-through visitors only. Limited occupancy. No winter use.	Ducted low-level heat. Summer cooling, on/off control, DX cooling, some humidification. Reheat capability.	Basic HVAC	B (if benign climate) C (if mild winter) D
Climate controlled	V	Insulated structures, double glazing, vapor retardant, double doors	Purpose-built museums, research libraries, galleries, exhibits, storage rooms	Education groups. Good open public facility. Unlimited occupancy.	Ducted heat, cooling, reheat, and humidification with control dead band.	Climate control, often with seasonal drift	AA (if mild winters) A B
	VI	Metal wall construction, interior rooms with sealed walls and controlled occupancy	Vaults, storage rooms, cases	No occupancy. Access by appointment.	Special heating, cooling, and humidity control with precision constant stability control.	Special constant environments	AA A Cool Cold Dry

Source: Adapted from Conrad (1995).

generate organic acids and aldehydes (see Table 2), which can be found at high levels in enclosures. Total VOCs are included; they can be used as a metric for the overall air quality in the area. Table 5 presents suggested concentration limits, action limits, air quality recommendations, natural background and urban concentrations, acute human toxicity levels, U.S. Environmental Protection Agency (EPA) Clean Air Act Limits, and World Health Organization time-weighted average (TWA) limits. Limits in the first four columns combine Tétreault's (2003) lowest observable adverse effect dose (LOAED) and no observable adverse effect level (NOAEL), which combine critical review of detailed in situ observations with labora-tory studies and provide substantial quantitative information on adverse effects of pollutants on materials. Extensive sets of LOAED and NOAEL are available in Tétreault (2003).

It is preferable to exclude pollutant sources by properly selecting construction, finishing materials, and cleaning products used inside collection storage and display spaces (e.g., silicon adhesive without acetic acid, not using vinegar-based cleaning products or amine compounds as corrosion inhibitors in humidification systems; see Table 2).

Controlling relative humidity below 45% for long-term preser-vation and maintaining temperature at about 65°F can minimize deterioration of collections from airborne pollutants. Cleanliness of the collection is also important, because fine particles, ultrafine par-ticulates, salts, fatty acids, or metallic dirt may initiate or accelerate some deterioration caused by gaseous pollutants.

For long-term preservation, low levels of airborne pollutants can be achieved in many ways, including building design, filtration, maintenance, and operations. If the HVAC system cannot provide the specified protection from contaminants, it may be necessary to place collections in appropriate protective enclosures or microclimates.

A building is a complex, dynamic environment that affects the indoor concentration and fate of airborne pollutants. Detailed discussion is beyond the scope of this chapter, but major factors influencing airborne pollutants include the following, and these should be considered during HVAC design:

- Outdoor pollutant types and load
- Visitor traffic and indoor activities
- Location of outdoor air intakes
- Location and type of air delivery vents in collection spaces
- Ratio of outside to recirculated air when the building is open to the public and staff, and when it is closed to the public and/or unoccupied
- Particle and gaseous filter efficiency and filter maintenance
- Location and fit of filters in HVAC system
- Janitorial, building, and grounds maintenance practices
- Nature of construction products and collections

The surface temperature of walls and artifacts can influence particle deposition. For these reasons, the HVAC engineer has con-siderable flexibility in controlling filtration, although some architec-tural, maintenance, and geographic factors are beyond control. Nevertheless, engineers should be aware of these issues and bring them into design discussion when appropriate.

Many institutions with a small or limited operating budget may prefer to use enclosures. Long-term preservation of collections in either rooms or enclosures must be discussed among the clients, design engineers, and conservation professionals. In addition to deci-sion factors already stated, ethical aspects of the exhibition, security, overall long-term preservation goals, object/pollutant interactions,

Table 5 Current Recommended Target Levels for Key Gaseous Pollutants[a] (in ppb, unless otherwise indicated)

Major Outdoor Pollutants in Museums	Suggested Pollutant Limits for Collections[b,c]		Action Limits[d]		Air Quality Recommendations		Reference Concentrations				
	Sensitive Materials[e]	General Collections	High	Extremely High	Archival Document Storage	Libraries, Archives, and Museums	Natural Background Levels	Urban Areas	Health: Acute Toxicity Level for 1 h Exposure[f]	EPA Clean Air Act Limits[g]	World Health Organization[h] TWA Limits
Nitrogen dioxide, NO₂	<0.05 to 2.6	2 to 10	26 to 104	>260	Canada: 2.6 USA: 2.6	2.6	0.05 to 4.9	1.6 to 68 USA: 22 to 52 Canada: 16 to 22 Europe: 2 to 34	244 OSHA: 5 ppm[i]	50 (1 y)	104 (1 h) 21 (annual) 62 (8 h)
Ozone, O₃	<0.05	0.5 to 5	25 to 60	75 to 250	Canada: 1.0 USA: 13	2.0	1 to 100	5 to 200 USA: 100 to 120 Canada: 17 to 21 Europe: 65 to 145	90 OSHA: 100	120 (1 h) 80	60 (8 h)
Sulfur dioxide, SO₂	<0.04 to 0.4	0.4 to 2	8 to 15	15 to 57	Canada: 0.4 USA: 0.4	1.0	0.04 to 11 Rural USA: 6 to 10 Europe: 1 to 14	2 to 380 USA: 4 to 6 Canada: 4 to 6 Europe: 2 to 94	251 OSHA: 5 ppm	30 (1 y) 140 (24 h)	190 (10 min) 10 (24 h) 19 (annual)

Major Indoor-Generated Pollutants in Museums	Suggested Pollutant Limits		Action Limits		Reference Concentrations				
	Sensitive Materials	General Collections	High	Extremely High	Natural Background Levels	Urban Areas	Health: Acute Toxicity Level for 1 h Exposure	EPA Clean Air Act Limits	World Health Organization TWA Limits
Hydrogen sulfide, H₂S	<0.010	<0.100	0.4 to 1.4	2.0 to 20	0.005 to 10	0.1 to 5	30 ppm OSHA: 10 ppm		107
Organic Carbonyl Pollutants									
Acetic acid,[j] CH₃COOH	<5	224 40 to 280	200 to 480	600 to 1000	0.1 to 4	0.1 to 16	OSHA: 10 ppm		
Formic acid,[k] HCOOH	<5	42 to 78	104 to 260	260 to 780	0.05 to 4	0.05 to 17	OSHA: 5 ppm		
Formaldehyde, HCHO	<0.1 to 5	10 to 20	16 to 120	160 to 480	0.4 to 1.6	1.6 to 24 New home: 50 to 60	75 OSHA: 750		80 (30 min)
Total VOCs (as hexane)[l]		<100	700	1700		New or renovated building 4500 to 9000			
Fine particles[m] (PM2.5)	<0.1 μg/m³	1 to 10 μg/m³	10 to 50	50 to 150	1 to 30	1 to 100			

Source: Grzywacz (2006)

Notes for Table 5

Note: This extrapolation of minimum risk for most collections over an extended period of exposition assumes temperature between 68 and 86°F, collection cleanliness, and less than 60% rh; hypersensitive objects such as lead, vulcanized natural rubber, silver, and most sensitive colorants are excluded and require special control measures.

See References:

1. World Health Organization 2000
2. Seinfeld 1986
3. Graedel 1984
4. Tétreault 2003
5. Grosjean 1988
6. Graedel, Kammlott, and Franey 1981
7. Sano 1999
8. Sano 2000
9. Bradley and Thickett 1999
10. Kawamura, Steinberg, and Kaplan 1996
11. Granby and Christensen 1997
12. Grosjean and Williams 1992
13. Hodgson et al. 2000
14. Rothweiler, Waeger, and Schlatter 1992
15. Lavédrine 2003
16. National Air Filtration Association (NAFA) 2004

[a]Current standards, based on best available sources; not meant to be absolute and final concentration recommendations. Concentration limits for materials and objects continue to be reviewed.

[b]Maximum levels allowed to ensure minimum risk to sensitive objects; assumes temperature between 59 and 77°F, clean collection, and less than 60% rh (ideally less than 50%) (Tétreault 2003).

[c]Temperature and relative humidity should always be minimized as well as pollutant concentration to reduce risk.

[d]Mitigation measures should be taken to protect objects in the collection.

[e]Sensitive materials are those that are at risk from the particular gaseous pollutant.

[f]Acute reference exposure levels (RELs) established by U.S. Office of Environmental Health Hazard Assessment (OEHHA 2003).

[g]U.S. Environmental Protection Agency Office of Air and Radiation Clean Air Act limits (EPA 2003).

[h]World Health Organization's maximum exposure recommendations (World Health Organization 2000).

[i]U.S. Department of Labor, Occupational Safety and Health Agency maximum permissible exposure limit (PEL) for 8 h work day.

[j]Acetic acid levels can be as high as 10,000 ppb inside enclosures made with inappropriate materials.

[k]Very little is known about effects of formic acid at various concentrations.

[l]Total VOCs are reported referenced to calibrated gas, such as hexane or toluene.

[m]Adapted from Tétreault (2003).

and IAQ performances of the HVAC system and enclosures should taken in consideration.

SYSTEM SELECTION AND DESIGN

A typical project for a new, purpose-built building consists of many different types of spaces, such as galleries, exhibition spaces, reading rooms, laboratories, conference rooms, stacks, storage rooms, restaurants or cafeterias, auditoriums, rare book vaults, offices, lounges, and study rooms. Areas housing objects or collections are considered to need a special environment, defined by the criteria in the previous section, for preventive conservation or preservation.

The HVAC system for a preservation environment must maintain relative humidity, maintain temperature and air movement, and filter air, evenly throughout the space, with minimum risk of damage to collections, at a cost the institution can support. Special HVAC system indications usually fall into two general categories: museums, including galleries and other spaces where environmentally sensitive objects are kept or displayed, and libraries, archives, and other spaces where primarily paper-based collections are stored and used, with some additional concern for film and other media.

One primary requirement is high performance with low or limited annual operating budgets. This shifts focus to capital investments in systems and features to minimize operating costs and problems.

Design Issues

Functional Organization. Maintaining an effective preservation environment should depend firstly on the basic architectural design (e.g., windows, vapor retardants), use of appropriate building materials, and the building's operation (e.g., hours of operation, availability of tempering sources). Ideally, the HVAC engineer should be involved early in project planning to ensure that space layout does not present unnecessary problems. In the best case, collections are isolated in microclimates (cases or storage areas) from visitors, users, staff, and all other functions. Where this is not possible, processes and activities that threaten collections should be physically and mechanically separated from the collections. Separate systems for collection and noncollection areas allow isolation of environments and can reduce project costs for noncollection areas.

A typical issue, particularly in fine art museums, is whether to treat executive offices as collection spaces. This should be considered carefully, not only for the added capital and operating cost, but also for the risks to the collection if offices are not so treated and are nonetheless used for collections display.

Frequently used entrances, such as the lobby and loading dock, are some of the most environmentally disruptive elements. The engineer should ensure that loads from these spaces are managed and isolated from primary collection areas.

Substantial holdings of film ideally should be housed separately from other collections; for details, see the section on Materials Damage Caused by Airborne Pollutants.

Humidity-Tolerant Building Envelope. Winter humidification should be a high priority in a heated building in temperate or cold climates, but the humidity tolerance of the building must be considered. Often, the building envelope has problem condensation (on single-glazed windows, on window frames, or in the exterior wall or roof) in winter at interior humidities as low as 25%. This condensation can cause cosmetic or substantive damage to the building. In such cases there are three alternatives: (1) keeping winter humidity stable and below the point where problem condensation occurs, (2) retrofitting the building envelope to tolerate higher humidity, or (3) reducing the space temperature.

Depending on the humidity level that can be maintained without problems, a collection may require that the envelope be modified to

support higher winter humidity. These changes benefit collections, but present considerable design challenges and expenses. If a collection needing higher humidity does not dominate a building, but only a manageable minority of the spaces, then separate humidified storage containers, cases, cabinets, or rooms are an alternative. These must be carefully sealed and isolated to preserve their internal environment and to protect the building from the moisture.

Reliability. Most collections can tolerate several hours of lost conditions without major damage, but some are at risk even with brief losses of control. The engineer, along with the conservator, should evaluate equipment failure scenarios against collection sensitivities and likely maintenance efforts to see what reasonable precautions can be taken to minimize downtime and damage to the collection. Spare equipment may need to be kept at the project site to allow timely repairs.

Loads. Certain load characteristics of collection buildings should be considered in system design. Usually, the HVAC system should operate 24 h/day. Galleries, exhibition spaces, and reading rooms tend to have high occupancy only at certain times. Some gallery occupancies are as high as 10 ft² per person, but stack or storage areas may have 1000 ft² per person or less. The system should be designed to handle this load as well as the more common part-loads. Many engineers design to 20 ft² per person because part of the room is never occupied. In other facilities, where the space is extensively used for receptions, openings, and other high-traffic activities, even higher density assumptions may be justified. Continual and close dialogue between the designer and client is therefore important.

Lighting loads vary widely from space to space and at different times of the day. The most common driver of sizing cooling in a museum is display lighting. The engineer should ensure that estimated lighting loads are realistic. Lighting typically varies from 2 to 8 W/ft² for display areas; figures as high as 15 W/ft² are sometimes requested by lighting designers, but are rarely needed. With growing awareness of damage caused by light, display areas for light-sensitive objects should have low illumination levels, and associated low lighting power densities.

Exhibit Cases. Exhibit cases should be designed to protect the collection from environmental extremes. Sealed or vented cases are typically used. Sealed cases rely on isolation from the ambient environment in the exhibition room and usually require passive or special conditioning systems independent of regular room air. Because sealed cases are subject to build-up of contaminants, they should be made of inert or low-emission materials, or those that emit gases benign to objects in the case. In a properly conditioned space, exhibit cases that are within, built into, or back up to this space can be vented.

Exhibit cases should not be conditioned by blowing supply air into the cases, or by drawing return air through the cases. Supply air temperature and humidity vary and can cause extremes if blown into a case. Even if temperature and humidity conditions are sensed inside the case as part of the control system, the typical high ratio of supply air to the case volume makes such treatments problematic. Acute temperature or humidity conditions in supply air are undiluted and immediately affect sensitive objects. Return air is more stable but tends to have higher levels of particulate contamination, leading to particle accumulation in the display case. Active conditioning of exhibit cases has been used successfully, but only by using purpose-designed equipment.

These systems are not configured like typical HVAC systems and have additional features, such as desiccant beds to stabilize supply air humidity. They are used primarily in cold climates, where conditioning large, historic galleries is problematic and case conditioning is often the only solution. Condition stability with these systems is less than ideal, and much less than sealed exhibit cases with passive conditioning agents. The primary value is where many cases need to be conditioned and rigorous sealing and reconditioning passive agents are impractical. Lights should always be housed in a separate ventilated compartment from the one housing the artifacts.

Cold Storage Vaults. Cold storage vaults extend the life of materials particularly sensitive to thermal deterioration, such as acetate films and color photographic materials. These vaults usually require special equipment; experienced turnkey or design-build vendors provide the most successful systems (Lavédrine 2003; Wilhelm 1993).

Primary Elements and Features

The following primary HVAC elements and features provide a good preservation environment for a museum or library, as in Figure 6 (Lull 1990).

Constant Air Volume. Air should be constantly circulated at sufficient volume, regardless of tempering needs, to ensure good circulation throughout the collection space. In general, perimeter radiation and other sensible-only heating or cooling elements should be avoided, because they can create local humidity extremes near collections.

Cooling System. Several systems are available, including direct expansion (DX) cooling, glycol, and central chilled water. Screw compressors are recommended to generate chilled water at 36°F for use in chilled-water coils, which generally have copper fins and tubes. Conventional, DX, and chilled-water systems have a limitation in producing 36°F chilled water; therefore, desiccant systems may be considered instead. Ice storage systems with glycol also could be used.

Heating System. Several systems are available to generate heat, including steam and oil with converter, modular boilers, and scotch marine boilers. Hot water is circulated through heating and reheat coils for temperature and humidity control in the space.

Humidification. Humidification should be provided by steam or deionized water introduced in the air system. Evaluate the moisture source for risks of pollutants. Often, heating steam is treated with compounds (especially amines) that can pose a risk to the collection (Volent and Baer 1985). Systems should be selected and designed to prevent standing pools of water, and should follow good humidification design as described in Chapters 1 and 21 of the 2008 *ASHRAE Handbook—HVAC Systems and Equipment*. Humidification methods include electronic steam humidifiers, clean steam humidifiers, evaporative pan humidifiers, spray-coil wetted-element systems, and ultrasonic humidification. All materials in humidification equipment should be selected to minimize microbial growth and degradation of system components.

Unlike most other applications, HVAC design for this building type is often more concerned with humidity control than temperature control. The averaging effect of a common mixed return air and common humidifier on a central system is preferred, but sometimes zone humidifiers have been necessary to recover from unsatisfactory conditions, even when the same humidity level is desired in each zone on the same system. Attempt to identify and correct the cause of the condition before taking drastic measures.

Maintaining widely different conditions in zones using the same air handler can be difficult to achieve, and wastes energy. If possible, different zone conditions should have the same absolute moisture content, using zone reheat to modify space humidity for different relative humidity requirements.

Dehumidification. The most common problem in museums and libraries is inadequate or ineffective dehumidification. Modest dehumidification can be achieved with most cooling systems, limited by the apparatus dew point at the cooling coil, and requiring adequate reheat. Most problems derive from compromises in the cooling medium temperature or lack of reheat. Some chilled-water systems may not reliably deliver water that is cold enough, or may have chilled-water temperature reset or cooling coils that are too shallow for dehumidification to occur. Zone reheat is essential to maintaining necessary conditions in the space.

Sebor (1995) suggests the following typical approaches to more aggressive dehumidification:

- **Low-temperature chilled water**, usually based on a glycol solution, offers familiar operation and stable control but requires glycol management.

- **DX refrigeration** tends to be better for small systems and has lower capital costs, but generally is less reliable, requires more energy, and may require a defrost cycle.

- **Desiccant dehumidifiers** can be quite effective if properly designed, installed, and maintained. Economy of operation is very sensitive to the cost of the regeneration heat source. Liquid desiccant systems eliminate (1) the need to cool the air below the dew point, and (2) reheat, both of which are very important cost factors for sustainability.

Desiccant systems (Figure 7) may be a good solution in many cases that require humidity between 30 and 35% rh year-round. Desiccant regeneration is required. Silica gel and rotary wheel dehumidifiers are commonly used.

Dehumidification systems should be additions to a typical cooling system; they cannot maintain comfort conditions by themselves. For libraries or archives requiring cool, dry conditions, a desiccant system may be required. Chapter 24 of the 2008 *ASHRAE Handbook—HVAC Systems and Equipment* has further information on desiccants.

Outdoor Air. Because the goal is to maintain a close-tolerance environment, excessive amounts of outdoor air for economizer cooling are problematic. Outdoor air is rarely at design temperature and humidity, and can introduce particles and gaseous pollution. Economizer cooling is often the primary cause of humidity fluctuations in museums and libraries and requires very large humidification systems; air-side economizers should not be used unless (1) bin analysis or other study shows outdoor air moisture content to be favorable for an economical number of hours, and (2) favorable outdoor air can be reliably selected by the control system. Outdoor air should be the minimum amount required to provide fresh air for occupants and to pressurize collection spaces. Relief air (see Figure 6) usually should not be used, unless required to support high outdoor-air requirements for high occupancy. Peak occupancies are rarely the norm, so outdoor air might be controlled by monitoring levels of carbon dioxide or other gases indicative of occupancy.

The added cost of preconditioning outdoor air, usually with a precooling coil, often benefits humidity control and economical operation. Cooling can reduce the amount of reheat needed for dehumidification, and may reduce static pressure on the primary fan by keeping the cooling coil dry.

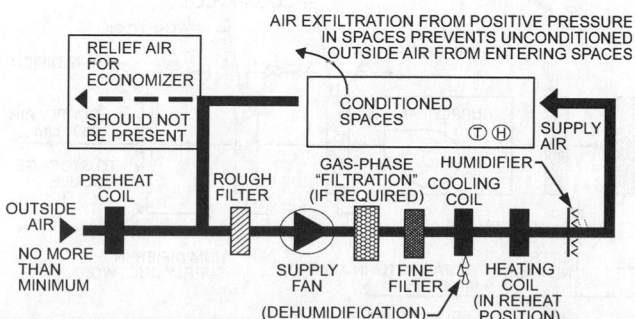

Fig. 6   Primary Elements of Preservation Environment HVAC System

Filtration

Particulate Filtration. Particulate filtration is essential for removal of contaminants that could foul the HVAC system, as well as particles that might degrade or deface artifacts being preserved. For this reason, particulate filtration is addressed here in two steps: prefiltration and fine-particulate filtration.

Prefiltration is required for to prevent fouling in cooling coils and build-up of dust in the fan, ductwork, or other HVAC components. It is also required to protect and prolong the functional service life of gas-phase filters and fine-particulate filters. These fouling-sized particles are generally considered to be the MERV E-3 (ASHRAE *Standard* 52.2) range particles or the 3 to 10 μm size range. Achieving at least 50% removal of the E-3 particle size range requires MERV 7 filtration.

Higher efficiencies may be possible with media configurations that operate at lower pressure loss. MERV 11 or higher prefilters can operate at similar or lower pressure drop and provide more protection for HVAC components and gas-phase filters.

Fine-particulate filtration protects artifacts and collections in the facility. This particle size is commonly referred to as the "accumulation" size and falls in the MERV E-1 range of particles (0.3 to 1 μm). Removal efficiencies of a minimum of 85% of the E-1 range are sufficient for preservation of most collections. MERV 15 filters are minimum 85% in the E-1 range, and minimum 90% in both the E-2 (1 to 3 μm) and E-3 (3 to 10 μm) ranges.

Some collections may require higher efficiencies than MERV 15 for long-term preservation. Options include microenclosures with minimal airflow and separate filtration, or HEPA (99.97% at 0.3 μm) filtration for the entire common area. Whenever HEPA filtration is used as the final filtration, serious consideration should be given to upgrading the prefiltration to protect the life of the HEPA filters.

Framing systems should be able to seal the air filters without bypass air leakage in the housing. Systems design with positive locking mechanisms for filters is a benefit.

High-voltage electrostatic air cleaners should be used with caution because of the potential to generate ozone, which can damage artifacts (see Table 2).

Gas-Phase Filtration. Outdoor air infiltration of gaseous pollutants, materials offgassing in new construction, and similar offgassing of furnishings and cleaning agents may threaten the stability of some collections. Sensitive collections of valuable holdings (e.g., low-fire ceramics, some metals and alloys, film, rare books) should use active control of gas-phase pollutants.

The primary compounds of concern include acetic acid, formaldehyde, hydrogen sulfide, nitrogen dioxide, ozone, and sulfur dioxide, all of which are removable with molecular filtration. The specific sorbent must be chosen for the various gaseous contaminants indigenous to the facility, because removal and retention properties are not all the same. Some gases are easily removed with activated carbon, whereas others may require treated carbon or potassium permanganate beds.

Careful thought should be given to the service life of molecular filtration media. Much service life testing has focused on the potential weight removal capacity of the sorbent when immersed in a challenge chemical; although this information is important, removal from an airstream may not be achievable even though the sorbent is not spent.

Air Distribution. High, monumental spaces are prone to thermal stratification. If this places collections at risk, then appropriate return and supply air may be required to ensure air motion across the entire space. Gallery and display area use and loads may change because of varied lighting and numbers of visitors. Although this can sometimes be addressed through temperature control zones, adjustable supply air may be more economical and more effective.

Supply air should not blow directly onto collections. Diffusing supply air along a wall can be a major problem in a gallery, where collections are displayed on walls. Floor supply should also be avoided because particles at foot level become entrained.

Controls. Control system design is critical for maintaining precise temperature and humidity control. Consideration should be given to industrial-grade controls for proper temperature and humidity. Control systems should be able to monitor and control humidity, temperatures, airflow, filter pressure drops, water alarms, capacity alarms, and failure scenarios.

Sensors, thermostats, and humidistats must be located in the collection space, not in the return airstream. Temperature variation is usually preferable to prolonged humidity swings. This strongly affects controls design, because conventional control treats temperature as the primary goal and humidity as supplementary. Where comfort conditions are not required, humidity-controlled heating, which modulates heating within a very broad temperature dead band to seek stable or moderated humidity conditions, might be used (LaFontaine 1982; Marcon 1987; Staniforth 1984). By tracking system performance using the temperature control system, operators can adjust zones based on actual operating history. Operating at lower system volume at night when lights are off, but maintaining more than 6 air changes per hour, can save energy.

Types of Systems

The type of HVAC system used is critical to achieving project environmental goals. Proper airflow filters the air, controls humidity,

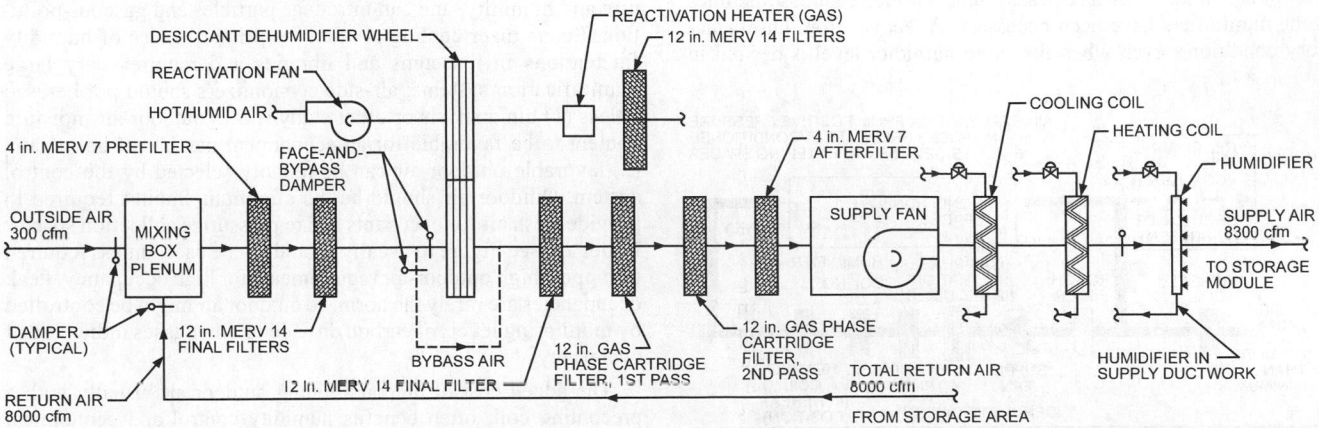

Fig. 7   Packaged Desiccant Dehumidification Unit
(Setty 2006)

and suppresses mold growth. Minimum airflow criteria vary from 6 to 8 air changes per hour (NBS 1983; Chapter 3 of this volume). These needs are usually best met with a constant-volume system.

The problems most often overlooked are maintenance access and risk to the collection from disruptions and leaks from overhead or decentralized equipment. Water or steam pipes over and in collection areas present the possibility of leaks, as do air-handling units. Some systems can provide full control without running any pipes to the zones, but others require two to six pipes to each zone, which often must be run over or in collection areas and are, unfortunately, the pipes most likely to leak. Leaks and maintenance can prevent effective use of spaces and result in lost space efficiency.

Central air-handling stations keep filtration, dehumidification, humidification, maintenance, and monitoring away from the collection. The investment in added space and the expense of the more elaborate duct system provide major returns in reduced disruption to the collection spaces and a dramatically extended service life for the distribution system. Unlike most commercial projects, space turnover in museums is low and rezoning is rarely needed. In some museums, 60-year-old multizone duct systems have been reused. Renovating the old system is economical, with most renovations confined entirely to the mechanical rooms. This is in comparison to common duct distribution systems (e.g., terminal reheat, dual-duct, and variable-air-volume), where renovations often require a new duct system and terminal equipment, involving major expense from demolishing the old ducts, installing the new duct system, and reinstalling architectural finishes.

Constant-Volume Reheat. A constant-volume reheat system can present problems if improperly applied. In many institutions, terminal reheat with steam or hot-water coils located near or over collection spaces cause chronic problems from steam and water leaks. Efficient zone-level humidification often suggests placing the humidifier downstream from the reheat coil; if the reheat coil is located near or over collection spaces, preventive maintenance on humidifiers further complicates maintenance problems. Constant-volume reheat systems are very effective when reheat coils and humidifiers are installed entirely within the mechanical space, instead of at the terminal, feeding through what is effectively a multizone distribution system.

Multizone System. A multizone air handler with zone reheat and zone humidification can be a stable and relatively energy-efficient solution. However, multizone systems without individual zone reheat and individual zone humidification have proved problematic in many institutions, requiring retrofit of zone equipment for stable humidity control. With proper layout and equipment complement, a multizone system can reduce the amount of reheat and be very energy-efficient.

Bovill (1988, Figure 10) shows preferences for constant-volume and multizone systems in collection spaces. Without the need for many temperature control zones, but with the need for high air quality, the choices are constant-volume, multizone with bypass, and dual-duct. When air handlers are outside collection areas (as recommended), the best choices are constant-volume and multizone with bypass, and dual-duct. When other systems are used, the client must be made fully aware of the possible compromises in performance, cost, and serviceability.

Dehumidification Coil. An important feature of multizone and dual-duct air handlers is a separate dehumidification coil upstream of both the hot and cold decks. This separate cooling coil, distinct from the one in the cold deck, is used during dehumidification demand. Air can be cooled to dew point even if it eventually flows through the hot deck. Without this feature, moist return air could be warmed in the hot deck and delivered back to the room without being dehumidified. An alternative is to locate a single cooling coil upstream of both decks, where the cold deck simply bypasses the hot deck, although this configuration can increase energy use.

Fan-Coil Units. Fan-coil units have been problematic when placed in and above collection areas. Fan-coil units expand and decentralize maintenance, requiring maintenance in collection areas and a net increase in overall facility maintenance. Because they cool locally, they need condensate drains, which can leak or back up over time. As all-water systems, they require four pressurized-water pipes to each unit, increasing the chance of piping leaks in collection areas.

Variable Air Volume (VAV). VAV, though appropriate for other types of buildings, tends to be inappropriate for collection housing because of poor humidity control, inadequate airflow, maintenance disruption, leaks in the collection spaces, and inflexibility to meet environmental needs. A VAV system may be chosen because of space and budget constraints, but almost always the cost and space required for a properly designed VAV system (full filtration, local humidification, local dehumidification, minimum air volume settings, well-planned piping and maintenance access, well-documented operating instructions) give no advantages over constant-volume systems. VAV systems can save energy compared to constant-volume systems, but usually at the collection's expense.

If used, a VAV system should look much like a constant-volume reheat system, with the minimum airflow to prevent mold growth, contamination buildup, and uneven conditions in the conditioned space. Terminal equipment should include reheat for each zone and be located in mechanical rooms or other spaces where access and service do not endanger a collection. If VAV performance becomes a problem, it can be easily converted to a constant-volume reheat system with only an adjustment of controls.

Fan-Powered Mixing Boxes. These are usually inappropriate for these facilities. Although fan-powered mixing boxes can help ensure air circulation to suppress mold growth, they do not allow effective air filtration for particles and gases. These fans also increase local maintenance requirements and present an added fire risk. If they include reheat, there is an added risk from leaks (with water or steam reheat) or fire (with electric reheat).

Energy and Operating Costs

Operating a preservation environment is often costly, but it is often necessary for long-term protection of a valuable collection. In most cases, the increase in usable life of a collection easily justifies the annual energy and operating costs to maintain the special environmental conditions. For institutions with small or limited operating budgets that cannot afford a major increase in annual energy cost, some compromises might be warranted or some initial capital investments made to reduce recurring annual costs. Careful commissioning using multiple instruments is helpful, and postconstruction tuning and off-hours volume reduction can reduce fan energy.

Scope of Special Environments. One of the best ways to reduce operating costs is to treat as little of the building as possible with the special environments. Spaces not needing preservation conditions should be on separate air systems that operate only when occupied.

Energy Efficiency. Using condenser heat to provide reheat for dehumidification can increase efficiency, and can substantially reduce dehumidification energy cost. Although an air-side economizer can cause problems, a water-side economizer can allow efficient winter cooling using condenser water. Because load varies between day and night operations, particularly in museums, night cooling loads are sometimes best met with a smaller off-hours chiller. Similarly, primary-secondary pumping with two-way control valves can be useful as loads vary over the day and across areas.

Daylighting. Using natural light is often proposed. Ayres et al. (1990) noted that this feature is always a net energy penalty. If used, the daylighting aperture should be minimized, and avoided as much as possible in and over collection areas. For lower risk of leaks and better-managed lighting, clerestories are preferred over skylights.

Humidistatically Controlled Heating. This specialized approach has limited application and must include safety controls, but is sometimes the only option that can handle envelope limitations in cold climates. In this approach, the heating system is controlled by a humidistat rather than a thermostat (LaFontaine and Michalski 1984); cold, damp air is heated until the relative humidity drops to 50%. Where interior temperatures drop consistently below 50°F, it solves the problem of humidity in a building that does not have an adequate envelope. Obviously, humidity-controlled heating does not provide human comfort in winter, but many small museums, historic buildings, and reserve collection buildings are essentially unoccupied in winter. A high-limit thermostat is necessary to stop overheating during warm weather, and a low-limit thermostat is optional if water pipe freezing is a concern. This approach has been used in Canada (LaFontaine 1982; Marcon 1987), the United States (Conrad 1994; Kerschner 1992, 2006), and in many historic buildings in Britain (Staniforth 1984). Maekawa and Toledo (2001) successfully applied humidistatic control in hot, humid climates to minimize microbial growth.

Some cautions apply. Foundations in a previously heated building may heave if the ground is waterlogged before freezing. Improving drainage, insulating the ground near the footings, and heating the basement reduce this risk. Problems have occurred in buildings with dense object storage and a very low infiltration rate, such as a specially sealed storage space (Padfield and Jensen 1996); a very slow supply of dehumidified air to the space can be helpful.

This approach is cost-effective in seasonal museums (especially for low-mass wood-frame buildings) in colder climates like the northern United States and Canada, and in maritime regions. Humidistat control does not work in warm, humid weather: without a high-limit thermostat it can heat hot, humid conditions to dangerous temperatures. In this case, domestic dehumidifiers can supplement humidistat control.

Hybrid (Load-Sharing) HVAC Systems. This approach involves enhancing preservation in museums and libraries by an optimum collocation of radiant and forced-convection systems. Most of the sensible loads are assigned to radiant panel systems, whereas latent loads and the remaining sensible loads are assigned to forced-convection systems. Decoupling the HVAC functions primarily into sensible and latent heat transfer components enables the designer to select better function-oriented HVAC components and to ensure higher accuracy and precision in control. Additionally, hybrid HVAC seems to satisfactorily balance the needs of preservation and human comfort, while keeping operating costs low. For example, in a library, human comfort could be maintained primarily by thermal radiation, while lowered air temperature increased the expected half-life of the books, with required fan power reduced by almost 50% (Kilkis et al. 1995). Moreover, because the forced-convection system is freed from satisfying most of the sensible heating and cooling loads, more accurate humidity control and faster response to humidity changes may be achieved. Multizoning in a hybrid HVAC system may be easier. Latent and sensible systems can also be integrated into a single HVAC unit called a **hybrid panel** (Kilkis 2002). When hydronic radiant panels, which require piping fluids into exhibition areas, are used, associated risks of damage from liquid leaks and the greater maintenance challenges must be balanced against expected savings. Liquid leakage risk might be minimized by using capillary tubing operated under negative pressure. To eliminate liquid leakage risks, all-air hybrid HVAC systems may be used. More information about panel heating and cooling and hybrid HVAC systems can be found in Chapter 6 of the 2008 *ASHRAE Handbook—HVAC Systems and Equipment.*

Maintenance and Ease of Operation. A common failing of designs is not accounting for ongoing operation and maintenance. Most designs work if properly adjusted and maintained, but many institutions do not have the staff, budget, or expertise to give the system the attention it needs. Maintenance needs for any system should be matched against the institution's staff capabilities. For large projects in larger cities, code-required staffing for the plant should be considered; sometimes, smaller reciprocating chillers can be used at night to preclude the licensed engineer needed to operate larger chillers. Small projects without HVAC maintenance staff may need packaged equipment that does not require daily attention.

Effects of maintenance activities on the collection must always be taken into account. For example, testing or accidental activation of a smoke removal system can radically change the collection environment. Transitions between winter and summer modes on economizers cause many operational problems. Tools and ladders in gallery and storage areas are a threat to the collection and special precautions need to be taken.

Contaminated air conveyance components (e.g., microbiological growth and other buildup) can contribute to pollution levels, lead to premature component failure, and affect heat transfer efficiency, resulting in higher utility costs. Regular inspection and cleaning is an important part of preventative maintenance.

REFERENCES

Allegrini, I., F. DeSantis, V. Di Palo, and A. Liberti. 1984. Measurement of particulate and gaseous ammonia at a suburban area by means of diffusion tubes (Denuders). *Journal of Aerosol Science* 15(4):465-471.

Ashley-Smith, J. 1999. *Risk assessment for object conservation.* Butterworth-Heinemann, Oxford.

ASHRAE. 2007. Method of testing general ventilation air-cleaning devices for removal efficiency by particle size. ANSI/ASHRAE *Standard* 52.2-2007.

ASHRAE. 2010. Thermal environmental conditions for human occupancy. ANSI/ASHRAE *Standard* 55-2010.

ASHRAE. 2010. Ventilation for acceptable indoor air quality. ANSI/ASHRAE *Standard* 62.1-2010.

Ayerst, G. 1968. Prevention of biodeterioration by control environmental conditions. In *Biodeterioration of Materials*, pp. 223-241. A.H. Walters and J.J. Elphick, eds. Elsevier, Amsterdam.

Ayres, J.M., H. Lau, and J.C. Haiad. 1990. Energy impact of various inside air temperatures and humidities in a museum when located in five U.S. cities. *ASHRAE Transactions* 96(2):100-111.

Bellan, L.M., L.G. Salmon, and G.R. Cass. 2000. A study on the human ability to detect soot deposition onto works of art. *Environmental Science & Technology* 34(10):1946-1952.

Beuchat, L.R. 1987. Influence of water activity on sporulation, germination, outgrowth, and toxin production. In *Water activity: Theory and applications to food*, pp. 137-152. L.B. Beuchat, and L.R. Rockland, eds. Marcel Dekker, New York.

Bogaty, H., K.S. Campbell, and W.D. Appel. 1952. The oxidation of cellulose by ozone in small concentrations. *Textile Research Journal* 22(2):81-83.

Bovill, C. 1988. Qualitative engineering. *ASHRAE Journal* 30(4):29-34.

CAGI. 2002. Standard air defined as in ISO standards. Compressed Air and Gas Institute, Cleveland. Available at http://www.cagi.org/toolbox/glossary.asp.

Cass, G.R., J.R. Druzik, D. Grosjean, W.W. Nazaroff, P.M. Whitemore, and C.L. Wittman. 1989. Protection of works of art from atmospheric ozone. *Research in Conservation Series* 5. Getty Conservation Institute, Los Angeles.

CCI/ICC. 2005. *Framework for preservation of museum collections.* Canadian Conservation Institute, Ottawa. Available at http://www.cci-icc.gc.ca/tools/framework/index_e.aspx.

Conrad, E. 1994. Balancing environmental needs of the building, the collection, and the user. *Preventive Conservation Practice, Theory and Research: Preprints of the Contributions to the Ottawa Congress*, A. Roy and P. Smith, eds. International Institute for Conservation of Historic and Artistic Works (IIC), London.

Conrad, E. 1995. *A table for classification of climatic control potential in buildings.* Landmark Facilities Group, CT.

Craft, M., C. Hawks, J. Johnson, M. Martin, and L. Mibach. 1996. *Preservation of collections: Assessment, evaluation and mitigation strategies.* American Institute for Conservation, Washington, D.C.

Dahlin, E., J.F. Henriksen, and O. Anda. 1997. Assessment of environmental risk factors in museums and archives. *European Cultural Heritage Newsletter on Research (ECHNR)* 10:94-97.

Daly, D. and S. Michalski. 1987. Methodology and status of the lining project, CCI. *ICOM Conservation Committee 8th Triennial Meeting*, Los Angeles, pp. 145-152.

Davies, T.D., B. Ramer, G. Kaspyzok, and A.C. Delany. 1984. Indoor/outdoor ozone concentrations at a contemporary art gallery. *Journal of the Air Pollution Control Association* 31(2):135-137.

Davis, N. 2006. Tracing the evolution of preservation environments in archives, museums, and libraries. *20th Annual Preservation Conference: Beyond the Numbers: Specifying and Achieving an Efficient Preservation Environment*. National Archives, Washington, D.C. Available at http://www.archives.gov/preservation/conferences/2006/presentations.html.

Erhardt, D. and M. Mecklenburg. 1994. Relative humidity re-examined. *Preventive Conservation Practice, Theory and Research: Preprints of the Contributions to the Ottawa Congress*, pp. 32-38. International Institute for Conservation of Historic and Artistic Works, London.

Graedel, T.E. 1984. Concentrations and metal interactions of atmospheric trace gases involved in corrosion. *Metallic Corrosion*, Toronto, pp. 396-401.

Graedel, T.E. and R. McGill. 1986. Degradation of materials in the atmosphere. *Environmental Science & Technology* 20(11):1093-1100.

Graedel, T.E., G.W. Kammlott, and J.P. Franey. 1981. Carbonyl sulfide: Potential agent of atmospheric sulfur corrosion. *Science* 212:663-665.

Granby, K. and C.S. Christensen. 1997. Urban and semi-rural observations of carboxylic acids and carbonyls. *Atmospheric Environment* 31(10): 1403-1415.

Groom, P. and T. Panisset. 1933. Studies in *Penicillium chrysogenum thom* in relation to temperature and relative humidity of the air. *Annals of Applied Biology* 20:633-660.

Grzywacz, C.M. 2006. Monitoring for gaseous pollutants in museum environments. In *Tools in Conservation*, E. Maggio. ed. Getty Conservation Institute, Los Angeles,

Grzywacz, C.M. and N.H. Tennent. 1994. Pollution monitoring in storage and display cabinets: Carbonyl pollutant levels in relation to artifact deterioration. *Preventive Conservation Practice, Theory and Research: Preprints of the Contributions to the Ottawa Congress*, pp. 164-170. International Institute for Conservation of Historic and Artistic Works (IIC), London.

Grzywacz, C. and N.H. Tennent. 1997. The threat of organic carbonyl pollutants to museum collections. *European Cultural Heritage Newsletter on Research* (ECHNR). European Commission DG XII Science, Research and Development, Brussels.

Hancock, R.P., N.A. Esman, and C.P. Furber. 1976. Visual response to dustiness. *Journal of the Air Pollution Control Association* 26(1):54-57.

Hatchfield, P.B. 2002. *Pollutants in the museum environment: Practical strategies for problem solving in design, exhibition, and storage*. Archetype Publications, London.

Hedley, G. 1988. Relative humidity and the stress/strain response of canvas painting: Uniaxial measurements of naturally aged samples. *Studies in Conservation* 33:133-148.

Hens, H.L.S.C. 1993. Mold risk: Guidelines and practice, commenting the results of the International Energy Agency EXCO on energy conservation in buildings and community systems, Annex 14: Condensation energy. In *Bugs, Mold and Rot III: Moisture Specifications and Control in Buildings*, pp. 19-28. W. Rose and A. Tenwolde, eds. National Institute of Building Sciences, Washington, D.C.

Hermance, H.W., C.A. Russell, E.J. Bauer, T.F. Egan, and H.V. Wadlow. 1971. Relation of airborne nitrate to telephone equipment damage. *Environmental Science & Technology* 5:781-789.

Hodgson, A.T., A.F. Rudd, D. Beal, and S. Chandra. 2000. Volatile organic compound concentrations and emission rates in new manufactured and site-built houses. *Indoor Air* 10(3):178-192.

IPI. 2006. *Preservation calculator*. Image Permanence Institute, Rochester Institute of Technology (RIT) College of Imaging Arts and Sciences. (Available at https://www.imagepermanenceinstitute.org/resources/calculators)

Kawamura, K., S. Steinberg, and I.R. Kaplan. 1996. Concentrations of monocarboxylic and dicarboxylic acids and aldehydes in southern California wet precipitations: Comparison of urban and nonurban samples and compositional changes during scavenging. *Atmospheric Environment* 30(7):1035-1052.

Kerschner, R.L. 1992. A practical approach to environmental requirements for collections in historic buildings. *Journal of the American Institute for Conservation* 31:65-76.

Kerschner, R.L. 2006. Providing safe and practical environments for cultural properties in historic buildings...and beyond. *20th Annual Preservation Conference: Beyond the Numbers: Specifying and Achieving an Efficient Preservation Environment*. National Archives, Washington, D.C. Available at http://www.archives.gov/preservation/conferences/2006/kerschner.pdf.

Kilkis, B.I. 2002. Modeling of a hybrid HVAC panel for library buildings. *ASHRAE Transactions* 108(2):693-698.

Kilkis, B.I., S.R. Suntur, and M. Sapci. 1995. Hybrid HVAC systems. *ASHRAE Journal* 37(12):23-28.

LaFontaine, R.H. 1979. Environmental norms for Canadian museums, art galleries, and archives. *CCI Technical Bulletin* 5, Canadian Conservation Institute, Ottawa.

LaFontaine, R.H. 1982. Humidistically-controlled heating: A new approach to relative humidity control in museum closed for the winter season. *Journal of the International Institute for Conservation: Canadian Group* 7(1,2):35-41.

LaFontaine, R.H., and S. Michalski. 1984. The control of relative humidity—Recent developments. *ICOM Conservation 7th Triennial Meeting*, Copenhagen, pp. 33-37.

Lavédrine, B. 2003. *A guide to the preventive conservation of photographic collections*. Getty Conservation Institute, Los Angeles.

Ligocki, M.P., H.I.H. Liu, G.R. Cass, and W. John. 1990. Measurements of particle deposition rates inside southern California museums. *Aerosol Science and Technology* 13:85-101.

Ligocki, M.P., L.G. Salmon, T. Fall, M.C. Jones, W.W. Nazaroff, and G.R. Cass. 1993. Characteristics of airborne particles inside southern California museums. *Atmospheric Environment, Part A: General Topics* 27(5): 697-711.

Lull, W. and L. Harriman. 2001. Museums, libraries and archives. In *Humidity Control Design Guide for Commercial and Institutional Buildings*, L. Harriman, G.W. Brundrett, and R. Kittler, eds. ASHRAE.

Maekawa, S. and F. Toledo. 2001. Sustainable climate control for historic buildings in hot and humid regions. *Renewable Energy for a Sustainable Development of the Built Environment: Proceedings of the 18th International Conference on Passive and Low Energy Architecture*, F.O.R. Pereira, R. Rüther, R.V.G. Souza, S. Afonso, and J.A.B. da Cunha Neto, eds.

Marcon, P.J. 1987. Controlling the environment within a new storage and display facility for the governor general's carriage. *Journal of the International Institute for Conservation: Canadian Group* 12:37-42.

Mecklenburg, M.F., C.S. Tumosa, and A. Pride. 2004. Preserving legacy buildings. *ASHRAE Journal* 46(6):S18-S23.

Meininghaus, R., L. Gunnarsen, and H.N. Knudsen. 2000. Diffusion and sorption of volatile organic compounds in building materials—Impact on indoor air quality. *Environmental Science & Technology* 34(15):3101-3108.

Michalski, S. 1991. Paintings, their response to temperature, relative humidity, shock and vibration. In *Works of Art in Transit*, pp. 223-248. M.F. Mecklenburg, ed. National Gallery, Washington, D.C.

Michalski, S. 1993. Relative humidity in museum, galleries and archives: Specification and control. *Bugs, Mold and Rot III: Moisture Specification and Control in Buildings*, pp. 51-61. W. Rose and A. Tenwolde, eds. National Institute of Building Science, Washington, D.C.

Michalski, S. 1995. *Wooden artifacts and humidity fluctuation: Different construction and different history mean different vulnerabilities*. Chart, Version 3.0. Canadian Conservation Institute, Ottawa.

Michalski, S. 1996a. *Environmental guidelines: Defining norms for large and varied collections*. American Institute for Conservation, Washington, D.C.

Michalski, S. 1996b. Quantified risk reduction in the humidity dilemma. *APT Bulletin* 37:25-30.

Michalski, S. 1998. Climate control priorities and solutions for collections in historic buildings. *Forum* 12(4):8-14.

Michalski, S. 1999. *Relative humidity and temperature guidelines for Canadian archives*. Canadian Council of Archives and Canadian Conservation Institute, Ottawa.

NAFA. 2004. *Recommended practice: Guidelines—Libraries, archives and museums*. National Air Filtration Association, Virginia Beach.

Nazaroff, W.W. and G.R. Cass. 1991. Protecting museum collections from soiling due to the deposition of airborne particles. *Atmospheric Environment* 25A(5/6):841-852.

Nazaroff, W.W., L.G. Salmon, and G.R. Cass. 1990. Concentration and fate of airborne particles in museums. *Environmental Science & Technology* 24(1):66-76.

NBS. 1983. *Air quality criteria for storage of paper-based records.* NBSIR 83-2795. National Institute of Standards and Technology, Gaithersburg, MD.

Nishimura, D.W. 1993. The IPI storage guide for acetate film. *Topics in Photographic Preservation* 5:123-137.

Office of Environmental Health Hazard Assessment. 2003. Air—Hot spots—Chronic RELS: All chronic reference exposure levels developed by the OEHHA. Available at http://www.oehha.org/air/chronic_rels/index.html.

Oreszczyn. T., M. Cassar, and K. Fernandez. 1994. Comparative studies of air-conditioned and non-air-conditioned museums: Preventive conservation practice, theory and research. *Preprints of the Contributions to the Ottawa Congress*, pp. 144-148. International Institute for Conservation of Historic and Artistic Works, London.

Padfield, T. and P. Jensen. 1996. Low energy climate control in stores: A postscript. *ICOM Conservation Committee, 9th Triennial Meeting*, Dresden, pp. 596-601.

Perera, D.Y. and D. Van den Eynde. 1987. Moisture and temperature induced stresses (hygrothermal stresses) in organic coatings. *Journal of Coatings Technology* 59(5):55-63.

Persily, A.K. 1999. Myths about building envelopes. *ASHRAE Journal* 41(3):39-45.

Raychaudhuri, M.R. and P. Brimblecombe. 2000. Formaldehyde oxidation and lead corrosion. *Studies in Conservation* 45(4):226-232.

Salmon, L.G., P.R. Mayo, G.R. Cass, and C.S. Christoforou. 2005. Airborne particles in new museum facilities. *Journal of Environmental Engineering* 131(10):1453-1461.

Sebor, A.J. 1995. Heating, ventilating, and air-conditioning systems. In *Storage of natural history collections: Ideas and practical solutions.* C.L. Rose, C.A. Hawks, and H.H. Genoways, eds. Society for the Preservation of Natural History Collections, Washington D.C.

Snow, D., M.H.G. Crichton, and N.C. Wright. 1944. Mould deterioration of feeding stuff in relation to humidity of storage. *Annals of Applied Biology* 31:102-110.

Staniforth, S. 1984. Environmental conservation. In *Manual of curatorship*, pp. 192-202. J.M.A. Thompson, ed. Butterworths, London.

Stock, T.H., and E.A. Venso. 1993. The impact of residential evaporative air cooling on indoor exposure to ozone. *Indoor Air '93: Proceedings of the 6th International Conference on Indoor Air Quality and Climate*, Helsinki, pp. 251-256. M. Jantunen, P. Kalliokoski, E. Kukkonen, K. Saarela, O. Seppänen, and H. Vuorelma, eds.

Tétreault, J. 1994. Display materials: The good, the bad, and the ugly. *Exhibition and Conservation SSCR*, pp. 21-22. Scottish Society for Conservation and Restoration (SSCR), Edinburgh.

Tétreault, J. 2003. *Airborne pollutants in museums, galleries and archives: Risk assessment, control strategies and preservation management.* Canadian Conservation Institute, Ottawa, ON.

Thickett, D., S.M. Bradley, and L.R. Lee. 1998. Assessment of the risks to metal artifacts posed by volatile carbonyl pollutants. *Metal 98: Proceedings of the International Conference on Metals Conservation*, pp. 260-264. W. Mourey and L. Robbiola, eds. James and James Science, London.

Thomson, G. 1994. *The museum environment (conservation and museology)*, 2nd ed. Butterworth-Heinemann, Oxford, U.K.

Toishi, K. and T. Kenjo. 1975. Some aspects of the conservation of works of art in buildings of new concrete. *Studies in Conservation* 20:118-122.

USDA. 1999. *Wood handbook.* U.S. Department of Agriculture, Washington, D.C.

Volent, P. and N.S. Baer. 1985. Volatile amines used as corrosion inhibitors in museum humidification systems. *International Journal of Museum Management and Curatorship* 4:359-364.

Waller, R. 1999. Internal pollutants, risk assessment and conservation priorities. *12th Triennial Meeting of ICOM Committee for Conservation*, Lyon, pp. 113-118. J. Bridgland, ed.

Waller, R., and S. Michalski. 2005. A paradigm shift for preventive conservation, and a software tool to facilitate the transition. *14th Triennial Meeting, ICOM Committee for Conservation*, The Hague, pp. 733-738. I. Verger, ed.

Watts, S. 1999. Hydrogen sulphide levels in museums: What do they mean? *Indoor Air Pollution: Detection and Prevention—Presentation Abstracts and Additional Notes*, pp. 14-16. Netherlands Institute for Cultural Heritage, Amsterdam. Available at http://www.iaq.dk/iap/iap1999/1999_04.htm.

Wilhelm, H. 1993. *The permanence and care of color photographs: Traditional and digital color prints, color negatives, slides, and motion pictures.* Preservation Publishing, Grinnell, Iowa.

World Health Organization. 2000. *Air quality guidelines for Europe*, 2nd ed. World Health Organization *European Series* 91, Regional Office for Europe, Copenhagen.

Yoon, Y.H. and P. Brimblecombe. 2001. The distribution of soiling by coarse particulate matter in the museum environment. *Indoor Air* 11(4): 232-240.

BIBLIOGRAPHY

ASHRAE. 1992. Gravimetric and dust-spot procedures for testing air-cleaning devices used in general ventilation for removing particulate matter. *Standard 52.1-1992.*

CAGI. 2006. *Toolbox: Metric to standard converter calculator.* Compressed Air and Gas Institute, Cleveland. (Available at http://www.cagi.org/toolbox/converter.htm)

Larsen, R. 1997. Deterioration and conservation of vegetable tanned leather. *European Cultural Heritage Newsletter on Research, EC Research Workshop—Effects of the Environment on Indoor Cultural Property* 10 (June):54-61.

Tennent, N.H., B.G. Cooksey, D. Littlejohn, B.J. Ottaway, S.E. Tarling, and M. Vickers. 1993. Unusual corrosion and efflorescence products on bronze and iron antiquities stored in wooden cabinets. *Conservation Science in the U.K.; Preprints of the Meeting held in Glasgow, May 1993*, pp. 60-66. N.H. Tennent, ed. James and James Science, London.

CHAPTER 24

ENVIRONMENTAL CONTROL FOR ANIMALS AND PLANTS

THE design of plant and animal housing is complicated because many environmental factors affect the production and well-being of living organisms. The financial constraint that equipment must repay costs through improved economic productivity must be considered by the designer. The engineer must balance costs of modifying the environment against economic losses of a plant or animal in a less-than-ideal environment.

Thus, design of plant and animal housing is affected by (1) economics, (2) concern for both workers and the care and well-being of animals, and (3) regulations on pollution, sanitation, and health assurance.

DESIGN FOR ANIMAL ENVIRONMENTS

Typical animal production plants modify the environment, to some degree, by housing or sheltering animals year-round or for parts of a year. The degree of modification is generally based on the expected increase in production. Animal sensible heat and moisture production data, combined with information on the effects of environment on growth, productivity, and reproduction, help designers select optimal equipment. Detailed information is available in a series of handbooks published by the MidWest Plan Service. These include *Mechanical Ventilating Systems for Livestock Housing* (MWPS 1990a), *Natural Ventilating Systems for Livestock Housing and Heating* (MWPS 1989), and *Cooling and Tempering Air for Livestock Housing* (MWPS 1990b). ASAE *Monograph* 6, Ventilation of Agricultural Structures (Hellickson and Walter 1983), also gives more detailed information.

Design Approach

Environmental control systems are typically designed to maintain thermal and air quality conditions within an acceptable range and as near the ideal for optimal animal performance as is practicable. Equipment is usually sized assuming steady-state energy and mass conservation equations. Experimental measurements confirm that heat and moisture production by animals is not constant and that there may be important thermal capacitance effects in livestock buildings. Nevertheless, for most design situations, the steady-state equations are acceptable.

Achieving the appropriate fresh air exchange rate and establishing the proper distribution within the room are generally the two most important design considerations. The optimal ventilation rate is selected according to the ventilation rate logic curve (Figure 1).

The preparation of this chapter is assigned to TC 2.2, Plant and Animal Environment.

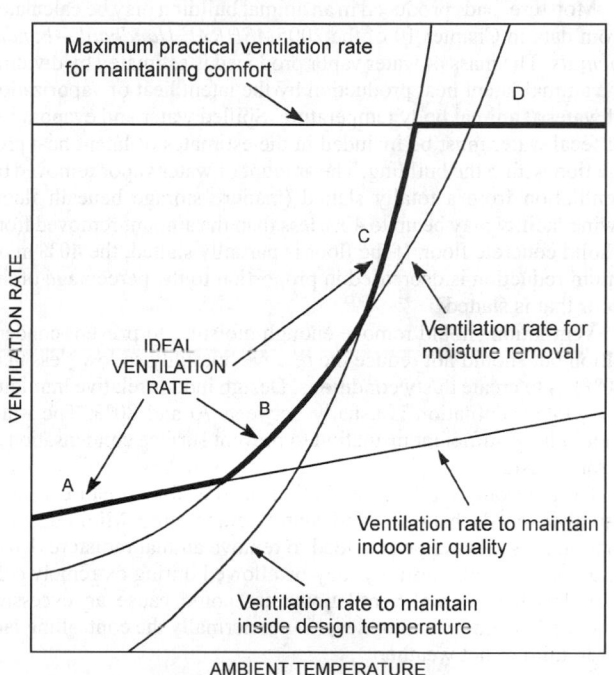

Fig. 1 Logic for Selecting Appropriate Ventilation Rate in Livestock Buildings
(Adapted from Christianson and Fehr 1983)

During the coldest weather, the ideal ventilation rate is that required to maintain indoor relative humidity at or below the maximum desired, and air contaminant concentrations within acceptable ranges (Rates A and B in Figure 1). Supplemental heating is often required to prevent the temperature from dropping below optimal levels.

In milder weather, the ventilation rate required for maintaining optimal room air temperature is greater than that required for moisture and air quality control (Rates C and D in Figure 1). In hot weather, the ventilation rate is chosen to minimize the temperature rise above ambient and to provide optimal air movement over animals. Cooling is sometimes used in hot weather. The maximum rate (D) is often set at 60 air changes per hour (ach) as a practical maximum.

Temperature Control

The temperature in an animal structure is computed from the sensible heat balance of the system, usually disregarding transient effects. Nonstandard buildings with low airflow rates and/or large thermal mass may require transient analysis. Steady-state heat transfer through walls, ceiling or roof, and ground is calculated as presented in Chapters 25 to 27 of the 2009 *ASHRAE Handbook—Fundamentals*.

Mature animals typically produce more heat per of unit floor area than do young stock. Chapter 10 of the 2005 *ASHRAE Handbook—Fundamentals* presents estimates of animal heat loads. Lighting and equipment heat loads are estimated from power ratings and operating times. Typically, the designer selects indoor and outdoor design temperatures and calculates the ventilation rate to maintain the temperature difference. Outdoor design temperatures are given in Chapter 14 of the 2009 *ASHRAE Handbook—Fundamentals*. The section on Recommended Practices by Species in this chapter presents indoor design temperature values for various livestock.

Moisture Control

Moisture loads produced in an animal building may be calculated from data in Chapter 10 of the 2005 *ASHRAE Handbook—Fundamentals*. The mass of water vapor produced is estimated by dividing the animal latent heat production by the latent heat of vaporization of water at animal body temperature. Spilled water and evaporation of fecal water must be included in the estimates of latent heat production within the building. The amount of water vapor removed by ventilation from a totally slatted (manure storage beneath floor) swine facility may be up to 40% less than the amount removed from a solid concrete floor. If the floor is partially slatted, the 40% maximum reduction is decreased in proportion to the percentage of the floor that is slatted.

Ventilation should remove enough moisture to prevent condensation but should not reduce the relative humidity so low (less than 40%) as to create dusty conditions. Design indoor relative humidity for winter ventilation is usually between 70 and 80%. The walls should have sufficient insulation to prevent surface condensation at 80% rh inside.

During cold weather, ventilation needed for moisture control usually exceeds that needed to control temperature. Minimum ventilation must always be provided to remove animal moisture. Up to a full day of high humidity may be allowed during extremely cold periods when normal ventilation rates could cause an excessive heating demand. Humidity level is not normally the controlling factor in mild or hot weather.

Air Quality Control

Contaminants. The most common and prevalent air contaminants in animal buildings are particulate matter (PM) and gases. In animal buildings, particulate matter originates mainly from feed, litter, fecal materials, and animals. Particulates include solid particles (or dust), liquid droplets, and microorganisms, can be deposited deep within the respiratory system. Particulates carry allergens that cause discomfort and health problems for workers in animal housing facilities. They also carry much of the odors in animal housing facilities, for potentially long distances from the facilities. Consequently, particulates pose major problems for animals, workers, and neighbors. Particulate levels in swine buildings have been measured to range from 0.028 to 0.43 mg/ft³. Dust has not been a major problem in dairy buildings; one two-year study found an average of only 0.014 mg/ft³ in a naturally ventilated dairy barn. Poultry building dust levels average around 0.057 to 0.20 mg/ft³, but levels up to 0.51 to 0.82 mg/ft³ have been measured during high activity.

The most common gas contaminants are ammonia, hydrogen sulfide, other odorous compounds, carbon dioxide, and carbon monoxide. High moisture levels can also aggravate other contaminant problems. Ammonia, which results from decomposition of manure, is the most important chronically present contaminant gas. Typical ammonia levels measured have been 10 to 50 ppm in poultry units, 0 to 20 ppm in cattle buildings, 5 to 30 ppm in swine units with liquid manure systems, and 10 to 50 ppm in swine units with

solid floors (Ni et al. 1998a). Up to 200 ppm have been measured in swine units in winter. Ammonia should be maintained below 25 ppm and, ideally, below 10 ppm.

Maghirang et al. (1995) and Zhang et al. (1992) found ammonia levels in laboratory animal rooms to be negligible, but concentrations could reach 60 ppm in cages. Weiss et al. (1991) found ammonia levels in rat cages of up to 350 ppm with four male rats per cage and 68 ppm with four female rats per cage. Hasenau et al. (1993) found that ammonia levels varied widely among various mouse microisolation cages; ammonia ranged from negligible to 520 ppm nine days after cleaning the cage.

Hydrogen sulfide, a by-product of microbial decomposition of stored manure, is the most important acute gas contaminant. During normal operation, hydrogen sulfide concentration is usually insignificant (i.e., below 1 ppm). A typical level of hydrogen sulfide in swine buildings is around 150 to 350 ppb (Ni et al. 1998b). However, levels can reach 200 to 330 ppm, and possibly up to 1000 to 8000 ppm during in-building manure agitation.

Odors from animal facilities are an increasing concern, both in the facilities and in the surrounding areas. Odors result from both gases and particulates; particulates are of primary concern because odorous gases can be quickly diluted below odor threshold concentrations in typical weather conditions, whereas particulates can retain odor for long periods. Methods that control particulate and odorous gas concentrations in the air also reduce odors, but controlling odor generation at the source appears to be the most promising method of odor control.

Barber et al. (1993), reporting on 173 pig buildings, found that carbon dioxide concentrations were below 3000 ppm in nearly all instances when the external temperature was above 32°F but almost always above 3000 ppm when the temperature was below 32°F. The report indicated that there was a very high penalty in heating cost in cold climates if the maximum allowed carbon dioxide concentration was less than 5000 ppm. Air quality control based on carbon dioxide concentrations was suggested by Donham et al. (1989). They suggested a carbon dioxide concentration of 1540 ppm as a threshold level, above which symptoms of respiratory disorders occurred in a population of swine building workers. For other industries, a carbon dioxide concentration of 5000 ppm is suggested as the time-weighted threshold limit value for 8 h of exposure (ACGIH 1998).

Other gas contaminants can also be important. Carbon monoxide from improperly operating unvented space heaters sometimes reaches problem levels. Methane is another occasional concern.

Control Methods. Three standard methods used to control air contaminant levels in animal facilities are

1. Reduce contaminant production at the sources.
2. Remove contaminants from the air by air cleaning.
3. Reduce contaminant concentration by dilution (ventilation).

The first line of defense is to reduce release of contaminants from the source, or at least to intercept and remove them before they reach workers and animals. Animal feces and urine are the largest sources of contaminants, but feed, litter, and the animals themselves are also a major source of contaminants, especially particulates. Successful operations effectively collect and remove all manure from the building within three days, before it decomposes enough to produce large quantities of contaminants. Removing ventilation air uniformly from manure storage or collection areas helps remove contaminants before they reach animal or worker areas.

Ammonia production can be minimized by removing wastes from the room and keeping floor surfaces or bedding dry. Immediately covering manure solids in gutters and pits with water also reduces ammonia, which is highly soluble in water. Because adverse effects of hydrogen sulfide on production begin to occur at 20 ppm, ventilation systems should be designed to maintain hydrogen sulfide levels below 20 ppm during agitation. When manure is agitated and removed from the storage, the building should be well ventilated and all animals and occupants evacuated to avoid potentially fatal concentrations of gases.

For laboratory animals, changing the bedding frequently and keeping the bedding dry with lower relative humidities and appropriate cage ventilation can reduce ammonia release. Individually ventilated laboratory animal cages or placing cages in mass air displacement units reduce contaminant production by keeping litter drier. Using localized contaminant containment work stations for dust-producing tasks such as cage changing may also help. For poultry or laboratory animals, the relative humidity of air surrounding the litter should be kept between 50 and 75% to reduce particulate and gas contaminant release. Relative humidities between 40 and 75% also reduce the viability of pathogens in the air. A moisture content of 25 to 30% (wet basis) in the litter or bedding keeps dust to a minimum. Adding 0.5 to 2% of edible oil or fat can significantly reduce dust emission from the feed. Respirable dust (smaller than 10 μm), which is most harmful to the health and comfort of personnel and animals, is primarily from feces, animal skins, and dead microorganisms. Respirable dust concentration should be kept below 0.0065 mg/ft³. Some dust control technologies are available. For example, sprinkling oil at 0.12 gal per 1000 ft² of floor area per day can reduce dust concentration by more that 80%. High animal activity levels release large quantities of particulates into the air, so management strategies to reduce agitation of animals are helpful.

Methods of removing contaminants from the air are essentially limited to particulate removal, because gas removal methods are often too costly for animal facilities. Some animal workers wear personal protection devices (appropriate masks) to reduce inhaled particulates. Room air filters reduce animal disease problems, but they have not proven practical for large animal facilities because of the large quantity of particulates and the difficulty in drawing particulates from the room and through a filter. Air scrubbers can remove gases and particulates, but the initial cost and maintenance make them impractical. Aerodynamic centrifugation is showing promise for removing the small particulates found in animal buildings.

Ventilation is the most prevalent method used to control gas contaminant levels in animal facilities. It is reasonably effective in removing gases, but not as effective in removing particulates. Pockets in a room with high concentrations of particulate contaminants are common. These polluted pockets occur in dead air spots or near large contaminant sources. Providing high levels of ventilation can be costly in winter, can create drafts on the animals, and can increase the release of gas contaminants by increasing air velocity across the source.

Disease Control

Airborne microbes can transfer disease-causing organisms among animals. For some situations, typically with young animals where there are low-level infections, it is important to minimize air mixing among animal groups. It is especially important to minimize air exchange between different animal rooms, so buildings need to be fairly airtight.

Poor thermal environments and air contaminants can increase stress on the animals, which can make them more susceptible to disease. Therefore, a good environmental control system is important for disease control.

Air Distribution

Air speed should be maintained below 50 fpm for most animal species in both cold and mild weather. Animal sensitivities to draft are comparable to those of humans, although some animals are more sensitive at different stages. Riskowski and Bundy (1988) documented that air velocities for optimal rates of gain and feed efficiencies can be below 25 fpm for young pigs at thermoneutral conditions.

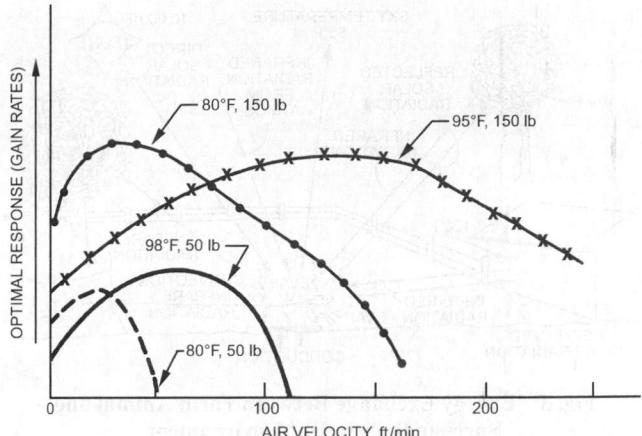

Fig. 2   Response of Swine to Air Velocity

Increased air movement during hot weather increases growth rates and improves heat tolerance. There are conflicting and limited data defining optimal air velocity in hot weather. Bond et al. (1965) and Riskowski and Bundy (1988) determined that both young and mature swine perform best when air speed is less than 200 fpm (Figure 2). Mount and Start (1980) did not observe performance penalties at air speeds increased to a maximum of 150 fpm.

Degree of Shelter

Livestock, especially young animals, need some protection from adverse climates. On the open range, mature cattle and sheep need protection during severe winter conditions. In winter, dairy cattle and swine may be protected from precipitation and wind with a three-sided, roofed shelter open on the leeward side. The windward side should also have approximately 10% of the wall surface area open to prevent negative pressure inside the shelter, which could cause rain and snow to be drawn into the building on the leeward side. These shelters do not protect against high temperature or high humidity.

In warmer climates, shades often provide adequate shelter, especially for large, mature animals such as dairy cows. Shades are commonly used in Arizona; research in Florida has shown an approximate 10% increase in milk production and a 75% increase in conception efficiency for shaded versus unshaded cows. The benefit of shades has not been documented for areas with less severe summer temperatures. Although shades for beef cattle are also common practice in the southwestern United States, beef cattle are somewhat less susceptible to heat stress, and extensive comparisons of various shade types in Florida have detected little or no differences in daily weight gain or feed conversion.

The energy exchange between an animal and various areas of the environment is illustrated in Figure 3. A well-designed shade makes maximum use of radiant heat sinks, such as the cold sky, and gives maximum protection from direct solar radiation and high surface temperature under the shade. Good design considers geometric orientation and material selection, including roof surface treatment and insulation material on the lower surface.

An ideal shade has a top surface that is highly reflective to solar energy and a lower surface that is highly absorptive to solar radiation reflected from the ground. A white-painted upper surface reflects solar radiation, yet emits infrared energy better than aluminum. The undersurface should be painted a dark color to prevent multiple reflection of shortwave energy onto animals under the shade.

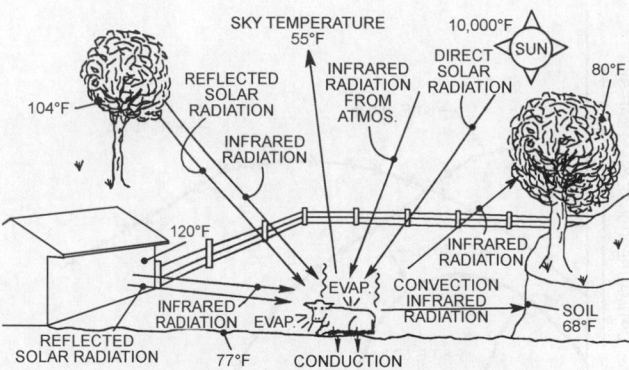

Fig. 3 Energy Exchange Between Farm Animal and Surroundings in Hot Environment

COOLING AND HEATING

Air Velocity

Increasing air velocity helps to facilitate the cooling of mature animals. It is especially beneficial when combined with skin wetting evaporative cooling. Mature swine benefit most with air velocities up to 200 fpm; cattle around 300 fpm; and poultry around 600 fpm. Air velocity can be increased with air circulation fans that blow air horizontally in circular patterns around the room, paddle fans that blow air downward, or tunnel cooling that moves air horizontally along the length of the building.

Evaporative Cooling

Supplemental cooling of animals in intensive housing conditions may be necessary during heat waves to prevent heat prostration, mortality, or serious losses in production and reproduction. Evaporative cooling, which may reduce ventilation air to 80°F or lower in most of the United States, is popular for poultry houses, and is sometimes used for swine and dairy housing.

Evaporative cooling is well suited to animal housing because the high air exchange rates effectively remove odors and ammonia, and increase air movement for convective heat relief. Initial cost, operating expense, and maintenance problems are all relatively low compared to other types of cooling systems. Evaporative cooling works best in areas with low relative humidity, but significant benefits can be obtained even in the humid southeastern United States.

Design. The pad area should be sized to maintain air velocities between 200 and 275 fpm through the pads. For most pad systems, these velocities produce evaporative efficiencies between 75 and 85%; they also increase pressures against the ventilating fans from 0.04 to 0.12 in. of water, depending on pad design.

The building and pad system must be airtight because air leaks caused by the negative-pressure ventilation reduce airflow through the pads, and hence reduce cooling effectiveness.

The most serious problem encountered with evaporative pads for agricultural applications is clogging by dust and other airborne particles. Whenever possible, fans should exhaust away from pads on adjacent buildings. Regular preventive maintenance is essential. Water bleed-off and the addition of algaecides to the water are recommended. When pads are not used in cool weather, they should be sealed to prevent dusty inside air from exhausting through them.

High-pressure fogging with water pressure of 500 psi is preferred to pad coolers for cooling air in broiler houses with built-up litter. The high pressure creates a fine aerosol, causing minimal litter wetting. Timers and/or thermostats control the cooling. Evaporative efficiency and installation cost are about one-half those of a well-designed evaporative pad. Foggers can also be used with naturally ventilated, open-sided housing. Low-pressure systems are not recommended for poultry, but may be used during emergencies.

Nozzles that produce water mist or spray droplets to wet animals directly are used extensively during hot weather in swine confinement facilities with solid concrete or slatted floors. Currently, misting or sprinkling systems with larger droplets that directly wet the skin surface of the animals (not merely the outer portion of the hair coat) are preferred. Timers that operate periodically, (e.g., 2 to 3 min on a 15 to 20 min cycle) help to conserve water.

Mechanical Refrigeration

Mechanical refrigeration can be designed for effective animal cooling, but it is considered uneconomical for most production animals. Air-conditioning loads for dairy housing may require 2.5 kW or more per cow. Recirculation of refrigeration air is usually not feasible because of high contaminant loads in the air in the animal housing. Sometimes, zone cooling of individual animals is used instead of whole-room cooling, particularly in swine farrowing houses, where a lower air temperature is needed for sows than for unweaned piglets. It is also beneficial for swine boars and gestating sows. Refrigerated air, 18 to 36°F below ambient temperature, is supplied through insulated ducts directly to the head and face of the animal. Air delivery rates are typically 20 to 40 cfm per animal for snout cooling, and 60 to 80 cfm per sow for zone cooling.

Earth Tubes

Some livestock facilities obtain cooling in summer and heating in winter by drawing ventilation air through tubing buried 6 to 13 ft below grade. These systems are most practical in the north central United States for animals that benefit from both cooling in summer and heating in winter.

Cooling and Tempering Air for Livestock Housing (MWPS 1990b) details design procedures for this method. A typical design uses 50 to 150 ft of 8 in. diameter pipe to provide 300 cfm of tempered air. Soil type and moisture, pipe depth, airflow, climate, and other factors affect the efficiency of buried pipe heat exchangers. The pipes must slope to drain condensation, and must not have dips that could plug with condensation.

Heat Exchangers

Ventilation accounts for 70 to 90% of the heat losses in typical livestock facilities during winter. Heat exchangers can reclaim some of the heat lost with the exhaust ventilating air. However, predicting fuel savings based on savings obtained during the coldest periods overestimates yearly savings from a heat exchanger. Estimates of energy savings based on air enthalpy can improve the accuracy of the predictions.

Heat exchanger design must address the problems of condensate freezing and/or dust accumulation on the heat-exchanging surfaces. If unresolved, these problems result in either reduced efficiency and/or the inconvenience of frequent cleaning.

Supplemental Heating

For poultry weighing 3.3 lb or more, for pigs heavier than 50 lb, and for other large animals such as dairy cows, body heat of animals at recommended space allocations is usually sufficient to maintain moderate temperatures (i.e., above 50°F) in a well-insulated structure. Combustion-type heaters are used to supplement heat for baby chicks and pigs. Supplemental heating also increases the moisture-holding capacity of the air, which reduces the quantity of air required for moisture removal. Various types of heating equipment may be included in ventilation, but they need to perform well in dusty and corrosive atmospheres.

Insulation Requirements

The amount of building insulation required depends on climate, animal space allocations, and animal heat and moisture production.

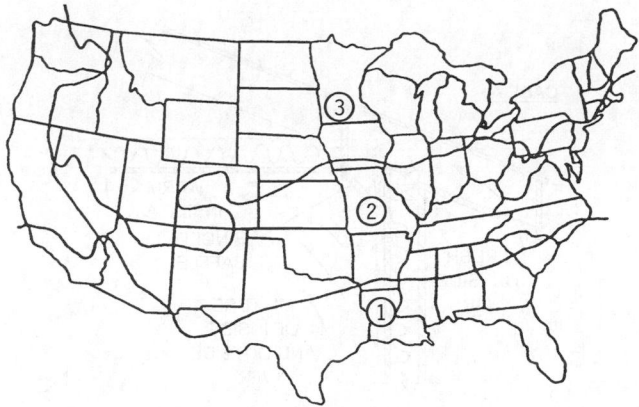

Fig. 4   Climatic Zones
(Reprinted with permission from ASAE *Standard* S401.2)

Table 1   Minimum Recommended Overall Coefficients of Heat Transmission *U* for Insulated Assemblies[a,b]

| Climatic Zone[d] | Recommended Minimum *U*, Btu/h·ft²·°F[c] | | | | | |
| | Cold | | Modified Environment | | Supplementally Heated | |
	Walls	Ceiling	Walls	Ceiling	Walls	Ceiling
1	—	0.17[e]	0.17[e]	0.071	0.071	0.045
2	—	0.17	0.17	0.059	0.071	0.040
3	—	0.17	0.083	0.040	0.050	0.030

[a]Use assembly U-factors that include framing effects, air spaces, air films, linings, and sidings. Determine assembly U-factors by testing the full assembly in accordance with ASTM *Standard* C1363 or calculate by the procedures presented in the 2009 *ASHRAE Handbook—Fundamentals*.
[b]Values shown are not the values necessary to provide a heat balance between heat produced by products or animals and heat transferred through the building.
[c]Current practice for poultry grow-out buildings uses a *U* of 0.11 to 0.14 Btu/h·ft²·°F in the roof and walls.
[d]Refer to Figure 4.
[e]Where ambient temperature and radiant heat load are severe, *U* = 0.83 Btu/h·ft²·°F.

Refer to Figure 4 and Table 1 for selecting insulation levels. In warm weather, ventilation between the roof and insulation helps reduce the radiant heat load from the ceiling. Insulation in warm climates can be more important for reducing radiant heat loads in summer than reducing building heat loss in winter.

Cold buildings have indoor conditions about the same as outside conditions. Examples are free-stall barns and open-front livestock buildings. Minimum insulation is frequently recommended in the roofs of these buildings to reduce solar heat gain in summer and to reduce condensation in winter.

Modified environment buildings rely on insulation, natural ventilation, and animal heat to remove moisture and to maintain the inside within a specified temperature range. Examples are warm free-stall barns, poultry production buildings, and swine finishing units.

Supplementary heated buildings require insulation, ventilation, and extra heat to maintain the desired inside temperature and humidity. Examples are swine farrowing and nursery buildings.

VENTILATION

Mechanical Ventilation

Mechanical ventilation uses fans to create a static pressure difference between the inside and outside of a building. Farm buildings use either positive pressure, with fans forcing air into a building, or negative pressure, with exhaust fans. Some ventilation systems use a combination of positive pressure to introduce air into a building and separate fans to remove air. These zero-pressure systems are particularly appropriate for heat exchangers.

Positive-Pressure Ventilation. Fans blow outside air into the ventilated space, forcing humid air out through any planned outlets and through leaks in walls and ceilings. If vapor barriers are not complete, moisture can condense within the walls and ceiling during cold weather. Condensation causes deterioration of building materials and reduces insulation effectiveness. The energy used by fan motors and rejected as heat is added to the building (an advantage in winter, but a disadvantage in summer).

Negative-Pressure Ventilation. Fans exhaust air from the ventilated space while drawing outside air in through planned inlets and leaks in walls, in ceilings, and around doors and windows. Air distribution in negative-pressure ventilation is often less complex and costly than positive- or neutral-pressure systems. Simple openings and baffled slots in walls control and distribute air in the building. However, at low airflow rates, negative pressure ventilation may not distribute air uniformly because of air leaks and wind pressure effects. Supplemental air mixing may be necessary.

Allowances should be made for reduced fan performance caused by dust, guards, and corrosion of louver joints (Person et al. 1979).

Totally enclosed fan motors are protected from exhaust air contaminants and humidity. Periodic cleaning helps prevent overheating. Negative-pressure ventilation is more commonly used than positive-pressure ventilation.

Ventilation should always be designed so that manure gases are not drawn into the building from manure storages connected to the building by underground pipes or channels.

Neutral-Pressure Ventilation. Neutral-pressure (push/pull) ventilation typically uses supply fans to distribute air down a distribution duct to room inlets, and exhaust fans to remove air from the room. Supply and exhaust fan capacities should be matched.

Neutral-pressure systems are often more expensive, but they achieve better control of the air. They are less susceptible to wind effects and to building leakage than positive- or negative-pressure systems. Neutral-pressure systems are most frequently used for young stock and for animals most sensitive to environmental conditions, primarily where cold weather is a concern.

Natural Ventilation

Either natural or mechanical ventilation is used to modify environments in livestock shelters. Natural ventilation is most common for mature animal housing, such as free-stall dairy, poultry growing, and swine finishing houses. Natural ventilation depends on pressure differences caused by wind and temperature differences. Well-designed natural ventilation keeps temperatures reasonably stable, if automatic controls regulate ventilation openings. Usually, a design includes an open ridge (with or without a rain cover) and openable sidewalls, which should cover at least 50% of the wall for summer operation. Ridge openings are about 2 in. wide for each 10 ft of house width, with a minimum ridge width of 6 in. to avoid freezing problems in cold climates. Upstand baffles on each side of the ridge opening greatly increase airflow (Riskowski et al. 1998). Small screens and square edges around sidewall openings can significantly reduce airflow through vents.

Openings can be adjusted automatically, with control based on air temperature. Some designs, referred to as flex housing, include a combination of mechanical and natural ventilation usually dictated by outside air temperature and/or the amount of ventilation required.

VENTILATION MANAGEMENT

Air Distribution

Pressure differences across walls and inlet or fan openings are usually maintained between 0.04 and 0.06 in. of water. (The exhaust fans are usually sized to provide proper ventilation at pressures up to 0.12 in. to compensate for wind effects.) This pressure difference creates inlet velocities of 600 to 1000 fpm, sufficient for effective air mixing, but low enough to cause only a small reduction in fan capacity. A properly planned inlet system distributes fresh air equally throughout the building. Negative pressure ventilation that relies on cracks around doors and windows does not distribute fresh air effectively. Inlets require adjustment, since winter airflow rates are typically less than 10% of summer rates. Automatic controllers and inlets are available to regulate inlet areas.

Positive pressure ventilation, with fans connected directly to perforated air distribution tubes, may combine heating, circulation, and ventilation in one system. Air distribution tubes or ducts connected to circulating fans are sometimes used to mix the air in negative pressure ventilation. Detailed design procedures for perforated ventilation tubes are described by Zhang (1994). However, dust in the ducts is of concern when air is recirculated, particularly when cold incoming air condenses moisture in the tubes.

Inlet Design. Inlet location and size most critically affect air distribution within a building. Continuous or intermittent inlets can be placed along the entire length of one or both outside walls. Building widths narrower than 20 ft may need only a single inlet along one wall. The total inlet area may be calculated by the system characteristic technique, which follows. Because the distribution of the inlet area is based on the geometry and size of the building, specific recommendations are difficult.

System Characteristic Technique. This technique determines the operating points for the ventilation rate and pressure difference across inlets. Fan airflow rate as a function of pressure difference across the fan should be available from the manufacturer. Allowances must be made for additional pressure losses from fan shutters or other devices such as light restriction systems or cooling pads.

Inlet flow characteristics are available for hinged baffle and center-ceiling flat baffle slotted inlets (Figure 5). Airflow rates can be calculated for the baffles in Figure 5 by the following:

For Case A:

$$Q = 285 W p^{0.5} \qquad (1)$$

For Case B:

$$Q = 183 W p^{0.5} \qquad (2)$$

For Case C (total airflow from sum of both sides):

$$Q = 320 W p^{0.5} (D/T)^{0.08} e^{(-0.867 \, W/T)} \qquad (3)$$

where

Q = airflow rate, cfm per foot length of slot opening
W = slot width, in.
p = pressure difference across the inlet, in. of water
D = baffle width, in.
T = width of slot in ceiling, in.

Zhang and Barber (1995) measured infiltration rates of five rooms in a newly built swine building at 0.12 cfm/ft² of surface area at 0.08 in. of water. Surface area included the area of walls and ceiling enclosing the room. It is important to include this infiltration rate into the ventilation design and management. For example, at 0.12 cfm/ft² of surface area, the infiltration represents 1.4 ach. In the heating season, the minimum ventilation is usually about 3 ach. Thus, large infiltration rates greatly reduce the airflow from the controlled inlet and adversely affect the air distribution.

Room Air Velocity. The average air velocity inside a slot-ventilated structure relates to the inlet air velocity, inlet slot width (or equivalent continuous length for boxed inlets), building width, and ceiling height. Estimates of air velocity within a barn, based on air exchange rates, may be very low because of the effects of jet velocity and recirculation. Conditions are usually partially turbulent, and there is no reliable way to predict room air velocity at

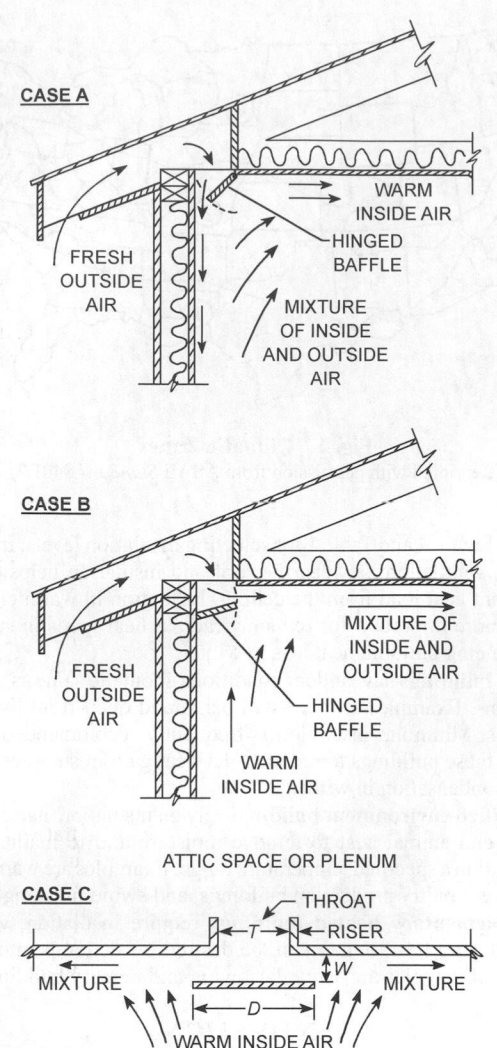

Fig. 5 Typical Livestock Building Inlet Configurations

animal level. General design guidelines keep the throw distance less than 20 ft from slots and less than 10 ft from perforated tubes.

Fans

Fans should not exhaust against prevailing winds, especially for cold-weather ventilation. If structural or other factors require installing fans on the windward side, fans rated to deliver the required capacity against at least 0.12 in. of water static pressure and with a relatively flat power curve should be selected. The fan motor should withstand a wind velocity of 30 mph, equivalent to a static pressure of 0.4 in. of water, without overloading beyond its service factor. Wind hoods on the fans or windbreak fences reduce the effects of wind.

Third-party test data should be used to obtain fan performance and energy efficiencies for fan selection (BESS Lab 1997). Fans should be tested with all accessories (e.g., louvers, guards, hoods) in place, just as they will be installed in the building. The accessories have a major effect on fan performance.

Flow Control. Because the numbers and size of livestock and climatic conditions vary, means to modulate ventilation rates are often required beyond the conventional off/on thermostat switch. The minimum ventilation rate to remove moisture, reduce air contaminant concentrations, and keep water from freezing should always be provided. Methods of modulating ventilation rates include (1) intermittent fan operation (fans operate for a percentage of the time controlled by a percentage timer with a 10 min cycle); (2) staging of fans using multiple units or fans with high/low-exhaust capability; (3) using multispeed fans [larger fans (1/2 hp and up) with two flow rates, the lower being about 60% of the maximum rate]; and (4) using variable-speed fans [split-capacitor motors designed to modulate fan speed smoothly from maximum down to 10 to 20% of the maximum rate (the controller is usually thermostatically adjusted)].

Generally, fans are spaced uniformly along the winter leeward side of a building. Maximum distance between fans is 115 to 165 ft. Fans may be grouped in a bank if this range is not exceeded. In housing with side curtains, exhaust fans that can be reversed or removed and placed inside the building in the summer are sometimes installed to increase air movement in combination with doors, walls, or windows being opened for natural ventilation.

Thermostats

Thermostats should be placed where they respond to a representative temperature as sensed by the animals. Thermostats need protection and should be placed to prevent potential physical or moisture damage (i.e., away from animals, ventilation inlets, water pipes, lights, heater exhausts, outside walls, or any other objects that will unduly affect performance). Thermostats also require periodic adjustment based on accurate thermometer readings taken in the immediate proximity of the animal.

Emergency Warning

Animals housed in a high-density, mechanically controlled environment are subject to considerable risk of heat prostration if a failure of power or ventilation equipment occurs. To reduce this danger, an alarm and an automatic standby electric generator are highly recommended. Many alarms detect failure of the ventilation. These alarms range from inexpensive power-off alarms to ones that sense temperature extremes and certain gases. Automatic telephone-dialing systems are effective as alarms and are relatively inexpensive. Building designs that allow some side wall panels (e.g., 25% of wall area) to be removed for emergency situations are also recommended.

RECOMMENDED PRACTICES BY SPECIES

Mature animals readily adapt to a broad range of temperatures, but efficiency of production varies. Younger animals are more temperature sensitive. Figure 6 illustrates animal production response to temperature.

Relative humidity has not been shown to influence animal performance, except when accompanied by thermal stress. Relative humidity consistently below 40% may contribute to excessive dustiness; above 80%, it may increase building and equipment deterioration. Disease pathogens also appear to be more viable at either low or high humidity. Relative humidity has a major influence on the effectiveness of skin-wetting cooling methods.

Dairy Cattle

Dairy cattle shelters include confinement stall barns, free stalls, and loose housing. In a stall barn, cattle are usually confined to stalls approximately 4 ft wide, where all chores, including milking and feeding, are conducted. Such a structure requires environmental modification, primarily through ventilation. Total space requirements are 50 to 75 ft^2 per cow. In free-stall housing, cattle are not confined to stalls but can move freely. Space requirements per cow are 75 to 100 ft^2. In loose housing, cattle are free to move within a fenced lot containing resting and feeding areas. Space required in sheltered loose housing is similar to that in free-stall housing. Shelters for resting and feeding areas are generally open-sided and require no air conditioning or mechanical ventilation,

but supplemental air mixing is often beneficial during warm weather. The milking area is in a separate area or facility and may be fully or partially enclosed, thus requiring some ventilation.

For dairy cattle, climate requirements for minimal economic loss are broad, and range from 35 to 75°F with 40 to 80% rh. Below 35°F, production efficiency declines and management problems increase. However, the effect of low temperature on milk production is not as extreme as are high temperatures, where evaporative coolers or other cooling methods may be warranted.

Ventilation Rates for Each 1100 lb Cow

Winter	Spring/Fall	Summer
36 to 47 cfm	142 to 190 cfm	230 to 470 cfm

Required ventilation rates depend on specific thermal characteristics of individual buildings and internal heating load. The relative humidity should be maintained between 50 and 80%.

Both loose housing and stall barns require an additional milk room to cool and hold the milk. Sanitation codes for milk production contain minimum ventilation requirements. The market being supplied should be consulted for all applicable codes. Some state codes require positive-pressure ventilation of milk rooms. Milk rooms are usually ventilated with fans at rates of 4 to 10 ach to satisfy requirements of local milk codes and to remove heat from milk coolers. Most milk codes require ventilation in the passageway (if any) between the milking area and the milk room.

Beef Cattle

Beef cattle ventilation requirements are similar to those of dairy cattle on a unit weight basis. Beef production facilities often provide only shade and wind breaks.

Swine

Swine housing can be grouped into four general classifications:

1. Farrowing pigs, from birth to 30 lb, and sows
2. Nursery pigs, from 30 to 75 lb
3. Growing/finishing pigs, from 75 lb to market weight
4. Breeding and gestation

In farrowing barns, two environments must be provided: one for sows and one for piglets. Because each requires a different temperature, zone heating and/or cooling is used. The environment within the nursery is similar to that within the farrowing barn for piglets. The requirements for growing barns and breeding stock housing are similar.

Currently recommended practices for **farrowing houses**:

- Temperature: 50 to 68°F, with small areas for piglets warmed to 82 to 90°F by brooders, heat lamps, or floor heat. Avoid cold drafts and extreme temperatures. Hovers are sometimes used. Provide supplemental cooling for sows (usually drippers or zone cooling) in extreme heat.
- Relative humidity: Up to 70% maximum
- Ventilation rate: 20 to 500 cfm per sow and litter (about 400 lb total weight). The low rate is for winter; the high rate is for summer temperature control.
- Space: 35 ft^2 per sow and litter (stall); 65 ft^2 per sow and litter (pens)

Recommendations for **nursery barns**:

- Temperature:
  80°F for first week after weaning. Lower room temperature 3°F per week to 72°F. Provide warm, draft-free floors. Provide supplemental cooling for extreme heat (temperatures 85°F and above).
- Ventilation rate:
  2 to 2.5 cfm per pig, 12 to 30 lb each
  3 to 35 cfm per pig, 30 to 75 lb each

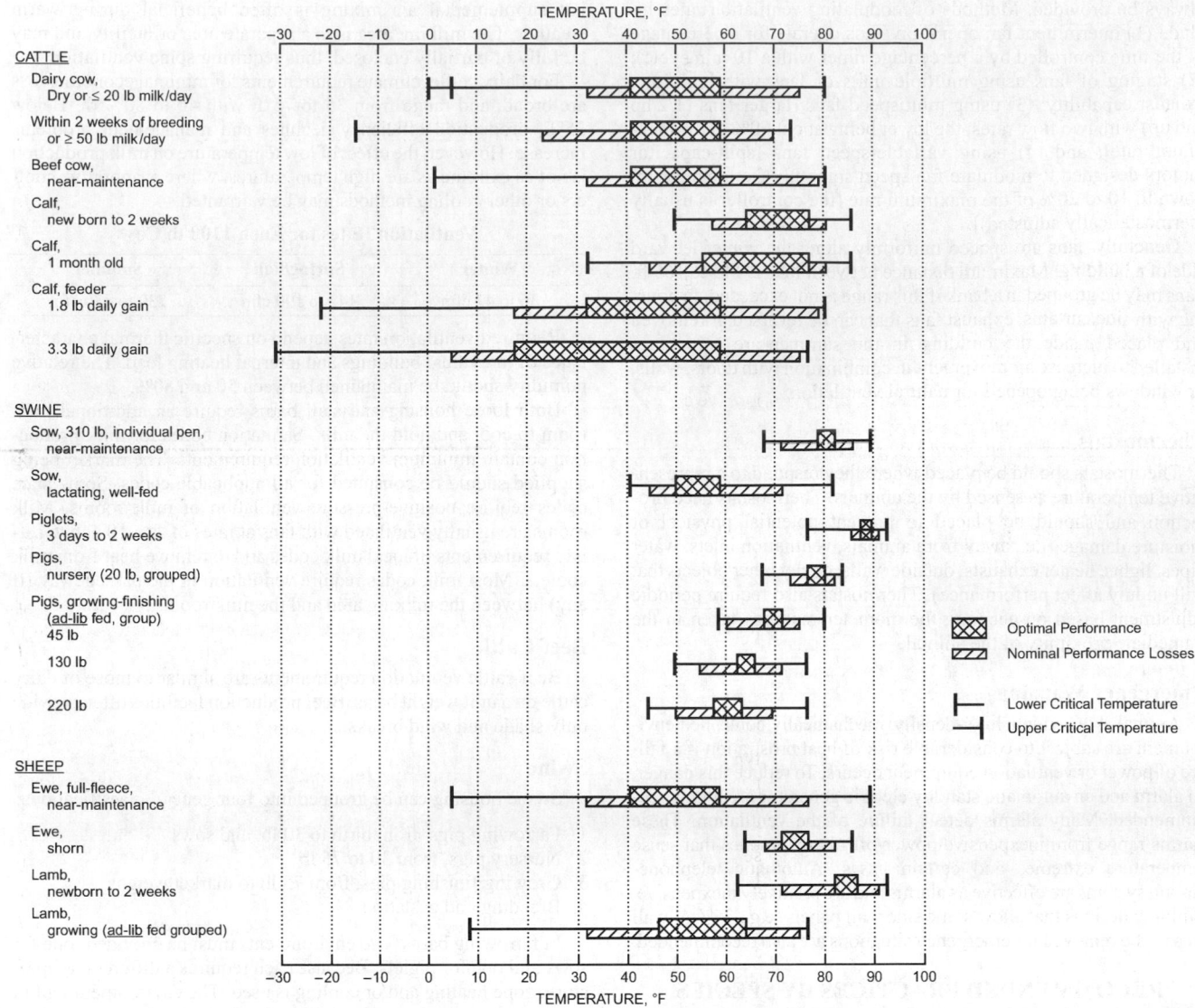

Fig. 6   Critical Ambient Temperatures and Temperature Zone for Optimum Performance and Nominal Performance Loss in Farm Animals

(Adapted from Hahn 1985, in *Stress Physiology in Livestock*, Vol. II, CRC Press)

- Space:
  2 to 2.5 ft² per pig, 12 to 30 lb each
  3 to 4 ft² per pig, 30 to 75 lb each

Recommendations for **growing** and **gestation barns**:

- Temperature:
  55 to 72°F preferred. Provide supplemental cooling (sprinklers or evaporative coolers) for extreme heat.
- Relative humidity:
  75% maximum in winter; no established limit in summer
- Ventilation rate:
  Growing pig (75 to 150 lb), 7 to 75 cfm
  Finishing pig (150 to 220 lb), 10 to 120 cfm
  Gestating sow (325 lb), 12 to 150 cfm
  Boar/breeding sow (400 lb), 14 to 300 cfm
- Space:
  6 ft² per pig, 75 to 150 lb each
  8 ft² per pig, 150 to 220 lb each
  14 to 24 ft² per sow, 240 to 500 lb each

Poultry

In broiler and brooder houses, growing chicks require changing environmental conditions, and heat and moisture dissipation rates increase as the chicks grow older. Supplemental heat, usually from brooders, is used until sensible heat produced by the birds is adequate to maintain an acceptable air temperature. At early stages of growth, moisture dissipation per bird is low. Consequently, low ventilation rates are recommended to prevent excessive heat loss. Litter is allowed to accumulate over 3 to 5 flock placements. Lack of low-cost litter material may justify the use of concrete floors. After each flock, caked litter is removed and fresh litter is added.

Housing for poultry may be open, curtain-sided or totally enclosed. Mechanical ventilation depends on the type of housing used. For open-sided housing, ventilation is generally natural airflow in warm weather, supplemented with stirring fans, and by fans with closed curtains in cold weather or during the brooding period. Mechanical ventilation is used in totally enclosed housing. Newer houses have smaller curtains and well-insulated construction to accommodate both natural hand mechanical ventilation operation.

Recommendations for **broiler houses**:

- Room temperature: 60 to 80°F
- Temperature under brooder hover: 86 to 91°F, reducing 5°F per week until room temperature is reached
- Relative humidity: 50 to 80%
- Ventilation rate: Sufficient to maintain house within 2 to 4°F of outside air conditions during summer. Generally, rates are about 0.1 cfm per lb live weight during winter and 1 to 2 cfm per lb for summer conditions.
- Space: 0.6 to 1.0 ft² per bird (for the first 21 days of brooding, only 50% of floor space is used)
- Light: Minimum of 10 lx or 1 footcandle to 28 days of age; 1 to 20 lx or 0.1 to 2 footcandles for growout (in enclosed housing).

Recommendations for **breeder houses** with birds on litter and slatted floors:

- Temperature: 50 to 86°F maximum; consider evaporative cooling if higher temperatures are expected.
- Relative humidity: 50 to 75%
- Ventilation rate: Same as for broilers on live weight basis.
- Space: 2 to 3 ft² per bird

Recommendations for **laying houses** with birds in cages:

- Temperature, relative humidity, and ventilation rate: Same as for breeders.
- Space: 50 to 65 in² per hen minimum
- Light: Controlled day length using light-controlled housing is generally practiced (January through June).

Laboratory Animals

The well-being and experimental response of laboratory animals depend greatly on the design of the facilities. Cage type, noise levels, light levels, air quality, and thermal environment can affect animal well-being and, in many cases, affect how the animal responds to experimental treatments (Clough 1982; Lindsey et al. 1978; McPherson 1975; Moreland 1975). If any of these factors vary across treatments or even within treatments, it can affect the validity of experimental results, or at least increase experimental error. Consequently, laboratory animal facilities must be designed and maintained to expose the animals to appropriate levels of these environmental conditions and to ensure that all animals in an experiment are in a uniform environment. See Chapter 16 for additional information on laboratory animal facilities.

In the United States, recommended environmental conditions within laboratory animal facilities are usually dictated by the *Guide for the Care and Use of Laboratory Animals* (ILAR 1996). Temperature recommendations vary from 61 to 84°F, depending on the species being housed. The acceptable range for relative humidity is 30 to 70%. For animals in confined spaces, daily temperature fluctuations should be minimized. Relative humidity must also be controlled, but not as precisely as temperature.

Ventilation recommendations are based on room air changes; however, cage ventilation rates may be inadequate in some cages and excessive in other cages, depending on cage and facility design. ILAR (1996) recommendations for room ventilation rates of 10 to 15 ach are an attempt to provide adequate ventilation for the room and cages. This recommendation is based on the assumption that adequate ventilation in the macroenvironment (room) provides sufficient ventilation to the microenvironment (cage). This may be a reasonable assumption when cages have a top of wire rods or mesh. However, several studies have shown that covering cages with filter tops, which provide a protective barrier for rodents and reduce airborne infections and diseases, especially neonatal diarrhea, can create significant differences in microenvironmental conditions.

Maghirang et al. (1995) and Riskowski et al. (1996) surveyed room and cage environmental conditions in several laboratory animal facilities and found that the animal's environmental needs may

not be met even though the facilities were designed and operated according to ILAR (1996). The microenvironments were often considerably poorer than the room conditions, especially in microisolator cages. For example, ammonia levels in cages were up to 60 ppm even though no ammonia was detected in a room. Cage temperatures were up to 7°F higher than room temperature and relative humidities up to 41% higher.

Furthermore, cage microenvironments in the same room were found to have significant variation (Riskowski et al. 1996): ammonia levels varied from 0 to 60 ppm, air temperature varied from 1 to 7°F higher than room temperature, relative humidity varied from 1 to 30% higher than room humidity, and average light levels varied from 2 to 337 lx. This survey found three identical rooms that had room ventilation rates from 4.4 to 12.5 ach but had no differences in room or cage environmental parameters.

A survey of laboratory animal environmental conditions in seven laboratory rat rooms was conducted by Zhang et al. (1992). They found that room air ammonia levels were under 0.5 ppm for all rooms, even though room airflow varied from 11 to 24 ach. Air exchange rates in the cages varied from less than 0.1 to 2.5 cfm per rat, and ammonia levels ranged from negligible to 60 ppm. Riskowski et al. (1996) measured several environmental parameters in rat shoebox cages in full-scale room mockups with various room and ventilation configurations. Significant variations in cage temperature and ventilation rates within a room were also found. Varying room ventilation rate from 5 to 15 ach did not have large effects on cage environmental conditions. These studies verify that designs based only on room air changes do not guarantee desired conditions in the animal cages.

In order to analyze the ventilation performance of different laboratory animal research facilities, Memarzadeh (1998) used **computational fluid dynamics** (CFD) to undertake computer simulation of over 100 different room configurations. CFD is a three-dimensional mathematical technique used to compute the motion of air, water, or any other gas or liquid. However, all conditions must be correctly specified in the simulation to produce accurate results. Empirical work defined inputs for such parameters as heat dissipation and surface temperature as well as the moisture, CO_2, and NH_3 mass generation rates for mice.

This approach compared favorably with experimentally measured temperatures and gas concentrations in a typical animal research facility. To investigate the relationships between room configuration parameters and the room and cage environments in laboratory animal research facilities, the following parameters were varied:

- Supply air diffuser type and orientation, air temperature, and air moisture content
- Room ventilation rate
- Exhaust location and number
- Room pressurization
- Rack layout and cage density
- Change station location, design, and status
- Leakage between the cage lower and upper moldings
- Room width

Room pressurization, change station design, and room width had little effect on ventilation performance. However, other factors found to affect either the macroenvironment or microenvironment or both led to the following observations:

- Ammonia production depends on relative humidity. Ten days after the last change of bedding, a high-humidity environment produced ammonia at about three times the rate of cages in a low-humidity environment.
- Acceptable room and cage ammonia concentrations after 5 days without changing cage bedding are produced by room supply airflow rates of around 0.85 cfm per 100 g of body mass of mice. This is equivalent to 5 ach for the room with single-density racks

considered in this study, and 10 ach for the room with double-density racks. The temperature of the supply air must be set appropriately for the heat load in the room. The room with single-density racks contained 1050 mice with a total mass of 21 kg and the room with double-density racks contained 2100 mice with a total mass of 42 kg.

- Increasing the room ventilation rate does not have a large effect on the cage ventilation. Increasing the supply airflow from 5 to 20 ach around single-density racks parallel to the walls reduces the CO_2 concentration from 1764 to 1667 ppm, a reduction of only 6%. For the double-density racks perpendicular to the walls, the reduction is larger, but still only from about 2300 to 1800 ppm (around 20%)

- Both the cage and the room ammonia concentrations can be reduced by increasing the supply air temperatures. This reduces the relative humidity for a given constant moisture content in the air, and the lower relative humidity leads to lower ammonia generation. Raising the supply discharge temperature from 66 to 72°F at 15 ach raises the room temperature by 5°F to around 73°F and the cages by 4°F to around 77°F. This can reduce ammonia concentrations by up to 50%.

- Using 72°F as the supply discharge temperature at 5 ach (the lowest flow rate considered) for double-density racks produces a room temperature around 79°F, with cage temperatures only slightly higher. Although this higher temperature provides a more comfortable environment for the mice (Gordon et al. 1997), the high room temperature may be unacceptable to the scientists working in the room.

- Ceiling or high-level exhausts tend to produce lower room temperatures (for a given supply air temperature, all CFD models were designed to have 72°F at the room exhaust) when compared to low-level exhausts. This indicates that low-level exhausts are less efficient at cooling the room.

- Low-level exhausts appear to ventilate the cages slightly better (up to 27% for the radial diffuser; much less for the slot diffuser) than ceiling or high-level exhausts when the cages are placed parallel to the walls, near the exhausts. Ammonia concentration in the cages decreased even further, although this is because of the higher temperatures in the low-level exhaust cases when compared to the ceiling and high-level exhausts. The room concentrations of CO_2 and ammonia do not show that any type of supply or exhaust is significantly better or worse than the other type.

DESIGN FOR PLANT FACILITIES

Greenhouses, plant growth chambers, and other facilities for indoor crop production overcome adverse outdoor environments and provide conditions conducive to economical crop production. The basic requirements of indoor crop production are (1) adequate light; (2) favorable temperatures; (3) favorable air or gas content; (4) protection from insects and disease; and (5) suitable growing media, substrate, and moisture. Because of their lower cost per unit of usable space, greenhouses are preferred over plant growth chambers for protected crop production.

This section covers greenhouses and plant growth facilities. Figure 7 shows the structural shapes of typical commercial greenhouses. Other greenhouses may have Gothic arches, curved glazing, or simple lean-to shapes. Glazing, in addition to traditional glass, now includes both film and rigid plastics. High light transmission by the glazing is usually important; good location and orientation of the house are important in providing desired light conditions. Location also affects heating and labor costs, exposure to plant disease and air pollution, and material handling requirements. As a general rule in the northern hemisphere, a greenhouse should be placed at a distance of at least 2.5 times the height of the object closest to it in the eastern, western, and southern directions.

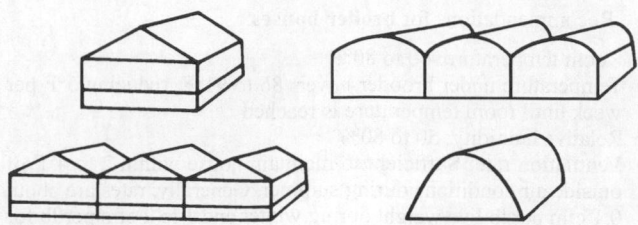

Fig. 7　Structural Shapes of Commercial Greenhouses

GREENHOUSES

Site Selection

Sunlight. Sunlight provides energy for plant growth and is often the limiting growth factor in greenhouses of the central and northern areas of North America during the winter. When planning greenhouses that are to be operated year-round, a designer should design for the greatest sunlight exposure during the short days of midwinter. The building site should have an open southern exposure, and if the land slopes, it should slope to the south.

Soil and Drainage. When plants are to be grown in the soil covered by the greenhouse, a growing site with deep, well-drained, fertile soil, preferably sandy loam or silt loam, should be chosen. Even though organic soil amendments can be added to poor soil, fewer problems occur with good natural soil. However, when good soil is not available, growing in artificial media should be considered. The greenhouse should be level, but the site can and often should be sloped and well-drained to reduce salt build-up and insufficient soil aeration. A high water table or a hardpan may produce water-saturated soil, increase greenhouse humidity, promote diseases, and prevent effective use of the greenhouse. If present, these problems can be alleviated by tile drains under and around the greenhouse. Ground beds should be level to prevent water from concentrating in low areas. Slopes within greenhouses also increase temperature and humidity stratification and create additional environmental problems.

Sheltered Areas. Provided they do not shade the greenhouse, surrounding trees act as wind barriers and help prevent winter heat loss. Deciduous trees are less effective than coniferous trees in midwinter, when the heat loss potential is greatest. In areas where snowdrifts occur, windbreaks and snowbreaks should be 100 ft or more from the greenhouse to prevent damage.

Orientation. Generally, in the northern hemisphere, for single-span greenhouses located north of 35° latitude, maximum transmission during winter is attained by an east-west orientation. South of 35° latitude, orientation is not important, provided headhouse structures do not shade the greenhouse. North-south orientation provides more light on an annual basis.

Gutter-connected or ridge-and-furrow greenhouses are oriented preferably with the ridge line north-south regardless of latitude. This orientation allows the shadow pattern caused by the gutter superstructure to move from the west to the east side of the gutter during the day. With an east-west orientation, the shadow pattern would remain north of the gutter, and the shadow would be widest and create the most shade during winter when light levels are already low. Also, the north-south orientation allows rows of tall crops, such as roses and staked tomatoes, to align with the long dimension of the house—an alignment that is generally more suitable to long rows and the plant support methods preferred by many growers.

The slope of the greenhouse roof is a critical part of greenhouse design. If the slope is too flat, a greater percentage of sunlight is reflected from the roof surface (Figure 8). A slope with a 1:2 rise-to-run ratio is the usual inclination for a gable roof.

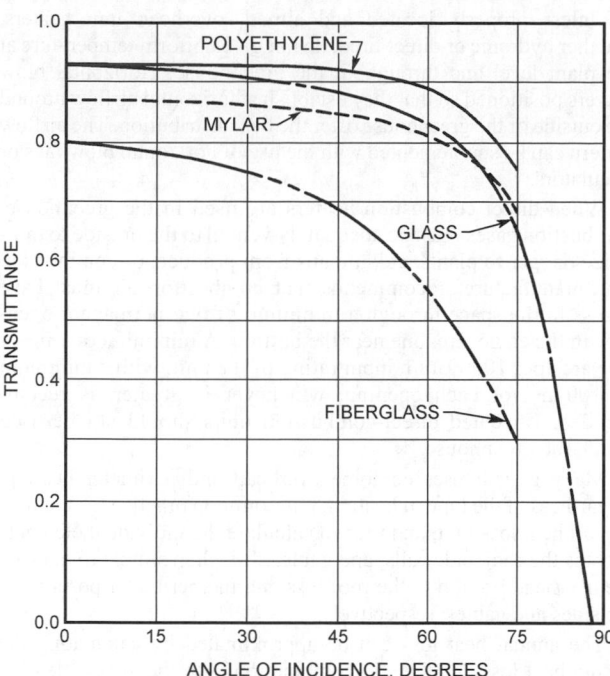

Fig. 8   Transmittance of Solar Radiation Through Glazing Materials for Various Angles of Incidence

Heating

Structural Heat Loss. Estimates for heating and cooling a greenhouse consider conduction, infiltration, and ventilation energy exchange. In addition, the calculations must consider solar energy load and electrical input, such as light sources, which are usually much greater for greenhouses than for conventional buildings. Generally, conduction q_c plus infiltration q_i are used to determine the peak requirements q_t for heating.

$$q_t = q_c + q_i \qquad (4)$$

$$q_c = UA(t_i - t_o) \qquad (5)$$

$$q_i = 0.018VN(t_i - t_o) \qquad (6)$$

where

U = overall heat loss coefficient, Btu/h·ft²·°F (Tables 2 and 3)
A = exposed surface area, ft²
t_i = inside temperature, °F
t_o = outside temperature, °F
V = greenhouse internal volume, ft³
N = number of air exchanges per hour (Table 4)

Type of Framing. The type of framing should be considered in determining overall heat loss. Aluminum framing and glazing systems may have the metal exposed to the exterior to a greater or lesser degree, and the heat transmission of this metal is higher than that of the glazing material. To allow for such a condition, the U-factor of the glazing material should be multiplied by the factors shown in Table 3.

Table 3   Construction U-Factor Multipliers

Metal frame and glazing system, 16 to 24 in. spacing	1.08
Metal frame and glazing system, 48 in. spacing	1.05
Fiberglass on metal frame	1.03
Film plastic on metal frame	1.02
Film or fiberglass on wood	1.00

Table 2   Suggested Heat Transmission Coefficients

		U, Btu/h·ft²·°F
Glass		
	Single glazing	1.13
	Double glazing	0.70
	Insulating	Manufacturers' data
Plastic film		
	Single film[a]	1.20
	Double film, inflated	0.70
	Single film over glass	0.85
	Double film over glass	0.60
Corrugated glass fiber		
	Reinforced panels	1.20
Plastic structured sheet[b]		
	16 mm thick	0.58
	8 mm thick	0.65
	6 mm thick	0.72

[a]Infrared barrier polyethylene films reduce heat loss; however, use this coefficient when designing heating systems because the structure could occasionally be covered with non-IR materials.
[b]Plastic structured sheets are double-walled, rigid plastic panels.

Table 4   Suggested Design Air Changes (N)

New Construction	
Single glass lapped (unsealed)	1.25
Single glass lapped (laps sealed)	1.0
Plastic film covered	0.6 to 1.0
Structured sheet	1.0
Film plastic over glass	0.9

Old Construction	
Good maintenance	1.5
Poor maintenance	2 to 4

Infiltration. Equation (6) may be used to calculate heat loss by infiltration. Table 4 suggests values for air changes N.

Radiation Energy Exchange. Solar gain can be estimated using the procedures outlined in Chapter 18 of the 2009 *ASHRAE Handbook—Fundamentals*. As a guide, when a greenhouse is filled with a mature crop of plants, one-half the incoming solar energy is converted to latent heat, and one-quarter to one-third, to sensible heat. The rest is either reflected out of the greenhouse or absorbed by the plants and used in photosynthesis.

Radiation from a greenhouse to a cold sky is more complex. Glass admits a large portion of solar radiation but does not transmit long-wave thermal radiation in excess of approximately 5000 nm. Plastic films transmit more of the thermal radiation but, in general, the total heat gains and losses are similar to those of glass. Newer plastic films containing infrared (IR) inhibitors reduce the thermal radiation loss. Plastic films and glass with improved radiation reflection are available at a somewhat higher cost. Some research greenhouses use a retractable horizontal heat curtain to reduce the effect of night sky losses. Normally, radiation energy exchange is not considered in calculating the design heat load.

Heating Systems. Greenhouses may have a variety of heaters. One is a convection heater that circulates hot water or steam through plain or finned pipe. The pipe is most commonly placed along walls and occasionally beneath plant benches to create desirable convection currents. A typical temperature distribution pattern created by perimeter heating is shown in Figure 9. More uniform temperatures can be achieved when about one-third the total heat comes from pipes spaced uniformly across the house. These pipes can be placed above or below the crop, but temperature stratification and shading are avoided when they are placed below. Outdoor weather conditions affect temperature distribution, especially on windy days in loosely constructed greenhouses. Manual or automatic overhead

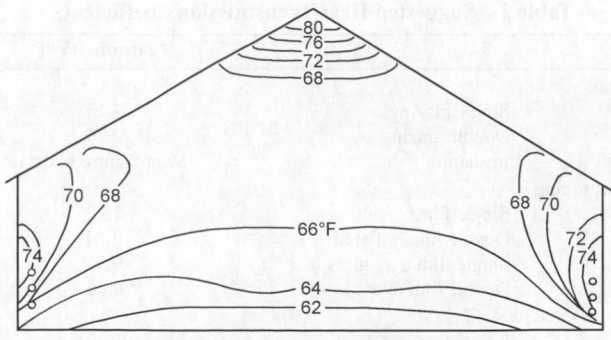

**Fig. 9   Temperature Profiles in a
Greenhouse Heated with Radiation Piping
along the Sidewalls**

pipes are also used for supplemental heating to prevent snow build-up on the roof. In a gutter-connected greenhouse in a cold climate, a heat pipe should be placed under each gutter to prevent snow accumulation.

An overhead tube heater consists of a unit heater that discharges into 12 to 30 in. diameter plastic film tubing perforated to provide uniform air distribution. The tube is suspended at 6 to 10 ft intervals and extends the length of the greenhouse. Variations include a tube and fan receiving the discharge of several unit heaters. The fan and tube system is used without heat to recirculate the air and, during cold weather, to introduce ventilation air. However, tubes sized for heat distribution may not be large enough for effective ventilation during warm weather.

Perforated tubing, 6 to 10 in. in diameter, placed at ground-level (underbench) heaters can also improve heat distribution. Ideally, the ground-level tubing should draw air from the top of the greenhouse for recirculation or heating. Tubes on or near the floor have the disadvantage of being obstacles to workers and reducing usable floor space.

Underfloor heating can supply up to 25% or more of the peak heating requirements in cold climates. A typical underfloor system uses 0.75 in. plastic pipe spaced 12 to 16 in. on center, and covered with 4 in. of gravel or porous concrete. Hot water, not exceeding 104°F, circulates at a rate of 2 to 2.5 gpm per loop. Pipe loops should generally not exceed 400 ft in length. This can provide 16 to 20 Btu·h·ft² from a bare floor, and about 75% as much when potted plants or seedling flats cover most of the floor.

Similar systems can heat soil directly, but root temperature must not exceed 77°F. When used with water from solar collectors or other heat sources, the underfloor area can store heat. This storage consists of a vinyl swimming pool liner placed on top of insulation and a moisture barrier at a depth of 8 to 12 in. below grade, and filled with 50% void gravel. Hot water from solar collectors or other clean sources enters and is pumped out on demand. Some heat sources, such as cooling water from power plants, cannot be used directly but require closed-loop heat transfer to avoid fouling the storage and the power plant cooling water.

Greenhouses can also be bottom-heated with 0.25 in. diameter EPDM tubing (or variations of that method) in a closed loop. The tubes can be placed directly in the growing medium of ground beds or under plant containers on raised benches. The best temperature uniformity is obtained by flow in alternate tubes in opposite directions. This method can supply all the greenhouse heat needed in mild climates.

Bottom heat, underfloor heating, and underbench heating are, because of the location of the heat source, more effective than overhead or peripheral heating, and can reduce energy loss by 20 to 30%.

Unless properly located and aimed, overhead unit heaters, whether hydronic or direct fired, do not give uniform temperature at the plant level and throughout the greenhouse. Horizontal blow heaters positioned so that they establish a horizontal airflow around the outside of the greenhouse offer the best distribution. The airflow pattern can be supplemented with the use of horizontal blow fans or circulators.

When direct combustion heaters are used in the greenhouse, combustion gases must be adequately vented to the outside to minimize danger to plants and humans from products of combustion. One manufacturer recommends that combustion air must have access to the space through a minimum of two permanent openings in the enclosure, one near the bottom. A minimum of 1 in² of free area per 1000 Btu/h input rating of the unit, with a minimum of 100 in² for each opening, whichever is greater, is recommended. Unvented direct-combustion units should not be used inside the greenhouse.

Many greenhouses combine overhead and perimeter heating. Regardless of the type of heating, it is common practice to calculate overall heat loss first, and then to calculate the individual elements such as the roof, sidewalls, and gables. It is then simple to allocate the overhead portion to the roof loss and the perimeter portions to the sides and gables, respectively.

The annual heat loss can be approximated by calculating the design heat loss and then, in combination with the annual degree-day tables using the 65°F base, estimating an annual heat loss and computing fuel usage on the basis of the rating of the particular fuel used. If a 50°F base is used, it can be prorated.

Heat curtains for energy conservation are becoming more important in greenhouse construction. Although this energy savings may be considered in the annual energy use, it should not be used when calculating design heat load; the practice is to open the heat curtains during snowstorms to facilitate snow melting, thereby nullifying its contribution to the design heat loss value.

Air-to-air and water-to-air heat pumps have been used experimentally on small-scale installations. Their usefulness is especially sensitive to the availability of a low-cost heat source.

Radiant (Infrared) Heating. Radiant heating is used in some limited applications for greenhouse heating. Steel pipes spaced at intervals and heated to a relatively high temperature by special gas heaters serve as the source of radiation. Because the energy is transmitted by radiation from a source of limited size, proper spacing is important to completely cover the heated area. Further, heavy-foliage crops can shade the lower parts of the plants and the soil, thus restricting the radiation from warming the root zone, which is important to plant growth.

Cogenerated Sources of Heat. Greenhouses have been built near or adjacent to power plants to use the heat and electricity generated by the facility. Although this energy may cost very little, an adequate standby energy source must be provided, unless the power supplier can assure that it will supply a reliable, continuous source of energy.

Cooling

Solar radiation is a considerable source of sensible heat gain; even though some of this energy is reflected from the greenhouse, some of it is converted into latent heat as the plants transpire moisture, and some is converted to plant material by photosynthesis. Natural ventilation, mechanical ventilation, shading, and evaporative cooling are common methods used to remove this heat. Mechanical refrigeration is seldom used to air-condition greenhouses because the cooling load and resulting cost is so high.

Natural Ventilation. Most older greenhouses and many new ones rely on natural ventilation with continuous roof sashes on each side of the ridge and continuous sashes in the sidewalls. The roof sashes are hinged at the ridge, and the wall sashes are hinged at the

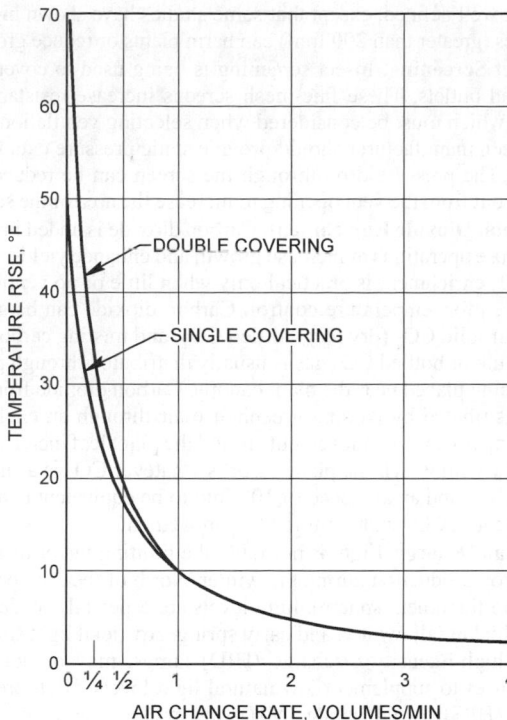

Fig. 10 Influence of Air Exchange Rate on Temperature Rise in Single- and Double-Covered Greenhouses

Table 5 Multipliers for Calculating Airflow for Fan-and-Pad Cooling

Elevation (Above Sea Level)		Max. Interior Light Intensity		Fan-to-Pad Temp. Difference	
ft	F_e	footcandles	F_l	°F	F_t
<1000	1.00	4000	0.80	10	0.70
1000	1.04	4500	0.90	9	0.78
2000	1.08	5000	1.00	8	0.88
3000	1.12	5500	1.10	7	1.00
4000	1.16	6000	1.20	6	1.17
5000	1.20	6500	1.30	5	1.40
6000	1.25	7000	1.40	4	1.75
7000	1.30	7500	1.50		
8000	1.36	8000	1.60		

Table 6 Velocity Factors for Calculating Airflow for Fan-to-Pad Cooling

Fan-to-Pad Distance, ft	F_v	Fan-to-Pad Distance, ft	F_v
20	2.24	65	1.24
25	2.00	70	1.20
30	1.83	75	1.15
35	1.69	80	1.12
40	1.58	85	1.08
45	1.49	90	1.05
50	1.41	95	1.03
55	1.35	100	1.00
60	1.29		

top of the sash. During much of the year, vents admit enough ventilating air for cooling without the added cost of running fans.

The principles of natural ventilation are explained in Chapter 16 of the 2009 *ASHRAE Handbook—Fundamentals*. Ventilation air is driven by wind and thermal buoyancy forces. Proper vent openings take advantage of pressure differences created by wind. Thermal buoyancy caused by the temperature difference between the inside and the outside of the greenhouse is enhanced by the area of the vent opening and the stack height (vertical distance between the center of the lower and upper opening). Within the limits of typical construction, the larger the vents, the greater the ventilating air exchanged. For a single greenhouse, the combined area of the sidewall vents should equal that of the roof vents. In ranges of several gutter-connected greenhouses, the sidewall area cannot equal the roof vent area.

Mechanical (Forced) Ventilation. Exhaust fans provide positive ventilation without depending on wind or thermal buoyancy forces. The fans are installed in the side or end walls of the greenhouse and draw air through vents on the opposite side or end walls. Air velocity through the inlets should not exceed 400 fpm.

Air exchange rates between 0.75 and 1 change per minute effectively control the temperature rise in a greenhouse. As shown in Figure 10, the temperature inside the greenhouse rises rapidly at lower airflow rates. At higher airflow rates, the reduction of the temperature rise is small, fan power requirements are increased, and plants may be damaged by the high air speed.

Shading. Shading compounds can be applied in varying amounts to the exterior of the roof of the greenhouse to achieve up to 50% shading. Durability of these compounds varies; ideally, the compound wears away during the summer and leaves the glazing clean in the fall, when shading is no longer needed. In practice, some physical cleaning is needed. Compounds used formerly usually contained lime, which corrodes aluminum and attacks some caulking. Most compounds used currently are formulated to avoid this problem.

Mechanically operated shade cloth systems with a wide range of shade levels are also available. They are mounted inside the greenhouse to protect them from the weather. Not all shading compounds or shade cloths are compatible with all plastic glazings, so the manufacturers' instructions and precautions should be followed.

Evaporative Cooling.

Fan-and-Pad Systems. Fans for fan-and-pad evaporative cooling are installed in the same manner as fans used for mechanical ventilation. Pads of cellulose material in a honeycomb form are installed on the inlet side. The pads are kept wet continuously when evaporative cooling is needed. As air is drawn through the pads, the water evaporates and cools the air. New pads cool the air by about 80% of the difference between the outdoor dry- and wet-bulb temperatures, or to 3 to 4°F above the wet-bulb temperature. The principles of applying evaporative cooling are explained in Chapter 40 of the 2008 *ASHRAE Handbook—HVAC Systems and Equipment* and in Chapter 52 of this volume.

The empirical base rate of airflow is 8 cfm per square foot of floor area. This flow rate is modified by multiplying it by factors for elevation (F_e), maximum interior light intensity (F_l), and allowable temperature rise between the pad and fans (F_t). These factors are listed in Table 5. The overall factor for the house is given by the following equation:

$$F_h = F_e F_l F_t \qquad (7)$$

The maximum fan-to-pad distance should be kept to 175 ft, although some greenhouses with distances of 225 ft have shown no serious reduction in effectiveness. With short distances, the air velocity becomes so low that the air feels clammy and stuffy, even though the airflow is sufficient for cooling. Therefore, a velocity factor F_v listed in Table 6 is used for distances less than 100 ft. For distance less than 100 ft, F_v is compared to F_h. The factor that gives the greatest airflow is used to modify the empirical base rate. For fan-to-pad distances greater than 100 ft, F_v can be ignored.

For best performance, pads should be installed on the windward side, and fans spaced within 25 ft of each other. Fans should not blow toward pads of an adjacent house unless it is at least 50 ft away. Fans in adjacent houses should be offset if they blow toward each other and are within 15 ft of each other.

Recommended air velocities through commonly used pads are listed in Table 7. Water flow and sump capacities are shown in Table 8. The system should also include a small, continuous bleed-off of water to reduce the build-up of dirt and other impurities.

Unit Evaporative Coolers. This equipment contains the pads, water pump, sump, and fan in one unit. Unit coolers are primarily used for small compartments. They are mounted 15 to 20 ft apart on the sidewall and blow directly into the greenhouse. They cool a distance of up to 50 ft from the unit. A side sash on the outside opposite wall is the best outlet, but roof vents may also work. The roof vent on the same side as the unit should be slightly open for better air distribution. If the roof vent on the opposite side is opened instead, air may flow directly out the vent and not cool the opposite side of the greenhouse.

Fog. In a direct-pressure atomizer, a high-pressure pump forces water at 800 to 1000 psi through a special fog nozzle. Fog is considered to be a water droplet smaller than 40 μm in diameter. The direct-pressure atomizer generates droplets of 35 μm or less. This requires a superior filter to minimize clogging of the very small nozzle orifices.

A line of nozzles placed along the top of the vent opening can cool the entering air nearly to its wet-bulb temperature. Additional lines in the greenhouse continue to cool the air as it absorbs heat in the space.

Fogging cools satisfactorily with less airflow than fan-and-pad systems, but the fan capacity must still be based on one air change per minute to ventilate the greenhouse when the cooler will be used without fog.

Other Environmental Controls

Humidity Control. At various times during the year, humidity may need to be controlled in the greenhouse. When the humidity is too high at night, it can be reduced by adding heat and ventilating simultaneously. When the humidity is too low during the day, it can be increased by turning on a fog or mist nozzle.

Winter Ventilation. During the winter, houses are normally closed tightly to conserve heat, but photosynthesis by the plants may lower the carbon dioxide level to such a point that it slows plant growth. Some ventilation helps maintain inside carbon dioxide levels. A normal rate of airflow for winter ventilation is 2 to 3 cfm per square foot of floor area.

Air Circulation. Continuous air circulation within the greenhouse reduces still-air conditions that favor plant diseases. Recirculating fans, heaters that blow air horizontally, and fans attached to polyethylene tubes are used to circulate air. The amount of recirculation has

Table 7 Recommended Air Velocity Through Various Pad Materials

Pad Type and Thickness	Air Face Velocity Through Pad,* fpm
Corrugated cellulose, 4 in. thick	250
Corrugated cellulose, 6 in. thick	350

*Speed may be increased by 25% where construction is limiting.

Table 8 Recommended Water Flow and Sump Capacity for Vertically Mounted Cooling Pad Materials

Pad Type and Thickness	Minimum Water Rate per Linear Foot of Pad, gpm	Minimum Sump Capacity per Unit Pad Area, gal/ft^2
Corrugated cellulose, 4 in. thick	0.5	0.8
Corrugated cellulose, 6 in. thick	0.8	1.0

not been well defined, except that some studies have shown high air velocities (greater than 200 fpm) can harm plants or reduce growth.

Insect Screening. Insect screening is being used to cover vent inlets and outlets. These fine-mesh screens increase resistance to airflow, which must be considered when selecting ventilation fans. The screen manufacturer should provide static pressure data for its screens. The pressure drop through the screen can be reduced by framing out from the vent opening to increase the area of the screen.

Carbon Dioxide Enrichment. Carbon dioxide is added in some greenhouse operations to increase growth and enhance yields. However, CO_2 enrichment is practical only when little or no ventilation is required for temperature control. Carbon dioxide can be generated from solid CO_2 (dry ice), bottled CO_2, and misting carbonated water. Bulk or bottled CO_2 gas is usually distributed through perforated tubing placed near the plant canopy. Carbon dioxide from dry ice is distributed by passing greenhouse air through an enclosure containing dry ice. Air movement around the plant leaf increases the efficiency with which the plant absorbs whatever CO_2 is available. One study found an air speed of 100 fpm to be equivalent to a 50% enrichment in CO_2 without forced air movement.

Radiant Energy. Light is normally the limiting factor in greenhouse crop production during the winter. North of the 35th parallel (in the northern hemisphere), light levels are especially inadequate or marginal in fall, winter, and early spring. Artificial light sources, usually high-intensity discharge (HID) lamps, may be added to greenhouses to supplement low natural light levels. High-pressure sodium (HPS), metal halide (MH), low-pressure sodium (LPS), and, occasionally, mercury lamps coated with a color-improving phosphor are currently used. Because differing irradiance or illuminance ratios are emitted by the various lamp types, the incident radiation is best described as radiant flux density (W/ft^2) between 400 and 850 nm, or as photon flux density between 400 and 700 nm, rather than in photometric terms of lux or footcandles.

To assist in relating irradiance to more familiar illuminance values, Table 9 shows constants for converting illuminance (lux) and photon flux density [$μmol/(s·m^2)$] of HPS, MH, LPS, and other lamps to the irradiance (W/m^2). One footcandle is approximately 10 lux.

Table 10 gives values of suggested irradiance at the top of the plant canopy, duration, and time of day for supplementing natural light levels for specific plants.

HID lamps in luminaires developed specifically for greenhouse use are often placed in a horizontal position, which may decrease both the light output and the life of the lamp. These drawbacks may be balanced by improved horizontal and vertical uniformity as compared to industrial parabolic reflectors.

Photoperiod Control. Artificial light sources are also used to lengthen the photoperiod during the short days of winter. Photoperiod control requires much lower light levels than those needed for photosynthesis and growth. Photoperiod illuminance needs to be only 0.6 to 1.1 W/ft^2. The incandescent lamp is the most effective light source for this purpose because of its higher far-red component. Lamps such as 150 W (PS-30) silverneck lamps spaced 10 to

Table 9 Constants to Convert to W/m^2

Light Source	klx	μmol/(s·m^2)
400 to 700 nm		
Incandescent (INC)	3.99	0.20
Fluorescent cool white (FCW)	2.93	0.22
Fluorescent warm white (FWW)	2.81	0.21
Discharge clear mercury (HG)	2.62	0.22
Metal halide (MH)	3.05	0.22
High-pressure sodium (HPS)	2.45	0.20
Low-pressure sodium (LPS)	1.92	0.20
Daylight	4.02	0.22

Table 10 Suggested Radiant Energy, Duration, and Time of Day for Supplemental Lighting in Greenhouses

Plant and Stage of Growth	W/ft²	Duration Hours	Duration Time
African violets	1 to 2	12 to 16	0600-1800
early-flowering			0600-2200
Ageratum	1 to 4.5	24	
early-flowering			
Begonias—fibrous rooted	1 to 2	24	
branching and early-flowering			
Carnation	1 to 2	16	0800-2400
branching and early-flowering			
Chrysanthemums	1 to 2	16	0800-2400
vegetable growth branching			
and multiflowering	1 to 2	8	0800-1600
Cineraria	0.6 to 1	24	
seedling growth (four weeks)			
Cucumber	1 to 2	24	
rapid growth and early-flowering			
Eggplant	1 to 4.5	24	
early-fruiting			
Foliage plants	0.6 to 1	24	
(Philodendron, Schefflera)			
rapid growth			
Geranium	1 to 4.5	24	
branching and early-flowering			
Gloxinia	1 to 4.5	16	0800-2400
early-flowering	0.6 to 1	24	
Lettuce	1 to 4.5	24	
rapid growth			
Marigold	1 to 4.5	24	
early-flowering			
Impatiens—New Guinea	1	16	0800-2400
branching and early-flowering			
Impatiens—Sultana	1 to 2	24	
branching and early-flowering			
Juniper	1 to 4.5	24	
vegetative growth			
Pepper	1 to 2	24	
early-fruiting, compact growth			
Petunia	1 to 4.5	24	
branching and early-flowering			
Poinsettia—vegetative growth	1	24	
branching and multiflowering	1 to 2	8	0800-1600
Rhododendron	1	16	0800-2400
vegetative growth (shearing tips)			
Roses (hybrid teas, miniatures)	1 to 4.5	24	
early-flowering and rapid regrowth			
Salvia	1 to 4.5	24	
early-flowering			
Snapdragon	1 to 4.5	24	
early-flowering			
Streptocarpus	1	16	0800-2400
early-flowering			
Tomato	1 to 2	16	0800-2400
rapid growth and early-flowering			
Trees (deciduous)	0.6	16	1600-0800
vegetative growth			
Zinnia	1 to 4.5	24	
early-flowering			

13 ft on centers and 13 ft above the plants provide a cost-effective system. Where a 13 ft height is not practical, 60 W extended service lamps on 6.5 ft centers are satisfactory. One method of photoperiod control is to interrupt the dark period by turning the lamps on at 2200 and off at 0200. The 4 h interruption, initially based on chrysanthemum response, induces a satisfactory long-day response in all photoperiodically sensitive species. Many species, however, respond to interruptions of 1 h or less. Demand charges can be reduced in large installations by operating some sections from 2000 to 2400 and others from 2400 to 0400. The biological response to these schedules, however, is much weaker than with the 2200 to 0200 schedule, so some varieties may flower prematurely. If the 4 h interruption period is used, it is not necessary to keep the light on throughout the interruption period. Photoperiod control of most plants can be accomplished by operating the lamps on light and dark cycles with 20% *on* times; for example, 12 s/min. The length of the dark period in the cycle is critical, and the system may fail if the dark period exceeds about 30 min. Demand charges can be reduced by alternate scheduling of the *on* times between houses or benches without reducing the biological effectiveness of the interruption.

Plant displays in places such as showrooms or shopping malls require enough light for plant maintenance and a spectral distribution that best shows the plants. Metal halide lamps, with or without incandescent highlighting, are often used for this purpose. Fluorescent lamps, frequently of the special phosphor plant-growth type, enhance color rendition, but are more difficult to install in aesthetically pleasing designs.

Design Conditions

Plant requirements vary from season to season and during different stages of growth. Even different varieties of the same species of plant may vary in their requirements. State and local cooperative extension offices are a good source of specific information on design conditions affecting plants. These offices also provide current, area-specific information on greenhouse operations.

Alternative Energy Sources and Energy Conservation

Limited progress has been achieved in heating commercial greenhouses with solar energy. Collecting and storing the heat requires a volume at least one-half the volume of the entire greenhouse. Passive solar units work at certain times of the year and, in a few localities, year-round.

If available, reject heat is a possible source of winter heat. Winter energy and solar (photovoltaic) sources are possible future energy sources for greenhouses, but the development of such systems is still in the research stage.

Energy Conservation. A number of energy-saving measures (e.g., thermal curtains, double glazing, and perimeter insulation) have been retrofitted to existing greenhouses and incorporated into new construction. Sound maintenance is necessary to keep heating system efficiency at a maximum level.

Automatic controls, such as thermostats, should be calibrated and cleaned at regular intervals, and heating-ventilation controls should interlock to avoid simultaneous operation. Boilers that can burn more than one type of fuel allow use of the most inexpensive fuel available.

Modifications to Reduce Heat Loss

Film covers that reduce heat loss are used widely in commercial greenhouses, particularly for growing foliage plants and other species that grow under low light levels. Irradiance (intensity) is reduced 10 to 15% per layer of plastic film.

One or two layers of transparent 4 or 6 mil continuous-sheet plastic is stretched over the entire greenhouse (leaving some vents uncovered), or from the ridge to the sidewall ventilation opening. When two layers are used, (outdoor) air at a pressure of 0.2 to 0.25 in. of water is introduced continuously between the layers of film to maintain the air space between them. When a single layer is used, an air space can be established by stretching the plastic over the glazing bars and fastening it around the edges, or a length of polyethylene tubing can be placed between the glass and the plastic and inflated (using outside air) to stretch the plastic sheet.

Double-Glazing Rigid Plastic. Double-wall panels are manufactured from acrylic and polycarbonate plastics, with walls separated by about 0.4 in. Panels are usually 48 in. wide and 96 in. or longer. Nearly all types of plastic panels have a high thermal expansion coefficient and require about 1% expansion space (0.12 in/ft). When a panel is new, light reduction is roughly 10 to 20%. Moisture accumulation between the walls of the panels must be avoided.

Double-Glazing Glass. The framing of most older greenhouses must be modified or replaced to accept double glazing with glass.

Light reduction is 10% more than with single glazing. Moisture and dust accumulation between glazings increases light loss. As with all types of double glazing, snow on the roof melts slowly and increases light loss. Snow may even accumulate enough to cause structural damage, especially in gutter-connected greenhouses.

Silicone Sealants. Transparent silicone sealant in the glass overlaps of conventional greenhouses reduces infiltration and may produce heat savings of 5 to 10% in older structures. There is little change in light transmission.

Precautions. The preceding methods reduce heat loss by reducing conduction and infiltration. They may also cause more condensation, higher relative humidity, lower carbon dioxide concentration, and an increase in ethylene and other pollutants. Combined with the reduced light levels, these factors may cause delayed crop production, elongated plants, soft plants, and various deformities and diseases, all of which reduce the marketable crop.

Thermal Blankets. Thermal blankets are any flexible material that is pulled from gutter to gutter and end to end in a greenhouse, or around and over each bench, at night. Materials ranging from plastic film to heavy cloth, or laminated combinations, have successfully reduced heat losses by 25 to 35% overall. Tightness of fit around edges and other obstructions is more important than the kind of material used. Some films are vaportight and retain moisture and gases. Others are porous and allow some gas exchange between the plants and the air outside the blanket. Opaque materials can control crop day length when short days are part of the requirement for that crop. Condensation may drip onto and collect on the upper sides of some blanket materials to such an extent that they collapse.

Multiple-layer blankets, with two or more layers separated by air spaces, have been developed. One such design combines a porous-material blanket and a transparent film blanket; the latter is used for summer shading. Another design has four layers of porous, aluminum foil-covered cloths, with the layers separated by air.

Thermal blankets may be opened and closed manually as well as automatically. The decision to open or close should be based on irradiance level and whether it is snowing, rather than on time of day. Two difficulties with thermal blankets are the physical problems of installation and use in greenhouses with interior supporting columns, and the loss of space from shading by the blanket when it is not in use during the day.

Other Recommendations. Although the foundation can be insulated, the insulating materials must be protected from moisture, and the foundation wall should be protected from freezing. All or most of the north wall can be insulated with opaque or reflective-surface materials. The insulation reduces the amount of diffuse light entering the greenhouse and, in cloudy climates, causes reduced crop growth near the north wall.

Ventilation fan cabinets should be insulated, and fans not needed in winter should be sealed against air leaks. Efficient management

and operation of existing facilities are the most cost-effective ways to reduce energy use.

PLANT GROWTH ENVIRONMENTAL FACILITIES

Controlled-environment rooms (CERs), also called plant growth chambers, include all controlled or partially controlled environmental facilities for growing plants, except greenhouses. CERs are indoor facilities. Units with floor areas less than 50 ft^2 may be moveable with self-contained or attached refrigeration units. CERs usually have artificial light sources, provide control of temperature and, in some cases, control relative humidity and CO_2 level.

CERs are used to study all aspects of botany. Some growers use growing rooms to increase seedling growth rate, produce more uniform seedlings, and grow specialized, high-value crops. The main components of the CER are (1) an insulated room or an insulated box with an access door; (2) a heating and cooling mechanism with associated air-moving devices and controls; and (3) a lamp module at the top of the insulated box or room. CERs are similar to walk-in cold storage rooms, except for the lighting and larger refrigeration system needed to handle heat produced by the lighting.

Location

The location for a CER must have space for the outside dimensions of the chamber, refrigeration equipment, ballast rack, and control panels. Additional space around the unit is necessary for servicing the various components of the system and, in some cases, for substrate, pots, nutrient solutions, and other paraphernalia associated with plant research. The location requires electricity, water, compressed air, and ventilation and exhaust air systems. For planning purposes, electrical densities of up to 140 W/ft^2 of controlled environment space) are possible, or 95 W/ft^2 of total space housing CERs.

Construction and Materials

Wall insulation should have a thermal conductance of less than 0.026 Btu/h·ft^2·°F. Materials should resist corrosion and moisture. The interior wall covering should be metal, with a high-reflectance white paint, or specular aluminum with a reflectivity of at least 80%. Reflective films or similar materials can be used, but require periodic replacement.

Floors and Drains

Floors that are part of the CER should be corrosion-resistant. Tar or asphalt waterproofing materials and volatile caulking compounds should not be used because they are likely to release phytotoxic gases into the chamber atmosphere. The floor must have a drain to remove spilled water and nutrient solutions. The drains should be trapped and equipped with screens to catch plant and substrate debris.

Plant Benches

Three bench styles for supporting the pots and other plant containers are normally encountered in plant growth chambers: (1) stationary benches; (2) benches or shelves built in sections that are adjustable in height; and (3) plant trucks, carts, or dollies on casters, which are used to move plants between chambers, greenhouses, and darkrooms. The bench supports containers filled with moist sand, soil, or other substrate, and is usually rated for loads of at least 50 lb/ft^2. The bench or truck top should be constructed of nonferrous, perforated metal, wire, or metal mesh to allow free passage of air around the plants and to let excess water drain from the containers to the floor and subsequently to the floor drain.

Normally, benches, shelves, or truck tops are adjustable in height so that small plants can be placed close to the lamps and thus receive a greater amount of light. As the plants grow, the shelf or bench is lowered so that the tops of the plants continue to receive the original radiant flux density.

Control

Environmental chambers require complex controls to provide the following:

- Automatic transfer from heating to cooling with 2°F or less dead zone and adjustable time delay.
- Automatic daily switching of the temperature set point for different day and night temperatures (setback may be as much as 10°F).
- Protection of sensors from radiation. Ideally, the sensors are located in a shielded, aspirated housing, but satisfactory performance can be attained by placing them in the return air duct.
- Control of the daily duration of light and dark periods. Ideally, this control should be programmable to change the light period each day to simulate the natural progression of day length. Photoperiod control, however, is normally accomplished with mechanical time clocks, which must have a control interval of 5 min or less for satisfactory timing.
- Protective control to prevent the chamber temperature from going more than a few degrees above or below the set point. Control should also prevent short-cycling of the refrigeration system, especially when condensers are remotely located.
- Control of the CO_2 level in enriched environment chambers.
- Audible and visual alarms to alert personnel of malfunctions.
- Maintenance of relative humidity to prescribed limits.

Data loggers, recorders, or recording controllers are recommended for monitoring daily operation. Solid-state, microprocessor-based controls are widely used for programming, controlling, and monitoring the CER conditions. Host systems are also used to program and monitor larger numbers of units in a common facility, and most offer remote access functions. Host systems tend to be vendor-specific in their use and application.

Heating, Air Conditioning, and Airflow

When the lights are on, cooling will normally be required, and the heater will rarely be called on to operate. When the lights are off, however, both heating and cooling may be needed. Conventional refrigeration is generally used with some modification. Direct-expansion units usually operate with a hot-gas bypass to prevent numerous on/off cycles, and secondary coolant may use aqueous ethylene glycol rather than chilled water. Heat is usually provided by electric heaters, but other energy sources can be used, including hot gas from the refrigeration.

The plant compartment is the heart of the growth chamber. The primary design objective, therefore, is to provide the most uniform, consistent, and regulated environmental conditions possible. Thus, airflow must be adequate to meet specified psychrometric conditions, but it is limited by the effects of high air speed on plant growth. As a rule, the average air speed in CERs is restricted to about 100 fpm.

To meet the uniform conditions required by a CER, conditioned air is normally moved through the space from bottom to top, although some CERs use top-to-bottom airflow. There is no apparent difference in plant growth between horizontal, upward, or downward airflow when the speed is less than 175 fpm. Regardless of the method, a temperature gradient is certain to exist, and should be kept as small as possible. Uniform airflow is more important than the direction of flow; thus, selection of properly designed diffusers or plenums with perforations is essential for achieving it.

The ducts or false sidewalls that direct air from the evaporator to the growing area should be small, but not so small that the noise increases appreciably more than acceptable building air duct noise. CER design should include some provision for cleaning the interior of the air ducts.

Air-conditioning equipment for relatively standard chambers provides temperatures that range from 45 to 90°F. Specialized CERs that require temperatures as low as −5°F need low-temperature refrigeration equipment and devices to defrost the evaporator without increasing the growing area temperature. Other chambers that require temperatures as high as 115°F need high-temperature components. The air temperature in the growing area must be controlled with the least possible variation about the set point. Temperature variation about the set point can be held to 0.5°F using solid-state controls, but in older facilities, the variation is 1 to 2°F.

The relative humidity in many CERs is simply an indicator of the existing psychrometric conditions and is usually between 50 and 80%, depending on the temperature. Relative humidity in the chamber can be increased by steam injection, misting, hot-water evaporators, and other conventional humidification methods. Steam injection causes the least temperature disturbance, and sprays or misting cause the greatest disturbance. Complete control of relative humidity requires dehumidification as well as humidification.

A typical humidity control includes a cold evaporator or steam injection to adjust the chamber air dew point. The air is then conditioned to the desired dry-bulb temperature by electric heaters, a hot-gas bypass evaporator, or a temperature-controlled evaporator. A dew point lower than about 40°F cannot be obtained with a cold-plate dehumidifier because of icing. Dew points lower than 40°F usually require a chemical dehumidifier in addition to the cold evaporator.

Lighting Environmental Chambers

The type of light source and number of lamps used in CERs are determined by the desired plant response. Traditionally, cool-white fluorescent plus incandescent lamps that produce 10% of the fluorescent illuminance are used. Nearly all illumination data are based on either cool-white or warm-white fluorescent, plus incandescent lamps. A number of fluorescent lamps have special phosphors hypothesized to be the spectral requirements of the plant. Some of these lamps are used in CERs, but there is little data to suggest that they are superior to cool-white and warm-white lamps. In recent years, high-intensity discharge lamps have been installed in CERs, either to obtain very high radiant flux densities, or to reduce the electrical load while maintaining a light level equal to that produced by the less efficient fluorescent-incandescent systems.

One method to design lighting for biological environments is to base light source output recommendations on photon flux density $\mu mol/(s \cdot m^2)$ between 400 and 700 nm, or, less frequently, as radiant flux density between 400 and 700 nm, or 400 and 850 nm. Rather than basing illuminance measurements on human vision, this allows comparisons between light sources as a function of plant photosynthetic potential. Table 9 shows constants for converting various measurement units to W/m^2. However, instruments that measure the 400 to 850 nm spectral range are generally not available, and some controversy exists about the effectiveness of 400 to 850 nm as compared to the 400 to 700 nm range in photosynthesis. The power conversion of various light sources is listed in Table 11.

The design requirements for plant growth lighting differ greatly from those for vision lighting. Plant growth lighting requires a greater degree of horizontal uniformity and, usually, higher light levels than vision lighting. In addition, plant growth lighting should have as much vertical uniformity as possible (a factor rarely important in vision lighting). Horizontal and vertical uniformity are much easier to attain with linear or broad sources, such as fluorescent lamps, than with point sources, such as HID lamps. Tables 12 and 13 show the type and number of lamps, mounting height, and spacing required to obtain several levels of incident energy. Because the data were taken directly under lamps with no reflecting wall surfaces nearby, the incident energy is perhaps one-half of what the plants would receive if the lamps had been placed in a small chamber with highly reflective walls.

Extended-life incandescents, which have a much longer life, lower lamp replacement requirements. These lamps have lower lumen output, but are nearly equivalent in the red portion of the spectrum. For safety, porcelain lamp holders and heat-resistant

Table 11   Input Power Conversion of Light Sources

Lamp Identification		Total Input Power, W	Radiation (400-700 nm), %	Radiation (400-850 nm), %	Other Radiation, %	Conduction and Convection, %	Ballast Loss, %
Incandescent	INC, 100A	100	7	15	75	10	0
Fluorescent							
Cool white	FCW	46	21	21	32	34	13
Cool white	FCW	225	19	19	34	35	12
Warm white	FWW	46	20	20	32	35	13
Plant growth A	PGA	46	13	13	35	39	13
Plant growth B	PGB	46	15	16	34	37	13
Infrared	FIR	46	2	9	39	39	13
Discharge							
Clear mercury	HG	440	12	13	61	17	9
Mercury deluxe	HG/DX	440	13	14	59	18	9
Metal halide	MH	460	27	30	42	15	13
High-pressure sodium	HPS	470	26	36	36	13	15
Low-pressure sodium	LPS	230	27	31	25	22	22

Note: Conversion efficiency is for lamps without luminaires. Values compiled from manufacturers' data, published information, and unpublished test data by R.W. Thimijan.

Table 12   Approximate Mounting Height and Spacing of Luminaires in Greenhouses

Lamp and Wattage	Irradiation, W/ft²			
	0.6	1.1	2.2	4.4
	Height and Spacing, in.			
HPS (400 W)	118	90	63	39
LPS (180 W)	94	67	47	31
MH (400 W)	106	79	55	35

lamp wiring should be used. Lamps used for CER lighting include fluorescent lamps (usually 1500 mA), 250, 400, and occasionally 1000 W HPS and MH lamps, 180 W LPS lamps, and various sizes of incandescent lamps. In many installations, the abnormally short life of incandescent lamps is caused by vibration from the lamp loft ventilation or from cooling fans. Increased incandescent lamp life under these conditions can be attained by using lamps constructed with a C9 filament.

Energy-saving lamps have approximately equal or slightly lower irradiance per input watt. Because the irradiance per lamp is lower, there is no advantage to using these lamps, except in tasks that can be accomplished with low light levels. Light output of all lamps declines with use, except perhaps for low-pressure sodium (LPS) lamps, which appear to maintain approximately constant output but require an increase in input power during use.

Fluorescent and metal halide designs should be based on 80% of the initial light level. Most CER lighting systems have difficulty maintaining a relatively constant light level over considerable periods of time. Combinations of MH and HPS lamps compound the problem, because the lumen depreciation of the two light sources is significantly different. Thus, over time, the spectral energy distribution at plant level shifts toward the HPS. Lumen output can be maintained in two ways: (1) individual lamps, or a combination of lamps, can be switched off initially and activated as the lumen output decreases; and (2) the oldest 25 to 33% of the lamps can be replaced periodically. Solid-state dimmer systems are commercially available only for low-wattage fluorescent lamps and for mercury lamps.

To maintain a constant distance from plant to light source, light fixtures in many CERs are mounted on movable, counterbalanced light banks. This design requirement precludes separation of the lamps from the plant chamber.

Large rooms, especially those constructed as an integral part of the building and retrofitted as CERs, rarely separate the lamps from the growing area with a transparent barrier. Rooms designed as CERs (at the time a building is constructed) and freestanding rooms or chambers usually separate the lamp from the growing area with a barrier of glass or rigid plastic. Light output from fluorescent

lamps is a function of the temperature of the lamp. Thus, the barrier serves a two-fold purpose: (1) to maintain optimum lamp temperature when the growing area temperature is higher or lower than optimum, and (2) to reduce the thermal radiation entering the growing area. Fluorescent lamps should operate in an ambient temperature and airflow environment that maintains the tube wall temperature at 104°F. Under most conditions, the light output of HID lamps is not affected by ambient temperature. The heat must be removed, however, to prevent high thermal radiation from causing adverse biological effects (Figure 11).

Transparent glass barriers remove nearly all radiation from about 350 to 2500 nm. Rigid plastic is less effective than glass; however, the lighter weight and lower breakage risk of plastic makes it a popular barrier material. Ultraviolet is also screened by both glass and plastic (more by plastic). Special UV-transmitting plastic (which degrades rapidly) can be obtained if the biological process requires UV light. When irradiance is very high, especially from HID lamps or large numbers of incandescent lamps or both, rigid plastic can soften from the heat and fall from the supports. Furthermore, very high irradiance and the resulting high temperatures can darken plastic, which can increase the absorptivity and temperature enough to destroy it. Under these conditions, heat-resistant glass may be necessary. The lamp compartment and barrier absolutely require positive ventilation regardless of the light source, and the lamp loft should have limit switches to shut down the lamps if the temperature rises to a critical level.

Phytotrons

A phytotron is a botanical laboratory comprising a series of chambers reproducing any condition of temperature, humidity, illumination, or other plant growth factor. They are typically found in plant-based research buildings. These facilities require substantial electrical and mechanical systems to generate light required for plant growth as well as to remove heat generated by lights and CER cooling systems.

Electrical Requirements. If the exact number and size of units is unknown, an electrical consumption of 0.2 kW/ft² may be assumed for lighting input to the CERs. If the CERs have a built-in refrigeration system, the compressor input is typically 80% of lighting input, because the units are designed to maintain the chamber at 50°F with lights on, creating a high latent load on the compressor at an inefficient operating point. Remote condensing units and remote air-cooled condensers require a separate electrical feed and interconnecting control wiring.

Heat Rejection. Most of the electrical input to the CERs is converted to heat. The heat rejection system must be able to remove that heat from the phytotron; this can be done in a number of ways.

Table 13   Height and Spacing of Luminaires

Light Source	Radiant Flux Density, W/ft²						
	0.03	0.08	0.28	0.84	1.67	2.5	4.6
Fluorescent—Cool White							
40 W single 4 ft lamp, 3.2 klm							
Radiant power, W/m², 400 to 700 nm	0.3	0.9	2.9	8.8			
Illumination, klx	0.10	0.30	1.0	3.0			
Lamps per 100 ft²	1.1	3.3	11	33			
Distance from plants, in.	114	67	36	21			
40 W 2-lamp fixtures (4 ft), 6.4 klm							
Radiant power, W/m², 400 to 700 nm	0.3	0.9	2.9	8.8			
Illumination, klx	0.10	0.30	1.0	3.0			
Fixtures per 100 ft²	0.6	1.7	5.5	16.7			
Distance from plants, in.	161	94	51	30			
215 W 2-8 ft lamps, 31.4 klm							
Radiant power, W/m², 400 to 700 nm	0.3	0.9	2.9	8.8	17.6	23.5	49.0
Illumination, klx	0.10	0.30	1.0	3.0	6.0	8.0	16.7
Lamps per 100 ft²	0.1+	0.4	1.2	3.6	7.1	9.3	20
Distance from plants, in.	346	201	110	63	43	39	28
High-Intensity Discharge							
Mercury-1 400 W parabolic reflector							
Radiant power, W/m², 400 to 700 nm	0.28	0.84	2.80	8.39	16.8	22.4	46.6
Illumination, klx	0.1	0.32	1.1	3.2	6.4	8.6	18.0
Lamps per 100 ft²	0.2	0.5	1.6	4.8	9.3	13.0	27
Distance from plants, in.	299	173	94	55	39	31	24
Metal halide-1 400 W							
Radiant power, W/m², 400 to 700 nm	0.77	0.80	2.68	8.03	16.1	21.4	44.6
Illumination, klx	0.09	0.26	0.88	2.6	5.3	7.0	15.0
Lamps per 100 ft²	0.09	0.2	0.7	2.2	4.4	5.8	12.0
Distance from plants, in.	445	256	142	83	59	51	34
High-pressure sodium 400 W							
Radiant power, W/m², 400 to 700 nm	0.22	0.65	2.18	6.52	13.0	17.4	36.2
Illumination, klx	0.09	0.27	0.89	2.7	5.3	7.1	15.0
Lamps per 100 ft²	0.05	0.14	0.5	1.4	2.8	3.6	7.6
Distance from plants, in.	559	323	177	102	71	63	43
Low-pressure sodium 180 W							
Radiant power, W/m², 400 to 700 nm	0.26	0.79	2.64	7.93	15.9	21.1	44.0
Illumination, klx	0.14	0.41	1.4	4.1	8.3	11.0	23.0
Lamps per 100 ft²	0.08	0.24	0.8	2.4	4.9	6.5	13.6
Distance from plants, in.	421	244	134	9	55	47	33
Incandescent							
Incandescent 100 W							
Radiant power, W/m², 400 to 700 nm	0.14	0.41	1.38	4.14	8.28	11.0	23.0
Illumination, klx	0.033	0.10	0.33	1.0	2.0	2.7	5.6
Lamps per 100 ft²	0.5	1.6	5.2	15.8	32	42	87
Distance from plants, in.	165	94	51	30	21	18	13
Incandescent 150 W flood							
Radiant power, W/m², 400 to 700 nm	0.14	0.41	1.38	4.14	8.28	11.0	23.0
Illumination, klx	0.033	0.098	0.33	1.0	2.0	2.6	5.5
Lamps per 100 ft²	0.3	0.9	3.3	9.3	19.5	26	54
Distance from plants, in.	212	122	67	39	28	24	16
Incandescent-Hg 160 W							
Radiant power, W/m², 400 to 700 nm	0.14	0.41	1.38	4.14	8.28	11.0	23.0
Illumination, klx	0.050	0.15	0.50	1.5	3.0	4.0	8.3
Lamps per 100 ft²	0.7	2.0	6.9	20.4	42	56	111
Distance from plants, in.	146	83	47	26	18	16	11
Sunlight							
Radiant power, W per 100 ft²	2.0	6.2	20.5	61.7	124	164	714
Illumination, klx	0.054	0.16	0.54	1.6	3.2	4.3	8.9

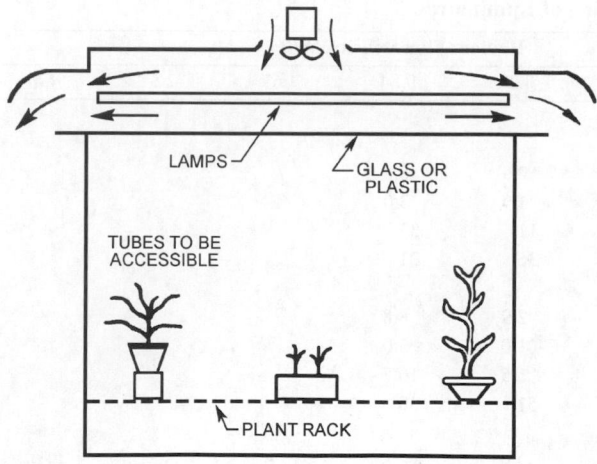

GROWTH CABINET – LIGHTS AIR-COOLED

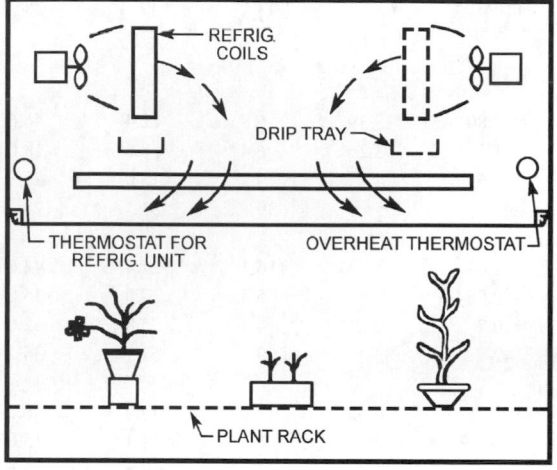

GROWTH CHAMBER – LIGHTS COOLED
BY REFRIGERATION

Fig. 11   Cooling Lamps in Growth Chambers

If the CERs are primarily self-contained air-cooled units, the room can be ventilated at a rate that maintains acceptable working conditions in the space (see Chapter 11). Because of the high ventilation rates needed, ensure that air returned to the space is properly filtered to limit the introduction of dust, pollen, insects, and bacteria from outside.

Self-contained CERs with water-cooled condensing units typically use a condenser water loop connected to a cooling tower or fluid cooler to reject heat. Chapter 13 of the 2008 *ASHRAE Handbook—HVAC Systems and Equipment* describes selection and design of these systems. Because phytotron facilities operate all year, operation of fluid coolers and cooling towers at ambient temperatures below freezing in cold climates must be considered. Sediment must also be removed from the condenser water, because the condenser on the CER is a relatively low-velocity point in the loop and will plug up with these solids.

Locations of remote condensing units or remote air-cooled condensers should be easily accessible, because they require servicing at all times of the year. Ensure good airflow around all air-cooled condensers so that discharge air from one unit is not reentrained into adjacent units. Locate equipment away from laboratory exhaust systems that could accelerate corrosion of metal on the units. Refrigerant piping must be carefully designed, sized, and installed to ensure

proper oil return and long-term operation of the compressors. Chapter 1 of the 2010 *ASHRAE Handbook—Refrigeration* details these requirements.

Central chilled water can be used for the CERs. The primary consideration is the chilled-water temperature to be provided to each unit. In practice, most CERs operate at internal temperatures between 68 to 77°F when lights are on. As a result, standard chilled-water supply temperatures of 45°F can be used successfully. When lights are off, temperatures of 50°F can be achieved using the same chilled-water temperature. Chambers that require cooler daytime temperatures can use water-cooled condensing units and reject their heat to the chilled-water loop. Some phytotrons use chilled-water supply temperatures of 14°F, but with a high failure rate on the compressors because of low suction temperatures and poor oil return.

Energy Conservation. Because of CERs' very high energy consumption and the predictable day/night cycle of the lighting load, consider balancing the units' schedule to limit electrical demand. Most plants require a 12 to 16 h daily photoperiod. By adjusting the day/night schedule, it is possible to reduce the phytotron's electrical demand by up to 25%.

Chilled-water CERs can have the lowest total energy consumption because of the economy of scale available by using large-capacity chillers versus small compressors. Large laboratory facilities can reject heat from the phytotron to preheat laboratory makeup air. In cold climates, chilled water can be produced without mechanical cooling at ambient temperatures below 28°F. If exposing chilled water to ambient air that could be below freezing, use an appropriate concentration of suitable antifreeze.

Condenser water can also be used to preheat fresh air or, because of its higher temperature, other process loads, such as domestic hot water.

Operating Considerations. CERs with self-contained compressors generate noise. When a large number of units are placed in a room, consideration should be given to attenuating this sound. Chilled-water and remote-condensing-unit CERs provide the quietest environment for workers, because the compressors are remotely located. The total installed cost of these systems may be higher because of the extra cost to remotely locate and energize the cooling systems.

Plants require CO_2 to grow. Many CERs in phytotrons have a central exhaust system to exhaust any chemicals used inside the chambers and to pull in a constant supply of air. Because the units are under a slight negative pressure, makeup air entering the unit must be filtered to limit uncontrolled spread of pollen, insect pests, and bacteria. The flow rate from units depends on the type of crop being grown. A normal rate of ventilation is 2 to 3 cfm per square foot of plant growth area.

Water is required for humidification, plant watering, and cleaning. This often means that three totally separate systems are used. High-purity water is often available in laboratory buildings, and can be used to directly humidify the chambers without introducing waterborne minerals into the chamber. Water for plants should be tempered to avoid root shock. A tempered-water loop with provision for introducing chemical fertilizer, supplied to designated hose stations in the phytotron, is common in larger installations, but normal municipal water supplies are all that is required. Cleaning of these areas is important.

CERs require drainage of cooling coil condensate and plant overwatering. It is important to provide good drainage near the units without excess use of drain lines running exposed across the floor. Similarly, any piping or ducts that operate below the room design dew-point temperature should be insulated to prevent condensation on those lines. These puddles of water are prime breeding grounds for plant pests, and could cause slip hazards for staff.

Keeping the phytotron clean is important for plants' health. Phytotrons typically have separate potting areas and harvest rooms, both of which generate a lot of dust and dirt. Potting areas must

Table 14 Mounting Height for Luminaires in Storage Areas

	Survival = 0.3 W/ft^2		Maintenance = 0.8 W/ft^2	
	Distance, ft	lux	Distance, ft	lux
Fluorescent (F)				
FCW two 40 W	3.0	1000	2.5	3000
FWW	3.0	1000	2.5	3000
FCW two 215 W	9.2	1000	5.2	3000
Discharge (HID)				
MH 400 W	10.8	800	6.6	2400
HPS 400 W	14.8	800	8.2	2400
LPS 180 W	11.2	1300	3.9	4000
Incandescent (INC)				
INC 160 W	4.3	350	1.0	1000
INC-HG 160 W	3.9	500	5.2	1500
DL	—	500	—	1500

remain sanitary to minimize contamination of seedlings and plantlets. In harvest rooms, mature plants may host insects that can damage young plants. Ventilation systems should keep harvest rooms at negative pressure relative to the cleaner potting areas and phytotron.

Genetically modified plants must be autoclaved once the plant is harvested. Provision should be made for an autoclave next to the harvest room, with a supply of steam or electricity. Odors and steam from the autoclave should be exhausted out of the building.

OTHER PLANT ENVIRONMENTAL FACILITIES

Plants may be held or processed in warehouse-type structures prior to sale or use in interior landscaping. Required temperatures range from slightly above freezing for cold storage of root stock and cut flowers, to 68 to 77°F for maintaining growing plants, usually in pots or containers. Provision must be made for venting fresh air to avoid CO_2 depletion.

Light duration must be controlled by a time clock. When they are in use, lamps and ballasts produce almost all the heat required in an insulated building. Ventilation and cooling may be required. Illumination levels depend on plant requirements. Table 14 shows approximate mounting heights for two levels of illumination. Luminaires mounted on chains permit lamp height to be adjusted to compensate for varying plant height.

The main concerns for interior landscape lighting are how it renders the color of plants, people, and furnishings, as well as how it meets the minimum irradiation requirements of plants. The temperature required for human occupancy is normally acceptable for plants. Light level and duration determine the types of plants that can be grown or maintained. Plants grow when exposed to higher levels, but do not survive below the suggested minimum. Plants may be grouped into three levels based on the following of irradiances:

Low (survival): A minimum light level of 0.07 W/ft^2 and a preferred level of 0.3 W/ft^2 irradiance for 8 to 12 h daily.

Medium (maintenance): A minimum of 0.3 W/ft^2 and a preferred level of 0.8 W/ft^2 irradiance for 8 to 12 h daily.

High (propagation): A minimum of 0.8 W/ft^2 and a preferred level of 2.2 W/ft^2 irradiance for 8 to 12 h daily.

Fluorescent (warm-white), metal halide, or incandescent lighting is usually chosen for public places. Table 13 lists the irradiance of various light sources.

REFERENCES

AGCIH. 1998. *Industrial ventilation: A manual of recommended practice*, 23rd ed. American Conference of Governmental Industrial Hygienists, Cincinnati, OH.

ASAE. 2003. Guidelines for use of thermal insulation in agricultural buildings. ANSI/ASAE *Standard* S401.2. American Society of Agricultural Engineers (now American Society of Agricultural and Biological Engineers), St. Joseph, MI.

ASTM. 2005. Test method for thermal performance of building materials and envelope assemblies by means of a hot box apparatus. *Standard* C1363-05. American Society for Testing and Materials, West Conshohocken, PA.

Barber, E.M., J.A. Dosman, C.S. Rhodes, G.I. Christison, and T.S. Hurst. 1993. Carbon dioxide as an indicator of air quality in swine buildings. *Proceedings of Third International Livestock Environment Symposium.* American Society of Agricultural Engineers (now American Society of Agricultural and Biological Engineers), St. Joseph, MI.

BESS Lab. 1997. *Agricultural ventilation fans—Performance and efficiencies.* Bioenvironmental and Structural Systems Laboratory, Department of Agricultural Engineering, University of Illinois at Urbana-Champaign.

Bond, T.E., H.H. Heitman, Jr., and C.F. Kelly. 1965. Effect of increased air velocities on heat and moisture loss and growth of swine. *Transactions of ASAE* 8(2):167-169, 174.

Christianson, L.L. and R.L. Fehr. 1983. Ventilation—Energy and economics. In *Ventilation of agricultural structures*, pp. 335-349. American Society of Agricultural Engineers (now American Society of Agricultural and Biological Engineers), St. Joseph, MI.

Clough, G. 1982. Environmental effects on animals used in biomedical research. *Biological Reviews* 57:487-523.

Donham, J.K., P. Haglind, Y. Peterson, R. Rylander, and L. Belin. 1989. Environmental and health studies of workers in Swedish swine confinement buildings. *British Journal of Industrial Medicine* 40:31-37.

Gordon, C.J., P. Becker, and J.S. Ali. 1997. *Behavioural thermoregulatory responses of single- and group-housed mice.* Neurotoxicology Division, National Health and Environmental Effects Research Laboratory, U.S. Environmental Protection Agency, Research Triangle Park, NC.

Hahn, G.L. 1985. Management and housing of farm animals in hot environments. In *Stress physiology in livestock*, vol. II, pp. 151-176. M. Yousef, ed. CRC Press, Boca Raton, FL.

Hasenau, J.J., R.B. Baggs, and A.L. Kraus. 1993. Microenvironments in cages using BALB/c and CD-1 mice. *Contemporary Topics* 32(1):11-16.

Hellickson, M.A. and J.N. Walker, eds. 1983. Ventilation of agricultural structures. ASAE *Monograph* 6. American Society of Agricultural Engineers (now American Society of Agricultural and Biological Engineers), St. Joseph, MI.

ILAR. 1996. *Guide for the care and use of laboratory animals.* National Institutes of Health, Bethesda, MD.

Lindsey, J.R., M.W. Conner, and H.J. Baker. 1978. Physical, chemical and microbial factors affecting biologic response. In *Laboratory Animal Housing*, pp. 31-43. Institute of Laboratory Animal Resources, National Academy of Sciences, Washington, D.C.

Maghirang, R.G., G.L. Riskowski, L.L. Christianson, and P.C. Harrison. 1995. Development of ventilation rates and design information for laboratory animal facilities—Part 1 field study. *ASHRAE Transactions* 101 (2):208-218.

McPherson, C. 1975. Why be concerned about the ventilation requirements of experimental animals. *ASHRAE Transactions* 81(2):539-541.

Memarzadeh, F. 1998. *Design handbook on animal research facilities using static microisolators*, vols. I and II. National Institutes of Health. Bethesda, MD.

Moreland, A.F. 1975. Characteristics of the research animal bioenvironment. *ASHRAE Transactions* 81(2):542-548.

Mount, L.E. and I.B. Start. 1980. A note on the effects of forced air movement and environmental temperature on weight gain in the pig after weaning. *Animal Production* 30(2):295.

MWPS. 1989. *Natural ventilating systems for livestock housing and heating.* MidWest Plan Service, Ames, IA.

MWPS. 1990a. *Cooling and tempering air for livestock housing.* MidWest Plan Service, Ames, IA.

MWPS. 1990b. *Mechanical ventilating systems for livestock housing.* MidWest Plan Service, Ames, IA.

Ni, J., A.J. Heber, T.T. Lim, R.K. Duggirala, B.L. Haymore, and C.A. Diehl. 1998a. Ammonia emission from a tunnel-ventilated swine finishing building. ASAE *Paper* 984051. American Society of Agricultural Engineers (now American Society of Agricultural and Biological Engineers), St. Joseph, MI.

Ni, J., A.J. Heber, T.T. Lim, R.K. Duggirala, B.L. Haymore, and C.A. Diehl. 1998b. Emissions of hydrogen sulfide from a mechanically-ventilated swine grow-finish unit. ASAE *Paper* 984050. American Society of Agricultural Engineers (now American Society of Agricultural and Biological Engineers), St. Joseph, MI.

Person, H.L., L.D. Jacobson, and K.A. Jordan. 1979. Effect of dirt, louvers and other attachments on fan performance. *Transactions of ASAE* 22(3): 612-616.

Riskowski, G.L. and D.S. Bundy. 1988. Effects of air velocity and temperature on weanling pigs. *Livestock environment III: Proceedings of the Third International Livestock Environment Symposium.* American Society of Agricultural Engineers (now American Society of Agricultural and Biological Engineers), St. Joseph, MI.

Riskowski, G.L., S.E. Ford, and K.O. Mankell. 1998. Laboratory measurements of wind effects on ridge vent performance. *ASHRAE Transactions* 104(1).

Riskowski, G.L., R.G. Maghirang, and W. Wang. 1996. Development of ventilation rates and design information for laboratory animal facilities—Part 2 laboratory tests. *ASHRAE Transactions* 102(2):195-209.

Weiss, J., G.T. Taylor, and W. Nicklas. 1991. *Ammonia concentrations in laboratory rat cages under various housing conditions.* American Association for Laboratory Animal Science, Cordova, TN.

Zhang, Y. 1994. *Swine building ventilation.* Prairie Swine Centre, Saskatoon, Saskatchewan, Canada.

Zhang, Y. and E.M. Barber. 1995. Air leakage and ventilation effectiveness for confinement livestock housing. *Transactions of ASAE* 38(5):1501-1504.

Zhang, Y., L.L. Christianson, G.L. Riskowski, B. Zhang, G. Taylor, H.W. Gonyou, and P.C. Harrison. 1992. A survey on laboratory rat environments. *ASHRAE Transactions* 98(2):247-253.

BIBLIOGRAPHY

ANIMALS

Handbooks and Proceedings

Albright, L.D. 1990. *Environment control for animals and plants, with computer applications.* American Society of Agricultural Engineers (now American Society of Agricultural and Biological Engineers), St. Joseph, MI.

ASAE. 1982. *Dairy housing II: Second National Dairy Housing Conference Proceedings.* American Society of Agricultural Engineers (now American Society of Agricultural and Biological Engineers), St. Joseph, MI.

ASAE. 1982. *Livestock environment II: Second International Livestock Environment Symposium.* American Society of Agricultural Engineers (now American Society of Agricultural and Biological Engineers), St. Joseph, MI.

ASAE. 1988. *Livestock environment III: Proceedings of the Third International Livestock Environment Symposium.* American Society of Agricultural Engineers (now American Society of Agricultural and Biological Engineers), St. Joseph, MI.

ASAE. 1993. *Livestock environment IV: Proceedings of the Fourth International Livestock Environment Symposium.* American Society of Agricultural Engineers (now American Society of Agricultural and Biological Engineers), St. Joseph, MI.

ASAE. 1993. Design of ventilation systems for livestock and poultry shelters. *Standard* EP270.5. American Society of Agricultural Engineers (now American Society of Agricultural and Biological Engineers), St. Joseph, MI.

Curtis, S.E. 1983. *Environmental management in animal agriculture.* Iowa State University Press, Ames.

Curtis, S.E., ed. 1988. *Guide for the care and use of agricultural animals in agricultural research and teaching.* Consortium for Developing a Guide for the Care and Use of Agricultural Animals in Agricultural Research and Teaching, Champaign, IL.

HEW. 1978. Guide for the care and use of laboratory animals. *Publication* (NIH)78-23. U.S. Department of Health, Education and Welfare, Washington, D.C.

Rechcigl, M., Jr., ed. 1982. *Handbook of agricultural productivity*, vol. II, *Animal productivity.* CRC Press, Boca Raton, FL.

Straub, H.E. 1989. *Building systems: Room air and air contaminant distribution.* ASHRAE.

Air Cooling

Canton, G.H., D.E. Buffington, and R.J. Collier. 1982. Inspired-air cooling for dairy cows. *Transactions of ASAE* 25(3):730-734.

Hahn, G.L. and D.D. Osburn. 1969. Feasibility of summer environmental control for dairy cattle based on expected production losses. *Transactions of ASAE* 12(4):448-451.

Hahn, G.L. and D.D. Osburn. 1970. Feasibility of evaporative cooling for dairy cattle based on expected production losses. *Transactions of ASAE* 12(3):289-291.

Heard, L., D. Froelich, L. Christianson, R. Woerman, and R. Witmer. 1986. Snout cooling effects on sows and litters. *Transactions of ASAE* 29(4): 1097-1101.

Morrison, S.R., M. Prokop, and G.P. Lofgreen. 1981. Sprinkling cattle for heat stress relief: Activation, temperature, duration of sprinkling, and pen area sprinkled. *Transactions of ASAE* 24(5):1299-1300.

Timmons, M.B. and G.R. Baughman. 1983. Experimental evaluation of poultry mist-fog systems. *Transactions of ASAE* 26(1):207-210.

Wilson, J.L., H.A. Hughes, and W.D. Weaver, Jr. 1983. Evaporative cooling with fogging nozzles in broiler houses. *Transactions of ASAE* 26(2): 557-561.

Air Pollution in Buildings

ACGIH. 2011. *TLVs® and BEIs®.* American Conference of Governmental Industrial Hygienists, Cincinnati.

Avery, G.L., G.E. Merva, and J.B. Gerrish. 1975. Hydrogen sulfide production in swine confinement units. *Transactions of ASAE* 18(1):149.

Bundy, D.S. and T.E. Hazen. 1975. Dust levels in swine confinement systems associated with different feeding methods. *Transactions of ASAE* 18(1):137.

Deboer, S. and W.D. Morrison. 1988. *The effects of the quality of the environment in livestock buildings on the productivity of swine and safety of humans—A literature review.* Department of Animal and Poultry Science, University of Guelph, Ontario.

Grub, W., C.A. Rollo, and J.R. Howes. 1965. Dust problems in poultry environment. *Transactions of ASAE* 8(3):338.

Effects of Environment on Production and Growth of Animals

Cattle

Anderson, J.F., D.W. Bates, and K.A. Jordan. 1978. Medical and engineering factors relating to calf health as influenced by the environment. *Transactions of ASAE* 21(6):1169.

Garrett, W.N. 1980. Factors influencing energetic efficiency of beef production. *Journal of Animal Science* 51(6):1434.

Gebremedhin, K.G., C.O. Cramer, and W.P. Porter. 1981. Predictions and measurements of heat production and food and water requirements of Holstein calves in different environments. *Transactions of ASAE* 24(3): 715.

Holmes, C.W. and N.A. McLean. 1975. Effects of air temperature and air movement on the heat produced by young Friesian and Jersey calves, with some measurements of the effects of artificial rain. *New Zealand Journal of Agricultural Research* 18(3):277.

Morrison, S.R., G.P. Lofgreen, and R.L. Givens. 1976. Effect of ventilation rate on beef cattle performance. *Transactions of ASAE* 19(3):530.

General

Hahn, G.L. 1982. Compensatory performance in livestock: Influences on environmental criteria. *Proceedings of the Second International Livestock Environment Symposium.* American Society of Agricultural Engineers (now American Society of Agricultural and Biological Engineers), St. Joseph, MI.

Hahn, G.L. 1981. Housing and management to reduce climatic impacts on livestock. *Journal of Animal Science* 52(1):175-186.

Pigs

Boon, C.R. 1982. The effect of air speed changes on the group postural behaviour of pigs. *Journal of Agricultural Engineering Research* 27(1): 71-79.

Christianson, L.L., D.P. Bane, S.E. Curtis, W.F. Hall, A.J. Muehling, and G.L. Riskowski. 1989. *Swine care guidelines for pork producers using environmentally controlled housing.* National Pork Producers Council, Des Moines, IA.

Close, W.H., L.E. Mount, and I.B. Start. 1971. The influence of environmental temperature and plane of nutrition on heat losses from groups of growing pigs. *Animal Production* 13(2):285.

Driggers, L.B., C.M. Stanislaw, and C.R. Weathers. 1976. Breeding facility design to eliminate effects of high environmental temperatures. *Transactions of ASAE* 19(5):903.

McCracken, K.J. and R. Gray. 1984. Further studies on the heat production and affective lower critical temperature of early-weaned pigs under commercial conditions of feeding and management. *Animal Production* 39:283-290.

Nienaber, J.A. and G.L. Hahn. 1988. Environmental temperature influences on heat production of ad-lib-fed nursery and growing-finishing swine. *Livestock Environment III: Proceedings of the Third International Livestock Environment Symposium*. American Society of Agricultural Engineers (now American Society of Agricultural and Biological Engineers), St. Joseph, MI.

Phillips, P.A., B.A. Young, and J.B. McQuitty. 1982. Liveweight, protein deposition and digestibility responses in growing pigs exposed to low temperature. *Canadian Journal of Animal Science* 62:95-108.

Poultry

Buffington, D.E., K.A. Jordan, W.A. Junnila, and L.L. Boyd. 1974. Heat production of active, growing turkeys. *Transactions of ASAE* 17(3):542.

Carr, L.E., T.A. Carter, and K.E. Felton. 1976. Low temperature brooding of broilers. *Transactions of ASAE* 19(3):553.

Riskowski, G.L., J.A. DeShazer, and F.B. Mather. 1977. Heat losses of white leghorn laying hens as affected by intermittent lighting schedules. *Transactions of ASAE* 20(4):727-731.

Siopes, T.D., M.B. Timmons, G.R. Baughman, and C.R. Parkhurst. 1983. The effect of light intensity on the growth performance of male turkeys. *Poultry Science* 62:2336-2342.

Sheep

Schanbacher, B.D., G.L. Hahn, and J.A. Nienaber. 1982. Photoperiodic influences on performance of market lambs. *Proceedings of the Second International Livestock Environment Symposium*. American Society of Agricultural Engineers (now American Society of Agricultural and Biological Engineers), St. Joseph, MI.

Vesely, J.A. 1978. Application of light control to shorten the production cycle in two breeds of sheep. *Animal Production* 26(2):169.

Modeling and Analysis

Albright, L.D. and N.R. Scott. 1974. An analysis of steady periodic building temperature variations in warm weather—Part I: A mathematical model. *Transactions of ASAE* 17(1):88-92, 98.

Albright, L.D. and N.R. Scott. 1974. An analysis of steady periodic building temperature variations in warm weather—Part II: Experimental verification and simulation. *Transactions of ASAE* 17(1):93-98.

Albright, L.D. and N.R. Scott. 1977. Diurnal temperature fluctuations in multi-air spaced buildings. *Transactions of ASAE* 20(2):319-326.

Bruce, J.M. and J.J. Clark. 1979. Models of heat production and critical temperature for growing pigs. *Animal Production* 28:353-369.

Christianson, L.L. and H.A. Hellickson. 1977. Simulation and optimization of energy requirements for livestock housing. *Transactions of ASAE* 20(2):327-335.

Ewan, R.C. and J.A. DeShazer. 1988. Mathematical modeling the growth of swine. *Livestock environment III*. Proceedings of the Third International Livestock Environment Symposium. American Society of Agricultural Engineers (now American Society of Agricultural and Biological Engineers), St. Joseph, MI.

Hellickson, M.L., K.A. Jordan, and R.D. Goodrich. 1978. Predicting beef animal performance with a mathematical model. *Transactions of ASAE* 21(5):938-943.

Teter, N.C., J.A. DeShazer, and T.L. Thompson. 1973. Operational characteristics of meat animals—Part I: Swine; Part II: Beef; Part III: Broilers. *Transactions of ASAE* 16:157-159, 740-742, 1165-1167.

Timmons, M.B. 1984. Use of physical models to predict the fluid motion in slot-ventilated livestock structures. *Transactions of ASAE* 27(2):502-507.

Timmons, M.B., L.D. Albright, R.B. Furry, and K.E. Torrance. 1980. Experimental and numerical study of air movement in slot-ventilated enclosures. *ASHRAE Transactions* 86(1):221-240.

Shades for Livestock

Bedwell, R.L. and M.D. Shanklin. 1962. Influence of radiant heat sink on thermally-induced stress in dairy cattle. Missouri Agricultural Experiment Station *Research Bulletin* 808.

Bond, T.E., L.W. Neubauer, and R.L. Givens. 1976. The influence of slope and orientation of effectiveness of livestock shades. *Transactions of ASAE* 19(1):134-137.

Roman-Ponce, H., W.W. Thatcher, D.E. Buffington, C.J. Wilcox, and H.H. VanHorn. 1977. Physiological and production responses of dairy cattle to a shade structure in a subtropical environment. *Journal of Dairy Science* 60(3):424.

Transport of Animals

Ashby, B.H., D.G. Stevens, W.A. Bailey, K.E. Hoke, and W.G. Kindya. 1979. *Environmental conditions on air shipment of livestock.* U.S. Department of Agriculture, SEA, Advances in Agricultural Technology, Northeastern Series 5.

Ashby, B.H., A.J. Sharp, T.H. Friend, W.A. Bailey, and M.R. Irwin. 1981. Experimental railcar for cattle transport. *Transactions of ASAE* 24(2): 452.

Ashby, B.H., H. Ota, W.A. Bailey, J.A. Whitehead, and W.G. Kindya. 1980. Heat and weight loss of rabbits during simulated air transport. *Transactions of ASAE* 23(1):162.

Grandin, R. 1988. *Livestock trucking guide.* Livestock Conservation Institute, Madison, WI.

Scher, S. 1980. Lab animal transportation receiving and quarantine. *Lab Animal* 9(3):53.

Stermer, R.A., T.H. Camp, and D.G. Stevens. 1982. Feeder cattle stress during handling and transportation. *Transactions of ASAE* 25(1):246-248.

Stevens, D.G., G.L. Hahn, T.E. Bond, and J.H. Langridge. 1974. *Environmental considerations for shipment of livestock by air freight.* U.S. Department of Agriculture, Animal and Plant Health Inspection Service.

Stevens, D.G. and G.L. Hahn. 1981. Minimum ventilation requirement for the air transportation of sheep. *Transactions of ASAE* 24(1):180.

Laboratory Animals

NIH. 1978. Laboratory animal housing. *Proceedings of a symposium held at Hunt Valley, MD, September 1976.* National Academy of Sciences, Washington, D.C.

McSheehy, T. 1976. *Laboratory animal handbook 7—Control of the animal house environment.* Laboratory Animals, Ltd.

Soave, O., W. Hoag, F. Gluckstein, and R. Adams. 1980. The laboratory animal data bank. *Lab Animal* 9(5):46-49.

Ventilation Systems

Albright, L.D. 1976. Air flows through hinged-baffle, slotted inlets. *Transactions of ASAE* 19(4):728, 732, 735.

Albright, L.D. 1978. Air flow through baffled, center-ceiling, slotted inlets. *Transactions of ASAE* 21(5):944-947, 952.

Albright, L.D. 1979. Designing slotted inlet ventilation by the systems characteristic technique. *Transactions of ASAE* 22(1):158.

Pohl, S.H. and M.A. Hellickson. 1978. Model study of five types of manure pit ventilation systems. *Transactions of ASAE* 21(3):542.

Randall, J.M. 1980. Selection of piggery ventilation systems and penning layouts based on the cooling effects of air speed and temperature. *Journal of Agricultural Engineering Research* 25(2):169-187.

Randall, J.M. and V.A. Battams. 1979. Stability criteria for air flow patterns in livestock buildings. *Journal of Agricultural Engineering Research* 24(4):361-374.

Timmons, M.B. 1984. Internal air velocities as affected by the size and location of continuous inlet slots. *Transactions of ASAE* 27(5):1514-1517.

Natural Ventilation

Bruce, J.M. 1982. Ventilation of a model livestock building by thermal buoyancy. *Transactions of ASAE* 25(6):1724-1726.

Jedele, D.G. 1979. Cold weather natural ventilation of buildings for swine finishing and gestation. *Transactions of ASAE* 22(3):598-601.

Timmons, M.B., R.W. Bottcher, and G.R. Baughman. 1984. Nomographs for predicting ventilation by thermal buoyancy. *Transactions of ASAE* 27(6):1891-1893.

PLANTS

Greenhouse and Plant Environment

Aldrich, R.A. and J.W. Bartok. 1984. *Greenhouse engineering.* Department of Agricultural Engineering, University of Connecticut, Storrs.

ASAE. 2002. Guidelines for measuring and reporting environmental parameters for plant experiments in growth chambers. ANSI/ASAE *Engineering Practice* EP411.2. American Society of Agricultural Engineers (now American Society of Agricultural and Biological Engineers), St. Joseph, MI.

Clegg, P. and D. Watkins. 1978. *The complete greenhouse book*. Garden Way Publishing, Charlotte, VT.

Downs, R.J. 1975. *Controlled environments for plant research*. Columbia University Press, New York.

Langhans, R.W. 1985. *Greenhouse management*. Halcyon Press, Ithaca, NY.

Mastalerz, J.W. 1977. *The greenhouse environment*. John Wiley & Sons, New York.

Nelson, P.V. 1978. *Greenhouse operation and management*. Reston Publishing, VA.

Pierce, J.H. 1977. *Greenhouse grow how*. Plants Alive Books, Seattle.

Riekels, J.W. 1977. *Hydroponics*. Ontario Ministry of Agriculture and Food, Fact Sheet 200-24, Toronto.

Riekels, J.W. 1975. *Nutrient solutions for hydroponics*. Ontario Ministry of Agriculture and Food, Fact Sheet 200-532, Toronto.

Sheldrake, R., Jr. and J.W. Boodley. *Commercial production of vegetable and flower plants*. Research Park, 1B-82, Cornell University, Ithaca, NY.

Tibbitts, T.W. and T.T. Kozlowski, eds. 1979. *Controlled environment guidelines for plant research*. Academic Press, New York.

Light and Radiation

Bickford, E.D. and S. Dunn. 1972. *Lighting for plant growth*. Kent State University Press, OH.

Carpenter, G.C. and L.J. Mousley. 1960. The artificial illumination of environmental control chambers for plant growth. *Journal of Agricultural Engineering Research* [U.K.] 5:283.

Campbell, L.E., R.W. Thimijan, and H.M. Cathey. 1975. Special radiant power of lamps used in horticulture. *Transactions of ASAE* 18(5):952.

Cathey, H.M. and L.E. Campbell. 1974. Lamps and lighting: A horticultural view. *Lighting Design & Application* 4:41.

Cathey, H.M. and L.E. Campbell. 1979. Relative efficiency of high- and low-pressure sodium and incandescent filament lamps used to supplement natural winter light in greenhouses. *Journal of the American Society for Horticultural Science* 104(6):812.

Cathey, H.M. and L.E. Campbell. 1980. Light and lighting systems for horticultural plants. *Horticultural Reviews* 11:491. AVI Publishing, Westport, CT.

Cathey, H.M., L.E. Campbell, and R.W. Thimijan. 1978. Comparative development of 11 plants grown under various fluorescent lamps and different duration of irradiation with and without additional incandescent lighting. *Journal of the American Society for Horticultural Science* 103:781.

Hughes, J., M.J. Tsujita, and D.P. Ormrod. 1979. *Commercial applications of supplementary lighting in greenhouses*. Ontario Ministry of Agriculture and Food, Fact Sheet 290-717, Toronto.

Kaufman, J.E., ed. 1981. IES *Lighting handbook: Application volume*. IES, New York.

Kaufman, J.E., ed. 1981. IES *Lighting handbook: Reference volume*. IES, New York.

Robbins, F.V. and C.K. Spillman. 1980. Solar energy transmission through two transparent covers. *Transactions of ASAE* 23(5).

Sager, J.C., J.L. Edwards, and W.H. Klein. 1982. Light energy utilization efficiency for photosynthesis. *Transactions of ASAE* 25(6):1737-1746.

Photoperiod

Heins, R.D., W.H. Healy, and H.F. Wilkens. 1980. Influence of night lighting with red, far red, and incandescent light on rooting of chrysanthemum cuttings. *HortScience* 15:84.

Carbon Dioxide

Bailey, W.A., et al. 1970. CO_2 systems for growing plants. *Transactions of ASAE* 13(2):63.

Gates, D.M. 1968. Transpiration and leaf temperature. *Annual Review of Plant Physiology* 19:211.

Holley, W.D. 1970. CO_2 enrichment for flower production. *Transactions of ASAE* 13(3):257.

Kretchman, J. and F.S. Howlett. 1970. Enrichment for vegetable production. *Transactions of ASAE* 13(2):252.

Tibbitts, T.W., J.C. McFarlane, D.T. Krizek, W.L. Berry, P.A. Hammer, R.H. Hodgsen, and R.W. Langhans. 1977. Contaminants in plant growth chambers. *Horticulture Science* 12:310.

Wittwer, S.H. 1970. Aspects of CO_2 enrichment for crop production. *Transactions of ASAE* 13(2):249.

Heating, Cooling, and Ventilation

ASAE. 2003. Heating, ventilating and cooling greenhouses. ANSI/ASAE *Engineering Practice* EP406.1. American Society of Agricultural Engineers (now American Society of Agricultural and Biological Engineers), St. Joseph, MI.

Albright, L.D., I. Seginer, L.S. Marsh, and A. Oko. 1985. In situ thermal calibration of unventilated greenhouses. *Journal of Agricultural Engineering Research* 31(3):265-281.

Buffington, D.E. and T.C. Skinner. 1979. Maintenance guide for greenhouse ventilation, evaporative cooling, and heating systems. *Publication* AE-17. Department of Agricultural Engineering, University of Florida, Gainesville.

Duncan, G.A. and J.N. Walker. 1979. Poly-tube heating ventilation systems and equipment. *Publication* AEN-7. Agricultural Engineering Department, University of Kentucky, Lexington.

Elwell, D.L., M.Y. Hamdy, W.L. Roller, A.E. Ahmed, H.N. Shapiro, J.J. Parker, and S.E. Johnson. 1985. *Soil heating using subsurface pipes*. Department of Agricultural Engineering, Ohio State University, Columbus.

Heins, R. and A. Rotz. 1980. Plant growth and energy savings with infrared heating. *Florists' Review* (October):20.

NGMA. 1989. *Greenhouse heat loss*. National Greenhouse Manufacturers' Association, Taylors, SC.

NGMA. 1989. *Standards for ventilating and cooling greenhouses*. National Greenhouse Manufacturers' Association, Taylors, SC.

NGMA. 1993. *Recommendation for using insect screens in greenhouse structures*. National Greenhouse Manufacturers' Association, Taylors, SC.

Roberts, W.J. and D. Mears. 1984. *Floor heating and bench heating extension bulletin for greenhouses*. Department of Agricultural and Biological Engineering, Cook College, Rutgers University, New Brunswick, NJ.

Roberts, W.J. and D. Mears. 1984. *Heating and ventilating greenhouses*. Department of Agricultural and Biological Engineering, Cook College, Rutgers University, New Brunswick, NJ.

Roberts, W.J. and D.R. Mears. 1979. *Floor heating of greenhouses*. Miscellaneous Publication, Rutgers University, New Brunswick, NJ.

Silverstein, S.D. 1976. Effect of infrared transparency on heat transfer through windows: A clarification of the greenhouse effect. *Science* 193:229.

Walker, J.N. and G.A. Duncan. 1975. Greenhouse heating systems. *Publication* AEN-31. Agricultural Engineering Department, University of Kentucky, Lexington.

Walker, J.N. and G.A. Duncan. 1979. Greenhouse ventilation systems. *Publication* AEN-30. Agricultural Engineering Department, University of Kentucky, Lexington.

Energy Conservation

Roberts, W.J., J.W. Bartok, Jr., E.E. Fabian, and J. Simpkins. 1985. *Energy conservation for commercial greenhouses*. NRAES-3. Department of Agricultural Engineering, Cornell University, Ithaca, NY.

Solar Energy Use

Albright, L.D., R.W. Langhans, and G.B. White. 1980. Passive solar heating applied to commercial greenhouses. *Publication* 115, Energy in Protected Civilization. Acta Horticultura.

Cathey, H.M. 1980. Energy-efficient crop production in greenhouses. *ASHRAE Transactions* 86(2):455.

Duncan, G.A., J.N. Walker, and L.W. Turner. 1979. *Energy for greenhouses*, Part I: Energy Conservation. Publication No. AEES-16. College of Agriculture, University of Kentucky, Lexington.

Duncan, G.A., J.N. Walker, and L.W. Turner. 1980. *Energy for greenhouses*, Part II: Alternative Sources of Energy. College of Agriculture, University of Kentucky, Lexington.

Gray, H.E. 1980. Energy management and conservation in greenhouses: A manufacturer's view. *ASHRAE Transactions* 86(2):443.

Roberts, W.J. and D.R. Mears. 1980. Research conservation and solar energy utilization in greenhouses. *ASHRAE Transactions* 86(2):433.

Short, T.H., M.F. Brugger, and W.L. Bauerle. 1980. Energy conservation ideas for new and existing commercial greenhouses. *ASHRAE Transactions* 86(2):448.

DRYING AND STORING SELECTED FARM CROPS

CONTROL of moisture content and temperature during storage is critical to preserving the nutritional and economic value of farm crops as they move from the field to the market. Fungi (mold) and insects feed on poorly stored crops and reduce crop quality. Relative humidity and temperature affect mold and insect growth, which is reduced to a minimum if the crop is kept cooler than 50°F and if the relative humidity of the air in equilibrium with the stored crop is less than 60%.

Mold growth and spoilage are a function of elapsed storage time, temperature, and moisture content above critical values. Approximate allowable storage life for cereal grains is shown in Table 1. For example, corn at 60°F and 20% wet basis (w.b.) moisture has a storage life of about 25 days. If it is dried to 18% w.b. after 12 days, half of its storage life has elapsed. Thus, the remaining storage life at 60°F and 18% w.b. moisture content is 25 days, not 50 days.

Insects thrive in stored grain if the moisture content and temperature are not properly controlled. At low moisture contents and temperatures below 50°F, insects remain dormant or die.

Most farm crops must be dried to, and maintained at, a suitable moisture content. For most grains, a suitable moisture content is in the range of 12 to 15% w.b., depending on the specific crop, storage temperature, and length of storage. Oilseeds such as peanuts, sunflower seeds, and flaxseeds must be dried to a moisture content of 8 to 9% w.b. Grain stored for more than a year, grain that is damaged, and seed stock should be dried to a lower moisture content. Moisture levels above these critical values lead to the growth of fungi, which may produce toxic compounds such as aflatoxin.

The maximum yield of dry matter can be obtained by harvesting when the corn has dried in the field to an average moisture content of 26% w.b. However, for quality-conscious markets, the minimum damage occurs when corn is harvested at 21 to 22% w.b. Wheat can be harvested when it has dried to 20% w.b., but harvesting at these moisture contents requires expensive mechanical drying. Although field drying requires less expense than operating drying equipment, total cost may be greater because field losses generally increase as the moisture content decreases.

The price of grain to be sold through commercial market channels is based on a specified moisture content, with price discounts for moisture levels above the specified amount. These discounts compensate for the weight of excess water, cover the cost of water removal, and control the supply of wet grain delivered to market. Grain dried to below the base moisture content set by the market (15.0% w.b. for corn, 13.0% w.b. for soybeans, and 13.5% w.b. for wheat) is not generally sold at a premium; thus, the seller loses the opportunity to sell water for the price of grain.

Grain Quantity

The **bushel** is the common measure used for marketing grain in the United States. Most dryers are rated in bushels per hour for a specified moisture content reduction. The use of the bushel as a measure causes considerable confusion. A bushel is a volume measure equal to 1.244 ft³. The bushel is used as a volume measure to estimate the holding capacity of bins, dryers, and other containers.

For buying and selling grain, for reporting production and consumption data, and for most other uses, the bushel weight is used. For example, the legal weight of a bushel is 56 lb for corn and 60 lb for wheat. When grain is marketed, bushels are computed as the load weight divided by the bushel weight. So, 56,000 lb of corn (regardless of moisture content) is 1000 bushels. Rice, grain sorghum, and sunflower are more commonly traded on the basis of the hundredweight (100 lb), a measure that does not connote volume. The relationship between bushel by volume and market bushel is the **bulk density** (listed for some crops in Table 2). For some crops, the market has defined a test weight parameter, lb/bu. Test weight is essentially the bulk density, with bushels and cubic feet related by the definition of 1 bushel = 1.244 cubic feet.

The terms **wet bushel** and **dry bushel** sometimes refer to the mass of grain before and after drying. For example, 56,000 lb of 25% moisture corn may be referred to as 1000 wet bushels or simply 1000 bushels. When the corn is dried to 15.5% moisture content (m.c.), only 49,704 lb or 49,704/56 = 888 bushels remain. Thus, a dryer rated on the basis of wet bushels (25% m.c.) shows a capacity 12.6% higher than if rated on the basis of dry bushels (15.5% m.c.).

Table 1 Approximate Allowable Storage Time (Days) for Cereal Grains

Moisture Content, % w.b.[a]	Temperature, °F					
	30	40	50	60	70	80
14	*	*	*	*	200	140
15	*	*	*	240	125	70
16	*	*	230	120	70	40
17	*	280	130	75	45	20
18	*	200	90	50	30	15
19	*	140	70	35	20	10
20	*	90	50	25	14	7
22	190	60	30	15	8	3
24	130	40	15	10	6	2
26	90	35	12	8	5	2
28	70	30	10	7	4	2
30	60	25	5	5	3	1

Based on composite of 0.5% maximum dry matter loss calculated on the basis of USDA research; *Transactions of ASAE* 333-337, 1972; and "Unheated Air Drying," Manitoba Agriculture Agdex 732-1, rev. 1986.

[a]Grain moisture content calculated as percent wet basis: (weight of water in a given amount of wet grain ÷ weight of the wet grain) × 100.

*Approximate allowable storage time exceeds 300 days.

The preparation of this chapter is assigned to TC 2.2, Plant and Animal Environment.

Table 2   Calculated Densities of Grains and Seeds Based on U.S. Department of Agriculture Data

	Bulk Density, lb/ft³
Alfalfa	48.0
Barley	38.4
Beans, dry	48.0
Bluegrass	11.2 to 24.0
Canola	40.2 to 48.2
Clover	48.0
Corn*	
Ear, husked	28.0
Shelled	44.8
Cottonseed	25.6
Oats	25.6
Peanuts, unshelled	
Virginia type	13.6
Runner, Southeastern	16.8
Spanish	19.8
Rice, rough	36.0
Rye	44.8
Sorghum	40.0
Soybeans	48.0
Sudan grass	32.0
Sunflower	
Nonoil	19.3
Oilseed	25.7
Wheat	48.0

*70 lb of husked ears of corn yield 1 bushel, or 56 lb of shelled corn. 70 lb of ears of corn occupy 2 volume bushels (2.5 ft³).

The percent of weight lost due to water removed may be calculated by the following equation:

$$\text{Moisture shrink, \%} = \frac{M_o - M_f}{100 - M_f} \times 100$$

where

M_o = original or initial moisture content, wet basis
M_f = final moisture content, wet basis

Applying the formula to drying a crop from 25% to 15%,

$$\text{Moisture shrink} = \frac{25 - 15}{100 - 15} \times 100 = 11.76\%$$

In this case, the moisture shrink is 11.76%, or an average 1.176% weight reduction for each percentage point of moisture reduction. The moisture shrink varies depending on the final moisture content. For example, the average shrink per point of moisture when drying from 20% to 10% is 1.111.

Economics

Producers generally have the choice of drying their grain on the farm before delivering it to market, or delivering wet grain with a price discount for excess moisture. The expense of drying on the farm includes both fixed and variable costs. Once a dryer is purchased, the costs of depreciation, interest, taxes, and repairs are fixed and minimally affected by volume of crops dried. The costs of labor, fuel, and electricity vary directly with the volume dried. Total drying costs vary widely, depending on the volume dried, the drying equipment, and fuel and equipment prices. Energy consumption depends primarily on dryer type. Generally, the faster the drying speed, the greater the energy consumption (Table 3).

DRYING EQUIPMENT AND PRACTICES

Contemporary crop-drying equipment depends on mass and energy transfer between the drying air and the product to be dried.

Table 3   Estimated Corn Drying Energy Requirement

Dryer Type	Btu/lb of Water Removed
Unheated air	1000 to 1200
Low temperature	1200 to 1500
Batch-in-bin, continuous-flow in-bin	1500 to 2000
High temperature	
Air recirculating	1800 to 2200
Without air recirculating	2000 to 3000
Combination drying, dryeration	1400 to 1800

Note: Includes all energy requirements for fans and heat.

The drying rate is a function of the initial temperature and moisture content of the crop, the air-circulation rate, the entering condition of the circulated air, the length of flow path through the products, and the time elapsed since the beginning of the drying operation. Outdoor air is frequently heated before it is circulated through the product. Heating increases the rate of heat transfer to the product, increases its temperature, and increases the vapor pressure of the product moisture. For more information on crop responses to drying, see Chapter 11 of the 2005 *ASHRAE Handbook—Fundamentals*.

Most crop-drying equipment consists of (1) a fan to move the air through the product, (2) a controlled heater to increase the ambient air temperature to the desired level, and (3) a container to distribute the drying air uniformly through the product. The exhaust air is vented to the atmosphere. Where climate and other factors are favorable, unheated air is used for drying, and the heater is omitted.

Fans

The fan selected for a given drying application should meet the same requirements important in any air-moving application. It must deliver the desired amount of air against the static resistance of the product in the bin or column, the resistance of the delivery system, and the resistance of the air inlet and outlet.

Foreign material in the grain can significantly change the required air pressure in the following ways:

- Foreign particles larger than the grain (straw, plant parts, and larger seeds) reduce airflow resistance. The airflow rate may be increased by 60% or more.
- Foreign particles smaller than the grain (broken grain, dust, and small seeds) increase the airflow resistance. The effect may be dramatic, decreasing the airflow rate by 50% or more.
- The method used to fill the dryer or the agitation or stirring of the grain after it is placed in the dryer can increase pressure requirements by up to 100%. In some grain, high moisture causes less pressure drop than does low moisture.

Vaneaxial fans are normally recommended when static pressures are less than 3 in. of water. Backward-curved centrifugal fans are commonly recommended when static pressures are higher than 4 in. of water column. Low-speed centrifugal fans operating at 1750 rpm perform well up to about 7 in. of water, and high-speed centrifugal fans operating at about 3500 rpm have the ability to develop static pressure up to about 10 in. of water. The in-line centrifugal fan consists of a centrifugal fan impeller mounted in the housing of an axial flow fan. A bell-shaped inlet funnels the air into the impeller. The in-line centrifugal fan operates at about 3450 rpm and has the ability to develop pressures up to 10 in. of water with 7.5 hp or larger fans.

After functional considerations are made, the initial cost of the dryer fan should be taken into account. Drying equipment has a low percentage of annual use in many applications, so the cost of dryer ownership per unit of material dried is sometimes greater than the energy cost of operation. The same considerations apply to other components of the dryer.

Heaters

Most crop dryer heaters are fueled by either natural gas, liquefied petroleum gas, or fuel oil, though some electric heaters are used. Dryers using coal, biomass (such as corn cobs, stubble, or wood), and solar energy have also been built.

Fuel combustion in crop dryers is similar to combustion in domestic and industrial furnaces. Heat is transferred to the drying air either indirectly, by means of a heat exchanger, or directly, by combining the combustion gases with the drying air. Direct combustion heating is generally limited to natural gas or liquefied petroleum (LP) gas heaters. Most grain dryers use direct combustion. Indirect heating is sometimes used in drying products such as hay because of its greater fire hazard.

Controls

In addition to the usual temperature controls for drying air, all heated air units must have safety controls similar to those found on space-heating equipment. These safety controls shut off the fuel in case of flame failure and stop the burner in case of overheating or excessive drying air temperatures. All controls should be set up to operate the machinery safely in the event of power failure.

SHALLOW-LAYER DRYING

Batch Dryers

The batch dryer cycles through the loading, drying, cooling, and unloading of the grain. Fans force hot air through columns (typically 12 in. wide) or layers (2 to 5 ft thick) of grain. Drying time depends on the type of grain and the amount of moisture to be removed. Some dryers circulate and mix the grain to prevent significant moisture content gradients from forming across the column. A circulation rate that is too fast or a poor selection of handling equipment may cause undue damage and loss of market quality. Batch dryers are suitable for farm operations and are often portable.

Continuous-Flow Dryers

This type of self-contained dryer passes a continuous stream of grain through the drying chamber. Some dryers use a second chamber to cool the hot, dry grain prior to storage. Handling and storage equipment must be available at all times to move grain to and from the dryers. These dryers have crossflow, concurrent flow, or counterflow designs.

Crossflow Dryers. A crossflow dryer is a column dryer that moves air perpendicular to the grain movement. These dryers commonly consist of two or more vertical columns surrounding the drying and cooling air plenums. The columns range in thickness from 8 to 16 in. Airflow rates range from 40 to 160 cfm per cubic foot of grain. The thermal efficiency of the drying process increases as column width increases and decreases as airflow rate increases. However, moisture uniformity and drying capacity increase as airflow rate increases and as column width decreases. Dryers are designed to obtain a desirable balance of airflow rate and column width for the expected moisture content levels and drying air temperatures. Performance is evaluated in terms of drying capacity, thermal efficiency, and dried product moisture uniformity.

As with the batch dryer, a moisture gradient forms across the column because the grain nearest the inside of the column is exposed to the driest air during the complete cycle. Several methods minimize the problem of uneven drying.

One method uses turnflow devices that split the grain stream and move the inside half of the column to the outside and the outside half to the inside. Although effective, turnflow devices tend to plug if the grain is trashy. Under these conditions, a scalper/cleaner should be used to clean the grain before it enters the dryer.

Another method is to divide the drying chamber into sections and duct the hot air so that its direction through the grain is reversed in alternate sections. This method produces about the same effect as the turnflow method.

A third method is to divide the drying chamber into sections and reduce the drying air temperature in each section consecutively. This method is the least effective.

Rack-Type Dryers. In this special type of crossflow dryer, grain flows over alternating rows of heated air supply ducts and air exhaust ducts (Figure 1). This action mixes the grain and alternates exposure to relatively hot drying air and air cooled by previous contact with the grain, promoting moisture uniformity and equal exposure of the product to the drying air.

Concurrent-Flow Dryers. In the concurrent-flow dryer, grain and drying air move in the same direction in the drying chamber. The drying chamber is coupled to a counterflow cooling section. Thus, the hottest air is in contact with the wettest grain, allowing the use of higher drying air temperatures (up to 450°F). Rapid evaporative cooling in the wettest grain prevents the grain temperature from reaching excessive levels. Because higher drying air temperatures are used, the energy efficiency is better than that obtained with a conventional crossflow dryer. In the cooling section, the coolest air initially contacts the coolest grain. The combination of drying and cooling chambers results in lower thermal stresses in the grain kernels during drying and cooling and, thus, a higher-quality product.

Counterflow Dryers. The grain and drying air move in opposite directions in the drying chamber of this dryer. Counterflow is common for in-bin dryers. Drying air enters from the bottom of the bin and exits from the top. The wet grain is loaded from overhead, and floor sweep augers can be used to bring the hot, dry grain to a center sump, where it is removed by another auger. The travel of the sweep is normally controlled by moisture- or temperature-sensing elements.

A drying zone exists only in the lower layers of the grain mass and is truncated at its lower edge so that the grain being removed is not overdried. As a part of the counterflow process, the warm, saturated or near-saturated air leaving the drying zone passes through the cool incoming grain. Some energy is used to heat the cool grain, but some moisture may condense on the cool grain if the bed is deep and the initial grain temperature is low.

Reducing Energy Costs

Recirculation. In most commercially available continuous-flow dryers, optional ducting systems recycle some of the exhaust air from the drying and cooling chambers back to the inlet of the drying chamber (Figure 2). Systems vary, but most make it possible to recirculate all of the air from the cooling chamber and from the lower

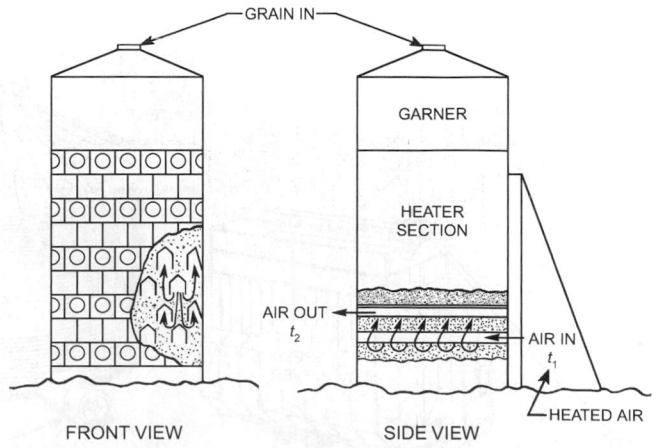

Fig. 1 Rack-Type Continuous-Flow Grain Dryer with Alternate Rows of Air Inlet and Outlet Ducts

two-thirds of the drying chamber. The relative humidity of this recirculated air for most crossflow dryers is less than 50%. Energy savings of up to 30% can be obtained from a well-designed system.

Dryeration. This is another means of reducing energy consumption and improving grain quality. In this process, hot grain with a moisture content one or two percentage points above that desired for storage is removed from the dryer (Figure 3). The hot grain is placed in a dryeration bin, where it tempers without airflow for at least 4 to 6 h. After the first grain delivered to the bin has tempered, the cooling fan is turned on as additional hot grain is delivered to the bin. The air cools the grain and removes 1 to 2% of its moisture before the grain is moved to final storage. If the cooling rate equals the filling rate, cooling is normally completed about 6 h after the last hot grain is added. The crop cooling rate should equal the filling rate of the dryeration bin. A faster cooling rate cools the grain before it has tempered. A slower rate may result in spoilage, since the allowable storage time for hot, damp grain may be only a few days. The required airflow rate is based on dryer capacity and crop density. An airflow rate of 12 cfm for each bushel per hour (bu/h) of dryer capacity provides the cooling capacity to keep up with the dryer when it is drying corn that weighs 56 lb/bu. Recommended airflow rates for some crops are listed in Table 4.

Combination Drying. This method was developed to improve drying thermal efficiency and corn quality. First, a high-temperature

dryer dries the corn to 18 to 20% moisture content. Then it is transferred to a bin, where the full-bin drying system brings the moisture down to a safe storage level.

Dryer Temperature. For energy savings, operating temperatures of batch and continuous-flow dryers are usually set at the highest level that will not damage the product for its end use.

DEEP-BED DRYING

A deep-bed drying system can be installed in any structure that holds grain. Most grain storage structures can be designed or adapted for drying if a means of distributing the drying air uniformly through the grain is provided. A perforated floor (Figure 4) and duct systems placed on the floor of the bin (Figure 5) are the two most common means.

Perforations in the floor should have a total area of at least 10% of the floor area. A perforated floor distributes air more uniformly and offers less resistance to airflow than do ducts, but a duct system is less expensive for larger floor area systems. Ducts can be removed after the grain is removed, and the structure can be cleaned and used for other purposes. Ducts should not be spaced farther apart than one-half times the depth of the grain. The amount of perforated area or the duct length will affect airflow distribution uniformity.

Air ducts and tunnels that disperse air into the grain should be large enough to prevent the air velocity from exceeding 2000 fpm; slower speeds are desirable. Sharp turns, obstructions, or abrupt changes in duct size should be eliminated, as they cause pressure loss. Operating methods for drying grain in storage bins are (1) full-bin

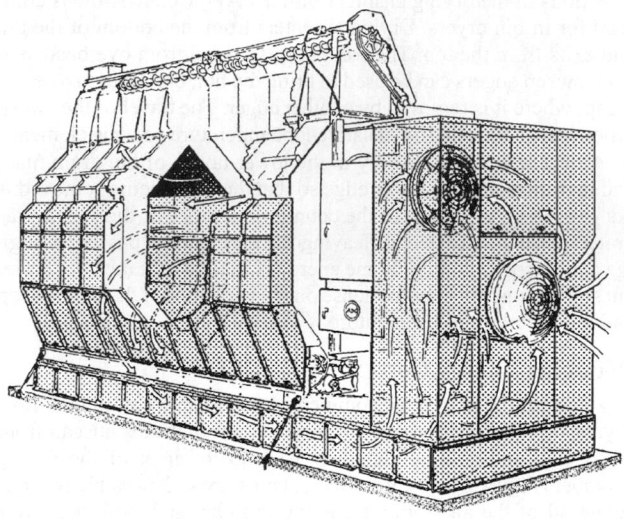

Fig. 2   Crop Dryer Recirculation Unit
(Courtesy Farm Fans, Inc., a division of The GSI Group)

Table 4   Recommended Airflow Rates for Dryeration

Crop	Weight, lb/bu	Recommended Dryeration Airflow Rate, cfm per bu/h
Barley	48	10
Corn	56	12
Durum	60	13
Edible beans	60	13
Flaxseeds	56	12
Millet	50	11
Oats	32	7
Rye	56	12
Sorghum	56	12
Soybeans	60	13
Nonoil sunflower seeds	24	5
Oil sunflower seeds	32	7
Hard red spring wheat	60	13

Note: Basic air volume is 12.9 ft³/lb.

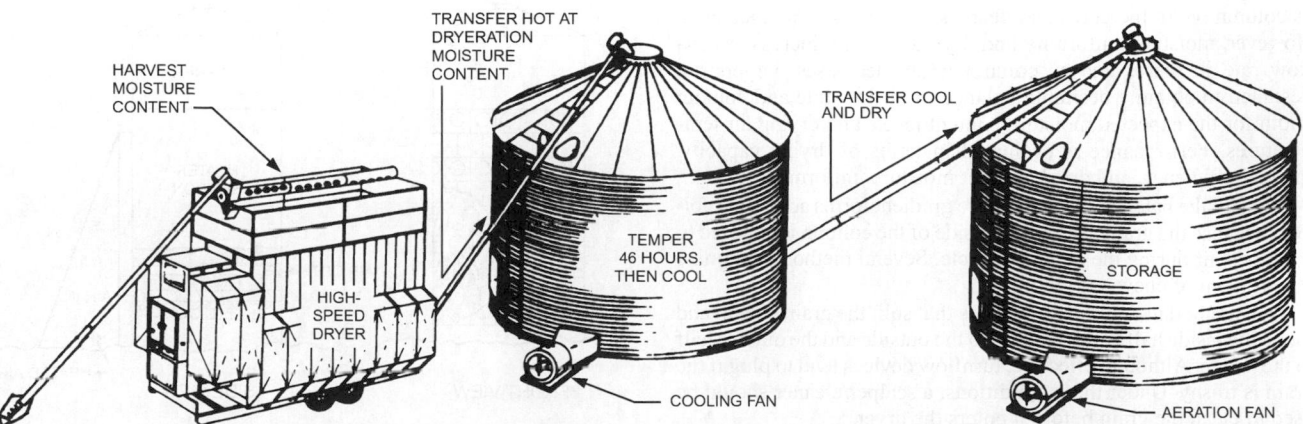

Fig. 3   Dryeration System Schematic

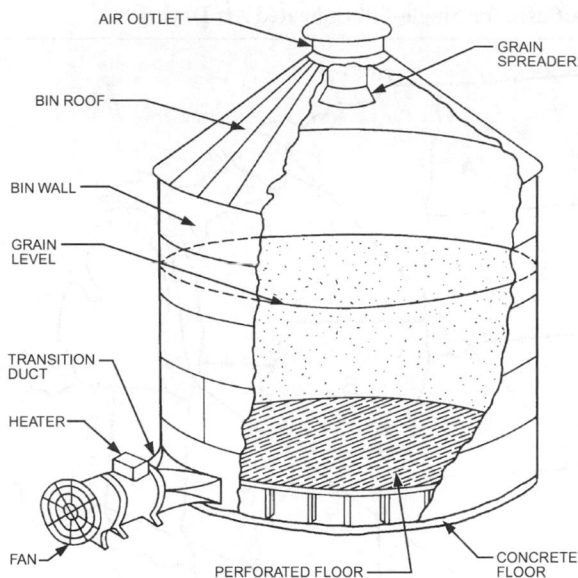

Fig. 4 Perforated Floor System for Bin Drying of Grain

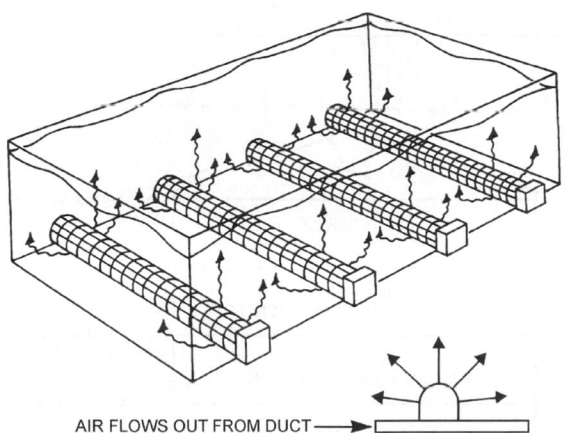

Fig. 5 Tunnel or Duct Air Distribution System

drying, (2) layer drying, (3) batch-in-bin drying, and (4) recirculating/continuous-flow bin drying.

Full-Bin Drying

Full-bin drying is generally performed with unheated air or air heated up to 10°F above ambient. A humidistat is frequently used to sense the humidity of the drying air and turn off the heater if the weather conditions are such that heated air would cause overdrying. A humidistat setting of 55% stops drying at approximately the 12% moisture level for most farm grains, assuming that the ambient relative humidity does not go below this point.

Airflow rate requirements for full-bin drying are generally calculated on the basis of cfm of air required per cubic foot or bushel of grain. The airflow rate recommendations depend on the weather conditions and on the type of grain and its moisture content. Airflow rate is important for successful drying. Because faster drying results from higher airflow rates, the highest economical airflow rate should be used. However, the cost of full-bin drying at high airflow rates may exceed the cost of using column dryers, or the electric power requirement may exceed the available capacity.

Recommendations for full-bin drying with unheated air are shown in Tables 5, 6, and 7. These recommendations apply to the

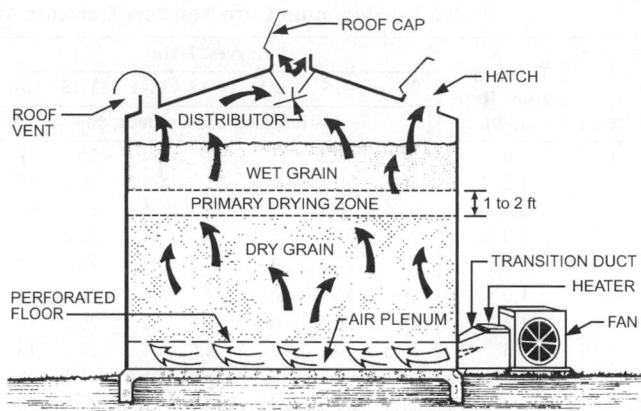

Fig. 6 Three Zones Within Grain During Full-Bin Drying

principal production areas of the continental United States and are based on experience under average conditions; they may not be applicable under unusual weather conditions or even usual weather conditions in the case of late-maturing crops. Full-bin drying may not be feasible in some geographical areas.

The maximum practical depth of grain to be dried (distance of air travel) is limited by the cost of the fan, motor, air distribution system, and power required. This depth seems to be 20 ft for corn and soybeans, and about 15 ft for wheat.

To ensure satisfactory drying, heated air may be used during periods of prolonged fog or rain. Burners should be sized to raise the temperature of the drying air by no more than 10°F above ambient. The temperature should not exceed about 80°F after heating. Overheating the drying air causes the grain to overdry and dry nonuniformly; heat is recommended only to counteract adverse weather conditions. Electric controllers can be applied to fan and heater operation to achieve the final desired grain moisture content.

Drying takes place in a drying zone, which advances upward through the grain (Figure 6). Grain above this drying zone remains at or slightly above the initial moisture content, while grain below the drying zone is at a moisture content in equilibrium with the drying air.

As the direction of air movement does not affect the rate of drying, other factors must be considered in choosing the direction. A pressure system moves the moisture-laden air up through the grain, and it is discharged under the roof. If there are insufficient roof outlets, moisture may condense on the underside of metal roofs. During pressure system ventilation, the wettest grain is near the top surface and is easy to monitor. Fan and motor waste heat enter into the airstream and contribute to drying.

A negative-pressure system moves air down through the grain. Moisture-laden air discharges from the fan to the outside; thus, roof condensation is not a problem. Also, the air picks up some solar heat from the roof. However, the wettest grain is near the bottom of the mass and is difficult to sample. Of the two systems, the pressure system is recommended because it is easier to manage.

The following management practices must be observed to ensure the best performance of the dryer:

1. Minimize foreign material. A scalper-cleaner is recommended for cleaning the grain to reduce air pressure and energy requirements and to help provide uniform airflow for elimination of wet spots.
2. Distribute the remaining foreign material uniformly by installing a grain distributor.
3. Place the grain in layers and keep it leveled.
4. Start the fan as soon as the floor or ducts are covered with grain.
5. Operate the fan continuously with unheated air unless it is raining heavily or there is a dense ground fog. Once all the grain is within 1% of desired storage moisture content, run the fans only when the relative humidity is below 70%.

Table 5 Maximum Corn Moisture Contents, Wet Mass Basis, for Single-Fill Unheated Air Drying

Zone	Full-Bin Airflow Rate, cfm/bu	9-1	9-15	10-1	10-15	11-1	11-15	12-1
				Initial Moisture Content, %				
A	1.0	18	19.5	21	22	24	20	18
	1.25	20	20.5	21.5	23	24.5	20.5	18
	1.5	20	20.5	22.5	23	25	21	18
	2.0	20.5	21	23	24	25.5	21.5	18
	3.0	22	22.5	24	25.5	27	22	18
B	1.0	19	20	20	21	23	20	18
	1.25	19	20	20.5	21.5	24	20.5	18
	1.5	19.5	20.5	21	22.5	24	21	18
	2.0	20	21	22.5	23.5	25	21.5	18
	3.0	21	22.5	23.5	24.5	26	22	18
C	1.0	19	19.5	20	21	22	20	18
	1.25	19	20	20.5	21.5	22.5	20.5	18
	1.5	19.5	20	21	22	23.5	21.5	18
	2.0	20	21	22	23	24.5	21.5	18
	3.0	21	22	23.5	24.5	25.5	22	18
D	1.0	19	19.5	20	21	22	20	18
	1.25	19	19.5	20.5	21	22.5	20.5	18
	1.5	19	19.5	21	22	23	21	18
	2.0	19.5	21	21.5	23	24	21.5	18
	3.0	20.5	21.5	23	24	25	22	18

Source: Midwest Plan Service, 1980. Reprinted with permission.

Table 6 Minimum Airflow Rate for Unheated Air Low-Temperature Drying of Small Grains and Sunflower in the Northern Plains of the United States

Airflow Rate		Maximum Initial Moisture Content, % Wet Basis	
cfm/bu	cfm/ft³	Small Grains	Sunflower
0.5	0.4	16	15
1.0	0.8	18	17
2.0	1.6	20	21

Layer Drying

In layer drying, successive layers of wet grain are placed on top of dry grain. When the top 6 in. has dried to within 1% of the desired moisture content, another layer is added (Figure 7). Compared to full-bin drying, layering reduces the time that the top layers of grain remain wet. Because the effective airflow rate is greater for lower layers, allowable harvest moisture content of grain in these levels can be greater than that in the upper layers. Either unheated air or air heated 10 to 20°F above ambient may be used, but using heated air controlled with a humidistat to prevent overdrying is most common. The first layer may be about 7 ft deep, with successive layers of about 3 ft.

Batch-in-Bin Drying

A storage bin adapted for drying may be used to dry several batches of grain during a harvest season, if the grain is kept to a shallow layer so that higher airflow rates and temperatures can be used. After the batch is dry, the bin is emptied, and the cycle is repeated. The drying capacity of the batch system (bu/yr) is greater than that of other in-storage drying systems. In a typical operation, batches of corn in 3 ft depths are dried from an initial moisture content of 25% with 130°F air at the rate of about 20 cfm per cubic foot. Considerable nonuniformity of moisture content may be present in the batch after drying is stopped; therefore, the grain should be well mixed as it is placed into storage. If the mixing is done well, grain that is too wet equalizes in moisture with grain that is too dry before spoilage can occur. Aeration of the grain in storage will facilitate the equalization of moisture.

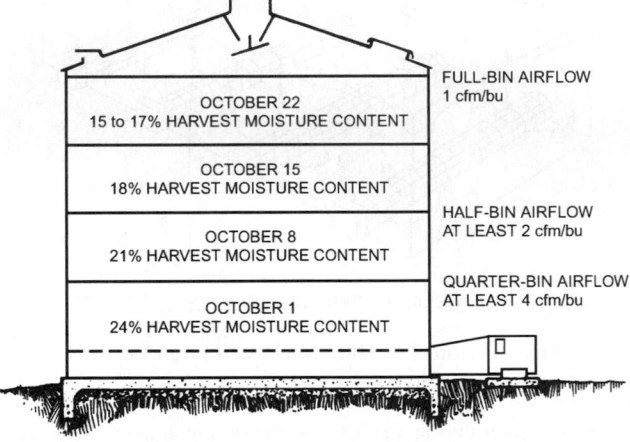

Fig. 7 Example of Layer Filling of Corn

Grain may be cooled in the dryer to ambient temperature before it is stored. Cooling is accomplished by operating the fan without the heater for about 1 h. Some additional drying occurs during the cooling process, particularly in the wetter portions of the batch.

Grain stirring devices are used with both full-bin and batch-in-bin drying systems. Typically, these devices consist of one or more open, 2 in. diameter, standard pitch augers suspended from the bin roof and extending to near the bin floor. The augers rotate and simultaneously travel horizontally around the bin, mixing the drying grain to reduce moisture gradients and prevent overdrying of the bottom grain. The augers also loosen the grain, allowing a higher airflow rate for a given fan. Stirring equipment reduces bin capacity by about 10%. Furthermore, commercial stirring devices are available only for round storage enclosures.

Recirculating/Continuous-Flow Bin Drying

This type of drying incorporates a tapered sweep auger that removes uniform layers of grain from the bottom of the bin as it dries

Table 7 Recommended Unheated Air Airflow Rate for Different Grains and Moisture Contents in the Southern United States

Type of Grain	Grain Moisture Content, %	Recommended Airflow Rate, cfm per ft³ of grain	cfm/bu
Wheat	25	4.8	6.0
	22	4.0	5.0
	20	2.4	3.0
	18	1.6	2.0
	16	0.8	1.0
Oats	25	2.4	3.0
	20	1.6	2.0
	18	1.2	1.5
	16	0.8	1.0
Shelled Corn	25	4.0	5.0
	20	2.4	3.0
	18	1.6	2.0
	16	0.8	1.0
Ear Corn	25	6.4	8.0
	18	3.2	4.0
Grain Sorghum	25	4.8	6.0
	22	4.0	5.0
	18	2.4	3.0
	15	1.6	2.0
Soybeans	25	4.8	6.0
	22	4.0	5.0
	18	2.4	3.0
	15	1.6	2.0

Compiled from USDA *Leaflet* 332 (1952) and Univ. of Georgia *Bulletin* NS 33 (1958).

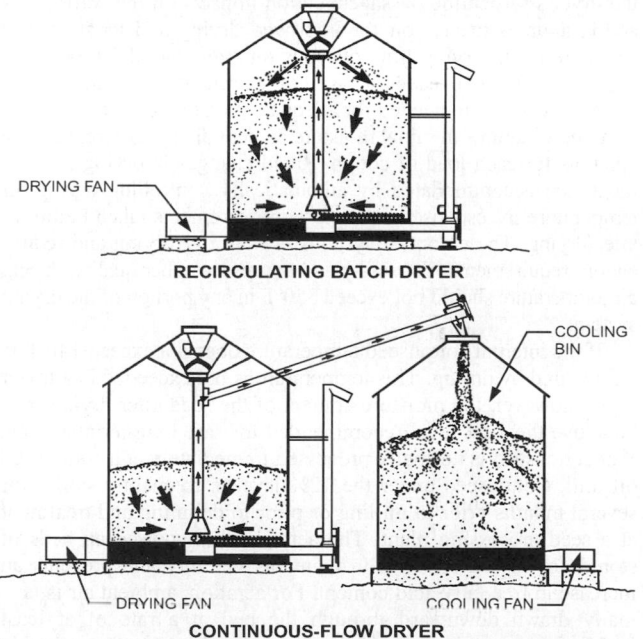

RECIRCULATING BATCH DRYER

DRYING FAN

COOLING BIN

DRYING FAN COOLING FAN

CONTINUOUS-FLOW DRYER

Fig. 8 Grain Recirculators Convert Bin Dryer to High-Speed Continuous-Flow Dryer

(Figure 8). The dry grain is then redistributed on top of the pile of grain or moved to a second bin for cooling. The sweep auger may be controlled by temperature or moisture sensors. When the desired condition is reached, the sensor starts the sweep auger, which removes a layer of grain. After a complete circuit of the bin, the sweep auger stops until the sensor determines that another layer is dry. Some drying takes place in the cooling bin. Up to two percentage points of moisture may be removed, depending on the management of the cooling bin.

DRYING SPECIFIC CROPS

SOYBEANS

Soybeans usually need drying only when there is inclement weather during the harvest season. Mature soybeans left exposed to rain or damp weather develop a dark brown color and a mealy or chalky texture. Seed quality deteriorates rapidly. Oil from weather-damaged beans costs more to refine and is often not of edible grade. In addition to preventing deterioration, the artificial drying of soybeans offers the advantage of early harvest, which reduces the chance of loss from bad weather and reduces natural and combine shatter loss. Soybeans harvested with a wet basis moisture content greater than 13.5% exhibit less damage.

Drying Soybeans for Commercial Use

Conventional corn-drying equipment can be used for soybeans, with some limitations on heat input. Soybeans for commercial use can be dried at 130 to 140°F; drying temperatures of 190°F reduce the oil yield. If the relative humidity of the drying air is below 40%, excessive seedcoat cracking occurs, causing many split beans in subsequent handling. Physical damage can cause fungal growth on the beans, storage problems, and a slight reduction in oil yield and quality. Flow-retarding devices should be used during handling, and beans should not be dropped more than 20 ft onto concrete floors.

Drying Soybeans for Seed and Food

The relative humidity of the drying air should be kept above 40%, regardless of the amount of heat used. The maximum drying temperature to avoid germination loss is 110°F. Natural air drying at a flow rate of 1.6 cfm per cubic foot is adequate for drying seed with an initial moisture content of up to 16% w.b.

If adding heat, raise the drying air temperature no more than 5°F above ambient. This drying method is slow, but it results in excellent quality and avoids overdrying. However, drying must be completed before spoilage occurs. At higher moisture contents, good results have been obtained using an airflow rate of 3.2 cfm per cubic foot with humidity control. Data on allowable drying time for soybeans are unavailable. Without better information, an estimate of storage life for oil crops can be made based on the values for corn, using an adjusted moisture content calculated by the following equation:

$$\text{Comparable moisture content} = \frac{\text{Oilseed moisture content}}{100 - \text{Seed oil content}} \times 100$$

A corn moisture content 2% greater than that of the soybeans should generally be used to estimate allowable drying time (e.g., 12% soybeans are comparable to 14% corn). Soybeans are dried from a lower initial moisture content than corn.

Dry high-moisture soybeans in a bin with the air temperature controlled to keep the relative humidity at 40% or higher. Airflow rates of 8.0 cfm per cubic foot are recommended, with the depth of the beans not to exceed 4 ft.

HAY

Hay normally contains 65 to 80% wet basis moisture at cutting. Field drying to 20% may result in a large loss of leaves. Alfalfa hay leaves average about 50% of the crop by weight, but they contain 70% of the protein and 90% of the carotene. The quality of hay can be increased and the risk of loss due to bad weather reduced if the hay is put under shelter when partially field dried (35% moisture content) and then artificially dried to a safe storage moisture content. In good drying weather, hay conditioned by mechanical means can be dried sufficiently in one day and placed

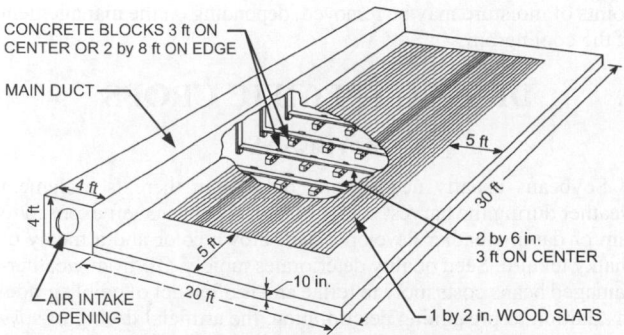

Fig. 9 Central Duct Hay-Drying System with Lateral Slatted Floor for Wide Mows

in the dryer. Hay may be long, chopped, or baled for this operation; unheated or heated air can be used.

In-Storage Drying

Unheated air is normally used for in-storage or mow drying. Hay is dried in the field to 30 to 40% moisture content before being placed in the dryer. For unheated air drying, airflow should be at least 200 cfm per ton. The fan should be able to deliver required airflow against static pressure of 1 to 2 in. of water.

Slotted floors, with at least 50% of the area open, are generally used for drying baled hay. For long or chopped hay in mows narrower than 36 ft wide, the center duct system is the most popular. A slotted floor should be placed on each side of the duct to within 5 ft of its ends and the outside walls (Figure 9). If the mow is wider than 36 ft, it should be divided crosswise into units of 28 ft or narrower. These should then be treated as individual dryers. If the storage depth exceeds about 13 ft, vertical flues and/or additional levels of ducts may be used. If tiered ducts are used, a vertical air chamber, about 75% of the probable hay depth, should be used. The supply ducts are then connected at 7 to 10 ft vertical intervals as the mow is filled. With either of these methods, hay in total depths up to 30 ft can be dried. The duct size should be such that the air velocity is less than 1000 fpm.

The maximum depth of wet hay that should be placed on a hay-drying system at any time depends on hay moisture content, weather conditions, the physical form of the hay, and the airflow rate. The maximum drying depth is about 16 ft for long hay, 13 ft for chopped hay, and 7 small rectangular bales deep for baled hay. Baled hay should have a density of about 8 lb/ft³. For best results, bales should be stacked tightly together on edge (parallel to the stems) to ensure that no openings exist between them.

For mow drying, the fan should run continuously during the first few days. Afterward, it should be operated only during low relative humidity weather. During prolonged wet periods, the fan should be operated only enough to keep the hay cool.

Batch Wagon Drying

Batch drying can be done on a slotted floor platform; however, because this method is labor-intensive, wagon dryers are more commonly used. With a wagon dryer system, hay is baled at about 45% moisture content to a density of about 11 lb/ft³. The hay is then stacked onto a wagon with tight, high sides and a slotted or expanded metal floor. Drying is accomplished most efficiently by forcing the heated air (up to 158°F) down the canvas duct of a plenum chamber secured to the top of the wagon. After 4 or 5 h of drying, the exhaust air is no longer saturated with moisture, and about 75% of it may be recirculated or passed through a second wagon of wet hay for greater drying efficiency.

In this method, the amount of hay harvested each day is limited by the capacity of the drying wagons. In this 24 h process, the hay

cut one day is stored the following day; only enough hay to load the drying wagons should be harvested each day.

The airflow rate in this method is normally much higher than when unheated air is used. About 40 cfm per square foot of wagon floor space is required. As with mow drying, the duct size should be such that the air velocity is less than 1000 fpm.

COTTON

Producers normally allow cotton to dry naturally in the field to 12% moisture content or less before harvest. Cotton harvested in this manner can be stored in trailers, baskets, or compacted stacks for extended periods with little loss in fiber or seed quality. Thus, cotton is not normally aerated or artificially dried prior to ginning. Cotton harvested during inclement weather and stored cotton exposed to precipitation must be dried at the cotton gin within a few days to prevent self-heating and deterioration of the fiber and seed.

Though cotton may be safely stored at moisture contents as high as 12%, moisture levels near the upper limit are too high for efficient ginning and for obtaining optimum fiber grade. The cleaning efficiency of cotton is inversely proportional to its moisture content, with the most efficient level being 5% fiber moisture content. However, fiber quality is best preserved when the fiber is separated from the seed at moisture contents between 6.5 and 8%. Therefore, if cotton comes into the system below this level, it can be cleaned, but moisture should be added prior to separating the fiber from the seed to improve the ginning quality. Dryers in the cotton gins are capable of drying the cotton to the desired moisture level.

The tower dryer is the most commonly used among several types of commercially available dryers. This device operates on a parallel flow principle: 14 to 24 cfm of drying air per pound of cotton also serves as the conveying medium. As it moves through the dryer's serpentine passages, cotton impacts on the walls. This action agitates the cotton for improved drying and lengthens its exposure time. Drying time depends on many variables, but total exposure seldom exceeds 15 s. For extremely wet cotton, two stages of drying are needed for adequate moisture control.

Wide variations in initial moisture content dictate different drying amounts for each load of cotton. Rapid changes in drying requirements are accommodated by automatically controlling drying air temperature in response to moisture measurements taken before or after drying. These control systems prevent overdrying and reduce energy requirements. For safety and to preserve fiber quality, drying air temperature should not exceed 350°F in any portion of the drying system.

If the internal cottonseed temperature does not exceed 140°F is unimpaired by drying. This temperature is not exceeded in a tower dryer; however, the moisture content of the seed after drying may be above the 12% level recommended for safe long-term storage. Wet cottonseed is normally processed immediately at a cottonseed oil mill. Cottonseed under the 12% level is frequently stored for several months prior to milling or prior to delinting and treatment at a seed processing plant. The aeration that cools deep beds of stored cottonseed effectively maintains viability and prevents an increase in free fatty acid content. For aeration, ambient air is normally drawn downward through the bed at a rate of at least 0.025 cfm per cubic foot of oil mill seed and 0.125 cfm per cubic foot of planting seed.

PEANUTS

Peanuts normally have a moisture content of about 50% at the time of digging. Allowing the peanuts to dry on the vines in the windrow for a few days removes much of this water. However, peanuts usually contain 20 to 30% moisture when removed from the vines, and some artificial drying is necessary. Drying should begin within 6 h after harvesting to keep the peanuts from self-heating. Both the maximum temperature and the rate of drying must be carefully controlled to maintain quality.

High temperatures result in an off flavor or bitterness. Drying too rapidly without high temperatures results in blandness or nuts that do not develop flavor during roasting. High temperatures, rapid drying, or excessive drying cause the skin to slip easily and the kernels to become brittle. These conditions result in high damage rates in the shelling operation but can be avoided if the moisture removal rate does not exceed 0.5% per hour. Because of these limitations, continuous-flow drying is not usually recommended for peanuts.

Peanuts can be dried in bulk bins using unheated air or air with supplemental heat. Under poor drying conditions, unheated air may cause spoilage, so supplemental heat is preferred. Air should be heated no more than 13 or 14°F to a maximum temperature of 95°F. An airflow rate of 10 to 25 cfm per cubic foot of peanuts should be used, depending on the initial moisture content.

The most common method of drying peanuts is bulk wagon drying. Peanuts are dried in depths of 5 to 6 ft, using airflow rates of 10 to 15 cfm per cubic foot of peanuts and air heated 11 to 14°F above ambient. This method retains quality and usually dries the peanuts in three to four days. Wagon drying reduces handling labor but may require additional investment in equipment.

RICE

Of all grains, rice is probably the most difficult to process without quality loss. Rice containing more than 13.5% moisture cannot be safely stored for long periods, yet the recommended harvest moisture content for best milling and germination ranges from 20 to 26%. When rice is harvested at this moisture content, drying must be started promptly to prevent souring. Normally, heated air is used in continuous-flow dryers, where large volumes of air are forced through 4 to 10 in. layers of rice. Temperatures as high as 130°F may be used, if (1) the temperature drop across the rice does not exceed 20 to 30°F, (2) the moisture reduction does not exceed two percentage points in a 0.5 h exposure, and (3) the rice temperature does not exceed 100°F During the tempering period following drying, the rice should be aerated to ambient temperature prior to the next pass through the dryer. This removes additional moisture and eliminates one to two dryer passes. It is estimated that full use of aeration following dryer passes could increase the maximum daily drying capacity by about 14%.

Unheated air or air with a small amount of added heat (13°F above ambient, but not exceeding 95°F) should be used for deep-bed rice drying. Too much heat overdries the bottom, resulting in checking (cracking), reduced milling qualities, and possible spoilage in the top. Because unheated air drying requires less investment and attention than supplemental heat drying, it is preferred when conditions permit. In the more humid rice-growing areas, supplemental heat is desirable to ensure that the rice dries. The time required for drying varies with weather conditions, moisture content, and airflow rate. In California, the recommended airflow rate is 0.2 to 2.4 cfm per cubic foot. Because of less favorable drying conditions in Arkansas, Louisiana, and Texas, greater airflow rates are recommended (e.g., a minimum of 2.0 cfm per cubic foot is recommended in Texas). Whether unheated air or supplemental heat is used, the fan should be turned on as soon as rice uniformly covers the air distribution system. The fan should then run continuously until the moisture content in the top 1 ft of rice is reduced to about 15%. At this point, the supplemental heat should be turned off. The rice can then be dried to a safe storage level by operating the fan only when the relative humidity is below 75%.

STORAGE PROBLEMS AND PRACTICES
MOISTURE MIGRATION

Redistribution of moisture generally occurs in stored grain when grain temperature is not controlled (Figure 10). Localized spoilage can occur even when the grain is stored at a safe moisture

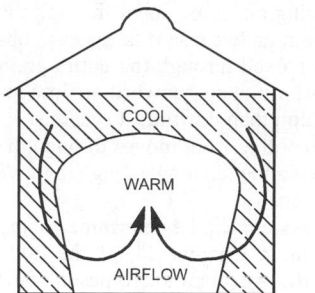 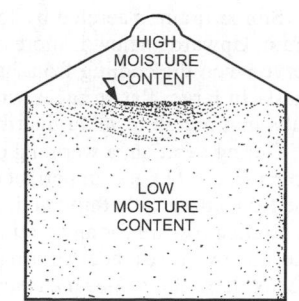

Fig. 10   Grain Storage Conditions Associated with Moisture Migration During Fall and Early Winter

level. Grain placed in storage in the fall at relatively high temperatures cools nonuniformly through contact with the outside surfaces of the storage bin as winter approaches. Thus, the grain near the outside walls and roof may be at cool outdoor temperatures while the grain nearer the center is still nearly the same temperature it was at harvest. These temperature differentials induce air convection currents that flow downward along the outside boundaries of the porous grain mass and upward through the center. When the cool air from the outer regions contacts the warm grain in the interior, the air is heated and its relative humidity is lowered, increasing its capacity to absorb moisture from the grain. When the warm, humid air reaches the cool grain near the top of the bin, it cools again and transfers vapor to the grain. Under extreme conditions, water condenses on the grain. The moisture concentration near the center of the grain surface causes significant spoilage if moisture migration is uncontrolled. During spring and summer, the temperature gradients are reversed. The grain moisture content increases most at depths of 2 to 4 ft below the surface. Daily variations in temperature do not cause significant moisture migration. Aside from seasonal temperature variations, the size of the grain mass is the most important factor in fall and winter moisture migration. In storages containing less than 1200 ft³, there is less trouble with moisture migration. The problem becomes critical in large storages and is aggravated by in-complete cooling of artificially dried grain. Artificially dried grain should be cooled to near ambient temperature soon after drying.

GRAIN AERATION

Aeration by mechanically moving ambient air through the grain mass is the best way to control moisture migration. Aeration systems are also used to cool grain after harvest, particularly in warmer climates where grain may be placed in storage at temperatures exceeding 100°F. After the harvest heat is removed, aeration may be continued in cooler weather to bring the grain to a temperature within 20°F of the coldest average monthly temperature. The temperature must be maintained below 50°F.

Aeration systems are not a means of drying because airflow rates are too low. However, in areas where the climate is favorable, carefully controlled aeration may be used to remove small amounts of moisture. Commercial storages may have pockets of higher-moisture grain if, for example, some batches of grain are delivered after a rain shower or early in the morning. Aeration can control heating damage in the higher-moisture pockets.

Aeration Systems Design

Aeration systems include fans capable of delivering the required amount of air at the required static pressure, suitable ducts or floors to distribute the air into the grain, and controls to regulate the operation of the fan. The airflow rate determines how many hours are required to cool the crop (Table 8). Most aeration systems are designed with airflow rates between 0.05 and 0.2 cfm/bu.

Stored grain is aerated by forcing air up or down through the grain. Upward airflow is more common because it is easier to observe when the cooling front has moved through the entire grain mass. In large, flat storages with long ducts, upward airflow results in more uniform air distribution than downdraft systems.

During aeration, a warming or cooling front moves through the crop (Figure 11); it is important to run the fan long enough to move the front completely through the crop.

Static pressure for an aeration system can be determined using the airflow resistance information in Chapter 11 of the 2005 *ASHRAE Handbook—Fundamentals*. All common types of fans are used in aeration systems. Attention should be given to noise levels with fans that are operated near residential areas or where people work for extended periods. The supply ducts connecting the fan to the distribution ducts in the grain should be designed and constructed according to the standards of good practice for any air-moving application. A maximum air velocity of 2500 fpm may be used, but 1600 to 2000 fpm is preferred. In large systems, one large fan may be attached to a manifold duct leading to several distribution ducts in one or more storages, or smaller individual fans may serve individual distribution ducts. Where a manifold is used, valves or dampers should be installed at each takeoff to allow adjustment or closure of airflow when part of the aerator is not needed.

Distribution ducts are usually perforated sheet metal with a circular or inverted U-shaped cross section, although many functional arrangements are possible. The area of the perforations should be at least 10% of the total duct surface. The holes should be uniformly spaced and small enough to prevent the passage of the grain into the duct (e.g., 0.1 in. holes or 0.08 in. wide slots do not pass wheat).

Since most problems develop in the center of the storage, and the crop cools naturally near the wall, the aeration system must provide good airflow in the center. Flush floor systems work well in storages with sweep augers and unloading equipment. Ducts should be easily removable for cleaning. Duct spacing should not exceed the depth of the crop; the distance between the duct and storage structure wall should not exceed one-half the depth of the crop for bins and flat storages. Common duct patterns for round bins are shown in Figure 12. Duct spacing for flat storages is shown in Figure 13.

When designing the distribution duct system for any type of storage, the following should be considered: (1) the cross-sectional area and length of the duct, which influences both the air velocity within the duct and the uniformity of air distribution; (2) the duct surface area, which affects the static pressure losses in the grain surrounding the duct; and (3) the distance between ducts, which influences the uniformity of airflow.

For upright storages where distribution ducts are relatively short, duct velocities up to 2000 fpm are permissible. Maximum recommended air velocities in ducts for flat storages are shown in Table 9. Furthermore, these velocities should not be exceeded in the air outlets from the storage; therefore, an air outlet area at least equal to the duct cross-sectional area should be provided.

The duct surface area that is perforated or otherwise open for air distribution must be great enough that the air velocity through the grain surrounding the duct is low enough to avoid excessive pressure loss. When a semicircular perforated duct is used, the entire surface area is effective; only 80% of the area of a circular duct resting on the floor is effective. For upright storages, the air velocity through the grain near the duct (duct face velocity) should be limited to 30 fpm or less; in flat storages, to 20 fpm or less.

Duct strength and anchoring are important. If ducts placed directly on the floor are to be held in place by the crop, the crop flow should be directly on top of the ducts to prevent movement and damage. Distribution ducts buried in the grain must be strong enough to withstand the pressure the grain exerts on them. In tall, upright storages, static grain pressures may reach 10 psi. When ducts are located

Table 8　Airflow Rates Corresponding to Approximate Grain Cooling Time

Airflow Rate, cfm/bu	Cooling Time, h
0.05	240
0.1	120
0.2	60
0.3	40
0.4	30
0.5	24
0.6	20
0.8	15
1.0	12

Table 9　Maximum Recommended Air Velocities Within Ducts for Flat Storages

Grain	Airflow Rate, cfm/bu	Air Velocity (fpm) within Ducts for Grain Depths of:				
		10 ft	20 ft	30 ft	40 ft	50 ft
Corn, soybeans, and other large grains	0.05	—	750	1000	1250	1250
	0.1	750	1000	1250	1500	1750
	0.2	1000	1250	—	—	—
Wheat, grain sorghum, and other small grains	0.05	—	1000	1500	1750	2000
	0.1	750	1500	2000	—	—
	0.2	1000	2000	—	—	—

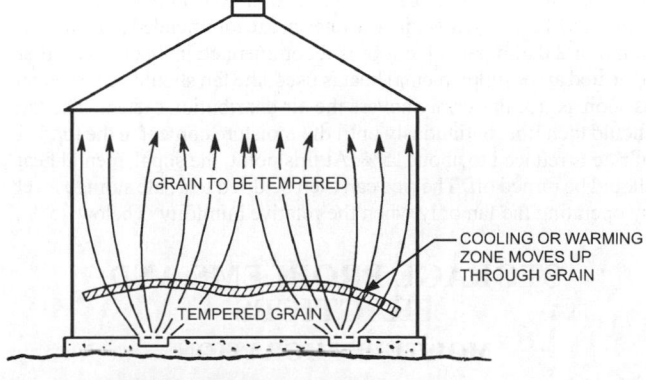

NEGATIVE-PRESSURE AERATION　　　　　　**POSITIVE-PRESSURE AERATION**

Fig. 11　Aerating to Change Grain Temperature

in the path of the grain flow, as in a hopper, they may be subjected to many times this pressure during grain unloading.

Operating Aeration Systems

The operation of aeration systems depends largely on the objectives to be attained and the locality. In general, cooling should be carried out any time the outdoor air temperature is about 15°F cooler than the grain. Stored grain should not be aerated when the air humidity is much above the equilibrium humidity of the grain because moisture will be added. The fan should be operated long enough to cool the crop completely, but it should then be shut off and covered, thus limiting the amount of grain that is rewetted.

Aeration to cool the grain should be started as soon as the storage is filled, and cooling air temperatures are available. Aeration to prevent moisture migration should be started whenever the average air temperature is 10 to 15°F below the highest grain temperature.

Aeration is usually continued as weather permits until the grain is uniformly cooled to within 20°F of the average temperature of the coldest month, or to 30 to 40°F.

Grain temperatures of about 32 to 50°F are desirable. In the northern corn belt, aeration may be resumed in the spring to equalize the grain temperature and raise it to between 40 and 50°F. This reduces the risk of localized heating from moisture migration. Storage problems are the only reason to aerate when air temperatures are above 60°F. Aeration fans and ducts should be covered when not in use.

In storages where fans are operated daily in fall and winter months, automatic controls work well when air is not too warm or humid. One thermostat usually prevents fan operation when the air temperature is too high, and another prevents operation when the air is too cold. A humidistat allows operation when the air is not too humid. Fan controllers that determine the equilibrium moisture content of the crop based on existing air conditions can regulate the fan based on entered information.

SEED STORAGE

Seed must be stored in a cool, dry environment to maintain viability. Most seed storages have refrigeration equipment to maintain a storage environment of 45 to 55°F. Seed storage conditions must be achieved before mold and insect damage occur.

BIBLIOGRAPHY

ASAE. 1993. Density, specific gravity, and mass-moisture relationships of grain for storage. ANSI/ASAE *Standard* D241.4. American Society of Agricultural Engineers (now American Society of Agricultural and Biological Engineers), St. Joseph, MI.

ASAE. 1995. Moisture relationship of plant-based agricultural products. ASAE *Standard* D245.5. American Society of Agricultural Engineers (now American Society of Agricultural and Biological Engineers), St. Joseph, MI.

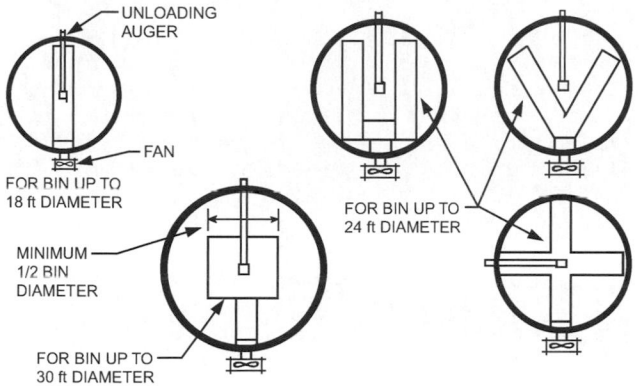

Fig. 12   Common Duct Patterns for Round Grain Bins

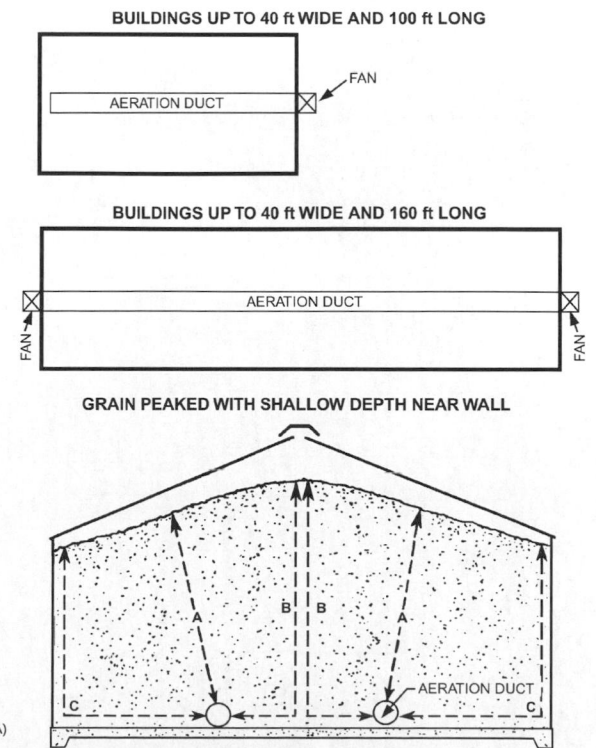

Fig. 13   Duct Arrangements for Large Flat Storages

ASAE. 1996. Resistance of airflow of grains, seeds, other agricultural products, and perforated metal sheets. ASAE *Standard* D272.2. American Society of Agricultural Engineers (now American Society of Agricultural and Biological Engineers), St. Joseph, MI.

Brooker, D.B., F. Bakker-Arkema, and C.W. Hall. 1992. *Drying and storage of grains and oilseeds.* Van Nostrand, Reinhold, NY.

MidWest Plan Service. 1988. *Grain drying, handling and storage handbook.* MWPS-13. Iowa State University, Ames.

MidWest Plan Service. 1980. *Low temperature and solar grain drying handbook.* MWPS-22. Iowa State University, Ames.

MidWest Plan Service. 1980. *Managing dry grain in storage.* AED-20. Iowa State University, Ames.

Hall, C.A. 1980. *Drying and storage of agricultural crops.* AVI Publishing, Westport, CT.

Hellevang, K.J. 1989. *Crop storage management.* AE-791. NDSU Extension Service, North Dakota State University, Fargo.

Hellevang, K.J. 1987. *Grain drying.* AE-701. NDSU Extension Service, North Dakota State University, Fargo.

Hellevang, K.J. 1983. *Natural air/low temperature crop drying.* EB-35. NDSU Extension Service, North Dakota State University, Fargo.

Saver, D.B. (ed.) 1992. *Storage of cereal grains and their products.* American Association of Cereal Chemists, St. Paul, MN.

Schuler, R.T, B.J. Holmes, R.J. Straub, and D.A. Rohweder. 1986. *Hay drying.* A3380. University of Wisconsin-Extension, Madison.

CHAPTER 26

AIR CONDITIONING OF WOOD AND PAPER PRODUCT FACILITIES

THIS chapter covers some of the standard requirements for air conditioning of facilities that manufacture finished wood products as well as for pulp and paper product process operations.

Special Warning: Certain industrial spaces may contain flammable, combustible, and/or toxic concentrations of vapors or dusts under either normal or abnormal conditions. In spaces such as these, there are life-safety issues that this chapter may not completely address. Special precautions must be taken in accordance with requirements of recognized authorities such as the National Fire Protection Association (NFPA), the Occupational Safety and Health Administration (OSHA), and the American National Standards Institute (ANSI). In all situations, engineers, designers, and installers who encounter conflicting codes and standards must defer to the code or standard that best addresses and safeguards life safety.

GENERAL WOOD PRODUCT OPERATIONS

In wood product manufacturing facilities, ventilation can be considered a part of the process. Metal ductwork should be used and grounded to prevent a buildup of static electricity. Hoods should be made of spark-free, noncombustible material. A pneumatic conveying system should be furnished to reduce the accumulation of wood dust in the collecting duct system. The airflow rate and velocity

---

The preparation of this chapter is assigned to TC 9.2, Industrial Air Conditioning.

should be able to maintain the air-dust mixture below the minimum explosive concentration level. If dampers are unavoidable in the system, they should be firmly fastened after balancing work. Dust collectors should be located outside the building. Fans or blowers should be placed downstream of the dust collector and air-cleaning equipment, and should be interlocked with the wood-processing equipment. When the fan or blower stops, the wood process should stop immediately and forward a signal to the alarm system.

Deflagration venting and suppression should be furnished for wood-processing workshops and wood-processing equipment such as vessels, reactors, mixers, blenders, mills, dryers, ovens, filters, dust collectors, storage equipment, material-handling equipment, and aerosol areas. The deflagration suppression system must be disarmed before performing any maintenance work to avoid possible injury from discharging the suppressant. Warning signs should be displayed prominently at all maintenance access points.

Finished lumber products to be used in heated buildings should be stored in areas that are heated 10 to 20°F above ambient. This provides sufficient protection for furniture stock, interior trim, cabinet material, and stock for products such as ax handles and glue-laminated beams. Air should be circulated within the storage areas. Lumber that is kiln-dried to a moisture content of 12% or less can be kept within a given moisture content range through storage in a heated shed. The moisture content can be regulated either manually or automatically by altering the dry-bulb temperature (Figure 1).

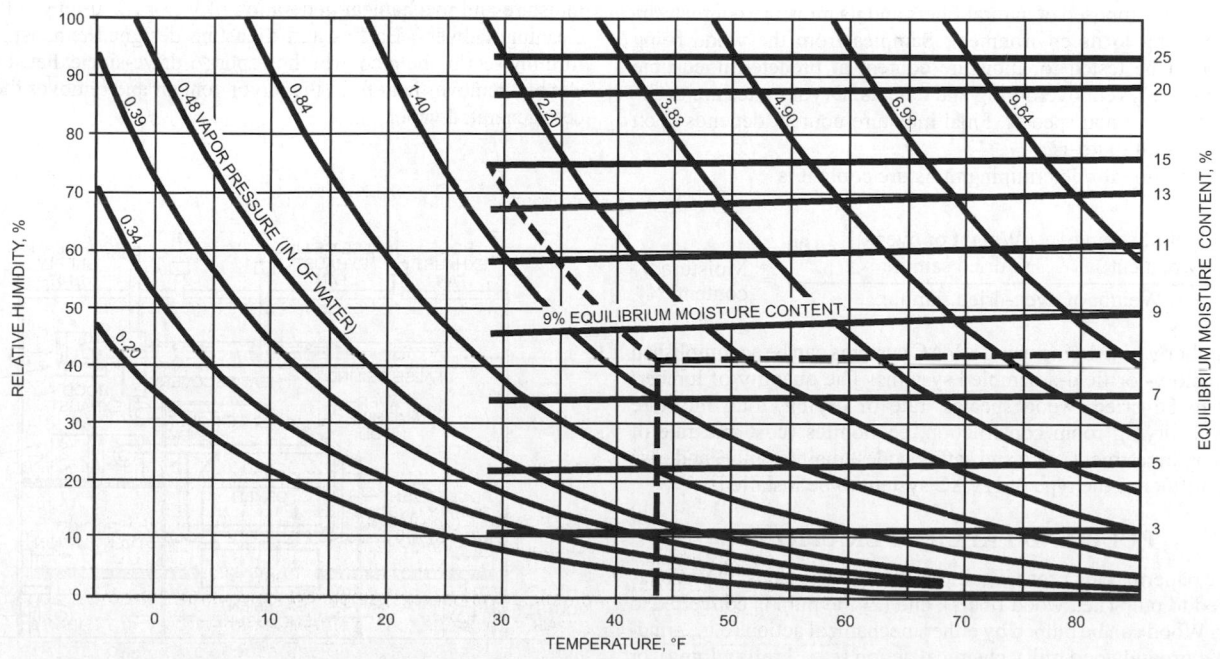

Fig. 1   Relationship of Temperature, Relative Humidity, and Vapor Pressure of Air and Equilibrium Moisture Content of Wood

Some special materials require close control of moisture content. For example, musical instrument stock must be dried to a given moisture level and maintained there because the moisture content of the wood affects the harmonics of most stringed wooden instruments. This control may require air conditioning, heating, and/or humidification, with or without reheating.

Process Area Air Conditioning

Temperature and humidity requirements in wood product process areas vary according to product, manufacturer, and governing code. For example, in match manufacturing, the match head must be cured (i.e., dried) after dipping. This requires careful control of humidity and temperature to avoid a temperature near the ignition point. Any process involving application of flammable substances should follow the ventilation recommendations of the National Fire Protection Association, the National Fire Code, and the U.S. Occupational Safety and Health Act.

Finished Product Storage

Finished lumber to be made into furniture, equipment parts, musical instruments, architectural woodwork, or other wood products of value is stored and/or manufactured under controlled temperature and humidity to maintain proper wood dryness. Improper drying can cause laminated or glued joints to fail. Finished wood that has changed dimension because of excess moisture gain or loss can cause fitting problems. Cracking, splitting, checking, warping, and discoloring are other problems with improperly dried and/or stored wood.

Green, rough, cut lumber is stacked end to end in layers, each layer being separated by wood strips to allow air circulation. Lumber can be stacked and left to dry naturally in open-sided sheds. Enclosed, heated kilns with steam coils and/or direct steam injection, forced air circulation, makeup air, and exhaust air vents could be used where faster, controlled drying is preferred. Drying (or addition of moisture) can be accomplished by HVAC systems using dehumidifying coils and/or desiccants, heating/reheat coils, humidifiers, makeup and exhaust air, distribution air ducts, and automatic controls. An insulated dehumidifying/humidifying room could be constructed and finished to minimize moisture migration from higher-humidity areas. Lumber can also be dried by solar kilns, microwaves, dielectric heating, superheated steam, and vacuum.

Wood is composed of natural fibers and its moisture content varies according to its environment. Samples from the wood being dried must be tested for moisture content at predetermined time intervals to prevent overdrying and defects. Drying rates are determined by the wood species. Final moisture content depends upon the wood's ultimate use.

The formula for determining moisture content is

$$\frac{\left[\left(\begin{array}{c}\text{Weight of sample}\\\text{when cut}\end{array}\right)-\left(\begin{array}{c}\text{Weight of oven-}\\\text{dried sample}\end{array}\right)\right] \times 100}{\text{Weight of oven-dried sample}} = \begin{array}{c}\text{Moisture}\\\text{content, \%}\end{array}$$

Lumber/wood drying using HVAC systems can be accomplished with factory- or field-assembled systems. The quantity of lumber/wood to be dried, wood species, rate of drying, total moisture removal, drying room construction, economics (cost and rate of return on investment), fire and safety codes, maintenance, and ease of use influence the type of HVAC system to be installed.

PULP AND PAPER OPERATIONS

The papermaking process comprises two basic steps: (1) wood is reduced to pulp (i.e., wood fibers), and (2) the pulp is converted to paper. Wood can be pulped by either mechanical action (e.g., grinding in a groundwood mill), chemical action (e.g., kraft pulping), or a combination of both.

Many different types of paper can be produced from pulp, ranging from the finest glossy finish to newsprint to bleached board to fluff pulp for disposable diapers. To make newsprint, a mixture of mechanical and chemical pulps is fed into the paper machine. To make kraft paper (e.g., grocery bags, corrugated containers), however, only unbleached chemical pulp is used. Disposable diaper material and photographic paper require bleached chemical pulp with a very low moisture content of 6 to 9%.

Paper Machine Area

In papermaking, extensive air systems are required to support and enhance the process (e.g., by preventing condensation) and to provide reasonable comfort for operating personnel. Radiant heat from steam and hot-water sources and mechanical energy dissipated as heat can result in summer temperatures in the machine room as high as 120°F. In addition, high paper machine operating speeds of 2000 to 4500 fpm and a stock temperature near 122°F produce warm vapor in the machine room.

Outdoor air makeup units and process exhausts absorb and remove room heat and water vapor released from the paper as it is dried (Figure 2). Makeup air is distributed to working areas above and below the operating floor. Part of the air delivered to the basement migrates to the operating floor through hatches and stairwells. Motor-cooling equipment distributes cooler basement air to the paper machine drive motors.

Wet and basement exhaust should be installed inside the room. Outdoor air intakes with insulated adjustable louvers should be installed on the outside wall to supplement the mechanical air supply. In facilities with no basement exterior wall, sufficient mechanical air intake should be provided. The exhaust, adjustable louver, or mechanical air intake should be furnished with modulating control. When the ambient temperature drops to near freezing, outdoor airflow must be reduced to a minimum and the appropriate heater started to prevent freezing.

The most severe ventilation demand occurs in the area between the wet-end forming section and press section and the dryer section. In the forming section, the pulp slurry, which contains about 90% water, is deposited on a traveling screen. Gravity, rolls, foils, vacuum, steam boxes, and three or more press roll nips are sequentially used to remove up to 50% of the water in the forming section and press section. The wet end is very humid because of evaporation of moisture and mechanical generation of vapor by turning rolls and cleaning showers. Baffles and a custom-designed exhaust in the forming section help control the vapor. A drive-side exhaust in the wet end removes heat from the motor vent air and removes the process generated vapor.

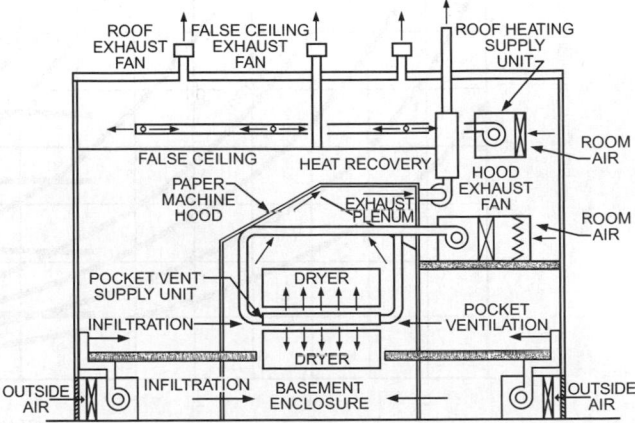

Fig. 2   Paper Machine Area

To prevent condensation or accumulated fiber from falling on the traveling web, a false ceiling is used with ducts connected to roof exhausters that remove humid air not captured at a lower point. At the wet end, heated inside air is usually circulated to scrub the underside of the roof to prevent condensation in cold weather. Additional roof exhaust may also remove accumulated heat from the dryer section and the dry end during warmer periods. Ventilation in the wet end should be predominantly by roof exhaust.

The large volume of moisture and vapor generated from the wet-end process rises and accumulates under the roof. To keep condensation from forming in winter, the roof is normally exhausted and hot air is distributed under the roof. Sufficient roof insulation should be installed to keep the inside surface temperature above the dew point. Heat transfer from the room to the interior surface is

$$\frac{t_r - t_{is}}{R_{r-is}} = \frac{t_{is} - t_o}{R_{is-o}} \qquad (1)$$

where

t_r = room air temperature, °F
t_{is} = roof interior surface temperature, °F
t_o = outdoor air temperature, °F
R_{r-is} = heat transfer resistance from room air to roof interior surface. In winter, $R_{r-is} = 0.61$ ft²·°F·h/Btu
R_{is-o} = required total R-value from roof interior surface to outdoor air, ft²·°F·h/Btu

For a given project, t_o and t_r have been determined and only t_{is} needs to be selected. For wet-end roof insulation and assuming 96% relative humidity, t_{is} can be shown on a psychrometric chart to be

$$t_{is} = t_r - 1.2°F \qquad (2)$$

Then Equation (1) can be simplified to find the required roof R-value as

$$R_{is-o} = \frac{0.61}{1.2}(t_r - t_o - 1.2) \qquad (3)$$

In the dryer section, the paper web is dried as it travels in a serpentine path around rotating steam-heated drums. Exhaust hoods remove heat from the dryers and moisture evaporated from the paper web. Most modern machines have enclosed hoods, which reduce the airflow required to less than 50% of that required for an open-hood exhaust. The temperature inside an enclosed hood ranges from 130 to 140°F at the operating floor to 180 to 200°F in the hood exhaust plenum at 70 to 90% rh, with an exhaust rate generally ranging from 300,000 to 400,000 cfm per machine.

Where possible, pocket ventilation air (see Figure 3) and hood supply air are drawn from the upper level of the machine room to take advantage of the preheating of makeup air by process heat as it rises. The basement of the dryer section is also enclosed to control infiltration of machine room air to the enclosed hood. The hood supply and the pocket ventilation air typically operate at 200°F; however, some systems run as high as 250°F. Enclosed hood exhaust is typically 300 cfm per ton of machine capacity. The pocket ventilation and hood supply are designed for 75 to 80% of the exhaust, with the balance infiltrated from the basement and machine room. Large volumes of air (500,000 to 800,000 cfm) are required to balance the paper machine's exhaust with the building air balance.

The potential for heat recovery from hood exhaust air should be evaluated. Most of the energy in steam supplied to the paper dryers is converted to latent heat in the hood exhaust as water evaporates from the paper web. Air-to-air heat exchangers are used where the air supply is located close to the exhaust. Air-to-liquid heat exchangers that recirculate water/glycol to heat remote makeup air units can also be used. Air-to-liquid systems provide more latent heat recovery, resulting in three to four times more total heat recovery than air-to-air units. Some machines use heat recovered from the exhaust air to heat process water. Ventilation in paper machine buildings in the United States ranges from 10 to 25 air changes per hour in northern mills to 20 to 50 in southern mills. In some plants, computers monitor the production rate and outdoor air temperature to optimize operation and conserve energy.

After fine, bond, and cut papers have been bundled and/or packaged, they should be wrapped in a nonpermeable material. Most papers are produced with less than 10% moisture by weight, the average being 7%. Dry paper and pulp are hygroscopic and begin to swell noticeably and deform permanently when the relative humidity exceeds 38%. Therefore, finished products should be stored under controlled conditions to maintain their uniform moisture content.

Finishing Area

To produce a precisely cut paper that stabilizes at a desirable equilibrium moisture content, the finishing areas require temperature and humidity control. Further converting operations such as printing and die cutting require optimum sheet moisture content for efficient processing. Finishing room conditions range from 70 to 75°F db and from 40 to 45% rh. Rooms should be maintained within reasonably close limits of the selected conditions. Without precise environmental control, the paper equilibrium moisture content varies, influencing dimensional stability, the tendency to curl, and further processing.

Process and Motor Control Rooms

In most pulp and paper applications, process control, motor control, and switchgear rooms are separate from the process environment. Air conditioning removes heat generated by equipment, lights, etc., and reduces the air-cleaning requirement. (See Chapter 19 for air conditioning in control rooms that include a computer, a computer terminal, or data processing equipment.) Ceiling grilles or diffusers should be located above access aisles to avoid the risk of condensation on control consoles or electrical equipment during start-up and recovery after an air-conditioning shutdown. Electrical rooms are usually maintained in the range of 75 to 80°F, with control rooms at 73°F; the humidity is maintained in the range of 45 to 55% in process control rooms and is not normally controlled in electrical equipment rooms.

Motor and electrical control rooms for process and electrical distribution control contain electronic equipment that is susceptible to corrosion. The typical pulp and paper mill environment contains both particulate and vapor-phase contaminants with sulfur- and chloride-based compounds. To protect equipment, multistage particulate and adsorbent filters should be used. They should have treated activated charcoal and potassium permanganate-impregnated alumina sections for vapor-phase contaminants, as well as fiberglass and cloth media for particulates.

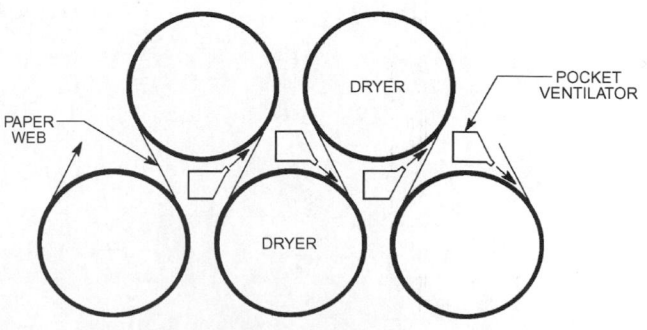

Fig. 3 Pocket Ventilation

To ensure normal operation of air-conditioning systems, redundancy of supply fans, fan motors, and fan power supply in air-handling units that serve process control rooms and motor control centers is strongly recommended in new-construction plants.

Switchgear and motor control centers are not as heat-sensitive as control rooms, but the moisture-laden air carries chemical residues onto the contact surfaces. Arcing, corrosion, and general deterioration can result. A minimum amount of filtered, outdoor air and air conditioning is used to protect these areas.

In most projects, the electric distribution control system (DCS) is energized before the room air conditioning is installed and started. If a temporary air conditioner is used in the DCS room, a condensate drain pan and temporary drain pipe should be installed to keep condensate from the cable channel beneath the DCS panels.

Paper Testing Laboratories

Design conditions in paper mill laboratories must be followed rigidly. The most recognized standard for testing environments for paper and paper products (paperboard, fiberboard, and containers) is TAPPI (the Technical Association of the Pulp and Paper Industry) *Standard* T402. Other standards include ASTM E171 and ISO/TC 125.

Standard pulp and paper testing laboratories have three environments: preconditioning, conditioning, and testing. The physical properties of a sample are different if it is brought to the testing humidity from a high humidity than if it is brought from a lower humidity. Preconditioning at lower relative humidity tends to eliminate hysteresis. For a preconditioning atmosphere, TAPPI *Standard* T402 recommends 10 to 35% rh and 72 to 104°F db. Samples are usually conditioned in a controlled, conditioned cabinet.

Conditioning and testing atmospheres should be maintained at 50 ± 2.0% rh and 73 ± 2°F db. However, a change of 2°F db at 73°F without starting a humidifier causes the relative humidity to fluctuate as much as 3%. A dry-bulb temperature tolerance of ±1°F must be held to maintain a ±2% rh. A well-designed temperature and humidity control system should be provided.

Miscellaneous Areas

The pulp digester area contains many components that release heat and contribute to dusty conditions. For batch digesters, the chip feeders are a source of dust and need hooded exhaust and makeup air. The wash and screen areas have numerous components with hooded exhausts that require considerable makeup air. Good ventilation controls fumes and humidity. The lime kiln feed-end releases extremely large amounts of heat and requires high ventilation rates or air conditioning.

Recovery-boiler and power-boiler buildings have conditions similar to those of power plants; the ventilation rates are also similar. The control rooms are generally air conditioned. The grinding motor room, in which groundwood is made, contains many large motors that require ventilation to keep the humidity low.

System Selection

The system and equipment selected for air conditioning a pulp and paper mill depends on many factors, including the plant layout and atmosphere, geographic location, roof and ceiling heights (which can exceed 100 ft), and degree of control desired. Chilled-water systems are economical and practical for most pulp and paper operations, because they have both the large cooling capacity needed by mills and the precision of control to maintain the proper temperature and humidity in laboratories and finishing areas. In the bleach plant, the manufacture of chlorine dioxide is enhanced by using water with a temperature of 45°F or lower; this water is often supplied by the chilled-water system. If clean plant or process water is available, water-cooled chillers are satisfactory and may be supplemented by water-cooled direct-expansion package units for small, remote areas. However, if plant water is not clean enough, a separate cooling tower and condenser water system should be installed for the air conditioning.

Most manufacturers prefer water-cooled over air-cooled systems because of the gases and particulates present in most paper mills. The most prevalent contaminants are chlorine gas, caustic soda, borax, phosphates, and sulfur compounds. With efficient air cleaning, the air quality in and about most mills is adequate for properly placed air-cooled chillers or condensing units that have properly applied coil and housing coatings. Phosphor-free brazed coil joints are recommended in areas where sulfur compounds are present.

Heat is readily available from processing operations and should be recovered whenever possible. Most plants have good-quality hot water and steam, which can be used for unit heater, central station, or reheat quite easily. Evaporative cooling should be considered. Newer plant air-conditioning methods, using energy conservation techniques such as temperature destratification and stratified air conditioning, have application in large structures. Absorption systems should be considered for pulp and paper mills because they provide some degree of energy recovery from the high-temperature steam processes.

BIBLIOGRAPHY

ACGIH. 2009. *Industrial ventilation: A manual of recommended practice*, 27th ed. American Conference of Governmental Industrial Hygienists, Cincinnati, OH.

ASTM. 2007. Standard specification for standard atmospheres for conditioning and testing flexible barrier materials. *Standard* E171-94 (2007). American Society for Testing and Materials, West Conshohocken, PA.

TAPPI. 1988. Conditioning and testing atmospheres for paper, board, pulp handsheets, and related products. *Test Method* T402. Technical Association of the Pulp and Paper Industry, Norcross, GA.

CHAPTER 27

POWER PLANTS

THIS chapter discusses HVAC systems for industrial facilities for the production of process heat and power and for electrical generating stations. Not every type of power plant is specifically covered, but the process areas addressed normally correspond to similar process areas in any plant. For example, wood-fired boilers are not specifically discussed, but the requirements for coal-fired boilers generally apply. Aspects of HVAC system design unique to nuclear power plants are covered in Chapter 28.

Special Warning: Certain industrial spaces may contain flammable, combustible, and/or toxic concentrations of vapors or dusts under either normal or abnormal conditions. In spaces such as these, there are life-safety issues that this chapter may not completely address. Special precautions must be taken in accordance with requirements of recognized authorities such as the National Fire Protection Association (NFPA), the Occupational Safety and Health Administration (OSHA), and the American National Standards Institute (ANSI). In all situations, engineers, designers, and installers who encounter conflicting codes and standards must defer to the code or standard that best addresses and safeguards life safety.

GENERAL DESIGN CRITERIA

Space-conditioning systems in power plant buildings are designed to maintain an environment for reliable operation of power generation systems and equipment and for the convenience and safety of plant personnel. A balance is achieved between the cost of the process systems designed to operate in an environment and the cost of providing that environment.

Environmental criteria for personnel safety and comfort are governed by several sources. The U.S. Occupational Safety and Health Administration (OSHA) defines noise, thermal environment, and air contaminant exposure limits. Chapters 14 and 31 of this volume and *Industrial Ventilation* by the American Conference of Governmental Industrial Hygienists (ACGIH 2010) also provide guidance for safety in work spaces, primarily in the areas of industrial ventilation and worker-related heat stress. The degree of worker comfort is somewhat subjective and more difficult to quantify. The plant owner or operator ordinarily establishes the balance between cost and worker comfort.

Exhaust vents are subject to regulation of the plant's air quality permit and local air pollution control board's requirements. For this reason, all exhaust vent locations should be properly identified and classified, and coordinated with the plant's environmental compliance department. Treatment of exhaust streams is discussed in Chapter 29 of the 2008 *ASHRAE Handbook—HVAC Systems and Equipment.*

Criteria should be clearly defined at the start of design, because they document an understanding between the process designer and the HVAC system engineer that is fundamental to achieving the environment required for the various process areas. Typical criteria for a coal-fired power plant are outlined in Table 1. They should be reviewed for compliance with local codes, the plant operator's experience and preferences, and the overall financial objectives of the facility. Additional discussion of criteria may be found in the sections on specific areas.

Temperature and Humidity

Selection of outdoor design temperatures is based on the operating expectations of the plant. If the power production facility is critical and must operate during severe conditions, then the effect of local extreme high and low temperatures on the systems should be evaluated. Electrical power consumption is usually highest under extreme outdoor conditions, so the plant should be designed to operate when needed the most. Other temperature ranges, indicated in Table 1, may be more appropriate for less-critical applications.

Indoor temperatures should match the specified operating temperatures of the equipment. Electrical equipment, such as switchgear, motor control centers, and motors, typically determines the design temperature limits in the plant; common temperature ratings are 104 or 122°F. Other areas such as elevator machine rooms may include electronic equipment with temperature restrictions.

In plant areas where compressed-gas containers are stored, the design temperature is according to the gas supplier. Typically, the minimum temperature should be high enough that the gas volume can be effectively released from the containers. If the gas is hazardous (e.g., chlorine), the minimum temperature does not apply during personnel occupancy periods, when high dilution ventilation rates are needed.

Practical ventilation rates for fuel-fired power plants provide indoor conditions 10 to 20°F above the outdoor ambient. Therefore, ventilation design criteria establish a temperature rise above the design outdoor temperature to produce an expected indoor temperature that matches the electrical equipment ratings. For example, an outdoor extreme design temperature of 112°F with a ventilation system designed for a 10°F rise would meet the requirements of 122°F-rated plant equipment unless a new record temperature occurred. However, the environment for workers should also be considered. Velocity (spot) cooling may be necessary in some areas to support work activities.

Low temperatures may affect plant reliability because of the potential for freezing. The selection of the low design temperature should be balanced by the selection of the heating design margin. If the record low temperature is used in the design, indoor design temperatures of 35 to 40°F may be used. In the heating system design, credit is generally not taken for heat generated from operating equipment.

The preparation of this chapter is assigned to TC 9.2, Industrial Air Conditioning.

Table 1 Design Criteria for Fuel-Fired Power Plant

Building/Area	Design Outdoor Cooling/ Heating Dry-Bulb[a]	Indoor Temperature, °F		Relative Humidity, %	Room Ventilation Rate, ach*	Filtration Efficiency, %	Pressur- ization	Redundancy[b]	Noise Criterion
		Maximum	Minimum						
Steam Turbine Area									
Suboperating level	0.4%/99.6%	Design outdoor + 10	45	None	30	None	None	Multiplicity	Background
Above operating floor	0.4%/99.6%	Design outdoor + 10	45	None	10	None	None	Multiplicity	Background
Combustion Turbine Area	0.4%/99.6%	Design outdoor + 18	45	None	20	None	None	Multiplicity	Background
Steam Generator Area									
Below burner elevation	0.4%/99.6%	Design outdoor + 10	45	None	30	None	None	Multiplicity	Background
Above operating floor	0.4%/99.6%	Design outdoor + 10	45	None	15	None	None	Multiplicity	Background
Other Non-Air-Conditioned Areas									
Shops	1%/99%	Design outdoor + 10	65	None	15	None	None	None	85 dBA
Air-Conditioned Areas[d]									
Control rooms and control equipment rooms containing instruments and electronics	Extreme (see text)	75 ± 2	72 ± 2	30 to 65	ASHRAE Std. 62.1	85 to 90 (see text)	Positive	100%	NC-40[c]
Offices	1%/99%	78	70	30 to 65	ASHRAE Std. 62.1	ASHRAE Std. 62.1	Positive	None	See text
Laboratories	1%/99%	78	70	30 to 65	ASHRAE Std. 62.1	High	Positive	None	See text
Locker rooms and toilets	1%/99%	78	70	None	ASHRAE Std. 62.1	ASHRAE Std. 62.1	Negative	None	See text
Shops (air-conditioned)	1%/99%	78	65	None	ASHRAE Std. 62.1	None	None	None	85 dBA
Mechanical Equipment									
Pumps, large power	0.4%/99.6%	Design outdoor + 10	45	None	30	None	None	Multiplicity	Background
Valve stations, miscellaneous	0.4%/99.6%	Design outdoor + 10	45	None	15	None	None	None	85 dBA
Elevator machine rooms	0.4%/99.6%	90	45	None	None	Low	Positive	None	85 dBA
Fire pump area	0.4%/99.6%	NFPA Std. 20	NFPA Std. 20	None	NFPA Std. 20	None	None	None	85 dBA
Diesel generator area	0.4%/99.6%	Design outdoor + 10	45	None	30	None	None	None	Background
Electrical Equipment[d]									
Enclosed transformer equipment areas	0.4%/99.6%	Design outdoor + 10	45	None	60	Low	Positive	100%	85 dBA
Critical equipment	Extreme (see text)	Design outdoor + 10	45	None	30	None	Positive	100%	85 dBA
Miscellaneous electrical equipment	0.4%/99.6%	Design outdoor + 10	45	None	20	None	None	Multiplicity	85 dBA
Water Treatment									
Chlorine equipment rooms									
When temporarily occupied	0.4%/99.6%	Design outdoor + 10	None	None	60	None	Negative	None	85 dBA
When unoccupied	0.4%/99.6%	Design outdoor + 10	60	None	15	None	Negative	None	85 dBA
Chemical treatment	0.4%/99.6%	Design outdoor + 10	60	None	10	None	None	None	85 dBA
Battery Rooms	0.4%/99.6%	77	77	None	As required for hydrogen dilution	None	Negative or neutral	50%	85 dBA

*Listed numbers are for estimating purposes only. When heat gain data are available, use Equation (1) to calculate required ventilation rate.

[a]See Chapter 14 of the 2009 *ASHRAE Handbook—Fundamentals* for design dry-bulb temperature data corresponding to given annual cumulative frequency of occurrence and specific geographic location of plant.

[b]Multiplicity indicates that the HVAC system should have multiple units.

[c]See Figure 6 in Chapter 8 of the 2009 *ASHRAE Handbook—Fundamentals* for noise criterion curves.

[d]See ASHRAE research project RP-1104 (White 2003) for heat release values.

The selection of outdoor design humidity levels affects the selection of cooling towers and evaporative cooling processes and the sizing of air-conditioning coils for outdoor air loads. When values from Chapter 14 of the 2009 *ASHRAE Handbook—Fundamentals* are used for design, the mean coincident wet bulb is appropriate. If extreme dry-bulb temperatures are selected for the design basis, the use of extreme wet bulbs is too restrictive because the extremes are not coin-

cident. It is prudent to use the wet bulb associated with the 1% dry bulb when extreme dry-bulb temperatures are used for the design.

Indoor design humidity is not a factor in ventilated areas unless the plant is in a harsh, corrosive environment. In this case, lower humidity reduces the potential for corrosion. In air-conditioned areas for personnel or electronic equipment, ASHRAE *Standard* 62.1, Instrumentation, Systems, and Automation Society (ISA)

Standard S71-04, and manufacturers' recommendations dictate the humidity criteria.

Ventilation Rates

Ventilation within plant structures provides heat removal and dilution of potentially hazardous gases. Ventilation rates for heat removal are calculated during HVAC system design to meet summer indoor design temperatures.

The numbers shown in Table 1 for air change rates are for estimating approximate ventilation needs. Actual heat emission rates should be obtained from equipment manufacturers or from the engineer's experience. American Boiler Manufacturer's Association (ABMA) heat loss curves (Stultz and Kitto 1992) can be used to approximate heat loads from boiler casings if better information is not available.

The ventilation rate for room heat removal is

$$Q = \frac{q}{(t_r - t_o)(60 \rho c_p)} \qquad (1)$$

where

> Q = ventilation rate, cfm
> q = room heat, Btu/h
> t_r = allowed room temperature from Table 1, °F
> t_o = outdoor air temperature, °F
> ρ = air density, lb_m/ft^3
> c_p = specific heat of air = 0.24 $Btu/lb_m \cdot °F$

Hazardous gases are mostly handled by the process system design functions. Natural gas and other combustible fuel gases are controlled by ignition safeties and may contain odorants for detection. Hydrogen and other gases used for generator and bus cooling are monitored for leakage by pressure loss or makeup rates. Escaped gases are diluted by outdoor air infiltration. For a building with very tight construction (i.e., very little natural infiltration), an analysis should be performed to verify that dilution rates are acceptable.

Flue gas is confined to the boiler and flue gas ductwork and generally poses no hazard. In some types of boilers and associated gas ducts, however, flue gas is at a higher pressure than the surroundings and can leak into occupied areas. Also, special treatment gases such as ammonia or sulfur compounds encountered in flue gas conditioning systems can leak into the boiler building, depending on the location of the treatment device in the flue gas stream. In these cases, gas detection monitors should be used.

Ventilation for areas with hazardous gases (e.g., chlorine) should be designed by specific gas industry standards such as *The Chlorine Manual* (CI 1997) or ACGIH (2010).

Infiltration and Exfiltration

Infiltration of outdoor air into boiler and power-generation structures or exfiltration of room air from these buildings is driven by thermal buoyancy of heated air. Both infiltration and exfiltration are beneficial; infiltration air dilutes fugitive fumes, whereas exfiltration air carries out excess heat during hot weather. However, infiltration adds to the cold-weather load on the heating system.

Filtration and Space Cleanliness

Filtration of ventilation air for process areas is usually not needed because some process areas are dirtier than the outdoor surroundings, and the process equipment is designed to operate in a dusty environment. However, the plant may be located in an area with sources of outdoor particulate contaminants that need to be managed to protect the process equipment. Power plants in dusty or sandy areas, or where there are seasonal nuisances such as airborne plant matter, may require filtration of ventilation air. Plants at industrial sites such as refineries and paper mills may need to address gaseous contaminants and corrosive gases, as well.

Indoor air cleanliness is a concern in control room HVAC system design. Even if the control center is in an independent building, remote from the boiler-turbine building, other operations such as coal transportation, coal crushing, fuel/air distribution and combustion, ash handling, fume heat recovery, fume/smoke exhaust diffusion, and so forth may contaminate the entire plant and its surroundings.

When potential outdoor contaminants are a factor, the quality of outdoor air may need to be evaluated. This may include collection of typical particulates and the use of corrosion coupons to quantify gaseous contaminants. The U.S. Environmental Protection Agency (EPA) is a source of data. Filtration requirements may include 30% dust-spot test efficiency prefilters, 65 to 90% efficiency final filters, and gas-phase filtration units.

Air-conditioned areas for people should meet ASHRAE *Standard* 62.1 requirements. Air-conditioned areas for control and electrical equipment should meet the requirements of the equipment manufacturer(s). Guidelines for reliability of electrical equipment are found in ISA *Standard* S71.04.

Redundancy

Maintaining design operating temperatures within the power plant is essential for reliable operation. Operating electrical equipment above its rated temperature reduces equipment life. Sensitive electronic equipment, such as in the main control center, may not function reliably at high temperatures. Low temperatures also affect plant availability; for example, low temperatures in batteries or freezing of pipes, instrument lines, or tanks could prevent normal plant operation.

The HVAC systems or components essential for plant operation should be designed with redundancy to ensure plant availability. Automatic switchover to the back-up system may be required in normally unoccupied areas. In areas where back-up systems are impractical, temperature monitoring and alarming systems should be considered to initiate temporary corrective measures.

HVAC systems that include multiple units (indicated as "multiplicity" in Table 1) also improve power plant reliability. A space ventilated by multiple fans, such as four at 1/4 capacity, may retain sufficient ventilation even if one fan is out of service.

Noise

Consideration should be given to noise levels produced by HVAC system equipment both inside and outside plant spaces. Indoor noise guidelines should be established for air-conditioned areas and ventilated areas with continuous occupancies. Outdoor noise levels are established by the environmental noise pollution concerns of adjacent areas.

Air-conditioned indoor spaces should meet the normal sound level guidelines for occupancies (such as offices) listed in Chapter 48. Special occupancies such as control rooms should follow the guidelines in Table 1.

Ventilated areas of the plant should be treated as other industrial areas following OSHA regulations. Sound levels indicated in Table 1 are suggested guidelines that may be appropriate in the absence of a specific engineered solution to meet the OSHA requirements. Where "background" is indicated in Table 1, noise generated by HVAC equipment is usually not a major noise source in comparison to the noise from processes or equipment such as turbine generators, motors, pumps, and relieving of process steam. In these areas, overall noise level criteria are established by the process equipment requirements.

HVAC system components contribute to the overall noise level outside the plant buildings either by generating noise or by having ventilation openings. HVAC designs for power plants in urban areas can be significantly influenced by outdoor noise level requirements. Equipment may have to include sound-absorbing materials or be located indoors in sound attenuation enclosures. Openings may require acoustical louvers.

VENTILATION APPROACH

Summer ventilation can be achieved by natural draft, forced mechanical supply and exhaust, or natural and mechanical combined systems. **Natural-draft systems** use a combination of adjustable inlet louvers and open doors or windows and relieve warmed air through roof or high side-wall openings. **Mechanical systems** use fans, power roof ventilators (PRVs), or air-handling units to move air. A **combined system** typically uses lower and upper wall and roof openings for natural ventilation while using mechanical ventilation as a supplement. With any ventilation arrangement, consideration should be given to physical separation of inlet and outlet openings to minimize recirculation, as discussed in Chapter 24 of the 2009 *ASHRAE Handbook—Fundamentals*.

APPLICATIONS

Large plants or units with layouts containing large ducts and equipment in the ventilation airflow path, possibly with limited separation between pieces of major equipment and/or exterior walls, imposing a pressure drop exceeding the capability of gravity ventilation systems, require mechanical assistance. The plant design may have cavities, such as the area under the boiler arch and between the casing and flue gas duct (see Figures 1 and 2), which require mechanical ventilation. Areas of the plant such as the conveyor gallery usually require mechanical assistance in the makeup air system to ensure pressurization control and proper functioning of dust collectors and associated equipment.

Natural or gravity systems are appropriate for facilities without basements and with relatively open airflow paths. The pressure drop of intake louvers, control dampers, powerhouse airflow path, and exhaust louvers/gravity vents should not exceed the minimum expected buoyancy forces. Care should also be taken to configure the system so as not to interfere with mechanical dust collection equipment.

In any system configuration, the engineer should ensure all areas of the plant are provided airflow. Air intakes and supply ducts/fans should be located to prevent short-circuiting with relief openings. Ventilation air should be supplied to electrical boards and other equipment located in upper elevations of the plant, because these areas are usually significantly hotter than the lower elevations.

Driving Forces

Natural ventilation systems use the thermal buoyancy of the air as the motive force for air movement through a building. Equations for determining differential pressures for natural ventilation are found in Chapter 16 of the 2009 *ASHRAE Handbook—Fundamentals*. With natural ventilation, air enters the enclosed space and is heated by the plant equipment. The difference in density between the inside air and the outdoor air causes air to be drawn into the building at low elevations and relieved at high elevations.

Mechanical ventilation depends on fans (or fans and buoyancy forces) to provide required ventilation regardless of the building configuration or the temperature difference.

Air Distribution

With natural ventilation, small differential pressures drive air movement. Accordingly, air is drawn into the building at low velocities; it penetrates a short distance into the building and then disperses.

Mechanical ventilation supplied from the walls or roof can distribute air more effectively throughout the structure.

A combined system uses both natural and mechanical ventilation to achieve effective air distribution and prevent air stagnation A

typical combined system uses lower-level sidewall openings as the primary natural air intake and roof or upper-level sidewall openings for hot air relief. The location and size of openings can prevent hot air accumulation under the roof naturally. Mechanical ventilation should be provided where sufficient airflow cannot be established.

Inlet and Exhaust Areas

Because of the low differential pressure driving the air, natural ventilation requires numerous large inlet louver and exhaust relief areas. Mechanical ventilation requires fewer openings.

Noise

Openings required for natural ventilation allow noise generated by inside plant equipment to pass more easily to the outdoors.

Mechanical equipment such as fans and PRVs generate noise directly, but the noise level can be managed by fan selection and acoustical treatment.

Impact on Plant Cleanliness

Natural ventilation creates negative pressure in the lower portions of the building, which may draw dust and fumes into the building through openings near ground level.

Mechanical ventilation can pressurize the building and can draw air from relatively clean sources at higher elevations.

Economics

The primary advantage to natural ventilation is that there are no operating costs for fan power. Because natural ventilation is passive, it is more reliable and has lower maintenance costs than a mechanical system. However, natural ventilation may not always be the most economical selection. The cost of louvers and inlet openings, architectural features, and gravity relief openings to achieve an acceptable ventilation rate may be higher than the first cost for mechanical ventilation.

Another consideration is the average building temperature. Because internal heat is the driving force, the naturally ventilated building is normally warmer than the power-ventilated building. This warmer average temperature may shorten the life of plant equipment such as expansion joints, seals, motors, electrical switchgear, and instrumentation. Warmer temperatures may also affect operator performance.

The large louver areas associated with natural ventilation may allow greater infiltration, thereby increasing the winter heating load. This additional heating cost may offset some of the summer energy savings of natural ventilation.

A combined system takes on the strengths of both natural and mechanical ventilation and can offer the advantages of reduced capital and operating costs.

STEAM GENERATOR BUILDINGS: INDUSTRIAL AND POWER FACILITIES

A steam generator is a device that uses heat energy to convert water to steam. The two basic subsystems of a steam generator are the heat energy system and the steam process system.

The heat energy system for a fueled (oil, gas, coal, etc.) steam generator includes fuel distribution piping or conveyors, preparation subsystems, and supply rate and ignition controls. Provision to supply and regulate combustion air is required at the combustion chamber; flue gas is handled downstream of the combustion area. With ash-producing fuels, bottom ash below the steam generator and fly ash entrained in the flue gas must be processed. Figure 1 shows a steam generator building with typical components.

Steam process components typically found in the steam generator building include an enclosure for the fire and heat transfer surfaces and feedwater equipment such as pumps, piping, and controls. Steam lines for primary and reheat steam are typically routed from

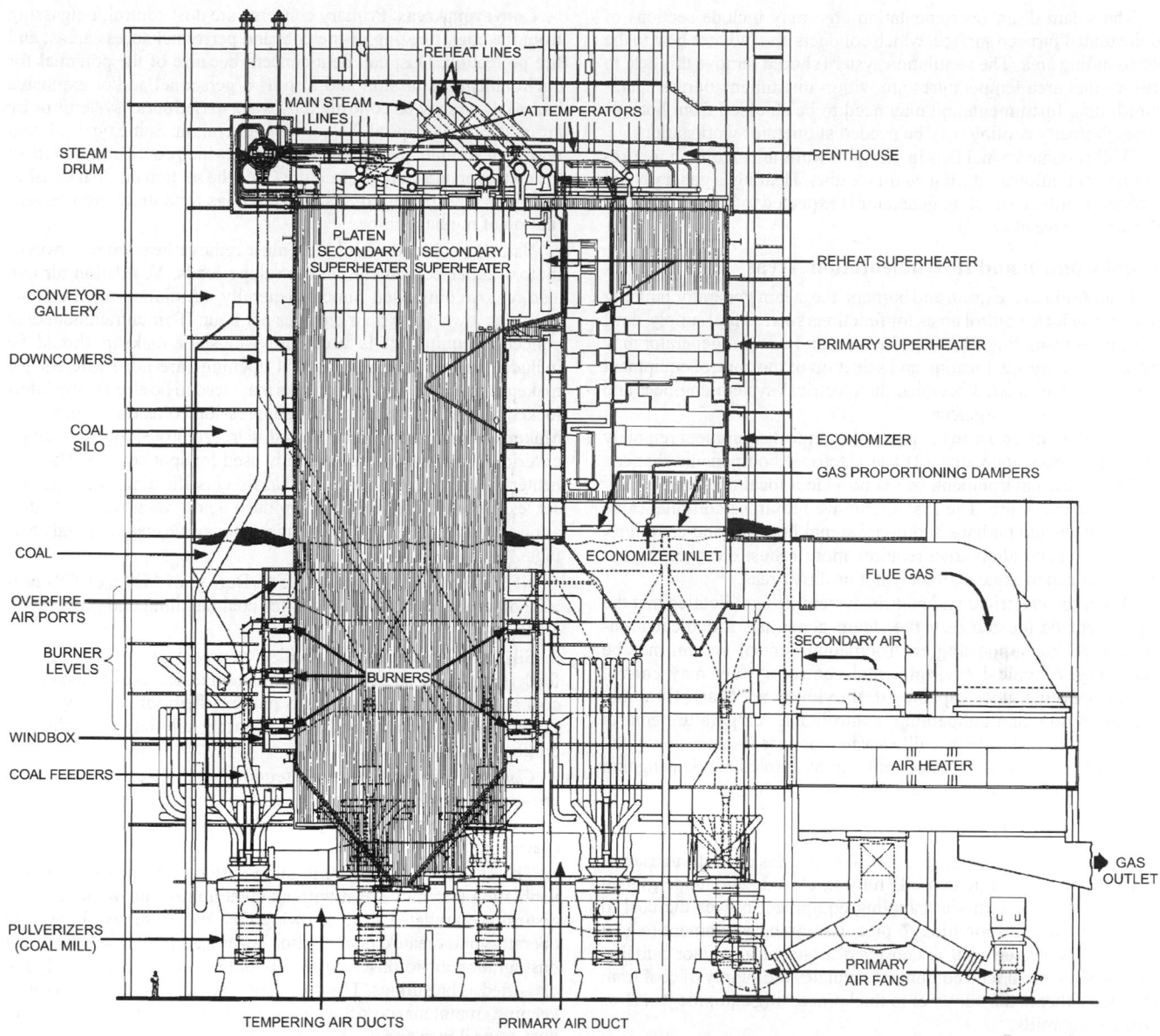

Fig. 1   Steam Generator Building
(Courtesy The Babcock & Wilcox Company)

the steam drum and reheat sections of the steam generator to the steam turbine or process systems.

The heat energy and steam process systems impose requirements on the HVAC systems for specific areas of the steam generator building.

Burner Areas

Fuel (gas, oil, coal, etc.) is transferred to the furnace, mixed with combustion air, and ignited in the burner area of the steam generator. Instrumentation must modulate the fuel in response to combustion needs. Viewports typically allow operators to monitor combustion. In many cases, these viewports are equipped with aspirating air systems to limit the amount of flue gas and heat escaping the boiler and entering the boiler room.

The burner area requires special attention for the steam generator building ventilation system. This area is often occupied by plant operators who monitor and inspect the controls and the combustion process. Heat is radiated and conducted to the adjacent spaces from

inspection ports, penetrations, and the steam generator. Leakage of fumes and combustion gases is also possible.

Both the burner area operator and the controls require ventilation with outdoor air. Outdoor air also provides dilution for fugitive fumes. Outdoor air can be ducted to burner areas and discharged by supply registers or blown directly into the area with wall-mounted fans, depending on the building arrangement. The flow rate is difficult to quantify; generally, 60 air changes per hour supplied to an area 15 to 20 ft around the steam generator provide adequate ventilation. Consider providing velocity cooling of personnel workstations. In cold climates, outdoor air may need to be tempered with indoor air.

Steam Drum Instrumentation Area

A typical steam generator has a steam drum at the top of the boiler that provides the water-to-steam interface. The water level in the drum is monitored to regulate the flow of steam and feedwater. This is a critical steam generator control function, so accurate and reliable process flow measurement is important.

The steam drum instrumentation area may include sections of uninsulated furnace surface, which conducts and radiates heat to the surrounding area. The ventilation system should remove this heat to ensure that area temperatures are within instrumentation temperature limits. Instrumentation may need to be shielded from hot surfaces. Velocity cooling may be needed at operator workstations.

Wall-mounted panel fans in the outer walls are an option for providing ventilation air during warm weather. Heating is generally not a concern unless the steam generator is expected to be out of service during cold weather.

Local Control and Instrumentation Areas

In addition to the drum and burners, the steam generator building may house local control areas for functions such as fuel supply, draft fans, or ash handling. Because areas around a steam generator may be hot and dirty, the location and selection of the control equipment should be coordinated between the electrical system engineer and the HVAC system engineer.

The alternatives are to (1) locate the control equipment remotely from the steam generator, (2) use electrical components that can withstand the environment, or (3) provide a local environmentally controlled enclosure. The first alternative requires additional cable and raceway and perhaps additional signal boosters and conditioners. The second alternative requires more robust electrical equipment that can tolerate extremely hot or dirty areas.

When the electrical and control system design dictates that the equipment be located near the steam generator, a dedicated enclosure with a supporting environmental control system may be necessary. A typical environmental control system may control an air-handling unit capable of providing adequate filtration, pressurization, and temperature control. The temperature control may be obtained with a chilled-water or direct-expansion (DX) coil with a remote condensing unit. An air-cooled condensing unit may be used if it is rated to match the surroundings.

Coal- and Ash-Handling Areas

Coal is typically stored on site, either in piles or in storage structures, including semienclosed, fully enclosed, and underground storage facilities. Material-handling equipment moves the coal to conveyors for transportation to preconditioning equipment (e.g., a crusher). Processed coal is conveyed to steam generator building storage silos. Coal feed equipment regulates the supply of coal from the silos either to the burner or to final processing equipment such as pulverizing mills.

Many new power plants are designed to burn low-cost coals from Wyoming's Powder River Basin and other similar coal seams. Many of these coals are extremely friable when dry, creating significant amounts of dust during handling. During new plant design, the material handling and ventilation systems should be designed to control dusting, preferably by using modern chutework and coal-handling machinery design to reduce or control dust emissions during fuel handling. This can reduce the effects of mechanical dust collectors and associated makeup air on plant HVAC systems.

Some coals spontaneously begin to burn during normal outdoor ambient weather and normal powerhouse interior conditions. Under some conditions, accumulated coal dust spontaneously smolders, then suddenly flashes if the surface crust of the pile is broken, exposing the smoldering coal to oxygen. It is important that equipment in areas subject to dust accumulation be designed to reduce dust accumulation with sloped or curved surfaces, and designed to facilitate manual cleaning and washdown. Spontaneous combustion must be addressed in ventilation and dust collector design and installation to facilitate emptying dust out of the collector without exposing plant personnel to fire and explosion hazards.

Coal-handling areas in the steam generator building that require special ventilation system consideration are the conveyor, silo, feeder, mill, and their transition areas, and ash-handling areas.

Conveyor Areas. Primary concerns are dust control, outgassing from the coal, freezing of the coal and personnel access areas, and fire protection. Dust can be a concern because of the potential for environmental emission and also as a personnel and/or explosive hazard. Dust may be controlled by water-based spray systems or by air induction pickups at the point of generation. Some types of coal may outgas small quantities of methane, which could accumulate in the conveyor and storage structures. See the section on Coal Crusher and Coal Transportation System Buildings for a discussion of ventilation of methane fumes.

Natural or forced ventilation must remove heat from conveyor motors, other equipment, and envelope loads. Ventilation air can also remove outgassed fumes. Generally, ventilation requirements can be as low as 2 to 5 air changes per hour. If air entrainment dust collection equipment is used, provisions for makeup should be included in the design. If natural openings are not sufficient for makeup air, supply ventilation fans may need to be electrically interlocked to operate with the dust suppression/collection equipment. Makeup air may have to be heated if freeze protection is a design criterion. Unit heaters are generally used for spot heating. The unit heater should be specified to the hazard classification for the area it serves. Because coal dust can produce acids when wet, consideration should be given to specifying noncorrosive materials and coatings.

NFPA *Standard* 120 and the U.S. Bureau of Mines (1978) provide other safety considerations for coal handling and preparation areas.

Silo and Feeder Areas. Coal is generally fully contained by feeders and silos, so no special ventilation is needed. Occasionally, coal systems include an inert gas purge system for fire prevention. Ventilation may be needed for life safety dilution ventilation of purge gases.

Coal Mill Areas. Coal mills require large power motors for the grinding process. These motors may have their own ventilation system, or the motor heat may be rejected directly to the surrounding space.

The challenge for the ventilation system is to provide enough ventilation to remove heat without creating high air velocities that disturb accumulated dust. Blowing dust can pose health risks to operators and create a dust ignition hazard. Although occurrence of dust ignition air-to-dust ratios is possible, this area is generally not classified as hazardous. The dust ignition risk is managed by housekeeping, maintenance of seals on the mill equipment, and other dust-control measures.

Forced supply ventilation is generally required for equipment cooling. Sidewall propeller fans work well if the mills are arranged near outer walls. Mills located in the interior of the building may require ducted supply air. Supply air velocities at the coal-handling equipment must be lower than the particulate entrainment velocity for the expected dust size. The maximum air velocity is established using the particle size distribution spectrum and the associated air settling velocities indicated in Figure 3 and Table 1 in Chapter 11 of the 2009 *ASHRAE Handbook—Fundamentals*.

Many utilities use carbon monoxide (CO) monitoring/trending to detect fires in coal bunkers, silos, and other areas. These systems are sometimes integrated with dust collector and other exhaust streams in these areas. Specification and design of the CO monitoring should be the responsibility of a special hazards fire protection engineer and coordinated with the industrial ventilation and HVAC design.

The National Fire Protection Association (NFPA) does not differentiate between coal types/seams in the codes and standards. The utility industry, however, has accumulated experience and established best practices in handling these fuels. This information is shared between member utilities through groups such as Edison Electric Institute and Electric Power Research Institute. Other

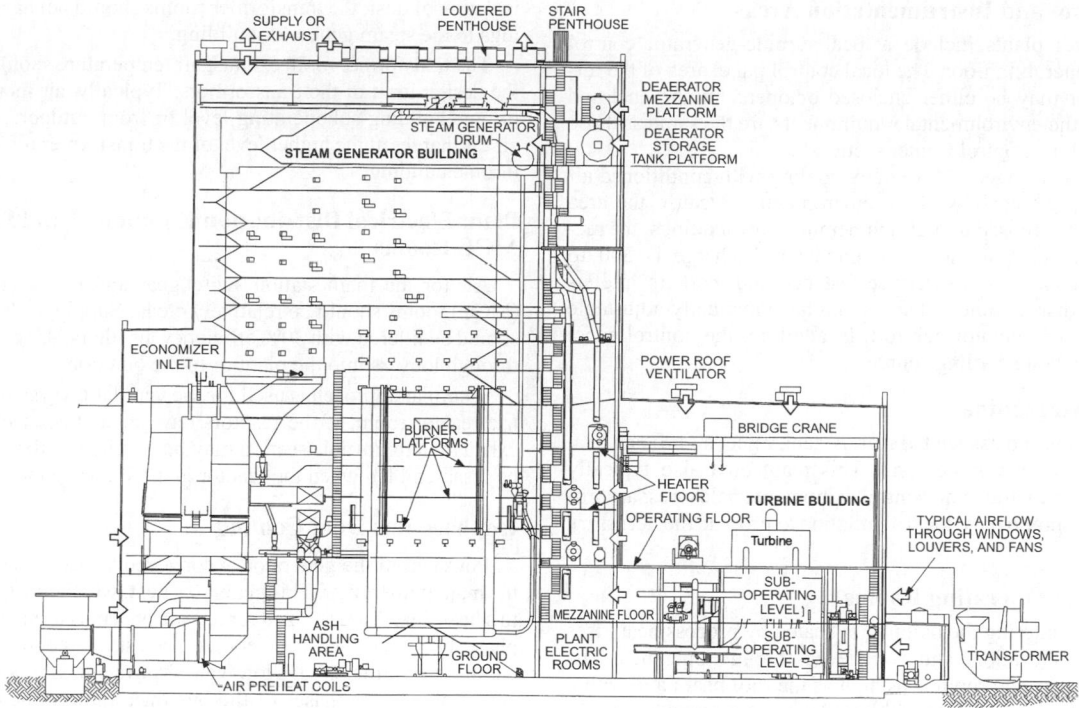

Fig. 2   Generation Building Arrangement
(Courtesy Black & Veatch LLP)

sources of specific requirements include local fire code authorities and the plant's insurance carrier.

Ash-Handling Areas. Ash is generated when coal or heavy fuel oil is burned. Fine ash particles carried by the flue gas from the top of the steam generator are called **fly ash**. Ash that accumulates as slag in the bottom of the steam generator is called **bottom ash**.

Ash-handling equipment generally demands no special HVAC system consideration. Although fly ash is captured in the flue gas stream by a baghouse or electrostatic precipitator, uncaptured (fugitive) fly ash can create problems in equipment mechanisms because of its abrasiveness. If fugitive fly ash is expected to be in the air, HVAC equipment in the ash-handling areas should include filters to capture the ash before it enters building areas.

Wet ash-handling systems have recently been installed in China. Powerful water sprinklers are furnished at the channel leaving the ash outlet from the bottom section of the steam generators, to thoroughly wash the passing ash stream. No dry ash remains as it leaves the channel. Wet ash forms a pulpy liquid and is transported by a special ash pulp pump station to a giant concrete pit, where it is stored and used for producing ash bricks.

Stack Effect

One consideration in HVAC system design for a steam generator building is the stack effect caused by buoyancy of heated air. A 300 ft tall steam generator building with 0°F outdoor air temperature and 100°F indoor air temperature may have 0.5 in. of water negative internal pressure at ground level. This high level of negative pressure causes abnormally large forces on doors, creating a hazard for operators.

Sources of Combustion Air

Large-draft fans supply combustion air for the steam generator. A positive-pressure steam generator is supplied by forced-draft fans, and a negative-pressure steam generator uses induced-draft fans. A balanced-draft steam generator, typical for a larger unit, uses both forced- and induced-draft fans. Because the air is heated to furnace temperatures in the combustion process, part of the fuel energy is used to heat the air. Forced-draft fans on a large steam generator can supply 100,000 cfm or more to the combustion process. A significant amount of energy is needed to preheat the combustion air.

Two prevailing methods of preheating combustion air are used. One method is to draw air in from outdoors and heat it using steam or hot-water coils using energy directly from the power cycle. Another method uses heat rejected from steam generator surfaces to the building space to heat combustion air. This method provides energy savings over heating outdoor air. Temperatures in the higher levels of the generator building can be 100°F or higher. Heat recovery is accomplished by locating the draft fan intake high in the building. Although the potential for savings is large in a cold climate, the total effect on the building heating and ventilation systems should be evaluated. One effect is that drawing the air from the building makes the building pressure more negative; this increases infiltration and adds to the building heating system load, possibly offsetting the potential power cycle thermal efficiency advantage. Increased negative pressure can also contribute to stack effect problems associated with negative pressure low in the building. The draft fan can also be used to supplement ventilation during warm outdoor conditions.

TURBINE GENERATOR BUILDING

As shown in Figure 2, a turbine generator building usually includes a high-bay operating level, a deaerator mezzanine, and one or more suboperating levels. Typically, the turbines and electric generators are located along the centerline of the building between the operating level and the first suboperating level and are the major heat sources in the building. Deaerators are another significant heat contributor; the deaerator mezzanine is commonly open to the turbine operating level. Other room heat sources are steam, steam condensate and hot-water piping, heat exchangers, steam valve stations and traps, motors, electric transformers, and other electrical equipment.

Local Control and Instrumentation Areas

Some power plants include a local turbine-generator control panel on the operating floor. The local control panel area of the turbine generator may be either enclosed or open; for the enclosed arrangement, the environmental requirements are the same as those given in the Main Control Center section.

For an open arrangement, velocity cooling with conditioned air improves the operator's working environment. Because the area may be directly exposed to high-temperature surroundings, the recommended velocity of the conditioned air discharge is 300 to 600 fpm, depending on its service distance and workers' preference. The air distribution system should have manually adjustable air deflectors for operator comfort. In addition, the control panel may need a separate cooling source.

Deaerator Mezzanine

The deaerator and associated storage tank reject significant heat at the deaerator mezzanine level. This plant area also typically includes instrumentation and control equipment enclosures; accordingly, the area should have local ventilation to provide the necessary cooling.

Bridge Crane Operating Rooms

Outdoor air entering the building is heated by process heat, rises toward higher elevation, and is relieved through openings. The bridge crane operating room is as high as the roof beam and within the building exhaust airstream. If the outdoor air temperature is 95°F, the crane operating room may be surrounded by 105°F or hotter air. Hence, the bridge crane operating room is normally air conditioned.

Because the bridge crane operating room moves within the building during its operation, through-the-wall mounted air conditioners are commonly used; to simplify the electrical work, an additional power plug in the crane for the air conditioner should be provided by the crane supplier. Provisions for the cooling-coil condensate drain should be included in the design.

Suboperating Level

The turbine generator is located on the operating floor, a large deck surface open to the turbine building roof. The deck may be 70% or more of the turbine building area. Below the operating level are one or more suboperating levels. Ventilation supply air should be provided to the suboperating levels and at the lower elevations of the operating floor. Air rising through the operating levels brings room heat to the roof area, where it is relieved through high-elevation openings (gravity vents) or exhausted by roof-mounted exhaust fans or PRVs.

The major heat sources in the suboperating levels are high-temperature mechanical and piping systems. Other contributors are electric transformer room exhaust, switchgear room exhaust, electric reactor room exhaust, electric motor heat, etc. Electrical equipment exhaust heat in the turbine generator building is small compared to heat emitted from mechanical and piping systems. Accordingly, ventilation air from the plant distribution electric room can generally be exhausted directly into the turbine building without ducting to the outdoors. Conditioned air may be supplied to local instrumentation panel areas.

For plants in cold climates (temperatures below 32°F), consider spot heating and/or exterior door heated-air curtains. Consideration should also be given to freeze protection of piping close to building walls; stack effect and wind-driven infiltration can increase local heating requirements at the building perimeter. This problem can be addressed by adding capacity to installed heating systems or by providing mobile, temporary heating.

Electric Transformer Rooms

Transformer rooms are typically located at a suboperating level between the turbine building and steam generator buildings. For isolation of dust, the transformer rooms should not have inlet openings to the steam generator building.

The transformer room exhaust air temperature should not exceed the design limit of the transformers. Typically, air intake is from a turbine building suboperating level or from outdoors, and exhaust air discharges at the higher level of the transformer room toward the turbine building.

Plant Electrical Distribution Equipment and Switchgear/MCC Rooms

Air for the main station switchgear and motor control center (MCC) rooms should be relatively clean. Supply air from outdoors should be filtered with 30% efficiency air filters. Air can be relieved through louvers into the plant or to the outdoors.

A similar approach is used for the ventilation system for an electric reactor room. If the reactors have ducted connections for the exhausts, a removable section may be required so that the ducts can be disassembled when the reactor is lifted during maintenance.

Isophase Bus Duct Cooling

Power from the generator is conveyed by isophase bus ducts to the main transformer. This generates heat, which must be dissipated to a heat sink. A specialized forced-air system is used to cool the isophase bus duct. The duct cooler consists of cooling coils, fans, dampers, and filters. For a low-velocity system, air is supplied to the cooler along two phases of the bus duct and returned in the third phase. For a higher-velocity system, air is supplied and returned midway between the transformer and generator, or is supplied at the generator end along bus ducts of all three phases and returned at the transformer end. Isophase bus cooling is essential for power production and delivery, so it is important that the designer specify sufficient redundancy (in the form of dual fans, dual cooling coils and a bypass duct to provide cooling with outdoor air, or another source) in the system design.

COMBUSTION TURBINE AREAS

Combustion turbines are adaptable for outdoor or indoor installation. The outdoor type is usually a skid-mounted structure, with support systems typically designed and furnished by the combustion turbine vendor. The indoor type is typically enclosed in a weatherproof and acoustically treated enclosure and may have indoor support systems, designed and installed separately.

Combustion turbine installations have some or all of the following support facilities: fuel oil handling facility, natural gas pressure-reducing facility, office or administration areas, maintenance shops, battery rooms, control rooms, distributed control system (DCS) control room, communication or computer room, and a water treatment facility. Heating and ventilation design issues associated with combustion turbines include but are not limited to airflow, combustion air source, hot duct and equipment surfaces, fuel supply, turbine inlet cooling, and noise.

The turbine manufacturer typically establishes HVAC requirements and, if required by the purchaser, provides HVAC equipment for the various compartments on the turbine skid. Turbine ventilation and equipment requirements should be coordinated with the building ventilation design.

Turbine and generator casings and the surface of the exhaust duct are large contributors to the heat removal requirements. Hot surfaces should be insulated with appropriate materials as much as is practical and with the approval of the turbine manufacturer, with appropriate airflow established to eliminate hot spots.

When heat recovery equipment is installed in the turbine exhaust, the designer must take into consideration the additional heat rejected into the building because of increased back pressure on the hot exhaust.

Airflow through the combustion turbine building should consider combustion air requirements, combustible gas dilution requirements based on design leakage rates, heat removal requirements, exhaust gas dilution, and electrical component cooling requirements. Combustion turbines draw combustion air through ducts from outside the building. The system design should ensure control of the building pressurization to keep the building under positive pressure, and building outdoor air should be filtered. This also offsets infiltration of dust, rain, snow, bugs, and other contaminants into the building. System design should address freeze protection for vulnerable components. For cold-climate applications during normal operation, mixing outdoor and building air to keep the outdoor supply air temperature above freezing is recommended. In electrically classified installations, the ventilation system must switch to 100% outdoor air during upset conditions.

Noise from combustion turbines is managed by a combination of site location considerations, acoustic enclosures; sound attenuation devices and engineered sound controls include those that are part of HVAC systems.

HVAC design requirements for various areas of the building or turbine skid may be obtained from Table 1. The design should use NFPA *Standards* such as 37, 70, and 90A; insurance carrier requirements; and applicable local codes and standards. HVAC system design and operation should be coordinated with the fire protection systems to ensure adequate concentration of fire suppressant and to prevent fire and smoke spread. HVAC systems also must shut down when fire or extremely high concentrations of gases are detected.

MAIN CONTROL CENTER

The main control center usually includes a control room, electronic and electric control panel and instrumentation rack room, computational equipment server room, automation process control system room, telecommunications equipment room, battery room, UPS room, engineer and operator training simulation room, and associated administration areas.

Because the control center usually contains temperature-sensitive electronic equipment critical to plant operation, it is generally provided with redundant air-handling units and refrigeration equipment. A back-up power supply may also be required. Passive components such as distribution ductwork and piping do not have to be duplicated. Controls should be designed so that failure of a component common to both the primary and back-up systems does not cause failure of both systems. Manual changeover is a simple solution to this problem. If the main control building is located in a fly-ash-contaminated area, room pressurization is highly recommended.

Control Rooms

The control room houses the computerized microprocessor, printer, electronic and emergency response controls, fire protection controls, communication and security systems, regional system networks, accessories, and relevant wiring and tubing systems. An air-conditioning system typical for office occupancy, with features to meet overall design requirements for reliability and the specific environmental needs of the control equipment, is generally appropriate.

Battery Rooms

Battery rooms should be maintained between approximately 70 and 80°F for optimum battery capacity and service life. Temperature variations are acceptable as long as they are accounted for in battery sizing calculations. The minimum room design temperature should be taken into account in determining battery capacity. Batteries produce hydrogen gas during charging, so the HVAC system must be designed to limit the hydrogen concentration to the lowest of the levels specified by IEEE *Standard* 484, ASHRAE guidelines, OSHA, and the lower explosive limit (LEL). The recommended hydrogen concentration in the battery room is 2% or less of the room volume. If battery design information is not available, it is recommended that a five air changes per hour be provided for the exhaust system.

TURBINE LUBRICATING OIL STORAGE

A typical power plant has a turbine lubricating oil storage tank and associated filtration equipment. If this storage area is inside the building, the tank should be vented to the outdoors or ventilation rates should provide for dilution of oil fumes. The ventilation systems should be coordinated with fire protection systems to ensure adequate fire suppressant concentration and to prevent spread of fire.

OIL STORAGE AND PUMP BUILDINGS

At a power plant, fuel oil may be the main source of energy for the steam generator, combustion turbine, or diesel generator. It may also be a back-up or supplemental fuel. Coal-fueled plants usually use oil or gas as the initial light-off fuel or for operation of an auxiliary steam generator. Auxiliary steam generators provide initial plant warm-up and building heating.

Oil for combustion is generally a light oil such as No. 2 fuel oil or a heavy oil such as No. 6. Light oils can be pumped at normal temperatures, but heavy oils are highly viscous and may need to be heated for pumping. Oils are usually received by rail or truck, transported by pipeline, and stored in tanks.

Enclosures for pumps, valves, heat exchangers, and associated equipment should be heated and ventilated to remove excess heat and to dilute hydrocarbon fumes. Tank ventilation is an integral part of the tank and piping system design, which is separate from the enclosure ventilation design. Fuel oils are classified in NFPA *Standard* 30 as either combustible or flammable, depending on their vapor pressure at the indoor design temperature. Flammable liquids are hazardous; combustible liquids are not.

The design of HVAC systems for areas containing combustible fuels involves following ventilation principles for heat removal and for good air mixing. Ventilation rates should dilute fumes expected from evaporation of spilled or leaking fuel, following ACGIH (2010) guidelines and material safety data sheets (MSDSs) provided by the material manufacturer. For fuel handling confined to piping systems, the expected leakage is nearly zero, so very low fresh air rates are required for ventilation (generally less than 1 air change per hour). If fuel is handled in open containers or hoses, higher rates are prudent.

If the fuel is flammable at temperatures expected in the room, NFPA *Standard* 30 and other safety and building codes should be followed. Electrical systems may need to be classified for operation in a hazardous location.

COAL CRUSHER AND COAL TRANSPORTATION SYSTEM BUILDINGS

Coal-handling facilities at a power plant receive and prepare coal and then transport it from the initial delivery point to the burners. Intermediate steps in the process may include long- or short-term storage, cleaning, and crushing. Receipt may include barge, rail car, or truck unloading. Storage may be in piles on the ground, underground, or in barns or silos. At the site, the coal is handled by mobile equipment or conveyor systems.

The following general HVAC considerations apply for the types of structures involved.

Potential for Dust Ignition Explosion

Most types of coal readily break down into dust particles when handled or conveyed. The dust can become fine enough and occur in the right particle size distribution and concentration to create a dust explosion. The design engineer should review and apply the referenced NFPA standards and guidelines to determine the dust ignition potential for each ventilation system application.

Ventilation of Conveyor and Crusher Motors in Coal Dust Environment

Heat from motors and process equipment should be removed through ventilation. The options are to use ducted, ventilated motors or to ventilate the building enclosures. Ventilation in enclosures containing coal should keep the velocity below the entrainment velocities of the expected particle sizes. Table 1 in Chapter 11 of the 2009 *ASHRAE Handbook—Fundamentals* has information on settling velocity. Generally, air should be mechanically exhausted to allow ventilation air to enter the building through louvers at low velocities.

Cooling or Ventilation of Electrical and Control Equipment

Electrical and control equipment may be located near coal piles or other coal-handling facilities. Air-conditioned control rooms should be pressurized with filtered outdoor air. Ventilated motor control or switchgear areas should also be pressurized with filtered air. Because of high dust concentrations in coal yards, ordinary filter media have a short life; a solution is to use inertial filters. For air-conditioned areas, inertial filters can be followed by higher-efficiency media filters.

For electrical equipment rooms adjacent to an area with the potential for a dust ignition explosion, NFPA *Standard* 496 should be followed. This standard recommends the flow of clean air away from electrical equipment into the dusty area.

Ventilation of Methane Fumes

Methane and other hydrocarbons are present in coal both as free gas in cracks and voids and as adsorbents within the coal. Although most of the methane is released from the interstitial coal structure during mining and handling, some methane or other potentially flammable gases may remain in the coal. Thus, flammable concentrations of methane can accumulate when large amounts of coal are stored. The design engineer should identify the potential for methane accumulation when designing for structures associated with silos or coal storage buildings. At the mine or mine mouth, methane gas emission rates as high as 5 ft^3/ton·day are possible; at other locations, the rate is usually less than 1 ft^3/ton·day. Dust collection air exhaust or natural ventilation is often sufficient to prevent the methane level from reaching the 1% explosion limit. The design engineer should apply guidelines from NFPA *Standards* 120, 123, 850, 8503, 8504, and 8505.

Underground Tunnels and Conveyors

Enclosed conveyors are generally of loose construction and require no ventilation. Smoke or gases in underground conveyor tunnels, hoppers, or conveyor transfer points could cause a personnel safety hazard. Ventilation systems should be coordinated with escape route passages to move fresh air from the direction of the egress. Ventilation rates in the range of 2 to 5 air changes per hour are generally appropriate for normal system operation.

Makeup of Dust Collection Air

Coal dust can be controlled by high-velocity air pickup at locations where coal is transferred. The airflow associated with these pickup points may be sufficient to meet the ventilation requirements. Air inlets must be provided. If additional ventilation is needed, the ventilation fan must coordinate with the dust collection system. For heated structures, makeup air may need to be heated.

HEATING/COOLING SYSTEMS

Selection of the heating and cooling systems in a power plant depends on several variables, including the geographical location and orientation of the plant and the type of fuel used. Most plants are ventilated with outdoor air, but it is customary in hot climates to air-condition many plant areas. Steam, hot water, gas, and electricity are alternative heating methods to be evaluated for the most economical choice. Electricity is the primary energy source for general-purpose cooling of various areas of the plant. Areas such as the main control room, office areas, and electrical switchgear/MCC rooms are air-conditioned for continuous human occupancy or for maintaining the operability of the electrical equipment and controls. Spot cooling may also be needed at local control panel areas in the power generation (turbine-electrical generator building) and steam generator (boiler) buildings.

Cooling

The cooling source may be either a centrally located system providing chilled water to various area coolers or individual direct-expansion area coolers with either air- or water-cooled condensing units. Selection depends on the layout of the areas to be cooled and the comparative costs of the two options. The condensing system of the water chiller may be cooled with either air or available water from the plant service water system. Air-cooled chillers are used when air near the proposed chiller location is moderately clean and no fly ash or coal dust problem is anticipated. For a water-cooled system, a closed-loop cooling tower is sometimes used if service water is poor or unavailable (e.g., during start-up or plant outages). To protect chillers from fouling and corrosion by the service water, a heat exchanger is sometimes used between the chiller condenser and the service water source. In a power plant with several operating units and individual self-supporting chilled-water systems, the chilled-water systems of each unit are sometimes interconnected to provide back-up and redundancy.

Heating

Heating in various areas of the power plant is usually provided by electric, steam, or hot-water unit heaters or heating coils in air-handling units. In a hot-water distribution system, glycol is usually added into the system for protection against freezing. Because the building's stack effect induces large quantities of infiltration air, heating requirements in the lower levels of the steam generator building may increase when the steam generator is operating. Pressurization fans directing cooler, outdoor air into the warmer upper elevations of the plant can offset this infiltration. Also, the design engineer may evaluate redistribution of hotter air from higher to lower elevations.

An alternative to heating the open areas of the plant is to use pipeline heat tracing and spot heating at personnel workstations. For this approach, the design engineer should consider all components that may require heat tracing, such as instrument lines, small and large pipes, traps, pumps, tanks, and other surfaces that may be subject to freezing temperatures. Often the large number of components and surfaces to be heat traced and insulated makes this impractical.

Hydroelectric Power Plants

Hydroelectric power plants consist of a dam structure, draft tubes with gates, and turbine generators. The facility may be arranged with the generation components within the dam structure or within structures attached to the dam. HVAC systems should be provided for reliable operation of the mechanical, electrical, and control equipment and office areas.

The system design should consider the geographical location, humidity, degree of automation needed, and potential for flooding. Dams are typically built in remote locations, so the design and arrangement of the equipment, air intakes, and exhaust ports should consider the potential for vandalism. Accordingly, intakes should include security grating or bars, should not provide a line of sight into the structure, and should be constructed of heavy-gage steel to thwart bullets. HVAC equipment should be indoors if practical, or made otherwise inaccessible.

Humidity is a major design consideration. The lake surface and outfall structure create humid outdoor conditions. Outdoor air used for ventilation may introduce humidity into indoor spaces that may cause corrosion of electrical and control equipment. In addition, dam structure and turbine components below the lake water level are usually at temperatures colder than the dew point of the outdoor air. Introducing unconditioned outdoor air can create a significant amount of condensation on structure and equipment surfaces.

Another consideration of the remote location of hydroelectric plants is that the HVAC systems need to be reliable and able to operate with minimum attention by operating and maintenance personnel. Thus, these systems may require seasonal changeover features and fully automatic functions. System controls should be integrated with the generation plant controls, such as communication of status and alarm of critical functions to an off-site monitoring facility.

HVAC design should consider the maximum and minimum lake water level. Intakes and exhausts should be above the maximum design flooding level. Suction points for cooling water sources should be below the minimum water level.

Because of the potential for introducing outdoor humidity, cooling of equipment areas, such as the turbine generator hall, turbine and wicker gate, and other mechanical equipment areas, should use a minimum amount of ventilation air. When practical, air should be dehumidified mechanically or with coils using cold lake water. Areas with heat-producing equipment such as for stop-log storage generally do not need ventilation air, which may introduce unwanted humidity. To minimize the effect of outdoor air humidity on electrical and control equipment areas, consideration should be given to water-cooled equipment and radiant cooling to cool structural surfaces. Battery rooms and oil storage areas should be designed for the specific hazards to minimize excessive ventilation rates.

ENERGY RECOVERY

Energy recovery should be considered in any new system design or system upgrade. Considerations in the power plant should include the following:

- **Interfaces with existing plant process systems.** Using waste steam or steam bled from a process stream must consider process system behavior during generating unit start-up, part load, full load, shutdown, and cold standby/shutdown conditions. Any water returned to the boiler steam cycle must consider effects on the generating unit boiler water chemistry/water treatment.
- **Operating environment,** including cross-contamination of clean airstreams from dirty airstreams, dusty environments, and corrosive conditions from substances such as flue gas. For details, see Chapter 25 of the 2008 *ASHRAE Handbook—HVAC Systems and Equipment.*
- **Safety,** including issues resulting from the presence of coal dust, flammable gases, etc. Fire codes and insurance underwriters should be consulted for guidance.
- **Control strategies.**
- **Constructability.**
- **Codes, standards, and local rules and regulations.**

Energy recovery systems should be evaluated by economic analysis of life-cycle costs, including

- Initial cost, including equipment and installation cost
- Fuel and station service power cost
- Operating cost savings
- Maintenance cost

Utility economic evaluations should follow the accounting guidelines and requirements as established by the Federal Energy Regulatory Commission (FERC). See Chapter 37 for further information on owing and operating costs.

REFERENCES

ACGIH. 2010. *Industrial ventilation: A manual of recommended practice*, 27th ed. American Conference of Governmental Industrial Hygienists, Cincinnati, OH.

ASHRAE. 2010. Ventilation for acceptable indoor air quality. ANSI/ASHRAE *Standard* 62.1-2010.

CI. 1997. *The chlorine manual*, 6th ed. *Pamphlet* 1. Chlorine Institute, Washington, D.C.

IEEE. 2002. Recommended practice for installation design and installation of vented lead-acid batteries for stationary applications. *Standard* 484-2002. Institute of Electrical and Electronics Engineers, Piscataway, NJ.

ISA. 1985. Environmental conditions for process measurement and control systems: Airborne contaminants. ANSI/ISA *Standard* ISA-71.04-1985. Instrumentation, Systems, and Automation Society, Research Triangle Park, NC.

NFPA. 2010. Installation of centrifugal fire pumps for fire protection. ANSI/NFPA *Standard* 20. National Fire Protection Association, Quincy, MA.

NFPA. 2008. Flammable and combustible liquids code. ANSI/NFPA *Standard* 30. National Fire Protection Association, Quincy, MA.

NFPA. 2010. Installation and use of stationary combustion engines and gas turbines. NFPA *Standard* 37. National Fire Protection Association, Quincy, MA.

NFPA. 2011. *National electrical code®*. ANSI/NFPA *Standard* 70. National Fire Protection Association, Quincy, MA.

NFPA. 2009. Installation of air conditioning and ventilating systems. NFPA *Standard* 90A. National Fire Protection Association, Quincy, MA.

NFPA. 2010. Coal preparation plants. ANSI/NFPA *Standard* 120. National Fire Protection Association, Quincy, MA.

NFPA. 1999. Fire prevention and control in underground bituminous coal mines. ANSI/NFPA *Standard* 123. National Fire Protection Association, Quincy, MA.

NFPA. 2008. Purged and pressurized enclosures for electrical equipment. ANSI/NFPA *Standard* 496. National Fire Protection Association, Quincy, MA.

NFPA. 2010. Recommended practice for fire protection for electric generating plants and high voltage direct current converter stations. ANSI/NFPA *Standard* 850. National Fire Protection Association, Quincy, MA.

NFPA. 1997. Pulverized fuel systems. ANSI/NFPA *Standard* 8503. National Fire Protection Association, Quincy, MA.

NFPA. 1996. Atmospheric fluidized-bed boiler operation. ANSI/NFPA *Standard* 8504. National Fire Protection Association, Quincy, MA.

NFPA. 1998. Stoker operation. ANSI/NFPA *Standard* 8505. National Fire Protection Association, Quincy, MA.

Stultz, S.C. and J.B. Kitto, eds. 1992. *Steam, its generation and use*, 40th ed. Babcock & Wilcox, Barberton, OH.

U.S. Bureau of Mines. 1978. Methane emissions from gassy coals in storage silos. *Report of Investigation* 8269.

White, W. 2003. Heat gain from electrical and control equipment (RP-1104). ASHRAE Research Project, *Final Report*.

BIBLIOGRAPHY

China Ministry of Power. 1975 and 1994. Technical code for designing fossil fuel power plants. People's Republic of China *Standard* DL 5000-94.

Copelin, W. and R. Foiles. 1995. Explosion protection from low sulfur coal. *Power Engineering* (November).

NFPA. 2003. *Fire protection handbook*, 19th ed., Section 9: Detection and alarm. National Fire Protection Association, Quincy, MA.

Shieh, C. 1966. *HVAC handbook*. China Ministry of Power.

NUCLEAR FACILITIES

THE HVAC requirements for facilities using radioactive materials are discussed in this chapter. Such facilities include nuclear power plants, fuel fabrication and processing plants, plutonium processing plants, hospitals, corporate and academic research facilities, and other facilities housing nuclear operations or materials. The information presented here should serve as a guide; however, careful and individual analysis of each facility is required for proper application.

BASIC TERMINOLOGY

Criticality, radiation fields, and regulation are three issues that are more important in the design of nuclear-related HVAC systems than in that of other special HVAC systems.

Criticality. Criticality considerations are unique to nuclear facilities. Criticality is the condition reached when the chain reaction of fissionable material, which produces extreme radiation and heat, becomes self-sustaining. Unexpected or uncontrolled conditions of criticality must be prevented at all cost. In the United States, only a limited number of facilities, including fuel-processing facilities, weapons facilities, naval shipboard reactors, and some national laboratories, handle special nuclear material subject to criticality concerns.

Radiation Fields. All facilities using nuclear materials contain radiation fields. They pose problems of material degradation and personnel exposure. Although material degradation is usually addressed by regulation, it must be considered in all designs. The personnel exposure hazard is more difficult to measure than the amount of material degradation because a radiation field cannot be detected without special instruments. It is the responsibility of the designer and of the end user to monitor radiation fields and limit personnel exposure.

Regulation. In the United States, the Department of Energy (DOE) regulates weapons-related facilities and national laboratories, and the Nuclear Regulatory Commission (NRC) regulates commercial nuclear plants. Other local, state, and federal regulations may also be applicable. For example, meeting an NRC requirement does not relieve the designer or operator of the responsibility of meeting Occupational Safety and Health Administration (OSHA) requirements. The design of an HVAC system to be used near radioactive materials must follow all guidelines set by these agencies and by the local, state, and federal governments.

For facilities outside the United States, a combination of national, local, and possibly some U.S. regulations apply. In Canada, the Canadian National Safety Commission (CNSC), formerly

the Atomic Energy Control Board (AECB), is responsible for nuclear regulation, whereas in the United Kingdom, the Nuclear Installations Inspectorate (NII) and the Environment Agency (EA), are involved in issuing operation licenses.

As Low as Reasonably Achievable (ALARA)

ALARA means that all aspects of a nuclear facility are designed to limit worker exposure and discharges to the environment to the minimum amount of radiation that is reasonably achievable. This refers not to meeting legal requirements, but rather to attaining the lowest cost-effective below-legal levels.

Design

HVAC requirements for a facility using or associated with radioactive materials depend on the type of facility and the specific service required. The following are design considerations:

- Physical layout of the HVAC system that minimizes the accumulation of material within piping and ductwork
- Control of the system so that portions can be safely shut down for maintenance and testing or in the case of any event, accident, or natural catastrophe that causes radioactivity to be released
- Modular design for facilities that change operations regularly
- Preservation of confinement integrity to limit the spread of radioactive contamination in the physical plant and surrounding areas

The design basis in existing nuclear facilities requires that safety-class systems and their components have active control for safe shutdown of the reactor, for mitigating a design basis accident (DBA) and for controlling radiation release to the environment as the result of an accident.

Advanced nuclear steam supply systems (NSSS) are being designed that incorporate more passive control to minimize dependence on mechanical equipment to mitigate the consequences of a DBA.

Normal or Power Design Basis

The normal or power design basis for nuclear power plants covers normal plant operation, including normal operation mode and normal shutdown mode. This design basis imposes no requirements more stringent than those specified for standard indoor conditions.

Safety Design Basis

The safety design basis establishes special requirements necessary for a safe work environment and public protection from exposure to radiation. Any system designated essential or safety related must mitigate the effect of a design basis accident, or natural catastrophe that may result in the release of radioactivity into the surroundings or the plant atmosphere. These safety systems must be

---

The preparation of this chapter is assigned to TC 9.2, Industrial Air Conditioning.

operable at all times unless allowed by a limited condition of operation (LCO). The degree to which an HVAC system contributes to safety determines which components must function during and after a DBA or specific combinations of such events as a safe shutdown earthquake (SSE), a tornado, a loss of coolant accident (LOCA), fuel-handling accident (FHA), control rod drop accident (CRDA), main steam line break (MSLB), and loss of off-site electrical power (LOSP). Non-safety-related systems are not credited in any design basis accident and are designed not to adversely affect safety-related systems.

Previously, safety classification of structures, systems, and components (SSC) was based on a deterministic approach, but will be changed to a risk-informed classification and classified as safety-significant (SS) or low safety-significant (LSS), and categorized in four groups. NRC *Regulatory Guide* 1.201 provides information on safety classification of systems, structures and components.

System Redundancy. Systems important to safety must be redundant and single-failure-proofed. Such a failure should not cause a failure in the back-up system. For additional redundancy requirements, refer to the section on Commercial Facilities.

Seismic Qualification. All safety-class components, including equipment, pipe, duct, and conduit, must be seismically qualified by testing or calculation to withstand and perform under the shock and vibration caused by an SSE or an operating-basis earthquake (OBE) (the largest earthquake postulated for the region). This qualification also covers any amplification by the building structure. In addition, any HVAC component that could, if it failed, jeopardize the essential function of a safety-related component, must be seismically qualified or restrained to prevent such failure.

Environmental Qualification. Safety-class components must be environmentally qualified; that is, the useful life of the component in the environment in which it operates must be determined through a program of accelerated aging. Environmental factors such as temperature, humidity, pressure, and cumulative radiation dose must be considered.

Quality Assurance. All designs and components of safety-class systems must comply with the requirements of a quality assurance (QA) program for design control, inspection, documentation, and traceability of material. For U.S. plant designs, refer to Appendix B of Title 10 of the U.S. *Code of Federal Regulations*, Part 50 (10CFR50) or ASME *Standard* NQA-1 for quality assurance program requirements.

Canadian plant designs use two related series of quality assurance standards: CAN3-286.0 and its six daughter standards, plus four standards in the Z299 series. Quality programs in the United Kingdom are based on ISO 9000. For other countries, refer to the applicable national regulations.

Emergency Power. All safety-class systems must have a backup power source such as an emergency diesel generator.

Outdoor Conditions

Chapters 14 and 15 of the 2009 *ASHRAE Handbook—Fundamentals*, the U.S. National Oceanic and Atmospheric Administration, national weather service of the site country, or site meteorology can provide information on outdoor conditions, temperature, humidity, solar load, altitude, and wind.

Nuclear facilities generally consist of heavy structures with high thermal inertia. Time and temperature lag should be considered in determining heat loads. For some applications, such as diesel generator buildings or safety-related pumphouses in nuclear power plants, the 24 h average temperature may be used as a steady-state value. For critical ventilation system design, site meteorological data should be evaluated.

Indoor Conditions

Indoor temperatures are dictated by occupancy, equipment or process requirements, and comfort requirements based on personnel activities. HVAC system temperatures are dictated by the environmental qualification of the safety-class equipment located in the space and by ambient conditions during the different operating modes of the equipment.

Indoor Pressures

Where control of airflow pattern is required, a specific building or area pressure relative to the outdoor atmosphere or to adjacent areas must be maintained. The effect of prevailing wind speed and direction, based on site meteorological information, should be considered. For process facilities with pressure zones, the pressure relationships are specified in the section on Confinement Systems.

In facilities where zoning is different from that in process facilities, and in cases where any airborne radioactivity must not spread to rooms within the same zone, this airborne radioactivity must be controlled by airflow.

Airborne Radioactivity

The level of airborne radioactivity within a facility and the amount released to the surroundings must be controlled to meet the requirements of 10CFR20, 10CFR50, 10CFR61, 10CFR100, 10CFR835, and U.S. DOE *Policy* P 441.1, or equivalent national regulations of the site country.

Tornado/Missile Protection

Protection from tornados and the objects or missiles launched by wind or other design basis events is normally required to prevent the release of radioactive material to the atmosphere. A tornado passing over a facility causes a sharp drop in ambient pressure. If exposed to this transient pressure, ducts and filter housings could collapse because the pressure inside the structure would still be that of the environment prior to the pressure drop. Protection is usually provided by tornado dampers and missile barriers in all appropriate openings to the outdoors. Tornado dampers are heavy-duty, low-leakage dampers designed for pressure differences in excess of 3 psi. They are normally considered safety-class and are environmentally and seismically qualified.

Fire Protection

Fire protection for HVAC and filtration systems must comply with applicable requirements of RG 1.189, Appendix R of 10CFR50, and NFPA, UL, and ANSI or equivalent standards of the site country. Design criteria should be developed for all building fire protection systems, including secondary sources, filter plenum protection, fire dampers, and systems for detection/suppression and smoke management. Fire protection systems may consist of a combination of building sprays, hoses and standpipes, and gaseous or foam suppression. The type of fire postulated in the Fire Hazard Analysis (FHA) or equivalent determines which kind of system is used.

A requirement specific to U.S. nuclear commercial facilities is protection of carbon filter plenums and ventilation ductwork. Manually activated water sprays (window nozzles, fog nozzles, or standard dry pipe/wet pipe system spray heads) are usually used for fire suppression in carbon filter plenums.

Heat detectors and fire suppression systems should be considered for special equipment such as glove boxes. Application of the two systems in combination allows the shutdown of one system at a time for repairs, modifications, or maintenance.

In a DOE facility, the exhaust system duct penetrating a fire-rated boundary does not need a fire damper for maintaining the integrity of the boundary if the duct is fire rated. The exhaust duct may be rated at up to two hours by either wrapping, spraying, or enclosing the duct in an approved material and qualifying it by an engineering analysis. Additional design guidance can be obtained from the *Nuclear Air Cleaning Handbook* (DOE-HDBK-1169-2003).

Smoke control criteria can be found in NFPA *Standards* 801, 803, 804, and 901, or equivalent standards of the site country.

SMOKE MANAGEMENT

The design objective for smoke management in a nuclear facility is to protect the plant operators and equipment from internally and externally generated smoke. Smoke management involves (1) use of materials with low smoke-producing characteristics, (2) prevention of smoke movement to areas where operators may be overcome, (3) use of differential pressures to contain smoke to fire areas, (4) smoke venting to permit access to selected areas, and (5) purging to permit access to areas after a fire.

Smoke control may be static, by prevention of smoke movement (NFPA 90A), or it may be dynamic, by controlling building pressure or air velocities (NFPA 92A). Ventilation systems in the affected areas should be shut down to prevent smoke from migrating and overcoming occupants in other areas. Smoke management for an *internal* fire source should allow the plant operator to shut down the reactor in a controlled manner and maintain shutdown condition. Smoke from an *external* fire should be isolated and appropriate measures provided to prevent smoke from entering the main control room envelope. This envelope includes the main control room and other necessary areas such as restrooms, kitchens, and offices. The location of the safe shutdown panels and the pathway to the safe shutdown panel must be such that, in case of abandonment of the main control room because of fire and smoke, safe egress is ensured.

Capabilities should be provided for purging smoke from fire areas to permit reentry into the areas after the fire is isolated and extinguished. Venting may be used to remove heat and smoke at the point of the fire to permit fire fighting and to control pressures generated by fires.

NFPA 90A, 204, and 92A and NUREG 800 SRP Branch Technical Position CMEB 9.5-1 provide guidance for smoke management and discuss the discharge of smoke and corrosive gases.

Control Room Habitability Zone

The HVAC system in a control room is a safety-related system that must fulfill the following requirements during all normal and postulated accident conditions:

- Maintain conditions comfortable to personnel, and ensure that control room equipment functions continuously and complies with its qualification limit
- Protect personnel from exposure to radiation or toxic chemicals, in the event of a design basis accident
- Protect personnel from combustion products (smoke) emitted from on-site and off-site fires

Additional information may be obtained from the NRC Standard Review Plans, Sections 6.4 and 9.4.1.

Air Filtration

HVAC filtration systems can be designed to remove either radioactive particles or radioactive gaseous iodine from the airstream. They filter potentially contaminated exhaust air prior to discharge to the environment and may also filter potentially contaminated makeup air for power plant control rooms and technical support centers.

The composition of the filter train is dictated by the type and concentration of the contaminant, the process air conditions, and the filtration levels required by the applicable regulations [e.g., NRC *Regulatory Guides* RG 1.52, RG 1.140; ASME AG-1, N509, and 510 (for equipment designed to N509), and N511 (for equipment designed to AG-1); 10CFR20, 10CFR100]. Filter trains may consist of one or more of the following components: prefilters, high-efficiency particulate air (HEPA) filters, carbon filters (adsorbers), heaters, demisters and associated ductwork, housings,

fans, dampers, and instrumentation. For nuclear-safety-related versions of this equipment, the latest edition of ASME AG-1 codifies rules for Materials; Design; Inspection and Testing; Fabrication; Packaging, Shipping, Receiving, Storage, and Handling; and Quality Assurance. Information common to all equipment is compiled in AG-1, Section AA: General Requirements. The AG-1 Code discusses specific rules for each of the major components in separate sections.

For DOE facilities, the *Nuclear Air Cleaning Handbook* (DOE-HDBK-1169-2003) recommends the design of systems and use of major components for nuclear process facilities and laboratories.

Demisters (Mist Eliminators). Demisters are required to protect HEPA and carbon filters if entrained moisture droplets are expected in the airstream. They should be fire resistant. For details, see AG-1, Section FA.

Heaters. Electric heating coils may be used to meet the relative humidity conditions requisite for carbon filters. For safety-class systems, electric heating coils should be connected to the emergency power supply. Interlocks should be provided to prevent heater operation when the exhaust fan is deenergized. For details, see AG-1, Section CA.

Prefilters/Postfilters. Extended-surface filters are selected for the efficiency required by the particular application. AG-1, Table FB-4200-1, lists the average atmospheric dust spot efficiency ranges. ARI 850 provides efficiency tolerances for the various classes. These types of filters are often used as prefilters for HEPA filters to prevent them from being loaded with atmospheric dust and to minimize replacement costs. High-efficiency (90 to 95%) filters are also often used as postfilters downstream of the carbon filter in lieu of downstream HEPAs. For details, see AG-1, Section FB.

European filter standards use efficiency tolerances from EN 779 in place of ARI 850.

HEPA Filters. HEPA filters are used where there is a risk of particulate airborne radioactivity. For details, see AG-1, Sections FC and FK. For DOE sites, the construction and quality assurance testing of HEPA filters are per DOE *Standards* 3020 and 3025.

Carbon Filters. Activated carbon adsorbers are used mainly to remove radioactive iodine in gaseous state. Bed depths are typically 2 or 4 in. Carbon filters have an efficiency of 99.9% for elemental iodine and 95 to 99% for organic iodine, although they lose efficiency as relative humidity increases. For this reason, they are often preceded by a heating element to keep the relative humidity of the entering air below 70%. Nuclear carbon filters can be either tray type (Type II), rechargeable (Type III), or modular (Type IV). For details on each type, see AG-1, Sections FD, FE, and FH. Carbon efficiency is tested in accordance with ASTM D3803-91 (2004) or its latest edition.

Both carbon and HEPA filters may be affected by exposure to paint solvent, chemicals, and fire, and thus should be tested upon exposure.

Design information for ventilation and air-conditioning system design, ductwork, housings, fans, dampers, and instrumentation are contained in AG-1, Sections CA, RA, SA, HA, BA, DA, and IA, respectively.

Sand Filters. Sand filtration is a passive air filtration system that consists of multiple layers of sand and gravel through which air is drawn and filtered. The air enters an inlet tunnel that runs the entire length of the filter. Smaller cross-sectional laterals running perpendicular to the inlet tunnel distribute air across the base of the sand. Air rises through several layers of various sizes of sand and gravel, typically at a facial velocity of 5 fpm. It is then collected in the outlet tunnel for discharge to the atmosphere. Sand filters require no maintenance, and the sand is not changed or replaced during its active service life. A detailed discussion of sand filters is given in Chapter 9 of the DOE's (2003) *Nuclear Air Cleaning Handbook*. At present, there are no national consensus codes and standards relating to sand filters, although ASME's Code of Nuclear and Air Gas Treatment

(CONAGT) Committee is developing a sand filter section for ASME *Code* AG-1.

DEPARTMENT OF ENERGY FACILITIES

The following discussion applies to U.S. National Laboratory facilities. Nonreactor nuclear HVAC systems must be designed in accordance with DOE *Order* O 420.1B. Critical items and systems in plutonium processing facilities are designed to confine radioactive materials under both normal and DBA conditions.

CONFINEMENT SYSTEMS

Zoning

Typical process facility confinement systems are shown in Figure 1. Process facilities comprise several zones.

Primary Confinement Zone. This zone includes the interior of the hot cell, canyon, glove box, or other means of containing radioactive material. Containment must prevent the spread of radioactivity within or from the building under both normal conditions and upset conditions up to and including a facility DBA. Complete isolation from neighboring facilities is necessary. Multistage HEPA filtration of the exhaust is required.

Secondary Confinement Zone. This zone is bounded by the walls, floors, roofs, and associated ventilation exhaust systems of the operating and maintenance areas or rooms surrounding the primary confinement zone.

Tertiary Confinement Zone. This zone is bounded by the walls, floors, roofs, and associated ventilation exhaust systems of the facility. They provide a final barrier against the release of hazardous material to the environment. Radiation monitoring may be required at exit points.

Uncontaminated Zone. This zone includes offices and cold shop areas.

Air Locks

Air locks in nuclear facilities are used as safety devices to maintain a negative differential pressure when a confinement zone is accessed. They are used for placing items in primary confinement areas and for personnel entry into secondary and tertiary confinement areas. Administrative controls ensure proper operation of the air lock doors.

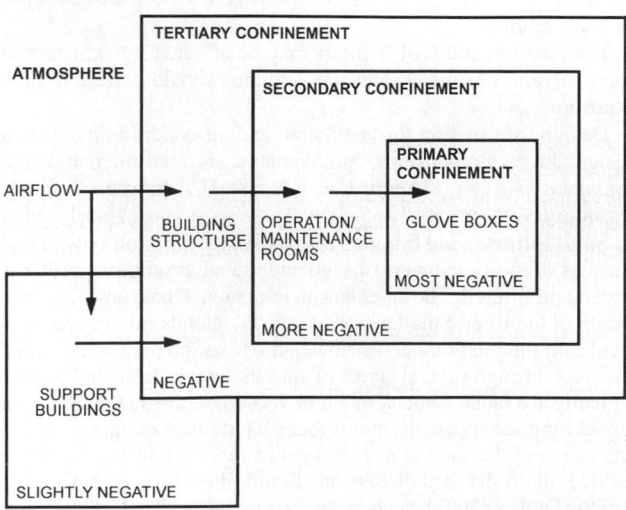

Fig. 1   Typical Process Facility Confinement Categories

There are three methods of ventilating personnel air locks (ventilated vestibules):

- The **clean conditioned supply air** method, where the air lock is at positive pressure with respect to the adjacent zones. For this method to be effective, the air lock must remain uncontaminated at all times.
- The **flow-through ventilation air** method, where no conditioned air is supplied to the air lock and the air lock stays at negative pressure with respect to the less contaminated zone.
- The **combined ventilation air** method, which is a combination of the other two methods. This may be the most effective method, when properly designed.

Zone Pressure Control

Negative static pressure increases (becomes more negative) from the uncontaminated zone to the primary confinement zone, causing any air leakage to be inward, toward areas of higher potential contamination. All zones should be maintained negative with respect to atmospheric pressure. Zone pressure control cannot be achieved through the ventilation system alone; confinement barrier construction must meet all applicable specifications.

Cascade Ventilation

Confinement barriers are enhanced by the use of a cascaded ventilation system, in which pressure gradients cause air to flow from areas of lower contamination to areas of higher contamination through engineered routes. In a cascade ventilation system, air is routed through areas or zones from lower contamination to higher contamination and then to highest contamination, thus reducing the number of separate ventilation systems and the amount of air required for contamination control.

Properly designed air locks should be provided for access between noncontaminated and contaminated areas. If there is a potential for development of differential pressure reversal, HEPA filters should be used at the inlet air openings between areas of higher and lower contamination levels to control the spread of contamination into less contaminated or cleaner areas. Appropriate sealing mechanisms should be used for doors or hatches leading into highly contaminated areas.

Differential Pressures

Differential pressures help ensure that air flows in the proper direction in case of a breach in a confinement zone barrier. The design engineer must incorporate the desired magnitudes of the differential pressures into the design early to avoid later operational problems. These magnitudes are normally specified in the design basis document of the safety analysis report (SAR). The following are approximate values for differential pressures between the three confinement zones.

Primary Confinement. With respect to the secondary confinement area, air-ventilated glove boxes are typically maintained at pressures of –0.3 to –1.25 in. of water, inert gas glove boxes at –0.3 to –1.5 in., and canyons and cells at a minimum of –1.0 in.

Secondary Confinement. Differential pressures of –0.03 to –0.15 in. of water with respect to the tertiary confinement area are typical.

Tertiary Confinement. Differential pressures of –0.01 to –0.15 in. of water with respect to the atmosphere are typical.

VENTILATION

Ventilation systems are designed to confine radioactive materials under normal and DBA conditions and to limit radioactive discharges to allowable levels. They ensure that airflows are, under all normal conditions, toward areas (zones) of progressively higher potential radioactive contamination. Air-handling equipment should be sized conservatively so that upsets in the airflow balance do not cause the airflow to reverse direction. Examples of upsets include improper use of an air lock, a credible breach in the confinement barrier, or excessive loading of HEPA filters.

HEPA filters at the ventilation inlets in all primary confinement zone barriers prevent movement of contamination toward zones of lower potential contamination in case of an airflow reversal. Ventilation system balancing helps ensure that the building air pressure is always negative with respect to the outdoor atmosphere.

Recirculating refers to the reuse of air in a particular zone or area. Room air recirculated from a space or zone may be returned to the primary air-handling unit for reconditioning and then, with the approval of health personnel, be returned to the same space (zone) or to a zone of greater potential contamination. All air recirculated from secondary and tertiary zones must be HEPA-filtered before reintroduction to the same space. Recirculating air is not permitted in primary confinement areas, except those with inert atmospheres.

A safety analysis is necessary to establish minimum acceptable response requirements for the ventilation system and its components, instruments, and controls under normal, abnormal, and accident conditions.

Analysis determines the number of exhaust filtration stages required in different areas of the facility to limit (in conformance with the applicable standards, policies, and guidelines) the amount of radioactive or toxic material released to the environment during normal and accident conditions. Consult DOE *Order* O 420.1B, *Standard* 1189, and Handbook 1169 (DOE 2003) for air-cleaning system criteria.

Ventilation Requirements

A partial recirculating ventilation system may be considered for economic reasons. However, it must be designed to prevent contaminated exhaust from entering the room air-recirculating systems.

The exhaust system is designed to (1) clean radioactive contamination from the discharge air, (2) safely handle combustion products, and (3) maintain the building under negative pressure relative to the outdoors.

Provisions may be made for independent shutdown of ventilation systems or isolation of portions of the systems to facilitate operations, filter change, maintenance, or emergency procedures such as fire fighting. All possible effects of partial shutdown on the airflows in interfacing ventilation systems should be considered. Positive means must be provided to control the backflow of air that might transport contamination. A HEPA filter installed at the interface between the enclosure and the ventilation system minimizes contamination in the ductwork; a prefilter reduces HEPA filter loading. These HEPA filters should not be considered the first stage of an airborne contamination cleaning system.

Ventilation Systems

The following is a partial list of elements that may be included in the overall air filtration and air-conditioning system:

- Air-sampling devices
- Pre- and/or postfilters (e.g., carbon adsorbers, multiple-bed sand, HEPA)
- Scrubbers
- Demisters
- Process vessel vent systems
- Condensers
- Distribution baffles
- Fire suppression systems
- Fire and smoke dampers
- Exhaust stacks
- Fans
- Coils
- Heat removal systems
- Pressure- and flow-measuring devices
- Duct test ports
- Radiation-measuring devices
- Criticality-safe drain systems
- Tornado dampers
- Smoke dampers

The ventilation system and associated fire suppression system are designed for fail-safe operation. The ventilation system is equipped with alarms and instruments that report and record its behavior through readouts in control areas and utility service areas.

Control Systems

Control systems for HVAC systems in nuclear facilities have some unique safety-related features. Because the exhaust system is to remain in operation during both normal and accident-related conditions, redundancy in the form of standby fans is often provided. These standby fans and their associated isolation dampers energize automatically upon a set reduction in either airflow rate or specific location pressure, as applicable. For DOE facilities, maintaining exhaust airflow is important, so fire dampers are excluded from all potentially contaminated exhaust ducts.

Pressure control in the facility interior maintains zones of increasing negative pressure in areas of increasing contamination potential. Care must be taken to prevent wind from unduly affecting the atmospheric control reference. Pulsations can cause the pressure control system to oscillate strongly, resulting in potential reversal of relative pressures. One alternative is to use a variety of balancing and barometric dampers to establish an air balance at the desired differential pressures, lock the dampers in place, and then control the exhaust air to a constant flow rate.

Air and Gaseous Effluents Containing Radioactivity

Air and all other gaseous effluents are exhausted through a ventilation system designed to remove radioactive particulates. Exhaust ducts or stacks located downstream of final filtration that may contain radioactive contaminants should have two monitors, one a continuous air monitor (CAM) and the other a fixed sampler. These monitors may be a combination unit. Exhaust stacks from nuclear facilities are usually equipped with an isokinetic sampling system that relies on a relatively constant airflow rate. The isokinetic sensing probe is a symmetrically arranged series of pickup tubes connected through sweeping bends to a stainless steel header that connects with capillary tubing to a sampling station (CAM). The air velocity through an isokinetic sampling system should be the same as the airstream being sampled. This ensures that particles captured by the isokinetic sampling probe and conveyed to the sampling station are the same size particles that are conveyed in the airstream being sampled. Typically, an exhaust system flow controller modulates the exhaust fan inlet dampers or motor speed to hold the exhaust airflow rate steady while the HEPA filters load.

CAMs can also be located in specific ducts where a potential for radiological contamination has been detected. These CAMs are generally placed beyond the final stage of HEPA filtration, as specified in ANSI *Standard* N13.1. Each monitoring system is connected to an emergency power supply.

The following are design considerations for CAM systems:

- Maintain fully developed turbulent flow at the nonisokinetic sampling point.
- Maintain fully developed laminar flow at the isokinetic sampling point.
- Ensure fully developed turbulent flow is developed between the final filtration and the isokinetic sampling point.
- For accurate CAM operation, heat tracing on the sampling air tubing may be required.
- Maintain the ratio of the sample airflow rate to total discharge airflow rate constant.

COMMERCIAL FACILITIES

OPERATING NUCLEAR POWER PLANTS

The two kinds of commercial light-water power reactors currently in operation in the United States, and in many other countries, are the pressurized water reactor (PWR) and the boiling water reactor (BWR). Heavy water (deuterium oxide) reactors are used in Canada and some other countries. Gas-cooled reactors constitute most of the installed base in Great Britain, but are in the process of being phased out. For all these types, the main objective of the HVAC systems, in addition to ensuring personnel comfort and reliable equipment operation, is protecting operating personnel and the general public from airborne radioactive contamination during normal and accident conditions. Radiation exposure limits are controlled by 10 CFR 20. The "as low as reasonably achievable" (ALARA) concept is the design objective of the HVAC system. The radiological dose is not allowed to exceed the limits as defined in 10CFR50 and 10CFR100. For other countries operating commercial nuclear plants, the specific national rules and regulations should be consulted.

NRC Regulatory Guides (RGs) delineate techniques of evaluating specific problems and provide guidance to applicants concerning the information the NRC needs for its review of the facility. The regulatory guides that relate directly to HVAC system design are RG 1.52, RG 1.78, RG 1.140, RG 1.194, RG 1.196, and RG 1.197. Deviations from RG criteria must be justified by the owner and approved by the NRC. Some countries also invoke NRC regulatory guides as part of the design requirements. Deviations from RG criteria must be requested through the applicable government agency in the country of construction.

The design of the HVAC systems for a U.S. nuclear power generating station must ultimately be approved by the NRC in accordance with Appendix A of 10CFR50. The NRC developed standard review plans (SRPs) as part of Regulatory Report NUREG-0800 to provide an orderly and thorough review. The SRP provides a good basis or checklist for the preparation of a safety analysis report (SAR). The SRP is the basis for information provided by an applicant in an SAR as required by Section 50.34 of 10CFR50. Technical specifications for nuclear power plant systems are developed by the owner and approved by the NRC as outlined in Section 50.36 of 10CFR50. Technical specifications define the safety limits, limiting conditions for operation (LCO), and surveillance requirements (SR) for all systems important to plant safety.

Minimum requirements for the performance, design, construction, acceptance testing, and quality assurance of equipment used in safety-related air and gas treatment systems in nuclear facilities are found in ASME N509, N510, N511, and AG-1.

Temperature and humidity conditions are dictated by the nuclear steam supply system (NSSS). For U.S. plants, the common modes of operation are normal, hot shutdown, cold shutdown, and refueling.

Normal Operation. NSSS temperature and humidity requirements are specified by the NSSS supplier. Some plants require recirculation filtration trains in the containment building to control the level of airborne contamination. In existing plants, containment cooling is necessary for maintaining the components in the containment and to ensure that design limits are not exceeded during accidents. Cooling is provided by a containment cooling system. Some next-generation plants are mostly of passive design and thus do not need active containment cooling systems.

Refueling Condition. The temperature in the refueling or fuel-handling area is determined by the need to perform refueling activities safely. Also, because personnel work in protective clothing, they can be vulnerable to heat stress. To prevent this, area cooling can be provided by a normal non-safety-related cooling system. Outdoor air should be provided for ventilation.

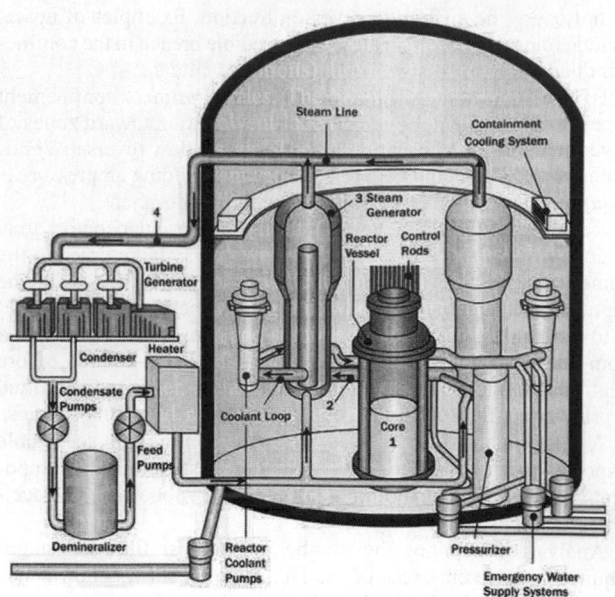

Fig. 2   Typical Pressurized-Water Reactor
(From NRC Web site: www.nrc.gov)

Accident Scenarios

Plants are analyzed for four types of accidents: (1) loss of coolant accident (LOCA), (2) fuel handling accident (FHA), (3) control rod drop accident (CRDA), and (4) a main steam line break (MSLB). For all plant types, it necessary to evaluate the limiting accident event and to conduct safety assessments, so measures can be taken to mitigate any accident's any accident's consequences.

Major NSSS Types

Pressurized-Water Reactors (PWRs). These reactors, widely used in the United States, use enriched uranium for fuel. The reactor, steam generators, and other components of the NSSS are housed in the containment structure. Other support systems are housed in the auxiliary building, control building, turbine building, and diesel building. In PWR design, the steam turbine is powered by nonradioactive steam for the generation of electricity. General design requirements of the PWR plant are contained in ANSI/ANS *Standard* 56.6-1986. Figure 2 shows a typical PWR.

Boiling-Water Reactor (BWRs). Also widely used in the United States, this type of design reactor uses enriched uranium for fuel. The reactor pressure vessel and related piping are housed within the primary containment, which is also referred to as the drywell (Figure 3). The drywell is a low-leakage, pressure-retaining structure designed to withstand the high temperature and pressure from a major break in the reactor coolant line. The drywell is housed within a concrete structure called the secondary containment or the reactor building. Other support systems are housed in the control building, turbine building, and diesel building. In BWR design, the steam turbine is powered by radioactive steam for the generation of electricity. General design requirements of the BWR plant are contained in ANSI/ANS *Standard* 56.7-1987.

Heavy Water Reactors. Canadian power reactors use natural uranium fuel and heavy water (deuterium oxide), which acts as a moderator and a coolant source. The reactor core is mounted in a large, horizontal steel vessel called a calandria, which is enclosed in a concrete containment structure. This design enables the reactors to be refueled while the unit is operating at full power.

Commercial Plant License Renewal and Power Uprate

Nuclear power plants were originally licensed for a 40 year term and a defined power output capacity. Most nuclear plants currently operating in the United States have applied for or will be applying for license renewal to extend the plant operating license from 40 to 60 years. As part of the license renewal process, passive plant components, such as HVAC ductwork and piping, are reviewed, and renewal licenses are granted based on the commitment to perform aging management review (AMR) of long-lived components. Credit is taken for the maintenance rule to address upkeep and replacement of active components. Maintenance rule requirements are outlined in 10CFR50.65, Nuclear Energy Institute documents, Nuclear Management Resources Council (NUMARC) 93-0, and NRC Regulatory Guide RG 1.160.

In addition to license renewal, most plants are increasing their power output capacity by as much as 5 to 15% by conducting power uprate evaluations. Power uprates are often done in steps, and are often categorized as (1) measurement uncertainty recapture power uprates, (2) stretch power uprates, and (3) extended-power uprates. As part of the power uprate, the effects of power output on plant and HVAC systems are evaluated, and these systems are upgraded as needed.

NEW NUCLEAR POWER PLANTS

Many new nuclear power generation plants are being considered for construction in the United States, as well as in other parts of the world. The U.S. NRC has streamlined the application and licensing process for the new reactors, and the *Code of Federal Regulations* 10CFR52 governs the issuance of combined construction and operating license. The new plants are being designed and licensed to satisfy the demand for additional generation capacity in a competitive environment, while contributing to sustainable growth. Advantages of the newer plants include standardized designs with shortened construction times; up to 60-year service life; and the possibility of using flexible fuel [including mixed oxide (MOX) fuel]. The status of design certification and combined construction and operating license (COL) applications for various new U.S. plants can be found at the NRC Web site (www.nrc.gov).

New reactors being considered in the United States are the advanced passive 1000 (AP1000), economic simplified boiling-

water reactor (ESBWR), U.S. evolutionary power reactor (USEPR), advanced boiling-water reactor (ABWR) and U.S. advanced pressurized water reactor (USAPWR). The AP1000 and the ESBWR use passive cooling systems, and the need for safety-related HVAC systems is limited. The ABWR and the USAPWR are enhanced designs of the boiling water reactor and pressurized water reactor, respectively, and HVAC systems for these reactors are not expected to be very much different than those at currently operating BWR and PWR plants. The USEPR is a pressurized water reactor of European design. Brief overviews of HVAC for AP1000, ESBWR, and USEPR plants are as follows.

Advanced Passive AP1000

HVAC systems at AP1000 plants have several differences from those at pressurized-water reactor plants. The main control room (MCR) has both a normal and an emergency HVAC system. The normal system maintains temperature and relative humidity during normal plant operation, and the emergency system uses passive cooling heat sinks to maintain habitability in the main control room during accident conditions. The normal HVAC system has supplemental air filtration that can be used to filter outdoor air with HEPA and charcoal filters, and maintains the main control room at a slight positive pressure to prevent infiltration of unfiltered air into the main control room envelope. On detection of high radiation in the supply air or an extended loss of ac power, the normal system is isolated and the safety-related emergency habitability system is activated. During emergency-mode operation, air is supplied to the MCR from pressurized storage tanks that are sized to meet the ventilation and pressurization requirements for a 72 h accident event. Passive heat sinks are used in the MCR, instrumentation and control (I&C), and dc equipment rooms to limit temperature rise in those rooms after loss of normal HVAC systems. The heat sinks primarily consist of the thermal mass of concrete in the ceilings and walls. Safety-related HVAC equipment is limited to containment isolation valves, control room isolation valves, and the emergency habitability system. Limiting active safety equipment and using passive safety features is part of the AP1000 design philosophy.

The AP1000 has a containment air filtration system, but it serves no safety-related function and is isolated during accident conditions. It is designed to provide intermittent venting of the containment to the atmosphere during normal plant operation. HVAC systems serving the AP1000 diesel generator, radioactive waste (radwaste), annex, and turbine buildings are similar to those in existing PWRs, with a few exceptions: (1) the containment HVAC recirculation system is non-safety-related; (2) the containment HVAC recirculation system uses chilled water to cool containment; and (3) the diesel generators (and thus the diesel generator building) are not safety related.

Economic Simplified Boiling-Water Reactor (ESBWR)

ESBWR HVAC systems have several differences from those at operating boiling-water reactor plants. The control building ventilation system has two subsystems: the control building general-area ventilation system (CBGAVS), which serves general areas in the control building, and the control room habitability-area (CRHA) ventilation system (CRHAVS), which serves the main control room. The CRHAVS provides cooling to MCR, which is served by two redundant recirculation air-handling units (AHUs). Recirculation air-handling units in the main control room draw air from the ceiling space plenum, condition it, and discharge it to an underfloor air distribution system. There is an outdoor air AHU for providing makeup to the MCR for normal habitability. During emergency mode, the CRHA is isolated at the boundary by isolation dampers. The recirculation AHU continues to cool while a battery-powered emergency filter unit (EFU) with HEPA and carbon filters provides filtered outdoor air. If power is lost, the EFU continues to operate for pressurization and to maintain the MCR at 0.125 in. of water positive

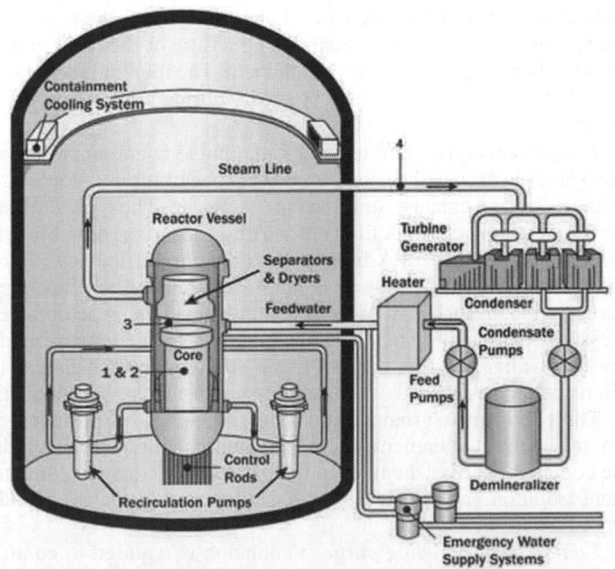

Fig. 3 Typical Boiling-Water Reactor
(From NRC Web site: www.nrc.gov)

relative to the surroundings for the 72 h accident coping period. During this time, the walls and boundary areas act as passive heat sinks to limit the temperature rise in the main control room.

The reactor building (RB) ventilation system (RBVS) has three non-safety-related subsystems: the contaminated-area ventilation subsystem (CONAVS), refueling and pool-area ventilation subsystem (REPAVS), and the reactor building clean-area ventilation subsystem (CLAVS). These systems are separated from the each other with isolation dampers. The CLAVS and CONAVS subsystems are split into separate trains for serving the two halves of the RB. The RBVS has two non-safety-related filter trains (with HEPA and carbon filters), which are backed up with power supply from the diesel generators. Those systems operate on loss of power, but are not credited in the accident analysis. The fuel building ventilation system (FBVS) has two non-safety-related subsystems: the fuel building general-area ventilation subsystem (FBGAVS) and the fuel building fuel-pool-area ventilation subsystem (FBFPVS). The FBGAVS and the FBFPVS subsystems are once-through systems, and room coolers provide supplementary cooling for selected rooms in the fuel building.

The ESBWR uses a centralized non-safety-related chilled-water plant and provides chilled water to various buildings, including the drywell, based on the primary/secondary loop design concept.

U.S. Evolutionary Power Reactor (USEPR)

Unlike the AP1000 and the ESBWR, the USEPR is not of passive design and is similar to PWR plants, except that the emergency (safety) systems have a four-loop or four-train design: there are four electrical trains of safety systems, four emergency diesels and four loops of safety chilled water. The main control room HVAC system has an active and diverse cooling system consisting of two trains of safety-related water-cooled chillers and two trains of safety-related air-cooled chillers. The main control room air-conditioning system (CRACS) is designed to maintain habitability in the main control room and adjoining rooms during normal operation and during accident conditions involving radiation and toxic gas releases. The CRACS maintains the control room envelope at a 0.125 in. of water (30 Pa) positive pressure relative to surrounding areas during accident conditions. The CRACS filtration and air-conditioning equipment and associated ductwork are located inside the control room pressure boundary, thus eliminating the potential for in leakage of unfiltered air. Smoke detectors and toxic gas sensors in the outdoor air supply duct actuate an alarm in the main control room and automatically place the cooling system in recirculation mode.

The containment building ventilation system (CBVS) is composed of three separate subsystems: the (1) non-safety-related full- and low-flow containment purge system, (2) non-safety-related containment filtration system, and (3) safety-related containment cooling system. The containment low-flow purge subsystem operates during normal plant operation to facilitate containment entry and to support outages activities, whereas the full-flow purge system operates only during outages to control the containment environment. The containment filtration subsystem consists of a filter unit with HEPA and carbon filters and a heater, and is used for cleanup of the containment environment. The containment cooling subsystem contains fans, cooling coils, and associated ductwork for cooling various areas of the containment.

PLANT HVAC&R SYSTEMS

PRESSURIZED-WATER REACTORS

Containment Building

Containment Cooling. The following systems are typical for containment cooling:

Reactor containment coolers. These units remove most of the heat load. Distribution of the air supply depends on the containment layout and the location of the major heat sources.

Reactor cavity air-handling units or fans. These units are usually transfer fans without coils that provide cool air to the reactor cavity.

Control rod or control element drive mechanism (CRDM or CEDM) air-handling units. The CRDM and CEDM are usually cooled by an induced-draft system using exhaust fans. Because the flow rates, pressure drops, and heat loads are generally high, the air should be cooled before it is returned to the containment atmosphere.

Essential containment cooling units. The containment air-cooling system, or a part of it, is normally designed to provide cooling during normal plant operation and after a postulated accident. The system must be able to perform at high temperature, pressure, humidity, and levels of radioactivity. Cooling coils are provided with essential plant service water.

System design must accommodate both normal and accident conditions. The ductwork must be able to endure the rapid pressure build-up associated with accident conditions, and fan motors must be sized to handle the high-density air.

In addition, the system must be analyzed to identify measures to mitigate the effect of water hammer as addressed in NRC's Generic Letter 96-06.

Radioactivity Control. Airborne radioactivity is controlled by the following means:

Essential air filtration units. Redundant filter units powered by two Class 1E buses are used to reduce the amount of airborne radioactivity. The typical system consists of a demister, a heater, a HEPA filter bank, and a carbon adsorber, possibly followed by a second HEPA filter bank or by a high-efficiency (90 to 95%) filter bank. The electric heater is designed to reduce the relative humidity of the incoming air from 100% to less than 70%. All the components of the air filtration unit are environmentally qualified (EQ), and designed to meet the requirements of a LOCA.

In the case of an accident and the subsequent operation of the filter train, the carbon can become loaded with radioactive iodine such that the decay heat could cause the carbon to self-ignite if the airflow stops. Heat build-up in the carbon adsorber bed should be evaluated, and an appropriate redundant decay heat removal mechanism should be provided.

Containment power access purge or minipurge. Ventilation is sometimes necessary during normal operation, when the reactor is under pressure, to control containment pressure or the level of airborne radioactivity within the containment. The maximum opening size allowed in the containment boundary during normal operation is 8 in.

The system consists of a supply fan, double containment isolation valves in each of the containment wall penetrations (supply and exhaust), and an exhaust filtration unit with a fan. The typical filtration unit contains a HEPA filter and a carbon adsorber, possibly followed by a second HEPA filter or by a medium-efficiency (90 to 95%) filter bank. The system is non-safety-related, and operates during personnel access into the containment with the reactor under pressure. During normal plant operation, the containment tends toward positive pressure because of air leaks and the exhaust fan is often operated as needed to control pressure inside the containment.

This **minipurge** system should not be connected to any duct system inside the containment. It should include a debris screen within the containment over the inlet and outlet ducts, so that the containment isolation valves can close even if blocked by debris or collapsed ducts.

Containment refueling purge. Ventilation is required to control the level of airborne radioactivity during refueling. Because the reactor is not under pressure during refueling, there are no restrictions on the size of the penetrations through the containment boundary. Large openings of 42 to 48 in., each protected by double containment isolation valves, may be provided. The required ventilation rate is typically based on 1 air change per hour.

The system consists of a supply air-handling unit, double containment isolation valves at each supply and exhaust containment penetration, and an exhaust fan. Filters are recommended.

Containment combustible gas control. In the case of a LOCA, when a strong solution of sodium hydroxide or boric acid is sprayed into the containment, various metals react and produce hydrogen. Also, if some of the fuel rods are not covered with water, the fuel rod cladding can react with steam at elevated temperatures to release hydrogen into the containment. Therefore, redundant hydrogen recombiners are needed to remove hydrogen from the containment atmosphere, recombine the hydrogen with oxygen, and return the air to the containment. The recombiners may be backed up by special exhaust filtration trains.

BOILING-WATER REACTORS

Primary Containment

The primary containment HVAC system consists of recirculating cooling units. It normally recirculates and cools the primary containment air to maintain the environmental conditions specified by the NSSS supplier. During an accident, the system may perform a safety-related function of recirculating the air to prevent stratification of any hydrogen that may be generated if the system is credited in the accident analysis. Depending on the specific plant design, the cooling function may or may not be safety related. Primary containment cooling is necessary for maintaining the life of the components inside the containment and for ensuring that the safety temperature limit of the primary containment is maintained during an accident.

Resistance temperature detectors (RTDs) measure containment temperatures at various locations, and provide input into the volumetric average temperature, which is used as the measure of the containment temperature. The plant's technical specification operating limit is established based on the volumetric average temperature. Temperature problems have been experienced in many BWR primary containments because of temperature stratification and underestimation of heat loads. The cooling system should be designed to adequately mix the air to prevent stratification. Heat load calculations should include a safety factor sufficient to allow for deficiencies and degradation in insulation.

Reactor Building

The reactor building completely encloses the primary containment, auxiliary equipment, and refueling area. Under normal conditions, the reactor building HVAC system maintains the design space conditions and minimizes the release of radioactivity to the environment. The HVAC system consists of a 100% outdoor air cooling system. Outdoor air is filtered, heated, or cooled as required before being distributed throughout the various building areas. The exhaust air flows from areas with the least potential contamination to areas of most potential contamination. Before exhausting to the environment, potentially contaminated air is filtered with HEPA filters and carbon adsorbers; all exhaust air is monitored for radioactivity. To ensure that no unmonitored exfiltration occurs during normal operations, the ventilation systems maintain the reactor building at a negative pressure relative to the atmosphere.

During an event involving a LOCA, MSLB, FHA, or high radiation in the ventilation exhaust, the HVAC system's safety-related function is to isolate the secondary containment consisting of the reactor building and the refueling area. Once isolated by fast-closing valves, the secondary containment boundary functions to contain any leakage from the primary containment or refueling area.

Once the secondary containment is isolated, a safety-related standby gas treatment system (SGTS) is started to reduce the ground level releases by drawing down the secondary containment pressure

to about –0.25 in. of water within 120 s. The SGTS exhausts air from the secondary containment to the environment at an elevated release location referred to as the main stack. The SGTS consists of redundant filtration trains, which consist primarily of HEPA filters and carbon adsorbers. The capacity of the SGTS is based on the amount of exhaust air needed to reduce the pressure in the secondary containment and maintain it at the design level, given the containment leakage rates and required drawdown times.

In addition to the SGTS, some designs include safety-related recirculating air systems within the secondary containment to mix, cool, and/or treat the air during accidents. These recirculation systems sometimes use portions of the normal ventilation system ductwork; therefore, if the ductwork is used for that purpose, then it must be classified as safety related.

Other than the emergency core cooling system (ECCS) pump rooms, the isolated secondary containment area is not cooled during accident events. All safety-related components in the secondary containment must be environmentally qualified to operate at the maximum temperature and the temperature profile for the accident event. Safety-related room coolers served by the plant service water provide cooling to the emergency core cooling system (ECCS) pumps during accident conditions.

Turbine Building

Only a BWR supplies radioactive steam directly to the turbine, which could cause a release of airborne radioactivity to the surroundings. Therefore, areas of the BWR turbine building in which release of airborne radioactivity is possible should be enclosed. These areas must be ventilated and the exhaust filtered to ensure that no radioactivity is released to the atmosphere. Filtration trains typically consist of a prefilter, a HEPA filter, and a carbon adsorber, possibly followed by a second HEPA filter bank or by a medium-efficiency (90 to 95%) filter bank. Filtration requirements are based on the plant and site configuration and commitments for 10CFR50, Appendix I requirements. Depending on outdoor air conditions at the location, the turbine building is cooled either with outdoor air or by area coolers served by a dedicated chilled-water system.

HEAVY WATER REACTORS

Containment Inlet Air-Conditioning/Exhaust Ventilation System

The production of heavy water in sufficient quantities for the needs of a heavy water reactor is complex and expensive. Once produced, however, deuterium oxide (D_2O) may be reused indefinitely as long as it does not become contaminated. Because heavy water reactor containments are vented and require makeup air, ordinary water (H_2O) is one contaminant that must be contained. This is normally accomplished by means of a non-nuclear safety desiccant air-conditioning unit mounted on the roof of the service building. This unit typically contains a rotary desiccant dryer, hot-water heating coils and chilled-water cooling coils both upstream and downstream of the desiccant wheel, a desiccant regeneration duct containing an electric heater, and a flow control system. The resulting inlet makeup air contains very little moisture. To prevent any radioactive contaminants from escaping up the stack, the containment exhaust ventilation unit typically contains a prefilter bank, a HEPA filter bank, a Type III carbon filter, a second HEPA filter bank, and an exhaust fan.

AREAS OUTSIDE PRIMARY CONTAINMENT

All areas located outside the primary containment are designed to the general requirements contained in ANSI/ANS *Standard* 59.2. These areas are common to any type of plant.

Auxiliary Building

The auxiliary building contains a large amount of support equipment, much of which handles potentially radioactive material. The building may be air conditioned for equipment protection, and the exhaust is filtered to prevent the release of potential airborne radioactivity. The filtration trains typically consist of a prefilter, a HEPA filter, and a carbon adsorber, possibly followed by a second HEPA filter bank or by a high-efficiency (90 to 95%) filter bank.

The HVAC system is a once-through system, as needed for general cooling. Ventilation is augmented by area or room coolers in the individual equipment rooms requiring additional cooling. The building is maintained at negative pressure relative to the outdoors.

If the equipment in these rooms is not safety related, the area is cooled by normal air-conditioning units. If they are safety related, the area is cooled by safety-related or essential area or room coolers units powered from the same Class 1E (according to IEEE *Standard* 323) power supply as the equipment in the room.

The normal and essential functions may be performed by one cooling unit having both a normal and an essential cooling coil and a safety-related fan served from a Class 1E bus. The normal coil can be a direct-expansion or chilled-water cooling coil served by a normal chilled-water system. The essential coil operates with chilled water from a safety-related chilled-water system or the plant service water, or a safety-related cooling water source.

Control Room

The control room HVAC system serves the control room habitability zone (those spaces that must be habitable following a postulated accident to allow orderly shutdown of the reactor) and performs the following functions:

- Controls indoor environmental conditions
- Provides pressurization to prevent infiltration
- Minimizes unfiltered inleakage to that credit in dose assessments
- Protects the zone from hazardous chemical fume intrusion
- Protects the zone from fire
- Removes noxious fumes, such as smoke

In defining the control room envelope or pressure boundary, it is necessary to ensure that, in the event of abandonment as a result of fire or smoke in the control room, access to the remote shutdown panel is safeguarded.

Design requirements for the control room HVAC system are outlined in 10CFR50, GDC 19 Appendix A, Standard Review Plan Sections 6.4 and 9.4.1, NCR *Regulatory Guides* RG 1.52, 1.78, 1.194, 1.196, and 1.197, and NUREG 0737. In 2003, NRC issued Generic Letter 2003-01 to address control room habitability findings at U.S. nuclear power plants, which suggested that licensees may not have been meeting the control room licensing and design basis, and applicable regulatory requirements, and that existing technical specification surveillance requirements may not have been adequate. As a result, all nuclear plants are required to conduct control room inleakage testing using the tracer gas (SF_6) to validate the integrity of the control room boundary and to ensure compliance with dose assessments per GDC 19, Appendix A. It is also necessary to develop and maintain a control room integrity program in accordance with the plant technical specifcation requirements to control and maintain boundary breaches. Plants are also required to conduct self-assessments of their control room habitability program every three years and inleakage testing every six years. Control room HVAC filter units are designed to filter radioactive contaminants and are fabricated, designed, and tested per ASME *Standards* N509, N510, N511, and AG-1.

Control Cable Spreading Rooms

These rooms are located directly above and below the control room. They are usually served by an independent ventilating or

cooling system or by the air-handling units that serve the electric switchgear room or the control room.

Diesel Generator Building

Nuclear power plants have an auxiliary or back-up power source for all essential and safety-related equipment in case of loss of off-site electrical power. The auxiliary power source consists of at least redundant diesel generators, each sized to meet the emergency power load. Heat released by the diesel generator and associated auxiliary systems is normally removed by a safety-related ventilation.

Emergency Electrical Switchgear Rooms

These rooms house the electrical switchgear that controls essential or safety-related equipment. The switchgear located in these rooms must be protected from excessive temperatures (1) to ensure that its qualified life, as determined by environmental qualification, is maintained and (2) to preserve power circuits required for proper operation of the plant, especially its safety-related equipment.

Battery Rooms

Battery rooms should be maintained at approximately 70 to 80°F for optimum battery capacity and service life. Temperature variations are acceptable as long as they are accounted for in battery sizing calculations. The minimum room design temperature should be taken into account in determining battery capacity. Because batteries produce hydrogen gas during charging periods, the HVAC system must be designed to limit the hydrogen concentration to the lowest of the levels specified by IEEE *Standard* 484, ASHRAE guidelines, OSHA, and the lower explosive limit (LEL) (see Chapter 11 of the 2009 *ASHRAE Handbook—Fundamentals* for more information). The recommended hydrogen concentration (by volume) in the battery room is 2% or less. If battery design information is not available, it is recommended that the exhaust system be designed to provide a minimum of five air changes per hour.

Fuel-Handling Building

New and spent fuel is stored in the fuel-handling building. The building is air conditioned for equipment protection and ventilated with a once-through air system to control potential airborne radioactivity. Normally, the level of airborne radioactivity is so low that the exhaust need not be filtered, although it should be monitored. If significant airborne radioactivity is detected, as can happen during an FHA, the normal building ventilation is isolated and the safety-related system started automatically. The safety-related ventilation system should exhaust through filtration trains powered by Class 1E buses.

Personnel Facilities

For nuclear power plants, these areas usually include decontamination facilities, laboratories, and medical treatment rooms.

Pumphouses

Cooling water pumps are protected by houses that are often ventilated by fans to remove the heat from the pump motors. If the pumps are essential or safety related, the ventilation equipment must also be considered safety related.

Radioactive Waste Building

The building is normally air conditioned for equipment protection and ventilated to control potential airborne radioactivity. The air may require filtration through HEPA filters and/or carbon adsorbers prior to release to the atmosphere.

Technical Support Center

The technical support center (TSC) is an outside facility located close to the control room. Although normally unoccupied, it is used by plant management and technical support personnel during training exercises and accident events.

The TSC HVAC system is designed to provide the same level of comfort (temperature and humidity) and radiological habitability conditions as provided for the control room. The TSC HVAC system is a non-safety-related system, but is augmented to the same level of importance as the main control room HVAC system. An air filtration system (HEPA carbon postfilter) provides the facility protection from radiological releases during an accident. Additional components, such as moisture separators, heaters, and prefilters, are sometimes also used. Because the operation and availability of the TSC is credited in the plant's emergency operating procedures, the TSC HVAC system should be designed with some redundancy such that maintenance on the HVAC system does not require declaration of the TSC as unavailable.

NONPOWER MEDICAL AND RESEARCH REACTORS

The requirements for HVAC and filtration systems for nuclear nonpower medical and research reactors are set by the NRC. The criteria depend on the type of reactor (ranging from a nonpressurized swimming pool type to a 10 MW or more pressurized reactor), the type of fuel, the degree of fuel enrichment, and the type of facility and environment. Many of the requirements discussed in the sections on various nuclear power plants apply to a certain degree to these reactors. It is therefore imperative for the designer to be familiar with the NRC requirements for the reactor under design.

LABORATORIES

Requirements for HVAC and filtration systems for laboratories using radioactive materials are set by the DOE and/or the NRC. Laboratories located at DOE facilities are governed by DOE regulations. All other laboratories using radioactive materials are regulated by the NRC. Other agencies may be responsible for regulating other toxic and carcinogenic material present in the facility.

Laboratory containment equipment for nuclear processing facilities is treated as a primary, secondary, or tertiary containment zone, depending on the level of radioactivity anticipated for the area and on the materials to be handled. For additional information see Chapter 16.

Glove Boxes

Glove boxes are windowed enclosures equipped with one or more flexible gloves for handling material inside the enclosure from the outside. The gloves, attached to a porthole in the enclosure, seal the enclosure from the surrounding environment. Glove boxes permit hazardous materials to be manipulated without being released to the environment.

Because the glove box is usually used to handle hazardous materials, the exhaust is filtered with a HEPA filter before leaving the box and prior to entering the main exhaust duct. In nuclear processing facilities, a glove box is considered primary confinement (see Figure 1), and is therefore subject to the regulations governing those areas. For nonnuclear processing facilities, the designer should know the designated application of the glove box and design the system according to the regulations governing that particular application.

Laboratory Fume Hoods

Nuclear laboratory fume hoods are similar to those used in nonnuclear applications. Air velocity across the hood opening must be sufficient to capture and contain all contaminants in the hood. Excessive hood face velocities should be avoided because they cause contaminants to escape when an obstruction (e.g., an operator) is positioned at the hood face. For information on fume hood testing, refer to ASHRAE *Standard* 110.

Radiobenches

A radiobench has the same shape as a glove box except that in lieu of the panel for the gloves, there is an open area. Air velocity across the opening is generally the same as for laboratory hoods. The level of radioactive contamination handled in a radiobench is much lower than that handled in a glove box.

DECOMMISSIONING OF NUCLEAR FACILITIES

The exhaust air filtration system for decontamination and decommissioning (D&D) activities in nuclear facilities depends on the type and level of radioactive material expected to be found during the D&D operations. The exhaust system should be engineered to accommodate the increase in dust loading, with more radioactive contamination than is generally anticipated, because the D&D activities dislodge previously fixed materials, making them airborne. Good housekeeping measures include chemical fixing and vacuuming the D&D area as frequently as necessary.

The following are some design considerations for ventilation systems required to protect the health and safety of the public and the D&D personnel:

- Maintain a higher negative pressure in the areas where D&D activities are being performed than in any of the adjacent areas.
- Provide an adequate capture velocity and transport velocity in the exhaust system from each D&D operation to capture and transport fine dust particles and gases to the exhaust filtration system.
- Exhaust system inlets should be as close to the D&D activity as possible to enhance the capture of contaminated materials and to minimize the amount of ductwork that is contaminated. A movable inlet capability is desirable.
- With portable enclosures, filtration of the enclosure inlet and exhaust air must sustain the correct negative internal pressure.

Low-Level Radioactive Waste

Requirements for the HVAC and filtration systems of low-level radioactive waste facilities are governed by 10CFR61. Each facility must have a ventilation system to control airborne radioactivity. The exhaust air is drawn through a filtration system that typically includes a demister, heater, prefilter, HEPA filter, and carbon adsorber, maybe followed by a second filter. Ventilation systems and their CAMs should be designed for the specific characteristics of the facility.

WASTE-HANDLING FACILITIES

The handling of radioactive waste requires inventory control of the different radioactive wastes. See the section on Codes and Standards for pertinent publications.

REPROCESSING PLANTS

A reprocessing plant is a specific-purpose facility. Spent nuclear fuel is opened and the contents dissolved in nitric acid to enable the constituents to be chemically separated and recovered. The offgas contains hazardous chemical and radioactive contaminants. Special cleanup equipment, such as condensers, scrubbers, cyclones, mist eliminators, and special filtration, is required to capture the vapors.

MIXED-OXIDE FUEL FABRICATION FACILITIES

The mixed-oxide (MOX) fuel fabrication facility (MFFF) is designed to produce MOX fuel for use in commercial nuclear power plants. In the United States, this facility is DOE operated and NRC licensed. MOX fuel fabrication consists of blending polished plutonium dioxide with uranium dioxide to form mixed-oxide pellets and loading them into fuel rods that form fuel assemblies for use as nuclear fuel. The MFFF HVAC design includes several important design features, and it complies with the performance criteria of 10CFR70, ASME AG-1, IEEE 323, and other applicable codes and standards.

In addition to maintaining the design environmental conditions, the HVAC systems are designed to maintain a pressure differential between the building confinement zones and between the building and the outdoors to ensure that cascaded airflow is from zones of lesser contamination potential to zones of greater contamination. The supply air system provides air to all confinement zones. Each confinement zone has its own exhaust system, and the pressure boundaries are maintained by cascaded airflow from tertiary confinement (negative) to secondary confinement (more negative), and then to primary confinement (most negative) zones. Primary and secondary confinement ventilation exhaust systems include both the intermediate and the final filter units in series. Tertiary confinement exhaust systems include only the final filter units. Intermediate filter units contain roughing filter(s) (i.e., stainless steel mesh filter) and HEPA filter(s) in series. Final filter units contain roughing filter(s), prefilter(s) (i.e., stainless steel/glass fiber mesh filter), and two banks of HEPA filter(s) in series. Air locks between confinement zones are designed to minimize personnel contamination exposure by maintaining pressures between various zones. Primary and secondary confinement exhaust systems are provided with on-site emergency diesel power, and battery back-up is provided for the primary confinement exhaust system.

Glove boxes (GB) are provided with room ventilation air, dry air, or inert gas, as determined by the process requirements. GB exhaust piping is connected to common headers for intermediate and final filtration. The ventilation system is designed to maintain a normal operating pressure of -1.2 to -2.0 in. of water. Pressure-relief valves and vacuum breakers are provided to keep GB pressures within their structural design limits under conditions such as loss of supply flow or failed open valves. Additionally, dump valves (high-volume relief valves connected to exhaust header) are provided to ensure that, in the event of a glove port or bag port breach, the capture velocity at the breach is at least 125 fpm.

Detailed analysis, using compressible-flow pipe network analysis software, is performed for the MFFF ventilation systems and GB ventilation system. and the results are used for airflow balancing and to confirm that confinement is maintained during normal and design-basis accident conditions.

Sprinkler systems for HEPA filtration units are eliminated by integrating ventilation exhaust system design with fire-safety design. To confirm this design approach, a prototype final filter unit is tested in a testing laboratory for higher soot loading and differential pressures than calculated design values. The test results should show no breach in any filter in the prototype unit.

Sheet metal thickness and reinforcement for galvanized steel and stainless steel duct systems, with pressures from $+30$ to -30 in. of water, round duct diameter up to 60 in., rectangular duct width up to 144 in., and temperature exposure up to 400°F, are designed using SMACNA standards. Seismic supports for the duct systems and redundant tornado dampers at outer wall openings are designed to protect the duct systems during abnormal events.

CODES AND STANDARDS

ANSI

Standard N13.1	Guide for Sampling Airborne Radioactive Materials in Nuclear Facilities

ASME

Standard AG-1	Code on Nuclear Air and Gas Treatment
Standard N509	Nuclear Power Plant Air-Cleaning Units and Components
Standard N510	Testing of Nuclear Air Treatment Systems
Standard N511	In-Service Testing of Nuclear Air Treatment, Heating, Ventilating, and Air-Conditioning Systems
Standard NQA-1	Quality Assurance Program Requirements for Nuclear Facility Applications

ANSI/ASHRAE

Standard 110	Method of Testing Performance of Laboratory Fume Hoods

ASTM

Standard 3803-1989	Standard Test Method for Nuclear-Grade Activated Carbon

DOE Orders

Order O 420.1B	Facility Safety

DOE Policy

Policy P 441.1	Radiological Health and Safety Policy

DOE Standards

Standard 1189	Integration of Safety into the Design Process
Standard 3020	Specification for HEPA Filters Used by DOE Contractors
Standard 3025	Quality Assurance Inspection and Testing of HEPA Filters

DOE Handbooks

DOE-HDBK-1169-2003	Nuclear Air Cleaning Handbook

ANSI/IEEE

Standard 323	Standard for Qualifying Class 1E Equipment for Nuclear Power Generating Stations
Standard 484	Recommended Practices for Installation Design and Installation of Vented Lead-Acid Batteries for Stationary Applications

NFPA

Standard 90A	Standard for the Installation of Air Conditioning and Ventilating Systems (1999)
Standard 90B	Standard for the Installation of Warm Air Heating and Air Conditioning Systems (1999)
Standard 92A	Recommended practice for smoke control systems (2000)
Standard 204	Standard for Smoke and Heat Venting (2002)
Standard 801	Standard for Facilities Handling Radioactive Materials
Standard 803	Standard for Fire Protection for Light Water Nuclear Power Plants
Standard 804	Standard for Fire Protection for Advanced Light Water Reactor Electric Generating Plants
Standard 805	Performance Based Standard for Fire Protection for Light Water Reactor Electric Generating Plants
Standard 806	Performance Based Standard for Fire Protection for Advanced Nuclear Reactor Electric Generating Plants
Standard 901	Classifications for Incident Reporting and Fire Protection Data

ARI

Standard 850	Commercial and Industrial Air Filter Equipment

Code of Federal Regulations

10CFR	Title 10 of the *Code of Federal Regulations*
Part 20	Standards for Protection Against Radiation
Part 50	Domestic Licensing of Production and Utilization Facilities

Part 52	Licenses, Certifications, and Approvals for Nuclear Power Plants
Part 61	Land Disposal of Radioactive Waste
Part 65	Requirements for Monitoring the Effectiveness of Maintenance at Nuclear Power Plants
Part 70	Domestic Licensing of Special Nuclear Materials
Part 100	Reactor Site Criteria
Part 835	Occupational Radiation Protection

NRC

NUREG-0696	Functional Criteria for Emergency Response Facilities
NUREG-0737	Clarification of TMI Action Plan Requirements
NUREG-0800	Standard Review Plans
SRP 6.4	Control Room Habitability Systems
SRP 9.4.1	Control Room Area Ventilation System
SRP 9.4.2	Spent Fuel Pool Area Ventilation System
SRP 9.4.3	Auxiliary and Radwaste Building Ventilation Systems
SRP 9.4.4	Turbine Area Ventilation System
SRP 9.4.5	Engineered Safety Feature Ventilation System
NUREG-CR-3786	A Review of Regulatory Requirements Governing Control Room Habitability

Regulatory Guides U.S. Nuclear Regulatory Commission

RG 1.52, Rev. 3	Design, Testing, and Maintenance Criteria for Engineered Safety Feature Atmospheric Cleanup System Air Filtration and Adsorption Units of LWR Nuclear Power Plants
RG 1.78, Rev. 2	Assumptions for Evaluating the Habitability of Nuclear Power Plant Control Room during a Postulated Hazardous Chemical Release
RG 1.140, Rev. 2	Design, Testing, and Maintenance Criteria for Normal Ventilation Exhaust System Air Filtration and Adsorption Units of LWR Nuclear Power Plants
RG 1.189	Fire Protection for Operating Nuclear Power Plants
RG 1.194	Atmospheric Relative Concentrations for Control Room Radiological Habitability Assessments at Nuclear Power Plants
RG 1.196	Control Room Habitability at Light-Water Nuclear Power Reactors
RG 1.197	Demonstrating Control Room Envelope Integrity at Nuclear Power Reactors
RG 1.201	Guidelines for Categorizing Structures, Systems, and Components in Nuclear Power Plants According to Their Safety Significance

Generic Letters

96-06	Assurance of Equipment Operability and Containment Integrity During Design-Basis Accident Conditions
2003-01	Control Room Habitability

Canadian Standards

CAN3-N286 Series	Quality Assurance for Nuclear Power Plants
CAN3-Z299	Quality Assurance Programs

European Standards

EN 779	Particulate Air Filters for General Ventilation—Requirements, Testing, Marking

MINE AIR CONDITIONING AND VENTILATION

IN underground mines, excess humidity, high temperatures, inadequate oxygen, and excessive concentrations of dangerous gases can lower worker productivity and can cause illness and death. Air cooling and ventilation are needed in deep underground mines to minimize heat stress and remove contaminants. As mines become deeper, heat removal and ventilation problems become more difficult and costly to solve.

Caution: This chapter presents only a very brief overview of the principles of mine ventilation planning. The person responsible for such planning should either be an experienced engineer, or work under the direct supervision of such an engineer. Seven English-language texts have been written on mine ventilation since 1980 (Bossard 1982; Hall 1981; Hartman et al. 1997; Hemp 1982; Kennedy 1996; McPherson 1993; Tien 1999). The ventilation engineer is strongly encouraged to study these references.

Special Warning: Certain industrial spaces may contain flammable, combustible, and/or toxic concentrations of vapors or dusts under either normal or abnormal conditions. In spaces such as these, there are life-safety issues that this chapter may not completely address. Special precautions must be taken in accordance with requirements of recognized authorities such as the National Fire Protection Association (NFPA), the Occupational Safety and Health Administration (OSHA), and the American National Standards Institute (ANSI). In all situations, engineers, designers, and installers who encounter conflicting codes and standards must defer to the code or standard that best addresses and safeguards life safety.

DEFINITIONS

Definitions specific to mine air conditioning and ventilation are as follows.

Heat stress is a qualitative assessment of the work environment based on temperature, humidity, air velocity, and radiant energy. Many heat stress indices have been proposed (see Chapter 9 of the 2009 *ASHRAE Handbook—Fundamentals* for a thorough discussion); the most common in the mining industry are effective temperature (Hartman et al. 1997), air cooling power (Howes and Nixon 1997), and wet-bulb temperature. The following wet-bulb temperature ranges were derived from experience at several deep western U.S. metal mines:

$t_{wb} \leq 80°F$ Worker efficiency 100%

$80 < t_{wb} \leq 85°F$ Economic range for acclimatized workers

$85 < t_{wb} \leq 91°F$ Safety factor range; corrective action required

$91°F < t_{wb}$ Only short-duration work with adequate breaks

Heat strain is the physiological response to heat stress. Effects include sweating, increased heart rate, fatigue, cramps, and progressively worsening illness up to heat stroke. Individuals have different tolerance levels for heat.

The preparation of this chapter is assigned to TC 9.2, Industrial Air Conditioning.

Reject temperature, based on the heat stress/strain relationship is the wet-bulb temperature at which air should be rejected to exhaust or recooled. Reject temperature ranges between 78 and 85°F wet bulb, depending on governmental regulation, air velocity, and expected metabolic heat generation rate of workers. Specifying the reject temperature is one of the first steps in planning air-conditioning systems. The ventilation engineer must be able to justify the reject temperature to management because of the economics involved. If too high, work productivity, health, safety, and morale suffer; if too low, capital and operating costs become excessive.

Critical ventilation depth is the depth at which the air temperature in the intake shaft rises to the reject temperature through autocompression and shaft heat loads. Work areas below the critical ventilation depth rely totally on air conditioning to remove heat. The critical ventilation depth is reached at about 8000 to 10,000 ft, depending on surface climate in the summer, geothermal gradient, and shaft heat loads such as pump systems.

Base heat load is calculated at an infinite airflow at the reject temperature passing through the work area. The temperature of an infinite airflow will not increase as air picks up heat. **Actual heat load** is measured or calculated at the average stope temperature. It is always greater than the base heat load because the average stope temperature is lower than the reject temperature. More heat is drawn from the wall rock. **Marginal heat load** is the difference between base and actual heat loads. It is the penalty paid for using less than an infinite airflow (i.e., the lower the airflow, the lower the inlet temperature required to maintain the reject temperature and the higher the heat load).

Temperature-dependent heat sources (TDHs) depend on the temperature difference between the source and air. Examples include wall rock, broken rock, and fissure water (in a ditch or pipe). **Temperature-independent heat sources (TIHs)** depend only on the energy input to a machine or device after the energy required to raise the potential energy of a substance, if any, is deducted. Examples include electric motors, lights, substation losses, and the calorific value of diesel fuel.

Passive thermal environmental control separates heat sources from ventilating airflows. Examples include insulating pipes and wall rock, and blocking off inactive areas. **Active thermal environmental control** removes heat via airflow and air conditioning quickly enough so that air temperature does not rise above the reject.

Positional efficiency, an important design parameter for mine cooling systems, is the cooling effect reaching the work area divided by the machine evaporator duty. The greater the distance between the machine and work area, the more heat that the cooling medium (air or water) picks up en route.

Percent utilization is the ratio of the evaporator duty of the refrigeration plant over a year in energy units to the duty if the plant had worked the entire year at 100% load. This consideration becomes important when evaluating surface versus underground plants.

Coefficient of performance (COP) usually is defined as the evaporator duty divided by the work of compression in similar units. In mines, the overall COP is used: the evaporator duty divided by all power-consuming devices needed to deliver cooling to the work sites. This includes pumps and fans as well as refrigeration machine compressors.

A **shaft** is a vertical opening or steep incline equipped with skips to hoist the ore, and cages (elevators) to move personnel and supplies. Electric cables and pipes for fresh water, compressed air, cooling water, and pump water are installed in shafts. **Drifts** and **tunnels** are both horizontal openings; a tunnel opens to daylight on both ends, whereas a drift, deep underground, does not. In metal mining, a **stope** is a production site where ore is actually mined. In coal mining, coal is usually produced by either longwall (one continuous production face hundreds of feet long) or room-and-pillar (multiple production faces in a grid of rooms with supporting pillars in between) methods.

SOURCES OF HEAT ENTERING MINE AIR

Adiabatic Compression

Air descending a shaft increases in temperature and pressure (because of the mass of air above it). As air flows down a shaft, it increases in temperature as if compressed in a compressor because of conversion of potential energy to internal energy, even if there is no heat interchange with the shaft and no evaporation of moisture.

For dry air at standard conditions (59°F at 14.696 psia), the specific heat at constant pressure c_p is 0.24 Btu/lb·°F. For most work, c_p can be assumed constant, but extreme conditions might warrant a more precise calculation: 1 Btu is added (for descending airflow) or subtracted (for exhaust) to each pound for every 778 ft. The dry-bulb temperature change is $1/(0.24 \times 778 \times 1) = 0.00535$°F per foot, or 1°F per 187 ft of elevation. The specific heat for water vapor is 0.45 Btu/lb·°F. So, for constant air-vapor mixtures, the change in dry-bulb temperature is $(1 + W)/(0.24 + 0.45W)$ per 778 ft of elevation, where W is the humidity ratio in pounds of water per pound of dry air.

The theoretical heat load imposed on intake air by adiabatic compression is given in Equation (1), which is a simplified form of the general energy equation:

$$q = 60Q\rho E \,\Delta d \qquad (1)$$

where

> q = theoretical heat of autocompression, Btu/h
> 60 = 60 min/h
> Q = airflow in shaft, cfm
> ρ = air density, lb/ft^3
> E = energy added per unit distance of elevation change, 1 Btu/778 ft·lb
> Δd = elevation change, ft

Example 1. What is the equivalent heat load from adiabatic compression of 300,000 cfm at 0.070 lb/ft^3 density flowing down a 5000 ft shaft?

Solution:

> $q = (60)(300,000)(0.070)(1/778)(5000) = 8,097,686$ Btu/h

The adiabatic compression process is seldom truly adiabatic: *autocompression* is a more appropriate term. Other heating or cooling sources, such as shaft wall rock, introduction of groundwater or water sprayed in the shaft to wet the guides, compressed-air and water pipes, or electrical facilities, often mask the effects of adiabatic compression. The actual temperature increase for air descending a shaft usually does not match the theoretical adiabatic temperature increase, for the following reasons:

- The effect of seasonal and daily surface temperature fluctuations, such as cool night air on the rock or shaft lining (rock exhibits thermal inertia, which absorbs and releases heat at different times of the day)
- The temperature gradient of ground rock related to depth

- Evaporation of moisture within the shaft, which suppresses the dry-bulb temperature rise while increasing the moisture content of the air

The wet-bulb temperature lapse rate varies, depending on the entering temperature and humidity ratio, and the pressure drop in the shaft. It averages about 2.5°F wet bulb per 1000 ft, and is much less sensitive to evaporation or condensation than the dry bulb.

Electromechanical Equipment

Electric motors and diesel engines transfer heat to the air. Loss components of substations, electric input to devices such as lights, and all energy used on a horizontal plane appear as heat added to the mine air. Energy expended in pumps, conveyors, and hoists to increase the potential energy of a material does not appear as heat, after losses are deducted.

Vehicles with electric drives, such as scoop-trams, trucks, and electric-hydraulic drill jumbos, release heat into the mine at a rate equivalent to the nameplate and a utilization factor. For example, a 150 hp electric loader operated at 80% of nameplate for 12 h a day liberates (150 hp)(42.4 Btu/min·hp)(0.80) (12 h/day)(60 min/h) = 3,663,360 Btu/day. Dividing by 24 h/day gives an average heat load over the day of 152,640 Btu/h. During the 12 h the loader is operating, the heat load is doubled to 305,280 Btu/h. The dilemma for the ventilation engineer is that, if heat loads are projected at the 152,640 rate, the stope temperature will exceed the reject temperature for half the day, and the stope will be overventilated for the other half; if projected at 305,380 Btu/h, the stope will be greatly overventilated when the loader is not present. Current practice is to accept the additional heat load while the loader is present. Operators get some relief when they leave the heading to dump rock, at which time the ventilation system can partially purge the heading.

Diesel equipment dissipates about 90% of the heat value of the fuel consumed, or 125,000 Btu/gal, to the air as heat (Bossard 1982). The heat flow rate is about three times higher for a diesel engine than for an equivalent electric motor. If the same 150 hp loader discussed previously were diesel-powered, the heat would average about 458,000 Btu/h over the day, and 916,000 Btu/h during actual loader operation. Both sensible and latent heat components of the air are increased because combustion produces water vapor. If a wet scrubber is used, exhaust gases are cooled by adiabatic saturation and the latent heat component increases even further.

Fans raise the air temperature about 0.45°F per in. of water static pressure. Pressures up to 10 in. of water are common in mine ventilation. This is detrimental only when fans are located on the intake side of work areas or circuits.

Groundwater

Transport of heat by groundwater has the largest variance in mine heat loads, ranging from essentially zero to overwhelming values. Groundwater usually has the same temperature as the virgin rock. Ventilating airflows can pick up more heat from hot drain water in an uncovered ditch than from wall rock. Thus, hot drain water should be stopped at its source or contained in pipelines or in covered ditches. Pipelines can be insulated, but the main goal is isolating the hot water so that evaporation cannot occur.

Heat release from open ditches increases in significance as airways age and heat flow from surrounding rock decreases. In one Montana mine, water in an open ditch was 40°F cooler than when it flowed out of the wall rock; the heat was transferred to the air. Evaporation of water from wall rock surfaces lowers the surface temperature of the rock, which increases the temperature gradient of the rock, depresses the dry-bulb temperature of the air, and allows more heat to flow from the rock. Most of this extra heat is expended in evaporation.

Table 1 Maximum Virgin Rock Temperatures

Mining District	Depth, ft	Temperature, °F
Kolar Gold Field, India	11,000	152
South Africa	12,000	135
Morro Velho, Brazil	8,000	130
North Broken Hill, Australia	3,530	112
Great Britain	4,000	114
Braloroe, BC, Canada	4,100	112.5
Kirkland Lake, Ontario	6,000	81
Falconbridge Mine, Ontario	6,000	84
Lockerby Mine, Ontario	4,000	96
Levac Borehold (Inco), Ontario	10,000	128
Garson Mine, Ontario	5,000	78
Lake Shore Mine, Ontario	6,000	73
Hollinger Mine, Ontario	4,000	58
Creighton Mine, Ontario	10,000	138
Superior, AZ	4,000	140
San Manuel, AZ	4,500	118
Butte, MT	5,200	150
Homestake Mine, SD	8,000	134
Ambrosia Lake, NM	4,000	140
Brunswick #12, New Brunswick, Canada	3,700	73
Belle Island Salt Mine, LA	1,400	88

Source: Fenton (1972).

Table 2 Thermal Properties of Rock Types

Rock Type	Thermal Conductivity, Btu/h·ft·°F	Diffusivity, ft²/h
Coal	1.27	0.050
Gabro	1.37	0.092
Granite	1.11	0.129
Pyritic shale	2.11	0.078
Quartzite	3.18	0.090
Sandstone	1.14	0.065
Shale	1.38	0.035
Rhyolite	2.00	0.043
Sudbury ore	1.50	0.049
North Idaho metamorphic	2.95	0.109

Source: Mine Ventilation Services Inc., Fresno, CA. Reprinted with permission.

Example 2. Water leaks from a rock fissure at 20 gpm and 125°F. If the water enters the shaft sump at 85°F, what is the rate of heat transfer to the air?

Solution:

$$\text{Heat rate} = (20 \text{ gpm})(60 \text{ min/h})(8.33 \text{ lb/gal})(1 \text{ Btu/lb·°F})$$
$$\times (125 - 85°F) = 399,840 \text{ Btu/h}$$

Wall Rock Heat Flow

Wall rock is the main heat source in most deep mines. Temperature at the earth's core has been estimated to be about 10,300°F. Heat flows from the core to the surface at an average of 0.022 Btu/h·ft². The implication for mine engineers is that a geothermal gradient exists: rock gets warmer as the mine deepens. The actual gradient varies from approximately 0.5 to over 4°F per 100 ft of depth, depending on the thermal conductivity of local rock. Table 1 gives depths and maximum **virgin rock temperatures (VRTs)** for various mining districts. Table 2 gives thermal conductivities and diffusivities for rock types commonly found in mining. These two variables are required for wall rock heat flow analysis.

Wall rock heat flow is unsteady-state: it decays with time because of the insulating effect of cooled rock near the rock/air boundary. Equations exist for both cylindrical and planar openings, but this section discusses cylindrical equations (Goch and Patterson 1940). The method can solve for either instantaneous or average heat flux rate. The instantaneous rate is recommended because it is better used for

older tunnels or drifts. For newer drifts, a series of instantaneous rates over short time periods is equivalent to the average rate. The Goch and Patterson calculations are easily performed on a computer using the following variables and equations:

$$\text{Fo} = \frac{\alpha\theta}{r^2} \tag{2}$$

$$\varepsilon = \{1.017 + 0.7288 \log_{10}(\text{Fo}) + 0.1459[\log_{10}(\text{Fo})]^2$$
$$- 0.01572[\log_{10}(\text{Fo})]^3 - 0.004525[\log_{10}(\text{Fo})]^4 \tag{3}$$
$$+ 0.001073[\log_{10}(\text{Fo})]^5\}^{-1}$$

$$\text{Heat Flux, Btu/h·ft}^2 = \frac{k(t_{vr} - t_a)(\varepsilon)}{r} \tag{4}$$

$$\text{Total Heat Flow, Btu/h} = (\text{Heat Flux})(L)(P) \tag{5}$$

where

Fo = Fourier number, dimensionless
k = thermal conductivity of rock, Btu/h·ft·°F
L = length of section, ft
P = perimeter of section, ft
r = radius of circular section, ft, or equivalent radius of rectangular section; $r = (A/\pi)^{1/2}$, where
$\quad A$ = cross-sectional area of section, ft²
t_a = air dry-bulb temperature, °F
t_{vr} = virgin rock temperature, °F
α = thermal diffusivity of rock (equals $k/\rho c$), ft²/h, where
$\quad \rho$ = rock density, lb/ft³
$\quad c$ = heat capacity, Btu/lb·°F
ε = function of Fourier number for instantaneous rate, dimensionless (Whillier and Thorpe 1982)
θ = average age of section, h

Example 3. A 500 ft long section of drift, 12 ft high by 15 ft wide, was driven in quartzite with a VRT of 110°F. The drift was started 20 days before the face was reached, and the face is 1 day old. One design criterion is keeping the average dry-bulb temperature of the air in the drift at 80°F. How much heat will flow into the section?

Solution: From Table 2, the thermal conductivity of quartzite is 3.18 Btu/h·ft·°F and the diffusivity is 0.090 ft²/h. The average age of the section is (20 + 1 days)/2 = 10.5 days, or 252 h. The cross-sectional area of the drift is 12 × 15 = 180 ft² and the perimeter is (12 + 15) × 2 = 54 ft. The equivalent radius of the drift is $(180/\pi)^{1/2}$ = 7.57 ft. The following equations are then applied:

Using Equation (2), $\text{Fo} = \dfrac{\alpha\theta}{r^2} = \dfrac{(0.090)(252)}{7.57^2} = 0.396$

Using Equation (3), $\varepsilon = 1.336$

Using Equation (4),

$$\text{Heat Flux} = \frac{k(t_{vr} - t_a)(\varepsilon)}{r} = \frac{(3.18)(110 - 80)(1.336)}{7.57}$$
$$= 16.84 \text{ Btu/h·ft}^2$$

Using Equation (5),

$$\text{Total Heat Flow} = (\text{Heat Flux})(L)(P)$$
$$= (16.84)(500)(54) = 454,700 \text{ Btu/h}$$

Thus, to keep the average temperature of the drift section at 80°F db, 37.9 tons of refrigeration are needed.

The Goch and Patterson method lacks a convective heat transfer coefficient at the rock/air boundary, and overestimates heat transfer in a dry drift by 8 to 15%. It also does not have a wetness factor. Because a drift with water on the perimeter draws more heat from wall rock, the method underestimates heat flow. Almost all drifts have some wetness on the floor, back, and side walls, though it may not be visible. Comparisons of the Goch and Patterson method with field measurements and results from commercial software under typical conditions (a drift with 20 to 60% of the perimeter wetted) indicate that the overestimate is nearly equal to the underestimate. When using Goch and Patterson for drift heat loads, keep drift section lengths under 200 ft and do not apply any contingency factor to the calculated heat load.

Heat load calculations for stoping require a large number of variables. Irregular shapes, sporadic advance rates, intermittent TIH sources, fissure water, and nonhomogeneous or anisotropic (with directionally differing heat conducting properties) rock are difficult to model (Duckworth and Mousset-Jones 1993; Marks and Shaffner 1993). For cut-and-fill stoping with a sand floor, measured heat loads are about 70% of the heat loads predicted by Goch and Patterson. Other stoping methods such as room-and-pillar or tabular reef mining are more amenable to planar heat load equations. Patterson (1992) gives empirical graphs relating heat load to productivity and depth.

Ventilation engineers needing to project heat loads for new mines or extensive tunnel projects can write their own computer program using the Goch and Patterson equations, or use a commercial software package. These programs account for convective heat transfer, wetness, elevation changes, and TIH sources that can make hand calculations tedious. However, program input must be carefully derived or the output will be misleading.

Heat from Broken Rock

Freshly blasted broken rock can liberate significant amounts of heat in a confined area. The broken rock's initial and final temperatures, and where the rock is cooled en route from the face to the hoisting facility, must be estimated.

$$\text{Heat, Btu} = (\text{mass})(\text{specific heat})(\text{VRT} - \text{final temperature}) \quad (6)$$

$$\text{Heat load, Btu/h} = \frac{\text{Heat}}{(\text{time, h})} \quad (7)$$

Example 4. A 12 ft high by 15 ft wide by 10 ft long drift round is blasted in quartzite where the VRT is 120°F. Quartzite has a 168 lb/ft³ density and a 0.2 Btu/lb·°F specific heat. By the time the rock is hoisted to the surface 4 h later, it has cooled to 90°F. What is the heat load imposed on the drift and shaft?

Solution:

$$\text{Heat} = (12 \text{ ft} \times 15 \text{ ft} \times 10 \text{ ft} \times 168 \text{ lb/ft}^3)(0.2 \text{ Btu/lb·°F})$$
$$\times (120 - 90°F) = 1.8144 \times 10^6 \text{ Btu}$$
$$\text{Heat load} = \frac{1.8144 \times 10^6 \text{ Btu}}{4h} = 453,600 \text{ Btu/h}$$

Heat from Other Sources

Heat produced by oxidation of timber and sulfide minerals can be locally significant and even cause mine fires. Fortunately, timber is seldom used for ground support in modern mines. Heat from blasting can also be appreciable. The typical heat potential in various explosives is similar to that of 60% dynamite, about 1800 Btu/lb. This heat is usually swept out of the mine between shifts and thus is not tallied in heat load projections. Body metabolism is only a concern in refuge chambers and is rarely if ever included in heat load projections. Although these heat sources are usually neglected, the ventilation engineer must remain vigilant for cases where local effects might be significant.

Summation of Mine Heat Loads

Mine cooling requirements should be estimated after mining methods, work sites, production rates, and equipment are specified, and heat sources identified. The time frame, during which the ventilation and cooling systems must provide an acceptable work environment, is normally 10 years, but can vary.

Total heat for a mine or mine section is the summation of all TIH and TDH sources. It helps to plot heat sources on a schematic.

The heat load from the surface to the entrance of the stope is assessed first, starting with TIH sources because they influence TDH sources. Shaft heat loads, autocompression, and drift heat loads are added to the air en route to stopes. The process should take only one iteration to find a stope entering temperature. Stope heat load is calculated by assuming that the wet bulb leaving the stope equals the design reject temperature. The air temperature entering the stope is estimated, and heat load equations are used to calculate the exit temperature. If this exit temperature exceeds the reject temperature, a lower stope entering temperature is assumed and a new exit temperature is calculated. The process is repeated with new stope entering temperatures until the calculated stope exit temperature equals the design reject temperature.

If the entering stope temperature calculated from the surface is greater than the entering stope temperature calculated from the reject temperature, higher airflow or air conditioning will be needed. Psychrometrics can determine the size of the airflow increase or cooling required.

HEAT EXCHANGERS

Underground heat exchangers can be water-to-refrigerant, air-to-refrigerant, water-to-water, air-to-water, or air-to-air. Brine can be used instead of water where freezing might occur. Heat exchangers can be direct (e.g., spray chambers) or indirect (e.g., conductive heat transfer through tubes or plates).

See Chapters 18 to 23, 25, 26, 28, 29, 36 to 39, and 41 to 47 of the 2008 *ASHRAE Handbook—HVAC Systems and Equipment* for general guidelines when designing large cooling plants for mine duty.

Shell-and-Tube and Plate Heat Exchangers

Shell-and-tube heat exchangers are the mainstay of refrigeration machines used in mines. Machines in the 200 to 400 ton range may use either direct expansion (DX) or flooded evaporators. In both cases, the working fluid (refrigerant or water) is circulated through the tubes.

South African mines often use plate-and-frame evaporators in large surface chilled-water plants (van der Walt and van Rensburg 1988). These machines can cool water to within a degree of freezing without danger of rupture. In contrast, shell-and-tube evaporators should not be expected to chill water below 38°F. Manufacturers must be consulted.

Shell-and-tube water-to-water heat exchangers have been used in mine cooling systems to avoid pumping return water against high heads (the U-tube effect). Chilled water from the surface is sent down to the high-pressure (tube) side of the exchanger. Water on the low-pressure side operates district chiller systems or spot coolers. The shell-and-tube water-to-water heat method is not very popular, perhaps because it requires high-pressure supply and return piping. The second law of thermodynamics limits the approach temperature of the outlet high-pressure water to the inlet low-pressure water temperature. This tends to limit heat removal in deep mines that would require at least three heat exchanger stations, in series, in the shaft.

Cooling Coils

Cooling coils can be DX or chilled-water coils. DX coils are used with spot coolers and typically range from 15 to 60 tons. Some modern spot coolers use dual coils in parallel for compactness. Chilled-water coils, used in district chiller systems, are also used in a wide range of sizes.

Air-side fouling is the main operational problem with cooling coils in mines. Coils with fin spacing tighter than 6 fins per inch are not recommended. Water-side fouling is minimal if the water is of fair quality, the circuit is closed, and a corrosion inhibitor is added to the circuit.

Small Spray Chambers

Small spray chambers can be used as an alternative to cooling coils. Heat transfer is direct, air-to-water. Spray chamber maintenance is minimal, and the amount of water sprayed is typically one-half to one-third that required for a cooling coil at the same duty. Some washing effect also occurs in the chamber.

Spray chambers are open systems that dump water into a ditch or collection pond after spraying. This water drains to the dewatering system or is pumped back to the chiller plant. Small spray chambers are still popular for small duties in mines that chill service water, but the pumping system must be able to handle the increased service water requirement.

Cooling Towers

When heat loads are large, the full capacity of a mine's heat removal system will probably be needed. A key component of the heat removal system is exhaust air. The ventilation engineer for a deep, hot mine must be proficient at designing underground cooling towers for condenser heat rejection, and spray chambers for cooling airflows.

Rather than using the standard HVAC&R method of assessing cooling tower performance, as described in Chapter 39 of the 2008 *ASHRAE Handbook—HVAC Systems and Equipment*, mining engineers use the South African factor-of-merit method, developed in the 1970s for designing direct-contact heat exchangers (Bluhm 1981; Burrows 1982; Whillier 1977). This method requires a psychrometric program and the following equations:

$$\Sigma - h \quad (c_{pw})(W)(t) \tag{8}$$

Note that t is either the wet-bulb temperature for air, or the water temperature, depending on whether Σ_{ai} or Σ_{wi} is calculated.

$$q = (M_w)(c_{pw})(t_{wi} - t_{wo}) \tag{9}$$

$$q = (M_a)(\Sigma_{ao} - \Sigma_{ai}) \tag{10}$$

$$n_w = \frac{t_{wi} - t_{wo}}{t_{wi} - t_{ai}} \tag{11}$$

$$R = \frac{(M_w)(c_{pw})(t_{wi} - t_{ai})}{(M_a)(\Sigma_{wi} - \Sigma_{ai})} \tag{12}$$

$$N = \frac{F}{(1 - F)(R^{0.4})} \tag{13}$$

$$n_w = \frac{\left(1 - e^{-N(1 - R)}\right)}{\left(1 - Re^{-N(1 - R)}\right)} \quad \text{(for counterflow towers)} \tag{14}$$

where

a = air
c_{pw} = specific heat of water at constant pressure, 1 Btu/lb·°F
F = factor of merit, roughly equivalent to UA factor in conductive heat transfer, dimensionless; ranges from 0 (no heat transfer) to 1 (as much as heat transfer as allowed by second law takes place)
h = enthalpy of moist air, Btu/lb
i = inlet
M = mass flow rate, water or air, lb/min
N = number of transfer units, an intermediate factor for calculating water efficiency
n_w = water efficiency, dimensionless
o = outlet
q = heat rate to be transferred in chamber, Btu/lb
R = tower capacity factor, dimensionless; ratio of heat capacity of water to heat capacity of air under limits of second law
t = temperature, °F
W = humidity ratio of moist air, lb water per lb air

Table 3 Factors of Merit

	Factor of Merit Range	Source
Vertical counterflow, open, unpacked	0.50 to 0.70	Hemp 1982
Horizontal cross-flow		
Single-stage	0.40 to 0.55	Hemp 1982
Two-stage	0.57 to 0.72	Marks 1988
Three-stage	0.69 to 0.81	Marks 1988
Four-stage	0.76 to 0.87	Marks 1988
Commercial packed cooling		
Counterflow tower	0.68 to 0.78	Patterson 1992
Cross-flow tower	0.55 to 0.65	Patterson 1992

w = water
Σ = energy of air, Btu/lb; total enthalpy minus enthalpy of liquid water evaporated into air [approximated by $(c_{pw})(W)(t)$, where t is wet bulb]; dependent only on wet bulb and barometric pressure

Designing an underground cooling tower requires the exhaust air mass flowrate M_a, the wet- and dry-bulb temperatures available at the tower, and the ambient barometric pressure. A psychrometric program is needed to calculate enthalpy, density, humidity ratio, and specific volume from the wet bulb, dry bulb, and barometric pressure. Tower airflow is taken from measurements or from the mine plan. The temperature, if not measurable (e.g., in a new mine), can be assumed to approach the reject temperature. Experience shows that air usually enters exhaust at about 82 to 83°F wb. Then,

1. Calculate heat rejection rate in the tower (evaporator duty × condenser heat rejection factor, typically between 1.2 and 1.4). Discuss with the manufacturer.
2. Select condenser water flow and use Equation (9) to calculate Δt_w in the tower and machine condensers.
3. Specify cooling tower diameter and tower air velocity.
4. Calculate M_w/M_a.
5. Calculate heat rejection rate per cfm in tower.
6. Calculate air enthalpy h and use Equation (8) to calculate Σ_{ai}.
7. Select a factor of merit for tower using Table 3.
8. Estimate tower capacity factor R (e.g., $R = 0.5$). Using $R = 1$ in Equation (14) will result in division by zero. Skip to Step 10 if $R = 1$, and use the value of F for n_w.
9. Use Equation (13) to calculate N.
10. Use Equation (14) to calculate water efficiency n_w.
11. Use Equation (11) to calculate inlet water temperature t_{wi}.
12. Use Equation (8) to calculate Σ_{wi}, the energy of air at inlet water temperature t_{wi}.
13. Use Equation (12) to calculate a new tower capacity factor R.
14. Compare the R calculated in Step 13 with the R estimated in Step 8. If different by more than 1%, return to Step 8 and re-estimate R. Repeat Steps 9 to 14 until the calculated R is within 1% of the estimated R.
15. Calculate air and water temperatures leaving the tower and the evaporation rate.

Keep the following empirical design criteria in mind during Steps 1 to 5:

- Realistic Δt_w in tower is 12 to 16°F
- Realistic water loading in tower is 6 to 18 gpm per square foot
- Optimum water velocity in machine condenser tubes (3 to 13 fpm, per manufacturer's recommendations) is based on tubing material and water quality
- Realistic maximum air velocity in tower is 1600 fpm
- Realistic ratios of the mass flows of water to air M_w/M_a range from 0.5 to 2.5
- A realistic heat rejection rate in tower is 32 to 65 Btu/h per cfm

Values outside these design parameters are sometimes used (especially when plant duty is increased at a future date), but the penalty paid is a higher condensing temperature and lower COP. Once q, F, M_a, t_{ai}, and M_w are specified, only one t_{wi}, t_{wo}, and t_{ao} will balance all equations.

Example 5. Design a cooling tower for a 1000 ton refrigeration plant planned for a deep, hot mine. Exhaust airflow for heat rejection is 250,000 cfm at 83°F saturated. Barometric pressure is 15.226 psia (31 in. of mercury, or 1000 ft below sea level). What size cooling tower is needed, how much condenser cooling water is required, what are the inlet and outlet air and water temperatures, and how much makeup water is needed?

Solution:

Step 1. For a refrigeration plant to produce 1000 tons of cooling, it must reject about 1000 tons × 12,000 Btu/h·ton × 1.25 condenser heat rejection factor = 15,000,000 Btu/h.

Step 2. Select a condenser water flow. For this example, start with 1 gpm per 12,000 Btu/h rejected. The condenser flow is thus 1250 gpm, or 10,413 lb/min at 8.33 lb/gal. The change in water temperature is calculated from Equation (9):

$$\Delta t_w = \frac{15,000,000 \text{ Btu/h}}{(10,413 \text{ lb/min})(60 \text{ min/h})(1 \text{ Btu/lb·°F})} = 24°F$$

That exceeds the realistic 12 to 16°F Δt_w, so arbitrarily increase the water flow to 2000 gpm and recalculate Δt_w. In practice, selecting the condenser water flow is anything but arbitrary. Generally, the higher the flow, the better, but higher flows require larger condensers to keep tube velocity within design limits, larger cooling towers with more nozzles, and significantly larger pumps. Actual condenser water flow is a compromise between machine and tower performance, capital cost, and overall plant COP (operating cost). At 2000 gpm for this example, $\Delta t_w = 15°F$, which is acceptable.

Step 3. Specify cooling tower diameter by using a midrange value to 12 gpm/ft^2.

$$(\pi/4)d^2 = \frac{2000 \text{ gpm}}{12 \text{ gpm/ft}^2} \quad \therefore \ d = 14.57 \text{ ft} \approx 15 \text{ ft}$$

$$\text{Air velocity} = \frac{250,000 \text{ cfm}}{(\pi/4)\left(15^2\right)} = 1415 \text{ fpm} \ (< 1600 \text{ fpm; acceptable})$$

Step 4. Calculate M_w/M_a.

$$M_w = (2000 \text{ gpm})(8.33 \text{ lb/gal}) = 16,660 \text{ lb}_w/\text{min}$$

The specific volume for 83°F saturated inlet air at 15.226 psia is 13.71 ft^3/lb$_a$. M_a is therefore

$$M_a = \frac{250,000 \text{ cfm}}{13.71 \text{ ft}^3/\text{lb}_a} = 18,235 \text{ lb}_a/\text{min}$$

$$M_w/M_a = \frac{16,660}{18,235} = 0.914 \quad (0.5 < M_w/M_a < 2.5; \text{ acceptable})$$

Step 5. The heat rejection rate in the tower is

$$\frac{15,000,000 \text{ Btu/h}}{250,000 \text{ cfm}} = 60 \text{ Btu/h per cfm}$$

This rate is approaching the upper acceptable limit. Consideration should be given to routing more air through the tower if possible. All design criteria have now been met.

Step 6. The enthalpy of air h_{ai} at 83°F saturated and 15.226 psia is 45.95 Btu/lb. $\Sigma_{ai} = 45.95 - (1)(0.0237)(83) = 43.98$ Btu/lb.

Step 7. Select a factor of merit for the tower. From Table 3, an open, unpacked, vertical counterflow cooling tower can conservatively be expected to have a 0.55 factor of merit. If the tower is well designed and actually has a higher factor, the tower will return cooler water to the plant and COP will increase.

Step 8. Estimate $R = 0.5$ (first pass).

Step 9. Calculate N from Equation (13):

$$N = \frac{0.55}{(1 - 0.55)\left(0.5^{0.4}\right)} = 1.613$$

Step 10. Calculate n_w from Equation (14):

$$n_w = \frac{1 - e^{-1.613(1 - 0.5)}}{1 - 0.5e^{-1.613(1 - 0.5)}} = 0.713$$

Step 11. Calculate t_{wi} from Equation (11) (after manipulation, and assuming that $t_{wi} - t_{wo} = \Delta t_w$):

$$t_{wi} = \frac{\Delta t_w}{n_w} + t_{ai} = \frac{15}{0.713} + 83 = 104.04°F$$

Step 12. Σ_{wi} at 104.04°F and 15.226 psia = 77.11 − (1)(0.047)(104.04) = 72.22 Btu/lb.

Step 13. Calculate the new R using Equation (12):

$$R = \frac{(16,660)(1)(104.04 - 83)}{(18,235)(72.22 - 43.98)} = 0.681$$

Step 14. The new R is higher than the 0.5 R estimated in Step 8. Return to Step 8 and iterate until the R calculated in Step 13 equals the R projected in Step 8. This occurs at $R = 0.662$.

Step 15. All other values can now be calculated.

$$\text{Per Step 11, } t_{wi} = 106.09°F$$

$$t_{wo} = 106.09 - 15 = 91.09°F$$

$$\Sigma_{ao} = \Sigma_{ai} + \frac{15,000,000 \text{ Btu/h}}{(18,235 \text{ lb/min})(60 \text{ min/h})}$$
$$= 43.98 + 13.71 = 57.69 \text{ Btu/lb}$$

$$t_{ao} = 94.5°F \text{ (via psychrometric iteration)}$$

The water evaporated in the tower is the difference in humidity ratios $\Delta W \times$ the mass flow of dry air. From psychrometric equations, $W_{83°F} = 0.0237 \text{ lb}_w/\text{lb}_a$ and $W_{94.5°F} = 0.0347 \text{ lb}_w/\text{lb}_a$.

$$\text{Evaporation rate} = (18,235 \text{ lb/min})\frac{0.0347 - 0.0237 \text{ lb}_w/\text{lb}_a}{8.33 \text{ lb}_w/\text{gal}}$$

$$= 24.1 \text{ gpm}$$

Total makeup water depends on evaporation rate, water carryover (if any), and blowdown used to control dissolved solids in the condenser circuit. Leakages and carryover can be deducted from the blowdown. Makeup water is usually planned at 1 to 3% of the condenser water flow, depending on the quality of the makeup water, allowable cycles of concentration of dissolved solids, and water treatment plan.

Vertical unpacked cooling towers in mines often use clog-resistant full-cone nozzles circling the top of the tower, at least 40 ft above the pond. South African mines tend to use ham-type sprayers. Nozzle pressure of 30 psig is typically specified: lower water pressures do not generate the fine water droplets preferred for heat transfer, and higher pressures increase pumping costs. Higher pressures can also impinge water drops into side walls, where the water runs in sheets down the sides. This drastically reduces the surface area of the water flow, which reduces heat transfer. Rings circling the tower are recommended to kick water running down the sides back into the airstream. Unpacked towers do not have as high a factor of merit as towers with film packing or splash bars, but they are virtually maintenance-free and have low resistance to airflow. Figure 1 shows a typical underground vertical counterflow cooling tower.

After a cooling tower begins operation, the actual factor of merit should be determined. This is accomplished by measuring air and water flow rates and temperatures at the tower inlet and outlet, and then working the cooling tower equations in reverse. The actual factor of merit can be used to determine performance at other inlet conditions. This applies to mine, industrial, and commercial cooling towers.

Large Spray Chambers (Bulk Air Coolers)

The procedure for designing spray coolers is the same as for cooling towers, with the following minor changes:

$$q = M_w c_{pw}(t_{wo} - t_{wi}) \tag{15}$$

$$q = M_a(\Sigma_{ai} - \Sigma_{ao}) \tag{16}$$

$$n_w = \frac{1 - e^{-R(1-X)}}{R} \tag{17}$$

where $X = e^N$ for horizontal cross-flow chambers.

A perfect counterflow tower has a factor of merit of 1.0, but the factor of merit for a single cross-flow chamber cannot exceed 0.63 (Bluhm 1981). Two-stage cross-flow chambers are most often specified. Counterflow performance is approximated, and the counterflow equation for water efficiency can be used. Three-stage chambers can be designed when water flow must be limited to control pumping costs. Four-stage chambers are rarely cost-effective in mining applications.

Spray chambers often use vee-jet nozzles at 30 psig, placed uniformly along the chamber length and designed to cover the cross section evenly. Sprays should just reach the back of the chamber. Mist eliminators are usually installed at the chamber exit. Whereas cooling towers need makeup water to replace evaporated water, bulk air coolers gain water through condensation. This water can be sent to the condenser side as makeup. Figure 2 shows a typical two-stage horizontal cross-flow spray chamber.

MINE-COOLING TECHNIQUES

A mine-cooling system typically sends air, water, or ice into the mine at a low enthalpy state and removes it at a higher one. In hot mines, heat is typically rejected to water being pumped to the surface, and to exhaust air being drawn from the mine. There are many combinations and variations on how this is accomplished. Economics and site-specific conditions determine the optimum methods.

Increasing Airflows

This alternative should be considered first: it is usually less expensive to moderately increase airflows than to install refrigeration if the mine is above the critical ventilation depth. Increasing airflows also helps remove diesel fumes, which is increasingly important for modern mining. However, in deep, usually older mines with small cross-sectional airways, airflow increases may not be practical because of the cube relationship between fan power and airflow increase through a given resistance. Circuit resistance reduction via new airways or stripping existing airways is very expensive.

Chilling Service Water

When a mine requires additional heat removal, and airflow increases are not practical, chilling service water should be considered. Most mining methods require that water be sprayed on rock immediately after blasting to control dust; chilled water can intercept rock heat before the heat escapes into the air. This is a very flexible method of heat removal because it is applied when and where it is needed the most. After blasted rock is removed, the water is turned off and routed elsewhere. Main water lines should be insulated when service water is chilled.

Water is usually chilled in surface plants. A single-pass system is used if regional water supplies are plentiful. If scarce, water is pumped to the surface, recooled, and returned underground. Recycling can ease discharge permit requirements and thus save on treatment costs. Some mines have zero discharge permits. Regions with low winter temperatures and low relative humidities during summer have a natural cooling capacity that is adaptable to water chilling on the surface. Warm mine water is precooled in an evaporative cooling tower before being sent to the refrigeration plant.

Reducing Water Pressure and Energy Recovery Systems

All mines send water underground for drilling, cleaning, suppressing dust, and wetting broken rock. Hot mines using extensive cooling systems often send large volumes of water underground solely for air conditioning. The pressure of descending water must be broken periodically. The most common methods are the open cascade system and pressure-reducing valves. Turbines can also break the pressure and recover a significant portion of the potential

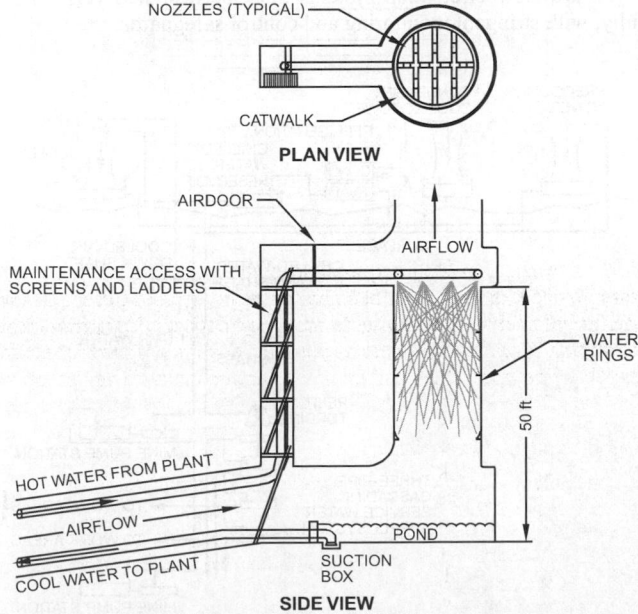

Fig. 1   Underground Open Counterflow Cooling Tower

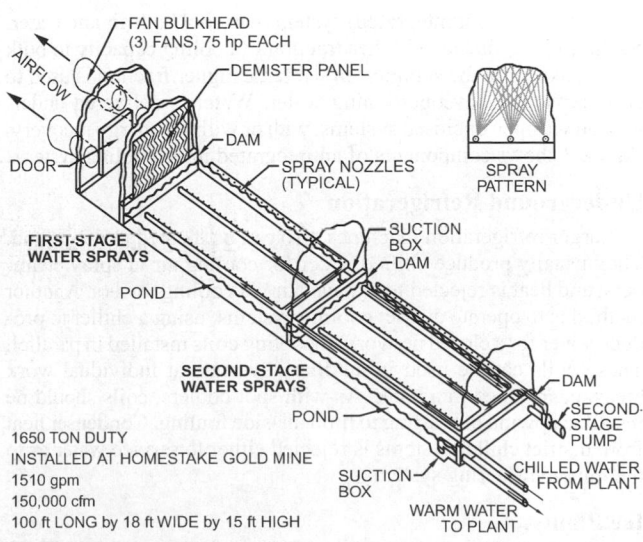

Fig. 2   Two-Stage Horizontal Spray Chamber

energy that would otherwise be lost. Two types of turbines are suitable for mine use: the Pelton wheel, which is most often used, and a centrifugal pump specially designed to run in reverse. The Pelton wheel rotor is shaped like the spokes of a wheel, with cups attached to the ends of the spokes. One or two nozzles shoot high-pressure water onto the cups, spinning the wheel. It is at least 80% efficient over a range of flows, simply constructed, and readily controlled. A wide operating range is important because water demand fluctuates. A turbine can turn either a generator or pump. Turning a generator is preferred because it separates service and cooling water from the mine dewatering system so that downtime in one system is less likely to disrupt the other.

Besides providing power to help return service water to the surface, turbines have another advantage: unrecovered potential energy is converted to heat at a rate of 1 Btu/lb per 778 ft of depth. If, for example, a 6000 ft deep mine uses 1000 gpm for air conditioning without energy recovery, the water will heat by 7.71°F. If 80% efficient turbines are used, the water temperature rise is about $0.2 \times 7.71 = 1.54$°F. The refrigerating effect lost is only 64 instead of 321 tons.

Other energy-recovery devices include hydrotransformers (large pistons transfer force from the high-pressure side to the lower-pressure side), and three-pipe feeder systems that deliver chilled water on one side while pumping out crushed ore on the other. These concepts have been tested in Europe and South Africa.

Bulk Cooling Versus Spot Cooling

Engineers must balance bulk cooling and spot cooling. **Bulk cooling** using a centrally located plant cools the entire mine, or a large section of it. Benefits are lower cost per ton installed, generally better maintenance, and lower temperatures in non-stoping areas such as haul drifts. Bulk cooling intake air is often done at warm-climate mines to provide winter-like or better conditions year round. Air is cooled in large direct-contact spray chambers adjacent to the shaft and then injected into the shaft below the main landing.

Cooling the entire mine draws more heat from surrounding wall rock, so a larger system must be designed to ensure proper stope cooling (i.e., positional efficiency suffers). When a multilevel mine is bulk cooled, cooling may be wasted on upper levels where heat load is low.

Spot cooling provides adequate temperature control in exploration and development headings, and in stopes on the fringes of mining activity. Total heat load is lower, but cost per ton is higher and temperatures in some areas might exceed design limits.

Combination (Integrated) Surface Systems

Combination (or integrated) systems can cool both air and water. Surface plants devote a higher fraction of cooling capacity to bulk cool intake air in the summer. In winter, a higher fraction is used to chill service or air-conditioning water. Water is delivered underground via open or closed systems, with or without energy recovery. Figure 3 shows components of an integrated mine cooling system.

Underground Refrigeration

Larger refrigeration machines also can be located underground. They usually produce chilled water for cooling air in spray chambers, and heat is rejected to exhaust air via cooling towers. Another method is to operate district cooling systems, using a chiller to produce water for a closed network of cooling coils installed in parallel. These coils can be used in auxiliary systems at individual work areas, or installed in a bank. As with spot coolers, coils should be installed upwind of blasting to limit air-side fouling. Condenser heat from district chiller systems is rejected either to service water or to the mine-dewatering system.

Ice Plants

For ultradeep mines (>12,000 ft), or those at the performance limits of existing water and airflow heat rejection systems, ice cooling should be considered. In going from 32 to 90°F before being pumped out of the mine, cooling water starting as ice can remove about 4.5 times the heat as the same mass flow of chilled water in going from 45 to 90°F:

Heat removal (Btu/lb) = Sensible + Latent

Heat removal of chilled water = (1 Btu/lb·°F)(90 − 45) = 45 Btu/lb

Heat removal of ice = (1 Btu/lb·°F)(90 − 32) + 144 Btu/lb = 202 Btu/lb

Heat removal factor increase of ice over water = 202/45 = 4.5

South African mines have been at the forefront in this application (Sheer et al. 2001). Both chunk and slurry delivery methods send ice to underground chambers, where it mixes with warm water returning from the mining area. The cold mixed water is then sent back to the mining area.

Several successful systems have been installed. Cost has dropped as technology improves; the overall COP of ice systems for ultradeep mines is now competitive with traditional cooling methods.

Thermal Storage

This Canadian innovation uses near-surface ice stopes or rock rubble to effectively and inexpensively heat intake air in the winter and cool it in the summer (Stachulak 1989).

Controlled Recirculation

This technique, used in conjunction with bulk air cooling, can reduce ventilation and air-conditioning requirements in older, deep mines, especially heavily mechanized ones (Tien 1999). Besides increasing air velocities in work areas without drawing more surface air through a high-resistance circuit, controlled recirculation reduces the heat load caused by autocompression. Using Equation (1), for every 100,000 cfm at standard density brought from the surface, the lost cooling capacity per 1000 ft of descent is

$$\text{Heat Load} = (100{,}000 \text{ ft}^3/\text{min})(60 \text{ min/h})(0.075 \text{ lb/ft}^3)$$
$$\times \left(\frac{1 \text{ Btu}}{778 \text{ ft/lb}} \right) (1000 \text{ ft})$$
$$= 578{,}406 \text{ Btu/h, or } 48.2 \text{ tons of cooling lost}$$
$$\text{per } 100{,}000 \text{ cfm, per } 1000 \text{ ft}$$

Controlled recirculation systems must be designed very carefully, with stringent monitoring and control safeguards.

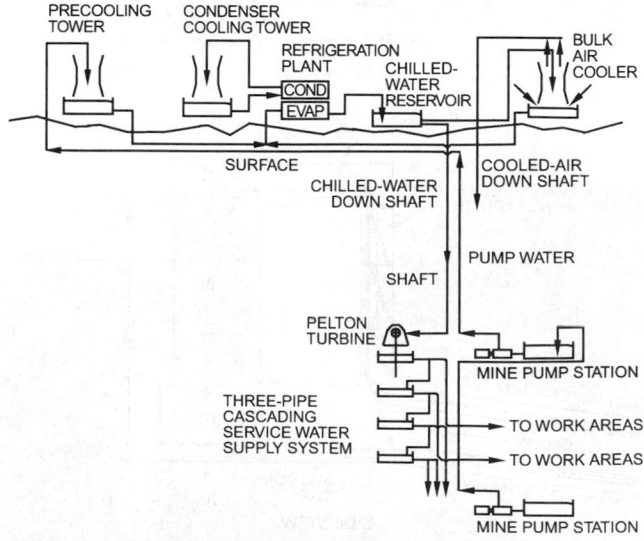

Fig. 3 Integrated Cooling System

Cabs and Vests

Mechanization can add significant heat, especially in confined auxiliary-ventilated spaces. Most noncoal mines have converted to diesel equipment in the last 20 years, although some use electric loaders and trucks. Coal mines are also mechanizing, but more slowly. The biggest problem with diesel engine heat is related to the greatest advantage of these vehicles: mobility. Heat and emissions from a diesel vehicle in a confined area can tax almost any ventilation system. Increasingly, cabs are being specified for large diesel vehicles. Cabs come equipped with window-type air conditioners and HEPA filters to capture diesel particulate and dust. After the diesel vehicle has left the heading, the mine ventilation and cooling system can provide an acceptable environment for other personnel. Cabs are expensive; however, because a loader often visits multiple headings in a week, the cost of a cab is much less than maintaining the design reject wet-bulb temperature in all headings at all times.

Cooling vests are not popular in mining. They are bulky, reduce mobility, and are time-consuming to prepare and use. Vests have limited application for mechanics, electricians, pipe-fitters, and others who must enter hot areas to set up ventilation and cooling. Vests using blue-ice packs or dry-ice last two to three hours; those using compressed-air venturi-type coolers require an umbilical cord.

Other Methods

Other methods being developed include the air cycle (air, compressed on the surface and sent underground to a turbine, turns a generator and exits at –40°F), and the ammonia cycle (sending down liquid ammonia, evaporating it, and sending the vapor back to surface condensers). These methods may be best suited to ultradeep mines where other cooling methods are already fully developed.

Transferring heat from current stopes to wall rock or rock rubble in previously worked-out stopes (the only method that does not remove heat from the mine) has also been considered. Refrigeration equipment would have to operate at a high condensing temperature to produce water hot enough to transfer heat to worked-out stopes.

SELECTING A MINE-COOLING METHOD

After mine-cooling and ventilation requirements have been projected, the designer must analyze and select the best method(s) for meeting those requirements. Cost-benefit analysis is the most widely used, but hardware reliability, dependency on outside factors, flexibility, safety, and technological level are just as important. Some factors to consider include the following:

- **Seasonal ambient conditions.** Warm-climate mines tend to bulk-cool air on the surface in industrial direct-contact spray chambers located close to the intake shaft.
- **Orebody and mining methods.** The more massive the orebody, the more ideal bulk cooling becomes. When stopes are scattered and continuously advanced into new areas, district or spot cooling might be better.
- **Mining rate.** This is critical: heat removal is energy related, not necessarily power related. A fast mining rate prompts a high instantaneous heat load (Btu/h), but less heat energy (Btu) per ton of production. This is because wall rock is covered by fill or isolated before its total heat energy has escaped into the airstream. The Btu per ton of production incurs the air-conditioning costs. Leave as much heat in the wall rock as possible.
- **Size and condition of major airways.** In older mines, small airways often limit airflow increases. This may prompt the need for air conditioning sooner than it normally would have been necessary.
- **Heat sources.** The contribution of TIH and TDH sources to the total can help determine the balance of passive to active thermal environmental controls, the ratio of airflow to air conditioning, and whether cabs should be specified.

Table 4 Basic Cooling Alternatives

	Warm Climate	Cool/Cold Climate
Massive ore-body (deep)	Large or medium airflow Chill service water Bulk-cool air on surface Bulk-cool air underground	Medium airflow Chill service water Bulk-cool air underground Thermal storage
Massive ore-body (shallow)	Large airflow Bulk-cool air on surface Chill service water Shell and tube	Large or medium airflow Chill service water on surface Shell and tube
Scattered orebody (multilevel)	Large airflow if not too deep Chill service water Bulk-cool air on surface District chiller systems Spot coolers/spray chambers	Large or medium airflow Chill service water District chiller systems Thermal storage
Ultradeep ore-body (massive and/or multilevel)	Limited airflow Chill service water District chiller systems Ice cooling Controlled recirculation	Limited airflow Chill service water District chiller systems Ice cooling Controlled recirculation
Small orebody	Bulk-cool air on surface Chill service water District chiller systems Spot coolers	District chiller systems Chill service water Spot coolers
Porous rock	District chiller systems Spot coolers	District chiller systems Spot coolers

- **Cost of power, water, labor, and supplies.** Knowing these costs is critical for assessing optimum capital expenditure to control operating costs. For example, if power cost is high, spending extra for a higher-COP system may be warranted.
- **Governmental regulations.** Heat stress standards can influence the size of the system. Other safety issues may constrain design, such as not using combustible pipe insulation or ammonia machines underground.

Basic cooling alternatives for specific cases are summarized in Table 4. Airflows are described as limited, medium, or large. One way to express airflow for a given mine is the ratio of tons of airflow per ton of ore. Limited airflow is defined as less than 8 tons of air per ton of ore, medium airflow is 8 to 16 tons per ton, and large airflow is over 16 tons per ton.

These ranges are based on an unpublished study of approximately 100 mines of all types, worldwide. The ranges discussed are for heat removal only. Additional airflow for methane or radon removal must be addressed separately.

MECHANICAL REFRIGERATION PLANTS

Surface Plants

Centrifugal or helical rotary screw machines are typically used in surface plants to chill water or bulk-cool air. Banks of machines are usually installed in parallel: plant design must accommodate one machine being down at any given time for maintenance while others operate. Shell-and-tube heat exchangers are standard, although plate-and-frame are used if water close to the freezing point is specified. The most common refrigerant for positive-displacement compression is HCFC-22. Ammonia is also commonly used in surface plants. Absorption machines can be considered if external waste heat is available.

Underground Plants

Large underground plants do the same work as surface plants, but are closer to work areas. Better positional efficiency and percent utilization are the advantages. Whereas surface plants use atmospheric air for heat rejection, underground plants use mine exhaust air, which raises natural ventilation pressure and aids circuit fans. Components for underground machines must be disassembled for transport down the shaft.

The main disadvantage of underground refrigeration is that heat rejection is limited by the amount of available exhaust air. Excavating underground refrigeration rooms and spray chambers is more costly than erecting prefabricated surface buildings. Maintenance is also more difficult because of shaft logistics. Power is more difficult to supply to an underground facility, and subject to more disruptions.

Spot Coolers

Spot coolers with 15 to 100 ton capacity allow driving long development headings, or cooling exploration sites before installing primary ventilation and cooling equipment. Development headings can be advanced more rapidly and under more comfortable conditions. Condenser heat is most often removed by service water, although some air-to-air condensers are used.

Spot coolers use reciprocating, scroll, or small screw compressors. Hermetic scroll compressors are becoming more popular because they handle liquid slugging better than reciprocating compressors and are less expensive. Spot coolers use direct exchange (DX) air cooling coils. The packaged unit includes a fan, which draws air through the coil (or coils, in a dual-coil unit) and then blows it through duct to the heading. Spot coolers must be compact and portable because they are moved often. The service water required is typically 1.1 gpm per ton, but can be less if the water temperature is under 55°F, or more if it is over 70°F. A return drain pipe is recommended to prevent contact between hot discharge water (often over 100°F) and ambient air. Coils sometimes receive dusty air immediately after blasting; if so, coils must be washed at least every other day.

Spot coolers are expensive, but often are the only choice for cooling exploration, development, and small-scale stoping on the fringes of mining activity.

Maintenance

Mines with extensive systems (e.g., a large chiller plant, or over 10 spot coolers) should employ a mechanic specializing in refrigeration. Mines with over 2000 tons probably need a second mechanic. These persons should be factory-trained and must be certified to handle refrigerants. Refrigeration specialists can be assisted periodically or full-time by apprentice mechanics. Another viable approach is a maintenance contract with the equipment manufacturer or supplier, or an independent HVAC&R shop. Some mines have a full-time person cleaning coils.

A fouling factor (ft²·h·°F/Btu) should be calculated from lab analysis of the condenser water, especially for district chillers using sump water. Planning a tube-cleaning regimen, either manual, acid circulation, or automatic with brushes and a flow reversal valve, is critical. Underground condensers can become plugged within a couple of weeks without cleaning, depending on the fouling factor. Water treatment is needed to control scale, corrosion, and organisms in surface or underground plants with cooling towers.

MINE AIR HEATING

Cold-climate mines typically heat intake air in the winter. In Canada, heating intake air can cost more than all other ventilation costs combined (Hall 1989). However, without heat, water in the shaft will freeze, disrupting hoisting operations and damaging shaft support members, cables, and pipes. Very cold air and icy floors are safety and health hazards; heavy gloves and other protective clothing required can make routine tasks difficult. Intake air is typically heated to just above the freezing point. Autocompression and shaft heat loads further temper the air as it downcasts into the mine.

Steam coils operated by boilers burning wood, coal, fuel oil, or natural gas often served as shaft heaters in the past. Electric resistance heaters have also been used, but they are expensive to operate. Waste heat from compressor stations has also been used.

When exhaust and intake shafts are located close together, a circulating glycol or heat pump system can be used to transfer heat from exhaust air to intake. For every degree of total heat (sensible plus latent) given up by warm saturated exhaust air, the same mass flow of cold intake air can be heated sensibly by 4 degrees. Either coils or a cooling tower extracts heat from exhaust air, and then coils transfer this heat to the intake air.

Controlled recirculation (up to 25% of total airflow) can also be applied to heat intake air (Hall 1989). The system is temporarily shut down during blasting.

Some cold-climate mines isolate the primary production shaft from the ventilation circuit. A slight upcast flow of uncontaminated air maintains good conditions in the shaft for hoisting ore and moving personnel and supplies. The disadvantage of this method is that ventilation duties of the production shaft must be transferred to one or more expensive stand-alone intake airways.

Natural gas and/or propane heaters are typically used at modern mines. Natural gas is preferred because it is less expensive and it burns more cleanly. Where natural gas is not available, propane must be trucked to the mine site. The same heater can burn either natural gas or propane; thus, propane can be used for back-up in case natural gas is cut off. Direct-fired heaters are usually preferred because the entire heat value of the fuel enters the intake airstream. If indirect heaters are used, roughly 15 to 25% of the heat is lost up the flue pipe.

Two types of natural gas or propane heaters have been used to heat intake air: (1) a grid of burner bars installed in a housing at the intake shaft, sometimes with louvers to adjust the flow of intake air and to mix air from the heaters with outdoor air; and (2) a crop dryer type of burner. Temperature sensors installed downstream can modulate both heater types to ensure that no more heat is applied than necessary to bring the temperature of the mixed intake air to 34°F.

Carbon monoxide sensors should also be installed downstream of the heaters. Experience at two mines in the western United States shows that the CO content of intake air heated by direct-fired burners can reach 10 to 20 parts per million.

Equation (18) is used to calculate the total heat required, assuming that the air has a low humidity ratio (which is the case for very cold air), and that no water is evaporated in the heater. Heating values for different fuels are given in Table 5.

$$\text{Heat, Btu/h} = (\text{Airflow, cfm})(60 \text{ min/h})(\text{density, lb/ft}^3)$$
$$\times (0.24 \text{ Btu/lb·°F})(\Delta t, \text{°F}) \tag{18}$$

where $\Delta t = 34$°F minus the intake air temperature.

Example 6. A mine is located where the atmospheric air temperature can drop to −20°F for two or more weeks per year. Occasionally the temperature drops to −30°F. An intake shaft handles 400,000 cfm, and the density of air entering the shaft in winter is 0.070 lb/ft³. What heating should be installed at the shaft intake to keep the shaft free of ice?

Solution: Sizing heaters is usually based on average cold periods, not extreme cold snaps. Here, a direct-fired heater is sized to raise −20°F

Table 5   Heating Values for Fuels

Fuel	Value	Source
Natural gas	1000 Btu/ft³	Kennedy 1996
Propane	90,000 Btu/gal	Kennedy 1996
Bituminous coal	12,300 to 14,400 Btu/lb	Abbeon Cal 2001
Fuel oil	143,000 Btu/gal	Abbeon Cal 2001
Wood	15,000,000 to 31,000,000 Btu/cord	Abbeon Cal 2001

air to 34°F. When the temperature drops to −30°F for short periods, intake airflow should be temporarily reduced. Using Equation (18),

Heat = (400,000)(60)(0.070)(0.24)[34 − (−20)] = 21,800,000 Btu/h

If natural gas is used, the volume required is

$$\frac{21,800,000 \text{ Btu/h}}{1000 \text{ Btu/ft}^3} = 21,800 \text{ ft}^3/\text{h}$$

If propane is used, the gallons required are

$$\frac{21,800,000 \text{ Btu/h}}{90,000 \text{ Btu/gal}} = 242 \text{ gal per hour}$$

MINE VENTILATION

Mine ventilation supplies oxygen to underground facilities, and removes dangerous or harmful contaminants such as methane, radon, strata gases, dust, blasting fumes, and diesel emissions. Ventilation also removes heat and helps control humidity in hot mines. Planning a ventilation system consists of five basic steps: (1) determining airflows, (2) planning the primary circuit, (3) specifying circuit fans and their installation, (4) determining auxiliary system requirements, and (5) assessing health and safety aspects.

Determining Airflows

Mining operations generate differing types and amounts of contaminants, and airflows dilute and remove these contaminants. The ventilation engineer must work closely with mine planning staff to understand where and how much production will take place, and what contaminants will be generated. The federal Mine Safety and Health Administration (MSHA) regulates contaminant concentrations to limits specified in the Federal Register, CFR 30. Controlling the most problematic contaminant normally keeps all others within their legal limits. For coal mines, contaminants of concern are typically methane and coal dust; for uranium mines, radon gas; for non-dieselized hard rock mines, usually silica dust and blasting fumes; for dieselized mines, typically diesel emissions. Design airflows for dieselized nonuranium metal mines range from 75 to 150 cfm per diesel horsepower, depending on the reference cited. With the current emphasis on controlling diesel emissions, start planning at 100 cfm per horsepower.

Total airflow is a summation of airflows for individual work areas, plus a leakage factor. Leakage is defined as airflow that does not ventilate any active work area or permanent site such as a pump room. A "tight" system minimizes leakage through well-constructed doors and seals, by minimizing the number of possible leakage paths, and by careful fan placement. Leakage can range from 10% of total airflow at a tight metal mine to 80% at some coal mines.

The ratio of tons of air per ton of ore production is about 2 to 4 for block cave mines, 6 to 8 for nondieselized cut-and-fill metal mines, and 9 to 16 for dieselized metal mines. Gassy coal and uranium mines can have significantly higher ratios, depending on the methane or radon generation rate.

Example 7. A new mechanized cut-and-fill gold mine is planned. Ore production is expected to be 1,200,000 tons per year. Intake air density is 0.070 lb/ft^3. What is the rough airflow required for ventilation?

Solution: The airflow range is 9 to 16 tons of air per ton of ore for dieselized metal mines. For a first-pass guess, assume an average 12.5 tons per ton. The total weight of the air through the mine in a year is

(1,200,000 tons ore per year) × (12.5 tons air per ton ore)
= 15,000,000 tons air per year

$$\text{Airflow cfm} = \frac{(15,000,000 \text{ tons/yr})(2000 \text{ lb/ton})}{\left(0.070 \text{ lb/ft}^3\right)(525,600 \text{ min/yr})} = 815,400 \text{ cfm}$$

Ratios provide a good first guess. However, the ventilation engineer should derive the total airflow by listing all operations, estimating leakages, and adding the specific airflows required to ventilate each operation (zero-based planning). As with reject temperature, the total airflow selected should be economically justifiable to management.

Airflow specification may change with time because of production, equipment, or mining method changes.

Planning the Circuit

With airflow specified and work sites plotted, the ventilation engineer must lay out the primary circuit. The three basic types of airways are intake, work area, and exhaust. Sizing airways is normally based on keeping velocity within acceptable limits: If velocity is too low, the airway is oversized and thus costs more than necessary; if it is too high, pressure drop is too large and raises operating costs. Air velocity should not exceed 1200 fpm in production shafts and haul drifts. Higher velocities can create dust problems and lead to employee discomfort. However, velocities in bare circular concrete exhaust shafts can approach 5000 fpm if necessary. Air velocity in vertical upcast exhaust shafts should avoid the 1400 to 2300 fpm range because water sheets can form, causing surging at the main fan.

Resistance to airflow is calculated using Atkinson's (1854) and McPherson's (1993) equations:

$$\Delta H = RQ^2 \frac{d}{0.075} \tag{19}$$

where

ΔH = pressure drop, in. of water
R = resistance, in. of water·min^2/ft^6
Q = airflow, cfm
d = actual air density, lb$_m$/ft^3
0.075 = standard air density, lb$_m$/ft^3

$$R = \frac{kLP}{5.2A^3} \tag{20}$$

where

k = friction factor, lb$_f$·min^2/ft^4 (includes effects of pipes, ground support, and rock surface roughness)
L = length, ft
P = perimeter of opening, ft
5.2 = conversion factor, lb$_f$/ft^2·in. of water
A = area of opening, ft^2

For rectangular drifts, the k factor can range from 22×10^{-10} for a smooth, concrete-lined straight drift to 90×10^{-10} for an unlined, irregular curved drift. For shafts, the k factor can range from 12×10^{-10} for a smooth-sided borehole to over 500×10^{-10} for a heavily timbered rectangular shaft. See Hartman (1997), McPherson (1993), and Tien (1999) for more precise airway resistance specification.

Example 8. Mine plans call for 65,000 cfm to be sent through 2000 ft of 10 ft wide by 10 ft high drift. The k factor from measurements of similar drifts is 50×10^{-10} lb·min^2/ft^4. The average temperature is 75°F wb and 80°F db. The barometric pressure is 13.8 psia. What is the resistance of this drift, and what is the air pressure drop?

Solution: Using psychrometric equations in Chapter 1 of the 2009 *ASHRAE Handbook—Fundamentals*, the density is 0.0683 lb/ft^3.

$$R = \frac{kLP}{5.2A^3} = \frac{\left(50 \times 10^{-10}\right)(2000)(40)}{(5.2)(100)^3}$$

$$= 7.69 \times 10^{-11} \text{ in. of water·min}^2/\text{ft}^6$$

$$\Delta H = RQ^2 \frac{d}{0.075} = (7.69 \times 10^{-11})(65,000)^2 \frac{0.0683}{0.075}$$

$$= 0.30 \text{ in. of water}$$

A mine ventilation circuit contains airways in series and in parallel. The overall resistance (Hartman 1997) is

For series: $\quad R_T = R_1 + R_2 + R_3 + \cdots + R_n \quad$ (21)

For parallel: $\quad \dfrac{1}{\sqrt{R_T}} = \dfrac{1}{\sqrt{R_1}} + \dfrac{1}{\sqrt{R_2}} + \dfrac{1}{\sqrt{R_3}} + \cdots + \dfrac{1}{\sqrt{R_n}} \quad$ (22)

Example 9. If Airway #1 has a resistance of 1×10^{-10}, Airway #2 has a resistance of 2×10^{-10}, and Airway #3 has a resistance of 3×10^{-10}, what is the resistance of these three branches in series and in parallel?

Solution:

Series:

$$R_T = 1 \times 10^{-10} + 2 \times 10^{-10} + 3 \times 10^{-10}$$
$$= 6.0 \times 10^{-10} \text{ in. of water·min}^2/\text{ft}^6$$

Parallel:

$$\frac{1}{\sqrt{R_T}} = \frac{1}{\sqrt{1 \times 10^{-10}}} + \frac{1}{\sqrt{2 \times 10^{-10}}} + \frac{1}{\sqrt{3 \times 10^{-10}}}$$
$$R_T = 1.92 \times 10^{-11} \text{ in. of water·min}^2/\text{ft}^6$$

Modern ventilation network computer analysis uses Kirchhoff's laws to balance airflows: (1) the summation of airflows into a junction equals the summation out, and (2) the summation of pressure drops around any enclosed mesh equals zero.

Computer simulation allows quick analysis of a wide range of scenarios. Most programs use a balancing algorithm based on the work of Hardy Cross in the 1960s and 1970s. The program iterates as it converges on final balanced airflows. Fan curves or regulators can be inserted in almost any branch.

Regulators or section booster fans control airflow in branches. Without regulation, too little or too much airflow may occur; nevertheless, circuits should be designed with a many free-split branches (branches without a fan or regulator) as possible to minimize overall resistance. Free-split branches are often located in circuit extremities.

A mine should have more intakes than exhausts. This enhances safety, because miners have more escape paths, and because more paths bring in fresh air if a fire occurs in one of the intakes. Also, exhaust shafts can generally handle greater air velocities and hence larger quantities, so fewer exhaust shafts are needed.

Metal mines often contain circuit booster fans. Underground boosters can create neutral points in the system where air short-circuits from intake to exhaust above the point, and recirculates from exhaust to intake below the point. Uncontrolled recirculation should be minimized.

Exhausting primary circuits are commonly used in both metal and coal mines (intakes do not have airlocks). Under normal operation, this produces a negative mine pressure gradient. If fans fail or are deactivated, barometric pressure in the mine rises, which temporarily helps keeps methane in coal mines from flowing away from gob (mining waste) areas (Kennedy 1996).

Specifying Circuit Fans

Primary fans are either centrifugal or axial. South African mines typically use large centrifugals, whereas most U.S. and Canadian mines use axials. Both types have advantages. Efficiency (up to 90%) is about the same with either type. Centrifugals are heavier duty, quieter, do not have a pronounced stall region, and can generate higher static pressures (over 30 in. of water). Axials are more compact, and airflows can be easily adjusted by blade angle changes. Primary fans range from 100 to over 3500 hp each. Surface installations with multiple fans are common for large airflows and for back-up

operation when one fan is turned off for maintenance. Circuit fans can also be installed underground, especially in metal mines.

Primary fans are specified while the circuit is designed. Engineers must often balance airway considerations (sizes and numbers) and fan specifications. A fan should be selected that will operate on an efficient part of its curve. Fan speed, quantity, pressure, and power are related in the fan laws equations, described in Chapter 20 of the 2008 *ASHRAE Handbook—HVAC Systems and Equipment*. Care must also be taken to anticipate future circuit changes as mining operations advance.

The fan installation must be designed after primary fans are selected. Consequences of fan downtime must be carefully considered, especially for coal mines, because methane concentrations can increase when circuit airflows decrease. Fans exhausting a mine are typically mounted horizontally near a vertical shaft or borehole. A 90° transition turns the air into the fan inlet. An isolation door is installed between the transition and fan. Coal mines require a blast door to dampen a shock wave caused by a possible methane or coal dust explosion. An evasé, or diffuser, is attached to the fan outlet to recover part of the velocity pressure exiting the mine. A silencer can be added if surface noise reduction is desired. Increasingly, variable-speed drives are used with electric motors to turn the fans. These drives provide soft start, and speeds from 50 to 100% of synchronous speed, and even to 110% for temporary emergency duty. The installation must be designed for accessibility and ease of maintenance.

Determining Auxiliary System Requirements

Auxiliary fan and duct systems deliver air into dead-end headings. These systems are generally not permitted in coal mines but are common in metal mines. A blowing system is most often used. A fan is set in a fresh air base at the start of a drift, and duct is installed as the drift advances. For drifts under 1000 ft, flexible brattice-cloth duct can be used. Longer drifts that require booster fans need rigid duct because duct gage pressure can drop below atmospheric. Rigid duct also offers less resistance than brattice-cloth duct, but it is about eight times as expensive.

The air quantity needed at the face is determined by the equipment used and the rate at which blasting and diesel fumes must be removed. Ducts are sized for the air quantity needed and for space limitations in the drift. Fans are selected to provide the specified airflow. In general, a single-stage axial fan can generate up to 10 in. of water static pressure, which should deliver required airflow up to 2500 ft through properly sized duct. A larger duct or a two-stage axial fan is needed if distance is much longer. For very long drifts, booster fans are needed about every 2500 ft.

An exhausting system is often used for drifts requiring quick ingress after blasting. Air flows to the face through the drift, captures fumes, and is blown back to the circuit through duct. This keeps the drift clear of fumes. Disadvantages include: (1) the air picks up heat and humidity en route to the face, (2) rigid duct is required, and (3) the face is not swept by air as with a blowing system. A face overlap fan and duct can be installed.

Chapter 21 of the 2009 *ASHRAE Handbook—Fundamentals* provides a friction chart for round duct. However, it is better to acquire the friction chart of the specific duct being considered from the supplier. Shock losses through couplings and bends must be tallied. One important consideration is leakage through couplings. This can be minimized by careful installation, keeping duct pressure under 10 in. of water static pressure, and installing longer pieces of duct (up to 100 ft for brattice cloth, or 20 ft for rigid).

Cassettes loaded with brattice-cloth duct are now used for drifts driven by rapidly advancing tunnel boring machines.

Duct damage is common in mines. Mobile equipment and fly rock from blasting can punch holes in the duct. These factors can drastically reduce airflow. Care must be taken to minimize damage, and to quickly repair or replace damaged pieces.

Assessing Health and Safety

Few aspects of underground mining have as direct an impact on health, safety, and morale as ventilation. No component of ventilation design should be undertaken without a rigorous review of health and safety aspects, including the risk of

- Fire and explosion
- Dangerous and toxic substances
- Heat
- Ventilation equipment usage

For metal mines, fire is the most significant potential ventilation hazard. Fuel, heat, and oxygen are required for combustion; removing any of these components will prevent combustion. Fuel sources such as oil, diesel fuel, and blasting agents are kept in special areas designed to keep out ignition sources. Sprinkler or chemical suppression systems can be installed in these areas as well as in repair shops. Mobile equipment fires are a special concern for modern mining: vehicles should be fitted with a dry chemical fire suppression system, triggered either automatically or by the operator. Electric substation and conveyor fires can be very dangerous.

Engineers should anticipate various scenarios in ventilation circuits. What are the fire risks in any given area? If a fire broke out in any location, how would circuits respond? Would fire-induced natural drafts change airflow quantities and directions? How would fire be detected, how would miners be notified, how would they escape, and how would the fire be fought? MSHA requires that refuge chambers be constructed if miners cannot be hoisted to the surface within 1 h of notification. The ventilation staff must work closely with the safety department and mine management in preplanning how to respond to different emergencies.

Spontaneous combustion is a problem for both metal and coal mines. Fortunately, timber is now seldom used for ground support, although many older mines have worked-out areas that contain timber. Coal, being combustible, can be particularly troublesome. Circuits must be designed so that spontaneous combustion fires will not contaminate active workings.

For coal mines, methane and coal dust explosions pose the greatest risk. Equipment must be rated "permissible," or non-sparking. Airways should be coated with rock dust to prevent a methane ignition from propagating. Methane is explosive in air from 5 to 15% concentration. Whenever methane reaches 0.25%, MSHA requires that changes be made to improve ventilation. At 0.5%, further steps must be taken, and no other work is permitted until the concentration drops. At 1%, all personnel except those working on ventilation must be evacuated from the affected area.

Ventilation is the first line of defense against toxic or asphyxiating gases. These can be generated by blasting (CO, CO_2, NH_3, and NO_x) and by diesel engines (CO, NO_x, SO_2, various hydrocarbon compounds, and soot). The rock itself can release CO_2 and H_2S.

The relationship between heat stress and accident frequency has been clearly established in South African mines (Stewart 1982). Work area temperatures should be kept under 85°F wb, especially where heavy physical work is performed.

Ventilation and air-conditioning equipment may also pose health and safety risks. All fan inlets require screens. Fans should be equipped with vibration sensors that can deactivate the fan if necessary. Silencers may be needed if personnel work nearby. Refrigeration rooms must be well ventilated in case of a sudden refrigerant release. Duct, pipe insulation, and other substances such as foam for

seals should be approved by MSHA in accordance with 30 CFR Part 7. Electrical systems must meet rigorous MSHA codes.

REFERENCES

Abbeon Cal, Inc. 2001. *Pocket ref*, 2nd ed. T.J. Glover, ed. Sequoia Publishing, Littleton, CO.

Atkinson, J.J. 1854. *On the theory of the ventilation of mines*. North of England Institute of Mining Engineering.

Bluhm, S.J. 1981. Heat transfer characteristics of direct-contact crossflow heat exchangers. Master's thesis presented to the University of the Witwatersrand. Republic of South Africa.

Bossard, F. 1982. Sources of heat entering mine air. *Mine Ventilation Design Practices*, 1st ed. Butte, MT.

Burrows, J. 1982. Refrigeration—Theory and operation. *Environmental engineering in South African mines*, pp. 631-637. The Mine Ventilation Society of South Africa.

Duckworth, I. and P. Mousset-Jones. 1993. A detailed heat balance study of a deep level mechanized cut-and-fill stope. *Proceedings of the 6th U.S. Mine Ventilation Symposium*, pp. 441-447. SME, Littleton, CO.

Fenton, J.L. 1972. Survey of underground mine heat sources. Master's thesis, Montana College of Mineral Science and Technology, Butte.

Goch, D.C. and H.S. Patterson. 1940. The heat flow into tunnels. *Journal of the Chemical, Metallurgical and Mining Society of South Africa* 41.

Hall, A.E., D.M. Mchaina, and S.G. Hardcastle. 1989. The use of controlled recirculation to reduce heating costs in Canada. *Proceedings of the 4th International Mine Ventilation Congress*. Australasia Institute of Mining and Metallurgy, Melbourne, Australia.

Hall, C.J. 1981. *Mine ventilation engineering*. Society of Mining Engineers and Lucas-Guinn Co., Hoboken, NY.

Hartman, H.L., J.M. Mutmansky, R.V. Ramami, and Y.J. Wang. 1997. *Mine ventilation and air conditioning*, 3rd ed. John Wiley & Sons, New York.

Howes, M.J., and C.A. Nixon. 1997. Development of procedure for safe working in hot conditions. *Proceedings of the 6th International Mine Ventilation Congress*, pp. 191-197. SME, Littleton, CO.

Kennedy, W.R. 1996. *Practical mine ventilation*. Intertec Publishing, Chicago.

Marks, J.R. 1988. Computer-aided design of large underground direct-contact heat exchangers. Master's thesis, University of Idaho.

Marks, J.R. and L.M. Shaffner. 1993. An empirical analysis of ventilation requirements for deep mechanized stoping at the Homestake Gold Mine. *Proceedings of the 6th U.S. Mine Ventilation Symposium*, Salt Lake City, pp. 381-385. SME, Littleton, CO.

McPherson, M.J. 1993. *Subsurface ventilation and environmental engineering*. Chapman & Hall, NY.

Patterson, A.M. ed. 1992. *The mine ventilation practitioner's data book*. The Mine Ventilation Society of South Africa.

Sheer, T.J., M.D. Butterworth, and R. Ramsden. 2001. Ice as a coolant for deep mines. *Proceedings of the 7th International Mine Ventilation Congress*. Research and Development Center for Electrical Engineering and Automation in Mining, Kracow.

Stachulak, J. 1989. Ventilation strategy and unique air conditioning at INCO Limited. *Proceedings of the 4th U.S. Mine Ventilation Symposium*, Berkeley, CA. Society for Mining Metallurgy and Exploration, Littleton, CO.

Stewart, J.M. 1982. Practical aspects of human heat stress. In *Environmental engineering in South African mines*, p. 574. Burrows, Hemp, Holding and Stroh, eds. The Mine Ventilation Society of South Africa.

Tien, J.C. 1999. Controlled recirculation. In *Practical mine ventilation engineering*, pp. 370-392. Intertec Publishing, Chicago.

van der Walt, J. and C.S.J. van Rensburg. 1988. The options for cooling deep mines. *Proceedings of the 4th International Mine Ventilation Congress*, p. 401. Australasian Institute of Mining and Metallurgy, Melbourne, Australia.

Whillier, A. 1977. Predicting the performance of forced-draught cooling towers. *Journal of the Mine Ventilation Society of South Africa* 30 (1):2-25.

Whillier and Thorpe. 1982. Sources of heat in mines. In *Environmental engineering in South African mines*, Burrows, Hemp, Holding and Stroh, eds. Mine Ventilation Society of South Africa.

INDUSTRIAL DRYING

DRYING removes water and other liquids from gases, liquids, and solids. The term is most commonly used, however, to describe removing water or solvent from solids by thermal means. **Dehumidification** refers to the drying of a gas, usually by condensation or by absorption with a drying agent (see Chapter 32 of the 2009 *ASHRAE Handbook—Fundamentals*). **Distillation**, particularly **fractional distillation**, is used to dry liquids.

It is cost-effective to separate as much water as possible from a solid using mechanical methods *before* drying using thermal methods. Mechanical methods such as filtration, screening, pressing, centrifuging, or settling require less power and less capital outlay per unit mass of water removed.

This chapter describes industrial drying systems and their advantages, disadvantages, relative energy consumption, and applications.

Special Warning: Certain industrial spaces may contain flammable, combustible, and/or toxic concentrations of vapors or dusts under either normal or abnormal conditions. In spaces such as these, there are life-safety issues that this chapter may not completely address. Special precautions must be taken in accordance with requirements of recognized authorities such as the National Fire Protection Association (NFPA), the Occupational Safety and Health Administration (OSHA), and the American National Standards Institute (ANSI). In all situations, engineers, designers, and installers who encounter conflicting codes and standards must defer to the code or standard that best addresses and safeguards life safety.

MECHANISM OF DRYING

When a solid dries, two processes occur simultaneously: (1) the transfer of heat to evaporate the liquid and (2) the transfer of mass as vapor and internal liquid. Factors governing the rate of each process determine the drying rate.

The principal objective in commercial drying is to supply the required heat efficiently. Heat transfer can occur by convection, conduction, radiation, or a combination of these. Industrial dryers differ in their methods of transferring heat to the solid. In general, heat must flow first to the outer surface of the solid and then into the interior. An exception is drying with high-frequency electrical currents, where heat is generated within the solid, producing a higher temperature at the interior than at the surface and causing heat to flow from inside the solid to the outer surfaces.

APPLYING HYGROMETRY TO DRYING

In many applications, recirculating the drying medium improves thermal efficiency. The optimum proportion of recycled air balances the lower heat loss associated with more recirculation against the higher drying rate associated with less recirculation.

Because the humidity of drying air is affected by the recycle ratio, the air humidity throughout the dryer must be analyzed to determine whether the predicted moisture pickup of the air is physically attainable. The maximum ability of air to absorb moisture corresponds to the difference between saturation moisture content at wet-bulb (or adiabatic cooling) temperature and moisture content at supply air dew point. The actual moisture pickup of air is determined by heat and mass transfer rates and is always less than the maximum attainable.

ASHRAE psychrometric charts for normal and high temperatures (No. 1 and No. 3) can be used for most drying calculations. The process does not exactly follow the adiabatic cooling lines because some heat is transferred to the material by direct radiation or by conduction from the metal tray or conveyor.

Example 1. A dryer has a capacity of 90.5 lb of bone-dry gelatin per hour. Initial moisture content is 228% bone-dry basis, and final moisture content is 32% bone-dry basis. For optimum drying, supply air is at 120°F db and 85°F wb in sufficient quantity that the exhaust air is 100°F db and 84.5°F wb. Makeup air is available at 80°F db and 65°F wb.

Find (1) the required amount of makeup and exhaust air and (2) the percentage of recirculated air.

Solution: In this example, the humidity in each of the three airstreams is fixed; hence, the recycle ratio is also determined. Refer to ASHRAE Psychrometric Chart No. 1 to obtain the humidity ratio of makeup air and exhaust air. To maintain a steady-state condition in the dryer, water evaporated from the material must be carried away by exhaust air. Therefore, the pickup (the difference in humidity ratio between exhaust air and makeup air) is equal to the rate at which water is evaporated from the material divided by the weight of dry air exhausted per hour.

Step 1. From ASHRAE Psychrometric Chart No. 1, the humidity ratios are as follows:

	Dry bulb, °F	Wet bulb, °F	Humidity ratio, lb/lb dry air
Supply air	120	85	0.018
Exhaust air	100	84.3	0.022
Makeup air	80	65.2	0.010

Moisture pickup is 0.022 − 0.010 = 0.012 lb/lb dry air. The rate of evaporation in the dryer is

$$90.5 (228 - 32)/100 = 177 \text{ lb/h}$$

The dry air required to remove the evaporated water is 177/0.012 = 14,750 lb/h.

Step 2. Assume x = percentage of recirculated air and (100 − x) = percentage of makeup air. Then

Humidity ratio of supply air =

 (Humidity ratio of exhaust and recirculated air)(x/100)

 + (Humidity ratio of makeup air)(100 − x)/100

Hence,

$$0.018 = 0.022(x/100) + 0.010(100 - x)/100$$

$$x = 66.7\% \text{ recirculated air}$$

$$100 - x = 33.3\% \text{ makeup air}$$

The preparation of this chapter is assigned to TC 9.2, Industrial Air Conditioning.

DETERMINING DRYING TIME

The following three methods of finding drying time are listed in order of preference:

- Conduct tests in a laboratory dryer simulating conditions for the commercial machine, or obtain performance data using the commercial machine.
- If the specific material is not available, obtain drying data on similar material by either of the above methods. This is subject to the investigator's experience and judgment.
- Estimate drying time from theoretical equations (see the Bibliography). Care should be taken in using the approximate values obtained by this method.

When designing commercial equipment, tests are conducted in a laboratory dryer that simulates commercial operating conditions. Samples used in the laboratory tests should be identical to the material found in the commercial operation. Results from several tested samples should be compared for consistency. Otherwise, test results may not reflect the drying characteristics of the commercial material accurately.

When laboratory testing is impractical, commercial drying data can be based on the equipment manufacturer's experience.

Commercial Drying Time

When selecting a commercial dryer, the estimated drying time determines what size machine is needed for a given capacity. If the drying time has been derived from laboratory tests, the following should be considered:

- In a laboratory dryer, considerable drying may result from radiation and heat conduction. In a commercial dryer, these factors are usually negligible.
- In a commercial dryer, humidity may be higher than in a laboratory dryer. In drying operations with controlled humidity, this factor can be eliminated by duplicating the commercial humidity condition in the laboratory dryer.
- Operating conditions are not as uniform in a commercial dryer as in a laboratory dryer.
- Because of the small sample used, the test material may not be representative of the commercial material.

Thus, the designer must use experience and judgment to modify the test drying time to suit the commercial conditions.

Dryer Calculations

To estimate preliminary cost for a commercial dryer, the circulating airflow rate, makeup and exhaust airflow rate, and heat balance must be determined.

Circulating Air. The required circulating or supply airflow rate is established by the optimum air velocity relative to the material. This can be obtained from laboratory tests or previous experience, keeping in mind that the air also has an optimum moisture pickup. (See the section on Applying Hygrometry to Drying.)

Makeup and Exhaust Air. The makeup and exhaust airflow rate required for steady-state conditions within the dryer is also discussed in the section on Applying Hygrometry to Drying. In a **continuously operating dryer**, the relationship between moisture content of the material and quantity of makeup air is given by

$$G_T(W_2 - W_1) = M(w_1 - w_2) \qquad (1)$$

where

G_T = dry air supplied as makeup air to the dryer, lb/h
M = stock dried in a continuous dryer, lb/h
W_1 = humidity ratio of entering air, lb water vapor per lb dry air
W_2 = humidity ratio of leaving air, lb water vapor per lb dry air (in a continuously operating dryer, W_2 is constant; in a batch dryer, W_2 varies during part of the cycle)
w_1 = dry basis moisture content of entering material, lb of water per lb

w_2 = dry basis moisture content of leaving material, lb of water per lb

In **batch dryers**, the drying operation is given as

$$G_T(W_2 - W_1) = (M_1) \frac{dw}{d\theta} \qquad (2)$$

where

M_1 = mass of material charged in a discontinuous dryer, lb per batch
$dw/d\theta$ = instantaneous time rate of evaporation corresponding to w

The makeup air quantity is constant and is based on the average evaporation rate. Equation (2) then becomes identical to Equation (1), where $M = M_1/\theta$. Under this condition, humidity in the batch dryer decreases during the drying cycle, whereas in the continuous dryer, humidity is constant with constant load.

Heat Balance. To estimate the fuel requirements of a dryer, a heat balance consisting of the following is needed:

- Radiation and convection losses from the dryer
- Heating of the commercial dry material to the leaving temperature (usually estimated)
- Vaporization of the water being removed from the material (usually considered to take place at the wet-bulb temperature)
- Heating of the vapor from the wet-bulb temperature in the dryer to the exhaust temperature
- Heating of the total water in the material from the entering temperature to the wet-bulb temperature in the dryer
- Heating of the makeup air from its initial temperature to the exhaust temperature

The energy absorbed must be supplied by the fuel. The selection and design of the heating equipment is an essential part of the overall design of the dryer.

Example 2. Magnesium hydroxide is dried from 82% to 4% moisture content (wet basis) in a continuous conveyor dryer with a fin-drum feed (see Figure 7). The desired production rate is 3000 lb/h. The optimum circulating air temperature for drying is 160°F, which is not limited by the existing steam pressure of the dryer.

Step 1. Laboratory tests indicate the following:

Specific heats
   air (c_a)   = 0.24 Btu/lb·°F
   material (c_m)   = 0.3 Btu/lb·°F
   water (c_w)   = 1.0 Btu/lb·°F
   water vapor (c_v)   = 0.45 Btu/lb·°F

Temperature of material entering dryer = 60°F

Temperature of makeup air
   dry bulb   = 70°F
   wet bulb   = 60°F

Temperature of circulating air
   dry bulb   = 160°F
   wet bulb   = 100°F

Air velocity through drying bed   = 250 fpm

Dryer bed loading   = 6.82 lb/ft²

Test drying time   = 25 min

Step 2. Previous experience indicates that the commercial drying time is 70% greater than the time obtained in the laboratory test. Thus, the commercial drying time is estimated to be $1.7 \times 25 = 42.5$ min.

Step 3. The holding capacity of the dryer bed is

   3000(42.5/60) = 2125 lb at 4% (wet basis)

The required conveyor area is 2125/6.82 = 312 ft². Assuming the conveyor is 8 ft wide, the length of the drying zone is 312/8 = 39 ft.

Step 4. The amount of water in the material entering the dryer is

   3000[82/(100 + 4)] = 2370 lb/h

The amount of water in the material leaving is

   3000[4/(100 + 4)] = 115 lb/h

Thus, the moisture removal rate is 2370 − 115 = 2255 lb/h.

Step 5. The air circulates perpendicular to the perforated plate conveyor, so the air volume is the face velocity times the conveyor area:

$$\text{Air volume} = 250 \times 312 = 78,000 \text{ cfm}$$

ASHRAE Psychrometric Charts 1 and 3 show these air properties:

Supply air (160°F db, 100°F wb)

| Humidity ratio | = 0.0285 lb per lb of dry air |
| Specific volume | = 16.33 ft³ per lb of dry air |

Makeup air (70°F db, 60°F wb)

| Humidity ratio W_1 | = 0.0086 lb per lb of dry air |

The mass flow rate of dry air is

$$(78,000 \times 60)/16.33 = 286,500 \text{ lb/h}$$

Step 6. The amount of moisture pickup is

$$2255/286,500 = 0.0079 \text{ lb per lb of dry air}$$

The humidity ratio of the exhaust air is

$$W_2 = 0.0285 + 0.0079 = 0.0364 \text{ lb per lb of dry air}$$

Substitute in Equation (1) and calculate G_T as follows:

$$G_T(0.0364 - 0.0086) = (3000/1.04)(82 - 4)/100$$

$$G_T = 81,000 \text{ lb dry air per hour}$$

Therefore,

| Makeup air | = $100 \times 81,000/286,500 = 28.3\%$ |
| Recirculated air | = 71.7% |

Step 7. Heat Balance

Sensible heat of material	= $M(t_{m2} - t_{m1})c_m$
	= (3000/1.04)(100 − 60)0.3
	= 34,600 Btu/h
Sensible heat of water	= $M_{w1}(t_w - t_{m1})c_w$
	= 2370(100 − 60)1.0
	= 94,800 Btu/h
Latent heat of evaporation	= $M(w_1 - w_2)H$
	= 2255 lb/h × 1037 Btu/lb
	= 2,338,400 Btu/h
Sensible heat of vapor	= $M(t_2 - t_w)c_v$
	= 2255(160 − 100)0.45
	= 60,900 Btu/h
Required heat for material	= 2,528,700 Btu/h

The temperature drop $(t_2 - t_3)$ through the bed is

$$\frac{\text{Required heat}}{\text{Supplied air, (lb/h)} \times c_a} = \frac{2,528,700}{286,500 \times 0.24} = 37°F$$

Therefore, the exhaust air temperature is 160 − 37 = 123°F.

Required heat for makeup air	= $G_T(t_3 - t_1)c_a$
	= 81,000(123 − 70)0.24
	= 1,030,000 Btu/h

The total heat required for material and makeup air is

$$2,528,700 + 1,030,000 = 3,559,000 \text{ Btu/h}$$

Additional heat that must be provided to compensate for radiation and convection losses can be calculated from the known construction of the dryer surfaces.

DRYING SYSTEM SELECTION

A general procedure for selecting a drying system is as follows:

1. Survey of suitable dryers.
2. Preliminary cost estimates of various types.
   (a) Initial investment
   (b) Operating cost
3. Drying tests conducted in prototype or laboratory units, preferably using the most promising equipment available. Sometimes a pilot plant is justified.
4. Summary of tests evaluating quality of samples of the dried products.

Factors that can overshadow the operating or investment cost include the following:

- Product quality, which should not be sacrificed
- Dusting, solvent, or other product losses
- Space limitation
- Bulk density of the product, which can affect packaging cost

Friedman (1951) and Parker (1963) discuss additional aids to dryer selection.

TYPES OF DRYING SYSTEMS

Radiant Infrared Drying

Thermal radiation may be applied by infrared lamps, gas-heated incandescent refractories, steam-heated sources, and, most often, electrically heated surfaces. Infrared heats only near the surface of a material, so it is best used to dry thin sheets.

Using infrared heating to dry webs such as uncoated materials has been relatively unsuccessful because of process control problems. Thermal efficiency can be low; heat transfer depends on the emitter's characteristics and configuration, and on the properties of the material to be dried.

Radiant heating is used for drying ink and other coatings on paper, textile fabrics, paint films, and lacquers. Inks have been specifically formulated for curing with tuned or narrow wavelength infrared radiation.

Ultraviolet Radiation Drying

Ultraviolet (UV) drying uses electromagnetic radiation. Inks and other coatings based on monomers are cure-dried when exposed to UV radiation. This method has superior properties (Chatterjee and Ramaswamy 1975): the print resists scuff, scratch, acid, alkali, and some solvents. Printing can also be done at higher speeds without damage to the web.

Major barriers to wider acceptance of UV drying include the high capital installation cost and the increased cost of inks. The cost and frequency of replacing UV lamps are greater than for infrared ovens.

Overexposure to radiation and ozone, which is formed by UV radiation's effect on atmospheric oxygen, can cause severe sunburn and possibly blood and eye damage. Safety measures include fitting the lamp housings with screens, shutters, and exhausts.

Conduction Drying

Drying rolls or drums (Figure 1), flat surfaces, open kettles, and immersion heaters are examples of direct-contact drying. The heating surface must have close contact with the material, and agitation may increase uniform heating or prevent overheating.

Conduction drying is used to manufacture and dry paper products. It (1) does not provide a high drying rate, (2) does not furnish uniform heat and mass transfer conditions, (3) usually results in a

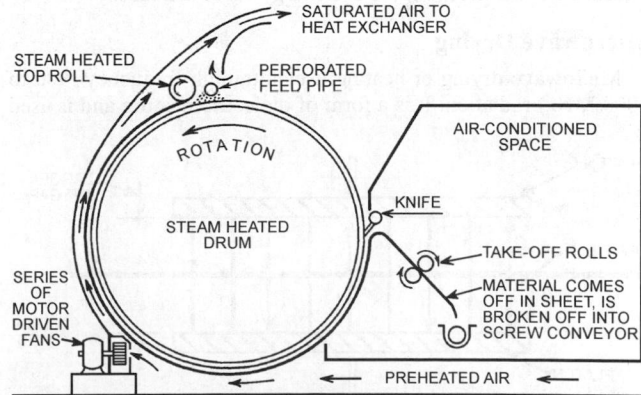

Fig. 1 Drum Dryer

poor moisture profile across the web, (4) lacks proper control, (5) is costly to operate and install, and (6) usually creates undesirable working conditions in areas surrounding the machine. Despite these disadvantages, replacing existing systems with other forms of drying is expensive. For example, Joas and Chance (1975) report that RF (dielectric) drying of paper requires approximately four times the capital cost, six times the operating (heat) cost, and five times the maintenance cost of steam cylinder conduction drying. However, augmenting conduction drying with dielectric drying sections offsets the high cost of RF drying and may produce savings and increased profits from greater production and higher final moisture content.

Further use of large conduction drying systems depends on reducing heat losses from the dryer, improving heat recovery, and incorporating other drying techniques to maintain quality.

Dielectric Drying

When wet material is placed in a strong, high-frequency (2 to 100 MHz) electrostatic field, heat is generated within the material. More heat is developed in the wetter areas than in the drier areas, resulting in automatic moisture profile correction. Water is evaporated without unduly heating the substrate. Therefore, in addition to its leveling properties, dielectric drying provides uniform heating throughout the web thickness.

Dielectric drying is controlled by varying field or frequency strength; varying field strength is easier and more effective. Response to this variation is quick, with neither time lag nor thermal lag in heating. The dielectric heater is a sensitive moisture meter.

Several electrode configurations are used. The platen type (Figure 2) is used for drying and baking foundry cores, heating plastic preforms, and drying glue lines in furniture. The rod or stray field types (Figure 3) are used for thin web materials such as paper and textile products. The double-rod types (over and under material) are used for thicker webs or flat stock, such as plywood.

Dielectric drying is popular in the textile industry. Because air is entrained between fibers, convection drying is slow and uneven. Because the yarn is usually transferred to large packages immediately after drying, however, even and correct moisture content can be obtained by dielectric drying. Knitting wool seems to benefit from internal steaming in hanks.

Warping caused by nonuniform drying is a serious problem for plywood and linerboard. Dielectric drying yields warp-free products.

Dielectric drying is not cost-effective for overall paper drying but has advantages when used at the dry end of a conventional steam drum dryer. It corrects moisture profile problems in the web without overdrying. This combination of conventional and dielectric drying is synergistic: the drying effect of the combination is greater than the sum of the two types of drying. This is more pronounced in thicker web materials, accounting for as much as a 16% line speed increase and a corresponding 2% energy input increase.

Microwave Drying

Microwave drying or heating uses ultrahigh-frequency (900 to 5000 MHz) radiation. It is a form of dielectric heating and is used for heating nonconductors. Because of its high frequency, microwave equipment is capable of generating extreme power densities.

Microwave drying is applied to thin materials in strip form by passing the strip through the gap of a split waveguide. Entry and exit shielding make continuous process applications difficult. Its many safety concerns make microwave drying more expensive than dielectric drying. Control is also difficult because microwave drying lacks the self-compensating properties of dielectrics.

Convection Drying (Direct Dryers)

Some convection drying occurs in almost all dryers. True convection dryers, however, use circulated hot air or other gases as the principal heat source. Each means of mechanically circulating air or gases has its advantages.

Rotary Dryers. These cylindrical drums cascade the material being dried through the airstream (Figure 4). The dryers are heated directly or indirectly, and air circulation is parallel or counterflow. A variation is the rotating-louver dryer, which introduces air beneath the flights to provide close contact.

Cabinet and Compartment Dryers. These batch dryers range from the heated loft (with only natural convection and usually poor and nonuniform drying) to self-contained units with forced draft and properly designed baffles. Several systems may be evacuated to dry delicate or hygroscopic materials at low temperatures. Material is usually spread in trays to increase the exposed surface. Figure 5 shows a dryer that can dry water-saturated products.

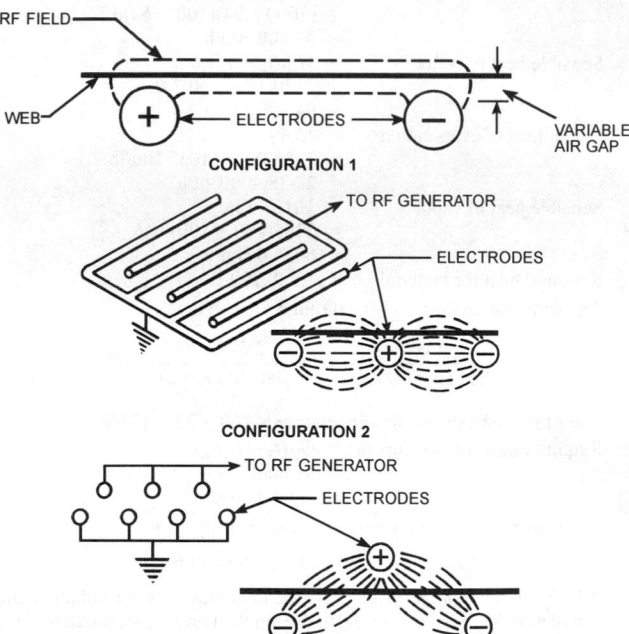

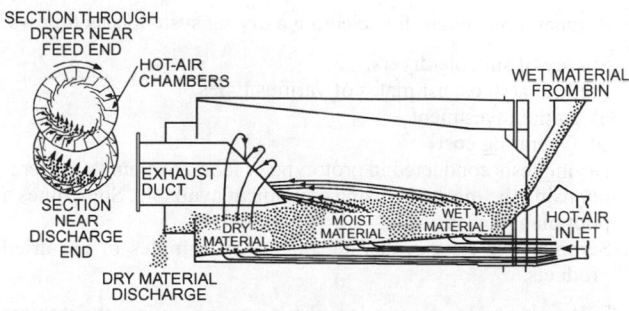

Fig. 3   Rod-Type Dielectric Dryers

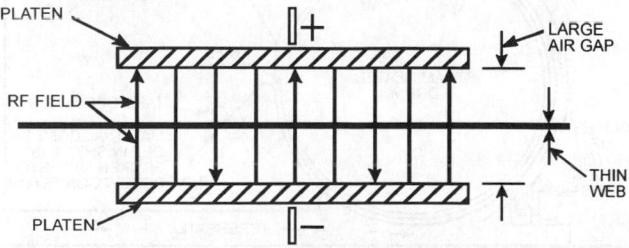

Fig. 2   Platen-Type Dielectric Dryer

Fig. 4   Cross Section and Longitudinal Section of Rotary

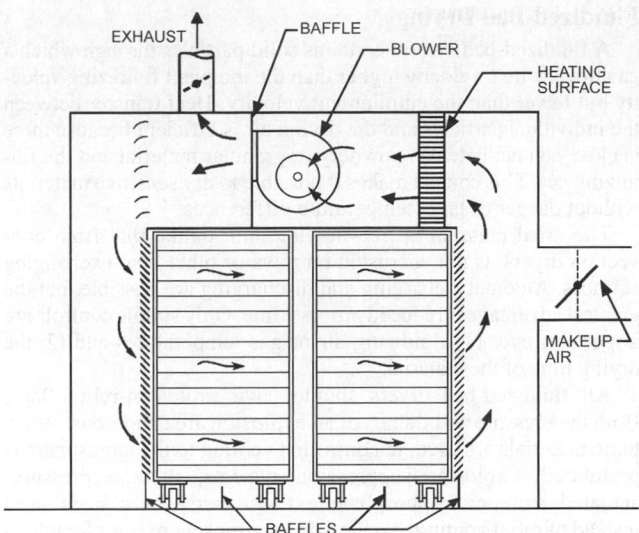

Fig. 5   Compartment Dryer Showing Trucks with Air Circulation

When designing dryers to process products saturated with solvents, special features must be included to prevent explosive gases from forming. Safe operation requires exhausting 100% of the air circulated during the initial drying period or any part of the drying cycle when solvent is evaporating at a high rate. At the end of the purge cycle, the air is recirculated and heat is gradually applied. To prevent explosions, laboratory dryers can be used to determine the amount of air circulated, cycle lengths, and rate that heat is applied for each product. In the drying cycle, dehumidified air, which is costly, should be recirculated as soon possible. The air *must not* be recirculated when cross-contamination of products is prohibited.

Dryers must have special safety features in case any part of the drying cycle fails. The following are some of the safety design features described in *Industrial Ovens and Driers* (FMEA 1990):

- Each compartment must have separate supply and exhaust fans and an explosion-relief panel.
- The exhaust fan blade tip speed should be 5000 fpm for forward-inclined blades, 6800 fpm for radial-tip, and 7500 fpm for backward-inclined. These speeds produce high static pressures at the fan, ensuring constant exhaust volumes under conditions such as negative pressures in the building or downdrafts in the exhaust stacks.
- Airflow failure switches in both the supply and exhaust ducts must shut off fans and the heating coil and must sound an alarm.
- A high-temperature limit controller in the supply duct must shut off the heat to the heating coil and must sound an alarm.
- An electric interlock on the dryer door must interrupt the drying cycle if the door is opened beyond a set point, such as that wide enough for a person to enter for product inspection.

Tunnel Dryers. Tunnel dryers are modified compartment dryers that operate continuously or semicontinuously. Fans circulate heated air or combustion gas is circulated by fans. The material is handled on trays or racks on trucks and moves through the dryer either intermittently or continuously. The airflow may be parallel, counterflow, or a combination obtained by center exhaust (Figure 6). Air may also flow across the tray surface, vertically through the bed, or in any combination of directions. By reheating or recirculating the air in the dryer, high saturation is reached before the air is exhausted, thus reducing the sensible heat loss.

The following problems with tunnel dryers have been experienced and should be considered in future designs:

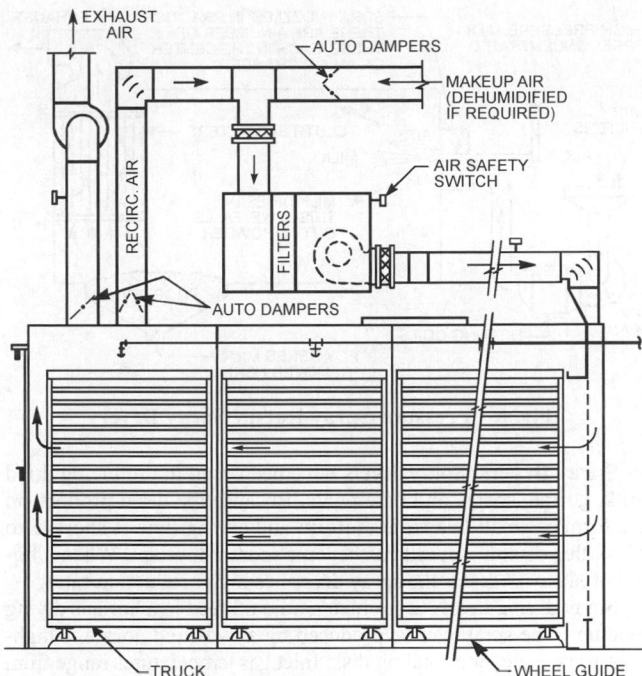

Fig. 6   Explosionproof Truck Dryer Showing Air Circulation and Safety Features

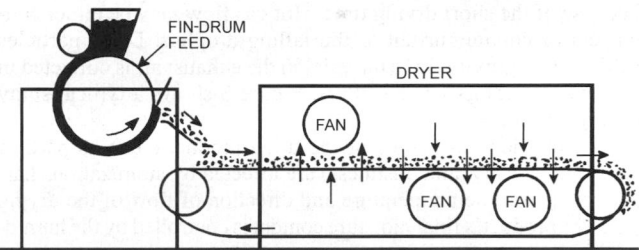

Fig. 7   Section of Blow-Through Continuous Dryer

- Operators may overload product trays to increase output, but this can overtax the system and increase drying time.
- Sometimes air from the drying tunnel is discharged into the production area, increasing the humidity. Air from the drying tunnel should be discharged to the drying system return or outdoors.
- Overloaded product trays add pressure drop, which decreases flow through the dryer. The control panel should indicate validated flow through the tunnel. High and low flow and high moisture levels should trigger alarms.
- Cycle times can be reduced by designing dryers for cross-flow rather than end-to-end flow.

A variation of the tunnel dryer is the strictly continuous dryer, which has one or more mesh belts that carry the product through it, as shown in Figure 7. Many combinations of temperature, humidity, air direction, and velocity are possible. Hot air leaks at the entrance and exit can be minimized by baffles or inclined ends, with the material entering and leaving from the bottom.

High-Velocity Dryers. High-velocity hoods or dryers have been used to supplement conventional cylinder dryers for drying paper. When used with conventional cylinder dryers, web instability and lack of process control result. Where internal diffusion is not the controlling factor in the drying rate, applications such as thin permeable webs offer more promise.

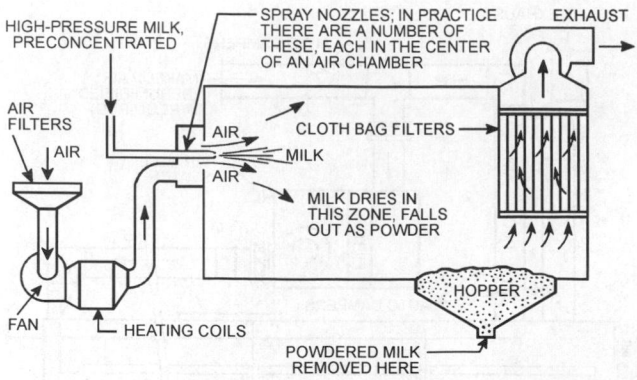

Fig. 8   Pressure-Spray Rotary Spray Dryer

Spray Dryers. Spray dryers have been used in producing dried milk, coffee, soaps, and detergents. Because the dried product (in the form of small beads) is uniform and drying time is short (5 to 15 s), this drying method has become more important. When a liquid or slurry is dried, the spray dryer has high production rates.

Spray drying involves atomizing a liquid feed in a hot-gas drying medium. The spray can be produced by a two-fluid nozzle, a high-pressure nozzle, or a rotating disk. Inlet gas temperatures range from 200 to 1400°F, with the high temperatures requiring special construction materials. Because thermal efficiency increases with the inlet gas temperature, high inlet temperatures are desirable. Even heat-sensitive products can be dried at higher temperatures because of the short drying time. Hot gas flow may be either concurrent or countercurrent to the falling droplets. Dried particles settle out by gravity. Fine material in the exhaust air is collected in cyclone separators or bag filters. Figure 8 shows a typical spray drying system.

The physical properties of the dried product (such as particle size, bulk density, and dustiness) are affected by atomization characteristics and the temperature and direction of flow of the drying gas. The product's final moisture content is controlled by the humidity and temperature of the exhaust gas stream.

Currently, pilot-plant or full-scale production operating data are required for design purposes. The drying chamber design is determined by the nozzle's spray characteristics and heat and mass transfer rates. There are empirical expressions that approximate mean particle diameter, drying time, chamber volume, and inlet and outlet gas temperatures.

Freeze Drying

Freeze drying has been applied to pharmaceuticals, serums, bacterial and viral cultures, vaccines, fruit juices, vegetables, coffee and tea extracts, seafoods, meats, and milk.

The material is frozen, then placed in a high-vacuum chamber connected to a low-temperature condenser or chemical desiccant. Heat is slowly applied to the frozen material by conduction or infrared radiation, allowing the volatile constituent, usually water, to sublime and condense or be absorbed by the desiccant. Most freeze-drying operations occur between 14 and −40°F under minimal pressure. Although this process is expensive and slow, it has advantages for heat-sensitive materials (see Chapter 29 of the 2010 *ASHRAE Handbook—Refrigeration*).

Vacuum Drying

Vacuum drying takes advantage of the decrease in the boiling point of water that occurs as the pressure is lowered. Vacuum drying of paper has been partially investigated. Serious complications arise if the paper breaks, and massive sections must be removed. Vacuum drying is used successfully for pulp drying, where lower speeds and higher weights make breakage relatively infrequent.

Fluidized-Bed Drying

A fluidized-bed system contains solid particles through which a gas flows with a velocity higher than the incipient fluidizing velocity but lower than the entrainment velocity. Heat transfer between the individual particles and the drying air is efficient because there is close contact between powdery or granular material and the fluidizing gas. This contact makes it possible to dry sensitive materials without danger of large temperature differences.

The dried material is free-flowing and, unlike that from convection dryers, is not encrusted on trays or other heat-exchanging surfaces. Automatic charging and discharging are possible, but the greatest advantage is reduced process time. Only simple controls are important: over (1) fluidizing air or gas temperatures and (2) the drying time of the material.

All fluidized-bed dryers should have explosion-relief flaps. Both the pressure and flames of an explosion are dangerous. When toxic materials are used, uncontrolled venting to the atmosphere is prohibited. Explosion suppression systems, such as pressure-actuated ammonium-phosphate extinguishers, have been used instead of relief venting. An inert dryer atmosphere is preferable to suppression systems because it prevents explosive mixtures from forming.

When organic and inflammable solvents are used in the fluidized-bed system, the closed system offers advantages other than explosion protection. A portion of the fluidizing gas is continuously run through a condenser, which strips the solvent vapors and greatly reduces air pollution problems, thus making solvent recovery convenient.

Materials dried in fluidized-bed installations include coal, limestone, cement rock, shales, foundry sand, phosphate rock, plastics, medicinal tablets, and foodstuffs. Leva (1959) and Othmer (1956) discuss the theory and methods of fluidization of solids. Clark (1967) and Vanecek et al. (1966) developed design equations and cost estimates.

Agitated-Bed Drying

Uniform drying is ensured by periodically or continually agitating a bed of preformed solids with a vibrating tray, a conveyor, or a vibrating mechanically operated rake, or, in some cases, by partial fluidization of the bed on a perforated tray or conveyor through which recycled drying air is directed. Drying and toasting cereals is an important application.

Drying in Superheated Vapor Atmospheres

When drying solids with air or another gas, the vaporized solvent (water or organic liquid) must diffuse through a stagnant gas film to reach the bulk gas stream. Because this film is the main resistance to mass transfer, the drying rate depends on the solvent vapor diffusion rate. If the gas is replaced by solvent vapor, resistance to mass transfer in the vapor phase is eliminated, and the drying rate depends only on the heat transfer rate. Drying rates in solvent vapor, such as superheated steam, are greater than those in air for equal temperatures and mass flow of the drying media.

This method also has higher thermal efficiency, easier solvent recovery, and a lower tendency to overdry, and it eliminates oxidation or other chemical reactions that occur when air is present. In drying cloth, superheated steam reduces the migration tendency of resins and dyes. Superheated vapor drying cannot be applied to heat-sensitive materials because of the high temperatures.

Commercial drying equipment with recycled solvent vapor as the drying medium is available. Installations have been built to dry textile sheeting and organic chemicals.

Flash Drying

Finely divided solid particles that are dispersed in a hot gas stream can be dried by flash drying, which is rapid and uniform. Commercial applications include drying pigments, synthetic resins,

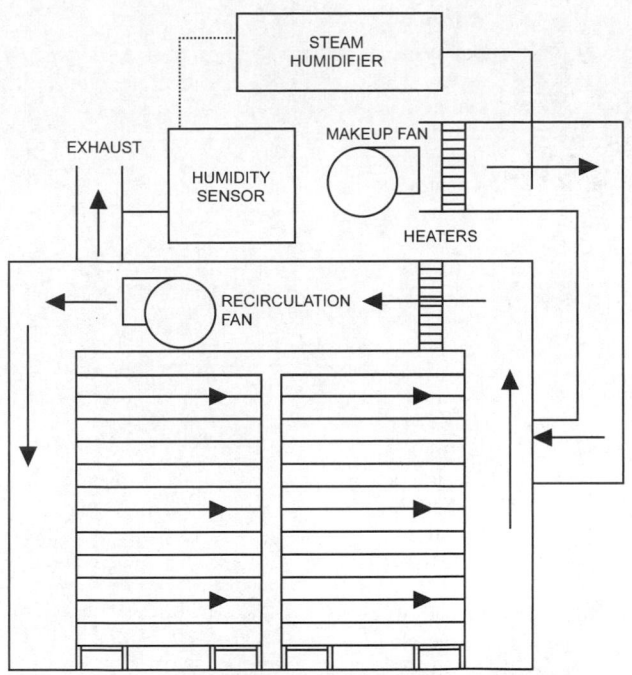

Fig. 9 Humidified Cross-Flow Tray Dryer

food products, hydrated compounds, gypsum, clays, and wood pulp.

Constant-Moisture Solvent Drying

In some cases it is desirable to dry organic solvents from a substance without changing moisture content. This is particularly true in drying pharmaceutical products, which are commonly bound with solvents such as isopropyl alcohol (IPA) or acetone. Loss of moisture content can affect the stability of some pharmaceutical compounds, which therefore must be liberated of bound solvents without changing the relative humidity. Most pharmaceutical facilities producing oral solid dosage (OSD) presentations manufacture in spaces that range from 30 to 45% rh. To maintain this level of humidity at the temperatures needed to liberate the organic solvents at the required rate, an oven must be fitted with an adequately sized humidifier. Figure 9 shows the compartment dryer from Figure 5 adapted for constant-moisture drying.

It is vital to size the preheat coil to the humidifier to allow enough "room" to provide the humidification level required. The amount of preheat is usually enough to provide makeup air to the oven at 90% rh. This allows the oven heaters to control drying temperature at the more common oven humidity levels of 30 to 45% rh. If electric heat is used, the ability to constantly vary the capacity of the heaters must be provided; silicon-controlled rectifiers (SCRs) are commonly used.

Additional insulation on the outside of the oven and the ductwork from the humidifier section onward should also be considered. Care must be taken to ensure that unwanted condensation does not form inside the oven. Dripping moisture can sometimes ruin the product, causing a substantial economic loss. If the product is pharmaceutical, then the humidifier makeup water source should be deionized or produced to comply with U.S. Pharmacopoeia 34—National Formulary—29 (effective August 2010).

REFERENCES

Chatterjee, P.C. and R. Ramaswamy. 1975. Ultraviolet radiation drying of inks. *British Ink Maker* 17(2):76.

Clark, W.E. 1967. Fluid bed drying. *Chemical Engineering* 74(March): 177.

FMEA. 1990. *Industrial ovens and driers*. Data Sheet No. 6-9. Factory Mutual Engineering Association, Norcross, GA.

Friedman, S.J. 1951. Steps in the selection of drying equipment. *Heating and Ventilating* (February):95.

Joas, J.G. and J.L. Chance. 1975. Moisture leveling with dielectric, air impingement and steam drying—A comparison. *Tappi* 58(3):112.

Leva, M. 1959. *Fluidization*. McGraw-Hill, New York.

Othmer, D.F. 1956. *Fluidization*. Reinhold Publishing, New York.

Parker, N.H. 1963. Aids to drier selection. *Chemical Engineering* 70(June 24):115.

U.S. Pharmacopoeia 34—National Formulary—29. 2010.

Vanecek, V., M. Markvart, and R. Drbohlav. 1966. *Fluidized bed drying*. Chemical Rubber Company, Cleveland, OH.

CHAPTER 31

VENTILATION OF THE INDUSTRIAL ENVIRONMENT

INDUSTRIAL environments require ventilation to reduce exposure to excess heat and contaminants that are generated in the workplace; in some situations, cooling may also be required. Ventilation is primarily used to control excess heat, odors, and hazardous particulate and chemical contaminants. These could affect workers' health and safety or, in some cases, become combustible or flammable when allowed to accumulate above their minimum explosible concentration (MEC) or lower flammable limit (LFL) [also called the lower explosive limit (LEL)] (Cashdollar 2009). Excess heat and contaminants can best be controlled by using local exhaust systems whenever possible. Local exhaust systems capture heated air and contaminants at their source and may require lower airflows than general (dilution) ventilation. See Chapter 32 for more information on the selection and design of industrial local exhaust systems.

General ventilation can be provided by mechanical (fan) systems, by natural draft, or by a combination of the two. Combination systems could include mechanically driven (fan-driven) supply air with air pressure relief through louvers or other types of vents, and mechanical exhaust with air replacement inlet louvers and/or doors.

Mechanical (fan-driven) supply systems provide the best control and the most comfortable and uniform environment, especially when there are extremes in local climatic conditions. The systems typically consist of an inlet section, filtration section, heating and/or cooling equipment, fans, ductwork, and air diffusers for distributing air in the workplace. When toxic gases or vapors are not present and there are no aerosol contaminants associated with adverse health effects, air cleaned in the general exhaust system or in packaged air filtration units can be recirculated via a return duct. When applied appropriately, air recirculation can be a major contributor to a sustainable industrial ventilation design and may reduce heating and cooling costs.

In addition, regardless of the method selected, any positive ventilation into an industrial space should be from a source that will be essentially free of any contaminants under both normal and abnormal conditions in the surrounding atmosphere. In many cases, this may require a sealed intake stack or ductwork, as opposed to a perimeter wall hood or other air intake device, wherein the source of intake should be from a point well above or beyond the veil of the hazardous space that may surround a ventilated space.

A general exhaust system, which removes air contaminated by gases, vapors, or particulates not captured by local exhausts, usually consists of one or more fans, plus inlets, ductwork, and air cleaners or filters. After air passes through the filters, it is either discharged outside or partially recirculated to the building workplace. The air filter system's cleaning efficiency should conform to environmental regulations and depends on factors such as building location, background contaminant concentrations in the atmosphere, type and toxicity of contaminants, and height and velocity of the building exhaust discharge.

Many industrial ventilation systems must handle simultaneous exposures to heat and hazardous substances. In these cases, the required ventilation can be provided by a combination of local exhaust, general ventilation air supply, and general exhaust systems. The ventilation engineer must carefully analyze supply and exhaust air requirements to determine the worst case. For example, air supply makeup for hood exhaust may be insufficient to control heat exposure. It is also important to consider seasonal climatic effects on ventilation system performance, especially for natural ventilation systems.

Most importantly, if the hazardous substances are ignitable gases or dusts, all electrical components of the ventilation system should be rated for the proper electrical classification in the absence of any ventilation, regardless of their locations in the ventilation system.

In specifying acceptable chemical contaminant and heat exposure levels, the industrial hygienist or industrial hygiene engineer must consult the appropriate occupational exposure limits that apply as well as any governing standards and guidelines. The legislated limits for the maximum airborne concentration of chemical substances to which a worker may be exposed are listed as (1) maximum average exposures to which a worker may be exposed over a given work day (generally assumes an 8 to 10 h work day and a traditional 40 h work week); (2) short-term exposure limits, which are the maximum average airborne concentration to which a worker may be exposed over any 15 min period; and (3) ceiling limits, which are the maximum airborne concentration to which a worker may be exposed at any time. However, occupational exposure limits for cold, heat, and contaminants are not lines of demarcation between safe and unsafe exposures. Rather, they represent conditions to which it is believed nearly all workers may be exposed day after day without adverse and/or long-term effects. Because a small percentage of workers may be affected by occupational exposure below the regulated limits, it is prudent to design for exposure levels well below the limits.

In the case of exposure to hazardous chemicals, the number of contaminant sources, their generation rates, and the effectiveness of exhaust hoods may not be known. Consequently, the ventilation engineer must rely on industrial hygiene engineering practices when designing toxic and/or hazardous chemical controls. Close cooperation among the industrial hygienist, process engineer, and ventilation engineer is required.

In the case of exposure to flammable or ignitable chemicals, the specific gravity of the contaminant source(s), their concentration, and the rating of all electrical devices within the space, along with any source or point of excessive heat, must be carefully considered to prevent possible loss of life or severe injury. As with all hazardous chemicals, cooperation of knowledgeable experts, including electrical engineers, is required.

This chapter describes principles of ventilation practice and includes other information on industrial hygiene in the industrial environment. Publications from the British Occupational Hygiene Society (1987), National Safety Council (2002), U.S. National

---

The preparation of this chapter is assigned to TC 5.8, Industrial Ventilation Systems.

Institute for Occupational Safety and Health (NIOSH 1986), and the U.S. Department of Health and Human Services (DHHS 1986) provide further information on industrial hygiene principles and their application.

VENTILATION DESIGN PRINCIPLES

Special Warning: Certain industrial spaces may contain flammable, combustible, and/or toxic concentrations of vapors or dusts under either normal or abnormal conditions. In spaces such as these, there are life safety issues that this chapter may not completely address. Special precautions must be taken in accordance with requirements of recognized authorities such as the National Fire Protection Association (NFPA), the Occupational Safety and Health Administration (OSHA), and the American National Standards Institute (ANSI). In all situations, engineers, designers, and installers who encounter conflicting codes and standards must defer to the code or standard that best addresses and safeguards life safety.

General Ventilation

General ventilation supplies and/or exhausts air to provide heat relief, dilute contaminants to an acceptable level, and replace exhaust air. Ventilation can be provided by natural or mechanical supply and/or exhaust systems. Industrial areas must comply with ASHRAE Standard 62.1-2010 and other standards as required [e.g., by NFPA]. Outdoor air is unacceptable for ventilation if it is known to contain any contaminant at a concentration above that given in ASHRAE Standard 62.1. If air is thought to contain any contaminant not listed in the standard, guidance on acceptable exposure levels should be obtained from relevant federal, state, provincial, or local jurisdictions. In addition to their role in controlling industrial contaminants, general ventilation rates must be sufficient to dilute the carbon dioxide produced by occupants.

For complex industrial ventilation problems, experimental scale models and computational fluid dynamics (CFD) models are often used in addition to field testing.

Makeup Air

When large volumes of air are exhausted to provide acceptable comfort and safety for personnel and acceptable conditions for process operations, this air must be replaced, either through intentional design strategy or through paths of least resistance. A safe and effective ventilation design should be strategic about the mechanism, locations, and physical parameters by which the makeup air enters the occupied space. Makeup air, consistently provided by good air distribution, allows more effective cooling in the summer and more efficient and effective heating in the winter. When makeup air design is not incorporated into the ventilation design scheme, it may lead to inefficient operation of local exhaust systems and/or combustion equipment and cross-drafts that affect occupant comfort and environmental control settings. Relying on windows or other air inlets that cannot function in year-round weather conditions is discouraged. Some factors to consider in makeup air design include the following:

- Makeup air must be sufficient to replace air being exhausted or consumed by combustion processes, local and general exhaust systems (see Chapter 32), or process equipment. (Large air compressors can consume a large amount of air and should be considered if air is drawn from within the building.)
- Makeup air systems should be designed to eliminate uncomfortable crossdrafts by properly arranging supply air outlets, and to prevent infiltration (through doors, windows, and similar openings) that may make hoods unsafe or ineffective, defeat environmental control, bring in or stir up dust, or adversely affect processes by cooling or disturbances. The design engineer needs to consider side drafts and other sources of air movement close to

the capture area of a local exhaust hood. Caplan and Knutson (1977, 1978) found that air movement in front of laboratory hoods can cause contaminants to escape from the hood and into the operator's breathing zone. In industrial applications, it is common to see large fans blowing air on workers in front of the hood. This can render the local exhaust hood ineffective to the point that no protection is provided for the worker.

- Makeup air should be obtained from a clean source with no more than trace amounts of any airborne contaminants or hazardous, ignitable substances. Supply air can be filtered, but infiltration air cannot. For transfer air use, see ASHRAE Standard 62.1.
- Makeup air for spaces contaminated by toxic, ignitable, or combustible chemicals may have to be acquired through carefully sealed ductwork from an area know to be free of contamination and be supplied at sufficient rates, pressures, and mixing efficiencies to (1) remove all contamination, and (2) prevent infiltration of similar contaminants from surrounding areas or adjacent spaces.
- Makeup air should be used to control building pressure and airflow from space to space to (1) avoid positive or negative pressures that make it difficult or unsafe to open doors, (2) minimize drafts, and (3) prevent infiltration.
- Makeup air should be used to reduce contaminant concentration, to control temperature and humidity, and minimize undesirable air movement.
- Makeup air systems should be designed to recover heat and conserve energy (see the section on Energy Conservation, Recovery, and Sustainability).

ACGIH (2010) provides information on adverse conditions that may result from specific negative pressure levels in buildings.

GENERAL COMFORT AND DILUTION VENTILATION

Effective air diffusion in ventilated rooms and the proper quantity of conditioned air are essential for creating an acceptable working environment, removing contaminants, and reducing installation and operating costs of a ventilation system. Ventilation systems must supply air at the proper velocity and temperature, with resulting contaminant concentrations within permissible occupational exposure limits (OELs). For the industrial environment, the most common objective is to provide tolerable (acceptable) working conditions rather than comfort (optimal) conditions.

General ventilation system design is based on the assumption that local exhaust ventilation, radiation shielding, and equipment insulation and encapsulation have been selected to minimize both heat load and contamination in the workplace (see the section on Heat Control). When work operations are generally restricted, such as with equipment operating stations or control booths, spot conditioning of the work environment with clean conditioned air (see the preceding section on Makeup Air) may further reduce the reliance on general ventilation for conditioning or contaminant dilution. In cold climates, infiltration and heat loss through the building envelope may need to be minimized by pressurizing buildings.

For more information on dilution ventilation, see ACGIH (2010).

Quantity of Supplied Air

Sufficient air must be supplied to replace air exhausted by process ventilation and local exhausts, dilute contaminants (gases, vapors, or airborne particles) not captured by local exhausts, prevent the entry of contaminants or hazardous (ignitable) substances from any surrounding atmosphere during ingress or egress, and provide the required thermal environment. The amount of supplied air should be the largest of the amounts needed for temperature control, dilution, and replacement.

Air Supply Methods

Air supply to industrial spaces can be by natural or mechanical ventilation systems. Although natural ventilation systems driven by gravity forces and/or wind effect are still widely used in industrial spaces (especially in hot premises in cold and moderate climates), they are inefficient in large buildings, may cause drafts, and may not solve air contamination problems because there is no practical filtration method available. Thus, most ventilation systems in industrial spaces are either mechanical (fan-driven) or a combination of mechanical supply with natural exhaust, using louvers or doors for air pressure relief (or for air replacement in exhaust systems).

The most common methods of air supply to industrial spaces are mixing, displacement, and localized.

Mixing Air Distribution. In mixing systems, air is normally supplied at velocities much greater than those acceptable in the occupied zone. Supply air temperature can be above, below, or equal to the air temperature in the occupied zone, depending on the heating/cooling load. The supply air diffuser jet mixes with room air by entrainment, which reduces air velocities and equalizes the air temperature. The occupied zone is ventilated either directly by the air jet or by reverse flow created by the jet. Properly selected and designed mixing air distribution creates relatively uniform air velocity, temperature, humidity, and air quality conditions in the occupied zone and over the room height.

Displacement Ventilation Systems. Conditioned air that is slightly cooler than the desired room air temperature in the occupied zone is supplied from air outlets at low air velocities (~100 fpm or less). Because of buoyancy, the cooler air spreads along the floor and floods the room's lower zone. Air close to the heat source is heated and rises upward as a convective air stream; in the upper zone, this stream spreads along the ceiling. The height of the lower zone depends on the air volume and temperature supplied to the occupied zone and on the amount of convective heat discharged by the sources.

Typically, outlets are located at or near the floor, and supply air is introduced directly into the occupied zone. In some applications (e.g., in computer rooms or hot industrial buildings), air may be supplied to the occupied zone through a raised floor. Exhaust or air returns are located at or close to the ceiling or roof.

Displacement ventilation is common in European countries. It is an option when contaminants are released in combination with surplus heat, and contaminated air is warmer (more buoyant) than the surrounding air. It is not a good choice when air turbulence can interfere with convective conveyance of heat and contaminants. Further information on displacement air distribution systems can be found in Goodfellow and Tahti (2001).

Localized Ventilation. Air is supplied locally for occupied regions or a few permanent work areas (Figure 1). Conditioned air is supplied toward the breathing zone of the occupants to create comfortable conditions and/or to reduce the concentration of pollutants. These zones may have air 5 to 10 times cleaner than the surrounding air. In localized ventilation systems, air is supplied through one of the following devices:

- Nozzles or grilles (e.g., for spot cooling), specially designed low-velocity/low-turbulence devices
- Perforated panels suspended on vertical duct drops and positioned close to the workstation

Local Area and Spot Cooling

In hot workplaces that have few work areas, it is likely impractical and energy-inefficient to maintain a comfortable environment in the entire space. However, environmentally controlled cabins, individual cooling, and spot cooling can improve working conditions in occupied areas.

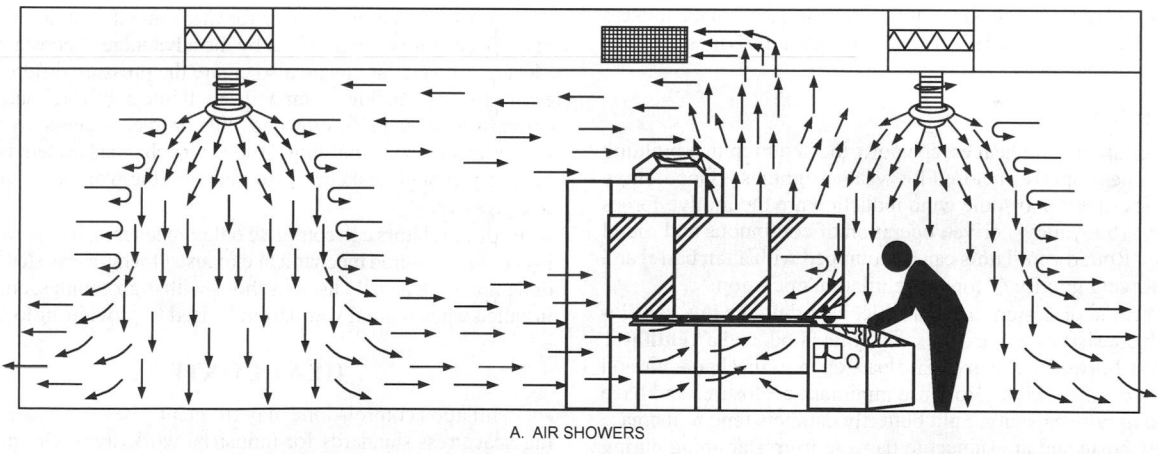

A. AIR SHOWERS

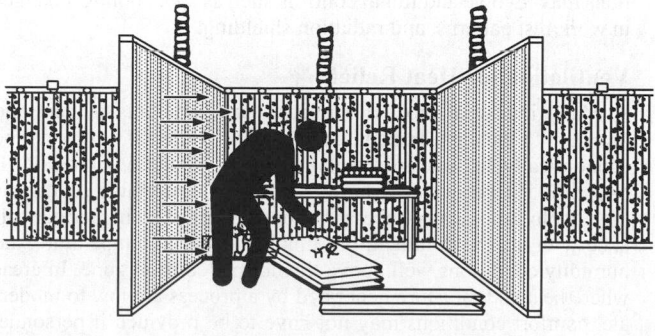

B. AIR OASIS WITH HORIZONTAL AIR SUPPLY

C. AIR OASIS WITH VERTICAL AIR SUPPLY

Fig. 1   Localized Ventilation Systems

Environmentally controlled cabins (e.g., operating cabs, pulpits, control rooms, enclosures) can not only provide thermal comfort, but when pressurized with a dedicated clean air supply (either dedicated source or through effective filtration), also can improve air quality in workers' occupational environments. There usually are significant economic benefits to properly designing, installing, and maintaining worker-protective environmental enclosures.

Spot cooling, probably the most popular method of improving the thermal environment, can be provided by radiation (changing mean radiant temperature), by convection (changing air velocity and/or air supply temperatures), or both. Spot-cooling equipment is fixed at the workstation, whereas in **individual cooling**, the worker wears the equipment.

Locker Room, Toilet, and Shower Space Ventilation

Ventilation of locker rooms, toilets, and shower spaces is important in industrial facilities to remove odor and reduce humidity. In some industries, adequate control of workroom contamination requires prevention of both ingestion and inhalation routes of exposure, so adequate hygienic facilities, including appropriate ventilation, may be required in locker rooms, changing rooms, showers, lunchrooms, and break rooms. State, provincial, and local regulations should be consulted early in design.

Supply air may be introduced through doors or wall grilles. In some cases, plant air may be so contaminated that filtration or (preferably) mechanical ventilation may be required. When control of workroom contaminants is inadequate or not feasible, minimizing the level of contamination in the locker rooms, lunchrooms, and break rooms by pressurizing these areas with excess supply air can reduce employee exposure.

When mechanical ventilation is used, the supply system should have adequate ducting and air distribution devices, such as diffusers or grilles, to distribute air throughout the area.

In locker rooms, exhaust should be taken primarily from the toilet and shower spaces as needed, and the remainder from the lockers and the room ceiling. ASHRAE *Standard* 62.1 provides requirements in this area.

Roof Ventilators

Roof ventilators are heat escape ports located high in a building and should be properly enclosed for weathertightness (Goodfellow 1985). Stack effect and some wind induction are the motive forces for gravity- (buoyancy-) driven operation of continuous and round ventilators. Round ventilators can be equipped with a fan barrel and motor, allowing gravity or forced ventilation operation.

Many ventilator designs are available, including the **low ventilator**, which consists of a stack fan with a rain hood, and a **ventilator with a split butterfly closure** that floats open to discharge air and closes by a counterweight. Both use minimum enclosures and have little or no gravity capacity. Split butterfly dampers tend to increase fan airflow noise and are subject to damage from slamming during strong winds. Because noise is frequently a problem in powered roof ventilators, the manufacturer's sound rating should be reviewed. Sound attenuators should be installed where required to meet the design sound ratings.

Continuous ventilation monitors remove substantial, concentrated heat loads most effectively. One type, the **streamlined continuous ventilator**, is efficient, weathertight, and designed to prevent backdraft; it usually has dampers that may be closed in winter to conserve building heat. Its capacity is limited only by the available roof area and the proper location and sizing of low-level air inlets. Continuous ventilation to achieve a slight pressure above the surrounding atmosphere (referred to as **pressurization** by NFPA) also can be used to reduce or declassify the electrical classification of enclosed spaces. Typically, reductions from class I, zone 1 or division 1 to class I, zone 2 or division 2 can be achieved by following the recommendations of NFPA *Standard* 496. This allows using general-purpose electrical devices instead of zone 2 or division 2 devices, or using zone 2 or division 2 electrical devices instead of zone 1 or division 1 devices, which (1) greatly reduces the cost of electrical equipment and (2) provides a sound alternative when particular devices are not available for the higher (more volatile) electrical area classifications.

Gravity ventilators, also highly effective, have low operating costs, do not generate noise, and are self-regulating (i.e., higher heat release increases airflow through the ventilators). Gravity ventilators can be affected by environmental conditions and thus should only be used for heat control rather than for the control of gaseous or aerosol contaminants. Care must be taken to ensure positive pressure at the ventilators, particularly during the heating season. Otherwise, outside air will enter the ventilators.

Next in order of heat removal capacity are (1) round gravity or wind-band ventilators, (2) round gravity ventilators with fan and motor added, (3) low-hood powered ventilators, and (4) vertical upblast powered ventilators. The shroud for the vertical upblast design has a peripheral baffle to deflect air upward instead of downward. Vertical discharge is highly desirable to reduce roof damage caused by hot air if it contains condensable oil or solvent vapor. Ventilators with direct-connected motors are desirable to avoid belt maintenance. Round gravity ventilators are applicable for warehouses with light heat loads and for manufacturing areas with high roofs and light loads.

Streamlined continuous ventilators must operate effectively without mechanical power. To ensure ventilator performance, sufficient low-level openings must be provided for incoming air; insufficient inlet area and significant space air currents are the most common reasons gravity roof ventilators malfunction. A positive supply of air around hot equipment may be necessary in large buildings where external wall inlets are remote from the equipment. Chapter 16 of the 2009 *ASHRAE Handbook—Fundamentals* has additional information on ventilation and infiltration.

The cost of electrical power for mechanical ventilation over that of roof ventilators can be offset by the advantage of constant airflow. Mechanical ventilation can also create the pressure differential necessary for good airflow, even with small inlets. Inlets should be sized correctly to avoid infiltration and other problems caused by high negative pressure in the building. Often, a mechanical system is justified to supply enough makeup air to maintain the work area under positive pressure.

Roof ventilators can comprise either mechanically operated openings or fan-powered mechanical exhaust. Operator-assisted openings or dampers are usually used in shops with high ceilings, and must be installed when natural ventilation is used to provide air to the space.

HEAT CONTROL

Ventilation control alone may frequently be inadequate for meeting heat stress standards for industrial work areas. Optimum solutions may involve additional controls such as spot cooling, changes in work/rest patterns, and radiation shielding.

Ventilation for Heat Relief

Many industrial processes release large amounts of heat and moisture to the environment. In such environments, it may not be economically feasible to maintain comfort conditions (ASHRAE *Standard* 55), particularly during summer. Comfortable conditions are not physiologically necessary: the body must be in thermal balance with the environment, but this can occur at temperature and humidity conditions well above the normal comfort zone. In areas where heat and moisture generated by a process are low to moderate, comfort conditions may not have to be provided if personnel exposures are infrequent and brief. In such cases, ventilation may be the only control necessary to prevent excessive physiological heat stress.

The engineer must distinguish between control needs for hot/dry industrial areas and warm/moist conditions. In hot/dry areas, a process gives off only sensible (primarily convective and radiant) heat without adding moisture to the air. This increases the heat load on exposed workers, but the rate of cooling by evaporation of perspiration may not be significantly reduced. Body heat equilibrium may be maintained, but could cause excessive perspiration. Hot/dry work situations occur around furnaces, forges, metal-extruding and rolling mills, glass-forming machines, etc.

In warm/moist conditions, a wet process may generate a significant latent heat load. The rise in sensible heat load on workers may be insignificant, but the increased moisture content of the air can seriously reduce cooling by evaporation of perspiration, making warm/moist conditions potentially more hazardous than hot/dry. Typical warm/moist operations are found in textile mills, laundries, dye houses, and deep mines, where water is used extensively for dust control.

Industrial heat load is also affected by local climate. Solar heat gain and elevated outdoor temperatures increase the heat load at the workplace, but may be insignificant compared to process heat generated locally. The moisture content of outdoor air is an important factor that can affect hot/dry work situations by restricting an individual's evaporative cooling. For warm/moist working environments, solar heat gain and elevated outdoor temperatures are even more important because moisture contributed by outdoor air is insignificant compared to that released by the process.

Both ASHRAE *Standard* 55 and International Organization for Standardization (ISO) *Standard* 7730 specify thermal comfort conditions for humans.

Methods for evaluating the general thermal state of the body both in comfort conditions and under heat and cold stress are based on analysis of the heat balance for the human body, as discussed in Chapter 9 of the 2009 *ASHRAE Handbook—Fundamentals*. A person may find the thermal environment unacceptable or intolerable because of local effects on the body caused by asymmetric radiation, air velocity, vertical air temperature differences, or contact with hot or cold surfaces (floors, machinery, tools, etc.).

Heat Stress—Thermal Standards

Another heat stress indicator for evaluating an environment's heat stress potential is the **wet-bulb globe temperature** (WBGT), defined as follows:

Outdoors with solar load

$$\text{WBGT} = 0.7t_{nwb} + 0.2t_g + 0.1t_{db} \qquad (1)$$

Indoors, or outdoors with no solar load

$$\text{WBGT} = 0.7t_{nwb} + 0.3t_g \qquad (2)$$

where

t_{nwb} = naturally ventilated wet-bulb temperature (no defined range of air velocity; different from saturation temperature or psychrometric wet-bulb temperature), °F
t_g = globe temperature (Vernon bulb thermometer, 6 in. diameter), °F
t_{db} = dry-bulb temperature (sensor shaded from solar radiation), °F

Coefficients in Equations (1) and (2) represent the fractional contributions of the component temperatures.

Exposure limits for heat stress for different levels of physical activity are shown in Figure 2 (NIOSH 1986), which depicts the allowable work regime (in terms of rest periods and work periods each hour) for different levels of work over a range of WBGT. When applying Figure 2, assume that the rest area has the same WBGT as the work area. The curves are valid for workers acclimatized to heat. ASHRAE *Standard* 62.1 provides some metabolic rates for different activities that can be used with Figure 2. Refer to NIOSH (1986) for recommended WBGT limits for nonacclimatized workers.

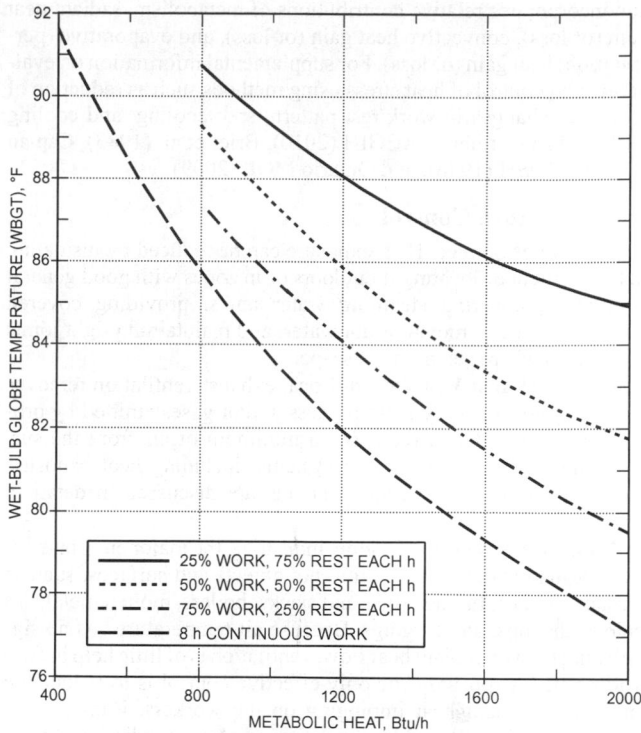

Fig. 2 Recommended Heat Stress Exposure Limits for Heat-Acclimatized Workers
[Adapted from NIOSH (1986)]

The **WBGT index** is an international standard (ISO *Standards* 7243 and 7730) for evaluating hot environments. The WBGT index and activity levels should be evaluated on 1 h mean values; that is, WBGT and activity are measured and estimated as time-weighted averages on a 1 h basis for continuous work, or on a 2 h basis when exposure is intermittent. Although recommended by NIOSH, the WBGT has not been accepted as the sole legal standard by the Occupational Safety and Health Administration (OSHA), and it may not apply for non-U.S. jurisdictions. The WBGT is generally used in conjunction with other methods to determine heat stress.

Although Figure 2 is useful for evaluating heat stress exposure limits, it is of limited use for control purposes or for evaluation of comfort. Air velocity and psychrometric wet-bulb measurements are usually needed to specify proper controls, and are only measured indirectly in WBGT determinations. Information on other useful tools, including the heat stress index (HSI), can be found in Chapter 9 of the 2009 *ASHRAE Handbook—Fundamentals* and in ISO *Standards* 7243, 7730, and 7933.

The thermal relationship between humans and their environment is determined by four independent environmental variables:

- Air temperature
- Radiant temperatures
- Moisture content of the air
- Air velocity

Together with the rate of internal heat production (metabolic rate) and clothing variables, these factors may combine in various ways to create different degrees of heat stress. The HSI is defined as the percent of the skin that is wetted by perspiration:

$$\text{HSI} = E_{sk}/E_{max} \times 100 \qquad (3)$$

where

E_{sk} = evaporative heat loss from the skin, Btu/h·ft²
E_{max} = maximum possible evaporative heat loss from the skin, Btu/h·ft²

and incorporates relative contributions of metabolism, radiant heat gain (or loss), convective heat gain (or loss), and evaporative (perspiration) heat gain (or loss). For supplemental information on evaluation and control of heat stress using methods such as reduction of radiation, changes in work/rest pattern, spot cooling, and cooling vests and suits, refer to ACGIH (2010), Brief et al. (1983), Caplan (1980), NIOSH (1986), and Ontario MOL (2009).

Heat Exposure Control

Control at Source. Heat exposure can be reduced by insulating hot equipment or locating it outdoors or in zones with good general ventilation, covering steaming water tanks, providing covered drains for direct removal of hot water, and maintaining tight joints and valves where steam may escape.

Local Exhaust Ventilation. Local exhaust ventilation removes heated air generated by a hot process and/or gases emitted by process equipment, while removing a minimum of air from the surrounding space. Local exhaust systems, including heat exposure control using overhead canopy hoods, are discussed in detail in Chapter 32.

Radiation Shielding. In some industries, the major environmental heat load is radiant heat from hot objects and surfaces, such as furnaces, ovens, furnace flues and stacks, boilers, molten metal, hot ingots, castings, and forgings. Because air temperature has no significant effect on radiant heat flow, ventilation is of little help in controlling such exposure. The only effective control is to reduce the amount of radiant heat impinging on the workers. Radiant heat exposure can be reduced by insulating or placing radiation shields around the source.

Radiation shields are effective in the following forms:

- **Reflective shielding.** Sheets of reflective material or insulating board are temporarily attached to the hot equipment or arranged in a semiportable floor stand.
- **Absorptive shielding (water-cooled).** These shields absorb and remove heat from hot equipment.
- **Transparent shields.** Heat-reflective tempered plate glass, reflective metal chain curtains, and close-mesh wire screens moderate radiation without obstructing the view of hot equipment.
- **Flexible shielding.** Aluminum-treated fabrics give a high degree of radiation shielding.
- **Protective clothing.** Reflective garments such as aprons, gauntlet gloves, and face shields provide moderate radiation shielding. For extreme radiation exposures, complete suits with vortex tube cooling may be required.

If the shield is a good reflector, it remains relatively cool in severe radiant heat. Bright or highly polished tinplate, stainless steel, and ordinary flat or corrugated aluminum sheets are efficient and durable. Foil-faced plasterboard, although less durable, reflects well on one side. To be efficient, however, the reflective shield must remain bright. Radiation shields are much more efficient when used in multiple layers; they should reflect the radiant heat back to the primary source, where the resulting hot gases can be removed by local exhaust. However, unless the shield completely surrounds the primary source, some infrared energy is reflected into the cooler surroundings and possibly into an occupied area. The direction of reflected heat should be studied to ensure proper shielding installation.

Spot Cooling. If the workplace is located near a source of radiant heat that cannot be entirely controlled by radiation shielding, spot cooling can be used. See Chapter 20 in the 2009 *ASHRAE Handbook—Fundamentals* and data from spot-cooling diffuser manufacturers for further information.

ENERGY CONSERVATION, RECOVERY, AND SUSTAINABILITY

Because of the large air volumes required to ventilate industrial plants, energy conservation and recovery should be practiced, and can provide substantial savings if this practice does not compromise overriding life safety concerns. Therefore, after all critical life safety issues have been adequately addressed, energy recovery should be incorporated into preliminary planning for an industrial plant wherever and whenever it is both safe and practical to deploy. When selecting energy recovery equipment, ensure that materials are compatible with all contaminants and hazardous substances that may be exhausted. Verify the acceptability of the energy recovery method with local codes.

In some cases, it is possible to provide unheated or partially heated makeup air to the building. Although most energy conservation and recovery methods in this section apply to heating, the savings possible with cooling systems are similar. The following are some methods of energy conservation and recovery:

- In the original design phase, process and equipment insulation and heat shields should be provided to minimize heat loads. Vapor-proofing and reducing the glass area may be required. Exhaust requirements for hoods and processes should be reviewed and kept to a practical, safe minimum; for more on local exhaust systems, see Chapter 32.
- Design the supply and exhaust general ventilation systems for optimal operation throughout the year. Air should be supplied as close to the occupied zone as possible. Clean, recirculated air can be used in winter makeup if does not increase the levels of contamination in the space (see the following bullet points for more details and restrictions) (ACGIH 2010).
- Supply air can be passed through air-to-air, liquid-to-air, or hot-gas-to-air heat exchangers to recover building or process heat. Rotary, regenerative, coil energy recovery (runaround), and air-to-air heat exchangers are discussed extensively in Chapter 25 of the 2008 *ASHRAE Handbook—HVAC Systems and Equipment*. Energy recovery is also discussed in Chapters 10 and 11 of ACGIH (2010).
- Operate the system for economy if this does not compromise life safety. Although CO_2-based demand control ventilation (DCV) is unsuitable for industrial spaces where human activity is not the main reason for ventilating the space (DOE 2004), industrial spaces may offer their own kind of demand control ventilation for providing makeup air to offset process and exhaust hood exhaust volumes. If the space does not contain a potential source of toxic contaminants or hazardous (ignitable) substances, shut such systems down at night, on weekends, or whenever possible, and operate makeup air in balance with the needs of process equipment and exhaust hoods. Keep heating supply air temperatures at the minimum, and cooling supply temperatures at the maximum, consistent with process needs and employee comfort. Keep the building in pressure balance so that uncomfortable drafts do not necessitate excessive heating.
- If the exhaust air has contaminants that do not pose an unacceptable health risk, and does not contain a potential source of hazardous (ignitable) substances, then recirculation can be considered. However, even then, contaminant concentrations in recirculated air must be determined so that allowable limits in the space are not exceeded. As recirculated air returns to the space, the concentration of contaminants in the partially filtered return air adds to the contaminant levels already existing in the space. It must be determined whether the concentration increases beyond the allowable time-weighted average (TWA) exposure limit during the period for which the worker is exposed. This period is usually assumed to be 8 h for an 8 h work shift, but could be any period of exposure. Once installed, real-time or periodic monitoring is likely required to support this determination. Depending on the contaminant's toxicity, the monitoring system may be re-quired to perform some form of corrective action or shut down once a target level (a percentage of the safe exposure limit) concentration is attained. Predicted energy cost savings from recirculation should be weighed

against the necessary costs of air cleaning and monitoring requirements (to include calibration and maintenance of monitoring equipment).

Assuming equilibrium has been established, the predicted TWA concentration at the workers' breathing zone can be calculated (ACGIH 2010):

$$C_B = \frac{Q_B}{Q_A}(C_G - C_M)(1 - f) + (C_O - C_M)f \\ + K_B C_R + (1 - K_B)C_M \tag{4}$$

where

C_B = TWA worker breathing zone contaminant concentration during recirculation, ppm
Q_B = total ventilation airflow without considering recirculation, cfm
Q_A = total ventilation airflow including recirculation, cfm
C_G = average space concentration if no recirculation, ppm
f = fraction of time worker spends at workstation
C_O = TWA contaminant concentration at breathing zone of workstation if no recirculation, ppm
K_B = fraction of worker breathing zone air that consists of recirculated air, 0 to 1.0
C_R = recirculated air (after air cleaner) discharge concentration, ppm, or

$$C_R = \frac{(1-\eta)(C_E - K_R C_M)}{1 - (1-\eta)K_R} \tag{5}$$

η = fractional air cleaner efficiency for contaminant
C_E = (local) exhaust concentration without recirculation, ppm
K_R = fraction of exhaust air that is recirculated air, 0 to 1.0
C_M = replacement air contaminant concentration, ppm

Other recirculation system examples are given in Chapter 8 of Goodfellow and Tahti (2001).

Example 1. An industrial space uses 10,000 cfm for ventilation, of which 5000 cfm is general exhaust and 5000 cfm local exhaust (ACGIH 2010). Local exhaust is recirculated through an air cleaner with an efficiency of 0.75. Recirculated air is directed toward the worker spaces, such that $K_B = 0.5$ and $K_R = 0.8$ (more of the recirculated air is locally exhausted than enters the worker's breathing zone). The worker is at the workstation 100% of the time ($f = 1$). The makeup air has a concentration of 5 ppm (C_M), the local exhaust has a concentration of 500 ppm (C_E), the space has an average concentration of 20 ppm (C_G), and without recirculation the worker's breathing zone is 35 ppm (C_O).

Solution: The concentration C_B at the breathing zone with recirculation is determined from

$$C_R = \frac{(1-0.75)[500 - 0.8(5)]}{1 - (1-0.75)(0.8)} = 155 \text{ ppm}$$

and

$$C_B = \frac{10,000}{5000}(20-5)(1-1) + (35-5)(1) + 0.5(155) \\ + (1-0.5)5 = 110 \text{ ppm}$$

which may or may not exceed the allowable TWA exposure limit of the worker space, depending on the specific contaminant.

REFERENCES

ACGIH. 2010. *Industrial ventilation: A manual of recommended practice for design*, 27th ed. American Conference of Governmental Industrial Hygienists, Cincinnati, OH.

ASHRAE. 2010. Thermal environmental conditions for human occupancy. ANSI/ASHRAE *Standard* 55-2010.

ASHRAE. 2010. Ventilation for acceptable indoor air quality. ANSI/ASHRAE *Standard* 62.1-2010.

Brief, R.S., S. Lipton, S. Amamnani, and R.W. Powell. 1983. Development of exposure control strategy for process equipment. *Annals of the American Conference of Governmental Industrial Hygienists* (5).

British Occupational Hygiene Society (BOHS). 1987. Controlling airborne contaminants in the workplace. *Technical Guide* 7. Science Review Ltd. and H&H Scientific Consultants, Leeds, U.K.

Caplan, K.J. 1980. Heat stress measurements. *Heating, Piping and Air Conditioning* (February):55-62.

Caplan, K.J. and G.W. Knutson. 1977. The effect of room air challenge on the efficiency of laboratory fume hoods. *ASHRAE Transactions* 83(1): 141-156.

Caplan, K.J. and G.W. Knutson. 1978. Laboratory fume hoods: Influence of room air supply. *ASHRAE Transactions* 84(1):522-537.

Cashdollar, K.L. 2000. Overview of dust explosibility characteristics. *Journal of Loss Prevention in the Process Industries* 13(3):183-199. http://www.cdc.gov/Niosh/mining/pubs/pdfs/odec.pdf.

DHHS. 1986. *Advanced industrial hygiene engineering.* PB87-229621. U.S. Department of Health and Human Services, Cincinnati, OH.

DOE. 2004. Demand-controlled ventilation using CO_2 sensors. *Federal Technology Alert* DOE/EE-0293. U.S. Department of Energy, Washington, D.C. http://www.energyinnovation.net/images/co2/dept_of_energy. pdf.

Goodfellow, H.D. 1985. *Advanced design of ventilation systems for contaminant control.* Elsevier Science B.V., Amsterdam.

Goodfellow, H. and E. Tahti, eds. 2001. *Industrial ventilation design guidebook.* Academic Press, New York.

ISO. 1989. Hot environments—Estimation of the heat stress on working man, based on the WBGT-index (wet bulb globe temperature). *Standard* 7243. International Organization for Standardization, Geneva.

ISO. 2005. Ergonomics of the thermal environment—Analytical determination and interpretation of thermal comfort using calculation of the PMV and PPD indices and local thermal comfort criteria. *Standard* 7730. International Organization for Standardization, Geneva.

ISO. 2004. Ergonomics of the environment—Analytical determination and interpretation of heat stress using calculation of the predicted heat strain. *Standard* 7933. International Organization for Standardization, Geneva.

National Safety Council. 2002. *Fundamentals of industrial hygiene*, 5th ed. Chicago.

NFPA. 2003. Purged and pressurized enclosures for electrical equipment. *Standard* 496-2003. National Fire Protection Association, Quincy, MA.

NIOSH. 1986. Criteria for a recommended standard: Occupational exposure to hot environments. CDC/NIOSH *Publication* 86-113. National Institute for Occupational Safety and Health, Washington, D.C. Available from http://www.cdc.gov/niosh/86-113.html.

Ontario MOL. 2009. *Heat stress (health and safety guidelines).* Ontario Ministry of Labor. Available from http://www.labour.gov.on.ca/english/hs/pubs/gl_heat.php.

BIBLIOGRAPHY

Alden, J.L. and J.M. Kane. 1982. *Design of industrial ventilation systems*, 5th ed. Industrial Press, New York.

Anderson, R. and M. Mehos. 1988. Evaluation of indoor air pollutant control techniques using scale experiments. ASHRAE Indoor Air Quality Conference.

Balchin, N.C., ed. 1991. *Health and safety in welding and allied processes*, 4th ed. Abington Publishing, Cambridge.

Bartknecht, W. 1989. *Dust explosions: Course, prevention, protection.* Springer-Verlag, Berlin.

Burgess, W.A., M.J. Ellenbecker, and R.D. Treitman. 1989. *Ventilation for control of the work environment.* John Wiley & Sons, New York.

Cawkwell, G.C. and H.D. Goodfellow. 1990. Multiple cell ventilation model with time-dependent emission sources. *Proceedings of the 2nd International Conference on Engineering Aero- and Thermodynamics of Ventilated Rooms*, AI-9, Oslo.

Chamberlin, L.A. 1988. *Use of controlled low velocity air patterns to improve operator environment at industrial work stations.* M.A. thesis, University of Massachusetts.

Cole, J.P. 1995. Ventilation systems to accommodate the industrial process. *Heating, Piping, Air Conditioning* (May).

Constance, J.D. 1983. *Controlling in-plant airborne contaminants.* Marcel Dekker, New York.

Cralley, L.V. and L.J. Cralley, eds. 1986. *Patty's industrial hygiene and toxicology*, vol. 3: *Industrial hygiene aspects of plant operations.* John Wiley & Sons, New York.

Fanger, P.O. 1982. *Thermal comfort*. Robert E. Krieger, Malabar, FL.

Flagan, R.C. and J.H. Seinfeld. 1988. *Fundamentals of air pollution engineering*. Prentice-Hall, Englewood Cliffs, NJ.

Godish, T. 1989. *Indoor air pollution control*. Lewis Publishers, Chelsea, MI.

Goldfield, J. 1980. *Contaminant concentration reduction: General ventilation versus local exhaust ventilation*. American Industrial Hygienists Association Journal 41(November).

Goodfellow, H.D. 1986. *Proceedings of Ventilation '85*. Elsevier Science B.V., Amsterdam.

Goodfellow, H.D. 1987. *Encyclopedia of physical science and technology*, vol. 14: *Ventilation, industrial*. Academic Press, San Diego, CA.

Goodfellow, H.D. and J.W. Smith. 1982. Industrial ventilation—A review and update. *American Industrial Hygiene Association Journal* 43 (March):175-184.

Harris, R.L. 1988. Design of dilution ventilation for sensible and latent heat. *Applied Industrial Hygiene* 3(1).

Hayashi, T., R.H. Howell, M. Shibata, and K. Tsuji. 1987. *Industrial ventilation and air conditioning*. CRC, Boca Raton, FL.

Heinsohn, R.J. 1991. *Industrial ventilation engineering principles*. John Wiley & Sons, New York.

Holcomb, M.L. and J.T. Radia. 1986. An engineering approach to feasibility assessment and design of recirculating exhaust systems. *Proceedings of Ventilation '85*. Elsevier Science B.V., Amsterdam.

Jackman, R. 1991. Displacement ventilation. Presented at CIBSE National Conference (April), University of Kent, Canterbury.

Laurikainen, J. 1995. Displacement ventilation system design method. Seminar presentations, Part 2. INVENT *Report* 46. FIMET, Helsinki.

Licht, W. 1988. *Air pollution control engineering*, 2nd ed. Marcel Dekker, New York.

McDermott, H.J. 1985. *Handbook of ventilation for contaminant control*, 2nd ed. Ann Arbor Science Publishers, Ann Arbor, MI.

Mehta, M.P., H.E. Ayer, B.E. Saltzman, and R. Ronk. 1988. Predicting concentration for indoor chemical spills. Presented at ASHRAE Indoor Air Quality Conference.

NFPA. [Annual.] *National fire codes*®. National Fire Protection Association, Quincy, MA

Olesen, B.W. and A.M. Zhivov. 1994. Evaluation of thermal environment in industrial work spaces. *ASHRAE Transactions* 100(2):623-635.

Pozin, G.M. 1993. Determination of the ventilating effectiveness in mechanically ventilated spaces. *Proceedings of the 6th International Conference on Indoor Air Quality (IAQ '93)*, Helsinki.

RoomVent '90. 1990. *Proceedings of the 2nd International Conference on Engineering Aero- and Thermodynamics of Ventilated Rooms*, Oslo.

RoomVent '92. 1992. *Proceedings of the 3rd International Conference on Engineering Aero- and Thermodynamics of Ventilated Rooms*, Aalborg, Denmark.

RoomVent '94. 1994. *Proceedings of the 4th International Conference on Engineering Aero- and Thermodynamics of Ventilated Rooms*, Krakow.

RoomVent '96. *Proceedings of the 5th International Conference on Air Distribution in Rooms*, Yokohama.

Schroy, J.M. 1986. A philosophy on engineering controls for workplace protection. *Annals of Occupational Hygiene* 30(2):231-236.

Shilkrot, E.O. and A.M. Zhivov. 1996. Zonal model for displacement ventilation design. *RoomVent '96, Proceedings of the 5th International Conference on Air Distribution in Rooms*, vol. 2. Yokohama.

Skaret, E. 1985. *Ventilation by displacement—Characterization and design applications*. Elsevier Science B.V., Amsterdam.

Skaret, E. and H.M. Mathisen. 1989. Ventilation efficiency—A guide to efficient ventilation. *ASHRAE Transactions* 89(2B):480-495.

Skistad, H. 1994. *Displacement ventilation*. Research Studies Press, John Wiley & Sons, West Sussex, U.K.

Stephanov, S.P. 1986. Investigation and optimization of air exchange in industrial halls ventilation. *Proceedings of Ventilation '85*. Elsevier Science B.V., Amsterdam.

Vincent, J.H. 1989. *Aerosol sampling, science and practice*. John Wiley & Sons, Oxford, U.K.

Volkavein, J.C., M.R. Engle, and T.D. Raether 1988. Dust control with clean air from an overhead air supply island (oasis). *Applied Industrial Hygiene* 3(August):8.

Wadden, R.A. and P.A. Scheff 1982. *Indoor air pollution: Characterization, prediction, and control*. John Wiley & Sons, New York.

Wilson, D.J. 1982. A design procedure for estimating air intake contamination from nearby exhaust vents. *ASHRAE Transactions* 89(2A):136-152.

Zhivov, A.M. and B.W. Olesen. 1993. Extending existing thermal comfort standards to work spaces. *Proceedings of the 6th International Conference on Indoor Air Quality (IAQ '93)*, Helsinki.

Zhivov, A.M. E.O. Shilkrot, P.V. Nielsen, and G.L. Riskowski. 1997. Displacement ventilation design. *Proceedings of the 5th International Symposium on Ventilation for Contaminant Control (Ventilation '97)*, vol. I, Ottawa.

INDUSTRIAL LOCAL EXHAUST

INDUSTRIAL exhaust ventilation systems collect and remove airborne contaminants consisting of particulate matter (dusts, fumes, smokes, fibers), vapors, and gases that can create a hazardous, unhealthy, or undesirable atmosphere. Exhaust systems can also salvage usable material, improve plant housekeeping, and capture and remove excessive heat or moisture. Often, industrial ventilation exhaust systems are considered life-safety systems and can contain hazardous gases and/or particles. Industrial exhaust systems also have to comply with ANSI/ASHRAE *Standard* 62.1 and other standards as required [e.g., by the National Fire Protection Agency (NFPA)].

Special Warning: Certain industrial spaces may contain flammable, combustible and/or toxic concentrations of vapors or dusts under either normal or abnormal conditions. In spaces such as these, there are life safety issues that this chapter may not completely address. Special precautions must be taken in accordance with requirements of recognized authorities such as the National Fire Protection Association (NFPA), Occupational Safety and Health Administration (OSHA), and American National Standards Institute (ANSI). In all situations, engineers, designers, and installers who encounter conflicting codes and standards must defer to the code or standard that best addresses and safeguards life safety.

Local Exhaust Versus General Ventilation

Local exhaust ventilation systems can be the most performance-effective and cost-effective method of controlling air pollutants and excessive heat. For many operations, capturing pollutants at or near their source is the only way to ensure compliance with occupational exposure limits that are measured within the worker's breathing zone. When properly designed, local exhaust ventilation optimizes ventilation exhaust airflow, thus optimizing system acquisition costs associated with equipment size and operating costs associated with energy consumption and makeup air tempering.

In some industrial ventilation designs, the emphasis is on filtering air captured by local exhausts before exhausting it to the outdoors or returning it to the production space. As a result, these systems are evaluated according to their filter efficiency or total particulate removal. However, if an insufficient percentage of emissions are captured, the degree of air-cleaning efficiency sometimes becomes irrelevant.

For a process exhaust system in the United States, the design engineer must verify if the system is permitted by the 1990 Clean Air Act. For more information, see the Environmental Protection Agency's Web site (www.epa.gov).

The pollutant-capturing efficiency of local ventilation systems depends on hood design, the hood's position relative to the source of contamination, temperature of the source being exhausted, and the induced air currents generated by the exhaust airflow. Selection and positioning of the hood significantly influence initial and operating costs of both local and general ventilation systems. In addition,

poorly designed and maintained local ventilation systems can cause deterioration of building structures and equipment, negative health effects, and decreased worker productivity.

No local exhaust ventilation system is 100% effective in capturing pollutants and/or excess heat. In addition, installation of local exhaust ventilation system may not be possible in some circumstances, because of the size, mobility, or mechanical interaction requirements of the process. In these situations, general ventilation is needed to dilute pollutants and/or excess heat (where pollutants are toxic or present a health risk to workers, local exhaust is required; dilution ventilation should be avoided). Air supplied by the general ventilation system is usually conditioned (heated, humidified, cooled, etc.). Supply air replaces air extracted by local and general exhaust systems and improves comfort conditions in the occupied zone.

Chapter 11 of the 2009 *ASHRAE Handbook—Fundamental* covers definitions, particle sizes, allowable concentrations, and upper and lower explosive limits of various air contaminants. Chapter 31 of this volume, Goodfellow and Tahti (2001), and Chapter 4 of *Industrial Ventilation: A Manual of Recommended Practice for Design* (American Conference of Governmental Industrial Hygienists [ACGIH] 2007) detail steps to determine air volumes necessary to dilute contaminant concentration using general ventilations.

Sufficient makeup air must be provided to replace air removed by the exhaust system. If replacement air is insufficient, building pressure becomes negative relative to atmospheric pressure and allows air to infiltrate through open doors, window cracks, and backfeed through combustion equipment vents. A negative pressure as little as 0.05 in. of water can cause drafts and might cause backdrafts in combustion vents, thereby creating a potential health hazard. From the sustainability perspective, a negative plant static pressure can also result in excessive energy use. If workers near the plant perimeter complain about cold drafts, unit heaters are often installed. Heat from these units often is drawn into the plant interior, overheating the interior. Too often, this overheating is addressed by exhausting more air from the interior, causing increased negative pressure and more infiltration. Negative plant pressure reduces the exhaust volumetric flow rate because of increased system resistance, which can also decrease local exhaust efficiencies or require additional energy to overcome the increased resistance. Wind effects on building balance may also play a role, and are discussed in Chapter 24 of the 2009 *ASHRAE Handbook—Fundamentals*.

Positive-pressure plants and balanced plants (those with equal exhaust and replacement air rates) use less energy. However, if there are clean and contaminated zones in the same building, the desired airflow direction is from clean to dirty, and zone boundary construction and pressure differentials should be designed accordingly.

Exhaust system discharge may be regulated under various federal, state, and local air pollution control regulations or ordinances. These regulations may require exhaust air treatment before discharge to the atmosphere. Chapter 29 of the 2008 *ASHRAE Handbook—HVAC Systems and Equipment* provides guidance and recommendations for discharge air treatment.

The preparation of this chapter is assigned to TC 5.8, Industrial Ventilation Systems.

LOCAL EXHAUST FUNDAMENTALS

System Components

Local exhaust ventilation systems typically consist of the following basic elements:

- Hood to capture pollutants and/or excessive heat
- Ducted system to transport polluted air to air cleaning device or building exhaust
- Air-cleaning device to remove captured pollutants from the airstream for recycling or disposal
- Air-moving device (e.g., fan or high-pressure air ejector), which provides motive power to generate the hood capture velocity plus overcome exhaust ventilation system resistance
- Exhaust stack, which discharges system air to the atmosphere

System Classification

Contaminant Source Type. Knowledge of the process or operation is essential before a local exhaust hood system can be designed.

Hood Type. Exhaust hoods are typically round, rectangular, or slotted to accommodate the geometry of the source. Hoods are either enclosing or nonenclosing (Figure 1). **Enclosing hoods** provide more effective and economical contaminant control because their exhaust rates and the effects of room air currents are minimal compared to those for nonenclosing hoods. Hood access openings for inspection and maintenance should be as small as possible and out of the natural path of the contaminant. Hood performance (i.e., how well it captures the contaminant) should ideally be verified by an industrial hygienist.

A **nonenclosing hood** can be used if access requirements make it necessary to leave all or part of the process open. Careful attention must be paid to airflow patterns and capture velocities around the process and hood (under dynamic conditions) and to the process characteristics to make nonenclosing hoods effective. The use of moveable baffles, curtains, strip curtains, and brush seals may allow the designer to increase the level of enclosure without interfering with the work process. The more of the process that can be enclosed, the less exhaust airflow required to control the contaminant(s).

System Mobility. Local exhaust systems with nonenclosing hoods can be **stationary** (i.e., having a fixed hood position), **moveable, portable,** or **built-in** (into the process equipment). Moveable hoods are used when process equipment must be accessed for repair and loading and unloading of materials (e.g., in electric ovens for melting steel).

The portable exhaust system shown in Figure 2 is commonly used for temporary exhausting of fumes and solvents in confined spaces or during maintenance. It has a built-in fan and filter and an exhaust hood connected to a flexible hose. Built-in local exhaust systems are commonly used to evacuate welding fumes, such as hoods built into stationary or turnover welding tables. Lateral exhaust hoods, which exhaust air through slots on the periphery of open vessels, such as those used for galvanizing metals, are another example of built-in local exhaust systems.

Effectiveness of Local Exhaust

The most effective hood design uses the minimum exhaust airflow rate to provide maximum contaminant control without compromising operator capability to complete the work task. **Capture effectiveness** should be high, but it is difficult and costly to develop hoods with efficiencies approaching 100%. Makeup air supplied by general ventilation to replace exhausted air can dilute contaminants that are not captured by the hood. Enclosing more of the process reduces the need to protect against contaminant escape through crossdrafts, convective currents, or process-generated contaminant momentum. In turn, this reduces the exhaust airflow required to control the contaminant(s).

Capture Velocity. Capture velocity is the air velocity required to entrain contaminants at the point of contaminant generation upstream of a hood. The contaminant enters the moving airstream near the point of generation and is carried along with the air into the hood. Designers use a designated capture velocity V_c to determine a volumetric flow rate to draw air into the hood. Table 1 shows ranges of capture velocities for several industrial operations. These figures are based on successful experience under ideal conditions. Once capture velocity upstream of the hood and hood position relative to the source are known, then the hood flow rate can be determined for the particular hood design. Velocity distributions for specific hoods must be known or determined.

Hood Volumetric Flow Rate. For a given hood configuration and capture velocity, the exhaust volumetric flow rate (the airflow rate that allows contaminant capture) can be calculated as

$$Q_o = V_o A_o \qquad (1)$$

where

Q_o = exhaust volumetric flow rate, cfm
V_o = average air velocity in hood opening that ensures capture velocity at point of contaminant release, fpm
A_o = hood opening area, ft^2

Low face velocities require that supply (makeup) air be as uniformly distributed as possible to minimize the effects of room air currents. This is one reason replacement air systems must be designed with exhaust systems in mind. Air should enter the hood

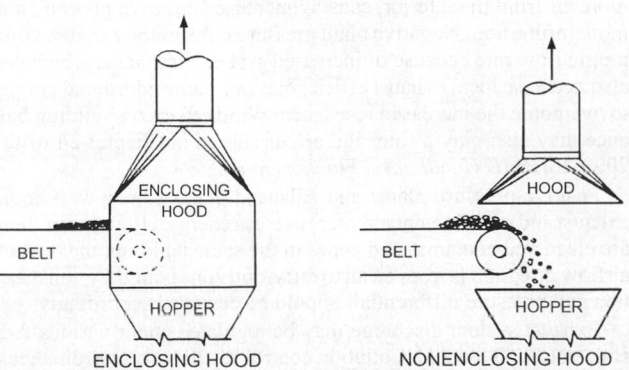

Fig. 1   Enclosing and Nonenclosing Hoods
(Adapted from ACGIH®, *Industrial Ventilation: A Manual of Recommended Practice*, 27th ed. Copyright 2010. Reprinted with permission.)

Fig. 2   Portable Fume Extractor with Built-in Fan and Filter

Table 1 Range of Capture (Control) Velocities

Condition of Contaminant Dispersion	Examples	Capture Velocity, fpm
Released with essentially no velocity into still air	Evaporation from tanks, degreasing, plating	50 to 100
Released at low velocity into moderately still air	Container filling, low-speed conveyor transfers, welding	100 to 200
Active generation into zone of rapid air motion	Barrel filling, chute loading of conveyors, crushing, cool shakeout	200 to 500
Released at high velocity into zone of very rapid air motion	Grinding, abrasive blasting, tumbling, hot shakeout	500 to 2000

Note: In each category above, a range of capture velocities is shown. The proper choice of values depends on several factors (Alden and Kane 1982):

Lower End of Range	Upper End of Range
1. Room air currents favorable to capture	1. Distributing room air currents
2. Contaminants of low toxicity or of nuisance value only	2. Contaminants of high toxicity
3. Intermittent, low production	3. High production, heavy use
4. Large hood; large air mass in motion	4. Small hood; local control only

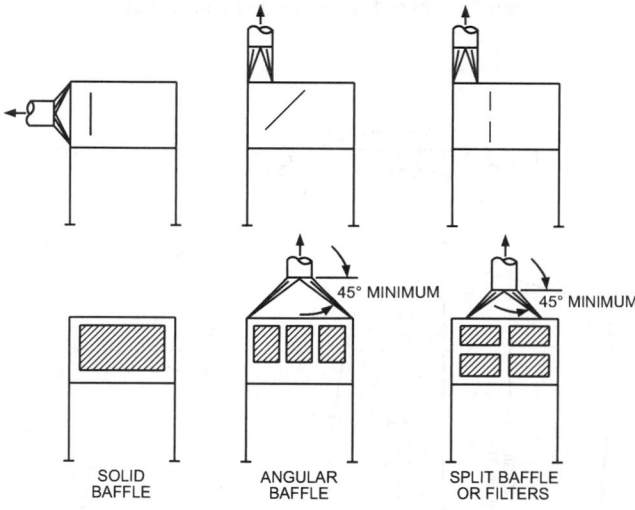

Fig. 3 Use of Interior Baffles to Ensure Good Air Distribution

uniformly. Hood flanges, side baffles, and interior baffles are sometimes necessary (Figure 3).

Airflow requirements for maintaining effective capture velocity at a contaminant source also vary with the distance between the source and hood. Chapter 6 of ACGIH (2007) provides methodology for estimating airflow requirements for specific hood configurations and locations relative to the contaminant source.

Airflow near the hood can be influenced by drafts from supply air jets (spot cooling jets) or by turbulence of the ambient air caused by jets, upward/downward convective flows, moving people, mobile equipment, and drafts from doors and windows. Process equipment may be another source of air movement. For example, high-speed rotating machines such as pulverizers, high-speed belt material transfer systems, falling granular materials, and escaping compressed air from pneumatic tools all produce air currents. These factors can significantly reduce the capturing effectiveness of local exhaust systems and should be accounted for in the exhaust system design.

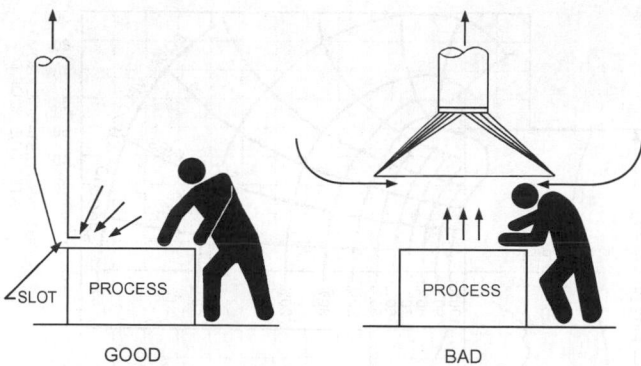

Fig. 4 Influence of Hood Location on Contamination of Air in the Operator's Breathing Zone
(Adapted from ACGIH®, *Industrial Ventilation: A Manual of Recommended Practice*, 27th ed. Copyright 2010. Reprinted with permission.)

Exhausted air may contain combustible pollutant/air mixtures. If it does, the amount by which the exhaust airflow rate should be increased to dilute combustible mixture must be verified to meet the requirements of National Fire Protection Association (NFPA) *Standard 86*.

Principles of Hood Design Optimization

Numerous studies of local exhaust systems and common practices have led to the following hood design principles:

- Hood location should be as close as possible to the source of contamination.
- The hood opening should be positioned so that it causes the contaminant to deviate the least from its natural path.
- The hood should be located so that the contaminant is drawn away from the operator's breathing zone.
- Hood size must be the same as or larger than the cross section of flow entering the hood. If the hood is smaller than the flow, a higher volumetric flow rate is required.
- Worker position with relation to contaminant source, hood design, and airflow path should be evaluated based on the principles given in Chapters 6 and 13 of ACGIH (2007).
- Canopy hoods (Figure 4) should not be used where the operator must bend over a tank or process (ACGIH 2007).

AIR MOVEMENT IN VICINITY OF LOCAL EXHAUST

Air capture velocities in front of the hood opening depend on the exhaust airflow rate, hood geometry, distance from hood face and surfaces surrounding the hood opening. Figure 5 shows velocity contours for an unflanged round duct hood. Studies have established the similarity of velocity contours (expressed as a percentage of the hood entrance velocity) for hoods with similar geometry (Dalla-Valle 1952). Figure 6 shows velocity contours for a rectangular hood with an **aspect ratio** (width divided by length) of 0.333. The profiles are similar to those for the round hood but are more elongated. If the aspect ratio is lower than about 0.2 (0.15 for flanged openings), the flow pattern in front of the hood changes from approximately spherical to approximately cylindrical. Velocity decreases rapidly with distance from the hood.

The design engineer should consider side drafts and other sources of air movement close to the capture area of a local exhaust hood. Caplan and Knutson (1977, 1978) found that air movement in front of laboratory hoods can cause contaminants to escape from the hood and into the operator's breathing zone. In industrial applications, it is common to see large fans blowing air onto workers who are located in front of an exhaust hood. This can render the local

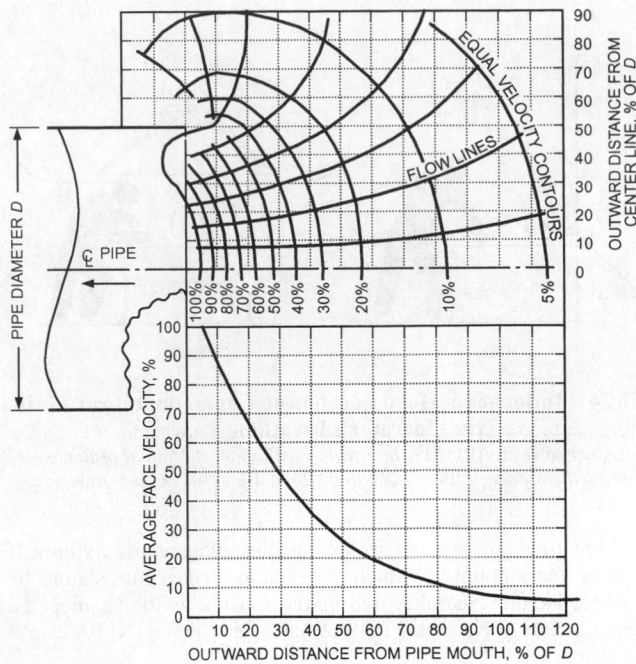

Fig. 5   Velocity Contours for Plain Round Opening
(Alden and Kane 1982; used by permission)

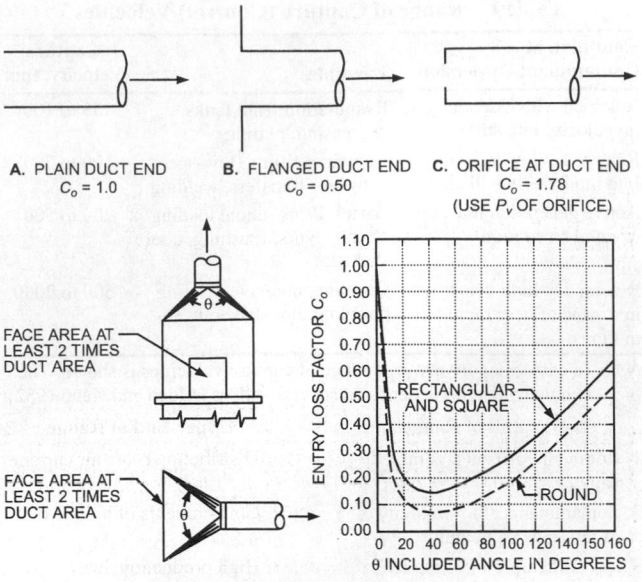

Fig. 7   Entry Losses for Typical Hoods

exhaust hood ineffective to the point that no protection is provided for the worker and/or their adjacent co-workers.

Pressure Loss in Hoods and Ducts

A vena contracta forms in the entrance of the hood or duct and produces a pressure loss, which can be described using pressure loss coefficient C_o or a static pressure entry loss (ACGIH 2007). When air enters a hood, the pressure loss, called **hood entry loss**, may have several components, depending on the hood's complexity. Simple hoods usually have a single pressure loss coefficient specified, defined as

$$C_L = \sqrt{\frac{P_v}{P_{s,h}}} \tag{2}$$

where

C_L = loss factor depending on hood type and geometry, dimensionless

$P_v = K\rho V^2/2g_c$, dynamic pressure inside duct (constant in duct after vena contracta), where K is proportionality constant, in. of water, ρ is air density in lb/ft^3, and g_c is the gravitational acceleration constant, 32.2 lb$_m$·ft/lb$_f$·s^2

$P_{s,h}$ = static pressure in hood duct because of velocity pressure increase and hood entry loss, in. of water

More information on loss factors and the design of exhaust ductwork is in Chapter 21 of the 2009 *ASHRAE Handbook—Fundamentals*, ACGIH (2007), and Brooks (2001).

The loss coefficient C_L is different from the hood entry loss coefficient. The entry loss coefficient C_o relates duct total pressure loss to duct velocity pressure. From Bernoulli's equation, hood total pressure is approximately zero at the entrance to the hood, and therefore the static pressure is equal to the negative of the velocity pressure:

$$P_s = -P_v \tag{3}$$

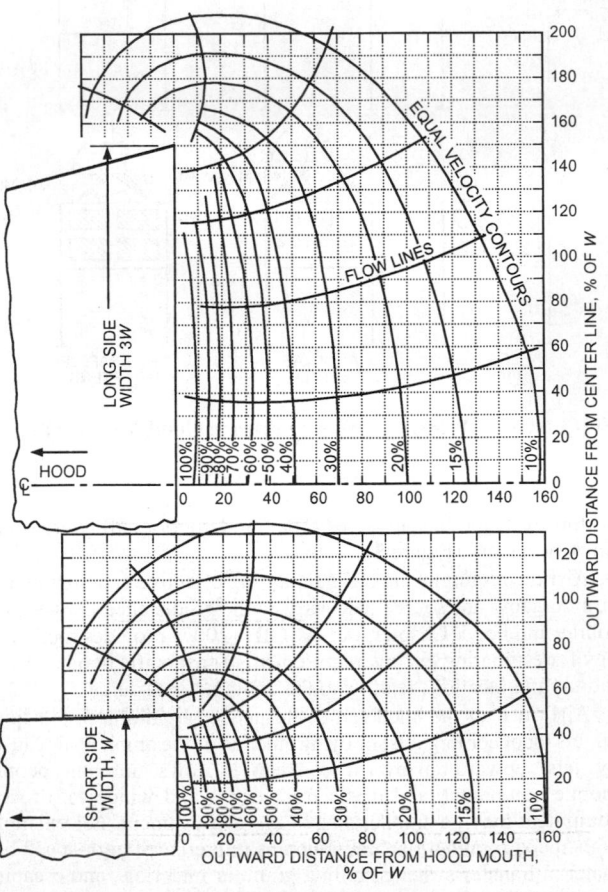

Fig. 6   Velocity Contours for Plain Rectangular Opening with Sides in a 1:3 Ratio
(Alden and Kane 1982; used by permission)

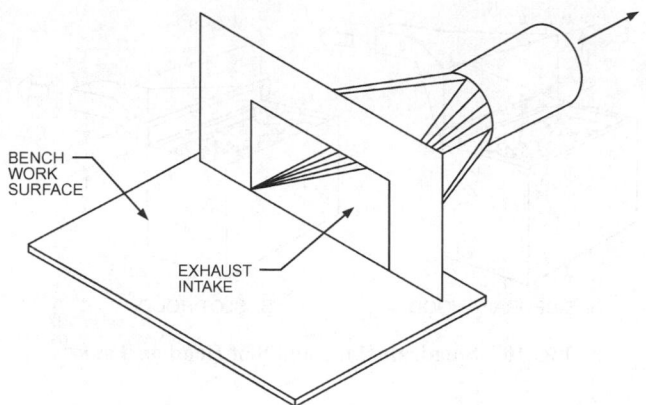

Fig. 8   Hood on Bench

Static pressure in the hood/duct is the static pressure (velocity pressure) plus the head loss, which is expressed as a fraction of the velocity pressure, as

$$P_{s,h} = P_v + C_o P_v \tag{4}$$

Rearranged, the hood/duct static pressure $P_{s,h}$ (hood suction) for hoods is

$$P_{s,h} = (1 + C_o)P_v \tag{5}$$

and the change in total pressure is

$$\Delta P_t = P_{s,h} - P_v = C_o P_v \tag{6}$$

Loss coefficients C_o for various hood shapes are given in Figure 7. For tapered hoods, Figure 5 shows that the optimum hood entry angle to minimize entry loss is 45°, but this may be impractical in many situations because of the required transition length. A 90° angle, with a corresponding loss factor of 0.25 (for rectangular openings), is typical for many tapered hoods.

Example 1. A nonenclosing side-draft flanged hood (Figure 8) with face dimensions of 1.5 by 4 ft rests on the bench. The required volumetric flow rate is 1560 cfm. The duct diameter is 9 in.; this gives a duct velocity of 3530 fpm. The hood is designed such that the largest angle of transition between the hood face and the duct is 90°. What is the suction pressure (static pressure) for this hood? Assume air density at 72°F.

Solution: The two transition angles cannot be equal. Whenever this is true, the larger angle is used to determine the loss factor from Figure 7. Because the transition piece originates from a rectangular opening, the curve marked "rectangular" must be used. This corresponds to a loss factor of 0.25. The duct velocity pressure is

$$P_v = \frac{\rho V^2}{2g_c} = \frac{(0.075)(3530)^2}{(2)(32.2)} \times \frac{12}{(62.4)(3600)} = 0.78 \text{ in. of water}$$

From Equation (5),

$$P_{s,h} = (1 + 0.25)(0.78) = 0.98 \text{ in. of water}$$

Compound Hoods. Losses for multislot hoods (Figure 9) or single-slot hoods with a plenum (compound hoods) must be analyzed somewhat differently. The slots distribute air over the hood face and do not influence capture efficiency. Slot velocity should be approximately 2000 fpm to provide required distribution at minimum energy cost; plenum velocities are typically 50% of slot velocities (approximately 1000 fpm). Higher velocities dissipate more energy and can cause hot spots in the face of the hood.

Losses occur when air passes through the slot and when air enters the duct. Because the velocities, and therefore the velocity pres-

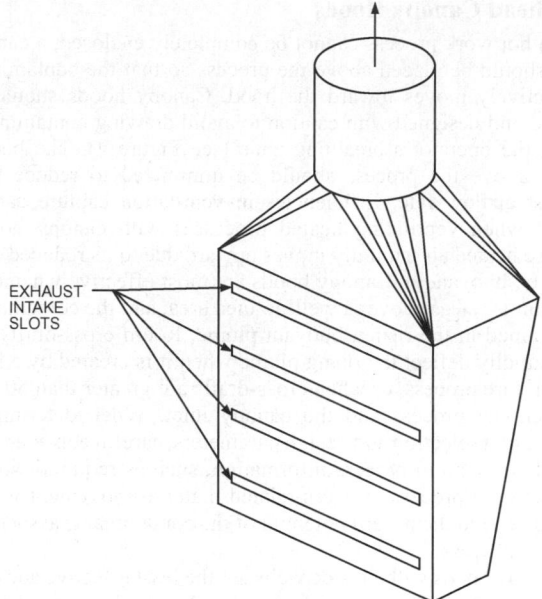

Fig. 9   Multislot Nonenclosing Hood

sures, can be different at the slot and at the duct entry locations, the hood suction must reflect both losses and is given by

$$P_{s,h} = P_v + (C_o P_v)_s + (C_o P_v)_d \tag{7}$$

where the first P_v is generally the higher of the two velocity pressures, s refers to the slot, and d refers to the duct entry location.

Example 2. A multislot hood has three slots, each 1 by 40 in. At the top of the plenum is a 90° transition into the 10 in. duct. The volumetric flow rate required for this hood is 1650 cfm. Determine the hood suction (static pressure). Assume air density at 72°F.

Solution: The slot velocity V_s is

$$V_s = \frac{(1650)(144)}{(3)(40)(1)} = 1980 \text{ fpm}$$

which is near the minimum slot velocity of 2000 fpm. Substituting this velocity,

$$P_v = \frac{\rho V^2}{2g} = \frac{(0.075)(1980)^2}{(2)(32.2)} \times \frac{(12)}{(62.4)(3600)} = 0.24 \text{ in. of water}$$

The duct area is 0.5454 ft². Therefore, duct velocity and velocity pressure are

$$V_d = Q/A$$

$$V_d = \frac{1650}{0.5454} = 3025 \text{ fpm}$$

Substituting this velocity,

$$P_v = \frac{(0.075)(3025)^2}{(2)(32.2)} \times \frac{(12)}{(62.4)(3600)} = 0.57 \text{ in. of water}$$

For a 90° transition into the duct, the loss factor is 0.25. For the slots, the loss factor is 1.78 (Figure 7). The duct velocity pressure is added to the sum of the two losses because it is larger than the slot velocity pressure. Using Equation (7),

$$P_{s,h} = 0.57 + (1.78)(0.24) + (0.25)(0.57) = 1.14 \text{ in. of water}$$

Exhaust volume requirements, minimum duct velocities, and entry loss factors for many specific operations are given in Chapter 13 of ACGIH (2007).

Overhead Canopy Hoods

If a hot work process cannot be completely enclosed, a canopy hood should be placed above the process so that the contaminant convectively moves toward the hood. Canopy hoods should be applied and designed with caution to avoid drawing contaminants across the operator's breathing zone (see Figure 4). The hood's height above the process should be minimized to reduce total exhaust airflow rate. Efficiencies in ventilation capture can be gained when ventilating heated processes with canopy hoods, because heated air naturally moves upward due to its reduced density (i.e., buoyancy). Canopy hoods are most effective when contaminant is released over a well-defined area, and the contaminant is entrained in the rising, buoyant plume. Room cross-drafts can substantially deflect the rising plume when it is created by a low-temperature process, or when cross-drafts are greater than 50 fpm between the process and the canopy inlet. When determining proper hood selection and design parameters, careful consideration should be given to process information, such as required worker access to the process, process-related material movement within the plume, and the hazard potential of the contaminants associated with the process.

Canopy hoods without side walls are the least effective and efficient method of controlling hot process plumes. The limitation of any hood design with distance between the hood face and surface of the source is the ability of cross-drafts to interfere with capturing contaminants rising from the hot process. Where cross-drafts greater than 50 fpm are present, hood designs should include side walls. At a minimum, one side wall should be included in the hood design on the side of the process where the cross-draft originates (upstream side).

Canopy Hoods with Sidewalls

When side walls are included, or when the process is close to a structural wall, the plume may attach to the wall. In this event, the plume entrainment volume is reduced compared to that in an unbounded plume, and the resulting flow in the plume is reduced to half the flow of an unbounded plume. If there are two walls attached at a right angle, the flow is reduced to 1/4 of the unbounded plume flow (Nielsen 1993).

Low Canopy Hoods

Whenever the distance between a canopy hood and the hot source is within 3 ft or the source diameter, whichever is smaller, this hood is considered to be a low canopy hood. Its close proximity to the source does not allow sufficient time for the plume to expand; thus, the diameter or cross section of the hot air column is approximately the same as the source. Under this design scenario, the diameter or side dimensions of the hood need only be about 1 ft larger than the source diameter at its widest cross-section (Hemeon 1963, 1999). For rectangular sources, rising plumes may be better controlled if the hood shape reflects that of the source. In this circumstance, perform the hood airflow and design calculations as for a circular source, once for the length and once for the width dimensions.

High Canopy Hood Use as Redundant Control Measure

The high canopy hood without side walls is the least favorable canopy design. The design can be used as a redundant measure for controlling large-volume process plumes. High canopy hoods are not recommended as a primary control measure for heated processes, because of the large volumes of air displaced to remove pollutants from the workplace. For example, arc furnace charging has a limited duration, and restricted canopy hood use while the furnace is being charged reduces the required volume of replacement air. Ideally, high canopy hood faces without walls should be round, because rising air from point sources and compact shapes (i.e., not line sources) becomes circular in cross section as it rises (Bill and

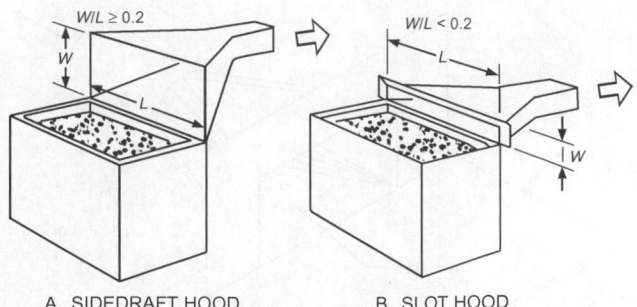

Fig. 10   Sidedraft Hood and Slot Hood on Tank

Gebhart 1975). This occurs because turbulence sweeps the plume edges inward to a minimal volume. However, it is more cost-effective to manufacture and install square or rectangular hoods. Baffles are recommended at the face of rectangular canopy hoods to approximate the area of a round hood face.

Ventilation Controls for Large-Scale Hot Processes

Equations to approximate the velocity, area, and volumetric flow of rising air above a large-scale cylindrical heated process with excess air temperatures ($\Delta T < 198°F$) are available from several sources (ACGIH 2007; Goodfellow 1985; Hemeon 1963, 1999; U.S. Public Health Service 1973). These equations derive from compilation of empirical research by Hemeon and others, and are useful for traditional large-scale, high-temperature processes (e.g., arc furnaces, tapping operations).

Ventilation Controls for Small-Scale Hot Processes

New equations to approximate the velocity, area, and volumetric flow of the rising air above a small-scale heated process have been developed, and validated within a range of excess air temperatures ($2°F < \Delta T < 54°F$) (McKernan and Ellenbecker 2007; McKernan et al. 2007a, 2007b). These equations are based on modern research applicable to designing engineering controls for heated processes, as well as historic work by Hemeon and others (Goodfellow 1985; Hemeon 1963, 1999; U.S. Public Health Service 1973). They are particularly useful for approximating the volumetric flow from discrete low-temperature sources. The historic equations of Hemeon and others continue to be useful for the traditional large-scale, high temperature processes (e.g., arc furnaces, tapping operations).

Sidedraft Hoods

Sidedraft hoods typically draw contaminant away from the operator's breathing zone. With a buoyant source, a sidedraft hood requires a higher exhaust volumetric flow rate than a low canopy hood. If a low canopy hood restricts the work process, a sidedraft hood may be more cost-effective than a high canopy hood. Examples of sidedraft hoods include multislotted "pickling" hoods near welding benches (Figure 9) and slot hoods on tanks (Figure 10).

OTHER LOCAL EXHAUST SYSTEM COMPONENTS

Duct Design and Construction

Duct Considerations. The second component of a local exhaust ventilation system is the duct through which contaminated air is transported from the hood(s). Round ducts are preferred because they (1) offer more uniform velocity to resist settling of material, (2) can withstand the higher static pressures normally found in industrial exhaust systems, and (3) are easier to seal. When design limitations require rectangular or flat oval ducts, the aspect ratio (height-to-width ratio) should be as close to unity as possible.

Minimum transport velocity is the velocity required to transport particles without settling. Table 2 lists some generally accepted

Table 2　Contaminant Transport Velocities

Nature of Contaminant	Examples	Minimum Transport Velocity, fpm
Vapor, gases, smoke	All vapors, gases, smoke	Usually 1000 to 2000
Fumes	Welding	2000 to 2500
Very fine light dust	Cotton lint, wood flour, litho powder	2500 to 3000
Dry dusts and powders	Fine rubber dust, molding powder dust, jute lint, cotton dust, shavings (light), soap dust, leather shavings	3000 to 4000
Average industrial dust	Grinding dust, buffing lint (dry), wool jute dust (shaker waste), coffee beans, shoe dust, granite dust, silica flour, general material handling, brick cutting, clay dust, foundry (general), limestone dust, asbestos dust in textile industries	3500 to 4000
Heavy dust	Sawdust (heavy and wet), metal turnings, foundry tumbling barrels and shakeout, sandblast dust, wood blocks, hog waste, brass turnings, cast-iron boring dust, lead dust	4000 to 4500
Heavy and moist dust	Lead dust with small chips, moist cement dust, asbestos chunks from transite pipe cutting machines, buffing lint (sticky), quicklime dust	4500 and up

Source: From American Conference of Governmental Industrial Hygienists (ACGIH®), *Industrial Ventilation: A Manual of Recommended Practice*, 27th ed. Copyright 2010. Reprinted with permission.

transport velocities as a function of the nature of the contaminants (ACGIH 2007). The values listed are typically higher than theoretical and experimental values to account for (1) damage to ducts, which increases system resistance and reduces volumetric flow and duct velocity; (2) duct leakage, which tends to decrease velocity in the duct system upstream of the leak; (3) fan wheel corrosion or erosion and/or belt slippage, which could reduce fan volume; and (4) reentrainment of settled particles caused by improper operation of the exhaust system. Design velocities can be higher than minimum transport velocities but should never be significantly lower.

When particle concentrations are low, the effect on fan power is negligible. Standard duct sizes and fittings should be used to cut cost and delivery time. Information on available sizes and cost of nonstandard sizes can be obtained from the contractor(s).

Duct Losses. Chapter 21 of the 2009 *ASHRAE Handbook—Fundamentals* covers the basics of duct design and design of metal-working exhaust systems. Loss coefficients are found in the ASHRAE *Duct Fitting Database* CD-ROM (ASHRAE 2008).

For systems conveying particles, elbows with a centerline radius-to-diameter ratio (r/D) greater than 1.5 are the most suitable. If $r/D \leq 1.5$, abrasion in dust-handling systems can reduce the life of elbows. Elbows, especially those with large diameters, are often made of seven or more gores. For converging flow fittings, a 30° entry angle is recommended to minimize energy losses and abrasion in dust-handling systems (Fitting ED5-1 in Chapter 21 of the 2009 *ASHRAE Handbook—Fundamentals*).

Where exhaust systems handling particles must allow for a substantial increase in future capacity, required transport velocities can be maintained by providing open-end stub branches in the main duct. Air is admitted through these stub branches at the proper pressure and volumetric flow rate until the future connection is installed. Figure 11 shows such an air bleed-in. Using outside air minimizes replacement air requirements, though care must be taken to consider potential adverse effects of temperature or humidity extremes associated with the two air streams. The size of the opening can be calculated by determining the pressure drop required across the orifice from the duct calculations. Then the orifice velocity pressure can be determined from one of the following equations:

$$P_{v,o} = \frac{\Delta P_{t,o}}{C_o} \qquad (8)$$

where

$P_{v,o}$ = orifice velocity pressure, in. of water

ΔT_o = total pressure to be dissipated across orifice, in. of water

C_o = orifice loss coefficient referenced to the velocity at the orifice cross-sectional area, dimensionless (see Figure 7)

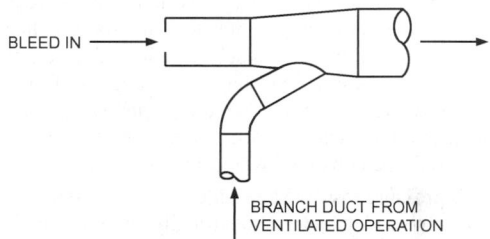

Fig. 11　Air Bleed-In

Once the velocity pressure is known, orifice velocity and size can be determined.

Occasionally, a counterweighted backdraft damper or spring-loaded air admittance valve, configured to allow airflow into the duct but not out, is used as an air bleed in lieu of an orifice in systems that operate under varying airflow conditions. This ensures the proper quantity of transport airflow inside the duct, helping to eliminate material fallout and subsequent duct blockage.

Integrating Duct Segments. Most systems have more than one hood. If the pressures are not designed to be the same for merging parallel airstreams, the system adjusts to equalize pressure at the common point; however, the resulting flow rates of the two merging airstreams will not necessarily be the same as designed. As a result, the hoods can fail to control the contaminant adequately, exposing workers to potentially hazardous contaminant concentrations. Two design methods ensure that the two pressures will be equal. The preferred design self-balances without external aids. This procedure is described in the section on Industrial Exhaust System Duct Design in Chapter 21 of the 2009 *ASHRAE Handbook—Fundamentals*. The second design, which uses adjustable balance devices such as blast gates or balancing dampers, is not recommended, especially when abrasive material is conveyed.

Duct Construction. Elbows and converging flow fittings should be made of thicker material than the straight duct, especially if abrasives are conveyed. Elbows with $r/D > 2$ with replaceable wear plates (wear backs) in the heel are often used where particulate loading is extremely heavy or the particles are very abrasive. When corrosive material is present, alternatives such as special coatings or different duct materials (fibrous glass or stainless steel) can be used. Cleanout openings should be located to allow access to the duct interior in the event of a blockage. Certain contaminants may require washdown systems and/or fire detection and suppression systems to comply with safety or fire prevention codes. These requirements should be verified with local code officials and insurance underwriters. NFPA standards provide guidance on fire safety. Industrial duct construction is described in Chapter 18 of the 2008 *ASHRAE Handbook—HVAC Systems and Equipment*, and in Sheet Metal and Air Conditioning Contractors' National Association (SMACNA) *Standard* 005-1999.

Air Cleaners

Air-cleaning equipment is usually selected to (1) conform to federal, state, or local emissions standards and regulations; (2) prevent reentrainment of contaminants to work areas; (3) reclaim usable materials; (4) allow cleaned air to recirculate to work spaces and/or processes; (5) prevent physical damage to adjacent properties; and (6) protect neighbors from contaminants.

Factors to consider when selecting air-cleaning equipment include the type of contaminant (number of components, particulate versus gaseous, moisture and heat in the airstream, and pollutant concentration), contaminant characteristics (e.g. volatility, reactivity), required contaminant removal efficiency, disposal method, and air or gas stream characteristics. Auxiliary systems such as instrument-grade compressed air, electricity, or water may be required and should be considered in equipment selection. Specific hazards such as explosions, fire, or toxicity must be considered in equipment selection, design, and location. See Chapters 28 and 29 of the 2008 *ASHRAE Handbook—HVAC Systems and Equipment* for information on equipment for removing airborne contaminants. An applications engineer should be consulted when selecting equipment.

The cleaner's pressure loss must be added to overall system pressure calculations. In some cleaners, specifically some fabric filters, loss increases as operation time increases. System design should incorporate the maximum pressure drop of the cleaner, or hood flow rates will be lower than designed during most of the duty cycle. Also, fabric collector losses are usually given only for a clean air plenum. A reacceleration to the duct velocity, with the associated entry losses, must be calculated during design. Most other cleaners are rated flange-to-flange with reacceleration included in the loss.

Air-Moving Devices

The type of air-moving device selected depends on the type and concentration of contaminant, the pressure rise required, and allowable noise levels. Fans are usually used. Chapter 20 of the 2008 *ASHRAE Handbook—HVAC Systems and Equipment* describes available fans; Air Movement and Control Association *Publication* 201 (AMCA 2002) describes proper connection of the fan(s) to the system. The fan should be located downstream of the air cleaner whenever possible to (1) reduce possible abrasion of the fan wheel blades and (2) create negative pressure within the air cleaner and the entire length of dirty duct so that air leaks into the exhaust system throughout its dirty side and control of the contaminant is maintained.

Fans handling flammable or explosive dusts should be specified as spark-resistant. AMCA provides three different spark-resistant fan construction specifications. The fan manufacturer should be consulted when handling these materials. Multiple NFPA standards give fire safety requirements for fans and systems handling explosive or flammable materials.

When possible, devices such as fans and pollution-control equipment should be located outside classified areas, and/or outside the building, to reduce the risk of fire or explosion.

In some instances, the fan is located upstream from the cleaner to help remove dust. This is especially true with cyclone collectors, for example, which are used in the woodworking industry. If explosive, corrosive, flammable, or sticky materials are handled, an injector (also known as an eductor) can transport the material to the air-cleaning equipment. Injectors create a shear layer that induces airflow into the duct. Injectors should be the last choice because their efficiency seldom exceeds 10%.

Energy Recovery to Increase Sustainability

Energy transfer from exhausted air to replacement air may be economically feasible, depending on the (1) location of the exhaust and replacement air ducts, (2) temperature of the exhausted gas, and (3) nature of the contaminants being exhausted. Heat transfer efficiency depends on the type of heat recovery system used.

If exhausted air contains particulate matter (e.g., dust, lint) or oil mist, the exhausted air should be filtered to prevent fouling the heat exchanger. If exhausted air contains gaseous and vaporous or volatile contaminants, such as hydrocarbons and water-soluble chemicals, their effect on the heat recovery device should be investigated. Chapter 25 of the 2008 *ASHRAE Handbook—HVAC Systems and Equipment* discusses air-to-air energy recovery systems.

When selecting energy recovery equipment for industrial exhaust systems, cross-contamination from the energy recovery device must be considered. Some types of energy recovery equipment may allow considerable cross-contamination (e.g., some heat wheels) from the exhaust into the supply airstream, whereas other types (e.g., run-around coils) do not. The exhaust side of the energy recovery device should be negatively pressured compared to the supply side, so that any leakage will be from the clean side into the contaminated side. This is not acceptable for some applications. The material of the energy recovery device must be compatible with the pollutants being exhausted. If the exhaust airstream destroys the heat exchanger, contamination can enter the supply airstream and cause additional equipment damage as well as increase exposure to workers.

Exhaust Stacks

The exhaust stack must be designed and located to prevent reentraining discharged air into supply system inlets. The building's shape and surroundings determine the atmospheric airflow over it. Chapter 24 of the 2009 *ASHRAE Handbook—Fundamentals* and Chapter 45 of this volume cover exhaust stack design. The typical code-required minimum stack height is intended to provide protection for workers near the stack, so discharged air will be above their breathing zone. The minimum required stack height does not protect against reentrainment of contaminated exhaust into any outside air intakes.

If rain protection is important, a no-loss stack head design (ACGIH 2007; SMACNA *Standard* 005) is recommended. Weather caps deflect air downward, increasing the chance that contaminants will recirculate into air inlets, have high friction losses, and provide less rain protection than a properly designed stack head. Weather caps should never be used with a contaminated or hazardous exhaust stream.

Figure 12 contrasts flow patterns of weather caps and stack heads. Loss data for stack heads are presented in the *Duct Fitting Database* CD-ROM (ASHRAE 2008). Losses in straight-duct stack heads are balanced by the pressure regain at the expansion to the larger-diameter stack head.

Instrumentation and Controls

Some industrial exhaust systems may require positive verification of system airflow. Indicators of performance failure may require both audible and visual warning indicators. Other instrumentation, such as dust collector level indication, rotary lock valve operation, or fire detection, may be required. Selection of electronic monitoring instruments should consider durability expectations, maintenance, and calibration requirements. Interfaces may be required with the process control system or with the balance of the plant ventilation system. Electrical devices in systems conveying flammable or explosive materials or in a hazardous location may need to meet certain electrical safety and code requirements. These requirements are determined by the owner, process equipment manufacturer, federal and state regulations, local codes, and/or insurance requirements.

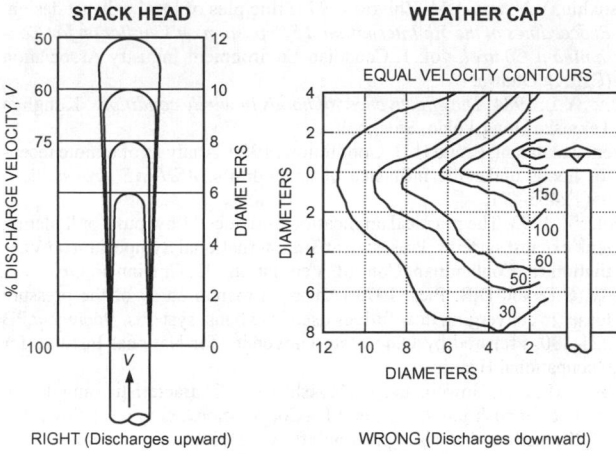

Fig. 12   Comparison of Flow Pattern for Stack Heads and Weather Caps

OPERATION

System Testing and Balancing

After installation, an exhaust system should be tested and balanced to ensure that it operates properly, with the required flow rates through each hood. If actual flow rates are different from design values, they should be corrected before the system is used. Testing is also necessary to obtain and document baseline data to determine (1) compliance with federal, state, and local codes; (2) by periodic inspections or real-time monitoring, whether maintenance on the system is needed to ensure design operation; (3) whether a system has sufficient capacity for additional airflow; (4) whether system leakage is acceptable; and (5) compliance with testing, adjusting, and balancing (TAB) standards. AMCA (1990) and Chapter 5 of ACGIH (2007) contain detailed information on preferred methods for testing systems.

Operation and Maintenance

Periodic inspection and maintenance are required for proper operation of exhaust systems. System designers should keep this requirement in mind and account for it through the installation of clean-out/inspection doors and through strategic placement of equipment that ensures access for maintenance activities. Systems are often changed or damaged after installation, resulting in low duct velocities and/or incorrect volumetric flow rates. Low duct velocities can cause contaminants to settle and plug the duct, reducing flow rates at affected hoods. Adding hoods to an existing system can change volumetric flow at the original hoods. In both cases, changed hood volumes can increase worker exposure and health risks. The maintenance program should include (1) inspecting ductwork for particulate accumulation and damage by erosion or physical abuse, (2) checking exhaust hoods for proper volumetric flow rates and physical condition, (3) checking fan drives, (4) maintaining air-cleaning equipment according to manufacturers' guidelines, and (5) confirming that the system continues to meet compliance with worker exposure and environmental pollution requirements.

REFERENCES

ACGIH. 2007. *Industrial ventilation: A manual of recommended practice for Operation and Maintenance.* Committee on Industrial Ventilation, American Conference of Governmental Industrial Hygienists, Cincinnati, OH.

ACGIH. 2010. *Industrial ventilation: A manual of recommended practice for Design,* 27th ed. Committee on Industrial Ventilation, American Conference of Governmental Industrial Hygienists, Cincinnati, OH.

Alden, J.L. and J.M. Kane. 1982. *Design of industrial ventilation systems,* 5th ed. Industrial Press, New York.

AMCA. 1990. Field performance measurement of fan systems. *Publication* 203-90. Air Movement and Control Association International, Arlington Heights, IL.

AMCA. 2002. Fans and systems. *Publication* 201-02. Air Movement and Control Association International, Arlington Heights, IL.

ASHRAE. 2010. Ventilation for acceptable indoor air quality. ANSI/ASHRAE *Standard* 62.1-2010.

ASHRAE. 2008. *Duct fitting database.*

Bill, R.G. and B. Gebhart. 1975. The transition of plane plumes. *International Journal of Heat and Mass Transfer* 18:513-526.

Brooks, P. 2001. Designing industrial exhaust systems. *ASHRAE Journal* 43(4):1-5.

Caplan, K.J. and G.W. Knutson. 1977. The effect of room air challenge on the efficiency of laboratory fume hoods. *ASHRAE Transactions* 83(1):141-156.

Caplan, K.J. and G.W. Knutson. 1978. Laboratory fume hoods: Influence of room air supply. *ASHRAE Transactions* 82(1):522-537.

DallaValle, J.M. 1952. *Exhaust hoods,* 2nd ed. Industrial Press, New York.

Goodfellow, H. 1985. Design of ventilation systems for fume control. In *Advanced design of ventilation systems for contaminant control,* pp. 359-438. Elsevier, New York.

Goodfellow, H. and E. Tahti, eds. 2001. *Industrial ventilation design guidebook.* Academic Press, New York.

Hemeon, W.C.L. 1963. Exhaust for hot processes. Ch. 8 in *Plant and process ventilation,* 2nd ed., pp. 160-196. Industrial Press, New York.

Hemeon, W.C.L. 1999. Exhaust for hot processes. Ch. 8 in *Hemeon's plant and process ventilation,* 3rd ed., pp. 117-147, D.J. Burton, ed. Lewis, New York.

McKernan, J.L. and M.J. Ellenbecker. 2007. Ventilation equations for improved exothermic process control. *Annals of Occupational Hygiene* 51:269-279.

McKernan, J.L., M.J. Ellenbecker, C.A Holcroft, and M.R. Petersen. 2007a. Evaluation of a proposed area equation for improved exothermic process control. *Annals of Occupational Hygiene* 51:725-738.

McKernan, J.L., M.J. Ellenbecker, C.A. Holcroft, and M.R. Petersen. 2007b. Evaluation of a proposed velocity equation for improved exothermic process control. *Annals of Occupational Hygiene* 51:357-369.

NFPA. 2003. Ovens and furnaces. ANSI/NFPA *Standard* 8603. National Fire Protection Association, Quincy, MA.

Nielsen, P.V. 1993. *Displacement ventilation: Theory and design.* Aalborg University, Aalborg, Denmark.

SMACNA. 1999. Round industrial duct construction standards, 2nd ed. ANSI/SMACNA/BSR *Standard* 005-1999. Sheet Metal and Air Conditioning Contractors' National Association, Chantilly, VA.

U.S. Public Health Service. 1973. Air pollution engineering manual. *Publication* 999-AP-40.

BIBLIOGRAPHY

Bastress, E., J. Niedzwocki, and A. Nugent. 1974. Ventilation required for grinding, buffing, and polishing operations. *Publication* 75107. U.S. Department of Health, Education, and Welfare. National Institute for Occupational Safety and Health, Washington, D.C.

Baturin, V.V. 1972. *Fundamentals of industrial ventilation,* 3rd English ed. Pergamon, New York.

Braconnier, R. 1988. Bibliographic review of velocity field in the vicinity of local exhaust hood openings. *American Industrial Hygiene Association Journal* 49(4):185-198.

Brandt, A.D., R.J. Steffy, and R.G Huebscher. 1947. Nature of airflow at suction openings. *ASHVE Transactions* 53:5576.

British Occupational Hygiene Society (BOHS). 1987. Controlling airborne contaminants in the workplace. *Technical Guide* 7. Science Review Ltd. and H&H Scientific Consultants, Leeds, U.K.

Burgess, W.A., M.J. Ellenbecker, and R.D. Treitman. 1989. *Ventilation for control of the work environment.* John Wiley & Sons, New York.

Chambers, D.T. 1993. *Local exhaust ventilation: A philosophical review of the current state-of-the-art with particular emphasis on improved worker protection.* DCE, Leicester, U.K.

Flynn, M.R. and M.J. Ellenbecker. 1985. The potential flow solution for airflow into a flanged circular hood. *American Industrial Hygiene Journal* 46(6):318-322.

Fuller, F.H. and A.W. Etchells. 1979. The rating of laboratory hood performance. *ASHRAE Journal* 21(10):49-53.

Garrison, R.P. 1977. *Nozzle performance and design for high-velocity/low-volume exhaust ventilation.* Ph.D. dissertation. University of Michigan, Ann Arbor.

Goodfellow, H.D. 1986. *Ventilation '85 (Conference Proceedings).* Elsevier, Amsterdam.

Hagopian, J.H. and E.K. Bastress. 1976. Recommended industrial ventilation guidelines. *Publication* 76162. U.S. Department of Health, Education, and Welfare, National Institute for Occupational Safety and Health, Washington, D.C.

Heinsohn, R.J. 1991. *Industrial ventilation: Engineering principles.* John Wiley & Sons, New York.

Heinsohn, R.J., K.C. Hsieh, and C.L. Merkle. 1985. Lateral ventilation systems for open vessels. *ASHRAE Transactions* 91(1B):361-382.

Hinds, W. 1982. *Aerosol technology: Properties, behavior, and measurement of airborne particles.* John Wiley & Sons, New York.

Huebener, D.J. and R.T. Hughes. 1985. Development of push-pull ventilation. *American Industrial Hygiene Association Journal* 46(5):262-267.

Kofoed, P. and P.V. Nielsen. 1991. Thermal plumes in ventilated rooms—Vertical volume flux influenced by enclosing walls. Presented at 12th Air Infiltration and Ventilation Centre Conference, Ottawa.

Ljungqvist, B. and C. Waering. 1988. Some observations on "modern" design of fume cupboards. *Proceedings of the 2nd International Symposium on Ventilation for Contaminant Control, Ventilation '88.* Pergamon, U.K.

Morton, B.R., G. Taylor, and J.S. Turner. 1956. Turbulent gravitational convection from maintained and instantaneous sources. *Proceedings of Royal Society* 234A:1.

Posokhin, V.N. and A.M. Zhivov 1997. Principles of local exhaust design. *Proceedings of the 5th International Symposium on Ventilation for Contaminant Control*, vol. 1. Canadian Environment Industry Association (CEIA), Ottawa.

Qiang, Y.L. 1984. *The effectiveness of hoods in windy conditions.* Kungliga Tekniska Hoggskolan, Stockholm.

Safemazandarani, P. and H.D. Goodfellow. 1989. Analysis of remote receptor hoods under the influence of cross-drafts. *ASHRAE Transactions* 95(1):465-471.

Sciola, V. 1993. The practical application of reduced flow push-pull plating tank exhaust systems. Presented at 3rd International Symposium on Ventilation for Contaminant Control, Ventilation '91, Cincinnati, OH.

Sepsy, C.F. and D.B. Pies. 1973. An experimental study of the pressure losses in converging flow fittings used in exhaust systems. *Document* PB 221 130. Prepared by Ohio State University for National Institute for Occupational Health.

Shibata, M., R.H. Howell, and T. Hayashi 1982. Characteristics and design method for push-pull hoods: Part I—Cooperation theory on airflow; Part 2—Streamline analysis of push-pull flows. *ASHRAE Transactions* 88(1): 535-570.

Silverman, L. 1942. Velocity characteristics of narrow exhaust slots. *Journal of Industrial Hygiene and Toxicology* 24 (November):267.

Sutton, O.G. 1950. The dispersion of hot gases in the atmosphere. *Journal of Meteorology* 7(5):307.

Zarouri, M.D., R.J. Heinsohn, and C.L. Merkle. 1983. Computer-aided design of a grinding booth for large castings. *ASHRAE Transactions* 89(2A):95-118.

Zarouri, M.D., R.J. Heinsohn, and C.L. Merkle. 1983. Numerical computation of trajectories and concentrations of particles in a grinding booth. *ASHRAE Transactions* 89(2A):119-135.

KITCHEN VENTILATION

KITCHEN ventilation is a complex web of interconnected HVAC systems. The main components typically include (1) cooling to address heat from cooking appliances, (2) makeup air to provide proper pressurization during cooking operations, and (3) exhaust to remove heat and effluent generated by cooking appliances. System design includes aspects of air conditioning, fire safety, ventilation, building pressurization, refrigeration, air distribution, and food service equipment. Kitchens are in many buildings, including restaurants, hotels, hospitals, retail malls, single- and multifamily dwellings, and correctional facilities. Each building type has special requirements for its kitchens, but many basic needs are common to all. This chapter provides an understanding of the different components of kitchen ventilation systems and where they can be applied. Additionally, background information is included to provide an understanding of the history and rationale behind these design decisions.

Kitchen ventilation has at least two purposes: (1) to provide a comfortable environment in the kitchen and (2) to ensure the safety of personnel working in the kitchen and of other building occupants. Comfort criteria often depend on the local climate, because some kitchens are not air conditioned. Kitchen ventilation ensures safety by providing a means to remove the heat, smoke, and grease (cooking effluent) produced during normal cooking operations.

HVAC system designers are most frequently involved in commercial kitchen applications, in which cooking effluent contains large amounts of grease or water vapor. Residential kitchens typically use a totally different type of hood. The amount of grease produced in residential applications is significantly less than in commercial applications, so the health and fire hazard is much lower.

The centerpiece of almost any kitchen ventilation system is an exhaust hood(s), used primarily to remove cooking effluent from kitchens. Effluent includes gaseous, liquid, and solid contaminants produced by the cooking process, and may also include products of fuel and even food combustion. These contaminants must be removed for both comfort and safety; effluent can be potentially life-threatening and, under certain conditions, flammable. Finally, it should be noted that the arrangement of food service equipment and its coordination with the hood(s) can greatly affect the energy used by these systems, which in turn affects kitchen operating costs. Quite often, the hood selection and appliance layout is determined by a kitchen facility designer. To minimize energy use and ensure a properly designed kitchen ventilation system, the HVAC engineer should reach out to the kitchen designer and share the practices and ideas presented in this chapter.

Sustainability Impact

Kitchens are some of the most intensive users of energy for a given area compared to other commercial or institutional occupancies. In addition to energy used during cooking, the kitchen ventilation system must address the large amount of heat emitted or convected into the kitchen from the cooking equipment, and supply and condition the ventilation air needed as part of the cooking effluent exhaust system. Given these factors, it is imperative that the kitchen ventilation system be designed with careful consideration of both first cost and operating costs. An additional factor to be considered is the cooking effluent, and any treatments that may be required before it is discharged into the atmosphere.

COMMERCIAL KITCHEN VENTILATION

ENERGY CONSIDERATIONS

Restaurants and commercial kitchens are the largest consumers of energy per unit of floor area when compared to other commercial or institutional occupancies (Itron and California Energy Commission 2006). Primary drivers of commercial kitchen energy use are the cooking appliances and the HVAC system. The kitchen exhaust ventilation system is often the largest energy-consuming component in a commercial food service facility. However, energy consumption associated with commercial kitchen ventilation (CKV) and HVAC systems, as well the conservation potential, can vary significantly (Fisher 2003). Beyond the design exhaust ventilation rate itself, the magnitude of energy consumption and cost of a CKV system is affected by factors such as geographic location (i.e., climate), system operating hours, static pressure and fan efficiencies, makeup air heating and cooling set points and level of dehumidification, efficiency of heating and cooling systems, level of interaction between kitchen and building HVAC system, appliances under the hood and associated radiant heat gain to space, and applied utility rates. Minimizing the exhaust airflow needed for cooking appliances and reducing radiant load from the appliances are primary considerations in optimizing CKV system design. Because climatic zones vary dramatically in temperature and humidity, energy-efficient designs have widely varying rates of returns on the investment. In new facilities, the designer can select conservation measures suitable for the climatic zone and the HVAC system to maximize the economic benefits.

The operating cost burden has stimulated energy efficiency design concepts and operating strategies discussed in this section and detailed in industry design guidelines (PG&E 2004). It has also impacted changes to ASHRAE *Standard* 90.1.

The updated Kitchen Exhaust Systems section of ASHRAE *Standard* 90.1 states that if a CKV system has a total exhaust airflow rate greater than 5000 cfm, the design must adhere to maximum exhaust rates specified for the different hood types. These rates apply to listed hoods, and are set 30% below the minimum values

---

The preparation of this chapter is assigned to TC 5.10, Kitchen Ventilation.

for unlisted hoods dictated by the *International Mechanical Code®* (IMC) (ICC 2009a); refer to Table 3 in this chapter for minimum unlisted hood airflow rates. If a kitchen or dining facility has a total kitchen hood exhaust airflow rate greater than 5000 cfm, it must have one of the following:

• At least 50% of all replacement air is transfer air that would otherwise be exhausted
• Demand ventilation system(s) on at least 75% of the exhaust air, capable of at least 50% reduction in exhaust and replacement air system airflow rates, including controls necessary to modulate airflow in response to appliance operation and to maintain full capture and containment of smoke, effluent, and combustion products during cooking and idle
• Listed energy recovery devices with a sensible heat recovery effectiveness of not less than 40% on at least 50% of the total exhaust airflow

Energy Conservation Strategies

Specifying Exhaust Hoods for Reduced Airflow. The type and style of exhaust hood selected depends on factors such as restaurant type, restaurant menu, and food service equipment installed, as well as flexibility for future kitchen upgrades. Exhaust flow rates are largely determined by the food service equipment and hood style. Wall-mounted canopy hoods function effectively at lower exhaust flow rates than single-island hoods. Single- and double-island canopy hoods are more sensitive to makeup air (MUA) supply and cross drafts than wall mounted canopy hoods (Swierczyna et al. 2009). Engineered backshelf (proximity) hoods may exhibit the lowest capture and containment flow rates. In some cases, a backshelf hood performs the same job as a wall-mounted canopy hood at one-third the exhaust rate. Cooking appliance type and duty rating must be included in the specification process, because not all hoods (particularly backshelf hoods) are rated or designed for all cooking appliance types or duty ratings.

Threshold exhaust rates for a specific hood and appliance configuration may be determined by laboratory testing under the specifications of ASTM *Standard* F1704-09. Similar in concept to the listed airflow rates derived from UL *Standard* 710, the threshold of containment and capture (C&C) for an ASTM *Standard* F1704 test is established under ideal laboratory conditions and is only a reference point for specifying the exhaust airflow rate for a CKV system.

Side Panels and Overhang. In many cases, side (or end) panels allow a reduced exhaust rate because they direct replacement airflow to the front of the hood and cooking equipment. They are a relatively inexpensive way to improve capture and containment and reduce the total exhaust rate. It is important to know that partial side panels can provide almost the same benefit as full panels. Although tending to defy its definition as an "island" canopy, end panels can improve the performance of a double- or single-island canopy hood. A significant benefit of end panels, when hoods are exposed to cross drafts, is mitigation of the negative effect those drafts have on hood performance. However, air distribution designs that eliminate cross drafts are preferred. Increasing overhang is another specification detail that can improve the hood's ability to capture and allow reduced exhaust rates. It is important that the engineer and food service designers work closely on appliance placement size and type, because they impact hood sizing, which in turn impacts lighting, HVAC, and most importantly hood performance. A hood that is too small (i.e., little or no overhang) may often not be capable of working properly at any airflow rate, whereas those with generous overhang may operate well at airflow rates reduced by 30% or more.

Custom-Designed Hoods. Hoods can be custom designed for specific cooking appliances or cooking processes. Customization often reduces exhaust and replacement air quantities and consequently reduces fan sizes, energy use, and energy costs. To operate at flow rates lower than required by code, custom-designed hoods must be either listed or approved by the local code official. The cost involved with custom design makes the process more applicable to chain restaurants, such as quick service, where a specific design may be installed repeatedly. Some single-site establishments may have architectural restrictions that demand a custom solution.

Transfer Air. ASHRAE *Standard* 62.1 specifies the quantity of outdoor air that must be provided to ventilate public spaces, such as dining rooms, in food service establishments. *Standard* 62.1 allows this ventilation air be reused by transfer from the ventilated public spaces to the kitchen, where it can be used to replace air exhausted by the hood system and assist kitchen comfort. By maximizing transfer of air from adjoining public spaces to the kitchen, the designer is able to minimize the quantity of dedicated outdoor air supplied to the kitchen as replacement air. This can reduce the energy load on the replacement air system, as well as improve thermal comfort in the kitchen. Most quick-service and fast-casual restaurants do not physically segregate the kitchen from the dining room, so conditioned air can be easily transferred from the dining area to the kitchen. In restaurants where the kitchen and dining room are physically segregated, ducts between the two areas may be required for proper flow of replacement air into the kitchen. Transfer fans may be required to overcome duct losses, because the differential pressure between spaces may be very low (<0.005 in. of water). This design can reduce kitchen replacement air requirements and enhance employee comfort, especially if the kitchen is not air conditioned. Codes may restrict transfer of air from adjoining spaces, other than public dining areas, in buildings such as hospitals. Adjoining spaces should not be overventilated to increase transfer airflow, because the heating and cooling conditions for dining and other public spaces are more energy-intense than for conditioning replacement air introduced directly into the kitchen.

Demand-Controlled Ventilation

Demand-controlled ventilation (DCV) refers to any engineered, automated method of modulating (i.e., variable reduction) the amount of air exhausted for a specific cooking operation in response to a part-load or no-load condition. In conjunction with this, the amount of replacement air (consisting of makeup, transfer, and outside air) is also modulated to maintain the same relative air ratios, airflow patterns, and pressurizations.

Use of DCV must be accounted for in the design of all involved ventilation systems to create a fully integrated system.

Complete capture and containment of all smoke and greasy vapor must be maintained when an exhaust system equipped with DCV is operated at less than 100% of design airflow.

Selection of all components, and design of the DCV system, must be such that stable operation can be maintained at all modulated and full-flow conditions.

When DCV is used as the method of compliance with ASHRAE *Standard* 90.1 (section 6.5.7.1.4 part b), it must meet all requirements of that section.

The exhaust system configuration and equipment to be served by a DCV system must be evaluated to determine the feasibility and cost benefits resulting from its use. An example of a situation where a DCV system may not be appropriate is when gas-underfired broilers are present.

Evaporative Cooling. Direct evaporative coolers are an alternative to mechanical cooling (or, for that matter, no cooling) of replacement air only in dry climates where dehumidification is not required. Indirect evaporative cooling has a wider range in geographical applications. Water costs, availability, and use restrictions should be considered.

Heat Recovery from Exhaust Hood Ventilation Air. High-temperature effluent, often in excess of 400°F, from the cooking equipment mixes with replacement room air, resulting in exhaust air temperatures well over 100°F. It is frequently assumed that this heated exhaust air is suitable for heat recovery; however, over time, smoke and grease in the exhaust air can foul the heat transfer surfaces. Under these conditions, the heat exchangers require regular maintenance (e.g., automatic washdown) to maintain heat recovery effectiveness and mitigate risk of fire. Because heat recovery systems are expensive, food service facilities with large ventilation rates and relatively light-duty cooking equipment are the best candidates for this equipment. Hospitals are a good example, with large exhaust rates and very low levels of grease production from the cooking equipment. An exhaust hood equipped with heat recovery is more likely to be cost-effective where the climate is extreme. A mild climate is not conducive to use of this conservation measure.

Optimized Heating and Cooling Set Points. IMC (ICC 2009a) requires that makeup air be conditioned to within 10°F of the kitchen space, except when makeup air is part of the air-conditioning system and does not adversely affect comfort conditions in the occupied space. The exception is important because it allows the design to be optimized to take advantage of the typically lower heating balance points of commercial kitchens. A commercial kitchen may be considered comfortable at temperatures of up to 85°F when space humidity is ≤60%. During heating seasons, space humidity is typically not an issue if the kitchen exhaust systems are properly designed and operated. During the heating season, space gains from unhooded appliances and radiant gains from hooded appliances, lighting, refrigeration units, and staff make it possible to maintain comfort using lower supply air temperatures. During cooling seasons, in climatic areas that require dehumidification, consideration must be given to all sources that may introduce moisture into commercial kitchens, including internal cooking, holding, and washing as well as local MUA systems or kitchen HVAC systems not designed for continuous dehumidification. Humidity control (≤60%) allows optimization of cooling set points at higher temperatures while maintaining a comfortable working environment.

A dedicated outdoor air system (DOAS) providing conditioned outdoor for kitchen heating, cooling and dehumidification also optimizes set points by using economizer operation when outdoor air alone (no conditioning) can maintain kitchen set points. The use of dedicated outdoor air to heat, cool, and dehumidify outdoor air to control space comfort and humidity and replace air exhausted through the hood has been demonstrated to be an energy-effective design for optimizing heating and cooling set points (Brown 2007).

Accordingly, when local MUA systems are used, it is essential that the heating set point of those units not be set higher than the forecasted heating/cooling balance point (e.g., 55°F) to avoid simultaneous replacement air heating and HVAC cooling. It may be more difficult to control comfort when local MUA (e.g., at 55°F) is introduced into a kitchen being heated by a conventional HVAC system introducing kitchen supply air at temperatures ≥80°F.

Reduced Exhaust and Associated Duct Velocities

Tempering outside replacement air can account for a large part of a food service facility's heating and cooling costs. By reducing exhaust flow rates (and the corresponding replacement air quantity) when little or no product is being cooked, energy cost can be significantly reduced when combined with a DVC system. Field evaluations by one large restaurant chain suggest that cooking appliances may operate under no-load conditions for 75% or more of an average business day (Spata and Turgeon 1995).

However, it has been difficult to reduce exhaust flow rates in a retrofit situation because of the minimum duct velocity restriction. National Fire Protection Association (NFPA) *Standard* 96 had historically required a minimum duct velocity of 1500 fpm. The common belief was that, if duct velocity were lowered, a higher percentage of grease would accumulate on the ductwork, which would then require more frequent duct cleaning. However, no data or research could be identified to support this assumption. Therefore, ASHRAE research project RP-1033 (Kuehn 2000) was undertaken to determine the true effect of duct velocity on grease deposition.

The project analyzed grease deposition as a function of mean duct velocity, using octanoic acid (commonly found in cooking oils and other foods). The results showed that, for design duct velocities below the traditional 1500 fpm threshold, grease deposition was not increased; in fact, in isothermal conditions, as duct velocity decreased, grease deposition on all internal sides of the duct also decreased. These results led to NFPA *Standard* 96 and the IMC changing their minimum duct velocity requirements from 1500 fpm to 500 fpm.

Another significant finding in the study was that, if there is a large temperature gradient between exhaust air inside the duct and the external duct wall, the rate of grease deposition increases significantly. Therefore, duct insulation should be considered where there are large temperature variations.

The primary benefit of these code-approved duct velocity changes is the potential for reduced food service energy consumption. By reducing excessive exhaust airflows, while maintaining necessary capture and containment, energy for fans as well as for heating, cooling and/or dehumidifying air that was previously wasted may now be saved. Previously, if a restaurant remodeled their cooking operation and the remodel resulted in reduced exhaust airflows, the owner had to install new, smaller-diameter ductwork to comply with the 1500 fpm duct velocity requirement. This is often too costly, if not also physically impractical. Now, if a system designed for heavy-duty equipment upgrades to more energy-efficient, lighter-duty equipment, exhaust airflows can be reduced without the expense of modifying ductwork.

Reduced code-approved duct velocity also facilitates the application of DVC systems with less resistance from local code authorities. From a new-facility design perspective, it is recommended that most kitchens be designed for an in-duct velocity between 1500 and 1800 fpm. This allows for reducing the airflows to 500 fpm if needed in the future or as part of a demand-ventilation control strategy.

SYSTEM INTEGRATION AND DESIGN

Ideally, system integration and balancing bring the many ventilation components together to provide the most comfortable, efficient, and economical performance of each component and of the entire system. In commercial kitchen ventilation, the replacement air system(s) must integrate and balance with the exhaust system and/or facility HVAC system(s). Even optimal system designs require field testing and balancing once installed. It is important to verify compliance with design, and equally important to confirm that the design meets the needs of the operating facility. Air balance is a critical step in any CKV commissioning process.

The following fundamentals should be considered and applied to all food service facilities, including restaurants, within the constraints of the particular facility and its location, equipment, and systems.

Principles

Although there are exceptions, the following are the fundamental principles of integrating and balancing food service facility systems for both comfort control and economical operation:

- In a freestanding building, the overall building should always be slightly positively pressurized (e.g. +0.005 in. of water) compared to atmosphere to minimize infiltration of unconditioned and/or unfiltered outdoor air. This pressurization helps prevent outdoor contamination such as dust and odors entering the building.
- In multiple-occupancy buildings, maintain the food preparation areas slightly negative to the dining and other adjacent areas but positive to atmosphere. This requires that the adjacent areas be maintained correspondingly higher to atmosphere. This helps prevent odor migration.
- Every kitchen should always be slightly negative (−0.001 in. of water) to adjacent rooms or areas immediately surrounding it to

help contain odors in the kitchen and to isolate the kitchen environment from the dining or customer environment.

- System HVAC design should limit cross-zoning airflow that would allow airflow supplied to food preparation areas being returned and resupplied to customer or other non-food-preparation areas. Odor contamination is an obvious potential problem. In addition, in conditions such as seasonal transitions, when adjacent zones may be in different modes (e.g., economizer versus air conditioning or heating), comfort may be adversely affected. Ideally, the kitchen HVAC system should be separate from all other zones' HVAC systems. Three situations to consider are the following:

  - During seasonal transitions, the kitchen zone may require air conditioning or may be served by ventilation air only, while dining areas require heating. Even in hot kitchens that may require cooling when the adjacent dining areas require heating, it is still important to maintain the pressure differential between these spaces and continue transfer of heated dining-area air into the kitchen.

  - If dedicated kitchen makeup air is heated, thermostatic control of the heating source should ideally be based on kitchen temperature rather than outside temperature. To limit kitchen personnel discomfort, it is important to control the low-temperature MUA set point and prevent drastic temperature variations between the kitchen space and MUA being introduced. Makeup air heating should be interlocked with kitchen HVAC cooling to prevent simultaneous heating and cooling. HVAC thermostat locations in kitchens must consider the potential conflicting temperatures.

  - Ideally, there should be no perceptible drafts in dining areas, with temperature variations of no more than 1°F. Kitchens might not be draft-free; however, velocities at or near exhaust hoods should be no greater than 75 fpm. Kitchen comfort is greatly impacted by radiant heat in work areas, but it is desirable to maintain sensible temperatures within 5°F. These conditions can be achieved with even distribution and thorough circulation of air in each zone by an adequate number of registers sized to preclude high air velocities. If there are noticeable drafts or temperature differences, dining customers will be uncomfortable and facility personnel are generally less comfortable and less productive.

Both design concepts and operating principles for proper integration and balance are involved in achieving desired results under varying conditions. The same principles are important in almost every aspect of food service ventilation.

In restaurants with multiple exhaust hoods, or hoods with demand control systems, exhaust airflow volume may vary throughout the day. Replacement air must be controlled to maintain proper building and kitchen differential pressures to ensure the kitchen remains negative to adjacent areas at all operating points. The more variable the exhaust, or the more numerous and smaller the zones involved, the more complex the design, but the overall pressure relationship principles must be maintained to provide optimum comfort, efficiency, and economy.

A different application is a kitchen with one side exposed to a larger building with common or remote dining. Examples are a food court in a mall or a small restaurant in a hospital, airport, or similar building. Positive pressure at the front of the kitchen might cause some cooking grease, vapor, and odors to spread into the common building space, which would be undesirable. In such a case, the kitchen area is held at a negative pressure relative to other common building areas as well as to its own back room storage or office space. Such spaces that include direct-vent appliances, such as gas-fired water heaters, must maintain the pressure required for safe appliance operation.

Multiple-Hood Systems

Single kitchen exhaust duct/fan systems serving multiple hoods present unique design and balancing challenges not encountered

with single-hood/duct/fan systems. Air balance is one of the main challenges. These systems may be designed with bleed ducts and/or balancing dampers. These dampers must be listed for this application. Additionally, most filters come in varying sizes to allow pressure loss equalization at varying airflows. Adjustable filters are available, but should not be used when they can be interchanged between hoods or within the same hood, because this interchange can disrupt the previously achieved balance. Balancing can also be accomplished by changing the number and/or size of filters in some hoods.

In some multitenant installations, the duct design may be completed and installed before the tenants have been identified. In cases such as master kitchen-exhaust systems, which are sometimes used in shopping center food courts, no single group is responsible for the entire design. The base building designer typically lays out ductwork to (or through) each tenant space, and each tenant selects a hood and lays out connecting ductwork. Often, the base building designer has incomplete information on tenant exhaust requirements. Therefore, one engineer must be responsible for defining criteria for each tenant's design and for evaluating proposed tenant work to ensure that tenant designs match the system's capacity. The engineer should also evaluate any proposed changes to the system, such as changing tenancy. Rudimentary computer modeling of the exhaust system may be helpful (Elovitz 1992). Given the unpredictability and volatility of tenant requirements, it may not be possible to balance the entire system perfectly. However, without adequate supervision, it is very probable the system will not achieve proper balance.

For greatest success with multiple-hood exhaust systems, minimize pressure losses in ducts by keeping velocities low, minimizing sharp transitions, and using hoods with relatively high pressure drops. When pressure loss in the ducts is low compared to the loss through the hood, changes in pressure loss in the ductwork because of field conditions or changes in design airflow have a smaller effect on total pressure loss and thus on actual airflow.

Minimum code-required air velocity (500 fpm) must be maintained in all parts of the exhaust ductwork at all times. If fewer or smaller hoods are installed than the design anticipated, resulting in low velocity in portions of the ductwork, the velocity must be brought up to the minimum. One way is to introduce outdoor air, preferably untempered, through a bleed duct system directly into the exhaust duct (Figure 1). The bypass duct should connect to the top or sides (at least 2 in. from the bottom) of the exhaust duct to prevent backflow of water or grease through the bypass duct when fans are off. This arrangement is also shown in NFPA *Standard* 96 and should be discussed with the authority having jurisdiction.

A fire damper is required in the bleed duct, located close to the exhaust duct. Bypass duct construction should be the same as the exhaust duct construction, including enclosure and clearance requirements, for at least several feet beyond the fire damper or as required by the local authority having jurisdiction (AHJ). Means to adjust the

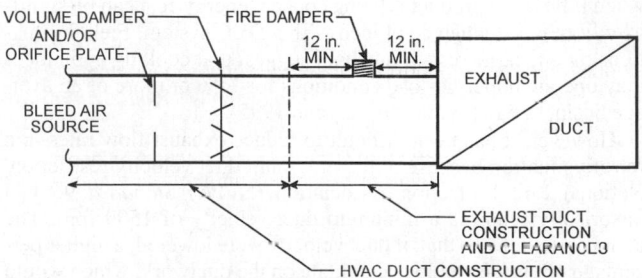

Fig. 1 Bleed Method of Introducing Outdoor Air Directly into Exhaust Duct
(Brohard et al. 2003)

bleed airflow must be provided upstream of the fire damper. All dampers must be in the clean bleed air duct so they are not exposed to grease-laden exhaust air. The difference in pressure between bleed and exhaust air duct may be great; the balancing device must be able to make a fine airflow adjustment against this pressure difference. It is best to provide two balancing devices in series, such as an orifice plate or blast gate for coarse adjustment followed by an opposed-blade damper for fine adjustment.

Directly measuring air velocities in the exhaust ductwork to assess exhaust system performance may be desirable. Velocity (pitot-tube) traverses may be performed in kitchen exhaust systems, but holes drilled for the pitot tube must be liquidtight to maintain the fire-safe integrity of the ductwork, per NFPA *Standard* 96. Holes should never be drilled in the bottom of a duct, where they may collect grease. Velocity traverses should not be performed when cooking is in progress because grease collects on the instrumentation.

Dynamic Volumetric Flow Rate Effects

Minimum exhaust flow rates for kitchen hoods are determined either by laboratory tests or by building code requirements. Energy codes specify maximum airflow rates. In either case, the installed system must ensure proper capture and containment under maximum cooking load conditions. The majority of kitchen exhaust systems use fixed-speed fans, which move the same volume of air at a given speed regardless of air density. Although the air volume remains constant, heat and moisture generated by the cooking process affect mass flow.

Exhaust fans for kitchen ventilation systems, like for other high-temperature exhaust processes, must be selected to provide adequate airflow at standard conditions to meet the mass flow needs of the actual cooking process. Testing for hood listing in accordance with UL *Standard* 710, for example, requires capture and containment testing using actual cooking, with cooking appliance heated to controlled surface temperatures and food product cooked, including flare-ups, to determine airflow. This airflow must be converted to standard air conditions to provide proper design ratings for fan selections.

COOKING EFFLUENT GENERATION AND CONTROL

Air quality, fire safety, labor cost, and maintenance costs are important concerns about emissions from a commercial cooking operation. Cooking emissions have also been identified as a major component of smog particulate. This has led to regulation in some major cities, requiring reduction of emissions from specific cooking operations.

In a fire, grease deposits in duct act as fuel. Reducing this grease can help prevent a small kitchen fire from becoming a major structural fire. In the past, the only control of grease build-up in exhaust ducts was frequent duct cleaning, which is expensive and disruptive to kitchen operation. It also depends on frequent duct inspections and regular cleaning. Grease build-up on fans, fire nozzles, roofs, and other ventilation equipment can be costly in additional maintenance and replacement costs. From an energy and sustainability perspective, it is desirable to reduce the atmospheric emissions and achieve the highest grease extraction or destruction with the lowest energy costs possible. For mechanical extractors, the pressure drop of the filters is the predominant driver for energy usage, whereas for other control systems there may be electrical components or water use that needs to be evaluated. Figure 2 presents some design guidance for what filtration may be desirable under various exhaust temperature and/or duty level situations.

Another issue that commonly comes up during kitchen design and operation is how often ductwork needs to be cleaned in restaurants. Table 1 presents inspection schedules adapted from Table 11.4 of NFPA *Standard* 96.

Table 1 Recommended Duct Cleaning Schedules

Type or Volume of Cooking	Inspection Frequency
Solid fuel	Monthly
High-volume cooking (gas charbroiler or wok cooking)	Quarterly
Moderate-volume	Semiannually
Low-volume (churches, day camps, seasonal businesses)	Annually

Source: NFPA *Standard* 96, Table 11.4.

Effluent Generation

During cooking operations on appliances, effluent is generated. This effluent includes water vapor and organic material (in both particulate and vapor form) released from the food. If the energy source for the appliance involves combustion additional contaminants may also be released.

Particle Size Comparisons (from Exhaust Systems). Effluent from five types of commercial cooking equipment has been measured under a typical exhaust hood (Kuehn et al. 1999). Foods that emit relatively large amounts of grease were selected. Figures 3A and 3B shows the measured amount of grease in the plume entering the hood above different appliances and the amount in the vapor phase, particles below 2.5 μm in size (PM 2.5), particles less than 10 μm in size (PM 10), and the total amount of particulate grease. Ovens and fryers generate little or no grease particulate emissions, whereas other processes generate significant amounts. However, gas underfired broilers (referred to as "gas broilers" in Figures 3A and 3B) generate much smaller particulates compared to the griddles and ranges, and these emissions depend on the broiler design. The amount of grease in the vapor phase is significant and varies from 30% to over 90% by mass; this affects the design approach for grease removal systems.

Carbon monoxide (CO) and carbon dioxide (CO_2) emissions are present in solid fuel and natural gas combustion processes but not in processes from electrical appliances. Additional CO and CO_2 emissions may be generated by gas underfired broilers when grease drippings land on extremely hot surfaces and burn. Nitrogen oxide (NO_x) emissions appear to be exclusively associated with gas appliances and related to total gas consumption.

Figure 3C shows the measured plume volumetric flow rate entering the hood. In general, gas appliances have slightly larger flow rates than electric because additional products of combustion must be vented. Gas underfired and electric broilers have plume flow rates considerably larger than the other appliances shown. Effluent flow rates from gas underfired and electric broilers are approximately 100 times larger than the actual volumetric flow rate created by vaporizing moisture and grease from food. The difference is caused by ambient air entrained into the effluent plume before it reaches the exhaust hood.

Thermal Plume Behavior

The most common method of contaminant control is to install an air inlet device (a hood) where the plume can enter it and be conveyed away by an exhaust system. The hood is generally located above or behind the heated surface to intercept normal upward flow. Understanding plume behavior is central to designing effective ventilation systems.

Effluent released from a noncooking cold process, such as metal grinding, is captured and removed by placing air inlets so that they catch forcibly ejected material, or by creating airstreams with sufficient velocity to induce the flow of effluent into an inlet. This technique has led to an empirical concept of capture velocity that is often misapplied to hot processes. Effluent (such as grease and smoke from cooking) released from a hot process and contained in a plume may be captured by locating an inlet hood so that the plume flows into it by buoyancy. Hood exhaust rate must equal or slightly exceed plume volumetric flow rate, but the hood need

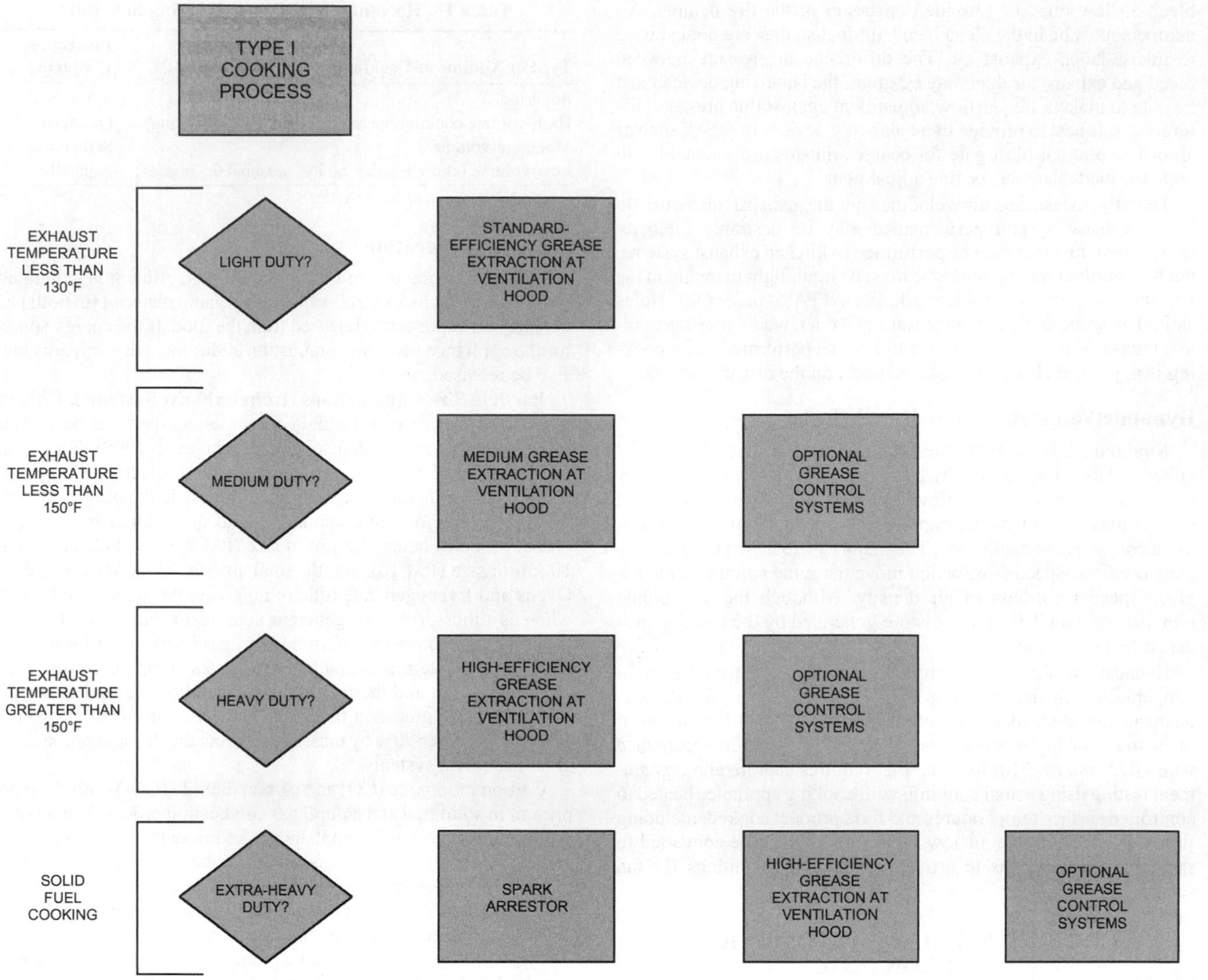

```
                    ┌─────────────┐
                    │   TYPE I    │
                    │  COOKING    │
                    │  PROCESS    │
                    └─────────────┘
```

EXHAUST
TEMPERATURE
LESS THAN
130°F

LIGHT DUTY?

STANDARD-
EFFICIENCY GREASE
EXTRACTION AT
VENTILATION
HOOD

EXHAUST
TEMPERATURE
LESS THAN
150°F

MEDIUM DUTY?

MEDIUM GREASE
EXTRACTION AT
VENTILATION
HOOD

OPTIONAL
GREASE
CONTROL
SYSTEMS

EXHAUST
TEMPERATURE
GREATER THAN
150°F

HEAVY DUTY?

HIGH-EFFICIENCY
GREASE
EXTRACTION AT
VENTILATION
HOOD

OPTIONAL
GREASE
CONTROL
SYSTEMS

SOLID
FUEL
COOKING

EXTRA-HEAVY
DUTY?

SPARK
ARRESTOR

HIGH-EFFICIENCY
GREASE
EXTRACTION AT
VENTILATION
HOOD

OPTIONAL
GREASE
CONTROL
SYSTEMS

Fig. 2　Typical Filter Guidelines Versus Appliance Duty and Exhaust Temperature

not actively induce capture of the effluent if the hood is large enough at its height above the cooking operation to encompass the plume as it expands during its rise. Additional exhaust airflow may be needed to resist cross currents that carry the plume away from the hood.

A heated plume, without cross currents or other interference, rises vertically, entraining additional air, which causes the plume to enlarge and its average velocity and temperature to decrease. If a surface parallel to the plume centerline (e.g., a back wall) is nearby, the plume will be drawn toward the surface by the Coanda effect. This tendency may also help direct the plume into the hood. Figure 4 illustrates a heated plume with and without cooking effluent as it rises from heated cooking appliances. Figure 4A shows two gas underfired broilers cooking hamburgers under a wall-mounted, exhaust-only, canopy hood. Note that the hood is mounted against a clear back wall to improve experimental observation. Figures 4B and 4C show the hot-air plume without cooking, visualized using a schlieren optical system, under full capture and spillage conditions, respectively

Effluent Control

Effluents generated by cooking include grease in particulate (solid or liquid) and vapor states, smoke particles, and volatile

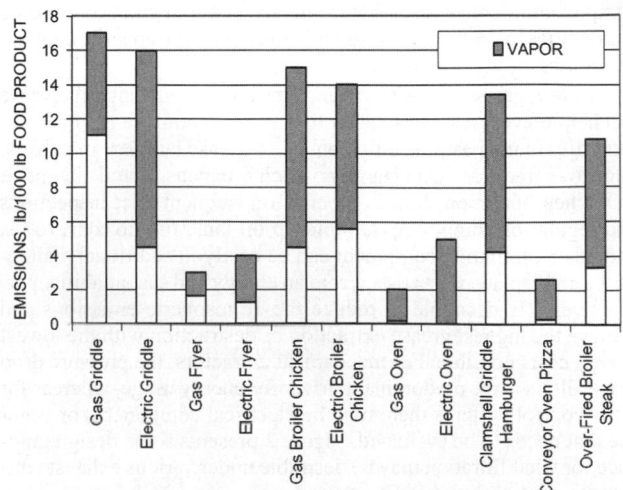

Fig. 3A　Grease in Particulate and Vapor Phases for Commercial Cooking Appliances with Total Emissions Less Than 50 lb/1000 lb of Food Cooked

organic compounds (VOCs or low-carbon aromatics, which are significant contributors to odor). Grease vapor is condensable and may condense into grease particulate in the exhaust airstream when diluted with room-temperature air or when it is exhausted into the cooler outdoor atmosphere..

Effluent controls in the vast majority of kitchen ventilation systems are limited to removing solid and liquid grease particles by mechanical grease removal devices in the hood. More effective devices reduce grease build-up downstream of the hood, lowering the frequency of duct cleaning and reducing the fire hazard.

The reported grease extraction efficiency of mechanical filtration systems (e.g., baffle filters and slot cartridge filters) may reflect the particulate removal performance of these devices. These devices are listed for their ability to limit flame penetration into the plenum and duct. Grease extraction performance can be evaluated using ASTM *Standard* F2519. Smaller aerodynamic particles (<2.5 μm) are not easily removed by mechanical extractors. If these particles must be removed, a pollution control unit is typically added, which removes

a large percentage of the grease that escaped the grease removal device in the hood, as well as smoke particles.

ASHRAE research project RP-745 (Gerstler et al. 1998) found that a significant proportion of grease effluent may be in vapor form (Figure 5), which is not removed by mechanical extractors.

Grease Extraction

The particulate range from cooking operations ranges from 0.01 to 100 μm. Different cooking operations have different ranges of particle sizes in the cooking plume and have been measured for many appliances (Gerstler et al. 1998; Kuehn et al. 2008). Grease particulates larger than 20 μm are too heavy to remain airborne and

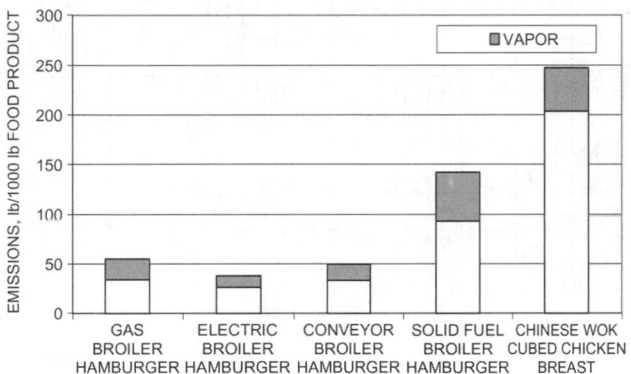

Fig. 3B   Grease in Particulate and Vapor Phases for Commercial Cooking Appliances with Total Emissions Greater Than 50 lb/1000 lb of Food Cooked

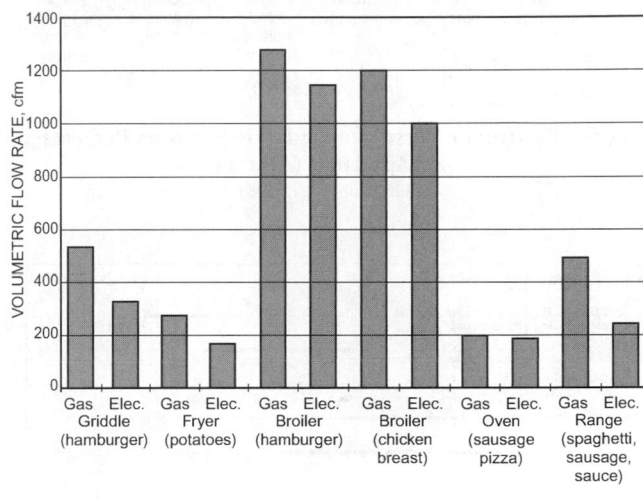

Fig. 3C   Plume Volumetric Flow Rate at Hood Entrance from Various Commercial Cooking Appliances
(Kuehn et al. 1999)

A. TEST SETUP: Two gas underfired broilers under 8 ft long wall-mounted canopy hood, cooking hamburgers.

B. Schlieren photo of capture and containment at 4400 cfm exhaust rate. Hot, clear air visualization, no cooking.

C. Schlieren photo of spillage and containment at 3300 cfm exhaust rate. Hot, clear air visualization, no cooking.

Fig. 4   Hot-Air Plume from Cooking Appliances under Wall-Mounted Canopy Hood

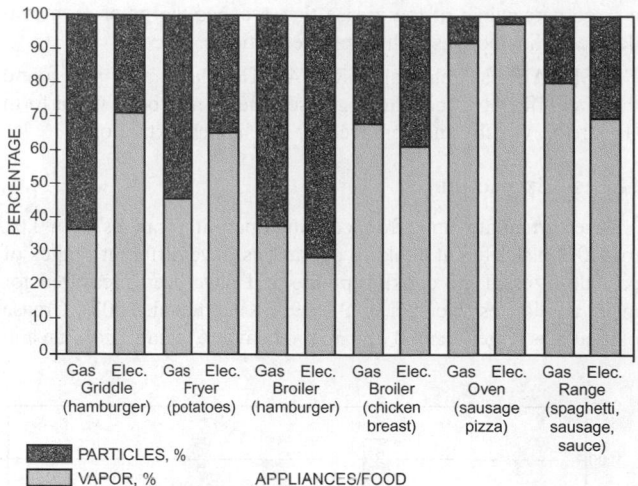

Fig. 5   Particulate Versus Vapor-Phase Emission Percentage per Appliance (Average)
(Gerstler et al. 1998)

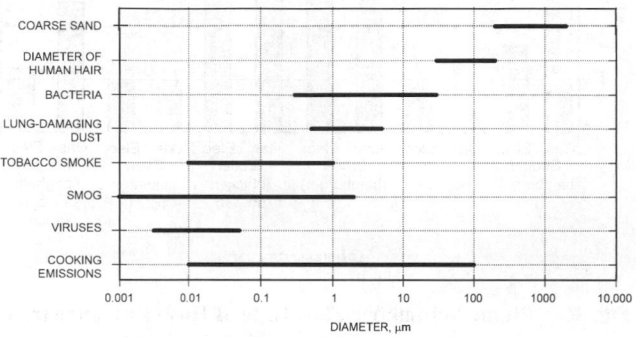

Fig. 6   Size Distribution of Common Particles

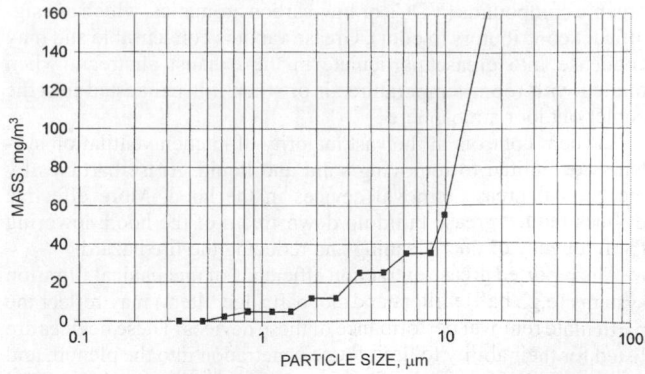

Fig. 7   Gas Griddle Mass Emission Versus Particle Size
(Kuehn et al. 1999)

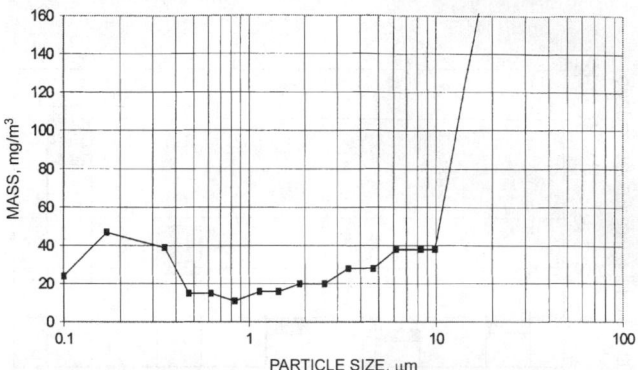

Fig. 8   Gas Underfired Broiler Mass Emission Versus Particle Size
(Kuehn et al. 1999)

drop out of the airstream. Figure 6 compares the size of particles from kitchen exhaust to common items.

Each combination of food product, cooking equipment, and cooking temperature creates a unique particle emissions profile, these profiles change over time during the cooking process. For example, the initial drop of French fries into a fryer gives off a short blast of large particles, whereas cooking a hamburger on a griddle gives off a continuous stream of particles and vapor. Burgers cooked on a broiler tend to burn and emit very small particles (< 1 μm in size).

Variations in the food product itself can also change the emissions of a cooking process. Hamburger with 23% fat content produces more grease than a 20% fat burger. Chicken breast may have a different effluent characteristic than chicken legs or thighs. Even cooking chicken with or without the skin changes the properties of emissions.

Figures 7 and 8 show typical particle emission profiles for a gas griddle and gas underfired broiler both cooking hamburgers (Kuehn et al. 1999).

ASTM *Standard* F2519-05 can be used to determine fractional filter efficiency for grease particulate. A fractional efficiency curve is a graph that gives a filter's efficiency over a range of particle sizes. Fractional efficiency curves are created by subjecting a test filter to a controlled distribution of particles and measuring the quantity of particles at each given size before and after the filter. The amount of reduction of particles is used to calculate the efficiency at each given

size. The fractional efficiency curve for a typical 20 by 20 in. baffle filter tested at 350 cfm/ft is shown in Figure 9.

Extraction efficiencies must be compared at the same airflow per linear length of filter. This gives a consistent way of comparing performance of extraction devices that may be built very differently, such as hoods with removable extractors and with stationary extractors. This is also consistent with the way exhaust flow rates for hoods are commonly specified. The airflow rate through a hood changes hood efficiency by changing the velocity at which the air travels through a filter.

To demonstrate what a filter fractional efficiency means with an actual cooking process, the gas underfired broilers (referred to as "charbroiler") emissions curve and the baffle filter efficiency curve have been plotted on one graph in Figure 10. The area under each emission curves is representative of the total particulate emissions for the gas underfired broiler. As can be seen by comparing the graph before and after the baffle filter, there is very little reduction in the amount of grease exhausted to the duct. The area under the "charbroiler after baffle" curve represents the amount of grease particulate exhausted into the duct.

The graphs and efficiencies shown here are only for particulate grease. There is also a vapor component of the grease that is exhausted, which cannot be removed by filtration. Some of the vapor condenses and is removed as particulate before reaching the filter. Some condenses in the duct and accumulates on the duct and fan. However, with elevated temperatures in the exhaust airstream, vapor may pass through and exit to the atmosphere.

Higher efficiency at a specific particle size may not be the only selection criteria for grease extraction. From an energy and sustainability standpoint, the ideal goal would be to have the highest grease

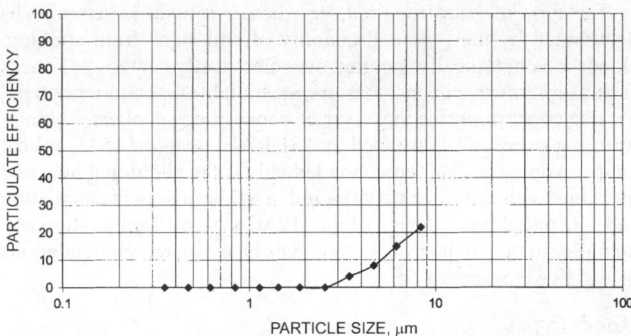

Fig. 9   Baffle Filter Particle Efficiency Versus Particle Size
(Kuehn et al. 1999)

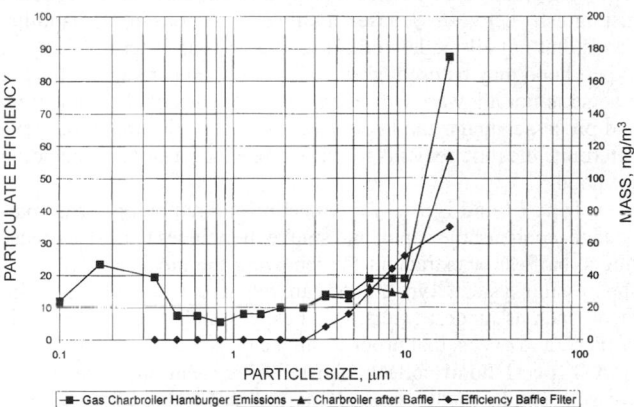

Gas Charbroiler Hamburger Emissions — Charbroiler after Baffle — Efficiency Baffle Filter

Fig. 10   Baffle Filter Particle Efficiency Versus Particle Size
(Kuehn et al. 1999)

extraction at the lowest pressure drop possible. Smaller particles can only be removed by shifting the efficiency curve towards the left.

More effective devices reduce grease build-up downstream of the hood, lowering the frequency of duct cleaning and reducing the fire hazard. Having higher-efficiency grease removal devices in the hood reduces the maintenance of downstream control equipment.

Concerns about air quality also emphasize the need for higher-efficiency grease extraction from the exhaust airstream than can be provided by filters or grease extractors in exhaust hoods. Cleaner exhaust discharge to outside may be required by increasingly stringent air quality regulations or where the exhaust discharge configuration is such that grease, smoke, or odors in discharge would create a nuisance. In some cases, exhaust air is cleaned so that it can be discharged inside (e.g., through recirculating systems). Several systems have been developed to clean the exhaust airstream, each of which presents special fire protection issues.

Where odor control is required in addition to grease removal, activated charcoal, other oxidizing bed filters, or deodorizing agents are used downstream of the grease filters. Because much cooking odor is gaseous and therefore not removed by air filtration, filtration upstream of the charcoal filters must remove virtually all grease in the airstream to prevent grease build-up on the charcoal filters.

The following technologies are applied to varying degrees for control of cooking effluent. They are listed by order of use in the exhaust stream after a mechanical filtration device, with particulate control upstream of VOC control. After the description of each technology are qualifications and concerns about its use. There is no consensus test protocol for evaluating these technologies in kitchen applications.

Electrostatic Precipitators (ESPs). Particulate removal is by high-voltage ionization, then collection on flat plates.

- Condensed grease can block airflow, especially when mounted outside.
- As the ionizer section becomes dirty, efficiency drops because the effective ionizer surface area is reduced.
- Under heavy loading, the unit may shut down because of voltage drop.

Ultraviolet (UV) Destruction. This system uses ultraviolet light to chemically convert the grease into an inert substance and ozone. Construction (not performance) is evaluated for safety in accordance with UL *Standard* 710C.

- Requires adequate exposure time for chemical reactions.
- Personnel should not look at light generated by high-intensity UV lamps.
- Exhaust fans should operate when UV lights are on because some forms of UV generate ozone.
- UV is more effective on very small particles and vapor.
- The required frequency of duct cleaning is reduced.
- Lamps need to be replaced periodically; as lamps become dirty, efficiency drops.

Water Mist, Scrubber, and Water Bath. Passage of the effluent stream through water mechanically entraps particulates and condenses grease vapor.

- High airflow can reduce efficiency of water baths.
- Water baths have high static pressure loss.
- Spray nozzles need much attention; water may need softening to minimize clogging.
- Drains tend to become clogged with grease, and grease traps require more frequent service. Mist and scrubber sections need significant length to maximize exposure time.

Pleated, Bag, and HEPA Filters. These devices are designed to remove very small particles by mechanical filtration. Some types also have an activated-carbon face coating for odor control.

- Filters become blocked quickly if too much grease enters.
- Static loss builds quickly with extraction, and airflow drops.
- Almost all filters are disposable and very expensive.

Activated-Carbon Filters. VOC control is through adsorption by fine activated charcoal particles.

- Require a large volume and thick bed to be effective.
- Are heavy and can be difficult to replace.
- Expensive to change and recharge. Many are disposable.
- Ruined quickly if they are grease-coated or subjected to water.
- Some concern that carbon is a source of fuel for a fire.

Oxidizing Pellet Bed Filters. VOC and odor control is by oxidation of gaseous effluent into solid compounds.

- Require a large volume and long bed to be effective.
- Are heavy to handle and can be difficult to replace.
- Expensive to change.
- Some concern about increased oxygen available in fire.

Incineration. Particulate, VOC, and odor control is by high-temperature oxidation (burning) into solid compounds.

- Must be at system terminus and clear of combustibles.
- Are expensive to install with adequate clearances.
- Can be difficult to access for service.
- Very expensive to operate.

Catalytic conversion. A catalytic or assisting material, when exposed to relatively high-temperature air, provides additional heat adequate to decompose (oxidize) most particulates and VOCs.

- Requires high temperature (450°F minimum).

- Expensive to operate because of high temperature requirement if integrated into the hood (can be cost-effective at the appliance level).

HEAT GAIN CALCULATIONS

Some special considerations must be taken into account when performing load calculations as part of the kitchen ventilation system design. The items listed are in addition to the standard HVAC load calculation gain and loss components (e.g., envelope, lighting, people).

Space Heat Gain

One of the challenges in performing cooling and heating load calculations for a kitchen is determining the space heat gains from the cooking appliances. Given that many of the largest cooking appliances include exhaust hoods to remove the smoke, grease, and heat, determining the heat gain can be challenge. There may also be a large number of smaller appliances that do not include exhaust hoods and reject all their heat directly to the space.

Tables 5A to 5E in Chapter 18 of the 2009 *ASHRAE Handbook—Fundamentals* list typical equipment and the heat rejected into the kitchen. The data contained in these tables were recently updated as part of ASHRAE research project RP-1362 (Swierczyna et al. 2008). For the majority of the appliances in the tables, the heat gain values were determined during the appliances in idle or standby condition: that is, the appliance was fully warmed up and in its ready-to-cook condition. (Typically, an appliance is in standby for as much as 70% of the day.) For appliances installed under an exhaust hood, the amount of heat emitted as radiation is listed, because this heat ends up heating nearby objects. For the appliances that are not installed under a hood because of their low energy consumption or lack of cooking effluent, the amount of both sensible and latent heat is listed, in addition to the radiation.

The greatest challenge with using these data is determining the diversity or usage factor of the appliances. It may be difficult to anticipate how often the appliance will be at full cooking or at some standby condition. Any assumptions made can be rendered incorrect by a change in kitchen throughput or sales. Determining the correct heat gain is an involved procedure that requires input from the entire kitchen design team.

The remaining heat gains from lighting, envelope, and people can be calculated following the procedure described in Chapter 18 of the 2009 *ASHRAE Handbook—Fundamentals*.

EXHAUST SYSTEMS

Exhaust systems remove effluent produced by appliances and cooking processes to provide fire and health safety, comfort, and aesthetics. Typical exhaust systems simultaneously incorporate fire prevention designs and fire suppression equipment. In most cases, these functions complement each other, but in other cases they may seem to conflict. Designs must balance the functions. For example, fire-actuated dampers may be installed to minimize the spread of fire to ducts, but maintaining an open duct might be better for removing the smoke of an appliance fire from the kitchen.

COMMERCIAL EXHAUST HOODS

The design, engineering, construction, installation, and maintenance of commercial kitchen exhaust hoods are governed by nationally recognized standards (e.g., NFPA *Standard* 96) and model codes [e.g., the IMC (ICC 2009a), *International Fuel Gas Code* (IFGC; ICC 2009b)]. In some cases, local codes may prevail. Before designing a kitchen ventilation system, the designer should identify governing codes and consult the AHJ. Local authorities with jurisdiction may have amendments or additions to these standards and codes.

The type of hood required, or whether a hood is required, is determined by the type and quantity of emissions from cooking. Hoods are not typically required over electrically heated appliances such as microwave ovens, toasters, steam tables, popcorn poppers, hot dog cookers, coffee makers, rice cookers, egg cookers, holding/ warming ovens (as mentioned in ASHRAE *Standard* 154), or heat lamps. Appliances can be unhooded only if the additional heat and moisture loads have been considered in a thorough load calculation and accounted for in design of the HVAC system. Temperature and humidity in the kitchen space should be based on recommendations of ASHRAE *Standard* 55.

Hood Types

Many types, categories, and styles of hoods are available, and selection depends on many factors. Hoods are classified by whether they are designed to handle grease; Type I hoods are designed for removing grease and smoke, and Type II hoods are not. Model codes distinguish between grease-handling and non-grease-handling hoods, but not all model codes use Type I/Type II terminology. A Type I hood may be used where a Type II hood is required, but the reverse is not allowed. However, characteristics of the equipment and processes under the hood, and not necessarily the hood type, determine the requirements for the entire exhaust system, including the hood.

A **Type I hood** is used for collecting and removing grease particulate, condensable vapor, and smoke. It includes (1) listed grease filters, baffles, or extractors for removing the grease and (2) fire-suppression system. Type I hoods are required over cooking equipment, such as ranges, fryers, griddles, gas underfired and electric broilers, and ovens, that produce smoke or grease-laden vapors.

A **Type II hood** collects and removes steam and heat where grease or smoke is not present. It may or may not have grease filters or baffles and typically does not have a fire-suppression system. It is usually used over dishwashers. A Type II hood is sometimes used over ovens, steamers, or kettles if they do not produce smoke or grease-laden vapor and as authorized by the AHJ.

Type I Hoods

Categories. Type I hoods fall into two categories: unlisted and listed. **Unlisted hoods** meet the design, construction, and performance criteria of applicable national and local codes and are not allowed to have fire-actuated exhaust dampers. **Listed hoods** are listed in accordance with Underwriters Laboratories (UL) *Standard* 710. Listed hoods are constructed in accordance with the terms of the hood manufacturer's listing, and are required to be installed in accordance with either NFPA *Standard* 96 or the model codes. Model codes include exceptions for listed hoods to show equivalency with the model code requirements.

The two subcategories of Type I listed hoods, as covered by UL *Standard* 710, are exhaust hoods with and without exhaust dampers. UL listings also distinguish between water-wash and dry hoods.

All listed hoods are subjected to electrical (if applicable), temperature, and cooking smoke and flare-up (capture and containment) tests. A listed exhaust hood with exhaust damper includes a fire-actuated damper, typically at the exhaust duct collar. In the event of a fire, the damper closes to prevent fire from entering the duct. Fire-actuated exhaust dampers are permitted only in listed hoods. Also, listed hoods that incorporate an integral supply air plenum include a fire-actuated damper in that plenum; the damper's location in the supply air plenum depends on plenum configuration. Refer to NFPA *Standard* 96 and UL *Standard* 710 for examples of the damper in exhaust hood supply air plenum.

Grease Removal. Most grease removal devices in Type I hoods operate on the same general principle: exhaust air passes through a series of baffles that create a centrifugal force to throw grease particles out of the airstream as the exhaust air passes around the baffles. The amount of grease removed varies with baffle design, air velocity, temperature, type of cooking, and other factors. NFPA *Standard* 96 does not allow use of mesh-filter as primary grease filter. To date, stand-alone mesh filters have not met the requirements of UL *Standard* 1046, and therefore cannot be used as primary grease filters. ASTM *Standard* F2519 provides a test method to determine the grease particle capture efficiency of grease filters and extractors. Grease removal devices generally fall into the following types:

- **Baffle filters** have a series of vertical baffles designed to capture grease and drain it into a container. The filters are arranged in a channel or bracket for easy insertion and removal for cleaning. Each hood usually has two or more baffle filters, which are typically constructed of aluminum, steel, or stainless steel and come in various standard sizes. Filters are cleaned by running them through a dishwasher or by soaking and rinsing. NFPA *Standard* 96 requires that grease filters be listed. Listed grease filters are tested and certified by a nationally recognized test laboratory in accordance with UL *Standard* 1046.

- **Removable extractors** (also called **cartridge filters**) have a single horizontal-slot air inlet. The filters are arranged in a channel or bracket for easy insertion and removal for cleaning. Each hood usually has two or more removable extractors, which are typically constructed of stainless steel and contain a series of horizontal baffles designed to remove grease and drain it into a container. Available in various sizes, they are cleaned by running them through a dishwasher or by soaking and rinsing. Removable extractors may be classified by a nationally recognized test laboratory in accordance with UL *Standard* 1046, or may be listed as part of the hood in accordance with UL *Standard* 710. Hoods that are listed with removable extractors cannot have those extractors replaced by other extractors.

- **Stationary extractors** are integral to the listed water-wash exhaust hoods and are typically constructed of stainless steel and contain a series of horizontal baffles that run the full length of the hood. The baffles are not removable for cleaning, though some have doors that can be removed to clean the extractors and plenum. The stationary extractor includes one or more water manifolds with spray nozzles that, when activated, wash the grease extractor with hot, detergent-injected water, removing accumulated grease. The wash cycle is typically activated at the end of the day, after cooking equipment and fans have been turned off; however, it can be activated more frequently. The cycle lasts 5 to 10 min, depending on the hood manufacturer, type of cooking, duration of operation, and water temperature and pressure. Most water-wash hood manufacturers recommend a water temperature of 130 to 180°F and water pressure of 30 to 80 psi. Average water consumption varies from 0.50 to 1.50 gpm per linear foot of hood, depending on manufacturer. Most water-wash hood manufacturers provide a manual and/or automatic means of activating the water-wash system in the event of a fire.

  Some water-wash hood manufacturers provide continuous cold water as an option. The cold water runs continuously during cooking and may or may not be recirculated, depending on the manufacturer. Typical cold-water usage is 1 gph per linear foot of hood. The advantage of this method is that it improves grease extraction and removal, partly through condensation of the grease. Many hood manufacturers recommend continuous cold water in hoods located over solid-fuel-burning cooking equipment, because the water also extinguishes hot embers that may be drawn up into the hood and helps cool the exhaust stream.

- **Multistage filters** use two or more stages of filtration to remove a larger percentage of grease. They typically consist of a baffle filter or removable extractor followed by a higher-efficiency filter, such as a packed bead bed. Each hood usually has two or more multistage filters, which are typically constructed of aluminum or stainless steel and are available in standard sizes. Filters are cleaned by

running them through a dishwasher or by soaking and rinsing. NFPA *Standard* 96 requires that grease filters be listed, so these multistage filters must be tested and certified by a nationally recognized test laboratory in accordance with UL *Standard* 1046.

UL *Standards* 710 and 1046 do not include grease extraction efficiency tests. Historically, grease extraction efficiency rates published by filter and hood manufacturers were usually derived from tests conducted by independent test laboratories retained by the manufacturer. Test methods and results therefore have varied greatly.

In 2005, however, a new grease filter and extractor test standard, ASTM *Standard* F2519, was published, which determines the grease particle capture efficiency of both removable filters and fixed extractors such as those used in water-wash hoods. The filters are evaluated by pressure drop as well as particulate capture efficiency. The test generates a controlled quantity of oleic acid particles in size ranging from 0.3 to 10 μm that are released into a hood to represent the cooking effluent. The particles are then sampled and counted downstream in the duct with an optical particle counter, with and without the filter or extractor in place. The difference in the counts is used to calculate the particulate capture efficiency graphed versus particle size. ASTM *Standard* F2519 measures particulate capture efficiency only, not vapor removal efficiency. A more detailed explanation is available in the Exhaust Systems section of this chapter.

Styles. Figure 11 shows the six basic styles for Type I hood applications. These style names are not used in all standards and codes but are well accepted in the industry. The styles are as follows:

- **Wall-mounted canopy,** used for all types of cooking equipment located against a wall.
- **Single-island canopy,** used for all types of cooking equipment in a single-line island configuration.
- **Double-island canopy,** used for all types of cooking equipment mounted back-to-back in an island configuration.
- **Back shelf/proximity,** used for counter-height equipment typically located against a wall, but possibly freestanding.
- **Eyebrow,** used for direct mounting to ovens and some dishwashers.
- **Pass-over,** used over counter-height equipment when pass-over configuration (from cooking side to serving side) is required.

Sizing. The size of the exhaust hood relative to cooking appliances is important in determining hood performance. Usually the hood must extend horizontally beyond the cooking appliances (on all open sides on canopy-style hoods and over the ends on back shelf and pass-over hoods) to capture expanding thermal currents rising from the appliances. For unlisted hoods, size and overhang requirements are dictated by the prevailing code; for listed hoods, by the terms of the manufacturer's listing. **Overhang** varies with hood style, distance between hood and cooking surface, and characteristics of cooking equipment. With back shelf and pass-over hoods, the front of the hood may be kept behind the front of the cooking equipment (**setback**) to allow head clearance for the cooks. These hoods may require a higher front inlet velocity to capture and contain expanding thermal currents. ASHRAE research (Swierczyna et al. 2006, 2010) indicates an appliance front overhang of 9 to 18 in. for canopy style and a 10 in. setback for back shelf/proximity style are preferable to current code minimums. All styles may have full or partial side panels to close the area between appliances and the hood. This may eliminate the side overhang requirement and generally reduces the exhaust flow rate requirement.

Exhaust Flow Rates. Exhaust flow rate requirements to capture, contain, and remove effluent vary considerably depending on hood style, overhang, distance from cooking surfaces to hood, presence and size of side panels, cooking equipment, food, and cooking processes involved. The hot cooking surfaces and product effluent create thermal air currents that are captured by the hood and then exhausted. The velocity of these currents depends largely on surface temperature and tends to vary from 15 fpm over steam equipment to 150 fpm over charcoal broilers. The required flow rate is determined by these thermal currents, a safety allowance to absorb cross-currents and flare-ups, and a safety factor for the style of hood.

Overhang and the presence or absence of side panels help determine the safety factor for different hood styles. Gas-fired cooking equipment may require an additional allowance for exhaust of combustion products and combustion air.

Because it is not practical to place a separate hood over each piece of equipment, general practice (reflected in ASHRAE *Standard* 154) is to categorize equipment into four groups, as shown in Table 2.

These categories apply to unlisted and listed Type I hoods. The exhaust flow rate requirement is based on the group of equipment under the hood. If there is more than one group, the flow rate is based on the heaviest-duty group unless the hood design allows different rates over different sections of the hood.

Though considered obsolete based on laboratory tests and research, some local codes may still require exhaust flow rates for unlisted canopy hoods to be calculated by multiplying the horizontal area of the hood opening by a specified air velocity. Some jurisdictions may use the length of the open perimeter of the hood times the vertical height between hood and appliance instead of the horizontal hood area. Swierczyna et al. (1997) found that these methods of calculation result in higher-than-necessary exhaust flow rates for deeper hoods, because the larger reservoirs of deeper hoods typically increase hood capture and containment performance.

Table 3 lists recommended exhaust flow rates by equipment duty category for unlisted hoods and typical exhaust flow rates for listed hoods. Rates for unlisted hoods are based on ASHRAE *Standard* 154. Typical design rates for listed hoods are based on published rates for listed hoods serving single categories of equipment, which vary from manufacturer to manufacturer. Rates are usually lower for listed hoods than for unlisted hoods, and it is generally advantageous to use listed hoods. Actual exhaust flow rates for hoods with internal short-circuit replacement air are typically higher than those in Table 3, although net exhaust rates (actual exhaust less internal replacement air quantity) are lower, which seriously compromises the hood's capture and containment performance (Brohard et al. 2003).

Table 2 Appliance Types by Duty Category

Light duty (400°F)	Electric or gas	Ovens (including standard, bake, roasting, revolving, retherm, convection, combination convection/steamer, conveyor, deck or deck-style pizza, pastry)
		Steam-jacketed kettles
		Compartment steamers (both pressure and atmospheric)
		Cheesemelters
		Rethermalizers
Medium duty (400°F)	Electric	Discrete element ranges (with or without oven)
	Electric or gas	Hot-top ranges
		Griddles
		Double-sided griddles
		Fryers (including open deep-fat fryers, donut fryers, kettle fryers, pressure fryers)
		Pasta cookers
		Conveyor (pizza) ovens
		Tilting skillets/braising pans
		Rotisseries
Heavy duty (600°F)	Gas	Open-burner ranges (with or without oven)
	Electric or gas	Gas underfired broilers
		Chain (conveyor) broilers
		Wok ranges
		Overfired (upright) salamander broilers
Extra-heavy duty (700°F)	Appliances using solid fuel such as wood, charcoal, briquettes, and mesquite to provide all or part of the heat source for cooking.	

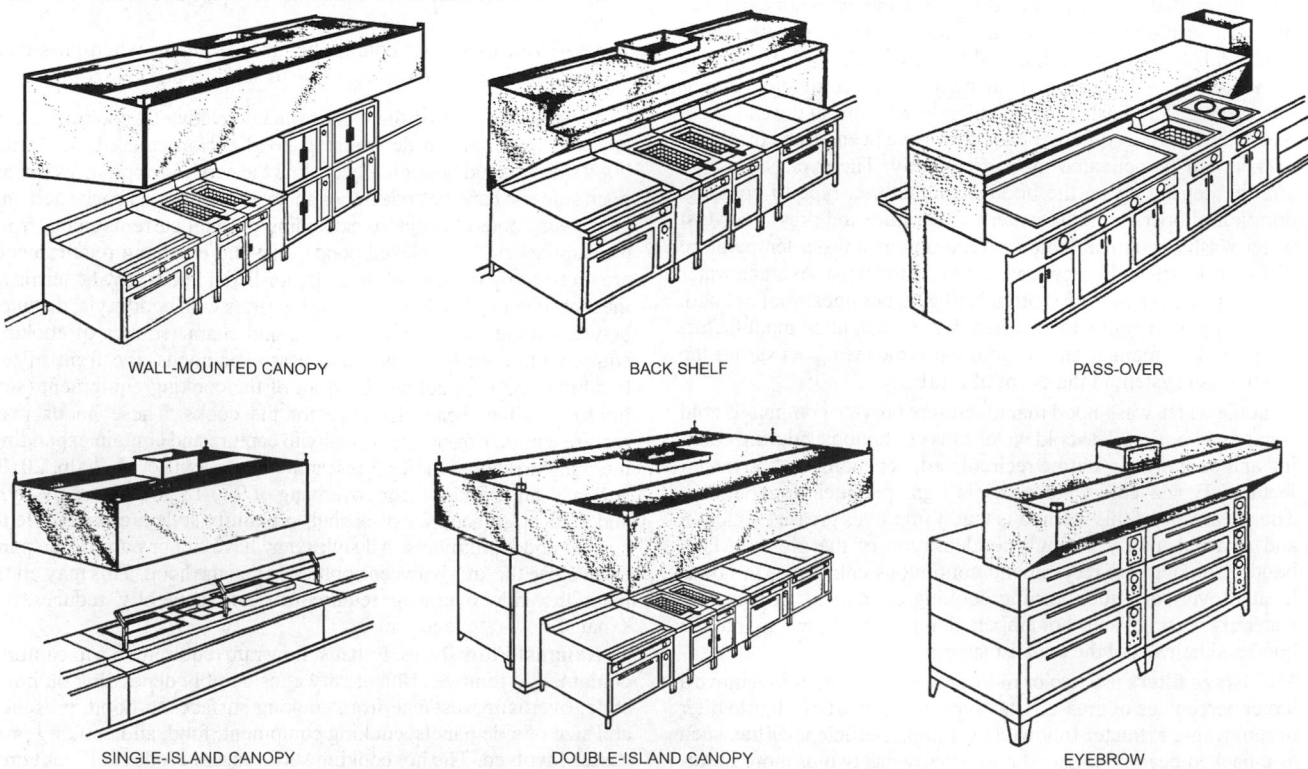

WALL-MOUNTED CANOPY   BACK SHELF   PASS-OVER

SINGLE-ISLAND CANOPY   DOUBLE-ISLAND CANOPY   EYEBROW

Fig. 11 Styles of Commercial Kitchen Exhaust Hoods

Table 3   Exhaust Flow Rates by Cooking Equipment Category for Unlisted and Listed Hood

Type of Hood	Minimum Exhaust Flow Rate, cfm per linear foot of hood			
	Light Duty	Medium Duty	Heavy Duty	Extra-Heavy Duty
Wall-mounted canopy, unlisted	200	300	400	550
listed	150 to 200	200 to 300	200 to 400	350+
Single-island, unlisted	400	500	600	700
listed	250 to 300	300 to 400	300 to 600	550+
Double-island (per side), unlisted	250	300	400	550
listed	150 to 200	200 to 300	250 to 400	500+
Eyebrow, unlisted	250	250	Not allowed	Not allowed
listed	150 to 250	150 to 250	—	—
Back shelf/proximity/pass-over, unlisted	300	300	400	Not allowed
listed	100 to 200	200 to 300	300 to 400	Not recommended

Source: ASHRAE Standard 154.

Listed hoods are allowed to operate at their listed exhaust flow rates by exceptions in the model codes. The exhaust flow rates for listed hoods are established by conducting tests per UL *Standard* 710. Typically, exhaust flow rates are much lower than those dictated by the model codes. Note that listed flow rate values are established under draft-free laboratory conditions, and actual operating conditions may compromise listed performance. Thus, manufacturers may recommend design values above their listed values.

Hoods listed in accordance with UL 710 cover one or more cooking equipment temperatures: 400, 600 and 700°F. In application, these temperature ratings correspond to duty ratings (see Table 2). The total exhaust flow rate is calculated by multiplying the hood exhaust flow rate by hood length.

ASTM *Standard* F1704 details a laboratory flow visualization procedure for determining the capture and containment threshold of an appliance/hood combination. This procedure can be applied to all hood types and configurations installed over any cooking appliances. ASTM *Standard* F2474 also provides a laboratory test procedure for determining heat gain of specific combinations of exhaust hood, cooking equipment, type of foods, and cooking processes. Results from a series of interlab heat gain tests (Fisher 1998; Swierczyna 2008) have been incorporated in Chapter 18 of the 2009 *ASHRAE Handbook—Fundamentals*.

Island Canopy Hoods

Island canopy hoods, particularly single-island style, have become popular in open cafeteria operations such as those found in university food service. In many cases, the food service consultant specifies gas underfired broilers and other heavy-duty cooking equipment as part of the design. For a given line of appliances, a single-island canopy hood requires significantly more exhaust than a wall-mounted canopy hood. Single-island canopy hoods present the most difficult capture and containment challenge in hood applications, and are often the source of the "hood" problem in a kitchen with display cooking. To address the lack of reliable performance data on island canopy hoods, ASHRAE research project RP-1480 (Swierczyna et al. 2010) was undertaken to determine appropriate exhaust airflow rates. The objective was to expand the database for the exhaust rates required for capture and containment of standardized cook lines under four island canopy hood configurations: rear filter single island, V-bank single island, and 8 ft deep and 10 ft deep double-island hoods. Four side panel designs, four supply air strategies, and two makeup air temperature set points were also evaluated to quantify the effects of these features on island hood performance.

Swierczyna et al. (2010) confirmed that single-island canopy hoods need significantly higher exhaust airflow rates than their wall-mounted counterparts to effectively ventilate cooking equipment for a given duty class. For example, although an exhaust rate

of 300 to 400 cfm/ft can be adequate for complete capture and containment with a wall-mounted canopy hood over a heavy-duty appliance line (ASHRAE 2003; PG&E 2010), a single-island canopy hood may require an exhaust rate in excess of 500 cfm/ft in many situations (measured along one side of the canopy hood). In fact, there were several test scenarios for single-island hoods where an exhaust rate in excess of 700 cfm/ft was required to achieve capture and containment. This contradicts common design practice, where the specified ventilation rates are often much closer to those for wall-canopy hoods.

Single-island hood performance was improved by the larger hood's V-bank filter configuration over the smaller hood's rear filter configuration for most test configurations. The plume was better aligned with the filters and was drawn towards the center, relative to the front and rear of the hood. The larger V-bank hood was found to be less sensitive to local air replacement. However, aggressive appliance plumes that focused on the flat bottom of the V-bank, or replacement air strategies that were focused at the side of the V-bank, proved challenging and indicated that a change of filter bank profile may improve hood performance.

The performance of a double-island canopy hood, with balanced replacement air, can be comparable to back-to-back wall-mounted canopy hoods for a given duty class of appliances. For example, a heavy-duty front line and a light-duty back line under the double-island hood required an exhaust airflow rate approximately 300 cfm/ft (measured along both sides of the hood). This rate is comparable to the ventilation rate for similar appliance duty classes under wall-mounted canopy hoods (Swierczyna et al. 2006). The double-island hood configuration performed as if a wall existed between them. Furthermore, the back-to-back appliance lines created a converging thermal plume that helped direct the plume toward the filter bank. However, without a wall between them, the double-island hood system was more susceptible to cross drafts than a wall-mounted hood configuration.

The configuration, volume, and temperature of makeup air introduced into the space was critical to the performance of the double-island canopy hood. Consistent with previous research (Brohard et al. 2003), reducing local makeup airflow rates and velocities corresponded with reduced capture and containment exhaust rates, in most cases. When the air volume and associated velocity and turbulence near the hood was minimized, the appliance plumes were more stable and the hood was able to capture and contain at a lower exhaust rate. However, when local makeup air was introduced aggressively through four-way diffusers, perforated diffusers, or a high-flow perforated perimeter supply system, hood performance degraded severely. For the double-island configurations, a perforated perimeter supply system operated at a low-flow, low-velocity condition was the best of the local makeup air configurations tested. When the perforated perimeter supply system delivered low-flow, low-velocity air adjacent to the hood (i.e., less than 60% of replacement air requirement), hood performance improved significantly over the high-flow, high-velocity introduction (i.e., greater than 60% of replacement air requirement), and in some cases, better than the exhaust-only configuration with displacement supply. Higher replacement air temperatures from ceiling diffusers also degraded the performance of island hoods. Unbalanced replacement air distribution was extremely detrimental to the performance of the double-island hoods.

Other research highlighted the advantages of using side panels for wall-mounted canopy hoods and a variety of makeup air conditions (Brohard et al. 2003; Swierczyna et al. 2006). However, results from the double-island canopy hood testing regarding side panels were inconclusive. A more extensive side panel (and center partition) investigation would need a larger laboratory where replacement air was introduced more uniformly around the hood to eliminate the effect of relatively high, directional local velocities.

A partition between the two appliances lines improved performance of a double-island hood when coupled with a balanced supply on both sides of the hood. However, if as little as 1000 cfm was exhausted from the side opposite from the supply air delivery, performance of the double-island hood degraded. This was contrary to the expectation that the partition would be more of a benefit with unbalanced replacement air and its ability to mitigate the effect of cross drafts.

Increased hood overhang was shown to be one of the most effective performance enhancements for island canopy hoods. With a heavy-duty three-broiler appliance line centered front-to-rear under the single island hoods, rather than at a minimum prescriptive front overhang dimension, a 14% exhaust reduction was possible for the smaller rear filter hood, and a 40% exhaust reduction was possible for the larger V-bank hood. Likewise, when side overhang was increased to 24 in. from the minimum of 6 in., a 41% exhaust rate reduction was found for both single-island hoods. However, the results did not show a significant performance difference between the 8 ft and 10 ft deep double-island hoods. Increased side overhang was found to be one of the most effective performance enhancements for double-island canopy hoods. Increasing the side overhang to 24 in. resulted in a 160 cfm/ft reduction in exhaust flow rate.

Tailored exhaust bias for double-island hoods may improve hood performance. With more exhaust volume focused over the more challenging appliances, the exhaust rate can be reduced for a given configuration. However, application of a specific bias for other applications or hood dimensions may yield different performance results and should be verified.

Specification of enhanced hood edge geometry should be considered by manufacturers and end-users. Although each design needs to be properly evaluated for its effect on hood performance, the design tested in this project was effective and was typical of edge design currently found in the industry.

Performance in the field should be verified to ensure proper hood capture and containment operation. As shown by RP-1480 (Swierczyna et al. 2010), many factors interact in the kitchen and affect hood performance. These interactions cannot be perfectly predicted for each installation. Therefore, a field test is best to verify proper kitchen ventilation and hood performance.

Wall Canopy Hoods, Appliance Positioning, and Diversity

ASHRAE research project RP-1202 (Swierczyna et al. 2006) quantified the effect of the position and/or combination of appliances under a wall canopy exhaust hood on the minimum C&C exhaust rate. Effects of side panels, front overhang, and rear seal were also investigated. The scope of this laboratory study was to investigate similar and dissimilar appliances under a 10 ft wall-mounted canopy hood. The appliances included three full-sized electric convection ovens, three two-vat gas fryers, and three 3 ft gas underfired broilers, representing the light, medium, and heavy-duty appliance categories, respectively. In addition to various physical appliance configurations, appliances were also varied in their usage: either off, at idle conditions, or at cooking conditions. A supplemental study investigated the effect of appliance accessories (including shelving and a salamander) and hood dimensions (including hood height, depth, and reservoir volume) on the minimum exhaust rate required for complete capture and containment.

The study demonstrated that subtle changes in appliance position and hood configuration could dramatically affect the exhaust rates required for complete capture and containment, regardless of appliance duty and/or usage. The wide range in C&C values for a given hood/appliance setup explains why a similar hood installed over virtually the same appliance line may perform successfully in one kitchen and fall short of expectations in another facility. The following conclusions are specific to the conditions tested by Swierczyna et al. (2006).

Airflow Requirements for Like-Duty Appliance Lines. Evaluation supported widely accepted commercial kitchen ventilation (CKV) design practices: higher ventilation rates are required for progressively heavier-duty appliances (Table 4). For a 10 ft wall-mounted canopy hood, at a defined median or good-case installation, the light-duty oven line required 1100 cfm (110 cfm/ft), the medium-duty fryer line required 2400 cfm (240 cfm/ft), and the heavy-duty broiler line required 4400 cfm (440 cfm/ft) to achieve C&C. Simply increasing front overhang as noted between the worst- and good-case installations in Table 4 reduced the C&C exhaust rate by 10 to 27%. Installing side panels in addition to the increased front overhang (best-case scenario) reduced the exhaust requirements by an additional 18 to 33%.

Diversity in Appliance Usage. Operation diversity was evaluated with cook lines of three similar appliances and included combinations of *cook* and *off* conditions. In most cases, operation

Table 4 Capture and Containment Exhaust Rates for Three Like-Duty Appliance Lines at Cooking Conditions with Various Front Overhang and Side Panel Configurations under 10 ft Wall-Mounted Canopy Hood

	Best Case	Good Case	Worst Case
Three electric full-sized convection ovens	9 in. front overhang full side panels 85 cfm/ft	9 in. front overhang 110 cfm/ft	6 in. front overhang 120 cfm/ft
Three two-vat gas fryers	18 in. front overhang partial side panels 160 cfm/ft	18 in. front overhang 240 cfm/ft	6 in. front overhang 330 cfm/ft
Three gas underfired broilers	12 in. front overhang partial side panels 330 cfm/ft*	12 in. front overhang 440 cfm/ft	0 in. front overhang (6 in. cook surface) 510 cfm/ft

*Adding a rear seal between back of appliance and wall to best-case configuration (6 in. of front overhang and partial side panels) further improved hood performance to an exhaust rate of 2800 cfm (280 cfm/ft).
Source: Swierczyna et al. (2006).

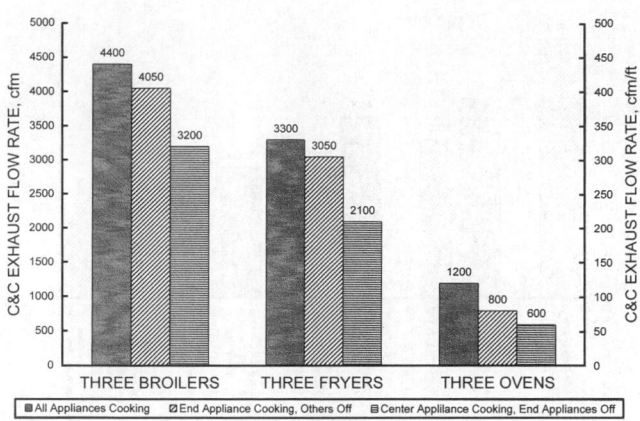

Fig. 12 Exhaust Capture and Containment Rates for One or Three Appliances Cooking from Like-Duty Classes under a 10 ft Wall-Canopy Hood
(Swierczyna et al. 2006)

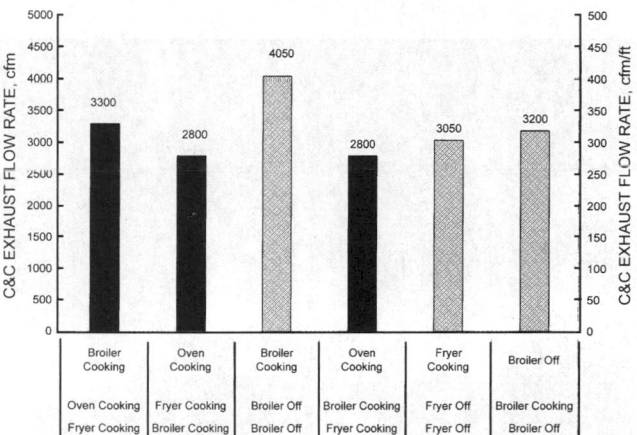

Fig. 13 Capture and Containment Exhaust Rates for Cooking Conditions on Multiduty Appliance Lines (Compared with Single-Duty Lines with Only One Appliance Operating) under 10 ft Wall Canopy Hood
(Swierczyna et al. 2006)

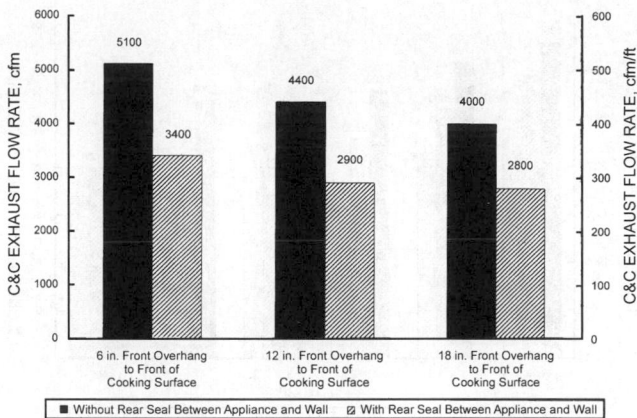

Fig. 14 Capture and Containment Exhaust Rates for Gas Underfired Broilers under 10 ft Wall Canopy Hood With and Without Rear Appliance Seal at Various Front Overhangs
(Swierczyna et al. 2006)

correlated directly with the required exhaust rate, with an emphasis on the operation of the end appliances (Figure 12). The capture and containment rate for the end appliance cooking and the other two like-duty appliances off was nearly the same rate as all three appliances cooking.

Changing the condition of end appliances from off to cooking had the greatest effect for medium-duty fryers, which required a 950 cfm increase in the exhaust rate. Because the fryers were thermostatically controlled, they responded to cooking operations by firing the burners. This, combined with an aggressive cooking plume, required a significantly increased exhaust rate for C&C. An aggressive thermal plume was present for the three heavy-duty gas underfired broilers; the exhaust rate increased 850 cfm. For the light-duty electric convection oven, there was a 200 cfm difference in turning the end appliances from off to cook. Figure 12 also shows that cooking with only the center appliance, with the two end appliances turned off, greatly reduced the exhaust requirement.

Diversity in Appliance Duty and Position (Side-to-Side). The study found that the capture and containment rate of a multiduty appliance line was less than the rate of the heaviest duty appliance in that line, applied over the length of the hood (Figure 13).

Appliance position testing confirmed the exhaust rate of an appliance line was most dependent on the duty of the end appliance. The end appliance drove the exhaust rate more than additional volume from the other two appliances, as they changed from off to cooking conditions or were varied in duty class. In most cases, the lowest exhaust requirements for particular appliance lines were achieved when the lowest-duty appliance was at the end of the appliance line. In other words, hood performance was optimized when the heaviest-duty appliance was in the middle of the appliance line.

Appliance Positioning (Front-to-Back) and Rear Seal. Increasing the front overhang by pushing appliances toward the back wall significantly decreased the required exhaust rates, not only because of the increased distance from the hood to the front of the appliance, but also because of the decreased distance between the back of the appliance and the wall. With a rear seal in place, some of the replacement air, which would have otherwise been drawn up from behind the appliances, was instead drawn in along the perimeter of the hood, helping guide the plume into the hood, as shown in Figure 14.

Hood Side Panels. Side panels installed on the 10 ft hood improved hood performance dramatically, by preventing the plume from spilling at the side of the hood and by increasing velocity along the front of the hood. Combining side panels (measuring 1 by 1 ft by 45°, 2 by 2 ft by 45°, 3 by 3 ft by 45°, 4 by 4 ft by 45°, or full) with the maximum hood overhang resulted in the lowest exhaust requirement for all cases tested. The example of the three two-vat gas fryer line is shown in Figure 15.

Effect of Shelving on Hood Capture and Containment Performance. Neither solid nor tubular shelving over the six-burner range required an increase in the exhaust rate. In fact, tubular shelving mounted to the back of the appliance showed a slight enhancement compared to having no shelving installed.

Effect of Hood Depth, Reservoir, and Mounting Height on Capture and Containment Performance. Comparing the 4 and 5 ft deep hoods, the deeper hood reduced capture and containment exhaust rates when appliances were positioned with maximum front overhang and minimum rear gap. The deeper hood had a negative effect when appliances remained in the minimum front overhang position. The effect of hood depth in conjunction with front overhang, side panels, and rear seal is shown in Figure 16.

Another advantage of the 5 ft over the 4 ft hood was its ability to capture and contain the plume when an oven door was opened. For a 4 ft hood and a 6 in. front overhang, an exhaust rate of 1200 cfm was required for the three electric ovens with the doors closed and

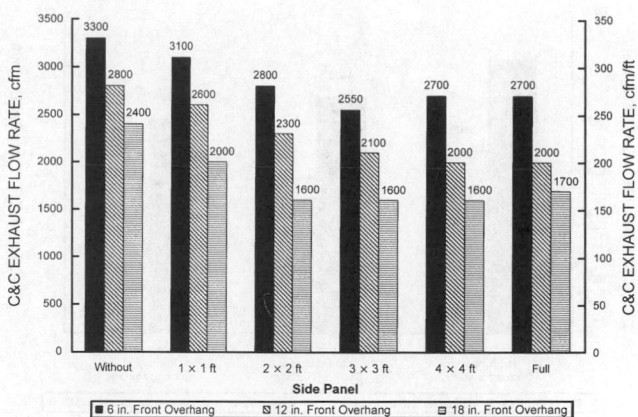

Fig. 15   Exhaust Capture and Containment Rates for Three Two-Vat Gas Fryers with Various Side Panel and Overhang Configurations under 10 ft Wall Canopy Hood
(Swierczyna et al. 2006)

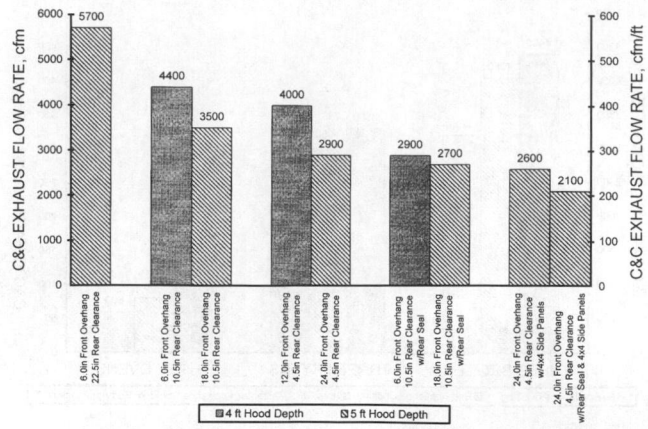

Fig. 16   Exhaust Capture and Containment Rates for a Heavy-Duty Gas Underfired Broiler Line under 10 ft Wall Canopy Hood with 4 and 5 ft Hood Depths and Front Various Front Overhangs
(Swierczyna et al. 2006)

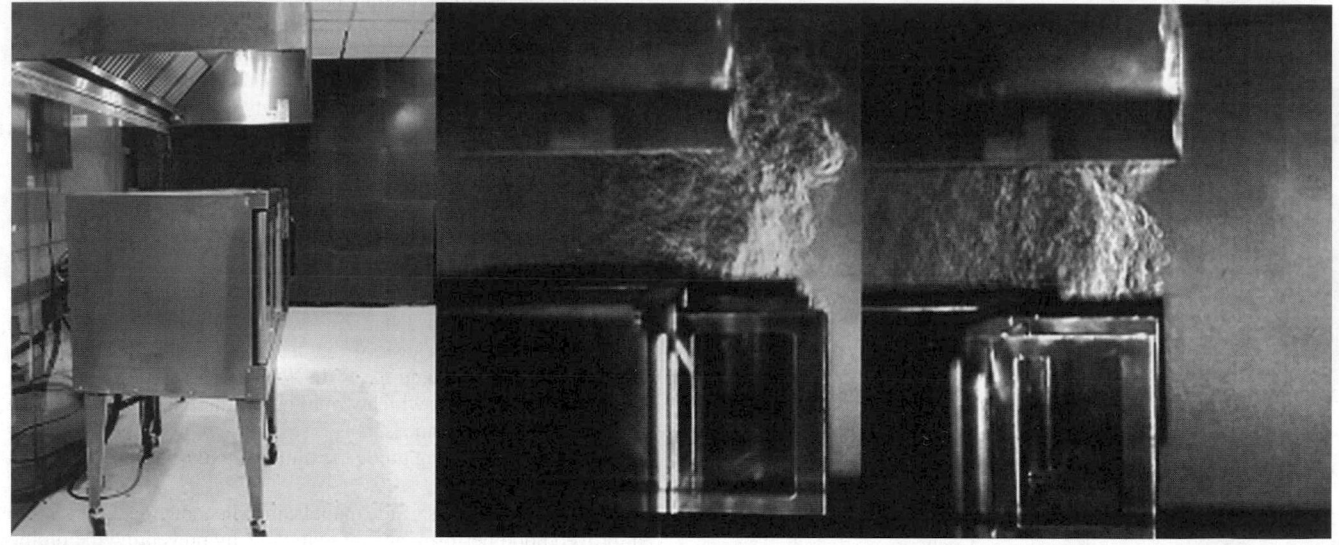

(A) NORMAL VIEW

(B) SCHLIEREN VIEW OF SPILLAGE WITH 4 ft DEEP HOOD WITH 6 in. OF FRONT OVERHANG

(C) SCHLIEREN VIEW DURING CAPTURE AND CONTAINMENT WITH 5 ft DEEP HOOD WITH 18 in. OF FRONT OVERHANG

Fig. 17   Three Ovens under Wall-Mounted Canopy Hood at Exhaust Rate of 3400 cfm
(Swierczyna et al. 2006)

5200 cfm with the doors open. Similarly, for a 5 ft deep hood with an 18 in. front overhang, 1200 cfm was required for three ovens with the doors closed and 3400 cfm with the doors open. The setup and schlieren views are shown in Figure 17.

The reservoir volume of the hood was increased by changing the hood height from 2 to 3 ft. When the gas underfired broiler was operated in the left appliance position, the increased hood volume marginally improved capture and containment performance. In contrast, a significant improvement was found for the appliance in the center position. This improvement indicated the plume was well located in the hood, and the increased hood volume may have allowed the plume to roll inside the hood and distribute itself more evenly along the length of the filter bank.

Minimizing hood mounting height had a positive effect on capture and containment performance. In most cases, a direct correlation could be made between the required exhaust rate and hood height for a given appliance line. The typical 6.5 ft mounting height

(for a canopy hood) was increased to 7 or 7.5 ft. For the gas underfired broiler installed at the end of the hood, increasing the hood height by 1 ft required a 14% increase in exhaust. However, when the broiler was in the center position, the increased hood height did not compromise capture and containment performance and required a reduced exhaust rate. The dramatic reduction in the exhaust requirement as the hood-to-appliance distance was reduced below the 6.5 ft mounting height illustrated the potential for optimizing CKV systems by using close-coupled or proximity-style hoods. This effect is shown in Figure 18.

Design Guidelines. Swierczyna et al. (2006) illustrated the potential for large variations in the airflow requirements for a specified appliance line and hood configuration. Best-practice design considerations that became evident included the following:

• Position heavy-duty appliances (e.g., broilers) in middle of the line.

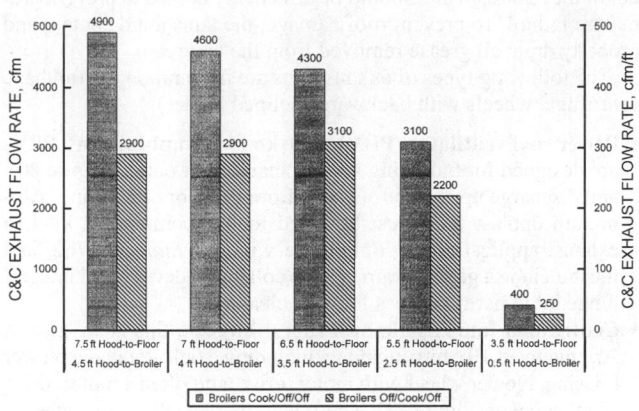

Fig. 18 Exhaust Capture and Containment Rates for a Gas Underfired Broiler under 10 ft Wall Canopy Hood at Various Mounting Heights
(Swierczyna et al. 2006)

Table 5 Exhaust Static Pressure Loss of Type I Hoods for Various Exhaust Airflows*

Type of Grease Removal Device	Hood Static Pressure Loss, in. of water			
	150 to 250 cfm/ft	250 to 350 cfm/ft	350 to 450 cfm/ft	500+ cfm/ft
Baffle filter	0.25 to 0.50	0.50 to 0.75	0.75 to 1.00	1.00 to 1.25
Extractor	0.80 to 1.35	1.30 to 1.70	1.70 to 3.00	2.90 to 4.20
Multistage	0.55 to 1.10	1.10 to 1.70	1.70 to 2.90	2.90 to 4.00

*Values based on 20 in. high filters and 1500 fpm through hood/duct collar.

- Position light-duty appliances (e.g., ovens) on the end of the line.
- Push back appliances (maximize front overhang, minimize rear gap).
- Seal area between rear of appliance and wall.
- Use side panels, end panels, and end walls.
- Installing shelving or ancillary equipment (e.g., salamander) behind or above a range should not negatively affect C&C performance, if other best practices (e.g., maximizing hood overhang) are observed.
- Use larger hoods, both deeper and taller.
- Installing hoods at lowest height practical (or allowed by code) to minimize distance from cooking surface to hood improves C&C performance.
- Introduce makeup air at low velocity. Do not locate four-way diffusers near hood, and minimize use of air curtains.

Replacement (Makeup) Air Options. Air exhausted from the kitchen must be replaced. Replacement air can be brought in through traditional methods, such as **ceiling diffusers**, or through systems built as an integral part of the hood. It may also be introduced using low-velocity displacement diffusers or transfer air from other zones. For further information, see the section on Replacement (Makeup) Air Systems.

Static Pressure. Static pressure drop through hoods depends on the type and design of the hood and grease removal devices, size of duct and duct connections, and flow rate. Table 5 provides a general guide for determining static pressure loss depending on the type of grease removal device and exhaust flow rate. Manufacturers' data should be consulted for actual values. Static pressure losses for exhaust ducts should be calculated for each installation.

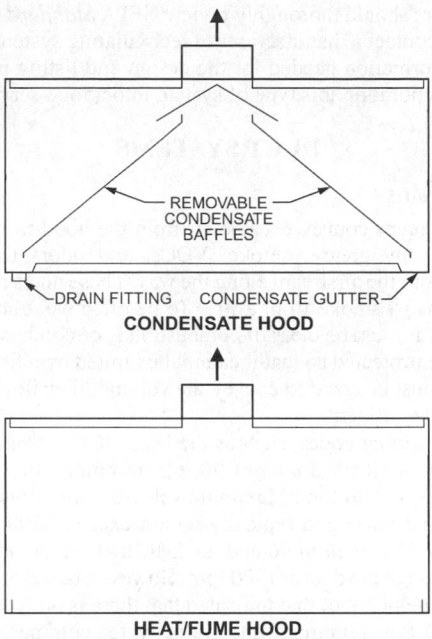

Fig. 19 Type II Hoods

Type II Hoods

Type II hoods (Figure 19) can be divided into the following two application categories:

- **Condensate hood.** For applications with high-moisture exhaust, condensate forms on interior surfaces of the hood. The hood is designed to direct the condensate toward a perimeter gutter for collection and drainage, allowing none to drip onto the appliance below. Flow rates are typically based on 300 to 500 cfm per foot of hood length. Hood material is usually noncorrosive, and filters are usually installed.
- **Heat/fume hood.** For hoods over equipment producing heat and fumes only, flow rates are typically based on 100 to 300 cfm per foot of hood. Filters are usually not installed.

Recirculating Systems

A recirculating system, previously called a **ductless hood**, consists of a cooking appliance/hood assembly designed to remove grease, smoke, and odor and to return the treated exhaust air directly back into the room. HVAC design must consider that recirculating systems discharge the total amount of heat and moisture generated by the cooking process back into the kitchen space, adding to the cooling load.

These hoods typically contain the following components in the exhaust stream: (1) a grease removal device such as a baffle filter, (2) a high-efficiency particulate air (HEPA) filter or an electrostatic precipitator (ESP), (3) some means of odor control such as activated charcoal, and (4) an exhaust fan. NFPA *Standard* 96, Chapter 13, is devoted entirely to recirculating systems and contains specific requirements such as (1) design, including interlocks of all critical components to prevent operation of the cooking appliance if any of the components are not operating; (2) fire extinguishing system, including specific nozzle locations; (3) maintenance, including a specific schedule for cleaning filters, ESP, hood, and fan; and (4) inspection and testing of the total operation and interlocks. In addition, NFPA *Standard* 96 requires that all recirculating systems be listed by a testing laboratory. The recognized standard for a recirculating system is UL *Standard* 710B. Recirculating systems should not be used over gas-fired or solid-fuel-fired cooking equipment.

Designers should thoroughly review NFPA *Standard* 96 requirements and contact a manufacturer of recirculating systems to obtain specific information needed for the design and listing information before incorporating this type of system into a food service design.

DUCT SYSTEMS

Duct Systems

Exhaust ducts convey exhaust air from the hood to the outside, along with any grease, smoke, VOCs, and odors that are not extracted from the airstream along the way. These ducts may also be used to exhaust smoke from a fire. To be effective, ducts must be greasetight; it must be clear of combustibles, or combustible material must be protected so that it cannot be ignited by a fire in a duct; and ducts must be sized to convey the volume of airflow necessary to remove the effluent.

Model building codes, such as the IMC (ICC 2009a), and standards, such as NFPA *Standard* 96, set minimum air velocity for exhaust ducts at 500 fpm. Maximum velocities are limited by pressure drop and noise and typically do not exceed 2500 fpm. Until recently, NFPA *Standard* 96 and the IMC had set the minimum air velocity through the duct at 1500 fpm. However, based on ASHRAE research (Kuehn 2000) that indicated that there is no basis for specifying 1500 fpm minimum duct velocity for commercial kitchen ventilation and that grease deposition in ducts does not increase when duct velocity is lowered to 500 fpm, NFPA and IMC requirements were changed to 500 fpm. This allows flexibility for design of variable-speed exhaust systems and retrofitting older systems, though because of spatial and cost constraints, current design practice for new single-speed systems generally is to design duct velocity between 1500 and 1800 fpm.

Ducts should have no traps that can hold grease, which would be an extra fuel source in the event of a fire, and ducts should pitch toward the hood or an approved reservoir for constant drainage of liquefied grease or condensates. On long duct runs, allowance must be made for possible thermal expansion because of fire, and the slope back to the hood or grease reservoir must conform to local code requirements.

Single-duct systems carry effluent from a single hood or section of a large hood to a single exhaust termination. In multiple-hood systems, several branch ducts carry effluent from several hoods to a single master duct that has a single termination. See the section on Multiple-Hood Systems for more information.

Ducts may be round or rectangular. Standards and model codes contain minimum specifications for duct materials and construction, including types and thickness of materials, joining methods, and minimum clearance of 18 in. to combustible materials. Listed factory-built modular grease duct systems are available as an alternative to code-prescribed welded systems. These listed systems typically incorporate stainless steel liners and double-wall, insulated construction, allowing reduced clearances to combustibles and nonwelded joint construction.

When fire-rated enclosures are required for grease ducts, either fired-rated enclosures are built around the duct or the newer listed, field-applied grease duct enclosures can be used directly on the grease duct, or the newer listed, factory-built, modular grease ducts with insulated construction can be used as an integral fire-rated enclosure. Most of these listed systems allow zero clearance to combustibles and also provide 1 h or 2 h fire resistance rating, and can be used in lieu of a fire-rated enclosure required in NFPA *Standard* 96 and IMC (ICC 2009a).

EXHAUST FANS

Types of Exhaust Fans

Exhaust fans for kitchen ventilation must be capable of handling hot, grease-laden air. The fan should be designed to keep the motor

out of the airstream and should be effectively cooled to prevent premature failure. To prevent roof damage, the fan should contain and properly drain all grease removed from the airstream.

The following types of exhaust fans are in common use (all have centrifugal wheels with backward-inclined blades):

- **Power roof ventilator (PRV).** Also known as **upblast** fans, PRVs are designed for mounting at the exhaust duct outlet (Figure 20), and discharge upward or outward from the roof or building. Aluminum upblast fans must be listed for the commercial kitchen exhaust application in compliance with UL *Standard* 762, and must include a grease drain, grease collection device, and integral hinge kit to permit access for duct cleaning.
- **Centrifugal fan.** Also known as a **utility set**, this is an AMCA Arrangement 10 centrifugal fan, including a field-rotatable blower housing, blower wheel with motor, drive, and often a motor/drive weather cover (Figure 21). These fans are typically constructed of steel and roof-mounted. Where approved, centrifugal fans can be mounted indoors and ducted to discharge outside. The inlet and outlet are at 90° to each other (single width, single inlet), and the outlet can usually be rotated to discharge at different angles around a vertical circle. The lowest part of the fan must drain to an approved container. When listed in accordance with UL *Standard* 762, a grease drain, grease collection device, and blower housing access panel are required.
- **Tubular centrifugal.** These fans, also known as **inline** fans, have the impeller mounted in a cylindrical housing discharging the gas in an axial direction (Figure 22). Where approved, these fans can be located in the duct inside a building if exterior fan mounting is not practical for wall or roof exhaust. They are always constructed of steel. The gasketed flange mounting must be greasetight yet removable for service. The lowest part of the fan must drain to an approved container. When listed in accordance with UL *Standard*

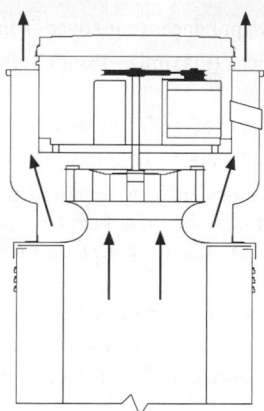

Fig. 20   Power Roof Ventilator (Upblast Fan)

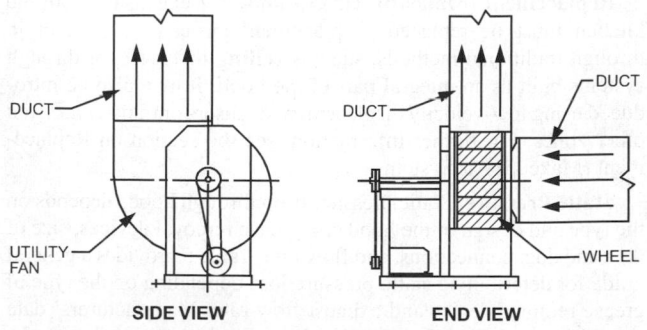

Fig. 21   Centrifugal Fan (Utility Set)

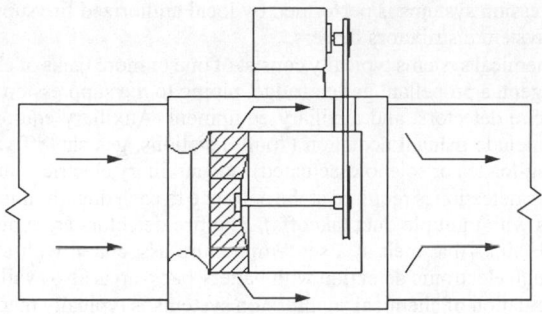

Fig. 22   Tubular Centrifugal (Inline) Fan

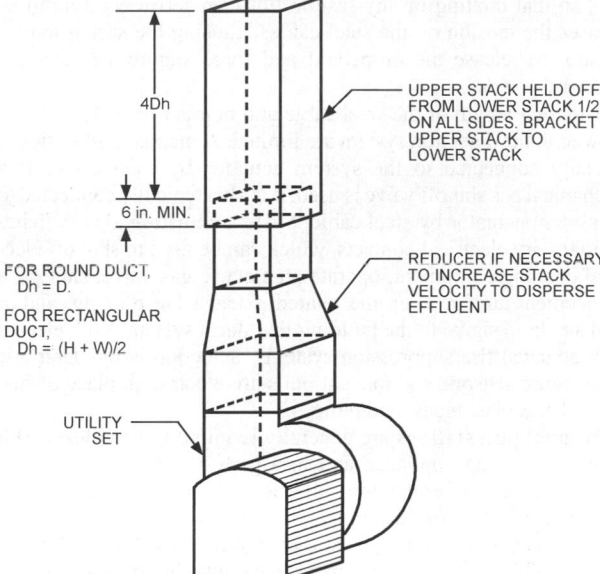

4Dh

6 in. MIN.

UPPER STACK HELD OFF
FROM LOWER STACK 1/2 in.
ON ALL SIDES. BRACKET
UPPER STACK TO
LOWER STACK

FOR ROUND DUCT,
Dh = D.

FOR RECTANGULAR
DUCT,
Dh = (H + W)/2

REDUCER IF NECESSARY
TO INCREASE STACK
VELOCITY TO DISPERSE
EFFLUENT

UTILITY
SET

**Fig. 23   Rooftop Centrifugal Fan (Utility Set) with
Vertical Discharge**

762, a grease drain, grease collection device, and blower housing access panel are required.

Exhaust Terminations

Rooftop. Rooftop terminations are preferred because discharge can be directed away from the building, the fan is at the end of the system, and the fan is accessible. Common concerns with rooftop terminations are as follows:

- Exhaust system discharge should be arranged to minimize reentry of effluent into any fresh-air intake or other opening to any building. This requires not only separating the exhaust from intakes, but also knowledge of the direction of the prevailing winds. Some codes specify a minimum distance to air intakes.
- In the event of a fire, neither flames, radiant heat, nor dripping grease should be able to ignite the roof or other nearby structures.
- All grease from the fan or duct termination should be collected and drained to a remote closed container to preclude ignition.
- Rainwater should be kept out of the exhaust system, especially out of the grease container. If this is not possible, then the grease container should be designed to separate water from grease and drain the water back onto the roof. Figure 23 shows a rooftop utility set with a stackhead fitting, which directs exhaust away from the roof and minimizes rain penetration. Discharge caps should not be used because they direct exhaust back toward the roof and can become grease-fouled.

Outside Wall. Wall terminations are less common today but are still occasionally used in new construction. The fan may or may not be the terminus of the system, located on the outside of the wall. Common concerns with wall terminations are as follows:

- Discharge from the exhaust system should not be able to enter any fresh-air intake or other opening to any building.
- Adequate clearance to combustibles must be maintained.
- To avoid grease draining down the side of the building, duct sections should pitch back to the hood inside, or a grease drain should be provided to drain grease back into a safe container inside the building.
- Discharge must not be directed downward or toward any pedestrian areas.
- Louvers should be designed to minimize their grease extraction and to prevent staining of the building facade.

Recirculating Systems. With these units, it is critical to keep components in good working order to maintain optimal performance. Otherwise, excessive grease, heat, and odors will accumulate in the premises.

As with other terminations, containing and removing grease and keeping the discharge as far as possible from combustibles are the main concerns. Some units are fairly portable and could be set in an unsafe location. The operator should be made aware of the importance of safety in locating the unit. These units are best for large, unconfined areas with a separate outside exhaust to keep the environment comfortable.

FIRE SAFETY

The combination of flammable grease vapor and particulates carried by kitchen ventilation systems and the potential of cooking equipment to be an ignition source creates a higher hazard level than normally found in HVAC systems. Design of an exhaust system serving commercial cooking equipment that may produce grease-laden vapors (i.e., a Type I exhaust system) must include a fire suppression system, as required by NFPA *Standard* 96 and the International Mechanical Code (IMC; ICC 2009a). The IMC further requires that the fire suppression system comply with the International Building Code (IBC; ICC 2009c) and the International Fire Code (IFC; ICC 2009d). By further reference, these codes and standards require that automatic fire suppression systems for Type I hoods must be listed to UL *Standard* 300.

Replacement air systems, air-conditioning systems serving a kitchen, and exhaust systems serving cooking equipment that does not produce grease-laden vapor have no specific fire protection requirements beyond those applicable to similar systems not located in kitchens. However, an exhaust system serving any grease-producing cooking equipment must be considered a grease exhaust system even if it also serves non-grease-producing equipment.

Fire safety starts with proper design, followed by proper operation and maintenance of the cooking equipment and the exhaust system, including frequent and thorough cleaning of grease deposits in the area of appliances, and exhaust filters, hoods, and ducts. After that, the three primary aspects of fire protection in a grease exhaust system are (1) to extinguish a fire quickly once it has started, (2) to prevent the spread of fire from or to the grease exhaust system, and (3) to prevent heat transfer to building components from a grease duct fire if the fire-extinguishing system fails. Additionally, UL *Standard* 300 requires that the fire suppression system not disperse burning grease outside the fire zone, and that, after a fire is suppressed by a fire suppression system, it must remain suppressed for at least 20 min.

Fire Suppression Systems

NFPA *Standard* 96 requires that exhaust systems serving grease-producing equipment must include a fire-extinguishing system that protects the cooking equipment, hood interior, hood filters or grease extractors, ducts, and any other grease-removal devices in the system.

Actuation of any fire-extinguishing system must not depend on building electricity. If actuation relies on electricity, it must be supplied with standby power, usually in the form of battery backup.

Listed fire suppression systems must also automatically shut off all supplies of fuel and energy to all equipment protected by that system. Any gas appliance not requiring protection but located under the same ventilating equipment must also be shut off. On operation of an extinguishing system, all electrical sources located under the ventilating equipment, if subject to exposure to discharge from the fire-extinguishing system, must be shut off. If the exhaust system is in a building with a fire alarm system, actuation of the fire-extinguishing system should send a signal to the fire alarm system.

Dry and Wet Chemical Systems. Wet chemical and combinations of wet chemical and water fire-extinguishing systems have comprised the majority of fire suppression systems since the publication of UL *Standard* 300 in 1994 and its subsequent citation by codes and standards. Dry chemical systems were popular through the early 1990s, but their use declined because they do not meet the requirements of UL *Standard* 300 and must be replaced with UL *Standard* 300 listed systems. Wet chemical systems are covered in NFPA *Standard* 17A, and though obsolete since UL *Standard* 300 was published, dry chemical systems are covered in NFPA *Standard* 17. Both standards provide detailed application information.

Fire suppression systems are tested for their ability to extinguish fires in cooking operations in accordance with UL *Standard* 300. Wet chemical systems extinguish fires by reacting with fats and grease to **saponify**, or form a soapy foam layer, which prevents oxygen from reaching the burning surface. This suppresses the fire and prevents reignition. Saponification is particularly important with deep fat fryers, where the frying medium may be hotter than its autoignition temperature for some time after the fire is extinguished. If the foam layer disappears or is disturbed before the frying medium has cooled below its autoignition temperature, the fat can reignite.

Frying media commonly used today, which contain a high percentage of vegetable oils, have autoignition points of about 685 to 710°F when new. Contamination and deterioration through normal use lowers the autoignition point. In addition to the formation of a foam blanket instead of a thin layer of powder, another advantage of wet chemical systems over dry chemical systems is that the former cools the frying media, bringing it below the autoignition point more quickly.

For a chemical system protecting the entire exhaust system, fire-extinguishing nozzles are located over the cooking equipment being protected, in the hood to protect grease-removal devices and the hood plenum, and at the duct collar (downstream from any fire dampers and pointing in the direction of effluent flow) to protect the grease duct.

Two types of nozzle arrangements are common for protecting appliances. Appliance specific coverage is provided by nozzles that are usually directed at the centers of individual appliances. Overlapping coverage is provided by a generally greater number of evenly spaced nozzles. Although overlapping coverage is slightly more expensive to install and maintain, this arrangement solves the common problem of appliances being periodically rearranged under hoods to meet operational needs.

The duct nozzle is rated to protect an unlimited length of duct, so additional nozzles are not required further downstream in the duct. Additional nozzles and piping in ducts would also make periodic duct cleaning more difficult.

Listed fire-extinguishing systems are available as preengineered (packaged) systems, either installed by authorized exhaust hood manufacturers or local authorized fire suppression system distributors/dealers. In either case, required periodic maintenance of fire

suppression systems is performed by local authorized fire suppression system distributors/dealers.

Chemical systems typically consist of one or more tanks of chemical agent, a propellant gas cartridge, piping to the suppression nozzles, fire detectors, and auxiliary equipment. Auxiliary equipment may include manual actuation ("pull") stations, gas shutoff valves (spring-loaded or solenoid-actuated), and auxiliary electric contacts.

Fire detection is required at the entrance to each duct (or ducts, in hoods with multiple duct takeoffs). The fire detectors are typically fusible links that melt at a set temperature associated with a fire, although electronic detection with battery back-up is also available.

Actuation of chemical suppression systems is typically mechanical, requiring no electric power, by means of a spring-loaded device that pierces the seal on a propellant canister. Fire detectors are typically interconnected with the system actuator by steel cables in tension, so that melting of any fusible links, in series configuration, releases the tension on the steel cables, causing the spring-loaded actuator to release the propellant and force suppressant through pipes and nozzles.

The total length of the steel cable and number of pulley elbows allowed in the detection system are limited. A manual pull station is typically connected to the system actuator by steel cable. If a mechanical gas shutoff valve is used, it is also typically connected to the system actuator by steel cable. System actuation also switches auxiliary dry electrical contacts, which can be used to shut off electrical cooking equipment, operate an electric gas valve, shut off a replacement air fan, keep the related exhaust fan running, and/or send an alarm signal to the building fire alarm system. With electrically actuated fire suppression systems, detection is by electronic temperature sensors, and manual pulls are electric, in place of fusible links, cables, pipes, and pulleys.

Manual pull stations are generally required to be at least 10 ft from the cooking appliance and in a path of egress. Some code authorities may prefer that the pull station be installed closer to the cooking equipment for faster response; however, if it is too close, it may not be possible to approach it once a fire has started. Refer to the applicable code requirements for each jurisdiction to determine specific requirements for location and mounting heights of pull stations.

Water Systems. Water can be used for protecting cooking equipment, hoods, and exhaust systems. Standard fire sprinklers may be used throughout the system, except over deep-fat fryers, where special automatic spray nozzles specifically listed for the application must be used. These nozzles must be aimed properly and supplied with the correct water pressure. Many hood manufacturers market a preengineered water spray system that typically includes a cabinet containing the necessary plumbing and electrical components to monitor the system and initiate fuel shutoff and building alarms.

Application of standard fire sprinklers for protection of cooking equipment, hoods, and exhaust systems is covered by NFPA *Standard* 13. NFPA *Standards* 25 and 96 cover maintenance of sprinkler systems serving an exhaust system. The sprinklers must connect to a wet-pipe building sprinkler system installed in compliance with NFPA *Standard* 13.

One advantage of a sprinkler system is that it has virtually unlimited capacity, whereas chemical systems have limited chemical supplies. Where sprinklers are used in ducts, the duct should be pitched to drain safely. NFPA *Standard* 13 requires that sprinklers used to protect ducts be installed every 10 ft on center in horizontal ducts, at the top of every vertical riser, and in the middle of any vertical offset. Any sprinklers exposed to freezing temperatures must be protected.

Combination Systems. Hoods that use water either for periodic cleaning (water-wash) or for grease removal (cold-water mist) can use this feature in conjunction with the fire-extinguishing system to protect the hood, grease-removal devices, and/or ducts in the event of a fire, if listed to UL *Standard* 300. The water supply for these systems may be from the kitchen water supply if flow and pressure requirements are met. Examples include (1) an approved water-wash or water-mist system to protect the hood in combination with a listed wet chemical system to protect ducts and the cooking appliances (2) a listed chemical fire suppression system in the hood backed up by water sprinklers in the duct, or (3) a listed wet chemical system for appliances, with simultaneous use of a hood water-wash system, with foam-forming chemical injected into the water, for hood plenum and duct.

Multiple-Hood Systems. All hoods connected to a multiple-hood exhaust system must usually meet several requirements. In the IMC (ICC 2009a), for example, the hoods must be on the same floor of the building, all interconnected hoods must be in the same room or in adjoining rooms, interconnecting ducts must not penetrate assemblies required to be fire-resistance rated, and the grease duct system must not serve solid-fuel-fired appliances.

The multiple-hood exhaust system must be designed to (1) prevent a fire in one hood or in the duct from spreading through the ducts to another hood and (2) protect against a fire starting in the common duct system. Of course, the first line of protection for the ducts is keeping them clean. Especially in a multiple-tenant system, a single entity must assume responsibility for cleaning the common duct frequently.

Each hood must have its own fire-extinguishing system to protect the hood and cooking surface. A single system might serve more than one hood, but in the event of fire under one hood, the system would discharge its suppressant under all hoods served, resulting in unnecessary cleanup expense and inconvenience. A water-mist system could serve multiple hoods if sprinkler heads were allowed to operate independently.

Because of the possibility of a fire spreading through ducts from one hood to another, the common duct must have its own fire extinguishing system. The appendices of NFPA *Standards* 17 and 17A present detailed examples of how common ducts can be protected, either by one system or by a combination of separate systems serving individual hoods. Different types of fire-extinguishing systems may be used to protect different portions of the exhaust system; however, in any case where two different types of system can discharge into the common duct at the same time, the agents must be compatible.

As mentioned earlier, actuation of the fire-extinguishing system protecting any hood must shut off fuel or power to all cooking equipment under that hood. When a common duct, or portion thereof, is protected by a chemical fire-extinguishing system that activates from a fire in a single hood, NFPA *Standards* 17 and 17A require shutoff of fuel or power to the cooking equipment under every hood served by that common duct, or every portion of it protected by the activated system, even if there is no fire in the other hoods served by that duct.

From an operational standpoint, it is usually most sensible to provide one or more fire-extinguishing systems to detect and protect against fire in common ducts and a separate system to protect each hood and its connecting ducts. This prevents a fire in the common duct from causing discharge of fire suppressant under an unaffected hood and it allows unaffected hoods to continue operation in the event of a fire under one hood unless the fire spreads to the common duct.

Preventing Fire Spread

The exhaust system must be designed and installed both to prevent a fire starting in the exhaust system from damaging the building or spreading to other building areas, and to prevent a fire in one building area from spreading to other parts of the building through the exhaust system. This protection has two main aspects: (1) maintaining clearance from the duct to other portions of the building, and (2) either enclosing the duct in a fire-resistance-rated enclosure, or wrapping the duct with a listed fire-rated product. These methods

are sometimes addressed by a listed insulated grease duct system that incorporates an integral fire resistance.

Clearance to Combustibles. A grease fire can generate gas temperatures of 2000°F or greater in the exhaust hood and duct. In such a grease fire, heat radiating from the hot surface can ignite combustible materials near the hood or duct. Additionally, if the hood or duct is not fully welded and liquidtight as required by codes and standards, grease liquid or vapor leaking from the hood or duct can ignite and spread fire to nearby combustible structure. Most codes require a minimum clearance of 18 in. from the hood and grease duct to any combustible material. However, even 18 in. may not be sufficient clearance to prevent ignition of combustibles in the case of a major grease fire, especially with large volumes of grease in larger ducts.

Several methods to protect combustible materials from the radiant heat of a grease fire and allow reduced clearance to combustibles are described in NFPA *Standard* 96 and the IMC (ICC 2009a). Based on testing and listing of grease ducts provided with integral insulation or wrapped with insulation, NFPA *Standard* 96 and other codes now allow listed insulation to be applied to the duct or a listed factory-built grease duct with integral insulation. For hoods, the clearance can be reduced as prescribed.

Listed grease ducts, typically with insulation between double walls or on the outside of single-wall ducts, may be installed with reduced clearance to combustibles in accordance with locally adopted codes and standards, if installed per manufacturers' instructions, which should include specific information regarding the listing. Listed grease ducts are tested and evaluated in accordance with UL *Standard* 1978.

NFPA *Standard* 96 requires a minimum clearance of 3 in. to "limited combustible" materials (e.g., gypsum wallboard on metal studs). The IMC (2009a) allows reduced clearance of ducts to 3 in. in proximity to noncombustibles on noncombustible structure, such as gypsum wallboard on metal studs. Clearance reduction is also available for hoods, but may differ by local code, so local codes and standards should be consulted accordingly.

Note that clearance-to-combustible issues are often seen in inspections of restaurant sites after grease fires. Many instances of inappropriate clearance reduction have been seen in which gypsum wallboard was mistakenly applied to wood studs and joists. In many of these cases, surrounding structure was ignited by heat from a grease fire, in spite of the gypsum wallboard barrier. The issue here is autoignition of the combustible material behind the gypsum wallboard from the high heat of the grease fire, even in cases where the gypsum wallboard layer is intact after the fire.

Enclosures. Normally, when a HVAC duct penetrates a fire-resistance-rated wall or floor, a fire damper is used to maintain the integrity of the wall or floor. Because fire dampers cannot be installed in a grease duct unless specifically approved for such use, there must be an alternative means of maintaining the integrity of rated walls or floors. Therefore, grease ducts that penetrate a fire-resistance-rated wall or floor/ceiling assembly must be continuously enclosed in a fire-rated enclosure from the point the duct penetrates the first fire barrier until the duct leaves the building. Listed grease ducts are also subject to these enclosure requirements. The requirements are similar to those for a vertical shaft (typically 1 h rating if the shaft penetrates fewer than three floors, 2 h rating if it penetrates three or more floors), except that the shaft can be both vertical and horizontal. In essence, the enclosure extends the room containing the hood through all the other compartments of the building without creating any unprotected openings to those compartments.

Where a duct is enclosed in a rated enclosure, whether vertical or horizontal, clearance must be maintained between the duct and the shaft. NFPA *Standard* 96 and the IMC (2009a) require a minimum 6 in. clearance and that the shaft be vented to the outside. IMC requires that each exhaust duct have its own dedicated enclosure.

Some listed grease ducts are designed and tested for use without shaft enclosure. Listed grease ducts of this type use fire barrier insulation and provide integral fire-rated resistance, which serves the same function as the shaft enclosure. These products are tested and listed in accordance with UL Standard 2221. They must be installed in accordance with the manufacturer's installation instructions.

Some insulation materials are listed to serve as a fire-resistance-rated enclosure for a grease duct when used to cover a duct. These insulations are tested and listed in accordance with ASTM Standard E2336. These listed insulations must be applied in accordance with the manufacturer's installation instruction.

Insulation materials that have not been specifically tested and approved for use as fire protection for grease ducts should not be used in lieu of rated enclosures or to reduce clearance to combustibles. Even insulation approved for other fire protection applications, such as to protect structural steel, may not be appropriate for grease ducts because of the high temperatures that may be encountered in a grease fire.

Exhaust and Supply Fire-Actuated Dampers. Because of the risk that the damper may become coated with grease and become a source of fuel in a fire, balancing and fire-actuated dampers are not allowed at any point in a exhaust system except where specifically listed for use or required as part of a listed device or system. Typically, fire dampers are found only at the hood collar and only if provided by the hood manufacturer as part of a listed hood.

Opinions differ regarding whether any fire-actuated dampers should be provided in the exhaust hood. On one hand, a fire-actuated damper at the exhaust collar may prevent a fire under the hood from spreading to the exhaust duct. However, like anything in the exhaust airstream, the fire-actuated damper and fusible link may become coated with grease if not properly maintained, which may impede damper operation. On the other hand, without the fire-actuated damper, the exhaust fan draws smoke and fire away from the hood. Although this cannot be expected to remove all smoke from the kitchen during a fire, it can help to contain smoke in the kitchen and minimize migration of smoke to other areas of the building.

A fire-actuated damper will generally close only in the event of a severe fire; most kitchen fires are extinguished before enough heat is released to trigger the fire-actuated damper. Thus, the hood fire-actuated damper remains open during relatively small fires, allowing the exhaust system to remove smoke, but can close in the event of a severe fire, helping to contain the fire in the kitchen area.

Fan Operations. If replacement air flow rates exceed 2000 cfm, the replacement air supply to the kitchen is generally required by codes and standards to be shut down during fire to avoid feeding air to the fire. However, if the exhaust system is intended to operate during a fire to remove smoke from the kitchen (as opposed to just containing it in the kitchen), the replacement air system must operate as well. If the hood has an integral replacement air plenum, a fire-actuated damper must be installed in the replacement air plenum to prevent a fire in the hood from entering the replacement air duct. NFPA *Standard* 96 details the instances where fire-actuated dampers are required in a hood replacement air plenum.

Regardless of whether fire-actuated dampers are installed in the exhaust system, NFPA *Standard* 96 calls for the exhaust fan to continue to run in the event of a fire unless fan shutdown is required by a listed component of the exhaust system or of the fire-extinguishing system. Listed fire-extinguishing systems protecting ducts are tested both with and without airflow, and exhaust airflow is not necessary for proper operation.

Control Systems. The IMC (2009a) requires that Type I (grease and smoke) hoods be designed and installed to automatically activate related exhaust fans whenever cooking operations occur. Further, the exhaust fan(s) must occur be activated through an interlock with the cooking appliances by means of a heat sensor or other approved means. An example of such an interlock is a listed electronic temperature sensor using a listed hood and duct accessory penetration in the duct collar and connecting the sensor electrically in parallel with fan control switches.

REPLACEMENT (MAKEUP) AIR SYSTEMS

In hood systems, where air exhausted through the hood is discharged to the outside, the volume of air exhausted must be replaced with uncontaminated outdoor air. Outdoor air must be introduced into the building through properly designed replacement air systems. Proper replacement air volume and distribution allow the hood exhaust fan to operate as designed and facilitate proper building pressurization, which is required for safe operation of direct-vent gas appliances (such as water heaters), prevention of kitchen odors migrating to adjacent building spaces, and/or maintaining a comfortable building environment. Proper pressurization enhances the building environment by preventing suction of unfiltered and/or unconditioned outdoor air into the building envelope through doors, windows, or air handlers. IMC (ICC 2009a) requires neutral or negative pressurization in rooms with mechanical exhaust. NFPA *Standard* 96 requires enough replacement air to prevent negative pressures from exceeding 0.02 in. of water, which may still be excessive for proper drafting of some direct vent appliances. To ensure pressure control, IMC also requires electrical interlock between exhaust and replacement air sources. This electrical interlock prevents excessive negative or positive pressures created by the exhaust fan or replacement air unit operating independently.

Indoor Air Quality

Traditionally, the primary purpose of replacement air has been to ensure proper operation of the hood. Kitchen thermal comfort and indoor air quality (IAQ) have been secondary. In some applications, thermal comfort and IAQ can be improved through adequate airflow and proper introduction of replacement air. In many of today's applications, outdoor air that meets IAQ standards is the most energy-efficient source for kitchen hood replacement air. ASHRAE *Standard* 62.1 requires that 10 cfm outdoor air per person, based on a maximum occupancy of 70 persons per 1000 ft^2, be supplied to nonsmoking restaurant dining areas. For nonsmoking cafeterias and fast food dining, the outdoor air requirement is 9 cfm per person, based on 100 persons per 1000 ft^2. The requirement is 9 cfm per person in nonsmoking bars/cocktail lounges at 100 persons per 1000 ft^2. These requirements may be increased or decreased in certain areas if approved by the authority having jurisdiction. Outdoor air requirements affect HVAC system sizing and may require another means of introducing outside air. A further requirement of *Standard* 62.1, that outdoor air be sufficient to provide for an exhaust rate of at least 0.70 cfm per square foot of kitchen space, is generally easily met.

Replacement Air Introduction

Replacement air may be introduced into the building through conventional HVAC apparatus, ventilators (no conditioning), or dedicated kitchen makeup air units and/or replacement air units. Replacement-air units are specifically designed to supply heated or cooled 100% outdoor air.

Conventional HVAC units used as replacement air sources may have fixed outdoor air intakes or economizer-controlled outside air dampers. HVAC units with economizers should have a barometric relief damper either in the return ductwork or in the HVAC unit itself. As the amount of outdoor air is increased, the increase in pressure in the system will open the relief damper, so that the return air volumetric rate is only enough to maintain approximately the amount of design supply air. The supply fan runs at a constant speed and thus moves a constant volume of air. The amount of air required for dedicated replacement air becomes the minimum set point for the economizer damper when the hoods are operating. Fixed outdoor air intakes must be set to allow the required amount of replacement air. Outdoor air dampers should be interlocked with hood controls to open to a preset minimum position when the hood system is energized. If the zone controls call for cooling, and outside conditions are within economizer range, the outside damper may be opened to allow greater amounts of outdoor air. The maximum setting for outdoor air dampers in unitary HVAC units is typically 25 to 30% of total unit air volume when compressors are running.

Operating in economizer mode should not change air discharge velocities or volumes, because the supply fan runs at a constant speed and thus moves a constant volume of air. However, field experience shows that large increases in air discharge velocities or volumes can occur at diffusers when HVAC units go into economizer mode. This is because the static loss through the fresh-air intake is considerably less than through the return air duct system, and thus a change from return air to fresh air reduces the overall static through the system, resulting in a relative increase in the total system flow. This can create air balance problems that negatively affect hood performance because of interference with capture and containment supply flow patterns at the hoods. A large increase in air velocity or volume from supply diffusers indicates a need for better balance between the fresh air and return air static losses. Some HVAC manufacturers state that a relief fan is required to ensure proper air balance if economizer controls call for outdoor air greater than 50% during economizer operation mode. A relief fan addresses static losses in the return duct system, thus helping minimize the static difference with the fresh-air intake. Lack of a barometric relief damper, or constrictions in the return ductwork, also may be the source of the problem.

In smaller commercial buildings, including restaurants and strip centers, individual unitary rooftop HVAC equipment is common. This unitary equipment may not be adequate to supply 100% of the replacement air volume. Outdoor air must be considered in the initial unit selection to obtain desired unit operation and space comfort. The space in which the hood is located should be maintained at a neutral or negative pressure relative to adjacent spaces. Therefore, HVAC economizers are not recommended for equipment supplying air directly to the space in which the hood is located, unless the economizer installation includes equipment and controls to maintain overall system air balance and to prevent excessive air discharge velocities or volumes.

Using enthalpy or temperature control is recommended. These controls cycle HVAC compressor(s), water supply, or heat source off when outdoor air conditions warrant and open outdoor air dampers if economizer controls are used. These additional controls provide kitchen comfort while conserving energy and saving money.

Replacement Air Categories

Three categories of replacement air have been defined for design of energy-efficient replacement air systems: supply, makeup, and transfer. IAQ engineers must design outdoor air systems to meet total building ventilation requirements. Replacement air for kitchen ventilation must integrate into the total building IAQ design. Total kitchen ventilation replacement air may consist of only dedicated makeup air; however, in many energy-efficient designs, outdoor air required for ventilating the kitchen or adjacent spaces is used as supply or transfer air to augment or even eliminate the need for dedicated makeup air. Typically, replacement air will be a combination of categories from multiple sources. The source of replacement air typically determines its category.

Supply air is outdoor air introduced through the HVAC or ventilating apparatus, dedicated to the comfort conditioning of the space in which the hood is located. In many cases this may be an ideal source of replacement air because it also provides comfort conditioning for the occupants.

Makeup air is outdoor air introduced through a system dedicated to provide replacement air specifically for the hood. It is typically delivered directly to or close to the hood. This air may or may not be ventilated. When conditioned, it may be heated only;

generally only in extreme environments will it be cooled. When included, makeup air typically receives less conditioning than space supply air. The IMC (ICC 2009a) requires makeup air be conditioned to within 10°F of the kitchen space, except when introducing replacement air that does not decrease kitchen comfort (see the section on Energy Considerations for additional information). This can be accomplished with proper distribution design. Typical sources of makeup air conditioning include electric resistance, direct and indirect gas-fired units, evaporative coolers, and water coils for cooling or heating (freeze protection required). Temperature of makeup air introduced varies with distribution system and type of operation.

Transfer air is outdoor air, introduced through the HVAC or ventilating apparatus, dedicated to comfort conditioning and ventilation requirements of a space adjacent to the hood. The device providing transfer air must be in operation and supplying outdoor air while the hood is operating. Air must not be transferred from spaces where airborne contaminants such as odors, germs, or dust may be introduced into the food preparation or serving areas. Air may be transferred through wall openings, door louvers, or ceiling grilles connected by duct above the ceiling. Depending on grille and duct pressure drop, a transfer fan(s) may be required to avoid drawing transfer air through lower-pressure-drop openings. When using openings through which food is passed, transfer velocities should not exceed 50 fpm to avoid excessive cooling of the food. Transfer air is an efficient source of replacement air because it performs many functions, including ventilating and/or conditioning the adjacent space, replacing air for the hood, and additional conditioning for the space in which the hood is located. Only the portion of air supplied to the adjacent space that originated as outdoor air may be transferred for replacement air. The IMC (ICC 2009a) recognizes the use of transfer air as a replacement (makeup) air source. In large buildings such as malls, supermarkets, and schools, adequate transfer air may be available to meet 100% of hood replacement air requirements. Malls and multiple-use-occupancy buildings may specify a minimum amount of transfer air to be taken from their space to keep cooking odors in the kitchen, or they may specify the maximum transfer air available to hold down the cost of conditioning outside air. Code restrictions may prevent the use of corridors as spaces through which transfer air may be routed. Conditions of transfer air are determined by conditioning requirements of the space into which the air is initially supplied.

Air Distribution

The design of a replacement air distribution system may enhance or degrade hood performance. Systems that use a combination of supply, makeup, and transfer air include various components of distribution. Distribution from each source into the vicinity of the hood, must be designed to eliminate high velocities, eddies, swirls, or stray currents that can interrupt the natural rising of the thermal plume from cooking equipment into the hood, thus degrading the performance of the hood. Methods of distribution may include conventional diffusers, compensating hood designs, transfer devices, and simple openings in partitions separating building spaces. Regardless of the method selected, it is important to always deliver replacement air to the hood (1) at proper velocity and (2) uniformly from all directions to which the hood is open. This minimizes excessive cross-currents that could cause spillage. Proper location and/or control of HVAC return grilles is therefore critical. The higher air velocities typically recommended for general ventilation or spot cooling with unconditioned air (75 to 200 fpm at worker) should be avoided around the hood. Hood manufacturers offer a variety of compensating hoods, plenums, and diffusers designed to introduce replacement air effectively.

Compensating Hoods. A common way of distributing replacement air is through compensating systems that are integral with the hood. Figure 24 shows four typical compensating hood configurations. Because actual flows and percentages may vary with hood

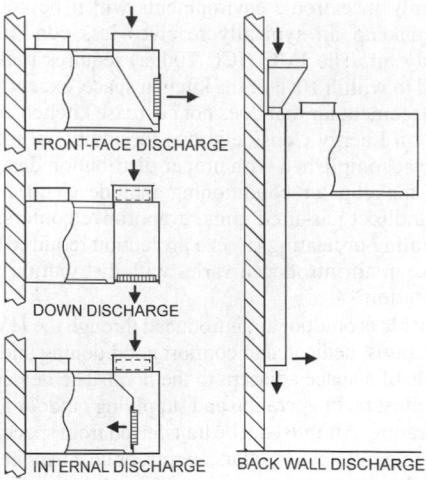

Fig. 24   Compensating Hood Configurations

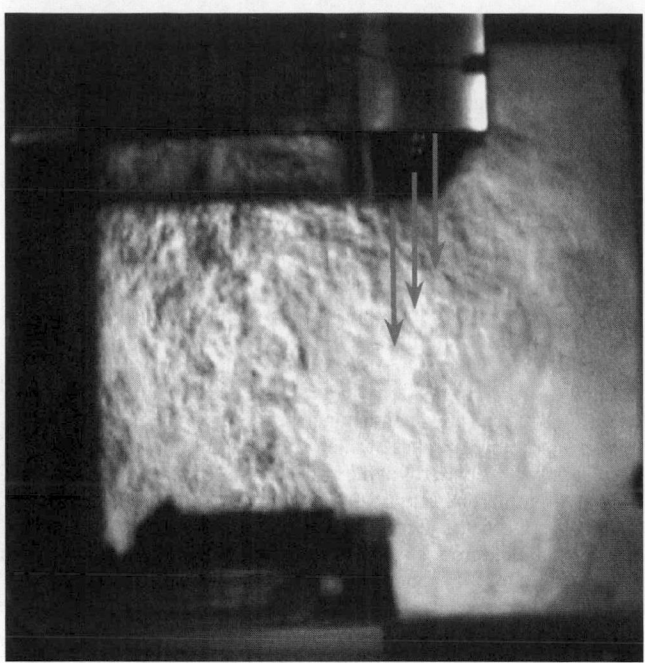

**Fig. 25   Schlieren Image Showing Thermal Plume Being
Pulled Outside Hood by Air Curtain**
(Brohard et al. 2003)

design, the manufacturer should be consulted about specific applications. The following are typical descriptions of configurations that include perimeter supply.

Brohard et al. (2003) investigated the effects of six methods of introducing replacement air on three hood styles, Three hood types were tested: (1) wall-mounted canopy, (2) island-mounted canopy, and (3) proximity (backshelf). Gas underfired broilers and gas griddles, respectively representing heavy-duty and medium-duty appliances, were tested. Idle and emulated cooking conditions also were tested. The MUA strategies included (1) displacement ventilation (base case), (2) ceiling diffuser, (3) hood face diffuser, (4) air curtain diffuser, (5) back wall supply, and (6) short-circuit supply. The influences of air mass disturbances (drafts) and tapered side panels were also investigated. Each makeup air strategy and specific configuration tested compromised the exhaust hood's ability to completely capture and contain the thermal plume and/or effluents at higher makeup airflow rates (expressed as a percentage of the threshold exhaust rate). Temperature of locally supplied makeup air also affected hood performance, because air density affects the dynamics of air movement around the hood. Generally, hotter makeup air temperatures (e.g., greater than 90°F) affect hood performance more adversely than cooler air (e.g., less than 75°F).

Air Curtain Supply. This method is typically used for spot-cooling the cooking staff to counter the severe radiant heat generated from equipment such as gas or electric broilers. The air must be heated and/or cooled, depending on local climate. Air curtain discharge can be along the length of the hood front only or along all open sides of the hood. When discharge velocity is too low, air tends to enter the hood directly and may have little effect on hood performance. When discharge velocity is too high, air entrains the cooking plume and spills it into the room. Ideal velocity and throw can improve hood performance and redirect the thermal plume toward the filters. Discharge velocities must be carefully selected to avoid discomfort to personnel and cooling of food.

Limit the percentage of makeup air supplied through an air curtain to less than 20% of the hood's exhaust flow. At these low air velocities, an air curtain may enhance capture and containment, depending on design details. However, at higher makeup airflow rates, the air curtain is one of the worst performing makeup air strategies. The negative effect of an air curtain is clearly illustrated in Figure 25 by the schlieren flow visualization recorded during a test of a wall-mounted canopy hood operating over two gas underfired broilers.

Introducing makeup air through an air curtain is a risky option. An air curtain (by itself or in combination with another pathway) is

not recommended, unless velocities are minimized and the designer has access to performance data on the actual air curtain configuration being specified. Typical air curtains are easily adjusted, which could cause cooking effluent to spill into the kitchen by inadvertently creating higher-than-specified discharge velocities.

Back-Wall Supply. A makeup air plenum is installed between the back of the hood and wall. The full-length plenum typically extends down the wall to approximately 6 in. below the cooking surface or 2 to 3 ft above the floor. The depth of the plenum is typically 6 in. Makeup air is discharged behind and below the cooking equipment. The bottom of the plenum is provided with diffusers and may also include a balancing damper. As with front-face discharge, air volume and discharge velocity dictate how far into the space the replacement air will travel. The amount of travel and local climate dictate the amount of heating and/or cooling needed. Support for wall shelves, salamander broilers, or cheesemelters mounted under the hood must be considered. The plenum structure typically does not provide sufficient support for mounting these items.

Back-wall supply can be an effective strategy for introducing makeup air (Figure 26). In most cases, it allows significant amounts of air to be locally supplied without a detrimental effect on hood C&C performance. Local makeup air mostly enters the kitchen space, rather than remaining contained in the cooking zone. This potentially creates an additional heat and moisture load on the kitchen, particularly because most makeup air supplied is mixed with room air before being exhausted.

To help ensure proper performance, the discharge of the back-wall supply should be at least 12 in. below cooking surfaces of appliances, to prevent the relatively high-velocity makeup air from interfering with gas burners and pilot lights. Back-wall plenums with larger discharge areas may provide increased airflow rates as long as discharge velocities remain below maximum thresholds. The quantity of air introduced through the back-wall supply should be no more than 60% of the hood's exhaust flow.

Front-Face Supply. Supplying air through the front face of the hood is a configuration recommended by many hood manufacturers. In theory, air exits the front-face unit horizontally into the kitchen space. However, a front-face discharge with louvers or

**Fig. 26   Schlieren Image Showing Thermal Plume Being
Captured with Back-Wall Supply**
(Brohard et al. 2003)

**Fig. 27   Schlieren Image Showing Thermal Plume Being
Pulled Outside Hood by Front Face**
(Brohard et al. 2003)

perforated face can perform poorly, if its design does not consider discharge air velocity and direction. Figure 27 presents a poorly designed perforated face supply, which can negatively affect hood capture performance in the same way as an air-curtain or four-way diffuser. To improve front-face performance, internal baffling and/or a double layer of perforated plates may be used improve the uniformity of airflow. In addition, greater distance between the lower capture edge of the hood and the bottom of the face discharge area may decrease the tendency of the makeup air supply to interfere with hood capture and containment. In general, face discharge velocities should not exceed 150 fpm (i.e., makeup air flow rate divided by gross discharge area) and should exit the front face in a horizontal direction.

Internal Makeup Air. This method, also known as **short-circuit,** introduces makeup air directly into the exhaust hood cavity. This design has limited application, and the amount of air that can be introduced varies considerably with the type of cooking equipment and exhaust flow rate. As noted previously, thermal currents from cooking equipment create a plume of a certain volume that the hood must remove. The hood must therefore draw at least this volume of air from the kitchen, in addition to any internal makeup. If the net exhaust flow rate (total exhaust less internal replacement air) is less than the plume volume, part of the plume may spill out of the hood. Internal makeup air is typically not conditioned; however, depending on local climate, manufacturer's design, type of cooking equipment, and local codes, conditioning may be required. Some local authorities approve internal discharge hoods, and some do not. For unlisted hoods, IMC (2009a) requires the net quantity of exhaust air to be calculated by subtracting any airflow supplied directly to a hood cavity from the total exhaust flow rate of a hood. Listed hoods are operated in accordance with the terms of the listing. All applicable codes must be consulted to ensure proper criteria are followed.

When short-circuit hoods are operated with excessive internal makeup air, they typically fail to capture and contain the cooking effluent (Figure 28). Additionally, the introduction of untempered makeup air results in uncomfortable kitchen conditions. Independent research (Brohard 2003) recommends not using this compensating hood design; therefore, there is no additional design information in this chapter.

**Fig. 28   Schlieren Image Showing Thermal Plume Being
Displaced by Short-Circuit Supply, Causing Hood to Spill**
(Brohard et al. 2003)

Multiple Discharge. This method may combine internal, perimeter, air curtain, and/or front face. Each may be served by a separate or common plenum. Balancing dampers may be provided for one or both discharge arrangements. These dampers may be used to fine-tune the amount of air discharged through the air curtain or front face. However, this method inherits the performance problems of each of the individual types, and the combining of them tends to compound these problems.

Fig. 29 Schlieren Image Shows Effective Plume Capture with Makeup Air Supplied Through 16 in. Wide Perforated Perimeter Supply, Shown with Additional Front Overhang
(Brohard et al. 2003)

Fig. 30 Schlieren Image Showing Thermal Plume Being Pulled Outside Hood by Air Discharged From Four-Way Diffuser
(Brohard et al. 2003)

Perforated Perimeter Supply. Perforated perimeter supply is similar to a front-face supply, but the air is directed downward, as in Figure 29, toward the hood capture area. This may be advantageous under some conditions, because air is directed downward into the hood capture zone.

For proper hood performance, discharge velocities should not exceed 150 fpm (i.e., makeup airflow rate divided by gross discharge area) from any section of the diffuser, and the distance to lower edge of the hood should be no less than 18 in., or the system begins to act like an air curtain. An increase in the plenum discharge area lowers the velocity for a given flow of makeup air and reduces the chance of it affecting capture and containment. If the perforated perimeter supply is extended along the sides of the hood as well as the front, the increased area allows proportionally more makeup air to be supplied. In all cases, the velocity downward 2 in. above the lower edge of the hood should not exceed 75 fpm.

Diffusers. There are various ways to distribute replacement air in the vicinity of the hood to avoid cross-currents that degrade hood performance. Nonaspirating diffusers are recommended, especially adjacent to the hood. Typical devices include the following (for more information on diffusers, see Chapter 20 of the 2009 *ASHRAE Handbook—Fundamentals* and Chapter 19 of the 2008 *ASHRAE Handbook—HVAC Systems and Equipment*).

Directional Ceiling Diffusers. Air from these two- or three-way diffusers should not be directed toward exhaust hoods, where it might disturb the thermal plume and adversely affect hood performance. The diffuser should be located so that the jet velocity at the lip of the hood does not exceed 75 fpm.

Four-Way Directional Ceiling Diffusers. Four-way directional diffusers located close to kitchen exhaust hoods (Figure 30) can have a detrimental effect on hood performance, particularly when flow through the diffuser approaches its design limit. They are not recommended within 15 ft of the hood.

Perforated Ceiling Diffusers. These nonaspirating, perforated-face diffusers may have internal deflecting louvers, but should not

be capable of directing the airflow toward the hood. The diffuser should be located so that the jet velocity at the lip of the hood does not exceed 75 fpm. In some code jurisdictions, when conventional ceiling diffusers are used, only perforated diffusers are allowed in commercial kitchens. Perforated ceiling diffusers can be used near the hood, although a greater number of these diffusers may be required to reduce air velocities for a given supply rate. To help ensure proper hood performance, air from a perforated diffuser near the hood should not be directed toward the hood. If ceiling-supplied air must be directed toward a hood, air discharge velocity at the diffuser face should be selected so that the terminal velocity does not exceed 75 fpm at the edge of the hood capture area.

Slot Diffusers. Because the slot opening of these devices is generally small compared to air volume, air velocity is often higher than that which would be obtained with two-, three-, and four-way diffusers. Also, because airflow is mostly downward, the potential for negatively affecting hood performance is quite high if outlets are near the hood. If used with relatively high ceilings, the potential for negative impact is less because the velocity diminishes as air diffuses downward. Slot diffusers are usually nonaspirating.

Displacement Diffusers. These devices, designed to provide low-velocity laminar flow over the diffuser surface, typically supply air from 50 to 70°F in a kitchen, depending on equipment loads. Hotter, stratified air is removed from the ceiling through exhaust ducts or returned to the HVAC system to be conditioned. In contrast with ceiling diffusers, which require complete mixing to be effective, stratification is the desired effect with displacement diffusers.

Displacement diffusers were used to determine the baseline for Brohard et al.'s (2003) makeup air study, because they provided a uniform, nearly laminar bulk airflow. This low-velocity bulk airflow is optimal for attaining C&C with the lowest exhaust rate. Therefore, supplying replacement air through displacement diffusers (Figure 31) may be an effective strategy for introducing replacement air. Adequate wall or floor space is required to accommodate displacement diffusers.

Fig. 31   Schlieren Image Showing Plume Being Effectively Captured when Makeup Air Is Supplied at Low Velocity From Displacement Diffusers
(Brohard et al. 2003)

Other Factors That Influence Hood Performance.

Hood Style. Wall-mounted canopy hoods function effectively with a lower exhaust flow rate than single-island hoods. Island canopy hoods are more sensitive to makeup air supply and cross drafts than wall-mounted canopy hoods. Back-shelf/proximity hoods generally exhibit lower capture and containment exhaust rates, and in some cases, perform the same job at one-third of the exhaust rate required by a wall-mounted canopy hood.

Cross Drafts. Cross drafts have a detrimental effect on all hood/appliance combinations. Cross drafts adversely affect island canopy hoods more than wall-mounted canopy hoods. A fan in a kitchen, especially pointing at the cooking area, severely degrades hood performance and may make capture impossible. Cross drafts required at least a 37% increase in exhaust flow rate; in some cases, C&C could not be achieved with a 235% increase in exhaust rate (Brohard et al. 2003). Cross drafts can result from portable fans, movement in the kitchen, or an unbalanced HVAC system, which may pull air from open drive-through windows or doors.

Side Panels. Side (or end) panels allow a reduced exhaust rate in most cases, because they direct replacement airflow to the front of the hood. Installing side panels improved C&C performance for static conditions an average 10 to 15% and up to 35% for dynamic (cross-draft) conditions. They are a relatively inexpensive way to enhance performance and reduce the total exhaust rate. Partial side panels can provide virtually the same benefit as full panels. One of the greatest benefits of side panels is to mitigate the negative effect of cross drafts.

The primary recommendation from the study (Brohard et al. 2003) was to reduce the impact that locally supplied makeup air may have on hood performance by minimizing makeup air velocity as it is introduced near the hood. This can be accomplished by minimizing the volume of makeup air through any single distribution system or by distributing through multiple configurations. The chances of makeup air affecting hood performance increase as the percentage of the locally supplied makeup air (relative to the total exhaust) is increased. In fact, the 80% rule of thumb for sizing airflow through a makeup air system may be a recipe for trouble.

Effective introduction of replacement air (whether supplied through displacement ventilation diffusers, perforated diffusers located in the ceiling, and/or as transfer air from adjacent spaces) should be designed to limit velocities approaching the hood to less than 75 fpm.

Design Recommendations. The first step to reducing the replacement air requirement is lowering the design exhaust rate, which can be accomplished by prudent selection and application of UL *Standard* 710 listed hoods. Using side panels on canopy hoods may increase effectiveness and mitigate cross drafts, and is highly recommended where applicable. The next step is to take credit for outdoor air that must be supplied by the HVAC system to meet code requirements for space or occupant ventilating. Depending on the architectural layout, it may be practical to transfer most of this air to the kitchen. Assuming the transfer air is conditioned and properly introduced, it may enhance hood performance and improve the kitchen environment.

AIR BALANCING

ASHRAE research (Kuehn 2010) has demonstrated that a high degree of correction is required to achieve accurate airflow measurements when using many of the instruments commonly used in the field to balance hood systems. Because of the level of correction required, hot-wire anemometers are not recommended. Therefore, balancing is best performed when the manufacturers of all system components provide a certified reference method of measuring the airflow of their equipment, rather than depending on generic measurements of duct flows or other forms of measurement in the field, which again can be erroneous. The equipment manufacturer should be able to develop a reference method of measuring airflow in a portion of the equipment that is dynamically stable in the laboratory as well as in the field. This method should relate directly to airflow by graph or formula.

Basic tools for balancing include the following:

- Volumetric flow hood
- Rotating vane anemometer
- Velocity grid
- Pitot tube/anemometers
- Manometer/pressure meter
- Voltage/amperage meter(s)
- Tachometer

Where applicable, field instruments should be new (i.e., less than one year old) or carry a current calibration certificate.

The general steps for air balancing in restaurants are as follows:

1. Verify all exhaust and HVAC equipment is installed correctly and operating correctly, including (but not limited to) verifying that exhaust ducts are fully welded and inspection doors are in place, HVAC and supply ducts are complete and sealed, fans are rotating the correct direction, and thermostats are set up correctly and set to *on* or occupied mode.
2. Tabulated results of measurements should be kept and used to create a balance chart to show the building's net exfiltration or infiltration.
3. Exhaust hoods should be set to their proper flow rates, with supply and exhaust fans on.
4. Next, supply airflow rate, whether part of combined HVAC units or separate replacement air units, should be set to design values through the coils and the design supply flows from each outlet, with approximately correct settings on the outdoor airflow rate. Then, correct outdoor and return airflow rates should be set proportionately for each unit, as applicable. These settings should be made with exhaust on, to ensure adequate relief for the

outdoor air. Where outdoor air and return air flows of a particular unit are expected to modulate, there should ideally be similar static losses through both airflow paths to preclude large changes in total supply air from the unit. Such changes, if large enough, could affect the efficiency of heat exchange and could also change airflows within and between zones, thereby upsetting air distribution and balance.

5. Next, outdoor air should be set with all fans (exhaust and supply) operating. Pressure difference between inside and outside should be checked to see that (1) nonkitchen zones of the building are at a positive pressure compared to outside and (2) kitchen-zone pressure is negative compared to the surrounding zones and positive or neutral compared to outside.

6. For applications with demand-controlled ventilation (DVC) systems, proper capture and containment, as well as differential pressures between zones and atmosphere, should be confirmed at minimum and at maximum flow rates. This requires that the replacement airflow rate compensate automatically with each increment of exhaust. It may require some adjustments in controls or in damper linkage settings to get the correct proportional response.

Performance Test

After initial airflows are verified to be at design values and the building is balanced, a performance evaluation of all exhaust hoods should be performed to verify capture and containment (C&C) at the design conditions.

Type I hood field testing should be conducted with all appliances under the hood at operating temperatures, with all sources providing replacement air for the hood operating, and with all sources of recirculated air in the space operating. C&C is verified visually by observing smoke or steam produced by actual cooking operation or by simulating cooking using devices such as smoke candles or smoke puffers.

Note that smoke bombs should not be used, because they typically create new effluent from a point source and do not necessarily show whether cooking effluent is being captured. Actual cooking at the normal production rate is the most reliable method of generating smoke.

Hood systems with demand control should be tested at both minimum and maximum airflow settings.

If hoods fail the performance test, examine the systems and correct any capture problems. Close attention should be given to the design considerations and guidelines in this chapter to correct any performance problems.

Follow-Up: Records

1. A punch list of any remaining issues encountered during installation, air balancing, or performance testing should be recorded and submitted to the facility management and any affected contractors so that these items can be corrected.

2. When the preceding steps are complete, the system is properly integrated and balanced. At this time, all fan speeds and damper settings (at all modes of operation) should be permanently marked on the equipment and in the test and balance report. Air balance records of exhaust systems, replacement air systems, HVAC supply and return serving the hood area, and individual diffuser and/or grille airflows must also be completed. Records of fan model(s) and size(s), fan wheel rpm (not motor rpm), and motor amp draw should all be recorded. These records should be kept by the food service facility for future reference.

3. For new facilities, after two or three days in operation, all belts in the system should be checked and readjusted because new belts wear in quickly and could begin slipping. This examination should take place no longer than a week after initial operation, and before the facility opens if possible. Obviously, direct-drive systems do not require this inspection or replacement.

4. Once the facility is operational, performance of the ventilation system should be checked to verify that the design is adequate for actual cooking operation, particularly at maximum cooking and at outside environmental extremes. Any necessary changes should be made, and all the records should be updated to show the changes.

5. Rechecking the air balance should not be necessary more than once every two years unless basic changes are made in facility operation. If there are any changes, such as adding a new type of cooking equipment or deleting exhaust connections, the system should be modified accordingly. Modifications should be recorded, added to owner's records, and marked on affected equipment.

OPERATIONS AND MAINTENANCE

Sustainability Impact

Proper operation and maintenance of all kitchen ventilation systems is an often overlooked requirement. Typically, most attention is focused on the production of food, which is the primary role of any commercial kitchen. Given the kitchen ventilation system's role in providing makeup air, heating and cooling, and cooking effluent extraction, ensuring it is properly operated and maintained is critical for minimizing both overall system energy use and any environmental impacts both inside and outside the restaurant caused by effluent produced during cooking. Systems that are not operated or maintained correctly are likely to consume excessive energy, may create uncomfortable conditions in the kitchen area, may create an environmentally hazardous condition in the kitchen (e.g., hoods that do not capture and contain cooking effluent), and may affect outdoor environmental conditions (e.g., when pollution control devices are not operating properly). Additionally, given the fire hazards associated with commercial cooking, improper operation and maintenance of a kitchen's ventilation system(s) can even create a life safety hazard. Maintaining a proper air balance is part of kitchen ventilation systems necessary maintenance.

Operation

All components of the kitchen's ventilation system, and in some instances the entire building's ventilation system, are designed to operate in balance with each other, even under variable loads, to properly capture, contain, and remove cooking effluent and heat and maintain proper space temperature control in the most efficient and economical manner. Deterioration in any of these components unbalances the system, affecting one or more of its design concepts. The ventilation system's design intent should be fully understood by the operator so that any deviations in operation can be noted and corrected. In addition to creating health and fire hazards, normal cooking effluent deposits can also unbalance the system, so they must be regularly removed.

All components of exhaust and replacement air systems affect proper capture, containment, and removal of cooking effluent. In the exhaust system, this includes the cooking equipment itself, exhaust hood, all filtration devices, ducts, exhaust fan, and any dampers. In the replacement air system, this includes the air-handling unit(s) with intake louvers, dampers, filters, fan wheels, heating and cooling coils, ducts, and supply registers. In systems that obtain their replacement air from the general HVAC system, this also includes return air registers and ducts.

When the system is first set up and balanced in new condition, these components are set to optimum efficiency. In time, all components become dirty; filtration devices, dampers, louvers, heating and cooling coils, and ducts become restricted; fan blades change shape as they accumulate dirt and grease; and fan belts loosen. In addition, dampers can come loose and change position, even closing, and ducts can develop leaks or be blocked if internal insulation sheets fall down.

All these changes deteriorate system performance. The operator should know how the system performed when it was new, to better recognize when it is no longer performing the same way. This knowledge allows problems to be found and corrected sooner and the peak efficiency and safety of system operation to better be maintained.

Maintenance

Maintenance may be classified as preventive or emergency (breakdown). **Preventive maintenance** keeps the system operating as close as possible to optimal performance, including maximum production and least shutdown. It is the most effective maintenance and is preferred.

Preventive maintenance can prevent most emergency shutdowns and emergency maintenance. It has a modest ongoing cost and fewer unexpected costs. Clearly the lowest-cost maintenance in the long run, it keeps the system components in peak condition, maximizes the system's energy efficiency, and extends the operating life of all components.

Emergency maintenance must be applied when a breakdown occurs. Sufficient staffing and money must be applied to the situation to bring the system back on line in the shortest possible time. Such emergencies can be of almost any nature. They are impossible to predict or address in advance, except to presume the type of component failures that could shut the system down and keep spares of these components on hand or readily accessible, so they can be quickly replaced. Preventive maintenance, which includes regular inspection of critical system components, is the most effective way to avoid emergency maintenance.

Following are brief descriptions of typical operations of various components of kitchen ventilation systems and the type of maintenance and cleaning required to bring the abnormally operating system back to normal. Many nontypical operations are not listed here.

Cooking Equipment

Normal Operation. Produces properly cooked product, of correct temperature, within expected time. Minimum smoke during cooking.

Abnormal Operation. Produces undercooked product, of lower temperature, with longer cooking times. Increased smoke during cooking.

Cleaning/Maintenance. Clean solid cooking surfaces between each cycle if possible, or at least once a day. Baked-on product insulates and retards heat transfer. Filter frying medium daily and change it on schedule recommended by supplier. Check that (1) fuel source is at correct rating, (2) thermostats are correctly calibrated, and (3) conditioned air is not blowing on cooking surface.

Solid-fuel appliances are listed as "Extra-Heavy Duty" (see Table 2) and require additional attention. A hood over a solid-fuel appliance must be individually vented and therefore not be combined at any point with another duct and fan system. Using a UL *Standard* 762 upblast, in-line, or utility set fan listed to 400 or 500°F is suggested because the airstream temperature may be hotter without cooler air combining from other, typically lower-temperature cooking appliances. Design, installation, and maintenance precautions for the use of and emissions from solid fuel include monthly duct cleaning with weekly inspections, spark arrestors, and additional spacing to fryers. Refer to NFPA *Standard* 96, Chapters 5 to 10 and 14, and IMC sections 507 and 906, for additional direction.

Exhaust Systems

Normal Operation. All cooking vapors are readily drawn into the exhaust hood, where they are captured and removed from the space. The environment immediately around the cooking operation is clear and fresh.

Abnormal Operation. Many cooking vapors do not enter the exhaust hood at all, and some that enter subsequently escape. The environment around the cooking operation, and likely in the entire kitchen, is contaminated with cooking vapors and a thin film of grease.

Cleaning/Maintenance. Clean all grease removal devices in the exhaust system. Hood filters should be cleaned at least daily. High-efficiency grease extractors may require frequent cleanings during each shift. For other devices, follow the minimum recommendations of the manufacturer; even these may not be adequate at very high flow rates or with products producing large amounts of effluent. Check that (1) all dampers are in their original position, (2) fan belts are properly tensioned, (3) the exhaust fan is operating at the proper speed and turning in the proper direction, (4) the exhaust duct is not restricted, and (5) the fan blades are clear.

NFPA *Standard* 96 design requirements for access to the system should be followed to facilitate cleaning the exhaust hood, ductwork, and fan. Cleaning should be done if the combustibles' depth is greater than 0.08 in. in any part of the system, and by a method that leaves no more than a 0.002 in. depth deposit of combustibles. Cleaning agents should be thoroughly rinsed off, and all loose grease particles should be removed, because they can ignite more readily. Agents should not be added to the surface after cleaning, because their textured surfaces merely collect more grease more quickly. Fire-extinguishing systems may only be disarmed by properly trained and qualified service personnel before cleaning, to prevent accidental discharge, and then reset by authorized personnel after cleaning. All access panels removed must be reinstalled after cleaning, with proper gasketing in place to prevent grease leaks and escape of fire.

Supply, Replacement, and Return Air Systems

Normal Operation. The environment in the kitchen area is clear, fresh, comfortable, and free of drafts and excessive air noise.

Abnormal Operation. The kitchen is smoky, choking, hot, and humid, and perhaps very drafty with excessive air noise.

Cleaning/Maintenance. Check that the replacement air system is operating and is providing the correct amount of air to the space. If it is not, the exhaust system cannot operate properly. Check that dampers are set correctly, filters and exchangers are clean, the belts are tight, the fan is turning in the correct direction, and supply and return ductwork and registers are open, with supply air discharging in the correct direction and pattern. If drafts persist, the system may need to be rebalanced. If noise persists in a balanced system, system changes may be required.

Filter cleaning or changing frequency varies widely depending on the quantity of airflow and contamination of local air. Once determined, the cleaning schedule must be maintained.

With replacement air systems, the air-handling unit, coils, and fan are usually cleaned in spring and fall, at the beginning of the seasonal change. More frequent cleaning or better-quality filtering may be required in some contaminated environments. Duct cleaning for the system is on a much longer cycle, but local codes should be checked as stricter requirements are invoked. Ventilation systems should be cleaned by professionals to ensure that none of the expensive system components are damaged. Cleaning companies should be required to carry adequate liability insurance. The Power Washers of North America (PWNA) and the International Kitchen Exhaust Cleaning Association (IKECA) provide descriptions of proper cleaning and inspection techniques and lists of their members.

RESIDENTIAL KITCHEN VENTILATION

Although commercial and residential cooking processes are similar, their ventilation requirements and procedures are different. Differences include exhaust airflow rate and installation height. In addition, residential kitchen ventilation is less concerned with replacement air, and energy consumption is comparatively insignificant because of lower airflow, smaller motors, and intermittent operation.

Equipment and Processes

Although the physics of cooking and the resulting effluent are about the same, residential cooking is usually done more conservatively. Heavy-duty and extra-heavy-duty equipment, such as upright broilers and solid-fuel-burning equipment (described in Table 2), is not used. Therefore, the high ventilation rates of commercial kitchen ventilation and equipment for delivering these rates are not often found in residential kitchens. However, some residential kitchens are designed to operate with commercial-type cooking equipment, with higher energy inputs rates than usually found. In these cases, the hood may be similar to a commercial hood, and the required ventilation rate may approach that required for small commercial facilities.

Cooking effluent and by-products of open-flame combustion must be more closely controlled in a residence than in a commercial kitchen, because any escaping effluent can be dispersed throughout a residence, whereas a commercial kitchen is designed to be negatively pressurized compared to surrounding spaces. A residence also has a much lower background ventilation rate, making escaped contaminant more persistent. This situation makes residential kitchen ventilation a different kind of challenge, because problems cannot be resolved by simply increasing the ventilation rate at the cooking process.

Residential cooking always produces a convective plume that carries with it cooking effluent, often including grease vapor and particles, as well as water vapor, and by-products of combustion when natural gas is the energy source. Sometimes there is spatter as well, but those particles are so large that they are not removed by ventilation. Residential kitchen hoods depend more on thermal buoyancy than mechanical exhaust to capture cooking effluent and by-products of combustion.

EXHAUST SYSTEMS

Hoods and Other Ventilation Equipment

Wall-mounted, conventional range hoods ventilate most residential kitchens. Overall, they do the best job at the lowest installed cost. There are unlimited style-based variations of the conventional range hood shape. Deep canopy hoods are somewhat more effective because of their capture volume. Other styles have less volume, or a more flat bottom, and may be somewhat less effective at capturing effluent. To the extent that residential range hoods are often mounted between cabinets, with portions of the cabinets extending below the sides of the hood, performance may be improved because the cabinet sides help contain and channel the exhaust flow into the hood.

An increasingly popular development in residential kitchen ventilation is using a ventilating microwave oven in place of the typical residential range hood. Microwave ovens used for this purpose typically include small mesh filters mounted on the bottom of the oven and an internal exhaust fan. Means are usually provided to direct the exhaust flow in two directions: back into the kitchen or upward to an exhaust duct leading outdoors. The latter is more expensive but highly preferred; otherwise, if directed back to the kitchen, walls, ceiling, and cabinet surfaces are likely to become coated with grease from condensed grease vapor, and grease residue can damage paint and varnish. Additionally, typical microwave oven ventilators do not include vertical surfaces that provide a reservoir volume to contain the convective plume during transient effects, such as removing the lid from a cooking vessel. Consequently, microwave oven ventilators often provide lower exhaust capture and containment performance than standard range hoods.

Downdraft range-top ventilators have also become more popular. Functionally, these are an exception, because they capture contaminants by producing velocities over the cooking surface greater than those of the convective plume. With enough velocity, their operation can be satisfactory; however, velocity may be limited to prevent adverse effects such as gas flame disturbance and cooking process cooling. Additionally, this method is more effective for exhaust from cooking near the range surface, and it is usually much less effective for capturing the convective plume from taller cooking vessels, because the convective plume is too far above the ventilator intake to be affected by it.

Ironically, many high-end kitchens have less efficient ventilation than standard range hoods. Inefficient methods include

- Mounting range tops in cooking islands with no exhaust hood or other means of ventilation
- Mounting ovens in cabinets, separate from rangetops, without any way to remove heat and effluents from the oven
- Using low-profile exhaust devices with insufficient overhang over the appliance and no reservoir to contain convective plume during dynamic effects
- Having duct runs, particularly in larger homes, with very high static pressure losses, so that the actual exhaust flow rate is much lower than the nominal exhaust fan rating

Whole-kitchen exhaust fans were more common in the past, but they are still used. Mounted in the kitchen wall or ceiling, they ventilate the entire kitchen volume rather than capturing contaminants at the source. For kitchen exhaust fans not above the cooking surface, and without a capturing hood, 15 air changes per hour (ach) is recommended; for ceiling-mounted fans, this is usually sufficient, but for wall-mounted fans, it may be marginal.

Residential exhaust hoods are often furnished with multiple-speed fans, so that users can match exhaust fan speeds (and noise) with the cooking process and resultant convective plume. Carrying this concept further, there are high-end residential exhaust hood manufacturers that provide an automatic two-speed control that increases fan speed when higher convective plume temperature is sensed.

Continuous low-level, whole-building ventilation is increasingly used to ensure good indoor air quality in modern, tightly built houses with less infiltration. ASHRAE *Standard* 62.2 requires kitchen ventilation in most residences. Some whole-building ventilation systems can intermittently increase airflow to achieve the needed reduction in cooking effluent. In that case, there must be provision to avoid introducing and accumulating grease and other cooking effluent that may cause undesirable growth of microorganisms.

Differences Between Commercial and Residential Equipment

Safety requirements covering residential cooking area fans are contained in UL *Standard* 507. These fans and accessories are intended for use in conjunction with residential gas and electric cooking appliances only, and are investigated to determine the effects of increased air temperature and grease on electrical components. The filters provided as a part of the fan are also checked for flammability and smoke propagation. Products include hood fans intended to mount directly over (but not directly on) ranges, separate hoods provided with lights or other wiring and intended for use over ranges in conjunction with a remote blower, downdraft fans, and oven ventilators for use over wall-insert ovens. Fans intended for mounting directly on cooking equipment are investigated in conjunction with the cooking appliances, and are typically listed as part of the accessory to the cooking appliance. Fans installed in close proximity to a stove, range, or oven where fumes, grease-laden air, or the like may be present and intended to discharge air away from the cooking area should be installed to discharge air to the exterior of the building and not into concealed walls or ceiling spaces or into the attic. Ductless fans intended for use in cooking areas are not required to discharge air to the building exterior.

Fire-actuated dampers are never part of the hood and are almost never used. Grease filters in residential hoods are much simpler, and grease collection channels are rarely used because inadequate maintenance could allow grease to pool, creating a fire and health hazard.

Conventional residential wall hoods usually have standard dimensions that match the standard 3 in. modular grid of residential cabinets. Heights of 6, 9, 12, and 24 in. are common, as are depths from 17 to 22 in. Width is usually the same as the cooking surface, with 30 in. width nearly standard in the United States. Current U.S. Housing and Urban Development (HUD) Manufactured Home Construction and Safety Standards call for 3 in. overhang per side.

Hood mounting height is usually 18, 24, or 30 in., and sometimes even higher with a sacrifice in collection efficiency. A lower-mounted hood captures more effectively because there is less opportunity for lateral air currents to disrupt the convective plume. Studies show 18 in. is the minimum height for cooking surface access. Some codes require a minimum of 30 in.) from the cooking surface to combustible cabinets. In that case, the bottom of a 6 in. hood can be 24 in. above the cooking surface.

A minimum airflow rate (exhaust capacity) of 40 cfm per linear foot of hood width has long been recommended by the Home Ventilating Institute (HVI 2004), and confirmed by field tests. Additional capacity, with speed control, is desirable for handling unusually vigorous cooking and cooking mistakes, because airflow can be briefly increased to clear the air, and speed can be reduced to a quieter level for normal cooking.

Exhaust Duct Systems

Residential hoods offer little opportunity for custom design of an exhaust system. The range hood has a built-in duct connector and the duct should be the same size, whether round or rectangular. A hood includes either an axial or a centrifugal fan. The centrifugal fan can develop higher pressure, but the axial fan is usually adequate for low-volume hoods. The great majority of residential hoods in the United States have HVI-certified airflow performance. In all cases, it is highly preferable to vent the exhaust hood outdoors through a roof cap, rather than venting back into the home, whether into the kitchen or elsewhere.

Replacement (Makeup) Air

The exhaust rate of residential hoods is generally low enough and natural infiltration sufficient to avoid the need for replacement air systems. Although this may cause slight negative pressurization of the residence, it is brief and is usually less than that caused by other equipment. Still, backdrafts through the flue of a combustion appliance should be avoided and residences with gas furnace and water heater should have the flue checked for adequate flow. NFPA *Standard* 54 provides a method of testing flues for adequate performance. Sealed-combustion furnaces and water heaters are of less concern.

Sometimes commercial-style cooking equipment approved for residential use is installed in residences. For the higher ventilation requirements, see earlier sections of this chapter, especially the section on Replacement (Makeup) Air Systems.

Energy Conservation

The energy cost of residential hoods is quite low because of the few annual running hours and the low rate of exhaust. For example, it typically costs less than $10 per heating season in Chicago to run a hood and heat replacement air, based on running at 150 cfm for an hour a day and using gas heat.

Fire Protection for Residential Hoods

Residential hoods must be installed with metal (preferably steel) duct, positioned to prevent grease pooling. Residential hood exhaust ducts are almost never cleaned, and there is no evidence that this causes fires.

There have been some attempts to make fire extinguishers available in residential hoods, but none has met with broad acceptance. However, grease fires on the residential cooking surface continue to occur, almost always the result of unattended cooking. There is no industry-accepted performance standard or consistency of design in residential fire-extinguishing equipment.

Maintenance

All listed hoods and kitchen exhaust fans are designed for cleaning, which should be done at intervals consistent with the cooking practices of the user. Although cleaning is sometimes thought to be for fire prevention, the health benefits of removing nutrients available for the growth of organisms can be more important.

RESEARCH

Research Overview

ASHRAE Technical Committee TC 5.10, Kitchen Ventilation, has been active in research related to kitchen ventilation as shown in Table 6. This research has tended to focus on answering questions related to field-related issues, such as how to measure exhaust airflow rates for hood and replacement air systems (RP-623 and RP-1376) and how much grease is produced by cooking appliances (RP-745 and RP-1375). Some of the research focused on design aspects of kitchen ventilation systems, from optimizing exhaust hood performance (RP-1202 and RP-1480), to evaluating the grease removal efficiency of filtering devices (RP-851 and RP-1151) and reducing the velocity of airflow in the exhaust ductwork (RP-1033). Other projects evaluated relationships between appliances and ventilation systems and the HVAC system in the space (RP-1362). Although historically projects have focused on mechanical systems in the space, there is a current project to evaluate the impact of the kitchen environment on thermal comfort in the space (RP-1469).

Benefits to the HVAC Industry

Many of the research projects that TC 5.10 sponsored have affected energy use and sustainability in the food service industry. RP-1033 data that showed that grease deposition on the walls of duct actually decreased when the duct velocity was lowered from 1500 fpm to 500 fpm. These data allowed both NFPA *Standard* 96 and the International Mechanical Code to allow lower duct velocities. These changes allow demand-controlled ventilation systems (in which airflow is lowered during noncooking periods of the day) to be used across the United States to achieve significant energy savings.

The two projects related to hood performance (RP-1202 and RP-1480) not only evaluated how wall canopy and island hoods perform with various appliances, but also evaluated methods of reducing the exhaust airflows required for the hoods to capture the cooking effluent more efficiently. These include items such as optimizing the appliance position underneath the hoods, installing side panels, and designing hoods to use larger overhangs if possible. If exhaust air is reduced, this also generally reduces how much conditioned air needs to be brought back into the space to replace the air that is exhausted, leading to large energy savings in restaurants. RP-1362 measured the heat gain from appliances underneath hoods, and these data can be used to more accurately size the HVAC equipment needed to condition the kitchen space.

Earlier projects related to grease emissions (RP-851, RP-745, and RP-1151) were used to help develop ASTM *Standard* F2519. Data from these research projects, along with *Standard* F2519 and data from RP-1375, revolutionized the kitchen ventilation industry with regard to how mechanical filters actually perform in the field. *Standard* F2519 provides a framework for making more efficient filters that help reduce the amount of grease built up in ductwork, on exhaust fans, and on the roof of buildings.

Table 6　Summary of TC 5.10 Research Projects

Year(s)	ASHRAE Project	Title
1993 to 1994	RP-623	A Field Test Method for Determining Exhaust Rates in Grease Hoods for Commercial Kitchens (Gordon and Parvin 1994)
1996 to 1997	RP-851	Determining the Efficiency of Grease-Removal Devices in Commercial Kitchen Applications (Schrock 1998)
1998 to 1999	RP-745	Identification and Characterization of Effluents from Various Cooking Appliances and Processes as Related to Optimum Design of Kitchen Ventilation Systems (Gerstler et al. 1998)
2000 to 2001	RP-1033	Effects of Air Velocity on Grease Deposition in Exhaust Ductwork (Kuehn 2000)
2001 to 2003	RP-1151	Development of a Draft Method of Test for Determining Grease Removal Efficiencies (Welch 2004)
2003 to 2005	RP-1202	Effect of Appliance Diversity and Position on Commercial Kitchen Hood Performance (Swierczyna et al. 2006)
2007 to 2008	RP-1375	Characterization of Effluents from Additional Cooking Appliances (Kuehn 2008)
2008 to 2009	RP-1362	Revised Heat Gain and Capture and Containment Exhaust Rates from Typical Commercial Cooking Appliances (Swierczyna 2008)
2008 to 2010	RP-1376	Method of Test to Evaluate Field Performance of Commercial Kitchen Ventilation Systems (Kuehn 2010)
2008 to 2009	RP-1480	Island Hood Energy Consumption and Energy Reduction Strategies (Swierczyna et al. 2010)
2010 to present	RP-1469	Thermal Comfort in Commercial Kitchens (Stoops [ongoing])

RP-623 and RP-1376 both examined how to accurately measure the exhaust and replacement air in food service establishments. By being able to more accurately measure the airflows, restaurants can be properly balanced to the design conditions so that excess energy is not consumed.

REFERENCES

ASHRAE. 2010. Thermal environmental conditions for human occupancy. ANSI/ASHRAE *Standard* 55-2010.

ASHRAE. 2010. Ventilation for acceptable indoor air quality. ANSI/ASHRAE *Standard* 62.1-2010.

ASHRAE. 2010. Ventilation and acceptable indoor air quality in low-rise residential buildings. ANSI/ASHRAE *Standard* 62.2-2010.

ASHRAE. 2010. Energy standard for buildings except low-rise residential buildings. ANSI/ASHRAE *Standard* 90.1-2010.

ASHRAE. 2003. Ventilation for commercial cooking operations. ANSI/ASHRAE *Standard* 154-2003.

ASTM. 2009. Test methods for fire resistive grease duct enclosure systems. *Standard* E2336-04(R 09). American Society for Testing and Materials, West Conshohocken, PA.

ASTM. 2009. Test method for capture and containment performance of commercial kitchen exhaust ventilation systems. *Standard* F1704-09. American Society for Testing and Materials, West Conshohocken, PA.

ASTM. 2009. Test method for heat gain to space performance of commercial kitchen ventilation/appliance systems. *Standard* F2474-09. American Society for Testing and Materials, West Conshohocken, PA.

ASTM. 2005. Test method for grease particle capture efficiency of commercial kitchen filters and extractors. *Standard* F2519-05. American Society for Testing and Materials, West Conshohocken, PA.

Brohard, G., D.R. Fisher, V.A. Smith, R.T. Swierczyna, and P.A. Sobiski. 2003. *Makeup air effects on kitchen exhaust hood performance.* California Energy Commission, Sacramento.

Brown, S.L. 2007. Dedicated outdoor air system for commercial kitchen ventilation. *ASHRAE Journal* (July).

Elovitz, G. 1992. Design considerations to master kitchen exhaust systems. *ASHRAE Transactions* 98(1):1199-1213.

Fisher, D.F. 1998. New recommended heat gains for commercial cooking equipment. *ASHRAE Transactions* 104(2).

Fisher, D.F. 2003. Predicting energy consumption: Clearing the air on kitchens. *ASHRAE Journal* (June).

Gerstler, W.D., T.H. Kuehn, D.Y.H. Pui, J.W. Ramsey, M.J. Rosen, R.R. Carlson, and S.D. Petersen. 1998. Identification and characterization of effluents from various cooking appliances and processes as related to optimum design of kitchen ventilation systems. ASHRAE Research Project RP-745 (Phase II), *Final Report*.

Gordon, E.B. and F.A. Parvin. 1994. A field test method for determining exhaust rates in grease hoods for commercial kitchens (RP-623). *ASHRAE Transactions* 100(2):412-419.

HUD. Manufactured home construction and safety standards. 24CFR3280. *Code of Federal Regulations*, U.S. Department of Housing and Urban Development, Washington, D.C.

HVI. 2008. *The guide to home ventilation and indoor air quality.* Home Ventilating Institute, Arlington Heights, IL.

ICC. 2009a. *International mechanical code.* International Code Council, Washington, D.C.

ICC. 2009b. *International fuel gas code.* International Code Council, Washington, D.C.

ICC. 2009c. *International building code.* International Code Council, Washington, D.C.

ICC. 2009d. *International fire code.* International Code Council, Washington, D.C.

Itron, Inc., and California Energy Commission. 2006. *California commercial end-use survey.* CEC-400-2006-005. http://www.energy.ca.gov/2006publications/CEC-400-2006-005/CEC-400-2006-005.PDF.

Kuehn, T.H. 2000. Effects of air velocity on grease deposition in exhaust ductwork (RP-1033). ASHRAE Research Project, *Final Report*.

Kuehn, T.H. 2008. Characterization of effluents from additional cooking appliances. ASHRAE Research Project RP-1375, *Final Report*.

Kuehn, T.H. 2010. Method of test to evaluate field performance of commercial kitchen ventilation systems. ASHRAE Research Project RP-1376, *Final Report*.

Kuehn, T.H., W.D. Gerstler, D.Y.H. Pui, and J.W. Ramsey. 1999. Comparison of emissions from selected commercial kitchen appliances and food products. *ASHRAE Transactions* 105(2):128-141.

NFPA. 2010. Installation of sprinkler systems. ANSI/NFPA *Standard* 13-02. National Fire Protection Association, Quincy, MA.

NFPA. 2009. Dry chemical extinguishing systems. ANSI/NFPA *Standard* 17-02. National Fire Protection Association, Quincy, MA.

NFPA. 2009. Wet chemical extinguishing systems. ANSI/NFPA *Standard* 17A-98. National Fire Protection Association, Quincy, MA.

NFPA. 2008. Inspection, testing, and maintenance of water-based fire protection systems. *Standard* 25. National Fire Protection Association, Quincy, MA.

NFPA. 2009. National fuel gas code. *Standard* 54. National Fire Protection Association, Quincy, MA.

NFPA. 2008. Ventilation control and fire protection of commercial cooking operations. *Standard* 96-08. National Fire Protection Association, Quincy, MA.

PG&E Food Service Technology Center. 2004. *Commercial kitchen ventilation design guide series.* PG&E Food Service Technology Center, San Ramon, CA. Available from http://www.fishnick.com/ventilation/design guides/.

PG&E Food Service Technology Center. 2010. *Wall-mounted canopy exhaust hood performance reports: Application of ASTM 1704, standard test method for capture and containment performance of commercial kitchen exhaust ventilation system.* PG&E Food Service Technology Center, San Ramon, CA.

Schrock, D.W. 1998. Determining the efficiency of grease-removal devices in commercial kitchen applications (RP-851). *ASHRAE Transactions* 104(2).

Spata, A.J. and S.M. Turgeon. 1995. Impact of reduced exhaust and ventilation rates at "no-load" cooking conditions in a commercial kitchen during winter operation. *ASHRAE Transactions* 101(2):606-610.

Swierczyna, R.T., V.A. Smith, and F.P. Schmid. 1997. New threshold exhaust flow rates for capture and containment of cooking effluent. *ASHRAE Transactions* 103(2):943-949.

Swierczyna, R.T., P. Sobiski, and D. Fisher. 2006. Effects of appliance diversity and position on commercial kitchen hood performance (RP-1202). *ASHRAE Transactions* 112(1).

Swierczyna, R., D. Fisher, and P. Sobiski. 2008. Revised heat gain and capture and containment exhaust rates from typical commercial cooking appliances. ASHRAE Research Project RP-1362, *Final Report*.

Swierczyna, R., P. Sobiski, and D. Fisher. 2010. Island hood energy consumption and energy consumption strategies. ASHRAE Research Project RP-1480, *Final Report*.

UL. 2005. Fire testing of fire extinguishing systems for protection of restaurant cooking areas, 3rd ed. *Standard* 300-05. Underwriters Laboratories, Northbrook, IL.

UL. 1999. Electric fans, 9th ed. ANSI/UL *Standard* 507-99. Underwriters Laboratories, Northbrook, IL.

UL. 1995. Exhaust hoods for commercial cooking equipment, 5th ed. *Standard* 710-95. Underwriters Laboratories, Northbrook, IL.

UL. 2010. Power roof ventilators for restaurant exhaust appliances. *Standard* 762-10. Underwriters Laboratories, Northbrook, IL.

UL. 2010. Grease filters for exhaust ducts, 3rd ed. *Standard* 1046-10. Underwriters Laboratories, Northbrook, IL.

UL. 2005. Grease ducts, 3rd ed. *Standard* 1978-05. Underwriters Laboratories, Northbrook, IL.

UL. 2009. Recirculating systems, 1st ed. *Standard* 710B-09. Underwriters Laboratories, Northbrook, IL.

UL. 2001. Fire resistive grease duct enclosure assemblies, 1st ed. *Standard* 2221-01. Underwriters Laboratories, Northbrook, IL.

Welch, W.A. 2004. Development of a draft method of test for determining grease removal efficiencies. ASHRAE Research Project RP-1151, *Final Report*.

BIBLIOGRAPHY

Bevirt, W.D. 1994. What engineers need to know about testing and balancing. *ASHRAE Transactions* 100(1):705-714.

Black, D.K. 1989. Commercial kitchen ventilation—Efficient exhaust and heat recovery. *ASHRAE Transactions* 95(1):780-786.

Claar, C.N., R.P. Mazzucchi, and J.A. Heidell. 1985. *The project on restaurant energy performance (PREP)—End use monitoring and analysis.* U.S. Department of Energy, Office of Building Energy Research and Development, Washington, D.C.

Farnsworth, C., A. Waters, R.M. Kelso, and D. Fritzsche. 1989. Development of a fully vented gas range. *ASHRAE Transactions* 95(1):759-768.

Frey, D.J., K.F. Johnson, and V.A. Smith. 1993. Computer modeling analysis of commercial kitchen equipment and engineered ventilation. *ASHRAE Transactions* 99(2):890-908.

Fritz, R.L. 1989. A realistic evaluation of kitchen ventilation hood designs. *ASHRAE Transactions* 95(1):769-779.

Fugler, D. 1989. Canadian research into the installed performance of kitchen exhaust fans. *ASHRAE Transactions* 95(1):753-758.

Gordon, E.B. and N.D. Burk. 1993. A two-dimensional finite-element analysis of a simple commercial kitchen ventilation system. *ASHRAE Transactions* 99(2):909-914.

Gordon, E.B., D.J. Horton, and F.A. Parvin. 1994. Development and application of a standard test method for the performance of exhaust hoods with commercial cooking appliances. *ASHRAE Transactions* 100(2):988-999.

Gordon, E.B., D.J. Horton, and F.A. Parvin. 1995. Description of a commercial kitchen ventilation (CKV) laboratory facility. *ASHRAE Transactions* 101(1):249-261.

Horton, D.J., J.N. Knapp, and E.J. Ladewski. 1993. Combined impact of ventilation rates and internal heat gains on HVAC operating costs in commercial kitchens. *ASHRAE Transactions* 99(2):877-883.

ICC. 2004. *Acceptance criteria for grease duct assemblies.* AC101-2004. International Code Council, Washington, D.C.

Kelso, R.M. and C. Rousseau. 1995. Kitchen ventilation. *ASHRAE Journal* 38(9):32-36.

Knapp, J.N. and W.A. Cheney. 1993. Development of high-efficiency air cleaners for grilling and deep-frying operations. *ASHRAE Transactions* 99(2):884-889.

Kuehn, T.H., J. Ramsey, H. Han, M. Perkovich, and S. Youssef. 1989. A study of kitchen range exhaust systems. *ASHRAE Transactions* 95(1):744-752.

Livchak, A., D. Schrock, and Z. Sun. 2005. The effect of supply air systems on kitchen thermal environment. *ASHRAE Transactions* 111(1):748-754.

Parikh, J.S. 1992. Testing and certification of fire and smoke dampers. *ASHRAE Journal* 34(11):30-33.

Pekkinen, J. and T.H. Takki-Halttunen. 1992. Ventilation efficiency and thermal comfort in commercial kitchens. *ASHRAE Transactions* 98(1):1214-1218.

Pekkinen, J.S. 1993. Thermal comfort and ventilation effectiveness in commercial kitchens. *ASHRAE Journal* 35(7):35-38.

Schmid, F.P., V.A. Smith, and R.T. Swierczyna. 1997. Schlieren flow visualization in commercial kitchen ventilation research. *ASHRAE Transactions* 103(2):937-942.

Shaub, E.G., A.J. Baker, N.D. Burk, E.B. Gordon, and P.G. Carswell. 1995. On development of a CFD platform for prediction of commercial kitchen ventilation flow fields. *ASHRAE Transactions* 101(2):581-593.

Smith, V.A., D.J. Frey, and C.V. Nicoulin. 1997. Minimum-energy kitchen ventilation for quick service restaurants. *ASHRAE Transactions* 103(2):950-961.

Smith, V.A., R.T. Swierczyna, and C.N. Claar. 1995. Application and enhancement of the standard test method for the performance of commercial kitchen ventilation systems. *ASHRAE Transactions* 101(2):594-605.

Smith, V.A. and D.R. Fisher. 2001. Estimating food service loads and profiles. *ASHRAE Transactions* 107(2).

Soling, S.P. and J. Knapp. 1985. Laboratory design of energy efficient exhaust hoods. *ASHRAE Transactions* 91(1B):383-392.

Sobiski, P.A., R.T Swierczyna, D.F. Fisher. 2005. 1202-RP supplemental: effects of range top usage, appliance accessories and hood dimensions on commercial kitchen hood performance. *ASHRAE Transactions* 112(1).

Swierczyna, R.T., D.R. Fisher, D.J. Horton. 2002. Effects of commercial kitchen pressure on exhaust system performance. *ASHRAE Transactions* 108(1).

VDI Verlag. 1999. *Raumlufttechnische Anlagen für Küchen (Ventilation equipment for kitchens).* VDI 2052.

Wolbrink, D.W. and J.R. Sarnosky. 1992. Residential kitchen ventilation—A guide for the specifying engineer. *ASHRAE Transactions* 91(1):1187-1198.

CHAPTER 34

GEOTHERMAL ENERGY

THE use of geothermal resources can be subdivided into three general categories: high-temperature (>300°F) electric power production, intermediate- and low-temperature (<300°F) direct-use applications, and ground-source heat pump applications (generally <90°F). This chapter covers only direct use (including wells, equipment, and applications) and ground-source heat pumps. Design aspects of the building heat pump loop may be found in Chapter 8 of the 2008 *ASHRAE Handbook—HVAC Systems and Equipment*.

RESOURCES

Geothermal energy is the thermal energy in the earth's crust: thermal energy in rock and fluid (water, steam, or water containing large amounts of dissolved solids) that fills the pores and fractures in the rock, sand, and gravel. Calculations show that the earth, originating from a completely molten state, would have cooled and become completely solid many thousands of years ago without an energy input beyond that of the sun. It is believed that the ultimate source of geothermal energy is radioactive decay within the earth (Bullard 1973).

Through plate motion and vulcanism, some of this energy is concentrated at high temperature near the surface of the earth. Energy is also transferred from deeper parts of the crust to the earth's surface by conduction and by convection in regions where geological conditions and the presence of water allow.

Because of variation in volcanic activity, radioactive decay, rock conductivities, and fluid circulation, different regions have different heat flows (through the crust to the surface), as well as different temperatures at a particular depth. The normal increase of temperature with depth (i.e., the normal geothermal gradient) is about 13.7°F per 1000 ft of depth, with gradients of 5 to 27°F per 1000 ft being common. Areas that have higher temperature gradients and/or higher-than-average heat flow rates constitute the most interesting and viable economic resources. However, areas with normal gradients may be valuable resources if certain geological features are present.

Geothermal resources of the United States are categorized into the following types:

Igneous point resources are associated with magma bodies, which result from volcanic activity. These bodies heat the surrounding and overlying rock by conduction and convection, as allowed by the rock permeability and fluid content in the rock pores.

Hydrothermal convection systems are hot fluids near the earth's surface that result from deep circulation of water in areas of high regional heat flow. A widely used resource, these fluids rise from natural convection between hotter, deeper formations and cooler formations near the surface. The passageway that provides for this deep convection must consist of adequately permeable fractures and faults.

Geopressured resources, present widely in the Gulf Coast of the United States, consist of regional occurrences of confined hot water in deep sedimentary strata, where pressures of greater than 10,000 psi are common. This resource also contains methane, which is dissolved in the geothermal fluid.

Radiogenic heat sources exist in various regions as granitic plutonic rocks that are relatively rich in uranium and thorium. These plutons have a higher heat flow than the surrounding rock; if the plutons are blanketed by sediments of low thermal conductivity, an elevated temperature at the base of the sedimentary section can result. This resource has been identified in the eastern United States.

Deep regional aquifers of commercial value can occur in deep sedimentary basins, even in areas of only normal temperature gradient. For deep aquifers to be of commercial value, (1) basins must be deep enough to provide usable temperature levels at the prevailing gradient, and (2) permeability in the aquifer must be adequate for flow.

Thermal energy in geothermal resources exists primarily in the rocks and only secondarily in the fluids that fill the pores and fractures. Thermal energy is usually extracted by bringing to the surface the hot water or steam that occurs naturally in the open spaces in the rock. Where rock permeability is low, the energy extraction rate is low. In permeable aquifers, fluid produced may be injected back into the aquifer at some distance from the production well to pass through the aquifer again and recover some of the energy in the rock. Figure 1 indicates geothermal resource areas in the United States.

Temperature

The temperature of fluids produced in the earth's crust and used for their thermal energy content varies from below 40°F to 680°F. As indicated in Figure 1, local gradients also vary with geologic conditions. The lower value represents fluids used as the low-temperature energy source for heat pumps, and the higher temperature represents an approximate value for the HGP-A well at Hilo, Hawaii.

The following classification by temperature is used in the geothermal industry:

High temperature	$t > 300°F$
Intermediate temperature	$195°F < t < 300°F$
Low temperature	$t < 195°F$

Electric generation is generally not economical for resources with temperatures below about 300°F, which is the reason for the division between high- and intermediate-temperature. However, binary (organic Rankine cycle) power plants, with the proper set of circumstances, have demonstrated that it is possible to generate electricity economically above 230°F. In 1988, there were 86 binary plants worldwide, generating a total of 126.3 MW (Di Pippo 1988).

The preparation of this chapter is assigned to TC 6.8, Geothermal Energy Utilization.

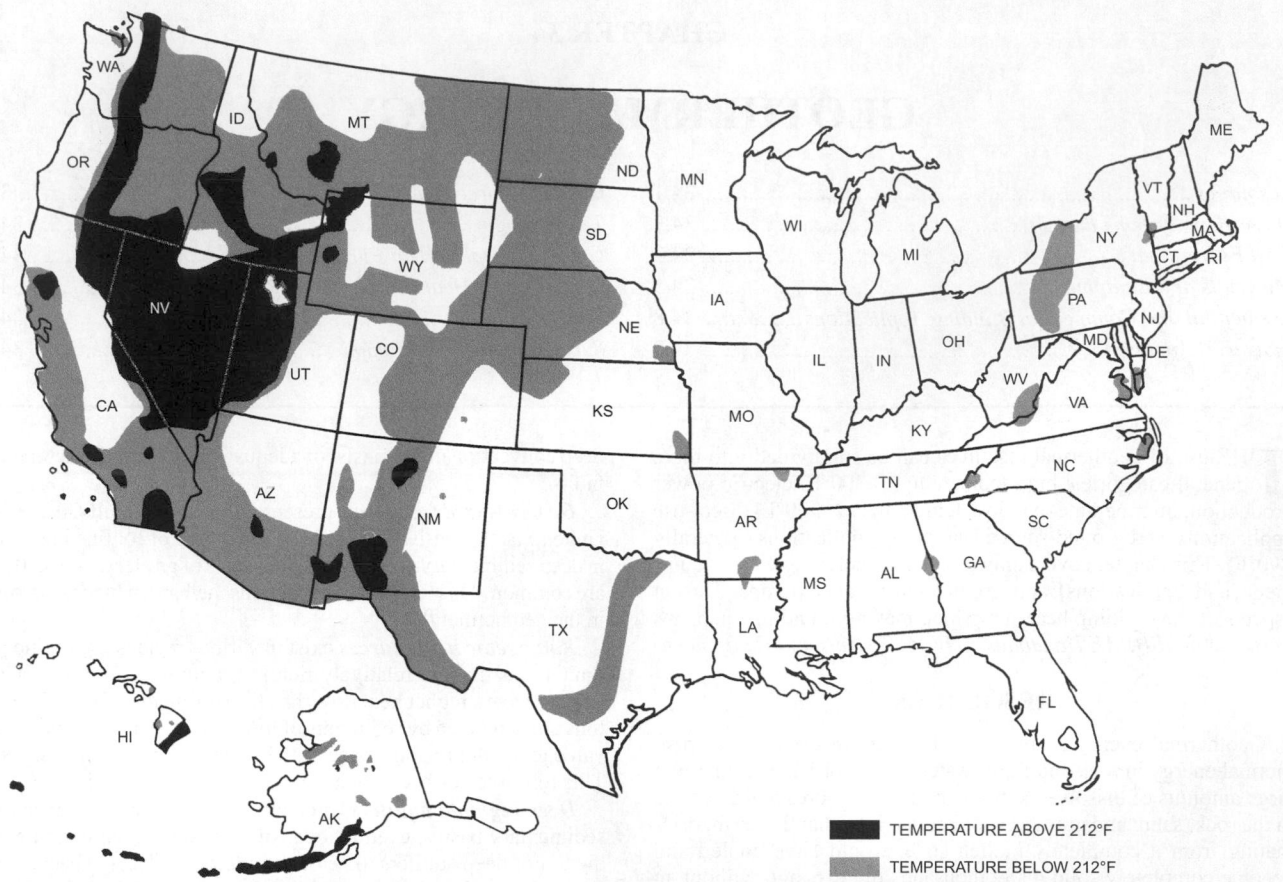

	TEMPERATURE ABOVE 212°F
	TEMPERATURE BELOW 212°F

Fig. 1   U.S. Hydrothermal Resource Areas
(Lienau et al. 1995)

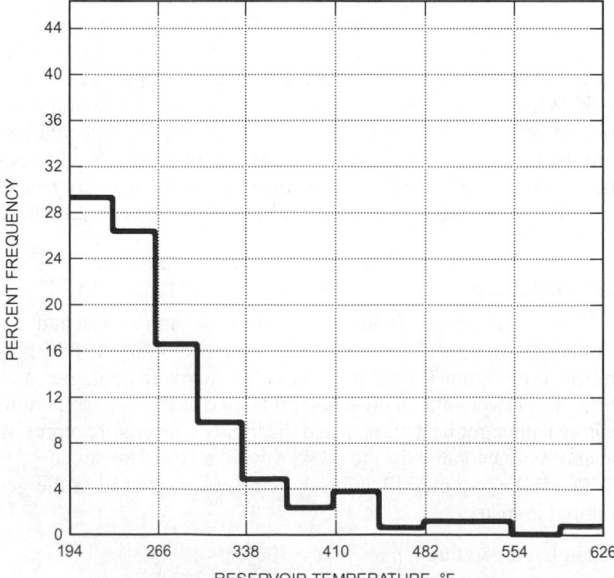

Fig. 2   Frequency of Identified Hydrothermal Convection Resources Versus Reservoir Temperature
(Muffler 1979)

Geothermal resources at lower temperatures are more common. The frequency by reservoir temperature of identified convective systems above 194°F is shown in Figure 2.

Fluids

Geothermal energy is extracted from the earth through naturally occurring fluids in rock pores and fractures. Fluids produced are steam, hot water, or a two-phase mixture of both. These may contain various amounts of impurities, notably dissolved gases and dissolved solids.

Geothermal resources that produce essentially dry steam are **vapor-dominated**. Although these are valuable resources, they are rare. Hot-water (**fluid-dominated**) resources are much more common and can be produced either as hot water or as a two-phase mixture of steam and hot water, depending on the pressure maintained on the production well. If pressure in the production casing or in the formation around the casing is reduced below the saturation pressure at that temperature, some of the fluid will flash, and a two-phase fluid will result. If pressure is maintained above the saturation pressure, the fluid remains single-phase. In fluid-dominated resources, both dissolved gases and dissolved solids are significant.

Geothermal fluid chemistry varies over a wide range. In the Imperial valley of California, some high-temperature geothermal fluids may contain up to 300,000 ppm of total dissolved solids (TDS). Fluids of this character are extremely difficult to accommodate in systems design and materials selection. In fact, most low-temperature fluids contain less than 3000 ppm and many meet drinking water standards. Despite this, even geothermal fluids of a few hundred ppm TDS can cause substantial problems with standard construction materials.

Present Use

Discoveries of concentrated radiogenic heat sources and deep regional aquifers in areas of near-normal temperature gradient indicate that 37 states in the United States have economically exploitable direct-use geothermal resources (Interagency Geothermal Coordinating Council 1980). The Geysers, in northern California, is the largest single geothermal development in the world.

The total electricity generated by geothermal development in the world was 7974 MW in 2000 (Lund et al. 2001). Direct application of geothermal energy for space heating and cooling, water heating, agricultural growth-related heating, and industrial processing represented about 51.6×10^9 Btu/h worldwide in 2000. In the United States in 2000, direct-use installed capacity amounted to 12.9×10^9 Btu/h, providing 19.3×10^{12} Btu/yr.

The major uses of geothermal energy in the United States are for heating greenhouse and aquaculture facilities. The principal industrial use is for food processing.

DIRECT-USE SYSTEMS DESIGN

A major goal in designing direct-use systems is capturing the most possible heat from each gallon of fluid pumped. System owning and operating costs are composed primarily of well pumping and well capitalization components; maximizing system Δt (i.e., minimizing flow requirements) minimizes well capital cost and pump operating cost. In many cases, system design can benefit from connecting loads in series according to temperature requirements. Direct-use system design is covered in detail in Anderson and Lund (1980) and Rafferty (1989a).

Direct-use systems can be divided into four subsystems: (1) production, including the producing wellbore and associated wellhead equipment; (2) transmission and distribution to transport geothermal energy from the resource site to the user site and then distribute it to the individual user loads; (3) user system; and (4) disposal, which can be either surface disposal or injection back into a formation.

In a typical direct-use system, geothermal fluid is produced from the production borehole by a lineshaft multistage centrifugal pump. When the geothermal fluid reaches the surface, it is delivered to the application site through the transmission and distribution system.

In the system in Figure 3, geothermal fluid is separated from the heating system by a heat exchanger. This secondary loop is especially desirable when the geothermal fluid is particularly corrosive and/or causes scaling. The geothermal fluid is pumped directly back into the ground without loss to the surrounding surface.

COST FACTORS

The following characteristics influence the cost of energy delivered from geothermal resources:

- Well depth

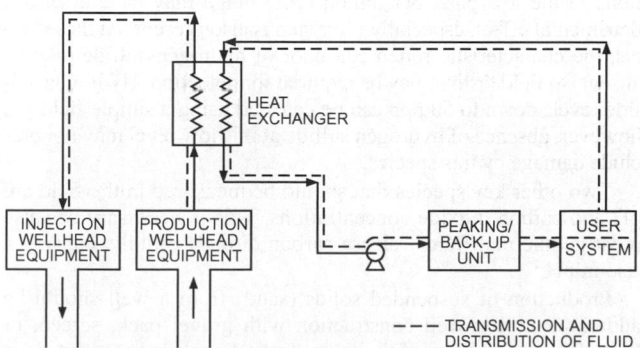

Fig. 3   Geothermal Direct-Use System with Wellhead Heat Exchanger and Injection Disposal

- Distance between resource location and application site
- Well flow rate
- Resource temperature
- Temperature drop
- Load factor
- Composition of fluid
- Ease of disposal

Many of these characteristics have a major influence because the cost of geothermal systems is primarily front-end capital cost; annual operating cost is relatively low.

Well Depth

The cost of the wells is usually one of the larger items in the overall cost of a geothermal system, and increases with resource depth. Compared to many geothermal areas worldwide, well depth requirements in the western United States are relatively shallow; most larger geothermal systems there operate with production wells of less than 2000 ft, and many at less than 1000 ft.

Distance Between Resource Location and Application Site

Direct use of geothermal energy must occur near the resource. The reason is primarily economic; although geothermal (or secondary) fluid could be transmitted over moderately long distances (greater than 60 miles) without great temperature loss, such transmission is generally not economically feasible. Most existing geothermal projects have transmission distances of less than 1 mile.

Well Flow Rate

Energy output from a production well varies directly with the fluid flow rate. The energy cost at the wellhead varies inversely with the well flow rate. A typical good resource has a production rate of 400 to 800 gpm per production well; however, geothermal direct-use wells have been designed to produce up to 2000 gpm.

Resource Temperature

The available temperature is fixed by the prevailing resource. The temperature can restrict applications. It often requires a reevaluation of accepted application temperatures, which were developed for uses served by conventional fuels for which the application temperature could be selected at any value in a relatively broad range. Most existing direct-use projects use fluids in the 130 to 230°F range.

Temperature Drop

Because well flow is limited, power output from a geothermal well is directly proportional to the temperature drop of the geothermal fluid connected to the system. Consequently, a larger temperature drop reduces operating (pumping) and capital (well and production pump) costs.

Cascading geothermal fluid to uses with lower temperature requirements can help achieve a large temperature difference (Δt). Most geothermal systems are designed for a Δt between 30 and 50°F, although one system was designed for a Δt of 100°F with a 190°F resource temperature.

Load Factor

Defined as the ratio of the average load to the design capacity of the system, the load factor effectively reflects the fraction of time that the initial investment in the system is working. Again, because geothermal cost is primarily initial rather than operating cost, this factor significantly affects a geothermal system's viability. As the load factor increases, so does the economy of using geothermal energy. The two main ways of increasing the load factor are (1) to select applications where it is naturally high, and (2) to use peaking equipment so that the geothermal design load is not the application peak load, but rather a reduced load that occurs over a longer period.

Table 1   Selected Chemical Species Affecting Fluid Disposal

Species	Reason for Control
Hydrogen sulfide (H_2S)	Odor
Boron (B^{3+})	Damage to agricultural crops
Fluoride (F^-)	Level limited in drinking water sources
Radioactive species	Levels limited in air, water, and soil

Source: Lunis (1989).

Composition of Fluid

The quality of the produced fluid is site specific and may vary from less than 1000 ppm TDS to heavily brined. Fluid quality influences two aspects of the design: (1) material selection to avoid corrosion and scaling effects, and (2) disposal or ultimate end use of the fluid.

Ease of Disposal

The costs associated with disposal, particularly when injection is involved, can substantially affect development costs. Historically, most geothermal effluent was disposed of on the surface, including discharge to irrigation, rivers, and lakes. This method of disposal is considerably less expensive than constructing injection wells.

Geothermal fluids sometimes contain chemical constituents that make surface disposal problematic. Some of these constituents are listed in Table 1.

Most new, large geothermal systems use injection for disposal to minimize environmental concerns and ensure long-term resource reliability. If injection is chosen, the depth at which the fluid can be injected affects well cost substantially. Many jurisdictions require the fluid be returned to the same or similar aquifers; thus, it may be necessary to bore the injection well to the same depth as the production well. Direct-use injection wells are considered Class V wells under the U.S. Environmental Protection Agency's Underground Injection Control (UIC) program. Water wells, along with terminology relating to the technology, are discussed in the section on Groundwater Heat Pumps.

Direct-Use Water Quality Testing

Low-temperature geothermal fluids commonly contain seven key chemical species that can significantly corrode standard materials of construction (Ellis 1989). These include

- Oxygen (generally from aeration)
- Hydrogen ion (pH)
- Chloride ion
- Sulfide species
- Carbon dioxide species
- Ammonia species
- Sulfate ion

The principal effects of these species are summarized in Table 2. Except as noted, the described effects are for carbon steel. Kindle and Woodruff (1981) present recommended procedures for complete chemical analysis of geothermal well water.

Two of these species are not reliably detected by standard water chemistry tests and deserve special mention. Dissolved oxygen does not occur naturally in low-temperature (120 to 220°F) geothermal fluids that contain traces of hydrogen sulfide. However, because of slow-reaction kinetics, oxygen from air inleakage may persist for some minutes. Once the geothermal fluid is produced, it is extremely difficult to prevent contamination, especially if pumps used to move the fluid are not downhole-submersible or lineshaft turbine pumps. Even if fluid systems are maintained at positive pressure, air inleakage at pump seals is likely, particularly with the low level of maintenance in many installations.

Hydrogen sulfide is ubiquitous in extremely low concentrations in geothermal fluids above 120°F. This corrosive species also occurs

Table 2   Principal Effects of Key Corrosive Species

Species	Principal Effects
Oxygen	• Extremely corrosive to carbon and low-alloy steels; 30 ppb shown to cause fourfold increase in carbon steel corrosion rate. • Concentrations above 50 ppb cause serious pitting. • In conjunction with chloride and high temperature, <100 ppb dissolved oxygen can cause chloride-stress corrosion cracking (chloride-SCC) of some austenitic stainless steels.
Hydrogen ion (pH)	• Primary cathodic reaction of steel corrosion in air-free brine is hydrogen ion reduction. Corrosion rate decreases sharply above pH 8. • Low pH (5) promotes sulfide stress cracking (SSC) of high-strength low-alloy (HSLA) steels and some other alloys coupled to steel. • Acid attack on cements.
Carbon dioxide species (dissolved carbon dioxide, bicarbonate ion, carbonate ion)	• Dissolved carbon dioxide lowers pH, increasing carbon and HSLA steel corrosion. • Dissolved carbon dioxide provides alternative proton reduction pathway, further exacerbating carbon and HSLA steel corrosion. • May exacerbate SSC. • Strong link between total alkalinity and corrosion of steel in low-temperature geothermal wells.
Hydrogen sulfide species (hydrogen sulfide, bisulfide ion, sulfide ion)	• Potent cathodic poison, promoting SSC of HSLA steels and some other alloys coupled to steel. • Highly corrosive to alloys containing both copper and nickel or silver in any proportions.
Ammonia species (ammonia, ammonium ion)	• Causes stress corrosion cracking (SCC) of some copper-based alloys.
Chloride ion	• Strong promoter of localized corrosion of carbon, HSLA, and stainless steel, as well as of other alloys. • Chloride-dependent threshold temperature for pitting and SCC. Different for each alloy. • Little if any effect on SSC. • Steel passivates at high temperature in 6070 ppm chloride solution (pH = 5) with carbon dioxide. 133,500 ppm chloride destroys passivity above 300°F.
Sulfate ion	• Primary effect is corrosion of cements.

Source: Ellis (1989).
Note: Except as indicated, described effects are for carbon steel.

naturally in many cooler groundwaters. For alloys such as cupronickel, which are strongly affected by it, hydrogen sulfide concentrations in the low parts per billion (10^9) range may have a serious detrimental effect, especially if oxygen is also present. At these levels, the characteristic rotten egg odor of hydrogen sulfide may be absent, so field testing may be required for detection. Hydrogen sulfide levels down to 50 ppb can be detected using a simple field kit; however, absence of hydrogen sulfide at this low level may not preclude damage by this species.

Two other key species that should be measured in the field are pH and carbon dioxide concentrations. This is necessary because most geothermal fluids release carbon dioxide rapidly, causing a rise in pH.

Production of suspended solids (sand) from a well should be addressed during well construction with gravel pack, screen, or both. Proper selection of the screen/gravel pack is based on sieve analysis of cutting samples from drilling. Surface separation is less desirable because a surface separator requires sand to pass first through the pump, reducing its useful life.

Biological fouling is largely a phenomenon of low-temperature (<90°F) wells. The most prominent organisms are various strains (*Galionella, Crenothrix*) of what are commonly referred to as iron bacteria. These organisms typically inhabit water with a pH range of 6.0 to 8.0, dissolved oxygen content of less than 5 ppm, ferrous iron content of less than 0.2 ppm, and a temperature of 46 to 61°F (Hackett and Lehr 1985). Iron bacteria can be identified microscopically. The most common treatment for iron bacteria infestation is chlorination, surging, and flushing; success depends on maintaining proper pH (less than 8.5), dosage, free residual chlorine content (200 to 500 ppm), contact time (24 h minimum), and agitation or surging. Precleaning (wire brushing) of the screen and redevelopment of the well after treatment are key to effectiveness. Hackett and Lehr (1985) provide additional detail on treatment.

MATERIALS AND EQUIPMENT

For system parts exposed to the fluid, materials selection is an important part of the design process. Chemical treatment of the geothermal fluid is not an effective strategy in most cases, because of economics and environmental (disposal) considerations. Corrosion and scaling in direct-use systems are generally addressed by isolating the fluid from the majority of the system using a plate heat exchanger.

Performance of Materials

Carbon Steel. The Ryznar index has traditionally been used to estimate the corrosivity and scaling tendencies of potable water supplies. However, one study found no significant correlation (at the 95% confidence level) between carbon steel corrosion and the Ryznar index (Ellis and Smith 1983). Therefore, the Ryznar and other indices based on calcium carbonate saturation should not be used to predict corrosion in geothermal systems, though they remain valid for scaling prediction.

In Class Va geothermal fluids [as described by Ellis (1989); <5000 ppm total key species (TKS), total alkalinity 207 to 1329 ppm as CaCO₃, pH 6.7 to 7.6], corrosion rates of about 5 to 20 mil/yr can be expected, often with severe pitting.

In Class Vb geothermal fluids [as described by Ellis (1989); <5000 ppm TKS, total alkalinity <210 ppm as CaCO₃, pH 7.8 to 9.85], carbon steel piping has given good service in a number of systems, as long as system design rigorously excluded oxygen. However, introduction of 30 ppb oxygen under turbulent flow conditions causes a fourfold increase in uniform corrosion. Saturation with air often increases the corrosion rate by at least 15 times. Oxygen contamination at the 50 ppb level often causes severe pitting. Chronic oxygen contamination causes rapid failure.

External surfaces of buried steel pipe must be protected from contact with groundwater. Groundwater is aerated and has caused pipe failures by external corrosion. Required external protection can be obtained by coatings, pipe-wrap, or preinsulated piping, provided the selected material resists the system operating temperature and thermal stress.

At temperatures above 135°F, galvanizing (zinc coating) does not reliably protect steel from either geothermal fluid or groundwater. Hydrogen blistering can be prevented by using void-free (killed) steels.

Low-alloy steels (steels containing not more than 4% alloying elements) have corrosion resistance similar, in most respects, to carbon steels. As with carbon steels, sulfide promotes entry of atomic hydrogen into the metal lattice. If the steel exceeds a hardness of Rockwell C22, sulfide stress cracking may occur.

Copper and Copper Alloys. Copper-tubed fan-coil units and heat exchangers have consistently poor performance because of traces of sulfide species found in geothermal fluids in the United States. Copper tubing rapidly becomes fouled with cuprous sulfide films more than 1 mm thick. Serious crevice corrosion occurs at

cracks in the film, and uniform corrosion rates of 2 to 6 mil/yr appear typical, based on failure analyses.

Experience in Iceland also indicates that copper is unsatisfactory for heat exchange service and that most brasses (Cu-Zn) and bronzes (Cu-Sn) are even less suitable. Cupronickel often performs more poorly than copper in low-temperature geothermal service because of trace sulfide.

Much less information is available regarding copper and copper alloys in non-heat-transfer service. Copper pipe shows corrosion behavior similar to copper heat exchange tubes under conditions of moderate turbulence (Reynolds numbers of 40,000 to 70,000). An internal inspection of yellow brass valves showed no significant corrosion. However, silicon bronze CA 875 (12-16Cr, 3-5Si, <0.05Pb, <0.05P), an alloy normally resistant to dealloying, failed in less than three years when used as a pump impeller. Leaded red brass (CA 836 or 838) and leaded red bronze (SAE 67) appear viable as pump internal parts. Based on a few tests at Class Va sites, aluminum bronzes have shown potential for corrosion in heavy-walled components (Ellis 1989).

Solder is yet another problem area for copper equipment. Lead-tin solder (50Pb, 50Sn) was observed to fail by dealloying after a few years' exposure. Silver solder (1Ag, 7P, Cu) was completely removed from joints in under two years. If the designer elects to accept this risk, solders containing at least 70% tin should be used.

Stainless Steel. Unlike copper and cupronickel, stainless steels are not affected by traces of hydrogen sulfide. Their most likely application is heat exchange surfaces. For economic reasons, most heat exchangers are probably of the plate-and-frame type, most of which are fabricated with one of two standard alloys, Type 304 and Type 316 austenitic stainless steel. Some pump and valve trim also are fabricated from these or other stainless steels.

These alloys are subject to pitting and crevice corrosion above a threshold chloride level, which depends on the chromium and molybdenum content of the alloy and on the temperature of the geothermal fluid. Above this temperature, the passivation film, which gives the stainless steel its corrosion resistance, is ruptured, and local pitting and crevice corrosion occur. Figure 4 shows the relationship between temperature, chloride level, and occurrence of localized corrosion of Type 304 and Type 316 stainless steel. This figure indicates, for example, that localized corrosion of Type 304 may occur in 80°F geothermal fluid if the chloride level exceeds approximately 210 ppm; Type 316 is resistant at that temperature

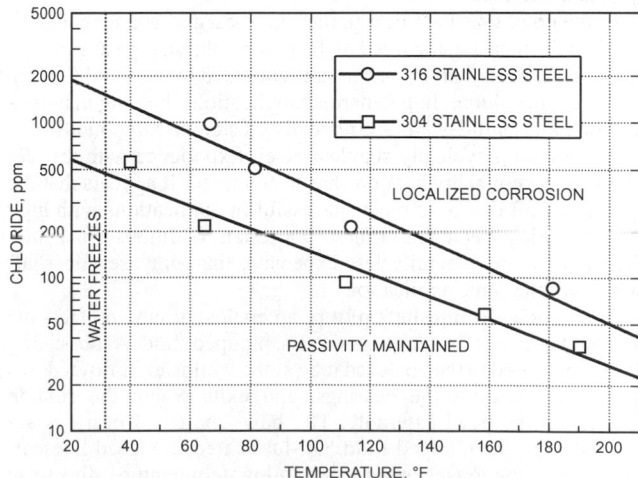

Fig. 4 Chloride Concentration Required to Produce Localized Corrosion of Stainless Steel as Function of Temperature
(Efrid and Moeller 1978)

until the chloride level reaches approximately 510 ppm. Because of its 2 to 3% molybdenum content, Type 316 is always more resistant to chlorides than is Type 304.

Aluminum. Aluminum alloys are not acceptable in most cases because of catastrophic pitting.

Titanium. This material has extremely good corrosion resistance and could be used for heat exchanger plates in any low-temperature geothermal fluid, regardless of dissolved oxygen content. Great care is required if acid cleaning is to be performed. The vendor's instructions must be followed. The titanium should not be scratched with iron or steel tools; this can cause pitting.

Chlorinated Polyvinyl Chloride (CPVC) and Fiber-Reinforced Plastic (FRP). These materials are easily fabricated and are not adversely affected by oxygen intrusion. External protection against groundwater is not required. The mechanical properties of these materials at higher temperatures may vary greatly from those at ambient temperature, and the materials' mechanical limits should not be exceeded. The usual mode of failure is creep rupture: strength decays with time. Manufacturer's directions for joining should be followed to avoid premature failure of joints.

Elastomeric Seals. Tests on O-ring materials in a low-temperature system in Texas indicated that a fluoroelastomer is the best material for piping of this nature; Buna-N is also acceptable (Ellis 1989). Neoprene, which developed extreme compression set, was a failure. Natural rubber and Buna-S should also be avoided. Ethylene-propylene terpolymer (EPDM) has been used successfully in gasket, O-ring, and valve seats in many systems. EPDM materials have swollen in some systems using oil-lubricated turbine pumps.

Pumps

Production well pumps are among the most critical components in a geothermal system and have been the source of much system downtime. Therefore, proper selection and design of the production well pump is extremely important. Well pumps are available for larger systems in two general configurations: lineshaft and submersible. The lineshaft type is most often used for direct-use systems (Rafferty 1989b).

Lineshaft Pumps. Lineshaft pumps are similar to those typically used in irrigation applications. An aboveground driver, typically an electric motor, rotates a vertical shaft extending down the well to the pump. The shaft rotates pump impellers in the pump bowl assembly, which is positioned at such a depth in the wellbore that adequate net positive suction head (NPSH) is available when the unit is operating. Two designs for the shaft/bearing portion of the pump are available: open and enclosed.

In the **open lineshaft pump**, the shaft bearings are supported in "spiders," which are anchored to the pump column pipe at 5 to 10 ft intervals. The shaft and bearings are lubricated by the fluid flowing up the pump column. In geothermal applications, bearing materials for open lineshaft designs are typically elastomer compounds. The shaft material is typically stainless steel. Experience with this design in geothermal applications has been mixed. It appears that the open lineshaft design is most successful in applications with high (<50 ft) static water levels or flowing artesian conditions. Open lineshaft pumps are generally less expensive than enclosed lineshaft pumps for the same application.

In an **enclosed lineshaft pump**, an enclosing tube protects the shaft and bearings from exposure to the pumped fluid. A lubricating fluid is admitted to the enclosed tube at the wellhead. It flows down the tube, lubricates the bearings, and exits where the column attaches to the bowl assembly. The bowl shaft and bearings are lubricated by the pumped fluid. Oil-lubricated, enclosed lineshaft pumps have the longest service life in low-temperature, direct-use applications.

These pumps typically include carbon or stainless steel shafts and bronze bearings in the lineshaft assembly, and stainless steel shafts and leaded red bronze bearings in the bowl assembly. Keyed-type

impeller connections (to the pump shaft) are superior to collet-type connections (Rafferty 1989b).

Because of the lineshaft bearings, lineshaft pump reliability decreases as pump-setting depth increases. Nichols (1978) indicates that, below about 800 ft, reliability is questionable, even under good pumping conditions.

Submersible Pumps. The electrical submersible pump consists of three primary components located downhole: the pump, the drive motor, and the motor protector. The pump is a vertical multistage centrifugal type. The motor is usually a three-phase induction type that is filled with oil for cooling and lubrication; it is cooled by heat transfer to the pumped fluid moving up the well. The motor protector is located between the pump and the motor and isolates the motor from the well fluid while allowing pressure equalization between the pump intake and the motor cavity.

The electrical submersible pump has several advantages over lineshaft pumps, particularly for wells requiring greater pump bowl setting depths. The deeper the well, the greater the economic advantage of the submersible pump. Moreover, it is more versatile, adapting more easily to different depths.

Submersible pumps have not demonstrated acceptable lifetimes in most geothermal applications. Although they are commonly used in high-temperature, downhole applications in the oil and gas industry, the acceptable overhaul interval in that industry is much shorter than in a geothermal application. In addition, most submersibles operate at 3600 rpm, resulting in greater susceptibility to erosion in aquifers that produce moderate amounts of sand. They have, however, been applied in geothermal projects where an existing well of relatively small diameter must be used. At 3600 rpm, they provide greater flow capacity for a given bowl size than an equivalent 1750 rpm lineshaft pump.

Standard "cold-water" submersible motors can be used at temperatures up to approximately 120°F with adequate precautions. These consist primarily of ensuring adequate water velocity past the motor (minimum 3 ft/s), which may require the use of a sleeve, and a small degree of motor oversizing (Franklin Electric 2001).

Well Pump Control. Well pumps serving variable loads are often controlled using variable-speed drives (VSDs). Submersible pumps can also be controlled using VSDs, but special precautions are required. Drive-rated motors are not commonly available for these applications, so external electronic protection should be used to prevent premature motor failure. In addition, the motor manufacturer must be aware that the motor will applied in a variable-speed application. Finally, because of the large static head in many well pump applications, controls should be configured to prevent the pump from operating at no-flow conditions.

Heat Exchangers

Geothermal fluids can be isolated with large central heat exchangers, as in the case of a district heating system, or with exchangers at individual buildings or loads. In both cases, the principle is to isolate the geothermal fluid from complicated systems or those that cannot readily be designed to be compatible with the geothermal fluid. The main types of heat exchangers used in transferring energy from the geothermal fluid are plate and downhole.

Plate Heat Exchangers. For all but the very smallest applications, plate-and-frame heat exchangers are the most commonly used design. Available in corrosion-resistant materials, easily cleanable, and able to accommodate increased loads by adding plates, these exchangers are well suited to geothermal applications. The high performance of plate heat exchangers is also an asset in many system designs. Because geothermal resource temperatures are often less than those used in conventional hot-water heating system design, minimizing temperature loss at the heat exchanger is frequently a design issue. Approach temperatures of 5°F and less are common.

Materials for plate heat exchangers in direct-use applications normally include Buna-N or EPDM gaskets and 316 or titanium plates. Plate selection is often a function of temperature and chloride content of the water. For applications characterized by chloride contents of >50 ppm at 200°F, titanium would be used. At lower temperatures, much higher chloride exposure can be tolerated (see Figure 4).

Downhole Heat Exchangers. The downhole heat exchanger (DHE) is an arrangement of pipes or tubes suspended in a wellbore (Culver and Reistad 1978). A secondary fluid circulates from the load through the exchanger and back to the load in a closed loop. The primary advantage of a DHE is that only heat is extracted from the earth, which eliminates the need to dispose of spent fluids. Other advantages are the elimination of (1) well pumps with their initial operating and maintenance costs, (2) the potential for depletion of groundwater, and (3) environmental and institutional restrictions on surface disposal. One disadvantage of a DHE is the limited amount of heat that can be extracted from or rejected to the well. The amount of heat extracted depends on the hydraulic conductivity of the aquifer and well design. Because of the limitations of natural convection, only about 10% of the heat output of the well is available from a DHE in comparison to pumping and using surface heat exchange (Reistad et al. 1979). With wells of approximately 200°F and depths of 500 ft, output under favorable conditions is sufficient to serve the needs of up to five homes.

The DHE in low- to moderate-temperature geothermal wells is installed in a casing, as shown in Figure 5.

Downhole heat exchangers with higher outputs rely on water circulation within the well, whereas lower-output DHEs rely on earth conduction. Circulation in the well can be accomplished by two methods: (1) undersized casing and (2) convection tube. Both methods rely on the difference in density between the water surrounding the DHE and that in the aquifer.

Circulation provides the following advantages:

- Water circulates around the DHE at velocities that, in optimum conditions, can approach those in the shell of a shell-and-tube exchanger.
- Hot water moving up the annulus heats the upper rocks and the well becomes nearly isothermal.

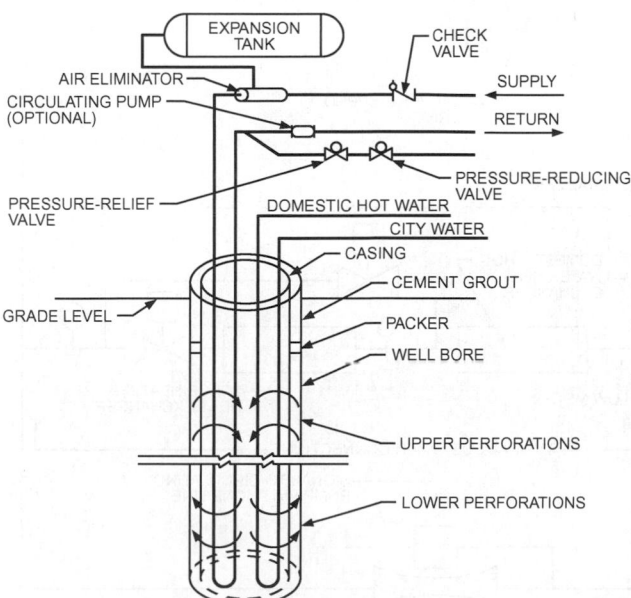

Fig. 5   Typical Connection of Downhole Heat Exchanger for Space and Domestic Hot-Water Heating
(Reistad et al. 1979)

- Some of the cool water, being denser than the water in the aquifer, sinks into the aquifer and is replaced by hotter water, which flows up the annulus.

Figure 5 shows well construction in competent formation (i.e., where the wellbore will stand open without a casing). An undersized casing with perforations at the lowest producing zone (usually near the bottom) and just below the static water level is installed. A packer near the top of the competent formation allows installation of an annular seal between it and the surface. When the DHE is installed and heat extracted, thermosiphoning causes cooler water inside the casing to move to the bottom, and hotter water moves up the annulus outside the casing.

Because most DHEs are used for space heating (an intermittent operation), heated rocks in the upper portion of the well store heat for the next cycle.

Where the well will not stand open without casing, a convection tube can be used. This is a pipe one-half the diameter of the casing either hung with its lower end above the well bottom and its upper end below the surface or set on the bottom with perforations at the bottom and below the static water level. If a U-bend DHE is used, it can be either inside or outside the convection tube. DHEs operate best in aquifers with a high hydraulic conductivity and that provide water movement for heat and mass transfer.

Valves

In large (>2.5 in.) pipe sizes, resilient-lined butterfly valves are preferred for geothermal applications. The lining material protects the valve body from exposure to the geothermal fluid. The rotary rather than reciprocating motion of the stem makes the valve less susceptible to leakage and build-up of scale deposits. For many direct-use applications, these valves are composed of Buna-N or EPDM seats, stainless steel shafts, and bronze or stainless steel disks. Where oil-lubricated well pumps are used, a seat material of oil-resistant material is recommended. Gate valves have been used in some larger geothermal systems but have been subject to stem leakage and seizure. After several years of use, they are no longer capable of 100% shutoff.

Piping

Piping in geothermal systems can be divided into two broad groups: pipes used inside buildings and those used outside, typically buried. Indoor piping carrying geothermal water is usually limited to the mechanical room. Carbon steel with grooved end joining is the most common material.

For buried piping, many existing systems use some form of nonmetallic piping, particularly asbestos cement (which is no longer available) and glass fiber. With the cost of glass fiber for larger sizes (>6 in.) sometimes prohibitive, ductile iron is frequently used. Available in sizes >2 in., ductile iron offers several positive characteristics: low cost, familiarity to installation crews, and wide availability. It requires no allowances for thermal expansion if push-on fittings are used.

Most larger diameter buried piping is preinsulated. The basic ductile iron pipe is surrounded by a layer of insulation (typically polyurethane), which is protected by an outer jacket of PVC or PE.

Standard ductile iron used for municipal water systems is sometimes modified for geothermal use. The seal coat used to protect the cement lining of the pipe is not suitable for the temperature of most geothermal applications; in applications where the geothermal water is especially soft or low in pH, the cement lining should be omitted, as well. Special high-temperature gaskets (usually EPDM) are used in geothermal applications. Few problems have been encountered in using ferrous piping with low-temperature geothermal fluids unless high chloride concentration, low (<7.0) pH, or oxygen is present in the fluid. Most cases of corrosion failure have resulted from external attack by soil moisture in buried applications.

RESIDENTIAL AND COMMERCIAL BUILDING APPLICATIONS

The primary applications for direct use of geothermal energy in the residential and commercial area are space and domestic water heating. Space cooling using the absorption process is possible but rarely applied.

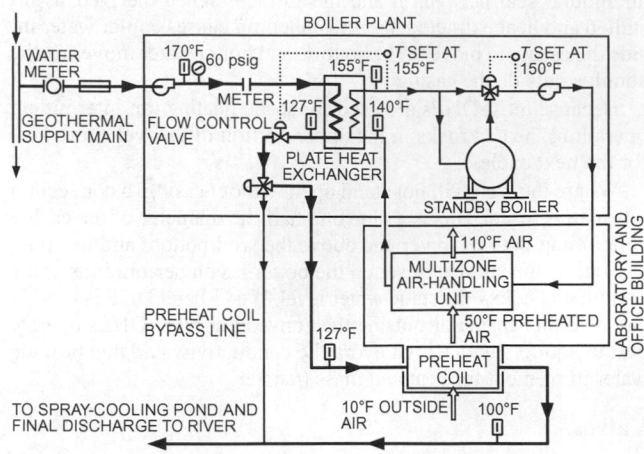

Fig. 6   Heating System Schematic

Space Heating

Figure 6 illustrates a system that uses geothermal fluid at 170°F (Austin 1978). The geothermal fluid is used in two main equipment components for heating the buildings: (1) a plate heat exchanger that supplies energy to a closed heating loop previously heated by a natural gas boiler (the boiler remains as a standby unit) and (2) a water-to-air coil used for preheating ventilation air. In this system, proper control is crucial for economical operation.

The average temperature of discharged fluid is 120 to 130°F. Geothermal fluid is used directly in the preheat coils in the buildings, which would probably not be the case if the system were designed today (Lienau 1979).

Figure 7 shows a geothermal district heating system that has a unique feature: its design is based on a peak load Δt of 100°F using a 190°F resource. It is of closed-loop design with central heat exchangers. The production well has an artesian shut-in pressure of 25 psi, so the system operates with no production pump for most of the year. During colder weather, a surface centrifugal pump located at the wellhead boosts the pressure.

Geothermal flow from the production well is initially controlled by a throttling valve on the supply line to the main heat exchanger, which responds to a temperature signal from the supply water on the closed-loop side of the heat exchanger. When the throttling valve reaches the full-open position, the production booster pump is enabled. The pump is controlled through a variable-speed drive that responds to the same supply-water signal as the throttling valve. The booster pump is designed for a peak flow rate of 300 gpm of 190°F water.

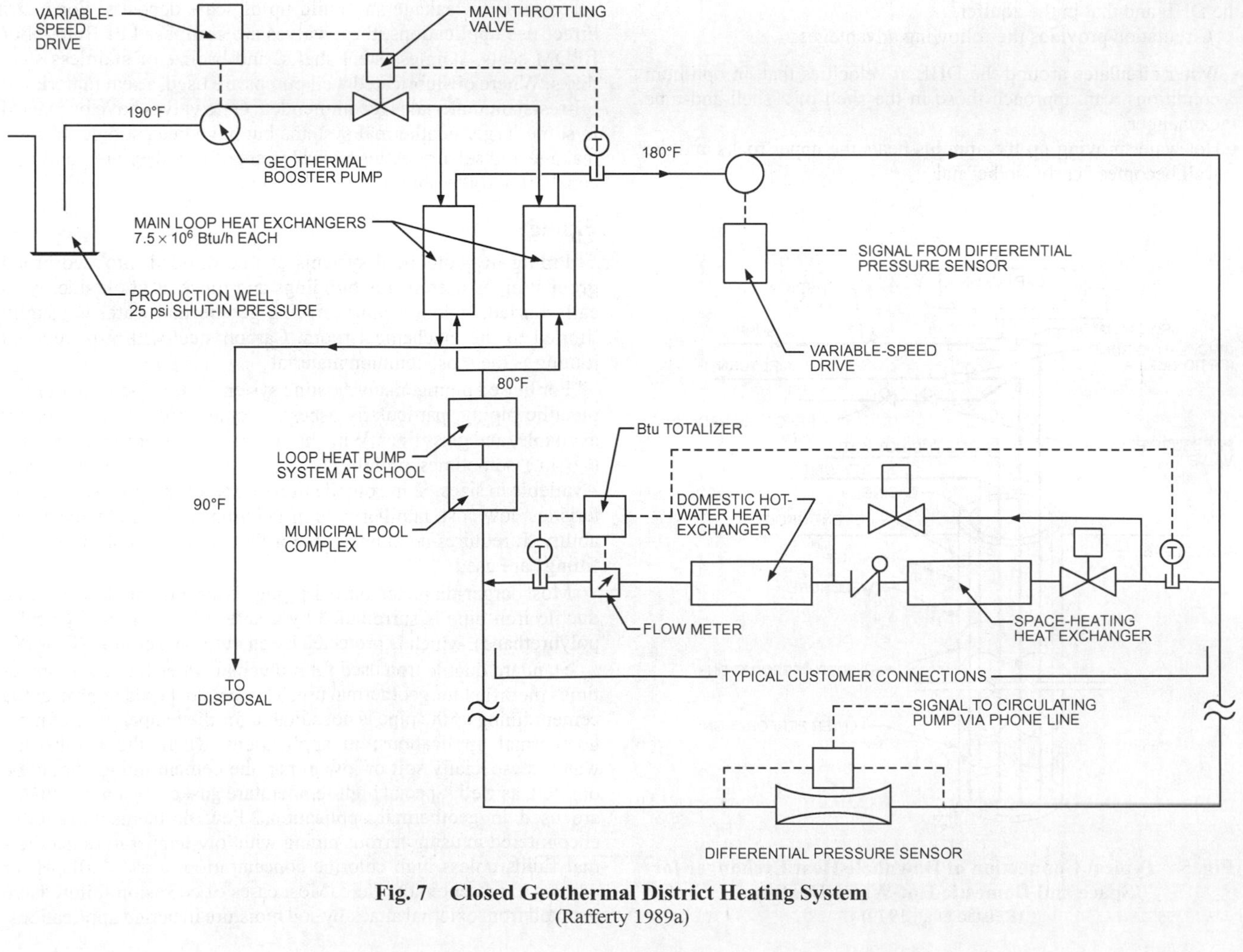

Fig. 7   Closed Geothermal District Heating System
(Rafferty 1989a)

A few district heating systems have also been installed using an open distribution system. In this design, central heat exchangers (as in Figure 7) are eliminated and the geothermal water is delivered to individual building heat exchangers. When more than a few buildings are connected to the system, using central heat exchangers is normally more cost-effective.

Terminal equipment used in geothermal systems is the same as that used in nongeothermal heating systems. However, certain types of equipment are better suited to geothermal design than others.

In many cases, buildings heated by low-temperature geothermal sources operate at lower supply water temperatures than conventional hydronic designs. Because many geothermal sources are designed to take advantage of a large Δt, proper selection of equipment for low flow and low temperature is important.

Finned-coil, forced-air systems generally function best in this low-temperature/high-Δt situation. One or two additional rows of coil depth compensate for the lower supply water temperature. Although an increased Δt affects coil circuiting, it improves controllability. This type of system should be able to use a supply water temperature as low as 110°F.

Radiant floor panels are well suited to very low water temperatures, particularly in industrial applications with little or no floor covering. In industrial settings, with a bare floor and a relatively low space-temperature requirement, the average water temperature could be as low as 95°F. For a higher space temperature and/or thick floor coverings, a higher water temperature may be required.

Baseboard convectors and similar equipment are the least capable of operating at low supply-water temperature. At 150°F average water temperatures, derating factors for this design load may be affected. This type of equipment can be operated at low temperatures from the geothermal source to provide base-load heating, with peak load supplied by a conventional boiler.

Domestic Water Heating

Domestic water heating in a district space-heating system is beneficial because it increases the overall size of the energy load, energy demand density, and load factor. For those resources that cannot heat water to the required temperature, preheating is usually possible. Whenever possible, the domestic hot-water load should be placed in series with the space-heating load to reduce system flow rates and increase Δt.

Space Cooling

Geothermal energy has seldom been used for cooling, although emphasis on solar energy and waste heat has created interest in cooling with thermal energy. The absorption cycle is most often used, and lithium bromide/water absorption machines are available in a wide range of capacities. Temperature and flow requirements for absorption chillers run counter to the general design philosophy for geothermal systems: they require high supply water temperatures and a small Δt on the hot-water side. Figure 8 illustrates the effect of reduced supply water temperature on machine performance. The machine is rated at a 240°F input temperature, so derating factors must be applied if the machine is operated below this temperature. For example, operation at a 200°F supply water temperature results in a 50% decrease in capacity, which seriously affects the economics of absorption cooling at a low resource temperature.

Coefficient of performance (COP) is less seriously affected by reduced supply water temperature. The nominal COP of a single-stage machine at 240°F is 0.65 to 0.70; that is, for each ton of cooling output, a heat input of 12,000 Btu/h divided by 0.65, or 18,460 Btu/h, is required.

Most absorption equipment is designed for steam input (an isothermal process) to the generator section. When this equipment is operated from a hot-water source, a relatively small Δt must be used. This creates a mismatch between building flow requirements

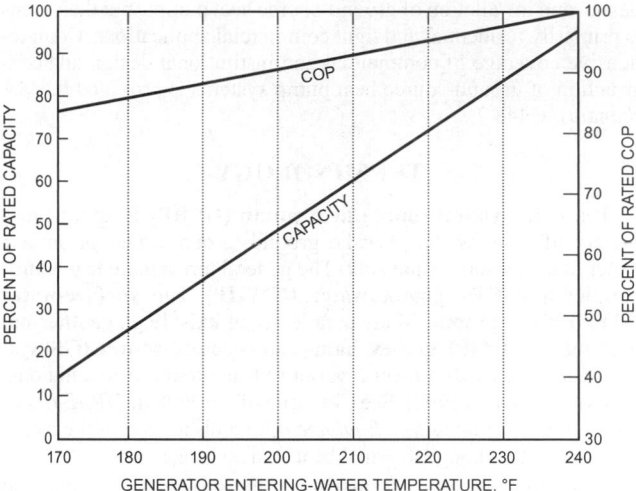

Fig. 8 Typical Lithium Bromide Absorption Chiller Performance Versus Temperature
(Christen 1977)

for space heating and cooling. For example, assume a 200,000 ft² building is to use a geothermal resource for heating and cooling. At 25 Btu/h·ft² and a design Δt of 40°F, the flow requirement for heating is 250 gpm. At 30 Btu/h·ft², a Δt of 15°F, and a COP of 0.65, the flow requirement for cooling is 1230 gpm.

Some small-capacity (3 to 25 ton) absorption equipment has been optimized for low-temperature operation in conjunction with solar heat. Although this equipment could be applied to geothermal resources, the prospects are questionable. Small absorption equipment generally competes with packaged direct-expansion units in this range; absorption equipment requires a great deal more mechanical auxiliary equipment for a given capacity. The cost of the chilled-water piping, pump, and coil; cooling-water piping, pump, and tower; and hot-water piping raises the capital cost of the absorption equipment substantially. Only in large sizes (>10 tons) and in areas with high electric rates and high cooling requirements (>2000 full-load hours) would this type of equipment offer an attractive investment to the owner (Rafferty 1989a).

INDUSTRIAL APPLICATIONS

Design philosophy for the use of geothermal energy in industrial applications, including agricultural facilities, is similar to that for space conditioning. However, these applications have the potential for much more economical use of the geothermal resource, primarily because they (1) operate year-round, which gives them greater load factors than possible with space-conditioning applications; (2) do not require extensive (and expensive) distribution to dispersed energy consumers, as is common in district heating; and (3) often require various temperatures and, consequently, may be able to make greater use of a particular resource than space conditioning, which is restricted to a specific temperature. In the United States, the primary non-space-heating applications of direct-use geothermal resources are dehydration (primarily vegetables), gold mining, and aquaculture.

GROUND-SOURCE HEAT PUMPS

Ground-source heat pumps were originally developed in the residential arena and are now widely applied in the commercial sector. Many of the installation recommendations and design guides appropriate to residential design must be amended for large buildings. Kavanaugh and Rafferty (1997) provide a more complete overview of design of ground-source heat pump systems. Kavanaugh (1991) and OSU (1988a, 1988b) provide a more detailed treatment of the design and installation of ground-source heat pumps, but their focus is primarily residential and light commercial applications. Comprehensive coverage of commercial and institutional design and construction of ground-source heat pump systems is provided in CSA *Standard* C448.2.

TERMINOLOGY

The term **ground-source heat pump (GSHP)** is applied to a variety of systems that use the ground, groundwater, or surface water as a heat source and sink. The general terms include **ground-coupled (GCHP)**, **groundwater (GWHP)**, and **surface-water (SWHP) heat pumps**. Many parallel terms exist [e.g., **geothermal heat pumps (GHP)**, **geo-exchange**, and **ground-source (GS) systems**] and are used to meet a variety of marketing or institutional needs (Kavanaugh 1992). See Chapter 8 of the 2008 *ASHRAE Handbook—HVAC Systems and Equipment* for a discussion of the merits of various other nongeothermal heat sources/sinks.

This chapter focuses primarily on the ground-loop portion of GSHP systems, although the heat pump units used in these systems are unique to GSHP technology as well. GSHP systems typically use extended-range water-source heat pump units, in most cases of water-to-air configuration. Extended-range units are specifically designed for operation at entering water temperatures between 23°F in heating mode and 104°F in cooling mode. Units not meeting the extended-range criteria are not suitable for use in GSHP systems. Some applications can include a free-cooling mode when water-loop temperatures fall near or below 55°F. This includes groundwater loops, deep-surface-water loops, and interior core zones of ground-coupled loops when perimeter zones require heating. This is typically accomplished by inserting a water coil in the return air stream before the refrigerant coil.

Ground-Coupled Heat Pump Systems

The GCHP is a subset of the GSHP and is often called a closed-loop heat pump. A GCHP system consists of a reversible vapor compression cycle that is linked to a closed ground heat exchanger buried in soil (Figure 9). The most widely used unit is a water-to-air heat pump, which circulates a water or a water/antifreeze solution through a liquid-to-refrigerant heat exchanger and a buried thermoplastic piping network. Heat pump units often include desuperheater heat exchangers (shown on the left in Figure 9). These devices use hot refrigerant at the compressor outlet to heat water. A second type of GCHP is the direct-expansion (DX) GCHP, which uses a buried copper piping network through which refrigerant is circulated.

The GCHP is further subdivided according to ground heat exchanger design: vertical and horizontal. **Vertical GCHPs** (Figure 10) generally consist of two small-diameter, high-density polyethylene (HDPE) tubes placed in a vertical borehole that is subsequently filled with a solid medium. The tubes are thermally fused at the bottom of the bore to a close return U-bend. Vertical tubes range from 0.75 to 1.5 in. nominal diameter. Bore depths range from 50 to 600 ft, depending on local drilling conditions and available equipment. Boreholes are typically 4 to 6 in. in diameter.

To reduce thermal interference between individual bores, a minimum borehole separation distance of 20 ft is recommended when loops are placed in a grid pattern. This distance may be reduced when bores are placed in a single row, the annual ground load is balanced (i.e., energy released in the ground is approximately equal to the energy extracted on an annual basis), or water movement or evaporation and subsequent recharge mitigates the effect of heat build-up in the loop field.

Advantages of the vertical GCHP are that it (1) requires relatively small plots of ground, (2) is in contact with soil that varies very little in temperature and thermal properties, (3) requires the smallest amount of pipe and pumping energy, and (4) can yield the most efficient GCHP system performance. Disadvantages are (1) typically higher cost because of expensive equipment needed to drill the borehole and (2) the limited availability of contractors to perform such work.

Hybrid systems are a variation of ground-coupled systems in which a smaller ground loop is used, augmented in cooling mode by a fluid cooler or a cooling tower. This approach can have merit in large cooling-dominated applications. The ground loop is sized to meet the heating requirements. The downsized loop is used in conjunction with the fluid cooler or cooling tower with an isolation heat

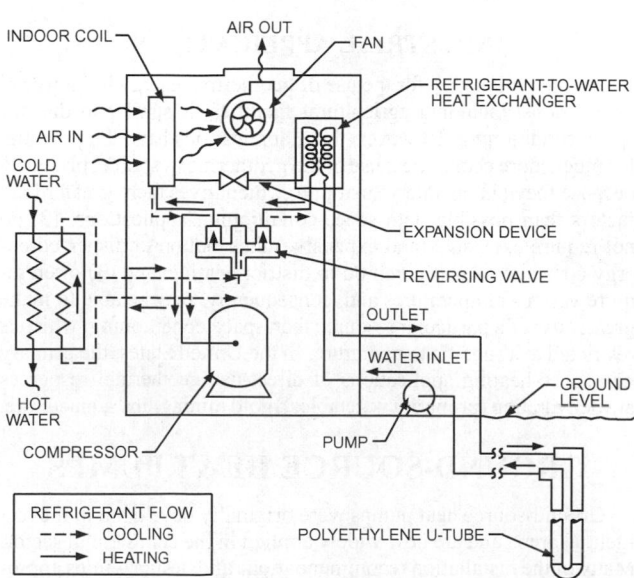

Fig. 9 Vertical Closed-Loop Ground-Coupled Heat Pump System

(Kavanaugh 1985)

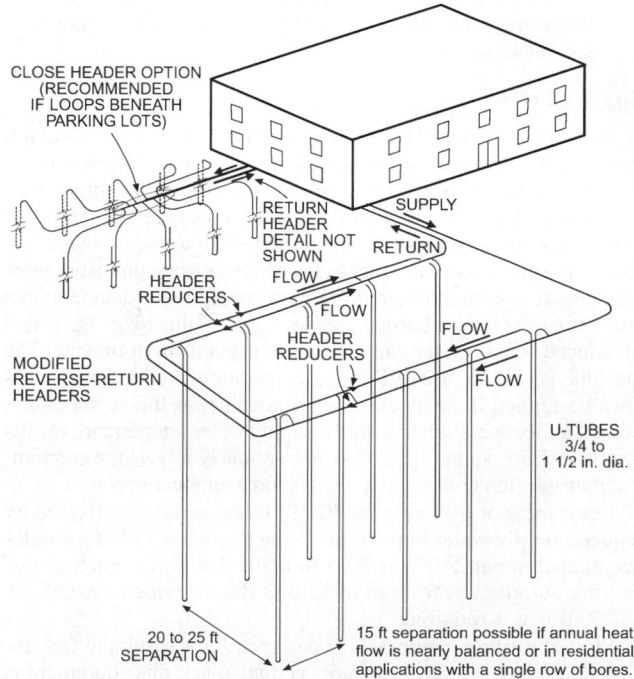

Fig. 10 Vertical Ground-Coupled Heat Pump Piping

exchanger to meet the heat rejection load. Using the cooler reduces the capital cost of the ground loop in such applications, but somewhat increases maintenance requirements. For heavily heating-dominant applications, a downsized loop also can be augmented with an auxiliary heat source such as electric resistance, solar collectors, or fossil fuel.

Horizontal GCHPs (Figure 11) can be divided into several subgroups, including single-pipe, multiple-pipe, spiral, and horizontally bored. Single-pipe horizontal GCHPs were initially placed in narrow trenches at least 4 ft deep. These designs require the greatest amount of ground area. Multiple pipes (usually two, four, or six), placed in a single trench, can reduce the amount of required ground area. Trench length is reduced with multiple-pipe GCHPs, but total pipe length must be increased to overcome thermal interference from adjacent pipes. The spiral coil is reported to further reduce required ground area. These horizontal ground heat exchangers are made by stretching small-diameter polyethylene tubing from the tight coil in which it is shipped into an extended coil that can be placed vertically in a narrow trench or laid flat at the bottom of a wide trench. Recommended trench lengths are much shorter than those of single-pipe horizontal GCHPs, but pipe lengths must be much longer to achieve equivalent thermal performance. When horizontally bored loops are grouted and placed in the deep earth, as shown in Figure 11, design lengths are near those for vertical systems, because annual temperature and moisture content variations approach deep-earth values.

Advantages of horizontal GCHPs are that (1) they are typically less expensive than vertical GCHPs because relatively low-cost installation equipment is widely available, (2) many residential applications have adequate ground area, and (3) trained equipment operators are more widely available. Disadvantages include, in addition to a larger ground area requirement, (1) greater adverse variations in performance because ground temperatures and thermal properties fluctuate with season, rainfall, and burial depth; (2) slightly higher pumping-energy requirements; and (3) lower system efficiencies. OSU (1988a, 1988b) and Svec (1990) cover the design and installation of horizontal GCHPs.

Groundwater Heat Pump Systems

The second subset of GSHPs is groundwater heat pumps (Figure 12). Until the development of GCHPs, they were the most widely used type of GSHP. In the commercial sector, GWHPs can be an attractive alternative because large quantities of water can be delivered from and returned to relatively inexpensive wells that require very little ground area. Whereas the cost per unit capacity of the ground heat exchanger is relatively constant for GCHPs, the cost per unit capacity of a well water system is much lower for a large GWHP system. A single pair of high-volume wells can serve an entire building. Properly designed groundwater loops with correctly developed water wells require no more maintenance than conventional air and water central HVAC. When groundwater is injected back into the aquifer by a second well, net water use is zero.

One widely used design places a central water-to-water heat exchanger between the groundwater and a closed water loop, which is connected to water-to-air heat pumps located in the building. A second possibility is to circulate groundwater through a heat recovery chiller (isolated with a heat exchanger), and to heat and cool the building with a distributed hydronic loop.

Both types and other variations may be suited for direct preconditioning in much of the United States. Groundwater below 60°F can be circulated directly through hydronic coils in series or in parallel with heat pumps. The cool groundwater can displace a large amount of energy that would otherwise have to be generated by mechanical refrigeration.

Advantages of GWHPs under suitable conditions are (1) they cost less than GCHP equipment, (2) the space required for the water well is very compact, (3) water well contractors are widely available, and

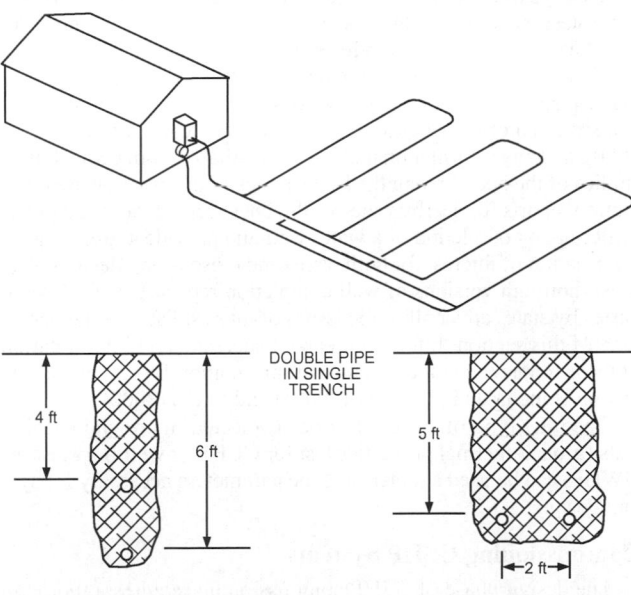

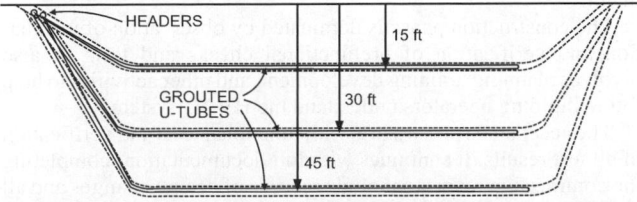

Note: If frost line is greater than 3 ft below grade, average depth of coils should be a minimum of 2 ft below frost line and upper pipe should be a minimum of 1 ft below frost line.

Fig. 11 Trenched Horizontal and Horizontally Bored Ground-Coupled Heat Pump Piping

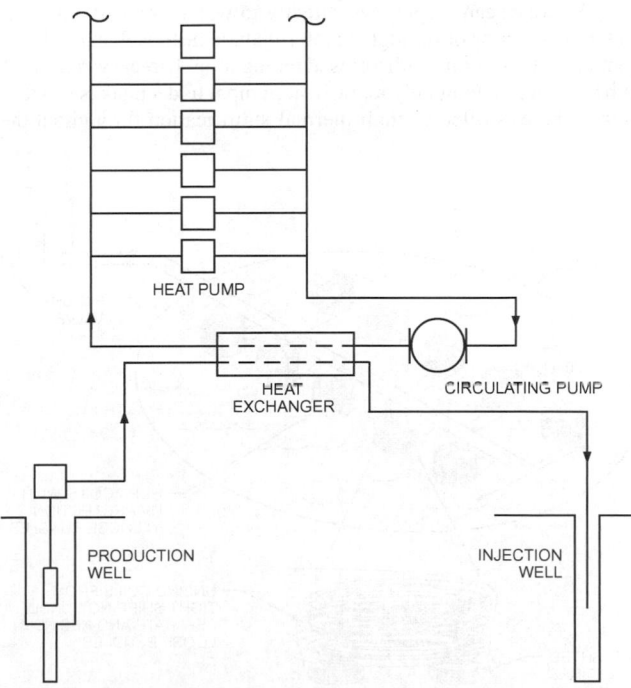

Fig. 12 Unitary Groundwater Heat Pump System

(4) the technology has been used for decades in some of the largest commercial systems.

Disadvantages are that (1) local environmental regulations may be restrictive, (2) water availability may be limited, (3) fouling precautions may be necessary if groundwater is used directly in the heat pumps and water quality is poor, and (4) pumping energy may be high if the system is poorly designed or draws from a deep aquifer.

Surface Water Heat Pump Systems

Surface water heat pumps have been included as a subset of GSHPs because of the similarities in applications and installation methods. SWHPs can be either closed-loop systems similar to GCHPs or open-loop systems similar to GWHPs. However, the thermal characteristics of surface water bodies are quite different than those of the ground or groundwater. Some unique applications are possible, though special precautions may be warranted.

Closed-loop SWHPs (Figure 13) consist of water-to-air or water-to-water heat pumps connected to a piping network placed in a lake, river, or other open body of water. A pump circulates water or a water/antifreeze solution through the heat pump water-to-refrigerant heat exchanger and the submerged piping loop, which transfers heat to or from the body of water. The recommended piping material is thermally fused HDPE tubing with ultraviolet (UV) radiation protection.

Advantages of closed-loop SWHPs are (1) relatively low cost (compared to GCHPs) because of reduced excavation costs, (2) low pumping-energy requirements, (3) low maintenance requirements, and (4) low operating cost. Disadvantages are (1) the possibility of coil damage in public lakes and (2) wide variation in water temperature with outdoor conditions if lakes are small and/or shallow. Such variation in water temperature would cause undesirable variations in efficiency and capacity, though not as severe as with air-source heat pumps.

Open-loop SWHPs can use surface water bodies the way cooling towers are used, but without the need for fan energy or frequent maintenance. In warm climates, lakes can also serve as heat sources during winter heating mode, but in colder climates where water temperatures drop below 45°F, closed-loop systems are the only viable option for heating.

Lake water can be pumped directly to water-to-air or water-to-water heat pumps or through an intermediate heat exchanger that is connected to the units with a closed piping loop. Direct systems tend to be smaller, having only a few heat pumps. In deep lakes (40 ft or more), there is often enough thermal stratification throughout the

year that direct cooling or precooling is possible. Water can be pumped from the bottom of deep lakes through a coil in the return air duct. Total cooling is possible if water is 50°F or below. Precooling is possible with warmer water, which can then be circulated through the heat pump units. Large-scale cooling-only systems have been deployed successfully in some locations, including Cornell University and the city of Toronto [Cornell University 2006; Enwave (no date)].

Site Characterization

Site characteristics influence the type of GSHP system most suitable for a particular location. Site characterization is the evaluation of a site's geology and hydrogeology with respect to its effect on GSHP system design. Important issues include presence or absence of water, depth to water, water (or soil/rock) temperature, depth to rock, rock type, and the nature and thickness of unconsolidated materials overlying the rock. Information about the nature of water resources at the site helps to determine whether an open-loop system may be possible. Depth to water affects pumping energy for an open-loop system and possibly the type of rig used for drilling closed-loop boreholes. Groundwater temperature in most locations is the same as the undisturbed ground temperature. These temperatures are key inputs to the design of GSHP systems. The types of soil and rock allow a preliminary evaluation of the range of thermal conductivity/diffusivity that might be expected. The thickness and nature of the unconsolidated (soil, gravel, sand, clay, etc.) materials overlying the rock influence whether casing is required in the upper portion of boreholes for closed-loop systems, a factor which increases drilling cost.

After the GSHP system type has been decided, specific details about the subsurface materials' (rock/soil) thermal conductivity and diffusivity, water well static and pumping levels, drawdown, etc., are necessary to design the system. Ways to obtain these more detailed data are discussed in other parts of this chapter.

There are many sources for gathering site characterization information, such as geologic and hydrologic maps, state geology and water regulatory agencies, the U.S. Geological Survey (USGS 2000), and any information that may be available from geotechnical studies of the site. Among the best sources of information are completion reports for nearby water wells. These reports are filed by the driller upon completion of a water well and provide a great deal of information of interest for both open- and closed-loop designs. The most thorough versions of well completion reports (level of detail varies by state) cover all of the issues of interest listed at the beginning of this section. Information about access to and interpretation of these reports and other sources of information for site characterization is included in Rafferty (2000a) and Sachs (2002).

Once the type of system has been selected, more site-specific tests (ground thermal properties test for GCHP or well flow test for GWHP) can be used to determine the parameters necessary for system design.

Commissioning GSHP Systems

The design phase of GSHP commissioning requires a thorough site survey and characterization, accurate load modeling, and ensuring that the design chosen (and its documentation) meets the design intent.

The construction phase is dominated by observation of installation and verification of prefunctional checks and tests. It also involves planning, training development, and other activities to help future building operators understand the HVAC system.

The acceptance phase starts with functional tests and verification of all test results. It continues with full documentation: completing the commission report to include records of design changes and all as-built plans and documents, and completing the operations and maintenance manual and system manual. Finally, after system testing and balancing is complete, the owner's operating staff are

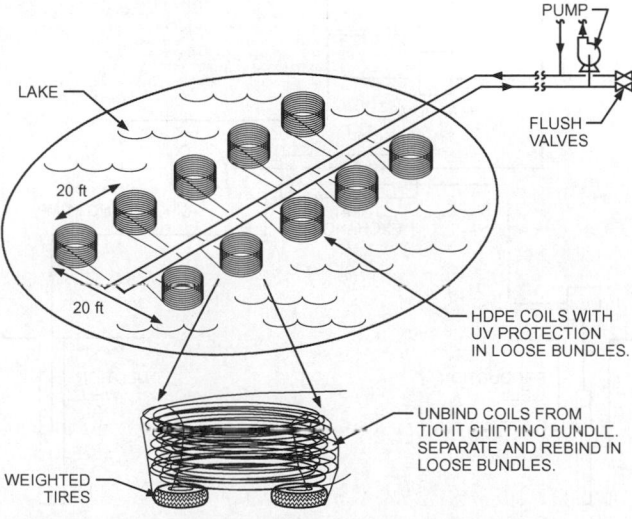

Fig. 13   Lake Loop Piping

PUMP

LAKE

FLUSH VALVES

20 ft

20 ft

HDPE COILS WITH UV PROTECTION IN LOOSE BUNDLES.

UNBIND COILS FROM TIGHT SHIPPING BUNDLE. SEPARATE AND REBIND IN LOOSE BUNDLES.

WEIGHTED TIRES

Table 3 Example of GSHP Commissioning Process for Mechanical Design

System	Function	Performed By	Witnessed By
Heat pump piping	Pressure test, clean, and fill	Contractor	A/E
Ground source piping	Pressure test, clean, fill, and purge air	Contractor Contractor	A/E
Pumps	Inspect, test, and start up	Contractor	—
Heat recovery unit	Inspect, test, and start up; provide clean set of filters, staff instruction	Manufacturer Contractor Manufacturer	CA — CA/owner
Heat pump units	Inspect, test, and start up; provide clean filters, staff instruction	Manufacturer Contractor Manufacturer	— — CA/owner
Chemical treatment	Flushing and cleaning, chemical treatment, staff instruction	Contractor Contractor/ manufacturer Manufacturer	A/E and CA — CA/owner
Balancing	Balancing, spot checking, follow-up site visits	TAB contractor TAB contractor TAB contractor	— A/E and CA CA
Controls	Installation/commissioning, staff instruction, performance testing, seasonal testing	Contractor CA CA CA	— CA/owner — —

Source: Caneta (2001).
A/E = Architect/engineer
CA = Commissioning authority
TAB = Testing, adjusting, and balancing

trained. The acceptance phase ends when "substantial completion" is reached. The warranty period begins from this date.

Table 3 provides information on tasks and participants involved in the GSHP commissioning process. Additional details on this topic, along with preventive maintenance and troubleshooting information, are included in Caneta (2001).

GROUND-COUPLED HEAT PUMPS

Vertical Design

This section provides an overview of a suggested design procedure; related information and equations are discussed in more detail in Kavanaugh and Rafferty (1997). Several public software programs are available for performing the repetitive computations necessary for system optimization. Shonder et al. (1999, 2000) tested the accuracy of these programs and found that only one program matched the results of field-measured data in the initial test. However, closer agreement was attained with several programs in subsequent evaluations.

A more recent publication (Kavanaugh 2008) updates the design recommendations for GCHP systems:

1. Calculate peak zone cooling and heating loads, and estimate off-peak loads.
2. Estimate annual heat rejection into and absorption from loop field to account for potential ground temperature change (see Tables 7 and 8).
3. Select preliminary loop operating temperatures and flow rate to begin optimization of first cost and efficiency (selecting temperatures near normal ground temperature results in high efficiencies but larger and more costly ground loops).
4. Correct heat pump performance at rated conditions to actual design conditions (see Table 13).
5. Select heat pumps to meet cooling and heating loads, and locate units to minimize duct cost and fan power and noise.
6. Arrange heat pump into ground loop circuits to minimize system cost, pump energy and electrical demand (see Figures 18 to 20).

7. Conduct site survey to determine ground thermal properties and drilling conditions (see following recommendations).
8. Determine and evaluate possible loop field arrangements that are likely to be optimum for the building and site (bore depth, separation distance, completion methods, annulus grout/fill, and header arrangements); include subheader circuits (typically 5 to 15 U-tubes on each) with isolation valves to allow air and debris flushing of sections of loop field through a set of full-port purge valves.
9. Determine optimum ground heat exchanger dimensions with Equations (2) to (5) or software; one or more alternatives (depth, number of bores, grout/fill material, etc.) that provide equivalent performance may yield more competitive bids.
10. Iterate to determine optimum operating temperatures, flows, loop field arrangement, depth, bores, grout/fill materials, etc.
11. Lay out interior piping and compute head loss through critical path.
12. Select pumps and control method, determine system efficiency, and consider modifying water distribution system if pump demand exceeds 8% of the system total demand or air distribution system if fan demand exceeds 12% of the system total.

Documents necessary to adequately describe a GCHP installation include, as a minimum,

- Heat pump specifications at rated conditions
- Pump(s) specifications, expansion tank size, and air separator
- Fluid specifications: system volume, inhibitors, antifreeze concentration (if required), water quality, etc.
- Design operating conditions: entering and leaving ground loop temperatures, return air temperatures (including wet bulb in cooling), airflow rates, and liquid flow rates
- Pipe header details with ground loop layout, including pipe diameters, spacing, and clearance from building and utilities
- Bore depth and approximate bore diameter
- Piping material specifications, and visual inspection and pressure testing requirements
- Grout/fill specifications: thermal conductivity and acceptable placement methods to eliminate voids
- Purge provisions and flow requirements to ensure removal of air and debris without reinjecting air when switching to adjacent subheader circuits
- Instructions on connecting to building loop(s) and coordinating building and ground loop flushing
- Sequence of operation for controls

In the design of vertical GCHPs, accurate knowledge of soil/rock formation thermal properties is critical. These properties can be estimated in the field by installing a loop of approximately the same size and depth as the heat exchangers planned for the site. Heat is added in a water loop at a constant rate, and data are collected as shown in Figure 14. Inverse methods are applied to find thermal conductivity, diffusivity, and temperature of the formation. These methods are based on the either the line source (Gehlin 1998; Mogensen 1983; Witte et al. 2002), the cylindrical heat source (Ingersoll and Zobel 1954), or a numerical algorithm (Austin et al. 2000; Shonder and Beck 1999; Spitler et al. 1999). More than one of these methods should be applied, when possible, to enhance reported accuracy. Recommended test specifications are as follows (Kavanaugh 2000, 2001):

- Thermal property tests should be performed for 36 to 48 h.
- Heat rate should be 15 to 25 W/ft of bore, which are the expected peak loads on the U-tubes for an actual heat pump system.
- Standard deviation of input power should be less than ±1.5% of the average value and peaks less than ±10% of average, or resulting temperature variation should be less than ±0.5°F from a straight trend line of a log (time) versus average loop temperature.

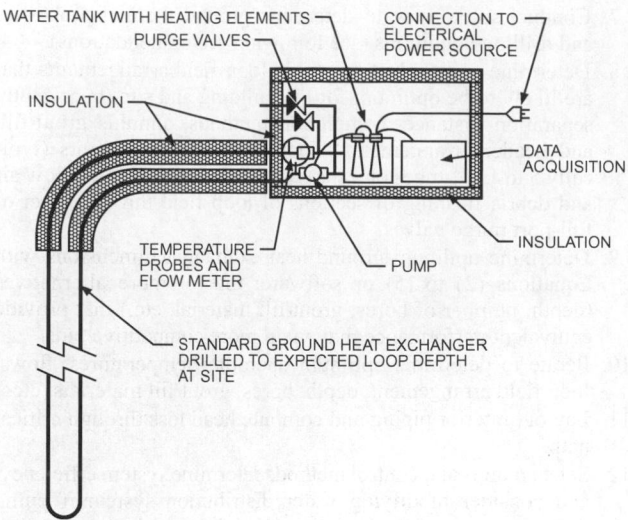

Fig. 14 Thermal Properties Test Apparatus

- Accuracy of temperature measurement and recording devices should be ±0.5°F.
- Combined accuracy of the power transducer and recording device should be ±2% of the reading.
- Flow rates should be sufficient to provide a differential loop temperature of 6 to 12°F. This is the temperature differential for an actual heat pump system.
- A waiting period of five days is suggested for low-conductivity soils ($k < 1.0$ Btu/h·ft·°F) after the ground loop has been installed and grouted (or filled) before the thermal conductivity test is initiated. A delay of three days is recommended for higher-conductivity formations ($k > 1.0$ Btu/h·ft·°F).
- The initial ground temperature measurement should be made at the end of the waiting period by directly inserting a probe inside a liquid-filled ground heat exchanger at three locations, representing the average, or by temperature measurement as liquid exits the loop during the period immediately after start-up.
- Data collection should be at least once every 10 min.
- All aboveground piping should be insulated with a minimum of 0.5 in. closed-cell insulation or equivalent. Test rigs should be enclosed in a sealed cabinet that is insulated with a minimum of 1.0 in. fiberglass insulation or equivalent.
- If retesting a bore is necessary, loop temperature should be allowed to return to within 0.5°F of the pretest initial ground temperature. This typically requires a 10 to 12 day delay in mid- to high-conductivity formations and 14 days in low-conductivity formations if a complete 48 h test has been conducted. Waiting periods can be proportionally reduced if tests were shorter.

The ground-loop design method uses a limited amount of information from commercial systems. A major missing component is long-term, field-monitored data, which are needed to further validate the design method so that the effects of water movement and long-term heat storage are more fully addressed. The conservative designer can assume no benefit from water movement; designers who assume maximum benefit must ignore annual imbalances in heat rejection and absorption.

One design method is based on the solution of the equation for heat transfer from a cylinder buried in the earth. This equation was developed and evaluated by Carslaw and Jaeger (1947) and was suggested by Ingersoll and Zobel (1954) as an appropriate method of sizing ground heat exchangers. Kavanaugh (1985) adjusted the method to account for the U-bend arrangement and hourly heat rate variations. Alternative design methods are described by Eskilson (1987), Morrison (1997), Spitler (2000), and Spitler et al. (2000).

The method of Ingersoll and Zobel (1954) can be used to handle these shorter-term variations. It uses the following steady-state heat transfer equation:

$$q = \frac{L(t_g - t_w)}{R} \qquad (1)$$

where

q = heat transfer rate, Btu/h
L = required bore length, ft
t_g = ground temperature, °F
t_w = liquid temperature, °F
R = effective thermal resistance of ground, ft·h·°F/Btu

The equation is rearranged to solve for the required bore length L. The steady-state equation is modified to represent the variable heat rate of a ground heat exchanger by using a series of constant-heat-rate "pulses." Thermal resistance of the ground per unit length is calculated as a function of time corresponding to the time span over which a particular heat pulse occurs. A term is also included to account for thermal resistance of the pipe wall and interfaces between the pipe and fluid and the pipe and the ground. The resulting equation takes the following form for cooling:

$$L_c = \frac{q_a R_{ga} + (q_{lc} - 3.41 W_c)(R_b + PLF_m R_{gm} + R_{gd} F_{sc})}{t_g - \dfrac{t_{wi} + t_{wo}}{2} - t_p} \qquad (2)$$

The required length for heating is

$$L_h = \frac{q_a R_{ga} + (q_{lh} - 3.41 W_h)(R_b + PLF_m R_{gm} + R_{gd} F_{sc})}{t_g - \dfrac{t_{wi} + t_{wo}}{2} - t_p} \qquad (3)$$

where

F_{sc} = short-circuit heat loss factor
L_c = required bore length for cooling, ft
L_h = required bore length for heating, ft
PLF_m = part-load factor during design month
q_a = net annual average heat transfer to ground, Btu/h
q_{lc} = building design cooling block load, Btu/h
q_{lh} = building design heating block load, Btu/h
R_{ga} = effective thermal resistance of ground (annual pulse), ft·h·°F/Btu
R_{gd} = effective thermal resistance of ground (peak daily pulse: 1 h minimum, 4 to 6 h recommended), ft·h·°F/Btu
R_{gm} = effective thermal resistance of ground (monthly pulse), ft·h·°F/Btu
R_b = thermal resistance of bore, ft·h·°F/Btu
t_g = undisturbed ground temperature, °F
t_p = temperature penalty for interference of adjacent bores, °F
t_{wi} = liquid temperature at heat pump inlet, °F
t_{wo} = liquid temperature at heat pump outlet, °F
W_c = system power input at design cooling load, W
W_h = system power input at design heating load, W

Note: Heat transfer rate, building loads, and temperature penalties are positive for heating and negative for cooling.

Equations (2) and (3) consider three different pulses of heat to account for long-term heat imbalances q_a, average monthly heat rates during the design month, and maximum heat rates for a short-term period during a design day. This period could be as short as 1 h, but a 4 to 6 h block is recommended.

The required bore is the larger of the two lengths L_c and L_h found from Equations (2) and (3). If L_c is larger than L_h, an oversized coil could be beneficial during the heating season. A second option is to install the smaller heating length along with a cooling tower to compensate for the undersized coil. If L_h is larger, the designer should

Table 4 Summary of Potential Completion Methods for Different Geological Regime Types

Geological Regime Type	Grout			Backfill with Cutting	Two-Fill with	
	$0.4 < k \leq 0.8$ Btu/h·ft·°F	$0.8 < k \leq 1.2$ Btu/h·ft·°F	$k > 1.2$ Btu/h·ft·°F		Cuttings Below Aquifers	Other* Below Aquifers
Clay or low-permeability rock,						
no aquifer	—	Yes	Yes	—	Yes	Yes
single-aquifer	—	Yes	Yes	—	—	Yes
multiple-aquifer	Yes	Yes	Yes	Yes	Yes	Yes
Permeable rock,						
no shallow aquifers	—	Yes	Yes	Yes	Yes	Yes
single-aquifer	—	Yes	Yes	Yes	Yes	Yes
multiple-aquifers	—	Yes	Yes	Yes	—	—
Karst terrains with secondary permeability	—	Yes	Yes	Yes	—	—
Fractured terrains with secondary permeability	—	Yes	Yes	Yes	Yes	Yes

*Use of backfill material that has thermal conductivity of $k \geq 1.4$ Btu/h·ft·°F   Yes = Recommended potentially viable backfill methods

install this length, and during cooling mode the efficiency benefits of an oversized ground coil could be used to compensate for the higher first cost.

Selection of the fill material for the borehole is a function of thermal, regulatory, and economic considerations. Historically, a relatively low-thermal-conductivity bentonite grout commonly used in the water well industry and, in some cases, drill cuttings have been used as fill. More recently, thermally enhanced materials have been developed. Nutter et al. (2001) contains a detailed evaluation of potential fills and grouts for vertical boreholes. Table 4 summarizes potential completion methods for various geological conditions. "Two-fill" refers to the practice of placing a low-permeability material in the upper portion of the hole and/or in intervals where it is required to separate individual aquifers, and a more thermally advantageous material in the remaining intervals.

Thermal resistance of the ground is calculated from ground properties, pipe dimensions, and operating periods of the representative heat rate pulses. Table 5 lists typical thermal properties for soils and fills for the annular region of the bore holes. Table 6 gives equivalent thermal resistance of the vertical high-density polyethylene (HDPE) U-tubes for two bore hole diameters d_b. Alternative methods of computing the thermal borehole resistance are presented by Bernier (2006), Hellström (1991), and Remund (1999).

The most difficult parameters to evaluate in Equations (2) and (3) are the equivalent thermal resistances of the ground. The solutions of Carslaw and Jaeger (1947) require that the time of operation, bore diameter, and thermal diffusivity of the ground be related in the dimensionless Fourier number (Fo):

$$Fo = \frac{4\alpha_g \tau}{d_b^2} \qquad (4)$$

where

α_g = thermal diffusivity of the ground, ft²/day
τ = time of operation, days
d_b = bore diameter, ft

The method may be modified to permit calculation of equivalent thermal resistances for varying heat pulses. A system can be modeled by three heat pulses, a 10 year (3650 day) pulse of q_a, a 1 month (30 day) pulse of q_m, and a 6 h (0.25 day) pulse of q_d. Three times are defined as

$$\tau_1 = 3650 \text{ days}$$
$$\tau_2 = 3650 + 30 = 3680 \text{ days}$$
$$\tau_f = 3650 + 30 + 0.25 = 3680.25 \text{ days}$$

The Fourier number is then computed with the following values:

$$Fo_f = 4\alpha\tau_f/d_b^2$$

Table 5 Thermal Properties of Selected Soils, Rocks, and Bore Grouts/Fills

	Dry Density, lb/ft³	Conductivity, Btu/h·ft·°F	Diffusivity, ft²/day
Soils			
Heavy clay, 15% water	120	0.8 to 1.1	0.45 to 0.65
5% water	120	0.6 to 0.8	0.5 to 0.65
Light clay, 15% water	80	0.4 to 0.6	0.35 to 0.5
5% water	80	0.3 to 0.5	0.35 to 0.6
Heavy sand, 15% water	120	1.6 to 2.2	0.9 to 1.2
5% water	120	1.2 to 1.9	1.0 to 1.5
Light sand, 15% water	80	0.6 to 1.2	0.5 to 1.0
5% water	80	0.5 to 1.1	0.6 to 1.3
Rocks			
Granite	165	1.3 to 2.1	0.9 to 1.4
Limestone	150 to 175	1.4 to 2.2	0.9 to 1.4
Sandstone		1.2 to 2.0	0.7 to 1.2
Shale, wet	160 to 170	0.8 to 1.4	0.7 to 0.9
dry		0.6 to 1.2	0.6 to 0.8
Grouts/Backfills			
Bentonite (20 to 30% solids)		0.42 to 0.43	
Neat cement (not recommended)		0.40 to 0.45	
20% bentonite/80% SiO₂ sand		0.85 to 0.95	
15% bentonite/85% SiO₂ sand		1.00 to 1.10	
10% bentonite/90% SiO₂ sand		1.20 to 1.40	
30% concrete/70% SiO₂ sand, s. plasticizer		1.20 to 1.40	

Source: Kavanaugh and Rafferty (1997).

Table 6 Thermal Resistance of Bores R_b for High-Density Polyethylene U-Tube Vertical Ground Heat Exchangers

U-Tube Diameter, in.	Bore Fill Conductivity,* Btu/h·ft·°F					
	4 in. Diameter Bore			6 in. Diameter Bore		
	0.5	1.0	1.5	0.5	1.0	1.5
3/4	0.19	0.09	0.06	0.23	0.11	0.08
1	0.17	0.08	0.06	0.20	0.10	0.07
1 1/4	0.15	0.08	0.05	0.18	0.09	0.06

*Based on DR 11, HDPE tubing with turbulent flow

Corrections for Other Tubes and Flows		
DR 9 Tubing	Re = 4000	Re = 1500
+0.02 Btu/h·ft·°F	+0.008 Btu/h·ft·°F	+0.025 Btu/h·ft·°F

Sources: Kavanaugh (2001) and Remund and Paul (2000).

$$\text{Fo}_1 = 4\alpha(\tau_f - \tau_1)/d_b^2$$

$$\text{Fo}_2 = 4\alpha(\tau_f - \tau_2)/d_b^2$$

An intermediate step in computing the ground's thermal resistance using the methods of Ingersoll and Zobel (1954) is to identify a G-factor, which is then determined from Figure 15 for each Fourier value.

$$R_{ga} = (G_f - G_1)/k_g \qquad (5a)$$

$$R_{gm} = (G_1 - G_2)/k_g \qquad (5b)$$

$$R_{gd} = G_2/k_g \qquad (5c)$$

Ranges of the ground thermal conductivity k_g are given in Table 6. State geological surveys are a good source of soil and rock data. However, geotechnical site surveys are highly recommended to determine load soil, rock types, and drilling conditions.

Performance degrades somewhat because of short-circuiting heat losses between the upward- and downward-flowing legs of a conventional U-bend loop. This degradation can be accounted for by introducing the short-circuit heat loss factor [F_{sc} in Equations (2) and (3)] in the table below. Normally U-tubes are piped in parallel to the supply and return headers. Occasionally, when bore depths are shallow, two or three loops can be piped in series. In these cases, short-circuit heat loss is reduced; thus, the values for F_{sc} are smaller than for a single bore piped in series.

	F_{sc}	
Bores per Loop	2 gpm/ton	3 gpm/ton
1	1.06	1.04
2	1.03	1.02
3	1.02	1.01

Temperature. The remaining terms in Equations (2) and (3) are temperatures. The local deep-ground temperature t_g can best be obtained from local water well logs and geological surveys. A second, less accurate source is a temperature contour map, similar to Figure 16, prepared by state geological surveys. A third source, which can yield ground temperatures within 4°F, is a map with contours, such as Figure 17. Comparison of Figures 16 and 17 indicates the complex variations that would not be accounted for without detailed contour maps.

Selecting the temperature t_{wi} of water entering the unit is critical in the design process. Choosing a value close to ground temperature results in higher system efficiency, but makes the required ground coil length very long and thus unreasonably expensive. Choosing a value far from t_g allows selection of a small, inexpensive ground coil, but the system's heat pumps will have both greatly reduced capacity during heating and high demand when cooling. Selecting t_{wi} to be 20 to 30°F higher than t_g in cooling and 10 to 20°F lower than t_g in heating is a good compromise between first cost and efficiency in many regions of the United States.

A final temperature to consider is the temperature penalty t_p resulting from thermal interferences from adjacent bores. The designer must select a reasonable separation distance to minimize required land area without causing large increases in the required bore length (L_c, L_h). Table 7 presents the temperature penalty for a 10 by 10 vertical grid of bores for various operating conditions after 10 years of operation in a nonporous soil where cooling effects from moisture evaporation or water movement do not mitigate temperature change (Kavanaugh 2003; Kavanaugh and Rafferty 1997). Correction factors are included to find the temperature penalty for four other grid patterns. Note that the higher the number of internal bores, the larger the correction factor.

In the table, adjustments are made to the number of equivalent full-load hours (EFLHs) in cooling and heating that correspond to

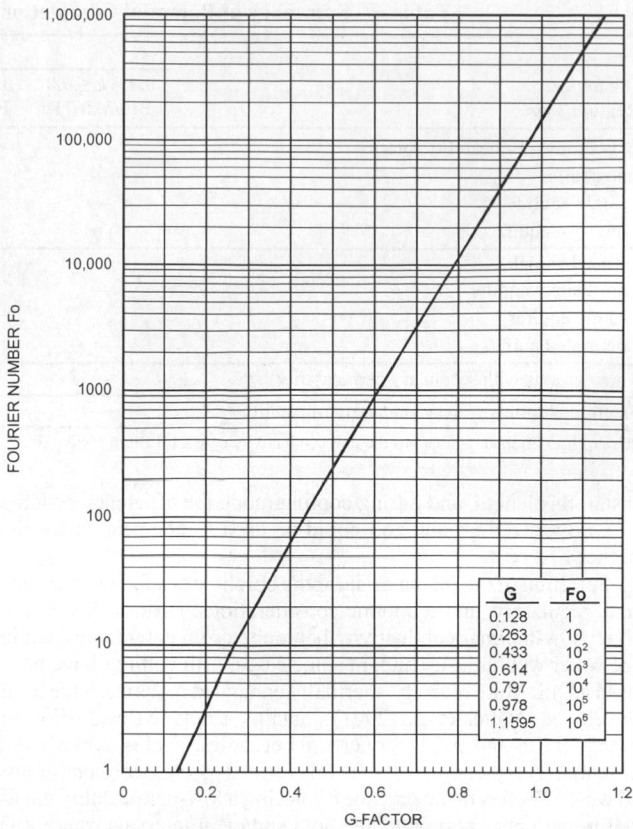

Fig. 15   Fourier/G-Factor Graph for Ground Thermal Resistance
(Kavanaugh and Rafferty 1997)

G	Fo
0.128	1
0.263	10
0.433	10^2
0.614	10^3
0.797	10^4
0.978	10^5
1.1595	10^6

values consistent with the local ground temperature. To mitigate long-term heat build-up for small separation distances, the required heat exchanger length is extended to maintain good system efficiencies. Larger separation distances result in shorter required lengths and smaller temperature changes, because there is greater thermal capacity available and greater area to diffuse heat to the far field.

The table applies only to a limited number of specific cases and is not intended for application to actual designs. It is intended to demonstrate trends for various ground temperatures, hours of operation in heating and cooling, and bore separation distances. Note that values of t_p in the table are significantly different from those obtained using the approach presented by Bernier et al. (2008).

Smaller bore lengths per ton of peak block load result in larger temperature changes; the relationship between bore length and temperature change is inverse and linear.

Values in this table represent worst-case scenarios, and the temperature change is usually mitigated by groundwater recharge (vertical flow), groundwater movement (horizontal flow), and evaporation (and condensation) of water in the soil.

Groundwater movement strongly affects the long-term temperature change in a densely packed ground loop field (Chiasson et al. 2000). A related factor is the evaporative cooling effect experienced with heat addition to the ground. Although thermal conductivity is somewhat reduced with lower moisture content (see Table 5), the net effect is beneficial in porous soils when water movement recharges the ground to original moisture levels. A similar effect may be experienced in cold climates when soil moisture freezes and the heat of solidification mitigates excessive temperature decline. Because these effects have not been thoroughly studied, the design engineer must establish a range of design lengths between one based on minimal groundwater movement, as in very tight clay soils with

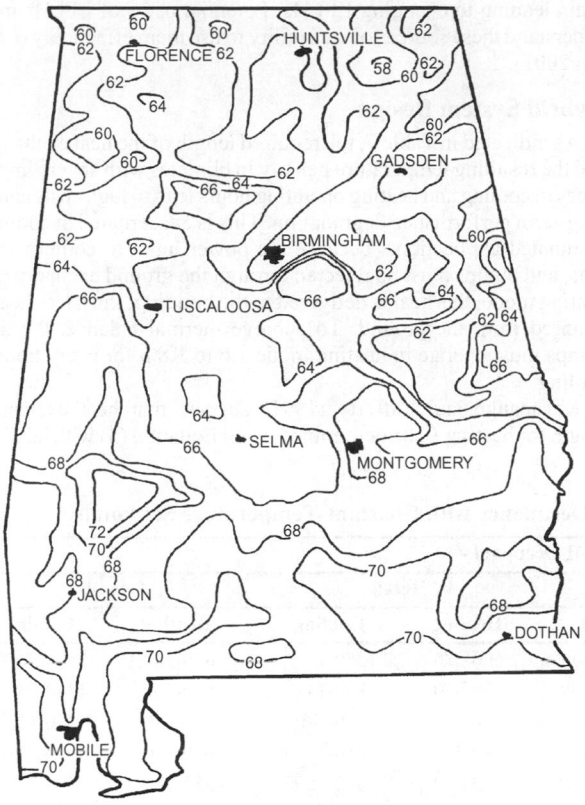

**Fig. 16   Water and Ground Temperatures in Alabama
at 50 to 100 ft Depth**
(Chandler 1987)

**Table 7   Long-Term Temperature Penalty for Worst-Case
Nonporous Formations for 10 × 10 grid and 100 ton Load**

$EFLH_c$, h/yr	$EFLH_h$, h/yr	EER, Btu/W·h	COP	T_g, °F	Bore Separation, ft	Bore Length, ft	$T_{penalty}$, °F
250	1250	17.6	3.6	42	15	230	−1.3
		17.6	3.6		20	221	−0.7
		17.6	3.6		25	217	−0.4
500	1000	16.8	3.7	45	15	218	−1.4
		16.8	3.7		20	210	−0.7
		16.8	3.7		25	206	−0.4
750	750	14.3	4.0	55	15	206	3.4
		14.3	4.0		20	195	1.8
		14.3	4.0		25	190	1.0
1000	500	13.3	4.4	65	15	284	6.9
		13.3	4.4		20	248	3.8
		13.3	4.4		25	231	2.0
1250	250	13.0	4.6	68	15	362	10.0
		13.0	4.6		20	289	5.7
		13.0	4.6		25	256	3.0
0	1500	Not recommended without solar or thermal regeneration					
1500	0	Not recommended without fluid cooler or cooling tower assist					

Note:
k_g = 1.4 Btu/h·ft·°F, k_{grout} = 0.85 Btu/h·ft·°F, rated EER/COP = 20.0/4.2 (GLHP).

Correction Factors for Other Grid Patterns:

1 × 10 grid	2 × 10 grid	5 × 5 grid	20 × 20 grid
C_f = 0.36	C_f = 0.45	C_f = 0.75	C_f = 1.14

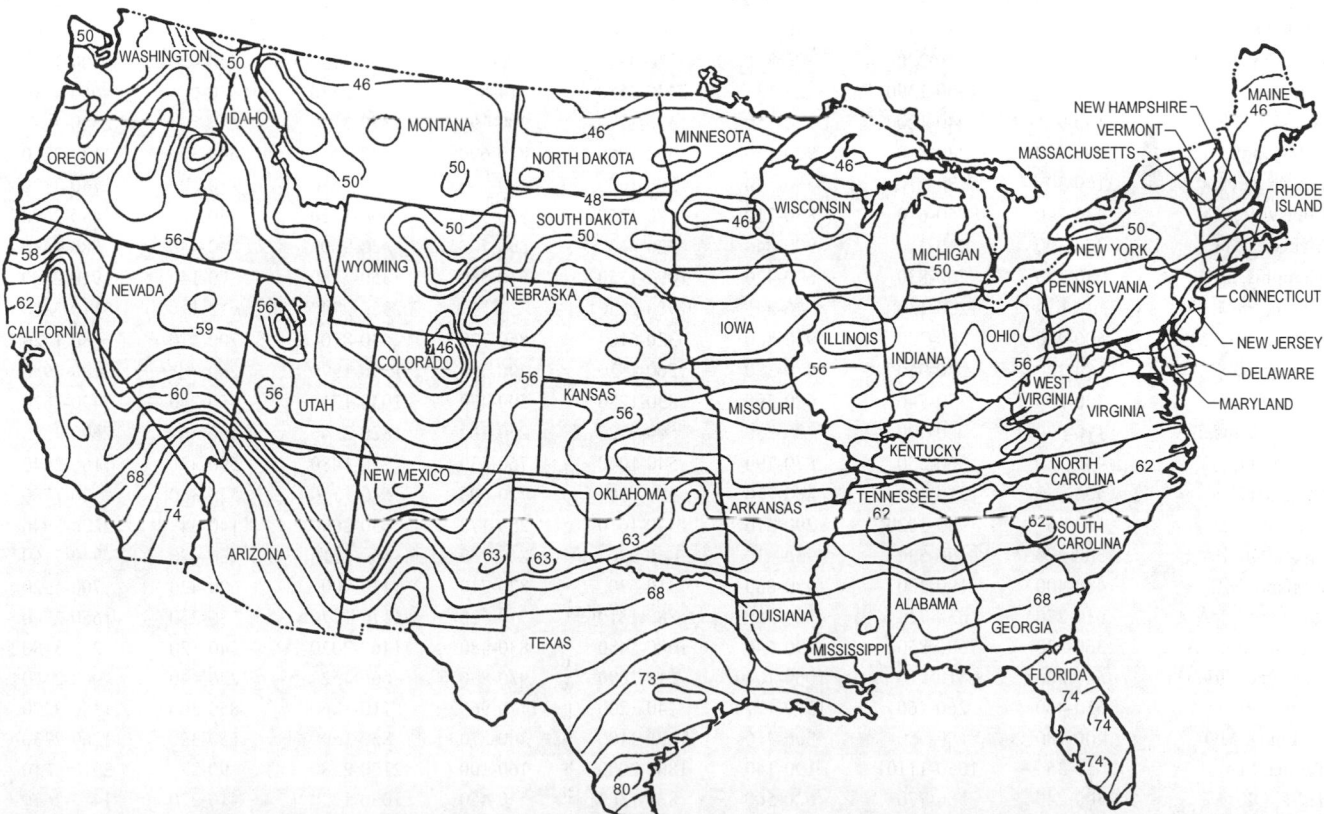

Fig. 17   Approximate Groundwater Temperatures (°F) in the Continental United States

poor percolation rates, and a second based on the higher rates characteristic of porous aquifers.

The long-term temperature change table and a commonly used software for GCHP system design both use EFLH calculations. To the extent that annual loads are proportional to peak loads, the equivalent full-load hours method provides a simple estimate of annual loads from peak loads. The EFLHs in Table 8 provide a quick means to estimate annual loads needed to size ground heat exchangers at the initial feasibility study phase of a project. Because EFLHs vary with changes in both annual loads and peak loads, not all building parameters effects are included in EFLHs. For instance, building operating hours change annual loads by increasing the amount of time that internal gains are at elevated levels, but they do not change the peak load. Occupancy hours can add load without increasing the installed capacity, thereby changing the EFLHs. Furthermore, changes in other parameters, such as internal gains, do not necessarily scale with system capacity in the same proportion as annual load,

again leading to changing EFLHs. Potential users of EFLHs must understand these sources of variability to use them effectively (Carlson 2001).

Hybrid System Design

As indicated in Table 7, the required length of the heat exchanger and the resulting temperature penalty in climates with severe imbalances in cooling and heating operating hours lead to high costs and/or long-term performance degradation. This is exacerbated in cooling-dominated applications, because the power input to compressors, fans, and pumps must be rejected through the ground exchanger. In heating mode, this heat is delivered to the building; thus, less heat is required from the ground. To achieve thermal balance, the heat pumps must operate in heating mode 1.6 to 1.8 h for every hour in cooling.

Kavanaugh and Rafferty (1997) suggest that heat exchanger length for heating L_h be determined using Equation (3) with heating-

Table 8   Equivalent Full-Load Hours (EFLH) for Typical Occupancy with Constant-Temperature Set Points

	EFLH Occupancy							
	School		Office		Retail		Hospital	
Location	Heating	Cooling	Heating	Cooling	Heating	Cooling	Heating	Cooling
Atlanta, GA	290-200	690-830	690-480	1080-1360	600-380	1380-1860	430-160	2010-2850
Baltimore, MD	460-320	500-610	890-720	690-1080	770-570	880-1480	590-300	1340-2340
Bismarck, ND	500-460	150-250	990-950	250-540	900-810	340-780	730-530	540-1290
Boston, MA	520-450	300-510	1000-960	450-970	870-760	610-1380	680-420	1020-2330
Charleston, WV	440-310	430-570	840-770	620-1140	730-620	820-1600	550-320	1260-2560
Charlotte, NC	320-200	650-730	780-530	1060-1340	670-420	1350-1830	490-180	1990-2820
Chicago, IL	470-390	280-410	920-820	420-780	810-670	550-1090	640-400	870-1780
Dallas, TX	200-120	830-890	520-340	1350-1580	440-280	1660-2090	310-100	2320-3100
Detroit, MI	480-400	230-360	1020-970	390-820	900-790	530-1170	710-460	870-1950
Fairbanks, AK	630-560	26-54	1170-1050	64-200	1090-930	110-320	930-690	210-600
Great Falls, MT	430-360	130-220	890-820	210-490	800-680	290-710	640-420	500-1210
Hilo, HI	1-0	1360-1390	23-13	2440-2580	14-8	2990-3370	0-0	4060-4910
Houston, TX	130-90	940-1000	350-250	1550-1770	300-190	1870-2290	200-70	2540-3320
Indianapolis, IN	480-400	380-560	920-840	560-1000	820-690	730-1410	640-390	1120-2250
Los Angeles, CA	160-80	780-910	580-370	1280-1670	440-250	1740-2350	180-20	2740-3770
Louisville, KY	430-290	550-670	830-710	770-1250	720-570	1000-1720	550-300	1480-2690
Madison, WI	470-390	210-310	900-840	320-640	800-700	420-900	640-440	680-1490
Memphis, TN	240-170	700-830	600-420	1090-1350	510-330	1350-1780	370-140	1910-2680
Miami, FL	12-6	1260-1300	46-34	1980-2150	37-25	2350-2740	12-1	3110-3890
Minneapolis, MN	500-420	200-300	950-860	320-610	860-720	430-870	700-470	680-1420
Montgomery, AL	180-120	840-910	470-330	1260-1510	400-250	1550-1990	260-90	2170-2950
Nashville, TN	320-250	570-740	680-590	830-1280	590-470	1030-1710	450-240	1490-2620
New Orleans, LA	110-67	920-990	320-230	1500-1720	260-160	1820-2240	160-46	2500-3280
New York, NY	440-350	360-550	870-790	540-1040	760-630	720-1480	590-330	1160-2440
Omaha, NE	400-330	310-440	800-720	480-820	720-600	610-1130	570-360	920-1780
Phoenix, AZ	110-65	950-1020	290-210	1340-1610	250-170	1630-2090	140-34	2220-3040
Pittsburgh, PA	500-470	300-530	950-910	440-920	840-750	600-1310	650-420	960-2160
Portland, ME	480-400	190-300	980-880	310-630	870-710	410-900	690-420	700-1520
Richmond, VA	410-270	630-730	820-660	880-1310	710-520	1110-1770	530-250	1650-2760
Sacramento, CA	360-220	680-850	990-640	1080-1430	830-480	1460-2020	540-120	2250-3180
Salt Lake City, UT	540-520	410-710	1060-1040	510-1090	930-830	660-1520	720-440	1060-2470
Seattle, WA	650-460	260-460	1370-1270	440-1200	1170-960	710-1860	850-360	1340-3270
St. Louis, MO	400-280	460-550	800-710	680-1100	700-570	850-1500	550-320	1260-2330
Tampa, FL	58-35	1050-1110	190-140	1800-2000	160-100	2170-2580	90-22	2910-3710
Tulsa, OK	300-240	580-770	620-560	830-1300	540-450	1030-1730	410-220	1470-2630

Notes: 1. The ranges in values are from internal gains at 0.6 and 2.5 W/ft². 2. Operating with large temperature setbacks during unoccupied periods (effectively turning off the system) reduces heating EFLHs by 20% and cooling EFLHs by 5%.

Equations relating EFLH to Heating and Cooling Degree Days allowing calculation of EFLH for locations other than those listed here can be found in Carlson (2001).

mode loop temperatures t_{wi} and t_{wo} as low as possible to minimize L_c. A fluid cooler or cooling tower with an isolation heat exchanger is sized to meet the capacity difference between the required cooling length L_c from Equation (2) and the heating length L_h:

$$q_{cooler} \text{ (tons)} = (L_c - L_h)/L_c \text{ per ton} \tag{6}$$

This strategy does not completely eliminate long-term ground temperature change. However, the fluid cooler or cooling tower can operate additional hours during off-peak periods to ensure no temperature change occurs. It is suggested that operation during periods of lower outdoor-air wet-bulb temperature substantially reduces the operating hours and energy required to achieve this balance.

A more detailed study (Hackel et al. 2009) included assumptions about typical installation and operating costs to demonstrate an optimized design strategy. A variety of cases were considered, and the simplified best-fit regression for the hybrid ground heat exchanger length L_{hyb} was found to be

$$L_{hyb} = 254 \text{ ft·h·°F/kBtu} \times q_h/(t_g - t_{wo}) \tag{7}$$

Hackel et al. (2009) explored the balance between theoretical installation and operating costs to find an optimized hybrid design strategy. Various cooling-dominated scenarios were considered, and the simplified best-fit regression for a theoretically optimal hybrid ground heat exchanger length L_{hyb} was found to be proportional to heating load:

$$L_{tot} = C_1 \frac{\dot{q}_{peak,heat}}{\Delta t_{ground}}$$

where $C_1 = 254$ ft·h·°F/kBtu, at $k = 1.4$ Btu/ft·h·°F. For other ground conductivities, the change in ground heat exchanger size is approximately inversely proportional to the change in conductivity. This basic strategy of sizing hybrids was found to be valid for a wide range of economic scenarios. Results of this study suggest that in most cases it is economically optimal to include antifreeze in the system and allow the entering fluid temperature to drop to 35°F or lower. If a cooling tower is used, its optimal size can be found according to the following equation. The tower should be controlled according to the Δt between the fluid temperature and the ambient (wet or dry bulb, depending on whether a tower or cooler is used).

$$C_{CCCT} = 1.3 \dot{q}_{unmet,cool} = 1.3 \left(\dot{q}_{peak,cool} - \frac{\dfrac{q_{GHX,cool}}{L_{tot}}}{C_1(t_{ground} - t_o)} \right)$$

where $C_1 = 4.72$ ft/ton·°F and $t_o = 0$°F.

Heating-Dominated Hybrids. Heating-dominated hybrids are needed in only the most severe cold climates, partially because of the constant heat creation of compressors. Most commercial buildings throughout the United States are cooling-dominated and would not benefit from supplemental heat, especially when ventilation heat recovery is used to temper cold winter air. In severely heating-dominated climates (i.e., approximately twice as much annual heating load as cooling load), however, a heating-dominated hybrid could decrease the size of the ground heat exchanger and balance the ground load. In an optimal system, the ground heat exchanger is sized to meet the cooling load and a supplemental device meets the remaining heating load.

A robust, efficient solution to a heating hybrid uses the heat sources to heat air or fluid directly (because boilers, electricity, and solar all provide high enough temperatures for this) and avoid operating the heat pump unnecessarily. Delivering this supplemental heat by coils separate from the geothermal fluid system increases system efficiency and eliminates the problems posed in the indirect approach. The separate heat source should be delivered through a

smaller, dedicated subsystem such as dedicated ventilation preheat, baseboard perimeter heating, or unit heaters. If the heat is not expected to operate often, another effective option is electric duct heaters placed at the heat pumps; in this case, ensure that controls do not leave this auxiliary heat on for extended periods.

In some situations, it may only be cost-effective to provide indirect heat through a supplemental heat coil on the geothermal fluid loop. Coil energy could be supplied by gas boiler, solar panels, electric resistance, or another source. However, there are several drawbacks to this approach:

- The heat coil's high potential temperatures are not covered by typical ground loop warranties and pose a danger to the loop.
- If controls are not maintained correctly, the boiler may dump heat to the ground, and only a fraction of that rejected heat would be recovered.
- The heat pumps operate whenever heating is needed, even though a high-temperature coil is also operating. This, coupled with the need for a heat exchanger in most of these systems, adds significant and unnecessary energy consumption to the system.

Pump and Piping System Options

Loop design can have a substantial effect on both pumping power requirements and system installed cost. A GSHP survey (Caneta 1995) reported that installed pumping power varied from 0.04 to 0.21 hp/ton of heat pump power. This represents 4 to 21% of the total demand of typical GSHP systems and up to 50% of the total energy for some pump control schemes. Table 9 gives a recommended set of guidelines for minimizing the power of closed-loop GSHPs and maximizing system efficiency.

Good Table 9 grades can be obtained by minimizing extensive piping arrangements with long interior and exterior piping runs, high-head-loss fittings, valves, and control devices. Designers must compare the costs and advantages of large central piping loops and larger pumps with those of multiple smaller loops and smaller pumps. Pumping rates greater than 3.0 gpm/ton in closed-loop systems result in marginal equipment capacity gains in modern water-to-air heat pumps, and typically decrease overall system efficiency.

Kavanaugh et al. (2002) found the total cost of vertical GCHP ground-loop systems (including headers) ranged from $3.00 to $16.60 per foot of vertical bore. The cost of headers is a significant portion of the total and in many cases exceeded the cost of the vertical bore. The savings in vertical loop costs because of central systems' load diversity often is not warranted because of the increased cost of large-diameter piping networks connecting equipment inside the building and below-grade circuits connecting exterior ground loops.

In low-rise buildings with large footprints, such as a school, multiple unitary loop systems (Figure 18) are recommended to offset the high cost of central interior piping and ground-loop header costs. Although the total length of vertical bore for unitary systems is greater than for central loop systems, the high cost of interior piping, exterior headers, and valve vaults often offsets the bore cost savings. Additionally, pump demand is substantially reduced in the unitary system because of the short header runs, so low-wattage on/off circulator pumps are suggested.

Table 9 Guidelines for Pump Power for GSHP Ground Loops

Installed Pump Power		Head at 3 gpm/ton,
hp$_{pump}$/100 tons	Grade	ft of water
< 5	A	< 46
5 to 7.5	B	46 to 69
7.5 to 10	C	69 to 92
10 to 15	D	92 to 138
> 15	F	> 138

Source: Kavanaugh and Rafferty (1997).

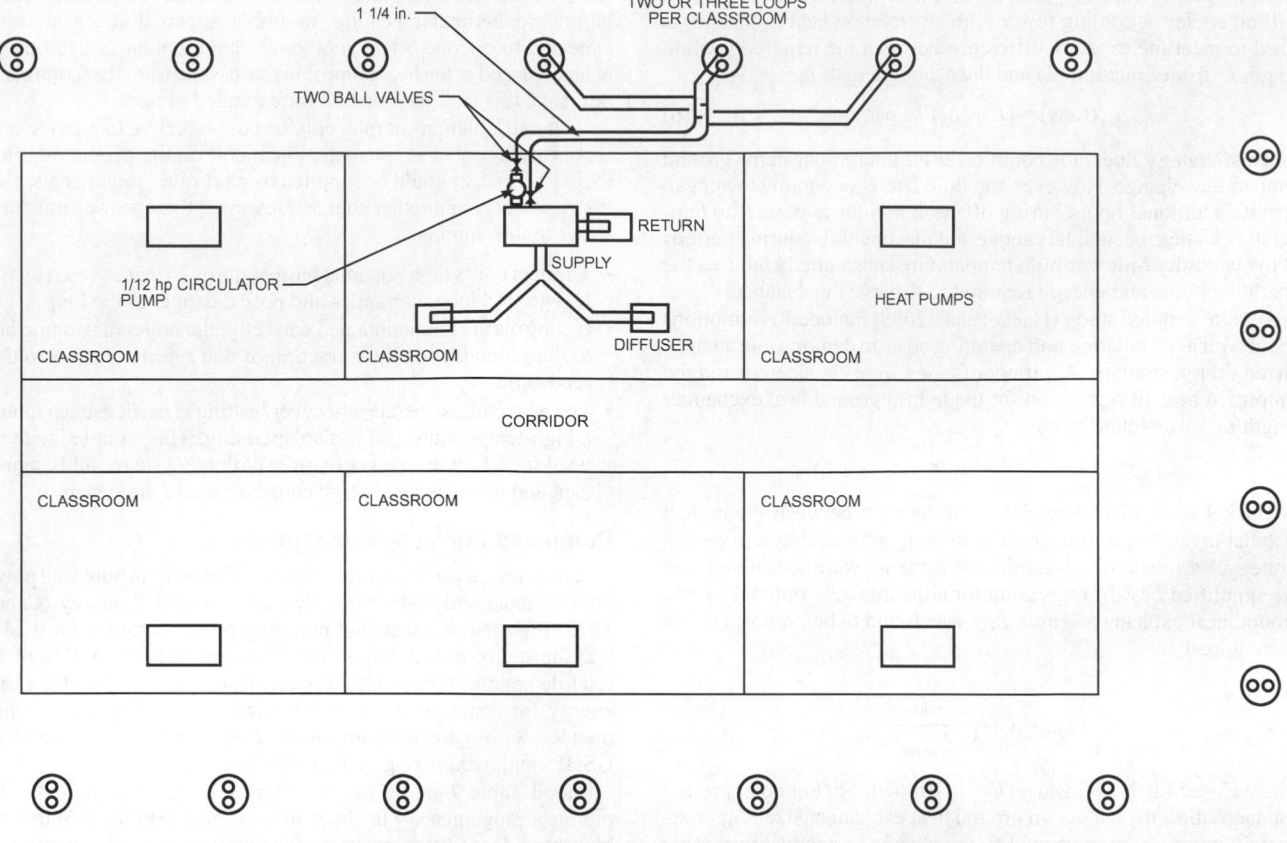

Fig. 18 Unitary GCHP Loops with On/Off Circulator Pumps

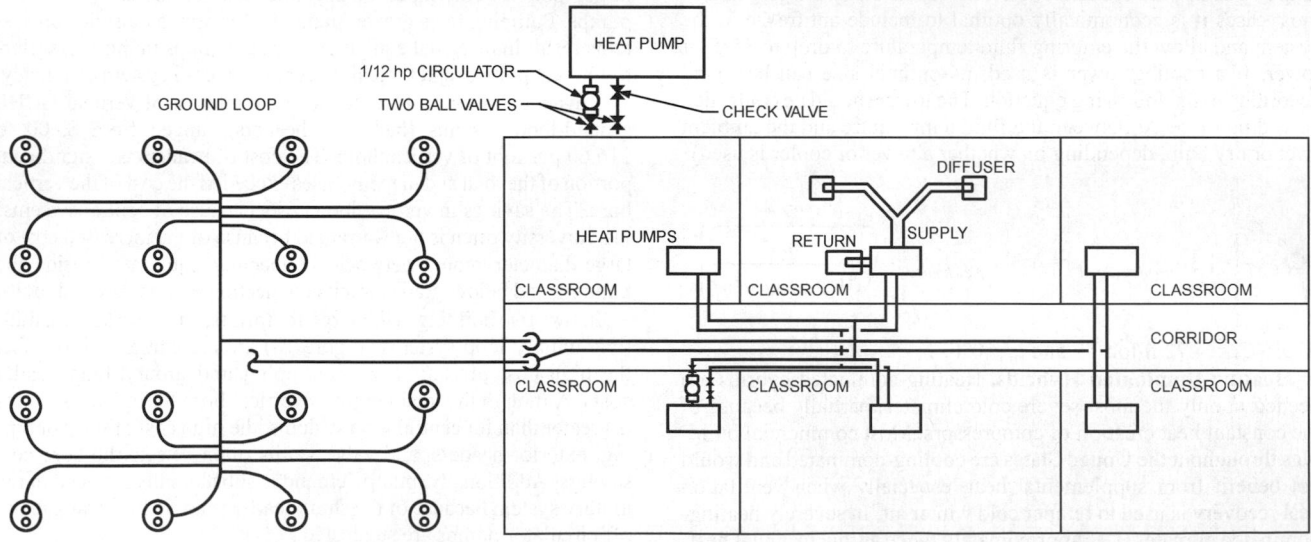

Fig. 19 Subcentral GCHP Loop with On/Off Circulator Pumps

A compromise in applications with significant load diversity is to group ground loops into multiple smaller subcentral loops in different areas of the building (Figure 19). Subcentral loops can be served by on/off circulator pumps located on each heat pump if a check valve is installed on each heat pump to prevent reverse water circulation through idle units.

Figure 20 is an example of a central loop that can effectively reduce the cost of the required vertical bore in buildings with higher load diversities. The central ground loop consists of several subheader sets, each having 6 to 20 vertical U-tube heat exchangers. The subheaders are gathered into a valve manifold located either near the center of the loop field in a below-grade vault or in the building equipment room. Each subheader set has isolation valves for independent purging of air and debris. Interior piping is similar to conventional water-source heat pump systems in which interior piping is routed to individual water-to-air heat pumps in each zone and/or heat pump water heaters and water-to-water heat pumps.

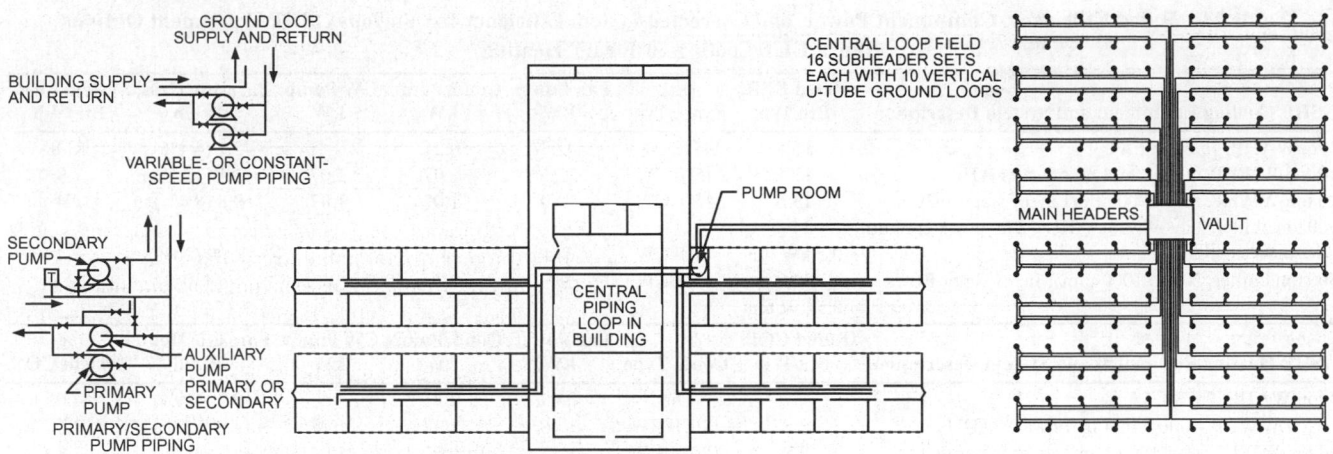

Fig. 20   Central Loop GCHP

Variable-speed drives (VSDs) are recommended for central loop systems because they offer substantial energy savings compared to primary/secondary pump schemes in GSHP applications. However, in buildings with primary occupancy of less than 60 h per week, measures should be incorporated to turn off the main VSD pump and provide some alternative means of pumping water to critical building zones during low-occupancy or unoccupied periods.

Table 10 is a summary of Tables 9 and 10 from Chapter 32 of the 2007 *ASHRAE Handbook—HVAC Applications*, which compared a central GCHP loop and a set of eight subcentral loops for a 40,000 ft² four-story office with a 100 ton cooling load. The total cost of the single central loop was higher ($2114 per ton, versus $1956 per ton). Although the vertical portion of the subcentral loop cost slightly more ($1343 per ton, versus $1327 per ton), the total system cost was projected to be lower because of the lower header cost.

Table 10 also compares a central loop with multiple unitary loops for a 57,000 ft² elementary school. The projected 17% reduction in total cost ($2099 per ton versus $2530 per ton) of the unitary system is more significant compared to the reduction in the four-story office because of the building's layout. The single-story, large-footprint design increases the cost of a central loop because of the long interior piping runs, even though the vertical portion of the central loop system is smaller. The unitary loop system benefits because the large footprint provides ample space for loops to be located near heat pump units.

Effect of GSHP Equipment Selection on Heat Exchanger Design

The ground heat exchanger must absorb the heat of compression and heat from auxiliary equipment (e.g., fans, pumps) in cooling mode. In heating mode, heat from auxiliary equipment reduces the amount of heat required from the ground. Therefore, the cooling- and heating-mode power values W_c and W_h in Equations (2) and (3) must include the auxiliary input.

Rated values published in compliance with ANSI/ARI/ASHRAE/ISO *Standard* 13256-1 do not include the auxiliary power required to circulate air and water through the distribution systems. Furthermore, the auxiliary power required to distribute chilled air (and water) can have a substantial negative effect on the equipment's cooling capacity.

Table 11 summarizes the air and water temperatures used to generate the rated performance of water-to-air heat pumps.

Table 12 summarizes the conditions for ANSI/ARI/ASHRAE/ISO *Standard* 13256-2, the rating standard for water-tower heat pumps.

Table 10   GCHP Piping Cost Comparison for Two Sample Buildings

	Four-Story Office Building, 40,000 ft²		Elementary School, 57,000 ft²	
	Central Loop	Sub-Central Loops (Eight)	Central Loop	Unitary Loops
Vertical loop cost per ft	$6.63	$6.63	$6.63	$6.63
Header and vertical loop cost per ft	$10.57	$9.67	$12.91	$10.71
Vertical loop cost per ton	$1327	$1343	$1300	$1430
Header and vertical loop cost per ton	$2114	$1956	$2530	$2099

Table 11   Rating Conditions for Water-to-Air Heat Pumps for Total Cooling (TC, Btu/h), Energy Efficiency Ratio (EER, Btu/W·h), Heating Capacity (HC, Btu/h) and Coefficient of Performance (COP, W/W)

Entering Liquid and Air	WLHP Water Loop	GWHP Ground-water	GLHP Ground Loop	GLHP-PL (Part-Load)
ELT (sink, cooling)	86°F	59°F	77°F	68°F
ELT (source, heating)	68°F	50°F	32°F	41°F
EAT (db/wb, cooling)	80.6/66.2°F	80.6/66.2°F	80.6/66.2°F	80.6/66.2°F
EAT (heating)	68°F	68°F	68°F	68°F

Source: ANSI/ARI/ASHRAE/ISO *Standard* 13256-1.
Required fan power to overcome external static pressure (ESP) and pump power to circulate liquid for piping loop not included in calculation of TC, EER, HC, and COP.

Table 12   Rating Conditions for Water-to-Water Heat Pumps for Total Cooling (TC, Btu/h), Energy Efficiency Ratio (EER, Btu/W·h), Heating Capacity (HC, Btu/h) and Coefficient of Performance (COP, W/W)

Entering Liquid and Air	WLHP (Water Loop)	GWHP (Ground-water)	GLHP (Ground Loop)	GLHP-PL (Part-Load)
ELT (sink)	86°F	59°F	77°F	68°F
ELT (source)	68°F	50°F	32°F	41°F
ELT (building)	53.6°F	53.6°F	53.6°F	53.6°F
ELT (building)	104°F	104°F	104°F	104°F

Source: ANSI/ARI/ASHRAE/ISO *Standard* 13256-2.
Pump power to circulate liquid for source/sink and building loops not included in calculation of TC, EER, HC, and COP.

Table 13 Rated Efficiency, Component Power, and Corrected System Efficiency for Various GSHP Equipment Options
(86°F ELT Cooling/50°F ELT Heating)

GSHP Cooling Equipment and System Description	Rated EER, Btu/Wh	Evap. Type	Fan Power, kW	Cond. Pump, kW	CW Pump, kW	Parasitic Heat, kBtu/h	System EER, Btu/Wh
4 ton WAHP, 75/63°F EAT	16.5	45°F DX	0.63	0.21	—	−2.1 (4.4%)	13.8
10 ton WWHP, 4000 cfm/4 in. of water AHU	13.6	45°F CW	3.36	1.07	1.07	−15.1 (12.6%)	7.9
10 ton WWHP, four 1000 cfm/1 in. of water FCUs	13.6	45°F CW	1.80	1.07	1.07	−9.8 (8.0%)	9.4
500 ton chiller, 4 in. of water AHUs, 2 in. of water return fans, series FPVAV	24.0 0.5 kW/ton	45°F CW	314	27	36	−1190 (20%)	7.4
500 ton chiller, 200 to 1000 cfm/1 in. of water FCUs	24.0 0.5 kW/ton	45°F CW	90	27	36	−430 (7.2%)	12.8

GSHP Heating Equipment and System Description	Rated COP, Btu/Wh	Cond. Type	Fan Power, kW	Cond. Pump, kW	CW Pump, kW	Parasitic Heat, kBtu/h	System COP
4 ton WAHP, 70°F EAT	4.7	Dir.	0.63	0.21	—	+2.1 (4.4%)	4.0
10 ton WWHP, 4000 cfm/4 in. of water AHU	3.8	120°F HW	3.36	1.07	1.07	+14.4 (12.0%)	2.5
10 ton WWHP, four 1000 cfm/1 in. of water FCUs	3.8	120°F HW	1.57	1.07	1.07	+8.3 (6.9%)	2.8
10 ton WWHP, 4000 cfm/4 in. of water AHU	3.8	100°F HW	3.36	1.07	1.07	+14.4 (12.0%)	3.1
10 ton WWHP, in-floor heat	3.8	100°F HW	0	1.07	1.07	+3.7 (3.1%)	3.7

Actual heat pump performance can be substantially different from rated conditions. Designers must convert rated performance to design conditions by accounting for the effect of auxiliary power input and for design ELTs and EWTs. When water-to-water heat pumps or chillers are used in this application, corrections should include power for the pump(s) of the source/sink loop and the chilled-/hot-water loop, and power for fans in the air distribution system. Table 13 demonstrates the difference between rated GSHP efficiency and actual system efficiencies for various options when the effect of auxiliary components is considered. Note that using a high-static-pressure air handler for air distribution significantly reduces cooling efficiency. In heating, 120°F hot water also lowers heating COP compared to direct condensation and hydronic systems (e.g., in-floor heating) that use lower-temperature water.

Horizontal and Small Vertical System Design

The buried pipe of a closed-loop GSHP may theoretically produce a change in temperature in the ground up to 16 ft away. For all practical purposes, however, the ground temperature is essentially unchanged beyond about 3 ft from the pipe loop. For that reason, the pipe can be buried relatively near the ground surface and still benefit from the moderating temperatures that the earth provides. Because the ground temperature may fluctuate as much as ±10°F at a depth of 6 ft, an antifreeze solution must be used in most heating-dominated regions. The critical design aspect of horizontal applications is to have enough buried pipe loop in the available land area to serve the equipment. The design guidelines for residential horizontal loop installations can be found in OSU (1988b).

A horizontal loop design has several advantages over a vertical loop design for a closed-loop ground source system:

- Ground-loop installation is usually less expensive than for vertical well designs because the capital cost of a backhoe or trencher is only a fraction of the cost of most drilling rigs.
- Most ground-source contractors must have a backhoe for construction of the header pits and trenches to the building, so they can perform the entire job without the need to schedule a special contractor.
- Large drilling rigs may not be able to get to all locations because of their size and weight.
- There is usually no potential for aquifer contamination because of the shallow depth of the trench.
- There is minimal residual temperature effect from unbalanced annual loads on the ground loop because the heat transfer to or from the ground loop is small compared to the normal heat transfer occurring at the ground surface.

Some limitations on selecting a horizontal loop design include the following:

- The minimum land area needed for most nonspiral horizontal loop designs for an average house is about 0.5 acre. Horizontal systems are not feasible for most urban houses, which are commonly built on smaller lots.
- The larger length of pipe buried relatively near the surface is more susceptible to being cut during excavations for other utilities.
- Soil moisture content must be properly accounted for in computing the required ground loop length, especially in sandy soils or on hilltops that may dry out in summer.
- Rocks and other obstructions near the surface may make excavation with a backhoe or trencher impractical.
- Multiple pipes are often placed in a single trench to reduce the land area needed for horizontal loop applications. Some common multiple-pipe arrangements are shown in Figure 21. When pipes are placed at two depths, the bottom row is placed first, and then the trench is partially backfilled before the upper row is put in place. Rarely are more than two layers of pipe used in a single trench because of the extra time needed for the partial backfilling. Higher pipe densities in the trench provide diminishing returns because thermal interference between multiple pipes reduces the heat transfer effectiveness of each pipe. The most common multiple-pipe applications are the two-pipe arrangement used with chain trenchers and the four- or six-pipe arrangements placed in trenches made with a wide backhoe bucket.

An overlapping spiral configuration, shown in Figure 22, has also been used with some success. However, it requires special attention during backfilling to ensure that soil fills all the pockets formed by the overlapping pipe. Large quantities of water must be added to compact the soil around the overlapping pipes. The backfilling must be performed in stages to guarantee complete filling around the pipes and good soil contact. The high pipe density (up to 10 ft of pipe per linear foot of trench) may cause problems in prolonged extreme weather conditions, either from soil drying during cooling or from freezing during heating. This spiral design has been used in vertical trenches cut with a chain trencher as well as in laying the coil flat on the bottom of a large pit excavated with a bulldozer. Installations using the horizontal spiral coil on the bottom of a pit have generally performed better than those with spiral coils that were stood upright in a vertical trench.

The extra time needed to backfill and the extra pipe length required make spiral configurations nearly as expensive to install as straight pipe configurations. However, the reduced land area needed for the more compact design may allow use on smaller residential

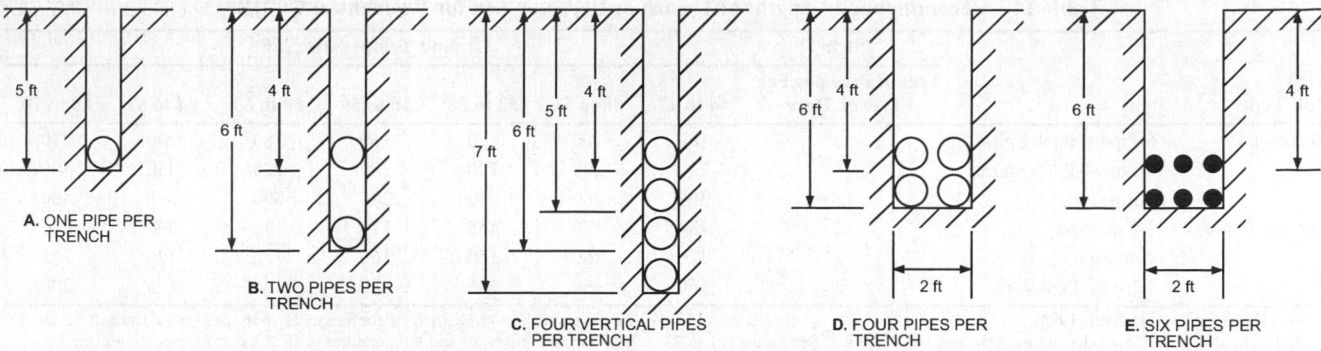

A. ONE PIPE PER TRENCH

B. TWO PIPES PER TRENCH

C. FOUR VERTICAL PIPES PER TRENCH

D. FOUR PIPES PER TRENCH

E. SIX PIPES PER TRENCH

Note: If frost line is greater than 3 ft below grade, average depth of coils should be a minimum of 2 ft below frost line and upper pipe should be a minimum of 1 ft below frost line.

Fig. 21   Horizontal Ground Loop Configurations

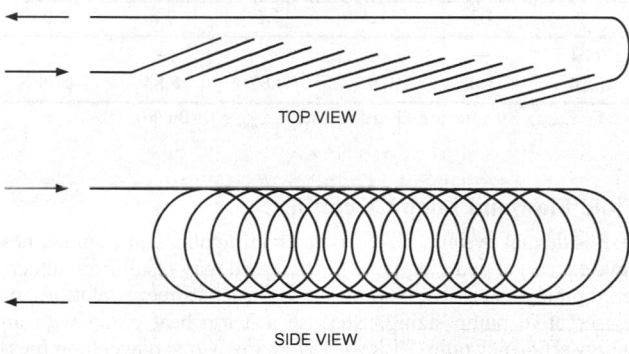

TOP VIEW

SIDE VIEW

Fig. 22   General Layout of Spiral Earth Coil

lots that would be too small for conventional horizontal-pipe ground-loop designs. The spiral pipe configuration laid flat in a horizontal pit arrangement is used commonly in the northern Midwest of the United States, where sandy soil causes vertical trenches to collapse. A large open pit is excavated by a bulldozer, and then the overlapping pipes laid flat on the bottom of the pit. The bulldozer is also used to cover the pipe, being careful to not run over them with the bulldozer tread.

Although most horizontal closed-loop systems are installed with either a chain trencher or a backhoe, horizontal boring machines are also now available for this application. Developed for buried utility applications such as electric or potable water service, these devices simply bore through the ground parallel to the ground surface. A detector at the surface can show the exact point where the boring head is located underground so that the bore does not penetrate other known utilities or cross over into a neighbor's lot.

Most horizontal loop installations place the pipe loops in a parallel rather than a single (series) loop to reduce pumping power (Figure 23). Parallel loops may require slightly more pipe, but may use smaller pipe and thus have smaller internal volumes, requiring less antifreeze (if needed). Also, smaller pipe is typically much less expensive for a given length, so total pipe cost should be less for parallel loops. An added benefit is that parallel loops can be flushed out with a smaller purge pump than is required for a larger single-pipe loop. A disadvantage of parallel loops is the potential for unequal flow in the loops and thus nonuniform heat exchange efficiency.

The time required to install a horizontal loop is not much different from that for a vertical system. For the arrangements described, a two-person crew can typically install the ground loop for an average house in a single day.

Soil characteristics are an important concern for any ground loop design. With horizontal loops, the soil type can be more easily

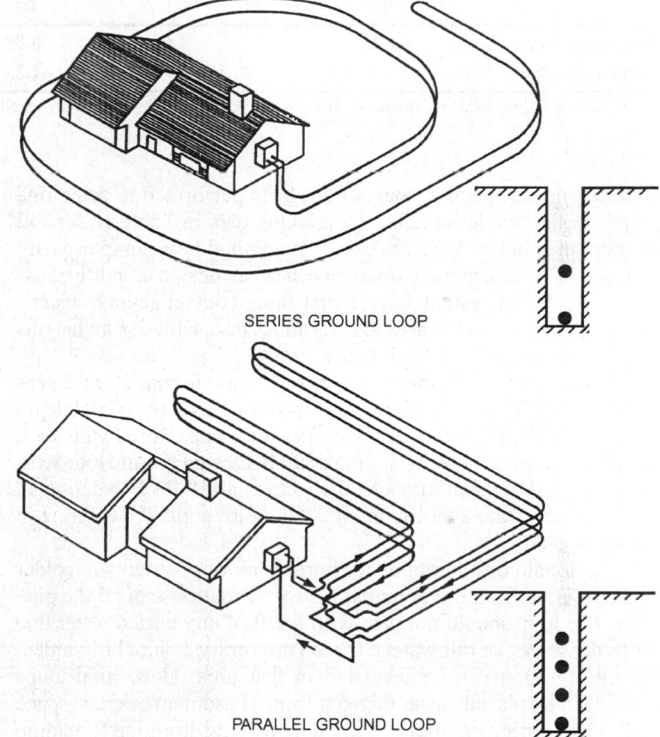

SERIES GROUND LOOP

PARALLEL GROUND LOOP

Fig. 23   Parallel and Series Ground Loop Configurations

determined because the excavated soil can be inspected and tested. EPRI (1990) lists criteria and simple test procedures that can be used to classify soil and rock for horizontal ground-loop design.

Although soil type and moisture content are important considerations in sizing the ground loop, some design guidelines have been developed based on extensive analysis of monitored systems in mostly southern climates (Kavanaugh and Calvert 1995). Table 14 gives recommended trench lengths for the various types of commonly used excavation methods. Heating-mode run times approaching 100% on a daily basis would be the norm at heating design conditions in heating-dominated climates. In contrast, daily run times of no more than 50% would be encountered at design cooling conditions in cooling-dominated climates. The combination of long run times and ice formation around the pipes makes performance of horizontal systems dependent on both the loop field design and how the system is matched to the building load. Though many thousands of these systems have been installed in heating

Table 14 Recommended Lengths of Trench or Bore per Ton for Residential GCHPs

Coil Type[a]		Pitch[b] Feet of Pipe per Feet Trench/Bore	Ground Temperature, °F						
			44 to 47	48 to 51	52 to 55	56 to 59	60 to 63	64 to 67	68 to 70
Horizontal	6-Pipe/6-Pitch Spiral	6	180	160	150	160	180	200	230
	4-Pipe/4-Pitch Spiral	4	220	200	190	200	220	250	300
	2-Pipe	2	300	280	250	280	300	340	400
Vertical U-tube	3/4 in. Pipe	2	180	170	155	170	180	200	230
	1 in. Pipe	2	170	160	150	160	170	190	215
	1 1/4 in. Pipe	2	160	150	145	150	160	175	200

Source: Kavanaugh and Calvert (1995).
[a]Lengths based on DR11 high-density polyethylene (HDPE) pipe. See Figures 21 to 23 for details.
[b]Multiply length of trench by pitch to find required length of pipe.

Note: Based on $k = 0.6$ Btu/h·ft·°F for horizontal loops and $k = 1.2$ Btu/h·ft·°F for vertical loops. Figures for soil temperatures < 56°F based on modeling using nominal heat pump capacity and assumption of auxiliary heat at design conditions.

Multiply Table 14 Values by Bold Values Below to Correct for Other Values of Ground Conductivity

	Ground Thermal Conductivity in Btu/h·ft·°F								
	0.4	0.6	0.8	1.0	1.2	1.4	1.6	1.8	2.0
Horizontal loop	1.22	1.0	0.89	0.82	—	—	—	—	—
Vertical loop*	—	—	1.23	1.10	1.0	0.93	0.87	0.83	0.79

*Vertical loop values based on an annular fill with $k = 0.85$ Btu/h·ft·°F. Multiply lengths by 1.2 for $k_{annulus} = 0.4$ Btu/h·ft·°F and 0.95 for $k_{annulus} = 1.1$ Btu/h·ft·°F.

climates, no comparable analysis has been performed to determine proper design guidelines. The loop length data in Table 14 for soil temperatures below 56°F are based on nominal heat pump capacity and use of supplemental resistance heat at design conditions. If installing such a system for the first time, contact several experienced contractors in the area to determine successful design lengths for the local climate and soil types.

Trench lengths in Table 14 are based on a minimum trench separation of 10 ft and minimum horizontal loop average burial depth of the greater of 5 ft or 2 ft below the frost line. Bore lengths are based on a vertical bore separation of 20 ft. Design ground loop temperatures are a maximum of 90°F return and 100°F entering in warm climates and a minimum of 28°F return and 22°F entering in cold climates.

Additional considerations for horizontal loop systems in colder climates arise from the potential for ice formation around the pipe loop. The loop should not pass within 2 ft of any buried water line (potable, sewer, or rainwater). If such proximity cannot be avoided, the GCHP loop can be insulated in that area. Horizontal loops should not be placed closer than 6 ft from a basement or crawl space wall when buried parallel to the wall. Heaving from ice formation could cause structural damage if placed in close proximity to the wall.

Leaks in heat-fused plastic pipe are rare when attention is paid to pipe cleanliness and proper fusion techniques. Should a leak occur, it is usually best to try to isolate the leaking parallel loop and abandon it in place. The effort required to find the source of the leak usually far outweighs the cost of replacing the defective loop. Because the loss of as little as 0.25 gal of water from the system causes the system to lose pressure and shut down, leaks cannot be located by looking for wet soil, as is commonly done with water lines.

Although leaks should be rare with properly thermally fused pipe, some states have adopted restrictions against the use of certain types of antifreeze mixtures in GCHP systems; check local water-quality regulations before selecting a mixture. Methanol has been used extensively because of its low cost and good physical properties when cold. Comprehensive studies by Heinonen and Tapscott (1996) and Heinonen et al. (1997) showed that propylene glycol is a good alternative when issues of flammability or environmental safety are important considerations. A more thorough discussion of antifreeze solutions is given in the Antifreeze Requirements section of this chapter.

Fluid Flow and Loop Circuiting

Residential systems, like commercial applications, sometimes have excessive pumping power. This trend may result from undersized piping, excessive amounts of viscous antifreeze solutions, or conservative pump sizing. Because a 3 ton heat pump with an energy efficiency ratio (EER) of 15 requires a total power (compressor and fan) of 2400 W, the addition of a second 1/6 hp pump (which draws 245 W) reduces system efficiency by 10%. Table 15 provides a guideline to ensure adequate liquid flow rate with the least possible number of pumps. The table is to be used in conjunction with Table 14 and applies to loops with 0 to 15% propylene glycol solutions (by volume). This solution has the reputation of being the most difficult of the commonly used solutions to pump when cold. However, it is no more difficult to pump than ethyl alcohol, and pumping penalties can be mitigated by adding only the required amount of antifreeze. Shorter loops may require higher levels of antifreeze solutions. See the section on Antifreeze Requirements for more details. Any exposed piping above the frost line must be insulated with closed-cell insulation with ultraviolet (UV) protection (paint or wrap).

Residential Design Example

Design the vertical ground coupling grid and the pumping loop for a 4 ton (48,000 Btu/h) heat pump system. The home is located in Nashville, Tennessee, and the header pipes can be brought into the equipment room where the unit will be located. The driller can bore 4.5 in. holes to a depth of 175 ft in the light limestone and clay at the site. The owner wants the drilling site to be located 75 ft from the house. Thermally enhanced grout with thermal conductivity of 0.85 Btu/h·ft·°F is used to fill the annular region between the U-tubes and borehole walls.

Solution:

The soil temperature is estimated to be 59°F in Nashville (see Figure 17). Table 14 suggests bore lengths of 170 ft/ton for 3/4 in. U-bends, 160 ft/ton for 1 in., and 150 ft/ton for 1 1/4 in.bores are deep, greater than 100 ft. However, 1 1/4 in. U-bends are very difficult to install into a 4.5 in. bore hole, and are not considered. Therefore, either 680 ft (170 ft/ton × 4 tons) of 3/4 in. U-bend coupling or 640 ft (160 × 4 tons) of 1 in. coupling is required. The latter is used in this example. Also, Table 14 is based on a soil conductivity of 1.2 Btu/h·ft·°F, which is an approximate average between limestone and clay, and a bore fill (or grout) conductivity of 0.85 Btu/h·ft·°F. If the ground conductivity is higher (i.e., more limestone than clay), the loops should be reduced as noted in Table 14; if lower, the loops should be lengthened. Loop

Table 15 Recommended Residential GCHP Piping Arrangements and Pumps

Coil Type*	Nominal Heat Pump Capacity, tons				
	2	3	4	5	6
	Required Flow Rate, gpm				
	5 to 6	7 1/2 to 9	10 to 12	12 to 15	15 to 18
	Number of Parallel Loops				
Spiral (10 pt.)	3 to 4	4 to 6	6 to 9	8 to 10	8 to 10
6-Pipe	3 to 4	4 to 6	6 to 9	8 to 10	8 to 10
4-Pipe	2 to 3	4 to 6	5 to 8	6 to 9	6 to 10
2-Pipe	2 to 4	3 to 5	4 to 6	5 to 8	6 to 10
Vertical 3/4 in. pipe	2 to 3	3 to 5	4 to 6	5 to 8	6 to 10
1 in. pipe	2 to 3	2 to 4	3 to 5	4 to 6	4 to 6
1 1/4 in. pipe	1 to 2	1 to 2	2 to 3	2 to 3	2 to 4

Trench Length	Header Diameter (PE Pipe), in.				
Less than 100 ft	1 1/4	1 1/4	1 1/2	1 1/2 to 2	1 1/2 to 2
100 to 200 ft	1 1/4	1 1/2	1 1/2	2	2

Size (No.) of Pumps Required					
1/12 hp (1)	1/6 hp (1)	1/12 hp (2)	1/6 hp (2)	1/6 hp (2)	

Source: Kavanaugh and Calvert (1995).
*Based on DR11 HDPE pipe.

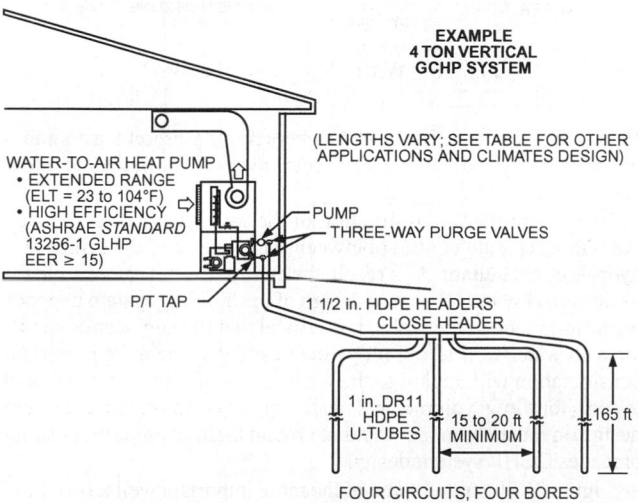

EXAMPLE
4 TON VERTICAL
GCHP SYSTEM

(LENGTHS VARY; SEE TABLE FOR OTHER APPLICATIONS AND CLIMATES DESIGN)

WATER-TO-AIR HEAT PUMP
• EXTENDED RANGE (ELT = 23 to 104°F)
• HIGH EFFICIENCY (ASHRAE *STANDARD* 13256-1 GLHP EER ≥ 15)

PUMP
THREE-WAY PURGE VALVES
P/T TAP
1 1/2 in. HDPE HEADERS
CLOSE HEADER
1 in. DR11 HDPE U-TUBES
15 to 20 ft MINIMUM
165 ft
FOUR CIRCUITS, FOUR BORES

Fig. 24 Residential Design Example

lengths also must be lengthened if the bore fill (or grout) conductivity is lower than 0.85 Btu/h·ft·°F, as noted in Table 14.

Layout is dictated by drilling conditions. The total length of 640 ft requires four bores, because the driller can only drill to 175 ft. This can be accomplished with four 160 to 165 ft holes. Table 15 suggests between three and five parallel circuits for the grid. Three and five circuits do not divide evenly into the four U-bends. Therefore, four circuits (one per U-bend) should be used in an arrangement similar to Figure 24.

GROUNDWATER HEAT PUMPS

A groundwater heat pump system (GWHP) removes groundwater from a well and delivers it to a heat pump (or an intermediate heat exchanger) to serve as a heat source or sink. Both unitary and central plant designs are used. In the unitary type, a large number of small water-to-air heat pumps are distributed throughout the building. The central plant design uses one or a small number of large-capacity

chillers supplying hot and chilled water to a two- or four-pipe distribution system. The unitary approach is more common and tends to be more energy-efficient.

Direct systems (in which groundwater is pumped directly to the heat pump without an intermediate heat exchanger) are not recommended except on the very smallest installations. Although some systems of this design have been successful, many have had serious difficulty even with groundwater of apparently benign chemistry. Thus, prudent design for commercial/industrial-scale projects isolates groundwater from the building system with a heat exchanger. The increased capital cost of installing the heat exchanger is only a small percentage of the total cost and, in view of these systems' greatly reduced maintenance requirements, is quickly recovered.

Past GWHP systems sometimes used surface disposal (to rivers, lakes, drainage ditches, etc.) of the groundwater. Injection, in general, should be the standard disposal method because it eliminates the potential for negative effects on the aquifer water level over time and preserves the positive environmental character associated with GSHP systems.

Regardless of the type of equipment installed in the building, the specific components for handling groundwater are similar. Primary items include (1) wells (supply and injection), (2) well pump and controls, and (3) groundwater heat exchanger. Some specifics of these items are discussed in the Direct-Use Systems section of this chapter. In addition to those comments, the following considerations apply specifically to unitary GWHP systems using a groundwater isolation heat exchanger.

Design Strategy

An open-loop system design must balance well pumping power with heat pump performance. As groundwater flow increases through a system, more favorable average temperatures are produced for the heat pumps. Higher groundwater flow rates, to a point, increase system EER or COP: increased well pump power is outweighed by decreased heat pump power requirements (because of the more favorable temperatures). At some point, additional increases in groundwater flow result in a greater increase in well pump power than the resulting decrease in heat pump power. The key strategy in open-loop system design is identifying the point of maximum system performance with respect to heat pump and well pump power requirements. Once this optimum relationship has been established for the design condition, the method of controlling the well pump determines the extent to which the relationship is preserved at off-peak conditions. This optimization process involves evaluating the performance of the heat pumps and well pump(s) over a range of groundwater flows. Key data necessary to make this calculation include well performance (flow and drawdown at various groundwater flows) and heat pump performance versus entering water temperatures at different flow rates. Well information is generally derived from well pump test results. Heat pump performance data are available from the manufacturer.

GWHP systems use the same type of extended-range unitary heat pumps as GCHP systems. Building loop pumping guidelines (see Table 9) in the GCHP portion of this chapter also apply to GWHP systems. In large commercial applications, the head loss associated with the isolation heat exchanger in a GWHP system is typically lower than that of an equivalently sized ground loop in a GCHP system. A guideline for building loop head loss in a GWHP system is

$$\text{Building loop head loss (ft of water)} = 28 + 0.1d$$

where d = pipeline distance from plate heat exchanger outlet to most distant heat pump unit inlet, in ft.

This calculation assumes a maximum head loss of 4 ft/100 ft fittings at 25% of total head loss, and a heat pump unit head loss of 12 ft. Because of their more extensive fittings, retrofit applications can sometimes exceed this value.

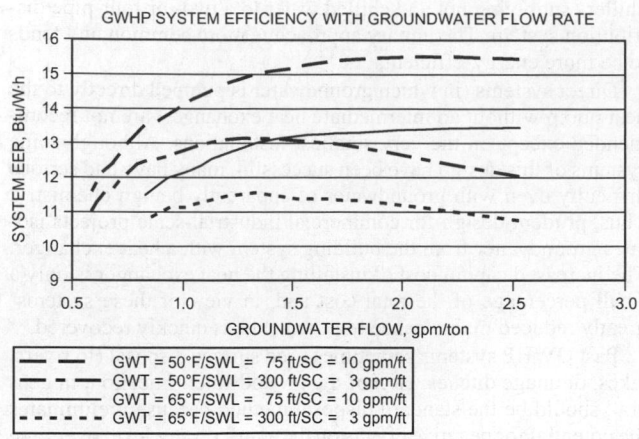

Fig. 25 Optimum Groundwater Flow for Maximum System EER
(Kavanaugh and Rafferty 1997)

Table 16 Example GWHP System* Design Data

Heat Pump EWT, °F	Heat Pump LWT, °F	Heat Pump EER	Ground-water LWT, °F	Ground-water Flow, gpm	Well Pump Head, ft	Well Pump kW	Loop Pump kW	System EER
61.0	72.4	17.6	68.4	289	256	23.7	4.8	11.8
63.0	74.5	17.3	70.5	233	229	17.5	4.8	12.5
65.0	76.5	16.9	72.5	196	210	13.7	4.8	12.9
67.0	78.6	16.5	74.6	169	197	11.4	4.8	13.0
69.0	80.6	16.1	76.6	149	186	9.7	4.8	13.1
71.0	82.7	15.7	78.7	133	179	8.5	4.8	13.0
73.0	84.7	15.3	80.7	120	172	7.5	4.8	12.9
75.0	86.7	15.1	82.7	110	167	6.7	4.8	12.9
77.0	88.8	14.9	84.8	101	163	6.0	4.8	12.8
79.0	90.8	14.6	86.8	94	159	5.5	4.8	12.6
81.0	92.3	14.2	88.9	88	156	5.1	4.8	12.4
83.0	94.9	13.4	90.9	82	153	4.7	4.8	12.2

*Block cooling load 85 tons, 60°F groundwater, 75 ft well static water level, 2 gpm/ft specific capacity, 37 ft surface head losses, 4°F heat exchanger approach, 213 gpm building loop flow at 65 ft head.

For moderate-efficiency heat pumps (EER of 14.2), efficient loop pump design (7.5 hp/100 tons), and a heat exchanger approach of 3°F, Figure 25 provides curves for two different groundwater temperatures [GWT = 50 and 65°F], two static water levels [SWL = 75 and 300 ft], and two well specific capacities [SC = 3 and 10 gpm/ft].

Although the four curves show a clear optimum flow, sometimes operating at a lower groundwater flow reduces well/pump capital cost and the problem of fluid disposal. These considerations are highly project specific, but do afford the designer some latitude in flow selection. Generally, an optimum design results in a groundwater flow rate that is less than the building loop flow rate.

Table 16 provides design data for a specific example system.

WATER WELLS

This section includes information on water wells that is generally common to both direct-use and groundwater heat pump (GWHP) systems.

An **aquifer** is a geologic unit that is capable of yielding groundwater to a well in sufficient quantities to be of practical use (UOP

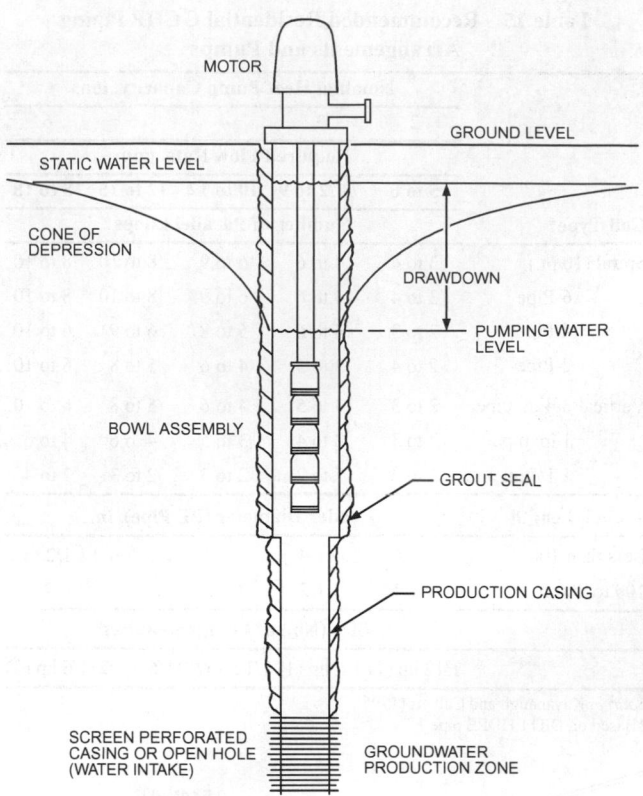

Fig. 26 Water Well Terminology

1975). Aquifers can exist in areas where water is present in conjunction with pore spaces in the subsurface materials sufficient to allow the water to move laterally.

In many projects, construction of the well (or wells) is handled through a separate contract between the owner and the driller or a hydrology consultant. As a result, the engineer is not responsible for its design. However, because design of the building system depends on performance of the wells, it is critical that the engineer be familiar with water well terminology and test data. The most important consideration with regard to the wells is that they be completed and tested (for flow volume and water quality) before final system design, in much the same way that ground thermal properties testing precedes GCHP system design.

Figure 26 illustrates some of the more important well terms. Several references (Anderson 1984; Campbell and Lehr 1973; EPA 1975; Roscoe Moss Company 1985) cover well drilling and well construction in detail.

Static water level (SWL) is the level that exists under static (nonpumping) conditions. In some cases, this level is much closer to the surface than that at which the driller encounters water during drilling. **Pumping water level (PWL)** is the level that exists under specific pumping conditions. Generally, this level is different for different pumping rates (higher pumping rates mean lower pumping levels). The difference between the SWL and the PWL is the **drawdown**. The **specific capacity** of a well is frequently quoted in gpm per foot of drawdown. For example, for a well with a static level of 50 ft that produces 150 gpm at a pumping level of 95 ft, drawdown = 95 − 50 = 45 ft; specific capacity = 150/45 = 3.33 gpm per foot.

Water entrance velocity (through the screen or perforated casing) can be an important design consideration. Velocity should be limited to a maximum of 0.1 fps (0.05 ft/s for injection wells) to avoid incrustation of the entrance openings. The **pump bowl assembly** (impeller housings and impellers) is always placed sufficiently below the expected pumping level to prevent cavitation at the peak production rate. For the previous example, this pump should be placed at least 115 ft below the casing top (pump setting depth = 115 ft) to allow for adequate submergence at peak flow. Along with any expected annual aquifer water level fluctuations, the specific **net positive suction head (NPSH) pressure** required for a pump varies with each application and should be carefully considered in selecting the setting depth.

For the well pump, **total pump head** is composed of four primary components: lift, column friction, surface requirements, and injection head (pressure). **Lift** is the vertical distance that water must be pumped to reach the surface. In the example, lift is 95 ft. The additional 20 ft of submergence imposes no static pump head (pressure).

Column friction, the friction loss in the pump column between the bowl assembly and the surface, is calculated from pump manufacturer data in a similar manner to other pipe friction calculations (see Chapter 22 of the 2009 *ASHRAE Handbook—Fundamentals*). **Surface pressure requirements** account for friction losses through piping, heat exchangers, and controls, and in many applications are between 25 and 35 ft. **Injection pressure requirements** are a function of well design, aquifer conditions, and water quality. In theory, an injection well penetrating the same aquifer as the production well experiences a water level rise (assuming equal flows) that mirrors the drawdown in the production well. Using the earlier example, an injection well with a 50 ft static level would experience a water level rise of 45 ft, resulting in a surface injection pressure of $45 - 50 = -5$ ft (i.e., a water level that remains 5 ft below the ground surface). Thus, no additional pump head is required for injection in the example.

In practice, injection pressure requirements usually exceed the theoretical value. With good (nonscaling) water quality, careful drilling, and little sand production, injection pressure should be near the theoretical value. For poor water quality, high sand production, or poor well construction, injection pressure may be 10 to 40% higher.

The well casing diameter depends on the diameter of the pump (bowl assembly) necessary to produce the required flow rate. Table 17 presents nominal casing sizes for a range of water flow rates.

In addition to the production well, most systems should include an injection well to dispose of the fluid after it has passed through the system. Injection stabilizes the aquifer from which the fluid is withdrawn by reducing or eliminating long-term drawdowns, and helps to ensure long-term productivity. Construction of injection wells differs from production wells primarily in the recommended screen velocity (0.05 ft/s, or 1/2 that of production wells) and well sealing design. Injection wells, particularly those likely to be subject to positive injection pressure, should be fully cased and sealed from the top of the injection zone to the surface.

Flow Testing

When possible, well testing should be completed before mechanical design. Only with actual flow test data and water chemical analysis information can accurate design proceed.

Flow testing can be divided into three different types of tests: rig, short-term, and long-term (Stiger et al. 1989). Rig tests are generally very short and are accomplished while the drilling rig is on site. The primary purpose of this test is to purge the well of remaining drilling fluids and cuttings and to get a preliminary indication of yield. The length of the test is generally governed by the time required for the water to run clean. The rate is determined by the available pumping equipment. Frequently, the well is "blown" or pumped with the drilling rig's air compressor. As a result, limited information about the well's production characteristics is available from a rig test. If the well is air lifted, it may not be useful to collect water samples for chemical analysis because certain chemical constituents may be oxidized by the compressed air.

Table 17 Nominal Well Surface Casing Sizes

Pump Bowl Diameter, in.	Suggested Casing Size, in.	Minimum Casing Size, in.	Submersible Flow Range (3450 rpm), gpm	Lineshaft Flow Range (1750 rpm), gpm
4	6	5	<80	<50
6	10	8	80 to 350	50 to 175
7	12	10	250 to 600	150 to 275
8	12	10	360 to 800	250 to 500
9	14	12	475 to 850	275 to 550
10	14	12	500 to 1000	
12	16	14	900 to 1300	

Properly conducted, short-term, single-well tests lasting 4 to 24 h yield information about well flow rate, temperature, drawdown, and recovery. These tests are used most frequently for direct-use and GWHP applications. The test is generally run with a temporary electric submersible pump or lineshaft turbine pump driven by an internal combustion engine and are often performed by a well pump contractor.

A step test (Table 18), the most common type, involves at least three production rates, the largest being equal to the design flow rate for the system served. The three points are the minimum required to determine a productivity curve for the well that relates production to drawdown (Stiger et al. 1989). The key parameters monitored during these tests are well water level and water flow. Water level and pumping rate should be stabilized at each point before flow is increased. In many cases, water level is monitored with a bubbler or an electric sounder, and flow is measured using an orifice meter. More sophisticated instrumentation (e.g., pressure transducers for water level, magnetic flow meters, data loggers) can also be used. Short-term testing is generally used for small projects and provides information on yield, drawdown, and specific capacity.

Test results should reflect stable flows rates and individual flow "steps" are extended until water level readings stabilize. In many cases brief intervals of turbidity may occur at flow changes but extensive periods of turbidity indicate instability in the near well formation.

Long-term tests of up to 30 days provide information on the reservoir. Normally these tests involve monitoring nearby wells to evaluate interference effects. The data are useful in calculating transmissivity and storage coefficient, reservoir boundaries, and recharge areas (Stiger et al. 1989) but are rarely used for direct-use and GWHP systems.

It is also important to collect background information before the test, and water level recovery data after pumping has ceased. Recovery data in particular can be used to evaluate skin effect, which is a type of well flow resistance caused by residual drilling fluids, insufficient screen or slotted liner area, or improper filter pack.

Groundwater Quality

The importance of groundwater quality depends on the system design. Systems using isolation heat exchangers commonly encounter no water quality issues (other than iron bacteria) that would prevent the application of a GWHP system operating under reasonable maintenance levels.

For systems that use groundwater directly in heat pump units (e.g., standing-column systems and small residential GWHP systems), several issues are of concern. The primary water quality problem in the United States is scaling, usually of calcium carbonate (lime). Because this type of scaling is partially temperature driven, the temperature of surfaces that groundwater contacts determines the extent to which scaling will occur. In these systems, peak temperatures in the refrigerant-to-water exchanger in cooling mode are likely to be over 160°F. For the same system using an isolation plate heat exchanger, the groundwater is unlikely to encounter temperatures over 90°F. Using the plate heat exchanger reduces

Table 18 Example Well Flow Test Results SWL 68 ft

Time Since Pump Start, min	Flow, gpm	Water Level, ft	Comments
5	125	78.3	clear
10	127	79.5	clear
15	125	81.1	clear
20	125	82	clear
25	125	83.1	clear
30	126	83.4	clear
40	125	83.3	clear
50	125	83.3	clear
60	125	83.3	clear
65	200	90.6	cloudy
70	200	96.8	clear
75	200	98.9	clear
80	200	99.5	clear
85	201	99.7	clear
90	200	100.1	clear
100	200	100.2	clear
120	200	100.2	clear
125	295	130.6	cloudy
130	300	135.8	cloudy
135	301	140.3	clear
140	300	141.5	clear
160	300	142.3	clear
170	300	145.7	clear
180	300	145.7	clear

the propensity for scaling and limits any scale that does occur to a single heat exchanger. Rafferty (2000b) provides information on water scaling potential on a state-by-state basis.

Although cupronickel, originally developed for seawater applications, is an excellent material in that role, it provides few benefits for most groundwater applications. It has been shown to perform more poorly than pure copper with hydrogen-sulfide-bearing waters.

Excessive iron, particularly ferrous iron, in the water can result in coating of heat transfer surfaces if the water is exposed to air (allowing the iron to oxidize to the ferric state, a form with much lower solubility in water). Periodically removing this iron from the plates of a single heat exchanger is much less labor-intensive than removing it from tens or hundreds of individual heat pump heat exchangers. Table 19 summarizes the minimum parameters that should be evaluated for a GWHP application.

Particulate matter (e.g., sand) in the groundwater stream, although usually not a problem in the mechanical system, can effectively plug injection wells. Sand production should be addressed in construction of the production well (screen/gravel pack/development). If it must be dealt with on the surface, a screen or strainer is preferable to a centrifugal separator. The strainer does not suffer from ineffectiveness at start-up, shutdown, and variable flow as does the centrifugal separator (Kavanaugh and Rafferty 1997). Perforation size selection is critical to a strainer's effectiveness. Sizing should be based on 90 to 100% removal of the particulate material. Particle size information can be based on the sieve analysis results of drill cuttings (used to size the well screen) or of a sample taken during well flow testing. In applications with very fine sand, multiple strainers in parallel may be necessary to control pressure drop (Rafferty 2008).

Well Pumps

Submersible pumps have not performed well in higher-temperature, direct-use projects. However, the submersible pump is a cost-effective option with normal groundwater temperatures, as encountered in heat pump applications. The low temperature eliminates the need to specify an industrial design for the motor/

Table 19 Water Chemistry Constituents

Quality	Comment
pH	Typical range: 6.5 to 9.0. Lower values typically associated with higher rates of general corrosion in ferrous and copper alloys; higher values associated with scaling.
TDS	Total dissolved solids: gross indicator of quantity of dissolved constituents. Higher levels associated with increased corrosion and/or scaling; used in calculation of scale index.
Fe	Iron: use care to prevent exposure to air; problems possible at >0.5 ppm.
Total M alkalinity	Ability of water to buffer acid; strongly linked to scale and used to calculate scaling index. Usually expressed as ppm $CaCO_3$.
Ca	Calcium ion: linked to scaling of water and used to calculate scaling index. Expressed in ppm Ca × 0.5 = ppm as $CaCO_3$.
CO_3/HCO_3	Carbonate/bicarbonate: varies in concentration with pH.
Hardness	Linked to scaling and used to calculate of scale index; at >100 ppm, scaling can occur. Expressed in ppm/17.1 = hardness in gr/gal.
Cl	Chloride: accelerates corrosion of carbon and stainless steels; may be elevated in coastal areas.
Mn	Manganese: causes black scale; possible deposits at >0.2 ppm.
O_2	Oxygen: dissolved gas; accelerates corrosion; promotes other reactions; test in field.
H_2S	Hydrogen sulphide: dissolved gas; rotten egg odor >0.5 ppm; attacks copper alloys; test in field.
CO_2	Carbon dioxide: dissolved gas, often present at pH < 7.5, test in field. GW pressurization keeps CO_2 in solution.
Stability index (Ryznar index)	Originally developed to predict corrosion but used in GWHP for scaling prediction; calculated from temperature, Ca, TDS, alkalinity, and hardness. Must use temperature reflective of application: 85°F for systems with plate heat exchanger, 150°F for nonisolated systems.
Saturation index (Langlier index)	Similar to stability index. Originally developed to predict corrosion but used in GWHP for scaling prediction; calculated from temperature, Ca, TDS, alkalinity, and hardness. Must use temperature reflective of application: 85°F for systems with plate heat exchanger, 150°F for nonisolated systems
BART	Bacteriological activity reaction test: broad indicator of various bacteria. Most common tests are for iron-reducing (IRB), slime-forming (SLYM), and sulfate-reducing (SRB) bacteria.

Source: Rafferty (2008).

protector, thereby greatly reducing the first cost relative to direct use. Caution should still be used for wells that are expected to produce moderate amounts of sand. The high speed (nominal 3600 rpm) of most submersible pumps makes them susceptible to erosion damage.

Small groundwater systems have frequently been identified with excessive well pump energy consumption. The reasons for excessive pump energy consumption (high water flow rate, coupling to the domestic pressure tank, and low efficiency of small submersible pumps) are generally not present in large, commercial groundwater systems. In large systems, the groundwater flow per unit capacity is frequently less than half that of residential systems. Pressure at the wellhead is not the 30 to 50 psi typical of domestic systems, but is rather a function only of pressure losses through the groundwater loop. Finally, large well pumps have efficiencies of up to 83% compared to the 35 to 40% range for small submersible pumps.

In GWHP system design, the control method for the well pump determines the extent to which the optimum relationship between well pump power and heat pump power is preserved at off-peak conditions. There are several ways the pump can be controlled. Multiple pumps can be staged to meet system loads, either with multiple wells or with multiple pumps installed in a single well. A dual set-point control similar to that used in boiler/tower systems energizes the well pump above a given temperature in cooling mode and below

Table 20   Controller Range Values for Dual Set-Point Well Pump Control*

	Building Loop Thermal Mass in Gallons per Ton of Peak Block Cooling Load						
	2	4	6	8	10	12	14
Cooling Range, °F	31	16	11	8	6	5	4
Heating Range, °F	18	9	6	4	3	3	3

*Table values for pumps > 5 hp. For pumps < 5 hp, three-phase range values may be reduced by 50%.

a given temperature in heating mode. Between those temperatures, the building loop floats without the addition of groundwater. To control well pump cycling, it is necessary to establish a temperature range (difference between pump-on and pump-off temperatures) over which the pump operates in both the heating and cooling modes. The size of this range is primarily a function of the building loop water volume in terms of gallons per peak per ton of peak block system load (Rafferty 2000c). Table 20 summarizes these data. In the example in Table 16, the optimum system building loop return temperature (at peak system EER) is 80.6°F. If this system had a water volume of 8 gal/ton, from Table 20, a range of 8°F in cooling mode would be required. This range would result in a well pump start temperature of 80.6 + (8/2) = 84.6°F and a well pump stop temperature of 80.6 − (8/2) = 76.6°F. A similar calculation can be made for heating mode. It is apparent from Table 20 that, for systems with very low thermal mass, the dual set-point method of control becomes impractical because of the very large temperature range required. For these applications, an alternative method of control (variable speed, staging, etc.) is required.

Well pumps may also be controlled using a variable-speed drive, which responds to building loop return temperature by varying groundwater flow to the exchanger to maintain the cooling or heating mode set point. Submersible-motor variable-speed applications are somewhat different than surface motor applications. Most manufacturers limit speed reduction to 50%, and other issues such as minimum water velocity for motor cooling, switching frequency, reactor requirement, and motor protection must be addressed. Additional information on VFD applications for submersible motors is available in Rafferty (2008).

Heat Exchangers

Design of a plate-and-frame heat exchanger is largely a tradeoff between pressure drop, which influences pumping (operating cost), and overall heat transfer coefficient, which influences surface area (capital cost). In general, exchangers in GWHP systems can be economically selected for approach temperatures (between loop return and groundwater leaving temperatures) as low as 3°F. Most selections involve an approach of between 3 and 7°F and a pressure drop of less than 10 psi on the building loop side. Excessive fouling factors (>0.0002 h·ft^2·°F/Btu) should not be specified when selecting plate heat exchangers, which can be easily disassembled and cleaned.

Heat exchanger cost may be reduced for groundwater applications by using Type 304 stainless steel plates rather than the Type 316 or titanium plates common in direct-use projects. The low temperature and generally low chloride content of heat pump fluids frequently make the less expensive Type 304 material acceptable. Chloride content of the groundwater, particularly in coastal areas, should always be compared to values in Figure 4 to determine plate material acceptability. Exchanger performance should be checked at minimum system flow rates to ensure adequate heat transfer. In some cases, very low design pressure drop selections can encounter inadequate heat transfer at minimum flows.

Central Plant Systems

Central plant systems, in which a conventional or heat recovery central chiller is connected to a four-pipe system, are the oldest type

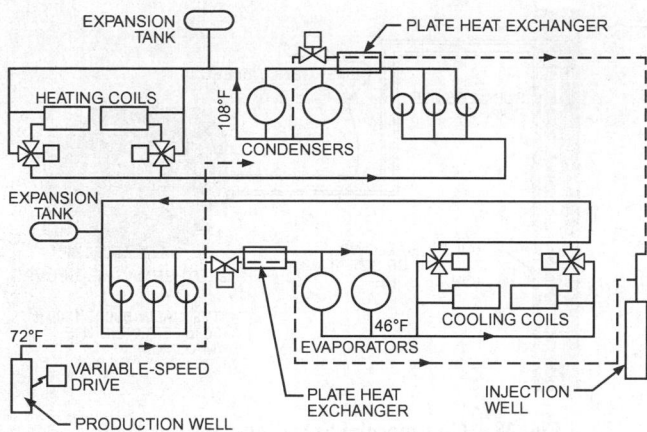

Fig. 27   Central Plant Groundwater System

of open-loop design, having first been installed in the late 1940s. Because of the cost and energy requirements of the central plant design, these systems typically do not result in the same level of energy efficiency as unitary GWHP systems.

For central plant groundwater systems, two heat exchangers are normally used: one in the chilled-water loop and one in the condenser water loop (Figure 27). The evaporator-loop exchanger provides a heat source for heating-dominated operation and the condenser-loop exchanger provides a heat sink for cooling-dominated operation.

Sizing of the **condenser-loop exchanger** is based on providing sufficient capacity to reject the condenser load in the absence of any building heating requirement.

Sizing of the **chilled-water-loop exchanger** must consider two loads. The primary criterion is the load required during heating-dominant operation. The exchanger must transfer sufficient heat (when combined with compressor heat) from the groundwater to the chilled-water loop to meet the building's space heating requirement. Depending on the relative groundwater and chilled-water temperatures and on the design temperature rise, exchangers may also provide some free cooling during cooling-dominant operation. If groundwater temperature is lower than that of chilled water returning to the exchanger, some chilled-water load can be met by the exchanger. This mode is most likely available in regions with groundwater temperatures below 60°F.

Central plant chiller controls must also allow for the unique operation with a groundwater source. Controls can be similar to those on a heat-recovery chiller with a tower, with one important difference. In a conventional heat-recovery chiller, waste heat is available only when there is a building chilled-water (or conditioning) load. In a groundwater system, a heat source (the groundwater) is available year-round. To take advantage of this source during the heating season, the chiller must be loaded in response to the heating load instead of the chilled-water load. That is, the control must include a heating-dominant mode and a cooling-dominant mode. Two general designs are available for this:

- Chiller capacity remains controlled by chilled-water (supply or return) temperature, and groundwater flow through the chilled-water exchanger is varied in response to the heating load
- Chiller capacity is controlled by the heating-water (condenser) loop temperature, and groundwater flow through the chilled-water exchanger is controlled by chilled-water temperature

For buildings with a significant heating load, the former may be more attractive, whereas the latter may be appropriate for conventional buildings in moderate to warm climates.

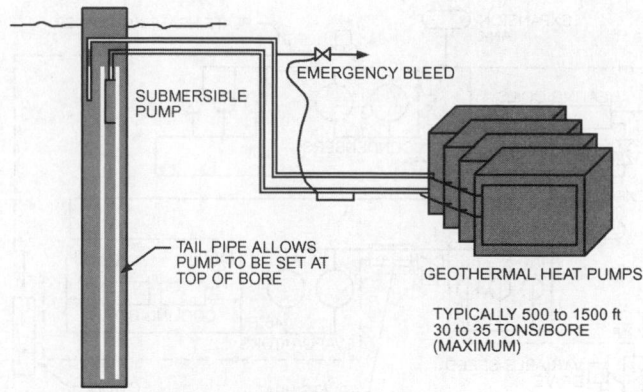

Fig. 28 Commercial Standing-Column Well

Standing-Column Systems

In practice, standing-column wells (SCWs) are a tradeoff between groundwater systems and ground-coupled systems. They do not require well flow testing of the sort necessary for groundwater systems, and conductive heat transfer can be based on existing closed-loop theory. Though standing-column systems have been applied mostly in the northeastern United States, approximately 60% of the country is underlain by near-surface (<150 ft) bedrock suitable for the systems.

A heating capacity of 350,000 to 420,000 Btu/h or cooling capacity of 30 to 35 tons can be expected from a 1500 ft deep standing-column well. Ideal spacing between SCWs is 50 to 75 ft (Orio et al. 2005) Typically, spacing between standing-column wells is greater than vertical closed-loop (GCHP) boreholes. These estimated capacities assume a 10% advective bleed flow, discussed later in this section. Closer spacing affects well field performance and can be evaluated with design software. Additional information on standing-column systems can be found in Spitler (2002).

The standing-column well combines supply and injection wells into one, and does not depend on the presence or flow of groundwater, beyond that of the typical bleed rate of 10 to 20% of total pumped flow. Standing-column wells are always augmented with a bleed circuit, to monitor the entering water temperature. Well water temperatures that are below or above design limits because of variations in rock conductivity, building anomalies, or nonstandard weather patterns can be restabilized (i.e., brought back to far-field temperatures by overflowing smaller amounts of water on command). This advective flow is a powerful short-term method of warming and cooling well columns that are beyond design limits.

Standing-column wells (Figure 28) consist of a borehole cased in steel or other material until competent bedrock is reached. The casing must be driven 25 to 50 ft into and sealed in competent bedrock. Bedrock sealing requirements vary by state. The remaining depth of the well is then self-supporting through bedrock. For deep commercial SCWs, a tail pipe (porter shroud) is inserted to form a conduit to draw up water, and an annulus to return water downward. This tail pipe is perforated at the bottom to form a diffuser. Water is drawn into the diffuser and up the central riser pipe to the submersible pump. The well pump must be located below the water table in line with the central riser pipe. The tail pipe allows a shorter, reduced-power wire size as well as more accessible well pump service.

The U.S. Environmental Protection Agency (EPA) Underground Injection Control program considers standing-column reinjection well water a Class V water use, type 5A7, noncontact cooling water for geothermal heating and cooling. The EPA and equivalent state agencies regard SCW reinjection as a beneficial use. Permitting or notice may be required, depending on average daily water flow rates. SCWs are serviced by qualified well contractors with minimal familiarization training.

SURFACE WATER HEAT PUMPS

Surface water bodies can be very good heat sources and sinks if properly used. In some cases, lakes can be the very best water supply for cooling. Various water circulation designs are possible; several of the more common are presented here.

In a **closed-loop system**, a water-to-air heat pump is linked to a submerged coil. Heat is exchanged to (cooling mode) or from (heating mode) the lake by the fluid (usually a water/antifreeze mixture) circulating inside the coil. The heat pump transfers heat to or from the air in the building.

In an **open-loop system**, water is pumped from the lake through a heat exchanger and returned to the lake some distance from the point at which it was removed. The pump can be located either slightly above or submerged below the lake water level. For heat pump operation in heating mode, this type is restricted to warmer climates. Entering lake water temperature must remain above 42°F to prevent freezing.

Thermal stratification of water often keeps large quantities of cold water undisturbed near the bottom of deep lakes. This water is cold enough to adequately cool buildings by simply being circulated through heat exchangers. A heat pump is not needed for cooling, and energy use is substantially reduced. Closed-loop coils may also be used in colder lakes. Heating can be provided by a separate source or with heat pumps in heating mode. As noted previously, precooling or supplemental total cooling are also allowed when water returning to the building is near or below 55°F.

Heat Transfer in Lakes

Heat is transferred to lakes by three primary modes: radiant energy from the sun, convective heat transfer from the surrounding air (when the air is warmer than the water), and conduction from the ground. Solar radiation, which can exceed 300 Btu/h per square foot of lake area, is the dominant heating mechanism, but it occurs primarily in the upper portion of the lake unless the lake is very clear. About 40% of solar radiation is absorbed at the surface (Pezent and Kavanaugh 1990). Approximately 93% of the remaining energy is absorbed at depths visible to the human eye.

Convection transfers heat to the lake when the lake surface is cooler than the air. Wind speed increases the rate at which heat is transferred to the lake, but maximum heat gain by convection is usually only 10 to 20% of maximum solar heat gain. Conduction gain from the ground is even less than convection gain (Pezent and Kavanaugh 1990).

Lakes are cooled primarily by evaporative heat transfer at the surface. Convective cooling or heating in warmer months contributes only a small percentage of the total because of the relatively small temperature difference between the air and lake surface. Back radiation typically occurs at night when the sky is clear, and can account for significant amount of cooling. The relatively warm water surface radiates heat to the cooler sky. For example, on a clear night, a cooling rate of up to 50 Btu/h·ft² is possible from a lake 25°F warmer than the sky. The last mode of heat transfer, conduction to the ground, does not play a major role in lake cooling (Pezent and Kavanaugh 1990).

To put these heat transfer rates in perspective, consider a 1 acre (43,560 ft²) lake used in connection with a 10 ton (120,000 Btu/h) heat pump. In cooling mode, the unit rejects approximately 150,000 Btu/h to the lake. This is 3.4 Btu/h·ft², or approximately 1% of the maximum heat gain from solar radiation in the summer. In winter, a 10 ton heat pump absorbs only about 90,000 Btu/h, or 2.1 Btu/h·ft², from the lake.

Thermal Patterns in Lakes

The maximum density of water occurs at 39.2°F, not at the freezing point of 32°F. This phenomenon, in combination with the normal modes of heat transfer to and from lakes, produces temperature profiles advantageous to efficient heat pump operation. In the winter, the coldest water is at the surface. It tends to remain at the surface and freeze. The bottom of a deep lake stays 5 to 10°F warmer than the surface. This condition is referred to as winter stagnation. The warmer water is a better heat source than the colder water at the surface.

As spring approaches, the surface water warms until the temperature approaches the maximum density point of 39.2°F. The winter stratification becomes unstable, and circulation loops begin to develop from top to bottom. This condition of spring overturn (Peirce 1964) causes the lake temperature to become fairly uniform.

Later in the spring, as water temperatures rise above 45°F, the circulation loops are in the upper portion of the lake. This pattern continues throughout the summer. The upper portion of the lake remains relatively warm, with evaporation cooling the lake and solar radiation warming it. The lower portion (hypolimnion) of the lake remains cold because most radiation is absorbed in the upper zone. Circulation loops do not penetrate to the lower zone, and conduction to the ground is quite small. The result is that, in deeper lakes with small or medium inflows, the upper zone is 70 to 90°F, the lower zone is 40 to 55°F, and the intermediate zone (thermocline) has a sharp change in temperature within a small change in depth. This condition is referred to as summer stagnation.

As fall begins, the water surface begins to cool by radiation and evaporation. With the approach of winter, the upper portion begins to cool toward the freezing point and the lower levels approach the maximum density temperature of 39.2°F. An ideal temperature-versus-depth chart is shown in Figure 29 for each of the four seasons (Peirce 1964).

Many lakes do exhibit near-ideal temperature profiles. However, (1) high inflow/outflow rates, (2) insufficient depth for stratification, (3) level fluctuation, (4) wind, and (5) lack of enough cold weather to establish sufficient amounts of cold water necessary for summer stratification can disrupt the profile. Therefore, a thermal survey of the lake should be conducted or existing surveys of similar lakes in similar geographic locations should be consulted (Hattemer and Kavanaugh 2005).

The thermal and environmental effects of heat rejection and absorption on larger lakes and streams has been studied to some degree (Hattemer et al. 2006). However, the impact of SWHPs on thermal stratification profiles is not well characterized. The relative importance of heat transfer modes is not well known, especially during heating mode. Design equations and tools that consider the many heat and mass flow modes of lakes, streams, and oceans are not publicly available, including data-gathering recommendations for determining the thermal response of bodies of water to SWHPs.

Closed-Loop Lake Water Heat Pump

The closed-loop lake water heat pump shown in Figure 30 has several advantages over the open loop:

- Fouling is reduced because clean water (or water/antifreeze solution) circulates through the heat pump
- Pumping-power requirement is lower because there is no elevation head from the lake surface to the heat pumps
- It is the only type recommended if a lake temperature below 40°F is possible: fluid outlet temperature is about 6°F below that of the inlet at a flow of 3 gpm per ton, and frosting occurs on heat exchanger surfaces when the bulk water temperature is in the 34 to 38°F range

Disadvantages of closed-loop systems include the following:

- Heat pump performance decreases slightly because circulation fluid temperature drops 4 to 12°F below lake temperature
- Coils may be damaged in public lakes; thermally fused polyethylene loops are much more resistant to damage than copper, glued plastic (PVC), or tubing with band-clamped joints
- Fouling can occur on the outside of the lake coil, particularly in murky lakes or where coils are located on or near the lake bottom.

High-density polyethylene (HDPE 3408) is recommended for in-lake piping. All connections must be either thermally socket-fused

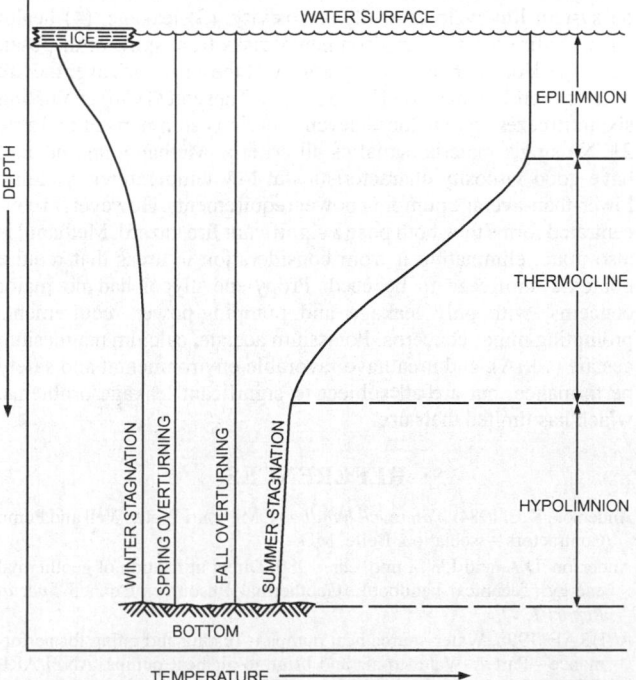

Fig. 29   Idealized Diagram of Annual Cycle of Thermal Stratification in Lakes

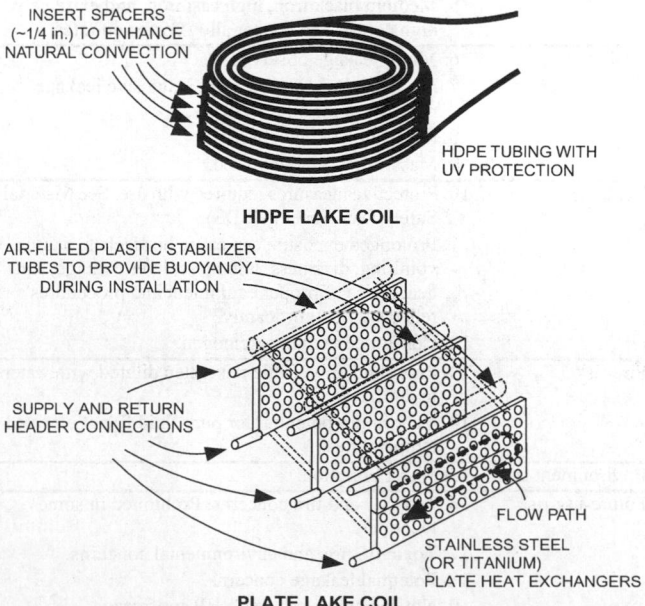

Fig. 30   Closed-Loop Lake Coils

or butt-fused. These plastic pipes should also have protection from UV radiation, especially when near the surface. Polyvinyl chloride (PVC) pipe and plastic pipe with band-clamped joints are not recommended.

Plate heat exchangers are also available, and manufacturers typically provide estimated capacities for specified conditions. However, use care in applying metal heat exchangers in cold climates. For an equal temperature difference between the reservoir and fluid inside the coil, the surface temperature of metal heat exchangers is closer to the freezing point of water. The higher thermal resistance of HDPE results in a larger required surface area and a much lower heat rate per unit of surface area. Thus, the surface temperature of an adequately designed HDPE coil tends to be higher than that of a metal tube or plate and less likely to develop ice on the coil exterior.

The piping networks of closed-loop systems resemble those used in ground-coupled heat pump systems. Both a large-diameter header between the heat pump and lake coil and several parallel loops of piping in the lake are required. Loops are spread out to limit thermal interference, hot spots, and cold pockets. Although this layout is preferred in terms of performance, installation is more time-consuming. Many contractors simply unbind plastic pipe coils and submerge them in a loose bundle. Some compensation for thermal interference is obtained by making bundled coils longer than the spread coils. A diagram of this type of installation is shown in Figure 30.

Copper coils have also been used successfully. Copper tubes have a very high thermal conductivity, so coils only one-fourth to one-third the length of plastic coils are required. However, copper pipe does not have the durability of PE 3408 or polybutylene, and if fouling is possible, coils must be significantly longer.

Antifreeze Requirements

Closed-loop horizontal and surface water heat exchanger systems often require antifreeze in the circulating water in locations with significant heating seasons. Antifreeze may not be needed in a comparable vertical borehole heat exchanger, because the deep ground temperature is essentially constant. At a depth of 6 ft, a typical value for horizontal heat exchangers, ground temperature varies by approximately ±10°F. Even if the mean ground temperature is 60°F in late winter, ground temperature at a 6 ft depth drops to 50°F. The heat extraction process lowers the temperature even further around the heat exchanger pipes, probably by an additional 10°F or more. Even with good heat transfer to the circulating water, the entering water temperature (leaving the ground heat exchanger) is around 40°F. Lakes that freeze at the surface in the winter approach 39°F at the bottom, yielding nearly the same margin of safety against freezing of the circulating fluid. An additional 10°F temperature difference is usually needed in the heat pump's refrigerant-to-water heat exchanger to transfer heat to the refrigerant. Having a refrigerant-to-water coil surface temperature below the freezing point of water risks growing a layer of ice on the water side of the heat exchanger. In the best case, coil icing restricts and may eventually block the flow of water and cause a shutdown. In the worst case, ice could burst the tubing in the coil and require a major service expense.

Several factors must be considered when selecting an antifreeze for a ground-loop heat exchanger; the most important are (1) effect on system life-cycle cost, (2) corrosivity, (3) leakage, (4) health risks, (5) fire risks, (6) environmental risks from spills or disposal, and (7) risk of future use (acceptability of the antifreeze over the life of the system). A study by Heinonen and Tapscott (1996) evaluating six antifreezes against these seven criteria is summarized in Table 21. No single material satisfies all criteria. Methanol and ethanol have good viscosity characteristics at low temperatures, yielding lower-than-average pumping power requirements. However, in concentrated forms they both pose a significant fire hazard. Methanol is also toxic, eliminating it from consideration in areas that require nontoxic antifreeze to be used. Propylene glycol had no major concerns, with only leakage and pumping-power requirements prompting minor concerns. Potassium acetate, calcium magnesium acetate (CMA), and urea have favorable environmental and safety performance, but are all subject to significant leakage problems, which has limited their use.

Table 21 Suitability of Selected GCHP Antifreeze Solutions

Category	Methanol	Ethanol	Propylene Glycol	Potassium Acetate	CMA	Urea
Life-cycle cost	***	***	**1	**1	**1	***
Corrosion	**2	**3	***	**	**4	*5
Leakage	***	**6	**6	*7	*8	*9
Health risk	*10,11	**10,12	***10	***10	***10	***10
Fire risk	*13	*13	***14	***	***	***
Environment risk	**15	**15	***	**15	**15	***
Future-use risk	*16	**17	***	**18	**19	**19

Key: *   Potential problems, caution in use required
      **   Minor potential for problems
      ***   Little or no potential for problems

Category	Notes
Life-cycle cost	1. Higher-than-average installation and energy costs.
Corrosion	2. High black iron and cast iron corrosion rates.
	3. High black iron cast iron, copper and copper alloy corrosion rates.
	4. Medium black iron, copper and copper alloy corrosion rates.
	5. Medium black iron, high cast iron, and extremely high copper and copper alloy corrosion rates.
Leakage	6. Minor leakage observed.
	7. Moderate leakage observed. Extensive leakage reported in installed systems.
	8. Moderate leakage observed.
	9. Massive leakage observed.
Health risk	10. Protective measures required with use. See Material Safety Data Sheet (MSDS).
	11. Prolonged exposure can cause headaches, nausea, vomiting, dizziness, blindness, liver damage, and death. Use of proper equipment and procedures reduces risk significantly.
	12. Confirmed human carcinogen.
Fire risk	13. Pure fluid only. Little risk when diluted with water in antifreeze.
	14. Very minor potential for pure fluid fire at elevated temperatures.
Environment risk	15. Water pollution.
Future-use risk	16. Toxicity and fire concerns. Prohibited in some locations.
	17. Toxicity, fire, and environmental concerns.
	18. Potential leakage concerns.
	19. Not currently used as GSHP antifreeze solution. May be difficult to obtain approval for use.

Source: Heinonen and Tapscott (1996).

REFERENCES

Anderson, K.E. 1984. *Water well handbook*. Missouri Water Well and Pump Contractors Association, Belle, MD.

Anderson, D.A. and J.W. Lund, eds. 1980. Direct utilization of geothermal energy: Technical handbook. Geothermal Resources Council *Special Report* 7.

ASHRAE. 1998. Water-source heat pumps—Testing and rating for performance—Part 1: Water-to-air and brine-to-air heat pumps. ANSI/ARI/ASHRAE/ISO *Standard* 13256-1:1998.

ASHRAE. 1998. Water-source heat pumps—Testing and rating for performance—Part 2: Water-to-water and brine-to-water heat pumps. ANSI/ARI/ASHRAE/ISO *Standard* 13256-2:1998.

Austin, J.C. 1978. A low temperature geothermal space heating demonstration project. *Geothermal Resources Council Transactions* 2(2).

Austin III, W.A., C. Yavuzturk, and J.D. Spitler. 2000. Development of an in-situ system for measuring ground thermal properties. *ASHRAE Transactions* 106(1):365-379.

Bernier, M.A. 2006. Closed-loop ground-coupled heat pump systems, *ASHRAE Journal* 48(9):12-19.

Bernier, M.A., A. Chahla, and P. Pinel. 2008. Long-term ground temperature changes in geo-exchange systems. *ASHRAE Transactions* 114(2):342-350.

Bullard, E. 1973. *Basic theories (Geothermal energy; Review of research and development).* UNESCO, Paris.

Campbell, M.D. and J.H. Lehr. 1973. *Water well technology.* McGraw-Hill, New York.

Caneta Research. 1995. *Commercial/institutional ground-source heat pump engineering manual.* ASHRAE.

Caneta Research. 2001. *Commissioning, preventative maintenance and troubleshooting guide for commercial GSHP systems.* ASHRAE SP-94.

Carlson, S. 2001. Development of equivalent full load heating and cooling hours for GCHPs applied in various building types and locations. ASHRAE TRP-1120, *Final Report.*

Carslaw, H.S. and J.C. Jaeger. 1947. *Heat conduction in solids.* Claremore Press, Oxford.

Chandler, R.V. 1987. Alabama streams, lakes, springs and ground waters for use in heating and cooling. *Bulletin* 129. Geological Survey of Alabama, Tuscaloosa.

Chiasson, A.D., S.J. Rees, and J.D. Spitler. 2000. A preliminary assessment of the effects of groundwater flow on closed-loop ground-source heat pump systems. *ASHRAE Transactions* 106(1):380-393.

Christen, J.E. 1977. *Central cooling—Absorption chillers.* Oak Ridge National Laboratories, Oak Ridge, TN.

Cornell University. 2006. *Lake source cooling.* http://www.utilities.cornell.edu/utl_ldlsc.html.

CSA. 2002. Design and installation of earth energy systems. *Standard* C448-02. Canadian Standards Association, Rexdale, ON.

Culver, G.G. and G.M. Reistad. 1978. Evaluation and design of downhole heat exchangers for direct applications. DOE *Report* RLO-2429-7.

Di Pippo, R. 1988. Industrial developments in geothermal power production. *Geothermal Resources Council Bulletin* 17(5).

Efrid, K.D. and G.E. Moeller. 1978. Electrochemical characteristics of 304 and 316 stainless steels in fresh water as functions of chloride concentration and temperature. *Paper* 87, Corrosion/78, Houston.

Ellis, P. 1989. *Geothermal direct use engineering and design guidebook,* Ch. 8, Materials selection guidelines. Oregon Institute of Technology, Geo-Heat Center, Klamath Falls.

Ellis, P. and C. Smith. 1983. *Addendum to material selection guidelines for geothermal energy utilization systems.* Radian Corporation, Austin.

Enwave. (no date). *Deep water lake cooling.* http://www.enwave.com/dlwc.php.

EPA. 1975. *Manual of water well construction practices.* EPA-570/9-75-001. U.S. Environmental Protection Agency, Washington, D.C. Available from http://www.epa.gov/nscep/.

EPRI. 1990. *Soil and rock classification for the design of ground-coupled heat pump systems.* International Ground Source Heat Pump Association. Electric Power Research Institute, National Rural Electric Cooperative Association, Oklahoma State University, Stillwater.

Eskilson, P. 1987. *Thermal analysis of heat extraction boreholes.* University of Lund, Sweden.

Franklin Electric. 2001. *Application manual for submersible pumps.* Franklin Electric, Bluffton, IN.

Gehlin, S. 1998. *Thermal response test, in-situ measurements of thermal properties in hard rock.* Licentiate thesis, Department of Environmental Engineering, Division of Water Resources Engineering, Luleå University of Technology, Sweden.

Hackel, S., G. Nellis, and S. Klein. 2009. Optimization of cooling dominated hybrid ground-coupled heat pump systems (RP-1384). *ASHRAE Transactions* 109(1).

Hackett, G. and J.H. Lehr. 1985. *Iron bacteria occurrence problems and control methods in water wells.* National Water Well Association, Worthington, OH.

Hattemer, B.G. and S.P. Kavanaugh. 2005. Design temperature data for surface water HVAC systems. *ASHRAE Transactions* 111(1).

Hattemer, B.G., S.P. Kavanaugh, and D. Williamson. 2006. Environmental impacts of surface water heat pump systems. *ASHRAE Transactions* 112(1).

Heinonen, E.W. and R.E. Tapscott. 1996. Assessment of anti-freeze solutions for ground-source heat pump systems. ASHRAE Research Project RP-863, *Report.*

Heinonen, E.W., R.E. Tapscott, M.W. Wildin, and A.N. Beall. 1997. Assessment of anti-freeze solutions for ground-source heat pump systems. ASHRAE BRP-90, *Report.*

Hellström, G. 1991. *Ground heat storage—Thermal analyses of duct storage systems.* Ph.D. dissertation. University of Lund, Sweden.

Ingersoll, L.R. and A.C. Zobel. 1954. *Heat conduction with engineering and geological application,* 2nd ed. McGraw-Hill, New York.

Interagency Geothermal Coordinating Council. 1980. Geothermal energy, research, development and demonstration program. DOE *Report* RA0050, IGCC-5. U.S. Department of Energy, Washington, D.C.

Kavanaugh, S.P. 1985. *Simulation and experimental verification of a vertical ground-coupled heat pump system.* Ph.D. dissertation, Oklahoma State University, Stillwater.

Kavanaugh, S.P. 1991. *Ground and water source heat pumps.* Oklahoma State University, Stillwater.

Kavanaugh, S.P. 1992. Ground-coupled heat pumps for commercial building. *ASHRAE Journal* 34(9):30-37.

Kavanaugh, S.P. 2000. Field tests for ground thermal properties—Methods and impact on GSHP system design. *ASHRAE Transactions* 106(1): DA00-13-4.

Kavanaugh, S.P. 2001. Investigation of methods for determining soil formation thermal characteristics from short term field tests. ASHRAE RP1118, *Final Report.*

Kavanaugh, S.P. 2003. Impact of operating hours on long-term heat storage and design of ground heat exchangers. *ASHRAE Transactions* 109(1).

Kavanaugh, S.P. 2008. A 12-step method for closed-loop ground-source heat-pump design. *ASHRAE Transactions* 114(2).

Kavanaugh, S.P. and T.H. Calvert. 1995. Performance of ground source heat pumps in North Alabama. *Final Report,* Alabama Universities and Tennessee Valley Authority Research Consortium. University of Alabama, Tuscaloosa.

Kavanaugh, S.P. and K. Rafferty. 1997. *Ground-source heat pumps—Design of geothermal systems for commercial and institutional buildings.* ASHRAE.

Kavanaugh, S.P., S. Lambert, and D. Messer. 2002. Development of guidelines for the selection and design of the pumping/piping subsystem for ground coupled heat pumps. ASHRAE TRP-1217, *Final Report.*

Kindle, C.H. and E.M. Woodruff. 1981. *Techniques for geothermal liquid sampling and analysis.* Battelle Pacific Northwest Laboratory, Richland, WA.

Lienau, P.J. 1979. Materials performance study of the OIT geothermal heating system. Geo-Heat Utilization Center *Quarterly Bulletin,* Oregon Institute of Technology, Klamath Falls.

Lienau, P., H. Ross, and P. Wright. 1995. Low temperature resource assessment. *Geothermal Resources Council Transactions* 19(1995).

Lund, J., T. Boyd, A. Sifford, and R. Bloomquist. 2001. Geothermal utilization in the United States—2000. *Proceedings of the 26th Annual Stanford Workshop—Reservoir Engineering.* Stanford University.

Lunis, B. 1989. *Geothermal direct use engineering and design guidebook,* Ch. 20, Environmental considerations. Oregon Institute of Technology, Geo-Heat Center, Klamath Falls.

Mogensen, P. 1983. Fluid to duct wall heat transfer in duct system heat storages. *Proceedings of the International Conference on Subsurface Heat Storage in Theory and Practice,* Stockholm, pp. 652-657.

Morrison, A. 1997. GS2000 software. *Proceedings of the Third International Heat Pumps in Cold Climates Conference,* Wolfville, Nova Scotia, pp. 67-76.

Muffler, L.J.P., ed. 1979. Assessment of geothermal resources of the United States—1978. U.S. Geological Survey *Circular* 790.

Nichols, C.R. 1978. Direct utilization of geothermal energy: DOE's resource assessment program. Direct Utilization of Geothermal Energy: A Symposium. Geothermal Resources Council.

Nutter, D., R. Couvillion, M.G. Sutton, K. Tan, R. Davis, and J. Hemphill. 2001. Investigation of borehole completion methods to optimize the environmental benefits of ground-coupled heat pumps (RP-1016). ASHRAE Research Project RP-1016, *Final Report.*

Orio, C., C.N. Johnson, S.J. Rees, A. Chiasson, D. Zheng, and J.D. Spitler. 2005. A survey of standing column well installations in North America (RP-1119). *ASHRAE Transactions* 111(2):109-121.

OSU. 1988a. *Closed-loop/ground-source heat pump systems installation guide.* International Ground Source Heat Pump Association, Oklahoma State University, Stillwater.

OSU. 1988b. *Closed loop ground source heat pump systems.* Oklahoma State University, Stillwater.

Peirce, L.B. 1964. Reservoir temperatures in north central Alabama. *Bulletin 8.* Geological Survey of Alabama, Tuscaloosa.

Pezent, M.C. and S.P. Kavanaugh. 1990. Development and verification of a thermal model of lakes used with water-source heat pumps. *ASHRAE Transactions* 96(1).

Rafferty, K. 1989a. *Geothermal direct use engineering and design guidebook,* Ch. 14, Absorption refrigeration. Oregon Institute of Technology, Geo-Heat Center, Klamath Falls.

Rafferty, K. 1989b. *A materials and equipment review of selected U.S. geothermal district heating systems.* Oregon Institute of Technology, Geo-Heat Center, Klamath Falls.

Rafferty, K. 2000a. A guide to online geologic information and publications for use in GSHP site characterization. *Transactions of the 2000 Heat Pumps in Cold Climates Conference,* Caneta Research, Missisauga, ON, Canada.

Rafferty, K. 2000b. *Scaling in geothermal heat pump systems.* Geo-Heat Center, Oregon Institute of Technology, Klamath Falls.

Rafferty, K. 2000c. Design aspects of commercial open loop heat pump systems. *Transactions of the 2000 Heat Pumps in Cold Climates Conference,* Caneta Research, Missisauga, ON, Canada.

Rafferty, K. 2008. Design issues in commercial open loop heat pump systems. *ASHRAE Transactions* 114(2).

Reistad, G.M., G.G. Culver, and M. Fukuda. 1979. Downhole heat exchangers for geothermal systems: Performance, economics and applicability. *ASHRAE Transactions* 85(1):929-939.

Remund, C. 1999. Borehole thermal resistance: Laboratory and field studies. *ASHRAE Transactions* 105(1).

Remund, C. and N. Paul. 2000. *Grouting for vertical geothermal heat pump systems: Engineering design and field procedures manual.* International Ground Source Heat Pump Association, Stillwater.

Roscoe Moss Company. 1985. *The engineers' manual for water well design.* Roscoe Moss Company, Los Angeles.

Sachs, H. 2002. *Geology and drilling methods for ground-source heat pump system installations: An introduction for engineers.* ASHRAE.

Shonder, J.A. and J.V. Beck. 1999. Determining effective soil formation properties from field data using a parameter estimation technique. *ASHRAE Transactions* 105(1):458-466.

Shonder, J.A., V.D. Baxter, J.W. Thornton, and P. Hughes. 1999. A new comparison of vertical ground heat exchanger design methods for residential applications. *ASHRAE Transactions* 105(2).

Shonder, J.A., V.D. Baxter, P.J. Hughes, and J.W. Thornton. 2000. A comparison of vertical ground heat exchanger design software for commercial applications. *ASHRAE Transactions* 106(1).

Spitler, J.D. 2000. GLHEPRO—A design tool for commercial building ground loop heat exchangers. *Proceedings of the Fourth International Heat Pumps in Cold Climates Conference,* Aylmer, Québec.

Spitler, J.D. 2002. R&D studies applied to standing column well design, ASHRAE RP-1119, *Report.*

Spitler, J.D., S.J. Rees, and C. Yavuzturk. 1999. More comments on in-situ borehole thermal conductivity testing. *The Source* 12(2):4-6.

Spitler, J.D, S.J. Rees, and C. Yavuzturk. 2000. Recent developments in ground source heat pump system design, modeling and applications. *Proceedings of the Dublin 2000 Conference.*

Stiger, S., J. Renner, and G. Culver. 1989. *Geothermal and direct use engineering and design guidebook,* Ch. 7, Well testing and reservoir evaluation. Oregon Institute of Technology, Geo-Heat Center, Klamath Falls.

Svec, O.J. 1990. Spiral ground heat exchangers for heat pump applications. *Proceedings of 3rd IEA Heat Pump Conference.* Pergamon Press, Tokyo.

UOP. 1975. *Ground water and wells.* Johnson Division, UOP Inc., St. Paul, MN.

USGS. 2000. *Ground water atlas of the United States.* U.S. Geological Survey, Reston, VA.

Witte, H., G. van Gelder, and J. Spitler. 2002. In-situ thermal conductivity testing: A Dutch perspective. *ASHRAE Transactions* 108(1).

BIBLIOGRAPHY

Allen, E. 1980. *Preliminary inventory of western U.S. cities with proximate hydrothermal potential.* Eliot Allen and Associates, Salem, OR.

Caneta Research. 1995. Operating experiences with commercial ground-source heat pumps. ASHRAE RP-863, *Report.*

Chiasson, A.D. and C. Yavusturk. 2003. Assessment of the viability of hybrid geothermal heat pump systems with solar thermal collectors. *ASHRAE Transactions* 109(2).

Cosner, S.R. and J.A. Apps. 1978. A compilation of data on fluids from geothermal resources in the United States. DOE *Report* LBL-5936. Lawrence Berkeley Laboratory, Berkeley, CA.

Kavanaugh, S. 1991. *Ground and water source heat pumps: A manual for the design and installation of ground coupled, ground water and lake water heating and cooling systems in southern climates.* Energy Information Services, Tuscaloosa.

Kavanaugh, S.P. and M.C. Pezent. 1990. Lake water applications of water-to-air heat pumps. *ASHRAE Transactions* 96(1):813-820.

Kavanaugh, S.P. and K.D. Rafferty, eds. 1995. *Commercial ground source heat pump systems—A collection of ASHRAE papers.* ASHRAE.

Lund, J., ed. 2000. *Geothermal direct use engineering and design guidebook.* Geo-Heat Center, Klamath Falls.

Lund, J.W., P.J. Lienau, G.G. Culver, and C.V. Higbee. 1979. Klamath Falls geothermal district heating. *Geothermal Resources Council Transactions* 3.

McCray, K., ed. 1997. *Guidelines for the construction of vertical boreholes for closed loop heat pump systems.* National Ground Water Association, Westerville, OH.

Mitchell, D.A. 1980. Performance of typical HVAC materials in two geothermal heating systems. *ASHRAE Transactions* 86(1):763-768.

Performance Pipe. 1998. *Polyethylene piping systems manual 10428-98 piping manual.* Performance Pipe Inc., Dallas.

SOLAR ENERGY USE

THE sun radiates considerable energy onto the earth. Putting that diffuse, rarely over 300 Btu/h·ft² energy to work has lead to the creation of many types of devices to convert that energy into useful forms, mainly heat and electricity. How that energy is valued economically drives the ebb and flow of the global solar industry. This chapter discusses several different types of solar equipment and system designs for various HVAC applications, as well as methods to determine the solar resource.

Worldwide, solar energy use varies in application and degree. In China and, to a lesser extent, Australasia, solar energy is widely used, particularly for water heating. In Europe, government incentives have fostered use of photovoltaic and thermal systems for both domestic hot-water and space heating. In the Middle East, solar power is used for desalination and absorption air conditioning. Solar energy use in the United States is relatively modest, driven by tax policy and utility programs that generally react to energy shortages or the price of oil.

Recent interest in sustainability and green buildings has led to an increased focus on solar energy devices for their nonpolluting and renewable qualities; replacing fossil fuel with domestic, renewable energy sources can also enhance national security by reducing dependence on imported energy.

For more information on the use of solar and other energy sources, see the Energy Information Administration (EAI) of the U.S. Department of Energy (www.eai.doe.gov) and the International Energy Agency (www.iea.org).

QUALITY AND QUANTITY OF SOLAR ENERGY

Solar Constant

Solar energy approaches the earth as electromagnetic radiation, with wavelengths ranging from 0.1 μm (x-rays) to 100 m (radio waves). The earth maintains a thermal equilibrium between the annual input of shortwave radiation (0.3 to 2.0 μm) from the sun and the outward flux of longwave radiation (3.0 to 30 μm). Only a limited band need be considered in terrestrial applications, because 99% of the sun's radiant energy has wavelengths between 0.28 and 4.96 μm. The current value of the solar constant (which is defined as the intensity of solar radiation on a surface normal to the sun's rays, just beyond the earth's atmosphere at the average earth-sun distance) is 433 Btu/h·ft². Chapter 15 of the 2009 *ASHRAE Handbook—Fundamentals* has further information on this topic.

Solar Angles

The axis about which the earth rotates is tilted at an angle of 23.45° to the plane of the earth's orbital plane and the sun's equator. The earth's tilted axis results in a day-by-day variation of the angle

The preparation of this chapter is assigned to TC 6.7, Solar Energy Utilization.

between the earth-sun line and the earth's equatorial plane, called the **solar declination** δ. This angle varies with the date, as shown in Table 1 for the year 1964 and in Table 2 for 1977. For other dates, the declination may be estimated by the following equation:

$$\delta = 23.45 \sin\left[360° \times \frac{284+N}{365}\right] \quad (1)$$

where N = year day, with January 1 = 1. For values of N, see Tables 1 and 2.

The relationship between δ and the date from year to year varies to an insignificant degree. The daily change in the declination is the primary reason for the changing seasons, with their variation in the distribution of solar radiation over the earth's surface and the varying number of hours of daylight and darkness. Note that the following sections are based in the northern hemisphere; sites in the southern hemisphere will be 180° from the examples (e.g., a solar panel should face north).

The earth's rotation causes the sun's apparent motion (Figure 1). The position of the sun can be defined in terms of its altitude β above the horizon (angle HOQ) and its azimuth ϕ, measured as angle HOS in the horizontal plane.

At solar noon, the sun is exactly on the meridian, which contains the south-north line. Consequently, the solar azimuth ϕ is 0°. The noon altitude $β_N$ is given by the following equation as

$$β_N = 90° - LAT + \delta \quad (2)$$

where LAT = latitude.

Because the earth's daily rotation and its annual orbit around the sun are regular and predictable, the solar altitude and azimuth may be readily calculated for any desired time of day when the latitude, longitude, and date (declination) are specified. Apparent solar time (AST) must be used, expressed in terms of the hour angle H, where

$$H = (\text{Number of hours from solar noon}) \times 15°$$
$$= \frac{\text{Number of minutes from solar noon}}{4} \quad (3)$$

Solar Time

Apparent solar time (AST) generally differs from local standard time (LST) or daylight saving time (DST), and the difference can be significant, particularly when DST is in effect. Because the sun appears to move at the rate of 360° in 24 h, its apparent rate of motion is 4 min per degree of longitude. The AST can be determined from the following equation:

Table 1 Date, Declination, and Equation of Time for the 21st Day of Each Month of 1964, with Data (A, B, C) Used to Calculate Direct Normal Radiation Intensity at the Earth's Surface

	Jan	Feb	Mar	Apr	May	June	July	Aug	Sept	Oct	Nov	Dec
Year Day	21	52	80	111	141	172	202	233	264	294	325	355
Declination δ, degrees	−19.9	−10.6	0.0	+11.9	+20.3	+23.45	+20.5	+12.1	0.0	−10.7	−19.9	−23.45
Equation of time, minutes	−11.2	−13.9	−7.5	+1.1	+3.3	−1.4	−6.2	−2.4	+7.5	+15.4	+13.8	+1.6
Solar noon		late			early			late			early	
A, Btu/h·ft^2	390	385	376	360	350	345	344	351	365	378	387	391
B, dimensionless	0.142	0.144	0.156	0.180	0.196	0.205	0.207	0.201	0.177	0.160	0.149	0.142
C, dimensionless	0.058	0.060	0.071	0.097	0.121	0.134	0.136	0.122	0.092	0.073	0.063	0.057

A = apparent solar irradiation at air mass zero for each month.
B = atmospheric extinction coefficient.
C = ratio of diffuse radiation on horizontal surface to direct normal irradiation.

Table 2 Solar Position Data for 1977

Date		Jan	Feb	Mar	Apr	May	June	July	Aug	Sept	Oct	Nov	Dec
1	Year Day	1	32	60	91	121	152	182	213	244	274	305	335
	Declination δ	−23.0	−17.0	−7.4	+4.7	+15.2	+22.1	+23.1	+17.9	+8.2	−3.3	−14.6	−21.9
	Eq of Time	−3.6	−13.7	−12.5	−4.0	+2.9	+2.4	−3.6	−6.2	+0.0	+10.2	+16.3	11.0
6	Year Day	6	37	65	96	126	157	187	218	249	279	310	340
	Declination δ	−22.4	−15.5	−5.5	+6.6	+16.6	+22.7	+22.7	+16.6	+6.7	−5.3	−16.1	−22.5
	Eq of Time	−5.9	−14.2	−11.4	−2.5	+3.5	+1.6	−4.5	−5.8	+1.6	+11.8	+16.3	+9.0
11	Year Day	11	42	70	101	131	162	192	223	254	284	315	345
	Declination δ	−21.7	−13.9	−3.5	−8.5	+17.9	+23.1	+22.1	+15.2	+4.4	−7.2	−17.5	−23.0
	Eq of Time	−8.0	−14.4	−10.2	−1.1	+3.7	+0.6	−5.3	−5.1	+3.3	+13.1	+15.9	+6.8
16	Year Day	16	47	75	106	136	167	197	228	259	289	320	350
	Declination δ	−20.8	−12.2	−1.6	+10.3	+19.2	+23.3	+21.3	+13.6	+2.5	−8.7	−18.8	−23.3
	Eq of Time	−9.8	−14.2	−8.8	+0.1	+3.8	−0.4	−5.9	−4.3	+5.0	+14.3	+15.2	+4.4
21	Year Day	21	52	80	111	141	172	202	233	264	294	325	355
	Declination δ	−19.6	−10.4	+0.4	+12.0	+20.3	+23.4	+20.6	+12.0	+0.5	−10.8	−20.0	−23.4
	Eq of Time	−11.4	−13.8	−7.4	+1.2	+3.6	−1.5	−6.2	−3.1	+6.8	+15.3	+14.1	+2.0
26	Year Day	26	57	85	116	146	177	207	238	269	299	330	360
	Declination δ	−18.6	−8.6	+2.4	+13.6	+21.2	+23.3	+19.3	+10.3	−1.4	−12.6	−21.0	−23.4
	Eq of Time	−12.6	−13.1	−5.8	+2.2	+3.2	−2.6	−6.4	−1.8	+8.6	+15.9	+12.7	−0.5

Source: ASHRAE *Standard* 93-1986 (RA91).
Notes: Units for declination are angular degrees; units for equation of time are minutes. Values of declination and equation of time vary slightly for specific dates in other years.

$$\text{AST} = \text{LST} + \text{Equation of time}$$
$$+ (4 \text{ min})(\text{LST meridian} - \text{Local longitude}) \qquad (4)$$

The longitudes of the seven standard time meridians that affect North America are Atlantic ST, 60°; Eastern ST, 75°; Central ST, 90°; Mountain ST, 105°; Pacific ST, 120°; Yukon ST, 135°; and Alaska-Hawaii ST, 150°.

The equation of time is the measure, in minutes, of the extent by which solar time, as determined by a sundial, runs faster or slower than local standard time (LST), as determined by a clock that runs at a uniform rate. Table 1 gives values of the declination of the sun and the equation of time for the 21st day of each month for the year 1964 (when the ASHRAE solar radiation tables were first calculated), and Table 2 gives values of δ and the equation of time for six days each month for the year 1977.

Example 1. Find AST at noon DST on July 21 for Washington, D.C., longitude = 77°, and for Chicago, longitude = 87.6°.

Solution: Noon DST is actually 11:00 AM LST. Washington is in the eastern time zone, and the LST meridian is 75°. From Table 1, the equation of time for July 21 is –6.2 min. Thus, from Equation (4), noon DST for Washington is actually

$$\text{AST} = 11:00 - 6.2 + 4(75 - 77) = 10:45.8 \text{ AST} = 10.76 \text{ h}$$

Chicago is in the central time zone, and the LST meridian is 90°. Thus, from Equation (4), noon central DST is

$$\text{AST} = 11:00 - 6.2 + 4(90 - 87.6) = 11:03.4 \text{ AST} = 11.06 \text{ h}$$

The hour angles H for these two examples (see Figure 2) are

for Washington, $H = (12.00 - 10.76) \, 15° = 18.6°$ east
for Chicago, $H = (12.00 - 11.06) \, 15° = 14.10°$ east

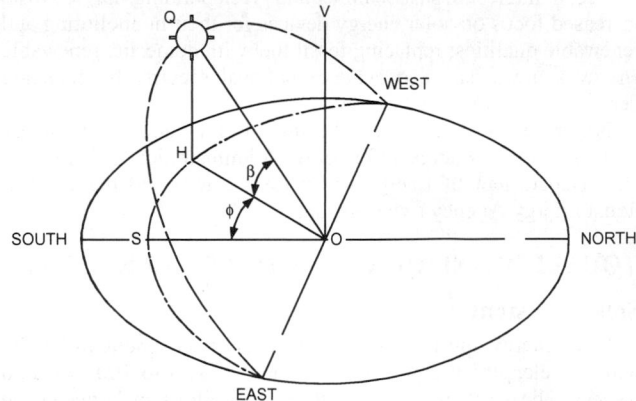

Fig. 1 Apparent Daily Path of the Sun Showing Solar Altitude (β) and Solar Azimuth (φ)

To find the solar altitude β and the azimuth φ when the hour angle H, latitude LAT, and declination δ are known, the following equations may be used:

$$\sin \beta = \cos(\text{LAT}) \cos \delta \cos H + \sin(\text{LAT}) \sin \delta \qquad (5)$$

$$\sin \phi = \cos \delta \sin H / \cos \beta \qquad (6)$$

or $$\cos \phi = (\cos H \cos \delta \sin \text{LAT} - \sin \delta \cos \text{LAT})/\cos \beta \qquad (7)$$

Tables 15 through 21 in Chapter 29 of the 1997 *ASHRAE Handbook—Fundamentals* give values for latitudes from 16 to 64° north. For any other date or latitude, interpolation between the tabulated values will give sufficiently accurate results.

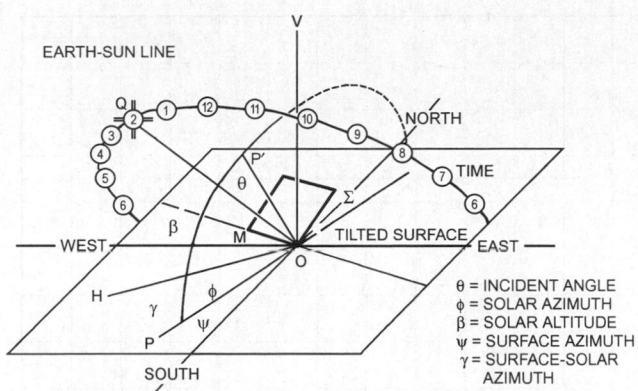

Fig. 2 Solar Angles with Respect to a Tilted Surface

Incident Angle

The angle between the line normal to the irradiated surface (OP′ in Figure 2) and the earth-sun line OQ is called the incident angle θ. It is important in solar technology because it affects the intensity of the direct component of solar radiation striking the surface and the surface's ability to absorb, transmit, or reflect the sun's rays.

To determine θ, the surface azimuth ψ and the surface-solar azimuth γ must be known. The surface azimuth (angle POS in Figure 2) is the angle between the south-north line SO and the normal PO to the intersection of the irradiated surface with the horizontal plane, shown as line OM. The surface-solar azimuth, angle HOP, is designated by γ and is the angular difference between the solar azimuth ϕ and the surface azimuth ψ. For surfaces facing east of south, $\gamma = \phi - \psi$ in the morning and $\gamma = \phi + \psi$ in the afternoon. For surfaces facing west of south, $\gamma = \phi + \psi$ in the morning and $\gamma = \phi - \psi$ in the afternoon. For south-facing surfaces, $\psi = 0°$, so $\psi = \phi$ for all conditions. The angles δ, β, and ϕ are always positive.

For a surface with a tilt angle Σ (measured from the horizontal), the angle of incidence θ between the direct solar beam and the normal to the surface (angle QOP′ in Figure 2) is given by

$$\cos\theta = \cos\beta\cos\gamma\sin\Sigma + \sin\beta\cos\Sigma \qquad (8)$$

For vertical surfaces, $\Sigma = 90°$, $\cos\Sigma = 0$, and $\sin\Sigma = 1.0$, so Equation (8) becomes

$$\cos\theta = \cos\beta\cos\gamma \qquad (9)$$

For horizontal surfaces, $\Sigma = 0°$, $\sin\Sigma = 0$, and $\cos\Sigma = 1.0$, so Equation (8) leads to

$$\theta_H = 90° - \beta \qquad (10)$$

Example 2. Find θ for a south-facing surface tilted upward 30° from the horizontal at 40° north latitude at 4:00 PM, AST, on August 21.

Solution: From Equation (3), at 4:00 PM on August 21,

$$H = 4 \times 15° = 60°$$

From Table 1,

$$\delta = 12.1°$$

From Equation (5),

$$\sin\beta = \cos 40° \cos 12.1° \cos 60° + \sin 40° \sin 12.1°$$
$$\beta = 30.6°$$

From Equation (6),

$$\sin\phi = \cos 12.1° \sin 60°/\cos 30.6°$$
$$\phi = 79.7°$$

The surface faces south, so $\phi = \gamma$. From Equation (8),

$$\cos\theta = \cos 30.6° \cos 79.7° \sin 30° + \sin 30.6° \cos 30°$$
$$\theta = 58.8°$$

ASHRAE *Standard* 93-2010 provides tabulated values of θ for horizontal and vertical surfaces and for south-facing surfaces tilted upward at angles equal to the latitude minus 10°, the latitude, the latitude plus 10°, and the latitude plus 20°. These tables cover the latitudes from 24° to 64° north, in 8° intervals.

Solar Spectrum

Beyond the earth's atmosphere, the effective blackbody temperature of the sun is 10,370°R. The maximum spectral intensity occurs at 0.48 μm in the green portion of the visible spectrum. Thekaekara (1973) presents tables and charts of the sun's extraterrestrial spectral irradiance from 0.120 to 100 μm, the range in which most of the sun's radiant energy is contained. The ultraviolet portion of the spectrum below 0.40 μm contains 8.73% of the total, another 38.15% is contained in the visible region between 0.40 and 0.70 μm, and the infrared region contains the remaining 53.12%.

Solar Radiation at the Earth's Surface

In passing through the earth's atmosphere, some of the sun's direct radiation I_D is scattered by nitrogen, oxygen, and other molecules, which are small compared to the wavelengths of the radiation; and by aerosols, water droplets, dust, and other particles with diameters comparable to the wavelengths (Gates 1966). This scattered radiation causes the sky to appear blue on clear days, and some of it reaches the earth as diffuse radiation I.

Attenuation of the solar rays is also caused by absorption, first by the ozone in the outer atmosphere, which causes a sharp cutoff at 0.29 μm of the ultraviolet radiation reaching the earth's surface (Figure 3). In the longer wavelengths, there are series of absorption bands caused by water vapor, carbon dioxide, and ozone. The total amount of attenuation at any given location is determined by (1) the length of the atmospheric path through which the rays travel and (2) the composition of the atmosphere. The path length is expressed in terms of the air mass m, which is the ratio of the mass of atmosphere in the actual earth-sun path to the mass that would exist if the sun were directly overhead at sea level ($m = 1.0$). For all practical purposes, at sea level, $m = 1.0/\sin\beta$. Beyond the earth's atmosphere, $m = 0$.

Before 1967, solar radiation data were based on an assumed solar constant of 419.7 Btu/h·ft² and on a standard sea-level atmosphere containing the equivalent depth of 2.8 mm of ozone, 20 mm of precipitable moisture, and 300 dust particles per cm³. Threlkeld and Jordan (1958) considered the wide variation of water vapor in the atmosphere above the United States at any given time, and particularly the seasonal variation, which finds three times as much moisture in the atmosphere in midsummer as in December, January, and February. The basic atmosphere was assumed to be at sea-level barometric pressure, with 2.5 mm of ozone, 200 dust particles per cm³, and an actual precipitable moisture content that varied throughout the year from 8 mm in midwinter to 28 mm in mid-July. Figure 4 shows the variation of the direct normal irradiation with solar altitude, as estimated for clear atmospheres and for an atmosphere with variable moisture content.

Stephenson (1967) showed that the intensity of the direct normal irradiation I_{DN} at the earth's surface on a clear day can be estimated by the following equation:

$$I_{DN} = Ae^{-B/\sin\beta} \qquad (11)$$

where A, the apparent extraterrestrial irradiation at $m = 0$, and B, the atmospheric extinction coefficient, are functions of the date and take into account the seasonal variation of the earth-sun distance and the air's water vapor content.

The values of the parameters A and B given in Table 1 were selected so that the resulting value of I_{DN} would be closely agree with the Threlkeld and Jordan (1958) values on average cloudless days. The values of I_{DN} given in Tables 15 through 21 in Chapter 29 of the 1997 *ASHRAE Handbook—Fundamentals* were obtained by using

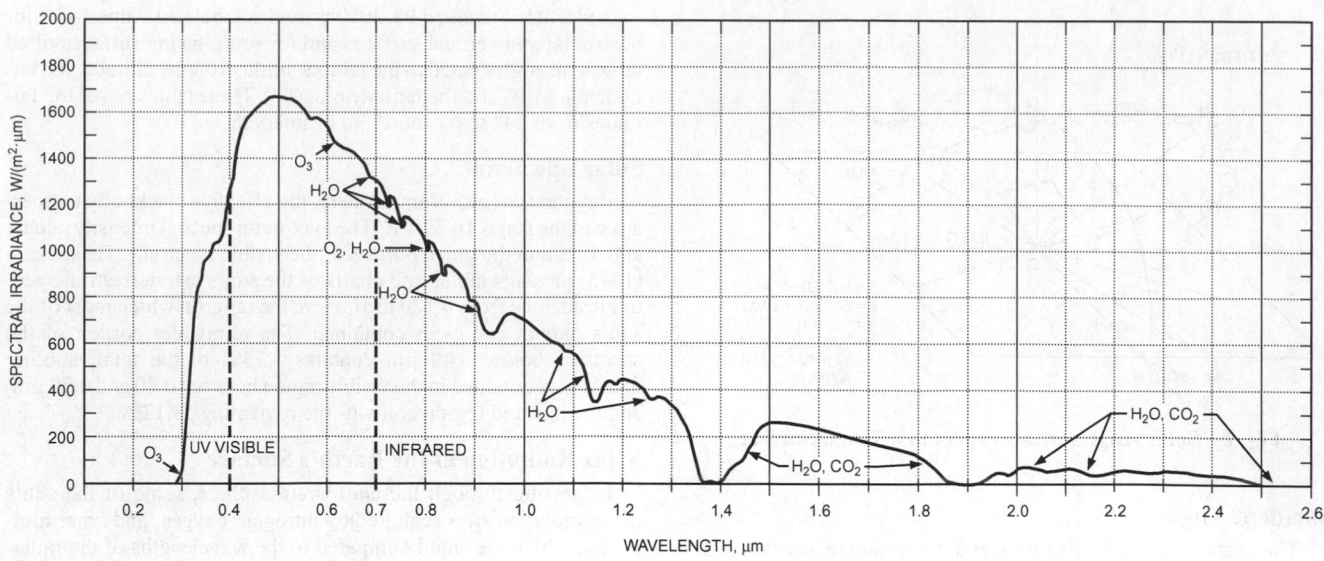

Fig. 3   Spectral Solar Irradiation at Sea Level for Air Mass = 1.0

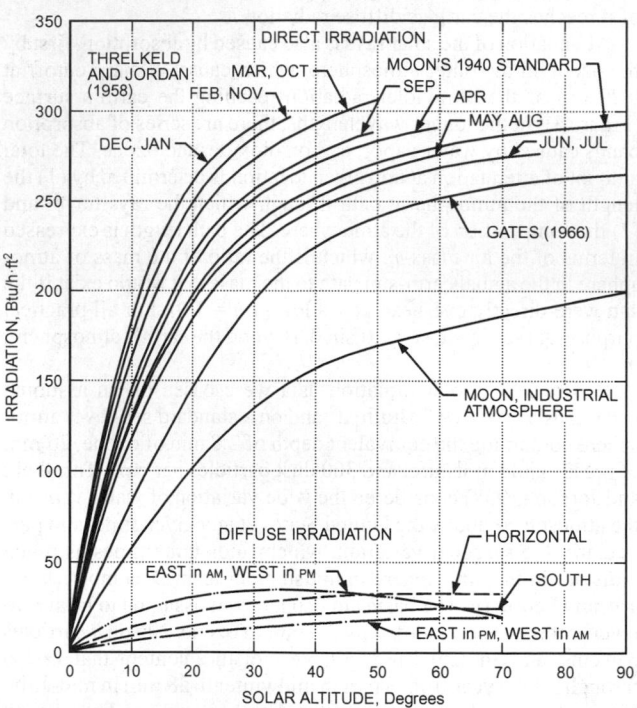

Fig. 4   Variation with Solar Altitude and Time of Year for Direct Normal Irradiation

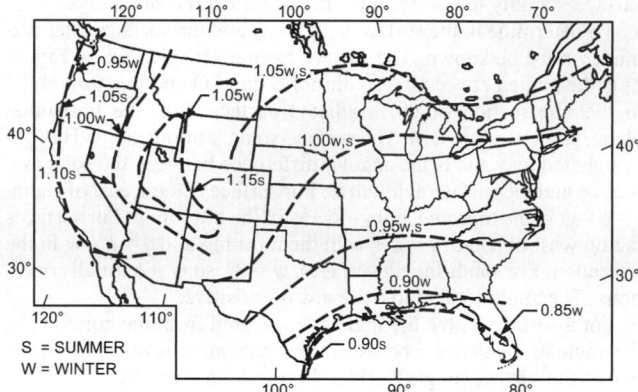

Fig. 5   Clearness Numbers for the United States

Equation (11) and data from Table 1. The values of solar altitude β and solar azimuth φ may be obtained from Equations (5) and (6). Additional information may be found in Chapters 14 and 15 of the 2009 *ASHRAE Handbook—Fundamentals*.

Because local values of atmospheric water content and elevation can vary markedly from the sea-level average, the concept of **clearness number** was introduced to express the ratio between the actual clear-day direct irradiation intensity at a specific location and the intensity calculated for the standard atmosphere for the same location and date.

Figure 5 shows the Threlkeld-Jordan map of winter and summer clearness numbers for the continental United States. Irradiation

values should be adjusted by the clearness numbers applicable to each particular location.

Design Values of Total Solar Irradiation

The total solar irradiation $I_{t\theta}$ of a terrestrial surface of any orientation and tilt with an incident angle θ is the sum of the direct component $I_{DN} \cos \theta$ plus the diffuse component $I_{d\theta}$ coming from the sky plus whatever amount of reflected shortwave radiation I_r may reach the surface from the earth or from adjacent surfaces:

$$I_{t\theta} = I_{DN} \cos \theta + I_{d\theta} + I_r \tag{12}$$

The diffuse component is difficult to estimate because of its non-directional nature and its wide variations. Figure 4 shows typical values of diffuse irradiation of horizontal and vertical surfaces. For clear days, Threlkeld (1963) derived a dimensionless parameter (designated as *C* in Table 1), which depends on the dust and moisture content of the atmosphere and thus varies throughout the year:

$$C = \frac{I_{dH}}{I_{DN}} \tag{13}$$

where I_{dH} is the diffuse radiation falling on a horizontal surface under a cloudless sky.

The following equation may be used to estimate the amount of diffuse radiation $I_{d\theta}$ that reaches a tilted or vertical surface:

$$I_{d\theta} = CI_{DN}F_{ss} \tag{14}$$

where

$$F_{ss} = \frac{1 + \cos \Sigma}{2} \tag{15}$$
= angle factor between the surface and the sky

The reflected radiation I_r from the foreground is given by

$$I_r = I_{tH}\rho_g F_{sg} \tag{16}$$

where

ρ_g = reflectance of the foreground
I_{tH} = total horizontal irradiation

$$F_{sg} = \frac{1 + \cos \Sigma}{2} \tag{17}$$
= angle factor between surface and earth

The intensity of reflected radiation that reaches any surface depends on the nature of the reflecting surface and on the incident angle between the sun's direct beam and the reflecting surface. Many measurements made of the reflection (albedo) of the earth under varying conditions show that clean, fresh snow has the highest reflectance (0.87) of any natural surface.

Threlkeld (1963) gives values of reflectance for commonly encountered surfaces at solar incident angles from 0 to 70°. Bituminous paving generally reflects less than 10% of the total incident solar irradiation; bituminous and gravel roofs reflect from 12 to 15%; concrete, depending on its age, reflects from 21 to 33%. Bright green grass reflects 20% at $\theta = 30°$ and 30% at $\theta = 65°$.

The maximum daily amount of solar irradiation that can be received at any given location is that which falls on a flat plate with its surface kept normal to the sun's rays so it receives both direct and diffuse radiation. For fixed flat-plate collectors, the total amount of clear-day irradiation depends on the orientation and slope. As shown by Figure 6 for 40° north latitude, the total irradiation of horizontal surfaces reaches its maximum in midsummer, whereas

vertical south-facing surfaces experience their maximum irradiation during the winter. These curves show the combined effects of the varying length of days and changing solar altitudes.

In general, flat-plate collectors are mounted at a fixed tilt angle Σ (above the horizontal) to give the optimum amount of irradiation for each purpose. Collectors intended for winter heating benefit from higher tilt angles than those used to operate cooling systems in summer. Solar water heaters, which should operate satisfactorily throughout the year, require an angle that is a compromise between the optimal values for summer and winter. Figure 6 shows the monthly variation of total day-long irradiation on the 21st day of each month at 40° north latitude for flat surfaces with various tilt angles.

Tables in ASHRAE *Standard* 93 give the total solar irradiation for the 21st day of each month at latitudes 24 to 64° north on surfaces with the following orientations: normal to the sun's rays (direct normal data do not include diffuse irradiation); horizontal; south-facing, tilted at (LAT–10), LAT, (LAT+10), (LAT+20), and 90° from the horizontal. The day-long total irradiation for fixed surfaces is highest for those that face south, but a deviation in azimuth of 15 to 20° causes only a small reduction.

Solar Energy for Flat-Plate Collectors

The preceding data apply to clear days. The irradiation for average days may be estimated for any specific location by referring to publications of the U.S. Weather Service. The *Climatic Atlas of the United States* (U.S. GPO 2005) gives maps of monthly and annual values of percentage of possible sunshine, total hours of sunshine, mean solar radiation, mean sky cover, wind speed, and wind direction. Chapter 14 in the 2009 *ASHRAE Handbook—Fundamentals* also provides several sources for obtaining solar data.

The total daily horizontal irradiation data reported by the U.S. Weather Bureau for approximately 100 stations before 1964 show that the percentage of total clear-day irradiation is approximately a linear function of the percentage of possible sunshine. The irradiation is not zero for days when the percentage of possible sunshine is reported as zero, because substantial amounts of energy reach the earth in the form of diffuse radiation. Instead, the following relationship exists for the percentage of possible sunshine:

$$\frac{\text{Day-long actual } I_{tH}}{\text{Clear day } I_{tH}} 100 = a + b \tag{18}$$

where a and b are constants for any specified month at any given location. See also Duffie and Beckman (1992) and Jordan and Liu (1977).

Longwave Atmospheric Radiation

In addition to the shortwave (0.3 to 2.0 μm) radiation it receives from the sun, the earth receives longwave radiation (4 to 100 μm, with maximum intensity near 10 μm) from the atmosphere. In turn, a surface on the earth emits longwave radiation q_{Rs} in accordance with the Stefan-Boltzmann law:

Table 3 Sky Emittance and Amount of Precipitable Moisture Versus Dew-Point Temperature

Dew Point, °F	Sky Emittance, e_{at}	Precipitable Water, in.
−20	0.68	0.12
−10	0.71	0.16
0	0.73	0.18
10	0.76	0.22
20	0.77	0.29
30	0.79	0.41
40	0.82	0.57
50	0.84	0.81
60	0.86	1.14
70	0.88	1.61

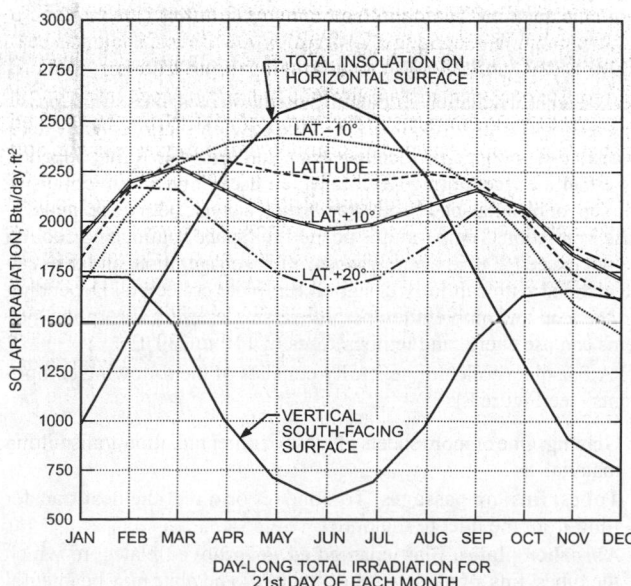

Fig. 6   Total Daily Irradiation for Horizontal, Tilted, and Vertical Surfaces at 40° North Latitude

$$q_{Rs} = e_s \sigma T_s^4 \qquad (19)$$

where

e_s = surface emittance
σ = Stefan-Boltzmann constant, 0.1712×10^{-8} Btu/h·ft²·°R⁴
T_s = absolute temperature of surface, °R

For most nonmetallic surfaces, the longwave hemispheric emittance is high, ranging from 0.84 for glass and dry sand to 0.95 for black built-up roofing. For highly polished metals and certain selective surfaces, e_s may be as low as 0.05 to 0.20.

Atmospheric radiation comes primarily from water vapor, carbon dioxide, and ozone (Bliss 1961); very little comes from oxygen and nitrogen, although they make up 99% of the air.

Approximately 90% of the incoming atmospheric radiation comes from the lowest 300 ft. Thus, air conditions at ground level largely determine the magnitude of the incoming radiation. Downward radiation from the atmosphere q_{Rat} may be expressed as

$$q_{Rat} = e_{at} \sigma T_{at}^4 \qquad (20)$$

The emittance of the atmosphere is a complex function of air temperature and moisture content. The dew point of the atmosphere near the ground determines the total amount of moisture in the atmosphere above the place where the dry-bulb and dew-point temperatures of the atmosphere are determined (Reitan 1963). Bliss (1961) found that the emittance of the atmosphere is related to the dew-point temperature, as shown by Table 3.

The apparent sky temperature is defined as the temperature at which the sky (as a blackbody) emits radiation at the rate actually emitted by the atmosphere at ground level temperature with its actual emittance e_{at}. Then,

$$\sigma T_{sky}^4 = e_{at} \sigma T_{at}^4 \qquad (21)$$

or

$$T_{sky}^4 = e_{at} T_{at}^4 \qquad (22)$$

Example 3. Consider a summer night condition when ground-level temperatures are 65°F dew point and 85°F dry bulb. From Table 3, e_{at} at 65°F dew point is 0.87, and the apparent sky temperature is

$$T_{sky} = 0.87^{0.25} (85 + 459.6) = 526.0°R$$

Thus, T_{sky} = 526.0 – 459.6 = 66.4°F, which is 18.6°F below the ground-level dry-bulb temperature.

For a winter night in Arizona, when temperatures at ground level are 60°F db and 25°F dp, from Table 3, the emittance of the atmosphere is 0.78, and the apparent sky temperature is 488.3°R or 28.7°F.

A simple relationship, which ignores vapor pressure of the atmosphere, may also be used to estimate the apparent sky temperature:

$$T_{sky} = 0.0411 T_{at}^{1.5} \qquad (23)$$

where T is in degrees Rankine.

If the temperature of the radiating surface is assumed to equal the atmospheric temperature, the heat loss from a black surface ($e_s = 1.0$) may be found from Figure 7.

Example 4. For the conditions in the previous example for summer, 85°F db and 65°F dp, the rate of radiative heat loss is about 23 Btu/h·ft². For winter, 60°F db and 25°F dp, the heat loss is about 27 Btu/h·ft².

Where a rough, unpainted roof is used as a heat dissipater, the rate of heat loss rises rapidly as the surface temperature goes up. For the summer example, a painted metallic roof, $e_s = 0.96$, at 100°F (559.6°R) will have a heat loss rate of

$$q_{Rs} = 0.96 \times 0.1712 \times 10^{-8}[559.6^4 - 526.0^4]$$
$$= 35.4 \text{ Btu/h·ft}^2$$

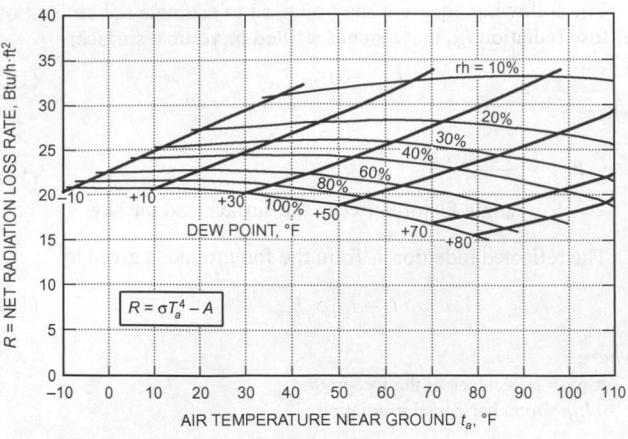

Fig. 7   Radiation Heat Loss to Sky from Horizontal Blackbody

This analysis shows that radiation alone is not an effective means of dissipating heat under summer conditions of high dew-point and ambient temperatures. In spring and fall, when both the dew-point and dry-bulb temperatures are relatively low, radiation becomes much more effective.

On overcast nights, when cloud cover is low, the clouds act much like blackbodies at ground-level temperature, and virtually no heat can be lost by radiation. Exchange of longwave radiation between the sky and terrestrial surfaces occurs in the daytime as well as at night, but the much greater magnitude of the solar irradiation masks the longwave effects.

SOLAR ENERGY COLLECTION

Solar energy can be converted by (1) chemical, (2) electrical, and (3) thermal processes. Photosynthesis is a chemical process that produces food and converts CO_2 to O_2. Photovoltaic cells convert solar energy to electricity. The section on Photovoltaic Applications discusses some applications for these devices. The thermal conversion process, the primary subject of this chapter, provides thermal energy for space heating and cooling, domestic water heating, power generation, distillation, and process heating.

Solar Heat Collection by Flat-Plate Collectors

The solar irradiation calculation methods presented in the previous sections may be used to estimate how much energy is likely to be available at any specific location, date, and time of day for collection by either a concentrating device, which uses only the direct rays of the sun, or by a flat-plate collector, which can use both direct and diffuse irradiation. Temperatures needed for space heating and cooling do not exceed 200°F, even for absorption refrigeration, and they can be attained with carefully designed flat-plate collectors. Depending on the load and ambient temperatures, single-effect absorption systems can use energizing temperatures of 110 to 230°F.

A flat-plate collector generally consists of the following components (see Figure 8):

- **Glazing.** One or more sheets of glass or other radiation-transmitting material.
- **Tubes, fins, or passages.** To conduct or direct the heat transfer fluid from the inlet to the outlet.
- **Absorber plates.** Flat, corrugated, or grooved plates, to which the tubes, fins, or passages are attached. The plate may be integral with the tubes.
- **Headers or manifolds.** To admit and discharge the heat transfer fluid.

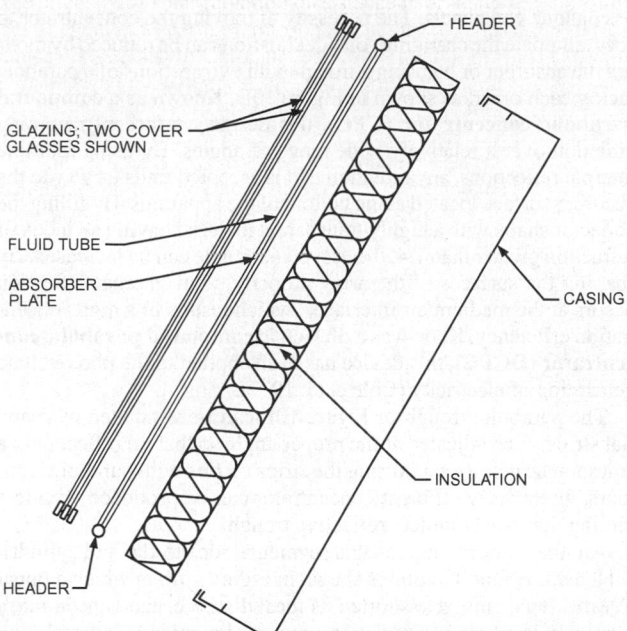

**Fig. 8   Exploded Cross Section Through Double-Glazed
Solar Water Heater**

- **Insulation.** To minimize heat loss from the back and sides of the collector.
- **Container or casing.** To surround the other components and protect them from dust, moisture, etc.

Flat-plate collectors have been built in a wide variety of designs from many different materials (Figure 9). They have been used to heat fluids such as water, water plus an antifreeze additive, or air. Their major purpose is to collect as much solar energy as possible at the lowest possible total cost. The collector should also have a long effective life, despite the adverse effects of the sun's ultraviolet radiation; corrosion or clogging because of acidity, alkalinity, or hardness of the heat transfer fluid; freezing or air-binding in the case of water, or deposition of dust or moisture in the case of air; and breakage of the glazing because of thermal expansion, hail, vandalism, or other causes. These problems can be minimized by using tempered glass.

Glazing Materials

Glass has been widely used to glaze flat-plate solar collectors because it can transmit as much as 90% of the incoming shortwave solar irradiation while transmitting very little of the longwave radiation emitted outward from the absorber plate. Glass with low iron content has a relatively high transmittance for solar radiation (approximately 0.85 to 0.90 at normal incidence), but its transmittance is essentially zero for the longwave thermal radiation (5.0 to 50 μm) emitted by sun-heated surfaces.

Plastic films and sheets also have high shortwave transmittance, but because most usable varieties also have transmission bands in the middle of the thermal radiation spectrum, their longwave transmittances may be as high as 0.40.

Plastics are also generally limited in the temperatures they can sustain without deteriorating or undergoing dimensional changes. Only a few kinds of plastics can withstand the sun's ultraviolet radiation for long periods. However, they are not broken by hail and other stones and, in the form of thin films, are completely flexible and have low mass.

The glass generally used in solar collectors may be either single-strength (0.085 to 0.100 in. thick) or double-strength (0.115 to 0.133 in. thick). Commercially available grades of window and

**Table 4   Variation with Incident Angle of Transmittance
for Single and Double Glazing and Absorptance
for Flat-Black Paint**

Incident Angle, Deg	Transmittance		Absorptance for Flat-Black Paint
	Single Glazing	Double Glazing	
0	0.87	0.77	0.96
10	0.87	0.77	0.96
20	0.87	0.77	0.96
30	0.87	0.76	0.95
40	0.86	0.75	0.94
50	0.84	0.73	0.92
60	0.79	0.67	0.88
70	0.68	0.53	0.82
80	0.42	0.25	0.67
90	0.00	0.00	0.00

0.87 and 0.85, respectively. For direct radiation, the transmittance varies markedly with the angle of incidence, as shown in Table 4, which gives transmittances for single and double glazing using double-strength clear window glass.

The 4% reflectance from each glass/air interface is the most important factor in reducing transmission, although a gain of about 3% in transmittance can be obtained by using water-white glass. Antireflective coatings and surface texture can also improve transmission significantly. The effect of dirt and dust on collector glazing may be quite small, and the cleansing effect of an occasional rainfall is usually adequate to maintain the transmittance within 2 to 4% of its maximum.

The glazing should admit as much solar irradiation as possible and reduce upward loss of heat as much as possible. Although glass is virtually opaque to the longwave radiation emitted by collector plates, absorption of that radiation causes an increase in the glass temperature and a loss of heat to the surrounding atmosphere by radiation and convection. This type of heat loss can be reduced by using an infrared-reflective coating on the underside of the glass; however, such coatings are expensive and reduce the effective solar transmittance of the glass by as much as 10%.

In addition to serving as a heat trap by admitting shortwave solar radiation and retaining longwave thermal radiation, the glazing also reduces heat loss by convection. The insulating effect of the glazing is enhanced by using several sheets of glass, or glass plus plastic. Loss from the back of the plate rarely exceeds 10% of the upward loss.

Collector Plates

The collector plate absorbs as much of the irradiation as possible through the glazing, while losing as little heat as possible up to the atmosphere and down through the back of the casing. The collector plates transfer retained heat to the transport fluid. The absorptance of the collector surface for shortwave solar radiation depends on the nature and color of the coating and on the incident angle, as shown in Table 4 for a typical flat-black paint.

By suitable electrolytic or chemical treatments, selective surfaces can be produced with high values of solar radiation absorptance α and low values of longwave emittance e. Essentially, typical selective surfaces consist of a thin upper layer, which is highly absorbent to shortwave solar radiation but relatively transparent to longwave thermal radiation, deposited on a substrate that has a high reflectance and a low emittance for longwave radiation. Selective surfaces are particularly important when the collector surface temperature is much higher than the ambient air temperature.

For fluid-heating collectors, passages must be integral with or firmly bonded to the absorber plate. A major problem is obtaining a good thermal bond between tubes and absorber plates without incurring excessive costs for labor or materials. Materials most frequently used for collector plates are copper, aluminum, and steel.

UV-resistant plastic extrusions are used for low-temperature application. If the entire collector area is in contact with the heat transfer fluid, the material's thermal conductance is not important.

Whillier (1964) concluded that steel tubes are as effective as copper if the bond conductance between tube and plate is good. Potential corrosion problems should be considered for any metals. Bond conductance can range from 1000 Btu/h·ft^2·°F for a securely soldered or brazed tube, to 3 Btu/h·ft^2·°F for a poorly clamped or badly soldered tube. Plates of copper, aluminum, or stainless steel with integral tubes are among the most effective types available. Figure 9 shows a few of the solar water and air heaters that have been used with varying degrees of success.

Concentrating Collectors

Temperatures far above those attainable by flat-plate collectors can be reached if a large amount of solar radiation is concentrated on a relatively small collection area. Simple **reflectors** can markedly increase the amount of direct radiation reaching a collector, as shown in Figure 10A.

Because of the apparent movement of the sun across the sky, conventional concentrating collectors must follow the sun's daily motion. There are two methods by which the sun's motion can be readily tracked. The altazimuth method requires the tracking device to turn in both altitude and azimuth; when performed properly, this method enables the concentrator to follow the sun exactly. **Paraboloidal solar furnaces** (Figure 10B) generally use this system. The polar, or equatorial, mounting points the axis of rotation at the North Star, tilted upward at the angle of the local latitude. By rotating the collector 15° per hour, it follows the sun perfectly (on March 21 and September 21). If the collector surface or aperture must be kept normal to the solar rays, a second motion is needed to correct for the change in the solar declination. This motion is not essential for most solar collectors.

The maximum variation in the angle of incidence for a collector on a polar mount is ±23.5° on June 21 and December 21; the incident angle correction then is cos 23.5° = 0.917.

Horizontal reflective parabolic troughs, oriented east and west (Figure 10C), require continuous adjustment to compensate for the changes in the sun's declination. There is inevitably some morning and afternoon shading of the reflecting surface if the concentrator has opaque end panels. The necessity of moving the concentrator to accommodate the changing solar declination can be reduced by moving the absorber or by using a trough with two sections of a parabola facing each other, as shown in Figure 10D. Known as a **compound parabolic concentrator (CPC)**, this design can accept incoming radiation over a relatively wide range of angles. By using multiple internal reflections, any radiation that is accepted finds its way to the absorber surface located at the bottom of the apparatus. By filling the collector shape with a highly transparent material having an index of refraction greater than 1.4, the acceptance angle can be increased. By shaping the surfaces of the array properly, total internal reflection occurs at the medium/air interfaces, which results in a high concentration efficiency. Known as a **dielectric compound parabolic concentrator (DCPC)**, this device has been applied to the photovoltaic generation of electricity (Cole et al. 1977).

The parabolic trough of Figure 10E can be simulated by many flat strips, each adjusted at the proper angle so that all reflect onto a common target. By supporting the strips on ribs with parabolic contours, a relatively efficient concentrator can be produced with less tooling than the complete reflective trough.

Another concept applies this segmental idea to flat and cylindrical lenses. A modification is shown in Figure 10F, in which a linear **Fresnel lens**, curved to shorten its focal distance, can concentrate a relatively large area of radiation onto an elongated receiver. Using the equatorial sun-following mounting, this type of concentrator has been used to attain temperatures well above those that can be reached with flat-plate collectors.

One disadvantage of concentrating collectors is that, except at low concentration ratios, they can use only the direct component of solar radiation, because the diffuse component cannot be concentrated by most types. However, an advantage of concentrating collectors is that, in summer, when the sun rises and sets well to the north of the east-west line, the sun-follower, with its axis oriented north-south, can begin to accept radiation directly from the sun long before a fixed, south-facing flat plate can receive anything other than diffuse radiation from the portion of the sky that it faces. At 40° north latitude, for example, the cumulative direct radiation available to a sun-follower on a clear day is 3180 Btu/ft^2, whereas the total radiation falling on the flat plate tilted upward at an angle equal to the latitude is only 2220 Btu/ft^2 each day. Thus, in relatively

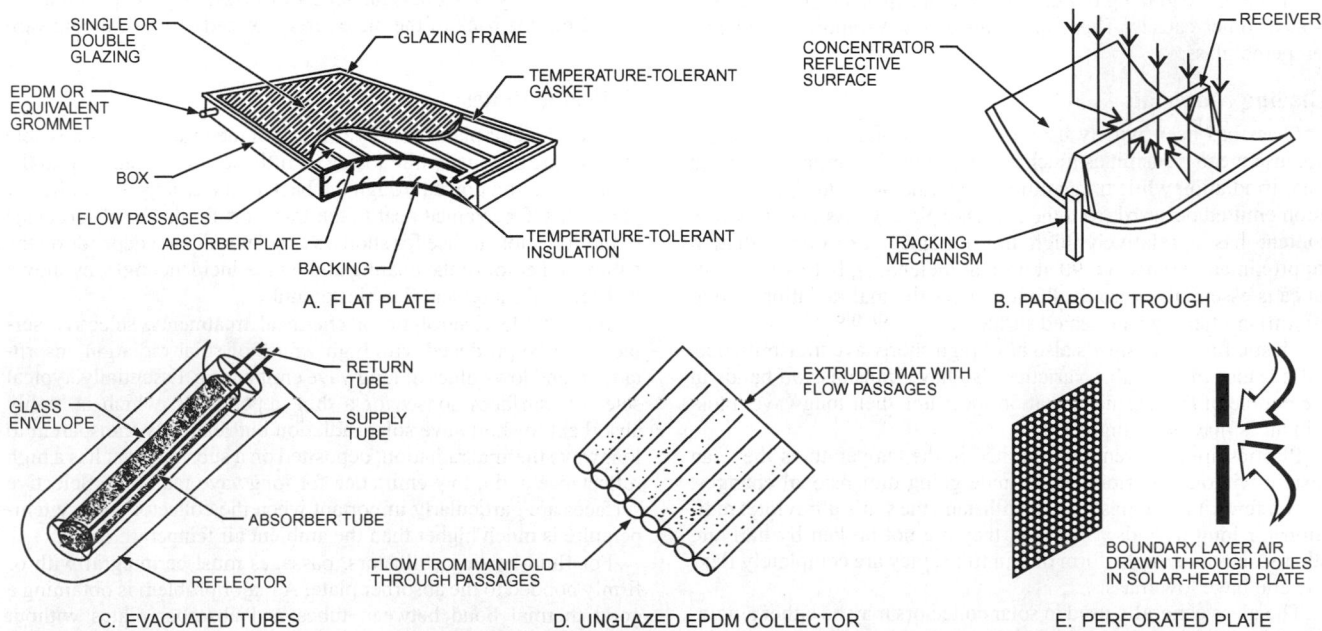

Fig. 9   **Various Types of Solar Collectors**

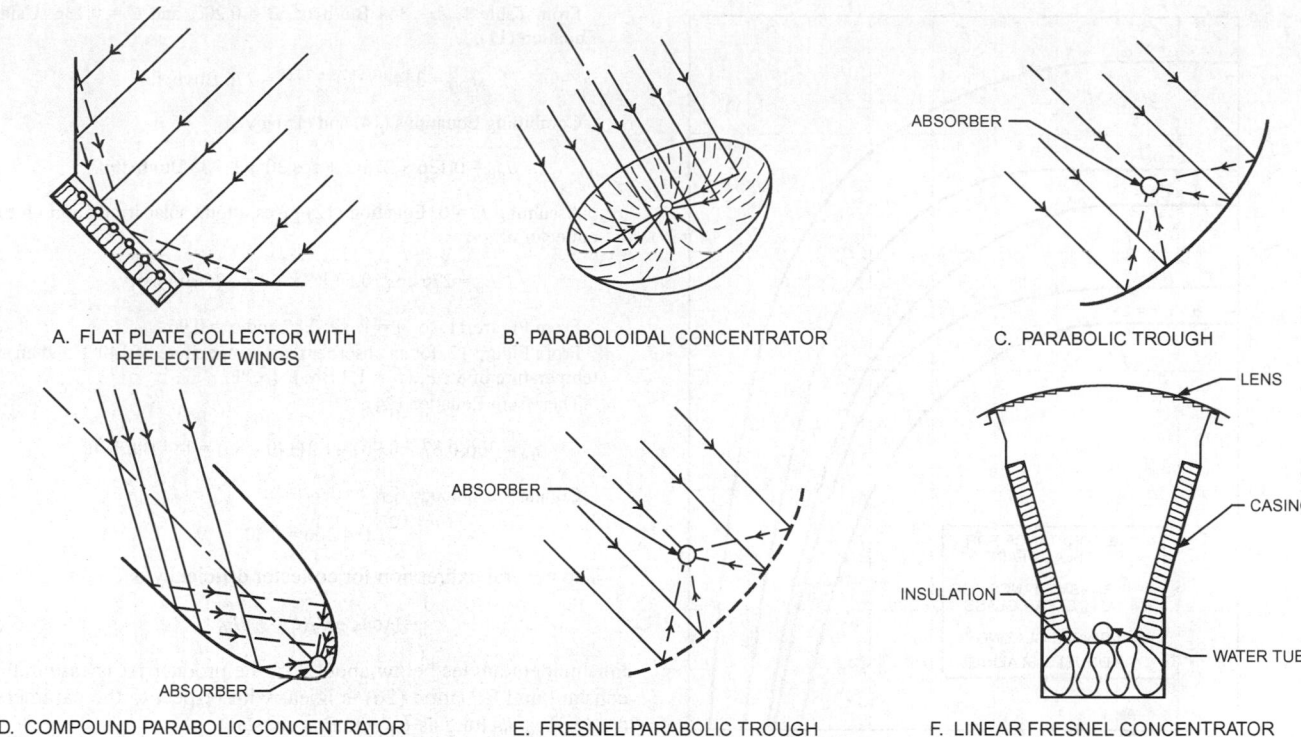

A. FLAT PLATE COLLECTOR WITH REFLECTIVE WINGS

B. PARABOLOIDAL CONCENTRATOR

C. PARABOLIC TROUGH

D. COMPOUND PARABOLIC CONCENTRATOR

E. FRESNEL PARABOLIC TROUGH

F. LINEAR FRESNEL CONCENTRATOR

Fig. 10   Types of Concentrating Collectors

cloudless areas, the concentrating collector may capture more radiation per unit of aperture area than a flat-plate collector.

To get extremely high inputs of radiant energy, many flat mirrors, or heliostats, using altazimuth mounts, can be used to reflect their incident direct solar radiation onto a common target. Using slightly concave mirror segments on the heliostats, large amounts of thermal energy can be directed into the cavity of a steam generator to produce steam at high temperature and pressure.

Collector Performance

The performance of collectors may be analyzed by a procedure originated by Hottel and Woertz (1942) and extended by Whillier (ASHRAE 1977). The basic equation is

$$q_u = I_{t\theta}(\tau\alpha)_\theta - U_L(t_p - t_{at}) = \dot{m}c_p(t_{fe} - t_{fi})/A_{ap} \qquad (24)$$

Equation (24) also may be adapted for use with concentrating collectors:

$$q_u = I_{DN}(\tau\alpha)_\theta(\rho\Gamma) - U_L(A_{abs}/A_{ap})(t_{abs} - t_a) \qquad (25)$$

where

q_u = useful heat gained by collector per unit of aperture area, Btu/h·ft²
$I_{t\theta}$ = total irradiation of collector, Btu/h·ft²
I_{DN} = direct normal irradiation, Btu/h·ft²
$(\tau\alpha)_\theta$ = transmittance τ of cover times absorptance α of plate at prevailing incident angle θ
U_L = upward heat loss coefficient, Btu/h·ft²·°F
t_p = temperature of absorber plate, °F
t_a = temperature of atmosphere, °F
t_{abs} = temperature of absorber, °F
$\dot{m}$ = fluid flow rate, lb/h
c_p = specific heat of fluid, Btu/lb·°F
t_{fe}, t_{fi} = temperatures of fluid leaving and entering collector, °F
$\rho\Gamma$ = reflectance of concentrator surface times fraction of reflected or refracted radiation that reaches absorber

A_{abs}, A_{ap} = areas of absorber surface and of aperture that admit or receive radiation, ft²

The total and direct normal irradiation for clear days may be found in ASHRAE *Standard* 93. The transmittance for single and double glazing and the absorptance for flat-black paint may be found in Table 4 for incident angles from 0 to 90°. These values, and the products of τ and α, are also shown in Figure 11. The solar-optical properties of the glazing and absorber plate change little until θ exceeds 30°, but, because all values reach zero when $\theta = 90°$, they drop off rapidly for values of θ beyond 40°.

For nonselective absorber plates, U_L varies with the temperature of the plate and the ambient air, as shown in Figure 12. For selective surfaces, which strongly reduce the emittance of the absorber plate, U_L is much lower than the values shown in Figure 12. Manufacturers of such surfaces should be asked for values applicable to their products, or test results that give the necessary information should be consulted.

Example 5. A flat-plate collector is operating in Denver, latitude = 40° north, on July 21 at noon solar time. The atmospheric temperature is 85°F, and the average temperature of the absorber plate is 140°F. The collector is single-glazed with flat-black paint on the absorber. The collector faces south, and the tilt angle is 30° from the horizontal. Find the rate of heat collection and collector efficiency. Neglect losses from the back and sides of the collector.

Solution: From Table 2, $\delta = 20.6°$.
From Equation (2),

$$\beta_N = 90° - 40° + 20.6° = 70.6°$$

From Equation (3), $H = 0$; therefore from Equation (6), $\sin \phi = 0$ and thus, $\phi = 0°$. Because the collector faces south, $\psi = 0°$, and $\gamma = \phi$. Thus $\gamma = 0°$. Then Equation (8) gives

$$\cos \theta = \cos 70.6° \cos 0° \sin 30° + \sin 70.6° \cos 30°$$
$$= (0.332)(1)(0.5) + (0.943)(0.866)$$
$$= 0.983$$
$$\theta = 10.6°$$

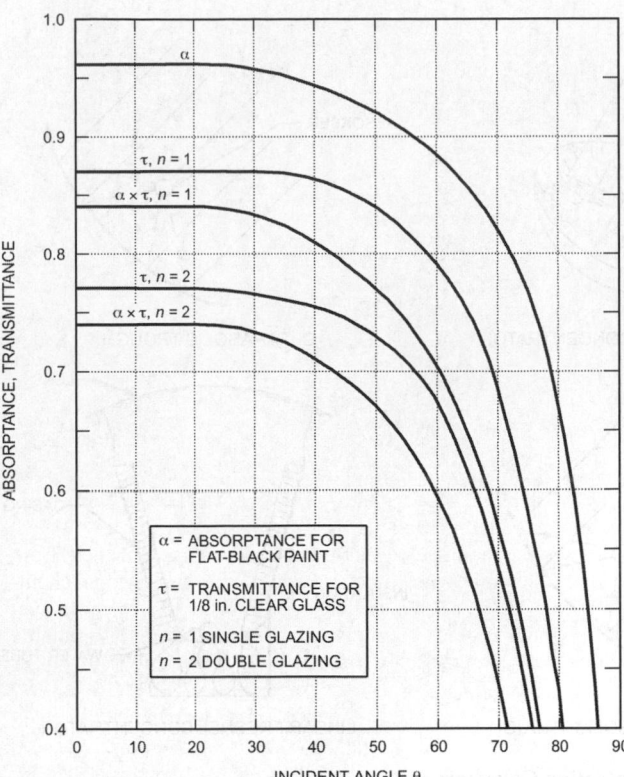

Fig. 11 Variation of Absorptance and Transmittance with Incident Angle

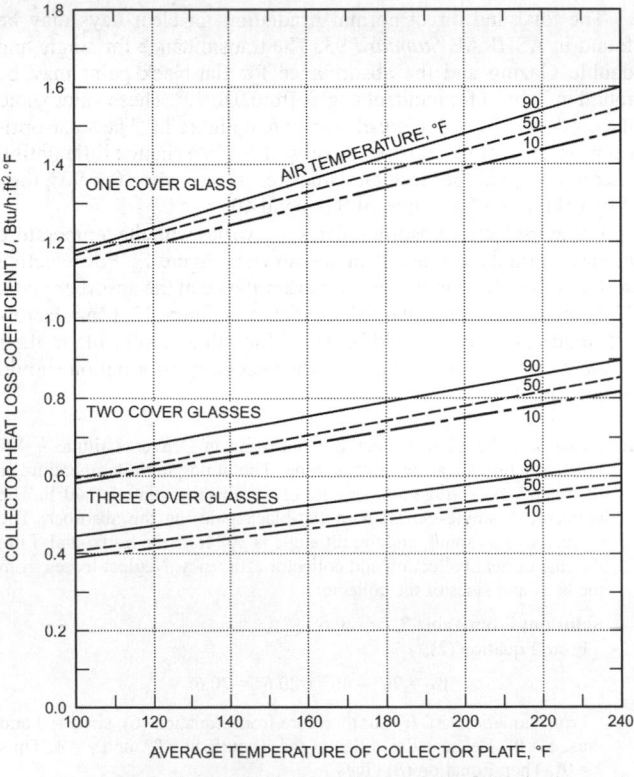

Note: Absorber plate is coated with nonselective flat-black paint.

Fig. 12 Variation of Upward Heat Loss Coefficient U_L with Collector Plate Temperature and Ambient Air Temperatures for Single-, Double-, and Triple-Glazed Collectors

From Table 1, $A = 344$ Btu/h·ft², $B = 0.207$, and $C = 0.136$. Using Equation (11),

$$I_{DN} = 344e^{-0.207/\sin 70.6°} = 276 \text{ Btu/h·ft}^2$$

Combining Equations (14) and (15) gives

$$I_{d\theta} = 0.136 \times 276(1 + \cos 30°)/2 = 35 \text{ Btu/h·ft}^2$$

Assuming $I_r = 0$, Equation (12) gives a total solar irradiation on the collector of

$$I_{t\theta} = 276 \cos 10.6° + 35 + 0 = 306 \text{ Btu/h·ft}^2$$

From Figure 11, for $n = 1$, $\tau = 0.87$ and $\alpha = 0.96$.

From Figure 12, for an absorber plate temperature of 140°F and an air temperature of 85°F, $U_L = 1.3$ Btu/h·ft²·°F.

Then from Equation (24),

$$q_u = 306(0.87 \times 0.96) - 1.3(140 - 85) = 184 \text{ Btu/h·ft}^2$$

Collector efficiency η is

$$184/306 = 0.60$$

The general expression for collector efficiency is

$$\eta = (\tau\alpha)_\theta - U_L(t_p - t_a)/I_{t\theta} \qquad (26)$$

For incident angles below about 35°, the product $\tau\alpha$ is essentially constant and Equation (26) is linear with respect to the parameter $(t_p - t_a)/I_{t\theta}$, as long as U_L remains constant.

ASHRAE (1977) suggested that an additional term, the **collector heat removal factor F_R**, be introduced to allow use of the fluid inlet temperature in Equations (24) and (26):

$$q_u = F_R[I_{t\theta}(\tau\alpha)_\theta - U_L(t_{fi} - t_a)] \qquad (27)$$

$$\eta = F_R(\tau\alpha)_\theta - F_R U_L(t_{fi} - t_a)/I_{t\theta} \qquad (28)$$

where F_R equals the ratio of the heat actually delivered by the collector to the heat that would be delivered if the absorber were at t_{fi}. F_R is found from the results of a test performed in accordance with ASHRAE *Standard* 93. The Solar Rating and Certification Corporation (SRCC) conducts this test in the United States and publishes the results, along with day-long performance outputs. Similar organizations around the world [e.g., the International Organization for Standardization (ISO), the European Committee for Standardization (CEN), Commonwealth Scientific and Industrial Research Organization (CISRO)] provide similar test results. For additional information on SRCC ratings, see Chapter 36 of the 2008 *ASHRAE Handbook—HVAC Systems and Equipment.*

The results of such a test are plotted in Figure 13. When the parameter is zero, because there is no temperature difference between fluid entering the collector and the atmosphere, the value of the y-intercept equals $F_R(\tau\alpha)$. The slope of the efficiency line equals the heat loss factor U_L multiplied by F_R. For the single-glazed, nonselective collector with the test results shown in Figure 13, the y-intercept is 0.82, and the x-intercept is 0.69 ft²·h·°F/Btu. This collector used high-transmittance single glazing, $\tau = 0.91$, and black paint with an absorptance of 0.97, so Equation (28) gives $F_R = 0.82/(0.91 \times 0.97) = 0.93$.

Assuming that the relationship between η and the parameter is actually linear, as shown, then the slope is $-0.82/0.69 = -1.190.12 = -6.78$; thus $U_L = 1.196.78/F_R = 1.196.78/0.93 = 1.28$ Btu/h·ft²·°F. The tests on which Figure 13 is based were run indoors. Wind speed and fluid velocity can affect measured efficiency.

Figure 13 also shows the efficiency of a double-glazed air heater with an unfinned absorber coated with flat-black paint. The y-intercept for the air heater B is considerably less than it is for water heater A because (1) transmittance of the double glazing used in B is lower than the transmittance of the single glazing used in A and

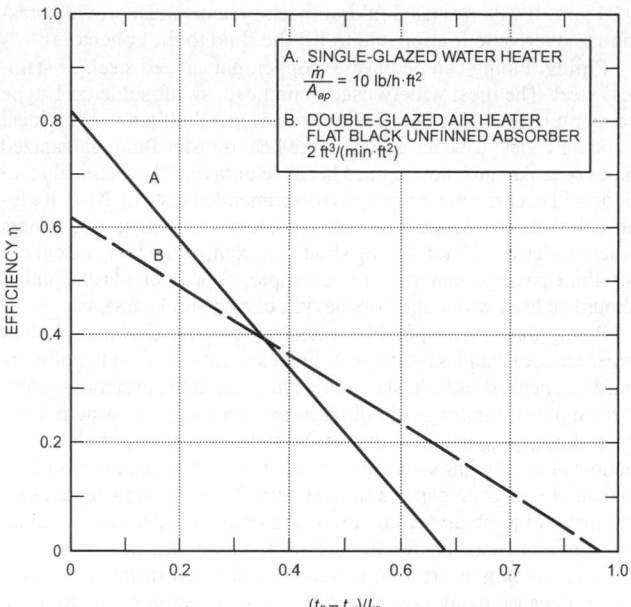

Fig. 13 Efficiency Versus $(t_{fi} - t_{at})/I_{t\theta}$ for Single-Glazed Solar Water Heater and Double-Glazed Solar Air Heater

(2) F_R is lower for B than for A because of the lower heat transfer coefficient between air and the unfinned metal absorber.

The x-intercept for air heater B is greater than it is for the water heater A because the upward loss coefficient U_L is much lower for the double-glazed air heater than for the single-glazed water heater. The data for both A and B were taken at near-normal incidence with high values of $I_{t\theta}$. For Example 5, using a single-glazed water heater, the value of the parameter would be close to $(140 - 85)/306 = 0.18$ ft²·h·°F/Btu, and the expected efficiency, 0.60, agrees closely with the test results.

As ASHRAE *Standard* 93 shows, the incident angles encountered with south-facing tilted collectors vary widely throughout the year. Considering a surface located at 40° north latitude with a tilt angle $\Sigma = 40°$, the incident angle θ depends on the time of day and the declination δ. On December 21, $\delta = 23.456°$; at 4 h before and after solar noon, the incident angle is 62.7°, and remains close to this value for the same solar time throughout the year. Total irradiation at these conditions varies from a low of 45 Btu/h·ft² on December 21 to approximately 140 Btu/h·ft² throughout most of the other months.

When irradiation is below about 100 Btu/h·ft², losses from the collector may exceed the heat that can be absorbed. This situation varies with the temperature difference between the collector inlet temperature and the ambient air, as suggested by Equation (27).

When the incident angle rises above 30°, the product of the glazing's transmittance and the collector plate's absorptance begins to diminish; thus, heat absorbed also drops. Losses from the collector are generally higher as time moves farther from solar noon, and consequently efficiency also drops. Thus, daylong efficiency is lower than near-noon performance. During early afternoon, efficiency is slightly higher than at the comparable morning time, because the ambient air temperature is lower in the morning than in the afternoon.

ASHRAE *Standard* 93 describes the **incident angle modifier**, which may be found by tests run when the incident angle is set at 30, 45, and 60°. Simon (1976) showed that for many flat-plate collectors, the incident angle modifier is a linear function of the quantity $(1/\cos \theta - 1)$. For evacuated tubular collectors, the incident angle modifier may grow with rising values of θ.

ASHRAE *Standard* 93 specifies that efficiency be reported in terms of the gross collector area A_g rather than the aperture area A_{ap}. The reported efficiency is lower than that given by Equation (28), but the total energy collected is not changed by this simplification:

$$\eta_g = \frac{\eta_{ap} A_{ap}}{A_g} \tag{29}$$

COMPONENTS

This section describes the major components involved in the collection, storage, transportation, control, and distribution of solar heat for a domestic hot-water system.

Collectors. Flat-plate collectors are most commonly used for water heating because of the year-round load requiring temperatures of 80 to 180°F. For discussions of other collectors and applications, see ASHRAE *Standard* 93, Chapter 36 of the 2008 *ASHRAE Handbook—HVAC Systems and Equipment*, and previous sections of this chapter. Collectors must withstand extreme weather (e.g., freezing, stagnation, high winds), as well as system pressures.

Heat Transfer Fluids. Heat transfer fluids transport heat from the solar collectors to the domestic water. There are potential chemical and mechanical problems with this transfer, primarily in systems in which a heat exchanger interface exists with the potable water supply. Both the chemical compositions of the heat transfer fluids (pH, toxicity, and chemical durability) and their mechanical properties (specific heat and viscosity) must be considered.

Except in unusual cases, or when potable water is being circulated, the energy transport fluid is nonpotable and could contaminate potable water. Even potable or nontoxic fluids in closed circuits are likely to become nonpotable because of contamination from metal piping, solder joints, and packing, or by the inadvertent installation of a toxic fluid at a later date.

Thermal Energy Storage. Heat collected by solar domestic and service water heaters is virtually always stored as a liquid in tanks. Storage tanks and bins should be well insulated. In domestic hot-water systems, heat is usually stored in one or two tanks. The hot-water outlet is at the top of the tank, and cold water enters the tank through a dip tube that extends down to within 4 to 6 in. of the tank bottom. The outlet on the tank to the collector loop should be approximately 4 in. above the tank bottom to prevent scale deposits from being drawn into the collectors. Water from the collector array returns to the middle to lower portion of the storage tank. This plumbing arrangement may take advantage of thermal stratification, depending on the delivery temperature from the collectors and the flow rate through the storage tank.

Single-tank electric auxiliary systems often incorporate storage and auxiliary heating in the same vessel. Conventional electric water heaters commonly have two heating elements: one near the top and one near the bottom. If a dual-element tank is used in a solar energy system, the bottom element should be disconnected and the top left functional to take advantage of fluid stratification. Standard gas- and oil-fired water heaters should not be used in single-tank arrangements. In gas and oil water heaters, heat is added to the bottom of the tanks, which reduces both stratification and collection efficiency in single-tank systems.

Dual-tank systems often use the solar domestic hot-water storage tank as a preheat tank. The second tank is normally a conventional domestic hot-water tank containing the auxiliary heat source. Multiple tanks are sometimes used in large institutions, where they operate similarly to dual-tank heaters. Although using two tanks may increase collector efficiency and the solar fraction, it increases tank heat losses. The water inlet is usually a dip tube that extends near the bottom of the tank.

Estimates for sizing storage tanks usually range from 1 to 2.5 gal per square foot of solar collector area. The estimate used most often is 1.8 gal per square foot of collector area, which usually provides enough heat for a sunless period of about a day. Storage volume should be analyzed and sized according to the project water requirements and draw schedule; however, solar applications typically require larger-than-normal tanks, usually the equivalent of the average daily load.

Heat Exchangers. Indirect solar water heaters require one or more heat exchangers. The potential exists for contamination in the transfer of heat energy from solar collectors to potable hot water. Heat exchangers influence the effectiveness of energy collected to heat domestic water. They also separate and protect the potable water supply from contamination when nonpotable heat transfer fluids are used. For this reason, various codes regulate the need for and design of heat exchangers.

Heat exchanger selection should consider the following:

- Heat exchange effectiveness
- Pressure drop, operating power, and flow rate
- Design pressure, configuration, size, materials, and location
- Cost and availability
- Reliable protection of potable water supply from contamination by heat transfer fluid
- Leak detection, inspection, and maintainability
- Material compatibility with other elements (e.g., metals and fluids)
- Thermal compatibility with design parameters such as operating temperature, and fluid thermal properties

Heat exchanger selection depends on characteristics of fluids that pass through the heat exchanger and properties of the exchanger itself. Fluid characteristics to consider are fluid type, specific heat, mass flow rate, and hot- and cold-fluid inlet and outlet temperatures. Physical properties of the heat exchanger to consider are the overall heat transfer coefficient of the heat exchanger and the heat transfer surface area.

For most solar domestic hot-water designs, only the hot and cold inlet temperatures are known; the other temperatures must be calculated using the heat exchanger's physical properties. Two quantities that are useful in determining a heat exchanger's heat transfer and a collector's performance characteristics when it is combined with a given heat exchanger are the (1) fluid capacitance rate, which is the product of the mass flow rate and specific heat of fluid passing through the heat exchanger, and (2) heat exchanger effectiveness, which relates the capacitance rate of the two fluids to the fluid inlet and outlet temperatures. Effectiveness is equal to the ratio of the actual heat transfer rate to the maximum heat transfer rate theoretically possible. Generally, a heat exchanger effectiveness of 0.4 or greater is desired.

Expansion Tanks. An indirect solar water heater operating in a closed collector loop requires an expansion tank to prevent excessive pressure. Fluid in solar collectors under stagnation conditions can boil, causing excessive pressure to develop in the collector loop, and expansion tanks must be sized for this condition. Expansion tank sizing formulas for closed-loop hydronic systems, found in Chapter 12 of the 2008 *ASHRAE Handbook—HVAC Systems and Equipment*, may be used for solar heater expansion tank sizing, but the expression for volume change caused by temperature increase should be replaced with the total volume of fluid in the solar collectors and of any piping located above the collectors, if significant. This sizing method provides a passive means for eliminating fluid loss through overtemperature or stagnation, common problems in closed-loop solar systems. This results in a larger expansion tank than typically found in hydronic systems, but the increase in cost is small compared to the savings in fluid replacement and maintenance costs (Lister and Newell 1989).

Pumps. Pumps circulate heat transfer liquid through collectors and heat exchangers. In solar domestic hot-water heaters, the pump is usually a centrifugal circulator driven by a motor of less than 300 W. The flow rate for collectors generally ranges from 0.015 to 0.04 gpm/ft^2. Pumps used in drainback systems must provide pressure to overcome friction and to lift the fluid to the collectors.

Piping. Piping can be plastic, copper, galvanized steel, or stainless steel. The most widely used is nonlead, sweat-soldered L-type copper tubing. M-type copper is also acceptable if allowed by local building codes. If water/glycol is the heat transfer fluid, galvanized pipes or tanks must not be used because unfavorable chemical reactions will occur; copper piping is recommended instead. Also, if glycol solutions or silicone fluids are used, they may leak through joints where water would not. Piping should be compatible with the collector fluid passage material; for example, copper or plastic piping should be used with collectors having copper fluid passages.

Piping that carries potable water can be plastic, copper, galvanized steel, or stainless steel. In indirect systems, corrosion inhibitors must be checked and adjusted no less than annually, preferably every three months. Inhibitors should also be checked if the system overheats during stagnation. If dissimilar metals are joined, dielectric or nonmetallic couplings should be used. The best protection is sacrificial anodes or getters in the fluid stream. Their location depends on the material to be protected, anode material, and electrical conductivity of the heat transfer fluid. Sacrificial anodes of magnesium, zinc, or aluminum are often used to reduce corrosion in storage tanks. Because many possibilities exist, each combination must be evaluated. A copper-aluminum or copper-galvanized steel joint is unacceptable because of severe galvanic corrosion. Aluminum, copper, and iron have a greater potential for corrosion.

Elimination of air, pipe expansion, and piping slope must be considered to avoid possible failures. Collector pipes (particularly manifolds) should be designed to allow expansion from stagnation temperature to extreme cold weather temperature. Expansion can be controlled with offset elbows in piping, hoses, or expansion couplings. Expansion loops should be avoided unless they are installed horizontally, particularly in systems that must drain for freeze protection. The collector array piping should slope 0.06 in. per foot for drainage (DOE 1978a).

Air can be eliminated by placing air vents at all piping high points and by air purging during filling. Flow control, isolation, and other valves in the collector piping must be chosen carefully so that these components do not restrict drainage significantly or back up water behind them. The collectors must drain completely.

Valves and Gages. Valves in solar domestic hot-water systems must be located to ensure system efficiency, satisfactory performance, and safety of equipment and personnel. Drain valves must be ball-type; gate valves may be used if the stem is installed horizontally. Check valves or other valves used for freeze protection or for reverse thermosiphoning must be reliable to avoid significant damage.

Auxiliary Heat Sources. On sunny days, a typical solar energy system should supply water at a predetermined temperature, and the solar storage tank should be large enough to hold sufficient water for a day or two. Because of the intermittent nature of solar radiation, an auxiliary heater must be installed to handle hot-water requirements. If a utility is the source of auxiliary energy, operation of the auxiliary heater can be timed to take advantage of off-peak utility rates. The auxiliary heater should be carefully integrated with the solar energy heater to obtain maximum solar energy use. For example, the auxiliary heater should not destroy any stratification that may exist in the solar-heated storage tank, which would reduce collector efficiency.

Ductwork, particularly in systems with air-type collectors, must be sealed carefully to avoid leakage in duct seams, damper shafts, collectors, and heat exchangers. Ducts should be sized using conventional air duct design methods.

Control. Controls regulate solar energy collection by controlling fluid circulation, activate system protection against freezing and overheating, and initiate auxiliary heating when it is required. The three major control components are sensors, controllers, and actuators.

Sensors detect conditions or measure quantities, such as temperature. Controllers receive output from the sensors, select a course of action, and signal a component to adjust the condition. Actuators, such as pumps, valves, dampers, and fans, execute controller commands and regulate the system.

Temperature sensors measure the temperature of the absorber plate near the collector outlet and near the bottom of the storage tank. The sensors send signals to a controller, such as a differential temperature thermostat, for interpretation.

The differential thermostat compares signals from the sensors with adjustable set points for high and low temperature differentials. The controller performs different functions, depending on which set points are met. In liquid systems, when the temperature difference between the collector and storage reaches a high set point, usually 12 to 15°F, the pump starts, automatic valves are activated, and circulation begins. When the temperature difference reaches a low set point, usually 4°F, the pump is shut off and the valves are deenergized and returned to their normal positions. To restart the system, the differential temperature set point must again be met. If the system has either freeze or overheat protection, the controller opens or closes valves or dampers and starts or stops pumps or fans to protect the system when its sensors detect conditions indicating that either freezing or overheating is about to occur.

Sensors must be selected to withstand high temperature, such as may occur during collector stagnation. Collector loop sensors should be located as close as possible to the outlet of the collectors, either in a pipe above the collector, on a pipe near the collector, or in the collector outlet passage.

The storage temperature sensor should be near the bottom of the storage tank to detect the temperature of fluid before it is pumped to the collector or heat exchanger. Storage fluid is usually coldest at that location because of thermal stratification and the location of the makeup water supply. The sensor should be either securely attached to the tank and well insulated, or immersed inside the tank near the collector supply.

The freeze protection sensor, if required, should be located so that it detects the coldest liquid temperature when the collector is shut down. Common locations are the back of the absorber plate at the bottom of the collector, the collector intake or return manifolds, or the center of the absorber plate. The center absorber plate location is recommended because reradiation to the night sky freezes the collector heat transfer fluid, even though the ambient temperature is above freezing. Some systems, such as the recirculation system, have two sensors for freeze protection; others, such as the draindown, use only one.

Control of on/off temperature differentials affects system efficiency. If the differential is too high, the collector starts later than it should; if it is too low, the collector starts too soon. The turn-on differential for liquid systems usually ranges from 10 to 30°F and is commonly lower in warmer climates and higher in cold climates. For air systems, the range is usually 25 to 45°F.

The turn-off temperature differential is more difficult to estimate. Selection depends on a comparison between the value of the energy collected and the cost of collecting it. It varies with individual systems, but a value of 4°F is typical and generally the fixed default value in the control.

Water temperature in the collector loop depends on ambient temperature, solar radiation, radiation from the collector to the night sky, and collector loop insulation. Freeze protection sensors should be set to detect 40°F.

Sensors are important but often overlooked control components. They must be selected and installed properly because no control can produce accurate outputs from unreliable sensor inputs. Sensors are used in conjunction with a differential temperature controller and are usually supplied by the controller manufacturer. Sensors must survive the anticipated operating conditions without physical damage or loss of accuracy. Low-voltage sensor circuits must be located away from high-voltage lines to avoid electromagnetic interference. Sensors attached to collectors should be able to withstand the stagnation temperature.

Sensor calibration, which is often overlooked by installers and maintenance personnel, is critical to system performance; a routine calibration maintenance schedule is essential.

Another control option is to use a photovoltaic (PV) panel that powers a pump. A properly sized and oriented PV panel converts sunlight into electricity to run a small circulating pump. No additional sensing is required because the PV panel and pump output increase with sunlight intensity and stop when no sunlight (collector energy) is available. Cromer (1984) showed that, with proper matching of pump and PV electrical characteristics, PV panel sizes as low as 5 W per 40 ft^2 of thermal panel may be used successfully. Difficulty with late starting and running too long in the afternoon can be alleviated by tilting the PV panel slightly to the east during commissioning of the installed system.

WATER HEATING

A solar water heater includes a solar collector that absorbs solar radiation and converts it to heat, which is then absorbed by a heat transfer fluid (water, a nonfreezing liquid, or air) that passes through the collector. The heat transfer fluid's heat is stored or used directly.

Portions of the solar energy system are exposed to the weather, so they must be protected from freezing. The system must also be protected from overheating caused by high insolation levels during periods of low energy demand.

In solar water heating, water is heated directly in the collector or indirectly by a heat transfer fluid that is heated in the collector, passes through a heat exchanger, and transfers its heat to the domestic or service water. The heat transfer fluid is transported by either natural or forced circulation. Natural circulation occurs by natural convection (thermosiphoning), whereas forced circulation uses pumps or fans. Except for thermosiphon systems, which need no control, solar domestic and service water heaters are controlled by differential thermostats.

Five types of solar energy systems are used to heat domestic and service hot water: thermosiphon, direct circulation, indirect, integral collector storage, and site built. Recirculation and draindown are two methods used to protect direct solar water heaters from freezing.

Thermosiphon Systems

Thermosiphon systems (Figure 14) heat potable water or a heat transfer fluid and rely on natural convection to transport it from the collector to storage. For direct systems, pressure-reducing valves are required when the city water pressure is greater than the working pressure of the collectors. In a thermosiphon system, the storage tank must be elevated above the collectors, which sometimes requires designing the upper level floor and ceiling joists to bear this additional load. Extremely hard or acidic water can cause scale deposits that clog or corrode the absorber fluid passages. Thermosiphon flow is induced whenever there is sufficient sunshine, so these systems do not need pumps.

Direct-Circulation Systems

A direct-circulation system (Figure 15) pumps potable water from storage to the collectors when there is enough solar energy available to warm it. It then returns the heated water to the storage tank until it is needed. Collectors can be mounted either above or below the storage tank. Direct-circulation systems are only feasible in areas where freezing is infrequent. Freeze protection is provided either by recirculating warm water from the storage tank or by flushing the collectors with cold water. Direct water-heating systems should not be used in areas where the water is extremely hard or

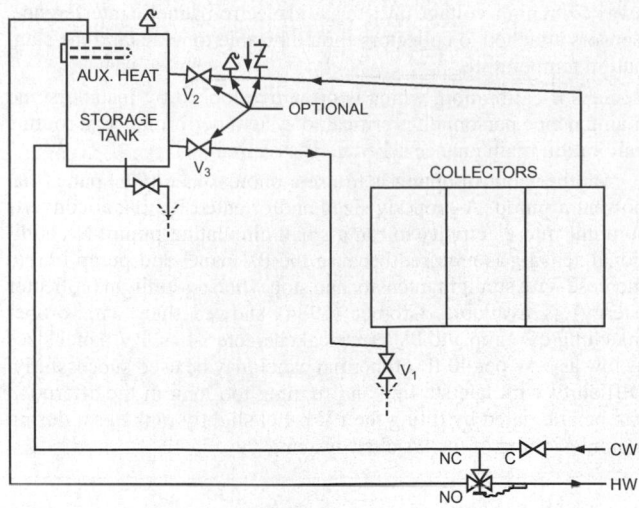

Fig. 14   Thermosiphon System

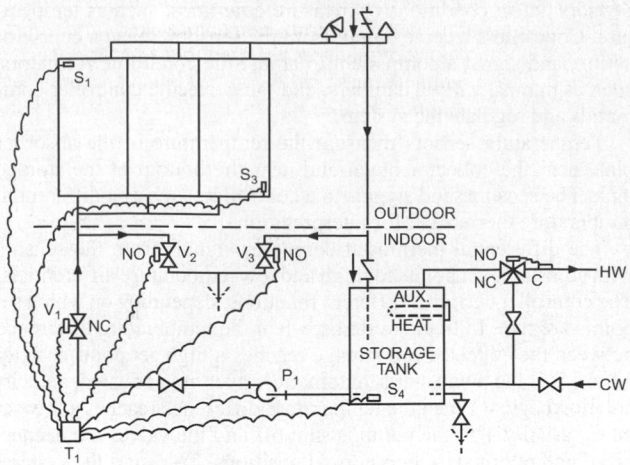

Fig. 16   Draindown System

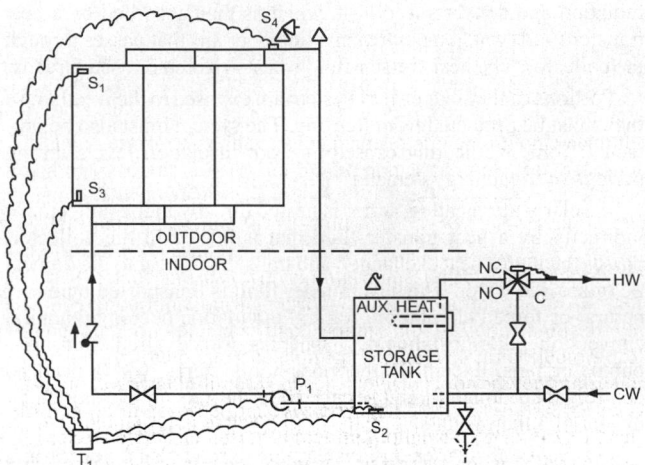

Fig. 15   Direct Circulation System

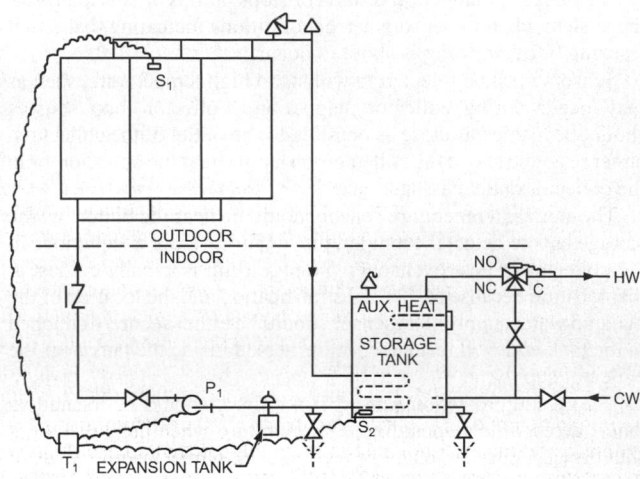

Fig. 17   Indirect Water Heating

acidic because scale deposits may clog or corrode the absorber fluid passages, rendering the system inoperable.

Direct-circulation systems are exposed to city water line pressures and must withstand pressures as required by local codes. Pressure-reducing valves and pressure-relief valves are required when city water pressure is greater than the working pressure of the collectors. Direct-circulation systems often use a single storage tank for both solar energy storage and the auxiliary water heater, but two-tank storage systems can be used.

Draindown Systems. Draindown systems (Figure 16) are direct-circulation water-heating systems in which potable water is pumped from storage to the collector array where it is heated. Circulation continues until usable solar heat is no longer available. When a freezing condition is anticipated or a power outage occurs, the system drains automatically by isolating the collector array and exterior piping from the city water pressure and using one or more valves for draining. Solar collectors and associated piping must be carefully sloped to drain the collector's exterior piping.

Indirect Water-Heating Systems

Indirect water-heating systems (Figure 17) circulate a freeze-protected heat transfer fluid through the closed collector loop to a heat exchanger, where its heat is transferred to the potable water. The most commonly used heat transfer fluids are water/ethylene glycol

and water/propylene glycol solutions, although other heat transfer fluids such as silicone oils, hydrocarbons, and refrigerants can also be used (ASHRAE 1983). These fluids are nonpotable, sometimes toxic, and normally require double-wall heat exchangers. The double-wall heat exchanger can be located inside the storage tank, or an external heat exchanger can be used. The collector loop is closed and therefore requires an expansion tank and a pressure-relief valve. A one- or two-tank storage can be used. Additional overtemperature protection may be needed to prevent the collector fluid from decomposing or becoming corrosive.

Designers should avoid automatic water makeup in systems using water/antifreeze solutions because a significant leak may raise the freezing temperature of the solution above the ambient temperature, causing the collector array and exterior piping to freeze. Also, antifreeze systems with large collector arrays and long pipe runs may need a time-delayed bypass loop around the heat exchanger to avoid freezing the heat exchanger on start-up.

Drainback Systems. Drainback systems are generally indirect water-heating systems that circulate treated or untreated water through the closed collector loop to a heat exchanger, where its heat is transferred to the potable water. Circulation continues until usable energy is no longer available. When the pump stops, the collector fluid drains by gravity to a storage or tank. In a pressurized system, the tank also serves as an expansion tank, so it must have a temperature- and pressure-relief valve to protect against excessive

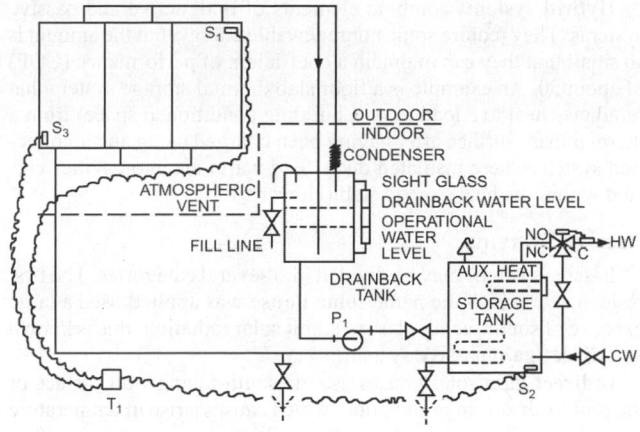

Fig. 18   Drainback System

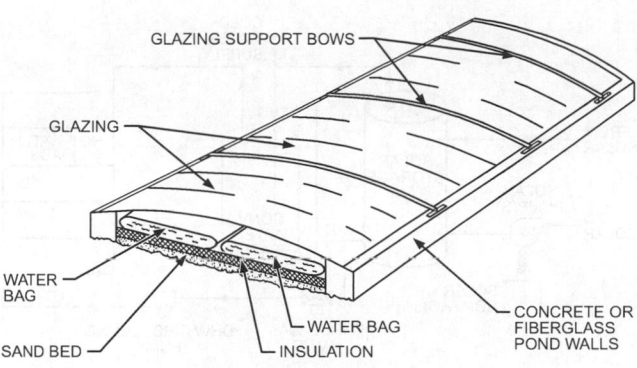

Fig. 19   Shallow Solar Pond

pressure. In an unpressurized system (Figure 18), the tank is open and vented to the atmosphere.

The collector loop is isolated from the potable water, so valves are not needed to actuate draining, and scaling is not a problem. The collector array and exterior piping must be sloped to drain completely, and the pumping pressure must be sufficient to lift water to the top of the collector array.

Integral Collector Storage Systems

Integral collector storage (ICS) systems use hot-water storage as part of the collector. Some types use the surface of a single tank as the absorber, and others use multiple long, thin tanks placed side by side horizontally to form the absorber surface. In this type of ICS, hot water is drawn from the top tank, and cold replacement water enters the bottom tank. Because of the greater nighttime heat loss from ICS systems, they are typically less efficient than pumped systems, and use of selective surfaces is strongly recommended. ICS systems are normally installed as a solar preheater without pumps or controllers. Flow through the ICS system occurs on demand, as hot water flows from the collector to a hot-water auxiliary tank in the structure.

SRCC provides annual performance results for these various types of systems at www.solar-rating.org.

Site-Built Systems

Site-built, large-volume solar air- or water-heating equipment is used in commercial and industrial applications. These site-built systems are based on a transpired solar collector for air heating and shallow solar pond technologies.

Transpired Solar Collector. This collector preheats outdoor air by drawing it through small holes in a metal panel. It is typically installed on south-facing walls and is designed to heat outdoor air for building ventilation or process applications (Kutscher 1996). The prefabricated panel, made of dark metal with thousands of small holes, efficiently heats and captures fresh air by drawing it through a perforated adsorber, eliminating the cost and the reflection losses associated with a glazing. The sun heats the metal panel, which in turn heats a boundary layer of air on its surface. Air is heated as it is drawn through the small holes into a ventilation system for delivery as ventilation air, for crop drying, or other process applications.

Shallow Solar Pond. The shallow solar pond (SSP) is a large-scale ICS solar water heater (Figure 19) capable of providing more than 5000 gal of hot water per day for commercial and industrial use. These ponds are built in standard modules and tied together to supply the required load. The SSP module can be ground mounted or installed on a roof. It is typically 16 ft wide and up to 200 ft long. The module contains one or two flat water bags similar to a water

bed. The bags rest on a layer of insulation inside concrete or fiberglass curbs. The bag is protected against damage and heat loss by greenhouse glazing. A typical pond filled to a 4 in. depth holds approximately 6000 gal of water.

Pool Heaters

Solar pool heaters do not require a separate storage tank, because the pool itself serves as storage. In most cases, the pool's filtration pump forces the water through the solar panels or plastic pipes. In some retrofit applications, a larger pump may be required to handle the needs of the solar heater, or a small pump may be added to boost pool water to the solar collectors.

Automatic control may be used to direct the flow of filtered water to the collectors when solar heat is available; this may also be accomplished manually. Normally, solar heaters are designed to drain down into the pool when the pump is turned off; this provides the collectors with freeze protection.

Four primary types of collector designs are used for swimming pool heat: (1) rigid black plastic panels (polypropylene), usually 4 by 10 ft or 4 by 8 ft; (2) tube-on-sheet panels, which usually have a metal deck (copper or aluminum) with copper water tubes; (3) an ethylene-propylene diene monomer (or ethylene-propylene terpolymer) (EPDM) rubber mat, extruded with water passages running its length; and (4) arrays of black plastic pipe, usually 1.5 in. diameter acrylonitrile butadiene styrene (ABS) plastic (Root et al. 1985).

Hot-Water Recirculation

Domestic hot-water (DHW) recirculation systems (Figures 20 and 21), which continuously circulate domestic hot water throughout a building, are found in motels, hotels, hospitals, dormitories, office buildings, and other commercial buildings. Recirculation heat losses in these systems are usually a significant part of the total water-heating load. In Figures 20 and 21, the three-way valve prevents heated water from returning to the solar storage tank when the return temperature is greater than the solar tank temperature. This ensures that heated water is used only when it is hot enough and prevents heating of the solar tank by the conventional heater. Using a cycle-timer to control the DHW circulating pump that is synchronized with the building hot-water consumption profile can also help reduce circulation losses. The return line on the makeup preheat system can go directly to the conventional water heater to eliminate the three-way valve and prevent the solar tank from being heated by auxiliary energy.

SOLAR HEATING AND COOLING SYSTEMS

The components and subsystems discussed previously may be combined to create a wide variety of solar heating and cooling systems. These systems fall into two principal categories: passive and active.

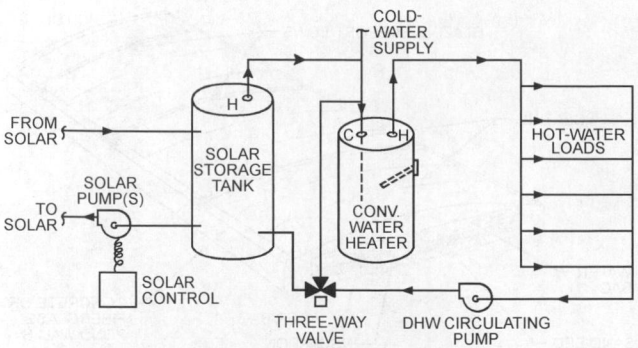

Fig. 20   DHW Recirculation System

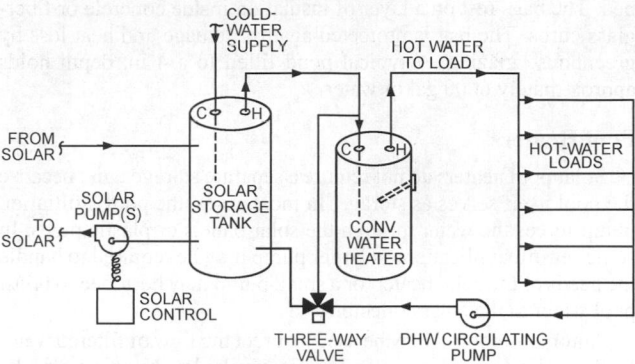

Fig. 21   DHW Recirculation System with Makeup Preheat

Passive solar systems require little, if any, nonrenewable energy to make them function (Yellott 1977; Yellott et al. 1976). Every building is passive in the sense that the sun tends to warm it by day, and it loses heat at night. Passive systems incorporate solar collection, storage, and distribution into the architectural design of the building and make minimal or no use of fans to deliver the collected energy to the structure. Passive solar heating, cooling, and lighting design must consider the building envelope and its orientation, the thermal storage mass, and window configuration and design. ASHRAE (1984), DOE (1980/1982), LBL (1981), and Mazria (1979) give estimates of energy savings resulting from the application of passive solar design concepts.

Active solar systems use either liquid or air as the collector fluid. Active systems must have a continuous availability of nonrenewable energy, generally in the form of electricity, to operate pumps and fans. A complete system includes solar collectors, energy storage devices, and pumps or fans for transferring energy to storage or to the load. The load can be space cooling, heating, or hot water. Although it is technically possible to construct a solar heating and cooling system to supply 100% of the design load, such a system would be uneconomical and oversized. The size of the solar system, and thus its ability to meet the load, is determined by life-cycle cost analysis that weighs the cost of energy saved against the amortized solar cost.

Active solar energy systems have been combined with heat pumps for water and/or space heating. The most economical arrangement in residential heating is a solar system in parallel with a heat pump, which supplies auxiliary energy when the solar source is not available. For domestic water systems requiring high water temperatures, a heat pump placed in series with the solar storage tank may be advantageous. Freeman et al. (1979) and Morehouse and Hughes (1979) present information on performance and estimated energy savings for solar-heat pumps.

Hybrid systems combine elements of both active and passive systems. They require some nonrenewable energy, but the amount is so small that they can maintain a coefficient of performance (COP) of about 50. An example is a floor slab thermal storage system that reradiates heat to a load (e.g., a building conditioned space) from a thermal mass surface after having been charged using an air collection system where insulated ducts feed warm air into cavities created within the heat storage slab (Howard 1986).

Passive Systems

Passive systems may be divided into several categories. The first residence to which the name **solar house** was applied used a large expanse of south-facing glass to admit solar radiation; this is known as a **direct-gain** passive system.

Indirect-gain solar houses use the south-facing wall surface or the roof to absorb solar radiation, which causes a rise in temperature that, in turn, conveys heat into the building in several ways. This principle was applied to the pueblos and cliff dwellings of the southwestern United States. Glass has led to modern adaptations of the indirect-gain principle (Balcomb et al. 1977; Trombe et al. 1977).

By glazing a large south-facing, massive masonry wall, solar energy can be absorbed during the day, and heat conduction to the inner surface provides radiant heating at night. The wall's mass and its relatively low thermal diffusivity delay the heat's arrival at the indoor surface until it is needed. The glazing reduces heat loss from the wall back to the atmosphere and increases the system's collection efficiency.

Openings in the wall near the floor and ceiling allow convection to transfer heat to the room. Air in the space between the glass and the wall warms as soon as the sun heats the outer surface of the wall. The heated air rises and enters the building through the upper openings. Cool air flows through the lower openings, and convective heat gain can be established as long as the sun is shining.

In another indirect-gain passive system, a metal roof/ceiling supports transparent plastic bags filled with water (Hay and Yellott 1969). Movable insulation above these water-filled bags is rolled away during the winter day to allow the sun to warm the stored water. The water then transmits heat indoors by convection and radiation. The insulation remains over the water bags at night or during overcast days. During the summer, the water bags are exposed at night for cooling by (1) convection, (2) radiation, and (3) evaporation of water on the bags. The insulation covers the water bags during the day to protect them from unwanted irradiation. Pittenger et al. (1978) tested a building for which water rather than insulation was moved to provide summer cooling and winter heating.

Attached greenhouses (sunspaces) can be used as solar attachments when the orientation and other local conditions are suitable. The greenhouse can provide a buffer between the exterior wall of the building and the outdoors. During daylight, warm air from the greenhouse can be introduced into the house by natural convection or a small fan.

In most passive systems, control is accomplished by moving a component that regulates the amount of solar radiation admitted into the structure. Manually operated window shades or venetian blinds are the most widely used and simplest controls.

Passive heating and cooling systems have been effective in field demonstrations A well-designed passive-solar-heated building may provide 45 to nearly 100% of daily heat requirements. Architectural design features can dramatically reduce air-conditioning loads through heat gain avoidance techniques and natural cooling, where climatically appropriate (Howard and Pollock 1982; Howard and Saunders 1989).

Passive solar daylighting has been shown to be cost-effective, providing dual benefits: it both reduces electric power demand and lowers cooling costs in properly designed interiors and atrium spaces.

COOLING BY NOCTURNAL
RADIATION AND EVAPORATION

Radiative cooling is a natural heat loss that causes formation of dew, frost, and ground fog. Because its effects are the most obvious at night, it is sometimes termed **nocturnal radiation**, although the process continues throughout the day. Thermal infrared radiation, which affects the surface temperature of a building wall or roof, may be estimated by using the **sol-air temperature** concept. Radiative cooling of window and skylight surfaces can be significant, especially under winter conditions when the dew-point temperature is low.

The most useful parameter for characterizing the radiative heat transfer between horizontal nonspectral emitting surfaces and the sky is the **sky temperature T_{sky}**. If S designates the total downward radiant heat flux emitted by the atmosphere, then T_{sky} is defined as

$$T_{sky}^4 = S/\sigma \qquad (30)$$

where $\sigma = 0.1712 \times 10^{-8}$ Btu/h·ft^2·°R^4.

The sky radiance is treated as if it originates from a blackbody emitter of temperature T_{sky}. The **net radiative cooling rate R_{net}** of a horizontal surface with absolute temperature T_{rad} and a nonspectral emittance ε is then

$$R_{net} = \varepsilon\sigma(T_{rad}^4 - T_{sky}^4) \qquad (31)$$

Values of ε for most nonmetallic construction materials are about 0.9.

Radiative building cooling has not been fully developed. Design methods and performance data compiled by Hay and Yellott (1969) and Marlatt et al. (1984) are available for residential roof ponds that use a sealed volume of water covered by sliding insulation panels as the combined rooftop radiator and thermal storage. Other conceptual radiative cooling designs have been proposed, but more developmental work is required (Givoni 1981; Mitchell and Biggs 1979).

The sky temperature is a function of atmospheric water vapor, the amount of cloud cover, and air temperature; the lowest sky temperatures occur under an arid, cloudless sky. The monthly average sky temperature depression, which is the average of the difference between the ambient air temperature and the sky temperature, typically lies between 9 and 43°F throughout the continental United States. Martin and Berdahl (1984) calculated this quantity using hourly weather data from 193 sites, as shown in the contour map for the month of July (Figure 22).

The sky temperature may be too high at night to effectively cool the structure. Martin and Berdahl (1984) suggest that the sky temperature should be less than 61°F to achieve reasonable cooling in July (Figure 23). In regions where sky temperatures fall below 61°F 40% or more of the month, all nighttime hours are effectively available for radiative cooling.

Clark (1981) modeled a horizontal radiator at various surface temperatures in convective contact with outdoor air for 77 U.S. locations. The average monthly cooling rates for a surface temperature of 76°F are plotted in Figure 24. If effective steps are taken to reduce the surface convection coefficient by modifying the radiator geometry or using an infrared-transparent glazing, it may be possible to improve performance beyond these values.

Active Systems

Active systems absorb solar radiation with collectors and convey it to storage using a suitable fluid. As heat is needed, it is obtained from storage via heated air or water. Control is exercised by several types of thermostats, the first being a differential device that starts the flow of fluid through the collectors when they have been sufficiently warmed by the sun. It also stops fluid flow when the collectors no longer gain heat. In locations where freezing occurs only

rarely, a low-temperature sensor on the collector controls a circulating pump when freezing is impending. This process wastes some stored heat, but it prevents costly damage to the collector panels. This system is not suitable for regions where freezing temperatures persist for long periods.

The space-heating thermostat is generally the conventional double-contact type that calls for heat when the temperature in the controlled space falls to a predetermined level. If the temperature in storage is adequate to meet the heating requirement, a pump or fan

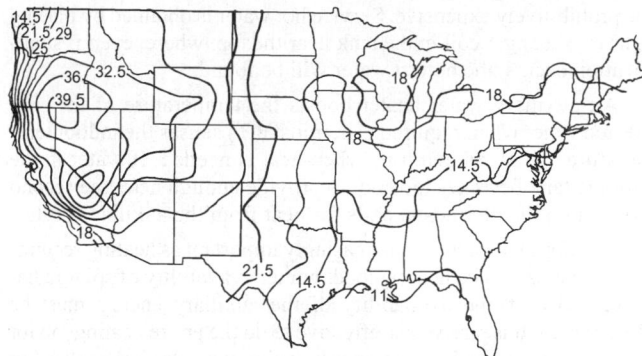

Fig. 22   Average Monthly Sky Temperature Depression
$(T_{air} - T_{sky})$ for July, °F
(Adapted from Martin and Berdahl 1984)

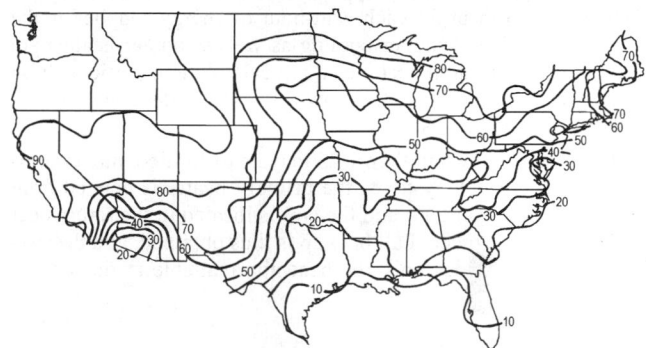

Fig. 23   Percentage of Monthly Hours when
Sky Temperature Falls below 61°F
(Adapted from Martin and Berdahl 1984)

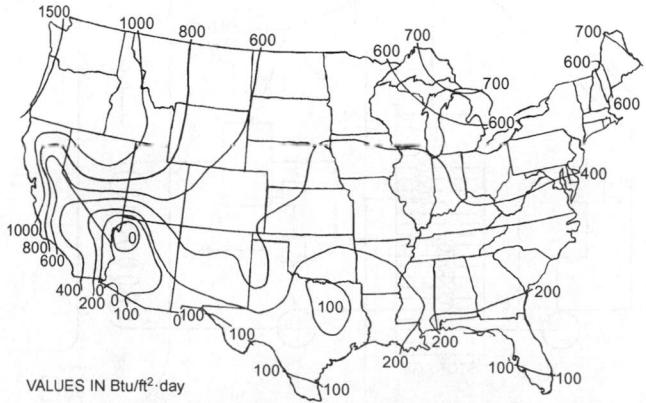

VALUES IN Btu/ft^2·day

Fig. 24   July Nocturnal Net Radiative Cooling Rate from
Horizontal Dry Surface at 76°F
(Adapted from Clark 1981)

is started to circulate the warm fluid. If the temperature in the storage subsystem is inadequate, the thermostat calls on the auxiliary or standby heat source.

Space Heating and Service Hot Water

Figure 25 shows one of the many systems for service hot water and space heating. In this case, a large, atmospheric pressure storage tank is used, from which water is pumped to the collectors by pump P_1 in response to the differential thermostat T_1. Drainback is used to prevent freezing, because the amount of antifreeze required would be prohibitively expensive. Service hot water is obtained by placing a heat exchanger coil in the tank near the top, where, even if stratification occurs, the hottest water will be found.

An auxiliary water heater boosts the temperature of the sun-heated water when required. Thermostat T_2 senses the indoor temperature and starts pump P_2 when heat is needed. If water in the storage tank becomes too cool to provide enough heat, the second contact on the thermostat calls for heat from the auxiliary heater.

Standby heat becomes increasingly important as heating requirements increase. The heating load, winter availability of solar radiation, and cost and availability of the auxiliary energy must be determined. It is rarely cost-effective to do the entire heating job for either space or service hot water by using the solar heat collection and storage system alone.

Electric resistance heaters have the lowest first cost, but often have high operating costs. Water-to-air heat pumps, which use sun-heated water from the storage tank as the evaporator energy source, are an alternative auxiliary heat source. The heat pump's COP is 10 to 14 Btu of heat for each watt-hour of energy supplied to the compressor. When summer cooling as well as winter heating are needed, the heat pump becomes a logical solution, particularly in large systems where a cooling tower is used to dissipate the heat withdrawn from the system.

The system shown in Figure 25 may be retrofitted into a warm-air furnace. In such systems, the primary heater is deleted from the space heating circuit, and the coil is located in the return duct of the existing furnace. Full backup is thus obtained, and the auxiliary heater provides only the heat not available at the storage temperature.

COOLING BY SOLAR ENERGY

Swartman et al. (1974) emphasize various absorption systems. Newton (in Jordan and Liu 1977) discusses commercially available water vapor/lithium bromide absorption refrigeration systems. Standard absorption chillers are generally designed to give rated capacity for activating fluid temperatures well above 200°F at full load and design condenser water temperature. Few flat-plate collectors can operate efficiently in this range; therefore, a lower hot-fluid temperature is used when solar energy provides the heat. Both condenser water temperature and percentage of design load are determinants of the optimum energizing temperature, which can be quite low, sometimes below 120°F. Proper control can raise the COP at these part-load conditions.

Many large commercial or institutional cooling installations must operate year-round, and Newton (in Jordon and Liu 1977) showed that the low-temperature cooling water available in winter enables the LiBr/H_2O to function well with a hot-fluid inlet temperature below 190°F. Residential chillers in sizes as low as 1.5 tons, with an inlet temperature in the range of 175°F, have been developed.

Solar Cooling with Absorption Refrigeration

When solar energy is used for cooling as well as for heating, the absorption system shown in Figure 26 or one of its many modifications, may be used. The collector and storage must operate at a temperature approaching 200°F on hot summer days when the water from the cooling tower exceeds 80°F, but considerably lower operating water temperatures may be used when cooler water is available from the tower. Controls for collection, cooling, and distribution are generally separated, with the circulating pump P_1 operating in response to the collector thermostat T_1, which is located in the air-conditioned space. When T_2 calls for heating, valves V_1 and V_2 direct water flow from the storage tank through the unactivated auxiliary heater to the fan-coil in the air distribution system. The fan F_1 in this unit may respond to the thermostat also, or it may have its own control circuit so that it can bring in outdoor air when a suitable temperature condition is present.

When thermostat T_2 calls for cooling, the valves direct hot water into the absorption unit's generator, and pumps P_3 and P_4 are

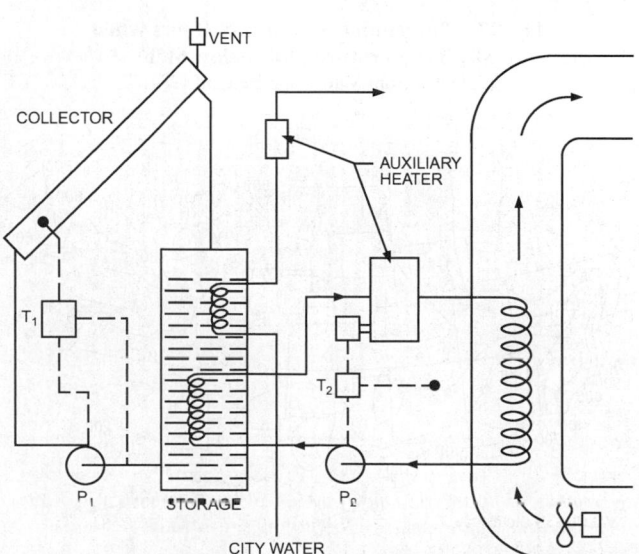

Fig. 25   Solar Collection, Storage, and Distribution System for Domestic Hot Water and Space Heating

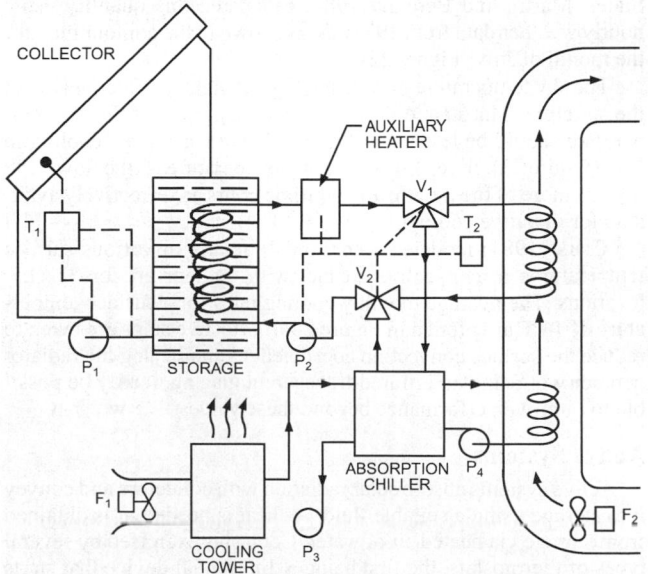

Fig. 26   Space Heating and Cooling System Using Lithium Bromide-Water Absorption Chiller

activated to pump cooling tower water through the absorber and condenser circuits and chilled water through the cooling coil in the air distribution system. A relatively large hot-water storage tank allows the unit to operate when no sunshine is available. A chilled-water storage tank (not shown) may be added so that the absorption unit can operate during the day whenever water is available at a sufficiently high temperature to make the unit function properly. The COP of a typical lithium bromide/water absorption unit may be as high as 0.75 under favorable conditions, but frequent on/off cycling of the unit to meet a high variable cooling load may cause significant loss in performance because the unit must be heated to operating temperature after each shutdown. Modulating systems are analyzed differently than on/off systems.

Water-cooled condensers are required with the absorption cycles, because the lithium bromide/water cycle operates with a relatively delicate balance among the temperatures of the three fluid circuits (cooling tower water, chilled water, and activating water). The steam-operated absorption systems, from which solar cooling systems are derived, customarily operate at energizing temperatures of 230 to 240°F, but these are above the capability of most flat-plate collectors. The solar cooling units are designed to operate at considerably lower temperature, but unit ratings are also lowered.

Smaller domestic units may operate with natural circulation, or **percolation**, which carries the lithium bromide/water solution from the generator (to which the activating heat is supplied) to the separator and condenser; there, the reconcentrated LiBr is returned to the absorber while the water vapor goes to the condenser before being returned to the evaporator, where cooling occurs. Larger units use a centrifugal pump to transfer the fluid.

SIZING SOLAR HEATING AND COOLING SYSTEMS: ENERGY REQUIREMENTS

Methods used to determine solar heating and/or cooling energy requirements for both active and passive/hybrid systems are described by Feldman and Merriam (1979) and Hunn et al. (1987). Descriptions of public- and private-domain methods are included. The following simulation techniques are suitable for active heating and cooling systems analysis, and for passive/hybrid heating, cooling, and lighting analysis.

Performance Evaluation Methods

The performance of any solar energy system is directly related to the (1) heating load, (2) amount of solar radiation available, and (3) solar energy system characteristics. Various calculation methods use different procedures and data when considering the available solar radiation. Some simplified methods consider only average annual incident solar radiation; complex methods may use hourly data.

Solar energy system characteristics, as well as individual component characteristics, are required to evaluate performance. The degree of complexity with which these systems and components are described varies from system to system.

The cost effectiveness of a solar domestic and service hot-water heating system depends on the initial cost and energy cost savings. A major task is to determine how much energy is saved. The **annual solar fraction** (the annual solar contribution to the water-heating load divided by the total water-heating load) can be used to estimate these savings. It is expressed as a decimal or percentage and generally ranges from 0.3 to 0.8 (30 to 80%), although more extreme values are possible.

Simplified Analysis Methods

A very simplified way to initially estimate the size of a solar heating system is to divide the average daily load by the daily output of a particular collector based on its SRCC rating for the application

and solar conditions. This requires knowledge of the average daily incident solar radiation for the site and the type of collector best suited to the application. See Chapter 36 of the 2008 *ASHRAE Handbook—HVAC Systems and Equipment* for more information.

Simplified analysis methods have the advantages of computational speed, low cost, rapid turnaround (especially important during iterative design phases), and ease of use by persons with little technical experience. Disadvantages include limited flexibility for design optimization, lack of control over assumptions, and a limited selection of systems that can be analyzed. Thus, if the application, configuration, or load characteristics under consideration are significantly nonstandard, a detailed computer simulation may be required to achieve accurate results. This section describes the *f*-Chart method for active solar heating and the solar load ratio method for passive solar heating (Dickinson and Cheremisinoff 1980; Klein and Beckman 1979; Lunde 1980).

Water-Heating Load

The amount of hot water required must be estimated accurately because it affects component selection. Oversized storage may result in low-temperature water that requires auxiliary heating to reach the desired supply temperature. Undersizing can prevent the collection and use of available solar energy. Chapter 50 gives methods to determine the load.

Active Heating/Cooling

Beckman et al. (1977) developed the *f*-Chart method using an hourly simulation program (Klein et al. 1976) to evaluate space heating and service water heating in many climates and conditions. The results of these analyses correlate the fraction *f* of the heat load met by solar energy. The correlations give the fraction *f* of the monthly heating load (for space heating and hot water) supplied by solar energy as a function of collector characteristics, heating loads, and weather. The standard error of the differences between detailed simulations in 14 locations in the United States and the *f*-Chart predictions was about 2.5%. Correlations also agree within the accuracy of measurements of long-term performance data. Beckman et al. (1977, 1981) and Duffie and Beckman (2006) discuss the method in detail.

The *f*-Chart method requires the following data:

- Monthly average daily radiation on a horizontal surface
- Monthly average ambient temperatures
- Collector thermal performance curve slope and intercept from standard collector tests; that is, $F_R U_L$ and $F_R(\tau\alpha)_n$ (see ASHRAE *Standard* 93 and Chapter 36 of the 2008 *ASHRAE Handbook—HVAC Systems and Equipment*)
- Monthly space- and water-heating loads

Standard Systems

The *f*-Chart assumes several standard systems and applies only to these liquid configurations. The standard **liquid heater** uses water, an antifreeze solution, or air as the heat transfer fluid in the collector loop and water as the storage medium (Figure 27). Energy is stored in the form of sensible heat in a water tank. A water-to-air heat exchanger transfers heat from the storage tank to the building. A liquid-to-liquid heat exchanger transfers energy from the main storage tank to a domestic hot-water preheat tank, which in turn supplies solar-heated water to a conventional water heater. A conventional furnace or heat pump is used to meet the space heating load when energy in the storage tank is depleted.

Figure 28 shows the assumed configuration for a **solar air heater** with a pebble-bed storage unit. Energy for domestic hot water is provided by heat exchange from air leaving the collector to a domestic water preheat tank, as in the liquid system. The hot water is further heated, if necessary, by a conventional water heater. During summer operation, a seasonal, manually operated storage

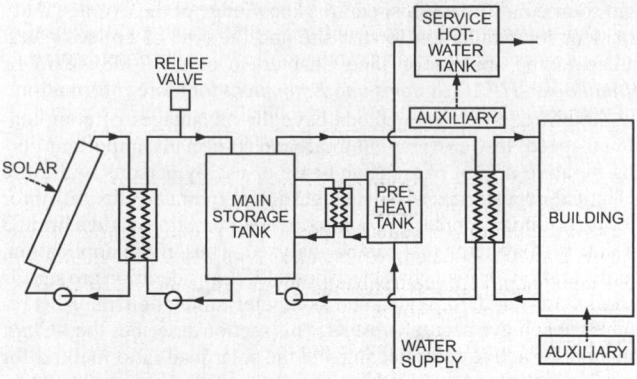

Fig. 27 Liquid-Based Solar Heating System
(Adapted from Beckman et al. 1977)

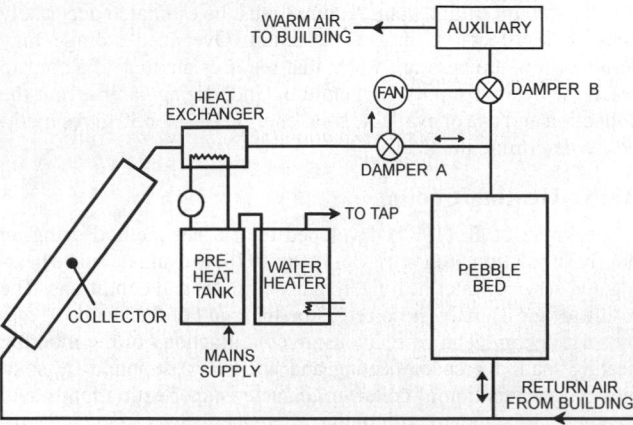

Fig. 28 Solar Air Heating System
(Adapted from Beckman et al. 1977)

bypass damper is used to avoid heat loss from the hot bed into the building.

The standard **solar domestic water heater** collector heats either air or liquid. Collected energy is transferred by a heat exchanger to a domestic water preheat tank that supplies solar-heated water to a conventional water heater. The water is further heated to the desired temperature by conventional fuel if necessary.

f-Chart Method

Computer simulations correlate dimensionless variables and the long-term performance of the systems. The fraction f of the monthly space- and water-heating loads supplied by solar energy is empirically related to two dimensionless groups. The first dimensionless group X is collector loss; the second Y is collector gain:

$$X = \frac{F_R U_L A_c \Delta\theta}{L}\left(\frac{F_r}{F_R}\right)(t_{ref} - \bar{t}_a) \tag{32}$$

$$Y = \frac{F_R(\tau\alpha)_n \overline{H}_T N A_c}{L}\left(\frac{F_r}{F_R}\right)\left[\frac{(\overline{\tau\alpha})}{(\tau\alpha)_n}\right] \tag{33}$$

where

A_c = area of solar collector, ft^2
F_r = collector heat exchanger efficiency factor
F_R = collector efficiency factor
U_L = collector overall energy loss coefficient, Btu/h·ft^2·°F
$\Delta\theta$ = total number of hours in month

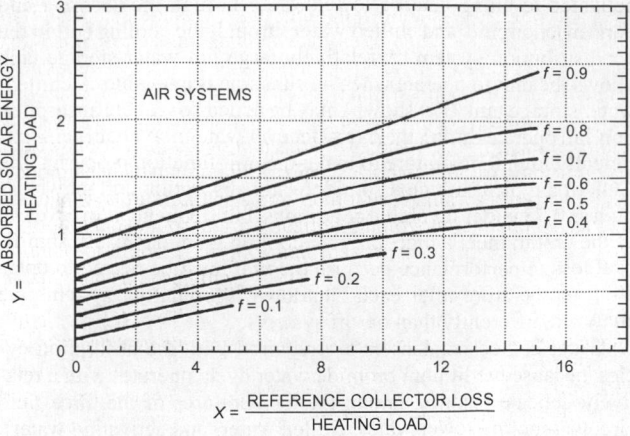

Fig. 29 Chart for Air System
(Adapted from Beckman et al. 1977)

$\bar{t}_a$ = monthly average ambient temperature, °F
L = monthly total heating load for space heating and hot water, Btu
$\overline{H}_T$ = monthly averaged daily radiation incident on collector surface per unit area, Btu/day·ft^2
N = number of days in month
$(\overline{\tau\alpha})$ = monthly average transmittance-absorptance product
$(\tau\alpha)_n$ = normal transmittance-absorptance product
t_{ref} = reference temperature, 212°F

$F_R U_L$ and $F_R(\tau\alpha)_n$ are obtained from collector test results. The ratios F_r/F_R and $(\overline{\tau\alpha})/(\tau\alpha)_n$ are calculated using methods given by Beckman et al. (1977). The value of $\bar{t}_a$ is obtained from meteorological records for the month and location desired. $\overline{H}_T$ is calculated from the monthly averaged daily radiation on a horizontal surface by the methods previously discussed in this chapter or in Duffie and Beckman (1992). The monthly load L can be determined by any appropriate load estimating method, including analytical techniques or measurements. Values of the collector area A_c are selected for the calculations. Thus, all the terms in these equations can be determined from available information.

Transmittance τ of the transparent collector cover and the absorptance α of the collector plate depend on the angle at which solar radiation nearly perpendicularly is incident on the collector surface. Collector tests are usually run with the radiation nearly perpendicularly incident on the collector. Thus, the value of $F_R(\tau\alpha)_n$ determined from these tests ordinarily corresponds to the transmittance and absorptance values for radiation at normal incidence. Depending on collector orientation and time of year, the monthly average values of transmittance and absorptance can be significantly lower. The f-Chart method requires knowledge of the ratio of the monthly average to normal incidence transmittance-absorptance.

The f-Chart method for liquid systems is similar to that for air systems. The fraction of the monthly total heating load supplied by the solar air heating system is correlated with the dimensionless groups X and Y, as shown in Figure 29. To determine the fraction of the heating load supplied by solar energy for a month, values of X and Y are calculated for the collector and heating load in question. The value of f is determined at the intersection of X and Y on the f-Chart, or from the following equivalent equations.

Air system:

$$f = 1.04Y - 0.065X - 0.159Y^2 + 0.00187X^2 - 0.0095Y^3 \tag{34}$$

Liquid system:

$$f = 1.029Y - 0.065X - 0.245Y^2 \\ + 0.0018X^2 + 0.025Y^3 \qquad (35)$$

This is done for each month of the year. The solar energy contribution for the month is the product of f and the total heating load L for the month. Finally, the fraction F of the annual heating load supplied by solar energy is the sum of the monthly solar energy contributions divided by the annual load:

$$F = \sum fL / \sum L$$

Example 6. Calculating the heating performance of a residence, assume that a solar heating system is to be designed for use in Madison, WI, with two-cover collectors facing south, inclined 58° with respect to the horizontal. The air heating collectors have the characteristics $F_R U_L = 0.50$ Btu/h·ft²·°F and $F_R(\tau\alpha)_n = 0.49$. The $\bar{t}_a$ is 19.4°F, the total space and water-heating load for January is 34.1 × 10⁶ Btu, and the solar radiation incident on the plane of the collector is 1.16 × 10³ Btu/day·ft². Determine the fraction of the load supplied by solar energy with a system having a collector area of 538.2 ft².

Solution: For air systems, there is no heat exchanger penalty factor and $F_r/F_R = 1$. The value of $(\overline{\tau\alpha})/(\tau\alpha)_n$ is 0.94 for a two-cover collector in January. Therefore, the values of X and Y are

$$X = 0.50(1)(212 - 19.4)(31)(24)(538.2)/(34.1 \times 10^6) = 1.13$$

$$Y = (0.49)(1)(0.94)(1.16 \times 10^3)(31)(538.2)/(34.1 \times 10^6) = 0.26$$

Then the fraction f of the energy supplied for January is 0.19. The total solar energy supplied by this system in January is

$$fL = (0.19)(34.1 \times 10^6) = 6.4 \times 10^6 \text{ Btu}$$

The annual system performance is obtained by summing the energy quantities for all months. The result is that 37% of the annual load is supplied by solar energy.

The collector heat removal factor F_R that appears in X and Y is a function of the collector fluid flow rate. Because of the higher cost of power for moving fluid through air collectors than through liquid collectors, the capacitance rate used in air heaters is ordinarily much lower than that in liquid heaters. As a result, air heaters generally have a lower F_R. Values of F_R corresponding to the expected airflow in the collector must be used to calculate X and Y.

Increased airflow rate tends to improve collector performance by increasing F_R, but it tends to decrease performance by reducing the degree of thermal stratification in the pebble bed (or water storage tank). The f-Chart for air systems is based on a collector airflow rate of 2 scfm per square foot of collector area. The performance with different collector airflow rates can be estimated by using the appropriate values of F_R in both X and Y. A further modification to the value of X is required to account for the change in degree of stratification in the pebble bed.

Air system performance is less sensitive to storage capacity than that of liquid systems for two reasons: (1) air systems can operate with air delivered directly to the building in which storage is not used, and (2) pebble beds are highly stratified and additional capacity is effectively added to the cold end of the bed, which is seldom heated and cooled to the same extent as the hot end. The f-Chart for air systems is for a nominal storage capacity. Performance of systems with other storage capacities can be determined by modifying the dimensionless group X as described in Beckman et al. (1977).

With modification, f-Charts can be used to estimate performance of solar water heating operating in the range of 120 to 160°F. The main water supply temperature and minimum acceptable hot-water temperature (i.e., the desired delivery temperature) both affect the performance of solar water heating. The dimensionless group X, which is related to collector energy loss, can be redefined to include these effects. If monthly values of X are multiplied by a correction factor, the f-Chart for liquid-based solar space- and water-heating systems can be used to estimate monthly values of f for water heating. Experiments and analysis show that the load profile for a well-designed heater has little effect on long-term performance. Although the f-Chart was originally developed for two-tank systems, it may be applied to single- and double-tank domestic hot-water systems with and without collector tank heat exchangers. Because of their low rates, use caution when modeling thermosiphon and ICS systems with the f-Chart.

For industrial process heating, absorption air conditioning, or other processes for which the delivery temperature is outside the normal f-Chart range, modified f-Charts are applicable (Klein et al. 1976). The concept underlying these charts is that of solar usability, which is the fraction of the total solar energy that is useful in the given process. This fraction depends on the required delivery temperature as well as collector characteristics and solar radiation. The procedure allows the energy delivered to be calculated in a manner similar to that for f-Charts. An example of the application of this method to solar-assisted heat pumps is presented in Svard et al. (1981).

Other Active Collector Methods

The **relative areas method**, based on correlations of the f-Chart method, predicts annual rather than monthly active heating performance (Barley and Winn 1978). An hourly simulation program was used to develop the **monthly solar-load ratio (SLR) method**, another simplified procedure for residential systems (Dickinson and Cheremisinoff 1980). Based on hour-by-hour simulations, a method was devised to estimate performance based on monthly values of horizontal solar radiation and heating degree-days. This SLR method has also been extended to nonresidential buildings for a range of design water temperatures (Dickinson and Cheremisinoff 1980; Schnurr et al. 1981).

Passive Heating

A widely accepted simplified passive space heating design tool is the solar-load ratio method (ASHRAE 1984; DOE 1980/1982). It can be applied manually, although like the f-Chart, software is also available. The SLR method for passive systems is based on correlating results of multiple hour-by-hour computer simulations, the algorithms of which have been validated against test cell data for the following generic passive heating types: direct gain, thermal storage wall, and attached sunspace. Monthly and annual performance, as expressed by the auxiliary heating requirement, is predicted by this method. The method applies to single-zone, envelope-dominated buildings. A simplified, annual-basis distillation of SLR results, the **load collector ratio (LCR) method**, and several simple-to-use rules have grown out of the SLR method. Several hand-held calculator and microcomputer programs have been written using the method (Nordham 1981).

The SLR method uses a single dimensionless correlating parameter (SLR), which Balcomb et al. (1982) define as a particular ratio of solar energy gains to building heating load:

$$\text{SLR} = \frac{\text{Solar energy absorbed}}{\text{Building heating load}} \qquad (36)$$

A correlation period of 1 month is used; thus the quantities in the SLR are calculated for a 1 month period.

The parameter that is correlated to the SLR, the **solar savings fraction (SSF),** is defined as

$$\text{SSF} = 1 - \frac{\text{Auxiliary heat}}{\text{Net reference load}} \qquad (37)$$

The SSF measures the energy saving expected from the passive solar building, relative to a reference nonpassive solar building.

In Equation (37), the net reference load is equal to the degree-day load DD of the nonsolar elements of the building:

$$\text{Net reference load} = (\text{NLC})(\text{DD}) \qquad (38)$$

where NLC is the net load coefficient, which is a modified *UA* coefficient computed by leaving out the solar elements of the building. The nominal units are Btu/°F·day. The term DD is the temperature departure in degree-days computed for an appropriate base temperature. A building energy analysis based on the SLR correlations begins with a calculation of the monthly SSF values. The monthly auxiliary heating requirement is then calculated by

$$\text{Auxiliary heat} = (\text{NLC})(\text{DD})(1 - \text{SSF}) \qquad (39)$$

Annual auxiliary heat is calculated by summing the monthly values.

By definition, SSF is the fraction of the heat load of the nonsolar portions of the building met by the solar element. If the solar elements of the building (south-facing walls and window in the northern hemisphere) were replaced by other elements so that the net annual flow of heat through these elements was zero, the building's annual heat consumption would be the net reference load. The savings achieved by the solar elements would therefore be the net reference load in Equation (38) minus the auxiliary heat in Equation (39), which gives

$$\text{Solar savings} = (\text{NLC})(\text{DD})(\text{SSF}) \qquad (40)$$

Although simple, in many situations and climates, Equation (40) is only approximately true because a normal solar-facing wall with a normal complement of opaque walls and windows has a near-zero effect over the entire heating season. In any case, the auxiliary heat estimate is the primary result and does not depend on this assumption.

The hour-by-hour simulations used as the basis for the SLR correlations are done with a detailed model of the building in which all the design parameters are specified. The only parameter that remains a variable is the solar collector area, which can be expressed in terms of the load collector ratio (LCR):

$$\text{LCR} = \frac{\text{NLC}}{A_p} = \frac{\text{Net load coefficient}}{\text{Projected collector area}} \qquad (41)$$

Performance variations are estimated from the correlations, which allow the user to account directly for thermostat set point, internal heat generation, glazing orientation, and configuration, shading, and other solar radiation modifiers. Major solar system characteristics are accounted for by selecting one of 94 reference designs. Other design parameters, such as thermal storage thickness and conductivity, and the spacing between glazings, are included in a series of sensitivity calculations obtained using hour-by-hour simulations. The results are generally presented in graphic form so that the designer can see the effect of changing a particular parameter.

Solar radiation correlations for the collector area have been determined using hour-by-hour simulation and typical meteorological year (TMY) weather data. These correlations are expressed as ratios of incident-to-horizontal radiation, transmitted-to-incident radiation, and absorbed-to-transmitted radiation as a function of the latitude minus mid-month solar declination and the atmospheric clearness index K_T.

The performance predictions of the SLR method have been compared to predictions made by the detailed hour-by-hour simulations for a variety of climates in the United States. The standard error in the prediction of the annual SSF, compared to the hour-by-hour simulation, is typically 2 to 4%.

The annual solar savings fraction calculation involves summing the results of 12 monthly calculations. For a particular city, the resulting SSF depends only on the LCR of Equation (41), the type of system, and the temperature base used in calculating the temperature departure. Thus, tables that relate SSF to LCR for various systems and for various degree-day base temperatures may be generated for a particular city. Such tables are easier for hand analysis than are the SLR correlations.

Annual SSF versus LCR tables have been developed for 209 locations in the United States and 14 cities in southern Canada for 94 reference designs and 12 base temperatures (ASHRAE 1984).

Example 7. [Abstracted from the detailed version in Balcomb et al. (1982).] Consider a small office building located in Denver, Colorado, with 3000 ft² of usable space and a sunspace entry foyer that faces due south; the projected collector area A_p is 420 ft². A sketch and preliminary plan are shown in Figure 30. Distribution of solar heat to the offices is primarily by convection through the doorways from the sunspace. The principal thermal mass is in the common wall that separates the sunspace from the offices and in the sunspace floor. Even though lighting and cooling are likely to have the greatest energy costs for this building, heating is a significant energy item and should be addressed by a design that integrates passive solar heating, cooling, and lighting.

Solution: Table 5 shows calculations of the net load coefficient. Then,

$$\text{NLC} = 24 \times 525 = 12{,}600 \text{ Btu/°F·day}$$

and the total load coefficient includes the solar aperture

$$\text{TLC} = 24 \times 676 = 16{,}224 \text{ Btu/°F·day}$$

From Equation (43), the load collector ratio is

$$\text{LCR} = 12{,}600/420 = 30 \text{ Btu/ft}^2\text{·°F·day}$$

If the daily internal heat is 130,000 Btu/day, and the average thermostat setting is 68°F, then

$$T_{base} = 68 - 130{,}000/16{,}224 = 60°F$$

(Note that solar gains to the space are not included in the internal gain term as they customarily are for nonsolar buildings.)

In this example, the solar system is type SSD1, defined in ASHRAE (1984) and Balcomb et al. (1982). It is a semiclosed sunspace with a 12 in. masonry common wall between it and the heated space (offices). The aperture is double-glazed, with a 50° tilt and no night insulation. To achieve a projected area of 420 ft², a sloped glazed area of 420/sin 50° = 548 ft² is required.

The SLR correlation for solar system type SSD1 is shown in Figure 31. Values of the absorbed solar energy S and heating degree-days DD to base temperature 60°F are determined monthly. For S, the solar radiation correlation presented in ASHRAE (1984) for tabulated Denver, CO, weather data is used. For January, the horizontal surface incident radiation is 840 Btu/ft²·day. The 60°F base degree-days are 933. Using the tabulated incident-to-absorbed coefficients found in

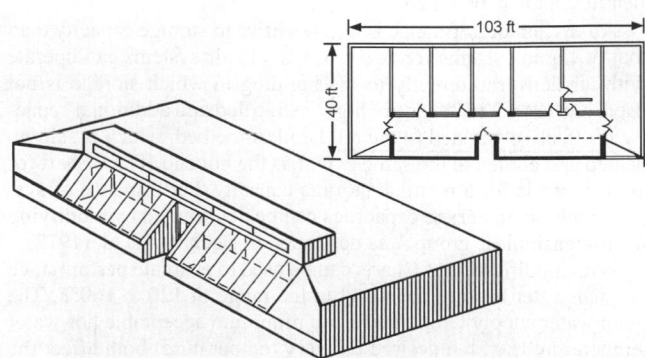

Fig. 30 Commercial Building in Example 7

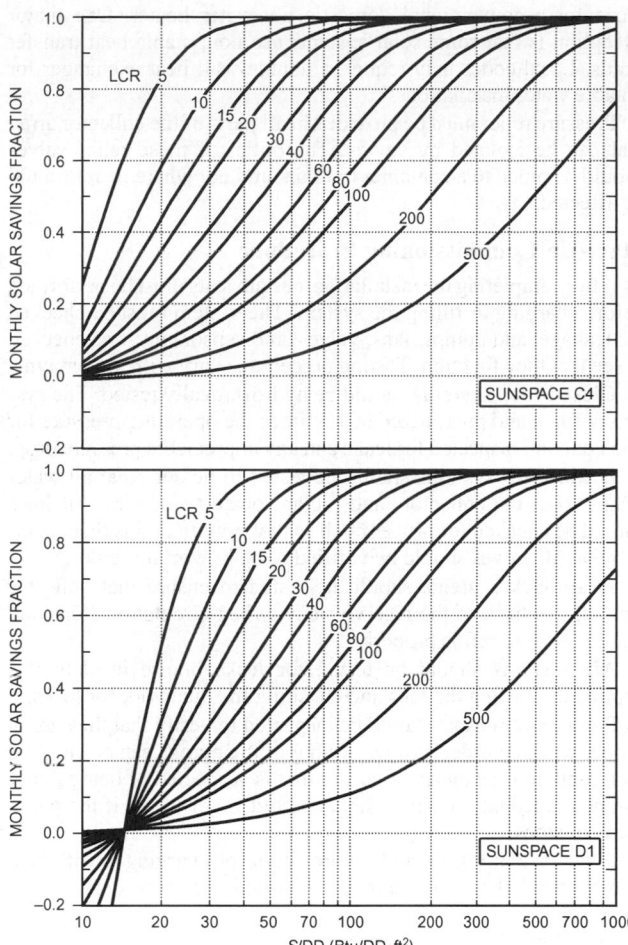

**Fig. 31   Monthly SSF Versus Monthly *S*/DD for
Various LCR Values**

Table 5   Calculations for Example 7

	Area *A*, ft²	U-Factor, Btu/h·ft²·°F	*UA*, Btu/ h·°F
Opaque wall	2000	0.04	80
Ceiling	3000	0.03	90
Floor (over crawl space)	3000	0.04	120
Windows (E, W, N)	100	0.55	55
		Subtotal	345
Infiltration			180
		Subtotal	525
Sunspace (treated as unheated space)			151
		Total	676

ASHRAE procedures are approximated with *V* = volume, ft³; *c* = heat capacity of Denver air, Btu/ft³·°F; ACH = air changes/h; equivalent *UA* for infiltration = *Vc*ACH. In this case, *V* = 24,000 ft³, *c* = 0.015 Btu/ft³·°F, and ACH = 0.5.

ASHRAE (1984), *S* = 43,681 Btu/ft². Thus *S*/DD = 46.8 Btu/ ft²·°F·day. From Figure 31, at an LCR = 30 Btu/ft²·°F·day, the SSF = 0.51. Therefore, for January,

$$\text{Net reference load} = (933)(12,600) = 11.76 \times 10^6 \text{ Btu}$$

$$\text{Solar savings} = (11.76 \times 10^6)(0.51) = 6.00 \times 10^6 \text{ Btu}$$

$$\text{Auxiliary heat} = (11.76 \times 10^6) - (6.00 \times 10^6) = 5.76 \times 10^6 \text{ Btu}$$

Repeating this calculation for each month and adding the results for the year yields an annual auxiliary heat of 19.94 × 10⁶ Btu.

Other Passive Heating Methods

The concept of usability has been applied to passive buildings. In this approach, the energy requirements of zero- and infinite-capacity buildings are calculated. The amount of solar energy that enters the building and exceeds the instantaneous load of the zero-capacity building is then calculated. This excess energy must be dumped in the zero-capacity building, but it can be stored to offset heating loads in a finite-capacity building. Methods are provided to interpolate between the zero- and infinite-capacity limits for finite-capacity buildings. Equations and graphs for direct gain and collector-storage wall systems are given in Monsen et al. (1981, 1982).

INSTALLATION GUIDELINES

Most solar components are the same as those in HVAC and hot-water systems (pumps, piping, valves, and controls), and their installation is not much different from a conventional installation. Solar collectors are the most unfamiliar component in a solar heater. They are located outdoors, which requires penetration of the building envelope. They also require a structural element to support them at the proper tilt and orientation toward the sun.

The site must taken into account. Collectors should be (1) located so that shading is minimized and (2) installed so that they are attractive both on and off site. They should also be located to minimize vandalism and to avoid a safety hazard.

Collectors should be placed near the storage tank to reduce piping cost and heat loss. The collector and piping must be installed so that they can be drained without trapping fluid in the system.

For best annual performance, collectors should be installed at a tilt angle above the horizontal that is appropriate for the local latitude. In the northern hemisphere they should be oriented toward true south, not magnetic south. Small variations in tilt (±10°) and orientation (±20°) do not reduce performance significantly.

Collector Mounting

Solar thermal collectors are usually mounted on the ground or on flat or pitched roofs. A roof location necessitates penetration of the building envelope by mounting hardware, piping, and control wiring. Ground or flat-roof-mounted collectors are generally rack mounted.

Pitched-roof mounting can be done several ways. Collectors can be mounted on structural **standoffs**, which support them at an angle other than that of the roof to optimize solar tilt. In another pitched-roof-mounting technique known as **direct mounting**, collectors are placed on a waterproof membrane on top of the roof sheeting. The finished roof surface together with the necessary collector structural attachments and flashing are then built up around the collector. A weatherproof seal between the collector and the roof must be maintained to prevent leakage, mildew, and rotting.

Integral mounting can be done for new pitched-roof or curtain-wall construction. The collector is attached to and supported by the structural framing members. The top of the collector then serves as the finished roof surface. Weathertightness is crucial to avoid damage and mildew. This building-integrated approach is often incorporated in green building design and generally is used for photovoltaic modules, rather than thermal collectors.

Collectors should support snow loads that occur on the roof area they cover. The collector tilt usually expedites snow sliding with only a small loss in efficiency. The roof structure should be free of objects that could impede snow sliding, and the collectors should be raised high enough to prevent snow buildup over them.

The mounting structure should be built to withstand winds of at least 100 mph, which impose a wind load of 40 lb/ft² on a vertical surface or an average of 25 lb/ft on a tilted roof (HUD 1977). Wind load requirements may be higher, depending on local building codes, especially in coastal areas exposed to hurricanes. Flat-plate collectors mounted flush with the roof surface should be constructed to withstand the same wind loads. See Chapter 55 for additional information.

The collector array becomes more vulnerable to wind gusts as the angle of the mount increases. This wind load, in addition to the equivalent roof area wind loads, should be determined according to accepted engineering procedures (ASCE *Standards* 7, 8, and 9).

Expansion and contraction of system components, material compatibility, and use of dissimilar metals must be considered. Collector arrays and mounting hardware (bolts, screws, washers, and angles) must be well protected from corrosion. Steel mounting hardware in contact with aluminum and copper piping in contact with aluminum hardware are both examples of metal combinations that have a high potential for corrosion. Dissimilar metals can be separated by washers made of fluorocarbon polymer, phenolic, or neoprene rubber.

Freeze Protection

Freeze protection is extremely important and is often the determining factor when selecting a system in the United States. Freezing can occur at ambient temperatures as high as 42°F because of radiation to the night sky. Manual freeze protection should not be used for commercial installations.

One simple way of protecting against freezing is to drain fluid from the collector array and interior piping when potential freezing conditions exist. Drainage may be automatic, as in draindown and drainback systems, or manual, as in direct thermosiphon systems. Automatic systems should be capable of fail-safe drainage operation, even in the event of pump failure or power outage. In some cases water may be designed to drain back through the pump, so the design must allow refilling without causing cavitation.

In areas where freezing is infrequent, recirculating water from storage to the collector array can be used as freeze protection. Freeze protection can also be provided by using fluids that resist freezing. Fluids such as water/glycol solutions, silicone oils, and hydrocarbon oils are circulated by pumps through the collector array and double wall heat exchanger. Draining the collector fluid is not required, because these fluids have freezing points well below the coldest anticipated outdoor temperature.

In mild climates where recirculation freeze protection is used, a second level of freeze protection should be provided by flushing the collector with cold supply water when the collector approaches near-freezing temperatures. This can be accomplished with a temperature-controlled valve that will automatically open a small port at a near-freezing temperature of about 40°F and then close at a slightly higher temperature.

Overheat Protection

During periods of high insolation and low hot-water demand, overheating can occur in the collectors or storage tanks. Protection against overheating must be considered for all portions of the solar hot-water system. Liquid expansion or excessive pressure can burst piping or storage tanks. Steam or other gases within a system can restrict liquid flow, making the system inoperable.

The most common methods of overheat protection stop circulation in the collection loop until the storage temperature decreases, discharge the overheated water and replace it with cold makeup water, or use a heat exchanger as a means of heat rejection. Some freeze-protection methods can also provide overheat protection by circulating the collector fluid at night to radiate excess heat.

For nonfreezing fluids such as glycol antifreezes, overheat protection is needed to limit fluid degradation at high temperatures during collector stagnation.

Safety

Safety precautions required for installing, operating, and servicing a solar domestic hot-water heater are essentially the same as those for a conventional domestic hot-water heater. One major exception is that some solar systems use nonpotable heat transfer fluids. Local codes may require a double-wall heat exchanger for potable water installations.

Pressure relief must be provided in all parts of the collector array that can be isolated by valves. The outlet of these relief valves should be piped to a container or drain, and not where people could be affected.

Start-Up Commissioning Procedure

After completing the installation, certain tests must be performed before charging or filling the system. The system must be checked for leakage, and pumps, fans, valves, and sensors must be checked to see that they function. Testing procedures vary with system type.

Closed-loop systems should be hydrostatically tested. The system is filled and pressurized to 1.5 times the operating pressure for one hour and inspected for leaks and any appreciable pressure drop.

Draindown systems should be tested to be sure that all water drains from the collectors and piping located outdoors. All lines should be checked for proper pitch so that gravity drains them completely. All valves should be verified to be in working order.

Drainback systems should be tested to ensure that collector fluid drains back to the reservoir tank when circulation stops and that the system refills properly.

Air systems should be tested for leaks before insulation is applied by starting the fans and checking the ductwork for leaks.

Pumps and sensors should be inspected to verify that they are in proper working order. Proper cycling of the pumps can be checked by a running time meter. A sensor that is suspected of being faulty can be dipped alternately in hot and cold water to see if the pump starts or stops.

After system testing and before filling or charging it with heat transfer fluid, the system should be flushed to remove debris.

Maintenance

All systems should be checked at least once a year in addition to any periodic maintenance that may be required for specific components. A log of all maintenance performed should be kept. The system designer and installer should provide the building owner with an operating manual that describes maintenance procedures and operating modes in sufficient detail to support ongoing performance.

The collectors' outer glazing should be hosed down periodically. Leaves, seeds, dirt, and other debris should be carefully swept from the collectors. Care should be taken not to damage plastic covers.

Without opening a sealed collector panel, the absorber plate should be checked for surface coating damage caused by peeling, crazing, or scratching. Also, the collector tubing should be inspected to ensure that it contacts the absorber. If the tubing is loose, the manufacturer should be consulted for repair instructions.

Heat transfer fluids should be tested and replaced at intervals suggested by the manufacturer. Also, the solar energy storage tank should be drained about every six months to remove sediment.

Performance Monitoring/Minimum Instrumentation

Temperature sensors and temperature differential controllers are required to operate most solar systems. However, additional instruments should be installed for monitoring, checking, and troubleshooting.

Thermometers should be located on the collector supply and return lines so that the temperature difference in the lines can be determined visually.

A pressure gage should be inserted on the discharge side of the pump. The gage can be used to monitor the pressure that the pump must work against and to indicate if the flow passages are blocked.

Running time meters on pumps and fans may be installed to determine if the system is cycling properly.

DESIGN, INSTALLATION, AND OPERATION CHECKLIST

The following checklist is for designers of solar heating and cooling systems. Specific values have not been included because these vary for each application. The designer must decide whether design figures are within acceptable limits for any particular project [see DOE (1978b) for further information]. The review order listed does not reflect their precedence or importance during design.

Collectors

- Check flow rate for compliance with manufacturer's recommendation.
- Check that collector area matches the application and claimed solar fraction of the load.
- Review collector instantaneous efficiency curve and check match between collector and system requirements.
- Relate collector construction to end use; two cover plates are not required for low-temperature collection in warm climates and may, in fact, be detrimental. Two cover plates are more efficient when the temperature difference between the absorber plate and outdoor air is high, such as in severe winter climates or when collecting at high temperatures for cooling. Radiation loss only becomes significant at relatively high absorber plate temperatures. Selective surfaces should be used in these cases. Flat-black surfaces are acceptable and sometimes more desirable for low collection temperatures.
- Check match between collector tilt angle, latitude, and collector end use.
- Check collector azimuth.
- Check collector location for potential shading and exposure to vandalism or accidental damage.
- Review provisions for high stagnation temperature. If not used, are liquid collectors drained or left filled in summer?
- Check for snow hang-up and ice formation. Will casing vents become blocked?
- Review precautions, if any, against outgassing.
- Check access for cleaning covers.
- Check mounting for stability in high winds.
- Check for architectural integration. Do collectors on roof present rainwater drainage or condensation problems? Do roof penetrations present potential leak problems?
- Check collector construction for structural integrity and durability. Will materials deteriorate under operating conditions? Will any pieces fall off?
- Are liquid collector passages organized in such a way as to allow natural fill and drain? Does mounting configuration affect this?
- Does air collector duct connection promote balanced airflow and even heat transfer? Are connections potentially leaky?

Heat Transfer Fluid

- Check that flow rate through the collector array matches system parameters.
- If antifreeze is used, check that flow rate has been modified to allow for the viscosity and specific heat.
- Review properties of proposed antifreeze. Some fluids are highly flammable. Check toxicity, vapor pressure, flash point, and boiling and freezing temperatures at atmospheric pressure.
- Check means of makeup into antifreeze system. An automatic water makeup system can result in freezing.
- Check that provisions are made for draining and filling the system (air vents at high points, drains at low points, pipes correctly graded in-between, drainback vented to storage or expansion tank).

- If system uses drainback freeze protection, check that
  - Provision is made for drainback volume and back venting
  - Pipes are graded for drainback
  - Solar primary pump is sized for lift head
  - Pump is self-priming if tank is below pump
- Check that collector pressure drop for drainback is slightly higher than static pressure between supply and return headers.
- Optimum pipe arrangement is reverse-return with collectors in parallel. Series collectors reduce flow rate and increase head. A combination of parallel/series can sometimes be beneficial, but check that equipment has been sized and selected properly.
- Cross-connections under different operating modes sometimes result in pumps operating in opposition or tandem, causing severe hydraulic problems.
- If heat exchangers are used, check that approach temperature differential has been recognized in the calculations.
- Check that adequate provisions are made for water expansion and contraction. Use specific volume/temperature tables for calculation. Each unique circuit must have its own provision for expansion and contraction.
- Three-port valves tend to leak through the closed port. This, together with reversed flows in some modes, can cause potential hydraulic problems. As a general rule, simple circuits and controls are better.

Airflow

- Check that flow rate through the collector array matches system design values.
- Check temperature rise across collectors using air mass flow and specific heat.
- Check that duct velocities are within design limits.
- Check that cold air or water cannot flow from collectors by gravity under "no-sun" conditions.
- Verify duct material and construction methods. Ductwork must be sealed to reduce loss.
- Check duct configuration for balanced flow through collector array.
- Check number of collectors in series. More than two collectors in series can reduce collection efficiency.

Thermal Storage

- Check that thermal storage capacity matches values of collector area, collection temperature, use temperature, and load.
- Verify that thermal inertia does not impede effective operation.
- Check provisions for promoting temperature stratification during both collection and use.
- Check that pipe and duct connections to storage are compatible with the control, ensuring that only the coolest air goes to the collectors and connections.
- If liquid storage is used for high-temperature (above 200°F) applications, check that tank material and construction can withstand the temperature and pressure.
- Check that storage location does not promote unwanted heat loss or gain and that adequate insulation is provided.
- Verify that liquid storage tanks are treated to resist corrosion. This is particularly important in tanks that are partially filled.
- Check that provision is made to protect liquid tanks from exposure to either overpressure or vacuum.

Uses

Domestic Hot Water

- Characteristics of domestic hot-water loads include short periods of high draw interspersed with long dormant periods. Check that domestic hot-water storage matches solar heat input.

- Check that provisions have been made to prevent reverse heating of the solar thermal storage by the domestic hot-water back-up heater.
- Check that the design allows cold makeup water preheating on days of low solar input.
- Verify that the antiscald valve limits domestic hot-water supply to a safe temperature during periods of high solar input.
- Depending on total dissolved solids, city water heated above 150°F may precipitate a calcium carbonate scale. If collectors are used to heat water directly, check provisions for preventing scale formation in absorber plate waterways.
- Check whether the heater is required to have a double-wall heat exchanger and that it conforms to appropriate codes if the collector uses nonpotable fluids.

Heating

- Warm-air heating systems can use solar energy directly at moderate temperatures. Check that air volume is sufficient to meet the heating load at low supply temperatures and that the limit thermostat has been reset.
- At times of low solar input, solar heat can still be used to meet part of the load by preheating return air. Check location of solar heating coil in system.
- Baseboard heaters require relatively high supply temperatures for satisfactory operation. Their output varies as the 1.5 power of the log mean temperature difference and falls off drastically at low temperatures. If solar is combined with baseboard heating, check that supply temperature is compatible with heating load.
- Heat exchangers imply an approach temperature difference that must be added to the system operating temperature to derive the minimum collection temperature. Verify calculations.
- Water-to-air heat pumps rely on a constant solar water heat source for operation. When the heat source is depleted, the back-up system must be used. Check that storage is adequate.

Cooling

- Solar-activated absorption cooling with fossil fuel back-up is currently the only commercially available active cooling. Be assured of all design criteria and a large amount of solar participation. Verify calculations.
- Storing both hot and chilled water may make better use of available storage capacity.

Controls

- Check that control design matches desired modes of operation.
- Verify that collector loop controls recognize solar input, collector temperature, and storage temperature.
- Verify that controls allow both the collector and usage loops to operate independently.
- Check that control sequences are reversible and always revert to the most economical mode.
- Check that controls are as simple as possible within the system requirements. Complex controls increase the frequency and possibility of breakdowns.
- Check that all controls are fail-safe.

Performance

- Check building heating, cooling, and domestic hot-water loads as applicable. Verify that building thermal characteristics are acceptable.
- Check solar energy collected on a monthly basis. Compare with loads and verify solar participation.

PHOTOVOLTAIC APPLICATIONS

Photovoltaic (PV) solar collectors, called PV modules, convert light directly into electricity. Chapter 36 of the 2008 *ASHRAE Handbook—HVAC Systems and Equipment* discusses the fundamentals of their operation. Photovoltaics are used to power remotely located communications equipment, remote monitoring, lighting, water pumping, battery charging, and cathodic protection, and for utility interactive power generation. The first step for any PV application is to analyze electric loads and then minimize them wherever possible. PV array size can be reduced substantially by using high-efficiency HVAC, appliances, motors, and other power-consuming devices. Using this strategy is almost always more cost-effective than adding more PV modules.

Communications. Photovoltaics can provide reliable power with little maintenance for communications systems, especially those in remote areas (away from the power grid) with extreme weather conditions such as high winds, heavy snows, and ice. Examples include communication relay towers, travelers' information transmitters, cellular telephones, mobile radio systems, emergency call boxes, and military test facilities. These systems range in size from a few watts for call boxes to several kilowatts for microwave repeater stations. For larger systems at remote sites, an engine generator is often combined with the photovoltaic-battery system. These hybrid systems with two or more generators can achieve nearly 100% availability.

Remote Site Electrification. Photovoltaics are used to provide power to rural residences, visitor centers in parks, vacation cabins, island and villages, remote research facilities, and military test areas. The varied loads include lighting, small appliances, water pumps (including circulators on solar water-heating systems), and communications equipment. The load demand varies from a few watts to tens of kilowatts. Deep-cycle battery back-up is most often used in this application for nighttime power, and is typically sized to meet the load for two to three days without PV input. Hybrid systems are also common, using wind or small hydro generation equipment.

Remote Monitoring. Photovoltaics provide power at remote sites to sensors, data loggers, and associated transmitters for meteorological monitoring, structural condition measurement, seismic recording, irrigation control, highway/traffic monitoring, security monitoring, and scientific research. Most of these applications require less than 200 W, and many can be powered by a single photovoltaic module. Vandalism may be a problem in some areas, so non-glass-covered modules are sometimes used. Mounting the modules on a tall pole or in an unobtrusive manner may also help avoid damage or theft. The batteries are often located in the same weather-resistant enclosure as the data acquisition/monitoring equipment. This enclosure is sometimes camouflaged or buried for protection. Some data loggers come with their own battery and charge regulator.

Signs and Signals. The most popular application for photovoltaics is warning signs. Typical devices include navigational beacons, audible signals such as sirens, highway warning signs, railroad signals, and aircraft warning beacons. Because these signals are critical to public safety, they must be operative at all times, and thus the reliability of the photovoltaic system is extremely important. High reliability can be achieved by using large-capacity batteries. Many of these systems operate in harsh environments. For maritime applications, special modules are used that are resistant to corrosion from salt water.

Water Pumping and Control. Photovoltaics are typically used for intermediate-sized water pumping applications (those larger than hand pumps and smaller than large engine-powered pumps). The range of sizes for photovoltaic-powered pumps is a few hundred watts to a few kilowatts. Applications include domestic use, water for campgrounds, irrigation, village water supplies, and livestock watering. Photovoltaics for livestock watering can be more economical than maintaining a distribution line to a remote pump on a ranch. Most pumping systems do not use batteries but store water in holding tanks. For this application, photovoltaic modules may be mounted on tracking frames that maximize energy production by tracking the sun each day.

Rest Room Facilities. Highway rest stops, public beach facilities, outdoor recreation parks, and public campgrounds may have photovoltaics to operate air circulation and ventilation fans, interior and exterior lights, and auxiliary water pumps for sinks and showers. For most of these applications, the initial cost for photovoltaic power is the least expensive option.

Charging Auto, Boat, and RV Batteries. Batteries self-discharge over time if they are not used. This is a problem for organizations that maintain a fleet of vehicles such as fire-fighting or snow removal equipment, some of which are infrequently used. Photovoltaic battery chargers can solve this problem by providing a trickle charging current that keeps the battery at a high state of charge. Often the PV module can be placed inside the windshield and plugged into the vehicle's power socket, thus using existing wiring and protection circuits and providing a quick disconnect for the module. Modules are installed on the roof or engine hood of larger vehicles. Another successful application is using PV modules to charge the batteries in electric vehicles.

Grid-Connected Systems. Grid-connected or utility-interactive PV systems use an inverter that converts dc output from the PV array to ac current. It is designed to operate in conjunction with the electric utility mains, synchronizing the ac output (phase, frequency, and voltage) with the utility. The PV output is directly connected to the inverter, and the ac output from the inverter is connected to the ac load circuit. Excess power generated during the day is fed back to the utility grid, and grid power is used at night. **Net metering** uses a single utility meter; when energy is fed back to the utility grid, the meter runs backwards.

SYMBOLS

a = constant for specified month at specified location

A = apparent extraterrestrial irradiation at $m = 0$, Btu/h·ft^2

A_{abs}, A_{ap} = areas of absorber surface and of aperture that admit or receive radiation, ft^2

A_{ap} = aperture area

A_c = area of solar collector, ft^2

ACH = air changes per hour

A_g = gross collector area

A_p = projected collector area

AST = apparent solar time

b = constant for specified month at specified location

B = atmospheric extinction coefficient

C = parameter depending on dust and moisture content of atmosphere

c_p = specific heat of fluid, Btu/lb·°F

DD = degree-day load of nonsolar elements

DST = daylight saving time

e_{at} = sky emittance

e_s = surface emittance

f = fraction of monthly space- and water-heating loads supplied by solar energy

F_r = collector heat exchanger efficiency factor

F_R = collector heat removal factor; ratio of heat actually delivered by collector to heat that would be delivered if absorber were at t_{fi}

F_{sg} = angle factor between surface and earth; Equation (17)

F_{ss} = angle factor between surface and sky; Equation (15)

H = hour angle

H_T = monthly averaged daily radiation incident on collector surface per unit area, Btu/day·ft^2

I_{dH} = diffuse radiation falling on horizontal surface under cloudless sky

I_{DN} = direct normal irradiation, Btu/h·ft^2

$I_{d\theta}$ = diffuse component of solar radiation from sky, Btu/h·ft^2

I_r = reflected shortwave radiation

I_{tH} = total horizon irradiation

$I_{t\theta}$ total solar irradiation of terrestrial surface or collector, Btu/h·ft^2

K_T = atmospheric clearness index

L = monthly total heating load for space heating and hot water, Btu

LCR = load collector ratio; net load coefficient divided by projected collector area

LST = local standard time

m = air mass; = 1.0/sin β at sea level and 0 outside earth's atmosphere; also fluid flow rate, lb/h

$\dot{m}$ = fluid flow rate, lb/h

N = year day, with January 1 = 1; also number of days in month

NLC = net load coefficient

q_{Rat} = downward radiation from atmosphere

q_{Rs} = longwave radiation emitted by surface on earth

q_u = useful heat gained by collector per unit of aperture area, Btu/h·ft^2

R_{net} = net radiative cooling rate of horizontal surface

S = total downward radiant heat flux from atmosphere

SSF = solar savings fraction

SLR = solar energy absorbed divided by building heating load

t_a = temperature of atmosphere, or monthly average ambient temperature, °F

$\bar{t}_a$ = monthly average ambient temperature, °F

t_{abs} = temperature of absorber, °F

T_{base} = base temperature

t_{fe}, t_{fi} = temperatures of fluid leaving and entering collector, respectively, °F

TLC = total load coefficient

t_p = temperature of absorber plate, °F

T_{rad} = absolute temperature of horizontal surface, °R

t_{ref} = reference temperature, 212°F

T_s = absolute temperature of surface, °R

T_{sky} = apparent sky temperature, °R

U_L = upward heat loss coefficient, Btu/h·ft^2·°F

X = reference collector loss divided by heating load

Y = absorbed solar energy divided by heating load

Greek

α = absorptance

β = sun's altitude above horizon

γ = surface-solar azimuth

δ = solar declination

$\Delta\theta$ = total number of hours in month

ε = nonspectral emittance of horizontal surface; about 0.9 for most nonmetallic construction materials

η = collector efficiency

θ = solar incident angle

ρ_g = reflectance of foreground

$\rho\Gamma$ = reflectance of concentrator surface times fraction of reflected or refracted radiation that reaches absorber

σ = Stefan-Boltzmann constant, 0.1712×10^{-8} Btu/h·ft^2·°R^4

Σ = tilt angle of surface, measured from horizontal

τ = transmittance

$(\overline{\tau\alpha})$ = monthly average transmittance-absorptance product

$(\tau\alpha)_n$ = normal transmittance-absorptance product

$(\tau\alpha)_\theta$ = transmittance τ of cover times absorptance α of plate at prevailing incident angle θ

ϕ = solar azimuth

ψ = surface azimuth

REFERENCES

ASCE. 2003. Minimum design loads for buildings and other structures. ANSI/ASCE *Standard* 7-2003. American Society of Civil Engineers, Reston, VA.

ASCE. 1991. Specification for the design of cold-formed stainless steel structural members. ANSI/ASCE *Standard* 8-91. American Society of Civil Engineers, Reston, VA.

ASCE. 1994. Standard practice for construction and inspection of composite slabs. ANSI/ASCE *Standard* 9-94. American Society of Civil Engineers, Reston, VA.

ASHRAE. 1977. *Applications of solar energy for heating and cooling of buildings.*

ASHRAE. 1983. *Solar domestic and service hot water manual.*

ASHRAE. 1984. *Passive solar heating analysis: A design manual.*

ASHRAE. 1986. Methods of testing to determine the thermal performance of solar collectors. ANSI/ASHRAE *Standard* 93-1986 (RA91).

ASHRAE. 2010. Methods of testing to determine the thermal performance of solar collectors. ANSI/ASHRAE *Standard* 93-2010.

Balcomb, J.D., J.C. Hedstrom, and R.D. McFarland. 1977. Thermal storage walls in New Mexico. *Solar Age* 2(8):20.

Balcomb, J.D., R.W. Jones, R.D. McFarland, and W.O. Wray. 1982. Expanding the SLR method. *Passive Solar Journal* 1:2.

Barley, C.D. and C.B. Winn. 1978. Optimal sizing of solar collectors by the method of relative areas. *Solar Energy* 21:4.

Beckman, W.A., S.A. Klein, and J.A. Duffie. 1977. *Solar heating design by the* f-*Chart method.* John Wiley, New York.

Beckman, W.A., S.A. Klein, and J.A. Duffie. 1981. Performance predictions for solar heating systems. In *Solar energy handbook,* J.F. Kreider and F. Kreith, eds. McGraw Hill, New York.

Bliss, R.W. 1961. Atmospheric radiation near the surface of the earth. *Solar Energy* 59(3):103.

Clark, G. 1981. Passive/hybrid comfort cooling by thermal radiation. *Proceedings of the International Passive and Hybrid Cooling Conference.* American Section of the International Solar Energy Society, Miami Beach, FL.

Cole, R.L., A.J. Gorski, R.M. Graven, W.R. McIntire, W.W. Schertz, R. Winston, and S. Zwerdling. 1977. Applications of compound parabolic concentrators to solar energy conversion. *Report* AMLw42. Argonne National Laboratory, Chicago.

Cromer, C.J. 1984. *Design of a DC-pump, photovoltaic-powered circulation system for a solar domestic hot water system.* Florida Solar Energy Center, Cocoa.

Dickinson, W.C. and P.N. Cheremisinoff, eds. 1980. *Solar energy technology handbook,* Part B: *Application, systems design and economics.* Marcel Dekker, New York.

DOE. 1978a. *SOLCOST—Solar hot water handbook; A simplified design method for sizing and costing residential and commercial solar service hot water systems,* 3rd ed. DOE/CS-0042/2. U.S. Department of Energy.

DOE. 1978b. *DOE facilities solar design handbook.* DOE/AD-0006/1. U.S. Department of Energy.

DOE. 1980/1982. *Passive solar design handbooks,* vols. 2 and 3, *Passive solar design analysis.* DOE Reports/CS-0127/2 and CS-0127/3. January, July. U.S. Department of Energy.

Duffie, J.A. and W.A. Beckman. 2006. *Solar thermal energy processes,* 3rd ed. Wiley Interscience, New York.

Feldman, S.J. and R.L. Merriam. 1979. Building energy analysis computer programs with solar heating and cooling system capabilities. Arthur D. Little, Inc. *Report* EPRIER-1146 (August) to the Electric Power Research Institute.

Freeman, T.L., J.W. Mitchell, and T.E. Audit. 1979. Performance of combined solar-heat pump systems. *Solar Energy* 22:2.

Gates, D.M. 1966. Spectral distribution of solar radiation at the earth's surface. *Science* 151(3710):523.

Givoni, B. 1981. Experimental studies on radiant and evaporative cooling of roofs. *Proceedings of the International Passive and Hybrid Cooling Conference.* American Section of the International Solar Energy Society, Miami Beach, FL.

Hay, H.R. and J.I. Yellott. 1969. Natural air conditioning with roof ponds and movable insulation. *ASHRAE Transactions* 75(1):165-177.

Hottel, H.C. and B.B. Woertz. 1942. The performance of flat-plate solar collectors. *Transactions of ASME* 64:91.

Howard, B.D. 1986. Air core systems for passive and hybrid energy-conserving buildings. *ASHRAE Transactions* 92(2B):815-830.

Howard, B.D. and E.O. Pollock. 1982. *Comparative report—Performance of passive solar heating systems.* Vitro Corp. U.S. DOE National Solar Data Program, Oak Ridge, TN.

Howard, B.D. and D.H. Saunders. 1989. Building performance monitoring—The thermal envelope perspective—Past, present, and future. Thermal Performance of the Exterior Envelopes of Buildings IV, ASHRAE.

HUD. 1977. *Intermediate minimum property standards supplement for solar heating and domestic hot water systems.* SD Cat. No. 0-236-648. U.S. Department of Housing and Urban Development.

Hunn, B.D., N. Carlisle, G. Franta, and W. Kolar. 1987. *Engineering principles and concepts for active solar systems.* SERI/SP-271-2892. Solar Energy Research Institute, Golden, CO.

Jordan, R.C. and B.Y.H. Liu, eds. 1977. Applications of solar energy for heating and cooling of buildings. ASHRAE *Publication* GRP 170.

Klein, S.A. and W.A. Beckman. 1979. A general design method for closed-loop solar energy systems. *Solar Energy* 22(3):269-282.

Klein, S.A., W.A. Beckman, and J.A. Duffie. 1976. TRNSYS—A transient simulation program. *ASHRAE Transactions* 82(1):623-633.

Kutscher, C.F. 1996. *Proceedings of the 19th World Energy Engineering Congress,* Atlanta, GA.

LBL. 1981. *DOE-2 reference manual version 2.1A.* Los Alamos Scientific Laboratory, LA-7689-M, Version 2.1A. Lawrence Berkeley Laboratory, LBL-8706 Rev. 2.

Lister, L. and T. Newell. 1989. *Expansion tank characteristics of closed loop, active solar energy collection systems; Solar engineering—1989.* American Society of Mechanical Engineers, New York.

Lunde, P.J. 1980. *Thermal engineering.* John Wiley & Sons, New York.

Marlatt, W., C. Murray, and S. Squire. 1984. Roofpond systems energy technology engineering center. Rockwell International, *Report* ETEC6.

Martin, M. and P. Berdahl. 1984. Characteristics of infrared sky radiation in the United States. *Solar Energy* 33(3/4):321-336.

Mazria, E. 1979. *The passive solar energy book.* Rodale, Emmaus, PA.

Mitchell, D. and K.L. Biggs. 1979. Radiative cooling of buildings at night. *Applied Energy* 5:263-275.

Monsen, W.A., S.A. Klein, and W.A. Beckman. 1981. Prediction of direct gain solar heating system performance. *Solar Energy* 27(2):143-147.

Monsen, W.A., S.A. Klein, and W.A. Beckman. 1982. The un-utilizability design method for collector-storage walls. *Solar Energy* 29(5):421-429.

Morehouse, J.H. and P.J. Hughes. 1979. Residential solar-heat pump systems: Thermal and economic performance. *Paper* 79-WA/SOL-25, ASME Winter Annual Meeting, New York.

Nordham, D. 1981. *Microcomputer methods for solar design and analysis.* Solar Energy Research Institute, SERI-SP-722-1127.

Pittenger, A.L., W.R. White, and J.I. Yellott. 1978. A new method of passive solar heating and cooling. *Proceedings of the Second National Passive Systems Conference,* Philadelphia, ISES and DOE.

Reitan, C.H. 1963. Surface dew point and water vapor aloft. *Journal of Applied Meteorology* 2(6):776.

Root, D.E., S. Chandra, C. Cromer, J. Harrison, D. LaHart, T. Merrigan, and J.G. Ventre. 1985. *Solar water and pool heating course manual,* 2 vols. Florida Solar Energy Center, Cocoa.

Schnurr, N.M., B.D. Hunn, and K.D. Williamson. 1981. The solar load ratio method applied to commercial buildings active solar system sizing. *Proceedings of the ASME Solar Energy Division Third Annual Conference on System Simulation,* Economic Analysis/Solar Heating and Cooling Operational Results, Reno, NV.

Simon, F.F. 1976. Flat-plate solar collector performance evaluation. *Solar Energy* 18(5):451.

Stephenson, D.G. 1967. Tables of solar altitude and azimuth; Intensity and solar heat gain tables. *Technical Paper* 243, Division of Building Research, National Research Council of Canada, Ottawa.

Svard, C.D., J.W. Mitchell, and W.A. Beckman. 1981. Design procedure and applications of solar-assisted series heat pump systems. *Journal of Solar Energy Engineering* 103(5):135.

Swartman, R.K., V. Ha, and A.J. Newton. 1974. Review of solar-powered refrigeration. *Paper* 73-WA/SOL-6. American Society of Mechanical Engineers, New York.

Thekaekara, M.P. 1973. Solar energy outside the earth's atmosphere. *Solar Energy* 14(2):109.

Threlkeld, J.L. 1963. Solar irradiation of surfaces on clear days. *ASHRAE Transactions* 69:24.

Threlkeld, J.L. and R.C. Jordan. 1958. Direct radiation available on clear days. *ASHRAE Transactions* 64:45.

Trombe, F., J.F. Robert, M. Caloanat, and B. Sesolis. 1977. Concrete walls for heat. *Solar Age* 2(8):13.

U.S. GPO. 2005. *Climate atlas of the United States,* 2nd ed. U.S. Government Printing Office, Washington, D.C. Available as CD or online at http://www.ncdc.noaa.gov/oa/ncdc.html.

Whillier, A. 1964. Thermal resistance of the tube-plate bond in solar heat collectors. *Solar Energy* 8(3):95.

Yellott, J.I. 1977. Passive solar heating and cooling systems. *ASHRAE Transactions* 83(2):429.

Yellott, J.I., D. Aiello, G. Rand, and M.Y. Kung. 1976. *Solar-oriented architecture.* Arizona State University Architecture Foundation, Tempe.

BIBLIOGRAPHY

Abdulla, S.H., S.A. Klein, and W.A. Beckman. 2000. *A new correlation for the prediction of the frequency distribution of daily solar radiation.* American Solar Energy Society, Boulder, CO.

ASHRAE. 1987. Methods of testing to determine the thermal performance of solar domestic water heating systems. *Standard* 95-1987.

ASHRAE. 1989. Methods of testing to determine the thermal performance of unglazed flat-plate liquid-type solar collectors. *Standard* 96-1980 (RA 1989).

ASHRAE. 1988. *Active solar heating systems design manual.*

ASHRAE. 1991. *Active solar heating systems installation manual.*

ASHRAE. 1995. *Bin and degree-hour weather data for simplified energy calculations.*

Colorado State University. 1980. *Solar heating and cooling of residential buildings: Design of systems.* U.S. Government Printing Office, Washington, D.C.

Colorado State University. 1980. *Solar heating and cooling of residential buildings: Sizing, insulation and operation of systems.* U.S. Government Printing Office, Washington, D.C.

Cook, J., ed. 2000. *Passive cooling*, 2nd ed. MIT, Cambridge, MA.

Diamond, S.C. and J.G. Avery. 1986. *Active solar energy system design, installation and maintenance: Technical applications manual.* LA-UR-86-4175.

DOE. 1980. *SOLCOST—Version 3.0; Solar design program for nonthermal specialists: User's guide.* DOE/ET-20108-T2, U.S. Department of Energy.

Edwards, D.K., J.T. Gier, K.E. Nelson, and R.D. Roddick. 1962. Spectral and directional thermal radiation characteristics of selective surfaces. *Solar Energy* 6(1):1.

Healey, H.M. 1988. Site-built large volume solar water heating systems for commercial and industrial facilities. *ASHRAE Transactions* 94(1): 1277-1286.

Hiller, M.D., J.W. Mitchell, and W.A. Beckman. 1997. *TRNSHD—A program for shading and insolation calculations.* ASES World Forum, Washington, D.C.

HUD. 1980. *Installation guidelines for solar DHW systems in one- and two-family dwellings*, 2nd ed. U.S. Department of Housing and Urban Development.

Kaplanis, S.N. 2006. New methodologies to estimate the hourly global solar radiation: Comparisons with existing models. *Renewable Energy* 31(6): 781-790.

Klein, S.A. and W.A. Beckman. 2001. *PV f-chart: Photovoltaic system analysis software; User's manual.* F-Chart Software, Middleton, WI.

Knapp, C.L., T.L. Stoffel, and S.D. Whitaker. 1980. *Insolation data manual.* SERI/SP-755-789. Solar Energy Research Institute, Golden, CO. (Supplement added 1982.)

Kreider, J.F. and F. Kreith. 1981. *Solar energy handbook.* McGraw-Hill, New York.

Kreider, J.F. and F. Kreith. 1982. *Solar heating and cooling: Active and passive design*, 2nd ed. McGraw-Hill, New York.

Lane, G.A. 1986. *Solar heat storage: Latent heat materials*, 2 vols. CRC, Boca Raton, FL.

Löf, G.O., J.A. Duffie, and C.D. Smith. 1966. World distribution of solar radiation. *Report* 21. Solar Energy Laboratory, University of Wisconsin, Madison.

Mueller Associates, Inc. 1985. *Active solar thermal design manual.* U.S. Department of Energy, Solar Energy Research Institute and ASHRAE.

Mumma, S.A. 1985. Solar collector tilt and azimuth charts for rotated collectors on sloping roofs. *Proceedings of the Joint ASME-ASES Solar Energy Conference*, Knoxville, TN.

NREL. 2003. *Solar resource information.* National Renewable Energy Laboratory. http://www.nrel.gov/rredc//solar_resource.html.

Solar Energy Research Institute. 1981. *Solar radiation energy resource atlas of the United States.* SERI/SP642-1037. Golden, CO.

U.S. Hydrographic Office. 1958. Tables of computed altitude and azimuth. *Hydrographic Office Bulletin* 214, vols. 2 and 3. U.S. Superintendent of Documents, Washington, D.C.

Weiss, W., ed. 2003. *Solar heating for houses: A design handbook for solar combisystems.* International Energy Agency, James & James, Ltd., London.

ENERGY USE AND MANAGEMENT

ENERGY management in buildings is the control of energy use and cost while maintaining indoor environmental conditions to provide comfort and to fully meet functional needs. This chapter provides guidance on establishing an effective, ongoing energy management program, as well as information on planning and implementing energy management projects. The energy manager should understand how energy is used in the building, to manage it effectively. There are opportunities for savings by reducing the unit price of purchased energy, and by improving the efficiency and reducing the use of energy-consuming systems.

Water/sewer costs and use may be included in the energy management activity. This could be called "utility management," but "energy management" is used in this chapter.

ENERGY MANAGEMENT

The specific processes by which building owners and operators control energy consumption and costs are as variable as their building types. Small buildings, such as residences and small commercial businesses, usually involve the efforts of one person. Energy management procedures should be as simple, specific, and direct as possible. General energy management advice, such as from utility energy surveys or state or provincial energy offices, can provide ideas, but these must be evaluated to determine whether they are applicable to the target building. Owners and operators of smaller buildings may only need advice on specific energy projects (e.g., boiler replacement, lighting retrofit). On the other hand, large or complex facilities, such as hospital or university campuses, industrial complexes, or large office buildings, usually require a team effort and process as represented in Figure 1.

Figure 1 is adapted from the ENERGY STAR® Web site (www. energystar.gov). On the ENERGY STAR Web site, each box in the flowchart refers the reader to numerous useful tips.

Energy management for existing buildings has these basic steps:

1. Appoint an energy manager to oversee the process and to ensure that someone is dedicated to the initiatives and accountable to the company.
2. Early communication to solicit feedback for other steps of the process.
3. Establish an energy accounting system that records energy and water consumption and associated costs. It should include comparisons with similar buildings, to benchmark and set performance goals.
4. Validate and analyze current and historical energy use data to help identify conservation energy-efficiency measures.
5. Carry out energy surveys and walk-through audits to identify low-cost/no-cost operations, maintenance, and energy-efficiency measures. Having a qualified energy professional do this is recommended.

The preparation of this chapter is assigned to TC 7.6, Building Energy Performance.

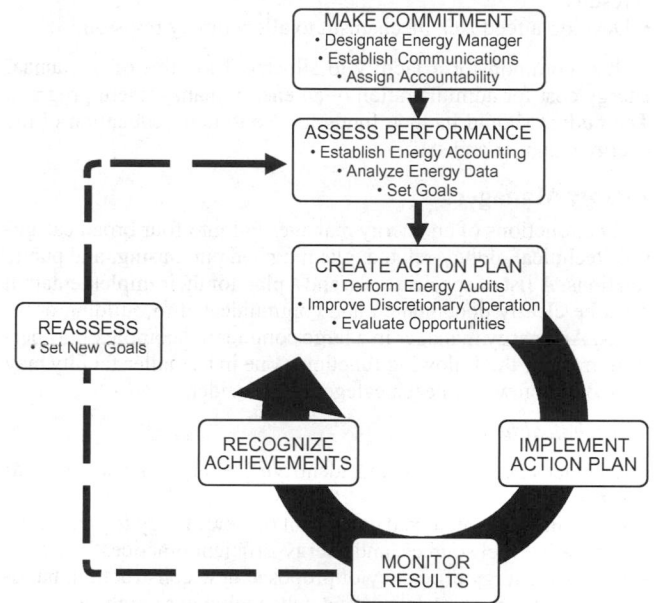

Fig. 1   An Energy Management Process

6. Using the survey results, change building operating procedures to eliminate energy waste.
7. Evaluate energy-efficiency measures for expected savings, estimated implementation costs, risks, and nonenergy benefits. Recommend a number of prioritized energy-efficiency projects for implementation.
8. Implement approved energy-efficiency measures (EEMs). Tender projects that must be outsourced.
9. Track results using the energy accounting system for overall performance, supplemented as needed by energy monitoring related to specific projects.
10. Compare results to past goals, revise as necessary, and develop new goals. Report to management and tenants. Return to step 7 and continue the process to maintain and continually improve building performance.

Each of these energy management program components is discussed in detail in the following sections.

ASHRAE *Standard* 100 gives details useful in energy management planning in existing buildings. Information on energy efficiency in new design can be found in all volumes of the *ASHRAE Handbook* and in ASHRAE *Standards* 90.1 and 90.2. The area most likely to be overlooked in new design is the ability to measure and monitor energy consumption and trends for each energy use category given in Chapter 41. Additional guidelines for this area can be found in Chapter 34 of the 2009 *ASHRAE Handbook—Fundamentals*.

Organizing for Energy Management

To be effective, energy management must be given the same emphasis as management of any other cost/profit center. Top management should

- Establish the energy cost/profit center
- Assign management responsibility for the program
- Assign an energy manager and provide training
- Allocate resources
- Clearly communicate the energy management program to all departments and personnel
- Set clear program goals
- Encourage ownership of the program by all levels of the organization
- Set up an ongoing reporting and analysis procedure to monitor results
- Develop a feedback mechanism to allow timely revisions

It is common for a facility to allocate 3 to 10% of the annual energy cost for administration of an energy management program. The budget should include funds for continuing education of the energy manager and staff.

Energy Managers

The functions of an energy manager fall into four broad categories: technical, policy-related, planning and purchasing, and public relations. A list of specific tasks and a plan for their implementation must be clearly documented and communicated to building occupants. An energy manager in a large commercial complex may perform most of the following functions; one in a smaller facility may have only a few from each category to consider.

Technical functions

- Conduct energy audits and identifying energy-efficiency measures
- Act as in-house technical consultant on new energy technologies, alternative fuel sources, and energy-efficient practices
- Evaluate energy efficiency of proposed new construction, building expansion, remodeling, and new equipment purchases
- Set performance standards for efficient operation and maintenance of equipment and facilities
- Review state-of-the-art energy management hardware
- Review building operation and maintenance procedures for optimal energy management
- Implement energy-efficiency measures (EEMs)
- Establish an energy accounting system
- Establish a baseline from which energy-saving improvements can be measured
- Measure and maintain effectiveness of EEMs
- Measure energy use in the field to verify design and operating conditions

Policy-related functions

- Fulfill energy policy established by top management
- Monitor federal and state (provincial) legislation and regulatory activities, and recommend policy/response
- Adhere to energy management building codes
- Represent the organization in energy associations
- Administer government-mandated reporting programs

Planning and purchasing functions

- Take advantage of fuel-switching and load management opportunities
- Purchase equipment based on life-cycle cost
- Take advantage of energy-efficiency programs offered by utilities and agencies
- Negotiate or advise on major utility contracts
- Develop contingency plans for supply interruptions or shortages

- Forecast and budget for short- and long-term energy requirements and costs
- Report regularly to top management and other stakeholders.

Public relations functions

- Make occupants aware of the benefits of efficient energy use
- Establish a mechanism to elicit and evaluate suggestions
- Recognize successful energy projects
- Establish an energy communications network
- Increase community awareness with press releases and appearances at civic group meetings

General qualifications

- A technical background, preferably in engineering
- Experience in energy-efficient design of building systems and processes
- Practical, hands-on experience with systems and equipment
- Goal-oriented management style
- Ability to work with people at all levels
- Technical report-writing and verbal communication skills

Desirable educational and professional qualifications

- Bachelor of science degree, preferably in mechanical, electrical, industrial, or chemical engineering
- Thorough knowledge of energy resource planning and conservation
- Ability to
  - Analyze and compile technical and statistical information and reports
  - Interpret plans and specifications for building facilities
- Knowledge of
  - Utility rates, energy efficiency, and planning
  - Automatic controls and systems instrumentation
  - Energy-related metering equipment and practices
  - Project management

If it is not possible to add a full-time manager, an existing employee with a technical background should be considered and trained. Energy management should not be a collateral duty of an employee who is already fully occupied. Another option is to hire a professional energy management consultant. Energy services companies (ESCOs) provide energy services as part of a contract, with payments based on realized savings. Other companies charge a fee to perform a variety of energy management functions.

COMMUNICATIONS

Energy management requires careful planning and help from all personnel that operate and use the facility. A communication plan should be regularly reviewed by both the energy manager and senior management. The initial communiqué should introduce the plan and express the support of top management for high-level goals. Providing early information to tenants and staff is important, because it takes time to change behaviors. Once the communication plan is launched, the energy manager should be prepared to answer a variety of questions from different areas of the company.

An effective communication strategy may include these tasks:

- Produce a regular newsletter
- Post energy-saving tips or reminders
- Hold annual seminars with maintenance and cleaning staff
- Meet with operations staff for training and feedback
- Report regularly to management and operations staff

Message content should be tailored to the specific audience. The more successful the communication is, the more quickly the energy management activities will become second nature. Diligent reporting promotes accountability and persistence of performance.

ENERGY ACCOUNTING SYSTEMS

An energy accounting system that tracks consumption and costs on a continuing basis is essential. It provides energy use data needed to confirm savings from energy-efficiency projects. The primary data source is utility bills, but other sources include

- Printouts from time-of-use meters
- Combustion efficiency, eddy current, and water quality tests
- Recordings of temperature and relative humidity
- Submetered energy use
- Event recordings
- Occupancy schedules and occupant activity levels
- Climate data
- Data from similar buildings in similar climates
- Infrared scans
- Production records
- Computer modeling

Energy Accounting Process

The energy manager establishes procedures for meter reading, monitoring, and tabulating facility energy use and profiles. The energy manager also periodically reviews utility rates, rate structures, and trends, and should subscribe to free utility mailing lists to track changes in their rate tariffs. The energy manager provides periodic reports to top management, summarizing the work accomplished, its cost-effectiveness, plans for future work, and projections of utility costs. Utility bill analysis software can be used to track avoided costs. If energy-efficiency measures are to be cost-effective, continued monitoring and periodic reauditing are necessary to ensure persistence. The procedures in ASHRAE *Guideline* 14 can be used for measurement and verification of energy savings.

Energy Accounting

Energy accounting means tracking utility bill data on a monthly basis to provide a current picture of building energy performance and to identify trends and instances of excess use. An Internet search for "energy accounting" will produce Web sites for the major commercial providers. In some cases software is sold for computer installation, or the accounting system is web-based and the user has a subscription. For many users, a simple spreadsheet is all that is needed. A comparison of the features of many available energy accounting software packages can be found at http://www.betterbricks.com/DetailPage.aspx?ID=518. Portfolio Manager, from the ENERGY STAR Web site, allows users to enter monthly energy usage, in kWh, therms, etc. The Portfolio Manager simultaneously calculates the facility's EUI and develops a normalized ENERGY STAR score (http://www.energystar.gov/benchmark). Portfolio Manager facilitates comparison of multiple buildings and goal setting, is useful for numerous building types, and is normalized by building type for weather.

Utility Rates

Because most energy management activities are dictated by economics, the energy manager must understand the utility rates that apply to each facility. Electric rates are more complex than gas or water rates and some rate structures make cost calculations difficult. In addition to general commercial or institutional electric rates, special rates may exist such as time of day, interruptible service, on peak/off peak, summer/winter, and peak demand. Electric rate schedules vary widely in North America; Chapters 37 and 56 discuss these in detail. Energy managers should work with local utility companies to identify the most favorable rates for their buildings, and must understand how demand is computed as well as the distinction between marginal and average costs (see the section on Improving Discretionary Operations). The utility representative can help develop the most cost-effective methods of metering and billing.

ANALYZING ENERGY DATA

Preparing for Cost and Efficiency Improvements

Opportunities for savings come in reducing (1) the cost per unit of energy, and then (2) energy consumption. Historically, energy users had little choice in selecting energy suppliers, and regulated tariffs applied based on certain customer characteristics. In recent years there has been a move in North America and other parts of the world to deregulate energy markets. and there is more flexibility in supply and pricing. Electric rate structures vary widely in North America; Chapter 37 discusses these in detail.

Electric utilities commonly meter both consumption and demand. **Demand** is the peak rate of consumption, typically averaged over a 15 or 30 min period. Electric utilities may also use a ratchet billing procedure for demand. Contact the local electric utility to fully understand the demand component.

Some utilities use **real-time pricing (RTP)**, in which the utility calculates the marginal cost of power per hour for the next day, determines the price, and sends this hourly price to customers. The customer can then determine the power consumption at different times of the day. A variation on RTP was introduced in some areas: **demand exchange and active load management** pays customers to shed loads during periods of high utility demand. Also called **demand reduction** or **demand response**, the utilities ask participating customers to reduce their consumption for a period of time on as little as a few hours' notice.

Caution is advised in designing or installing systems that take advantage of utility rate provisions, because the structure or provisions of utility rates cannot be guaranteed for the life of the system. Provisions that change include on-peak times, declining block rates, and demand ratchets. Chapter 56 has additional information on billing rates.

Analyzing Energy Use Data

Any reliable utility data should be examined. Utilities often provide metered data with measurement intervals as short as 15 min. Data from shorter time intervals make anomalies more apparent. High consumption at certain periods may reveal opportunities for cost reduction (Haberl and Komor 1990a, 1990b). If monthly data are used, they should be analyzed over several years.

A base year should be established as a reference point. Record the dates of meter readings so that energy use can be normalized for the number of days in a billing period. Any periods in which consumption was estimated rather than measured should be noted.

If energy data are available for more than one building or department, each should be tabulated separately. Initial tabulations should include both energy and cost per unit area (in an industrial facility, this may be energy and cost per unit of goods produced). Document variables such as heating or cooling degree-days, percent occupancy, quantity of goods produced, building occupancy, hours of operation, or daily weather conditions (see Chapter 14, Climatic Design Information, in the 2009 *ASHRAE Handbook—Fundamentals*). Because these variables may not be directly proportional to energy use, it is best to plot information separately or to superimpose one plot over another. Examples of ways to normalize energy consumption for temperature and other variations are provided in ASHRAE *Guideline* 14.

Potential savings areas can be identified by separating base energy consumption from weather-dependent energy consumption. **Base-load energy use** is the amount of energy consumed independent of weather, such as for lighting, motors, domestic hot water, and miscellaneous office equipment. When a building has electric cooling and no electric heating, the base-load electric energy use is normally the energy consumed during the winter. The annual base-load energy use may also be estimated by taking the average monthly use during nonheating or noncooling months and multiplying by 12. For many buildings, subtracting the base-load energy use

Table 1 Electricity Consumption for Atlanta Example Building

| | | Occupancy Factor | 32.7% | | | | Building Area: 30,700 ft^2 | | | |
| | | Summer ELF 2002 | 82.5% | Summer ELF 2003 | 37.4% | Summer ELF 2004 | 54.7% | | | |

Year	Month	Bill Start	Bill End	Billing Period	Billed Use, kWh	Actual Demand, kW	Billed Demand, kW	LF	Daily Use, kWh	Daily Base Use, kWh	Monthly Base Use, kWh	Percent Excess Use, kWh
2002	Jan-02	1/2/2002	1/31/2002	29	54,600	166	166	47.3%	1883	1665	48,285	11.6%
2002	Feb-02	1/31/2002	2/28/2002	28	46,620	148	166	46.9%	1665^a	1665	46,620	0.0%
2002	Mar-02	2/28/2002	4/1/2002	32	60,900	140^{b,c}	166	56.6%	1903	1665	53,280	12.5%
2002	Apr-02	4/1/2002	4/29/2002	28	56,340	166	166	50.5%	2012	1665	46,620	17.3%
2002	May-02	4/29/2002	5/31/2002	32	65,520	159	166	53.7%	2048	1665	53,280	18.7%
2002	Jun-02	5/31/2002	6/28/2002	28	63,540	180	180	52.5%	2269	1665	46,620	26.6%
2002	Jul-02	6/28/2002	7/31/2002	33	76,860	158	171	61.4%	2329	1665	54,945	28.5%
2002	Aug-02	7/31/2002	8/30/2002	30	82,620	192	192	59.8%	2754^a	1665	49,950	39.5%
2002	Sep-02	8/30/2002	9/30/2002	31	66,780	195^b	195^b	46.0%	2154	1665	51,615	22.7%
2002	Oct-02	9/30/2002	10/29/2002	29	60,720	193	185	45.2%	2094	1665	48,285	20.5%
2002	Nov-02	10/29/2002	12/2/2002	34	62,100	151	185	50.4%	1826	1665	56,610	8.8%
2002	Dec-02	12/2/2002	1/2/2003	31	60,180	166	185	48.7%	1941	1665	51,615	14.2%
2003	Jan-03	1/2/2003	1/31/2003	29	57,120	178	185	46.1%	1970	1704	49,429	13.5%
2003	Feb-03	1/31/2003	3/3/2003	31	61,920	145	185	57.4%	1997	1704	52,838	14.7%
2003	Mar-03	3/3/2003	4/1/2003	29	60,060	140	185	61.6%	2071	1704	49,429	17.7%
2003	Apr-03	4/1/2003	4/30/2003	29	62,640	154	185	58.4%	2160	1704	49,429	21.1%
2003	May-03	4/30/2003	6/2/2003	33	73,440	161	185	57.6%	2225^a	1704	56,247	23.4%
2003	Jun-03	6/2/2003	6/28/2003	26	53,100	171	185	49.8%	2042	1704	44,316	16.5%
2003	Jul-03	6/28/2003	7/30/2003	32	67,320	180^b	185^b	48.7%	2104	1704	54,542	19.0%
2003	Aug-03	7/30/2003	8/29/2003	30	66,000	170	185	53.9%	2200	1704	51,133	22.5%
2003	Sep-03	8/29/2003	9/30/2003	32	63,960	149	171	55.9%	1999	1704	54,542	14.7%
2003	Oct-03	9/30/2003	10/30/2003	30	55,260	122	171	62.9%	1842	1704	51,133	7.5%
2003	Nov-03	10/30/2003	11/26/2003	27	46,020	140	171	50.7%	1704^a	1704	46,020	0.0%
2003	Dec-03	11/26/2003	12/30/2003	34	61,260	141	171	53.2%	1802	1704	57,951	5.4%
2004	Jan-04	12/30/2003	1/30/2004	31	59,040	145	171	54.7%	1905	1676	51,960	12.0%
2004	Feb-04	1/30/2004	2/28/2004	29	54,240	159	171	49.0%	1870	1676	48,608	10.4%
2004	Mar-04	2/28/2004	3/19/2004	20	37,080	122	171	63.3%	1854	1676	33,523	9.6%
2004	Apr-04	3/19/2004	3/31/2004	12	22,140	133	171	57.8%	1845	1676	20,114	9.2%
2004	May-04	3/31/2004	5/4/2004	34	64,260	148	171	53.2%	1890	1676	56,988	11.3%
2004	Jun-04	5/4/2004	6/2/2004	29	63,720	148	171	61.9%	2197	1676	48,608	23.7%
2004	Jul-04	6/2/2004	7/2/2004	30	69,120	169	169	56.8%	2304	1676	50,284	27.3%
2004	Aug-04	7/2/2004	8/3/2004	32	73,800	170^b	170^b	56.5%	2306^a	1676	53,636	27.3%
2004	Sep-04	8/3/2004	9/1/2004	29	64,500	166^b	166^b	55.8%	2224	1676	48,608	24.6%
2004	Oct-04	9/1/2004	10/1/2004	30	60,060	152	161	54.9%	2002	1676	50,284	16.3%
2004	Nov-04	10/1/2004	11/2/2004	32	65,760	128	161	66.9%	2055	1676	53,636	18.4%
2004	Dec-04	11/2/2004	12/3/2004	31	51,960	132	161	52.9%	1676^a	1676	51,960	0.0%

	kWh·y/ft^2			Days	Total kWh	Peak kW	Billed kW	Avg LF	Daily Base Use, kWh	Total Base Use, kWh
2002	24.65			365	756,780	195	195	51.6%	1665	607,725
2003	23.72			362	728,100	180	185	51.5%	1704	617,009
2004	22.33			339	685,680	170	171	52.4%	1676	568,208

^{a}Maximum or minimum value for year. ^{b}Peak demand for year. ^{c}Minimum demand used in seasonal ELF calculation.

from total annual energy use yields a good estimate of heating or cooling energy consumption. This approach is not valid when building operation differs from summer to winter, when cooling operates year-round, or when space heating is used during summer (e.g., for reheat). Base-load analysis can be improved by using hourly load data. **Electric load factors (ELFs)** and occupancy factors can also be used instead of hourly energy profiles (Haberl and Komor 1990a, 1990b).

Although it can be difficult to relate heating and cooling energy directly to weather, several authors, including Fels (1986) and Spielvogel (1984), suggest that this is possible using a curve-fitting method to calculate the balance point of a building (discussed in Chapter 19 of the 2009 *ASHRAE Handbook—Fundamentals*). For this method, building use must be regular, and actual rather than estimated data must be used, along with accurate dates and weather data.

More detailed breakdown of energy use requires that some metered data be collected daily (winter versus summer days, week-

days versus weekends) and that some hourly information be collected to develop profiles for night (unoccupied), morning warmup, day (occupied), and shutdown. Submetering of energy end uses is recommended for optimal energy management. For more information, see Chapter 41.

An example spreadsheet using three years of electricity bill data for a two-story office building in Atlanta, Georgia, is presented in Table 1. (See Chapter 18 of the 2009 *ASHRAE Handbook—Fundamentals* for floor plans and elevations of the building.)

Electrical Use Profile

The **electrical use profile (EUP)** report, shown in Figure 2, divides electrical consumption into base and weather-dependent consumption. The average daily consumption for each month appears in the daily use column in Table 1, and is plotted in the EUP

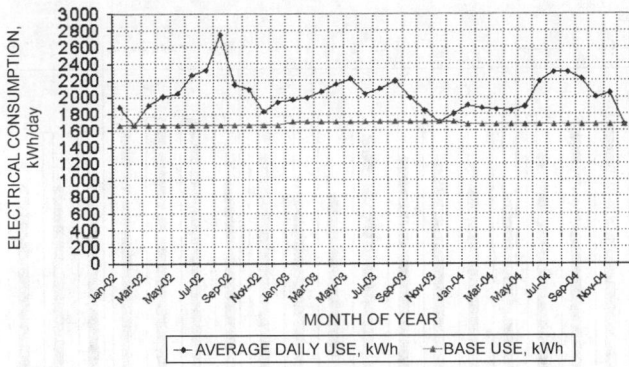

Fig. 2 Electrical Use Profile for Atlanta Example Building

graph. The average daily consumption is calculated by dividing the consumption for a particular month by its billing days.

The lowest value in the daily-use column is used to plot the facility's base electrical consumption (shown as the base use line) in Figure 2. Where a facility uses electricity only for cooling or heating, or in an all-electric facility where there is no overlap between cooling and heating, the difference between these two lines represents the weather-dependent electrical consumption.

Weather-dependent energy consumption (either electric or other fuels) may then be compared to the **cooling degree-days (CDD)** or **heating degree-days (HDD)** totals for the same time period (see Chapter 14 of the 2009 *ASHRAE Handbook—Fundamentals*). This comparison shows how the building performs from month to month or year to year. The HDDs stop and CDDs start at the balance point, defined as the outdoor temperature at which, for a specified interior temperature, the total heat loss is equal to the heat gain from the sun, occupants, lights, etc. Note that all-electric buildings may have periods of overlap between heating and cooling, causing the base load to be overestimated and the heating and cooling estimates to be conservative.

Examine the average daily use line to see whether it follows the expected seasonal curve. For example, the shoulders of the curve for an electrically cooled, gas-heated hospital should closely follow the base electrical use line in the winter. As summer approaches, this curve should rise steadily to reflect the increased cooling load. Errors in meter readings, reading dates, or consumption variances appear as unusual peaks or valleys. Reexamine the data and correct errors as necessary.

If an unusual profile remains after correcting any errors, an area of potential energy savings may exist. For example, if the average daily use line for the facility is running near summer levels during March, April, May, October, and November, simultaneous heating and cooling may be occurring. This situation is illustrated in Figure 2, and often occurs with dual-duct systems.

Simultaneous heating and cooling is also indicated in the percent excess use column of Table 1. The values show the percent difference between the value appearing in the monthly base use column and the billed consumption for the month. In Figure 2, note how the excess consumption for spring and fall months runs close to the summer percentages. The monthly base use is the lowest value from the daily use column multiplied by the number of billing days for each month.

For electrically cooled, gas-heated facilities, weather-dependent consumption is the difference between the totals of the monthly base use column and the billed use column.

For an all-electric facility, subtract the total monthly consumption from total billed use for the cooling months, then do the same calculations for heating months to determine the electric cooling and heating loads, respectively.

Calculating Electrical Load and Occupancy Factors

Another method for detecting potential energy savings is to compare the facility's electrical load factor to its occupancy factor. An ELF exceeding its occupancy factor indicates a higher-than-expected electric use occurring outside normal occupancy (e.g., lights or fans are left on or air conditioning is not shut off as early in the day as possible in summer). Setback thermostats, direct digital control (DDC) strategies, time-of-day scheduling, and lighting controls can address this.

The ELF is the ratio between the average daily use and the maximum possible use if peak demand operated for a 24 h period. The occupancy factor is the ratio between the hours a building actually is occupied and 24 h/day occupancy.

To calculate the ELF, find the month with the lowest demand on the utility data analysis spreadsheet. This value represents the base monthly peak demand, and is usually found in the same or adjacent month as the month with the lowest consumption. From the EUP report, find the lowest value in the daily use column. For example, the lowest average daily use for the office building in Table 1 is 1704 kWh (in November 2003), and the lowest monthly demand from the spreadsheet is 122 kW (in October 2003). The ELF is calculated as follows:

$$\text{ELF} = \frac{\text{Lowest average daily use}}{\text{Lowest monthly demand} \times 24} = \frac{1704}{122 \times 24} = 58\%$$

The office is normally occupied from 7:30 AM to 6:30 PM, Monday to Friday. Therefore, the occupancy factor is calculated as

$$\begin{matrix}\text{Occupancy} \\ \text{factor}\end{matrix} = \frac{\text{Actual weekly occupied hours}}{24 \text{ h} \times 7 \text{ days}} = \frac{55}{168} = 33\%$$

Calculating Seasonal ELFs

ELFs can also be calculated for cooling and heating seasons. Typical defaults are May to August as cooling months, and the rest of the year as heating months, but these change based on climate.

The steps for calculating a seasonal ELF are as follows:

1. The daily base consumption is determined from the daily use column of the EUP report. Subtract the lowest value of the year from the highest value of the season.
2. The base demand is determined by subtracting the lowest monthly demand for the year from the demand recorded for the month with the highest daily use. These calculations can be refined further if on- and off-peak data are available.

For example, because the electrically cooled Atlanta example building operates year-round, the summer ELF must also be calculated. The daily base consumption (1089) is determined by subtracting the lowest value (1665) from the highest cooling-season value (2754) in the daily use column of the EUP report.

From the spreadsheet, take the demand from September 2002 (the month with the peak cooling-season actual demand) and subtract the lowest monthly demand from the spreadsheet (195 − 140) to determine the cooling-season base demand (55). Thus, the summer ELF is

$$\text{Summer ELF} = \frac{1089}{55 \times 24} = 82\% \text{ (for 2002)}$$

These calculations show that the cooling equipment is operating beyond building occupancy (82% versus 33%) Therefore, excessive equipment run times should be investigated. Note that comparing the ELF to the occupancy factor is meaningless for buildings occupied 24 h a day, such as hospitals.

Similar tables and charts may be created for natural gas, water, and other utilities.

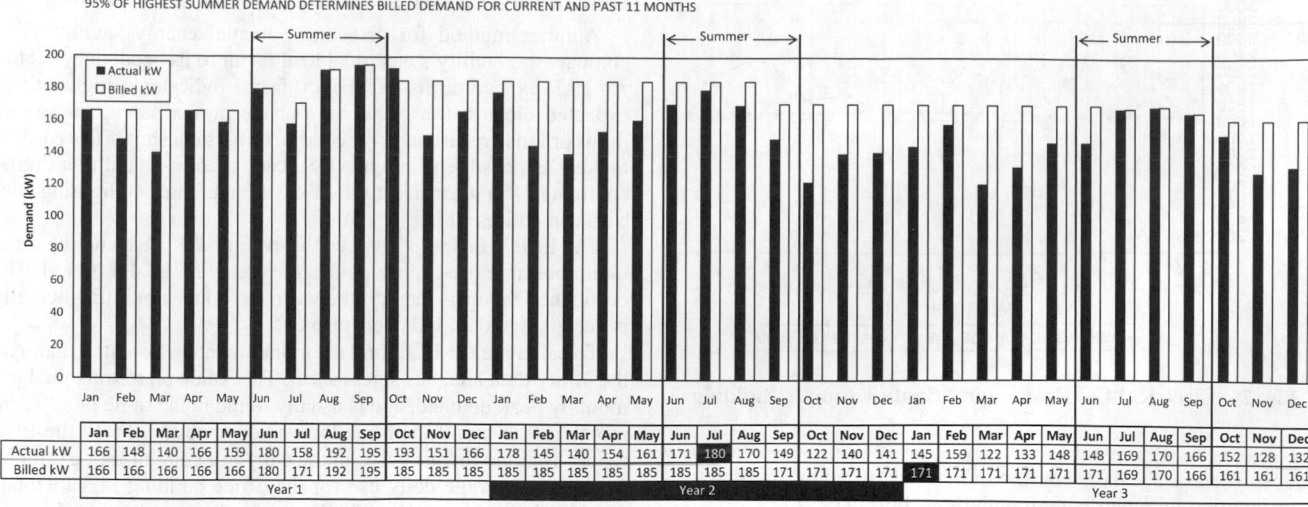

95% OF HIGHEST SUMMER DEMAND DETERMINES BILLED DEMAND FOR CURRENT AND PAST 11 MONTHS

	Jan	Feb	Mar	Apr	May	Jun	Jul	Aug	Sep	Oct	Nov	Dec	Jan	Feb	Mar	Apr	May	Jun	Jul	Aug	Sep	Oct	Nov	Dec	Jan	Feb	Mar	Apr	May	Jun	Jul	Aug	Sep	Oct	Nov	Dec
Actual kW	166	148	140	166	159	180	158	192	195	193	151	166	178	145	140	154	161	171	180	170	149	122	140	141	145	159	122	133	148	148	169	170	166	152	128	132
Billed kW	166	166	166	166	166	180	171	192	195	185	185	185	185	185	185	185	185	185	185	185	171	171	171	171	171	171	171	171	171	169	170	166	166	161	161	161
	Year 1												Year 2												Year 3											

Fig. 3 Comparison Between Actual and Billed Demand for Atlanta Example Building

Electric Demand Billing

The Atlanta example building has a ratchet-type demand rate (see Chapter 56), and billed demand is determined as a percentage of actual demand in the summer months. The ratchet is illustrated in Figure 3, where billed demand is the greater of the measured demand or 95% of the highest measured demand within the past 12 months. The billed demand for January of year 3 was 171 kW ($171 = 0.95 \times 180$), or 95% of the actual demand from July of year 2.

In Table 1, the actual demand in the first six months of 2003 had no effect on the billed demand, and therefore no effect on the dollar amount of the bill; the same is true for the last three months of the year. Because of the demand ratchet, the billed demand in January 2004 (171 kW) was set in July 2003. This means that any conservation measures that reduce peak demand will not affect billed demand until the following summer (e.g., June to September 2004); however, consumption savings begin at the next billing cycle. The effect of demand ratchet rates is that any conservation measures implemented have a longer initial payback period simply because of the utility rate structure. The energy manager should investigate other rate structures, such as a time-of-use (TOU) or seasonal rates. Rate structures for smaller buildings may not include demand charges.

Benchmarking Energy Use

Benchmarking (comparing a building's normalized energy consumption to that of similar buildings) can be a useful first measure of energy efficiency. Relative energy use is commonly expressed in **energy utilization index (EUI**; energy use per unit area per year) and **cost utilization index (CUI**; energy cost per unit area per year). The Atlanta example building is 30,700 ft^2 in size, so its 2004 EUI is 76,200 Btu/ft^2 and its CUI is \$1.47/ft^2.

Two sources of benchmarking data for U.S. buildings are ENERGY STAR (www.energystar.gov) and the U.S. Department of Energy's Energy Information Administration (DOE/EIA). Data on U.S. buildings in all sectors are summarized in periodic reports by the DOE/EIA. Tables 2 to 4 present DOE/EIA CBECS data in a combined format. Table 2 lists EUI input data and EUI distributions for the buildings surveyed in 2003. Table 3 lists the 2003 *Commercial Buildings Energy Consumption Survey* (CBECS) electricity per unit of floor area, and Table 4 shows CUI distributions. More complete and up-to-date information on the CBECS is available at www.eia.doe.gov/emeu/cbecs. When referring to these tables, keep in mind the facility's operating or occupied hours of facility and current utility rates.

Databases. Compiling a database of past energy use and cost is important. All reliable utility data should be examined. ASHRAE *Standard* 105 contains information that allows uniform, consistent expressions of energy consumption in new and existing buildings.

The energy use database for a new building may consist solely of typical data for similar buildings, as in Table 2. This may be supplemented by energy simulation data developed during design. A new building should be commissioned to ensure proper operation of all systems, including any energy-efficiency features (see ASHRAE *Guideline* 1.1 and Chapter 43).

All the data presented in these tables come from detailed reports of consumption patterns, and it is important to understand how they were derived. When using the data, verify correct use with the original EIA documents.

Mazzucchi (1992) lists data elements useful for normalizing and comparing utility billing information. Metered energy consumption and cost data are also published by trade associations, such as the Building Owners and Managers Association International (BOMA), the National Restaurant Association (NRA), and the American Hotel and Lodging Association (AH&LA). In some cases, local energy consumption data may be available from local utility companies or state or provincial energy offices.

Additional energy use information for homes and commercial buildings in Canada can be found at the Office of Energy Efficiency at http://www.oee.nrcan.gc.ca/corporate/statistics/neud/dpa/data_e /publications.cfm. In Europe, benchmarking data are defined on a national basis in the frame of the European Directive on the Energy Performance of Buildings (EPBD) (EC 2010). Balaras et al. (2007) provides an overview of relevant data for residential buildings, although detailed data for commercial buildings are rather limited (Gaglia et al 2007).

SURVEYS AND AUDITS

This section provides guidance on conducting building surveys and describes the levels of intensity of investigation.

Energy Audits

The objective of an energy audit is to identify opportunities to reduce energy use and/or cost. The results should provide the information needed by an owner/operator to decide which recommendations to implement. Energy audits may include the following:

1. Collect and analyze historical energy use
   - Review more than one year of energy bills (preferably three years)

Table 2   2003 Commercial Sector Floor Area and EUI Percentiles

Building Use	Calculated, Weighted Number of Buildings, Hundreds	Calculated, Weighted Floor Area, 10^9 ft^2	Actual Number of Buildings, N	Calculated, Weighted Energy Use Index (EUI) Values Site Energy, kBtu/yr per gross square foot Percentiles 10th	25th	50th	75th	90th	Mean
Administrative/professional office	442	6.63	555	28.1	41	62	93	138	75
Bank/other financial	104	1.10	75	55.7	67	87	117	184	106
Clinic/other outpatient health	66	0.75	100	28.7	41	66	97	175	84
College/university	34	1.42	88	14.1	67	108	178	215	122
Convenience store	57	0.16	28	68.6	156	232	352	415	274
Convenience store with gas station	72	0.28	32	82.2	135	211	278	409	225
Distribution/shipping center	155	5.25	231	8.7	17	33	54	91	45
Dormitory/fraternity/sorority	16	0.51	37	36.3	65	74	100	154	90
Elementary/middle school	177	4.75	331	21.1	35	54	93	127	76
Entertainment/culture	27	0.50	50	1.7	29	46	134	418	95
Fast food	78	0.26	95	176.3	268	418	816	933	534
Fire station/police station	53	0.38	47	6.9	24	82	112	137	78
Government office	84	1.55	150	31.5	52	77	103	149	85
Grocery store/food market	86	0.71	117	98.1	138	185	239	437	213
High school	68	2.52	126	19.8	44	65	99	130	75
Hospital/inpatient health	8	1.90	217	108.1	169	196	279	355	227
Hotel	20	1.90	86	39.7	51	73	116	183	95
Laboratory	9	0.65	43	98.0	165	270	505	925	362
Library	20	0.56	36	35.0	67	92	121	197	104
Medical office (diagnostic)	54	0.50	58	14.1	25	44	100	137	60
Medical office (nondiagnostic)	37	0.22	33	25.7	40	52	66	109	59
Mixed-use office	84	2.30	172	20.0	38	71	106	158	88
Motel or inn	70	1.05	109	23.9	37	67	102	197	87
Nonrefrigerated warehouse	229	3.05	172	2.3	6	19	46	87	34
Nursing home/assisted living	22	0.98	73	41.6	77	116	184	205	124
Other	70	1.08	68	5.5	29	69	96	118	74
Other classroom education	51	0.71	60	4.3	23	40	64	108	51
Other food sales	10	0.10	10	31.5	37	58	190	343	126
Other food service	58	0.33	56	39.6	71	125	309	548	242
Other lodging	16	0.65	28	31.2	54	71	83	146	76
Other office	73	0.41	52	15.3	41	57	84	146	69
Other public assembly	32	0.42	31	9.9	30	42	73	155	65
Other public order and safety	17	0.71	38	44.0	58	93	160	308	127
Other retail	47	0.24	42	32.7	65	92	146	205	120
Other service	139	0.48	171	28.0	50	86	164	303	168
Post office/postal center	19	0.50	23	7.2	58	64	76	97	64
Preschool/daycare	56	0.48	46	18.8	35	59	112	121	75
Recreation	96	1.28	99	13.4	24	40	88	152	68
Refrigerated warehouse	15	0.53	20	6.5	13	143	190	257	127
Religious worship	370	3.75	313	9.3	17	33	63	88	46
Repair shop	76	0.65	51	7.0	13	30	54	72	37
Restaurant/cafeteria	161	1.06	212	51.8	117	207	462	635	302
Retail store	347	3.48	460	14.2	25	45	93	170	72
Self-storage	198	1.26	84	2.1	4	7	10	15	9
Social/meeting	101	1.18	78	7.9	15	41	71	93	52
Vacant	182	2.57	178	1.4	3	12	31	77	26
Vehicle dealership/showroom	50	0.60	40	24.5	40	82	110	248	110
Vehicle service/repair shop	212	1.21	131	10.1	16	37	86	137	58
Vehicle storage/maintenance	176	1.21	99	0.9	4	21	53	152	54
SUM or Mean for sector	4645	64.78	5451	9.8	26	56	108	207	97

Source: Calculated based on DOE/EIA preliminary 2003 CBECS microdata.

- Review billing rate class options with utility
- Review monthly patterns for irregularities
- Derive target goals for energy, demand, and cost indices for a building with similar characteristics and climate
2. Study the building and its operational characteristics
   - Acquire a basic understanding of the mechanical and electrical systems
   - Perform a walk-through survey to become familiar with its construction, equipment, operation, and maintenance

- Meet with owner/operator and occupants to learn of special problems or needs
- Identify any required repairs to existing systems and equipment
3. Identify potential modifications to reduce energy use or cost
   - Identify low-cost/no-cost changes to the facility or to operating and maintenance procedures
   - Identify potential equipment retrofit opportunities
   - Identify training required for operating staff

Table 3   Electricity Index Percentiles from 2003 Commercial Survey

Building Use	Weighted Electricity Use Index Values, kWh/yr per gross square foot					
	Percentiles					
	10th	25th	50th	75th	90th	Mean
Administrative/professional office	3.54	6.7	11.0	15.0	24.1	12.7
Bank/other financial	6.23	14.5	22.2	29.5	33.3	22.5
Clinic/other outpatient health	4.94	9.4	15.2	20.7	27.3	16.6
College/university	4.13	10.5	15.0	24.0	42.3	17.7
Convenience store	20.09	43.3	65.3	78.7	107.4	69.6
Convenience store with gas station	24.09	37.7	48.1	79.0	120.0	62.0
Distribution/shipping center	1.77	2.9	4.5	7.4	9.9	5.7
Dormitory/fraternity/sorority	2.16	3.3	5.1	6.6	16.6	6.7
Elementary/middle school	3.45	5.7	9.3	14.0	19.7	12.1
Entertainment/culture	0.49	1.0	7.4	16.9	122.5	20.9
Fast food	27.97	48.0	81.8	131.2	168.1	95.5
Fire station/police station	1.14	3.8	6.6	12.6	22.0	9.8
Government office	3.96	8.1	10.8	19.3	26.0	14.3
Grocery store/food market	26.12	32.2	42.4	54.4	100.6	51.7
High school	3.50	4.5	7.5	12.8	19.3	9.7
Hospital/inpatient health	15.24	21.8	24.0	35.6	45.9	28.7
Hotel	6.73	11.6	14.3	18.3	27.4	16.4
Laboratory	11.43	25.5	39.2	54.6	95.6	44.1
Library	6.34	8.7	15.5	23.2	34.3	17.3
Medical office (diagnostic)	2.21	4.1	7.6	13.8	18.3	9.6
Medical office (nondiagnostic)	2.41	4.5	7.4	12.1	15.3	8.6
Mixed-use office	3.40	5.5	11.1	18.0	28.9	14.3
Motel or inn	4.95	7.5	10.8	18.1	26.3	13.6
Nonrefrigerated warehouse	0.38	1.0	2.9	5.9	10.7	5.4
Nursing home/assisted living	6.33	8.1	14.9	21.0	25.9	15.9
Other	1.60	3.0	5.8	12.2	24.7	9.5
Other classroom education	1.27	2.8	4.9	9.2	15.7	6.6
Other food sales	9.22	9.2	10.8	12.6	58.5	22.0
Other food service	8.85	15.4	27.2	60.1	89.5	40.3
Other lodging	2.86	3.7	14.0	21.0	22.7	12.0
Other office	3.04	4.5	9.4	16.2	18.3	10.8
Other public assembly	1.13	2.6	3.4	12.3	13.8	7.5
Other public order and safety	5.45	14.4	16.7	20.7	42.1	18.9
Other retail	4.87	6.7	22.4	27.2	38.3	19.8
Other service	4.13	7.5	13.4	19.6	28.6	16.3
Post office/postal center	2.10	3.2	7.2	13.3	21.3	9.9
Preschool/daycare	3.34	5.5	8.8	12.1	28.9	11.6
Recreation	1.59	2.9	5.1	10.8	19.3	8.8
Refrigerated warehouse	1.89	3.8	35.2	51.1	55.7	28.5
Religious worship	1.06	1.9	3.5	6.0	8.6	4.5
Repair shop	1.88	2.6	6.1	7.6	14.2	6.8
Restaurant/cafeteria	9.76	15.2	28.7	49.9	88.2	37.9
Retail store	2.41	3.9	8.1	15.2	27.3	12.5
Self-storage	0.63	1.2	2.1	2.8	3.8	2.2
Social/meeting	1.01	1.8	2.9	7.5	12.8	6.2
Vacant	0.29	0.4	1.7	3.8	7.8	3.2
Vehicle dealership/showroom	2.50	7.2	13.8	21.9	33.9	15.7
Vehicle service/repair shop	1.96	3.3	5.6	9.8	18.6	8.2
Vehicle storage/maintenance	0.27	1.2	3.3	6.4	10.4	5.2
SUM or Mean for sector	1.59	3.6	8.3	17.1	35.4	15.7

Source: Calculated based on DOE/EIA preliminary 2003 CBECS microdata.

- Perform a rough estimate of the breakdown of energy consumption for significant end-use categories
4. Perform an engineering and economic analysis of potential modifications
   - For each practical measure, determine resultant savings
   - Estimate effects on building operations and maintenance costs
   - Prepare a financial evaluation of estimated total potential investment
5. Prepare a rank ordered list
   - List all possible energy savings modifications
   - Select those that may be considered practical by the building owner
   - Assume that modifications with highest operational priority and/or best return on investment will be implemented first

- Provide preliminary implementation costs and savings estimates
- Assume that modifications with highest operational priority and/or best return on investment will be implemented first
6. Report results
   - Provide description of building, operating requirements, and major energy-using systems
   - Clearly state savings from each modification and assumptions on which each is based
   - Review list of practical modifications with the owner
   - Prioritize modifications in recommended order of implementation
   - Recommend measurement and verification methods

Table 4　Electricity Index Percentiles from 2003 Commercial Survey

| Building Use | Weighted Electricity Use Index Values, kWh/yr per gross square foot | | | | | |
| | Percentiles | | | | | |
	10th	25th	50th	75th	90th	Mean
Administrative/professional office	3.54	6.7	11.0	15.0	24.1	12.7
Bank/other financial	6.23	14.5	22.2	29.5	33.3	22.5
Clinic/other outpatient health	4.94	9.4	15.2	20.7	27.3	16.6
College/university	4.13	10.5	15.0	24.0	42.3	17.7
Convenience store	20.09	43.3	65.3	78.7	107.4	69.6
Convenience store with gas station	24.09	37.7	48.1	79.0	120.0	62.0
Distribution/shipping center	1.77	2.9	4.5	7.4	9.9	5.7
Dormitory/fraternity/sorority	2.16	3.3	5.1	6.6	16.6	6.7
Elementary/middle school	3.45	5.7	9.3	14.0	19.7	12.1
Entertainment/culture	0.49	1.0	7.4	16.9	122.5	20.9
Fast food	27.97	48.0	81.8	131.2	168.1	95.5
Fire station/police station	1.14	3.8	6.6	12.6	22.0	9.8
Government office	3.96	8.1	10.8	19.3	26.0	14.3
Grocery store/food market	26.12	32.2	42.4	54.4	100.6	51.7
High school	3.50	4.5	7.5	12.8	19.3	9.7
Hospital/inpatient health	15.24	21.8	24.0	35.6	45.9	28.7
Hotel	6.73	11.6	14.3	18.3	27.4	16.4
Laboratory	11.43	25.5	39.2	54.6	95.6	44.1
Library	6.34	8.7	15.5	23.2	34.3	17.3
Medical office (diagnostic)	2.21	4.1	7.6	13.8	18.3	9.6
Medical office (nondiagnostic)	2.41	4.5	7.4	12.1	15.3	8.6
Mixed-use office	3.40	5.5	11.1	18.0	28.9	14.3
Motel or inn	4.95	7.5	10.8	18.1	26.3	13.6
Nonrefrigerated warehouse	0.38	1.0	2.9	5.9	10.7	5.4
Nursing home/assisted living	6.33	8.1	14.9	21.0	25.9	15.9
Other	1.60	3.0	5.8	12.2	24.7	9.5
Other classroom education	1.27	2.8	4.9	9.2	15.7	6.6
Other food sales	9.22	9.2	10.8	12.6	58.5	22.0
Other food service	8.85	15.4	27.2	60.1	89.5	40.3
Other lodging	2.86	3.7	14.0	21.0	22.7	12.0
Other office	3.04	4.5	9.4	16.2	18.3	10.8
Other public assembly	1.13	2.6	3.4	12.3	13.8	7.5
Other public order and safety	5.45	14.4	16.7	20.7	42.1	18.9
Other retail	4.87	6.7	22.4	27.2	38.3	19.8
Other service	4.13	7.5	13.4	19.6	28.6	16.3
Post office/postal center	2.10	3.2	7.2	13.3	21.3	9.9
Preschool/daycare	3.34	5.5	8.8	12.1	28.9	11.6
Recreation	1.59	2.9	5.1	10.8	19.3	8.8
Refrigerated warehouse	1.89	3.8	35.2	51.1	55.7	28.5
Religious worship	1.06	1.9	3.5	6.0	8.6	4.5
Repair shop	1.88	2.6	6.1	7.6	14.2	6.8
Restaurant/cafeteria	9.76	15.2	28.7	49.9	88.2	37.9
Retail store	2.41	3.9	8.1	15.2	27.3	12.5
Self-storage	0.63	1.2	2.1	2.8	3.8	2.2
Social/meeting	1.01	1.8	2.9	7.5	12.8	6.2
Vacant	0.29	0.4	1.7	3.8	7.8	3.2
Vehicle dealership/showroom	2.50	7.2	13.8	21.9	33.9	15.7
Vehicle service/repair shop	1.96	3.3	5.6	9.8	18.6	8.2
Vehicle storage/maintenance	0.27	1.2	3.3	6.4	10.4	5.2
SUM or Mean for sector	1.59	3.6	8.3	17.1	35.4	15.7

Source: Calculated based on DOE/EIA preliminary 2003 CBECS microdata.

ASHRAE (2004) identifies the following four levels of effort in the audit process.

Preliminary Energy Use Analysis. This involves analysis of historic utility use and cost and development of the energy utilization index (EUI) of the building. Compare the building's EUI to similar buildings to determine if further engineering study and analysis are likely to produce significant energy savings.

Level I: Walk-Through Analysis. This assesses a building's current energy cost and efficiency by analyzing energy bills and briefly surveying the building. The auditor should be accompanied by the building operator. Level I analysis identifies low-cost/no-cost measures and capital improvements that merit further consideration, along with an initial estimate of costs and savings. The level of detail depends on the experience of the auditor and the client's specifications. The Level I audit is most applicable when there is some doubt about the energy savings potential of a building, or when an owner wishes to establish which buildings in a portfolio have the greatest potential savings. The results can be used to develop a priority list for a Level II or III audit.

Level II: Energy Survey and Analysis. This includes a more detailed building survey and energy analysis, including a breakdown of energy use in the building, a savings and cost analysis of all practical measures that meet the owner's constraints, and a discussion of any effect on operation and maintenance procedures. It also lists potential capital-intensive improvements that require more thorough data collection and analysis, along with an initial judgment of potential costs and savings. This level of analysis is adequate for most buildings.

Level III: Detailed Analysis of Capital-Intensive Modifications. This focuses on potential capital-intensive projects identified during Level II and involves more detailed field data gathering and engineering analysis. It provides detailed project cost and savings information with a level of confidence high enough for major capital investment decisions.

The levels of energy audits do not have sharp boundaries. They are general categories for identifying the type of information that can be expected and an indication of the level of confidence in the results. In a complete energy management program, Level II audits should be performed on all facilities.

A thorough systems approach produces the best results. This approach has been described as starting at the end rather than at the beginning. For example, consider a factory with steam boilers in constant operation. An expedient (and often cost-effective) approach is to measure the combustion efficiency of each boiler and to improve boiler efficiency. Beginning at the end requires finding all or most of the end uses of steam in the plant, which could reveal considerable waste by venting to the atmosphere, defective steam traps, uninsulated lines, and lines through unused heat exchangers. Eliminating end-use waste can produce greater savings than improving boiler efficiency.

A detailed process for conducting audits is outlined in ASHRAE (2004).

IMPROVING DISCRETIONARY OPERATIONS

Basic Energy Management

Control Energy System Use. The most effective method to reduce energy costs is through discretionary operations, such as turning off equipment when not needed. Ways to conserve energy include the following:

- Shut down HVAC&R systems when not required
- Reduce air leakage
- Reduce ventilation rates during periods of low occupancy
- Shut down exhaust fans when not required
- Seal or repair leaks in ducts and pipes
- Reduce water leakage
- Turn off lighting: remove unnecessary lighting, add switched circuits, use motion sensors and light-sensitive controls
- Use temperature setup and setback
- Cool with outside air
- Seal unused vents and ducts to the outside
- Tune up systems before heating and cooling seasons
- Take transformers offline during idle periods

Purchase Lower-Cost Energy. This is the second most effective method for reducing energy costs. Building operators and managers must understand all the options in purchasing energy and design systems to take advantage of changing energy costs. The following options should be considered:

- Choosing or negotiating lower-cost utility rates
- Procuring electricity or fuels through brokers
- Correcting power factor penalties
- Controlling peak electric billing demand
- Utility-sponsored demand response programs
- Transportation and interruptible natural gas rates
- Cogeneration
- Lower-cost liquid fuels
- Increasing volume for onsite storage
- Avoiding sales or excise taxes where possible
- Incentive rebates from utilities and manufacturers

Optimize Energy Systems Operation. The third most effective method for reducing energy costs is to tune energy systems to optimal performance, an ongoing process combining training, preventive maintenance, and system adjustment. Tasks for optimizing performance include

- Training operating personnel
- Tuning combustion equipment
- Adjusting gas burners to optimal efficiency
- Following an established maintenance program
- Cleaning or replacing filters
- Cleaning fan blades and ductwork
- Cycling ventilation systems to coincide with occupied spaces
- Using water treatment

Purchase Efficient Replacement Systems. This method is more expensive than the other three, presents energy managers with the greatest liability, and may be less cost-effective. It is critical to ensure that possible equipment or system replacements are objectively evaluated to confirm both the replacement costs and benefits to the owner. The optimum time for replacing less-efficient equipment is near the end of its expected life or when major repairs are needed. Systems commonly replaced include

- Lighting systems and lamps
- Heating and cooling equipment
- Energy distribution systems (pumps and fans)
- Motors
- Thermal envelope components
- Controls and energy management systems

Optimizing More Complex System Operation

As the complexity of building systems increases, additional strategies are needed to optimize energy systems. According to ASHRAE *Guideline* 0-2005, approaches include **recommissioning** (applied to a project that has been delivered using the commissioning process), **retrocommissioning** (applied to an existing facility that was not previously commissioned), and **ongoing commissioning** (continuation of the commissioning process well into the occupancy and operations phase to verify that a project continues to meet current and evolving owner's project requirements). See Chapter 43 for more information.

These approaches typically require a strong team effort of the facility staff and third-party consultants to identify and fix comfort problems as well as aggressively optimize HVAC operation and control. Some important measures typically implemented include

- Optimizing hot and cold deck reset schedules
- Optimizing duct static pressure reset schedules
- Optimizing pump control
- Optimizing terminal box settings/control
- Optimizing sequencing and water temperature reset schedules of boilers and chillers
- Identifying and repairing stuck or leaky valves and dampers
- Training operating personnel in optimum operating strategies
- Setting up monitoring and reporting of key system performance indicators

Implementing these measures has been found to reduce energy use by an average of about 20% (Claridge et al. 1998). Approaches to commissioning and optimizing operation of existing buildings can be found in ASHRAE *Guideline* 1.1-2007, Claridge and Liu (2000), Haasl and Sharp (1999), Kurt et al. (2003), Liu et al. (1997), Poulos (2007), and Tseng (2005).

ENERGY-EFFICIENCY MEASURES

Identifying Energy-Efficiency Measures

Various energy-efficiency measures (EEMs) can be quantitatively evaluated from end-use energy profiles. Important considerations in this process are as follows:

- System interaction
- Utility rate structure

- Payback
- Alignment with corporate goals
- Installation requirements
- Life of the measure
- Energy measurement and verification requirements
- Maintenance costs
- Tenant/occupant comfort
- Effect on building operation and appearance

Accurate energy savings calculations can be made only if system interaction is allowed for and fully understood. Annual simulation models may be necessary to accurately estimate the interactions between various EEMs.

Using average costs per unit of energy in calculating the energy cost avoidance of a particular measure is likely to result in incorrect energy costs and cost avoidance, because actual energy cost avoidance may not be proportional to the energy saved, depending on the billing method for energy used.

PNNL (1990) discusses 118 EEMs, including the following:

Boilers	Condensate systems
Envelope infiltration	Water treatment
Weather-stripping	Caulking
Fuel systems	Vestibules
Chillers	Steam distribution
Vapor barrier	Hydronic systems
Glazing	Pumps
Piping insulation	Steam traps
Instrumentation	Domestic water heating
Shading	Fixtures
Thermal shutters	Swimming pools
Surface color	Cooling towers
Roof covering	Condensing units
Lamps	Air-handling units
Ballasts	Unitary equipment
Light switching options	Outside air control
Photo cell controls	Balancing
Demand limiting	Shutdown
Power factor correction	Minimizing reheat
Energy recovery	Power distribution
Filters	Cooking practices
Humidification	Refrigeration
Dishwashing	System air leakage
Vending machines	System interaction
Heat/cool storage	Space segregation
Time-of-day rates	Computer controls
Cogeneration	Heat pumps
Active solar systems	Staff training
Occupant indoctrination	Documentation
Controls	Thermostats
Setback	Space planning
T5 lighting	Variable-frequency drives

In addition, previously implemented energy-efficiency measures should be evaluated to (1) ensure that devices are in good working order and measures are still effective, and (2) consider revising them to reflect changes in technology, building use, and/or energy cost.

Evaluating Energy-Efficiency Measures

In establishing EEM priorities, the capital cost, cost-effectiveness, effect on indoor environment, and resources available must be considered. Factors involved in evaluating the desirability of energy-efficiency measures are as follows:

- Rate of return (simple payback, life-cycle cost, net present value)
- Total savings (energy, cost avoidance)
- Initial cost (required investment)
- Other benefits (safety, comfort, improved system reliability, improved productivity)

- Liabilities (increased maintenance costs, potential obsolescence)
- Risk of failure (confidence in predicted savings, rate of increase in energy costs, maintenance complications, success of others with the same measures)

Project success also depends on the availability of

- Management attention, commitment, and follow-through
- Technical expertise
- Personnel
- Investment capital

Some owners are reluctant to implement EEMs because of bad experiences with energy projects. To reduce the risk of failure, documented performance of EEMs in similar situations should be obtained and evaluated. One common problem is that energy consumption for individual end uses is overestimated, and the predicted savings are not achieved. When doubt exists about energy consumption, temporary monitoring or spot measurements should be made and evaluated.

Heating Effects of Electrical Equipment

Electrical equipment and appliances, from lighting systems and office equipment to motors and water heaters, provide useful services; however, the electrical energy they use eventually appears as heat within the building, which can either be useful or detrimental, depending on the season. In cold weather, heat produced by electrical equipment can help reduce the load on the building's heating system. In contrast, during warm weather, it adds to the air-conditioning load.

Energy-efficient equipment and appliances consume less energy to produce the same useful work, but they also produce less heat. As a result, efficient electrical equipment increases the load on heating systems in winter and reduces the load on air-conditioning systems in summer. Effects of energy-efficient equipment and appliances on energy use for building heating and air conditioning systems are commonly called **interactive effects** or **cross effects**.

When considering the overall net savings of an energy-efficiency measure, it is important to consider its interactive effects on building heating, cooling, and refrigeration systems. Weighing the interactive effects results in better-informed decisions and realistic expectations of savings.

The percentage of heat that is useful in a specific building or room depends on several factors, including the following:

- Location of light fixtures
- Location of heaters and their thermostats or other sensors
- Type of ceiling
- Size of building
- Whether room is an interior or exterior space
- Extent of heating and cooling seasons
- Type of heating, ventilation, and air-conditioning system used in each room

Unfortunately, interactive effects are often quite complex and may require assessment by a specialist; for details, see Rundquist et al. (1993).

Exploring Financing Options

Financing alternatives also need to be considered. When evaluating proposed energy management projects, particularly those with a significant capital cost, it is important to include a life-cycle cost analysis. This not only provides good information about the financial attractiveness (or otherwise) of a project, but also assures management that the project has been carefully considered and evaluated before presentation.

Several life-cycle cost procedures are available. Chapter 37 contains details on these and other factors that should be considered in such an analysis.

Capital for energy-efficiency improvements is available from various public and private sources, and can be accessed through a wide and flexible range of financing instruments. There are variations and combinations, but the five general mechanisms for financing investments in energy efficiency are the following:

- **Internal funds**, or direct allocations from an organization's own internal capital or operating budget
- **Debt financing**, with capital borrowed directly by an organization from private lenders
- **Lease** or **lease-purchase agreements**, in which equipment is acquired through an operating or financing lease with little or no up-front costs, and payments are made over five to ten years
- **Energy performance contracts**, in which improvements are financed, installed, and maintained by a third party, which guarantees savings and payments based on those savings
- **Utility (or other) incentives**, such as rebates, grants, or other financial assistance offered by an energy utility or public benefits fund for design and purchase of energy-efficient systems and equipment

An organization may use several of these financing mechanisms in various combinations. The most appropriate set of options depends on the type of organization (public or private), size and complexity of a project, internal capital constraints, in-house expertise, and other factors (Turner 2001).

IMPLEMENTING ENERGY-EFFICIENCY MEASURES

When all desirable EEMs have been considered and a list of recommendations is developed, a report should be prepared for management. Each recommendation should include the following:

- Present condition of the system or equipment to be modified
- Recommended action
- Who should accomplish the action
- Necessary documentation or follow-up required
- Measurement and verification protocol to be used
- Potential interferences to successful completion
- Disruption to workplace or production
- Staff effort and training required
- Risk of failure
- Interactions with other end uses and EEMs
- Economic analysis (including payback, investment cost, and estimated savings figures) using corporate economic evaluation criteria
- Schedule for implementation

The energy manager must be prepared to sell the plans to upper management. Energy-efficiency measures must generally be financially justified if they are to be adopted. Every organization has limited funds available that must be used in the most effective way. The energy manager competes with others in the organization for the same funds. A successful plan must be presented in a form that is easily understood by the decision makers. Finally, the energy manager must present nonfinancial benefits, such as improved product quality or the possibility of postponing other expenditures.

After approval by management, the energy manager directs the completion of energy-efficiency measures. If utility rebates are used, the necessary approvals should be acquired before proceeding with the work. Some measures require that an architect or engineer prepare plans and specifications for the retrofit. The package of services required usually includes drawings, specifications, assistance in obtaining competitive bids, evaluation of the bids, selection of contractors, construction observation, final check-out, and assistance in training personnel in the proper application of the revisions.

MONITORING RESULTS

Once energy-efficiency measures are under way, procedures need to be established to record, frequently and regularly, energy consumption and costs for each building and/or end-use category in a manner consistent with functional cost accountability. Turner et al. (2001) found that consumption increased by more than 5% over two years because of component failures and controls changes after implementing optimum practices in a group of 10 buildings. Data may be obtained from the utility, but additional metering may be needed to monitor energy consumption accurately. Metering can use devices that automatically read and transmit data to a central location, or less expensive metering devices that require regular readings by building maintenance and/or security personnel. Costs for automatic metering devices, such as adding points to a DDC system, must be weighed against the benefits. Many energy managers find it helpful to collect energy consumption information hourly.

The energy manager should review data while they are current and take immediate action if profiles indicate a trend in the wrong direction. These trends could be caused by uncalibrated controls, changes in operating practices, or mechanical system failure, which should be isolated and corrected as soon as possible.

EVALUATING SUCCESS AND ESTABLISHING NEW GOALS

Comparing facility performance before and after implementing EEMs helps keep operating staff on track with their energy-efficiency efforts, ensuring that performance is maintained. Evaluating and reporting energy performance involves four steps:

1. Establishing key performance indicators
2. Tracking performance
3. Developing new goals
4. Reporting

Establishing Key Performance Indicators

It is important to determine performance factors of the energy management program. These are expressed in terms of key performance indicators (KPIs). The definition of key performance indicators determines what data need to be collected, how often to collect it, and how to present it to senior management. Suggested basic key performance indicators are

- Energy use index (EUI), total energy use per unit of gross floor area
- Cost utilization index (CUI), total energy cost per unit of total gross floor area
- Electrical energy use per unit of total gross floor area

Energy Policy Act. The Energy Policy Act (EPAct 2005) set goals for federal buildings to decrease their energy consumption by 2% per year between 2006 and 2015, compared to a baseline of 2004 consumption. Thus, by 2010, for example, the target percentage reduction from 2004 values was 10%. For this initiative, the following sample KPI definitions could be used:

- 2004 benchmark measurement (energy use per unit area) reduced by 4% to set 2007 target, and by 10% to set the 2010 target, and by 16% to set the 2013 target
- Energy use data, summed monthly and annually for reporting against targets

Executive Order 13514 October 2009. Executive Order 13514 further set goals for U.S. federal agencies to develop and implement strategic energy sustainability plans for 2011 through 2021 to reduce buildings' energy use intensity (EUI), increase renewable energy use, obtain net-zero-energy buildings by 2030, and ensure that all products and services are ENERGY STAR or Federal Energy Management Program (FEMP) designated.

ENERGY STAR Tools. The U.S. Environmental Protection Agency's (EPA) ENERGY STAR web site offers the free online benchmarking tool, Target Finder (I-P units only; accessible from www.energystar.gov/index.cfm?c=new_bldg_design.bus_target_finder). This tool compares actual building performance to target values, and to other similar buildings. Figure 4 shows sample results for the Atlanta example building's general office space (omitting the computer center's floor space and electricity use). ENERGY STAR also offers an online Portfolio Manager (www.energystar.gov/index.cfm?c=evaluate_performance.bus_portfoliomanager), which provides secure performance data management and benchmarking for multiple buildings. Annual benchmarking with these (or similar) tools helps track improvements, both over time and in comparison with other buildings.

Building Energy Labels

The ASHRAE Building Energy Quotient (eQ) labeling program rates new and existing buildings (Jarnagin 2009). Like the EPA's ENERGY STAR program, Building eQ focuses solely on energy, but provides additional features, including potential side-by-side comparison of operational and asset (as-designed) ratings; peak-demand reduction and demand management opportunities; on-site renewable energy; indoor environmental quality indicators; and a list of operational features, including commissioning activities, energy-efficiency improvements, and information on improving performance. The Building eQ scale allows differentiation among buildings at the highest levels of performance and encourages the design and operation of net-zero-energy buildings.

The Building eQ program provides an easily understood scale to convey a building's energy use to the public. Through an on-site assessment, the building owner is provided with building-specific information that can be used to improve the building. Documentation on previous energy-efficiency upgrades and commissioned systems is also included. With procedures for both an asset and operational rating, building owners can make side-by-side comparisons that could further reconcile differences between designed and measured energy use.

The label itself is the most visible aspect of the program (Figure 5). It is simple to understand and is targeted at the general public. It could be posted in a building lobby and could satisfy compliance with many of the programs being developed at the state and local level requiring display of energy use. The certificate contains technical information that explains the score on the label and provides information useful to the building owner, prospective owners and tenants, and operations and maintenance personnel. This includes many of the value-added features described previously. The documentation accompanying the label and certificate provides background information useful for engineers, architects, and technically savvy building owners or prospective owners in determining the current state of the building and opportunities for improving its energy use. More information is available at http://buildingeq.com/.

Throughout the European Union, the European Commission's directive on the energy performance of buildings (EPBD) has been in effect since January 4, 2006. Despite difficulties, all EU member states have brought into force national laws, regulations, and administrative provisions for setting minimum requirements on the energy performance of new buildings and for existing buildings that are being renovated, as well as energy performance certification of buildings. Additional requirements include regular inspection of building systems and installations, assessment of existing facilities, and provision of advice on possible improvements and alternative solutions. The objective is to properly design new buildings and renovate existing buildings in a manner that will use the minimum nonrenewable energy, produce minimum air pollution as a result of the building operating systems, and minimize construction waste, all with acceptable investment and operating costs, while improving the indoor environment for comfort, health, and safety.

An energy performance certificate (EPC) is issued when buildings are constructed, sold, or rented out. The EPC documents the energy performance of the building, expressed as a numeric indicator that allows benchmarking. The certificate includes recommendations for cost-effective improvement of the energy performance, and it is valid for up to 10 years.

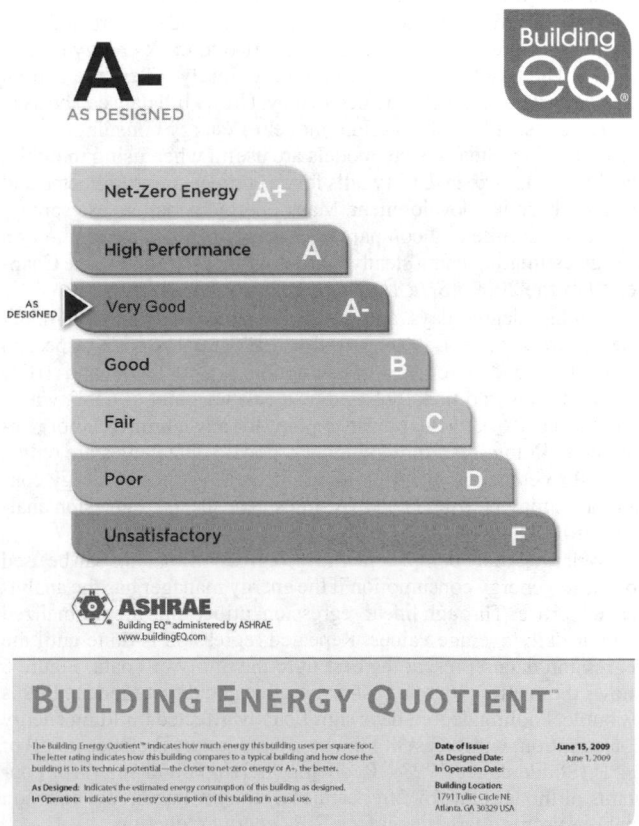

Target Energy Performance Results (estimated)			
Energy	Design	Target	Top 10%
Score	67	75	90
Site Energy Use Intensity (kBtu/Sq. Ft./yr)	69.8	64.7	49.0
Estimated Total Annual Energy (kBtu)	2,143,077.2	1,986,976.8	1,505,328.2
Total Annual Energy Cost ($)	$ 40,826	$ 37,852	$ 28,677

Facility Information   Edit

ASHRAE HQ
Atlanta, GA
United States

Facility Characteristics	Edit		Design Energy			Edit
Space Type	Gross Floor Area (Sq. Ft.)		Energy Source	Units	Estimated Total Annual Energy Use	Energy Rate ($/Unit)
Open Parking Lot[a]	78,242					
Office (General)	30,690		Electricity	kWh	628,100	$ 0.065/kWh
Total Gross Floor Area	30,690		Design Energy default values - Source: DOE-EIA			
[a] not included in total						

Fig. 4   ENERGY STAR Rating for Atlanta Building

Fig. 5   ASHRAE Building eQ Label

According to the EPBD, minimum energy performance requirements are set for new buildings and for major renovations of large existing buildings in each EU member state. Energy performance should be upgraded to meet minimum requirements that are technically, functionally, and economically feasible. In the case of large new buildings, alternative energy supply systems should be considered (e.g., decentralized energy supply systems based on renewable energy, combined heat and power, district or block heating or cooling, heat pumps). The concerted action (CA) EPBD that was launched by the European Commission provides updated information on the implementation status in the various European countries (www.epbd-ca.org).

Tracking Performance

The next step is to create a tracking mechanism to provide high-level KPI views, giving an overall indication of energy performance. Daily monitoring can be a valuable, proactive tool. Most DDC systems can monitor energy performance and notify the energy engineer when energy usage is off track.

For example, using the data presented in Table 1, a daily target usage/day could be determined based on outside air temperature and building occupancy schedule. If the daily use rises above the target use by a predetermined amount, the DDC system can indicate an alarm and send a notification. The energy manager can then investigate the cause of the discrepancy and correct any operational errors before long-term performance is affected. When implementing this type of performance-monitoring strategy, it is important that the measurement and verification plan provide standard operating procedures (SOPs) to facilitate troubleshooting of energy performance alarms. Procedures are discussed in ANSI/ASHRAE *Standard* 105.

Establishing New Goals

Implementing the baseline model is a three-step process: (1) the baseline period is selected, (2) the baseline model is created, and (3) one or more target models are identified to track energy performance. The baseline period should most closely reflect the current or expected building use and occupancy. Utility bill data can be used to create a steady-state baseline model of energy consumption for each building. Steady-state models are useful when using monthly, weekly, or daily data. Utility bills for an entire year are collected and used for baseline development. Many energy managers use spreadsheets to compile and compare the data. For more information on energy estimating using steady-state, data-driven models, see Chapter 19 of the 2009 *ASHRAE Handbook—Fundamentals*.

Cooling degree-days and heating degree-days are commonly used to track successes compared to EEM targets with respect to weather-dependent energy consumption. Local CDD and HDD information is traditionally based on a balance point of 65°F, which is not typically the actual balance point for any commercial or residential building; therefore, regional or local HDD values are only a general reference point. A building's weather-affected energy consumption may be calculated by using spreadsheets, regression analysis, or building energy modeling software.

For larger, more complex facilities, regression analysis can be used to analyze energy consumption if the energy manager has the analytical expertise. Through linear regression, utility bills are normalized to their daily average values. Repeated regression is done until the regression data represent the best fit to the utility bill data. Figure 6 shows the scatter plot of a best-fit baseline and target models. In this example, cooling degree-days significantly affected building energy consumption, with a best fit for a base temperature (balance point) of 54°F (Sonderegger 1998). Reducing the slope and intercept constants of the baseline by 20% creates a straight-line model equation that represents a target goal for a 20% energy reduction.

The utility bill data steady-state model is also referred to as whole-building measurement and verification. More information

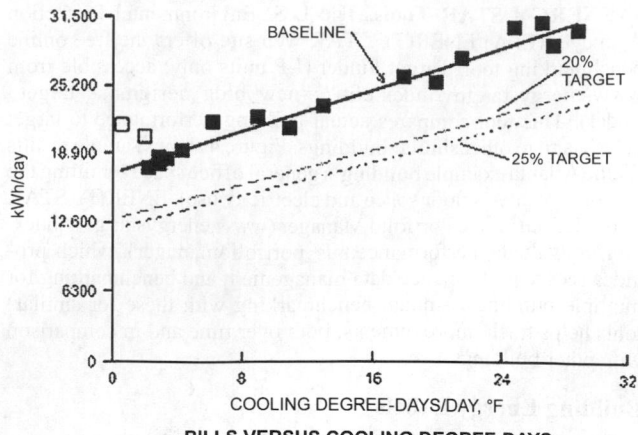

BILLS VERSUS COOLING DEGREE-DAYS

Fig. 6   Scatter Plot, Showing Best-Fit Baseline Model and Target Models

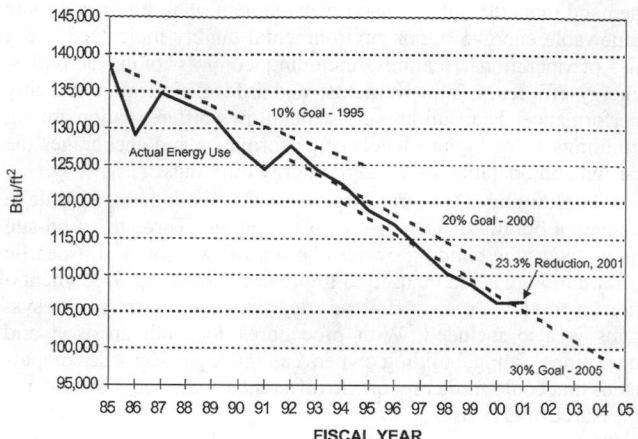

Fig. 7   Progress Toward Energy Reduction Goals for Federal Standard Buildings

about this process can be found in ASHRAE *Guideline* 14 and EVO (2002).

Reporting

When developing presentation materials to document energy performance, make sure that report content shows performance as related to key performance indicators (KPIs) used by the organization. Reports should be pertinent to the audience. Whereas a report to the company's administration would show how the energy management program affects operating and maintenance costs, a separate report to the operations staff might show how their daily decisions and actions change daily load profiles.

Figure 7 shows progress toward energy reduction goals for federal buildings presented to the U.S. Congress for fiscal year 2001 (DOE 2004). The figure compares energy performance against energy goals established in 1999.

Reports must be easy to understand by their readers. Keep management aware of the progress of changes to resource consumption, utility costs, and any effects (positive or negative) on the indoor environment as perceived by staff. Provide information on any major activities, savings to date, and future planned activities. Provide narrative reports with pie charts or bar graphs of cost per resource. Figure 8 shows an example of monthly gas use in a facility from year to year.

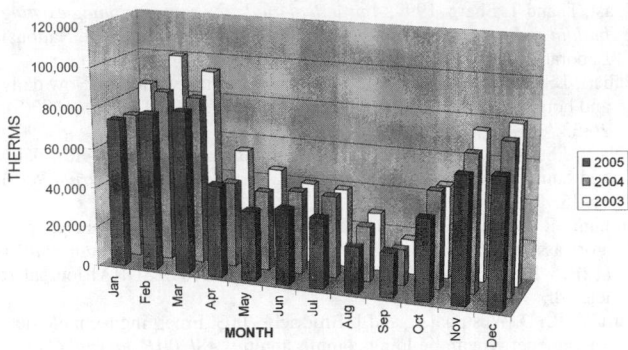

Fig. 8   Monthly Comparison of Natural Gas Use by Year

BUILDING EMERGENCY ENERGY USE REDUCTION

This section provides information to help building owners and operators maintain the best operating condition for the facilities during various energy emergencies. The need for occasional short-term reductions in energy use has increased because of rising energy costs and supply reductions (voluntary or mandatory) or equipment failures. In limited instances, utilities have implemented rolling blackouts, requested voluntary reductions, and asked users to operate emergency generators.

Implementing Emergency Energy Use Reductions

Each building manager or operator should use the energy team approach and identify an individual with the necessary authority and knowledge to review and fit recommendations into a building energy management plan. Because energy reduction requirements may occur with little or no advance notice, contingency plans should be developed and reviewed by the energy team. Each type of energy emergency requires a specific plan to reduce building energy use and still maintain the best possible building environment. The plan should include measures to reduce specific types of energy use in the building, as well as provisions for both slight and major energy use reduction. In some cases, existing building energy management systems can be used to implement demand shedding. The plan should be tested regularly. The following steps should be taken in developing a building emergency energy use reduction plan:

1. Develop a list of measures applicable to each building.
2. Estimate the amount and type of energy savings for each measure and appropriate combination of measures (e.g., account for air-conditioning savings from reduced lighting and other internal loads). Tabulate demand and usage savings separately for response to different types of emergencies.
3. For various levels of possible energy emergency, develop a plan that maintains the best building environment under the circumstances. Develop the plan so that actions taken can be energy-source-specific. That is, group actions to be taken to reduce energy consumption for each type of energy used in the building. Include both short- and long-term measures in the plan. Operational changes may be implemented quickly and prove adequate for short-term emergencies.
4. Experiment with the plan; record energy consumption and demand reduction data, and revise the plan as necessary. Much of the experimentation may be done on weekends to minimize disruption.
5. Meet with the local utility provider(s) and back-up fuel suppliers to review the plan.
6. Meet with building occupants annually to review the plan to ensure that actions taken do not cause major disruptions or compromise life safety or security provisions. Establish a procedure for notification of building occupants before actions are taken.

7. Be certain that there is a plan to minimize entrapment of occupants in elevators in case of emergency disruptions.
8. Review the plan annually with building security and the fire department to ensure that emergency efforts are not hindered by the plan and that security or emergency people know what to expect (reduced lighting, lower temperatures, elevators out of operation, etc.).
9. When preparing the plan, **do not**
   - Take lighting fixtures out of service that are on night lighting circuits, provide lighting for security cameras, or provide egress lighting during a power failure
   - Remove elevators or lifts from service that will be required for emergency or ADA purposes
   - Reduce ventilation or exhaust in laboratories or other areas where hazardous conditions exist

Some measures can be implemented permanently. Depending on the level of energy emergency and the building priority, the following actions may be considered in developing the plan for emergency energy reduction:

General
- Change operating hours
- Move personnel into other building areas (consolidation)
- Ensure that emergency generators are tuned up and run frequently enough to increase dependability, service the expected electrical load, and keep alternative fuel supply at optimal level
- Shut off nonessential equipment
- Review the amount of uninterruptible power supply (UPS) time available for critical equipment, and upgrade if necessary

Thermal Envelope
- Use all existing blinds, draperies, and window coverings
- Install interior window insulation
- Caulk and seal around unused exterior doors and windows (but do not seal doors required for emergency egress or that may be required by the fire department in an emergency).
- Install solar shading devices in summer
- Seal all unused vents and ducts to outside

HVAC Systems and Equipment
- Modify controls or control set points to raise and lower temperature and humidity as necessary
- Shut off or isolate all nonessential equipment and spaces
- Lower thermostat set points in winter
- Raise chilled-water temperature
- Lower hot-water temperature (*Note*: Keep hot-water temperature higher than 145°F if a noncondensing gas boiler is used)
- Reduce or eliminate reheat and recool
- Reduce (and eliminate during unoccupied hours) mechanical ventilation and exhaust airflow
- Raise thermostat set points in summer or turn cooling equipment off

Lighting Systems
- Evaluate overlit areas and remove lamps or reduce lamp wattage
- Use task lighting where appropriate
- Move building functions to exterior or daylit areas
- Turn off electric lights in areas with adequate natural light
- Revise building cleaning and security procedures to minimize lighting periods
- Consolidate parking and turn off unused parking security lighting

Special Equipment
- Take transformers offline during periods of nonuse
- Shut off or regulate the use of vertical transportation systems

- Shut off unused or unnecessary equipment, such as photocopiers, music systems, and computers
- Reduce or turn off potable hot-water supply

Building Operation Demand Reduction

- Sequence or interlock heating or air-conditioning systems
- Disconnect or turn off all nonessential loads
- Reduce lighting levels
- Preheat or precool, if possible, before utility-imposed emergency periods

When Power Is Restored

- To prevent overloading the system, turn equipment back on gradually
- Test and verify proper operation of critical equipment, security, and fire and smoke alarms
- Check monitors on temperature-sensitive equipment
- Discuss lessons learned with staff and make any necessary changes to emergency plan
- Restock whatever emergency supplies were used, including alternative fuels

REFERENCES

ASHRAE. 2004. *Procedures for commercial building energy audits.*
ASHRAE. 2005. The commissioning process. *Guideline* 0-2005.
ASHRAE. 2007. The HVAC&R technical requirements for the commissioning process. *Guideline* 1.1-2007.
ASHRAE. 2002. Measurement of energy and demand savings. *Guideline* 14-2002.
ASHRAE. 2007. Energy standards for buildings except low-rise residential buildings. ANSI/ASHRAE/IESNA *Standard* 90.1-2007.
ASHRAE. 2007. Energy-efficient design of low-rise residential buildings. ANSI/ASHRAE *Standard* 90.2-2007.
ASHRAE. 2006. Energy conservation in existing buildings. ANSI/ASHRAE/IESNA *Standard* 100-2006.
ASHRAE. 2007. Standard methods of measuring, expressing, and comparing building energy performance. ANSI/ASHRAE *Standard* 105-2007.
ASHRAE. 2004. *Procedures for commercial building energy audits.*
Balaras, C.A., A.G. Gaglia, E. Georgopoulou, S. Mirasgedis, Y. Sarafidis, and D.P. Lalas. 2007. European residential buildings and empirical assessment of the Hellenic building stock, energy consumption, emissions & potential energy savings. *Building and Environment* 42(3):1298-1314.
Claridge, D.E. and M. Liu. 2000. HVAC system commissioning. In *Handbook of heating, ventilation, and air conditioning*, pp. 7.1-7.25. J.F. Kreider, ed. CRC Press, Boca Raton, FL.
Claridge, D.E., M. Liu, W.D. Turner, Y. Zhu, M. Abbas, and J.S. Haberl. 1998. Energy and comfort benefits of continuous commissioning in buildings. *Proceedings of the International Conference Improving Electricity Efficiency in Commercial Buildings*, Amsterdam, pp. 12.5.1-12.5.17.
DOE. 2004. *Annual report to Congress on federal government energy management and conservation programs, fiscal year 2001.* U.S. Department of Energy, Washington, D.C. http://www1.eere.energy.gov/femp/pdfs/annrep01.pdf.
DOE/EIA. 2003. *Nonresidential buildings energy consumption survey: 2003 commercial buildings energy consumption survey (CBECS) public use files.* http://www.eia.doe.gov/emeu/cbecs/cbecs2003/public_use_2003/cbecs_pudata2003.html.
EC. 2010. *Directive on the energy performance of buildings.* COM 2010/31/EU. European Commission.
EPAct. 2005. *Energy Policy Act.* U.S. Department of Energy, Washington, D.C.
EVO. 2002. *International performance measurement and verification protocol (IPMVP), vol. I: Concepts and options for determining savings.* Efficiency Value Organization, San Francisco.
Fels, M. 1986. Special issue devoted to the Princeton Scorekeeping Method (PRISM). *Energy and Buildings* 9(1 and 2).
Gaglia, A.G., C.A. Balaras, S. Mirasgedis, E. Georgopoulou, Y. Sarafidis, and D.P. Lalas. 2007. Empirical assessment of the Hellenic non-residential building stock, energy consumption, emissions and potential energy savings. *Energy Conversion and Management* 48(4):1160-1175.

Haasl, T. and T. Sharp. 1999. *A practical guide for commissioning existing buildings.* Portland Energy Conservation, Inc., and Oak Ridge National Laboratory for U.S. DOE, ORNL/TM-1999/34.
Haberl, J.S. and P.S. Komor. 1990a. Improving energy audits—How daily and hourly consumption data can help, part 1. *ASHRAE Journal* 90(8):26-33.
Haberl, J.S. and P.S. Komor. 1990b. Improving energy audits—How daily and hourly consumption data can help, part 2. *ASHRAE Journal* 90(9):26-36.
Jarnagin, R. 2009. ASHRAE Building eQ program will help owners, operators assess buildings, and guide good decisions. *ASHRAE Journal* 51(12):18-19. http://www.buildingeq.com/files/122009ASHRAEJournalarticle.pdf.
Kurt, W.R., D. Westphalen, and J. Brodrick. 2003. Emerging technologies: Saving energy with building commissioning. *ASHRAE Journal* 45(11):65-66.
Liu, M., D.E. Claridge, J.S. Haberl, and W.D. Turner. 1997. Improving building energy systems performance by continuous commissioning. *Proceedings of the Thirty-Second Intersociety Energy Conversion Engineering Conference*, Honolulu, vol. 3.
Mazzucchi, R.P. 1992. A guide for analyzing and reporting building characteristics and energy use in commercial buildings. *ASHRAE Transactions* 92(1):1067-1080.
NRC. 2000. *Commercial and institutional building energy use survey (CIBEUS): Detailed statistical report.* Natural Resources Canada, Office of Energy Efficiency, Ottawa.
PNNL. 1990. *Architect's and engineer's guide to energy conservation in existing buildings*, vol. 2, ch.1. DOE/RL/01830 P-H4. Pacific Northwest National Laboratories, Richland, WA.
Poulos, J. (2007). Existing building commissioning. *ASHRAE Journal* 49(9):66-78.
Rundquist, R.A., K.F. Johnson, and D.J. Aumann. 1993. Calculating lighting and HVAC interactions. *ASHRAE Journal* 35(11):28-37.
Sonderegger, R.C. 1998. A baseline model for utility bill analysis using both weather and non-weather-related variables. *ASHRAE Transactions* 104(2):859-870.
Spielvogel, L.G. 1984. One approach to energy use evaluation. *ASHRAE Transactions* 90(1B):424-435.
Tseng, P.C. 2005. Commissioning sustainable buildings. *ASHRAE Journal* 47(9):S20-S24.
Turner, W.C. 2001. *Energy management handbook*, 4th ed. Fairmount Press, Lilburn, GA.
Turner, W.D., D. Claridge, S. Deng, S. Cho, M. Liu, T. Hassl, C. Dethell, Jr., and H. Bruner, Jr. 2001. Persistence of savings from continuous commissioning. 9th National Conference on Building Commissioning, Cherry Hill, NJ.

BIBLIOGRAPHY

ASHRAE. 2010. *Greenguide: The design, construction, and operation of sustainable buildings*, 3rd ed.
Duff, J.M. 1999. A justification for energy managers. *ASHRAE Transactions* 105(1):988-992.
EPA. (no date). Portfolio manager overview. U.S. Environmental Protection Agency and U.S. Department of Energy ENERGY STAR program, Washington, D.C. http://www.energystar.gov/benchmark.
Hay, J.C. and I. Sud. 1997. Evaluation of proposed ASHRAE energy audit form and procedures. *ASHRAE Transactions* 103(2):90-120.
Langley, G., et.al. 2009. *The improvement guide: A practical approach to enhancing organizational performance.* Jossey-Bass, San Francisco.
MacDonald, J.M. and D.M. Wasserman. 1989. *Investigation of metered data analysis methods for commercial and related buildings.* ORNL/CON-279. Oak Ridge National Laboratories, Oak Ridge, TN.
Mendell, M.J. and A.G. Mirer. 2009. Indoor thermal factors and symptoms in office workers: Findings from the US EPA BASE study. *Indoor Air* 19:291-302.
Miller, W. 1999. Resource conservation management. *ASHRAE Transactions* 105(1):993-1002.
Mills, E. and P. Matthew. 2009. Monitoring-based commissioning: Benchmarking analysis of 24 UC/CSU/IOU projects. Lawrence Berkeley National Laboratory *Report* 1972E.
Mills, E. 2009. Building commissioning: A golden opportunity for reducing energy costs and greenhouse gas emissions. *Report* for California Energy Commission Public Interest Energy Research. http://cx.lbl.gov/2009-assessment.html.

Russell, C. 2006. Energy management pathfinding. *Strategic Planning for Energy and the Environment* 25(3).

Sikorski, B.D. and B.A. O'Donnell. 1999. Savings impact of a corporate energy manager. *ASHRAE Transactions* 105(1):977-987.

Waltz, J.P. 2000. *Computerized building energy simulation.* Fairmont Press, Lilburn, GA.

ONLINE RESOURCES

ENERGY STAR financial evaluation tools: www.energystar.gov/index.cfm?c=assess_value.financial_tools
- Building upgrade value calculator
- Cash flow opportunity calculator
- Financial value calculator

End-use energy survey spreadsheet tool: www.focusonenergy.com/files/Document_Management_System/Business_Programs/equipmentusage_spreadsheet.xls

Building energy software tools directory: http://apps1.eere.energy.gov/buildings/tools_directory/

This directory provides information on almost 400 building software tools for evaluating energy efficiency, renewable energy, and sustainability in buildings. The energy tools listed in this directory include databases, spreadsheets, component and systems analyses, and whole-building energy performance simulation programs. A short description is provided for each tool along with other information, including expertise required, users, audience, input, output, computer platforms, programming language, strengths, weaknesses, technical contact, and availability.

2008 Buildings Energy Data Book (March 2009) on uses of energy in buildings: http://buildingsdatabook.eren.doe.gov/docs%5CDataBooks%5C2008_BEDB_Updated.pdf

U.S. Energy Information Administration's commercial buildings energy consumption survey (commercial energy uses and costs): www.eia.doe.gov/emeu/cbecs/

Emissions associated with energy generation (eGRID): www.epa.gov/cleanenergy/energy-resources/egrid/index.html

Climate zone information: http://resourcecenter.pnl.gov/cocoon/morf/ResourceCenter/article/1420

CHAPTER 37

OWNING AND OPERATING COSTS

OWNING and operating cost information for the HVAC system should be part of the investment plan of a facility. This information can be used for preparing annual budgets, managing assets, and selecting design options. Table 1 shows a representative form that summarizes these costs.

A properly engineered system must also be economical, but this is difficult to assess because of the complexities surrounding effective money management and the inherent difficulty of predicting future operating and maintenance expenses. Complex tax structures and the time value of money can affect the final engineering decision. This does not imply use of either the cheapest or the most expensive system; instead, it demands intelligent analysis of financial objectives and the owner's requirements.

Certain tangible and intangible costs or benefits must also be considered when assessing owning and operating costs. Local codes may require highly skilled or certified operators for specific types of equipment. This could be a significant cost over the life of the system. Similarly, intangible items such as aesthetics, acoustics, comfort, safety, security, flexibility, and environmental impact may vary by location and be important to a particular building or facility.

OWNING COSTS

The following elements must be established to calculate annual owning costs: (1) initial cost, (2) analysis or study period, (3) interest or discount rate, and (4) other periodic costs such as insurance, property taxes, refurbishment, or disposal fees. Once established, these elements are coupled with operating costs to develop an economic analysis, which may be a simple payback evaluation or an in-depth analysis such as outlined in the section on Economic Analysis Techniques.

Initial Cost

Major decisions affecting annual owning and operating costs for the life of the building must generally be made before completing contract drawings and specifications. To achieve the best performance and economics, alternative methods of solving the engineering problems peculiar to each project should be compared in the early stages of design. Oversimplified estimates can lead to substantial errors in evaluating the system.

The evaluation should lead to a thorough understanding of installation costs and accessory requirements for the system(s) under consideration. Detailed lists of materials, controls, space and structural requirements, services, installation labor, and so forth can be prepared to increase accuracy in preliminary cost estimates. A reasonable estimate of capital cost of components may be derived from cost records of recent installations of comparable design or from quotations submitted by manufacturers and contractors, or by

The preparation of this chapter is assigned to TC 7.8, Owning and Operating Costs.

Table 1 Owning and Operating Cost Data and Summary

OWNING COSTS

I. Initial Cost of System	_____
II. Periodic Costs	
A. Income taxes	_____
B. Property taxes	_____
C. Insurance	_____
D. Rent	_____
E. Other periodic costs	_____
Total Periodic Costs	_____
III. Replacement Cost	_____
IV. Salvage Value	_____
Total Owning Costs	_____

OPERATING COSTS

V. Annual Utility, Fuel, Water, etc., Costs	
A. Utilities	
1. Electricity	_____
2. Natural gas	_____
3. Water/sewer	_____
4. Purchased steam	_____
5. Purchased hot/chilled water	_____
B. Fuels	
1. Propane	_____
2. Fuel oil	_____
3. Diesel	_____
4. Coal	_____
C. On-site generation of electricity	_____
D. Other utility, fuel, water, etc., costs	_____
Total	_____
VI. Annual Maintenance Allowances/Costs	
A. In-house labor	_____
B. Contracted maintenance service	_____
C. In-house materials	_____
D. Other maintenance allowances/costs	_____
(e.g., water treatment)	
Total	_____
VII. Annual Administration Costs	_____
Total Annual Operating Costs	_____

TOTAL ANNUAL OWNING AND OPERATING COSTS _____

Table 2 Initial Cost Checklist

Energy and Fuel Service Costs

Fuel service, storage, handling, piping, and distribution costs
Electrical service entrance and distribution equipment costs
Total energy plant

Heat-Producing Equipment

Boilers and furnaces
Steam-water converters
Heat pumps or resistance heaters
Makeup air heaters
Heat-producing equipment auxiliaries

Refrigeration Equipment

Compressors, chillers, or absorption units
Cooling towers, condensers, well water supplies
Refrigeration equipment auxiliaries

Heat Distribution Equipment

Pumps, reducing valves, piping, piping insulation, etc.
Terminal units or devices

Cooling Distribution Equipment

Pumps, piping, piping insulation, condensate drains, etc.
Terminal units, mixing boxes, diffusers, grilles, etc.

Air Treatment and Distribution Equipment

Air heaters, humidifiers, dehumidifiers, filters, etc.
Fans, ducts, duct insulation, dampers, etc.
Exhaust and return systems
Heat recovery systems

System and Controls Automation

Terminal or zone controls
System program control
Alarms and indicator system
Energy management system

Building Construction and Alteration

Mechanical and electric space
Chimneys and flues
Building insulation
Solar radiation controls
Acoustical and vibration treatment
Distribution shafts, machinery foundations, furring

Table 3 Median Service Life

Equipment Type	Median Service Life, Years	Total No. of Units	No. of Units Replaced
DX air distribution equipment	>24	1907	284
Chillers, centrifugal	>25	234	34
Cooling towers, metal	>22	170	24
Boilers, hot-water, steel gas-fired	>22	117	24
Controls, pneumatic	>18	101	25
electronic	>7	68	6
Potable hot-water heaters, electric	>21	304	36

updated and current. The database was seeded with information gathered from a sample of 163 commercial office buildings located in major metropolitan areas across the United States. Abramson et al. (2005) provide details on the distribution of building size, age, and other characteristics. Table 3 presents estimates of median service life for various HVAC components in this sample.

Median service life in Table 3 is based on analysis of survival curves, which take into account the units still in service and the units replaced at each age (Hiller 2000). Conditional and total survival rates are calculated for each age, and the percent survival over time is plotted. Units still in service are included up to the point where the age is equal to their current age at the time of the study. After that point, these units are censored (removed from the population). Median service life in this table indicates the highest age at which the survival rate remains at or above 50% while the sample size is 30 or more. There is no hard-and-fast rule about the number of units needed in a sample before it is considered statistically large enough to be representative, but usually the number should be larger than 25 to 30 (Lovvorn and Hiller 2002). This rule-of-thumb is used because each unit removal represents greater than a 3% change in survival rate as the sample size drops below 30, and that percentage increases rapidly as the sample size gets even smaller.

The database initially developed and seeded under research project TRP-1237 (Abramson et al. 2005) is now available online, providing engineers with equipment service life and annual maintenance costs for a variety of building types and HVAC systems. The database can be accessed at www.ashrae.org/database.

As of the end of 2009 this database contained more than 300 building types, with service life data on more than 38,000 pieces of equipment.

The database allows users to access up-to-date information to determine a range of statistical values for equipment owning and operating costs. Users are encouraged to contribute their own service life and maintenance cost data, further expanding the utility of this tool. Over time, this input will provide sufficient service life and maintenance cost data to allow comparative analysis of many different HVAC systems types in a broad variety of applications. Data can be entered by logging into the database and registering, which is free. With this, ASHRAE is providing the necessary methods and information to assist in using life-cycle analysis techniques to help select the most appropriate HVAC system for a specific application. This system of collecting data also greatly reduces the time between data collection and when users can access the information.

Figure 1 presents the survival curve for centrifugal chillers, based on data in Abramson et al. (2005). The point at which survival rate drops to 50% based on all data in the survey is 31 years. However, because the sample size drops below the statistically relevant number of 30 units at 25 years, the median service life of centrifugal chillers can only be stated with confidence as >25 years.

Table 4 compares the estimates of median service life in Abramson et al. (2005) with those developed with those in Akalin (1978). Most differences are on the order of one to five years.

Estimated service life of new equipment or components of systems not listed in Table 3 or 4 may be obtained from manufacturers,

consulting commercially available cost-estimating guides and software. Table 2 shows a representative checklist for initial costs.

Analysis Period

The time frame over which an economic analysis is performed greatly affects the results. The analysis period is usually determined by specific objectives, such as length of planned ownership or loan repayment period. However, as the length of time in the analysis period increases, there is a diminishing effect on net present-value calculations. The chosen analysis period is often unrelated to the equipment depreciation period or service life, although these factors may be important in the analysis.

Service Life

For many years, this chapter included estimates of service lives for various HVAC system components, based on a survey conducted in 1976 under ASHRAE research project RP-186 (Akalin 1978). These estimates have been useful to a generation of practitioners, but changes in technology, materials, manufacturing techniques, and maintenance practices now call into question the continued validity of the original estimates. Consequently, ASHRAE research project TRP-1237 developed an Internet-based data collection tool and database on HVAC equipment service life and maintenance costs, to allow equipment owning and operating cost data to be continually

Table 4   Comparison of Service Life Estimates

Equipment Item	Median Service Life, Years		Equipment Item	Median Service Life, Years		Equipment Item	Median Service Life, Years	
	Abramson et al. (2005)	Akalin (1978)		Abramson et al. (2005)	Akalin (1978)		Abramson et al. (2005)	Akalin (1978)
Air Conditioners			**Air Terminals**			**Condensers**		
Window unit	N/A*	10	Diffusers, grilles, and registers	N/A*	27	Air-cooled	N/A	20
Residential single or split package	N/A*	15	Induction and fan-coil units	N/A*	20	Evaporative	N/A*	20
Commercial through-the-wall	N/A*	15	VAV and double-duct boxes	N/A*	20	**Insulation**		
Water-cooled package	>24	15	**Air washers**	N/A*	17	Molded	N/A*	20
Heat pumps			**Ductwork**	N/A*	30	Blanket	N/A*	24
Residential air-to-air	N/A*	15[b]	**Dampers**	N/A*	20	**Pumps**		
Commercial air-to-air	N/A*	15	**Fans**	N/A*		Base-mounted	N/A*	20
Commercial water-to-air	>24	19	Centrifugal	N/A*	25	Pipe-mounted	N/A*	10
Roof-top air conditioners			Axial	N/A*	20	Sump and well	N/A*	10
Single-zone	N/A*	15	Propeller	N/A*	15	Condensate	N/A*	15
Multizone	N/A*	15	Ventilating roof-mounted	N/A*	20	**Reciprocating engines**	N/A*	20
Boilers, Hot-Water (Steam)			**Coils**			**Steam turbines**	N/A*	30
Steel water-tube	>22	24 (30)	DX, water, or steam	N/A*	20	**Electric motors**	N/A*	18
Steel fire-tube		25 (25)	Electric	N/A*	15	**Motor starters**	N/A*	17
Cast iron	N/A*	35 (30)	**Heat Exchangers**			**Electric transformers**	N/A*	30
Electric	N/A*	15	Shell-and-tube	N/A*	24	**Controls**		
Burners	N/A*	21	**Reciprocating compressors**	N/A*	20	Pneumatic	N/A*	20
Furnaces			**Packaged Chillers**			Electric	N/A*	16
Gas- or oil-fired	N/A*	18	Reciprocating	N/A*	20	Electronic	N/A*	15
Unit heaters			Centrifugal	>25	23	**Valve actuators**		
Gas or electric	N/A*	13	Absorption	N/A*	23	Hydraulic	N/A*	15
Hot-water or steam	N/A*	20	**Cooling Towers**			Pneumatic	N/A*	20
Radiant heaters			Galvanized metal	>22	20	Self-contained		10
Electric	N/A*	10	Wood	N/A*	20			
Hot-water or steam	N/A*	25	Ceramic	N/A*	34			

*N/A: Not enough data yet in Abramson et al. (2005). Note that data from Akalin (1978) for these categories may be outdated and not statistically relevant. Use these data with caution until enough updated data are accumulated in Abramson et al.

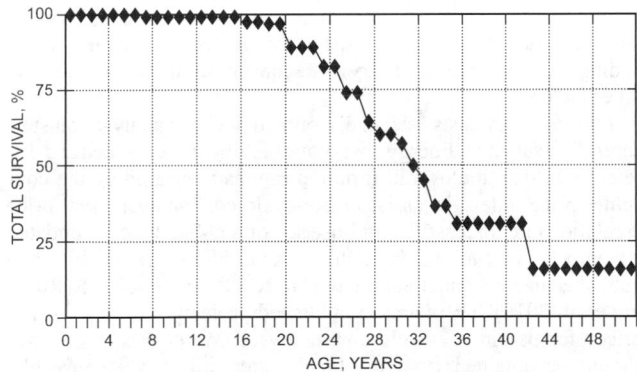

Fig. 1   Survival Curve for Centrifugal Chillers
[Based on data in Abramson et al. (2005)]

associations, consortia, or governmental agencies. Because of the proprietary nature of information from some of these sources, the variety of criteria used in compiling the data, and the diverse objectives in disseminating them, extreme care is necessary in comparing service life from different sources. Designs, materials, and components of equipment listed in Tables 3 and 4 have changed over time and may have altered the estimated service lives of those equipment categories. Therefore, establishing equivalent comparisons of service life is important.

As noted, service life is a function of the time when equipment is replaced. Replacement may be for any reason, including, but not limited to, failure, general obsolescence, reduced reliability, excessive maintenance cost, and changed system requirements (e.g., building characteristics, energy prices, environmental considerations).

Service lives shown in the tables are based on the age of the equipment when it was replaced, regardless of the reason it was replaced.

Locations in potentially corrosive environments and unique maintenance variables affect service life. Examples include the following:

- **Coastal and marine environments**, especially in tropical locations, are characterized by abundant sodium chloride (salt) that is carried by sea spray, mist, or fog.

  Many owners require equipment specifications stating that HVAC equipment located along coastal waters will have corrosion-resistant materials or coatings. Design criteria for systems installed under these conditions should be carefully considered.

- **Industrial** applications provide many challenges to the HVAC designer. It is very important to know if emissions from the industrial plant contain products of combustion from coal, fuel oils, or releases of sulfur oxides (SO_2, SO_3) and nitrogen oxides (NO_x) into the atmosphere. These gases typically accumulate and return to the ground in the form of acid rain or dew.

  Not only is it important to know the products being emitted from the industrial plant being designed, but also the adjacent upwind or downwind facilities. HVAC system design for a plant located downwind from a paper mill requires extraordinary corrosion protection or recognition of a reduced service life of the HVAC equipment.

- **Urban** areas generally have high levels of automotive emissions as well as abundant combustion by-products. Both of these contain elevated sulfur oxide and nitrogen oxide concentrations.

- **Maintenance** factors also affect life expectancy. The HVAC designer should temper the service life expectancy of equipment with a **maintenance factor**. To achieve the estimated service life values in Table 3, HVAC equipment must be maintained properly, including good filter-changing practices and good maintenance procedures. For example, chilled-water coils with more than four rows and close fin spacing are virtually impossible to clean even using extraordinary methods; they are often replaced with multiple coils in series, with a maximum of four rows and lighter fin spacing.

Depreciation

Depreciation periods are usually set by federal, state, or local tax laws, which change periodically. Consult applicable tax laws for more information on depreciation.

Interest or Discount Rate

Most major economic analyses consider the opportunity cost of borrowing money, inflation, and the time value of money. **Opportunity cost** of money reflects the earnings that investing (or lending) the money can produce. **Inflation** (price escalation) decreases the purchasing or investing power (value) of future money because it can buy less in the future. **Time value** of money reflects the fact that money received today is more useful than the same amount received a year from now, even with zero inflation, because the money is available earlier for reinvestment.

The cost or value of money must also be considered. When borrowing money, a percentage fee or interest rate must normally be paid. However, the interest rate may not necessarily be the correct cost of money to use in an economic analysis. Another factor, called the **discount rate**, is more commonly used to reflect the true cost of money [see Fuller and Petersen (1996) for detailed discussions]. Discount rates used for analyses vary depending on individual investment, profit, and other opportunities. Interest rates, in contrast, tend to be more centrally fixed by lending institutions.

To minimize the confusion caused by the vague definition and variable nature of discount rates, the U.S. government has specified particular discount rates to be used in economic analyses relating to federal expenditures. These discount rates are updated annually (Rushing et al. 2010) but may not be appropriate for private-sector economic analyses.

Periodic Costs

Regularly or periodically recurring costs include insurance, property taxes, income taxes, rent, refurbishment expenses, disposal fees (e.g., refrigerant recycling costs), occasional major repair costs, and decommissioning expenses.

Insurance. Insurance reimburses a property owner for a financial loss so that equipment can be repaired or replaced. Insurance often indemnifies the owner from liability, as well. Financial recovery may include replacing income, rents, or profits lost because of property damage.

Some of the principal factors that influence the total annual insurance premium are building size, construction materials, amount and size of mechanical equipment, geographic location, and policy deductibles. Some regulations set minimum required insurance coverage and premiums that may be charged for various forms of insurable property.

Property Taxes. Property taxes differ widely and may be collected by one or more agencies, such as state, county, or local governments or special assessment districts. Furthermore, property taxes may apply to both real (land, buildings) and personal (everything else) property. Property taxes are most often calculated as a percentage of assessed value, but are also determined in other ways, such as fixed fees, license fees, registration fees, etc. Moreover, definitions of assessed value vary widely in different geographic areas. Tax experts should be consulted for applicable practices in a given area.

Income Taxes. Taxes are generally imposed in proportion to net income, after allowance for expenses, depreciation, and numerous other factors. Special tax treatment is often granted to encourage certain investments. Income tax professionals can provide up-to-date information on income tax treatments.

Other Periodic Costs. Examples of other costs include changes in regulations that require unscheduled equipment refurbishment to eliminate use of hazardous substances, and disposal costs for such substances.

Replacement Costs and Salvage Value. Replacement costs and salvage value should be evaluated when calculating owning cost. Replacement cost is the cost to remove existing equipment and install new equipment. Salvage value is the value of equipment or its components for recycling or other uses. Equipment's salvage value may be negative when removal, disposal, or decommissioning costs are considered.

OPERATING COSTS

Operating costs are those incurred by the actual operation of the system. They include costs of fuel and electricity, wages, supplies, water, material, and maintenance parts and services. Energy is a large part of total operating costs. Chapter 19 of the 2009 *ASHRAE Handbook—Fundamentals* outlines how fuel and electrical requirements are estimated. Because most energy management activities are dictated by economics, the facility manager must understand the utility rates that apply to each facility. Electric rates are usually more complex than gas or water rates. In addition to general commercial or institutional electric rates, special rates may exist such as time of day, interruptible service, on-peak/off-peak, summer/winter, and peak demand. Electric rate schedules vary widely in North America. The facility manager should work with local utility companies to identify the most favorable rates and to understand how to qualify for them. The local utility representative can help the facility manager develop the most cost-effective methods of metering and billing. The facility manager must understand the utility rates, including the distinction between marginal and average costs and, in the case of demand-based electric rates, how demand is computed.

Note that, in general, total energy consumption cannot be multiplied by a per-unit energy cost to arrive at a correct annual utility cost, because rate schedules (especially for electricity) often have a sliding scale of prices that vary with consumption, time of day, and other factors.

Future energy costs used in discounted payback analyses must be carefully evaluated. Energy costs have historically escalated at a different rate than the overall inflation rate as measured by the consumer price index. To assist in life-cycle cost analysis, fuel price escalation rate forecasts by end-use sector and fuel type are updated annually by the National Institute of Standards and Technology and published in the Annual Supplement to NIST *Handbook* 135 (Rushing et al. 2010). There are no published projection rates for water prices for use in life-cycle cost analyses. Water escalation rates should be obtained from the local water utility when possible. Building designers should use energy price projections from their local utility in place of regional forecasts whenever possible, especially when evaluating alternative fuel types.

Deregulation in some areas may allow increased access to nontraditional energy providers and pricing structures; in other areas, traditional utility infrastructures and practices may prevail. The amount and profile of the energy used by the facility will also determine energy cost. Unbundling energy services (having separate contracts for energy and for its transportation to point of use) may dictate separate agreements for each service component or may be packaged by a single provider. Contract length and price stability are factors in assessing nontraditional versus traditional energy suppliers when estimating operating costs. The degree of energy supply and system reliability and price stability considered necessary by the owner/occupants of a building may require considerable deliberation. The sensitivity of a building's functionality to energy-related variables should dictate the degree of attention allocated in evaluating these factors.

Electrical Energy

The total cost of electricity is determined by a rate schedule and is usually a combination of several components: consumption (kilowatt-hours), demand (kilowatts) fuel adjustment charges, special allowances or other adjustments, and applicable taxes. Of these, consumption and demand are the major cost components and the ones the owner or facility manager may be able to affect.

Electricity Consumption Charges. Most electric rates have step-rate schedules for consumption, and the cost of the last unit consumed may be substantially different from that of the first. The last unit is usually cheaper than the first because the fixed costs to the utility may already have been recovered from earlier consumption costs. Because of this, the energy analysis cannot use average costs to accurately predict savings from implementation of energy conservation measures. Average costs will overstate the savings possible between alternative equipment or systems; instead, marginal (or incremental) costs must be used.

To reflect time-varying operating costs or to encourage peak shifting, electric utilities may charge different rates for consumption according to the time of use and season, with higher costs occurring during the peak period of use.

Fuel Adjustment Charge. Because of substantial variations in fuel prices, electric utilities may apply a fuel adjustment charge to recover costs. This adjustment may not be reflected in the rate schedule. The fuel adjustment is usually a charge per unit of consumption and may be positive or negative, depending on how much of the actual fuel cost is recovered in the energy consumption rate. The charge may vary monthly or seasonally.

Allowances or Adjustments. Special discounts or rates may be available for customers who can receive power at higher voltages or for those who own transformers or similar equipment. Special rates or riders may be available for specific interruptible loads such as domestic water heaters.

Certain facility electrical systems may produce a low power factor [i.e., ratio of real (active) kilowatt power to apparent (reactive) kVA power], which means that the utility must supply more current on an intermittent basis, thus increasing their costs. These costs may be passed on as an adjustment to the utility bill if the power factor is below a level established by the utility.

When calculating power bills, utilities should be asked to provide detailed cost estimates for various consumption levels. The final calculation should include any applicable special rates, allowances, taxes, and fuel adjustment charges.

Demand Charges. Electric rates may also have demand charges based on the customer's peak kilowatt demand. Whereas consumption charges typically cover the utility's operating costs, demand charges typically cover the owning costs.

Demand charges may be formulated in a variety of ways:

- *Straight charge.* Cost per kilowatt per month, charged for the peak demand of the month.
- *Excess charge.* Cost per kilowatt above a base demand (e.g., 50 kW), which may be established each month.
- *Maximum demand (ratchet).* Cost per kilowatt for maximum annual demand, which may be reset only once a year. This established demand may either benefit or penalize the owner.
- *Combination demand.* Cost per hour of operation of demand. In addition to a basic demand charge, utilities may include further demand charges as demand-related consumption charges.

The actual demand represents the peak energy use averaged over a specific period, usually 15, 30, or 60 min. Accordingly, high electrical loads of only a few minutes' duration may never be recorded at the full instantaneous value. Alternatively, peak demand is recorded as the average of several consecutive short periods (i.e., 5 min out of each hour).

The particular method of demand metering and billing is important when load shedding or shifting devices are considered.

Table 5 Electricity Data Consumption and Demand for Atlanta Example Building, 2003 to 2004

	Billing Days	Consumption, kWh	Actual Demand, kW	Billing Demand, kW	Total Cost, US$
Jan. 2003	29	57,120	178	185	4,118
Feb. 2003	31	61,920	145	185	4,251
Mar. 2003	29	60,060	140	185	4,199
Apr. 2003	29	62,640	154	185	4,271
May. 2003	33	73,440	161	185	4,569
Jun. 2003	26	53,100	171	185	4,007
Jul. 2003	32	67,320	180	185	4,400
Aug. 2003	30	66,000	170	185	4,364
Sep. 2003	32	63,960	149	171	4,127
Oct. 2003	30	55,260	122	171	3,865
Nov. 2003	27	46,020	140	171	3,613
Dec. 2003	34	61,260	141	171	4,028
Total 2003	**362**	**670,980**			**49,812**
Jan. 2004	31	59,040	145	171	3,967
Feb. 2004	29	54,240	159	171	3,837
Mar. 2004	20	37,080	122	171	2,584
Apr. 2004	12	22,140	133	171	1,547
May. 2004	34	64,260	148	171	4,110
Jun. 2004	29	63,720	148	171	4,321
Jul. 2004	30	69,120	169	169	4,458
Aug. 2004	32	73,800	170	170	4,605
Sep. 2004	29	64,500	166	166	4,281
Oct. 2004	30	60,060	152	161	3,866
Nov. 2004	32	65,760	128	161	4,018
Dec. 2004	31	51,960	132	161	3,646
Total 2004	**339**	**685,680**			**45,240**

The portion of the total bill attributed to demand may vary greatly, from 0% to as high as 70%.

- *Real-time* or *time-of-day rates.* Cost of electricity at time of use. An increasing number of utilities offer these rates. End users who can shift operations or install electric load-shifting equipment, such as thermal storage, can take advantage of such rates. Because these rates usually reflect a utility's overall load profile and possibly the availability of specific generating resources, contact with the supplying utility is essential to determine whether these rates are a reasonable option for a specific application.

Understanding Electric Rates. To illustrate a typical commercial electric rate with a ratchet, electricity consumption and demand data for an example building are presented in Table 5.

The example building in Table 5 is on a ratcheted rate, and bill demand is determined as a percentage of actual demand in the summer. How the ratchet operates is illustrated in Figure 2.

Table 5 shows that the actual demand in the first six months of 2004 had no effect on the billing demand, and therefore no effect on the dollar amount of the bill. The same is true for the last three months of the year. Because of the ratchet, the billing demand in the first half of 2004 was set the previous summer. Likewise, billing demand for the last half of 2004 and first half of 2005 was set by the peak actual demand of 180 kW in July 2003. This tells the facility manager to pay attention to demand in the summer months (June to September) and that demand is not a factor in the winter (October to May) months for this particular rate. (Note that Atlanta's climate is hot and humid; in other climates, winter electric demand is an important determinant of costs.) Consumption must be monitored all year long.

Understanding the electric rates is key when evaluating the economics of energy conservation projects. Some projects save electrical demand but not consumption; others save mostly consumption but have little effect on demand. Electric rates must be correctly

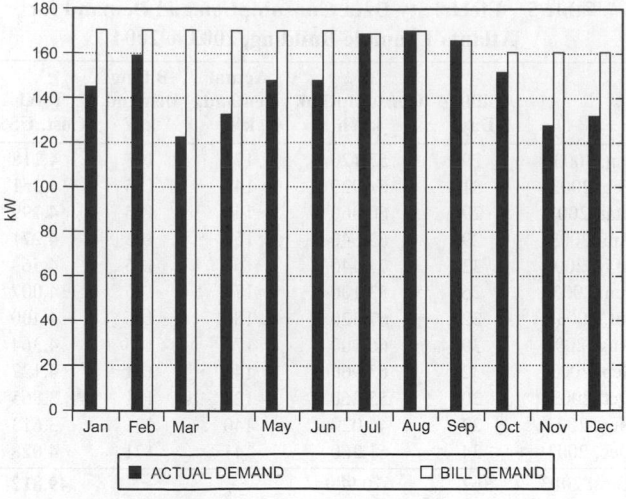

**Fig. 2   Bill Demand and Actual Demand for
Atlanta Example Building, 2004**

applied for economic analyses to be accurate. Chapter 56 contains a thorough discussion of various electric rates.

Natural Gas

Rates. Conventional natural gas rates are usually a combination of two main components: (1) utility rate or base charges for gas consumption and (2) purchased gas adjustment (PGA) charges.

Although gas is usually metered by volume, it is often sold by energy content. The utility rate is the amount the local distribution company charges per unit of energy to deliver the gas to a particular location. This rate may be graduated in steps; the first 100 units of gas consumed may not be the same price as the last 100 units. The PGA is an adjustment for the cost of the gas per unit of energy to the local utility. It is similar to the electric fuel adjustment charge. The total cost per unit is then the sum of the appropriate utility rate and the PGA, plus taxes and other adjustments.

Interruptible Gas Rates and Contract/Transport Gas. Large industrial plants usually have the ability to burn alternative fuels and can qualify for special interruptible gas rates. During peak periods of severe cold weather, these customers' supply may be curtailed by the gas utility, and they may have to switch to propane, fuel oil, or some other back-up fuel. The utility rate and PGA are usually considerably cheaper for these interruptible customers than they are for firm-rate (noninterruptible) customers.

Deregulation of the natural gas industry allows end users to negotiate for gas supplies on the open market. The customer actually contracts with a gas producer or broker and pays for the gas at the source. Transport fees must be negotiated with the pipeline companies carrying the gas to the customer's local gas utility. This can be a very complicated administrative process and is usually economically feasible only for large gas users. Some local utilities have special rates for delivering contract gas volumes through their system; others simply charge a standard utility fee (PGA is not applied because the customer has already negotiated with the supplier for the cost of the fuel itself).

When calculating natural gas bills, be sure to determine which utility rate and PGA and/or contract gas price is appropriate for the particular interruptible or firm-rate customer. As with electric bills, the final calculation should include any taxes, prompt payment discounts, or other applicable adjustments.

Other Fossil Fuels

Propane, fuel oil, and diesel are examples of other fossil fuels in widespread use. Calculating the cost of these fuels is usually much simpler than calculating typical utility rates.

The cost of the fuel itself is usually a simple charge per unit volume or per unit mass. The customer is free to negotiate for the best price. However, trucking or delivery fees must also be included in final calculations. Some customers may have their own transport trucks, but most seek the best delivered price. If storage tanks are not customer-owned, rental fees must be considered. Periodic replacement of diesel-type fuels may be necessary because of storage or shelf-life limitations and must also be considered. The final fuel cost calculation should include any of these costs that are applicable, as well as appropriate taxes.

It is usually difficult, however, to relate usage of stored fossil fuels (e.g., fuel oil) with their operating costs. This is because propane or fuel oil is bought in bulk and stored until needed, and normally not metered or measured as it is consumed, whereas natural gas and power are metered and billed for as they are used.

Energy Source Choices

In planning for a new facility, the designer may undertake **energy master planning**. One component of energy master planning is choice of fuels. Typical necessary decisions include, for example, whether the building should be heated by electricity or natural gas, how service hot water should be produced, whether a hybrid heating plant (i.e., a combination of both electric and gas boilers) should be considered, and whether emergency generators should be fueled by diesel or natural gas.

Decision makers should consider histories or forecasts of price volatility when selecting energy sources. In addition to national trending, local energy price trends from energy suppliers can be informative. These evaluations are particularly important where relative operating costs parity exists between various fuel options, or where selecting more efficient equipment may help mitigate utility price concerns.

Many sources of historic and projected energy costs are available for reference. In addition to federal projections, utility and energy supplier annual reports and accompanying financial data may provide insight into future energy costs. Indicators such as constrained or declining energy supply or production may be key factors in projecting future energy pricing trends. Pricing patterns that suggest unusual levels of energy price volatility should be carefully analyzed and tested at extreme predicted price levels to assess potential effects on system operating costs.

Under conditions of rapidly evolving energy prices or new pricing options, imminent technological improvements, or pending environmental standards and mandates, the adaptability of design options must be carefully evaluated. Where appropriate, contingency planning for accommodating foreseeable alterations to building systems may be prudent. Using diverse energy sources or suppliers in lieu of single sourcing may reduce cost of shifting energy use in the event that single-source pricing becomes volatile, and may even provide negotiating leverage for facility owners.

Water and Sewer Costs

Water and sewer costs have risen in many parts of the country, and should not be overlooked in economic analyses. Fortunately, these rates are usually very simple and straightforward: commonly, a charge per hundred cubic feet (CCF) for water and a different charge per CCF for sewer. Because water consumption is metered and sewage is not, most rates use the water consumption quantity to compute the sewer charge. If an owner uses water that is not returned to sewer, there may be an opportunity to receive a credit or refund. Owners frequently use irrigation meters for watering grounds when the water authority has a special irrigation rate with no sewer charge. Another opportunity that is sometimes overlooked is to separately meter makeup water for cooling towers. This can be done with an irrigation meter if the costs of setting the meter can be justified; alternatively, it may be done by installing an in-line water meter for the cooling tower, in which case the owner reports the usage annually and applies for a credit or refund.

Because of rising costs of water and sewer, water recycling and reclamation is becoming more cost effective. For example, it may now be cost effective in some circumstances to capture cooling coil condensate and pump it to a cooling tower for makeup water.

MAINTENANCE COSTS

The quality of maintenance and maintenance supervision can be a major factor in overall life-cycle cost of a mechanical system. The maintenance cost of mechanical systems varies widely depending upon configuration, equipment locations, accessibility, system complexity, service duty, geography, and system reliability requirements. Maintenance costs can be difficult to predict, because each system or facility is unique.

Dohrmann and Alereza (1986) obtained maintenance costs and HVAC system information from 342 buildings located in 35 states in the United States. In 1983 U.S. dollars, data collected showed a mean HVAC system maintenance cost of $0.32/ft^2 per year, with a median cost of $0.24/ft^2 per year. Building age has a statistically significant but minor effect on HVAC maintenance costs. Analysis also indicated that building size is not statistically significant in explaining cost variation. The type of maintenance program or service agency that building management chooses can also have a significant effect on total HVAC maintenance costs. Although extensive or thorough routine and preventive maintenance programs cost more to administer, they usually extend equipment life; improve reliability; and reduce system downtime, energy costs, and overall life-cycle costs.

Some maintenance cost data are available, both in the public domain and from proprietary sources used by various commercial service providers. These sources may include equipment manufacturers, independent service providers, insurers, government agencies (e.g., the U.S. General Services Administration), and industry-related organizations [e.g., the Building Owners and Managers Association (BOMA)] and service industry publications. More traditional, widely used products and components are likely to have statistically reliable records. However, design changes or modifications necessitated by industry changes, such as alternative refrigerants, may make historical data less relevant.

Newer HVAC products, components, system configurations, control systems and protocols, and upgraded or revised system applications present an additional challenge. Care is required when using data not drawn from broad experience or field reports. In many cases, maintenance information is proprietary or was sponsored by a particular entity or group. Particular care should be taken when using such data. It is the user's responsibility to obtain these data and to determine their appropriateness and suitability for the application being considered.

ASHRAE research project TRP-1237 (Abramson et al. 2005) developed a standardized Internet-based data collection tool and database on HVAC equipment service life and maintenance costs. The database was seeded with data on 163 buildings from around the country. Maintenance cost data were gathered for total HVAC system maintenance costs from 100 facilities. In 2004 dollars, the mean HVAC maintenance cost from these data was $0.47/ft^2, and the median cost was $0.44/ft^2. Table 6 compares these figures with estimates reported by Dohrmann and Alereza (1983), both in terms of contemporary dollars, and in 2004 dollars, and shows that the cost per square foot varies widely between studies.

Table 6   Comparison of Maintenance Costs Between Studies

Survey	Cost per ft^2, as Reported		Consumer Price Index	Cost per ft^2, 2004 Dollars	
	Mean	Median		Mean	Median
Dohrmann and Alereza (1983)	$0.32	$0.24	99.6	$0.61	$0.46
Abramson et al. (2005)	$0.47	$0.44	188.9	$0.47	$0.44

Estimating Maintenance Costs

Total HVAC maintenance cost for new and existing buildings with various types of equipment may be estimated several ways, using several resources. Equipment maintenance requirements can be obtained from the equipment manufacturers for large or custom pieces of equipment. Estimating in-house labor requirements can be difficult; BOMA (2003) provides guidance on this topic. Many independent mechanical service companies provide preventative maintenance contracts. These firms typically have proprietary estimating programs developed through their experience, and often provide generalized maintenance costs to engineers and owners upon request, without obligation.

When evaluating various HVAC systems during design or retrofit, the absolute magnitude of maintenance costs may not be as important as the relative costs. Whichever estimating method or resource is selected, it should be used consistently throughout any evaluation. Mixing information from different resources in an evaluation may provide erroneous results.

Applying simple costs per unit of building floor area for maintenance is highly discouraged. Maintenance costs can be generalized by system types. When projecting maintenance costs for different HVAC systems, the major system components need to be identified with a required level of maintenance. The potential long-term costs of environmental issues on maintenance costs should also be considered.

Factors Affecting Maintenance Costs

Maintenance costs are primarily a measure of labor activity. System design, layout, and configuration can significantly affect the amount of time and effort required for maintenance and, therefore, the maintenance cost. Factors to consider when evaluating maintenance costs include the following:

- **Quantity and type of equipment.** Each piece of equipment requires a core amount of maintenance and time, regardless of its size or capacity. A greater number of similar pieces of equipment are generally more expensive to maintain than larger but fewer units. For example, one manufacturer suggests the annual maintenance for centrifugal chillers is 24 h for a nominal 1000 ton chiller and 16 h for a nominal 500 ton chiller. Therefore, the total maintenance labor for a 1000 ton chiller plant with two 500 ton chillers would be 32 h, or 1/3 more than a single 1000 to chiller.

- **Equipment location and access.** The ability to maintain equipment in a repeatable and cost-effective manner is significantly affected by the equipment's location and accessibility. Equipment that is difficult to access increases the amount of time required to maintain it, and therefore increases maintenance cost. Equipment maintenance requiring erection of ladders and scaffolding or hydraulic lifts increases maintenance costs while likely reducing the quantity and quality of maintenance performed. Equipment location may also dictate an unusual working condition that could require more service personnel than normal. For example, maintenance performed in a confined space (per OSHA definitions) requires an additional person to be present, for safety reasons.

- **System run time.** The number of hours of operation for a HVAC system affects maintenance costs. Many maintenance tasks are dictated by equipment run time. The greater the run time, the more often these tasks need to be performed.

- **Critical systems.** High-reliability systems require more maintenance to ensure uninterrupted system operation. Critical system maintenance is also usually performed with stringent shutdown and failsafe procedures that tend to increase the amount of time required to service equipment. An office building system can be turned off for a short time with little effect on occupants, allowing maintenance almost any time. Shutdown of a hospital operating room or pharmaceutical manufacturing HVAC system, on the other hand, must be coordinated closely with the operation of the facility to eliminate risk to patients or product. Maintenance on critical systems may sometimes incur labor premiums because of unusual shutdown requirements.

- **System complexity.** More complex systems tend to involve more equipment and sophisticated controls. Highly sophisticated systems may require highly skilled service personnel, who tend to be more costly.

- **Service environment.** HVAC systems subjected to harsh operating conditions (e.g., coastal and marine environments) or environments like industrial operations may require more frequent and/or additional maintenance.

- **Local conditions.** The physical location of the facility may require additional maintenance. Equipment in dusty or dirty areas or exposed to seasonal conditions (e.g., high pollen, leaves) may require more frequent or more difficult cleaning of equipment and filters. Additional maintenance tasks may be needed.

- **Geographical location.** Maintenance costs for remote locations must consider the cost of getting to and from the locations. Labor costs for the number of anticipated trips and their duration for either in-house or outsourced service personnel to travel to and from the site must be added to the maintenance cost to properly estimate the total maintenance cost.

- **Equipment age.** The effect of age on equipment repair costs varies significantly by type of HVAC equipment. Technologies in equipment design and application have changed significantly, affecting maintenance costs.

- **Available infrastructure.** Maintenance costs are affected by the availability of an infrastructure that can maintain equipment, components, and systems. Available infrastructure varies on a national, regional, and local basis and is an important consideration in the HVAC system selection process.

REFRIGERANT PHASEOUTS

Production phaseout of many commonly used refrigerants has required building owners to decide between, replacing existing equipment or retrofitting for alternative refrigerants. Several factors must be considered, including

- **Initial Cost.** New equipment may have a significantly higher installed cost than retrofitting existing equipment. For example, retrofitting an existing centrifugal chiller to operate on R-123 may cost 50% of the cost for a new chiller, making the installation cost of a new chiller seem a prudent alternative. Conversely, the cost of rigging a new unit may significantly raise the installed cost, improving the first-cost advantage of refrigerant conversion.

- **Operating Costs.** The overall efficiency of new equipment is often substantially better than that of existing equipment, depending on age, usage, and level of maintenance performed over the life of the existing unit. In addition, conversion to alternative refrigerants may reduce capacity and/or efficiency of the existing equipment.

- **Maintenance Costs.** The maintenance cost for new equipment is generally lower than that for existing equipment. However, the level of retrofit required to attain compatibility between existing equipment and new refrigerant often includes replacement or remanufacture of major unit components, which can bring the maintenance and repair costs in line with those expected of new equipment.

- **Equipment Useful Life.** The effect of a retrofit on equipment useful life is determined by the extent of modification required. Complete remanufacture of a unit should extend the remaining useful life to a level comparable to that of new equipment.

Replacing existing equipment or converting to alternative refrigerants can improve overall system efficiency. Reduced capacity requirements and introduction of new technologies such as variable-speed drives and microprocessor-based controllers can substantially reduce annual operating costs and significantly improve a project's economic benefit.

Information should be gathered to complete Table 1 for each alternative. The techniques described in the section on Economic Analysis Techniques may then be applied to compare the relative values of each option.

Other Sources

The DOE's Federal Energy Management Program (FEMP) Web site (www.eere.energy.gov/femp/procurement) has up-to-date information on energy-efficient federal procurement. Products that qualify for the EPA/DOE ENERGY STAR label are listed, as are efficiency recommendations, cost effectiveness examples, and purchasing guidance. FEMP also provides Web-based cost-calculator tools that simplify the energy cost comparison between products with different efficiencies.

The General Services Administration (GSA) has a basic ordering agreement (BOA) that offers a streamlined procurement method for some HVAC products based on lowest life-cycle cost. For chillers purchased through commercial sources, the BOA can still be used as a guide in preparing specifications.

OTHER ISSUES

Financing Alternatives

Alternative financing is commonly used in third-party funding of projects, particularly retrofit projects, and is variously called privatization, third-party financing, energy services outsourcing, performance contracting, energy savings performance contracting (ESPC), or innovative financing. In these programs, an outside party performs an energy study to identify or quantify attractive energy-saving retrofit projects and then (to varying degrees) designs, builds, and finances the retrofit program on behalf of the owner or host facility. These contracts range in complexity from simple projects such as lighting upgrades to more detailed projects involving all aspects of energy consumption and facility operation.

Alternative financing can be used to accomplish any or all of the following objectives:

- Upgrade capital equipment
- Provide for maintenance of existing facilities
- Speed project implementation
- Conserve or defer capital outlay
- Save energy
- Save money

The benefits of alternative financing are not free. In general terms, these financing agreements transfer the risk of attaining future savings from the owner to the contractor, for which the contractor is paid. In addition, these innovative owning and operating cost reduction approaches have important tax consequences that should be investigated on a case-by-case basis.

There are many variations of the basic arrangements and nearly as many terms to define them. Common nomenclature includes guaranteed savings (performance-based), shared savings, paid from savings, guaranteed savings loans, capital leases, municipal leases, and operating leases. For more information, refer to the U.S.

Department of Energy's Web site and to the DOE (2007). A few examples of alternative financing techniques follow.

Leasing. Among the most common methods of alternative financing is the lease arrangement. In a true lease or lease-purchase arrangement, outside financing provides capital for construction of a facility. The institution then leases the facility at a fixed monthly charge and assumes responsibility for fuel and personnel costs associated with its operation. Leasing is also commonly available for individual pieces of equipment or retrofit systems and often includes all design and installation costs. Equipment suppliers or independent third parties retain ownership of new equipment and lease it to the user.

Outsourcing. For a cogeneration, steam, or chilled-water plant, either a lease or an energy output contract can be used. An energy output contract enables a private company to provide all the capital and operating costs, such as personnel and fuel, while the host facility purchases energy from the operating company at a variable monthly charge.

Energy Savings. Retrofit projects that lower energy usage create an income stream that can be used to amortize the investment. In **paid-from-savings** programs, utility payments remain constant over a period of years while the contractor is paid out of savings until the project is amortized. In **shared savings** programs, the institution receives a percentage of savings over a longer period of years until the project becomes its property. In a **guaranteed savings** program, the owner retains all the savings and is guaranteed that a certain level of savings will be attained. A portion of the savings is used to amortize the project. In any type of energy savings project, building operation and use can strongly affect the amount of savings actually realized.

Low-Interest Financing. In this arrangement, the supplier offers equipment with special financing arrangements at below-market interest rates.

Cost Sharing. Several variations of cost-sharing programs exist. In some instances, two or more groups jointly purchase and share new equipment or facilities, thereby increasing use of the equipment and improving the economic benefits for both parties. In other cases, equipment suppliers or independent third parties (such as utilities) who receive an indirect benefit may share part of the equipment or project cost to establish a market foothold for the product.

District Energy Service

District energy service is increasingly available to building owners; district heating and cooling eliminates most on-site heating and cooling equipment. A third party produces treated water or steam and pipes it from a central plant directly to the building. The building owner then pays a metered rate for the energy that is used.

A cost comparison of district energy service versus on-site generation requires careful examination of numerous, often site-specific, factors extending beyond demand and energy charges for fuel. District heating and cooling eliminates or minimizes most costs associated with installation, maintenance, administration, repair, and operation of on-site heating and cooling equipment. Specifically, costs associated with providing water, water treatment, specialized maintenance services, insurance, staff time, space to house on-site equipment, and structural additions needed to support equipment should be considered. Costs associated with auxiliary equipment, which represent 20 to 30% of the total plant annual operating costs, should also be included.

Any analysis that fails to include all the associated costs does not give a clear picture of the building owner's heating and cooling alternatives. In addition to the tangible costs, there are a number of other factors that should be considered, such as convenience, risk, environmental issues, flexibility, and back-up.

On-Site Electrical Power Generation

On-site electrical power generation covers a broad range of applications, from emergency back-up to power for a single piece of equipment to an on-site power plant supplying 100% of the facility's electrical power needs. Various system types and fuel sources are available, but the economic principles described in this chapter apply equally to all of them. Other chapters (e.g., Chapters 7 and 36 of the 2008 *ASHRAE Handbook—HVAC Systems and Equipment*) may be helpful in describing system details.

An economic study of on-site electrical power generation should include consideration of all owning, operating, and maintenance costs. Typically, on-site generation is capital intensive (i.e., high first cost) and therefore requires a high use rate to produce savings adequate to support the investment. High use rates mean high run time, which requires planned maintenance and careful operation.

Owning costs include any related systems required to adapt the building to on-site power generation. Additional equipment is required if the building will also use purchased power from a utility. Costs associated with shared equipment should also be considered. For example, if the power source for the generator is a steam turbine, and a hot-water boiler would otherwise be used to meet the HVAC demand, the boiler would need to be a larger, high-pressure steam boiler with a heat exchanger to meet the hot-water needs. Operation and maintenance costs for the boiler also are increased because of the increased operating hours.

Costs of an initial investment and ongoing inventory of spare parts must also be considered. Most equipment manufacturers provide a recommended spare parts list as well as recommended maintenance schedules, typically daily, weekly, and monthly routine maintenance and periodic major overhauls. Major overhaul frequency depends on equipment use and requires taking the equipment off-line. The cost of either lost building use or the provision of electricity from an alternative source during the shutdown should be considered.

ECONOMIC ANALYSIS TECHNIQUES

Analysis of overall owning and operating costs and comparisons of alternatives require an understanding of the cost of lost opportunities, inflation, and the time value of money. This process of economic analysis of alternatives falls into two general categories: simple payback analysis and detailed economic analyses (life-cycle cost analyses).

A simple payback analysis reveals options that have short versus long paybacks. Often, however, alternatives are similar and have similar paybacks. For a more accurate comparison, a more comprehensive economic analysis is warranted. Many times it is appropriate to have both a simple payback analysis and a detailed economic analysis. The simple payback analysis shows which options should not be considered further, and the detailed economic analysis determines which of the viable options are the strongest. The strongest options can be accepted or further analyzed if they include competing alternatives.

Simple Payback

In the simple payback technique, a projection of the revenue stream, cost savings, and other factors is estimated and compared to the initial capital outlay. This simple technique ignores the cost of borrowing money (interest) and lost opportunity costs. It also ignores inflation and the time value of money.

Example 1. Equipment item 1 costs $10,000 and will save $2000 per year in operating costs; equipment item 2 costs $12,000 and saves $3000 per year. Which item has the best simple payback?

Item 1 $10,000($2000/yr) = 5-year simple payback

Item 2 $12,000/($3000/yr) = 4-year simple payback

Because analysis of equipment for the duration of its realistic life can produce a very different result, the simple payback technique should be used with caution.

More Sophisticated Economic Analysis Methods

Economic analysis should consider details of both positive and negative costs over the analysis period, such as varying inflation rates, capital and interest costs, salvage costs, replacement costs, interest deductions, depreciation allowances, taxes, tax credits, mortgage payments, and all other costs associated with a particular system. See the section on Symbols for definitions of variables.

Present-Value (Present Worth) Analysis. All sophisticated economic analysis methods use the basic principles of present value analysis to account for the time value of money. Therefore, a good understanding of these principles is important.

The total present value (present worth) for any analysis is determined by summing the present worths of all individual items under consideration, both future single-payment items and series of equal future payments. The scenario with the highest present value is the preferred alternative.

Single-Payment Present-Value Analysis. The cost or value of money is a function of the available interest rate and inflation rate. The future value F of a present sum of money P over n periods with compound interest rate i per period is

$$F = P(1 + i)^n \qquad (1)$$

Conversely, the present value or present worth P of a future sum of money F is given by

$$P = F/(1 + i)^n \qquad (2)$$

or

$$P = F \times \text{PWF}(i,n)_{sgl} \qquad (3)$$

where the single-payment present-worth factor $\text{PWF}(i,n)_{sgl}$ is defined as

$$\text{PWF}(i,n)_{sgl} = 1/(1 + i)^n \qquad (4)$$

Example 2. Calculate the value in 10 years at 10% per year interest of a system presently valued at $10,000

$$F = P(1 + i)^n = \$10{,}000\,(1 + 0.1)^{10} = \$25{,}937.42$$

Example 3. Using the present-worth factor for 10% per year interest and an analysis period of 10 years, calculate the present value of a future sum of money valued at $10,000. (Stated another way, determine what sum of money must be invested today at 10% per year interest to yield $10,000 10 years from now.)

$$P = F \times \text{PWF}(i,n)_{sgl}$$
$$P = \$10{,}000 \times 1/(1 + 0.1)^{10}$$
$$= \$3855.43$$

Series of Equal Payments. The present-worth factor for a series of future equal payments (e.g., operating costs) is given by

$$\text{PWF}(i,n)_{ser} = \frac{(1 + i)^n - 1}{i(1 + i)^n} \qquad (5)$$

The present value P of those future equal payments (PMT) is then the product of the present-worth factor and the payment [i.e., $P = \text{PWF}(i,n)_{ser} \times \text{PMT}$].

The number of future equal payments to repay a present value of money is determined by the capital recovery factor (CRF), which is the reciprocal of the present-worth factor for a series of equal payments:

$$\text{CRF} = \text{PMT}/P \qquad (6)$$

$$\text{CRF}(i,n)_r = \frac{i(1 + i)^n}{(1 + i)^n - 1} = \frac{i}{1 - (1 + i)^{-n}} \qquad (7)$$

The CRF is often used to describe periodic uniform mortgage or loan payments.

Note that when payment periods other than annual are to be studied, the interest rate must be expressed per appropriate period. For example, if monthly payments or return on investment are being analyzed, then interest must be expressed per month, not per year, and n must be expressed in months.

Example 4. Determine the present value of an annual operating cost of $1000 per year over 10 years, assuming 10% per year interest rate.

$$\text{PWF}(i,n)_{ser} = [(1 + 0.1)^{10} - 1]/[0.1(1 + 0.1)^{10}] = 6.14$$
$$P = \$1000(6.14) = \$6140$$

Example 5. Determine the uniform monthly mortgage payments for a loan of $100,000 to be repaid over 30 years at 10% per year interest. Because the payment period is monthly, the payback duration is $30(12) = 360$ monthly periods, and the interest rate per period is $0.1/12 = 0.00833$ per month.

$$\text{CRF}(i,n) = 0.00833(1 + 0.00833)^{360}/[(1 + 0.00833)^{360} - 1]$$
$$= 0.008773$$
$$\text{PMT} = P(\text{CRF})$$
$$= \$100{,}000(0.008773)$$
$$= \$877.30 \text{ per month}$$

Improved Payback Analysis. This somewhat more sophisticated payback approach is similar to the simple payback method, except that the cost of money (interest rate, discount rate, etc.) is considered. Solving Equation (7) for n yields the following:

$$n = \frac{\ln[\text{CRF}/(\text{CRF} - i)]}{\ln(1 + i)} \qquad (8)$$

Given known investment amounts and earnings, CRFs can be calculated for the alternative investments. Subsequently, the number of periods until payback has been achieved can be calculated using Equation (8).

Example 6. Compare the years to payback of the same items described in Example 2 if the value of money is 10% per year.

Item 1

$$
\begin{aligned}
\text{cost} &= \$10{,}000 \\
\text{savings} &= \$2000/\text{year} \\
\text{CRF} &= \$2000/\$10{,}000 = 0.2 \\
n &= \ln[0.2/(0.2 - 0.1)]/\ln(1 + 0.1) = 7.3 \text{ years}
\end{aligned}
$$

Item 2

$$
\begin{aligned}
\text{cost} &= \$12{,}000 \\
\text{savings} &= \$3000/\text{year} \\
\text{CRF} &= \$3000/\$12{,}000 = 0.25 \\
n &= \ln[0.25/(0.25 - 0.1)]/\ln(1 + 0.1) = 5.4 \text{ years}
\end{aligned}
$$

If years to payback is the sole criteria for comparison, Item 2 is preferable because the investment is repaid in a shorter period of time.

Accounting for Inflation. Different economic goods may inflate at different rates. Inflation reflects the rise in the real cost of a commodity over time and is separate from the time value of money. Inflation must often be accounted for in an economic evaluation. One way to account for inflation is to substitute effective interest rates that account for inflation into the equations given in this chapter.

The effective interest rate i', sometimes called the real rate, accounts for inflation rate j and interest rate i or discount rate i_d; it can be expressed as follows (Kreider and Kreith 1982):

$$i' = \frac{1+i}{1+j} - 1 = \frac{i-j}{1+j} \qquad (9)$$

Different effective interest rates can be applied to individual components of cost. Projections for future fuel and energy prices are available in the *Annual Supplement to NIST Handbook 135* (Rushing et al. 2010).

Example 7. Determine the present worth P of an annual operating cost of $1000 over 10 years, given a discount rate of 10% per year and an inflation rate of 5% per year.

$$i' = (0.1 - 0.05)/(1 + 0.05) = 0.0476$$

$$\text{PWF}(i',n)_{ser} = \frac{(1 + 0.0476)^{10} - 1}{0.0476(1 + 0.0476)^{10}} = 7.813$$

$$P = \$1000(7.813) = \$7813$$

The following are three common methods of present-value analysis that include life-cycle cost factors (life of equipment, analysis period, discount rate, energy escalation rates, maintenance cost, etc., as shown in Table 1). These comparison techniques rely on the same assumptions and economic analysis theories but display the results in different forms. They also use the same definition of each term. All can be displayed as a single calculation or as a cash flow table using a series of calculations for each year of the analysis period.

Savings-to-Investment Ratio. Most large military-sponsored work and many other U.S. government entities require a savings-to-investment-ratio (SIR) method. Simply put, SIR is the ratio of an option's savings to its costs. This ratio defines the relative economic strength of each option. The higher the ratio, the better the economic strength. If the ratio is less than 1, the measure does not pay for itself within the analysis period. The escalated savings on an annual and a special (nonannual) basis is calculated and discounted. Costs are shown on an annual and special basis for each year over the life of the system or option. Savings and investments are both discounted separately on an annual basis, and then the discounted total cumulative savings is divided by the discounted total cumulative investments (costs). The analysis period is usually the life of the system or equipment being considered.

The SIR is the sum of a series of operation-related savings from a project alternative divided by the sum of its additional investment-related costs. Typically, this is over a period of years (5, 10, or 20 years, or the typical expected life span).

The general equation for the SIR simply rearranges these two terms as a ratio:

$$\text{SIR}_{A:BC} = \frac{\displaystyle\sum_{t=0}^{N} S_t/(1+d)^t}{\displaystyle\sum_{t=0}^{N} I_t/(1+d)^t} \qquad (10)$$

where

$\text{SIR}_{A:BC}$ = ratio of PV savings to additional PV investment costs of (mutually exclusive) alternative A to base case BC

S_t = savings in year t in operational costs attributable to alternative

I_t = investment-related costs in year t attributable to alternative

t = year of occurrence (where 0 is base date)

d = discount rate

N = length of study

A more practical SIR base-case equation for buildings is as follows:

$$\text{SIR}_{A:BC} = \frac{E + W + \text{OM\&R}}{I_o + \text{Repl} - \text{Res}} \qquad (11)$$

where

$\text{SIR}_{A:BC}$ = ratio of operational savings to investment-related additional costs computed for alternative A to base case BC

E = $(E_{BC} - E_A)$, savings in energy costs attributable to alternative relative to base case

W = $(W_{BC} - W_A)$, savings in water costs attributable to alternative

$\text{OM\&R}$ = difference in OM&R costs; $\text{OM\&R}_{BC} - \text{OM\&R}_A$

I_o = additional initial investment cost required for alternative relative to base case; $(I_A - I_{BC})$

Repl = difference in capital replacement costs; $(\text{Repl}_A - \text{Repl}_{BC})$

Res = difference in residual value; $(\text{Res}_A - \text{Res}_{BC})$

where all amounts are in present values.

Example 8: SIR Computation. For this example, the numerator and denominator are defined as follows:

Numerator:
 PV of operational savings attributable to the alternative = $91,030

Denominator:
 PV of additional investment costs required for the alternative = $7,239

Thus,

$$\text{SIR}_{A:BC} = \frac{\$91,030}{\$7,239} = 12.6$$

A ratio of 12.6 means that the energy-conserving design generates an average return of $12.6 for every $1 invested, over and above the minimum required rate of return imposed by the discount rate. The project alternative in this example is clearly cost effective. A ratio of 1.0 indicates that the cost of the investment equals its savings; a ratio of less than 1.0 indicates an uneconomic alternative that would cost more than it would save.

Summary of SIR Method

- An investment is cost effective if its SIR is greater than 1.0; this is equivalent to having net savings greater than zero.
- The SIR is a relative measure; it must be calculated with respect to a designated base case.
- When computing the SIR of an alternative relative to its base case, the same study period and the same discount rate must be used.
- The SIR is useful for evaluating a single project alternative against a base case or for ranking independent project alternatives; it is not useful for evaluating multiple mutually exclusive alternative.

Internal Rate of Return. The internal rate of return (IRR) method calculates a return on investment over the defined analysis period. The annual savings and costs are not discounted, and a cash flow is established for each year of the analysis period, to be used with an initial cost (or value of the loan). Annual recurring and special (nonannual) savings and costs can be used. The cash flow is then discounted until a calculated discount rate is found that yields a net present value of zero. This method assumes savings are reinvested at the same calculated rate of return; therefore, the calculated rates of return can be overstated compared to the actual rates of return.

Another version of this is the **modified** or **adjusted internal rate of return (MIRR or AIRR)**. In this version, reinvested savings are assumed to have a given rate of return on investment, and the financed moneys a given interest rate. The cash flow is then discounted until a calculated discount rate is found that yields a net present value of zero. This method gives a more realistic indication of expected return on investment, but the difference between alternatives can be small.

The most straightforward method of calculating the AIRR requires that the SIR for a project (relative to its base case) be calculated first. Then the AIRR can be computed easily using the following equation:

$$\text{AIRR} = (1 + r)(\text{SIR})^{1/N} - 1 \qquad (12)$$

Table 7   Two Alternative LCC Examples

Alternative 1: Purchase Chilled Water from Utility

					Year					
---	0	1	2	3	4	5	6	7	8	9	10
First costs		—	—	—	—	—	—	—	—	—	—
Chilled-water costs		$65,250	$66,881	$68,553	$70,267	$72,024	$73,824	$75,670	$77,562	$79,501	$81,488
Replacement costs		—	—	—	—	—	—	—	—	—	—
Maintenance costs		—	—	—	—	—	—	—	—	—	—
Net annual cash flow		65,250	66,881	68,553	70,267	72,024	73,824	75,670	77,501	79,501	81,488
Present value of cash flow		60,417	57,340	54,420	51,648	49,018	46,522	44,153	41,904	39,770	37,745

	Year									
---	11	12	13	14	15	16	17	18	19	20
Financing annual payments	—	—	—	—	—	—	—	—	—	—
Chilled-water costs	$83,526	$85,614	$87,754	$89,948	$92,197	$94,501	$96,864	$99,286	$101,768	$104,312
Replacement costs	—	—	—	—	—	—	—	—	—	—
Maintenance costs	—	—	—	—	—	—	—	—	—	—
Net annual cash flow	83,526	85,614	87,754	89,948	92,197	94,501	96,864	99,286	101,768	104,312
Present value of cash flow	35,823	33,998	32,267	30,624	29,064	27,584	26,179	24,846	23,581	22,380

20-year life-cycle cost $769,823

Alternative 2: Install Chiller and Tower

					Year					
---	0	1	2	3	4	5	6	7	8	9	10
First costs	$220,000	—	—	—	—	—	—	—	—	—	—
Energy costs		$18,750	$19,688	$20,672	$21,705	$22,791	$23,930	$25,127	$26,383	$27,702	$29,087
Replacement costs		—	—	—	—	—	—	—	—	—	90,000
Maintenance costs		15,200	15,656	16,126	16,609	17,108	17,621	18,150	18,694	19,255	19,833
Net annual cash flow	220,000	33,950	35,344	36,798	38,315	39,898	41,551	43,276	45,077	46,957	138,920
Present value of cash flow	220,000	31,435	30,301	29,211	28,163	27,154	26,184	25,251	24,354	23,490	64,347

	Year									
---	11	12	13	14	15	16	17	18	19	20
Financing annual payments	—	—	—	—	—	—	—	—	—	—
Energy costs	$30,542	$32,069	$33,672	$35,356	$37,124	$38,980	$40,929	$42,975	$45,124	$47,380
Replacement costs	—	—	—	—	—	—	—	—	—	—
Maintenance costs	20,428	21,040	21,672	22,322	22,991	23,681	24,392	25,123	25,877	26,653
Net annual cash flow	50,969	53,109	55,344	57,678	60,115	62,661	65,320	68,099	71,001	74,034
Present value of cash flow	21,860	21,090	20,350	19,637	18,951	18,290	17,654	17,042	16,452	15,884

20-year life-cycle cost $717,100

where r is the reinvestment rate and N is the number of years in the study period. Using the SIR of 12.6 from Equation (10) and a reinvestment rate of 3% [the minimum acceptable rate of return (MARR)], the AIRR is found as follows:

$$\text{AIRR}_{A:BC} = (1 + 0.03)(12.6)^{1/20} - 1 = 0.1691$$

Because an AIRR of 16.9% for the alternative is greater than the MARR, which in this example is the FEMP discount rate of 3%, the project alternative is considered to be cost effective in this application.

Life-Cycle Costs. This method of analysis compares the cumulative total of implementation, operating, and maintenance costs. The total costs are discounted over the life of the system or over the loan repayment period. The costs and investments are both discounted and displayed as a total combined life-cycle cost at the end of the analysis period. The options are compared to determine which has the lowest total cost over the anticipated project life.

Example 9. A municipality is evaluating two different methods of providing chilled water for cooling a government office building: purchasing chilled water from a central chilled-water utility service in the area, or installing a conventional chiller plant. Because the municipality is not a tax-paying entity, the evaluation does not need to consider taxes, allowing for either a current or constant dollar analysis.

The first-year price of the chilled-water utility service contract is $65,250 per year, and is expected to increase at a rate of 2.5% per year.

The chiller and cooling tower would cost $220,000, with an expected life of 20 years. A major overhaul ($90,000) of the chiller is expected to occur in year ten. Annual costs for preventative maintenance ($1400), labor ($10,000), water ($2000) and chemical treatments ($1800) are all expected to keep pace with inflation, which is estimated to average 3% annually over the study period. The annual electric cost ($18,750) is expected to increase at a rate of 5% per year. The municipality uses a discount rate of 8% to evaluate financial decisions.

Which option has the lowest life-cycle cost?

Solution. Table 7 compares the two alternatives. For the values provided, alternative 1 has a 20-year life cycle cost of $769,283 and alternative 2 has a 20-year life cycle cost of $717,100. If LCC is the only basis for the decision, alternative 2 is preferable because it has the lower life-cycle cost.

Computer Analysis

Many computer programs are available that incorporate economic analysis methods. These range from simple macros developed for popular spreadsheet applications to more comprehensive, menu-driven computer programs. Commonly used examples of the

Table 8 Commonly Used Discount Formulas

Name	Algebraic Form[a,b]	Name	Algebraic Form[a,b]
Single compound-amount (SCA) equation	$F = P \cdot [(1 + d)^n]$	Uniform compound-amount (UCA) equation	$F = A \cdot \left[\dfrac{(1 + d)^n - 1}{d} \right]$
Single present-value (SPV) equation	$P = F \cdot \left[\dfrac{1}{(1 + d)^n} \right]$	Uniform present-value (UPV) equation	$P = A \cdot \left[\dfrac{(1 + d)^n - 1}{d(1 + d)^n} \right]$
Uniform sinking-fund (USF) equation	$A = F \cdot \left[\dfrac{d}{(1 + d)^n - 1} \right]$	Modified uniform present-value (UPV*) equation	$P = A_0 \cdot \left(\dfrac{1 + e}{d - e} \right) \cdot \left[1 - \left(\dfrac{1 + e}{1 + d} \right)^n \right]$
Uniform capital recovery (UCR) equation	$A = P \cdot \left[\dfrac{d(1 + d)^n}{(1 + d)^n - 1} \right]$		

where

A = end-of-period payment (or receipt) in a uniform series of payments (or receipts) over n periods at d interest or discount rate

A_0 = initial value of a periodic payment (receipt) evaluated at beginning of study period

$A_t = A_0(1 + e)^t$, where $t = 1, \ldots, n$

d = interest or discount rate

e = price escalation rate per period

Source: NIST *Handbook* 135 (Fuller and Petersen 1996).

[a]Note that the USF, UCR, UCA, and UPV equations yield undefined answers when $d = 0$. The correct algebraic forms for this special case would be as follows: USF formula, $A = F/N$; UCR formula, $A = P/N$; UCA formula, $F = An$. The UPV* equation also yields an undefined answer when $e = d$. In this case, $P = A_0 \cdot n$.

[b]The terms by which known values are multiplied are formulas for the factors found in discount factor tables. Using acronyms to represent the factor formulas, the discounting equations can also be written as $F = P \times$ SCA, $P = F \times$ SPV, $A = F \times$ USF, $A = P \times$ UCR, $F = $ UCA, $P = A \times$ UPV, and $P = A_0 \times$ UPV*.

latter include Building Life-Cycle Cost (BLCC) and PC-ECON-PACK.

BLCC was developed by the National Institute of Standards and Technology (NIST) for the U.S. Department of Energy (DOE). The program follows criteria established by the Federal Energy Management Program (FEMP) and the Office of Management and Budget (OMB). It is intended for evaluation of energy conservation investments in nonmilitary government buildings; however, it is also appropriate for similar evaluations of commercial facilities.

PC-ECONPACK, developed by the U.S. Army Corps of Engineers for use by the DOD, uses economic criteria established by the OMB. The program performs standardized life-cycle cost calculations such as net present value, equivalent uniform annual cost, SIR, and discounted payback period.

Macros developed for common spreadsheet programs generally contain preprogrammed functions for various life-cycle cost calculations. Although typically not as sophisticated as the menu-driven programs, the macros are easy to install and learn.

Reference Equations

Table 8 lists commonly used discount formulas as addressed by NIST. Refer to NIST *Handbook* 135 (Fuller and Petersen 1996) for detailed discussions.

SYMBOLS

AIRR = modified or adjusted internal rate of return (MIRR or AIRR)
c = cooling system adjustment factor
C = total annual building HVAC maintenance cost
C_e = annual operating cost for energy
$C_{s,assess}$ = assessed system value
$C_{s,init}$ = initial system cost
$C_{s,salv}$ = system salvage value at end of study period
C_y = uniform annualized mechanical system owning, operating, and maintenance costs
CRF = capital recovery factor
CRF(i,n) = capital recovery factor for interest rate i and analysis period n
CRF(i',n) = capital recovery factory for interest rate i' for items other than fuel and analysis period n
CRF(i'',n) = capital recovery factor for fuel interest rate i'' and analysis period n
CRF(i_m,n) = capital recovery factor for loan or mortgage rate i_m and analysis period n
d = distribution system adjustment factor

D_k = depreciation during period k
$D_{k,SL}$ = depreciation during period k from straight-line depreciation method
$D_{k,SD}$ = depreciation during period k from sum-of-digits depreciation method
F = future value of sum of money
h = heating system adjustment factor
i = compound interest rate per period
i_d = discount rate per period
i_m = market mortgage rate
i' = effective interest rate for all but fuel
i'' = effective interest rate for fuel
I = insurance cost per period
ITC = investment tax credit
j = inflation rate per period
j_e = fuel inflation rate per period
k = end of period(s) during which replacement(s), repair(s), depreciation, or interest are calculated
M = maintenance cost per period
n = number of periods under analysis
P = present value of a sum of money
P_k = outstanding principle on loan at end of period k
PMT = future equal payments
PWF = present worth factor
PWF(i_d,k) = present worth factor for discount rate i_d at end of period k
PWF(i',k) = present worth factor for effective interest rate i' at end of period k
PWF(i,n)$_{sgl}$ = single payment present worth factor
PWF(i,n)$_{ser}$ = present worth factor for a series of future equal payments
R_k = net replacement, repair, or disposal costs at end of period k
SIR = savings-to-investment ratio
T_{inc} = net income tax rate
T_{prop} = property tax rate
T_{salv} = tax rate applicable to salvage value of system

REFERENCES

Abramson, B., D. Herman, and L. Wong. 2005. Interactive Web-based owning and operating cost database (TRP-1237). ASHRAE Research Project, *Final Report*.

Akalin, M.T. 1978. Equipment life and maintenance cost survey (RP-186). *ASHRAE Transactions* 84(2):94-106.

BOMA. 2003. *Preventive maintenance and building operation efficiency.* Building Owners and Managers Association, Washington, D.C.

DOE. 2007. The international performances measurement and verification protocol (IPMVP). *Publication* No. DOE/EE-0157. U.S. Department of Energy. http://www.ipmvp.org/.

Dohrmann, D.R. and T. Alereza. 1986. Analysis of survey data on HVAC maintenance costs (RP-382). *ASHRAE Transactions* 92(2A):550-565.

Fuller, S.K. and S.R. Petersen. 1996. Life-cycle costing manual for the federal energy management program, 1995 edition. NIST *Handbook* 135. National Institute of Standards and Technology, Gaithersburg, MD. http://fire.nist.gov/bfrlpubs/build96/art121.html.

Hiller, C.C. 2000. Determining equipment service life. *ASHRAE Journal* 42(8):48-54.

Kreider, J. and F. Kreith. 1982. *Solar heating and cooling: Active and passive design*. McGraw-Hill, New York.

Lovvorn, N.C. and C.C. Hiller. 2002. Heat pump life revisited. *ASHRAE Transactions* 108(2):107-112.

NIST. Building life-cycle cost computer program (BLCC 5.2-04). Available from the U.S. Department of Energy Efficiency and Renewable Energy Federal Energy Management Program, Washington, D.C. http://www.eere.energy.gov/femp/information/download_blcc.cfm#blcc5.

OMB. 1992. Guidelines and discount rates for benefit-cost analysis of federal programs. *Circular* A-94. Office of Management and Budget, Washington, D.C. Available at http://www.whitehouse.gov/OMB/circulars/a094/a094.html.

Rushing, A.S., Kneifel, J.D., and B.C. Lippiatt. 2010. Energy price indices and discount factors for life-cycle cost analysis—2010. *Annual Supplement to NIST Handbook* 135 and *NBS Special Publication* 709. NISTIR 85-3273-25. National Institute of Standards and Technology, Gaithersburg, MD. http://www1.eere.energy.gov/femp/pdfs/ashb10.pdf

BIBLIOGRAPHY

ASHRAE. 1999. HVAC maintenance costs (RP-929). ASHRAE Research Project, *Final Report*.

ASTM. 2004. Standard terminology of building economics. *Standard* E833-04. American Society for Testing and Materials, International, West Conshohocken, PA.

Easton Consultants. 1986. Survey of residential heat pump service life and maintenance Issues. AGA S-77126. American Gas Association, Arlington, VA.

Haberl, J. 1993. Economic calculations for ASHRAE *Handbook*. ESL-TR-93/04-07. Energy Systems Laboratory, Texas A&M University, College Station. http://repository.tamu.edu/bitstream/handle/1969.1/2113/ESL-TR-93-04-07.pdf?sequence=1.

Kurtz, M. 1984. *Handbook of engineering economics: Guide for engineers, technicians, scientists, and managers*. McGraw-Hill, New York.

Lovvorn, N.C. and C.C. Hiller. 1985. A study of heat pump service life. *ASHRAE Transactions* 91(2B):573-588.

Quirin, D.G. 1967. *The capital expenditure decision*. Richard D. Win, Inc., Homewood, IL.

U.S. Department of Commerce, Bureau of Economic Analysis. (Monthly) *Survey of current business*. U.S. Department of Commerce Bureau of Economic Analysis, Washington, D.C. http://www.bea.gov/scb/index.htm.

USA-CERL. 1998. Life cycle cost in design computer program (WinLCCID 98). Available from Building Systems Laboratory, University of Illinois, Urbana-Champaign.

U.S. Department of Labor. 2005. *Annual percent changes from 1913 to present*. Bureau of Labor Statistics. Available from http://www.bls.gov/cpi.

USACE. PC-ECONPACK computer program (ECONPACK 4.0.2). U.S. Army Corps of Engineers, Huntsville, AL. http://www.hnd.usace.army.mil/paxspt/econ/download.aspx.

TESTING, ADJUSTING, AND BALANCING

S YSTEMS that control the environment in a building change with time and use, and must be rebalanced accordingly. The designer must consider initial and supplementary testing and balancing requirements for commissioning. Complete and accurate operating and maintenance instructions that include intent of design and how to test, adjust, and balance the building systems are essential. Building operating personnel must be well-trained, or qualified operating service organizations must be employed to ensure optimum comfort, proper process operations, and economical operation.

This chapter does not suggest which groups or individuals should perform a complete testing, adjusting, and balancing procedure. However, the procedure must produce repeatable results that meet the design intent and the owner's requirements. Overall, one source must be responsible for testing, adjusting, and balancing all systems. As part of this responsibility, the testing organization should check all equipment under field conditions to ensure compliance.

Testing and balancing should be repeated as systems are renovated and changed. Testing boilers and other pressure vessels for compliance with safety codes is not the primary function of the testing and balancing firm; rather, it is to verify and adjust operating conditions in relation to design conditions for flow, temperature, pressure drop, noise, and vibration. ASHRAE Standard 111 details procedures not covered in this chapter.

TERMINOLOGY

Testing, adjusting, and balancing (TAB) is the process of checking and adjusting all environmental systems in a building to produce the design objectives. This process includes (1) balancing air and water distribution systems, (2) adjusting the total system to provide design quantities, (3) electrical measurement, (4) establishing quantitative performance of all equipment, (5) verifying automatic control system operation and sequences of operation, and (6) sound and vibration measurement. These procedures are accomplished by checking installations for conformity to design, measuring and establishing the fluid quantities of the system as required to meet design specifications, and recording and reporting the results.

The following definitions are used in this chapter. Refer to ASHRAE Terminology of HVAC&R (1991) for additional definitions.

Test. Determine quantitative performance of equipment.

Adjust. Regulate the specified fluid flow rate and air patterns at terminal equipment (e.g., reduce fan speed, adjust a damper).

Balance. Proportion flows in the distribution system (submains, branches, and terminals) according to specified design quantities.

Balanced System. A system designed to deliver heat transfer required for occupant comfort or process load at design conditions. A minimum heat transfer of 97% should be provided to the space or

load served at design flow. The flow required for minimum heat transfer establishes the system's flow tolerance. The fluid distribution system should be designed to allow flow to maintain the required tolerance and verify its performance.

Procedure. An approach to and execution of a sequence of work operations to yield repeatable results.

Report forms. Test data sheets arranged in logical order for submission and review. They should also form the permanent record to be used as the basis for any future TAB work.

Terminal. A point where the controlled medium (fluid or energy) enters or leaves the distribution system. In air systems, these may be variable- or constant-volume boxes, registers, grilles, diffusers, louvers, and hoods. In water systems, these may be heat transfer coils, fan-coil units, convectors, or finned-tube radiation or radiant panels.

GENERAL CRITERIA

Effective and efficient TAB requires a systematic, thoroughly planned procedure implemented by experienced and qualified staff. All activities, including organization, calibration of instruments, and execution of the work, should be scheduled. Air-side work must be coordinated with water-side and control work. Preparation includes planning and scheduling all procedures, collecting necessary data (including all change orders), reviewing data, studying the system to be worked on, preparing forms, and making preliminary field inspections.

Air leakage in a conduit (duct) system can significantly reduce performance, so conduits (ducts) must be designed, constructed, and installed to minimize and control leakage. During construction, all duct systems should be sealed and tested for air leakage. Water, steam, and pneumatic piping should be tested for leakage, which can harm people and equipment.

Design Considerations

TAB begins as design functions, with most of the devices required for adjustments being integral parts of the design and installation. To ensure that proper balance can be achieved, the engineer should show and specify a sufficient number of dampers, valves, flow measuring locations, and flow-balancing devices; these must be properly located in required straight lengths of pipe or duct for accurate measurement. Testing depends on system characteristics and layout. Interaction between individual terminals varies with pressures, flow requirements, and control devices.

The design engineer should specify balancing tolerances. Minimum flow tolerances are ±10% for individual terminals and branches in noncritical applications and ±5% for main air ducts. For critical water systems where differential pressures must be maintained, tolerances of ±5% are suggested. For critical air systems, recommendations are the following:

The preparation of this chapter is assigned to TC 7.7, Testing and Balancing.

Positive zones:
Supply air	0 to +10%
Exhaust and return air	0 to −10%

Negative zones:
Supply air	0 to −10%
Exhaust and return air	0 to +10%

Balancing Devices. Balancing devices should be used to provide maximum flow-limiting ability without causing excessive noise. Flow reduction should be uniform over the entire duct or pipe. Single-blade dampers or butterfly balancing valves are not good balancing valves because of the uneven flow pattern at high pressure drops. Pressure drop across equipment is not an accurate flow measurement but can be used to determine whether the manufacturer design pressure is within specified limits. Liberal use of pressure taps at critical points is recommended.

AIR VOLUMETRIC MEASUREMENT METHODS

General

The pitot-tube traverse is the generally accepted method of measuring airflow in ducts; ways to measure airflow at individual terminals are described by manufacturers. The primary objective is to establish repeatable measurement procedures that correlate with the pitot-tube traverse.

Laboratory tests, data, and techniques prescribed by equipment and air terminal manufacturers must be reviewed and checked for accuracy, applicability, and repeatability of results. Conversion factors that correlate field data with laboratory results must be developed to predict the equipment's actual field performance.

Air Devices

All flow-measuring instruments should be field verified by comparing to pitot-tube traverses to establish correction and/or density factors.

Generally, correction factors given by air diffuser manufacturers should be checked for accuracy by field measurement and by comparing actual flow measured by pitot-tube traverse to actual measured velocity. Air diffuser manufacturers usually base their volumetric test measurements on a deflecting vane anemometer. The velocity is multiplied by an empirical effective area to obtain the air diffuser's delivery. Accurate results are obtained by measuring at the vena contracta with the probe of the deflecting vane anemometer. Methods advocated for measuring airflow of troffer-type terminals are similar to those for air diffusers.

A capture hood is frequently used to measure device airflows, primarily of diffusers and slots. Loss coefficients should be established for hood measurements with varying flow and deflection settings. If the air does not fill the measurement grid, the readings will require a correction factor.

Rotating vane anemometers are commonly used to measure airflow from sidewall grilles. Effective areas (correction factors) should be established with the face dampers fully open and deflection set uniformly on all grilles. Correction factors are required when measuring airflow in open ducts [i.e., damper openings and fume hoods (Sauer and Howell 1990)].

Duct Flow

The preferred method of measuring duct volumetric flow is the pitot-tube traverse average. The maximum straight run should be obtained before and after the traverse station. To obtain the best duct velocity profile, measuring points should be located as shown in Chapter 36 of the 2009 *ASHRAE Handbook—Fundamentals* and ASHRAE *Standard* 111. When using factory-fabricated volume-measuring stations, the measurements should be checked against a pitot-tube traverse.

Power input to a fan's driver should be used as only a guide to indicate its delivery; it may also be used to verify performance determined by a reliable method (e.g., pitot-tube traverse of system's main) that considers possible system effects. For some fans, the flow rate is not proportional to the power needed to drive them. In some cases, as with forward-curved-blade fans, the same power is required for two or more flow rates. The backward-curved-blade centrifugal fan is the only type with a flow rate that varies directly with power input.

If an installation has an inadequate straight length of ductwork or no ductwork to allow a pitot-tube traverse, the procedure from Sauer and Howell (1990) can be followed: a vane anemometer reads air velocities at multiple points across the face of a coil to determine a loss coefficient.

Mixture Plenums

Approach conditions are often so unfavorable that the air quantities comprising a mixture (e.g., outdoor and return air) cannot be determined accurately by volumetric measurements. In such cases, the mixture's temperature indicates the balance (proportions) between the component airstreams. Temperatures must be measured carefully to account for stratification, and the difference between outdoor and return temperatures must be greater than 20°F. The temperature of the mixture can be calculated as follows:

$$Q_t t_m = Q_o t_o + Q_r t_r \qquad (1)$$

where

Q_t = total measured air quantity, %
Q_o = outdoor air quantity, %
Q_r = return air quantity, %
t_m = temperature of outdoor and return mixture, °F
t_o = outdoor temperature, °F
t_r = return temperature, °F

Pressure Measurement

Air pressures measured include barometric, static, velocity, total, and differential. For field evaluation of air-handling performance, pressure should be measured per ASHRAE *Standard* 111 and analyzed together with manufacturers' fan curves and system effect as predicted by AMCA *Standard* 210. When measured in the field, pressure readings, air quantity, and power input often do not correlate with manufacturers' certified performance curves unless proper correction is made.

Pressure drops through equipment such as coils, dampers, or filters should not be used to measure airflow. Pressure is an acceptable means of establishing flow volumes only where it is required by, and performed in accordance with, the manufacturer certifying the equipment.

Stratification

Normal design minimizes conditions causing air turbulence, to produce the least friction, resistance, and consequent pressure loss. Under some conditions, however, air turbulence is desirable and necessary. For example, two airstreams of different temperatures can stratify in smooth, uninterrupted flow conditions. In this situation, design should promote mixing. Return and outdoor airstreams at the inlet side of the air-handling unit tend to stratify where enlargement of the inlet plenum or casing size decreases air velocity. Without a deliberate effort to mix the two airstreams (e.g., in cold climates, placing the outdoor air entry at the top of the plenum and return air at the bottom of the plenum to allow natural mixing), stratification can be carried throughout the system (e.g., filter, coils, eliminators, fans, ducts). Stratification can freeze coils and rupture tubes, and can affect temperature control in plenums, spaces, or both.

Stratification can also be reduced by adding vanes to break up and mix the airstreams. No solution to stratification problems is guaranteed; each condition must be evaluated by field measurements and experimentation.

BALANCING PROCEDURES FOR AIR DISTRIBUTION

No one established procedure is applicable to all systems. The bibliography lists sources of additional information.

Instruments for Testing and Balancing

The minimum instruments necessary for air balance are

- Manometer calibrated in 0.005 in. of water divisions
- Combination inclined/vertical manometer (0 to 10 in. of water)
- Pitot tubes in various lengths, as required
- Tachometer (direct-contact, self-timing) or strobe light
- Clamp-on ammeter with voltage scales [root-mean-square (RMS) type]
- Rotating vane anemometer
- Deflecting vane anemometer
- Thermal anemometer
- Capture hood
- Digital thermometers (0.1°F increments as a minimum) and glass stem thermometers (0.1°F graduations minimum)
- Sound level meter with octave band filter set, calibrator, and microphone
- Vibration analyzer capable of measuring displacement velocity and acceleration
- Water flowmeters (0 to 50 in. of water and 0 to 400 in. of water ranges)
- Compound gage
- Test gages (100 psi and 300 psi)
- Sling psychrometer
- Etched-stem thermometer (30 to 120°F in 0.1°F increments)
- Hygrometers
- Relative humidity and dew-point instruments

Instruments must be calibrated periodically to verify their accuracy and repeatability before use in the field.

Preliminary Procedure for Air Balancing

1. Before balancing, all pressure tests (duct leakage) of duct and piping systems must be complete and acceptable.
2. Obtain as-built design drawings and specifications, and become thoroughly acquainted with the design intent.
3. Obtain copies of approved shop drawings of all air-handling equipment, outlets (supply, return, and exhaust), and temperature control diagrams, including performance curves. Compare design requirements with shop drawing capacities.
4. Compare design to installed equipment and field installation.
5. Walk the system from air-handling equipment to terminal units to determine variations of installation from design.
6. Check dampers (both volume and fire) for correct and locked position and temperature control for completeness of installation before starting fans.
7. Prepare report test sheets for both fans and outlets. Obtain manufacturer's outlet factors and recommended test procedure. A summation of required outlet volumes allows cross-checking with required fan volumes.
8. Determine the best locations in the main and branch ductwork for the most accurate duct traverses.
9. Place all outlet dampers in full open position.
10. Prepare schematic diagrams of system as-built ductwork and piping layouts to facilitate reporting.
11. Check filters for cleanliness and proper installation (no air bypass). If specifications require, establish procedure to simulate dirty filters.
12. For variable-volume systems, develop a plan to simulate diversity (if required).

Equipment and System Check

1. All fans (supply, return, and exhaust) must be operating before checking the following items:
   - Motor amperage and voltage to guard against overload
   - Fan rotation
   - Operability of static pressure limit switch
   - Automatic dampers for proper position
   - Air and water controls operating to deliver required temperatures
   - Air leaks in the casing and around the coils and filter frames must be stopped. Note points where piping enters the casing to ensure that escutcheons are right. Do not rely on pipe insulation to seal these openings (insulation may shrink). In prefabricated units, check that all panel-fastening holes are filled.
2. Traverse the main supply ductwork whenever possible. All main branches should also be traversed where duct arrangement allows. Traverse points and method of traverse should be selected as follows:
   - Traverse each main or branch after the longest possible straight run for the duct involved.
   - For test hole spacing, refer to Chapter 36 of the 2009 *ASHRAE Handbook—Fundamentals*.
   - Traverse using a pitot tube and manometer where velocities are over 600 fpm. Below this velocity, use either a micromanometer and pitot tube or an electronic multimeter and a pitot tube.
   - Note temperature and barometric pressure and correct for standard air quantity if needed.
   - After establishing the total air being delivered, adjust fan speed to obtain design airflow, if necessary. Check power and speed to confirm motor power and/or critical fan speed are not exceeded.
   - Proportionally adjust branch dampers until each has the proper air volume.
   - With all dampers and registers in the system proportioned and with supply and return fans operating at or near design airflow, set the minimum outdoor and return air ratio. If duct traverse locations are not available, this can be done by measuring the mixture temperature in the return air, outdoor air louver, and filter section. The mixture temperature may be approximated from Equation (1).

   The greater the temperature difference between outdoor and return air, the easier it is to get accurate damper settings. Take the temperature at many points in a uniform traverse to be sure there is no stratification.

   After the minimum outdoor air damper has been set for the proper percentage of outdoor air, run another traverse of mixture temperatures and install baffling if variation from the average is more than 5%. Remember that stratified mixed-air temperatures vary greatly with outdoor temperature in cold weather, whereas return air temperature has only a minor effect.
3. Balance terminal outlets in each control zone in proportion to each other, as follows:
   - Once the preliminary fan quantity is set, proportion the terminal outlet balance from the outlets into the branches to the fan. Concentrate on proportioning the flow rather than the absolute quantity. As fan settings and branch dampers change, the outlet terminal quantities remain proportional. Branch dampers should be used for major adjusting and terminal dampers for trim or minor adjustment only. It may be necessary to install additional sub-branch dampers to decrease the use of terminal dampers that create objectionable noise.
   - Normally, several passes through the entire system are necessary to obtain proper outlet values.
   - The total tested outlet air quantity compared to duct traverse air quantities may indicate duct leakage.
   - With total design air established in the branches and at the outlets, (1) take new fan motor amperage readings, (2) find static pressure across the fan, (3) read and record static pressure

across each component (intake, filters, coils, mixing dampers), and (4) take a final duct traverse.

Multizone Systems

Balancing should be accomplished as follows:

1. When adjusting multizone constant-volume systems, establish the ratio of the design volume through the cooling coil to total fan volume to achieve the desired diversity factor. Keep the proportion of cold to total air constant during the balance. However, check each zone on full cooling. If the design calls for full flow through the cooling coil, the entire system should be set to full flow through the cooling side while making tests. Perform the same procedure for the hot-air side.
2. Check leaving air temperature at each zone to verify that hot and cold damper inlet leakage is not greater than the established maximum allowable leakage.
3. Check apparatus and main trunks, as outlined in the section on Equipment and System Check.
4. Proportionately balance diffusers or grilles.
5. Change control settings to full heating, and ensure that controls function properly. Verify airflow at each diffuser. Check for stratification.
6. If the engineer included a diversity factor in selecting the main apparatus, it will not be possible to get full flow to all zones simultaneously, as outlined in item 3 under Equipment and System Check. Zones equaling the diversity should be set to heating.

Dual-Duct Systems

Balancing should be accomplished as follows:

1. When adjusting dual-duct constant-volume systems, establish the ratio of the design volume through the cooling coil to total fan volume to achieve the desired diversity factor. Keep the proportion of cold to total air constant during the balance. If the design calls for full flow through the cooling coil, the entire system should be set to full flow through the cooling side while making tests. Perform the same procedure for the hot-air side.
2. Check leaving air temperature at the nearest terminal to verify that hot and cold damper inlet leakage is not greater than the established maximum allowable leakage.
3. Check apparatus and main trunks, as outlined in the section on Equipment and System Check.
4. Determine whether static pressure at the end of the system (the longest duct run) is at or above the minimum required for mixing box operation. Proceed to extreme end of the system and check the static pressure with an inclined manometer. Pressure should exceed the minimum static pressure recommended by the mixing box manufacturer.
5. Proportionately balance diffusers or grilles on the low-pressure side of the box, as described for low-pressure systems in the previous section.
6. Change control settings to full heating, and ensure that the controls and dual-duct boxes function properly. Spot-check airflow at several diffusers. Check for stratification.
7. If the engineer has included a diversity factor in selecting the main apparatus, it will not be possible to get full flow from all boxes simultaneously, as outlined in item 3 under Equipment and System Check. Mixing boxes closest to the fan should be set to full heating to force the air to the end of the system.

VARIABLE-VOLUME SYSTEMS

Many types of variable air volume (VAV) systems have been developed to conserve energy. They can be categorized as pressure-dependent or pressure-independent.

Pressure-dependent systems incorporate air terminal boxes with a thermostat signal controlling a damper actuator. Air volume to the space varies to maintain space temperature; air temperature supplied

to terminal boxes remains constant. The balance of this system constantly varies with loading changes; therefore, any balancing procedure will not produce repeatable data unless changes in system load are simulated by using the same configuration of thermostat settings each time the system is tested (i.e., the same terminal boxes are fixed in the minimum and maximum positions for the test). Each terminal box requires a balancing damper upstream of its inlet.

In a pressure-dependent system with pneumatic controls, setting minimum airflows to the space (other than at no flow) is not suggested unless the terminal box has a normally closed damper and the manufacturer of the damper actuator provides adjustable mechanical stops.

Pressure-independent systems incorporate air terminal boxes with a thermostat signal used as a master control to open or close the damper actuator, and a velocity controller used as a sub-master control to maintain the maximum and minimum amounts of air to be supplied to the space. Air volume to the space varies to maintain the space temperature; air temperature supplied to the terminal remains constant. Take care to verify the operating range of the damper actuator as it responds to the velocity controller to prevent dead bands or overlap of control in response to other system components (e.g., double-duct VAV, fan-powered boxes, retrofit systems). Care should also be taken to verify the action of the thermostat with regard to the damper position, as the velocity controller can change the control signal ratio or reverse the control signal.

The pressure-independent system requires verifying that the velocity controller is operating properly; it can be adversely affected by inlet duct configurations if a multipoint sensor is not used (Griggs et al. 1990). The primary difference between the two systems is that the pressure-dependent system supplies a different amount of air to the space as pressure upstream of the terminal box changes. If the thermostats are not calibrated properly to meet the space load, zones may overcool or overheat. When zones overcool and receive greater amounts of supply air than required, they decrease the amount of air that can be supplied to overheated zones. The pressure-independent system is not affected by improper thermostat calibration in the same way as a pressure-dependent system, because minimum and maximum airflow limits may be set for each zone.

Static Control

Static control saves energy and prevents overpressurizing the duct system. The following procedures and equipment are some of the means used to control static pressure.

Fan Control. Consult ASHRAE *Standard* 90.1.

Discharge Damper. Losses and noise should be considered.

Vortex Damper. Losses from inlet air conditions are a problem, and the vortex damper does not completely close. The minimum expected airflow should be evaluated.

Variable Inlet Cones. System loss can be a problem because the cone does not typically close completely. The minimum expected airflow should be evaluated.

Varying Fan Speed Mechanically. Slippage, loss of belts, cost of belt replacement, and the initial cost of components are concerns.

Variable Pitch-in-Motion Fans. Maintenance and preventing the fan from running in the stall condition must be evaluated.

Varying Fan Speed Electrically. Varying voltage or frequency to the fan motor is usually the most efficient method. Some motor drives may cause electrical noise and affect other devices.

In controlling VAV fan systems, location of the static pressure sensors is critical and should be field-verified to give the most representative point of operation. After the terminal boxes have been proportioned, static pressure control can be verified by observing static pressure changes at the fan discharge and the static pressure sensor as load is simulated from maximum to minimum airflow (i.e., set all terminal boxes to balanced airflow conditions and determine whether any changes in static pressure occur by placing one terminal box at a time to minimum airflow, until all terminals are placed at the minimal airflow setting). The maximum to minimum air volume changes should be within the fan curve performance (speed or total pressure).

Diversity

Diversity may be used on a VAV system, assuming that total airflow is lower by design and that all terminal boxes never fully open at the same time. Duct leakage should be avoided. All ductwork upstream of the terminal box should be considered medium-pressure, whether in a low- or medium-pressure system.

A procedure to test the total air on the system should be established by setting terminal boxes to the zero or minimum position nearest the fan. During peak load conditions, care should be taken to verify that adequate pressure is available upstream of all terminal boxes to achieve design airflow to the spaces.

Outdoor Air Requirements

Maintaining a space under a slight positive or neutral pressure to atmosphere is difficult with all variable-volume systems. In most systems, the exhaust requirement for the space is constant; thus, outdoor air used to equal the exhaust air and meet minimum outdoor air requirements for building codes must also remain constant. Because of the location of the outdoor air intake and pressure changes, this does not usually happen. Outdoor air should enter the fan at a point of constant pressure (i.e., supply fan volume can be controlled by proportional static pressure control, which can control the return air fan volume). Makeup air fans can also be used for outdoor air control.

Return Air Fans

If return air fans are required in series with a supply fan, control and sizing of the fans is most important. Serious over- and under-pressurization can occur, especially during economizer cycles.

Types of VAV Systems

Single-Duct VAV. This system uses a pressure-dependent or -independent terminal and usually has reheat at a minimal setting on the terminal unit, or a separate heating system. Consult ASHRAE *Standard* 90.1 when considering this system.

Bypass. This system uses a pressure-dependent damper, which, on demand for heating, closes the damper to the space and opens to the return air plenum. Bypass sometimes uses a constant bypass airflow or a reduced amount of airflow bypassed to the return plenum in relation to the amount supplied to the space. No economic value can be obtained by varying fan speed with this system. A control problem can exist if any return air sensing is done to control a warm-up or cool-down cycle. Consult ASHRAE *Standard* 90.1 when considering this system.

VAV Using Single-Duct VAV and Fan-Powered, Pressure-Dependent Terminals. This system has a primary source of air from the fan to the terminal and a secondary powered fan source that pulls air from the return air plenum before the additional heat source. In some fan-powered boxes, backdraft dampers allow duct leakage when the system calls for the damper to be fully closed. Typical applications include geographic areas where the ratio of heating hours to cooling hours is low.

Double-Duct VAV. This type of terminal uses two single-duct variable terminals. It is controlled by velocity controllers that operate in sequence so that both hot and cold ducts can be opened or closed. Some controls have a downstream flow sensor in the terminal unit. The total-airflow sensor is in the inlet and controlled by the thermostat. As this inlet damper closes, the downstream controller opens the other damper to maintain set airflow. Low pressure in the decks controlled by the thermostat may cause unwanted mixing of air, which results in excess energy use or discomfort in the space.

Balancing the VAV System

The general procedure for balancing a VAV system is

1. Determine the required maximum air volume to be delivered by the supply and return air fans. Load diversity usually means that the volume will be somewhat less than the outlet total.
2. Obtain fan curves on these units, and request information on surge characteristics from the fan manufacturer.
3. If inlet vortex damper control is used, obtain the fan manufacturer's data on deaeration of the fan when used with the damper. If speed control is used, find the maximum and minimum speed that can be used on the project.
4. Obtain from the manufacturer the minimum and maximum operating pressures for terminal or variable-volume boxes to be used on the project.
5. Construct a theoretical system curve, including an approximate surge area. The system curve starts at the boxes' minimum inlet static pressure, plus system loss at minimum flow, and terminates at the design maximum flow. The operating range using an inlet vane damper is between the surge line intersection with the system curve and the maximum design flow. When variable-speed control is used, the operating range is between (1) the minimum speed that can produce the necessary minimum box static pressure at minimum flow still in the fan's stable range and (2) the maximum speed necessary to obtain maximum design flow.
6. Position the terminal boxes to the proportion of maximum fan air volume to total installed terminal maximum volume.
7. Proportion outlets, and verify design volume with the VAV box on maximum flow. Verify minimum flow setting.
8. Set the fan to operate at approximate design speed.
9. Verify flow at all terminal boxes. Identify which boxes are not in control and the inlet static pressures, if any. If all terminal boxes are in control, reduce the fan flow until the remote terminals are just in control and within the tolerances of filter loading.
10. Run a total air traverse with a pitot tube.
11. Run steps (8) through (10) with the return or exhaust fan set at design flow as measured by a pitot-tube traverse and with the system set on minimum outdoor air.
12. Set terminals to minimum, and adjust the inlet vane or speed controller until minimum static pressure and airflow are obtained.
13. Temperature control personnel, balancing personnel, and the design engineer should agree on final placement of the sensor for the static pressure controller. This sensor must be placed in a representative location in the supply duct to sense average maximum and minimum static pressures in the system.
14. Check return air fan speed or its inlet vane damper, which tracks or adjusts to the supply fan airflow, to ensure proper outdoor air volume.
15. Operate the system on 100% outdoor air (weather permitting), and check supply and return fans for proper power and static pressure.

Induction Systems

Most induction systems use high-velocity air distribution. Balancing should be accomplished as follows:

1. For apparatus and main trunk capacities, perform general VAV balancing procedures.
2. Determine primary airflow at each terminal unit by reading the unit plenum pressure with a manometer and locating the point on the charts (or curves) of air quantity versus static pressure supplied by the unit manufacturer.
3. Normally, about three complete passes around the entire system are required for proper adjustment. Make a final pass without adjustments to record the end result.

4. To provide the quietest possible operation, adjust the fan to run at the slowest speed that provides sufficient nozzle pressure to all units with minimum throttling of all unit and riser dampers.

5. After balancing each induction system with minimum outdoor air, reposition to allow maximum outdoor air and check power and static pressure readings.

Report Information

To be of value to the consulting engineer and owner's maintenance department, the air-handling report should consist of at least the following items:

1. *Design*
   - Air quantity to be delivered
   - Fan static pressure
   - Motor power installed or required
   - Percent of outdoor air under minimum conditions
   - Fan speed
   - Input power required to obtain this air quantity at design static pressure

2. *Installation*
   - Equipment manufacturer (indicate model and serial numbers)
   - Size of unit installed
   - Arrangement of air-handling unit
   - Nameplate power and voltage, phase, cycles, and full-load amperes of installed motor

3. *Field tests*
   - Fan speed
   - Power readings (voltage, amperes of all phases at motor terminals)
   - Total pressure differential across unit components
   - Fan suction and fan discharge static pressure (equals fan total pressure)
   - Plot of actual readings on manufacturer's fan performance curve to show the installed fan operating point
   - Measured airflow rate

   It is important to establish initial static pressures accurately for the air treatment equipment and duct system so that the variation in air quantity caused by filter loading can be calculated. It enables the designer to ensure that the total air quantity never is less than the minimum requirements. Because the design air quantity for peak loading of the filters has already been calculated, it also serves as a check of dirt loading in coils.

4. *Terminal Outlets*
   - Outlet by room designation and position
   - Manufacture and type
   - Size (using manufacturer's designation to ensure proper factor)
   - Manufacturer's outlet factor [where no factors are available, or field tests indicate listed factors are incorrect, a factor must be determined in the field by traverse of a duct leading to a single outlet; this also applies to capture hood readouts (see ASHRAE *Standard* 111)]
   - Adjustment pattern for every air terminal

5. *Additional Information (if applicable)*
   - Air-handling units
     - Belt number and size
     - Drive and driven sheave size
     - Belt position on adjusted drive sheaves (bottom, middle, and top)
     - Motor speed under full load
     - Motor heater size
     - Filter type and static pressure at initial use and full load; time to replace
     - Variations of velocity at various points across face of coil
     - Existence of vortex or discharge dampers, or both

- Distribution system
  - Unusual duct arrangements
  - Branch duct static readings in double-duct and induction system
  - Ceiling pressure readings where plenum ceiling distribution is used; tightness of ceiling
  - With wind conditions outside less than 5 mph, relationship of building to outdoor pressure under both minimum and maximum outdoor air
  - Induction unit manufacturer and size (including required air quantity and plenum pressures for each unit) and test plenum pressure and resulting primary air delivery from manufacturer's listed curves
- All equipment nameplates visible and easily readable

Many independent firms have developed detailed procedures suitable to their own operations and the area in which they function. These procedures are often available for information and evaluation on request.

PRINCIPLES AND PROCEDURES FOR BALANCING HYDRONIC SYSTEMS

Heat Transfer at Reduced Flow Rate

The typical heating-only hydronic terminal (200°F, 20°F Δt) gradually reduces heat output as flow is reduced (Figure 1). Decreasing water flow to 20% of design reduces heat transfer to 65% of that at full design flow. The control valve must reduce water flow to 10% to reduce heat output to 50%. This relative insensitivity to changing flow rates is because the governing coefficient for heat transfer is the air-side coefficient; a change in internal or water-side coefficient with flow rate does not materially affect the overall heat transfer coefficient. This means (1) heat transfer for water-to-air terminals is established by the mean air-to-water temperature difference, (2) heat transfer is measurably changed, and (3) a change in mean water temperature requires a greater change in water flow rate.

Tests of hydronic coil performance show that when flow is throttled to the coil, the water-side differential temperature of the coil increases with respect to design selection. This applies to both

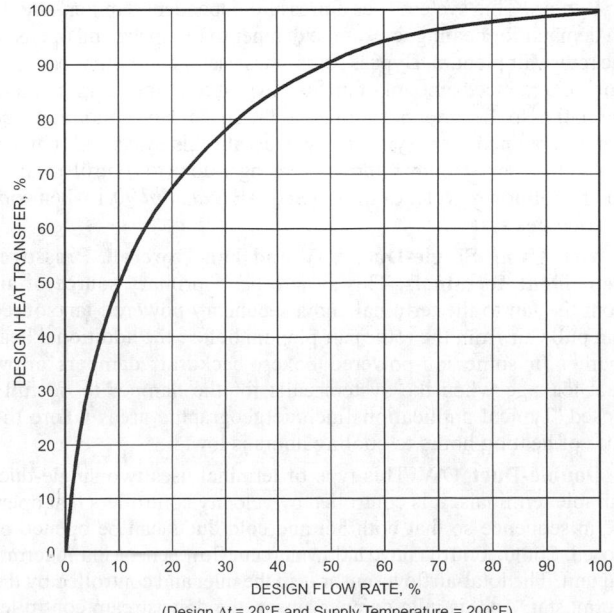

Fig. 1   Effects of Flow Variation on Heat Transfer from a Hydronic Terminal

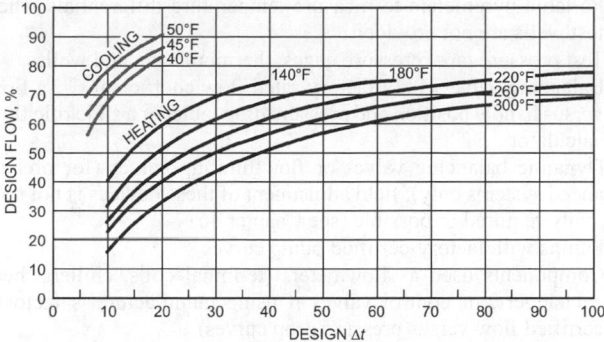

Fig. 2 Percent of Design Flow Versus Design Δt to Maintain 90% Terminal Heat Transfer for Various Supply Water Temperatures

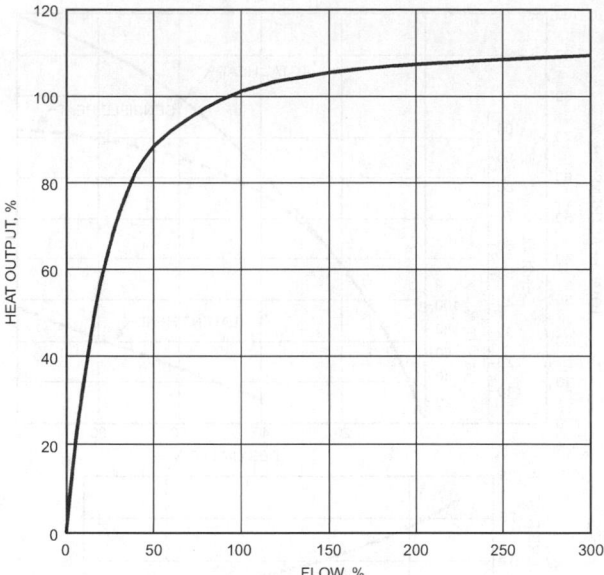

Fig. 3 Typical Heating Coil Heat Transfer Versus Water Flow

Table 1 Load Flow Variation

	% Design Flow at 90% Load	Other Load, Order of %		
Load Type		**Sensible**	**Total**	**Latent**
Sensible	65	90	84	58
Total	75	95	90	65
Latent	90	98	95	90

Note: Dual-temperature systems are designed to chilled flow requirements and often operate on a 10°F temperature drop at full-load heating.

constant-volume and variable-volume air-handling units. In constantly circulated coils that control temperature by changing coil entering water temperature, decreasing source flow to the circuit decreases the water-side differential temperature.

A secondary concern applies to heating terminals. Unlike chilled water, hot water can be supplied at a wide range of temperatures. Inadequate terminal heating capacity caused by insufficient flow can sometimes be overcome by raising supply water temperature. Design below the 250°F limit (ASME low-pressure boiler code) must be considered.

Figure 2 shows the flow variation when 90% terminal capacity is acceptable. Note that heating tolerance decreases with temperature and flow rates and that chilled-water terminals are much less tolerant of flow variation than hot-water terminals.

Dual-temperature heating/cooling hydronic systems are sometimes first started during the heating season. Adequate heating ability in the terminals may suggest that the system is balanced. Figure 2 shows that 40% of design flow through the terminal provides 90% of design heating with 140°F supply water and a 10°F temperature drop. Increased supply water temperature establishes the same heat transfer at terminal flow rates of less than 40% design.

Sometimes, dual-temperature water systems have decreased flow during the cooling season because of chiller pressure drop; this could cause a flow reduction of 25%. For example, during the cooling season, a terminal that heated satisfactorily would only receive 30% of the design flow rate.

Although the example of reduced flow rate at $\Delta t = 20°F$ only affects heat transfer by 10%, this reduced heat transfer rate may have the following negative effects:

- Object of the system is to deliver (or remove) heat where required. When flow is reduced from design rate, the system must supply heating or cooling for a longer period to maintain room temperature.
- As load reaches design conditions, the reduced flow rate is unable to maintain room design conditions.
- Control valves with average range ability (30:1) and reasonable authority ($\beta = 0.5$) may act as on/off controllers instead of throttling flows to the terminal. The resultant change in riser friction loss may cause overflow or underflow in other system terminals. Attempting to throttle may cause wear on the valve plug or seat because of higher velocities at the vena contracta of the valve. In extreme situations, cavitations may occur.

Terminals with lower water temperature drops have greater tolerance for unbalanced conditions. However, larger water flows are necessary, requiring larger pipes, pumps, and pumping cost.

System balance becomes more important in terminals with a large temperature difference. Less water flow is required, which reduces the size of pipes, valves, and pumps, as well as pumping costs.

A more linear emission curve gives better system control. If flow varies by more than 5% at design flow conditions, heat transfer can fall off rapidly, ultimately causing poorer control of the wet-bulb temperature and potentially decreasing system air quality.

Heat Transfer at Excessive Flow

Increasing the flow rate above design in an effort to increase heat transfer requires careful consideration. Figure 3 shows that increasing the flow to 200% of design only increases heat transfer by 6% but increases resistance or pressure drop four times and power by the cube of the original power (pump laws) for a lower design Δt. In coils with larger water-side design Δt, heat transfer can increase.

Generalized Chilled Water Terminal— Heat Transfer Versus Flow

Heat transfer for a typical chilled-water coil in an air duct versus water flow rate is shown in Figure 4. The curves are based on ARI rating points: 45°F inlet water at a 10°F rise with entering air at 80°F db and 67°F wb. The basic curve applies to catalog ratings for lower dry-bulb temperatures providing a consistent entering-air moisture content (e.g., 75°F db, 65°F wb). Changes in inlet water temperature, temperature rise, air velocity, and dry- and wet-bulb temperatures cause terminal performance to deviate from the curves. Figure 4 is only a general representation and does not apply to all chilled-water terminals. Comparing Figure 4 with Figure 1 indicates the similarity of nonlinear heat transfer and flow for both the heating and cooling terminals.

Table 1 shows that if the coil is selected for the load and flow is reduced to 90% of load, three flow variations can satisfy the reduced load at various sensible and latent combinations. Note that the reduction in flow will not maintain a chilled-water design differen-

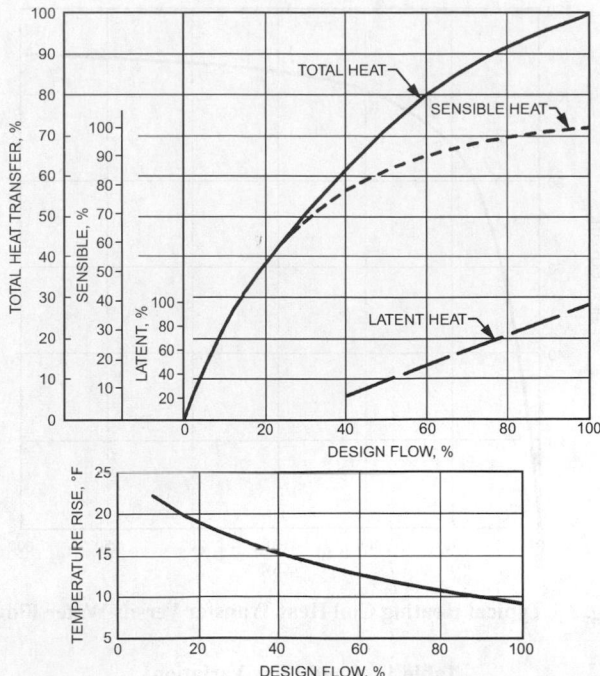

Fig. 4   Chilled Water Terminal Heat Transfer Versus Flow

tial when coil velocity drops below 1.5 fps. This affects chiller loading and unloading.

Flow Tolerance and Balance Procedure

The design procedure rests on a design flow rate and an allowable flow tolerance. The designer must define both the terminal's flow rates and feasible flow tolerance, remembering that the cost of balancing rises with tightened flow tolerance. Any overflow increases pumping cost, and any flow decrease reduces the maximum heating or cooling at design conditions.

WATER-SIDE BALANCING

Water-side balancing adjustments should be made with a thorough understanding of piping friction loss calculations and measured system pressure losses. It is good practice to show expected losses of pipes, fittings, and terminals and expected pressures in operation on schematic system drawings.

The water side should tested by direct flow measurement. This method is accurate because it deals with system flow as a function of differential pressures, and avoids compounding errors introduced by temperature difference procedures. Measuring flow at each terminal enables proportional balancing and, ultimately, matching pump head and flow to actual system requirements by trimming the pump impeller or reducing pump motor power. Often, reducing pump operating cost pays for the cost of water-side balancing.

Equipment

Proper equipment selection and preplanning are needed to successfully balance hydronic systems. Circumstances sometimes dictate that flow, temperature, and pressure be measured. The designer should specify the water flow balancing devices for installation during construction and testing during hydronic system balancing. The devices may consist of all or some of the following:

- Flowmeters (ultrasonic stations, turbines, venturi, orifice plate, multiported pitot tubes, and flow indicators)
- Manometers, ultrasonic digital meters, and differential pressure gages (analog or digital)
- Portable digital meter to measure flow and pressure drop

- Portable pyrometers to measure temperature differentials when test wells are not provided
- Test pressure taps, pressure gages, thermometers, and wells.
- Balancing valve with a factory-rated flow coefficient C_v, a flow versus handle position and pressure drop table, or a slide rule flow calculator
- Dynamic balancing valves or flow-limiting valves (for prebalanced systems only); field adjustment of these devices is not normally required or possible (see Chapter 46)
- Pumps with factory-certified pump curves
- Components used as flowmeters (terminal coils, chillers, heat exchangers, or control valves if using manufacturer's factory-certified flow versus pressure drop curves)

Record Keeping

Balancing requires accurate record keeping while making field measurements. Dated and signed field test reports help the designer or customer in work approval, and the owner has a valuable reference when documenting future changes.

Sizing Balancing Valves

A balancing valve is placed in the system to adjust water flow to a terminal, branch, zone, riser, or main. It should be located on the leaving side of the hydronic branch. General branch layout is from takeoff to entering service valve, then to the coil, control valve, and balancing/service valve. Pressure is thereby left on the coil, helping keep dissolved air in solution and preventing false balance problems resulting from air bind.

A common valve sizing method is to select for line size; however, balancing valves should be selected to pass design flows when near or at their fully open position with 12 in. of water minimum pressure drop. Larger Δp is recommended for accurate pressure readings. Many balancing valves and measuring meters give an accuracy of ±5% of range down to a pressure drop of 12 in. of water with the balancing valve wide open. Too large a balancing valve pressure drop affects the performance and flow characteristic of the control valve. Too small a pressure drop affects its flow measurement accuracy as it is closed to balance the system. Equation (2) may be used to determine the flow coefficient C_v for a balancing valve or to size a control valve.

The flow coefficient C_v is defined as the number of gallons of water per minute that flows through a wide-open valve with a pressure drop of 1 psi at 60°F. This is shown as

$$C_v = Q \sqrt{s_f / \Delta p} \qquad (2)$$

where

    Q = design flow for terminal or valve, gpm
    s_f = specific gravity of fluid
    Δp = pressure drop, psi
    Δh = pressure drop, ft of water

If pressure drop is determined in feet of water Δh, Equation (2) can be shown as

$$C_v = 1.5 Q \sqrt{s_f / \Delta h} \qquad (3)$$

HYDRONIC BALANCING METHODS

Various techniques are used to balance hydronic systems. Balance by temperature difference and water balance by proportional method are the most common.

Preparation. Minimally, preparation before balancing should include collecting the following:

1. Pump submittal data; pump curves, motor data, etc.
2. Starter sizes and overload protection information
3. Control valve C_v ratings and temperature control diagrams

4. Chiller, boiler, and heat exchanger information; flow and head loss
5. Terminal unit information; flow and head loss data
6. Pressure relief and reducing valve setting
7. Flowmeter calibration curves
8. Other pertinent data

System Preparation for Static System

1. Examine piping system: Identify main pipes, risers, branches and terminals on as-built drawings. Check that flows for all balancing devices are indicated on drawings before beginning work. Check that design flows for each riser equal the sum of the design flows through the terminals.
2. Examine reducing valve
3. Examine pressure relief valves
4. Examine expansion tank
5. For pumps, confirm
   • Location and size
   • Vented volute
   • Alignment
   • Grouting
   • Motor and lubrication
   • Nameplate data
   • Pump rotational direction
6. For strainers, confirm
   • Location and size
   • Mesh size and cleanliness
7. Confirm location and size of terminal units
8. Control valves:
   • Confirm location and size
   • Confirm port locations and flow direction
   • Set all valves open to coil
   • Confirm actuator has required force to close valve under loaded conditions
9. Ensure calibration of all measuring instruments, and that all calibration data are known for balancing devices
10. Remove all air from piping; all high points should have air vents

Pump Start-Up

1. Start pump and confirm rotational direction; if rotation is incorrect, have corrected.
2. Read differential head and apply to pump curve to observe flow approximates design.
3. Slowly close pump (if pump is under 25 hp) throttle valve to shutoff. Read pump differential head from gages.
   • If shutoff head corresponds with published curve, the previously prepared velocity head correction curve can be used as a pump flow calibration curve.
   • A significant difference between observed and published shutoff head can be caused by an unvented volute, a partially plugged impeller, or by an impeller size different from that specified.

Confirmation of System Venting

1. Confirm tank location and size.
2. Shut off pump; record shutoff gauge pressure at tank junction.
3. Start pump and record operating pressure at tank junction.
4. Compare operating to shutoff pressures at tank junction. If there is no pressure change, the system is air-free.
5. Eliminate free air.
   • No air separation: Shut off pump and revent. Retest and revent until tank junction pressure is stable.
   • Air separation: Operate system until free air has been separated out, indicated by stable tank junction pressure.

Balancing

For single-, multiple-, and parallel pump systems, after pump start-up and confirmation of system venting,

1. Adjust pump throttle until pump head differential corresponds to design.
2. Record pump motor voltage and amperage, and pump strainer head, at design flow.
3. Balance equipment room piping circuit so that pumped flow remains constant over alternative flow paths.
4. Record chiller and boiler circuits (for multiple-pump systems, requires a flowmeter installed between header piping).

For multiple-pump systems only,

5. Check for variable flow in source circuits when control valves are operated.
6. Confirm
   • Pump suction pressure remains above cavitations range for all operating conditions.
   • Pump flow rates remain constant.
   • Source working pressures are unaffected.

For parallel-pump systems, follow steps (1) to (4), then shut off pumps alternately and

5. Record head differential and flow rate through operating pump, and operating pump motor voltage and current.
6. Confirm that operational point is satisfactory (no overload, cavitation potential, etc.).

Balance by Temperature Difference

This common balancing procedure is based on measuring the water temperature difference between supply and return at the terminal. The designer selects the cooling and/or heating terminal for a calculated design load at full-load conditions. At less than full load, which is true for most operating hours, the temperature drop is proportionately less. Figure 5 demonstrates this relationship for a heating system at a design Δt of 20°F for outdoor design of 10°F and room design of 70°F.

For every outdoor temperature other than design, the balancing technician should construct a similar chart and read off the Δt for balancing. For example, at 50% load, or 30°F outdoor air, the Δt required is 10°F, or 50% of the design drop.

This method is a rough approximation and should not be used where great accuracy is required. It is not accurate enough for cooling systems.

Water Balance by Proportional Method

Preset Method. A thorough understanding of the pressure drops in the system riser piping, branches, coils, control valves, and bal-

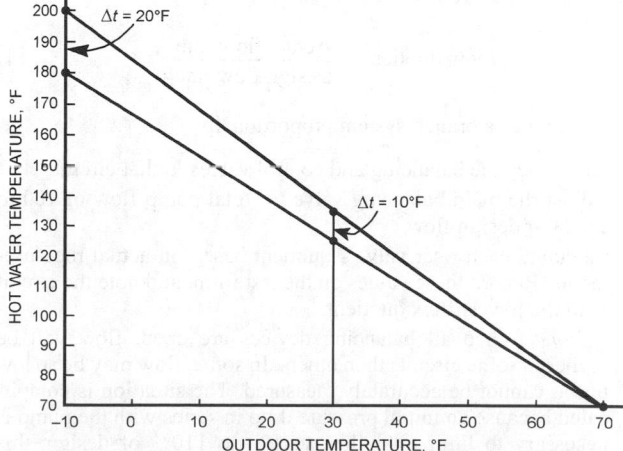

Fig. 5   Water Temperature Versus Outdoor Temperature Showing Approximate Temperature Difference

ancing valves is needed. Generally, several pipe and valve sizes are available for designing systems with high or low pressure drops. A flow-limiting or trim device will be required. Knowing system pressure losses in design allows the designer to select a balancing device to absorb excess system pressures in the branch, and to shift pressure drop (which might be absorbed by a balancing device nearly close to achieve balance) to the pipes, coils, and valves so the balancing device merely trims these components' performance at design flow. It may also indicate where high-head-loss circuits can exist for either relocation in the piping network, or hydraulic isolation through hybrid piping techniques. The installed balancing device should never be closed more than 40 to 50%; below this point flow reading accuracy falls to ±20 to 30%. Knowing a starting point for setting the valve (preset) allows the designer to iterate system piping design. This may not always be practical in large systems, but minimizing head and flow saves energy over the life of the facility and allows for proper temperature control. In this method,

1. Analyze the piping network for the largest hydraulic loss based on design flow and pipe friction loss. The pump should be selected to provide the total of all terminal flows, and the head required to move water through the hydraulically greatest circuit. Balance devices in this circuit should be sized only for the loss required for flow measurement accuracy. Trimming is not required.
2. Analyze differences in pressure drop in the pumping circuit for each terminal without using a balancing device. The difference between each circuit and the pump head (which represents the drop in the farthest circuit) is the required drop for the balancing device.
3. Select a balancing device that will achieve this drop with minimum valve throttling. If greater than two pipe sizes smaller, shift design drop into control valve or coil (or both), equalizing pressure drop across the devices.
4. Monitor system elevations and pressure drops to ensure air management, minimizing pocket collections and false pressure references that could lead to phantom balancing problems.
5. Use proportional balancing methods as outlined for field testing and adjustment.

Proportional Balancing

Proportional water-side balancing may use design data, but relies most on as-built conditions and measurements and adapts well to design diversity factors. This method works well with multiple-riser systems. When several terminals are connected to the same circuit, any variation of differential pressure at the circuit inlet affects flows in all other units in the same proportion. Circuits are proportionally balanced to each other by a flow quotient:

$$\text{Flow quotient} = \frac{\text{Actual flow rate}}{\text{Design flow rate}} \quad (4)$$

To balance a branch system proportionally,

1. Fully open the balancing and control valves in that circuit.
2. Adjust the main balancing valve for total pump flow of 100 to 110% of design flow.
3. Calculate each riser valve's quotient based on actual measurements. Record these values on the test form, and note the circuit with the lowest flow quotient.

Note: When all balancing devices are open, flow will be higher in some circuits than others. In some, flow may be so low that it cannot be accurately measured. The situation is complicated because an initial pressure drop in series with the pump is necessary to limit total flow to 100 to 110% of design; this decreases the available differential pressure for the distribution system. After all other risers are balanced, restart analysis of risers with unmeasurable flow at step (2).

4. Identify the riser with the highest flow ratio. Begin balancing with this riser, then continue to the next highest flow ratio, and so on. When selecting the branch with the highest flow ratio,
   - Measure flow in all branches of the selected riser.
   - In branches with flow higher than 150% of design, close the balancing valves to reduce flow to about 110% of design.
   - Readjust total pump flow using the main valve.
   - Start balancing in branches with a flow ratio greater than or equal to 1. Start with the branch with the highest flow ratio.

The reference circuit has the lowest quotient and the greatest pressure loss. Adjust all other balancing valves in that branch until they have the same quotient as the reference circuit (at least one valve in the branch should be fully open).

When a second valve is adjusted, the flow quotient in the reference valve also changes; continued adjustment is required to make their flow quotients equal. Once they are equal, they will remain equal or in proportional balance to each other while other valves in the branch are adjusted or until there is a change in pressure or flow.

When all balancing valves are adjusted to their branches' respective flow quotients, total system water flow is adjusted to the design by setting the balancing valve at the pump discharge to a flow quotient of 1.

Pressure drop across the balancing valve at pump discharge is produced by the pump that is not required to provide design flow. This excess pressure can be removed by trimming the pump impeller or reducing pump speed. The pump discharge balancing valve must then be reopened fully to provide the design flow.

As in variable-speed pumping, diversity and flow changes are well accommodated by a system that has been proportionately balanced. Because the balancing valves have been balanced to each other at a particular flow (design), any changes in flow are proportionately distributed.

Balancing the water side in a system that uses diversity must be done at full flow. Because components are selected based on heat transfer at full flow, they must be balanced to this point. To accomplish full-flow proportional balance, shut off part of the system while balancing the remaining sections. When a section has been balanced, shut it off and open the section that was open originally to complete full balance of the system. When balancing, care should be taken if the building is occupied or if load is nearly full.

Variable-Speed Pumping. To achieve hydronic balance, full flow through the system is required during balancing, after which the system can be placed on automatic control and the pump speed allowed to change. After the full-flow condition is balanced and the system differential pressure set point is established, to control the variable-speed pumps, observe the flow on the circuit with the greatest resistance as the other circuits are closed one at a time. The flow in the observed circuit should remain equal to, or more than, the previously set flow. Water flow may become laminar at less than 2 fps, which may alter the heat transfer characteristics of the system.

Other Balancing Techniques

Flow Balancing by Rated Differential Procedure. This procedure depends on deriving a performance curve for the test coil, comparing water temperature difference Δt_w to entering water temperature t_{ew} minus entering air temperature t_{ea}. One point of the desired curve can be determined from manufacturer's ratings, which are published as ($t_{ew} - t_{ea}$). A second point is established by observing that the heat transfer from air to water is zero when ($t_{ew} - t_{ea}$) is zero (consequently, $\Delta t_w = 0$). With these two points, an approximate performance curve can be drawn (Figure 6). Then, for any other ($t_{ew} - t_{ea}$), this curve is used to determine the appropriate Δt_w. The basic curve applies to catalog ratings for lower dry-bulb temperatures, providing a consistent entering air moisture content (e.g., 75°F db, 65°F wb). Changes in inlet water temperature, temperature rise, air velocity, and dry- and wet-bulb temperatures cause terminal

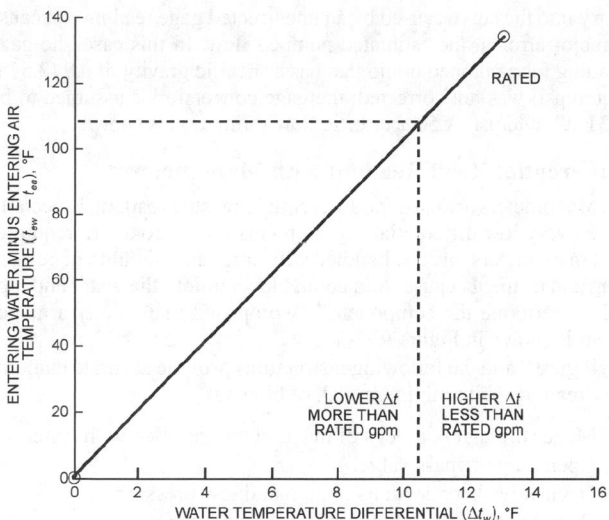

Fig. 6   Coil Performance Curve

performance to deviate from the curves. The curve may also be used for cooling coils for sensible transfer (dry coil).

Flow Balancing by Total Heat Transfer. This procedure determines water flow by running an energy balance around the coil. From field measurements of airflow, wet- and dry-bulb temperatures up- and downstream of the coil, and the difference Δt_w between entering and leaving water temperatures, water flow can be determined by the following equations:

$$Q_w = q/500\ \Delta t_w \qquad (5)$$

$$q_{cooling} = 4.5 Q_a (h_1 - h_2) \qquad (6)$$

$$q_{heating} = 1.08 Q_a (t_1 - t_2) \qquad (7)$$

where

Q_w = water flow rate, gpm
q = load, Btu/h
$q_{cooling}$ = cooling load, Btu/h
$q_{heating}$ = heating load, Btu/h
Q_a = airflow rate, cfm
h = enthalpy, Btu/lb
t = temperature, °F

Example 1. Find the water flow for a cooling system with the following characteristics:

Test data

t_{ewb} = entering wet-bulb temperature = 68.5°F
t_{lwb} = leaving wet-bulb temperature = 53.5°F
Q_a = airflow rate = 22,000 cfm
t_{lw} = leaving water temperature = 59.0°F
t_{ew} = entering water temperature = 47.5°F

From psychrometric chart

h_1 = 32.84 Btu/lb
h_2 = 22.32 Btu/lb

Solution: From Equations (5) and (6),

$$Q_w = \frac{4.5 \times 22,000(32.84 - 22.32)}{500(59.0 - 47.5)} = 181\ \text{gpm}$$

The desired water flow is achieved by successive manual adjustments and recalculations. Note that these temperatures can be greatly influenced by the heat of compression, stratification, bypassing, and duct leakage.

General Balance Procedures

All the variations of balancing hydronic systems cannot be listed; however, the general method should balance the system and minimize operating cost. Excess pump pressure (operating power) can be eliminated by trimming the pump impeller. Allowing excess pressure to be absorbed by throttle valves adds a lifelong operating-cost penalty to the operation.

The following is a general procedure based on setting the balance valves on the site:

1. Develop a flow diagram if one is not included in the design drawings. Illustrate all balance instrumentation, and include any additional instrument requirements.
2. Compare pumps, primary heat exchangers, and specified terminal units, and determine whether a design diversity factor can be achieved.
3. Examine the control diagram and determine the control adjustments needed to obtain design flow conditions.

Balance Procedure—Primary and Secondary Circuits

1. Inspect the system completely to ensure that (1) it has been flushed out, it is clean, and all air is removed; (2) all manual valves are open or in operating position; (3) all automatic valves are in their proper positions and operative; and (4) the expansion tank is properly charged.
2. Place controls in position for design flow.
3. Examine flow diagram and piping for obvious short circuits; check flow and adjust the balance valve.
4. Take pump suction, discharge, and differential pressure readings at both full and no flow. For larger pumps, a no-flow condition may not be safe. In any event, valves should be closed slowly.
5. Read pump motor amperage and voltage, and determine approximate power.
6. Establish a pump curve, and determine approximate flow rate.
7. If a total flow station exists, determine the flow and compare with pump curve flow.
8. If possible, set total flow about 10% high using the total flow station first and the pump differential pressure second; then maintain pumped flow at a constant value as balance proceeds by adjusting the pump throttle valve.
9. Any branch main flow stations should be tested and set, starting by setting the shortest runs low as balancing proceeds to the longer branch runs.
10. With primary and secondary pumping circuits, a reasonable balance must be obtained in the primary loop before the secondary loop can be considered. The secondary pumps must be running and terminal units must be open to flow when the primary loop is being balanced, unless the secondary loop is decoupled.

FLUID FLOW MEASUREMENT

Flow Measurement Based on Manufacturer's Data

Any component (terminal, control valve, or chiller) that has an accurate, factory-certified flow/pressure drop relationship can be used as a flow-indicating device. The flow and pressure drop may be used to establish an equivalent flow coefficient as shown in Equation (3). According to the Bernoulli equation, pressure drop varies as the square of the velocity or flow rate, assuming density is constant:

$$Q_1^2/Q_2^2 = \Delta h_1/\Delta h_2 \qquad (8)$$

For example, a chiller has a certified pressure drop of 25 ft of water at 100 gpm. The calculated flow with a field-measured pressure drop of 30 ft is

$$Q_2 = 100 \sqrt{30/25} = 109.5 \text{ gpm}$$

Flow calculated in this manner is only an estimate. The accuracy of components used as flow indicators depends on the accuracy of (1) cataloged information concerning flow/pressure drop relationships and (2) pressure differential readings. As a rule, the component should be factory-certified flow tested if it is to be used as a flow indicator.

Pressure Differential Readout by Gage

Gages are used to read differential pressures. Gages are usually used for high differential pressures and manometers for lower differentials. Accurate gage readout is diminished when two gages are used, especially when the gages are permanently mounted and, as such, subject to malfunction.

A single high-quality gage should be used for differential readout (Figure 7). This gage should be alternately valved to the high- and low-pressure side to establish the differential. A single gage needs no static height correction, and errors caused by gage calibration are eliminated.

Differential pressure can also be read from differential gages, thus eliminating the need to subtract outlet from inlet pressures to establish differential pressure. Differential pressure gages are usually dual gages mechanically linked to read differential pressure. The differential pressure gage readout can be stated in terms of psi or in feet of head of 60°F water.

Conversion of Differential Pressure to Head

Pressure gage readings can be restated to fluid head, which is a function of fluid density. The common hydronic system conversion factor is related to water density at about 60°F; 1 psi equals 2.31 ft. Pressure gages can be calibrated to feet of water head using this conversion. Because the calibration only applies to water at 60°F, the readout may require correction when the gage is applied to water at a significantly higher temperature.

Pressure gage conversion and correction factors for various fluid specific gravities (in relation to water at 60°F) are shown in Table 2. The differential gage readout should only be defined in terms of the head of the fluid actually causing the flow pressure differential. When this is done, the resultant fluid head can be applied to the C_v to determine actual flow through any flow device, provided the manufacturer has correctly stated the flow to fluid head relationship.

For example, a manufacturer may test a boiler or control valve with 100°F water. If the test differential pressure is converted to head at 100°F, a C_v independent of test temperature and density may be calculated. Differential pressures from another test made in the field at 250°F may be converted to head at 250°F. The C_v calculated with this head is also independent of temperature. The manufacturer's data can then be directly correlated with the field test to establish flow rate at 250°F.

A density correction must be made to the gage reading when differential heads are used to estimate pump flows as in Figure 8. This is because of the shape of the pump curve. An incorrect head difference

entry into the curve caused by an uncorrected gage reading can cause a major error in the estimated pumped flow. In this case, the gage reading for a pumped liquid that has a specific gravity of 0.9 (2.57 ft liquid/psi) was not corrected; the gage conversion is assumed to be 2.31 ft liquid/psi. A 50% error in flow estimation is shown.

Differential Head Readout with Manometers

Manometers are used for differential pressure readout, especially when very low differentials, great precision, or both, are required. But manometers must be handled with care; they should not be used for field testing because fluid could blow out into the water and rapidly deteriorate the components. A proposed manometer arrangement is shown in Figure 9.

Figure 9 and the following instructions provide accurate manometer readings with minimum risk of blowout.

1. Make sure that both legs of manometer are filled with water.
2. Open purge bypass valve.
3. Open valved connections to high and low pressure.
4. Open bypass vent valve slowly and purge air here.
5. Open manometer block vents and purge air at each point.
6. Close needle valves. The columns should zero in if the manometer is free of air. If not, vent again.
7. Open needle valves and begin throttling purge bypass valve slowly, watching the fluid columns. If the manometer has an adequate available fluid column, the valve can be closed and the differential reading taken. However, if the fluid column reaches the top of the manometer before the valve is completely closed, insufficient manometer height is indicated and further throttling will blow fluid into the blowout collector. A longer manometer or the single gage readout method should then be used.

An error is often introduced when converting inches of gage fluid to feet of test fluid. The conversion factor changes with test fluid temperature, density, or both. Conversion factors shown in Table 2 are to a water base, and the counterbalancing water height H (Figure 9) is at room temperature.

Orifice Plates, Venturi, and Flow Indicators

Manufacturers provide flow information for several devices used in hydronic system balance. In general, the devices can be classified as (1) orifice flowmeters, (2) venturi flowmeters, (3) velocity impact meters, (4) pitot-tube flowmeters, (5) bypass spring impact flowmeters, (6) calibrated balance valves, (7) turbine flowmeters, and (8) ultrasonic flowmeters.

The **orifice flowmeter** is widely used and is extremely accurate. The meter is calibrated and shows differential pressure versus flow. Accuracy generally increases as the pressure differential across the

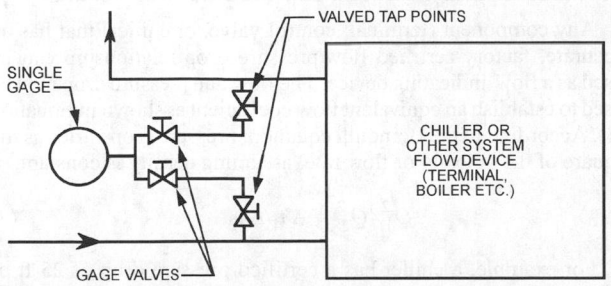

Fig. 7   Single Gage for Reading Differential Pressure

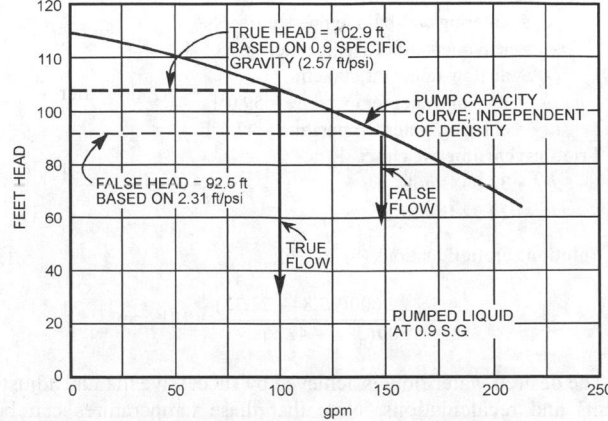

Fig. 8   Fluid Density Correction Chart for Pump Curves

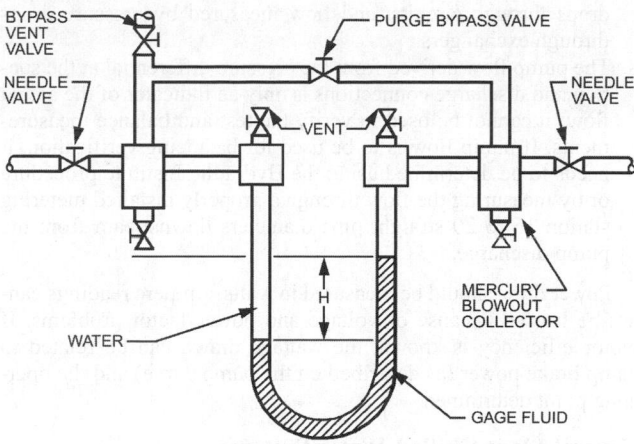

Fig. 9 Fluid Manometer Arrangement for Accurate Reading and Blowout Protection

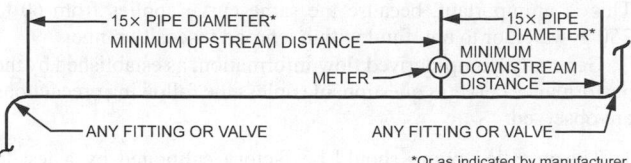

Fig. 10 Minimum Installation Dimensions for Flowmeter

Table 2 Differential Pressure Conversion to Head

Fluid Specific Gravity	Corresponding Water Temperature, °F	Foot Fluid Head Equal to 1 psi[a]	Correction Factor When Gage is Stated to Feet of Water (60°F)[b]
1.5		1.54	
1.4		1.65	
1.3		1.78	
1.2		1.93	
1.1		2.10	
1.0	60	2.31	1.00
0.98	150	2.36	1.02
0.96	200	2.41	1.04
0.94	250	2.46	1.065
0.92	300	2.51	1.09
0.90	340	2.57	1.11
0.80		2.89	
0.70		3.30	
0.60		3.85	
0.50		4.63	

[a]Differential psi readout is multiplied by this number to obtain feet fluid head when gage is calibrated in psi.
[b]Differential feet water head readout is multiplied by this number to obtain feet fluid head when gage calibration is stated to feet head of 60°F water.

meter increases. The differential pressure readout instrument may be a manometer, differential gage, or single gage (see Figure 7).

The **venturi flowmeter** has lower pressure loss than the orifice plate meter because a carefully formed flow path increases velocity head recovery. The venturi flowmeter is placed in a main flow line where it can be read continuously.

Velocity impact meters have precise construction and calibration. The meters are generally made of specially contoured glass or plastic, which allows observation of a flow float. As flow increases, the flow float rises in the calibrated tube to indicate flow rate. Velocity impact meters generally have high accuracy.

A special version of the velocity impact meter is applied to hydronic systems. This version operates on the velocity head difference between the pipe side wall and the pipe center, which causes fluid to flow through a small flowmeter. Accuracy depends on the location of the impact tube and on a velocity profile that corresponds to theory and the laboratory test calibration base. Generally, the accuracy of this **bypass flow impact** or differential velocity head flowmeter is less than a flow-through meter, which can operate without creating a pressure loss in the hydronic system.

The **pitot-tube flowmeter** is also used for pipe flow measurement. Manometers are generally used to measure velocity head differences because these differences are low.

The **bypass spring impact flowmeter** uses a defined piping pressure drop to cause a correlated bypass side branch flow. The side branch flow pushes against a spring that increases in length with increased flow. Each individual flowmeter is calibrated to relate extended spring length position to main flow. The bypass spring impact flowmeter has, as its principal merit, a direct readout. However, dirt on the spring reduces accuracy. The bypass is opened only when a reading is made. Flow readings can be taken at any time.

The **calibrated balance valve** is an adjustable orifice flowmeter. Balance valves can be calibrated so that a flow/pressure drop relationship can be obtained for each incremental setting of the valve. A ball, rotating plug, or butterfly valve may have its setting expressed in percent open or degree open; a globe valve, in percent open or number of turns. The calibrated balance valve must be manufactured with precision and care to ensure that each valve of a particular size has the same calibration characteristics.

The **turbine flowmeter** is a mechanical device. The velocity of the liquid spins a wheel in the meter, which generates a 4 to 20 mA output that may be calibrated in units of flow. The meter must be well maintained, because wear or water impurities on the bearing may slow the wheel, and debris may clog or break the wheel.

The **ultrasonic flowmeter** senses sound signals, which are calibrated in units of flow. The ultrasonic metering station may be installed as part of the piping, or it may be a strap-on meter. In either case, the meter has no moving parts to maintain, nor does it intrude into the pipe and cause a pressure drop. Two distinct types of ultrasonic meter are available: (1) the transit time meter for HVAC or clear-water systems and (2) the Doppler meter for systems handling sewage or large amounts of particulate matter.

If any of the above meters are to be useful, the minimum distance of straight pipe upstream and downstream, as recommended by the meter manufacturer and flow measurement handbooks, must be adhered to. Figure 10 presents minimum installation suggestions.

Using a Pump as an Indicator

Although the pump is not a meter, it can be used as an indicator of flow together with the other system components. Differential pressure readings across a pump can be correlated with the pump curve to establish the pump flow rate. Accuracy depends on (1) accuracy of readout, (2) pump curve shape, (3) actual conformance of the pump to its published curve, (4) pump operation without cavitation, (5) air-free operation, and (6) velocity head correction.

When a differential pressure reading must be taken, a single gage with manifold provides the greatest accuracy (Figure 11). The pump suction to discharge differential can be used to establish pump differential pressure and, consequently, pump flow rate. The single gage and manifold may also be used to check for strainer clogging by measuring the pressure differential across the strainer.

If the pump curve is based on fluid head, pressure differential, as obtained from the gage reading, needs to be converted to head, which is pressure divided by the fluid weight per cubic foot. The pump differential head is then used to determine pump flow rate (Figure 12). As long as the differential head used to enter the pump curve is expressed as head of the fluid being pumped, the pump curve shown by the manufacturer should be used as described. The pump curve may state that it was defined by test with 85°F water.

This is unimportant, because the same curve applies from 60 to 250°F water, or to any fluid within a broad viscosity range.

Generally, pump-derived flow information, as established by the performance curve, is questionable unless the following precautions are observed:

1. The installed pump should be factory calibrated by a test to establish the actual flow/pressure relationship for that particular pump. Production pumps can vary from the cataloged curve because of minor changes in impeller diameter, interior casting tolerances, and machine fits.

2. When a calibration curve is not available for a centrifugal pump being tested, the discharge valve can be closed briefly to establish the no-flow shutoff pressure, which can be compared to the published curve. If the shutoff pressure differs from that published, draw a new curve parallel to the published curve. Though not exact, the new curve usually fits the actual pumping circumstance more accurately. Clearance between the impeller and casing minimizes the danger of damage to the pump during a no-flow test, but manufacturer verification is necessary.

3. Differential head should be determined as accurately as possible, especially for pumps with flat flow curves.

4. The pump should be operating air-free and without cavitation. A cavitating pump will not operate to its curve, and differential readings will provide false results.

5. Ensure that the pump is operating above the minimum net positive suction head.

6. Power readings can be used (1) as a check for the operating point when the pump curve is flat or (2) as a reference check when there is suspicion that the pump is cavitating or providing false readings because of air.

7. The flow determined by the pump curve should be compared to the flow measured at the flowmeters, flow measured by pressure drops through circuits, and flow measured by pressure drops through exchangers.

8. The pump flow derived from the pressure differential at the suction and discharge connections is only an indicator of the actual flow; it cannot be used to verify the test and balance measurements. If pump flow is to be used for balancing verification, it needs to be determined using the Hydraulic Institute procedure or by measuring the flow through a properly installed metering station 15 to 20 straight pipe diameters downstream from the pump discharge.

Power draw should be measured in watts. Ampere readings cannot be trusted because of voltage and power factor problems. If motor efficiency is known, the wattage drawn can be related to pump brake power (as described on the pump curve) and the operating point determined.

Central Plant Chilled-Water Systems

For existing installations, establishing accurate thermal load profiles is of prime importance because it establishes proper primary chilled-water supply temperature and flow. In new installations, actual load profiles can be compared with design load profiles to obtain valid operating data.

To perform proper testing and balancing, all interconnecting points between the primary and secondary systems must be designed with sufficient temperature, pressure, and flow connections so that adequate data may be indicated and/or recorded.

Water Flow Instruments

As indicated previously, proper location and use of instruments is vital to accurate balancing. Instruments for testing temperature and pressure at various locations are listed in Table 3. Flow-indicating devices should be placed in water systems as follows:

- At each major heating coil bank (10 gpm or more)
- At each major cooling coil bank (10 gpm or more)
- At each bridge in primary/secondary systems
- At each main pumping station
- At each water chiller evaporator
- At each water chiller condenser

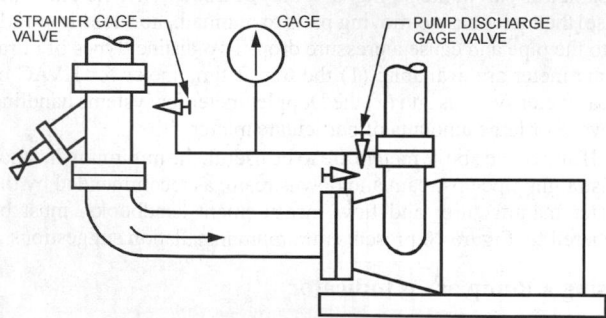

Fig. 11 Single Gage for Differential Readout Across Pump and Strainer

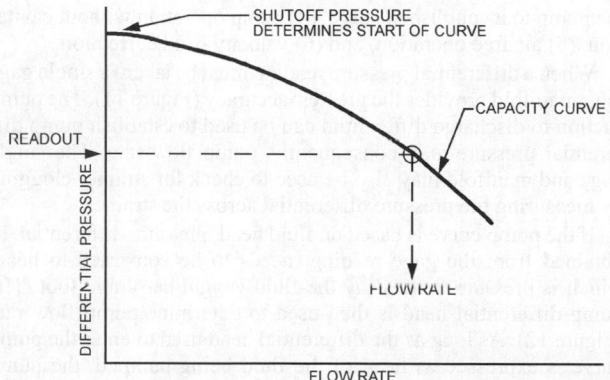

Fig. 12 Differential Pressure Used to Determine Pump Flow

Table 3 Instruments for Monitoring a Water System

Point of Information	Manifold Gage	Single Gage	Thermometer	Test Well	Pressure Tap
Pump—Suction, discharge	x				
Strainer—In, out					x
Cooler—In, out		x	x		
Condensers—In, out		x	x		
Concentrator—In, out		x	x		
Absorber—In, out		x	x		
Tower cell—In, out				x	x
Heat exchanger—In, out	x		x		
Coil—In, out				x	x
Coil bank—In, out		x	x		
Booster coil—In, out					x
Cool panel—In, out					x
Heat panel—In, out				x	x
Unit heater—In, out					x
Induction—In, out					x
Fan coil—In, out					x
Water boiler—In, out		x			
Three-way valve—All ports					x
Zone return main		x			
Bridge—In, out		x			
Water makeup		x			
Expansion tank		x			
Strainer pump					x
Strainer main	x				
Zone three-way—All ports				x	x

- At each water boiler outlet
- At each floor takeoff to booster reheat coils, fan coil units, induction units, ceiling panels, and radiation (do not exceed 25 terminals off of any one zone meter probe)
- At each vertical riser to fan coil units, induction units, and radiation
- At the point of tie-in to existing systems

STEAM DISTRIBUTION

Procedures for Steam Balancing Variable Flow Systems

Steam distribution cannot be balanced by adjustable flow-regulating devices. Instead, fixed restrictions built into the piping in accordance with carefully designed pipe and orifice sizes are used to regulate flow.

It is important to have a balanced distribution of steam to all portions of the steam piping at all loads. This is best accomplished by properly designing the steam distribution piping, which includes carefully considering steam pressure, steam quantities required by each branch circuit, pressure drops, steam velocities, and pipe sizes. Just as other flow systems are balanced, steam distribution systems are balanced by ensuring that the pressure drops are equalized at design flow rates for all portions of the piping. Only marginal balancing can be done by pipe sizing. Therefore, additional steps must be taken to achieve a balanced performance.

Steam flow balance can be improved by using spring-type packless supply valves equipped with precalibrated orifices. The valves should have a tight shutoff between 25 in. of Hg and 60 psig. These valves have a nonrising stem, are available with a lockshield, and have a replaceable disk. Orifice flanges can also be used to regulate and measure steam flow at appropriate locations throughout the system. The orifice sizes are determined by the pressure drop required for a given flow rate at a given location. A schedule should be prepared showing (1) orifice sizes, (2) valve or pipe sizes, (3) required flow rates, and (4) corresponding pressure differentials for each flow rate. It may be useful to calculate pressure differentials for several flow rates for each orifice size. Such a schedule should be maintained for future reference.

After the appropriate regulating orifices are installed in the proper locations, the system should be tested for tightness by sealing all openings in the system and applying a vacuum of 20 in. of Hg, held for 2 hours. Next, the system should be readied for warm-up and pressurizing with steam following the procedures outlined in Section VI of the ASME *Boiler and Pressure Vessel Code*. After the initial warm-up and system pressurization, evaluate system steam flow, and compare it to system requirements. The orifice schedule calculated earlier will now be of value should any of the orifices need to be changed.

Steam Flow Measuring Devices

Many devices are available for measuring flow in steam piping: (1) steam meters, (2) condensate meters, (3) orifice plates, (4) venturi fittings, (5) steam recorders, and (6) manometers for reading differential pressures across orifice plates and venturi fittings. Some of these devices are permanently affixed to the piping system to facilitate taking instantaneous readings that may be necessary for proper operation and control. A surface pyrometer used in conjunction with a pressure gage is a convenient way to determine steam saturation temperature and the degree of superheat at various locations in the system. This information can be used to evaluate performance characteristics.

COOLING TOWERS

Field-testing cooling towers is demanding and difficult. ASME *Standard* PTC 23 and CTI *Standard Specification* ATC-105 establish procedures for these tests. Certain general guidelines for testing cooling towers are as follows:

Conditions at Time of Test
- Water flow within 15% of design

- Heat load within 30% of design and stabilized
- Entering wet bulb within 12°F of design

Using these limitations and field readings that are as accurate as possible, a projection to design conditions produces an accuracy of ±5% of tower performance.

Conditions for Performing Test
- Water-circulating systems serving tower should be thoroughly cleaned of all dirt and foreign matter. Samples of water should be clear and indicate clear passage of water through pumps, piping, screens, and strainers.
- Fans serving cooling tower should operate in proper rotation. Foreign obstructions should be removed. Permanent obstruction should be noted.
- Interior filling of cooling tower should be clean and free of foreign materials such as scale, algae, or tar.
- Water level in tower basin should be maintained at proper level. Visually check basin sump during full flow to verify that centrifugal action of the water is not causing entrainment of air, which could cause pump cavitation.
- Water-circulating pumps should be tested with full flow through tower. If flow exceeds design, it should be valved down until design is reached. The flow that is finally set should be maintained throughout the test period. All valves, except necessary balancing valves, should be fully open.
- If makeup and blowdown have facilities to determine flow, set them to design flow at full flow through the tower. If flow cannot be determined, shut off both.

Instruments

Testing and balancing agencies provide instruments to perform the required tests. Mechanical contractors provide and install all components such as orifice plates, venturis, balancing valves, thermometer wells, gage cocks, and corporation cocks. Designers specify measuring point locations.

Instruments used should be recently calibrated as follows:

Temperature
- Thermometer with divisions of 0.2°F in a proper well for water should be used.
- Thermometer with solar shield and 0.2°F divisions or thermocouple with 0.2°F readings having mechanical aspiration for wet-bulb readings should be used.
- Sling psychrometer may be used for rough checks.
- Thermometer with 0.2°F divisions should be used for dry-bulb readings.

Water Flow
- Orifice, venturi, and balancing valve pressure drops can be read using a manometer or recently calibrated differential pressure gage.
- Where corporation cocks are installed, a pitot tube and manometer traverse can be made by trained technicians.

Test Method

1. Conduct water flow tests to determine volume of water in tower, and of makeup and blowdown water.
2. Conduct water temperature tests, if possible, in suitable wells as close to tower as possible. Temperature readings at pumps or condensing element are not acceptable in tower evaluation. If there are no wells, surface pyrometer readings are acceptable.
3. Take makeup water volume and temperature readings at point of entry into tower.
4. Take blowdown volume and temperature readings at point of discharge from tower.
5. Take inlet and outlet dry- and wet-bulb temperature readings using prescribed instruments.

- Use wet-bulb entering and leaving temperatures to determine tower actual performance against design.
- Use wet- and dry-bulb entering and leaving temperatures to determine evaporation involved.

6. If tower has a ducted inlet or outlet where a reasonable duct traverse can be made, use this air volume as a cross-check of tower performance.
7. Take wet- and dry-bulb temperature readings 3 to 5 ft from the tower on all inlet sides, halfway between the base and the top of inlet louvers at no more than 5 ft spacing horizontally the readings. Note any unusual inlet conditions.
8. Note wind velocity and direction at time of test.
9. Take test readings continually with a minimum time lapse between readings.
10. If the first test indicates a tower deficiency, perform two additional tests to verify the original readings.

TEMPERATURE CONTROL VERIFICATION

The test and balance technician should work closely with the temperature control installer to ensure that the project is completed correctly. The balancing technician needs to verify proper operation of the control and communicate findings back to the agency responsible for ensuring that the controls have been installed correctly. This is usually the HVAC system designer, although others may be involved. Generally, the balancing technician does not adjust, relocate, or calibrate the controls. However, this is not always the case, and differences do occur with VAV terminal unit controllers. The balancing technician should be familiar with the specifications and design intent of the project so that all responsibilities are understood.

During the design and specification phase of the project, the designer should specify verification procedures for the controls and responsibilities for the contractor who installs the temperature controls. It is important that the designer specify the (1) degree of coordination between the installer of the control and the balancing technician and (2) testing responsibilities of each.

Verification of control operation starts with the balancing technician reviewing the submitted documents and shop drawings of the control system. In some cases, the controls technician should instruct the balancing technician in the operation of certain control elements, such as digital terminal unit controllers. This is followed by schedule coordination between the control and balancing technicians. In addition, the balancing and controls technicians need to work together when reviewing the operation of some sections of the HVAC system, particularly with VAV systems and the setting of the flow measurement parameters in digital terminal unit controllers.

Major mechanical systems should be verified after testing, adjusting, and balancing is completed. The control system should be operated in stages to prove it can match system capacity to varying load conditions. Mechanical subsystem controllers should be verified when balancing data are collected, considering that the entire system may not be completely functional at the time of verification. Testing and verification should account for seasonal variations; tests should be performed under varying outdoor loads to ensure operational performance. Retesting a random sample of terminal units may be desirable to verify the control technician's work.

Suggested Procedures

The following verification procedures may be used with either pneumatic or electrical controls:

1. Obtain design drawings and documentation, and become well acquainted with the design intent and specified responsibilities.
2. Obtain copies of approved control shop drawings.
3. Compare design to installed field equipment.
4. Obtain recommended operating and test procedures from manufacturers.

5. Verify with the control contractor that all controllers are calibrated and commissioned.
6. Check location of transmitters and controllers. Note adverse conditions that would affect control, and suggest relocation as necessary.
7. Note settings on controllers. Note discrepancies between set point for controller and actual measured variable.
8. Verify operation of all limiting controllers, positioners, and relays (e.g., high- and low-temperature thermostats, high- and low-differential pressure switches, etc.).
9. Activate controlled devices, checking for free travel and proper operation of stroke for both dampers and valves. Verify normally open (NO) or normally closed (NC) operation.
10. Verify sequence of operation of controlled devices. Note line pressures and controlled device positions. Correlate to air or water flow measurements. Note speed of response to step change.
11. Confirm interaction of electrically operated switch transducers.
12. Confirm interaction of interlock and lockout systems.
13. Coordinate balancing and control technicians' schedules to avoid duplication of work and testing errors.

Pneumatic System Modifications

1. Verify main control supply air pressure and observe compressor and dryer operation.
2. For hybrid systems using electronic transducers for pneumatic actuation, modify procedures accordingly.

Electronic Systems Modifications

1. Monitor voltages of power supply and controller output. Determine whether the system operates on a grounded or nongrounded power supply, and check condition. Although electronic controls now have more robust electronic circuits, improper grounding can cause functional variation in controller and actuator performance from system to system.
2. Note operation of electric actuators using spring return. Generally, actuators should be under control and use springs only upon power failure to return to a fail-safe position.

Direct Digital Controllers

Direct digital control (DDC) offers nontraditional challenges to the balancing technician. Many control devices, such as sensors and actuators, are the same as those in electronic and pneumatic systems. Currently DDC is dominated by two types of controllers: fully programmable or application-specific. Fully programmable controllers offer a group of functions linked together in an applications program to control a system such as an air-handling unit. Application-specific controllers are functionally defined with the programming necessary to carry out the functions required for a system, but not all adjustments and settings are defined. Both types of controllers and their functions have some variations. One of the functions is adaptive control, which includes control algorithms that automatically adjust settings of various controller functions.

The balancing technician must understand controller functions so that they do not interfere with the test and balance functions. Literacy in computer programming is not necessary, although it does help. When testing the DDC,

1. Obtain controller application program. Discuss application of the designer's sequence with the control programmer.
2. Coordinate testing and adjustment of controlled systems with mechanical systems testing. Avoid duplication of efforts between technicians.
3. Coordinate storage (e.g., saving to central DDC database and controller memory) of all required system adjustments with control technician.

In cases where the balancing agency is required to test discrete points in the control system,

1. Establish criteria for test with the designer.
2. Use reference standards that test the end device through the entire controller chain (e.g., device, wiring, controller, communications, and operator monitoring device). An example would be using a dry block temperature calibrator (a testing device that allows a temperature to be set, monitored, and maintained in a small chamber) to test a space temperature sensor. The sensor is installed with extra wire so that it may be removed from the wall and placed in the calibrator chamber. After the system is thermally stabilized, the temperature is read at the controller and the central monitor, if installed.
3. Report findings of reference and all points of reading.

FIELD SURVEY FOR ENERGY AUDIT

An energy audit is an organized survey of a specific building to identify and measure all energy uses, determine probable sources of energy losses, and list energy conservation opportunities. This is usually performed as a team effort under the direction of a qualified energy engineer. The field data can be gathered by firms employing technicians trained in testing, adjusting, and balancing.

Instruments

To determine a building's energy use characteristics, existing conditions must be accurately measured with proper instruments. Accurate measurements point out opportunities to reduce waste and provide a record of the actual conditions in the building before energy conservation measures were taken. They provide a compilation of installed equipment data and a record of equipment performance before changes. Judgments will be made based on the information gathered during the field survey; that which is not accurately measured cannot be properly evaluated.

Generally, instruments used for testing, adjusting, and balancing are sufficient for energy conservation surveying. Possible additional instruments include a power factor meter, light meter, combustion testing equipment, refrigeration gages, and equipment for recording temperatures, fluid flow rates, and energy use over time. Only high-quality instruments should be used.

Observation of system operation and any information the technician can obtain from the operating personnel pertaining to the operation should be included in the report.

Data Recording

Organized record keeping is extremely important. A camera is also helpful. Photographs of building components and mechanical and electrical equipment can be reviewed later when the data are analyzed.

Data sheets for energy conservation field surveys contain different and, in some cases, more comprehensive information than those used for testing, adjusting, and balancing. Generally, the energy engineer determines the degree of fieldwork to be performed; data sheets should be compatible with the instructions received.

Building Systems

The most effective way to reduce building energy waste is to identify, define, and tabulate the energy load by building system. For this purpose, load is defined as the quantity of energy used in a building, or by one of its subsystems, for a given period. By following this procedure, the most effective energy conservation opportunities can be achieved more quickly because high priorities can be assigned to systems that consume the most energy.

A building can be divided into nonenergized and energized systems. Nonenergized systems do not require outside energy sources such as electricity and fuel. Energized systems (e.g., mechanical and electrical systems) require outside energy. Energized and nonenergized systems can be divided into subsystems defined by function. Nonenergized subsystems are (1) building site, envelope, and interior; (2) building use; and (3) building operation.

Building Site, Envelope, and Interior. These subsystems should be surveyed to determine how they can be modified to reduce the building load that the mechanical and electrical systems must meet without adversely affecting the building's appearance. It is important to compare actual conditions with conditions assumed by the designer, so that mechanical and electrical systems can be adjusted to balance their capacities to satisfy actual needs.

Building Use. These loads can be classified as people occupancy or operation loads. People occupancy loads are related to schedule, density, and mixing of occupancy types (e.g., process and office). People operation loads are varied and include (1) operation of manual window shading devices; (2) setting of room thermostats; and (3) conservation-related habits such as turning off lights, closing doors and windows, turning off energized equipment when not in use, and not wasting domestic hot or chilled water.

Building Operation. This subsystem consists of the operation and maintenance of all the building subsystems. The load on the building operation subsystem is affected by factors such as (1) the time at which janitorial services are performed, (2) janitorial crew size and time required to clean, (3) amount of lighting used to perform janitorial functions, (4) quality of equipment maintenance program, (5) system operational practices, and (6) equipment efficiencies.

Building Energized Systems

The energized subsystems of the building are generally plumbing, heating, ventilating, cooling, space conditioning, control, electrical, and food service. Although these systems are interrelated and often use common components, logical organization of data requires evaluating the energy use of each subsystem as independently as possible. In this way, proper energy conservation measures for each subsystem can be developed.

Process Loads

In addition to building subsystem loads, the process load in most buildings must be evaluated by the energy field auditor. Most tasks not only require energy for performance, but also affect the energy consumption of other building subsystems. For example, if a process releases large amounts of heat to the space, the process consumes energy and also imposes a large load on the cooling system.

Guidelines for Developing a Field Study Form

A brief checklist follows that outlines requirements for a field study form needed to conduct an energy audit.

Inspection and Observation of All Systems. Record physical and mechanical condition of the following:

- Fan blades, fan scroll, drives, belt tightness, and alignment
- Filters, coils, and housing tightness
- Ductwork (equipment room and space, where possible)
- Strainers
- Insulation ducts and piping
- Makeup water treatment and cooling tower

Interview of Physical Plant Supervisor. Record answers to the following survey questions:

- Is the system operating as designed? If not, what changes have been made to ensure its performance?
- Have there been modifications or additions to the system?
- If the system has had a problem, list problems by frequency of occurrence.
- Are any systems cycled? If so, which systems and when, and would building load allow cycling systems?

Recording System Information. Record the following system/equipment identification:

- Type of system: single-zone, multizone, double-duct, low- or high-velocity, reheat, variable-volume, or other

- System arrangement: fixed minimum outdoor air, no relief, gravity or power relief, economizer gravity relief, exhaust return, or other
- Air-handling equipment: fans (supply, return, and exhaust) manufacturer, model, size, type, and class; dampers (vortex, scroll, or discharge); motors manufacturer, power requirement, full load amperes, voltage, phase, and service factor
- Chilled- and hot-water coils: area, tubes on face, fin spacing, and number of rows (coil data necessary when shop drawings are not available)
- Terminals: high-pressure mixing box manufacturer, model, and type (reheat, constant-volume, variable-volume, induction); grilles, registers, and diffusers manufacturer, model, style, and loss coefficient to convert field-measured velocity to flow rate
- Main heating and cooling pumps, over 5 hp: manufacturer, pump service and identification, model, size, impeller diameter, speed, flow rate, head at full flow, and head at no flow; motor data (power, speed, voltage, amperes, and service factor)
- Refrigeration equipment: chiller manufacturer, type, model, serial number, nominal tons, brake horsepower, total heat rejection, motor (horsepower, amperes, volts), chiller pressure drop, entering and leaving chilled water temperatures, condenser pressure drop, condenser entering and leaving water temperatures, running amperes and volts, no-load running amperes and volts
- Cooling tower: manufacturer, size, type, nominal tons, range, flow rate, and entering wet-bulb temperature
- Heating equipment: boiler (small through medium) manufacturer, fuel, energy input (rated), and heat output (rated)

Recording Test Data. Record the following test data:

- Systems in normal mode of operation (if possible): fan motor running amperes and volts and power factor (over 5 hp); fan speed, total air (pitot-tube traverse where possible), and static pressure (discharge static minus inlet total); static profile drawing (static pressure across filters, heating coil, cooling coil, and dampers); static pressure at ends of runs of the system (identifying locations)
- Cooling coils: entering and leaving dry- and wet-bulb temperatures, entering and leaving water temperatures, coil pressure drop (where pressure taps permit and manufacturer's ratings can be obtained), flow rate of coil (when other than fan), outdoor wet and dry bulb, time of day, and conditions (sunny or cloudy)
- Heating coils: entering and leaving dry-bulb temperatures, entering and leaving water temperatures, coil pressure drop (where pressure taps permit and manufacturer's ratings can be obtained), and flow rate through coil (when other than fan)
- Pumps: no-flow head, full-flow discharge pressure, full-flow suction pressure, full-flow differential pressure, motor running amperes and volts, and power factor (over 5 hp)
- Chiller (under cooling load conditions): chiller pressure drop, entering and leaving chilled water temperatures, condenser pressure drop, entering and leaving condenser water temperatures, running amperes and volts, no-load running amperes and volts, chilled water on and off, and condenser water on and off
- Cooling tower: water flow rate in tower, entering and leaving water temperatures, entering and leaving wet bulb, fan motor [amperes, volts, power factor (over 5 hp), and ambient wet bulb]
- Boiler (full fire): input energy (if possible), percent CO_2, stack temperature, efficiency, and complete Orsat test on large boilers
- Boiler controls: description of operation
- Temperature controls: operating and set point temperatures for mixed air controller, leaving air controller, hot-deck controller, cold-deck controller, outdoor reset, interlock controls, and damper controls; description of complete control system and any malfunctions
- Outdoor air intake versus exhaust air: total airflow measured by pitot-tube traverses of both outdoor air intake and exhaust air systems, where possible. Determine whether an imbalance in the exhaust system causes infiltration. Observe exterior walls to determine whether outdoor air can infiltrate return air (record outdoor air, return air, and return air plenum dry- and wet-bulb temperatures). The greater the differential between outdoor and return air, the more evident the problem will be.

REPORTS

The report is a record of the HVAC system balancing and must be accurate in all respects. If the data are incomplete, it must be listed as a deficiency for correction by the installing contractor.

Reports should comply with ASHRAE *Standard* 111, be complete, and include the location of test drawings. An instrument list including serial numbers and current and future calibration dates should also be provided.

TESTING FOR SOUND AND VIBRATION

Testing for sound and vibration ensures that equipment is operating satisfactorily and that no objectionable noise and vibration are transmitted to the building structure and occupied space. Although sound and vibration are specialized fields that require expertise not normally developed by the HVAC engineer, the procedures to test HVAC are relatively simple and can be performed with a minimum of equipment by following the steps outlined in this section. Although this section provides useful information for resolving common noise and vibration problems, Chapter 48 should be consulted for information on problem solving or the design of HVAC.

Testing for Sound

Present technology does not test whether equipment is operating with desired sound levels; field tests can only determine sound pressure levels, and equipment ratings are almost always in terms of sound power levels. Until new techniques are developed, the testing engineer can only determine (1) whether sound pressure levels are within desired limits and (2) which equipment, systems, or components are the source of excessive or disturbing transmission.

Sound-Measuring Instruments. Although an experienced listener can often determine whether systems are operating in an acceptably quiet manner, sound-measuring instruments are necessary to determine whether system noise levels are in compliance with specified criteria, and if not, to obtain and report detailed information to evaluate the cause of noncompliance. Instruments normally used in field testing are as follows.

The **precision sound level meter** is used to measure sound pressure level. The most basic sound level meters measure overall sound pressure level and have up to three weighted scales that provide limited filtering capability. The instrument is useful in assessing outdoor noise levels in certain situations and can provide limited information on the low-frequency content of overall noise levels, but it provides insufficient information for problem diagnosis and solution. Its usefulness in evaluating indoor HVAC sound sources is thus limited.

Proper evaluation of HVAC sound sources requires a sound level meter capable of filtering overall sound levels into frequency increments of one octave or less.

Sound analyzers provide detailed information about sound pressure levels at various frequencies through filtering networks. The most popular sound analyzers are the octave band and center frequency, which break the sound into the eight octave bands of audible sound. Instruments are also available for 0.33, 0.1, and narrower spectrum analysis; however, these are primarily for laboratory and research applications. Sound analyzers (octave midband or center frequency) are required where specifications are based on noise criteria (NC) and room criteria (RC) curves or similar frequency criteria and for problem jobs where a knowledge of frequency is necessary to determine proper corrective action.

Personal computers are a versatile sound-measuring tool. Software used on portable computers has all the functional capabilities described previously, plus many that previously required a fully equipped acoustical laboratory. This type of sound-measuring system is many times faster and much more versatile than conventional sound level meters. With suitable accessories, it can also be used to evaluate vibration levels. Accuracy and calibration to applicable standards are of concern for software.

Regardless of which sound-measuring system is used, it should be calibrated before each use. Some systems have built-in calibration, while others use external calibrators. Much information is available on the proper application and use of sound-measuring instruments.

Air noise, caused by air flowing at a velocity of over 1000 fpm or by winds over 12 mph, can cause substantial error in sound measurements because of wind effect on the microphone. For outdoor measurements or in drafty places, either a wind screen for the microphone or a special microphone is required. When in doubt, use a wind screen on standard microphones.

Sound Level Criteria. Without specified values, the testing engineer must determine whether sound levels are within acceptable limits (Chapter 47). Note that complete absence of noise is seldom a design criterion, except for certain critical locations such as sound and recording studios. In most locations, a certain amount of noise is desirable to mask other noises and provide speech privacy; it also provides an acoustically pleasing environment, because few people can function effectively in extreme quiet. Table 1 in Chapter 8 of the 2009 *ASHRAE Handbook—Fundamentals* lists typical sound pressure levels. In determining allowable HVAC equipment noise, it is as inappropriate to demand 30 dB for a factory where the normal noise level is 75 dB as it is to specify 60 dB for a private office where the normal noise level might be 35 dB.

Most field sound-measuring instruments and techniques yield an accuracy of ±3 dB, the smallest difference in sound pressure level that the average person can discern. A reasonable tolerance for sound criteria is 5 dB; if 35 dBA is considered the maximum allowable noise, the design engineer should specify 30 dBA.

The measured sound level of any location is a combination of all sound sources present, including sound generated by HVAC equipment as well as sound from other sources such as plumbing systems and fixtures, elevators, light ballasts, and outdoor noises. In testing for sound, all sources from other than HVAC equipment are considered background or ambient noise.

Background sound measurements generally have to be made (1) when the specification requires that the sound levels from HVAC equipment only, as opposed to the sound level in a space, not exceed a certain specified level; (2) when the sound level in the space exceeds a desirable level, in which case the noise contributed by the HVAC system must be determined; and (3) in residential locations where little significant background noise is generated during the evening hours and where generally low allowable noise levels are specified or desired. Because background noise from outdoor sources such as vehicular traffic can fluctuate widely, sound measurements for residential locations are best made in the normally quiet evening hours. Procedures for residential sound measurements can be found in ASTM *Standard* E1574, Measurement of Sound in Residential Spaces.

Sound Testing. Ideally, a building should be completed and ready for occupancy before sound level tests are taken. All spaces in which readings will be taken should be furnished with whatever drapes, carpeting, and furniture are typical because these affect the room absorption, which can affect sound levels and the subjective quality of the sound. In actual practice, because most tests must be conducted before the space is completely finished and furnished for final occupancy, the testing engineer must make some allowances. Because furnishings increase the absorption coefficient and reduce by about 4 dB the sound pressure level that can be expected between most live and dead spaces, the following guidelines should suffice for measurements made in unfurnished spaces. If the sound pressure level is 5 dB or more over the specified or desired criterion, it can be

assumed that the criterion will not be met, even with the increased absorption provided by furnishings. If the sound pressure level is 0 to 4 dB greater than the specified or desired criterion, recheck when the room is furnished to determine compliance.

Follow this general procedure:

1. Obtain a complete set of accurate, as-built drawings and specifications, including duct and piping details. Review specifications to determine sound and vibration criteria and any special instructions for testing.
2. Visually check for noncompliance with plans and specifications, obvious errors, and poor workmanship. Turn system on for aural check. Listen for noise and vibration, especially duct leaks and loose fittings.
3. Adjust and balance equipment, as described in other sections, so that final acoustical tests are made with the HVAC as it will be operating. It is desirable to perform acoustical tests for both summer and winter operation, but where this is not practical, make tests for the summer operating mode, as it usually has the potential for higher sound levels. Tests must be made for all mechanical equipment and systems, including standby.
4. Check calibration of instruments.
5. Measure sound levels in all areas as required, combining measurements as indicated in step 3 if equipment or systems must be operated separately. Before final measurements are made in any particular area, survey the area using an A-weighted scale reading (dBA) to determine the location of the highest sound pressure level. Indicate this location on a testing form, and use it for test measurements. Restrict the preliminary survey to determine location of test measurements to areas that can be occupied by standing or sitting personnel. For example, measurements would not be made directly in front of a diffuser located in the ceiling, but would be made as close to the diffuser as standing or sitting personnel might be situated. In the absence of specified sound criteria, the testing engineer should measure sound pressure levels in all occupied spaces to determine compliance with criteria indicated in Chapter 48 and to locate any sources of excessive or disturbing noise. With octave band sound level measurements, overall NC and RC values can be determined. Measurements of 63 to 8000 Hz should be considered the minimum when sound levels in the 16 and 31.5 Hz octave band are desired, from completeness and evaluation of the possibility of noise-induced vibration as evaluated by RC and NCB methods.
6. Determine whether background noise measurements must be made.
   - If specification requires determining sound level from HVAC equipment only, background noise readings must be taken with HVAC equipment turned off.
   - If specification requires compliance with a specific noise level or criterion (e.g., sound levels in office areas not to exceed 35 dBA), ambient noise measurements must be made only if the noise level in any area exceeds the specified value.
   - For residential locations and areas requiring very low noise, such as sound recording studios and locations used during the normally quieter evening hours, it is usually desirable to take sound measurements in the evening and/or take ambient noise measurements.
7. For outdoor noise measurements to determine noise radiated by outdoor or roof-mounted equipment such as cooling towers and condensing units, the section on Sound Control for Outdoor Equipment in Chapter 48, which presents proper procedure and necessary calculations, should be consulted.

Noise Transmission Problems. Regardless of precautions taken by the specifying engineer and installing contractors, situations can occur where the sound level exceeds specified or desired levels, and there will be occasional complaints of noise in completed installations. A thorough understanding of Chapter 48 and the section on

Testing for Vibration in this chapter is desirable before attempting to resolve any noise and vibration transmission problems. The following is intended as an overall guide rather than a detailed problem-solving procedure.

All noise transmission problems can be evaluated in terms of the source-path-receiver concept. Objectionable transmission can be resolved by (1) reducing noise at the source by replacing defective equipment, repairing improper operation, proper balancing and adjusting, and replacing with quieter equipment; (2) attenuating paths of transmission with silencers, vibration isolators, and wall treatment to increase transmission loss; and (3) reducing or masking objectionable noise at the receiver by increasing room absorption or introducing a nonobjectionable masking sound. The following discussion includes ways to identify actual noise sources using simple instruments or no instruments and possible corrections.

When troubleshooting in the field, the engineer should listen to the offending sound. The best instruments are no substitute for careful listening, because the human ear has the remarkable ability to identify certain familiar sounds such as bearing squeak or duct leaks and can discern small changes in frequency or sound character that might not be apparent from meter readings only. The ear is also a good direction and range finder; noise generally gets louder as one approaches the source, and direction can often be determined by turning the head. Hands can also identify noise sources. Air jets from duct leaks can often be felt, and the sound of rattling or vibrating panels or parts often changes or stops when these parts are touched.

In trying to locate noise sources and transmission paths, the engineer should consider the location of the affected area. In areas remote from equipment rooms containing significant noise producers but adjacent to shafts, noise is usually the result of structure-borne transmission through pipe and duct supports and anchors. In areas adjoining, above, or below equipment rooms, noise is usually caused by openings (acoustical leaks) in the separating floor or wall or by improper, ineffective, or maladjusted vibration isolation systems.

Unless the noise source or path of transmission is quite obvious, the best way to identify it is by eliminating all sources systematically as follows:

1. Turn off all equipment to make sure that the objectionable noise is caused by the HVAC. If the noise stops, the HVAC components (compressors, fans, and pumps) must be operated separately to determine which are contributing to the objectionable noise. Where one source of disturbing noise predominates, the test can be performed starting with all equipment in operation and turning off components or systems until the disturbing noise is eliminated. Tests can also be performed starting with all equipment turned off and operating various component equipment singularly, which permits evaluation of noise from each individual component.

   Any equipment can be termed a predominant noise source if, when the equipment is shut off, the sound level drops 3 dBA or if, when measurements are taken with equipment operating individually, the sound level is within 3 dBA of the overall objectionable measurement.

   When a sound level meter is not used, it is best to start with all equipment operating and shut off components one at a time because the ear can reliably detect differences and changes in noise but not absolute levels.

2. When some part of the HVAC system is established as the source of objectionable noise, try to further isolate the source. By walking around the room, determine whether the noise is coming through air outlets or returns, hung ceiling, or floors or walls.

3. If the noise is coming through the hung ceiling, check that ducts and pipes are isolated properly and not touching the hung ceiling supports or electrical fixtures, which would provide large noise radiating surfaces. If ducts and pipes are the source of noise and are isolated properly, possible remedies to reduce noise include

changing flow conditions, installing silencers, and/or wrapping the duct or pipe with an acoustical barrier or lagging such as a lead blanket or other materials suitable for the location (see Chapter 48).

4. If noise is coming through the walls, ceiling, or floor, check for any openings to adjoining shafts or equipment rooms, and check vibration isolation systems to ensure that there is no structure-borne transmission from nearby equipment rooms or shafts.

5. Noise traced to air outlets or returns usually requires careful evaluation by an engineer or acoustical consultant to determine the source and proper corrective action (see Chapter 48). In general, air outlets can be selected to meet any acoustical design goal by keeping the velocity sufficiently low. For any given outlet, the sound level increases about 2 dB for each 10% increase in airflow velocity over the vanes, and doubling the velocity increases the sound level by about 12 to 15 dB. Approach conditions caused by improperly located control dampers or improperly sized diffuser necks can increase sound levels by 10 to 20 dB. Using variable-frequency drive (VFD) speed controllers on air-handling units can help evaluate air velocity concerns. Dampers used to limit airflow often influence overall sound levels.

   A simple, effective instrument that aids in locating noise sources is a microphone mounted on a pole. It can be used to localize noises in hard-to-reach places, such as hung ceilings and behind heavy furniture.

6. If noise is traced to an air outlet, measure the A-weighted sound level close to it but with no air blowing against the microphone. Then, remove the inner assembly or core of the air outlet and repeat the reading with the meter and the observer in exactly the same position as before. If the second reading is more than 3 dB below the first, a significant amount of noise is caused by airflow over the vanes of the diffuser or grille. In this case, check whether the system is balanced properly. As little as 10% too much air increases the sound generated by an air outlet by 2.5 dB. As a last resort, a larger air outlet could be substituted to obtain lower air velocities and hence less turbulence for the same air quality. Before this is considered, however, the air approach to the outlet should be checked.

   Noise far exceeding the normal rating of a diffuser or grille is generated when a throttled damper is installed close to it. Air jets impinge on the vanes or cones of the outlet and produce **edge tones** similar to the hiss heard when blowing against the edge of a ruler. The material of the vanes has no effect on this noise, although loose vanes may cause additional noise from vibration.

   When balancing air outlets with integral volume dampers, consider the static pressure drop across the damper, as well as the air quantity. Separate volume dampers should be installed sufficiently upstream from the outlet that there is no jet impingement. Plenum inlets should be brought in from the side, so that jets do not impinge on the outlet vanes.

7. If air outlets are eliminated as sources of excessive noise, inspect the fan room. If possible, change fan speed by about 10%. If resonance is involved, this small change can make a significant difference.

8. Sometimes fans are poorly matched to the system. If a belt-driven fan delivers air at a higher static pressure than is needed to move the design air quantity through the system, reduce fan speed by changing sheaves. If the fan does not deliver enough air, consider increasing fan speed only after checking the duct system for unavoidable losses. Turbulence in the air approach to the fan inlet increases fan sound generation and decreases its air capacity. Other parts that may cause excessive turbulence are dampers, duct bends, and sudden enlargements or contractions of the duct. When investigating fan noise, seek assistance from the fan supplier or manufacturer.

9. If additional acoustical treatment is to be installed in the duct-work, obtain a frequency analysis. This involves the use of an octave-band analyzer and should generally be left to a trained engineer or acoustic analysis consultant.

Testing for Vibration

Vibration testing is necessary to ensure that (1) equipment is operating within satisfactory vibration levels and (2) objectionable vibration and noise are not transmitted to the building structure. Although these two factors are interrelated, they are not necessarily interdependent. A different solution is required for each, and it is essential to test both the isolation and vibration levels of equipment.

General Procedure.

1. Make a visual check of all equipment for obvious errors that must be corrected immediately.
2. Make sure all isolation is free-floating and not short-circuited by obstruction between equipment or equipment base and building structure.
3. Turn on the system for an aural check of any obviously rough op-eration. Checking bearings with vibration measurement instru-mentation is especially important because bearings can become defective in transit and/or if equipment was not properly stored, installed, or maintained. Defective bearings should be replaced immediately to avoid damage to the shaft and other components.
4. Adjust and balance equipment and systems so that final vibration tests are made on equipment as it will actually be operating.
5. Test equipment vibration.

Instruments. Although instruments are not required to test vi-bration isolation systems, they are essential to test equipment vibra-tion properly.

Sound-level meters and **computer-driven sound-measuring systems** are the most useful instruments for measuring and evaluat-ing vibration. Usually, they are fitted with accelerometers or vibra-tion pickups for a full range of vibration measurement and analysis. Other instruments used for testing vibration in the field are de-scribed as follows.

Reed vibrometers are relatively inexpensive and are often used for testing vibration, but their relative inaccuracy limits their usefulness.

Vibrometers are moderately priced and measure vibration am-plitude by means of a light beam projected on a graduated scale.

Vibrographs are moderately priced mechanical instruments that measure both amplitude and frequency. They provide a chart recording amplitude, frequency, and actual wave form of vibration. They can be used for simple, accurate determination of the natural frequency of shafts, components, and systems by a **bump test**.

Reed vibrometers, vibrometers, and vibrographs have largely been supplanted by electronic meters that are more accurate and have become much more affordable.

Vibration meters are moderately priced, relatively simple-to-use modern electronic instruments that measure the vibration ampli-tude. They provide a single broadband (summation of all frequen-cies) number identifying the magnitude of the vibration level. Both analog and digital readouts are common.

Vibration analyzers are relatively expensive electronic instru-ments that measure amplitude and frequency, usually incorporating a variable filter.

Strobe lights are often used with many of the other instruments for analyzing and balancing rotating equipment.

Stethoscopes are available as inexpensive mechanic's type (basi-cally, a standard stethoscope with a probe attachment), relatively in-expensive models incorporating a tunable filter, and moderately priced powered types that electronically amplify sound and provide some type of meter and/or chart recording.

The choice of instruments depends on the test. Vibrometers and vibration meters can be used to measure vibration amplitude as an acceptance check. Because they cannot measure frequency, they

cannot be used for analysis and primarily function as a go/no-go instrument. The best acceptance criteria consider both amplitude and frequency. Anyone seriously concerned with vibration testing should use an instrument that can determine frequency as well as amplitude, such as a vibrograph or vibration analyzer.

Vibration measurement instruments (both meters and analyzers) made specifically for measuring machinery vibration typically use **moving coil velocity transducers**, which are sizable and rugged. These are typically limited to a lower frequency of 500 cycles per minute (cpm) [8.33 Hz] with normal calibration. If measuring very-low-speed machinery such as large fans, cooling towers, or compressors operating below this limit, use an adjustment factor provided by the instrument manufacturer or use an instrument with a lower low-frequency limit, which typically uses a smaller accel-erometer as the vibration pickup transducer.

Testing Vibration Isolation.

1. Ensure that equipment is **free-floating** by applying an unbal-anced load, which should cause the equipment to move freely and easily. On floor-mounted equipment, check that there are no obstructions between the base or foundation and the building structure that would cause transmission while still permitting equipment to rock relatively free because of the application of an unbalanced force (Figure 13). On suspended equipment, check that hanger rods are not touching the hanger. Rigid connections such as pipes and ducts can prohibit mounts from functioning properly and from providing a transmission path. Note that the fact that the equipment is free floating does not mean that the isolators are functioning properly. For example, a 500 rpm fan on isolators with a natural frequency of 500 cpm (8.33 Hz) could be free-floating but would actually be in resonance, resulting in transmission to the building and excessive movement.
2. Determine whether isolators are adjusted properly and providing desired isolation efficiency. All isolators supporting a piece of equipment should have approximately the same deflection (i.e., they should be compressed the same under the equipment). If not, they have been improperly adjusted, installed, or selected; this should be corrected immediately. Note that isolation effi-ciency cannot be checked by comparing vibration amplitude on equipment to amplitude on the structure (Figure 14).

The only accurate check of isolation efficiencies is to compare vibration measurements of equipment operating with isolators to measurements of equipment operating without isolators. Be-cause this is usually impractical, it is better to check whether the isolator's deflection is as specified and whether the specified or desired isolation efficiency is being provided. Figure 15 shows natural frequency of isolators as a function of deflection and in-dicates the theoretical isolation efficiencies for various frequen-cies at which the equipment operates.

Although it is easy to determine the deflection of spring mounts by measuring the difference between the free heights with a ruler (information as shown on submittal drawings or available from a manufacturer), these measurements are difficult with most pad or rubber mounts. Further, most pad and rubber mounts do not lend themselves to accurate determination of natural frequency as a func-tion of deflection. For these mounts, the most practical approach is to

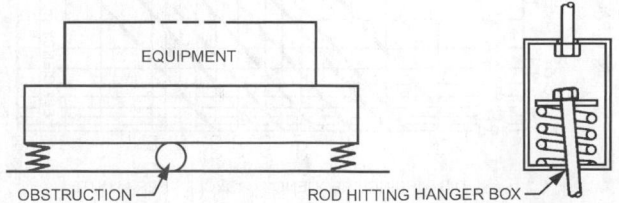

Fig. 13 Obstructed Isolation Systems

check that there is no excessive vibration of the base and no notice-able or objectionable vibration transmission to the building structure.

If isolators are in the 90% efficiency range and there is transmission to the building structure, either the equipment is operating roughly or there is a flanking path of transmission, such as connecting piping or obstruction, under the base.

Testing Equipment Vibration. Testing equipment vibration is necessary as an acceptance check to determine whether equipment is functioning properly and to ensure that objectionable vibration and noise are not transmitted. Although a person familiar with equipment can determine when it is operating roughly, instruments are usually required to determine accurately whether vibration levels are satisfactory.

Vibration Tolerances. Vibration tolerance criteria are listed in Table 45 of Chapter 48. These criteria are based on equipment installed on vibration isolators and can be met by any reasonably smoothly running equipment. Note that values in Chapter 48 are based on root-mean-square (RMS) values; other sources often use peak-to-peak or peak values, especially for displacements. For sinusoid responses, it is simple to obtain peak from RMS values, but in application the relationship may not be so straightforward. The main advantage of RMS is that the same instrumentation can be used for both sound and vibration measurements by simply changing the transducer. Also, there is only one recognized reference level

for the decibel used for sound-pressure levels, but for vibration levels there are several recognized ones but no single standard. A common mistake in interpreting vibration data is misunderstanding the reference level and whether the vibration data are RMS, peak-to-peak, or peak values. Use great care in publishing and interpreting vibration data and converting to and from linear absolute values and levels in decibels.

Procedure for Testing Equipment Vibration.

1. Determine operating speeds of equipment from nameplates, drawings, or a speed-measuring device such as a tachometer or strobe, and indicate them on the test form. For any equipment where the driving speed (motor) is different from the driven speed (fan wheel, rotor, impeller) because of belt drive or gear reducers, indicate both driving and driven speeds.
2. Determine acceptance criteria from specifications, and indicate them on the test form. If specifications do not provide criteria, use those shown in Chapter 48.
3. Ensure that the vibration isolation system is functioning properly (see the section on Testing Vibration Isolation).
4. Operate equipment and make visual and aural checks for any apparent rough operation. Any defective bearings, misalignment, or obvious rough operation should be corrected before proceeding further. If not corrected, equipment should be considered unacceptable.
5. Measure and record vibration at bearings of driving and driven components in horizontal, vertical, and, if possible, axial directions. At least one axial measurement should be made for each rotating component (fan motor, pump motor).
6. Evaluate measurements.

Evaluating Vibration Measurements.

Amplitude Measurement. When specification for acceptable equipment vibration is based on amplitude measurements only, and measurements are made with an instrument that measures only amplitude (e.g., a vibration meter or vibrometer),

- No measurement should exceed specified values or values shown in Tables 45 or 46 of Chapter 48, taking into consideration reduced values for equipment installed on inertia blocks
- No measurement should exceed values shown in Tables 45 or 46 of Chapter 48 for driving and driven speeds, taking into consideration reduced values for equipment installed on inertia blocks. For example, with a belt-driven fan operating at 800 rpm and having an 1800 rpm driving motor, amplitude measurements at fan bearings must be in accordance with values shown for 800 cpm (13.3 Hz), and measurements at motor bearings must be in accordance with values shown for 1800 cpm (30 Hz). If measurements at motor bearings exceed specified values, take measurements of the motor only with belts removed to determine whether there is feedback vibration from the fan.
- No axial vibration measurement should exceed maximum radial (vertical or horizontal) vibration at the same location.

Amplitude and Frequency Measurement. When specification for acceptable equipment vibration is based on both amplitude and frequency measurements, and measurements are made with instruments that measure both amplitude and frequency (e.g., a vibrograph or vibration analyzer),

- Amplitude measurements at driving and driven speeds should not exceed specified values or values shown in Tables 45 or 46 of Chapter 48, taking into consideration reduced values for equipment installed on inertia blocks. Measurements that exceed acceptable amounts may be evaluated as explained in the section on Vibration Analysis.
- Axial vibration measurements should not exceed maximum radial (vertical or horizontal) vibration at the same location.
- The presence of any vibration at frequencies other than driving or driven speeds is generally reason to rate operation unacceptable;

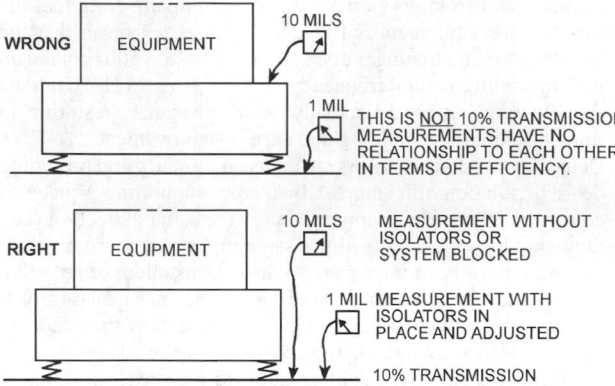

Fig. 14   Testing Isolation Efficiency

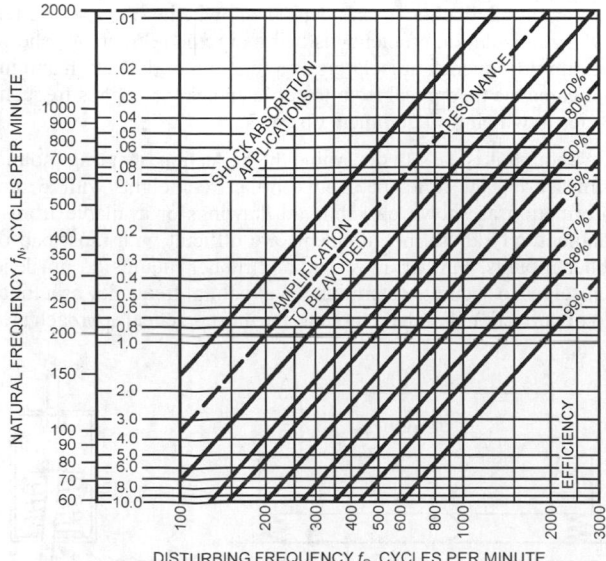

Fig. 15   Isolator Natural Frequencies and Efficiencies

this vibration should be analyzed as explained in the section on Vibration Analysis.

Vibration Analysis. The following guide covers most vibration problems that may be encountered.

Axial Vibration Exceeds Radial Vibration. When the amplitude of axial vibration (parallel with shaft) at any bearing exceeds radial vibration (perpendicular to shaft, vertical or horizontal), it usually indicates misalignment, most common on direct-driven equipment because flexible couplings accommodate parallel and angular misalignment of shafts. This misalignment can generate forces that cause axial vibration, which can cause premature bearing failure, so misalignment should be checked carefully and corrected promptly. Other possible causes of large-amplitude axial vibration are resonance, defective bearings, insufficient rigidity of bearing supports or equipment, and loose hold-down bolts.

Vibration Amplitude Exceeds Allowable Tolerance at Rotational Speed. The allowable vibration limits established by Table 41 of Chapter 48 are based on vibration caused by rotor imbalance, which results in vibration at rotational frequency. Although vibration caused by imbalance must be at the frequency at which the part is rotating, a vibration at rotational frequency does not have to be caused by imbalance. An unbalanced rotating part develops centrifugal force, which causes it to vibrate at rotational frequency. Vibration at rotational frequency can also result from other conditions such as a bent shaft, an eccentric sheave, misalignment, and resonance. If vibration amplitude exceeds allowable tolerance at rotational frequency, the following steps should be taken before performing field balancing of rotating parts:

1. Check vibration amplitude as equipment goes up to operating speed and as it coasts to a stop. Any significant peaks at or near operating speed, as shown in Figure 16, indicate probable resonance (i.e., some part having a natural frequency close to the operating speed, resulting in greatly amplified levels of vibration).

   A bent shaft or eccentricity usually causes imbalance that results in significantly higher vibration amplitude at lower speeds, as shown in Figure 17, whereas vibration caused by imbalance generally increases as speed increases.

   If a bent shaft or eccentricity is suspected, check the dial indicator. A bent shaft or eccentricity between bearings as shown in Figure 18A can usually be compensated for by field balancing, although some axial vibration might remain. Field balancing cannot correct vibration caused by a bent shaft on direct-connected equipment, on belt-driven equipment where the shaft is bent at the location of sheave, or if the sheave is eccentric (Figure 18B). This is because the center-to-center distance of the sheaves fluctuates, each revolution resulting in vibration.

2. For belt- or gear-driven equipment where vibration is at motor driving frequency rather than driven speed, it is best to disconnect the drive to perform tests. If the vibration amplitude of the motor operating by itself does not exceed specified or allowable values, excessive vibration (when the drive is connected) is probably a function of bent shaft, misalignment, eccentricity, resonance, or loose holddown bolts.

3. Vibration caused by imbalance can be corrected in the field by firms specializing in this service or by testing personnel if they have appropriate equipment and experience.

Vibration at Other than Rotational Frequency. Vibration at frequencies other than driving and driven speeds is generally considered unacceptable. Table 4 shows some common conditions that can cause vibration at other than rotational frequency.

Resonance. If resonance is suspected, determine which part of the system is in resonance.

Isolation Mounts. The natural frequency of the most commonly used spring mounts is a function of spring deflection, as shown in Figure 19, and it is relatively easy to calculate by determining the difference between the free and operating height of the mount, as

Table 4 Common Causes of Vibration Other than Unbalance at Rotation Frequency

Frequency	Source
0.5 × rpm	Vibration at approximately 0.5 rpm can result from improperly loaded sleeve bearings. This vibration will usually disappear suddenly as equipment coasts down from operating speed.
2 × rpm	Equipment is not tightly secured or bolted down.
2 × rpm	Misalignment of couplings or shafts usually results in vibration at twice rotational frequency and generally a relatively high axial vibration.
Many × rpm	Defective antifriction (ball, roller) bearings usually result in low-amplitude, high-frequency, erratic vibration. Because defective bearings usually produce noise rather than any significantly measurable vibration, it is best to check all bearings with a listening device.

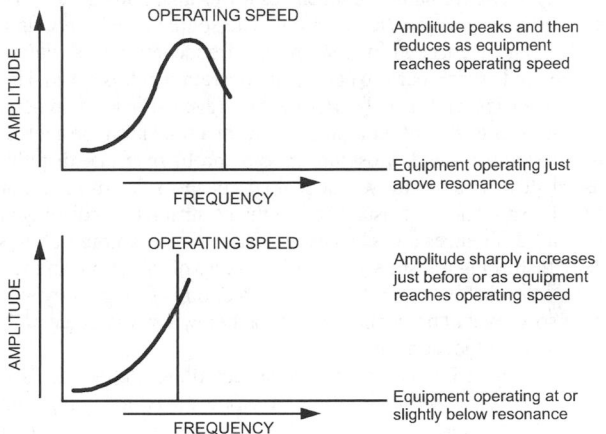

Fig. 16 Vibration from Resonant Condition

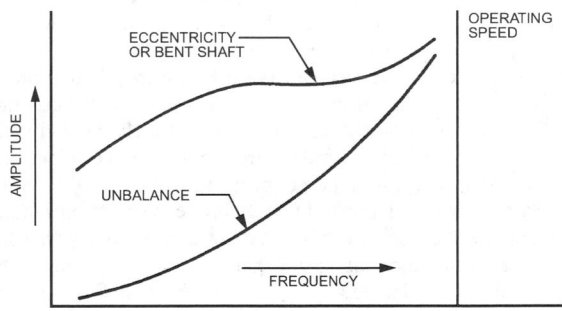

Fig. 17 Vibration Caused by Eccentricity

Fig. 18 Bent Shafts

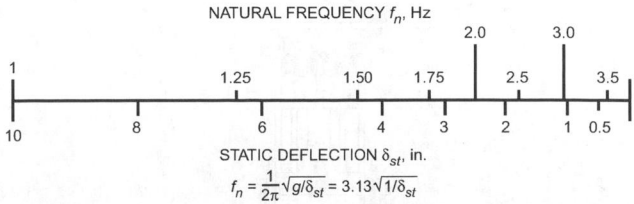

$$f_n = \frac{1}{2\pi}\sqrt{g/\delta_{st}} = 3.13\sqrt{1/\delta_{st}}$$

Fig. 19 Natural Frequency of Vibration Isolators

explained in the section on Testing Vibration Isolation. This technique cannot be applied to rubber, pad, or fiberglass mounts, which have a natural frequency in the 300 to 3000 cpm (5 to 50 Hz) range. Natural frequency for such mounts is determined by a bump test. Any resonance with isolators should be immediately corrected because it results in excessive movement of equipment and more transmission to the building structure than if equipment were attached solidly to the building (installed without isolators).

Components. Resonance can occur with any shaft, structural base, casing, and connected piping. The easiest way to determine natural frequency is to perform a bump test with a vibration spectrum analyzer. This test consists of bumping the part and measuring; the part will vibrate at its natural frequency, which will show up as a response peak on the analyzer. However, most of these instruments are restricted to frequencies above 500 cpm (8.3 Hz). They therefore cannot be used to determine natural frequencies of most isolation systems, which usually have natural frequencies lower than 500 cpm (8.3 Hz).

Checking for Vibration Transmission. The source of vibration transmission can be checked by determining frequency with a vibration analyzer and tracing back to equipment operating at this speed. However, the easiest and usually best method (even if test equipment is being used) is to shut off components one at a time until the source of transmission is located. Most transmission problems cause disturbing noise; listening is the most practical approach to determine a noise source because the ear is usually better than instruments at distinguishing small differences and changes in character and amount of noise. Where disturbing transmission consists solely of vibration, an instrument will probably be helpful, unless vibration is significantly above the sensory level of perception. Vibration below sensory perception is generally not objectionable.

If equipment is located near the affected area, check isolation mounts and equipment vibration. If vibration is not being transmitted through the base, or if the area is remote from equipment, the probable cause is transmission through connected piping and/ or ducts. Ducts can usually be isolated by isolation hangers. However, transmission through connected piping is very common and presents many problems that should be understood before attempting to correct them (see the following section).

Vibration and Noise Transmission in Piping. Vibration and noise in connected piping can be generated by either equipment (e.g., pump or compressor) or flow (velocity). Mechanical vibration from equipment can be transmitted through the walls of pipes or by a water column. Flexible pipe connectors, which provide system flexibility to permit isolators to function properly and protect equipment from stress caused by misalignment and thermal expansion, can be useful in attenuating mechanical vibration transmitted through a pipe wall. However, they rarely suppress flow vibration and noise and only slightly attenuate mechanical vibration as transmitted through a water column.

Tie rods are often used with flexible rubber hose and rubber expansion joints (Figure 20). Although they accommodate thermal movements, they hinder vibration and noise isolation. This is because pressure in the system causes the hose or joint to expand until resilient washers under tie rods are virtually rigid. To isolate noise adequately with a flexible rubber connector, tie rods and anchor piping should not be used. However, this technique generally

cannot be used with pumps on spring mounts, which would still permit the hose to elongate. Flexible metal hose can be used with spring-isolated pumps because wire braid serves as tie rods; metal hose controls vibration but not noise.

Problems of transmission through connected piping are best resolved by changes in the system to reduce noise (improve flow characteristics, reduce impeller size) or by completely isolating piping from the building structure. Note, however, that it is almost impossible to isolate piping completely from the structure, because the required resiliency is inconsistent with rigidity requirements of pipe anchors and guides. Chapter 48 contains information on flexible pipe connectors and resilient pipe supports, anchors, and guides, which should help resolve any piping noise transmission problems.

REFERENCES

AMCA. 1999. Laboratory methods of testing fans for rating. *Standard* 210-99/ASHRAE *Standard* 51-1999. Air Movement and Control Association, Arlington Heights, IL.

ASHRAE. 2010. Energy standard for buildings except low-rise residential buildings. ANSI/ASHRAE/IESNA *Standard* 90.1-2010.

ASHRAE. 2008. Measurement, testing, adjusting and balancing of building HVAC systems. *Standard* 111-2008.

ASHRAE. 1991. *Terminology of HVAC&R*, 2nd ed.

ASME. 2003. Atmospheric water cooling equipment. *Standard* PTC 23-03. American Society of Mechanical Engineers, New York.

ASME. 2010. *Boiler and pressure vessel code*, Section VI. American Society of Mechanical Engineers, New York.

ASTM. 2006. Standard test method for measurement of sound in residential spaces. *Standard* E1574-98 (RA 2006). American Society for Testing and Materials, West Conshohocken, PA.

CTI. 1997. Standard specifications for thermal testing of wet/dry cooling towers. *Standard Specification* ATC-105. Cooling Tower Institute, Houston.

Griggs, E.I., W.B. Swim, and H.G. Yoon. 1990. Duct velocity profiles and the placement of air control sensors. *ASHRAE Transactions* 96(1):523-541.

Sauer, H.J. and R.H. Howell. 1990. Airflow measurements at coil faces with vane anemometers: Statistical correction and recommended field measurement procedure. *ASHRAE Transactions* 96(1):502-511.

BIBLIOGRAPHY

AABC. 2002. *National standards for total system balance*, 6th ed. Associated Air Balance Council, Washington, D.C.

AABC. 1997. *Test and balance procedures*. Associated Air Balance Council, Washington, D.C.

AMCA. 2007. *Fan application manual*. Air Movement and Control Association, Arlington Heights, IL.

Armstrong Pump. 1986. *Technology of balancing hydronic heating and cooling systems*. Armstrong Pump, North Tonawanda, NY.

ASA. 2006. Specification for sound level meters. *Standard* 1.4-83 (R 2006). Acoustical Society of America, New York.

ASHRAE. 2007. The HVAC commissioning process. *Guideline* 1-2007.

Coad, W.J. 1985. Variable flow in hydronic systems for improved stability, simplicity and energy economics. *ASHRAE Transactions* 91(1B):224-237.

Eads, W.G. 1983. Testing, balancing and adjusting of environmental systems. In *Fan Engineering*, 8th ed. Buffalo Forge Company, Buffalo, NY.

Gladstone, J. 1981. *Air conditioning—Testing and balancing: A field practice manual*. Van Nostrand Reinhold, New York.

Gupton, G. 1989. *HVAC controls, operation and maintenance*. Van Nostrand Reinhold, New York.

Haines, R.W. 1987. *Control systems for heating, ventilating and air conditioning*, 4th ed. Van Nostrand Reinhold, New York.

Hansen, E.G. 1985. *Hydronic system design and operation*. McGraw-Hill, New York.

Miller, R.W. 1983. *Flow measurement engineering handbook*. McGraw-Hill, New York.

NEBB. 2005. *Procedural standards for testing, balancing and adjusting of environmental systems*, 7th ed. National Environmental Balancing Bureau, Vienna, VA.

NEBB. 1986. *Testing, adjusting, balancing manual for technicians*, 1st ed. National Environmental Balancing Bureau, Vienna, VA.

SMACNA. 2002. *HVAC systems—Testing, adjusting and balancing*, 3rd ed. Sheet Metal and Air Conditioning Contractors' National Association, Merrifield, VA.

SMACNA. 1985. *HVAC air duct leakage test manual*, 1st ed. Sheet Metal and Air Conditioning Contractors' National Association, Merrifield, VA.

Trane Company. 1996. *Trane air conditioning manual*. Trane, LaCrosse, WI.

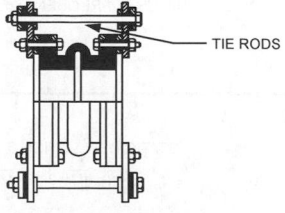

Fig. 20　Typical Tie Rod Assembly

CHAPTER 39

OPERATION AND MAINTENANCE MANAGEMENT

CAPITAL investments provide most of the built environment humankind enjoys today. To derive the greatest return on the capital investment, and to ensure future generations will continue to enjoy these benefits, the built environment must be sustainable. Significant components of the sustainable facility are the ways in which the structure and its systems are operated and preserved for the long term. This chapter presents several strategies, methods, procedures, and techniques for building operation and maintenance management programs that minimize asset failure and preserve system function to deliver its intended purpose.

Evolving building system complexity and increasing operating costs demand that equipment and systems providing thermal comfort and beneficial indoor air quality be properly maintained to achieve energy efficiency and building owner reliability requirements. These factors clearly imply that a highly organized, systematic approach for properly and effectively functioning building assets is necessary to achieve a successful maintenance program. Maintenance management is the formal effort required to plan, design, and implement a maintenance program tailored to the specific needs of the facility.

Traditionally, considerable focus has been devoted to minimizing first costs (i.e., capital investment) of construction. However, choices made regarding operation and maintenance (O&M) can have a greater impact on costs of ownership over the entire life of a facility.

OPERATION AND MAINTENANCE AS PART OF BUILDING LIFE-CYCLE COSTS

Operation and maintenance are major contributors to the total costs of ownership over the life of the facility. It is useful to compare maintenance costs to the total costs of facility ownership. In general, the major categories of the life-cycle cost of facility ownership include design and construction; operations and maintenance; acquisition, renewal, and disposal; and employee salaries and benefits. Life-cycle costs of a facility are distributed as shown in Figures 1, 2, and 3.

Figure 1 represents the major categories of facility life-cycle costs as pillars supporting a building. Within each category, examples of the typical elements that make up the foundation of the category are shown.

Figure 2 (Christian and Pandeya 1997) illustrates the relative financial resources required to sustain each phase of facility ownership when the major categories are defined as design and construction; operations and maintenance; and acquisition, renewal, and disposal. Over a nominal 30-year term, O&M comprises the largest segment of facility ownership cost: between 60 and 85% of the life-cycle cost. Figure 3 (NIBS 1998) categorizes life-cycle costs

and design and construction, operations and maintenance, and employee salaries and benefits.

From Figures 2 and 3, it is clear that operation and maintenance costs contribute a considerable amount to the total cost of ownership over the life cycle of any facility. Figure 4 (BOMA 2008) further defines operations and maintenance costs for a typical office building and shows that operations and maintenance costs most directly related to HVAC&R (i.e., maintenance and repair and utilities) make up over 50% of the operations costs for a building. Given the life-cycle cost of a facility, the operations and maintenance staff should be involved with the predesign, design, and construction

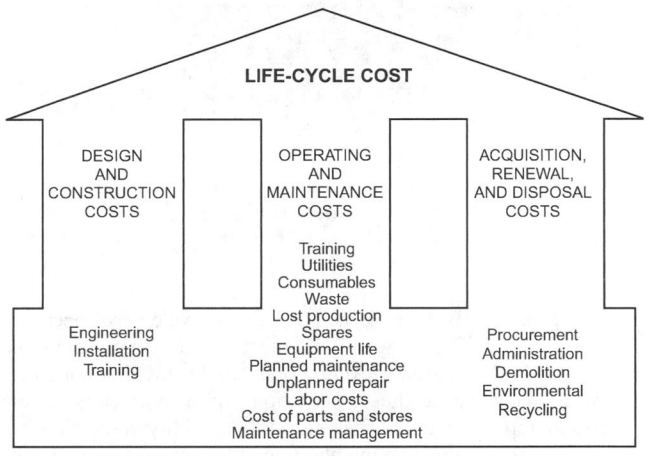

Fig. 1 Three Pillars of Typical Life-Cycle Cost with Cost Elements

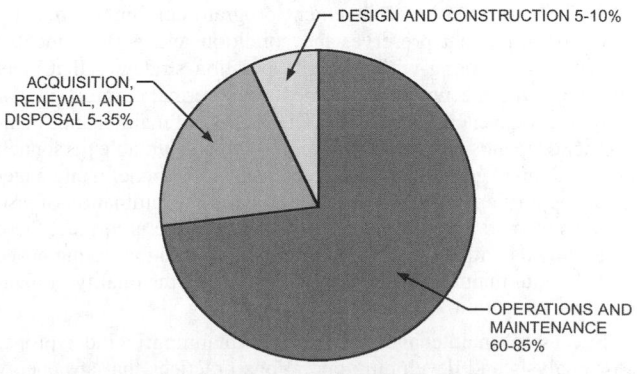

Fig. 2 Life-Cycle Cost Elements for Developing, Operating, and Maintaining Nonresidential Buildings
(Adapted from Christian and Pandeya 1997)

The preparation of this chapter is assigned to TC 7.3, Operation and Maintenance Management. Significant content on fault detection and diagnostics was provided by TC 7.5, Smart Building Systems.

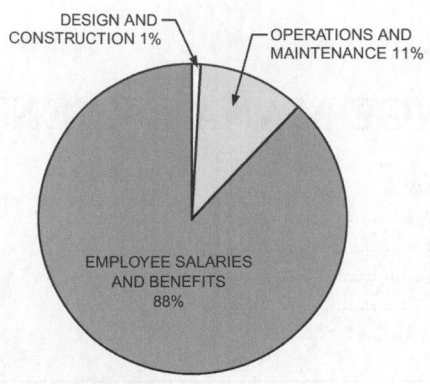

Fig. 3   Life-Cycle Cost Elements: Business Costs for Nonresidential Buildings, Including Salaries and Benefits to Occupants
(Adapted from NIBS 1998)

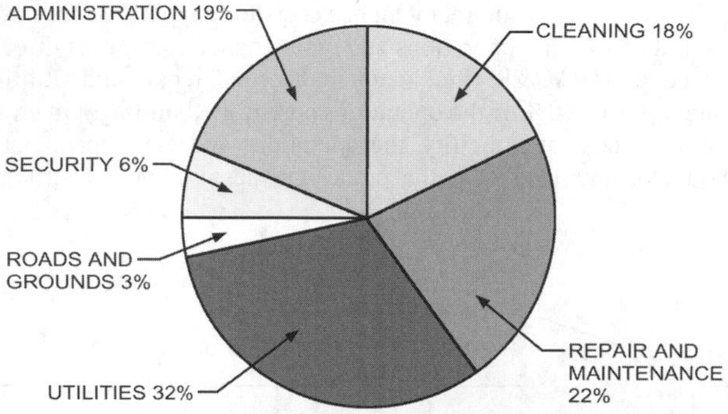

Fig. 4   Operations and Maintenance Cost Elements for Typical Office Building
(Adapted from BOMA 2008)

phases of the facility. During each phase, life-cycle costs, occupant comfort, energy efficiency, and maintenance strategies should be considered in all decision-making. For example, any first-cost compromises made during the construction phase must consider the long-term impacts on both life-cycle cost and ability to satisfy occupant thermal comfort while maintaining indoor air quality with efficient energy use.

ELEMENTS OF SUCCESSFUL PROGRAMS

A successful O&M management program combines a maintenance program that preserves the condition and performance of building assets along with prudent operation strategies that meet building owner/occupant requirements for thermal comfort, indoor air quality, and energy efficiency. The success of maintenance management depends on dedicated, trained, and accountable personnel; clearly defined goals and objectives; measurable benefits; management support; and constant examination and reexamination of process results and goal achievement. The challenge is to meet these requirements with fixed or declining budgets and increasing costs. A successful maintenance program depends on the quality of planning and execution.

Successful maintenance management planning includes proper cost analysis and developing operations practices that are energy efficient and minimize equipment downtime and disruption to building occupants and/or processes. With increasing equipment and system complexity, appropriate technical expertise (whether

in-house or contracted) is key to keeping systems running in optimal fashion.

Successful facility ownership requires investing appropriately in acquisition, design, and construction practices that consider future O&M expenses. Low first cost of capital projects or replacement of obsolete equipment must be weighed against the long-term effects of potentially more frequent and/or more time-consuming maintenance and repair, resulting in higher operating costs. In addition, maintenance management should consider an optimum mix of techniques to minimize maintenance costs and mitigate risk of equipment and system failure.

Critical components of a facilities maintenance program plan include an inventory of assets to be maintained, program performance objectives, defined condition indicators, inspection and work tasks created to achieve the performance objectives, documentation, and a process to review changes in asset condition and task effectiveness.

Creating and Implementing an Effective O&M Program

An effective operation and maintenance program is created from the preceding elements. In addition to the inventory of maintained assets, a maintenance plan should be developed to describe the goals, objectives, and implementation procedures of the maintenance program. The maintenance plan is written for the size, scope, age, and complexity of the specific systems and equipment serving the facility. The plan describes all required tasks, their frequency, salient condition indicators and the parties responsible for performing the work, documentation, and program monitoring tasks.

The following is a list of important terms to understand when developing a maintenance program:

Maintenance management is the planning, implementation, and review of maintenance activities. Specific levels of maintenance rigor should be established, ideally determined by cost-effectiveness, health, and safety concerns. Cost-effectiveness is the balance among system effectiveness, including maintenance levels, equipment and system availability, reliability, maintainability, performance, and life-cycle costs.

Performance objectives can be addressed in many ways. Desired outcomes can be measured in terms of equipment and system deliverables such as thermal comfort, energy efficiency, and indoor air quality. Other measures include up-time, mean time between failure, mean time to repair, and normalized cost data. The plan should include the source of the objectives.

Condition indicators rely on descriptions of unacceptable equipment and system conditions, which can be based on physical condition and/or performance output. In general, condition indicators are measurements and observations of conditions known to lead to equipment and system failure or performance degradation. Presence of these indicators is a signal that remedial actions should be expedited to avoid failures or larger repair costs later on.

Inspection and maintenance tasks are developed to minimize the risk of equipment and system failure to achieve operating performance goals for the facility. Inspection tasks are devoted to asset condition assessment. When abnormal conditions are found, the cause must be investigated and remedied. Maintenance tasks in general comprise a series of activities such as cleaning, adjustment, service, or replacement as conditions warrant. Arguably, unacceptable conditions require more significant repair work.

Task frequency for inspection and maintenance of new equipment and systems is initially established based on the manufacturer's recommendations. Based on the rate of asset condition or performance degradation, the task frequency can be adjusted to provide cost-effective assurance that the asset is in acceptable condition. The monitoring and review elements of a maintenance program are important for keeping maintenance costs in control. For maintenance programs established for equipment and systems in service past the warranty period, task frequencies based on O&M staff experience or advice from a maintenance engineer may be used.

Documentation is a critical element of any maintenance program and includes identifying the asset's capacity, its location, the list of inspection and maintenance tasks and their frequencies, results of inspections and maintenance, and verification that inspection and maintenance tasks have been carried out. Archived results of prior conditions and maintenance are key factors in determining whether task frequencies can be adjusted.

Setting Clearly Defined Maintenance Goals

Maintenance management goals are closely related and interdependent with facility operating goals and objectives. The maintenance program must account for building operating strategy and procedures, and for the criticality of the equipment and system in question. In general, the maintenance program is established to mitigate asset degradation and failure while enabling these assets to deliver the required indoor environment. Goals must account for expected funding and human capital resources, age of equipment and systems, and capital projects planned for the near term.

The facility's operating strategy dictates when equipment and systems can be shut down for inspection and maintenance. In some systems for which continuous operation is a critical requirement, outages must be scheduled in advance or workarounds must be developed. In extreme cases, there may need to be redundant capacity to allow off-line maintenance without interrupting operation of the equipment or system. In general, maintenance goals are focused on equipment reliability or uptime. Either task intervals or procedures are developed to eliminate or minimize asset downtime.

Clearly defined, measurable goals provide targets against which to measure progress and serve to align the facility maintenance and maintenance management staff with the facility operating strategy, required resources and acceptable levels of risk. It is helpful to understand the interrelationship between several terms that are related to asset performance when establishing program goals. Common descriptors of desired outcomes include the following:

- **Availability** is the amount of time that machinery or equipment is operable when needed. Often referred to as *uptime*, availability improvements lead to increased production for manufacturing and industrial companies. Availability is often confused with reliability, which is a component of availability.
- **Capability** is a system's ability to satisfactorily provide required service. It is the probability of meeting functional requirements when operating under a designated set of conditions. An example of capability is the ability of a heating system to meet the heating load at the winter design temperature. Capability must be verified when the system is first commissioned and whenever functional requirements change.
- **Deliverability** is the total amount of production delivered to users per unit of time. High deliverability is clearly supported by levels of the other concepts.
- **Dependability** is the measure of a system's condition. Assuming the system was operative at the beginning of its service life, dependability is the probability of its operating at any other given time until the end of its life. For systems that cannot be repaired during use, dependability is the probability that there will not be a failure during use. For systems that can be repaired, dependability is governed by how easily and quickly repairs can be made.
- **Durability** is the average expected service life of a system or facility. Table 3 of Chapter 37 lists median years of equipment service life. Individual manufacturers quantify durability as design life, which is the average number of hours of operation before failure, extrapolated from accelerated life tests and from stressing critical components to economic destruction.
- **Maintainability** is the ease, accuracy, safety, and economy of performing maintenance. The purpose of maintainability is to improve the effectiveness and efficiency of maintenance so that up time may be optimized. Maintainability is an important design consideration and contributes to availability.

For some industries, maintainability is quantitative, corresponding to the probability of performing a maintenance action or repair in a specified period of time using prescribed procedures in a specified environment. For others, maintainability is simply the ease with which maintenance actions can be performed.

- **Operability** is the efficient conversion of labor, raw materials, and energy into product, in which the value of the ratio of output to input is optimal. Maintainability and reliability contribute to availability. High levels of availability minimize the input required for a given output, thereby contributing to high levels of operability.
- **Reliability** is the probability that a system or facility will perform its intended function for a specified period of time when used under specific conditions and environment. Issues affecting reliability include operating practices, equipment and system design, installation, and maintenance practices. Reliability contributes to availability.
- **Sustainability** is "providing for the needs of the present without detracting from the ability to fulfill the needs of the future" (ASHRAE 2010). Sustainable maintenance management includes identifying and reducing a building's detrimental environmental impacts during its operating life. Sustainability in buildings cannot be achieved simply through sustainable design practices, but also requires sustainable operation and maintenance.

There are multiple perspectives of the interrelationship among maintainability, reliability, and availability. Maintainability and reliability are often grouped together, because they deal with essentially the same design concepts from different perspectives. Reliability analyses often serve as input to maintainability analyses. Here, maintainability and reliability are considered independent concepts, both of which contribute to availability. Increased availability and operability can significantly improve profitability. Figure 5 shows these relationships: reliability and maintainability combine to yield availability, which then combines with operability to yield deliverability. Improvements in any area (all else being equal) ultimately lead to improved results or profitability, because these concepts are interrelated.

For example, good reliability and availability may minimize or eliminate the need for installed spare capacity, reducing the facility's footprint and resource consumption. Good operability increases efficiency and reduces energy use.

The facility operating strategy should embody the preceding concepts, which can be described in measurable outcomes that support achieving the goals of the maintenance program. A maintenance program's effectiveness can be measured by its impact on profitability. Cost-effective improvements to any of the concepts shown in Figure 5 will contribute to profitability and the maintenance program's effectiveness. In addition, goals may be established to indicate how efficiently the maintenance program is implemented. Good maintenance management monitors progress toward achieving goals over time. The trends are more important than the incremental values.

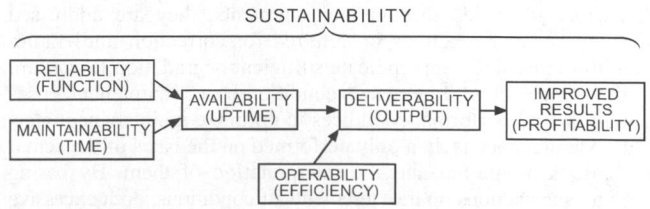

Fig. 5 General Interrelationship of Concepts

Comparing recent with historical data on condition and performance provides valuable input for an effective maintenance program.

Choosing the Best Combination of O&M Strategies

Operation is the processes and methods used when the building is working correctly, in contrast to maintenance, which is the processes and methods necessary to repair and replace equipment. Operation of a building includes a set of procedures and standards to keep systems and equipment operating as designed. Daily or weekly walkthroughs of mechanical rooms, roofs with mechanical equipment, and other locations with mechanical equipment should be performed to help identify equipment conditions that adversely affect equipment operation.

A building automation system (BAS) is an important tool to measure, benchmark, and analyze energy consumption and other operational data over the building's life. To use a BAS to track whole-building, system, and equipment performance, it is important to make sure the BAS has been properly commissioned, has an appropriate number of alarms set, and has properly set-up trends logs to collect necessary data. For more information about benchmarking and analyzing energy use, see Chapter 36.

There are three basic maintenance strategies. In **run-to-failure maintenance**, minimal or no resources are invested in maintenance until equipment or systems break down (i.e., fail). **Preventive maintenance** is scheduled, either by run time or by the calendar. **Condition-based and predictive maintenance** uses predictions of future equipment condition to optimize maintenance actions. Within each group, the skills and management tools required can vary from simple to complex. Maintenance programs may incorporate features of all three approaches into a single program. Many arguments can be made about the cost-effectiveness of each of these programs.

Operation and maintenance costs represent a significant portion of a facility's total life-cycle cost (see Figure 2). Therefore, the cost-effectiveness of maintenance management is paramount.

Run-to-Failure and Repair. Failure is the inability of a system or equipment to perform its intended function at an acceptable level. Run-to-failure is applied when the cost of maintenance or repair may exceed the cost of replacement or losses in the event of failure. Only minimum maintenance, such as cleaning or filter change, is performed. The equipment may or may not be monitored for proper operation, depending on the consequences of failure. This is a highly reactive approach to dealing with abnormal conditions. For example, a window air conditioner may be run although it is vibrating and making noise, then be replaced rather than repaired (a failure-triggered response in which operation is fully restored without embellishment).

Preventive Maintenance. Preventive maintenance classifies available resources to ensure proper operation of a system or equipment under the maintenance program. Durability, reliability, efficiency, and safety are the principal objectives. Preventive maintenance is scheduled, based on run time or by the calendar.

Condition-Based and Predictive Maintenance. Many repairs are a direct result of maintenance-induced faults occurring during scheduled maintenance. Condition-based maintenance uses measurements of the performance of equipment and systems to guide maintenance activities. As performance degradation or operational faults are identified through measurements, they are addressed through corrective actions or deferred for correction until future condition measurements indicate sufficient degradation to warrant maintenance action. In essence, condition-based maintenance uses condition and performance indices to optimize maintenance intervals. Maintenance is then only performed on the basis of the actual conditions monitored and the interpretation of them. By basing maintenance actions on measurements of conditions, both excessive and too-infrequent maintenance can be avoided, thereby optimizing plant reliability and maximizing maintenance personnel productiv-

ity. There can, however, be added costs involved in supplementary training and instrumentation that must be evaluated for cost effectiveness relative to the benefits of other maintenance strategies.

Measurements can be performed continuously or periodically (e.g., weekly, monthly). Continuous monitoring provides continual information, enabling identification of catastrophic failures (e.g., failure of a motor or compressor) as they occur. Faults and performance degradation are detected by comparing the monitored indicator of performance with expected values, based on benchmarks (e.g., manufacturer specifications, historical performance, values from models). In some cases, alarms based on measured values of conditions exceeding known threshold values (e.g., maximum acceptable temperature, tabulated values of acceptable vibration) are used to notify operations and maintenance staff that action may be warranted to correct conditions. For complex systems, it may be necessary to monitor several conditions (e.g., temperatures, vibration, and load) to assess the overall system condition. Trending over a period of time, either using measurements of conditions at intervals over time or continuous measurements can be used to track gradual deterioration and remedial work can be performed when deterioration reaches a critical level or planned in advance based on the rate of deterioration (which is a predictive approach).

In the case of air filters in a variable-air-volume (VAV) system in which the supply fan draws air through the filter and is controlled to a fixed static pressure, the criterion for replacing the filter is a function of the maximum differential pressure the fully loaded filter can withstand without bursting and the energy consumption of the fan as the filter loading increases. It may be more economical to change the filter before the bursting pressure is reached if the rate of loading is slow. Continuous monitoring with a building automation system rather than changing filters on a fixed time schedule makes it possible to detect rapidly accelerated dust loading (e.g., by nearby construction work), so that the system alerts building staff before filters are overloaded and could potentially burst.

In a constant-volume system (again, assuming the main fan draws air through the filter), the filter change criterion is a function of the maximum differential pressure the filter can withstand without bursting and the drop in flow rate that can be tolerated by the system. As the filter becomes dirty, energy consumption increases and filter efficiency decreases. A fixed time interval for filter bank changing/cleaning may not be optimum. Changing the filter based on a monitored condition optimizes filter life and minimizes labor costs.

Routine operating plant inspection conducted by the technician during regularly scheduled plant tours can be an effective condition-monitoring practice. A technician's knowledge, experience, and familiarity with the plant can be valuable tools in plant diagnostics. Plant familiarity, however, is valuable only when based on a solid knowledge of the underlying physical principles, and it is lost with frequent technician staff changes. Many physical parameters or conditions can be measured objectively using both special equipment and conventional building automation system sensors. One such condition is vibration on rolling element bearings. Vibration data are captured by computers. Special software is used to analyze the data to determine whether shaft alignment is correct, whether there are excessively unbalanced forces in the rotating mass, the state of bearing lubrication, and/or faults with the fixed or moving bearing surfaces or rolling elements. Not only can this technique be used to diagnose faults and determine repair requirements at an early stage, but it can also be used after completing a repair to ensure that the underlying cause of fault has been removed.

Other techniques include (1) using thermal infrared images of electrical connections to determine whether mechanical joints are tight, (2) analyzing oil and grease for contamination (e.g., water in fuel oil on diesel engines), (3) analyzing electrical current to diagnose motor winding faults, (4) measuring differential pressure across filter banks and heat exchangers to determine optimum change/cleaning frequency, and (5) measuring temperature differences for correct control valve response, chiller operation, and air handling.

In some cases, multiple parameters are required. For example, to determine the degree of contamination of an air filter bank on a VAV system, it is necessary to measure the differential pressure across the filter and interpret it in terms of the actual flow rate through the filter. This can be done either by forcing the fan on to high speed and then measuring the differential pressure, or by combining a flow rate signal with a differential pressure signal.

With predictive maintenance, indicators of performance degradation are extrapolated into the future to predict when an unacceptable degree of degradation will be reached and repair or maintenance should be performed. The extrapolation is based on a statistically valid approach that accounts for uncertainty in the future projections. The resulting projections of the time to failure (or unacceptable performance) can be used to plan maintenance in advance. Predictions can be based on nearly any indicator of performance degradation (e.g., pressure drop across a filter) that changes gradually over time. Several techniques are frequently associated with predictive maintenance, including nondestructive testing, chemical analysis, vibration and noise monitoring, and routine visual inspection and logging.

Automated Fault Detection and Diagnosis (AFDD). Whereas the previously described techniques and methods generally require human expertise to interpret quantitative information (e.g., a time-series plot of the values of a measured variable, such as electric power use) to reach conclusions about the condition of equipment or a system, AFDD interprets values and trends in measured parameters to automatically reach conclusions about the presence of faults or degree of performance degradation. This enables performance monitoring to be performed continuously for a large inventory of equipment, which is not possible manually.

AFDD uses a set of software-based systematic procedures that automatically compare measured and expected system performance and determine the causes of discrepancies (i.e., symptoms or indicators of faults, caused by the faults themselves). **Fault detection** is the process of determining whether the monitored system deviates from normal operation, and **fault diagnosis** is the process of isolating the detected fault(s) from other possible faults. This procedure is based on a thorough knowledge of the physical principles underlying the operation of HVAC systems, equipment, and components. Measurements provide data on actual performance at the time they are taken, and models that capture normal operational behavior are generally used to provide values of the same variables if the system were operating properly. Models used include engineering first-principles-based models, purely empirical models based on past performance, gray-box models that combine some engineering modeling with empirical data, and various statistical approaches. These methods differ primarily by the modeling technique used, methods for detecting differences between measured and expected performance, and methods used to distinguish among diagnostic outcomes. These methods are then coded in software to automate execution on measured data as they become available.

The ability to detect unacceptable conditions in HVAC&R systems has existed for some time and has been used primarily in safety devices intended to protect expensive equipment from catastrophic failure and for single point alarms tied to occupant comfort. These techniques were generally based on a parameter exceeding a predefined threshold (e.g., a maximum acceptable pressure above which a relief valve opens or equipment shuts down). They have also been used for alarms in building automation systems to alert operators to an unacceptable condition (e.g., a chilled-water temperature that is too high). These alarms are usually provided for binary devices that, when activated, indicate an abnormal condition or discrepancy. More recent motivating factors for development and use of AFDD include increasing energy efficiency, improving indoor air

quality, and reducing unscheduled equipment downtime (Braun 1999). At the same time, AFDD capabilities have been expanded well beyond single-point alarms, providing the ability to detect and diagnose faults based on many different variables. AFDD has the potential to be a significant element of a maintenance program by enabling prolonged equipment life for everything from chillers and other large equipment to individual actuators.

The benefits of AFDD capabilities have been validated in part by surveys and site measurements that have documented a wide variety of operating faults in common HVAC equipment. Numerous studies of maintenance records and independent field studies of faults show that the prevalence of faults in HVAC equipment is substantial (Breuker and Braun 1998; Breuker et al. 2000; Comstock 2002; House et al. 2001a, 2003; Jacobs et al. 2003; Katipamula et al. 1999; Proctor 2004; Rossi 2004; Seem et al. 1999). These faults included packaged air conditioners and heat pumps with economizers not operating properly; low (and high) refrigerant charges; condenser and filter fouling; faulty sensors; electrical problems; chillers with faulty controls, condensers, compressors, lubrication, piping, and evaporators; and air-handling units with too little or too much outdoor-air ventilation, poor economizer control, stuck outdoor-air dampers and other problems.

Types of AFDD Tools. Portable service tools exist to evaluate the performance of packaged and unitary vapor-compression equipment systems and perform corrective measures to address problems. Self-contained, microprocessor-based portable hardware is used during a service visit for data acquisition and analysis. The sensors for making measurements and evaluating system performance may be installed temporarily or permanently. Data are usually collected for a relatively small period of time (minutes) while the equipment is operating at steady-state conditions.

Application of these tools generally involves using connected sensors to collect values of several performance parameters (e.g., superheat or subcooling) that have corresponding performance expectations based on system characteristics (e.g., expansion device) and operating conditions. The measured and expected performance are compared using rules defined in the analysis software to detect and diagnose system faults. Data and messages are displayed in real time to guide system service or repair. Several different methods are used for analysis and detection among these tools.

Local controllers with embedded AFDD include fault detection and diagnostic algorithms as part of the software code. Integrated diagnostic and control logic gives the AFDD code access to data at the controller's short sampling interval; these higher-frequency data may enable detection of problems (e.g., unstable control loops) that might be difficult to detect using data collected at typical trending intervals. Embedded AFDD tools can also reduce network traffic, if applicable, by executing algorithms locally and propagating only key parameters or results to a display or central control-system computer. Embedding algorithms in the controllers can also facilitate integration of the output from the AFDD tool with the alarming capabilities of the equipment control system. Computational and memory limits, however, may place practical restrictions on the complexity of the algorithms that are embedded in local controllers.

Central workstation AFDD tools use dedicated software to detect and diagnose HVAC system faults using data from a building automation system (where applicable) and, in some cases, from other dedicated stand-alone sensors. This software usually resides on a computer that is part of a building automation system, or has access to stored data from a BAS. The BAS may serve one or several buildings. Data acquisition and analysis may be near-real time or periodic (e.g., hourly or daily), depending on the software features. Because the algorithms are implemented on a computer that has significant computational resources, analytical methods and historical data can be used in ways that are not possible with hand-held devices and local controllers. A key strength of workstation AFDD software is its ability to detect system-level faults arising from interactions among components.

Workstation AFDD software may require extensive effort for configuration. In particular, mapping points from the building automation system to the AFDD tool can be a significant programming task, depending on the number of measurement and control points used by the AFDD tool.

Web-based AFDD software may obtain data from a BAS, independent data acquisition system, or controller-embedded AFDD software. In this case, the Internet is used to remotely acquire and display results. In some cases, wireless data acquisition systems are used to communicate between equipment and a remote network operations center, where data are managed and a Web-based user interface is hosted. Using the Internet or long-distance wireless communication for data acquisition allows gathering data from many buildings, virtually on a coincident basis, and supports enterprise-wide reporting. AFDD processing and analysis may be done locally at the building or remotely. Updating software remotely is another advantage of Web-based AFDD. Web-based systems are emerging for detecting and diagnosing faults in individual equipment and whole-building energy consumption (Brambley et al. 2005). A significant challenge for Web-based AFDD is Internet security, which may require additional hardware and software administration, even if the access is periodic and not continuous.

Characteristics of AFDD Systems. Inherent characteristics of AFDD systems can be adjusted to suit a variety of applications and users.

Sensitivity is the lowest fault severity level required to trigger the correct detection and diagnosis of a fault. This characteristic is vital for monitoring safety-critical systems where early identification of small malfunctions could prevent loss of life, costly loss of product, and/or catastrophic damage to equipment. In this case, the fault severity threshold should be established at a low level. For non-safety-critical applications (which covers most HVAC equipment), the sensitivity threshold can be higher so that faults have significant performance or cost impacts before operators are alerted.

False alarm rate is the rate at which an AFDD system reports faults when they do not actually exist (i.e., when operation is normal). A high false alarm rate could result in significant economic loss from unnecessary service inspections or stoppage of equipment operation.

The challenge is to establish a balance between adequate AFDD system sensitivity and minimizing false alarms. One technique used for critical systems, installing redundant sensors, has been found to mitigate excessive false alarms while maintaining desired fault sensitivity. Installation of redundant sensors, however, increases equipment costs and has been resisted for most HVAC equipment. Anecdotal evidence based on traditional building automation system alarming capabilities suggests that a method to disable or adjust AFDD system sensitivity is required for HVAC systems so operations and maintenance staff may cope with excessive false alarms. Without this capability, AFDD tools may be removed from service for the nuisance they create.

Sensitivity and false alarm rate are useful for quantifying the performance on an AFDD tool; however, AFDD tools have numerous other characteristics that impact these performance criteria and also affect the cost of implementing a particular method. House et al. (2001b) identified the following additional characteristics that should be considered when selecting an AFDD tool:

- Number of sensors and control signals used
- Amount of design data used
- Training data required
- User-selected parameters

Generally, increases in any of these factors lead to increased complexity, time, and cost associated with using the AFDD method or tool, but often with increased quality of the results. Therefore,

users should seek a balance between these considerations that is appropriate for the intended application.

AFDD in Practice. The primary objective of an AFDD system is early detection of faults and diagnosis of their causes, enabling correction of the faults before additional damage to the system, loss of service, or measurable energy waste occurs. This is accomplished by continuously monitoring the operations of a system, using AFDD procedures to detect and diagnose abnormal fault conditions, evaluating the significance of the detected faults, and deciding how to respond. Figure 6 shows an example of an operation and maintenance process using AFDD comprising four distinct functional processes. The first two steps in this process are fault detection and, if a fault is detected, fault diagnosis. After diagnosis, the software can analyze the various consequences of the faults detected. System operators can then evaluate the results from the automated fault impact analysis to arrive at an integrated assessment of overall fault significance (based on operational requirements for safety, availability, cost, energy use, comfort, health, environmental impacts, or effects on other performance indicators). Once fault evaluation is completed, system operators determine how to respond (e.g., by taking a corrective action or possibly even no action). Together, these four steps can alert operations and maintenance staff to problems, help identify the root causes of problems so that they can be properly corrected, and help prioritize maintenance activities to ensure that the most important problems are attended to first. This forms the basis for condition-based maintenance.

BENEFITS OF DETECTING AND DIAGNOSING EQUIPMENT FAULTS

Most O&M studies conducted from 1995 to 2007 demonstrated that maintenance performed on HVAC systems to preserve intended equipment performance and condition also accrued beneficial energy and operating cost savings. However, the magnitude of the benefit may not offset the effort to achieve the savings. A study of rooftop air-handling units by the Electric Power Research Institute (Krill 1997) concluded that the cost of annual maintenance would likely exceed utility savings but could result in improved occupant comfort.

A preponderance of other studies, however, shows that savings on operating exceed the cost of AFDD and servicing of equipment, particularly when account is taken for optimizing service task scheduling, reducing unnecessary on-site inspections, and decreasing the peaks and valleys of seasonal work (Li and Braun 2007a, 2007b, 2007c; Rossi 2004). Indications to date for specific applications are that use of AFDD can contribute to avoiding operating and maintenance costs. Building owners and operators must carefully evaluate the cost and benefits of implementing and administering AFDD technology in their specific situation.

Effective Maintenance

Effective maintenance provides the required reliability and availability at the lowest cost by identifying and implementing actions that reduce the probability of failure to an acceptable level.

Effective maintenance management provides reliability and availability at the lowest cost. This strategy involves identifying and implementing actions that cost-effectively reduce the probability of failure. In general, investments are made to promote longevity of mission-critical assets to help the organization succeed over the long term. The intent of effective maintenance management is to minimize operation and maintenance costs through proactive efforts in all three segments.

Maintenance management is effective when integrating proactive maintenance effectiveness and efficiency into the total cost of facility ownership. If adequate measures for cost-effective maintainability are not integrated into the design and construction phases of a project, reliability and/or uptime are likely to be adversely

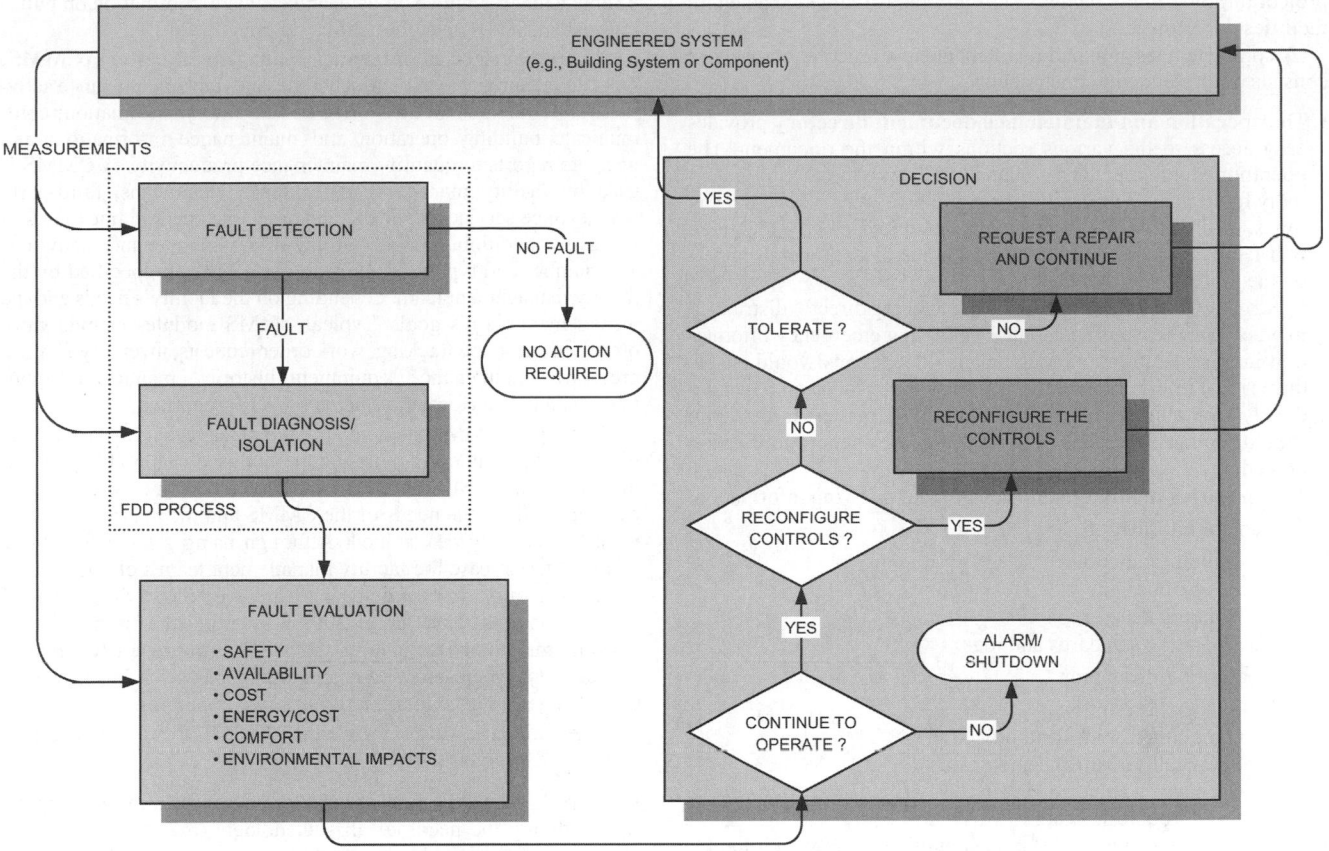

Fig. 6 Generic Application of AFDD to Operation and Maintenance of Engineered Systems
[Adapted from Katipamula and Brambley (2005)]

affected and total life-cycle costs increase significantly. Appropriate levels of maintainability seldom occur by chance. They require up-front planning, setting objectives, disciplined design implementation, and feedback from prior projects. It is vital to identify critical maintainability and production reliability issues and integrate them into facility project designs to achieve long-term facility owning and operating benefits.

Dirty evaporator and condenser coils and filters require additional energy to operate, because inefficient heat transfer creates higher compressor discharge pressures, so more energy is required for a given load. The system must also operate longer to satisfy space conditions. Although these items are variable and difficult to quantify, proper maintenance improves system operation and equipment life, regardless of system size.

Additional Maintenance Strategies

Corrective maintenance classifies resources (expended or reserved) for predicting and correcting conditions of impending failure. Corrective action is strictly remedial and always performed before failure occurs. An identical procedure performed in response to failure is classified as a repair. Corrective action may be taken during a shutdown caused by failure, if the action is optional and unrelated.

Planned maintenance classifies resources invested in selected functions at specified intervals. All functions and resources in this classification must be planned, budgeted, and scheduled. Planned maintenance includes preventive and corrective maintenance.

Unplanned maintenance classifies resources expended or reserved to handle interruptions in the operation or function of systems or equipment covered by the maintenance program. This classification is defined by a repair response.

DOCUMENTATION

Information on the facilities, collateral equipment, and intended operation procedures is essential for planning facilities maintenance actions, efficiently performing facilities maintenance, documenting maintenance histories, following up on maintenance performance, energy reporting, and management reporting. For these reasons, documentation is a critical element in a successful maintenance program. Operation and maintenance documentation should be prepared as outlined in ASHRAE *Guideline* 4.

For new construction, operation and maintenance documentation requirements should be established as part of the project requirements. Deliverables should support the expected maintenance strategy, skills of the maintenance and operations staff, and anticipated resources to be committed to performing operations and maintenance. For maintenance programs being developed for existing facilities, the requirements are the same; however, the operations and maintenance staff may be more involved with sourcing and compiling the documents.

Operation and Maintenance Documents

Information should be compiled into a manual as soon as it becomes available. This information can be used to support design and construction activities, systems commissioning, training of operation and maintenance staff, start-up, and troubleshooting. In addition to the operation and maintenance manual being available to the construction team, an appropriate number of manuals should be set aside for the building owner's staff after construction turnover. It is critical that all information required operating the systems and maintaining the equipment must be compiled prior to project turnover to the owner's staff and be available to the entire facilities department.

A complete operation and maintenance documentation package consists of the following documents:

- The **operation and maintenance document directory** provides easy access to the various sections within the document. The operation and maintenance manual will serve the facility potentially for decades. During this time staff turnover will occur many times. A directory that is well organized and clearly identified will facilitate quick reference by technicians and operators new on the job.
- **Emergency information.** In addition to being directly distributed to emergency response personnel, including emergency information in the operation and maintenance documents would enable this critical information to be kept in a single place and be immediately available during emergencies. Emergency information should include emergency and staff and/or agency notification procedures.
- The **operating manual** should contain the following information:

  I. General information

     A. Building function
     B. Basis of design
     C. Building description
     D. Operating standards and logs

  II. Technical information

     A. System description
     B. Operating routines and procedures
     C. Seasonal start-up and shutdown
     D. Special procedures
     E. Basic troubleshooting

- The **maintenance manual** should contain the following information:

  I. Equipment data sheets (specific to installed equipment)

     A. Operating and nameplate data
     B. Warranty information

  II. Maintenance program and procedure information

     A. Manufacturer's installation, operation, and maintenance instructions
     B. Spare parts information
     C. Corrective, preventive, and predictive maintenance actions, as applicable
     D. Schedule of actions, including frequency
     E. Action descriptions
     F. History

- **Test reports** provide a record of observed performance during start-up and commissioning. The records should be compiled throughout the service life of the facility.
- Copies of **construction documents** ("as builts").

These documents should be available to the entire facilities department.

Initial system maintenance procedures should be detailed in the maintenance manual (with individual equipment maintenance frequency detailed in manufacturers' literature) that was furnished during the initial installation or developed after installation. The maintenance program should be tailored to each specific facility and system type.

Documentation Methods

There are two basic methods for collecting and archiving operation and maintenance documents: (1) a bound written document that is executed and updated by hand, and (2) an electronic, computer-based database management system. The chosen method should be congruent with facility and maintenance program complexity and scope and with the skill level of maintenance staff. The object is to be able to enter, archive, update, and evaluate information on building systems and assets efficiently and effectively.

A **computerized maintenance management system (CMMS)** is a software program to plan, schedule, and track maintenance activities; store maintenance histories and inventory information; communicate building operation and maintenance information; and generate reports to quantify maintenance productivity. A CMMS is used by facility managers, maintenance technicians, third-party maintenance service providers, and asset managers to track the status, asset condition, and cost of day-to-day maintenance activities. The number and type of modules in the CMMS is specified by the facility management team, depending on the facility's needs and the management team's goals. Typical CMMS modules include work order generator and tracking, work order requests, inventory control, preventive maintenance, equipment histories, maintenance contracts, and key performance indicator (KPI) reporting.

Although a CMMS is not required to manage maintenance activities, they are becoming more commonly used (Sapp 2008). When implementing a CMMS in a new or existing facility, or upgrading an existing CMMS, the needs of the CMMS and the planning process must be carefully determined. Although using a CMMS has the potential to increase the facility management team's efficiency and serve as a historical maintenance archive, more than 50% of implementations fail (Berger 2009). Selecting the right software does not guarantee that using a CMMS will improve maintenance productivity: when selecting or upgrading a CMMS, business process changes are as important, if not more important.

When implementing or upgrading a CMMS, the following is recommended (Berger 2009):

- Determine the maintenance process and software requirements, considering the needs of all stakeholders (maintenance, operations, engineering, IT, materials management, purchasing, and finance).
- Allow 3 to 6 months to design new processes and develop a set of system requirements, using a participatory approach to include all stakeholders.

STAFFING

Skills, Responsibilities, and Training

To be effective and efficient, operation and maintenance management staff must have both technical and managerial skills. Technical skills include the ability to understand how mechanical systems operate and how maintenance of mechanical equipment is performed, as well as analytical problem-solving expertise of the physical plant engineer. Physical plant engineers require a variety of skills to effectively operate HVAC systems, implement a maintenance program, and manage operating and maintenance staff and meet requirements of the investment plan. Good physical plant engineering solutions are developed while the investment plan is being formulated, and continue throughout the life of the facility.

Managerial skills include overseeing the facility in life cycle terms and on a day-to-day basis. Facility management at this level requires the development of maintenance strategies; determination of program goals and objectives; and administration of contracts with tenants, service providers, and labor unions. Even when specialized contract maintenance companies provide service, the facility manager requires these skills.

An effective facility manager should be able to manage and train staff, and plan and control a facility's operation and maintenance with the cooperation of senior management and all departments. The facility manager's responsibilities include administering the O&M budget and protecting the life-cycle objectives. There are nine core competency areas of facility management (IFMA 2009), seven of which apply to operations and maintenance:

- **Operations and maintenance**: Oversee acquisition; installation; operation; maintenance; and disposition of building systems, furniture and equipment, grounds, and exterior elements
- **Human and environmental factors**: Develop and implement practices that promote and protect health, safety, security, quality of work life, the environment, and organizational effectiveness
- **Planning and project management**: Develop facility plans; plan and manage each phase of the project, including construction and relocations
- **Leadership and management**: Plan, organize, administer, and manage the facility's functions; manage personnel assigned to each function
- **Quality assessment and innovation**: Manage the process of assessing the quality of services and facility effectiveness; manage benchmarking process and audits
- **Communication**: Communicate effectively
- **Technology**: Plan, direct, and manage facility management business and operational technologies, and support the organization's technological infrastructure

Training should be an important part a facility management department. The pace of technology used in buildings continues to increase, making training for all facilities even more important. Training can be done in-house or by a contracted third party who provides training as a business. When developing an in-house training program or evaluating training by a third-party, make sure these key components are addressed:

- **Personnel**: Who will attend the training; create a list of maintenance managers, supervisors, and trades workers
- **Course content**: Course requirements, type of training, and media required
- **Schedule**: When the training will take place and how it will be coordinated with day-to-day assignments
- **Master training plan**: A written document to record personnel, course content, and time schedule requirements (Westerkamp 1997)

In-House Versus Contracted Staff

In most situations, an on-site maintenance staff changes filters, belts and motors and lubricates bearings. However, tasks such as cleaning condenser and evaporator coils; assessing refrigeration and building automation systems; and assessing and repairing heating components, such as gas and oil burners and electronic elements; may require a specialized technician. Additionally, tasks that may have previously been done by the on-site maintenance staff may now require certification and special equipment: for example, the U.S. Environmental Protection Agency (EPA) certifies technicians and recovery units for refrigerant handling. In most small facilities, maintenance contractors provide services requiring specialized technicians. In larger facilities, maintenance may be contracted out for complex equipment when staff does not have adequate technical expertise. Whenever the operator cannot service and repair the systems or components installed, the owner should ensure that qualified contractors and technicians perform the work.

The frequency of maintenance depends in part on system run time and type of operation in the facility. ASHRAE *Standard* 180 provides maintenance frequencies for air distribution systems, chillers, boilers, condensing units, building automation systems, cooling towers, dehumidification and humidification, engines, microturbines, fan coils, pumps, rooftop units, and other HVAC systems.

Commissioning and training on how to commission and retrocommission mechanical systems are increasingly important with these types of systems to ensure optimum comfort, efficient system operation, and minimal operation and maintenance costs. Without detailed commissioning documentation, operation staff cannot effectively consider factors such as energy budgets when addressing building occupants' comfort complaints.

Criticality and Complexity

The criticality and complexity of a system or piece of equipment must be considered when developing the maintenance strategy. Building complexity and criticality ranges from residential, commercial, institutional, and industrial, to mission-critical facilities, such as cleanrooms and operating rooms.

The operations and maintenance strategy selected must balance system and equipment up-time, customer needs, energy efficiency, cost, and the use of proactive and reactive maintenance approaches. Criticality and complexity of systems and equipment also vary in a facility. In many facilities, larger, more complex systems and equipment (e.g., chillers and boilers) are more critical than smaller equipment, such as toilet exhaust fans and vestibule cabinet unit heaters.

Customer Service

Customer service here means responding to requests and complaints of the building occupants. In nonmanufacturing facilities, the level of customer service provided, or the perceived level of service provided, is a key component of a successful maintenance organization. All people in the maintenance organization, from technicians to management, should work together to deliver high-quality customer service to building occupants.

MANAGING CHANGES IN BUILDINGS

Renovation and Retrofit Projects

Design for Ease of Operation and Maintenance. Operation, maintenance, and maintainability of all HVAC&R systems should be considered during building design. Any successful operation and maintenance program must include proper documentation of design intent and criteria. ASHRAE *Guideline* 4 provides a methodology to properly document HVAC systems. Newly installed systems should be commissioned according to the methods and procedures in ASHRAE *Guideline* 1.1 to ensure that they are functioning as designed. It is then the responsibility of management and operational staff to maintain design functionality throughout the life of the building.

For new construction or renovation, the building owner should work with the designer to clearly define facility requirements. In addition to meeting the owner's project requirements, the designer must provide a safe and efficient facility with adequate space to inspect and repair components. The designer must reach agreement with the owner on the criticality of each system to establish issues such as access, redundancy, and component isolation requirements.

Commissioning Throughout Turnover to Operation and Maintenance. Existing systems may need to be reconfigured and recommissioned to accommodate changes. See Chapter 43 for additional information on commissioning.

Retrocommissioning or Ongoing Commissioning. Retrocommissioning (or ongoing commissioning; see Chapter 43) applies to buildings that have been in operation for some time, regardless of whether they were initially commissioned. This process needs to be applied any time major building retrofits or usage changes are contemplated by the owner/manager. It is also advisable to consider this process if there has been a significant changeover in operational personnel, to ensure proper facility operation as originally intended.

Conversion to New Technology

Building systems and equipment are based on the technology available to planners and designers at the time of preparation, construction, and installation. Maintenance, repairs, and operating schemes should be adhered to throughout the required service life of the facility or system. During the service life, new technology (e.g., high-efficiency equipment, smart technology systems, sustainability advances) may become available to increase overall efficiency and affect maintenance, repair, and replacement programs. Conversion from existing to new technology must be assessed in life-cycle terms. Existing technology must be assessed for the degree of loss from a shorter return on investment. The new technology must be assessed for (1) all initial, operation, and maintenance costs; (2) the correlation between service life and the facility's remaining service life; and (3) the cost of conversion, including revenue losses from associated downtime. As facilities and systems are upgraded to new technologies, facility operations and maintenance personnel must be trained to ensure that new equipment operates efficiently and effectively throughout its service life to maintain the anticipated benefits of technology improvements.

Document and Communicate

As a facility is planned, designed, constructed, and occupied, the documentation and information about the facility increases. Facility owners are increasingly aware of the need to document and communicate from the early stages of planning through the facility's life to ensure minimal information is lost. During planning and design, documentation must include detailed plans and specifications identifying all aspects of the equipment, including physical size, location, system interactions, operating set points (e.g., temperature, humidity, airflow, energy usage, relative pressure), flow diagrams, instrumentation, and control sequences. The equipment supplier must provide equipment meeting the specifications, and should also recommend alternatives that offer lower life-cycle costs for the owner and designer to evaluate. The installing contractor must turn over the newly installed system to the owner in an organized and comprehensive manner, including complete documentation with O&M manuals and commissioning reports (e.g., air balance, fume hood certification). The design and commissioning team should ensure that a comprehensive systems manual as outlined in ASHRAE *Guideline* 0 is provided to the owner, along with training on equipment and systems. The design team and the contractor should also ensure that complete and correct as-built or conformance drawings are provided showing the actual built facility, including all changes required during the construction process.

Technological advances now allow much of this information to be provided electronically. Software and hardware systems are available to create building information models (BIM) during facility planning and construction that can become a resource for the owner throughout the facility's service life. Building information may be linked into the CMMS, building automation, and other smart building systems to allow building owners, operating engineers, technicians, and other facility personnel immediate access to the information required to maintain, operate, and service the building systems. During the building's life cycle, as modifications and changes are made to the facility, this information must be continuously updated to ensure accurate records continue to be maintained.

REFERENCES

ASHRAE. 2005. The commissioning process. *Guideline* 0-2005.

ASHRAE. 2007. HVAC&R technical requirements for the commissioning process. *Guideline* 1.1-2007.

ASHRAE. 2008. Preparation of operating and maintenance documentation for building systems. *Guideline* 4-2008

ASHRAE. 2008. Standard practice for inspection and maintenance of commercial building HVAC systems. ANSI/ASHRAE/ACCA *Standard* 180-2008.

ASHRAE. 2010. *ASHRAE greenguide: The design, construction, and operation of sustainable buildings*, 3rd ed.

Berger, D. 2009. *2009 CMMS/EAM review: Power up a winner—How to find the right asset management system for your plant*. www.plantservices.com/articles/2009/066.html.

BOMA. 2008. *Experience exchange report (EER)*. Building Owners and Managers Association International, Washington D.C.

Brambley, M.R., S. Katipamula, and P. O'Neill. 2005. Facility energy management via a commercial web service. Ch. 18 in *Information technology for energy managers*, vol. II: *Web based energy information and control systems case studies and applications*, pp. 229-240, B.L. Capehart and L.C. Capehart, eds. Fairmont/CRC, Lilburn, GA.

Braun, J.E. 1999. Automated fault detection and diagnostics for the HVAC&R industry. *HVAC&R Research* 5(2):85-86.

Breuker, M.S. and J.E. Braun. 1998. Common faults and their impact for rooftop air conditioners. *HVAC&R Research* 4(3):303-318.

Breuker, M.S., T. Rossi, and J. Braun. 2000. Smart maintenance for rooftop units. *ASHRAE Journal* 42(11):41-47.

Christian, J. and A. Pandeya. 1997. Cost predictions of facilities. *Journal of Management in Engineering* 13(1):52-61.

Comstock, M.C., J.E. Braun, and E.A. Groll. 2002. A survey of common faults for chillers. *ASHRAE Transactions* 108(1):819-825.

House, J.M., H. Vaezi-Nejad, and J.M. Whitcomb. 2001a. An expert rule set for fault detection in air-handling units. *ASHRAE Transactions* 107(1):858-871.

House, J.M., J.E. Braun, T.M. Rossi, and G.E. Kelly. 2001b. Section D: Evaluation of FDD tools. In *Final report: Demonstrating automated fault detection and diagnosis methods in real buildings*, A. Dexter and J. Pakanen, eds. International Energy Agency on Energy Conservation in Buildings and Community Systems, Annex 34. VTT Technical Research Centre of Finland, Espoo, Finland.

House, J.M., K.D. Lee, and L.K. Norford. 2003. Controls and diagnostics for air distribution systems. *ASME Journal of Solar Energy Engineering* 125(3):310-317.

IFMA. 2009. *Facility management competency areas*. http://www.ifma.org/learning/fm_credentials/competencies.cfm.

Jacobs, P., V. Smith, C. Higgins, and M. Brost. 2003. Small commercial rooftops: Field problems, solutions and the role of manufacturers. *Proceedings of the 2003 National Conference on Building Commissioning*, Portland Energy Conservation, Portland, OR.

Katipamula, S. and M.R. Brambley. 2005. Methods for fault detection, diagnostics and prognostics for building systems—a review, part I. *HVAC&R Research* 11(1):3-25.

Katipamula, S., R.G. Pratt, D.P. Chassin, Z.T. Taylor, K. Gowri, and M.R. Brambley. 1999. Automated fault detection and diagnosis for outdoor-air ventilation systems and economizers: Methodology and results from field testing. *ASHRAE Transactions* 105(1):555-567.

Krill, W. 1997. The impact of maintenance on package unitary equipment. *Final Report*, TR-107273 3831, Electric Power Research Institute, Palo Alto, CA.

Li, H. and J.E. Braun. 2007a. An overall performance index for characterizing the economic impact of faults in direct expansion cooling equipment. *International Journal of Refrigeration* 30(2):299-310

Li, H. and J.E. Braun. 2007b. An economic evaluation of the benefits associated with application of automated fault detection and diagnosis in rooftop air conditioners. *ASHRAE Transactions* 113(2):200-210.

Li, H. and J.E. Braun. 2007c. A methodology for diagnosing multiple-simultaneous faults in vapor compression air conditioners. *HVAC&R Research* 13(2):369-395.

NIBS. 1998. Excellence in facility management: Five federal case studies. *Publication* 5600-1. Facility Maintenance and Operations Committee, National Institute of Building Sciences, Washington, D.C.

Proctor, J. 2004. Residential and small commercial air conditioning—Rated efficiency isn't automatic. ASHRAE Public Session, Winter Meeting, Anaheim, CA.

Rossi, T.M. 2004. Unitary air conditioning field performance. *Proceedings of the Tenth International Refrigeration and Air Conditioning Conference at Purdue*, pp. R146.1-R146.9.

Sapp, D. 2008. *Computerized maintenance management systems (CMMS)*. www.wbdg.org/om/cmms.php.

Seem, J.E., J.M. House and R.H. Monroe. 1999. On-line monitoring and fault detection of control system performance. *ASHRAE Journal* 41(7):21-26.

Westerkamp, T. 1997. *Maintenance manager's standard manual*. Prentice Hall, Paramus, NJ.

BIBLIOGRAPHY

Blanchard, B.S., D. Verma, and E.L. Peterson. 1995. *Maintainability: A key to effective serviceability and maintenance management*. John Wiley & Sons, New York.

Campbell, J.D. 1995. *Uptime, strategies for excellence in maintenance management.* Productivity Press, Portland, OR.

Criswell, J.W. 1989. *Planned maintenance for productivity and energy conservation*, 3rd ed. Prentice Hall, Englewood Cliffs, NJ.

Fuchs, S.J. 1992. *Complete building equipment maintenance desk book.* Prentice Hall, Englewood Cliffs, NJ.

Katipamula, S., and M.R. Brambley. 2004. Methods for fault detection, diagnostics and prognostics for building systems—A review, part II. *International Journal of HVAC&R Research* (now *HVAC&R Research*) 11(2):169-187.

Mobely, R.K. 1989. *An introduction to predictive maintenance.* Van Nostrand Reinhold, New York.

Moubray, J. 1997. *Reliability-centered maintenance*, 2nd ed. Industrial Press, New York.

NCMS. 1999. *Reliability and maintainability guideline for manufacturing machinery and equipment*, 2nd ed. National Center for Manufacturing Sciences, Ann Arbor, MI.

National Academy of Sciences. 1998. *Stewardship of federal facilities: A proactive strategy for managing the nation's public assets.* National Academy, Washington, D.C.

Petrocelly, K.L. 1988. *Physical plant operations handbook.* Prentice Hall, Englewood Cliffs, NJ.

Smith, A.M. 1993. *Reliability-centered maintenance.* McGraw-Hill, New York.

Stum, K. 2000. Compilation and evaluation of information sources for HVAC&R system operations and maintenance procedures. *Report*, ASHRAE Research Project RP-1025.

COMPUTER APPLICATIONS

COMPUTERS are used in a wide variety of applications in the HVAC industry. Rapid technological advances and the decreasing cost of computing power, memory, and secondary storage have changed many aspects of the HVAC industry. New HVAC design tools allow optimal solutions to be found in engineering applications. Building operations benefit from low-cost networking to achieve multivendor control system interoperability. Consulting engineers can search manufacturers' equipment and specifications interactively over the Internet. Designers can collaborate from remote locations. More powerful applications that are also easier to use are now affordable by a wider segment of the industry; many HVAC calculations, such as heating and cooling loads, can be performed easily and automatically.

Business applications and infrastructure have also positively affected the HVAC industry. Open communications standards and internetworking allow fast, efficient communications throughout company and industry circles. HVAC design, manufacture, installation, and maintenance functions benefit as businesses build computing infrastructure through corporate information services (IS) or information technology (IT). Many advances in the business community have been adopted by the HVAC industry as de facto standards.

COMPUTER SYSTEM COMPONENTS AND TECHNOLOGIES

Because of rapid advances in computer technology, computers offer tremendous power at very low cost. However, the total cost of ownership of a personal computer is not limited to the cost of the computer hardware. Software, network connectivity, support, and maintenance expenses quickly surpass the initial cost of hardware.

Selecting a computer platform includes a wide variety of issues, such as

- Analysis of application needs
- Corporate computer support architecture and standards
- Vendor support, including guaranteed response time
- Central processing unit (CPU) power, random-access memory (RAM), and secondary or hard disk storage capacity
- System compatibility and interoperability between vendors
- Ease of use and required training
- Data backup strategy: how will the data on the computer be restored in the event of a system failure?
- Security issues: what data will be accessible, and how will data be protected from unauthorized access?
- Network communications capability and compatibility
- Technical support live and/or online, including driver update support
- Cost/benefits comparison
- Information system infrastructure/interoperability requirements
- Technology reliability and obsolescence
- Total cost of ownership

The preparation of this chapter is assigned to TC 1.5, Computer Applications.

Virtually all brand-name personal computers now have more than enough speed for basic business applications such as word processing and spreadsheets. Personal computers differ in their ability to handle multimedia graphics and sound, which are useful in advanced software applications.

An information system comprised of computer networks depends on the ability of computers to communicate with each other in a standard architecture. Technological advances have resulted in computer systems becoming obsolete in less than five years. Therefore, it is important to plan for compatibility with future business computer requirements. Software and hardware must match the needs of the user, and be consistent with the business system architecture.

System architecture standards from organizations such as the Institute of Electrical and Electronics Engineers (IEEE), the American National Standards Institute (ANSI), and the International Organization for Standardization (ISO) have resulted in well-defined standards such as Ethernet, TCP/IP, and HTTP, further described in the section on Networking Components. The combination of popular standards has resulted in network and Internet accessibility from local computers.

COMPUTER HARDWARE

Hardware is the physical equipment (electronic parts, cables, printed circuit boards, etc.) that provides the foundation for the software to run. Hardware processing power has increased dramatically, such that the average business system provides much more capacity than the average user needs. Specialized tasks, however, often require specialized hardware.

Mainframes are large, expensive computer systems that serve the majority of large-scale business and technical needs. Mainframe computers best serve large nationwide databases, weather prediction, and scientific modeling applications. They support many users simultaneously, and are usually accessed over a network from a personal computer (PC). More powerful mainframes are called **supercomputers**.

Personal computers (PCs) are desktop computers tailored to individual user needs, and provide much more convenient access to computing resources than the traditional mainframe. PCs are used in a wide variety of information-gathering, communication, organization, computation, and scheduling roles.

PC servers allow access to files and data on a network. PC servers are a special class of computers that support large amounts of memory, fast connection speed to the network, large disk storage for access of many files, and applications designed to be usable from other computer systems. Because system reliability is critical, there are additional features such as better construction, component redundancy, and removable hard drives.

Workstations are application-specific computers serving specialized needs such as computer-aided design (CAD) and manufacturing (CAM), product design and simulation, mapping geographic information system (GIS), or specialized graphics design. Workstations have greater processing power and graphical display capabilities than personal computers, and may also have other improvements found in PC servers.

Clusters are several computers organized to work together. Clusters of PCs can approach the computing power of a mainframe at a fraction of the cost, if efficient application software is available. Clusters of workstations can provide more computing power than even the largest mainframe. Clusters of mainframes can be extremely powerful tools.

Minicomputers are medium-sized business computer systems that serve medium to large application needs. Minicomputers have been used to economically scale applications that have grown too large for PC servers. Many small business e-commerce applications have moved to the minicomputer platform.

A **Web server** is a PC server, minicomputer, or mainframe that stores and distributes Web-based information; it is typically used to distribute documents and other information across the Internet.

Laptop computers are smaller, battery-powered personal computers optimized for light weight and portability. Compared to a similarly priced personal computer, laptops are generally slower and have fewer features, though they generally include most basic features found in a desktop computer. A smaller, lighter laptop computer is called a **notebook** or **netbook**.

Palmtop computers serve a specialized niche, allowing access to some of the same information available from a laptop, but in a much smaller package. Palmtops may have no keys, or very small keys that are much smaller than the standard keyboard.

Personal digital assistants (PDAs) support limited functionality of laptop computers in a notecard-sized format. A typical interface method is a stylus with a graphical touch-sensitive screen; standard functions include a calendar, address book, e-mail, and a calculator. Options include synchronizing hardware to upload address lists or e-mail to a PC, and a modem or wireless card for portable Internet access. Some models combine the functionality of a PDA with a cell phone and a digital camera.

Smartphones are mobile phones that offer advanced computing ability and connectivity. Smartphones run complete operating system software, providing a platform for application developers, and are able to run many different types of software applications, including those specific to HVAC engineering. The advantages of smartphone applications over desktop applications are that they are simple to use and can easily be deployed in the field for quick calculations.

Calculators and organizers are available usually for fixed arithmetic functions and, on the high end, programmable applications. This hardware format is typically inexpensive and portable, but software is typically proprietary. Laptops, palmtops, and PDAs have challenged the advantages of calculators and organizers.

COMPUTER SOFTWARE

Software is computer instructions or data that can be stored electronically in storage hardware, such as random access memory (RAM), or secondary storage such as hard disk drives, DVDs, or CDs.

Software is often divided into two categories:

- *Systems software* includes the operating system and utilities associated with the operating system, such as file copying.
- *Applications software* includes the programs used to do the user's work, such as word processors, spreadsheets, or databases. HVAC-specific programs belong to the applications software category.

Operating Systems

The operating system (OS) is the basic control program that allows execution and control of application software. Operating systems (system software) also provide a common environment for applications to use shared resources such as memory, secondary storage (e.g., hard disk storage), processing time, network access, the keyboard, graphic display environments, desktops, file services, and printer services. Application software must be selected first; that decision will typically limit the hardware computer platform and OS.

Utility Software

Utility programs perform a wide variety of organizing and data-handling operations such as copying files, printing files, file management, network analysis, disk size utilities, CPU speed tests, improving security, and other related operations. Most utilities perform one or two specific functions.

Compression Software. These utilities compress and archive data files in a standard format, such as ZIP. Compression utilities allow files and file systems to be compressed, making the transfer of information and files much more efficient without any loss of content or data. These utilities are valuable when file size is important, such as transmitting files over a limited-bandwidth communication system, and for archival purposes. A typical compression ratio for text files is 3:1. E-mail attachments are often compressed.

Antivirus Software. An important software package for most computer systems is antivirus software (see the section on E-mail, Viruses, and Hoaxes). Antivirus software detects, quarantines, repairs, and eliminates files that contain viruses. A computer virus may take various forms and is commonly attached invisibly to a word-processed or e-mailed document. Antivirus software is imperative when a computer is accessing Internet-based information. Antivirus software must be updated frequently to be able to catch new viruses. Updates are often available for free after purchasing antivirus software.

Antispyware Software. Spyware is a type of program that sends information from your computer to another computer, usually with malicious intent. It may send passwords, credit card numbers, or any other information that is typed into your computer. Antispyware software will detect spyware and remove it.

Firewall Software. A firewall is a utility that controls and limits all network communications going from or to your computer. With a firewall, only programs with permission are allowed to communicate over the network. This should stop spyware and other similar malicious programs. Also, a firewall limits the communications it accepts over the network, which acts against other types of attacks. A large organization should have a firewall separating their entire network from the Internet, but it is still important to have a firewall on each PC.

Back-Up Software. The ability to recover from a massive failure such as a successful virus attack, accidental deletion of files, flood, fire, hardware (especially hard disk) failure, and other inevitable disasters is extremely important. Typically, information service (IS) computers use a rotating digital tape back-up strategy that performs massive back-ups of network resources. Individual computers may not be included in these back-ups, so regularly scheduled back-ups must be performed. A relatively inexpensive way to back up small amounts of data is to store data files and documents in a compact disk recordable (CD-R) format, which allows roughly up to 700 MB of storage at low cost. Digital versatile disks (DVDs) promise roughly 10 times the storage capacity of CD-R. Tape back-up provides an extremely low cost per megabyte stored and is useful if quick access and retrieval of individual back-up files is not an issue. Back-ups should be stored in a secure location away from the computer system. There are two basic types of back-ups: complete, in which all files are backed up, or incremental, in which only files changed since the last back-up are stored. A typical strategy is to perform a complete back-up once a week or month, and an incremental back-up every day.

Application Software

Most useful application software falls into one of these categories:

Word Processing. Word processing applications are used for tasks that range from creating simple printable documents to providing products for advanced enterprise-level electronic distribution and publishing systems. Features may include automatic text layout, spellchecking, printing, index and table of contents generation, simple graphics, and the ability to save files in a variety of formats such as rich text format (RTF), hypertext markup language (HTML), plain text, or American Standard Code for Information Interchange (ASCII). Most word processors allow the user to view the output before printing.

Specification Writing. A useful extension to word processors is application software to create specifications for the consulting engineer. Features include ensuring selectable sections for creating desired features, and consistent naming of materials, specifications, and processes or sequences.

Specification software can save time by allowing the reuse of sections from a master specification, along with previously developed specification details, sometimes with small changes. Specification tools are most useful in situations that do not require completely new specifications from project to project.

Desktop Publishing. This is a specialized class of word processors with high-end, professional text and graphics layout features. Desktop publishing systems have been extended to incorporate professional-quality graphical images, many different styles of text and fonts, and special effects such as color fading, shadowing, and artistic effects. Using inexpensive, high-quality printers, the quality of output from desktop publishing software can be remarkable. In general, the limiting factor in the use of desktop publishing software is the time and training investment to use it efficiently.

Web Publishing. Word processors are best suited for printing documents on paper; Web publishing tools are designed to format information and transfer the documents to Web sites. In many regards, Web publishing software is similar to word processing software, allowing text, graphics, and tables to be formatted for display. Unique extensions from Web software include the ability to manage links, or connections within a related collection of documents.

Web publishing software saves information in HTML, a standard format used by Web browsers. Web browsers can display information in many different screen resolutions and sizes, which makes a standard layout of documents especially difficult.

Other features of Web publishing software include tools to enforce consistency of Web page format and style, lower resolution of pictures and graphics to reduce loading time, and preview what the final Web page will look like to the end user.

Standardized Web Document Formats. Created as a result of special formatting requirements for printing and display on Web pages, this software converts word processing documents to a standard format. This standard format requires the use of a Web browser software extension called a "plug-in," typically offered at no charge as a viewing tool. The viewing tool allows users to see and print the document in exactly the same format. The disadvantage of standardized formats is the inability to modify the information.

Spreadsheets. Spreadsheet applications allow the user to create tables for adding and subtracting columns of numbers in one complete view. Spreadsheets contain thousands of cells into which the user may enter labels, numbers, formulas, or even programs ("macros") to execute such functions as subroutines for a specific calculation, database sorting and searching, conversions, and other calculations. Automatic recalculations allow displayed information to be evaluated quickly. Spreadsheets are useful for importing and formatting data into a graphical display, trend analysis, and/or statistical computations. Users can input different variable values into a mathematical representation of the process or system, and evaluate the results immediately or save them as separate files of graphs for later analysis.

Databases. Database applications are designed to store, organize, and sort information in a specialized format optimized for speed and utility. Simple PC database applications are typically designed for single users, whereas database management systems (DBMSs) are designed for larger applications involving multiple-user access. DBMSs support security hierarchies, record and file locking, and mission-critical network-accessible applications such as financial and banking applications. The standard language for accessing the database information is structured query language (SQL).

Presentations. Software optimized for group presentations displays information on digital projection systems. Presentation software is able to present simple text and graphics in an attractive screen format package for seminars, demonstrations, and training. Additionally, most presentation software allows the creation of notes for the speaker to consult during the presentation, and slide handouts. Popular extensions to presentation software include sound effects and multimedia effects such as video and audio clips. Libraries of vendor and third-party graphics are available at extra charge or from the Internet.

Accounting. Accounting software for accounts receivable, accounts payable, payroll, and taxes is available for many hardware platforms. Many accounting software packages contain extra modules and applications that allow extensions for unique accounting requirements. Financial reports for projects can be easily generated from standard report packages. Large accounting systems can be integrated with databases and automatic tax-reporting functions. Individual accounts and corrections can be monitored and updated automatically. Large accounting systems also allow integration and reporting from individual employees via Web browser applications.

Project Scheduling. Meeting project requirements and efficiently coordinating project resources are difficult tasks; project scheduling software helps solve these problems. Most project scheduling software can take resources (such as individual workers), length of task, percent allocation of resources, and task interdependencies, and present the information as a critical path method (CPM) chart or as a program evaluation and review technique (PERT) chart to evaluate the schedule. Many applications can perform resource leveling, resulting in an efficient allocation of resources to complete the project.

As project scheduling software is updated to include real data, the predicted completion date and the total amount of resources used (e.g., hours worked on individual tasks, material requirements) can be used as predictive record for future projects.

Management. Management software can review past and current data on inventory, accounting, purchasing, and project control, and predict future performance. As more data are accumulated, statistical analysis can predict performance with a specified level of confidence.

Much of the data required for these techniques may require access to standard industry data in industry-wide databases, financial forecasts, code standards, and demographic and other resources. In many cases, this information can be obtained on the Internet for a nominal fee.

Web Meetings. Similar to teleconferences, Web meetings coordinate demonstrations and meetings over the Internet. Participants use a common Web site to view and update a slide show presentation, with optional voice commentary handled either through traditional teleconferencing or through streaming audio or video feeds. Some types allow participants to view the desktop of the moderator's computer, allowing them to share a live view of any application the moderator chooses.

Graphics and Imaging. Graphics software packages readily produce images, x-y charts, graphs, pie charts, and custom graphics. Libraries of graphics and images exist for use in word processing, presentation, desktop publishing, and Web page applications. Many graphics programs can produce output in numerous popular computer graphics formats such as Joint Photographic Experts Group (JPG), graphics interchange format (GIF), tagged information file format (TIFF), bitmap (BMP), and other standards. Specialty programs allow graphics files to be converted between graphics formats. Multimedia tools allow video and sound to be combined with traditional text and graphics.

Images require large amounts of data to be represented accurately. Graphics are usually stored in either bitmap or vector format. The **bitmap** format is commonly used for picture and photo-oriented graphics, whereas **vector** formats are commonly used in CAD/CAM and architectural drawings. Although the output of bitmap images is simple on devices such as graphic printers, rescaling these images does not always produce satisfactory output. Conversions between formats give varying results. Bitmap images are frequently converted to JPG (typically for photographic images) and GIF (typically for simple drawings) file formats.

Web Browsers. Most Internet-enabled computers have Web browsers that support text and graphical Web page displays. The underlying language of the World Wide Web is HTML, and navigation is by clicking on links to access different computer systems and pages on the Web.

Browsers are a preferred interface for accessing vastly different computer systems. Full-feature browsers also support Java® applets (small programs) and a host of extension applications that are available at nominal or no charge. These include streaming sound, streaming multimedia, and graphics formats other than GIF and JPG, which are natively supported by browsers. Some other available features include access to e-mail, file transfer protocol (FTP; for transferring files to and from servers over the Internet), chat rooms, newsgroups and forums, Telnet access to computers, and access to secure e-commerce sites that use encryption routines to protect sensitive data such as credit card information, bank account balances, and stock transfer information.

Communications Applications. Communication over the Internet has become widespread, for both business and personal use. Voice over Internet protocol (VOIP) services allow users to place phone calls over a data network, such as the Internet or a local intranet. Calls can be placed to another user on the network, or to users on the traditional telephone (land-line) network. Text messaging sends short messages (less than 160 characters), usually from a digital cell phone. Instant messaging (IM) allows interactive text-based messaging, usually from a PC to another user on a PC. It is also possible to send messages between text messaging and IM systems.

SOFTWARE AVAILABILITY

Freeware is copyrighted software that is free. The user can run the software, but must obey the copyright restrictions.

Public domain software is available to the public for little or no charge. Public domain software, which is not copyrighted, is often confused with freeware, which is copyrighted. Public domain software can be used without the restrictions associated with copyrighted software.

Shareware is software available for users to try before buying it; after the trial period, users are expected to register or stop using the application. There is usually a nominal registration fee.

Software can be **purchased** from manufacturers, distributors, representatives, discount stores, and computer specialty stores. Software price, support, return policies, and distribution vary from vendor to vendor. During installation, most software displays a detailed software license granting *use* rather than *ownership* of the software program. This license gives specific restrictions on how the software is to be used. Most high-volume software does not offer direct support, but many software companies offer pay-per-incident and other fee arrangements if desired. Many applications have discussion groups or message boards on the Internet, where solutions to problems may be found. Most off-the-shelf, shrinkwrapped software is nonreturnable once opened.

CUSTOM PROGRAMMING

Using an existing software package is usually preferable to designing a custom application, which is much more expensive and potentially riskier, especially for large, mission-critical software

such as client-server applications. If, after careful evaluation, prepackaged software will not suffice, the choice must be made between in-house and outside development. Custom software can be (1) contracted out to a specialized firm, (2) developed solely by internal staff, or (3) developed by internal staff with consultation help by an outside firm.

Contracting to an outside firm involves hiring a specialized outside party to define, estimate, schedule, and create the software. Funds should be budgeted for the outside organization to support modifications or enhancements not specifically covered in the contract. Contracting outside is suitable for an organization that does not have the resources or expertise to accomplish the project. Disadvantages of this approach include the expense and lack of control over the program. Licensing and ownership issues must be defined in the contract.

Developing the program internally is viable only if the skills and resources are available. Internal projects are easier to control because the people involved are co-located.

Consultation help from an outside firm involves a skilled outside party assisting internal staff, who may have insufficient skill in the area on their own. Outside firms can provide expertise to get a project going quickly. Long-term support and maintenance of the software are done in-house.

Specifications are key to the success of a software project. Calculations, human interface, reports, user documents, and testing procedures should be carefully detailed and agreed on by all parties before development begins. Software testing should be specified at the beginning of the project to avoid the common problem of low quality because of hasty and inadequate testing. Design testing should address the human interface, a wide range of input values including improper input, the various functions, and output to screen, disk, or other media. Field tests should include conditions experienced by the final users of the software.

With any development approach, good, understandable documentation is required. If for any reason the software cannot be adequately supported, the program will have to be replaced at substantial cost.

SOFTWARE LANGUAGES

Application software is written in a variety of computer languages. Most commercial software today is written in C++ or C to allow fast speed and efficient use of resources.

Visual Basic® is an extremely popular and easy-to-use graphical programming language for Windows® that has proprietary extensions of the original BASIC language. High-performance software modeling tools such as finite element analysis (FEA) and graphics modeling use languages such as FORTRAN and C.

Several modern languages, such as Java® and Smalltalk, use object-oriented programming. Java is a favored programming language for applications that run over computer networks such as the Internet. C++ mixes both traditional procedural C language programming and object-oriented extension concepts. The major disadvantage of Java and C++ is the high level of programming skill required to create programs.

Java is a powerful programming language and environment with extensions to support Internet, security, and multiplatform implementation. Java applets are small, secure, self-contained programs that need a host environment, such as a Web browser, to run. Many new applications use standard Web browsers with applets to create useful displays and interfaces. JavaBeans are modular components that provide users with standard modular and graphical interfaces of components. Enterprise JavaBeans (EJB) are server-based programs that allow interoperability and language standardization to be scalable and interchangeable between server vendors.

Relational system databases use structured query language (SQL). Object databases are still relatively new, but fully support the object-oriented paradigm of object references, instead of the traditional relational model.

INPUT/OUTPUT (I/O) DEVICES

Basic computer peripherals (e.g., monitor, keyboard, mouse, etc.) allow interaction with the computer and are sufficient for many applications. More advanced I/O capability is available through digital cameras, scanners, bar code readers, stylus pads, voice recognition devices, and fax machines. Common devices include the following.

Printer types include laser and ink-jet. Laser printers offer a lower cost per sheet, double-sided printing, faster throughput, and large storage capacity. Ink-jet printers offer color printing and lower initial cost but have a higher operating cost.

CD and DVD drives can be used to read data from a CD or DVD. CD and DVD burners can be used to write data to a new CD or DVD, or to copy an existing one. A CD typically holds 700 MB of data, whereas a DVD holds 4.7 GB. There are two types of writeable CDs: CD-R, and CD-RW. Once data is written to a CD-R, it cannot be changed. Data written to a CD-RW can be modified or deleted, like data on a floppy disk. DVD+R and DVD-R are similar to CD-R, and DVD+RW and DVD-RW are similar to CD-RW. There is also a double-layer (DL) DVD that holds 8.5 GB. Newer types of disks can hold up to 50 GB.

External storage can be used to store files on a location other than the local computer. USB flash drives are a simple way to save files or to move files between computers. They are typically very small (1.5 to 2.75 in.), but offer less capacity than other methods (usually between 128 MB and 64 GB). External hard drives offer a wide range of capacity (typically 2 to 1000 GB), but are often larger and less convenient to move between computers. External hard drives are often used to supplement the storage located on a PC's internal hard drive, or to store a back-up of data on an internal hard drive.

Digital cameras are useful for photographs and digital storage of images. These digitized images can be transmitted electronically and viewed or incorporated into Web or traditional documents. Most digital cameras can store dozens or hundreds of pictures, depending on the amount of memory they have. Some can also store up to several minutes of video. PC video cameras (Web cams) allow a live image such as live conferencing to be viewed and recorded. A cell phone with a digital camera may allow users to transmit (e-mail) pictures or video from any location where there is cell phone service.

Scanners digitize printed photographs and images, and have a characteristic image quality expressed in dots per inch (dpi). The image files can be large and should preferably be stored using a compression format such as JPG. Special scanner software can also read text from a scanned document through a process called optical character recognition (OCR). The scanned text can then be viewed or edited in a word processor.

Bar code readers speed database input by converting a manual scan of a bar code to the alphanumeric input in a database field.

Stylus pads combine the navigational functionality of a mouse with the ability to enter handwritten notes and sketches and run automated scripts based on movements on the pad.

Voice recognition software automatically converts speech to typed text. These programs take some tuning to learn the accent, diction, and idioms of the user. Many voice recognition software packages also enable automated scripts or macros to run on verbal commands.

Fax software emulates a printer driver. It allows the user to print a document to a target fax machine through a modem. It also allows the user to receive faxes as electronic documents. Other features include integration with electronic contact databases, templates for transmittal pages, and automated tracking of faxes sent and received.

Convergence

Many devices serve more than one purpose. A common example is devices that combine the functions of a scanner, printer, fax machine, and copier. Many new cell phones also have cameras and MP3 players, and can also function like a USB flash drive or a modem. This trend toward multifunctionality is expected to continue.

NETWORKING COMPONENTS

Computers use resources that are commonly arranged into networks comprised of hardware and software components. Hardware network components include hubs, switches, routers, bridges, and repeaters.

Software network components usually take one of two forms: (1) dynamic link libraries (DLLs) to control network hardware, or (2) applications to communicate between computers according to adopted networking standards (e.g., IEEE 802.3 Ethernet).

Numerous computer network configurations exist for intra-, inter-, and extra-office data communication. Some common examples include local area networks (LANs), wide-area networks (WANs), metropolitan area networks (MANs), intranets, extranets, and the Internet.

Figure 1 is a diagram of large business system architecture with Internet connections. The system has a large number of personal computers connected via Ethernet to a central wiring closet device called a hub. This hub simplifies the wiring infrastructure by allowing centralized wiring termination for modular utility connections to individual computers. The firewall, a computer security barrier, protects against unauthorized access from the outside Internet to individual nodes inside the firewall. Users may connect to the Internet either through a corporate router or indirectly from home through an Internet service provider (ISP), which may charge a fee for the service. Higher-bandwidth connections typically cost more as speed increases.

LAN PROTOCOLS

The Ethernet protocol is the most widely used and accepted local area network (LAN) standard, and is used extensively in industry. Because of its pervasive acceptance in industry, HVAC building automation protocols such as BACnet® and LonWorks® allow their protocols to use Ethernet in their physical layer. Ethernet is a multi-access, packet-switching network where each node can request access to the network on an equal basis with any other node. Ethernet uses the collision sense, multiple access/collision detect (CSMA/CD) technique: if two nodes transmit at the same time, a collision occurs; both nodes then wait a random period of time and try again. This allows efficient allocation of bandwidth based on need.

An Ethernet LAN can use different media, such as the following:

- 10Base5, also known as Thick Net, which uses the large coaxial cable RG-11
- 10Base2, also known as Thin Net, which uses the smaller coaxial cable RG-58
- Fiber-optic cable
- 10Base-T (Twisted Pair), the most popular type, which has a wiring limitation length of 30 m and is typically used in centralized wiring systems. These cables typically use a RJ-45 connector.

Faster versions of Ethernet include 100Base-T or Fast Ethernet, 1000Base-T or Gigabit Ethernet, and 10GBase-T or 10-Gigabit Ethernet, which support data rates of 100 Mbps, 1000 Mbps (1 Gbps), and 10 Gbps respectively. 1000Base-T and 10GBase-T have variants that use fiber or twisted pair connections.

802.11, also known as Wi-Fi, is a wireless LAN standard created and maintained by the IEEE LAN/MAN Standards Committee (IEEE 802). The base current version is IEE *Standard* 802.11-2007, and the older protocols are 802.11a, 802.11b, and 801.11g; the latest

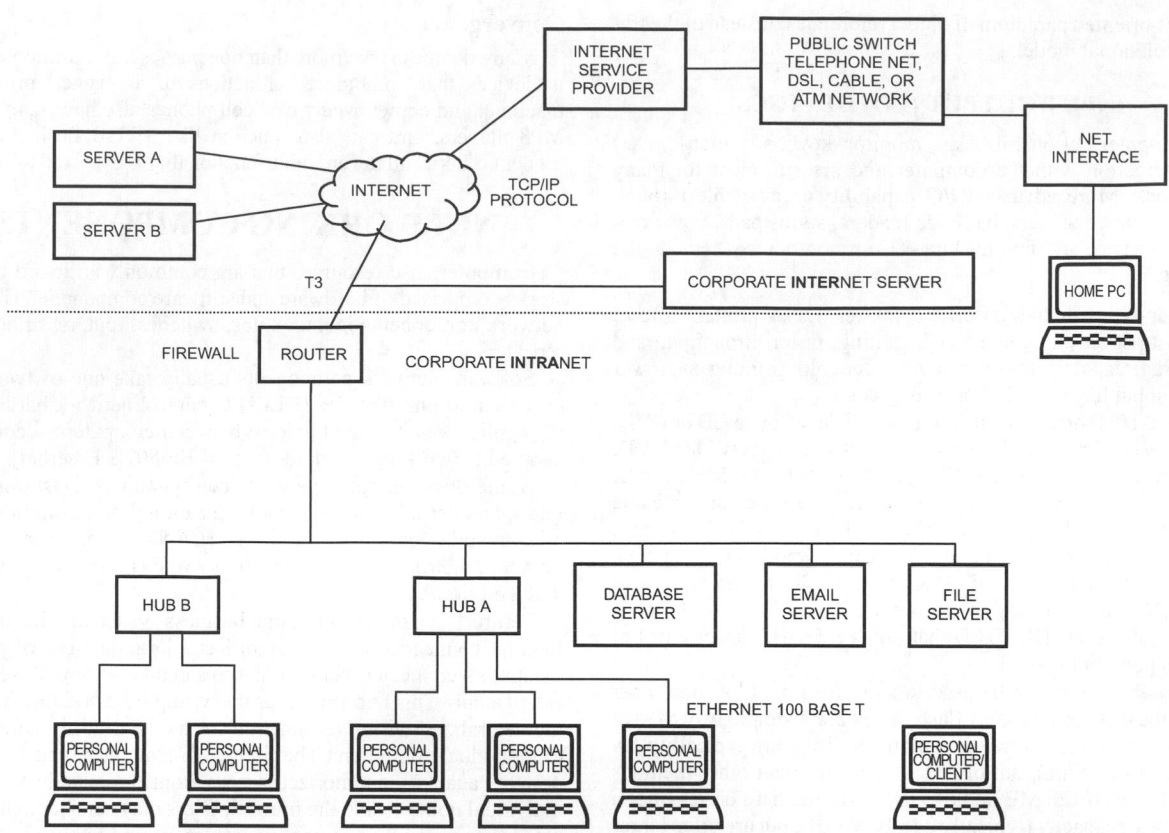

Fig. 1   Example of Business System Architecture

protocol is 802.11n. Instead of a cable, the computer and hub communicate using low-power radio waves.

802.16, also known as WiMAX, is a wireless protocol that can be used over a much longer distance than Wi-Fi. It may be configured to act as an element of a LAN, a WAN, or a MAN.

TRANSMISSION CONTROL PROTOCOL/ INTERNET PROTOCOL

Transmission control protocol/Internet protocol (TCP/IP, RFC1122) is a set of protocols that allow computers to communicate with each other. TCP/IP is the language of the Internet, and is also used in local networks, intranets, and extranets. As communication protocols such as TCP/IP have gained acceptance, availability of components based on these protocols has increased and driven networking costs down. A complementary protocol, UDP/IP (RFC768), is often used by HVAC controllers, as well as by many streaming audio, video, and VOIP sites.

WIRELESS SERVICE

One of several immediately applicable new technologies is wireless connections for computers and equipment (e.g., to the Internet or to a LAN). Wireless communications to remote sensors can provide a low-cost solution where the cost of running conventional wiring is prohibitive or inappropriate.

Wireless monitoring components include a wireless transmitter and a Web site. The transmitter is mounted on or near the equipment being monitored. The Web site determines what alerts are sent to whom, and the method of delivery (e.g., e-mail, pager, telephone). Alert messages specify the condition that triggered the alert and the location, make, model, and serial number of the equipment. Also, field service technicians can inspect service procedures, service

requests, and maintenance from a remote computer at the job site to determine repair history on a piece of equipment.

Issues such as reliability, battery replacement alerts, maintenance, and reliability because of electromagnetic noise and interference should be considered when using wireless communications.

TELEPHONY STANDARDS

Regular analog phone service offers limited bandwidth and can support analog phone modems up to 56 kilobits per second (kbps). Digital phone standards include digital subscriber line (DSL), which can support bandwidth from 640 kbps to 3 Mbps but is limited to a 10,000 ft wire length from the telephone switching office. Cable modems provide high-speed access from traditional cable TV service companies. Cable networks generally have download speeds roughly between 500 kbps and 6 Mbps, but are often limited to about 768 kbps upload speed. Fiber-optic Internet connections are available in some areas, offering download bandwidth up to 50 Mbps and upload bandwidth up to 20 Mbps. As with DSL and cable modems, fiber-optic users may not have the maximum bandwidth available at all times. Note that, for most Internet users, even the slowest bandwidth available from a DSL, cable, or fiber-optic connection is sufficient. Higher bandwidth is most useful for users who frequently download music, video, or other large files. In addition to the wired connection types, wireless connection standards with a range and bandwidth suitable to compete with DSL, cable, and fiber connections are being developed.

T1 and T3 are high-speed digital communication lines used by main communications trunks between servers, or for access to the Internet from an Internet service provider or company connection line. T1 connection speed is a maximum of 1.5 Mbps; T3 maximum speed is 44.736 Mbps. Asynchronous transfer mode (ATM) digital service provides high-speed communication at a typical maximum data rate of 155 Mbps.

FIREWALLS

Firewalls (computer security barriers), frequently used to prevent outside individuals from accessing internal, proprietary networked resources, can be designed to block certain messages from coming into or leaving the network. Firewalls may be implemented in a high-speed router, or as a separate device connected to a router. They can efficiently block all communication to or from a specific IP address. All messages that enter and leave the intranet pass through the firewall, which examines each message and halts progress of illegal communications. Firewalls are an essential part of corporate computer security systems, and are also commonly implemented as software on a PC.

INTERNET AND APPLICATIONS

The Internet originated in 1969 as an experimental project of the Advanced Research Project Agency (ARPA), and was called ARPANET. From its inception, the network was designed to be a decentralized, self-maintaining series of redundant links between computers and computer networks, capable of rapidly transmitting communications without direct human involvement or control, and with the automatic ability to reroute communications if one or more individual links were damaged or otherwise unavailable. Among other goals, this redundancy of linked computers was designed to allow vital research and communications to continue even if portions of the network were damaged by a war or other catastrophic event.

Messages between computers on the Internet do not necessarily travel entirely along the same path. The Internet uses packet-switching communication protocols that allow individual messages to be subdivided into smaller packets. These are then sent independently to the destination and automatically reassembled by the receiving computer. Although all packets of a given message often travel along the same path to the destination, if computers along the route become overloaded, packets can be rerouted to less loaded computers.

No single entity (academic, corporate, governmental, or nonprofit) administers the Internet. It exists and functions because hundreds of thousands of separate operators of computers and computer networks independently decided to use common data transfer protocols to exchange communications and information with other computers. No centralized storage location, control point, or communications channel exists for the Internet, and no single entity can control all of the information conveyed on the Internet.

INTERNET COMMUNICATION METHODS

The most common methods of communications on the Internet (as well as within the major on-line services) can be grouped into the following categories and are described in the following sections:

- One-to-one messaging (e.g., e-mail or text messaging)
- One-to-many messaging (e.g., mailing list servers)
- Distributed message databases (e.g., USENET news groups)
- Real-time communication (e.g., Internet Relay Chat, IM, VOIP)
- Real-time remote computer use (e.g., Telnet)
- Remote information retrieval (e.g., ftp, gopher, the World Wide Web)

Most of these methods of communication can be used to transmit text, data, computer programs, sound, visual images (i.e., pictures), and moving video images.

Servers

Network servers are fast becoming an important part of building automation and computer services systems. A Web page runs on a generic Web server. Files available for central storage and shared resources among users of a network are referred to as a file server. FTP servers allow users outside a company to store and manage large files such as drawing and construction documents. Mail servers are central storage servers that provide e-mail services to users. Database servers provide core database functionality, and can be added to application servers to access specific application services.

E-mail

Electronic mail (e-mail) is an extremely useful communications tool for use with local area networks, wide area networks, and the Internet. Using e-mail software, a message can be sent to individuals or to groups. Unlike postal mail, simple e-mail generally is not "sealed" or secure, and can be accessed or viewed on intermediate computers between the sender and recipient unless the message is encrypted. Modern e-mail software can attach graphics, executable programs, and other types of files. These features should be used with caution because of the large network resources required to support the file transfer.

Computer Viruses, Worms, and Trojan Horses. Computer viruses are programs that replicate themselves and typically do damage to the host system; they are frequently received through e-mail attachments. A **worm** is a self-contained program that spreads working copies of itself to other computer systems, typically through network connections. A **Trojan horse** is a program that performs undocumented activities not desired by the program user.

Antivirus software is vital to maintaining system security. Some PC operating systems are more vulnerable to viruses and worms than others.

Some e-mail programs automatically execute script files attached to e-mail; this leaves the user open to viruses and other types of attack. As a precaution, e-mail programs that automatically execute script files should not be used, or, if no other program is available, automatic script execution should be turned off. Only attachments from reputable or trusted sources should be opened, and then only after scanning the attachment with antivirus software.

Hoaxes. A common problem associated with e-mail is the hoax or chain letter. A hoax is commonly spread as a warning about serious consequences of a new virus, offers of free money or prizes, or dubious information about a person, product, or organization. A common request from the hoax is to send it to everyone you know, in effect overloading the system with useless information. Although hoaxes and chain letters do not damage the host computer, they tax network resources, spread misinformation, and distract users from real work. Users should not pass hoaxes along, but should simply delete the message. Legitimate warnings about viruses come from IT staff or antivirus vendors. Individual hoaxes can be investigated through Web sites devoted to hoaxes and virus myths (e.g., Lawrence Livermore National Laboratory 2002).

Mailing Lists

The Internet also contains automatic mailing list services (e.g., LISTSERV) that allow communication about particular subjects of interest to a group of people. Users can subscribe to a mailing list on a particular topic of interest to them. The subscriber can submit messages on the topic to the mailing list server that are forwarded, either automatically or through a human moderator overseeing the list, to the e-mail accounts of all list subscribers. A recipient can reply to the message and have the reply also distributed to everyone on the mailing list. Most mailing lists automatically forward all incoming messages to all mailing list subscribers (i.e., they are unmoderated).

Distributed Message Databases

Distributed message databases, such as USENET newsgroups, are similar to mailing lists but differ in how communications are transmitted. User-sponsored newsgroups are among the most popular and widespread applications of Internet services, and cover all imaginable topics of interest. Like mailing lists, newsgroups are open discussions and exchanges on particular topics. Users can view the entire database of messages at any time. Some newsgroups are moderated, but most are open. One common variation of a user-generated message database is called a **wiki**.

Real-Time Communication

Internet users can also engage in an immediate dialogue, in real time, with other people on the Internet, ranging from one-to-one communications to Internet Relay Chat (IRC), which allows two or more to type messages to each other that almost immediately appear on the receivers' computer screens. IRC is analogous to a telephone party line, using a computer and keyboard rather than a telephone. With IRC, however, at any one time there are thousands of different party lines available, with collectively tens of thousands of users discussing a huge range of subjects. Moreover, a user can create a new party line to discuss a different topic at any time. Some IRC conversations are moderated or have channel operators. Commercial online services often have their own chat systems to allow their members to converse.

Real-Time Remote Computer Use

Another method to access and control remote computers in real time is using Telnet, the main Internet protocol for connecting remote computers. For example, using Telnet, a researcher can use the computing power of a supercomputer located at a different university, or a student can connect to a remote library to access the library's online catalog program.

Remote Information Retrieval

The most well-known use of the Internet is searching for and retrieving information located on remote computers. The three primary methods to locate and retrieve information on the Internet are ftp, Gopher, and HTTP (the World Wide Web).

A simple method, **file transfer protocol** (ftp) allows the user to transfer files between the local computer and a remote computer. The program and format named **Gopher** guides a search through the resources available on a remote computer.

World Wide Web. The World Wide Web (WWW) serves as the platform for a global, online store of knowledge, containing information from diverse sources and accessible to Internet users around the world. Though information on the Web is contained in individual computers, the fact that each of these computers is connected to the Internet allows all of the information to become part of a single body of knowledge. It is currently the most advanced information system developed on the Internet, and encompasses most information in previous networked information systems such as ftp, Gopher, wide-area information server (WAIS) protocol (used to search indexed databases on remote servers), and USENET.

Basic Operation. The WWW is a series of documents stored in different computers all over the Internet. An essential element of the Web is that each document has an address (like a telephone number).

Most organizations now have home pages on the Web, which guide users through links to information about or relevant to that organization. Links may also take the user from the original Web site to another Web site on another computer connected to the Internet. These links from one computer to another, from one document to another across the Internet, are what unify the Web into a single body of knowledge, and what makes the Web unique.

Publishing. The WWW allows people and organizations to communicate easily and quickly. Publishing information on the Web requires only a computer connected to the Internet and running WWW server software. The computer can be anything from an inexpensive personal computer to several state-of-the-art computers filling a small building. Many Web publishers choose instead to lease disk storage space from a company with the necessary facilities, eliminating the need to own any equipment.

Information to be published on the Web must be formatted according to Web standards, which ensure that all users who want to view the material will be able to do so.

Web page design can significantly affect user satisfaction. Pages with content limited to one screen tend be easier to use than long pages that require scrolling. Large graphics files or special software requirements for viewing may discourage low-bandwidth-connection users (e.g., Internet access over regular analog phone lines) because of long waits. Maintaining and updating content is important: regular changes encourage repeat users. Web pages that give current information available nowhere else create value for users and encourage new and repeat visitors. However, care must be taken to obtain written permission for use of copyrighted material. Web sites can be open to all Internet users, or closed (an **intranet**), accessible only to those with advance authorization. Many publishers choose to keep their sites open to give their information the widest potential audience.

Searching the Web. Systems known as **search engines** allow users to search by category or by key words for particular information. The engine then searches the Web and presents a list linked to sites that contain the search term(s). Users can then follow individual links, browsing through the information on each site, until the desired material is found. There are also Web-based directories maintained by humans that can often give more precise results for a search in a particular area.

Search engines results can be narrowed by adding more key words: for example, a search for the word "HVAC" may end up with a list several hundred pages long, but a search for "HVAC VAV" could narrow the focus.

Common Standards. The Web links disparate information on Internet-linked computers by using a common information storage format called **hypertext markup language (HTML)** and a common language for the exchange of Web documents, **hypertext transfer protocol (HTTP)**. HTML documents can contain text, images, sound, animation, moving video, and links to other resources. Although the information itself may be in many different formats and stored on computers that are not otherwise compatible, these Web standards provide a basic set of standards that allow communication and exchange of information.

Many specialized programs exist to create and edit HTML files. Like HTML, **extensible markup language (XML)**, is a subset of the standard generalized markup language (SGML) protocol. XML applies the same document presentation principles of HTML to data interchange beyond simply displaying the data. XML separates data content from its presentation. This allows for different visual renditions of the same message, for displays on different devices.

Supervisory Groups. No single authority controls the entire Internet, but it is guided by the Internet Society (ISOC). ISOC is a voluntary membership organization intended to promote global information exchange through the Internet. The Internet Architecture Board (IAB) defines Internet communication standards. The IAB also keeps track of information that must remain unique. For example, each computer on the Internet has a unique 32-bit address. The IAB does not actually assign the addresses, but it makes the rules about how to assign them.

Collaborative Design

The Internet is revolutionizing how project teams collaborate. Interactions such as sharing or jointly developing drawings, documents, or software can now take place efficiently with collaborators around the world.

Collaborative design raises many coordination issues. Methods of sharing data, such as distributed databases, revision and developmental accounting, and security concerns associated with the transmission of design information must be carefully considered.

SECURITY ISSUES

Computer security requires sound planning and continuous maintenance and monitoring. Mail servers can scan for viruses before users open e-mail, and individual virus software loaded on individual machines serves as a second layer of protection (see also the section on Computer Viruses, Worms, and Trojan Horses). Always save attached files to a safe directory and scan with antivirus software before using. Typically, IS and IT staff have a dedicated support person and strategy to maintain security.

A critical area for computer security is user names and passwords. Many systems require frequent changes of passwords with minimum length, number, case, and other requirements. Passwords should never be released to anyone.

HVAC SOFTWARE APPLICATIONS

Over the past decade, the use of computer software designing mechanical systems has increased. The advantages of software-assisted design over traditional manual methods include the following:

- Computers can better preserve design assumptions and document the design intent through consolidating and organizing resources (e.g., central databases, Web sites with design documents)
- It is often faster and easier to modify assumptions used in a calculation and recalculate the results using a computer
- The computational speed of computers allows designers to tackle more difficult design problems and to use more accurate (but computationally intensive) methods
- Computers assist designers in analyzing design alternatives.

In the future, interoperable software design and application of expert systems could reduce the burden of design. Interoperable applications automatically pass design assumptions and calculation results between applications, reducing both the burden of user input and the opportunity for user error. Expert and rule-based systems could automate design standards to the point where simple system selection or distribution system sizing can be automated.

There is a wide variety of software used in the HVAC industry. This section covers applications that assist mechanical engineers in design, analysis, and specification of mechanical systems, including

- HVAC design software
- HVAC simulation software
- CAD/graphics software
- Miscellaneous utilities and applets

Much of this software is also used by HVAC equipment manufacturers to design their products, and by controls contractors to develop control systems.

In addition, sources for these applications and the fundamentals of interoperable computer applications for the HVAC&R industry are discussed.

The distinction between HVAC design and simulation is subtle. The convention used here is that design software is used for static analysis: evaluating equipment and distribution systems under a specific design condition. In contrast, HVAC simulation is used for dynamic system analysis over a range of time or operating conditions. Many applications provide both functions through a single interface.

HVAC DESIGN CALCULATIONS

HVAC design software includes programs for the following tasks:

- HVAC heating and cooling load calculations
- Duct system sizing and analysis
- Pipe system sizing and analysis

- Acoustical analysis
- Equipment selection

Although computers are now widely used in the design process, most programs perform analysis, not design, of a specified system. That is, the engineer proposes a design and the program calculates the consequences of that design. When the engineer can define alternatives, a program may aid design by analyzing the performance of the alternatives and then selecting the best according to predetermined criteria. Thus, a program to calculate the cooling load of a building requires specifications for the building and its systems; it then simulates that building's performance under certain conditions of weather, occupancy, and scheduling. Although most piping and duct design tools can size pipe or ductwork based on predetermined criteria, most engineers still decide duct dimensions, air quantities, duct routing, and so forth and use software to analyze the performance of that design.

Some interoperable software can pass values from one design phase to be used as inputs in the next. For example, a load program may be used to determine the design air quantities at the zone and space level. Through interoperability, this information can be preserved in the building model and used as an input for the duct sizing program. Interoperable software is further described in the section on Interoperable Computer Applications for the HVAC&R Industry.

Because computer programs perform repetitive calculations rapidly, the designer can explore a wide range of alternatives and use selection criteria based on annual energy costs or life-cycle costs.

Heat and Cooling Load Design

Calculating design thermal loads in a building is a necessary step in selecting HVAC equipment for virtually any building project. To ensure that heating equipment can maintain satisfactory building temperature under all conditions, peak heating loads are usually calculated for steady-state conditions without solar or internal heat gains. This relatively simple calculation can be performed by hand or with a computer.

Peak cooling loads are more transient than heating loads. Radiative heat transfer within a space and thermal storage cause thermal loads to lag behind instantaneous heat gains and losses. This lag can be important with cooling loads, because the peak is both reduced in magnitude and delayed in time compared to the heat gains that cause it. A further complication is that loads peak in different zones at different times (e.g., the solar load in a west-facing zone will peak in the afternoon, and an east-facing zone will usually peak in the morning). To properly account for thermal lag and the noncoincidence of zone loads, a computer program is generally required to calculate cooling loads for a multiple-zone system. These programs simulate loads for each hour of the day for every day of the year (or a representative subset of the days of the year), and report peak loads at both the zone and system level, recording the date and time of their occurrence. Because the same input data are used for heating and cooling loads, these programs report peak heating loads, as well.

Various calculation methods are used to account for the transience of cooling loads; the most common methods are described in detail in Chapters 17 and 18 of the 2009 *ASHRAE Handbook— Fundamentals*. The most accurate but computationally intensive method, the heat-balance (HB) method, is a fundamental calculation of heat transfer from all sources with the construction elements of the building (walls, floor, windows, skylights, furniture, ceiling, slab, and roof). Only the computational speed of present-day personal computers allows these calculations to be performed for large multizone buildings in a reasonable amount of time (ranging from several minutes to an hour or more). Several simplified methods or approximations of thermal lag are available to speed computation time, including the radiant time-series (RTS), transfer function (TFM), total equivalent temperature differential with time averaging (TETD/TA) and cooling load temperature differential with cooling load factors (CLTD/CLF) methods. These approximations are less accurate than the heat balance method, but can be sufficient if they are appropriately applied.

Both the TETD/TA and the TFM methods require a history of thermal gains and loads. Because histories are not initially known, they are assumed zero and the building under analysis is taken through a number of daily weather and occupancy cycles to establish a proper 24 h load profile. Thus, procedures required for the TFM and TETD/TA methods involve so many individual calculations that noncomputerized calculations are highly impractical. The CLTD/CLF method, on the other hand, is meant to be a manual calculation method and can be implemented by a spreadsheet program. The CLTD and CLF tables presented in the 2001 *ASHRAE Handbook—Fundamentals* are, in fact, based on application of the TFM to certain geometries and building constructions. An automated version of TFM is preferred to CLTD/CLF because it is more accurate. The automated TETD/TA method, however, can provide a good approximation with significantly less computation than the TFM.

Characteristics of a Loads Program. In general, a loads program requires user input for most or all of the following:

• Full building description, including the construction and other heat transfer characteristics of the walls, roof, windows, and other building elements; size; orientation; geometry of the rooms, zones, and building; and possibly also shading geometries
• Sensible and latent internal loads from lights and equipment, and their corresponding operating schedules
• Sensible and latent internal loads from people
• Indoor and outdoor design conditions
• Geographic data such as latitude and elevation
• Ventilation requirements and amount of infiltration
• Number of zones per system and number of systems

With this input, loads programs calculate both the heating and cooling loads and perform a psychrometric analysis. Output typically includes peak room and zone loads, supply air quantities, and total system (coil) loads.

Selecting a Loads Program. Beyond general characteristics, such as hardware and software requirements, type of interface (icon-based, menu-driven versus command-driven), availability of manuals and support, and cost, some program-specific characteristics should be considered when selecting a loads program. The following are among items to be considered:

• Ease of use and compatibility with the existing operating system
• Type of building to be analyzed: residential versus commercial (residential-only loads programs tend to be simpler to use than more general-purpose programs meant for commercial and industrial use, but residential-only programs generally have limited abilities)
• Method of calculation for the cooling load (note that some programs support more than one method of calculation)
• Program limits on such items as number of systems, zones, rooms, and surfaces per room
• Sophistication of modeling techniques (e.g., capability of handling exterior or interior shading devices, tilted walls, daylighting, skylights, etc.)
• Units of input and output
• Program complexity (in general, the more sophisticated and flexible programs require more input and are somewhat more difficult to use than simpler programs)
• Capability of handling the system types under investigation
• Certification of the program for energy code compliance
• Ability to share data with other programs, such as CAD and energy analysis

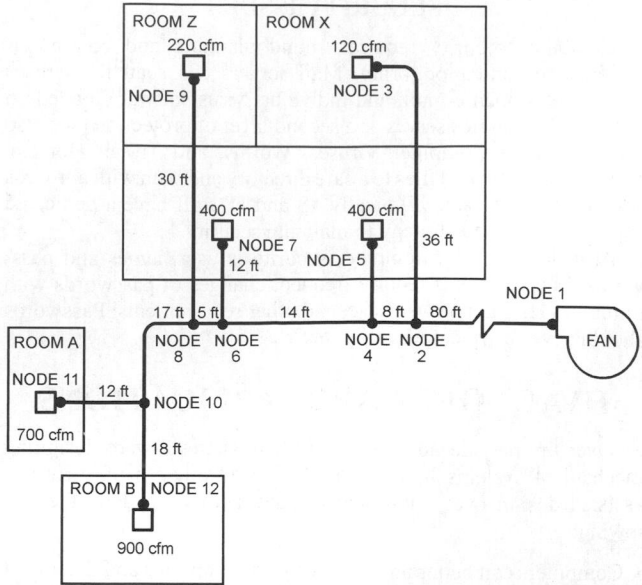

Fig. 2   Example of Duct System Node Designation

Duct Design

Chapter 21 of the 2009 *ASHRAE Handbook—Fundamentals* describes several methods for duct design. Major considerations in duct design include the physical space required by the duct, cost of the ductwork and fittings, energy required to move air through the duct, acoustics, minimum velocities to maintain contaminants in suspension (fume exhaust), flexibility for future growth, zoning, and constructability. Duct software can help with the layout and analysis of new and/or existing distribution systems.

Duct design is more an art than a science, which complicates automation of design. Duct design programs and calculation methods are limited in their ability to (1) calculate actual operating pressures because of the impact of fan and duct system effects, and (2) take into consideration many unmeasurable factors in the process of laying out and sizing ducts (e.g., provision for space limitations, local duct construction preferences and practices, local labor and material costs, acoustics, and accommodating future remodeling).

Selecting and Using a Program. Duct design involves routing or laying out ductwork, selecting fittings, and sizing ducts. Computer programs can rapidly analyze design alternatives and determine critical paths in long and complicated distribution systems. Furthermore, programs and printouts can preserve the assumptions used in the design: any duct calculation requires accurate estimates of pressure losses in duct sections and the definition of the interrelations of velocity heads, static pressures, total pressures, and fitting losses.

The general computer procedure is to designate nodes (the beginning and end of duct sections) by number. Details about each node (e.g., divided flow fitting and terminal) and each section of duct between nodes (maximum velocity, flow rate, length, fitting codes, size limitation, insulation, and acoustic liner) are used as input data (Figure 2).

Characteristics of a duct design program include the following:

• Calculations for supply, return, and exhaust
• Sizing by constant friction, velocity reduction, static regain, and constant velocity methods
• Analysis of existing ducts
• Inclusion of fitting codes for a variety of common fittings
• Identification of the duct run with the highest pressure loss, and tabulation of all individual losses in each run

- Printout of all input data for verification
- Provision for error messages
- Calculation and printout of airflow for each duct section
- Printout of velocity, fitting pressure loss, duct pressure loss, and total static pressure change for each duct section
- Graphics showing schematic or line diagrams indicating duct size, shape, flow rate, and air temperature
- Calculation of system heat gain/loss and correction of temperatures and flow rates, including possible system resizing
- Specification of maximum velocities, size constraints, and insulation thicknesses
- Consideration of insulated or acoustically lined duct
- Bill of materials for sheet metal, insulation, and acoustic liner
- Acoustic calculations for each section
- File-sharing with other programs, such as spreadsheets and CAD

Because many duct design programs are available, the following factors should be considered in program selection:

- Maximum number of branches that can be calculated
- Maximum number of terminals that can be calculated
- Types of fittings that can be selected
- Number of different types of fittings that can be accommodated in each branch
- Ability of the program to balance pressure losses in branches
- Ability to handle two- and three-dimensional layouts
- Ability to size a double-duct system
- Ability to handle draw-through and blow-through systems
- Ability to prepare cost estimates
- Ability to calculate fan motor power
- Provision for determining acoustical requirements at each terminal
- Ability to update the fitting library

Optimization Techniques for Duct Sizing. For an optimized duct design, fan pressure and duct cross sections are selected by minimizing the life-cycle cost, which is an objective function that includes initial and energy costs. Many constraints, including constant pressure balancing, acoustic restrictions, and size limitations, must be satisfied. Duct optimization is a mathematical problem with a nonlinear objective function and many nonlinear constraints. The solution must be taken from a set of standard diameters and standard equipment. Numerical methods for duct optimization exist, such as the T method (Tsal et al. 1988), coordinate descent (Tsal and Chechik 1968), Lagrange multipliers (Kovarik 1971; Stoecker et al. 1971), dynamic programming (Tsal and Chechik 1968), and reduced gradient (Arkin and Shitzer 1979).

Piping Design

Many computer programs are available to size or calculate the flexibility of piping systems. Sizing programs normally size piping and estimate pump requirements based on velocity and pressure drop limits. Some consider heat gain or loss from piping sections. Several programs produce a bill of materials or cost estimates for the piping system. Piping flexibility programs perform stress and deflection analysis. Many piping design programs can account for thermal effects in pipe sizing, as well as deflections, stresses, and moments.

Numerous programs are available to analyze refrigerant piping layouts. These programs aid in design of properly sized pipes or in troubleshooting existing systems. Some programs are generic, using the physical properties of refrigerants along with system practices to give pipe sizes that may be used. Other programs "mix and match" evaporators and condensers from a particular manufacturer and recommend pipe sizes. See Chapters 2 and 3 of the 2010 *ASHRAE Handbook—Refrigeration* for specifics on refrigerant pipe sizing.

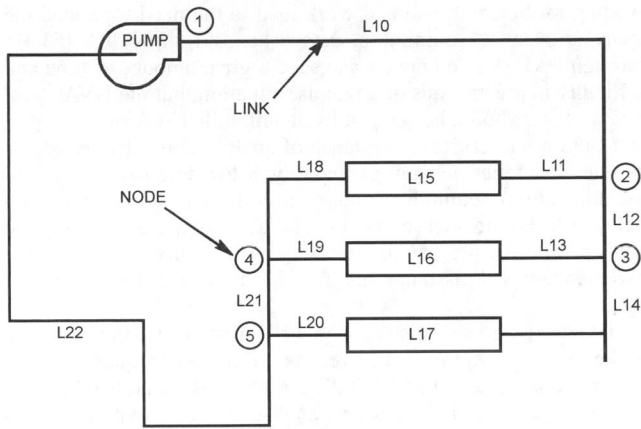

Fig. 3   Examples of Nodes for Piping System

The general technique for computerizing piping design problems is similar to that for duct design. A typical piping problem in its nodal representation is shown in Figure 3.

Useful piping programs do the following:

- Provide sufficient design information
- Perform calculations for both open and closed systems
- Calculate flow, pipe size, and pressure drop in each section
- Handle three-dimensional piping systems
- Cover a wide selection of common valves and fittings, including specialized types (e.g., solenoid and pressure-regulating valves)
- Consider different piping materials such as steel, copper, and plastic by including generalized friction factor routines
- Accommodate liquids, gases, and steam by providing property information for a multiplicity of fluids
- Calculate pump capacity and head required for liquids
- Calculate available terminal pressure for nonreturn pipes
- Calculate required expansion tank size
- Estimate heat gain or loss for each portion of the system
- Prepare a cost estimate, including costs of pipe and insulation materials and associated labor
- Print out a bill of materials
- Calculate balance valve requirements
- Perform a pipe flexibility analysis
- Perform a pipe stress analysis
- Print out a graphic display of the system
- Allow customization of specific design parameters and conditions, such as maximum and minimum velocities, maximum pressure drops, condensing temperature, superheat temperature, and subcooling temperature
- Allow piping evaluation for off-design conditions
- Provide links to other programs, such as equipment simulation programs

Limiting factors to consider in piping program selection include the following:

- Maximum number of terminals the program can accommodate
- Maximum number of circuits the program can handle
- Maximum number of nodes each circuit can have
- Maximum number of nodes the program can handle
- Compressibility effects for gases and steam
- Provision for two-phase fluids

Acoustic Calculation

Chapter 48 summarizes sound generation and attenuation in HVAC systems. Applying these data and methodology often requires a large amount of computation. All sound-generation mechanisms and transmission paths are potential candidates for analysis.

Adding to the computational work load is the need to extend the analysis over, at a minimum, octave bands 1 through 8 (63 Hz through 8 kHz). A computer can save a great amount of time and difficulty in the analysis of any noise situation, but the HVAC system designer should be wary of using unfamiliar software.

Caution and critical acceptance of analytical results are necessary at all frequencies, but particularly at low frequencies. Not all manufacturers of equipment and sound control devices provide data below 125 Hz; in such cases, the HVAC designer conducting the analysis and the programmer developing the software must make experience-based assumptions for these critical low-frequency ranges.

The designer/analyst should be well satisfied if predictions are within 5 dB of field-measured results. In the low-frequency "rumble" regions, results within 10 dB are often as accurate as can be expected, particularly in areas of fan discharge. Conservative analysis and application of results is necessary, especially acoustically critical spaces.

Several easy-to-use acoustics programs are currently available, but they are often less detailed than custom programs developed by acoustic consultants for their own use. Acoustics programs are designed for comparative sound studies and allow the design of comparatively quiet systems. Acoustic analysis should address the following key areas of the HVAC system:

- Sound generation by HVAC equipment
- Sound attenuation and regeneration in duct elements
- Wall and floor sound attenuation
- Ceiling sound attenuation
- Sound breakout or break-in in ducts or casings
- Room absorption effect (relation of sound power criteria to sound pressure experienced)

Algorithm-based programs are preferred because they cover more situations (see Chapter 48); however, algorithms require assumptions. Basic algorithms, along with sound data from the acoustics laboratories of equipment manufacturers, are incorporated to various degrees in acoustics programs. The HVAC equipment sound levels in acoustics programs should come from the manufacturer and be based on measured data, because there is a wide variation in the sound generated by similar pieces of equipment. Some generic equipment sound generation data, which may be used as a last resort in the absence of specific measured data, are found in Chapter 48. Whenever possible, equipment sound power data by octave band (including 32 Hz and 63 Hz) should be obtained for the path under study. A good sound prediction program relates all performance data.

Many more specialized acoustics programs are also available. Various manufacturers provide equipment selection programs that not only select optimum equipment for a specific application, but also provide associated sound power data by octave bands. Data from these programs should be incorporated in the general acoustic analysis. For example, duct design programs may contain sound predictions for discharge airborne sound based on the discharge sound power of fans, noise generation/attenuation of duct fittings, attenuation and end reflections of VAV terminals, attenuation of ceiling tile, and room effect. VAV terminal selection programs generally contain subprograms that estimate the space NC level near the VAV unit in the occupied space. However, projected space NC levels alone may not be acceptable substitutes for octave-band data. The designer/analyst should be aware of assumptions, such as room effect, made by the manufacturer in presenting acoustical data.

Predictive acoustic software allows the designer to examine HVAC-generated sound in a realistic, affordable time frame. HVAC-oriented acoustic consultants generally assist designers by providing cost-effective sound control ideas for sound-critical applications. A well-executed analysis of the various components and sound paths enables the designer to assess the relative importance of each and to direct corrective measures, where necessary, to the most critical areas. However, computer-generated results should supplement the designer's skills, not replace them.

Equipment Selection and Simulation

Equipment-related computer applications include programs for equipment selection, equipment optimization, and equipment simulation.

Equipment selection programs are basically computerized catalogs. The program locates an existing equipment model that satisfies the entered criteria. The output is a model number, performance data, and sometimes alternative selections.

Equipment optimization programs display all possible equipment alternatives and let the user establish ranges of performance data or first cost to narrow the selection. The user continues to narrow performance ranges until the best selection is found. The performance data used to optimize selections vary by product family.

Equipment simulation programs calculate the full- and part-load performance of specific equipment over time, generally one year. The calculated performance is matched against an equipment load profile to determine energy requirements. Utility rate structures and related economic data are then used to project equipment operating cost, life-cycle cost, and comparative payback. Before accepting output from an equipment simulation program, the user must understand the assumptions made, especially concerning load profile and weather.

Some advantages of equipment programs include the following:

- High speed and accuracy of the selection procedure
- Pertinent data presented in an orderly fashion
- More consistent selections than with manual procedures
- More extensive selection capability
- Multiple or alternative solutions
- Small changes in specifications or operating parameters easily and quickly evaluated
- Data-sharing with other programs, such as spreadsheets and CAD

Simulation programs have the advantages of (1) projecting part-load performance quickly and accurately, (2) establishing minimum part-load performance, and (3) projecting operating costs and offering payback-associated higher-performance product options.

There are programs for nearly every type of HVAC equipment, and industry standards apply to the selection of many types. The more common programs and their optimization parameters include the following:

Air distribution units	Pressure drop, first cost, sound, throw
Air-handling units, rooftop units	Power, first cost, sound, filtration, footprint, heating and cooling capacity
Boilers	First cost, efficiency, stack losses
Cooling towers	First cost, design capacity, power, flow rate, air temperatures
Chillers	Power input, condenser head, evaporator head, capacity, first cost, compressor size, evaporator size, condenser size
Coils	Capacity, first cost, fluid pressure drop, air pressure drop, rows, fin spacing
Fan-coils	Capacity, first cost, sound, power
Fans	Volume flow, power, sound, first cost, minimum volume flow
Heat recovery equipment	Capacity, first cost, air pressure drop, water pressure drop (if used), effectiveness
Pumps	Capacity, head, impeller size, first cost, power
Air terminal units (variable- and constant-volume flow)	Volume flow rate, air pressure drop, sound, first cost

Some extensive selection programs have evolved. For example, coil selection programs can select steam, hot-water, chilled-water, and refrigerant (direct-expansion) coils. Generally, they select coils according to procedures in AHRI *Standards* 410 and 430.

Chiller and refrigeration equipment selection programs can choose optimal equipment based on factors such as lowest first cost, highest efficiency, best load factor, and best life-cycle performance. In addition, some manufacturers offer modular equipment for customization of their product. This type of equipment is ideal for computerized selection.

However, equipment selection programs have limitations. The logic of most manufacturers' programs is proprietary and not available to the user. All programs incorporate built-in approximations or assumptions, some of which may not be known to the user. Equipment selection programs should be qualified before use.

HVAC SIMULATION

Energy Simulation

Unlike peak load calculation programs, building energy simulation programs integrate loads over time (usually a year), consider the systems serving the loads, and calculate the energy required by the equipment to support the system. Most energy programs simulate the performance of already designed systems, although programs are now available that make selections formerly left to the designer, such as equipment sizes, system air volume, and fan power. Energy programs are necessary for making decisions about building energy use and, along with life-cycle costing routines, quantify the effect of proposed energy conservation measures during the design phase. In new building design, energy programs help determine the appropriate type and size of building systems and components; they can also be used to explore the effects of design tradeoffs and evaluate the benefits of innovative control strategies and the efficiency of new equipment.

Energy programs that track building energy use accurately can help determine whether a building is operating efficiently or wastefully. They have also been used to allocate costs from a central heating/cooling plant among customers of the plant. However, such programs must be adequately calibrated to measured data from the building under consideration.

Characteristics of Building Energy Simulation. Most programs simulate a wide range of buildings, mechanical equipment, and control options. However, computational results differ substantially across programs. For example, the shading effect from overhangs, side projections, and adjacent buildings frequently affects a building's energy consumption; however, the diverse approaches to load calculation result in a wide range of answers.

The choice of weather data also influences the load calculation. Depending on the requirements of each program, various weather data are used:

- Typical hourly data for one year only, from averaged weather data
- Typical hourly data for one year, as well as design conditions for typical design days
- Reduced data, commonly a typical day or days per month for the year
- Typical reduced data, nonserial or bin format
- Actual hourly data, recorded on site or nearby, for analysis where the simulation is being compared to actual utility billing data or measured hourly data

Simulation programs differ significantly in the methods they use to simulate the mass effects of buildings, ground, and furniture. How accurate these methods are and how well they delay peak heating and cooling can lead to significant uncertainty in predicting building heating and cooling needs, sizing equipment to meet those loads, and predicting system energy needs.

Both air-side and energy conversion simulations are required to handle the wide variations among central heating, ventilation, and air-conditioning systems. To properly estimate energy use, simulations must be performed for each combination of system design, operating scheme, and control sequence.

Simulation Techniques. Two methods are used in computer simulation of energy systems: the fixed schematic technique and the component relation technique.

The **fixed-schematic-with-options technique**, the first and most prevalent method, involves writing a calculation procedure that defines a given set of systems. The schematic is then fixed, with the user's options usually limited to equipment performance characteristics, fuel types, and the choice of certain components.

The **component relation technique** is organized around components rather than systems. Each component is described mathematically and placed in a library. User input includes the definition of the schematic, as well as equipment characteristics and capacities. Once all components have been identified and a mathematical model for each has been formulated, they may be connected and information may be transferred between them. Although its generality leads to certain inefficiencies, the component relation technique does offer versatility in defining system configurations.

Selecting an Energy Program. In selecting an energy analysis program, factors such as cost, availability, ease of use, technical support, and accuracy are important. However, the fundamental consideration is whether the program will do what is required of it. It should be sensitive to the parameters of concern, and its output should include necessary data. For other considerations, see Chapter 19 of the 2009 *ASHRAE Handbook—Fundamentals*.

Time can be saved if the initial input file for an energy program can also be used for load calculations. Some programs interface directly with computer-aided design (CAD) files, greatly reducing the time needed to create an energy program input file.

Comparisons of Energy Programs. Because energy analysis programs use different calculation methods, results vary significantly. Many comparisons, verifications, and validations of simulation programs have been made and reported (see the Bibliography). Conclusions from these reports can be summarized as follows:

- Results obtained by using several programs on the same building range from good agreement to no agreement at all. The degree of agreement depends on interpretations by the user and the ability of the programs to model the building.
- Several people using several programs on the same building will probably not agree on the results of an energy analysis.
- The same person using different programs on the same building may or may not find good agreement, depending on the complexity of the building and its systems and on the ability of the programs to model specific conditions in that building.
- *Forward* computer simulation programs, which calculate the performance of a building given a set of descriptive inputs, weather conditions, and occupancy conditions, are best suited for design.
- Calibrating hourly forward computer simulation programs is possible, but can require considerable effort for a moderately complex commercial office building. Details on scheduled use, equipment set points, and even certain on-site measurements may be necessary for a closely calibrated model. Special-purpose graphic plots are useful in calibrating a simulation program with data from monthly, daily, or hourly measurements.
- *Inverse, empirical,* or *system parameter identification* models may also be useful in determining the characteristics of building energy usage. Such models can determine relevant building parameters from a given set of actual performance data. This is the inverse of the traditional building modeling approach, hence the name.

Energy Programs to Model Existing Buildings. Computer energy analysis of existing buildings can accommodate complex situations; evaluate the energy effects of many alternatives, such as changes in control settings, occupancy, and equipment performance; and predict relative magnitudes of energy use. There are many programs available, varying widely in cost, degree of complexity, and ease of use.

A general input-data acquisition procedure should be followed in computer energy analysis of existing buildings. First, **energy consumption data** must be obtained for a one- to two-year period. For electricity, these data usually consist of metered electrical energy consumption and demand on at least a month-by-month basis. For natural gas, the data are in a similar form and are almost always on a monthly basis. For both electricity and natural gas, it is helpful to record the dates when the meters were read; these dates are important in weather normalization for determining average billing period temperatures. For other types of fuel, such as oil and coal, the only data available may be delivery amounts and dates.

Unless fuel use is metered or measured daily or monthly, consumption for any specific period shorter than one season or year is difficult to determine. Data should be converted to per-day usage or adjusted to account for differences in length of metering periods. The data tell how much energy went into the building on a gross basis. Unless extensive submetering is used, it is nearly impossible to determine when, how, and for what that energy was used. It may be necessary to install such meters to determine energy use.

The **thermal and electricity usage characteristics** of a building and its energy-consuming systems as a time-varying function of ambient conditions and occupancy must also be determined. Most computer programs can use as much detailed information about the building and its mechanical and electrical system as is available. Where the energy implications of these details are significant, it is worth the effort of obtaining them. Testing fan systems for air quantities, pressures, control set points, and actions can provide valuable information on deviations from design conditions. Test information on pumps can also be useful.

Data on **building occupancy** are among the most difficult to obtain. Because most energy analysis programs simulate the building on an hourly basis for a one-year period, it is necessary to know how the building is used for each of those hours. Frequent observation of the building during days, nights, and weekends shows which energy-consuming systems are being used and to what degree. Measured, submetered hourly data for at least one week are needed to begin to understand weekday/weekend schedule-dependent loads.

Weather data, usually one year of hour-by-hour weather data, are necessary for simulation (see Chapter 14 of the 2009 *ASHRAE Handbook—Fundamentals*). The actual weather data for the year in which energy consumption data were recorded significantly improves simulation of an existing building. Where the energy-consuming nature of the building is related more to internal than to external loads, selection of weather data is less important; however, for residential buildings or buildings with large outside air loads, the selection of weather data can significantly affect results. The purpose of the simulation should also be considered when choosing weather data: either specific-year data, data representative of long-term averages, or data showing temperature extremes may be needed, depending on the goal of the simulation.

Usually, the results of the first computer runs do not agree with actual metered energy consumption data. The following are possible reasons for this discrepancy:

- Insufficient understanding of energy-consuming systems that create the greatest use
- Inaccurate information on occupancy and time of building use
- Inappropriate design information on air quantities, set points, and control sequences

The input building description must be adjusted and trial runs continued until the results approximate actual energy use. Matching the metered energy consumption precisely is difficult; in any month, results within 10% are considered adequate.

The following techniques for calibrating a simulation program to measured data from a building should be considered:

- Matching submetered loads or 24 h day profiles of simulated whole-building electric loads to measured data
- Matching *x-y* scatter plots of simulated daily whole-building thermal loads versus average daily temperature to measured data
- Matching simulated monthly energy use and demand profiles to utility billing data

Having a simulation of the building as it is being used permits subsequent computer runs to evaluate the energy effects of various alternatives or modifications. The evaluation may be accomplished simply by changing the input parameters and running the program again. The effect of various alternatives may then be compared and an appropriate one selected.

Computational Fluid Dynamics

Computational fluid dynamics (CFD) is a technique for simulation, study, analysis, and prediction of fluid flow and heat transfer in well-defined, bounded spaces. It is based on equations (Navier-Stokes, thermal energy, and species equations, with the appropriate equation of state) that govern the physical behavior of a flow/thermal system, under the premise that mass, momentum, thermal energy, and species concentration are conserved locally and globally within the model.

Usually these equations are in the form of nonlinear partial differential equations relating velocities, pressure, temperature, and some scalar variables to space directions and time. Initial and boundary conditions are specified for each domain, and numerical methods such as finite difference, finite volume, or finite element analysis are utilized to solve these equations. The domain (typically space) is divided (discretized) into cells or elements, the nodal points are defined, then equations are solved for each discrete cell. After solving these equations, the dependent variables (velocities, pressure, temperature, and scalar) are made available at the nodes.

CFD analysis can be applied to a variety of tasks, including location of airflow inlets/outlets in a space, velocities required for system performance, flow pattern determination, temperature distribution in space, and smoke movement and build-up in a space with fire.

HVAC GRAPHICS, CAD, AND BUILDING DATA MODELS

Computer-Aided Design

Computer graphics of buildings and their systems help coordinate interdisciplinary designs and simplify modifications. CAD (referring to computer-aided drafting at times and to computer-aided design at other times) is a subset of CADD (computer-aided design and drafting). CADD encompasses the creation of **building data models** and the storing and use of attributes of graphic elements, such as automated area takeoff or the storage of design characteristics in a database. Drafting, or the creation of construction documents, is quickly becoming a mere by-product of the design process. The distinction between engineering and drafting in engineering firms is decreasing as design and production work become unified. In this section, CADD refers to all aspects of design.

Building data models express not only true dimensions, but also how different elements are connected and where they are located. Some "intelligent" models can even calculate how different components interact with each other, and how all the components act as a system. It is easy to extract lengths, areas, and volumes, as well as the orientation and connectedness (topology) of building parts from computerized drawings of buildings. Drawings can be changed easily and, in most cases, do not require redrawing construction documents. Building walls or ductwork, for example, can be added, removed, enlarged, reduced, and relocated. The user can rotate an entire building on the site plan without redrawing the building or destroying the previous drawing. Section and elevation views can also be specified and drawn almost instantly, as can fabrication drawings. Building data models can also be exported using standard formats and then imported to other programs, such as those for calculating heating and cooling loads, duct sizing, or operating sheet metal plasma cutting machines.

CAD systems automate (1) material and cost estimating, (2) design and analysis of HVAC systems, (3) visualization and interference checking, and even (4) manufacturing of HVAC components such as ductwork. These systems are now customized to HVAC applications; for instance, double- and single-line duct layout can be re-represented with a single command and automatically generate three-dimensional (3D) building data models.

Computer-generated drawings of building parts can be linked to nongraphic characteristics, which can be extracted for reports, schedules, and specifications. For instance, an architectural designer can draw wall partitions and windows, and then have the computer automatically tabulate the number and size of windows, the areas and lengths of walls, and the areas and volumes of rooms and zones. Similarly, the building data model can store information for later use in reports, schedules, design procedures, or drawing notes. For example, the airflow, voltage, weight, manufacturer, model number, cost, and other data about a fan displayed on a drawing can be stored and associated with the fan in the data model. The link between the graphics and attributes makes it possible to review the characteristics of any item or to generate schedules of items in a certain area on a drawing. Conversely, the graphics-to-data links allow the designer to enhance graphic items having a particular characteristic by searching for that characteristic in the data files and then making the associated graphics brighter, flashing, bolder, or different-colored in the display.

Computer-generated data models can also help the designer visualize the building and its systems and check for interferences, such as structure and ductwork occupying the same space. Some software packages now perform automatic interference checking. Layering features also help an HVAC designer coordinate designs with other disciplines. Mistakes in design can be corrected even before the drawings are plotted to eliminate costly field rework or, worse, at-fault litigation. For complex projects, interference software applications are available that assist in managing interferences throughout the building data model.

HVAC system models and associated data can be used by building owners and maintenance personnel for ongoing facilities management, strategic planning, and maintenance. For instance, an air-handling unit can be associated with a preventative maintenance schedule in a separate facility management application. Models are useful for computer-aided manufacturing such as duct construction. The 3D building data model can also aid in developing sections and details for building drawings. CAD programs can also automate cross-referencing of drawings, drawing notations, and building documents.

In addition to the concept of the single-building data model, the Internet is has greatly influenced engineering workflows. As bandwidth continues to widen, the speed of the Internet has allowed engineering firms to expand their clients, partners, and services worldwide. Software applications are available that allow users to design projects and share building model data globally over the Internet, and can be used for simple viewing and redlining as well. Project managers can view all aspects of the design process live and redline drawings simply by using a Web browser.

Computer Graphics and Modeling

Computer graphics programs may be used to create and manipulate pictorial information. Information can be assimilated more easily when presented as graphical displays, diagrams, and models. Historically, building scale models has improved the design engineer's understanding and analysis of early prototype designs. The psychrometric chart has also been invaluable to HVAC engineers for many years.

Combining alphanumerics (text) with computer graphics lets the design engineer quickly evaluate design alternatives. Computers can simulate design problems with three-dimensional color images that can predict the performance of mechanical systems before they are constructed. Simulation graphics software is also used to evaluate design conditions that cannot ordinarily be tested with scale models because of high costs or time constraints. Product manufacturability and economic feasibility may thus be determined without constructing a working prototype. Whereas historically only a few dozen scale-model tests could be performed before full-scale manufacturing, combined computer and scale-model testing allows hundreds or thousands of tests to be performed, improving reliability in product design. Graphic modeling and simulation of complex fluid flow fields are also used to simulate airflow fields in unidirectional cleanrooms, and may eventually make cleanroom mock-ups obsolete (Busnaina et al. 1988). Computer models of particle trajectories, transport mechanisms, and contamination propagation are also commercially available (Busnaina 1987).

Integrated CADD and expert systems can help the building designer and planner with construction design drawings and with construction simulation for planning and scheduling complex building construction scenarios (Potter 1987).

Architects use software that creates a 3D model of the building. **Building information modeling (BIM)** is a building design and documentation methodology that relies on the creation and collection of interrelated computable information about a building project so that reliable, coordinated, and internally consistent digital representations of the building are available for design decision making, production of high-quality construction documents, construction planning, and predicting performance in various ways.

BIM is about (1) integrating design and construction processes, (2) making them interoperable, and (3) the software tools needed to achieve that. ASHRAE's (2009) *Introduction to Building Information Modeling (BIM)* is intended to serve as a starting point for members considering adopting BIM tools and applications as part of their business practices. It explores the benefits, costs, risks and rewards associated with BIM, interoperability, and integration. In addition, for those already applying BIM and BIM-related technologies, it may provide ideas to help them unearth new opportunities and expand their services into new markets.

Engineers using BIM today are working with software where objects carry information that describes their individual characteristics as well as how they interact with other objects to create a system. Systems created with these "intelligent" objects allow creation of drawings that are a virtual representation of building systems exactly as they would be installed during construction. Because these systems then inherit the intelligence of the objects that compose them, powerful modification capabilities allow these systems to propagate design changes throughout the system. As required systems are created, a full 3D systems model is assembled that accurately represents what is to be constructed in the field. It is this 3D systems model that enables quick and accurate generation of construction documents. The 3D model can simply be presented from various orientations to create the necessary views and plans rather than needing to reproduce a representation from a separate perspective. Section and elevation views can be generated instantly through a few simple steps to identify the view required to section and the plane that cuts through the design to create these views directly from the systems model. As a design evolves and design changes are made, these views can be updated automatically. By taking advantage of the design model, tools such as automated collision detection can quickly identify conflicts in designs in both 2D and 3D before construction begins. Schedules can be produced that are directly linked to the design and, as additions or changes are made, these are automatically updated. Schedules can also include formulas to calculate values; therefore, even detailed design information, such as heating or electrical load requirements, can be captured in the schedules or extracted for use in other applications, such as spreadsheets.

Models can be exported in standard file formats to quickly calculate heating and cooling loads, size duct systems using equal friction or static regain methods, as well as carry out full-building energy simulation using equipment performance parameters established in the BIM design.

Estimating software imports data directly for sheet metal, piping, plumbing, and mechanical cost estimating and generates a comprehensive labor and materials estimate, eliminating manual data entry and the potential for error.

Fabrication drawings can be created directly from design documents, eliminating the need to redraw. The fabrication software automatically produces an actual cost estimate of the system as it is to be fabricated without time-consuming manual measurements and data entry through "takeoff" of the system. The ductwork model can be exported to an application that controls the sheet metal plasma cutting machine to directly manufacture the ductwork, reducing the time required to complete construction projects while eliminating errors.

A simple, helpful tool for the HVAC engineer is graphic representation of **thermodynamic properties and thermodynamic cycle analysis**. These programs may be as simple as computer-generated psychrometric charts with cross-hair cursor retrieval of properties and computer display magnification of chart areas for easier data retrieval. More complex software can graphically represent thermodynamic cyclic paths overlaid on 2D or 3D thermodynamic property graphs. With simultaneous graphic display of the calculated results of Carnot efficiency and COP, the user can understand more readily the cycle fundamentals and the practicality of the cycle synthesis (Abtahi et al. 1986).

Typical **piping and duct system simulation** software produces large amounts of data to be analyzed. Pictorially enhanced simulation output of piping system curves, pump curves, and load curves speeds the sizing and selection process. Hypothetical scenarios with numeric/graphic output increase the designer's understanding of the system and help avoid design problems that may otherwise become apparent only after installation (Chen 1988). Graphic-assisted fan and duct system design and analysis programs are also available (Mills 1989).

Airflow analysis of flow patterns and air streamlines is done by solving fundamental equations of fluid mechanics. Finite-element and finite-volume modeling techniques are used to produce two- or three-dimensional pictorial-assisted displays, with velocity vectors and velocity pressures at individual nodes solved and numerically displayed.

An example of pictorial output is Figure 5 in Chapter 18, where calculated airflow streamlines have been overlaid on a graphic of the cleanroom. With this display, the cleanroom designer can see potential problems in the configuration, such as the circular flow pattern in the lower left and right.

Major features and benefits associated with most computer flow models include the following:

- Two- or three-dimensional modeling of simple cleanroom configurations
- Modeling of both laminar and turbulent airflows
- CAD of air inlets and outlets of varying sizes and of room construction features
- Allowance for varying boundary conditions associated with walls, floors, and ceilings
- Pictorial display of aerodynamic effects of process equipment, workbenches, and people

- Prediction of specific airflow patterns in all or part of a cleanroom
- Reduced costs of design verification for new cleanrooms
- Graphical representation of flow streamlines and velocity vectors to assist in flow analysis

A graphical representation of simulated particle trajectories and propagation is shown in Figure 6 of Chapter 18. For this kind of graphic output, the user inputs a concentrated particle contamination and the program simulates the propagation of particle populations. In this example, the circles represent particle sources, and the diameter is a function of the concentration present. Such simulations help the cleanroom designer determine protective barrier placement.

Although research shows excellent correlation between flow modeling done by computer and in simple mock-ups, modeling software should not be considered a panacea for design; simulation of flow around complex shapes is still being developed, as are improved simulations of low-Reynolds-number flow.

Simulations range from two-dimensional and black and white to three-dimensional, color, and fully animated. Three-dimensional color outputs are often used not only for attractive presentations, but to increase user productivity (Mills 1989).

HVAC UTILITIES

Utility programs and modules are calculation tools that can be used both alone and integrated into other design tools.

Unit Conversion Programs

Many small utilities automatically convert between different sets of measurement units. Many of these are available as freeware or shareware. Chapter 38 of the 2009 *ASHRAE Handbook—Fundamentals* provides factors for unit conversions between SI and I-P units.

Psychrometric Utilities

Chapter 1 of the 2009 *ASHRAE Handbook—Fundamentals* provides equations to calculate psychrometric properties. Most of the utilities and Visual Basic® functions that are available to automate these calculations determine psychrometric properties from any two of the properties and a barometric pressure. Some also calculate entering and leaving properties from entering state-point(s) and a defined process (e.g., mixing of air streams, dehumidification), and plot to a psychrometric chart. Charting can be performed by scanning a psychrometric chart and determining the x and y axes by using the linear properties of dry-bulb temperature and humidity ratio.

Thermal Comfort Modules

An ASHRAE utility (ASHRAE 1997) calculates the predicted mean vote (PMV) for thermal comfort from the inputs for clothing, metabolic rate, mean radiant temperature, air speed, dry-bulb temperature, and relative humidity. This utility uses the calculations presented in ASHRAE *Standard* 55-1992.

Refrigeration Properties and Design

The program REFPROP (NIST; see http://www.nist.gov/srd/nist23.htm) allows the user to examine thermodynamic and transport properties for 109 pure refrigerants and blends as of version 8.0. It may be used in an interactive mode. Because source code is included, it may also be used as part of a program that requires refrigerant properties. Other refrigerant property programs are also available from sources such as universities. Programs should be able to address the following refrigerant properties:

- Enthalpy
- Entropy
- Viscosity
- Thermal conductivity

Ventilation

Many ventilation programs are available to help designers meet the requirements of ASHRAE *Standard* 62, Ventilation for Acceptable Indoor Air Quality. As with any computer program, users should evaluate any program's technical capabilities, such as

- Ventilation requirements by application
- Application of the Multiple Space equation in *Standard* 62
- Use of ventilation effectiveness
- Application for spaces with intermittent or variable occupancy

In addition to technical capabilities, the units of input and output and the ability to interface with programs for input or output should also be examined.

SOURCES OF HVAC SOFTWARE

Many applications have been developed as the result of ASHRAE-funded research, ranging from database retrieval for use in HVAC&R calculations to demonstration versions of advanced tools.

The ASHRAE Web site (www.ashrae.org) and the ASHRAE Online Bookstore are good sources for obtaining this software, and provide information about these programs and their system requirements, as well as other sources of HVAC-related software.

INTEROPERABLE COMPUTER APPLICATIONS FOR THE HVAC&R INDUSTRY

HVAC&R computer applications have historically been stand-alone tools that require manual reentry of data to accomplish multidisciplinary analysis. In most projects, these data must then be manually transcribed for use in other professional software, for work such as HVAC calculations and design, cost estimating, building energy performance simulation, and energy code compliance checking. This process requires time and resources. Furthermore, error detection and correction can be just as resource consuming; undetected errors can lead to serious problems. Any changes to the design require corresponding changes in transcription for other applications. Each data element in a building design may typically be independently recreated seven times in the course of design. Much of this information may be lost and/or outdated later in the building life cycle during construction, renovation, and facility maintenance.

The goal of automatically sharing data between different tools is becoming possible because of software interoperability, the ability to share the same information among software applications used by practitioners performing different industry tasks. National and international efforts to develop standardized data models that allow software interoperability are well under way. A commonly used format for interoperability between BIM tools and HVAC analysis software today is Green Building XML (gbXML), an open schema that can store all information about a building needed by building analysis software applications. gbXML is supported by an international consortium of software vendors. Another key product of these efforts is the Industry Foundation Classes (IFC) data model developed by the International Alliance for Interoperability (IAI). Members of the IAI, such as ASHRAE, work to develop cross-platform and cross-application communication standards.

Interoperability also requires that software vendors incorporate standard data models in their software applications. As shown in Figure 4, this can be done either by directly accessing and updating an IFC or gbXML data file or by indirect access through an IFC- or gbXML-compliant data server. Several CAD tools now have implementations based on the evolving IFC and gbXML data model standard, allowing building geometry input in these tools to be easily shared with other IFC- and gbXML-compliant tools. Nongeometric building data (e.g., construction material properties, HVAC&R equipment characteristics) are also included in the IFC and gbXML

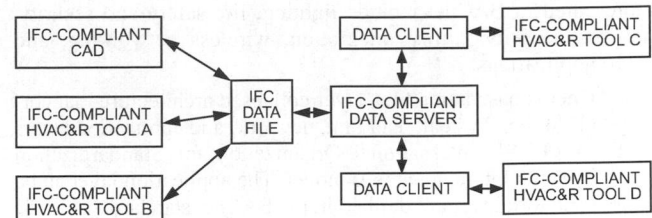

Fig. 4   Software Interoperability Based on IFC Data Model Standard

data models. In addition to CAD tools, many other IFC- and gbXML-based HVAC&R software applications are available. Current information on state-of-the-art software interoperability is available in the section on Further Internet Resources at the end of the chapter.

MONITORING AND CONTROL

A control system performs two primary functions: monitoring (also referred to as data acquisition or data gathering) and control (also referred to as device control). A control consists of a set of measured (monitored) parameters, a set of controlled parameters, and a control function that translates the measurement data into control signals applied to controlled parameters.

Supervisory control includes (1) **total system monitoring** (functions such as alarm reporting, energy measurement and calculation, logs, and trend reports) and (2) **overall control** (functions such as manual overrides, optimizing modification or discharge local loop set points, optimizing start-stop of subsystems, and controlled interaction between subsystems).

Types of supervisory control, depending on focus, include the following:

- Building automation system (BAS): automating monitoring and control
- Energy monitoring and control system (EMCS): conserving energy by both automatic and manual control with the aid of energy monitoring
- Energy management system (EMS): conserving energy by specific automatic control programs
- Facility management system (FMS): HVAC control of a subset of multiple subsystems or buildings, including fire, security, elevator, or manufacturing systems

BAS has become the most popular term for description of a computerized control system that may provide one or more of these functions.

INTEROPERABILITY

Standard communication protocols are developed in committee by professional societies, *open* protocols are created by manufacturers but available for all to use, and *proprietary* protocols are developed by manufacturers but not freely distributed. User needs should be determined before selecting a particular protocol for a given application.

There are two major standard protocols used in HVAC building automation today: BACnet® and LonWorks®.

BACnet is the ASHRAE *Standard* 135 Building Automation and Control Networks Protocol. It provides mechanisms by which computerized equipment for a variety of building control functions may exchange information, regardless of the particular building service it performs. As a result, the BACnet protocol may be used by head-end computers, general-purpose direct digital controllers, and application-specific or unitary controllers. Working groups represented in BACnet include lighting, life safety and security, network security, utility integration, wireless networking, and XML applications.

BACnet is based on a four-layer collapsed architecture that corresponds to the physical, data link, network, and application layers of the ISO/OSI (International Organization for Standardization Open Systems Interconnection) model. The application layer and a simple network layer are defined in the BACnet standard.

The **physical layer** provides a means of connecting devices and transmitting the electronic signals that convey the data. BACnet devices often use Ethernet networking, and can coexist with PCs on the same network.

The **data link layer** organizes the data into frames or packets, regulates access to the medium, provides addressing, and handles some error recovery and flow control.

Functions provided by the **network layer** include translation of global addresses to local addresses, routing messages through one or more networks, accommodating differences in network types and in the maximum message size permitted by those networks, sequencing, flow control, error control, and multiplexing. BACnet is designed so that there is only one logical path between devices, thus eliminating the need for optimal path routing algorithms.

The **presentation layer** provides a way for communicating partners to negotiate the transfer syntax used to conduct the communication. This transfer syntax is a translation from the abstract user view of data at the application layer to sequences of octets treated as data at the lower layers.

The **application layer** of the protocol provides the communication services required by the applications to perform their functions, in this case monitoring and control of the HVAC&R and other building functions.

LONWORKS defines a protocol for interoperability between control and automation devices. Task groups represented in LONMARK include HVAC, fire, industrial, lighting, vertical transportation (elevators), automated food service equipment, home/utility, network tools, refrigeration, router, security, semiconductor, sunblinds, system integration, and transportation.

Building automation system networks are local area networks, even though some applications must exchange information with devices in a building that is very far away. This long-distance communication is typically done through telephone networks. Using the Internet is becoming a popular, low-cost alternative to telephones.

DIRECT DIGITAL CONTROL (DDC) APPLICATIONS

Building management systems use standard workstation technology to connect or view system information. Standardization of building and vendor control protocol will allow future systems to access BACnet® and LONWORKS® systems. Although there are many approaches to communicating building information, the Web-browser-based client/server approach is becoming a standard interface.

Graphical representation of data from automation systems (e.g., operating conditions, trend data) is effective for enhancing operation and corrective action.

Direct digital control (DDC) allows precise and repeatable control using vendor hardware and software platforms. Many vendors allow access to DDC and set-point parameters, or are supporting common application standards such as distributed component object model (DCOM) ActiveX® and Enterprise JavaBeans®/common object request broker architecture (EJB/CORBA). Purchasing outside standards from companies specializing in certain components allows HVAC workstation providers to use common components for analysis, trending, and graphical presentation.

CONTROL COMPONENTIZATION

HVAC manufacturers use DDC microprocessors to create powerful, low-cost controls and feature-rich equipment. In the past, many of the control devices were proprietary and information was not shared between manufacturers. With standard communication protocols, diverse applications such as HVAC, energy, security, lighting, and fire controls in large buildings can use a single integrated control system.

Standard protocols allow application-specific component control systems designed for particular HVAC equipment to be included in building-wide strategies. For example, low-cost componentized DDCs provide operation, safety, maintenance, and even self-diagnostic information and control functions. Making a chiller or a variable-speed drive controller a part of a cohesive, building-wide control system is now possible because of standard protocols.

In the past, because all major HVAC equipment required some type of control system, control points were duplicated. Acceptance of communication standards has greatly reduced total control costs and increased functionality by providing information (e.g., part-load performance criteria) to users that was previously only available from manufacturer's testing.

As the functionality of component control increases and associated cost decreases, component-level interaction makes building-wide system integration less critical.

INTERNET-ENABLED BUILDING AUTOMATION SYSTEMS

With Internet-enabled building automation systems, additional information outside the traditional control system (e.g., weather data, real-time energy data, pricing rates, manufacturers' equipment data) can be combined with real-time equipment data to create valuable analysis information. Commercial applications allow device integration to view real-time data, command equipment, and trend data. Users can manage energy consumption and costs, correct comfort problems, and solve equipment issues from distributed locations. Access control systems are available through the Internet using existing Internet standards such as TCP/IP and HTTP/XML.

CONVERGENCE OF INFORMATION SYSTEMS AND INFORMATION TECHNOLOGY

As the information system (IS) and information technology (IT) industry has advanced, so have applications that can be used in building automation. Internet browser technology is a standard application, allowing information to be shared across nontraditional computer boundaries.

Because there is such demand in business for IT products, expertise and knowledge formerly reserved for IT applications can be applied to, for instance, user interface and building automation applications. A critical concern about integrating IT systems and building automation is information security. New security standards, such as the BACnet Network Security specification, should help address these concerns. Another common concern is that building automation systems will use a large amount of network bandwidth, but experience shows that bandwidth use by a properly configured system is not appreciable.

Web-based building controls that use the IT approach should create an open, interoperable structure for control automation systems. Implementing common field-bus standards and integration with Web browsers can be helpful for legacy system support. Rapid adoption of the Internet and standardization efforts in the field bus and building automation protocols have made developing a single standard for both Internet and control a goal for many vendors.

Combining building automation with Internet functionality will provide functionality in a user-friendly form.

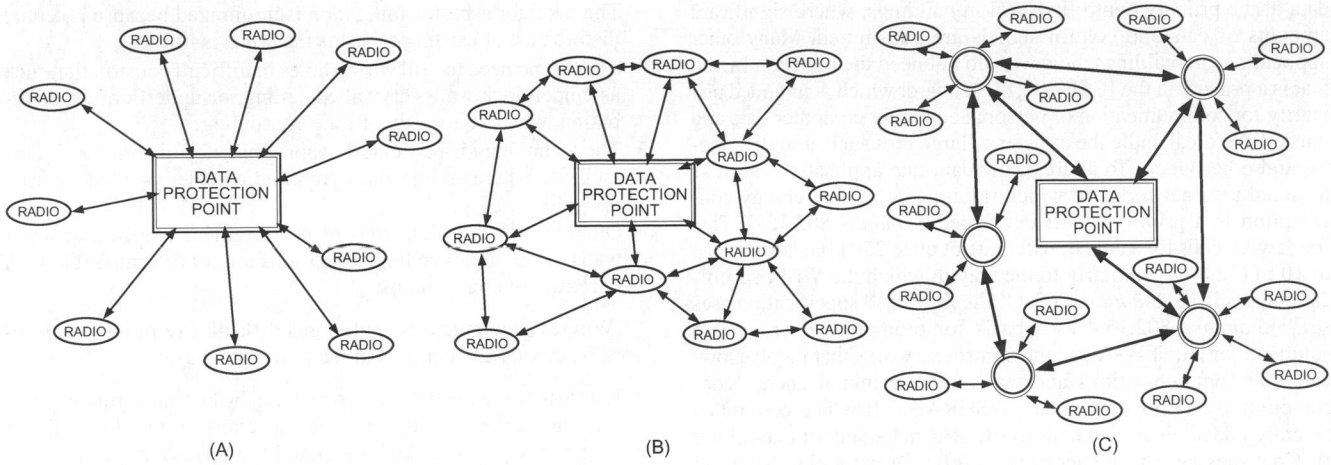

Fig. 5 (A) Star, (B) Mesh, and (C) Hybrid Wireless Network Topologies

WIRELESS COMMUNICATION

Wireless communication technologies present many opportunities for enhanced information exchange in buildings. By removing the cost and time associated with wiring sensors, controllers, actuators, data acquisition systems, and computers to each other, wireless systems open new possibilities for designing and operating buildings to perform efficiently while maintaining a comfortable indoor environment. This section gives an overview of wireless communication technologies relevant to building operations, discusses applications where wireless equipment can be used, and provides guidance on selecting wireless systems for HVAC&R applications.

TECHNOLOGIES

Radiofrequency Data Transfer

In this section, *wireless* refers to the technique by which data are transferred between devices, as opposed to how power is delivered to the devices. A wireless transmitter and receiver must both consist of a radio, antenna, and microprocessor. For a transmitter, the microprocessor converts the data to a signal that is sent over the air through the antenna; for a receiver, it converts the radiofrequency (RF) signal received at the antenna to data. Radios can often be configured to serve as both transmitters and receivers. Data can be encoded in RF signals in a number of ways; for details, see Rappaport (2002). These data transmission design schemes allow for timing and/or frequency modulation as well as error checking, ensuring that the signal is properly sent and received.

Data can theoretically be transmitted at any frequency in the electromagnetic spectrum, but the radio communications spectrum is generally considered to be from 3 kHz to 300 GHz. Most practical uses, however, occur in the lower portion of that range (U.S. General Accounting Office 2004). The industrial, scientific, and medical (ISM) bands are reserved for license-free operations, and other segments of the spectrum are dedicated to licensed operation. The International Telecommunication Union defined 12 ISM bands, with the most popular bands for communications for these applications centered at 434 MHz, 915 MHz, 2.45 GHz, and 5.80 GHz. The 434 MHz band is valid only in Europe, Africa, the Middle East, and the former Soviet Union, and the 915 MHz band is allotted only in the Americas, Greenland, and parts of the Pacific Islands. Consequently, designers often select the 2.45 and 5.80 GHz bands because of their worldwide acceptance.

System Design

Collections of wireless sensors, controllers, and actuators form a network. Figure 5 shows three main topologies for wireless networks. In a star network, all wireless devices communicate directly to a base station. In a mesh network, each wireless device, or node, can communicate with all other wireless devices in the network, with each node serving as a transmitter, receiver and repeater of signals. The networks can be designed to relay messages from one device to another, and messages can bypass a central receiving point to allow direct communication between wireless devices, such as a sensor and an actuator. Mesh networks are often designed to set themselves up in an ad-hoc manner and to be self-healing. Ad-hoc configuration allows the user to simply physically place the wireless equipment, and software then determines the best path through the available nodes for messages from a wireless device to reach their destinations. These ad-hoc networks also allow wireless nodes to be added automatically to an existing network. Self-healing mesh networks automatically find alternative routes for messages to be passed should a radio node in the network fail or should a link between two nodes become lost. This failure can happen for a number of reasons such as physical obstacles introduced between nodes, new sources of electromagnetic interference between nodes, and often difficult-to-determine causes for communication loss between specific sets of nodes.

The sophistication of mesh networks allows for robust networks that are easily expanded. One drawback of the mesh network compared to star networks is the added complexity and, hence, cost of the equipment. A second downside of these networks is that radio nodes are expected to relay (or repeat) data and commands from neighboring nodes in addition to their own information, thereby using more energy and decreasing battery life.

Hybrid networks combine the concepts of both mesh and star networks. An example would be a series of data collection points connected in a mesh network to which sensors and actuators are each attached in a star configuration.

Standards

Various standards have been developed for wireless data communications, the most familiar of which is the IEEE 802.11 family of standards for wireless local area networks (IEEE *Standard* 802.11). The most recent addendum (802.11g) allows for data transfer at a rate of 54 Mbps, although the next-generation standard (802.11n) promises data transfer at up to 600 Mbps. The 802.11 standard forms the basis of products that conform to the Wi-Fi™ specification, an industry-based specification developed by the Wi-Fi Alliance. Products that comply with this standard can be used to transfer data at the primary controller level in buildings, where significant amounts of data and control signals are transmitted. Many other applications in buildings, however, do not need the high bandwidth that is a priority in the IEEE 802.11 standard, which is intended primarily for communication of personal and other computer data and must, therefore, handle the transfer of large files such as audio, photos and other video. To address low-data-rate applications such as personal area networks, in which minimizing radios' energy consumption is a priority, IEEE developed *Standard* 802.15.4. This framework calls for a data transfer rate of up to 250 kbits/s at a range of 10 m (32.8 ft). Similarly to the way in which the Wi-Fi specification uses IEEE *Standard* 802.11, the ZigBee™ specification uses IEEE *Standard* 802.15.4 as a basis for promoting interoperable solutions for wireless sensor and control networks that involve low-data-rate communications and that require minimal energy consumption by radio equipment. ASHRAE's BACnet committee recently added an amendment to the BACnet standard that allows BACnet messages to be sent over ZigBee™ networks (Martocci 2008).

Other standards are also in place for wireless data communications. The Bluetooth™ standard is based on IEEE 802.15.1 standard for wireless personal area networks (IEEE *Standard* 802.15.1). This standard focuses on higher-data-rate applications than those addressed by 802.15.4 and has gained acceptance in short-distance applications such as wireless headsets for phones and wireless peripherals for computers. The WiMedia Alliance promotes open standards based on Ultra-Wideband data communications. The target bandwidth enabled by these standards is 480 Mbps. WiMedia™ is not to be confused with WiMAX™, which is based on IEEE *Standard* 802.16 and is designed for broadband Ethernet transmission as part of a metropolitan area network.

APPLICATIONS

RF wireless data communication can be used in the same building automation applications as wired data communication. Wireless equipment can be used at any point in the building control network as well as in monitoring applications that are separate from the control system. The major advantage of wireless technology over wired methods is the ease of installation, which leads to labor cost savings, greater flexibility in installation scheduling, less damage to existing construction, and less interference with occupants of existing buildings.

In new and existing construction, advantages include the following:

- Installation may be quicker because it is not necessary to pull communication wires and/or power wiring (in some cases), or to run conduit
- No need for unattractive conduits or wire molding.
- Systems can be used in buildings with aesthetic restrictions on visible wires.
- Deployment involves less complicated coordination between construction trades (e.g., no need to wait for electricians to run wires, or concern if drywall goes up before the electrician arrives).
- Wi-Fi can be used to set up monitoring and control networks before structured Ethernet cabling is installed (e.g., during construction).

Advantages when adding sensing and control systems to existing buildings include the following:

- There is less disruption to the physical facility and occupants during installation.
- Installation does not require opening up walls and ceilings.
- Less dust and wire clippings are generated (especially important in health care setting).
- Structural integrity is never compromised by drilling holes to run wires.

- The need for asbestos mitigation is minimized because potential disturbance of existing asbestos materials is limited.
- There is no need to drill wiring holes in difficult construction such as cinderblock walls, drywalled ceiling, or underfloor air distribution, where it is costly or time consuming.
- The technology is particularly appropriate for historic or vintage facilities where architectural preservation rules preclude tearing up walls.
- Otherwise inaccessible sites of measurement like behind water walls or up on tall ceilings or atriums are more compatible with wireless communications.

Wireless equipment provides added flexibility in locating sensors, actuators, and controls in the following ways:

- Wireless devices can be moved without the burden of running new communication cabling. This is especially useful for building operators who reconfigure their facility's spaces frequently.
- Wireless devices can be placed in the optimal location for sensing and control. They are not restricted only to the places where cabling can be run. Open floor plans can easily be accommodated.
- Wireless devices can also easily be added to an existing wireless network in an open floor plan area because there is no concern about finding a communication cable to tap into.

Finally, wireless technology affords the ability to incrementally modify a building automation system and may add a degree of redundancy that is not possible in wired applications:

- Incremental deployment of devices can proceed based on the application needs instead of being dictated by the location of the nearest live communication cable.
- Old wired networks can be upgraded incrementally to wireless networks without concern about disrupting communication to the rest of the wired devices caused by removal of data wiring or power wiring on the old devices.
- Wireless technology eliminates dependency on unreliable, previously installed network cabling that may suffer from deterioration due to age, years of miswiring, or simple incompatibilities.
- A wireless network can result in reduced single points of failure. If a wired communication cable gets cut, all the downstream devices are lost. In a wireless mesh network, redundant paths of communication exist that automatically correct for failures in the network.

Examples of Applications of Wireless Systems in Buildings

Several documented cases of the use of wireless communication systems in buildings have been presented (Healy 2005; Kintner-Meyer and Brambley 2002; Raimo 2006; Ruiz 2007; Wills 2004). These articles discuss the application of wireless sensor networks for equipment and environmental monitoring in both commercial and residential installations, the use of wireless systems for HVAC controls, and wireless architectures for buildings.

SELECTION OF WIRELESS SYSTEMS

This section is based on information in Brambley et al. (2005).

Challenges

The primary issues in applying wireless technologies inside buildings are associated with (1) powering devices (mainly sensors) where line power is not available, (2) the maximum distance over which devices can communicate, (3) interference caused by signals from other radio transmitters (such as wireless LANs) and microwave ovens that leak electromagnetic energy, (4) attenuation as the RF signal travels from the transmitter through walls, furnishings, and even air to reach the receiver, and (5) security.

Powering Sensors. Many different wireless technologies are available, ranging from cellular phone networks to wireless temperature sensors. In building automation applications, line power may not be conveniently available at all locations where sensors should be placed. These sensors need (1) a source of electrical power with an acceptably long life (usually at least 5 years), and (2) a low rate of power consumption by the wireless nodes to minimize the amount of power required. Most wireless sensor nodes use batteries; to maximize battery life, communication protocols for wireless sensor networks must minimize power use. Although not widely available today for wireless sensor applications, environmental power harvesting shows promise to power wireless sensors in the future. A limited number of products using power harvesting are available today, such as a light switch that uses electric power generated when a person pushes the switch to power transmission of the signal to an actuator on the light circuit remote from the switch, creating a movable electric switch that would simplify remodeling in cases where switches must be relocated.

Communication Range. A radio that has a maximum line-of-sight range of 300 ft outdoors may be limited to 60 ft or even less indoors. The range depends on several factors, including the frequency at which the radio communicates (lower frequencies penetrate materials better), the specific materials used in building construction (e.g., wood absorbs more signal power per unit path length than air), and layout of walls and spaces. Therefore, the wireless technology and communication protocol for indoor sensor networks must provide adequate communication ranges in less-than-ideal indoor environments.

Interference. Interference is usually caused by electromagnetic noise from other wireless devices or random thermal noise. Spread spectrum techniques increase resistance to interference from a single-frequency source by spreading the signal over a defined spectrum. Multiple transmitters usually can communicate clearly in a common frequency band, although spread spectrum is not immune to interference (particularly with heavily loaded frequency bands used by large numbers of wireless devices). Early tests in buildings with 30 to 100 wireless sensors did not identify problems with crosstalk or data loss in transmission, but communications may become less reliable as frequency band crowding increases (e.g., hundreds or thousands of devices sharing the band). Field testing under these usage conditions is needed.

Attenuation. RF signals weaken over distance; the amount of attenuation depends on how far the signal travels, and through what materials. Different materials have different levels of attenuation per unit of path through the material (Table 1).

Signal attenuation can be compensated for by using repeaters. Repeaters are usually placed close enough to transmission sources that a high percentage of the signals will successfully reach the repeater. The repeater receives signals, amplifies them, and retransmits them (often at higher power) to extend the transmission range. In a wireless mesh network, every node serves as a repeater.

Security. Growing threats to networking infrastructure have increased security requirements for facility automation systems. Wireless networks are particularly vulnerable because no direct "hard" physical link is required to connect to them. Data encryption (encoding data in a format that is not readable except by someone with the "key") has successfully defended against intrusion and provided security for wireless local area networks (LANs). Encryption, however, requires additional computational power on each wireless device, which raises the cost of wireless networks and the power requirements of individual devices. Research is ongoing by developers, vendors, and standards committees to provide security without additional power requirements.

Practical Design and Installation Considerations

Laying out a wireless network indoors is probably as much art as it is science. Every building is unique, if not in its construction and

Table 1 Signal Attenuation for Selected Building Materials for 900 to 908 MHz Band

Construction Material		Attenuation, dB
Drywall,	1/4 in.	0.2
	1/2 in.	0.3
	3/4 in.	0.5
Plywood, dry,	1/4 in., dry	0.5
	1/4 in., wet	1.7
	1/2 in., dry	0.6
	1/2 in., wet	2.0
Glass,	1/4 in.	0.8
	1/2 in.	2
	3/4 in.	3
Lumber,	1.5 in.	3
	3 in.	3
	6.75 in.	6
Brick,	3.5 in.	4
	10.5 in.	7
Reinforced concrete, 8 in., with 1% rebar mesh		27

Source: Stein (2004)

floor plan, then at least in the type and layout of its furnishings. Predicting wireless signal strength throughout a building requires characterizing the structure and its layout, furnishings, and equipment in great detail and using that information to model RF signal propagation. No tools are available to do this accurately, although equipment can test RF signal propagation in buildings, which can help with sensor placement. Changes in space use or furnishings can make these data obsolete, however. The following are some practical points to keep in mind when designing a wireless network.

Determining Receiver Location. The number and locations of receivers strongly influences the design of a wireless sensor network for in-building monitoring. A stand-alone wireless network (not connected to a wired control network) may have some flexibility in choosing the location of the primary receiver that provides data to users (e.g., a monitoring workstation). From a communications perspective, the best location is open and provides the best line-of-sight pathways between the most wireless sensors and the receiver. Convenient connection of the receiver to a computer where data will be processed and viewed is another important consideration. These factors must be balanced. If the design requires integration of the wireless sensor network with an existing building automation system (BAS), then receivers must be located near points of connection to the BAS. Locations are constrained somewhat in this case, but there are typically still many options. Frequently, a convenient integration point is a control panel that provides easy access to communication cables as well as electricity to power the receiver and integration devices. In commercial buildings, BAS network wires are often laid in cabling conduits (open or closed) above the ceiling panel, and are relatively easily accessible. Often, though, the lack of electric power in the ceiling space makes a control panel more convenient.

Signal Attenuation and Transmitter Range. Estimating the range of the transmitting devices is important from a cost point of view. If a single transmission path cannot connect a transmitting device to the ultimate end node, additional hardware will be required for signal repeating or amplification, adding to the installation's total cost. The following discussion provides an overview of how to estimate how many repeater or amplification devices an installation requires. It does not replace a thorough RF survey of a facility to determine the exact number and locations of receivers, repeaters, or intermediate nodes necessary to ensure robust communication.

A transmitter's range depends on the three variables: (1) attenuation because of distance between wireless devices, (2) attenuation caused by the signals traveling through construction material along

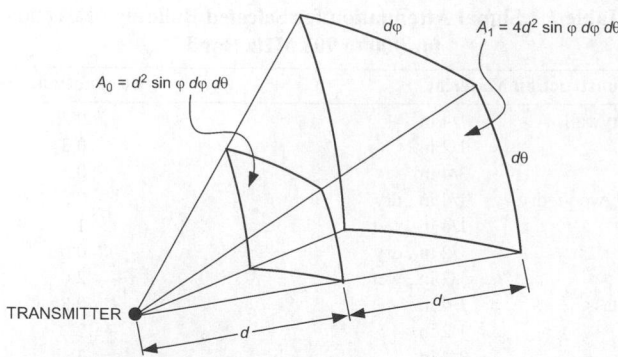

Fig. 6 Example of RF Propagation Area Increasing with Distance from Transmitter

Table 2 Attenuation of RF Signal in Free Air for Selected Distances

Distance, ft	0.2	3	8	16	33	66	131	262
Signal strength, dB	0	−26	−34	−40	−46	−52	−58	−64
Attenuation along propagation path, dB	0	26	34	40	46	52	58	64

the signal pathways, and (3) overall electromagnetic noise levels in the facility.

Attenuation of signal strength caused by distance between the transmitter and receiver (free path loss) depends on the relation of the transmitter's electromagnetic energy per unit area to the distance of the receiving surface (see Figure 6). The energy per unit area at a distance d from the transmitter decreases in proportion to $1/d^2$. Therefore, every time d doubles, the energy density or signal strength received decreases by a factor of 1/4.

This relationship accounts only for signal dispersion across a large area distant from the source. In practice, other factors also affect the strength of signal received, even for an unobstructed path: absorption by moisture in the air or by the ground, partial signal cancellation by waves reflected by the ground, and other reflections. In general, this causes signal strength at distance d from the transmitter to decrease in proportion to $1/d^m$, where $2 < m < 4$ (Su et al. 2004).

The following example illustrates signal attenuation with distance from the transmitter in free air for a 900 MHz transmitter, and shows how simple relations can be used to estimate potential transmission range.

Example 1. Assume that the measured signal strength of a small transmitter is 645 mW/in² at a distance of 2 in. from the transmitter's antenna. The transmission path efficiency or transmission loss is customarily expressed in decibels, a logarithmic measure of a power ratio, and is defined as

$$dB = 10 \log_{10} (p_1/p_0)$$

where p_1 is the power density in W/cm² and p_0 is the power density at a reference point in W/cm².

For p_0, select the power density measured 5 cm from the transmitter's antenna. Table 2 shows the emitted signal's attenuation (traveling through air only) as a function of distance from the transmitter. For every doubling of the distance, the signal strength decreases by 6 dB (i.e., the attenuation increases by 6 dB).

Further, assume that the measured ambient noise is −75 dB. For a signal to be detectable, its strength should be at least 10 dB above the noise level (i.e., a signal margin of 10 dB or greater is recommended) (Inovonics 1997). Using Table 2, the transmission range for this system that meets the 10 dB signal margin requirements is about 262 ft, because −75 dB +10 dB = −65 dB, which is less than −64 dB at 262 ft.

If the receiver is placed in a mechanical room of a building, the signal from the most distant transmitter must go through two brick walls and two layers of drywall. Using signal attenuation estimates from Table 1, the combined attenuation of the brick and drywall is 14.6 dB [2 × 0.3 (for the 1/2 in. drywall) + 2 × 7 (for the 10.5 in. brick wall) = 14.6]; for practical purposes, round this to 15 dB. Adding the material-related attenuation of 15 dB to the −65 dB signal strength requirement yields −50 dB as the new indoor signal strength requirement for the free air transmission segment. Table 2 shows that the transmission range is between 33 and 66 ft, only 1/8 to 1/4 of the range in open air. This example illustrates how significantly radio signals can be attenuated

simply by the structure itself. Furniture further adds to attenuation and complicates prediction of signal strength as a function of location in buildings. Therefore, to characterize indoor environments with respect to RF signal propagation, empirical surveying is recommended.

RF Surveying. The purpose of an RF facility survey is to determine the actual attenuation of RF signal strength throughout the facility. This information, together with knowledge of the locations at which sensors will be positioned, is used to lay out the wireless network. The layout will include the number of repeaters and receivers in the network and their locations. For instance, for a multistory facility there may be good reasons for placing one receiver on each floor, provided the data are needed only on each floor (e.g., one user per floor for that floor) or there is another means to communicate the data between floors (such as a wired BAS connection on each floor). If the data are needed at a computer located on a specific floor (such as a control room in the basement), a repeater might be used on each floor to transmit signals to the location of a central receiver located close to where the data are needed. If communication between receivers on different floors is not sufficient, there may be opportunities to route signals inside an elevator shaft, stair case, or on the exterior of the building. The most cost-effective solution is usually determined by the difference between the cost of repeaters and receivers for a completely wireless network and the cost of interfacing the receivers for separate wireless networks on each floor to a preexisting wired network that would communicate the data from all wireless networks to the control room. The layout with the lowest total cost that provides sufficiently reliable communication is generally optimal.

Most vendors of wireless sensor networks offer RF survey kits that are specific for the vendors' technologies. These kits consist of a transmitter and a receiver. The transmitter is often a modified sensor transmitter that is programmed to transmit at frequent time intervals. The receiver generally is connected to (or part of) an indicator of signal strength, together making a wireless signal-strength meter. These meters may simply give an indication whether the signal strength is adequate or provide numerical values of signal strength and background noise levels from which the adequacy of signal strength can be determined.

Before the RF facility survey is performed, potential receiver and sensor locations need to be known. The survey is then performed by placing the transmitter in anticipated locations for the receivers, then moving the signal-strength meter to locations where sensors will be positioned. By taking measurements throughout the facility, the limits of transmission range where the signal can no longer be detected (or is not of sufficient strength) can be identified. Repeaters will then need to be located in the layout within the transmission range to extend the range further.

Radiofrequency surveying is generally done by the wireless technology vendor or installer. Depending on the diversity of noise level in the facility and the complexity of the interior layout, an RF survey can be performed for office buildings with a floor space of 100,000 ft² in 2 to 4 h.

Although RF surveys are critical for successfully designing and installing a wireless network that uses a star topology, systems using a mesh network topology with sufficient sensor density will ultimately not require RF surveys for installation. With sufficient densities of sensors (i.e., relatively short distances between sensors and multiple neighboring sensors within the communication range of each node), these networks are self-configuring and the multiple potential transmission paths ensure reliable, consistent communications. In the near term, use caution when assuming that mesh networks will perform reliably for every application, especially in cases where high sensor density is not anticipated. For low sensor density installations, communication over long distances may require a higher-power repeater to connect a local mesh network to the point where the data are needed (or a daisy-chain of nodes for this communication). In these cases, the advantages of mesh networking are lost in the region where individual devices carry all data communicated and those devices become potential single points of failure for the entire mesh that they connect to the point of data use.

Other Practical Considerations. Several other factors should be considered in deciding to use wireless sensing in buildings. Stein (2004) provides a nice summary of practical considerations for monitoring with wireless sensor networks. In addition to communication range, some of the key considerations that need to be assessed when selecting a wireless sensing network are

- Component prices
- Availability of support
- Compatibility with different types of sensors with different outputs
- Battery backup for line powered devices
- Low-battery indicators for battery-powered devices
- On-board memory
- Proper packaging and technical specifications for the environment where devices will be located
- Battery life and factors that affect it
- Frequency of data collection and its relationship to battery life (where applicable)
- Need for and availability of integration boxes or gateways to connect wireless sensor networks to BASs, other local area networks, or the Internet
- Availability of software for viewing or processing the data for the intended purpose
- Compatibility among products from different vendors; this is rare today but will improve as manufacturers adopt new standards such as IEEE *Standard* 802.15.4 (Gutierrez et al. 2003) and Zignsbee (Kinney 2003)
- Tools for configuring, commissioning, repairing, and adding nodes to the sensor network
- Software to monitor network performance

Most important is ensuring the selected wireless network meets the requirements of the intended application. All factors need to be considered and assessed with respect to satisfying the requirements of the application and the specific facility. Each installation is unique.

REFERENCES

Abtahi, H., T.L. Wong, and J. Villanueva, III. 1986. Computer aided analysis in thermodynamic cycles. *Proceedings of the 1986 ASME International Computers in Engineering Conference* 2.

AHRI. 2001. Forced-circulation air-cooling and air-heating coils. *Standard* 410-2001. Air-Conditioning, Heating, and Refrigeration Institute, Arlington, VA.

AHRI. 1999. Central station air-handling units. *Standard* 430-99. Air-Conditioning, Heating, and Refrigeration Institute, Arlington, VA.

Arkin, H. and A. Shitzer. 1979. Computer aided optimal life-cycle design of rectangular air supply duct systems. *ASHRAE Transactions* 85(1):197-213.

ASHRAE. 1997. *Thermal comfort tool CD.*

ASHRAE. 2008. BACnet®: A data communications protocol for building automation and control networks. ANSI/ASHRAE *Standard* 135-2008.

ASHRAE. 2009. *An introduction to building information modeling (BIM): A guide for ASHRAE members.* Available from http://www.ashrae.org/bimguide.

Berners-Lee, T. 1996. *Presentation to CDA challenge by CDT et al.* http://www.w3.org/People/Berners-Lee/9602affi.html.

Brambley, M.R., M. Kintner-Meyer, S. Katipamula, and P. O'Neill. 2005. Wireless sensor applications for building operation and management. Chapter 27 in *Information technology for energy managers*, vol. II: *Web based energy information and control systems case studies and applications*, B.L. Capehart and L.C. Capehart, eds., pp. 341-367. Fairmont/CRC, Lilburn, GA.

Busnaina, A.A. 1987. Modeling of clean rooms on the IBM personal computer. *Proceedings of the Institute of Environmental Sciences*, pp. 292-297.

Busnaina, A.A., S. Abuzeid, and M.A.R. Sharif. 1988. Three-dimensional numerical simulation of fluid flow and particle transport in a clean room. *Proceedings of the Institute of Environmental Sciences*, pp. 326-330.

Chen, T.Y.W. 1988. Optimization of pumping system design. *Proceedings of the 1988 ASME International Computers in Engineering Conference.*

Gutierrez, J.A., E. Callaway, and R. Barrett, eds. 2003. *Low-rate wireless personal area networks: Enabling wireless sensors with IEEE 802.15.4.* Institute of Electrical and Electronics Engineers, Inc., Piscataway, NJ.

Healy, W.M. 2005. Lessons learned in wireless monitoring. *ASHRAE Journal* 47(10):54-58.

IEEE. 2007. Wireless LAN medium access control (MAC) and physical layer (PHY) specifications. *Standard* 802.11. Institute of Electrical and Electronics Engineers, Inc., Piscataway, NJ.

IEEE. 2005. Wireless medium access control (MAC) and physical layer (PHY) specifications for wireless personal area networks (WPANs). *Standard* 802.15.1. Institute of Electrical and Electronics Engineers, Inc., Piscataway, NJ.

IEEE. 2003. Wireless medium access control (MAC) and physical layer (PHY) specifications for low-rate wireless personal area networks (LR-WPANs). *Standard* 802.15.4. Institute of Electrical and Electronics Engineers, Inc., Piscataway, NJ.

IEEE. 2004. Air interface for fixed broadband wireless access systems. *Standard* 802.16. Institute of Electrical and Electronics Engineers, Inc.

Inovonics. 1997. *FA116 executive programmer, user manual for FA416, FA426 and FA464 Frequency Agile™ receivers.* Inovonics Corporation, Louisville, CO.

Kintner-Meyer, M. and M. Brambley. 2002. Pros & cons of wireless. *ASHRAE Journal* 44(11):54-61.

Kinney, P. 2003. *ZigBee technology: Wireless control that simply works.* ZigBee Alliance, Inc. Available at http://www.zigbee.org/imwp/idms/popups/pop_download.asp?contentID=5438.

Kovarik, M. 1971. Automatic design of optimal duct systems. Use of computers for environmental engineering related to buildings. National Bureau of Standards *Building Science Series* 39(October).

Kuehn, T.H. 1988. Computer simulation of airflow and particle transport in cleanrooms. *Journal of Environmental Sciences* 31.

Lawrence Livermore National Laboratory. 2002. Hoaxbusters. http://www.hoaxbusters.org.

Martocci, J. 2008. BACnet unplugged. *ASHRAE Journal* 50(6):42-46.

McGowan, J.J. 2002. *DDC's future.* Available at http://www.automatedbuildings.com/news/jan01/articles/mcg/mcg.htm.

Mills, R.B. 1989. Why 3D graphics? In *Computer Aided Engineering*. Penton Publishing, Cleveland, OH.

Potter, C.D. 1987. *CAD in construction.* Penton Publishing, Cleveland, OH.

Raimo, J. 2006. Wireless mesh controller networks. *ASHRAE Journal* 48(10):34-38.

Rappaport, T.S. 2002. *Wireless communications.* Prentice Hall PTR, Upper Saddle River, NJ.

Ruiz, J. 2007. Going wireless. *ASHRAE Journal* 49(6):33-43.

Sinclair, K. 2002. The componentization era is here! http://www.automatedbuildings.com/news/mar01/articles/component/component.htm.

Stein, P. 2004. Practical considerations for environmental monitoring with wireless sensor networks. *Remote Site & Equipment Management* (June/July).

Stoecker, W.F., R.C. Winn, and C.O. Pedersen. 1971. Optimization of an air supply duct system. Use of computers for environmental engineering related to buildings. National Bureau of Standards *Building Science Series* 39(October).

Su, W., O.B. Akun, and E. Cayirici. 2004. Communication protocols for sensor networks. In *Wireless sensor networks*, C.S. Raghavendra, K.M. Sivalingam, and T. Znati, eds., pp. 21-50. Kluwer Academic, Boston, MA.

Tsal, R.J. and E.I. Chechik. 1968. *Use of computers in HVAC systems.* Budivelnick Publishing, Kiev. Available from Library of Congress, Service-TD153.T77 (Russian).

Tsal, R.J., H.F. Behls, and R. Mangel. 1988. T-method duct design: Part I, optimization theory; part II, calculation procedure and economic analysis. ASHRAE *Technical Data Bulletin* (June).

U.S. General Accounting Office. 2004. Spectrum management: Better knowledge needed to take advantage of technologies that may improve spectrum efficiency. *Report* GAO-04-666.

Wills, J. 2004. Will HVAC control go wireless? *ASHRAE Journal* 46(7): 46-52.

BIBLIOGRAPHY

BLIS-Project. 2002. *Building Lifecycle Interoperable Software Project home page.* http://www.blis-project.org/.

Brothers, R.W. and K.R. Cooney. 1989. A knowledge-based system for comfort diagnostics. *ASHRAE Journal* 31(9).

Diamond, S.C., C.C. Cappiello, and B.D. Hunn. 1985. User-effect validation tests of the DOE-2 building energy analysis computer program. *ASHRAE Transactions* 91(2B):712-724.

Diamond, S.C. and B.D. Hunn. 1981. Comparison of DOE-2 computer program simulations to metered data for seven commercial buildings. *ASHRAE Transactions* 87(1):1222-1231.

Fitzgerald, N. 2002. Virus-L/comp.virus FAQ v2.00. http://www.faqs.org/faqs/computer-virus/faq/.

Haberl, J.S. and D. Claridge. 1985. Retrofit energy studies of a recreation center. *ASHRAE Transactions* 91(2B):1421-1433.

Hsieh, E. 1988. Calibrated computer models of commercial buildings and their role in building design and operation. *Report* 230, Center for Energy and Environmental Studies, Princeton University, Princeton, NJ.

International Alliance for Interoperability. 2002. *Home page.* http://www.buildingsmart.com.

Judkoff, R. 1988. International energy agency design tool evaluation procedure. Solar Energy Research Institute *Report* SERI/TP-254-3371.

Judkoff, R., D. Wortman, and B. O'Doherty. 1981. A comparative study of four building energy simulations, phase II: DOE-2. 1, BLAST-3.0, SUN-CAT-2.4 and DEROB. Solar Energy Research Institute *Report* SERI/TP-721-1326 (July).

Kaplan, M.B., J. McFerran, J. Jansen, and R. Pratt. 1990. Reconciliation of a DOE2.IC model with monitored end-use data for a small office building. *ASHRAE Transactions* 96(1):981-993.

Kusuda, T. and J. Bean. 1981. Comparison of calculated hourly cooling load and indoor temperature with measured data for a high mass building tested in an environmental chamber. *ASHRAE Transactions* 87(1):1232-1240.

Kuehn, T.H. 1988. Computer simulation of airflow and particle transport in cleanrooms. *Journal of Environmental Sciences* 31.

Lemmon, E.W., M.L. Huber, and M.O. McLinden. 2007. *NIST standard reference database 23: Reference fluid thermodynamic and transport properties—REFPROP*, v. 8.0. National Institute of Standards and Technology, Gaithersburg, MD.

Li, K.W., W.K. Lee, and J. Stanislo. 1986. Three-dimensional graphical representation of thermodynamic properties. *Proceedings of the 1986 ASME International Computers in Engineering Conference* 1.

McQuiston, F.C. and J.D. Spitler. 1992. *Cooling and heating load calculation manual.* ASHRAE.

Milne, M. and S. Yoshikawa. 1978. Solar-5: An interactive computer-aided passive solar building design system. *Proceedings of the Third National Passive Solar Conference*, Newark, DE.

Olgyay, V. 1963. *Design with climate: Bioclimatic approach to architectural regionalism.* Princeton University Press, Princeton, NJ.

Rabl, A. 1988. Parameter estimation in buildings: Methods for dynamic analysis of measured energy use. *ASME Journal of Solar Energy, Engineering* 110.

Rankin, J.R. 1989. *Computer graphics software construction.* Prentice Hall, New York.

Robertson, D.K. and J. Christian. 1985. Comparison of four computer models with experimental data from test buildings in New Mexico. *ASHRAE Transactions* 91(2B):591-607.

Sharimugavelu, I., T.H. Kuehn, and B.Y.H. Liu. 1987. Numerical simulation of flow fields in clean rooms. *Proceedings of the Institute of Environmental Sciences*, pp. 298-303.

Sorrell, F., T. Luckenback, and T. Phelps. 1985. Validation of hourly building energy models for residential buildings. *ASHRAE Transactions* 91(2).

Spielvogel, L.G. 1975. Computer energy analysis for existing buildings. *ASHRAE Journal* 7(August):40.

Spitler, J.D. 2008. *Load calculation applications manual.* ASHRAE.

Sturgess, G.J., W.P.C. Inko-Tariah, and R.H. James. 1986. Postprocessing computational fluid dynamic simulations of gas turbine combustor. *Proceedings of the 1986 ASME International Computers in Engineering Conference* 3.

Subbarao, K. 1988. PSTAR Primary and secondary terms analysis and renormalization, a unified approach to building energy simulations and short-term monitoring. Solar Energy Research Institute *Report* SERM254-3175.

Tanenbaum, A.S. 1996. *Computer networks.* Prentice Hall, New York.

Yuill, G. 1985. Verification of the BLAST computer program for two houses. *ASHRAE Transactions* 91(2B):687-700.

FURTHER INTERNET RESOURCES

BACnet Web site:	www.bacnet.org
BACnet Manufacturers Association:	www.bacnetassociation.org
FIATECH:	www.fiatech.org
International Alliance for Interoperability:	www.iai-international.org
LonMark Interoperability Association:	www.lonmark.org
DOE Building Energy Software Tools Directory:	www.eere.energy.gov/buildings/tools_directory
OnGuard Online:	www.onguardonline.gov
Green Building XML:	www.gbxml.org

BUILDING ENERGY MONITORING

BUILDING energy monitoring was conducted on a large scale in the 1980s and 1990s, and the need to capture lessons learned and document project requirements that often were not addressed adequately in these large projects led to the development of this chapter. The intent of such projects is to provide realistic, empirical information from field data to enhance understanding of actual building energy performance and help quantify changes in performance over time. Although different building energy monitoring projects can have different objectives and scopes, all have several issues in common that allow methodologies and procedures (monitoring protocols) to be standardized.

This chapter provides guidelines for developing building monitoring projects that provide the necessary measured data at acceptable cost. The intended audience comprises building owners, building energy monitoring practitioners, and data end users such as energy and energy service suppliers, energy end users, building system designers, public and private research organizations, utility program managers and evaluators, equipment manufacturers, and officials who regulate residential and commercial building energy systems. A new section has been added on small projects to show how the methodology can be simplified.

Monitoring projects can be **uninstrumented** (i.e., no additional instrumentation beyond the utility meter) or **instrumented** (i.e., billing data supplemented by additional sources, such as an installed instrumentation package, portable data loggers, or building automation system). Uninstrumented approaches are generally simpler and less costly, but they can be subject to more uncertainty in interpretation, especially when changes made to the building represent a small fraction of total energy use. It is important to determine (1) the accuracy needed to meet objectives, (2) the type of monitoring needed to provide this accuracy, and (3) whether the desired accuracy justifies the cost of an instrumented approach.

Instrumented field monitoring projects generally involve a data acquisition system (DAS), which typically comprise various sensors and data-recording devices (e.g., data loggers) or a suitably equipped building automation system. Projects may involve a single building or hundreds of buildings and may be carried out over periods ranging from weeks to years. Most monitoring projects involve the following activities:

- Project planning
- Site installation and calibration of data acquisition equipment (if required)
- Ongoing data collection and verification
- Data analysis and reporting

These activities often require support by several professional disciplines (e.g., engineering, data analysis, management) and construction trades (e.g., electricians, controls technicians, pipe fitters).

Useful building energy performance data cover whole buildings, lighting, HVAC equipment, water heating, meter readings, utility demand and load factors, excess capacity, controller actuation, and building and component lifetimes. Current monitoring practices vary considerably. For example, a utility load research project may tend to characterize the average performance of buildings with relatively few data points per building, whereas a test of new technology performance may involve monitoring hundreds of parameters in a single facility. Monitoring projects range from broad research studies to very specific, contractually required savings verification carried out by performance contractors. However, all practitioners should use accepted standards of monitoring practices to communicate results. Key elements in this process are (1) classifying the types of project monitoring and (2) developing consensus on the purposes, approaches, and problems associated with each type (Haberl et al. 1990; Misuriello 1987). For example, energy savings from energy service performance contracts can be specified on either a whole-building or component basis. Monitoring requirements for each approach vary widely and must be carefully matched to the specific project. Procedures in ASHRAE *Guideline* 14-2002 and the IPMVP (2007) can be used to determine monitoring requirements.

REASONS FOR ENERGY MONITORING

Monitoring projects can be broadly categorized by their goals, objectives, experimental approach, level of monitoring detail, and uses (Table 1). Other factors, such as resources available, data validation and analysis procedures, duration and frequency of data collection, and instrumentation, are common to most, if not all, projects.

Energy End Use

Energy end-use projects typically focus on individual energy systems in a particular market sector or building type. Monitoring usually requires separate meters or data collection channels for each end use, and analysts must account for all factors that may affect energy use. Examples of this approach include detailed utility load research efforts, evaluation of utility incentive programs, and end-use calibration of computer simulations. Depending on the project objectives, the frequency of data collection may range from one-time measurements of full-load operation to continuous time-series measurements.

Specific Technology Assessment

Specific technology assessment projects monitor field performance of particular equipment or technologies that affect building energy use, such as envelope retrofit measures, major end-use system loads or savings from retrofits (e.g., lighting), or retrofits to or performance of mechanical equipment.

The typical goal of retrofit performance monitoring projects is to estimate savings resulting from the retrofit despite potentially significant variation in indoor/outdoor conditions, building characteristics, and occupant behavior unrelated to the retrofit. The frequency and complexity of data collection depend on project objectives and site-specific conditions. Projects in this category assess variations in

The preparation of this chapter is assigned to TC 7.6, Building Energy Performance.

Table 1   Characteristics of Major Monitoring Project Types

Project Type	Goals and Objectives	General Approach	Level of Detail	Uses
Energy end use	Determine characteristics of specific energy end uses in building.	Often uses large, statistically designed sample. Monitor energy demand or use profile of each end use of interest.	Detailed data on end uses metered. Collect building and operating data that affect end use.	Load forecasting by end use. Identify and confirm energy conservation or demand-side management opportunities. Simulation calculations. Rate design.
Specific technology assessment	Measure field performance of building system technology or retrofit measure in individual buildings.	Characterize individual building or technology, occupant behavior, and operation. Account and correct for variations.	Uses detailed audit, submetering, indoor temperature, on-site weather, and occupant surveys. May use weekly, hourly, or short-term data.	Technology evaluation. Retrofit performance. Validate models and predictions.
Energy savings measurement and verification	Estimate the impact of retrofit, commissioning, or other building alteration to serve as basis for payments or benefits calculation.	Preretrofit consumption is used to create baseline model. Postretrofit consumption is measured; the difference between the two is savings.	Varies substantially, including verification of potential to provide savings, retrofit isolation, or whole-building or calibrated simulation.	Focused on specific campus, building, component, or system. Amount and frequency of data varies widely between projects.
Building operation and diagnostics	Solve problems. Measure physical or operating parameters that affect energy use or that are needed to model building or system performance.	Typically uses one-time and/or short-term measurement with special methods, such as infrared imaging, flue gas analysis, blower door, or coheating.	Focused on specific building component or system. Amount and frequency of data vary widely between projects.	Energy audit. Identify and solve operation and maintenance, indoor air quality, or system problems. Provide input for models. Building commissioning.

performance between different buildings or for the same building before and after the retrofit.

Field tests of end-use equipment are often characterized by detailed monitoring of all critical performance parameters and operational modes. In evaluating equipment performance or energy efficiency improvements, it is preferable to measure in situ performance. Although manufacturers' data and laboratory performance measurements can provide excellent data for sizing and selecting equipment, installed performance can vary significantly from that at design conditions. The project scope may include reliability, maintenance, design, energy efficiency, sizing, and environmental effects (Phelan et al. 1997a, 1997b).

Savings Measurement and Verification (M&V)

Accountability is increasingly necessary in energy performance retrofits, whether they are performed as part of energy savings performance contracting (ESPC) or performed directly by the owner. In either case, savings measurement and verification (M&V) is an important part of the project. Because the actual energy savings cannot be measured directly, the appropriate role of energy monitoring methodology is to

- Ensure that appropriate data are available, including preretrofit data if retrofits are installed
- Accurately define baseline conditions and assumptions
- Confirm that proper equipment and systems were installed and have the potential to generate the predicted energy savings
- Take postretrofit measurements
- Estimate the energy savings achieved

Proper assessment of an energy retrofit involves comparing before and after energy use, and adjusting for all nonretrofit changes that affected energy use. Weather and occupancy are examples of factors that often change. To assess the effectiveness of the retrofit alone, the influence of these other complicating factors must be removed as best possible. Relationships must be found between energy use and these factors to remove the influence of the factors from the energy savings measurement. These relationships are usually determined through data analysis, not textbook equations. Because data analysis can be conducted in an infinite number of ways, there can be no absolute certainty about the relationship chosen. The need for certainty must be carefully balanced with measurement and analysis costs, recognizing that absolute certainty is not achievable. Among the numerous sources of uncertainty are instrumentation or

measurement error, normalization or model error, sampling error, and errors of assumptions. Each source can be minimized to varying degrees by using more sophisticated measurement equipment, analysis methods, sample sizes, and assumptions. However, more certain savings determinations generally follow the law of diminishing returns, where further increases in certainty come at progressively greater expense. Total certainty is seldom achievable, and even less frequently cost-effective (ASHRAE *Guideline* 14).

Other resources are also available. One of the widest known is the *International Performance Measurement and Verification Protocol* (IPMVP 2007). The IPMVP is more general than ASHRAE *Guideline* 14 but provides important background for understanding the larger context of M&V efforts.

Building Diagnostics

Diagnostic projects measure physical and operating parameters that determine the energy use of buildings and systems. Usually, the project goal is to determine the cause of problems, model or improve energy performance of a building or system(s), or isolate effects of components. Diagnostic tests frequently involve one-time measurements or short-term monitoring. To give insight, the frequency of measurement must be several times faster than the rate of change of the effect being monitored. Some diagnostic tests require intermittent, ongoing data collection.

The most basic energy diagnostic for buildings is determining rate of energy use or power for a specific period, from essentially a single point in time to a few weeks. The scope of measurement may include the whole building or only one component. The purpose can range from measurement system parameter estimation to verification of nameplate information. Daily or weekly profiles may also be of interest.

A large number of diagnostic measurement procedures are used for energy measurements in residential buildings, particularly single-family. Typical measurements for single-family residences include (1) flue gas and other analysis procedures to determine steady-state furnace combustion efficiency and the efficiency of other end uses, such as air conditioners, refrigerators, and water heaters; (2) fan pressurization tests to measure and locate building envelope air leakage (ASTM *Standard* E779) and tests to measure airtightness of air distribution systems (Modera 1989; Robison and Lambert 1989); and (3) infrared thermography to locate thermal defects in the building envelope and other methods to determine overall building envelope parameters (Subbarao 1988).

Table 2 Comparison of Small Projects to Overall Methodology

Project Characteristic	Small Project Approach	Overall Methodology Coverage
Project problem areas	• Project goals and resources are iteratively evaluated in short time. Only one to possibly a few people involved. Small group allows more informal procedures and high interaction as needed. • Data products loosely defined. Data collection starts. Initial and ongoing analysis indicates any data management or quality control issues. • Data products refined over time as needed, based on analysis. • Accuracy evaluated on the fly as data are collected. • Commitment is simply needing to finish the work. • Advice still sought where needed.	Project goals, project costs and resources, data products, data management, data quality control, commitment, accuracy requirements, advice.
Building and occupant characteristics	Typically, only one to few buildings included; work is reasonably local. Characteristic data collected on site, at convenience of project person(s), depending on project location. Return trips likely already needed for simplified data collection approach; supporting data can be collected on return trips.	Fairly extensive data structure (e.g., a characteristic database) and definition of levels of detail may be needed to handle possibly many buildings and improve ability to report results. With many buildings, only one trip per building may be acceptable.
Project design	Project personnel usually know what they want to measure and report, and may not want to be confused by complex approach in this chapter. Knowledge of experimental approaches may also be understood minimally but still applied successfully, without specific declaration, to small project.	Three higher-level general approaches: (1) fewer buildings or systems with more detailed measurements, (2) many buildings or systems with less detailed measurements, or (3) many buildings or systems with more detailed measurements. Six major experimental design approaches: on/off, before/after, test/reference, simulated occupancy, nonexperimental reference, engineering field test.
Reporting	Reporting is informal to somewhat formal (more like straightforward engineering project than research project). Reported results are often minimal but provide key information sought.	Research project report is likely required, possibly hundreds of pages long, with multiple appendices. Extensive databases likely generated and must be quality checked and corrected for use by others. Data user access procedures may have to be developed. Databases must be maintained over years in many cases and must be well documented. Extensive research results may have been generated and should be reported. For example, Fracastoro and Lyberg (1983) discussion of guiding principles for residential projects is 300 pages long.

Energy systems in multifamily buildings can be much more complex than those in single-family homes, but the types of diagnostics are similar: combustion equipment diagnostics, air leakage measurements, and infrared thermography to identify thermal defects or moisture problems (DeCicco et al. 1995). Some techniques are designed to determine the operating efficiency of steam and hot-water boilers and to measure air leakage between apartments.

Diagnostic techniques have been designed to measure the overall airtightness of office building envelopes and the thermal performance of walls (Armstrong et al. 2001; Persily et al. 1988; Sellers et al. 2004). Practicing engineers also use a host of monitoring techniques to aid in diagnostics and analysis of equipment energy performance. Portable data loggers are often used to collect time-synchronized distributed data, allowing multiple data sets (e.g., chiller performance and ambient conditions) to be collected and quickly analyzed. Similar short-term monitoring procedures are used to provide more detailed and complete commercial building system commissioning. Short-term, in situ tests have also been developed for pumps, fans, and chillers (Phelan et al. 1997a, 1997b).

Diagnostics are also well suited to support development and implementation of building energy management programs (see Chapter 36). Long-term diagnostic measurements have even supported energy improvements (Liu et al. 1994). Diagnostic measurement projects can generally be designed using procedures adapted to specific project requirements (see the section on Steps for Project Design and Implementation).

Equipment for diagnostic measurement may be installed temporarily or permanently to aid energy management efforts. Designers should consider providing permanent or portable check metering of major electrical loads in new building designs. Building automation systems also can be used to collect the data required for diagnostics. The same concept can be extended to fuel and thermal energy use.

SMALL PROJECTS

Most energy metering projects are done on a small scale and will have the project steps described in this chapter simplified and compacted. This section describes briefly how to use the information in this chapter for a small project.

Small projects will still be potentially impacted by the issues described here, but if only a small group of people are doing all the work, they can choose what to address and how to address the project requirements. Table 2 compares small-project approaches and the material in this chapter.

How to Use This Chapter for Small Projects

- Skim through the chapter and make a few notes on items that may apply for the project in question.
- Generate a brief project plan to make sure major issues that may cause problems are not overlooked.
- Generate a brief checklist from the item notes for final check-off during or at the end of project.
- Clarify what building or site characteristics data may be needed, and be sure to collect those data.
- Analyze data from the start, to make sure there are no quality or data issues.
- Consider whether any data should be made available to others at the end of the project, and if so, develop a data format for exchange.
- Skim through the chapter again when the report is being prepared to gather ideas about what to include in the report.

PROTOCOLS FOR PERFORMANCE MONITORING

Examples of procedures (**protocols**) for evaluating energy savings for projects involving retrofit of existing building energy systems are presented here. These protocols should also be useful to those interested in more general building energy monitoring.

Building monitoring has been significantly simplified and made more professional in recent years by the development of fairly standardized monitoring protocols. Although there may be no way to define a protocol to encompass all types of monitoring applications, repeatable and understandable methods of measuring and verifying retrofit savings are needed. However, following a protocol does not replace adequate project planning and careful assessment of project objectives and constraints.

Residential Retrofit Monitoring

Protocols for residential building retrofit performance can answer specific questions associated with actual measured performance. For example, Ternes (1986) developed a single-family retrofit monitoring protocol, a data specification guideline that identifies important parameters to be measured. Both one-time and time-sequential data parameters are covered, and parameters are defined carefully to ensure consistency and comparability between experiments. Discrepancies between predicted and actual performance, as measured by the energy bill, are common. This protocol improves on billing data methods in two ways: (1) internal temperature is monitored, which eliminates a major unknown variable in data interpretation; and (2) data are taken more frequently than monthly, which potentially shortens monitoring duration. Utility bill analysis generally requires a full season of pre- and post-retrofit data. The single-family retrofit protocol may require only a single season.

Ternes (1986) identified both a minimum set of data, which must be collected in all field studies that use the protocol, and optional extensions to the minimum data set that can be used to study additional issues. See Table 3 for details. Szydlowski and Diamond (1989) developed a similar method for multifamily buildings.

The single-family retrofit monitoring protocol recommends a before/after experimental design, and the minimum data set allows performance to be measured on a normalized basis with weekly time-series data. (Some researchers recommend daily.) The protocol also allows hourly recording intervals for time-integrated parameters, an extension of the basic data requirements in the minimum data set. The minimum data set may also be extended through optional data parameter sets for users seeking more information.

Data parameters in this protocol have been grouped into four data sets: basic, occupant behavior, microclimate, and distribution system (Table 3). The minimum data set consists of a weekly option of the basic data parameter set. Time-sequential measurements are monitored continuously during the field study. These are all time-integrated parameters (i.e., appropriate average values of a parameter over the recording period, rather than instantaneous values).

This protocol also addresses instrumentation installation, accuracy, and measurement frequency and expected ranges for all time-sequential parameters (Table 4). The minimum data set (weekly option of the basic data) must always be collected. At the user's discretion, hourly data may be collected, which allows two optional parameters to be monitored. Parameters from the optional data sets may be chosen, or other data not described in the protocol added, to arrive at the final data set.

This protocol standardizes experimental design and data collection specifications, enabling independent researchers to compare project results more readily. Moreover, including both minimum and optional data sets and two recording intervals accommodates projects of varying financial resources.

Commercial Retrofit Monitoring

Several related guidelines have been created for the particular application of retrofit savings (M&V). ASHRAE *Guideline* 14 provides methods for effectively and reliably measuring the energy and demand savings due to building energy projects. The guideline defines a minimum acceptable level of performance in measuring

Table 3   Data Parameters for Residential Retrofit Monitoring

	Recording Period	
	Minimum	**Optional**
Basic Parameters		
House description		Once
Space-conditioning system description		Once
Entrance interview information		Once
Exit interview information		Once
Pre- and post-retrofit infiltration rates		Once
Metered space-conditioning system performance		Once
Retrofit installation quality verification		Once
Heating and cooling equipment energy consumption	Weekly	Hourly
Weather station climatic information	Weekly	Hourly
Indoor temperature	Weekly	Hourly
House gas or oil consumption	Weekly	Hourly
House electricity consumption	Weekly	Hourly
Wood heating use	—	Hourly
Domestic hot water energy consumption	Weekly	Hourly
Optional Parameters		
Occupant behavior		
Additional indoor temperatures	Weekly	Hourly
Heating thermostat set point	—	Hourly
Cooling thermostat set point	—	Hourly
Indoor humidity	Weekly	—
Microclimate		
Outdoor temperature	Weekly	Hourly
Solar radiation	Weekly	Hourly
Outdoor humidity	Weekly	Hourly
Wind speed	Weekly	Hourly
Wind direction	Weekly	Hourly
Shading		Once
Shielding		Once
Distribution system		
Evaluation of ductwork infiltration		Once

Source: Ternes (1986).

energy and demand savings from energy conservation measures in residential, commercial, or industrial buildings. These measurements can serve as the basis for commercial transactions between energy services providers and customers who rely on measured energy savings as the basis for financing payments. Three approaches are discussed: whole building, retrofit isolation, and calibrated simulation. The guideline includes an extensive resource on physical measurement, uncertainty, and regression techniques. Example M&V plans are also provided.

The *International Performance Measurement and Verification Protocol* (IPMVP 2007) provides guidance to buyers, sellers, and financiers of energy projects on quantifying energy savings performance of energy retrofits. The Federal Energy Management Program has produced guidelines specific to federal projects, which include many procedures usable for calculating retrofit savings in nonfederal buildings (FEMP 2008).

On a more detailed level, ASHRAE Research Project RP-827 resulted in separate guidelines for in situ testing of chillers, fans, and pumps to evaluate installed energy efficiency (Phelan et al. 1997a, 1997b). The guidelines specify the physical characteristics to be measured; number, range, and accuracy of data points required; methods of artificial loading; and calculation equations with a rigorous uncertainty analysis.

In addition to these specialized protocols for particular monitoring applications, a number of specific laboratory and field measurement standards exist (see Chapter 57), and many monitoring source books are in circulation.

Finally, MacDonald et al. (1989) developed a protocol for field monitoring studies of energy improvements (retrofits) for commercial buildings. Similar to the residential protocol, it addresses data

Table 4 Time-Sequential Parameters for Residential Retrofit Monitoring

Data Parameter	Accuracy[a]	Range	Stored Value per Recording Period	Scan Rate[b] Option 1	Option 2
Basic Parameters					
Heating/cooling equipment energy consumption	3%		Total consumption	15 s	15 s
Indoor temperature	1.0°F	50 to 95°F	Average temperature	1 h	1 min
House gas or oil consumption	3%		Total consumption	15 s	15 s
House electricity consumption	3%		Total consumption	15 s	15 s
Wood heating use	1.0°F	50 to 800°F	Average surface temperature or total use time		1 min
Domestic hot water	3%		Total consumption	15 s	15 s
Optional Data Parameter Sets					
Occupant behavior					
Additional indoor temperatures	1.0°F	50 to 95°F	Average temperature	1 h	1 min
Heating thermostat set point	1.0°F	50 to 95°F	Average set point		1 min
Cooling thermostat set point	1.0°F	50 to 95°F	Average set point		1 min
Indoor humidity	5% rh	10 to 95% rh	Average humidity	1 h	
Microclimate					
Outdoor temperature	1.0°F	−40 to 120°F	Average temperature	1 h	1 min
Solar radiation	10 Btu/h·ft²	0 to 350 Btu/h·ft²	Total horizontal radiation	1 min	1 min
Outdoor humidity	5% rh	10 to 95% rh	Average humidity	1 h	1 min
Wind speed	0.5 mph	0 to 20 mph	Average speed	1 min	1 min
Wind direction	5°	0 to 360°	Average direction	1 min	1 min

Source: Ternes (1986). [a]All accuracies are stated values. [b]Applicable scan rates if nonintegrating instrumentation is used.

Table 5 Performance Data Requirements of Commercial Retrofit Protocol

Projects with Submetering		
	Before Retrofit	**After Retrofit**
Utility billing data (for each fuel)	12 month minimum	3 month minimum (12 months if weather normalization required)
Submetered data (for all recording intervals)	All data for each major end use, up to 12 months	All data for each major end use, up to 12 months
	Type	**Recording Interval**	**Period Length**
Temperature data (daily maximum and minimum must be provided for any periods without integrated averages)	Maximum and minimum —or— Integrated averages	Daily —or— Same as for submetered data but not longer than daily	Same as billing data length —or— Length of submetering

Projects Without Submetering		
	Before Retrofit	**After Retrofit**
Utility billing data (for each fuel)	12 month minimum	12 month minimum
	Type	**Recording Interval**	**Period Length**
Temperature data	Maximum and minimum —or— Integrated averages	Daily	Same as billing data length

requirements for monitoring studies. Commercial buildings are more complex, with a diverse array of potential efficiency improvements. Consequently, the approach to specifying measurement procedures, describing buildings, and determining the range of analysis must differ.

The strategy used for this protocol is to specify data requirements, analysis, performance data with optional extensions, and a building core data set that describes the field performance of efficiency improvements. This protocol requires a description of the approach used for analyzing building energy performance. The necessary performance data, including identification of a minimum data set, are outlined in Table 5.

Commercial New Construction Monitoring

New building construction offers the potential for monitoring building subsystem energy consumption at a reasonable cost. Information obtained by monitoring operating hours or direct energy consumption can benefit building owners and managers by

- Verifying design intent
- Alerting them to inefficient or improper operation of equipment

- Providing data that can be useful in determining benefits of alternative operating strategies or replacement equipment
- Evaluating costs of operation for extending occupancy hours for special conditions or event
- Demonstrating effects of poor maintenance or identifying when maintenance procedures are not followed
- Diagnosing and fixing comfort and other space condition problems
- Diagnosing power quality problems
- Submetering tenants
- Verifying/improving savings of a performance contract
- Maintaining persistence of energy savings

To provide data necessary to improve building systems operation, monitoring should be considered for boilers, chillers, cooling towers, heat pumps, air-handling unit fans, large fan-coil units, major exhaust fans, major pumps, comfort cooling compressors, lighting panels, electric heaters, receptacle panels, substations, motor control centers, major feeders, service water heaters, process loads, and computer rooms.

Guidance on energy monitoring to determine energy savings for new construction design modifications is available in IPMVP (2006).

Construction documents may include provisions for various meters to monitor equipment and system operation. Some equipment can be specified to have factory-installed hour meters that record actual operating hours of the equipment. Hour meters can also be easily field-installed on any electrical motor.

More sophisticated power-monitoring systems, with electrical switchgear, substations, switchboards, and motor control centers, can be specified. These systems can monitor energy demand, energy consumption, power factor, neutral current, etc., and can be linked to a computer. These same systems can be installed on circuits to existing or retrofit fans, chillers, lighting panels, etc. Some equipment commonly used for improving system efficiency, such as variable-frequency drives, can be provided with capability to monitor kilowatt output, kilowatt-hours consumed, and other variables.

Using direct digital control (DDC) or building automation systems for monitoring is particularly appropriate in new construction. These systems can monitor, calculate, and record system status, water use, energy use at the main meter or of particular end-use systems, demand, and hours of operation, as well as start and stop building systems, control lighting, and print alarms when systems do not operate within specified limits. Initial specification of the new control system should include specific requirements for sensors, calculations, and trend logging and reporting functions. Special issues related to sensors and monitoring approaches can be found in Piette et al. (2000).

COMMON MONITORING ISSUES

Field monitoring projects require effective management of various professional skills. Project staff must understand the building systems being examined, quality control of data, data management, data acquisition, and sensor technology. In addition to data collection, processing, and analysis, the logistics of field monitoring projects require coordinating equipment procurement, delivery, and installation.

Key issues include the **accuracy** and **reliability** of collected data. Projects have been compromised by inaccurate or missing data, which could have been avoided by periodic sensor calibration and ongoing data verification.

Planning

Many common problems in monitoring projects can be avoided by effective and comprehensive planning.

Project Goals. Project goals and data requirements should be established before hardware is selected. Unfortunately, projects are often driven by hardware selection rather than by project objectives, either because monitoring hardware must be ordered several months before data collection begins or because project initiation procedures are lacking. As a result, the hardware may be inappropriate for the particular monitoring task, or critical data points may be overlooked.

Project Costs and Resources. After goal setting, the feasibility of the anticipated project should be reviewed in light of available resources. Projects to which significant resources can be devoted usually involve different approaches from those with more limited resources. This issue should be addressed early on and reviewed throughout the course of the project. Although it is difficult at this early stage to assess with certainty the cost of an anticipated project, rough estimates can be quite helpful.

Data Products. It is important to establish the type and format of the final results calculated from data before selecting data points. Failure to plan these data products first may lead to failure to answer critical questions.

Data Management. Failure to anticipate the typically large amounts of data collected can lead to major difficulties. The computer and personnel resources needed to verify, retrieve, analyze, and archive data can be estimated based on experience with previous projects.

Data Quality Control. It is also important to check and validate the quality and reasonableness of data before use. Failure to use some type of quality control typically results in data errors and invalid results.

Commitment. Many projects require long-term commitment of personnel and resources. Project success depends on long-term, daily attention to detail and on staff continuity.

Accuracy Requirements. The required accuracy of data products and accuracy of the final data and experimental design needed to meet these requirements should be determined early on. After the required accuracy is specified, the sample size (number of buildings, control buildings, or pieces of equipment) must be chosen, and the required measurement precision (including error propagation into final data products) must be determined. Because tradeoffs must usually be made between cost and accuracy, this process is often iterative. It is further complicated by a large number of independent variables (e.g., occupants, operating modes) and the stochastic nature of many variables (e.g., weather).

Advice. Expert advice should be sought from others who have experience with the type of monitoring envisioned.

Implementation and Data Management

The following steps can facilitate smooth project implementation and data management:

- Calibrate sensors before installation. Spot-check calibration on site. During long-term monitoring projects, recalibrate sensors periodically. Appropriate procedures and standards should be used in all calibration procedures [see ASHRAE *Guideline* 14 and IPMVP (2007)].
- Track sensor performance regularly. Quick detection of sensor failure or calibration problems is essential. Ideally, this should be an automated or a daily task. The value of data is high because they may be difficult or impossible to reconstruct.
- Generate and review data on a timely, periodic basis. Problems that often occur in developing final data products include missing data from failed sensors, data points not installed because of planning oversights, and anomalous data for which there are no explanatory records. If data products are specified as part of general project planning and produced periodically, production problems can be identified and resolved as they occur. Automating the process of checking data reliability and accuracy can be invaluable in keeping the project on track and in preventing sensor failure and data loss.

Data Analysis and Reporting

For most projects, the collected data must be analyzed and put into reports. Because the objective of the project is to translate these data into information and ultimately into knowledge and action, the importance of this step cannot be overemphasized. Clear, convenient, and informative formats should be devised in the planning stages and adhered to throughout the project.

Close attention must be paid to resource allocation to ensure that adequate resources are dedicated to verification, management, and analysis of data and to ongoing maintenance of monitoring equipment. As a quality control procedure and to make data analysis more manageable, these activities should be ongoing. Data analysis should be carefully defined before the project begins.

STEPS FOR PROJECT DESIGN AND IMPLEMENTATION

This section describes methodology for designing effective field monitoring projects that meet desired goals with available project resources. The task components and relationships among the nine activities constituting this methodology are identified in Figure 1.

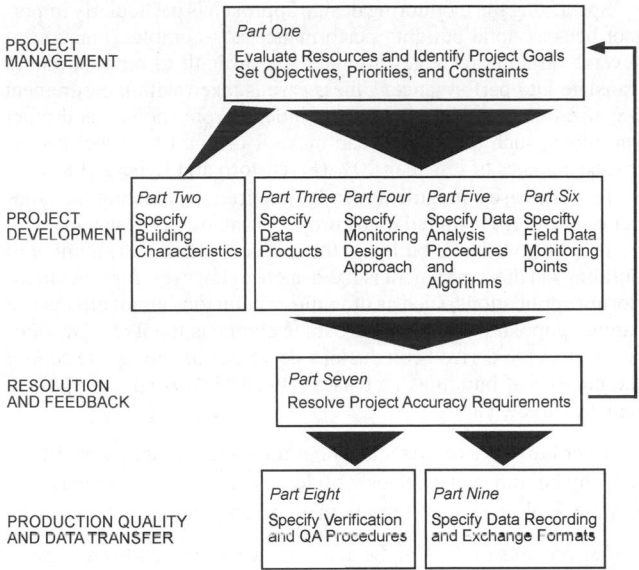

PROJECT MANAGEMENT

Part One
Evaluate Resources and Identify Project Goals
Set Objectives, Priorities, and Constraints

PROJECT DEVELOPMENT

Part Two
Specify Building Characteristics

Part Three
Specify Data Products

Part Four
Specify Monitoring Design Approach

Part Five
Specify Data Analysis Procedures and Algorithms

Part Six
Specify Field Data Monitoring Points

RESOLUTION AND FEEDBACK

Part Seven
Resolve Project Accuracy Requirements

PRODUCTION QUALITY AND DATA TRANSFER

Part Eight
Specify Verification and QA Procedures

Part Nine
Specify Data Recording and Exchange Formats

Fig. 1   Methodology for Designing Field Monitoring Projects

The activities fall into four categories: project management, project development, resolution and feedback, and production quality and data transfer. Field monitoring projects vary in terms of resources, goals and objectives, data product requirements, and other variables, affecting how methodology should be applied. Nonetheless, the methodology provides a proper framework for advance planning, which helps minimize or prevent implementation problems.

An iterative approach to planning activities is best. The scope, accuracy, and techniques can be adjusted based on cost estimates and resource assessments. The initial design should be performed simply and quickly to estimate cost and evaluate resources. If costs are out of line with resources, such as when desired levels of instrumentation exceed the resources available for the project, adjustments are needed. Planning should identify and resolve any tradeoffs necessary to execute the project within a given budget. Examples include reducing the scope of the project versus relaxing instrumentation specifications or accuracy requirements. These decisions often depend on what questions the project must answer and which questions can be eliminated, simplified, or narrowed.

One frequent oversight in project planning is failing to reserve sufficient time and resources for later analysis and reporting of data. Unanticipated additional costs associated with data collection and problem resolution should not jeopardize these resources.

Documenting the results of project planning should cover all nine parts of the process. This report can be a useful part of an overall project plan that may document other important project information, such as resources to be used, schedule, etc.

Part One: Identify Project Objectives, Resources, and Constraints

Start with a clear understanding of the decision to be made or action to be taken as a result of the project. The goals and objectives statement determines the overall direction and scope of the data collection and analysis effort. The statement should also list questions to be answered by empirical data, noting the error or uncertainty associated with the desired result. Realistic assessment of error is needed because requiring too small an uncertainty leads to an overly complex and expensive project. It is important in monitoring projects that a data acquisition plan be developed and followed with a clear idea of the **research questions** to be answered.

Resource requirements for equipment, personnel, and other items must feed into budget estimates to determine expected funding needed for different project objectives. Scheduling requirements must also be considered, and any project constraints defined and considered. Tradeoffs on budget and objectives require that priorities be established.

Even if a project is not research-oriented, it is attempting to obtain information, which can be stated in the form of questions. Research questions can have varying scopes and levels of detail, addressing entire systems or specific components. Some examples of research questions follow:

- Measurement and verification: Have contractors fulfilled their responsibilities of installing equipment and improving systems to achieve the agreed-upon energy savings?
- Classes of buildings: To an accuracy of 20%, how much energy has been saved by using a building construction/performance standard mandated in the jurisdiction?
- Particular buildings: Has a lapse in building maintenance caused energy performance to degrade?
- Particular components: What is the average reduction in demand charges during summer peak periods because of the installation of an ice storage system in this building?

Research questions vary widely in technical complexity, generally taking one of the following three forms:

- How does the building/component perform?
- Why does the building/component perform as it does?
- Which building/component should be targeted to achieve optimal cost-effectiveness?

The first form of question can sometimes be answered generically for a class of typical buildings without detailed monitoring and analysis, although detailed planning and thorough analysis are still required. The second and third forms usually require detailed monitoring and analysis and, thus, detailed planning.

In general, more detailed and precise goal statements are better. They ensure that the project is constrained in scope and developed to meet specific accuracy and reliability requirements. Usually, projects attempt to answer more than one research question, and often consider both primary and secondary questions. All data collected should have a purpose of helping to answer a project question; the more specific the questions, the easier identifying required data becomes.

Part Two: Specify Building and Occupant Characteristics

Measured energy data will not be meaningful later to people who were not involved in the project unless the characteristics of the building being monitored and its use have been documented. To meet this need, a data structure (e.g., a characteristic database) can be developed to describe the buildings.

Building characteristics can be collected at many levels of detail, depending on the type of monitoring project and the parameters that affect results. For projects that determine whole-building performance, it is important to provide at least enough detail to document the following:

- General building type, use, configuration, and envelope (particularly energy-related aspects)
- Building occupant information (number, occupancy schedule, activities)
- Internal loads
- HVAC system descriptive information characterizing key parameters affecting HVAC system performance
- Type and quantity of other energy-using systems
- Any building changes that occur during the monitoring project
- Entrance interview information focusing on energy-related behavior of building occupants before monitoring

• Exit interview information documenting physical or lifestyle changes at the test site that may affect data analysis

The minimum level of detail is known as **summary characteristics** data. **Simulation-level characteristics** (detailed information collected for hourly simulation model input) may be desirable for some buildings. Regardless of the level of detail, the data should provide a context for analysts, who may not be familiar with the project, to understand the building and its energy use.

Part Three: Specify Data Products and Project Output

The objective of a monitoring project is typically not to produce data, but to answer a question. However, the data must be of high quality and must be presented to key decision makers and analysts in a convenient, informative format. The specific **data products** (format and content of data needed to meet project goals and objectives) must be identified and evaluated for feasibility and usefulness in answering project questions identified in Part One. Final data products must be clearly specified, together with the minimum acceptable data requirements for the project. It is important to clearly define an **analysis path** showing what will be calculated and what data are necessary to achieve desired results. Clear communication is critical to ensure that project requirements are satisfied and factors contributing to monitoring costs are understood.

Evaluation results can be presented in many forms, often as interim and final reports (possibly by heating and/or cooling season), technical notes, or technical papers. These documents must convey specific results of the field monitoring clearly and concisely. They should also contain estimates of the accuracy of the results.

The composition of data presentations and analysis summaries should be determined early to ensure that all critical parameters are identified (Hough et al. 1987). For instance, mock-ups of data tables, charts, and graphs can be used to identify requirements. Previously reported results can be used to provide examples of useful output. Data products should also be prioritized to accommodate possible cost tradeoffs or revisions resulting from other steps in the process such as error analysis (see Part Seven).

Although requirements for the minimum acceptable data results can often be specified during planning, data analysis typically reveals further requirements. Thus, budget plans should include allowances and optional data product specifications to handle additional or unique project output requirements uncovered during data analysis.

Longer-term goals and future information needs should be anticipated and explained to project personnel. For example, a project may have short- and long-term data needs (e.g., demonstrating reductions in peak electrical demand versus demonstrating cost-effectiveness or reliability to a target audience). Initial results on demand reduction may not be the ultimate goal, but rather a step toward later presentations on cost reductions achieved. Thus, it is prudent to consider long-term and potential future data needs so that additional supporting information, such as photographs or testimonials, may be identified and obtained.

Part Four: Specify Monitoring Design Approach

A general monitoring design must be developed that defines three interacting factors: the number of buildings in the study, the monitoring approach (or experimental design), and the level of detail in the data being measured. A less detailed or precise approach can be considered if the number of buildings is increased, and vice versa. If the goal is related to a specific product, the monitoring design must isolate the effects of that product. Haberl et al. (1990) discuss monitoring designs. For example, for retrofit M&V, protocols have been written allowing a range of different monitoring methods, from retrofit verification to retrofit isolation (ASHRAE *Guideline* 14; FEMP 2008; IPMVP 2007). Some monitoring approaches are more suited than others to larger numbers of buildings.

Specifying the monitoring design approach is particularly important because total building performance is a complex function of several variables, changes in which are difficult to monitor and to translate into performance. Unless care is taken with measurement organization and accuracy, uncertainties, errors (noise), and other variations, such as weather, can make it difficult to detect performance changes of less than 20% (Fracastoro and Lyberg 1983).

In some cases, judgment may be required in selecting the number of buildings involved in the project. If an owner seeks information about a particular building, the choice is simple: the number of buildings in the experiment is fixed at one. However, for other monitoring applications, such as drawing conclusions about effects in a sample population of buildings, some choice is involved. Generally, error in the derived conclusions decreases as the square root of the number of buildings increases (Box 1978). A specific project may be directed at

• Fewer buildings or systems with more detailed measurements
• Many buildings or systems with less detailed measurements
• Many buildings or systems with more detailed measurements

For projects of the first type, accuracy requirements are usually resolved initially by determining expected variations of measured quantities (dependent variables) about their average values in response to expected variations of independent variables. For buildings, a typical concern is the response of heating and cooling loads to changes in temperature or other weather variables. The response of building lighting energy use to daylighting is another example of the relationship between dependent and independent variables. Fluctuations in response are caused by (1) outside influences not quantified by measured energy use data and (2) limitations and uncertainties associated with measurement equipment and procedures. Thus, accuracy must often be determined using statistical methods to describe mean tendencies of dependent variables.

For projects of the second and third types, the increased number of buildings improves confidence in the mean tendencies of the dependent response(s) of interest. Larger sample sizes are also needed for experimental designs with control groups, which are used to adjust for some outside influences. For more information, see Box (1978), Fracastoro and Lyberg (1983), and Hirst and Reed (1991). Projects can also use more complex, multilevel measurement and modeling approaches to handle an array of technologies or help improve confidence in results (Hughes and Shonder 1998).

Most monitoring procedures use one or more of the following general experimental approaches:

Before/After. Building, system, or component energy consumption is monitored before and after a new component or retrofit improvement is installed. Changes in factors not related to the retrofit, such as the weather and building operation during the two periods, must be accounted for, often requiring a model-based analysis (Fels 1986; Hirst et al. 1983; Kissock et al. 1992; Robison and Lambert 1989; Sharp and MacDonald 1990). This experimental design is the primary concern of most current building energy monitoring documents (ASHRAE *Guideline* 14; FEMP 2008; IPMVP 2002).

Test/Reference. The building energy consumption data of two "identical" buildings, one with the product or retrofit being investigated, are compared. Because buildings cannot be absolutely identical (e.g., different air leakage distributions, insulation effectiveness, temperature settings, and solar exposure), measurements should be taken before installation as well, to allow calibration. Once the product or retrofit is installed, any deviation from the calibration relationship can be attributed to the product or retrofit (Fracastoro and Lyberg 1983; Levins and Karnitz 1986).

On/Off. If the retrofit or product can be activated or deactivated at will, energy consumption can be measured in a number of repeated on/off cycles. On-period consumption is then compared to off-period consumption (Cohen et al. 1987; Woller 1989).

Table 6 Advantages and Disadvantages of Common Experimental Approaches

Mode	Advantages	Disadvantages
Before/after	No reference building required. Same occupants implies smaller occupant variations. Modeling processes are mostly identical before/after.	Weather different before/after. More than one heating/cooling season may be needed. Model is required to account for weather and other changes.
Test/reference	One season of data may be adequate. Small climate difference between buildings.	Reference building required. Calibration phase required (may extend testing to two seasons). Occupants in either or both buildings can change behavior.
On/off	No reference building required. One season may be adequate. Modeling processes are mostly identical before/after. Most occupancy changes are small.	Requires reversible product. Cycle may be too long if time constants are large. Model is required to account for weather differences in cycles. Dynamic model accounting for transients may be needed.
Simulated occupancy	Noise from occupancy is eliminated. A variety of standard schedules can be studied.	Not "real" occupants. Expensive apparatus required. Extra cost of keeping building unoccupied.
Nonexperimental reference	Cost of actual reference building eliminated. With simulation, weather variation is eliminated.	Database may be lacking in strata entries. Simulation errors and definition of reference problematic. With database, weather changes usually not possible.
Engineering field test	Information focused on product of interest. Minimal number of buildings required. Same occupants during test.	Extensive instrumentation of product processes required. Models required to extrapolate to other buildings and climates. Occupancy effects not determined.

Source: Partially based on Fracastoro and Lyberg (1983).

Simulated Occupancy. In some cases, the desire to reduce noise can lead the experimenter to postulate certain standard profiles for temperature set points, internal gains, moisture release, or window manipulation and to introduce this profile into the building by computer-controlled devices. The reference is often given by the test/reference design. In this case, both occupant and weather variations are nearly eliminated in the comparison (Levins and Karnitz 1986).

Nonexperimental Reference. A reference for assessing the performance of a building can be derived nonexperimentally using (1) a normalized, stratified performance database, such as energy use per unit area classified by building type (MacDonald and Wasserman 1989) or (2) a reasonable standard building, simulated by a calculated hourly or bin-method calibrated building energy performance model subject to the same weather, equipment type, and occupancy as the monitored building.

This design, also called **calibrated simulation**, is a secondary concern of current building energy monitoring documents (ASHRAE *Guideline* 14; FEMP 2008; IPMVP 2007).

Engineering Field Test. When an experiment focuses on testing a particular piece of equipment, actual performance in a building is often of interest. The building provides a realistic environment for testing the equipment for reliability, maintenance requirements, and comfort and noise levels, as well as energy usage. Energy consumption of mechanical equipment is significantly affected by the system control strategy. Testing procedures related to determining energy impacts should be designed to incorporate the control strategy of the equipment and its system (Phelan et al. 1997a, 1997b). This type of monitoring and testing can also be used to calibrate computer simulation models of as-built and as-operated buildings, which can then be used to evaluate whole-building energy consumption. The equipment may be extensively instrumented.

Some of the general advantages and disadvantages of these approaches are listed in Table 6 (Fracastoro and Lyberg 1983). Combining monitoring design choices has been successful (e.g., the before/after and test/reference approaches). Questions to be considered in choosing a monitoring approach include the following:

- Can the building alteration being investigated be turned on and off at will? The on/off design offers considerable advantages.
- Are occupancy and occupant behavior critical? Changes in building tenants, use schedules, internal gains, temperature set points, and natural or forced ventilation practices should be considered

because any one of these variables can ruin an experiment if it is not constant or accounted for.

- Are actual baseline energy performance data critical? In before/after designs, time must be allotted to characterize the before case as precisely as the after case. For instances in which heating and cooling systems are evaluated, data may be required for a wide range of anticipated ambient conditions.
- Is it a test of an individual technology, or are multiple technologies installed as a package being tested? If the effects of individual technologies are sought, detailed component data and careful model-based analyses are required.
- Does the technology have a single mode or multiple modes of operation? Can the modes be controlled to suit the experiment? If many modes are involved, it is necessary to test over a variety of conditions and conduct model-based analysis (Phelan et al. 1997a, 1997b).

Part Five: Specify Data Analysis Procedures and Algorithms

Data are useless unless they are distilled into meaningful products that allow conclusions to be drawn. Too often, data are collected and never analyzed. This planning step focuses on specifying the minimum acceptable data analysis procedures and algorithms and detailing how collected data will be processed to produce desired data products. In this step, monitoring practitioners should do the following:

- Determine the independent variables and analysis constants to be measured in the field (e.g., fan power, lighting and receptacle power, indoor air temperature).
- Develop engineering calculations and equations (algorithms) necessary to convert field data to end products. This may include use of statistical methods and simulation modeling.
- Specify detailed items, such as the frequency of data collection, the required range of independent variables to be captured in the data set, and the reasons certain data must be obtained at different intervals. For example, 15 min interval demand data are assembled into hourly data streams to match utility billing data.

Determine proper NIST-traceable calibration standards for each sensor type to be used. For details, see the references in ASHRAE *Guideline* 14 and IPMVP (2007) for specific types of sensors. However, it is often impractical to implement standards in the field. For

Table 7 Whole-Building Analysis Guidelines

Project Goal	Class of Method		
	Empirical (Billing Data)*	**Time-Integrated Model***	**Dynamic Model**
Building evaluation	Yes, but expect fluctuations in 20 to 30% range.	Yes, extra care needed beyond 15% uncertainty.	Yes, extra care needed beyond 10% uncertainty.
Building retrofit evaluation	Not generally applicable using monthly data, unless large samples are used. Requires daily data and various normalization techniques for reasonable accuracy.	Yes, but difficult beyond 15% uncertainty. Method cannot distinguish multiple retrofit effects.	Yes, can resolve 5% change with short-term tests. Can estimate multiple retrofit effects.
Component evaluation	Not applicable.	Not applicable unless submetering is done to supplement.	Yes, about 5% accuracy, but best with submetering.

Note: Error figures are approximate for total energy use in a single building. All methods improve with selection of more buildings.

*Accuracy can be improved by decreasing time step to weekly or daily. These methods are of little use when outdoor temperature approaches balance temperature.

example, maintaining the length of straight ductwork required for an airflow sensor is usually difficult, requiring compromise.

Algorithm inputs can be assumed values (e.g., energy value of a unit volume of natural gas), one-time measurements (e.g., leakage area of a house), or time-series measurements (e.g., fuel consumption and outdoor and indoor temperatures at the site). The algorithms may pertain to (1) utility level aggregates of buildings, (2) particular whole-building performance, or (3) performance of instrumented components.

Chapter 19 of the 2009 *ASHRAE Handbook—Fundamentals* contains a lengthy discussion on modeling procedures, and readers should consult this material for more information on modeling. In this chapter, the discussion is categorized differently, with a view toward procedures and issues related to field energy monitoring projects.

Table 7 provides a guide to selecting an analysis method. The error quotations are rough estimates for a single-building scenario.

Empirical Methods. Although empirical methods are the simplest, they can have large uncertainty and may generate little or no information for small sample sizes. The simplest empirical methods are based on annual consumption values, tracking annual numbers and looking for degradation. Questions about building performance relative to other buildings are based on comparing certain performance indices between the building and an appropriate reference. The ENERGY STAR tools (www.energystar.gov) are probably the best-known performance comparison tools.

For commercial buildings, the most common simple index is the **energy use intensity (EUI)**, which is annual consumption, either by fuel type or summed over all fuel types, divided by the gross floor area [see MacDonald and Wasserman (1989) for a discussion of indices]. Comparison is often made only on the basis of general building type, which can ignore potentially large variations in how much floor area is heated or cooled, climate, number of workers in a building, number and type of computers in a building, and HVAC systems. Variations can be accommodated somewhat by stratifying the database from which the reference EUI is chosen. Computer simulations are often used to set reasonable comparison values.

The Commercial Buildings Energy Consumption Survey (CBECS) database (summarized in Chapter 36) has been used to develop ENERGY STAR energy use benchmarking methods for several building types in the United States. Many of the building tools have been updated to 2003 data. Building types covered as of June 2009 include the following:

- Offices (general offices, financial centers, bank branches, and courthouses)
- K to 12 schools
- Supermarkets/grocery stores
- Hospitals (acute care and children's)
- Medical offices and clinics
- Hotels/motels
- Residence halls/dormitories

- Retail stores
- Warehouses (refrigerated and nonrefrigerated)

For some of these building types, some secondary spaces are allowed:

- Computer data centers
- Garages and parking lots
- Swimming pools

Initial work in this area covered only electricity use for office buildings (Sharp 1996), but the methods have been extended to cover all fuels for the listed building types. Results show that electricity use of office buildings is most significantly explained by the number of workers in the building, number of computers, whether the building is owner-occupied, and number of operating hours each week. Only a subset of these parameters might be used to determine a benchmark within a specific census division. ENERGY STAR documentation on current tools and the performance normalization factors can be found at www.energystar.gov/buildings.

Simple empirical methods applied to retrofit applications should include at least some periods of data on daily energy use and average daily temperature (recorded locally) to account for variations in occupancy and building schedules. Monthly EUI or billing data provide more information for empirical analysis and can be used for extended analysis of energy impacts of retrofit applications, for example, in **conditional demand analysis** (Hirst and Reed 1991). Monthly data can also be used to detect billing errors, improper equipment operation during unoccupied hours, and seasonal space condition problems (Haberl and Komor 1990a, 1990b). Daily data are often used in these analyses, and raw hourly total building consumption data, when available, provide more detailed information on occupied versus unoccupied performance. Hourly, daily, monthly, and annual EUI across buildings can be directly compared when reduced to average power per unit area (power density). To avoid false correlations, the method of analysis should have statistical significance that can be traced to realistic parameters (Haberl et al. 1996).

Model-Based Methods. These techniques allow a wide range of additional data normalization to potentially improve the accuracy of comparisons and provide estimates of cause-and-effect relations. The analyst must carefully define the system and postulate a useful form of the governing energy balance/system performance equation or system of equations. Explicit terms are retained for equipment or processes of particular interest. As part of the data analysis, whole-building data (driving forces and thermal or energy response) are used to determine the model's significant parameters. The parameters themselves can provide insight, although parameter interpretation can be difficult, particularly with time-integrated billing data methods. The model can then be used for normalization processes as well as future diagnostic and control applications. Two general classes of models are used in analysis methods: time-integrated methods and dynamic techniques (Balcomb et al. 1993). Simulation modeling results used for design of new buildings typically do not cover all energy used in a building and thus are usually difficult to compare with empirical data.

Time-Integrated Methods. Based on algebraic calculation of the building energy balance, time-integrated methods are often used before data comparison to correct annual consumption for variations in outdoor temperature, internal gains, and internal temperature (ASHRAE 2002; Fels 1986; Haberl and Claridge 1987). This type of correction is essential for most retrofit applications.

Time-integrated methods can be used with whole-building energy consumption data (billing data) or with submetered end-use data. For example, standard time-integrated methods are often used to separately integrate end-use consumption data on heating, cooling (Ternes and Wilkes 1993), domestic water heating, and others for comparison and analysis. Time-integrated methods are generally reliable, as long as the following three conditions are accounted for:

- *Appropriate time step.* Generally, the time step should be as long as or longer than the response time of the building or building system for which energy use is being integrated. For example, the response of daylighting controls to natural illumination levels can be rapid, allowing short time steps for data integration. In contrast, the response of cooling system energy use to changes in cooling load can be comparatively slow. In this instance, either a time step long enough to average over these slow variations or a dynamic model should be used. In general, an appropriate time step should account for the physical behavior of the energy system(s) and the expression of this behavior in model parameters.
- *Linearity of model results.* Generally, time-integrated models should not be applied to data used to estimate nonlinear effects. Air infiltration, for example, is nonlinear when estimated using wind speed and indoor/outdoor temperature difference data in certain models. Estimation errors result if these parameters are independently time-integrated and then used to calculate air infiltration. These nonlinear effects should be modeled at each time step (each hour, for example).
- *End-use uniformity within data set.* End-use data sets should be **uniform** (i.e., should not inadvertently contain observations with measurements of end uses other than those intended). During mild weather, for example, HVAC systems may provide both heating and cooling over the course of a day, creating data observations of both heating and cooling measurements. In a time-integrated model of heating energy use, these cooling energy observations lead to error. These observations should be identified or otherwise flagged by their true end use.

For whole-building energy consumption data (billing data), reasonable results can be expected from heating analysis models when the building is dominantly responsive to indoor/outdoor temperature differences. Billing data analysis yields little of interest when internal gains are large compared to skin loads, as in large commercial buildings and industrial applications. Daily, weekly, and monthly whole-building heating season consumption integration steps have been used (Claridge et al. 1991; Fels 1986; Sharp and MacDonald 1990; Ternes 1986). Cooling analysis results have been less reliable because cooling load is not strictly proportional to variable-base cooling degree-days (Fels 1986; Haberl et al. 1996; Kissock et al. 1992). Problems also arise when solar gains are dominant and vary by season.

Dynamic Techniques. Dynamic models, both **macrodynamic** (whole-building) or **microdynamic** (component-specific), offer great promise for reducing monitoring duration and increasing conclusion accuracy. Furthermore, individual effects from multiple measures and system interactions can be examined explicitly. Dynamic whole-building analysis is generally accompanied by detailed instrumentation of specific technologies.

Dynamic techniques create a dynamic physical model for the building, adjusting model parameters to fit experimental data (Duffy et al. 1988; Subbarao 1988). In residential applications, computer-controlled electric heaters can be used to maintain a steady interior temperature overnight, extracting from these data an experimental value for the building steady-state load coefficient. A cooldown period can also be used to extract information on internal building mass thermal storage. Daytime data can be used to renormalize the building response (computed from a microdynamic model) to solar radiation, which is particularly appropriate for buildings with glazing areas over 10% of the building floor area (Subbarao et al. 1986). Once the data with electric heaters have been taken, the building can be used as a dynamic calorimeter to assess the performance of auxiliary heating and cooling systems.

Similar techniques have been applied to commercial buildings (Burch et al. 1990; Norford et al. 1985). In these cases, delivered energy from the HVAC system must be monitored directly in lieu of using electric heaters. Because ventilation is a major variable in the building energy balance, outdoor airflow rate should also be monitored directly. Simultaneous heating and cooling (common in large buildings) requires a multizone treatment, which has not been adequately tested in any of the dynamic techniques.

Equipment-specific monitoring guidelines using dynamic modeling have been successfully tested in a variety of applications. For fans and pumps, relatively simple regression techniques from short-term monitoring provided accurate estimates of annual energy consumption when combined with an annual equipment load profile. For chillers, a thermodynamic model used with short-term monitoring captured the most important operating parameters for estimating installed annual energy performance. In all cases, the key to accurate model results was capturing a wide enough range of the independent load variable in monitored data to reflect annual operating characteristics (Phelan et al. 1997a, 1997b).

Part Six: Specify Field Data Monitoring Points

Careful specification of field monitoring points is critical to identifying variables that need to be monitored or measured in the field to produce required data.

The analysis method determines the data to be measured in the field. The simplest methods require no onsite instrumentation. As methods become more complex, data channels increase. For engineering field tests conducted with dynamic techniques, up to 100 data channels may be required.

Because metering projects are often conducted in buildings with changing conditions, special consideration must be given to identifying and monitoring significant changes in climate, systems, and operation during the monitoring period. Additional monitoring points may be required to measure variables that are assumed to be constant, insignificant, or related to other measured variables to draw sound conclusions from the measurements. Because the necessary data may be obtained in several ways, data analysts, equipment installers, and data acquisition system engineers should work together to develop tactics that best suit the project requirements. It is important to anticipate the need for supplemental measurements in response to project needs that may not become apparent until actual equipment installation occurs.

The cost of data collection is a nonlinear function of the number, accuracy, and duration of measurements that must be considered while planning within budget constraints. Costs per data point typically decrease as the number of points increases, but increase as accuracy requirements increase. Duration of monitoring can have many different effects. If the extent of data applications is unknown, such as in research projects, the value of other concurrent measurements should be considered because the incremental cost of alternative analyses may be small.

For any project involving large amounts of data, data quality verification (see Part Eight) should be automated (Lopez and Haberl

1992). Although this may require adding monitoring points to facilitate energy balances or redundancy checks, the added costs are likely to be offset by savings in data verification for large projects.

If multiple sites are to be monitored, common protocols for selecting and describing all field monitoring points should be established so data can be more readily verified, normalized, compared, and averaged. Protocols also add consistency in selecting monitoring points. Pilot installations should be conducted to provide data for a test of the system and to ensure that the necessary data points have been properly specified and described.

Monitoring Equipment. General considerations in selecting monitoring equipment include the following:

- Evaluate equipment thoroughly under actual test conditions before committing to large-scale procurement. Particular attention should be paid to any sensitivity to power outages and to protection against power surges and lightning.
- Consider local setup and testing of complex data acquisition systems that are to be installed in the field.
- Avoid unproven data acquisition equipment (Sparks et al. 1992). Untested equipment, even if donated, may not be a good value.
- Consider costs and benefits of remote data interrogation and programming.
- Evaluate quality and reliability of data loggers and instrumentation; these issues may be more important than cost, particularly when data acquisition sites are distant.
- Verify vendor claims by calling references or obtaining performance guarantees.
- Consider portable battery-powered data loggers in lieu of hardwired loggers if the monitoring budget is limited and the length of the monitoring period is less than a few months.
- Ensure that monitoring equipment and installation methods are consistent with prevailing laws, building codes, and standards of good practice.

Using a building energy management system or direct digital control systems for data acquisition may decrease costs. This should be considered only when the sensors and their accuracy and limitations (scan rate, etc.) are thoroughly understood. When merging data from two data acquisition systems, problems may arise such as differing reliability and low data resolution (e.g., 1 kW resolution of a circuit that draws 10 kW fully loaded). These problems can often be avoided, however, by adding appropriate sensors and setting up custom logging or calculations with point, memory, and programming capacity.

Once the required field data monitoring points are specified, these requirements should be clearly communicated to all members of the project team to ensure that the actual monitoring points are accurately described. This can be accomplished by publishing handbooks for measurement plan development and equipment installation and by outlining procedures for diagnostic tests and technology assessments.

Because hardware needs vary considerably by project, specific selection guidelines are not provided here. However, general characteristics of data acquisition hardware components are shown in Table 8. Some typical concerns for selecting data acquisition hardware are outlined in Table 9. In general, data logger and instrumentation hardware should be standardized, with replacements available in the event of failure. Also consider redundant measurements for critical data components that are likely to fail, such as modems, flowmeters, shunt resistors on current transformers, and devices with moving parts (O'Neal et al. 1993). Some measurements that are more difficult to obtain accurately because of instrumentation limitations are summarized in Table 10.

Safety must be considered in equipment selection and installation. Installation teams of two or more individuals reduce risks. When contemplating thermal metering, the presence of asbestos insulation on water piping should be determined. Properly licensed trades personnel, such as an electrician or welder, should be a fundamental part of any team installing electrical monitoring equipment.

To prevent inadvertent tampering, occupants and maintenance personnel should be carefully briefed on what is being done and the purpose of sensors and equipment. Data loggers should have a dedicated (non-occupant-switchable) hard-wired power supply to prevent accidental power loss.

Sensors. Sensors should be selected to obtain each measurement on the field data list. Next, conversion and proportion constants should be specified for each sensor type, and the accuracy, resolution, and repeatability of each sensor should be noted. Sensors should be calibrated before they are installed in the field, preferably with a NIST-traceable calibration procedure. They should be checked periodically for drift, recalibrated, and then postcalibrated at the conclusion of the experiment (Haberl et al. 1996). Instrument calibration is particularly important for flow and power measurement.

Particular attention must be paid to sensor location. For example, if the method requires an average indoor temperature, examine the potential for internal temperature variation; data from several temperature sensors must often be averaged. Alternatively, temperature

Table 8   General Characteristics of Data Acquisition System (DAS)

Types of DAS	Typical Use	Typical Data Retrieval	Comments
Manual readings	Total energy use	Monthly or daily written logs	Human factors may affect accuracy and reading period. Data must be manually entered for computer analysis.
Pulse counter, solid state (1, 4, or 8 channels)	Total energy use (some end use)	Polled by telephone to mainframe or minicomputer	Computer hardware and software are needed for transfer and conversion of pulse data. Can be expensive. Can handle large numbers of sites. User-friendly.
Stick-on battery powered logger (1 to 8 channels)	Diagnostics, technology assessment, end use	Monthly manual download to PC	Very useful for remote sites. Can record pulse counts, temperature, etc., up to thousands of records.
Plug-in A/D boards for PCs	Diagnostics, technology assessment, control	On-site real-time collection and storage	Usually small-quantity, unique applications. PC programming capability needed to set up data software and configure boards.
Simple field DAS (usually 16 to 32 channels)	Technology assessment, residential end use (some diagnostics)	Phone retrieval to host computer for primary storage (usually daily to weekly)	Can use PCs as hosts for data retrieval. Good A/D conversion available. Low cost per channel. Requires programming skills to set up field unit and configure communications for data transfer.
Advanced field DAS (usually >40 channels/ units)	Diagnostics, energy control systems, commercial end use	On-site real-time collection and data storage, or phone retrieval	Usually designed for single buildings. Can be PC-based or stand-alone unit. Can run applications/diagnostic programs. User-friendly.
Direct digital control or building automation system	On-site diagnostics, energy measurement and verification	Proprietary data collection procedures, manual or automated export to spreadsheet.	Requires significant coordination with building operation personnel. Sensor accuracy, calibration, and installation require confirmation. Good for projects with limited instrumentation budget.

Table 9 Practical Concerns for Selecting and Using Data Acquisition Hardware

Components	Field Application Concerns
Data logger unit and peripherals	• Select equipment for field application. Flexible or adaptable input capabilities desirable. • Equipment should store data electronically for easy transfer to a computer. • Remote programming capability should be available to minimize on-site software modifications. • Avoid equipment with cooling fans. • Use high-quality, reliable communication devices or methods. • Make sure logger/computer and communication reset after power outage.
Cabling and interconnection hardware	• Use only signal-grade cable: shielded, twisted-pair with drain wire for analog signals. • Mitigate sources of common mode and normal mode signal noise.
Sensors	• Use rugged, reliable sensors rated for field application. • Use a signal splitter if sharing existing sensors or signals with other recorders or energy management control system (EMCS). • Select ranges so sensors operate at 50 to 75% of full scale. • Choose sensors that do not require special signal conditioning or power supply, if possible. • Precalibrate sensors and recalibrate periodically. • When possible, use redundant channels to cross-check critical channels that can drift.

Table 10 Instrumentation Accuracy and Reliability

Instrument	Problems
Hygrometers	Drift, saturation, and accuracy over time; need for calibration to remove temperature dependence; aspirated systems need to be cleaned periodically. Chilled-mirror systems require frequent maintenance.
Flowmeters	Need for calibration, reliability. Moving parts prone to failure. Pipe size must be verified before calibration or installation.
Btu meters	60 Hz noise from surroundings, calibration.
Single-ended voltage	Grounding problems, spurious line voltages, 60 Hz noise.
Outdoor air temperature sensor	Must be properly shielded from solar radiation. Aspiration may reduce solar radiation effects but decrease long-term reliability.
RTD sensors	Signal wire length affects readings.
Power meters	Polarity of current transformers (CTs) often marked incorrectly, problems with shunt resistors and CT output. Devices should be checked before installation.

sensors adjacent to HVAC thermostats detect the temperature to which the HVAC equipment reacts.

Scanning and Recording Intervals. Measurement frequency and data storage can affect the accuracy of results. Scanning differs from storage in that data channels may be read (scanned) many times per second, for example, whereas average data may be recorded and stored every 15 min. Most data loggers maintain temporary storage registers, accumulating an integrated average of channel readings from each scan. The average is then recorded at the specified interval.

After the channel list is compiled and sensor accuracy requirements established, scan rates should be assigned. Some sensors, such as indoor and outdoor temperature sensors, may require low scan rates (once every 5 min). Others, such as total electric sensors,

may contain high-frequency transients that require rapid sampling (many times per second). The scan rate must be fast enough to ensure that all significant effects are monitored.

The maximum sampling rate is usually programmed into the logger, and averages are stored at a specified time step (e.g., hourly). Some loggers can scan different channels at different rates. The logger's interrupt capability can also be used for rapid, infrequent transients. Interrupt channels signal the data logger to start monitoring an event only once it begins. In some cases, online computation of derived quantities must be considered. For example, if heat flow in an air duct is required, it can be computed from a differential temperature measurement multiplied by an air mass flow rate determined from a one-time measurement. However, it should be computed and totaled only when the fan is operating.

Part Seven: Resolve Data Product Accuracies

Data collected by monitoring equipment are usually used for the following purposes:

- Direct reporting of primary measurement data
- Reporting of secondary or deduced quantities (e.g., thermal energy consumed by a building, found by multiplying mass flow rate and temperature difference
- Subsequent interpretation and analyses (e.g., to develop a statistical model of energy used by a building versus outdoor dry-bulb temperature)

In all three cases, the value of the measurements is dramatically increased if the associated uncertainty can be quantified.

Basic Concepts. Uncertainty can be better understood in terms of confidence limits, which define the range of values that can be expected to include the true value with a stated probability (ASHRAE *Guideline* 2). Thus, a statement that the 95% confidence limits are 5.1 to 8.2 implies that the true value is between 5.1 and 8.2 in 19 out of 20 predictions or, more loosely, that we are 95% confident that the true value lies between 5.1 and 8.2. For a given set of n observations with normal (Gaussian) error distribution, the total variance about the mean predicted value $\overline{X}'$ provides a direct indication of the confidence limits. Thus the "true" mean value X' of the random variable is bounded as follows:

$$\overline{X}' \pm \left(t_{\alpha/2,\,n-1} \sqrt{\sigma^2/n} \right) \qquad (1)$$

where

α = level of significance
$\overline{X}'$ = mean predicted value of random variable X
$t_{\alpha/2,n-1}$ = t-statistic with probability of $1 - \alpha/2$ and $n - 1$ degrees of freedom (tabulated in most statistical textbooks)
n = number of observations, with Gaussian error distribution
σ^2 = estimated measurement variance

The terms **accuracy** and **precision** are often used to distinguish between bias errors and random errors. A set of measurements with small bias errors is said to have high accuracy, and a set of measurements with small random errors is said to have high precision. In repeated measurements of a given sample by the same technique (single-sample data), each measurement has the same bias. Bias errors include those that are (1) known and can be calibrated out, (2) negligible and are ignored, and (3) estimated and are included in the uncertainty analysis. It is usually difficult to estimate bias limits, and this effect is often overlooked. However, a proper error analysis should include bias error, which is usually written as a plus-minus error. ASME *Standard* PTC 19.1 has a more complete discussion.

Because bias errors b_m and random errors ε_m are usually uncorrelated, measurement variance σ^2 can be expressed as

$$\sigma_{meas}^2(b_m, \varepsilon_m) = \sigma^2(b_m) + \sigma^2(\varepsilon_m) \qquad (2)$$

For further information on uncertainty, see Chapter 36 of the 2009 *ASHRAE Handbook—Fundamentals*. For more information on uncertainty calculation methods related specifically to building energy savings monitoring, see ASHRAE *Guideline* 14-2002, Annex B.

Primary Measurement Uncertainty. Sensor and measuring equipment manufacturers usually specify measurement variances; frequent recalibration minimizes bias errors. As indicated by Equation (1), increasing the number of measurements n reduces the uncertainty bounds.

Uncertainty in Derived Quantities. Once a specific algorithm or equation for obtaining final data from physical measurements has been established, standard techniques can be used to incorporate primary measurement uncertainties into the final data product. For random errors, the Kline and McClintock (1953) error propagation method, based on a first-order Taylor series expansion, is widely used to determine measurement uncertainties in derived variables in single-sample experiments. Bias errors are difficult to account for; the usual practice is to calibrate them out and exclude them from the uncertainty analysis.

Uncertainty in Statistical Regression Models. Statistical regression models developed from measured data are usually used for predictive purposes. Measurement errors are much smaller than model errors, which arise because the regression model is imperfect; that is, it is unable to explain the entire variation in the regressor variable (Box 1978). Measurement error is inherently contained in the identified model, so total prediction variance is simply given by the model prediction uncertainty.

Determining prediction errors from regression models is subject to different types of problems. The various sources of error can be classified into three categories (Reddy et al. 1998):

- Model *misspecification* errors occur because the functional form of the regression model is usually an approximation of the true driving function of the response variable.
- Model *prediction* errors occur because a model is never perfect.
- Model *extrapolation* errors occur when a model is used for prediction outside the region covered by the data from which the model was developed. Models developed from short data sets, which do not satisfactorily represent the annual behavior of the system, are subject to this error. This error cannot be quantified in statistical terms alone, but certain experimental conditions are likely to lead to accurate predictive models. This falls under the purview of experimental design (Box 1978).

Misspecification and extrapolation errors are likely to introduce bias and random error. If ordinary least-squares regression is used for parameter estimation, and if the model is subsequently used for prediction, model prediction will be purely random. Thus, models identified from short data sets and used to predict seasonal or annual energy use are affected by misspecification and extrapolation errors.

The least-squares method of calculating linear regression coefficients cannot produce unbiased estimators of slope and intercept if there are errors associated with measuring the predictor variable. The uncertainty analysis methodology developed for in situ equipment testing uses standard linear regression practices to find the functional relationship and then estimates the increased uncertainty in the regression prediction because of random and bias errors in both variable measurements (Phelan et al. 1996).

Experimental Design. Errors can also be estimated based on historical experience (e.g., using results from previous similar projects). Alternatively, a pilot study can obtain an estimate of potential errors in a proposed analysis. Some estimate of potential error must be available to determine whether project goals and objectives are reasonable.

Estimating data uncertainty is one part of the iterative procedure associated with proper experimental design. If the final data product uncertainty determined using the given evaluation procedure is unacceptable, uncertainty can be reduced in one or more of the following ways:

- Reducing overall measurement uncertainty (improving sensor precision)
- Increasing the duration of monitoring to average out stochastic variations
- Increasing the number of buildings tested (sample size)

On the other hand, if simulations or estimates indicate that the expected bias in the final data products is unacceptable, the bias may be reduced by one or more of the following steps:

- Adding sensors to get an unbiased measurement of the quantity
- Using more detailed models and analysis procedures
- Increasing data acquisition frequency, combined with a more detailed model, to address biases from sensor or system nonlinearities.

Accuracy Versus Cost. The need for accuracy must be carefully balanced against measurement and analysis costs. Accuracy loss can stem from instrumentation or measurement error, normalization or model error, sampling or statistical error, and errors of assumptions. Each of these sources can be controlled to varying degrees. However, in general, more accurate methods follow the law of diminishing returns, in which further reductions in error come at progressively greater expense.

Because of this tradeoff, the optimal measurement solution is usually found by an iterative approach, where incremental improvements in accuracy are assessed relative to the increase in measurement cost. Such optimization requires that a value be placed on increasing levels of accuracy. One method of evaluating the uncertainty of a proposed method is to calculate results using the highest

Table 11   Quality Assurance Elements

Time Frame	Hardware	Engineering Data	Characteristics Data
Initial start-up	Bench calibration (1)	Installation verification (1)	Field verification (1)
	Field calibration (1)	Collection verification (1, 2)	Completeness check (1)
	Installation verification (1)	Processing verification (1, 2)	Reasonableness check (1, 2)
		Result production (1, 2)	Result production (1, 2)
Ongoing	Functional testing (1)	Quality checking (2)	Problem diagnosis (3)
	Failure mode diagnosis (3)	Reasonableness checking (2)	Data reconstruction (4)
	Repair/maintenance (4)	Failure mode diagnosis (3)	Change control (1)
	Change control (1)	Data reconstruction (4)	
		Change control (1)	
Periodic	Preventive maintenance (1)	Summary report preparation and review (2)	Scheduled updates/resurveys (1)
	Calibration (1)		Summary report preparation and review (2)

(1) Actions to ensure good data.   (2) Actions to check data quality.   (3) Actions to diagnose problems.   (4) Actions to repair problems.

and lowest values in the confidence interval. The difference between these values can be translated into a monetary amount that is at risk. The question that must be answered is whether further measurement investment is warranted to reduce this risk.

Part Eight: Specify Verification and Quality Assurance Procedures

Establishing and using data quality assurance (QA) procedures can be very important to the success of a field monitoring project. The amount and importance of the data to be collected help determine the extent and formality of QA procedures. For most projects, the entire data path, from sensor installation to procedures that generate results for the final report, should be considered for verification tests. In addition, the data flow path should be checked routinely for failure of sensors or test equipment, as well as unexpected or unauthorized modifications to equipment.

QA often requires complex data handling. Building energy monitoring projects collect data from sensors and manipulate those data into results. Data handling in a project with only a few sensors and required readings can consist of a relatively simple data flow on paper. Computers, which are generally used in one or more stages of the process, require a different level of process documentation because much of what occurs has no direct paper trail.

Computers facilitate collection of large data sets and increase project complexity. To maximize automation, computers require development of specific software. Often, separate computers are involved in each step, so passing information from one computer to another must be automated in large projects. To move data as smoothly as possible, an automated **data pipeline** should be developed; this minimizes the delay from data collection to results production and maximizes the cost-effectiveness of the entire project.

Because collected data are valuable, data back-up procedures must also be part of QA. At a minimum, the basic data, either raw or first-level processed, should be stored in at least one and preferably two different back-up locations, apart from the main data storage location, to allow data recovery in case of hardware failure, fire, vandalism, etc. Back-ups should occur at regular intervals, probably not less than weekly for larger projects.

Automated data verification should be used when possible. Frequent data acquisition (preferably automated) with a quality control review of summarized and plotted data is essential to ensure that reliable data are collected. Verification procedures should be performed at frequent intervals (daily or weekly), depending on the importance of missing data. This minimizes data loss because of equipment failure and/or changes at a building site. It also allows processed information to be applied quickly.

The following QA actions should take place:

- Calibrate hardware and establish a good control procedure for collection of data. Use NIST-traceable calibration methods.
- Verify data, check for reasonableness, and prepare a summary report to ensure data quality after collection.
- Perform initial analysis of data. Significant findings may lead to changes in procedures for checking data quality.
- Thoroughly document and control procedures applied to remedy problems. These procedures may entail changes in hardware or collected data (such as data reconstruction), which can have a fundamental effect on the results reported.
- Archive raw data obtained from the site to ensure project integrity.

Three aspects of a monitoring project that require QA are shown in Table 11: hardware, engineering data, and characteristics data. Three QA reviews are necessary for each aspect: (1) initial QA confirms that the project starts correctly; (2) ongoing QA confirms that information collected by the project continues to satisfy quality requirements; and (3) periodic QA involves additional checks, established at the beginning of the project, to ensure continued performance at an acceptable quality level.

Information about data quality and the QA process should be readily available to data users. Otherwise, significant analytical resources may be expended to determine data quality, or the analyses may never be performed because of uncertainties.

Part Nine: Specify Recording and Data Exchange Formats

This step specifies the formats in which data are supplied to the end user or other data analysts. Both raw and processed (adjusted for missing data or anomalous readings) data formats should be specified. In addition, if supplemental analyses are planned, the medium and format to be used (type of disk, possibly magnetic tape type, spreadsheet, character encoding standard) should be specified. These requirements can be determined by analyzing the software data format specifications. Common formats for raw data are comma- and blank-delimited American Standard Code for Information Exchange (ASCII), which do not require data conversion.

Data documentation is essential for all monitoring projects, especially when several organizations are involved. Data usability is improved by specifying and adhering to data recording and exchange formats. Most data transfer problems are related to inadequate documentation. Other problems include hardware or software incompatibility, errors in electronic storage media, errors or inconsistencies in the data, and transmittal of the wrong data set. The following precautions can prevent some of these problems:

- Provide documentation to accompany the data transfer (Table 12). Because these guidelines apply to general data, models, programs, and other types of information, items listed in Table 12 may not apply to every case.
- Provide documentation of transfer media, including the computer operating system, software used to create the files, media format (e.g., ASCII, application-specific), and media characteristics (note that many CD-ROM formats exist, some of which are proprietary and not widely used, so care should be taken to use commonly available methods).

Table 12 Documentation Included with Computer Data to Be Transferred

1. Title and/or acronym

2. Contact person (name, address, phone number)

3. Description of file (file format, number of records, geographic coverage, spatial resolution, time period covered, temporal resolution, sampling methods, uncertainty/reliability)

4. Definition of data values (variable names, units, codes, missing value representation, location, method of measurement, variable derivation, variable formats)

5. Original uses of file

6. Size of file (number of records, bytes)

7. Original source (person, agency, citation)

8. Pertinent references (complete citation) on materials providing additional information

9. Appropriate reference citation for the file

10. Credit line (for use in acknowledgments)

11. Restrictions on use of data/program

12. Disclaimer, such as

- Unverified data; use at your own risk.

- Draft data; use with caution.

- Clean data to the best of our knowledge. Please let us know of any possible errors or questionable values.

- Program under development.

- Program tested under the following conditions (conditions specified by author).

- Provide procedures to check the accuracy and completeness of data transfer, including statistics or frequency counts for variables and hard-copy versions of the file. Test input data and corresponding output results for models on other programs.
- Keep all raw data, including erroneous records.
- Convert and correct data; save routines for later use.
- Limit equipment access to authorized individuals.
- Check incoming data soon after they are collected, using simple time-series and x-y inspection plots.
- Automate as many routines as possible to avoid operator error.

REFERENCES

Armstrong, P.R., D.L. Hadley, R.D. Stenner, and M.C. Janus. 2001. Whole-building airflow network characterization by a many-pressure-states (MPS) technique. *ASHRAE Transactions* 107(2):645-657.

ASHRAE. 2005. Engineering analysis of experimental data. ASHRAE *Guideline* 2-2005.

ASHRAE. 2002. Measurement of energy and demand savings. ASHRAE *Guideline* 14-2002.

ASME. 1998. Measurement uncertainty: Instruments and apparatus. ANSI/ASME *Standard* PTC 19.1-98 (R2004). American Society of Mechanical Engineers, New York.

ASTM. 2003. Test method for determining air leakage rate by fan pressurization. *Standard* E779-03. American Society for Testing and Materials, West Conshohocken, PA.

Balcomb, J.D., J.D. Burch, and K. Subbarao. 1993. Short-term energy monitoring of residences. *ASHRAE Transactions* 99(2):935-944.

Box, G.E.P. 1978. *Statistics for experimenters: An introduction to design, data analysis and model-building.* John Wiley & Sons, New York.

Burch, J.D., K. Subbarao, A. Lekov, M. Warren, L. Norford, and M. Krarti. 1990. Short-term energy monitoring in a large commercial building. *ASHRAE Transactions* 96(1):1459-1477.

Claridge, D.E., J.S. Haberl, W.D. Turner, D.L. O'Neal, W.M. Heffington, C. Tombari, M. Roberts, and S. Jaeger. 1991. Improving energy conservation retrofits with measured savings. *ASHRAE Journal* 33(10):14-22.

Cohen, R.R., P.W. O'Callaghan, S.D. Probert, N.M. Gibson, D.J. Nevrala, and G.F. Wright. 1987. Energy storage in a central heating system: Spa school field trail. *Building Service Engineering Research Technology* 8:79-84.

DeCicco, J., R. Diamond, S.L. Nolden, J. DeBarros, and T. Wilson. 1995. Improving energy efficiency in apartment buildings. *Proceedings ACEEE*, Washington, D.C., pp. 234-236.

Duffy, J.J., D. Saunders, and J. Spears. 1988. Low-cost method for evaluation of space heating efficiency of existing homes. *Proceedings of the 12th Passive Solar Conference*, ASES, Boulder, CO.

Fels, M.F., ed. 1986. Measuring energy savings: The scorekeeping approach. *Energy and Buildings*, vol. 9.

FEMP. 2008. *M&V guidelines: Measurement and verification for federal energy projects.* Federal Energy Management Program, U.S. Department of Energy, Washington D.C. http://www1.eere.energy.gov/femp/pdfs/mv_guidelines.pdf.

Fracastoro, G.V. and M.D. Lyberg. 1983. *Guiding principles concerning design of experiments, instruments, instrumentation, and measuring techniques.* Swedish Council for Building Research, Stockholm. Available from www.ecbcs.org/docs/annex_03_guiding_principles.pdf.

Haberl, J.S. and D.E. Claridge. 1987. An expert system for building energy consumption analysis: Prototype results. *ASHRAE Transactions* 93(1):979-998.

Haberl, J.S. and P.S. Komor. 1990a. Improving energy audits: How annual and monthly consumption data can help. *ASHRAE Journal* 32(8):26-33.

Haberl, J.S. and P.S. Komor. 1990b. Improving energy audits: How daily and hourly data can help. *ASHRAE Journal* 32(9):26-36.

Haberl, J., D.E. Claridge, and D. Harrje. 1990. The design of field experiments and demonstration. *Proceedings of the IEA Field Monitoring for a Purpose Workshop* (Chalmers University, Gothenburg, Sweden, April), pp. 33-58 (Available from Energy Systems Lab, Texas A&M University).

Haberl, J.A., Reddy, D. Claridge, D. Turner, D. O'Neal, and W. Heffington. 1996. Measuring energy-saving retrofits: Experiences from the Texas LoanSTAR Program. ORNL *Report* ORNL/Sub/93-SP090-1. Oak Ridge National Laboratory, Oak Ridge, TN.

Hirst, E. and J. Reed, eds. 1991. Handbook of evaluation of utility DSM programs. ORNL *Report* ORNL/CON-336, Oak Ridge National Laboratory, Oak Ridge, TN.

Hirst, E., D. White, and R. Goeltz. 1983. Comparison of actual electricity savings with audit predictions in the BPA residential weatherization pilot program. ORNL *Report* ORNL/CON-142. Oak Ridge National Laboratory, Oak Ridge, TN.

Hough, R.E., P.J. Hughes, R.J. Hackner, and W.E. Clark. 1987. Results-oriented methodology for monitoring HVAC equipment in the field. *ASHRAE Transactions* 93(1):1569-1579.

Hughes, P.J. and J.A. Shonder. 1998. The evaluation of a 4000-home geothermal heat pump retrofit at Fort Polk, Louisiana: Final report. ORNL *Report* ORNL/CON-460. Oak Ridge National Laboratory, Oak Ridge, TN.

IPMVP. 2006. *International performance measurement and verification protocol*, vol. III: *Concepts and practices for determining energy savings in new construction.* EVO 30000-1.2006. Efficiency Valuation Organization, Washington, D.C. Available from www.evo-world.org.

IPMVP. 2007. *International performance measurement and verification protocol*, vol. I, *Concepts and options for determining energy and water savings.* EVO 10000-1.2007. Efficiency Valuation Organization, Washington, D.C. Available from www.evo-world.org.

Kissock, J.K., D.E. Claridge, J.S. Haberl, and T.A. Reddy. 1992. Measuring retrofit savings for the Texas LoanSTAR program: Preliminary methodology and results. *Solar Engineering 1992—Proceedings of the 1992 ASME-JSES-KSES International Solar Engineering Conference*, Maui, pp. 299-308.

Kline, S. and F.A. McClintock, 1953. Describing uncertainties in single-sample experiments, *Mechanical Engineering* 75:2-8.

Levins, W.P. and M.A. Karnitz. 1986. Cooling-energy measurements of unoccupied single-family houses with attics containing radiant barriers. ORNL *Report* ORNL/CON-200. Oak Ridge National Laboratory, Oak Ridge, TN (See also ORNL *Report* ORNL/CON-213).

Liu, M., J. Houcek, A. Athar, A. Reddy, D. Claridge, and J. Haberl. 1994. Identifying and implementing improved operation and maintenance measures in Texas LoanSTAR buildings. *Proceedings of the 1994 ACEEE Summer Study on Energy Efficiency in Buildings* 5:153.

Lopez, R. and J.S. Haberl. 1992. Data processing routines for monitored building energy data. *Solar Engineering 1992—Proceedings of the 1992 ASME-JSES-KSES International Solar Engineering Conference*, Maui, pp. 329-336.

MacDonald, J.M. and D.M. Wasserman. 1989. Investigation of metered data analysis methods for commercial and related buildings. ORNL *Report* ORNL/CON-279. Oak Ridge National Laboratory, Oak Ridge, TN. Available from eber.ed.ornl.gov/commercialproducts/CON-279.htm.

MacDonald, J.M., T.R. Sharp, and M.B. Gettings. 1989. A protocol for monitoring energy efficiency improvements in commercial and related buildings. ORNL *Report* ORNL/CON-291. Oak Ridge National Laboratory, Oak Ridge, TN.

Misuriello, H. 1987. A uniform procedure for the development and dissemination of monitoring protocols. *ASHRAE Transactions* 93(1):1619-1629.

Modera, M.P. 1989. Residential duct system leakage: Magnitude, impacts, and potential for reduction. *ASHRAE Transactions* 95(2):561-569.

Norford, L.K, A. Rabl, and R.H. Socolow. 1985. Measurement of thermal characteristics of office buildings. ASHRAE Conference on the Thermal Performance of Building Envelopes III, Clearwater Beach, FL.

O'Neal, D.L., J. Bryant, C. Boecker, and C. Bohmer. 1993. *Instrumenting buildings to determine retrofit savings: Murphy's law revisited.* ESL-PA93/03-02. Energy Systems Lab, Texas A&M University.

Persily, A.K., R. Grot, J.B. Fan, and Y.M. Chang. 1988. Diagnostic techniques for evaluating office building envelopes. *ASHRAE Transactions* 94(1):987-1006.

Phelan, J., M. Brandemuehl, and M. Krarti. 1996. *Final report: Methodology development to measure in-situ chiller, fan and pump performance.* JCEM TR/96/3. Joint Center for Energy Management, University of Colorado, Boulder.

Phelan, J., M. Brandemuehl, and M. Krarti. 1997a. In-situ performance testing of chillers for energy analysis. *ASHRAE Transactions* 103(1):290-302.

Phelan, J., M. Brandemuehl, and M. Krarti. 1997b. In-situ performance testing of fans and pumps for energy analysis. *ASHRAE Transactions* 103(1):318-332.

Reddy, T.A., J.K. Kissock, and D.K. Ruch. 1998. Regression modelling in determination of retrofit savings. *ASME Journal of Solar Energy Engineering* 120:185.

Robison, D.H. and L.A. Lambert. 1989. Field investigation of residential infiltration and heating duct leakage. *ASHRAE Transactions* 95(2): 542-550.

Sellers, D., H. Friedman, L. Luskay, and T. Haasl. 2004. Commissioning and envelope leakage: Using HVAC operating strategies to meet design and construction challenges. *Proceedings of the 2004 ACEEE Summer Study on Energy Efficiency in Buildings*, pp. 3.287-3.299. http://www.eceee.org/conference_proceedings/ACEEE_buildings/2004/Panel_3/p3_24/Paper/&rct=j&q=sellers commissioning envelope leakage.

Sharp, T. 1996. Energy benchmarking in commercial office buildings. *Proceedings of the 1996 ACEEE Summer Study on Energy Efficiency in Buildings*, pp. 4.321-4.329. http://www.energy.ca.gov/greenbuilding/documents/background/13-ORNL_COM_BLDG_BENCHMARKING.PDF.

Sharp, T.R. and J.M. MacDonald. 1990. Effective, low-cost HVAC controls upgrade in a small bank building. *ASHRAE Transactions* 96(1):1011-1017.

Sparks, R., J.S. Haberl, S. Bhattacharyya, M. Rayaprolu, J. Wang, and S. Vadlamani. 1992. Use of simplified system models to measure retrofit energy savings. *Solar Engineering 1992—Proceedings of the 1992 ASME-JSES-KSES International Solar Engineering Conference*, Maui, pp. 325-328.

Subbarao, K. 1988. *PSTAR—A unified approach to building energy simulations and short-term monitoring*. SERI/TR-254-3175. Solar Energy Research Institute, Golden, CO.

Subbarao, K., J.D. Burch, and H. Jeon. 1986. *Building as a dynamic calorimeter: Determination of heating system efficiency*. SERI/TR-254-2947. Solar Energy Research Institute, Golden, CO.

Szydlowski, R.F. and R.C. Diamond. 1989. *Data specification protocol for multifamily buildings*. LBL-27206. Lawrence Berkeley Laboratory, Berkeley.

Ternes, M.P. 1986. *Single-family building retrofit performance monitoring protocol: Data specification guideline*. ORNL *Report* ORNL/CON-196. Oak Ridge National Laboratory, Oak Ridge, TN.

Ternes, M.P. and K.E. Wilkes. 1993. Air-conditioning electricity savings and demand reductions from exterior masonry wall insulation applied to Arizona residences. *ASHRAE Transactions* 99(2):843-854.

Woller, B.E. 1989. Data acquisition and analysis of residential HVAC alternatives. *ASHRAE Transactions* 95(1):679-686.

BIBLIOGRAPHY

ASHRAE/CIBSE/USGBC. 2010. *Performance measurement protocols for commercial buildings*. ASHRAE, in collaboration with the Chartered Institute of Building Engineers, London, U.K., and U.S. Green Building Council, Washington, D.C.

ASTM. 2005. Guide for developing energy monitoring protocols for commercial or institutional buildings or facilities. ASTM *Standard* E1464-92 (RA2005).

Atif, M.R., J.A. Love, and P. Littlefair. 1997. Daylighting monitoring protocols & procedures for buildings. National Research Council Canada *Report* NRCC 41369. Available from nrc.ca/irc/fulltext/nrcc41369/nrcc41369.pdf.

EIA. 1998. *A look at commercial buildings in 1995: Characteristics, energy consumption, and energy expenditures*. U.S. Department of Energy, Energy Information Administration. DOE/EIA-0625(95).

IPMVP. 2002. *International performance measurement and verification protocol*, vol. II, *Concepts and practices for improved indoor environmental quality*. Available from www.evo-world.org.

MacDonald, J.M. 2004. Commercial sector and energy use. In *Encyclopedia of energy*. Elsevier.

CHAPTER 42

SUPERVISORY CONTROL STRATEGIES AND OPTIMIZATION

COMPUTERIZED energy management and control systems provide an excellent means of reducing utility costs associated with maintaining environmental conditions in commercial buildings. These systems can incorporate advanced control strategies that respond to changing weather, building conditions, and utility rates to minimize operating costs.

HVAC systems are typically controlled using a two-level control structure. Lower-level **local-loop control** of a single set point is provided by an actuator. For example, the supply air temperature from a cooling coil is controlled by adjusting the opening of a valve that provides chilled water to the coil. The upper control level, **supervisory control**, specifies set points and other time-dependent modes of operation.

The performance of large, commercial HVAC systems can be improved through better local-loop and supervisory control. Proper tuning of local-loop controllers can enhance comfort, reduce energy use, and increase component life. Set points and operating modes for cooling plant equipment can be adjusted by the supervisor to maximize overall operating efficiency. Dynamic control strategies for ice or chilled-water storage systems can significantly reduce on-peak electrical energy and demand costs to minimize total utility costs. Similarly, thermal storage inherent in a building's structure can be dynamically controlled to minimize utility costs. In general, strategies that take advantage of thermal storage work best when forecasts of future energy requirements are available.

This chapter focuses on the opportunities and control strategies associated with using computerized control for centralized cooling and heating systems. The chapter is divided into three major sections: (1) background information on the effects of and opportunities for adjusting control variables, terminology, and descriptions of the systems and control variables considered in this chapter; (2) basic optimization methods used throughout in the application sections of the chapter (intended for researchers and developers of advanced control strategies); and (3) applications of both static and dynamic optimization methods (e.g., cooling plants with and without storage) as well as supervisory control strategies that can be implemented in computerized control systems, intended for practitioners.

TERMINOLOGY

Air distribution system: includes terminal units [variable-air-volume (VAV) boxes, etc.], air-handling units (AHUs), ducts, and controls. In each AHU, ventilation air is mixed with return air from the zones and fed to the cooling/heating coil.

The preparation of this chapter is assigned to TC 7.5, Smart Building Systems.

Air-side economizer control: used to select between minimum and maximum ventilation air, depending on the condition of the outdoor air relative to the conditions of the return air. Under certain outdoor air conditions, AHU dampers may be modulated to provide a mixed air condition that can satisfy the cooling load without the need for mechanical cooling.

Building thermal mass storage: storing energy in the form of sensible heat in building materials, interior equipment, and furnishings.

CAV systems: air-handling systems that have fixed-speed fans and provide no feedback control of airflow to the zones. Zone temperature is controlled to a set point using a feedback controller that regulates the amount of local reheat applied to the air entering each zone.

Charging: storing cooling capacity by removing heat from a cool storage device; or storing heating capacity by adding heat to a heat storage device.

Chilled-/hot-water/steam loop: consists of pumps, pipes, valves, and controls. Two different types of pumping systems are considered in this chapter: primary and primary/secondary. With a **primary pumping** system, a single piping loop is used and water that flows through the chiller or boiler also flows through the cooling or heating coils. When steam is used, a steam piping loop sends steam to a hot-water converter, returning hot condensate back to the boiler. Another piping loop then carries hot water through the heating coils. Often, fixed-speed pumps are used with their control dedicated to chiller or boiler control. **Dedicated control** means that each pump is cycled on and off with the chiller or boiler that it serves. Systems with fixed-speed pumps and two-way cooling or heating coil valves often incorporate a **water bypass valve** to maintain relatively constant flow rates and reduce system pressure drop and pumping costs at low loads. The valve is typically controlled to maintain a fixed pressure difference between the main supply and return lines. This set point is termed the **chilled-** or **hot-water loop differential pressure**. Sometimes, primary systems use one or more variable-speed pumps to further reduce pumping costs at low loads. In this case, water bypass is not used and pumps are controlled directly to maintain a water loop differential pressure set point.

Chiller or boiler plant: one or more chillers or boilers, typically arranged in parallel with dedicated pumps, provide the primary source of cooling or heating for the system. Individual feedback controllers adjust the capacity of each chiller or boiler to maintain a specified supply water temperature (or steam header pressure for steam boilers). Additional control variables include the number of chillers or boilers operating and the relative loading for each. For a given total cooling or heating requirement, individual chiller or boiler loads can be controlled by using different water supply set points for constant individual flow or by adjusting individual flows for identical set points.

Chiller priority: control strategy for partial storage systems that uses the chiller to directly meet as much of the load as possible, normally by operating at full capacity most of the time. Thermal storage is used to supplement chiller operation only when the load exceeds the chiller capacity.

Condenser water loop: consists of cooling towers, pumps, piping, and controls. Cooling towers reject heat to the environment through heat transfer and possibly evaporation (for wet towers) to the ambient air. Typically, large towers incorporate multiple cells sharing a common sump, with individual fans having two or more speed settings. Often, a feedback controller adjusts tower fan speeds to maintain a temperature set point for water leaving the cooling tower, termed the **condenser water supply temperature.** Typically, condenser water pumps are dedicated to individual chillers (i.e., each pump is cycled on and off with the chiller it serves).

Demand limiting: a partial storage operating strategy that limits the capacity of the cooling system during the on-peak period. The cooling system capacity may be limited based on its cooling capacity, its electric demand, or the facility demand.

Discharge capacity: maximum rate at which cooling can be supplied from a cool storage device.

Discharging: using stored cooling capacity by adding thermal energy to a cool storage device or removing thermal energy from a heat storage device.

Ice storage: types of ice storage systems include the following. An **ice harvester** is a machine that cyclically forms a layer of ice on a smooth cooling surface, utilizing the refrigerant inside the heat exchanger, then delivers it to a storage container by heating the surface of the cooling plate, normally by reversing the refrigeration process and delivering hot gases inside the heat exchanger. In **ice-on-coil-external melt**, tubes or pipes (coil) are immersed in water and ice is formed on the outside of the tubes or pipes by circulating colder secondary medium or refrigerant inside the tubing or pipes, and is melted externally by circulation the unfrozen water outside the tubes or pipes to the load. **Ice-on-coil-internal melt** is similar, except the ice is melted internally by circulating the same secondary coolant or refrigerant to the load.

Load leveling: a partial storage sizing strategy that minimizes storage equipment size and storage capacity. The system operates with the refrigeration equipment running at full capacity for 24 h to meet the normal cooling minimum load profile and, when the load is less than the chiller output, the excess cooling is stored. When the load exceeds the chiller capacity, the additional cooling requirement is obtained from the thermal storage system.

Load profile: compilation of instantaneous thermal loads over a period of time, normally 24 hours.

Nominal chiller capacity: (1) chiller capacity at standard ARI (*Standard* 550/590) rating conditions, or (2) chiller capacity at a given operating condition selected for the purpose of quick chiller sizing selections.

Partial storage: a cool storage sizing strategy in which only a portion of the on-peak cooling load is met from thermal storage, with the remainder of the load being met by operating the chilling equipment.

Precooling: a thermal energy storage (TES) strategy that allows a properly designed chiller system to operate more efficiently with lower condensing temperatures (low night ambient temperature) higher evaporating temperatures (60 to 69°F chilled water) and at or near its most efficient part-load point. Precooling can be applied to the conditioned space, or directly to mass by passing chilled air or water through building elements such as concrete floor decks.

Primary/secondary chilled- or hot-water systems: systems designed specifically for variable-speed pumping. In the primary loop, fixed-speed pumps provide a relatively constant flow of water to the chillers or boilers. This design ensures good chiller or boiler performance and, for chilled-water systems, reduces the risk of freezing on evaporator tubes. The secondary loop incorporates one or more variable-speed pumps that are controlled to maintain a water loop differential pressure set point. The primary and secondary loops may be separated by a heat exchanger. However, it is more common to use direct coupling with a common pipe.

Storage capacity: the maximum amount of cooling (or heating) that can be achieved by the stored medium in the thermal storage device. **Nominal storage capacity** is a theoretical capacity of the thermal storage device. In many cases, this may be greater than the usable storage capacity. This measure should not be used to compare usable capacities of alternative storage systems.

Storage cycle: a period (usually one day) in which a substantial charge and discharge of a thermal storage device has occurred, beginning and ending at the same state or same time of day.

Storage efficiency: the ratio of useful heating or cooling extracted during the discharge cycle to that imparted to storage during the charging cycle. One may also define an exergic efficiency that accounts for mixing and thermal destratification as well as conduction losses.

Storage inventory: the amount of usable heating or cooling capacity remaining in a thermal storage device.

Storage priority: a control strategy that uses stored cooling to meet as much of the load as possible. Chillers are operated only if the load exceeds the storage system's available cooling capacity.

Supply air temperature: temperature of air leaving an AHU or package unit. The temperature supplied to the zones may differ from the supply air temperature due to heat transfer in the ductwork and local reheat. A local-loop controller adjusts the flow of water through the air handler cooling/heating coil using a two- or three-way valve to maintain a specified set-point temperature for the air downstream of the cooling/heating coil. The value of the supply air temperature set-point will affect energy consumption and can be adjusted by the supervisory control system.

System COP: ratio of cooling rate ($kW_{thermal}$) to system input power ($kW_{electric}$) required to operate all distribution fans and pumps as well as compressors and condenser or cooling tower fans and pumps.

Temperature set point: zone temperature set points are typically fixed values within the comfort zone during the occupied time, and zone humidity is allowed to "float" within a range dictated by the system design and choice of the supply air set-point temperature. **Night setup** is often used in summer to raise the zone temperature set points during unoccupied times and reduce the cooling requirements. Similarly, **night setback** is often used in winter to lower the zone temperature set points during unoccupied times and reduce the heating requirements. Setup and setbacks can also be used during occupied periods to temporarily curtail energy consumption.

Thermal energy storage (TES): thermal energy storage systems are of three general types. **Discrete (or active) TES** uses a tank of water (usually stratified), packed bed of rock or phase-change material, or a tank of ice and water with immersed coil. **Intrinsic (or passive) TES** uses the thermal capacitance of the building fabric, such as concrete columns, beams and decks, or wall board, possibly augmented by phase-change material. **Thermally active building systems (TABS)** cool floors or other elements directly with embedded pipes or ducts, giving storage capacity and efficiency substantially greater than can achieved by passive TES.

System capacity: maximum amount of cooling that can be supplied by the entire cooling system, which may include the chillers and the thermal storage. **Usable storage capacity** is the total amount of beneficial cooling able to be discharged from a thermal storage device. (This may be less than the nominal storage capacity, because the distribution header piping may not allow discharging the entire cooling capacity of the thermal storage device.)

Total cooling load: The integrated thermal load that must be met by the cooling plant over a given period of time.

VAV systems have a feedback controller that regulates the airflow to each zone to maintain zone temperature set point. The zone airflows are regulated using dampers located in VAV boxes in each zone. VAV systems also incorporate feedback control of the primary airflow through various means of fan capacity control. Typically, inputs to a fan outlet damper, inlet vanes, blade pitch, or variable speed motor are adjusted to maintain a **duct static pressure** set point in the supply duct, as described in Chapter 47.

CONTROL VARIABLES

Systems and Controls

Figures 1 and 2 show schematics of typical centralized cooling and heating systems for which control strategies are presented in this chapter. For cooling systems, the strategies generally assume that the equipment is electrically driven and that heat is rejected to the environment by cooling towers. For heating systems, boilers may be fired by a variety of fuels or powered by electricity, but are typically fired from either natural gas or #2 or #6 fuel oil. However, some strategies apply to any type of system (e.g., return from night setup or setback). For describing different systems and controls, it is useful to divide the system into the subsystems depicted in Figures 1 and 2: air distribution system, chilled-/hot-water loop, chiller/boiler plant, condenser water loop. A variant of the chilled-water system shown in Figure 1 uses chilled water directly for sensible cooling by radiant panels, chilled beams, or radiant floors.

Control of a VAV cooling system (Figure 1) responds to changes in building cooling requirements. As the cooling load increases, the zone temperature rises as energy gains to the zone air increase. The zone controller responds to higher temperatures by increasing local

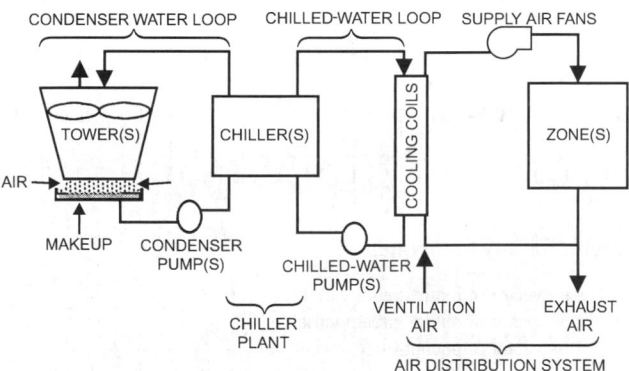

Fig. 1   Schematic of Chilled-Water Cooling System

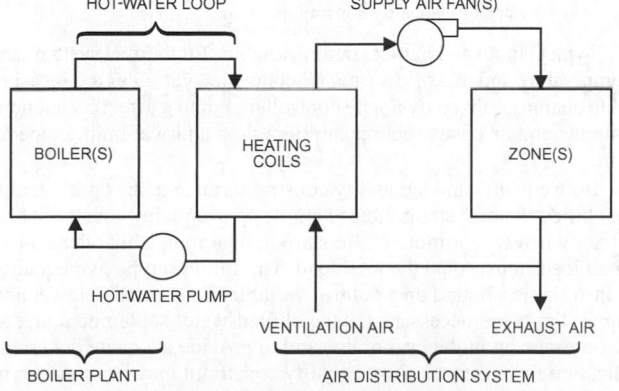

Fig. 2   Schematic of Hot-Water Heating System

flow of cool air by opening a damper. Opening a damper reduces static pressure in the primary supply duct, which causes the fan controller to create additional airflow. With greater airflow, the supply air temperature of the cooling coils increases, which causes the air handler feedback controller to increase the water flow by opening the cooling coil valves. This increases the chilled-water flow and heat transfer to the chilled water (i.e., the cooling load).

Control of a radiant cooling system (a variant of Figure 1) responds directly to zone load by increasing the chilled-water flow rate. The chilled-water temperature may be controlled by the neediest zone or by an open-loop control.

Control of a hot-water heating system (Figure 2) is similar. As the heating load increases, the zone temperature falls as energy gains to the zone air decrease. The zone controller responds to lower temperatures by opening a control valve and increasing the flow of hot water through the local reheat coil. The supply airflow rate is usually maintained at its minimum value when a VAV system is in heating mode. Increasing water flow through the reheat coils reduces the temperature of the water returned to the boiler. With lower return water temperature, the supply water temperature drops, which causes the feedback controller to increase the boiler firing rate to maintain the desired supply water temperature.

For both heating and cooling, an increase in the building load results in an increase in water flow rate, which is ultimately propagated through the central system. For fixed-speed chilled- or hot-water pumps, the differential pressure controller closes the chilled- or hot-water bypass valve and keeps the overall flow relatively constant. For variable-speed pumping, the differential pressure controller increases pump speed. In a chilled-water system, the return water temperature and/or flow rate to the chillers increases, leading to an increase in the chilled-water supply temperature. The chiller controller responds by increasing the chiller cooling capacity to maintain the chilled-water supply set point (and match the cooling coil loads). The increased energy removed by the chiller increases the heat rejected to the condenser water loop, which increases the temperature of water leaving the condenser. The increased water temperature entering the cooling tower increases the water temperature leaving the tower. The tower controller responds to the higher condenser water supply temperature and increases the tower airflow. At some load, the current set of operating chillers is not sufficient to meet the load (i.e., maintain the chilled-water supply set points) and an additional chiller is brought online. For a hot-water system, the return water temperature and/or flow rate to the boilers decreases, leading to a decrease in the hot-water supply temperature. The boiler controller responds by increasing the boiler heating capacity to maintain the hot-water supply set point (and match the heating coil loads).

For all-electric cooling without thermal storage, minimizing power at each point in time is equivalent to minimizing energy costs. Therefore, supervisory control variables should be chosen to maximize the **coefficient of performance (COP)** of the system at all times while meeting the building load requirements. The COP is defined as the ratio of total cooling load to total system power consumption. In addition to the control variables, the COP depends primarily on the cooling load and the ambient wet- and dry-bulb temperatures. Often, the cooling load is expressed in a dimensionless form as a part-load ratio (PLR), which is the cooling load under a given condition divided by the design cooling capacity.

For cooling or heating systems with thermal storage, performance depends on the time history of charging and discharging. In this case, controls should minimize operating costs integrated over the billing period or storage cycle. In addition, safety features that minimize the risk of prematurely depleting storage capacity may be important.

For any of these scenarios, several local-loop controllers respond to load change to maintain specified set points. A supervisory controller establishes modes of operation and chooses (or resets) values of set points. At any given time, cooling or heating needs can be met with various combinations of modes of operation and set points. This chapter discusses several methods for determining supervisory control variables that provide good overall performance.

Sampling Intervals for Reset Controls

Proper sampling intervals are required when resetting the set point for proportional-integral (PI) feedback control loops to prevent oscillation of the process variable in those loops. In general, the sampling time interval between reset commands should be greater than the settling time for the loop. For example, resetting both the chilled-water supply temperature set point and cooling coil discharge air temperature set point is necessary for optimal control. In this case, resetting the set points for chilled-water supply temperature and cooling coil discharge air temperature should not occur simultaneously, but at staggered intervals; the interval of reset for either loop should be greater than the settling time for the coil (the time for the discharge air temperature of the coil to reach a new steady-state temperature).

For cascaded loops, the sampling interval between reset commands should be greater than the settling time for the slower loop. An example of a cascaded loop is a VAV box controller with its flow set point determined from the space temperature of its associated zone, and the box damper controlled to maintain the flow set point. For example, in resetting static pressure set point on a variable-speed air-handler fan based on VAV box damper position, the interval between resets should be greater than the settling time of the flow control loop in a pressure-independent VAV box. (Settling time in this example is the time for the flow rate to reach a new steady-state value.)

OPTIMIZATION METHODS

STATIC OPTIMIZATION

Optimal supervisory control of cooling equipment involves determining the control that minimizes the total operating cost. For an all-electric system without significant storage, cost optimization leads to minimization of power at each instant in time. Optimal control depends on time, albeit indirectly through changing cooling requirements and ambient conditions. Static optimization techniques applied to a general simulation can be used to determine optimal supervisory control variables. The simulation may be based on physical (Hiller and Glicksman 1977; Stoecker 1980) or empirical models. However, for control variable optimization, empirical and semiempirical models (where the model parameters are estimated from measurements) are often used. This section presents a framework for determining optimal control and a simplified approach for estimating control laws for cooling plants.

General Static Optimization Problem

Figure 3 depicts the general nature of the static optimization problem for a system of interconnected components. Each component in a system is represented as a separate set of mathematical relationships organized into a computer model. Its output variables and operating cost are functions of parameter, input, output, uncontrolled, and controlled variables. The structure of the complete set of equations to be solved for the entire system is dictated by the manner in which the components are interconnected.

The problem is formally stated as the minimization of the sum of the operating costs of each component J_i with respect to all discrete and continuous controls, or

Minimize

$$J(\mathbf{f}, \mathbf{M}, \mathbf{u}) = \sum_{i=1}^{n} J_i(\mathbf{x}_i, \mathbf{y}_i, \mathbf{f}_i, \mathbf{M}_i, \mathbf{u}_i) \qquad (1)$$

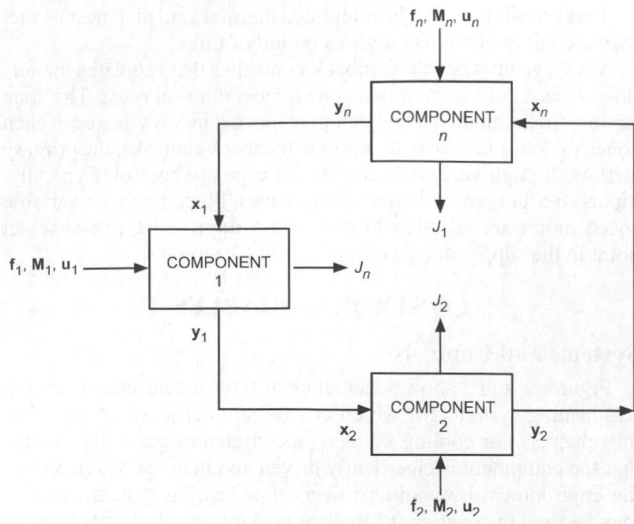

Fig. 3 Schematic of Modular Optimization Problem

with respect to $\mathbf{M}$ and $\mathbf{u}$, subject to equality constraints of the form

$$\mathbf{g}(\mathbf{f}, \mathbf{M}, \mathbf{u}) = \begin{bmatrix} \mathbf{g}_1(\mathbf{f}_1, \mathbf{M}_1, \mathbf{u}_1, \mathbf{x}_1, \mathbf{y}_1) \\ \mathbf{g}_2(\mathbf{f}_2, \mathbf{M}_2, \mathbf{u}_2, \mathbf{x}_2, \mathbf{y}_2) \\ \cdot \\ \mathbf{g}_n(\mathbf{f}_n, \mathbf{M}_n, \mathbf{u}_n, \mathbf{x}_n, \mathbf{y}_n) \end{bmatrix} = 0 \qquad (2)$$

and inequality constraints of the form

$$\mathbf{h}(\mathbf{f}, \mathbf{M}, \mathbf{u}) = \begin{bmatrix} h_1(\mathbf{f}_1, \mathbf{M}_1, \mathbf{u}_1, \mathbf{x}_1, \mathbf{y}_1) \\ h_2(\mathbf{f}_2, \mathbf{M}_2, \mathbf{u}_2, \mathbf{x}_2, \mathbf{y}_2) \\ \cdot \\ h_n(\mathbf{f}_n, \mathbf{M}_n, \mathbf{u}_n, \mathbf{x}_n, \mathbf{y}_n) \end{bmatrix} \geq 0 \qquad (3)$$

where, for any component i,

$\mathbf{x}_i$ = vector of input stream variables
$\mathbf{y}_i$ = vector of output stream variables
$\mathbf{f}_i$ = vector of uncontrolled variables
$\mathbf{M}_i$ = vector of discrete control variables
$\mathbf{u}_i$ = vector of continuous control variables
J_i = operating cost
$\mathbf{g}_i$ = vector of equality constraints
h_i = vector of inequality constraints

Typical input and output stream variables for thermal systems are temperature and mass flow rate. Uncontrolled variables are measurable quantities that may not be controlled, but that affect component outputs and/or costs, such as ambient dry- and wet-bulb temperature.

Both equality and inequality constraints arise in the optimization of chilled-water systems. One example of an equality constraint that arises when two or more chillers are in operation is that the sum of their loads must equal the total load. The simplest type of inequality constraint is a bound on a control variable. For example, lower and upper limits are necessary for the chilled-water set temperature, to avoid freezing in the evaporator and to provide adequate dehumidification for the zones. Any equality constraint may be rewritten in the form of Equation (2) such that when it is satisfied, the constraint equation is equal to zero. Similarly, inequality constraints may be expressed as Equation (3), so that the constraint equation is greater than or equal to zero to avoid violation.

Braun (1988) and Braun et al. (1989a) presented a component-based nonlinear optimization and simulation tool and used it to investigate optimal performance. Each component is represented as a separate subroutine with its own parameters, controls, inputs, and outputs. The optimization problem is solved efficiently by using second-order representations for costs that arise from curve-fits or Taylor-series approximations. Application of the component-based optimization led to many guidelines for control and a simplified system-based optimization methodology. In particular, the results showed that optimal set points could be correlated as a linear function of load and ambient wet-bulb temperature.

Zakula et al. (2010) show that, for chillers with high part-load efficiency over a wide range of compressor speed, optimal condenser fan speed and chilled-water specific flow rate are highly nonlinear functions of cooling load, and without optimal control, transport power can be much greater than compressor power at part-load and low-lift conditions. Gayeski et al. (2010a) demonstrated an empirical method for deriving near-optimal control laws from measured performance.

Cumali (1988, 1994) presented a method for real-time global optimization of HVAC systems, including the central plant and associated piping and duct networks. The method uses a building load model for the zones, based on a coupled weighting-factor method similar to that used in DOE-2 (Winkelman et al. 1993). Variable time steps are used to predict loads over a 5 to 15 min period. The models are based on thermodynamic, heat transfer, and fluid mechanics fundamentals and are calibrated to match actual performance, using data obtained from the building and plant. Pipe and duct networks are represented as incidence and circuit matrices, and both dynamic and static losses are included. Fluid energy transfers are coupled with zone loads using custom weighting factors calibrated for each zone. The resulting equations are grouped to represent feasible equipment allocations for each range of building loads, and solved using a nonlinear solver. The objective function is the cost of delivering or removing energy to meet the loads, and is constrained by the comfort criteria for each zone. The objective function is minimized using the reduced gradient method, subject to constraints on comfort and equipment operation. Optimization starts with the feasible points as determined by a nonlinear equation solver for each combination of equipment allocation. The values of set points that minimize the objective function are determined; the allocation with the least cost is the desired operation mode. Results obtained from this approach have been applied to high-rise office buildings in San Francisco with central plants, VAV, dual-duct, and induction systems. Electrical demand reductions of 8 to 12% and energy savings of 18 to 23% were achieved.

Simplified System-Based Optimization Approach

The component-based optimization method presented by Braun et al. (1989a) was used to develop a simpler method for determining optimal control. The method involves correlating overall cooling plant power consumption using a quadratic functional form. Minimizing this function leads to linear control laws for control variables in terms of uncontrolled variables. The technique may be used to tune parameters of the cooling tower and chilled-water reset strategies presented in the section on Supervisory Control Strategies and Tools. It may also be used to define strategies for supply air temperature reset for VAV systems and flow control for variable-speed condenser water pumps.

In the vicinity of any optimal control point, plant power consumption may be approximated as a quadratic function of the continuous control variables for each of the operating modes (i.e., discrete control mode). A quadratic function also correlates power consumption in terms of uncontrolled variables (i.e., load, ambient temperature) over a wide range of conditions. This leads to the following general functional relationship between overall cooling plant power and the controlled and uncontrolled variables:

$$J(\mathbf{f}, \mathbf{M}, \mathbf{u}) = \mathbf{u}^T \mathbf{A} \mathbf{u} + \mathbf{b}^T \mathbf{u} + \mathbf{f}^T \mathbf{C} \mathbf{f} + \mathbf{d}^T \mathbf{f} + \mathbf{f}^T \mathbf{E} \mathbf{u} + g \quad (4)$$

where J is the total plant power, $\mathbf{u}$ is a vector of continuous and free control variables, $\mathbf{f}$ is a vector of uncontrolled variables, $\mathbf{M}$ is a vector of discrete control variables, and the superscript T designates the transpose vector. $\mathbf{A}$, $\mathbf{C}$, and $\mathbf{E}$ are coefficient matrices, $\mathbf{b}$ and $\mathbf{d}$ are coefficient vectors, and g is a scalar. The empirical coefficients of this function depend on the operating modes so that these constants must be determined for each feasible combination of discrete control modes.

A solution for the optimal control vector that minimizes power may be determined analytically by applying the first-order condition for a minimum. Equating the Jacobian of Equation (4) with respect to the control vector to zero and solving for the optimal control set points gives

$$\mathbf{u}^* = \mathbf{k} + \mathbf{K}\mathbf{f} \quad (5)$$

where

$$\mathbf{k} = -\mathbf{A}^{-1}\mathbf{b}/2 \quad (6)$$

$$\mathbf{K} = -\mathbf{A}^{-1}\mathbf{E}/2 \quad (7)$$

The cost associated with the unconstrained control of Equation (5) is

$$J^* = \mathbf{f}^T \theta \mathbf{f} + \sigma^T \mathbf{f} + \tau \quad (8)$$

where

$$\theta = \mathbf{K}^T \mathbf{A}\mathbf{k} + \mathbf{E}\mathbf{K} + \mathbf{C} \quad (9)$$

$$\sigma = 2\mathbf{K}\mathbf{A}\mathbf{k} + \mathbf{K}\mathbf{b} + \mathbf{E}\mathbf{k} + \mathbf{d} \quad (10)$$

$$\tau = \mathbf{K}^T \mathbf{A}\mathbf{k} + \mathbf{b}^T \mathbf{k} + g \quad (11)$$

The control defined by Equation (5) results in a minimum power consumption if $\mathbf{A}$ is positive definite. If this condition holds and if the system power consumption is adequately correlated with Equation (4), then Equation (5) dictates that the optimal continuous, free control variables vary as a nearly linear function of the uncontrolled variables. However, a different linear relationship applies to each feasible combination of discrete control modes. The minimum cost associated with each mode combination must be computed from Equation (8) and compared to identify the minimum.

Uncontrolled Variables. As discussed in the background section, optimal control variables primarily depend on ambient wet-bulb temperature (or dry-bulb temperature in the case of air-cooled chillers) and total chilled-water load. The load affects the heat transfer requirements for all heat exchangers, whereas the wet-bulb temperature affects chilled- and condenser water temperatures necessary to achieve a given heat transfer rate. As discussed in the section on Supervisory Control Strategies and Tools, cooling coil heat transfer depends on the coil entering wet-bulb temperature. However, this reduces to an ambient wet-bulb temperature dependence for a given ventilation mode (e.g., minimum outside air or economizer) and fixed zone conditions. Thus, separate cost functions are necessary for each ventilation mode, with load and ambient wet bulb as uncontrolled variables. Alternatively, for a specified ventilation strategy (e.g., the economizer strategy from the section on Supervisory Control Strategies and Tools), three uncontrolled variables could be used for all ventilation modes: load, ambient wet-bulb temperature, and average cooling-coil inlet wet-bulb temperature.

For radiant cooling systems, if latent cooling is handled by package dehumidifying equipment (e.g., a dedicated outdoor air system), the chilled-water set point can be raised well above what is needed for dehumidification and the main chiller will operate more efficiently. Gayeski et al. (2010a) and Katipamula et al. (2010b) show that part-load efficiency may benefit greatly from using optimal chilled-water flow rate and temperature in response to part-load ratio.

Additional uncontrolled variables that could be important if varied over a wide range are the individual-zone latent-to-sensible load ratios and the ratios of individual sensible zone loads to the total sensible loads for all zones. However, these variables are difficult to determine from measurements and are of secondary importance.

Free Control Variables. The number of independent or "free" control variables in the optimization can be reduced significantly by using the simplified strategies presented in the section on Supervisory Control Strategies and Tools. For instance, the optimal static pressure set point for a VAV system should keep at least one VAV box fully open and should not be considered as a free optimization variable. Similarly, supply air temperature for a constant-air-volume (CAV) system should be set to minimize reheat. Additional near-optimal guidelines were presented for sequencing of cooling tower fans, sequencing of chillers, loading of chillers, reset of pressure differential set point for variable-speed pumping, and chilled-water reset with fixed-speed pumping. Furthermore, Braun et al. (1989b) showed that using identical supply air set points for multiple air handlers gives near-optimal results for VAV systems.

For all variable-speed auxiliary equipment (i.e., pumps and fans), the free set-point variables to use in Equation (1) could be reduced to the following: (1) supply air set temperature, (2) chilled-water set temperature, (3) tower airflow relative to design capacity, and (4) condenser water flow relative to design capacity. All other continuous supervisory control variables are dependent on these variables with the simplified strategies presented in the section on Supervisory Control Strategies and Tools.

Some of the dependent control variables may be discrete control variables. For instance, variable-flow pumping may be implemented with multiple fixed- and variable-speed pumps, where the number of operating pumps is a discrete variable that changes when a variable-speed pump reaches its capacity. These discrete changes could lead to discrete changes in cost because of changes in overall pump efficiency. However, this has a relatively small effect on overall power consumption and may be neglected in fitting the overall cost function to changes in the control variables.

Some discrete control variables may also be independent variables. In general, different cost functions arise for all operating modes consisting of each possible combination of discrete control variables. With all variable-speed pumps and fans, the only significant discrete control variable is the number of operating chillers. Then, optimization involves determining optimal values of only four continuous control variables for each of the feasible chiller modes. A chiller mode defines which of the available chillers are to be online. The chiller mode giving the minimum overall power consumption represents the optimum. For a chiller mode to be feasible, the specified chillers must operate safely within their capacity and surge limits. In practice, abrupt changes in the chiller modes should also be avoided. Large chillers should not be cycled on or off except when the savings associated with the change are significant.

Using fixed-speed equipment reduces the number of free continuous control variables. For instance, supply air temperature is removed as a control variable for CAV systems, and chilled-water temperature is not included for fixed-speed chilled-water pumping. However, for multiple chilled-water pumps not dedicated to chillers, the number of operating pumps can become a free discrete control variable. Similarly, for multiple fixed-speed cooling tower fans and condenser water pumps, each of the discrete combinations can be considered as a separate mode. However, for multiple cooling tower cells with multiple fan speeds, the number of possible combinations may be large. A simpler approach that works satisfactorily is to treat relative flows as continuous control variables during the optimization and to select the discrete relative flow that is closest to the optimal value. At least three relative flows (discrete flow modes) are necessary for each chiller mode to fit the quadratic cost function. The number of possible sequencing modes for fixed-speed pumps is generally much more limited than that for cooling tower fans, with two or three possibilities (at most) for each chiller mode. In fact, with many current designs, individual pumps are physically coupled with chillers, and it is impossible to operate more or fewer pumps than the number of operating chillers. Thus, it is generally best to treat the control of fixed-speed condenser water pumps with a set of discrete control possibilities rather than use a continuous control approximation.

Training. The coefficients of Equation (4) must be determined empirically, and a variety of approaches have been proposed. One approach is to apply regression techniques directly to measurements of total power consumption. Because the cost function is linear with respect to the empirical coefficients, linear regression techniques may be used. A set of experiments can be performed over the expected range of operating conditions. Large amounts of data that include the entire range must be taken to account for measurement uncertainty. The regression could possibly be performed online using least-squares recursive parameter updating (Ljung and Söderström 1983). However, precautions should be taken to ensure that the matrix **A** is positive definite, which guarantees a minimum. If system power is relatively "flat," automated methods could generate coefficients that produce a maximum in power consumption rather than a minimum (Brandemuehl and Bradford 1998).

Rather than fitting empirical coefficients of the system-cost function of Equation (4), the coefficients of the optimal control Equation (5) and the minimum-cost function of Equation (8) may be estimated directly. At a limited set of conditions, optimal values of the continuous control and free variables may be estimated through trial-and-error variations. Only three independent conditions are necessary to determine coefficients of the linear control law given by Equation (5) if the load and wet bulb are the only uncontrolled variables. The coefficients of the minimum cost function can then be determined from system measurements with the linear control law in effect. The disadvantage of this approach is that there is no direct way to handle physical constraints on the controls.

Summary and Constraint Implementation. The methodology for determining the near-optimal control of a chilled-water system may be summarized as follows:

1. Change the chiller operating mode if system operation is at the limits of chiller operation (near surge or maximum capacity).
2. For the current set of conditions (load and wet bulb), estimate the feasible modes of operation that avoid operating the chiller and condenser pump at their limits.
3. For the current operating mode, determine optimal values of the continuous controls using Equation (5).
4. Determine a constrained optimum if controls exceed their bounds.
5. Repeat steps 3 and 4 for each feasible operating mode.
6. Change the operating mode if the optimal cost associated with the new mode is significantly less than that associated with the current mode.
7. Change the values of the continuous control variables. When treating multiple-speed fan control with a continuous variable, use the discrete control closest to the optimal continuous value.

If the linear optimal control Equation (5) is directly determined from optimal control results, then the constraints on controls may be handled directly. Otherwise, a simple solution is to constrain the individual control variables as necessary and neglect the effects of the constraints on the optimal values of the other controls and the minimum cost function. The variables of primary concern for constraints are the chilled-water and supply air set temperatures.

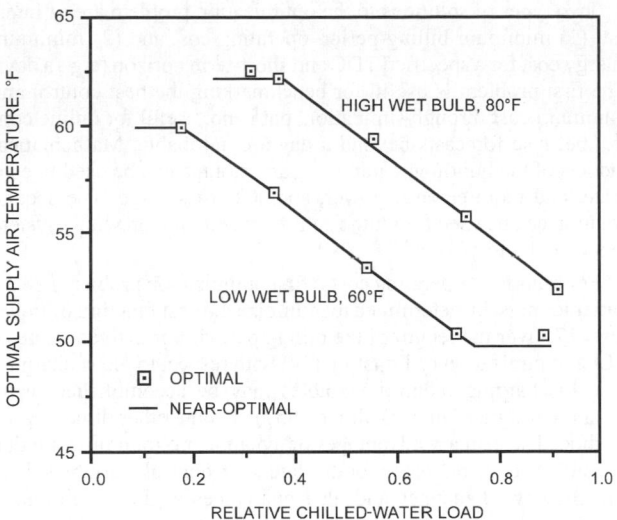

Fig. 4 Comparisons of Optimal Supply Air Temperature

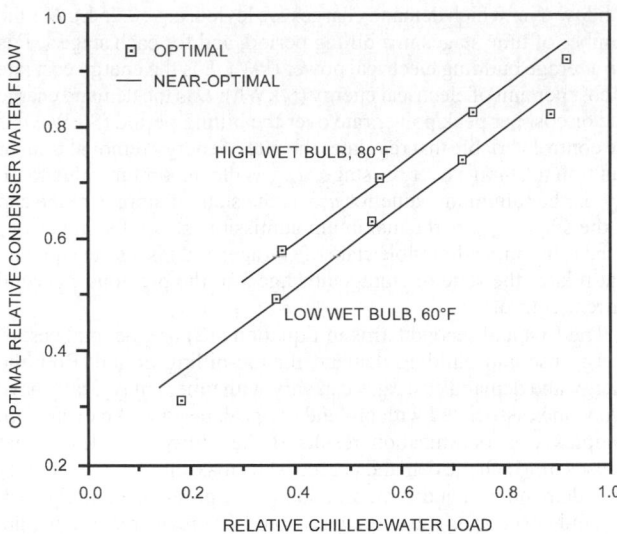

Fig. 5 Comparisons of Optimal Condenser Pump Control

These controls must be bounded for proper comfort and safe operation of the equipment. On the other hand, cooling tower fans and condenser water pumps should be sized so the system performs efficiently at design loads, and constraints on control of this equipment should only occur under extreme conditions.

The optimal value of the chilled-water supply temperature is coupled to the optimal value of the supply air temperature, so decoupling these variables in evaluating constraints is generally not justified. However, optimization studies indicate that when either control is operated at a bound, the optimal value of the other free control is approximately bounded at a value that depends only on the ambient wet-bulb temperature. The optimal value of this free control (either chilled-water or supply air set point) may be estimated at the load at which the other control reaches its limit. Coupling between optimal values of the chilled-water and condenser water loop controls is not as strong; interactions between constraints on these variables may be neglected.

Case Studies. Braun et al. (1987) correlated the power consumption of the Dallas-Ft. Worth airport chiller, condenser pumps, and cooling tower fans with the quadratic cost function given by Equation (4) and showed good agreement with data. Because the chilled-water loop control was not considered, the chilled-water set point was treated as a known uncontrolled variable. The discrete control variables associated with the four tower cells with two-speed fans and the three condenser pumps were treated as continuous control variables. The optimal control determined by the near-optimal Equation (5) also agreed well with that determined using a nonlinear optimization applied to a detailed simulation of the system.

In subsequent work, Braun et al. (1989a) considered complete system simulations (cooling plant and air handlers) to evaluate the performance of the quadratic, system-based approach. A number of different system characteristics were considered. Figures 4, 5, 22, and 23 show comparative results between the controls as determined with the component-based and system-based methods for a range of loads, for a relatively low and high ambient wet-bulb temperature (60°F and 80°F).

In Figures 4 and 22, optimal values of the chilled-water and supply air temperatures are compared for a system with variable air and water flow. The near-optimal control equation provides a good fit to the optimization results for all conditions considered. The chilled-water temperature was constrained between 38 and 55°F, while the supply air set point was allowed to float freely. Figures 4 and 22 show that, for the conditions where the chilled-water temperature is constrained, the optimal supply air temperature is

also nearly bounded at a value that depends on the ambient wet bulb.

Optimal relative cooling tower air and condenser water flow rates are compared in Figures 5 and 23 for a system with variable-speed cooling tower fans and condenser water pumps. Although the optimal controls are not exactly linear functions of the load, the linear control equation provides an adequate fit. The differences in these controls result in insignificant differences in overall power consumption, because, as discussed in the background section, the optimum is extremely flat with respect to these variables. The nonlinearity of the condenser loop controls is partly due to the constraints imposed on the chilled-water set temperature. However, this effect is not very significant. Figures 5 and 23 also suggest that the optimal condenser loop control is not very sensitive to ambient wet-bulb temperature.

DYNAMIC OPTIMIZATION

Dynamic Optimization with Discrete Storage

The optimal supervisory control for storage is a complex function of such factors as utility rates, load profile, chiller characteristics, storage characteristics, and weather. For a utility rate structure that includes both time-of-use energy and demand charges, the optimal strategy can depend on variables that extend over a monthly time scale. The overall problem of minimizing the utility cost over a billing period (e.g., a month) can be mathematically described as follows:

Minimize

$$J = \sum_{k=1}^{N} (E_k P_k \Delta\tau) + \text{Max}_{1 \leq k \leq N}(D_k P_k) \tag{12}$$

with respect to the control variables $(u_1, u_2, \ldots, u_N)$ and subject to the following constraints for each stage k:

$$u_{min,k} \leq u_k \leq u_{max,k} \tag{13}$$

$$x_k = f(x_{k-1}, u_k, k) \tag{14}$$

$$x_{min} \leq x_k \leq x_{max} \tag{15}$$

$$x_N = x_0 \tag{16}$$

where J is the utility cost associated with the billing period (e.g., a month); Δt is the stage time interval (typically equal to the time window over which demand charges are levied, e.g., 0.25 h); N is the number of time stages in a billing period, and for each stage k, P is the average building electrical power (kW); E is the energy cost rate or cost per unit of electrical energy ($/kWh); D is the demand charge rate or cost per peak power rate over the billing period ($/kW); u is the control variable that regulates the rate of energy removal from or addition to storage over the stage; u_{max} is the maximum value for u; u_{min} is the minimum value for u; x is the state of storage at the end of the stage; x_{max} is the maximum admissible state of storage; x_{min} is the minimum admissible state of storage; and f is a state equation that relates the state of storage at stage k to the previous state and current control.

The first and second terms in Equation (12) are the total cost of energy use and building demand for the billing period. Both the energy and demand cost rates can vary with time, but typically have two values associated with on- and off-peak periods. An even more complex cost optimization results if the utility includes ratchet clauses in which the demand charge is the maximum of the monthly peak demand cost and some fraction of the previous monthly peak demand cost during the cooling season. With real-time pricing, the demand charge, which is the second term in Equation (12), might not exist and the hourly energy rates would vary over time according to the generation costs.

For ice or chilled-water storage systems, the control variable could be the rate at which energy is added or removed from storage. In this case, the constraint given by Equation (13) arises from limits that depend on the chiller and storage heat exchanger and can also depend on the state of storage. For use of building thermal mass, the control variable could be the zone temperature(s) and the constraint of Equation (13) would be associated with comfort considerations or capacity constraints. Different comfort limits would probably apply for occupied and unoccupied periods.

In this general formulation the state equation, Equation (14), is treated as an equality constraint. The state of storage at any stage k is a function of the previous state (x_{k-1}), the control (u_k), and other time-dependent factors (e.g., ambient temperature). For lumped storage systems (e.g., ice), the state of storage can be characterized with a single-state variable. However, for a distributed storage (e.g., a building structure), multiple-state equations may be necessary to properly characterize the dynamics. The state of storage may also be constrained to be between states associated with full discharge and full charge [Equation (15)]. The constraint of Equation (16) forces a steady-periodic solution to the problem. This constraint becomes less important as the length of analysis increases.

To determine a control strategy for charging and discharging storage that minimizes utility cost, Equation (12) must be minimized over the entire billing period because of the influence of the demand charge. Alternatively, the optimization problem can be posed as a series of shorter-term (e.g., daily or weekly) optimizations with a constraint on the peak demand charge according to

Minimize

$$J = \sum_{k=1}^{N} \{E_k P_k \Delta\tau\} + \text{TDC} \qquad (17)$$

with respect to the control variables ($u_1, u_2,..., u_N$) and a billing period target demand cost (TDC) and subject to the constraints of Equations (13) through (16) and the following additional constraint:

$$D_k P_k \leq \text{TDC} \qquad (18)$$

which arises from the form of the cost function chosen for Equation (17). At each stage, the demand cost must be less than or equal to the peak demand cost for the billing period. The peak or target demand cost TDC is an optimization variable that affects both energy and demand costs. Using Equation (17) rather than Equation (12) simplifies the numerical solution.

Two types of solutions to the optimization problem are of interest: (1) minimum billing-period operating cost and (2) minimum energy cost for a specified TDC and short-term horizon (e.g., a day). The first problem is useful for benchmarking the best control and minimum cost through simulation, but is not useful for online control because forecasts beyond a day are unreliable. Mathematical models of the building, equipment, and storage can be used to estimate load requirements, power, and state of storage. The second solution can be used for online control in conjunction with a system model and a forecaster.

For minimum operating costs (first optimization problem), $N + 1$ variables must be determined to minimize the cost function of Equation (17) over the length of the billing period. For a given value of TDC, minimization of Equation (17) with respect to the N charging (and discharging) control variables may be accomplished using dynamic programming (Bellman 1957) or some other direct search method. The primary advantages of dynamic programming are that it handles constraints on both state and control variables in a straightforward manner and also guarantees a global minimum. However, the computation becomes excessive if more than one state variable is needed to characterize storage. The N-variable optimization problem is resolved at each iteration of an outer loop optimization for TDC. Brent's algorithm (1973) is a robust method for solving the one-dimensional optimization for the demand target because it does not require derivative information. This is important because TDC appears as an inequality constraint in the dynamic programming solution and may not always be triggered.

For shorter-term optimizations (second optimization problem), dynamic programming can still be used to minimize Equation (17) with respect to the N charging (and discharging) control variables for a specified TDC. However, an optimal value for TDC cannot be determined when demand charges are imposed. For ice storage, Drees and Braun (1996) found that a simple and near-optimal approach is to set TDC to zero at the beginning of each billing period. Therefore, the optimizer minimizes the demand cost for the first optimization period (e.g., a day) and then uses this demand as the target for the billing period unless it is exceeded. For online optimization, the optimization problem can be resolved at regular intervals (e.g., 1 h) during each day's operation.

Ice Storage Control Optimization. Several researchers have studied optimal supervisory control of ice storage systems. Braun (1992) solved daily optimization problems for two limiting cases: minimum energy (i.e., no demand charge) and minimum demand (no energy charge). Results of the optimizations for different days and utility rates were compared with simple chiller-priority and load-limiting control strategies (see the section on Supervisory Control Strategies and Tools). For the ice-on-pipe system considered, load-limiting control was found to be near optimal for both energy and demand costs with on-peak to off-peak energy cost ratios greater than about 1.4.

Drees and Braun (1996) solved both daily and monthly optimization problems for a range of systems with internal-melt area-constrained ice storage tanks. The optimization results were used to develop rules that became part of a rule-based, near-optimal controller presented in the section on Supervisory Control Strategies and Tools. For a range of partial-storage systems, load profiles, and utility rate structures, the monthly electrical costs for the rule-based control strategy were, on average, within about 3% of the optimal costs.

Henze et al. (1997a) developed a simulation environment that determines the optimal control strategy to minimize operating cost, including energy and demand charges, over the billing period. A modular cooling plant model was used that includes three compressor types (screw, reciprocating, and centrifugal), three ice storage media (internal melt, external melt, and ice harvester), a water-cooled condenser, central air handler, and all required fans and pumps. The simulation tool was used to compare the performance of chiller-priority, constant-proportion, storage-priority, and optimal control.

Henze et al. (1997b) presented a predictive optimal controller for use with **real-time pricing (RTP)** structures. For the RTP structure considered, the demand term of Equation (17) disappears and the optimization problem only involves a 24 h period. The controller calculates the optimal control trajectory at each time step (e.g., 30 min), executes the first step of that trajectory, and then repeats that process at the next time step. The controller requires a model of the plant and storage, along with a forecast of the future cooling loads.

To apply the optimization approach described in the previous section, models for storage, system power consumption, and building loads are needed. For online optimization, simple empirical models that can be trained using system measurements are appropriate. However, physically based models would be best for simulation studies.

The optimization studies that have been performed for ice storage assumed that the state of storage could be represented with a single-state variable. Assuming negligible heat gains from the environment, the relative state of charge (i.e., fraction of the maximum available storage capacity) for any stage k is

$$ x_k = x_{k-1} + \frac{u_k \Delta t}{C_s} \qquad (19) $$

where C_s is the maximum change in internal energy of the storage tank that can occur during a discharge cycle and u_k is the storage charging rate. The state of charge defined in this manner must be between zero and one.

The charging rate for storage depends on the storage heat exchanger area, secondary fluid flow rate and inlet temperature, and the thickness of ice. At any stage, the maximum charging rate can be expressed as

$$ u_{k,max} = \varepsilon_{c,k,max} \dot{m}_{f,max} c_f (t_s - t_{f,i}) \qquad (20) $$

where $\varepsilon_{c,k,max}$ is the heat transfer effectiveness for charging at the current state of storage if the secondary fluid flow rate were at its maximum value of $m_{f,max}$, c_f is the secondary fluid specific heat, $t_{f,i}$ is the temperature of secondary fluid inlet to the tank, and t_s is the temperature at which the storage medium melts or freezes (e.g., 32°F).

The minimum charging rate is actually the negative of the maximum discharging rate and can be given by

$$ u_{k,min} = \varepsilon_{d,max} \dot{m}_{f,max} c_f (t_s - t_{f,i}) \qquad (21) $$

where $\varepsilon_{d,k,max}$ is the heat transfer effectiveness for discharging at the current state of storage if the secondary fluid flow rate were at its maximum value of $m_{f,max}$.

In general, the heat transfer effectiveness for charging and discharging at the design flow can be correlated as a function of state of charge using manufacturers' data (e.g., Drees and Braun 1995).

A model for the total building power is also needed. At any time

$$ P = P_{noncooling} + P_{plant} + P_{dist} \qquad (22) $$

where $P_{noncooling}$ is the building electrical use that is not associated with the cooling system (e.g., lights), P_{plant} is the power needed to operate the cooling plant, and P_{dist} is the power associated with the distribution of secondary fluid and air through the cooling coils. The models used by Henze et al. (1997a) predict cooling plant and distribution system power with a component-based simulation that is appropriate for simulation studies. Alternatively, for online optimization, plant and distribution system power can be represented with empirical correlations. Drees (1994) used curve-fits of plant power

consumption in terms of cooling load and ambient wet-bulb temperature. At any time, the cooling requirement for the chiller is the difference between the building load requirement and the storage discharge rate. The chiller supply temperature is then determined from an energy balance on the chiller and used to evaluate the limits on the storage charging and discharging rates in Equations (21) and (22). The chiller cooling rate must be greater than a minimum value for safe operation and less than the chiller capacity. Drees (1994) correlated the maximum cooling capacity as a function of the ambient wet-bulb temperature and the chiller supply temperature. For simulation studies, a building model may be used to estimate building cooling loads. For online optimization, a forecaster would provide estimates of future building cooling loads.

Dynamic Optimization with TABS

Optimal supervisory control for precooling building mass usually requires transient thermal response models of the conditioned zones, because the state of charge cannot be measured and is not well defined in any case. For passive TES, Equation (12) is retained as the objective function and Equation (22) also applies, with P_{plant} a function of capacity u_k and conditions that affect system COP. One of these conditions is chilled-water return temperature, which is in turn a function of current and past capacities and zone temperatures and past return temperatures: essentially, a conduction transfer function. The constraint on plant capacity [Equation (13)] is retained as well.

Equation (14) is replaced by a transient thermal response model that expresses room temperature x_k in terms of current and past weather, direct heating and cooling (including direct solar and internal gains), and control actions $u_k \ldots u_{k-n}$, and in terms of past room temperatures $x_{k-1} \ldots x_{k-n}$. The comprehensive room transfer function [CRTF (Seem 1989a)] is such a model. Methods and results of training a CRTF have been reported by Armstrong (2000, 2006) and Gayeski et al. (2010b). Equation (15) serves to constrain room temperature during occupied hours and constraint (16) is not normally needed if timestep 0 is assigned to an occupied hour. Note that for TABS the supervisory control cannot generally be expressed in terms of zone temperature set points.

Equation (12), may involve only the chiller power used to meet sensible load, $P = u/COP(u,t)$. It is possible to, with, e.g., liquid desiccant storage or ice storage, to shift latent load. However with enthalpy recovery (see Chapter 44 of the 2008 *ASHRAE Handbook—HVAC Systems and Equipment*), latent loads are a small fraction of total load and not usually considered to be attractive for peak shifting; in this case, DOAS system power may be treated as part of the noncooling power term in Equation (22). Setting $E = 1$ and $D = 0$ in Equation (12) will produce the highest energy savings.

APPLICATIONS OF STATIC OPTIMIZATION

Controls for Boilers

Almost two-thirds of fossil fuel consumption in the United States involves use of a boiler, furnace, or other fired system (Parker et al. 1997). Boiler systems in many heating applications can be operated automatically by energy management and control systems to reduce utility costs associated with maintaining proper environmental conditions. Boiler efficiency depends on many factors, such as combustion airflow rate, load factor, and water temperature in hot-water boilers (or pressure for steam boilers). Opportunities for energy and cost reduction in boiler plants include excess air control, sequencing and loading of multiple boilers, and resetting the hot-water supply temperature set point (for hot-water boilers) or the steam pressure set point (for steam boilers) (Dyer and Maples 1981).

Excess Air in the Combustion Process. Combustion would occur with greatest efficiency if air and fuel could be mixed in the exact proportions indicated in the chemical reaction equation. This is called **stoichiometric combustion**. The ratio of the volume of air needed to burn completely one unit volume of the fuel is known as the **stoichiometric air/fuel ratio**. The heat released when the fuel burns completely is known as the **heat of combustion**.

In practice, it is impossible to achieve stoichiometric combustion because burners cannot mix air and fuel perfectly. In combustion processes, **excess air** is generally defined as air introduced above the stoichiometric or theoretical amount required for complete combustion of the fuel. Only the minimum amount of excess air to ensure complete combustion should be supplied to the burner; more than this amount increases the heat rejected to the stack and reduces efficiency. Combustion efficiency depends on the amount of excess air (or O_2) in the flue gas, stack temperature rise above burner inlet air temperature, and amount of unburned hydrocarbons. As shown in Figure 6, combustion efficiency drops sharply when deficient air is supplied to the burner because the amount of unburned hydrocarbons rises sharply, thereby wasting fuel. Operating a boiler with high excess air also heats the air unnecessarily, resulting in lower combustion efficiency. Combustion efficiency is optimized when excess air is reduced to the minimum.

To determine the minimum excess air for a particular boiler, flue gas combustible content as a function of excess O_2 should be charted as shown in Figure 7. For a gas-fueled boiler, carbon monoxide should be monitored; for liquid or solid fuel, monitor the smoke spot number (SSN). Different firing rates should be considered because the excess air minimum varies with the firing rate (percent load). Figure 7 shows curves for high and low firing rates. As shown, low firing rates generally produce a more gradual curve; high-rate curves are steeper. For burners and firing rates with a steep combustible content curve, small changes in the amount of excess

O_2 may cause unstable operation. The optimal control set point for excess air should generally be 0.5 to 1% above minimum, to allow for slight variations in fuel composition, intake air temperature and humidity, barometric pressure, and control system characteristics. Table 1 lists typical, normally attainable optimum excess air levels, classified by fuel type and firing method.

Carbon monoxide upper control limits vary with the boiler fuel used. The CO limit for gas-fired boilers may be set typically at 400, 200, or 100 ppm. For No. 2 fuel oil, the maximum SSN is typically 1; for No. 6 fuel oil, SSN = 4. However, for any fuel used, local environmental regulations may require lower limits.

To maintain safe unit output conditions, excess air requirements may be greater than the levels indicated in this table. This condition may arise when operating loads are substantially less than the design rating. Where possible, the vendor's predicted performance curves should be checked. If they are unavailable, excess air should be reduced to minimum levels consistent with satisfactory output.

Oxygen Trim Control. An oxygen trim control system adjusts the airflow rate using an electromechanical actuator mounted on the boiler's forced-draft fan damper linkage, and measures excess oxygen using a zirconium oxide sensor mounted in the boiler stack. The oxygen sensor signal is compared with a set point value obtained

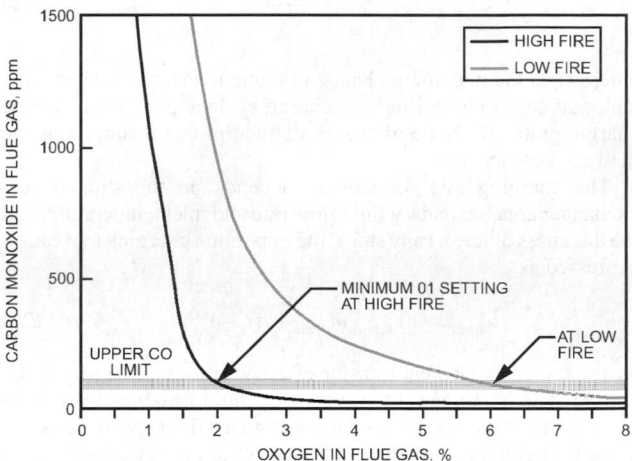

Fig. 7 Hypothetical CO-O_2 Characteristic Combustion Curves for a Gas-Fired Industrial Boiler
Parker et al. (1997). Copyright 1997 by Fairmont Press, Inc.
700 Indian Trail, Lilburn, GA 30047, www.fairmontpress.com
Reprinted by permission from *Energy Management Handbook*.

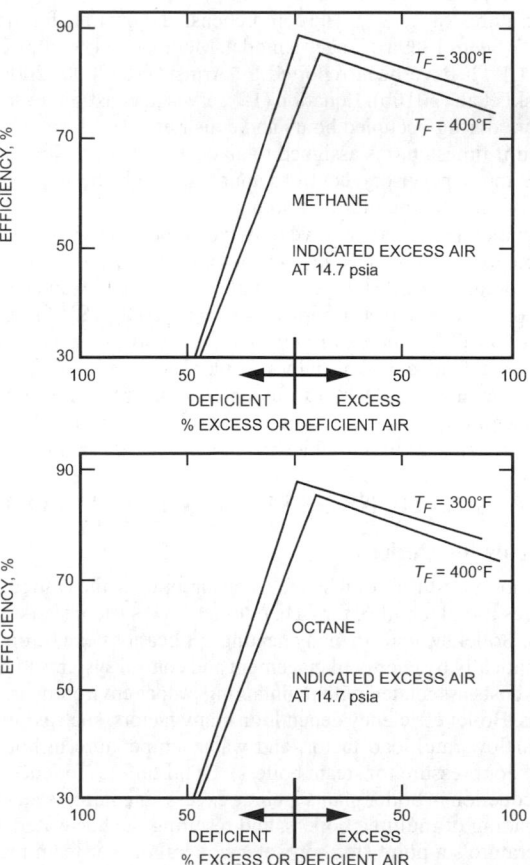

Fig. 6 Effect of Percent of Excess Air on Combustion Efficiency
BEI (1991). Copyright 1991 by the Boiler Efficiency Institute.
Reprinted by permission from *Boiler Efficiency Improvement*.

Table 1 Typical Optimum Excess Air for Various Boiler Types

Fuel Type	Firing Method	Optimum Excess Air, %	Equivalent O_2 (by Volume)
Natural gas	Natural draft	20-30	4-5
	Forced draft	5-10	1-2
	Low excess air	0.4-2.0	0.1-0.5
Propane	—	5-10	1-2
Coke oven gas	—	5-10	1-2
No. 2 oil	Rotary cup	15-20	3-4
	Air-atomized	10-15	2-3
	Steam-atomized	10-15	2-3
No. 6 oil	Steam-atomized	10-15	2-3
Coal	Pulverized	15-20	3-3.5
	Stoker	20-30	3.5-5
	Cyclone	7-15	1.5-3

Source: Parker et al. (1997). Copyright 1997 by Fairmont Press, Inc., 700 Indian Trail, Lilburn, GA 30047, www.fairmontpress.com. Reprinted by permission from *Energy Management Handbook*.

from the boiler's excess air set point curve for the given firing rate. The oxygen trim controller adjusts ("trims") the damper setting to regulate the oxygen level in the boiler stack at this set point. In the event of an electronic failure, the boiler defaults to the air setting determined by the mechanical linkages.

Carbon Monoxide Trim Control. Carbon monoxide trim control systems are also used to control excess air, and offer several advantages over oxygen trim systems. In carbon monoxide trim systems, the amount of unburned fuel (in the form of carbon monoxide) in the flue gas is measured directly by a carbon monoxide sensor and the air/fuel ratio control is set to actual combustion conditions rather than preset oxygen levels. Thus, the system continuously controls for minimum excess air. Carbon monoxide trim systems are also independent of fuel type and are virtually unaffected by combustion air temperature, humidity, and barometric pressure conditions. However, they cost more than oxygen trim systems because of the expense of the carbon monoxide sensor. Also, the carbon monoxide level in the boiler stack is not always a measure of excess air. A dirty burner, poor atomization, flame chilling, flame impingement on the boiler tubes, or poor fuel mixing can also raise the carbon monoxide level in the boiler stack (Taplin 1998).

Sequencing and Loading of Multiple Boilers. Generally, boilers operate most efficiently at a 65 to 85% full-load rating. Boiler efficiencies fall off at higher and lower load points, with the decrease most pronounced at low load conditions. Boiler efficiency can be calculated by means of stack temperature and percent O_2 (or percent excess air) in the boiler stack for a given fuel type. Part-load curves of boiler efficiency versus hot-water or steam load should be developed for each boiler. These curves should be dynamically updated at discrete load levels based on the hot-water or steam plant characteristics to allow the control strategy to continuously predict the input fuel requirement for any given heat load. When the hot-water temperature or steam pressure drops below set point for the predetermined time interval (e.g., 5 min), the most efficient combination of boilers must be selected and turned on to meet the load. The least efficient boiler should be shut down and banked in hot standby if its capacity drops below the spare capacity of the current number of boilers operating (or, for primary/secondary hot-water systems, if the flow rate of the associated primary hot-water pump is less than the difference between primary and secondary hot-water flow rates) for a predetermined time interval (e.g., 5 min). The spare capacity of the current online boilers is equal to their full-load capacity minus the current hot water load.

Resetting Supply Water Temperature and Pressure. Standby losses are reduced and overall efficiencies enhanced by operating hot-water boilers at the lowest acceptable temperature. Condensing boilers achieve significantly higher combustion efficiencies at water temperatures below the flue gas dew point when they are operating in condensing mode (see Chapter 31 in the 2008 *ASHRAE Handbook—HVAC Systems and Equipment*). Hot-water boilers of this type are very efficient at part-load operation when a high water temperature is not required. Energy savings are therefore possible if the supply water temperature is maintained at the minimum level required to satisfy the largest heating load. However, to minimize condensation of flue gases and consequent boiler damage from acid, water temperature should not be reset below that recommended by the boiler manufacturer (typically 140°F).

Similarly, energy can be saved in steam heating systems by maintaining supply pressure at the minimum level required to satisfy the largest heating load.

In practice, reset control is only possible if boiler controls interface with the energy management and control system.

Operating Constraints. Note that there are practical limitations on the extent of automatic operation if damage to the boiler is to be prevented. Control strategies to reduce boiler energy consumption can also conflict with recommended boiler operating practice. For example, in addition to the flue gas condensation concerns mentioned previously, rapid changes in boiler jacket temperature (**thermal shock**) brought about by abrupt changes in boiler water temperature or flow, firing rate, or air temperature entering the boiler should be avoided. The repeated occurrence of such transient conditions may weaken the metal and lead to cracking and/or loose tubes. It is therefore important to follow all of the recommendations of the American Boiler Manufacturers Association (ABMA 1998).

Controls for Cooling Without Storage

Figure 1 depicts multiple chillers, cooling towers, and pumps providing chilled water to air-handling units to cool air that is supplied to building zones. At any given time, cooling needs may be met with different modes of operation and set points. However, one set of control set points and modes results in minimum power consumption. This optimal control point results from tradeoffs between the energy consumption of different components. For instance, increasing the number of cooling tower cells (or fan speeds) increases fan power but reduces chiller power because the temperature of the water supplied to the chiller's condenser is decreased. Similarly, increasing condenser water flow by adding pumps (or increasing pump speed) decreases chiller power but increases pump power.

Similar tradeoffs exist for the chilled-water loop variables of systems with variable-speed chilled-water pumps and air handler fans. For instance, increasing the chilled-water set point reduces chiller power but increases pump power because greater flow is needed to meet the load. Increasing the supply air set point increases fan power, but decreases pump power.

Figure 8 illustrates the sensitivity of the total power consumption to condenser water-loop controls (from Braun et al. 1989b) for a single chiller load, ambient wet-bulb temperature, and chilled-water supply temperature. Contours of constant power consumption are plotted versus cooling tower fan and condenser water pump speed for a system with variable-speed fans and pumps. Near the optimum, power consumption is not sensitive to either of these control variables, but increases significantly away from the optimum. The rate of increase in power consumption is particularly large at low condenser pump speeds. A minimum pump speed is necessary to overcome the static pressure associated with the height of the water discharge in the cooling tower above the sump. As the pump speed approaches this value, condenser flow approaches zero and chiller power increases dramatically. A pump speed that is too high is generally better than one that is too low. The broad area near the optimum indicates that, for a given load, the optimal setting does not need to be accurately determined. However, optimal settings change significantly when there are widely varying chiller loads and ambient wet-bulb temperature.

Figure 9 illustrates the sensitivity of power consumption to chilled-water and supply air set-point temperatures for a system with variable-speed chilled-water pumps and air handler fans (Braun et al. 1989b). Within about 3°F of the optimum values, power consumption is within 1% of the minimum. Outside this range, sensitivity to the set points increases significantly. The penalty associated with operation away from the optimum is greater in the direction of smaller differences between the supply air and chilled-water set points. As this temperature difference is reduced, the required flow of chilled water to this coil increases and the chilled-water pumping power is greater. For a given chilled-water or supply air temperature, the temperature difference is limited by the heat transfer characteristics of the coil. As this limit is approached, the required water flow and pumping power would become infinite if the pump speed were not constrained. It is generally better to have too large rather than too small a temperature difference between the supply air and chilled-water set points.

For constant chilled-water flow, tradeoffs in energy use with chilled-water set point are very different than for variable-flow systems. Increasing the chilled-water set point reduces chiller

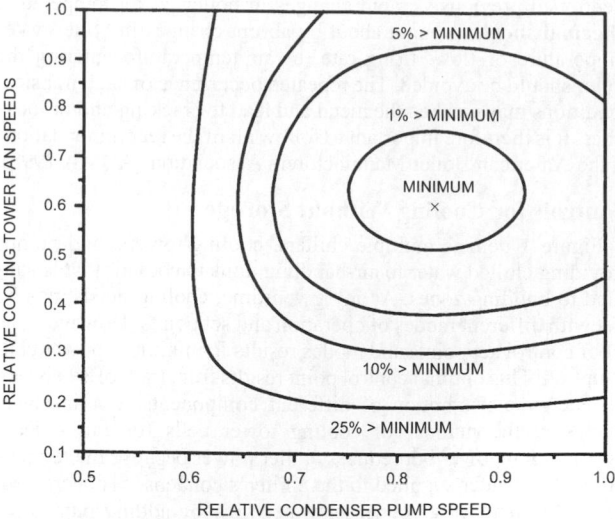

Fig. 8　Example Power Contours for Condenser-Loop Control Variables

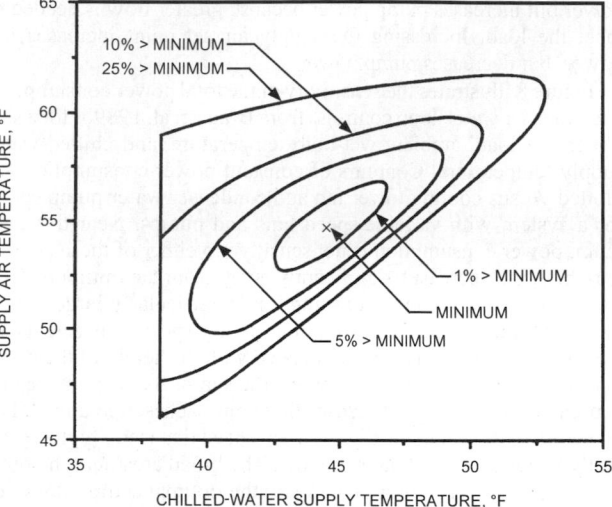

Fig. 9　Example Power Contours for Chilled-Water and Supply Air Temperatures

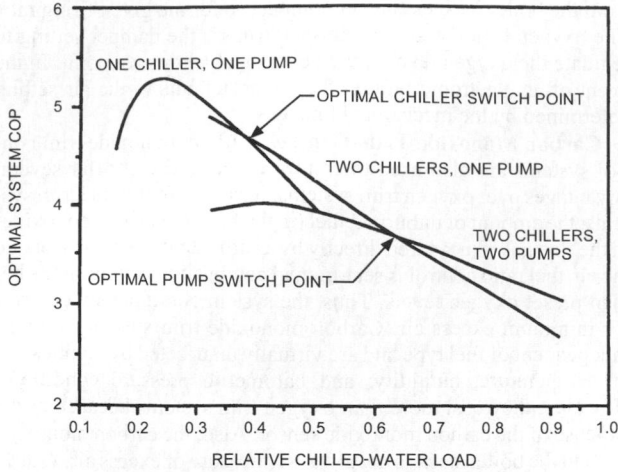

Fig. 10　Example of Effect of Chiller and Pump Sequencing on Optimal Performance

power consumption, but has little effect on chilled-water pumping energy. Therefore, the benefits of chilled-water temperature reset are more significant than for variable-flow systems (although variable-flow systems use less energy). For constant chilled-water flow, the minimum-cost strategy is to raise the chilled-water set point to the highest value that will keep all discharge air temperatures at their set points and keep zone humidities within acceptable bounds.

For constant-volume (CAV) air-handling systems, tradeoffs in energy use with supply air set point are also very different than for variable-air-volume systems. Increasing the supply air set point for cooling reduces both the cooling load and reheat required, but does not change fan energy. Again, the benefits of supply air temperature reset for CAV systems are more significant than for VAV systems (although VAV systems use less energy). In general, the set point for a CAV system should be at the highest value that will keep all zone temperatures at their set points and all humidities within acceptable limits.

In addition to the set points used by local-loop controllers, a number of operational modes can affect performance. For instance, significant energy savings are possible when a system is properly switched over to an economizer cycle. At the onset of economizer operation, return dampers are closed, outdoor air dampers are opened, and the maximum possible outdoor air is supplied to cooling coils. Two different types of switchover are typically used: (1) dry-bulb and (2) enthalpy. With a dry-bulb economizer, the switchover occurs when the ambient dry-bulb temperature is less than a specified value, typically between 55 and 65°F. With an enthalpy economizer, the switchover typically happens when the outdoor enthalpy (or wet-bulb temperature) is less than the enthalpy (or wet-bulb temperature) of the return air. Although the enthalpy economizer yields lower overall energy consumption, it requires wet-bulb temperature or dry-bulb and relative humidity measurements.

Another important operation mode is the sequencing of chillers and pumps. Sequencing defines the order and conditions associated with bringing equipment online or offline. Optimal sequencing depends on the individual design and part-load performance characteristics of the equipment. For instance, more-efficient chillers should generally be brought online before less-efficient ones. Furthermore, the conditions where chillers and pumps should be brought online depend on their performance characteristics at part-load conditions.

Figure 10 shows an example of optimal system performance (i.e., optimal set-point choices) for different combinations of chillers and fixed-speed pumps in parallel as a function of load relative to the design load for a given ambient wet-bulb temperature. For this system (from Braun et al. 1989b), each component (chillers, chilled-water pumps, and condenser water pumps) in each parallel set is identical and sized to meet half of the design requirements. The best performance occurs at about 25% of the design load with one chiller and pump operating. As load increases, system COP decreases because of decreasing chiller COP and a nonlinear increase in the power consumption of cooling tower and air handler fans. A second chiller should be brought online at the point where the overall COP of the system is the same with or without the chiller. For this system, this optimal switch point occurs at about 38% of the total design load or about 75% of the individual chiller's capacity. The optimal switch point for bringing a second condenser and chilled-water pump online occurs at a much higher relative chilled load (0.62) than the switch point for adding or removing a chiller (0.38). However, pumps are typically sequenced with chillers (i.e., they are brought online together). In this case, Figure 10 shows that the optimal switch point for bringing a second chiller online (with

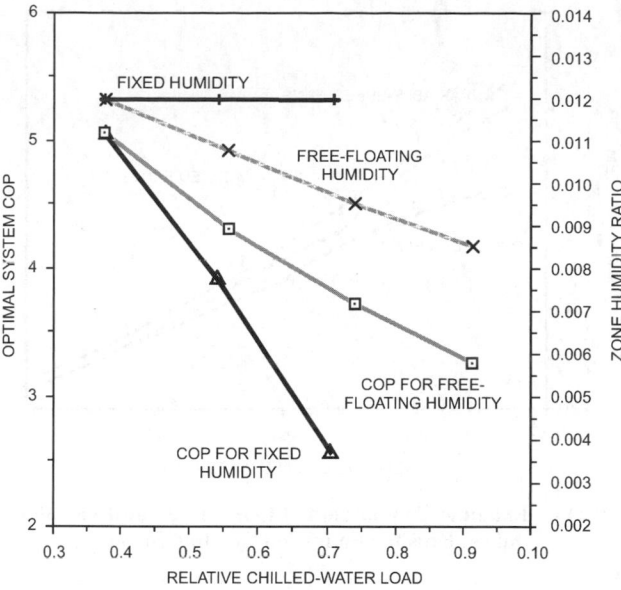

**Fig. 11   Example Comparison of Free-Floating
and Fixed Humidity**

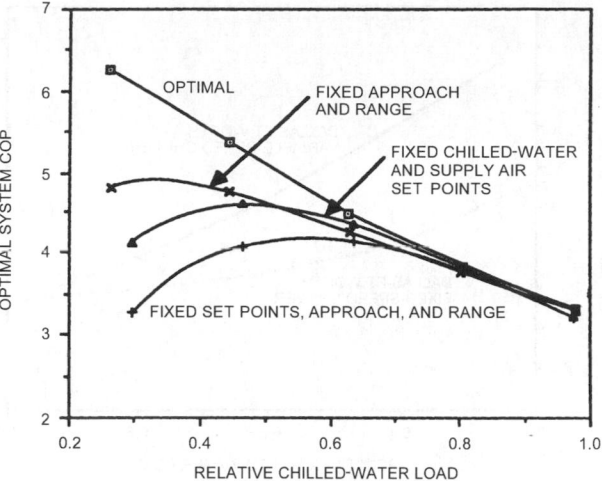

**Fig. 12   Comparisons of Optimal Control with
Conventional Control Strategies**

pumps) is about 50% of the overall design load or at the design capacity of the individual chiller. This is generally the case for sequencing chillers with dedicated pumps.

In most cases, zone humidities are allowed to float between upper and lower limits dictated by comfort (see Chapter 8 of the 2009 *ASHRAE Handbook—Fundamentals*). However, VAV systems can control zone humidity and temperature simultaneously. For a zone being cooled, the equipment operating costs are minimized when the zone temperature is at the upper bound of the comfort region. However, operating simultaneously at the upper limit of humidity does not minimize operating costs. Figure 11 shows an example comparison of system COP and zone humidity associated with fixed and free-floating zone humidity as a function of the relative load (from Braun et al. 1989b). Over the range of loads, allowing the humidity to float within the comfort zone produces a lower cost and zone humidity than setting the humidity at the highest acceptable value. The largest differences occur at the highest loads. Operation with the zone at the upper humidity bound results in lower latent loads than with a free-floating humidity, but this humidity control constraint requires a higher supply air temperature, which in turn results in greater air handler power consumption. For minimum energy costs, the humidity should be allowed to float freely within the bounds of human comfort.

Effects of Load and Ambient Conditions on Optimal Supervisory Control. When the ratio of individual zone loads to total load does not change significantly with time, the optimal control variables are functions of the total sensible and latent gains to the zones and of the ambient dry- and wet-bulb temperatures. For systems with wet cooling towers and climates where moisture is removed from conditioned air, the effect of the ambient dry-bulb temperature alone is small because air enthalpy depends primarily on wet-bulb temperature, and the performance of wet-surface heat exchangers is driven primarily by the enthalpy difference. Typically, zone latent gains are on the order of 15 to 25% of the total zone gains, and changes in latent gains have a relatively small effect on performance for a given total load. Consequently, in many cases optimal supervisory control variables depend primarily on ambient wet-bulb temperature and total chilled-water load. However, load distributions between zones may also be important if they change significantly over time.

Generally, optimal chilled-water and supply air temperatures decrease with increasing load for a fixed ambient wet-bulb temperature and increase with increasing ambient wet-bulb temperature for a fixed load. Furthermore, optimal cooling tower airflow and condenser water flow rates increase with increasing load and ambient wet-bulb temperature.

Performance Comparisons for Supervisory Control Strategies. Optimization of plant operation is most important when loads vary and when operation is far from design conditions for a significant period. Various strategies are used for chilled-water systems at off-design conditions. Commonly, the chilled-water and supply air set-point temperatures are changed only according to the ambient dry-bulb temperature. In some systems, cooling tower airflow and condenser water flow are not varied in response to changes in the load and ambient wet-bulb temperature. In other systems, these flow rates are controlled to maintain constant temperature differences between cooling tower outlet and ambient wet-bulb temperature (approach) and between cooling tower inlet and outlet (range), regardless of load and wet-bulb temperature. Although these strategies seem reasonable, they do not generally minimize operating costs.

Figure 12 shows a comparison of the COPs for optimal control and three alternative strategies as a function of load for a fixed ambient wet-bulb temperature. This system (from Braun et al. 1989b) incorporated the use of variable-speed pumps and fans. The three strategies are

- Fixed chilled-water and supply air temperature set points (40 and 52°F, respectively), with optimal condenser-loop control
- Fixed tower approach and range (5 and 12°F, respectively), with optimal chilled-water loop control
- Fixed set points, approach, and range

Because the fixed values were chosen to be optimal at design conditions, the differences in performance for all strategies are minimal at high loads. However, at part-load conditions, Figure 12 shows that the savings associated with the use of optimal control can become significant. Optimal control of the chilled-water loop results in greater savings than that for the condenser loop for part-load ratios less than about 50%. The overall savings over a cooling season depend on the time variation of the load. If the cooling load is relatively constant and near the design load, fixed values of temperature set points, approach, and range could be chosen to give near-optimal performance. However, for typical building loads with significant daily and seasonal variations, the penalty for using a

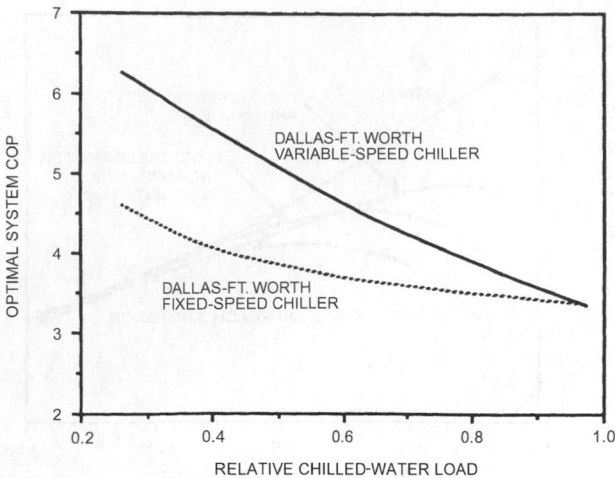

**Fig. 13   Example of Optimal Performance for
Variable- and Fixed-Speed Chillers**

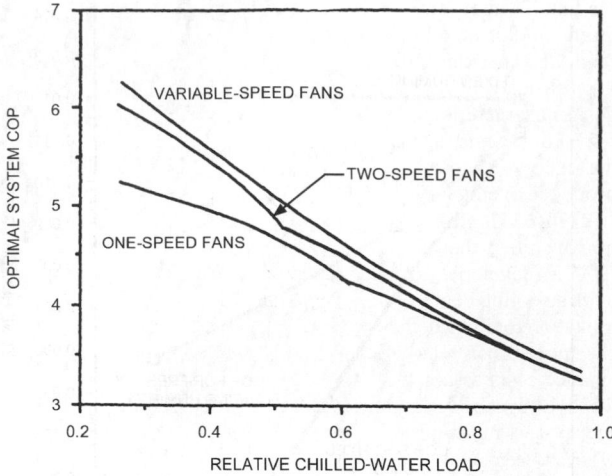

**Fig. 14   Example Comparison of One-, Two-, and Variable-
Speed Fans for Four-Cell Cooling Tower**

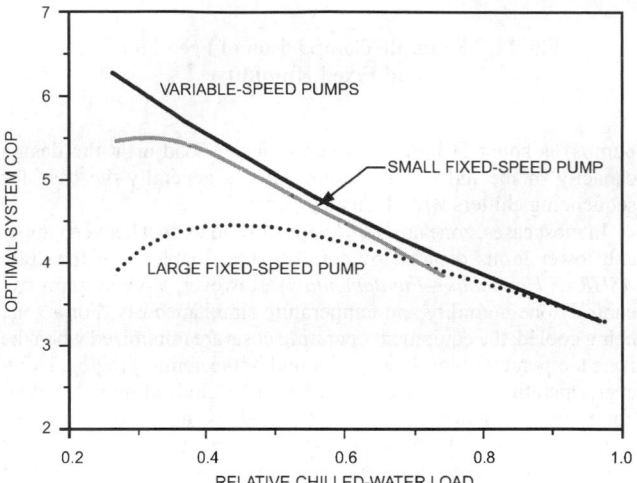

**Fig. 15   Example of Optimal Performance for
Variable- and Fixed-Speed Chillers**

fixed set-point control strategy is typically in the range of 5 to 20% of the cooling system energy.

Even greater energy savings are possible with economizer control and discharge air temperature reset with constant-volume systems. Kao (1985) investigated the effect of different economizer and supply air reset strategies on both heating and cooling energy use for CAV, VAV, and dual-duct air handling systems for four different buildings. The results indicated that substantial improvements in a building's energy use may be obtained.

Variable- Versus Fixed-Speed Equipment. Using variable-speed motors for chillers, fans, and pumps can significantly reduce energy costs but can also complicate the problem of determining optimal control. The overall savings from using variable-speed equipment over a cooling season depend on the time variation of the load. Typically, using variable-speed drives reduces equipment operating costs 20 to 50% compared to equipment with fixed-speed drives.

Figure 13 gives the overall optimal system performance for a cooling plant with either variable- or fixed-speed, variable-vane control of a centrifugal chiller. At part-load conditions, the system COP associated with using a variable-speed chiller is improved as much as 25%. However, the power requirements are similar at conditions associated with peak loads, because at full load the vanes are wide open and the speed under variable-speed control and fixed-speed operation is the same. The results in Figure 13 are from a single case study of a large chilled-water facility at the Dallas/Ft. Worth Airport (Braun et al. 1989b), constructed in the mid-1970s, where the existing chiller was retrofitted with a variable-speed drive. Differences in performance between variable- and fixed-speed chillers may be smaller for current equipment.

The most common design for cooling towers places multiple tower cells in parallel with a common sump. Each tower cell has a fan with one, two, or possibly three operating speeds. Although multiple cells with multiple fan settings offer wide flexibility in control, using variable-speed tower fans can provide additional improvements in overall system performance. Figure 14 shows an example comparison of optimal performance for single-speed, two-speed, and variable-speed tower fans as a function of load for a given wet-bulb temperature for a system with four cells (Braun et al. 1989b). The variable-speed option results in higher COP under all conditions. In contrast, for discrete fan control, the tower cells are isolated when their fans are off and the performance is poorer. Below about 70% of full-load conditions, there is a 15% difference in total energy consumption between single-speed and variable-speed fans. Between two-speed and variable-speed fans, the differences are much smaller, about 3 to 5% over the entire range.

Fixed-speed pumps that are sized to give proper flow to a chiller at design conditions are oversized for part-load conditions. Thus, the system will have higher operating costs than with a variable-speed pump of the same design capacity. Multiple pumps with different capacities have increased flexibility in control, and using a smaller fixed-speed pump for low loads can reduce overall power consumption. The optimal performance for variable-speed and fixed-speed pumps applied to both the condenser and chilled-water flow loops is shown in Figure 15 (Braun et al. 1989b). Large fixed-speed pumps were sized for design conditions; the small pumps were sized to have one-half the flow capacity of the large pumps. Below about 60% of full-load conditions, a variable-speed pump showed a significant improvement over the use of a single, large fixed-speed pump. With the addition of a small fixed-speed pump, improvements with the variable-speed pump were significant at about 40% of the maximum load.

Fan energy consumed by VAV systems is strongly influenced by the device used to vary the airflow. Centrifugal fans with variable-speed drives typically provide the most energy efficient performance. Brothers and Warren (1986) compared the fan energy consumption for a typical office building in various U.S. locations. The analysis focused on centrifugal and vaneaxial fans with three typical flow modulation devices: (1) dampers on the outlet side of the fan, (2) inlet vanes on the fan, and (3) variable-speed control of the fan motor. In all locations, the centrifugal fan used less energy than the vaneaxial fan. Vaneaxial fans have higher efficiencies at the full-load design point, but centrifugal fans have better off-design characteristics that lead to lower annual energy consumption. For a centrifugal fan, inlet vane control saved about 20% of the energy compared to damper control. Variable-speed control produced average savings of 57% compared to inlet vane control.

Hybrid Cooling Plants. Hybrid cooling plants use a combination of chillers that are powered by electricity and natural gas. Braun (2007a) developed a set of near-optimal operating strategies for hybrid cooling plants to reduce operating costs. Operating cost minimization for hybrid plants must account for effects of electrical and gas energy costs, electrical demand costs, and differences in maintenance costs associated with different chillers. Control strategies for hybrid cooling plants were developed by separating hourly energy cost minimization from the problem of determining trade-offs between monthly energy and demand costs. A demand constraint was set for each month, based on a heuristic strategy, and energy cost optimal strategies that attempted to satisfy the demand constraint were applied for cooling tower and chiller control at each decision interval. Simulated costs associated with the individual control strategies compared well with costs for optimal control.

Moreover, Braun (2007b) presented an algorithm for determining cooling tower fan settings in hybrid plants in response to loadings on individual chillers. Parameters of the algorithm were evaluated using design information for the chillers and cooling tower fans. In addition to reducing operating costs, use of the open-loop control strategy simplifies the control and improves the stability of tower control compared with using a constant condenser water supply or approach to wet-bulb. Simulated plant cooling costs associated with the algorithm were compared with costs for optimized settings, and were within 1% of the minimum costs. The developed control method is general, in the sense that it also applies to cooling plants that have all-electric or all-natural-gas chillers.

APPLICATIONS OF DYNAMIC OPTIMIZATION

Controls for Cooling Systems With Discrete Storage

Using a thermal storage system allows part or all of the cooling load to be shifted from on-peak to off-peak hours. As described in Chapter 50 of the 2008 *ASHRAE Handbook—HVAC Systems and Equipment*, there are several possible storage media and system configurations. Figure 16 depicts a generic storage system coupled to a cooling system and a building load. The storage medium could be ice, chilled water, or the building structure itself (**building thermal mass**). In ice storage, the cooling equipment **charges** (operates at low temperatures to make ice) during unoccupied periods when the cost of electricity is low. During times of occupancy and higher electric rates, ice is melted (**storage discharging**) as the storage meets part of the building load in combination with the primary cooling equipment. In building structure storage, the building is the storage medium; charging and discharging are accomplished by adjusting space temperatures over a relatively narrow range.

Utility incentives encouraging use of thermal storage are generally in the form of time-varying energy and peak demand charges. The commercial consumer is charged more for energy during the daytime and is also levied an additional charge each month based on the peak power consumption during the on-peak period. These incentives can be significant, depending on location, and are often the most important factor affecting an optimal control strategy for systems with thermal storage.

The primary control variables for the thermal storage systems depicted in Figure 16 are the rate of (1) energy removal from storage by the cooling system (charging rate) and (2) energy addition because of the load (discharging rate). Determining the optimal charging and discharging rates differs considerably from determining

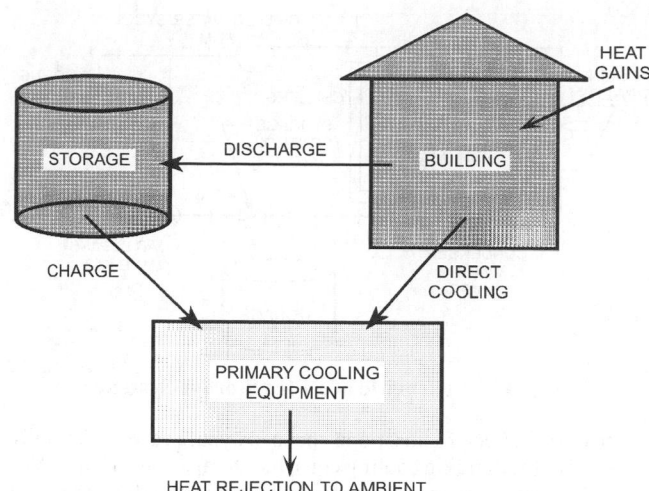

Fig. 16   Generic Storage System for Cooling

optimal set points for cooling plants that do not have storage. With thermal storage, control decisions (i.e., charging and discharging rates) determined for the current hour affect costs and control decisions for several hours in the future. Optimal control of thermal storage systems involves finding a sequence of charging and discharging rates that minimizes the total cost of providing cooling over an extended period of time, such as a day, and requires forecasting and application of dynamic optimization techniques. Constraints include limits on charging and discharging rates. The optimal control sequence results from tradeoffs between the costs of cooling the storage during off-peak hours and the cost of meeting the load during on-peak hours. In the absence of any utility incentives to use electricity at night, optimal control generally minimizes the use of ice storage because the cooling equipment operates less efficiently while charging at low temperatures. For building thermal mass systems, precooling increases heat gains from the ambient to the building. Close attention to system design and supervisory control is essential if this penalty is to be offset by potentially more efficient operation of the chiller at higher chilled-water and lower average heat rejection (night operation) temperatures.

Online optimal control of thermal storage, like other HVAC optimal control schemes, is rarely implemented because of the high initial costs associated with sensors (e.g., power) and software implementation. However, heuristic control strategies have been developed that provide near-optimal performance under most circumstances. The following sections provide background on developing control strategies for ice storage and building thermal mass. Detailed descriptions of some specific control strategies are given in the Supervisory Control Strategies and Tools section of this chapter.

Ice Storage. This section emphasizes ice storage applications, although much of the information applies to chilled-water storage as well. Figure 17 shows a schematic of a typical ice storage system. The system consists of one or more chillers, cooling tower cells, condenser water pumps, chilled-water/glycol distribution pumps, ice storage tanks, and valves for controlling charging and discharging modes of operation. Ice is made at night and used during the day to provide part of a building's cooling requirements; the storage is not sized to handle the full on-peak load requirement on the design day. Typically, in a load-leveling scheme, the storage and chiller capacity are sized such that chiller operates at full capacity during the on-peak period on the design day.

Typical modes of operation for the system in Figure 17 are as follows:

• **Storage charging mode:** Typically, charging (i.e., ice making) only occurs when the building is unoccupied and off-peak electric

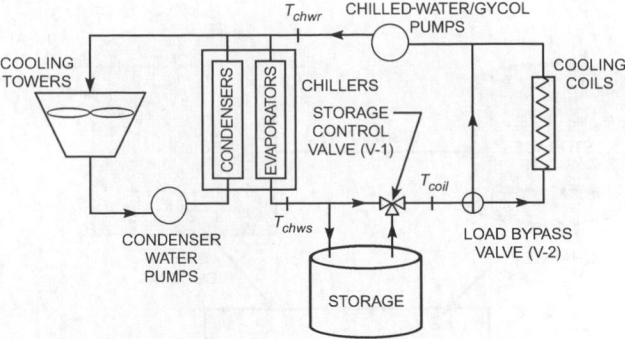

Fig. 17   Schematic of an Ice Storage System

rates are in effect. In this mode, the load bypass valve V-2 is fully closed to the building cooling coils, the storage control valve V-1 is fully open to the ice storage tank (the total chilled-water/glycol flow is through the tank), and the chiller produces low temperatures (e.g., 20°F) sufficient to make ice within the tank.

- **Storage discharging mode:** Discharging of storage (i.e., ice melting) only occurs when the building is occupied. In this mode, valve V-2 is open to the building cooling coils and valve V-1 modulates the mixture of flows from the storage tank and chiller to maintain a constant supply temperature to the building cooling coils (e.g., 38°F). Individual valves at the cooling coils modulate their chilled-water/glycol flow to maintain supply air temperatures to the zones.
- **Direct chiller mode:** The chiller may operate to meet the load directly without using storage during the occupied mode (typically when off-peak electric rates are in effect). In this mode, valve V-1 is fully closed with respect to the storage tank.

For a typical partial-storage system, the storage meets only a portion of the on-peak cooling loads on the design day and the chiller operates at capacity during the on-peak period. Thus, the peak power is limited by the capacity of the chiller. For off-design days, there are many different control strategies that meet the building's cooling requirements. However, each method has a different overall operating cost.

The best control strategy for a given day is a function of several factors, including utility rates, load profile, chiller characteristics, storage characteristics, and weather. For a utility rate structure that includes both time-of-use energy and demand charges, the optimal strategy can depend on variables that extend over a monthly time scale. Consider the charges typically associated with electrical use within a building. The first charge is the total cost of energy use for the building over the billing period, which is usually a month. Typically, the energy cost rate varies according to time of use, with high rates during the daytime on weekdays and low costs at night and on weekends. The second charge, the building demand cost, is the product of the peak power consumption during the billing period and the demand cost rate for that stage. The demand cost rate can also vary with time of day, with higher rates for on-peak periods. To determine a control strategy for charging and discharging storage that minimizes utility costs for a given system, it is necessary to perform a minimization of the total cost over the entire billing period because of the demand charge. An even more complicated cost optimization results if the utility rate includes ratchet clauses, whereby the demand charge is the maximum of the monthly peak demand cost and some fraction of the previous monthly peak demand cost within the cooling season. In either case, it is not worthwhile to perform an optimization over time periods longer than those for which reliable forecasts of cooling requirements or ambient conditions could be performed (e.g., 1 day). It is therefore important to have simple control strategies for charging and discharging storage over a daily cycle.

The following control strategies for limiting cases provide further insight:

- If the demand cost rate is zero and the energy cost rate does not vary with time, minimizing cost is equivalent to minimizing total electrical energy use. In general, cooling plant efficiency is lower when it is being used to make ice than when it is providing cooling for the building. Thus, in this case, the optimal strategy for minimum energy use minimizes the use of storage. Although this may seem like a trivial example, the most common control strategy in use today for partial ice storage systems, chiller-priority control, attempts to minimize the use of storage.
- If the demand cost rate is zero but energy costs are higher during on-peak than off-peak periods, minimizing cost then involves tradeoffs between energy use and energy cost rates. For relatively small differences between on-peak and off-peak rates of less than about 30%, energy penalties for ice making typically outweigh the effect of reduced rates, and chiller-priority control is optimal for many cases. However, with higher differentials between on-peak and off-peak energy rates or with chillers having smaller charging-mode energy penalties, the optimal strategy might maximize the use of storage. A control strategy that attempts to maximize the load-shifting potential of storage is called storage-priority control; in this scheme, the chiller operates during the off-peak period to fully charge storage. During the on-peak period, storage is used to cool the building in a manner that minimizes use of the chiller(s). Partial-storage systems that use storage-priority control strategies require forecasts for building cooling requirements to avoid prematurely depleting storage.
- If only on-peak demand costs are considered, then the optimal control strategy tends to maximize the use of storage and controls the discharge of storage in a manner to always minimize the peak building power. A storage-priority, demand-minimization control strategy for partial-storage systems requires both cooling load and noncooling electrical use forecasts.

A number of control strategies based on these three simple limiting cases have been proposed for ice storage systems (Braun 1992; Drees and Braun 1996; Grumman and Butkus 1988; Rawlings 1985; Spethmann 1989; Tamblyn 1985). Braun (1992) appears to have been the first to evaluate the performance of chiller-priority and storage-priority control strategies as compared with optimal control. The storage-priority strategy was termed load-limiting control because it attempts to minimize the peak cooling load during the on-peak period. For the system considered, the load-limiting strategy provided near-optimal control in terms of demand costs in all cases and worked well with respect to energy costs when time-of-day energy charges were available. However, the scope of the study was limited in terms of the systems considered.

Krarti et al. (1996) evaluated chiller-priority and storage-priority control strategies as compared with optimal control for a wide range of systems, utility rate structures, and operating conditions. Similar to Braun (1992), they concluded that load-limiting, storage-priority control provides near-optimal performance when there are significant differentials between on-peak and off-peak energy and demand charges. However, optimal control provides superior performance in the absence of time-of-day incentives. In general, the monthly utility costs associated with chiller-priority control were significantly higher than optimal and storage-priority control. However, without time-of-use energy charges, chiller-priority control did provide good performance for individual days when the daily peak power was less than the monthly peak. General guidance based on the work is presented by Henze et al. (2003). Drees and Braun (1996) developed a simple rule-based control strategy that combines elements of storage- and chiller-priority strategies in a way that results in near-optimal performance under all conditions. The strategy was derived from heuristics obtained through both daily and monthly optimization results for several simulated systems. A modified version of this strategy is presented in the Supervisory Control Strategies and Tools section of this chapter.

Braun (2007c, 2007d) also developed a near-optimal control method for charging and discharging of cool storage systems when RTP electric rates are available The algorithm requires relatively low-cost measurements (cooling load and storage state of charge) and very little plant specific information, is computationally simple, and ensures that building cooling requirements are always met (e.g., storage is not prematurely depleted). The control method was evaluated for ice storage systems using a simulation tool for different combinations of cooling plants, storage sizes, buildings, locations, and RTP rates.

Controls for Precooling of Building Thermal Mass

For conventional night setup strategies, it is well known that building mass works to increase operating costs. A massless building would require no time for precooling or preheating and would have lower overall cooling or heating loads than an actual building. However, under proper circumstances, using a building's thermal storage for load shifting can significantly reduce operational costs and energy use, even though the total zone loads may increase. This is especially true for cooling of high-performance buildings, in which internal loads may dominate.

At any given time, the cooling requirement for a space is caused by convection from internal gains (lights, equipment, and people) and interior surfaces. Because a significant fraction of the internal gain is radiated to interior surfaces, the state of a building's thermal storage and the convective coupling dictates the cooling requirement. Precooling the building during unoccupied times reduces the overall convection from exposed surfaces during the occupied period as compared with night setup control and can reduce daytime cooling requirements. The potential for storing thermal energy in the structure and furnishings of conventional commercial buildings is significant when compared to the load requirements. Typically, internal gains are about 3 to 7 W per square foot of floor space. The thermal capacity for typical concrete building structures is approximately 2 to 4 W·h/°F per square foot of floor area. Thus, for an internal space, the energy storage can handle the load for about 1 h for every 2°F of precooling of the thermal mass.

Opportunities for reducing operating costs by using building thermal mass for cooling derive from four effects: (1) reduction in demand costs, (2) use of low-cost off-peak electrical energy, (3) reduced mechanical cooling from the use of cool nighttime air for ventilation precooling, and (4) improved mechanical cooling efficiency from increased operation at more favorable part-load and ambient conditions. However, these benefits must be balanced with the increase in total cooling requirement that occurs with precooling the thermal mass. Therefore, the savings associated with load shifting and demand reductions depend on both the method of control and the specific application.

Several simulation studies have been performed that demonstrate a substantial benefit to precooling buildings in terms of cost savings and peak cooling load reduction (Andresen and Brandemuehl 1992; Braun 1990; Rabl and Norford 1991; Snyder and Newell 1990). Possible energy savings ranged from 0 to 25%; possible reductions in total building peak electrical demand ranged from 15 to 50% compared with conventional control. The results can be sensitive to the convective coupling between the air and the thermal mass, and the mass of the furnishings may be important (Andresen and Brandemuehl 1992).

Determining the optimal set of building temperatures over time that minimizes operating costs is complex. Keeney and Braun (1996) developed a simplified approach for determining optimal control of building thermal mass using two optimization variables for the precool period and a set of rules for the occupied period of each day. This approach significantly reduces the computation required for determining the optimal control as compared with considering

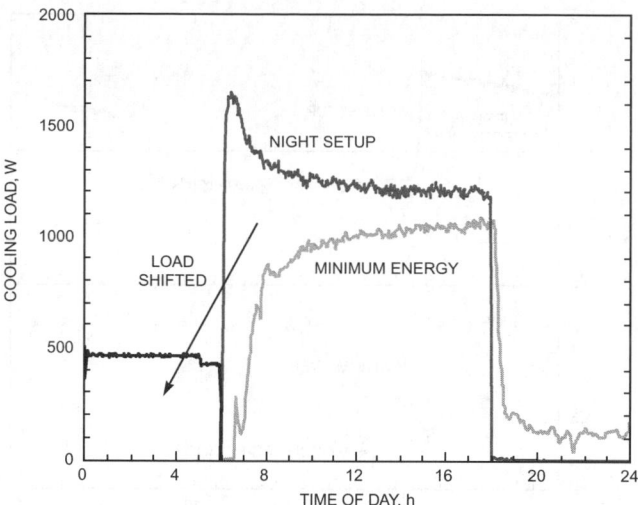

Fig. 18 Comparison of Cooling Requirements for Minimum Energy and Night Setup Control
(Morris et al. 1994)

hourly zone set points as optimization variables. Results of the simplified approach compared well with those of detailed optimizations for a range of systems (over 1000 different combinations of building types, weather conditions, cooling plants, and utility rates).

Morris et al. (1994) performed a set of experiments using a test facility at the National Institute of Standards and Technology (NIST) to demonstrate the potential for load shifting and load leveling when control was optimized. Two different control strategies were considered: (1) minimum cooling system energy use and (2) minimum peak cooling system electrical demand. The two strategies were implemented in the test facility and compared with night setup control. Figure 18 shows the 24 h time variation in the cooling requirement for the test facility allowed to reach a steady-periodic condition for both the minimum energy use strategy and conventional night setup control. The results indicate a significant load-shifting potential for the optimal control. Overall, the cooling requirements during the occupied period were approximately 40% less for optimal than for night setup control.

Comfort conditions were also monitored for the tests. Figure 19 gives the time variation of predicted mean vote (PMV) for the two control strategies as determined from measurements at the facility. A PMV of zero is a thermally neutral sensation, positive is too warm, and negative too cool. In the region of ±0.5, comfort is not compromised to any significant extent. Figure 19 shows that the comfort conditions were essentially identical for the two control methods during the occupied period. The space temperature, which has the dominant effect on comfort, was maintained at 75°F during the occupied period for both control methods. During the unoccupied period, the cooling system was off for night setup control and the temperature floated to warm comfort conditions. On the other hand, the optimal controller precooled the space, resulting in cool comfort conditions before occupancy. During these tests, the minimum space temperature during precooling was 68°F, and the space temperature set point was raised to 75°F just before occupancy.

Figure 20 shows the 24 h time variation in the cooling requirement for the test facility for both the minimum peak demand strategy and conventional night setup. Optimal control involved precooling the structure and adjusting the space temperatures within the comfort zone (–0.5 < PMV < 0.5) during the occupied period to achieve the minimum demand. Although the true minimum was not achieved during the tests, the peak cooling rate during the occupied period was approximately 40% less for minimum peak demand control than for night setup control.

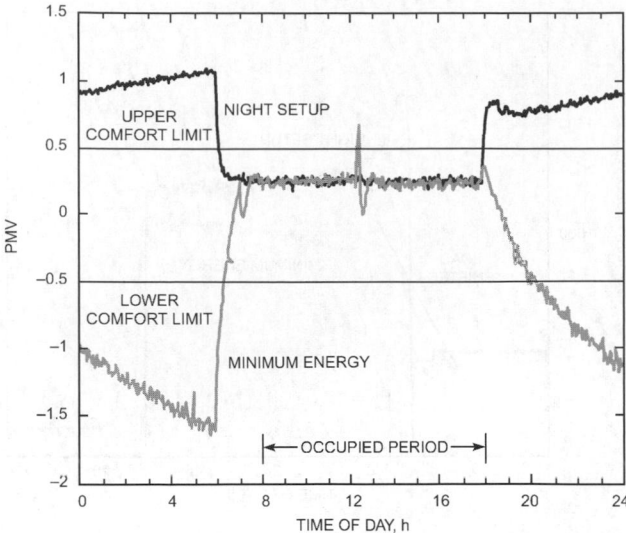

Fig. 19   Comparison of Predicted Mean Vote (PMV) for Minimum Energy and Night Setup Control
(Morris et al. 1994)

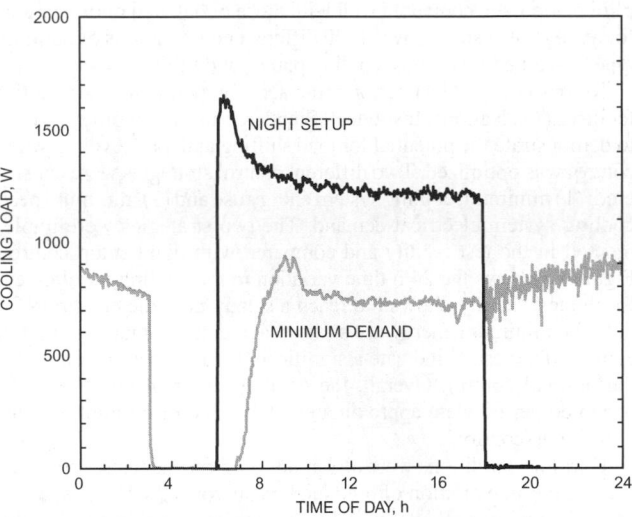

Fig. 20   Comparison of Cooling Requirements for Minimum Demand and Night Setup Control
(Morris et al. 1994)

Morris et al. (1994) demonstrated significant savings potential for control of building thermal mass; however, they also showed that the cost savings are very sensitive to the application, operating conditions, and method of control. For example, an investigation into the effect of precooling on the on-peak cooling requirements for an existing building (which may not have been a good candidate for use of building thermal storage) showed only a 10% reduction in the cooling energy required during the occupied period, with a substantial increase in the total cooling required and no reduction in the peak cooling requirement (Ruud et al. 1990). System simulations can be used to identify (1) whether the system is a good candidate for using building thermal mass and (2) an effective method for control, before implementing a strategy in a particular building.

Keeney and Braun (1997) used system simulation to develop a control strategy that was then tested in a large commercial building located northwest of Chicago. The goal of the control strategy was to use building thermal mass to limit the peak cooling load for

Table 2   Cooling Season Energy, Demand, and Total Costs and Savings Potential of Different Building Mass Control Strategies

Strategy	Costs in U.S. Dollars			Savings, %
	Energy	Demand	Total	
Night setup	$90,802	$189,034	$279,836	0.0
Light precool	$84,346	$147,581	$231,928	17.1
Moderate precool	$83,541	$143,859	$227,400	18.7
Extended precool	$81,715	$134,551	$216,266	22.7
Maximum discharge	$72,638	$ 91,282	$163,920	41.4
Two-hour linear rise	$72,671	$ 91,372	$164,043	41.4
Four-hour linear rise	$73,779	$115,137	$188,916	32.5
Nine-hour linear rise	$77,095	$141,124	$218,219	22.0

Source: Braun et al. (2001).
Note: Building located in Chicago, Illinois.

continued building operation in the event of the loss of one of the four central chiller units. The algorithm was tested using two nearly identical buildings separated by a large, separately cooled entrance area. The east building used the existing building control strategy; the west building used the precooling strategy developed for this project. The precooling control strategy successfully limited the peak load to 75% of the cooling capacity for the west building, whereas the east building operated at 100% of capacity. Details of the strategy and case study results are presented in the Supervisory Control Strategies and Tools section of this chapter.

Braun et al. (2001) used on-site measurements from the same building used by Keeney and Braun (1997) to train site-specific models that were then used to develop site-specific control strategies for using building thermal mass and to evaluate the possible cost savings of these strategies. The building is an excellent candidate for using building thermal mass because it has (1) a large differential between on-peak and off-peak energy rates (about a 2-to-1 ratio), (2) a large demand charge (about $16/kW), (3) a heavy structure with significant exposed mass, and (4) cooling loads that are dominated by internal gains, leading to a high storage efficiency. The model underpredicted the total HVAC bill by about 5% but worked well enough to be used in comparing the performance of alternative control strategies.

Table 2 gives estimates of cooling-related costs and savings over the course of three summer months for different control strategies. The light precool and moderate precool strategies are simple strategies that precool the building at a fixed set point of 67°F before occupancy and then maintain a fixed discharge set point in the middle of the comfort range (73°F) during occupancy. The light precool begins at 3 AM, whereas moderate precool starts at 1 AM. The extended precool strategy attempts to maintain the cooled thermal mass until the onset of the on-peak period. In this case, the set point at occupancy is maintained at the lower limit of comfort (69°F) until the on-peak period begins at 10 AM. At this point, the set point is raised to the middle of the comfort range (73°F). The other strategies use the extended precooling, but the entire comfort range is used throughout the on-peak, occupied period. The maximum discharge strategy attempts to discharge the mass as quickly as possible after the on-peak period begins. In this case, the set point is raised to the upper limit of comfort within an hour after the on-peak period begins. The slow linear rise strategy raises the set point linearly over the entire on-peak, occupied period (9 h in this case), whereas the fast linear rise strategy raises the set point over 4 h.

The strategies that do not use the entire comfort range during the occupied period (light precool, moderate precool and extended precool) all provided about 20% savings compared to night setup. Each of these strategies reduced both energy and demand costs, but the demand costs and reductions were significantly greater than the energy costs and savings. The decreases in energy costs were due to favorable on-to-off peak energy rate ratios of about 2 to 1. The high on-peak demand charges provided even greater incentives for precooling. The savings increased with the length of the precooling period, particularly when precooling was performed close to the onset of on-peak rates. The maximum discharge strategy, which maximizes discharge of the thermal storage within the structure, provided the largest savings (41%). Much of the additional savings came from reduced demand costs. The linear rise strategies also provided considerable savings with greater savings associated with faster increases in the set point temperature.

Morgan and Krarti (2006) performed both simulation analyses and field testing to evaluate various precooling strategies. They found that energy cost savings associated with precooling thermal mass depends on several factors, including thermal mass level, climate, and utility rate. For time-of-use (TOU) utility rates, they found that energy cost savings are primarily affected by the ratio of on-peak to off-peak demand charges as well as the ratio of on-peak to off-peak energy charges (refer to Chapter 50 of the 2008 *ASHRAE Handbook—HVAC Systems and Equipment*).

More recently, an extensive study of building thermal mass control was conducted in the context of ASHRAE research project RP-1313 (Henze et al. 2007), in which optimal building thermal mass control strategies were investigated for time-of-use electric utility rates structures, including demand charges, with the help of a newly developed integrated optimization and building simulation tool (Henze et al. 2008). Cheng et al. (2008) identified the primary factors that influence optimal control of passive thermal storage, where optimal control strategies are determined with the objective of minimizing total energy and demand costs. A fractional factorial analysis was used to investigate how cost savings are affected by several building and system characteristics, utility rate structures, and climates. Utility rates, internal loads, building mass level, and equipment efficiency were found to have the largest impacts on cost savings, whereas building envelope characteristics did not have a significant impact. Although the magnitude of savings is affected by climate, the relative impacts of each of these factors are largely independent of weather.

Using the same simulation and optimization environment, Henze et al. (2009) presented advances toward near-optimal building thermal mass control derived from full factorial analyses of the important parameters influencing passive thermal storage for a range of buildings and climate/utility rate structure combinations. In response to the actual utility rates imposed in the investigated cities, insights and control simplifications were derived from those buildings deemed suitable candidates. The near-optimal strategies were derived from the optimal control trajectory, consisting of four variables, and then tested for effectiveness and validated with respect to uncertainty regarding building parameters and climate variations. Although no universally applicable control guideline could be found, a significant number of cases (i.e., combinations of buildings, weather, and utility rate structure) were investigated and offer both insight into and recommendations for simplified control strategies. These recommendations are a good starting point for experimentation with building thermal mass control for a substantial range of building types, equipment, climates, and utility rates.

The cost savings potential of optimal passive thermal storage controls were examined by Greensfelder et al. (2010) for the case of day-ahead, real-time electricity rate structures. The operational strategies of three office building models were optimized in four U.S. cities (Chicago, New York, Houston, and Los Angeles) using price and weather data for the summer of 2008. Building thermal mass was optimized using a predictive optimal controller to define supervisory control strategies in terms of building global cooling temperature set points. A global minimization algorithm determined optimal set-point trajectories for each day divided into four distinct time periods (called building modes). Cost savings were found to range from 0 to 14%, depending on the building, climate, and characteristics of the rate signal. The best cost savings occurred in the presence of price spikes or cool nighttime temperatures.

Moreover, it was found that low internal gains favored a more flexible precooling strategy, whereas high internal gains coupled with low thermal mass resulted in poor precooling performance.

Controls for Thermally Active Building Systems

One way to reduce cooling energy further along the path to net-zero-energy buildings is **thermally active building systems (TABS)** or **active-core cooling**: direct cooling of building mass by chilled water in conjunction with chillers designed for very high part-load efficiency in low-lift operation and enthalpy-recovery dedicated outdoor air systems (DOAS).

TABS differs from passive storage by cooling the mass directly from the inside, thus eliminating charging-mode convective or radiative coupling resistances. By directly precooling the mass, instead of the occupied space, substantially better storage efficiency and larger effective diurnal storage capacity are achieved. Precooling energy percentage savings may be 5 to 35% higher because the condenser-evaporator temperature difference is lower to begin with and because, with the elimination of supply fans and mechanical cooling, energy use is dominated by compressor operation at very low pressure ratios (Armstrong et al. 2009; Katipamula et al. 2010a). Economizer-mode energy, which involves pumps only rather than fan transport energy, is extremely low as well.

Katipamula et al. (2010) simulated idealized thermal energy storage (TES) to approximate TABS, and a variable-speed air-cooled chiller with variable-speed distribution pump. The TES and distribution options are shown schematically in Figure 21.

The chiller/distribution system was modeled with compressor, condenser fan, and chilled-water pump capable of wide 10:1 speed ranges and statically optimized control, resulting in evaporator temperatures of up to 64°F at 10% of design load. Chiller performance was measured by an independent testing laboratory over a wide range of part-load fraction, outdoor temperatures, and chilled-water temperatures (Katipamula et al. 2010b). Performance curves fit to the data are shown in Figure 22. The performance map on the left is for a standard chilled-water reset schedule recommended for VAV systems (ASHRAE *Standard* 90.1), and the right-hand map represents performance of the chiller with radiant ceiling panels (RCPs) or chilled-beam distribution system serving a conditioned space at $T_{op} = 75°F$. Compressor, condenser fan, and chilled-water pump are operated at optimal speed under any given condition and part-load fraction.

A flat electricity rate was assumed, to achieve minimum annual energy use, and the performance of standard and high-performance versions of 12 building types was simulated in 16 climates. The load shifting achieved by TES is illustrated in Figure 23. The joint distribution of chiller output is expressed in full-load operating hours for each bin cell of 5°F in outdoor temperature by 10% in rated capacity. Compared to the base-case load distribution, operation hours for the TES system are significantly shifted in two respects. The bin in which the full-load equivalent operating hours (FLEOH) peak occurs is typically 15°F lower than in the baseline chiller load distribution. Although the chiller continues to operate at high outdoor temperatures, it does so at much lower part-load ratios. Conversely, the FLEOH of operation at low outdoor temperature and low part-load ratio significantly increased.

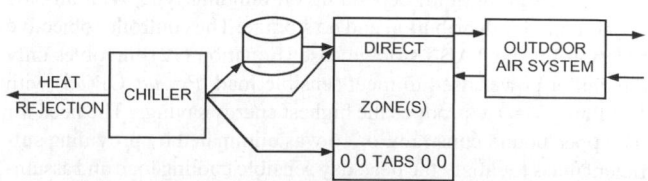

Fig. 21   Schematic of an Active-Core Cooling System

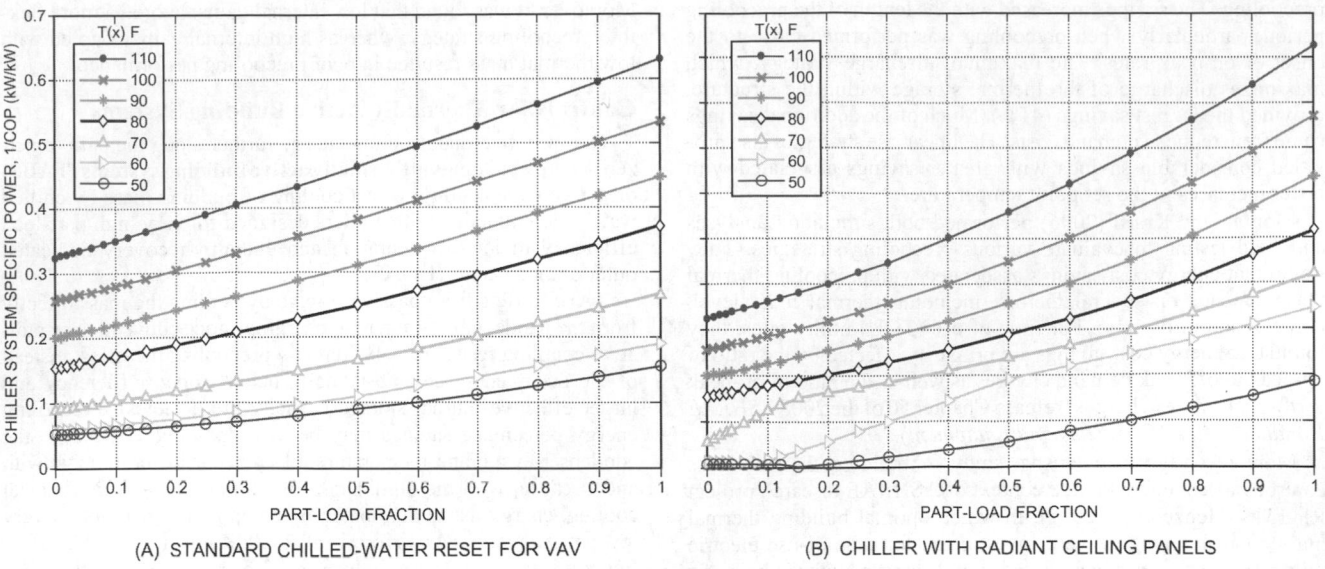

Fig. 22   Performance of an Optimally Controlled Chiller for Two Different Load-Side Boundary Conditions

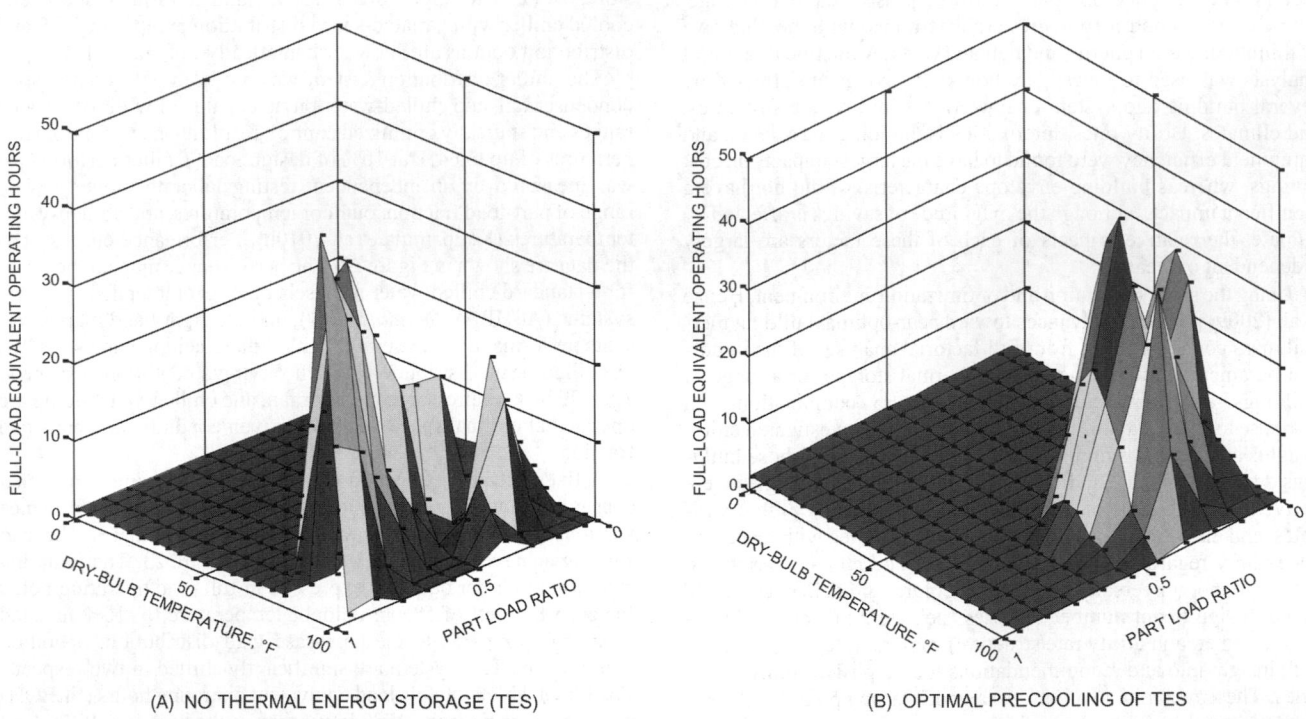

Fig. 23   Chiller Load Distributions for Chicago

The base case for savings used a two-speed chiller built from the same components as the statically optimized chiller and VAV or CAV air-handling unit, depending on building type, with air-side economizer for distribution and no storage. The controller objective function for the TABS storage case [Equation (12)] involves only the chiller power used to meet sensible load, $P = u/COP(u,t)$ with $E = 1$ and $D = 0$ to produce the highest energy savings. The need for the upper-bound constraint of 16 was eliminated by providing sufficient mass to satisfy the peak day sensible cooling load and assuming zero charge carryover from day to day, In the simulations, perfect 24 h forecasts of load and outdoor temperature were used; in

practice, a model-based predictive control using the forecast methods from this chapter and publicly or commercially available forecasts of weather would result in some loss of performance (Krarti et al. 2007). The results in Table 3 indicate substantial energy savings potential for most of the building types in all 16 climates. For high-performance buildings, the savings were substantial in most building types and climates where economizer cooling potential is modest and the annual sensible load is dominant. Capital costs of the base and low-lift active-core systems were estimated and compared. For large office buildings, the TABS cooling system capital cost was estimated to be less than that of the VAV system.

Table 3 Energy Savings Potential for Precooling with High Part-Load Efficiency Chiller

Building Type	Standard Building			High-Performance Building		
	Mini-mum	Maxi-mum	Average	Mini-mum	Maxi-mum	Average
Office, small	59%	77%	70%	−19%	43%	25%
medium	13%	52%	37%	7%	50%	25%
large	21%	61%	40%	−4%	36%	13%
Retail, standalone	56%	73%	66%	31%	50%	41%
strip mall	45%	63%	56%	−7%	41%	16%
Primary school	46%	55%	51%	22%	46%	34%
Secondary school	32%	49%	43%	10%	37%	26%
Hotel, large	16%	57%	44%	−11%	47%	29%
Supermarket	59%	78%	68%	35%	63%	51%
Warehouse	50%	81%	69%	−5%	69%	40%
Outpatient	65%	83%	78%	34%	67%	53%
Hospital	48%	76%	64%	10%	49%	37%

These results are based on simulations in which the control for charging of the TABS is assumed to be modeled exactly. One problem with TABS is that precooling can no longer be reliably controlled by room temperature set-point adjustments. A further problem is that the state of charge cannot be readily measured. For these reasons, supervisory control of a real TABS implementation is considered to require some form of model-based predictive control.

A full-scale laboratory test of TABS cooling with model-based predictive control was conducted by Gayeski (2010b). The test results showed energy savings of 25 to 30% compared to an SEER-16 all-air system for typical Atlanta and Phoenix summer conditions. In each two-week-long test, the same variable-speed compressor/condenser unit was connected to a conventional indoor unit for the baseline case and to a hydronic evaporator supplying chilled water to a 6 in. concrete slab for the optimal precooling case. The results were comparable to the simulation results of Katipamula et al. (2010a) for a medium office building in Atlanta and Phoenix.

Control of Combined Thermal Energy Storage Systems

Combined Passive and Active Thermal Storage. Most investigations into the optimal control of combined active and passive building thermal storage inventory rely on a detailed white-box or gray-box model of building thermal response and equipment performance [see, e.g., Henze et al. (2005)]. However, Liu and Henze (2006a, 2006b) describe a novel approach to optimally control commercial building passive and active thermal storage inventory simultaneously: a hybrid control scheme that combines features of model-based optimal control and model-free reinforcement learning control.

Theoretically, the reinforcement learning algorithm, based on Watkins and Dayan (1992), approximates dynamic-programming-based optimal control by sampling the cost space and can reach the true optimum, given properly selected learning parameters and long enough learning time. The amount of required training is not yet realistic if the controller is directly implemented in a commercial building application. This constitutes the major drawback of the reinforcement learning control approach; contextual information in some form needs to be introduced to expedite learning of the fundamental features of the problem, while reinforcement learning fine-tunes the controller. This realization inspired the development of the hybrid learning control scheme (Liu and Henze 2006a).

Liu and Henze (2006b) analyzed the performance of this hybrid controller installed in a full-scale laboratory facility. Operating cost savings from the application of both active and passive TES were attained with the hybrid control approach compared with conventional building control; however, the savings were lower than for the case of model-based predictive optimal control. As for the case of model-based predictive control, the hybrid controller's performance is largely affected by the quality of the training model, and extensive real-time learning is required for the learning controller to eliminate any false cues it receives during the initial training period. Nevertheless, compared with standard reinforcement learning, Liu and Henze's proposed hybrid controller is much more readily implemented in a commercial building.

Combination of TABS and Direct Sensible Cooling. A control problem similar to that of the combined passive and active TES is raised by the TABS cooling system. If piping is embedded deep in a floor or ceiling mass to increase storage capacity, the ability to control peak zone temperatures is seriously degraded and it becomes necessary to augment the TABS with fast-responding terminal units such as RCPs or chilled beams. Further research is needed, not only to develop appropriate control strategies, but to understand the relation between TABS piping depth and annual cooling system performance.

SUPERVISORY CONTROL STRATEGIES AND TOOLS

COOLING TOWER FAN CONTROL

Figure 24 shows a schematic of the condenser loop for a typical chilled-water unit consisting of centrifugal chillers, cooling towers, and condenser water pumps. Typically, the condenser water pump control is directed by the chiller control to provide relatively constant flow for individual chillers. However, the cooling tower cells may be independently controlled to maximize system efficiency.

Typically, cooling tower fans are controlled using a feedback controller that attempts to maintain a temperature set point for the water supplied to the chiller condensers. Often, the condenser water supply temperature set point is held constant. However, a better strategy is to maintain a constant temperature difference between the condenser water supply and the ambient wet bulb (constant approach). Additional savings are possible through optimal control.

With a single feedback controller, the controller output signal must be converted to a specific fan sequence that depends on the number of operating cells and the individual fan speeds. Typically, with the discrete control associated with one- or two-speed tower fans, the set point cannot be realized, resulting in the potential for oscillating tower fan control. Fan cycling can be reduced by using deadbands, "sluggish" control parameters, and/or lower limits for on and off periods.

Braun and Diderrich (1990) demonstrated that feedback control for cooling tower fans could be eliminated by using an open-loop supervisory control strategy. This strategy requires only measuring chiller loading to specify the control and is inherently stable. The tower fan control is separated into two parts: tower sequencing and optimal airflow. For a given total tower airflow, general rules for optimal tower sequencing are used to specify the number of operating cells and fan speeds that give the minimum power consumption for both the chillers and tower fans. The optimal tower airflow is estimated with an open-loop control equation that uses design information for the cooling tower and chiller. The computational procedure is presented in this section, and the control strategy is summarized in a set of steps and sample calculations.

Near-Optimal Tower Fan Sequencing

For variable-speed fans, minimum power consumption results when all cooling tower cells are operated under all conditions. Tower airflow varies almost linearly with fan speed, whereas the fan power consumption varies approximately with the cube of the speed. Thus, for the same total airflow, operating more cells in parallel allows for lower individual fan speeds and lower overall fan power consumption. An additional benefit associated with full-cell

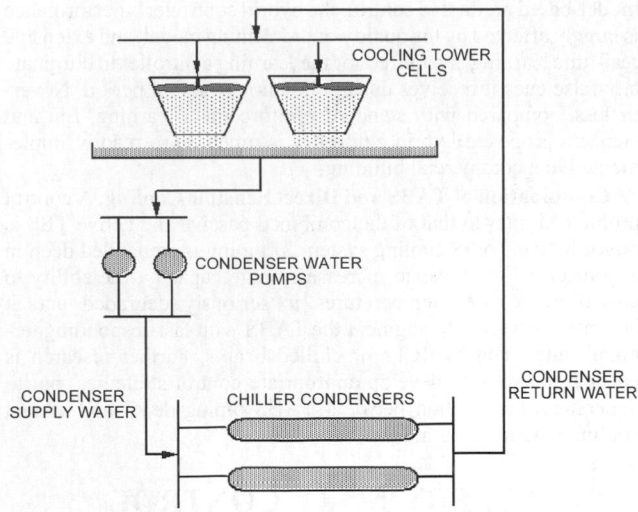

Fig. 24   Condenser Water Loop Schematic

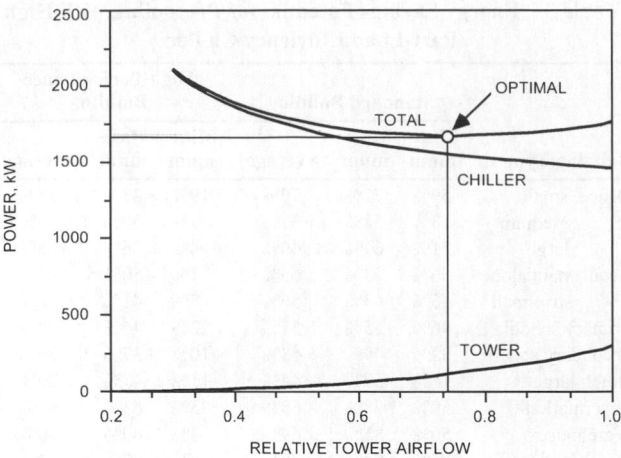

Fig. 25   Tradeoffs Between Chiller Power and Fan Power with Tower Airflow

operation is lower water pressure drops across the spray nozzles, which results in lower pumping power requirements. However, at very low pressure drops, inadequate spray distribution may adversely affect the thermal performance of the cooling tower.

Most cooling towers use multiple-speed rather than continuously adjustable variable-speed fans. In this case, it is not optimal to operate all tower cells under all conditions. The optimal number of cells operating and individual fan speeds depend on the system characteristics and ambient conditions. However, simple relationships exist for the best sequencing of cooling tower fans as capacity is added or removed. When additional tower capacity is required, Braun et al. (1989b) showed that, in almost all practical cases, the speed of the tower fan operating at the lowest speed (including fans that are off) should be increased first. The rules for bringing cell fans online are as follows:

Sequencing Rules

- **All variable-speed fans:** Operate all cells with fans at equal speeds.
- **Multiple-speed fans:** Activate lowest-speed fans first when adding tower capacity. Reverse for removing capacity.
- **Variable/multiple-speed fans:** Operate all cells with variable-speed fans at equal speeds. Activate lowest-speed fans first when adding tower capacity with multiple-speed fans. Add multiple-speed fan capacity when variable-speed fan speeds match the fan speed associated with the next multiple-speed fan increment to be added.

Similarly, for removing tower capacity, the highest fan speeds are the first to be reduced and sequences defined here are reversed.

These guidelines were derived by evaluating incremental power changes for fan sequencing. For two-speed fans, the incremental power increase associated with adding a low-speed fan is less than that for increasing one to high speed if the low speed is less than 79% of the high fan speed. In addition, if the low speed is greater than 50% of the high speed, then the incremental increase in airflow is greater (and therefore thermal performance is better) for adding the low-speed fan. Most commonly, the low speed of a two-speed cooling tower fan is between one-half and three-quarters of full speed. In this case, tower cells should be brought online at low speed before any operating cells are set to high speed. Similarly, the fan speeds should be reduced to low speed before any cells are brought offline.

For three-speed fans, low speed is typically greater than or equal to one-third of full speed, and the difference between the high and

intermediate speeds is equal to the difference between the intermediate and low speeds. In this situation, the best sequencing strategy is to activate the lowest fan speeds first when adding tower capacity and deactivate the highest fan speeds first when removing capacity. Typical three-speed combinations that satisfy these criteria are (1) one-third, two-thirds, and full speed or (2) one-half, three-quarters, and full speed.

Another issue related to control of multiple cooling tower cells with multiple-speed fans concerns the distribution of water flow to the individual cells. Typically, water flow is divided equally among the operating cells. Even though the overall thermal performance of the cooling tower is best when the flow is divided such that the ratio of water-to-airflow rates is identical for all cooling tower cells, equal water flow distribution results in near-optimal performance.

Near-Optimal Tower Airflow

Figure 25 illustrates the tradeoff between the chiller and cooling tower fan power associated with increasing tower airflow for variable-speed fans. As airflow increases, fan power increases with a cubic relationship. At the same time, there is a reduction in the temperature of the water supplied to the condenser of the chiller, resulting in lower chiller power consumption. The minimum total power occurs at a point where the rate of increase in fan power with airflow is equal to the rate of decrease in chiller power. Near the optimum, the total power consumption is not very sensitive to the control. This "flat" optimum indicates extreme accuracy is not needed to determine the optimum control. In general, it is better to have too high rather than too low a fan speed.

Braun et al. (1989b) showed that the tower control that minimizes the instantaneous power consumption of a cooling plant varies as a near-linear function of the load over a wide range of conditions. Although optimal control depends on the ambient wet-bulb temperature, this dependence is small compared to the load effect. Figure 26 shows an example of how the optimal tower control varies for a specific plant. The tower airflow as a fraction of the design capacity is plotted as a function of load relative to design load for two different wet-bulb temperatures. For a 20°F change in wet-bulb temperature, the optimal control varies only about 5% of the tower capacity. This difference in control results in less than a 1% difference in the plant power consumption. Figure 26 also shows that linear functions work well in correlating the optimal control over a wide range of loads for the two wet-bulb temperatures. Given the insensitivity to wet-bulb temperature and the fact that the load is highly correlated with wet bulb, a single linear relationship is adequate in correlating the optimal tower control in terms of load.

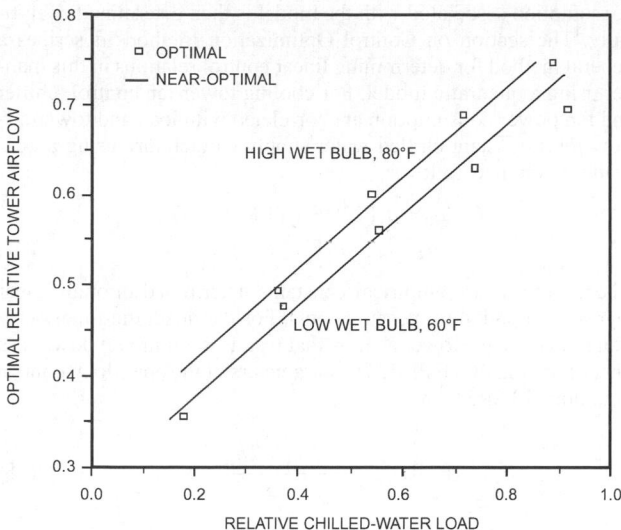

Fig. 26   Example of Optimal Tower Fan Control

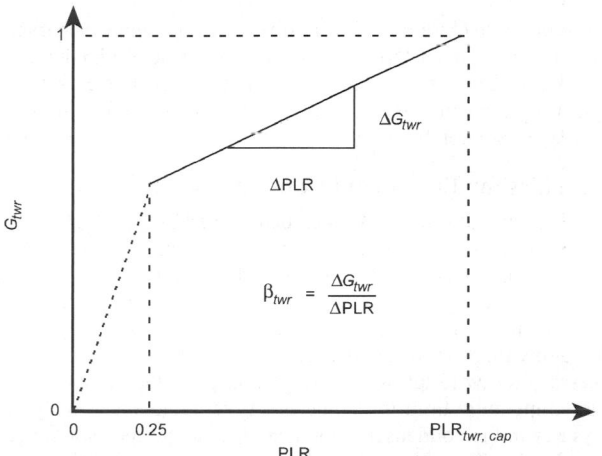

Fig. 27   Fractional Tower Airflow Versus Part-Load Ratio

Figure 27 depicts the general form to determine tower airflow as a function of load. The (unconstrained) relative tower airflow is computed as a linear function of the part-load ratio as

$$G_{twr} = 1 - \beta_{twr} (PLR_{twr,cap} - PLR) \text{ for } 1.0 < PLR < 0.25 \quad (23)$$

where

G_{twr} = tower airflow divided by maximum airflow with all cells operating at high speed

PLR = chilled-water load divided by design total chiller plant cooling capacity (part-load ratio)

$PLR_{twr,cap}$ = part-load ratio (value of PLR) at which tower operates at its capacity ($G_{twr} = 1$)

β_{twr} = slope of relative tower airflow (G_{twr}) versus part-load ratio (PLR) function

The linear relationship between airflow and load is only valid for loads greater than about 25% of the design load. For many installations, chillers do not operate at these small loads. However, for those situations in which chiller operation is necessary below 25% of full load, the tower airflow should be ramped to zero as the load goes to zero according to

$$G_{twr} = 4PLR \left[1 - \beta_{twr} (PLR_{twr,cap} - 0.25)\right] \text{ for } PLR < 0.25 \quad (24)$$

Table 4   Parameter Estimates for Near-Optimal Tower Control Equation

Parameter	One-Speed Fans	Two-Speed Fans	Variable-Speed Fans
$PLR_{twr,cap}$	PLR_0	$\sqrt{2}\,PLR_0$	$\sqrt{3}\,PLR_0$
β_{twr}	$\dfrac{1}{PLR_{twr,cap}}$	$\dfrac{2}{3PLR_{twr,cap}}$	$\dfrac{1}{2PLR_{twr,cap}}$

Note:
$$PLR_0 = \frac{1}{\sqrt{\dfrac{P_{ch,des}}{P_{twr,des}} S_{cwr,des}(a_{twr,des} + r_{twr,des})}}$$

The results of either Equation (23) or (24) must be constrained between 0 and 1. This fraction of tower capacity is then converted to a tower control using the sequencing rules of the section on Near-Optimal Tower Fan Sequencing.

The variables of the open-loop linear control Equation (23) that yield near-optimal control depend on the system's characteristics. Detailed measurements may be taken over a range of conditions and used to accurately estimate these variables. However, this requires measuring component power consumption along with considerable time and expertise, and may not be cost effective unless performed by on-site plant personnel. Alternatively, simple estimates of these parameters may be obtained using design data.

Open-Loop Parameter Estimates Using Design Data. Good estimates for the parameters of Equation (23) may be determined analytically using design information as summarized in Table 4. These estimates were derived by Braun and Diderrich (1990) by applying optimization theory to a simplified mathematical model of the chiller and cooling tower, assuming that the tower fans are sequenced in a near-optimal manner. In general, these estimates are conservative in that they should provide greater rather than less than the optimal tower airflow. The results given in Table 4 for variable-speed fans should also provide adequate estimates for three-speed fans.

Design factors that affect the parameter estimates given in Table 4 are the (1) ratio of chiller power to cooling tower fan power at design conditions $P_{ch,des}/P_{twr,des}$, (2) sensitivity of chiller power to changes in condenser water return temperature at design conditions $S_{cwr,des}$, and (3) sum of the tower approach and range at design conditions $(a_{twr,des} + r_{twr,des})$. Chiller power consumption at design conditions is the total power consumption of all plant chillers operating at their design cooling capacity. Likewise, the design tower fan power is the total power associated with all tower cells operating at high speed. As the ratio of chiller power to tower fan power increases, it becomes more beneficial to operate the tower at higher airflows. This is reflected in a decrease in the part-load ratio at which the tower reaches its capacity, $PLR_{twr,cap}$. If the tower airflow were free (i.e., zero fan power), then $PLR_{twr,cap}$ would go to zero, and the best strategy would be to operate the towers at full capacity independent of the load. A typical value for the ratio of the chiller power to the cooling tower fan power at design conditions is 10.

The chiller sensitivity factor $S_{cwr,des}$ is the incremental increase in chiller power for each degree increase in condenser water temperature as a fraction of the power or

$$S_{cwr,des} = \frac{\text{Change in chiller power}}{\text{Change in cond. water return temp.} \times \text{Chiller power}} \quad (25)$$

If the chiller power increases by 2% for a 1°F increase in condenser water temperature, $S_{cwr,des}$ is equal to 0.02/°F. A large sensitivity factor means that the chiller power is very sensitive to the cooling tower control favoring operation at higher airflow rates (low $PLR_{twr,cap}$). The sensitivity factor should be evaluated at design conditions using chiller performance data. Typically, the sensitivity factor is between 0.01 and 0.03/°F. For multiple chillers with different performance characteristics, the sensitivity factor at design conditions is estimated as

$$S_{cwr,des} = \frac{\sum\limits_{i=1}^{N_{ch}} S_{cwr,des,i} P_{ch,des,i}}{\sum\limits_{i=1}^{N_{ch}} P_{ch,des,i}} \qquad (26)$$

where $S_{cwr,des,i}$ is the sensitivity factor and $P_{ch,des,i}$ is the power consumption for the ith chiller at the design conditions, and N_{ch} is the total number of chillers.

The design approach to wet bulb $a_{twr,des}$ is the temperature difference between the condenser water supply and the ambient wet bulb for the tower, operating at its air and water flow capacity at plant design conditions. The design range $r_{twr,des}$ is the water temperature difference across the tower at these same conditions (condenser water return minus supply temperature). The sum of $a_{twr,des}$ and $r_{twr,des}$ is the temperature difference between the tower inlet and the ambient wet bulb and represents a measure of the tower's capability to reject heat to ambient relative to the system requirements. A small temperature difference (tower approach plus range) results from a high tower heat transfer effectiveness or high water flow rate and yields lower condenser water temperatures with lower chiller power consumption. Typical values for the design approach and range are 7 and 10°F.

The part-load ratio associated with the tower operating at full capacity, $\text{PLR}_{twr,cap}$ may be greater than or less than one. Values less than unity imply that, from an energy point of view, the tower is not sized for optimal operation at design load conditions and that it should operate at its capacity for a range of loads less than the design load. Values of $\text{PLR}_{twr,cap}$ greater than one imply that the tower is oversized for the design load and that it should never operate at its capacity.

For multiple chillers with very different performance characteristics, different open-loop parameters may be used for any combination of operating chillers. The sensitivity factors and chiller design power used to determine the open-loop control parameters in Table 4 should be estimated for each combination of operating chillers, and the part-load ratio used in Equation (23) should be determined using the design capacity for the operating chillers (not all chillers). In this case, N_{ch} in Equation (26) represents the number of operating chillers.

Open-Loop Parameter Estimates Using Plant Measurements. Energy consumption can be reduced slightly by determining the open-loop control parameters from plant measurements. However, this results in additional complexity associated with implementation. One method for estimating the open-loop control parameters of Equation (23) from plant measurements involves performing a set of one-time trial-and-error experiments. At a given set of conditions (i.e., cooling load and ambient conditions), the optimal tower control is estimated by varying the fan settings and monitoring the total chiller and fan power consumption. Each tower control setting and load condition must be maintained for a sufficient time for the power consumption to approach steady-state and to hold the chilled-water supply temperature constant. The control setting that produces the minimum total power consumption is deemed optimal. This set of experiments is performed for a number of chilled-water cooling loads and the best-fit straight line through the resulting data points is used to estimate the parameters of Equation (23). As initial control settings for each load, Equation (23) may be used with estimates from design data as summarized in the previous section.

Another method for estimating the variables of Equation (23) uses an empirical model for total power consumption that is fit to plant measurements. The control that minimizes the power

consumption associated with the model is then determined analytically. The section on Control Optimization Methods describes a general method for determining linear control relations in this manner using a quadratic model. For cooling tower fan control, chiller and fan power consumption are correlated with load and tower airflow for a constant chilled-water supply temperature using a quadratic function as follows:

$$P = a_0 + a_1 \text{PLR} + a_2 \text{PLR}^2 + a_3 G_{twr}$$
$$+ a_4 G^2_{twr} + a_5 \text{PLR} \times G_{twr} \qquad (27)$$

where a_0 to a_5 are empirical constants determined through linear regression applied to measurements. For the quadratic function of Equation (27), the tower airflow that results in minimum power is a linear function of the PLR. The parameters of the open-loop control Equation (23) are then

$$\text{PLR}_{twr,cap} = -\frac{a_3 + 2a_4}{a_5} \qquad (28)$$

$$\beta_{twr} = -\frac{a_5}{2a_4} \qquad (29)$$

For multiple chillers with very different performance characteristics, different open-loop parameters can be determined for any combination of operating chillers. In this case, separate correlations for near-optimal airflow or power consumption must be determined for each chiller combination.

Overrides for Equipment Constraints

The fractional tower airflow as determined by Equations (23) or (24) must be bounded between 0 and 1 according to the physical constraints of the equipment. Additional constraints on the temperature of the supply water to the chiller condensers are necessary to avoid potential chiller maintenance problems. Many (older) chillers have a low limit on the condenser water supply temperature that is necessary to avoid lubrication migration from the compressor. A high-temperature limit is also necessary to avoid excessively high pressures in the condenser, which can lead to compressor surge in centrifugal chillers. If condenser water temperature falls below the low limit, then it is necessary to override the open-loop tower control and reduce tower airflow to go above this limit. Similarly, if the high limit is exceeded, then tower airflow should be increased as required.

Implementation

Before commissioning, the parameters of the open-loop control Equation (23) must be specified. These parameters are estimated using Table 4. After the system is in operation, these parameters may be fine-tuned with measurements as outlined previously. If multiple chillers have significantly different performance characteristics, it may be advantageous to determine different parameters for Equation (23) depending on the combination of operating chillers.

The relative tower airflow must be converted to a specific set of tower fan settings using the sequencing rules defined previously. This involves defining a relationship (i.e., table) for fan settings as a function of tower airflow. The table is constructed by defining the best fan settings for each possible increment of airflow. The conversion process between the continuous output of Equations (23) or (24) and the fan control involves choosing the set of discrete fan settings from the table that produces a tower airflow closest to the desired flow. However, in general, it is better to have greater rather than less than the optimal airflow. A good general rule is to choose the set of discrete fan controls that results in a relative airflow that is closest to, but not more than 10% less than, the output of Equations (23) or (24).

With the parameters of Equation (23) specified, the following procedure is applied at each decision interval (e.g., 15 min) to determine the tower control:

1. If the temperature of supply water to the chiller condenser is less than the low limit, then reduce tower airflow by one increment according to the near-optimal sequencing rules and exit the algorithm. Otherwise go to step 2.
2. If the temperature of supply water to the chiller condenser is greater than the high limit, then increase tower airflow by one increment according to the near-optimal sequencing rules and exit the algorithm. Otherwise go to step 3.
3. Determine the chilled-water load relative to the design load.
4. If the chilled-water load has changed by a significant amount (e.g., 10%) since the last control change, then go to step 5. Otherwise, exit the algorithm.
5. If the part-load ratio is greater than 0.25, then compute the near-optimal tower airflow as a fraction of tower capacity G_{twr} with Equation (23). Otherwise, determine G_{twr} with Equation (24).
6. Limit G_{twr} to keep the change from the previous decision interval less than a minimum value (e.g., less than 0.1 change).
7. Restrict the value of G_{twr} between 0 and 1.
8. Convert the value of G_{twr} to a specific set of control functions for each of the tower cell fans according to the near-optimal sequencing rules.

This procedure requires some estimate of the chilled-water load, along with a measurement of the condenser water supply temperature. However, the accuracy of the load estimates is not extremely critical. In general, near-optimal control determined with load estimates that are accurate to within 5 to 10% results in total power consumption that is within 1% of the minimum. The best method for determining the chilled-water load is from the product of the measured chilled-water flow rate and the temperature difference between the chilled-water return and supply. For systems that use constant flow pumping to the chillers, the flow rates may be estimated from design data for the pumps and system pressure-drop characteristics.

Example 1. Consider an example plant consisting of four 550 ton chillers with four cooling tower cells, each having two-speed fans. Each chiller consumes approximately 330 kW at the design capacity, and each tower fan uses 40 kW at high speed. At design conditions, the chiller power increases approximately 6.6 kW for a 1°F increase in condenser water temperature, giving a sensitivity factor of 6.6/330 or 0.02/°F. The tower design approach and range from manufacturer's data are 7 and 10°F.

Solution:

The first step in applying the open-loop control algorithm to this problem is determining the parameters of Equation (23) from the design data. From Table 4, the part-load ratio at which operation of the tower is at its capacity is estimated for the two-speed fans as

$$PLR_{twr,cap} = \frac{1}{\sqrt{\frac{1}{2}(0.02/°F)\frac{4 \times 330 \text{ kW}}{4 \times 40 \text{ kW}}(7 + 10°F)}} = 0.84$$

and the slope of the fractional airflow versus part-load ratio is estimated to be

$$\beta_{twr} = \frac{2}{3 \times 0.84} = 0.79$$

Given these parameters and the part-load ratio, the fractional tower airflow is estimated as

```
IF (PLR > 0.25) THEN
    Gtwr = 1 – βtwr(PLRtwr,cap – PLR)
ELSE
    Gtwr = 4PLR[1 – βtwr(PLRtwr,cap – 0.25)]
    Gtwr = MIN[1, MAX(0, Gtwr)]
```

To convert G_{twr} into a specific tower control, the tower sequencing must be defined. The following table gives this information in a form that specifies the relationship between G_{twr} and tower control for this example.

Cooling Tower Fan Sequencing for Example 1

Sequence No.	G_{twr}	Tower Fan Speeds			
		Cell #1	Cell #2	Cell #3	Cell #4
1	0.125	Low	Off	Off	Off
2	0.250	Low	Low	Off	Off
3	0.375	Low	Low	Low	Off
4	0.500	Low	Low	Low	Low
5	0.625	High	Low	Low	Low
6	0.750	High	High	Low	Low
7	0.875	High	High	High	Low
8	1.000	High	High	High	High

For a specific chilled-water load, the fan control should be the sequence of tower fan settings from the table that results in a value of G_{twr} that is closest to, but not more than 10% less than, the output of Equations (23) or (24). Note that this example assumes that proper water flow can be maintained over all cooling tower cells.

CHILLED-WATER RESET WITH FIXED-SPEED PUMPING

Figure 28 shows a common configuration using fixed-speed chilled-water pumps with two-way valves at the cooling coils. A two-way bypass valve controlled to maintain a fixed pressure difference between the main supply and return lines is used to ensure relatively constant flow through chiller evaporators and reduce pressure drop and pumping costs at low loads. However, additional pump and chiller power savings can be realized by adjusting the chilled-water supply temperature to keep some cooling coil valves open and thereby minimize the bypass flow.

Ideally, the chilled-water temperature should be adjusted to maintain all discharge air temperatures with a minimal number of cooling-coil control valves in a saturated (fully open) condition. The procedure described in this section is designed to accomplish this goal in a reliable and stable manner that reacts quickly to changing conditions.

Pump Sequencing

In plants where one chilled-water pump is dedicated to each chiller, the sequencing of chilled-water pumps is defined by the sequencing of chillers. In some installations, the chiller pumps are not

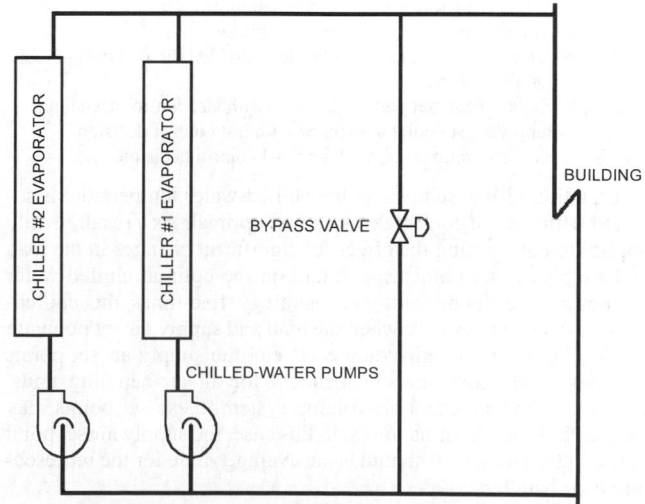

Fig. 28 Typical Chilled-Water Distribution for Fixed-Speed Pumping

dedicated to chillers, but instead are arranged in parallel, sharing common headers. In this case, the order for bringing pumps online and offline and the conditions for adding or removing chilled-water pump capacity must be specified. For pumps of different capacities, the logical order for bringing pumps online is from small to large. For pumps of similar capacity, the most efficient pumps should be brought online first and taken offline last.

Optimal Chilled-Water Temperature

One method for determining the optimal chilled-water temperature is to monitor the water control valve positions of "representative" air handlers and to adjust the set temperature incrementally at fixed decision intervals until a single control valve is fully open. The representative air handlers should be chosen to include load diversity at all times and ensure reliable data. One difficulty of this control approach is that valve position data are often unreliable. The valve could be stuck open or the saturation indicator could be faulty. This problem can be overcome by also monitoring discharge air temperatures, using them as a consistency check on valve position data. If a valve is unsaturated, this implies that the coil has sufficient capacity to maintain the discharge air temperature near the set point. Conversely, if a valve remains saturated at 100% open, the discharge air temperature should ultimately increase above the set point. These considerations lead to the following simple rules for increasing or decreasing the chilled-water set point in response to valve position and discharge air temperature data.

- If all water valves are unsaturated or the discharge air temperatures associated with all saturated valves are lower than the set point, increase the chilled-water set temperature.
- If more than one valve is saturated at 100% open and their corresponding discharge air temperatures are greater than their set points, decrease the chilled-water temperature.

In implementing these rules, a fixed increment for increasing or decreasing the chilled-water temperature must be chosen. A small increment results in more stable control, but also results in a slow response to sudden changes in load or supply air temperature set points. Using a first-order approximation, the chilled-water temperature can be reset in response to sudden changes in load and supply air temperature set point according to

$$t_{chws} = t_{as} - \frac{\text{PLR}}{\text{PLR}_o}(t_{as,o} - t_{chws,o}) \qquad (30)$$

where

t_{chws} = new chilled-water set-point temperature
t_{as} = current supply air set-point temperature
PLR = current part-load ratio (chiller load divided by total design load for all chillers)
$t_{chws,o}$ = chilled-water set point associated with last control decision
$t_{as,o}$ = supply air set point associated with last control decision
PLR_o = part-load ratio associated with last control decision

Equation (30) assumes that the chilled-water temperature associated with the last control decision was optimal. As a result, it only applies to anticipating the effects of significant changes in the load and supply air set-point temperature on the optimal chilled-water set point. The "bump-and-wait" strategy fine-tunes the chilled-water supply temperature when the load and supply air set point are stable. For a variable-air-volume system, the supply air set points are most often constant and identical for all air-handling units. However, for a constant-air-volume system, these set points may vary with different air handlers. In this case, the supply air set point to use in Equation (30) should be an average value for the representative air handlers.

Equation (30) indicates that the optimal chilled-water supply temperature increases with increasing supply air temperature and decreasing load. This is because these changes cause the cooling-coil

valves to close; optimal control involves keeping at least one valve open. Increasing supply air temperature causes the cooling-coil valves to close somewhat because of a larger average temperature difference for heat transfer between the water and air. A lower load requires smaller air-to-water temperature differences, which also leads to control valves closing.

Overrides for Equipment and Comfort Constraints

For a given chiller load, the chilled-water temperature has both upper and lower limits. The lower limit is necessary to avoid ice formation on the evaporator tubes of the chiller. This limit depends primarily on the load in relation to the size of the evaporator or, in other words, the temperature difference between the chilled water and refrigerant. At small temperature differences (large area or small load), the evaporator can tolerate a lower chilled-water temperature to avoid freezing than at large temperature differences. The lower limit on the chilled-water set point should be evaluated at the design load, because the overall system performance is improved by increasing chilled-water temperature above this limit for loads less than design. This lower limit can range from 38 to 44°F when chilled water is used for dehumidification, to 55°F for TABS, and as high as 60°F for chilled water serving radiant or chilled-beam cooling systems.

An upper limit on the chilled-water temperature arises from comfort constraints associated with the zones and the possibility of microbial growth associated with high humidities. For the available flows, the chilled-water temperature should be low enough to provide discharge air at a temperature and humidity sufficient to maintain all zones in the comfort region and avoid microbial growth. This upper limit varies with both load and entering air conditions and is accounted for by monitoring the zone conditions to ensure that they are in the comfort zone. If zone temperatures or humidities are not within reasonable bounds, then the discharge air temperature set point should be lowered. For radiant panel, chilled-beam, or TAB systems, there is no specific upper chilled-water limit as long as the load is satisfied.

Implementation

At each decision interval (e.g., 5 min), the following algorithm would be applied for determining the optimal chilled-water set-point temperature:

1. Determine the time-averaged total chilled-water load for the previous decision interval.
2. If the chilled-water load or supply air set-point temperature has changed by a significant amount (e.g., 10%) since the last control change, then estimate a new optimal chilled-water set point with Equation (30) and go to step 6. Otherwise, go to step 3.
3. Determine the time-averaged position of (or controller output for) the cooling-coil water valves and corresponding discharge air temperatures for representative air handlers.
4. If more than one valve is saturated at 100% open and their corresponding supply air temperatures are greater than set point (e.g., 1°F), then decrease the chilled-water temperature by a fixed amount (e.g., 0.5°F) and go to step 6. Otherwise, go to step 5.
5. If all water valves are unsaturated or the supply air temperatures associated with all valves that are saturated are lower than the set point, then raise the chilled-water set temperature by a fixed amount (e.g., 0.5°F). Otherwise, exit the algorithm with the chilled-water set point unchanged.
6. Limit the chilled-water set-point temperature between the upper and lower limits dictated by comfort, humidity, and equipment safety.

Implementing this algorithm requires some estimate of the chilled-water load, along with a measurement of the discharge air temperatures and control valve positions. However, a highly accurate estimate of the load is not necessary.

CHILLED-WATER RESET WITH VARIABLE-SPEED PUMPING

Figure 29 shows a common configuration for systems using variable-speed chilled-water pumps with primary/secondary water loops. The primary pumps are fixed speed and are generally sequenced with chillers to provide a relatively constant flow of water through the chiller evaporators. The secondary chilled-water pumps are variable speed and are typically controlled to maintain a specified set point for pressure difference between supply and return flows for the cooling coils.

Although variable-speed pumps are usually used with primary/secondary chilled-water loops, they may also be applied to systems with a single chilled-water loop. In either case, variable-speed pumps offer the potential for a significant operating cost saving when both chilled-water and pressure differential set points are optimized in response to changing loads. This section presents an algorithm for determining near-optimal values of these control variables.

Optimal Differential Pressure Set Points

In practically all variable-speed chilled-water pumping applications, pump speed is controlled to maintain a constant pressure differential between the main chilled-water supply and return lines. However, this approach is not optimal. To maintain a constant pressure differential with changing flow, the control valves for the air-handling units must close as the load (i.e., flow) is reduced, resulting in an increase in the flow resistance. The best strategy for a given chilled-water set point is to reset the differential pressure set point to maintain all discharge air temperatures with at least one control valve in a saturated (fully open) condition. This results in a relatively constant flow resistance and greater pump savings at low loads. With variable differential pressure set points, optimizing the chilled-water loop is described in terms of finding the chilled-water temperature that minimizes the sum of chiller and pumping power, with pump control dependent on set point and load.

Near-Optimal Chilled-Water Set Point

The optimal chilled-water supply temperature at a given load results from a tradeoff between chiller and pumping power, as illustrated in Figure 30. As the chilled-water temperature increases, chiller power is reduced due to a reduction in the lift requirements of the chiller. For a higher set temperature, more chilled-water flow is necessary to meet the load requirements, and the pumping power requirements increase. The minimum total power occurs at a point where the rate of increase in pumping power with chilled-water

temperature is equal to the rate of decrease in chiller power. This optimal set point moves to lower values as the load increases.

Braun et al. (1989b) demonstrated that the optimal chilled-water set point varies as a near-linear function of both load and wet-bulb temperature over a wide range of conditions. Figure 31 shows an example of how the optimal set point varies for a specific plant. The set point is plotted as a function of load relative to design load for two different wet-bulb temperatures. In general, the optimal chilled-water temperature decreases with load because pump power becomes a larger fraction of total power. A lower set-point limit is set to avoid conditions that could form ice on evaporator tubes or too high a chiller "lift," and an upper limit is established to ensure adequate cooling-coil dehumidification. For a given load, the chilled-water set point increases with wet-bulb temperature because energy transfer across each cooling coil is proportional to the difference between its entering air wet-bulb temperature and the entering water temperature (the chilled-water set point). For a constant load, this temperature difference is constant and the chilled-water supply temperature increases linearly with entering air wet-bulb temperature.

The results in Figure 31 were obtained for a system where both the chilled-water supply and supply air set points to the zones were optimized. For this case, the supply air temperatures varied between

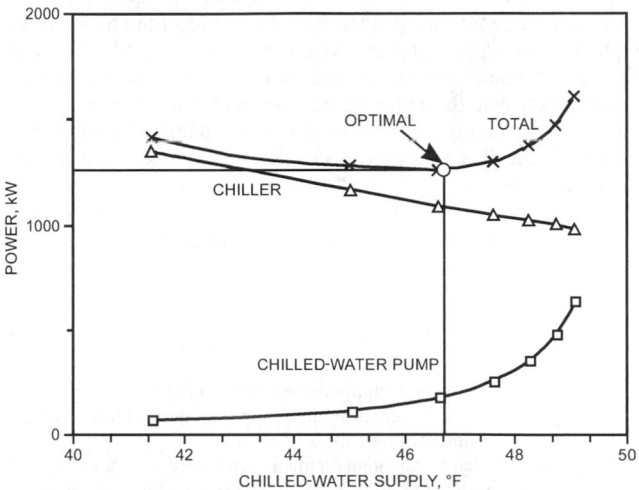

Fig. 30 Tradeoff of Chiller and Pump Power with Chilled-Water Set Point

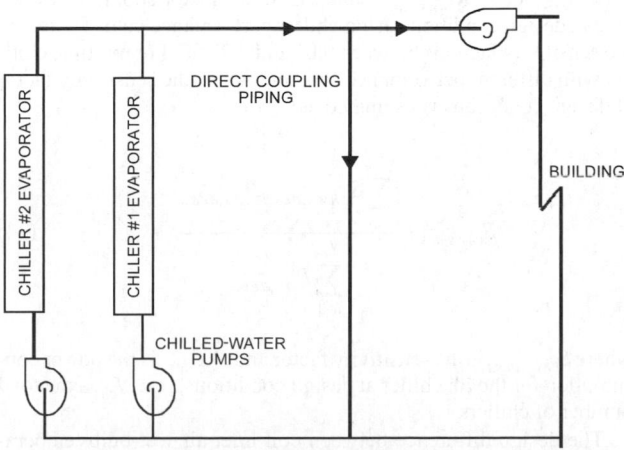

Fig. 29 Typical Chilled-Water Distribution for Primary/Secondary Pumping

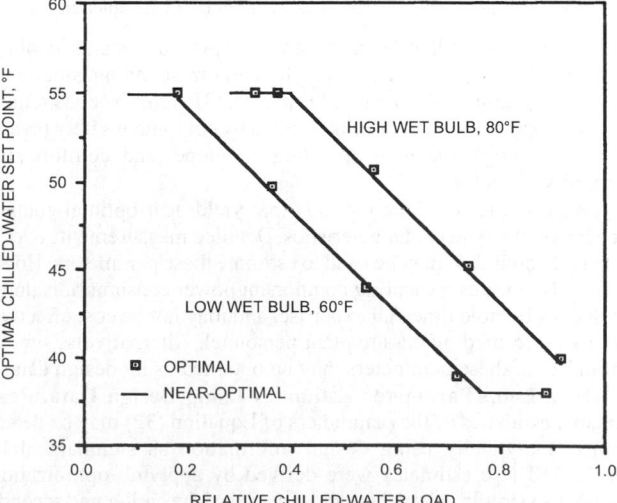

Fig. 31 Comparisons of Optimal Chilled-Water Temperature

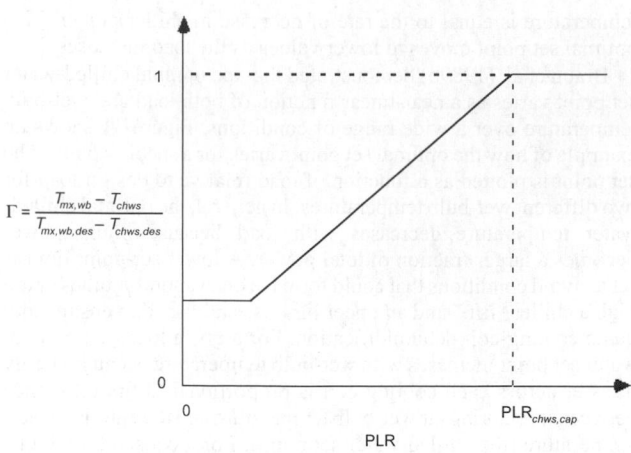

Fig. 32   Dimensionless Chilled-Water Set Point Versus Part-Load Ratio

Table 5   Parameter Estimates for Near-Optimal Chilled-Water Set Point Equation

Parameter	Estimate
$PLR_{chws,cap}$	$\sqrt{\dfrac{1}{3} \times \dfrac{P_{ch,des}}{P_{chwp,des}}} S_{chws,des}(t_{mx,wb,des} - t_{chws,des})$
β_{chws}	$\dfrac{0.5}{PLR_{chws,cap}}$

55°F at high loads and 60°F at low loads. More typically, supply air temperatures are constant at 55°F, and the variation in chilled-water supply temperature is smaller than that shown in Figure 30.

Figure 32 depicts the general form for an algorithm to determine chilled-water supply set points as a function of load and the average wet-bulb temperature entering the cooling coils. A normalized difference between the entering air wet-bulb temperature and the chilled-water supply temperature is shown as a linear function of the part-load ratio. The (unconstrained) chilled-water set point is determined as

$$t_{chws} = t_{mx,wb} - \Gamma(t_{mx,wb,des} - t_{chws,des}) \qquad (31)$$

where

$$\Gamma = 1 - \beta_{chws}(PLR_{chws,cap} - PLR) \qquad (32)$$

t_{chws} = chilled-water supply temperature set point
$t_{mx,wb}$ = average or "representative" wet-bulb temperature of air entering cooling coils
$t_{chws,des}$ = chilled-water supply temperature at design conditions
$t_{mx,wb,des}$ = wet-bulb temperature of air entering cooling coils at design conditions
PLR = chilled-water load divided by the total chiller cooling capacity (part-load ratio)
$PLR_{chws,cap}$ = part-load ratio (value of PLR) at which $\Gamma = 1$
β_{chws} = slope of the Γ versus part-load ratio (PLR) function

For RCP and chilled-beam systems that provide sensible cooling only, replace $t_{mx,wb,des}$ and $t_{mx,wb,des}$ by the corresponding zone operative temperatures. The result of Equation (31) must be constrained between upper and lower limits dictated by equipment safety (evaporator freezing), machine operating envelope, and comfort and humidity concerns.

The variables of Equation (32) that yield near-optimal control depend on the system characteristics. Detailed measurements over a range of conditions may be used to estimate these parameters. However, this requires measuring component power consumption along with considerable time and expertise, and may not be cost effective unless performed by on-site plant personnel. Alternatively, simple estimates of these parameters may be obtained using design data.

Open-Loop Parameter Estimates Using Design Data. Reasonable estimates of the parameters of Equation (32) may be determined analytically using design information as summarized in Table 5. These estimates were derived by applying optimization theory to a simplified mathematical model of the chiller and secondary-loop water pumps, assuming that a differential pressure reset strategy is used, pump efficiencies are constant, and the supply air

temperature is not varied in response to changes in chilled-water supply temperature. In general, these parameter estimates are conservative in that they should provide a relatively low estimate of the optimal chilled-water set point.

The design factors that affect the parameter estimates given in Table 5 are the (1) ratio of the chiller power to chilled-water pump power at design conditions $P_{ch,des}/P_{chwp,des}$, (2) chiller power's sensitivity to changes in chilled-water temperature at design conditions $S_{chws,des}$, and (3) difference between design entering air wet-bulb temperature to the cooling coil and the chilled-water supply temperature ($t_{mx,wb,des} - t_{chws,des}$).

Chiller power consumption at design conditions is the total power consumption of all plant chillers operating at their design cooling capacity. Likewise, the design pump power is the total power associated with all secondary chilled-water supply pumps operating at high speed. As the ratio of chiller power to pump power increases, it becomes more beneficial to operate the chillers at higher chilled-water temperatures and the pumps at higher flows. This is reflected in an increase in $PLR_{chws,cap}$. If chiller power were free, then $PLR_{chws,cap}$ would go to zero, and the best strategy would be to operate the chillers at the minimum possible set point, resulting in low chilled-water flow rates. Typical values for the ratio of chiller power to pump power at design conditions are between 10 and 20, depending primarily on whether primary/secondary pumping is used.

Chiller sensitivity factor $S_{chws,des}$ is the incremental increase in chiller power for each degree decrease in chilled-water temperature as a fraction of the power:

$$S_{cwr,des} = \frac{\text{Increase in chiller power}}{\text{Decrease in chilled-water temp.} \times \text{Chiller power}} \qquad (33)$$

If chiller power increases by 2% for a 1°F decrease in chilled-water temperature, then $S_{chws,des}$ is equal to 0.02/°F. A large sensitivity factor means that chiller power is very sensitive to the set point control favoring operation at higher set point temperatures and flows (higher $PLR_{chws,cap}$). The sensitivity factor should be evaluated at design conditions using chiller performance data. Typically, the sensitivity factor is between 0.01 and 0.03/°F. For multiple chillers with different performance characteristics, the sensitivity factor at design conditions is estimated as

$$S_{chws,des} = \frac{\displaystyle\sum_{i=1}^{N_{ch}} S_{chws,des,i} P_{ch,des,i}}{\displaystyle\sum_{i=1}^{N_{ch}} P_{ch,des,i}} \qquad (34)$$

where $S_{chws,des,i}$ is the sensitivity factor and $P_{ch,des,i}$ is the power consumption for the ith chiller at design conditions, and N_{ch} is the total number of chillers.

The design difference between coil inlet air wet-bulb temperature and entering water temperature should be evaluated for a typical air handler operating at design load and flows. A small temperature difference results from a high coil heat transfer effectiveness or high water flow rate, allowing higher chilled-water temperatures with lower chiller power consumption. This is evident from Equation (31), where chilled-water set point decreases linearly with $(t_{mx,wb,des} - t_{chws,des})$ for a given Γ, and Γ is inversely related to the square root of $(t_{mx,wb,des} - t_{chws,des})$. Typically, this temperature difference is about 20°F.

Example 2. Consider an example plant with primary/secondary chilled-water pumping. There are four 550 ton chillers, each with a dedicated primary pump. Each chiller consumes approximately 330 kW at design capacity. At design conditions, chiller power increases approximately 6.6 kW for a 1°F decrease in chilled-water temperature, giving a sensitivity factor of 6.6/330 or 0.02/°F. The design chilled-water set point is 42°F, and the coil entering wet-bulb temperature is 62°F at design conditions. The secondary loop uses three identical 60 hp chilled-water pumps: one with a variable-speed and two with fixed-speed motors.

Solution:

The first step in applying the open-loop control algorithm to this problem is determining the parameters of Equation (32) from the design data. From Table 5, the part-load ratio at which the chilled-water temperature reaches a minimum (with the design entering wet-bulb temperature to the coils) is

$$\text{PLR}_{chws,cap} = \sqrt{\frac{1}{3} \times \frac{4 \times 330 \text{ kW}}{3 \times 60 \text{ hp} \times 0.75 \text{ kW/hp}} (0.02/°\text{F})(20°\text{F})} = 1.14$$

and the slope of the set point versus part-load ratio is estimated to be

$$\beta_{chws} = \frac{0.5}{1.14} = 0.44$$

Given these parameters and the part-load ratio, the unconstrained chilled-water set-point temperature from Equations (9) and (10) is then

$$t_{chws} = t_{mx,wb} - [1 - 0.44(1.14 - \text{PLR})]20$$

Pump Sequencing

Variable-speed pumps are sometimes used in combination with fixed-speed or other variable-speed pumps. Pump sequencing involves determining both the order and point that pumps should be brought online and offline.

Pumps should be brought online in an order that allows a continuous variation in flow rate and maximized operating efficiency of the pumps at each switch point for the specific pressure loss characteristic. For a combination of fixed- and variable-speed pumps, at least one variable-speed pump should be brought online before any fixed-speed pumps. For single-loop systems (i.e., no secondary loop) with variable-speed pumps, the pressure drop characteristics change when chillers are added or removed and the optimal sequencing of pumps depends on the sequencing of chillers.

An additional pump should be brought online whenever the current set of pumps is operating at full capacity and can no longer satisfy the differential pressure set point. This situation can be detected by monitoring the differential pressure or the controller output signal. Insufficient pump capacity leads to extended periods with differential pressures that are less than the set point and a controller output that is saturated at 100%. A pump may be taken offline whenever the remaining pumps have sufficient capacity to maintain the differential pressure set point. This condition can be determined by comparing the current (time-averaged) controller output with the controller output (time-averaged) at the point just after the last pump was brought online. The pump can be brought offline when the current output is less than the switch point value by a specified dead band (e.g., 5%).

Overrides for Equipment and Comfort Constraints

The chilled-water temperature is bounded by upper and lower limits dictated by comfort, humidity, and equipment safety concerns. However, within these bounds, the chilled-water temperature

may not always be low enough to maintain supply air set-point temperatures for the cooling coils. This situation might occur at high loads when the chilled-water flow is at a maximum and is detectable by monitoring the coil discharge air temperatures. Limits on the pressure differential set point might also be imposed to ensure adequate controllability of the cooling-coil control valves.

Implementation

Before commissioning, the parameters of the open-loop control for chilled-water set point [Equation (32)] must be estimated using the results of Table 5. After the system is in operation, these parameters may be fine-tuned with measurements as outlined in the section on Control Optimization Methods. With the parameters specified, the control algorithm is separated into two reset strategies: chilled-water temperature and pressure differential.

Chilled-Water Temperature Reset. The chilled-water supply temperature set point is reset at fixed decision intervals (e.g., 15 min) using the following procedure:

1. Determine the time-averaged position of (or controller output for) the cooling-coil water valves and corresponding discharge air temperatures for "representative" air handlers over the previous decision interval.
2. If more than one valve is saturated at 100% open and their corresponding discharge air temperatures are greater than set point (e.g., 1°F), then decrease the chilled-water temperature by a fixed amount (e.g., 0.5°F) and go to step 5. Otherwise, go to step 3.
3. Determine the total chilled-water flow and load.
4. Estimate an optimal chilled-water set point with Equations (9) and (10). Increase or decrease the actual set point in the direction of the near-optimal value by a fixed amount (e.g., 1°F).
5. Limit the new set point between upper and lower constraints dictated by comfort and equipment safety.

Pump Sequencing. Secondary pumps should be brought online or offline at fixed decision intervals (e.g., 15 min) with the following logic:

1. Evaluate the time-averaged pump controller output over the previous decision interval.
2. If the pump controller is saturated at 100%, then bring the next pump online. Otherwise, go to step 3.
3. If the pump control output is significantly less (e.g., 5%) than the value associated with the first time interval after the last pump was brought online, then bring that pump offline.

Differential Pressure Reset. The set point for differential pressure between supply and return lines should be reset at smaller time intervals than the supply water temperature reset and pump sequencing strategies (e.g., 5 min) using the following procedure:

1. Check the water valve positions (or controller output) for "representative" air handlers and determine the time-averaged values over the last decision interval.
2. If more than one valve has been saturated at 100% open, then increase the differential pressure set point by a fixed value (e.g., 5% of the design value) and go to step 4. Otherwise, go to step 3.
3. If none of the valves have been saturated, then decrease the differential pressure set point by a fixed value (e.g., 5% of the design value).
4. Limit the differential pressure set point between upper and lower constraints.

SEQUENCING AND LOADING MULTIPLE CHILLERS

Multiple chillers are normally configured in parallel and typically controlled to give identical chilled-water supply temperatures. In most cases, controlling for identical set temperatures is the best and simplest strategy. With this approach, the relative loading on operating chillers is controlled by the relative chilled-water flow rates. Typically, the distribution of flow rates to heat exchangers for both chilled and condenser water are dictated by chiller pressure drop characteristics and may be adjusted through flow balancing, but are not controlled using a feedback controller. In addition to the distribution of chilled and condenser water flow rates, the chiller sequencing affects energy consumption. Chiller sequencing defines the conditions under which chillers are brought online and offline. Simple guidelines may be established for each of these controls to provide near-optimal operation.

Near-Optimal Condenser Water Flow Distribution

In general, the condenser water flow to each chiller should be set to give identical leaving condenser water temperatures. This condition approximately corresponds to relative condenser flow rates equal to the relative loads on the chillers, even if the chillers are loaded unevenly. Figure 33 shows results for four sets of two chillers operated in parallel. The curves represent data from chillers at three different installations: (1) a 5500 ton variable-speed chiller at the Dallas-Ft. Worth airport, Texas (Braun et al. 1989b); (2) a 550 ton fixed-speed chiller at an office building in Atlanta (Hackner et al. 1984, 1985); and (3) a 1250 ton fixed-speed chiller at a large office building in Charlotte, North Carolina (Lau et al. 1985). The capacities of the chillers in the two office buildings were scaled up for comparison with the Dallas-Ft. Worth airport chiller.

The overall chiller coefficient of performance (COP) is plotted versus the difference between the condenser water return temperatures for equal chiller loading. For multiple chillers having similar performance characteristics (either variable- or fixed-speed), it is best to distribute the condenser water flow rates so that each chiller has the same leaving condenser water temperature. For situations where chillers do not have identical performance, equal leaving condenser water temperatures result in chiller performance that is close to the optimum. Even for variable- and fixed-speed chiller combinations that have very different performance characteristics, the penalty associated with using identical condenser leaving-water temperatures is small. To achieve equal condenser leaving-water temperatures, it is necessary to properly balance the condenser water flow rates at design operating conditions.

Optimal Chiller Load Distribution

Assuming identical chilled-water return and chiller supply temperatures, the relative chilled-water load for each parallel chiller

(load divided by total load) that is operating could be controlled by its relative chilled-water flow rate (flow divided by total flow). To change the relative loadings in response to operating conditions, the individual flow rates must be controlled. However, this is typically not done and it is probably sufficient to establish the load distributions based on design information and then balance the flow rates to achieve these load distributions. Alternatively, the individual chiller loads can be precisely controlled through variation of individual chiller supply water set points.

Chillers with Similar Performance Characteristics. Braun et al. (1989b) showed that, for chillers with identical design COPs and part-load characteristics, a minimum or maximum power consumption occurs when each chiller is loaded according to the ratio of its capacity to the total capacity of all operating chillers. This is equivalent to each chiller operating at equal part-load ratios (load divided by cooling capacity at design conditions). For the ith chiller, the optimal chiller loading is then

$$\dot{Q}_{ch,i}^{*} = \frac{\dot{Q}_{load}}{\sum_{i=1}^{N} \dot{Q}_{ch,des,i}} \dot{Q}_{ch,des,i} \tag{35}$$

where $\dot{Q}_{load}$ is the total chiller load, $\dot{Q}_{ch,des,i}$ is the cooling capacity of the ith chiller at design conditions, and N is the number of chillers operating.

The loading determined with Equation (35) could result in either minimum or maximum power consumption. However, this solution gives a minimum when the chillers are operating at loads greater than the point at which the maximum COP occurs (i.e., chiller COP decreases with increased loading). Typically, the maximum COP occurs at loads that are less than the nominal design capacity.

Figure 34 shows the effect of relative loading on chiller COP for different sets of identical chillers loaded at approximately 70% of their total capacities. Three of the chiller sets have maximum COPs when evenly loaded [matching the criterion of Equation (35)], whereas the fourth (Dallas-Ft. Worth fixed-speed) obtains a minimum at that point. The part-load characteristic of the Dallas-Ft. Worth fixed-speed chiller is unusual in that the maximum overall COP occurs at its maximum capacity. This chiller was retrofitted with a different refrigerant and drive motor, which derated its capacity from 8700 to 5500 tons. As a result, the evaporators and

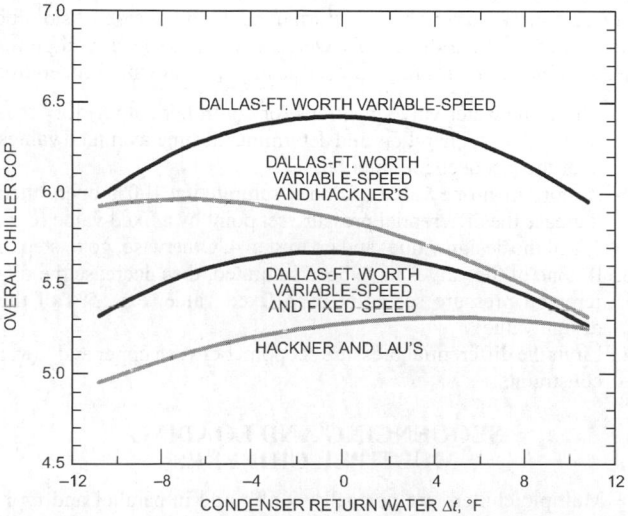

Fig. 33 Effect of Condenser Water Flow Distribution for Two Chillers In Parallel

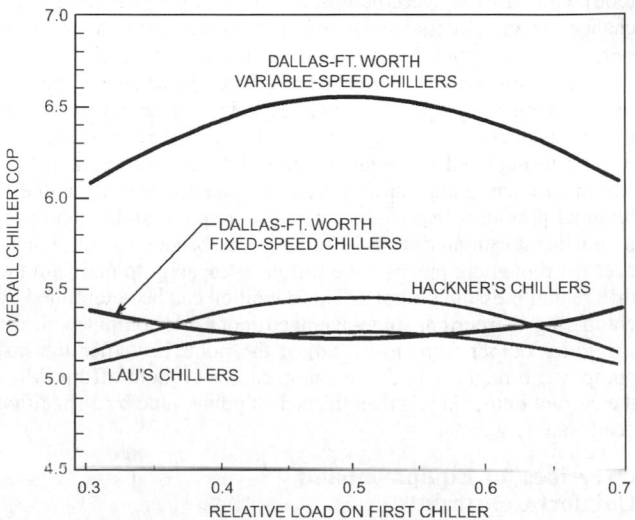

Fig. 34 Effect of Relative Loading for Two Identical Parallel Chillers

condensers are oversized for its current capacity. Overall, the penalty associated with equally loading the Dallas-Ft. Worth fixed-speed chillers is small compared with optimal loading, and this strategy is probably appropriate. However, a slight reduction in energy consumption is possible if one of the two chillers operates at full capacity. The loading criterion of Equation (35) also works well for many combinations of chillers with different performance characteristics.

To achieve specified relative chiller loadings with equal chilled-water set points, chilled-water flow rates must be properly balanced. The relative loadings of Equation (35) only depend on design information, and flow balancing can be achieved through proper design and commissioning.

Chillers with Different Performance Characteristics. For the general case of chillers with significantly different part-load characteristics, a point of minimum or maximum overall power occurs where the partial derivatives of the individual chiller's power consumption with respect to their loads are equal:

$$\frac{\partial P_{ch,i}}{\partial \dot{Q}_{ch,i}} = \frac{\partial P_{ch,j}}{\partial \dot{Q}_{ch,j}} \qquad \text{for all } i \text{ and } j \qquad (36)$$

and subject to the constraint that

$$\sum_{i=1}^{N} \dot{Q}_{ch,i} = \dot{Q}_{load} \qquad (37)$$

where $\dot{Q}_{ch,i}$ is the cooling load for the ith chiller and $\dot{Q}_{load}$ is the total cooling load.

In general, the power consumption of a chiller can be correlated as a quadratic function of cooling load and difference between the leaving condenser water and chilled-water supply temperatures according to

$$P_{ch,i} = a_{0,i} + a_{1,i}(t_{cwr,i} - t_{chws,i}) + a_{2,i}(t_{cwr,i} - t_{chws,i})^2 \qquad (38)$$
$$+ a_{3,i}\dot{Q}_{ch,i} + a_{4,i}\dot{Q}_{ch,i}^2 + a_{5,i}(t_{cwr,i} - t_{chws,i})\dot{Q}_{ch,i}$$

where, for the ith chiller, t_{cwr} is the leaving condenser water temperature and t_{chws} is the chilled-water supply temperature. The coefficients of Equation (38) ($a_{0,i}$ to $a_{5,i}$) can be determined for each chiller through regression applied to measured or manufacturers' data.

If each chiller has identical leaving condenser and chilled-water supply temperatures, the criterion of Equation (36) applied to the correlation of Equation (38) leads to

$$a_{3,i} + 2a_{4,i}\dot{Q}_{ch,i}^* + a_{5,i}(t_{cwr} - t_{chws})$$
$$= a_{3,j} + 2a_{4,j}\dot{Q}_{ch,j}^* + a_{5,j}(t_{cwr} - t_{chws}) \qquad \text{for } i \neq j \qquad (39)$$

where $\dot{Q}_{ch,i}^*$ is the optimal load for the ith chiller.

Equations (39) and (37) represent a system of N linear equations in terms of N chiller loads that can be solved to give minimum (or possibly maximum) power consumption. For a given combination of chillers, the solution depends on the operating temperatures and total load. However, the individual chiller loads must be constrained to be less than the maximum chiller capacity at these conditions. If an individual chiller load determined from these equations is greater than its cooling capacity, then this chiller should be fully loaded and Equations (39) and (37) should be resolved for the remaining chillers [Equation (39) should only include unconstrained chillers].

To control individual chiller loads with identical chilled-water supply temperatures, individual chilled-water flow rates need to be controlled with two-way valves, which is not typical. However, the distribution of chiller loads could be changed for a fixed-flow distribution by using different chilled-water set-point temperatures.

For a given flow and load distribution, the individual chiller set point for parallel chillers is determined according to

$$t_{chws,i} = t_{chwr} - \frac{\dot{Q}_{ch,i}}{f_{F,i}\dot{Q}_{load}}(t_{chwr} - t_{chws}^*) \qquad (40)$$

where $f_{F,i}$ is the flow for the ith chiller divided by total flow, t_{chws}^* is the chilled-water supply temperature set point for the combination of chillers determined using the previously defined reset strategies, and t_{chwr} is the temperature of water returned to chillers from the building.

Substituting Equation (40) into Equation (38) and then applying the criterion of Equation (36) leads to the following:

$$A_i + B_i\dot{Q}_{ch,i}^* = A_j + B_j\dot{Q}_{ch,j}^* \qquad \text{for } i \neq j \qquad (41)$$

where

$$A_i = a_{3,i} + [a_{1,i} + 2a_{2,i}(t_{cwr} - t_{chwr})]\frac{t_{chwr} - t_{chws}^*}{f_{F,i}\dot{Q}_{load}}$$
$$+ a_{5,i}(t_{cwr} - t_{chws}^*)$$

$$B_i = 2\left[a_{4,i} + a_{5,i}\frac{t_{chwr} - t_{chws}^*}{f_{F,i}\dot{Q}_{load}} + a_{2,i}\left(\frac{t_{chwr} - t_{chws}^*}{f_{F,i}\dot{Q}_{load}}\right)^2\right]$$

Optimal chiller loads are determined by solving the linear system of equations represented by Equations (41) and (37). The individual chiller set points are then evaluated with Equation (40). If any set points are less than the minimum set point or greater than the maximum set point, then the set point should be constrained and Equations (41) and (37) should be resolved for the remaining chillers [Equation (41) should only include unconstrained chillers].

Example 3. Determine the optimal loading for two chillers using the three methods outlined in this section. Table 6 gives the design cooling capacities and coefficients of the curve-fit of Equation (38) for the two chillers. The chillers are operating with a total cooling load of 1440 tons, condenser water return temperature of 85°F, an overall chilled-water supply temperature set point of 45°F, and a chilled-water return temperature of 55°F. Figure 35 shows the COPs for the two chillers as a function load relative to their design loads for the given operating temperatures. Chiller 1 is more efficient at higher part-ratios and less efficient at lower part-load ratios as compared with chiller 2.

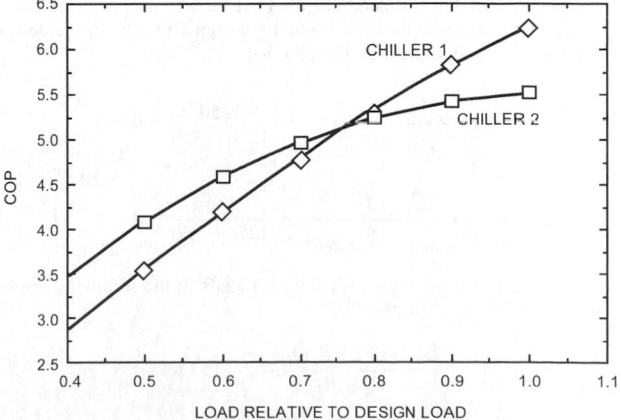

Fig. 35 Chiller COP for Two Chillers

**Table 6　Chiller Characteristics for
Optimal Loading Example 3**

Variable	Units	Chiller 1	Chiller 2
$\dot{Q}_{ch,des,i}$	tons	1250	550
$a_{0,i}$	kW	106.4	119.7
$a_{1,i}$	kW/°F	6.147	0.1875
$a_{2,i}$	kW/°F^2	0.1792	0.04789
$a_{3,i}$	kW/ton	−0.0735	−0.3673
$a_{4,i}$	kW/ton^2	0.0001324	0.0005324
$a_{5,i}$	kW/ton·°F	−0.001009	0.008526

Solution:

First, consider operating the chillers at equal part-load ratios. The ratio of the cooling load to the cooling capacity of the operating chillers is 1440/(1250 + 550) = 0.8. From Equation (35), the individual chiller loads are

$$\dot{Q}^*_{ch,1} = 0.8(1250 \text{ tons}) = 1000 \text{ tons}$$

$$\dot{Q}^*_{ch,2} = 0.8(550 \text{ tons}) = 440 \text{ tons}$$

The power for each chiller is computed for the specified operating conditions with Equation (38) and the coefficients of Table 6. For the case of equal part-load ratios, the total chiller power consumption is

$$P_{ch} = P_{ch,1} + P_{ch,2} = 657.5 \text{ kW} + 295.3 \text{ kW} = 952.8 \text{ kW}$$

A second solution is determined for optimal chiller loads for the case of equal chilled-water temperature set points and controllable flow for each chiller. In this case, algebraic manipulation of Equations (37) and (39) produces the following results for the individual chiller loads:

$$\dot{Q}^*_{ch,1} = \frac{(a_{3,2} - a_{3,1}) + 2a_{4,2}\dot{Q}_{load} + (a_{5,2} - a_{5,1})(t_{cwr} - t_{chws})}{2(a_{4,1} + a_{4,2})}$$

$$= 1219 \text{ tons}$$

$$\dot{Q}_{ch,2} = \dot{Q}_{load} - \dot{Q}_{ch,1} = 221 \text{ tons}$$

The resulting power consumption is then

$$P_{ch} = P_{ch,1} + P_{ch,2} = 696.9 \text{ kW} + 224 \text{ kW} = 920.9 \text{ kW}$$

Optimal loading of the chillers reduces the overall chiller power consumption by about 4% through heavier loading of chiller 1 and lighter loading of chiller 2 (see Figure 35).

Finally, optimal chiller loading is determined for the case where the individual loadings are controlled by using different chilled-water temperature set points (individual flow is not controllable). To apply Equations (40) and (41), the relative chilled-water flow rate for each chiller must be known. For this example, the relative flow for the ith chiller is assumed to be equal to the ratio of its design capacity to the design capacity for the operating chillers, so that

$$f_{F,1} = \frac{\dot{Q}_{ch,des,1}}{\dot{Q}_{ch,des,1} + \dot{Q}_{ch,des,2}} = \frac{1250}{1250 + 550} = 0.694$$

$$f_{F,2} = \frac{\dot{Q}_{ch,des,2}}{\dot{Q}_{ch,des,1} + \dot{Q}_{ch,des,2}} = \frac{550}{1250 + 550} = 0.306$$

Then, solving Equations (37) and (41) leads to the following results for the individual chiller loads:

$$\dot{Q}^*_{ch,1} = \frac{(A_2 - A_1) + B_2\dot{Q}_{load}}{B_1 + B_2} = 1153 \text{ tons}$$

$$\dot{Q}_{ch,2} = \dot{Q}_{load} - \dot{Q}_{ch,1} = 287 \text{ tons}$$

These loads lead to a total chiller power consumption of

$$P_{ch} = P_{ch,1} + P_{ch,2} = 713.8 \text{ kW} + 218.1 \text{ kW} = 931.9 \text{ kW}$$

Individual chilled-water set points are determined from Equation (40) and are 43.5 and 48.5°F for chillers 1 and 2, respectively. Power consumption has increased slightly from the case of identical chiller set points and variable flow.

Note that changing either flows or chilled-water set points complicates overall system control as compared with loading the chillers with fixed part-load ratios, and leads to relatively small savings.

Order for Bringing Chillers Online and Offline

For chillers with similar efficiencies, the order in which chillers are brought online and offline may be dictated by their cooling capacities and the desire to provide even runtimes. However, whenever beneficial and possible, chillers should be brought online in an order that minimizes the incremental increase in energy consumption. At a given condition, the power consumption of any chiller can be evaluated using the correlation given by Equation (38), where the coefficients are determined using manufacturers' data or in-situ measurements. Then, the overall power consumption for all operating chillers is

$$P_{ch} = \sum_{i=1}^{N} P_{ch,i} \tag{42}$$

When additional chiller capacity is required (see next section), the projected power of all valid chiller combinations should be evaluated using Equation (38), with the projected load determined by Equation (35) for chillers with similar performance characteristics, or the solution of Equations (36) subject to the constraint in Equation (37) for chillers with significantly different performance characteristics. Valid chiller combinations involve chillers that are not in alarm or locked out, and with load ratios between a low limit (e.g., 30%) and 100%. The best chiller combination to bring on line should result in the smallest increase (or largest decrease) in overall chiller power consumption as estimated with Equations (42) and (38) with chiller loading determined as outlined in the previous section. For systems with dedicated chilled-water and condenser water pumps and/or cooling towers, the associated power of this equipment should be added to Equation (38) to estimate the load of the chiller plus its auxiliary equipment. It is recommended that these estimates be performed online using in-situ measurements of each chiller's discharge chilled-water temperature and entering condenser water temperature, and the projected load of the chiller. A chiller should be shut down when its load drops below the spare capacity load of the current number of online chillers; for a primary/secondary chilled-water system, the primary chilled-water flow will remain above the secondary chilled-water flow once the chiller is shut down. (The spare capacity load is equal to the rated capacity of the online chillers minus the actual measured load of the online chillers.)

For chillers with similar design cooling capacities, a simpler (although suboptimal) approach can be used for determining the order for bringing chillers online and offline. In this case, the chiller with the highest peak COP can be brought online first, followed by the second most efficient chiller, etc., and then brought offline in reverse order. The maximum COP for each chiller can be evaluated using manufacturers' design and part-load data or from curve-fits to in-situ performance.

The chiller load associated with maximum COP for each chiller can be determined by applying a first-order condition for a maximum, using Equation (38) and the definition of COP. For this functional form, the maximum (or possibly minimum) COP occurs for

$$\dot{Q}^*_{ch,i} = \sqrt{\frac{a_{0,i} + a_{1,i}(t_{cwr} - t_{chws}) + a_{2,i}(t_{cwr} - t_{chws})^2}{a_{4,i}}} \tag{43}$$

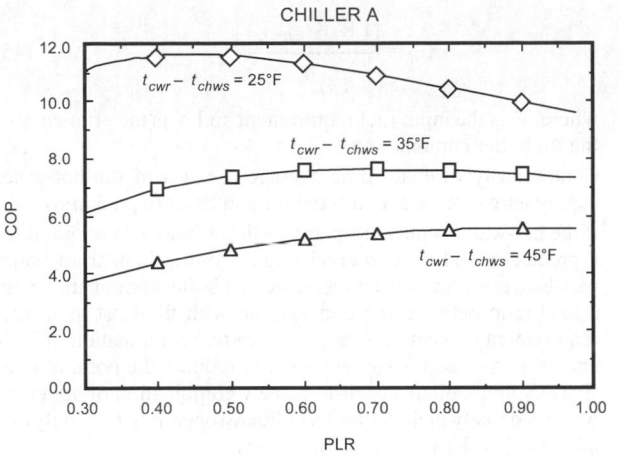

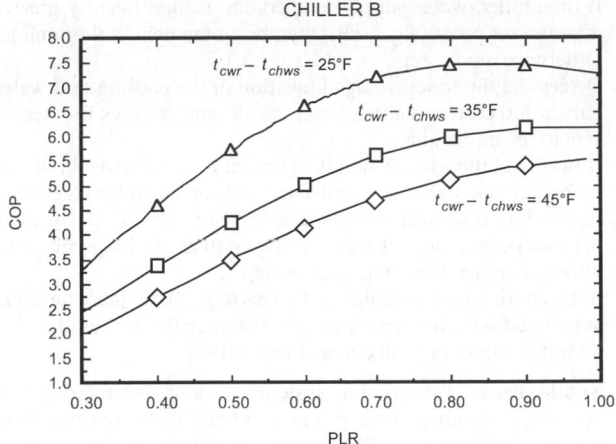

Fig. 36 Chiller A and B Performance Characteristics for Maximum COP, Example 4

The load determined from Equation (43) yields a maximum COP whenever it is real and bounded between upper and lower limits. Otherwise, it can be assumed that the maximum COP occurs at full load conditions. Typically, the maximum COP for centrifugal chillers occurs between about 40 and 80% and, for small multicompressor or inverter drive chillers, between 15 and 40% of design load. COP increases as the temperature difference between the condenser leaving water and chilled-water supply decreases. Equation (43) could be applied online to determine the rank ordering of chillers to bring online as a function of operating temperatures. However, it is often sufficient to use Equation (43) at the design temperature difference and establish a chiller sequencing order at the design or commissioning stage.

Example 4. Determine the loads for maximum COP for two different chillers at a chilled-water set point of 45°F and a condenser water return temperature of 80°F. Table 7 gives the design cooling capacities and coefficients of the curve-fit of Equation (38) for the two chillers. Figure 36 shows the COPs of the two chillers determined from the correlations as a function of relative load (PLR) and temperature difference ($t_{cwr} - t_{chws}$). These chillers have identical performance at design conditions, but very different part-load characteristics because of different methods used for capacity control.

Solution:

Loading associated with the maximum COP for each chiller is determined using Equation (43) and the coefficients of Table 7. Power for each chiller is then determined using Equation (38) and the COP follows directly. Results of the calculations are given in Table 8. The maximum COP for chiller A is about 20% greater than that for chiller B at the specified operating temperatures and should be brought online first.

Load Conditions for Bringing Chillers Online or Offline

In general, chillers should be brought online at conditions where the total power (including pumps and tower or condenser fans) of operating with the additional chiller would be less than without it. Conversely, a chiller should be taken offline when the total power of operating with that chiller would be less than with it. In practice, the switch point for bringing a chiller online should be greater than that for bringing that same chiller offline (e.g., 10%), to ensure stable control. The optimal sequencing of chillers depends primarily on their part-load characteristics and the manner in which the chiller pumps are controlled.

Dedicated Pumps. Where individual condenser and chilled-water pumps are dedicated to the chiller, Braun et al. (1989b) and Hackner et al. (1985) showed that a chiller should be brought online when the operating chillers reach their capacity. This conclusion is the result of considering both the chiller and pumping power in determining optimal control. If pumping power is ignored, the optimal

Table 7 Chiller Characteristics for Maximum COP, Example 4

Variable	Unit	Chiller A	Chiller B
$\dot{Q}_{ch,des,i}$	tons	5421	5421
$a_{0,i}$	kW	262.6	187.2
$a_{1,i}$	kW/°F	−25.36	96.19
$a_{2,i}$	kW/°F^2	0.9718	−0.4314
$a_{3,i}$	kW/ton	−0.02568	−0.4314
$a_{4,i}$	kW/ton^2	0.00004046	0.0001106
$a_{5,i}$	kW/ton·°F	0.005289	−0.004537

Table 8 Results for Maximum COP, Example 4

Variable	Chiller A	Chiller B
$\dot{Q}^*_{ch,i}$	3738 ton	5229 ton
PLR_i	0.690	0.965
$P_{ch,i}$	1727 kW	2964 kW
COP_i	7.61	6.21

chiller sequencing occurs when chiller efficiency is maximized at each load. Because maximum efficiency often occurs at part-load conditions, the optimal point for adding or removing chillers may occur when chillers are operating at less than their capacity. However, the additional pumping power required for bringing additional pumps online with the chiller usually offsets any reductions in overall chiller power consumption associated with part-load operation.

When pumps are dedicated to chillers, situations may arise where chillers are operating at less than their capacity but chilled-water flow to the cooling coils is insufficient to meet the building load. This generally results from inadequate design or improper maintenance. Under these circumstances, either some zone conditions need to float to reduce the chilled-water set point (if possible), or an additional chiller needs to be brought online. Monitoring the zone air-handler conditions is one way to detect this situation. If (1) the chilled-water set point is at its lower limit, (2) any air-handler water control valves are saturated at 100% open, and (3) their corresponding discharge air temperatures are significantly greater (e.g., 2°F) than set point, then the chilled-water flow is probably insufficient and an additional chiller/pump combination could be brought online. One advantage of this approach is that it is consistent with the reset strategies for both fixed- and variable-speed chilled-water systems.

Chillers can be brought online or offline with the following logic:

1. Evaluate the time-averaged values of the chilled-water supply temperature and overall cooling load over a fixed time interval (e.g., 5 min).

2. If the chilled-water supply temperature is significantly greater than the set point (e.g., 1°F), then bring the next chiller online. Otherwise, go to step 3.

3. Determine the time-averaged position of the cooling-coil water valves and corresponding discharge air temperatures for "representative" air handlers.

4. If (a) the chilled-water supply set point is at its lower limit, (b) more than one valve is saturated at 100% open, and (c) their corresponding discharge air temperatures are significantly greater than set point (e.g., 1°F), then bring another chiller/pump combination online. Otherwise, go to step 5.

5. If the cooling load is significantly less (e.g., 10%) than the value associated with the first time interval after the last chiller was brought online, then take that chiller offline.

Nondedicated Pumps. For systems without dedicated chiller pumps (e.g., variable-speed primary systems), the optimal load conditions for bringing chillers online or offline do not generally occur at full chiller capacity. In determining optimal chiller switch points, ideally both chiller and pumping power should be considered because pressure drop characteristics and pumping change when a chiller is brought online or offline. However, simple estimates of optimal switch points may be determined by considering only chiller power.

A chiller should be brought online whenever it would reduce the overall chiller power or if the current chillers can no longer meet the load (see previous section). A chiller should be added if the power consumption associated with ($N + 1$) chillers is significantly less (e.g., 5%) than the current N chillers, with both conditions evaluated using Equation (42) with correlations of the form given in Equation (38), and sequencing and loading determined as outlined in previous sections. Conversely, a chiller should be removed if the power consumption associated with the ($N - 1$) chillers is significantly less (e.g., 5%) than the current N chillers. The decision to add or remove chillers is readily determined using the current load and operating temperatures.

STRATEGIES FOR BOILERS

Load Conditions for Bringing Boilers Online or Offline

The specifics of the strategy for bringing boilers online depend on the type of boiler. Hot-water boilers have dedicated or nondedicated hot-water pumps; steam boilers do not have hot-water pumps, but rely on differences in steam pressure between the boiler steam header discharge and the point of use to distribute steam throughout the system.

Hot-Water Boilers with Dedicated Pumps. The strategy for hot-water boilers with dedicated hot-water pumps is similar to that for chillers with dedicated chilled-water and condenser water pumps: another boiler should be brought online when operating boilers reach capacity, because the efficiency of the boiler should include the power to drive its associated hot-water pump. This can be determined when hot-water temperature drops below its set point for a predetermined time interval (e.g., 5 min).

Hot-water boilers with dedicated pumps can be brought online and offline with the following logic:

1. Continuously calculate the load ratio of each boiler or boiler combination. For the ith boiler,

$$LR_i = \frac{\dot{Q}_{load}}{\dot{Q}_{blr,des,i}} \qquad (44)$$

where LR_i is load ratio of the ith boiler combination, $\dot{Q}_{load}$ is the total boiler plant load, and $\dot{Q}_{blr,des,i}$ is the design (rated) output of the ith boiler combination.

2. Every sampling interval (e.g., 60 s), calculate the predicted input fuel requirement for each boiler combination as

$$IF_i = \frac{(LR_i)\dot{Q}_{blr,des,i}}{\eta_i} \qquad (45)$$

where IF_i is the input fuel requirement and η_i is the efficiency of the ith boiler combination.

3. Continuously evaluate time-averaged values of the hot-water supply temperature over a fixed time interval (e.g., 5 min).

4. If the hot-water supply temperature drops below its set point for a predetermined time interval (e.g., 5 min), then, from boiler part-load performance curves, select the boiler combination with a load ratio between 0.5 and 1.0 and with the least input fuel requirement to meet the load, and turn this combination of boilers on. Note that this strategy greatly reduces the possibility of short-cycling boilers because the new combination of boilers to be started likely includes boilers already operating (i.e., only one additional boiler is likely to be added).

5. If the capacity of the least efficient online boiler drops below the spare capacity of the current number of boilers operating (or for a primary/secondary hot-water system, if the flow rate of the associated primary hot-water pump is less than the difference between primary and secondary hot-water flow rates) for a predetermined time interval (e.g., 5 min), then shut down and bank this boiler in hot standby.

Hot-Water Boilers with Nondedicated Pumps or Steam Boilers. For hot-water systems without dedicated hot-water pumps or for steam systems, the optimal load conditions for bringing boilers online or offline do not generally occur at the full capacity of the online boilers. For these systems, a new boiler combination should be brought online whenever the hot-water supply temperature or steam pressure falls below set point for a predetermined time interval (e.g., 5 min) *and* the part-load efficiency curves of the boiler combination predicts that the new combination of boilers can meet the required load using significantly less (e.g., 5%) input fuel.

Optimal Boiler Load Distribution

Optimal load distribution strategies for boilers are similar to those for chillers. For boilers with similar performance characteristics, the optimal boiler loading is similar to Equation (35) for chillers:

$$\dot{Q}_{blr,i}^* = \frac{\dot{Q}_{load}}{\displaystyle\sum_{i=1}^{N} \dot{Q}_{blr,des,i}} \dot{Q}_{blr,des,i} \qquad (46)$$

where $\dot{Q}_{blr,i}^*$ is the optimal load for the ith boiler.

For boilers with significantly different performance characteristics, the criterion for optimal boiler loading is similar to Equations (36) and (37) for chillers, except that boiler cost of operation is used:

$$\frac{\partial C_{blr,i}}{\partial \dot{Q}_{blr,i}} = \frac{\partial C_{blr,j}}{\partial \dot{Q}_{blr,j}} \qquad \text{for all } i \text{ and } j \qquad (47)$$

and subject to the constraint that

$$\sum_{i=1}^{N} \dot{Q}_{blr,i} = \dot{Q}_{load} \qquad (48)$$

where $C_{blr,i}$ and $C_{blr,j}$ are the operating costs of boiler i and j, respectively, $\dot{Q}_{blr,i}$ is heating load for the ith boiler, and Q_{load} is total heating load.

In general, the boiler operating-cost curve can be calculated as a quadratic function of heating load only. For the ith boiler,

$$C_{blr,i} = b_{0,i} \dot{Q}_{blr,i}^2 + b_{1,i} \dot{Q}_{blr,i} + b_{2,i} \qquad (49)$$

Applying the criterion of Equation (47) to Equation (49),

$$2b_{0,i} \dot{Q}_{blr,i}^* + b_{1,i} = 2b_{0,j} \dot{Q}_{blr,j}^* + b_{1,j} \qquad (50)$$

where $\dot{Q}_{blr,i}^*$ is the optimal load for the ith boiler.

Maintaining Boilers in Standby Mode

It is generally more economical to run fewer boilers at a high rating. However, the integrity of the steam or hot-water supply must be maintained in the event of a forced outage of one of the operating boilers or if the facility experiences highly diverse load swings throughout the heating season. Both conditions can often be satisfied by maintaining a boiler in standby or "live bank" mode. For example, in this mode, a steam boiler is isolated from the steam system at no load but is kept at system operating pressure by periodic firing of either the igniters or a main burner to counteract ambient heat losses.

Supply Water and Supply Pressure Reset for Boilers

Simple control strategies can be used to generate a suboptimal hot-water temperature (for hot-water boilers) or steam pressure (for steam boilers). An energy management and control system must be interfaced to the boiler controls and be capable of monitoring the position of the valve controlling the flow of hot water to the heating coils or steam pressure at the most critical zone. For a hot-water system,

1. Continuously monitor the hot-water valve position of the various heating zones.
2. If none of the hot-water valves are greater than 95% open, lower boiler hot-water supply temperature by a small increment (e.g., 1°F) each reset time interval (a predetermined interval established by system thermal lag characteristics, e.g., 15 to 20 min).
3. Once one hot-water valve opens beyond 95%, stop downward resets of boiler hot-water temperature set point.
4. If two or more hot-water valves open beyond 95%, raise boiler hot-water temperature set point by a small increment (e.g., 1°F) each reset interval.

For a steam system, the steam header pressure should be lowered to a value that just satisfies the highest pressure demand. *Caution*: for nongravity condensate return systems, steam pressure reset could impede condensate return (see Chapter 10 of the 2008 *ASHRAE Handbook—HVAC Systems and Equipment*).

STRATEGIES FOR AIR-HANDLING UNITS

Air Handler Sequencing and Economizer Cooling

Traditional air handler sequencing strategies use a single PI controller to control heating, cooling with outdoor air, mechanical cooling with 100% outdoor air, and mechanical cooling with minimum outdoor air. Sequencing between these different modes is accomplished by splitting the controller output into different regions of operation, as shown in Figure 37.

Figure 37 depicts the relationship between the control signal to the valves and dampers and the feedback controller output. The controller adjusts its output to maintain the supply air temperature set point. If output is between 100 and 200%, mechanical cooling is used. When outdoor conditions are suitable, the outdoor air dampers switch from minimum position (minimum ventilation air) to fully open. For a dry-bulb economizer, this switch point occurs when

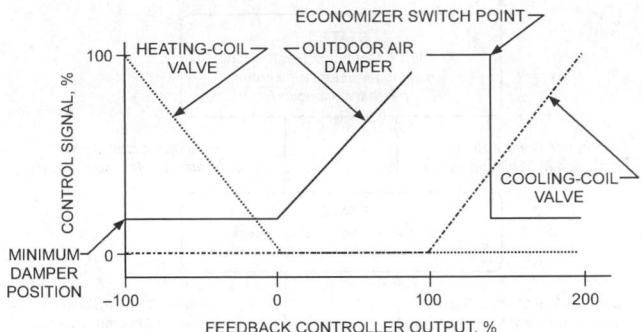

Fig. 37 AHU Sequencing Strategy with Single Feedback Controller

ambient air is less than a specified value. This switch point should be less than the switch point to return to minimum outdoor air, to ensure stable control. The economizer switchover temperature may be significantly lower than the return air temperature (e.g., 10°F lower) in humid climates where latent ventilation loads are significant. However, in dry climates, the switchover temperature may be close to the return temperature (e.g., 75°F). An enthalpy (or wet-bulb) economizer compares outdoor and return air enthalpies (or wet-bulb temperatures) to initiate or terminate economizer operation. In general, enthalpy economizers yield lower energy costs than dry-bulb economizers, but require measurements of outdoor and return air humidity. Humidity sensors require regular maintenance to ensure accurate readings. When controller output is between 0 and 100% (see Figure 37), the cooling coil valve is fully closed and cooling is provided by ambient air only. In this case, the controller output modulates the position of the outdoor air dampers to maintain the set point. If the controller output signal is between −100 and 0%, the heating coil is used to maintain set point and the outdoor air dampers are set at their minimum position.

A single feedback controller is difficult to tune to perform well for all four modes of operation associated with an AHU. An alternative to the traditional sequencing strategy is to use three separate feedback controllers, as described by Seem et al. (1999). This approach can improve temperature control, reduce actuator usage, and reduce energy costs. Figure 38 shows a state transition diagram for implementing a sequencing strategy that incorporates separate feedback controllers.

In state 1, a feedback controller adjusts the heating valve to maintain the supply air set-point temperature with minimum outdoor air. The transition to state 2 occurs after the control signal has been saturated at the no-heating position for a period equal to a specified state transition delay (e.g., 3 min). In state 2, a second feedback controller adjusts the outdoor and return air dampers to achieve set point with heating and cooling valves closed. Transition back to state 1 only occurs after the damper control signal is saturated at its minimum value for the state transition delay, whereas transition to state 3 is associated with saturation at the maximum damper position for the state transition delay. In state 3, the outdoor air damper remains fully open and a third feedback controller is used to adjust the flow of cooling water to maintain the supply air temperature at set point. Transition back to state 2 occurs if the controller output is saturated at its minimum value for the state transition delay. For a dry-bulb economizer, transition to state 4 occurs when the ambient dry-bulb temperature is greater than the switchover temperature by a dead band (e.g., 2°F). The feedback controller continues to modulate the cooling-coil valve to achieve set point. Transition back to state 3 occurs when the ambient dry-bulb is less than the switchover temperature (e.g., 65°F). For an enthalpy economizer, the ambient enthalpy is compared with return air enthalpy to initiate transitions between states 3 and 4.

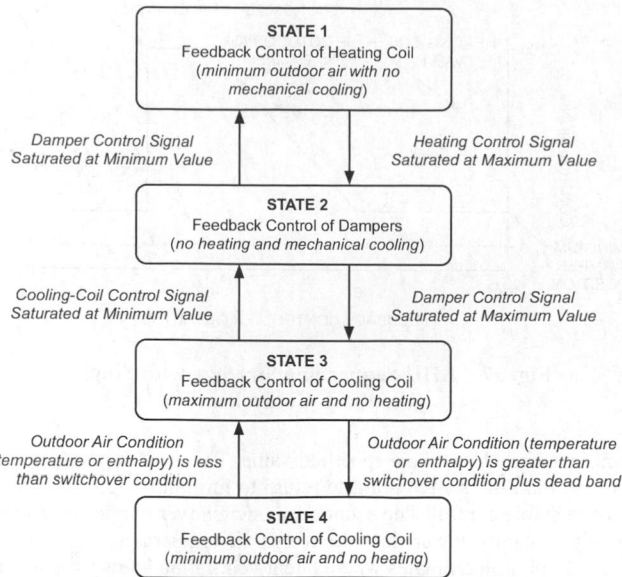

Fig. 38   AHU Sequencing Strategy with Multiple Feedback Controllers

Supply Air Temperature Reset for Constant Air Volume (CAV)

The benefits of resetting supply air temperature set points for CAV systems are significant. Increasing the supply air set point for cooling reduces both the cooling and reheat required, but does not change fan energy. In general, the set point for a CAV system could be set at the highest value that will keep all zone temperatures at their set points and all humidities within acceptable limits. A simple reset strategy based on this concept follows.

At each the decision interval (e.g., 5 min), the following logic can be applied:

1. Check controller outputs for representative zone reheat units and determine time-averaged values over the last decision interval.
2. If any controller output is less than a threshold value (e.g., 5%), then decrease the supply air set point by a fixed value (e.g., 0.5°F) and go to step 4. Otherwise, go to step 3.
3. If all zone humidities are acceptable and all controller outputs are greater than a threshold value (e.g., 10%), then increase the discharge air set point by a fixed value (e.g., 0.5°F) and go to step 4. Otherwise, do not change the set point.
4. Limit the set point between upper and lower limits based on comfort considerations.

Static Pressure Reset for Variable Air Volume (VAV)

Flow may be modulated in a VAV system by using dampers on the outlet side of the fan, inlet vanes on the fan, vane-axial fans with controllable pitch fan blades, or variable-speed control of the fan motor. Typically, inputs to any of these controlled devices are modulated to maintain a duct static pressure set point as described in Chapter 47. In a single-duct VAV system, the duct static pressure set point is typically selected by the designer to be a fixed value. The sensor is located at a point in the ductwork such that the established set point will ensure proper operation of the zone VAV boxes under varying load (supply airflow) conditions. A shortcoming of this approach is that static pressure is controlled based on a single sensor intended to represent the pressure available to all VAV boxes. Poor

location or malfunction of this sensor will cause operating problems.

For a fixed static pressure set point, all of the VAV boxes tend to close as zone loads and flow requirements decrease. Therefore, flow resistance increases with decreasing load. Significant fan energy savings are possible if the static pressure set point is reset so that at least one of the VAV boxes remains open. With this approach, flow resistance remains relatively constant. Englander and Norford (1992), Hartman (1993), Warren and Norford (1993), and Wei et al. (2004) proposed several different strategies based on this concept. Englander and Norford used simulations to show that either static pressure or fan speed can be controlled directly using a flow error signal from one or more zones and simple rules. Their technique forms the basis of the following reset strategy.

At each decision interval (e.g., 5 min), the following logic can be applied:

1. Check the controller outputs for representative VAV boxes and determine time-averaged values over the last decision interval.
2. If any of the controller outputs are greater than a threshold value (e.g., 98%), then increase the static pressure set point by a fixed value (e.g., 5% of the design range) and go to step 4. Otherwise, go to step 3.
3. If all controller outputs are less than a threshold value (e.g., 90%), then decrease the static pressure set point by a fixed value (e.g., 5% of the design range) and go to step 4. Otherwise, do not change the set point.
4. Limit the set point between upper and lower limits based on upper and lower flow limits and duct design.

STRATEGIES FOR BUILDING ZONE TEMPERATURE SET POINTS

Recovery from Night Setback or Setup

For buildings that are not continuously occupied, significant savings in operating costs may be realized by raising the building set-point temperature for cooling (setup) and by lowering the set point for heating (setback) during unoccupied times. Bloomfield and Fisk (1977) showed energy savings of 12% for a heavyweight building and 34% for a lightweight building.

An optimal controller for return from night setback or setup returns zone temperatures to the comfort range precisely when the building becomes occupied. Seem et al. (1989b) compared seven different algorithms for minimum return time. Each method requires estimating parameters from measurements of the actual return times from night setback or setup.

Seem et al. (1989b) showed that the optimal return time for cooling was not strongly influenced by the outdoor temperature. The following quadratic function of the initial zone temperature was found to be adequate for estimating the return time:

$$\tau = a_0 + a_1 t_{z,i} + a_2 t_{z,i}^2 \qquad (51)$$

where τ is an estimate of optimal return time, $t_{z,i}$ is initial zone temperature at the beginning of the return period, and a_0, a_1, and a_2 are empirical parameters. The parameters of Equation (51) may be estimated by applying linear least-squares techniques to the difference between the actual return time and the estimates. These parameters may be continuously corrected using recursive updating schemes as outlined by Ljung and Söderström (1983).

For heating, Seem et al. found that ambient temperature has a significant effect on the return time and that the following relationship works well in correlating return times:

$$\tau = a_0 + (1 - w)(a_1 t_{z,i} + a_2 t_{z,i}^2) + w a_3 t_a \qquad (52)$$

where t_a is the ambient temperature, a_0, a_1, a_2, and a_3 are empirical parameters, and w is a weighting function given by

$$w = 1000^{-(t_{z,i} - t_{unocc})/(t_{occ} - t_{unocc})} \qquad (53)$$

where t_{unocc} and t_{occ} are the zone set points for unoccupied and occupied periods. Within the context of Equation (52), this function weights the outdoor temperature more heavily when the initial zone temperature is close to the set-point temperature during the unoccupied time. Again, the parameters of Equation (52) may be estimated by applying linear least-squares techniques to the difference between the actual return time and the estimates.

Ideally, separate equations should be used for zones that have significantly different return times. Equipment operation is initiated for the zone with the earliest return time. In a building with a central cooling system, the equipment should be operated above some minimum load limit. With this constraint, some zones need to be returned to their set points earlier than the optimum time.

Optimal start algorithms often use a measure of the building mass temperature rather than the space temperature to determine return time. Although use of space temperature results in lower energy costs (i.e., shorter return time), the mass temperature may result in better comfort conditions at the time of occupancy.

Emergency Strategy to Limit Peak Cooling Requirements

Keeney and Braun (1997) developed a simple control strategy that uses building thermal mass to reduce peak cooling requirements in the event of a loss of a chiller. This emergency strategy is used only on days where cooling capacity is not sufficient to keep the building in the comfort range using night setup control. It involves precooling the building during unoccupied times and allowing the temperature to float through the comfort zone during occupancy.

The precooling control strategy is depicted in Figure 39 along with conventional night setup control. Precooling is controlled at a constant temperature set point t_{pre}. The warm-up period is used to reset the zone air temperature set point so that the cooling system turns off without calling for heating. During this time, the zone air is warmed by lighting and equipment loads. The occupied set point t_{occ} is set at the low end of the comfort region so that the building mass charge is held as long as cooling capacity is available. This set point is maintained until the limit on cooling capacity is reached. After this point, temperatures in the zones float up and the building thermal mass provides additional cooling. If the precooling and occupied set points have been chosen properly and the cooling capacity is sufficient, zone conditions will remain comfortable throughout the occupied period. The peak cooling requirement can be reduced by as much as 25% using this strategy as compared with night setup control. Thus, the loss of one of four identical chillers

could be tolerated. This strategy could also be used for spaces such as auditoriums that have a high occupancy density for a short period.

The length of time and temperature for precooling and the occupied temperature set point chosen for this strategy strongly influence the capacity reduction and could affect occupant comfort. A reasonable strategy is to precool at 68°F beginning at midnight, allow a 30 min warm-up period before occupancy, and then adjust the occupied set point to 70°F. The zone temperature will then rise above this set point when the chillers are operating at capacity.

Case Study. The control strategy was tested in a 1.4 million square foot office building located near Chicago, Illinois. The facility has two identical buildings with very similar internal gains and solar radiation loads, connected by a large, separately cooled entrance area. During tests, the east building used the existing building control strategy and the west building used the precooling strategy.

Four 900 ton vapor compression chillers normally provide chilled water to the air-handling units. Loss of one chiller results in a 25% reduction of the total capacity. This condition was simulated by limiting the vane position of the two chiller units that cool the west building to 75%. The capacity limitation was imposed directly at the chiller control panels. Set points were provided to local zone controllers from a modern energy management and control system. Chiller cooling loads and zone thermal comfort conditions were monitored throughout the tests.

Consistent with simulation predictions, the precooling control strategy successfully limited the peak load to 75% of the cooling capacity for the west building, whereas the east building operated at 100% of capacity. Figure 40 shows the total chiller coil load for the east and west buildings for a week of testing in the middle of August 1995. The cooling-coil load profile on Monday is the most dramatic example of load shifting during this test period. The peak cooling load for this facility often occurs on Monday morning. The cooling limit was achieved on Monday during a period in which a heat emergency had been declared in the city. The severe ambient conditions were compounded by a power outage that caused a loss of the west-side chiller units for approximately 20 min. Under these demanding conditions, the precooling strategy maintained occupant comfort while successfully limiting cooling demand of the west side of the building to less than 75% of that for the east side.

The east-side cooling requirement was at or below the 75% chiller capacity target for Tuesday through Friday, so the emergency precooling strategy was not necessary. For these off-design days, the emergency strategy was not effective in reducing the on-peak cooling requirements because discharge of the mass was not initiated when capacity was below the target. The thermal mass

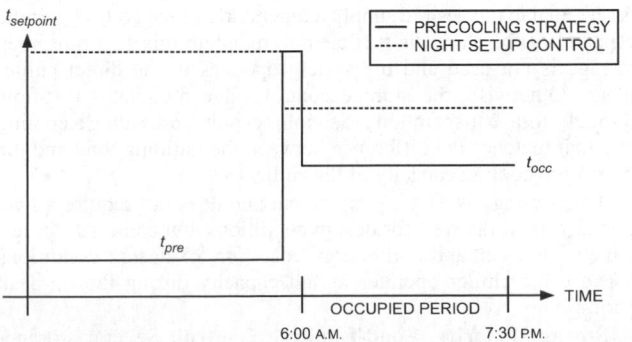

Fig. 39   Zone Air Temperature Set Points

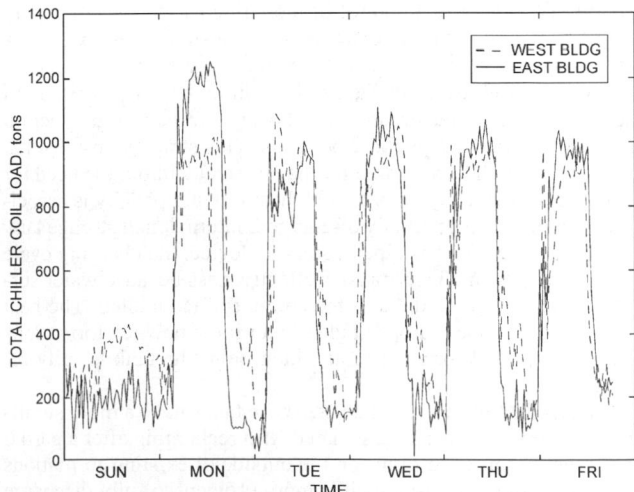

Fig. 40   Total Coil Load for East and West Chiller Units

remained charged so that peak reduction would occur if the target value on the off-design days were reset to a lower value.

Precooling the top floor of the facility had already been implemented into the conventional control strategy used for the east building. This was necessary to maintain comfort conditions with full cooling capacity on hot days. As a result, even greater peak reduction would have been recorded if the precool strategy had been compared with conventional night setup control. The total electrical use was greater for the precooled west building; however, the strategy was designed for an emergency and does not attempt to minimize costs.

This emergency strategy should only be applied on days when the available cooling capacity is not sufficient to maintain comfort conditions when using night setup control. Otherwise, the costs of providing cooling could increase significantly. ASHRAE research project RP-1313 (Cheng et al. 2008; Henze et al. 2007, 2008) developed simplified strategies for optimal control of building zone set points to minimize operating costs.

CONTROL OF DISCRETE COOL THERMAL STORAGE

The choice of a control strategy for a thermal storage system results from a tradeoff between performance (i.e., operating costs) and ease of implementation (i.e., initial costs). Chiller-priority control has the lowest implementation costs, but generally leads to the highest operating costs. Storage-priority strategies provide superior performance, but require the use of a forecaster and a measurement of state of charge for storage. This section presents details of chiller-priority, load-limiting, and rule-based control applied to ice storage systems. Each of these strategies shares the same procedures for charging storage, but differs in the manner in which storage is discharged. In general, the control strategies presented in this section are appropriate for systems with utility rate structures that include time-of-use energy and demand charges, but would not be appropriate in conjunction with real-time pricing. Additional information on control strategies for cool storage systems can be found in Chapter 50 of the 2008 *ASHRAE Handbook—HVAC Systems and Equipment*.

Charging Strategies

Ice making should be initiated when both the building is unoccupied and off-peak electrical rates are in effect. During the ice-making period, the chiller should operate at full capacity. Cooling plants for ice storage generally operate most efficiently at full load because of the auxiliaries and the characteristics of ice-making chillers. With feedback control of the chilled-water/glycol supply temperature, full capacity control is accomplished by establishing a low enough set point to ensure this condition (e.g., 20°F).

Internal Melt Storage Tanks. The chiller should operate until the tank reaches its maximum state of charge or the charging period (i.e., off-peak, unoccupied period) ends. This strategy ensures that sufficient ice will be available for the next day without the need for a forecaster. Typically, only a small heat transfer penalty is associated with restoring a partially discharged, internal melt storage tank to a full charge. For this type of storage device, the charging cycle always starts with a high transfer effectiveness because water surrounds the tubes regardless of the amount of ice melted. The heat transfer effectiveness drops gradually until the new ice formations intersect with old formations, at which point the tank is fully recharged.

External Melt Storage Tanks. These tanks have a more significant heat transfer penalty associated with recharging after a partial discharge, because ice forms on the outside of existing formations during charging. In this case, it is more efficient to fully discharge the tank each day and only recharge as necessary to meet the next day's cooling requirements. To ensure that adequate ice is available,

the maximum possible storage capacity needed for the next day must be forecast. The storage requirements for the next day depend on the discharge strategy used and the building load. In general, the state of charge for storage necessary to meet the next day's load can be estimated according to

$$X_{chg} = \sum_{k=1}^{\substack{occupied \\ period}} \frac{\hat{Q}_{load,k} - \hat{Q}_{ch,k}}{C_s} \quad (54)$$

where X_{chg} is the relative state of charge at the end of the charging period, C_s is the maximum change in internal energy of the storage tank that can occur during a normal discharge cycle, and $\hat{Q}_{load,k}$ and $\hat{Q}_{ch,k}$ are forecasts of the building load and chiller cooling requirement for the kth stage (e.g., hour) of the occupied period. The relative state of charge is defined in terms of two reference states: the fully discharged and fully charged conditions that correspond to values of zero and one. These conditions are defined for a given storage based on its particular operating strategy (ASHRAE *Standard* 150; Elleson 1996). The fully charged condition exists when the control stops the charge cycle as part of its normal sequence. Similarly, the fully discharged condition is the point where no more usable cooling is recovered from the tank. Typically, zero state of charge corresponds to a tank of water at a uniform temperature of 32°F and a complete charge is associated with a tank having maximum ice build at 32°F. (The fully discharged and fully charged conditions are arbitrarily selected reference states.) In abnormal circumstances, a storage tank can be discharged or charged beyond these conditions, resulting in relative states of charge below zero or above one.

Hourly forecasts of the next day's cooling requirement can be determined using the algorithm described in the section on Forecasting Diurnal Energy Requirements. However, long-term forecasts are highly uncertain and a safety factor based on previous forecast errors is appropriate (e.g., uncertainty of two or three times the standard deviation of the errors of previous forecasts). Estimates of hourly chiller requirements should be determined using the intended discharge strategy (described in the next section) and building load forecasts.

Discharging Strategies

Three discharge strategies are presented for use with utility structures having on-peak and off-peak energy and demand charges: (1) chiller-priority control, (2) storage-priority, load-limiting control, and (3) a rule-based strategy that uses both chiller-priority and load-limiting strategies.

Chiller-Priority Discharge. During the storage discharge mode, the chiller operates at full cooling capacity (or less if sufficient to meet the load) and storage matches the difference between the building requirement and chiller capacity. For the example system illustrated in Figure 17, the chiller supply temperature set point t_{chws} is set equal to the desired supply temperature for the coils t_{coil}. If the capacity of the chiller is sufficient to maintain this set point, then storage is not used and the system operates in the direct chiller mode. Otherwise, the storage control valve modulates the flow through storage to maintain the supply set point, providing a cooling rate that matches the difference between the building load and the maximum cooling capacity of the chillers.

This strategy is easy to implement and does not require a load forecast. It works well for design conditions, but can result in relatively high demand and energy costs for off-design conditions because the chiller operates at full capacity during the on-peak period.

Storage-Priority, Load-Limiting Control. Several storage-priority control approaches ensure that storage is not depleted prematurely. Braun (1992) presented a storage-priority control strategy, termed *load-limiting* control, which tends to minimize the peak cooling plant power demand. The operation of equipment for load-limiting control during different parts of the occupied period can be described as follows:

- **Off-Peak, Occupied Period.** During this period, the goal is to minimize use of storage, and the chiller-priority described in the previous section should be applied.
- **On-Peak, Occupied Period.** During this period, the goal is to operate the chillers at a constant load while discharging the ice storage such that the ice is completely melted when the off-peak period begins. This requires using a building cooling-load forecaster. At each decision interval (e.g., 15 min), the following steps are applied:

    (1) Forecast the total integrated building cooling requirement until the end of the discharging period.
    (2) Estimate the state of charge of the ice storage tank from measurements.
    (3) At any time, the chiller loading for load-limiting control is determined as

$$\dot{Q}_{LLC} = \text{Max}\left[\frac{\hat{Q}_{load,occ} - (X - X_{min})C_s}{\Delta t_{on}}, \hat{Q}_{ch,min}\right] \quad (55)$$

where $\hat{Q}_{load,occ}$ is a forecast of the integrated building load for the rest of the on-peak period, Δt_{on} is the time remaining in the on-peak period, X is the current state of charge defined as the fraction of the maximum storage capacity, X_{min} is a minimum allowable state of charge, C_s is the maximum possible energy that could be added to storage during discharge, and $\hat{Q}_{ch,min}$ is the minimum allowable chiller cooling capacity. If the chiller does not need to be operated during the remainder of the occupied, on-peak period, the minimum allowable cooling capacity could be set to zero. Otherwise, the cooling capacity should be set to the minimum at which the chiller can safely operate.

    (4) Determine the chiller set-point temperature necessary to achieve the desired loading as

$$t_{chws} = t_{chwr} - \frac{\dot{Q}_{LLC}}{C_{chw}} \quad (56)$$

where t_{chwr} is the temperature of water/glycol returned to the chiller and C_{chw} is the capacitance rate (mass flow times specific heat) of the flow stream.

Hourly forecasts of cooling loads can be determined using the algorithm described in the section on Forecasting Diurnal Energy Requirements. The hourly forecasts are then integrated to give a forecast of the total cooling requirement. To ensure sufficient cooling capacity, a worst-case forecast of cooling requirements could be estimated as the sum of the best forecast and two or three times the standard deviation of the errors of previous forecasts.

Rule-Based Controller. Drees and Braun (1996) presented a rule-based controller that combines elements of chiller-priority and storage-priority strategies, along with a demand-limiting algorithm to achieve near-optimal control. The demand-limiting algorithm requires a measurement of the total building electrical use. A simpler strategy is described here that does not require this measurement and yields equivalent performance whenever the peak demand for the billing period is coincident with the peak cooling load.

Figure 41 shows a flowchart for the discharge strategy that is applied during each decision interval (e.g., 15 min) during the occupied period. Block 1 determines whether storage use should be maximized or minimized. Block 2 is used if storage use lowers daily

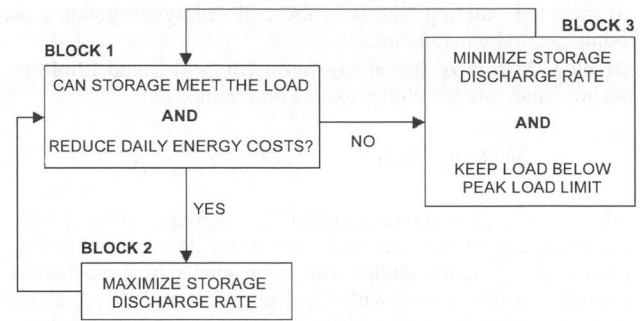

**Fig. 41 Flowchart for Rule-Based
Controller Discharge Strategy**

energy costs and storage is sufficient to meet the remainder of the load for the occupied period without operating the chillers. Otherwise, the goal of the strategy in block 3 is to minimize storage use while keeping peak load below a limit. This strategy tends to keep the chiller(s) heavily loaded (and therefore, if part-load efficiency is poor, operating efficiently) until they are no longer needed. The logic in each block is as follows:

- **Block 1: Discharge Strategy Selection.** The discharge of storage will not reduce the energy cost whenever the cost of replenishing the ice is greater than the cost of providing direct cooling by the chiller(s). This situation is always the case during the off-peak, occupied period because the electricity rates are the same as those associated with the charging period and chillers are less efficient in ice-making mode than when providing direct cooling. Furthermore, during the on-peak, occupied period, using storage generally reduces energy costs whenever the following criterion holds:

$$\text{ECR} > \text{COP}_d/\text{COP}_c \quad (57)$$

where ECR is the ratio of on-peak to off-peak energy charges and COP_d and COP_c are coefficients of performance for the cooling plant (including chiller, pumps, and cooling tower fans) during discharging and charging of the tank. The COPs should be evaluated at the worst-case charging and discharging conditions associated with the design day. Typically, this ratio is between about 1.2 and 1.8 for systems with cooling towers. However, the ratio can be lower because of the effect of cool nighttime temperatures, especially for systems with air-cooled condensers in dry climates.

If the criterion of Equation (57) is satisfied, control switches from chiller-priority to storage-priority strategy whenever storage capacity is greater than the remaining integrated load. Therefore, storage-priority control is enabled whenever

$$(X - X_{min})C_s \geq \hat{Q}_{load,occ} \quad (58)$$

Hourly forecasts of cooling loads can be determined using the algorithm described in the section on Forecasting Diurnal Energy Requirements and then integrated to give a forecast of the total cooling requirement. To ensure that adequate ice is available, worst-case hourly forecasts can be determined by adding the expected value of the hourly forecasts and the forecast errors associated with a specified confidence interval (e.g., two standard deviations for a 95% confidence interval). The worst-case hourly forecasts can then be integrated to give a worst-case integrated forecast.

- **Block 2: Maximum Use of Storage.** In this mode, the chillers are turned off and storage is used to meet the entire load throughout the remainder of the occupied period. However, a chiller may need to be turned on if the storage discharge rate is not sufficient to meet the building load (i.e., the coil supply temperature set point cannot be maintained).

- **Block 3: Minimize Use of Storage with Peak Load Limiting.** At any time, a target chiller load is determined as

$$\dot{Q}_{ch} = \text{Min}[\text{Max}(\dot{Q}_{ch,peak}, \dot{Q}_{LCC}), \dot{Q}_{load}] \qquad (59)$$

where $\dot{Q}_{ch,peak}$ is the peak chiller cooling requirement that has occurred during the on-peak period for the current billing period, $\dot{Q}_{LCC}$ is the chiller load associated with load-limiting control and determined with Equation (55), and $\dot{Q}_{load}$ is the current building load. The chiller set-point temperature necessary to achieve the desired loading is determined as

$$t_{chws} = \text{Max}\left(t_{chwr} - \frac{\dot{Q}_{ch}}{C_{chw}}, t_{coil}\right) \qquad (60)$$

On the first day of each billing period, $\dot{Q}_{ch,peak}$ is set to zero. For this first day, applying Equation (59) leads to the load-limiting control strategy described in the previous section. On subsequent days, load-limiting control is used only if the current peak limit would lead to premature depletion of storage. Whenever the current load is less than $\dot{Q}_{ch,peak}$ and $\dot{Q}_{LCC}$, Equations (59) and (60) lead to chiller-priority control.

FORECASTING DIURNAL DEMAND PROFILES

As discussed previously, forecasts of cooling requirements and electrical use in buildings are often necessary for the control of thermal storage to shift electrical use from on-peak to off-peak periods. In addition, forecasts can help plant operators anticipate major changes in operating modes, such as bringing additional chillers online.

In most methods, predictions are estimated as a function of time-varying input variables that affect cooling requirements and electrical use. Examples of inputs that affect building energy use include (1) ambient dry-bulb temperature, (2) ambient wet-bulb temperature, (3) solar radiation, (4) building occupancy, and (5) wind speed. Methods that include time-varying measured input variables are often termed **deterministic methods**.

Not all inputs affecting cooling requirements and electric use are easily measured. For instance, building occupancy is difficult to determine and solar radiation measurements are expensive. In addition, the accuracy of forecast models that use deterministic inputs depends on the accuracy of future predictions of the inputs. As a result, most inputs that affect building energy use are typically not used.

Much of the time-dependent variation in cooling loads and electrical use for a building can be captured with time as a deterministic input. For instance, building occupancy follows a regular schedule that depends on time of day and time of year. In addition, variations in ambient conditions follow a regular daily and seasonal pattern. Many forecasting methods use time in place of unmeasured deterministic inputs in a functional form that captures the average time dependence of the variation in energy use.

A deterministic model has limited accuracy for forecasts because of both unmeasured and unpredictable (random) input variables. Short-term forecasts can be improved significantly by adding previous values of deterministic inputs and previous output measurements (cooling requirements or electrical use) as inputs to the forecasting model. The time history of these inputs provides valuable information about recent trends in the time variation of the forecasted variable and the unmeasured input variables that affect it. Most forecasting methods use historical variables to predict the future.

Any forecasting method requires that a functional form is defined and the parameters of the model are learned based on measured data. Either offline or online methods can be used to estimate parameters. Offline methods involve estimating parameters from a batch of collected data. Typically, parameters are determined by minimizing the sum of squares of the forecast errors. The parameters of the process are assumed to be constant over time in the offline methods. Online methods allow the parameters of the forecasting model to vary slowly with time. Again, the sum of squares of the forecast errors is minimized, but sequentially or recursively. Often, a forgetting factor is used to give additional weight to the recent data. The ability to track time-varying systems can be important when forecasting cooling requirements or electricity use in buildings, because of the influence of seasonal variations in weather.

Forrester and Wepfer (1984) presented a forecasting algorithm that uses current and previous ambient temperatures and previous loads to predict future requirements. Trends on an hourly time scale are accounted for with measured inputs for a few hours preceding the current time. Day-to-day trends are considered by using the value of the load that occurred 24 h earlier as an input. One of the major limitations of this model is its inability to accurately predict loads when an occupied day (e.g., Monday) follows an unoccupied (e.g., Sunday) or when an unoccupied day follows an occupied day (e.g., Saturday). The cooling load for a particular hour of the day on a Monday depends very little on the requirement 24 h earlier on Sunday. Forrester and Wepfer (1984) described a number of methods for eliminating this 24 h indicator. MacArthur et al. (1989) also presented a load profile prediction algorithm that uses a 24 h regressor.

Armstrong et al. (1989) presented a very simple method for forecasting either cooling or electrical requirements that does not use the 24 h regressor; Seem and Braun (1991) further developed and validated this method. The "average" time-of-day and time-of-week trends are modeled using a lookup table with time and type of day (e.g., occupied versus unoccupied) as the deterministic input variables. Entries in the table are updated using an exponentially weighted, moving-average model. Short-term trends are modeled using previous hourly measurements of cooling requirements in an autoregressive (AR) model. Model parameters adapt to slow changes in system characteristics. The combination of updating the table and modifying model parameters works well in adapting the forecasting algorithm to changes in season and occupancy schedule.

Kreider and Wang (1991) used artificial neural networks (ANNs) to predict energy consumption of various HVAC equipment in a commercial building. Data inputs to the ANN included (1) previous hour's electrical power consumption, (2) building occupancy, (3) wind speed, (4) ambient relative humidity, (5) ambient dry-bulb temperature, (6) previous hour's ambient dry-bulb temperature, (7) two hour's previous ambient temperature, and (8) sine and cosine of the hour number to roughly represent the diurnal change of temperature and solar insolation. The primary purpose in developing these models was to detect changes in equipment and system performance for monitoring purposes. However, the authors suggested that an ANN-based predictor might be valuable when used to predict energy consumption with a network based on recent historical data. Forecasts of all deterministic input variables are necessary to apply this method.

Gibson and Kraft (1993) used an ANN to predict building electrical consumption as part of the operation and control of a thermal energy storage (TES) cooling system. The ANN used the following inputs: (1) electric demand of occupants (lighting and other loads), (2) electric demand of TES cooling tower fans, (3) outdoor ambient temperature, (4) outdoor ambient temperature/inside target temperature, (5) outdoor ambient relative humidity, (6) on/off status for building cooling, (7) cooling system on/off status, (8) chiller #1 direct-cooling mode on/off status, (9) chiller #2 direct-cooling mode on/off status, (10) ice storage discharging mode on/off status,

(11) ice storage charging mode on/off status, (12) chiller #1 charging mode on/off status, and (13) chiller #2 charging mode on/off status. To use this forecaster, values of each of these inputs must be predicted. Although the authors suggest that average occupancy demand profile be used as an input, they do not state how the other input variables should be forecast.

A Forecasting Algorithm

This section presents an algorithm for forecasting hourly cooling requirements or electrical use in buildings that is based on the method developed by Seem and Braun (1991). At a given hour n, the forecast value is

$$\hat{E}(n) = \hat{X}(n) + \hat{D}(h,d) \qquad (61)$$

where

$\hat{E}(n)$ = forecast cooling load or electrical use for hour n

$\hat{X}(n)$ = stochastic or probabilistic part of forecast for hour n

$\hat{D}(h,d)$ = deterministic part of forecast at hour n associated with hth hour of day and current day type d

The deterministic part of the forecast is simply a lookup table for the forecasted variable in terms of hour of day h and type of day d. Seem and Braun recommend using three distinct day types: unoccupied days, occupied days following unoccupied days, and occupied days following occupied days. The three day types account for differences between the building responses associated with return from night setup and return from weekend setup. The building operator or control engineer must specify the number of day types and a calendar of day types.

Given the hour of day and day type, the deterministic part of the forecast is simply the value stored in that location in the table. Table entries are updated when a new measurement becomes available for that hour and day type. Updates are accomplished through the use of an exponentially weighted, moving-average (EWMA) model as

$$\hat{D}(h,d) = \hat{D}(h,d)_{old} + \lambda[E(n) - \hat{D}(h,d)_{old}] \qquad (62)$$

where

$E(n)$ = measured value of cooling load or electrical use for current hour n

λ = exponential smoothing constant, $0 < \lambda < 1$

$\hat{D}(h,d)_{old}$ = previous table entry for $\hat{D}(h,d)$

As λ increases, the more recent observations have more influence on the average. As λ approaches zero, the table entry approaches the average of all data for that hour and day type. When λ equals one, the table entry is updated with the most recent measured value. Seem and Braun recommend using a value of 0.30 for λ in conjunction with three day types and 0.18 with two day types.

The stochastic portion of the forecast is estimated with a third-order autoregressive model, AR(3), of forecasting errors associated with the deterministic model. With this model, an estimate of the next hour's error in the deterministic model forecast is given by the following equation:

$$\hat{X}(n+1) = \phi_1 X(n) + \phi_2 X(n-1) + \phi_3 X(n-2) \qquad (63)$$

where

$X(n)$ = difference between measurement and deterministic forecast of cooling load or electrical use at any hour n

ϕ_1, ϕ_2, ϕ_3 = parameters of AR(3) model that must be learned

The error in the deterministic forecast at any hour is simply

$$X(n) = E(n) - \hat{D}(h,d) \qquad (64)$$

For forecasting more than one hour ahead, conditional expectation is used to estimate the deterministic model forecast errors using the AR(3) model as follows:

$$\hat{X}(n+2) = \phi_1 \hat{X}(n+1) + \phi_2 X(n) + \phi_3 X(n-1)$$

$$\hat{X}(n+3) = \phi_1 \hat{X}(n+2) + \phi_2 \hat{X}(n+1) + \phi_3 X(n)$$

$$\vdots$$

$$\hat{X}(n+k) = \phi_1 \hat{X}(n+k+1) + \phi_2 \hat{X}(n+k-2)$$
$$+ \phi_3 \hat{X}(n+k-3) \qquad \text{for } k > 3 \qquad (65)$$

Online estimation of the AR(3) model parameters is accomplished by minimizing the following time-dependent cost function:

$$J(\phi) = \sum_{k=1}^{n} \alpha^{n-k}[X(k) - \hat{X}(k)]^2 \qquad (66)$$

where the constant α is called the forgetting factor and has a value between 0 and 1. With this formulation, the residual for the current time step has a weight of one and the residual for k time steps back has a weight of α_k. By choosing a value of α that is positive and less than one, recent data have greater influence on the parameter estimates. In this manner, the model can track changes due to seasonal or other effects. Seem and Braun recommend using a forgetting factor of 0.99. Parameters of the AR(3) model should be updated at each hour when a new measurement becomes available.

Ljung and Söderström (1983) describe online estimation methods for determining coefficients of an AR model. The parameter estimates should be evaluated for stability. If an AR model is not stable, then the forecasts will grow without bound as the time of forecasts increases. Ljung and Söderström discuss methods for checking stability.

Seem and Braun compared forecasts of electrical usage with both simulated and measured data. Figure 42 shows the standard deviation of the 1 h through 24 h errors in electrical use forecasts for annual simulation results. Results are given for the deterministic model alone, deterministic plus AR(2), and deterministic plus AR(3). For the combined models, the standard deviation of the residuals increases as the forecast length increases. For short time

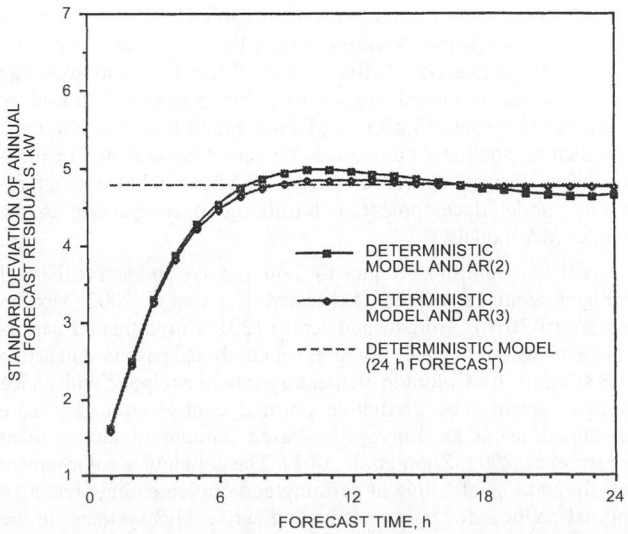

**Fig. 42   Standard Deviation of Annual Errors for
1 to 24 h Forecasts**

BUILDING ELECTRICAL CONSUMPTION PROFILES

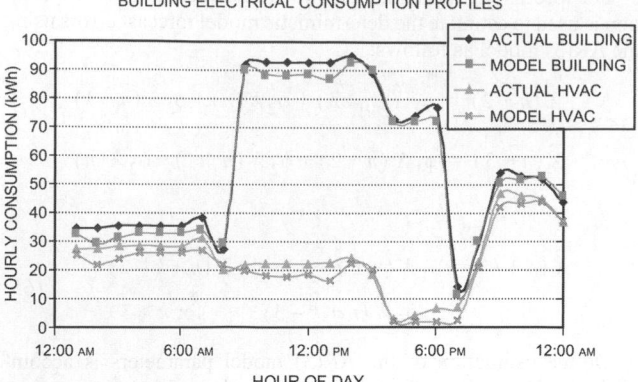

Fig. 43   Building Electricity Use Profiles for 6 h Predictive Optimal Control

BUILDING ELECTRICAL CONSUMPTION PROFILES

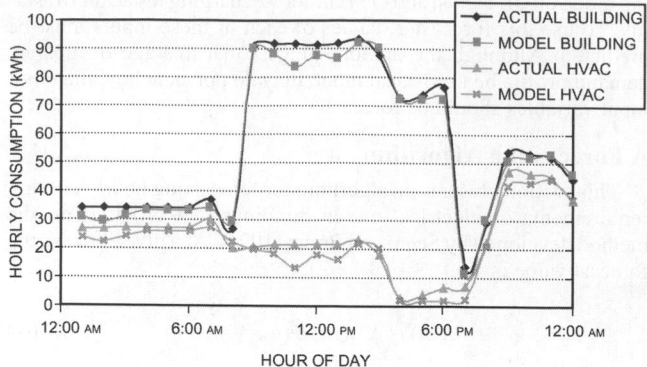

Fig. 44   Building Electricity Use Profiles for 24 h Predictive Optimal Control

steps (i.e., less than 6 h), the combined deterministic and stochastic models provide much better forecasts than the purely deterministic model (i.e., lookup table).

Seem and Braun also investigated a method for adjusting the deterministic forecast based on using weather service forecasts of maximum daily ambient temperature as an input. For short periods (i.e., less than 4 h), forecasts for the temperature-dependent model were nearly identical to forecasts for the temperature-independent model. For longer periods, the temperature-dependent model provided better forecasts than the temperature-independent model.

PREDICTIVE CONTROL STRATEGIES

Model predictive controls applied to commercial buildings require short-term weather forecasts to optimally adjust set points in a supervisory control environment. Review of the literature reveals that several researchers are convinced that nonlinear forecasting models based on neural networks (NNs) provide superior performance over traditional time-series analysis. Florita and Henze (2009) sought to identify the complexity required for short-term weather forecasting in the context of a model predictive control environment. Moving average (MA) models with various enhancements and neural network models were used to predict weather variables seasonally in numerous geographic locations. Their performance was statistically assessed using the coefficient of variation (CV) and mean bias error (MBE) values. When a cyclical two-stage model predictive control process of policy planning followed by execution was used, Florita and Henze found that even the most complicated nonlinear autoregressive neural network with exogenous input does not appear to warrant the additional efforts in forecasting model development and training in comparison to the simpler MA models.

Predictive control strategies for both passive and active thermal storage systems have been field-tested (Krarti et al. 2007; Morgan and Krarti 2010). Morgan and Krarti (2010) investigated performance of various control strategies for combined passive and active TES systems in a Colorado elementary school equipped with an ice storage system. The predictive optimal control strategies were developed using an EnergyPlus-based simulation environment (Krarti et al. 2007; Zhou et al. 2005). The simulation environment was found to be effective in defining and implementing predictive optimal controls for both passive and active TES systems for the buildings. Figures 43 and 44 show examples of the field testing results for the building energy use performed by Morgan and Krarti (2010).

REFERENCES

ABMA. 1998. *Guideline for the integration of boilers and automated control systems in heating applications.* American Boiler Manufacturers Association, Arlington, VA.

Andresen, I. and M.J. Brandemuehl. 1992. Heat storage in building thermal mass: A parametric study. *ASHRAE Transactions* 98(1).

Armstrong, P.R., T.N. Bechtel, C.E. Hancock, S.E. Jarvis, J.E. Seem, and T.E. Vere. 1989. *Environment for structured implementation of general and advanced HVAC controls—Phase II.* Ch. 7: Small business innovative research program. DOE Contract DE-AC02-85ER 80290.

Armstrong, P.R., S. Katipamula, W. Jiang, D. Winiarski, and L.K. Norford. 2009. Efficient low-lift cooling with radiant distribution, thermal storage and variable-speed chiller controls, Part II: Annual energy use and savings. *HVAC&R Research* 15(2):402-432.

ASHRAE. 2010. Energy standard for buildings except low-rise residential buildings. ANSI/ASHRAE/IESNA *Standard* 90.1-2010.

ASHRAE. 2004. Method of testing the performance of cool storage systems. ANSI/ASHRAE *Standard* 150-2000 (RA 2004).

BEI. 1991. *Boiler efficiency improvement.* Boiler Efficiency Institute, Auburn, AL.

Bellman, R. 1957. *Dynamic programming.* Princeton University Press, Princeton, NJ.

Bloomfield, D.P. and D.J. Fisk. 1977. The optimization of intermittent heating. *Buildings and Environment* 12:43-55.

Brandemuehl, M.J. and J. Bradford. 1998. Implementation of on-line optimal supervisory control of cooling plants without storage (RP-823). ASHRAE Research Project, *Final Report.*

Braun, J.E. 1988. *Methodologies for the design and control of central cooling plants.* Ph.D. dissertation, University of Wisconsin-Madison.

Braun, J.E. 1990. Reducing energy costs and peak electrical demands through optimal control of building thermal storage. *ASHRAE Transactions* 96(1).

Braun, J.E. 1992. A comparison of chiller-priority, storage-priority, and optimal control of an ice-storage system. *ASHRAE Transactions* 98(1): 893-902.

Braun, J.E. 2007a. A near-optimal control strategy for cool storage systems with dynamic electric rates. *HVAC&R Research* 13(4):557-580.

Braun, J.E. 2007b. Impact of control on operating costs for cool storage systems with dynamic electric rates. *ASHRAE Transactions* 113.

Braun, J.E. 2007c. Near-optimal control strategies for hybrid cooling plants. *HVAC&R Research* 13(4):599-622.

Braun, J.E. 2007d. A general control algorithm for cooling towers in cooling plants with electric and/or gas-driven chillers. *HVAC&R Research* 13(4):581-598.

Braun, J.E. and G.T. Diderrich. 1990. Near-optimal control of cooling towers for chilled-water systems. *ASHRAE Transactions* 96(2):806-813.

Braun, J.E., J.W. Mitchell, S.A. Klein, and W.A. Beckman. 1987. Performance and control characteristics of a large central cooling system. *ASHRAE Transactions* 93(1):1830-1852.

Braun, J.E., S.A. Klein, J.W. Mitchell, and W.A. Beckman. 1989a. Applications of optimal control to chilled water systems without storage. *ASHRAE Transactions* 95(1):663-675.

Braun, J.E., S.A. Klein, J.W. Mitchell, and W.A. Beckman. 1989b. Methodologies for optimal control to chilled water systems without storage. *ASHRAE Transactions* 95(1):652-662.

Braun, J.E., K.W. Montgomery, and N. Chaturvedi. 2001. Evaluating the performance of building thermal mass control strategies. *International Journal of HVAC&R Research* (now *HVAC&R Research*) 7(4):403-428.

Brent, R.P. 1973. *Algorithms for minimization without derivatives*, Ch. 5. Prentice Hall.

Brothers, P.W. and M.L. Warren. 1986. Fan energy use in variable air volume systems. *ASHRAE Transactions* 92(2B):19-29.

Cheng, H., M.J. Brandemuehl, G.P. Henze, A.R. Florita, and C. Felsmann. 2008. Evaluation of the primary factors impacting the optimal control of passive thermal storage (RP-1313). *ASHRAE Transactions* 114(2):57-64.

Cumali, Z. 1988. Global optimization of HVAC system operations in real time. *ASHRAE Transactions* 94(1):1729-1744.

Cumali, Z. 1994. Application of real-time optimization to building systems. *ASHRAE Transactions* 100(1).

Drees, K.H. 1994. *Modeling and control of area-constrained ice storage systems*. M.S. thesis, Purdue University, West Lafayette, IN.

Drees, K.H. and J.E. Braun. 1995. Modeling of area-constrained ice storage tanks. *International Journal of HVAC&R Research* (now *HVAC&R Research*) 1(2):143-159.

Drees, K.H. and J.E. Braun. 1996. Development and evaluation of a rule-based control strategy for ice storage systems. *International Journal of HVAC&R Research* (now *HVAC&R Research*) 2(4):312-336.

Dyer, D.F. and G. Maples. 1981. *Boiler efficiency improvement*. Boiler Efficiency Institute. Auburn, AL.

Elleson, J.S. 1996. *Successful cool storage projects: From planning to operation*. ASHRAE.

Englander, S.L. and L.K. Norford. 1992. Saving fan energy in VAV systems, Part 2: Supply fan control for static pressure minimization using DDC zone feedback. *ASHRAE Transactions* 98(1):19-32.

Florita, A.R. and G.P. Henze. 2009. Comparison of short-term weather forecasting models for model predictive control. *HVAC&R Research* 15(5):835-853.

Forrester, J.R. and W.J. Wepfer. 1984. Formulation of a load prediction algorithm for a large commercial building. *ASHRAE Transactions* 90(2B):536-551.

Gayeski, N.T., P.R. Armstrong, and L.K. Norford. 2010a. Experimental characterization of a variable-speed heat pump with optimal control over a wide range of conditions and load. In preparation for submission to *HVAC&R Research*.

Gayeski, N.T., P.R. Armstrong, and L.K. Norford. 2010b. Predictive precooling control with passive thermal energy storage: implementation and measured performance. In preparation for submission to *HVAC&R Research*.

Gibson, G.L. and T.T. Kraft. 1993. Electric demand prediction using artificial neural network technology. *ASHRAE Journal* 35(3):60-68.

Greensfelder, E.M., G.P. Henze, and C. Felsmann. 2010. An investigation of optimal control of passive building thermal storage with real time pricing. Submitted to *Journal of Building Performance Simulation*.

Grumman, D.L. and A.S. Butkus, Jr. 1988. Ice storage application to an Illinois hospital. *ASHRAE Transactions* 94(1):1879-1893.

Hackner, R.J., J.W. Mitchell, and W.A. Beckman. 1984. HVAC system dynamics and energy use in buildings—Part I. *ASHRAE Transactions* 90(2B):523-535.

Hackner, R.J., J.W. Mitchell, and W.A. Beckman. 1985. HVAC system dynamics and energy use in buildings—Part II. *ASHRAE Transactions* 91(1B):781.

Hartman, T. 1993. Terminal regulated air volume (TRAV) systems. *ASHRAE Transactions* 99(1):791-800.

Henze, G.P., M. Krarti, and M.J. Brandemuehl. 1997a. A simulation environment for the analysis of ice storage controls. *International Journal of HVAC&R Research* (now *HVAC&R Research*) 3(2):128-148.

Henze, G.P., R.H. Dodier, and M. Krarti. 1997b. Development of a predictive optimal controller for thermal energy storage systems. *International Journal of HVAC&R Research* (now *HVAC&R Research*) 3(3):233-264.

Henze, G.P., M. Krarti, and M.J. Brandemuehl. 2003. Guidelines for improved performance of ice storage systems. *Energy and Buildings* 35(2):111-127.

Henze, G.P., D. Kalz, S. Liu, and C. Felsmann. 2005. Experimental analysis of model-based predictive optimal control for active and passive building thermal storage inventory. *HVAC&R Research* 11(2):189-214.

Henze, G.P., M.J. Brandemuehl, C. Felsmann, A. Florita, and H. Cheng. 2007. Evaluation of building thermal mass savings. ASHRAE Research Project RP-1313, *Final Report*.

Henze, G.P., C. Felsmann, A.R. Florita, M.J. Brandemuehl, H. Cheng, and C.E. Waters. 2008. Optimization of building thermal mass control in the presence of energy and demand charges (RP-1313). *ASHRAE Transactions* 114(2):75-84.

Henze, G.P., A.R. Florita, M.J. Brandemuehl, C. Felsmann and H. Cheng. 2009. Advances in near-optimal control of passive building thermal storage. *Proceedings of the ASME 3rd International Conference on Energy Sustainability*, San Francisco.

Hiller, C.C. and L. Glicksman. 1977. Heat pump improvement using compressor flow modulation. *ASHRAE Transactions* 83(2)

Kao, J.Y. 1985. Control strategies and building energy consumption. *ASHRAE Transactions* 91(2B):510-817.

Katipamula, S., P.R. Armstrong, W. Wang, N. Fernandez, H. Cho, W. Goetzler, J. Burgos, R. Radhakrishnan, and C. Ahlfeldt. 2010a. *Cost-effective integration of efficient low-lift baseload cooling equipment: FY08 final report*. PNNL-19114.

Katipamula, S., P.R. Armstrong, W. Wang, N. Fernandez, H. Cho, W. Goetzler, J. Burgos, R. Radhakrishnan, and C. Ahlfeldt. 2010b. *Development of high-efficiency low-lift vapor compression system—Final report*. PNNL-19227.

Keeney, K.R. and J.E. Braun. 1996. A simplified method for determining optimal cooling control strategies for thermal storage in building mass. *International Journal of HVAC&R Research* (now *HVAC&R Research*) 2(1):1-20.

Keeney, K.R. and J.E. Braun. 1997. Application of building precooling to reduce peak cooling requirements. *ASHRAE Transactions* 103(1):463-469.

Krarti, M., M.J. Brandemuehl, and G.P. Henze. 1996. Evaluation of optimal control for ice systems (RP-809). ASHRAE Research Project, *Report*.

Krarti, M., G. Henze, G. Zhou, P. Ihm, S. Liu, and S. Morgan. 2007. *Real-time predictive optimal control of active and passive thermal energy storage systems: Final report*. Department of Energy Contract DE-FC-36-03G)13026.

Kreider, J.F. and X.A. Wang. 1991. Artificial neural networks demonstration for automated generation of energy use predictors for commercial buildings. *ASHRAE Transactions* 97(2):775-779.

Lau, A.S., W.A. Beckman, and J.W. Mitchell. 1985. Development of computer control—Routines for a large chilled water plant. *ASHRAE Transactions* 91(1B):766-780.

Liu, S. and G.P. Henze. 2006a. Experimental analysis of simulated reinforcement learning control for active and passive building thermal storage inventory—Part 1: Theoretical foundation. *Energy and Buildings* 38(2):142-147.

Liu, S. and G.P. Henze. 2006b. Experimental analysis of simulated reinforcement learning control for active and passive building thermal storage inventory—Part 2: Results and analysis. *Energy and Buildings* 38(2):148-161.

Ljung, L. and T. Söderström. 1983. *Theory and practice of recursive identification*. MIT Press, Cambridge, MA.

MacArthur, J.W., A. Mathur, and J. Zhao. 1989. On-line recursive estimation for load profile prediction. *ASHRAE Transactions* 95(1):621-628.

Morgan, S. and M. Krarti. 2006. Impact of electricity rate structures on energy cost savings of pre-cooling controls for office buildings. *Building and Environment* 42(8):2810-2818.

Morgan S. and M. Krarti. 2010. Field testing of optimal controls of passive and active thermal storage. *ASHRAE Transactions* 116(1).

Morris, F.B., J.E. Braun, and S. Treado. 1994. Experimental and simulated performance of optimal control of building thermal storage. *ASHRAE Transactions* 100(1):402-414.

Parker, S.A., R.B. Scollon, and R.D. Smith. 1997. Boilers and fired systems. *Energy Management Handbook*. Fairmont Press, Lilburn, GA.

Rabl, A. and L.K. Norford. 1991. Peak load reduction by preconditioning buildings at night. *International Journal of Energy Research* 15:781-798.

Ruud, M.D., J.W. Mitchell, and S.A. Klein. 1990. Use of building thermal mass to offset cooling loads. *ASHRAE Transactions* 96(2).

Rawlings, L.K. 1985. Strategies to optimize ice storage. *ASHRAE Journal* 27(5):39-44.

Seem, J.E. and J.E. Braun. 1991. Adaptive methods for real-time forecasting of building electrical demand. *ASHRAE Transactions* 97(1):710-721.

Seem, J.E., S.A. Klein, W.A. Beckman, and J.W. Mitchell. 1989a. Comprehensive room transfer functions for efficient calculation of the transient heat transfer process in buildings. *Journal of Heat Transfer* 111:264-273.

Seem, J.E., P.R. Armstrong, and C.E. Hancock. 1989b. Comparison of seven methods for forecasting the time to return from night setback. *ASHRAE Transactions* 95(2).

Seem, J.E., S.A. Klein, W.A. Beckman, and J.W. Mitchell. 1990. Model reduction of transfer functions using a dominant root method. *ASME Journal of Heat Transfer* 112:547-554.

Seem, J.S., C. Park, and J.M. House. 1999. A new sequencing control strategy for air-handling units. *International Journal of HVAC&R Research* (now *HVAC&R Research*) 5(1):35-58.

Snyder, M.E. and T.A. Newell. 1990. Cooling cost minimization using building mass for thermal storage. *ASHRAE Transactions* 96(2):830-838.

Spethmann, D.H. 1989. Optimal control for cool storage. *ASHRAE Transactions* 95(1):1189-1193.

Stoecker, W.F. 1980. *Design of thermal systems.* McGraw-Hill, New York.

Tamblyn, R.T. 1985. Control concepts for thermal storage. *ASHRAE Transactions* 91(1B):5-11.

Taplin, H.R. 1998. *Boiler plant and distribution system optimization manual.* Fairmont Press, Lilburn, GA.

Warren, M. and L.K. Norford. 1993. Integrating VAV zone requirements with supply fan operation. *ASHRAE Journal* 35(4):43-46.

Watkins, C. and P. Dayan. 1992. Q-learning. *Machine Learning* 8:279-292.

Wei, G., M. Liu, and D.E. Claridge. 2004. Integrated damper and pressure reset for VAV supply air fan control. *ASHRAE Transactions* 110(2):309-313.

Winkelman, F.C., W.F. Buhl, B. Birdsall, A.E. Erdem, K.L. Ellington, and Hirsch & Associates. 1993. DOE-2 BDL summary, version 21.E. LBNL *Report* 34946, Lawrence Berkeley National Laboratory, Berkeley, CA.

Zakula, T., N.T. Gayeski, P.R. Armstrong, and L.K. Norford. 2010. Variable-speed heat pump performance model, model validation, and optimal control. In preparation for submission to *HVAC&R Research.*

Zhou, G., P. Ihm, M. Krarti, S. Liu, and G.P. Henze. 2005. Integration of an internal optimization module within EnergyPlus. *IBPSA Proceedings.*

HVAC COMMISSIONING

COMMISSIONING implements a quality-oriented process for achieving, verifying, and documenting that the performance of facilities, systems, and assemblies meets defined objectives and criteria. The defined objectives and criteria are often referred to as the **owner's project requirements (OPR)**, which involve achieving, verifying, and documenting the performance of each assembly or system to meet the building's operational needs. The commissioning process uses the owner's project requirements as the reference to determine acceptance of the design and/or construction. Commissioning includes verifying and documenting that the project operational and maintenance documentation and training of operation and maintenance personnel occur. The result should be fully functional systems that can be properly operated and maintained throughout the life of the building.

This chapter gives an overview of the general commissioning process as covered in ASHRAE *Guideline* 0-2005, developed for the National Institute of Building Sciences' total building commissioning program, as well as the best practices for applying the process from ASHRAE *Guideline* 1.1-2007. Although this chapter provides less detail and is less prescriptive, it provides more narrative discussion on some issues than these two guidelines.

Recommissioning applies commissioning to a project that has been previously delivered using the commissioning process. This may be a scheduled recommissioning developed as part of ongoing commissioning, or it may be triggered by use change, operational problems, or other needs. **Existing building commissioning** (often called **retrocommissioning**) applies commissioning to an existing facility that may or may not have been previously commissioned. It consists of systematically investigating, analyzing, and adjusting operations of existing building equipment, systems, and assemblies, as well as training and documentation for operators to ensure that required performance (including energy, comfort, and IAQ) is achieved. Buildings require maintenance and tuning to prevent performance degradation. Existing building commissioning should be performed as part of ongoing efforts to maintain a comfortable and efficient environment within the building. It has broad application to virtually every building type and vintage with excellent cost/benefit results and payback ratios. Existing building commissioning starts with development of the owner's current facility requirements, reviews the existing design, and then tests existing systems. Any major retrofits required follow the process for new building commissioning as defined in this chapter.

Applicability

The commissioning process described here applies to new construction, major renovations, and all systems and assemblies. Although this chapter focuses on HVAC, commissioning can be applied to the building as a total system, which includes structural

The preparation of this chapter is assigned to TC 7.9, Building Commissioning.

elements, building envelope, life safety features, electrical systems, communication systems, plumbing, irrigation, controls, and HVAC systems (ASHRAE *Guideline* 0). Based on owners' preference and project contract scope, total commissioning can include industrial process and process equipment, systems, piping, instrumentation, electrical, and related control, or these topics may be treated as an independent phase of project commissioning.

Which building systems should be commissioned varies with the systems and assemblies used, building size, project type, and objectives. Owners and commissioning providers often focus on systems and assemblies under the commissioning umbrella that have (1) historically not performed well at turnover (e.g., outside air economizers and variable-speed drives), (2) are mission-critical (e.g., air cleanliness in a cleanroom, emergency power in a hospital), (3) will be costly to fix during occupancy if they fail (e.g., chilled-water piping, window flashing assemblies), or (4) present a life-safety risk if they fail (e.g., fire alarm, smoke control, moisture penetration). Recommendations in this chapter should be appropriately modified for each project. Although commissioning may begin at any time during the project life cycle, owners obtain the highest benefits when commissioning begins at the conceptual or predesign phase.

Background

Equipment, components, systems, and assemblies have become more complex. More specialization has occurred in the disciplines and trades, with increased interactions between all elements. This increased specialization and interaction requires increased integration between disciplines and specialized systems by the delivery team. Owners often use low-bid policies, and scopes of design professionals are often narrowed. The result has been buildings that do not meet owner expectations and often do not work as intended because of programming, design, and construction deficiencies. Commissioning is a value-added service that helps overcome these infrastructure inadequacies and fundamentally improve the performance of building systems and living conditions for occupants.

Benefits

The primary benefits of commissioning include improvements in all of the following areas:

- **Predesign and design**
  - Owners develop better understanding of what they want and need through clear, documented owner's project requirements (OPR)
  - Designers understand better what owner is requesting
  - Designers reduce their risk with better communication and input from owner
  - Owners understand better what designers are proposing through a clear, documented basis of design (BOD)
  - Experts review and improve commissioning documents

- **Construction (including system and assembly performance)**
  - Improved specifications and drawings, resulting in improved coordination between all groups

- Specifying systems that can be properly commissioned and tested, and are within owner's ability to maintain
- Tools to help contractors perform better installations (e.g., construction checklists)
- Performance accountability through construction observation, issue management, and testing
- Documented verification of system and assembly performance
- Thorough training requirements in construction documents
- Verifying training completion
- Formal acceptance testing at completion
- **Occupancy and operations (including maintenance)**
  - Thorough documentation in construction contract
  - Verifying documentation submittals

Commissioning also reduces potential change orders, contractor callbacks, and time required to fine-tune and debug systems during occupancy, and smooths turnover. Building performance improvements give better building and system control, improve energy efficiency, enhance indoor environmental quality, and contribute to increased occupant productivity.

Key Contributors

- Owner
- Engineer/architect
- Commissioning authority (CA)
- Operations and maintenance personnel
- Occupants and users
- Design professionals
- Contractors
- Suppliers
- Independent test and balancing company

Definitions

Basis of Design (BOD). A document that records the concepts, calculations, decisions, and product selections used to meet the OPR and to satisfy applicable regulatory requirements, standards, and guidelines. The document includes both narrative descriptions and lists of individual items that support the design process.

Commissioning Authority (CA). An entity identified by the owner who leads, plans, schedules, and coordinates the commissioning team to implement the commissioning process.

Commissioning Plan (CP). A document that outlines the organization, schedule, allocation of resources, and documentation requirements of the commissioning process.

Construction Checklist. A form used by the contractor to verify that appropriate components are on site, ready for installation, correctly installed, and functional.

Owner's Project Requirements (OPR). A document that details the functional requirements of a project and the expectations of how it will be used and operated. These include project goals, measurable performance criteria, cost considerations, benchmarks, success criteria, and supporting information. (The term **project intent** is used by some owners for their commissioning-process OPR.)

Systems Manual. A system-focused, composite document that includes the operation manual, maintenance manual, and additional information of use to the owner during occupancy and operations.

Test Procedures. A written protocol that defines methods, personnel, and expectations for tests conducted on components, equipment, assemblies, systems, and interfaces among systems.

COMMISSIONING OBJECTIVE

The commissioning objective focuses on documented confirmation that a facility fulfills the specified performance requirements for the building owner, occupants, and operators. To reach this objective, it is necessary to (1) clearly document the owner's project requirements, including performance and maintainability; and (2) verify and document compliance with these criteria throughout design, construction, acceptance, and initial operation phases.

Specific goals for commissioning include

- Providing documentation and tools to improve quality of deliverables (e.g., forms, tracking software, performance calculation tools)
- Verifying and documenting that systems and assemblies perform according to OPR by end of construction with building occupancy
- Providing a uniform and effective process for delivery of construction projects
- Using quality-based sampling techniques to detect systemic problems
- Verifying proper coordination among all contractors, subcontractors, vendors, and manufacturers of all furnished equipment and assemblies in the completed systems
- Verifying that adequate and accurate system and assembly documentation is provided to owner
- Verifying that operation and maintenance personnel and occupants are properly trained

MANAGEMENT AND RESPONSIBILITIES

Management Strategies

In each project, a qualified party should be designated as the commissioning authority.

Predesign and Design. Commissioning during predesign and design is most often managed by an independent CA who is not part of the formal designer-of-record team. An independent, objective view is critical. The CA normally provides input to the owner and designers but does not have ultimate authority over design decisions. The CA should also coordinate, conduct, or approve activities such as assisting in development of the OPR, conducting statistical sampling reviews, and developing commissioning specifications and test procedures. The CA may also review plan designs. In some projects, commissioning is the designer's responsibility, using either their own staff or a consultant.

Construction. During construction, because of the variety of players, construction management scenarios, and the owner's objectives, numerous methods are used to manage the commissioning process. To maintain objectivity, the CA should be independent. If the contractor or designer hires the CA, the potential conflict of interest must be carefully managed. The two primary methods to manage commissioning during construction are commissioning-authority-managed and contractor-managed. In the **commissioning-authority-managed approach**, the CA performs many of the planning and technical tasks, such as developing the commissioning plan and test procedures and directing, witnessing, and documenting execution of tests, performed by either the contractor or themselves. In the **contractor-managed approach**, the contractor may develop the commissioning plan, write test procedures, and direct and document testing, with the CA reviewing and approving the plan, witnessing selected tests, and reviewing completed test reports. The CA should report to the owner on the adequacy of a contractor-managed commissioning plan. The contractor may assign staff, subcontractor, or subconsultant to manage and coordinate commissioning responsibilities. This approach gives the contractor more responsibility. Some view this method as less objective, but others consider it more integrated into the building delivery process than the CA-managed approach.

Some project plans use both management approaches, particularly when a substantial amount of electrical equipment is being tested. HVAC and controls follow the commissioning-authority-managed approach, and electrical system commissioning follows the contractor-managed approach, but the entire process is still overseen by the single CA.

Team Members

Effective building commissioning requires a team effort. The size and makeup of the team depends on the size and complexity of the project and the owner's desire for quality assurance. Team members include the owner, occupants, design professionals, construction manager, general contractor, subcontractors, operation and maintenance (O&M), suppliers, equipment manufacturers, and the CA. All members, particularly the O&M manager, need to be brought into the commissioning process early, preferably during predesign.

The level of effort of team members changes during the different project phases. For example, during design, the designer is a key player in the commissioning process, whereas the contractor may not have been selected. During construction, the general contractor's and installing subcontractors' roles increase.

The scope of work of the CA, design professionals, and contractors should be clearly and completely identified in their contracts. Without this, change orders, incomplete or missed tasks, and otherwise dysfunctional commissioning may result.

Roles and Responsibilities

The commissioning team's responsibilities are to conduct commissioning activities in a logical, sequential, and efficient manner using consistent protocols and forms, centralized documentation, clear and regular communications and consultations with all necessary parties, frequently updated timelines and schedules, and appropriate technical expertise. The following sections summarize the responsibilities of each party. Additional detail is found in the Commissioning Process section.

Commissioning Authority. Specific responsibilities vary with the management scenario and the CA's specific scope of services. Ideally, the same party or firm acts as CA through all project phases. The CA organizes and leads the commissioning team throughout the project.

Design Phase (Including Predesign). During predesign, the CA develops the predesign and design-phase commissioning plan and ensures the OPR is developed.

During design, the CA develops detailed commissioning activities. The core CA responsibilities are

- Reviewing designer's BOD, plans, and specifications, ensuring they meet the OPR
- Developing initial construction-phase commissioning plan
- Review owner's request for quotation (RFQ) before issuing for construction bid, ensuring that the commissioning, training, and documentation requirements for all contractors and suppliers are reflected in construction contract documents.

Construction and Acceptance. During construction, the CA is in charge of the commissioning process and makes the final recommendations to the owner about functional performance of commissioned building systems and assemblies. The CA directs commissioning activities, possibly performing many of them, depending on the management scenario in place. The CA is an independent and objective advocate for the owner. The core commissioning activities during construction involve

- Reviewing selected construction submittals to ensure conformance with OPR, with updates in commissioning plan as needed
- Observing installations, start-up, and functional performance tests, including documenting any conditions that require correction
- Co-organizing with the discipline design engineer, and planning, developing, reviewing, approving, and executing or observing testing
- Codeveloping or assisting with systems manual
- Reviewing O&M manual submissions
- Verifying operator and maintenance personnel training and documentation

- Submitting documented results to owner on all commissioning performed

These tasks may vary (e.g., some commissioning scopes involve preparing the O&M or electronic facility's manuals, preparing detailed maintenance management plans, or conducting operator and maintenance personnel training).

Occupancy and Operations. During occupancy and operations, the CA helps resolve commissioning issues and directs opposite-season testing. Often, the CA participates in a near-warranty-end review of system and assembly performance.

Independence. If the CA's firm has other project responsibilities, a potential conflict of interest exists. Wherever this occurs, the CA should disclose in writing the nature of the conflict and the means by which it will be managed. If the CA is not under direct contract to the owner, the owner's interests need to be protected through appropriate oversight of the CA's work.

Qualifications. The CA should fully understand commissioning, design, and construction processes and have technical design, operations, maintenance, and troubleshooting knowledge of the systems and assemblies being commissioned. Excellent written and verbal communication skills are critical. The CA may represent an individual or a team of commissioning experts, depending on system complexity, the number of disciplines involved, and commissioning scope. Thus, the ability to manage diverse disciplines over long timelines is also important.

Construction Manager. The construction manager's role varies with construction responsibilities. When they have significant oversight for the owner (e.g., schedule management, submittal review, change order authority), their commissioning role is more like the owner's: they ensure the contractor is executing their commissioning responsibilities according to the commissioning plan and help resolve issues.

General Contractors.

Design. The general contractor (if yet selected) reviews commissioning requirements and performance criteria for coordination, schedule, and cost implications.

Construction and Acceptance. The contractor's role and responsibilities are

- Ensuring subcontractors' commissioning work is completed and cooperating with CA in executing the commissioning plan
- Providing input into commissioning plan for CA's review and approval
- Integrating commissioning schedule into overall project schedule
- Participating in commissioning meetings
- Responding to questions and issues raised by CA
- Resolving issues identified during commissioning and coordinating correction of identified deficiencies
- Providing equipment, system, and assembly data needed by CA
- Performing specified training
- Submitting required portions of systems manuals

In the contractor-managed approach, the general contractor is often required to hire a third party with direct commissioning skills to manage and execute the contractor commissioning requirements.

Trade Contractors.

Design. Trade contractors of specialty or complex systems or designs should review commissioning requirements and performance criteria of their systems for coordination, schedule, and cost implications.

Construction and Acceptance. The responsibilities of the installing trade contractors (and vendors, as appropriate) include

- Participating with CA (and the contractor's commissioning manager, when applicable) in executing commissioning plan
- Providing input into commissioning plan for CA's review and approval

- Coordinating with other trades as necessary to facilitate a smooth and complete commissioning process
- Participating in commissioning meetings
- Responding to questions and issues concerning their work raised by CA
- Executing and documenting tasks in construction checklist and start-up process
- Performing and documenting tests when in their scope
- Participating in resolving issues within their scope identified during commissioning
- Correcting identified deficiencies within their scope
- Providing required documentation for systems manuals and commissioning reports

Commissioning-related activities of trade contractors are to prepare O&M manuals and submissions to the systems manual and provide training on commissioned systems and assemblies. To avoid confusion, the OPR should specify which commissioning activities are the trade contractor's responsibility, and which are the CA's.

Architect and Engineers (Designers).

Design. The design professionals should develop complete basis-of-design (BOD) documentation, including design narratives, rationale, and criteria, according to their scopes of services, and update this document with each new design submission. They provide input to the commissioning plan, respond to questions and concerns by the CA and others, respond to design review comments, and incorporate commissioning requirements in construction contract documents.

Construction and Acceptance. During construction, designers

- Review the commissioning plan
- Attend selected commissioning meetings
- Answer questions about system design and intended operation
- Update design narratives in the BOD to reflect as-built conditions
- Respond to or incorporate CA comments on construction submittals and O&M manuals
- Help resolve design-related issues raised during commissioning
- Perform specified training
- Submit required portions of systems manuals

Additional tasks sometimes required are to present system description overviews for primary systems during O&M staff training, review and approve testing plans and procedures, review completed test forms, or witness selected tests.

Owner's Project Management Staff. The owner's project management staff's ultimate responsibility is to see that the commissioning plan is executed. The owner, with guidance from the CA, should include specific responsibilities in all commissioning team members' scopes of services, make sure there is sufficient time for commissioning in the project schedule, ensure the CA is receiving cooperation from other team members, and ensure that other owner responsibilities (e.g., developing the OPR, having O&M staff participate during construction) are fulfilled. The owner ensures that all design review and construction-phase issues identified through commissioning are resolved in a timely manner.

Owner's Representatives. The owner's representatives are individuals or firms hired to represent the owner's interest during specified phases of the building process. The owner typically retains the project architect or project engineer responsible for HVAC design and the CA as a team of owner's representatives.

Owner's Operations Staff.

Predesign. The owner's O&M staff should participate in the development of the OPR during predesign.

Design. During design, O&M staff may contribute to reviews of the designer's BOD, plans, and specifications.

Construction and Acceptance. During construction, the owner's O&M staff should

- Assist in reviewing selected submittals
- Assist in construction observation, verifying completion of construction checklists and observing start-up
- Participate in or witness testing, within pre-established lines of responsibility and authority
- Review O&M and systems manual
- Participate in training

COMMISSIONING PROCESS

Commissioning should begin during predesign, and formally continue through the first year of occupancy and operations. Although circumstances may require owners to begin commissioning at the design or construction stage of a project, this later implementation should, when possible, capture the same information and verifications developed when commissioning begins at project inception.

PREDESIGN-PHASE COMMISSIONING

Objectives

The primary activities and objectives of commissioning during predesign are to

- Develop owner's project requirements (OPR)
- Identify scope and budget for commissioning process
- Develop initial commissioning plan
- Review and accept predesign-phase commissioning-process activities
- Review and use lessons learned from previous projects

Activities

Commissioning Team and Management. During the predesign phase, a team is formed to oversee and accomplish commissioning. Responsibility for leadership of the commissioning team should be defined and assigned to the CA at the beginning of predesign.

Owner's Project Requirements (OPR). The OPR forms the basis from which all design, construction, acceptance, and operational decisions are made. It describes the functional requirements of the facility and expectations of how it will be used and operated. It includes project and design goals, budgets, limitations, schedules, owner directives, and supporting information, as well as necessary information for all disciplines to properly plan, design, construct, operate, and maintain systems and assemblies (ASHRAE *Guideline* 0).

The OPR is generally a set of concise objective qualitative statements, each with one or more quantitative performance metrics or criteria. The following information should be included:

- Functional requirements, needs and goals for building use, operation, maintenance, renovation, and expansion, including user's requirements and space temperature requirements
- Occupancy schedules and space plan requirements, including zone-based control areas
- Sustainability, reliability, durability, safety, and aesthetic goals
- Quality of materials and construction
- Warranty, project documentation, and training requirements
- Goals for the process and outcome of design and construction (e.g., budgets, schedules, change orders, safety, aesthetics, effects on adjacent or integral occupied spaces and tenants)
- General commissioning scope and objectives
- General statements about codes, standards, and regulations to be followed
- Limitations likely to affect design decisions
- Specific features, systems, assemblies, or brands the owner requires (these will be repeated in the design narrative)
- Instructions to designers on types of design tools and aids expected to be used

The CA ensures that the OPR is developed and is clear and complete. The CA may develop or help develop the OPR with the owner or provide direction and review of the OPR developed by others. Facilitated workshops, surveys, and questionnaires are useful for developing the OPR. Later during design, additional OPR statements with performance criteria may be added to the formal list, as desired by the owner and commissioning team. The OPR should still be developed, even if not originally generated in predesign, and included in the systems manual.

Scope and Budget for Commissioning. During predesign, the owner, with assistance from the CA, develops a scope and a rough budget for commissioning. At minimum, design-phase activities should be initially scoped. Once a design-phase commissioning plan is developed, the scope and budget may need to be adjusted. The scope and budget should reflect the commissioning objectives in the OPR.

Selecting areas to commission is typically based on the budget, systems or assemblies with which the owner has experienced problems on previous projects, complexity of systems and assemblies, and criticality of the system or assembly in meeting the OPR. During predesign and design, the list of areas to be commissioned may be general (e.g., electrical lighting controls, emergency power, general electrical equipment, HVAC, domestic water system, and envelope fenestration, etc.). Later in design but before scoping construction-phase commissioning, additional detail should be added to each of these categories, and others added as needed to ensure that the scope of commissioning is clear. Adding this detail increases the cost of commissioning, and needs to be specified early in the design phase.

Historically, commissioning focused on HVAC. Owners are now asking for more systems to be commissioned, including lighting controls, fire and life safety systems, vertical and horizontal transport systems, envelope, plumbing, landscaping, sustainability features, structural elements, many electrical equipment components, security, data, and communications. Refer to the section on Commissioning Costs for budgeting guidelines.

Predesign-Phase Commissioning Plan

One predesign-phase commissioning task should be drafting the commissioning plan for the design phase. The CA develops this plan with review and comment by the owner and designer, and the plan is updated as the project progresses. The design-phase commissioning plan should include the following:

- Objectives and scope of commissioning
- Overview of the process
- Detailed commissioning-process activities for design phase
- General commissioning-process activities for construction and operations/occupancy phases
- Roles and responsibilities
- Deliverables
- Communication protocols
- Schedule
- Checklist of requirements and formats
- Verification and acceptance procedures

Acceptance of Predesign Commissioning

The owner's project requirements and commissioning plan should be formally accepted during predesign, after review and comment by the CA.

DESIGN-PHASE COMMISSIONING

Objectives

Design-phase commissioning objectives include the following:

- Update the owner's project requirements (OPR)
- Verify basis of design (BOD) document against OPR

- Update the design-phase commissioning plan developed during predesign
- Develop and incorporate commissioning requirements into project specifications
- Develop commissioning plan for construction and occupancy/operations phases, including draft construction checklist
- Verify plans and specifications against BOD and OPR
- Begin codeveloping with relevant discipline design engineer for systems manual
- Define training requirements for O&M personnel
- Perform commissioning-focused design reviews
- Accept design-phase commissioning

Activities

Update Design-Phase Commissioning Plan. The initial design-phase commissioning plan is developed during predesign. As more becomes known about systems and assemblies likely to be a part of the project and as project objectives are clarified, the commissioning plan may need to be updated with additional details. The CA must participate in value engineering and constructability review sessions to ensure that commissioning can be performed. The owner and designer then review and comment on the updated plan, which then becomes the guide for the rest of the design phase.

Update the OPR. As design progresses, additional OPR and performance criteria are likely to be identified. Other criteria may need to be altered as more detailed budget and design data become available.

Verify the Basis of Design. All BOD elements can be grouped under one of two terms: design narrative or design criteria. These two terms provide a useful separation when writing the design basis.

The **design narrative** is the written description and discussion of the concepts and features the designers *intend* (during schematic design phase) to incorporate into the design or what they *have* incorporated (during the balance of design) to meet the OPR and associated performance criteria as well as codes, standards, and regulations. This narrative should be understandable by all parties of the building construction and operation process, though it may address fairly technical and specialized issues. It includes a brief section on what systems were considered and why they were accepted or rejected, along with the rationale for the system selected. The design narrative should be updated with each phase of design.

The **design criteria** are the project-specific information, including underlying assumptions for calculations, calculation methodology, codes and standards followed, equipment used as the basis of design, and design assumptions needed to make design calculations and other decisions, such as

- Diversity and safety factors used in sizing
- Classes of systems and components (duct class, cleanroom class, explosive or other hazardous classifications, etc.)
- Level of redundancy
- Occupant density
- Limitations and restrictions of systems and assemblies
- Inside and outside conditions (space temperature; relative humidity; lighting power density; glazing fraction; U-value and shading coefficient; roof, wall, and ceiling R-values; ventilation and infiltration rates; etc.)
- Fire and life safety issues
- Summary of primary HVAC load calculations and the methods used

Development and Use. The BOD is written by the designer and increases in details as design progresses. The CA may need to obtain this explanatory information from the designer. An updated BOD with increased detail should be submitted with each new design submission. Each submission is reviewed by the owner and CA as part of design reviews.

Develop Commissioning Plan for Construction and Occupancy/Operations Phases. The commissioning plan (CP) is a document that outlines the organization, schedule, allocation of resources, and documentation requirements of the commissioning process. This is an overall plan, developed during the predesign, design, and construction phases, that provides the structure, schedule, and coordination planning for commissioning. The CP includes specifications detailing the scope, objectives, and process of commissioning during the construction and occupancy/operations phases of the project. The CP must specify the scope of work, roles, responsibilities, and requirements of the construction contractor. The commissioning plan for the construction and occupancy/operations phases describes the following:

- Commissioning process
- Scope of commissioning effort, including systems, assemblies, and components being commissioned
- Rigor of commissioning
- Roles and responsibilities of each team member
- Team contact information
- Communication protocols between team members, including documentation requirements
- Commissioning overview and details of submittal activities
- Construction observation, checklisting, and start-up activities
- Preliminary schedule for commissioning activities
- Process for dealing with deficiencies
- Test procedure development and execution
- Prefunctional/functional test procedures
- Operation and maintenance (O&M) manual review
- Warranty-period activities
- Operation training procedures
- Systems manual development
- Description of summary report, progress and reporting logs, and initial schedule (including phasing, if applicable)
- Procedures for documenting commissioning activities and resolving issues

The commissioning plan developed during predesign is updated to include construction-phase activities. At the beginning of the design phase, the plan is general and is used primarily to guide development of commissioning specifications. The owner and designer review and comment on the plan. As design progresses, the CA updates and finalizes the plan when the construction documents are completed. The commissioning plan can be issued with the bid documents for reference.

Develop and Incorporate Commissioning Requirements into Project Specifications. The specifications in the CP are needed by contractors so they can include commissioning responsibilities in pricing and understand how to execute the work. Because commissioning is still relatively new to the building industry, descriptive process language should be included, rather than just delineating requirements. Frequently, for reference, the responsibilities of other team members not bound by the specifications (e.g., owner, CA, construction manager, architect) are given in the commissioning specifications to ensure clarity and put the contractor's responsibilities in context.

The specification should include definitions, a list of equipment and systems to be commissioned, submittal, construction checklist, testing and documentation requirements, and sample checklists and test forms. If the project uses contractor-managed commissioning, the specification should identify skills and qualifications required of the contractor's commissioning lead.

The OPR, along with as much BOD information as possible, should be included in the construction documents and labeled as "Informational Purposes Only" to differentiate from the contractor's contractual obligations. Training and O&M manual requirements of the contractor also should be included.

It is critical that the project specifications in the CP clearly define how the quality control and testing functions that have traditionally been a part of many construction projects (e.g., fire alarm, elevator, duct pressure, room pressurization, emergency power testing) will be integrated with HVAC commissioning. Responsibility for checkout and test procedures, including test procedure review, direction, execution, witnessing, documentation, and approval, must all be clearly described. The acceptance criteria for the test should be included in the specifications. Acceptance criteria should be based on the OPR and the systems selected. For example, a project may require tight temperature or humidity tolerances to meet certification criteria. Systems designed for these projects should be able to control to those tolerances. A system with staged cooling (direct expansion with compressor staging) may not be able to meet a $\pm1°F$ level of control. This should be taken into consideration when selecting the systems for the project.

The CA ensures that contractor responsibilities for commissioning are appropriately incorporated into the project specifications. Placing the general commissioning requirements, process descriptions, and specifications in a single section ensures that all parties know where to look for their responsibilities and find common terminology.

Often, the commissioning authority writes the commissioning specifications and then works with the designer to integrate them into the project specifications. Alternatively, the designer can develop the commissioning specifications, with the CA reviewing and recommending revisions.

Begin Developing Systems Manual. During design, the systems manual contains the OPR, BOD, and drawings and specifications, updated at each design submission and during and after construction. The CA is often responsible for assembling and maintaining the systems manual; however, the contract documents for the CA or design professionals should delineate who is responsible for assembling the systems manual.

The systems manual differs significantly from traditional O&M manuals. This manual expands the scope to include other project information developed and gathered during commissioning, such as traditional equipment O&M data, design and construction documents (OPR, BOD, plans, specifications, and approved construction submittals), system schematics, final commissioning report, training records, commissioning test procedures (filled-in and blank), and optimization and diagnostic data (which can include operational procedures for specific emergency situations, seasonal changeover procedures, fire and emergency power response matrix, smoke management system operation during and after fire, energy efficiency recommendations, troubleshooting guide, recommissioning frequency, and diagnostic building automation system trend logs). Scopes of work should clearly identify whether the systems manual includes all project systems and assemblies or just commissioned ones. For more information, see ASHRAE *Guideline* 4-2008.

The owner, designer, contractor, and commissioning authority each have development responsibilities for parts of the systems manual. Construction documents should list the contents and requirements for the systems manual and the responsible party for generating, compiling, and finishing each part of the required documentation. Systems manuals should be available for and used in operator training. Much of the systems manual can be put into electronic media format. The ability to search and auto-update enhances the usability and accessibility of the data.

Define Training Requirements. During the design phase, the training requirements of O&M personnel and occupants are identified relative to the systems and assemblies to be installed in the facility. O&M personnel must have the knowledge and skills required to operate the facility to meet the OPR. Occupants also need to understand their effect on the use of the facility and the ability to meet project requirements. Both groups require training.

Training needs can be identified using a group-technique workshop, interviews, or surveys with the owner and occupant representatives after the systems and assemblies have been specified, and before issuing the construction documents. The contractor's training responsibilities need to be incorporated into the project specifications and should include requirements for the number of training hours for each item of equipment or assembly and submittals of training plans and qualifications of trainers. Training likely requires participation of the designer (for system overviews), the CA (for system overviews, recommissioning, optimization, diagnostics, and using and maintaining the systems manual), and possibly the contractor, and should be included in their scopes of work. Because turnover in O&M and occupants will occur, training materials should be reusable (e.g., video, written manuals, computer presentations).

Perform Design Reviews. Design review by parties not part of the formal designer-of-record team should be conducted to provide an independent perspective on performance, operations, and maintenance. These document reviews, conducted by experts in the field, should start as early as possible, when options and issues can be more easily resolved. The reviews may be coordinated by the CA and should include the owner's technical staff. The CA may attend some design team meetings, and formally reviews and comments on the design at various stages of development [ideally, at least once during schematic design (predesign), design development, and construction document phases]. The CA's design review is not intended to replace peer-to-peer design reviews that check for accuracy and completeness of the design and calculations.

A targeted design review may cover the following:

- General quality review of documents, including legibility, consistency, and level of completeness
- Coordination between disciplines
- Specification applicability to project and consistency with drawings
- Verification that BOD assumptions and rationale are reasonable
- Verification that system and assembly narrative descriptions are clear and consistent with OPR and the BOD is updated with resolved issues
- Verification that plans and specifications are consistent with BOD and OPR, and plans and specifications are updated with resolved issues

Potential system performance problems, issues likely to result in change orders, areas where correct installation is difficult, energy efficiency improvements, environmental sustainability, indoor environmental quality issues, fire and life safety issues, operation and maintenance issues, and other issues may be addressed in these design reviews, depending on the owner's desires and CA's scope. Required reviews ensure that training and systems manual requirements are adequately reflected in construction documents.

Some reviews use sampling, giving 10 to 20% of the drawings and specifications for an in-depth review; if only minimal issues are identified, the owner accepts the submission. If significant issues are identified in the sample, either the submittal is sent back to the designer for revamping and a thorough review, or the CA may perform a thorough review, depending upon the scope of work defined in the CA's contract. After the design team has addressed the issues, the CA performs a new review. In this type of review, the design team is still responsible for their traditional peer review of construction documents for accuracy. The CA makes recommendations to facilitate commissioning and improve building performance, without approving or disapproving either design or documents. The design team is ultimately responsible for design. The CA should be able to justify all of the recommendations made. It is the responsibility of the owner or project manager to evaluate all review findings with the design team and see that the responsible team member

implements the approved ones. All issues are tracked to resolution and verified in later reviews to have been incorporated as agreed.

Accept Design-Phase Commissioning Activities. Commissioning should include the owner's formal acceptance of the BOD, updated OPR, CP, and the design, after review and comment by the CA.

Additional Commissioning Team Tasks. Additional design-phase responsibilities of the commissioning team (led by the CA, who is frequently responsible for these requirements) include the following:

- Build and maintain cohesiveness and cooperation among the project team
- Assist owner in preparing requests for project services that outline commissioning roles and responsibilities developed in the commissioning plan
- Ensure that commissioning activities are clearly stated in all project scopes of work
- Develop scope and budget for project-specific commissioning-process activities
- Identify specialists responsible for commissioning specific systems and assemblies
- Conduct and document commissioning team meetings
- Inform all commissioning team members of decisions that result in modifications to the OPR
- Integrate commissioning into the project schedule
- Track and document issues and deviations relating to the OPR and document resolutions
- Write and review commissioning reports

CONSTRUCTION-PHASE COMMISSIONING

Objectives

Commissioning activities should take place throughout the construction phase and include verification and documentation that

- All acceptance testing requirements are documented
- All systems and assemblies are provided and installed as specified
- All systems and assemblies are started and function properly
- All acceptance testing requirements are documented
- The systems manual is updated and provided to facility staff
- Facility staff and occupants receive specified training and orientation
- Acceptance testing occurs

Activities

The following primary commissioning activities (in approximately sequential order) address commissioning objectives. The CA coordinates and ensures that all activities occur and perform successfully.

Bidding and Contract Negotiation. A member of the commissioning team (usually the CA) may attend the prebid conference to present an overview of commissioning requirements and answer questions. Changes that occur during bidding and contract negotiations related to commissioned systems and assemblies are also reviewed to ensure they agree with the OPR.

Commissioning Planning and Kickoff Meetings. The CA coordinates construction-phase planning and kickoff meetings. The planning meeting held with the contractor, owner, designer, and CA focuses on reviewing requirements and establishing specific communication and reporting protocols. The commissioning plan is updated from this meeting. The kickoff meeting is held with additional construction team members, who generally include the mechanical, controls, electrical, and test and balancing contractors. At this meeting, the commissioning provider outlines the roles and responsibilities of each project team member, specifies procedures for documenting activities and resolving issues, and reviews the preliminary construction commissioning plan and schedule. Team members provide comments on the plan and schedule, and the CA uses these suggestions to help finalize the commissioning plan and schedule.

Commissioning Plan Update. The planning and kickoff meetings usually result in an updated commissioning plan. Later, any project phasing or other schedule and scope-related issues (e.g., testing and training plans and schedules) are clarified in further updates.

Submittal Reviews.

Construction Submittals. The CA reviews equipment and material submittals of commissioned systems and assemblies to obtain information needed to develop construction checklists, make meaningful observations of construction progress, and aid in developing comprehensive tests. Submittals are also reviewed to identify construction-related performance issues before construction progress makes them more difficult and expensive to address. Submittals should be reviewed concurrently by the design team to allow any discrepancies to be identified and communicated to the design team before formal approval.

Controls Submittal and Integration Meeting. Before the contractor develops the controls submittal, the CA coordinates a controls integration meeting to discuss and resolve methods for implementing performance specifications or strategies, interlocks between systems, priority of control between packaged controls and the central control system, the control system database, point names, graphic details and layout, access levels, etc.

Coordination Drawings. The CA may help the owner monitor the development and coordination of shop drawings to ensure synchronization between trades.

Early O&M Data. Information beyond typical construction submittals requested by the CA includes installation and start-up procedures, operation and maintenance information, equipment performance data, and control drawings before formal O&M manual submittals. This information allows the CA to become familiar with systems and assemblies to develop construction checklists, start-up plans, and test procedures.

Contract Modifications Review. Construction documentation issued during this phase, including requests for information, construction field directives, and change orders, should be reviewed by the CA to identify issues that may affect commissioning and compliance with construction documents, BOD, or OPR.

Schedule Commissioning Field Activities. The CA works with the contractors and construction manager to coordinate the commissioning schedule and ensure that commissioning activities are integrated into the master construction schedule.

Construction and Commissioning Meetings. The CA attends periodic planning and job-site meetings to stay informed on construction progress and to update parties involved in commissioning. During initial construction, the CA may attend regular construction meetings and hold a line item on the agenda. Later, the CA may convene entire meetings devoted to commissioning issues, with more frequent meetings as construction progresses. Attendees vary with the purpose of the meeting. Team members should be represented at meetings by parties with technical expertise who are authorized to make commitments and decisions for their respective organizations. The CA should distribute minutes from these meetings.

Progress Reports. The CA provides periodic progress reports to the owner and contractor with increasing frequency as construction progresses. These reports indicate current progress, next steps, and critical issues affecting progress and construction schedule.

Update Owner's Project Requirement and Basis of Design. When contract negotiations and/or changes and clarifications made during construction alter or add to the OPR or BOD, these documents should be updated. Normally, the CA updates the OPR and the designer updates the BOD. Final construction updates to these documents are made at the end of testing, typically a few months into occupancy.

Coordinate Owner's Representatives Participation. The commissioning plan should describe participation of the owner's representatives in work such as submittal review, construction checklist verification, construction observation, test procedure review and execution, and O&M manual review. The CA normally coordinates this participation with the contractors.

Construction Observation. The CA should make planned, systematic visits to the site to observe installation of systems and assemblies. The owner's staff may assist in construction observation. The CA should verify that the first few of any large-quantity items (e.g., variable-air-volume terminal units) are installed properly and used as a mock-up or standard to judge the rest of the installation. Any conditions not in compliance with the construction documents or BOD or that may affect system performance, commissioning, operation, or other project requirements should be documented. These observations normally focus on areas where observers have found problems before, or spot-check items on construction checklists. Less often, practitioners are tasked with validations or detailed inspections verifying that equipment or assemblies have been installed properly in every detail. Some practitioners make formal construction observation reports, whereas others merge findings into the regular issue logs and progress reports. Site visits should be used to verify completion of construction checklists.

The CA normally witnesses many of the contractor's start-up activities for major equipment to ensure checklists and start-up are documented properly and to gain additional feature and function information from installing technicians.

Construction Checklists and Start-up. At the beginning of construction, construction checklists are developed (usually by the CA in cooperation with the discipline engineer, but sometimes by the contractor or equipment manufacturer) for most commissioned systems and equipment. They are attached to or integrated with manufacturer's installation and start-up procedures. In most projects, contractors fill out the checklists during installation, during normal checkout of equipment and systems, and before and during system start-up, though some commissioning practitioners fill out the checklists themselves. The contractor fully documents start-up and initial checkout, including the construction checklists, and submits them to the CA, who reviews the forms and spot-checks selected items in the field later in the project, to ensure systems are ready for testing.

Some CA practitioners statistically sample items on checklists to verify proper completion (typically random or targeted sampling 2 to 20%). If an inordinate fraction of the sampled items are deficient (typically more than 10%), the contractor is required to check and document all remaining items. The contractual documents need to contain details of the sampling and actions based on the results.

Commissioning Issues Management. The CA keeps a record of all commissioning issues that require action by the design team, contractor, or owner. The issues should remain uniquely identified, be tied to equipment and systems, and prioritized relative to performance, cost, and schedule. Issues are tracked to resolution and completely documented. The CA distributes the updated log to the owner, contractor, construction manager, and HVAC design engineer at construction and commissioning meetings. This log can also be placed on project Web sites. In the contractor-managed scenario, the contractor's commissioning manager or subconsultant may manage the contractor's issues log. In that case, to minimize conflicts of interest, the commissioning authority is often required to report all issues simultaneously to the contractor and to the owner.

Developing Test Procedures. Step-by-step test procedures and project-specific documentation formats are used for all commissioned equipment and assemblies. **Manual tests** evaluate systems with immediate results. **Monitoring testing** uses the building automation system or data loggers to record system parameters over time and analyze the data days or weeks later. **Automated testing** gathers or analyzes system performance data completely electronically, or with significant help from software.

Test procedure writing begins immediately after the submittal, because test procedures need to be reviewed and approved before testing occurs, which is generally scheduled about three to six weeks after the submittal review. Test procedures may be based on specifications, applicable standards and codes, submittal data, O&M data, data shipped with the equipment, approved control drawings, and existing test procedures of similar equipment or components. Tests should cover all functions and modes.

Procedural documents clearly describe the test prerequisites, required test conditions, individual systematic test procedures, expected system response and acceptance criteria for each procedure, actual response or findings, and any pertinent discussion. Test procedures differ from **testing requirements** found in the specifications, which describe *what* modes and features are to be tested and verified and under what conditions. Test procedures describe the step-by-step method of *how* to test. Simple checklists may be appropriate for testing simple components, but dynamic testing of interacting components requires more detailed procedures and forms.

The responsible HVAC design engineer should organize the preparation of HVAC system testing, adjusting, and balancing (TAB) procedure together with the test and balancing professional and the commissioning authority, depending on their scopes of work. The CA is responsible for verifying that the test procedures are written and appropriate for determining that equipment, assemblies, and systems function correctly. All parties should have input into the final test procedures to ensure that equipment, assemblies, systems, or people will not be endangered or warranties voided. Industry standard test procedures [e.g., ASHRAE, Air-Conditioning and Refrigeration Institute (ARI), American Composites Manufacturers Association (ACMA)] should be referenced whenever possible.

Testing and Verification.

Responsibilities and Management. Not all testing and verification falls under HVAC commissioning. Traditional air and water testing, adjusting, and balancing is often the sole responsibility of the contractor or by independent contract to the owner. Building envelope, elevators, and electrical system testing are also generally excluded from HVAC commissioning. There is some movement in the industry to centralize coordination for quality assurance/quality control (QA/QC) functions under the commissioning team. Each project is unique, and different approaches can be warranted.

Critical issues include ensuring that

- Appropriate testing rigor is applied
- Technically qualified parties execute and document the testing
- Objectivity is maintained
- Testing is well documented

For systems not usually thoroughly tested by the contractor [e.g., HVAC systems and controls, lighting controls, specialty plumbing, and envelope and interfaces between systems (security, communications, controls, HVAC, fire protection, emergency power)], the CA may write test procedures that go beyond HVAC tests. The CA then directs, witnesses, and documents each test executed by the contractor. For these systems, the controls subcontractor usually executes the tests, although the CA may test some equipment with or without the contractor present.

Testing that has traditionally been conducted by the contractor (e.g., fire alarm, fire protection, elevator, duct and pipe tests, emergency power, some electrical equipment) ideally should be centrally coordinated. This can be the responsibility of the contractor or of the CA. The specifications should clearly establish testing and documentation requirements and define the responsible party. The level of confidence and objectivity can be increased by requiring experts in specific disciplines to witness tests, particularly in some electrical system and envelope assembly field testing. Increasing the required amount of field witnessing by the CA also improves the confidence that commissioning was correctly performed.

Within a given discipline, there may be differing levels of autonomy. For example, in tests of electrical equipment (e.g., circuit breakers), the contractor may conduct and document the bolt-torque tests, and also be required to hire an independent certified testing agency to conduct other necessary tests that require more specialized expertise and test equipment.

The owner's technical staff can assist in and benefit from participation in any of the above scenarios. The designer and owner's project management staff may witness selected tests.

Verification Testing Scheduling. Verification testing should be performed after equipment and assemblies are complete and started up, construction checklists checked out and submitted, and air and water balancing completed. The contractor is then ready to turn the system over to the owner. Most projects require a certificate of readiness from the contractor certifying that the system has been thoroughly checked out and verified to be completely functional. Ideally, manual testing occurs before substantial completion, but schedule slippage may require testing to occur after this milestone. Some short-term monitoring may be completed with manual testing, but sometimes is postponed until early occupancy. Opposite-season and other deferred testing should be conducted during seasonal changes or peak seasonal conditions.

Testing Scope. At a minimum, testing includes observing and documenting system operation and function during normal operation, through each of their sequences of operation, and all other modes of operation and conditions, including manual, bypass, emergency, standby, high and low load, and seasonal extremes, and comparing actual performance to that specified in the construction documents. Testing may also be conducted to verify performance criteria found in the BOD and OPR, including system optimization, though deficiencies in these areas are not normally the contractor's responsibility.

Manual Testing Methods. Testing includes observing normal operation; changing set points, schedules, and timers; and exercising power disconnects, speed controls, overwriting sensor values, etc., to cause perturbations in the system. System response and results are recorded on test procedure forms, and any issues are documented. Small corrections are often made during testing. Less pressing corrections or issues with unknown solutions are investigated later, corrected, and retested.

Building automation systems (BAS), when present, can be the backbone for conducting much of the testing, collecting, and archiving data. Before using the BAS, critical sensors, actuators, and features should be verified as calibrated so the system readouts are reliable (although all sensors and actuators should have been calibrated by the contractor and documented on construction checklists). The results are viewed on the building automation system screen or at the equipment. Other tests may require hand-held instruments or visual verification (e.g., evaluating caulking and flashings on window installations).

Monitoring. Some testing requires monitoring (trending) system operation over time through the BAS or data loggers (when the BAS does not monitor desired points). Monitoring can be used to document that systems are performing properly during test conditions over the monitoring period. However, this is not a substitute for manual testing, which can cover a wide range of conditions. Monitoring provides a view of system interactions over the course of normal, start-up, shutdown, and weekend operation. Normally, the CA analyzes monitored data and submits a report, with any concerns added to the issues log.

Automated Testing. Various semiautomated testing is conducted in permanent onboard equipment controllers. Currently, most truly automated testing focuses on identifying electrical faults in controller components and is used during vendor start-up and troubleshooting activities. Some use logic to identify parameters outside limits, which indicate component malfunctions such as hunting and calibration issues. Different types of automated testing intended to help commissioning are under development. Some are primarily tools to gather and display monitored data; others help the analyzer make diagnoses. Equipment manufacturers often integrate automated commissioning testing capabilities into onboard controllers on their equipment.

Training. Training should include, as appropriate, (1) the general purpose of the system; (2) use and management of the systems manual; (3) review of control drawings and schematics; (4) start-up, shutdown, seasonal changeover, and normal, unoccupied, and manual operation; (5) controls set-up and programming; (6) diagnostics, troubleshooting, and alarms; (7) interactions with other systems; (8) adjustments and optimizing methods for energy conservation; (9) relevant health and safety issues; (10) special maintenance and replacement sources; (11) tenant interaction issues; and (12) discussion of why specific features are environmentally sustainable. Occupants may also need orientation on certain systems, assemblies, and features in the building, particularly sustainable design features that can be easily circumvented.

The CA helps the owner ensure that adequate training plans are used by the contractor and that training is completed according to the construction documents. (See the discussion of defining training requirements in the section on Commissioning During Design.) Some CAs conduct testing with a sample of trainees to verify the efficacy of the training.

Most training should be accomplished during construction, before substantial completion. However, for complex systems (e.g., control systems), multiple training sessions should occur before and after substantial completion. Training for systems that will not come into operation until the next season may be delayed. A meaningful training program typically includes using the operation and maintenance components of the systems manual, which must be submitted before training begins. Selected training materials can be video-recorded as desired by the owner.

Commissioning Record. The CA compiles all commissioning documentation and project data, which are submitted and become part of the systems manual. The commissioning record contains the salient documentation of commissioning, including the commissioning final report, issues log, commissioning plan, progress reports, submittal and O&M manual reviews, training record, test schedules, construction checklists, start-up reports, tests, and trend log analysis, grouped by equipment.

Final Commissioning Report. The CA should write (or review) and submit a final commissioning report detailing, for each piece of commissioned equipment or assembly, the adequacy of equipment or assemblies meeting contract documents. The following areas should be covered: (1) installation, including procedures used for testing equipment with respect to specifications; (2) functional performance and efficiency, including test results; (3) O&M manual documentation; and (4) operator training. Noncompliance items should be specifically listed. A brief description of the verification method used (manual testing, trend logs, data loggers, etc.) and observations and conclusions from the testing should be included. The CA updates the final commissioning report after occupancy/operations-phase commissioning. The commissioning documents also should include, among other things,

- Certificates and warranties of system completion with a complete set of as-built drawings submitted from mechanical, electrical, piping, plumbing, control, and fire protection contractors
- Complete records of all problems and solutions occurred during start-up, testing, and adjustments submitted by every individual contractor or subcontractor
- Certified system testing and balancing report from the licensed TAB company, with verified major equipment models, capacities,

and all tested performance records conforming to system design criteria
- If room pressurization is required, a complete room-to-room pressurization map in the TAB report
- If room cleanliness is required, a certified as-built room cleanliness report of testing during completion of construction and installation

Systems Manual Submittal. The CA usually compiles the systems manual and provides it to the owner. At the end of construction, the designer, contractor, owner, and CA provide elements of the systems manual generated during the construction phase. The systems manual should include commissioning test procedures, results of commissioning tests, issue logs and resolution, system schematics, O&M information, record drawings, construction checklists, start-up reports, and trend log analysis, grouped by equipment. The CA normally reviews and approves systems manual submissions by the contractor and designer, similar to traditional O&M manual reviews. Electronic systems manuals, now developed occasionally, will likely become standard in the future.

OCCUPANCY- AND OPERATIONS-PHASE COMMISSIONING

Occupancy- and operations-phase commissioning typically begins with resolving the findings from performance monitoring over the first month or two into occupancy, and ends with the completion of the first year of occupancy.

Objectives

Commissioning during this phase should ensure the following:

- Initial maintenance and operator training is complete.
- Systems and assemblies received functional opposite-season verification.
- Outstanding performance issues are identified and resolved before warranty expiration.
- Commissioning process evaluation is conducted and satisfactorily resolved.

Activities

Verifying Initial Training Completion. The CA ensures that any remaining training is conducted according to the contract documents, either by reviewing documentation of the training or through witnessing portions of the training. This normally applies to control systems and training on major systems for which peak season is not near the end of the construction phase.

Seasonal Testing. Seasonal testing verifies proper operation of those systems for which peak-load conditions are not available before substantial completion. Additionally, intermediate-season testing may be required for part-load, and changeover testing may be required. For example, when completion occurs in winter, final full-load cooling system testing must wait until the following summer. Intermediate-season testing verifies system changeover controls and ability to maintain space conditions per OPR. Testing should be performed by the appropriate contractor and witnessed by the CA and building operators. However, the owner's operations staff and the CA, if sufficiently proficient with the controls system, can execute the tests and recall contractors only if there are problems.

Near-Warranty-End Review. The CA may also be asked to return a few months before the contractor's one-year warranty expires, to interview facility staff and review system operation. By acting as the owner's technical representative, the CA assists facility staff to address any problems or warranty issues.

Documentation Update. Any identified operations-phase concerns are added to the issues log and the final commissioning report is amended to include occupancy/operations-phase commissioning activities. Changes to the BOD, OPR, or record documents are documented by updating the systems manual near the end of the warranty. Changes to sequences of operation require particular care in ensuring that these updates occur.

Commissioning Process Evaluation. The CA should meet briefly with the owner; general, controls, mechanical, and electrical contractors; and mechanical and electrical designers to discuss the commissioning process for this project. Topics to be addressed include what went well, what could be improved, what would best be done differently next time, etc. This will benefit all parties in commissioning future projects. The CA will submit a report on this meeting to the owner.

The occupancy/operations phase typically begins with resolving the findings from monitoring a month or two into occupancy, and ends when the one-year equipment warranties expire.

Additional Activities. The CA may also be given other responsibilities during the warranty period, such as helping develop a maintenance management program, optimizing system performance, and developing electronic facility manuals.

Ongoing or Recommissioning. Ongoing monitoring and periodic retesting and calibration of selected systems and assemblies are recommended to ensure they comply with the OPR, operating and functioning optimally throughout their life. This is sometimes called recommissioning. Some recommissioning methods rely more on semicontinuous monitoring of primary system performance parameters with periodic analysis. Other approaches consist of recalibrating and retesting targeted systems and components on a regular schedule, including both manual testing and monitoring. Calibration and test frequency vary with equipment and its application.

COMMISSIONING COSTS

Commissioning costs vary considerably with project size and building type, equipment type, scope, and traveling requirements (Mills et al. 2004; Wilkenson 2000). Historically, commissioning focused on HVAC and controls, and started during construction. However, QA/QC for increasing numbers of systems is included in commissioning, and the process now frequently begins in the design phase. Currently, the commissioning industry is not mature; budget estimates, even for relatively detailed scopes of work, vary widely.

Clear definition of tasks, deliverables, systems and components to be commissioned, rigor, and testing methods must be provided for comparative pricing. The costing guidelines that follow must be used with great caution and are provided only for rough planning purposes. Understanding what is and is not included in each cost number is critical. *Owners should consult commissioning providers with their planned projects to obtain budget estimates, and practitioners should use detailed cost breakdowns for their pricing.*

DESIGN-PHASE COSTS
(INCLUDING PREDESIGN AND DESIGN)

Predesign-phase costs include the CA's efforts in attending predesign meetings and design reviews with the architect's consulting team and owner's representatives. This portion of work may range from 8 to 12% of the CA's contract. Design-phase costs include the CA's reviewing design submittals, coordinated with the designer, and developing sections of the systems manual (design intent and basic operations from the control submittal). This portion of the work may range from 15 to 20% of the CA contract.

For a project that includes the discussed tasks for all HVAC and controls components, a moderate level of electrical systems commissioning, and minor plumbing and envelope commissioning, the total commissioning costs (CA cost plus the additional work of the designers) may range from 0.2 to 0.6% of the total construction cost for a typical office building. This estimate assumes two moderate design reviews. Different types of buildings or more complex

Table 1 Estimated Commissioning Authority Costs to Owner for Construction and Occupancy/Operations Phases

Commissioned System	Total Commissioning Cost
HVAC and controls[a]	2.0 to 3.0% of mechanical
Electrical system[a]	1.0 to 2.0% of electrical
HVAC, controls, and light electrical[b]	0.5 to 1.5% of construction

Sources:
[a]Wilkinson (2000).
[b]PECI (2000).

buildings with larger scopes of design review may cost considerably more.

CONSTRUCTION- AND OCCUPANCY/
OPERATIONS-PHASE COSTS

Table 1 estimates the CA's costs for the construction and occupancy/operation phases under the CA-managed approach. It includes construction- and occupancy/operations-phase commissioning for the HVAC system (including fire and life safety controls, changeover season, and opposite season) and electrical system (including lighting controls, emergency power, and limited connection and grounding checks). It does not include specialty testing such as full infrared scanning, power quality, switchgear, transformer, or low-voltage-system testing. Complex systems and critical applications have higher costs. *For a given building type and complexity,* larger buildings tend to come in at the lower end of the range and smaller buildings at the higher.

The listed costs cover only the CA fees; there are also costs to the contractor, designers, and owner's staff. For the CA-managed approach, costs for the mechanical contractor attending meetings, documenting construction checklists, and assisting with testing approximate 10 to 20% of the CA's mechanical commissioning costs. The electrical contractor's costs may equal the CA's electrical commissioning costs for electrical commissioning (because contractors are usually responsible for hiring their own electrical testing company to perform electrical tests). International Electrical Testing Association (NETA) tests are often already part of the normal construction program, and the only additional commissioning costs are for the CA to coordinate testing, spot-witness, and review reports.

Commissioning costs for the contractor-managed approach are similar overall, but more costs are shifted from the CA to the contractor and their commissioning manager and staff.

EXISTING BUILDINGS

Existing building commissioning (also called retrocommissioning, sometimes written as "RCx") involves commissioning building HVAC equipment after the equipment has been installed, and the facility is running and occupied. HVAC equipment performance normally degrades with use and time, at a rate depending on the quality of maintenance and operations and hours of operation. Quality of maintenance also affects equipment life expectancy. A facility retrocommissioning effort should include developing an owner's current facility requirements or owner's intent, documentation of the existing system, a survey identifying operational inefficiencies for the facility, quantifying and prioritizing the inefficiencies found, determining how best to optimize the equipment or operation, implementing the change, training operating personnel, documenting operations, and then reverifying with ongoing measurements that the retrocommissioning activities produced and continue to produce the desired effect (Claridge et al. 2000). Other definitions specify it as a one-time event with a different set of project phases (Haasl and Sharp 1999; Thorne and Nadel 2003). All of these approaches provide methodologies to improve operation, improve work environment, safely improve process productivity, and optimize a facility's energy use with direct consideration given to current operational requirements.

Existing building commissioning activities include reviewing utility bills and building documents, optimizing chiller and boiler systems, implementing various equipment operational scheduling, optimizing air delivery systems, setting up temperature resets for water- and air-side operations, optimizing indoor air quality control, and verifying control systems are functioning as needed. Training and documentation for operators should be included to ensure that the required performance can be achieved. When existing systems may not have the capacity to meet the owner's project requirements, system deficiencies need to be documented with a decision on when (or whether) upgrading will be done. For example, indoor air quality objectives may not be met because a system was designed under an older standard or code. Temperature objectives may not be met because additional computer equipment loads have been added to some spaces, and the original system was designed to handle a lower load. In each case, a documented recommendation should be provided to the owner on the options available.

RCx has been shown to be a very cost-effective way to improve occupant comfort and productivity and optimize operational costs. Energy savings of over 20% with a two-year payback have been reported (Claridge et al. 2004). Also, Kats et al. (1996) showed that RCx typically achieved 40% or more savings than was estimated during a range of audits.

Buildings with systems ranging from older pneumatic controls to newer building automation systems (BASs) have been successfully retrocommissioned. Pneumatic controls limit the number of RCx options that can be implemented, and also require separate data logging for monitoring parameters used to calculate energy savings. Modern BASs enable lower-cost RCx and also allow trend logging of various parameters to sustain the achieved savings if the systems are verified to be functioning properly and calibrated.

CERTIFICATION

Several groups offer certification of commissioning authorities and providers, including the following.

- ASHRAE's Commissioning Process Management Professional (CPMP) program targets individuals who manage and oversee the commissioning process and commissioning team members. Recipients are usually design/consulting professionals and technologists.

- The AABC Commissioning Group (ACG) offers a certification program for TAB engineers.

- The Association of Energy Engineers (AEE) offers a Certified Building Commissioning Professional (CBCP®) certification.

- The Building Commissioning Association (BCA) offers certification for a Certified Commissioning Professional™ (CCP™).

- The National Environmental Balancing Bureau (NEBB) offers certification for commissioning providers by system type (e.g., HVAC, plumbing, fire protection).

- The University of Wisconsin offers three levels of certification: professional, managerial, and technical support.

REFERENCES

ASHRAE. 2005. The commissioning process. ASHRAE *Guideline* 0-2005.

ASHRAE. 2007. HVAC&R technical requirements for the commissioning process. ASHRAE *Guideline* 1.1-2007.

ASHRAE. 2008. Preparation of operating and maintenance documentation for building systems. ASHRAE *Guideline* 4-2008.

Claridge, D.E., C.H. Culp, M. Liu, S. Deng, W.D. Turner, and J.S. Haberl. 2000. Campus-wide continuous commissioning of university buildings. *Proceeding of the ACEEE 2000 Summer Study on Energy Efficiency in Buildings*, Pacific Grove, CA, pp. 3.101-3.112.

Claridge, D.E., W.D. Turner, M. Liu, S. Deng, G. Wei, C.H. Culp, H. Chen, and S.Y. Cho. 2004. Is commissioning once enough? *Energy Engineering* 101(4):7-19.

Haasl, T. and T. Sharp. 1999. *A practical guide for commissioning existing buildings*. ORNL/TM-1999/34. Portland Energy Conservation, OR, and Oak Ridge National Laboratory, Oak Ridge, TN.

Kats, G.H., A.H. Rosenfeld, T.A. McIntosh, and S.A. McGaraghan. 1996. *Energy efficiency as a commodity: The emergence of an efficiency secondary market for savings in commercial buildings*. U.S. Department of Energy Protocol. http://www.eceee.org/conference_proceedings/eceee/1997/Panel_2/p2_26/Paper.

Mills, E., H. Friedman, T. Powell, N. Bourassa, D. Claridge, T. Haasl, and M.A. Piette. 2004. The cost-effectiveness of commercial-buildings commissioning: A meta-analysis of energy and non-energy impacts in existing buildings and new construction in the United States. Lawrence Berkeley National Laboratory *Report* LBNL-56637.

PECI. 2000. *The National Conference on Building Commissioning Proceedings*. Portland Energy Conservation, OR.

Thorne, J. and S. Nadel. 2003. Retro-commissioning: Program strategies to capture energy savings in existing buildings. *Report* A035. American Council for an Energy-Efficient Economy, Washington, D.C.

Wilkinson, R. 2000. Establishing commissioning fees. *ASHRAE Journal* 42(2):41-47.

BIBLIOGRAPHY

ASHRAE. 2001. Commissioning smoke management systems. ASHRAE *Guideline* 5-1994 (RA 2001).

ASHRAE. 2002. Measurement of energy and demand savings. ASHRAE *Guideline* 14-2002.

ASHRAE. 2002. *Laboratory design guide*.

DOE. 2002. *Continuous commissioning guidebook*. http://www1.eere.energy.gov/femp/operations_maintenance/om_ccguide.html.

Idaho Department of Administration. 1999. *State of Idaho retrocommissioning guidelines*. http://adm.idaho.gov/pubworks/archengr/app7rcg.pdf.

Idaho Department of Administration. 2000. *New-building commissioning guidelines*. http://adm.idaho.gov/pubworks/archengr/app7rcg.pdf.

MDAE. 2000. *Best practices in commissioning in the state of Montana*. Montana Division of Architecture and Engineering.

NEBB. 1993. *Procedural standards for building systems commissioning*. National Environmental Balancing Bureau, Gaithersburg, MD.

SMACNA. 1995. *HVAC systems commissioning manual*. Sheet Metal and Air Conditioning Contractors' National Association, Chantilly, VA.

U.S. DOC. 1992. HVAC functional inspection and testing guide. NTIS *Technical Report* PB92-173012. U.S. Department of Commerce, Washington, D.C., and General Services Administration, Washington, D.C.

U.S. DOE/PECI. 1997. Model commissioning plan and guide commissioning specifications. NTIS *Technical Report* DE97004564. U.S. Department of Energy, Washington, D.C., and Portland Energy Conservation, OR.

CHAPTER 44

BUILDING ENVELOPES

PROPER building envelope design requires knowledge of the physics governing building performance as well as of building materials and how they are assembled. This chapter provides practical information for designing new building envelopes and retrofits to existing envelopes, always with the notion that the envelope must work well in concert with the building's surroundings and the HVAC system. The information can also be useful for those involved with building envelope investigation and analysis.

This chapter was developed with the integrated design approach in mind and assumes that the architect, HVAC designer, building envelope designer, and others involved in envelope design and construction communicate and understand the interrelationships between the building enclosure and mechanical systems. Integrated design requires a clear statement of the owner's project requirements (OPR) and design intent, and is described in greater detail in Chapter 58. That chapter may be used as a basis for finding common agreement among designers and engineers using the integrated design approach. The growing use of integrated design in project delivery highlights the building envelope as the principal site where architectural design and mechanical engineering meet.

A successful building envelope design requires that the team be knowledgeable about and responsible for the performance requirements described in this chapter. This chapter does not distinguish the individual responsibilities of each team member, but rather is intended to serve the team as a whole.

Buildings are designed and constructed to provide shelter from the weather and normally house conditioned, habitable spaces for occupants. The **building envelope** is an assembly of components and materials that separate the conditioned indoor environment from the outdoor environment. The envelope typically includes the **foundation**, **walls**, **windows**, **doors**, and **roof**. Partitions between interior building zones that have substantially different environmental conditions (such as a swimming pool compared to an office area) are often required to function similarly to building envelopes.

Performance requirements for the building envelope include the following (Handegord and Hutcheon 1989; Hendriks and Hens 2000; Hutcheon 1963):

- Control heat flow
- Control airflow, including airborne contaminants
- Control liquid water penetration with rain as the most important source
- Control water vapor flow
- Control light, solar, and other radiation
- Control noise
- Control fire
- Provide strength and rigidity against outside influences (sometimes structural)
- Be durable

The preparation of this chapter is assigned to TC 4.4, Building Materials and Building Envelope Performance.

- Be constructable, maintainable, and repairable
- Be aesthetically pleasing
- Be economical
- Be sustainable

These performance requirements and their effects on one another must be understood by the project team. Building envelopes should be designed for good overall performance. The first eight listed items arise from the envelope's function of separating the conditioned and unconditioned environments. Parties responsible for HVAC and building envelope design must be knowledgeable about how each system affects the performance of the other. Review of the heat, air, and moisture characteristics of the proposed envelope is needed for appropriate design of HVAC systems. The building envelope must also be designed with an understanding of the interior and exterior environmental design conditions; consequently, the architect or principal designer needs to provide the specific performance requirements to the HVAC designer, including provisions to achieve minimum airtightness, interior occupancy criteria, and special-use considerations. With a building envelope design suited to the operating requirements, the space-conditioning (HVAC) system generally is smaller in capacity and may have simpler control and distribution systems, normally resulting in a system with greater efficiency.

This chapter applies information in Chapters 25 to 27 of the 2009 *ASHRAE Handbook—Fundamentals* to building envelope design. It incorporates much of the material from previous versions (until 2005) of Chapter 24 in that volume.

TERMINOLOGY

For definitions related to the physics of heat air and moisture transport, see the Terminology section in Chapter 25 of the 2009 *ASHRAE Handbook—Fundamentals*.

An **air barrier** is a component or set of components in a building envelope that forms a continuous barrier that controls airflow across the envelope or assembly.

A **building assembly** is any part of the building envelope (e.g., wall, window, roof) that has boundary conditions at the conditioned space and the exterior.

A **building envelope** or **building enclosure** is the overall physical structure that provides separation between conditioned spaces and the outdoor environment or any indoor environment that is substantially different from the conditioned one.

A **(building) component** is any physical element or material within a building assembly.

Moisture **condensation** is the change in phase from vapor to liquid water. Condensation occurs typically on materials such as glass or metal that are not porous or hygroscopic and on capillary porous materials that are capillary saturated. Condensation should be distinguished from phase change between vapor and bound water in capillary or open porous materials (see **moisture content**).

Durability is the ability of a building or any of its components to perform its required functions in its service environment over a period of time without unforeseen cost for maintenance or repair (CSA 1995).

Fenestration includes all areas (including the frames) in the building envelope that let in light. Fenestration includes windows, curtain walls (vision areas), clerestories, skylights, and glazed doors. Fenestration excludes insulated spandrels and solid doors. **Fenestration area** is the total area of fenestration measured using the rough opening, including the rough opening for doors.

Hygrothermal design analysis is a set of calculation procedures that uses building design and component physical properties to predict heat, air, and moisture performance of envelopes and assemblies under design conditions. See Chapters 25 to 27 in the 2009 *ASHRAE Handbook—Fundamentals*.

Infiltration is uncontrolled inward air leakage through open, porous materials, cracks, and crevices in any building component and around windows and doors caused by pressure differences.

Exfiltration is uncontrolled outward air leakage through open, porous materials, cracks, and crevices in any building component and around windows and doors caused by pressure differences.

Wind washing is uncontrolled wind-induced flow of outside air in and behind insulation layers

Air intrusion is uncontrolled pressure-induced flow of indoor air in and in front of air-permeable insulation layers, caused by wind pressures, stack effect, or HVAC systems.

Convective loop is uncontrolled stack-induced convective flow of cavity air in and around insulation layers

Thermal insulation is any material specifically designed to decrease heat flow by equivalent conduction through a building envelope or envelope assembly.

Moisture content is the ratio of mass of water to volume of dry material in porous and hygroscopic materials, in lb/ft^3. **Bound water** describes the phase of water bound in hygroscopic materials. **Sorption** (and **desorption**) describes the change in phase between vapor and bound water.

A **plenum** is a compartment or chamber to which one or more ducts are connected, that forms part of an air distribution system, and is not used for occupancy or storage. A plenum is often under positive or negative air pressure relative to adjacent spaces.

The **R-value** of a material is the thermal resistance for a given thickness of that material, as provided by the manufacturer or listed in Table 4 of Chapter 26 in the 2009 *ASHRAE Handbook—Fundamentals*. The **system R-value** (R_S) is the sum of the individual R-values for each material, excluding air films. The **total R-value** (R_T) is the sum of the system R-value and the interior and exterior air-film resistances (see Chapter 25 in the 2009 *ASHRAE Handbook—Fundamentals*).

A **thermal break** is a thermally resistive element that decreases heat conduction through an assembly.

A **thermal bridge** is a thermally conductive element through an otherwise thermally resistive assembly.

U-factor or **thermal transmittance** is the rate of heat transfer per unit surface of an assembly per unit temperature difference between the environments at both sides of the assembly. The clear value only considers the surface film resistances and R-values of the material layers comprising the assembly. The U-factor is $1/R_T$. A **whole-wall** or **effective U-factor** also takes into account thermal bridging, convective loops, wind washing, and indoor air washing effects (see Chapter 25 in the 2009 *ASHRAE Handbook—Fundamentals*).

A **vapor retarder** or **vapor barrier** is any component in a building envelope with a low permeance to moisture flow by diffusion.

A **water-resistive barrier (WRB)** is a building envelope component designed to prevent inward movement of liquid water.

GOVERNING PRINCIPLES

The building envelope is the key element in managing the environmental loads on the building. These loads are a function of the climate and the indoor conditions, such as air temperature, relative humidity, and air pressure differential. There is a strong interdependence between a building HVAC system and envelope that must be considered when designing or modifying a building. This interdependence centers on controlling heat flow, airflow (including control of airborne contaminants), and water and water vapor flow. Design parameters involved are as follows.

Design Parameters

Heat. The type and amount of insulation to be provided depends on the climate, governing codes, and building use. The insulation should be continuous, while considering the limitations of the materials and systems. Discontinuities (or thermal bridges) are the sites of unwanted heat transfer that reduce energy efficiency which may result in premature soiling (e.g., ghosting), surface condensation, and/or mold growth. In heating- and cooling-dominated climates, reduced thermal performance can affect indoor conditions and increase HVAC loads. The thermal conductivities and R-values of insulating materials allow them to be compared for their effect on heat transfer, though their properties related to air and moisture transfer vary widely.

Air. Some buildings are designed for natural ventilation when building use and climate allow, and for mechanical space conditioning (with ventilation) at other times. During periods of space conditioning, the building envelope should show minimal air leakage. This allows better control of (1) HVAC, (2) inflow of airborne contaminants in the building, and (3) noise transmission. The HVAC system can generate pressure differentials across the envelope that increase air leakage and may create moisture and thermal problems. It is important to review the interaction of the HVAC system and envelope at the design stage.

Moisture. Building envelopes should be designed to shed rainwater, prevent accumulation of moisture in moisture-sensitive materials, and allow draining and drying of water that accumulates. A secondary means of moisture transport through building envelopes is airflow through openings in the envelopes. Liquid water and frost can accumulate on cold materials in a wall assembly along air movement paths. Vapor diffusion can play a role in wetting and especially in drying of building materials.

Although vapor diffusion control in an envelope assembly is important, field experience shows that most moisture problems are associated with bulk water penetration and moisture accumulation caused by air leakage. Despite the historic and code emphasis on vapor barriers, their effect is often secondary.

A beneficial exercise during building envelope design is to trace the continuity of the elements providing thermal, air, and water protection to the envelope as assembly details are refined. Continuity of the WRB and the air barrier is essential to their performance. Absolute continuity of a vapor barrier is not essential to its performance.

Hygrothermal analysis can be used to predict envelope performance and compare these results with the requirements. ASHRAE *Standard* 160 provides guidance for performing a hygrothermal design analysis. Inputs for hygrothermal modeling include assembly configuration, material properties, initial conditions, and indoor and outdoor climate conditions. Analysis tools generate outputs that may include heat and moisture flux and material moisture content. Because of the number of assumptions and limitations of calculations required to complete an analysis, results should be used for guidance to supplement the designer's understanding of envelope performance. They should not be considered an absolute prediction of actual hygrothermal performance.

Other Important Performance Criteria

Strength and Rigidity. The air barrier system must be able to withstand air pressures to which the building will be subjected. These pressures can often be large, and strength is critical in severe-weather areas such as hurricane zones.

Noise. For most occupancies, building envelopes should be designed and constructed to reduce noise transfer between conditioned and unconditioned spaces. Sound insulation can be particularly important near noisy areas such as airports, railways, and highways, especially for occupancies that require indoor quiet (e.g., hospitals, hotels, residences, theaters).

Constructability. During design, how a building envelope will be built must be considered. If there are practical limitations to how the envelope elements are physically put together, the design intent will be lost and problems are likely to develop during construction. Simplicity in design is an effective way to improve the chances of construction in accordance with the design intent. Construction of the various components must be sequenced so that all components can be correctly assembled, particularly when coordination between multiple trades is required. Investigation of many building failures demonstrates that poor sequencing between trades is a common contributing factor. Integrating constructability early in project design minimizes the number of failures and helps maximize the potential to achieve the results described in this chapter. During construction review, specific attention should be given to areas of the building where multiple systems are connected and multiple trades are involved. Use of mockups, design reviews, modeling, and other methods can enhance constructability.

Maintainability and Repairability. Building envelopes comprise many parts and components, with different anticipated service lives. Exterior cladding materials and fenestration may need replacement during the expected life of a building. Foundations and framing are the core elements of a building and should last for the entire building service life. Care should be taken not to cover shorter-lived building components with components having a longer anticipated lifespan.

Maintenance of the envelope and HVAC system is important to ensure a functional building. A poorly maintained HVAC system can result in substantial energy loss and have a detrimental effect on the building envelope by subjecting it to unexpected pressures and/or moisture loads. Verification of airflows and pressure differentials has been incorporated into many sustainable design tools as a check to ensure the building mechanical system is functioning as intended long after commissioning.

Sustainability. ASHRAE (2006) defines sustainability as "providing for the needs of the present without detracting from the ability to fulfill the needs of the future." In this chapter, *sustainability* refers primarily to durability and energy performance. Durability is essential for sustainable buildings, and moisture control is essential to durability. Additional information on sustainability can be found in Chapter 35 of the 2009 *ASHRAE Handbook—Fundamentals*.

With regard to durability, thermal insulation keeps interior and exterior materials near their respective temperatures. In a well-insulated building in a cold climate, the exterior materials are subjected to harsher conditions (lower temperatures, wetter conditions, slower drying, and longer periods at subfreezing temperatures) than in one poorly insulated. Exterior materials for well-insulated buildings should therefore be sufficiently robust to withstand the conditions to which they will be exposed.

From an energy perspective, completeness of the air and thermal barrier is critical to achieve good performance. For details on sustainable energy use in buildings, see ASHRAE *Standard* 90.1.

Quality Control. Ensuring good envelope performance demands a well-established quality control program: the design must minimize damage risk while maximizing thermal efficiency (quality assurance). For that purpose, redundancies should be incorporated that allow for imperfections in construction (e.g., providing for drainage or drying to remove moisture from wall cavities). The building should be designed to enable construction in a manner so it can be effectively maintained and repaired over its anticipated service life. Quality control methods such as construction review and the use of site mock-ups can be invaluable tools to improve the probability of good construction.

The building envelope differs greatly from mechanical and lighting systems in terms of inspections and commissioning. The building envelope is normally inspected during key phases of construction to check compliance with the construction documents and design intent, before many of the elements are enclosed within a wall system. Once enclosed, it is often very difficult to return to these areas to make repairs. At these key phases of completion, inspection measures are critical to ensure that any changes in design maintain the intent. Mechanical and lighting systems, on the other hand, are normally commissioned at the end of a project, meaning that their full-service testing for compliance does not occur until the building is operational and occupied.

DESIGN PRINCIPLES

Air conditioning and humidification can substantially change the moisture loads on the building envelope. New building materials may have significantly different thermal and moisture characteristics than traditional materials. The interdependency between the building envelope and the HVAC system has greater consequences to building durability and performance under problematic heat, air, and moisture conditions.

Heat Flow Control

A building envelope must adequately reduce heat flow to maintain energy efficiency and ensure indoor thermal comfort. Generally, heat flow control is achieved by installing thermal insulation as part of a wall, floor, or roof assembly.

The most common insulation materials used in building envelopes are glass fiber, mineral wool, cellulose, foam boards, and spray-applied foams. All these materials have exposure and performance limitations (e.g., fire, noise, moisture, ultraviolet) and should be selected carefully to promote long-term performance (see Table 4 in Chapter 26 of the 2009 *ASHRAE Handbook—Fundamentals* for common insulation materials and their properties).

Depending on the type of envelope assembly, the location of insulation in the wall can have a direct effect on thermal performance. For example, placing continuous insulation, such as rigid foam or mineral fiber board, outboard of the exterior sheathing on stud walls reduces conductive heat transfer through the studs and improves overall thermal performance.

As with all building envelope assemblies, correct installation of insulation is important in maintaining good thermal performance. For example, small voids left in insulation can result in an appreciable increase in heat flow, with the voids having a greater significance in more highly insulated assemblies. Verschoor (1977) found that convective air currents around thin wall insulation installed vertically with air spaces on both sides increased heat loss by 60%. Lecompte (1990) found losses up to 300% depending on the size and distribution of openings around insulation materials. Other factors, including vibration, temperature cycling, and other mechanical forces, can affect thermal performance by causing settling or other dimensional changes.

The thermal barrier should be continuous around the building envelope. This means aligning insulation planes in walls with thermal breaks in windows or providing continuity of the thermal plane around corners or at wall/roof connections.

Thermal Performance

Table 4 in Chapter 26 of the 2009 *ASHRAE Handbook—Fundamentals* gives thermal resistances of building materials. The thermal resistance of building assemblies is usually less than the sum of the material resistances, and may be significantly less. Data for clear-wall areas [summarized by James and Goss (1993)] do not include the effects of intersections with floors, roofs, and partitions, and do not account for thermal bridges at framing and partitions, air leakage, or convective air loops.

To test the validity of applying clear-wall data to the wall system, a series of three-dimensional heat conduction simulations was performed on a single-family, detached, one-story house, assuming no air leakage (Kosny and Desjarlais 1994). These simulations showed that, for a conventional wood-frame stud wall system with studs installed on 16 in. centers and no continuous outboard insulation, the average area-weighted whole-wall R-value was 91% of the clear-wall R-value. For a similar wall using 3.5 in. steel studs, the whole-wall R-value was only 83% of the clear-wall R-value. A similar two-dimensional analysis of an attached, two-story, steel-stud house by Tuluca et al. (1997) showed the R-value for the wall system to be 40 to 50% of the clear-wall R-value. Thermal bridging occurred through framing, metal ties, and exposed slab edges. Simply using the published insulation R-values alone as the R-value for the whole wall overestimates the real thermal performance. For constructions containing steel studs, for example, use either the R-value zone method or the modified zone method to determine real thermal performance that considers framing effects (see Chapter 25 of the 2009 *ASHRAE Handbook—Fundamentals*).

Thermal Mass

Thermal mass describes the ability of a material layer to store thermal energy and the ability of an opaque envelope component to dampen and delay transfer of heat. That damping, if combined with moderate glazing, effective solar shading, a correct ventilation strategy, and inside partitioning with high thermal storage, can help moderate indoor temperature fluctuation under outdoor temperature swings (Brandemuehl et al. 1990). Increased thermal mass may also positively affect energy efficiency (Kosny et al. 1998; Newell and Snyder 1990; Wilcox et al. 1985). Finally, increased thermal mass can help to shift demand for heating and cooling to off-peak periods. Thermal mass as a design tool is effective where the outdoor diurnal temperature fluctuates around the indoor comfort range. In areas where this does not occur, thermal mass has little effect.

Damping and time delay are defined by the order in which opaque envelope components are arranged. Best results are achieved when the thermal insulation faces outside and layers with large heat capacity face the inside, as confirmed by an in-depth study of six wall configurations by Kosny et al. (1998). Damping capability of such walls also increases with thermal resistance (Van Geem 1986).

Hourly-based computer simulations using transient energy simulation tools may provide a good estimate of thermal mass effects.

Thermal Bridges

A thermal bridge is an envelope area with significantly higher rate of heat transfer than the contiguous enclosure. Primary causes for thermal bridging are (1) parts with low thermal resistance perforating layers with high thermal resistance, (2) geometries that create zones where large exterior surfaces connect to much smaller interior surfaces, and (3) chilled or warmed edges at the edge of insulation as a result of discontinuities.

Thermal bridges increase energy use. They may lead to moisture condensation in and on the envelope, result in possible mold growth, accelerate surface fouling (ghosting), increase crack risk, and create surfaces with nonuniform temperatures that can result in indoor comfort problems.

Slab edges, perimeter beams, balconies, and decks protruding from the building envelope are common areas for thermal bridging.

The effect of thermal bridges in envelopes can be assessed using the zone method, modified zone method, acceptable sources of test data, or computer simulation tools. Refer to Chapter 27 of the 2009 *ASHRAE Handbook—Fundamentals* for details.

Thermal bridges created by webs of concrete masonry units (CMUs) dictate the maximum thermal efficiency a CMU can attain. To reduce their effect, blocks containing two webs instead of the usual three have been used, and web cross section has been reduced by up to 40%. However, even such changes in block design have not significantly improved wall R-value. Instead, applying low-density concrete with significantly lower thermal conductivity effectively improved thermal performance of these masonry units (Kosny and Christian 1995a, 1995b). The same strategy can be used for CMUs containing core-insulating inserts or insulation fill.

A comparison between the uninsulated slab edge detail of Figure 1A and the insulated version in Figure 1B illustrates the importance of designing to reduce thermal bridging. [See Steven Winter Associates (1988) for numerical examples summarized in this section.]

Air inside a masonry cavity often is at or near the outdoor air temperature. Figure 1A shows that, without cavity insulation, the concrete slab edge and steel beam are exposed to that outdoor air temperature. Both of these elements are made of relatively thermally conductive materials, steel being considerably more conductive than concrete. The result is significant heat loss and a low floor surface temperature near the slab edge, which can be a source of occupant discomfort and possible condensation if indoor humidity is sufficiently high. Adding insulation on the outside of these elements (Figure 1B) keeps them inside the thermal envelope. It keeps the structural steel warm, and prevents cold-weather condensation that could lead to corrosion, damage to the masonry below, and moisture damage to the surrounding finishes that could lead to interior mold growth.

Insulation is sometimes specified for the interior surface of the perimeter beam to increase the thermal resistance of this wall segment. Site conditions rarely allow thorough installation, so the top and bottom flange edges remain exposed, creating a strong thermal bypass and marginalizing the insulation's effectiveness. Moreover, use of a vapor-permeable insulation makes vapor condensation more likely because it decreases the temperature of the beam chord but does not stop water vapor migration.

Concrete balcony decks are often formed by slab extensions that pass through the building envelope. As the exposed exterior surface exchanges heat with the outside, the result is extra heat loss at the interior floor and low floor temperatures during cold weather. These low temperatures may extend to the top and bottom of close-by interior walls, with condensation each time their surface temperature drops below the dew point of the interior air. The same mechanism can increase energy use during hot weather because building mechanical systems compensate for additional heat gain. Thermal break elements between slab and balcony or a careful addition of insulation panels can moderate the thermal bridge effect. One- or two-dimensional heat transfer models can be used to analyze more complicated assemblies.

Air Leakage Control

Uncontrolled air leakage in the form of infiltration, exfiltration, intrusion, and wind washing increases space conditioning costs and may cause moisture problems (see Chapter 25 of the 2009 *ASHRAE Handbook—Fundamentals*). Air leakage is much more effective than diffusion for transporting water vapor in the building envelope and causing interstitial condensation. Forensic field observations support these research results. Uncontrolled air leakage also short-circuits the transient response of envelope assemblies. It degrades sound insulation of the envelope and may cause draft and related thermal discomfort. These drawbacks underline the need to plan for airflow control in building envelope design by installing an air barrier.

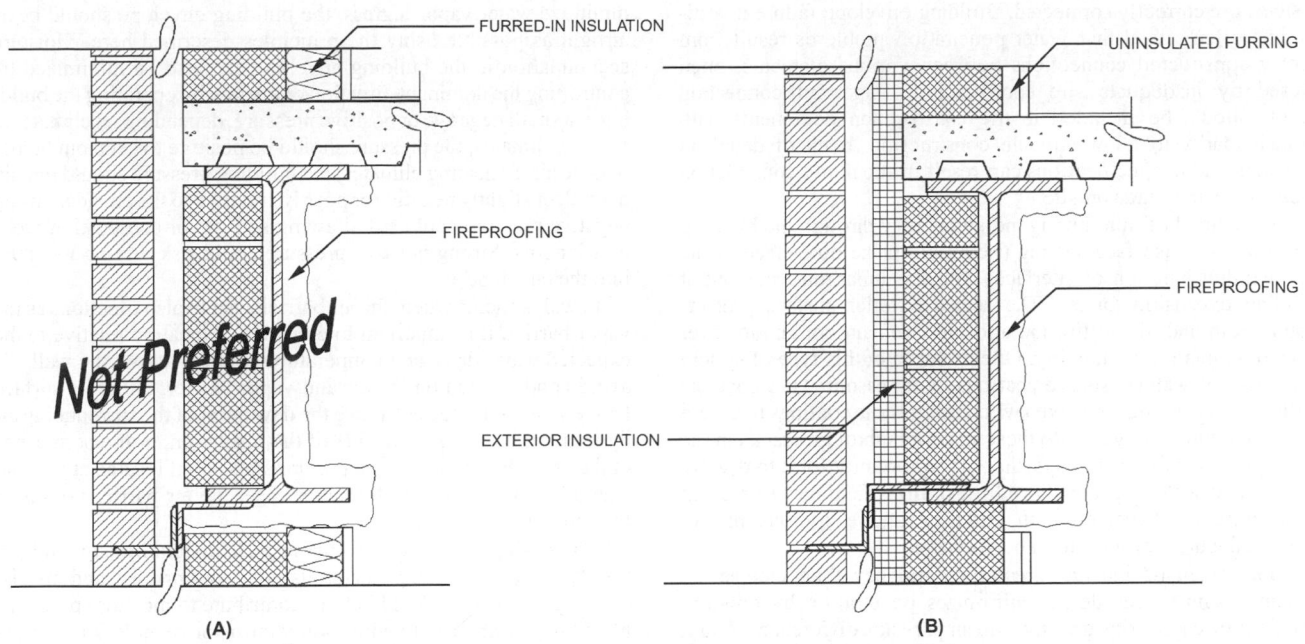

FURRED-IN INSULATION

FIREPROOFING

Not Preferred

EXTERIOR INSULATION

(A)

UNINSULATED FURRING

FIREPROOFING

EXTERIOR INSULATION

(B)

Air barriers, vapor barriers, drainage planes, weeps, and flashings omitted for clarity.

Fig. 1   Schematic Detail of (A) Uninsulated and (B) Insulated Slab Edge and Metal Shelf Angle.

Most published literature on air barrier requirements stipulates that an air barrier's air leakage permeance should not exceed 0.004 cfm/ft² at 0.3 in. of water when tested in accordance with ASTM *Standard* E2178 (NRCC 2005; U.S. ACE 2009).

The air barrier must be sufficiently supported on both sides and be strong enough to resist expected loads from stack effect, mechanical-system-induced pressures, and wind. For example, a sheet-membrane air barrier material installed using staples against a sheathing in a cavity wall is well supported by the sheathing in a positive-pressure direction, but may be subject to tearing at the fastener points under negative pressures, which tend to pull the membrane away from the wall.

The air barrier must be continuous across the entire building envelope. To ensure continuity at windows and other penetrations, components creating the barrier must be connected with an airtight, durable joint. It is particularly important to ensure continuity of air barrier systems at junctions that create complicated geometries where two or more elements in different planes intersect (e.g., roof/wall intersections, wall/floor intersections, corners). Maintaining continuity at these assemblies can be further complicated because they often involve different construction trades, requiring coordination between workers.

Penetrations (e.g., electrical outlets, light fixtures, plumbing stacks) should be minimized and, if unavoidable, sealed carefully. Airtight electrical boxes are available. Maintaining air barrier system integrity throughout the building's life should also be considered. Cutting holes in gypsum board assemblies during renovations, for example, can result in widespread air barrier failure if the board is intended to be the primary air barrier in the assembly.

The best location for the air barrier is generally where it is easiest to assemble into a continuous and durable system. The location of the air barrier in a wall assembly may affect overall hygrothermal performance. Permeance properties of the barrier material may be important; see the following section on Water Vapor Control.

Although the intent should be to construct an airtight building envelope, not all cracks and openings can be sealed in existing buildings, nor can an absolutely tight construction be achieved in new buildings. The objective is to provide an enclosure that is as tight as possible. Calling out a vapor barrier (see the section on Water

Vapor Control) as "continuous" is not a sufficient specification for an air barrier.

Moisture Control

Building envelopes are subject to several moisture loads (moisture entry mechanisms), including liquid water and water vapor from air leakage and/or diffusion. Historically, the primary moisture control strategy for walls was to restrict water entry by storing and redistributing moisture before evaporation. Today, good envelope design still requires minimizing accumulation and maximizing removal of moisture that gets in. Although weather-resistive components may be designed with the goal of eliminating water infiltration, some redundancy is obtained when assemblies are also designed to accommodate drainage and drying of incidental moisture intrusion. Hygrothermal modeling tools, as described in Chapters 25 and 27 of the 2009 *ASHRAE Handbook—Fundamentals*, can assist designers in understanding the drying potential of assemblies based on the assumed loads; however, current models cannot effectively predict drainage.

Liquid Water Control

Field observations indicate that moisture problems in buildings are most frequently caused by exterior liquid water penetrating or passing through the building envelope.

Rain is the most significant moisture source for buildings. Strategies to reduce the rain load on exterior walls include using building overhangs to prevent wetting, flashings, drip edges, and other water-shedding elements. During rain, poor flashing details at the interface of dissimilar materials, incomplete terminations of cladding systems, and other discontinuities in the moisture barrier may result in water entry, which can cause severe damage and loss of durability. For roofs and walls, cladding and building envelope components must be integrated to prevent water infiltration. At grade, rainwater should be carried away from the foundation through gutters, downspouts, and positive grading (i.e., sloping the surrounding grade to direct water away from the building).

An important consideration when designing an envelope for liquid water control is sequencing during construction and coordination between trades. Proper sequencing is essential to ensure that systems are correctly connected. Building envelope failure investigations often reveal that water penetration problems result from poorly constructed connections between various elements, often caused by inadequate site coordination. Important connection details should be included in the construction documents with enough clarity to allow suitable construction. Lack of detail on drawings and in specifications can result in too many construction decisions being made on site.

One method of minimizing moisture entry through the building envelope is to use face sealing (i.e., sealing the outer face of the wall/window junction, of interfaces with dissimilar materials, and at building expansion joints). The sealed exterior surface protects against rain and air infiltration and must remain continuous over time to maintain functionality. One example of this type of system is the use of water-resistive coatings over masonry and concrete walls. Great care must be taken when using these coatings to ensure that, when moisture gets into these materials through cracks in the coatings or by other pathways, there is opportunity for it to dry. By nature, face-sealed systems have little or no redundancy to prevent water ingress and accumulation, and they require rigorous maintenance schedules for long-term performance.

Rain screen design has greater redundancy than face-sealed systems. Rain screen design minimizes penetration by raindrop momentum, capillarity, gravity, and air pressure differences. A rain screen wall contains several components from inside to outside: an air barrier, a WRB (which may perform as air, moisture, and vapor barrier), the air space, and a rain screen. The air space may be an empty air cavity or a cavity filled with a material that drains freely. The air space must be vented to the outside through the rain screen and flashed to drain to the outside water that penetrates the rain screen. The rain screen's airflow resistance must be much lower than that of the air barrier, so that it acts as a deterrent for water penetration but is not a watertight seal. It is prudent to protect the WRB and air barrier from temperature extremes and direct exposure to ultraviolet light. With little pressure difference across the rain screen and with good cavity drainage detailing of the cavity, the potential for rain entry into the wall is significantly reduced. Interfaces of the exterior wall air barrier with fenestration air barriers, floors, and inside partition walls must be carefully considered, and may require site mock-ups to adequately determine the best solution.

For greater liquid water control, the air cavity may be designed as a pressure-moderation chamber, which involves making the WRB airtight. The cavity should also be compartmented to avoid lateral airflow, especially around corners of the building.

Water Vapor Control

Water vapor entry into the building envelope can be limited by airflow control and water vapor barriers. As described previously, air barriers are intended to restrict air leakage and control convective water vapor ingress, whereas vapor barriers are designed to restrict vapor flow caused by diffusion. It is important for building designers to understand the difference between the two mechanisms and how they are controlled.

Moisture deposition caused by air leakage is a point-load problem: a large volume of water can be deposited in a discrete location, often near the air leakage point, and can result in substantial damage. For that reason, an air barrier has to be continuous and sealed. This differs from moisture deposition cause by vapor diffusion, which is an area function that is directly related to the vapor drive and the vapor permeability of the materials that separate the two zones. A vapor barrier should therefore be continuous, but does not necessarily have to be sealed. The only time a vapor barrier is required to be sealed is when it also functions as the air barrier.

Water Vapor Transport Through Air Movement. Air leakage is more effective than diffusion for transporting water vapor in the building envelope, and therefore it is more important to control. To minimize water vapor ingress, the building envelope should be as airtight as possible using the principles described here. Moisture accumulation in the building envelope can also be minimized by controlling the dominant direction of airflow by operating the building at a small negative or positive pressure, depending on climate. In cooling climates, the pressure should be positive to keep out humid outside air. In heating climates, the building pressure should remain neutral, or slightly negative or positive relative to the outside. Strong negative pressure could risk drawing soil gas or combustion products indoors. Strong positive pressure could risk driving moisture into the envelope.

In wall systems where the air barrier system also functions as the vapor barrier, it is important to consider its location relative to the expected vapor drive and temperature gradient across the wall. To avoid condensation on the air and vapor barrier, either its surface temperature must be kept above the dew point of the surrounding air by locating it on the warm side of the insulation, or the permeance of the assembly must allow vapor transmission if located at the cold side of the insulation. In the latter case, the air barrier no longer functions as a vapor barrier.

Water Vapor Transport Through Diffusion. Moisture migration by diffusion through materials is a slow process and, as discussed previously, is less likely to contribute to moisture problems in buildings compared to liquid water intrusion or air leakage. However, diffusion can cause moisture problems in special occupancy types or in buildings that experience high moisture loads.

The overall diffusion performance of a building is a product of the diffusion characteristics of the envelope materials. A vapor barrier is not necessarily a sheet of plastic; many different materials or combination of materials can be used to control vapor diffusion, depending on design conditions. Many building envelope components, such as some peel-and-stick membranes, metal panels, and glass, have very low vapor permeance. Design and selection of building materials should be based on analysis to verify the desired performance of the assembly under the applicable loads.

Many common building and finish materials with low vapor permeance can have undesirable effects when placed in high-moisture-risk environments. One example is vinyl wallpaper placed at the interior wall surfaces. Under large inward vapor drives or excessive water penetration, this vapor-impermeable layer can lead to moisture accumulation through condensation, or can prevent drying by limiting vapor flows, both of which can lead to significant moisture damage. Careful attention must be paid to the type of materials selected for a wall construction, and where they are located in the wall assembly.

Use of vapor-impermeable layers at both the interior and exterior should generally be avoided so that the assembly can dry. Heat, air, and moisture calculations and modeling can be used to analyze an assembly, its climatic exposure, and desired indoor operating conditions to determine what methods of vapor control are appropriate. For details, see Chapters 25 to 27 of the 2009 *ASHRAE Handbook—Fundamentals*.

Common Envelope Problems

Wall/Window Interface. Air infiltration at the wall/window interface can reduce window performance and damage surrounding building materials and even remote materials, depending on the leakage path. In cold climates, warm, humid indoor air can increase moisture content in the wall cavity around the window. Excess moisture may damage the interior finish, seals of glazing units, insulation, exterior cladding, and possibly structural elements. Uncontrolled cold-air infiltration through the interface in turn can affect occupants' health and comfort by creating a dry indoor environment, cold drafts, and surface condensation on the window frame and glass edges. In warm, humid climates, leakage at the wall/window interface can result in interior fungal growth, distortion of interior window trim, and deterioration of the interior gypsum wallboard, particularly in air-conditioned buildings (because their interior surfaces are colder).

Control of air and water leakage at the wall/window interface is often difficult because that is where multiple systems intersect. Each system may incorporate a different approach for air and water control and must be integrated to provide continuity. Additionally, the different systems are commonly installed by different trades, which require deliberate sequencing to achieve the intended result. These intersections are often complex, making it difficult to inspect and test for performance compliance. Water penetration can result in moisture damage in any climate, although it is generally most severe in climates that impose a low drying potential on a building.

Control of Surface Condensation

To reduce the potential for condensation on the interior surface of glazing and the window frame, as well as on the surrounding interior wall finishes, the inside surface temperature can be controlled in the following ways:

- Select windows with appropriate condensation resistance for new construction or retrofit.
- Seal the wall/window interface and between the sash and frame of operable windows to minimize air leakage.
- Make the area of window frame exposed to the interior larger than the area exposed to the outside. Window and curtain wall systems with metal frame extensions on the inside have a higher resistance to condensation but contribute to heat loss.
- Reduce excessive interior humidity levels.
- Keep thermal breaks in the window system in the same plane as the wall insulation.

Continuity of the plane of thermal insulation between wall and window maximizes a window's thermal potential and reduces the potential for condensation on interior surfaces of the window frame, glass, and surrounding finish. Insulation in the joint between wall and window frame also compensates for the expected differential movement between the frame and the wall rough opening.

Interzonal Environmental Loads

Indoor partition walls separating zones such as indoor swimming pools, ice rinks, and freezers from zones conditioned as normal environments should be treated as envelope assemblies. Apart from weather-related loads, these partition walls must perform as building envelopes and need to control airflow, heat flow, liquid water (e.g., swimming pools, industrial settings), and water vapor flow.

Interstitial Spaces

Building envelope design must consider that modern buildings comprise a collection of interconnected internal chambers and cavities that provide potential pathways for unplanned airflows throughout the building. Concrete masonry cavity walls, wood- and steel-stud gypsum board partitions, chases, soffits, shafts, utility service penetrations, and many other details contain cavities that connect to adjacent elements to varying degrees wherever holes and openings have not been closed and sealed. The amount of air leakage depends on pressure differences, number and size of openings, and length and tortuosity of paths. Pressure differences generally derive from HVAC operation, differences in air density, and wind. These unplanned airflows often have no obvious adverse effects on occupants or the building, but in many cases they can significantly negatively affect energy demand, moisture deposition, and indoor air quality. Uncontrolled air leakage paths also add to the risk of undesirable sound transmission between rooms. A full discussion of all possible sources of unplanned airflows is beyond the scope of this chapter; see the References and Bibliography for additional information sources.

A common practice in commercial buildings is to use the space above a dropped ceiling as the HVAC system's return air plenum (Figure 2). The complex three-dimensional assemblies where the exterior wall adjoins a roof or floor at the building perimeter can be difficult to seal against air leakage and thermal bridging if the unique conditions of each design situation are not deliberately addressed. The interior gypsum board in the occupied space often terminates above the visual sight line of the suspended ceiling without extending to the roof or floor deck above. Unless carefully detailed and constructed, this can result in a leakage point at the building perimeter where negative pressure in the plenum can draw outside air into the return air system. If it is cold outdoors, an accidental sensible load penalty is imposed. For hot and humid climates, accidental latent and sensible loads are added that not only affect the energy required to condition the building, but also may lead to moisture and indoor air quality problems from condensation on interior surfaces that are cooler than the dew-point temperature of the incoming outdoor air. Sustained elevated relative humidity in the ceiling space can perpetuate mold growth. In general, but especially for demanding applications such as health-care and laboratory buildings, ducted returns should be used rather than depressurizing the entire ceiling plenum. Air pressure between the ceiling plenum and the occupied space can be equalized by installing ceiling grates in the suspended ceiling. Other configurations that may be more appropriate to the specific building design conditions can also be effective.

Similar considerations regarding energy loss and moisture deposition also apply to underfloor air supply systems that can force conditioned air out through the building envelope.

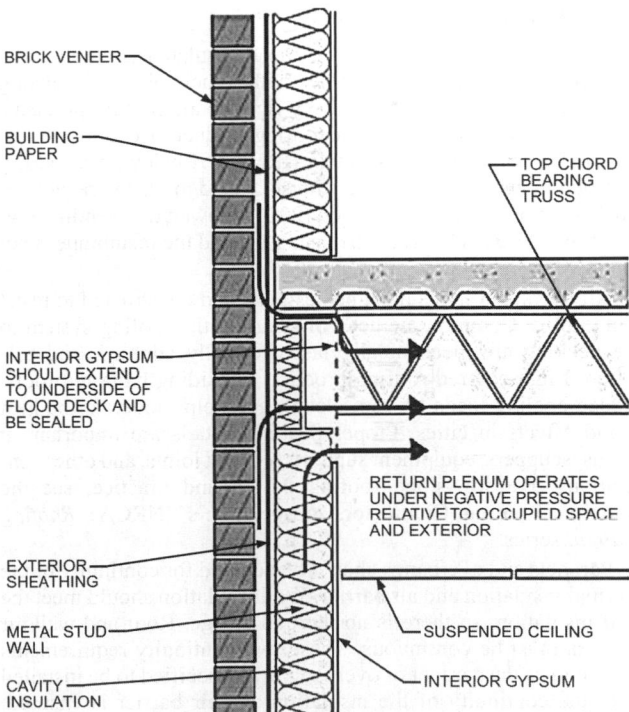

BRICK VENEER

BUILDING PAPER

TOP CHORD BEARING TRUSS

INTERIOR GYPSUM SHOULD EXTEND TO UNDERSIDE OF FLOOR DECK AND BE SEALED

RETURN PLENUM OPERATES UNDER NEGATIVE PRESSURE RELATIVE TO OCCUPIED SPACE AND EXTERIOR

EXTERIOR SHEATHING

METAL STUD WALL

SUSPENDED CEILING

CAVITY INSULATION

INTERIOR GYPSUM

Negative-pressure field in the dropped-ceiling return plenum extends to exterior, accidentally coupling the HVAC system to the building enclosure.

Fig. 2   Dropped-Ceiling Return Plenum
Lstiburek 2007

QUICK DESIGN GUIDE FOR HIGH-PERFORMANCE BUILDING ENVELOPES

1. Avoid excessive glazing. Follow energy standard and energy code requirements that set maximum glazed areas. Refer to the current version of ASHRAE *Standard* 90.1 for maximum roof and wall glazing areas.
2. Provide appropriate amounts of thermal insulation in foundation and above-grade walls and roofs. See current versions of ASHRAE *Standards* 90.1 and 90.2.
3. The completed building envelope should be airtight. Determine the level of airtightness to be achieved. Design airtight connections at all junctures in the air barrier system. Construction sequence should allow visual review and ensure performance with the selected criterion.
4. Resolve vapor barrier and vapor control issues using simplified or full hygrothermal analysis (ASHRAE *Standard* 160). The necessity and requirements for vapor control are specific to building usage, materials, and climate, and should be included in the design process.
5. Provide for effective shedding of rainwater away from the exterior wall. Rain screen principles to drain water away from the building should be used where possible.
6. Provide sound insulation appropriate for the building application. Some specialized facilities such as hospitals, libraries, and theaters require additional care to control sound transmission.
7. Refer to the ASHRAE *Advanced Energy Design Guide* series for additional information on building design.

COMMON ENVELOPE ASSEMBLIES

ROOFS

Low-Slope Roof Assemblies

Low-slope roofs are typically compact insulated assemblies of a waterproofing membrane together with structural and insulating board material. Insulation is usually rigid board or foam products. The insulation may be below the roofing product(s), or above for an inverted roof membrane assembly (IRMA). If below, it is usually installed in two or more layers with staggered joints to prevent airflow at the panel joint. Tobiasson (2010) showed that venting low-slope roof systems between the insulation and the membrane is not effective for moisture control.

All roofing systems must be designed and constructed to resist wind uplift. Common methods of securing the roofing system to the decking are mechanically fastened, fully adhered, and ballasted. Light-colored roofing products, including ballasted roofs, reduce cooling loads (if kept clean), and help moderate the heat island effects in cities. Proper flashing details are important at drains, scuppers, equipment supports, control joints, and other penetrations. For low-slope roofing design and practice, see the National Roofing Contractors Association's (NRCA) *Roofing Manual* series.

Parapets and overhangs should be detailed for continuity of the thermal insulation and air barrier. Wall insulation should meet the roof insulation, so there is no thermal bridge. Roof and wall air barriers must be continuous. These two continuity requirements may require the parapet or overhang to be specified to be installed after the continuity of the insulation and air barrier is ensured. Some parapet details, such as upper termination of interior drywall, flutes in steel decks, three-dimensional conditions, and fireproofing sequence for steel, are often overlooked. The roof/wall junction requires coordination and proper sequencing between trades to ensure proper continuity of the air and thermal control layers.

Steep-Roof Assemblies

There are two main insulation locations in steep-roof assemblies: at the roof (cathedral) and in the ceiling. For both types, exterior sheathing and roof materials should be able to accommodate wide temperature swings caused by radiant exchange with the sun and sky.

Insulated Sloped-Roof (ISR) Assembly. An insulated sloped roof assembly (e.g., cathedral ceiling) is a compact system with insulation parallel to the roof and no air cavity. Principles used in the design and construction of low-slope roofs may be applied in steep-sloped roof construction. According to Hens and Janssens (1999), moisture control is ensured only if airtightness is effective and can be maintained. Air entry and wind washing in insulated cathedral ceilings lead to degraded thermal and moisture (durability) performance. TenWolde and Carll (1992) showed that ventilation of ISR assemblies may increase air leakage, and that the net moisture effect depends on whether the principal source of makeup air is from indoors or outdoors. Use of vapor-permeable versus low-permeance thermal insulation in ISR assemblies can be an important factor in assembly design and performance. Their selection depends on the expected direction of vapor flow within the roofing system, drying potential, and other design considerations.

Attics. Standard North American attic construction provides for insulation at the ceiling plane level, leaving the attic space as unconditioned. The ceiling should be made airtight (Jordan et al. 1948). Air exchange with the outdoors provided by vents typically reduces attic air temperature on sunny afternoons. Other effects of ventilation, such as wood moisture content and roof surface temperature attenuation, depend more on factors other than the presence or absence of ventilation. TenWolde and Rose (1999) critically review four commonly cited reasons for attic ventilation: (1) preventing moisture damage, (2) enhancing the service life of temperature-sensitive roofing materials, (3) preventing ice dams, and (4) reducing cooling load.

The following additional design and construction elements should be considered for attics and insulated sloped-roof construction:

- Valleys are areas of high water concentration and are common sites of leaks. They should be designed to channel high volumes of water and for ease of repair.
- Mechanical equipment and ductwork should be placed in conditioned spaces. Their placement in unconditioned spaces leads to excess heat loss, energy consumption, and potential condensation on cold ductwork surfaces.
- Ice dams are typically caused by snow melting on the roof. Heat sources in the attic that could cause ice dams (e.g., chimneys, air leakage through the ceiling, attic-mounted equipment) should be identified and addressed.

Vegetated Roofing

There has been significant interest in vegetated roofing in recent years in an attempt to save energy or to control building rainwater outflow. On vegetated roofs, both growth medium and vegetation are installed on top of the roof. These assemblies normally require additional precautions and materials to protect the roof membrane from the plants. Quality control is even more critical during installation of a vegetated roof, because repair and eventual replacement are normally considerably more expensive and resource-intensive because of the presence of growing medium.

Storing rainwater on the roof can reduce the load on a municipality's storm system. There are many methods and products to achieve this. Storage of water and placement of a growing medium can add a considerable amount of mass to a roofing system. Checks to ensure structural capacity are essential for existing buildings, and new buildings must include structural provisions to support the additional weight. Some structural designers include provisions for the weight of a possible future vegetated roofing system because the incremental costs to do so are very small.

Vegetated roofing is often referred to as **green roofing**. Green roofing also refers to roofing systems that have other sustainability attributes such as high albedo.

WALLS

Curtain Walls

There has been tremendous growth in the use of curtain-wall systems for new building construction in recent years. A modern curtain-wall system is a highly engineered product based on mass production, standardization, and precise manufacturing (CMHC 2004). The systems generally consist of lightweight metal framing components connected to form a matrix to contain transparent and opaque wall areas. The framing is typically anchored to the building structural columns or floor slabs. Window wall systems are types of curtain walls that are typically installed floor to floor. Many older systems left slab edges exposed. New window wall systems often include drop-down panels that cover and provide thermal protection to the slab edge.

There are two basic kinds of systems: **stick built**, which are assembled on site from horizontal (rails) and vertical members (mullions), and **unitized systems**, which are largely assembled in a shop and delivered in sections that are then connected to form the wall. Both types are generally field glazed. Custom systems are also available for specialized applications. Glazing and opaque elements of the curtain-wall frame are fastened using exterior battens (pressure plates), structural sealant, or both.

Detailed information on the various types of curtain wall systems and glazing methods can be found in Canada Mortgage and Housing Corporation's *Best Practice Guide—Glass and Metal Curtain Walls* (CMHC 2004), or the National Institute of Building Sciences' *Whole Building Design Guide* (NIBS 2010).

Curtain-wall assemblies form the entire exterior wall where installed. They need to perform all the functions of a building envelope, though they have some significant differences from other walls. Curtain-wall assemblies are made of materials that have no moisture storage capabilities, whereas most other wall types (masonry, concrete, etc.) can store and release moisture over time. They nevertheless can retain significant volumes of water that may lead to penetration if watertightness and drainage are not provided.

There are two common methods for a curtain-wall system to manage exterior moisture: face-sealed and rain-screen design. **Face-sealed systems** are more susceptible to water penetration because they have no redundancy. Once water penetrates the exterior moisture barrier, there is no way for it to drain back to the exterior. **Rain screen designs** provide redundancy by draining moisture that bypasses the primary outer seal to the outside. This provides more protection from rain penetration than face-sealed systems.

Because curtain-wall systems are highly engineered and assembled out of precisely manufactured components, it is essential that they are assembled in accordance with manufacturers' specifications, with all recommended accessories installed. Omission of any components, such as corner blocks or other drainage elements, can result in reduced drainage capacity, water storage, and eventual penetration to the inside.

Curtain-wall systems can be used to cover floor slab edges to reduce thermal bridging. They bring in lots of natural light and can provide great occupant views. However, heavy use of glass often results in poorer thermal performance than for traditional wall assemblies, or even spandrel sections in a curtain-wall system. The large glazing areas in curtain walls should be considered in terms of solar gain and sizing of mechanical equipment. In cold climates, the large areas of glass and framing can have a radiative cooling effect on a space, and adversely affect thermal comfort. Conversely, in hot climates, large areas of glass can result in heat gains that can overwhelm cooling equipment, or necessitate cooling equipment in areas where mechanical cooling is not typically required. Often, thermal performance of curtain-wall systems is overstated by reporting center-of-glass or center-of-spandrel-panel thermal performance values, rather than the overall thermal performance (U-value) of the assembly.

Whereas traditional wall assemblies are often built by a number of trades, curtain-wall systems are often constructed by one installer. Connecting the curtain-wall system to surrounding systems, such as roof assemblies or wall, requires coordination of trades. Common problems found in the field include

- Improper or poor connection of curtain wall to roofing system
- Missing corner blocks or drainage elements, resulting in reduced moisture performance
- Missing or blocked vertical drainage channels, resulting in stored water at the glazing head

Sloped glazing is similar to curtain-wall systems, although the importance of using a rain-screen design to prevent water penetration to a building interior is enhanced. Properly flashed head joints and detailed drainage at the base of the sloped glazing are essential for long-term performance.

Precast Concrete Panels

Insulated precast wall (sandwich) panels are often constructed with solid concrete at the perimeter that encloses the insulation (Figure 3A). This concrete has a much greater thermal conductivity than the insulation, resulting in appreciable heat loss through these areas, particularly when added up across an entire building envelope. By changing how panels are assembled (e.g., connecting inner and outer panel sections with plastic tie-rods), a much greater thermal performance is achieved and the excessive perimeter heat loss is eliminated (Figure 3B). These types of panels are more susceptible to water penetration, however, so proper panel joint design and execution are even more critical with these types of panels.

Precast concrete wall assemblies should be connected using two-stage jointing system. Two-stage joints consist of an inner and outer sealant joint. The exterior joint forms an outside weather barrier that keeps most exterior moisture from entering the joint. The inner joint is normally sloped at the bottom of each panel joint to drain water

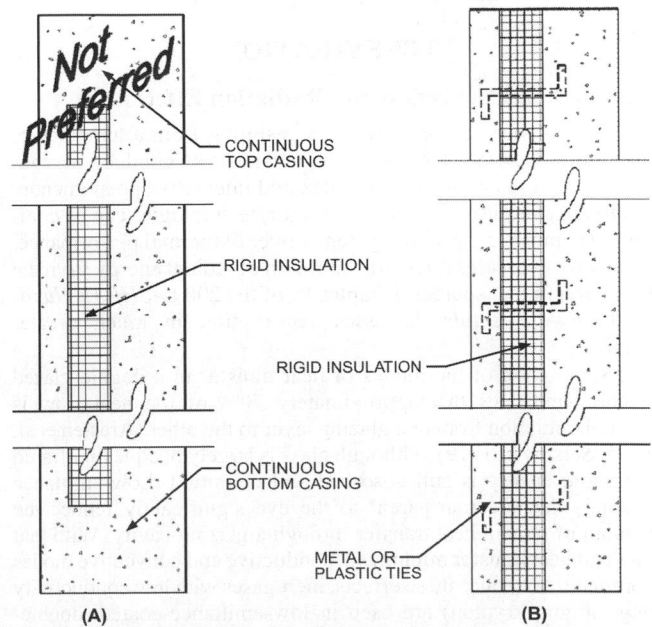

Fig. 3   Sandwich Panel with Insulation Encased in Concrete

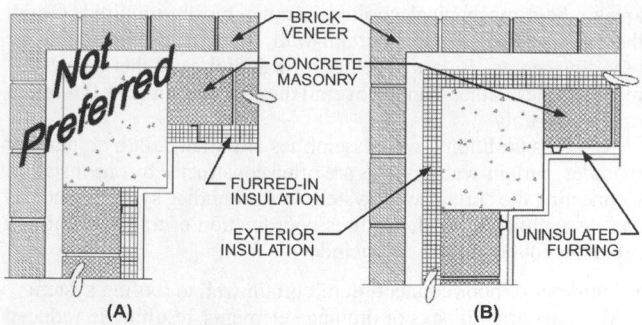

Fig. 4   Details of Insulation Around Column in Masonry Wall

that bypasses the outer joint back to the outside, and forms part of the wall's air-barrier assembly. For more information on precast panel joinery, see CMHC (2002).

Steel-Stud Wall Assemblies

Steel-stud wall assemblies are commonly used in multifamily, commercial, and industrial construction. These wall assemblies represent a particular challenge because of the studs' high thermal conductivity. Adding thermal breaks can prevent excessive heat loss and problems associated with condensation in the wall assembly and on interior finishes.

Adding insulated sheathing or insulation outboard of the steel studs significantly reduces heat loss and helps avoid cold surfaces at the interior and in the wall cavity; however, system limitations and manufacturer's instructions should be considered.

Wall Geometry with High Thermal Conductivity

Concrete columns in masonry walls are often left uninsulated. Figure 4 illustrates a wall system with a column at the junction of two exterior brick walls with CMU back-up. The assembly in Figure 4A represents a significant thermal bridge. In Figure 4B, the walls are insulated with polystyrene on metal furring. The amount of heat flow is greatly reduced in this assembly, and no condensation or subsequent mold growth would be expected on interior surfaces of the column under normal operating conditions.

FENESTRATION

Conduction/Convection and Radiation Effects

Heat transfer through a window resulting from a temperature differential between the inside and outside (i.e., conduction, convection, and radiation) is a complex and interactive phenomenon. Although glass itself is a poor insulator, technologies can be combined to improve a glazing system's overall thermal performance. Glass also decreases direct transmission of radiant energy from the room or ambient sources. Chapter 15 of the 2009 *ASHRAE Handbook—Fundamentals* discusses fenestration in much greater detail.

Examination of the modes of heat transfer in a double-glazed window indicates that approximately 70% of the heat flow is through radiation from one glazing layer to the other (Arasteh et al. 1985; Selkowitz 1979). Although glass is largely opaque to infrared radiation, energy is still absorbed and reemitted. Low-emittance coatings that are transparent to the eye significantly reduce the amount of radiant heat transfer through a glazing cavity. With that mode of heat transfer minimized, conductive and convective modes dominate. To reduce these effects, inert gases with low conductivity (e.g., argon, krypton) are used in low-emittance-coated, double-glazed windows. Inert-gas-filled triple or quadruple glazing layers with low-emittance coatings can additionally reduce heat transfer,

although overall thermal performance normally remains significantly less than a common insulated wall assembly. See Chapter 15 of the 2009 *ASHRAE Handbook—Fundamentals* for design information.

Air Infiltration Effects

Air infiltration at fenestration is influenced by the pressure difference between the inside and outside environments (a function of wind speed, inside/outside temperature differences, and mechanical air balance) as well as window-sealing characteristics (Klems 1983; Weidt and Weidt 1980). The infiltration rate of a fenestration product is a function of its method of operating (if any), weatherstripping material used, and construction quality. See Chapter 16 of the 2009 *ASHRAE Handbook—Fundamentals* for design information.

Solar Gain

Solar gain through windows can play a significant role in a building's energy balance. Glazings transmit, reflect, or absorb a given wavelength of solar radiation, depending on glazing characteristics. Transmitted solar radiation contributes heat to a space. Absorbed solar radiation is reemitted and/or conducted either to the inside or outside (depending on the glazing system configuration). Solar radiation that is reflected away from a building through the use of reflective glass does not contribute heat to a space (Arasteh et al. 1989), but the designer should be aware of the potential for secondary solar gains reflected from an adjacent structure. Clear glass, the most common glazing material, transmits fairly evenly across the solar spectrum. Tinted or heat-absorbing glass absorbs solar radiation and gives the glass a specific color. Some tints exhibit a significant degree of spectral sensitivity (i.e., they do not transmit, absorb, and reflect evenly across the solar spectrum). These types of glazing elements offer great flexibility and can be tailored for specific climates or uses (e.g., provide ample daylighting without overheating an interior space).

Interactions Between Thermal Loss and Solar Gain

In heating-dominated applications, solar gain provides a significant amount of heat. In some cases, heat supplied by the window can offset that lost through the window. The amount depends on characteristics of the site (e.g., how much solar gain is available, how cold the climate is) and the window (e.g., its U-factor and how much incident solar radiation is transmitted).

Typical passive solar applications try to maximize the amount of solar heat gain by installing significant areas of southeast- to southwest-facing glass, which receives the most solar radiation during the winter in the northern hemisphere. However, high-performance windows facing north in a heating-dominated but sunny climate can provide more solar gain to a space than heat loss (Arasteh et al. 1989; Dubrous and Wilson 1992; Sullivan et al. 1992).

Control of Rain Entry

Applying the rain screen principle at the wall/window interface requires the same features as applying it to the wall, including (1) an airflow retarder at the inside, (2) a rain deflector on the outside of the interface, and (3) a drainage path to the outside.

The line of airtightness is on the inside of the assembly, so it is protected from water, ultraviolet rays, and extremes of temperature. The rain deflector on the outside acts as a rain deterrent only, not as a watertight/airtight seal. A nonairtight rain deflector does not threaten system weathertightness, because pressure differences across the rain deflector are small. The key is to maintain airtightness on the inside of the joint. With little pressure difference across the rain deflector and with good detailing for outward drainage of the cavity, rain entry in the wall should be minimal. The interface between an exterior wall air barrier and an interior fenestration air barrier must be carefully considered during the design phase, and may require site mock-ups to adequately determine the best solution.

FOUNDATIONS

Three common types of below- and at-grade constructions are **basements**, **crawlspaces**, and **slabs**. Many buildings have combinations of these types. Key concerns for these assemblies are heat transfer and moisture management.

Heat Transfer

Heat transfer in below- and at-grade constructions is complex. Factors affecting it include thermal conductivity and wetness of the soil, height and temperature of the water table, amount and profile of insulation, and geometry. For a simplified means of estimating heat transfer in these constructions, see Chapter 18 of the 2009 *ASHRAE Handbook—Fundamentals*.

Measured heat transfer through slabs on grade with 10 different insulation profiles can be found in Bareither et al. 1948. Kusuda and Bean (1987) and Mitalas (1983) provide slab shape factors for steady-state calculations of heat loss. Transient calculations are provided in Hagentoft (1988) and Shipp (1982). An overview is provided in Labs et al. (1988). The following general principles may be applied:

- Heat loss through slabs is concentrated at the slab edge. Insulation is more critical there than in the center of the slab.
- For sections of basements or crawlspaces that are above grade or shallow below grade, insulation on the outside is more effective than insulation at the interior. For deeper parts of a basement, interior insulation is more effective because of the thermal bridge effect of the footing, which is typically uninsulated.
- In crawlspaces and basements, insulation may be applied in the foundation walls or in the floor system above. If insulation is applied in the floor system, the space below may be cold and wet. Some form of moisture protection on the underside of an insulated floor system may be necessary.

Moisture

Below- and at-grade constructions are affected by rainwater surrounding the building and by the below-grade water table. For all construction types, ensure that the soil or other finish outside the building is sloped away from the building. Discharge from scuppers, gutters, and downspouts should be conducted far enough from the building that discharge water cannot saturate soil that is in contact with the below- or at-grade assembly (Rose 2010).

Foundation depth should correspond to the hydrologic conditions of the building site, so that the underground water table is not allowed to encroach on the below-grade space. If it does, then a sump pump must be used to locally lower the water table.

Basements and crawlspaces may encounter some flooding in the course of their service lives. Consequently, basement finishes should be selected so that they can withstand water loading and are readily cleaned afterwards.

Crawlspaces have been known to be sources of moisture to the building since their introduction in the mid-1900s. Britton (1947) found that the use of low-vapor-permeance ground covers plus ventilation resulted in drier conditions. His recommendation for ventilation was picked up by many codes and guidelines. Venting crawlspaces, however, creates an unconditioned space beneath the building, which may lead to energy penalties and high relative humidity during warm weather. It is therefore recommended that crawlspaces should be treated like basements with insulated perimeter walls, insulated bottom floors, and protection from moisture ingress. The crawlspace should be accessible, well illuminated, and clean.

In cold climates, building foundations should resist frost heave. This is commonly done by ensuring footings below frost depth. The frost-protected shallow foundation uses below-grade insulation outboard of the foundation to ensure nonfreezing conditions at the slab edge. Attachment frost heaving may occur (Labs et al. 1988); it is prevented by avoiding puddles of water in contact with the foundation during freezing weather by draining water away from the building, as described previously.

EXISTING AND HISTORIC BUILDINGS

In recent years, there has been a noticeable shift in emphasis from new construction to work on existing buildings. There can be tremendous advantages in materials use, embodied energy, carbon dioxide emissions, and other environmental issues to renovating existing buildings rather than replacing them with new structures, or adding new structures. In the United States, work on buildings of historic significance is addressed by the Secretary of the Interior's Standards for the Treatment of Historic Properties (1977). For more information, see Park (2009).

Determining what materials were used in the original construction and their actual properties can be a challenge. Many materials are hidden within the structure and require some disassembly to determine underlying structure and components. Necessary investigative work may range from review of accurate as-built drawings, to field testing, disassembly, and historical research. Depending on the age of the building, multiple renovations may have created many different assemblies in the building, so a review of available plans and permits issued for the building helps avoid unknown conditions during construction. Properties of older materials can be markedly different from their modern equivalents, so research may be necessary to approximate their properties: in some cases, actual testing of the materials may be the only way to discover how the materials will react. Chemical compatibility between reused and newer materials may be a concern. For instance, new materials for roofing, waterproofing, and sealant or barrier systems may not be compatible with materials used in the past.

Existing buildings provide the benefit of having a performance history. This can be helpful in understanding the reality of energy consumption as well as moisture storage, air leakage, and water vapor movement through materials and assemblies. The building history can be used to identify defects or weak points in the assembly or interactions between the HVAC system and envelope. The building operates in an equilibrium that depends on indoor and outdoor conditions and building material properties. Changes to operating conditions or materials can affect everything and alter this equilibrium.

A review of changes such as HVAC upgrades or building envelope improvements should be incorporated into the design. Some sustainable practices are rapidly evolving, and applicable codes and guidelines are still developing. As a result, there can be challenges to following the letter of the code on renovation projects without fully understanding the physical phenomena behind envelope performance and interactions with HVAC system effects. This can result in problems ranging from water leakage to advanced degradation of wall components. Sometimes, novel approaches must be sought that allow compliance with the intent of the code while optimizing the durability and performance of existing materials/assemblies.

An additional concern for existing buildings involves a common practice of owners: some large-scale renovation projects are phased over a period of time. If an overall plan is not developed that includes understanding of the interdependence between the building envelope and HVAC systems, then problems and/or extra expense are very likely to occur over the multiple years that it takes to complete the entire renovation plan. The order in which changes are done (e.g., envelope first versus HVAC system first) could also greatly affect overall costs as well as the end result. For example, in hot, humid climates, if the HVAC system is replaced before an envelope-tightening project, extra capacity may have to be installed to meet the interim needs. This extra capacity could cause problems with overcooling and high relative humidity after the envelope tightening occurs.

Successful renovation projects begin with documenting existing conditions and careful analysis of the effects of potential changes.

New materials and systems are selected based on compatibility with existing materials and systems as well as durability and long-term performance characteristics.

Building Materials

Specific issues to be considered for material changes and selections include addition of new materials, removal of old materials, and replacement of existing materials with a modern equivalent; a full understanding of the purpose of specific layers and materials in the original design is also required. An example involves the use of stone or brick masonry. In older buildings, solid stone or brick masonry walls were designed to perform as barrier systems by absorbing and gradually releasing moisture. These materials relied on heat flow to keep them dry and prevent freeze/thaw damage in cold climates.

Reuse of wall systems during energy retrofits poses particular challenges in cold climates. Adding exterior insulation has the desirable attributes of providing continuous insulation and moisture control while insulating structural elements from thermal and moisture extremes, but it may conflict with preservation aesthetics as well as zoning requirements. Adding interior insulation increases thermal stresses while reducing the drying potential of the exterior facade, and may lead to increased freeze/thaw cycling in cold climates. It is also difficult to achieve continuous thermal insulation and air barrier around existing interior wall elements. Unavoidable discontinuities create thermal bridges that could become condensation sites. Floors supported by exterior masonry are one example. See Chapter 27 of the 2009 *ASHRAE Handbook—Fundamentals* for examples of calculating thermal resistance values for complex wall assemblies.

Hygrothermal analysis and understanding of the potential consequences of measures planned on the durability of facades and structural elements must be part of the renovation design process, because this may impose limits on the insulation strategy. Guidance for performing hygrothermal analysis can be found in ASHRAE *Standard* 160.

Changing HVAC Equipment and/or Control Strategy

When upgrading or replacing existing mechanical systems, the effect of the new HVAC system on the building envelope must be considered. This is particularly important with the addition of humidification. In cold, and even some mild, climates, humidifying a previously nonhumidified space may lead to damaging moisture accumulation in walls and roofs unless the building envelope can withstand the loads. For a typical nonhumidified building, the interior relative humidity is normally lowest when the exterior temperature is also at a minimum. This is beneficial to the building, because condensation risk on the interior is moderated during the coldest period of the year. Adding humidification (by mechanical means or occupant activities) reduces this benefit, and in cold climates may result in surface condensation and/or mold growth for several months out of the year. Durability often is negatively impacted when humidification is added to an existing building unless changes are also made to the envelope to resist condensation.

The building envelope requires more attention than just estimating its properties to calculate heating and cooling loads for new equipment. In older buildings with little to no insulation, successful HVAC system design needs to counteract the large envelope losses and solar gains; otherwise, occupant comfort can suffer. This can be particularly noticeable adjacent to windows with single-pane glazing or thermally inefficient walls where additional heating or cooling maybe required to compensate for excessive heat loss. Localized use of space heaters or fans in an existing building indicates deficiencies in the previous HVAC system. HVAC system design may have to include perimeter heaters, specialized VAV distribution, or localized heating to prevent low temperatures or moisture accumulation in concealed spaces. Additionally, information on existing drawings used to determine thermal performance properties for equipment sizing calculations may not reflect changes already made to the building envelope. Important changes that can affect performance can include window replacements or reflective coatings, adding insulation during reroofing or interior renovation projects, etc. Changing the pressure distribution in building zones or across the exterior envelope also significantly affects the building envelope's moisture performance.

Envelope Modifications Without Mechanical System Upgrades

Building envelope systems are often upgraded without corresponding upgrades to mechanical systems. Failure to modify the existing mechanical system to account for changes in the dynamic performance of the building envelope can result in problems such as excess humidity or lack of interior environmental control. Retrofitting a building for improved thermal insulation or solar control at glazing systems can effectively reduce cooling loads. If the mechanical systems are not modified accordingly, they become oversized for the renovated conditions. This can result in numerous problems, such as poor interior temperature and relative humidity control, or more subtle issues such as inefficient operation. Retaining an older mechanical system may also prevent the full energy savings of an enclosure upgrade from being realized.

Improved airtightness should be a criterion of any building envelope retrofit project. For buildings in cold climates that previously relied on incidental air leakage to provide ventilation, reducing leakage rates may result in high interior air moisture levels. In the previously leaky building, incidental leakage was sufficient to dilute interior moisture with dry outdoor air and maintain reasonable humidity levels. These buildings should use a ventilation system using appropriate energy recovery techniques to maintain adequate fresh air.

Designers should assess the relationship between the building envelope and mechanical system, and design appropriate modifications. Design considerations for converting existing buildings to high-performance buildings include the following:

- Investigate existing building envelope conditions, taking care to note sites of damage and repair that may affect performance. In many cases, it may be beneficial to gather information on the building's historical performance, including utility bills and past occupancy types.

- Provide documentation of existing conditions in the building, building envelope, and mechanical systems. For historic buildings, complete a historic structures report.

- Identify major air and/or water leakage sites, and provide remedial measures to correct flow through these sites. If appropriate, adopt a strategy to reach an airtightness performance target.

- Consider improving thermal performance. Exterior insulation is often preferred over interior or cavity insulation, because cavity or interior insulation typically allows structural and other members to act as thermal bridges. Local preservation requirements may govern the location of insulation.

- Review possible changes to envelope operating characteristics that result from adding thermal insulation or changing HVAC operation. A change in the moisture and/or heat flow function of the envelope can have a significant effect on durability.

- With improved envelope performance, consider downsized mechanical equipment.

REFERENCES

Arasteh, D.K., M.S. Reilly, and M.D. Rubin. 1989. A versatile procedure for calculating heat transfer through windows. *ASHRAE Transactions* 95(2): 755-765.

Arasteh, D.K., S. Selkowitz, and J. Hartmann. 1985. Detailed thermal performance data on conventional and highly insulating window systems. *Thermal Performance of the Exterior Envelopes of Buildings III*, pp. 830-845.

ASHRAE. *Advanced Energy Design Guides.*

ASHRAE. 2006. *ASHRAE greenguide: The design, construction, and operation of sustainable buildings*, 2nd ed.

ASHRAE. 2007. Energy standard for buildings except low-rise residential buildings. ANSI/ASHRAE/IESNA *Standard* 90.1-2007.

ASHRAE. 2007. Energy-efficient design of low-rise residential buildings. ANSI/ASHRAE *Standard* 90.2-2007.

ASHRAE. 2009. Criteria for moisture-control design analysis in buildings. ANSI/ASHRAE *Standard* 160-2009.

ASTM. 2003. Standard test method for air permeance of building materials. *Standard* E2178. American Society for Testing and Materials, West Conshohocken, PA.

Bareither, H.D., A.N. Fleming, and B.E. Alberty. 1948. Temperature and heat-loss characteristic of concrete floors laid on the ground. University of Illinois Small Home Council-Building Research Council Research *Report* 48-1.

Brandemuehl, M.J., J.L. Lepore, and J.F. Kreider. 1990. Modeling and testing the interaction of conditioned air with building thermal mass. *ASHRAE Transactions* 96(2):871-875.

Britton, R. 1947. Crawl spaces: Their effect on dwellings—An analysis of causes and results—Suggested good practice requirements. HHFA *Technical Bulletin* 2. Housing and Home Finance Agency. Washington, D.C.

CMHC. 2002. *Best practice guide: Architectural precast concrete—Walls and structure.* Canada Mortgage and Housing Corporation, Ottawa, ON.

CMHC. 2004. *Best practice guide: Glass and metal curtain walls.* Canada Mortgage and Housing Corporation, Ottawa, ON.

CSA. 2007. Guideline on durability in buildings. *Standard* S478-95 (R2007). Canadian Standards Association, Mississauga, ON.

Dubrous, F. and A.G. Wilson. 1992. A simple method for computing energy performance for different locations and orientations. *ASHRAE Transactions* 98(1):841-849.

Hagentoft, C.-E. 1988. *Heat loss to the ground from a building: Slab on the ground and cellar.* Department of Building Technology, Lund Institute of Technology, Sweden.

Handegord, G. and N. Hutcheon. 1989. *Building science for a cold climate.* Construction Technology Centre Atlantic, Inc., Fredericton, NB.

Hendriks, L. and H. Hens. 2000. Building envelopes in a holistic perspective. IEA ECBCS Annex 32, *Final Report*, vol. 1. International Energy Agency, Energy Conservation in Buildings and Community Systems, Leuven, Belgium.

Hens, H. and A. Janssens. 1999. Heat and moisture response of vented and compact cathedral ceilings: A test house evaluation. *ASHRAE Transactions* 105(1).

Hutcheon, N. 1963. Requirements for exterior walls. *Canadian Building Digest* 48.

James, T.B. and W.P. Goss. 1993. *Heat transmission coefficients for walls, roofs, ceilings, and floors.* ASHRAE.

Jordan, C.A., E.C. Peck, F.A. Strange, and L.V. Teesdale. 1948. Attic condensation in tightly built houses. Housing and Home Finance Agency *Technical Bulletin* 6.

Klems, J. 1983. Methods of estimating air infiltration through windows. *Energy and Buildings* 5:243-252.

Kosny J. and J.E. Christian. 1995a. Steady-state thermal performance of concrete masonry unit wall systems. *Thermal Envelopes VI Conference*, Clearwater, FL.

Kosny, J. and J. Christian. 1995b. Reducing the uncertainties associated with using the ASHRAE-zone method for R-value calculations of metal frame walls. *ASHRAE Transactions* 101(2):779-788.

Kosny, J. and A.O. Desjarlais. 1994. Influence of architectural details on the overall thermal performance of residential wall systems. *Journal of Thermal Insulation and Building Envelopes* 18:53-69.

Kosny, J., E. Kossecka, A.O. Desjarlais, and J.E. Christian. 1998. Dynamic thermal performance of concrete and masonry walls. *Thermal Envelopes VII Conference*, Clearwater, FL.

Kusuda, T. and J.W. Bean. 1987. Design heat loss factors for basement and slab floors. *Thermal insulation: Materials and systems*. ASTM STP 922. F.J. Powell and S.L. Matthews, eds. American Society for Testing and Materials, West Conshohocken, PA.

Labs, K., J. Carmody, R. Sterling, L. Shen, Y.J. Huang, and D. Parker. 1988. *Building foundation design handbook.* ORNL/Sub/86-72143/1. Oak Ridge National Laboratory, Oak Ridge, TN.

Lecompte, J. 1990. Influence of natural convection in an insulated cavity on the thermal performance of a wall. *Insulation Materials, Testing and Applications*, pp. 397-420. ASTM STP 1030. American Society for Testing and Materials, West Conshohocken, PA.

Lstiburek, J. 2007. Building sciences: The hollow building. *ASHRAE Journal* 49(6):56-58.

Mitalas, G.P. 1983. Calculation of basement heat loss. *ASHRAE Transactions* 89(1B):420-437. DRB Paper No. 1198. Division of Building Research. National Research Council Canada, Ottawa, ON. Available at http://www.nrc-cnrc.gc.ca/obj/irc/doc/pubs/nrcc23378/nrcc23378.pdf.

Newell, T.A. and M.E. Snyder. 1990. Cooling cost minimization using building mass for thermal storage. *ASHRAE Transactions* 96(2):830-838.

NIBS. 2010. *Whole building design guide.* National Institute of Building Sciences, Washington, D.C. http://www.wbdg.org.

NRCA. *Roofing and Waterproofing Manual.* National Roofing Contractors Association, Rosemont, IL.

NRCC. 2005. *National building code of Canada.* National Research Council Canada.

Park, S. 2009. Moisture in historic buildings and preservation guidance. ASTM *Manual* 18: *Moisture control in buildings: The key factor in mold prevention*, H. Trechsel and M. Bomberg, eds. American Society for Testing and Materials, West Conshohocken, PA.

Rose, W. 2010. Recommendations for remedial and preventive actions for existing residential buildings. ASTM *Manual* 18: *Moisture control in buildings: The key factor in mold prevention*, H. Trechsel and M. Bomberg, eds. American Society for Testing and Materials, West Conshohocken, PA.

Secretary of the Interior. 1977. Standards for the treatment of historic properties. Available from www.nps.gov/history/hps/tps.

Selkowitz, S.E. 1979. Thermal performance of insulating window systems. *ASHRAE Transactions* 85(2):669-685.

Shipp, P. 1982. Basement, crawl space and slab-on-grade performance. *Thermal Performance of the Exterior Envelopes of Buildings II.* ASHRAE.

Steven Winter Associates. 1988. Catalog of thermal bridges in commercial and multi-family residential constructions. *Report* 88-SA407/1 for Oak Ridge National Laboratory, Oak Ridge, TN.

Sullivan, R., B. Chin, D. Arasteh, and S. Selkowitz. 1992. A residential fenestration performance design tool. *ASHRAE Transactions* 98(1):832-840.

TenWolde, A. and C. Carll. 1992. Effect of cavity ventilation on moisture in walls and roofs. *Thermal Performance of the Exterior Envelopes of Buildings V*, pp. 555-562. ASHRAE.

TenWolde, A. and W. Rose. 1999. Issues related to venting of attics and cathedral ceilings. *ASHRAE Transactions* 105(1).

Tobiasson, W. 2010. Roofs. ASTM *Manual* 18: *Moisture control in buildings: The key factor in mold prevention*, H. Trechsel and M. Bomberg, eds. American Society for Testing and Materials, West Conshohocken, PA.

Tuluca, A., D. Lahiri, and J. Zaidi. 1997. Calculation methods and insulation techniques for steel stud walls in low-rise multifamily housing. *ASHRAE Transactions* 103(1):550-562.

U.S. ACE. 2009. *Engineering and Construction Bulletin 2009-29.* U.S. Army Corps of Engineers.

Van Geem, M.G. 1986. Summary of calibrated hot box test results for twenty-one wall assemblies. *ASHRAE Transactions* 92(2B):584-602.

Verschoor, J.D. 1977. Effectiveness of building insulation applications. USN/CEL *Report* CR78.006-NTIS AD-A053 452/9ST.

Weidt, J.L. and J. Weidt. 1980. Field air leakage of newly installed residential windows. LBL *Report* 11111. Lawrence Berkeley Laboratory, CA.

Wilcox, B., A. Gumerlock, C. Barnaby, R. Mitchell, and C. Huizenza. 1985. The effects of thermal mass exterior walls on heating and cooling loads in commercial buildings. *Thermal Performance of the Exterior Envelopes of Buildings III*, pp. 1187-1224.

BIBLIOGRAPHY

IEA. 1991. *Condensation and energy: Guidelines and practice*. IEA ECBCS Annex 14. International Energy Agency, Energy Conservation in Buildings and Community Systems, Leuven, Belgium. Available at http://www.ecbcs.org/docs/annex_14_guidelines_and_practice.pdf.

Korsgaard, V. and C.R. Pedersen. 1992. Laboratory and practical experience with a novel water-permeable vapor retarder. *Thermal Performance of the Exterior Envelopes of Buildings V*, pp. 480-490. ASHRAE.

Sanders, C. 1996. Environmental conditions. IEA ECBCS Annex 24, *Final Report*, vol. 2. International Energy Agency, Energy Conservation in Buildings and Community Systems, Leuven, Belgium.

Trechsel, H.R., ed. 1994. Moisture control in buildings. ASTM *Manual* 18: *Moisture control in buildings: The key factor in mold prevention*, H. Trechsel and M. Bomberg, eds. American Society for Testing and Materials, West Conshohocken, PA.

BUILDING AIR INTAKE AND EXHAUST DESIGN

FRESH air enters a building through its air intake. Likewise, building exhausts remove air contaminants from a building so wind can dilute the emissions. If the intake or exhaust system is not well designed, contaminants from nearby outside sources (e.g., vehicle exhaust, emergency generator, laboratory fume hoods on nearby buildings) or from the building itself (e.g., laboratory fume hood exhaust) can enter the building with insufficient dilution. Poorly diluted contaminants may cause odors, health impacts, and reduced indoor air quality. This chapter discusses proper design of exhaust stacks and placement of air intakes to avoid adverse air quality impacts. Chapter 24 of the 2009 *ASHRAE Handbook—Fundamentals* more fully describes wind and airflow patterns around buildings. Related information can also be found in Chapters 8, 17, 32, 33, and 34 of this volume, Chapters 11 and 12 of the 2009 *ASHRAE Handbook—Fundamentals*, and Chapters 28, 29, and 34 of the 2008 *ASHRAE Handbook—HVAC Systems and Equipment*.

EXHAUST STACK AND AIR INTAKE DESIGN STRATEGIES

Stack Design Strategies

The dilution a stack exhaust can provide is limited by the dispersion capability of the atmosphere. Before discharge, exhaust contamination can be reduced by filters, collectors, and scrubbers if needed to maintain acceptable air quality. The ultimate goal of the stack design is to specify the lowest flow, exhaust velocity, and stack height that ensures acceptable air quality at all locations of concern. This also ensures that energy consumption is minimized.

Central exhausts that combine flows from many collecting stations should always be used where safe and practical. By combining several exhaust streams, central systems dilute intermittent bursts of contamination from a single station. Also, the combined flow forms an exhaust plume that rises a greater distance above the emitting building. If necessary for air quality or architectural reasons, additional air volume can be added to the exhaust near the exit with a makeup air unit to increase initial dilution and exhaust plume rise. This added air volume does not need heating or cooling, and the additional energy cost is lower. A small increase in stack height may also achieve the same benefit but without any added energy cost.

In some cases, separate exhaust systems are mandatory. The nature of the contaminants to be combined, recommended industrial hygiene practice, and applicable safety codes need to be considered. Separate exhaust stacks could be grouped in a tight cluster to take advantage of the larger plume rise of the resulting combined jet. Also, a single stack location for a central exhaust system or a tight cluster of stacks allows building air intakes to be positioned as far as possible from the exhaust. Petersen and Reifschneider (2008) provide guidelines on optimum arrangements for ganged stacks. In general, for a tight cluster to be considered as a single stack (i.e., to add stack momentums together) in dilution calculations, the stacks must be uncapped and nearly be touching the middle stack of the group.

The preparation of this chapter is assigned to TC 4.3, Ventilation Requirements and Infiltration.

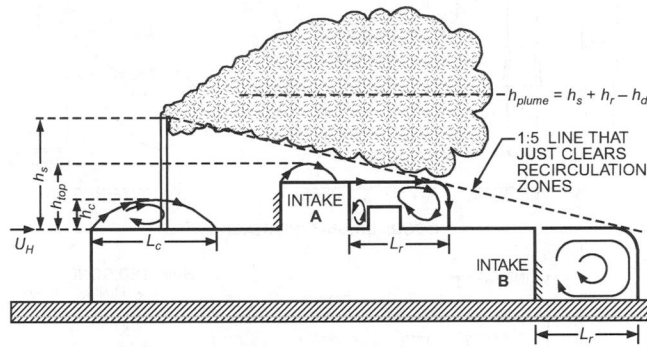

Fig. 1   Flow Recirculation Regions and Exhaust Parameters
(Wilson 1982)

As shown in Figure 1, stack height h_s is measured above the roof level on which the air intake is located. Wilson and Winkel (1982) demonstrated that stacks terminating below the level of adjacent walls and architectural enclosures frequently do not effectively reduce roof-level exhaust contamination. To take full advantage of their height, stacks should be located on the highest roof of a building.

Architectural screens used to mask rooftop equipment adversely affect exhaust dilution, depending on porosity, relative height, and distance from the stack. Petersen et al. (1999) found that exhaust dispersion improves with increased screen porosity.

Large buildings, structures, and terrain close to the emitting building can adversely affect stack exhaust dilution, because the emitting building can be within the recirculation flow zones downwind of these nearby flow obstacles (Wilson et al. 1998a). In addition, an air intake located on a nearby taller building can be contaminated by exhausts from the shorter building. Wherever possible, facilities emitting toxic or highly odorous contaminants should not be located near taller buildings or at the base of steep terrain.

As shown in Figure 2, stacks should be vertically directed and uncapped. Stack caps that deflect the exhaust jet have a detrimental effect on exhaust plume rise. Small conical stack caps often do not completely exclude rain, because rain does not usually fall straight down; periods of heavy rainfall are often accompanied by high winds that deflect raindrops under the cap and into the stack (Changnon 1966). A stack exhaust velocity V_e of about 2500 fpm prevents condensed moisture from draining down the stack and keeps rain from entering the stack. For intermittently operated systems, protection from rain and snow should be provided by stack drains, as shown in Figure 2F to 2J, rather than stack caps.

Recommended Stack Exhaust Velocity

High stack exhaust velocity and temperatures increase plume rise, which tends to reduce intake contamination. Exhaust velocity V_e should be maintained above 2000 fpm (even with drains in the stack) to provide adequate plume rise and jet dilution. Velocities above 2000 fpm provide still more plume rise and dilution, but above 3000 to 4000 fpm, noise, vibration, and energy costs can become an important concern. An exit nozzle (Figure 2B) can be

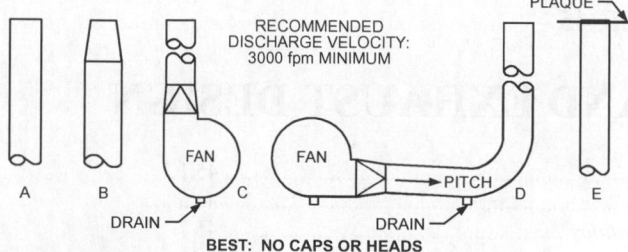

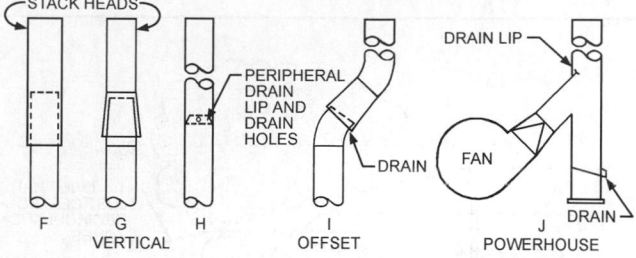

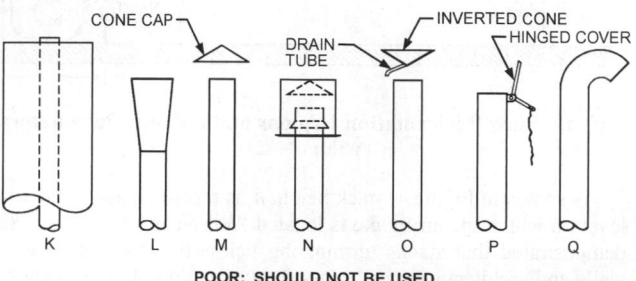

**Fig. 2   Stack Designs Providing Vertical Discharge
and Rain Protection**

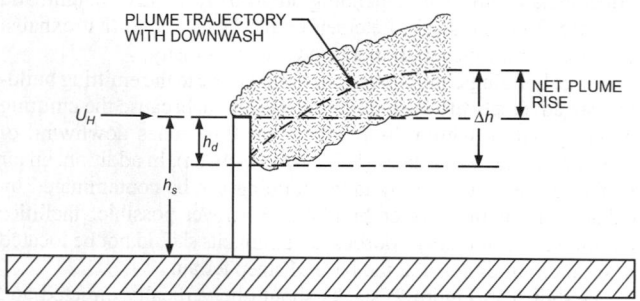

**Fig. 3   Reduction of Effective Stack Height
by Stack Wake Downwash**

used to increase exhaust velocity and plume rise. Many laboratory fume hood systems use variable-volume fans that reduce flow from hoods when they are closed. Stack exhaust velocity calculation must be based on the minimum total flow rate from the system, not the maximum.

Exceptions to these exhaust velocity recommendations include when corrosive condensate droplets are discharged. In this case, a velocity of 1000 fpm in the stack and a condensate drain are recommended to reduce droplet emission. At this low exhaust velocity, a taller stack may be needed to counteract downwash caused by low exit velocity. Another exception is when a detailed dispersion modeling analysis is conducted. Such an analysis can determine the minimum exit velocity needed to maintain acceptable dilution versus stack height. Generally, the taller the stack, the lower the required exit velocity and hence fan energy.

Stack wake downwash occurs where low-velocity exhausts are pulled downward by negative pressures immediately downwind of the stack, as shown in Figure 3. V_e should be at least 1.5 times the design speed U_H at roof height in the approach wind to avoid stack wake downwash. A meteorological station design wind speed U_{met} that is exceeded less than 1% of the time can be used. This value can be obtained from Chapter 14 of the 2009 *ASHRAE Handbook—Fundamentals*, or estimated by applying Table 2 of Chapter 24 of that volume to annual average wind speed. Because wind speed increases with height, a correction for roof height should be applied for buildings significantly higher than 30 ft, using Equation (4) and Table 1 of Chapter 24 of the 2009 *ASHRAE Handbook—Fundamentals*.

Other Stack Design Standards

Minimum heights for chimneys and other flues are discussed in the *International Building Code* (ICC 2006). For laboratory fume hood exhausts, American Industrial Hygiene Association (AIHA) *Standard* Z9.5 recommends a minimum stack height of 10 ft above the adjacent roof line, an exhaust velocity V_e of 3000 fpm, and a stack height extending one stack diameter above any architectural screen; National Fire Protection Association (NFPA) *Standard* 45 specifies a minimum stack height of 10 ft to protect rooftop workers. Toxic chemical emissions may also be regulated by federal, state, and local air quality agencies.

Contamination Sources

Some contamination sources that need consideration in stack and intake design include the following.

Toxic Stack Exhausts. Boilers, emergency generators, and laboratory fume hoods are some sources that can seriously affect building indoor air quality because of toxic air pollutants. These sources, especially diesel-fueled emergency generators, can also produce strong odors that may require administrative measures, such as generator testing during low building occupancy or temporarily closing the intakes.

Automobile and Truck Traffic. Heavily traveled roads and parking garages emit carbon monoxide, dust, and other pollutants. Diesel trucks and ambulances are common sources of odor complaints (Smeaton et al. 1991). Intakes near vehicle loading zones should be avoided. Overhead canopies on vehicle docks do not prevent hot vehicle exhaust from rising to intakes above the canopy. When the loading zone is in the flow recirculation region downwind from the building, vehicle exhaust may spread upwind over large sections of the building surface (Ratcliff et al. 1994). Garbage containers may also be a source of odors, and garbage trucks may emit diesel exhaust with strong odors.

Kitchen Cooking Hoods. Kitchen exhaust can be a source of odors and cause plugging and corrosion of heat exchangers. Grease hoods have stronger odors than other general kitchen exhausts. Grease and odor removal equipment beyond that for code requirements may be needed if air intakes cannot be placed far away.

Evaporative Cooling Towers. Outbreaks of Legionnaires' disease have been linked to bacteria in cooling tower drift droplets being drawn into the building through air intakes (Puckorius 1999). ASHRAE *Guideline* 12 gives advice on cooling tower maintenance for minimizing the risk of Legionnaires' disease, and suggests keeping cooling towers as far away as possible from intakes, operable windows, and outside public areas. No specific minimum separation distance is provided or available. Prevailing wind directions should also be considered to minimize risk. Evaporative cooling towers can have several other effects: water vapor can increase air-conditioning loads, condensing and freezing water vapor can damage equipment, and ice can block intake grilles and filters. Chemicals added to retard scaling and biological contamination may be emitted from the cooling tower, creating odors or health effects, as discussed by Vanderheyden and Schuyler (1994).

Building General Exhaust Air. General indoor air that is exhausted will normally contain elevated concentrations of carbon dioxide, dust, copier toner, off-gassing from materials, cleaning agents, and body odors. General exhaust air should not be allowed to reenter the building without sufficient dilution.

Stagnant Water Bodies, Snow, and Leaves. Stagnant water bodies can be sources of objectionable odors and potentially harmful organisms. Poor drainage should be avoided on the roof or ground near the intake. Restricted airflow from snow drifts, fallen leaves, and other debris can be avoided in the design stage with elevated louvers above ground or roof level.

Rain and Fog. Direct intake of rain and fog can increase growth of microorganisms in the building. AMCA (2009) recommends selecting louvers and grilles with low rain penetration and installing drains just inside the louvers and grilles. In locations with chronic fog, some outdoor air treatment is recommended. One approach is to recirculate some part of the indoor air to evaporate entrained water droplets, even during full air-side economizer operation (maximum outdoor air use).

Environmental Tobacco Smoke. Outdoor air intakes should not be placed close to outside smoking areas.

Plumbing Vents. Codes frequently require a minimum distance between plumbing vents and intakes to avoid odors.

Smoke from Fires. Smoke from fires is a significant safety hazard because of its direct health effects and from reduced visibility during evacuation. NFPA *Standard* 92A discusses the need for good air intake placement relative to smoke exhaust points.

Construction. Construction dust and equipment exhaust can be a significant nuisance over a long period. Temporary preconditioning of outdoor air is necessary in such situations, but is rarely provided. A simple solution is to provide room and access to the outdoor air duct for adding temporary air treatment filters or other devices, or a sufficient length of duct so that such equipment could be added when needed. Intake louvers and outdoor air ducts also require more frequent inspections and cleaning when construction occurs nearby.

Vandalism and Terrorism. Acts of vandalism and terrorism are of increasing concern. Louvers and grilles are potential points of illegal access to buildings, so their placement and construction are important. Intentional introduction of offensive or potentially harmful gaseous substances is also of concern. Some prudent initial design considerations might be elevating grilles and louvers away from easy pedestrian access and specifying security bars and other devices. Also, unlocked stair tower doors required for roof access during emergency evacuations may limit use of rooftop air intakes in sensitive applications because individuals would have ready access to the louvers. For more information, see ASHRAE's (2003) *Risk Management Guidance for Health, Safety, and Environmental Security under Extraordinary Incidents.*

General Guidance on Intake Placement

Carefully placed outdoor air intakes can reduce stack height requirements and help maintain acceptable indoor air quality. Rock and Moylan (1999) review recent literature on air intake locations and design. Petersen and LeCompte (2002) also showed the benefit of placing air intakes on building sidewalls. ASHRAE *Standard* 62.1-2010 highlights the need to locate makeup air inlets and exhaust outlets to avoid contamination.

Experience provides some general guidelines on air intake placement. Unless the appropriate dispersion modeling analysis is conducted, intakes should never be located in the same architectural screen enclosure as contaminated exhaust outlets. This is especially the case for low-momentum or capped exhausts (which tend to be trapped in the wind recirculation zone within the screen). For more information, see the section on Influence of Architectural Screens on Exhaust Dilution.

If exhaust is discharged from several locations on a roof, intakes should be sited to minimize contamination. Typically, this means maximizing separation distance. Where all exhausts of concern are emitted from a single, relatively tall stack or tight cluster of stacks, a possible intake location might be close to the base of this tall stack, if this location is not adversely affected by other exhaust locations, or is not influenced by tall adjacent structures creating downwash. However, contaminant leakage from the side of the stack has been observed in positively pressurized areas between the exhaust fans and stack exit (Hitchings 1997; Knutson 1997), so air intakes should not be placed very close to highly toxic or odorous exhaust stacks regardless of stack height.

Intakes near vehicle loading zones should be avoided. Overhead canopies on vehicle docks do not effectively protect air intakes, and vehicle exhaust may spread over large sections of the building surface. Loading zones also may have garbage and solid waste receptacles that create odors; trucks that serve the receptacles also produce odors. Air intakes should also not be placed near traffic or truck waiting areas. General building exhausts should also not be placed near outside contamination sources because flow reversal and ingestion of air through exhaust outlets can occur under some conditions (Seem et al. 1998).

Examining airflow around a building can help determine air intake placement. When wind is perpendicular to the upwind wall, air flows up and down the wall, dividing at about two-thirds up the wall (Figures 4 and 5). The downward flow creates ground-level swirl (shown in Figure 4) that stirs up dust and debris. To take advantage of the natural separation of wind over the upper and lower half of a building, toxic or nuisance exhausts should be located on the roof and intakes located on the lower one-third of the building, but high enough to avoid wind-blown dust, debris, and vehicle exhaust. If ground-level sources (e.g., wind-blown dust and vehicle exhaust) are major sources of contamination, rooftop intake is desirable.

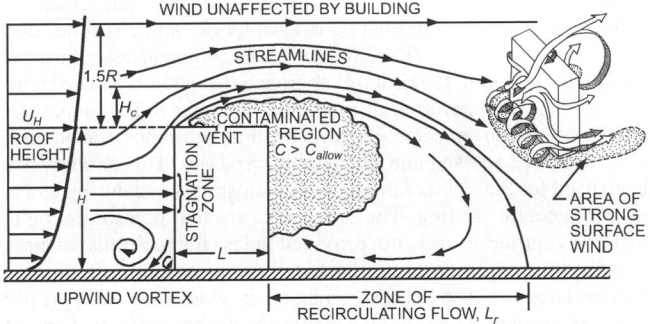

Fig. 4   Flow Patterns Around Rectangular Building

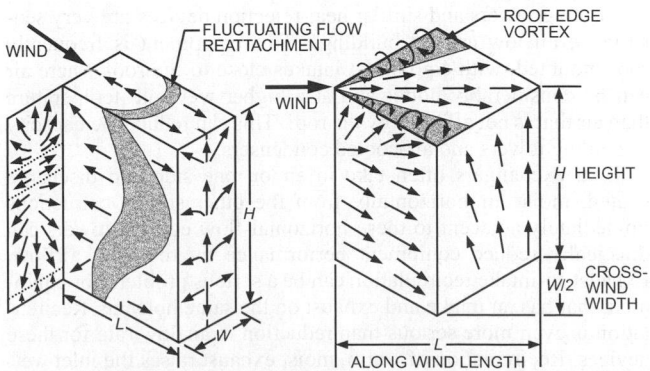

Fig. 5   Surface Flow Patterns and Building Dimensions

Code Requirements for Air Intakes

Many model building codes exist, and local governments adopt and amend codes as needed. Architects and building systems designers need to be familiar with local and national codes applicable to each project. Mechanical and plumbing codes typically give minimum required separation distances for some situations. However, maintaining these separation distances does not necessarily guarantee that intake contamination will not occur.

One example of a model building code is the *Uniform Mechanical Code* (UMC) (IAPMO 1997a), which has been widely adopted in the United States. The UMC requires that exhausts be at least 3 ft from property lines and 3 ft from openings into buildings. Makeup air intakes should be placed to avoid recirculation. Grease and explosives-bearing ducts, combustion vents, and refrigeration equipment have special requirements: intakes should be at least 10 ft from combustion vents, plumbing vents, and exhaust air outlets, and be at least 10 ft above a road. Cooling towers should be 5 ft above or 20 ft away from intakes.

The *Uniform Plumbing Code* (UPC) (IAPMO 1997b), requires that exhaust vents from domestic water heaters be 3 ft or more above air inlets. Sanitary vents must be 10 ft or more from or 3 ft above air intakes. When UPC and UMC requirements conflict, the UPC provisions govern. However, local jurisdictions may modify codes, so the adopted versions may have significantly different requirements than the model codes.

Treatment and Control Strategies

When available intake/exhaust separation does not provide the desired dilution factor, or intakes must be placed in undesirable locations, ventilation air requires some degree of treatment, as discussed in Section 6.2.1 of ASHRAE *Standard* 62.1. Fibrous media, inertial collectors, and electrostatic air cleaners, if properly selected, installed, and maintained, can effectively treat airborne particles. Reducing gaseous pollutants requires scrubbing, absorptive, adsorptive, or incinerating techniques. Biological hazards require special methods such as using high-efficiency particulate air (HEPA) filters and ultraviolet light. Chapters 28 and 29 of the 2008 *ASHRAE Handbook—HVAC Systems and Equipment* describe these treatments in detail. One control approach that should be used with care is selective operation of intakes. If a sensor in the intake airstream detects an unacceptable level of some substance, the outdoor air dampers are closed until the condition passes. This strategy has been used for helicopter landing pads at hospitals and during emergency generator testing. The drawbacks are that pressurization is lost and ventilation air is not provided unless the recirculated air is heavily treated. In areas of chronically poor outdoor air quality, such as large urban areas with stagnant air, extensive and typically costly treatment of recirculated air may be the only effective option when outdoor air dampers are closed for extended periods.

Intake Locations for Heat-Rejection Devices

Cooling towers and similar heat-rejection devices are very sensitive to airflow around buildings. This equipment is frequently roof-mounted, with equipment intakes close to the roof where air can be considerably hotter and at a higher wet-bulb temperature than air that is not affected by the roof. This can reduce the capacity of cooling towers and air-cooled condensers.

Heat exchangers often take in air on one side and discharge heated, moist air horizontally from the other side. Obstructions immediately adjacent to these horizontal-flow cooling towers can drastically reduce equipment performance by reducing airflow. Exhaust-to-intake recirculation can be a serious problem for equipment that has an intake and exhaust on the same housing. Recirculation is even more serious than reduction in airflow rate for these devices. Recirculation of warm, moist exhaust raises the inlet wet-bulb temperature, which reduces performance. Recirculation can be caused by adverse wind direction, local disturbance of the airflow

by an upwind obstruction, or by a close downwind obstruction. Vertical exhaust ducts may need to be extended to reduce recirculation and improve equipment effectiveness.

Wind Recirculation Zones on Flat-Roofed Buildings

Stack height design must begin by considering the wind recirculation regions (Figure 6). To avoid exhaust reentry, the stack plume must avoid rooftop air intakes and wind recirculation regions on the roof and in the wake downwind of the building. Where stacks or exhaust vents discharge within this region, gases rapidly diffuse to the roof and may enter ventilation intakes or other openings. Figures 4 and 6 show that exhaust gas from an improperly designed stack is entrained into the recirculating flow zone behind the downwind face and is brought back into contact with the building.

Wilson (1979) found that, for a flat-roofed building, the upwind roof edge recirculation region height H_c at location X_c and its recirculation length L_c (shown in Figures 1 and 6) are proportional to the building size scale R:

$$H_c = 0.22R \tag{1}$$

$$X_c = 0.5R \tag{2}$$

$$L_c = 0.9R \tag{3}$$

and the wind recirculation cavity length L_r on the downwind side of the building is approximately

$$L_r = R \tag{4}$$

where R is the building scaling length:

$$R = B_s^{0.67} B_L^{0.33} \tag{5}$$

where B_s is the smaller of the building upwind face dimensions (height or width) and B_L is the larger. These equations are approximate but are recommended for use. The dimensions of flow recirculating zones depend on the amount of turbulence in the approaching wind. High levels of turbulence from upwind obstacles can decrease the coefficients in Equations (1) to (4) by up to a factor of two. Turbulence in the recirculation region and in the approaching wind also causes considerable fluctuation in the position of flow reattachment locations (Figures 5 and 6).

Rooftop obstacles such as penthouses, equipment housings, and architectural screens are accounted for in stack design by calculating the scale length R for each of these rooftop obstacles from Equation (5) using the upwind face dimensions of the obstacle. The

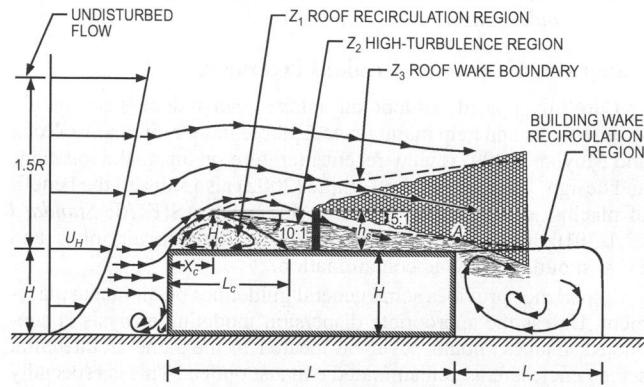

Fig. 6   Design Procedure for Required Stack Height to Avoid Contamination
(Wilson 1979)

recirculation regions for each obstacle are then calculated from Equations (1) to (4). When a rooftop obstacle is close to the upwind edge of a roof or another obstacle, the flow recirculation zones interact. Wilson (1979) gives methods for dealing with these situations.

Building-generated turbulence is confined to the roof wake region, whose upper boundary Z_3 in Figure 6 is

$$Z_3/R = 0.28(x/R)^{0.33} \qquad (6)$$

where x is the distance from the upwind roof edge where the recirculation region forms. Building-generated turbulence decreases with increasing height above roof level. At the edge of rooftop wake boundary Z_3, turbulence intensity is close to the background level in the approach wind. The high levels of turbulence in the air below the boundary Z_2 in Figure 6 rapidly diffuse exhaust gases downward to contaminate roof-level intakes. As shown in Figure 6, the boundary Z_2 of this high-turbulence region downwind of a wind recirculation region is approximated by a straight line sloping at 10:1 downward from the top of the wind recirculation zone to the roof. The stack in Figure 6 may be inadequate because at point A the plume intersects the high turbulence boundary Z_2. The geometric method for stack height is discussed in more detail in the next section.

GEOMETRIC METHOD FOR ESTIMATING STACK HEIGHT

This section presents a method of specifying stack height h_s so that the lower edge of the exhaust plume lies above air intakes and wind recirculation zones on the roof and downwind of the emitting building, based on flow visualization studies (Wilson 1979). This method does not calculate exhaust dilution in the plume; instead, it estimates the size of recirculation and high turbulence zones, and the stack height to avoid contamination is calculated from the shape of the exhaust plume. High vertical exhaust velocity is accounted for with a plume rise calculation that shifts the plume upward. Low vertical exhaust velocity that allows stack wake downwash of the plume (see Figure 3) is accounted for by reducing the effective stack height.

This stack height should prevent reentry of exhaust gas into the emitting building most of the time, provided no large buildings, structures, or terrain are nearby to disturb the approaching wind. The geometric method considers only intakes on the emitting building. Additional stack height or an exhaust-to-intake roof-level dilution calculation is often required if the exhaust plume can impinge on a nearby building (Wilson et al. 1998b). Dilution calculations should be used if this method produces an unsatisfactorily high stack, or if exhaust gases are highly toxic releases from fume hood exhaust.

Rooftop obstacles can significantly alter dispersion from exhaust stacks immediately downwind of the obstacles and of similar height to the obstacles (Saathoff et al. 2002). The goal of the geometric stack method is to ensure that the exhaust plume is well above the recirculation zones associated with these obstacles.

Step 1. Use Equations (1) to (5) to calculate the height and location of flow recirculation zones 1 and 2 and the recirculation zone downwind of the building (see Figure 6). All zones associated with rooftop obstacles up- and downwind of the stack location should be included. Note that zone 3 is not used in the geometric design method.

Step 2. Draw the recirculation regions on the top and downwind sides of penthouses, equipment housings, architectural screens, and other rooftop obstacles up- and downwind of the stack location. If there are intakes on the downwind wall of the building, include the building recirculation region L_r on this wall. Now, calculate the height h_{sc} of a stack with a rain cap (i.e., no plume rise), draw a line sloping down at 1:5 in the wind direction above the roof. Slide this line down toward the building as shown in Figure 1 until it contacts

any one of the recirculation zones on any obstacle up- or downwind of the stack (or until the line contacts any portion of the building if there are no rooftop zones or sidewall intakes). With the line in this position, its height at the stack location is the smallest allowable plume height h_{sc} for that wind direction. Repeat for other wind directions to find the worst-case (highest) required plume height.

This estimated h_{sc} is based on an assumption that the plume spreads up and down from h_{sc} with a 1:5 slope (11.3°), as shown in Figure 6. (This slope represents a downward spread of approximately two standard deviations of a bell-curve Gaussian plume concentration distribution in the vertical direction.)

Step 3. Reduce the stack height to give credit for plume rise from uncapped stacks. Only jet momentum rise is used; buoyancy rise is neglected as a safety factor. For an uncapped stack of diameter d_e, plume rise h_r from the vertical jet momentum of the exhaust is estimated versus downwind distance from Briggs (1984) as

$$h_r = \min\{\beta h_x, \beta h_f\} \qquad (7)$$

where the plume rise versus downwind distance, in ft, is

$$h_x = \left\{ \left(\frac{3F_m x}{\beta_j^2 U_H^2} \right)^{1/3} \right\}$$

momentum flux, in ft^4/s^2, is

$$F_m = V_e^2 \left(\frac{d_e^2}{4} \right)$$

the jet entrainment coefficient is

$$\beta_j = \frac{1}{3} + \frac{U_H}{V_e}$$

the final plume rise, in ft, is

$$h_f = \frac{0.9[F_m U_H / U_*]^{1/2}}{U_H \beta_j}$$

and the logarithmic wind profile equation is

$$U_H/U_* = 2.5 \ln(H/z_o)$$

where

β = stack capping factor: 1.0 without cap, 0 with cap
x = distance downwind of stack, ft
V_e = stack exit velocity, fpm
U_H = wind speed at building top, fpm
H = building height above ground level, ft
U_* = friction velocity, ft
z_o = surface roughness length, ft

Table 1 describes various z_o values for a range of sites. For example if z_o equals 2.13 ft and $H = 49.2$ ft, substituting into the logarithmic wind profile equation gives $U_H/U^* = 7.9$.

$$d_e = \sqrt{4A_e/\pi} \qquad (8)$$

For an uncapped stack, the capping factor is $\beta = 1.0$. For a capped stack, $\beta = 0$, so $h_r = 0$, and no credit is given for plume rise. U_H is the maximum design wind speed at roof height for which air intake contamination must be avoided. This maximum design speed must be at

Table 1   Atmospheric Boundary Layer Parameters

Terrain Category	z_o, ft	a	δ, ft
Flat, water, desert	0.03	0.10	700
Flat, airport, grassland	0.16	0.14	900
Suburban	2.1	0.22	1200
Urban	6.0	0.33	1500

least as large as the hourly wind speed exceeded 1% of the time. This 1% design speed is listed for many cities in Chapter 14 of the 2009 *ASHRAE Handbook—Fundamentals* (on the CD-ROM). For cities not on this list, set U_H equal to 2.5 times the annual average hourly wind speed as recommended in Table 2 of Chapter 24 of the 2009 *ASHRAE Handbook—Fundamentals*.

The plume rise of Equation (7) plus the physical stack height should not be considered equivalent to an effective stack height. A real stack of that height has better performance for two reasons: the effective height is achieved immediately instead of somewhere downstream, and the plume is higher than the effective height because of exhaust momentum. Stack height plus plume rise are additive in the geometric method as a simplification, but there are other conservatisms built into the geometric method to offset this approach.

Step 4. Increase stack height, if necessary, to account for stack wake downwash caused by low exhaust velocity, as described in the section on Recommended Stack Exhaust Velocity. For a vertically directed jet from an uncapped stack ($\beta = 1.0$), Wilson et al. (1998b) recommend a stack wake downwash adjustment h_d of

$$h_d = d_e(3.0 - \beta V_e/U_H) \qquad (9)$$

for $V_e/U_H < 3.0$. For $V_e/U_H > 3.0$, there is no downwash and $h_d = 0$. Rain caps and louvers are frequently used on stacks of gas- and oil-fired furnaces and packaged ventilation units, for which $\beta = 0$ and $h_d = 3.0d_e$.

The final adjusted stack height h_s recommended is

$$h_s = h_{sc} - h_r + h_d \qquad (10)$$

The advantage of using an uncapped stack instead of a capped stack is considerable. If the minimum recommended exhaust velocity V_e of $1.5U_H$ is maintained for an uncapped stack ($\beta = 1.0$), plume downwash $h_d = 1.5d_e$ and $h_r = 4.5d_e$. For a capped stack ($\beta = 0$), $h_d = 3.0d_e$ and $h_r = 0$. Using these values in Equation (10), an uncapped stack can be made $6.0d_e$ shorter than a capped stack.

Example 1. The stack height h_s of the uncapped vertical exhaust on the building in Figure 1 must be specified to avoid excessive contamination of intakes A and B by stack gases. The stack has a diameter d_e of 1.64 ft and an exhaust velocity V_e of 1770 fpm. It is located 52.5 ft from the upwind edge of the roof. The penthouse's upwind wall (with intake A) is located 98.4 ft from the upwind edge of the roof, 13.1 ft high, and 23.0 ft long in the wind direction. The top of intake A is 6.56 ft below the penthouse roof. The building has a height H of 49.2 ft and a length of 203 ft. The top of intake B is 19.7 ft below roof level. The width (measured into the page) of the building is 164 ft, and the penthouse is 29.5 ft wide. The annual average hourly wind speed is 7.95 mph at a nearby airport with an anemometer height H_{met} of 32.8 ft. The building is in suburban terrain (see Table 1). Calculate the required stack height h_s by the geometrical method using the lowest allowable design wind speed.

Solution: The first step is to set the height h_{sc} of a capped stack by projecting lines with 1:5 slopes so that recirculation zones are covered, as shown in Figure 1. The only influence of intake location is that the downwind recirculation zone must be considered if there is an intake on the downwind wall, which is true for intake B in this example.

First, check the rooftop recirculation zone associated with the penthouse. To find the height of this recirculation zone, use Equation (5):

$$R = (13.1)^{0.67}(29.5)^{0.33} = 17.1 \text{ ft}$$

Then use Equations (1) and (2):

$$H_c = (0.22)(17.1) = 3.76 \text{ ft}$$

$$X_c = (0.5)(17.1) = 8.55 \text{ ft}$$

With the 1:5 slope of the lower plume boundary shown in Figure 6, the capped stack height in Figure 1 (measured from the main roof) must be

$$h_{sc} = 0.2(98.4 - 52.5 + 8.55) + 3.76 + 13.1 = 27.8 \text{ ft}$$

to avoid the recirculation zone above the penthouse.

Next, check the building wake recirculation zone downwind of the building. The plume must also avoid this region because intake B is located there. The length of this recirculation region is found using Equation (4):

$$L_r = (49.2)^{0.67}(164)^{0.33} = 73.2 \text{ ft}$$

Projecting the downwind corner of this recirculation region upwind with a 1:5 slope to the stack location gives the required height of a no-downwash capped stack above the main roof level as

$$h_{sc} = 0.2(203 + 73.2 - 52.5) = 44.7 \text{ ft}$$

for the plume to avoid the recirculation zone on the downwind side of the building.

The design stack height is set by the condition of avoiding contamination of the building wake, because avoiding the penthouse roof recirculation requires only a 27.8 ft capped stack. Credit for plume rise h_r from the uncapped stack requires calculation of the building wind speed U_H at $H = 49.2$ ft. The minimum allowable design wind speed is the speed that is exceeded 1% of the time at the meteorological station. In this case, for $H_{met} = 32.8$ ft at the airport meteorological station, this 1% wind speed is $U_{met} = 2.5(7.95) = 19.9$ mph $= 1751$ fpm. With the airport in open terrain (see Table 1), and the building in suburban terrain, the wind speed adjustment parameters are $a_{met} = 0.14$ and $\delta_{met} = 900$ ft at the airport, and $a = 0.22$ and $\delta = 1200$ ft at the building. Using Equation (4) in Chapter 24 of the 2009 *ASHRAE Handbook—Fundamentals*, with building height $H = 49.2$ ft,

$$U_H = 1751\left(\frac{900}{32.8}\right)^{0.14}\left(\frac{49.2}{1200}\right)^{0.22} = 1379 \text{ fpm}$$

Because $V_e/U_H = 1770/1379 = 1.28$ is less than 3.0, there is some plume downwash as shown in Figure 3. From Equation (9),

$$h_d = 1.64[3 - 1.0(1.28)] = 2.8 \text{ ft}$$

Then, using Equation (7), the plume rise at the design wind speed and the distance to the closest intake A is calculated to be

$$
\begin{aligned}
x &= 46.4 \text{ ft} \\
H &= 49.2 \text{ ft} \\
d &= 1.64 \text{ ft} \\
V_e &= 1770 \text{ fpm} \\
U_H &= 1379 \text{ fpm} \\
z_o &= 2 \text{ ft} \\
F_m &= 2,106,562 \text{ ft}^4/\text{s}^2 \\
B_j &= 1.11 \\
h_x &= 5.00 \text{ ft} \\
U_H/U_* &= 7.9 \\
h_f &= 2.42 \text{ ft} \\
h_r &= 2.42 \text{ ft}
\end{aligned}
$$

Using these values in Equation (10), the uncapped height h_{sc} is, with the height reduction credit for the 6.3 ft rise and the height addition to account for the 2.8 ft downwash,

$$h_s = 44.7 - 2.4 + 2.8 = 45.1 \text{ ft}$$

As shown in Figure 1, this stack height is measured above the main roof. If stack height is higher than desirable, an alternative is to use dilution calculations. The geometric method does not directly account for dilution within the plume.

EXHAUST-TO-INTAKE DILUTION OR CONCENTRATION CALCULATIONS

Worst-Case Critical Dilution or Maximum Concentration

The geometric stack design procedure does not give a quantitative estimate of the worst-case critical dilution factor D_{crit} at an air intake. If a required dilution can be specified with known stack emissions and required health limits, odor thresholds, or air quality regulations, computing critical dilutions is the preferred method for specifying stack heights. Petersen et al. (2002, 2004) and Smeaton et al. (1991) discuss use of emission information and formulation of dilution requirements in more detail. Exhaust from a single-source dedicated stack may require more atmospheric dilution than a single stack with the same exhausts combined, because emissions are diluted in the exhaust manifold.

This section describes the methods for computing outside dilution of exhausts emitted from a rooftop stack due to atmospheric dispersion processes. The resulting dilution can be converted to contaminant concentration for comparison to odor thresholds or health limits. Dispersion of pollutants from building exhaust depends on the combined effect of atmospheric turbulence in the wind approaching the building and turbulence generated by the building itself. This building-generated turbulence is most intense in and near the flow recirculation zones that occur on the upwind edges of the building (Figures 1 and 5). Dilution of exhaust gas is estimated using design procedures developed for tall isolated stacks (EPA 1995, 2004), with modifications to include the high turbulence levels experienced by a plume diffusing over a building roof in an urban area (Schulman et al. 2000). Halitsky (1982), Hosker (1984), Meroney (1982), and Wilson and Britter (1982) are good references regarding gas diffusion near buildings.

Dispersion models (including physical models) ranging from the simple to very complex are intended to help the designer investigate how pollutants will be distributed in the atmosphere, around the building, and around adjacent buildings and areas. These models identify potential problems that could result in too-high pollution concentrations by air intakes, entrances, or other sensitive areas that can fairly easily be corrected during design by changing design parameters such as exhaust exit velocity, stack location or height, etc. Identifying these problems during the design phase allows for a less expensive, more efficient solution than trying to correct a real problem after the building is completed, when, for example, changing the location and/or height of the stack can be very costly.

Dilution and Concentration Definitions

A building exhaust system releases a mixture of building air and pollutant gas at concentration C_e (mass of pollutant per volume of air) into the atmosphere through a stack or vent on the building. The exhaust mixes with atmospheric air to produce a pollutant concentration C, which may contaminate an air intake or receptor if the concentration is larger than some specified allowable value C_{allow} (see Figure 4). The dilution factor D between source and receptor mass concentrations is defined as

$$D = C_e/C \qquad (11)$$

where

C_e = contaminant mass concentration in exhaust, lb/ft³
C = contaminant mass concentration at receptor, lb/ft³

The dilution increases with distance from the source, starting from its initial value of unity. If C is replaced by C_{allow} in Equation (11), the atmospheric dilution D_{req} required to meet the allowable concentration at the intake (receptor) is

$$D_{req} = C_e/C_{allow} \qquad (12)$$

The exhaust (source) concentration is given by

$$C_e = \dot{m}/Q_e = \dot{m}/(A_e V_e) \qquad (13)$$

where

$\dot{m}$ = contaminant mass release rate, lb/s
$Q_e = A_e V_e$ = total exhaust volumetric flow rate, ft³/s
A_e = exhaust face area, ft²
V_e = exhaust face velocity, ft/s

The concentration units of mass per mixture volume are appropriate for gaseous pollutants, aerosols, dusts, and vapors. The concentration of gaseous pollutants is usually stated as a volume fraction f (contaminant volume/mixture volume), or as ppm (parts per million) if the volume fraction is multiplied by 10^6. The pollutant volume fraction f_e in the exhaust is

$$f_e = Q/Q_e \qquad (14)$$

where Q is the volumetric release rate of the contaminant gas. Both Q and Q_e are calculated at exhaust temperature T_e.

The volume concentration dilution factor D_v is

$$D_v = f_e/f \qquad (15)$$

where f is the contaminant volume fraction at the receptor. If the exhaust gas mixture has a relative molecular mass close to that of air, D_v may be calculated from the mass concentration dilution D by

$$D_v = (T_e/T_a)D \qquad (16)$$

where

T_e = exhaust air absolute temperature, °R
T_a = outdoor ambient air absolute temperature, °R

Many building exhausts are close enough to ambient temperature that volume fraction and mass concentration dilutions D_v and D are equal.

Roof-Level Dilution Estimation Method

This section presents equations for predicting worst-case roof-level dilution of exhaust from a vertical stack on a roof. The equations assume a bell-shaped Gaussian concentration profile in both the vertical and cross-wind horizontal directions. Gaussian profiles have been used in many atmospheric dispersion models, such as those used by the U.S. Environmental Protection Agency. Considering their simplicity, bell-shaped Gaussian concentration profiles in the cross-wind y and vertical z directions at a given horizontal distance x represent atmospheric dispersion remarkably well (Brown et al. 1993).

The dilution equations predict the roof-level dilution D_r, which is the ratio of contaminant concentration C_e at the exit point of the exhaust to the maximum concentration C_r on the plume centerline at roof level, giving $D_r = C_e/C_r$. The centerline of the plume is defined in the x direction, with y the lateral (cross-wind) distance off the plume centerline (axis), and z the vertical. Dilution is affected by three processes:

- Wind carries the plume downwind. The higher the wind speed, the greater the dilution on the plume axis. Wind speed U_H carrying the plume is the wind speed in the undisturbed flow approaching the top of the building.
- Wind turbulence spreads the plume vertically and laterally (cross-wind). Plume spreads in the cross-wind y direction and the vertical z direction are σ_y and σ_z. These plume spreads increase with downwind distance.
- The plume is carried vertically by the initial buoyancy and vertical momentum of the exhaust at the stack exit. The higher the plume,

the greater the dilution at the roof surface. The stronger the wind, the less the plume rises, which may produce less dilution.

Thus, wind speed has two influences: (1) at very low wind speed, the exhaust jet from an uncapped stack rises high above roof level, producing a large exhaust dilution D_r at a given intake location; and (2) at high wind speed, the plume rise is low but the dilution is large because of longitudinal stretching of the plume by the wind. Between these extremes is the critical wind speed $U_{H,crit}$, at which the smallest amount of dilution occurs for a given exhaust and intake location.

Before performing Gaussian dilution calculations on a rooftop, the effect of rooftop obstacles, wind recirculation zones, and intake location(s) must be considered. Dilution depends on the vertical separation ζ between the plume centerline h_{plume} and h_{top}, defined as the highest of the intake, all active obstacles (discussed later), and recirculation zones defined in Equation (1). Vertical separation ζ is defined as

$$\zeta = h_{plume} - h_{top}$$
$$= 0 \text{ if } h_{plume} < h_{top} \qquad (17)$$

Plume centerline h_{plume} versus downwind distance defined as

$$h_{plume} = h_s + h_r - h_d \qquad (18)$$

where h_s is the physical stack height above the roof, h_r is the plume rise versus downwind distance defined in Equation (7), and h_d is the stack wake downwash defined in Equation (9) (see Figure 1).

To determine which rooftop obstacles are considered active in defining h_{top}, start by drawing a line in plan view through the stack location and the intake of interest. All obstacles along this line or one obstacle width laterally (y-direction) from the line are considered active. Obstacles and recirculation zones upstream of the stack and downstream of the intake should also be considered. The value of h_{top} is the higher of the active obstacles, recirculation zones (including zones on top of active obstacles) defined in Equation (1), and the height of the intakes. All of these heights should be referenced to the same roof level used to determine h_s. Once h_{top} is defined and ζ is calculated, dilution at the intake is calculated from

$$D_r(x) = \frac{4U_H \sigma_y \sigma_z}{V_e d_e^2} \exp\left(\frac{\zeta^2}{2\sigma_z^2}\right) \qquad (19)$$

When plume height is less than the height of intakes, active obstacles, and recirculation zones (i.e., $h_{plume} < h_{top}$), then $\zeta = 0$ and dilution should be calculated using Equation (23).

If exhaust gases are hot, buoyancy increases the rise of the exhaust gas mixture and produces lower concentrations (higher dilutions) at roof level. For all exhausts except very hot flue gases from combustion appliances, it is recommended that plume rise from buoyancy be neglected in dilution calculations and stack design on buildings. By neglecting buoyant plume rise, the predicted dilution has an inherent safety factor, particularly at low wind speed, where buoyancy rise is significant.

Cross-Wind and Vertical Plume Spreads for Dilution Calculations

Close to the stack, dispersion is governed by mechanical mixing, and the following equations for lateral and vertical plume spread can used (Cimorelli et al. 2005):

$$\sigma_y = (i_y^2 x^2 + \sigma_o^2)^{1/2} \qquad (20)$$

$$\sigma_z = (i_z^2 x^2 + \sigma_o^2)^{1/2} \qquad (21)$$

where i_y is lateral turbulence intensity, i_z is vertical turbulence intensity, x is distance downwind from the stack, and σ_o is initial source size, normally set equal to $0.35d_e$. The lateral and vertical turbulence intensity can be calculated using the following equations:

$$i_y = 0.75i_x; \quad i_z = 0.5i_x \qquad (22)$$

where

$$i_x = n \ln\left(\frac{90.4}{z_o}\right) / \ln\left(\frac{z}{z_o}\right)$$

$$n = 0.19 + 0.096 \log_{10}z_o + 0.016(\log_{10}z_o)^2$$

where inputs z and z_o are in feet.

The averaging time over which exhaust gas concentration exposures are predicted is important in determining roof-level dilution. The preceding equations provide dilution estimates for a 10 to 15 min averaging time, which corresponds to the averaging time for ACGIH short-term exposure limits. If odors are a concern, peak minimum dilution may be needed; the formula below can be used to estimate dilution for shorter averaging times.

$$(D_r)_s = (D_r)_{15} \times \left(\frac{t_s}{15}\right)^{0.2}$$

where $(D_r)_s$ is the estimate for a shorter averaging time t_s and $(D_r)_{15}$ is the estimate for the 15 min averaging time. For example, assume a predicted dilution of $(D_r)_{15} = 100$. The dilution for a 1 min averaging time would be $100 \times (1/15)^{0.2} = 58$. If estimates for longer averaging times are needed, the preceding estimates should be assumed to be hourly averages and the following conservative multiplication factors (EPA 1992) should be used to obtain the estimated maximum concentrations for 3 h, 8 h, 24 h, or annual averaging times:

Averaging Time	Scaling or Multiplying Factors
3 h	1.1
8 h	1.4
24 h	2.5
Annual	12.5

Equations (17), (18), (19), and (20) imply that dilution does not depend on the location of either the exhaust or intake, only on the horizontal distance x between them and the vertical separation ζ. This is reasonable when both exhaust and intake locations are on the same building wall or on the roof. Dilution can increase if the intake is below roof level on the sidewall of a building and the exhaust stack is located on the roof. Petersen et al. (2004) provide methods for estimating the increased dilution on sidewalls.

Stack Design Using Dilution Calculations

Example 2. In general, a spreadsheet should be designed using the preceding equations to calculate dilution from an exhaust at any specified location on the roof. This example shows the results for a calculation at intake A using the information provided in Example 1. The distance from stack to intake is 46.26 ft, and highest point on intake A is 6.56 ft above the roof. Assume a 20 ft stack height above the roof. Figure 7 provides the calculated dilution versus wind speed. The recommended method involves the following steps: (1) specify the site conditions using Table 1; (2) carry out the calculations outlined in Figure 7 using surface roughness in Table 1 for range of wind speeds; (3) repeat the calculations using a surface roughness half this value and 1.5 times this value; (4) find the lowest dilution for the range of wind speeds of interest for each surface roughness; and (5) determine the overall minimum dilution and use this value to determine the acceptability of the exhaust design.

Dilution from Flush Exhaust Vents with No Stack

For exhaust grilles and louvers on the roof or walls of a building or penthouse, vertical separation $\zeta = 0$ in Equation (19). Combining Equations (19), (20), and (21) gives

$$D_s(x) = \frac{4U_H\sigma_y\sigma_z}{V_e d_e^2} \tag{23}$$

The subscript s in the dilution D_s from a surface exhaust distinguishes it from the roof-level dilution D_r from a stack [Equation (19)].

Minimum critical dilutions for flush exhausts can be calculated using the approximate value $U_{H,crit} = 400$ fpm based on the observation that, for wind speeds less than 400 fpm at roof height, the atmosphere tends to develop high levels of turbulence that increase exhaust-to-intake dilution. For flush roof exhausts with no stack, and for wall intakes, the experiments of Petersen et al. (2004) suggest that D_s at the intake is at least a factor of 2 larger than the dilution at a roof-level intake at the same stretched-string distance S from the stack.

Example 3. The exhaust flow of $Q_e = 3740$ cfm in Example 1 comes from a louvered grille at location A. The exhaust grille is 2.3 ft high and 2.3 ft wide. What is the critical exhaust-to-intake dilution factor at intake B on the downwind wall of the building for an averaging time of 15 min?

Solution: Exhaust grille A has a face area of $A_e = 5.29$ ft^2 and an exhaust velocity V_e of 707 fpm (3.59 m/s). From Equation (8), the effective exhaust diameter is $d_e = [(4)(5.29)/\pi]^{0.5} = 2.60$ ft. The stretched-string distance S from exhaust A to intake B is the sum of the 6.56 ft from the top of A to the top of the penthouse, plus the 23 ft length of the penthouse, plus the sloped line of horizontal length 81.6 ft from the downwind edge of the penthouse roof, and a vertical drop of 13.1 ft to the roof, plus the 19.7 ft to intake B:

$$S = 6.56 + 23.0 + \sqrt{81.6^2 + (13.1 + 19.7)^2} = 117.5 \text{ ft}$$

The critical wind speed is assumed to be $U_{H,crit} = 400$ fpm for capped stacks and flush vents. The critical dilution of exhaust from A at intake

B is calculated from Equation (23). Using a similar method as outlined in Example 2, Figure 8 shows the calculation table results where S and x are assumed equal.

These are conservative estimates because they represent the dilution at a low critical wind speed of 400 fpm, which is 4.5 mph.

Dilution at a Building Sidewall (Hidden) Intakes

Petersen et al. (2004) provided results of an ASHRAE research study that outlined methods for estimating dilution or concentration for visible versus hidden intakes. A hidden intake is typically on a building side wall or on the side wall of a roof obstruction opposite the exhaust source. A visible intake is at roof level or on top of an obstruction, directly above the hidden intake. The basic approach starts following the method for estimating dilution at a rooftop intake. Dilution is calculated at the rooftop location above the sidewall receptor; dilution at this distance is then increased by the factors given in Petersen et al. (2004). A conservative dilution increase factor for most building configurations is 2.0.

EPA Models

In late 2005, the EPA recommended that the AERMOD modeling system (Cimorelli et al. 2005) be used instead of the previously preferred ISC (EPA 1995). The new model includes state-of-the-art boundary layer parameterization techniques, convective dispersion, plume rise formulations, and complex terrain/plume interactions, as well as a building downwash algorithm. AERMOD can be used to calculate short-term (hourly) exposure and long-term (monthly and annual) exposure. Both the short- and long-term models are divided into three source classifications: (1) point source, (2) line source, and (3) area source. For exhaust stack design, the point source is the model of interest. The EPA (2006) guideline also describes a short- and a long-term dry deposition model. AERMOD uses the Gaussian equation to calculate the concentration of the contaminant concentration downwind of the source. The models consider the wind speed profile, use plume rise formulas, calculate dispersions factors (which take into consideration different landscapes, building wakes and downwash, and buoyancy), calculate the vertical distribution, and consider decay of the contaminant. More information on AERMOD and other EPA models can be found at http://www.epa.gov/scram001/. Remember that the EPA models are primarily designed to predict concentration (or dilution) values downwind of the building on which the exhausts are located. For predicting the impact at

Zo =	2.13	ft			n =	0.22									
U_H	σ_θ	i_x	$i_y x$	$i_z x$	σ_y	σ_z	h_d	β_i	U_H/U_*	h_x	h_f	h_r	h_{plume}	ζ	D_r
(fpm)	(ft)	(-)	(ft)	(ft)	(ft)	(ft)	(ft)	(ft)	(-)	(-)	(ft)	(ft)	(ft)	(ft)	(ft)	(-)
984	0.57	0.27	9.43	6.28	9.44	6.28	1.97	0.89	7.85	7.26	4.18	4.18	22.21	15.67	1100
1378	0.57	0.27	9.43	6.28	9.44	6.28	2.81	1.11	7.85	5.00	2.39	2.39	19.58	13.04	591
1771	0.57	0.27	9.43	6.28	9.44	6.28	3.28	1.33	7.85	3.74	1.55	1.55	18.27	11.73	504
													Min		504
Zo =	1.066	ft			n =	0.20									
U_H	σ_θ	i_x	$i_y x$	$i_z x$	σ_y	σ_z	h_d	β_j	U_H/U_*	h_x	h_f	h_r	h_{plume}	ζ	D_r
(fpm)	(ft)	(-)	(ft)	(ft)	(ft)	(ft)	(ft)	(-)	(-)	(ft)	(ft)	(ft)	(ft)	(ft)	(-)
984	0.57	0.23	8.07	5.38	8.09	5.38	1.50	0.89	9.58	7.26	4.62	4.62	22.65	16.11	3194
1378	0.57	0.23	8.07	5.38	8.09	5.38	1.50	1.11	9.58	5.00	2.64	2.64	19.83	13.29	1065
1771	0.57	0.23	8.07	5.38	8.09	5.38	1.50	1.33	9.58	3.74	1.71	1.71	18.43	11.89	745
													Min		745
Zo =	3.198	ft			n =	0.24									
U_H	σ_θ	i_x	$i_y x$	$i_z x$	σ_y	σ_z	h_d	β_j	U_H/U_*	h_x	h_f	h_r	h_{plume}	ζ	D_r
(fpm)	(ft)	(-)	(ft)	(ft)	(ft)	(ft)	(ft)	(-)	(-)	(ft)	(ft)	(ft)	(ft)	(ft)	(-)
984	0.57	0.30	10.39	6.93	10.41	6.93	1.50	0.89	6.83	7.26	3.90	3.90	21.93	15.39	704
1378	0.57	0.30	10.39	6.93	10.41	6.93	1.50	1.11	6.83	5.00	2.23	2.23	19.42	12.88	470
1771	0.57	0.30	10.39	6.93	10.41	6.93	1.50	1.33	6.83	3.74	1.45	1.45	18.17	11.63	438
													Min		438
													Overall Minimum		438

Fig. 7   Spreadsheet for Example 2

the building intakes, operable windows, and entrances, alternative modeling methods are required.

Wind Tunnel Modeling

Wind tunnel modeling is often the preferred method for predicting maximum concentrations for stack designs and locations of interest when energy and equipment optimization is desired. It is the recommended approach because it gives the most accurate estimates of concentration levels in complex building environments. A wind tunnel modeling study is like a full-scale field study, except it is conducted before a project is built. Typically, a scale model of the building under evaluation, along with the surrounding buildings and terrain within a 1000 ft radius, is placed in an atmospheric boundary layer wind tunnel. A tracer gas is released from the exhaust sources of interest, and concentration levels of this gas are then measured at receptor locations of interest (e.g., air intakes, operable windows) and converted to full-scale concentration values. Next, these values are compared against the appropriate health or odor design criteria to evaluate the acceptability of the exhaust design. Snyder (1981) and Petersen and Cochran (2008) provide more information on scale-model simulation and testing methods. Scale modeling is also discussed in Chapter 24 of the 2009 *ASHRAE Handbook—Fundamentals*.

Wind-tunnel studies are highly technical, so care should be taken when selecting a dispersion modeling consultant. Factors such as past experience and staff technical qualifications are extremely important.

Computer Simulations Using Computational Fluid Dynamics (CFD)

CFD models are used successfully to model internal flow paths in areas such as vivariums and atriums, as well as in external aerodynamics for the aerospace industry. Aerospace CFD turbulence models, however, are ill suited for modeling atmospheric turbulence in complex full-scale building environments because of the differing geometric scales. More information on CFD modeling is in Chapter 24 of the 2009 *ASHRAE Handbook—Fundamentals*.

Based on the current state of the art, CFD models should be used with extreme caution when modeling exhaust plumes from laboratory pollutant sources. Currently, CFD models can both over- and underpredict concentration levels by orders of magnitude, leading to potentially unsafe designs. If a CFD study is conducted for such an application, supporting full-scale or wind tunnel validation studies

should be carried out. Various commercial software packages are available for CFD-driven airflow analysis. Most have advanced user interfaces and resulting visualization capabilities, as well as sophisticated physical models and solver options. Usually, commercial software includes advanced technical user support provided by vendor specialists. Several open-source research codes are available, as well, but require a much greater user insight into the underlying solution methods and hardware platforms. Normally, no user support or problem-specific validation data are available. Regardless of the software package choice, obtaining an accurate numerical solution requires expertise, training, and understanding of the fundamental aspects of CFD algorithm construction and implementation.

OTHER CONSIDERATIONS

Annual Hours of Occurrence of Highest Intake Contamination

To assess the severity of the hazard caused by intake contamination, it is useful to know the number of hours per year that exhaust-to-intake dilution is likely is lower than some allowable minimum dilution. The first step in making a frequency-of-occurrence estimate is to use the method outlined in Examples 2 and 3 to estimate the minimum dilution versus wind speed at the intake or receptor location of interest. Next, the wind speed range at which the dilution is unacceptable is specified. Weather data are then used to calculate the number of hours per year that wind speeds fall in this range for the 22.5° wind direction sector centered on a line joining the exhaust and intake.

Combined Exhausts

When exhaust from several collecting stations is combined in a single vent (as recommended in the section on Exhaust Stack and Air Intake Design Strategies), the plume rise increases due to the higher mass flow in the combined jet and results in significantly lower roof-level intake concentration C_r compared to that from separate exhausts. Where possible, exhausts should be combined before release to take advantage of this increase in overall dilution. For example, consider a single fume hood exhaust stack at 1000 cfm with a 3000 fpm exit velocity ($A_e = 0.33$ ft) versus 10 such fume hoods combined into a single stack at 10,000 cfm () and 3000 fpm ($A_e = 3.3$ ft). The F_m term in Equation (7) is proportional to the exhaust velocity times the exhaust area. For this example, F_m would

$Z_o =$	2.13	ft		$n =$	0.22					
U_H	σ_0	i_x	$i_y\,x$	$i_z\,x$	σ_y	σ_z	h_d	ζ	D_r
(fpm)	(ft)	(-)	(ft)	(ft)	(ft)	(ft)	(ft)	(ft)	(-)
394	0.91	0.27	23.93	15.96	23.95	15.96	3.15	-	126

$Z_o =$	1.066	ft		$n =$	0.20					
U_H	σ_0	i_x	$i_y\,x$	$i_z\,x$	σ_y	σ_z	h_d	ζ	D_r
(fpm)	(ft)	(-)	(ft)	(ft)	(ft)	(ft)	(ft)	(ft)	(-)
394	0.91	0.23	20.48	13.66	20.50	13.66	2.37	-	93

$Z_o =$	3.198	ft		$n =$	0.24					
U_H	σ_0	i_x	$i_y\,x$	$i_z\,x$	σ_y	σ_z	h_d	ζ	D_r
(fpm)	(ft)	(-)	(ft)	(ft)	(ft)	(ft)	(ft)	(ft)	(-)
394	0.91	0.30	26.38	17.59	26.40	17.59	2.37	-	153
								Overall Minimum	93

Fig. 8   Spreadsheet for Example 3

increase by a factor 10. Based on Equation (7), the plume rise would increase by a factor of $(10)^{0.33}$, or 2.1.

Ganged Exhausts

Greater plume rise and dilution can be achieved by grouping individual stacks close together. Petersen and Reifschneider (2008) summarized an ASHRAE research study on this topic and provided a method for calculating the plume rise and subsequent dilution for ganged stacks, to optimize stack arrangements. They found that, when stacks are nearly touching, the momentum terms for the individual stacks can add together and Equation (7) can be used to compute plume rise for ganged arrangements. If stacks are situated up- and downwind of each other, the F_m terms can also be added for quite large separation distances. If the stacks are not up- or downwind of each other, the paper presents an equation that can used to estimate the fraction of the momentum that can be added. This is a particularly advantageous strategy when a dedicated low-flow exhaust is needed that has toxic chemicals. This stack can be placed next to any high-flow stack and achieve the same plume rise and resulting high dilution.

Influence of Architectural Screens on Exhaust Dilution

Architectural screens are often placed around rooftop equipment to reduce noise or hide equipment. Unfortunately, these screens interact with windflow patterns on the roof and can adversely affect exhaust dilution and thermal efficiency of equipment inside the screen. This section describes a method to account for these screens by modifying the physical stack height h. Architectural screens generate flow recirculation regions similar to those shown downwind of the building and penthouse in Figure 1. These screens are often made of porous materials with mesh or louvers, which influence the height of the recirculation cavity above the screens.

To incorporate the effect of architectural screens into existing dilution prediction equations, a stack height reduction factor F_h is introduced. Using the equation developed by Petersen et al. (1999), stack height h_s above the roof must be multiplied by F_h when the stack is enclosed within a screen. Effective stack height $h_{s,eff}$ measured above roof level is

$$h_{s,eff} = F_h h_s \qquad (24)$$

The stack height reduction factor F_h is directly related to screen porosity P_s. For stack heights above the top of the screen that are less than 2.5 times the height of the screen,

$$F_h = 0.81 P_s + 0.20 \qquad (25)$$

where porosity is

$$P_s = \frac{\text{Open area}}{\text{Total area}} \qquad (26)$$

F_h is applied directly to h_s after the required stack height has been calculated, and should be included where the physical stack height is less than 2.5 times the height of the architectural screen.

Example 4. Calculate the required stack height for an uncapped stack with a height of 15.5 ft above roof level, surrounded by a 10 ft high, 50% porous architectural screen.

Solution: To determine the effect of the 50% porous screen, use Equation (25) with $P_s = 0.5$

$$F_h = 0.81(0.5) + 0.2 = 0.605$$

The screen has reduced the effective stack height from its actual height of 15.5 ft to

$$h_{s,eff} = 0.605(15.5) = 9.38 \text{ ft}$$

When the effect of the 50% porous screen is added, the 15.5 ft stack height is found to behave like a 9.38 ft stack. The actual stack height h_s must be increased to account for the screen's effect. This is most easily done by dividing h_s by the stack height reduction factor:

$$h_{s,corrected} = \frac{h_s}{F_h} = \frac{15.5}{0.605} \, 25.6 \text{ ft}$$

The corrected 25.6 ft high stack effectively behaves like a 15.5 ft tall stack, and should produce the same dilution at downwind air intakes. The correct height should be used for input when estimating dilution.

Emissions Characterization

Typical exhaust sources of concern are fume hoods, emergency generators, kitchens, vivariums, loading docks, traffic, cooling towers, and boilers. Chemical emissions from each source should be characterized to determine a design criterion, or critical concentration. Three types of information are needed to characterize the emissions: (1) a list of the toxic or odorous substances that may be emitted, (2) the health limits and odor thresholds for each emitted substance, and (3) the maximum potential emission rate for each substance.

Recommended health limits C_{health} are based on ANSI/AIHA *Standard* Z9.5, which specifies air intake concentrations no higher than 20% of acceptable indoor concentrations for routine emission and 100% of acceptable indoor concentrations for accidental releases. Acceptable indoor concentrations are frequently taken to be the minimum short-term exposure limits (STEL) from the American Conference of Governmental Industrial Hygienists (ACGIH), the Occupational Safety and Health Administration (OSHA), and the National Institute of Occupational Safety and Health (NIOSH), as listed in ACGIH. ACGIH also provides odor thresholds.

For laboratories, emission rates are typically based on small-scale accidental releases in fume hoods or in room, either liquid spills or emptying of a lecture bottle of compressed gas. Evaporation from liquid spills is computed from equations in EPA (1992) based on a worst-case spill in a fume hood or a room. Compressed gas leaks typically assume a fractured lecture bottle empties in one minute. For other sources, such as emergency generators, boilers, and vehicles, chemical emissions rates are often available from the manufacturer.

For general laboratory design purposes, Chapter 16 provides an example emission characterization (i.e., design criterion). A 16 cfm chemical emission rate (e.g., from a liquid spill or lecture bottle fracture) is assumed, along with a limiting concentration of 3 mg/kg or less at an intake. For dispersion modeling purposes, the emission characterization can be expressed as in SI units as 400 μg/m³ per g/s, or in dilution units of 1:5300 per 1000 cfm of exhaust flow. Chapter 16 includes the following disclaimers regarding this design criterion: (1) laboratories using extremely hazardous substances should conduct a chemical-specific analysis based on published health limits, (2) a more lenient limit may be justified for laboratories with low levels of chemical usage, and (3) project-specific requirements must be developed in consultation with the safety officer.

Chapter 16's criterion may be put into perspective by considering the as-manufactured and as-installed chemical hood containment requirements outlined in ANSI/AIHA *Standard* Z9.5-2003 (i.e., a concentration at a manikin outside the chemical hood of 0.05 ppm or less for as-manufactured, and 0.10 ppm or less for as-installed, with a 0.14 cfm accidental release in the hood as measured using the ANSI/ASHRAE *Standard* 110-1995 test method). The as-manufactured requirement is equivalent to a design criterion of 750 μg/m³ per g/s, and the as-installed requirement is equivalent to a design criterion of 1500 μg/m³ per g/s. Hence, the ASHRAE criterion for a manikin representing a worker outside the chemical hood is 1.9 to 3.8 times less restrictive than that for the air intake or other outdoor locations. It seems reasonable that the air intake has more strict criteria, because the worker at the chemical hood can shut the hood or walk away to avoid adverse exposure. Also, the ANSI/ASHRAE *Standard* 110-1995 test is not necessarily a worst-case exposure scenario for the worker.

SYMBOLS

A_e = stack or exhaust exit face area, ft^2

B_L = larger of two upwind building face dimensions H and W, ft

B_s = smaller of two upwind building face dimensions H and W, ft

C = contaminant mass concentration at receptor at ambient air temperature T_e, Equation (11), lb/ft^3

C_{allow} = allowable concentration of contaminant at receptor, Equation (12)

C_e = contaminant mass concentration in exhaust at exhaust temperature T_e, Equation (11), lb/ft^3

D = dilution factor between source and receptor mass concentrations, Equation (11)

D_{crit} = critical dilution factor at roof level for uncapped vertical exhaust at critical wind speed $U_{H,crit}$ that produces smallest value of D_r for given exhaust-to-intake distance S and stack height h_s

D_r = roof-level dilution factor D at given wind speed for all exhaust locations at same fixed distance S from intake, Equation (19)

D_{req} = atmospheric dilution required to meet allowable concentration of contaminant C_{allow}, Equation (12)

D_S = dilution at a wall or roof intake from a flush exhaust grille or louvered exhaust, Equation (23)

D_v = dilution factor between source and receptor using volume fraction concentrations, Equation (15)

d_e = effective exhaust stack diameter, Equation (8), ft

F_h = stack height adjustment factor to adjust existing stack height above screen for influence of screen of exhaust gas dilution, Equation (25)

F_m = momentum flux, ft^4/s^2

f = contaminant volume concentration fraction at receptor; ratio of contaminant gas volume to total mixture volume, Equation (15), ppm × 10^{-6}

f_e = contaminant volume concentration fraction in exhaust gas; ratio of contaminant gas volume to total mixture volume, Equation (14), ppm × 10^{-6}

H_c = maximum height above roof level of upwind roof edge flow recirculation zone, Equation (1), ft

H_{met} = anemometer height, ft

h_{crit} = height of plume above roof level with plume rise and downwash calculated at the critical wind speed U_{crit}, ft

h_d = downwash correction to be subtracted from stack height, Equation (9), ft

h_f = final plume rise, ft

h_{plume} = final plume height, Equation (18), ft

h_r = plume rise of uncapped vertical exhaust jet, Equation (7), ft

h_s = physical exhaust stack height (typically above roof unless otherwise specified), ft

h_{sc} = required height of capped exhaust stack to avoid excessive intake contamination, Equation (10), ft

$h_{s,eff}$ = effective exhaust stack height above roof on which it is located, corrected for an architectural screen surrounding the stack, Equation (24), ft

h_{top} = height of highest of intake, active obstacle, or recirculation zone on a rooftop between the stack and intake, Equation (17), ft

h_x = plume rise at downwind distance x, ft

L = length of building in wind direction, Figure 5, ft

L_c = length of upwind roof edge recirculation zone, Equation (3), ft

L_r = length of flow recirculation zone behind rooftop obstacle or building, Equation (4), ft

$\dot{m}$ = contaminant mass release rate, Equation (13), lb/s

P_s = porosity of an architectural screen near a stack, Equation (26)

Q = contaminant volumetric release rate, Equation (14), ft^3/s

Q_e = total exhaust volumetric flow rate, Equation (13), ft^3/s

R = scaling length for roof flow patterns, Equation (5), ft

S = stretched string distance from exhaust to intake, ft

T_a = outdoor ambient air absolute temperature, Equation (16), °R

t_{avg} = time interval over which receptor (intake) concentrations are averaged in computing dilution, Equation (20), min

T_e = exhaust air mixture absolute temperature, Equation (16), °R

U_H = mean wind speed at height H of upwind wall in undisturbed flow approaching building, Equation (7), ft/s

$U_{H,crit}$ = critical wind speed that produces smallest roof-level dilution factor D_{crit} for uncapped vertical exhaust at given X and h_s, ft/s

V_e = exhaust gas velocity, Equation (13), ft/s

W = width of upwind building face, ft

X_c = distance from upwind roof edge to H_c, Equation (2), ft

x = horizontal distance from upwind roof edge where recirculation region forms in direction of wind, ft

x = downwind horizontal distance from center of stack, Equations (20) and (21), ft

y = cross-wind distance off the plume centerline, ft

z = vertical distance, ft

Z_1 = height of flow recirculation zone boundary above roof, Figure 6, ft

Z_2 = height of high-turbulence zone boundary above roof, Figure 6, ft

Z_3 = height of roof edge wake boundary above the roof, Equation (6) and Figure 6, ft

Greek

β = capping factor; 1.0 for vertical uncapped roof exhaust; 0 for capped, louvered, or downward-facing exhaust

β_j = jet entrainment coefficient

ζ = vertical separation above h_{top}, Equation (17), ft

σ_o = standard deviation of initial plume spread at the exhaust used to account for initial dilution, Equation (20), ft

σ_y = standard deviation of cross-wind plume spread, Equation (20), ft

σ_z = standard deviation of vertical plume spread, Equation (21), ft

REFERENCES

AIHA. 2003. Laboratory ventilation. ANSI/AIHA *Standard* Z9.5-2003. American Industrial Hygiene Association, Fairfax, VA.

AMCA. 2009. Application manual for air louvers. AMCA *Publication* 501-09. Air Movement and Control Association, Arlington Heights, IL.

ASHRAE. 2000. Minimizing the risk of Legionellosis associated with building water systems. ASHRAE *Guideline* 12-2000.

ASHRAE. 2003. *Risk management guidance for health, safety, and environmental security under extraordinary incidents.*

ASHRAE. 2010. Ventilation for acceptable indoor air quality. ANSI/ASHRAE *Standard* 62.1-2010.

Briggs, G.A. 1984. Plume rise and buoyancy effects. In *Atmospheric Science and Power Production*, D. Randerson, ed. U.S. Department of Energy DOE/TIC-27601 (DE 84005177), Washington, D.C.

Brown, M., S.P. Arya, and W.H. Snyder. 1993. Vertical dispersion from surface and elevated releases: An investigation of a non-Gaussian plume model. *Journal of Applied Meteorology* 32:490-505.

Changnon, S.A. 1966. Selected rain-wind relations applicable to stack design. *Heating, Piping, and Air Conditioning* 38(3):93.

Cimorelli, A.J., S.G. Perry, A. Venkatram, J.C. Weil, R.J. Paine, R.B. Wilson, R.F. Lee, W.D. Peters, and R.W. Brode. 2005. AERMOD: A dispersion model for industrial source applications. Part I: General model formulation and boundary layer characterization. *Journal of Applied Meteorology* 44:682-693.

EPA. 1992. *Workbook of screening techniques for assessing impacts of toxic air pollutants (revised).* EPA-454/R-92-024. U.S. Environmental Protection Agency, Office of Air Quality, Planning and Standards, Research Triangle Park, NC.

EPA. 1995. *User's guide for the Industrial Source Complex (ISC3) dispersion models,* vol. 2: *Description of model algorithms.* EPA-454/B-95003B. U.S. Environmental Protection Agency, Research Triangle Park, NC.

EPA. 2004. *AERMOD: Description of model formulation.* EPA-454/R-03-004, September. U.S. Environmental Protection Agency, Research Triangle Park, NC.

EPA. 2006. *Addendum to user's guide for the AMS/EPA Regulatory Model—AERMOD,* U.S. Environmental Protection Agency, Washington, D.C.

Halitsky, J. 1982. Atmospheric dilution of fume hood exhaust gases. *American Industrial Hygiene Association Journal* 43(3):185-189.

Hitchings, D.T. 1997. Laboratory fume hood and exhaust fan penthouse exposure risk analysis using the ANSI/ASHRAE *Standard* 110-1995 and other tracer gas methods. *ASHRAE Transactions* 103(2).

Hosker, R.P. 1984. Flow and diffusion near obstacles. In *Atmospheric Science and Power Production*, D. Randerson, ed. U.S. Department of Energy DOE/TIC-27601 (DE 84005177).

IAPMO. 1997a. *Uniform mechanical code.* International Association of Plumbing and Mechanical Officials, Ontario, California.

IAPMO. 1997b. *Uniform plumbing code.* International Association of Plumbing and Mechanical Officials, Ontario, California.

ICC. 2006. *International building code.* International Code Council, Falls Church, VA.

Knutson, G.W. 1997. Potential exposure to airborne contamination in fan penthouses. *ASHRAE Transactions* 103(2).

Meroney, R.N. 1982. Turbulent diffusion near buildings. *Engineering Meteorology* 48:525.

NFPA. 2004. Fire protection for laboratories using chemicals. ANSI/NPPA *Standard* 45-04. National Fire Protection Association, Quincy, MA.

NFPA. 2006. Recommended practice for smoke-control systems. NPPA *Standard* 92A-2006. National Fire Protection Association, Quincy, MA.

Petersen, R.L. and B.C. Cochran. 2008. Wind tunnel modeling of pollutant dispersion. Ch. 24A in *Air quality modeling.* EnviroComp Institute and Air and Waste Management Association.

Petersen, R.L. and J.D. Reifschneider. 2008. The effect of ganging on pollutant dispersion from building exhaust stacks. *ASHRAE Transactions* 114(1).

Petersen, R.L. and J.W. LeCompte. 2002. Exhaust contamination of hidden versus visible air intakes. *Final Report,* ASHRAE RP-1168.

Petersen, R.L., M.A. Ratcliff, and J.J. Carter. 1999. Influence of architectural screens on rooftop concentrations due to effluent from short stacks. *ASHRAE Transactions* 105(1).

Petersen, R.L., B.C. Cochran, and J.J. Carter. 2002. Specifying exhaust and intake systems. *ASHRAE Journal* 44(8):30-35.

Petersen, R.L., J.J. Carter, and J.W. LeCompte. 2004. Exhaust contamination of hidden vs. visible air intakes. *ASHRAE Transactions* 110(1).

Puckorius, P.R. 1999. Update on Legionnaires' disease and cooling systems: Case history reviews—What happened/what to do and current guidelines. *ASHRAE Transactions* 105(2).

Ratcliff, M.A., R.L. Petersen, and B.C. Cochran. 1994. Wind tunnel modeling of diesel motors for fresh air intake design. *ASHRAE Transactions* 100(2):603-611.

Rock, B.A. and K.A. Moylan. 1999. Placement of ventilation air intakes for improved IAQ. *ASHRAE Transactions* 105(1).

Saathoff, P., L. Lazure, T. Stathopoulos, and H. Peperkamp. 2002. The influence of a rooftop structure on the dispersion of exhaust from a rooftop stack. Presented at the 2002 ASHRAE Meeting, Honolulu.

Schulman, L., D. Strimaitis, and J. Scire. 2000. Development and evaluation of the PRIME plume rise and building downwash model. *Journal of the Air and Waste Management Association* 50:378-390.

Seem, J.E., J.M. House, and C.J. Klaassen. 1998. Volume matching control: Leave the outside air damper wide open. *ASHRAE Journal* 40(2): 58-60.

Smeaton, W.H., M.F. Lepage, and G.D. Schuyler. 1991. Using wind tunnel data and other criteria to judge acceptability of exhaust stacks. *ASHRAE Transactions* 97(2):583-588.

Snyder, W.H. 1981. Guideline for fluid modeling of atmospheric diffusion. EPA 600/8-81-009. U.S. Environmental Protection Agency, Environmental Sciences Research Laboratory, Office of Research and Development, Research Triangle Park, NC.

Vanderheyden, M.D. and G.D. Schuyler. 1994. Evaluation and quantification of the impact of cooling tower emissions on indoor air quality. *ASHRAE Transactions* 100(2):612-620.

Wilson, D.J. 1979. Flow patterns over flat roofed buildings and application to exhaust stack design. *ASHRAE Transactions* 85:284-295.

Wilson, D.J. 1982. Critical wind speeds for maximum exhaust gas reentry from flush vents at roof level intakes. *ASHRAE Transactions* 88(1):503-513.

Wilson, D.J. and R.E. Britter. 1982. Estimates of building surface concentrations from nearby point sources. *Atmospheric Environment* 16: 2631-2646.

Wilson, D.J. and G. Winkel. 1982. The effect of varying exhaust stack height on contaminant concentration at roof level. *ASHRAE Transactions* 88(1):513-533.

Wilson, D.J., I. Fabris, and M.Y. Ackerman. 1998a. Measuring adjacent building effects on laboratory exhaust stack design. *ASHRAE Transactions* 104(2):1012-1028.

Wilson, D.J., I. Fabris, J. Chen, and M.Y. Ackerman. 1998b. Adjacent building effects on laboratory fume hood exhaust stack design. *Final Report,* ASHRAE RP-897.

BIBLIOGRAPHY

Chui, E.H. and D.J. Wilson. 1988. Effects of varying wind direction on exhaust gas dilution. *Journal of Wind Engineering and Industrial Aerodynamics* 31:87-104.

EPA. 2003. *AERMOD: Latest feature and evaluation results.* EPA-454/R-03-003. U.S. Environmental Protection Agency, Research Triangle Park, NC.

Gregoric, M., L.R. Davis, and D.J. Bushnell. 1982. An experimental investigation of merging buoyant jets in a crossflow. *Journal of Heat Transfer, Transactions of ASME* 104:236-240.

Li, W.W. and R.N. Meroney. 1983. Gas dispersion near a cubical building. *Journal of Wind Engineering and Industrial Aerodynamics* 12:15-33.

McElroy, J.L. and F. Pooler. 1968. *The St. Louis dispersion study.* U.S. Public Health Service, National Air Pollution Control Administration.

Petersen, R.L., B.C. Cochran, and J.J. Carter. 2002. Specifying exhaust and intake systems. *ASHRAE Journal* 44(8):30-35.

Petersen, R.L., J.J. Carter, and B.C. Cochran. 2005. Modeling exhaust dispersion for specifying acceptable exhaust/intake designs. *Laboratories for the 21st Century Best Practices Guide,* DOE/GO-102005-2104. U.S. Environmental Protection Agency, Washington, D.C. http://www.nrel.gov/docs/fy05osti/37601.pdf.

Snyder, William H., and R.E. Lawson. 1991. Fluid modeling simulation of stack-tip downwash for neutrally buoyant plumes. *Atmospheric Environment* 25A.

Wollenweber, G.C. and H.A. Panofsky. 1989. Dependence of velocity variance on sampling time. *Boundary Layer Meteorology* 47:205-215.

CONTROL OF GASEOUS INDOOR AIR CONTAMINANTS

THE purpose of gas phase filtration is to remove from the air contaminants that would adversely affect the occupants, processes, or contents of a space. The effects are problematic at different concentration levels for different contaminants. There are four categories of harmful effects: toxicity, odor, irritation, and material damage. In most cases, contaminants become annoying through irritation or odor before they reach levels toxic to humans, but this is not always true. For example, the potentially deadly contaminant carbon monoxide has no odor.

Indoor gaseous contaminants can sometimes be controlled with ventilation air drawn from outdoors, diluting the contaminants to acceptable levels. However, available outdoor air may contain undesirable gaseous contaminants at unacceptable concentrations. If so, it requires treatment by gaseous contaminant removal equipment before being used for ventilation. In addition, minimizing outdoor airflow by using a high recirculation rate and filtration is an attractive means of energy conservation. However, recirculated air cannot be made equivalent to fresh outdoor air by removing only particulate contaminants. Noxious, odorous, and toxic contaminants must also be removed by gaseous contaminant control equipment, which is frequently different from particulate filtration equipment.

This chapter covers design procedures for gaseous contaminant control for occupied spaces only. Procedures discussed are appropriate to control odors and gaseous irritants. Control of contaminants for the express purpose of protecting building occupants (whether against deliberate attack or industrial accidents) or to protect artifacts (such as in museums) requires application of the same design principles, but applied more rigorously and with great emphasis on having specific design and performance data, providing redundancy, and added engineering safety factors. Design for protection is not a focus of this chapter, although published design guidance is included and referenced; for more detail, see Chapter 59. Aspects of air-cleaning design for museums, libraries, and archives are included in Chapter 23 of this volume, and control of gaseous contaminants from industrial processes and stack gases is covered in Chapter 29 of the 2008 *ASHRAE Handbook—HVAC Systems and Equipment*.

TERMINOLOGY

The terminology used in control of gaseous air contaminants is specific to the field, and the meaning of some terms familiar from particle filtration is slightly different. In particular, gaseous contaminant technology performance is a function of (1) the specific contaminant, (2) its concentration, (3) airflow rate, and (4) environmental conditions. Several methods of measuring the performance of a gaseous control device, some unique to this application, are defined in the following.

Absorption. Process of one material being retained by another. Usually it is the physical dissolving of a gas, liquid, or solid in a liquid. It is important to distinguish absorption from the surface phenomenon of **adsorption**, which is one of the most important processes in operation of air cleaners that remove gaseous contaminants.

Adsorption, physical. Attraction of a contaminant to the surface, both outer surface and inner pore surface, of adsorbent media by physical forces (Van der Waals forces).

Activity. Mass of contaminant contained in a physical adsorbent at saturation, expressed as a percentage or fraction of the adsorbent mass (i.e., grams contaminant/grams adsorbent). Activity is an equilibrium property under particular challenge conditions, and is not a function of airflow. (In most cases, commercial bed filters are changed for efficiency reasons well before the adsorbent is saturated.) If a saturated adsorbent bed is then exposed to clean air, some of the adsorbed contaminant will desorb. Activity is generally greater than retentivity.

Breakthrough. While removing gaseous contaminants from an airstream passing through a media bed, the point at which downstream contaminant concentration is measurable and begins to rise rapidly.

Breakthrough curve. Plot of contaminant penetration versus time.

Breakthrough time. Operating time (at constant operating conditions) before a certain penetration is achieved. For instance, the 10% breakthrough time is the time between beginning to challenge a physical adsorbent or chemisorbent and the time at which air discharged contains 10% of the contaminant feed concentration. Continued operation leads to 50% and eventually to 100% breakthrough, at which point a physical adsorbent is saturated. For a chemisorbent, the media is exhausted. (Some commercial devices are designed to allow some of the challenge gas to bypass the adsorbent. These devices break through immediately, and breakthrough time, as defined here, does not apply.)

Catalyst. Any substance of which a small proportion notably affects the rate of a chemical reaction without itself being consumed or undergoing a chemical change. Most catalysts accelerate reactions, but a few retard them (negative catalysts, or inhibitors).

Channeling. Disproportionate or uneven flow of fluid (gas or liquid) through passages of lower resistance, which can occur in fixed beds or columns of granular media because of nonuniform packing, irregular sizes and shapes of media, gas pockets, wall effects, and other causes.

Challenge. Airstream containing contaminant(s) of interest that is fed to the air cleaner.

Chemisorption (chemical adsorption). Binding of a contaminant to the surface of a solid by forces with energy levels approximately those of a chemical bond.

Concentration. Quantity of one substance dispersed in a defined amount of another.

The preparation of this chapter is assigned to TC 2.3, Gaseous Air Contaminants and Gas Contaminant Removal Equipment.

Density, apparent (bulk density). Mass under specified conditions of a unit volume of a solid physical adsorbent or chemisorbent, including its pore volume and interparticle voids.

Efficiency. (1 – Penetration); usually expressed as a percentage or decimal fraction.

Efficiency curve. Plot of contaminant removal efficiency against time for a particular challenge concentration and airflow.

HEPA filter. High-efficiency particle air filter.

Mass transfer zone. Depth of physical adsorption or chemisorption media required to remove essentially all of an incoming contaminant; dependent on type of media, media granule size, contaminant nature, contaminant inlet concentration, and environmental conditions.

Mean particle diameter. Weighted average particle size, in millimeters, of a granular adsorbent; computed by multiplying the percent retained in a size fraction by the respective mean sieve openings, summing these values, and dividing by 100.

Media. Granular or pelletized physical adsorbent (or chemisorbent) used in gaseous contaminant control equipment. Also used to refer to a material (e.g., a nonwoven) that contains a physical adsorbent or chemisorbent.

Penetration. Ratio of breakthrough (downstream) concentration to challenge (inlet) concentration, usually expressed as a percentage or decimal fraction. Unlike particulate filters, physical adsorbents and chemisorbents both decline in efficiency as they load. The decline can be very sudden, and is usually not linear with time.

Pressure drop. Difference in pressure between two points in an airflow system, caused by frictional resistance to airflow in a duct, filter, or other system component such as a media bed or air-cleaning device.

Removal efficiency. Measure of amount of challenge gas removed at a given time by physical and/or chemical means.

Residence time. Theoretical time period that a contaminant molecule is within the boundaries of the media bed of a physical adsorbent or chemisorbent. The longer the residence time, the higher the efficiency, and the longer the bed life. For gaseous contaminant control equipment, residence time is computed as

$$\text{Residence time} = \frac{\text{Bed area exposed to airflow} \times \text{Bed depth}}{\text{Airflow rate}} \quad (1)$$

For commercial gaseous contaminant air cleaners, residence time computation neglects the fact that a significant fraction of the volume of the bed is occupied by the media. For example, a unitary adsorber containing trays totaling 40 ft^2 media in a 1 in. deep bed, challenged at 2000 cfm, has a residence time of 0.1 s. Given this definition, a deeper media bed, lower airflow rate, or media beds in series increase residence time and thus performance. Because gaseous contaminant air cleaners all tend to have approximately the same granule size, residence time is a generally useful indicator of performance. In some engineering disciplines, the media volume is subtracted from the nominal volume of packed beds when calculating residence time. This gives a shorter residence time value and is not normally used for HVAC.

Different ways of arranging the media, different media, or different media granule sizes all change the residence time. The geometry and packaging of some technologies makes computation of residence time difficult. For example, the flow pattern in pleated fiber-carbon composite media is difficult to specify, making residence time computation uncertain. Therefore, although residence times can be computed for partial-bypass filters, fiber-adsorbent composite filters, or fiber-bonded filters, they cannot be compared directly and may serve more as a rating than as an actual residence time. Manufacturers might publish equivalent residence time values that say that a particular physical adsorbent or chemisorbent performs the same as a traditional deep-bed air cleaner, but no standard test exists to verify such a rating.

Retentivity. Amount of a particular contaminant remaining in a physical adsorbent after a saturated bed reaches equilibrium in clean air, usually stated as a percentage or fraction of the adsorbent mass. Retentivity represents the ability of an adsorbent to resist desorption of the contaminant. Retentivity is generally less than activity.

Saturation. State of a physical adsorbent when it contains all the contaminant it can hold at the challenge concentration, temperature, and humidity of operation.

Vapor (vapor-phase contaminant). Substance in gas form that naturally occurs as a solid or liquid at the temperature and pressure of the location where it is present.

VOCs. Volatile organic compounds.

GASEOUS CONTAMINANTS

Ambient air contains nearly constant amounts of nitrogen (78% by volume), oxygen (21%), and argon (0.9%), with varying amounts of carbon dioxide (about 0.04%) and water vapor (up to 3.5%). In addition, trace quantities of inert gases (neon, xenon, krypton, helium, etc.) are always present.

Gases and vapors other than these natural constituents of air are usually considered to be gaseous contaminants. Their concentrations are almost always small, but they may have serious effects on building occupants, construction materials, or contents. Removing these gaseous contaminants is often desirable or necessary.

Sources of nonindustrial contaminants are discussed in Chapter 11 in the 2009 *ASHRAE Handbook—Fundamentals*. However, for convenience, data on some of the contaminants in cigarette smoke (Table 1), and some common contaminants emitted by building materials (Table 2), indoor combustion appliances (Table 3), and occupants (Table 4) are provided here.

Table 5 gives typical outdoor concentrations for gaseous contaminants at urban sites; however, these values may be exceeded if the building under consideration is located near a fossil fuel power plant, refinery, chemical production facility, sewage treatment plant, municipal refuse dump or incinerator, animal feed lot, or other major source of gaseous contaminants. If such sources have a significant influence on the intake air, a field survey or dispersion model must be run. Many computer programs have been developed to expedite such calculations.

Table 1 Major Contaminants in Typical Cigarette Smoke

Contaminant	Weighted Mean ETS Generation Rate, µg/cigarette	Weighted Standard Error, µg/cigarette	Method
Carbon monoxide	55,101	1,064	Nondispersive IR
Ammonia	4,148	107	Cation exchange cartridge
Acetaldehyde	2,500	54	DNPH cartridge
Formaldehyde	1,330	34	
Summary VOC measurements			
Total HC by FID	27,810	83	FID
Total sorbed and IDed VOC	11,270		Sorbent tube/GC
Total sorbed VOC	1,907.1	525	Sorbent tube/GC
Respirable particles	13,674	411	Gravimetric

DNPH = 2,4-dinitrophenylhydrazine; IR = infrared; VOC = volatile organic compound; ETS = environmental tobacco smoke; FID = flame ionization detector; GC = gas chromatography.
Source: Martin et al. (1997).

Table 2   Example Generation of Gaseous Contaminants by Building Materials

	Emission Factor Averages (ranges), μg/(h·m²)					
Contaminant	Acoustic Ceiling Panels	Carpets	Fiberboards	Gypsum Boards	Paints on Gypsum Board	Particle Boards
4-Phenylcyclo-hexene (PCH)		8.4 (n.d.- 85)				
Acetaldehyde		2.8 (n.d.- 37)	9.0 (n.d.-32)			28 (n.d.-55)
Acetic acid			8.4 (n.d.-26)			
Acetone	12 (n.d.-33)		35 (n.d.-67)	37 (n.d.-110)	35 (n.d.-120)	
Ethylene glycol			140 (n.d.-290)		19 (n.d.-190)	160 (140-200)
Formaldehyde	5.8 (n.d.-25)	3.6 (n.d.- 41)	220 (n.d.-570)	6.8 (n.d.-19)		49 (n.d.-97)
Naphthalene		11 (n.d.-59)	3.0 (n.d.-8.2)			
n-Heptane			21 (n.d.-53)			
Nonanal	4.9 (1.7-11)	11 (n.d.-68)		10 (n.d.-28)	3.7 (n.d.-24)	
Toluene			19 (n.d.-46)			
TVOC*	32 (3.2-150)	1900 (270-9100)	400 (52-850)	15 (n.d.-61)	2500 (170-6200)	420 (240-510)

	Emission Factor Averages (ranges) in μg/(h·m²)					
Contaminant	Plastic Laminates and Assemblies	Non-Rubber-Based Resilient Flooring	Rubber-Based Resilient Flooring	Tackable Wall Panels	Thermal Insulations	Wall Bases (Rubber-Based)
1,2,4-Trimethylbenzene			210 (n.d.-590)			
2-Butoxy-ethanol		2.7 (n.d.- 24)	1.6 (n.d.-24)			
Acetaldehyde		11 (n.d.- 49)				
Acetone	75 (4.8-150)	120 (n.d.- 830)			12 (1.8-21)	220 (30-400)
Butyric acid		0.51 (n.d. - 5.1)				
Dodecane			1.3 (n.d.-20)			
Ethylene glycol		38 (n.d.- 210)				
Formaldehyde	13 (n.d.-29)	6.8 (n.d.- 79)			5.9 (0.35-14)	32 (3.6-61)
Naphthalene		3.4 (n.d.- 14)	5.6 (n.d.-28)	6.6 (6.6)		
n-Butanol						100 (n.d.-200)
Nonanal		5.7 (n.d.- 19)	1.4 (n.d.-11)		1.8 (0.57-4)	
Octane						150 (n.d.-300)
Phenol	9.4 (4.4-19)	35 (n.d.- 310)				340 (n.d.-680)
Toluene		5.1 (n.d.- 12)				
Undecane						140 (13-270)
TVOC*	160 (6.3-310)	680 (100-2100)	15000 (1500-100000)	270 (100-430)	7.5 (0.57-26)	7100 (1200-13000)

Source: CIWMB (2003).

n.d. = nondetectable

*TVOC concentrations calculated from total ion current (TIC) from GC/MS analysis by adding areas of integrated peaks with retention times greater than 5 min, subtracting from sum of area of internal standard chlorobenzene-d5, and using response factor of chlorobenzene-d5 as calibration.

Using Source Data to Predict Indoor Concentrations

Source data such as those in Tables 1 to 5 provide the type of raw information on which control system designs can be based. Outdoor air contaminants enter buildings through the outdoor air intake and through infiltration. The indoor sources enter the occupied space air and are distributed through the ventilation system. If measurements are not available, source data can be used to predict the contaminant challenge to air-cleaning systems using building air quality models. The following relatively simple published model is intended as an introduction to the topic.

Meckler and Janssen (1988) described a model for calculating the effect of outdoor pollution on indoor air quality, which is outlined in this section.

A recirculating air-handling schematic is shown schematically in Figure 1. In this case, mixing is not perfect; the horizontal dashed line represents the boundary of the region close to the ceiling through which air passes directly from the inlet diffuser to the return air intake. Ventilation effectiveness E_v is the fraction of total air supplied to the space that mixes with room air and does not bypass the room along the ceiling. Meckler and Janssen suggest a value of 0.8 for E_v. Any people in the space are additional sources and sinks for gaseous contaminants. In the ventilated space, the steady-state contaminant concentration results from the summation of all processes adding contaminants to the space a divided by the total ventilation and other processes removing amounts b. The steady-state concen-

tration C_{ss} for a single component can be expressed as (Meckler and Janssen 1988)

$$C_{ss} = a/b \qquad (2)$$

where

$$a = C_x(Q_i + 0.01PE_vQ_v/f) + 0.5885(G_i + NG_O) \qquad (3)$$

$$b = Q_e + Q_h + Q_L + k_dA + NQ_O(1 - 0.01P_O) + (E_vQ - Q_v)(1 - 0.01P)/f \qquad (4)$$

and

A = surface area inside ventilated space on which contaminant can be adsorbed, ft²
C_{ss} = steady-state indoor concentration of contaminant, μg/m³
C_x = outdoor concentration of contaminant, μg/m³
E_v = ventilation effectiveness, fraction
$f = 1 - 0.01P(1 - E_v)$
G_i = generation rate for contaminant by nonoccupant sources, μg/h
G_O = generation rate for contaminant by an occupant, μg/h
k_d = deposition velocity on a for contaminant, fpm
N = number of occupants
P = filter penetration for contaminant, %
P_O = penetration of contaminant through human lung, %
Q = total flow, cfm
Q_e = exhaust airflow, cfm
Q_h = hood flow, cfm

Table 3 Example Generation of Gaseous Contaminants by Indoor Combustion Equipment

	Generation Rates, µg/Btu					Typical Heating Rate, 1000 Btu/h	Typical Use, hour/day	Vented or Unvented	Fuel
	CO_2	CO	NO_2	NO	HCHO				
Convective heater	53,500	88	13	18	1.45	31	4	U	Natural gas
Controlled-combustion wood stove		14	0.04	0.07		13	10	V	Oak, pine
Range oven		210	11	23		32	1.0*	U	Natural gas
Range-top burner		68	11	18	1.1	9.5/burner	1.7	U	Natural gas

*Sterling and Kobayashi (1981) found that gas ranges are used for supplemental heating by about 25% of users in older apartments. This increases the time of use per day to that of unvented convective heaters.

Sources: Cole (1983); Leaderer et al. (1987); Moschandreas and Relwani (1989); Sterling and Kobayashi (1981); Traynor et al. (1985); and Wade et al. (1975).

Table 4 Example Total-Body Emission of Some Gaseous Contaminants by Humans

Contaminant	Typical Emission, µg/h	Contaminant	Typical Emission, µg/h
Acetaldehyde	35	Methane	1,710
Acetone	475	Methanol	6
Ammonia	15,600	Methylene chloride	88
Benzene	16	Propane	1.3
2-Butanone (MEK)	9,700	Tetrachloroethane	1.4
Carbon dioxide	32×10^6	Tetrachloroethylene	1
Carbon monoxide	10,000	Toluene	23
Chloroform	3	1,1,1-Trichloroethane	42
Dioxane	0.4	Vinyl chloride monomer	0.4
Hydrogen sulfide	15	Xylene	0.003

Sources: Anthony and Thibodeau (1980); Brugnone et al. (1989); Cohen et al. (1971); Conkle et al. (1975); Gorban et al. (1964); Hunt and Williams (1977); and Nefedov et al. (1972).

Q_i = infiltration flow, cfm
Q_L = leakage (exfiltration) flow, cfm
Q_O = average respiratory flow for a single occupant, cfm
Q_v = ventilation (makeup) airflow, cfm

Flow continuity allows the expression for b to be simplified, which may make it easier to determine flows:

$$b = Q_i + Q_v + k_d A + NQ_O(1 - 0.01P_O) \qquad (5)$$

The parameters for this model must be determined carefully so that nothing significant is ignored. Leakage flow Q_L, for example, may include flow up chimneys or toilet vents.

The steady-state concentration is of interest for design. It may also help to know how rapidly concentration changes when conditions change suddenly. The dynamic equation for the building in Figure 1 is

$$C_I = C_{ss} + (C_0 - C_{ss})e^{-b\theta/V} \qquad (6)$$

where

V = volume of the ventilated space, ft^3
C_0 = concentration in space at time $\theta = 0$
C_I = concentration in space θ minutes after a change of conditions

C_{ss} is given by Equation (2), and b by Equation (5), with the parameters for the new condition inserted.

Reducing air infiltration, leakage, and ventilation air to reduce energy consumption raises concerns about indoor contaminant build-up. A low-leakage structure may be simulated by letting $Q_i = Q_L = Q_h = 0$. Then

$$C_{ss} = \frac{\dfrac{0.01 P E_v Q_v C_x}{f} + 2119(G_i + NG_O)}{Q_e + K_d A + NQ_O(1 - 0.01P_O) + \dfrac{(E_v Q - Q_v)(1 - 0.01P)}{f}} \qquad (7)$$

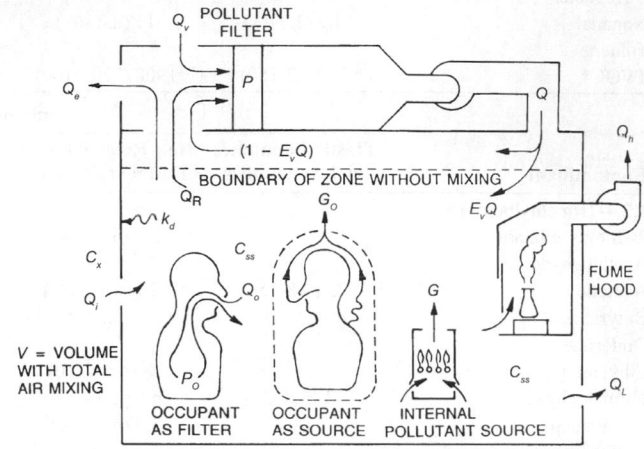

Fig. 1 Recirculatory Air-Handling and Gaseous Contaminant Schematic

Even if ventilation airflow $Q_v = 0$, a low-penetration (high-efficiency) gaseous contaminant filter and a high recirculation rate help lower the internal contaminant concentration. In commercial structures, infiltration and exfiltration are never zero. The only inhabited spaces operating on 100% recirculated air are space capsules, undersea structures, and structures with life-support to eliminate carbon dioxide and carbon monoxide and supply oxygen.

Real buildings generally have many rooms, with multiple and varying sources of gaseous contaminants and complex room-to-room air changes. In addition, mechanisms other than adsorption may eliminate gaseous contaminants on building interior surfaces. Nazaroff and Cass (1986) provide estimates for contaminant deposition velocity k_d in Equations (2) to (5) that range from 0.0006 to 0.12 fpm for surface adsorption only. A worst-case analysis, yielding the highest estimate of indoor concentration, is obtained by setting $k_d = 0$. Nazaroff and Cass (1986) and Sparks (1988) describe computer programs to handle these calculations. Details on multizone modeling can be found in Chapter 13 of the 2009 *ASHRAE Handbook—Fundamentals*.

The assumption of bypass and mixing used in the models presented here approximates the multiple-room case, because gaseous contaminants are readily dispersed by airflows. Also, a gaseous contaminant diffuses from a zone of high concentration to one of low concentration, even with low rates of turbulent mixing.

Quantities appropriate for the flows in Equations (2) to (7) are discussed in the sections on Local Source Control and Dilution Through General Ventilation. Infiltration flow can be determined approximately by the techniques described in Chapter 16 of the 2009 *ASHRAE Handbook—Fundamentals* or, for existing buildings, by tracer or blower-door measurements. ASTM *Standard* E741 defines procedures for tracer-decay measurements. Tracer and blower-door techniques are given in ASTM (1980); DeFrees and

Table 5 Typical U.S. Outdoor Concentration of Selected Gaseous Air Contaminants

Inorganic Air Contaminants[a]

Inorganic Name	CAS Number	Period of Average	Arithmetic Mean Concentration	
			$\mu g/m^3$	ppb
Carbon monoxide	630-08-0	1 year (2008)	2000	2
Nitrogen dioxide	10102-44-0	1 year (2008)	29	15
Ozone	10028-15-6	3 years (2006-08)	149	76

Organic Air Contaminants[b]

VOC Name	CAS number	Number of Sites Tested	Frequency Detected (% of Sites)	Arithmetic Mean Concentration	
				($\mu g/m^3$)	(ppb)
Chloromethane	74-87-3	87	99	2.6	1.3
Benzene	71-43-2	67	99	3.0	0.94
Acetone	67-64-1	67	98	8.6	3.6
Acetaldehyde	75-07-0	86	98	3.4	1.9
Toluene	108-88-3	69	96	5.1	1.4
Formaldehyde	50-00-0	99	95	3.9	3.2
Phenol	108-95-2	40	93	1.6	0.42
m- and p-Xylenes	1330-20-7	69	92	3.2	0.74
Ethanol	64-17-5	13	92	32	17
Dichlorodifluoromethane	75-71-8	87	91	7.1	1.4
o-Xylene	95-47-6	69	89	1.2	0.28
Nonanal	124-19-6	40	89	1.1	0.19
2-Butanone	78-93-3	66	88	1.4	0.48
1,2,4-Trimethylbenzene	95-63-6	69	87	1.2	0.24
Ethylbenzene	100-41-4	69	84	0.9	0.21
n-Decane	124-18-5	69	80	0.97	0.17
n-Hexane	110-54-3	38	75	1.7	0.48
Tetrachloroethene	127-18-4	69	73	1.1	0.16
4-Ethyltoluene	622-96-8	69	72	0.53	0.11
n-Undecane	1120-21-4	69	70	0.6	0.094
Nonane	111-84-2	69	66	0.59	0.11
1,1,1-Trichloroethane	71-55-6	66	65	0.88	0.16
Styrene	100-42-5	69	61	0.39	0.092
Ethyl acetate	141-78-6	66	58	0.43	0.12
Octane	111-65-9	68	56	0.44	0.094
1,3,5-Trimethylbenzene	108-67-8	69	56	0.41	0.083
Hexanal	66-25-1	40	53	0.65	0.16

[a]*Source*: EPA (2009). Note that only statistically viable datasets were used to calculate the national average concentrations, so the numbers may not be fully representative.
[b]*Source*: EPA (1997a).

Amberger (1987) describe a variation on the blower-door technique useful for large structures.

PROBLEM ASSESSMENT

Consensus design criteria (allowable upper limit of C_{ss} for any contaminant) do not exist for most nontoxic chemicals. Chapter 10 of the 2009 *ASHRAE Handbook—Fundamentals* discusses health effects of gaseous contaminants and provides some guidance on acceptable indoor concentrations, and Chapter 11 of that volume discusses the nature and non-health-related effects of gaseous contaminants, as well as providing some guidance on assessment.

Ideally, design for control of gaseous contaminants is based on accurate knowledge of the identity and concentration (as a function of time) of the contaminants to be controlled and other chemical species that are present, as well as the sources of each contaminant. This knowledge may come from estimates of source strength and modeling or direct measurement of the sources, or from direct measurements of the contaminants in the indoor air. Unfortunately, definitive assessment is seldom possible, so often careful observation, experience, and judgment must supplement data as the basis for design. For instance, certain molecular contaminants have distinctive odors, or have known sources in different geographical regions.

Two general cases exist: (1) new ventilation systems in new buildings for which contaminant loads must be estimated or measured, and (2) modification of existing ventilation systems to solve particular problems. For the first case, models such as described previously must be used. Identify contaminant-generating activities, estimate and sum the building sources, and identify outside air contaminants. Gaps in contaminant load data must be filled with estimates or measurements. Once contaminants and loads are identified, design can begin.

For control of a particular problem, measurements may also be required to identify the contaminant. Assessing the problem can become an indoor air quality investigation, including building inspection, occupant questionnaires, and local sampling and analysis. The *Building Air Quality Guide* (EPA 1991) is a useful basic guide for such investigations. Again, once the contaminants and loads are understood, design can begin.

Contaminant Load Estimates

Valuable guidance on estimating contaminant loads in industrial situations is given by Burton (2003). In the 2009 *ASHRAE Handbook—Fundamentals*, Chapter 11 discusses sampling and measurement techniques for industrial and nonindustrial environments, and Chapter 12 covers evaluating odor levels.

Results of sampling and analysis identify contaminants and their concentrations at particular places and times or over known periods. Several measurements, which may overlap or have gaps in the contaminants analyzed and times of measurement, are usually used to estimate the overall contaminant load. Measurements are used to develop a time-dependent estimate of contamination in the building, either formally through material balance or informally through experience with similar buildings and contamination. The degree of formality applied depends on the perceived severity of potential effects.

CONTROL STRATEGIES

Four control strategies may be used to improve the indoor air quality in a building: (1) elimination of sources, (2) local hooding with exhaust or recirculated air cleaning, (3) dilution with increased general ventilation, and (4) general ventilation air cleaning with or without increased ventilation rates. Usually, the first three are favored because of cost considerations. Control by general air cleaning is more difficult because it is applied after the contaminants are fully dispersed and at their lowest concentration.

Elimination of Sources

This strategy is the most effective and often the least expensive. For instance, prohibiting smoking in a building or isolating it to limited areas greatly reduces indoor pollution, even when rules are poorly enforced (Elliott and Rowe 1975; Lee et al. 1986). Radon gas can be controlled by installing traps in sewage drains and sealing and venting leaky foundations and crawlspaces to prevent entry of the gas (EPA 1986, 1987). Using waterborne materials instead of those requiring organic solvents may reduce VOCs, although Girman et al. (1984) show that the reverse is sometimes true. Substituting carbon dioxide for halocarbons in spray-can propellants is an example of using a relatively innocuous contaminant instead of a more troublesome one. Growth of mildew and other organisms that emit odorous contaminants can be restrained by controlling condensation and applying fungicides and bactericides, provided they are registered for the use and carefully chosen to have low off-gassing potential.

Local Source Control

Local source control is more effective than control by general ventilation when discrete sources in a building generate substantial amounts of gaseous contaminants. If these contaminants are toxic, irritating, or strongly odorous, local control and outdoor exhaust is essential. Bathrooms and kitchens are the most common examples. Some office equipment benefits from direct exhaust. Exhaust rates are sometimes set by local codes. The minimum transport velocity required for capturing large particles is larger than that required for gaseous contaminants; otherwise, the problems of capture are the same for both gases and particles.

Hoods are normally provided with exhaust fans and stacks that vent to the outdoors. Hoods use large quantities of tempered makeup air, which requires a great deal of fan energy, so hoods waste heating and cooling energy. Makeup for air exhausted by a hood should be supplied so that the general ventilation balance is not upset when a hood exhaust fan is turned on. Back diffusion from an open hood to the general work space can be eliminated by surrounding the work space near the hood with an isolation enclosure, which not only isolates the contaminants, but also keeps unnecessary personnel out of the area. Glass walls for the enclosure decrease the claustrophobic effect of working in a small space.

Increasingly, codes require filtration of hood exhausts to prevent toxic releases to the outdoors. Hoods should be equipped with controls that decrease their flow when maximum protection is not needed. Hoods are sometimes arranged to exhaust air back into the occupied space, saving heating and cooling that air. This practice must be limited to hoods exhausting the most innocuous contaminants because of the risk of filter failure. Design of effective hoods is described in *Industrial Ventilation: A Manual of Recommended Practice for Design* (ACGIH 2010), and in Chapter 32 of this volume.

Dilution Through General Ventilation

In residential and commercial buildings, the chief use of local source control and hooding occurs in kitchens, bathrooms, and occasionally around specific point sources such as diazo printers. Where there is no local control of contaminants, the general ventilation distribution system can sometimes provide contaminant control through dilution. These systems must meet both thermal load requirements and contaminant control standards. Complete mixing and a relatively uniform air supply per occupant are desirable for both purposes. The air distribution guidelines in Chapters 20 and 21 of the 2009 *ASHRAE Handbook—Fundamentals* are appropriate for contamination control by general ventilation. Airflow requirements set by ASHRAE *Standard* 62.1 must be met.

When local exhaust is combined with general ventilation, a proper supply of makeup air must balance the exhaust flow for any hoods present to maintain the desired over- or underpressure in the building or in specific rooms. Supply fans may be needed to provide enough pressure to maintain flow balance. For instance, clean spaces are designed so that static pressure forces air to flow from cleaner to less clean spaces, and the effects of doors opening and wind pressure, etc., dictate the need for backdraft dampers. Chapter 18 covers clean spaces in detail.

CONTROL BY VENTILATION AIR CLEANING

If eliminating sources, local hooding, or dilution cannot control contaminants, or are only partially effective, the ventilation air must be cleaned. Designing such a system requires understanding of the capabilities and limitations of various control processes.

Design goals are discussed at greater length in the section on Air Cleaner System Design, but it is appropriate to mention at this point that complete and permanent removal of every contaminant is often not necessary. Intermittent nuisance odors, for instance, can often be controlled satisfactorily and economically using a design that "shaves the peak" to below the odor threshold and then slowly releases the contaminant back into the air, still below the odor threshold. On the other hand, such an approach would be inappropriate for a contaminant that affected occupants' health.

Gas Contaminant Control Processes

Many chemical and physical processes remove gases or vapors from air, but those of highest current commercial interest to the HVAC engineer are physical adsorption and chemisorption. The operational parameters of greatest interest are removal efficiency, pressure drop, operational lifetime, first cost, and operating and maintenance cost. Other removal processes have been proposed, but currently have limited application in HVAC work, and are only briefly discussed.

Physical Adsorption. Physical adsorption is a surface phenomenon similar in many ways to condensation. Contaminant gas molecules strike a surface and remain bound to it (adsorbed) for an appreciable time by molecular attraction (van der Waals forces). Therefore, high surface area is crucial for effective adsorbents. Surfaces of gaseous contamination control adsorption media are expanded in two ways to enhance adsorption. First, the media are provided in granular, pelletized, or fibrous form to increase the gross surface exposed to an airstream. Second, the media's surface is treated or activated to develop microscopic pores, greatly increasing the area available for molecular contact. These internal pores account for the majority of available surface area in most commercially available adsorbents. Typical activated alumina has a surface

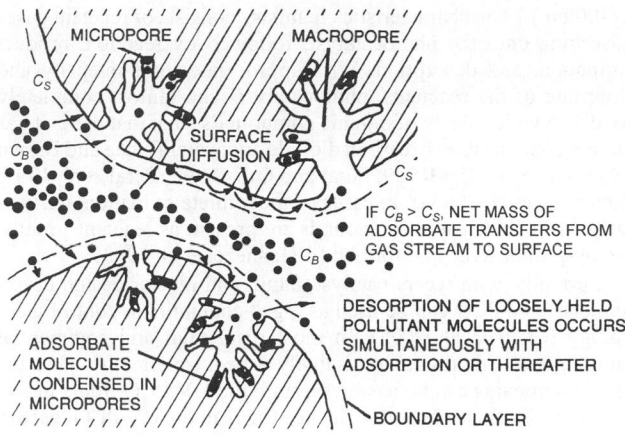

Fig. 2 Steps in Contaminant Adsorption

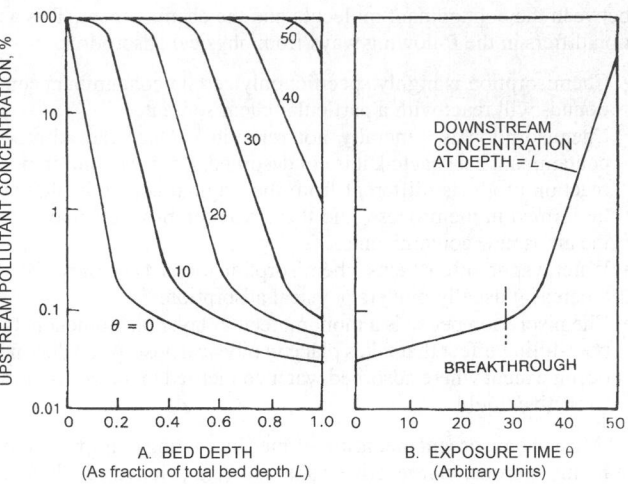

Fig. 3 Dependence of Contaminant Concentration on Bed Depth and Exposure Time

area of 1 to 1.6×10^6 ft^2 per pound; typical activated carbon has a surface area from 4 to 8×10^6 ft^2/lb. Pores of various microscopic sizes and shapes form minute traps that can fill with condensed contaminant molecules.

The most common adsorbent granules are millimeter-sized, and the granules are used in the form of packed beds. In general, packed beds composed of larger monodispersed particles have lower pressure drops per unit depth of sorbent than those composed of smaller monodispersed particles. On the other hand, the surface area of the adsorbent is more accessible to the contaminant with the smaller particles.

Several steps must occur in physical adsorption of a molecule (Figure 2):

1. The molecule is transported from the carrier gas stream across the boundary layer surrounding the adsorbent granule. This occurs randomly, with molecular movement both to and from the surface; the net flow of molecules is toward the surface when the concentration of contaminant in the gas flow is greater than at the granule surface. For this reason, adsorption decreases as contaminant load on the adsorbent surface increases. Very low concentrations in the gas flow also result in low adsorption rates.
2. The molecules of the contaminant diffuse into the pores to occupy that portion of the surface. Diffusion distances are lower and adsorption rates higher for smaller particles of adsorbent.
3. The contaminant molecules are bound to the surface. (Adsorption is exothermic, releasing energy. At the low concentrations and adsorption rates generally found in HVAC applications, adsorbents operate nearly isothermally.)

Any of these steps may determine the rate at which adsorption occurs. In general, step 3 is very fast for physical adsorption, but reversible: adsorbed molecules can be desorbed later, either when cleaner air passes through the adsorbent bed or when another contaminant arrives that either binds more tightly to the adsorbent surface or is present at a much higher concentration. Complete desorption usually requires adding thermal energy to the bed.

When a contaminant is fed at constant concentration and constant gas flow rate to an adsorbent bed of sufficient depth L, the gas stream concentration within the bed varies with time θ and bed depth, as shown in Figure 3A. When bed loading begins ($\theta = 0$), the contaminant concentration decreases logarithmically with bed depth; deeper into the bed, the slope of the concentration-versus-bed depth curve flattens at a very low value. Later, the entrance of the adsorbent bed becomes loaded with contaminant, so contaminant concentrations in the gas stream are higher at each bed depth.

Distribution of contaminant in an adsorbent bed is often described in terms of an idealized **mass transfer zone (MTZ)**. Conceptually,

all contaminant adsorption takes place in the MTZ. Upstream, the adsorbent is spent and the concentration is equal to the inlet concentration. Downstream, all contaminant has been adsorbed and the concentration is zero. The movement of the MTZ through the media bed is known as the **adsorption wave**. Though in actuality the front and back of the zone are not sharply defined, for many media/contaminant combinations the MTZ provides a very useful picture of media performance.

For the same constant contaminant feed, the pattern of downstream concentration versus time for an adsorbent bed of depth L is shown in Figure 3B. Usually, downstream concentration is very low until breakthrough time θ_{BT}, when the concentration rises rapidly until downstream concentration is the same as upstream. The logarithmic scale makes the downstream concentration appear higher than it is. At the breakthrough point, the downstream concentration is at less than 0.1% of the upstream, just as the slope of the curve increases. For purposes of defining the MTZ, choosing this point is reasonable. For protection purposes, any contaminant downstream might be too much; for nuisance odors, staying below the odor threshold might be adequate. The interval between these various breakthrough times could be very short or significant, depending on contaminant and media. However, not all adsorbent/contaminant combinations show as sharp a breakthrough as in Figure 3B.

Multiple contaminants produce more complicated penetration patterns: individually, each contaminant might behave as shown in Figure 3B, but each has its own time scale. The better-adsorbing contaminants are captured in the upstream part of the bed, and the poorer are adsorbed further downstream. As the challenge continues, the better-adsorbing compound progressively displaces the other, meaning the displaced component can leave the adsorbent bed at a higher concentration than it entered.

Underhill et al. (1988) and Yoon and Nelson (1988) discuss the effect of relative humidity on physical adsorption. Water vapor acts as a second contaminant, generally present at much higher concentrations than typical indoor contaminants, altering adsorption parameters by reducing the amount of the first contaminant that can be held by the bed and shortening breakthrough times. For solvent-soluble VOCs adsorbed on carbon, relative humidity's effect is modest up to about 50%, and greater at higher percentages. On the other hand, chemicals that dissolve in water may experience increased adsorption into the water layer at high relative humidities.

Chemisorption. The three steps described for physical adsorbents also apply to chemisorption. However, the third step in chemisorption involves chemical reaction with electron exchange between the contaminant molecule and the chemisorbent. This action differs in the following ways from physical adsorption:

- Chemisorption is highly specific; only certain contaminant compounds will react with a particular chemisorbent.
- Chemisorption is generally not reversible. Once the adsorbed contaminant has reacted, it is not desorbed. However, one or more reaction products, different from the original contaminant, may be formed in the process, and these reaction products may enter the air as new contaminants.
- Water vapor often helps chemisorption or is necessary for it, whereas it usually hinders physical adsorption.
- Chemisorption per se is a monomolecular layer phenomenon; the pore-filling effect that takes place in physical adsorption does not occur, except where adsorbed water condensed in the pores forms a reactive liquid.

Most chemisorbent media are formed by coating or impregnating a highly porous, nonreactive substrate (e.g., activated alumina, zeolite, or carbon) with a chemical reactant. The reactant will eventually become exhausted, but the substrate may have physical adsorption ability that remains active when chemisorption ceases.

General Considerations. Physical adsorption and chemisorption are the removal processes most commonly used in gaseous contaminant filtration. In most cases, the processes for both involve media supplied as granules or pellets, which are held in a retaining structure that allows air being treated to pass through the media with an acceptable pressure drop at the operating airflow. Granular media are traditionally a few millimetres in all dimensions, typically on the order of 4×6 or 4×8 U.S. mesh pellets or flakes.

Other Processes. Although physical adsorption and chemisorption are the most frequently used, the following processes are used in some applications.

Liquid absorption devices (scrubbers) and **combustion devices** are used to clean exhaust stack gases and process gas effluent. They are not commonly applied to indoor air cleanup. Additional information may be found in Chapter 29 of the 2008 *ASHRAE Handbook—HVAC Systems and Equipment*.

Catalysts can clean air by stimulating a chemical reaction on the surface of the media. **Catalytic combustion** or **catalytic oxidation (CatOx)** oxidizes moderate concentrations of unburned hydrocarbons in air. In general, the goal with catalytic oxidation is to achieve an adequate reaction rate (contaminant destruction rate) at ambient temperature. Reaction products are a concern, because oxidation of nonhydrocarbon VOCs or other reactions can produce undesirable by-products. This technology has been used industrially for years, but its potential use for indoor air cleaning is relatively new. Equipped with custom catalysts and operated at elevated pressures and temperatures, CatOx can be extremely effective at the removal of indoor contaminants, but is not currently cost-competitive in commercial indoor air or HVAC applications. Availability of waste heat significantly improves CatOx cost competitiveness. CatOx systems have potential application in security and protection applications.

Photocatalysis [or **photocatalytic oxidation (PCO)**] uses light [usually ultraviolet (UV)] and a photocatalyst to perform **reduction-oxidation (redox)** chemistry on the catalyst's surface as first observed and reported by Fujishima and Honda (1972). The device admits reactant gases, notably air contaminants, in a feed stream and emits product species. The photocatalyst can be granular, bulk, or unsupported, or it can be supported as a thin film on media such as glass, polymer, ceramic, or metal. The light sources must emit photons of energy greater than that of the intrinsic band gap energy E_g of the photocatalyst. For example, the photocatalyst titanium dioxide (TiO_2) has band-gap energy of 3.1 eV. For this material, ultraviolet light with wavelengths less than 380 nm has sufficient energy to overcome the E_g of TiO_2. [Note that numerous groups are developing materials that can be activated with visible light

(>400 nm).] The characteristic chemistry consists of reactant gases adsorbing onto the photocatalyst, followed by reaction, product formation, and desorption. With appropriate light intensity and flow rate of the reactants, photocatalysis can almost completely oxidize a wide variety of organic compounds such that the exit gas stream contains mostly carbon dioxide and water (Obee and Brown 1995; Obee and Hay 1999; Peral and Ollis 1992; Peral et al. 1997; Tompkins et al. 2005a). In cases of incomplete oxidation, particularly when chlorinated compounds are present as reactants, multiple by-products may be formed (d'Hennezel et al. 1998).

Currently, with recent catalyst, lamp, and reactor design developments, UV-PCO can be used as a gas-contaminant control technology (Chen et al. 2005). In most residential and commercial building applications, control is likely to be most effective when the air contaminants can be passed through the UV-PCO filters multiple times. Because of short residence times in the UV-PCO reactor, the single-pass removal efficiency is between a few percent and 10 to 20% (or higher), depending on the gas contaminant (i.e., gas contaminant removal efficiency is higher for larger molecules and those with lower vapor pressure, compared to smaller molecules or those with a higher vapor pressure). Therefore, in an HVAC application, the preferred location for a UV-PCO filter is in the return air or mixed air, where gas contaminants pass through the filter many times in a given time period. UV-PCO can be an attractive gas contaminant control technology because of its promise of reduced maintenance (no filters and/or adsorption media to maintain and dispose of periodically) and ability to treat a wide variety of airborne chemicals. ASHRAE research project RP-1134 exhaustively reviewed the literature on UV photocatalysis (Tompkins et al. 2005a, 2005b). Ongoing ASHRAE research project RP-1457, By-Product Production from Photocatalytic Oxidation Associated with Indoor Air Cleaning Devices, is currently investigating by-product formation from typical indoor air contaminants using several commercially available UV-PCO air cleaners.

Sometimes, the UV-PCO unit is followed by a gas-phase media section that can adsorb any partially oxidized molecules to prevent them from recirculating back into the occupied space. A further extension of UV-PCO being studied is use of UVV (i.e., low-energy UV light with wavelength close to the visible limit of 400 nm) along with UVC to generate radicals such as ozone, hydroxyls, and peroxides, which increase destruction efficiency and hence single-pass efficiency. A downstream gas-phase media section is necessary in this case to destroy these radicals and prevent them from passing into occupied spaces.

Biofiltration is effective for low concentrations of many VOCs found in buildings (Janni et al. 2001). It is suitable for exhaust air cleaning, and is used in a variety of applications, including plastics, paper, and agricultural industries and sewage treatment plants. Operating costs are low, and installation is cost-competitive. However, concerns over using uncharacterized mixtures of bacteria in the filter, possible downstream emissions of microbials or chemicals, and the risk of unexpected or undetected failure make it unsuitable for cleaning air circulated to people.

Ozone is sometimes touted as a panacea for removing gas-phase contaminants from indoor air. However, considerable controversy surrounds its use in indoor air. Ozone is a criteria pollutant and its maximum allowable concentration [8 h time-weighted average (TWA)] is regulated in both indoor (OSHA 1994) and outdoor air (EPA 1997b). Some ozone generators can quickly produce hazardous levels of ozone (Shaughnessy and Oatman 1991). Furthermore, the efficacy of ozone at low concentrations for removing gaseous pollutants has not been documented in the literature (Boeniger 1995). Human sensory results obtained in conjunction with a study by Nelson et al. (1993) showed that an ozone/negative ion generator used in a tobacco-smoke environment (1) produced unacceptable ozone levels at the manufacturer's recommended settings, and (2) when adjusted to produce acceptable ozone levels, produced more

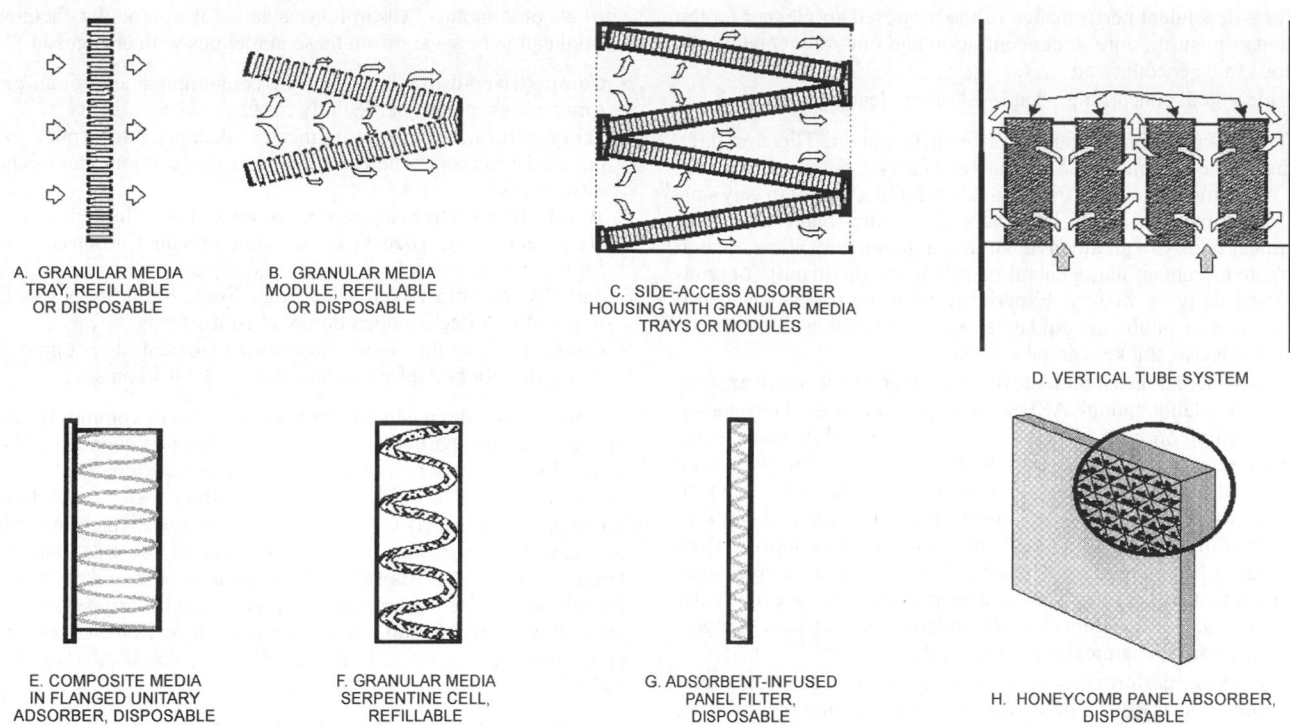

A. GRANULAR MEDIA
TRAY, REFILLABLE
OR DISPOSABLE

B. GRANULAR MEDIA
MODULE, REFILLABLE
OR DISPOSABLE

C. SIDE-ACCESS ADSORBER
HOUSING WITH GRANULAR MEDIA
TRAYS OR MODULES

D. VERTICAL TUBE SYSTEM

E. COMPOSITE MEDIA
IN FLANGED UNITARY
ADSORBER, DISPOSABLE

F. GRANULAR MEDIA
SERPENTINE CELL,
REFILLABLE

G. ADSORBENT-INFUSED
PANEL FILTER,
DISPOSABLE

H. HONEYCOMB PANEL ABSORBER,
DISPOSABLE

Fig. 4   Sectional and Schematic Views of Typical Adsorber and Chemisorber Configurations

odor and eye irritation over time than environmental tobacco smoke (ETS). Other work by Nelson (unpublished) has shown the rapid oxidation of NO to NO_2 by ozone and only a minor decrease in nicotine concentrations when ozone is used to "clean" the air. In light of the potential for generating hazardous ozone levels indoors and the lack of scientific data supporting its efficacy, using ozone to combat ETS in indoor air is not recommended.

Odor counteractants and **odor masking** are not truly control methods; they may apply only to specific odors and have limited effectiveness. They also add potential contaminants to the air.

EQUIPMENT

The purpose of gas-phase filtration equipment is to expose the chosen filtration media or device to the air to be filtered. In most cases, the filtration method uses granular media material, supplied either in bulk, or incorporated into a filter device that can be refillable or disposable. Typically, the gas filter has a particulate prefilter and afterfilter.

The most common retaining structure places the granular media between perforated retaining sheets or screens, as shown in Figures 4A, 4B, 4C, 4D, 4F, and 4H. The perforated retainers or screens must have holes smaller than the smallest particle of the media, and are typically made of aluminum; stainless, painted, plated, or coated steel; plastics; and kraftboard.

Media may also be retained in fibrous filter media or other porous support structures, and very fine media can be attached to the surface or within the structure of some particulate filter media, as shown in Figures 4E and 4G.

Effect of Media Size. Filtration devices that use small-diameter media generally have higher initial efficiency than the same media in larger particles, because of the larger exposed surface area. Devices that use larger-diameter media generally provide more overall filtration capacity because of the greater mass of media exposed to the air to be filtered.

Equipment Configurations. The typical media-holding devices shown in Figure 4 vary in thickness from 0.625 in. to as much as

6 in. Though they can be mounted perpendicular to the airflow (Figure 4A), they more often hold the media at an angle to the airstream (Figures 4B to 4F) to reduce pressure drop and increase residence time.

The honeycomb panels in Figure 4H hold the media in small channels formed by a corrugated spacer material, faced with a plastic mesh material. Though they usually have a low pressure drop, these are often in a holding frame configured at an angle to the airflow, as in Figure 4C.

Most of these devices are removed from the airstream, to either be refilled, or replaced. An exception is the vertical tube system, shown in Figure 4D. It is filled with bulk media, usually from plastic pails or large bags, fed in through top-access hatch(es). Expended media is removed by vacuuming from the same hatches, or from bottom hatches or hoppers.

Bypass. Performance of any air cleaner installation is limited by the airflow integrity of the total installation. A 100% efficient filter mounted in a housing that allows 10% bypass is a 90% efficient installation. Designers should consider the desired overall efficiency and ensure that the housing and filter together meet performance goals. One method of detecting significant bypass is to measure whether the filter achieves its rated pressure drop at full flow; if the pressure drop is low, bypass is likely.

AIR CLEANER SYSTEM DESIGN

Air cleaner system design consists of determining and sizing the air cleaning technology to be applied, and then choosing equipment with characteristics (size and pressure drop) that can be incorporated into the mechanical design.

The gaseous contaminant control designer ideally should have the following information:

- Exact chemical identity of the contaminants present in significant concentrations (not just the ones of concern)
- Rates at which contaminants are generated in the space
- Rates at which contaminants are brought into the space with outdoor air

- Time-dependent performance of the proposed air cleaner for the contaminant mixture at concentration and environmental conditions to be encountered
- A clear goal concerning what level of air cleaning is needed

This information is usually difficult to obtain. The first three items can be obtained by sampling and analysis, but funding is usually not sufficient to carry out adequate sampling except in very simple contamination cases. Designers must often make do with a chemical family (e.g., aldehydes). Investigation may allow a rough estimate of contaminant generation rate based on quantity of product used daily or weekly. Experience with the particular control application or published guidance [e.g., Rock (2006) for environmental tobacco smoke] can be very helpful.

Experimental measurements of air cleaner performance are usually not available, though ASHRAE is in the process of developing the upcoming *Standard* 145.2 that will define such measurements for individual contaminant gases. In the meantime, performance can be estimated using Equations (2) to (7) when the exact chemical identity of a contaminant is known. The chemical and physical properties influencing a contaminant's collection by control devices can usually be obtained from handbooks and technical publications. Contaminant properties of special importance are relative molecular mass, normal boiling point (i.e., at standard pressure), heat of vaporization, polarity, chemical reactivity, and diffusivity.

Air cleaner performance with mixtures of chemically dissimilar compounds is very difficult to predict. Some gaseous contaminants, including ozone, radon, and sulfur trioxide, have unique properties that require design judgment and experience.

Finally, design goals must be considered. For a museum or archive, the ideal design goal is total removal of the target contaminants with no subsequent desorption or release of by-products. For any chemical that may affect health, the design goal is to reduce the concentration to below the level of health effects. Again, desorption back into the space must be minimized. For odor control, however, 100% removal may be unnecessary and desorption back into the space at a later time with a lower concentration may be economical and acceptable.

The first step in design is selecting an appropriate physical or chemical adsorption medium. Next, the air cleaner's location in the HVAC system must be decided and any HVAC concerns addressed. Then the air cleaner must be sized so that sufficient media is used to achieve design efficiency and capacity goals and to estimate media replacement requirements. Finally, commercial equipment that most economically meets the needs of the application can be selected. These steps are not completely independent.

Media Selection

Media selection is clear for many general applications, but for others is a matter of judgment. In general, gaseous contaminants that have the same or higher boiling point as water can be removed by physical adsorption using standard activated carbon. Those with a lower boiling point usually require chemisorption for complete removal. Figure 5 shows the media and equipment selection process.

In practice, different examples of the same application may be served well by somewhat different media selections. Any guidance must be tempered by consideration of the specifics of a particular location, and guidance given by different manufacturers may differ somewhat. Table 6 consolidates general guidance for numerous commercial applications from multiple manufacturers. Within each media group, the applications are listed alphabetically; similar applications appear in more than one list, because some applications may be well served with either a single medium or by a blend, with the best choice determined by the specific contaminants present (both chemical identity and concentration.) Acceptability may hinge on a specific, hard-to-control chemical that is present at one site but not at another. Adsorption capacity for a particular chemical or application may vary from these guidelines with changes in

- **Competitive adsorption.** Multiple contaminants confound performance estimates, particularly for physical adsorbents.
- **Temperature.** A temperature increase decreases adsorption in a physical adsorbent, whereas it increases the reaction rates of chemisorbents.
- **Humidity.** For physical adsorbents, the effect of humidity (generally for rh > 50%) depends on the contaminant. Carbon capacity for water-miscible solvents increases; capacity for immiscible or partially miscible solvents decreases. Some humidity is usually required for effective operation of chemisorbents.
- **Concentration.** Increased contaminant concentration improves adsorption for both physical and chemical adsorbents.

Table 7 provides a general guide to selection of commonly used media to control particular chemicals or types of chemicals. The media covered are permanganate-impregnated alumina (PIA), activated carbon (AC), acid-impregnated carbon (AIC), and base-impregnated carbon (BIC). The numeral 1 indicates the best media to use, and 2 the second choice. As was true of Table 6, some difference in opinion exists as to which media is best, and chemicals for which there is disagreement are tagged with an exclamation point. Where information is unavailable, media can be evaluated for their ability to remove specific gases using ASHRAE *Standard* 145.1.

Air Cleaner Location and Other HVAC Concerns

Outside Air Intakes. Proper location of the outside air intake is especially important for applications requiring gaseous contaminant filters because outside contaminants can load the filters and reduce their operating lifetime. Outside air should not be drawn from areas where point sources of gaseous contaminants are likely: building exhaust discharge points, roads, loading docks, parking decks and spaces, etc. See Chapter 45 for more information on air inlets.

To further help reduce the amount of contaminants from outside air, at least on days of high ambient pollution levels, the quantity of outside air should be minimized.

Air Cleaner Locations. The three principal uses for gaseous contaminant control equipment in an HVAC system are

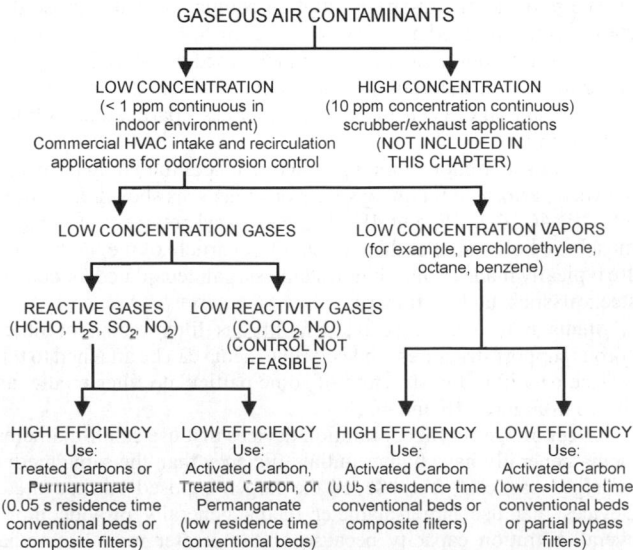

Fig. 5   Media and Equipment Selection Schematic
(Adapted with permission. ©1992 Extraction Systems, Inc.)

Table 6 Media Selection by Commercial Application

Commercial Application	Contaminants/Species
Activated Carbon or Carbon/Permanganate-Impregnated Alumina Blend	
Airport terminals (air side and non-air side), art studios, athletic clubs, auditoriums, banks (customer area), banquet rooms, beauty salons, bus terminals, clinics, darkrooms, decal application, dentists'/doctors' offices, dry cleaners (dust area), factories (office area), florists, grocery stores, kitchen exhausts, locker rooms, office buildings, painted rooms, pharmacies, photo stores, photographic studios, physiotherapy, recreation halls, rendering plants, stores	Multiple volatile organic gases/solvent vapors and inorganic gases; possibly some gases poorly adsorbed by carbon. Multiple organics and inorganics, fumes, food odors, body odors, floral scents, odorous fumes, moldy odors, paint fumes, furniture, ETS, NO_x, SO_x, O_3, mercaptans, valeric acid, formaldehyde
Activated Carbon/Permanganate-Impregnated Alumina Blend	
Bars, bingo halls, brasseries, cafeterias, casinos, cocktail lounges, conference rooms, correctional facilities, funeral homes, geriatrics, hospitals, hotels (smoking, renovation), ICUs, libraries, lounges, lunch rooms, motels, museums, night clubs, nurseries, paint shops (office), penal institutions, projection booths, psychiatric institutions and wards, public toilets, restaurants, segregated smoking rooms, storage rooms, theaters, waiting rooms	Mixed gases/ETS; good possibility of volatile organic gases and/ or solvent vapors. ETS, body odors, urine, excreta, perfume, multiple odors, food odors, kitchen fumes, food, furniture/furnishings offgassing, multiple organics and inorganics, paint
Activated Carbon or Permanganate-Impregnated Alumina	
Barber shops, dining rooms	Mixed gases, ETS, food odors
Carbon/Permanganate Blend or Permanganate-Impregnated Alumina	
Embalming rooms, fruit/vegetable storage, greenhouses	Multiple organics; organic gases poorly sorbed by activated carbon. Multiple organics, formaldehyde, ethylene
Activated Carbon or Permanganate-Impregnated Alumina/Acid-Impregnated Carbon Blend	
Garbage disposal areas	Mixture of volatile organics and inorganics with ammonia
Permanganate-Impregnated Alumina	
Autopsy rooms, banks (vault area), fish markets, hospitals (autopsy), morgues	Volatile organic gases poorly adsorbed by activated carbon. Formaldehyde, trimethyl amine
Permanganate-Impregnated Alumina/Acid-Impregnated Carbon Blend	
Pet shops, animal holding rooms, veterinary hospitals	Mixed organic gases with significant ammonia urine, excreta, animal odors
Activated Carbon/Acid-Impregnated Carbon Blend	
Printing plants	Mixed hydrocarbons and ammonia
Acid-Impregnated Carbon	
Fertilizer plants (office)	Largely ammonia

Notes: Permanganate impregnant is potassium permanganate. Acid impregnants vary.   ETS = environmental tobacco smoke

- **Outside air treatment.** Air-cleaning equipment can be located at the outside air intake to treat outside air only. This treatment is used principally when indoor gaseous contaminants are adequately controlled by outdoor air ventilation, but the outdoor air needs to be cleaned to achieve satisfactory air quality.
- **Bypass or partial supply air treatment.** Bypass can be achieved with a bypass duct and control damper or by installing an air cleaner that allows substantial bypass. Partial supply air treatment may be appropriate where a specific threshold contamination level is targeted, when outside and inside contamination rates are known, and the required level of reduction is small to moderate.
- **Full supply air treatment.** Full treatment achieves the best contaminant control, but with the highest cost and largest equipment volume. This approach is most often used in ventilation strategies that reduce outdoor air while maintaining good indoor air quality.

When outdoor air quality is adequate, treatment of recirculated ventilation air alone may be adequate to control indoor contaminants such as bioeffluents. Full or bypass treatment of the supply air may be appropriate, depending on the source strength.

Sizing Gaseous Contaminant Control Equipment

A number of issues need to be taken into account during equipment sizing. These include

- Nature of contaminant(s) to be controlled
- Average and peak concentrations of contaminant(s)
- Efficiency of the media used

- HVAC considerations, including layout and space available for equipment installation, and air velocity through the section where the air cleaner is to be installed, as well as those concerns discussed in previous sections

The more undesirable the contaminant of concern, and the greater its concentration, the greater the quantity of media required for removal, and the larger the air cleaner installation. Media efficiency depends on the residence time of contaminant(s) in the air cleaner, which in turn depends on media bed depth and HVAC air velocity.

There are currently two common sizing approaches: testing and calculations with Equations (2) to (7); or relying on manufacturers' guidance.

Available Test Methods and Equations. ASHRAE *Standard* 145.1 provides a method for testing granular media at small scale in a laboratory, and can be used to compare media performances against specific contaminants. ASHRAE *Draft Standard* 145.2 provides a method of testing a 2 × 2 ft filter under laboratory conditions and can supply more direct evaluation of potential air cleaning installations. Equations (2) to (7) can be used to size equipment, though the contaminant concentration data required to use the equations effectively may be difficult to obtain.

General procedures for developing a specification for an air cleaner installation are as follows.

1. Choose a physical or chemical adsorbent suited to the contaminant(s) using testing or taking guidance from Tables 6 and 7.

Table 7 Media Selection by Contaminant

Gaseous Contaminant	PIA	AC	AIC	BIC	Gaseous Contaminant	PIA	AC	AIC	BIC	Gaseous Contaminant	PIA	AC	AIC	BIC
Acetaldehyde	1	2			Dichlorofloromethane			1		Methyl formate	2	1		
Acetic acid (!)	1	2		2,1	R-114 (see note)			1		Methyl isobutyl ketone	2	1		
Acetic anhydride (!)	1,2	1		2	Diethylamine	2	1			Methyl sulfide	1			
Acetone (!)	1	2			Dimethylamine			1	2	Methyl vinyl ketone	2	1		
Acetylene	1				Dioctyl phthalate			1		Naphtha		1		
Acrolein	1	2			Dioxane	1	2			Naphthalene		1		
Acrylic acid (!)	1	1		2	Ethanol	1	2			Nicotine	1	2		
Allyl sulfide	1	2			Ethyl acetate	2	1			Nitric acid				1
Ammonia (NH$_3$)			1		Ethyl chloride (!)	1,2	2,1			Nitric oxide (NO)	1			2
Aniline	2	1			Ethylene (C$_2$H$_4$)	1				Nitrobenzene		1		
Arsine	1				Ethylene oxide	1	2			Nitrogen dioxide	1			2
Benzene		1			Ethyl ether	2	1			Nitromethane	1			
Borane (!)	1	2,2			Ethyl mercaptan (!)	1,1	2		2	Nitrous oxide				1
Bromine		1			Formaldehyde	1				Octane (!)	2	1,1		
1,3 Butadiene	1	2			Gasoline	1				Ozone (O$_3$) (!)	2	1,1		
Butane		1			General halocarbons			1		Perchloroethylene	2	1		
2-Butanone	1	2			General hydrocarbons	2	1			Peroxy acetyl nitrate (PAN)		1		
2-Butoxyethanol	2	1			General VOC	2	1			Phenol	2	1		
Butyl acetate (!)	1,2	2,1			Heptane		1			Phosgene	2	1		
Butyl alcohol	2	1			Hydrogen bromide		2		1	Phosphine	1			
Butyl mercaptan	2	1			Hydrogen chloride		2		1	Putrescine	1	2		
Butylene	2	1			Hydrogen cyanide	1				Pyridine (!)	1	1		
Butyne	2	1			Hydrogen fluoride	1			1	Skatole	2	1		
Butyraldehyde	2	1			Hydrogen iodide	2				Silane	1			
Butyric acid		1		2	Hydrogen selenide		1			Stoddard solvent		1		
Cadaverine	2	1			Hydrogen sulfide	1			1	Stibine	1			
Camphor		1			Iodine			1		Styrene (!)	2	1,1		
Carbon dioxide (CO$_2$)	Carbon w/catalyst				Iodoform	2	1			Sulfur dioxide	1			1
Carbon disulfide	2	1			Isopropanol	2	1			Sulfur trioxide	1			1
Carbon monoxide (CO)	Carbon w/catalyst				Kerosene		1			Sulfuric acid		2		1
Carbon tetrachloride		1			Lactic acid		1			Toluene		1		
Chlorine (Cl$_2$)				1	Menthol	2	1			Triethylamine		2	1	
Chloroform		1			Mercury vapor	Impreg. AC				Trichlorethylene		1		
Creosote (!)	1,2	2,1			Methanol	2	1			1,1,1, trichloroethane (!)	1	2,1		
Cyclohexane		1			Methyl acrylate	2	1			R-11 (see below)		1		
Cyclohexanol	2	1			Methyl bromide (!)	2,1	1			Turpentine	2	1		
Cyclohexanone	2	1			Methyl butyl ketone (!)	1,2	2,1			Urea (!)	2	1,1		
Cyclohexene		1			Methyl cellosolve acetate	2	1			Uric acid (!)	1	1		2,2
Decane		1			Methylchloroform		1			Vinyl chloride		1		
Diborane	1				Methylcyclohexane		1			Xylene		1		
Dichlorobenzene		1			Methylene chloride		1							

1 = primary media selection for contaminant; 2 = secondary media selection.
PIA = permanganate-impregnated alumina; AC = activated carbon;
AIC = acid-impregnated carbon; BIC = base-impregnated carbon
R-114 is dichlorotetrafluoroethane; R-11 is trichlorofluoromethane.

Comments: Some contaminant molecules have isomers that, because they have different physical properties (boiling point, vapor pressures), require different treatment methods. For some contaminants, preferred treatment is ion exchange or another (nonlisted) impregnated carbon. For some contaminants, manufacturer recommendations differ. "!" is used to identify these cases.

2. Pick an appropriate efficiency for the adsorbent (complete removal or partial bypass), depending on the contaminant(s).

3. Choose a desired operating adsorbent end point of 10%, 50%, or other breakthrough, depending on the application and allowable steady-state concentration. A building ventilation performance model, with the adsorber appropriately positioned, allows calculation of the expected indoor concentration at various breakthroughs and efficiencies.

4. Obtain a measurement or estimate of breakthrough time at adsorbent use conditions as developed in step 3.

5. Determine the change-out rate for the adsorbent as set by the breakthrough time.

6. Match the computed design requirements to available air cleaning equipment and specify.

Manufacturers' Design Guidance. Most manufacturers of control system components offer selection guidance. Some of the approaches for traditional granular beds are summarized here. Note that inclusion of this information or exclusion of other approaches does not imply acceptance or endorsement by ASHRAE, but is meant to be an abbreviated overview of present-day practice. The general expectation is of an HVAC application service life of 9 to 18 months for a 3120 h air-conditioning year. These values are simply a summary of conventional wisdom directed at meeting that goal, and they can be substantially in error.

Manufacturers with laboratory testing facilities may evaluate and specify filters by measuring the initial removal efficiency of the whole filter installed in its frame (similar to the ASHRAE *Standard* 145.2 test), and measure the amount of contaminant removed as adsorbent capacity is consumed during the test. Curves of efficiency versus capacity are used to guide the customer through the selection process.

In situations where multiple contaminants at low concentrations occur, such as in most IAQ investigative work and applications, neither the total load nor specific contaminant can typically be determined. In these case, a broad-based control design approach is usually recommended, consisting of two media banks: activated carbon, followed by permanganate-impregnated alumina. Sometimes these two media are combined into one bank because of space or pressure drop limitations.

- For light-duty applications, the recommendation is to use particulate filter media infused with carbon and/or permanganate media and pleated into a traditional filter design.
- For medium-duty applications, granular media are used in 1 to 3 in. deep refillable or disposable bulk-fill modules for increased efficiency and service life.

A useful approach for carbon-based air cleaners divides HVAC and IAQ applications into three categories and recommends specific equipment selection based on efficiency and activated carbon weight, as follows:

1. Heavy-duty outdoor air or mixed air IAQ applications with a relatively constant VOC generation rate and relatively constant moderate to severe outdoor air pollution:

   - For cleaned-air-equivalent air quality (i.e., air cleaned well enough to substitute for outdoor air), use equipment with >90% efficiency and 75 to 100 lb of high-grade carbon per 2000 cfm For a 1 in. bed in common commercial unitary adsorbers, this corresponds to a 0.1 s residence time.
   - With severe outdoor air pollution from nearby sources, it may be necessary to size equipment at 90 lb of carbon per 1000 to 1400 cfm, which corresponds to 0.14 to 0.2 s residence time.

2. For medium-duty return or mixed air IAQ applications with constant low to moderate VOC generation and cleaned-air-equivalent air quality, use equipment efficiencies of 20 to 90% and 8 to 75 lb of carbon per 2000 cfm. This corresponds to partial bypass equipment at the low-efficiency end and ranges up to about 0.08 s residence time at the high-efficiency end.

3. For light-duty mixed air IAQ applications with intermittent low to high VOC generation and an intermittent low to high outdoor air pollution load, use equipment efficiencies of >75% for odor control, which corresponds to a partial bypass design.

Another form of manufacturer's guidance of activated carbons is shown in Table 8, which gives suggested packed-bed residence time ranges, developed for the 4 × 6 or 4 × 8 mesh coconut shell carbon typically used in packed beds, for various applications. Residence times in Table 8 are appropriate for moderate to thick beds of large-particle carbons, but do not apply universally to commercial adsorbents. Different ways of arranging the carbon, different adsorbents, or different carbon granule sizes change the residence time required to get a particular result. This is especially true of very finely dispersed activated carbon, which has very fast adsorption kinetics. In addition, the geometry and packaging of some adsorbent technologies make computation of residence time difficult. For instance, cylindrical beds with radial flow have air velocities that decrease from the center to the periphery, so special computational techniques are needed to put residence time on the same basis as for a flat bed. Similarly, the flow pattern in pleated fiber/carbon composite media is difficult to specify, making residence time computation uncertain. Therefore, although residence times can be computed for partial-bypass filters, fiber-adsorbent composite filters, or fiber-bonded filters, they cannot be compared directly with those in Table 8, and serve more as a rating than as an actual residence time. By applying residence time as a rating, manufacturers may publish equivalent residence time values that say, in effect, that this adsorber performs the same as a traditional deep bed adsorber. No standard test exists to verify such a rating.

Table 8 Suggested 4 × 6 or 4 × 8 Mesh Coconut Shell Carbon Residence Time Ranges

Application	Residence Time,[a] s
HVAC odor control applications for indoor air quality	
Light to medium duty	0.03 to 0.07
Heavy duty	0.06 to 0.14
Cleanroom corrosion control	
Recirculation	0.03 to 0.07
Intakes	0.03 to 0.14
Industrial corrosion control (refineries, wastewater, pulp, etc.)	0.12 to 0.28+
Corrosive/reactive low-level exhaust applications	0.12 to 0.28+
Toxic gas control[b]	0.28+
Museums	
Standard applications	0.06 to 0.12
Recirculation applications	0.03 to 0.07
Critical air intake	0.28+
Critical recirculation application	0.12 to 0.28
Nuclear applications	0.12 to 0.28+

Notes: [a]All residence times given are rules of thumb using 4 × 6 or 4 × 8 mesh carbon in granular beds for the application indicated. Other carbon packages, such as pleated filters containing finely divided carbon, may be dramatically different. Particularly at low (ppb) contaminant concentrations, well-designed pleated carbon filters can be very efficient and have adequate capacity.
[b]Design for toxic gas control must include appropriate overdesign and fail-safe provisions for all equipment used.
Source: Adapted and used with permission. ©1992 Extraction Systems, Inc.

Special Cases

Ozone reaches an equilibrium concentration in a ventilated space without a filtration device. It does so partly because ozone molecules react in air to form oxygen, but also because they react with people, plants, and materials in the space. This oxidation is harmful to all three, and therefore natural ozone decay is not a satisfactory way to control ozone except at low concentrations (<0.1 ppmv). Fortunately, activated carbon adsorbs ozone readily, both reacting with it and catalyzing its conversion to oxygen.

Radon is a radioactive gas that decays by alpha-particle emission, eventually yielding individual atoms of polonium, bismuth, and lead. These atoms form ions, called radon daughters or radon progeny, which are also radioactive; they are especially toxic, lodging deep in the lung, where they emit cancer-producing alpha and beta particles. Radon progeny, both attached to larger aerosol particles and unattached, can be captured by particulate air filters. Radon gas itself may be removed with activated carbon, but in HVAC systems this method costs too much for the benefit derived. Control of radon emission at the source and ventilation are the accepted methods of radon control.

Museums, libraries, archives, and similar applications are special cases of air cleaner design for protection, and may require very efficient air cleaning; see Chapter 23 for specifics.

Building protection applications, whether to protect occupants against industrial accidents or deliberate acts, cannot reasonably stand alone. It makes little sense to design a complex system that can be easily overcome by physical acts. Air cleaner design alone is not enough. Air cleaning must be part of a complete and in-depth security program. The designer must have a scenario or series of scenarios against which to design protective systems.

Because a given air-cleaning technology may not protect against all challenges, protection against deliberate acts requires a robust design. The air-cleaning technology can be chosen to protect against the most challenging contaminant. Once the challenge contaminants are identified, a more rigorous application of the design method described previously is applied. Military specification hardware is generally suitable, although high-end commercial designs can provide significant protection. Testing of the installed filters is generally required, maintenance costs are significant, and the cost in space allocated to the installation, energy, and capital is high.

U.S. government guidance for the designer is available online (FEMA 2003a, 2003b; NIOSH 2002, 2003). The FEMA Web site also includes a number of other applicable documents. For additional guidance, see Chapter 59.

Energy Concerns

Pressure drop across the contaminant filter directly affects energy use. Data on the resistance of the filter as a function of airflow and on the resistance of the heating/cooling coils must be provided by the manufacturer. Currently, no standard test of pressure drop across a full-scale gaseous air cleaner is specified, but users can require that the initial pressure drop measurement from the particulate test (ASHRAE *Standard* 52.2) be conducted and reported. In addition to the gaseous contaminant filter itself, pressure drop through the housing, any added duct elements, and any particulate filters required up- and/or downstream of the gaseous contaminant filter must be included in the energy analysis.

Choosing between using outside air only and outside air plus filtered recirculated air is complex, but can be based on technical or maintenance factors, convenience, economics, or a combination of these. An energy-consumption calculation is useful. Replacing outdoor air with filtered indoor air reduces the amount of air that must be conditioned at an added expense in recirculation pressure drop. Outdoor air or filtered recirculated air may be used in any ratio, provided the air quality level is maintained. Janssen (1989) discusses the logic of these requirements.

Where building habitability can be maintained with ventilation alone, an economizer cycle is feasible under appropriate outdoor conditions. However, economizer mode may not be feasible at high humidities, because high humidity degrades the performance of carbon adsorbents.

Economic Considerations

Capital and operating costs for each competing system should be identified. Chapter 37 provides general information on performing an economic analysis. Table 9 is a checklist of filtration items to be considered in an analysis. It is important that the fan maintain adequate flow with an in-line air cleaner in place. If a larger blower is required, space must be available. Modifying unitary equipment that was not designed to handle the additional pressure drop through air-cleaning equipment can be expensive. With built-up designs, the added initial cost of providing air cleaners and their pressure drop can be much less because the increases may be only a small fraction of the total.

The life of the adsorbent media is very important. The economic benefits of regenerating spent carbon should be evaluated in light of the cost and generally reduced activity levels of regenerated material. Regeneration of impregnated carbon or any carbon containing

Table 9   Items Included in Economic Comparisons Between Competing Gaseous Contaminant Control Systems

Capital Costs	Operating Costs
Added filtration equipment	Replacement or reactivation of gaseous
Fan	contaminant filter media
Motor	Disposal of spent gaseous contaminant
Controls	filter media
Plenum	Added electric power
Spare media holding units	Maintenance labor
Floor space	

hazardous contaminants is never permitted. Spent alumina- or zeolite-based adsorbents also cannot be regenerated.

SAFETY

Gaseous contaminant removal equipment generally has a low hazard potential. Contaminant concentrations are low, temperature is moderate, and the equipment is normally not closed in. Alumina- or zeolite-based media do not support combustion, but carbon filter banks have been known to catch fire, usually from an external source such as a welder's torch. Check local codes and fire authorities for their regulations on carbon. One authority requires automatic sprinklers in the duct upstream and downstream of carbon filter banks. As a minimum, a smoke detector should be installed downstream of the filter bank to shut down the fan and sound an alarm in case of fire.

Access for safe maintenance and change-out of adsorbent beds must be provided. Physical and chemical adsorbents are both much heavier than particulate filters. Suitable lifting equipment must be available during installation and removal to prevent injury.

If adsorbent trays are to be refilled on site, safety equipment must be provided to deal with the dust this generates. Hooding, dust masks, and gloves are all required to refill adsorber trays from bulk containers.

INSTALLATION, START-UP, AND COMMISSIONING

This section provides general guidance on installing gaseous contaminant removal equipment. Most manufacturers can also provide complete details and drawings for design.

Particulate Filters. A minimum efficiency reporting value (MERV) 7 particulate filter (per ASHRAE *Standard* 52.2) should be installed ahead of the adsorbent bed. Higher efficiency particulate filtration is often desirable. Physical adsorbents and chemisorbents cannot function properly if their surfaces are covered and their pores clogged with dirt. UV-PCO system performance is also degraded unless protected by prefilters. If the air is extremely dirty (e.g., from diesel exhaust), the filter should have a much higher efficiency. One manufacturer requires a MERV 14 particulate filter for such applications. Weschler et al. (1994) report that carbon service life for ozone control was lengthened by using improved prefiltration.

Afterfilters are often used in critical applications where dust from media at start-up is likely, or where vibration of the adsorbent bed may cause granular media to shed particles. These filters are frequently MERV 7, but higher efficiencies may be needed in some applications.

Equipment Weight. Physical and chemical adsorption equipment is much heavier than particulate filtration equipment, so supporting structures and frames must be designed accordingly. A typical 24 by 24 in. unit consisting of a permanent holding frame and adsorbent-loaded trays has an installed weight of approximately 200 lb.

Minimize or Eliminate Bypass. Adsorbent beds and ducts in the outside air supply and in exhaust from hoods must be tightly sealed to prevent bypass of contaminants. Bypass leakage is not critical in most recirculating indoor air systems, but it is good practice to caulk all seams between individual holding frames. Granular media settles and compacts, and media retainers, such as trays or modules, must be loaded with media following manufacturers' recommendations to eliminate bypass through the media bed.

When to Install Media. When to install adsorbent beds in their holding frames depends on building circumstances. If they are installed at the same time as their holding frames and if the HVAC is turned on during the latter phases of construction, the adsorbers will adsorb paint and solvent vapors and other contaminants before the building is ready for beneficial occupancy. In some situations, adsorbing vapors and gases in the ventilation system before official start-up may be desired or needed. However, adsorbent life will be reduced correspondingly. If adsorbent beds are not loaded until the building is ready for occupancy, the unremoved contaminants may seriously reduce the initial indoor air quality of the building. Thus, shortened life is an acceptable trade-off for the quality of air at the time of occupancy (NAFA 1997). If the media are not in place during fan testing, the test and balance contractor must be instructed to place blank-offs or restrictions in the frames to simulate adsorbent bed pressure drop. The HVAC designer's job specifications must clearly state when media are to be installed.

Pressure Gages. If prefiltration is adequate, adsorber pressure drop will not increase during normal operation. A pressure-drop-measuring device (gage or manometer) is thus not required as it is for a particulate filter bank. However, a gage may be useful to detect fouling or unintentional bypass. If the prefilters or afterfilters are installed immediately adjacent to the adsorbers, it may be more feasible to install the gage across the entire assembly.

Provision for Testing. At any time after installation of new media, determining the remaining adsorbent capacity or operating life may be required. (See the section on When to Change Media under Operation and Maintenance.) The installation should provide access ports to the fully mixed airstream both up- and downstream of the air cleaner. If media samples will be removed to determine remaining life, access must be provided to obtain those samples. No standard method for field evaluation of media life currently exists.

Start-Up and Commissioning

Special procedures are not required during start-up of an air handler with installed gaseous contaminant air cleaners. The testing and balancing contractor normally is required to measure and record resistance of all installed filter banks, including adsorbers, for comparison with design conditions.

The commissioning authority may require an activity test on a random sample of media to determine if the new media suffered prior exposure that reduced its life or if it meets specifications. An in situ air sampling test may also be required on the adsorbent; however, no standard method for this test exists. See Chapter 43 for more on commissioning.

OPERATION AND MAINTENANCE

Bypass units and filters with adsorbent-infused fibrous media require frequent changing to maintain even low efficiency, but frequent maintenance is not required for complete removal units. Complete removal media units usually have a replaceable cell that cannot be regenerated or reactivated. This section covers maintenance of complete removal equipment with refillable trays or modules only.

When to Change Media

The changeout point of physical and chemical adsorbent is difficult to determine. Sometimes media are changed when breakthrough occurs and occupants complain; but if the application is sensitive, tests for estimated residual activity may be made periodically. A sample of the media in use is pulled from the adsorbent bed or from a pilot cell placed in front of the bed. The sample is sent to the manufacturer or an independent test laboratory for analysis, and the changeout time is estimated knowing the time in service and the life remaining in the sample.

In corrosion control installations, specially prepared metal "coupons" are placed in the space being protected by the adsorbent. After some time, usually a month, the coupons are sent to an analytical lab for measurement of corrosion thickness, which indicates the effectiveness of the gaseous contaminant control and provides an indication of system life. A standardized methodology for these tests is described in ISA *Standard* 71.04.

Replacement and Reactivation

Replacing granular media in permanent trays or modules is not the same as reactivation (regeneration), which is the process of restoring spent activated carbon media to its original efficiency (or close to it). In some unit operations in some industrial applications (e.g., pressure swing adsorption), spent carbon is regenerated in special high-temperature vessels in the absence of oxygen to drive off contaminants. Chemisorber modules can be replaced (or media changed), but chemisorber media, including impregnated carbon, cannot be regenerated.

Building operating personnel may choose to dump and refill trays and modules at the site after replacing those removed with a spare set already loaded with fresh media. They may also choose to dump the trays locally and send the empty trays to a filter service company for refilling, or they may simply exchange their spent trays for fresh ones. Disposing of spent sorbent by dumping must be limited to building air quality applications where no identifiable hazardous chemicals have been collected.

ENVIRONMENTAL INFLUENCES ON AIR CLEANERS

Environmental conditions, particularly temperature and humidity, affect the performance of most gaseous contaminant control equipment. Physical adsorbents such as activated carbon are particularly susceptible. The user should confirm performance for any control device at the expected normal environmental conditions as well as at extremes that might be encountered during equipment outages. The following information is an overview.

High relative humidity in the treated airstream lowers efficiency of physical adsorbers, such as carbon, because of competition for adsorption sites from the much more numerous water molecules. Often, performance is relatively stable up to 40 to 50% rh, but some compounds can degrade at higher humidities. The chemical nature of the contaminant(s) and the concentration both affect performance degradation as a function of relative humidity. On the other hand, very low relative humidities may make some chemisorption impossible. Therefore, media performance must be evaluated over the expected range of operation, and the relative humidity and temperature of the gaseous contaminant control should be held within design limits.

The effect of relative humidity swings can be better understood by considering a hypothetical physical adsorbent with a saturation capacity for a contaminant of 10% at 50% rh and 5% at 70% rh. Over an extended period at its normal operating condition of 50% rh, the sorbent might reach a loading of 2%. At this point a humidity swing to 70% rh would not cause a problem, and the adsorbent could load up to 5% capacity. Should the humidity then swing back to 50%, the adsorbent could continue to adsorb up to 10% by weight of the contaminant. However, if the adsorbent were loaded to 8% by weight at 50% rh and the humidity rose to 70% rh, the carbon would be above its equilibrium capacity and desorption would occur until equilibrium was reached.

Similarly, swings in temperature and contaminant concentration can affect physical adsorbent performance. Increasing temperature reduces capacity, and increasing concentration increases capacity. Additionally, changes in the identity of the contaminant in the airstream can affect overall performance as strongly adsorbed contaminants displace weakly held contaminants.

All physical and chemical adsorption media have a modest ability to capture dust particles and lint, which eventually plug the openings in and between media granules and cause a rapid rise in the pressure drop across the media or a decrease in airflow. All granular gaseous adsorption beds need to be protected against particle buildup by installing particulate filters upstream. A prefilter with a minimum ASHRAE *Standard* 52.2 MERV of 7 is recommended.

Vibration breaks up the granules to some degree, depending on the granule hardness. ASTM *Standard* D3802 describes a test for measuring the resistance of activated carbon to abrasion. Critical systems using activated carbon require hardness above 92%, as described by *Standard* D3802.

Physical adsorption and chemisorption media sometimes accelerate corrosion of metals they touch. Consequently, media holding cells, trays, and modules should not be constructed of uncoated aluminum or steel. Painted steel or ABS plastic are common and exhibit good material service lives in many applications. Coated or stainless steel components may be required in more aggressive environments.

TESTING MEDIA, EQUIPMENT, AND SYSTEMS

Testing may be conducted in the laboratory with small-scale media beds or small pieces of treated fabric or composite material; on full-scale air cleaners in a laboratory test rig capable of generating the test atmosphere; or in the field. Laboratory tests with specific challenge gases are generally intended to evaluate media for developmental, acceptance, or comparative purposes. Full-scale tests using specific challenge contaminants are required to evaluate a complete adsorber as constructed and sold, and ultimately are needed to validate performance claims. Field tests under actual job conditions are used to ensure that the air cleaners were properly installed and to evaluate remaining media life.

Laboratory Tests of Media and Complete Air Cleaners

Small granular media samples have been tested in a laboratory for many years, and most manufacturers have developed their own methods. ASTM *Standard* D5228 describes a test method, but it is not entirely applicable to HVAC work because indoor air tends to have a wide range of contaminants at concentrations several orders of magnitude lower than used in testing.

Fundamental Media Properties Evaluation. The test used to evaluate fundamental properties of physical adsorption media is static: measurement of the adsorption isotherm. In this test, a small sample of the adsorbent media is exposed to the pollutant vapor at successively increasing pressures, and the mass of pollutant adsorbed at each pressure is measured. The low-pressure section of an isotherm can be used to predict kinetic behavior, although the calculation is not simple. For many years, the test outlined in ASTM *Standard* D3467, which measured a single point on the isotherm of an activated carbon using carbon tetrachloride (CCl_4) vapor, was widely used for specifying performance of activated carbons. The test has been replaced by one described in ASTM *Standard* D5228, which uses butane as a test contaminant, because of carbon tetrachloride's toxicity. A correlation has been developed between the results of the two procedures so that users accustomed to CCl_4 numbers can recognize the performance levels given by ASTM D5228. It is a qualitative measure of performance at other conditions, and a useful quality control procedure. Another qualitative measure of performance is the Brunauer-Emmett-Teller (BET) method (ASTM *Standard* D4567), in which the surface area is determined by measuring the mass of an adsorbed monolayer of nitrogen. The results of this test are reported in square metres per gram of sorbent or catalyst. This number is often used as an index of media quality, with high numbers indicating high quality.

Small-Scale Dynamic Media Testing. ASHRAE *Standard* 145.1 provides a flow-through test of physical and chemical adsorption media at small scale (about 2 in. diameter test bed) and relatively high concentration (100 ppm). The test is intended to provide data for meaningful comparisons of media, provided that the same contaminant gas challenge is used for each. *Standard* 145.1 was developed based on several publications describing similar test procedures. Steady concentrations of a single contaminant are fed

to a media bed, and the downstream concentration is determined as a function of the total contaminant captured by the filter (ASTM 1989; Mahajan 1987, 1989; Nelson and Correia 1976).

Because physical adsorbent performance is a function of concentration, testing at high concentrations does not directly predict performance at low concentrations. The corrections described in the section on Physical Adsorption must be applied. If attempts are made to speed the test by high-concentration loading, pollutant desorption from the filter may confuse the results (Ostojic 1985). An adsorber cannot be tested for every pollutant, and there is no general agreement on which contaminants should be considered typical. Nevertheless, tests run according to the previously mentioned references do give useful measures of filter performance on single contaminants, and they do give a basis for estimates of filter penetrations and filter lives. Filter penetration data thus obtained can be used as *P* in Equations (3), (4), and (7) to estimate steady-state indoor concentrations.

VanOsdell et al. (2006) presented data from ASHRAE Research Project RP-792 showing that, for particular VOCs adsorbed on activated carbon, tests at high concentration could be extrapolated to indoor levels. In addition, tests of a carbon with one chemical, toluene, were used to predict breakthough times for four other chemicals with modest success. The breakthrough times correlated well on a log-log plot of breakthrough time versus challenge concentration, and the predictions, based only on chemical properties and toluene and carbon performance data, were within approximately 100% of measured.

Because physical adsorbers, chemisorbers, and catalysts are affected by the temperature and relative humidity of the carrier gas and the moisture content of the filter bed, they should be tested over the range of conditions expected in the application. Contaminant capture and reaction product generation need to be evaluated by sweeping the test filter with unpolluted air and measuring downstream concentrations. Reaction products may be as toxic, odorous, or corrosive as captured pollutants.

Full-Scale Laboratory Tests of Complete Air Cleaners. Full-scale tests of in-duct air cleaners are the system test analogs of the media tests described previously. ASHRAE is developing *Standard* 145.2, which details a full-scale performance test for in-duct air cleaners in a laboratory setting. In critical applications, such as chemical warfare protective devices and nuclear safety applications, sorption media are evaluated in a small canister, using the same carrier gas velocity as in the full-scale unit. The full-scale unit is then checked for leakage through gaskets, structural member joints, and thin spots or gross open passages in the sorption media by feeding a readily adsorbed contaminant to the filter and probing for its presence downstream (ASME *Standard* AG-1). Some HVAC filter manufacturers perform full-scale laboratory testing routinely for development and testing.

Chamber Decay Laboratory Tests of Complete Air Cleaners. Gaseous contaminant control devices can be tested in sealed chambers by recirculating contaminated air through them and measuring the decay of an initial contaminant concentration over time. Chamber decay tests are generally used for devices that physically cannot be tested in a duct, or that have single-pass efficiencies so low that they cannot be reliably measured using up- and downstream measurements. The procedures used are gaseous contaminant analogs to the Association of Home Appliance Manufacturers (AHAM) particulate air cleaner test method, but no consensus test standard exists and test methods vary between laboratories. Decay tests can provide valuable data, but the results are affected by factors extraneous to the air cleaner itself, such as errors introduced by adsorption on the chamber surfaces, leaks in the chamber, and the drawing of test samples. Robust test quality assurance and quality control are required to obtain meaningful data. Daisey and Hodgson (1989) compared the pollutant decay rate with and without the control device to overcome these uncertainties.

Field Tests of
Installed Air Cleaners

Gaseous contaminant air cleaners are expensive enough to place a premium on using their full capacity (service life). For odor control applications, the most reliable measure of the continued usefulness of gaseous contaminant air cleaners may be a lack of complaints, and complaints often serve as early indicators of exhausted granular beds. This approach is not acceptable for toxic contaminants, for which more formal procedures must be used. Unfortunately, and despite significant effort, there is not a simple, accepted standard for field-testing the capacity of gaseous air-cleaning equipment. Two approaches have been used: (1) laboratory testing of samples of media removed from the field and (2) in situ upstream/downstream gas measurements using the ambient contaminant(s) as the challenge.

The classic and still widely used technique of evaluating the status of a granular air cleaner in the field is to remove a small sample and ship it to a laboratory. Media manufacturers offer this service. The sample needs to be representative of the air cleaner, so should be obtained from near the center of the air cleaner. Consolidated multiple small samples are more representative of the air cleaner than one larger sample. The sample needs to be handled and shipped so that it remains representative when it reaches the laboratory; laboratories can suggest procedures. Given a representative sample and good handling, sampled media test results give a good indication of the state of a media bed.

Unfortunately, media sampling as a field-testing procedure has disadvantages. First, it applies only to granular media that can be sampled; cutting a hole in a bonded or pleated media product to obtain a sample destroys the filter. In addition, opening a granular media air cleaner to obtain samples is sufficiently disagreeable to prevent its frequent application.

These disadvantages have led to attempts to find more convenient ways to evaluate filters in the field. The most widely used alternative approach is to sample the air up- and downstream of the filter and use the ratio to estimate the remaining filter capacity.

Field tests of gaseous contaminant air cleaners are conducted using the same general techniques discussed under contaminant measurement and analysis. Depending on the contaminant, type of air cleaner, and application, field testing can be accomplished with active or passive sampling techniques. Liu (1998) discussed the relative merits of the various techniques. Either may be superior in any given case. For indoor air applications with relatively constant contaminant sources, passive samplers have advantages by capturing an integrated sample and being more economical. (Real-time samplers are used infrequently in this role.)

Up- and downstream measurements are evaluated by converting them to efficiency or fractional penetration and comparing them to measurements made at the time of installation. Because gaseous challenge contaminants cannot be injected into the HVAC system in occupied buildings, the up- and downstream samplers are exposed only to the ambient contaminants, which usually vary in nature and concentration. This complicates interpretation of the data, because air cleaner efficiency varies with concentration and nature of the contaminant. Efficiency from field samples is most directly interpreted if there is a single contaminant (or relatively consistent group of contaminants evaluated as TVOC) with a relatively constant challenge concentration. For multiple contaminants at multiple concentrations, judgment and experience are needed to interpret downstream measurements. Bayer et al. (2005) attempted to use single-component analyses to evaluate air cleaners in the field. That is, the up- and downstream samples were analyzed not for TVOC, but for particular chemicals (heptane, toluene, ethyl benzene, and formaldehyde). Bayer et al. found that the time variations in the challenge made the efficiency determinations so variable that the procedure could not be used reliably.

REFERENCES

ACGIH. 2010. *Industrial ventilation: A manual of recommended practice*, 27th ed. American Conference of Governmental Industrial Hygienists, Cincinnati, OH.

Anthony, C.P. and G.A. Thibodeau. 1980. *Textbook of anatomy and physiology.* C.V. Mosby, St. Louis.

ASHRAE. 2007. Method of testing general ventilation air-cleaning devices for removal efficiency by particle size. ANSI/ASHRAE *Standard* 52.2-2007.

ASHRAE. 2010. Ventilation for acceptable indoor air quality. ANSI/ASHRAE *Standard* 62.1-2010.

ASHRAE. 2008. Laboratory test method for assessing the performance of gas-phase air-cleaning media: Loose granular media. ANSI/ASHRAE *Standard* 145.1-2008.

ASHRAE. (in progress). Method of testing gaseous contaminant air cleaning devices for removal efficiency. ASHRAE *Draft Standard* 145.2.

ASME. 2009. Code on nuclear air and gas treatment. ANSI/ASME *Standard* AG-1-2009. American Society of Mechanical Engineers, New York.

ASTM. 1980. Building air change rate and infiltration measurements. *STP* 719. American Society for Testing and Materials, West Conshohocken, PA.

ASTM. 2009. Standard test method for carbon tetrachloride activity of activated carbon. *Standard* D3467-04 (R2009). American Society for Testing and Materials, West Conshohocken, PA.

ASTM. 2008. Standard practice for sampling atmospheres to collect organic compound vapors (activated charcoal tube adsorption method). *Standard* D3686-08. American Society for Testing and Materials, West Conshohocken, PA.

ASTM. 2010. Standard test method for ball-pan hardness of activated carbon. *Standard* D3802-10. American Society for Testing and Materials, West Conshohocken, PA.

ASTM. 2008. Standard test method for single-point determination of specific surface area of catalysts and catalyst carriers using nitrogen adsorption by continuous flow method. *Standard* D4567-03 (R2008). American Society for Testing and Materials, West Conshohocken, PA.

ASTM. 2010. Standard test method for determination of butane working capacity of activated carbon. *Standard* D5228-92 (R2010). American Society for Testing and Materials, West Conshohocken, PA.

ASTM. 2010. Standard test method for determination of butane activity of activated carbon. *Standard* D5742-95 (R2010). American Society for Testing and Materials, West Conshohocken, PA.

ASTM. 2006. Standard test method for determining air change in a single zone by means of a tracer gas dilution. *Standard* E741-00 (R2006). American Society for Testing and Materials, West Conshohocken, PA.

Bayer, C.W. and R.J. Hendry. 2005. Field test methods to measure contaminant removal effectiveness of gas phase air filtration equipment. *ASHRAE Transactions* 111(2):285-298.

Boeniger, M.F. 1995. Use of ozone generating devices to improve indoor air quality. *American Industrial Hygiene Association Journal* 56:590-598.

Brugnone, R., L. Perbellini, R.B. Faccini, G. Pasini, G. Maranelli, L. Romeo, M. Gobbi, and A. Zedde. 1989. Breath and blood levels of benzene, toluene, cumene and styrene in nonoccupational exposure. *International Archives of Environmental Health* 61:303-311.

Burton, D.J. 2003. *Industrial ventilation workbook*, 6th ed. American Conference of Governmental Industrial Hygienists, Cincinnati, OH.

Chen, W., J.S. Zhang, and Z. Zhang. 2005. Performance of air cleaners for removing multiple volatile organic compounds in indoor air. *ASHRAE Transactions* 111:1101-1114.

CIWMB. 2003. Building material emissions study. California Integrated Waste Management Board, *Publication* 433-03-015. Available from http://www.calrecycle.ca.gov/Publications/default.asp?pubid=1027.

Cohen, S.I., N.M. Perkins, H.K. Ury, and J.R. Goldsmith. 1971. Carbon monoxide uptake in cigarette smoking. *Archives of Environmental Health* 22(1):55-60.

Cole, J.T. 1983. Constituent source emission rate characterization of three gas-fired domestic ranges. APCA *Paper* 83-64.3. Air and Waste Management Association, Pittsburgh.

Conkle, J.P., B.J. Camp, and B.E. Welch. 1975. Trace composition of human respiratory gas. *Archives of Environmental Health* 30(6):290-295.

Daisey, J.M. and A.T. Hodgson. 1989. Initial efficiencies of air cleaners for the removal of nitrogen dioxide and volatile organic compounds. *Atmospheric Environment* 23(9):1885-1892.

DeFrees, J.A. and R.F. Amberger. 1987. Natural infiltration analysis of a residential high-rise building. *IAQ '87—Practical Control of Indoor Air Problems*, pp. 195-210. ASHRAE.

d'Hennezel, O., P. Pichat, and D.F. Ollis. 1998. Benzene and toluene gas-phase photocatalytic degradation over H_2O and HCl pretreated TiO2: By-products and mechanisms. *Journal of Photochemistry and Photobiology A: Chemistry* 118(3):197-204.

Elliott, L.P. and D.R. Rowe. 1975. Air quality during public gatherings. *Journal of the Air Pollution Control Association* 25(6):635-636.

EPA/NIOSH. 1991. *Building air quality—A guide for building owners and facility managers*. U.S. Environmental Protection Agency.

EPA. 1986. *Radon reduction techniques for detached houses: Technical guidance*, 2nd ed. EPA/625/5-86/O19. U.S. Environmental Protection Agency, Center for Environmental Research Information, Cincinnati, OH.

EPA. 1987. *Radon reduction methods: A homeowner's guide*, 2nd ed. EPA-87-010. U.S. Environmental Protection Agency, Center for Environmental Research Information, Cincinnati, OH.

EPA. 1997a. *Data from the building assessment survey and evaluation (BASE) study*. Available from http://www.epa.gov/iaq/base/summarized_data.html.

EPA. 1997b. National ambient air quality standards for ozone. *Federal Register* 62(138):38856. U.S. Environmental Protection Agency, Washington, D.C.

EPA. 2009. *Data from the air quality system measurements*. Available from http://www.epa.gov/ttn/airs/airsaqs/detaildata/.

FEMA. 2003a. Reference manual to mitigate potential terrorist attacks against buildings. FEMA *Publication* 426. Available from http://www.fema.gov/library/viewRecord.do?id=1559.

FEMA. 2003b. Primer for design of commercial buildings to mitigate terrorist attacks. FEMA *Publication* 427. Available from http://www.fema.gov/library/viewRecord.do?id=1560.

Fujishima, A. and K. Honda. 1972. Electrochemical photolysis of water at a semiconductor electrode. *Nature* 238:37-38.

Girman, J.R., A.P. Hodgson. A.F. Newton, and A.R Winks. 1984. Emissions of volatile organic compounds from adhesives for indoor application. *Report* LBL 1 7594. Lawrence Berkeley Laboratory, Berkeley, CA.

Gorban, C.M., I.I. Kondratyeva, and L.Z. Poddubnaya. 1964. Gaseous activity products excreted by man in an airtight chamber. In *Problems of Space Biology*. JPRS/NASA. National Technical Information Service, Springfield, VA.

Hunt, R.D. and D.T. Williams. 1977. Spectrometric measurement of ammonia in normal human breath. *American Laboratory* (June):10-23.

ISA. 1985. Environmental conditions for process measurement and control systems: Airborne contaminants. *Standard* 71.04-1985. Instrument Society of America, Research Triangle Park, NC.

Janssen, J.H. 1989. Ventilation for acceptable indoor air quality. *ASHRAE Journal* 31(10):40-45.

Janni, K.A., W.J. Maier, T.H. Kuehn, C.-H. Yang, B.B. Bridges, D. Vesley, and M.A. Nellis. 2001. Evaluation of biofiltration of air—An innovative air pollution control technology (RP-880). *ASHRAE Transactions* 107(1):198-214.

Leaderer, B.P., R.T. Zagranski, M. Berwick, and J.A.J. Stolwijk. 1987. Predicting NO_2 levels in residences based upon sources and source uses: A multi-variate model. *Atmospheric Environment* 21(2):361-368.

Lee, H.K., T.A. McKenna, L.N. Renton, and J. Kirkbride. 1986. Impact of a new smoking policy on office air quality. *Proceedings of Indoor Air Quality in Cold Climates*, pp. 307-322. Air and Waste Management Association, Pittsburgh.

Liu, R.-T. 1998. Measuring the effectiveness of gas-phase air filtration equipment—Field test methods and applications. *ASHRAE Transactions* 104(2):25-35.

Mahajan, B.M. 1987. *A method of measuring the effectiveness of gaseous contaminant removal devices*. NBSIR 87-3666. National Institute of Standards and Technology, Gaithersburg, MD.

Mahajan, B.M. 1989. *A method of measuring the effectiveness of gaseous contaminant removal filters*. NBSIR 89-4119. National Institute of Standards and Technology, Gaithersburg, MD.

Martin, P., D.L Heavner, P.R. Nelson, K.C. Maiolo, C.H. Risner, P.S. Simmons, W.T. Morgan, and M.W. Ogden. 1997. Environmental tobacco smoke (ETS): A market cigarette study. *Environment International* 23(1):75-89.

Meckler, M. and J.E. Janssen. 1988. Use of air cleaners to reduce outdoor air requirements. *IAQ '88: Engineering Solutions to Indoor Air Problems*, pp. 130-147. ASHRAE.

Moschandreas, D.J. and S.M. Relwani. 1989. Field measurement of NO_2 gas-top burner emission rates. *Environment International* 15:499-492.

NAFA. 1997. *Installation, operation and maintenance of air filtration systems*. National Air Filtration Association, Washington, D.C.

Nazaroff, W.W. and G.R. Cass. 1986. Mathematical modeling of chemically reactive pollutants in indoor air. *Environmental Science and Technology* 20:924-934.

Nefedov, I.G., V.P. Savina, and N.L. Sokolov. 1972. Expired air as a source of spacecraft carbon monoxide. Presented at 23rd International Aeronautical Congress, International Astronautical Federation, Paris.

Nelson, G.O. and A.N. Correia. 1976. Respirator cartridge efficiency studies: VIII. Summary and Conclusions. *American Industrial Hygiene Association Journal* 37:514-525.

Nelson, P.R., Sears, S.B., and D.L. Heavner. 1993. Application of methods for evaluating air cleaner performance. *Indoor Environment* 2:111-117.

NIOSH. 2002. Guidance for protecting building environments from airborne chemical, biological, or radiological attacks. DHHS (NIOSH) *Publication* 2002-139. Department of Health and Human Services, National Institute for Occupational Safety and Health, Cincinnati.

NIOSH. 2003. Guidance for filtration and air-cleaning systems to protect building environments from airborne chemical, biological, or radiological attacks. DHHS (NIOSH) *Publication* 2003-136. (Available at http://www.cdc.gov/niosh/docs/2003-136, and in PDF at http://www.cdc.gov/niosh/docs/2003-136/pdfs/2003-136.pdf). Department of Health and Human Services, National Institute for Occupational Safety and Health, Cincinnati.

Obee, T.N. and R.T. Brown. 1995. TiO_2 photocatalysis for indoor air applications: Effects of humidity and trace contaminant levels on the oxidation rates of formaldehyde, toluene and 1,3-butadiene. *Environmental Science and Technology* 29(5):1223-1231.

Obee, T.N. and S.O. Hay. 1999. The estimation of photocatalytic rate constants based on molecular structure: Extending to multi-component systems. *Journal of Advanced Oxidation Technologies* 4(2):147-152.

OSHA. 1994. Asbestos. 29CFR1910.1000. *Code of Federal Regulations*, U.S. Occupational Safety and Health Administration, Washington, D.C.

Ostojic, N. 1985. Test method for gaseous contaminant removal devices. *ASHRAE Transactions* 91(2):594-614.

Peral, J. and D.F. Ollis. 1992. Heterogeneous photocatalytic oxidation of gas-phase organics for air purification: Acetone, 1-butanol, butyraldehyde, formaldehyde and *m*-xylene oxidation. *Journal of Catalysis* 136(2):554-565.

Peral, J., X. Domenech, and D.F. Ollis. 1997. Heterogeneous photocatalysis for purification, decontamination and deodorization of air. *Journal of Chemical Technology and Biotechnology* 70(2):117-140.

Rivers, R.D. 1988. Practical test method for gaseous contaminant removal devices. *Proceedings of the Symposium on Gaseous and Vaporous Removal Equipment Test Methods* (NBSIR 88-3716). National Institute of Standards and Technology, Gaithersburg, MD.

Rock, B.A. 2006. *Ventilation for environmental tobacco smoke—Controlling ETS irritants where smoking is allowed*. Elsevier, New York.

Shaughnessy, R.J. and L. Oatman. 1991. The use of ozone generators for the control of indoor air contaminants in an occupied environment. *IAQ '91: Healthy Buildings*, pp. 318-324. ASHRAE.

Sparks, L.E. 1988. *Indoor air quality model*, version 1.0. EPA-60018-88097a. U.S. Environmental Protection Agency, Research Triangle Park, NC.

Sterling, T.D. and D. Kobayashi. 1981. Use of gas ranges for cooking and heating in urban dwellings. *Journal of the Air Pollution Control Association* 31(2):162-165.

Tompkins, D.T., B.J. Lawnicki, W.A. Zeltner, and M.A. Anderson. 2005a. Evaluation of photocatalysis for gas-phase air cleaning—Part 1: Process, technical, and sizing considerations (RP-1134). *ASHRAE Transactions* 111(2):60-84.

Tompkins, D.T., B.J. Lawnicki, W.A. Zeltner, and M.A. Anderson. 2005b. Evaluation of photocatalysis for gas-phase air cleaning—Part 2: Economics and utilization (RP-1134). *ASHRAE Transactions* 111(2):85-95.

Traynor, G.W. I.A. Nitschke, W.A. Clarke, G.P. Adams, and J.E. Rizzuto. 1985. A detailed study of thirty houses with indoor combustion sources. *Paper* 85-30A.3. Air and Waste Management Association, Pittsburgh.

Underhill, D.T., G. Mackerel, and M. Javorsky. 1988. Effects of relative humidity on adsorption of contaminants on activated carbon. *Proceedings of the Symposium on Gaseous and Vaporous Removal Equipment Test Methods* (NBSIR 88-3716). National Institute of Standards and Technology, Gaithersburg, MD.

VanOsdell, D.W., C.E. Rodes, and M.K. Owen. 2006. Laboratory testing of full-scale in-duct gas air cleaners. *ASHRAE Transactions* 112(2):418-429.

Wade, W.A., W.A. Cote, and J.E. Yocum. 1975. A study of indoor air quality. *Journal of the Air Pollution Control Association* 25(9):933-939.

Weschler, C.J., H.C. Shields, and D.V. Naik. 1994. Ozone-removal efficiencies of activated carbon filters after more than three years of continuous service. *ASHRAE Transactions* 100(2):1121-1129.

Yoon, Y.H. and J.H. Nelson. 1988. A theoretical study of the effect of humidity on respirator cartridge service life. *American Industrial Hygiene Association Journal* 49(7):325-332.

BIBLIOGRAPHY

ATC. 1990. *Technical assistance document for sampling and analysis of toxic organic compounds in ambient air.* EPA/600/8-90-005. Environmental Protection Agency, Research Triangle Park, NC.

Berglund, B., U. Berglund, and T. Lindvall. 1986. Assessment of discomfort and irritation from the indoor air. *IAQ '86: Managing Indoor Air for Health and Energy Conservation*, pp. 138-149. ASHRAE.

Cain, W.S., C.R. Shoaf, S.F. Velasquez, S. Selevan, and W. Vickery. 1992. *Reference guide to odor thresholds for hazardous air pollutants listed in the Clean Air Act Amendments of 1990.* EPA/600/R-92/047. Environmental Protection Agency.

Chung, T-W., T.K. Ghosh, A.L. Hines, and D. Novosel. 1993. Removal of selected pollutants from air during dehumidification by lithium chloride and triethylene glycol solutions. *ASHRAE Transactions* 99(1):834-841.

Fanger, P.O. 1989. The new comfort equation for indoor air quality. *IAQ '89: The Human Equation: Health and Comfort*, pp. 251-254. ASHRAE.

Foresti, R., Jr. and O. Dennison. 1996. Formaldehyde originating from foam insulation. *IAQ '86: Managing Indoor Air for Health and Energy Conservation*, pp. 523-537. ASHRAE.

Freedman, R.W., B.I. Ferber, and A.M. Hartstein. 1973. Service lives of respirator cartridges versus several classes of organic vapors. *American Industrial Hygiene Association Journal* (2):55-60.

Gully, A.J., R.M. Bethea, R.R. Graham, and M.C. Meador. 1969. *Removal of acid gases and oxides of nitrogen from spacecraft cabin atmospheres.* NASA-CR-1388. National Aeronautics and Space Administration. National Technical Information Service, Springfield, VA.

Idem, S. 2002. Leakage of ducted air terminal connections. ASHRAE *Research Project RP-1132.*

Jonas, L.A., E.B. Sansone, and T.S. Farris. 1983. Prediction of activated carbon performance for binary vapor mixtures. *American Industrial Hygiene Association Journal* 44:716-719.

Kelly, T.J. and D.H. Kinkead. 1993. Testing of chemically treated adsorbent air purifiers. *ASHRAE Journal* 35(7):14-23.

Lodge, J.E., ed. 1988. *Methods of air sampling and analysis*, 3rd ed. Lewis Publishers, Chelsea, MI.

Mehta, M.R, H.E. Ayer, B.E. Saltzman, and R. Romk. 1988. Predicting concentrations for indoor chemical spills. *IAQ '88: Engineering Solutions to Indoor Air Problems*, pp. 231-250. ASHRAE.

Miller, G.C. and P.C. Reist. 1977. Respirator cartridge service lives for exposure to vinyl chloride. *American Industrial Hygiene Association Journal* 38:498-502.

Moschandreas, D.J. and S.M. Relwani. 1989. Field measurement of NO_2 gas-top burner emission rates. *Environment International* 15:499-492.

NAFA. 1996. *Guide to air filtration*, 2nd ed. National Air Filtration Association, Washington, D.C.

Nagda, N.L. and H.E. Rector. 1983. *Guidelines for monitoring indoor-air quality.* EPA 600/4-83-046. Environmental Protection Agency, Research Triangle Park, NC.

NIOSH. 1994. *NIOSH manual of analytical methods*, 4th ed. 2 vols. U.S. Department of Health and Human Services, National Institute for Occupational Safety and Health, Cincinnati.

NIOSH. (yearly) *Annual registry of toxic effects of chemical substances.* U.S. Department of Health and Human Services, National Institute for Occupational Safety and Health, Washington, D.C.

NIOSH. (intermittent). *Criteria for recommended standard for occupational exposure to* (compound). U.S. Department of Health and Human Services, National Institute for Occupational Safety and Health, Washington, D.C.

Nirmalakhandan, N.N. and R.E. Speece. 1993. Prediction of activated carbon adsorption capacities for organic vapors using quantitative structure activity relationship methods. *Environmental Science and Technology* 27:1512-1516.

Perry, Chilton, and Kirkpatrick, 1997. *Perry's chemical engineers' handbook*, 7th ed. McGraw-Hill, New York.

Revoir, W.H. and J.A. Jones. 1972. Superior adsorbents for removal of mercury vapor from air. AIHA conference paper. American Industrial Hygiene Association, Akron, OH.

Riggen, R.M., W.T. Wimberly, and N.T. Murphy. 1990. *Compendium of methods for determination of toxic organic compounds in ambient air.* EPA/600/D 89/186. Environmental Protection Agency, Research Triangle Park, NC. Also second supplement, 1990, EPA/600/4-89/017.

Surgeon General. 1979. *Smoking and health.* U.S. Department of Health and Human Services, National Institute for Occupational Safety and Health, Washington, D.C.

Taylor, D.G., R.E. Kupel, and J.M. Bryant. 1977. *Documentation of the NIOSH validation tests.* U.S. Department of Health and Human Services, National Institute for Occupational Safety and Health, Washington, D.C.

Turk, A. 1954. Odorous atmospheric gases and vapors: Properties, collection, and analysis. *Annals of the New York Academy of Sciences* 58:193-214.

Weschler, C.J. and H.C. Shields. 1989. The effects of ventilation, filtration, and outdoor air on the composition of indoor air at a telephone office building. *Environment International* 15:593-604.

White, J.B., J.C. Reaves, R.C. Reist, and L.S. Mann. 1988. A data base on the sources of indoor air pollution emissions. *IAQ '88: Engineering Solutions to Indoor Air Problems.* ASHRAE.

Winberry, W.T., Jr., L. Forehand, N.T. Murphy, A. Ceroli, B. Phinney, and A. Evans. 1990. *EPA compendium of methods for the determination of air pollutants in indoor air.* EPA/600/S4-90/010. Environmental Protection Agency. (NTIS-PB 90-200288AS).

Woods, J.E., J.E. Janssen, and B.C. Krafthefer. 1996. Rationalization of equivalence between the ventilation rate and air quality procedures in ASHRAE *Standard 62. IAQ '86: Managing Indoor Air for Health and Energy Conservation.* ASHRAE.

Yu, H.H.S. and R.R. Raber. 1992. Air-cleaning strategies for equivalent indoor air quality. *ASHRAE Transactions* 98(1):173-181.

DESIGN AND APPLICATION OF CONTROLS

AUTOMATIC control of HVAC systems and equipment usually includes control of temperature, humidity, pressure, and flow rates of air and water. Automatic controls can sequence equipment operation to meet load requirements and to provide safe equipment operation using pneumatic, mechanical, electrical, electronic, and/or direct digital control (DDC) devices. Automatic controls are only fully effective when applied to well-designed mechanical systems; they cannot compensate for misapplied systems, excessive under- or oversizing, or highly nonlinear processes.

This chapter addresses control of typical HVAC systems, design of controls for system coordination and for energy conservation, and control system commissioning. Chapter 7 of the 2009 *ASHRAE Handbook—Fundamentals* covers details of component hardware and the basics of control.

SYSTEM TYPES

A pneumatic, electronic, or direct digital control system has several physical control loops, with each loop including a controlled variable (e.g., temperature), controlled device (e.g., actuator), and the process to be controlled (e.g., heating coil). DDC systems can more easily share a sensor value with several control loops or have multiple control loops selectively activate an actuator.

DDC systems allow information such as system status or alarms to be collected in central controllers and shared between HVAC systems, enabling advanced, energy-saving, system-level applications through a common communication protocol. ASHRAE *Guideline* 13 and *Standard* 135 have more detailed discussions of networking and interoperability.

HEATING SYSTEMS

Heating systems include boilers, fired by either fuel combustion or electric resistance, direct flame-to-air furnaces, and electric resistance air heaters. Load affects the required rate of heat input to a heating system. The rate is controlled by cycling a fixed-intensity energy source on and off, or by modulating the intensity of the heating process. Flame cycling and modulation can be handled by the boiler control package, or the DDC can send commands to the boiler controls. The control designer decides under what circumstances to turn boilers on and off in sequence and, for hot-water boilers, at what temperature set point to maintain the boiler supply water.

Hot-Water and Steam Boilers

Hot-water distribution control includes temperature control at hot-water boilers or the converter, reset of heating water temperature, and control of pumps and distribution systems. Other controls to be considered include (1) minimum water flow through boilers, (2) protecting boilers from temperature shock and condensation on the heat exchanger, and (3) coil low-temperature detection. If multiple or alternative heating sources (e.g., condenser heat recovery, solar storage) are used, the control strategy must also include a way

to sequence hot-water sources or select the most economical source.

Figure 1 shows a system for load control of a fossil-fuel-fired boiler. Boiler safety controls, usually factory installed with the boiler, include flame-failure, high-temperature, and other cutouts. Field-installed operating controls must allow safety controls to function in all modes of operation. Intermittent burner firing usually controls capacity, although burner modulation is common in larger systems. In most cases, the boiler is controlled to maintain a constant water temperature, although an outdoor air thermostat or other control strategies can reset the temperature if the boiler is not used for domestic water heating. A typical outdoor air reset schedule is shown in Figure 1. With DDC, reset can be controlled from zone demand, which can improve energy performance and ensure all zones are satisfied. To minimize condensation of flue gases and boiler damage, water temperature should not be reset below that recommended by the manufacturer, typically 140°F entering water temperature, or condensation may occur and lead to corrosion-related failure. Condensing boilers are specifically designed to allow flue gases to condense, and should operate at lower water temperatures to harness latent energy in the flue gas. Aggressive reset of hot-water temperatures improves the efficiency of condensing boilers, because efficiency is a strong function of boiler entering water temperature. Systems with sufficiently high pump operating costs can use variable-speed pump drives to reduce secondary pumping capacity to match the load and conserve energy. ASME (2004) requires that, for boilers above a certain size, a manually operated remote shutdown switch be located outside the boiler room door to disconnect power to burner controls.

Hot-water heat exchangers or steam-to-water converters are sometimes used instead of boilers as hot-water generators. Converters typically do not include a control package; therefore, the engineer must design the control scheme. The schematic in Figure 2 can be used with either low-pressure steam or boiler water from 200 to 360°F. The supply water temperature sensor controls two modulating two-way valves in a 1/3 and 2/3 arrangement in a steam or high-temperature hot-water supply line. An outdoor temperature sensor (or zone demand for an integrated DDC system) can be used to reset the supply water temperature downward as load decreases

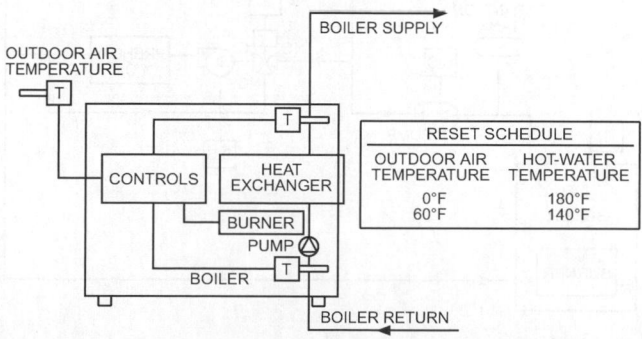

Fig. 1 Boiler Control

The preparation of this chapter is assigned to TC 1.4, Control Theory and Application.

to improve the controllability of heating valves at low load and to reduce piping losses. A flow or differential pressure switch interlock should close the two-way valve when the hot-water pump is not operating. Ensure that the flow switch will operate as expected at minimum flow rate on variable-flow systems. With an integrated DDC system, feedback from zone heating valves can be used to control starting and stopping of the hot-water pumps. When shutting down a steam converter or high-temperature hot-water system, close the steam valves and allow the water to circulate long enough to remove residual heat in the converter and prevent the pressure relief valve from opening.

Hot-Water Distribution Systems

Hot water is distributed using variable flow (primarily two-way valves at coils) or constant flow (three-way valves at coils). An example constant flow system is shown in Figure 3. Variable-flow systems are similar to the chilled-water distribution systems shown in Figures 10 and 11. Some boilers require constant flow or very high minimum flow rates. They typically are piped using a primary/ secondary system (see Figure 11). These boilers usually are required by their listing to have flow switches to enable the boiler only when flow is proven. Boilers that require small (or zero) minimum flow rates are usually piped in a primary-only configuration with a bypass to maintain minimum flow (see Figure 10). A flow meter in the boiler circuit is usually installed to control the bypass valve. The bypass can also be controlled to maintain minimum boiler entering water temperatures for noncondensing boilers.

Heating Coils

Heating coils that are not subject to freezing can be controlled by simple two- or three-way modulating valves (Figure 4). Steam-distributing coils are required to ensure proper steam coil control.

(For information on air-side coils, see the section on Air Systems.) The modulating valve is controlled by coil discharge air temperature or by space temperature, depending on the HVAC system. In cold regions, valves are set to open to allow heating if control power fails. In many systems, the outdoor air temperature resets the heating discharge air controller.

Heating coils in central air-handling units preheat, reheat, or heat, depending on the climate and the amount of minimum outdoor air needed. They can also provide morning warm-up on systems with limited zone heating capacity.

Heating coils using steam or hot water must have protection against freezing in cold climates, unless (1) the minimum outdoor air quantity is small enough to keep the mixed air temperature above freezing and (2) enough mixing occurs to prevent stratification. Even when the average mixed-air temperature is above freezing, inadequate air mixing may allow freezing air to impinge on small areas of the coil, causing localized freezing. This blocks flow and, without a heat source, the rest of the coil and equipment downstream is at risk. Preheating coils, those that heat 100% outdoor air, will always need protection against freezing in cold climates.

Steam preheat coils should have two-position valves and vacuum breakers to prevent condensate build-up in the coil. The valve should be fully open when outdoor (or mixed) air temperature is below freezing. This causes unacceptably high coil discharge temperatures at times, necessitating face-and-bypass dampers for final temperature control (Figure 5). The bypass damper should be sized to provide the same pressure drop at full bypass airflow as the combination of face damper and coil does at full airflow. When the outdoor air temperature is safely above freezing (roughly 35°F), the bypass damper is full open to the coil face and the coil valve can be modulated to improve controllability.

Hot-water coils must maintain a minimum water velocity in the tubes (on the order of 3 fps) to prevent freezing. A two-position valve combined with face-and-bypass dampers (Figure 5) or a coil

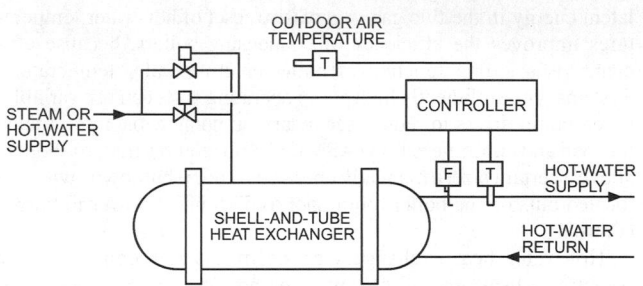

Fig. 2   Steam-to-Water Heat Exchanger Control

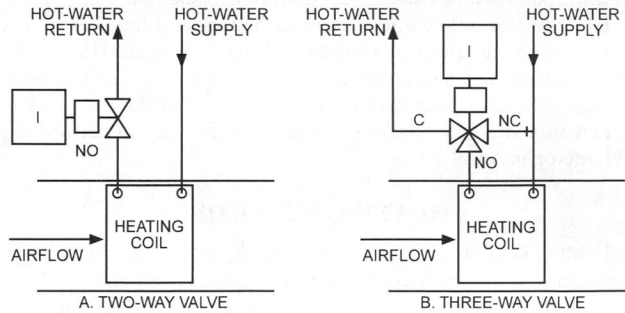

Fig. 4   Control of Hot-Water Coils

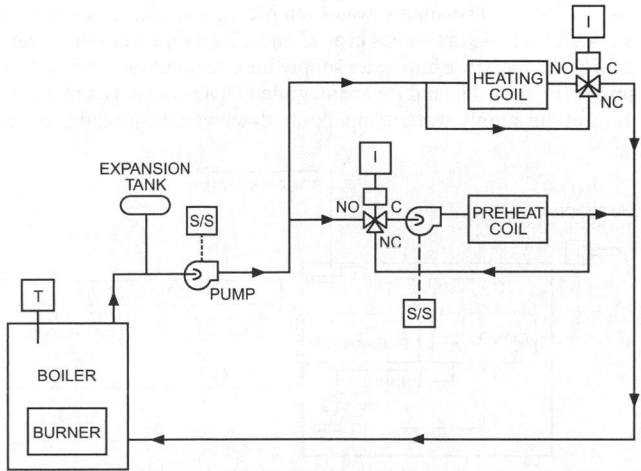

Fig. 3   Load and Zone Control in Constant Flow System

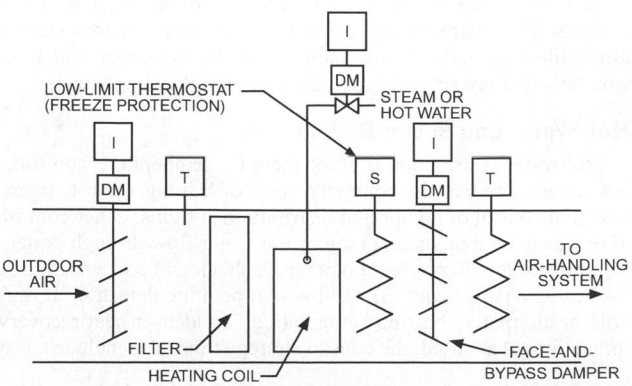

Fig. 5   Preheat with Face-and-Bypass Dampers

pump can be used. There are many coil pump piping schemes; the most common are shown in Figures 6 and 7. In each scheme, the control valve modulates to maintain the desired coil air discharge temperature and the pump maintains the minimum tube water velocity when the outdoor air is below freezing. Pumped coils can still freeze in very cold regions, so additional low-temperature detection measures such as glycol-based fluids should be used. Another protective device, a low-temperature detector (commonly called a freezestat), is a long, refrigerant-filled capillary tube used as a low-temperature sensing switch. If any short section of the tube is exposed to a low temperature (typically 38°F), it can provide an alarm or a hardwired interlock to shut the outdoor damper and open the return damper, or shut down the fan.

Figure 6 shows the conventional primary/secondary (or secondary/tertiary) arrangement where the coil pump and the pumps feeding the coil are hydraulically independent. It results in constant flow through the coil and in either variable flow through the primary loop, if a two-way valve is used, or in constant flow through the primary loop, if a three-way valve is used (shown dashed in the figure).

In Figure 7A, the coil pump is in series with the primary pumps, which can affect flow through parallel coils that do not have pumps when the three-way valve is open to the system. The pump

is decoupled when the three-way valve is closed to the system. The three-way valve must be oriented with the common port connected to the pump discharge so the pump is never deadheaded.

Figure 7B shows the coil pump piped in parallel with the primary pumps. This design has the advantage that hot-water flow can be achieved through the coil if either the primary pump or the coil pump fails. This design results in coil flow that varies from the pump design flow rate (when the control valve is closed) up through the sum of the pump flow rate plus the primary system flow rate (when the valve is wide open). Unlike the options in the previous two figures, the primary pump must be sized for the pressure drop through the coil at this high flow rate. Flow through the primary circuit may be variable, if a two-way valve is used, or constant, if a three-way valve is used.

Some systems may use a glycol solution in combination with any of these methods; however, glycol affects control valve sizing (see Chapter 46 of the 2008 *ASHRAE Handbook—HVAC Systems and Equipment*).

Steam Coils. Modulating steam coils are controlled in much the same way as water coils. Control valve size and characteristics are important to achieve proper control (see Chapter 46 of the 2008 *ASHRAE Handbook—HVAC Systems and Equipment*). Because the entering steam is hotter than the entering temperature of most water coils, a steam coil typically responds more rapidly than a comparable water coil. In low-temperature applications, two-position control should be used, as discussed previously. For large coils, control valves should be in a 1/3 and 2/3 arrangement.

Electric heating coils (duct heaters) are controlled in either two-position or modulating mode. Two-position operation uses power relays with contacts sized to handle the amperage of the heating coil. Timed two-position control requires a timer and contactors. The timer can be electromechanical, but it is usually electronic and provides a time base of 1 to 5 min. Step controllers provide cam-operated sequencing control of up to 10 stages of electric heat. Each stage may require a contactor, depending on the step controller contact rating. Thermostat demand determines the percentage of on-time. Because rapid cycling of mechanical or mercury contactors can cause maintenance problems, solid-state controllers are preferred. These devices make cycling so rapid that control appears proportional; therefore, face-and-bypass dampers are not used. Use of electric heating coils is restricted in some areas by energy standards. Code compliance should be checked before using this application. A control system with a solid-state controller and safety controls is shown in Figure 8.

Current in individual elements of electric duct heaters is normally limited to a maximum safe value established by the *National Electrical Code*® (NFPA 2005) or local codes. Two safety devices in addition to the airflow interlock device are usually applied to duct

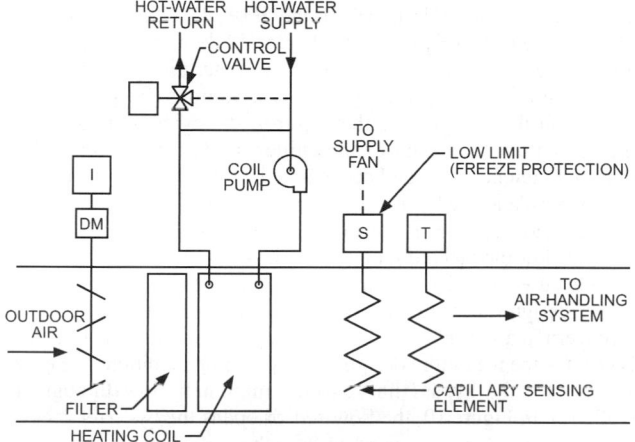

Fig. 6   Coil Pump Piped Primary/Secondary

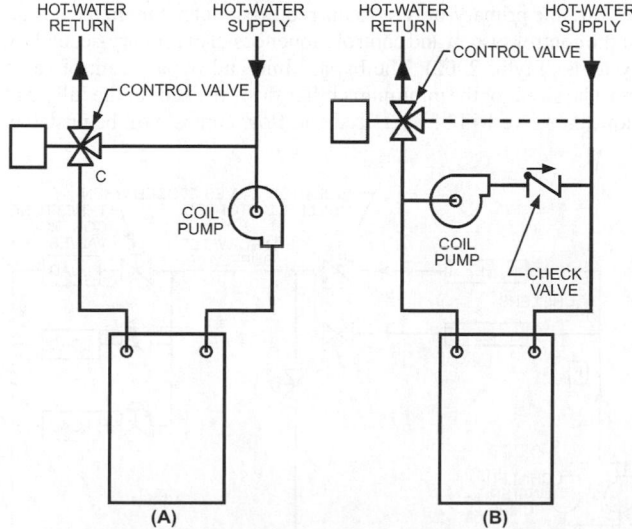

Fig. 7   Pumped Hot-Water Coil Variations:
(A) Series and (B) Parallel

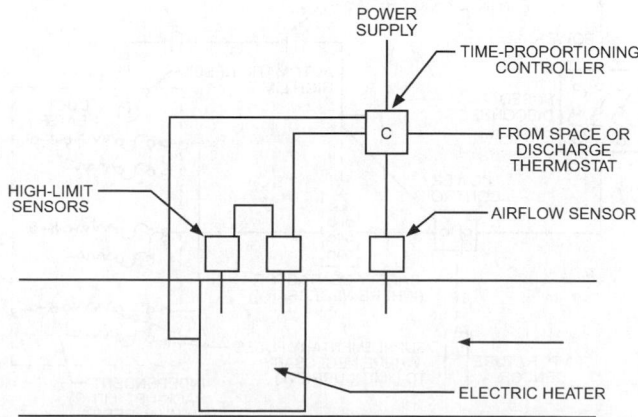

Fig. 8   Electric Heat: Solid-State Controller

heaters (Figure 9). The automatic reset high-limit thermostat normally turns off the control circuit. If the control circuit has an inherent time delay or uses solid-state switching devices, a separate safety contactor may be desirable. The manual reset back-up high-limit safety device is generally set independently to interrupt all current to the heater if other control devices fail. An electric heater must have a minimum airflow switch and two high-temperature limit sensors: one with manual reset and one with automatic reset. If still energized, electric coils and heaters can be damaged through overheating when air stops flowing around them. Therefore, control and power circuits must interlock with heat transfer devices (pumps and fans) to shut off electrical energy when the device shuts down. Flow or differential pressure switches may be used for this purpose; however, they should be calibrated to energize only when there is airflow. This precaution shuts off power if a fire damper closes or duct lining blocks the air passage. Limit thermostats should also be installed to turn off heaters when temperatures exceed safe operating levels.

Radiant Heating and Cooling

Radiators (more accurately called **convectors**) can be used either alone or to supplement another heat source. The control strategy depends on the function performed. For a radiation-only heating application, rooms are usually controlled individually; each radiator and convector is equipped with an automatic control valve. Depending on room size, one thermostat may control one valve or several valves in unison. Unit-mounted thermostats and packaged controls allow lower component cost and better assembly quality, and avoid the cost and coordination of a second trade for remote sensor installation. Wall-mounted thermostats give the best results when controlling the space for the comfort of seated occupants.

For supplemental heating applications, where perimeter radiation is used to offset perimeter heat losses (the zone or space load is handled separately by a zone air system), the radiation control should be sequenced with the main zone system to ensure there is no simultaneous heating and cooling. In the past, it was common to control the radiant system based on outdoor air temperature reset of the water temperature perhaps zoned by exposure with a solar compensating outdoor sensor, but this can result in "fighting" between the radiator and the main zone system and is disallowed by energy standards such as ASHRAE *Standard* 90.1.

Radiant panels combine controlled-temperature room surfaces with central air conditioning and ventilation. The radiant panel can be in the floor, walls, or ceiling. Panel temperature is maintained by circulating water or air, or by electric resistance. The central air system can be a basic one-zone, constant-temperature, constant-volume system, with the radiant panel operated by individual room

control thermostats, or it can include some or all the features of dual-duct, reheat, multizone, or VAV systems, with the radiant panel operated as a one-zone, constant-temperature system. Where hydronic tubing or electric heating elements are embedded in concrete, the rate of slab temperature change must be limited to prevent thermal expansion from cracking the concrete.

Radiant panels for both heating and cooling require controls similar to those for a four-pipe heating/cooling fan coil. To prevent condensation, ventilation air supplied to the space during the cooling cycle should have a dew point below that of the radiant panel surface. The dew point should be actively controlled to prevent condensation; as internal latent loads increase, the chilled-water temperature for the radiant cooling panels should be reset upward if the dew point becomes too high.

COOLING SYSTEMS

Chillers

The manufacturer almost always supplies chillers with an automatic control package installed. Control functions fall into two categories: capacity and safety.

Because of the wide variety of chiller types, sizes, drives, manufacturers, piping configurations, pumps, cooling towers, distribution systems, and loads, most central chiller plants, including their controls, are custom designed. In the 2008 *ASHRAE Handbook—HVAC Systems and Equipment*, Chapter 42 describes various chillers (e.g., centrifugal, reciprocating, screw, scroll), and Chapter 12 covers variations in various piping configurations and some associated control concepts. Chiller control strategies should always include an understanding of the chiller limits for minimum flow, minimum temperature, and acceptable rate of change for either.

Chiller plants are generally one of two types: variable flow (Figures 10 and 11) or constant flow (Figure 12). The figures show a parallel-flow piping configuration. Series-flow chiller configurations are often used in variable-primary-flow applications (Figure 10). The higher design water pressure drop of a series configuration is less of a concern in a variable-primary-flow application, because little time is spent at the maximum design flow operating condition. In Figures 10 and 11, the bypass line ensures minimum flow through the chiller(s). In Figure 10, the flow sensor opens the by-pass valve as needed to maintain the minimum flow the chiller requires.

Control of the remote load determines which type should be used. Throttling two-way coil valves vary flow in response to load, and are used with variable-flow systems.

Variable primary-only systems require greater care in the design of the control system and control sequences than primary-secondary systems (Taylor 2002). The bypass line and bypass control valve must be sized for the minimum chiller flow; if sized for the full plant flow, the valve will be oversized and flow control will be unstable,

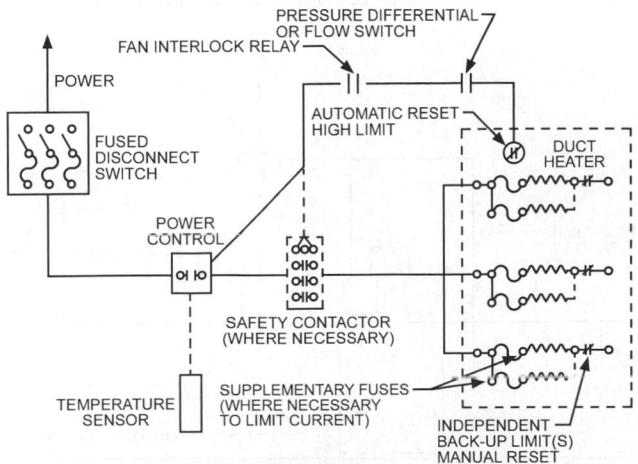

Fig. 9   Duct Heater Control

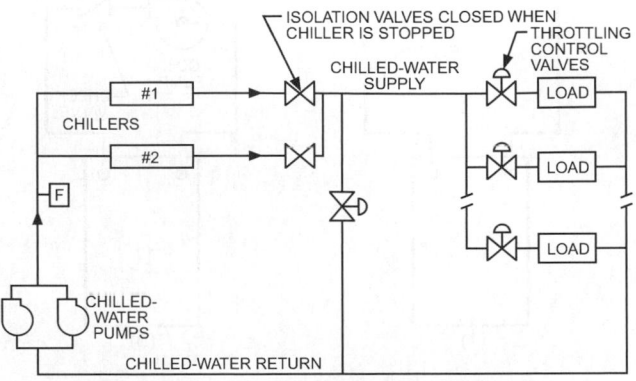

Fig. 10   Variable-Flow Chilled-Water System (Primary Only)

possibly causing chiller trips. Staging up from one chiller to two (or more) must be gradual to avoid a sudden drop in flow from active chillers.

The decision to reset chilled-water supply temperature set point in a variable-flow chiller system should be based on facility requirements. Chilled-water reset may not be appropriate in variable-flow systems with tight space temperature and humidity requirements, such as hospitals, computer data centers, museums, or some manufacturing facilities. Chilled-water reset always improves efficiency in constant-flow systems, and almost always improves efficiency in variable-flow systems, despite some increase in pump energy offsetting chiller energy savings. To ensure that no loads are starved, the chilled-water set point should be reset from the zone or system valve with the greatest load (load reset).

The constant-flow system (Figure 12) is only constant flow under each combination of chillers on line; chillers and pumps can be staged if all loads experience similar load profiles. Staging may be prevented even at low plant loads if some coils are at high load while others are at low load, causing the chillers to operate at low loads. Use of constant-flow systems may be limited to smaller systems by energy codes.

Chiller Plant Operation Optimization

Chapter 42 contains an extensive discussion on optimized control of chiller plants and a detailed description of the control strategies that can be applied. This section highlights the general conclusions that can be drawn from the specific optimization strategy in the section on Sequencing and Loading of Multiple Chillers of that chapter.

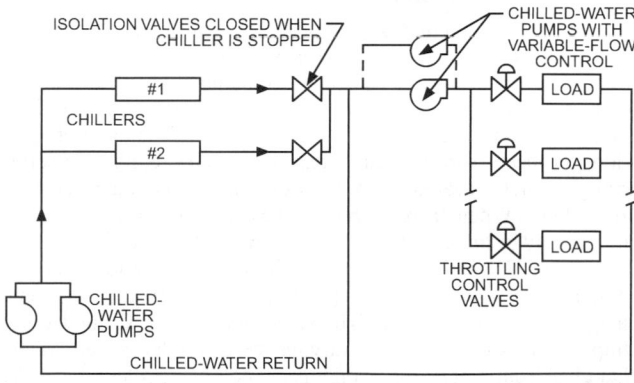

Fig. 11   Variable-Flow Chilled-Water System (Primary/Secondary)

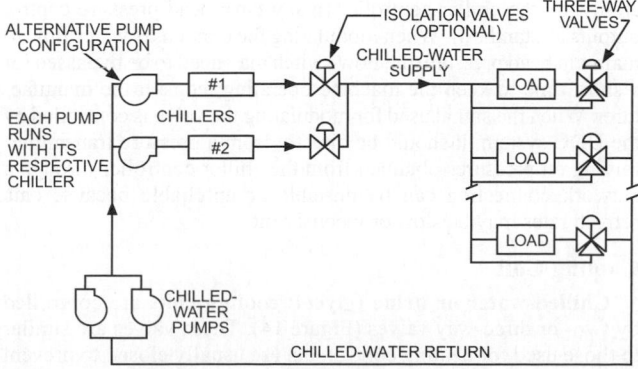

Fig. 12   Constant-Flow Chilled-Water System (Primary Only)

Multiple chillers should be operated at a point that minimizes overall plant power consumption, considering the power consumption of the auxiliary chilled- and condenser water pumps and cooling tower fans as well as the consumption of the chillers. For the general case of chillers with different part-load characteristics, multiple chillers should be operated at equal marginal coefficients of performance. That is, the ratio of the incremental change in thermal load to the incremental change in input power should be identical for each online chiller. For constant-speed-drive chillers, sequencing is straightforward. It is best to run one chiller until fully loaded before bringing on a second machine. This dynamic changes for chillers with variable-speed drives, or if the chilled- or condenser water pumps are piped in a parallel (headered) arrangement or have variable-speed drives; for plants of this configuration, often overall plant power consumption is minimized when more chillers are operating at reduced load compared to fewer chillers operating at higher loads. In general, the optimal point of operation for each plant component (chillers, pumps, cooling tower fans, and air handler fans) that results in minimum plant power consumption occurs when each component is operating at equal marginal performance.

Chiller efficiency is a function of the percent of full load on the chiller and the difference in refrigerant pressure between the condenser and evaporator. In practice, the pressure is represented by condenser water exit temperature minus chilled-water supply temperature. To reduce the refrigerant pressure differential, the chilled-water supply temperature must be increased and/or the condenser water temperature decreased. An energy saving of 1 to 2% is obtained for each 1°F reduction in condenser water temperature supplied to a chiller. The savings are even greater when using variable-speed drives on centrifugal chillers.

The following methods are used to reduce the refrigerant pressure differential:

• Use chilled-water load reset to raise chilled-water supply set point as load decreases based on valve demand, as discussed previously. Note that additional pump power for variable-flow systems must be considered in calculating net energy savings.
• Lower condenser water temperature to lowest safe temperature (use manufacturer's recommendations), operate at full condenser water pump capacity, and maintain water temperature within design cooling tower approach (see Chapter 39 of the 2008 *ASHRAE Handbook—HVAC Systems and Equipment*).

Because cooling tower performance is tightly coupled to chiller performance, the cooling tower fans should be controlled to minimize the sum of the chiller and tower power consumption. The section on Supervisory Control Strategies and Tools in Chapter 42 describes a near-optimal algorithm for optimizing control of the tower fans according to this criterion. In this algorithm, condenser water temperature is allowed to float between high and low limits, and tower fans are controlled according a mathematical formula involving the part-load ratio (normalized load) measured on the chilled-water side of the plant and design parameters from the cooling tower and chiller manufacturer.

Cooling Tower

Cooling tower fans are typically controlled to maintain condenser water supply (CWS) temperature set point, as described previously. The CWS set point may be reset based on chiller load or, in the case of hydronic free cooling, space conditions.

Figure 13 shows a typical cooling tower control schematic.

When the system includes large condenser water sumps, the temperature sensor's location must be considered. Large sumps introduce significant time delays into the system that must be accounted for. Often, condenser water supply piping does not run full at all times, particularly when draining to a sump. In this case, placing the temperature sensor so that is in contact with the water is important. This may necessitate locating the sensor on the

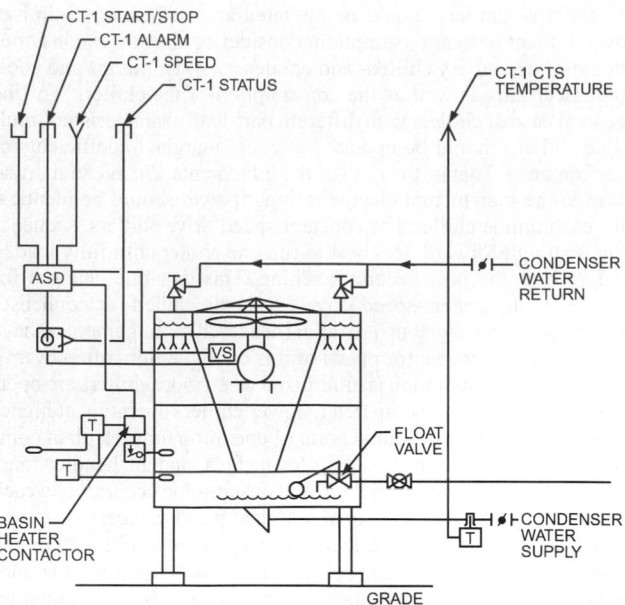

Fig. 13 Cooling Tower

bottom of an elbow or angled into the lower half of the pipe (to avoid mounting at the pipe bottom, where it could be susceptible to moisture collection).

Two-speed motors or variable-speed drives can reduce fan power consumption at part-load conditions and stabilize condenser water temperature. Variable-speed drives (VSDs) improve control because fan speed can be better matched to the cooling load. When there are multiple towers, efficiency is maximized when as many cells as possible are active, which increases mass transfer area and reduces required fan speed. It may be necessary, however, to shut off flow to some cells to maintain minimum tower flow rate, as recommended by the tower manufacturer to minimize scaling. Fan speed should be controlled as follows.

Two-Speed Motors. The lowest fan speed should be used. For three two-speed towers, staging should be as follows:

Tower	Stage 1	Stage 2	Stage 3	Stage 4	Stage 5	Stage 6	Stage 7
1	Off	Low	Low	Low	High	High	High
2	Off	Off	Low	Low	Low	High	High
3	Off	Off	Off	Low	Low	Low	High

Provisions should be made to decelerate the fan when switching from high to low speeds.

Variable-Speed Drives. All of the operating tower fans should operate at the same speed.

Tower fans should also have a vibration switch hard-wired to shut down the fan if excessive vibration is sensed.

Most tower manufacturers do not recommend bypassing water around the tower as a normal temperature control function, because this causes localized drying of the tower media and spot scaling. Bypass valves may be required for some low-temperature operations or hydronic economizer operations. Check with the tower manufacturer for bypass valve applications.

Where year-round tower operation is required in cold regions, cooling towers may require sump heating and/or continuous full flow over the tower to prevent ice formation, and may also require deicing. The cooling tower sump thermostat controls an electric heating element or hot-water or steam valve to keep water in the sump from freezing. Typically, sump heating is locked out during tower operation and when outdoor air temperatures are above freezing. When operating the tower in cold weather, full water flow is needed to prevent localized ice formation. If towers build up ice, reversing the fan rotation and sending air backward through the tower can deice them, either manually or automatically. Many operating personnel prefer to do this operation manually so they can observe the towers while deicing is under way. If deicing cycles are needed, the fan starter or VSD must be able to reverse the fan direction. There should be deceleration time interlocks so that the fans spin to a stop before engaging reverse operation. Automatic deicing control strategies are available from tower manufacturers. Towers that have internal balancing piping or chambers must be allowed to drain into the basin when the tower is not operating in cold weather.

Auxiliary Control Functions. The control system may be required to cycle a fill valve to maintain sump level, or to monitor water consumption or water treatment systems. If the tower does not have sump heaters, the control system may be required to drain the tower and piping when outdoor air approaches freezing, and refill the tower on the next call for cooling.

Air-Cooled Chillers

Air-cooled chillers are controlled similarly to other chillers. If the chiller is to operate during cold weather, it must be equipped with a low-ambient kit. Typically this is a modulating damper or variable-speed fan that limits airflow across the condenser, usually provided as part of the chiller package. In very cold conditions, additional equipment is required. With variable-flow evaporators, careful attention to the manufacturer's recommended minimum flow must be observed. Chilled-water supply temperature can be reset as described above. The chiller may be equipped with a barrel heater, usually controlled by the chiller's packaged controls, to prevent the evaporator from freezing in cold weather.

Water-Side Economizers

Water-side economizers are typically flat-plate heat exchangers where one side is cold tower water and the other side is chilled water. The heat exchanger prevents contamination of the chilled water by debris and chemicals found in tower water. The heat exchangers can be piped in series or in parallel. With parallel operation, the heat exchanger functions like another chiller. In series arrangement, the heat exchanger precools the chilled-water return to the chillers. When there is enough heat exchanger capacity, the chillers may be turned off and a bypass opened to direct flow around the chillers. The series arrangement allows for integrated (simultaneous) economizer and chiller operation, which is required by some energy standards such as ASHRAE *Standard* 90.1. Chilled-water temperature is controlled by varying the tower fan speed. When changing from water-side economizer mode to chiller mode, the condenser water is cold, which requires that chiller head pressure control be addressed. Consult the chiller manufacturer for the requirements of each specific machine. To maintain condenser head pressure, many manufacturers recommend self-contained modulating valves or control valves modulated by the DDC system or, preferably, by the chiller controller (many have head pressure control outputs as standard). When modulating the condenser water flow to maintain head pressure, the flow switch may need to be bypassed for a short time to keep the machine operating; consult the manufacturer. When the signal used for modulating the valve is controlled by the DDC system, it should be directly from a pressure transmitter; relying on pressures obtained from the chiller controller through a network connection can be unstable or unreliable because data refresh rates may be slow or inconsistent.

Cooling Coil

Chilled-water or brine (glycol) cooling coils are controlled by two- or three-way valves (Figure 14). These valves are similar to those used for heating control, but are usually closed to prevent cooling when the fan is off. The valve typically modulates to maintain discharge temperature or space temperature set point.

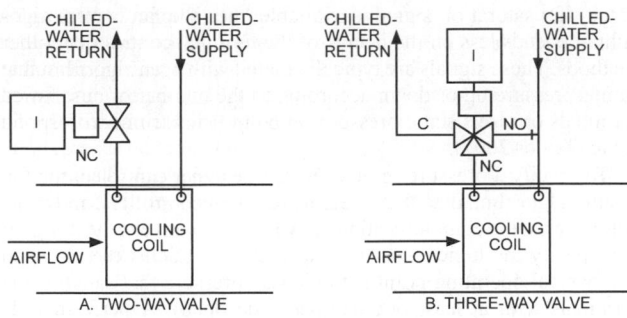

Fig. 14   Control of Chilled-Water Coils

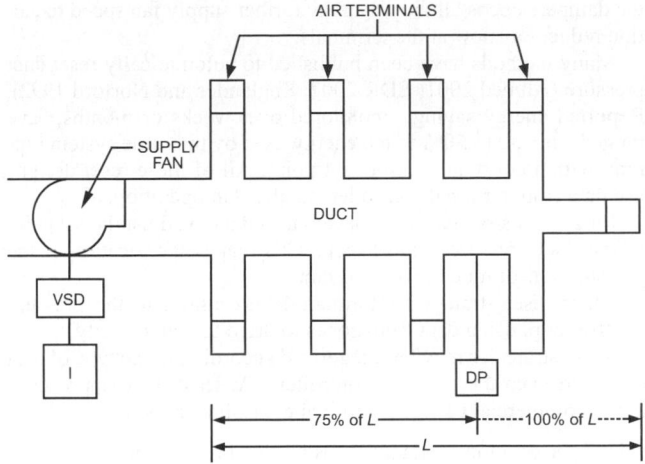

Fig. 15   Duct Static-Pressure Control

Direct-expansion (DX) cooling equipment is usually controlled by an air, space, or coil discharge temperature feedback loop in discrete stages, by starting and stopping compressors and by applying mechanical unloaders or hot-gas bypass valves. Most DX systems for commercial application have one to six stages.

When set up properly, a cycling DX system under a steady load operates like a two-position control loop (see Chapter 7 of the 2009 *ASHRAE Handbook—Fundamentals*). Some stages run steadily, and one stage cycles on and off. The behavior of the closed loop can be described by the cycling rate and corresponding swing in controlled air temperature. Most DDC algorithms for DX control address both temperature control (e.g., set points, feedback gains, staging dead bands) and equipment cycling restriction (e.g., minimum-on timer, minimum-off timer, interstage delay.) These two characteristics are inextricably linked by thermal sizing and loading conditions. Either characteristic can be affected by adjusting the control algorithm, but is not possible to affect both characteristics independently: any reduction in temperature swing is accompanied by an increase in cycling rate. Overtightening temperature control will conflict with cycle controls and be rendered irrelevant.

When set up improperly, a DX system may cycle through multiple stages as the temperature oscillates around the set point, rather than having only one stage cycle on and off. Compared to proper operation, both the cycle rate and temperature swing are excessive. This operation can usually be corrected by adjusting parameters in the feedback algorithm.

Some staging systems are arranged so that it takes a greater temperature error to activate the higher stages. The result is that, at higher loads, the system operates at higher temperatures. This is usually not desired; it is usually intended that the system operate in the same temperature range, regardless of loading on the stages. This can be accomplished in many ways, including a proportional-plus-integral (PI) controller output driving a staging module.

If the DX system serves a single zone, the feedback signal is usually the space temperature, which usually varies by 1 to 4°F. If the DX equipment serves multiple zones, as in a VAV air handler, the feedback signal is usually the coil discharge temperature. Measured at this point, temperature swings appear much larger, though the effect on zone comfort is the same. When adjusting or specifying a DX system for discharge temperature control, it is important to allow the wider range of temperatures.

AIR SYSTEMS

Variable Air Volume (VAV)

Supply Fan Control. Most VAV systems have pressure-independent terminals, which means a separate airflow control loop operates each terminal damper. In normal operation, these loops combine to set the airflow though the supply fan. So, within limits,

the fan controller does not directly control airflow to the zones. The VAV fan controller

- Ensures the pressure in the duct is enough to serve the terminals
- Prevents excessive pressure disrupting the terminal flow loops or causing damage to duct systems
- Avoids unnecessary energy consumption at the fan
- Keeps the fan in a stable region of the pressure-flow curve

Historically, a variety of mechanisms (e.g., bypass damper, variable inlet vanes) have been used to regulate fan output. These methods vary widely in efficiency and energy consumption. Currently, variable-speed drives (VSDs) are the most common because their low energy consumption and falling first cost makes them cost effective. They are mandated for most VAV systems by energy standards such as ASHRAE *Standard* 90.1.

The most common variable-airflow method is a closed-loop proportional-with-integral (PI) control, using the pressure measured at a selected point in the duct system. Historically, the set point was a constant, selected by the designer and confirmed by the balancer during system commissioning. However, this control strategy is based on the readings of a single sensor that is assumed to represent the pressure available to all VAV boxes. Choosing duct pressure sensor location can be difficult: if it malfunctions or is placed in a non-representative location, operating problems will result; if it is located too close to the fan, the sensor will not sufficiently indicate service of the terminals. This usually leads to excessive energy consumption. Some have reported that placing the sensor at the far end of the duct system couples fan control too closely with the action of a single terminal, making it difficult to stabilize the system. Experience indicates that performance is satisfactory when the sensor is located at 75 to 100% of the distance from the first to the most remote terminal (Figure 15). ASHRAE *Standard* 90.1 requires that the location result in a set point no higher than one-third of the total system static pressure drop.

Even with a good sensor location, fixed-pressure set point uses more energy than necessary. There are many operating hours when the fan pushes air through a system full of partly closed dampers. Many energy standards such as ASHRAE *Standard* 90.1 require automatically adjusting duct pressure based on zone demand as system load varies for systems with DDC at the zone level integrated with the air handler control. Airflow to zones is still regulated by flow loops in the terminal controllers and is unaffected, but all else being equal the system meets the load more efficiently with the terminal dampers closer to open. This reduces energy consumption at the fan. Ideally, pressure is reduced to the point that at least one of the dampers opens all the way. Any further supply fan speed reduction reduces airflow at the terminals.

Many methods have been published to automatically reset duct pressure (Ahmed 2001; EDR 2007; Englander and Norford 1992). Reported energy savings, monitored over weeks or months, have ranged from 30 to 50% of fan energy used by the same system running with a constant-pressure set point. All of these reset designs use data from terminal controllers to alter fan operation.

Most reset strategies use zone control data to adjust the set point of the duct pressure control loop. This makes the location of the pressure sensor much less important.

Other reset strategies (Hartman 1993) eliminate the pressure control loop, using data from zones to drive the fan directly.

Reset strategies may be categorized according to the type of data collected from the terminal controllers. At least three approaches are in commercial use. The terminal controllers may deliver

- Damper position (or damper position and flow error)
- Flow set point
- Saturation signal (terminal indicates that the pressure is insufficient)

Data available for coordinating a fan control system vary with model of terminal controller. Most have both a flow set point and a damper position value, though the suitability of the data for coordination varies. Control system designers should ensure that the data available from terminal controllers, the fan control strategy, and network data capacity are compatible.

The signal selected for coordination can determine the data communication load that the fan control strategy places on the network. Flow set points and saturation signals tend to change less often than damper position or flow measurements, so using them may be more practical with lower available bandwidth, especially in systems with many terminals. Saturation signals are binary, so they do not indicate their distance from the critical point. This can affect reset algorithm design.

One approach (using **damper position data**) is based on the idea that the desired mechanical operating point occurs when at least one damper is fully (or almost fully) open, and all terminals deliver the required flow. The fan controller processes damper position data from each terminal and adjusts duct pressure (or fan speed) to drive one damper open. To ensure that the open box is not starved, the reference may be set a little lower (95% open, for example), or the controller may check flow (or flow error) data from the terminal controllers. Floating actuator application methods may result in unreliable damper position values for some terminal controllers. It is important to take this into account when selecting a reset method.

Another approach (using **flow set points**) is based on the fact that the required pressure depends on the distribution system (ducts, terminals, diffusers) and required flow. One way is to add the flow set points from each terminal and then use an empirically determined function to set the pressure. A more exact approach puts the individual flow set points into a calibrated model of the duct network, and calculates the pressure needed at fan discharge to drive the required flow to each terminal (Kalore et al. 2003). This online optimization applies the same calculations used to size a fan in real time. The pressure control loop then adjusts the fan speed to maintain the calculated pressure, which results in all terminals being satisfied, with one critical damper fully open (Ahmed 2001, 2002). This method is now in commercial use. In contrast to a reset based on damper position or saturation signals, a reset based on flow set point is open loop; this means that performance depends on careful calibration, but is inherently stable.

A third approach (using a **saturation signal**) distributes more of the logic. Each terminal controller uses flow data, damper position data, timing, or other information to decide whether its local loop is sufficiently supplied by the fan. If not, the saturation signal is activated. If a saturation signal is available, then the fan control algorithm depends less on the details of the terminal control than other methods. These signals are typically mated with a fan algorithm that ramps pressure up or down according to the number of unsatisfied terminals or resets static pressure set point using trim-and-respond logic (Taylor 2007).

To specify a pressure reset system, a designer can select the fan control algorithm, data that integrate terminal controllers, and characteristics of the communication network. Alternatively, the designer can specify the logic in performance terms (i.e., that the intended mechanical operating point is the lowest pressure that satisfies the terminals with at least one damper wide open). A performance-based specification allows proposals from vendors with a wider variety of equipment and algorithms.

Duct Static Pressure Limit Control. In larger fan systems, or where fire or fire/smoke dampers could close off a significant percentage of airflow, static pressure limit controls are recommended. When the high limit set point is reached (or low limit on the suction side of the fans for systems with economizer dampers), the fan is deenergized. Limit controls should be manually reset. On large fans, inertia of the fan wheel could damage the ductwork even after the fan is deenergized. Additional protection for the ductwork (e.g., mechanical relief dampers) is needed in these situations.

Space Pressure Control. Differential static-pressure control, differential airflow, and directional bleed airflow are methods used to control pressurization of a space relative to adjacent spaces or the outdoors. Typical applications include pressure barriers for any occupied space to maintain interior comfort conditions, to prevent infiltration of moist unfiltered and untreated air. Applications requiring higher-performance controls include cleanrooms (positive pressure to prevent infiltration; see Chapter 18), laboratories and health care infection control (positive or negative, depending on use; see Chapter 8), and various manufacturing processes, such as spray-painting rooms (see Chapters 31 and 32 for industrial applications). The pressure controller usually modulates fan speed or dampers to maintain the desired pressure relationship or bleed airflow direction as exhaust volumes change. An alternative is to supply sufficient makeup air and to modulate a separate exhaust system to maintain space pressurization flow as auxiliary exhausts in the space are turned on or off.

Health Care Pressurization Codes, Regulations, and Application Design Guides. FGI (2010) incorporated ASHRAE/ASHE *Standard* 170-2008 into their guidelines. These standards include requirements for differential pressure or differential flow control for rooms such as positive and negative isolation rooms. Refer to the standards for details.

Building Pressurization. A slight positive building pressure (0.005 to 0.08 in. of water) is generally desired to reduce infiltration of unconditioned outdoor air. Pressure results from the development of a pressurization flow between adjacent pressure zones. A zone is positive to an adjacent zone if the pressurization flow across the zone barrier is positive. Generally, outdoor air is required to pressurize the building as a whole.

Static pressure control is one method for control of the relief or exhaust fan; this requires direct measurement of the space and outdoor static pressures. The inside static pressure measuring location must be selected carefully, away from openings to the outdoors, elevator lobbies, and other locations where it can be affected by wind pressure and drafts. Stack effect also impacts the reading for tall buildings in hot or cold weather; multiple pressure zones with independent sensors controls may be required to maintain positive pressure on all floors without overpressurizing some. The outdoor static pressure measuring location must also be selected carefully, typically 10 to 15 ft above the building and oriented to minimize wind effects from all directions. Even with good sensor port locations, pressure readings can fluctuate and should be buffered before using for control. If multiple fan systems serve

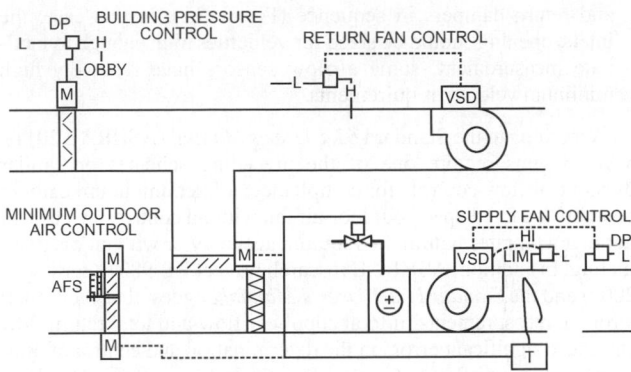

Fig. 16   Supply/Return Fan Control

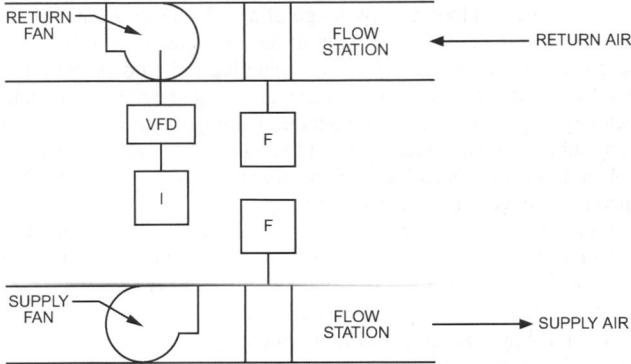

Fig. 17   Airflow Tracking Control

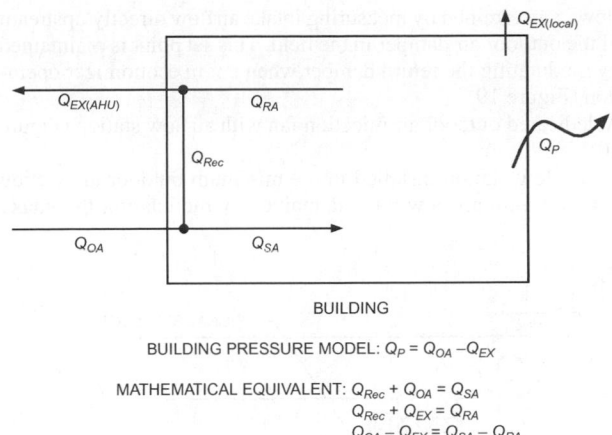

BUILDING

BUILDING PRESSURE MODEL: $Q_P = Q_{OA} - Q_{EX}$

MATHEMATICAL EQUIVALENT: $Q_{Rec} + Q_{OA} = Q_{SA}$
$Q_{Rec} + Q_{EX} = Q_{RA}$
$Q_{OA} - Q_{EX} = Q_{SA} - Q_{RA}$

Fig. 18   Building Pressure Model

areas that are open to one another, a single pressure control loop should be used to prevent instability.

The amount of minimum outdoor air for pressurization varies with building permeability and relief or exhaust fan operation. Control of building pressurization can affect the amount of outdoor air entering the building.

Proper return fan control for VAV systems is required for building pressurization. In one approach, outlined in ASHRAE *Guideline* 16, the return fan is controlled to maintain the return air plenum pressure while exhaust (relief) air dampers are controlled to maintain building static pressure (see Figure 16). For relief fan systems, the relief fan speed is generally directly controlled by building pressure.

Direct-measurement pressurization flow compares an interior static pressure location to an outdoor reference to modulate relief fan speed or relief dampers. This control allows for greater operational repeatability, and improved energy savings potential where there are natural relief paths such as operable windows.

Indirect building pressure control uses duct or fan airflow measurements to control a fixed differential air volume by modulating dampers, fan speed, or discharge rates (Figure 17). Because return air is typically the controlled variable and because its rate is set to track the normal changes in VAV supply at a fixed rate, this method is referred to as return fan or airflow tracking. The airflow differential set point is often determined empirically during commissioning as that needed to maintain a slight positive pressure with doors and windows closed.

Fixed-differential air volume to maintain pressurization flow, rather than measured space static pressure, results in very stable control. It avoids the instabilities described previously for direct pressure control caused by fluctuating pressures from gusts of wind, opening doors and windows, and multiple air-handling systems serving interconnected areas that interact. However, the control is indirect, so actual space pressure varies (e.g., with stack effect as outdoor air temperature changes).

Airflow quantity is indicated in Figure 18 by Q. Q_P is leakage in or out of the room, driven by the net pressure differential. Note that each surface may have a different ΔP because this value is relative to the pressure in the space on the other side of the wall.

When the control strategy changes from occupied (ventilation air required) to unoccupied warm-up, which does not require ventilation but needs thermal control to change the air-balancing requirements, warm-up is accomplished by setting return airflow equal to or just slightly less than the supply fan airflow, with toilet and other exhaust fans turned off and limiting supply fan volume to return fan capability. If exhaust fans remain running, then the supply fan must deliver sufficient outdoor air to make up the exhaust and still have a slightly pressurized space. During night cooldown, when using large quantities of outdoor air, the return fan operates in the normal mode (Kettler 1995).

Unstable fan operation in VAV systems can usually be avoided by proper fan sizing. However, if airflow reduction is large (typically over 60%), fan sequencing is often required to maintain airflow in the fan's stable range. Zone-based static pressure set point reset, described previously, also allows the fan system with variable-speed drives to almost completely avoid the unstable region of its operating curves until fan speed is so low that instabilities are minor. This logic can allow very large VAV fans to serve very small airflows, during off-hours for instance.

Supply air temperature reset can be used to improve energy performance in most multiple-zone systems, including VAV systems, and is required by energy standards such as ASHRAE *Standard* 90.1 for systems that have simultaneous heating and cooling at the zone level. In cool weather, supply air temperature can be reset upward based on zone demand, similar to static pressure reset. This reduces reheat energy losses and extends economizer operation, reducing mechanical cooling energy. In warmer weather, when space heat is not needed, supply air temperature should be reduced to reduce fan energy (EDR 2007).

Minimum Outdoor Air Control. Fixed minimum outdoor airflow control provides dilution air for ventilation, pressurization flow (usually exfiltration), combustion air for processes converting fuel to heat, and makeup air for exhaust fans.

Several variations of minimum outdoor airflow control for VAV systems are possible (ASHRAE 2011; Felker and Felker 2010; Kettler 2000):

• Differential pressure is measured across the outdoor air intake louver or two-position minimum outdoor air damper. The differential pressure set point correlating to the minimum outdoor airflow is determined by measuring intake airflow directly upstream of the outdoor air damper in the field. This set point is maintained by modulating the return damper when not in economizer operation (Figure 19).

- A dedicated outdoor air injection fan with airflow station (Figure 20).
- An airflow station installed in the minimum outdoor air section with a minimum flow rate maintained by modulating the intake

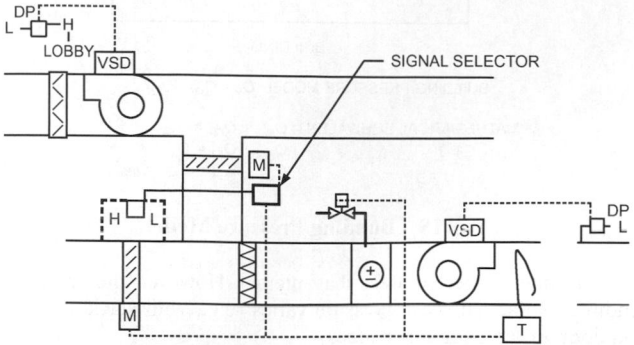

Fig. 19   Minimum Outdoor Air Control Using Differential Pressure Controls

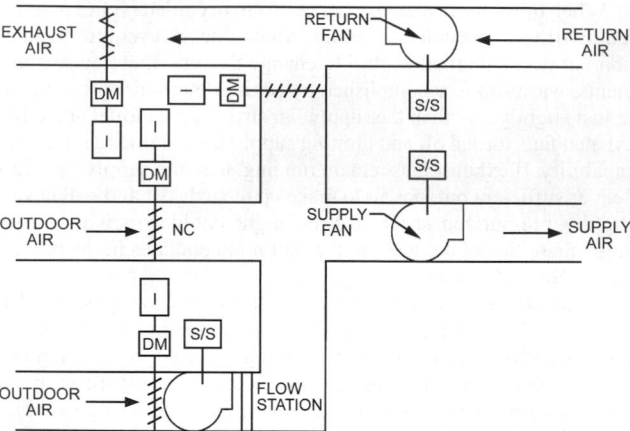

Fig. 20   Minimum Outdoor Air Control with Outdoor Air Injection Fan

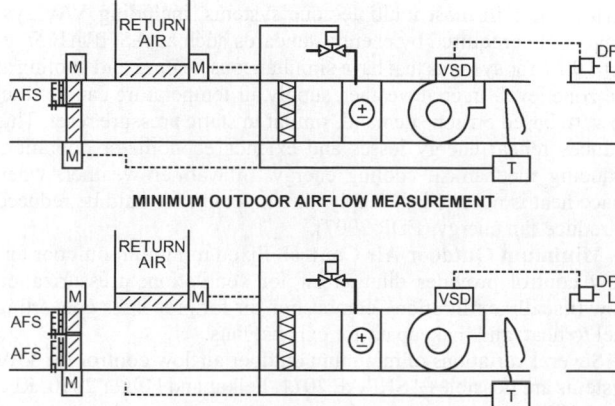

MINIMUM OUTDOOR AIRFLOW MEASUREMENT

TOTAL OUTDOOR AIRFLOW MEASUREMENT USING STAGED AIRFLOW STATIONS

Fig. 21   Outdoor Air Control with Airflow Measuring Stations

and return dampers in sequence (Figure 21). In this case, the intake opening should be sized for velocities high enough to facilitate measurement; some airflow sensors have relatively high minimum velocity requirements.

According to the Standard *62.1 User's Manual* (ASHRAE 2011), VAV systems require one of the preceding schemes or similar dynamic airflow controls for compliance; a fixed minimum damper position or a fixed-speed outdoor-air fan without control devices will not maintain rates within the required accuracy or without overventilating. In addition, ASHRAE research project RP-980 (Krarti et al. 2000) and the *Standard 62.1 User's Manual* suggest that even small errors in measurements of total supply airflow and total return flow can cause significant errors in the determination and control of minimum outdoor airflow rates, making return fan or airflow tracking unsatisfactory for minimum outdoor ventilation control. (It may be a reasonable means of building pressurization control, but minimum outdoor air must be controlled independently.)

If the total outdoor airflow range of a VAV design from a minimum of less than 50% to maximum design capacity is to be measured, the selected measurement technology should provide the needed reliability across the entire anticipated temperature and velocity range. One method for accomplishing this with pitot arrays is by subdividing the intake and using dual airflow stations sized for 1/4 and 3/4, or 1/3 and 2/3, of the maximum opening size. This increases the velocity pressure for the pitot array to ensure accurate measurement at minimum pressure drop (Kettler 2000). Thermal velocity sensors, which have a much lower minimum velocity than pitot devices, may be used without creating damper sections.

Regardless of the type of system, pressurization flow rate and outdoor airflow rate are controlled separately: the two functions are related but must be independently controlled.

The outdoor airflow set point for dilution ventilation should be established using ASHRAE *Standard* 62.1. In addition, the outdoor air set point for pressurization should be established by adding the pressurization flow requirement to the sum of the local exhausts in the zones served by the air-handling system. The greater of the two dictates the outdoor air set point.

Traditional economizer controls call for the outdoor air and recirculation dampers to be modulated inversely to maintain set point: one opens as the other closes. A more energy-efficient approach for VAV systems is to decouple the outdoor air and recirculation dampers by individually actuating each. The outdoor airflow rate is then controlled by sequencing the dampers (ASHRAE *Guideline* 16). This reduces pressure drop and thus reduces fan energy.

Dynamic Reset of Intake Rates. Demand-controlled ventilation (DCV) is a control scheme designed to reduce the amount of outdoor air when the spaces served have less than design occupancy. The most common scheme is to use CO_2 concentration to reset the occupant component of the minimum outdoor air rate required by ASHRAE *Standard* 62.1. **Ventilation reset control (VRC)** is a related control scheme for resetting outdoor air and minimum supply air rates as system ventilation efficiency changes because of operational changes in the system. Both control schemes are required by ASHRAE *Standard* 90.1 for many applications and are described in detail the user's manual (ASHRAE 2011) for ASHRAE *Standard* 62.1.

When implementing dynamic ventilation reset schemes that reduce outdoor air intake, provisions must be made to ensure that pressurization flow is maintained (i.e., the relationship between the outdoor air and the exhaust air is maintained). When outdoor air dew point approaches or exceeds 60°F, a net positive pressurization flow is required to prevent transport of water and outdoor air contaminants into the building or its envelope (ASHRAE *Standard* 62.1).

Air-Side Economizer Cycle. Economizer-cycle control reduces cooling costs when outdoor air is cool and dry enough to be used as a cooling medium. The economizer is enabled by a high-limit

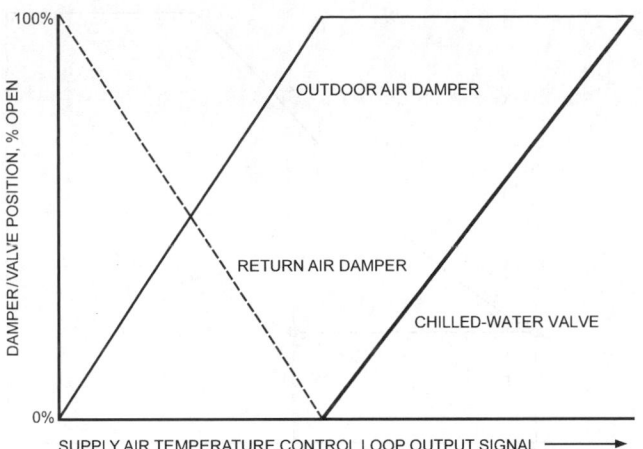

Fig. 22 "Integrated" Economizer Cycle Control

Table 1 Economizer Damper Type and Sizing

Relief System	Damper	Blade Type	Face Velocity, fpm
Return Fan	Relief/exhaust	Opposed	1000 to 1500
	Outdoor air	Parallel	400 to 1000
	Return air	Parallel	Per ΔP across damper ~1500
Relief Fan or Barometric	Outdoor air	Parallel	400 to 1000
	Return air	Parallel	800 to 1000

device. When enabled, the economizer return and outdoor air dampers modulate to maintain a supply air temperature in sequence with the mechanical cooling. Typically, the economizer is controlled in sequence with the mechanical cooling, using the same supply air temperature control loop. Figure 22 shows integrated control, in which the economizer and mechanical cooling can be active at the same time. This is mandated in most applications by ASHRAE *Standard* 90.1. The outdoor (and exhaust/relief) dampers would fully close, and the return air dampers open, when the supply fan is not operating. When the outdoor air temperature exceeds the economizer high-limit set point, the economizer is disabled and only minimum outdoor air is supplied.

ASHRAE *Guideline* 16 addresses the sizing and selection of dampers for outdoor air economizer systems. Table 1 summarizes the guideline's recommendations as a function of the relief air system. The guideline should be referenced for additional details and rationale.

High-limit controls are intended to enable the economizer when supplying outdoor air uses less energy than recirculating air. Common high-limit controls are

- Fixed dry-bulb temperature (compares outdoor air dry bulb to a fixed set point)
- Differential dry-bulb temperature (compares outdoor air dry bulb to return air dry bulb)
- Fixed enthalpy (compares outdoor air enthalpy to a fixed set point)
- Differential enthalpy (compares outdoor air enthalpy to return air enthalpy)
- Electronic enthalpy (compares outdoor air temperature and humidity to a set point that is a curve on the psychrometric chart)
- Combinations of these controls

ASHRAE *Standard* 90.1 includes some limitations on which controllers can be used and controller set points based on climate zone. The most energy-efficient high limit theoretically is a combination of differential enthalpy and differential dry-bulb temperature. However, it effectively requires four sensors (temperature and

humidity in both outdoor air and return airstreams), all of which have inaccuracy and can get out of calibration, in particular the humidity sensors. Sensor error may result in increased energy usage relative to other, less expensive high-limit controls. In practice, the simplest, least expensive, and most reliable high-limit control is a fixed outdoor air dry-bulb temperature sensor set to the set point required by ASHRAE *Standard* 90.1.

The relief air system should be enabled during economizer operation because the large quantities of outdoor air should leave the building along a planned path of flow and not an unplanned path, such as entry doors that may be pushed open.

VAV warm-up control during unoccupied periods requires no outdoor air if exhaust fans are off; typically, outdoor and exhaust dampers remain closed. The supply fan and return fan airflow are offset to maintain zero differential airflow.

Where outdoor conditions allow, night cooldown control (**night purge**) provides 100% outdoor air for cooling during unoccupied periods. The space is cooled to the space set point, typically 9°F above outdoor air temperature. Limit controls prevent operation if outdoor air is above space dry-bulb temperature, if outdoor dew-point temperature is excessive, or if outdoor dry-bulb temperature is too cold (typically 50°F or below). When outdoor air conditions are acceptable and the space requires cooling, the cooldown cycle is the first phase of the optimum start sequence.

Constant-Volume (CV) Systems

In a constant-volume system, supply and return fan airflow rates are manually set to meet the maximum airflow requirements for thermal load and ventilation. The air-handling unit's heating coil (where applicable), economizer dampers, and cooling coil are controlled in sequence to maintain the supply air temperature at a set point that is reset to satisfy the zone with the maximum load. Reheat coils on constant-volume terminal boxes are controlled by individual space thermostats to establish the final space temperatures. If warm-up mode is used, the supply air set point is adjusted upward to the desired value.

Control strategies for economizers, demand-controlled ventilation, morning warm-up, and night cooldown are the same as for VAV systems discussed previously.

Changeover/Bypass Zoning Systems

Changeover/bypass zoning systems, often referred to as variable-volume/variable-temperature systems, are typically used to convert a single-zone constant-volume air-handling unit into multiple variable-volume temperature-controlled zones (Figure 23). The unit fan usually operates continuously in occupied mode, and cycles as required to maintain setback/setup zone temperatures in unoccupied mode. The unit remains a constant-volume unit and, as zones close off airflow to a particular zone, a static pressure sensor in the supply duct typically modulates a damper to bypass air back to the return duct to keep airflow across the unit's coils relatively constant. A controller on the unit turns on stages of heat or cooling based on zone thermostat requirements. Zone control is typically pressure-dependent, and the zone controllers modulate a damper between maximum and minimum position based on the zone's temperature requirements and supply air temperature. Pressure-independent control is an option, and is recommended if supply air distribution systems are large or nonsymmetrical; the zone controller measures airflow to the zone and resets the air volume to the zone between maximum and minimum airflow limits based on zone temperature requirements and supply air temperature. Digital controllers on the zone dampers "vote" for heating or cooling based on zone temperature and deviation from set point. When the vote favors heating, the unit controller stages heating on to satisfy the zone requiring the most heat and broadcasts to the zone controllers that warm air is coming down the duct. Any zone requiring heat opens its damper, and any zone requiring cooling closes its damper to minimum

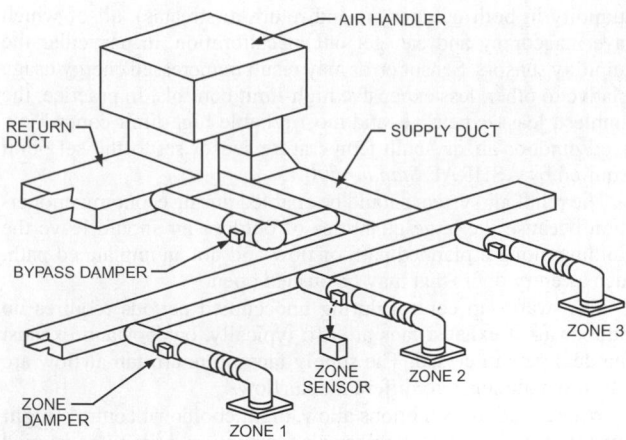

Fig. 23 Changeover/Bypass Zoning System

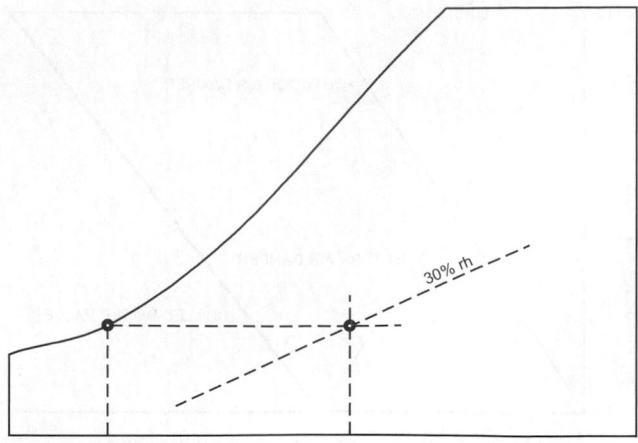

Fig. 24 Psychrometric Chart: Cooling and Dehumidifying, Practical Low Limit

position or flow. Similarly, when the vote is for cooling, the unit turns on stages of cooling to satisfy the zone requiring the most cooling and broadcasts to the zones that cool air is coming down the duct. If no zone requires heat or cooling, the supply fan remains on in occupied mode, but either keeps the heating and cooling off or cycles heating and cooling as required to maintain a supply temperature approximately equal to the zone temperature ±5°F.

Terminal Units

A system is considered to be variable-volume if primary airflow to the space varies. Total airflow to the space (primary air + plenum air) may be constant for some terminal units, even in a variable-volume system.

These systems typically serve fewer than 15 zones per air-handling unit, and should serve zones with similar thermal loads (e.g., all internal zones or all zones on the same exterior exposure), so that the unit is not continually switching between heating and cooling. To ensure minimum outdoor air ventilation is maintained, outdoor airflow must be controlled like a VAV system (see the section on Minimum Outdoor Air Control). A simple fixed damper position typically is not adequate.

Humidity Control

Humidity control relies on the output of a humidity sensor located either in the space or in the return air duct. Most comfort cooling involves some natural but uncontrolled dehumidification. The amount of dehumidification is a function of the effective coil surface temperature and is limited by the coolant's freezing point. If water condensing out of the airstream freezes on the coil surface, airflow is restricted and, in severe cases, may be shut off. The practical limit is about 40°F dew point on the coil surface. As indicated in Figure 24, this results in a relative humidity of no less than 30% at a space temperature of 75°F. When lower humidity is needed for a process application (e.g., dryroom), a desiccant dehumidifier is required.

Although simple cooling by refrigeration typically provides dehumidification as a byproduct of the cooling process, without additional equipment, it does not directly control space humidity. **Dehumidification** can be directly controlled in several ways. One way is to override control of the cooling coil. The supply air temperature leaving the coil is lowered until sufficient moisture is removed from the supply air to maintain the humidity set point. When a maximum relative humidity limit is required, a space or return air humidistat is provided in addition to the space thermostat. To limit maximum humidity, a control function selects the higher of the output signals from the two devices and controls the cooling coil valve accordingly. A reheat coil is required to maintain the space

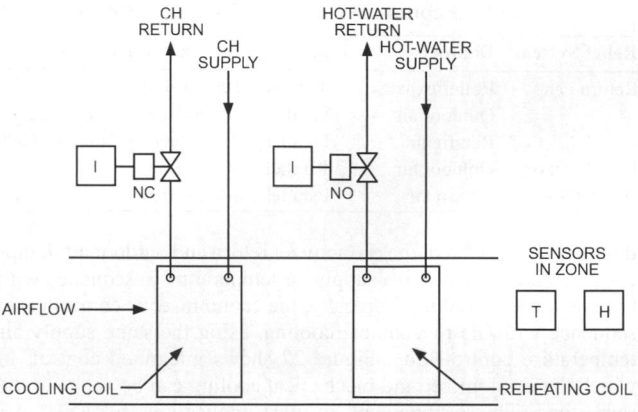

Fig. 25 Cooling and Dehumidifying with Reheat

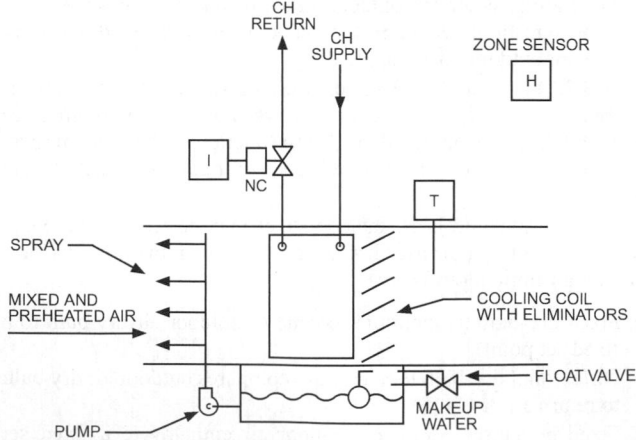

Fig. 26 Sprayed-Coil Dehumidifier

temperature if moisture removal results in too low a supply air temperature (Figure 25).

Sprayed-coil dehumidifiers (Figure 26) have been used for dehumidification. Space relative humidity ranging from 35 to 55% at 75°F can be obtained with this equipment; however, the costs of maintenance, reheat, and removal of solid deposits on the coil make the sprayed-coil dehumidifier less desirable than other methods.

A **desiccant-based dehumidifier** can lower space humidity below that possible with cooling/dehumidifying coils. This device adsorbs moisture using silica gel or a similar material. For continuous operation, heat is added to regenerate the material. The adsorption also generates heat (Figure 27). Figure 28 shows a typical control.

Humidification can be achieved by adding moisture to supply air, using evaporative pans (usually heated), steam injection, or atomizing spray tubes. A space or return air humidity sensor provides the necessary signal for the controller. A humidity sensor in the duct should be used to minimize moisture carryover or condensation in the duct (Figure 29). With proper use and control, humidifiers can achieve high space humidity, although they more often are used to maintain design minimum humidity during the heating season.

Terminal Units

A system is considered to be variable-volume if primary airflow to the space varies. Total airflow to the space (primary air + plenum

air) may be constant for some terminal units, even in a variable-volume system.

Single-Duct, Constant-Volume. Reheat terminals use a single constant-volume fan system that serves multiple zones (Figure 30). All of the system's supply air is cooled to satisfy the greatest zone cooling load. Air delivered to other zones is then reheated with heating coils (hot-water, steam, or electric) in individual zone ducts. The reheat coil valve (or electric heating element) is reset as required to maintain the space condition. Because these systems consume more energy than VAV systems, they are generally limited by energy standards to applications with fixed ventilation needs, such as hospitals and special processes or laboratories.

No fan control is required because the design, selection, and adjustment of fan components determine the air volume and duct static pressure. The same temperature air is supplied to all zones. However, the controller can vary the supply temperature to respond to demand from the zone with the greatest cooling load, thus conserving energy.

Single-Duct Variable-Volume. A **throttling VAV terminal** has an inlet damper that controls the flow of supply air (Figure 31). For spaces requiring heating, a reheat coil can be installed in the discharge. With pressure-independent controls, the space temperature sensor does not control the inlet damper directly. The space temperature control loop output is used to reset the primary airflow delivered to the space between a maximum and minimum rate. Direct control of airflow makes the VAV box independent of variations in duct static pressure. A common control sequence for this system is the single maximum scheme depicted in Figure 31. In this sequence, as temperature in the space drops below the set point, the damper begins to close and reduce airflow to the space. When airflow reaches the minimum limit, the valve on the reheat coil begins to open.

One disadvantage of this sequence is that the minimum flow set point must be high enough to meet the design heating load at a supply air temperature that is low enough to prevent stratification (e.g., less than 90°F). Therefore, the minimum flow set point typically must be 30 to 50% of the maximum flow set point, as limited by energy standards such as ASHRAE *Standard* 90.1. This wastes a great deal of reheat and fan energy, particularly for zones that are very conservatively sized.

A more energy-efficient sequence is the dual maximum sequence in Figure 32. As the space goes from design cooling load to design heating load, the airflow set point is first reset from the cooling maximum to the minimum. Then the supply air temperature is reset from

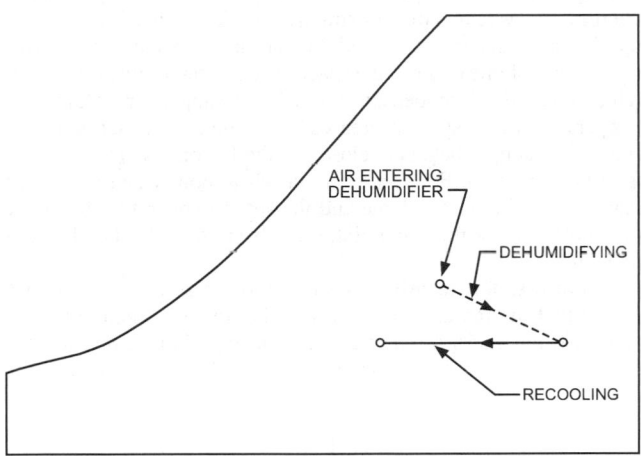

Fig. 27 Psychrometric Chart: Desiccant-Based Dehumidification

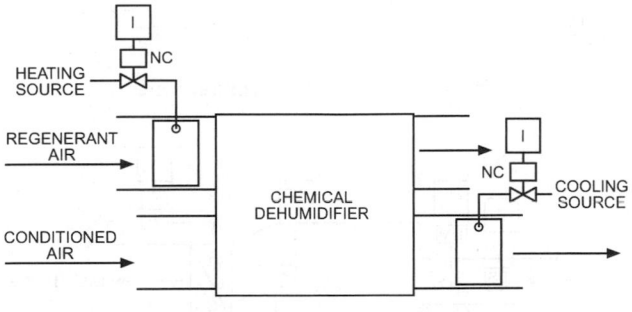

Fig. 28 Desiccant Dehumidifier

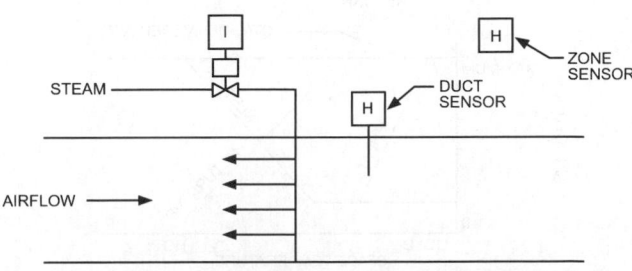

Fig. 29 Steam Injection Humidifier

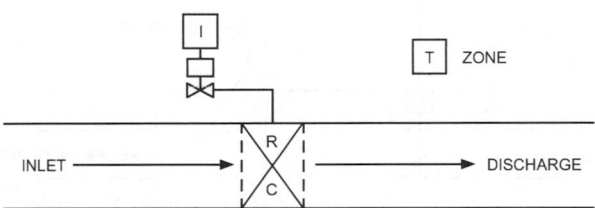

Fig. 30 Single-Duct Constant-Volume Zone Reheat

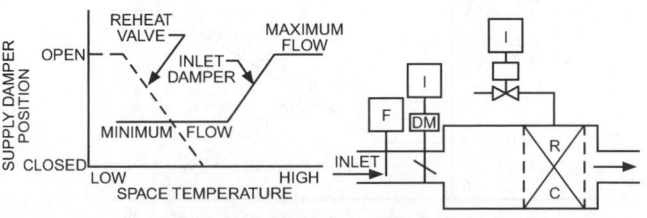

Fig. 31 Throttling VAV Terminal Unit: Single Minimum Control Sequence

minimum (e.g., 55°F) to maximum (e.g., 90°F), and the reheat coil is modulated to maintain the supply air temperature at set point. Lastly, the airflow set point is reset from the minimum up to the heating maximum. One of the advantages of the dual maximum sequence is that the minimum flow set point is not limited by stratification (as described for the single maximum) and can be set as low as 10 to 20% of the maximum flow, depending on ventilation requirements and the lowest nonzero controllable flow. Thus, the dual maximum sequence can greatly reduce wasted reheat and fan energy. This logic is mandated by some energy standards wherever DDC zone controls are used.

An **induction VAV terminal** controls space temperature by reducing supply airflow to the space and by inducing return air from the plenum into the airstream for the space (Figure 33). Both dampers are controlled simultaneously, so as the primary air opening decreases, the return air opening increases. When space temperature drops below the set point, the supply air damper begins to close and the return air damper begins to open.

A **bypass VAV terminal** has a damper that diverts part of the supply air into the return plenum (Figure 34). Control of the diverting damper is based on output of the space temperature sensor. When temperature in the space drops below the set point, the bypass damper begins to open, routing some supply air to the plenum, which

reduces the amount of supply air entering the space. When the bypass is fully open, the reheat coil's control valve opens as required to maintain space temperature. A manual balancing damper in the bypass is adjusted to match the resistance in the discharge duct. In this way, the supply of air from the primary system remains at a constant volume. The maximum airflow through the bypass must be restricted to maintain minimum airflow into the space. Although airflow to the space is reduced, the fan's total airflow remains constant, so fan power and associated energy cost are not reduced. These terminals can be added to a single-zone constant-volume system to provide zoning without the energy penalty of a conventional reheat system. However, if a return or exhaust fan is not used and a majority of the terminals go to bypass, the return plenum may become positive in relation to the space, forcing return air back into the space.

A **series fan-powered VAV terminal unit** has an integral fan in series with the primary supply air VAV damper that supplies a constant volume of air to the space (Figure 35). In addition to enhancing air distribution in the space, a reheat coil can be added for space heating and to maintain a minimum temperature in the space when the primary system is off, for strategies such as setback and warmup. When the space is occupied, the fan runs constantly to provide a constant volume of air to the space. The fan can draw air from the return plenum to compensate for reduced supply air volume. As temperature in the space decreases below the cooling set point, the supply air damper begins to close and the fan draws more air from the return plenum. For zones with a reheat coil, when supply air reaches its minimum volume and the space temperature begins to drop below the heating set point, the valve to the reheat coil begins to open.

A **parallel fan terminal** is similar to the series fan terminal, except that the fan is in parallel with the primary supply air VAV damper (Figure 36). A reheat coil may be placed in the discharge to the space or in the return plenum opening. The fan is intended to

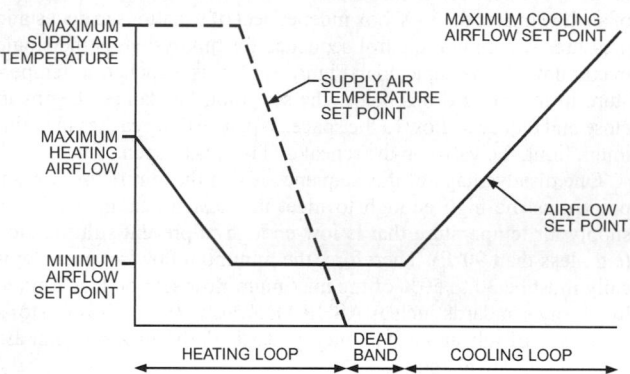

**Fig. 32 Throttling VAV Terminal Unit:
Dual Minimum Control Sequence**

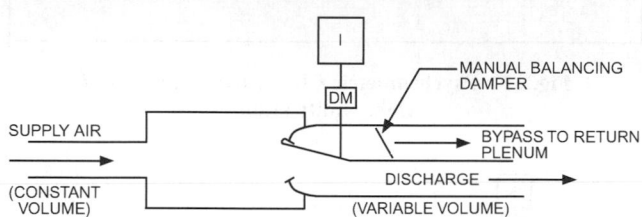

Fig. 34 Bypass VAV Terminal Unit

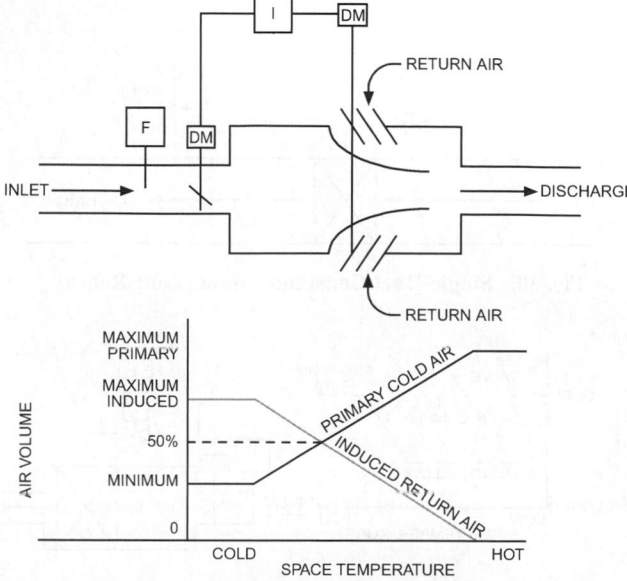

Fig. 33 Induction VAV Terminal Unit

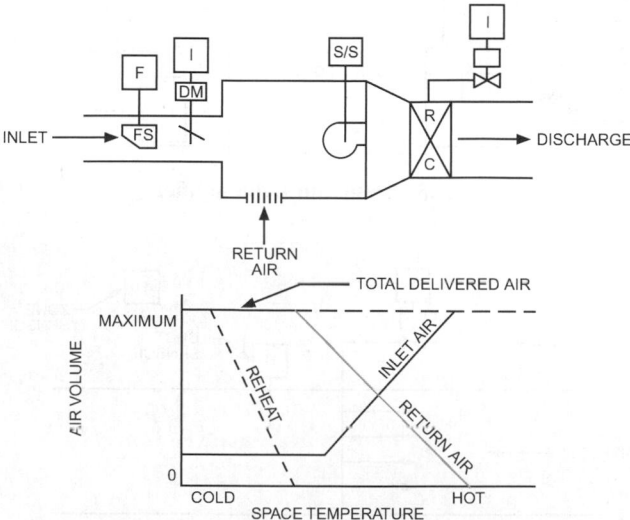

Fig. 35 Series Fan-Powered VAV Terminal Unit

operate primarily in heating mode, but may also operate to maintain a minimum airflow to the space, allowing reduced primary airflow rates. Total airflow to the space is the sum of the fan output and supply air quantity. When space temperature drops below the cooling set point, the supply air damper begins to reduce the quantity of supply air entering the terminal. Once the supply damper reaches its minimum position and the space temperature begins to drop below the heating set point, the reheat coil valve starts to open. When the space is unoccupied and requires heating for setback or warm-up, the supply air damper is closed, the fan turns on, and the reheat coil valve modulates to maintain the unoccupied set point.

Variable-volume, dual-duct terminal units (Figure 37) have inlet dampers (with individual damper actuators and airflow controllers) on the cooling and heating supply ducts and no total airflow volume damper. The space thermostat resets the airflow controller set points in sequence as the space load changes. The airflow controllers maintain adjustable minimum flows for ventilation. If the heating supply has sufficient ventilation air, there need not be any overlap of damper operations (one snaps closed and the other snaps open at the heat/cool changeover point), resulting in no simultaneous heating

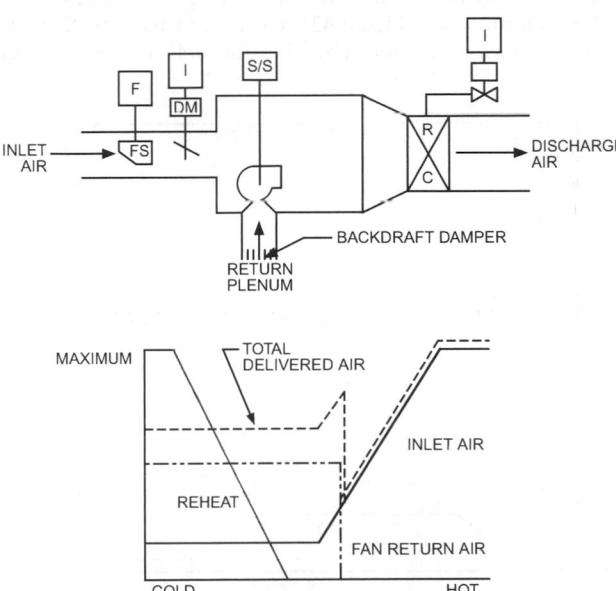

Fig. 36   Parallel Fan Terminal Unit

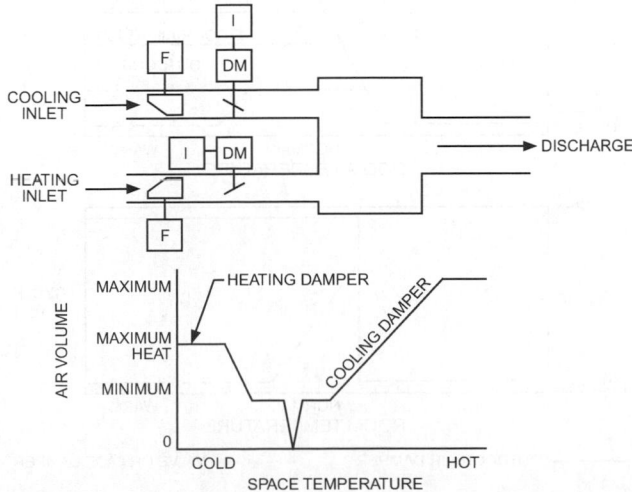

Fig. 37   Variable-Volume Dual-Duct Terminal Unit

and cooling in the terminal unit. On systems where the heating supply does not have sufficient ventilation air (e.g., on some dual-fan dual-duct applications), the cooling damper can be controlled to a minimum for ventilation.

Single-Zone Systems

A single-zone system (Figure 38) is an air handler (usually factory-packaged) serving a single zone. Single-zone systems do not require terminal boxes, because zone temperature can be maintained by modulating the heating and cooling control valves (and optional economizer dampers) in sequence. The fan is typically constant volume, but modern designs include two-speed or variable-speed fans, which are now a requirement for many applications of energy standards such as ASHRAE *Standard* 90.1. To control the system, a supply air temperature sensor must be added with the supply air temperature set point and fan speed controlled by zone temperature control loops, as indicated in Figure 39. The supply air temperature is maintained at set point like a conventional VAV system.

A **unit ventilator** is designed to heat, ventilate, and cool a space by introducing up to 100% outdoor air. Optionally, it can cool and dehumidify with a cooling coil (either chilled-water or direct-expansion). Heating can be by hot water, steam, or electric resistance. Control of these coils can be by valves or by face-and-bypass dampers. Consequently, controls applied to unit ventilators are many and varied. The four most commonly used control schemes are Cycle I, Cycle II, Cycle III, and Cycle W.

Cycle I Control. Except during the warm-up stage, Cycle I (Figure 40) supplies 100% outdoor air at all times. During warm-up, the heating valve is open, the outdoor air (OA) damper is closed, and the return air (RA) damper is open. As temperature rises into the operating range of the space thermostat, the OA damper opens fully, and the RA damper closes. The heating valve is positioned to maintain space temperature. The airstream thermostat can override space thermostat action on the heating valve to prevent discharge air from dropping below a minimum temperature. Figure 40 shows positions of the heating valve and ventilation dampers in relation to space temperature.

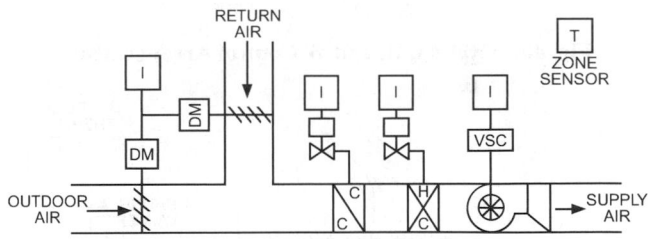

Fig. 38   Single-Zone Fan System

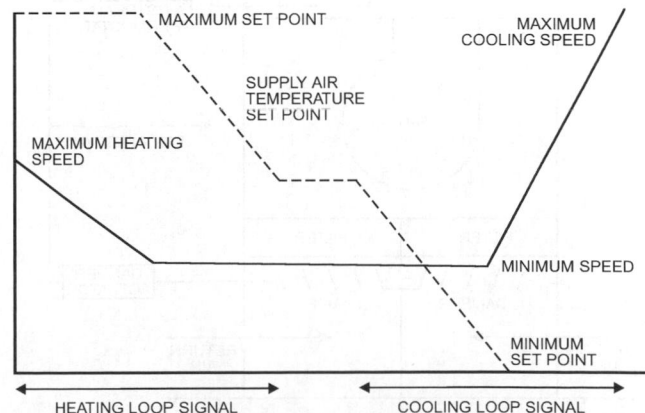

Fig. 39   Single-Zone VAV Control

Cycle II Control. During the heating stage, Cycle II (Figure 40) supplies a set minimum quantity of outdoor air. Outdoor air is gradually increased as required for cooling. During warm-up, the heating valve is open, the OA damper is closed, and the RA damper is open. As space temperature rises into the operating range of the space thermostat, ventilation dampers move to their set minimum ventilation positions. The heating valve and ventilation dampers are operated in sequence as required to maintain space temperature. The airstream thermostat can override space thermostat action on the heating valve and ventilation dampers to prevent discharge air from dropping below a minimum temperature. Figure 40 shows the relative positions of the heating valve and ventilation dampers with respect to space temperature.

Cycle III Control. During heating, ventilating, and cooling stages, Cycle III (Figure 41) supplies a variable amount of outdoor air as required to maintain the air entering the heating coil at a fixed temperature (typically 55°F). When heat is not required, this air is used for cooling. During warm-up, the heating valve is open, the OA air damper is closed, and the RA damper is open. As the space temperature rises into the operating range of the space thermostat, ventilation dampers control the air entering the heating coil at the set temperature. Space temperature is controlled by positioning the heating valve as required. Figure 41 shows the relative positions of the heating valve and ventilation dampers with respect to space temperature.

Day/night thermostats are frequently used with any of these control schemes to maintain a lower space temperature during unoccupied periods by cycling the fan with the outdoor air damper closed. Another common option is a low-temperature limit control placed next to the heating coil to turn off the unit when near-freezing temperatures are sensed.

Cycle W Control. Cycle W is similar to Cycle II, except that the heating valve is controlled by the room thermostat and the dampers are controlled by the low-limit thermostat (Figures 40 and 42).

Makeup air units (Figure 43) replace air exhausted from the building through exfiltration or by laboratory or industrial processes.

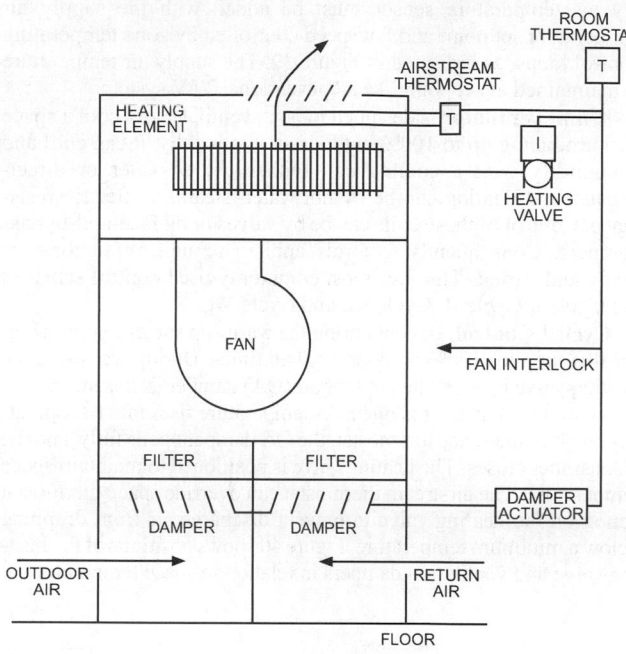

Fig. 40   Cycles I, II, and W Control Arrangements

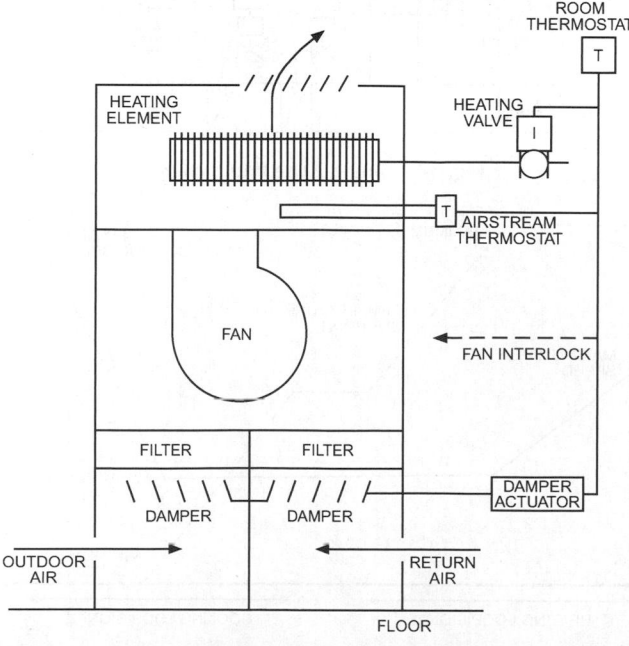

Fig. 41   Cycle III Control Arrangement

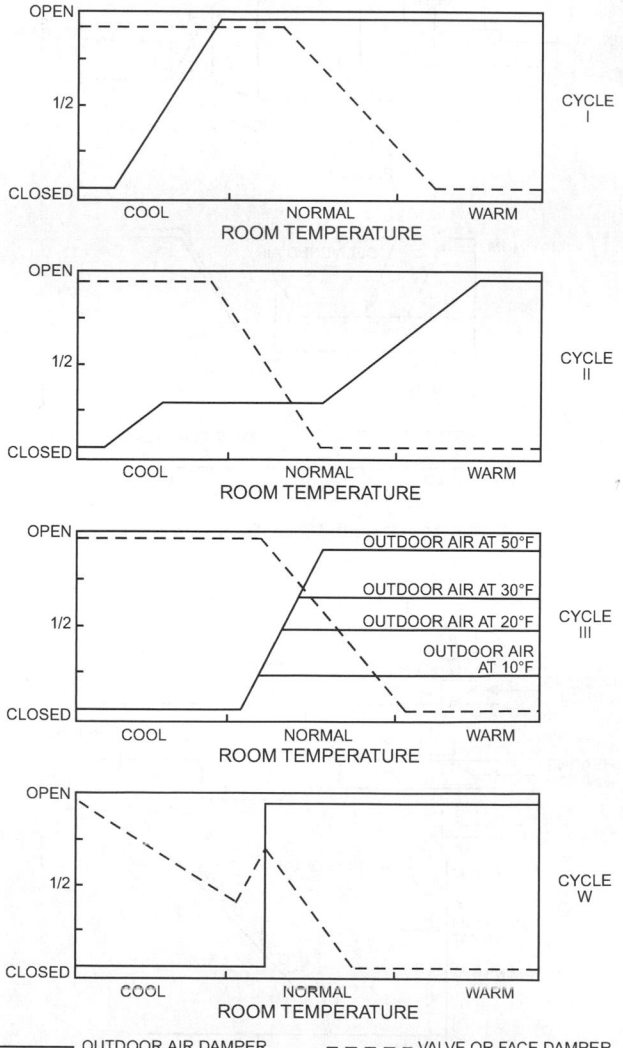

Fig. 42   Valve and Damper Positions with Respect to Room Temperature

Makeup air is often supplied at or near space conditions, but may also provide space heating and cooling. The makeup air fan is usually turned on, either manually or automatically, whenever exhaust fans are turned on. However, the fan should not start until the outdoor air damper is fully open as proven by an end switch. The two-position outdoor air damper remains closed when the makeup fan is not in operation. The outdoor air limit control opens the preheat coil valve when outdoor air temperature drops to the point where the air requires heating to raise it to the desired supply air temperature. Feed-forward control should be used to open the heating coil quickly

on start-up to prevent low-temperature shutdowns. A capillary element freezestat located adjacent to the coil shuts the fan down for low-temperature detection if air temperature approaches freezing at any spot along the sensing element.

Fuel-fired makeup air units have staged or modulating fuel-fired direct or indirect furnaces. Units have manufacturer-supplied controls and safeties for flame proving, airflow proving, and discharge air low limit to meet the ANSI *Standard* Z83.4/CSA *Standard* 3.7 combined safety standard. The manufacturer's controls either include temperature controls or provide an interface for the control contractor's temperature controls.

Multiple-Zone, Single-Duct System

This system (Figure 44) serves single-duct terminal units such as VAV and constant-volume terminals and series and parallel fan terminals, described previously. The fan is either variable volume or constant volume, depending on terminal type.

Multiple-Zone, Dual-Duct Systems

A **single-fan, dual-duct system** uses a single fan to supply separate heating and cooling ducts (Figure 45). Dual-duct terminal mixing boxes are used to control the zone temperature. For VAV terminals, static control is similar to that in VAV single-duct systems except that static pressure sensors are needed in each supply duct. A controller allows the sensor detecting the lowest pressure to control the fan output, thus ensuring that there is adequate static pressure to supply the necessary air for all zones.

The hot deck has its own heating coil, and the cold deck has its own cooling coil. Each coil is controlled by its own **discharge air**

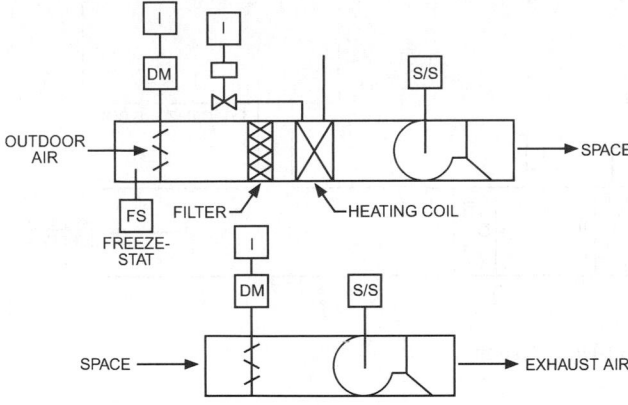

Fig. 43   Makeup Air Unit

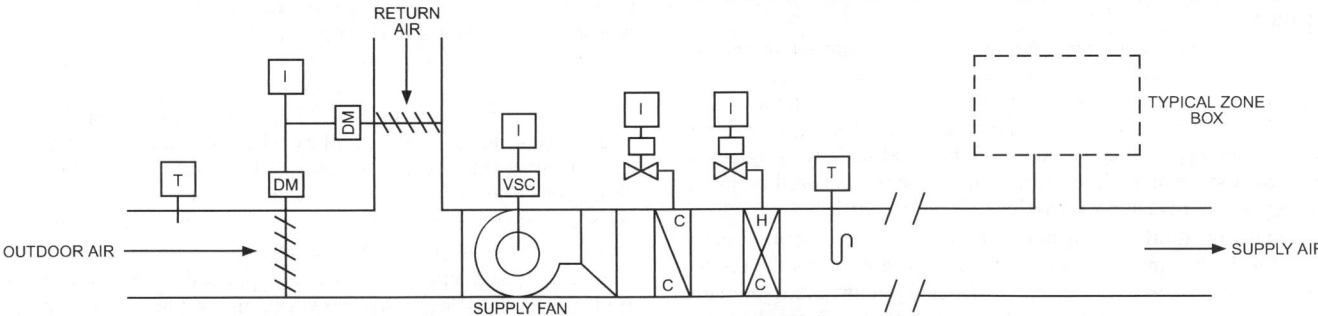

Fig. 44   Multiple-Zone, Single-Duct System

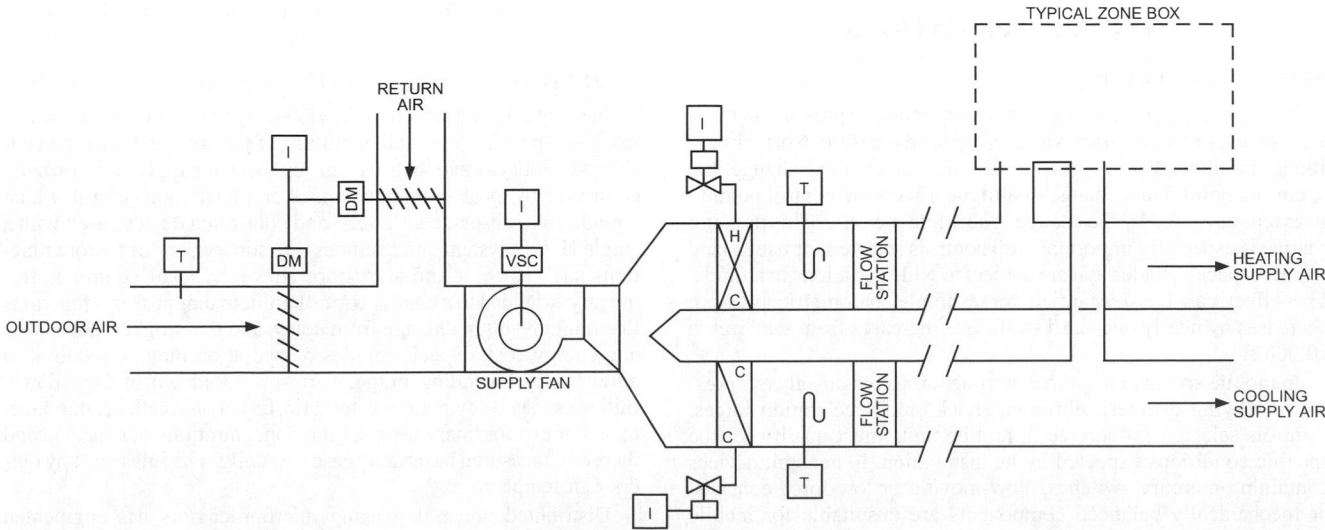

Fig. 45   Single-Fan, Dual-Duct System

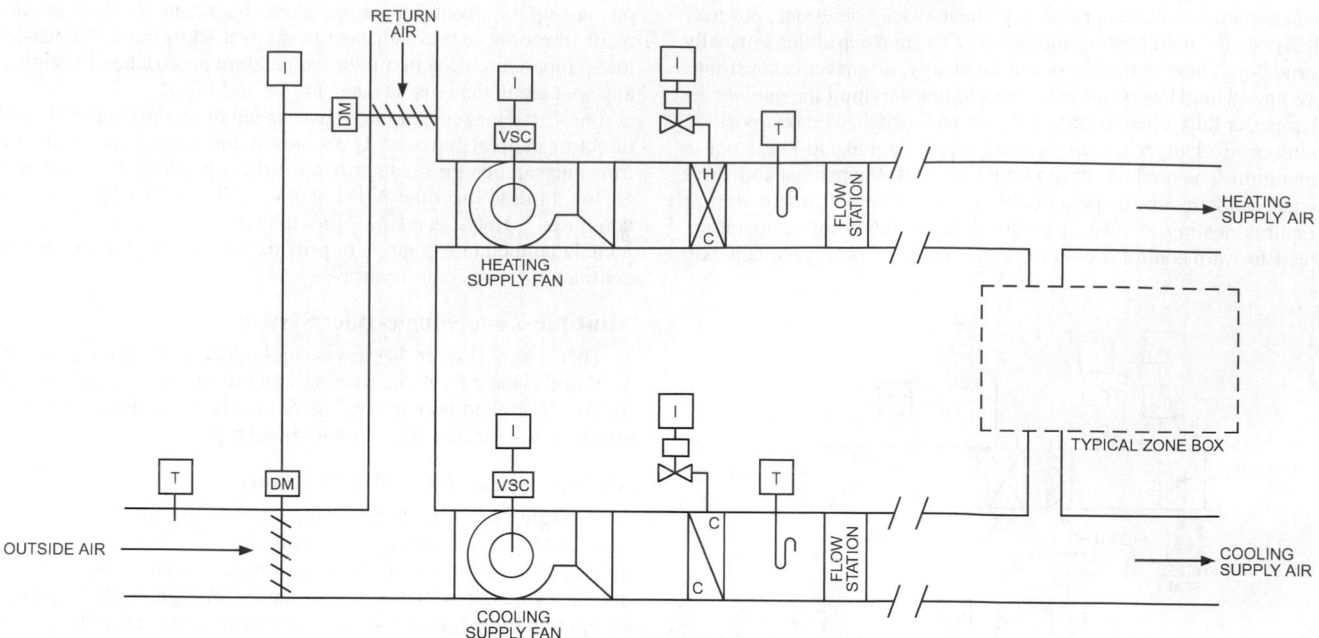

Fig. 46   Dual-Fan, Dual-Duct System

temperature controller. The hot deck set point may be reset from the zone with the greatest heating demand, and the cold deck set point may be reset from the zone with the greatest cooling demand.

Cooling supply air temperature control is similar to that in single-duct systems, with economizer dampers sequenced with the cooling coil. The economizer causes the supply air temperature entering the hot deck to be as cold as the cold deck, increasing heating energy usage. Because of this inefficiency, single-fan, dual-duct systems with air economizers are not allowed by many energy standards such as ASHRAE *Standard* 90.1.

Dual-fan, dual-duct systems (Figure 46) use separate supply fans for the heating and cooling ducts. This eliminates the economizer inefficiency of single-fan, dual-duct systems. Static-pressure control is similar to that for VAV dual-duct, single-fan systems, except that each supply fan has its own static pressure sensor and control.

SPECIAL APPLICATIONS

Mobile Unit Control

The operating point of any control that relies on pressure to operate a switch or valve varies with atmospheric pressure. Normal variations in atmospheric pressure do not noticeably change the operating point, but a change in altitude affects the control point to an extent governed by the change in absolute pressure. This pressure change is especially important with controls selected for use in land and aerospace vehicles that are subject to wide variations in altitude. The effect can be substantial; for example, barometric pressure decreases by nearly one-third as altitude increases from sea level to 10,000 ft.

In mobile applications, three detrimental factors are always present in varying degrees: vibration, shock, and acceleration forces. Controls selected for service in mobile units must qualify for the specific conditions expected in the installation. In general, devices containing mercury switches, slow-moving or low-force contacts, or mechanically balanced components are unsuitable for mobile applications; electronic solid-state devices are generally less susceptible to these three factors.

Explosive Atmospheres

Sealed-in-glass contacts are not considered explosionproof; therefore, other means must be provided to eliminate the possibility of a spark in an explosive atmosphere.

When using electric control, the control case and contacts can be surrounded with an explosionproof case, allowing only the capsule and the capillary tubing to extend into the conditioned space. It is often possible to use a long capillary tube and mount the instrument case in a nonexplosive atmosphere. This method can be duplicated with an electronic control by placing an electronic sensor in the conditioned space and feeding its signal to an electronic transducer located in the nonexplosive atmosphere.

Because pneumatic control uses compressed air, it is safe in otherwise hazardous locations. However, many pneumatic controls interface with electrical components. All electrical components require appropriate explosionproof protection.

Sections 500 to 503 of the *National Electrical Code* (NFPA *Standard* 70) include detailed information on electrical installation protection requirements for various types of hazardous atmospheres.

DESIGN CONSIDERATIONS AND PRINCIPLES

In designing and selecting the HVAC system for the entire building, the type, size, use, and operation of the structure must be considered. Subsystems such as fan and water supply are normally controlled by local automatic control or a local loop control, which includes the sensors, controllers, and controlled devices used with a single HVAC system and excludes any supervisory or remote functions such as reset and start/stop. However, local control is frequently extended to a central control point to diagnose malfunctions that might result in damage from delay, and that might reduce labor and energy costs. Special modes of operation may be required to allow for load shedding, purge, warm-up, cooldown, and lockdown. Initiators may be manual or automatic, based on weather, announcements of extraordinary events, high concentrations of expected and therefore measured hazardous gases, or daily schedules reset by outdoor air temperature.

Distributed processing using microprocessors has augmented computer use at many locations other than the central control point. The local loop controller is now more commonly DDC instead of a pneumatic or electric thermostat, integrated with energy management functions performed by upper level DDCs forming a complete building automation system (BAS).

Because HVAC systems are designed to meet maximum design conditions, they nearly always function at partial capacity. The system must be adjusted and operated for many years, so the simplest control that produces the necessary results is usually the best.

Extraordinary Incidents

Building owners and design engineers are sometimes interested in applying the building automation system (BAS) to implement strategies that protect occupants from airborne attack. It is crucial that the engineer does not approach this complex topic as a control system issue. The BAS may include protective features, but only in the context of a comprehensively designed ventilation system. A protective ventilation strategy only makes sense in the context of a thorough risk assessment and an overall security plan. If a protective ventilation strategy is attempted, it is crucial to consider every air movement device and pathway, not just the main fan(s) and damper(s). It is also necessary to consider possible interaction of a protective operation with other emergency control operations, such as the response to a fire/life safety device (e.g., a smoke detector).

Many references are available to guide an engineer or building owner in organizing a comprehensive plan, including ASHRAE (2003b), FEMA (2003), and NIOSH (2002). Also see Chapter 59 for more information on this topic.

Mechanical and Electrical Coordination

Even a pneumatic control system includes wiring, conduit, switchgear, and electrical distribution for many electrical devices. The mechanical designer must inform the electrical designer of the total electrical requirements if controls are to be wired by the electrical contractor. Requirements include (1) the devices to be furnished and/or connected, (2) electrical load, (3) location of electrical items, (4) a description of each control function, and (5) whether the control system needs to be on emergency power or UPS.

Coordination is essential. Proper coordination should produce a control diagram that shows the interface with other control elements to form a complete and usable system. As an option, the control engineer may develop a complete performance specification and require the control contractor to install all wiring related to the specified sequence. The control designer must run the final checks of drawings and specifications. Both mechanical and electrical specifications must be checked for compatibility and uniformity.

Sequences of Operation

DDC systems require that the engineer define how the system is to be controlled. These sequences are then programmed into the system by the DDC system installer. Writing clear, unambiguous, concise, yet comprehensive sequences of controls is key to the energy and comfort performance of the HVAC system, yet it is very difficult to do well. It first requires a clear understanding of how controls work, the limitations of the specific DDC hardware specified and the HVAC system design, and a knack for clear thinking and writing. Techniques for writing successful control sequences are discussed in ASHRAE *Guideline* 13. Sequences for typical HVAC systems can be found in the ASHRAE (2005) CD-ROM, *Sequences of Operation for Common HVAC Systems*.

Energy-Efficient Controls

The U.S. Green Building Council's (USGBC) Leadership in Energy Efficient Design (LEED) program for new construction and existing buildings provides guidance on designing and building commercial, institutional, and government buildings to produce quantifiable benefits for occupants, owners, and the environment.

During design and construction, LEED provides many opportunities for using a building automation system (BAS). LEED awards points, based on design criteria, to determine at which level a building is certified: platinum, gold, silver, or certified. (For specifics, see http://www.usgbc.org.) Many of these points depend on the sophistication, flexibility, and power of the BAS, not only for comfort and energy efficiency, but also for sustainability verification throughout the facility's lifetime.

Applications. It is recommended that the LEED design team choose the controls consultant/contractor and to have them on board with the team early in design, allowing for controls input on the control schemes.

For any facility, the BAS provides thermal comfort and routine programming (e.g., occupied/unoccupied). Under LEED, the controls consultant or contractor uses the BAS for benchmarking and alarming when reference points are exceeded (e.g., monitoring facility electrical use). Once a baseline is established, programs with established alarm limits can alert the owner if the facility exceeds the baseline by 10% or more, so corrections can be made and continued higher utility expenses avoided. Depending on the facility, this may be monitored for hourly, daily, weekly, or monthly comparison.

The BAS can also

- Turn solar panels for optimum sun exposure
- Adjust blinds and awnings
- Keep indoor air quality (IAQ) acceptable by adjusting outdoor air (OA) without overventilating
- Monitor and adjust indoor and outdoor lighting
- Control irrigation based on weather
- Collect weather data
- Track scheduling hours for occupancy
- Monitor equipment run times and set points
- Track electricity use profiles
- Monitor efficiency of large equipment performance [e.g., chillers, variable-frequency-drive (VFD) pumping, boilers)

All of these factors affect building sustainability.

Because the BAS operates in real time, and can compare real-time data to previously defined baselines, it is a valuable tool. Open protocols allow the BAS to monitor, control, and provide critical alarming for non-HVAC equipment (e.g., for power monitoring, chiller performance, VFDs for VAV systems or variable-speed pumping, water efficiency, emergency generators, indoor and outdoor lighting, boilers) for a minimal investment.

The BAS provides an excellent tool for commissioning both HVAC and other equipment, as required by LEED. Its data acquisition abilities allow comparisons of daily performance as building use changes, thereby allowing the commissioning agent to determine whether equipment is operating properly, and allowing the owner to compare real-time data to previously defined benchmarks in **measurement and verification (M&V)**. Typical operation sequences, such as optimizing outdoor air, chiller efficiency, and multiple sensing points for providing thermal comfort, are, however, still used, and are critical for maintaining sustainability.

Perhaps the most important aspect of a successful control system is training the owner. Many owners do not fully realize the capabilities of BAS, many of which are intangible and somewhat obscure, so it is critical that the owner be given proper training in using the controls system. The order in which training takes place is equally important: mechanical equipment training should come first, then operation and maintenance (O&M) and controls layout configurations, use of the controls system to provide thermal comfort and maintain equipment, and, finally, written M&V procedures to maintain sustainability.

If the facility cannot justify a full-time energy manager, then the owner should consider third-party contracting to ensure the facility maintains its design energy efficiency.

CONTROL PRINCIPLES FOR ENERGY CONSERVATION

After the general needs of a building have been established and the building and system subdivision has been made, the mechanical system and its control approach can be considered. Designing systems that conserve energy requires knowledge of (1) the building, (2) its operating schedule, (3) the systems to be installed, and (4) ASHRAE *Standard* 90.1. The principles or approaches that conserve energy are as follows:

- *Run equipment only when needed.* Schedule HVAC unit operation for occupied periods. Run heat at night only to maintain internal temperature at around 55 to 60°F to prevent freezing. Start morning warm-up as late as possible to achieve design internal temperature by occupancy time, considering residual space temperature, outdoor temperature, and equipment capacity (optimum start control). Under most conditions, heating and cooling equipment can be shut down some time before the end of occupancy, depending on internal and external load and space temperature (optimum stop control). Calculate shutdown time so that space temperature does not drift out of the selected comfort zone before occupancy ends and ensure that ventilation is provided throughout occupancy.
- *Sequence heating and cooling.* Do not supply heating and cooling simultaneously unless it is required for humidity control. Central fan systems should use cool outdoor air, if available, in sequence between heating and cooling. Zoning and system selection should eliminate, or at least minimize, simultaneous heating and cooling. Also, humidification and dehumidification should not take place concurrently.
- *Provide only the heating or cooling actually needed.* Reset the supply temperature of hot and cold air (or water). In air systems that support variable-speed fans, reset duct static pressure to provide thermal comfort at the lowest possible fan speed and energy consumption.
- *Supply heating and cooling from the most efficient source.* Use free or low-cost energy sources first, then higher-cost sources as necessary.
- *Apply outdoor air control.* When on minimum outdoor air, supply no less than that recommended by ASHRAE *Standard* 62.1 and, on VAV systems, include controls that ensure that outdoor air rates are maintained under all expected supply air operating conditions.

System Selection

The mechanical system significantly affects the control of zones and subsystems. System type and number and location of zones influence the amount of simultaneous heating and cooling that occurs. For perimeter areas, heating and cooling should be controlled in sequence to minimize simultaneous heating and cooling. In general, this sequencing must be accomplished by the control system because only a few mechanical systems (e.g., two-pipe and single-coil) have the ability to prevent simultaneous heating and cooling. Systems that require engineered control systems to minimize simultaneous heating and cooling include the following:

- *VAV cooling with zone reheat.* Reduce cooling energy and/or air volume to a minimum before applying reheat.
- *Four-pipe heating and cooling for unitary equipment.* Sequence heating and cooling.
- *Dual-duct systems.* Condition only one duct (either hot or cold) at a time. The other duct should supply a mixture of outdoor and return air.
- *Single-zone heating/cooling.* Sequence heating and cooling.

Some exceptions exist, such as dehumidification with reheat.

Control zones are determined by the location of the thermostat or temperature sensor that sets the requirements for heating and cooling supplied to the space. Typically, control zones are for a room or an open area of a floor.

Energy standards such as ASHRAE *Standard* 90.1 no longer allow constant-volume systems that reheat cold air or that mix heated and cooled air, except in special applications such as hospitals. If used, they should be designed for minimal use of reheat through zoning to match actual dynamic loads and resetting cold and warm air temperatures based on the zone(s) with the greatest demand. Heating and cooling supply zones should be structured to cover areas of similar load. Areas with different exterior exposures should have different supply zones.

Systems that provide changeover switching between heating and cooling prevent simultaneous heating and cooling. Some examples are hot or cold secondary water for fan-coils or single-zone fan systems. They usually require small operational zones, which have low load diversity, to allow changeover from warm to cold water without occupant dissatisfaction.

Systems for building interiors usually require year-round cooling and are somewhat simpler to control than exterior systems. These interior areas normally use all-air systems with a constant supply air temperature, with or without VAV control. Proper control techniques and operational understanding can reduce the energy used to treat these areas. General load characteristics of different parts of a building may lead to selecting different systems for each.

Load Matching

With individual room control, the environment in a space can be controlled more accurately and energy can be conserved if the entire system can be controlled in response to the major factor influencing the load. Thus, water temperature in a water-heating system, steam temperature or pressure in a steam-heating system, or delivered air temperature in a central fan system can be varied as building load varies. Control of the entire system relieves individual space controls of part of their burden and provides more accurate space control. Also, modifying the basic rate of heating or cooling input in accordance with the entire system load reduces losses in the distribution system.

The system must always satisfy the area with the greatest demand. Individual controls handle demand variations in the area the system serves. The more accurate the system zoning, the greater the control, the smaller the distribution losses, and the more effectively space conditions are maintained by individual controls.

Buildings or zones with a modular arrangement can be designed for subdivision to meet occupant needs. Before subdivision, operating inefficiencies can occur if a zone has more than one thermostat. In an area where one thermostat activates heating while another activates cooling, the terminals should be controlled from a single thermostat until the area is properly subdivided.

Size of Controlled Area

No individually controlled area should exceed about 5000 ft² because the difficulty of obtaining good distribution and of finding a representative location for space control increases with zone area. Each individually controlled area must have similar load characteristics throughout. Equitable distribution, provided through competent engineering design, careful equipment sizing, and proper system balancing, is necessary to maintain uniform conditions throughout a controlled area. The control can measure conditions only at its location; it cannot compensate for nonuniform conditions caused by improper distribution or inadequate design. Areas or rooms having dissimilar load characteristics or different conditions to be maintained should be controlled individually. The smaller the controlled area, the better the control and the performance and flexibility.

Location of Space Sensors

Space sensors and controllers must be located where they accurately sense the variables they control and where the condition is representative of the area (zone) they serve. In large open areas having more than one zone, thermostats should be located in the middle of their zones to prevent them from sensing conditions in surrounding zones. Typically, space temperature controllers or sensors are placed in the following locations.

- **Wall-mounted thermostats** or **sensors** are usually placed on inside walls or columns in the space they serve. Avoid outdoor wall locations. Mount thermostats at generally accessible heights according to the Americans with Disabilities Act (ADA) (USDOJ 1994) (usually 48 in.) and in locations where they will not be affected by heat from sources such as direct sun rays, wall pipes or ducts, convectors, or direct air currents from diffusers or equipment (e.g., copy machines, coffeemakers, refrigerators). The wall itself should be sealed tightly if it penetrates a pressurized supply air plenum either under the floor or overhead. Air circulation should be ample and unimpeded by furniture or other obstructions, and the thermostat should be protected against mechanical injury. Thermostats in spaces such as corridors, lobbies, or foyers should be used to control those areas only.
- **Return air thermostats** can control floor-mounted unitary conditioners such as induction or fan-coil units and unit ventilators. On induction and fan-coil units, the sensing element is behind the return air grille. On classroom unit ventilators that use up to 100% outdoor air for natural cooling, however, a forced-flow sampling chamber should be provided for the sensing element, which should be located carefully to avoid radiant effect and to ensure adequate air velocity across the element.

If return air sensing is used with a central fan system, locate the sensing element as near as possible to the space being controlled to eliminate any influence from other spaces and the effect of any heat gain or loss in the duct. Where supply/return light fixtures are used to return air to a ceiling plenum, the return air sensing element can be located in the return air opening. Be sure to offset the set point to compensate for heat from the light fixtures.

- **Diffuser-mounted thermostats** usually have sensing elements mounted on circular or square ceiling supply diffusers and depend on aspiration of room air into the supply airstream. They should be used only on high-aspiration diffusers adjusted for a horizontal air pattern. The diffuser on which the element is mounted should be in the center of the occupied area of the controlled zone.
- **CO_2 sensors** for DCV are usually located in spaces with high occupant densities (e.g., conference rooms, auditoriums, courtrooms). Locating the sensor in return air ducts/plenums that serve multiple spaces measures average concentrations and does not provide information on CO_2 levels in rooms with the highest concentrations. CO_2 sensors should be located in the breathing zone of the occupied space [see ASHRAE (2011) and USGBC (2009)].

Commissioning

Commissioning is the process of ensuring that systems are designed, installed, functionally tested, and capable of being operated and maintained in conformity with the design intent. Commissioning HVAC systems begins with planning and includes design, construction, start-up, acceptance, and training, and can be applied throughout the life of the building.

For HVAC systems, **functional performance testing (FPT)** is an important part of the commissioning process. FPT is the process of determining the ability of HVAC system to deliver heating, ventilating, and air conditioning in accordance with the final design intent. Commissioning is team-oriented and generally involves cooperation of various parties, including the owner, design engineers, and contractors and subcontractors. A commissioning authority (the designated person, company, or agent who implements the overall commissioning process) generally leads the process. Each commissioning process must have a plan that defines the commissioning process and is developed in increasing detail as the project progresses. Phases include plan, design, construction, and acceptance. The most useful tool used to challenge (simulate changes to) systems operation is the control system itself. A DDC system provides the added convenience of central execution of test steps, and the ability to record responses. Commissioning the DDC system must occur before it can be used to validate the HVAC systems. Commissioning DDC systems is discussed in ASHRAE GPC 11. Commissioning HVAC systems is recommended for construction of new buildings, and should be repeated periodically in existing buildings. See Chapter 43 and ASHRAE *Guidelines* 0 and 1 for more information.

REFERENCES

Ahmed, O. 2001. A model-based control for VAV distribution systems. Presented at CLIMA Conference, Napoli.

Ahmed, O. 2002. *Variable-volume, variable-pressure system.* International Facilities Management Association, Research and Development Council, Orlando, FL.

ANSI. 2003. Non-recirculating direct gas-fired industrial air heaters. ANSI *Standard* Z83.4-2003/CSA *Standard* 3.7-2003. American National Standards Institute, Washington, D.C., and Canadian Standards Association, Toronto.

ASHRAE. 2003. *Risk management guidance for health, safety, and environmental security under extraordinary incidents.* Presidential Ad Hoc Committee for Building Health and Safety Under Extraordinary Incidents.

ASHRAE. 2005. *Sequences of operation for common HVAC systems.*

ASHRAE. 2011. *Standard 62.1-2010 user's manual.*

ASHRAE. 2005. The commissioning process. *Guideline* 0-2005.

ASHRAE. 2008. HVAC&R technical requirements for the commissioning process. *Guideline* 1.1.

ASHRAE. 2007. Specifying direct digital control systems. *Guideline* 13-2007

ASHRAE. 2003. Selecting outdoor, return, and relief dampers for air-side economizer systems. ASHRAE *Guideline* 16.

ASHRAE. 2010. Ventilation for acceptable indoor air quality. ANSI/ ASHRAE *Standard* 62.1-2010.

ASHRAE. 2010. Energy standard for buildings except low-rise residential buildings. ANSI/ASHRAE/IES *Standard* 90.1-2010.

ASHRAE 2008. BACnet—A data communication protocol for building automation and control networks. ANSI/ASHRAE *Standard* 135-2008,

ASME. 2004. *Controls and safety devices for automatically fired boilers.* American Society of Mechanical Engineers, New York.

Avery, G. 1992. The instability of VAV systems. *HPAC Engineering* (February):47-50.

EDR. 2007. *Advanced variable air volume system design guide.* Energy Design Resources, Pacific Gas and Electric Company.

Englander, S.L. and L.K. Norford. 1992. Saving fan energy in VAV systems—Part 2: Supply fan control for static pressure minimization. *ASHRAE Transactions* 98(1):19-32.

FEMA. 2003. Reference manual to mitigate potential terrorist attacks against buildings. *Publication* 426, Federal Emergency Management Agency, U.S. Department of Homeland Security, Washington, D.C.

Felker, L.G. and T.L. Felker. 2010. *Dampers and airflow control.* ASHRAE.

FGI. 2010. *Guidelines for design and construction of health care facilities,* 2010 ed. Facilities Guideline Institute.

Hartman, T. 1993. Terminal regulated air volume (TRAV) systems. *ASHRAE Transactions* 99(1):791-800.

Kalore, P., O. Ahmed, and M. Cascia. 2003. Dynamic control of a building fluid distribution system. Presented at IEEE Conference on Control Applications, Istanbul.

Kettler, J. 1995. Minimum ventilation control fan tracking vs. workable solutions. *ASHRAE Transactions* 101(2):625-630.

Kettler, J. 2000. Measuring and controlling outdoor airflow. *ASHRAE IAQ Applications* (Winter).

Kettler, J. 2004. Return fans or relief fans: How to choose? *ASHRAE Journal* 46(4):28-32.

Krarti, M., C.C. Schroeder, E. Jeanette, and M.J. Brandemuehl. 2000. Experimental analysis of measurement and control techniques of outside air intake rates in VAV systems (RP-980). *ASHRAE Transactions* 106(1): 39-52.

NFPA. 2005. National electrical code®. ANSI/NFPA *Standard* 70-05. National Fire Protection Association, Quincy, MA.

NIOSH. 2002. Guidance for protecting building environments from airborne chemical, biological, or radiological attacks. DHHS *Publication* 2002-139. National Institute for Occupational Safety and Health, Department of Health and Human Services, Cincinnati.

Taylor, S. 2002. Primary-only vs. primary-secondary variable flow systems. *ASHRAE Journal* 44(2):25-29.

Taylor, S. 2007. Increasing efficiency with VAV system static pressure setpoint reset. *ASHRAE Journal* 49(6):24-32.

USDOJ. 1994. *ADA standards for accessible design.* Title III Regulations, Appendix A. 28CFR36. U.S. Department of Justice, Washington, D.C.

USGBC. 2009. *LEED® reference guide for green building design and construction.* U.S. Green Building Council, Washington, D.C.

BIBLIOGRAPHY

Federspiel, C.C. 2005. Detecting critical supply duct pressure, *ASHRAE Transactions* 111(1):957-963.

Salsbury, T. and B. Chen. 2002. A new sequence controller for multistage systems of known relative capacities. *International Journal of HVAC&R Research* (now *HVAC&R Research*) 8(4):403-428.

Seem, J., M. House, G. Kelly, and C. Klaassen. 2000. A damper control system for preventing reverse airflow through exhaust air dampers. *International Journal of HVAC&R Research* (now *HVAC&R Research*) 6(2): 135-148.

Shadpour, F. 2000. *Fundamentals of HVAC direct digital control: Practical applications and design.* Hacienda Blue, Escondido, CA.

NOISE AND VIBRATION CONTROL

HVAC equipment for a building is one of the major sources of interior noise, and its effect on the acoustical environment is important. Further, noise from equipment located outdoors often propagates to the community. Therefore, mechanical equipment must be selected, and equipment spaces designed, with an emphasis on both the intended uses of the equipment and the goal of providing acceptable sound levels in occupied spaces of the building and in the surrounding community. Operation of HVAC equipment can also induce mechanical vibration that propagates into occupied spaces through structureborne paths such as piping, ductwork, and mounts. Vibration can cause direct discomfort and also create secondary radiation of noise from vibrating walls, floors, piping, etc.

In this chapter, *sound* and *noise* are used interchangeably, although only *unwanted* sound is considered to be noise.

System analysis for noise control uses the source-path-receiver concept. The source of the sound is the noise-generating mechanism. The sound travels from the source via a path, which can be through the air (airborne) or through the structure (structureborne), or a combination of both paths, until it reaches the receiver (building occupant or outdoor neighbor).

Components of the mechanical system (e.g., fans, dampers, diffusers, duct junctions) all may produce sound by the nature of the airflow through and around them. As a result, almost all HVAC components must be considered. Because sound travels effectively in the same or opposite direction of airflow, downstream and upstream paths are often equally important.

This chapter provides basic sound and vibration principles and data needed by HVAC system designers. Many of the equations associated with sound and vibration control for HVAC may be found in Chapter 8 of the 2009 *ASHRAE Handbook—Fundamentals*. Additional technical discussions along with detailed HVAC component and system design examples can be found in the references.

DATA RELIABILITY

Data in this chapter come from both consulting experience and research studies. Use caution when applying the data, especially for situations that extrapolate from the framework of the original research. Test data tolerances and cumulative system effects lead to a typical uncertainty of ±2 dB. However, significantly greater variations may occur, especially in low frequency ranges and particularly in the 63 Hz octave band, where experience suggests that even correctly performed estimates may disagree with actual measured levels by 5 dB, so conservative design practices should be followed.

ACOUSTICAL DESIGN OF HVAC SYSTEMS

For most HVAC systems, sound sources are associated with the building's mechanical and electrical equipment. As shown in Figure 1, there are many possible paths for airborne and structureborne sound and vibration transmission between a sound source and receiver. Noise control involves (1) selecting a quiet source, (2) optimizing room sound absorption, and (3) designing propagation paths for minimal noise transmission.

Different sources produce sounds that have different frequency distributions, called **spectral characteristics**. For example, as shown in Figure 2, fan noise generally contributes to sound levels in

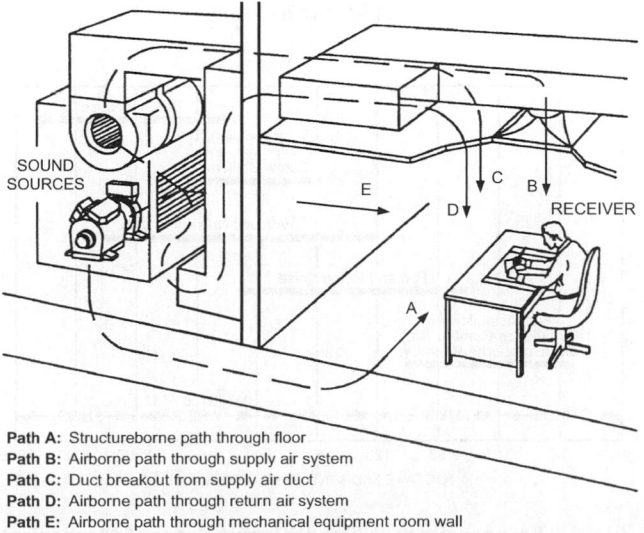

Path A: Structureborne path through floor
Path B: Airborne path through supply air system
Path C: Duct breakout from supply air duct
Path D: Airborne path through return air system
Path E: Airborne path through mechanical equipment room wall

Fig. 1 Typical Paths of Noise and Vibration Propagation in HVAC Systems

The preparation of this chapter is assigned to TC 2.6, Sound and Vibration Control.

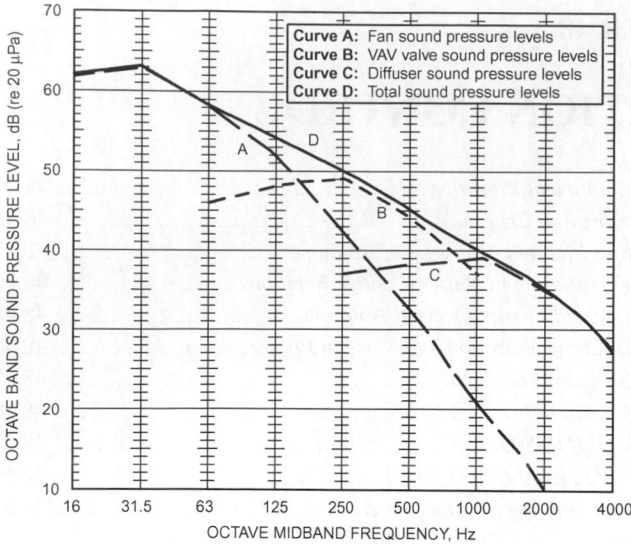

Fig. 2   HVAC Sound Spectrum Components for Occupied Spaces

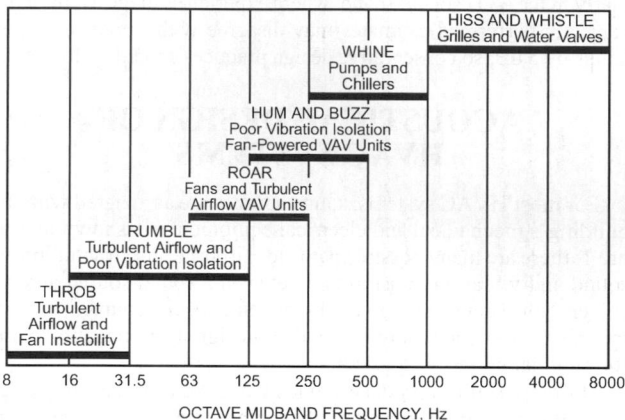

Fig. 3   Frequency Ranges of Likely Sources of Sound-Related Complaints
(Schaffer 2005)

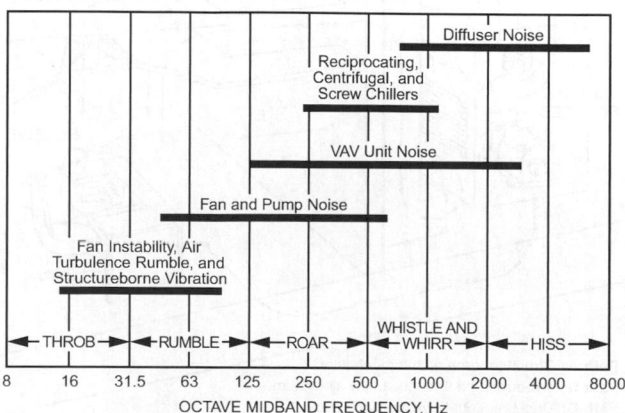

Fig. 4   Frequencies at Which Different Types of Mechanical Equipment Generally Control Sound Spectra
(Schaffer 2005)

the 16 to 250 Hz octave bands (curve A). [Frequencies that designate the octave bands are often called **octave midband** (or **center**) **frequencies**.] Variable-air-volume (VAV) valve noise usually contributes to sound levels in the 63 to 1000 Hz octave bands (curve B). Diffuser noise usually contributes to the overall HVAC noise in the 250 to 8000 Hz octave bands (curve C). The overall sound pressure level associated with all of these sound sources combined is shown as curve D.

Figure 3 (Schaffer 2005) shows the frequency ranges and descriptive terminology of the most likely sources of HVAC sound-related complaints. Figure 4 (Schaffer 2005) shows the frequencies at which different types of mechanical equipment generally control the sound spectra in a room. Occupant complaints may occur, however, despite a well-designed sound spectrum in the room. Criteria specified in this chapter do not necessarily correspond with all individuals' acceptability criteria.

RECEIVER CONSIDERATIONS

Indoor Sound Criteria

Whether an occupant considers the background noise acceptable generally depends on two factors. First is the **perceived loudness** of the noise relative to that of normal activities; if it is clearly noticeable, it is likely to be distracting and cause complaint. Second is the **sound quality** of the background noise; if the noise is perceived as a rumble, throb, roar, hiss, or tone, this may result in complaints of annoyance and stress. The frequency spectrum is then said to be unbalanced.

The acoustical design must ensure that HVAC noise is of sufficiently low level and unobtrusive quality so as not to interfere with occupancy use requirements. If background noise reduces speech intelligibility, for example, complaints of lost productivity can result. Accordingly, methods of rating HVAC-related noise ideally assess both perceived loudness and sound quality.

Design Guidelines for HVAC-Related Background Sound in Rooms. Table 1 presents recommended goals for indoor background noise levels in various types of unoccupied rooms served by HVAC systems. Perceived loudness and task interference are factored into the numerical part of the rating. The sound quality design target is assumed to be a neutral-sounding spectrum, although some spectral imbalance is probably tolerable within limits for most users. The criteria used are described in the next section.

An acceptable noise level depends on the specific use of the space, so each number rating typically represents a range of ±5 dB for the design target. For example, private offices and conference rooms are listed as NC/RC 30. This means that unless there are extenuating circumstances, the background noise level should be less than NC/RC 35, but in some locations (e.g., executive offices or specialty conference rooms), a noise criterion of as low as NC/RC 25 might be warranted. On the other hand, there is not necessarily a benefit to achieving the lower number in regular offices, as some background noise maintains a minimum level of acoustic privacy between adjacent offices.

The NC/RC designations relate to reference curves with octave band sound pressure levels that are (1) selected based on appropriate loudness in the speech interference range (500-2000 Hz) and (2) show contours for high and low frequencies that are balanced at the same loudness level. Acoustical evaluation based on octave bands and target balanced contours is recommended, because overall dBA ratings do not reflect undesirable contributions of excessive low-frequency noise. The dBA and dBC levels are listed only as approximate references in the case of simplistic measurements, where dBA indicates relative loudness and dBC indicates prevalence of low-frequency noise. Exact specifications should be established by acoustical experts considering occupant sensitivity.

Criteria Descriptions. This section presents ways to rate or measure the sound to determine acceptability. The information

Table 1 Design Guidelines for HVAC-Related Background Sound in Rooms

Room Types		Octave Band Analysis[a] NC/RC[b]	Approximate Overall Sound Pressure Level[a]	
			dBA[c]	dBC[c]
Rooms with Intrusion from Outdoor Noise Sources[d]	Traffic noise	N/A	45	70
	Aircraft flyovers	N/A	45	70
Residences, Apartments, Condominiums	Living areas	30	35	60
	Bathrooms, kitchens, utility rooms	35	40	60
Hotels/Motels	Individual rooms or suites	30	35	60
	Meeting/banquet rooms	30	35	60
	Corridors and lobbies	40	45	65
	Service/support areas	40	45	65
Office Buildings	Executive and private offices	30	35	60
	Conference rooms	30	35	60
	Teleconference rooms	25	30	55
	Open-plan offices	40	45	65
	Corridors and lobbies	40	45	65
Courtrooms	Unamplified speech	30	35	60
	Amplified speech	35	40	60
Performing Arts Spaces	Drama theaters, concert and recital halls	20	25	50
	Music teaching studios	25	30	55
	Music practice rooms	30	35	60
Hospitals and Clinics	Patient rooms	30	35	60
	Wards	35	40	60
	Operating and procedure rooms	35	40	60
	Corridors and lobbies	40	45	65
Laboratories	Testing/research with minimal speech communication	50	55	75
	Extensive phone use and speech communication	45	50	70
	Group teaching	35	40	60
Churches, Mosques, Synagogues	General assembly with critical music programs[e]	25	30	55
Schools[f]	Classrooms	30	35	60
	Large lecture rooms with speech amplification	30	35	60
	Large lecture rooms without speech amplification	25	30	55
Libraries		30	35	60
Indoor Stadiums, Gymnasiums	Gymnasiums and natatoriums[g]	45	50	70
	Large-seating-capacity spaces with speech amplification[g]	50	55	75

N/A = Not applicable

[a]Values and ranges are based on judgment and experience, and represent general limits of acceptability for typical building occupancies.

[b]NC: this metric plots octave band sound levels against a family of reference curves, with the number rating equal to the highest tangent line value.

RC: when sound quality in the space is important, the RC metric provides a diagnostic tool to quantify both the speech interference level and spectral imbalance.

[c]dBA and dBC: these are overall sound pressure level measurements with A- and C-weighting, and serve as good references for a fast, single-number measurement. They are also appropriate for specification in cases where no octave band sound data are available for design.

[d]Intrusive noise is addressed here for use in evaluating possible non-HVAC noise that is likely to contribute to background noise levels.

[e]An experienced acoustical consultant should be retained for guidance on acoustically critical spaces (below RC 30) and for all performing arts spaces.

[f]Some educators and others believe that HVAC-related sound criteria for schools, as listed in previous editions of this table, are too high and impede learning for affected groups of all ages. See ANSI/ASA *Standard* S12.60 (ASA 2009, 2010) for classroom acoustics and a justification for lower sound criteria in schools. The HVAC component of total noise meets the background noise requirement of that standard if HVAC-related background sound is approximately NC/RC 25. Within this category, designs for K-8 schools should be quieter than those for high schools and colleges.

[g]RC or NC criteria for these spaces need only be selected for the desired speech and hearing conditions.

should help the design engineer select the most appropriate background noise rating method for a specific project. Current methods described here and in other references include the traditional A-weighted sound pressure level (dBA) and tangent Noise Criteria (NC), the Room Criterion (RC) and more recent RC Mark II, the Balanced Noise Criterion (NCB), and the Room Noise Criteria (RNC). Each method was developed based on data for specific applications; hence, not all are equally suitable for rating HVAC-related noise in the variety of applications encountered. The preferred sound rating methods generally comprise two distinct parts: a family of criterion curves (specifying sound levels by octave bands), and a procedure for rating the calculated or measured sound data relative to the criterion curves with regard to sound quality.

Ideally, HVAC-related background noise should have the following characteristics:

- Balanced contributions from all parts of the sound spectrum with no predominant frequency bands of noise
- No audible tones such as hum or whine
- No fluctuations in level such as throbbing or pulsing

dBA and dBC: A- and C-Weighted Sound Level. The A-weighted sound level (described in Chapter 8 of the 2009 *ASHRAE Handbook—Fundamentals*) has been used for more than 60 years as a single-number measure of the relative loudness of noise, especially for outdoor environmental noise standards. The rating is expressed as a number followed by dBA (e.g., 40 dBA).

A-weighted sound levels can be measured with simple sound level meters. The ratings correlate fairly well with human judgments of relative loudness but take no account of spectral balance or sound quality. Thus, two different spectra can result in the same numeric value, but have quite different subjective qualities.

Along with dBA, there is also a C-weighted sound level, denoted as dBC, which is more sensitive to low-frequency sound contributions to the overall sound level than is dBA. When the quantity (dBC – dBA) is large (e.g., greater than 25 dB), significant low-frequency sound is present. It is recommended that when specifying background sound levels in dBA, the dBC is also included in the specification and does not exceed the dBA reading by more than 20 dB.

NC: Noise Criteria Method. The NC method for rating noise (described in Chapter 8 of the 2009 *ASHRAE Handbook—Fundamentals*) has been used for more than 50 years. It is a single-number rating that is somewhat sensitive to the relative loudness and speech interference properties of a given noise spectrum. The method consists of a family of criterion curves, shown in Figure 5, extending from 63 to 8000 Hz, and a **tangency rating procedure**. The criterion curves define the limits of octave band spectra that must not be exceeded to meet occupant acceptance in certain spaces. The rating is expressed as NC followed by a number (e.g., NC 40). The octave midband frequency of the point at which the spectrum is tangent to the highest NC curve should also be reported [e.g. NC 40 (125 Hz)]. The NC values are formally defined only in 5 dB increments, with intermediate values determined by discretionary interpolation.

Widely used and understood, the NC method is sensitive to level but has the disadvantage that the tangency method used to determine the rating does not require that the noise spectrum precisely follow the balanced shape of the NC curves. Thus, sounds with different frequency content can have the same numeric rating, but rank differently on the basis of sound quality. With the advent of VAV systems, low-frequency content (i.e., below the 63 Hz octave band) is prevalent, and the NC rating method fails to properly address this

issue (Ebbing and Blazier 1992). Consequently, if the NC method is chosen, sound levels at frequencies below 63 Hz must be evaluated by other means.

In HVAC systems that do not produce excessive low-frequency noise and strong discernable pure tones, the NC rating correlates relatively well with occupant satisfaction if sound quality is not a significant concern. NC rating is often used because of its simplicity.

RC/RC Mark II: Room Criteria Method. ASHRAE previously recommended the Room Criterion (RC) curves (beginning in Chapter 43 in the 1995 *ASHRAE Handbook—HVAC Systems and Equipment*; Blazier 1981a, 1981b) as an enhanced method for rating HVAC system related noise. The revised RC Mark II method is now preferred.

The RC method is a family of criterion curves and a rating procedure. The shape of these curves represents a well-balanced, bland-sounding spectrum, including two additional octave bands (16 and 31.5 Hz) to deal with excessive low-frequency noise. This rating procedure assesses background noise in spaces on the basis of its effect on speech, and on subjective sound quality. The rating value is expressed as RC followed by a number that represents the level of noise in the speech interference region of the spectrum, and a letter to indicate the quality [e.g., RC 35(N), where N denotes the desirable neutral rating]. The RC method includes evaluation of the potential for noise-induced vibration from excessive airborne sound levels at and below 63 Hz.

Based on experience and ASHRAE-sponsored research (Broner 1994), the RC method was revised to the RC Mark II method (Blazier 1997). Like its predecessor, the RC Mark II method is intended for use as a diagnostic tool for analyzing noise problems in the field. The RC Mark II method is complicated, but computerized spreadsheets and HVAC system analysis programs are available to perform the calculations and graphical analysis.

The RC Mark II method has three parts: (1) a family of criterion curves (Figure 6), (2) a procedure for determining the RC numerical rating and the noise spectral balance (quality), and (3) a procedure for estimating occupant satisfaction when the spectrum does not have the shape of an RC curve (quality assessment index) (Blazier 1995). The rating is expressed as RC followed by a number and a letter [e.g., RC 35(N)]. The number is the arithmetic average rounded to the nearest integer of sound pressure levels in the 500, 1000, and 2000 Hz octave bands (the main speech frequency region) and is known as the preferred speech interference level (PSIL). The letter is a qualitative descriptor that identifies the sound's perceived character: (N) for neutral, (LF) for low-frequency rumble, (MF) for midfrequency roar, and (HF) for high-frequency hiss. There are also two subcategories of the low-frequency descriptor: (LF$_B$), denoting a moderate but perceptible degree of sound-induced ceiling/wall vibration, and (LF$_A$), denoting a noticeable degree of sound-induced vibration.

Each reference curve in Figure 6 identifies the shape of a neutral, bland-sounding spectrum, indexed to a curve number corresponding to the sound level in the 1000 Hz octave band. The shape of these curves is based on Blazier (1981a, 1981b), modified at 16 Hz following recommendations of the research in Broner (1994). Regions A and B denote levels at which sound can induce vibration in light wall and ceiling construction, which can potentially cause rattles in light fixtures, furniture, etc. Curve T is the octave band threshold of hearing as defined by ANSI *Standard* 12.2.

Procedure for Determining the RC Mark II Rating for a System.

Step 1. Obtain the arithmetic average of the sound levels in the principal speech frequency range represented by the levels in the 500, 1000, and 2000 Hz octave bands [preferred speech interference level (PSIL)]. [This is not to be confused with the ANSI-defined "speech-interference level" (SIL), a four-band average obtained by including the 4000 Hz octave band as used with the NCB method.]

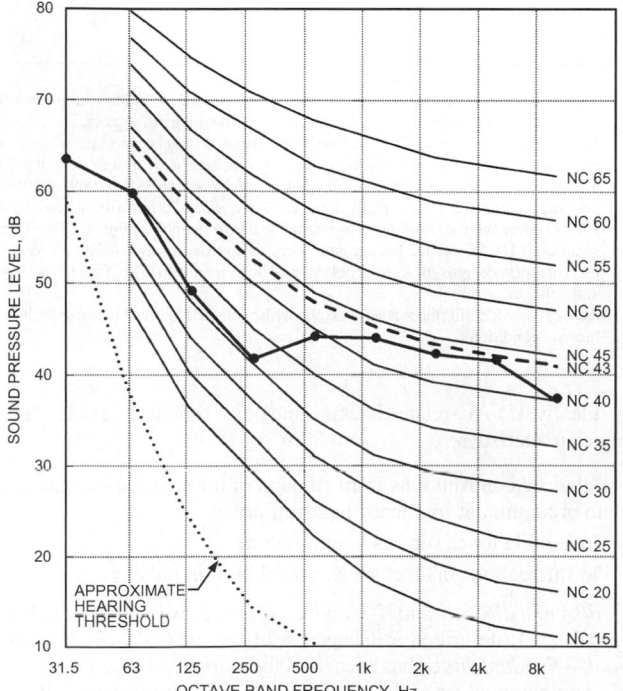

Fig. 5 Noise Criteria Curves

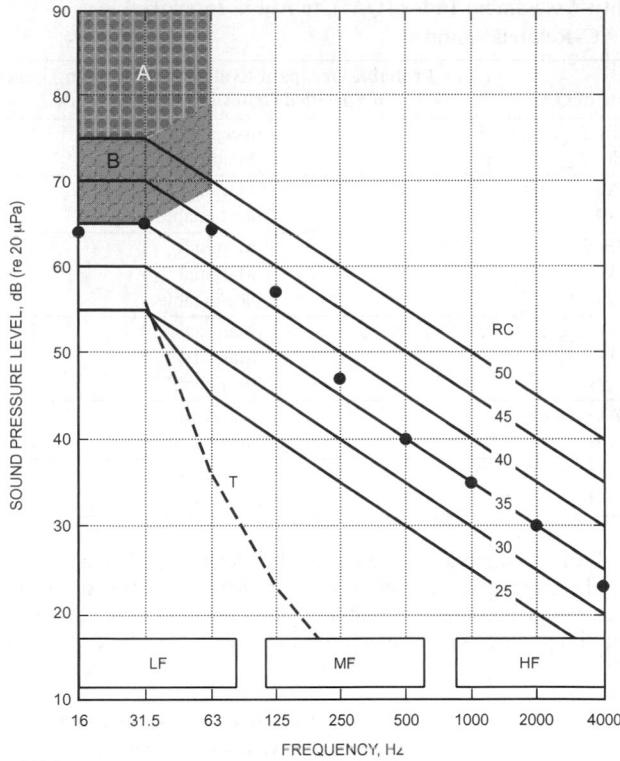

SOUND PRESSURE LEVEL, dB (re 20 µPa)

RC

50

45

40

35

30

25

LF   MF   HF

FREQUENCY, Hz

Note:
- Noise levels for lightweight wall and ceiling constructions:
  – In shaded region B are likely to generate vibration that may be perceptible. There is a slight possibility of rattles in light fixtures, doors, windows, etc.
  – In shaded region A have a high probability of generating easily perceptible noise-induced vibration. Audible rattling in light fixtures, doors, windows, etc. may be anticipated.
- Regions LF, MF, and HF are explained in the text.
- Solid dots are sound pressure levels for the example discussed in the text.

Fig. 6   Room Criterion Curves, Mark II

The RC reference curve is chosen to be that which has the same value at 1000 Hz as the calculated average value.

Step 2. Calculate the quality assessment index (QAI) (Blazier 1995), which measures the deviation of the spectrum under evaluation from the shape of the RC reference curve. Calculate the *energy-averaged* spectral deviations from the RC reference curve in each of three frequency groups: low (LF; 16 to 63 Hz), medium (MF; 125 to 500 Hz), and high (HF; 1000 to 4000 Hz). (A simple arithmetic average of these deviations is often adequate for most engineering purposes.) Equation (1) gives the procedure for the LF region; repeat for the MF and HF regions by substituting the corresponding values at each frequency.

$$LF = 10 \log[(10^{0.1\Delta L_{16}} + 10^{0.1\Delta L_{31.5}} + 10^{0.1\Delta L_{63}})/3] \qquad (1)$$

The ΔL terms are the differences between the spectrum being evaluated and the RC reference curve in each frequency band. In this way, three specific spectral deviation factors, expressed in dB with either positive or negative values, are associated with the spectrum being rated. QAI is the *range* in dB between the highest and lowest values of the spectral deviation factors.

If QAI ≤ 5 dB, the spectrum is assigned a *neutral* (N) rating. If QAI *exceeds* 5 dB, the sound quality descriptor of the RC rating is the letter designation of the frequency region of the deviation factor having the highest *positive* value.

Example 1. The spectrum plotted in Figure 6 indicated by large dots is processed in Table 2. The arithmetic average of the sound levels in the 500, 1000, and 2000 Hz octave bands is 35 dB, so the RC 35 curve is selected as the reference for spectrum quality evaluation.

Table 2   Example 1 Calculation of RC Mark II Rating

	Frequency, Hz								
	16	**31**	**63**	**125**	**250**	**500**	**1000**	**2000**	**4000**
Spectrum levels	64	65	64	57	47	40	35	30	23
Average of 500 to 2000 Hz levels							35		
RC contour	60	60	55	50	45	40	35	30	25
Levels—RC contour	4	5	9	7	2	0	0	0	−2
		LF			MF			HF	
Spectral deviations		6.6			4.0			−0.6	
QAI					7.2				
RC Mark II rating				**RC 35(LF)**					

The spectral deviation factors in the LF, MF, and HF regions are 6.6, 4.0, and −0.6 respectively, giving a QAI of 7.2. The maximum *positive* deviation factor occurs in the LF region and QAI exceeds 5; therefore, the rating of the spectrum is RC 35(LF). An average room occupant should perceive this spectrum as rumbly in character.

Estimating Occupant Satisfaction Using QAI.

The QAI estimates the probable reaction of an occupant when system design does not produce optimum sound quality. The basis for estimating occupant satisfaction is that changes in sound level of less than 5 dB do not cause subjects to change their ranking of sounds of similar spectral content. However, level changes greater than 5 dB do significantly affect subjective judgments. A QAI of 5 dB or less corresponds to a generally acceptable condition, provided that the perceived level of the sound is in a range consistent with the given type of space occupancy as recommended in Table 2. (An exception to this rule occurs when sound pressure levels in the 16 or 31 Hz octave bands exceed 65 dB. In such cases, there is potential for acoustically induced vibration in typical lightweight office construction. Levels above 75 dB in these bands indicate a significant problem with induced vibration.)

A QAI that exceeds 5 dB but is less than or equal to 10 dB represents a marginal situation, in which acceptance by an occupant is questionable. However, a QAI greater than 10 dB will likely be objectionable to the average occupant. Table 3 lists sound quality descriptors and QAI values and relates them to probable occupant reaction to the noise.

The numerical part of the RC rating may sometimes be less than the specified maximum for the space use, but with a sound quality descriptor other than the desirable (N). For example, a maximum of RC 40(N) is specified, but the actual noise environment turns out to be RC 35(MF). There is insufficient knowledge in this area to decide which spectrum is preferable.

Even at moderate levels, if the dominant portion of the background noise occurs at a very low frequency, some people can experience a sense of oppressiveness or depression in the environment (Persson-Wayne et al. 1997). Such a complaint may result after exposure to that environment for several hours, and thus may not be noticeable during a short exposure period.

NCB: Balanced Noise Criteria Method. The NCB method (ANSI *Standard* S12.2; Beranek 1989) is used to specify or evaluate room noise, including that from occupant activities. The NCB criterion curves (Figure 7) are intended as an improvement over the NC curves by including the two low-frequency octave bands (16 and 31.5 Hz), and by lowering permissible noise levels at high frequencies (4000 and 8000 Hz). Rating is based on the speech interference level (SIL = the average of the four sound pressure levels at octave midband frequencies of 500, 1000, 2000, and 4000 Hz) with additional tests for rumble and hiss compliance. The rating is expressed as NCB followed by a number (e.g., NCB 40).

The NCB method is better than the NC method in determining whether a noise spectrum has an unbalanced shape sufficient to

Table 3 Definition of Sound-Quality Descriptor and Quality-Assessment Index (QAI), to Aid in Interpreting RC Mark II Ratings of HVAC-Related Sound

Sound-Quality Descriptor	Description of Subjective Perception	Magnitude of QAI	Probable Occupant Evaluation, Assuming Level of Specified Criterion is Not Exceeded
(N) Neutral (Bland)	Balanced sound spectrum, no single frequency range dominant	QAI ≤ 5 dB, $L_{16}, L_{31} ≤ 65$	Acceptable
		QAI ≤ 5 dB, $L_{16}, L_{31} > 65$	Marginal
(LF) Rumble	Low-frequency range dominant (16 to 63 Hz)	5 dB < QAI ≤ 10 dB	Marginal
		QAI > 10 dB	Objectionable
(LFV$_B$) Rumble, with moderately perceptible room surface vibration	Low-frequency range dominant (16 to 63 Hz)	QAI ≤ 5 dB, $65 < L_{16}, L_{31} < 75$	Marginal
		5 dB < QAI ≤ 10 dB	Marginal
		QAI > 10 dB	Objectionable
(LFV$_A$) Rumble, with clearly perceptible room surface vibration	Low-frequency range dominant (16 to 63 Hz)	QAI ≤ 5 dB, $L_{16}, L_{31} > 75$	Marginal
		5 dB < QAI ≤ 10 dB	Marginal
		QAI > 10 dB	Objectionable
(MF) Roar	Mid-frequency range dominant (125 to 500 Hz)	5 dB < QAI ≤ 10 dB	Marginal
		QAI > 10 dB	Objectionable
(HF) Hiss	High-frequency range dominant (1000 to 4000 Hz)	5 dB < QAI ≤ 10 dB	Marginal
		QAI > 10 dB	Objectionable

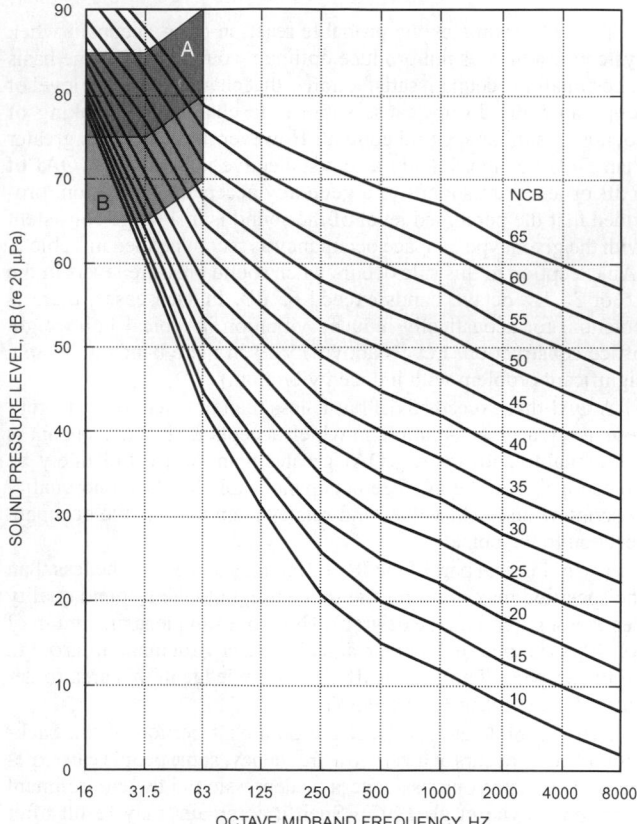

Fig. 7 NCB Noise Criterion Curves

demand corrective action, and it addresses the issue of low-frequency noise. Rating is, however, more complicated than the familiar NC tangency method. The NCB method can still be used as a tangency method; if so used, the point of tangency, which sets the rating, must be cited.

RNC: Room Noise Criteria Method. This rating method has been recently introduced and is described in detail in the American National Standards Institute (ANSI) *Standard* S12.2-2008. It is mentioned here for reference only and, at present, ASHRAE has no formal position on the use of this method.

Table 4 summarizes the essential differences, advantages, and disadvantages of rating methods used to characterize HVAC-related background noise. Unfortunately, at this time there is no acceptable and simple process to characterize the effects of audible tones and level fluctuations, so none of these rating methods address these issues.

Noise Criteria for Plumbing Systems. Acceptable noise levels from plumbing fixtures and piping have not been previously identified in the literature. Continuous noise from plumbing fixtures and piping systems with circulating fluids should meet the same noise criteria as HVAC systems. However, many sounds from plumbing fixtures and piping are of short duration or are transient, and typically have a somewhat higher threshold of acceptance. Examples of these sources include water flow noise associated with typical restroom fixtures; noise from waste lines connected to restroom, kitchen, and/or laundry drains; and noise from jetted bathtubs.

Table 5 presents suggested maximum A-weighted sound pressure levels for various transient plumbing noise sources in buildings with multiple occupancies. These criteria are minimum standards and are intended to apply to plumbing systems serving adjacent and nearby units in multifamily housing projects (apartments and condominiums), hospitals, educational facilities, and office buildings. Plumbing noise levels in high-end luxury condominiums or private homes should be 5 to 10 dB lower than levels shown in Table 5.

Achieving the recommended plumbing noise criteria in the finished space usually requires special attention to pipe installation details, selection of suitable piping materials, design flow velocities, and selection of appropriate fixtures.

Determining Compliance. When taking field measurements to determine whether a space complies with the guidelines presented in Table 1, the following precautions must be taken:

- Measure the noise with an integrating sound level meter with a real-time frequency analyzer meeting type 1 or 2 specifications, as defined in ANSI *Standards* S1.4, S1.11, and S1.43. The meter should have been calibrated by an accredited calibration laboratory, with some assurance that the calibration accuracy has been maintained.
- Set the meter to display and save the equivalent energy sound pressure level (L_{eq}) with the desired frequency filtering (e.g., octave bands, A-weighted, etc.). Each measurement should be a minimum of 15 s long.
- Place the measurement microphone in potential listening locations at least 3.2 ft from room boundaries and noise sources and at least 1.6 ft from furniture. More than one location may be measured, and the microphone may be moved during measurement; movement should not exceed 6 in/s.

Table 4   Comparison of Sound Rating Methods

Method	Overview	Considers Speech Interference Effects	Evaluates Sound Quality	Components Presently Rated by Each Method
dBA	No quality assessment Frequently used for outdoor noise ordinances	Yes	No	Cooling towers Water chillers Condensing units
NC	Can rate components Limited quality assessment Does not evaluate low-frequency rumble	Yes	Somewhat	Air terminals Diffusers
RC Mark II	Used to evaluate systems Should not be used to evaluate components Evaluates sound quality Provides improved diagnostics capability	Yes	Yes	Not used for component rating
NCB	Can rate components Some quality assessment	Yes	Somewhat	See NC
RNC	Some quality assessment Attempts to quantify fluctuations	Yes	Somewhat	Not used for component rating

Table 5   Plumbing Noise Levels

Receiving (listening) room	L_{max} (slow response)
Residential bedroom/living room/dining room	35
Hospital patient room/classroom	40
Private office/conference room	40
Residential bathroom/kitchen	45
Open office/lobby/corridor	50

- Note the operational conditions of the HVAC system at the time of the test. Turn off all non-HVAC system noises during the test. If possible, measure in a normally furnished, unoccupied room.
- The test may be repeated with the entire HVAC system turned off, to determine whether the room's ambient noise level from non-HVAC sources is contributing to the results.
- Record the sound level meter make, model, and serial number; measured sound pressure levels for each microphone location; HVAC system's operating conditions; and microphone location(s).

When these levels are used as a basis for compliance verification, the following additional information must be provided:

- What sound metrics are to be measured (specify L_{eq} or L_{max} levels, etc., in each octave frequency band)
- Where and how the sound levels are to be measured (specify the space average over a defined area or specific points for a specified minimum time duration, etc.)
- What type(s) of instruments are to be used to make the sound measurements (specify ANSI or IEC Type 1 or Type 2 sound level meters with octave band filters, etc.)
- How sound measurements instruments are to be calibrated or checked (specify that instruments are to be checked with an acoustical calibrator both before and after taking sound level measurements, etc.)
- How sound level measurements are to be adjusted for the presence of other sound sources (specify that background sound level measurements be performed without other sound sources under consideration operating; if background sound levels are within 10 dB of operational sound levels, then corrections should be performed; etc.)
- How results of sound measurements are to be interpreted (specify whether octave band sound levels, NC, RC, dBA, dBC or other values are to be reported)

Unless these six points are clearly stipulated, the specified sound criteria may be unenforceable.

When applying the levels specified in Table 1 as a basis for design, sound from non-HVAC sources, such as traffic and office equipment, may establish the lower limit for sound levels in a space.

Outdoor Sound Criteria

Acceptable outdoor sound levels are generally specified by local noise ordinances or other government codes, which almost always use the A-weighted noise level (dBA) as their metric. The usual metric is either L_{max} (maximum noise level over a period), L_{eq} (average noise level over a period), or L_p (no indication of the measure). The time constant (FAST or SLOW) used for L_{max} or L_p depends on the code.

Some communities have no ordinance and depend on state regulations that often use the day/night noise level descriptor L_{DN}, which is a combination of the daytime (7:00 AM to 10:00 PM) and nighttime (10:00 PM to 7:00 AM) average noise levels (L_{eq}) with a 10 dB penalty for nighttime. Other descriptors also exist; specific requirements should be identified at the outset of each project. In some cases, regulatory agencies may also impose project-specific noise conditions on the basis of community reaction and for maintaining an appropriate acoustic environment at the project vicinity.

Measurement or estimation of community noise is based on a location, often at the receiver's property line, from a height of approximately 4 ft that represents ear height for a typical person seated at ground level to any height to address upper floor elevations, but can be anywhere within the property line, and often near the façade of the closest dwelling unit. Alternatively, the measurement may be made at the property line of the noise source.

In the absence of a local noise ordinance, county or state laws or codes or those of a similar community should be used. Even if activity noise levels do not exceed those specified by an ordinance, community acceptance is not ensured. Very low ambient levels or a noise source with an often-repeated, time-varying characteristic or strong tonal content may increase the likelihood of complaints. In the absence of local ordinances, noise levels between 45 and 55 dBA may be considered in residential zones and 55 to 65 dBA in commercial zones. These are for outdoor use areas and, with standard building constructions, they also typically result in acceptable interior noise levels. Often, daytime noise levels (the period of daytime to be defined) are 10 dB higher than nighttime levels.

Although most ordinances are given as A-weighted pressure level, attenuation by distance, barriers, buildings, and atmosphere are all frequency-dependent. Thus, A-weighted levels do not give an accurate estimation of noise levels at distances from the source. If A-weighted sound levels of sources must be determined by means other than measurement, then octave band or one-third octave band measurements of source sound pressure level at a distance, or (preferably) sound power level, must be obtained before calculating the attenuation.

BASIC ACOUSTICAL DESIGN TECHNIQUES

When selecting fans and other related mechanical equipment and when designing air distribution systems to minimize sound transmitted from system components to occupied spaces, consider the following:

- Design the air distribution system to minimize flow resistance and turbulence. High flow resistance increases required fan pressure, which results in higher noise being generated by the fan, especially at low frequencies. Turbulence also increases flow noise generated by duct fittings and dampers, especially at low frequencies.
- Select a fan to operate as near as possible to its rated peak efficiency when handling the required airflow and static pressure. Also, select a fan that generates the lowest possible noise at required design conditions. Using an oversized or undersized fan that does not operate at or near rated peak efficiency can substantially increase noise levels.
- Design duct connections at both fan inlet and outlet for uniform and straight airflow. Both turbulence (at fan inlet and outlet) and flow separation at the fan blades can significantly increase fan-generated noise. Turning vanes near fan outlets can also increase turbulence and noise, especially if airflow is not sufficiently uniform.
- Select duct silencers that do not significantly increase the required fan total static pressure. Selecting silencers with static pressure losses of 0.35 in. of water or less can minimize regenerated noise from silencer airflow.
- Place fan-powered mixing boxes associated with variable-volume-air distribution systems away from noise-sensitive areas.
- Minimize flow-generated noise by elbows or duct branch takeoffs whenever possible by locating them at least four to five duct diameters from each other. For high-velocity systems, it may be necessary to increase this distance to up to 10 duct diameters in critical noise areas. Using flow straighteners or honeycomb grids, often called "egg crates," in the necks of short-length takeoffs that lead directly to grilles, registers, and diffusers is preferred to using volume extractors that protrude into the main duct airflow.
- Keep airflow velocity in ducts serving sound-sensitive spaces as low as possible by increasing the duct size to minimize turbulence and flow-generated noise (see Tables 8 and 9, in the section on Aerodynamically Generated Sound in Ducts).
- Duct transitions should not exceed an included expansion angle of 15°, or the resulting flow separation may produce rumble noise.
- Use turning vanes in large 90° rectangular elbows and branch takeoffs. This provides a smoother directional transition, thus reducing turbulence.
- Place grilles, diffusers, and registers into occupied spaces as far as possible from elbows and branch takeoffs.
- Minimize use of volume dampers near grilles, diffusers, and registers in acoustically critical situations.
- Vibration-isolate all reciprocating and rotating equipment connected to structure. Also, it is usually necessary to vibration-isolate mechanical equipment in the basement of a building as well as piping supported from the ceiling slab of a basement, directly below tenant space. It may be necessary to use flexible piping connectors and flexible electrical conduit between rotating or reciprocating equipment and pipes and ducts that are connected to the equipment.
- Vibration-isolate ducts and pipes, using spring and/or neoprene hangers for at least the first 50 ft from vibration-isolated equipment.
- Use barriers near outdoor equipment when noise associated with the equipment will disturb adjacent properties. In normal practice, barriers typically produce no more than 15 dB of sound attenuation in the midfrequency range. To be effective, the noise barrier must at least block the direct "line of sight" between the source and receiver.

Table 6 lists several common sound sources associated with mechanical equipment noise. Anticipated sound transmission paths and recommended noise reduction methods are also listed. Airborne and/or structureborne sound can follow any or all of the transmission paths associated with a specified sound source. Schaffer (2005) has more detailed information in this area.

SOURCE SOUND LEVELS

Accurate acoustical analysis of HVAC systems depends in part on reliable equipment sound data. These data are often available from equipment manufacturers in the form of sound pressure levels at a specified distance from the equipment or, preferably, equipment sound power levels. Standards used to determine equipment and component sound data are listed at the end of this chapter.

When reviewing manufacturers' sound data, obtain certification that the data have been obtained according to one or more of the relevant industry standards. If they have not, the equipment should be rejected in favor of equipment for which data have been obtained according to relevant industry standards. See Ebbing and Blazier (1998) for further information.

Fans

Prediction of Fan Sound Power. The sound power generated by a fan performing at a given duty is best obtained from manufacturers' test data taken under approved test conditions (AMCA *Standard* 300 or ASHRAE *Standard* 68/AMCA *Standard* 330). Applications of air-handling products range from stand-alone fans to systems with various modules and attachments. These appurtenances and modules can have a significant effect on air-handler sound power levels. In addition, fans of similar aerodynamic performance can have significant acoustical differences.

Predicting air-handling unit sound power from fan sound levels is difficult. Fan sound determined by tests may be quite different once the fan is installed in an air handler, which in effect creates a new acoustical environment. Proper testing to determine resulting sound power levels once a fan is installed is essential. Fan manufacturers are in the best position to supply information on their products, and should be consulted for data when evaluating the acoustic performance of fans for an air handler application. Similarly, air handler manufacturers are in the best position to supply acoustic information on air handlers.

Air handler manufacturers typically provide discharge, inlet, and casing-radiated sound power levels for their units based on one of two methods. A common method is the **fan-plus-algorithm** method: the fan is tested as a stand-alone item, typically using AMCA *Standard* 300, and an algorithm is used to predict the effect of the rest of the air-handling unit on the sound as it travels from the fan to the discharge and intake openings or is radiated through a casing with known transmission loss values. Another method is described in **AHRI *Standard* 260**, in which the entire unit is tested as an assembly, including fans, filters, coils, plenums, casing, etc., and the sound power level at the inlet and discharge openings, as well as the radiated sound power, is measured in a qualified reverberant room. Whenever possible, data obtained by the AHRI 260 method should be used because it eliminates much of the uncertainty present in the fan-plus-algorithm method. For a detailed description of fan operations, see Chapter 20 in the 2008 *ASHRAE Handbook—HVAC Systems and Equipment*. Different fan types have different noise characteristics and within a fan type, several factors influence noise.

Point of Fan Operation. The point of fan operation has a major effect on acoustical output. Fan selection at the calculated point of maximum efficiency is common practice to ensure minimum power consumption. In general, for a given design, fan sound is at a minimum near the point of maximum efficiency. Noise increases as the operating point shifts to the right, as shown in Figure 8 (higher airflow and lower static pressure). Low-frequency

Table 6   Sound Sources, Transmission Paths, and Recommended Noise Reduction Methods

Sound Source	Path No.
Circulating fans; grilles; registers; diffusers; unitary equipment in room	1
Induction coil and fan-powered VAV mixing units	1, 2
Unitary equipment located outside of room served; remotely located air-handling equipment, such as fans, blowers, dampers, duct fittings, and air washers	2, 3
Compressors, pumps, and other reciprocating and rotating equipment (excluding air-handling equipment)	4, 5, 6
Cooling towers; air-cooled condensers	4, 5, 6, 7
Exhaust fans; window air conditioners	7, 8
Sound transmission between rooms	9, 10

No.	Transmission Paths	Noise Reduction Methods
1	Direct sound radiated from sound source to ear Reflected sound from walls, ceiling, and floor	Direct sound can be controlled only by selecting quiet equipment. Reflected sound is controlled by adding sound absorption to room and to equipment location.
2	Air- and structureborne sound radiated from casings and through walls of ducts and plenums is transmitted through walls and ceiling into room	Design duct and fittings for low turbulence; locate high-velocity ducts in noncritical areas; isolate ducts and sound plenums from structure with neoprene or spring hangers.
3	Airborne sound radiated through supply and return air ducts to diffusers in room and then to listener by Path 1	Select fans for minimum sound power; use ducts lined with sound-absorbing material; use duct silencers or sound plenums in supply and return air ducts.
4	Noise transmitted through equipment room walls and floors to adjacent rooms	Locate equipment rooms away from critical areas; use masonry blocks or concrete for mechanical equipment room walls; use floating floors in mechanical rooms.
5	Vibration transmitted via building structure to adjacent walls and ceilings, from which it radiates as noise into room by Path 1	Mount all machines on properly designed vibration isolators; design mechanical equipment room for dynamic loads; balance rotating and reciprocating equipment.
6	Vibration transmission along pipes and duct walls	Isolate pipe and ducts from structure with neoprene or spring hangers; install flexible connectors between pipes, ducts, and vibrating machines.
7	Noise radiated to outside enters room windows	Locate equipment away from critical areas; use barriers and covers to interrupt noise paths; select quiet equipment.
8	Indoor noise follows Path 1	Select quiet equipment.
9	Noise transmitted to an air diffuser in a room, into a duct, and out through an air diffuser in another room	Design and install duct attenuation to match transmission loss of wall between rooms; use crosstalk silencers in ductwork.
10	Sound transmission through, over, and around room partition	Extend partition to ceiling slab and tightly seal all around; seal all pipe, conduit, duct, and other partition penetrations.

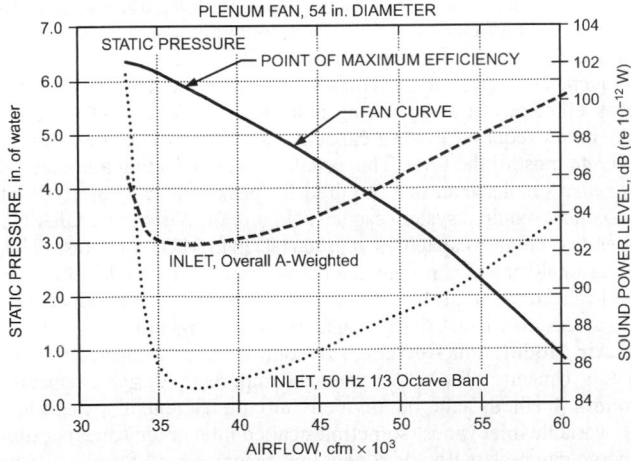

Note that A-weighted sound power level and 50 Hz 1/3 octave band sound power level rise as operating point moves away from maximum efficiency point.

Fig. 8   Test Data for Plenum Fan, Comparing Operating Point (Static Pressure and Airflow), A-Weighted Sound Power Level

noise can increase substantially at operating points to the left of maximum efficiency (lower airflow and higher static pressure). These operating points should be avoided.

Blade-Pass Frequency. The blade-pass frequency is represented by the number of times per second a fans impeller passes a stationary item: f_{bp} = (rpm × number of impeller blades)/60. All fans generate a tone at this frequency and its multiples (harmonics). Whether this tone is objectionable or barely noticeable depends on the type and design of the fan and the point of operation.

Housed Centrifugal Fans. Forward-curved (FC) fans are commonly used in a wide range of standard air-handler products. The blade-pass of FC fans is typically less prominent and is at a higher frequency than other fans. The most distinguishing acoustical concern of FC fans is the prevalent occurrence of low-frequency rumble from airflow turbulence generated at blade tips, which can be exacerbated by nonideal discharge duct conditions (less than five diameters of straight duct). FC fans are commonly thought to have 16, 31.5, and 63 Hz (full octave band) rumble, particularly when operating to the left of the maximum efficiency point.

Backward-inclined (BI) fans and airfoil (AF) fans are generally louder at the blade-pass frequency than a given FC fan selected for the same duty, but are much more energy-efficient at higher pressures and airflow. The blade-pass tone generally increases in prominence with increasing fan speed and is typically in a frequency range that is difficult to attenuate. Below the blade-pass frequency, these fans generally have lower sound amplitude than FC fans and are often quieter at high frequencies.

Care should be taken with all types of housed fans to allow adequate clearance around the inlets. Also, note that belt guards and inlet screens may decrease airflow and increase sound generation.

Plenum Fans. A plenum fan has no housing around the fan impeller and discharges directly into the chamber, pressurizing the plenum, and forcing air through the attached ductwork. Air flows into the fan impeller through an inlet bell located in the chamber wall. These fans can substantially lower discharge sound power levels if the fan plenum is appropriately sized and acoustically treated with sound-absorptive material.

The plenum discharge should be located away from the fan's air blast, because blowing directly into the duct can aggravate the blade-pass sound. Avoid obstructing the inlet or crowding the coils or filters.

Vaneaxial Fans. Generally thought to have the lowest amplitudes of low-frequency sound of any of the fan types, axial fans are often used in applications where the higher-frequency noise can be managed with attenuation devices. In the useful operating range, noise from axial fans is a strong function of the inlet airflow symmetry and blade tip speed.

Propeller Fans. Sound from propeller fans generally has a low-frequency-dominated spectrum shape; the blade-pass frequency is typically prominent and occurs in the low-frequency bands because of the small number of blades. Propeller fan blade-pass frequency noise is very sensitive to inlet obstructions. For some propeller fan designs, the shape of the fan venturi (inlet) is also a very important parameter that affects sound levels. In some applications, noise of a propeller fan is described as sounding like a helicopter. Propeller fans are most commonly used on condensers and for power exhausts.

Minimizing Fan Noise. To minimize the required air distribution system sound attenuation, proper fan selection and installation are vital. The following factors should be considered:

- Design the air distribution system for minimum airflow resistance. High system resistance requires fans to operate at a higher brake horsepower, which generates higher sound power levels.
- Carefully analyze system pressure losses. Higher-than-expected system resistance may result in higher sound power levels than originally estimated.
- Examine the sound power levels of different fan types and designs. Select a fan (or fans) that generates the lowest sound power levels while meeting other fan selection requirements.
- Many fans generate tones at the blade-pass frequency and its harmonics that may require additional acoustical treatment of the system. Amplitude of these tones can be affected by resonance within the duct system, fan design, and inlet flow distortions caused by poor inlet duct design, or by operation of an inlet volume control damper. When possible, use variable-speed volume control instead of volume control dampers.
- Design duct connections at both fan inlet and outlet for uniform and straight airflow. Avoid unstable, turbulent, and swirling inlet airflow. Deviation from acceptable practice can severely degrade both aerodynamic and acoustic performance of any fan and invalidate manufacturers' ratings or other performance predictions.

Variable-Air-Volume (VAV) Systems

General Design Considerations. As in other aspects of HVAC system design, ducts for VAV systems should be designed for the lowest practical static pressure loss, especially ductwork closest to the fan or air-handling unit (AHU). High airflow velocities and convoluted duct routing with closely spaced fittings can cause turbulent airflow that results in excessive pressure drop and fan instabilities that can cause excessive noise, fan stall, or both.

Many VAV noise complaints have been traced to control problems. Although most problems are associated with improper installation, many are caused by poor design. The designer should specify high-quality fans or air handlers within their optimum ranges, not at the edge of their operation ranges where low system tolerances can lead to inaccurate fan flow capacity control. Also, in-duct static pressure sensors should be placed in duct sections having the lowest possible air turbulence (i.e., at least three equivalent duct diameters from any elbow, takeoff, transition, offset, or damper).

Balancing. VAV noise problems have also been traced to improper air balancing. For example, air balance contractors commonly balance an air distribution system by setting all damper positions without considering the possibility of reducing fan speed. The result is a duct system in which no damper is completely open and the fan delivers air at a higher static pressure than would otherwise be necessary. If the duct system is balanced with at least one balancing damper wide open, fan speed and corresponding fan noise could be reduced. Lower sound levels occur if most balancing dampers are

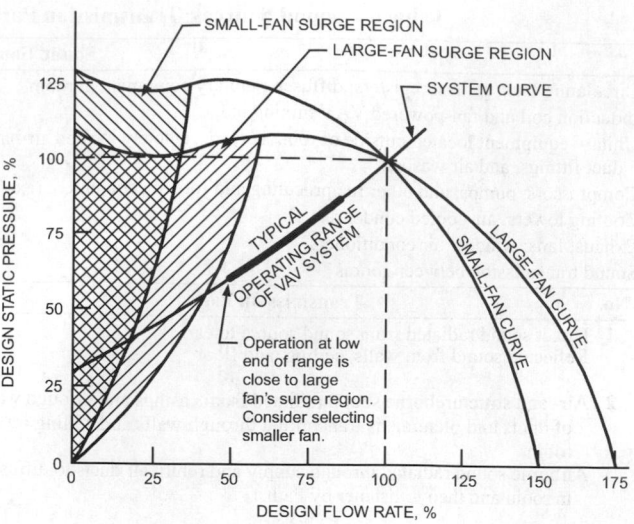

Fig. 9   Basis for Fan Selection in VAV Systems

wide open or eliminated. The specified goal should be to balance the system at the lowest static pressure required to operate the box located at the farthest point in the system.

Fan Selection. For constant-volume systems, fans should be selected to operate at maximum efficiency at design airflow. However, VAV systems must be selected to operate with efficiency and stability throughout the operating range. For example, a fan selected for peak efficiency at full output may aerodynamically stall at an operating point of 50% of full output, resulting in significantly increased low-frequency noise and unstable airflow. A stalling fan can indicate operation in the **surge region**, a region of operational instability where airflow reverses direction at the fan blade because of insufficient air entering the fan wheel. Similarly, a fan selected to operate most efficiently at the 50% output point may be very inefficient at full output, resulting in substantially increased fan noise at all frequencies. In general, a fan for a VAV system should be selected for peak efficiency at an operating point between 70 and 80% of the maximum required system capacity, which is where the fan will operate most of the time. This usually means selecting a fan that is one size smaller than that required for peak efficiency at 100% of maximum required system capacity (Figure 9). When the smaller fan operates at higher capacities, it produces up to 5 dB more noise. This occasional increase in sound level is usually more tolerable than stall-related sound problems that can occur with a larger fan operating at less than 100% design capacity most of the time.

Air Modulation Devices. The control method selected to vary the air capacity of a VAV system is important. Variable-capacity control methods can be divided into three general categories: (1) variable inlet vanes (sometimes called inlet guide vanes) or discharge dampers that yield a new fan system curve at each vane or damper setting, (2) variable-pitch fan blades (usually used on axial fans) that adjust the blade angle for optimum efficiency at varying capacity requirements, and (3) variable-speed motor drives in which motor speed is varied by modulation of the power line frequency or by mechanical means such as gears or continuous belt adjustment. Inlet vane and discharge damper volume controls can add noise to a fan system at reduced capacities, whereas variable-speed motor drives and variable-pitch fan blade systems are quieter at reduced air output than at full air output.

Variable-Inlet Vanes and Discharge Dampers. Variable-inlet vanes vary airflow capacity by changing inlet airflow to a fan wheel. This type of air modulation varies the total air volume and pressure at the fan, but fan speed remains constant. Although fan pressure and air volume reductions at the fan reduce duct system noise by reducing air velocities and pressures in the duct work, there is an associated increase in fan noise caused by airflow turbulence and flow distortions at the inlet vanes. Fan manufacturers' test data show that, on airfoil centrifugal fans, as vanes mounted inside the fan inlet (nested inlet vanes) close, the sound level at the blade-pass frequency of the fan increases by 2 to 8 dB, depending on the percent of total air volume restricted. For externally mounted inlet vanes, the increase is on the order of 2 to 3 dB. The increase for forward-curved fan wheels with inlet vanes is about 1 to 2 dB less than that for airfoil fan wheels. In-line axial fans with inlet vanes generate increased noise levels of 2 to 8 dB in the low-frequency octave bands for a 25 to 50% closed vane position.

Discharge dampers, typically located immediately downstream of the supply air fan, reduce airflow and increase pressure drop across the fan while fan speed remains constant. Because of air turbulence and flow distortions created by the high pressure drop across discharge dampers, there is a high probability of duct rumble near the damper location. If the dampers are throttled to a very low flow, a stall condition can occur at the fan, resulting in an increase in low-frequency noise.

Variable-Pitch Fans for Capacity Control. Variable-pitch fan blade controls vary the fan blade angle to reduce airflow. This type of system is predominantly used in axial fans. As air volume and pressure are reduced at the fan, there is a corresponding noise reduction. In the 125 to 4000 Hz octave bands, this reduction usually varies between 2 to 5 dB for a 20% reduction in air volume, and between 8 to 12 dB for a 60% reduction in air volume.

Variable-Speed-Motor-Controlled Fan. Three types of electronic variable-speed control units are used with fans: (1) current source inverter, (2) voltage source inverter, and (3) pulse-width modulation (PWM). The current source inverter and third-generation PWM control units are usually the quietest of the three controls. In all three types, matching motors to control units and the quality of the motor windings determine the motor's noise output. The motor typically emits a pure tone with an amplitude that depends on the smoothness of the waveform from the line current. The frequency of the motor tone depends on the motor type, windings, and speed, but is typically at the drive's switching frequency. Some drives allow adjustment to a higher frequency that does not carry as well, but at a cost of lower drive efficiency. Both inverter control units and motors should be enclosed in areas, such as mechanical rooms or electrical rooms, where the noise effect on surrounding rooms is minimal. The primary acoustic advantage of variable-speed fans is reduction of fan speed, which translates into reduced noise; dB reduction is approximately equal to 50 × log (higher speed/lower speed). Because this speed reduction generally follows the fan system curve, a fan selected at optimum efficiency initially (lowest noise) does not lose efficiency as the speed is reduced. When using variable-speed controllers,

- Select fan vibration isolators on the basis of the lowest practical speed of the fan. For example, the lowest rotational speed might be 600 rpm for a 1000 rpm fan in a commercial system.
- Select a controller with a feature typically called "critical frequency jump band." This feature allows a user to program the controller to avoid certain fan or motor rpm settings that might excite vibration isolation system or building structure resonance frequencies, or correspond to speeds of other fans in the same system.
- Check the intersection of the fan's curve at various speeds against the duct system curve. When selecting a fan controlled by a variable-speed motor controller, keep in mind that the system curve does not go to zero static pressure at no flow. The system curve is asymptotic at the static pressure control set point, typically 1 to 1.5 in. of water. An improperly selected fan may be forced to operate in its stall range at slower fan speeds.

Terminal Units. Fans and pressure-reducing valves in VAV units should have manufacturer-published sound data indicating sound power levels that (1) are discharged from the low-pressure end of the unit and (2) radiate from the exterior shell of the unit. These sound power levels vary as a function of valve position and fan point of operation. Sound data for VAV units should be obtained according to the procedures specified by the latest ARI *Standard* 880. In critical situations, a mock-up test should be conducted of a production terminal box under project conditions and space finishes. The test is required because minor changes in box motor, fan, or valve components can affect the noise generated by such equipment.

If the VAV unit is located in noncritical areas (e.g., above a storeroom or corridor), sound radiated from the shell of the unit may be of no concern. If, however, the unit is located above a critical space and separated from the space by a ceiling with little or no sound transmission loss at low frequencies, sound radiated from the shell into the space below may exceed the desired noise criterion. In this case, it may be necessary to relocate the unit to a noncritical area or to enclose it with a high-transmission-loss construction. Room sound levels can be estimated using attenuation factors detailed in AHRI *Standard* 885. In general, fan-powered VAV units should not be placed above or near any room with a required sound criterion rating of less than RC 40(N) (Schaffer 2005). For further information, see the section on Indoor Sound Criteria.

Full shutoff of VAV units can produce excessive duct system pressure at low flow, sometimes causing a fan to go into stall, resulting in accompanying roar, rumble, and surge. Systems providing more than 30% of their air to VAV devices should be provided with a means of static pressure control. Variable-frequency drives are preferred, but in the case of constant-volume air handlers, some means of bypass pressure control should be used to relieve system pressure as VAV devices close down (Schaffer 2005).

Rooftop-Mounted Air Handlers

Rooftop air handlers can have unique noise control requirements because these units are often integrated into a lightweight roof construction. Large roof openings are often required for supply and return air duct connections. These ducts run directly from noise-generating rooftop air handlers to the building interior. Generally, there is insufficient space or distance between roof-mounted equipment and the closest occupied spaces below the roof to apply standard sound control treatments. Rooftop units should be located above spaces that are not acoustically sensitive and should be placed as far as possible from the nearest occupied space. This measure can reduce the amount of sound control treatment necessary to achieve an acoustically acceptable installation.

The common sound transmission paths associated with rooftop air handlers (Figure 10) are

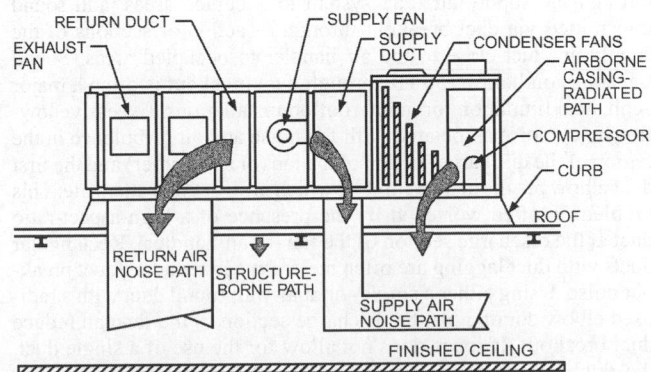

Fig. 10   Sound Paths for Typical Rooftop Installations

- Flanking-path-borne sound from condenser fans, or compressors breaking in through lightweight roofs or through windows
- Airborne through bottom of rooftop unit to spaces below
- Structureborne from vibrating equipment in rooftop unit to building structure
- Ductborne through supply air duct from air handler
- Ductborne through return air duct to air handler
- Duct breakout noise (see the section on Sound Radiation Through Duct Walls)

Flanking-path noise can enter through lightweight roof structures, adjacent walls, and windows. Avoid placing rooftop units on light structure over sensitive spaces or close to higher sidewalls with windows or other lightly constructed building elements. If it is necessary to place the rooftop unit over sensitive spaces or lightly constructed walls, then lagging with additional layers of gypsum board or other similar material may be required in these areas.

Using proper vibration isolation can minimize structureborne sound and vibration from vibrating equipment in a rooftop unit. Special curb mounting bases are available to support and provide vibration isolation for rooftop units. For roofs constructed with open web joists, thin long-span slabs, wooden construction, and any unusually light construction, evaluate all equipment weighing more than 300 lb to determine the additional deflection of the structure at mounting points caused by the equipment. Isolator deflection should be a minimum of 10 times the additional static deflection. If the required spring isolator deflection exceeds commercially available products, stiffen the supporting structure or change the equipment location.

Airborne paths are associated with casing-radiated sound that passes through the air-handler enclosure and roof structure to the spaces below. Airborne sound can result from air-handler noise or from other equipment components in the rooftop unit. Rooftop units should not be placed on open curbs or over a large opening in the roof structure through which both supply and return air ducts pass. Roof penetrations should be limited to two openings sized to accommodate only the supply and return air ducts. These openings should be properly sealed after installation of the ducts. If a large single opening exists under the rooftop unit, it should be structurally, acoustically, and flexibly sealed with one or more layers of gypsum board or other similar material around the supply and return air ducts. Airborne sound transmission to spaces below a rooftop unit can be greatly reduced by placing the rooftop unit on a structural support extending above the roof structure, and running supply and return air ducts horizontally along the roof for several duct diameters before the ducts turn to penetrate the roof. The roof deck/ceiling system below the unit can be constructed to adequately attenuate sound radiated from the bottom of the unit.

Ductborne transmission of sound through the supply air duct consists of two components: sound transmitted from the air handler through the supply air duct system to occupied areas, and sound transmitted via duct breakout through a section or sections of the supply air duct close to the air handler to occupied areas. Sound transmission below 250 Hz through duct breakout is often a major acoustical limitation for many rooftop installations. Excessive low-frequency noise associated with fan noise and air turbulence in the region of the discharge section of the fan (or air handler) and the first duct elbow results in duct rumble, which is difficult to attenuate. This problem is often worsened by the presence of a high-aspect-ratio duct at the discharge section of the fan (or air handler). Rectangular ducts with duct lagging are often ineffective in reducing duct breakout noise. Using either a single- or dual-wall round duct with a radiused elbow coming off the discharge section of the fan can reduce duct breakout. If space does not allow for the use of a single duct, the duct can be split into several parallel round ducts. Another effective method is using an acoustic plenum chamber constructed of a minimum 2 in. thick, dual-wall plenum panel, lined with fiber-glass and with a perforated inner liner, at the discharge section of the fan. Either round or rectangular ducts can be taken off the plenum as necessary for the rest of the supply air distribution system. Table 7 shows 12 possible rooftop discharge duct configurations with their associated low-frequency noise reduction potential (Beatty 1987; Harold 1986, 1991).

Ductborne transmission of sound through the return air duct of a rooftop unit is often a problem because there is generally only one short return air duct section between the plenum space above a ceiling and the return air section of the air handler. This does not allow for adequate sound attenuation between the fan inlet and spaces below the air handler. Sound attenuation through the return air duct system can be improved by adding at least one (more if possible) branch division where the return air duct is split into two sections that extend several duct diameters before they terminate into the plenum space above the ceiling. The inside surfaces of all return air ducts should be lined with a minimum of 1 in. thick duct liner. If conditions permit, duct silencers in duct branches or an acoustic plenum chamber at the air-handler inlet section give better sound conditions.

Aerodynamically Generated Sound in Ducts

Aerodynamic sound is generated when airflow turbulence occurs at duct elements such as duct fittings, dampers, air modulation units, sound attenuators, and room air devices. For details on air modulation units and sound attenuators, see the sections on Variable-Air-Volume Systems and Duct Silencers.

Although fans are a major source of sound in HVAC systems, aerodynamically generated sound can often exceed fan sound because of close proximity to the receiver. When making octave-band fan sound calculations using a source-path-receiver analysis, aerodynamically generated sound must be added in the path sound calculations at the location of the element.

Duct Velocities. The extent of aerodynamic sound is related to the airflow turbulence and velocity through the duct element. The sound amplitude of aerodynamically generated sound in ducts is proportional to the fifth, sixth, and seventh power of the duct airflow velocity in the vicinity of a duct element (Bullock 1970; Ingard et al. 1968). Therefore, reducing duct airflow velocity significantly reduces flow-generated noise. Tables 8 (Schaffer 2005) and 9 (Egan 1988) give guidelines for recommended airflow velocities in duct sections and duct outlets to avoid problems associated with aerodynamically generated sound in ducts.

Fixed Duct Fittings. Fixed duct fittings include elbows, tees, transitions, fixed dampers, and branch takeoffs. In all cases, less generated air turbulence and lower airflow velocities result in less aerodynamic sound. Figures 11 and 12 show typical frequency spectra for specific sizes of elbows and transitions. Data in these figures are based on empirical data obtained from ASHRAE RP-37 (Ingard et al. 1968). Normalized data from ASHRAE RP-37 and others, which can apply to all types of duct fittings and dampers, have been published (Bullock 1970) and presented in ASHRAE RP-265 (Ver 1983a). When multiple duct fittings are installed adjacent to each other, aerodynamic sound can increase significantly because of the added air turbulence and increased velocity pressures. Note that the magnitude of the field-measured static pressure drop across fixed duct fittings does not relate to the aerodynamic generated sound. However, total pressure drop across a duct fitting, which includes the velocity pressure change resulting from air turbulence, does affect aerodynamically generated sound.

Operable Volume Dampers. Operable damper aerodynamic sound is created because the damper is an obstacle in the airstream, and air turbulence increases as the damper closes. Because total pressure drop across the damper also increases with closure, the aerodynamic sound is related to the total pressure drop. Both single-blade and multiblade dampers, used to balance and control the airflow in a duct system and at room air devices, have similar

Table 7   Duct Breakout Insertion Loss—Potential Low-Frequency Improvement over Bare Duct and Elbow

Discharge Duct Configuration, 12 ft of Horizontal Supply Duct	Duct Breakout Insertion Loss at Low Frequencies, dB			Side View	End View
	63 Hz	125 Hz	250 Hz		
Rectangular duct: no turning vanes (reference)	0	0	0		22 GAGE
Rectangular duct: one-dimensional turning vanes	0	1	1		TURNING VANES
Rectangular duct: two-dimensional turning vanes	0	1	1		TURNING VANES
Rectangular duct: wrapped with foam insulation and two layers of lead	4	3	5	SEE END VIEW	FOAM INSULATION WITH TWO LAYERS LEAD
Rectangular duct: wrapped with glass fiber and one layer 5/8 in. gypsum board	4	7	6	SEE END VIEW	GLASS FIBER PRESSED FLAT AGAINST DUCT
Rectangular duct: wrapped with glass fiber and two layers 5/8 in. gypsum board	7	9	9	SEE END VIEW	GYPSUM BOARD SCREWED TIGHT
Rectangular plenum drop (12 ga.): three parallel rectangular supply ducts (22 ga.)	1	2	4	12 GAGE	22 GAGE
Rectangular plenum drop (12 ga.): one round supply duct (18 ga.)	8	10	6	12 CACE	18 GAGE
Rectangular plenum drop (12 ga.): three parallel round supply ducts (24 ga.)	11	14	8	12 GAGE	24 GAGE
Rectangular (14 ga.) to multiple drop: round mitered elbows with turning vanes, three parallel round supply ducts (24 ga.)	18	12	13	24 GAGE	14 GAGE
Rectangular (14 ga.) to multiple drop: round mitered elbows with turning vanes, three parallel round lined double-wall, 22 in. OD supply ducts (24 ga.)	18	13	16	24 GAGE	14 GAGE
Round drop: radiused elbow (14 ga.), single 37 in. diameter supply duct	15	17	10	14 GAGE, 18 GAGE	

frequency spectra. Figure 13 shows the frequency spectrum for a 45° damper in a 24 by 24 in. duct (Ingard et al. 1968).

Depending on its location relative to a room air device, a damper can generate sound that is transmitted down the duct to the room air device, or radiate sound through the ceiling space into the occupied space below. When an operable control damper is installed close to an air device to achieve system balance, the acoustic performance of the air outlet must be based not only on the air volume handled, but also on the magnitude of the air turbulence generated at the damper. The sound level produced by closing the damper is accounted for by adding a correction to the air device sound rating. As the damper is modulated for air balance, this quantity is proportional to the pressure ratio (PR), that is, the throttled total pressure drop across the damper divided by the minimum total pressure drop across the damper. Table 10 provides decibel corrections to determine the effect of damper location on linear diffuser sound ratings.

Volume dampers in sound-critical spaces should always be a minimum of 5 to 10 duct diameters from air device, with an acoustically lined duct between the damper and air device. Acoustically lined plenums may also be used between the damper and room air device to reduce damper sound. Linear air devices with a round duct connected to an insulated plenum have been successfully used for damper sound control. However, acoustical lining in this type of plenum does not minimize the sound generated by air flowing through a short section of the linear air device. If multiple inlets/outlets are used to spread airflow uniformly over the lined plenum and air device, then the linear slot generates less sound.

Proper air balancing of a fan/duct system directly affects aerodynamically generated sound even in a correctly designed and installed duct system. Primary volume dampers in the longest duct from a fan should always be nearly wide-open. If the primary damper in the longest duct run is more than 20% closed, the duct system has not been properly air balanced, and the fan may operate at a higher speed than required for the duct system. The result is an increase in air velocities and turbulence throughout the entire duct system, with excessive aerodynamic sound generated at all duct elements.

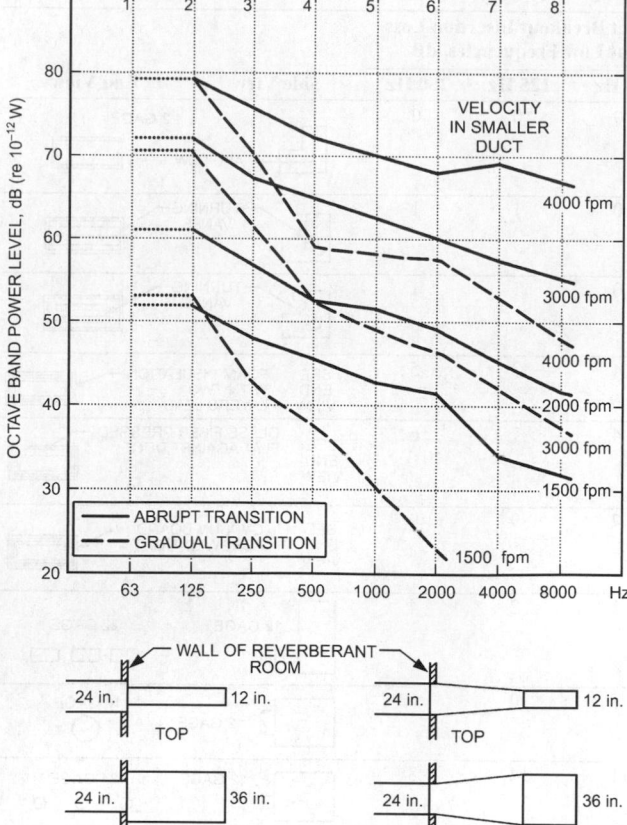

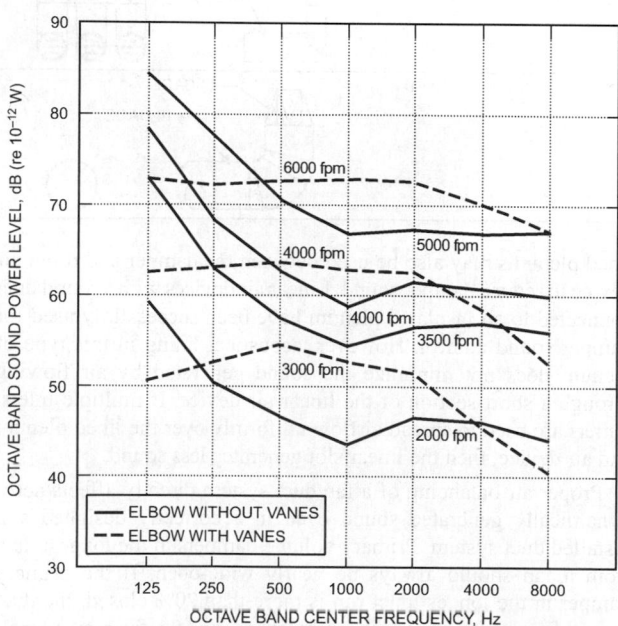

Fig. 11   Velocity-Generated Sound of Duct Transitions

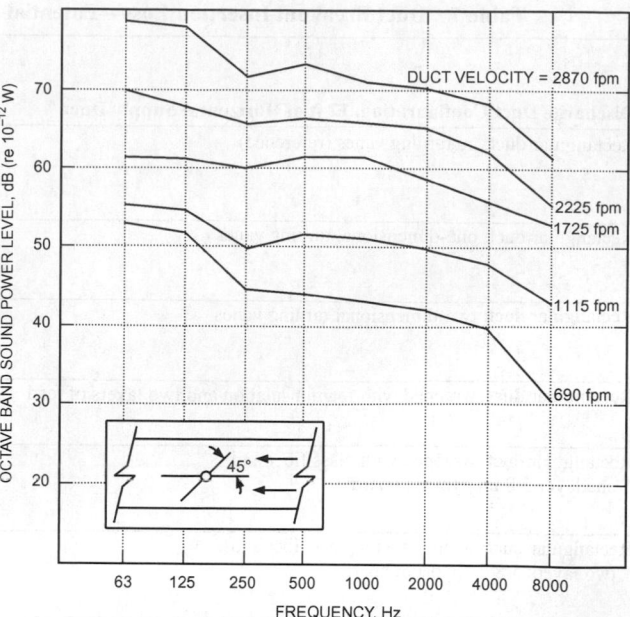

Fig. 13   Velocity-Generated Sound of 24 by 24 in.
Volume Damper

Table 8   Maximum Recommended Duct Airflow Velocities to Achieve Specified Acoustic Design Criteria

		Maximum Airflow Velocity, fpm	
Main Duct Location	Design RC(N)	Rectangular Duct	Circular Duct
In shaft or above drywall ceiling	45	3500	5000
	35	2500	3500
	25	1700	2500
Above suspended acoustic ceiling	45	2500	4500
	35	1750	3000
	25	1200	2000
Duct located within occupied space	45	2000	3900
	35	1450	2600
	25	950	1700

Notes:
1. Branch ducts should have airflow velocities of about 80% of values listed.
2. Velocities in final runouts to outlets should be 50% of values or less.
3. Elbows and other fittings can increase airflow noise substantially, depending on type. Thus, duct airflow velocities should be reduced accordingly.

Table 9   Maximum Recommended Air Velocities at Neck of Supply Diffusers or Return Registers to Achieve Specified Acoustical Design Criteria

Type of Opening	Design RC(N)	"Free" Opening Airflow Velocity, fpm
Supply air outlet	45	625
	40	560
	35	500
	30	425
	25	350
Return air opening	45	750
	40	675
	35	600
	30	500
	25	425

Note: Table intended for use when no sound data are available for selected grilles or diffusers, or no diffuser or grille is used. The number of diffusers or grilles increases sound levels, depending on proximity to receiver. Allowable outlet or opening airflow velocities should be reduced accordingly in these cases.

Note: Comparison of octave band sound power levels produced by airflow through
8 × 8 in. rectangular elbow with and without 7 circular arc turning vanes.

Fig. 12   Velocity-Generated Sound of Elbows

Table 10 Decibels to Be Added to Diffuser Sound Rating to Allow for Throttling of Volume Damper

Location of Volume Damper	Damper Pressure Ratio					
	1.5	2	2.5	3	4	6
	dB to Be Added to Diffuser Sound Rating					
In neck of linear diffuser	5	9	12	15	18	24
In inlet of plenum of linear diffusers	2	3	4	5	6	9
In supply duct at least 5 ft from inlet plenum of linear diffuser	0	0	0	2	3	5

Room Air Devices (Grilles, Registers, Diffusers). Manufacturers' test data should be obtained in accordance with ASHRAE *Standard* 70 or ARI *Standard* 890 for room air devices such as grilles, registers, diffusers, air-handling light fixtures, and air-handling suspension bars. Devices should be selected to meet the noise criterion required or specified for the room. However, the manufacturer's sound power rating is obtained with a uniform velocity distribution throughout the air device neck or grille collar; this is often not met in practice when a duct turn, sharp transition, or a balancing damper immediately precedes the entrance to the diffuser. In these cases, airflow is turbulent and noise generated by the device can be substantially higher than the manufacturer's published data (by as much as 12 dB). In some cases, placing an equalizer grid in the neck of the air device can substantially reduce this turbulence. The equalizer grid can help provide a uniform velocity gradient within the neck of the device, so the sound power generated in the field will be closer to that listed in the manufacturer's catalog.

At present, air devices are rated by manufacturers in terms of noise criterion (NC) levels, which usually includes a receiver room effect sound correction of 10 dB. The NC ratings may be useful for comparison between different air devices, but are not helpful for source-path-receiver calculations in terms of octave bands. For a complete analysis, the designer should request the component sound power level data in octave bands from the manufacturer. Whether using NC levels or sound power levels, the designer should also correct manufacturer's data for actual room effect, location of air devices, and number of air devices used in a specific design. The acoustical room effect is the reduction in sound level caused by distance from the sound source (e.g., air outlet); the room volume and amount of acoustical absorption present also affect the value. For more information, see the section on Receiver Room Sound Correction. For example, in a small room with an actual calculated room effect of 6 dB, and given a manufacturer's room effect correction of 10 dB, the discrepancy (in this case, 4 dB) must be added to the manufacturer's data. When an air device is located at the intersection of the ceiling and vertical wall, 6 dB should be added, and in the corner of a room, 9 dB should be added to manufacturer's data. When multiple room air devices are located in a small room or grouped together in a large room, the sound of air devices is additive by up to 10 × log(number of air devices).

A flexible duct connection between a branch air duct and an air device provides a convenient means to align the air device with the ceiling grid. The resulting misalignment in this connection, as shown in Figure 14, can cause as much as 12 to 15 dB higher sound levels in the air device's aerodynamically generated sound.

Avoiding Aerodynamically Generated Noise. Aerodynamic noise in duct systems can be avoided by

- Sizing ductwork and duct elements for low air velocities
- Avoiding abrupt changes in duct cross-sectional area or direction
- Providing smooth airflow at all duct elements, including branches, elbows, tees, transitions, and room air devices
- Providing straight ductwork (preferably 5 to 10 duct diameters) between duct elements

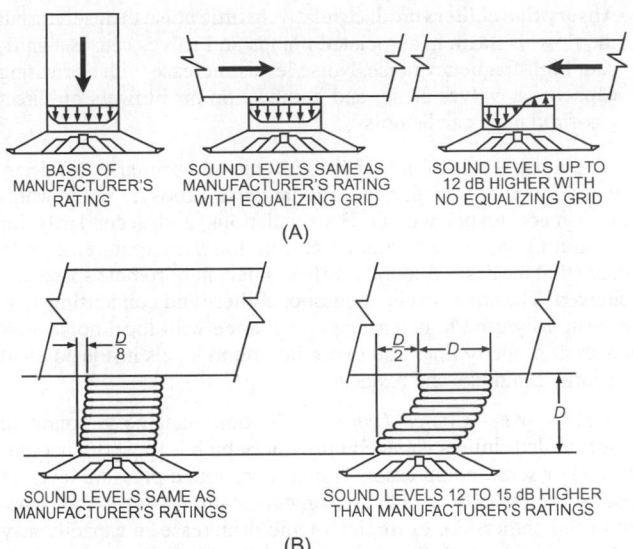

Fig. 14 (A) Proper and Improper Airflow Condition to an Outlet; (B) Effect of Proper and Improper Alignment of Flexible Duct Connector

- Air-balancing duct system for lowest reasonable fan speed with dampers generally open
- Locating volume control dampers a minimum of 3 (preferably 5 to 10) duct diameters away from room air devices (Schaffer 2005)

Chillers and Air-Cooled Condensers

Chillers and air-cooled condensers have components such as compressors, motors, gears, and fans that can produce significant amounts of both broadband and tonal noise. The broadband noise is typically caused by flows of refrigerant, water, and air, whereas the tonal noise is caused by rotation of compressors, motors, gears, and fans (in fan-cooled equipment). Chiller and condenser noise is significant in the octave bands from 63 to 4000 Hz and depends primarily on the type of compressor used.

Noise from Compressors and Chillers. All compressors produce tonal noise to varying degrees. Acoustical differences among compressors relate in large part to their tonal content:

- **Centrifugal compressor** tonal noise comes from rotation of the impeller and gears (if present). Impeller blade-related tonal content is typically not very strong but radiates from the condenser shell. Centrifugal compressor sound levels typically increase at reduced chiller capacity, because of the extra turbulence induced in the refrigerant circuit by the compressor inlet vanes, as well as rotating stall noise generated in the compressor diffuser. If capacity is reduced using motor-speed control, the resulting compressor sound levels generally decrease with decreasing capacity.
- **Reciprocating compressor** noise has a low-frequency drumming quality, caused by the oscillatory motion of pistons. The tonal content is high, and the sound level decreases very little with decreasing capacity.
- **Scroll compressors** tend to produce relatively weak tones.
- **Screw compressors** (sometimes called helical rotor or rotary compressors) generate very strong tones in the 250 to 2000 Hz octave bands. Rotor-induced tones can be amplified by resonances in the oil separation circuit, the refrigerant lines, and by efficient sound radiation from the condenser and evaporator shells connected to the compressor via these components. Screw compressors have been a source of chiller-noise complaints in many installations where their tonal characteristics have not been properly accounted for in the building design process.

- **Absorption chillers** produce relatively little noise themselves, but the flow of steam in associated pumps and valves causes significant high-frequency noise. Noise levels increase with decreasing capacity as valves close, and combustion air blowers on direct gas-fired units can be noisy.

The noise levels of indoor chillers are used primarily for determining compliance with occupational noise exposure in the workplace (in accordance with OSHA regulations) and, secondarily, for determining equipment room transmission loss requirements to ensure that the desired sound levels in adjacent or remote spaces are achieved. The noise levels of outdoor chillers and condensing units are primarily used to determine compliance with local noise ordinances at property lines and to predict sound levels inside adjacent or nearby buildings and residences.

Indoor Water-Cooled Chillers. The dominant noise source in water-cooled chillers is the compressor, which is most often a centrifugal or screw compressor. The average sound pressure levels at distances close to the chiller are sometimes insensitive to the capacity of the chiller. For example, a tenfold increase in capacity may only result in a 2 to 3 dBA increase in the published sound pressure levels. Even though physical sizes of chillers differ greatly, adjacent sound pressure levels may be comparable. However, as physical sizes of chillers increase, their radiated sound power levels increase significantly. Therefore, two chillers that have similar loudness or sound pressure levels could have much different sound power levels because of the surface area [see Equations (2) and (3)].

Factory-provided sound data for indoor chillers are typically obtained using AHRI *Standard* 575, which requires measuring the A-weighted and octave band sound pressure level (L_p) values at many locations 3.28 ft from the chiller and 4.92 ft above the floor. AHRI 575 sound pressure levels are generally available at operating points of 25, 50, and 100% of a chiller's nominal full capacity. The average A-weighted sound pressure levels can be used directly along with exposure times to determine OSHA compliance in the machinery room. The ranges of AHRI 575 values for typical centrifugal and screw chillers are shown in Figures 15 and 16, respectively. The spread of data includes both the effects of capacity and operating condition.

AHRI 575 measurements for factory-provided ratings are often made in very large rooms with large amounts of sound absorption. For that reason, assessment of sound pressure levels in situ should typically be adjusted for each chiller installation to account for the mechanical room's size and surface treatment. For a given chiller at a given operating point, a small equipment room (or one with mostly hard surface finishes) has higher L_p values than one that is large or has sound-absorbing treatments on its ceiling and walls. Figure 17 shows maximum typical adjustment factors that should be added to factory-provided AHRI 575 values to estimate the L_p values in specific installations due to reverberant (reflective) sound effects. The adjustment for each octave band requires knowing the size of an imaginary box that is circumscribed 3.28 ft away from the top and sides of the chiller (the AHRI 575 measurement surface), the dimensions of the equipment room, and the average sound absorption coefficient of the room surfaces. The adjustment in each octave band depends on the ratio of the areas of the equipment room and the imaginary box as well as the average sound absorption of the room finishes. Each curve in Figure 17 is for a different value of the average sound absorption, with the higher curves being for lower values.

Example 2. Estimate the reverberant L_p values in a 45 by 40 by 20 ft tall mechanical equipment room (MER) that houses a 360 ton centrifugal chiller. The room has a concrete floor and gypsum board walls and ceiling; all surfaces have an average absorption coefficient of 0.1. The chiller dimensions are 60 in. wide, 80 in. tall, and 120 in. long.

Solution:

The AHRI 575 measurement surface area S_M is determined by adding 3.28 ft to the chiller height and 6.56 ft to both its length and width. The floor area is not included in this calculation. The result is a box that has dimensions of 140 in. wide, 200 in. long and 120 in. tall. The surface area of this box is approximately 751 ft². The surface area of the equipment room (floor included) S_R is 7000 ft². Therefore, the ratio of

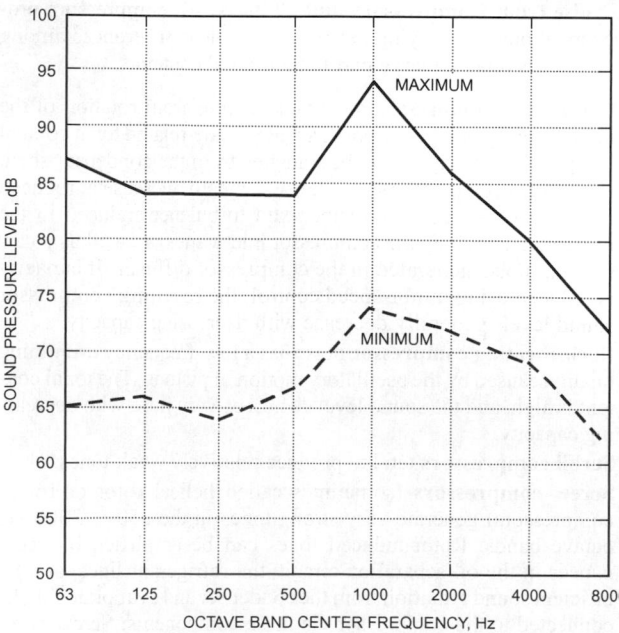

Fig. 15   **Typical Minimum and Maximum AHRI 575 L_p Values for Centrifugal Chillers (130 to 1300 Tons)**

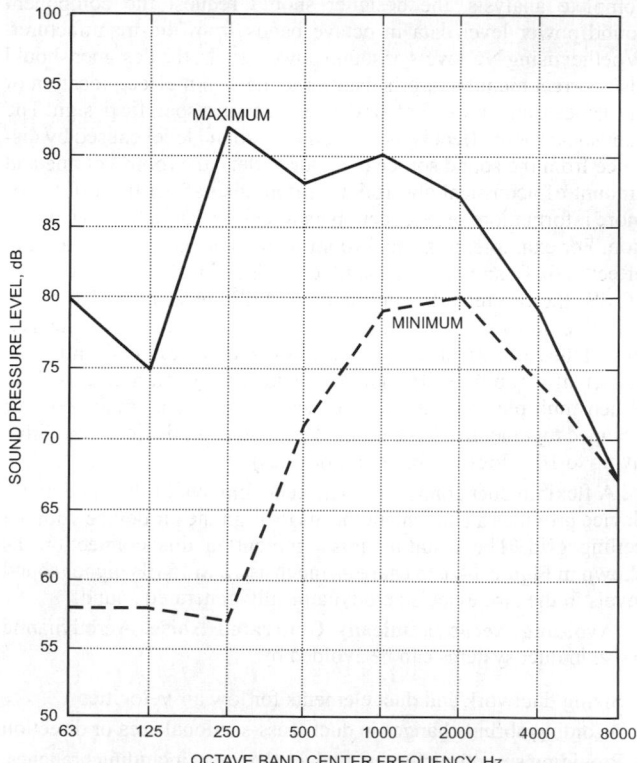

Fig. 16   **Typical Minimum and Maximum AHRI 575 L_p Values for Screw Chillers (130 to 400 Tons)**

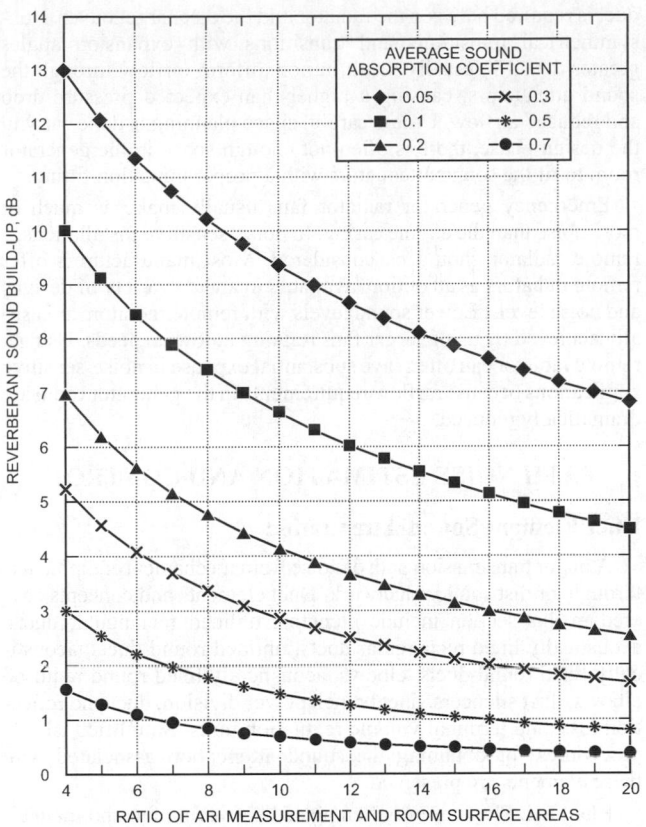

Fig. 17   Estimated Sound Level Build-Up in Mechanical Room for AHRI 575 Chiller Sound Levels

Table 11   Calculations for Reverberation Build-Up

	Octave Midband Frequency, Hz							
	63	125	250	500	1000	2000	4000	8000
AHRI 575 L_p values	73	74	73	72	74	72	69	63
Adjustment from Figure 17	7	7	7	7	7	7	7	7
Approximate revised L_p in MER	80	81	80	79	81	79	76	70

the areas S_R/S_M is 7000/751 = 9.3. Because the average absorption coefficient value for room is 0.1 for all octave bands, see Figure 17 for the adjustment factor and Table 11 for calculations.

The approximate reverberant L_p values in the last line of Example 2 can be used with sound transmission loss data of the construction to estimate transmitted L_p values in rooms adjacent to a chiller room.

An alternative approach to this method for estimating L_p in adjacent rooms is to use an estimate of the sound power levels of the chiller from the factory-provided AHRI 575 values (Stabley 2006) in conjunction with sound transmission loss data. A conversion factor (CF) is determined and used to convert the AHRI 575 sound pressure values to sound power L_w values. The conversion factor is calculated using

$$CF = 10\log(S/S_o) \tag{2}$$

where

S = area of measurement parallelepiped (excluding the top) used in AHRI 575
  = $2(L \times H) + 2(W \times H)$

$L, W,$
and H = length, width, and height of measurement parallelepiped, ft
S_o = 10.72 ft^2

and

$$L_w = L_p + CF \tag{3}$$

where

L_w = sound power level (A-weighted or octave band)
L_p = sound pressure levels per AHRI 575

This approach assumes that the factory-provided data were obtained in a "free-field" environment.

Indoor chillers are often offered with various types of factory noise-reduction options. These options can include variable-speed drives, or variable-geometry diffusers on centrifugal compressors, that reduce the strength of the noise sources internal to the machine. They may also include various external noise-attenuation devices ranging from compressor, refrigerant line and heat exchanger blankets (typically providing overall noise reduction of 2 to 6 dBA), to complete enclosures with sound-absorbing inner surfaces (which may reduce the overall noise by as much as 18 dBA). The amount of compressor noise reduction achieved by external attenuation approaches is usually limited by structureborne transmission of compressor vibration into the equipment frame and heat exchanger shells, which act as sounding boards. Attenuation options for chiller-noise control vary widely, depending on the application and the type of compressor used. Typically, they either reduce the sound radiating from the source (using acoustic enclosures or blankets) or reduce the internal sound-generating mechanisms of the source (using variable-speed drives on compressors and variable-geometry diffusers for centrifugal compressors). The effectiveness of each approach is affected by variables such as the type of compressor and its behavior with load, heat exchanger design, and type of prime mover used.

Field-installed noise-control options include full-sized sheet metal housings with specially treated openings for piping, electrical conduit, and ventilation. This option may require upgraded building construction. For more information, refer to the section on Mechanical Equipment Room Sound Isolation.

Outdoor Air-Cooled Chillers and Condensers. Outdoor units often use either reciprocating, scroll, or screw compressors. They are also used as the chiller portion of rooftop packaged units. The dominant noise sources in outdoor air-cooled chillers are the compressors and the condenser fans, which are typically low-cost, high-speed propeller fans. For air-cooled condensing units, propeller fans are the only significant noise source.

Factory sound data for outdoor equipment are obtained in accordance with AHRI *Standard* 370, which requires the determination of the equipment's octave band sound power levels (L_w), the A-weighted overall sound power level (L_{wa}), and the tone-adjusted A-weighted overall sound power level (L_{wat}). Because AHRI 370 is a sound power measurement technique, it provides certifiable sound data that can be compared across chiller manufacturers with greater certainty than is possible using the sound-pressure-based AHRI 575. The range of AHRI 370 L_w values for outdoor chillers in the 20 to 380 ton range is given in Figure 18.

Factory-supplied noise reduction options for outdoor equipment include compressor enclosures, component sound blankets, over-sized condenser fans, and variable-speed condenser fans. Because air-cooled equipment needs a free flow of cooling air, full enclosures are not feasible. However, strategically placed barriers can help reduce noise propagation on a selective basis. For more information, see the section on Sound Control for Outdoor Equipment.

Emergency Generators

Emergency or standby generators create very high sound levels and require special consideration, especially if used inside an

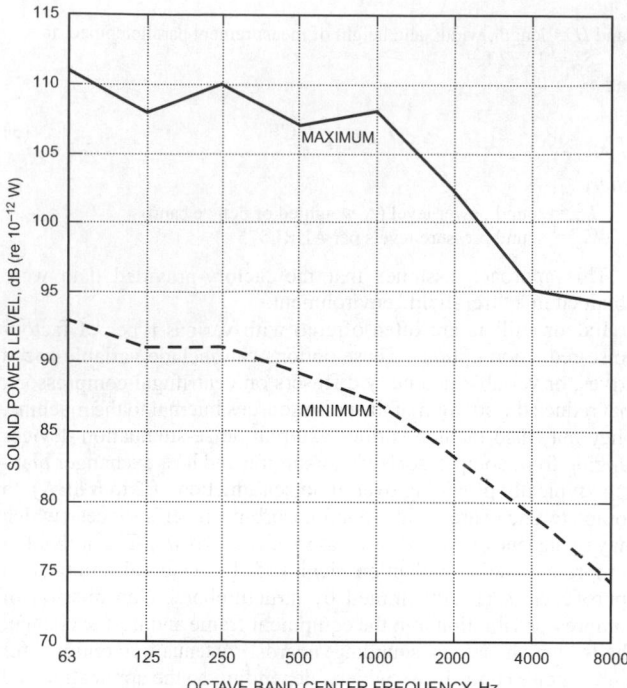

Fig. 18 Typical AHRI 370 L_w Values for Outdoor Chillers (20 to 380 Tons)

occupied building. The primary noise sources include the engine casing, radiator, and engine exhaust, and must be considered separately if the generator is installed inside a building. Sound power levels for these units depend on the power rating, fuel type, engine speed, exhaust muffler design, and radiator system. Overall sound power levels can be as high as 130 dBA (re 10^{-12} W) for larger (1.5 to 2.0 MW) diesel-powered units with standard mufflers. Noise levels inside generator rooms almost always exceed 100 dBA if the power rating of the unit is greater than 50 kW.

Noise from the generator casing is broadband with a relatively uniform spectrum. Octave band noise data are usually available from the generator manufacturer. Casing noise does not vary much with load. Conversely, exhaust noise typically contains strong tones at the engine shaft's running frequency and at the engine firing rate. Standard engine exhaust mufflers reduce exhaust noise by 20 to 25 dBA (compared to unsilenced exhaust), but even with this noise reduction, strong tones still radiate from the exhaust outlet in most cases. High-performance (critical and supercritical grade) mufflers are available but are larger and more expensive than standard units. Exhaust pipes should be routed away from noise-sensitive areas, and the exhaust outlet should be located and oriented to ensure that community noise levels are not excessive. In occupied buildings, the entire exhaust pipe should be suspended from the structure above with spring hangers.

Engine casing noise is best controlled by enclosing the generator in a sound-rated enclosure. The biggest problem with generator room design is finding adequate space for ventilation air. Generators require a substantial volume of air for engine cooling, and controlling engine noise transmission using air intake and exhaust paths can be difficult. In most cases, air intake and exhaust openings require sound attenuators. Because engine radiators usually use propeller fans to move air across the radiator core, the ventilation system cannot always handle the added pressure drop created by sound attenuators. In some cases, auxiliary fans are needed to draw fresh air into the generator room through the intake silencers. Sound attenuators at the discharge opening should be located between the radiator and exhaust louver. A smooth, slowly expanding transition

duct is required between the radiator and the discharge louver. Non-symmetrical transitions and transitions with expansion angles greater than 15° usually result in nonuniform airflow through the sound attenuators, causing a higher-than-expected pressure drop and reduced airflow. Unless careful space planning is done early in the design phase, there is often not enough space in the generator room to fit the sound attenuators with a proper transition fitting.

Emergency generator radiator fans usually make as much or more noise than the engine casing. In noise-sensitive installations, a remote radiator should be considered. Most manufacturers offer remote radiators as an option, available in a wide variety of designs and noise levels. Lower sound levels with remote radiators are usually achieved by using larger fans running at lower speeds. Using a remote radiator can often save substantial expense in noise-sensitive applications because airflow requirements in the generator room are dramatically reduced.

PATH NOISE ESTIMATION AND CONTROL

Duct Element Sound Attenuation

A major transmission path of noise from mechanical equipment is through air distribution ductwork. Duct elements and concepts covered in this section include plenums, unlined rectangular ducts, acoustically lined rectangular ducts, unlined round ducts, acoustically lined round ducts, elbows, acoustically lined round radiused elbows, duct silencers, duct branch power division, duct end reflection loss, and terminal volume regulation units. Simplified tabular procedures for obtaining the sound attenuation associated with these elements are presented.

Plenums. Plenums are often placed between a fan and main air distribution ducts to smooth turbulent airflow. They are typically lined with acoustically absorbent material to reduce fan and other mechanical noise. Plenums are usually large rectangular enclosures with an inlet and one or more outlets.

Based on experience, ASHRAE-sponsored research (Mouratidis and Becker 2004), and earlier work (Wells 1958), transmission loss associated with a plenum can be expressed using the following considerations:

- Frequency range (based on the cutoff frequency described in the following paragraphs), which is defined as the upper limit for plane wave sound propagation
- In-line inlet and outlet openings
- End-in/end-out versus end-in/side-out orientation (i.e., in-line versus elbow configuration)

At frequencies above the cutoff frequency, as defined by the plenum's inlet duct dimensions, the wavelength of sound is small compared to the characteristic dimensions of the plenum. Plane wave propagation in a duct exists at frequencies below the cutoff, creating a need to consider two frequency ranges, where

$$f_{co} = \frac{c}{2a} \qquad \text{or} \qquad f_{co} = 0.586\frac{c}{d} \qquad (4)$$

where

f_{co} = cutoff frequency, Hz
c = speed of sound in air, ft/s
a = larger cross-sectional dimension of rectangular duct, ft
d = diameter of round duct, ft

The **cutoff frequency f_{co}** is the frequency above which plane waves no longer propagate in a duct. At these higher frequencies, waves that propagate in the duct create **cross** or **spinning modes**. The **transmission loss (TL)** in this higher frequency range may be predicted using the following relationship:

$$TL = b\left[\frac{S_{out}Q}{4\pi r^2} + \frac{S_{out}(1-\alpha_a)}{S\alpha_a}\right]^n + OAE \quad (5)$$

where

TL = transmission loss, dB
b = 3.505
n = –0.359
S_{out} = area of plenum outlet, ft^2
S = total inside surface area of plenum minus inlet and outlet areas, ft^2
r = distance between centers of inlet and outlet of plenum, ft
Q = directivity factor; taken as 2 for opening near center of wall, or 4 for opening near corner of plenum
α_a = average absorption coefficient of plenum lining [see Equation (8)]
OAE = offset angle effect; additional attenuation found in Tables 14 and 15, which tabulate frequency-dependent sound transmission properties that are manifested when inlet and outlet of plenum are not in a direct line; 90° angle is referred to as elbow effect

The average absorption coefficient α_a of plenum lining is given by

$$\alpha_a = \frac{S_1\alpha_1 + S_2\alpha_2}{S_1 + S_2} \quad (6)$$

where

α_1 = sound absorption coefficient of any bare or unlined inside surfaces of plenum
S_1 = surface area of any bare or unlined inside surfaces of plenum, ft^2
α_2 = sound absorption coefficient of acoustically lined inside surfaces of plenum
S_2 = surface area of acoustically lined inside surfaces of plenum, ft^2

In many situations, inside surfaces of a plenum chamber are lined with a sound-absorbing material. For these situations, $\alpha_a = \alpha_2$. Table 12 gives sound absorption coefficients for selected common plenum materials.

Note: transmission loss (TL) of a plenum is the difference between the duct sound power level at the outlet and inlet of the plenum, unlike **insertion loss (IL)** ratings for silencers, which represent the difference (at a downstream measurement location) between the duct sound pressure levels with the silencer and with no silencer (replaced with an empty duct). For purposes here, both TL and IL are interpreted as attenuation, or the net reduction in propagating duct sound power.

For frequencies that correspond to plane wave propagation in the duct (below the cutoff frequency), the following relationship applies, with a lower frequency limit of 50 Hz:

$$TL = A_fS + W_e + OAE \quad (7)$$

where

A_f = surface area coefficient, dB/ft^2 (see Table 13 for small and large plenum size ranges)
W_e = wall effect, dB (see Table 13 for common HVAC plenum wall types)

The maximum TL predicted by Equation (7) should be limited to 20 dB at $f < f_{co}$.

For an end-in/end-out plenum configuration, where the openings are not in-line, the offset angle θ must be considered in the TL calculation. The value of θ is obtained from the following relationship:

$$\cos\theta = \frac{l}{r} = \frac{l}{\sqrt{l^2 + r_v^2 + r_h^2}} \quad (8)$$

where (refer to Figure 19)
θ = offset angle representing r to long axis l of duct

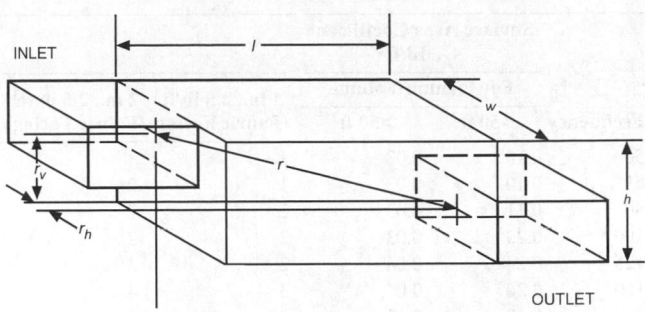

Fig. 19   Schematic of End-In/End-Out Plenum

Table 12   Sound Absorption Coefficients α of Selected Plenum Materials

	Octave Midband Frequency, Hz						
	63	125	250	500	1000	2000	4000
Non-sound-absorbing material							
Concrete	0.01	0.01	0.01	0.02	0.02	0.02	0.03
Bare sheet metal	0.04	0.04	0.04	0.05	0.05	0.05	0.07
Sound-absorbing material (fiberglass insulation board)							
1 in., 3.0 lb/ft^3	0.05	0.11	0.28	0.68	0.90	0.93	0.96
2 in., 3.0 lb/ft^3	0.10	0.17	0.86	1.00	1.00	1.00	1.00
3 in., 3.0 lb/ft^3	0.30	0.53	1.00	1.00	1.00	1.00	1.00
4 in., 3.0 lb/ft^3	0.50	0.84	1.00	1.00	1.00	1.00	0.97

Note: 63 Hz values estimated from higher-frequency values.

l = length of plenum, ft
r_v = vertical offset between axes of plenum inlet and outlet, ft
r_h = horizontal offset between axes of plenum inlet and outlet, ft

For a given offset angle, apply the applicable effects on TL (decibel addition or subtraction) for angles up to 45° (Table 14).

For an end-in/side-out plenum configuration, where openings are perpendicular to each other, the elbow effect must be considered in the TL calculation. For any plenum configuration involving an elbow condition, apply the applicable effects on TL (decibel addition or subtraction) for the two frequency ranges, both above and below the cutoff (Table 15).

For plenum applications within a practical size envelope of 20 to 1100 ft^3 volume or 50 to 650 ft^2 surface area, using duct sizes in the range $12 < d < 48$ in., this model may be applied with an anticipated standard deviation of ±3.5 to 5.0 dB for 50 Hz $< f \leq f_c$ and ±1.5 to 3.0 for $f_c < f \leq 5000$ Hz. Use caution when applying these prediction methods for plenum configurations where either the width or height dimension is $<1.5d$. In this case, the plenum may not perform as an expansion chamber, thus changing its broadband TL characteristics significantly.

Example 3. A small plenum with acoustically lined surfaces is 5.9 ft high, 4.0 ft wide, and 5.9 ft long. The inlet and outlet are each 3.0 ft wide by 2.0 ft high. The horizontal offset between centers of the plenum inlet and outlet is 1.0 ft. The vertical offset is 4.0 ft. The inside of the plenum is completely lined with 1 in. thick fiberglass insulation board, with sound absorption values as shown in Table 8. Determine the transmission loss TL associated with this plenum.

Solution:

The areas of the inlet section, outlet section, and overall surfaces are

$$S_{in} = 3.0 \quad 2.0 = 6.0 \text{ ft}^2$$

$$S_{out} = 3.0 \quad 2.0 = 6.0 \text{ ft}^2$$

Table 13   Low-Frequency Characteristics of Plenum TL

Frequency	Surface Area Coefficient A_f, dB/ft² For Plenum Volume <50 ft³	>50 ft³	Wall Effect W_e, dB Added 1 in., 2.5 lb/ft³ (Fabric Facing)	2 in., 2.5 lb/ft³ (Fabric Facing)	4 in., 2.5 lb/ft³ (Perf. Facing)	8 in., 2.5 lb/ft³ (Perf. Facing)	4 in. (Tuned, No Media)	4 in., 2.5 lb/ft³ (Double Solid Metal)
50	0.14	0.03	1	1	0	1	0	0
63	0.10	0.03	1	2	3	7	1	3
80	0.11	0.03	2	2	3	9	2	7
100	0.23	0.03	2	2	4	12	1	6
125	0.24	0.04	2	3	6	12	1	4
160	0.20	0.04	3	4	11	11	0	2
200	0.10	0.03	4	10	16	15	4	3
250	0.22	0.04	5	9	13	12	1	1
315	0.07	0.03	6	12	14	14	5	2
400	0.07	0.02	8	13	13	14	7	1
500	0.11	0.02	9	13	12	13	8	0

Source: Mouratidis and Becker (2004).

Table 14   Offset Angle Effects on TL for End-Outlet Plenum

Offset Angle Effects on TL for $f \le f_c$

Frequency, Hz	Offset Angle θ 0	15.0	22.5	30.0	37.5	45.0
50	0	0	0	0	0	0
63	0	0	0	0	0	0
80	0	0	−1	−3	−4	−6
100	0	1	0	−2	−3	−6
125	0	1	0	−2	−4	−6
160	0	0	−1	−2	−3	−4
200	0	0	−1	−2	−3	−5
250	0	1	2	3	5	7
315	0	4	6	8	10	14
400	0	2	4	6	9	13
500	0	1	3	6	10	15
≥630	N/A	N/A	N/A	N/A	N/A	N/A

Offset Angle Effects on TL for $f > f_c$

Frequency, Hz	Offset Angle θ 0	15.0	22.5	30.0	37.5	45.0
≤160	N/A	N/A	N/A	N/A	N/A	N/A
200	0	1	4	9	14	20
250	0	2	4	8	13	19
315	0	1	2	3	4	5
400	0	1	2	3	4	6
500	0	0	1	2	4	5
630	0	1	2	3	5	7
800	0	1	2	2	3	3
1000	0	1	2	4	6	9
1250	0	0	2	4	6	9
1600	0	0	1	1	2	3
2000	0	1	2	4	7	10
2500	0	1	2	3	5	8
3150	0	0	2	4	6	9
4000	0	0	2	5	8	12
5000	0	0	3	6	10	15

N/A = not applicable

S = Total surface area (all walls with lining)
$$= 2(5.9 \times 4.0) + 2(5.9 \times 5.9) + 2(4.0 \times 5.9) - 6.0 - 6.0$$
$$= 152.0 \text{ ft}^2$$

with l = 5.9 ft, r_v = 4.0 ft, and r_h = 1.0 ft,

$$r = (5.9^2 + 4.0^2 + 1.0^2)^{1/2} = 7.2 \text{ ft}$$

$$\theta = \cos^{-1}(5.9/7.2) = 35°$$

Table 15   Elbow Effect, dB

Frequency, Hz	$>f_c$	$\le f_c$
50	0	2
63	0	3
80	0	6
100	0	5
125	0	3
160	0	0
200	3	−2
250	6	−3
315	3	−1
400	3	0
500	2	0
630	2	0
800	3	0
1000	2	0
1250	2	0
1600	2	0
2000	2	0
2500	2	0
3150	2	0
4000	2	0
5000	1	0

N/A = not applicable

The cutoff frequency f_c is

$$f_c = 1132/(2 \times 3.0) = 189 \text{ Hz}$$

where 1132 ft/s is the approximate speed of sound in standard air.

Frequency Range #1 (1/3-octave TL in 50 Hz $\le f \le f_c$ range)

$$TL = A_f \times S + W_e + OAE \qquad (9)$$

(Consult Table 12 for A_f and W_e and Table 14 for offset angle effect.)

Frequency Range #2 (1/3-octave TL in $f_c < f \le$ 5000 Hz range)

$$TL = b\left[\frac{S_{out}Q}{4\pi r^2} + \frac{S_{out}(1 - \alpha_a)}{S\alpha_a}\right]^n + OAE \qquad (10)$$

where

b = 3.505
n = −0.359
Q = 4 (directivity factor for inlet opening close to adjacent wall or bihedral corner of plenum)
α_a = 1/3-octave average absorption values for 1 in. fiberglass lining (see Table 13)
OAE = see Table 14

Note: for angles between tabulated values in Table 14, use linear interpolation.

The results are tabulated as follows:

	(1)	(2)	(3)	(4)	(5)	(6)
1/3-Octave TL in	Freq., Hz	A_f, dB/ft^2	W_e, dB	OAE dB	TL for Frequency Range 1,[a] dB	TL for Frequency Range 2,[b] dB	Net TL,[c] dB

1/3-Octave TL in	Freq., Hz	A_f, dB/ft^2	W_e, dB	OAE dB	TL Range 1 dB	TL Range 2 dB	Net TL dB
$50 \le f \le$ 189 Hz	50	0.03	1	0	6		6
	63	0.03	1	0	6		6
	80	0.03	2	−4	3		3
	100	0.03	2	−3	4		4
	125	0.04	2	−4	4		4
	160	0.04	3	−3	6		6
$f_c < f \le$ 5000 Hz	200			12		20	20
	250			11		19	19
	315			4		13	13
	400			4		14	14
	500			3		13	13
	630			4		14	14
	800			3		14	14
	1000			5		16	16
	1300			5		16	16
	1600			2		13	13
	2000			6		17	17
	2500			4		15	15
	3200			5		16	16
	4000			7		18	18
	5000			8		19	19

OAE = offset angle effect
*Column 1 × S + column 2 + column 3
[b]Includes OAE value from column 3 per calculation from Equation (10).
[c]From column 4 or 5, depending on appropriate frequency range.

Unlined Rectangular Sheet Metal Ducts. Straight, unlined rectangular sheet metal ducts provide a fairly significant amount of low-frequency sound attenuation. Table 16 shows the results of selected unlined rectangular sheet metal ducts (Cummings 1983; Reynolds and Bledsoe 1989a; Ver 1978; Woods Fan Division 1973). Attenuation values in Table 16 apply only to rectangular sheet metal ducts with the lightest gages allowed by Sheet Metal and Air Conditioning Contractors' National Association, Inc. (SMACNA) HVAC duct construction standards. Attenuation for lengths greater than 10 ft is not well documented.

Sound energy attenuated at low frequencies in rectangular ducts may manifest itself as breakout noise along the duct. Low-frequency breakout noise should therefore be checked. For additional information on breakout noise, see the section on Sound Radiation Through Duct Walls.

Acoustically Lined Rectangular Sheet Metal Ducts. Internal duct lining for rectangular sheet metal ducts can be used to provide both thermal insulation and sound attenuation. The thickness of duct linings for thermal insulation usually varies from 0.5 to 2 in; density of fiberglass lining usually varies between 1.5 and 3.0 lb/ft^3, but may be as low as 0.75 lb/ft^3. For fiberglass duct lining to attenuate fan sound effectively, it should have a minimum thickness of 1 in. Tables 17 and 18 give attenuation values of selected rectangular sheet metal ducts for 1 and 2 in. duct lining, respectively (Kuntz 1986; Kuntz and Hoover 1987; Machen and Haines 1983; Reynolds and Bledsoe 1989a). Note that attenuation values shown in these tables are based on laboratory tests using 10 ft lengths of duct; for designs incorporating other distances, actual values will be different. The total attenuated noise will never be below the generated noise level in the duct.

Insertion loss values in Tables 17 and 18 are the difference in the sound pressure level measured in a reverberation room with sound

Table 16   Sound Attenuation in Unlined Rectangular Sheet Metal Ducts

		Attenuation, dB/ft Octave Midband Frequency, Hz			
Duct Size, in.	P/A, 1/ft	63	125	250	>250
6 × 6	8.0	0.30	0.20	0.10	0.10
12 × 12	4.0	0.35	0.20	0.10	0.06
12 × 24	3.0	0.40	0.20	0.10	0.05
24 × 24	2.0	0.25	0.20	0.10	0.03
48 × 48	1.0	0.15	0.10	0.07	0.02
72 × 72	0.7	0.10	0.10	0.05	0.02

Table 17   Insertion Loss for Rectangular Sheet Metal Ducts with 1 in. Fiberglass Lining

	Insertion Loss, dB/ft Octave Midband Frequency, Hz					
Dimensions, in.	125	250	500	1000	2000	4000
6 × 6	0.6	1.5	2.7	5.8	7.4	4.3
6 × 10	0.5	1.2	2.4	5.1	6.1	3.7
6 × 12	0.5	1.2	2.3	5.0	5.8	3.6
6 × 18	0.5	1.0	2.2	4.7	5.2	3.3
8 × 8	0.5	1.2	2.3	5.0	5.8	3.6
8 × 12	0.4	1.0	2.1	4.5	4.9	3.2
8 × 16	0.4	0.9	2.0	4.3	4.5	3.0
8 × 24	0.4	0.8	1.9	4.0	4.1	2.8
10 × 10	0.4	1.0	2.1	4.4	4.7	3.1
10 × 16	0.4	0.8	1.9	4.0	4.0	2.7
10 × 20	0.3	0.8	1.8	3.8	3.7	2.6
10 × 30	0.3	0.7	1.7	3.6	3.3	2.4
12 × 12	0.4	0.8	1.9	4.0	4.1	2.8
12 × 18	0.3	0.7	1.7	3.7	3.5	2.5
12 × 24	0.3	0.6	1.7	3.5	3.2	2.3
12 × 36	0.3	0.6	1.6	3.3	2.9	2.2
15 × 15	0.3	0.7	1.7	3.6	3.3	2.4
15 × 22	0.3	0.6	1.6	3.3	2.9	2.2
15 × 30	0.3	0.5	1.5	3.1	2.6	2.0
15 × 45	0.2	0.5	1.4	2.9	2.4	1.9
18 × 18	0.3	0.6	1.6	3.3	2.9	2.2
18 × 28	0.2	0.5	1.4	3.0	2.4	1.9
18 × 36	0.2	0.5	1.4	2.8	2.2	1.8
18 × 54	0.2	0.4	1.3	2.7	2.0	1.7
24 × 24	0.2	0.5	1.4	2.8	2.2	1.8
24 × 36	0.2	0.4	1.2	2.6	1.9	1.6
24 × 48	0.2	0.4	1.2	2.4	1.7	1.5
24 × 72	0.2	0.3	1.1	2.3	1.6	1.4
30 × 30	0.2	0.4	1.2	2.5	1.8	1.6
30 × 45	0.2	0.3	1.1	2.3	1.6	1.4
30 × 60	0.2	0.3	1.1	2.2	1.4	1.3
30 × 90	0.1	0.3	1.0	2.1	1.3	1.2
36 × 36	0.2	0.3	1.1	2.3	1.6	1.4
36 × 54	0.1	0.3	1.0	2.1	1.3	1.2
36 × 72	0.1	0.3	1.0	2.0	1.2	1.2
36 × 108	0.1	0.2	0.9	1.9	1.1	1.1
42 × 42	0.2	0.3	1.0	2.1	1.4	1.3
42 × 64	0.1	0.3	0.9	1.9	1.2	1.1
42 × 84	0.1	0.2	0.9	1.8	1.1	1.1
42 × 126	0.1	0.2	0.9	1.7	1.0	1.0
48 × 48	0.1	0.3	1.0	2.0	1.2	1.2
48 × 72	0.1	0.2	0.9	1.8	1.0	1.0
48 × 96	0.1	0.2	0.8	1.7	1.0	1.0
48 × 144	0.1	0.2	0.8	1.6	0.9	0.9

Table 18 Insertion Loss for Rectangular Sheet Metal Ducts with 2 in. Fiberglass Lining

Dimensions, in.	Insertion Loss, dB/ft Octave Midband Frequency, Hz					
	125	250	500	1000	2000	4000
6 × 6	0.8	2.9	4.9	7.2	7.4	4.3
6 × 10	0.7	2.4	4.4	6.4	6.1	3.7
6 × 12	0.6	2.3	4.2	6.2	5.8	3.6
6 × 18	0.6	2.1	4.0	5.8	5.2	3.3
8 × 8	0.6	2.3	4.2	6.2	5.8	3.6
8 × 12	0.6	1.9	3.9	5.6	4.9	3.2
8 × 16	0.5	1.8	3.7	5.4	4.5	3.0
8 × 24	0.5	1.6	3.5	5.0	4.1	2.8
10 × 10	0.6	1.9	3.8	5.5	4.7	3.1
10 × 16	0.5	1.6	3.4	5.0	4.0	2.7
10 × 20	0.4	1.5	3.3	4.8	3.7	2.6
10 × 30	0.4	1.3	3.1	4.5	3.3	2.4
12 × 12	0.5	1.6	3.5	5.0	4.1	2.8
12 × 18	0.4	1.4	3.2	4.6	3.5	2.5
12 × 24	0.4	1.3	3.0	4.3	3.2	2.3
12 × 36	0.4	1.2	2.9	4.1	2.9	2.2
15 × 15	0.4	1.3	3.1	4.5	3.3	2.4
15 × 22	0.4	1.2	2.9	4.1	2.9	2.2
15 × 30	0.3	1.1	2.7	3.9	2.6	2.0
15 × 45	0.3	1.0	2.6	3.6	2.4	1.9
18 × 18	0.4	1.2	2.9	4.1	2.9	2.2
18 × 28	0.3	1.0	2.6	3.7	2.4	1.9
18 × 36	0.3	0.9	2.5	3.5	2.2	1.8
18 × 54	0.3	0.8	2.3	3.3	2.0	1.7
24 × 24	0.3	0.9	2.5	3.5	2.2	1.8
24 × 36	0.3	0.8	2.3	3.2	1.9	1.6
24 × 48	0.2	0.7	2.2	3.0	1.7	1.5
24 × 72	0.2	0.7	2.0	2.9	1.6	1.4
30 × 30	0.2	0.8	2.2	3.1	1.8	1.6
30 × 45	0.2	0.7	2.0	2.9	1.6	1.4
30 × 60	0.2	0.6	1.9	2.7	1.4	1.3
30 × 90	0.2	0.5	1.8	2.6	1.3	1.2
36 × 36	0.2	0.7	2.0	2.9	1.6	1.4
36 × 54	0.2	0.6	1.9	2.6	1.3	1.2
36 × 72	0.2	0.5	1.8	2.5	1.2	1.2
36 × 108	0.2	0.5	1.7	2.3	1.1	1.1
42 × 42	0.2	0.6	1.9	2.6	1.4	1.3
42 × 64	0.2	0.5	1.7	2.4	1.2	1.1
42 × 84	0.2	0.5	1.6	2.3	1.1	1.1
42 × 126	0.1	0.4	1.6	2.2	1.0	1.0
48 × 48	0.2	0.5	1.8	2.5	1.2	1.2
48 × 72	0.2	0.4	1.6	2.3	1.0	1.0
48 × 96	0.1	0.4	1.5	2.1	1.0	1.0
48 × 144	0.1	0.4	1.5	2.0	0.9	0.9

Table 19 Sound Attenuation in Unlined Straight Round Ducts

Diameter, in.	Attenuation, dB/ft Octave Midband Frequency, Hz						
	63	125	250	500	1000	2000	4000
$D \leq 7$	0.03	0.03	0.05	0.05	0.10	0.10	0.10
$7 < D \leq 15$	0.03	0.03	0.03	0.05	0.07	0.07	0.07
$15 < D \leq 30$	0.02	0.02	0.02	0.03	0.05	0.05	0.05
$30 < D \leq 60$	0.01	0.01	0.01	0.02	0.02	0.02	0.02

Table 20 Insertion Loss for Acoustically Lined Round Ducts with 1 in. Lining

Diameter, in.	Insertion Loss, dB/ft Octave Midband Frequency, Hz							
	63	125	250	500	1000	2000	4000	8000
6	0.38	0.59	0.93	1.53	2.17	2.31	2.04	1.26
8	0.32	0.54	0.89	1.50	2.19	2.17	1.83	1.18
10	0.27	0.50	0.85	1.48	2.20	2.04	1.64	1.12
12	0.23	0.46	0.81	1.45	2.18	1.91	1.48	1.05
14	0.19	0.42	0.77	1.43	2.14	1.79	1.34	1.00
16	0.16	0.38	0.73	1.40	2.08	1.67	1.21	0.95
18	0.13	0.35	0.69	1.37	2.01	1.56	1.10	0.90
20	0.11	0.31	0.65	1.34	1.92	1.45	1.00	0.87
22	0.08	0.28	0.61	1.31	1.82	1.34	0.92	0.83
24	0.07	0.25	0.57	1.28	1.71	1.24	0.85	0.80
26	0.05	0.22	0.53	1.24	1.59	1.14	0.79	0.77
28	0.03	0.19	0.49	1.20	1.46	1.04	0.74	0.74
30	0.02	0.16	0.45	1.16	1.33	0.95	0.69	0.71
32	0.01	0.14	0.42	1.12	1.20	0.87	0.66	0.69
34	0	0.11	0.38	1.07	1.07	0.79	0.63	0.66
36	0	0.08	0.35	1.02	0.93	0.71	0.60	0.64
38	0	0.06	0.31	0.96	0.80	0.64	0.58	0.61
40	0	0.03	0.28	0.91	0.68	0.57	0.55	0.58
42	0	0.01	0.25	0.84	0.56	0.50	0.53	0.55
44	0	0	0.23	0.78	0.45	0.44	0.51	0.52
46	0	0	0.20	0.71	0.35	0.39	0.48	0.48
48	0	0	0.18	0.63	0.26	0.34	0.45	0.44
50	0	0	0.15	0.55	0.19	0.29	0.41	0.40
52	0	0	0.14	0.46	0.13	0.25	0.37	0.34
54	0	0	0.12	0.37	0.09	0.22	0.31	0.29
56	0	0	0.10	0.28	0.08	0.18	0.25	0.22
58	0	0	0.09	0.17	0.08	0.16	0.18	0.15
60	0	0	0.08	0.06	0.10	0.14	0.09	0.07

propagating through an unlined section of rectangular duct minus the corresponding sound pressure level measured when the unlined section of rectangular duct is replaced with a similar section of acoustically lined rectangular duct. The net result is the attenuation resulting from adding duct liner to a sheet metal duct.

Insertion loss and attenuation values discussed in this section apply only to rectangular sheet metal ducts made with the lightest gages allowed by SMACNA HVAC duct construction standards. Attenuation for lengths greater than 10 ft is not well documented.

Unlined Round Sheet Metal Ducts. As with unlined rectangular ducts, unlined round ducts provide some natural sound attenuation that should be taken into account when designing a duct system. Compared to rectangular ducts, round ducts are much more rigid and thus do not absorb as much sound energy. Because of this, round ducts only provide about one-tenth the sound attenuation at low frequencies as rectangular ducts. However, breakout from round ducts is significantly less than that from other shapes. Table 19 lists sound

attenuation values for unlined round ducts (Kuntz and Hoover 1987; Woods Fan Division 1973).

Acoustically Lined Round Sheet Metal Ducts. The literature provides very little data on insertion loss for acoustically lined round ducts; usually only manufacturers' product data are available. Tables 20 and 21 give insertion loss values for dual-wall round sheet metal ducts with 1 and 2 in. acoustical lining, respectively (Reynolds and Bledsoe 1989b). The acoustical lining is a 0.75 lb/ft^3 density fiberglass blanket, which is covered by an internal liner of perforated galvanized sheet metal with an open area of 25%. The data in Tables 20 and 21 were collected from 20 ft duct sections. Because there are many options available for round ducts, attenuation may significantly vary from the data provided in the tables.

Rectangular Sheet Metal Duct Elbows. Table 22 displays insertion loss values for unlined and lined square elbows without turning vanes (Beranek 1960). For lined square elbows, duct lining must extend at least two duct widths w beyond the elbow. Table 22 applies only where the duct is lined before and after the elbow. Table 23 gives insertion loss values for unlined radiused elbows. Table 24 gives insertion loss values for unlined and lined square elbows with turning vanes. The quantity fw in Tables 22 to 24 is the midfrequency of the octave band times the width of the elbow (Figure 20) (Beranek 1960; Ver 1983b).

Table 21 Insertion Loss for Acoustically Lined Round Ducts with 2 in. Lining

Diameter, in.	Insertion Loss, dB/ft Octave Midband Frequency, Hz							
	63	125	250	500	1000	2000	4000	8000
6	0.56	0.80	1.37	2.25	2.17	2.31	2.04	1.26
8	0.51	0.75	1.33	2.23	2.19	2.17	1.83	1.18
10	0.46	0.71	1.29	2.20	2.20	2.04	1.64	1.12
12	0.42	0.67	1.25	2.18	2.18	1.91	1.48	1.05
14	0.38	0.63	1.21	2.15	2.14	1.79	1.34	1.00
16	0.35	0.59	1.17	2.12	2.08	1.67	1.21	0.95
18	0.32	0.56	1.13	2.10	2.01	1.56	1.10	0.90
20	0.29	0.52	1.09	2.07	1.92	1.45	1.00	0.87
22	0.27	0.49	1.05	2.03	1.82	1.34	0.92	0.83
24	0.25	0.46	1.01	2.00	1.71	1.24	0.85	0.80
26	0.24	0.43	0.97	1.96	1.59	1.14	0.79	0.77
28	0.22	0.40	0.93	1.93	1.46	1.04	0.74	0.74
30	0.21	0.37	0.90	1.88	1.33	0.95	0.69	0.71
32	0.20	0.34	0.86	1.84	1.20	0.87	0.66	0.69
34	0.19	0.32	0.82	1.79	1.07	0.79	0.63	0.66
36	0.18	0.29	0.79	1.74	0.93	0.71	0.60	0.64
38	0.17	0.27	0.76	1.69	0.80	0.64	0.58	0.61
40	0.16	0.24	0.73	1.63	0.68	0.57	0.55	0.58
42	0.15	0.22	0.70	1.57	0.56	0.50	0.53	0.55
44	0.13	0.20	0.67	1.50	0.45	0.44	0.51	0.52
46	0.12	0.17	0.64	1.43	0.35	0.39	0.48	0.48
48	0.11	0.15	0.62	1.36	0.26	0.34	0.45	0.44
50	0.09	0.12	0.60	1.28	0.19	0.29	0.41	0.40
52	0.07	0.10	0.58	1.19	0.13	0.25	0.37	0.34
54	0.05	0.08	0.56	1.10	0.09	0.22	0.31	0.29
56	0.02	0.05	0.55	1.00	0.08	0.18	0.25	0.22
58	0	0.03	0.53	0.90	0.08	0.16	0.18	0.15
60	0	0	0.53	0.79	0.10	0.14	0.09	0.07

Table 22 Insertion Loss of Unlined and Lined Square Elbows Without Turning Vanes

	Insertion Loss, dB	
	Unlined Elbows	Lined Elbows
$fw < 1.9$	0	0
$1.9 \leq fw < 3.8$	1	1
$3.8 \leq fw < 7.5$	5	6
$7.5 \leq fw < 15$	8	11
$15 \leq fw < 30$	4	10
$fw > 30$	3	10

Note: f = center frequency, kHz, and w = width, in.

Table 23 Insertion Loss of Radiused Elbows

	Insertion Loss, dB
$fw < 1.9$	0
$1.9 \leq fw < 3.8$	1
$3.8 \leq fw < 7.5$	2
$fw > 7.5$	3

Note: f = center frequency, kHz, and w = width, in.

Nonmetallic Insulated Flexible Ducts. Nonmetallic insulated flexible ducts can significantly reduce airborne noise. Insertion loss values for specified duct diameters and lengths are given in Table 25 and in Appendix D of ARI *Standard* 885. Recommended duct lengths are normally 3 to 6 ft. Take care to keep flexible ducts straight; bends should have as long a radius as possible. Although an abrupt bend may provide some additional insertion loss, the airflow-generated noise associated with airflow in the bend may be unacceptably high. Because of potentially high breakout sound levels associated with flexible ducts, care should be taken when using flexible ducts above sound-sensitive spaces.

Table 24 Insertion Loss of Unlined and Lined Square Elbows with Turning Vanes

	Insertion Loss, dB	
	Unlined Elbows	Lined Elbows
$fw < 1.9$	0	0
$1.9 \leq fw < 3.8$	1	1
$3.8 \leq fw < 7.5$	4	4
$7.5 \leq fw < 15$	6	7
$fw > 15$	4	7

Note: f = center frequency, kHz, and w = width, in.

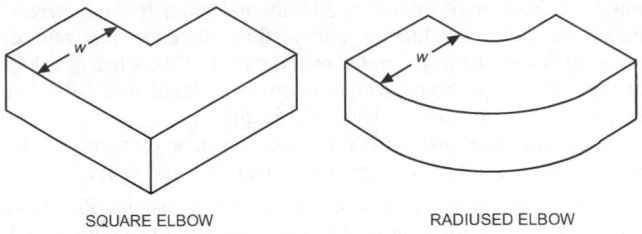

SQUARE ELBOW   RADIUSED ELBOW

Fig. 20 Rectangular Duct Elbows

Duct Branch Sound Power Division. When sound traveling in a duct encounters a junction, the sound power contained in the incident sound waves in the main feeder duct is distributed between the branches associated with the junction (Ver 1982, 1983b). This division of sound power is called branch sound power division. The corresponding attenuation of sound power transmitted down each branch of the junction is comprised of two components. The first is associated with reflection of the incident sound wave if the sum of the cross-sectional areas of individual branches ΣS_{Bi} differs from the cross-sectional area of the main feeder duct. The second and more dominant component is associated with energy division according to the ratio of the cross-sectional area of an individual branch S_{Bi} divided by ΣS_{Bi}. Values for the attenuation of sound power ΔL_{Bi} are given in Table 26.

Duct Silencers. Silencers, sometimes called sound attenuators, sound traps, or mufflers, are designed to reduce the noise transmitted from a source to a receiver. For HVAC applications, the most common silencers are duct silencers, installed on the intake and/or discharge side of a fan or air handler. Also, they may be used on the receiver side of other noise generators such as terminal boxes, valves, dampers, etc.

Duct silencers are available in varying shapes and sizes to fit project ductwork. Generally, a duct silencer's outer appearance is similar to a piece of ductwork. It consists of a sheet metal casing with length commonly ranging from 3 to 10 ft. Common shapes include rectangular, round, elbow, tee, and transitional. Figure 21 shows some duct silencer configurations.

All silencers can be rated for (1) insertion loss, (2) dynamic insertion loss, (3) pressure drop, and (4) self-generated noise in accordance with ASTM E477 test standards. As such, the performance is under rather ideal conditions as seen in Figure 22.

Insertion loss is the reduction in the sound power level at the receiver after the silencer is installed ("inserted") in the system. Insertion loss is measured as a function of frequency and commonly published in full octave bands ranging from 63 to 8000 Hz.

Dynamic insertion loss is insertion loss with given airflow direction and velocity. A silencer's insertion loss varies depending on whether sound is traveling in the same or opposite direction as airflow. Silencer performance changes with absolute duct velocity. However, airflow velocity generally does not significantly affect silencers giving a pressure drop of 0.35 in. of water or less, including system effects.

Pressure drop is measured across the silencer at a given velocity. Good flow conditions are required for accurate measurements at both the inlet and discharge of the silencer. The measuring points are usually 2.5 to 5 duct diameters upstream and downstream of the silencer to avoid turbulent flow areas near the silencer and to allow for any static pressure regain. For nonideal installations, with duct elbows or transitions closer than 2.5 to 5 duct diameters, the total system effect will be larger than the laboratory test data.

Airflow-generated self noise is the sound power generated on the receiving side by the silencer when quiet air flows through it. This represents the **noise floor**, or the lowest level achievable regardless of high insertion loss values. A silencer's self-generated noise is a function of frequency and internal geometry, and is referenced to specific velocities and airflow direction (forward or reverse). The airflow-generated sound power of the silencer is logarithmically proportional to silencer cross-sectional area. Self noise generally does not vary with silencer length.

There are three types of HVAC duct silencers: dissipative (with acoustic media), fiber-free reactive (no media), and active.

Dissipative Silencers. Dissipative silencers use sound-absorptive media such as fiberglass as the primary means of attenuating sound; mineral wool can be used in high-temperature applications but may contain too much contamination ("shot") for commercial HVAC

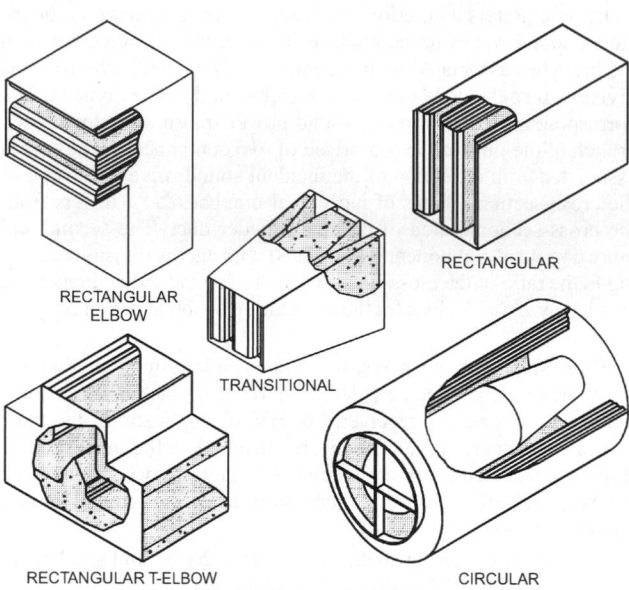

Fig. 21　Duct Silencer Configurations

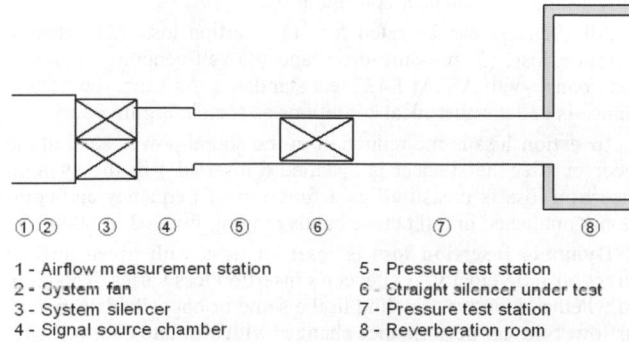

1 - Airflow measurement station   5 - Pressure test station
2 - System fan   6 - Straight silencer under test
3 - System silencer   7 - Pressure test station
4 - Signal source chamber   8 - Reverberation room

**Fig. 22　Typical Facility for Rating Straight Duct Silencers
With of Without Airflow**

Table 25　Insertion Loss for Lined Flexible Duct

Diameter, in.	Length, ft	Insertion Loss, dB Octave Midband Frequency, Hz						
		63	125	250	500	1000	2000	4000
4	12	6	11	12	31	37	42	27
	9	5	8	9	23	28	32	20
	6	3	6	6	16	19	21	14
	3	2	3	3	8	9	11	7
5	12	7	12	14	32	38	41	26
	9	5	9	11	24	29	31	20
	6	4	6	7	16	19	21	13
	3	2	3	4	8	10	10	7
6	12	8	12	17	33	38	40	26
	9	6	9	13	25	29	30	20
	6	4	6	9	17	19	20	13
	3	2	3	4	8	10	10	7
7	12	9	12	19	33	37	38	25
	9	6	9	14	25	28	29	19
	6	4	6	10	17	19	19	13
	3	2	3	5	8	9	10	6
8	12	8	11	21	33	37	37	24
	9	6	8	16	25	28	28	18
	6	4	6	11	17	19	19	12
	3	2	3	5	8	9	9	6
9	12	8	11	22	33	37	36	22
	9	6	8	17	25	28	27	17
	6	4	6	11	17	19	18	11
	3	2	3	6	8	9	9	6
10	12	8	10	22	32	36	34	21
	9	6	8	17	24	27	26	16
	6	4	5	11	16	18	17	11
	3	2	3	6	8	9	9	5
12	12	7	9	20	30	34	31	18
	9	5	7	15	23	26	23	14
	6	3	5	10	15	17	16	9
	3	2	2	5	8	9	8	5
14	12	5	7	16	27	31	27	14
	9	4	5	12	20	23	20	11
	6	3	4	8	14	16	14	7
	3	1	2	4	7	8	7	4
16	12	2	4	9	23	28	23	9
	9	2	3	7	17	21	17	7
	6	1	2	5	12	14	12	5
	3	1	1	2	6	7	6	2

Note: 63 Hz insertion loss values estimated from higher-frequency insertion loss values.

Table 26　Duct Branch Sound Power Division

$S_{Bi}/\sum S_{Bi}$	ΔL_{Bi}	$S_{Bi}/\sum S_{Bi}$	ΔL_{Bi}
1.00	0	0.10	10
0.80	1	0.08	11
0.63	2	0.063	12
0.50	3	0.050	13
0.40	4	0.040	14
0.32	5	0.032	15
0.25	6	0.025	16
0.20	7	0.020	17
0.16	8	0.016	18
0.12	9	0.012	19

applications. Usually, the absorptive medium is covered by perforated metal to protect it from erosion by airflow. If internal silencer velocities are high (faster than 4500 fpm), media erosion may be further reduced by a layer of material such as fiberglass cloth or polymer film liner placed between the absorptive media and the perforated metal. Dissipative silencers may be supplied as hospital-grade or as film-lined silencers that include special polymer film linings to prevent contamination of the airstream by acoustical media fibers and prevent particles from the airstream from getting into the media. These silencers are commonly used in hospitals, pharmaceutical facilities, cleanrooms, and other places where indoor air quality is of paramount concern. Consult manufacturers for construction and testing performance details.

Dissipative silencer performance is primarily a function of silencer length; airflow constriction; number, thickness, and shape of splitters or centerbodies; and type and density of absorptive media. The absorptive media allows dissipative silencers to provide significant insertion loss performance over a wide frequency range.

Insertion loss performance does not necessarily increase linearly with silencer length; for a given length, silencer designs can produce varying insertion loss and pressure drop data. Even at the same pressure drop and length, silencers can be configured to provide varying insertion loss performance across the frequency spectrum.

Reactive Silencers. Reactive silencers are constructed only of metal, both solid and perforated, with chambers of specially designed shapes and sizes behind the perforated metal that are tuned as resonators to react with and reduce sound power at certain frequencies. The outward appearance of reactive silencers is similar to that of their dissipative counterparts. However, because of tuning, insertion loss over a wide frequency range is more difficult to achieve. Longer lengths may be required to achieve similar insertion loss performance as dissipative silencers. Airflow generally increases the insertion loss of reactive silencers.

Figure 23 compares insertion loss of dissipative silencers, with and without protective film materials, against that of a reactive silencer for the same pressure drop.

Active Silencers. Active duct silencers, sometimes called noise canceling systems, produce inverse sound waves that cancel noise primarily at low frequencies. An input microphone measures noise in the duct and converts it to electrical signals, which are processed digitally to generate opposite, "mirror-image" sound signals of equal amplitude. A secondary noise source destructively interferes with the primary noise and cancels a significant portion of it. An error microphone measures residual sound beyond the silencer and provides feedback to adjust the computer model for improved performance.

Because components are mounted outside the airflow, there is no pressure loss or airflow-generated noise. Performance is limited, however, if excessive turbulence is detected by the microphones. Manufacturers recommend using active silencers where duct velocities are less than 1500 fpm and where duct configurations are conducive to smooth, evenly distributed airflow.

Active silencers have significant low-frequency insertion loss, and are self-regulating because, if fan noise levels increase, an active silencer can increase performance to compensate for the increased source noise. Mid- and high-frequency insertion loss is minimal, however, so if required, combinations of active (for low-frequency components) and passive (for mid- and high-frequency components) can be used to achieve insertion loss over a wide frequency range.

Test Standard. Data for dissipative and reactive silencers should be obtained from tests consistent with the procedures outlined in ASTM *Standard* E477. (This standard has not been verified for determining performance of active silencers.) Because insertion loss measurements use a substitution technique, reasonably precise insertion loss values can be achieved (±3 dB down to 125 Hz, and ±5 dB at lower frequencies). Airflow-generated noise values can be obtained with similar accuracy. Round-robin tests performed at several manufacturers' and independent testing laboratories showed that airflow-generated sound power data has an expected standard deviation of ±3 to 6 dB over the octave band frequency range of 125 to 8000 Hz. (For normal distribution, uncertainty with a 95% confidence interval is about two standard deviations.)

Silencer Selection Issues. When selecting a duct silencer, consider the following:

- Insertion loss required to achieve required room sound criteria
- Allowable pressure drop (if no specific requirement, then keep under 0.35 in. of water, including system effects; when system effects are unknown, keep under 0.20 in. of water, excluding system effects) at system duct velocity
- Silencer location and available space
- Amount of airflow-generated noise that can be tolerated
- Indoor air quality concerns
- Duct configuration

Insertion Loss Issues

To determine the insertion loss required, analyze the duct system, summing noise-generating mechanisms and subtracting attenuation elements (not including the silencer). The silencer's required insertion loss is the amount by which the estimated resultant sound pressure level in the space exceeds the room criteria for the space. The user should consider both the sound path through the ductwork and outlets as well as potential locations where sound may break out of the ductwork.

Allowable Pressure Drop Issues

Care should be taken in applying test data to actual project installations. Adverse aerodynamic system effects can significantly affect silencer performance. That is, if the silencer is located where less-than-ideal conditions exists on the inlet and/or the discharge of the silencer (3 to 5 duct diameters of straight duct), then the silencer's effective pressure drop (PD) is increased (total silencer PD = silencer PD per ASTM E477 + system effect losses). In some situations, the added system effect losses can be greater than the silencer's pressure drop. Some manufacturers give guidelines for estimated pressure loss increases from varying silencer inlet and

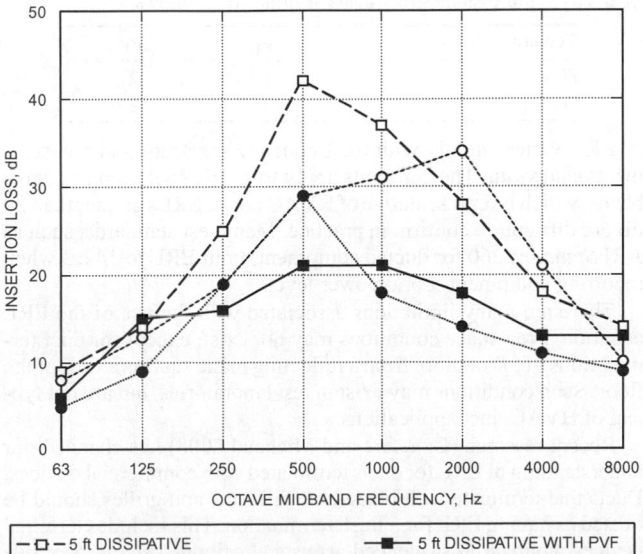

Fig. 23   Comparison of 5 ft Long Dissipative and Reactive Silencer Performance

Table 27 Approximate Silencer System Effect Factors

Silencer Condition	Pressure Drop Factor*
Inlet (within 3 to 4 duct diameters)	
Straight unobstructed duct	1.0
Free air/plenum with smooth inlet	1.05
Radiused elbow, with turning vanes	1.05
no turning vanes	1.1
Miter elbow	1.3
Free air/plenum with sharp inlet	1.1 to 1.30
Fan	1.1 to 1.3
Outlet (within 3 to 4 duct diameters)	
Straight unobstructed duct	1.0
Duct doubles area abruptly	1.4
Radiused elbow, with turning vanes	1.5
no turning vanes	1.9
Miter elbow	2.0
Abrupt expansion/plenum	2.0
Fan	1.2 to 1.4

*Silencer pressure drop (including system effects) = silencer pressure drop per test code × pressure drop factor (inlet) × pressure drop factor (outlet).

discharge configurations (Table 27); these should be considered as general guidelines. Substantial variations can occur depending on the type of silencer, its internal geometry, size of silencer, size of duct, airflow turbulence, etc. For example, an elbow fitting located immediately after a silencer prevents regain of the silencer's leaving velocity pressure. In addition, local velocities in the elbow fitting are greater than the average duct velocity that produces higher overall static pressure losses.

Silencer Location Issues

Silencers should generally be located as close to the noise source as possible but far enough away to allow a uniform flow profile to develop. This helps contain noise at the source and limits potential points where unsilenced noise may break out. However, because turbulent airflow usually exists close to noise sources such as fans, valves, dampers, etc., the user should carefully evaluate aerodynamic system effects.

A straight silencer has a lower first cost than a transitional or elbow silencer. If space limitations prohibit effective use of a straight silencer, or if pressure drop (including system effects) is greater than the loss allowed, use of elbow or transitional silencers should be evaluated. Special fan inlet and discharge silencers, including cone and inlet box silencers, minimize aerodynamic system effects, and contain noise at the source.

Airflow-Generated Noise Issues

In most installations, airflow-generated noise is much less than, and does not contribute to, the reduced noise level on the receiver side of the silencer. This is especially true if the silencer is properly located close to the source. In general, airflow-generated noise should be evaluated if pressure drops exceed 0.35 in. of water (including system effects), the noise criterion is below NC/RC 35, or if the silencer is located very close to or in the occupied space.

To evaluate airflow-generated noise, sum the noise-generating mechanisms (from noise source to silencer) and subtract the attenuation elements (including silencer) in the order they occur to determine the resultant sound power level on the quiet side of the silencer. This resultant level must be summed logarithmically with the silencer's generated noise (referenced to actual duct velocities, inlet and discharge configurations, and cross-sectional area). If the generated noise is more than 10 dB below the residual sound, then the silencer's generated noise will have no effect on system noise levels.

Table 28 Duct End Reflection Loss (ERL): Duct Terminated Flush with Wall

Duct Diameter, in.	End Reflection Loss, dB Octave Midband Frequency, Hz				
	63	125	250	500	1000
6	18	12	7	3	1
8	15	10	5	2	1
10	14	8	4	1	0
12	12	7	3	1	0
16	10	5	2	1	0
20	8	4	1	0	0
24	7	3	1	0	0
28	6	2	1	0	0
32	5	2	1	0	0
36	4	2	0	0	0
48	3	1	0	0	0
72	1	0	0	0	0

Duct End Reflection Loss. When low-frequency sound waves encounter the end of a duct that is terminated into a large room, some of the incident sound energy is reflected back into the duct. Duct end reflection loss (ERL) values for a duct terminated flush with a wall are shown in Table 28.

To use Table 28 for a rectangular duct, calculate the effective duct diameter D by

$$D = \sqrt{4A/\pi} \qquad (11)$$

where A is the cross-sectional area of the rectangular duct (ft^2). For the frequency range and duct sizes of interest to HVAC designers, the duct ERL may be accurately computed using a simplified equation (Cunefare and Michaud 2008) of the form

$$ERL = 10 \log10 \left[1 + \left(\frac{a_1 c_o}{\pi f D} \right)^{a_2} \right] \qquad (12)$$

where

c_o = speed of sound (dimensionally consistent with D), ft/s
f = frequency [Hz]
a_1 and a_2 = dimensionless constants determined as follows:

Termination	a_1	a_2
Flush	0.7	2
Free space	1	2

ERL varies slightly with the frequency spectrum and measurement bandwidth. The constants apply to a pink spectrum in octave bands, which is representative of HVAC noise. ERLs greater than 20 dB are difficult to confirm in practice. Many test standards, such as ARI *Standard* 260 for ducted equipment, limit ERL to 14 dB when reporting equipment sound power levels.

There are many limitations associated with the use of the ERL equation. Free-space conditions may not exist, except for duct terminations of $5D$ or more from a reflecting plane such as a wall or the floor. Such conditions may exist in test laboratories, but are not typical of HVAC duct applications.

Recent research (Cunefare and Michaud 2008) has changed our understanding of ERL for ducts terminated with commercial devices. Ducts that terminate with blade-type diffusers and grilles should be treated as having ERL for a flush termination. This includes terminal devices mounted in suspended acoustical ceiling systems. Slot diffusers characterized by high aspect ratios and mounted in a rigid baffle have frequency-independent ERL that may be determined by the analytical expression for the area ratio of the diffuser to duct cross-sectional area. Finally, using flexible duct upstream of diffusers, grilles and other terminal devices reduces ERL to near zero above 63 Hz for all terminal devices. This research suggests that a significant amount of the low-frequency sound that would normally be reflected back into the duct from an open termination is either transmitted through the flexible duct or radiated by the termination. There is however a frequency-independent ERL associated with the area change in the transition to the flexible duct.

Finally, ERL values are based on analytical assumptions and empirical data for long and straight duct sections. Many air distribution systems do not have long straight sections (greater than 3D) before they terminate into a room. Many duct sections between a main feed branch and a diffuser may be curved or short. The effects of these configurations on duct end reflection loss are not known. Table 28 can be used with reasonable accuracy for many diffuser configurations. However, caution should be used when a diffuser configuration differs from the conditions used to derive these ERL values.

Sound Radiation Through Duct Walls

Duct Rumble. Duct rumble is low-frequency sound generated by vibration of a flat duct surface. The vibration is caused when an HVAC fan and its connected ductwork act as a semiclosed, compressible-fluid pumping system; both acoustic and aerodynamic air pressure fluctuations at the fan are transmitted to other locations in the duct system. Rumbling occurs at the duct's resonance frequencies (Ebbing et al. 1978), and duct rumble levels of 65 to 95 dB in the 16 to 100 Hz frequency range have been measured in occupied spaces. With belt-driven fans, the rumble sound level fluctuates

above and below the mean dB level by 5 to 25 dB at a rate of 2 to 10 "beats" per second (Blazier 1993). The most common beat frequency occurs at the difference between the fan rpm and twice the belt frequency (belt rpm = fan sheave diameter × sheave rpm × π/belt length). As shown in Figure 24, duct rumble is dependent on the level of duct vibration. The very low resonant frequencies at which duct rumble occurs means that the sound wavelengths are very long (10 to 70 ft), and the rumble can exert sound energy over long distances. Lightweight architectural structures such as metal frame and drywall systems near a source of duct rumble can easily vibrate and rattle in sympathy to the rumble.

Case histories indicate that duct vibration is much more prevalent when there is a dramatic change in airflow direction near the fan, and at large, flat, unreinforced duct surfaces (usually greater than 48 in. in any dimension) near the fan. Problems can occur with dimensions as small as 18 in. if high noise levels are present. Figure 25 shows duct configurations near a centrifugal fan. Good to optimum designs of fan discharge transitions minimize potential for duct rumble; however, this may not completely eliminate the potential for duct rumble, which also depends heavily on flow turbulence at the fan wheel, duct stiffness, air velocity in the duct, and duct resonant characteristics.

Duct liner, sound attenuators, and duct lagging with mass-loaded vinyl over fiberglass do not reduce duct rumble. One approach to eliminate or reduce rumble is to alter the fan or motor speed, which changes the frequency of air pressure fluctuations so that they differ from duct wall resonance frequencies. Another method is to apply rigid materials, such as duct reinforcements and drywall, directly to the duct wall to change the wall resonance frequencies (Figure 26). Noise reductions of 5 to 11 dB in the 31.5 and 63 Hz octave frequency bands are possible using this treatment.

Mass-loaded materials applied in combination with absorptive materials do not alleviate duct rumble noise unless both materials are completely decoupled from the duct by a large air separation (preferably greater than 6 in.). The mass-loaded material should have a surface density greater than 4 lb/ft^2. An example of this type of construction, using two layers of drywall, is shown in Figure 27. Because the treatment is decoupled from the duct wall, it provides the greatest noise reduction. Mass-loaded/absorptive material directly attached to a round duct can be an effective noise control treatment for high-frequency noise above the duct rumble frequency range of 16 to 100 Hz. In addition, the stiffness of round ductwork prevents flexure of the duct wall. Where space allows, round ductwork is an effective method to prevent duct rumble (Harold 1986). However, unless round ducts are used throughout the primary duct system, duct rumble can be still generated at a remote point where round duct is converted to rectangular or flat oval.

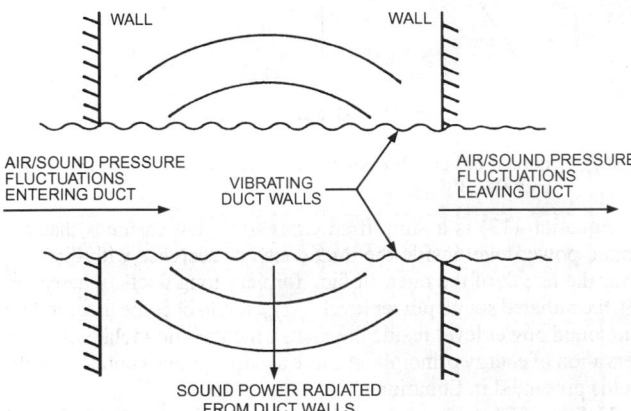

Fig. 24   Transmission of Rumble Noise Through Duct Walls

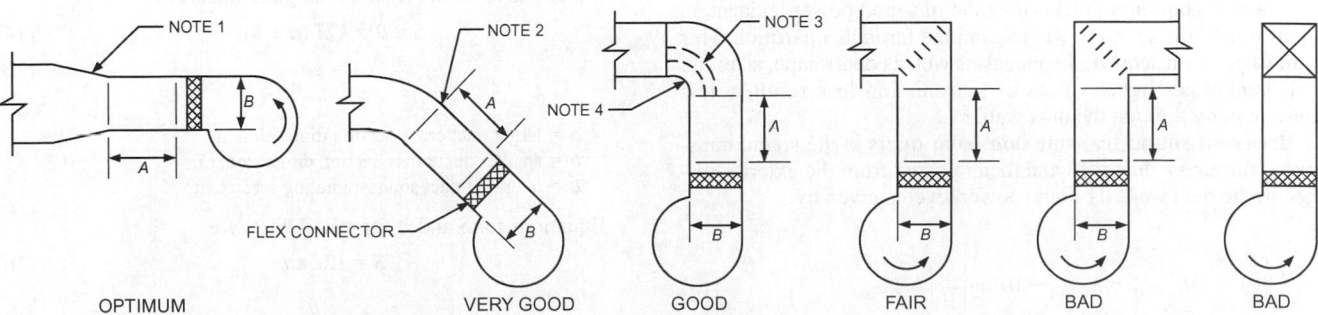

Notes:
1. Slopes of 1 in 7 preferred. Slopes of 1 in 4 permitted below 2000 fpm.
2. Dimension A should be at least 1.5 times B, where B is largest discharge duct dimension.
3. Rugged turning vanes should extend full radius of elbow.
4. Minimum 6 in. radius required.

Fig. 25   Various Outlet Configurations for Centrifugal Fans and Their Possible Rumble Conditions

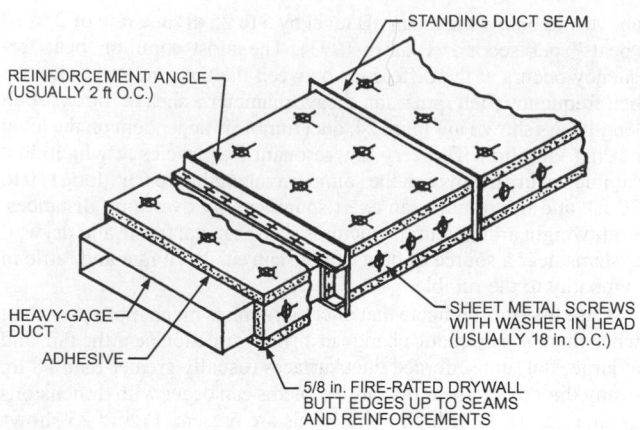

Fig. 26 Drywall Lagging for Duct Rumble

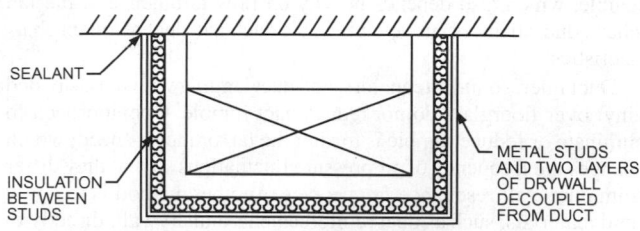

Fig. 27 Decoupled Drywall Enclosure for Duct Rumble

Round ducts can have a resonant ring resonance frequency, which depends on duct material and diameter. The ring frequency is a resonance frequency that occurs where the circumference of the duct is equal to the wavelength of the bending waves in the duct wall. On rare occasions, loud in-duct noise, such as blade-pass frequency noise from a centrifugal or axial fan, can excite this resonance. In all cases, this resonance causes an increase in radiated noise in the frequency region close to the ring frequency.

Sound Breakout and Break-In from Ducts. Breakout is sound associated with fan or airflow noise inside a duct that radiates through duct walls into the surrounding area (Figure 28). Breakout can be a problem if it is not adequately attenuated before the duct runs over an occupied space (Cummings 1983; Lilly 1987). Sound that is transmitted into a duct from the surrounding area is called **break-in** (Figure 29). The main factors affecting breakout and break-in sound transmission are the transmission loss of the duct, total exposed surface area of the duct, and presence of any acoustical duct liner.

Transmission loss (TL) is the ratio of sound power incident on a partition to the sound power transmitted through a partition. This ratio varies with acoustic frequency as well as duct shape, size, and wall thickness. Higher values of transmission loss result in less noise passing through the duct wall.

Breakout sound transmission from ducts is the sound transmitted through a duct wall and then radiated from the exterior surface of the duct wall. Its sound power level is given by

$$L_{w(out)} = L_{w(in)} + 10\log\left(\frac{S}{A}\right) - \text{TL}_{out} \qquad (13)$$

where

$L_{w(out)}$ = sound power level of sound radiated from outside surface of duct walls, dB
$L_{w(in)}$ = sound power level of sound inside duct, dB
S = surface area of outside sound-radiating surface of duct, in²
A = cross-section area of inside of duct, in²

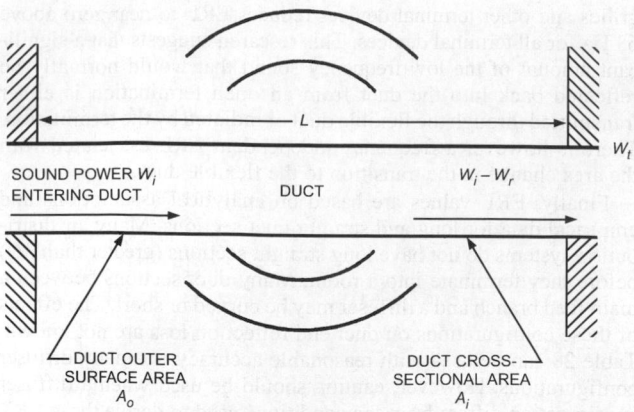

Fig. 28 Breakout Noise

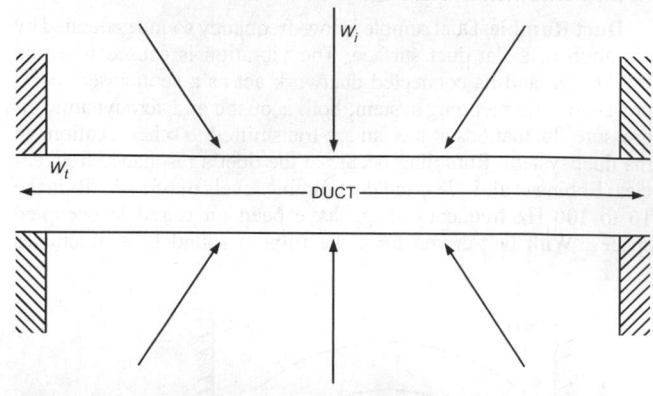

Fig. 29 Break-In Noise

TL_{out} = normalized duct breakout transmission loss (independent of S and A), dB

Equation (13) is a simplified expression that assumes that the sound power level inside the duct does not decrease with distance over the length of the duct. In fact, for very long ducts (when $S \gg A$), the radiated sound power level $L_{w(out)}$ could become greater than the sound power level inside the duct, which would violate the conservation of energy principle. A more accurate expression for breakout is presented in Equation (20).

Values of TL_{out} for rectangular ducts are given in Table 29, for round ducts in Table 30, and for flat oval ducts in Table 31 (Cummings 1983, 1985; Lilly 1987).

Equations for S and A for rectangular ducts are

$$S = 2 \times 12L(a + b) \qquad (14)$$
$$A = ab \qquad (15)$$

where

a = larger duct cross-section dimension, in.
b = smaller duct cross-section dimension, in.
L = length of duct sound-radiating surface, ft

Equations for S and A for round ducts are

$$S = 12L\pi d \qquad (16)$$
$$A = \pi\frac{d^2}{4} \qquad (17)$$

where

d = duct diameter, in.
L = length of duct sound-radiating surface, ft

For flat oval ducts,

Table 29 TL_out Versus Frequency for Rectangular Ducts

Duct Size, in.	Gage	TL_out, dB — Octave Midband Frequency, Hz							
		63	125	250	500	1000	2000	4000	8000
12 × 12	24	21	24	27	30	33	36	41	45
12 × 24	24	19	22	25	28	31	35	41	45
12 × 48	22	19	22	25	28	31	37	43	45
24 × 24	22	20	23	26	29	32	37	43	45
24 × 48	20	20	23	26	29	31	39	45	45
48 × 48	18	21	24	27	30	35	41	45	45
48 × 96	18	19	22	25	29	35	41	45	45

Note: Data are for duct lengths of 20 ft, but values may be used for cross section shown regardless of length.

Table 30 Experimentally Measured TL_out Versus Frequency for Round Ducts

Diameter, in.	Length, ft	Gage	Octave Midband Frequency, Hz						
			63	125	250	500	1000	2000	4000
Long Seam Ducts									
8	15	26	>45	(53)	55	52	44	35	34
14	15	24	>50	60	54	36	34	31	25
22	15	22	>47	53	37	33	33	27	25
32	15	22	(51)	46	26	26	24	22	38
Spiral Wound Ducts									
12	12	26*	52	51	53	51	50	46	36
24	24	24	51	53	51	44	36	26	29
	24	24*	51	51	54	44	39	33	47
	10	16	>48	53	36	32	32	28	41
36	24	20	51	51	52	46	36	32	55

*Ducts internally lined with 1 in. thick 1.5 pcf fiberglass with 24 ga perforated sheet metal inner liner.

$$S = 12L[2(a-b) + \pi b] \tag{18}$$

$$A = b(a-b) + \frac{\pi b^2}{4} \tag{19}$$

where

a = length of major axis, in.
b = length of minor axis, in.
L = length of duct sound-radiating surface, ft

Equation (13) assumes no decrease in the internal sound power level with distance along the length of the duct. Thus, it is valid only for relatively short lengths of unlined duct. For long ducts or ducts that have internal acoustic lining, one approach is to divide the duct into sections, each of which is short enough to be modeled as a section of duct with constant internal sound power level over the length of each section. The recommended maximum length of each section is the length that would result in a 1 dB reduction in the internal sound power level at the frequency of interest. Alternatively, the total sound power radiated from any duct of any length (including an internally lined duct) can be calculated in a single step with a modified version of Equation (13) (Lilly 1987):

$$L_{w(out)} = L_{w(in)} + 10 \log\left(\frac{S^*}{A}\right) - TL_{out} \tag{20}$$

where S^* is the effective surface area of the duct. $S^* = PL^*$, where P = duct perimeter, and L^* = effective length. The effective length L^* is calculated as

$$L^* = \frac{\gamma^L - 1}{\ln \gamma} \tag{21}$$

where

$$\gamma = 10^{(-\alpha/10)} \tag{22}$$

Table 31 TL_out Versus Frequency for Flat Oval Ducts

Duct Size, in.	Gage	TL_out, dB — Octave Midband Frequency, Hz						
		63	125	250	500	1000	2000	4000
12 × 6	24	31	34	37	40	43	—	—
24 × 6	24	24	27	30	33	36	—	—
24 × 12	24	28	31	34	37	—	—	—
48 × 12	22	23	26	29	32	—	—	—
48 × 24	22	27	30	33	—	—	—	—
96 × 24	20	22	25	28	—	—	—	—
96 × 48	18	28	31	—	—	—	—	—

Note: Data are for duct lengths of 20 ft, but values may be used for cross section shown regardless of length.

where α = duct attenuation rate, dB/ft (see Tables 16 to 21). For lined rectangular ducts, Tables 17 and 18 do not have data at 63 Hz. For rough approximations, use Table 16 values.

In most rooms where the listener is close to the duct, an estimate of the breakout sound pressure level can be obtained from

$$L_p = L_{w(out)} - 10 \log(\pi r L) + 10 \tag{23}$$

where

L_p = sound pressure level at a specified point in the space, dB
$L_{w(out)}$ = sound power level of sound radiated from outside surface of duct walls, given by Equation (13) or Equation (20), dB
r = distance between duct and position for which L_p is calculated, ft
L = length of the duct sound-radiating surface, ft

Note that Equation (23) gives sound pressure from a duct that is in a wide-open ceiling plenum space. If the duct is in a tight space between floor slab and ceiling, it may be up to 6 dB louder.

Example 4. A 24 in. by 24 in. by 25 ft long rectangular supply duct is constructed of 22 ga sheet metal. Given the following sound power levels in the duct, what are the breakout sound pressure levels 5 ft from the surface of the duct?

Solution: Using Equations (13) and (23),

	Octave Midband Frequency, Hz						
	63	125	250	500	1000	2000	4000
$L_{w(in)}$	90	85	80	75	70	65	60
$-TL_{out}$ (Table 29)	−20	−23	−26	−29	−32	−37	−43
$10 \log(S/A)$	17	17	17	17	17	17	17
$L_{w(out)}$	89	79	71	63	55	45	34
$-10 \log(\pi r L) + 10$	−16	−16	−16	−16	−16	−16	−16
L_p, dB	71	62	55	47	39	29	18

Using Equations (21) to (23),

	Octave Midband Frequency, Hz						
	63	125	250	500	1000	2000	4000
$L_{w(in)}$	90	85	80	75	70	65	60
$-TL_{out}$ (Table 29)	−20	−23	−26	−29	−32	−37	−43
α, dB/ft (Table 16)	0.25	0.2	0.1	0.03	0.03	0.03	0.03
γ	0.94	0.95	0.98	0.99	0.99	0.99	0.99
L^*, ft	13	15	19	23	23	23	23
$10 \log(S^*/A)$	14	15	16	17	17	17	17
$L_{w(out)}$	84	77	70	63	55	45	34
$-10 \log(\pi r L) + 10$	−16	−16	−16	−16	−16	−16	−16
L_p, dB	68	61	54	47	39	29	18

Example 5. Repeat Example 4, but with 2 in. thick internal duct liner.

Solution: Using Equations (13) and (23),

	Octave Midband Frequency, Hz						
	63	125	250	500	1000	2000	4000
$L_{w(in)}$	90	85	80	75	70	65	60
$-TL_{out}$ (Table 29)	−20	−23	−26	−29	−32	−37	−43
$10 \log(S/A)$	17	17	17	17	17	17	17
$L_{w(out)}$	87	79	71	63	55	45	34
$-10 \log(\pi r L) + 10$	−16	−16	−16	−16	−16	−16	−16
L_p, dB	71	63	55	47	39	29	18

Using Equations (21) to (23),

	Octave Midband Frequency, Hz						
	63	125	250	500	1000	2000	4000
$L_{w(in)}$	90	85	80	75	70	65	60
$-TL_{out}$ (Table 29)	−20	−23	−26	−29	−32	−37	−43
α, dB/ft (Table 18)	0.25	0.3	0.9	2.5	3.5	2.2	1.8
γ	0.94	0.93	0.81	0.56	0.45	0.60	0.66
L^*, ft	13	12	5	2	1	2	2
$10 \log(S^*/A)$	14	14	10	5	4	6	7
$L_{w(out)}$	84	76	64	51	42	34	24
$-10 \log(\pi r L) + 10$	−16	−16	−16	−16	−16	−16	−16
L_p, dB	68	60	48	35	26	18	8

Example 6. Repeat Example 5 using 24 in. diameter spiral round duct, 24 ga, 25 ft long with 1 in. thick acoustical duct lining.

Solution: Using Equations (13) and (23),

	Octave Midband Frequency, Hz						
	63	125	250	500	1000	2000	4000
$L_{w(in)}$	90	85	80	75	70	65	60
$-TL_{out}$ (Table 30)	−51	−51	−54	−44	−39	−33	−47
$10 \log(S/A)$	17	17	17	17	17	17	17
$L_{w(out)}$	56	51	43	48	48	49	30
$-10 \log(\pi r L) + 10$	−16	−16	−16	−16	−16	−16	−16
L_p, dB	40	35	27	32	32	33	14

Using Equations (δ1) to (23) yields

	Octave Midband Frequency, Hz						
	63	125	250	500	1000	2000	4000
$L_{w(in)}$	90	85	80	75	70	65	60
$-TL_{out}$ (Table 30)	−51	−51	−54	−44	−39	−33	−47
α, dB/ft (Table 20)	0.7	0.5	0.57	1.28	1.71	1.24	0.85
γ	0.98	0.94	0.88	0.74	0.67	0.75	0.82
L^*, ft	21	13	7.3	3.4	2.5	3.5	5.1
$10 \log(S^*/A)$	16	14	12	8.3	7.1	8.5	10
$L_{w(out)}$	55	48	38	39	38	40	23
$-10 \log(\pi r L) + 10$	−16	−16	−16	−16	−16	−16	−16
L_p, dB	39	32	22	24	22	25	7

Using round duct eliminates the low-frequency rumble present with rectangular ducts but introduces some mid- and high-frequency noise that can be reduced by adding duct liner as shown.

When sound is not transmitted through the wall of a round duct, it propagates down the duct and may become a problem at another point in the duct system. Round flexible and rigid fiberglass ducts do not have high transmission loss properties because they lack the mass or stiffness associated with round sheet metal ducts.

Table 32 TL_{in} Versus Frequency for Rectangular Ducts

Duct Size, in.	Gage	TL_{in}, dB Octave Midband Frequency, Hz							
		63	125	250	500	1000	2000	4000	8000
12 × 12	24	16	16	16	25	30	33	38	42
12 × 24	24	15	15	17	25	28	32	38	42
12 × 48	22	14	14	22	25	28	34	40	42
24 × 24	22	13	13	21	26	29	34	40	42
24 × 48	20	12	15	23	26	28	36	42	42
48 × 48	18	10	19	24	27	32	38	42	42
48 × 96	18	11	19	22	26	32	38	42	42

Note: Data are for duct lengths of 20 ft.

Table 33 Experimentally Measured TL_{in} Versus Frequency for Circular Ducts

Diameter, in.	Length, ft	Gage	TL_{in}, dB Octave Midband Frequency, Hz						
			63	125	250	500	1000	2000	4000
Long Seam Ducts									
8	15	26	>17	(31)	39	42	41	32	31
14	15	24	>27	43	43	31	31	28	22
22	15	22	>28	40	30	30	30	24	22
32	15	22	(35)	36	23	23	21	19	35
Spiral Wound Ducts									
8	10	26	>20	>42	>59	>62	53	43	26
14	10	26	>20	>36	44	28	31	32	22
26	10	24	>27	38	20	23	22	19	33
26	10	16	>30	>41	30	29	29	25	38
32	10	22	>27	32	25	22	23	21	37

Note: In cases where background sound swamped sound radiated from duct walls, a lower limit on TL_{in} is indicated by >. Parentheses indicate measurements in which background sound produced greater uncertainty than usual.

Table 34 TL_{in} Versus Frequency for Flat Oval Ducts

Duct Size, in.	Gage	TL_{in}, dB Octave Midband Frequency, Hz						
		63	125	250	500	1000	2000	4000
12 × 6	24	18	18	22	31	40	—	—
24 × 6	24	17	17	18	30	33	—	—
24 × 12	24	15	16	25	34	—	—	—
48 × 12	22	14	14	26	29	—	—	—
48 × 24	22	12	21	30	—	—	—	—
96 × 24	20	11	22	25	—	—	—	—
96 × 48	18	19	28	—	—	—	—	—

Note: Data are for duct lengths of 20 ft.

Whenever duct sound breakout is a concern, fiberglass or flexible round duct should not be used; these ducts have little or no transmission loss, and are essentially transparent to sound.

Break-in sound transmission into ducts is sound transmitted into a duct through the duct walls from the space outside the duct. Its sound power level is given by

$$L_{w(in)} = L_{w(out)} - TL_{in} - 3 \qquad (24)$$

where

$L_{w(in)}$ = sound power level of sound transmitted into duct and then transmitted upstream or downstream of point of entry, dB

$L_{w(out)}$ = sound power level of sound incident on outside of duct walls, dB

TL_{in} = duct break-in transmission loss, dB

Values for TL_{in} for rectangular ducts are given in Table 32, for round ducts in Table 33, and for flat oval ducts in Table 34 (Cummings 1983, 1985).

RECEIVER ROOM SOUND CORRECTION

The sound pressure level at a given location in a room caused by a particular sound source is a function of the sound power level and sound radiation characteristics of the sound source, acoustic properties of the room (surface treatments, furnishings, etc.), room volume, and distance between the sound source and point of observation. Two types of sound sources are typically encountered in HVAC system applications: **point** and **line**. Typical point sources are grilles, registers and diffusers; air-valve and fan-powered air terminal units and fan-coil units located in ceiling plenums; and return air openings. Line sources are usually associated with sound breakout from air ducts and long slot diffusers.

For a point source in an enclosed space, classical diffuse-field theory predicts that as the distance between the source and point of observation is increased, the sound pressure level initially decreases at the rate of 6 dB per doubling of distance. At some point, the reverberant sound field begins to dominate and the sound pressure level remains at a constant level.

For point sound sources in **reflective unfurnished rooms**, the classic diffuse equation for converting sound power to pressure could be used:

$$L_p = L_W + 10\log(Q/4\pi r^2 + 4/R) + 10.3 \qquad (25)$$

where

L_p = sound pressure level, dB (re 20 µPa)
L_W = sound power level, dB (re 10^{-12} W)
Q = directivity of sound source, dimensionless; see Figure 30
r = distance from source, ft
R = room constant = $[S\alpha/(1-\alpha)]$ = sum of all surface areas and their corresponding absorption coefficients, ft^2

A further discussion of assumptions used in converting power to pressure is available in Chapter 8 of the 2009 *ASHRAE Handbook—Fundamentals*.

However, investigators have found that diffuse-field theory does not apply in rooms with furniture or other sound-scattering objects (Schultz 1985; Thompson 1981). Instead, sound pressure levels decrease at the rate of around 3 dB per doubling of distance between sound source and point of observation. Generally, a true reverberant sound field does not exist in small rooms (room volumes less than 15,000 ft^3). In larger rooms reverberant fields usually exist, but typically at distances from the sound sources that are significantly greater than those predicted by diffuse-field theory.

Most **normally furnished rooms** of regular proportions have acoustic characteristics that range from *average* to *medium dead*. These usually include carpeted rooms with sound-absorptive ceilings. If such a room has a volume less than 15,000 ft^3 and the sound source is a single point source, sound pressure levels associated with the sound source can be obtained from (Schultz 1985).

$$L_p = L_w + A - B \qquad (26)$$

where

L_p = sound pressure level at specified distance from sound source, dB (re 20 µPa)
L_w = sound power level of sound source, dB (re 10^{-12} W)

Values for A and B are given in Tables 35 and 36.

For rooms larger than 15,000 ft^3, the following equation may be used:

$$L_p = L_w - 10\log r - 5\log V - 3\log f + 25 \qquad (27)$$

In another alternative calculation for a normally furnished room with volume greater than 15,000 ft^3 and a single point sound source, the sound pressure levels associated with the sound source can be obtained from

Table 35   Values for A in Equation (26)

Room Volume, ft^3	Value for A, dB Octave Midband Frequency, Hz						
	63	125	250	500	1000	2000	4000
1500	4	3	2	1	0	−1	−2
2500	3	2	1	0	−1	−2	−3
4000	2	1	0	−1	−2	−3	−4
6000	1	0	−1	−2	−3	−4	−5
10,000	0	−1	−2	−3	−4	−5	−6
15,000	−1	−2	−3	−4	−5	−6	−7

Table 36   Values for B in Equation (26)

Distance from Sound Source, ft	Value for B, dB
3	5
4	6
5	7
6	8
8	9
10	10
13	11
16	12
20	13

$$L_p = L_w - C - 5 \qquad (28)$$

Values for C are given in Table 37. Equation (28) can be used for room volumes of up to 150,000 ft^3, with accuracy typically within 2 to 5 dB.

Distributed Array of Ceiling Sound Sources

In many office buildings, air supply outlets are located flush with the ceiling of the conditioned space and constitute an array of distributed ceiling sound sources. The geometric pattern depends on the floor area served by each outlet, ceiling height, and thermal load distribution. In interior zones of a building where thermal load requirements are essentially uniform, air delivery per outlet is usually the same throughout the space; thus, these outlets emit nominally equal sound power levels. One way to calculate sound pressure levels in a room with a distributed array is to use Equation (26) or (28) to calculate the sound pressure levels for each individual air outlet at specified locations in the room and then logarithmically add the sound pressure levels for each diffuser at each observation point. This procedure can be very tedious for a room with a large number of ceiling air outlets.

For a distributed array of ceiling sound sources (air outlets) of nominally equal sound power, room sound pressure levels tend to be uniform in a plane parallel to the ceiling. Although sound pressure levels decrease with distance from the ceiling along a vertical axis, they are nominally constant along any selected horizontal plane. Equation (29) simplifies calculation for a distributed ceiling array. For this case, use a reference plane 5 ft above the floor.

Thus, $L_{p(5)}$ is obtained from

$$L_{p(5)} = L_{W(s)} - D \qquad (29)$$

where

$L_{p(5)}$ = sound pressure level at distance of 5 ft above floor, dB (re 20 µPa)
$L_{W(s)}$ = sound power level of single diffuser in array, dB (re 10^{-12} W)

Values of D are given in Table 38.

Nonstandard Rooms

The previous equations assume that the acoustical characteristics of a room range from average to medium dead, which is generally

Table 37 Values for C in Equation (28)

Distance from Sound Source, ft	Value for C, dB Octave Midband Frequency, Hz						
	63	125	250	500	1000	2000	4000
3	5	5	6	6	6	7	10
4	6	7	7	7	8	9	12
5	7	8	8	8	9	11	14
6	8	9	9	9	10	12	16
8	9	10	10	11	12	14	18
10	10	11	12	12	13	16	20
13	11	12	13	13	15	18	22
16	12	13	14	15	16	19	24
20	13	15	15	16	17	20	26
25	14	16	16	17	19	22	28
32	15	17	17	18	20	23	30

Table 38 Values for D in Equation (29)

Floor Area per Diffuser, ft²	Value for D, dB Octave Midband Frequency, Hz						
	63	125	250	500	1000	2000	4000
Ceiling height 8 to 9 ft							
100 to 150	2	3	4	5	6	7	8
200 to 250	3	4	5	6	7	8	9
Ceiling height 10 to 12 ft							
150 to 200	4	5	6	7	8	9	10
250 to 300	5	6	7	8	9	10	11
Ceiling height 14 to 16 ft							
250 to 300	7	8	9	10	11	12	13
350 to 400	8	9	10	11	12	13	14

true of most rooms. However, some rooms may be acoustically *medium live* to *live* (i.e., they have little sound absorption). These rooms may be sports or athletic areas, concert halls, or other rooms designed to be live, or they may be rooms that are improperly designed from an acoustic standpoint. The previous equations should not be used for acoustically live rooms because they can overestimate the decrease in sound pressure levels associated with room sound correction by as much as 10 to 15 dB. When these or other types of nonstandard rooms are encountered, it is best to use the services of an acoustical engineer.

Line Sound Sources

Sound from breakout from air ducts or long slot diffusers may be modeled as line sources. To convert sound power levels to the corresponding sound pressure levels in a room for such cases, the following equation may be used:

$$L_p = L_W + 10\log(1/\pi rL + 4/R) + 10.3 \qquad (30)$$

where

L_p = sound pressure level, dB (re 20 μPa)
L_W = sound power level, dB (re 10^{-12} W)
r = distance from source, ft
L = length of line source, ft
R = room constant = $[S\alpha/(1 - \alpha)]$ = sum of all surface areas and their corresponding absorption coefficients, ft²

This is the classic diffuse room equation for a line source, and may not produce accurate results for standard nondiffuse rooms. Unfortunately, no information is available at this time on how to correct more accurately for the effect of the receiver room on line sources.

Room Noise Measurement

Measuring HVAC system noise in a room is complicated by several factors, including the spatial and temporal variability of the noise, variable HVAC system operating conditions, modal characteristics of the room, and intrusion of noise from exterior sources. How the noise measurements should be taken depends to some extent on the purpose of the measurement. Is the purpose of the measurement to verify that the noise level in the room meets a specific criteria, or is it meant to troubleshoot an alleged problem? The specific measurement requirements vary depending on the intent.

For commissioning purposes, there are two levels of assessment: (1) a survey method may be used to make a quick assessment of a space and (2) an engineering method for a more detailed and accurate assessment. The survey approach is typically used to assess whether there may be a noise problem in the room. The survey method requires a Type 1 integrating sound level meter equipped with octave band filters if octave band levels are specified by the applicable noise criterion. The measurements can be taken at a single point or at several points, but all measurement points must be at a likely location for the listener's ears. No measurement locations may be less than 3 ft from a room boundary or less than 18 in. from any object in the room. The measurement microphone must be fixed (or slowly moving) for each measurement, and the minimum duration of each measurement is 15 s. It is recognized that HVAC noise is a time-varying signal, so the energy average sound pressure level L_{eq} must be compared against the noise criterion, not the maximum sound level recorded during the measurement.

If the survey method detects a potential noise problem or if a complaint has been registered by an occupant of the space, the engineering method may be implemented if compliance with a noise level specification is required. This method uses the same instrumentation but requires a minimum of 4 separate measurement locations, uniformly distributed throughout the room. For larger rooms (greater than 215 ft²) additional measurement points must be added, proportional to the floor area of the room. Unless specified otherwise, the energy average L_{eq} of all measurement locations in the room is compared against the noise criterion.

If the purpose of the noise measurement is troubleshooting a known problem, more sophisticated instrumentation (e.g., narrow band analyzers, vibration sensors, intensity probes, etc.) may be required. Troubleshooting work should be provided by a competent acoustical consultant with specific experience in this field of work. Contact the Institute of Noise Control Engineers (www.inceusa.org) or the National Council of Acoustical Consultants (www.ncac.com) for a list of experts.

In any case, it is important for the operating conditions of the HVAC system to be known at the time of the measurements. If the system contains compressors that cycle on and off during normal operation, the measurements must be taken while the compressors are running. For variable-volume systems, the measurements should be taken at design (maximum) volume. If the condition rarely operates under design flow conditions, measurements must also be taken at a more typical operating condition.

It is also important to make sure that noise from extraneous (non-HVAC) sources does not contaminate the measurements. Room noise measurements may be corrected for these sounds by taking one set of measurements with the HVAC system operating under test conditions and additional measurements with the HVAC system shut down entirely. This correction can only be applied if the ambient noise is shown to be relatively constant with time. If the energy average of two independent ambient noise level measurements (one obtained before and the other obtained after the HVAC system noise measurement) is more than 6 dB below the HVAC noise level in any octave band, then the ambient adjusted HVAC noise level in that octave band may be computed using the following equation:

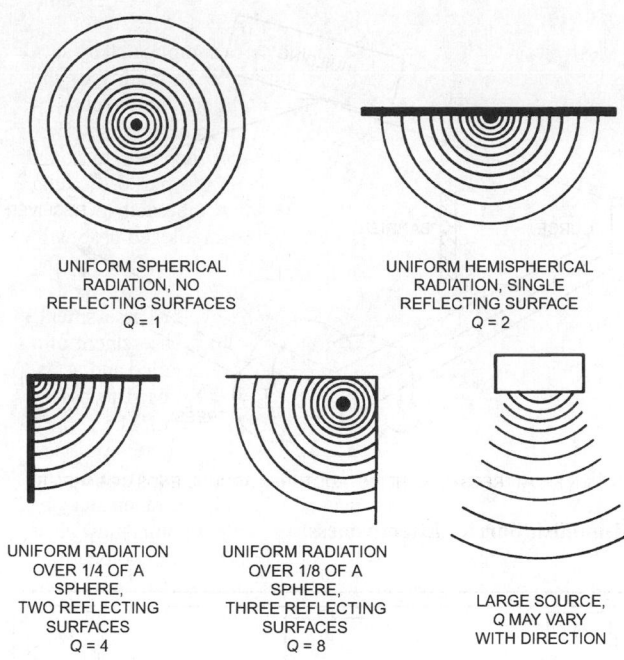

Fig. 30   Directivity Factors for Various Radiation Patterns

$$L_{p\ ambient\ adjusted} = 10 \log \left[10^{(L_{p\ HVAC}/10)} - 10^{(L_{p\ ambient}/10)} \right]$$

where

$L_{p\ HVAC}$ = sound pressure level with HVAC system operating
$L_{p\ ambient}$ = energy average ambient sound pressure level with HVAC system off

The ambient noise correction cannot be allowed if the difference between the two ambient noise levels in any frequency band is more than 3 dB. If this occurs, the ambient noise is not constant with time and the entire set of measurements should be repeated. It should be emphasized that that the ambient noise correction is not required.

For more information, see the section on Troubleshooting.

SOUND CONTROL FOR OUTDOOR EQUIPMENT

Outdoor mechanical equipment should be carefully selected, installed, and maintained to minimize sound radiated by the equipment, and to comply with local noise codes. Equipment with strong tonal components is more likely to provoke complaints than equipment with a broadband noise spectrum.

Sound Propagation Outdoors

If the equipment sound power level spectrum and ambient sound pressure level spectrum are known, the contribution of the equipment to the sound level at any location can be estimated by analyzing the sound transmission paths involved. When there are no intervening barriers and no attenuation because of berms, ground absorption, or atmospheric effects, the principal factor in sound pressure level reduction is distance. The following equation may be used to estimate the sound pressure level of equipment at a distance from it and at any frequency when the sound power level is known:

$$L_p = L_w + 10 \log Q - 20 \log d - 0.7 \qquad (31)$$

where

d = distance from acoustic center of source to distant point, ft
L_p = sound pressure level at distance d from sound source, dB
L_w = sound power level of sound source, dB

Table 39   Insertion Loss Values of Ideal Solid Barrier

Path-Length Difference, ft	Insertion Loss, dB Octave Midband Frequency, Hz							
	31	63	125	250	500	1000	2000	4000
0.01	5	5	5	5	5	6	7	8
0.02	5	5	5	5	5	6	8	9
0.05	5	5	5	5	6	7	9	10
0.1	5	5	5	6	7	9	11	13
0.2	5	5	6	8	9	11	13	16
0.5	6	7	9	10	12	15	18	20
1	7	8	10	12	14	17	20	22
2	8	10	12	14	17	20	22	23
5	10	12	14	17	20	22	23	24
10	12	15	17	20	22	23	24	24
20	15	18	20	22	23	24	24	24
50	18	20	23	24	24	24	24	24

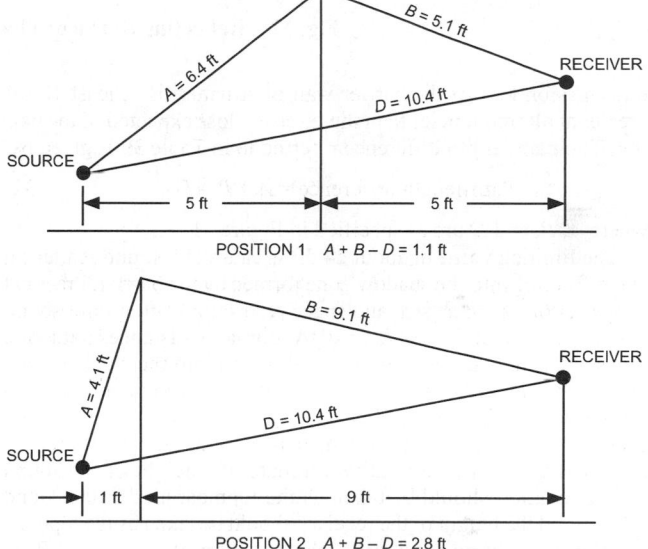

Fig. 31   Noise Barrier

Q = directivity factor associated with way sound radiates from sound source (see Figure 30)

Equation (31) does not apply where d is less than twice the maximum dimension of the sound source. L_p may be low by up to 5 dB where d is between two and five times the maximum sound source dimension. Also, if the distance is greater than about 500 ft, wind, thermal gradients, and atmospheric sound absorption need to be considered.

For complex cases, refer to texts on acoustics (e.g., Beranek 1971) and international standards such as ISO *Standard* 9613-2.

Sound Barriers

A sound barrier is a solid structure that intercepts the direct sound path from a sound source to a receiver. It reduces the sound pressure level within its shadow zone. Figure 31 illustrates the geometrical aspects of an outdoor barrier where no extraneous surfaces reflect sound into the protected area. Here the barrier is treated as an intentionally constructed noise control structure. If a sound barrier is placed between a sound source and receiver location, the sound pressure level L_p in Equation (26) is reduced by the **insertion loss (IL)** associated with the barrier.

Table 39 gives the insertion loss of an outdoor ideal solid barrier when no surfaces reflect sound into the shadow zone, and the sound

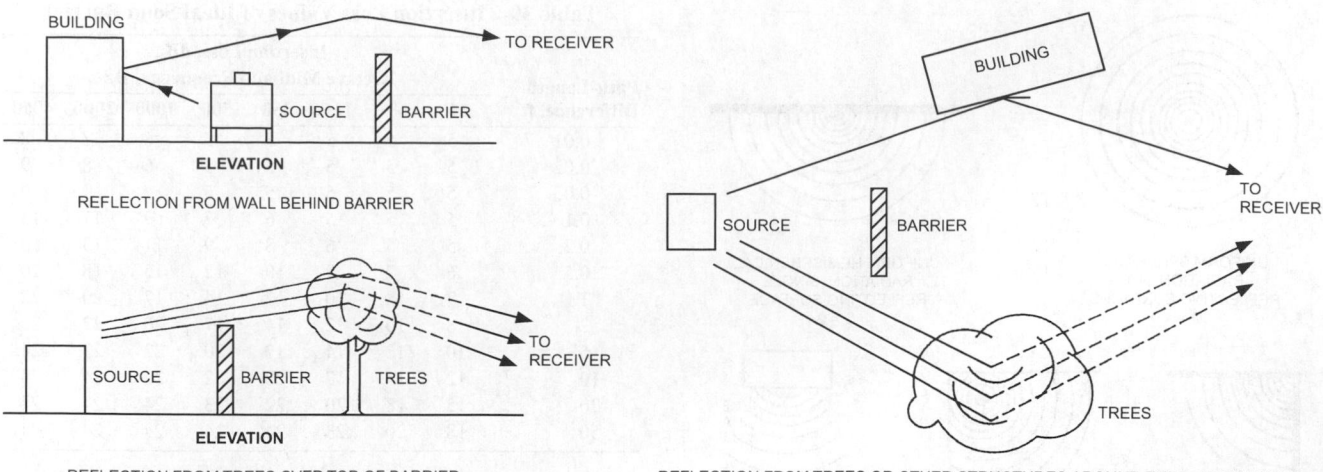

Fig. 32   Reflecting Surfaces That Can Diminish Barrier Effectiveness

transmission loss of the barrier wall or structure is at least 10 dB greater at all frequencies than the insertion loss expected of the barrier. The path-length difference referred to in Table 39 is given by

$$\text{Path-length difference} = A + B - D \qquad (32)$$

where A, B, and D are as specified in Figure 31.

The limiting value of about 24 dB is caused by sound scattering and refracting into the shadow zone formed by the barrier. Practical constructions such as size and space restrictions often limit sound barrier performance to 10 to 15 dBA. For large distances outdoors, this scattering and bending of sound waves into the shadow zone reduces barrier effectiveness. At large distances, atmospheric conditions can significantly affect sound path losses by amounts even greater than those provided by the barrier, with typical differences of 10 dBA. For a conservative estimate, the height of the sound source location should be taken as the topmost part of the sound source, and the height of the receiver should be taken as the topmost location of a sound receiver, such as the top of the second-floor windows in a two-floor house or at a height of 5 ft for a standing person.

Reflecting Surfaces. No other surfaces should be located where they can reflect sound around the ends or over the top of the barrier into the barrier shadow zone. Figure 32 shows examples of reflecting surfaces that can reduce the effectiveness of a barrier wall.

Width of Barrier. Each end of the barrier should extend horizontally beyond the line of sight from the outer edge of the source to the outer edge of the receiver position by a distance of at least three times the path-length difference. Near the ends of the barrier, the effectiveness of the noise isolation is reduced because some sound is diffracted over the top and around the ends. Also, some sound is reflected or scattered from various nonflat surfaces along the ground near the ends of the barrier. In critical situations, the barrier should completely enclose the sound source to eliminate or reduce the effects of reflecting surfaces.

Reflection from a Barrier. A large, flat reflecting surface, such as a barrier wall, may reflect more sound toward the source than there would have been with no wall present. If the wall produces no special focusing effect, reflections from the wall will produce levels on the side of the barrier facing the source that are 2 to 3 dB higher. Using acoustical absorption on the barrier surface (source side) reduces this increase.

FUME HOOD DUCT DESIGN

Fume hood exhaust systems are often the major sound source in a laboratory. The exhaust system may consist of individual exhaust

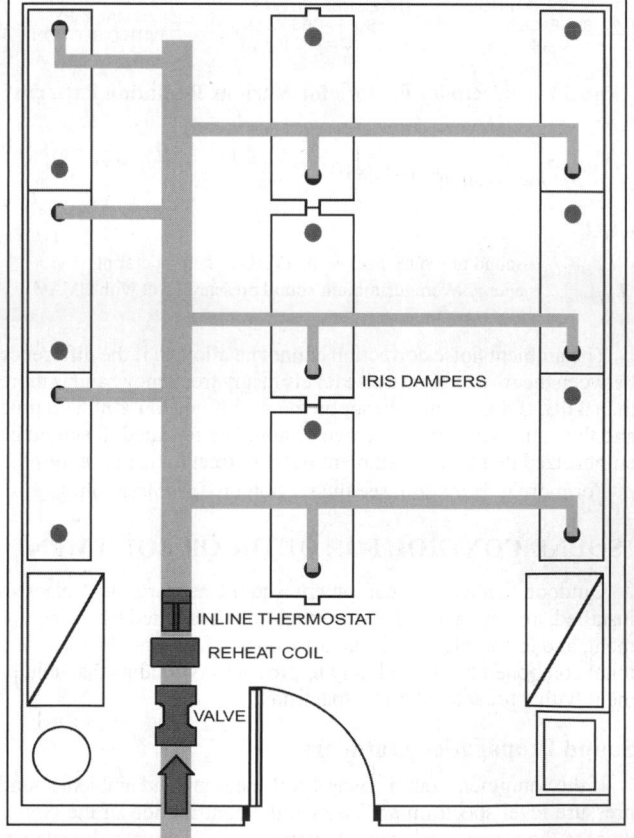

Fig. 33   Typical Manifold Lab Exhaust Layout

fans ducted to separate fume hoods, or a central exhaust fan connected through a collection duct system (commonly known as a manifold) to a large number of hoods, as shown in Figure 33. In either case, a redundancy system consisting of two fans might be used. In addition to fan noise, other sound sources are the air terminal unit serving the hood and aerodynamically generated noise from airflow in the ducts and control valves. Sound pressure levels produced in the laboratory space should be estimated using procedures described in this section and manufacturer-supplied noise emission

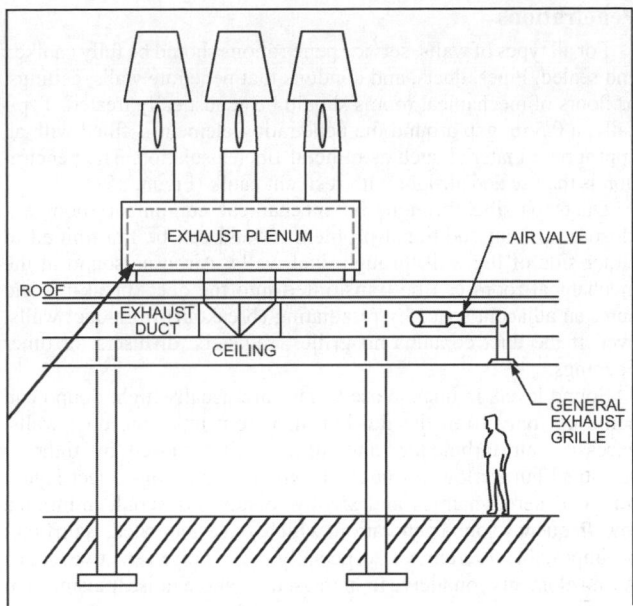

Fig. 34   Inlet Plenum for Multiple Exhaust Fans

data. Recommended noise level design criteria for laboratory spaces using fume hoods are given in Table 1.

To minimize static pressure loss and fan power consumption in a duct system, fume hood ducts should be sized to allow rated airflow at no greater than 2000 fpm or at a velocity consistent with regulatory requirements. Duct velocities over 2000 fpm should be avoided for acoustical reasons and to conserve energy. Above this speed, the design criteria given in Table 1 are unlikely to be met, even with a silenced fan(s).

Noise control measures for fume hood systems include the following:

Fan(s)
- Where conditions allow, use backward-inclined, airfoil, or forward-curved centrifugal fans instead of radial-blade fans, and use caution if applying axial-blade fans.
- Select fan(s) to operate at a low tip speed and maximum efficiency.
- Try to run redundant fans at reduced capacity instead of operating one fan at full capacity.

Manifold
- Design the manifold upstream of the exhaust fan(s) to double as an acoustic plenum as shown in Figure 34 with sound-absorbent sidewall panels, which can be constructed with nonporous lining or packless cavities with perforated inner wall. Fan low-frequency noise can be reduced when the manifold is parallel-piped, with large surface area compared to the cross-sectional area of the fume hood ducts connecting to it.

Duct silencers
- Use prefabricated duct silencers or sections of lined ducts where conditions allow. In addition to galvanized steel, silencers can be fabricated with stainless steel, aluminum, or plastic.
- Silencers should be packless design, or have nonporous fill.

Exhaust silencers
- Where outdoor noise is an issue, round silencers may be required between the fan and the discharge cone. These should be packless or nonporous-fill design.

Duct design
- Design duct elements such as elbows and junctions with low friction to minimize duct pressure loss and aerodynamically

generated noise. Use round ducts, because rectangular ducts can have a noise breakout issue.

Laboratory flow control valves and air terminal units
- Allow 2 to 3 ft minimum of straight duct upstream and downstream from the terminal unit to reduce aerodynamically generated noise at the unit.
- Additional straight length may be required on the room side of valves to accommodate high-performance silencers.
- Noise generation in flow control valves increases exponentially with pressure loss, so system supply or exhaust pressure should be set at the level necessary to achieve design flow rate with minimal safety factor.

Fume hood location
- Where possible, locate fume hoods in private alcoves to reduce lab occupant's noise exposure.

All potential noise control measures should be carefully evaluated for compliance with applicable codes, safety requirements, and corrosion resistance requirements of the specific fume hood system. In addition, vibration isolation for fume hood exhaust fans is generally required. For some laboratory facilities, particularly those with highly vibration-sensitive instruments such as electron microscopes, vibration control can be critical, and a vibration specialist should be consulted.

MECHANICAL EQUIPMENT ROOM SOUND ISOLATION

Location

Locating HVAC equipment in a common room allows the designer to control noise affecting nearby spaces. Often, these spaces have background noise level criteria that dictate the type of construction and treatment necessary to achieve sufficient reduction in equipment noise transmitted to other spaces.

The most effective noise control measure for indoor mechanical equipment rooms is to locate them as far away as possible from noise-sensitive areas. In some cases, this requires a separate structure, such as a central chiller plant, to house equipment. Subterranean basement locations are typically best for noisy equipment because the basement usually affects the fewest adjacent locations. Penthouse equipment rooms are common but can create significant challenges for noise and vibration isolation. Rooms containing air-handling units should provide sufficient room for the equipment and associated ductwork to allow smooth transitions and full-radius curved elbows. A building corner location can work well by reducing the number of adjacent interior spaces and the amount of associated outside-air ductwork. Using adjacent spaces such as corridors, closets, and storage rooms as buffer zones can provide effective noise control. A common mistake in locating mechanical equipment rooms is to position the room in the core of the building between a stairway, an elevator shaft, and a telecommunications closet, leaving only one wall where supply and return air ductwork can enter and leave the room. This leads to high-velocity air in the ductwork and high static pressures for fans to overcome, leading to higher noise levels.

Once the mechanical equipment room location has been established, the amount of noise created in the room should be assessed and appropriate constructions selected for walls, ceilings, and floors. Concrete masonry units of various available thickness and densities are often used for their durability and effectiveness in reducing low-frequency noise levels. Typically, heavier and thicker materials contain more sound within the space. Special masonry units that provide a limited amount of acoustical absorption using slotted openings and resonator cavities can also be used. The sound isolation of a masonry wall can be significantly improved by using furred-out gypsum wallboard and insulation in cavities. Chillers and other equipment with very high noise levels are best situated in rooms with concrete masonry unit walls.

Wall Design

Often, because of structural issues and weight limitations, mechanical equipment room walls are built from gypsum wallboard on metal or wood studs. To adequately attenuate low-frequency noise, sufficient mass and thickness must be provided in the wall partition construction. This typically entails using multiple layers of gypsum wallboard on both sides of the wall with batt insulation in the cavities. Where greater levels of noise reduction are required, walls are built on double, staggered-stud construction using two separate rows of studs on separate tracks with multiple layers of gypsum wallboard and batt insulation in the cavities (see Table 40).

Doors

Doors into mechanical equipment rooms are frequently the weak link in the enclosure. Where noise control is important, the doors should be as heavy as possible, gasketed around the perimeter, have no grilles or other openings, and be self-closing. If they lead to sensitive spaces, two doors separated by a 3 to 10 ft corridor may be necessary.

Penetrations

For all types of walls, service penetrations should be fully caulked and sealed. Pipes, ducts, and conduits that penetrate walls, ceilings, or floors of mechanical rooms should be acoustically treated. Typically, a 0.5 in. gap around the penetrating element is filled with an appropriate material such as mineral fiber insulation. The penetration is then sealed airtight with resilient caulk (Figure 35).

Ducts passing through the mechanical equipment room enclosure pose an additional problem. Sound can be transmitted to either side of the wall through duct walls. Airborne sound in the mechanical room can be transmitted into the duct (break-in) and enter an adjacent space by reradiating (breakout) from duct walls, even if the duct contains no grilles, registers, diffusers, or other openings.

Sound levels in ducts close to fans are usually high. Sound can come not only from the fan but also from pulsating duct walls, excessive air turbulence, and air buffeting caused by tight or restricted fan airflow entrance or exit configurations. Duct layout for good aerodynamics and airflow conditions should minimize low-frequency sound generation, which, once generated, is difficult or impossible to remove, especially near noise-sensitive areas. Avoid elements conducive to increased breakout noise transmission

Table 40   Sound Transmission Class (STC) and Transmission Loss Values of Typical Mechanical Equipment Room Wall, Floor, and Ceiling Types, dB

Room Construction Type	STC	Octave Midband Frequency, Hz						
		63	125	250	500	1000	2000	4000
8 in. CMU*	50	35	35	41	44	50	57	64
8 in. CMU with 5/8 in. GWB* on furring strips	53	33	32	44	50	56	59	65
5/8 in. GWB on both sides of 3 5/8 in. metal studs	38	18	16	33	47	55	43	47
5/8 in. GWB on both sides of 3 5/8 in. metal studs with fiberglass insulation in cavity	49	16	23	44	58	64	52	53
2 layers of 5/8 in. GWB on both sides of 3 5/8 in. metal studs with fiberglass insulation in cavity	56	19	32	50	62	67	58	63
Double row of 3 5/8 in. metal studs, 1 in. apart, each with 2 layers of 5/8 in. GWB and fiberglass insulation in cavity	64	23	40	54	62	71	69	74
6 in. solid concrete floor/ceiling	53	40	40	40	49	58	67	76
6 in. solid concrete floor with 4 in. isolated concrete slab and fiberglass insulation in cavity	72	44	52	58	73	87	97	100
6 in. solid concrete floor with two layers of 5/8 in. GWB hung on spring isolators with fiberglass insulation in cavity	84	53	63	70	84	93	104	105

Note: Actual material composition (e.g., density, porosity, stiffness) affects transmission loss and STC values.
*CMU = concrete masonry unit; GWB = gypsum wallboard.

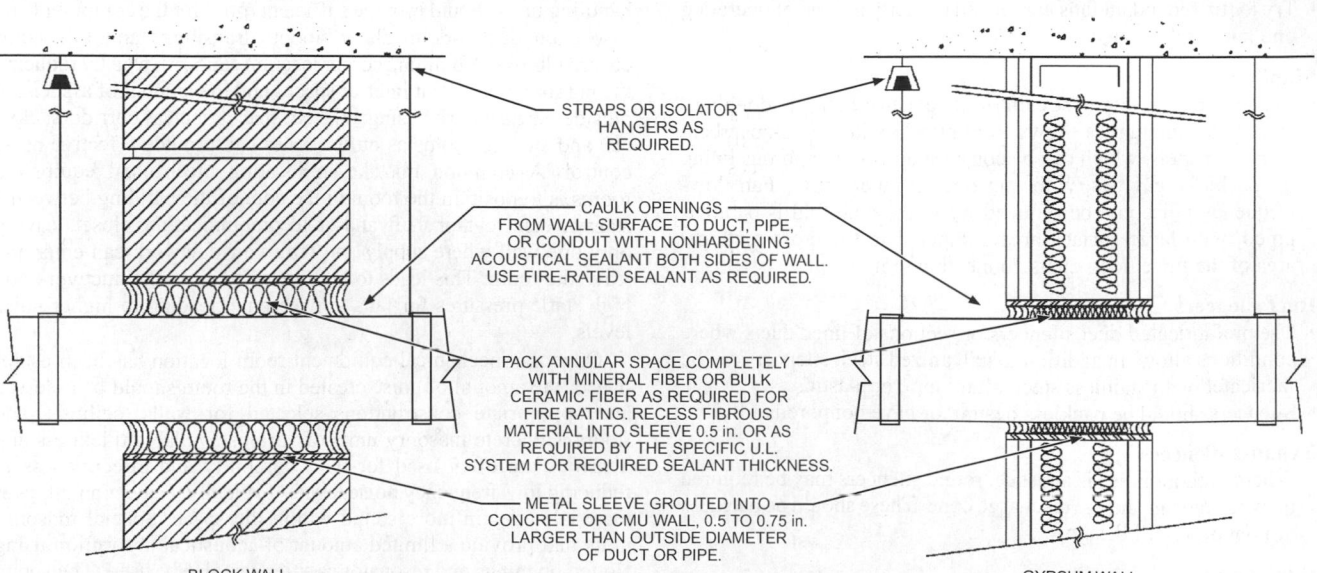

STRAPS OR ISOLATOR HANGERS AS REQUIRED.

CAULK OPENINGS FROM WALL SURFACE TO DUCT, PIPE, OR CONDUIT WITH NONHARDENING ACOUSTICAL SEALANT BOTH SIDES OF WALL. USE FIRE-RATED SEALANT AS REQUIRED.

PACK ANNULAR SPACE COMPLETELY WITH MINERAL FIBER OR BULK CERAMIC FIBER AS REQUIRED FOR FIRE RATING. RECESS FIBROUS MATERIAL INTO SLEEVE 0.5 in. OR AS REQUIRED BY THE SPECIFIC U.L. SYSTEM FOR REQUIRED SEALANT THICKNESS.

METAL SLEEVE GROUTED INTO CONCRETE OR CMU WALL, 0.5 TO 0.75 in. LARGER THAN OUTSIDE DIAMETER OF DUCT OR PIPE.

BLOCK WALL   GYPSUM WALL

Fig. 35   Duct, Conduit, and Pipe Penetration Details

and/or with a tendency to vibrate at low frequencies because of non-laminar airflow. Round ductwork is most resistant to these problems, followed by square and rectangular ducts with aspect ratios less than 2:1. Heavier-than-normal gage metal ductwork, such as 16 ga within the mechanical room and over noise-sensitive spaces, can also be used.

Mechanical Chases

Mechanical chases and shafts should be acoustically treated the same way as mechanical equipment rooms, especially if they contain noise-producing ductwork, pipes, and equipment such as fans and pumps. The shaft should be closed at the mechanical equipment room, and shaft wall construction must provide sufficient reduction of mechanical noise from the shaft to noise-sensitive areas to obtain acceptable noise levels. Chases should not be allowed to become "speaking tubes" between spaces requiring different acoustical environments. Crosstalk through the shaft must be prevented. Pipes, ducts, conduits, or equipment should be vibration-isolated so that mechanical vibration and structureborne noise is not transmitted to the shaft walls and into the building structure.

When mechanical equipment rooms are used as supply or return plenums, all openings into the equipment room plenum space may require noise control treatment, especially if any sound-critical space is immediately adjacent. This is particularly true if the space above an acoustical tile ceiling just outside the equipment room is used as a return air plenum. Most acoustical tile ceilings are almost acoustically transparent at low frequencies.

Often, supply ducts are run inside a chase that is also used for return air. It is best to attenuate supply and return paths at the fan rather than let duct breakout noise require additional noise control at the chase return air inlets. Care should be used to prevent turbulent noise generation in the supply duct through proper supply duct design.

Special Construction Types

Sound transmission loss values for some typical constructions are given in Table 40. These data are compiled from controlled laboratory tests and represent a condition typically superior to that found in field installations, because the in situ acoustical performance of any wall, floor, or ceiling is adversely affected by flanking paths, holes, penetrations, and other anomalies. Flanking paths include intersections of the wall, floor, or ceiling surface with another wall, floor, or ceiling that is structurally connected. Higher levels of sound isolation can be achieved by decoupling the surfaces and using double-walled, floating floor, or barrier-ceiling constructions.

Floating Floors and Barrier Ceilings

Correctly installed floating floors and barrier ceilings can provide very high levels of sound isolation, allowing mechanical equipment rooms to be placed adjacent to noise-sensitive spaces. Like double-walled construction, these configurations decouple two surfaces by providing separate supporting structures. However, these types of construction can add significant cost and coordination complexity and should be carefully evaluated.

In a floating floor, the upper floor slab is typically a 4 in. concrete slab resting on spring, neoprene, and/or fiberglass vibration isolators supported by the subfloor (typically another concrete slab of appropriate thickness). An air gap is maintained between the two slabs and resilient materials are used around the upper slab's perimeter to decouple it from surrounding walls. Air gap and upper slab thickness both impact noise isolation performance and should be considered. Any heavy equipment should be properly supported to account for additional loading and possible short circuiting.

Because natural frequencies for floating floor systems are limited by the dynamic response of the air trapped between the floating and structural slabs, these types of systems are not recommended to control structureborne vibration. It is extremely difficult on a floating floor to achieve a natural frequency of less than 15 Hz. While this is low enough to have a significant impact on audible frequencies, it is not low enough to control the vibrations generated by common equipment types.

Mounting equipment directly to a floating floor can reduce flanking paths and result in some acoustic benefits. However, caution should be exercised to ensure adequate damping is provided in the floating floor and that there are no common natural frequencies present between the floating floor isolation system and any equipment isolation mounted to it. Improper selection can increase the transmitted vibration rather than reduce it.

Similarly, barrier ceilings are typically composed of multiple layers of gypsum wallboard attached to a frame suspended from the structure above with vibration-isolating hangers. A sound barrier ceiling's construction is influenced by its purpose. If the ceiling isolates two vertically adjacent spaces, mechanical equipment should be placed below the ceiling to ensure a continuous drywall barrier. If noise levels in the occupied space below the sound barrier ceiling are critical, mechanical equipment should be placed above with minimal ceiling penetrations that are properly sealed. Regardless, ductwork, piping, and other equipment require careful coordination of hangers and supports to ensure no rigid contact with the ceiling. Ceiling penetrations should be minimized, because these will reduce barrier ceiling performance. In the case of recessed lighting, it is often necessary to use a gypsum wallboard enclosure around the entire fixture so that it does not serve as a flanking path for noise.

Design and selection of the floating floor or ceiling should be carefully considered to properly support the dead and live loads it must carry. The floating system (floor or ceiling) *is not meant as a means of equipment vibration isolation and serves primarily to control airborne sound transmission.* Acoustical performance of the floating system depends greatly on construction quality, which requires careful coordination between all trades. All penetrations and intersections with other surfaces must allow the floor or ceiling to float without any rigid connections. This typically entails maintaining clearances of at least 0.25 in. and filling all gaps with resilient materials such as nonhardening caulk.

Sound Transmission in Return Air Systems

The fan return air system provides a sound path (through ducts or a ceiling plenum) between a fan and occupied rooms. Where there is a direct opening to the mechanical equipment room from the ceiling plenum, sound levels in adjacent spaces can be high, originating from the fan and other sources in the mechanical equipment room. Low system attenuation between the mechanical equipment room and adjacent spaces exacerbates the problem.

Fan intake sound power levels control sound in ducted return air systems; sound power levels of the fan intake and casing-radiated noise components affect plenum return air systems. In some installations, sound from other equipment located in the mechanical equipment room may also radiate through the wall opening and into adjacent spaces. Good design yields room return air system sound levels that are approximately 5 dB below the corresponding room supply air system sound levels.

When sound levels in spaces adjacent to mechanical equipment rooms are too high, noise control measures must be provided. The controlling sound paths between the mechanical equipment room and adjacent spaces must be identified. Ducted return air systems can be modified using methods applicable to ducted supply systems. Ceiling plenum return systems should only be used for spaces that are remote from mechanical equipment rooms.

Ceiling plenum systems may require additional modifications. Prefabricated silencers can be effective when installed at the mechanical equipment room wall opening or at the suction side of the fan. Improvements in ceiling transmission loss are often limited by typical ceiling penetrations and lighting fixtures. Modifications to the mechanical equipment room wall can be effective for some

Table 41 Environmental Correction to Be Subtracted from Device Sound Power

	Octave Band Frequency, Hz					
63	125	250	500	1000	2000	4000
4	2	1	0	0	0	0

Table 42 Compensation Factors for Source Area Effect

Area range, ft²			
63 Hz	125 Hz	250 Hz	Adjustment, dB
<2.6	<2.2		–3
2.8 to 4.9	2.4 to 4.6	<2.3	–2
5.1 to 7.2	4.9 to 7.1	2.7 to 6.3	–1
7.4 to 9.4	7.3 to 9.5	6.7 to 10.3	0
9.7 to 11.7	9.8 to 12.0	10.7 to 14.3	1
11.9 to 14.0	12.2 to 14.4	14.7 to 18.3	2
14.2 to 16.3	14.6 to 16.8	18.7 to 22.3	3
16.5 to 18.5	17.1 to 19.3		4
18.8 to 20.8	19.5 to 21.7		5
21.0 to 23.1			6

Note: Find correct area in each frequency column and read adjustment from last column on right.

constructions. Adding acoustical absorption in the mechanical equipment room reduces build-up of reverberant sound energy in this space; however, this typically reduces high-frequency noise by a maximum of 4 dB and low-frequency noise only slightly in areas near the return opening.

Sound Transmission Through Ceilings

When terminal units, fan-coil units, air-handling units, ducts, or return air openings to mechanical equipment rooms are located in a ceiling plenum above a room, sound transmission through the ceiling system can be high enough to cause excessive noise levels in that room. There is no standard test procedure for measuring direct transmission of sound through ceilings from sources close to the ceiling. As a result, ceiling product manufacturers rarely publish data that can be used in calculations. The problem is complicated by the presence of light fixtures, diffusers, grilles, and speakers that reduce the ceiling's transmission loss. Experiments have shown that, for ceiling panels supported in a T-bar grid system, leakage between the panels and grid is the major transmission path; differences among panel types are small, and light fixtures, diffusers, etc., have only a localized effect.

To estimate room sound pressure levels associated with sound transmission through the ceiling, sound power levels in the ceiling plenum must be adjusted to account for the transmission loss of the ceiling system and plenum. Measured data must also be adjusted to account for sound absorption in the room. The procedures given here are based on ASHRAE research (Warnock 1998):

1. Obtain octave band radiated sound power levels of device.
2. Subtract environmental correction from Table 41.
3. Calculate surface area of bottom panel of source closest to ceiling tiles (ft²).
4. From Table 42, find adjustment to be subtracted from sound power values at three frequencies given there.
5. Select ceiling/plenum attenuation from Table 43 according to ceiling type in use. Note that these values include a typical room effect, so when using these data in analysis, there is no additional line item from Equation (20) or (30).
6. Subtract the three sets of values, taking account of sign where necessary, from sound power values. The result is the average sound pressure level in the room.
7. The sound field in the room may be assumed as uniform up to distances of 16 ft from the source.

Example 7. A terminal unit with an area of 14 ft² and a known sound power level is to be used above a standard 5/8 in. thick mineral fiber ceiling system in a T-bar grid. What room sound pressure levels can be expected?

Solution:

Step		Octave Band Frequency, Hz						
		63	125	250	500	1000	2000	4000
1	Sound power	71	71	65	55	54	53	45
2	Environmental (Table 41)	–4	–2	–1	0	0	0	0
4	Area adjustment (Table 42)	–2	–2	–1	0	0	0	0
5	Ceiling/plenum (Table 43)	–13	–15	–17	–19	–25	–30	–33
6	Room sound pressure levels, dB	52	52	46	36	29	23	12

HVAC NOISE-REDUCTION DESIGN PROCEDURES

These HVAC system design procedures address the 63 to 4000 Hz octave band midfrequency range. Although it is desirable to extend this frequency range down into the 31.5 Hz octave band, acoustical calculations below the 125 Hz octave band are generally not reliable. With a few exceptions, if acoustical design criteria are met at 4000 Hz, then the 8000 Hz requirements are also met. Guidelines in this chapter and other guides maximize the probability of meeting acoustical design criteria in the 31.5 to 8000 Hz octave bands.

There is reasonable probability that the acoustical design criteria will be met when the following requirements are satisfied:

- Systems are designed in accordance with the equipment selection, placement, and integration guidelines in this chapter, other ASHRAE guides, and manufacturers' application notes and bulletins.
- Acoustical calculations based on the information included in this chapter and the information provided by the equipment manufacturer indicate that the system will not exceed the selected acoustical design criteria values in the 63 to 4000 Hz octave band frequency range.

The following suggested design procedure uses the NC method, which is the most commonly used. Other criteria, such as NCB or RC, may be used. However, it is often difficult to acquire low-frequency sound data, and low-frequency acoustical calculations for HVAC system components are not reliable.

1. Determine the design goal for HVAC system noise for each critical area according to its use and construction. Choose the desirable NC criterion from Table 1.
2. Select equipment and fittings (e.g., air inlet and outlet grilles, registers, diffusers, and air terminal and fan-coil units that radiate sound directly into a room) that are operating comfortably with their specified duty and are quiet for the class of equipment in question. The appropriate selection of equipment and fitting will generally result in an efficient acoustic design to meet design goals.
3. Complete initial design and layout of the HVAC system. Include typical duct lining where appropriate. Provide space for duct sound attenuators. Confirm that the airflow velocities are compliant with the specified rates in Table 8 and 9 of the Aerodynamically Generated Sound in Ducts section of this Chapter.
4. Calculate sound pressure level in the room of interest:
   (a) Acquire sound power data from manufacturers for equipment such as air-handling units, packaged rooftop units, exhaust fans, variable-air-volume terminal units, fan-powered terminal units, etc. If manufacturers' data are not available, estimate sound power level based on methods in this chapter or other authoritative sources.
   (b) Calculate sound attenuation and regenerated sound power of duct elements in the air distribution system of interest.

Table 43 Ceiling/Plenum/Room Attenuations in dB for Generic Ceiling in T-Bar Suspension Systems

Tile Type	Approximate Density, lb/ft²	Tile Thickness, in.	Octave Midband Frequency, Hz						
			63	125	250	500	1000	2000	4000
Mineral fiber	1.0	5/8	13	16	18	20	26	31	36
	0.5	5/8	13	15	17	19	25	30	33
Glass fiber	0.1	5/8	13	16	15	17	17	18	19
	0.6	2	14	17	18	21	25	29	35
Glass fiber with TL backing	0.6	2	14	17	18	22	27	32	39
Gypsum board tiles	1.8	1/2	14	16	18	18	21	22	22
Solid gypsum board ceiling	1.8	1/2	18	21	25	25	27	27	28
	2.3	5/8	20	23	27	27	29	29	30
Double layer of gypsum board	3.7	1	24	27	31	31	33	33	34
	4.5	11/4	26	29	33	33	35	35	36
Mineral fiber tiles, concealed spline mount.	0.5 to 1	5/8	20	23	21	24	29	33	34

Source: Warnock (1998)

(c) Tabulate sound power and attenuation for each component in each sound transmission path. Start at the supply air fan or packaged air-conditioning unit and end at the room. Investigate both the supply and return air paths in similar ways. Investigate possible duct sound breakout when fans are adjacent to, or roof-mounted fans are above, the room of interest. Combine sound power levels from all paths. See Example 8 for calculation procedures for supply and return air paths, including duct breakout noise contributions. Include a placeholder for the duct sound attenuator so that it is a simple matter to include in calculations later.

(d) Convert sound power levels to corresponding sound pressure levels in the room using the ASHRAE room correction procedure.

5. If the mechanical equipment room is adjacent to the room of interest, determine sound pressure levels in the room of interest that are associated with sound transmitted through the mechanical equipment room wall. Air-handling units, ventilation and exhaust fans, chillers, pumps, electrical transformers, and in-strument air compressors are typical equipment to consider. Make sure that noise transmission from adjacent external spaces outside the room in question, such as cooling towers or air-cooled chillers, is also considered. Also consider the vibration isolation requirements for equipment, piping, and ductwork. [See Egan (1988) or Reynolds and Bledsoe (1991) for calculation procedure.]

6. Combine on an energy basis the sound pressure levels in the room of interest that are associated with all sound paths between the mechanical equipment room or roof-mounted unit(s) and the room of interest. Establish the controlling noise-transmission paths.

7. Determine the corresponding NC level associated with the calculated total sound pressure levels in the room of interest. Take special note of unbalanced sound spectra and tonal characteristics.

8. If the NC level satisfies the criteria established in step 1, analysis is complete. If the NC level exceeds the design goal, determine the octave frequency bands in which the corresponding sound pressure levels are exceeded and the sound paths associated with these octave frequency bands as determined in step 6. If the resulting noise levels are high enough to cause perceivable vibration, consider both airborne and structureborne noise.

9. Redesign the system:
   (a) Reselect the offending noise source. This is typically the least costly, most energy-efficient, and most effective change, but is not always possible.
   (b) Add sound attenuation to paths that contribute to excessive sound pressure levels in the room of interest. This may be achieved by using thicker internal acoustic-grade insulation

or proprietary silencers. Note that silencers preferably should be inserted at the penetration of MER walls or external building elements to minimize breakout before the silencer.

   (c) Consider increasing the length of ductwork or introducing bends or a plenum to increase the sound attenuation. Care needs to be taken to ensure that the breakout noise path(s) are still acceptable where additional ductwork is introduced.
   (d) Increase the sound transmission loss properties of building elements where this is the controlling noise path. This may be achieved by installing additional mass (i.e., thicker walls or filled CMUs) or by introducing an air gap with a secondary layer (i.e., double glazing).
   (e) Consider installing noise barriers around external plant to minimize the direct line of sight between plant and critical spaces. The manufacturers' requirements for access and airflow around equipment must be carefully considered where noise barriers are used.
   (f) If resultant noise levels are high enough to cause perceivable vibration, then major redesign and possibly use of supplemental vibration isolation for equipment and building systems are often required.
   (g) Reference should also be made to this chapter's sections on Acoustical Design of HVAC Systems and Basic Acoustical Design Techniques.

10. Repeat steps 4 through 9 until the desired design goal is achieved. Involve the complete design team where major problems are found. Often, simple design changes to building architectural and equipment selection can eliminate potential problems once the problems are identified. Ensure that all valid noise-transmission paths are assessed.

11. Repeat steps 3 through 10 for every room that is to be analyzed.

12. Make sure that environmental noise radiated by outdoor equipment such as air-cooled chillers, exhaust fans, condensers, and cooling towers does not disturb adjacent properties or interfere with criteria established in step 1 or any applicable building or zoning noise ordinances.

Example 8. This example illustrates step 4 in the design process. Previous examples demonstrate how to calculate equipment- and air distribution system airflow-generated sound power levels and attenuation values. Here, the individual elements are combined to determine sound pressure levels associated with a specific HVAC system. Only a summary of tabulated results is listed rather than showing complete calculations for each element. Calculations for each element are strictly based on the methods in this chapter or manufacturers' data. Noise transmission via the roof structure has not been considered in this example.

Air is supplied to the HVAC system by the rooftop unit shown in Figure 36. The receiver room is directly below the unit. The room has the following dimensions: length = 20 ft, width = 20 ft, and height = 9 ft. Assume that the roof penetrations for supply and return air ducts

Path 1: Ducted Supply Air

Element ID	Description	Sound Power Attenuation, dB							Regenerated Sound Power, dB							Path Sound Power, dB						
		63	125	250	500	1000	2000	4000	63	125	250	500	1000	2000	4000	63	125	250	500	1000	2000	4000
01	Supply air fan, 7000 cfm, 2.5 in. of water	0	0	0	0	0	0	0	92	86	80	78	78	74	71	92	86	80	78	78	74	71
02	22 in. dia., 90° rad. unlined elbow	0	1	2	3	3	3	3	0	0	0	0	0	0	0	92	85	78	75	75	71	68
03	22 × 44 in. long sound attenuator	4	7	19	31	38	38	27	68	79	69	60	59	59	55	88	82	69	60	59	59	55
04	22 in. dia., 8 ft long unlined duct	0	0	0	0	0	0	0	0	0	0	0	0	0	0	88	82	69	60	59	59	55
05.2	10 in. dia. branch, 22 in. dia. main, branch path	8	8	8	8	8	8	8	0	0	0	0	0	0	0	80	74	61	52	51	51	47
06	10 in. dia., 6 ft long unlined duct	0	0	0	0	0	0	0	0	0	0	0	0	0	0	80	74	61	52	51	51	47
07	VAV terminal	0	0	0	0	0	0	0	0	74	70	65	63	60	55	80	77	71	65	63	61	56
08	10 in. dia., 2 ft long unlined duct	0	0	0	0	0	0	0	0	0	0	0	0	0	0	80	77	71	65	63	61	56
09	10 in. dia., 90° rad. unlined elbow	0	0	1	2	3	3	3	0	0	0	0	0	0	0	80	77	70	63	60	58	53
10	10 in. dia. diffuser end reflection	14	8	4	1	0	0	0	0	0	0	0	0	0	0	66	69	66	62	60	58	53
11	15 × 15 in. rectangular diffuser	0	0	0	0	0	0	0	31	36	39	40	39	36	30	66	69	66	62	60	58	53
12	ASHRAE room correction: point source*	4	5	6	7	8	9	10	0	0	0	0	0	0	0	62	64	60	55	52	49	43

Path 2: Breakout Noise from 22 in. Main Duct

Element ID	Description	Sound Power Attenuation, dB							Regenerated Sound Power, dB							Path Sound Power, dB						
		63	125	250	500	1000	2000	4000	63	125	250	500	1000	2000	4000	63	125	250	500	1000	2000	4000
01	Supply air fan, 7000 cfm, 2.5 in. of water	0	0	0	0	0	0	0	92	86	80	78	78	74	71	92	86	80	78	78	74	71
02	22 in. dia., 90° rad. unlined elbow	0	1	2	3	3	3	3	0	0	0	0	0	0	0	92	85	78	75	75	71	68
03	22 × 44 in. long sound attenuator	4	7	19	31	38	38	27	68	79	69	60	59	59	55	88	82	69	60	59	59	55
04	22 in. dia., 8 ft long unlined duct	0	0	0	0	0	0	0	0	0	0	0	0	0	0	88	82	69	60	59	59	55
05.1	10 in. dia. branch, 22 in. dia. main, main path	1	1	1	1	1	1	1	0	0	0	0	0	0	0	87	81	68	59	58	58	54
13	22 in. dia., 20 ft long 24 ga. duct breakout	35	37	35	28	20	10	13	0	0	0	0	0	0	0	52	44	34	32	38	48	42
14	2 ft × 4 ft × 5/8 in. lay-in ceiling	13	15	17	19	25	30	33	0	0	0	0	0	0	0	39	29	17	13	13	18	9

Path 3: Return Air

Element ID	Description	Sound Power Attenuation, dB							Regenerated Sound Power, dB							Path Sound Power, dB						
		63	125	250	500	1000	2000	4000	63	125	250	500	1000	2000	4000	63	125	250	500	1000	2000	4000
16	Return air fan, 7000 cfm, 2.5 in. of water	0	0	0	0	0	0	0	82	79	80	78	78	74	71	82	79	80	78	78	74	71
17	36 × 72 in., 90° mitered unlined elbow	1	5	8	4	3	3	3	0	0	0	0	0	0	0	81	74	72	74	75	71	68
18	36 × 72 in., 8 ft long lined duct	0	1	2	8	16	10	10	0	0	0	0	0	0	0	81	73	70	66	59	61	58
19	36 × 72 in. end reflection loss	4	1	0	0	0	0	0	0	0	0	0	0	0	0	77	72	70	66	59	61	58
14	2 ft × 4 ft × 5/8 in. lay-in ceiling	13	15	17	19	25	30	33	0	0	0	0	0	0	0	64	57	53	47	34	31	25

*Based on a location 5 ft above floor at a receiver 4 ft from source

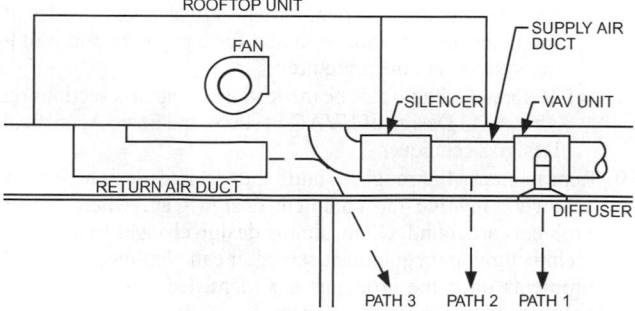

Fig. 36   Sound Paths Layout for Example 8

are well sealed and there are no other roof penetrations. In this example, it is assumed that breakout noise (upstream of the supply air silencer) is negligible. The supply side of the rooftop unit is ducted to a VAV terminal control unit serving the room in question. Although these units can create both ductborne and radiated noise, only the ductborne noise has been considered in this example. A return air grille conducts air to a common ceiling return air plenum. The return air is then directed to the rooftop unit through a short rectangular return air duct.

The following three sound paths are to be examined. Note that in this example, neither the noise transmission via the roof structure nor any other breakout noise upstream of the silencer has been considered. Those paths, plus VAV unit-radiated noise and other potential breakout noise paths, have been excluded from the analysis for simplification. In critical applications, each of those separate elements must also be considered.

Path 1. Fan airborne supply air sound that enters the room from the supply air system through the ceiling diffuser

Path 2. Fan airborne supply air sound that breaks out through the wall of the main supply air duct into the plenum space above the room

Path 3. Fan airborne return air sound that enters the room from the inlet of the return air duct

The tabulated calculations for each path follow:

Path Description	Path Sound Pressure Level, dB							
	63	125	250	500	1000	2000	4000	NC
1 Ducted supply air path	62	64	60	55	52	49	43	52
2 Breakout noise from 22 in. main duct	39	29	17	12	13	18	9	19
3 Return air path	64	57	53	47	34	31	25	44
Total L_p	66	65	60	56	52	49	43	52

Calculation Procedure

Analysis for each path begins at the rooftop unit and proceeds through the different system elements to the receiver room. The element numbers in the tables correspond to those in Figure 37. The source of each element calculation is listed in Table 44. Sound data for the rooftop unit (supply and return openings), VAV terminal, diffuser, and duct sound attenuator are manufacturers' data.

A spreadsheet was used to perform the calculations associated with this example. This type of calculation is often performed iteratively, as described in the preceding design procedure, but using a well-crafted spreadsheet increases the speed and accuracy of calculations.

Calculation tables for paths 1, 2, and 3 are organized with an element in each row. Three spectra (63 Hz to 4000 Hz) are shown for each element. The first and second spectra (sound power attenuation and regenerated sound power) are either calculated based on the

Table 44   Path Element Sound Calculation Reference

ID	Description	Data Source Reference
01	Supply air fan, 7000 cfm, 2.5 in. of water	Manufacturer's data
02	22 in. dia., 90° rad. unlined elbow	Attenuation: Table 23
03	22 × 44 in. long sound attenuator	Manufacturer's data
04	22 in. dia., 8 ft long unlined duct	Attenuation: Table 27
05.2	10 in. dia. branch, 22 in. dia. main, branch path	Attenuation: Table 27
05.1	10 in. dia. branch, 22 in. dia. main, main path	Attenuation: Table 27
06	10 in. dia., 6 ft long unlined duct	Attenuation: Table 16
07	VAV terminal	Manufacturer's data
08	10 in. dia., 2 ft long unlined duct	Attenuation: Table 16
09	10 in. dia., 90° rad. unlined elbow	Attenuation: Table 23
10	10 in. dia. diffuser, end reflection	Attenuation: Table 28
11	15 × 15 in. rectangular diffuser	Manufacturer's data
12	ASHRAE room correction: point source	Equation (26), Tables 35 and 36
13	22 in. dia., 20 ft long, 24 ga. duct breakout	Attenuation: Equation (20), Table 30
14	2 × 4 ft × 5/8 in. lay-in ceiling	Attenuation: Table 43
15	ASHRAE room correction: line source	Equation (30)
16	Return air fan, 7000 cfm, 2.5 in. of water	Manufacturer's data
17	36 × 72 in., 90° mitered unlined elbow	Attenuation: Tables 22 and 24
18	36 × 72 in., 8 ft long lined duct	Attenuation: Table 18; assume 0 dB at 63 Hz
19	36 × 72 in. end reflection loss	Attenuation: Table 27, $D = 43.5$ in.
20	ASHRAE room correction	Point source: Equation (26), Tables 35 and 36

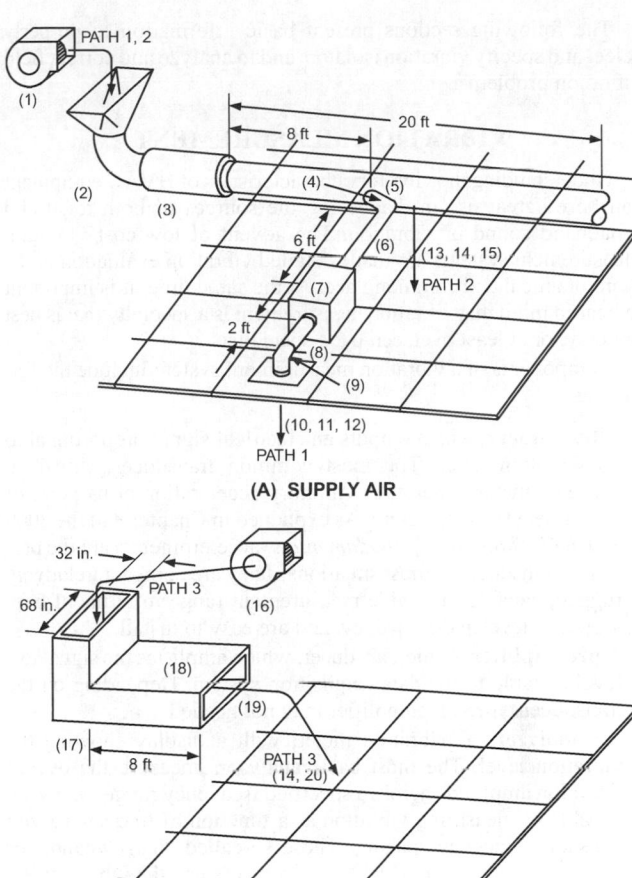

(A) SUPPLY AIR

(B) RETURN AIR

Fig. 37   (A) Supply and (B) Return Air Layout for Example 8

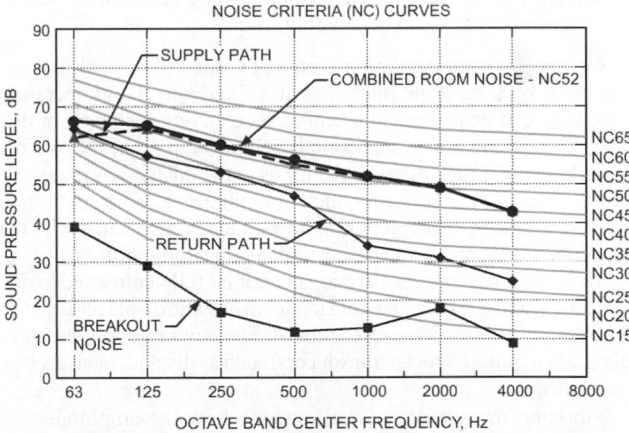

Fig. 38   NC Rating Calculated

equations and tables in this chapter or acquired from a manufacturer. It is important to note that sound power data and not sound pressure data must be used in these calculations.

The spreadsheet subtracts the sound attenuation spectrum, band by band, from the path sound power spectrum in the previous row. Then, the resultant sound power is logarithmically added per band to the element sound power. This calculation is performed for each element (row).

$$L_w = 10 \log\left(10^{L_{w1}/10} + 10^{L_{w2}/10}\right) \qquad (33)$$

The last element in each path is either the ASHRAE room sound correction or the ceiling/plenum/room attenuations outlined in Table 43, which is entered as an attenuation spectrum as it is subtracted directly from the final path sound power spectrum.

VIBRATION ISOLATION AND CONTROL

Mechanical vibration and vibration-induced noise are common sources of occupant complaints in modern buildings. Lightweight construction in buildings provides conditions that can result in vibration-related problems. Mandates for energy conservation have resulted in many buildings being designed with variable air volume systems with variable-speed equipment. As rotating equipment spins slower, its forcing frequency approaches the structure's resonant frequency, which can amplify vibration-induced noise. Mechanical equipment is often located in penthouses or on the roof, where structures are typically the most susceptible to inducing vibration-related problems. Mechanical equipment rooms are typically located on intermediate level floors, close to the occupied areas they serve.

Occupant complaints associated with building vibration typically take one or more of three forms:

- The level of vibration perceived by building occupants is of sufficient magnitude to cause concern or alarm
- Vibration energy from mechanical equipment, which is transmitted to the building structure, is transmitted to various parts of the building and then is radiated as structureborne noise
- Vibration in a building may interfere with proper operation of sensitive equipment or instrumentation

The following sections present basic information to properly select and specify vibration isolators and to analyze and correct field vibration problems.

VIBRATION MEASUREMENT

Understanding the vibratory characteristics of HVAC equipment can be of great use in diagnosing the sources of both tonal and broadband sound or vibration. The advent of low-cost vibration measurement systems has made detailed vibration evaluation much more practical and commonplace. At the same time, it is important to bear in mind that vibration measurement is a specialty that is best done by, or at least overseen by, a specialist.

Components of a vibration measurement system include the following:

- A **transducer**, which outputs an electrical signal proportional to its vibration level. The most common transducer, called an "accelerometer," measures vibratory acceleration at its point of attachment to the structure. As explained in Chapter 8 of the 2009 *ASHRAE Handbook—Fundamentals*, accelerometers are the preferred transducer in most situations. They are compact, relatively rugged, capable of a wide measurement range in terms of both vibration level and frequency, and are easy to install.
- A **preamplifier** for the transducer, which amplifies the signal to a level suitable to the data acquisition system. Depending on the transducer type, a preamplifier may not needed.
- An **analyzer**, or vibration meter, with a display showing the vibration level. The most basic analyzers measure the overall vibration amplitude across a specified frequency range. Many are capable of measuring vibration as a function of frequency, with constant frequency spacing. These so-called "narrowband," or fast Fourier transform (FFT), analyzers display the vibration frequency spectrum with a very high degree of resolution, typically at hundreds or even thousands of frequencies. Alternative analyzers of the constant-percentage bandwidth type measure the vibration spectrum across a relatively small number of frequency bands, the widths of which increase proportionally to the center frequency of each band. These often display vibration at octave, 1/3rd octave, or 1/12th octave frequencies.

Any steady-state vibration frequency spectrum, such as that generated by a machine operating at a fixed speed and operating condition, can be expressed as either acceleration, velocity, or displacement. In this situation, a simple relationship makes it possible to easily convert each of these quantities to the others. It is important to note that many transducers, including all accelerometers, cannot measure vibration below a minimum frequency associated with the transducer. The practical implication of that limitation is that, in many cases, measured acceleration cannot be fully converted to displacement, and can never be used to quantify static displacement. For that reason, in cases where very-low-frequency vibration measurements are required, special transducers, such as displacement probes, are needed.

Vibration measurements must specify how the amplitudes are expressed. These can be either peak (the maximum level), peak-to-peak (the range between minima and maxima), or rms (root mean square). The peak-to-peak value is twice the peak, and the rms is the peak divided by the square root of two.

Several factors must be considered when making vibration measurements. One of these is transducer attachment to the vibrating object. An extremely rigid attachment method, such as dental cement, or a screwed connection with oil between the surfaces, is required for accurate measurement at very high frequency (about 5 kHz). Epoxies or other high-quality glues tend to be somewhat more limited but are acceptable in nearly all situations. Using magnetic attachments, though convenient and fully acceptable in many cases, limits the upper frequency range of accurate data (typically about 150 Hz). Another common but frequency-limited method of attachment is wax. In any case, it is essential to validate that the attachment used in a given application is capable of measuring vibration to the needed degree of accuracy.

Several data processing factors must be considered for narrowband (FFT) measurement. These significantly affect the quality of spectral data and how they are interpreted. Among them are the window type (especially for tonal sources), the number of averages, the window overlap, the frequency resolution (the inverse of the total sample time), and the maximum frequency (the inverse of the sampling frequency). Again, guidance from a specialist should be sought when establishing these factors for a given measurement.

Typical applications of vibration measurement include

- Comparison of overall vibration levels (the total across a defined frequency range) with general guidelines representing typical levels to be expected from various classes of machinery. This most basic measurement is often used in connection with routine machinery maintenance or monitoring.
- Comparison of vibration spectral values with either equipment specifications, building specifications, or general guidelines. These more complete data, typically defined in terms of octave or 1/3 octave frequencies, provide more detailed guidance for machinery health monitoring, equipment qualification, or building certification.
- Comparison of vibration spectral values above and below vibration isolators, such as pads or springs, to determine if they are providing the anticipated vibration reduction. Note that, as explained in Chapter 8 of the 2009 *ASHRAE Handbook—Fundamentals*, interpretation of the results of these measurements may not be straightforward.
- Using a narrowband measurement system, determination of exact frequencies of tonal vibration sources. This information can be critical in identifying the specific machine or vibration component responsible for excessive vibration or noise. In some cases, a high degree of measurement resolution is required to separate closely spaced tones. For example, in 60 Hz applications, twice the motor or compressor running speeds are typically close to 118 Hz, while twice the electrical line frequency is 120 Hz. Clearly, while the difference between these frequencies is inaudible, knowing which source is responsible for a problem is essential to developing a solution.

Finally, it is noted that many specialized applications of vibration and dynamic measurement require advanced data acquisition equipment, data analysis software, and associated training. Examples are

- Transient vibration measurement
- Impact/frequency response measurement
- Modal testing
- Rotating equipment balancing
- Direct displacement measurement (e.g., rotating shaft orbit analysis)

EQUIPMENT VIBRATION

Any vibrating, reciprocating, or rotating equipment should be mounted such that it does not transmit significant levels of vibration into the surrounding or supporting structure. Vibrations transmitted via machine mounts or attached piping, ductwork, or electrical connections can result in vibrating walls, floors, and/or ceilings, which in turn radiate sound and/or vibration. Hence, it is important to provide vibration isolation for all attachments to a vibrating machine, including structural mounts and the connections to piping, ductwork, and the electrical system. It is also important to mitigate residual vibrations in attached piping and ductwork, even when equipment is properly isolated. It takes very little vibration energy to produce audible noise.

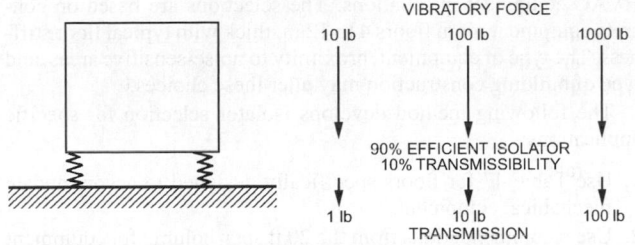

Fig. 39   Transmission to Structure Varies as Function of Magnitude of Vibration Force

Table 45   Human Comfort and Equipment Vibration Criteria

Human Comfort	Time of Day	8 to 80 Hz Curve,[a] μin/s
Workshops	All	32,000
Office areas	All[b]	16,000
Residential (good environmental standards)	0700-2200[b]	8000
	2200-0700[b]	5600
Hospital operating rooms and critical work areas	All	4000

Equipment Requirements	Curve[a]
Adequate for computer equipment, probe test equipment, and microscopes less than 40×	8000
Bench microscopes up to 100× magnification; laboratory robots	4000
Bench microscopes up to 400× magnification; optical and other precision balances; coordinate measuring machines; metrology laboratories; optical comparators; microelectronics manufacturing equipment; proximity and projection aligners, etc.	2000
Microsurgery, eye surgery, neurosurgery; bench microscopes at magnification greater than 400×; optical equipment on isolation tables; microelectronic manufacturing equipment, such as inspection and lithography equipment (including steppers) to 3 mm line widths[c]	1000
Electron microscopes up to 30,000× magnification; microtomes; magnetic resonance imagers; microelectronics manufacturing equipment, such as lithography and inspection equipment to 1 mm detail size[c]	500
Electron microscopes at magnification greater than 30,000×; mass spectrometers; cell implant equipment; microelectronics manufacturing equipment, such as aligners, steppers, and other critical equipment for photolithography with line widths of 1/2 μm; includes electron beam systems[c]	250
Unisolated laser and optical research systems; microelectronics manufacturing equipment, such as aligners, steppers, and other critical equipment for photolithography with line widths of 1/4 μm; includes electron beam systems[c]	125

[a]See Figure 41 for corresponding curves.
[b]In areas where individuals are sensitive to vibration, use Residential Day curve.
[c]Classes of microelectronics manufacturing equipment:

Vibration can be isolated or reduced to a fraction of the original force with resilient mounts between the equipment and the supporting structure, provided that the supporting structure has sufficient stiffness and mass. **Isolation efficiency** is the percentage of vibratory force *not* transmitted to the support structure. Figure 39 shows that 90% efficiency results in 10% of the vibration force being transmitted. In this case, the magnitude of transmission to the building is a function of the magnitude of the vibration force. Figure 40 shows the effect of different efficiency levels.

VIBRATION CRITERIA

For the HVAC designer, vibration criteria are specified relative to three areas: (1) human response to vibration, (2) vibration levels

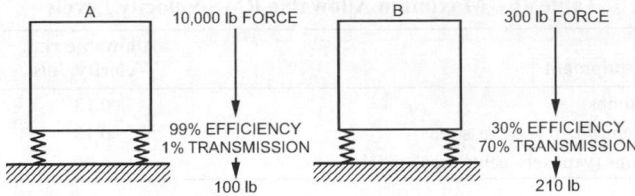

Fig. 40   Interrelationship of Equipment Vibration, Isolation Efficiency, and Transmission

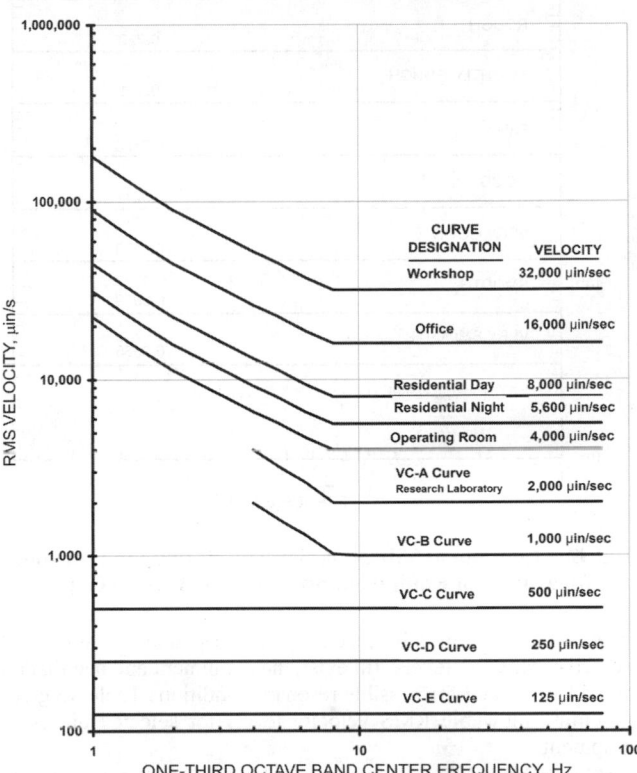

Fig. 41   Building Vibration Criteria for Vibration Measured on Building Structure

in a building, and (3) vibration severity of an operating machine. Figure 41 and Table 45 present recommended acceptable criteria for vibration in a building structure (IEST 2005; Murray et al. 1997). The vibration values in Figure 41 are measured in one-third octave bands using vibration transducers (usually accelerometers) placed on the building structure near vibrating equipment or in areas containing occupants or sensitive equipment. Occupant vibration criteria are based on guidelines recommended in ANSI S2.71-1983 (R2006) and ISO *Standard* 2631-2. For sensitive equipment, acceptable vibration values specified by equipment manufacturers should be used. If none are available, then criteria from IEST (2005), as reflected in Figure 41 and Table 45, can be used.

If acceptable vibration values are not available from equipment manufacturers, the values specified in Figure 42 can be used. This figure gives recommended equipment vibration severity ratings based on measured RMS velocity values (IRD 1988). The vibration values in Figure 42 are measured by vibration transducers (usually accelerometers) mounted directly on equipment, equipment structures, or bearing caps. Vibration levels measured on equipment and components can be affected by equipment unbalance, misalignment of equipment components, and resonance interaction between a vibrating piece of equipment and the floor on which it is placed. If

Table 46 Maximum Allowable RMS Velocity Levels

Equipment	Allowable rms Velocity, in/s
Pumps	0.13
Centrifugal compressors	0.13
Fans (vent sets, centrifugal, axial)	0.09

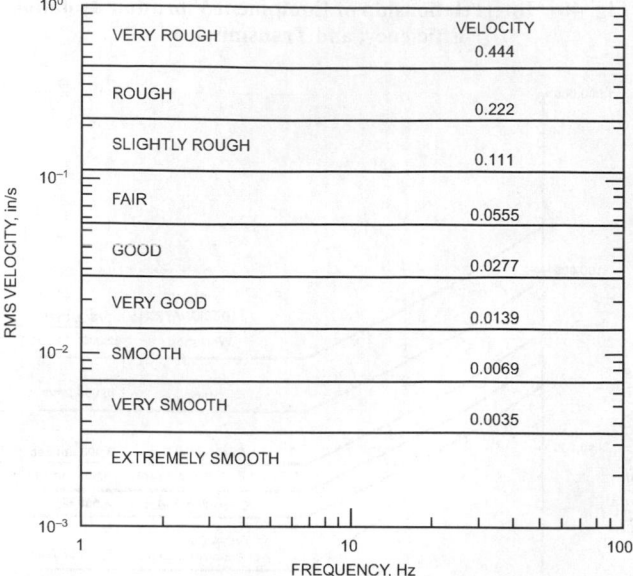

Fig. 42 Equipment Vibration Severity Rating for Vibration Measured on Equipment Structure or Bearing Caps

a piece of equipment is balanced within acceptable tolerances and excessive vibration levels still exist, the equipment and installation should be checked for possible resonant conditions Table 46 gives maximum allowable RMS velocity levels for selected pieces of equipment.

Vibration levels measured on equipment structures should be in or below the "Good" region in Figure 42. Machine vibration levels in the "Fair" or "Slightly Rough" regions may indicate potential problems requiring maintenance. Machines with vibration levels in these regions should be monitored to ensure problems do not arise. Machine vibration levels in the "Rough" and "Very Rough" regions indicate a potentially serious problem; immediate action should be taken to identify and correct the problem.

SPECIFICATION OF VIBRATION ISOLATORS

Vibration isolators must be selected not only to provide required isolation efficiency but also to compensate for floor stiffness. Longer floor spans can be more flexible and thus more easily set into vibration. Floor spans, equipment operating speeds, equipment power, damping, and other factors are considered in Table 47.

In specifying isolator deflection rather than isolation efficiency or transmissibility, a designer can compensate for floor stiffness and building resonances by selecting isolators that have more deflection than the supporting floor. To apply the information from Table 47, base type, isolator type, and minimum deflection columns are added to the equipment schedule. These isolator specifications are then incorporated into mechanical specifications for the project.

Minimum deflections in the table are based on the experience of acoustical and mechanical consultants and vibration control manufacturers. Recommended isolator type, base type, and minimum static deflection are reasonable and safe recommendations for most

HVAC equipment installations. The selections are based on concrete equipment room floors 4 to 12 in. thick with typical floor stiffness. The type of equipment, proximity to noise-sensitive areas, and type of building construction may alter these choices.

The following method develops isolator selection for specific applications:

1. Use Table 47 for floors specifically designed to accommodate mechanical equipment.
2. Use recommendations from the 20 ft span column for equipment on ground-supported slabs adjacent to noise-sensitive areas.
3. For roofs and floors constructed with open web joists; thin, long span slabs; wooden construction; and any unusual light construction, evaluate all equipment weighing more than 300 lb to determine the additional deflection of the structure caused by the equipment. Isolator deflection should be 15 times the additional deflection or the deflection shown in Table 47, whichever is greater. If the required spring isolator deflection exceeds commercially available products, consider air springs, stiffen the supporting structure, or change the equipment location.
4. When mechanical equipment is adjacent to noise-sensitive areas, it is important to not only provide adequate vibration isolation but also to coordinate the construction of the surrounding floors, ceilings, and walls to isolate mechanical equipment room noise.

Selecting Vibration Isolators to Meet Isolator Deflection Requirements

An overview of the procedure to select vibration isolators is as follows:

1. Establish total weight of equipment to be supported. This includes all equipment and support framework. The weight of piping connected to equipment may also need to be considered, because this may be partly supported from the equipment.
2. Establish operating weight (e.g., weight of water in a chiller or cooling tower).
3. Determine the location of supporting springs.
4. Calculate the distribution of weight onto each of the supporting springs using static force distribution methods.
5. Consider any dynamic forces that may change the weight distribution over the supporting springs.
6. Select vibration isolators to achieve the minimum deflection based on the vibration-isolator spring constant as advised by the manufacturer.

Note that the preceding procedure does not satisfy seismic requirements, which must be considered in vibration isolator selection to meet applicable codes and standards.

Where requested or for sensitive projects, the following calculations may be presented for review:

1. Calculation of dry and operating weights (including any thrust forces)
2. Calculation of operating weights at each of the support points, considering the operating condition of the equipment
3. Calculation of isolator deflection at each of the support points, given the selected vibration isolator spring constant

VIBRATION- AND NOISE-SENSITIVE FACILITIES

Vibration-sensitive facilities identified in the section on Vibration Criteria are likely to require detailed assessment. Table 47 reflects typical application of vibration isolators in buildings to satisfy human comfort requirements. A specialist should be engaged to design vibration isolators for facilities with sensitive noise and vibration requirements, such as concert halls or facilities with electron microscopes. The specialist will select vibration isolators based on the proximity to sensitive areas, structural design of the facility,

Table 47 Selection Guide for Vibration Isolation

Equipment Type	Horsepower and Other	RPM	Slab on Grade Base Isolator Type	Slab on Grade Isolator Type	Slab on Grade Min. Defl., in.	Up to 20 ft Base Isolator Type	Up to 20 ft Isolator Type	Up to 20 ft Min. Defl., in.	20 to 30 ft Base Isolator Type	20 to 30 ft Isolator Type	20 to 30 ft Min. Defl., in.	30 to 40 ft Base Isolator Type	30 to 40 ft Isolator Type	30 to 40 ft Min. Defl., in.	Reference Notes
Refrigeration Machines and Chillers															
Reciprocating	All	All	A	2	0.25	A	4	0.75	A	4	1.50	A	4	2.50	2,3,12
Centrifugal, scroll	All	All	A	1	0.25	A	4	0.75	A	4	1.50	A	4	1.50	2,3,4,8, 12
Screw	All	All	A	1	1.00	A	4	1.5	A	4	2.50	A	4	2.50	2,3,4,12
Absorption	All	All	A	1	0.25	A	4	0.75	A	4	1.50	A	4	1.50	
Air-cooled recip., scroll	All	All	A	1	0.25	A	4	1.50	A	4	1.50	A	4	2.50	2,4,5,12
Air-cooled screw	All	All	A	4	1.00	A	4	1.50	B	4	2.50	B	4	2.50	2,4,5,8,12
Air Compressors and Vacuum Pumps															
Tank-mounted horiz.	≤10	All	A	3	0.75	A	3	0.75	A	3	1.50	A	3	1.50	3,15
	≥15	All	C	3	0.75	C	3	0.75	C	3	1.50	C	3	1.50	3,15
Tank-mounted vert.	All	All	C	3	0.75	C	3	0.75	C	3	1.50	C	3	1.50	3,15
Base-mounted	All	All	C	3	0.75	C	3	0.75	C	3	1.50	C	3	1.50	3,14,15
Large reciprocating	All	All	C	3	0.75	C	3	0.75	C	3	1.50	C	3	1.50	3,14,15
Pumps															
Close-coupled	≤7.5	All	B	2	0.25	C	3	0.75	C	3	0.75	C	3	0.75	16
	≥10	All	C	3	0.75	C	3	0.75	C	3	1.50	C	3	1.50	16
Inline	5 to 25	All	A	3	0.75	A	3	1.50	A	3	1.50	A	3	1.50	
	≥30	All	A	3	1.50	A	3	1.50	A	3	1.50	A	3	2.50	
End suction and double-	≤40	All	C	3	0.75	C	3	0.75	C	3	1.50	C	3	1.50	16
suction/split case	50 to 125	All	C	3	0.75	C	3	0.75	C	3	1.50	C	3	2.50	10,16
	≥150	All	C	3	0.75	C	3	1.50	C	3	2.50	C	3	3.50	10,16
Packaged pump systems	All	All	A	3	0.75	A	3	0.75	A	3	1.50	C	3	2.50	
Cooling Towers	All	Up to 300	A	1	6.4	A	4	89	A	4	89	A	4	89	5,8,18
		301 to 500	A	1	6.4	A	4	64	A	4	64	A	4	64	5,18
		501 and up	A	1	6.4	A	4	19	A	4	19	A	4	38	5,18
Boilers															
Fire-tube	All	All	A	1	0.25	B B	4	0.75	B	4	1.50	B	4	2.50	4
Water-tube, copper fin	All	All	A	1	0.12	A	1	0.12	A	1	0.12	B	4	0.25	
Axial Fans, Plenum Fans, Cabinet Fans, Fan Sections, Centrifugal Inline Fans															
Up to 22 in. diameter	All	All	A	2	0.25	A	3	0.75	A	3	0.75	C	3	0.75	4,9,8
24 in. diameter and up	≤2 in. SP	Up to 300	B	3	2.50	C	3	3.50	C	3	3.50	C	3	3.50	9,8
		301 to 500	B	3	0.75	B	3	1.50	C	3	2.50	C	3	2.50	9,8
		501 and up	B	3	0.75	B	3	1.50	B	3	1.50	B	3	1.50	9,8
	≥2.1 in. SP	Up to 300	C	3	2.50	C	3	3.50	C	3	3.50	C	3	3.50	3,8,9
		301 to 500	C	3	1.50	C	3	1.50	C	3	2.50	C	3	2.50	3,8,9
		501 and up	C	3	0.75	C	3	1.50	C	3	1.50	C	3	2.50	3,8,9
Centrifugal Fans															
Up to 22 in. diameter	All	All	B	2	0.25	B	3	0.75	B	3	0.75	B	3	1.50	9,19
24 in. diameter and up	≤40	Up to 300	B	3	2.50	B	3	3.50	B	3	3.50	B	3	3.50	8,19
		301 to 500	B	3	1.50	B	3	1.50	B	3	2.50	B	3	2.50	8,19
		501 and up	B	3	0.75	B	3	0.75	B	3	0.75	B	3	1.50	8,19
	≥50	Up to 300	C	3	2.50	C	3	3.50	C	3	3.50	C	3	3.50	2,3,8,9,19
		301 to 500	C	3	1.50	C	3	1.50	C	3	2.50	C	3	2.50	2,3,8,9,19
		501 and up	C	3	1.00	C	3	1.50	C	3	1.50	C	3	2.50	2,3,8,9,19
Propeller Fans															
Wall-mounted	All	All	A	1	0.25	A	1	0.25	A	1	0.25	A	1	0.25	
Roof-mounted	All	All	A	1	0.25	A	1	0.25	B	4	1.50	D	4	1.50	
Heat Pumps, Fan-Coils, Computer Room Units	All	All	A	3	0.75	A	3	0.75	A	3	0.75	A/D	3	1.50	
Condensing Units	All	All	A	1	0.25	A	4	0.75	A	4	1.50	A/D	4	1.50	
Packaged AH, AC, H, and V Units															
All	≤10	All	A	3	0.75	A	3	0.75	A	3	0.75	A	3	0.75	19
	≤15, ≤4 in. SP	Up to 300	A	3	0.75	A	3	3.50	A	3	3.50	C	3	3.50	2,4,8,19
		301 to 500	A	3	0.75	A	3	2.50	A	3	2.50	A	3	2.50	4,19
		501 and up	A	3	0.75	A	3	1.50	A	3	1.50	A	3	1.50	4,19
	>15, >4 in. SP	Up to 300	B	3	0.75	C	3	3.50	C	3	3.50	C	3	3.50	2,3,4,8,9
		301 to 500	B	3	0.75	C	3	1.50	C	3	2.50	C	3	2.50	2,3,4,9
		501 and up	B	3	0.75	C	3	1.50	C	3	1.50	C	3	2.50	2,3,4,9

Table 47   Selection Guide for Vibration Isolation (*Continued*)

Equipment Type	Horsepower and Other	RPM	Slab on Grade Base Type	Slab on Grade Isolator Type	Slab on Grade Min. Defl., in.	Up to 20 ft Base Type	Up to 20 ft Isolator Type	Up to 20 ft Min. Defl., in.	20 to 30 ft Base Type	20 to 30 ft Isolator Type	20 to 30 ft Min. Defl., in.	30 to 40 ft Base Type	30 to 40 ft Isolator Type	30 to 40 ft Min. Defl., in.	Reference Notes
Packaged Rooftop Equipment	All	All	A/D	1	0.25	D	3	0.75			See Reference Note 17				5,6,8,17
Ducted Rotating Equipment															
Small fans, fan-powered	≤600 cfm		A	3	0.50	A	3	0.50	A	3	0.50	A	3	0.50	7
boxes	≥601 cfm		A	3	0.75	A	3	0.75	A	3	0.75	A	3	0.75	7
Engine-Driven Generators	All	All	A	3	0.75	C	3	1.50	C	3	2.50	C	3	3.50	2,3,4

Piping and Ducts (See sections on Isolating Vibration and Noise in Piping Systems and Isolating Duct Vibration for isolator selection.)

Base Types:
A. No base, isolators attached directly to equipment (Note 28)
B. Structural steel rails or base (Notes 29 and 30)
C. Concrete inertia base (Note 31)
D. Curb-mounted base (Note 32)

Isolator Types:
1. Pad, rubber, or glass fiber (Notes 20 and 21)
2. Rubber floor isolator or hanger (Notes 20 and 25)
3. Spring floor isolator or hanger (Notes 22, 23, and 26)
4. Restrained spring isolator (Notes 22 and 24)
5. Thrust restraint (Note 27)
6. Air spring (Note 25)

Notes for Table 47: Selection Guide for Vibration Isolation

These notes are keyed to the column titled *Reference Notes* in Table 47 and to other reference numbers throughout the table. Although the guide is conservative, cases may arise where vibration transmission to the building is still excessive. If the problem persists after all short circuits have been eliminated, it can almost always be corrected by altering the support path (e.g., from ceiling to floor), increasing isolator deflection, using low-frequency air springs, changing operating speed, improving rotating component balancing, or, as a last resort, changing floor frequency by stiffening or adding more mass. Assistance from a qualified vibration consultant can be very useful in resolving these problems.

Note 1. Isolator deflections shown are based on a reasonably expected floor stiffness according to floor span and class of equipment. Certain spaces may dictate higher levels of isolation. For example, bar joist roofs may require a static deflection of 1.5 in. over factories, but 2.5 in. over commercial office buildings.

Note 2. For large equipment capable of generating substantial vibratory forces and structureborne noise, increase isolator deflection, if necessary, so isolator stiffness is less than one-tenth the stiffness of the supporting structure, as defined by the deflection due to load at the equipment support.

Note 3. For noisy equipment adjoining or near noise-sensitive areas, see the section on Mechanical Equipment Room Sound Isolation.

Note 4. Certain designs cannot be installed directly on individual isolators (type A), and the equipment manufacturer or a vibration specialist should be consulted on the need for supplemental support (base type).

Note 5. Wind load conditions must be considered. Restraint can be achieved with restrained spring isolators (type 4), supplemental bracing, snubbers, or limit stops. Also see Chapter 55.

Note 6. Certain types of equipment require a curb-mounted base (type D). Airborne noise must be considered.

Note 7. See section on Resilient Pipe Hangers and Supports for hanger locations adjoining equipment and in equipment rooms.

Note 8. To avoid isolator resonance problems, select isolator deflection so that resonance frequency is 40% or less of the lowest normal operating speed of equipment (see Chapter 8 in the 2009 *ASHRAE Handbook—Fundamentals*). Some equipment, such as variable-frequency drives, and high-speed equipment, such as screw chillers and vaneaxial fans, contain very-high-frequency vibration. This equipment creates new technical challenges in the isolation of high-frequency noise and vibration from a building's structure. Structural resonances both internal and external to the isolators can significantly degrade their performance at high frequencies. Unfortunately, at present no test standard exists for measuring the high-frequency dynamic properties of isolators, and commercially available products are not tested to determine their effectiveness for high frequencies. To reduce the chance of high-frequency vibration transmission, add a 1 in. thick pad (type 1, Note 20) to the base plate of spring isolators (type 3, Note 22, 23, 24). For some sensitive locations, air springs (Note 25) may be required. If equipment is located near extremely noise-sensitive areas, follow the recommendations of an acoustical consultant.

Note 14. Compressors: When using Y, W, and multihead and multicylinder compressors, obtain the magnitude of unbalanced forces from the equipment manufacturer so the need for an inertia base can be evaluated.

Note 15. Compressors: Base-mounted compressors through 5 hp and horizontal tank-type air compressors through 10 hp can be installed directly on spring isolators (type 3) with structural bases (type B) if required, and compressors 15 to 100 hp on spring isolators (type 3) with inertia bases (type C) weighing 1 to 2 times the compressor weight.

Note 16. Pumps: Concrete inertia bases (type C) are preferred for all flexible-coupled pumps and are desirable for most close-coupled pumps, although steel bases (type B) can be used. Close-coupled pumps should not be installed directly on individual isolators (type A) because the impeller usually overhangs the motor support base, causing the rear mounting to be in tension. The primary requirements for type C bases are strength and shape to accommodate base elbow supports. Mass is not usually a factor, except for pumps over 75 hp, where extra mass helps limit excess movement due to starting torque and forces. Concrete bases (type C) should be designed for a thickness of one-tenth the longest dimension with minimum thickness as follows: (1) for up to 30 hp, 6 in.; (2) for 40 to 75 hp, 8 in.; and (3) for 100 hp and up, 12 in.

Pumps over 75 hp and multistage pumps may exhibit excessive motion at start-up ("heaving"); supplemental restraining devices can be installed if necessary. Pumps over 125 hp may generate high starting forces; a vibration specialist should be consulted.

Note 17. Packaged Rooftop Air-Conditioning Equipment: This equipment is usually installed on lightweight structures that are susceptible to sound and vibration transmission problems. The noise problems are compounded further by curb-mounted equipment, which requires large roof openings for supply and return air.

The table shows type D vibration isolator selections for all spans up to 20 ft, but extreme care must be taken for equipment located on spans of over 20 ft, especially if construction is open web joists or thin, lightweight slabs. The recommended procedure is to determine the additional deflection caused by equipment in the roof. If additional roof deflection is 0.25 in. or less, the isolator should be selected for 10 times the additional roof deflection. If additional roof deflection is over 0.25 in., supplemental roof stiffening should be installed to bring the roof deflection down below 0.25 in., or the unit should be relocated to a stiffer roof position.

For mechanical units capable of generating high noise levels, mount the unit on a platform above the roof deck to provide an air gap (buffer zone) and locate the unit away from the associated roof penetration to allow acoustical treatment of ducts before they enter the building.

Some rooftop equipment has compressors, fans, and other equipment isolated internally. This isolation is not always reliable because of internal short-circuiting, inadequate static deflection, or panel resonances. It is recommended that rooftop equipment over 300 lb be isolated externally, as if internal isolation was not used.

Note 9. To limit undesirable movement, thrust restraints (type 5) are required for all ceiling-suspended and floor-mounted units operating at 2 in. of water or more total static pressure.

Note 10. Pumps over 75 hp may need extra mass and restraints.

Note 11. See text for full discussion.

Isolation for Specific Equipment

Note 12. Refrigeration Machines: Large centrifugal, screw, and reciprocating refrigeration machines may generate very high noise levels; special attention is required when such equipment is installed in upper-story locations or near noise-sensitive areas. If equipment is located near extremely noise-sensitive areas, follow the recommendations of an acoustical consultant.

Note 13. Compressors: The two basic reciprocating compressors are (1) single- and double-cylinder vertical, horizontal or L-head, which are usually air compressors; and (2) Y, W, and multihead or multicylinder air and refrigeration compressors. Single- and double-cylinder compressors generate high vibratory forces requiring large inertia bases (type C) and are generally not suitable for upper-story locations. If this equipment must be installed in an upper-story location or at-grade location near noise-sensitive areas, the expected maximum unbalanced force data must be obtained from the equipment manufacturer and a vibration specialist consulted for design of the isolation system.

Note 18. Cooling Towers: These are normally isolated with restrained spring isolators (type 4) directly under the tower or tower dunnage. High-deflection isolators proposed for use directly under the motor-fan assembly must be used with extreme caution to ensure stability and safety under all weather conditions. See Note 5.

Note 19. Fans and Air-Handling Equipment: Consider the following in selecting isolation systems for fans and air-handling equipment:

1. Fans with wheel diameters of 22 in. and less and all fans operating at speeds up to 300 rpm do not generate large vibratory forces. For fans operating under 300 rpm, select isolator deflection so the isolator natural frequency is 40% or less than the fan speed. For example, for a fan operating at 275 rpm, $0.4 \times 275 = 110$ rpm. Therefore, an isolator natural frequency of 110 rpm or lower is required. This can be accomplished with a 3 in. deflection isolator (type 3).
2. Flexible duct connectors should be installed at the intake and discharge of all fans and air-handling equipment to reduce vibration transmission to air duct structures.
3. Inertia bases (type C) are recommended for all class 2 and 3 fans and air-handling equipment because extra mass allows the use of stiffer springs, which limit heaving movements.
4. Thrust restraints (type 5) that incorporate the same deflection as isolators should be used for all fan heads, all suspended fans, and all base-mounted and suspended air-handling equipment operating at 2 in. or more total static pressure. Restraint movement adjustment must be made under normal operational static pressures.

Vibration Isolators: Materials, Types, and Configurations

Notes 20 through 32 include figures to assist in evaluating commercially available isolators for HVAC equipment. The isolator selected for a particular application depends on the required deflection, life, cost, and compatibility with associated structures.

RUBBER PADS (Type 1)

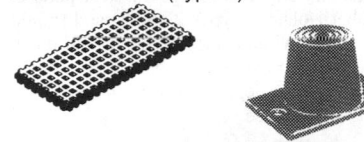

RUBBER MOUNTS (Type 2)

Note 20. Rubber isolators are available in pad (type 1) and molded (type 2) configurations. Pads are used in single or multiple layers. Molded isolators come in a range of 30 to 70 durometer (a measure of stiffness). Material in excess of 70 durometer is usually ineffective as an isolator. Isolators are designed for up to 0.5 in. deflection, but are used where 0.3 in. or less deflection is required. Solid rubber and composite fabric and rubber pads are also available. They provide high load capacities with small deflection and are used as noise barriers under columns and for pipe supports. These pad types work well only when they are properly loaded and the weight load is evenly distributed over the entire pad surface. Metal loading plates can be used for this purpose.

GLASS FIBER PADS (Type 1)

Note 21. Glass fiber with elastic coating (type 1). This type of isolation pad is precompressed molded fiberglass pads individually coated with a flexible, moisture-impervious elastomeric membrane. Natural frequency of fiberglass vibration isolators should be essentially constant for the operating load range of the supported equipment. Weight load is evenly distributed over the entire pad surface. Metal loading plates can be used for this purpose.

SPRING ISOLATOR (Type 3)

Note 22. Steel springs are the most popular and versatile isolators for HVAC applications because they are available for almost any deflection and have a virtually unlimited life. Spring isolators may have a rubber acoustical barrier to reduce transmission of high-frequency vibration and noise that can migrate down the steel spring coil. They should be corrosion-protected if installed outdoors or in a corrosive environment. The basic types include the following:

Note 23. *Open spring isolators* (type 3) consist of top and bottom load plates with adjustment bolts for leveling equipment. Springs should be designed with a horizontal stiffness of at least 80% of the vertical stiffness (k_x/k_y) to ensure stability. Similarly, the springs should have a minimum ratio of 0.8 for the diameter divided by the deflected spring height.

RESTRAINED SPRING ISOLATOR (Type 4)

Note 24. *Restrained spring isolators* (type 4) have hold-down bolts to limit vertical as well as horizontal movement. They are used with (a) equipment with large variations in mass (e.g., boilers, chillers, cooling towers) to restrict movement and prevent strain on piping when water is removed, and (b) outdoor equipment, such as condensing units and cooling towers, to prevent excessive movement due to wind loads. Spring criteria should be the same as open spring isolators, and restraints should have adequate clearance so that they are activated only when a temporary restraint is needed.

Closed mounts or *housed spring isolators* consist of two telescoping housings separated by a resilient material. These provide lateral snubbing and some vertical damping of equipment movement, but do not limit the vertical movement. Care should be taken in selection and installation to minimize binding and short-circuiting.

AIR SPRINGS

ROLLING LOBE   BELLOWS

Note 25. Air springs can be designed for any frequency, but are economical only in applications with natural frequencies of 1.33 Hz or less (6 in. or greater deflection). They do not transmit high-frequency noise and are often used to replace high-deflection springs on problem jobs (e.g., large transformers on upper-floor installations). A constant air supply (an air compressor with an air dryer) and leveling valves are typically required.

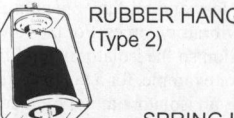
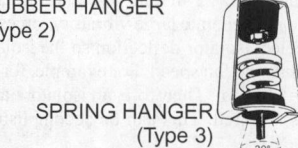

RUBBER HANGER
(Type 2)

SPRING HANGER
(Type 3)

Note 26. Isolation hangers (types 2 and 3) are used for suspended pipe and equipment and have rubber, springs, or a combination of spring and rubber elements. Criteria should be similar to open spring isolators, though lateral stability is less important. Where support rod angular misalignment is a concern, use hangers that have sufficient clearance and/or incorporate rubber bushings to prevent the rod from touching the housing. Swivel or traveler arrangements may be necessary for connections to piping systems subject to large thermal movements.

Precompressed spring hangers incorporate some means of precompression or preloading of the isolator spring to minimize movement of the isolated equipment or system. These are typically used on piping systems that can change weight substantially between installation and operation.

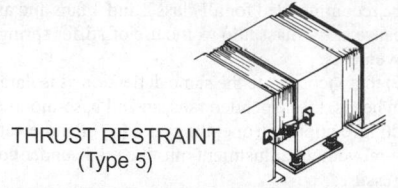

THRUST RESTRAINT
(Type 5)

Note 27. Thrust restraints (type 5) are similar to spring hangers or isolators and are installed in pairs to resist the thrust caused by air pressure. These are typically sized to limit lateral movement to 0.25 in. or less.

DIRECT ISOLATION (Type A)

Note 28. Direct isolation (type A) is used when equipment is unitary and rigid and does not require additional support. Direct isolation can be used with large chillers, some fans, packaged air-handling units, and air-cooled condensers. If there is any doubt that the equipment can be supported directly on isolators, use structural bases (type B) or inertia bases (type C), or consult the equipment manufacturer.

STRUCTURAL BASES (Type B)

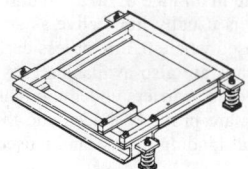

Note 29. Structural bases (type B) are used where equipment cannot be supported at individual locations and/or where some means is necessary to maintain alignment of component parts in equipment. These bases can be used with spring or rubber isolators (types 2 and 3) and should have enough rigidity to resist all starting and operating forces without supplemental hold-down devices. Bases are made in rectangular configurations using structural members with a depth equal to one-tenth the longest span between isolators. Typical base depth is between 4 and 12 in., except where structural or alignment considerations dictate otherwise.

STRUCTURAL RAILS (Type B)

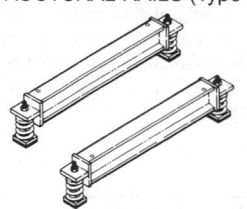

Note 30. Structural rails (type B) are used to support equipment that does not require a unitary base or where the isolators are outside the equipment and the rails act as a cradle. Structural rails can be used with spring or rubber isolators and should be rigid enough to support the equipment without flexing. Usual practice is to use structural members with a depth one-tenth of the longest span between isolators, typically between 4 and 12 in., except where structural considerations dictate otherwise.

CONCRETE BASES (Type C)

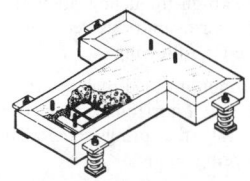

Note 31. Concrete bases (type C) are used where the supported equipment requires a rigid support (e.g., flexible-coupled pumps) or excess heaving motion may occur with spring isolators. They consist of a steel pouring form usually with welded-in reinforcing bars, provision for equipment hold-down, and isolator brackets. Like structural bases, concrete bases should be sized to support piping elbow supports, rectangular or T-shaped, and for rigidity, have a depth equal to one-tenth the longest span between isolators. Base depth is typically between 6 and 12 in. unless additional depth is specifically required for mass, rigidity, or component alignment.

CURB ISOLATION (Type D)

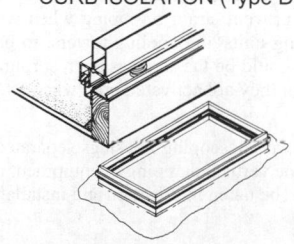

Note 32. Curb isolation systems (type D) are specifically designed for curb-supported rooftop equipment and have spring isolation with a watertight, and sometimes airtight, assembly. *Rooftop rails* consist of upper and lower frames separated by nonadjustable springs and rest on top of architectural roof curbs. *Isolation curbs* incorporate the roof curb into their design as well. Both kinds are designed with springs that have static deflections in the 1 to 3 in. range to meet the design criteria described in type 3. Flexible elastomeric seals are typically most effective for weatherproofing between the upper and lower frames. A continuous sponge gasket around the perimeter of the top frame is typically applied to further weatherproof the installation.

and type and operating duty of vibration sources. Vibration propagation through soil may also need to be considered.

INTERNAL VERSUS EXTERNAL ISOLATION

Vibration isolators are most effective if the isolator base is attached directly to the building structure at a support point possessing high stiffness compared to the stiffness of the isolator. In many cases, the vibrating equipment (e.g., internal components of air-handling units) can be effectively isolated with internal vibration isolators, where only the moving parts (e.g., fan/motor assembly) are supported by the isolators. This approach can reduce the load supported by the isolators and thus can reduce the cost of isolation. The other primary advantage of internal isolation is reduction of vibration and structureborne noise into the air-handling unit housing. Disadvantages of internal isolation can include the following:

- The isolator is often not easily visible in the field to verify that it is functioning properly.
- The isolator support point may be near the middle of a beam, which often provides inadequate stiffness for optimum isolator performance.
- Short-circuiting of housed isolators, caused by horizontal thrust of the fan, can occur.
- Commonly provided isolators are not selected based on building support structure or noise criteria, and may not provide sufficient vibration control.
- Internal isolation does nothing to reduce vibration in equipment casing and structure caused by air movement.

Typically, vibration isolation devices should be applied to either the internal components or the external casing, but not both.

It is possible to use both internal and external vibration isolation on the same air-handling unit, but isolator stiffness selection must avoid resonances at or near normal fan and motor rotational speeds. There are multiple resonance frequencies to consider: if the fan or motor operates at or near one of these frequencies, vibration levels could become excessive. The probability of such an interaction increases significantly if the fan and motor are driven by a variable-frequency drive. Implementing both internal and external vibration isolation on the same unit should only be attempted with the guidance of an experienced vibration consultant.

ISOLATING VIBRATION AND NOISE IN PIPING SYSTEMS

All piping systems have mechanical vibration generated by the equipment and impeller-generated and flow-induced vibration and noise, which is transmitted by the pipe wall and the water column. In addition, equipment supported by vibration isolators exhibits some motion from pressure thrusts during operation. Vibration isolators have even greater movement during start-up and shutdown as equipment vibration passes through the isolators' resonance frequency. The piping system must be flexible enough to (1) reduce vibration transmission along the connected piping, (2) allow equipment movement without reducing the performance of vibration isolators, and (3) accommodate equipment movement or thermal movement of the piping at connections without imposing undue strain on the connections and equipment.

Flow noise and vibration in piping can be reintroduced by turbulence, sharp pressure drops, and entrained air; however, this can be minimized by sizing pipe so that velocities are 4 fps maximum for pipe 2 in. and smaller and using a pressure drop limitation of 4 ft of water per 100 ft of pipe length with a maximum velocity of 10 fps for larger pipe sizes. Care should be taken not to exceed these limits.

Resilient Pipe Hangers and Supports

Resilient pipe hangers and supports may be used to prevent vibration and noise transmission from the piping system to the building structure and to provide flexibility in the piping.

Suspended Piping. Isolation hangers described in Note 26 of Table 47 should be used for all piping in equipment rooms and up to 50 ft from vibration-isolated equipment and PRV stations. To avoid reducing the effectiveness of equipment isolators, at least the first three hangers from the equipment should provide the same deflection as the equipment isolators, with a maximum limitation of 2 in. deflection; the remaining hangers should be spring or combination spring and rubber with 0.75 in. deflection.

The first two hangers adjacent to the equipment should be the positioning or precompressed type, to prevent load transfer to equipment flanges when the piping system is filled. The positioning hanger aids in installing large pipe, and many engineers specify this type for all isolated pipe hangers for piping 8 in. and larger.

Piping over 2 in. in diameter that is suspended below or within 50 ft noise-sensitive areas should be hung with isolation hangers. Hangers adjacent to noise-sensitive areas should be the spring and rubber combination type 3.

Floor-Supported Piping. Floor supports for piping in equipment rooms and adjacent to isolated equipment should use vibration isolators as described in Table 47. They should be selected according to the guidelines for hangers. The first two adjacent floor supports should be the restrained spring type, with a blocking feature that prevents load transfer to equipment flanges as the piping is filled or drained. Where pipe is subjected to large thermal movement, a slide plate (PTFE, graphite, or steel) should be installed on top of the isolator, and a thermal barrier should be used when rubber products are installed directly beneath steam or hot-water lines.

Riser Supports, Anchors, and Guides. Many piping systems have anchors and guides, especially in the risers, to permit expansion joints, bends, or pipe loops to function properly. Anchors and guides are designed to eliminate or limit (guide) pipe movement and must be rigidly attached to the structure; this is inconsistent with the resiliency required for effective isolation. The engineer should try to locate the pipe shafts, anchors, and guides in noncritical areas, such as next to elevator shafts, stairwells, and toilets, rather than adjoining noise-sensitive areas. Where concern about vibration transmission exists, some type of vibration isolation support or acoustical support is required for pipe supports, anchors, and guides.

Because anchors or guides must be rigidly attached to the structure, the isolator cannot deflect in the sense previously discussed, and the primary interest is that of an acoustical barrier. Heavy-duty rubber pads that can accommodate large loads with minimal deflection can provide such an acoustical barrier. Figure 43 shows some arrangements for resilient anchors and guides. Similar resilient supports can be used for the pipe.

Resilient supports for pipe, anchors, and guides can attenuate noise transmission, but they do not provide the resiliency required to isolate vibration. Vibration must be controlled in an anchor guide system by designing flexible pipe connectors and resilient isolation hangers or supports.

Completely spring-isolated riser systems that eliminate the anchors and guides have been used successfully in many instances and give effective vibration and acoustical isolation. In this type of isolation system, the springs are sized to accommodate thermal growth as well as to guide and support the pipe. These systems provide predictable load transfer because of thermal expansion and contraction, but require careful engineering to accommodate movements encountered not only in the riser but also in the branch takeoff to avoid overstressing the piping.

Piping Penetrations. HVAC systems typically have piping that must penetrate floors, walls, and ceilings. If these penetrations are not properly treated, they provide a path for airborne noise, which can destroy the acoustical integrity of the occupied space. Seal open-

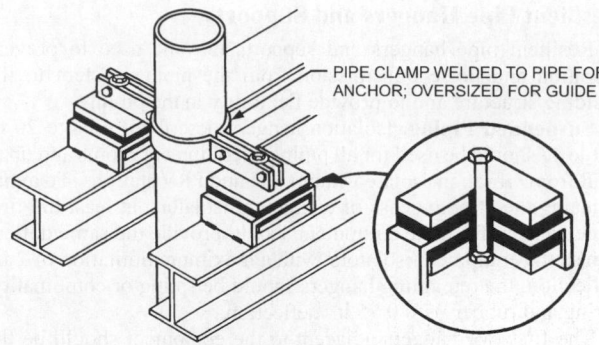

PIPE CLAMP WELDED TO PIPE FOR ANCHOR; OVERSIZED FOR GUIDE

RESILIENTLY ISOLATED PIPE ANCHORS AND GUIDES

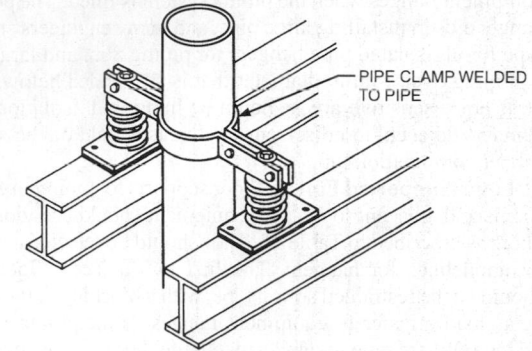

PIPE CLAMP WELDED TO PIPE

SPRING-ISOLATED RISER SYSTEM

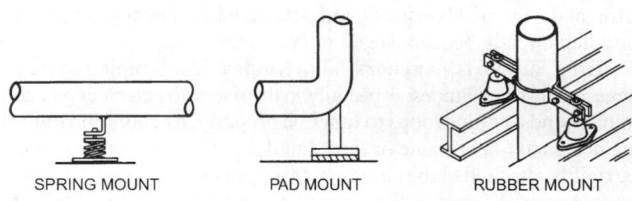

SPRING MOUNT   PAD MOUNT   RUBBER MOUNT

CONVENTIONAL ISOLATORS AS PIPE SUPPORTS FOR LINES WITH EXPANSION JOINTS

Fig. 43   Resilient Anchors and Guides for Pipes

Table 48   Recommended Live Lengths[a] of Flexible Rubber and Metal Hose

Nominal Diameter, in.	Length,[b] in.	Nominal Diameter, in.	Length,[b] in.
0.75	12	4	18
1	12	5	24
1.5	12	6	24
2	12	8	24
2.5	12	10	24
3	18	12	36

[a]Live length is end-to-end length for integral flanged rubber hose and is end-to-end less total fitting length for all other types.
[b]Per recommendations of Rubber Expansion Division, Fluid Sealing Association.

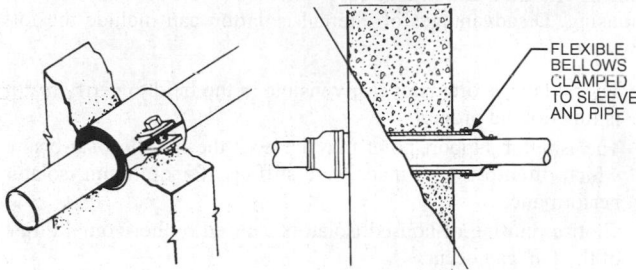

FLEXIBLE BELLOWS CLAMPED TO SLEEVE AND PIPE

Fig. 44   Acoustical Pipe Penetration Seals

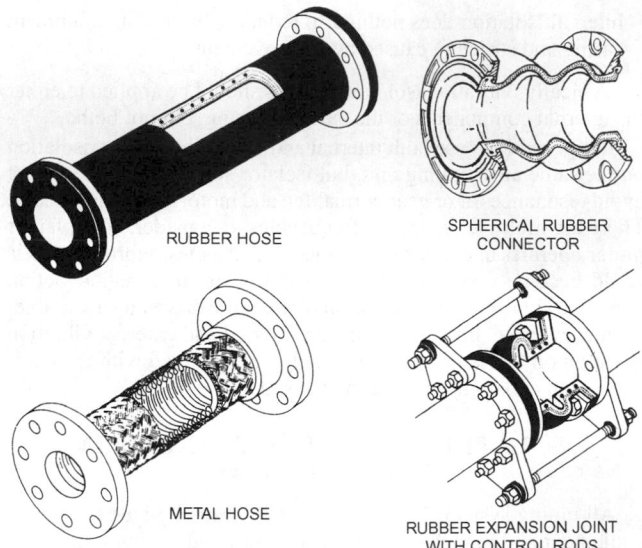

RUBBER HOSE   SPHERICAL RUBBER CONNECTOR

METAL HOSE   RUBBER EXPANSION JOINT WITH CONTROL RODS

Fig. 45   Flexible Pipe Connectors

ings in pipe sleeves between noisy areas, such as equipment rooms, and occupied spaces with an acoustical barrier such as fibrous material and resilient acoustical caulking, or with engineered pipe penetration seals as shown in Figure 44.

Flexible Pipe Connectors. Flexible pipe connectors (1) provide piping flexibility to permit isolators to function properly, (2) protect equipment from strain caused by misalignment and expansion or contraction of piping, and (3) attenuate noise and vibration transmission along the piping (Figure 45). Connectors are available in two configurations: (1) hose type, a straight or slightly corrugated wall construction of either rubber or metal; and (2) arched or expansion-joint type, a short-length connector with one or more large-radius arches, of rubber, PTFE, or metal. Metal expansion joints are acoustically ineffective and are seldom successfully used for vibration and sound isolation in HVAC systems; they should not be expected to substitute for conventional pipe vibration isolators.

To accommodate pressure thrust, flexible connectors require an end restraint, which is either (1) added to the connector, (2) incorporated by its design, (3) added to the piping system (anchoring), or (4) built in by the stiffness of the system. Connector extension caused by pressure thrust on isolated equipment should also be considered when flexible connectors are used. Overextension causes failure. Manufacturers' recommendations on restraint, pressure, and temperature limitations must be strictly observed.

Hose Connectors. Hose connectors accommodate lateral movement perpendicular to length and have very limited or no axial movement capability. Rubber hose connectors can have molded or hand-wrapped construction with wire reinforcing, and are available with metal-threaded end fittings or integral rubber flanges. Application of threaded fittings should be limited to 3 in. and smaller pipe diameter. The fittings should be the mechanically expanded type to minimize the possibility of pressure thrust blowout. Flanged types are available in larger pipe sizes. Table 48 lists recommended lengths.

Metal hose is constructed with a corrugated inner core and a braided cover, which helps attain a pressure rating and provides end restraints that eliminate the need for supplemental control assemblies. Short lengths of metal hose or corrugated metal bellows, or pump connectors, are available without braid and have built-in control assemblies. Metal hose is used to control misalignment and vibration rather than noise and is used primarily where temperature or pressure of flow media precludes the use of other material. Table 48 provides recommended lengths.

Expansion Joint or Arched Connectors. Expansion joint or arched connectors have one or more convolutions or arches and can accommodate all modes of axial, lateral, and angular movement and misalignment. When made of rubber, they are commonly called expansion joints, spool joints, or spherical connectors; in PTFE, they are known as couplings or expansion joints.

Rubber expansion or spool joints are available in two basic types: (1) hand-wrapped with wire and fabric reinforcing, and (2) molded with fabric and wire or with high-strength fabric only (instead of metal) for reinforcing. The handmade type is available in a variety of materials and lengths for special applications. Rubber spherical connectors are molded with high-strength fabric or tire cord reinforcing instead of metal. Their distinguishing characteristic is a large-radius arch. The shape and construction of some designs allow use without control assemblies in systems operating to 150 psi, and are the most effective for minimizing transmission of vibration. Where thrust restraints are not built in, they must be used as described for rubber hose joints.

PTFE expansion joints and couplings are similar in construction to rubber expansion joints with reinforcing metal rings.

In evaluating these devices, consider temperature, pressure, and service conditions as well as each device's ability to attenuate vibration and noise. Metal hose connections can accommodate misalignment and attenuate mechanical vibration transmitted through the pipe wall, but do little to attenuate noise. This type of connector has superior resistance to long-term temperature effects. Rubber hose, expansion joints, and spherical connectors attenuate vibration and impeller-generated noise transmitted through the pipe wall. Because rubber expansion joint and spherical connector walls are flexible, they have the ability to grow volumetrically and attenuate noise and vibration at blade-pass frequencies. This is a particularly desirable feature in uninsulated piping systems, such as for condenser or domestic water, which may run adjacent to noise-sensitive areas. However, high pressure has a detrimental effect on the ability of the connector to attenuate vibration and noise.

Because none of the flexible pipe connectors control flow or velocity noise or completely isolate vibration and noise transmission to the piping system, resilient pipe hangers and supports should be used; these are shown in Note 26 for Table 47 and are described in the Resilient Pipe Hangers and Supports section.

Isolating Duct Vibration

Flexible canvas and rubber duct connections should be used at fan intake and discharge. However, they are not completely effective because they become rigid under pressure, allowing the vibrating fan to pull on the duct wall. To maintain a slack position of the flexible duct connections, thrust restraints (see Note 27, Table 47) should be used on all equipment as indicated in Table 47.

Although vibration transmission from ducts isolated by flexible connectors is not common, flow pulsations within the duct can cause mechanical vibration in the duct walls, which can be transmitted through rigid hangers. Spring or combination spring and rubber hangers are recommended wherever ducts are suspended below or near a noise-sensitive area.

SEISMIC PROTECTION

Seismic restraint requirements are specified by applicable building codes that define design forces to be resisted by the mechanical system, depending on building location and occupancy, location of the system in the building, and whether it is used for life safety. Where required, seismic protection of resiliently mounted equipment poses a unique problem, because resiliently mounted systems are much more susceptible to earthquake damage from overturning forces, the impact limits of bare restraints, and resonances inherent in vibration isolators.

A deficiency in seismic restraint design or anchorage may not become apparent until an earthquake occurs, with possible catastrophic consequences. Adequacy of the restraint system and anchorage to resist code design forces must be verified before the event, by either equipment tests, calculations, or dynamic analysis, depending on the item, with calculations or dynamic analysis performed under the direction of a professional engineer. These analysis items may be supplied as a package by the vibration isolation vendor.

Restraints for floor-mounted equipment should be designed with adequate clearances so that they are not engaged during normal operation of the equipment. Contact surfaces (snubbers) should be protected with resilient pad material to limit shock during an earthquake, and restraints should be strong enough to resist the forces in any direction. The integrity of these devices can be verified by a comprehensive analysis, but is more frequently verified by laboratory tests.

Calculations or dynamic analyses should have an engineer's seal to verify that input forces are obtained in accordance with code or specification requirements. Additionally, a professional engineer should make the anchorage calculations in accordance with accepted standards. For more information, see Chapter 55.

VIBRATION INVESTIGATIONS

Theoretically, a vibration-isolation system can be designed to mitigate even the most extreme sources of mechanical vibration. However, isolators should not be used to mask a condition that should be corrected before it damages the equipment and its operation. High vibration levels can indicate a faulty equipment operating condition in need of correction, or they can be a symptom of a resonance interaction between a vibrating piece of equipment and the structure(s) on which it is supported or to which it is attached.

Vibration investigations can include

- Measurement of vibration levels on vibrating equipment (see Figure 42 for recommended vibration criteria)
- Measurement of vibration levels in building structures to which vibrating equipment is connected, such as a building floor, piping systems, etc. (see Figure 41 and Table 45 for recommended building vibration criteria)
- Examination of equipment vibration generated by system components, such as bearings, drives, pumps, etc.
- Measurement of the natural frequencies (resonances) of vibrating equipment or connected structure(s)
- Examination of equipment installation factors, such as equipment alignment, vibration isolator placement, etc. (see Table 47)
- Measurement of the unbalance of reciprocating or rotating equipment components

COMMISSIONING

In the initial design and final commissioning phases of an HVAC system, sound criteria are needed to determine the degree of noise impact and the amount of noise and vibration reduction required for acceptable background sound and vibration levels based on occupancy usage. This chapter is intended primarily to assist with the design phase and provide limited assistance with diagnosing problems. Detailed diagnosis of problems may require the assistance of an acoustical consultant or an engineer experienced in HVAC system noise and vibration analysis. The section on Testing for Sound and Vibration in Chapter 38 should be consulted for the commissioning phase.

TROUBLESHOOTING

Despite all precautions, situations may arise where there is disturbing noise and vibration. Problems can be identified and corrected by

- Determining which equipment or system is the source of the problem
- Determining whether the problem is one of airborne sound, vibration (structureborne noise), or a combination of both
- Applying appropriate solutions

Troubleshooting can be time-consuming, expensive, and difficult, and use of an experienced acoustical consultant is often warranted. Proper diagnosis of the problem is most critical to allow for developing the right solution. Once a noise or vibration transmission problem exists, occupants become more sensitive and require greater reduction of the sound and vibration levels than would initially have been satisfactory. The need for troubleshooting should be minimized by properly designing, installing, and testing the system as soon as it is operational and before the building is occupied.

DETERMINING PROBLEM SOURCE

The system or equipment that is the source of the problem can often be determined without instrumentation. Vibration and noise levels are usually well above the sensory level of perception and are readily felt or heard.

A simple, accurate method of determining the problem source is to turn individual pieces of equipment on and off until the vibration or noise is eliminated. Because the source of the problem is often more than one piece of equipment or the interaction of two or more systems, it is always good practice to double check by shutting off the system and operating the equipment individually. Reynolds and Bevirt (1994) and Schaffer (2005) provide practical information on the measurement and assessment of sound and vibration in buildings.

DETERMINING PROBLEM TYPE

Once the source is identified, the next step is to determine whether the problem is one of noise or vibration. Clearly perceptible vibration is often a clue that vibration transmission is the major cause of the problem. The possibility that lightweight wall or ceiling panels are excited by airborne noise should be considered. However, even if the vibration is not readily perceptible, the problem may still be one of vibration transmission causing structureborne noise. This can be checked using the following procedure:

- If a sound level meter is available, some readings should be taken. If the difference between C-weighted and overall (unweighted or linear) readings is greater than 6 dB, or if the slope of the acoustic spectrum is steeper than 6 dB per octave at low frequencies (below 63 Hz), vibration is likely a contributing factor.
- If excessive noise is found close to the equipment and/or main ductwork, airborne noise is probably the main contributor.
- If the affected area is remote from source equipment, there is no problem in intermediary spaces, and noise does not appear to be coming from the duct system or diffusers, structureborne noise is probably the cause.

One important step in diagnosing many noise or vibration problems, particularly if the affected area is close to the mechanical equipment room, is to check the equipment's vibration isolation system. A simple test is to have one person listen in the affected area while another shouts loudly in the equipment room. If the voice cannot be heard, the problem is likely one of structureborne noise. If the voice can be heard, check for openings in the wall or floor separating the areas. If no such openings exist, the structure separating the areas does not provide adequate transmission loss. In

these situations, see the section on Mechanical Equipment Room Sound Isolation for possible solutions.

Noise Problems

If ductborne sound (i.e., noise from grilles or diffusers or duct breakout noise) appears to be the problem, measure the sound-pressure levels and compare them with the design criteria (NC, RC, etc.). It is often helpful to obtain sound data with and without terminal devices installed. Comparison of the two results shows how much noise a given terminal device contributes. If this reveals the responsible components, the engineer can analyze each sound source using the procedures presented in this chapter to determine whether sufficient attenuation has been provided.

If the sound source is a fan, pump, or similar rotating equipment, an important question is whether it is operating near the most efficient part of its operating curve, where most equipment operates best and generates predictable levels of sound and vibration as published by equipment manufacturers and used in the building design. Excessive vibration and noise can occur if a fan or pump is trying to move too little or too much air or water. Check that vanes, dampers, and valves are in the correct operating position and that the system has been properly balanced.

Vibration Problems

Vibration and structureborne noise problems can be caused by

- Equipment improperly specified or installed, poorly balanced, misaligned, or operating outside of design conditions
- Equipment with inadequate or improper vibration isolation
- Flanking transmission paths such as rigid pipe or duct connections, obstructions under the base of vibration-isolated equipment, improperly installed equipment seismic restraints shorting vibration isolation, or shipping blocks not removed after the equipment has been installed and in operation
- Excessive floor flexibility indicative of improper structural support conditions for equipment or inadequate or improper vibration isolation
- Resonances in equipment, vibration isolation system, building structure, or connected structures (e.g., piping)

Most field-encountered problems result from improperly selected or installed isolators and flanking paths of transmission, which can be simply evaluated and corrected. If the equipment lacks vibration isolators, in many cases it is possible to add isolators (see Table 47) without altering connected ducts or piping by using structural brackets. Floor flexibility and resonance problems are sometimes encountered and usually require analysis by experts. However, the procedures in the following sections can help identify such problems.

Testing Vibration Isolation Systems. Improperly functioning vibration isolation systems are the cause of most field-encountered problems and can be evaluated and corrected by the following procedures:

1. Ensure that the system is free-floating by bouncing the base, which should cause the equipment to move up and down freely and easily. On floor-mounted equipment, check that there are no obstructions between the base and the floor that would short-circuit the isolation system. This is best accomplished by passing a rod under the equipment. A small obstruction might allow the base to rock, giving the impression that it is free-floating when it is not. On suspended equipment, make sure that rods are not touching the hanger box. Rigid connections such as pipes and ducts can prevent equipment from floating freely, prohibit isolators from functioning properly, and provide flanking paths for vibration transmission.
2. Determine whether isolator static deflection is as specified or required, changing it if necessary, as recommended in Table 47. A common problem is inadequate deflection caused by underloaded

isolators. Overloaded isolators are not generally a problem as long as the system is free-floating and there is space between the spring coils.

With most commonly used spring isolators, static deflection can be determined by measuring the operating height and comparing it to free-height information available from the manufacturer. Once the actual isolator deflection is known, determine its adequacy by comparing it with the recommended deflection in Table 47.

The efficiency of a vibration isolator depends on the ratio of the forcing frequency to the natural frequency of the isolator. If the natural frequency of the isolator is less than 25% of the forcing frequency (usually considered the operating speed of the equipment), the isolator will operate at an efficiency of 95% or greater, and generally provides sufficient vibration isolation except in solutions where heavy equipment is installed on extremely long-span floors or very flexible floors or roofs. If a transmission problem exists, it may be caused by (1) excessively rough equipment operation, (2) the system not being free-floating or flanking path transmission, or (3) a resonance or floor stiffness problem.

It is easy to determine the natural frequency of spring isolators from the static deflection determined by spring height measurements, but these measurements are difficult with pad and elastomeric isolators and are often not accurate in determining their natural frequencies. Although such isolators can theoretically provide natural frequencies as low as 4 Hz, they typically provide higher natural frequencies and generally do not provide the desired isolation efficiencies for upper floor equipment locations. Therefore, it is recommended to avoid using elastomeric mounts in general for (1) equipment on elevated floors, (2) major equipment, (3) critical applications, and (4) equipment on variable-speed operation; in all such cases, spring isolation should be considered and properly specified.

In general, it is very difficult to determine whether vibration isolation efficiencies intended in design have been achieved in field installations using field vibration measurements. However, vibration measurements can readily be made on vibrating equipment, equipment supports, floors supporting vibration-isolated equipment, and floors in adjacent areas to determine whether vibration criteria specified in Table 45 or in Figures 37 and 38 have been achieved.

Floor Flexibility Problems. Floor flexibility problems can occur with heavy equipment installed on long-span floors or thin slabs and with rooftop equipment installed on light structures of open web joist construction. If floor flexibility is suspected, the isolators should be one-tenth or less as stiff as the floor to eliminate the problem. Floor stiffness can be determined by calculating the additional floor deflection caused by a specific piece of equipment.

For example, if a 10,000 lb piece of equipment causes floor deflection of an additional 0.1 in., floor stiffness is 100,000 lb/in., and an isolator combined stiffness of 10,000 lb/in. or less must be used. Note that floor stiffness or spring rate, not total floor deflection, is determined. In this example, the total floor deflection might be 1 in., but if the problem equipment causes 0.1 in. of that deflection, 0.1 in. is the factor that identifies floor stiffness of 100,000 lb/in.

As a general guideline, limiting the additional floor deflection (not total deflection) due to the weight of the equipment to 0.3 in. is advisable, even when the equipment is provided with proper vibration isolation. This may need to be further reduced for vibration in acoustically critical adjacencies.

Resonance Problems. These problems occur when the equipment's operating speed is the same as or close to the resonance frequency of (1) an equipment component such as a fan shaft or bearing support pedestal, (2) the vibration isolation system, or (3) the resonance frequency of the floor or other building component, such as a wall. Vibration resonances can cause excessive equipment vibration levels, as well as objectionable and possibly destructive

vibration transmission in a building. These conditions must always be identified and corrected.

When vibrating mechanical equipment is mounted on vibration isolators on a flexible floor, there are two resonance frequencies that must be considered: that of the floor and that of the isolated equipment. The lower frequency should be controlled by the stiffness (and consequently the static deflection) of the vibration isolators. This frequency should be significantly less than the normal operating speed (or frequency) of the mechanical equipment and is generally not a problem. The higher resonance frequency is associated with and primarily controlled by the stiffness of the supporting structure. This resonance frequency is usually not affected by increasing or decreasing the static deflection of the mechanical equipment vibration isolators.

Sometimes, when the floor under mechanical equipment is flexible (as occurs with some long-span floor systems and with roof systems supporting rooftop packaged units), the operating speed of the mechanical equipment can coincide with the floor resonance frequency. When this occurs, changing the static deflection of the vibration isolators may not solve the problem. Alternatives include changing the rotating speed of the equipment, stiffening the structure, or adjusting a variable-frequency drive to avoid the resonant frequency.

Vibration Isolation System Resonance. Always characterized by excessive equipment vibration, vibration isolation system rigid-body resonance (characterized by the mass of the equipment vibrating on the stiffness of the isolators) usually results in objectionable transmission to the building structure. However, transmission might not occur if the equipment is on grade or on a stiff floor. Vibration isolation system rigid-body resonances can be measured with instrumentation or, more simply, by determining the isolator natural frequency as described in the section on Testing Vibration Isolation Systems and comparing this figure to the operating speed of the equipment.

When a vibration isolation system resonance problem exists, the system natural frequency must be changed using the following guidelines:

- If equipment is installed on excessively stiff pad or rubber elastomeric mounts, isolators with the deflection recommended in Table 47 should be installed.

- If equipment is installed on spring isolators and there is objectionable vibration or noise transmission to the structure, determine whether the isolator is providing the designed static deflection. For example, an improperly selected or installed nominal 2 in. deflection isolator could be experiencing only 1/8 in. deflection under its static load, which would be in resonance with equipment operating at 500 rpm. If this is the case, the isolators should be replaced with ones having enough capacity to provide the requisite 2 in. deflection. However, if there is no transmission problem with the isolators, it is not necessary to use greater-deflection isolators than can be conveniently installed.

- If equipment is installed on spring isolators and there is objectionable noise or vibration transmission, replace the isolators with springs of the deflection recommended in Table 47.

- If equipment is installed on spring isolators of the recommended stiffness and there is objectionable high-frequency (200 Hz or greater) noise or vibration, it is possible that resonances internal to the spring are the culprit. These resonances, sometimes called surge frequencies, can be important in applications where equipment (e.g., screw compressors, inverters) generates high-frequency noise. To control their adverse effects, many isolator designs incorporate an elastomeric pad under the spring. It may also be possible to identify an elastomeric mount that can provide the desired static deflection; these typically have better high-frequency characteristics than springs.

Building Resonances. These problems occur when some part of the structure has a resonance frequency coincident with the disturbing frequency (often the operating speed) of some of the equipment. These problems can exist even if the isolator deflections recommended in Table 48 are used. The resulting objectionable noise or vibration should be evaluated and corrected. Often, the resonance problem is associated with the floor on which the equipment is installed, but it can also occur in a remotely located floor, wall, or other building component. If a noise or vibration problem has a remote source that cannot be associated with piping or ducts, building resonance must be suspected.

Building resonance problems can be resolved by the following:

- Reduce the vibration force by balancing the equipment. This is not a practical solution for a true resonance problem. However, it is effective when the disturbing frequency is close to the floor's natural frequency, as evidenced by the equal displacement of the floor and the equipment, especially when the equipment is operating with excessive vibration.
- Change the disturbing frequency by changing the equipment operating speed. This is practical only for belt-driven equipment, or equipment driven by variable-frequency drives.
- Modify the structure to shift the structural response. Although this requires upsizing the structure and can be costly, if feasible it is often the most effective means of resolving vibration issues.

REFERENCES

AMCA. 2005. Reverberant room method for sound testing of fans. *Standard* 300-2005. Air Movement and Control Association International, Arlington Heights, IL.

AMCA. 1997. Laboratory method of testing to determine the sound power in a duct. ANSI/AMCA *Standard* 330-97. Air Movement and Control Association International, Arlington Heights, IL.

ANSI. 2006. Specification for sound level meters. *Standard* S1.4-1983 (R2006). American National Standards Institute, Washington, D.C.

ANSI. 2004. Specification for octave-band and fractional-octave-band analog and digital filters. *Standard* S1.11-2004. American National Standards Institute, Washington, D.C.

ANSI. 2002. Specifications for integrating-averaging sound level meters. *Standard* S1.43-1997 (R2002). American National Standards Institute, Washington, D.C.

ANSI. 1995. Criteria for evaluating room noise. *Standard* S12.2-1995. American National Standards Institute, Washington, D.C.

ANSI. 1983. Guide to the evaluation of human exposure to vibration in buildings. *Standard* S2.71-1983. American National Standards Institute, Washington, D.C.

AHRI. 2001. Sound rating of ducted air moving and conditioning equipment. *Standard* 260-2001. Air-Conditioning, Heating, and Refrigeration Institute, Arlington, VA.

AHRI. 2001. Sound rating of large outdoor refrigerating and air-conditioning equipment. *Standard* 370-2001. Air-Conditioning, Heating, and Refrigeration Institute, Arlington, VA.

AHRI. 1994. Method of measuring machinery sound within an equipment space. *Standard* 575-1994. Air-Conditioning, Heating, and Refrigeration Institute, Arlington, VA.

AHRI. 1998. Air terminals. *Standard* 880-1998. Air-Conditioning, Heating, and Refrigeration Institute, Arlington, VA.

AHRI. 1998. Procedure for estimating occupied space sound levels in the application of air terminals and air outlets. *Standard* 885-1998. Air-Conditioning, Heating, and Refrigeration Institute, Arlington, VA.

AHRI. 2001. Rating of air diffusers and air diffuser assemblies. *Standard* 890-2001. Air-Conditioning, Heating, and Refrigeration Institute, Arlington, VA.

ASA. 2009. Acoustical performance criteria, design requirements, and guidelines for schools, part 2. Relocatable classroom factors. ANSI/ASA *Standard* S12.60-2009. Acoustical Society of America, Melville, NY.

ASA. 2010. Acoustical performance criteria, design requirements, and guidelines for schools, part 1: Permanent schools. ANSI/ASA *Standard* S12.60-2010. Acoustical Society of America, Melville, NY.

ASHRAE. 1997. Laboratory method of testing to determine the sound power in a duct. ANSI/ASHRAE *Standard* 68-1997.

ASHRAE. 2006. Method of testing for rating the performance of air outlets and inlets. ANSI/ASHRAE *Standard* 70-2006.

ASTM. 2006. Test method for measuring acoustical and airflow performance of duct liner materials and prefabricated silencers. *Standard* E477-2006a. American Society for Testing and Materials, West Conshohocken, PA.

Beatty, J. 1987. Discharge duct configurations to control rooftop sound. *Heating/Piping/Air Conditioning* (July).

Beranek, L.L. 1960. *Noise reduction.* McGraw-Hill, New York.

Beranek, L.L. 1971. *Noise and vibration control.* McGraw-Hill, New York.

Beranek, L.L. 1989. Balanced noise criterion (NCB) curves. *Journal of the Acoustic Society of America* (86):650-654.

Blazier, W.E., Jr. 1981a. Revised noise criteria for design and rating of HVAC systems. *ASHRAE Transactions* 87(1):647-657.

Blazier, W.E., Jr. 1981b. Revised noise criteria for application in the acoustical design and rating of HVAC systems. *Noise Control Engineering Journal* 16(2):64-73.

Blazier, W.E., Jr. 1993. Control of low frequency noise in HVAC air-handling equipment and systems. *ASHRAE Transactions* 99(2):1031-1036.

Blazier, W.E., Jr. 1995. Sound quality considerations in rating noise from heating, ventilating and air-conditioning (HVAC) systems in buildings. *Noise Control Engineering Journal* 43(3).

Blazier, W.E., Jr. 1997. RC Mark II; a refined procedure for rating the noise of heating, ventilating and air-conditioning (HVAC) systems in buildings. *Noise Control Engineering Journal* 45(6).

Broner, N. 1994. Low-frequency noise assessment metrics—What do we know? (RP-714). *ASHRAE Transactions* 100(2):380-388.

Bullock, C.E. 1970. Aerodynamic sound generation by duct elements. *ASHRAE Transactions* 76(2):97-109.

Cummings, A. 1983. Acoustic noise transmission through the walls of air-conditioning ducts. *Final Report*, Department of Mechanical and Aerospace Engineering, University of Missouri, Rolla.

Cummings, A. 1985. Acoustic noise transmission through duct walls. *ASHRAE Transactions* 91(2A):48-61.

Cunefare, K. and A. Michaud. 2008. Reflection of airborne noise at duct terminations. ASHRAE Research Project RP-1314, *Final Report*.

Ebbing, C.E. and W.E. Blazier, Jr. 1992. HVAC low frequency noise in buildings. *Proceedings INTER-NOISE* 92(2):767-770.

Ebbing, C.E. and W.E. Blazier, Jr. 1998. *Application of manufacturer's sound data.* ASHRAE.

Ebbing, C.E., D. Fragnito, and S. Inglis. 1978. Control of low frequency duct-generated noise in building air distribution systems. *ASHRAE Transactions* 84(2):191-203.

Egan, M.D. 1988. *Architectural acoustics.* McGraw-Hill, New York.

Harold, R.G. 1986. Round duct can stop rumble noise in air-handling installations. *ASHRAE Transactions* 92(2A):189-202.

Harold, R.G. 1991. Sound and vibration considerations in rooftop installations. *ASHRAE Transactions* 97(1):445-451.

IEST. 2005. Considerations in clean room design. IEST *Recommended Practice* RP-CC012.1. Institute of Environmental Sciences and Technology, Rolling Meadows, IL.

Ingard, U., A. Oppenheim, and M. Hirschorn. 1968. Noise generation in ducts (RP-37). *ASHRAE Transactions* 74(1):V.1.1-V.1.10.

IRD. 1988. *Vibration technology,* vol. 1. IRD Mechanalysis, Columbus, OH.

ISO. 2003. Mechanical vibration and shock—Evaluation of human exposure to whole body vibration, Part 2: Vibration in buildings. *Standard* 2631-2-2003. International Organization for Standardization, Geneva, Switzerland.

ISO. 1966. Acoustics—Attenuation of sound during propagation outdoors—Part 2: General method of calculation. *Standard* 9613-2-1966. International Organization for Standardization, Geneva, Switzerland.

Kuntz, H.L. 1986. The determination of the interrelationship between the physical and acoustical properties of fibrous duct liner materials and lined duct sound attenuation. *Report* 1068. Hoover Keith and Bruce, Houston, TX.

Kuntz, H.L. and R.M Hoover. 1987. The interrelationships between the physical properties of fibrous duct lining materials and lined duct sound attenuation. *ASHRAE Transactions* 93(2):449-470.

Lilly, J. 1987. Breakout in HVAC duct systems. *Sound & Vibration* 21(10).

Machen, J. and J.C. Haines. 1983. Sound insertion loss properties of linacoustic and competitive duct liners. *Report* 436-T-1778. Johns-Manville Research and Development Center, Denver, CO.

Mouratidis, E. and J. Becker. 2004. The acoustic properties of common HVAC plena (RP-1026). *ASHRAE Transactions* 110(2):597-696.

Murray, T.M., D.E. Allen, and E.E. Ungar. 1997. *Steel design guide series 11: Floor vibrations due to human activity.* American Institute of Steel Construction, Chicago.

Persson-Wayne, K., S. Benton, H.G. Leventhall, and R. Rylander. 1997. Effects on performance and work quality due to low frequency ventilation noise. *Journal of Sound and Vibration* 205(4):467-474.

Reynolds, D.D. and W.D. Bevirt. 1994. *Procedural standards for the measurement and assessment of sound and vibration.* National Environmental Balancing Bureau, Rockville, MD.

Reynolds, D.D. and J.M. Bledsoe. 1989a. Sound attenuation of unlined and acoustically lined rectangular ducts. *ASHRAE Transactions* 95(1):90-95.

Reynolds, D.D. and J.M. Bledsoe. 1989b. Sound attenuation of acoustically lined circular ducts and radiused elbows. *ASHRAE Transactions* 95(1):96-99.

Reynolds, D.D. and J.M. Bledsoe. 1991. *Algorithms for HVAC acoustics.* ASHRAE.

Schaffer, M.E. 2005. *A practical guide to noise and vibration control for HVAC systems,* 2nd ed. ASHRAE.

Schultz, T.J. 1985. Relationship between sound power level and sound pressure level in dwellings and offices. *ASHRAE Transactions* 91(1A): 124-153.

Stabley, R.E. 2006. Sound ratings of water-cooled chillers—Sound pressure or power levels? In *Proceedings from Inter-Noise 2006,* Honolulu. Institute of Noise Control Engineering of the USA, Indianapolis, IN.

Thompson, J.K. 1981. The room acoustics equation: Its limitation and potential. *ASHRAE Transactions* 87(2):1049-1057.

Ver, I.L. 1978. A review of the attenuation of sound in straight lined and unlined ductwork of rectangular cross section. *ASHRAE Transactions* 84(1):122-149.

Ver, I.L. 1982. A study to determine the noise generation and noise attenuation of lined and unlined duct fittings. *Report* 5092. Bolt, Beranek and Newman, Boston.

Ver, I.L. 1983a. Noise generation and noise attenuation of duct fittings— A review: Part I (RP-265). *ASHRAE Transactions* 90(2A):354-382.

Ver, I.L. 1983b. Noise generation and noise attenuation of duct fittings—A review: Part II (RP-265). *ASHRAE Transactions* 90(2A):383-390.

Warnock, A.C.C. 1998. Transmission of sound from air terminal devices through ceiling systems. *ASHRAE Transactions* 104(1A):650-657.

Wells, R.J. 1958. Acoustical plenum chambers. *Noise Control* (July).

Woods Fan Division. 1973. *Design for sound.* English Electric Company.

BIBLIOGRAPHY

AIA. 2001. *Guidelines for design and construction of hospital and health care facilities.* American Institute of Architects, Washington, D.C.

ANSI. 2002. Guide for the evaluation of human exposure to whole body vibration. *Standard* S3.18-2002. American National Standards Institute, Washington, D.C.

ANSI. 2002. Determination of sound power levels of noise sources using sound pressure—Precision method for reverberation rooms. *Standard* S12.51-2002. American National Standards Institute, Washington, D.C.

ANSI. 1999. Determination of sound power levels of noise sources using sound pressure—Engineering methods in an essentially free field over a reflecting plane. *Standard* S12.54-1999. American National Standards Institute, Washington, D.C.

AHRI. 2001. Performance and calibration of reference sound sources. *Standard* 250-2001. Air-Conditioning, Heating, and Refrigeration Institute, Arlington, VA.

AHRI. 1995. Sound rating of outdoor unitary equipment. *Standard* 270-1995. Air-Conditioning, Heating, and Refrigeration Institute, Arlington, VA.

AHRI. 1997. Application of sound rating levels of outdoor unitary equipment. *Standard* 275-1997. Air-Conditioning, Heating, and Refrigeration Institute, Arlington, VA.

AHRI. 1995. Requirements for the qualification of reverberant rooms in the 63 Hz octave band. *Standard* 280-1995. Air-Conditioning, Heating, and Refrigeration Institute, Arlington, VA.

AHRI. 2000. Sound rating and sound transmission loss of packaged terminal equipment. *Standard* 300-2000. Air-Conditioning, Heating, and Refrigeration Institute, Arlington, VA.

AHRI. 2000. Sound rating of non-ducted indoor air-conditioning equipment. *Standard* 350-2000. Air-Conditioning, Heating, and Refrigeration Institute, Arlington, VA.

AHRI. 2005. Method of measuring sound and vibration of refrigerant compressors. *Standard* 530-2005. Air-Conditioning, Heating, and Refrigeration Institute, Arlington, VA.

AHRI. 1997. Assessing the impact of air-conditioning outdoor sound levels in the residential community. *Guideline* L-1997. Air-Conditioning, Heating, and Refrigeration Institute, Arlington, VA.

ASHRAE. 2004. Ventilation for acceptable indoor air quality. *Standard* 62.1-2004.

Cowan, J.P. 1994. *Handbook of environmental acoustics.* Van Nostrand Reinhold, New York.

DHHS. 2005. *Eleventh annual report on carcinogens.* U.S. Department of Health and Human Services, Public Health Service, Washington, D.C.

Environment Canada. 1994. Mineral fibres: Priority substances list assessment report. *Canadian Environmental Protection Act,* Ottawa.

Morey, P.R. and C.M. Williams. 1991. Is porous insulation inside an HVAC system compatible with healthy building? ASHRAE IAQ Symposium.

Reynolds, D.D. and W.D. Bevirt. 1989. *Sound and vibration design and analysis.* National Environmental Balancing Bureau, Rockville, MD.

Sandbakken, M., L. Pande, and M.J. Crocker. 1981. Investigation of end reflection of coefficient accuracy problems with AMCA *Standard* 300-67. *HL* 81-16. Ray W. Herrick Laboratories, Purdue University, West Lafayette, IN.

Schultz, T.J. 1972. *Community noise ratings.* Applied Science Publishers, London.

SMACNA. 1990. *HVAC systems duct design,* 3rd ed. Sheet Metal and Air Conditioning Contractors' National Association, Vienna, VA.

Ungar, E.E., D.H. Sturz, and C.H. Amick. 1990. Vibration control design of high technology facilities. *Sound and Vibration* 24(7):20-27.

Ver, I.L. 1983. Prediction of sound transmission through duct walls: Breakout and pickup. *ASHRAE Transactions* 90(2A):391-413.

RESOURCES

Acoustical Society of America (ASA)...............................www.asa.aip.org

The Air-Conditioning, Heating, and Refrigeration Institute (AHRI)... www.ahrinet.org

Institute of Noise Control Engineers (INCE)....................www.inceusa.org

National Council of Acoustical Consultants (NCAC).........www.ncac.com

National Environmental Balancing Bureau (NEBB).............www.nebb.org

Noise Pollution Clearinghouse .... www.nonoise.org

North American Insulation Manufacturers Association (NAIMA) ....www.naima.org

Testing, Adjusting and Balancing Bureau (TABB) .... www.tabbcertified.org

Vibration Institute ....www.vibinst.org

Sheet Metal and Air Conditioning Contractors' National Association (SMACNA) ....www.smacna.org

WATER TREATMENT

THIS chapter covers the fundamentals of water treatment and some of the common problems associated with water in heating and air-conditioning equipment.

WATER CHARACTERISTICS

Chemical Characteristics

When rain falls, it dissolves carbon dioxide and oxygen in the atmosphere. The carbon dioxide mixes with the water to form carbonic acid (H_2CO_3). When carbonic acid contacts soil that contains limestone ($CaCO_3$), it dissolves the calcium to form calcium carbonate. Calcium carbonate in water used in heating or air-conditioning applications can eventually become scale, which can increase energy costs, maintenance time, equipment shutdowns, and could eventually lead to equipment replacement.

The following paragraphs discuss typical chemical and physical properties of water used for HVAC applications.

Alkalinity is a measure of the capacity to neutralize strong acids. In natural waters, the alkalinity almost always consists of bicarbonate, although some carbonate may also be present. Borate, hydroxide, phosphate, and other constituents, if present, are included in the alkalinity measurement in treated waters. Alkalinity also contributes to scale formation.

Alkalinity is measured using two different end-point indicators. The **phenolphthalein alkalinity** (P alkalinity) measures the strong alkali present; the **methyl orange alkalinity** (M alkalinity), or **total alkalinity**, measures the total alkalinity in the water. Note that the total alkalinity includes the phenolphthalein alkalinity. For most natural waters, in which the concentration of phosphates, borates, and other noncarbonated alkaline materials is small, the actual chemical species present can be estimated from the two alkalinity measurements (Table 1).

Alkalinity or acidity is often confused with pH. Such confusion may be avoided by keeping in mind that the pH is a measure of hydrogen ion concentration expressed as the logarithm of its reciprocal.

Chlorides have no effect on scale formation but do contribute to corrosion because of their conductivity and because the small size of the chloride ion permits the continuous flow of corrosion current when surface films are porous. The amount of chlorides in the water is a useful measuring tool in evaporative systems. Virtually all other constituents in the water increase or decrease when common treatment chemicals are added or because of chemical changes that take place in normal operation. With few exceptions, only evaporation affects chloride concentration, so the ratio of chlorides in a water sample from an operating system to those of the makeup water provides a measure of how much the water has been concentrated. (*Note*: Chloride levels will change if the system is continuously chlorinated.)

Dissolved solids consist of salts and other materials that combine with water as a solution. They can affect the formation of corrosion and scale. Low-solids waters are generally corrosive because they have less tendency to deposit protective scale. If a high-solids

water is nonscaling, it tends to produce more intensive corrosion because of its high conductivity. Dissolved solids are often referred to as total dissolved solids (TDS).

Conductivity or **specific conductance** measures the ability of a water to conduct electricity. Conductivity increases with the total dissolved solids. Specific conductance can be used to estimate total dissolved solids.

Silica can form particularly hard-to-remove deposits if allowed to concentrate. Fortunately, silicate deposition is less likely than other deposits.

Soluble iron in water can originate from metal corrosion in water systems or as a contaminant in the makeup water supply. The iron can form heat-insulating deposits by precipitation as iron hydroxide or iron phosphate (if a phosphate-based water treatment product is used or if phosphate is present in the makeup water).

Sulfates also contribute to scale formation in high-calcium waters. Calcium sulfate scale, however, forms only at much higher concentrations than the more common calcium carbonate scale. High sulfates also contribute to increased corrosion because of their high conductivity.

Suspended solids include both organic and inorganic solids suspended in water (particularly unpurified water from surface sources or those that have been circulating in open equipment). Organic matter in surface supplies may be colloidal. Naturally occurring compounds such as lignins and tannins are often colloidal. At high velocities, hard suspended particles can abrade equipment. Settled suspended matter of all types can contribute to concentration cell corrosion.

Turbidity can be interpreted as a lack of clearness or brilliance in a water. It should not be confused with color. A water may be dark in color but still clear and not turbid. Turbidity is due to suspended matter in a finely divided state. Clay, silt, organic matter, microscopic organisms, and similar materials are contributing causes of turbidity. Although suspended matter and turbidity are closely related they are not synonymous. Suspended matter is the quantity of material in a water that can be removed by filtration. The turbidity of water used in HVAC systems should be as low as possible. This is particularly true of boiler feedwater. The turbidity can concentrate in the boiler and may settle out as sludge or mud and lead to deposition. It can also cause increased boiler blowdown, plugging, overheating, priming, and foaming.

Biological Characteristics

Bacteria, algae, and fungi can be present in water systems, and their growth can cause operating, maintenance, and health problems.

Table 1 Alkalinity Interpretation for Waters[a]

If	Then, Carbonate	Bicarbonate	Free CO_2
P Alk = 0	0	M Alk	Present
P Alk < 0.5M Alk	2P Alk	M Alk – 2P Alk	0
P Alk = 0.5M Alk	2P Alk = M Alk	0	0
P Alk > 0.5M Alk [b]	2(M Alk – P Alk)	0	0

[a] P Alk = Phenolphthalein alkalinity. M Alk = Methyl orange (total) alkalinity.
[b] Treated waters only. Hydroxide also present.

The preparation of this chapter is assigned to TC 3.6, Water Treatment.

Microorganism growth is affected by temperature, food sources, and water availability. Biological growth can occur in most water systems below 150°F. Problems caused by biological materials range from green algae growth in cooling towers to slime formation from bacteria in secluded and dark areas. The result can be plugging of equipment where flow is essential. Dead algae are often characterized by foul odors and mud-like deposits, which are often high in silica from the cell walls of the diatoms. These problems can be eliminated, or at least reduced to reasonable levels, by mechanical or chemical treatment.

CORROSION CONTROL

Corrosion is the destruction of a metal or alloy by chemical or electrochemical reaction with its environment. In most instances, this reaction is electrochemical in nature, much like that in an electric battery. For corrosion to occur, a corrosion cell consisting of an anode, a cathode, an electrolyte, and an electrical connection must exist. Metal ions dissolve into the electrolyte (water) at the anode. Electrically charged particles (electrons) are left behind. These electrons flow through the metal to other points (cathodes) where electron-consuming reactions occur. The result of this activity is the loss of metal and often the formation of a deposit.

Types of Corrosion

Corrosion can often be characterized as general, localized or pitting, galvanic, caustic cracking, corrosion fatigue, erosion corrosion, and microbiological corrosion.

General corrosion is uniformly distributed over the metal surface. The considerable amount of iron oxide produced by generalized corrosion contributes to fouling.

Localized or pitting corrosion exists when only small areas of the metal corrode. Pitting is the most serious form of corrosion because the action is concentrated in a small area. Pitting may perforate the metal in a short time.

Galvanic corrosion can occur when two different metals are in contact. The more active (less noble) metal corrodes rapidly. Common examples of galvanic corrosion in water systems are steel and brass, aluminum and steel, zinc and steel, and zinc and brass. If galvanic corrosion occurs, the metal named first corrodes.

Corrosion fatigue may occur by one of two mechanisms. In the first mechanism, cyclic stresses (for example, those created by rapid heating and cooling) are concentrated at points where corrosion has roughened or pitted the metal surface. In the second type, cracks often originate where metal surfaces are covered by a dense protective oxide film, and cracking occurs from the action of applied cyclic stresses.

Caustic cracking occurs when metal is stressed, water has a high caustic content, a trace of silica is present, and a mechanism of concentration occurs.

Stress corrosion cracking can occur in alloys exposed to a characteristic corrosive environment and subject to tensile stress. The stress may be either applied, residual (from processing), or a combination. The resulting crack formation is generally intergranular.

Erosion corrosion occurs as a result of flow or impact that physically removes the protective metal oxide surface. This type of corrosion usually takes place due to altered flow patterns or a flow rate that is above design.

Microbiologically influenced corrosion occurs when bacteria secrete corrosive acids on the metal surface. Often bacteria commingle with dirt and silt that protect the bacteria from biocides.

Factors That Contribute to Corrosion

Moisture. Corrosion does not occur in dry environments. However, some moisture is present as water vapor in most environments. In pure oxygen, almost no iron corrosion occurs at relative humidities up to 99%. However, when contaminants such as sulfur dioxide or solid particles of charcoal are present, corrosion can proceed at relative humidities of 50% or more. The corrosion reaction proceeds on surfaces of exposed metals such as iron and unalloyed steel as long as the metal remains wet. Many alloys develop protective corrosion-product films or oxide coatings and are thus unaffected by moisture.

Oxygen. In electrolytes consisting of water solutions of salts or acids, the presence of dissolved oxygen accelerates the corrosion rate of ferrous metals by depolarizing the cathodic areas through reaction with hydrogen generated at the cathode. In many systems, such as boilers or water heaters, most of the dissolved oxygen and other dissolved gases are removed from the water by deaeration to reduce the potential for corrosion.

Solutes. In ferrous materials such as iron and steel, mineral acids accelerate the corrosion rate, whereas alkalies decrease it. The likelihood of cathodic polarization, which slows down corrosion, decreases as the concentration of hydrogen ions (i.e., the acidity of the environment) increases. Relative acidity or alkalinity of a solution is defined as pH; a neutral solution has a pH of 7. The corrosivity of most salt solutions depend on their pH. Because alkaline solutions are generally less corrosive to ferrous systems, it is practical in many closed water systems to minimize corrosion by adding alkali or alkaline salt to raise the pH to 9 or higher.

Differential Solute Concentration. For the corrosion reaction to proceed, a potential difference between anode and cathode areas is required. This potential difference can be established between different locations on a metal surface because of differences in solute concentration in the environment at these locations. Corrosion caused by such conditions is called **concentration cell corrosion**. These cells can be metal ions or oxygen concentration cells.

In the metal ion cell, the metal surface in contact with the higher concentration of dissolved metal ion becomes the cathodic area, and the surface in contact with the lower concentration becomes the anode. The metal ions involved may be a constituent of the environment or may come from the corroding surface itself. The concentration differences may be caused by the sweeping away of the dissolved metal ions at one location and not at another.

In an oxygen concentration cell, the surface area in contact with the surface environment of higher oxygen concentration becomes the cathodic area, and the surface in contact with the surface environment of lower oxygen concentration becomes the anode. Crevices or foreign deposits on the metal surface can create conditions that contribute to corrosion. The anodic area, where corrosion proceeds, is in the crevice or under the deposit. Although they are manifestations of concentration cell corrosion, **crevice corrosion** and **deposit attack** are sometimes referred to as separate forms of corrosion.

Galvanic or Dissimilar Metal Corrosion. Another factor that can accelerate the corrosion process is the difference in potential of dissimilar metals coupled together and immersed in an electrolyte. The following factors control the severity of corrosion resulting from such dissimilar metal coupling:

- Relative differences in position (potential) in the galvanic series, with reference to a standard electrode. The greater the difference, the greater the force of the reaction. The galvanic series for metals in flowing aerated seawater is shown in Table 2.

- Relative area relationship between anode and cathode areas. Because the amount of current flow and, therefore, total metal loss is determined by the potential difference and resistance of the circuits, a small anodic area corrodes more rapidly; it is penetrated at a greater rate than a large anodic area.

- Polarization of either the cathodic or anodic area. Polarization can reduce the potential difference and thus reduce the rate of attack of the anode.

- The mineral content of water. As mineral content increases, the resulting higher electrical conductivity increases the rate of galvanic corrosion.

Table 2 Galvanic Series of Metals and Alloys in Flowing Aerated Seawater at 40 to 80°F

Corroded End (Anodic or Least Noble)
Magnesium alloys
Zinc
Beryllium
Aluminum alloys
Cadmium
Mild steel, wrought iron
Cast iron, flake or ductile
Low-alloy high-strength steel
Ni-Resist, Types 1 & 2
Naval Brass (CA464), yellow brass (CA268), Al brass (CA687), red brass (CA230), admiralty brass (CA443), Mn Bronze
Tin
Copper (CA102, 110), Si Bronze (CA655)
Lead-tin solder
Tin bronze (G & M)
Stainless steel, 12 to 14% Cr (AISI types 410,416)
Nickel silver (CA 732, 735, 745, 752, 764, 770, 794)
90/10 Copper-nickel (CA 706)
80/20 Copper-nickel (CA 710)
Stainless steel, 16 to 18% Cr (AISI Types 430)
Lead
70/30 Copper-nickel (CA 715)
Nickel-aluminum bronze
Silver braze alloys
Nickel 200
Silver
Stainless steel, 18% Cr, 8% Ni (AISI Types 302, 304, 321, 347)
Stainless steel, 18% Cr, 12% Ni-Mo (AISI Types 316, 317)
Titanium
Graphite, graphitized cast iron
Protected End (Cathodic or Most Noble)

Stress. Stresses in metallic structures rarely have significant effects on the uniform corrosion resistance of metals and alloys. Stresses in specific metals and alloys can cause corrosion cracking when the metals are exposed to specific corrosive environments. The cracking can have catastrophic effects on the usefulness of the metal.

Almost all metals and alloys exhibit susceptibility to stress corrosion cracking in at least one environment. Common examples are steels in hot caustic solutions, high zinc content brasses in ammonia, and stainless steels in hot chlorides. Metal manufacturers have technical details on specific materials and their resistance to stress corrosion.

Temperature. According to studies of chemical reaction rates, corrosion rates double for every 18°F rise in temperature. However, such a ratio is not necessarily valid for nonlaboratory corrosion reactions. The effect of temperature on a particular system is difficult to predict without specific knowledge of the characteristics of the metals involved and the environmental conditions.

An increase in temperature may increase the corrosion rate, but only to a point. Oxygen solubility decreases as temperature increases and, in an open system, may approach zero as water boils. Beyond a critical temperature level, the corrosion rate may decrease due to a decrease in oxygen solubility. However, in a closed system, where oxygen cannot escape, the corrosion rate may continue to increase with an increase in temperature.

For those alloys, such as stainless steel, that depend on oxygen in the environment for maintaining a protective oxide film, the reduction in oxygen content due to an increase in temperature can accelerate the corrosion rate by preventing oxide film formation.

Temperature can affect corrosion potential by causing a salt dissolved in the environment to precipitate on the metal surface as a protective layer of scale. One example is calcium carbonate scale in hard waters. Temperature can also affect the nature of the corrosion product, which may be relatively stable and protective in certain temperature ranges and unstable and non protective in others. An example of this is zinc in distilled water; the corrosion product is non protective from 140 to 190°F but reasonably protective at other temperatures.

Pressure. Where dissolved gases such as oxygen and carbon dioxide affect the corrosion rate, pressure on the system may increase their solubility and thus increase corrosion. Similarly, a vacuum on the system reduces the solubility of the dissolved gas, thus reducing corrosion. In a heated system, pressure may rise with temperature. It is difficult and impractical to control system corrosion by pressure control alone.

Flow Velocity. The effect of flow velocity on the corrosion rate of systems depends on several factors, including

- Amount of oxygen in the water
- Type of metal (iron and steel are most susceptible)
- Flow rate

In metal systems where corrosion products retard corrosion by acting as a physical barrier, high flow velocities may cause the removal of those protective barriers and increase the potential for corrosion. A turbulent environment may cause uneven attack, from both erosion and corrosion. This corrosion is called **erosion corrosion**. It is commonly found in piping with sharp bends where the flow velocity is high. Copper and softer metals are more susceptible to this type of attack.

Preventive and Protective Measures

Materials Selection. Any piece of heating or air-conditioning equipment can be made of metals that are virtually corrosion-proof under normal and typical operating conditions. However, economics usually dictate material choices. When selecting construction materials the following factors should be considered:

- Corrosion resistance of the metal in the operating environment
- Corrosion products that may be formed and their effects on equipment operation
- Ease of construction using a particular material
- Design and fabrication limitations on corrosion potential
- Economics of construction, operation, and maintenance during the projected life of the equipment, (i.e., expenses may be minimized in the long run by paying more for a corrosion-resistant material and avoiding regular maintenance).
- Use of dissimilar metals should be avoided. Where dissimilar materials are used, insulating gaskets and/or organic coatings must be used to prevent galvanic corrosion.
- Compatibility of chemical additives with materials in the system

Protective Coatings. The operating environment has a significant role in the selection of protective coatings. Even with a coating suited for that environment, the protective material depends on the adhesion of the coating to the base material, which itself depends on the surface preparation and application technique.

Maintenance. Defects in a coating are difficult to prevent. These defects can be either flaws introduced into the coating during application or mechanical damage sustained after application. In order to maintain corrosion protection, defects must be repaired.

Cycles of Concentration. Some corrosion control may be achieved by optimizing the cycles of concentration (the degree to which soluble mineral solids in the makeup water have increased in the circulating water due to evaporation). Generally, adjustment of the blowdown rate and pH to produce a slightly scale-forming condition (see section on Scale Control) will result in an optimum condition between excess corrosion and excess scale.

Chemical Methods. Chemical protective film-forming chemical inhibitors reduce or stop corrosion by interfering with the corrosion mechanism. Inhibitors usually affect either the anode or the cathode.

Anodic corrosion inhibitors establish a protective film on the anode. Though these inhibitors can be effective, they can be dangerous—if insufficient anodic inhibitor is present, the entire corrosion

Table 3 Typical Corrosion Inhibitors

Anodic Corrosion Inhibitors
   Molybdate
   Nitrite
   Orthophosphate
   Silicate
Mainly Cathodic Corrosion Inhibitors
   Bicarbonate
   Polyphosphate
   Phosphonate
   Zinc
   Polysilicate
General
   Soluble oils
   Other organics, such as azole or carboxylate

potential occurs at the unprotected anode sites. This causes severe localized (or pitting) attack.

Cathodic corrosion inhibitors form a protective film on the cathode. These inhibitors reduce the corrosion rate in direct proportion to the reduction of the cathodic area.

General corrosion inhibitors protect by forming a film on all metal surfaces whether anodic or cathodic.

Table 3 lists typical corrosion inhibitors. The most important factor in an effective corrosion inhibition program is the consistent control of both the corrosion inhibition chemicals and the key water characteristics. No program will work without controlling these factors.

Cathodic Protection. Sacrificial anodes reduce galvanic attack by providing a metal (usually zinc, but sometimes magnesium) that is higher on the galvanic series than either of the two metals that are coupled together. The sacrificial anode thereby becomes anodic to both metals and supplies electrons to these cathodic surfaces. Proper design and placement of these anodes are important. When properly used, they can reduce loss of steel from the tube sheet of exchangers using copper tubes. Sacrificial anodes have helped supplement chemical programs in many cooling water and process water systems.

Impressed-current protection is a similar corrosion control technique that reverses the corrosion cell's normal current flow by impressing a stronger current of opposite polarity. Direct current is applied to an anode—inert (platinum, graphite) or expendable (aluminum, cast iron)—reversing the galvanic flow and converting the steel from a corroding anode to a protective cathode. The method is very effective in protecting essential equipment such as elevated water storage tanks, steel tanks, or softeners.

White Rust on Galvanized Steel Cooling Towers

White rust is a zinc corrosion product that forms on galvanized surfaces. It appears as a white, waxy or fluffy deposit composed of loosely adhering zinc carbonate. The loose crystal structure allows continued access of the corrosive water to exposed zinc. Unusually rapid corrosion of galvanized steel, as evidenced by white rust, can affect galvanized steel cooling towers under certain conditions.

Before chromates in cooling tower water were banned, the common treatment system consisted of chromates for corrosion control and sulfuric acid for scale control. This control method generally has been replaced by alkaline treatment involving scale inhibitors at a higher pH. Alkaline water chemistry is naturally less corrosive to steel and copper, but create an environment where white rust on galvanized steel can occur. Also, some scale prevention programs soften the water to reduce hardness, rather than use acid to reduce alkalinity. The resulting soft water is corrosive to galvanized steel.

Prevention. White rust can be prevented by promoting the formation of a nonporous surface layer of basic zinc carbonate. This barrier layer is formed during a process called **passivation** and normally protects the galvanized steel for many years. Passivation is best accomplished by controlling pH during initial operation of the cooling tower. Control of the cooling water pH in the range of 7 to 8 for 45 to 60 days usually allows passivation of galvanized surfaces to occur. In addition to pH control, operation with moderate hardness levels of 100 to 300 ppm as $CaCO_3$ and alkalinity levels of 100 to 300 ppm as $CaCO_3$ will promote passivation. Where pH control is not possible, certain phosphate-based inhibitors may help protect galvanized steel. A water treatment company should be consulted for specific formulations.

SCALE CONTROL

Scale is a dense coating of predominantly inorganic material formed from the precipitation of water-soluble constituents. Some common scales are

- Calcium carbonate
- Calcium phosphate
- Magnesium salts
- Silica

The following principal factors determine whether or not a water is scale forming:

- Temperature
- Alkalinity or acidity (pH)
- Amount of scale-forming material present
- Influence of other dissolved materials, which may or may not be scale-forming

As any of these factors changes, scaling tendencies also change. Most salts become more soluble as temperature increases. However, some salts, such as calcium carbonate, become less soluble as temperature increases. Therefore, they often cause deposits at higher temperatures.

A change in pH or alkalinity can greatly affect scale formation. For example, as pH or alkalinity increases, calcium carbonate—the most common scale constituent in cooling systems—decreases in solubility and deposits on surfaces. Some materials, such as silica (SiO_2), are less soluble at lower alkalinities. When the amount of scale-forming material dissolved in water exceeds its saturation point, scale may result. In addition, other dissolved solids may influence scale-forming tendencies. In general, a higher level of scale-forming dissolved solids results in a greater chance for scale formation. Indices such as the Langelier Saturation Index (Langelier 1936) and the Ryznar Stability Index (Ryznar 1944) can be useful tools to predict the calcium carbonate scaling tendency of water. These indices are calculated using the pH, alkalinity, calcium hardness, temperature, and total dissolved solids of the water, and indicate whether the water will favor precipitating or dissolving of calcium carbonate.

Methods used to control scale formation include

- Limit the concentration of scale-forming minerals by controlling cycles of concentration or by removing the minerals before they enter the system (see the section on External Treatments later in this section). **Cycles of concentration** is the ratio of makeup rate to the sum of blowdown and drift rates. The cycles of concentration can be monitored by calculating the ratio of chloride ion, which is highly soluble, in the system water to that in the makeup water.
- Make mechanical changes in the system to reduce the chances for scale formation. Increased water flow and exchangers with larger surface areas are examples.
- Feed acid to keep the common scale forming minerals (e.g., calcium carbonate) dissolved.
- Treat with chemicals designed to prevent scale. Chemical scale inhibitors work by the following mechanisms:

1. **Threshold inhibition chemicals** prevent scale formation by keeping the scale forming minerals in solution and not allowing a deposit to form. Threshold inhibitors include organic phosphates, polyphosphates, and polymeric compounds.
2. **Scale conditioners** modify the crystal structure of scale, creating a bulky, transportable sludge instead of a hard deposit. Scale conditioners include lignins, tannins, and polymeric compounds.

Nonchemical Methods

Equipment based on magnetic, electromagnetic, or electrostatic technology has been used for scale control in boiler water, cooling water, and other process applications.

Magnetic systems are designed to cause scale-forming minerals to precipitate in a low-temperature area away from heat exchanger surfaces, thus producing nonadherent particles (e.g., aragonite form of calcium carbonate versus the hard, adherent calcite form). The precipitated particles can then be removed by blowdown, mechanical means, or physical flushing.

The objective of **electrostatics** is to prevent scale-forming reactions by imposing a surface charge on dissolved ions that causes them to repel.

Results of side-by-side comparative tests with conventional water treatment have been mixed. A Federal Technology Alert report regarding these technologies (DOE 1998) stresses that success of the application depends largely on the experience of the installer. The report includes a discussion of the potential benefits achieved and the necessary precautions to consider when applying these systems.

ASHRAE Research Project RP-1155 (Cho 2002) studied physical water treatment (PWT). For this study, a PWT device was defined as a nonchemical method of water treatment for scale prevention or mitigation. Bulk precipitation was proposed as the mechanism of scale prevention. Three different devices described as permanent magnets, a solenoid coil device, and a high-voltage electrode were tested under laboratory conditions. Fouling resistance data obtained in a heat transfer test section supported the benefit of all three devices when configured in optimum conditions.

External Treatments

Minerals may also be removed by various external pretreatment methods such as reverse osmosis and ion exchange. Zeolite softening, demineralization, and dealkalization are examples of ion exchange processes.

BIOLOGICAL GROWTH CONTROL

Biological growth (algae, bacteria, and fungi) can interfere with a cooling operation due to fouling or corrosion, and may present a health hazard if present in aerosols produced by the equipment. Heating equipment operates above normal biological limits and therefore has fewer microbial problems. When considering biological growth in a cooling system, it is important to distinguish between free-living planktonic organisms and sessile (attached) organisms. Sessile organisms cause the majority of the problems, though they may have entered and multiplied as planktonic organisms.

Biological fouling can be caused by a wide variety of organisms that produce biofilm and slime masses. Slimes can be formed by bacteria, algae, yeasts, or molds and frequently consist of a mixture of these organisms combined with organic and inorganic debris. Organisms such as barnacles and mussels may cause fouling when river, estuarine, or sea water is used. Biological fouling can significantly reduce the efficiency of cooling by reducing heat transfer, increasing back pressure on recirculation pumps, disrupting flow patterns over cooling media, plugging heat exchangers, and blocking distribution systems. In extreme cases the additional mass of slime has caused the cooling media to collapse.

Microorganisms can dramatically enhance, accelerate or, in some cases, initiate localized corrosion (pitting). Microorganisms can influence localized corrosion directly by their metabolism or indirectly by the deposits they form. Indirect influence may not be mediated by simply killing the microorganisms; deposit removal is usually necessary, while direct influence can be substantially mediated by inhibiting microorganism metabolism.

Algae use energy from the sun to convert bicarbonate or carbon dioxide into biomass. Masses of algae can block piping, distribution holes, and nozzles. A distribution deck cover, which drastically reduces the sunlight reaching the algae, is one of the most cost-effective control devices for a cooling tower. Biocides are also used to assist in the control of algae.

Algae can also provide nutrients for other microorganisms in the cooling system, increasing the biomass in the water. Bacteria can grow in systems even when nutrient levels are relatively low. Yeasts and fungi are much slower growing than bacteria and find it difficult to compete in bulk waters for the available food. Fungi do thrive in partially wetted and high humidity areas such as the cooling media. Wood-destroying species of fungi can be a major concern for wooden cooling towers as the fungi consume the cellulose and/or lignin in the wood, reducing its structural integrity.

Most waters contain organisms capable of producing biological slime, but optimal conditions for growth are poorly understood. Equipment near nutrient sources or that has process leaks acting as a food source is particularly susceptible to slime formation. Even a thin layer of biofilm significantly reduces heat transfer rates in exchangers.

Control Measures

Eliminating sunlight from wetted surfaces such as distribution troughs, cooling media, and sumps significantly reduces algae growth. Eliminating deadlegs and low flow areas in the piping and the cooling loop reduces biological growth in those areas. Careful selection of materials of construction can remove nutrient sources and environmental niches for growth. Maintaining a high quality makeup water supply with low bacteria counts also helps minimize biological growth. Equipment should also be designed with adequate access for inspection, sampling, and manual cleaning.

Sometimes the effective control of slime and algae requires a combination of mechanical and chemical treatments. For example, when a system already contains a considerable accumulation of slime, a preliminary mechanical cleaning makes the subsequent application of a biocidal chemical more effective in killing the growth and more effective in preventing further growth. A build-up of scale deposits, corrosion product, and sediment in a cooling system also reduces the effectiveness of chemical biocides. Routine manual cleaning of cooling towers, including the use of high level chlorination and a biodispersant (surfactant), helps control Legionella bacteria as well as other microorganisms.

Microbiocides. Chemical biocides used to control biological growth in cooling systems fall into two broad categories: oxidizing and nonoxidizing biocides.

Oxidizing biocides (chlorine, chlorine-yielding compounds, bromine, bromochlorodimethylhydantoin (BCDMH, or BCD), ozone, iodine, and chlorine dioxide) are among the most effective microbiocidal chemicals. However, they are not always appropriate for control in cooling systems with a high organic loading. In air washers, the odor may become offensive; in wooden cooling towers, excessive concentrations of oxidizing biocides can cause delignification; and overdosing of oxidizing biocides may cause corrosion of metallic components. In systems large enough to justify the cost of equipment to control feeding of oxidizing biocides accurately, the application may be safe and economical. The most effective use of oxidizing biocides is to maintain a constant low level residual in the system. However, if halogen-based oxidizing biocides are fed intermittently (slug dosed), a pH near 7 is advantageous because, at this neutral pH, halogens are present as the hypohalous acid (HOR, where R represents the halogen) form over the hypohalous ion (OR⁻) form. The effectiveness of this shock feeding is enhanced due to the faster killing action of hypohalous acid over that of hypohalous ion. The residual biocide concentration should be tested, using a field test kit, on a routine basis. Most halogenation programs can benefit from the use of dispersants or surfactants (chlorine helpers) to break up microbiological masses.

Chlorine has been the oxidizing biocide of choice for many years, either as chlorine gas or in the liquid form as sodium hypochlorite. Other forms of chlorine, such as powders or pellets, are also available. Use of chlorine gas is declining due to the health and safety concerns involved in handling this material and in part due to environmental pressures concerning the formation of chloramines and trihalomethane.

Bromine is produced either by the reaction of sodium hypochlorite with sodium bromide on site, or by release from pellets. Bromine has certain advantages over chlorine: it is less volatile, and bromamines break down more rapidly than chloramines in the environment. Also, when slug feeding biocide in high pH systems, hypobromous acid may have an advantage because its dissociation constant is lower than that of chlorine. This effect is less important when biocides are fed continuously.

Ozone has several advantages compared to chlorine: it does not produce chloramines or trihalomethane, it breaks down to nontoxic compounds rapidly in the environment, it controls biofilm better, and it requires significantly less chemical handling. The use of ozone-generating equipment in an enclosed space however, requires care be taken to protect operators from the toxic gas. Also, research by ASHRAE has shown that ozone is only marginally effective as a scale and corrosion inhibitor (Gan et al. 1996; Nasrazadani and Chao 1996).

Water conditions should be reviewed to determine the need for scale and corrosion inhibitors and then, as with all oxidizing biocides, inhibitor chemicals should be carefully selected to ensure compatibility. To maximize the biocidal performance of the ozone, the injection equipment should be designed to provide adequate contact of the ozone with the circulating water. In larger systems, care should be taken to ensure that the ozone is not depleted before the water has circulated through the entire system.

Iodine is provided in pelletized form, often from a rechargeable cartridge. Iodine is a relatively expensive chemical for use on cooling towers and is probably only suitable for use on smaller systems.

Nonoxidizing Biocides. When selecting a nonoxidizing microbiocide, the pH of the circulating water and the chemical compatibility with the corrosion and/or scale inhibitor product must be considered. The following list, while not exhaustive, identifies some of these products:

• Quaternary ammonium compounds
• Methylene bis(thiocyanate) (MBT)
• Isothiazolones
• Thiadiazine thione
• Dithiocarbamates
• Decyl thioethanamine (DTEA)
• Glutaraldehyde
• Dodecylguanidine
• Benzotriazole
• Tetrakis(hydroxymethyl)phosphonium sulfate (THPS)
• Dibromo-nitrilopropionamide (DBNPA)
• Bromo-nitropropane-diol
• Bromo-nitrostyrene (BNS)
• Proprietary blends

The manner in which nonoxidizing biocides are fed is important. Sometimes the continuous feeding of low dosages is neither effective nor economical. Slug feeding large concentrations to achieve a toxic level of the chemical in the water for a sufficient time to kill the organisms present can show better results. Water blowdown rate and biocide hydrolysis (chemical degradation) rate affect the required dosage. The hydrolysis rate of the biocide is affected by the type of biocide, along with the temperature and pH of the system water. Dosage rates are proportional to system volume; dosage concentrations should be sufficient to ensure that the contact time of the biocide is long enough to obtain a high kill rate of microorganisms before the minimum inhibitory concentration of the biocide is reached. The period between nonoxidizing biocide additions should be based on the system half life, with sequential additions timed to prevent regrowth of bacteria in the water.

Handling Microbiocides. All microbiocides must be handled with care to ensure personal safety. In the United States, cooling water microbiocides are approved and regulated through the EPA and, by law, must be handled in accordance with labeled instructions. Maintenance staff handling the biocides should read the material safety data sheets and be provided with all the appropriate safety equipment to handle the substance. Automatic feed systems should be used that minimize and eliminate the handling of biocides by maintenance personnel.

Other Biocides. Ultraviolet irradiation deactivates the microorganisms as the water passes through a quartz tube. The intensity of the light and thorough contact with the water are critical in obtaining a satisfactory kill of microorganisms. Suspended solids in the water or deposits on the quartz tube significantly reduce the effectiveness of this treatment method. Therefore, a filter is often installed upstream of the lamp to minimize these problems. Because the ultraviolet light leaves no residual material in the water, sessile organisms and organisms that do not pass the light source are not affected by the ultraviolet treatment. Ultraviolet irradiation may be effective on humidifiers and air washers where the application of biocidal chemicals is unacceptable and where 100% of the recirculating water passes the lamp. Ultraviolet irradiation is less effective where all the microorganisms cannot be exposed to the treatment, such as in cooling towers. Ultraviolet lamps require replacement after approximately every 8000 h of operation.

Metallic ions, namely copper and silver, effectively control microbial populations under very specific circumstances. Either singularly or in combination, copper and silver ions are released into the water via electrochemical means to generate 1 to 2 ppm of copper and/or 0.5 to 1.0 ppm of silver. The ions assist in the control of bacterial populations in the presence of a free chlorine residual of at least 0.2 ppm. Copper, in particular, effectively controls algae.

Liu et al. (1994) reported control of *Legionella pneumophila* bacteria in a hospital hot water supply using copper-silver ionization. In this case, *Legionella* colonization decreased significantly when copper and silver concentrations exceeded 0.4 and 0.04 ppm, respectively. Also, residual disinfection prevented *Legionella* colonization for two months after the copper-silver unit was inactivated.

Significant limitations exist in the use of copper and silver ion for cooling systems. Many states are restricting the discharge of these ions to surface waters, and if the pH of the system water rises above 7.8, the efficacy of the treatment is significantly reduced. Systems that have steel or aluminum heat exchangers should not be treated by this method, as the potential for the deposition of the copper ion and subsequent galvanic corrosion is significant.

Legionnaires' Disease

Like other living things, *Legionella pneumophila*, the bacterium that causes Legionnaires' disease (legionellosis), requires moisture for survival. *Legionella* bacteria are widely distributed in natural water systems and are present in many drinking water supplies. Potable hot water systems between 80 and 120°F, cooling towers, certain types of humidifiers, evaporative condensers, whirlpools and spas, and the various components of air conditioners are considered to be amplifiers. These bacteria are killed in a matter of minutes when exposed to temperatures above 140°F.

Legionellosis can be acquired by inhalation of *Legionella* organisms in aerosols. Aerosols can be produced by cooling towers, evaporative condensers, decorative fountains, showers, and misters. It has been reported that the aerosol from cooling towers can be transmitted over a distance of up to 2 miles. If air inlet ducts of nearby air conditioners draw the aerosol from contaminated cooling towers into the building, the air distribution system itself can transmit the disease. When an outbreak of Legionnaires' disease occurs, cooling towers are often the suspected source. However, other water systems may produce an aerosol and should not be neglected. Amplification of *Legionella* within protozoans has been demonstrated, and *Legionella* bacteria are thought to be protected from biocides while growing intracellularly. Amplification of *Legionella* bacteria in biofilm and slime masses has been shown by a number of researchers. Microbial control programs should consider the effectiveness of the products against slimes as part of the *Legionella* control program.

Humidifiers. Units that generate a water aerosol for humidity control can become amplifiers of bacteria if their reservoirs are poorly maintained. Manufacturers' recommendations on cleaning and maintenance should be followed closely. Steam humidifiers should not pose a bacterial problem because the steam does not contain bacteria.

Whirlpools and Spas. These units pose a potential hazard to users if not properly maintained. The complexity of some units makes cleaning difficult, so a firm that specializes in cleaning these devices may need to be hired. Manufacturers' instructions should be followed carefully.

Decorative Fountains. If these systems become contaminated, *Legionella* may multiply and pose a hazard to those nearby. The recirculating water should be kept clean and clear with proper filtration. Continuous chlorination or use of another biocide with EPA approval for decorative fountains is recommended. Indoor decorative fountains with ponds containing fish and other aquatic life cannot be chlorinated or treated; steps should be taken to ensure that water does not become airborne via sprays or cascading waterfalls.

Roof Ponds. These are used to lower the roof temperature and thus the demand on the air conditioning. If contaminated, such ponds can pose a risk to the staff and general public. Roof ponds should be monitored and treated in the same manner as recirculating cooling water systems (see the section on Selection of Water Treatment).

Safety Showers. Safety showers are mandated for safety, but they are used infrequently. Standing reservoirs of nonsterile water at room temperature should be avoided. A preventative maintenance program that includes flushing is suggested.

Vegetable Misters or Sprays. Although an older-style ultrasonic mist machine has been implicated in at least one case of this disease, once-through cool-water sprayers, (directly connected to a cold, potable water line with no reservoir) do not appear to be a problem.

Machine Shop Cooling. Water used in machining operations may be mixed with oil or other ingredients to improve cooling and cutting efficiency. It is typically stored in holding tanks that are subject to bacterial contamination, including by *Legionella*. Bacterial activity should be monitored and controlled.

Ice-Making Machines. While *Legionella* bacteria are not likely to amplify in this cold environment, bacterial amplification can occur prior to freezing the water. Graman et al. (1997) described a case in which *Legionella* were presumed to have been transmitted by aspiration of ice or ice water from an ice machine.

Prevention and Control. The *Legionella* count required to cause illness has not been firmly established because many factors are involved, including (1) virulence and number of *Legionella* in the air, (2) rate at which the aerosol dries, (3) wind direction, and (4) susceptibility to the disease of the person breathing the air. The organism is often found in sites not associated with an outbreak of the disease. It has been shown that it is feasible to operate cooling systems with *Legionella* bacteria below the limit of detection and that the only method to prove that a system is operating at these levels is to specifically test for *Legionella* bacteria, rather than to infer from total bacteria count measurements.

Periodic monitoring of circulating water for total bacteria count and *Legionella* count can be accomplished using culture methods. However, routine culturing of samples from building water systems may not yield an accurate prediction of the risk of *Legionella* transmission. Monitoring system cleanliness and using a microbial control agent that has proven efficacy or is generally regarded as effective in controlling *Legionella* populations are also important. Other measures to decrease risk include optimizing cooling tower design to minimize drift, eliminating deadlegs or low flow areas, selecting materials that do not promote the growth of *Legionella*, and locating the tower so that drift is not injected into the air handlers. The *Legionellosis Position Statement*, an ASHRAE Position Paper (1998) has further information on this topic.

SUSPENDED SOLIDS AND DEPOSITATION CONTROL

Mechanical Filtration

Strainers, filters, and separators may be used to reduce suspended solids to an acceptable low level. Generally, if the screen is 200 mesh, equivalent to about 0.003 in., it is called a strainer; if it is finer than 200 mesh, it is called a filter.

Strainers. A strainer is a closed vessel with a cleanable screen designed to remove and retain foreign particles down to 0.001 in. diameter from various flowing fluids. Strainers extract material that is not wanted in the fluid, and allow saving the extracted product if it is valuable. Strainers are available as single-basket or duplex units, manual or automatic cleaning units, and may be made of cast iron, bronze, stainless steel, copper-nickel alloys, or plastic. Magnetic inserts are available where microscopic iron or steel particles are present in the fluid.

Cartridge Filters. These are typically used as final filters to remove nearly all suspended particles from about 0.004 in. down to 0.00004 in. or less. Cartridge filters are typically disposable (i.e., once plugged, they must be replaced). The frequency of replacement, and thus the economical feasibility of their use, depends on the concentration of suspended solids in the fluid, the size of the smallest particles to be removed, and the removal efficiency of the cartridge filter selected.

In general, cartridge filters are favored in systems where contamination levels are less than 0.01% by mass (< 100 ppm). They are available in many different materials of construction and configurations. Filter media materials include yarns, felts, papers, nonwoven materials, resin-bonded fabric, woven wire cloths, sintered metal, and ceramic structures. The standard configuration is a cylinder with an overall length of approximately 10 in., an outside diameter of approximately 2.5 to 2.75 in., and an inside diameter of about 1 to 1.5 in., where the filtered fluid collects in the perforated internal core. Overall lengths from 4 to 40 in. are readily available.

Cartridges made of yarns, resin-bonded, or melt-blown fibers normally have a structure that increases in density towards the center. These depth-type filters capture particles throughout the total media thickness. Thin media, such as pleated paper (membrane types), have a narrow pore size distribution design to capture particles at or near the surface of the filter. Surface-type filters can normally handle higher flow rates and provide higher removal efficiency than equivalent depth filters. Cartridge filters are rated according to manufacturers' guidelines. Surface-type filters have an absolute rating, while depth-type filters have a nominal rating that reflects their general classification function. Higher efficiency melt-blown depth filters are available with absolute ratings as needed.

Sand Filters. A downflow filter is used to remove suspended solids from a water stream. The degree of suspended solids removal depends on the combinations and grades of the medium being used in the vessel. During the filtration mode, water enters the top of the filter vessel. After passing through a flow impingement plate, it enters the quiescent (calm) freeboard area above the medium.

In multimedia downflow vessels, various grain sizes and types of media are used to filter the water. This design increases the suspended solids holding capacity of the system, which in turn increases the backwashing interval. Multimedia vessels might also be used for low suspended solids applications, where chemical additives are required. In the multimedia vessel, the fluid enters the top layer of anthracite media, which has an effective size of 0.04 in. This relatively coarse layer removes the larger suspended particles, a substantial portion of the smaller particles, and small quantities of free oil. Flow continues down through the next layer of fine garnet material, which has an effective size of 0.012 in. A more finely divided range of suspended solids is removed in this polishing layer. The fluid continues into the final layer, a coarse garnet material that has an effective size of 0.08 in. Contained in this layer is the header/lateral assembly that collects the filtered water.

When the vessel has retained enough suspended solids to develop a substantial pressure drop, the unit must be backwashed either manually or automatically by reversing the direction of flow. This operation removes the accumulated solids out through the top of the vessel.

Centrifugal-Gravity Separators. In this type of separator, liquids/solids enter the unit tangentially, which sets up a circular flow. Liquids/solids are drawn through tangential slots and accelerated into the separation chamber. Centrifugal action tosses the particles heavier than the liquid to the perimeter of the separation chamber. Solids gently drop along the perimeter and into the separator's quiescent collection chamber. Solids-free liquid is drawn into the separator's vortex (low-pressure area) and up through the separator's outlet. Solids are either purged periodically or continuously bled from the separator by either a manual or automatic valve system.

Bag-Type Filters. These filters are composed of a bag of mesh or felt supported by a removable perforated metal basket, placed in a closed housing with an inlet and outlet. The housing is a welded, tubular pressure vessel with a hinged cover on top for access to the bag and basket. Housings are made of carbon or stainless steel. The inlet can be in the cover, in the side (above the bag), or in the bottom (and internally piped to the bag). The side inlet is the simplest type. In any case, the liquid enters the top of the bag. The outlet is located at the bottom of the side (below the bag). Pipe connections can be threaded or flanged. Single-basket housings can handle up to 220 gpm, multibaskets up to 3500 gpm.

The support basket is usually of 304 stainless steel perforated with 1/8 in. holes. (Heavy wire mesh baskets also exist.) The baskets can be lined with fine wire mesh and used by themselves as strainers, without adding a filter bag. Some manufacturers offer a second, inner basket (and bag) that fits inside the primary basket. This provides for two-stage filtering: first a coarse filtering stage, then a finer one. The benefits are longer service time and possible elimination of a second housing to accomplish the same function.

The filter bags are made of many materials (cotton, nylon, polypropylene, and polyester) with a range of ratings from 0.00004 to 0.033 in. Most common are felted materials because of their depth-filtering quality, which provides high dirt-loading capability, and their fine pores. Mesh bags are generally coarser, but are reusable and, therefore, less costly. The bags have a metal ring sewn into their opening; this holds the bag open and seats it on top of the basket rim.

In operation, the liquid enters the bag from above, flows out through the basket, and exits the housing cleaned of particulate down to the desired size. The contaminant is trapped inside the bag, making it easy to remove without spilling any downstream.

Special Methods. Localized areas frequently can be protected by special methods. Thus, pump-packing glands or mechanical shaft seals can be protected by fresh water makeup or by circulating

water from the pump casing through a cyclone separator or filter, then into the lubricating chamber.

In smaller equipment, a good dirt-control measure is to install backflush connections and shutoff valves on all condensers and heat exchangers so that accumulated settled dirt can be removed by backflushing with makeup water or detergent solutions. These connections can also be used for acid cleaning to remove calcium carbonate scale.

In specifying filtration systems, third-party testing by a qualified university or private test agency should be requested. The test report documentation should include a description of methods, piping diagrams, performance data, and certification.

Filters described in this section may also be used where industrial process cooling water is involved. For this type of service, consultation with the filtration equipment manufacturer is essential to ensure proper application.

START-UP AND SHUTDOWN OF COOLING TOWER SYSTEMS

The following guidelines are for start-up (or recommissioning) and shutdown of cooling tower systems:

Start-Up and Recommissioning for Drained Systems

- Clean all debris, such as leaves and dirt from the cooling tower.
- Close building air intakes in the area of the cooling tower to prevent entrainment of biocide and biological aerosols in the building air handling systems.
- Fill the system with water. While operating the condensing water pump(s) and *prior to operating the cooling tower fans*, execute one of the following two biocidal treatment programs:

  1. Resume treatment with the biocide that had been used prior to shutdown. Use the services of the water treatment supplier. Maintain the maximum recommended biocide residual (for the specific biocide) for a period sufficient to bring the system under good biological control (residual and time varies with the biocide).
  2. Treat the system with sodium hypochlorite at a level of 4 to 5 ppm free chlorine residual at a pH of 7.0 to 7.6. The residual level of free chlorine must be held at 4 to 5 ppm for 6 h. Commercially available test kits can be used to measure the residual of free chlorine.

- Once one of the two biocidal treatment programs has been successfully completed, turn on the fan and the put the system in service. The standard water treatment program (including biocidal treatment) should be resumed at this time.

Start-Up and Recommissioning for Undrained (Stagnant) Systems

- Remove accessible solid debris from bulk water storage vessel.
- Close building air intakes in the area of the cooling tower to prevent entrainment of biocide and biological aerosols in the building air handlers.
- Perform one of the two biocide pretreatment procedures (described in the section on Start-Up for Drained Systems) directly to the bulk water storage vessel (cooling tower sump, draindown tank, etc.). *Do not circulate stagnant bulk cooling water over cooling tower fill or operate cooling tower fans during pretreatment.*
- Stagnant cooling water may be circulated with condenser water pumps if tower fill is bypassed. Otherwise, add approved biocide directly to the bulk water source and mix with manual or sidestream flow methods. Take care to prevent the creation of aerosol spray from the stagnant cooling water from any point in the cooling water system.
- When one of the two biocidal pretreatments has been successfully completed, the cooling water should be circulated over the tower

fill. If biocide residual is maintained at a satisfactory level for at least 6 h, the cooling tower fans may then be operated safely.

Shutdown

When the system is to be shut down for an extended period, the entire system (cooling tower, system piping, heat exchangers, etc.) should be flushed and drained using the following procedure:

- Add a dispersant and biocide to the system and recirculate for 12 to 24 h. Confer with a water treatment consultant for suitable chemicals and dosage levels.
- Shut down pumps and completely drain all water distribution piping and headers, as well as the cooling loop. Remove water and debris from dead heads and low areas in the piping, which may not have completely drained.
- Rinse silt and debris from the sump. Pay special attention to corners and crevices. Add a mild solution of detergent and disinfectant to the sump and rinse. If the sump does not completely drain, pump out the remaining water and residue.
- If the equipment cannot be completely drained and is exposed to cold temperatures, freeze protection may be required.

SELECTION OF WATER TREATMENT

As discussed in the previous sections, many methods are available to prevent or correct water-caused problems. The selection of the proper water treatment method, and the chemicals and equipment necessary to apply that method, depends on many factors. The chemical characteristics of the water, which change with the operation of the equipment, are important. Other factors contributing to the selection of proper water treatment are

- Economics
- Chemistry control mechanisms
- Dynamics of the operating system
- Design of major components (e.g., the cooling tower or boiler)
- Number of operators available
- Training and qualifications of personnel
- Preventive maintenance program

Once-Through Systems

Economics is an overriding concern in treating water for once-through systems (in which a very large volume of water passes through the system only once). Protection can be obtained with relatively little treatment per unit mass of water because the water does not change significantly in composition while passing through equipment. However, the quantity of water to be treated is usually so large that any treatment other than simple filtration or the addition of a few parts per million of a polyphosphate, silicate, or other inexpensive chemical may not be practical or affordable. Intermittent treatment with polyelectrolytes can help maintain clean conditions when the cooling water is sediment-laden. In such systems, it is generally less expensive to invest more in corrosion-resistant construction materials than to attempt to treat the water.

Open Recirculating Systems

In an open recirculating system with chemical treatment, more chemical must be present because the water composition changes significantly by evaporation. Corrosive and scaling constituents are concentrated. However, treatment chemicals also concentrate by evaporation; therefore, after the initial dosage, only moderate dosages maintain the higher level of treatment needed. The selection of a water treatment program for an open recirculating system depends on the following major factors:

- Economics
- Water quality
- Performance criteria (e.g., corrosion rate, bacteria count, etc.)
- System metallurgy
- Available staffing

- Automation capabilities
- Environmental requirements
- Water treatment supplier (some technologies are superior to others in terms of economics, ease of use, safety, and impact on the environment)

An open recirculating system is typically treated with a scale inhibitor, corrosion inhibitor, oxidizing biocide, nonoxidizing biocide, and possibly a dispersant. The exact treatment program depends on the previously mentioned conditions.

A water treatment control scheme for a cooling tower might include

- Chemistry and cycles of concentration control using a conductivity controller
- Alkalinity control using automatic injection of sulfuric acid based on pH
- Scale control using contacting water meters, proportional feed, or traced control technology
- Oxidizing biocide control using an ORP (oxidation-reduction potential) controller
- Nonoxidizing biocide control using timers and pump systems

Air Washers and Sprayed-Coil Units

A water treatment program for an air washer or a sprayed-coil unit is usually complex and depends on the purpose and function of the system. Some systems, such as sprayed coils in office buildings, are used primarily to control the temperature and humidity. Other systems are intended to remove dust, oil vapor, and other airborne contaminants from an airstream. Unless the water is properly treated, the fouling characteristics of the contaminants removed from the air can cause operating problems.

Scale control is important in air washers or sprayed coils providing humidification. The minerals in the water may become concentrated (by evaporation) enough to cause problems. Inhibitor/dispersant treatment commonly used in cooling towers are often used in air washers to control scale formation and corrosion.

Suitable dispersants and surfactants are often needed to control oil and dust removed from the airstream. The type of dispersant depends on the nature of the contaminant and the degree of contamination. For maximum operating efficiency, dispersants should produce minimal amounts of foam.

Control of slime and bacterial growth is also necessary in the treatment of air washers and sprayed coils. The potential for biological growth is enhanced, especially if the water contains contaminants that are nutrients for the microorganisms. Due to variations in conditions and applications of air-washing installations and the possibility of toxicity problems, individual treatment options should be discussed with water treatment experts before a program is chosen. All microbiocides applied in air washers must have specific regulatory approval.

Ice Machines

Lime scale formation, cloudy or "milky" ice, objectionable taste and odor, and sediment are the most frequently encountered water problems in ice machines. Lime scale formation is probably the most serious problem because it interferes with the harvest cycle by forming on the freezing surfaces, preventing the smooth release of ice from the surface to the harvest bin.

Scale is caused by dissolved minerals in the water. Water freezes in a pure state and dissolved minerals concentrate in the unfrozen water; some eventually deposit on the machine's freezing surfaces as scale. As shown in Table 4, the probability of scale formation in an ice machine varies directly with the concentrations of carbonate and bicarbonate in the water circulating in the machine.

Dirty or scaled-up ice makers should be thoroughly cleaned before the water treatment program is started. Water distributor

Table 4   Amount of Scale in an Ice Machine

Total Alkalinity as Bicarbonate, ppm	Hardness as Calcium Carbonate, ppm			
	0 to 49	50 to 99	100 to 199	200 and up
0 to 49	None	Very light	Very light	Very light
50 to 99	Very light	Moderate	Moderate	Moderate
100 to 199	Very light	Troublesome	Troublesome	Heavy
200 and up	Very light	Troublesome	Heavy	Very heavy

holes should be cleared, and all loose sediment and other material should be flushed from the system. Existing scale can be removed by circulating an acid solution through the system.

Slowly soluble food-grade polyphosphates inhibit scale formation on freezing surfaces during normal operation by keeping hardness in solution. Although polyphosphates can inhibit lime scale in the ice-making section and help prevent sludge deposits of loose particles of lime scale in the sump, they do not eliminate the soft, milky, or white ice caused by a high concentration of dissolved minerals in the water.

Even with proper chemical treatment, recirculating water having a mineral content above 500 to 1000 ppm cannot produce clear ice. Increasing bleed off or reducing the thickness of the ice slab or the size of the cubes may mitigate the problem. However, demineralizing or distillation equipment is usually needed to prevent white ice production.

Another problem frequently encountered with ice machines is objectionable taste or odor. When water containing a material having an offensive taste or odor is used in an ice machine, the taste or odor is trapped in the ice. An activated carbon filter on the makeup water line can remove the objectionable material from the water. Carbon filters must be serviced or replaced regularly to avoid organic buildup in the carbon bed.

Occasionally, slime growth causes an odor problem in an ice machine. This problem can be controlled by regularly cleaning the machine with a food-grade acid. If slime deposits persist, sterilization of the ice machine may be helpful.

Feedwater often contains suspended solids such as mud, rust, silt, and dirt. To remove these contaminants, a sediment filter of appropriate size can be installed in the feed lines.

Closed Recirculating Systems

In a closed recirculating system, water composition remains fairly constant with very little loss of either water or treatment chemical. Closed systems are often defined as those requiring less than 5% makeup per year. The need for water treatment in such systems (i.e., water heating, chilled water, combined cooling and heating, and closed loop condenser water systems) is often ignored based on the rationalization that the total amount of scale from the water initially filling the system would be insufficient to interfere significantly with heat transfer, and that corrosion would not be serious. However, leakage losses are common, and corrosion products can accumulate sufficiently to foul heat transfer surfaces. Therefore, all systems should be adequately treated to control corrosion. Systems with high makeup rates should be treated to control scale as well.

The selection of a treatment program for closed systems should consider the following factors:

- Economics
- System metallurgy
- Operating conditions
- Makeup rate
- System size

Possible treatment technologies include

- Buffered nitrite
- Molybdate

- Silicates
- Polyphosphates
- Oxygen scavengers
- Organic blends

Before new systems are treated, they must be cleaned and flushed. Grease, oil, construction dust, dirt, and mill scale are always present in varying degrees and must be removed from the metallic surfaces to ensure adequate heat transfer and to reduce the opportunity for localized corrosion. Detergent cleaners with organic dispersants are available for proper cleaning and preparation of new closed systems.

Water Heating Systems

Secondary and Low-Temperature. Closed, chilled-water systems that are converted to secondary water heating during winter and primary low-temperature water heating, both of which usually operate in the range of 140 to 250°F, require sufficient inhibitors to control corrosion to less than 0.005 in. per year. Ethylene glycol or propylene glycol may be used as antifreeze in secondary hot water systems. Such glycols are available commercially, with inhibitors such as sodium nitrite, potassium phosphate, and organic inhibitors for nonferrous metals added by the manufacturer. These require no further treatment, but softened water should be used for all filling and makeup requirements. Samples should be checked periodically to ensure that the inhibitor has not been depleted. Analytical services are available from the glycol manufacturers and others for this purpose.

Environment- and High-Temperature. Environment-temperature water heating systems (250 to 350°F) and high-temperature, high-pressure hot water systems (above 350°F) require careful consideration of treatment for corrosion and deposit control. Makeup water for such systems should be demineralized or softened to prevent scale deposits. For corrosion control, oxygen scavengers such as sodium sulfite can be added to remove dissolved oxygen.

Electrode boilers are sometimes used to supply low- or high-temperature hot water. Such systems use heat generated due to the electrical resistance of the water between electrodes. The conductivity of the recirculating water must be in a specific range depending on the voltage used. Treatment of this type of system for corrosion and deposit control varies. In some cases oil-based corrosion inhibitors that do not contribute to the conductivity of the recirculating water are used.

Brine Systems

Systems containing brine, a strong solution of sodium chloride or calcium chloride, must be treated to control corrosion and deposits. Sodium nitrite at a minimum 3000 ppm in calcium brines or 4000 ppm in sodium brines, and a pH between 7.0 and 8.5 should provide adequate protection. Organic inhibitors are available that may provide adequate protection where nitrites cannot be used. Molybdates should not be used with calcium brines because insoluble calcium molybdate will precipitate.

Boiler Systems

Many treatment methods are available for steam-producing boilers; the method selected depends on

- Makeup water quality
- Makeup water quantity (or percentage condensate return)
- Pretreatment equipment
- Boiler operating conditions
- Steam purity requirements
- Economics

Pretreatment for makeup water could consist of

- Water softeners (for removal of calcium and magnesium hardness)
- Deaerators (for removal of dissolved gases; especially oxygen and carbon dioxide)

- Dealkalizers (to remove alkalinity in systems with high makeup rates and high alkalinity makeup water)
- Demineralizers (to remove almost all the hardness, alkalinity, and solids, depending on the application)
- Reverse osmosis (can remove up to 99% of minerals and dissolved solids from the water leaving almost pure H_2O)

Contaminants that are not removed mechanically by pretreatment equipment must be treated chemically. Once in the feedwater, dissolved gases, hardness, and dissolved minerals must be treated to prevent deposition and corrosion. ASME (1994) and ABMA (1995) have published recommended water chemistry limits for boiler water.

Treatment of the boiler water is often determined by the end use of the steam. Treatment regimens may include reduction of alkalinity; hardness removal; silica removal; oxygen reduction; and the feed of scale and corrosion inhibitors, oxygen scavengers, condensate treatment chemicals, and antifoams. These regimens are described in the following paragraphs.

Prevention of Scale. After the feedwater is pretreated, scale is controlled with phosphates, acrylates, polymers, chelates, and coagulation programs. Chelates, polymers, and acrylates work by binding the hardness, thereby preventing precipitation and scale formation. Phosphates and coagulation programs work in combination with sludge conditioners (tannins, lignins, starches, and synthetic polymers) to produce a softened precipitate that is removed by blowdown of the boiler.

Prevention of Corrosion and Oxygen Pitting. While boilers can corrode as the result of low boiler water pH or misuse of certain chemicals, corrosion is primarily caused by oxygen. After mechanical deaeration, boiler feedwater must be treated chemically. Effective deaeration removes most of the dissolved oxygen. (Most properly operating deaerators can remove oxygen down to 0.007 ppm and deaerating heaters can remove oxygen down to 0.04 ppm.) Oxygen scavengers such as catalyzed sodium sulfite and proprietary organic oxygen scavengers should then be fed to react with the residual oxygen in feedwater after deaeration. Oxygen scavengers will not only provide added protection to the boiler but to the steam and condensate system as well. Oxygen at levels as low as 0.005 ppm can cause oxygen pitting in the steam and condensate system if not chemically reduced by oxygen scavengers.

Most of the corrosion damage to boilers and associated equipment occurs during idle periods. The corrosion is caused by the exposure of wet metal to oxygen in the air or water. For this reason, special precautions must be taken to prevent corrosion while boilers are out of service.

Wet Boiler Lay-Up. This is a method of storing boilers full of water so that they can be returned to service. It involves adding extra chemicals (usually something to increase alkalinity, an oxygen scavenger, and a dispersant) to the boiler water. The water level is raised in the idle boiler to eliminate air spaces, and the boiler is kept completely full of treated water. Superheaters require special protection. Nitrogen gas can also be used on airtight boilers to maintain a positive pressure on the boiler, thereby preventing oxygen in-leakage.

Dry Boiler Lay-Up. This method of lay-up is usually for longer boiler outages. It involves draining, cleaning, and drying the boiler. A material that absorbs moisture, such as hydrated lime or silica gel, is placed in trays inside the boiler. The boiler is then sealed carefully to keep out air. Periodic inspection and replacement of the drying chemical are required during long storage periods.

Steam and Condensate Systems

Two problems associated with steam and condensate systems include general corrosion and pitting corrosion. In order to prevent condensate corrosion, the systems must be protected from acidic conditions, which lead to general corrosion, and oxygen, which leads to pitting corrosion. Protection can be by mechanical or chemical means or a combination of both. The following methods are commonly used for condensate system protection.

Protection from General Corrosion

Mechanical Protection. Reduce alkalinity from boiler feedwater to minimize the amount of carbon dioxide in the system. Carbon dioxide reacts with water (condensate) and forms carbonic acid that will corrode condensate system metal. Alkalinity can be reduced by dealkalization, demineralization, and reverse osmosis. Note: In many cases mechanical reduction of alkalinity is not needed due to low alkalinity makeup water and/or feedwater.

Chemical Protection. Use volatile amines, such as morpholine, diethylaminoethanol (DEAE), and cyclohexylamine, to neutralize carbonic acid and keep the condensate pH between 8.0 and 9.0.

Protection from Oxygen Corrosion

Mechanical Protection. Reduce oxygen from all boiler feedwater to prevent oxygen carryover to the steam and condensate system (via mechanical methods and chemical oxygen scavengers).

Chemical Protection. Feed filming amines, such as octadecylamine, to the steam to form a thin, hydrophobic film on the condensate system surfaces.

Chemical Protection. Feed a volatile oxygen scavenger to the steam to scavenge the oxygen.

The need for chemical treatment can be reduced by designing and maintaining tight return systems so that the condensate is returned to the boiler and less makeup is required in the boiler feedwater. The greater the amount of makeup, the more the system requires increased chemical treatment.

TERMINOLOGY

The following terms are commonly used in the water treatment industry as they pertain to corrosion, scale formation and fouling.

Alkalinity. The sum of bicarbonate, carbonate, and hydroxide ions in water. Other ions, such as borate, phosphate, or silicate, can also contribute to alkalinity.

Anion. A negatively charged ion of an electrolyte that migrates toward the anode influenced by an electric potential gradient.

Anode. The electrode of an electrolytic cell at which oxidation occurs.

Biological deposits. Water-formed deposits of biological organisms or the products of their life processes, such as barnacles, algae, or slimes.

Cathode. The electrode of an electrolytic cell at which reduction occurs.

Cation. A positively charged ion of an electrolyte that migrates toward the cathode influenced by an electric potential gradient.

Corrosion. The deterioration of a material, usually a metal, by reaction with its environment.

Corrosivity. The capacity of an environment or environmental factor to bring about destruction of a specific metal by the process of corrosion.

Electrolyte. A solution through which an electric current can flow.

Filtration. Process of passing a liquid through a porous material in such a manner as to remove suspended matter from the liquid.

Galvanic corrosion. Corrosion resulting from the contact of two dissimilar metals in an electrolyte or from the contact of two similar metals in an electrolyte of nonuniform concentration.

Hardness. The sum of the calcium and magnesium ions in water; usually expressed in ppm as $CaCO_3$.

Inhibitor. A chemical substance that reduces the rate of corrosion, scale formation, fouling, or slime production.

Ion. An electrically charged atom or group of atoms.

Passivity. The tendency of a metal to become inactive in a given environment.

pH. The logarithm of the reciprocal of the hydrogen ion concentration of a solution. pH values below 7 are increasingly acidic; those above 7 are increasingly alkaline.

Polarization. The deviation from the open circuit potential of an electrode resulting from the passage of current.

ppm. Parts per million by mass. In water, ppm are essentially the same as milligrams per liter (mg/L); 10,000 ppm (mg/L) = 1%.

Scale. 1. The formation at high temperature of thick corrosion product layers on a metal surface. 2. The precipitation of water-insoluble constituents on a surface.

Sludge. A sedimentary water-formed deposit, either of biological origin or suspended particles from the air.

Tuberculation. The formation over a surface of scattered, knob-like mounds of localized corrosion products.

Water-formed deposit. Any accumulation of insoluble material derived from water or formed by the reaction with water on surfaces in contact with it.

REFERENCES

ABMA. 2005. Boiler water quality requirements and associated steam quality for industrial/commercial and institutional boilers. *Publication* 402. American Boiler Manufacturers Association, Arlington, VA.

ASHRAE. 1998. Legionellosis position statement and Legionellosis position paper.

ASME. 1994. Consensus on operating practices for the control of feedwater and boiler water chemistry in modern industrial boilers. Research Committee on Water in Thermal Power Systems, Industrial Boiler Subcommittee. American Society of Mechanical Engineers, New York.

Cho, Y.I. 2002. Efficiency of physical water treatments in controlling calcium scale accumulation in recirculating open cooling water system. ASHRAE Research Project RP-1155, *Final Report*.

DOE. 1998. Non-chemical technologies for scale and hardness control. *Federal Technology Alert* DOE/EE-0162. U.S. Department of Energy.

Gan, F., D.-T. Chin, and A. Meitz. 1996. Laboratory evaluation of ozone as a corrosion inhibitor for carbon steel, copper, and galvanized steel in cooling water. *ASHRAE Transactions* 102(1).

Graman, P.S., G.A. Quinlan, and J.A. Rank. 1997. Nosocomial legionellosis traced to a contaminated ice machine. *Infection Control and Hospital Epidemiology* 18(9):637-640.

Langelier, W.F. 1936. The analytical control of anticorrosion water treatment. *Journal of the American Water Works Association* 28:1500.

Liu, Z., J.E. Stout, L. Tedesco, M. Boldin, C. Hwang, W.F. Diven, and V.L Yu. 1994. Controlled evaluation of copper-silver ionization in eradicating *Legionella pneumophila* from a hospital water distribution system. *The Journal of Infectious Diseases* 169:919-922.

Nasrazadani, S. and T.J. Chao. 1996. Laboratory evaluations of ozone as a scale inhibitor for use in open recirculating cooling systems. *ASHRAE Transactions* 102(2).

Ryznar, J.W. 1944. A new index for determining amount of calcium carbonate scale formed by a water. *Journal of the American Water Works Association* 36:472.

CHAPTER 50

SERVICE WATER HEATING

WATER HEATING energy use is second only to space conditioning in most residential buildings, and is also significant in many commercial and industrial settings. In some climates and applications, water heating is the largest energy use in a building. Moreover, quick availability of adequate amounts of hot water is an important factor in user satisfaction. Both water and energy waste can be significant in poorly designed service water-heating systems: from over- or undersizing pipes and equipment, from poor building layout, and from poor system design and operating strategies. Good service water-heating system design and operating practices can often reduce first costs as well as operating costs. The information in this chapter is thus critical for the sustainable design and operation of many buildings.

SYSTEM ELEMENTS

A service water-heating system has (1) one or more heat energy sources, (2) heat transfer equipment, (3) a distribution system, and (4) terminal hot-water usage devices.

Heat energy sources may be (1) fuel combustion; (2) electrical conversion; (3) solar energy; (4) geothermal, air, or other environmental energy; and/or (5) recovered waste heat from sources such as flue gases, ventilation and air-conditioning systems, refrigeration cycles, and process waste discharge.

Heat transfer equipment is direct, indirect, or a combination of the two. For direct equipment, heat is derived from combustion of fuel or direct conversion of electrical energy into heat and is applied within the water-heating equipment. For indirect heat transfer equipment, heat energy is developed from remote heat sources (e.g., boilers; solar energy collection; air, geothermal, or other environmental source; cogeneration; refrigeration; waste heat) and is then transferred to the water in a separate piece of equipment. Storage tanks may be part of or associated with either type of heat transfer equipment.

Distribution systems transport hot water produced by water-heating equipment to terminal hot-water usage devices. Water consumed must be replenished from the building water service main. For locations where constant supply temperatures are desired, circulation piping or a means of heat maintenance must be provided.

Terminal hot-water usage devices are plumbing fixtures and equipment requiring hot water that may have periods of irregular flow, constant flow, and no flow. These patterns and their related water usage vary with different buildings, process applications, and personal preference.

In this chapter, it is assumed that an adequate supply of building service water is available. If this is not the case, alternative strategies such as water accumulation, pressure control, and flow restoration should be considered.

WATER-HEATING TERMINOLOGY

Recovery efficiency. Heat absorbed by the water divided by heat input to heating unit during the period that water temperature is raised from inlet temperature to final temperature (includes heat losses from water heater jacket and/or tank).

Recovery rate. The amount of hot water that a residential water heater can continually produce, usually reported as flow rate in gallons per hour that can be maintained for a specified temperature rise through the water heater.

Fixture unit. A number, on an arbitrarily chosen scale, that expresses the load-producing effects on the system of different kinds of fixtures.

Thermal efficiency. Heat in water flowing from the heater outlet divided by the energy input to the heating unit over a specific period of steady-state conditions (includes heat losses from the water heater jacket and/or tank).

Input efficiency. Heat entering water in the heating device divided by energy input to the heating unit over a specific period of steady-state conditions, or while heating from cold to hot, depending on how stated (steady-state versus average input efficiency); it does not include heat losses from the water heater jacket and/or tank. When used with fossil-fuel-fired equipment, this is commonly called **combustion efficiency**.

Energy factor. The delivered efficiency of a residential water heater when operated as specified in U.S. Department of Energy (DOE) test procedures (DOE 2001). See also ASHRAE *Standard* 118.2.

First-hour rating. An indicator of the maximum amount of hot water a residential water heater can supply in 1 h. This rating is used by the Federal Trade Commission (FTC) for comparative purposes. Because peak draws taken over periods less than 1 h frequently drive residential equipment sizing, first-hour rating alone should not be used for equipment sizing. As for larger systems, storage tank volume and heating rate also play important roles.

Standby loss. As applied to a tank water heater (under test conditions with no water flow), the average hourly energy consumption divided by the average hourly heat energy contained in stored water, expressed as a percent per hour. This can be converted to the average Btu/h energy consumption required to maintain any water/air temperature difference by taking the percent times the temperature difference, times 8.25 Btu/gal·°F (a nominal specific heat for water), times the tank capacity, and then dividing by 100.

Standby loss coefficient. The heat input (in Btu/h·°F) into a storage water heater when operated as specified in U.S. Department of Energy (DOE) test procedures (DOE 2001). This value is essentially the standby loss divided by the difference in temperature between the average stored water temperature and the surrounding air temperature, and also includes the effect of recovery efficiency for providing the necessary heat energy to offset the heat loss.

The preparation of this chapter is assigned to TC 6.6, Service Water Heating Systems.

Hot-water distribution efficiency. Heat contained in the water at points of use divided by heat delivered at the heater outlet at a given flow.

Heater/system efficiency. Heat contained in the water at points of use divided by the heat input to the heating unit at a given flow rate (thermal efficiency times distribution efficiency).

Heat trap. A device to counteract the natural convection of heated water in a vertical pipe. Commercially available heat traps for large equipment are generally 360° loops of tubing; heat traps can also be constructed of pipes connected to the water heater (inlet or outlet) that direct flow downward before connecting to the vertical supply or hot-water distribution system. Tubing or piping heat traps should have a loop diameter or length of downward piping of at least 12 in. Various prefabricated check-valve-like heat traps are available for residential-sized equipment, using balls, flexible flaps, or moving disks.

Overall system efficiency. Heat energy in the water delivered at points of use divided by the total energy supplied to the heater for any selected period.

System standby loss. The amount of heat lost from the water-heating system and the auxiliary power consumed during periods of nonuse of service hot water.

SYSTEM PLANNING

The goals of system planning are to (1) size the system properly; (2) optimize system efficiency; and (3) minimize first, operating, and overall life-cycle costs. It is important to design systems so that they perform well from both functional and energy-use perspectives. Flow rate, temperature, and total flow over specific time periods are the primary factors to be determined in the hydraulic and thermal design of a water-heating and piping system. Operating pressures, time of delivery, and water quality are also factors to consider. Separate procedures are used to select water-heating equipment and to design the piping system. However, water-heating equipment sizing and piping system design should be considered together for best system design. Oversized or excessively long piping exacerbates delivery delay and/or energy waste.

Water-heating equipment, storage facilities, and piping should (1) have enough capacity to provide the required hot water while minimizing waste of energy or water and (2) allow economical system installation, maintenance, and operation.

Water-heating equipment types and designs are based on the (1) energy source, (2) application of the developed energy to heat the water, and (3) control method used to deliver the necessary hot water at the required temperature under varying water demand conditions. Application of water-heating equipment within the overall design of the hot-water system is based on (1) location of the equipment within the system, (2) related temperature requirements, (3) volume of water to be used, and (4) flow rate. Consideration of peak demand effects on utilities is also of growing importance.

Additional planning is required when the system providing the potable hot water is also used for space heating or other purposes. Some special water heater designs are made for this purpose, but this end can also be achieved by combining needed components in an appropriate manner.

Energy Sources

The choice of energy source(s) is influenced by equipment type and location. These decisions should be made only after evaluating purchase, installation, operating, and maintenance costs. A life-cycle analysis is highly recommended.

In making energy conservation choices, consult the ANSI/ASHRAE/IESNA *Standards* 90.1 and 90.2, or the sections on Service Water Heating of ANSI/ASHRAE/IESNA *Standard* 100, as well as the section on Design Considerations in this chapter.

WATER-HEATING EQUIPMENT

Gas-Fired Systems

Automatic storage water heaters incorporate the burner(s), storage tank, outer jacket, and controls such as a thermostat in a single unit and have an input-to-storage capacity ratio of less than 4000 Btu/h per gallon.

Automatic instantaneous water heaters are produced in two distinctly different types. Tank-type instantaneous heaters have an input-to-storage capacity ratio of 4000 Btu/h per gallon or more and a thermostat to control energy input to the heater. Water-tube instantaneous heaters have minimal water storage capacity. They usually have a flow switch that controls the burner, and may have a modulating fuel valve that varies fuel flow as water flow changes.

Tankless water heaters have almost no storage capacity, and heat water as it flows once through the water heater. Heating rate required varies with water flow rate and needed temperature rise. Most modern gas-fired tankless water heaters have a flow switch or equivalent to confirm flow before the burner activates. Some have advanced multistage or modulating burners to better control outlet temperature. Some also incorporate fixed or modulating water flow rate controls to ensure that water temperature reaches at least a minimum outlet temperature (i.e., it restricts flow rate, possibly below that which the user requested, to avoid undesirably cool outlet water temperature if burners are already operating at maximum heating rate). Most advanced designs also incorporate electronic ignition controls, thus minimizing standby energy losses compared to having a tank and continuously burning pilot light. Properly applied tankless water heaters thus have lower overall energy use and higher efficiency compared to minimum-efficiency tank types serving the same loads. Note, however, that tankless gas water heaters have on/off cycling-rate-related energy losses that, under some draw patterns, may reduce their apparent efficiency advantage. They may also have minimum flow rate requirements before they activate, which may require users to modify their behavior (e.g., use a higher flow rate than they normally would and/or leave water running when they would normally turn it off) to obtain hot water. Sometimes, it may be beneficial to reduce the hot-water delivery temperature to reduce point-of-use mixing with cold water, thereby increasing hot-water flow rate.

Circulating tank water heaters are classified in two types: (1) automatic, in which the thermostat is located in the water heater, and (2) nonautomatic, in which the thermostat is located within an associated storage tank.

Hot-water supply boilers are capable of providing service hot water. They are typically installed with separate storage tanks and applied as an alternative to circulating tank water heaters. Outdoor models are wind- and rain-tested. They are available in most of the classifications previously listed.

Direct-vent models are to be installed inside, but are not vented through a conventional chimney or gas vent and do not use ambient air for combustion. They must be installed with the means specified by the equipment manufacturer for venting (typically horizontal) and for supplying combustion air from outside the building.

Power vent equipment uses a powered fan or blower to move combustion products, allowing horizontal as well as vertical venting.

Direct-fired equipment passes cold water through a stainless steel or other heat exchange medium, which breaks up the water into very small droplets. These droplets then come into direct contact with heat rising from a flame, which heats the water directly.

Residential water-heating equipment is usually the automatic storage type, although increasing numbers of tankless water heaters are being installed. For industrial and commercial applications, commonly used types of heaters are (1) automatic storage, (2) circulating tank, (3) instantaneous/tankless and (4) hot-water supply boilers.

Installation guidelines for gas-fired water heaters can be found in the National Fuel Gas Code, NFPA *Standard* 54/ANSI *Standard* Z223.1. This code also covers sizing and installation of venting equipment and controls.

Oil-Fired Systems

Oil-fired water heaters are generally the storage tank type. Models with a storage tank of 50 gal or less with an input rating of 105,000 Btu/h or less are usually considered residential models. Commercial models are offered in a wide range of input ratings and tank sizes. There are models available with combination gas/oil burners, which can be switched to burn either fuel, depending on local availability.

Installation guidelines for oil-fired water heaters can be found in NFPA *Standard* 31/ANSI *Standard* Z95.1.

Electric

Electric water heaters are generally the storage type, consisting of a tank with one or more immersion heating elements. The heating elements consist of resistance wire embedded in refractories having good heat conduction properties and electrical insulating values. Heating elements are fitted into a threaded or flanged mounting for insertion into a tank. Thermostats controlling heating elements may be of the immersion or surface-mounted type.

Residential storage tank water heaters range up to 120 gal with input up to 12 kW. They have a primary resistance heating element near the bottom and often a secondary element located in the upper portion of the tank. Each element is controlled by its own thermostat. In dual-element heaters, the thermostats are usually interlocked so that the lower heating element cannot operate if the top element is operating. Thus, only one heating element operates at a time to limit the current draw.

Commercial storage tank water heaters are available in many combinations of element quantity, wattage, voltage, and storage capacity. Storage tanks may be horizontal or vertical. Compact, low-volume models are used in point-of-use applications to reduce hot-water piping length. Locating the water heater near the point of use makes recirculation loops unnecessary.

Instantaneous *or* **tankless electric water heaters** have almost no storage capacity and heat water as it flows once through the water heater. Heating rate required varies with water flow rate and needed temperature rise. Tankless electric water heaters for residential applications are available in heating capacities from a low of about 1.5 kW to a high of about 38 kW. Smaller-capacity units (typically 12 kW or less, but this varies with geographic location and entering cold-water temperature) are sometimes used in lavatory (sink) and other point-of-use applications such as remote low-use showers, small hot tubs, whirlpool baths, and other low-flow-rate applications. Larger sizes (above 18 kW) can sometimes be used in whole-house applications, depending on geographic location (and hence entering cold water temperature) and site hot-water use profiles (see Table 15). Tankless water heaters can, if equipped with appropriate controls, be used in booster and/or recirculating water-heating systems. Note that not all models can be used to heat already partially warmed water: this capability varies among models.

Heat pump water heaters (HPWHs) use a vapor-compression refrigeration cycle to extract energy from an air, ground, or water source to heat water. HPWHs may be add-on or integrated devices. Integrated units include a storage tank and supplemental heating. Add-on units can be connected to a storage tank. The heat pump cycle of some HPWHs has a maximum output temperature of 125 to 140°F, although other cycles can deliver temperatures in the range of 180°F. Where a higher delivery temperature is required than the HPWH can produce, a supplemental or booster water heater downstream of the storage tank should be used. HPWHs function most efficiently where inlet water temperature is low and source temperature is warm. Systems should be sized to allow high HPWH run time, which in some cases can require additional storage tank capacity above that normal for the application. As the HPWH collects heat, it provides a potentially useful cooling effect and, in air-source units, also dehumidifies the air. The effect of HPWH cooling output on the building's energy balance should be considered. Cooling output should be directed to provide occupant comfort and avoid interfering with temperature-sensitive equipment (EPRI 1990).

Demand-controlled water heating can significantly reduce the cost of heating water electrically. Demand controllers operate on the principle that a building's peak electrical demand exists for a short period, during which heated water can be supplied from storage rather than through additional energy applications. Shifting the use of electricity for service water heating from peak demand periods allows water heating at the lowest electric energy cost in many electric rate schedules. The building electrical load must be detected and compared with peak demand data. When the load is below peak, the control device allows the water heater to operate. Some controllers can program deferred loads in steps as capacity is available. The priority sequence may involve each of several banks of elements in (1) a water heater, (2) multiple water heaters, or (3) water-heating and other equipment having a deferrable load, such as pool heating and snow melting. When load controllers are used, hot-water storage must be sized appropriately.

Electric off-peak storage water heating is a water-heating equipment load management strategy whereby electrical demand to a water-heating system is time-controlled, primarily in relation to the building or utility electrical load profile. This approach may require increased tank storage capacity and/or stored-water temperature to accommodate water use during peak periods.

Sizing recommendations in this chapter apply only to water heating without demand or off-peak control. When demand control devices are used, the storage and recovery rate may need to be increased to supply all the hot water needed during the peak period and during the ensuing recovery period. Manian and Chackeris (1974) include a detailed discussion on load-limited storage heating system design.

Indirect Water Heating

In indirect water heating, the heating medium is steam, hot water, or another fluid that has been heated in a separate generator or boiler. The water heater extracts heat through an external or internal heat exchanger.

When the heating medium is at a higher pressure than the service water, the service water may be contaminated by leakage of the heating medium through a damaged heat transfer surface. In the United States, some national, state, and local codes require double-wall, vented tubing in indirect water heaters to reduce the possibility of cross-contamination. When the heating medium is at a lower pressure than the service water, other jurisdictions allow single-wall tubing heaters because any leak would be into the heating medium.

If the heating medium is steam, high rates of condensation occur, particularly when a sudden demand causes an inflow of cold water. The steam pipe and condensate return pipes should be of ample size. Condensate may be cooled by preheating the cold-water supply to the heater.

Corrosion is minimized on the heating medium side of the heat exchanger because no makeup water, and hence no oxygen, is brought into that system. The metal temperature of the service water side of the heat exchanger is usually less than that in direct-fired water heaters. This minimizes scale formation from hard water.

Storage water heaters are designed for service conditions where hot-water requirements are not constant (i.e., where a large volume of heated water is held in storage for periods of peak load). The amount of storage required depends on the load's nature and water heater's recovery capacity. An individual tank or several tanks joined by a manifold may be used to provide the required storage.

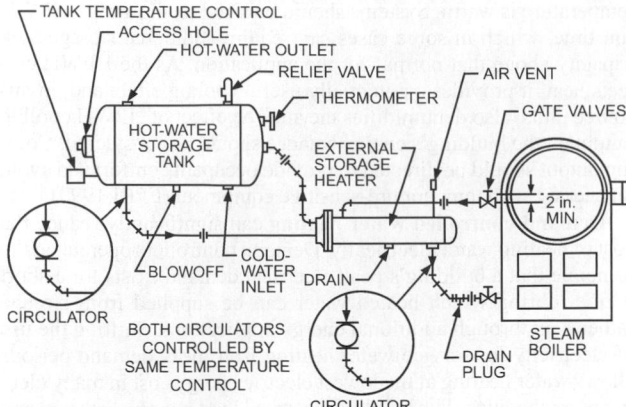

Fig. 1　Indirect, External Storage Water Heater

External storage water heaters are designed for connection to a separate tank (Figure 1). Boiler water circulates through the heater shell, while service water from the storage tank circulates through the tubes and back to the tank. Circulating pumps are usually installed in both the boiler water piping circuit and the circuits between the heat exchanger and the storage tank. Steam can also be used as the heating medium in a similar scheme.

Instantaneous indirect water heaters (tankless coils) are best used for a steady, continuous supply of hot water. In these units, the water is heated as it flows through the tubes. Because the heating medium flows through a shell, the ratio of hot-water volume to heating medium volume is small. As a result, variable flow of the service water causes uncertain temperature control unless a thermostatic mixing valve is used to maintain the hot-water supply to the plumbing fixtures at a more uniform temperature.

Some indirect instantaneous water heaters are located inside a boiler. The boiler is provided with a special opening through which the coil can be inserted. Although the coil can be placed in the steam space above the water line of a steam boiler, it is usually placed below the water line. The water heater transfers heat from the boiler water to the service water. The gross output of the boiler must be sufficient to serve all loads.

Semi-Instantaneous

These water heaters have limited storage to meet the average momentary surges of hot-water demand. They usually consist of a heating element and control assembly devised for close control of the temperature of the leaving hot water.

Circulating Tank

These water heaters are instantaneous or semi-instantaneous types used with a separate storage tank and a circulating pump. The storage acts as a flywheel to accommodate variations in the demand for hot water.

Blending Injection

These water heaters inject steam or hot water directly into the process or volume of water to be heated. They are often associated with point-of-use applications (e.g., certain types of commercial laundry, food, and process equipment). *Caution*: Cross-contamination of potable water is possible.

Solar

Availability of solar energy at the building site, efficiency and cost of solar collectors, system installation costs, and availability and cost of other fuels determine whether solar energy collection units should be used as a primary heat energy source. Solar energy equipment can also be included to supplement other energy sources and conserve fuel or electrical energy.

The basic elements of a solar water heater include solar collectors, a storage tank, piping, controls, and a transfer medium. The system may use natural or forced circulation. Auxiliary heat energy sources may be added, if needed.

Collector design must allow operation in below-freezing conditions, where applicable. Antifreeze solutions in a separate collector piping circuit arrangement are often used, as are systems that allow water to drain back to heated areas when low temperatures occur. Uniform flow distribution in the collector or bank of collectors and stratification in the storage tank are important for good system performance.

Application of solar water heaters depends on (1) auxiliary energy requirements; (2) collector orientation; (3) temperature of the cold water; (4) general site, climatic, and solar conditions; (5) installation requirements; (6) area of collectors; and (7) amount of storage. Chapter 35 has more detailed design information.

Waste Heat Use

Waste heat recovery can reduce energy cost and the energy requirement of the building heating and service water-heating equipment. Waste heat can be recovered from equipment or processes by using appropriate heat exchangers in the hot gaseous or liquid streams. Heat recovered is frequently used to preheat water entering the service water heater. A conventional water heater is typically required to augment the output of a waste heat recovery device and to provide hot water during periods when the host system is not in operation.

Refrigeration Heat Reclaim

These systems heat water with heat that would otherwise be rejected through a refrigeration, air-conditioning, or heat pump condenser. Refrigeration heat reclaim uses refrigerant-to-water heat exchangers connected to the refrigeration circuit between the compressor and condenser of a host refrigeration or air-conditioning system to extract heat. Water is heated only when the host is operating. Because many simple systems reclaim only superheat energy from the refrigerant, they are often called **desuperheaters**. However, some units are also designed to provide partial or full condensing. The refrigeration heat reclaim heat exchanger is generally of vented, double-wall construction to isolate potable water from refrigerant. Some heat reclaim devices are designed for use with multiple refrigerant circuits. Controls are required to limit high water temperature, prevent low condenser pressure, and provide for freeze protection. Refrigeration systems with higher run time and lower efficiency provide more heat reclaim potential. Most systems are designed with a preheat water storage tank connected in series with a conventional water heater (EPRI 1992). In all installations, care must be taken to prevent inappropriately venting refrigerants.

Combination Heating

A **combo system** provides hot water for both space heating and domestic use. A space-heating coil and a space-cooling coil are often included with the air handler to provide year-round comfort. Combo systems also can use other types of heat exchangers for space heating, such as baseboard convectors or floor heating coils. A method of testing combo systems is given in ASHRAE *Standard* 124. The test procedures allow the calculation of combined annual efficiency (CAE), as well as space- and water-heating efficiency factors. Kweller (1992), Pietsch and Talbert (1989), Pietsch et al. (1994), Subherwal (1986), and Talbert et al. (1992) provide additional design information on these heaters.

DESIGN CONSIDERATIONS

Hot-water system design should consider the following:

- Water heaters of different sizes and insulation may have different standby losses, recovery efficiency, thermal efficiency, or energy factors.

- A distribution system should be properly laid out, sized, and insulated to deliver adequate water quantities at temperatures satisfactory for the uses served. This reduces standby loss and improves hot-water distribution efficiency. Locating fixtures or usage devices close to each other and to the water-heating equipment is particularly important for minimizing piping lengths and diameters, and thus reducing wait times as well as water and energy waste.

- Heat traps between recirculation mains and infrequently used branch lines reduce convection losses to these lines and improve heater/system efficiency. In small residential systems, heat traps can be applied directly to the water heater for the same purpose.

- Controlling circulating pumps to operate only as needed to maintain proper temperature at the end of the main reduces losses on return lines.

- Provision for shutdown of circulators during building vacancy reduces standby losses.

DISTRIBUTION

Piping Material

Traditional piping materials include galvanized steel used with galvanized malleable iron screwed fittings. Copper piping and copper water tube types K, L, or M have been used with brass, bronze, or wrought copper water solder fittings. Legislation or plumbing code changes have banned the use of lead in solders or pipe-jointing compounds in potable water piping because of possible lead contamination of the water supply. See ASHRAE *Standard* 90 series for pipe insulation requirements.

Today, most potable water supplies require treatment before distribution; this may cause the water to become more corrosive. Therefore, depending on the water supply, traditional galvanized steel piping or copper tube may no longer be satisfactory, because of accelerated corrosion. Galvanized steel piping is particularly susceptible to corrosion (1) when hot water is between 140 and 180°F and (2) where repairs have been made using copper tube without a nonmetallic coupling. Note that plumbing can be either piping (relatively thick wall) or tubing (relatively thin wall), although *piping* is used in this chapter for both.

Before selecting any water piping material or system, consult the local code authority. The local water supply authority should also be consulted about any history of water aggressiveness causing failures of any particular material.

Alternative piping materials that may be considered are (1) stainless steel tube and (2) various plastic piping and tubes. Particular care must be taken to ensure that the application meets the design limitations set by the manufacturer, particularly regarding temperature and pressure limits, and that the correct materials and methods of joining are used. These precautions are easily taken with new projects, but become more difficult during repairs of existing work. Using incompatible piping, fittings, and jointing methods or materials must be avoided, because they can cause severe problems, such as corrosion or leakage caused by differential thermal expansion.

Pipe Sizing

Sizing hot-water supply pipes from a hydraulic (pressure drop) perspective involves the same principles as sizing cold-water supply pipes (see Chapter 22 of the 2009 *ASHRAE Handbook—Fundamentals*). The water distribution system must be correctly sized for the total hot-water system to function properly. Hot-water demand varies with the type of establishment, usage, occupancy, and time of day. The piping system should be able to meet peak demand at an acceptable pressure loss. It is important not to oversize hot-water supply pipes, because this adversely affects system heat loss and overall energy use.

Supply Piping

Table 16, Figures 25 and 26, and manufacturers' specifications for fixtures and appliances can be used to determine hot-water demands. These demands, together with procedures given in Chapter 22 of the 2009 *ASHRAE Handbook—Fundamentals*, are used to size the mains, branches, and risers.

Allowance for pressure drop through the heater should not be overlooked when sizing hot-water distribution systems, particularly where instantaneous water heaters are used and where the available pressure is low.

Pressure Differential

Sizing both cold- and hot-water piping requires that the pressure differential at the point of use of blended hot and cold water be kept to a minimum. This is particularly important for tubs and showers, because sudden changes in flow at fixtures cause discomfort and a possible scalding hazard. Pressure-compensating devices are available.

Piping Heat Loss and Hot-Water Delivery Delays

Good hot-water distribution system layout is very important, for both user satisfaction and energy use. This has become increasingly important with the mandated use of low-flow fixtures, which can cause lengthy delays and increased water waste while waiting for hot water to arrive at fixtures compared to higher-flow designs. In general, it is desirable to put fixtures close to each other and close to the water heater(s) that serve them. This minimizes both the diameter and length of the hot-water piping required. Recent work has shown that energy loss from hot-water piping due to both heat loss and water waste waiting for hot water to arrive at fixtures can be a significant percentage of total water-heating system energy use (Hiller 2005a; Klein 2004a, 2004b, 2004c; Lutz 2005). Energy losses from hot-water distribution systems usually amount to at least 10-20% of total hot-water system energy use in most potable water-heating systems (Hiller 2005a), and are often as high as 50%; losses of over 90% have been found in some installations (Hiller and Miller 2002; Hiller et al. 2002).

Hiller (2005a, 2005b, 2006a, 2006b) measured both piping heat loss and time, water, and energy waste while waiting for hot water to arrive at fixtures. This research measured piping heat loss *UA* factors for several commonly used piping sizes, types, and insulation levels. *UA*$_{flowing}$ values are a slight function of water flow rate and temperature difference between the hot water and the surroundings. However, for many practical calculation purposes, *UA* can be considered constant at the values shown in Table 1.

Hiller (2008) found that bare copper piping buried in damp sand (typical of under-slab piping) exhibited heat loss rates over eight times higher than the same pipe in air. This much higher heat loss rate is believed to be caused by moisture in the sand near the pipe behaving like a heat pipe by evaporating, recondensing (thus transferring heat to sand particles a short distance away much faster than conduction would), and then wicking back to the pipe. Adding insulation to buried piping dramatically reduced the heat-pipe effect by lowering the surface temperature seen by the moisture. Hence, as can be seen in Table 1, adding 3/4 in. foam pipe insulation to copper piping reduces the heat loss rate in air to around one-half of the uninsulated value, but adding the same insulation to pipe buried in damp sand reduces the heat loss rate to only around 6% of its uninsulated value, a reduction by a factor of around 16. Thus adding pipe insulation is highly beneficial for buried piping, and is recommended.

Table 1 also shows that all of the plastic pipes tested to date exhibit moderately to significantly higher heat loss rates than comparably sized copper pipes when tested uninsulated in air. However, when insulated, they exhibit moderately to significantly lower heat loss rates than comparably sized copper pipes with the same insulation. Adding 3/4 in. foam reduces plastic pipe heat loss rates to

Table 1 Piping Heat Loss Factors for Foam In Piping Heat Loss Factors for Foam Insulation with Thermal Conductivity of 0.02 Btu/h·ft²·°F

Nominal Pipe Size	Foam Insulation Thickness, in.	$UA_{zero\,flow}$, Btu/h·ft·°F	High-Value $UA_{flowing}$, Btu/h·ft·°F
1/2 in. rigid copper	0	0.226	0.36
	0.5	0.128	0.20
	0.75	0.116	0.19
3/4 in. rigid copper	0	0.388	0.44
	0.5	0.150	0.25
	0.75	0.142	0.24
3/4 in. rolled copper	0	0.334	0.334
	0.75	0.138	0.16
3/4 in. roll CU-sand	0	1.2	2.82
	0.75	0.155	0.177
3/4 in. PEX-AL-PEX[a]	0	0.550	0.546
	0.5	0.199	0.199
	0.75	0.158	0.18
3/4 in. PEX[b]	0	0.535	0.585
	0.75	0.159	0.19.
1/2 in. PEX	0	0.438	0.438
	0.75	0.13	0.13
3/8 in. PEX	0		
3/4 in. CPVC	0	0.44	0.52
	0.75	0.148	0.17

Note: Results are for horizontal in-air tests unless otherwise noted.

Sources: Hiller (2005a, 2005b, 2006b, 2008, 2009).

[a]High-density cross-linked polyethylene, aluminum, high-density cross-linked polyethylene multilayer pipe.

[b]High-density cross-linked polyethylene.

around 30% of their uninsulated values when tested in air. This is a reduction in heat loss rate by a factor of three, compared to a factor of two for insulation on copper piping. This result suggests that plastic pipes have higher emissivity for radiation heat loss from the piping than does copper. Theoretical analysis suggests that, for the pipe sizes tested, radiation heat loss from the pipes represents between 30% and 70% of total heat loss rate from the pipes, depending on pipe type and size. It has been suggested that the emissivity of copper pipe may increase with age as the outer surface oxidizes to its normal dull-brown appearance from its original bright, shiny surface. Repeat tests on aged copper pipe have not yet been performed.

The *UA* factors of Table 1 are used in Equations (1) to (8) to determine heat loss rates from piping during both flowing and zero-flow (cooldown) conditions, and to find temperature drop while water is flowing through pipe, and pipe temperature at any time during cooldown. Note that piping heat loss and pipe temperature drop are not constant with length under flowing conditions, because the temperature of each successive length of pipe is less than the one before it. The same is true for zero-flow pipe cooldown with respect to time, because the pipe is at a progressively lower temperature at each successive time interval. The result is that pipe temperatures decay inverse-exponentially with length under flowing conditions and with time under cooldown conditions. This is why log-mean temperature difference must be used in heat loss calculations instead of a simple linear temperature difference (Rohsenow and Choi 1961).

Under flowing conditions,

$$Q = mc_p(T_{hot\,in} - T_{hot\,out}) \quad (1)$$

and

$$Q = UA_{flowing}\Delta T_{lm} \quad (2)$$

For water flowing in pipes in a constant-air-temperature environment,

$$\Delta T_{lm} = \frac{[(T_{hot\,in} - T_{air}) - (T_{hot\,out} - T_{air})]}{\ln[(T_{hot\,in} - T_{air})/(T_{hot\,out} - T_{air})]} \quad (3)$$

When $UA_{flowing}$, water flow rate, air temperature, and entering water temperature are known, Equations (1) to (3) can be combined and rearranged to determine pipe-exiting water temperature as follows:

$$T_{hot\,out} = T_{air} + (T_{hot\,in} - T_{air})e^{-\left[\frac{(UA_{flowing})(L_{pipe})}{(mc_{p_{water}})}\right]} \quad (4)$$

where

ΔT_{lm} = log mean temperature difference, °F
Q = heat loss rate, Btu/h
m = water flow rate, lb$_m$/h
c_p = specific heat of water, 1 Btu/lb$_m$·°F
$T_{hot\,in}$ = water temperature entering pipe, °F
$T_{hot\,out}$ = water temperature leaving pipe, °F
$UA_{flowing}$ = flowing heat loss factor per foot of pipe, Btu/h·ft·°F
L_{pipe} = length of hot-water pipe, ft

Note that the quantity $(UA_{flowing})(L_{pipe})/(mc_p)_{water}$ must be nondimensional, so appropriate units must be used.

Under zero-flow cooldown conditions,

$$Q = (Mc_p)_{w,p,i}(T_{hot\,t_1} - T_{hot\,t_2})/(t_2 - t_1) \quad (5)$$

$$Q = UA_{zero\text{-}flow}(\Delta T_{lm}) \quad (6)$$

And for pipe in a constant-air-temperature environment:

$$\Delta T_{lm} = \frac{[(T_{hot\,t_1} - T_{air}) - (T_{hot\,t_2} - T_{air})]}{\ln[(T_{hot\,t_1} - T_{air})/(T_{hot\,t_2} - T_{air})]} \quad (7)$$

$$T_{hot\,t_2} = T_{air} + (T_{hot\,t_1} - T_{air})e^{-\left[\frac{(UA_{zero\text{-}flow})(t_2 - t_1)}{(Mc_p)_{w,p,i}}\right]} \quad (8)$$

where

t_1 = initial time
t_2 = final time
Q = average heat loss rate from time t_1 to time t_2, Btu/h
$(Mc_p)_{w,p,i}$ = sum of mass times specific heat for water, pipe, and insulation, Btu/ft·°F
$T_{hot\,t_1}$ = pipe temperature at t_1, °F
$T_{hot\,t_2}$ = pipe temperature at t_2, °F
$UA_{zero\text{-}flow}$ = zero-flow heat loss factor per foot of pipe, Btu/h·ft·°F

Note that the quantity $(UA_{zero\text{-}flow})(t_2 - t_1)/(Mc_p)_{w,p,i}$ must be nondimensional, so appropriate units must be used.

Pipe temperature at any time during the cooldown process is determined by Equation (8). Total energy lost from piping during zero-flow cooldown is determined by calculating the pipe temperature at time t_2 and multiplying the average heat loss rate between t_1 and t_2 determined by Equation (5) times the duration of the cooldown period $(t_2 - t_1)$. An alternative is to calculate heat loss over short time periods using Equation (6) and sum the results.

Table 2 contains earlier piping heat loss data, and shows computed piping *UA* values based on those data.

Hiller (2005a, 2005b, 2006b) also produced tables of water/energy wasted while waiting for hot water to arrive at fixtures. Waste is a strong function of pipe material, interior finish, diameter, fittings present, flow rate, initial pipe temperature, and entering hot-water

Table 2 Approximate Heat Loss from Piping at 140°F Inlet, 70°F Ambient

Nominal Size, in.	Bare Copper Tubing, Btu/h·ft	Bare Copper UA, Btu/h·ft·°F	0.5 in. Glass Fiber Insulated Copper Tubing, Btu/h·ft	0.5 in. Glass Fiber Insulated Copper UA, Btu/h·ft·°F
0.75	30	0.43	17.7	0.25
1	38	0.54	20.3	0.29
1.25	45	0.64	23.4	0.33
1.5	53	0.76	25.4	0.36
2	66	0.94	29.6	0.42
2.5	80	1.14	33.8	0.48
3	94	1.34	39.5	0.56
4	120	1.71	48.4	0.69

temperature. The amount of water wasted to drain is generally an amount greater than pipe volume because temperature of some of the first hot water traveling through the pipe is degraded to below a usable temperature.

Initial flow of hot water into a pipe full of cooler water often does not behave as predicted by steady-state flow theory, because both hot and cold water are flowing simultaneously in the same pipe (a non-steady-state condition). At least three different flow regimes were identified: (1) stratified flow (at low flow rates in horizontal pipes, hot water flows farther along the top side of the pipe than on the bottom side; this can happen even in small-diameter pipes), (2) normal turbulent flow, and (3) shear flow (a relatively sharp hot/cold interface with little turbulence-induced mixing of hot and cold water because the normal boundary layer is slow to develop under some conditions). These flow regimes are important because each causes different amounts of temperature degradation as hot water flows through the pipe.

For detailed information on time, water, and energy waste while waiting for hot water to arrive at fixtures, see Hiller (2005b). Simply summarized here, the amount of water waste can be expressed as the ratio of the actual amount of water (actual flow or AF) wasted while waiting for hot-enough-to-use water to arrive at fixtures (defined as 105°F by Hiller) divided by pipe volume (PV). When the pipe cools below a usable temperature, AF/PV ratios are usually in the range of 1.0 to 2.0, but can go to infinity at low flow rates in long, uninsulated pipe in cold or otherwise adverse (e.g., damp) heat transfer environments. The critical length of pipe at which AF/PV goes to infinity can be calculated for any flow rate and temperature conditions, using the piping $UA_{flowing}$ factors and Equations (1) to (4).

For preliminary engineering design and energy use calculations, Hiller recommends assuming AF/PV values of 1.25 to 1.75. For more refined analyses, accounting better for temperature effects on AF/PV ratio, the data tables in the original reference should be consulted. More such data on a larger variety of pipe sizes, types, and environments would be beneficial, but are not currently available.

Examples 11 to 14 demonstrate how to use piping heat loss and delivery water waste information to calculate hot-water system energy use.

Hot-Water Recirculation Loops and Return Piping

Hot-water recirculation loops are commonly used where piping lengths are long and hot water is desired immediately at fixtures. In recirculation-loop systems, return piping and a circulation device are provided. Some recirculation-loop systems use buoyancy-driven natural convection forces to circulate flow, but most are equipped with circulating pumps to force water through the piping and back to the water heater, thus keeping water in the piping hot.

The water circulation pump may be controlled by a thermostat (in the return line) set to start and stop the pump over an acceptable temperature range. This thermostat can significantly reduce both

heat loss and pumping energy in some applications. An automatic time switch or other control should turn water circulation off when hot water is not required. Other, more advanced circulating pump control schemes, such as on-demand types using manual initiation, flow switches, or occupancy sensors, are also available. Because hot water is corrosive, circulating pumps should be made of corrosion-resistant material.

For small installations, a simplified pump sizing method is to allow 1 gpm for every 20 fixture units in the system, or to allow 0.5 gpm for each 3/4 or 1 in. riser; 1 gpm for each 1 1/4 or 1 1/2 in. riser; and 2 gpm for each riser 2 in. or larger.

Dunn et al. (1959) and Werden and Spielvogel (1969a, 1969b) discuss heat loss calculations for large systems. For larger installations, piping heat losses become significant. A quick method to size the pump and return for larger systems is as follows:

1. Determine total length of all hot-water supply and return piping.
2. Choose an appropriate value for piping heat loss from Tables 1 or 2 or other engineering data (usually supplied by insulation companies, etc.). Multiply this value by the total length of piping involved.

   A rough estimation can be made by multiplying the total length of covered pipe by 30 Btu/h·ft or uninsulated pipe by 60 Btu/h·ft. Table 2 gives actual heat losses in pipes at a service water temperature of 140°F and ambient temperature of 70°F. The values of 30 or 60 Btu/h·ft are only recommended for ease in calculation.
3. Determine pump capacity as follows:

$$Q_p = \frac{q}{60\rho c_p \, \Delta t} \qquad (9)$$

where

Q_p = pump capacity, gpm
q = heat loss, Btu/h
ρ = density of water = 8.25 lb/gal (120°F)
c_p = specific heat of water = 1 Btu/lb·°F
Δt = allowable temperature drop, °F

For a 20°F allowable temperature drop,

$$Q_p(\text{gpm}) = \frac{q}{60 \times 8.25 \times 1 \times 20} = \frac{q}{9900} \qquad (10)$$

Caution: This calculation assumes that a 20°F temperature drop is acceptable at the last fixture.

4. Select a pump to provide the required flow rate, and obtain from the pump curves the pressure created at this flow.
5. Multiply the head by 100 and divide by the total length of hot water return piping to determine the allowable friction loss per 100 ft of pipe.
6. Determine the required flow in each circulating loop, and size the hot water return pipe based on this flow and the allowable friction loss from Step 5.

Where multiple risers or horizontal loops are used, balancing valves with means of testing are recommended in the return lines. A swing-type check valve should be placed in each return to prevent entry of cold water or reversal of flow, particularly during periods of high hot-water demand.

Three common methods of arranging circulation lines are shown in Figure 2. Although the diagrams apply to multistory buildings, arrangements (A) and (B) are also used in residential designs. In circulation systems, air venting, pressure drops through the heaters and storage tanks, balancing, and line losses should be considered. In Figures 2A and 2B, air is vented by connecting the circulating line below the top fixture supply. With this arrangement, air is eliminated from the system each time the top fixture is opened. Generally, for small installations, a nominal pipe size (NPS) 1/2 or 3/4 in. hot-water return is ample.

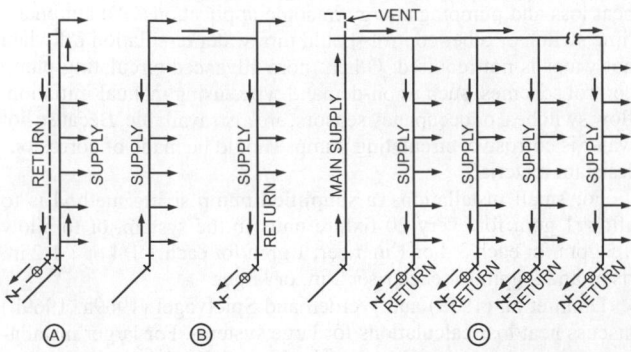

Fig. 2   Arrangements of Hot-Water Circulation Lines

All storage tanks and piping on recirculating systems should be insulated as recommended by the ASHRAE *Standard* 90 series and *Standard* 100.

Heat-Traced, Nonreturn Piping

In this system, the fixtures can be as remote as in the hot-water recirculation loops and return piping section. The hot-water supply piping is heat traced with electric resistance heating cable preinstalled under the pipe insulation. Electrical energy input is self-regulated by the cable's construction to maintain the required water temperature at the fixtures. No return piping system or circulation pump is required.

Multiple Water Heaters

Depending on fixture spacing, required pipe lengths, and draw spacing, it may be more energy-efficient (and sometimes provide lower first cost) to use more than one water heater rather than using extensive piping runs. Energy losses from high-efficiency water heaters can be lower than recirculation-loop piping heat losses if the distance from water heaters to fixtures exceeds 30 to 60 ft (Hiller 2005a). Although there are considerations beyond energy use, such as installation, maintenance, and space requirements, using more than one water heater should always be evaluated when designing water heating systems, even in residences, because of the potentially large energy savings.

Special Piping—Commercial Dishwashers

Adequate flow rate and pressure must be maintained for automatic dishwashers in commercial kitchens. To reduce operating difficulties, piping for automatic dishwashers should be installed according to the following recommendations:

- The cold-water feed line to the water heater should be no smaller than NPS 1.
- The supply line that carries 180°F water from the water heater to the dishwasher should not be smaller than NPS 3/4.
- No auxiliary feed lines should connect to the 180°F supply line.
- A return line should be installed if the source of 180°F water is more than 5 ft from the dishwasher.
- Forced circulation by a pump should be used if the water heater is installed on the same level as the dishwasher, if the length of return piping is more than 60 ft, or if the water lines are trapped.
- If a circulating pump is used, it is generally installed in the return line. It may be controlled by (1) the dishwasher wash switch, (2) a manual switch located near the dishwasher, or (3) an immersion or strap-on thermostat located in the return line.
- A pressure-reducing valve should be installed in the low-temperature supply line to a booster water heater, but external to a recirculating loop. It should be adjusted, with the water flowing, to the value stated by the washer manufacturer (typically 20 psi).

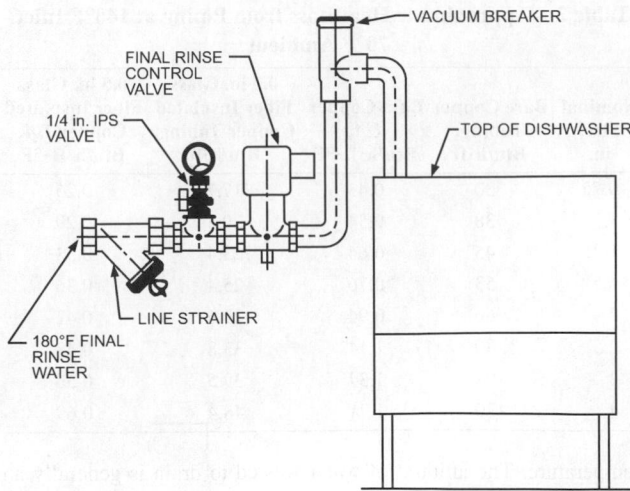

Fig. 3   National Sanitation Foundation (NSF) Plumbing Requirements for Commercial Dishwasher

- A check valve should be installed in the return circulating line.
- If a check-valve water meter or a backflow prevention device is installed in the cold-water line ahead of the heater, it is necessary to install a properly sized diaphragm-type expansion tank between the water meter or prevention device and the heater.
- National Sanitation Foundation (NSF) standards require an NPS 1/4 IPS connection for a pressure gage mounted adjacent to the supply side of the control valve. They also require a water-line strainer ahead of any electrically operated control valve (Figure 3).
- NSF standards do not allow copper water lines that are not under constant pressure, except for the line downstream of the solenoid valve on the rinse line to the cabinet.

Water Pressure—Commercial Kitchens

Proper flow pressure must be maintained to achieve efficient dishwashing. NSF standards for dishwasher water flow pressure are 15 psig minimum, 25 psig maximum, and 20 psig ideal. Flow pressure is the line pressure measured when water is flowing through the rinse arms of the dishwasher.

Low flow pressure can be caused by undersized water piping, stoppage in piping, or excess pressure drop through heaters. Low water pressure causes an inadequate rinse, resulting in poor drying and sanitizing of the dishes. If flow pressure in the supply line to the dishwasher is below 15 psig, a booster pump or other means should be installed to provide supply water at 20 psig.

Flow pressure over 25 psig causes atomization of the 180°F rinse water, resulting in excessive temperature drop (which can be as much as 15°F between rinse nozzle and dishes). A pressure regulator should be installed in the supply water line adjacent to the dishwasher and external to the return circulating loop (if used). The regulator should be set to maintain a pressure of 20 psig.

Two-Temperature Service

Where multiple temperature requirements are met by a single system, the system temperature is determined by the maximum temperature needed. Where the bulk of the hot water is needed at the higher temperature, lower temperatures can be obtained by mixing hot and cold water. Automatic mixing valves reduce the temperature of the hot water available at certain outlets to prevent injury or damage (Figure 4). Applicable codes should be consulted for mixing valve requirements.

Where predominant use is at a lower temperature, the common design heats all water to the lower temperature and then uses a separate booster heater to further heat the water for the higher-

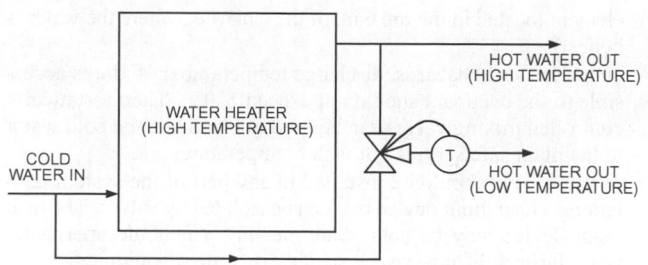

Fig. 4 Two-Temperature Service with Mixing Valve

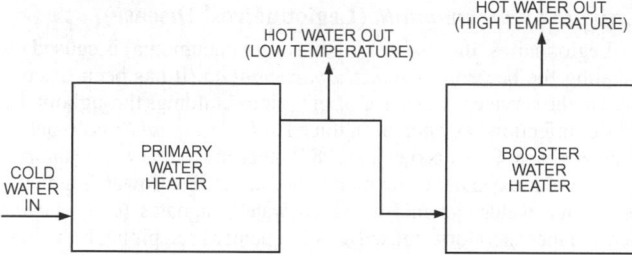

Fig. 5 Two-Temperature Service with Primary Heater and Booster Heater in Series

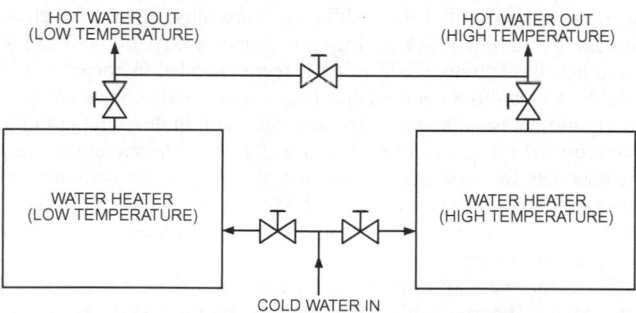

Fig. 6 Two-Temperature Service with Separate Heater for Each Service

temperature service (Figure 5). This method offers better protection against scalding.

A third method uses separate heaters for the higher-temperature service (Figure 6). It is common practice to cross-connect the two heaters, so that one heater can serve the complete installation temporarily while the other is valved off for maintenance. Each heater should be sized for the total load unless hot-water consumption can be reduced during maintenance periods.

Manifolding

Where one heater does not have sufficient capacity, two or more water heaters may be installed in parallel. If blending is needed, a single mixing valve of adequate capacity should be used. It is difficult to obtain even flow through parallel mixing valves.

Heaters installed in parallel should have similar specifications: the same input and storage capacity, with inlet and outlet piping arranged so that an equal flow is received from each heater under all demand conditions.

An easy way to get balanced, parallel flow is to use reverse/return piping (Figure 7). The unit having its inlet closest to the cold-water supply is piped so that its outlet is farthest from the hot-water supply line. Quite often this results in a hot-water supply line that reverses direction (see dashed line, Figure 7) to bring it back to the first unit in line; hence the name reverse/return.

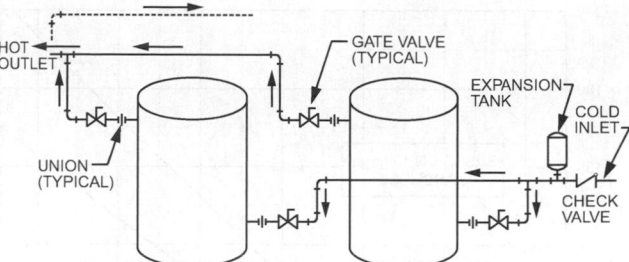

Fig. 7 Reverse/Return Manifold System

TERMINAL HOT-WATER USAGE DEVICES

Details on the vast number of devices using service hot water are beyond the scope of this chapter. Nonetheless, they are important to a successful overall design. Consult the manufacturer's literature for information on required flow rates, temperature and pressure limits, and/or other operating factors for specific items.

WATER QUALITY, SCALE, AND CORROSION

A complete water analysis and an understanding of system requirements are needed to protect water-heating systems from scale and corrosion. Analysis shows whether water is hard or soft. Hard water, unless treated, causes scaling or liming of heat transfer and water storage surfaces; soft water may aggravate corrosion problems and sacrificial anode consumption (Talbert et al. 1986).

Scale formation is also affected by system requirements and equipment. As shown in Figure 8, the rate of scaling increases with temperature and use because calcium carbonate and other scaling compounds lose solubility at higher temperatures. In water tube-type equipment, scaling problems can be offset by increasing water velocity over the heat transfer surfaces, which reduces the tube surface temperature. Also, flow turbulence, if high enough, works to keep any scale that does precipitate off the surface. When water hardness is over 8 gr/gal, water softening or other water treatment is often recommended.

Corrosion problems increase with temperature because corrosive oxygen and carbon dioxide gases are released from the water. Electrical conductivity also increases with temperature, enhancing electrochemical reactions such as rusting (Taborek et al. 1972). A deposit of scale provides some protection from corrosion; however, this deposit also reduces the heat transfer rate, and it is not under the control of the system designer (Talbert et al. 1986).

Steel vessels can be protected to varying degrees by galvanizing or by lining with copper, glass, cement, electroless nickel-phosphorus, or other corrosion-resistant material. Glass-lined vessels are almost always supplied with electrochemical protection. Typically, one or more anode rods of magnesium, aluminum, or zinc alloy are installed in the vessel by the manufacturer. This electrochemically active material sacrifices itself to reduce or prevent corrosion of the tank (the cathode). Higher temperature, softened water, and high water use may lead to rapid anode consumption. Manufacturers recommend periodic replacement of the anode rod(s) to prolong the life of the vessel. Some waters have very little electrochemical activity. In this instance, a standard anode shows little or no activity, and the vessel is not adequately protected. If this condition is suspected, consult the equipment manufacturer on the possible need for a high-potential anode, or consider using vessels made of nonferrous material.

Water heaters and hot-water storage tanks constructed of stainless steel, copper, or other nonferrous alloys are protected against oxygen corrosion. However, care must still be taken, as some stain-

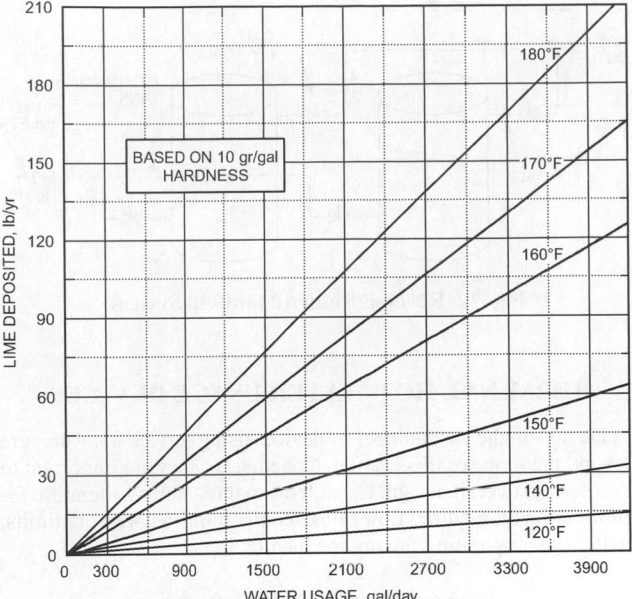

Fig. 8 Lime Deposited Versus Temperature and Water Use
(Based on data from Purdue University Bulletin No. 74)

less steel may be adversely affected by chlorides, and copper may be attacked by ammonia or carbon dioxide.

SAFETY DEVICES FOR HOT-WATER SUPPLIES

Regulatory agencies differ as to the selection of protective devices and methods of installation. It is therefore essential to check and comply with the manufacturer's instructions and the applicable local codes. In the absence of such instructions and codes, the following recommendations may be used as a guide:

- Water expands when it is heated. Although the water-heating system is initially under service pressure, the pressure rises rapidly if backflow is prevented by devices such as a check valve, pressure-reducing valve, or backflow preventer in the cold-water line or by temporarily shutting off the cold-water valve. When backflow is prevented, the pressure rise during heating may cause the safety relief valve to weep to relieve the pressure. However, if the safety relief valve is inadequate, inoperative, or missing, pressure rise may rupture the tank or cause other damage. Systems having this potential problem must be protected by a properly sized expansion tank located on the cold-water line downstream of and as close as practical to the device preventing backflow.
- Temperature-limiting devices (energy cutoff/high limit) prevent water temperatures from exceeding 210°F by stopping the flow of fuel or energy. These devices should be listed and labeled by a recognized certifying agency.
- Safety relief valves open when pressure exceeds the valve setting. These valves are typically applied to water-heating and hot-water supply boilers. The set pressure should not exceed the maximum allowable working pressure of the boiler. The heat input pressure steam rating (in Btu/h) should equal or exceed the maximum output rating for the boiler. The valves should comply with current applicable standards or the ASME *Boiler and Pressure Vessel Code*.
- Temperature and pressure safety relief valves also open if the water temperature reaches 210°F. These valves are typically applied to water heaters and hot-water storage tanks. The heat input temperature/steam rating (in Btu/h) should equal or exceed the heat input rating of the water heater. Combination temperature- and pressure-relief valves should be installed with the temperature-sensitive

element located in the top 6 in. of the tank (i.e., where the water is hottest).
- To reduce scald hazards, discharge temperature at fixtures accessible to the occupant should not exceed 120°F. Thermostatically controlled mixing valves can be used to blend hot and cold water to maintain safe service hot-water temperatures.
- A relief valve should be installed in any part of the system containing a heat input device that can be isolated by valves. The heat input device may be solar water-heating panels, desuperheater water heaters, heat recovery devices, or similar equipment.

SPECIAL CONCERNS

Legionella pneumophila (Legionnaires' Disease)

Legionnaires' disease (a form of severe pneumonia) is caused by inhaling the bacteria *Legionella pneumophila*. It has been discovered in the service water systems of various buildings throughout the world. Infection has often been traced to *L. pneumophila* colonies in shower heads. Ciesielski et al. (1984) determined that *L. pneumophila* can colonize in hot water maintained at 115°F or lower. Segments of service water systems in which water stagnates (e.g., shower heads, faucet aerators, unused or infrequently used piping branches, some sections of storage water heaters) provide ideal breeding locations.

Service water temperature in the 140°F range is recommended to limit the potential for *L. pneumophila* growth. This high temperature increases the potential for scalding, so care must be taken such as installing an anti-scald or mixing valve. Supervised periodic flushing of fixture heads with 170°F water is recommended in hospitals and health care facilities because the already weakened patients are generally more susceptible to infection. Although higher temperatures are required for eradication of *Legionella*, lower temperatures may be adequate for prevention. Note, too, that *Legionella* colonies can grow on the cold-water side of the system in stagnant flow areas, especially if those areas can rise in temperature because of heat input from the surroundings.

Note that susceptibility to Legionnaire's disease varies among individuals. People with compromised immune systems (organ transplant patients or otherwise on immunosuppressant drugs, AIDS patients, smokers, elderly, those with other chronic health conditions or injuries) are at greater risk of contracting the disease at lower exposure levels.

More information on this subject can be found in ASHRAE *Guideline* 12-2000.

Scalding

Scalding is an important concern in design and operation of potable hot-water systems. Figure 9 (Moritz 1947) shows plots of exposure time versus water temperature that results in both first-degree (pain, redness, swelling, minor tissue damage) and full-thickness third-degree (permanent damage, scarring) skin burns in adults. Children burn even more rapidly. Note that, at the high temperatures required to eradicate *Legionella* bacteria (140°F and above), burns can occur almost instantaneously (3 s or less exposure). Safety dictates some tradeoffs to limit scalding injuries (e.g., during pressure transients that may inhibit proper operation of temperature regulating valves) while minimizing risk of Legionnaire's disease.

Temperature Requirement

Typical temperature requirements for some services are shown in Table 3. A 140°F water temperature minimizes flue gas condensation in the equipment.

Hot Water from Tanks and Storage Systems

With storage systems, 60 to 80% of the hot water in a tank is assumed to be usable before dilution by cold water lowers the temperature below an acceptable level. However, better designs can

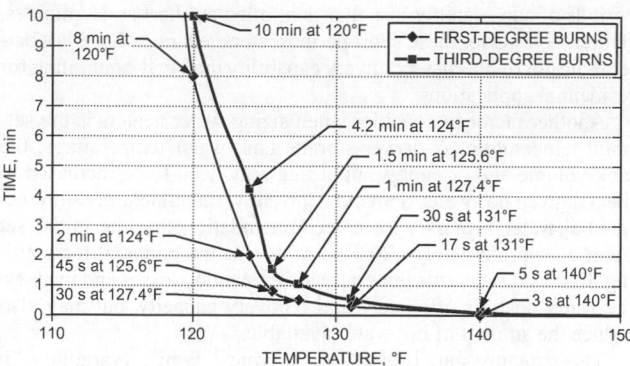

Fig. 9 Time for Adult Skin Burns in Hot Water

Table 3 Representative Hot-Water Temperatures

Use	Temperature, °F
Lavatory	
Hand washing	105
Shaving	115
Showers and tubs	110
Therapeutic baths	95
Commercial or institutional laundry, based on fabric	up to 180
Residential dish washing and laundry	140
Surgical scrubbing	110
Commercial spray-type dish washing[a]	
Single- or multiple-tank hood or rack type	
Wash	150 min.
Final rinse	180 to 195
Single-tank conveyor type	
Wash	160 min.
Final rinse	180 to 195
Single-tank rack or door type	
Single-temperature wash and rinse	165 min.
Chemical sanitizing types[b]	140
Multiple-tank conveyor type	
Wash	150 min.
Pumped rinse	160 min.
Final rinse	180 to 195
Chemical sanitizing glass washer	
Wash	140
Rinse	75 min.

[a]As required by NSF.
[b]See manufacturer for actual temperature required.

exceed 90%. Thus, the maximum hot water available from a self-contained storage heater is usually

$$V_t = Rd + MS_t \qquad (11)$$

where

V_t = available hot water, gal
R = recovery rate at required temperature, gph
d = duration of peak hot-water demand, h
M = ratio of usable water to storage tank capacity
S_t = storage capacity of heater tank, gal

However, Equation (11) only applies if the water draw rate is less than the available reheat rate. Otherwise, the tank cannot heat the flowing water to a usable temperature during the draw, and V_t drops to the same as an unfired tank. For example, a fossil-fuel-fired heater with a fuel input rate of 44,000 Btu/h and an input efficiency of 80% can raise the temperature of water being drawn through a storage tank at a rate of 3 gpm by approximately 23°F. If the

entering cold-water temperature is 60°F, the water will be heated to only 83°F, too cold to be useful, so the heating rate cannot contribute to effective storage tank capacity under a prolonged draw at this flow rate. In reality, draw rates are rarely constant during peak draw or other times. Computer simulation models allow equipment sizing under these more realistic conditions (Hiller 1992).

Maximum usable hot water from an unfired tank is

$$V_a = MS_a \qquad (12)$$

where

V_a = usable water available from unfired tank, gal
S_a = capacity of unfired tank, gal

Note: Assumes tank water at required temperature.

Hot water obtained from a water heater using a storage heater with an auxiliary storage tank is

$$V_z = V_t + V_a \qquad (13)$$

where V_z = total hot water available during one peak, in gallons.

Placement of Water Heaters

Many types of water heaters may be expected to leak at the end of their useful life. They should be placed where leakage will not cause damage. Alternatively, suitable drain pans piped to drains must be provided.

Water heaters not requiring combustion air may generally be placed in any suitable location, as long as relief valve discharge pipes open to a safe location.

Water heaters requiring ambient combustion air must be located in areas with air openings large enough to admit the required combustion/dilution air (see NFPA *Standard* 54/ANSI Z223.1).

For water heaters located in areas where flammable vapors are likely to be present, precautions should be taken against ignition. For water heaters installed in residential garages, additional precautions should be taken. Consult local codes for additional requirements or see sections 5.1.9 through 5.1.12 of NFPA *Standard* 54/ANSI Z223.1.

Outdoor models with a weather-proofed jacket are available. Direct-vent gas- and oil-fired models are also available; they are to be installed inside, but are not vented through a conventional chimney or gas vent. They use outdoor air for combustion. They must be installed with the means specified by the manufacturer for venting (typically horizontal) and for supplying air for combustion from outside the building.

Air-source heat pump water heaters require access to an adequate air supply from which heat can be extracted. For residential units, a room of at least 800 ft³ or ducted air is recommended. See manufacturer's literature for more information.

HOT-WATER REQUIREMENTS AND STORAGE EQUIPMENT SIZING

Methods for sizing storage water heaters vary. Those using recovery versus storage curves are based on extensive research. All methods provide adequate hot water if the designer allows for unusual conditions. To serve a hot-water load adequately, the needs of both the peak energy withdrawal rate and total integrated energy delivery for end uses must be met. Meeting these needs can be done either by providing a heating rate large enough to meet the peak energy withdrawal rate of the system (and modulating that heating input for smaller loads), or by providing a lower heating rate combined with storage (from which the peak rates can be satisfied). Lower costs are usually achieved by using at least some storage. A variety of different heating rate/storage volume combinations can be used to meet the needs of a given water-heating load profile (Hiller 1998).

Load Diversity

The greatest difficulty in designing water-heating systems comes from uncertainty about design hot-water loads, especially for buildings not yet built. Although it is fairly simple to test maximum flow rates of various hot-water fixtures and appliances, actual flow rates and durations are user-dependent. Moreover, the timing of different hot-water use events varies from day to day, with some overlap, but almost never will all fixtures be used simultaneously. As the number of hot-water-using fixtures and appliances grows, the percent of those fixtures used simultaneously decreases.

Some of the hot-water load information in this chapter is based on limited-scale field testing combined with statistical analysis to estimate load demand or **diversity** factors (percent of total possible load that is ever actually used at one time) versus number of end use points, number of people, etc. Much of the work to provide these diversity factors dates from the 1930s to the 1960s; it remains, however, the best information currently available (with a few exceptions, as noted). Of greatest concern is the fact that most of the data from those early studies were for fixtures that used water at much higher flow rates than modern energy-efficient fixtures (e.g., low-flow shower heads and sink aerators, energy-efficient washing machines and dishwashers). Some research has provided limited information on hot-water use by more modern fixtures, and on their use diversity (Becker et al. 1991; Goldner 1994a, 1994b; Goldner and Price 1999; Hiller 1998; Hiller and Lowenstein 1996, 1998; Thrasher and DeWerth 1994), but much more information in a variety of applications is needed before the design procedures can be updated. Using the older load diversity information usually results in a water-heating system that adequately serves the loads, but often results in substantial oversizing. Oversizing can be a deterrent to using modern high-efficiency water-heating equipment, which may have higher first cost per unit of capacity than less efficient equipment. Sustainable design must consider these effects.

Residential

Table 4 shows typical hot-water usage in a residence, including usage rates of modern ultralow-use appliances and fixtures. It is more difficult to show typical values for newer devices, because some automatically adjust the amount of hot water they use based on sensed load or cycle setting. In its *Minimum Property Standards for Housing*, the U.S. Department of Housing and Urban Development (HUD 1994) established minimum permissible water heater sizes (Table 5). Storage water heaters may vary from the sizes shown if combinations of recovery and storage are used that produce the required 1 h draw.

The first-hour rating (FHR) is the theoretical maximum amount of hot water that the water heater can supply in 1 h of operation under specific test conditions (DOE 1998). The linear regression lines shown in Figure 10 represent the FHR for 1556 electric heaters and 2901 gas heaters [GAMA (Continuous Maintenance); Hiller 1998]. Regression lines are not included for oil-fired and heat-pump water heaters because of limited data. The FHR represents water-heater performance characteristics that are similar to those represented by the 1 h draw values listed in Table 5. Residential water-heating equipment sizing is frequently driven by amounts of water used over

periods of considerably less than 1 h, often as short as 15 minutes. (Hiller 1998) Over these short periods, storage tank volume is a better indicator of hot-water delivery capability than first-hour rating for residential applications.

Another factor to consider when sizing water heaters is the set-point temperature. At lower storage tank water temperatures, the tank volume and/or energy input rate may need to be increased to meet a given hot-water demand. Currently, manufacturers ship residential water heaters with a recommendation that the initial set point be approximately 120°F to minimize the potential for scalding. Reduced set points generally lower standby losses and increase the water heater's efficiency and recovery capacity, but may also reduce the amount of hot water available.

The structure and lifestyle of a typical family (variations in family size, age of family members, presence and age of children, hot-water use volume and temperature, and other factors) cause

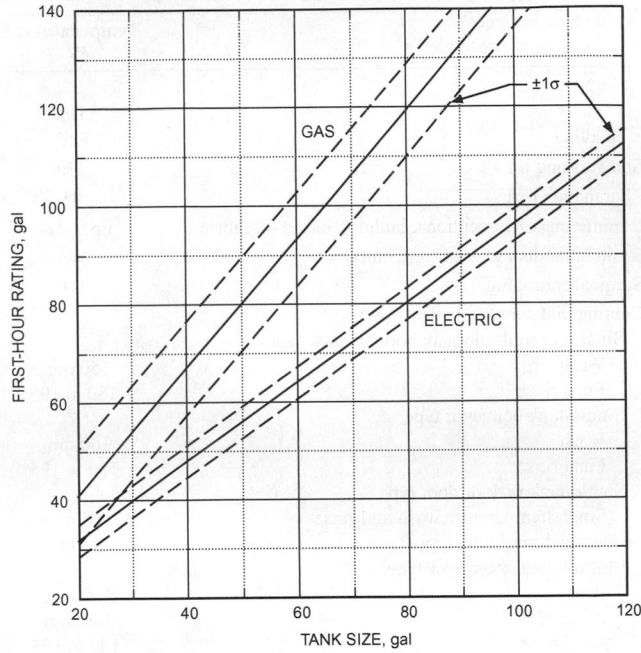

Fig. 10   First-Hour Rating (FHR) Relationships for Residential Water Heaters

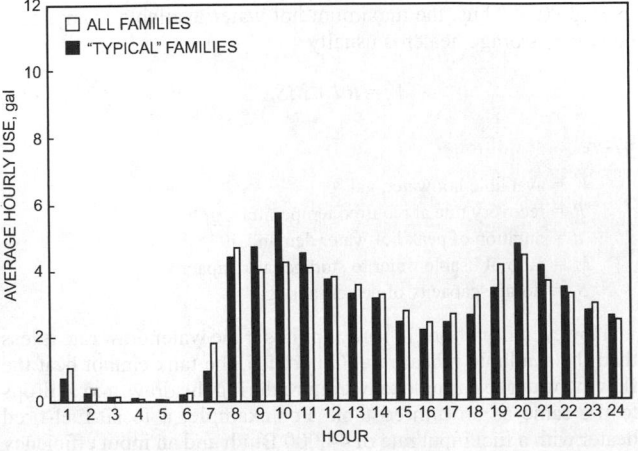

Fig. 11   Residential Average Hourly Hot-Water Use

Table 4   Typical Residential Use of Hot Water

Use	High Flow, Gallons/Task	Low Flow (Water Savers Used), Gallons/Task	Ultralow Flow, Gallons/Task
Food preparation	5	3	3
Hand dish washing	4	4	3
Automatic dishwasher	15	15	3 to 10
Clothes washer	32	21	5 to 15
Shower or bath	20	15	10 to 15
Face and hand washing	4	2	1 to 2

Table 5 HUD-FHA Minimum Water Heater Capacities for One- and Two-Family Living Units

Number of Baths	1 to 1.5			2 to 2.5				3 to 3.5			
Number of Bedrooms	1	2	3	2	3	4	5	3	4	5	6
Gas[a]											
Storage, gal	20	30	30	30	40	40	50	40	50	50	50
1000 Btu/h input	27	36	36	36	36	38	47	38	38	47	50
1 h draw, gal	43	60	60	60	70	72	90	72	82	90	92
Recovery, gph	23	30	30	30	30	32	40	32	32	40	42
Electric[a]											
Storage, gal	20	30	40	40	50	50	66	50	66	66	80
kW input	2.5	3.5	4.5	4.5	5.5	5.5	5.5	5.5	5.5	5.5	5.5
1 h draw, gal	30	44	58	58	72	72	88	72	88	88	102
Recovery, gph	10	14	18	18	22	22	22	22	22	22	22
Oil[a]											
Storage, gal	30	30	30	30	30	30	30	30	30	30	30
1000 Btu/h input	70	70	70	70	70	70	70	70	70	70	70
1 h draw, gal	89	89	89	89	89	89	89	89	89	89	89
Recovery, gph	59	59	59	59	59	59	59	59	59	59	59
Tank-type indirect[b,c]											
I-W-H-rated draw, gal in 3 h, 100°F rise		40	40		66	66[e]	66	66	66	66	66
Manufacturer-rated draw, gal in 3 h, 100°F rise		49	49		75	75[e]	75	75	75	75	75
Tank capacity, gal		66	66		66	66[e]	82	66	82	82	82
Tankless-type indirect[c,d]											
I-W-H-rated draw, gpm, 100°F rise		2.75	2.75		3.25	3.25[e]	3.75	3.25	3.75	3.75	3.75
Manufacturer-rated draw, gal in 5 min, 100°F rise		15	15		25	25[e]	35	25	35	35	35

Note: Applies to tank-type water heaters only

[a]Storage capacity, input, and recovery requirements indicated are typical and may vary with manufacturer. Any combination of requirements to produce stated 1 h draw is satisfactory.

[b]Boiler-connected water heater capacities (180°F boiler water, internal or external connection).

[c]Heater capacities and inputs are minimum allowable. Variations in tank size are permitted when recovery is based on 4 gph/kW at 100°F rise for electrical, AGA recovery ratings for gas, and IBR ratings for steam and hot-water heaters.

[d]Boiler-connected heater capacities (200°F boiler water, internal or external connection).

[e]Also for 1 to 1.5 baths and 4 bedrooms for indirect water heaters.

hot-water consumption demand patterns to fluctuate widely in both magnitude and time distribution.

Perlman and Mills (1985) developed the overall and peak average hot-water use volumes shown in Table 6. Average hourly patterns and 95% confidence level profiles are illustrated in Figures 11 and 12. Samples of results from the analysis of similarities in hot-water use are given in Figures 13 and 14.

Commercial and Institutional

Most commercial and institutional establishments use hot or warm water. The specific requirements vary in total volume, flow rate, duration of peak load period, and temperature. Water heaters and systems should be selected based on these requirements.

This section covers sizing recommendations for central storage water-heating systems. Hot-water usage data and sizing curves for dormitories, motels, nursing homes, office buildings, food service establishments, apartments, and schools are based on EEI-sponsored research (Werden and Spielvogel 1969a, 1969b). Caution must be taken in applying these data to small buildings. Also, within any given category there may be significant variation. For example, the motel category encompasses standard, luxury, resort, and convention motels.

When additional hot-water requirements exist, increase the recovery and/or storage capacity accordingly. For example, if there is food service in an office building, the recovery and storage capacities required for each additional hot-water use should be added when sizing a single central water-heating system.

Peak hourly and daily demands for various categories of commercial and institutional buildings are shown in Table 7. These demands for central-storage hot water represent the maximum flows metered in this 129-building study, excluding extremely high and very infrequent peaks. Table 7 also shows average hot-water consumption figures for these buildings. Averages for schools and food service establishments are based on actual days of operation; all others are based on total days. These averages can be used to estimate monthly consumption of hot water, but are not intended

Table 6 Overall (OVL) and Peak Average Hot-Water Use

Group	Average Hot-Water Use, gal							
	Hourly		Daily		Weekly		Monthly	
	OVL	Peak	OVL	Peak	OVL	Peak	OVL	Peak
All families	2.6	4.6	62.4	67.1	436	495	1897	2034
"Typical" families	2.6	5.8	63.1	66.6	442	528	1921	2078

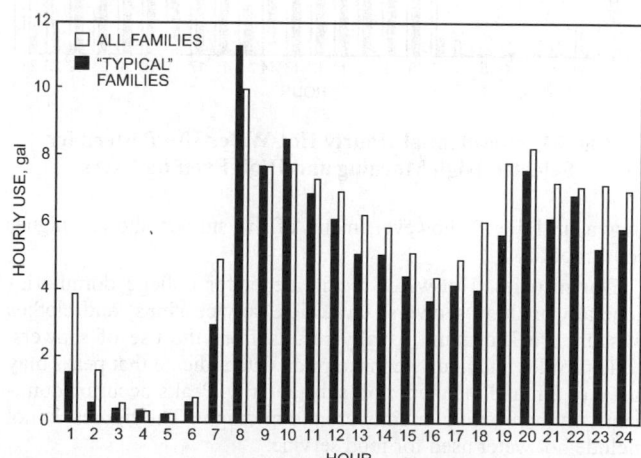

Fig. 12 Residential Hourly Hot-Water Use, 95% Confidence Level

for sizing purposes because they do not show the time distribution of draws.

Research conducted for ASHRAE (Becker et al. 1991; Thrasher and DeWerth 1994) and others (Goldner 1994a, 1994b) included a compilation and review of service hot-water use information in commercial and multifamily structures along with new monitoring data. Some of this work found consumption comparable to those

Table 7　Hot-Water Demands and Use for Various Types of Buildings*

Type of Building	Maximum Hourly	Maximum Daily	Average Daily
Men's dormitories	3.8 gal/student	22.0 gal/student	13.1 gal/student
Women's dormitories	5.0 gal/student	26.5 gal/student	12.3 gal/student
Motels: Number of units[a]			
20 or less	6.0 gal/unit	35.0 gal/unit	20.0 gal/unit
60	5.0 gal/unit	25.0 gal/unit	14.0 gal/unit
100 or more	4.0 gal/unit	15.0 gal/unit	10.0 gal/unit
Nursing homes	4.5 gal/bed	30.0 gal/bed	18.4 gal/bed
Office buildings	0.4 gal/person	2.0 gal/person	1.0 gal/person
Food service establishments			
Type A: Full-meal restaurants and cafeterias	1.5 gal/max meals/h	11.0 gal/max meals/day	2.4 gal/average meals/day[b]
Type B: Drive-ins, grills, luncheonettes, sandwich, and snack shops	0.7 gal/max meals/h	6.0 gal/max meals/day	0.7 gal/average meals/day[b]
Apartment houses: Number of apartments			
20 or less	12.0 gal/apartment	80.0 gal/apartment	42.0 gal/apartment
50	10.0 gal/apartment	73.0 gal/apartment	40.0 gal/apartment
75	8.5 gal/apartment	66.0 gal/apartment	38.0 gal/apartment
100	7.0 gal/apartment	60.0 gal/apartment	37.0 gal/apartment
200 or more	5.0 gal/apartment	50.0 gal/apartment	35.0 gal/apartment
Elementary schools	0.6 gal/student	1.5 gal/student	0.6 gal/student[b]
Junior and senior high schools	1.0 gal/student	3.6 gal/student	1.8 gal/student[b]

*Data predate modern low-flow fixtures and appliances.　　　[a]Interpolate for intermediate values.　　　[b]Per day of operation.

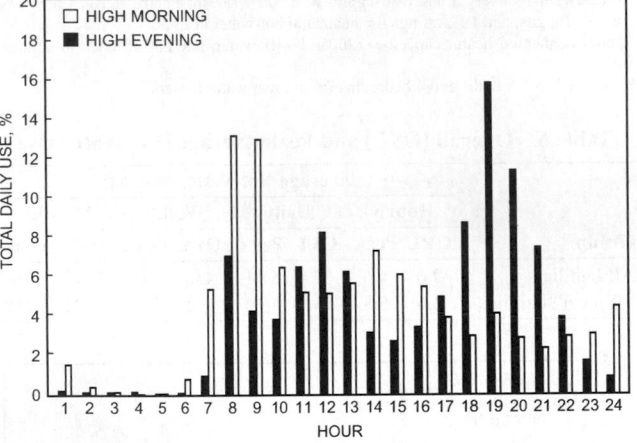

Fig. 13　Residential Hourly Hot-Water Use Pattern for Selected High Morning and High Evening Users

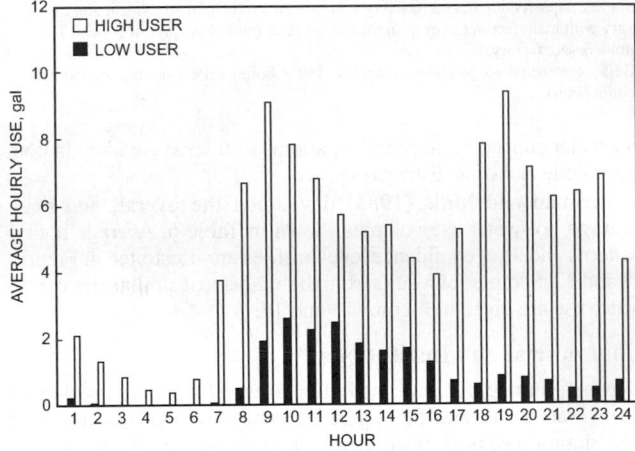

Fig. 14　Residential Average Hourly Hot-Water Use Patterns for Low and High Users

shown in Table 7; however, many of the studies showed higher consumption.

Dormitories. Hot-water requirements for college dormitories generally include showers, lavatories, service sinks, and clothes washers. Peak demand usually results from the use of showers. Load profiles and hourly consumption data indicate that peaks may last 1 or 2 h and then taper off substantially. Peaks occur predominantly in the evening, mainly around midnight. The figures do not include hot water used for food service.

Military Barracks. Design criteria for military barracks are available from the engineering departments of the U.S. Department of Defense. Some measured data exist for hot-water use in these facilities. For published data, contact the U.S. Army Corps of Engineers or Naval Facilities Engineering Command.

Motels. Domestic hot-water requirements are for tubs and showers, lavatories, and general cleaning purposes. Recommendations are based on tests at low- and high-rise motels located in urban, suburban, rural, highway, and resort areas. Peak demand, usually from shower use, may last 1 or 2 h and then drop off sharply. Food service, laundry, and swimming pool requirements are not included.

Nursing Homes. Hot water is required for tubs and showers, wash basins, service sinks, kitchen equipment, and general cleaning. These figures include hot water for kitchen use. When other equipment, such as that for heavy laundry and hydrotherapy purposes, is to be used, its hot-water requirement should be added.

Office Buildings. Hot-water requirements are primarily for cleaning and lavatory use by occupants and visitors. Older office buildings often use hot-water recirculation-loop systems and are thus good candidates for water-heating distribution system efficiency upgrades through more modern controls and/or addition of point-of-use water heaters. Hot-water use for food service in office buildings is not included.

Food Service Establishments. Hot-water requirements are primarily for dish washing. Other uses include food preparation, cleaning pots and pans and floors, and hand washing for employees and customers. Recommendations are for establishments serving food at tables, counters, booths, and parked cars. Establishments that use disposable service exclusively are not covered in Table 7.

Dish washing, as metered in these tests, is based on normal practice of dish washing after meals, not on indiscriminate or continuous use of machines irrespective of the flow of soiled dishes. Recommendations include hot water supplied to dishwasher booster heaters.

Apartments. Hot-water requirements for both garden-type and high-rise apartments are for one- and two-bath apartments, for showers, lavatories, kitchen sinks, dishwashers, clothes washers, and general cleaning purposes. Clothes washers can be either in individual apartments or centrally located. These data apply to central water-heating systems only.

Elementary Schools. Hot-water requirements are for lavatories, cafeteria and kitchen use, and general cleaning purposes. When showers are used, their additional hot-water requirements should be added. Recommendations include hot water for dishwashers but not for extended school operation such as evening classes.

High Schools. Senior high schools, grades 9 or 10 through 12, require hot water for showers, lavatories, dishwashers, kitchens, and general cleaning. Junior high schools, grades 7 through 8 or 9, have requirements similar to those of the senior high schools. Junior high schools without showers follow the recommendations for elementary schools.

Requirements for high schools are based on daytime use. Recommendations do not take into account hot-water use for additional activities such as night school. In such cases, the maximum hourly demand remains the same, but the maximum and average daily use increase, usually by the number of additional people using showers and, to a lesser extent, eating and washing facilities.

Additional Data.

Fast Food Restaurants. Hot water is used for food preparation, cleanup, and rest rooms. Dish washing is usually not a significant load. In most facilities, peak usage occurs during the cleanup period, typically soon after opening and immediately before closing. Hot-water consumption varies significantly among individual facilities. Fast food restaurants typically consume 250 to 500 gal per day (EPRI 1994).

Supermarkets. The trend in supermarket design is to incorporate food preparation and food service functions, substantially increasing the usage of hot water. Peak usage is usually associated with cleanup periods, often at night, with a total consumption of 300 to 1000 gal per day (EPRI 1994).

Apartments. Table 8 shows cumulative hot-water use over time for apartment buildings, taken from a series of field tests by Becker et al. (1991), Goldner (1994a, 1994b), Goldner and Price (1999), and Thrasher and DeWerth (1994). These data include use diversity information, and enable use of modern water-heating equipment sizing methods for this building type, making it easy to understand the variety of heating rate and storage volume combinations that can serve a given load profile (see Example 1). Unlike Table 7, Table 8 presents low/medium/high (LMH) guidelines rather than specific singular volumes, and gives better time resolution of peak hot-water use information. The same information is shown graphically in Figure 15. Note that these studies showed that occupants on average use more hot water when water-heating costs are included in the rent, than if the occupants pay directly for water-heating energy use.

The low-use peak hot-water consumption profile represents the lowest peak profile seen in the tests, and is generally associated with apartment buildings having mostly a mix of the following occupant demographics:

- All occupants working
- One person working, while one stays at home
- Seniors
- Couples
- Middle income
- Higher population density

The medium-use peak hot-water consumption profile represents the overall average highest peak profile seen in the tests, and is generally associated with apartment buildings having mostly a mix of the following occupant demographics:

Table 8 Hot-Water Demand and Use Guidelines for Apartment Buildings (Gallons per Person at 120°F Delivered to Fixtures)

Guideline	Peak Minutes						Maximum Daily	Average Daily
	5	15	30	60	120	180		
Low	0.4	1.0	1.7	2.8	4.5	6.1	20	14
Medium	0.7	1.7	2.9	4.8	8.0	11.0	49	30
High	1.2	3.0	5.1	8.5	14.5	19.0	90	54

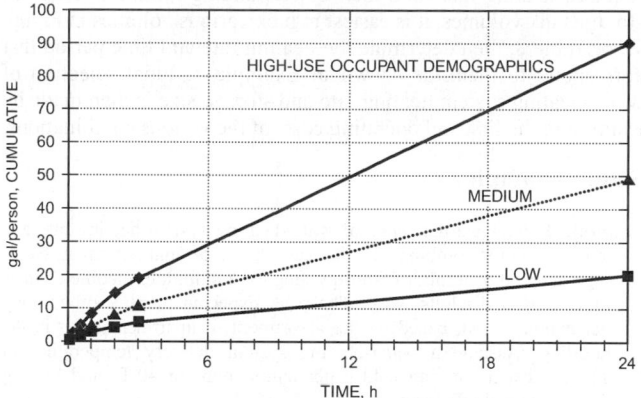

Fig. 15 Apartment Building Cumulative Hot-Water Use Versus Time (from Table 8)

- Families
- Singles
- On public assistance
- Single-parent households

The high-use peak hot-water consumption profile represents the highest peak profile seen in the tests, and is generally associated with apartment buildings having mostly a mix of the following occupant demographics:

- High percentage of children
- Low income
- On public assistance
- No occupants working
- Families
- Single-parent households

In applying these guidelines, the designer should note that a building may outlast its current use. This may be a reason to increase the design capacity for domestic hot water or allow space and connections for future enhancement of the service hot-water system. Building management practices, such as the explicit prohibition (in the lease) of apartment clothes washers or the existence of bath/kitchen hook-ups, should be factored into the design process. A diversity factor that lowers the probability of coincident consumption should also be used in larger buildings.

The information in Table 8 and Figure 15 generates a water-heating equipment sizing method for apartment buildings. The cumulative total hot-water consumption versus time (which includes all necessary load diversity information) can be used to select a range of heating rate and storage volume options, all of which will satisfy the load. The key is that plots of cumulative total hot-water consumption versus time as shown in Figure 15 also represent, by the slope of a line drawn from zero time through the cumulative volume used at any given time, the average hot-water flow rate up to that point in time. Up to any point in time, the minimum average heating rate needed to satisfy the load is one that can heat the average hot-water flow rate through that time from the local entering cold-water temperature to the water-heating system delivery temperature. (More accurately, the heating rate needed is determined by the local slope of the hot-water use versus time curve, not the average slope. This is because storage can supply the hot water supplied up to a selected time, and the heating rate only needs to provide the additional energy after storage is depleted. Eventually, however, storage needs to be reheated, which must also be considered. See the two methods illustrated in Example 1.) The storage volume needed for that heating rate is the total cumulative flow through that time (Hiller 1998). To evaluate the range of minimum required heating rates and their corresponding minimum required storage tank volumes, it is easiest to pick various volumes in Figure 15 or Table 8, then determine the heating rate and time period that correspond to them, as shown in Example 1. Final selection of water-heating system heating rate and storage size is then made by examining the first and operating costs of the various combinations.

Sizing Examples

Example 1. Evaluate the range of water-heating system heating-rate and storage volume combinations that can serve a 58-unit apartment building occupied by a mix of families, singles, and middle-income couples in which most adults work. The peak expected number of building occupants is 198, based on the assortment of apartment sizes in the building. Assume a water-heating system delivery temperature of 120°F, design entering cold-water temperature of 40°F, and heating device thermal efficiency of 80%.

Simplified Method.

Solution: The stated occupant demographics represent a medium load. Multiplying the volume per person versus time from the medium values in Table 8 by the number of occupants gives the cumulative amount needed at any point in time and the average flow rate (and hence heating rate) required through that time.

At 5 min, the peak design cumulative volume is (0.7 gal) × (198 people) = 138.6 gal. The average flow rate over 5 min is 138.6 gal/5 min = 27.72 gal/min. The required heating rate is thus, from Equation (1) and dividing by the input efficiency,

$$q = (27.72 \text{ gal/min})(60 \text{ min/h})(8.4 \text{ lb}_m/\text{gal})(1 \text{ Btu/lb}_m \cdot °F)$$
$$\times (120 - 40°F)/0.8 = 1,397,088 \text{ Btu/h}$$

Assuming 70% of the storage tank volume can be extracted at a useful temperature (the other 30% being degraded by mixing in the tank), the required tank volume for this heating rate is

$$V = 138.6 \text{ gal}/0.7 = 198 \text{ gal}$$

Note that, because the heating rate divided by storage capacity (7056 Btu/h·gal) exceeds 4000 Btu/h·gal, this system is considered an instantaneous water heater.

At 60 min, (4.8 gal/person)(198 people) = 950.4 gal. Average flow rate = 950.4 gal/60 min = 15.8 gal/min.

$$q = (15.8 \text{ gal/min})(60 \text{ min/h})(8.4 \text{ lb}_m/\text{gal})(1 \text{ Btu/lb}_m \cdot °F)$$
$$\times (120 - 40°F)/0.8 = 798,336 \text{ Btu/h}$$
$$V = 950.4 \text{ gal}/0.7 = 1358 \text{ gal}$$

Doing these calculations at other volumes and times yields the combinations of heating rate and storage volume that can serve the load (Table 9A).

More Accurate Method.

The preceding simplified method calculates the needed heating rate by computing the average water flow rate from the beginning of all draws for the day. In reality, because some storage is present, the water-heating device only needs to provide a heating rate computed from the local slope of the hot-water use curve, not the average slope. In other words for example, the flow over the first 5 min could have been provided entirely from storage without any heat input at all. The water heater only needs to heat in real time the amount of hot water needed over succeeding time periods. Consequently, the simplified heating rate

computational method works, but results in some degree of heating rate oversizing.

Solution: Using the more accurate heating rate sizing method is similar to using the simplified method, except the local slope of the hot-water use curve versus time must be found at each time interval to determine the necessary heating rate.

At 5 min, the peak design cumulative volume (Table 9A) is 139 gal. At 15 min, the peak design cumulative volume (Table 9A) is 337 gal. The incremental flow rate (representing the local slope of the hot water use line) is hence (337 gal – 139 gal)/10 min = 19.8 gal/min. The needed heating rate is thus more accurately computed as

$$Q = (19.8 \text{ gal/min})(60 \text{ min/h})(8.4 \text{ lb}_m/\text{gal})(1 \text{ Btu/lb}_m \cdot °F)$$
$$(120°F – 40°F)/0.8 = 997,920 \text{ Btu/h}$$

Note that heating rate divided by storage capacity (5040 Btu/h·gal) exceeds 4000 Btu/h·gal, so the more accurately sized system is still considered an instantaneous water heater.

From Table 9A, the peak design cumulative volume at 120 min is 1584 gal, and is 2178 gal at 180 min. The incremental flow rate slope is thus (2178 gal – 1584 gal)/(180 min – 120 min) = 9.9 gal/min. The heating rate needed when using 1584 gal of storage is more accurately computed as

$$Q = (9.9 \text{ gal/min})(60 \text{ min/h})(8.4 \text{ lb}_m/\text{gal})(1 \text{ Btu/lb}_m \cdot °F)$$
$$(120°F – 40°F)/0.8 = 498,960 \text{ Btu/h}$$

From Table 9A, the peak design cumulative volume at 180 min is 2178 gal, and at 1440 min is 9702 gal. Consequently, the incremental flow rate slope is (9702 gal – 2178 gal)/(1440 min – 180 min) = 5.97 gal/min. The heating rate needed when using 2178 gal of storage is thus more accurately computed as

$$Q = (5.97 \text{ gal/min})(60 \text{ min/h})(8.4 \text{ lb}_m/\text{gal})(1 \text{ Btu/lb}_m \cdot °F)$$
$$(120°F – 40°F)/0.8 = 300,888 \text{ Btu/h}$$

It is important to recognize, however, when using this more accurate heating rate sizing method, that storage must eventually be reheated. The minimum heating rate used should therefore not be less than that computed using the 24 h average flow rate.

Doing these calculations at other volumes and times yields the more accurate combinations of heating rate and storage volume that can serve the load, as shown in Table 9B.

Table 9A Example 1, Simplified Method: Heating Rate and Storage Volume Options

Time, min	Gallons per Person	Total Gallons for 198 People	Average Gallons per Minute	Heating Rate, Btu/h	Storage Volume, gal
5	0.7	139	28	1,397,088	198
15	1.7	337	22	1,130,976	481
30	2.9	574	19	964,656	820
60	4.8	950	16	798,336	1358
120	8	1584	13	665,280	2263
180	11	2178	12	609,840	3111
1440	49	9702	7	339,570	13,860

Table 9B Example 1, More Accurate Method: Heating Rate and Storage Volume Options

Time, min	Gallons per Person	Total Gallons for 198 People	Local Slope of Incremental Gallons per Minute	Heating Rate, Btu/h	Storage Volume, gal
5	0.7	139	19.8	997,920	198
15	1.7	337	15.8	796,320	481
30	2.9	574	12.5	630,000	820
60	4.8	950	10.6	534,240	1,358
120	8	1,584	9.9	498,960	2,263
180	11	2,178	5.97	339,570	3,111
1440	49	9,702	7	339,570	13,860

*Heating rate should not be lower than that needed to satisfy load on a 24 h basis.

There are several techniques to size water-heating systems using the more limited draw profile information in older data. Figures 16 to 23 show relationships between recovery and storage capacity for various building categories. Any combination of storage and recovery rate that falls on the proper curve satisfies building requirements. Using the minimum recovery rate and maximum storage capacity on the curves yields the smallest hot-water capacity able to satisfy the building requirement. The higher the recovery rate, the greater the 24 h heating capacity and the smaller the storage capacity required. Note that the data in Figures 16 to 23 predate modern low-flow fixtures and appliances.

These curves can be used to select recovery and storage requirements to accommodate water heaters that have fixed storage or recovery rates. Where hot-water demands are not coincident with peak electric, steam, or gas demands, greater heater inputs can be selected if they do not create additional energy system demands, and the corresponding storage tank size can be selected from the curves.

Ratings of gas-fired water-heating equipment are based on sea-level operation and apply up to 2000 ft. For operation above 2000 ft, and in the absence of specific recommendations from the local authority, equipment ratings should be reduced by 4% for each 1000 ft above sea level before selecting appropriately sized equipment.

Recovery rates in Figures 16 to 23 represent the actual hot water required without considering system heat losses. Heat losses from storage tanks and recirculating hot-water piping should be calculated and added to the recovery rates shown. Storage tanks and hot-water piping must be insulated.

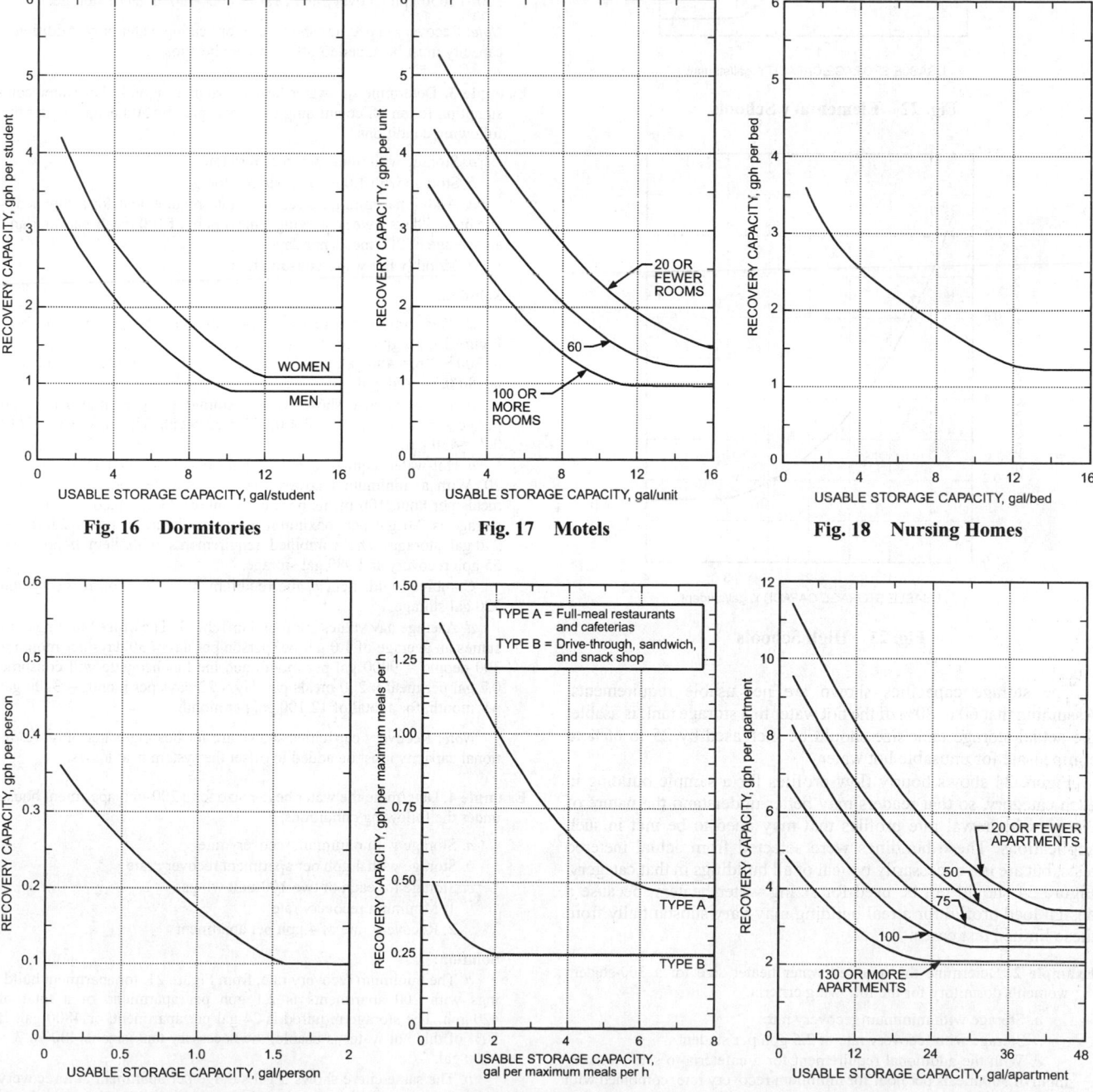

Fig. 16   Dormitories   Fig. 17   Motels   Fig. 18   Nursing Homes

Fig. 19   Office Buildings   Fig. 20   Food Service   Fig. 21   Apartments

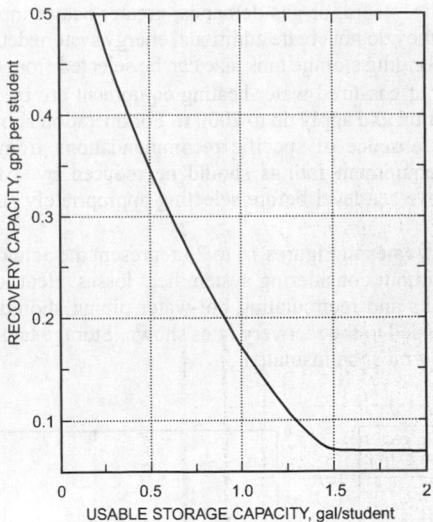

Fig. 22 Elementary Schools

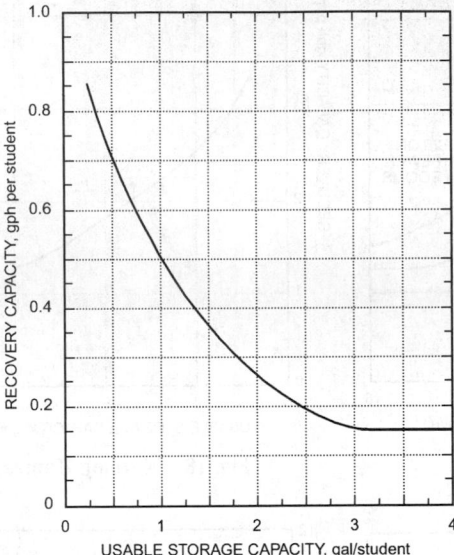

Fig. 23 High Schools

The storage capacities shown are net usable requirements. Assuming that 60 to 80% of the hot water in a storage tank is usable, the actual storage tank size should be increased by 25 to 66% to compensate for unusable hot water.

Figure 24 shows hourly flow profiles for a sample building in each category, so that readers may better understand the nature of energy withdrawal rate profiles that may need to be met in such applications. These buildings were selected from actual metered tests, but are not necessarily typical of all buildings in that category. Figure 24 should not be used for sizing water heaters, because a design load profile for a real building may vary substantially from these limited test cases.

Example 2. Determine the required water heater size for a 300-student women's dormitory for the following criteria:

    a. Storage with minimum recovery rate

    b. Storage with recovery rate of 2.5 gph per student

    c. With the additional requirement for a cafeteria to serve a maximum of 300 meals per hour for minimum recovery rate, combined with item *a;* and for a recovery rate of 1.0 gph per maximum meals per hour, combined with item *b*

Solution:

    a. The minimum recovery rate from Figure 16 for women's dormitories is 1.1 gph per student, or 330 gph total. At this rate, storage required is 12 gal per student or 3600 gal total. On a 70% net usable basis, the necessary tank size is 3600/0.7 = 5150 gal.

    b. The same curve shows 5 gal storage per student at 2.5 gph recovery, or $300 \times 5 = 1500$ gal storage with recovery of $300 \times 2.5 = 750$ gph. The tank size is 1500/0.7 = 2150 gal.

    c. Requirements for a cafeteria can be determined from Figure 20 and added to those for the dormitory. For the case of minimum recovery rate, the cafeteria (Type A) requires $300 \times 0.45 = 135$ gph recovery rate and $300 \times 7/0.7 = 3000$ gal of additional storage. The entire building then requires $330 + 135 = 465$ gph recovery and $5150 + 3000 = 8150$ gal of storage.

    With 1 gph recovery at the maximum hourly meal output, the recovery required is 300 gph, with $300 \times 2.0/0.7 = 860$ gal of additional storage. Combining this with item *b,* the entire building requires $750 + 300 = 1050$ gph recovery and $2150 + 860 = 3010$ gal of storage.

Note: Recovery capacities shown are for heating water only. Additional capacity must be added to offset system heat losses.

Example 3. Determine the water-heater size and monthly hot-water consumption for an office building to be occupied by 300 people under the following conditions:

    a. Storage with minimum recovery rate

    b. Storage with 1.0 gal per person storage

    c. Additional minimum recovery rate requirement for a luncheonette open 5 days a week, serving a maximum of 100 meals per hour and an average of 200 meals per day

    d. Monthly hot-water consumption

Solution:

    a. With minimum recovery rate of 0.1 gph per person from Figure 19, 30 gph recovery is required; storage is 1.6 gal per person, or $300 \times 1.6 = 480$ gal. If 70% of the hot water is usable, the tank size is 480/0.7 = 690 gal.

    b. The curve also shows 1.0 gal storage per person at 0.175 gph per person recovery, or $300 \times 0.175 = 52.5$ gph. The tank size is 300/0.7 = 430 gal.

    c. Hot-water requirements for a luncheonette (Type B) are in Figure 20. With a minimum recovery capacity of 0.25 gph per maximum meals per hour, 100 meals per hour requires 25 gph recovery, and the storage is 2.0 gal per maximum meals per hour, or $100 \times 2.0/0.7 = 290$ gal storage. The combined requirements with item *a* are then 55 gph recovery and 980 gal storage.

    Combined with item *b,* the requirement is 77.5 gph recovery and 720 gal storage.

    d. Average day values are found in Table 7. The office building consumes an average of 1.0 gal per person per day × 30 days per month × 300 people = 9000 gal per month and the luncheonette will consume 0.7 gal per meal × 200 meals per day × 22 days per month = 3100 gal per month, for a total of 12,100 gal per month.

Note: Recovery capacities shown are for heating water only. Additional capacity must be added to offset the system heat losses.

Example 4. Determine the water heater size for a 200-unit apartment house under the following conditions:

    a. Storage with minimum recovery rate

    b. Storage with 4 gph per apartment recovery rate

    c. Storage for each of two 100-unit wings

        1. Minimum recovery rate

        2. Recovery rate of 4 gph per apartment

Solution:

    a. The minimum recovery rate, from Figure 21, for apartment buildings with 200 apartments is 2.1 gph per apartment, or a total of 420 gph. The storage required is 24 gal per apartment, or 4800 gal. If 70% of this hot water is usable, the necessary tank size is 4800/0.7 = 6900 gal.

    b. The same curve shows 5 gal storage per apartment at a recovery rate of 4 gph per apartment, or $200 \times 4 = 800$ gph. The tank size is $200 \times 5/0.7 = 1400$ gal.

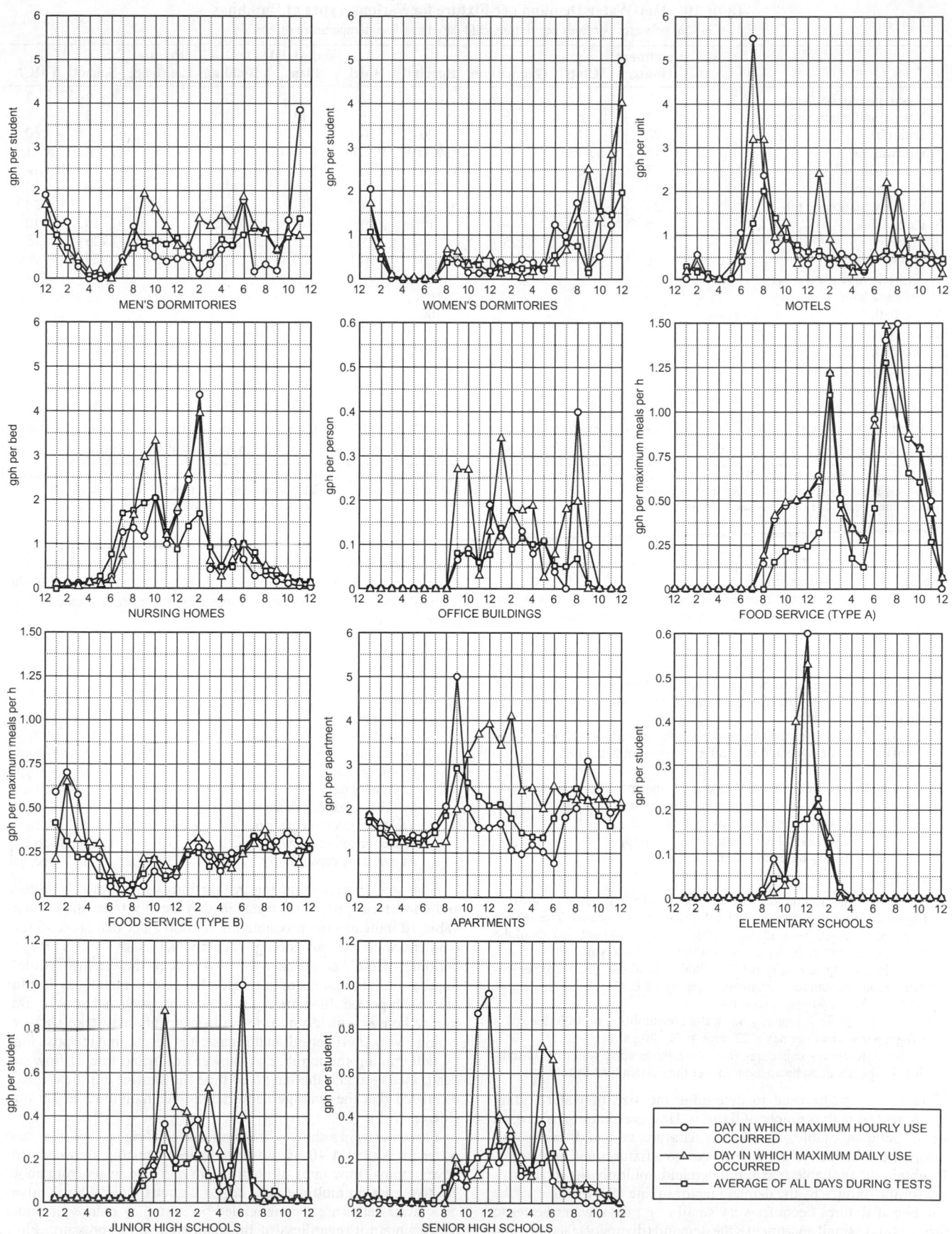

Fig. 24 Hourly Flow Profiles for Various Building Types

Table 10 Hot-Water Demand per Fixture for Various Types of Buildings
(Gallons of water per hour per fixture, calculated at a final temperature of 140°F)

	Apartment House	Club	Gymnasium	Hospital	Hotel	Industrial Plant	Office Building	Private Residence	School	YMCA
1. Basin, private lavatory	2	2	2	2	2	2	2	2	2	2
2. Basin, public lavatory	4	6	8	6	8	12	6	—	15	8
3. Bathtub[c]	20	20	30	20	20	—	—	20	—	30
4. Dishwasher[a]	15	50-150	—	50-150	50-200	20-100	—	15	20-100	20-100
5. Foot basin	3	3	12	3	3	12	—	3	3	12
6. Kitchen sink	10	20	—	20	30	20	20	10	20	20
7. Laundry, stationary tub	20	28	—	28	28	—	—	20	—	28
8. Pantry sink	5	10	—	10	10	—	10	5	10	10
9. Shower	30	150	225	75	75	225	30	30	225	225
10. Service sink	20	20	—	20	30	20	20	15	20	20
11. Hydrotherapeutic shower				400						
12. Hubbard bath				600						
13. Leg bath				100						
14. Arm bath				35						
15. Sitz bath				30						
16. Continuous-flow bath				165						
17. Circular wash sink				20	20	30	20		30	
18. Semicircular wash sink				10	10	15	10		15	
19. DEMAND FACTOR	0.30	0.30	0.40	0.25	0.25	0.40	0.30	0.30	0.40	0.40
20. STORAGE CAPACITY FACTOR[b]	1.25	0.90	1.00	0.60	0.80	1.00	2.00	0.70	1.00	1.00

Note: Data sources predate low-flow fixtures and appliances.
[a]Dishwasher requirements should be taken from this table or from manufacturers' data for model to be used, if known.

[b]Ratio of storage tank capacity to probable maximum demand/h. Storage capacity may be reduced where unlimited supply of steam is available from central street steam system or large boiler plant.
[c]Whirlpool baths require specific consideration based on capacity. They are not included in the bathtub category.

c. Solution for a 200-unit apartment house having two wings, each with its own hot-water system.

1. With minimum recovery rate of 2.5 gph per apartment (see Figure 21), a 250 gph recovery is required, and the necessary storage is 28 gal per apartment, or 100 × 28 = 2800 gal. The required tank size is 2800/0.7 = 4000 gal for each wing.

2. The curve shows that, for a recovery rate of 4 gph per apartment, storage is 14 gal per apartment, or 100 × 14 = 1400 gal, with recovery of 100 × 4 = 400 gph. The necessary tank size is 1400/0.7 = 2000 gal in each wing.

Note: Recovery capacities shown are for heating water only. Additional capacity must be added to offset the system heat loss.

Example 5. Determine the water-heater size and monthly hot-water consumption for a 2000-student high school under the following conditions:
a. Storage with minimum recovery rate
b. Storage with 4000 gal maximum storage capacity
c. Monthly hot-water consumption

Solution:
a. With the minimum recovery rate of 0.15 gph per student (from Figure 23) for high schools, 300 gph recovery is required. The storage required is 3.0 gal per student, or 2000 × 3.0 = 6000 gal. If 70% of the hot water is usable, the tank size is 6000/0.7 = 8600 gal.

b. Net storage capacity is 0.7 × 4000 = 2800 gal, or 1.4 gal per student. From the curve, a recovery capacity of 0.37 gph per student or 2000 × 0.37 = 740 gph is required.

c. From Table 7, monthly hot-water consumption is 2000 students × 1.8 gal per student per day × 22 days = 79,000 gal.

Note: Recovery capacities shown are for heating water only. Additional capacity must be added to offset the system heat loss.

Table 10 can be used to determine the size of water-heating equipment from the number of fixtures. However, caution is advised when using this table, because its data are very old, taken well before the introduction of modern low-flow fixtures and appliances. To obtain the probable maximum demand, multiply the total quantity for the fixtures by the demand factor in line 19. Note that, as the number of fixtures becomes very small (e.g., for a water heater to serve a single small apartment), the demand (diversity) factors listed in Table 10 are no longer valid. In all cases, total demand is never less than the demand for the largest single fixture. The heater or coil

should have a water-heating capacity equal to this probable maximum demand. The storage tank should have a capacity equal to the probable maximum demand multiplied by the storage capacity factor in line 20.

Example 6. Determine heater and storage tank size for an apartment building from a number of fixtures.

Solution:

60 lavatories	×	2 gph	=	120 gph
30 bathtubs	×	20 gph	=	600 gph
30 showers	×	30 gph	=	900 gph
60 kitchen sinks	×	10 gph	=	600 gph
15 laundry tubs	×	20 gph	=	300 gph
Possible maximum demand			=	2520 gph
Probable maximum demand	=	2520 × 0.30	=	756 gph
Heater or coil capacity			=	756 gph
Storage tank capacity	=	756 × 1.25	=	945 gal

Showers. In many housing installations such as motels, hotels, and dormitories, peak hot-water load is usually from shower use. Table 10 indicates the probable hourly hot-water demand and recommended demand and storage capacity factors for various types of buildings. Hotels could have a 3 to 4 h peak shower load. Motels require similar volumes of hot water, but peak demand may last for only a 2 h period. In some types of housing, such as barracks, fraternity houses, and dormitories, all occupants may take showers within a very short period. In this case, it is best to find the peak load by determining the number of shower heads and rate of flow per head; then estimate the length of time the shower will be on. It is estimated that the average shower time per individual is 7.5 min (Meier 1985).

Flow rate from a shower head varies depending on type, size, and water pressure. At 40 psi water pressure, available shower heads have nominal flow rates of blended hot and cold water from about 2.5 to 6 gpm. In multiple-shower installations, flow control valves on shower heads are recommended because they reduce flow rate and maintain it regardless of fluctuations in water pressure. Flow can usually be reduced to 50% of the manufacturer's maximum flow rating without adversely affecting the spray pattern of the shower head. Flow control valves are commonly available with capacities from 1.5 to 4.0 gpm.

If the manufacturer's flow rate for a shower head is not available and no flow control valve is used, the following average flow rates may serve as a guide for sizing the water heater:

Small shower head	2.5 gpm
Medium shower head	4.5 gpm
Large shower head	6 gpm

Note that the maximum flow rate allowed by U.S. federal energy efficiency standards is 2.5 gpm, as of 1992. However, higher-flow-rate models are still sold.

Food Service. In restaurants, bacteria are usually killed by rinsing washed dishes with 180 to 195°F water for several seconds. In addition, an ample supply of general-purpose hot water, usually 140 to 150°F, is required for the wash cycle of dishwashers. Although a water temperature of 140°F is reasonable for dish washing in private dwellings, in public places, the NSF (e.g., *Standard* 3) or local health departments require 180 to 195°F water in the rinsing cycle. However, the NSF allows a lower temperature when certain types of machines and chemicals are used. The two-temperature hot-water requirements of food service establishments present special problems. The lower-temperature water is distributed for general use, but the 180°F water should be confined to the equipment requiring it and should be obtained by boosting the temperature. It would be dangerous to distribute 180°F water for general use. ANSI/NSF *Standard* 3-2001 covers the design of dishwashers and water heaters used by restaurants. The American Gas Association's (Dunn et al. 1959) recommended procedure for sizing water heaters for restaurants consists of determining the following:

- Types and sizes of dishwashers used (manufacturers' data should be consulted to determine the initial fill requirements of the wash tanks)
- Required quantity of general-purpose hot water
- Duration of peak hot-water demand period
- Inlet water temperature
- Type and capacity of existing water-heating system
- Type of water heating system desired

After the quantity of hot water withdrawn from the storage tank each hour has been taken into account, the following equation may be used to size the required heater(s). The general-purpose and 180 to 195°F water requirements are determined from Tables 11 and 12.

$$q_i = Q_h c_p \rho \Delta t / \eta \qquad (14)$$

where

q_i = heater input, Btu/h
Q_h = flow rate, gph
c_p = specific heat of water = 1.00 Btu/lb·°F
ρ = density of water = 8.33 lb/gal
Δt = temperature rise, °F
η = heater efficiency

To determine the quantity of usable hot water from storage, the duration of consecutive peak demand must be estimated. This peak usually coincides with the dishwashing period during and after the main meal and may last from 1 to 4 h. Any hour in which the dishwasher is used at 70% or more of capacity should be considered a peak hour. If the peak demand lasts for 4 h or more, the value of a storage tank is reduced, unless especially large tanks are used. Some storage capacity is desirable to meet momentary high draws.

NSF *Standard* 5 recommendations for hot-water rinse demand are based on 100% operating capacity of the machines, as are the data provided in Table 11. NSF *Standard* 5 states that 70% of operating rinse capacity is all that is normally attained, except for rackless-type conveyor machines.

Examples 7, 8, and 9 demonstrate the use of Equation (14) in conjunction with Tables 11 and 12.

Table 11 NSF Final Rinse Water Requirement for Dishwashers[a]

Type and Size of Dishwasher	Flow Rate, gpm	180 to 195°F Hot-Water Requirements	
		Heaters Without Internal Storage,[b] gph	Heaters with Internal Storage to Meet Flow Demand,[c] gph
Door type:			
16 × 16 in.	6.94	416	69
18 × 18 in.	8.67	520	87
20 × 20 in.	10.4	624	104
Undercounter	5	300	70
Conveyor type:			
Single tank	6.94	416	416
Multiple tank (dishes flat)	5.78	347	347
Multiple tank (dishes inclined)	4.62	277	277
Silver washers	7	420	45
Utensil washers	8	480	75
Makeup water requirements	2.31	139	139

Note: Values are extracted from a previous version of ANSI/NSF *Standard* 3. The current version of ANSI/NSF *Standard* 3-2001 is performance-based and no longer lists minimum flow rates.
[a]Flow pressure at dishwashers assumed to be 20 psig.
[b]Based on flow rate in gpm.
[c]Based on dishwasher operation at 100% of mechanical capacity.

Table 12 General-Purpose Hot-Water (140°F) Requirement for Various Kitchens Uses[a,b]

Equipment	gph
Vegetable sink	45
Single pot sink	30
Double pot sink	60
Triple pot sink	90
Prescrapper (open type)	180
Preflush (hand-operated)	45
Preflush (closed type)	240
Recirculating preflush	40
Bar sink	30
Lavatories (each)	5

Source: Dunn et al. (1959).
[a]Supply water pressure at equipment assumed to be 20 psig.
[b]Dishwasher operation at 100% of mechanical capacity.

Example 7. Determine the hot-water demand for water heating in a cafeteria kitchen with one vegetable sink, five lavatories, one prescrapper, one utensil washer, and one two-tank conveyor dishwasher (dishes inclined) with makeup device. The initial fill requirement for the tank of the utensil washer is 85 gph at 140°F. The initial fill requirement for the dishwasher is 20 gph for each tank, or a total of 40 gph, at 140°F. The maximum period of consecutive operation of the dishwasher at or above 70% capacity is assumed to be 2 h. Supply water temperature is 60°F.

Solution: The required quantities of general-purpose (140°F) and rinse (180°F) water for the equipment, from Tables 11 and 12 and given values, are shown in the following tabulation:

Item	Quantity Required at 140°F, gph	Quantity Required at 180°F, gph
Vegetable sink	45	—
Lavatories (5)	25	—
Prescrapper	180	—
Dishwasher	—	277
Initial tank fill	40	—
Makeup water	—	139
Utensil washer	—	75
Initial tank fill	85	—
Total requirements	375	491

The total consumption of 140°F water is 375 gph. The total consumption at 180°F depends on the type of heater to be used. For a heater that has enough internal storage capacity to meet the flow demand, the total consumption of 180°F water is (277 + 139 + 75) = 491 gph based on the requirements taken from Table 11, or approximately 350 gph (0.70 × 491 = 344 gph). For an instantaneous heater without internal storage capacity, the total quantity of 180°F water consumed must be based on the flow demand. From Table 11, the quantity required for the dishwasher is 277 gph; for the makeup, 139 gph; and for the utensil washer, 480 gph. The total consumption of 180°F water is 277 + 139 + 480 = 896 gph, or approximately 900 gph.

Example 8. Determine fuel input requirements (assume 75% heater efficiency) for heating water in the cafeteria kitchen described in Example 7, by the following systems, which are among many possible solutions:

   a. Separate, self-contained, storage-type heaters

   b. Single instantaneous-type heater having no internal storage, to supply both 180°F and 140°F water through a mixing valve

   c. Separate instantaneous-type heaters having no internal storage

Solution:

   a. The temperature rise for 140°F water is 140 – 60 = 80°F. From Equation (14), the fuel input required to produce 375 gph of 140°F water with an 80°F temperature rise at 75% efficiency is about 333,000 Btu/h. One or more heaters may be selected to meet this total requirement.

   From Equation (14), the fuel input required to produce 491 gph of 180°F water with a temperature rise of 180 – 60 = 120°F at 75% efficiency is 654,000 Btu/h. One or more heaters with this total requirement may be selected from manufacturers' catalogs.

   b. Correct sizing of instantaneous-type heaters depends on the flow rate of the 180°F rinse water. From Example 7, the consumption of 180°F water based on the flow rate is 900 gph; consumption of 140°F water is 375 gph.

   Fuel input required to produce 900 gph (277 + 480 + 139) of 180°F water with a 120°F temperature rise is 1,200,000 Btu/h. Fuel input to produce 375 gph of 140°F water with a temperature rise of 80°F is 333,000 Btu/h. Total heater requirement is 1,200,000 + 333,000 = 1,533,000 Btu/h. One or more heaters meeting this total input requirement can be selected from manufacturers' catalogs.

   c. Fuel input required to produce 140°F water is the same as for solution *b*, 333,000 Btu/h. One or more heaters meeting this total requirement can be selected.

   Fuel input required to produce 180°F water is also the same as in solution *b*, 1,200,000 Btu/h. One or more heaters meeting this total requirement can be selected.

Example 9. A luncheonette has purchased a door-type dishwasher that handles 16 by 16 in. racks. The existing hot-water system can supply the necessary 140°F water to meet all requirements for general-purpose use and for the booster heater that is to be installed. Determine the size of the following booster heaters operating at 75% thermal efficiency required to heat 140°F water to provide sufficient 180°F rinse water for the dishwasher:

   a. Booster heater with no storage capacity

   b. Booster heater with enough storage capacity to meet flow demand

Solution:

   a. Because the heater is the instantaneous type, it must be sized to meet the 180°F water demand at a rated flow. From Table 11, this rated flow is 6.94 gpm, or 416 gph. From Equation (14), the required fuel input with a 40°F temperature rise is 185,000 Btu/h. A heater meeting this input requirement can be selected from manufacturers' catalogs.

   b. In designing a system with a booster heater having storage capacity, the dishwasher's hourly flow demand can be used instead of the flow demand used in solution *a*. The flow demand from Table 11 is 69 gph when the dishwasher is operating at 100% mechanical capacity. From Equation (14), with a 40°F temperature rise, the fuel input required is 30,700 Btu/h. A booster heater with this input can be selected from manufacturers' catalogs.

Estimating Procedure. Hot-water requirements for kitchens are sometimes estimated on the basis of the number of meals served (assuming eight dishes per meal). Demand for 180°F water for a dishwasher is

$$D_1 = C_1 N/\theta \tag{15}$$

where

   D_1 = water for dishwasher, gph
   N = number of meals served
   θ = hours of service
   C_1 = 0.8 for single-tank dishwasher, 0.5 for two-tank dishwasher

Demand for water for a sink with gas burners is

$$D_2 = C_2 V \tag{16}$$

where

   D_2 = water for sink, gph
   C_2 = 3
   V = sink capacity (15 in. depth), gal

Demand for general-purpose hot water at 140°F is

$$D_3 = C_3 N/(\theta + 2) \tag{17}$$

where

   D_3 = general-purpose water, gph
   C_3 = 1.2

Total demand is

$$D = D_1 + D_2 + D_3 \tag{18}$$

For soda fountains and luncheonettes, use 75% of the total demand. For hotel meals or other elaborate meals, use 125%.

Schools. Service water heating in schools is needed for janitorial work, lavatories, cafeterias, shower rooms, and sometimes swimming pools.

Hot water used in cafeterias is about 70% of that usually required in a commercial restaurant serving adults and can be estimated by the method used for restaurants. Where NSF sizing is required, follow *Standard* 5.

Shower and food service loads are not ordinarily concurrent. Each should be determined separately, and the larger load should determine the size of the water heater(s) and the tank. Provision must be made to supply 180°F sanitizing rinse. The booster must be sized according to the temperature of the supply water. If feasible, the same water can be used for both needs. If the distance between the two points of need is great, a separate water heater should be used.

A separate heater system for swimming pools can be sized as outlined in the section on Swimming Pools/Health Clubs.

Domestic Coin-Operated Laundries. Small domestic machines in coin laundries or apartment house laundry rooms have a wide range of draw rates and cycle times. Domestic machines provide a wash water temperature (normal) as low as 120°F. Some manufacturers recommend a temperature of 160°F; however, the average appears to be 140°F. Hot-water sizing calculations must ensure a supply to both the instantaneous draw requirements of a number of machines filling at one time and the average hourly requirements.

The number of machines drawing at any one time varies widely; the percentage is usually higher in smaller installations. One or two customers starting several machines at about the same time has a much sharper effect in a laundry with 15 or 20 machines than in one with 40 machines. Simultaneous draw may be estimated as follows:

1 to 11 machines	100% of possible draw
12 to 24 machines	80% of possible draw
25 to 35 machines	60% of possible draw
36 to 45 machines	50% of possible draw

Possible peak draw can be calculated from

$$F = NPV_f/T \qquad (19)$$

where

F = peak draw, gpm
N = number of washers installed
P = number of washers drawing hot water divided by N
V_f = quantity of hot water supplied to machine during hot-wash fill, gal
T = wash fill period, min

Recovery rate can be calculated from

$$R = 60NPV_f/(\theta + 10) \qquad (20)$$

where

R = total hot water (machines adjusted to hottest water setting), gph
θ = actual machine cycle time, min

Note: $(\theta + 10)$ is the cycle time plus 10 min for loading and unloading.

Commercial Laundries. Commercial laundries generally use a storage water heater. The water may be softened to reduce soap use and improve quality. The trend is toward installing high-capacity washer-extractor wash wheels, resulting in high peak demand.

Sizing Data. Laundries can normally be divided into five categories. The required hot water is determined by the weight of the material processed. Average hot-water requirements at 180°F are

Institutional	2 gal/lb·h
Commercial	2 gal/lb·h
Linen supply	2.5 gal/lb·h
Industrial	2.5 gal/lb·h
Diaper	2.5 gal/lb·h

Total weight of the material times these values give the average hourly hot-water requirements. The designer must consider peak requirements; for example, a 600 lb machine may have a 20 gpm average requirement, but the peak requirement could be 350 gpm.

In a multiple-machine operation, it is not reasonable to fill all machines at the momentary peak rate. Diversity factors can be estimated by using 1.0 of the largest machine plus the following balance:

	Total number of machines				
	2	3 to 5	6 to 8	9 to 11	12 and over
1.0 +	0.6	0.45	0.4	0.35	0.3

For example, four machines have a diversity factor of 1.0 + 0.45 = 1.45.

Types of Systems. Service water-heating systems for laundries are pressurized or vented. The pressurized system uses city water pressure, and the full peak flow rates are received by the softeners, reclaimer, condensate cooler, water heater, and lines to the wash wheels. Flow surges and stops at each operation in the cycle. A pressurized system depends on an adequate water service.

The vented system uses pumps from a vented (open) hot-water heater or tank to supply hot water. The tank's water level fluctuates from about 6 in. above the heating element to a point 12 in. from the top of the tank; this fluctuation defines the working volume. The level drops for each machine fill, and makeup water runs continuously at the average flow rate and water service pressure during the complete washing cycle. The tank is sized to have full working volume at the beginning of each cycle. Lines and softeners may be sized for the average flow rate from the water service to the tank, not the peak machine fill rate as with a closed, pressurized system.

Waste heat exchangers have continuous flow across the heating surface at a low flow rate, with continuous heat reclamation from the wastewater and flash steam. Automatic flow-regulating valves on the inlet water manifold control this low flow rate. Rapid fill of

machines increases production (i.e., more batches can be processed).

Heat Recovery. Commercial laundries are ideally suited for heat recovery because 135°F wastewater is discharged to the sewer. Fresh water can be conservatively preheated to within 15°F of the wastewater temperature for the next operation in the wash cycle. Regions with an annual average temperature of 55°F can increase to 120°F the initial temperature of fresh water going into the hot-water heater. For each 1000 gph or 8330 lb per hour of water preheated 65°F (55 to 120°F), heat reclamation and associated energy savings is 540,000 Btu/h.

Flash steam from a condensate receiving tank is often wasted to the atmosphere. Heat in this flash steam can be reclaimed with a suitable heat exchanger, to preheat makeup water to the heater by 10 to 20°F above the existing makeup temperature.

Swimming Pools/Health Clubs. The desirable temperature for swimming pools is 80°F. Most manufacturers of water heaters and boilers offer specialized models for pool heating; these include a pool temperature controller and a water bypass to prevent condensation. The water-heating system is usually installed before the return of treated water to the pool. A circulation rate to generate a change of water every 8 h for residential pools and 6 h for commercial pools is acceptable. An indirect heater, in which piping is embedded in the walls or floor of the pool, has the advantage of reduced corrosion, scaling, and condensation because pool water does not flow through the pipes, but its disadvantage is the high initial installation cost.

The installation should have a pool temperature control and a water pressure or flow safety switch. The temperature control should be installed at the inlet to the heater; the pressure or flow switch can be installed at either the inlet or outlet, depending on the manufacturer's instructions. It affords protection against inadequate water flow.

Sizing should be based on four considerations:

- Conduction through the pool walls
- Convection from the pool surface
- Radiation from the pool surface
- Evaporation from the pool surface

Except in aboveground pools and in rare cases where cold groundwater flows past the pool walls, conduction losses are small and can be ignored. Because convection losses depend on temperature differentials and wind speed, these losses can be greatly reduced by installing windbreaks such as hedges, solid fences, or buildings.

Radiation losses occur when the pool surface is subjected to temperature differentials; these frequently occur at night, when the sky temperature may be as much as 80°F below ambient air temperature. This usually occurs on clear, cool nights. During the daytime, however, an unshaded pool receives a large amount of radiant energy, often as much as 100,000 Btu/h. These losses and gains may offset each other. An easy method of controlling nighttime radiation losses is to use a floating pool cover; this also substantially reduces evaporative losses.

Evaporative losses constitute the greatest heat loss from the pool (50 to 60% in most cases). If it is possible to cut evaporative losses drastically, the pool's heating requirement may be cut by as much as 50%. A floating pool cover can accomplish this.

A pool heater with an input great enough to provide a heat-up time of 24 h would be the ideal solution. However, it may not be the most economical system for pools that are in continuous use during an extended swimming season. In this instance, a less expensive unit providing an extended heat-up period of as much as 48 h can be used. Pool water may be heated by several methods. Fuel-fired water heaters and boilers, electric boilers, tankless electric circulation water heaters, air-source heat pumps, and solar heaters have all been used successfully. Air-source heat pumps and solar heating systems are often used to extend a swimming season rather than to allow intermittent use with rapid pickup.

The following equations provide some assistance in determining the area and volume of pools.

Elliptical

Area = $3.14AB$
A = Short radius
B = Long radius
Volume = 7.5 gal/ft^3 × Area × Average Depth

Kidney Shape

Area = $0.45L (A+B)$ (approximately)
L = Length
A = Width at one end
B = Width at other end
Volume = 7.5 gal/ft^3 × Area × Average Depth

Oval (for circular, set $L = 0$)

Area = $3.14R^2 + LW$
L = Length of straight sides
W = Width or $2R$
R = Radius of ends
Volume = 7.5 gal/ft^3 × Area × Average Depth

Rectangular

Area = LW
L = Length
W = Width
Volume = 7.5 gal/ft^3 × Area × Average Depth

The following is an effective method for heating outdoor pools. Additional equations can be found in Chapter 5.

1. Obtain pool water capacity, in gallons.
2. Determine the desired heat pickup time in hours.
3. Determine the desired pool temperature. If not known, use 80°F.
4. Determine the average temperature of the coldest month of use.

The required heater output q_t can now be determined by the following equations:

$$q_1 = \rho c_p V(t_f - t_i)/\theta \tag{21}$$

where

q_1 = pool heat-up rate, Btu/h
ρ = density of water = 8.33 lb/gal
c_p = specific heat of water = 1.00 Btu/lb·°F
V = pool volume, gal
t_f = desired temperature (usually 80°F)
t_i = initial temperature of pool, °F
θ = pool heat-up time, h

$$q_2 = UA(t_p - t_a) \tag{22}$$

where

q_2 = heat loss from pool surface, Btu/h
U = surface heat transfer coefficient = 10.5 Btu/h·ft^2·°F
A = pool surface area, ft^2
t_p = pool temperature, °F
t_a = ambient temperature, °F

$$q_t = q_1 + q_2 \tag{23}$$

Notes: These heat loss equations assume a wind velocity of 3 to 5 mph. For pools sheltered by nearby fences, dense shrubbery, or buildings, an average wind velocity of less than 3.5 mph can be assumed. In this case, use 75% of the values calculated by Equation (22). For a velocity of 5 mph, multiply by 1.25; for 10 mph, multiply by 2.0.

Because Equation (22) applies to the coldest monthly temperatures, results calculated may not be economical. Therefore, a value of one-half the surface loss plus the heat-up value yields a more

viable heater output figure. Heater input then equals output divided by fuel source efficiency.

Whirlpools and Spas. Hot-water requirements for whirlpool baths and spas depend on temperature, fill rate, and total volume. Water may be stored separately at the desired temperature or, more commonly, regulated at the point of entry by blending. If rapid filling is desired, provide storage at least equal to the volume needed; fill rate can then be varied at will. An alternative is to establish a maximum fill rate and provide an instantaneous water heater that can handle the flow.

Industrial Plants. Hot water (potable) is used in industrial plants for cafeterias, showers, lavatories, gravity sprinkler tanks, and industrial processes. Employee cleanup load is usually heaviest and not concurrent with other uses. Other loads should be checked before sizing, however, to be certain that this is true.

Employee cleanup load includes (1) wash troughs or standard lavatories, (2) multiple wash sinks, and/or (3) showers. Hot-water requirements for employees using standard wash fixtures can be estimated at 1 gal of hot water for each clerical and light-industrial employee per work shift and 2 gal for each heavy-industrial worker.

For sizing purposes, the number of workers using multiple wash fountains is disregarded. Hot-water demand is based on full flow for the entire cleanup period. This usage over a 10 min period is indicated in Table 13. The shower load depends on the flow rate of the shower heads and their length of use. Table 13 may be used to estimate flow based on a 15 min period.

Water heaters used to prevent freezing in gravity sprinkler or water storage tanks should be part of a separate system. The load depends on tank heat loss, tank capacity, and winter design temperature.

Process hot-water load must be determined separately. Volume and temperature vary with the specific process. If the process load occurs at the same time as the shower or cafeteria load, the system must be sized to reflect this total demand. In some cases, it may be preferable to use separate systems, depending on the various load sizes and distance between them.

Ready-Mix Concrete. In cold weather, ready-mix concrete plants need hot water to mix the concrete so that it will not be ruined by freezing before it sets. Operators prefer to keep the mix at about 70°F by adding hot water to the cold aggregate. Usually, water at about 150°F is considered proper for cold weather. If the water temperature is too high, some of the concrete will flash set.

Generally, 30 gal of hot water per cubic yard of concrete mix is used for sizing. To obtain the total hot-water load, this number is multiplied by the number of trucks loaded each hour and the capacity of the trucks. The hot water is dumped into the mix as quickly as possible at each loading, so ample hot-water storage or large heat exchangers must be used. Table 14 shows a method of sizing water heaters for concrete plants.

Part of the heat may be obtained by heating the aggregate bin by circulating hot water through pipe coils in the walls or sides of the

Table 13 Hot-Water Usage for Industrial Wash Fountains and Showers

	Multiple Wash Fountains		Showers	
Type	**Gal of 140°F Water Required for 10 min Period**[a]		**Flow Rate, gpm**	**Gal of 140°F Water Required for 15 min Period**[b]
36 in. Circular	40		3	29.0
Semicircular	22		4	39.0
54 in. Circular	66		5	48.7
Semicircular	40		6	58.0

[a]Based on 110°F wash water and 40°F cold water at average flow rates.
[b]Based on 105°F shower water and 40°F cold water.

Table 14 Water Heater Sizing for Ready-Mix Concrete Plant
(Input and Storage Tank Capacity to Supply 150°F Water
at 40°F Inlet Temperature)

Truck Capacity, yd³	Water Heater Storage Tank Volume, gal	Time Interval Between Trucks, min*					
		50	35	25	10	5	0
		Water Heater Capacity, 1000 Btu/h					
6	430	458	612	785	1375	1830	2760
7.5	490	527	700	900	1580	2100	3150
9	560	596	792	1020	1790	2380	3580
11	640	687	915	1175	2060	2740	4120

*This table assumes 10 min loading time for each truck. Thus, for a 50 min interval between trucks, it is assumed that 1 truck/h is served. For 0 min between trucks, it is assumed that one truck loads immediately after the truck ahead has pulled away. Thus, 6 trucks/h are served. It also assumes each truck carries a 120 gal storage tank of hot water for washing down at the end of dumping the load. This hot water is drawn from the storage tank and must be added to the total hot-water demands. This has been included in the table.

bin. This can allow a lower mixing-water temperature, and the aggregate flows easily from the bins. When aggregate is not heated, it often freezes into chunks, which must be thawed before they will pass through the dump gates. If hot water is used for thawing, too much water accumulates in the aggregate, and control of the final product may vary beyond allowable limits. Therefore, jets of steam supplied by a small boiler and directed on the large chunks are often used for thawing.

Sizing Tankless Water Heaters

Although tankless water heaters are sometimes also referred to as instantaneous water heaters, in this chapter the two types are distinct. Larger instantaneous water heaters for bigger commercial, institutional, and industrial applications may still have some water storage tank volume, even though their ratio of heating rate divided by storage volume is large. Smaller commercial and residential systems only contain a volume of water sufficient to fill the chambers or tubing where the heating is done; they do not incorporate storage tanks, and are truly tankless as the term is used here.

Tankless water heaters offer potential efficiency advantages over tank-type units for several reasons. Because they do not store heated water, they have low standby energy loss (typically, only a small amount of electricity to run controls). This energy savings potential can be significant for low-use applications. Another potential advantage is that the lack of a storage tank means they are much smaller than tank-type units and can more easily be located close to points of use (especially electric tankless units). Locating units close to points of use reduces energy losses in the hot-water distribution system, sometimes substantially. This ease of positioning may also make it easier to use more than one water heater, reducing hot-water distribution system heat losses still further by eliminating even more piping.

There are many good applications of both electric and fossil-fired tankless water heaters in residences, commercial, institutional, and industrial settings. Tankless water heaters are especially useful for providing more localized heating in point-of-use or near-point-of-use applications because they do not take up much space. In general, tankless water heaters are designed to completely heat cold water in one pass through the heater. There are exceptions, however, because some models with advanced controls can also heat pre-warmed water by controllable amounts. See the discussion below about modulating heat input rates.

Tankless water heaters generally have some sort of flow detection method (e.g., a flow switch or method of differential temperature measurement that indicates flow is occurring). Water heating only begins once water flow is confirmed. Outlet temperature from tankless water heaters is determined by the flow rate, entering cold-water temperature, and applied heating rate. Simpler systems do not actively control outlet temperature, other than to turn off the heat

input if exit temperature exceeds a set value. These systems are more likely to specify the use of water flow restrictors to restrict flow through the units to minimize undesirably cool water exiting the units.

Systems with more advanced controls continuously monitor the exit water temperature and modulate the heat input and/or water flow rate to maintain the specified outlet temperature. Advanced electric tankless water heaters modulate power to the heating elements, either in steps (multiple heating elements) or by varying the voltage and/or current supplied to the heating elements, or both. Advanced fossil-fired tankless water heaters, which are available in both natural-gas- and propane-fired versions, modulate the heating rate by either modulating heat input in steps (e.g. using multiple burners), or by modulating gas flow rate to the burner(s), or some combination of the two. These designs can be used as booster heaters or in recirculated heating systems (i.e., they can work well with prewarmed entering water temperatures) because they can better control exit temperature.

One of the most important tankless water heater sizing considerations is having adequate heat input rate to heat the desired flow rate of water by a temperature rise needed to make the water warm enough to use. Table 15 shows the necessary heat input rate [not considering heat input efficiency: divide table values by heat input efficiency in decimal form (e.g., 0.8 for some fossil-fired heaters) to determine total energy input rate required for tankless water heaters versus flow rate and needed temperature rise. The heating rates shown are computed using Equation (1).

Note that 105°F is about the minimum acceptable temperature for human use at fixtures. Accounting for heat loss in piping and/or when atomizing droplets in a showerhead, 110°F is a more typical requirement. The needed temperature rise in a cold climate where the entering cold-water temperature may be 35°F would thus be 110°F – 35°F = 75°F; in a warm climate where the entering cold water temperature may be 85°F, the temperature rise would be 110°F – 85°F = 25°F. For comparison, the temperature rise specified in the U.S. federal water heater testing and rating procedure is 135°F – 58°F = 77°F. For reference, typical flow rate ranges are as follows:

- Hand-washing sinks: 0.2 to 1.0 gpm
- Showers: 0.8 to 2.5 gpm
- Bathtub fill rates: 1.0 to 6.0 gpm
- Dishwasher fill rates: 1.0 to 3.0 gpm
- Clothes-washing machine fill rates: 1.0 to 6.0 gpm
- Residential whole-house recurring peak rates: around 3.0 to 4.0 gpm
- Residential whole-house severe-peak flow rates: 6.0 to 8.0 gpm

As can be seen from Table 15, whole-house tankless water heaters need to be able to provide heating rates on the order of 75,000 to 150,000 Btu/h in all but the warmest climates. Note, however, that in single-family residential applications, users have the opportunity to learn what works and what does not, and are likely to adjust their hot-water use habits somewhat to obtain adequately hot water from whatever water-heating system is used. They could do this for example, by avoiding hot-water use from multiple fixtures simultaneously, and reducing demanded flow rates.

An important issue in the sizing of tankless water heaters is thus what peak hot-water energy rate load to design for. It is generally acceptable to design the water-heating system to meet a peak hot-water load (in terms of energy rate needed, not just water flow rate needed) that is not exceeded by 97.5% of all draws. The difficulty in sizing whole-house tankless water heaters comes in predicting how draws will coincide to create the peak energy demand rate. This peak coincident energy demand rate must be estimated by the person sizing the system, because ASHRAE does not currently have a statistically valid amount of data on peak residential water/energy flow rates with which to make recommendations. Sizing recommendations are easier with storage-type water heaters because their

Table 15   Needed Tankless Water Heater Output Heat Rates, Btu/h*

Flow Rate, gpm	Temperature Rise						
	10°F	25°F	50°F	55°F	75°F	77°F	100°F
0.1	504	1,260	2,520	2,772	3,780	3,881	5,040
0.5	2,520	6,300	12,600	13,860	18,900	19,404	25,200
1.0	5,040	12,600	25,200	27,720	37,800	38,808	50,400
1.5	7,560	18,900	37,800	41,580	56,700	58,212	75,600
2.0	10,080	25,200	50,400	55,440	75,600	776,196	100,800
2.5	12,600	31,500	63,000	69,300	94,500	97,020	126,000
3.0	15,120	37,800	75,600	83,160	113,400	116,424	151,200
3.5	17,640	44,100	88,200	97,020	132,300	135,828	176,400
4.0	20,160	50,400	100,800	110,880	151,200	155,232	201,600
4.5	22,680	56,700	113,400	124,740	170,100	174,636	226,800
5.0	25,200	63,000	126,000	138,600	189,000	194,040	252,000
6.0	30,240	75,600	151,200	166,320	226,800	232,848	302,400
7.0	35,280	88,200	176,400	194,040	264,600	271,656	352,800
8.0	40,320	100,800	201,600	221,760	302,400	310,464	403,200
9.0	45,360	113,400	226,800	249,480	340,200	349,272	453,600
10.0	50,400	126,000	252,000	277,200	378,000	388,080	504,000

*Divide table values by input efficiency to determine required heat input rate. Values in Btu/h apply to fossil fuels, and those in kW apply to electric tankless heaters.

Table 16   Hot-Water Demand in Fixture Units (140°F Water)

	Apartments	Club	Gymnasium	Hospital	Hotels and Dormitories	Industrial Plant	Office Building	School	YMCA
Basin, private lavatory	0.75	0.75	0.75	0.75	0.75	0.75	0.75	0.75	0.75
Basin, public lavatory	—	1	1	1	1	1	1	1	1
Bathtub	1.5	1.5	—	1.5	1.5	—	—	—	—
Dishwasher*	1.5	Five fixture units per 250 seating capacity							
Therapeutic bath	—	—	—	5	—	—	—	—	—
Kitchen sink	0.75	1.5	—	3	1.5	3	—	0.75	3
Pantry sink	—	2.5	—	2.5	2.5	—	—	2.5	2.5
Service sink	1.5	2.5	—	2.5	2.5	2.5	2.5	2.5	2.5
Shower	1.5	1.5	1.5	1.5	1.5	3.5	—	1.5	1.5
Circular wash fountain	—	2.5	2.5	2.5	—	4	—	2.5	2.5
Semicircular wash fountain	—	1.5	1.5	1.5	—	3	—	1.5	1.5

Note: Data predate modern low-flow fixtures and appliances.   *See Water-Heating Terminology section for definition of fixture unit.

sizing is done more based on integrated total energy requirements, and is not highly dependent on knowledge of peak flow rates.

An issue related to proper sizing of tankless water heaters is the size of fuel piping and electrical service needed. Because gas-fired tankless water heaters must have significantly higher fuel burn rates than typical tank-types, larger gas piping may be required. The same is true for electric tankless water heaters, where a whole-house unit may require larger wiring and often additional (multiple) circuit breakers. Consequently, large tankless water heaters, both gas and electric, can in some cases require a service entrance upgrade. Notably, diversified electrical demand for large numbers of electric tankless water heaters is not much different (generally a little lower) than tank types, because of the lower number of tankless water heaters that are on at any point in time compared to tank types. However, as number of users on an electrical line decreases, demand diversity decreases, which can result in increased electrical demand compared to tank types as the number of users on the line decreases to fairly few. The number that "few" represents varies with size of the tankless units.

Sizing Instantaneous and Semi-Instantaneous Water Heaters

The methods for sizing storage water-heating equipment should not be used for instantaneous and semi-instantaneous heaters. The following is based on the Hunter (1941) method for sizing hot- and cold-water piping, with diversity factors applied for hot water and various building types.

Fixture units (Table 16) are assigned to each fixture using hot water and totalled. Maximum hot-water demand is obtained from

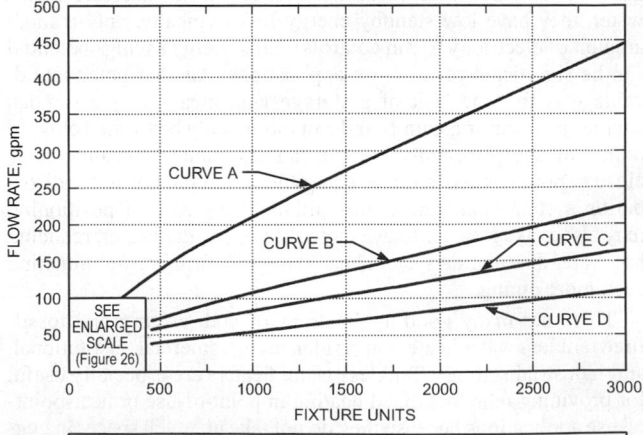

Fig. 25   Modified Hunter Curve for Calculating Hot-Water Flow Rate

(Data predate modern low-flow fixtures and appliances)

Figures 25 or 26 by matching total fixture units to the curve for the type of building. Special consideration should be given to applications involving periodic use of shower banks, process equipment, laundry machines, etc., as may occur in field houses, gymnasiums, factories, hospitals, and other facilities. Because these applications could have all equipment on at the same time, total hot-water capacity should be determined and added to the maximum hot-water demand from the modified Hunter curves. Often, the temperature of hot water arriving at fixtures is higher than is needed, and hot and cold water are mixed together at the fixture to provide the desired temperature. Equation (24), derived from a simple energy balance on mixing hot and cold water, shows the ratio of hot-water flow to desired end-use flow for any given hot, cold, and mixed end-use temperatures.

$$\text{Hot-water flow rate} = \frac{(\text{Mixed-temperature flow rate})(T_{mixed} - T_{cold})}{(T_{hot} - T_{cold})} \quad (24)$$

Once the actual hot-water flow rate is known, the heater can then be selected for the total demand and total temperature rise required. For critical applications such as hospitals, multiple heaters with 100% reserve capacity are recommended. Consider multiple heaters for buildings in which continuity of service is important. The minimum recommended size for semi-instantaneous heaters is 10 gpm, except for restaurants, for which it is 15 gpm. When system flow is not easily determined, the heater may be sized for full flow of the piping system at a maximum speed of 600 fpm. Heaters with low flows must be sized carefully, and care should be taken in the estimation of diversity factors. Unusual hot-water requirements should be analyzed to determine whether additional capacity is required. One example is a dormitory in a military school, where all showers and lavatories are used simultaneously when students return from a drill. In this case, the heater and piping should be sized for full system flow.

Whereas the fixture count method bases heater size of the diversified system on hot-water flow, hot-water piping should be sized for full flow to the fixtures. Recirculating hot-water systems are adaptable to instantaneous heaters.

To make preliminary estimates of hot-water demand when the fixture count is not known, use Table 17 with Figure 25 or Figure 26. The result is usually higher than the demand determined from the actual fixture count. Actual heater size should be determined from Table 16. Hot-water consumption over time can be assumed to be

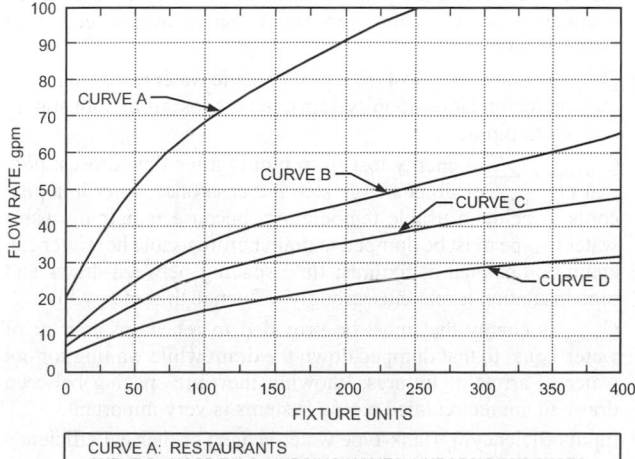

Fig. 26   Enlarged Section of Figure 25
(Modified Hunter Curve)
(Data predate modern low-flow fixtures and appliances)

the same as that in the section on Hot-Water Requirements and Storage Equipment Sizing.

Example 10. A 600-student elementary school has the following fixture count: 60 public lavatories, 6 service sinks, 4 kitchen sinks, 6 showers, and 1 dishwasher at 8 gpm. Determine the hot-water flow rate for sizing a semi-instantaneous heater based on the following:

a. Estimated number of fixture units

b. Actual fixture count

Solution:

a. Use Table 17 to find the estimated fixture count: 600 students × 0.3 fixture units per student = 180 fixture units. As showers are not included, Table 16 shows 1.5 fixture units per shower × 6 showers = 9 additional fixture units. The basic flow is determined from curve D of Figure 26, which shows that the total flow for 189 fixture units is 23 gpm.

b. To size the unit based on actual fixture count and Table 16, the calculation is as follows:

60	public lavatories	× 1.0 FU =	60 FU
6	service sinks	× 2.5 FU =	15 FU
4	kitchen sinks	× 0.75 FU =	3 FU
6	showers	× 1.5 FU =	9 FU
	Subtotal		87 FU

At 87 fixture units, curve D of Figure 26 shows 16 gpm, to which must be added the dishwasher requirement of 8 gpm. Thus, the total flow is 24 gpm.

Comparing the flow based on actual fixture count to that obtained from the preliminary estimate shows the preliminary estimate to be slightly lower in this case. It is possible that the preliminary estimate could have been as much as twice the final fixture count. To prevent oversizing of equipment, use the actual fixture count method to select the unit.

Sizing Refrigerant-Based Water Heaters

Refrigerant-based water heaters such as heat pump water heaters and refrigeration heat reclaim systems cannot be sized like conventional systems to meet peak loads. The seasonal and instantaneous efficiency and output of these systems vary greatly with operating conditions. Computer software that performs detailed performance simulations taking these factors into account should be used for sizing and analysis. The capacities of these systems and any related supplemental water-heating equipment should be selected to achieve a high average daily run time (typically 12 to 24 h) and the lowest combination of operating and equipment cost. Heat pump water heater installations often benefit from having a greater ratio of storage tank capacity per unit of heating capacity than for conventional water-heating equipment. Because the heat input efficiency of heat pump water heaters is dramatically higher (200 to 300%) than conventional resistance and fossil-fuel-fired equipment, energy penalties from storage tank heat loss are substantially lower for heat pump water heater systems compared to conventional systems. Larger storage capacity allows use of smaller heat pumps, often reducing total system costs. For heat pump water heaters, adequacy of the heat source and potential effect of the cooling output must be addressed.

Table 17   Preliminary Hot-Water Demand Estimate

Type of Building	Fixture Units
Hospital or nursing home	2.50 per bed
Hotel or motel	2.50 per room
Office building	0.15 per person
Elementary school	0.30 per student*
Junior and senior high school	0.30 per student*
Apartment house	3.00 per apartment

*Plus shower load (in fixture units).

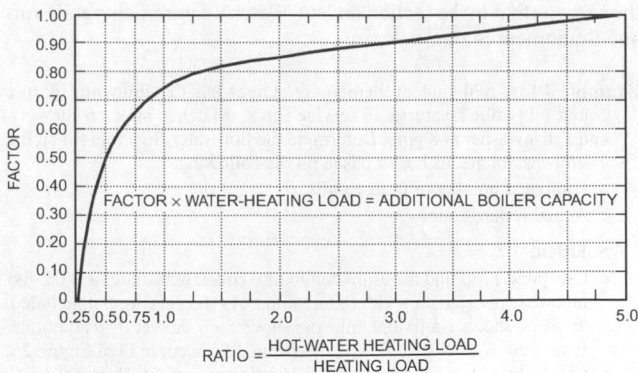

Fig. 27   **Sizing Factor for Combination Heating and Water-Heating Boilers**

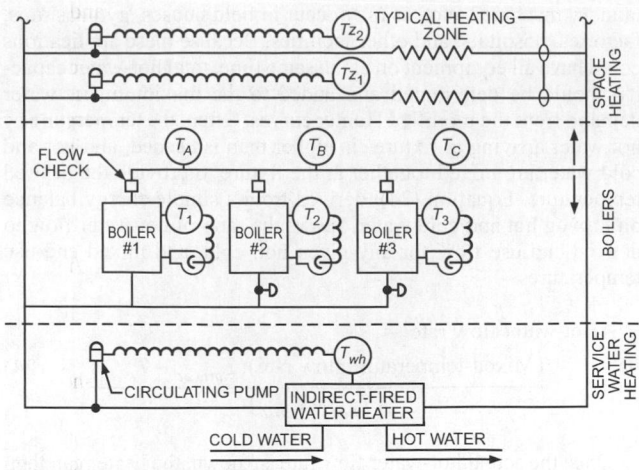

Fig. 28   **Typical Modular Boiler for Combined Space and Water Heating**

BOILERS FOR INDIRECT WATER HEATING

When service water is heated indirectly by a space heating boiler, Figure 27 may be used to determine the additional boiler capacity required to meet the recovery demands of the domestic water-heating load. Indirect heaters include immersion coils in boilers as well as heat exchangers with space-heating media.

Because the boiler capacity must meet not only the water supply requirement but also the space heating loads, Figure 27 indicates the reduction of additional heat supply for water heating if the ratio of water-heating load to space-heating load is low. This reduction is possible because

- Maximum space-heating requirements do not occur at the time of day when the maximum peak hot-water demands occur.
- Space-heating requirements are based on the lowest outdoor design temperature, which may occur for only a few days of the total heating season.
- An additional heat supply or boiler capacity to compensate for pickup and radiation losses is usual. The pickup load cannot occur at the same time as the peak hot-water demand because the building must be brought to a comfortable temperature before the occupants use hot water.

The factor obtained from Figure 27 is multiplied by the peak water-heating load to obtain the additional boiler output capacity required.

For reduced standby losses in summer and improved efficiency in winter, step-fired modular boilers may be used. Units not in operation cool down and reduce or eliminate jacket losses. Heated boiler water should not pass through an idle boiler. Figure 28 shows a typical modular boiler combination space- and water-heating arrangement.

Typical Control Sequence

1. Any control zone or indirectly fired water heater thermostat (e.g., T_{z_1} or T_{wh} in Figure 28) starts its circulating pump and supplies power to boiler no. 1 control circuit.
2. If T_1 is not satisfied, burner is turned on, boiler cycles as long as any circulating pump is on.
3. If after 5 min T_A is not satisfied, V_1 opens and boiler no. 2 comes on line.
4. If after 5 min T_B is not satisfied, V_2 opens and boiler no. 3 comes on line.
5. If T_C is satisfied and two boilers or fewer are firing for a minimum of 10 min, V_2 closes.
6. If T_B is satisfied and only one boiler is firing for a minimum of 10 min, V_1 closes.
7. If all circulating pumps are off, boiler no. 1 shuts down.

ASHRAE/IES *Standards* 90.1 and 90.2 discuss combination service water-heating/space-heating boilers and establish restrictions on their use. The ASHRAE/IES *Standard* 100 section on Service Water Heating also has information on this subject.

WATER-HEATING ENERGY USE

Energy use in water-heating systems includes the following factors, not all of which apply in a given type of system (Hiller 2006c):

- Q_{water} is energy content in water actually used, relative to entering cold-water temperature.
- $Q_{tank\,loss}$ is standby heat loss from water heater storage tank; it is proportional to time and temperature difference between water in tank and surroundings.
- $Q_{cycling\,loss}$ is energy loss from on/off cycling of heat input device, where energy invested in mass of heating device (e.g., heat exchanger) and water in it is lost to surroundings after device turns off; loss is proportional to number of heating cycles (e.g., in a tankless instantaneous water heater). Some fossil-fuel-fired tankless water heaters have pre- and/or postfiring combustion air blower operation to purge combustion products from combustion chamber, which can cause very rapid loss of invested energy in heat exchanger.
- Q_{piping} is heat energy lost from piping while water is flowing; note that, on recirculation-loop systems, heat is lost from both supply and return piping.
- $Q_{cooldown}$ is heat energy lost from piping after flow ceases; note that $Q_{cooldown}$ exhibits a large step increase once water in a pipe cools to below a usable temperature, because remaining warm water in pipe must be dumped to drain before usable hot water can again be obtained at fixtures; time spacing between draws and pipe insulation levels thus strongly influence this energy loss.
- Q_{dump} is energy that must be provided to reheat an amount of water equal to that dumped down the drain while waiting for hot water to arrive at fixtures; knowing the time spacing between draws in nonrecirculated piping systems is very important.
- Input efficiency η_i (tank-type water heater) or thermal efficiency η_t (tankless water heater or heating device external to tank) of heating device must be considered when calculating total water-heating system energy use.
- $Q_{circulating\,pump}$ is energy used to move water within system, if done with pumps. There are often multiple circulating pumps in system (e.g., to circulate water from storage tanks to heating devices, recirculation-loop pumps).

- $Q_{parasitics}$ is energy to operate fans, blowers, controls, and other devices.
- Q_{supply} is energy used to deliver potable water to system and force it through system. Includes pumping energy for well pumps or city water supply pumps, and water treatment system energy.
- $Q_{disposal}$ is energy used to treat and dispose of waste water, including pumping energy and other treatment system energy.

Total piping system energy use is thus

$$Q_{total} = Q_{water}/\eta_i + Q_{tank\,loss}/\eta_i + Q_{cycling\,loss} + Q_{piping}/\eta_i$$
$$+ Q_{cooldown}/\eta_i + Q_{dump}/\eta_i + Q_{circulating\,pump}$$
$$+ Q_{parasitics} + Q_{supply} + Q_{disposal}$$

Additional energy use terms may apply in some water-heating systems.

The following simple examples demonstrate how to compute water-heating system energy use for different system types and draw patterns.

Assumptions for Examples 11 to 14 include the following:

- Two fixtures 100 ft apart
- Six 5 min long draws per day of 1.0 gal/min, 105°F water at each fixture, spaced 3 min apart compared to 4 h apart
- Water heater output temperature of 120°F
- Tank-type fossil-fuel-fired water heater with input efficiency $\eta_i = 0.80$ and an energy factor of 0.59, yielding $UA_{tank} = 11.27$ Btu/h·°F, including energy input efficiency (note that tank heat loss rate is a function of energy factor rating, not tank size. However, for equal amounts of insulation, smaller tanks have higher energy factor rating)
- $T_{air} = 67.5$°F for both piping and tank
- $T_{cold} = 58$°F entering cold-water temperature
- Supply piping is 3/4 in. rigid copper with 1/2 in. thick foam insulation $(Mc_p)_{w,p,i} = 0.2542$ Btu/ft·°F, (from Table 1) pipe volume = 0.02514 gal/ft
- Return piping is 1/2 in. rigid copper with 1/2 in. thick foam insulation (RL system only)
- For simplicity, neglect short lengths of piping between fixture branch piping and main recirculation-loop piping (or tank if at location of fixture)
- For simplicity, neglect supply and disposal energy, recirculating pump energy, and other parasitics.

Example 11. Assume a continuously running hot-water recirculation-loop system with an allowed loop temperature drop to the farthest fixture of 5°F, and assuming one fixture is near the water heater. Note that because this is a continuously running recirculation-loop system, time spacing between draws is unimportant because the supply and return piping are always hot.

First, compute the recirculation loop flow rate needed to prevent temperature dropping below 115°F, using Equations (1) to (4) and (9).

$$\Delta T_{lm} = \frac{(120 - 67.5°F) - (115 - 67.5°F)}{\ln[(120 - 67.5°F)/(115 - 67.5°F)]} = 49.96°F$$

$UA_{flowing\,supply} = 0.25$ Btu/h·ft·°F (from Table 1)

$$Q_{piping\,supply} = (0.25\text{ Btu/h·ft·°F})(100\text{ ft})(49.96°F)/0.8$$
$$= (1561\text{ Btu/h})(24\text{ h/day})$$
$$= 37,470\text{ Btu/day}$$

$$m_{circulating\,pump} = \frac{(1561\text{ Btu/h})(0.8)}{(60\text{ min/h})(8.25\text{ lb}_m/\text{gal})(1\text{ Btu/lb}_m\cdot°F)(5°F)}$$
$$= 0.50\text{ gal/min}$$

$UA_{flowing\,return} = 0.20$ Btu/h·ft·°F (from Table 1)

$$T_{hot\,return\,out} = 67.5°F + (115 - 67.5°F)$$
$$\times e^{-\left\{\frac{(0.20\text{ Btu/h·ft·°F})(100\text{ ft})}{(0.5\text{ gal/min})(60\text{ min/h})(8.25\text{ lb}_m/\text{gal})(1\text{ Btu/lb}_m\cdot°F)}\right\}}$$
$$= 111.31°F$$

$$Q_{piping\,return} = (0.50\text{ gal/min})(60\text{ min/h})(8.25\text{ lb}_m/\text{gal})$$
$$\times (1\text{ Btu/lb}_m\cdot°F)(115 - 111.31°F)/0.8$$
$$= (1142\text{ Btu/h})(24\text{ h/day})$$
$$= 27,408\text{ Btu/day}$$

Next, determine the amount of hot water mixed with cold water to deliver the 105°F fixture delivery temperature, from Equation (24). For the fixture at the water heater,

$$m_{hot\,near} = (1\text{ gal/min})(105 - 58°F)/(120 - 58°F) = 0.758\text{ gal/min}$$

And for the far fixture,

$$m_{hot\,far} = (1\text{ gal/min})(105 - 58°F)/(115 - 58°F) = 0.825\text{ gal/min}$$

Consequently,

$$Q_{water\,near} = (0.758\text{ gal/min})(8.33\text{ lb}_m/\text{gal})(1\text{ Btu/lb}_m\cdot°F)$$
$$\times (120 - 58°F)(5\text{ min/draw})(6\text{ draws/day})/0.8$$
$$= 14,680\text{ Btu/day}$$

$$Q_{water\,far} = (0.825\text{ gal/min})(8.33\text{ lb}_m/\text{gal})(1\text{ Btu/lb}_m\cdot°F)$$
$$\times (115 - 58°F)(5\text{ min/draw})(6\text{ draws/day})/0.8$$
$$= 14,689\text{ Btu/day}$$

This is the same as for the near fixture, as it should be, because piping heat loss is separately computed.

$$Q_{tank\,heat\,loss} = (11.27\text{ Btu/h·°F})(120 - 67.5°F)(24\text{ h/day})$$
$$= 14,200\text{ Btu/day}$$

Thus,

$$Q_{total\,RL\,system} = 37,470 + 27,408 + 14,680 + 14,689 + 14,200$$
$$= 108,447\text{ Btu/day}$$

With the recirculation system, energy use is the same regardless of draw spacing.

Example 12. Assume a nonrecirculated piping system, one fixture at water heater, draws 3 min and 4 h apart.

First, determine the steady-state delivery temperature at the far fixture, and the actual hot-water flow rate to that fixture. This requires iteration: guessing an initial piping outlet temperature, calculating an estimated hot-water flow rate using Equation (24), and then calculating a new piping outlet temperature based on the calculated flow rate.

Guess $T_{hot\,out\,1} = 120$°F. Then,

$$m_{hot\,1} = \frac{(1.0\text{ gal/min})(105 - 58°F)}{(120 - 58°F)}$$
$$= 0.758\text{ gal/min [from Equation (24)]}$$

$T_{hot\,out\,2} = 116.6$°F [from Equation (2), where

$$UA_{flowing} = 0.25\text{ Btu/h·ft·°F, }L = 100\text{ ft}$$

$$m_{hot\,2} = \frac{(1.0\text{ gal/min})(105 - 58°F)}{(116.6 - 58°F)} = 0.80\text{ gal/min}$$

$$T_{hot\,out\,3} = 116.8°F$$

$$m_{hot\,3} = \frac{(1.0\text{ gal/min})(105 - 58°F)}{(116.8 - 58°F)} = 0.7993\text{ gal/min}$$

$$T_{hot\,out\,4} = 116.79°F$$

Thus,

$$Q_{water\,far} + Q_{piping} = (0.7993\ \text{gal/min})(8.33\ \text{lb}_m/\text{gal})(1\ \text{Btu/lb}_m \cdot {}^\circ\text{F})$$

$$\times (120 - 58\,{}^\circ\text{F})(5\ \text{min/draw})(6\ \text{draws/day})/0.8$$

$$= 15{,}480\ \text{Btu/day}$$

Note that, in this computation, water energy and piping flowing heat loss energy are calculated together for simplicity.

$$Q_{water\,near} = \text{same as in Example 11} = 14{,}680\ \text{Btu/day}$$

Next, compute the pipe temperature at the end of both the 3 min and 4 h cooldown (cd) periods, accounting for the different draw spacing scenarios. For simplicity, base the heat loss calculations on an average pipe temperature of $(120 + 116.79\,{}^\circ\text{F})/2 = 118.4\,{}^\circ\text{F}$.

Using $UA_{zero\,flow} = 0.15\ \text{Btu/h} \cdot \text{ft} \cdot {}^\circ\text{F}$ from Table 1, and Equation (8),

$$T_{pipe\,3\,min} = 67.5\,{}^\circ\text{F} + (118.4 - 67.5\,{}^\circ\text{F})$$

$$\times\, e^{-\left\{ \dfrac{(0.15\ \text{Btu/h} \cdot \text{ft} \cdot {}^\circ\text{F})(3\ \text{min})}{(60\ \text{min/h})(0.2542\ \text{Btu/ft} \cdot {}^\circ\text{F})} \right\}}$$

$$= 116.92\,{}^\circ\text{F}$$

and

$$T_{pipe\,4\,h} = 72.3\,{}^\circ\text{F}$$

The pipe does not cool below a usable temperature with the 3 min draw spacing, but it does with the 4 h draw spacing. This means that, for the 3 min draw spacing, there are five draws with small amounts of piping cooldown between draws plus one complete cooldown for the last draw of the day, whereas for the 4 h draw spacing, there are six complete cooldowns that result in dumping water in the pipe to drain at the next draw. Because pipe length to the fixture at the water heater is essentially zero under the assumptions here, only draws at the far fixture result in piping energy loses.

From Equation (5),

$$Q_{cd\,3\,min} = (0.2542\ \text{Btu/ft} \cdot {}^\circ\text{F})(100\ \text{ft})(118.4 - 116.92\,{}^\circ\text{F})$$

$$\times 5\ \text{cd/day}/0.8 + 1\ \text{cd (lumped into } Q_{dump})$$

$$= 235\ \text{Btu/day}$$

To estimate Q_{dump} and the amount of water waste, assume an AF/PV ratio of 1.5. Thus, each time the pipe cools below a usable temperature, $(1.5)(0.02514\ \text{gal/ft})(100\ \text{ft}) = 3.77$ gal of water must be dumped to drain.

$$Q_{dump\,3\,min} = (1\ \text{dump/day})(3.77\ \text{gal/dump})(8.33\ \text{lb}_m/\text{gal})$$

$$\times (1\ \text{Btu/lb}_m \cdot {}^\circ\text{F})(120 - 58\,{}^\circ\text{F})/0.8$$

$$= 2434\ \text{Btu/day}$$

$$Q_{dump\,4\,h} = (6\ \text{dumps/day})(2434\ \text{Btu/dump}) = 14{,}604\ \text{Btu/day}$$

and

$$Q_{cd\,4\,h} = 0$$

because all cooldown energy is lumped into Q_{dump}.

$$Q_{tank\,heat\,loss} = 14{,}200\ \text{Btu/day, as in Example 11}$$

To simplify calculation of total water-heating system energy use, it is convenient to add the cooldown energy term computed as shown to the Q_{water} term calculated as if all hot water were delivered to the fixture at a constant flow rate and the steady-state temperature. In reality, the hot-water flow rate to the fixture varies during the initial part of a draw as the cooled but still usable water temperature increases to the steady-state value as flow progresses. The energy use thus computed will be mathematically correct either way.

$$Q_{total\,non\text{-}RL,\,3\,min\,spacing} = 15{,}480 + 14{,}680 + 235 + 2434 + 14{,}200$$

$$= 47{,}033\ \text{Btu/day}$$

$$Q_{total\,non\text{-}RL,\,4\,h\,spacing} = 15{,}480 + 14{,}680 + 0 + 14{,}604 + 14{,}200$$

$$= 58{,}968\ \text{Btu/day}$$

Note the large increase in energy use when draws are spaced far enough apart for the pipe to cool to below a usable temperature between draws. Also, the time spent waiting for hot water to arrive at the far fixture is

$$t_{wait} = (3.77\ \text{gal})/(1\ \text{gal/min}) = 3.77\ \text{min}$$

Example 13. Assume two full-sized water heaters, one at each fixture; no piping.

In this case, tank heat loss is doubled, but piping heat loss is eliminated.

$$Q_{water} = (2)(14{,}680\ \text{Btu/day}) = 29{,}360\ \text{Btu/day}$$

$$Q_{tank\,heat\,loss} = (2)(14{,}200\ \text{Btu/day}) = 28{,}400\ \text{Btu/day}$$

$$Q_{total\,2\text{-}tank} = 29{,}360 + 28{,}400 = 57{,}760\ \text{Btu/day}$$

Draw spacing is irrelevant to the two-tank system because there is no piping.

Example 14. Assume two smaller water heaters, one at each fixture; no piping.

When two separate water heaters are used, each can be smaller than if one water heater were used. Assuming a smaller tank-type fossil-fuel-fired water heater with input efficiency $\eta_i = 0.80$ and an energy factor of 0.61, yielding $UA_{tank} = 9.86\ \text{Btu/h} \cdot {}^\circ\text{F}$, including energy input efficiency,

$$Q_{water} = (2)(14{,}680\ \text{Btu/day}) = 29{,}360\ \text{Btu/day}$$

$$Q_{tank\,heat\,loss} = (2)(9.86\ \text{Btu/h} \cdot {}^\circ\text{F})(120 - 67.5\,{}^\circ\text{F})(24\ \text{h/day})$$

$$= 24{,}847\ \text{Btu/day}$$

$$Q_{total\,2\text{-}tank} = 29{,}360 + 24{,}847 = 54{,}207\ \text{Btu/day}$$

Again, draw spacing is irrelevant to the two-tank system because there is no piping.

Table 18 compares water and energy use of Examples 11 to 14, and shows that the continuously running recirculation-loop system uses substantially more energy than the other approaches (on the order of twice as much). This is not uncommon. Also note that, in these examples, the two-tank approach saves both water waste and energy. The multiple-water-heater approach has at worst only a small negative energy effect if done properly, and under real water draw scenarios usually uses less energy than other options. This is why multiple-water-heater design options should always be considered.

Table 18   Results Comparisons for Examples 11 to 14

System Type	3 min Draw Spacing			4 h Draw Spacing		
	Energy Use, Btu/day	Energy Use Compared to One Tank, %	Water Waste, gal/day	Energy Use, Btu/day	Energy Use Compared to One Tank, %	Water Waste, gal/day
Recirculation loop	108,447	231	0	108,447	184	0
One-tank	47,033	100	3.8	58,968	100	22.6
Two-tank (large)	57,760	123	0	57,760	98	0
Two-tank (small)	54,207	115	0	54,207	92	0

In some cases, multiple-water-heater systems can have lower first costs than alternatives. Note that multiple-water-heater systems can use different types of water heaters for different parts of the system: fossil-fuel-fired or heat pump water heaters can be used to serve larger loads, whereas electric resistance water heaters may be preferred for serving smaller loads. In some cases, space limitations, life and maintenance issues, and other factors may make multiple-water-heater systems unattractive.

Both simplified and detailed computer models (Hiller 1992, 2000) are available to help calculate water heater energy use. These are especially useful for analyzing the energy used by heat pump water heaters, where efficiency and heating capacity vary strongly with both source (e.g., air, water) and sink (water) temperature. Computer models are also under development to compute water and energy waste associated with hot-water distribution systems.

REFERENCES

ANSI/NSF. 2003. Commercial warewashing equipment. *Standard* 3-2003. American National Standards Institute, Washington, D.C. and NSF International, Ann Arbor, MI.

ASHRAE. 2004. Energy standard for buildings except low-rise residential buildings. ANSI/ASHRAE/IESNA *Standard* 90.1-2004.

ASHRAE. 2004. Energy-efficient design of low-rise residential buildings. ANSI/ASHRAE/IESNA *Standard* 90.2-2004.

ASHRAE. 1995. Energy conservation in existing buildings. ANSI/ASHRAE/IESNA *Standard* 100-1995.

ASHRAE. 2003. Methods of testing for rating commercial gas, electric, and oil water heaters. *Standard* 118.1-2003.

ASHRAE. 2006. Methods of testing for rating residential water heaters. *Standard* 118.2-2006.

ASHRAE. 1991. Methods of testing for rating combination space-heating and water-heating appliances. *Standard* 124-1991.

ASHRAE. 2000. Minimizing the risk of Legionellosis associated with building water systems. *Guideline* 12-2000.

ASME. 1998. *Boiler and pressure vessel code.* Section IV-98: Rules for construction of heating boilers; Section VIII D1-98: Rules for construction of pressure vessels. ASME International, New York.

Becker, B.R., W.H. Thrasher, and D.W. DeWerth. 1991. Comparison of collected and compiled existing data on service hot water use patterns in residential and commercial establishments. *ASHRAE Transactions* 97(2):231-239.

Ciesielski, C.A., M.J. Blaser, and W.L. Wang. 1984. Role of stagnation and obstruction of water flow in isolation of *Legionella pneumophila* from hospital plumbing. *Applied and Environmental Microbiology* (November):984-987.

DOE. 2001. *Final rule regarding test procedures and energy conservation standards for water heaters.* 10CFR430, Federal Register 55(201). U.S. Department of Energy.

DOE. 1998. *Uniform test method for measuring the energy consumption of water heaters.* 10CFR430, Subpart B, Appendix E. U.S. Department of Energy.

Dunn, T.Z., R.N. Spear, B.E. Twigg, and D. Williams. 1959. Water heating for commercial kitchens. *Air Conditioning, Heating and Ventilating* (May):70. Also published as a bulletin, *Enough hot water—Hot enough.* American Gas Association (1959).

EPRI. 1990. *Commercial heat pump water heaters applications handbook.* CU-6666. Electric Power Research Institute, Palo Alto, CA.

EPRI. 1992. *Commercial water heating applications handbook.* TR-100212. Electric Power Research Institute, Palo Alto, CA.

EPRI. 1994. *High-efficiency electric technology fact sheet: Commercial heat pump water heaters.* BR-103415. Electric Power Research Institute, Palo Alto, CA.

GAMA. (Continuous Maintenance). *Consumers' directory of certified efficiency ratings for heating and water heating equipment.* Gas Appliance Manufacturers Association, Arlington, VA. Available from http://www.gamanet.org.

Goldner, F.S. 1994a. Energy use and domestic hot water consumption: Final report—Phase 1. *Report* 94-19. New York State Energy Research and Development Authority, Albany, NY.

Goldner, F.S. 1994b. DHW system sizing criteria for multifamily buildings. *ASHRAE Transactions* 100(1):963-977.

Goldner, F.S. and D.C. Price. 1996. DHW modeling: System sizing and selection criteria, Phase 2. *Interim Project Research Report* 1. New York State Energy Research and Development Authority, Albany.

Hiller, C.C. 1992. *WATSIM® 1.0: Detailed water heating analysis code.* Applied Energy Technology Co., Davis, CA.

Hiller, C.C. 1998. New hot water consumption analysis and water-heating system sizing methodology. *ASHRAE Transactions* 104(1B):1864-1877.

Hiller, C.C. 2000. *WATSMPL® 2.0: Simplified water heating analysis code.* Applied Energy Technology Co., Davis, CA.

Hiller, C.C. 2005a. Comparing water heater vs. hot water distribution system energy losses. *ASHRAE Transactions* 111(2):407-418.

Hiller, C.C. 2005b. Hot water distribution system research—Phase I final report. California Energy Commission *Report* CEC-500-2005-161.

Hiller, C.C. 2005c. Rethinking school potable water heating systems. *ASHRAE Journal* 47(5):48-56.

Hiller, C.C. 2006a. Hot water distribution system piping time, water, and energy waste—Phase I test results. *ASHRAE Transactions* 112(1):415-425.

Hiller, C.C. 2006b. Hot water distribution system piping heat loss factors—Phase I test results. *ASHRAE Transactions* 112(1):436-446.

Hiller, C.C. 2008. Hot water distribution system piping heat loss factors, both in air and buried—Phase II test results. *ASHRAE Transactions* 114(2).

Hiller, C.C. 2009. *Hot water distribution system research—Phase III interim report.* California Energy Commission, October.

Hiller, C.C. and A.I. Lowenstein. 1996. Disaggregating residential hot water use. ASHRAE *Paper* AT-96-18-1.

Hiller, C.C. and A.I. Lowenstein. 1998. Disaggregating residential hot water use—Part II. ASHRAE *Paper* SF-98-31-2.

HUD. 1994. Minimum property standards for housing. *Directive* 4910.1. U.S. Department of Housing and Urban Development, Washington, D.C.

Hunter, R.B. 1941. Water distributing systems for buildings. National Bureau of Standards *Report* BMS 79.

Klein, G. 2004a. Hot-water distribution systems: Part 1. *Plumbing Systems & Design* (Mar/Apr):36-39.

Klein, G. 2004b. Hot-water distribution systems: Part 2. *Plumbing Systems & Design.* (May/June):16-18.

Klein, G. 2004c. Hot-water distribution systems: Part 3. *Plumbing Systems & Design.* (Sept/Oct):14-17.

Kweller, E.R. 1992. Derivation of the combined annual efficiency of space/water heaters in ASHRAE 124-1991. *ASHRAE Transactions* 98(1):665-675.

Lutz, J.E. 2005. Estimating energy and water losses in residential hot water distribution systems. *ASHRAE Transactions* 111(2):418-422.

Manian, V.S. and W. Chackeris. 1974. Off peak domestic hot water systems for large apartment buildings. *ASHRAE Transactions* 80(1):147-165.

Meier, A. 1985. Low-flow showerheads, family strife, and cold feet. *Home Energy.*

Moritz, A.R. and R.C. Henriques. 1947. Studies of thermal injury: The relative importance of time and surface temperature in the causation of cutaneous burns. *American Journal of Pathology* 23:695.

NFPA. 2006. Installation of oil-burning equipment. *Standard* 31-2006. National Fire Protection Association, Quincy, MA.

NFPA. 2006. National fuel gas code. *Standard* 54/ANSI *Standard* Z223.1. National Fire Protection Association, Quincy, MA.

NSF. 2000. Water heaters, hot water supply boilers, and heat recovery equipment. NSF *Standard* 5-2000. NSF International, Ann Arbor, MI.

Perlman, M. and B. Mills. 1985. Development of residential hot water use patterns. *ASHRAE Transactions* 91(2A):657-679.

Pietsch, J.A. and S.G. Talbert. 1989. Equipment sizing procedures for combination space-heating/water-heating systems. *ASHRAE Transactions* 95(2):250-258.

Pietsch, J.A., S.G. Talbert, and S.H. Stanbouly. 1994. Annual cycling characteristics of components in gas-fired combination space/water systems. *ASHRAE Transactions* 100(1):923-934.

Rohsenow, W.M. and H.Y. Choi. 1961. *Heat, mass and momentum transfer.* Prentice Hall, New York.

Subherwal, B.R. 1986. Combination water-heating/space-heating appliance performance. *ASHRAE Transactions* 92(2B):415-432.

Taborek, J., T. Aoku, R.B. Ritter, J.W. Paeln, and J.G. Knudsen. 1972. Fouling—The major unresolved problem in heat transfer. *Chemical Engineering Progress* 68(2):59.

Talbert, S.G., G.H. Stickford, D.C. Newman, and W.N. Stiegelmeyer. 1986. The effect of hard water scale buildup and water treatment on residential water heater performance. *ASHRAE Transactions* 92(2B):433-447.

Talbert, S.G., J.G. Murray, R.A. Borgeson, V.P. Kam, and J.A. Pietsch. 1992. Operating characteristics and annual efficiencies of combination space/water-heating systems. *ASHRAE Transactions* 98(1):655-664.

Thrasher, W.H. and D.W. DeWerth. 1994. New hot-water use data for five commercial buildings (RP-600). *ASHRAE Transactions* 100(1):935-947.

Werden, R.G. and L.G. Spielvogel. 1969a. Sizing of service water heating equipment in commercial and institutional buildings, Part I. *ASHRAE Transactions* 75(I):81.

Werden, R.G. and L.G. Spielvogel. 1969b. Sizing of service water heating equipment in commercial and institutional buildings, Part II. *ASHRAE Transactions* 75(II):181.

BIBLIOGRAPHY

AGA. Comprehensive on commercial and industrial water heating. Catalog No. R-00980. American Gas Association, Washington, D.C.

AGA. Sizing and equipment data for specifying swimming pool heaters. Catalog No. R-00995. American Gas Association, Washington, D.C.

AGA. 1962. Water heating application in coin operated laundries. Catalog No. C-10540. American Gas Association, Washington, D.C.

AGA. 1965. *Gas engineers handbook.* American Gas Association, Washington, D.C.

ANSI/AGA. 2004. Gas water heaters, vol. I: Storage water heaters with input ratings of 75,000 Btu [22 kW] per hour or less. *Standard* Z21.10.1-2004. American National Standards Institute and American Gas Association, Washington, D.C.

ANSI/AGA. 2004. Gas water heaters, vol. III: Storage, with input ratings above 75,000 Btu [22 kW] per hour, circulating, and instantaneous water heaters. *Standard* Z21.10.3-2004. American National Standards Institute and American Gas Association, Washington, D.C.

ANSI. 2000. Relief valves for hot water supply systems. *Standard* Z21.22-2000. American National Standards Institute, Washington, D.C.

ANSI. 2001. Gas-fired pool heaters. *Standard* Z21.56-2001. American National Standards Institute, Washington, D.C.

ANSI. 2000. Automatic gas shutoff devices for hot water supply systems. ANSI *Standard* Z21.87-2000. American National Standards Institute, Washington, D.C.

Brooks, F.A. *Use of solar energy for heating water.* Smithsonian Institution, Washington, D.C.

Carpenter, S.C. and J.P. Kokko. 1988. Estimating hot water use in existing commercial buildings. *ASHRAE Transactions* 94(2):3-12

Coleman, J.J. 1974. Waste water heat reclamation. *ASHRAE Transactions* 80(2):370.

EPRI. 1992. *WATSMPL® 1.0: Detailed water heating simulation model user's manual.* TR-101702. Electric Power Research Institute, Palo Alto, CA.

EPRI. 1993. *HOTCALC 2.0: Commercial water heating performance simulation tool,* v. 2.0. SW-100210-R1. Electric Power Research Institute, Palo Alto, CA.

GRI. 1993. TANK computer program user's manual with diskettes. GRI-93/0186 *Topical Report,* available only to licensees. Gas Research Institute.

Hebrank, E.F. 1956. Investigation of the performance of automatic storage-type gas and electric domestic water heaters. *Engineering Experiment Bulletin* 436. University of Illinois.

Hiller, C.C. 2008. Hot water distribution system piping time, water, and energy waste—Phase II test results. *ASHRAE Transactions* 114(2).

Hiller, C.C. 2009. *Hot water distribution system research—Phase III interim report.* California Energy Commission, October.

Hiller, C.C. and J. Miller. 2002. Field test comparison of a potable hot water recirculation-loop system vs. point-of-use electric resistance water heaters in a high school. EPRI *Report* 1007022.

Hiller, C.C., J. Miller, and D. Dinse. 2002. Field test comparison of a potable hot water recirculation-loop system vs. point-of-use electric resistance water heaters in a high school. *ASHRAE Transactions* 108(2):771-779.

Jones, P.G. 1982. The consumption of hot water in commercial building. *Building Services Engineering, Research and Technology* 3:95-109.

Olivares, T.C. 1987. Hot water system design for multi-residential buildings. *Report* 87-239-K. Ontario Hydro Research Division.

Schultz, W.W. and V.W. Goldschmidt. 1978. Effect of distribution lines on stand-by loss of service water heater. *ASHRAE Transactions* 84(1):256-265.

Smith, F.T. 1965. Sizing guide for gas water heaters for in-ground swimming pools. Catalog No. R-00999. American Gas Association, Cleveland, OH.

UL. 1996. Household electric storage tank water heaters. UL *Standard* 174-1996. Underwriters Laboratories, Northbrook, IL.

UL. 1995. Oil-fired unit heaters. UL *Standard* 731-1995. Underwriters Laboratories, Northbrook, IL.

UL. 2001. Electric water heaters for pools and tubs. UL *Standard* 1261-2001. Underwriters Laboratories, Northbrook, IL.

UL. 1995. Electric booster and commercial storage tank water heaters. UL *Standard* 1453-1995. Underwriters Laboratories, Northbrook, IL.

Vine, E., R. Diamond, and R. Szydlowski. 1987. Domestic hot water consumption in four low income apartment buildings. *Energy* 12(6).

Wetherington, T.I., Jr. 1975. Heat recovery water heating. *Building Systems Design* (December/January).

CHAPTER 51

SNOW MELTING AND FREEZE PROTECTION

THE practicality of melting snow or ice by supplying heat to the exposed surface has been demonstrated in many installations, including sidewalks, roadways, ramps, bridges, access ramps, and parking spaces for the handicapped, and runways. Melting eliminates the need for snow removal by chemical means, provides greater safety for pedestrians and vehicles, and reduces the labor and cost of slush removal. Other advantages include eliminating piled snow, reducing liability, and reducing health risks of manual and mechanized shoveling.

This chapter covers three types of snow-melting and freeze protection systems:

1. Hot fluid circulated in slab-embedded pipes (**hydronic**)
2. Embedded **electric** heater cables or wire
3. Overhead high-intensity **infrared** radiant heating

Detailed information about slab heating can be found in Chapter 6 of the 2008 *ASHRAE Handbook—HVAC Systems and Equipment*. More information about infrared heating can be found in Chapter 15 of the same volume.

Components of the system design include (1) heat requirement, (2) slab design, (3) control, and (4) hydronic or electric system design.

SNOW-MELTING HEAT FLUX REQUIREMENT

The heat required for snow melting depends on five atmospheric factors: (1) rate of snowfall, (2) snowfall-coincident air dry-bulb temperature, (3) humidity, (4) wind speed near the heated surface, and (5) apparent sky temperature. The dimensions of the snow-melting slab affect heat and mass transfer rates at the surface. Other factors such as back and edge heat losses must be considered in the complete design.

Heat Balance

The processes that establish the heat requirement at the snow-melting surface can be described by terms in the following equation, which is the steady-state energy balance for required total heat flux (heat flow rate per unit surface area) q_o at the upper surface of a snow-melting slab during snowfall.

$$q_o = q_s + q_m + A_r(q_h + q_e) \quad (1)$$

where

q_o = heat flux required at snow-melting surface, Btu/h·ft^2
q_s = sensible heat flux, Btu/h·ft^2
q_m = latent heat flux, Btu/h·ft^2
A_r = snow-free area ratio, dimensionless
q_h = convective and radiative heat flux from snow-free surface, Btu/h·ft^2
q_e = heat flux of evaporation, Btu/h·ft^2

The preparation of this chapter is assigned to TC 6.5, Radiant Heating and Cooling.

Sensible and Latent Heat Fluxes. The sensible heat flux q_s is the heat flux required to raise the temperature of snow falling on the slab to the melting temperature plus, after the snow has melted, to raise the temperature of the liquid to the assigned temperature t_f of the liquid film. The snow is assumed to fall at air temperature t_a. The latent heat flux q_m is the heat flux required to melt the snow. Under steady-state conditions, both q_s and q_m are directly proportional to the snowfall rate s.

Snow-Free Area Ratio. Sensible and latent (melting) heat fluxes occur on the entire slab during snowfall. On the other hand, heat and mass transfer at the slab surface depend on whether there is a snow layer on the surface. Any snow accumulation on the slab acts to partially insulate the surface from heat losses and evaporation. The insulating effect of partial snow cover can be large. Because snow may cover a portion of the slab area, it is convenient to think of the insulating effect in terms of an effective or equivalent snow-covered area A_s, which is perfectly insulated and from which no evaporation and heat transfer occurs. The balance is then considered to be the equivalent snow-free area A_f. This area is assumed to be completely covered with a thin liquid film; therefore, both heat and mass transfer occur at the maximum rates for the existing environmental conditions. It is convenient to define a dimensionless **snow-free area ratio** A_r:

$$A_r = \frac{A_f}{A_t} \quad (2)$$

where

A_f = equivalent snow-free area, ft^2
A_s = equivalent snow-covered area, ft^2
$A_t = A_f + A_s$ = total area, ft^2

Therefore,

$$0 \leq A_r \leq 1$$

To satisfy $A_r = 1$, the system must melt snow rapidly enough that no accumulation occurs. For $A_r = 0$, the surface is covered with snow of sufficient thickness to prevent heat and evaporation losses. Practical snow-melting systems operate between these limits. Earlier studies indicate that sufficient snow-melting system design information is obtained by considering three values of the free area ratio: 0, 0.5, and 1.0 (Chapman 1952).

Heat Flux because of Surface Convection, Radiation, and Evaporation. Using the snow-free area ratio, appropriate heat and mass transfer relations can be written for the snow-free fraction of the slab A_r. These appear as the third and fourth terms on the right-hand side of Equation (1). On the snow-free surface, maintained at film temperature t_f, heat is transferred to the surroundings and mass is transferred from the evaporating liquid film. Heat flux q_h includes convective losses to the ambient air at temperature t_a and radiative losses to the surroundings, which are at mean radiant temperature T_{MR}. The convection heat transfer coefficient is a function of wind speed and a characteristic dimension of the snow-melting surface. This heat transfer coefficient is also a function of the thermodynamic properties of the air, which vary slightly over the temperature range for various snowfall events. The mean radiant temperature depends on air temperature, relative humidity, cloudiness, cloud altitude, and whether snow is falling.

The heat flux q_e from surface film evaporation is equal to the evaporation rate multiplied by the heat of vaporization. The evaporation rate is driven by the difference in vapor pressure between the wet surface of the snow-melting slab and the ambient air. The evaporation rate is a function of wind speed, a characteristic dimension of the slab, and the thermodynamic properties of the ambient air.

Heat Flux Equations

Sensible Heat Flux. The sensible heat flux q_s is given by the following equation:

$$q_s = \rho_{water} s [c_{p,ice}(t_s - t_a) + c_{p,water}(t_f - t_s)]/c_1 \qquad (3)$$

where

$c_{p,ice}$ = specific heat of ice, Btu/lb·°F
$c_{p,water}$ = specific heat of water, Btu/lb·°F
s = snowfall rate water equivalent, in/h
t_a = ambient temperature coincident with snowfall, °F
t_f = liquid film temperature, °F
t_s = melting temperature, °F
ρ_{water} = density of water, lb/ft³
c_1 = 12 in/ft

The density of water, specific heat of ice, and specific heat of water are approximately constant over the temperature range of interest and are evaluated at 32°F. The ambient temperature and snowfall rate are available from weather data. The liquid film temperature is usually taken as 33°F.

Melting Heat Flux. The heat flux q_m required to melt the snow is given by the following equation:

$$q_m = \rho_{water} s h_{if}/c_1 \qquad (4)$$

where h_{if} = heat of fusion of snow, Btu/lb.

Convective and Radiative Heat Flux from a Snow-Free Surface. The corresponding heat flux q_h is given by the following equation:

$$q_h = h_c(t_s - t_a) + \sigma \varepsilon_s (T_f^4 - T_{MR}^4) \qquad (5)$$

where

h_c = convection heat transfer coefficient for turbulent flow, Btu/h·ft²·°F
T_f = liquid film temperature, °R
T_{MR} = mean radiant temperature of surroundings, °R
σ = Stefan-Boltzmann constant = 0.1712×10^{-8} Btu/h·ft²·°R⁴
ε_s = emittance of surface, dimensionless

The convection heat transfer coefficient over the slab on a plane horizontal surface is given by the following equations (Incropera and DeWitt 1996):

$$h_c = 0.037 \left(\frac{k_{air}}{L} \right) Re_L^{0.8} Pr^{1/3} \qquad (6)$$

where

k_{air} = thermal conductivity of air at t_a, Btu·ft/h·ft²·°F
L = characteristic length of slab in direction of wind, ft
Pr = Prandtl number for air, taken as Pr = 0.7
Re_L = Reynolds number based on characteristic length L

and

$$Re_L = \frac{VL}{\nu_{air}} c_2 \qquad (7)$$

where

V = design wind speed near slab surface, mph
ν_{air} = kinematic viscosity of air, ft²/h
c_2 = 5280 ft/mile

Without specific wind data for winter, the extreme wind data in Chapter 14 of the 2009 *ASHRAE Handbook—Fundamentals* may be used; however, it should be noted that these wind speeds may not correspond to actual measured data. If the snow-melting surface is not horizontal, the convection heat transfer coefficient might be different, but in many applications, this difference is negligible.

From Equations (6) and (7), it can be seen that the turbulent convection heat transfer coefficient is a function of $L^{-0.2}$. Because of this relationship, shorter snow-melting slabs have higher convective heat transfer coefficients than longer slabs. For design, the shortest dimension should be used (e.g., for a long, narrow driveway or sidewalk, use the width). A snow-melting slab length $L = 20$ ft is used in the heat transfer calculations that resulted in Tables 1, 2, and 3.

The **mean radiant temperature** T_{MR} in Equation (5) is the equivalent blackbody temperature of the surroundings of the snow-melting slab. Under snowfall conditions, the entire surroundings are approximately at the ambient air temperature (i.e., $T_{MR} = T_a$). When there is no snow precipitation (e.g., during idling and after snowfall operations for $A_r < 1$), the mean radiant temperature is approximated by the following equation:

$$T_{MR} = [T_{cloud}^4 F_{sc} + T_{sky\,clear}^4 (1 - F_{sc})]^{1/4} \qquad (8)$$

where

F_{sc} = fraction of radiation exchange that occurs between slab and clouds
T_{cloud} = temperature of clouds, °R
$T_{sky\,clear}$ = temperature of clear sky, °R

The equivalent blackbody temperature of a clear sky is primarily a function of the ambient air temperature and the water content of the atmosphere. An approximation for the clear sky temperature is given by the following equation, which is a curve fit of data in Ramsey et al. (1982):

$$T_{sky\,clear} = T_a - (1.99036 \times 10^3 - 7.562 T_a$$
$$+ 7.407 \times 10^{-3} T_a^2 - 56.325 \phi + 26.25 \phi^2) \qquad (9)$$

where

T_a = ambient temperature, °R
ϕ = relative humidity of air at elevation for which typical weather measurements are made, decimal

The cloud-covered portion of the sky is assumed to be at T_{cloud}. The height of the clouds may be assumed to be 10,000 ft. The temperature of the clouds at 10,000 ft is calculated by subtracting the product of the average lapse rate (rate of decrease of atmospheric temperature with height) and the altitude from the atmospheric temperature T_a. The average lapse rate, determined from the tables of U.S. Standard Atmospheres (COESA 1976), is 3.5°F per 1000 ft of elevation (Ramsey et al. 1982). Therefore, for clouds at 10,000 ft,

$$T_{cloud} = Ta - 35 \qquad (10)$$

Under most conditions, this method of approximating the temperature of the clouds provides an acceptable estimate. However, when the atmosphere contains a very high water content, the temperature calculated for a clear sky using Equation (9) may be warmer than the cloud temperature estimated using Equation (10). When that condition exists, T_{cloud} is set equal to the calculated clear sky temperature $T_{sky\,clear}$.

Evaporation Heat Flux. The heat flux q_e required to evaporate water from a wet surface is given by

$$q_e = \rho_{dry\ air} h_m (W_f - W_a) h_{fg} \qquad (11)$$

where

h_m = mass transfer coefficient, ft/h
W_a = humidity ratio of ambient air, lb_{vapor}/lb_{air}
W_f = humidity ratio of saturated air at film surface temperature, lb_{vapor}/lb_{air}
h_{fg} = heat of vaporization (enthalpy difference between saturated water vapor and saturated liquid water), Btu/lb
$\rho_{dry\ air}$ = density of dry air, lb/ft^3

Determination of the mass transfer coefficient is based on the analogy between heat transfer and mass transfer. Details of the analogy are given in Chapter 5 of the 2009 *ASHRAE Handbook—Fundamentals*. For external flow where mass transfer occurs at the convective surface and the water vapor component is dilute, the following equation relates the mass transfer coefficient h_m to the heat transfer coefficient h_c [Equation (6)]:

$$h_m = \left(\frac{Pr}{Sc}\right)^{2/3} \frac{h_c}{\rho_{dry\ air} c_{p,air}} \qquad (12)$$

where Sc = Schmidt number. In applying Equation (11), the values Pr = 0.7 and Sc = 0.6 were used to generate the values in Tables 1 through 4.

The humidity ratios both in the atmosphere and at the surface of the water film are calculated using the standard psychrometric relation given in the following equation (from Chapter 1 of the 2009 *ASHRAE Handbook—Fundamentals*):

$$W = 0.622 \left(\frac{p_v}{p - p_v}\right) \qquad (13)$$

where

p = atmospheric pressure, psi
p_v = partial pressure of water vapor, psi

The atmospheric pressure in Equation (13) is corrected for altitude using the following equation (Kuehn et al. 1998):

$$p = p_{std} \left(1 - \frac{Az}{T_o}\right)^{5.265} \qquad (14)$$

where

p_{std} = standard atmospheric pressure, psi
A = 0.00356°R/ft
z = altitude of the location above sea level, ft
T_o = 518.7°R

Altitudes of specific locations are found in Chapter 14 of the 2009 *ASHRAE Handbook—Fundamentals*.

The vapor pressure p_v for the calculation of W_a is equal to the saturation vapor pressure p_s at the dew-point temperature of the air. Saturated conditions exist at the water film surface. Therefore, the vapor pressure used in calculating W_f is the saturation pressure at the film temperature t_f. The saturation partial pressures of water vapor for temperatures above and below freezing are found in tables of the thermodynamic properties of water at saturation or can be calculated using appropriate equations. Both are presented in Chapter 1 of the 2009 *ASHRAE Handbook—Fundamentals*.

Heat Flux Calculations. Equations (1) to (14) can be used to determine the required heat fluxes of a snow-melting system. However, calculations must be made for coincident values of snowfall rate, wind speed, ambient temperature, and dew-point temperature (or another

other measure of humidity). By computing the heat flux for each snowfall hour over a period of several years, a frequency distribution of hourly heat fluxes can be developed. Annual averages or maximums for climatic factors should never be used in sizing a system because they are unlikely to coexist. Finally, it is critical to note that the preceding analysis only describes what is happening at the upper surface of the snow-melting surface. Edge losses and back losses have not been taken into account.

Example 1. During the snowfall that occurred during the 8 PM hour on December 26, 1985, in the Detroit metropolitan area, the following simultaneous conditions existed: air dry-bulb temperature = 17°F, dew-point temperature = 14°F, wind speed = 19.7 mph, and snowfall rate = 0.10 in. of liquid water equivalent per hour. Assuming L = 20 ft, Pr = 0.7, and Sc = 0.6, calculate the surface heat flux q_o for a snow-free area ratio of A_r = 1.0. The thermodynamic and transport properties used in the calculation are taken from Chapters 1 and 33 of the 2009 *ASHRAE Handbook—Fundamentals*. The emittance of the wet surface of the heated slab is 0.9.

Solution:

By Equation (3),

$$q_s = 62.4 \times \frac{0.10}{12} [0.49(32 - 17) + 1.0(33 - 32)] = 4.3 \text{ Btu/h·ft}^2$$

By Equation (4),

$$q_m = 62.4 \times \frac{0.10}{12} \times 143.3 = 74.5 \text{ Btu/h·ft}^2$$

By Equation (7),

$$Re_L = \frac{19.7 \times 20 \times 5280}{0.49} = 4.24 \times 10^6$$

By Equation (6),

$$h_c = 0.037 \left(\frac{0.0135}{20}\right) (4.24 \times 10^6)^{0.8} (0.7)^{1/3} = 4.44 \text{ Btu/h·ft}^2 \cdot °F$$

By Equation (5),

$$q_h = 4.44(33 - 17) + (0.1712 \times 10^{-8})(0.9)(493^4 - 477^4)$$
$$= 83.5 \text{ Btu/h·ft}^2$$

By Equation (12),

$$h_m = \left(\frac{0.7}{0.6}\right)^{2/3} \frac{4.44}{0.083 \times 0.24} = 247 \text{ ft/h}$$

Obtain the values of the saturation vapor pressures at the dew-point temperature 14°F and the film temperature 33°F from Table 3 in Chapter 1 of the 2009 *ASHRAE Handbook—Fundamentals*. Then, use Equation (13) to obtain W_a = 0.00160 lb_{vapor}/lb_{air} and W_f = 0.00393 lb_{vapor}/lb_{air}. By Equation (11),

$$q_e = 0.083 \times 247(0.00393 - 0.00160) \times 1075 = 51.3 \text{ Btu/h·ft}^2$$

By Equation (1),

$$q_o = 4.3 + 74.5 + 1.0(83.5 + 51.3)$$
$$= 214 \text{ Btu/h·ft}^2 \times 0.2931 = 62.7 \text{ W/ft}^2$$

Note that this is the heat flux needed at the snow-melting surface of the slab. Back and edge losses must be added as discussed in the section on Back and Edge Heat Losses.

Weather Data and Heat Flux Calculation Results

Table 1 shows frequencies of snow-melting loads for 46 cities in the United States (Ramsey et al. 1999). For the calculations, the temperature of the surface of the snow-melting slab was taken to be 33°F. Any time the ambient temperature was below 32°F and it was

Table 1   Frequencies of Snow-Melting Surface Heat Fluxes at Steady-State Conditions[a]

Location	Snowfall Hours per Year	Snow-Free Area Ratio, A_r	Heat Fluxes Not Exceeded During Indicated Percentage of Snowfall Hours from 1982 Through 1993, Btu/h·ft² [b]					
			75%	90%	95%	98%	99%	100%
Albany, NY	156	1	89	125	149	187	212	321
		0.5	60	86	110	138	170	276
		0	37	62	83	119	146	276
Albuquerque, NM	44	1	70	118	168	191	242	393
		0.5	51	81	96	117	156	229
		0	30	46	61	89	92	194
Amarillo, TX	64	1	113	150	168	212	228	318
		0.5	71	88	108	124	142	305
		0	24	46	62	89	115	292
Billings, MT	225	1	112	164	187	212	237	340
		0.5	64	89	102	116	128	179
		0	22	33	45	60	68	113
Bismarck, ND	158	1	151	199	231	275	307	477
		0.5	83	107	124	148	165	243
		0	16	30	39	60	73	180
Boise, ID	85	1	58	79	100	126	146	203
		0.5	38	52	66	80	89	164
		0	22	31	40	53	62	164
Boston, MA	112	1	96	137	165	202	229	365
		0.5	65	95	112	149	190	365
		0	37	75	93	121	172	365
Buffalo, NY	292	1	115	166	210	277	330	570
		0.5	68	97	127	164	188	389
		0	23	39	55	93	112	248
Burlington, VT	204	1	91	130	154	184	200	343
		0.5	58	78	92	113	128	343
		0	23	40	55	78	94	343
Cheyenne, WY	224	1	119	172	201	229	261	354
		0.5	70	97	111	132	149	288
		0	16	37	52	77	100	285
Chicago, IL, O'Hare International Airport	124	1	96	126	153	186	235	521
		0.5	58	77	94	113	137	265
		0	23	38	53	75	83	150
Cleveland, OH	188	1	85	124	157	195	230	432
		0.5	52	73	92	118	147	235
		0	23	37	47	69	92	225
Colorado Springs, CO	159	1	89	135	167	202	219	327
		0.5	57	82	99	124	140	218
		0	23	45	61	87	112	165
Columbus, OH, International Airport	92	1	71	101	123	149	175	328
		0.5	45	60	71	87	95	184
		0	15	30	45	60	62	135
Des Moines, IA	127	1	120	174	208	255	289	414
		0.5	74	102	120	149	180	310
		0	24	46	69	94	108	231
Detroit, MI, Metro Airport	153	1	92	130	156	192	212	360
		0.5	57	77	94	118	134	227
		0	23	38	47	75	89	194
Duluth, MN	238	1	123	171	201	238	250	370
		0.5	71	97	114	131	142	213
		0	22	32	46	68	77	196
Ely, NV	153	1	67	97	116	134	162	242
		0.5	44	66	83	111	129	241
		0	23	45	67	97	112	240
Eugene, OR	18	1	59	110	139	165	171	224
		0.5	47	77	93	119	122	164
		0	30	53	70	102	120	164
Fairbanks, AK	288	1	91	121	144	174	202	391
		0.5	52	68	78	94	108	200
		0	15	23	31	40	48	87
Baltimore, MD, BWI Airport	56	1	87	139	172	235	282	431
		0.5	69	108	147	200	238	369
		0	46	84	119	181	214	306
Great Falls, MT	233	1	123	171	193	233	276	392
		0.5	71	93	107	129	144	210
		0	17	31	45	60	75	143
Indianapolis, IN	96	1	95	134	158	194	215	284
		0.5	58	80	96	116	124	209
		0	23	38	52	83	99	209

[a]Heat fluxes are at the snow-melting surface only. See text for calculation of back and edge heat loss fluxes.   [b]Multiply values by 0.2931 to convert to W/ft².

Table 1 Frequencies of Snow-Melting Surface Heat Fluxes at Steady-State Conditions[a] (*Continued*)

Location	Snowfall Hours per Year	Snow-Free Area Ratio, A_r	Heat Fluxes Not Exceeded During Indicated Percentage of Snowfall Hours from 1982 Through 1993, Btu/h·ft²[b]					
			75%	90%	95%	98%	99%	100%
Lexington, KY	50	1	81	108	123	150	170	233
		0.5	49	65	74	85	95	197
		0	16	30	39	46	55	162
Madison, WI	161	1	99	138	164	206	241	449
		0.5	61	82	98	129	163	245
		0	23	39	60	91	113	194
Memphis, TN	13	1	106	141	172	200	206	213
		0.5	75	96	115	118	130	157
		0	40	75	76	90	97	123
Milwaukee, WI	161	1	101	135	164	196	207	431
		0.5	62	83	101	128	147	246
		0	23	46	68	98	120	239
Minneapolis-St. Paul, MN	199	1	119	169	193	229	254	332
		0.5	73	99	114	138	154	287
		0	23	45	61	91	113	245
New York, NY, JFK Airport	61	1	91	134	164	207	222	333
		0.5	63	93	118	145	164	325
		0	38	68	86	113	133	316
Oklahoma City, OK	35	1	117	168	215	248	260	280
		0.5	72	101	123	133	144	208
		0	24	46	68	78	113	190
Omaha, NE	94	1	108	148	189	222	259	363
		0.5	65	89	105	128	135	186
		0	23	38	60	90	100	136
Peoria, IL	91	1	95	139	166	201	227	436
		0.5	58	83	99	119	130	250
		0	23	38	53	76	92	228
Philadelphia, PA, International Airport	56	1	94	129	154	208	246	329
		0.5	65	90	112	162	185	267
		0	38	63	79	111	150	225
Pittsburgh, PA, International Airport	168	1	83	125	159	194	219	423
		0.5	51	75	94	111	129	216
		0	16	31	46	68	77	136
Portland, ME	157	1	120	168	195	234	266	428
		0.5	76	108	132	168	199	376
		0	39	67	90	130	152	324
Portland, OR	15	1	50	78	102	177	239	296
		0.5	39	55	81	114	130	199
		0	23	45	60	78	102	128
Rapid City, SD	177	1	139	203	252	312	351	482
		0.5	78	111	132	164	183	245
		0	16	30	38	53	65	179
Reno, NV	63	1	50	72	89	116	137	191
		0.5	36	55	75	105	115	172
		0	23	45	68	91	113	159
Salt Lake City, UT	142	1	52	77	89	110	120	171
		0.5	39	62	76	96	104	171
		0	30	60	75	89	104	171
Sault Ste. Marie, MI	425	1	112	153	183	216	249	439
		0.5	66	88	104	125	142	239
		0	23	37	47	68	83	188
Seattle, WA	27	1	56	107	138	171	205	210
		0.5	45	72	97	122	133	175
		0	37	52	75	96	123	151
Spokane, WA	144	1	67	98	116	141	159	227
		0.5	45	61	73	84	95	145
		0	23	37	45	54	67	112
Springfield, MO	58	1	110	155	179	215	224	292
		0.5	70	95	117	142	171	240
		0	32	54	76	115	129	227
St. Louis, MO, International Airport	62	1	97	147	170	193	227	344
		0.5	66	90	105	126	144	269
		0	31	53	68	97	104	194
Topeka, KS	61	1	102	153	192	234	245	291
		0.5	64	92	110	132	139	185
		0	23	39	52	68	84	167
Wichita, KS	60	1	115	163	209	248	285	326
		0.5	71	96	116	137	153	168
		0	24	45	57	75	83	158

[a]Heat fluxes are at the snow-melting surface only. See text for calculation of back and edge heat loss fluxes.　　[b]Multiply values by 0.2931 to convert to W/ft².

not snowing, it was assumed that the system was idling (i.e., that heat was supplied to the slab so that melting would start immediately when snow began to fall).

Weather data were taken for the years 1982 through 1993. These years were selected because of their completeness of data. The weather data included hourly values of the precipitation amount in equivalent depth of liquid water, precipitation type, ambient dry-bulb and dew-point temperatures, wind speed, and sky cover. All weather elements for 1982 to 1990 were obtained from the *Solar and Meteorological Surface Observation Network 1961 to 1990 (SAMSON), Version 1.0* (NCDC 1993). For 1991 to 1993, all weather elements except precipitation were taken from *DATSAV2* data obtained from the National Climatic Data Center as described in Colliver et al. (1998). The precipitation data for these years were taken from NCDC's *Hourly Cooperative Dataset* (NCDC 1990).

All wind speeds used were taken directly from the weather data. Wind speed V_{met} is usually measured at height of approximately 33 ft. As indicated in the section on Heat Balance, the heat and mass transfer coefficients are functions of a characteristic dimension of the snow-melting slab. The dimension used in generating the values of Table 1 was 20 ft. Sensitivity of the load to both wind speed and the characteristic dimension is included in Table 2. During snowfall, the sky temperature was taken as equal to the ambient temperature.

The first data column in Table 1 presents the average number of snowfall hours per year for each location. All surface heat fluxes were computed for snow-free area ratios of 1, 0.5, and 0, and the frequencies of snow-melting loads are presented. The frequency indicates the percentage of time that the required snow-melting surface heat flux does not exceed the value in the table for that ratio.

This table is used to design a snow-melting system for a given level of customer satisfaction depending on criticality of function. For example, although a heliport at the rooftop of a hospital may require almost 100% satisfactory operation at a snow-free ratio of 1, a residential driveway may be considered satisfactory at 90% and $A_r = 0.5$ design conditions. To optimize cost, different percentiles may be applied to different sections of the slab. For example, train station slab embark/disembark areas may be designed for a higher percentile and A_r than other sections.

Figures 1 and 2 show the distribution in the United States of snow-melting surface heat fluxes for snow-free area ratios of 1.0 and 0, respectively. The values presented satisfy the loads 99% of the time (as listed in the 99% column of Table 1). Local values can be approximated by interpolating between values given on the figure; however, extreme care must be taken because special local climatological conditions exist for many areas (e.g., lake effect snow). Both altitude and geography should be considered in making interpolations. Generally, locations in the northern plains of the United States require the maximum snow-melting heat flux (Chapman 1999).

Example for Surface Heat Flux Calculation Using Table 1

Example 2. Consider the design of a system for Albany, New York, which has an installed heat flux capacity at the top surface of approximately 149 Btu/h·ft². Based on data in Table 1, this system will keep the surface completely free of snow 95% of the time. Because there are 156 snowfall hours in an average year, this design would have some accumulation of snow approximately 8 h per year (the remaining 5% of the 156 h). This design will also meet the load for more than 98% of the time (i.e., more than 153 of the 156 snowfall hours) at an area ratio of 0.5 and more than 99% of the time for an area ratio of 0. $A_r = 0.5$ means that there is a thin layer of snow over all or part of the slab such that it acts as though half the slab is insulated by a snow layer; $A_r = 0$ means that the snow layer is sufficient to insulate the surface from heat and evaporation losses, but that snow is melting at the base of this layer at the same rate that it is falling on the top of the layer.

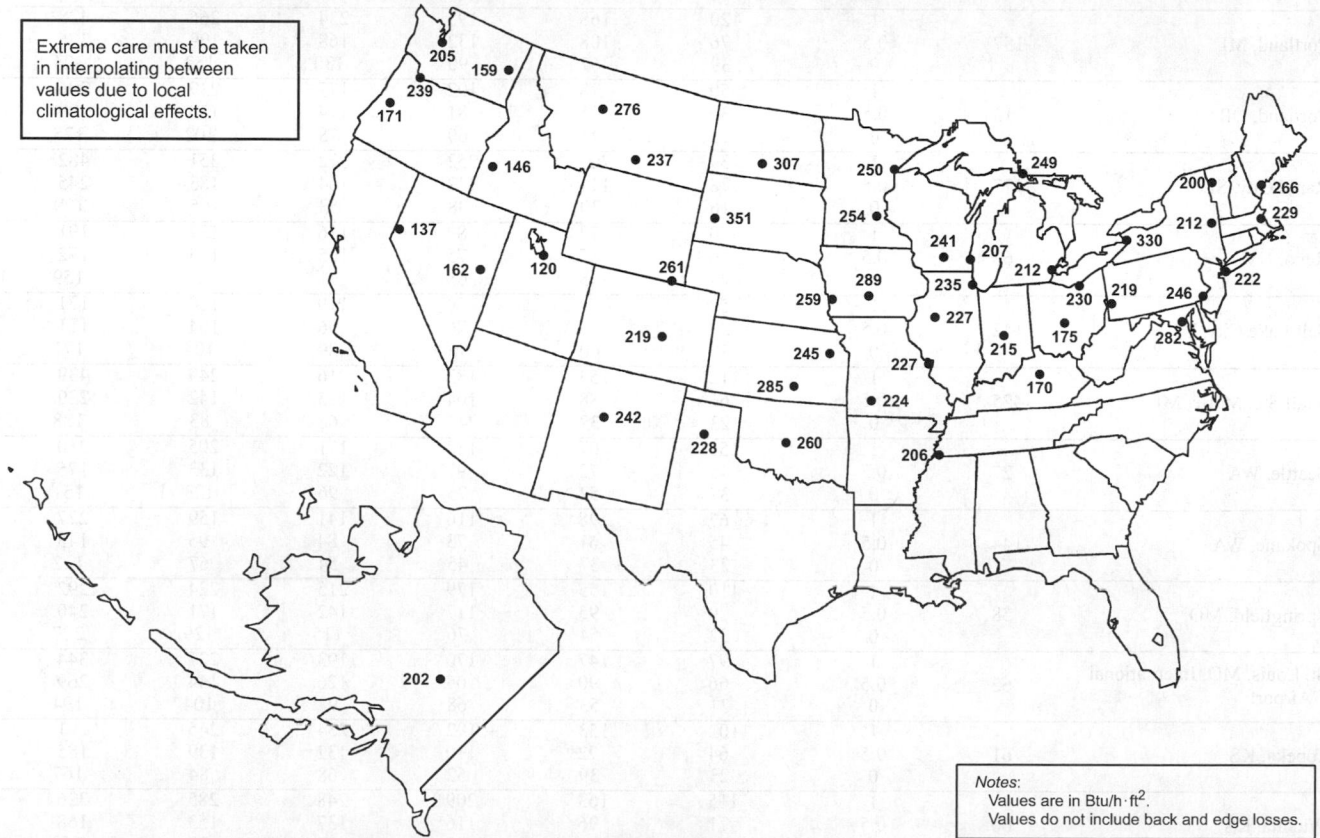

Extreme care must be taken in interpolating between values due to local climatological effects.

Notes:
Values are in Btu/h·ft².
Values do not include back and edge losses.

Fig. 1   Snow-Melting Surface Heat Fluxes Required to Provide Snow-Free Area Ratio of 1.0 for 99% of Time

Therefore, the results for this system can be interpreted to mean the following (all times are rounded to the nearest hour):

(a) For all but 8 h of the year, the slab will be snow-free.

(b) For the less than 5 h between the 95% and 98% nonexceedance values, there will be a thin build-up of snow on part of the slab.

(c) For the less than 2 h between the 98% and 99% nonexceedance values, snow will accumulate on the slab to a thickness at which the snow blanket insulates the slab, but the thickness will not increase beyond that level.

(d) For less than 2 h, the system cannot keep up with the snowfall.

An examination of the 100% column shows that to keep up with the snowfall the last 1% of the time, in this case less than 2 h for an average year, would require a system capacity of approximately 276 Btu/h·ft²; to attempt to keep the slab completely snow-free the entire season requires a capacity of 321 Btu/h·ft². Based on this interpretation, the designer and customer must decide the acceptable operating conditions. Note that the heat flux values in this example do not include back or edge losses, which must be added in sizing energy source and heat delivery systems.

Sensitivity of Design Surface Heat Flux to Wind Speed and Surface Size

Some snow-melting systems are in sheltered areas, whereas others may be in locations where surroundings create a wind tunnel effect. Similarly, systems vary in size from the baseline characteristic length of 20 ft. For example, sidewalks exposed to a crosswind may have a characteristic length on the order of 5 ft. In such cases, the wind speed will be either less than or greater than the meteorological value V_{met} used in Table 1. To establish the sensitivity of the surface heat flux to wind speed and characteristic length, calculations were performed at combinations of wind speeds $0.5V_{met}$, V_{met}, and $2V_{met}$ and L values of 5 ft and 20 ft for area ratios A_r of 1.0 and 0.5. Wind speed and system size do not affect the load values for $A_r = 0$ because

the calculations assume that no heat or mass transfer from the surface occurs at this condition. Table 2 presents a set of mean values of multipliers that can be applied to the loads presented in Table 1; these multipliers were established by examining the effect on the 99% nonexceedance values. The ratio of V to V_{met} in a given design problem may be determined from information given in Chapter 24 of the 2009 *ASHRAE Handbook—Fundamentals*. The closest ratio in Table 2 can then be selected. The designer is cautioned that these are to be used only as guidelines on the effect of wind speed and size variations.

Back and Edge Heat Losses

The surface heat fluxes in Table 1 do not account for heat losses from the back and edges of the slab. Adlam (1950) demonstrated that these back and edge losses may vary from 4 to 50%, depending on factors such as slab construction, operating temperature, ground temperature, and back and edge insulation and exposure. With the construction shown in Figure 3 or Figure 5 and ground temperature of 40°F at a depth of 24 in., back losses are approximately 20%. Higher losses occur with (1) colder ground, (2) more cover over the slab, or (3) exposed back, such as on bridges or parking decks.

Transient Analysis of System Performance

Determination of snow-melting surface heat fluxes as described in Equations (1) to (14) is based on steady-state analysis. Snow-melting systems generally have heating elements embedded in material of significant thermal mass. Transient effects, such as occur when the system is started, may be significant. A transient analysis method was developed by Spitler et al. (2002) and showed that particular storm conditions could change the snow-melting surface heat flux requirement significantly, depending on the precipitation rate at a particular time in the storm. It is therefore

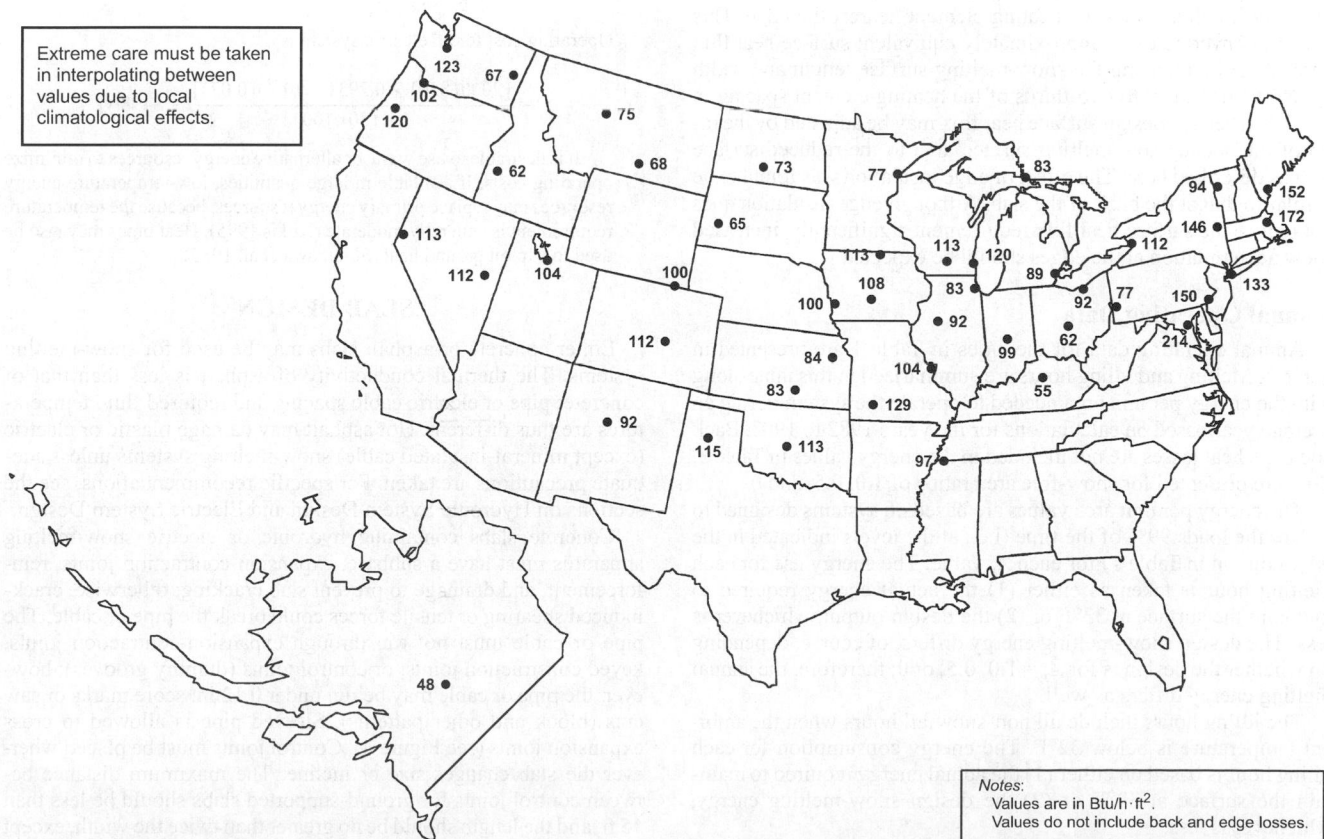

Extreme care must be taken in interpolating between values due to local climatological effects.

Notes:
Values are in Btu/h·ft².
Values do not include back and edge losses.

Fig. 2   Snow-Melting Surface Heat Fluxes Required to Provide Snow-Free Area Ratio of 0 for 99% of Time

Table 2 Mean Sensitivity of Snow-Melting Surface Heat Fluxes to Wind Speed and Slab Length

For loads not exceeded during 99% of snowfall hours, 1982 through 1993

Snow-Free Area Ratio, A_r	Ratio of Flux at Stated Condition to Flux at $L = 20$ ft and $V = V_{met}$				
	$L = 20$ ft		$L = 5$ ft		
	$V = 0.5V_{met}$	$V = 2V_{met}$	$V = V_{met}$	$V = 0.5V_{met}$	$V = 2V_{met}$
1	0.7	1.6	1.2	0.8	2.0
0.5	0.8	1.4	1.2	0.9	1.7
0	1.0	1.0	1.0	1.0	1.0

Note: Based on data from U.S. locations.
L = characteristic length
V_{met} = meteorological wind speed from NCDC

difficult to find simple general design rules that account for transient effects. To keep slab surfaces clear from snow during the first hour of the snowstorm, when the system is just starting to operate, heat fluxes up to five times greater than those indicated by steady-state analysis could be required. In general, greater heat input is required for greater spacing and greater depth of the heating elements in the slab; however, because of transient effects, this is not always true.

Transient analysis (Spitler et al. 2002) has also been used to examine back and edge losses. Heat fluxes because of back losses in systems without insulation ranged from 10 to 30% of the surface heat flux, depending primarily on the particular storm data, though higher losses occurred where heating elements were embedded deeper in the slab. In cases where 2 in. of insulation was applied below the slab, back losses were significantly reduced, to 1 to 4%. Although peak losses were reduced, the surface heat fluxes required to melt the snow were not significantly affected by the presence of insulation, because transient effects at the start of system operation drive the peak design surface heat fluxes. Edge losses ranged from 15 to 35% of the heat delivered by the heating element nearest the edge. This may be converted to an approximately equivalent surface heat flux percentage by reducing the snow-melting surface length and width by an amount equal to two-thirds of the heating element spacing in the slab. Then the design surface heat flux may be adjusted by the ratio of the actual snow-melting surface area to the reduced surface area as described here. The effect of edge insulation was found to be similar to that at the back of the slab. Although edge insulation does not reduce the surface heat flux requirement significantly, increased snow accumulation at the edges should be expected.

Annual Operating Data

Annual operating data for the cities in Table 1 are presented in Table 3. Melting and idling hours are summarized in this table along with the energy per unit area needed to operate the system during an average year based on calculations for the years 1982 to 1993. Back and edge heat losses are not included in the energy values in Table 3. Data are presented for snow-free area ratios of 1.0, 0.5, and 0.

The energy per unit area values are based on systems designed to satisfy the loads 99% of the time (i.e., at the levels indicated in the 99% column in Table 1) for each A_r value. The energy use for each melting hour is taken as either (1) the actual energy required to maintain the surface at 33°F or (2) the design output, whichever is less. The design snow-melting energy differs, of course, depending on whether the design is for A_r = 1.0, 0.5, or 0; therefore, the annual melting energy differs as well.

The idling hours include all non-snowfall hours when the ambient temperature is below 32°F. The energy consumption for each idling hour is based on either (1) the actual energy required to maintain the surface at 32°F or (2) the design snow-melting energy, whichever is less.

In Table 3, the column labeled "2% Min. Snow Temp." is the temperature below which only 2% of the snowfall hours occur. This

table may only be used to predict annual operating costs. Use Table 1 for system sizing.

Annual Operating Cost Example

Example 3. A snow-melting system of 2000 ft² is to be installed in Chicago. The application is considered critical enough that the system is designed to remain snow-free 99% of the time. Table 3 shows that the annual energy requirement to melt the snow is 8501 Btu/ft². Assuming a fossil fuel cost of $8 per 10^6 Btu, an electric cost of $0.07 per kWh, and back loss at 30%, find and compare the annual cost to melt snow with hydronic and electric systems. For the hydronic system, boiler combustion efficiency is 0.85 and energy distribution efficiency is 0.90.

Solution: Operating cost O may be expressed as follows:

$$O = \frac{A_t Q_a F}{[1-(B/100)](\eta_b \eta_d)} \tag{15}$$

where

O = annual operating cost $/yr
A_t = total snow-melting area, ft²
Q_a = annual snow-melting or idling energy requirement, Btu/ft² or kWh/ft²
F = primary energy cost $/Btu or $/kWh
B = back heat loss percentage, %
η_b = combustion efficiency of boiler (or COP of a heat pump in heating mode), dimensionless. If a waste energy source is directly used for snow-melting or idling purposes, the combustion efficiency term is neglected.
η_d = energy distribution efficiency, dimensionless (in an electric system, efficiencies may be taken to be 1)

Operating cost for a hydronic system is

$$O = \frac{(2000)(8501)(8 \times 10^{-6})}{\left[1-\left(\dfrac{30}{100}\right)\right][(0.85)(0.90)]} = \$254/\text{yr}$$

Operating cost for an electric system is

$$O = \frac{[2000(8501 \times 0.2931 \times 10^{-3})(0.07)]}{1-(30/100)} = \$498/\text{yr}$$

It is desirable to use waste or alternative energy resources to minimize operating costs. If available in large quantities, low-temperature energy resources may replace primary energy resources, because the temperature requirement is generally moderate (Kilkis 1995). Heat pipes may also be used to exploit ground heat (Shirakawa et al. 1985).

SLAB DESIGN

Either concrete or asphalt slabs may be used for snow-melting systems. The thermal conductivity of asphalt is less than that of concrete; pipe or electric cable spacing and required fluid temperatures are thus different. Hot asphalt may damage plastic or electric (except mineral-insulated cable) snow-melting systems unless adequate precautions are taken. For specific recommendations, see the sections on Hydronic System Design and Electric System Design.

Concrete slabs containing hydronic or electric snow-melting apparatus must have a subbase, expansion-contraction joints, reinforcement, and drainage to prevent slab cracking; otherwise, crack-induced shearing or tensile forces could break the pipe or cable. The pipe or cable must not run through expansion-contraction joints, keyed construction joints, or control joints (dummy grooves); however, the pipe or cable may be run under 0.12 in. score marks or saw cuts (block and other patterns). Sleeved pipe is allowed to cross expansion joints (see Figure 4). Control joints must be placed wherever the slab changes size or incline. The maximum distance between control joints for ground-supported slabs should be less than 15 ft, and the length should be no greater than twice the width, except for ribbon driveways or sidewalks. In ground-supported slabs, most cracking occurs during the early cure. Depending on the amount of

Table 3   Annual Operating Data at 99% Satisfaction Level of Heat Flux Requirement

City	Time, h/yr Melting	Time, h/yr Idling	2% Min. Snow Temp., °F	System Designed for $A_r = 1$ Melting	System Designed for $A_r = 1$ Idling	System Designed for $A_r = 0.5$ Melting	System Designed for $A_r = 0.5$ Idling	System Designed for $A_r = 0$ Melting	System Designed for $A_r = 0$ Idling
Albany, NY	156	1,883	9.3	10,132	109,230	7,252	109,004	4,371	108,420
Albuquerque, NM	44	954	16.3	2,455	38,504	1,729	38,495	984	38,332
Amarillo, TX	64	1,212	6.8	5,276	62,557	3,314	62,136	1,357	61,170
Billings, MT	225	1,800	−10.8	17,299	116,947	10,526	111,803	3,716	91,360
Bismarck, ND	158	2,887	−8.8	16,295	207,888	9,321	201,565	2,300	157,503
Boise, ID	85	1,611	5.3	3,543	74,724	2,449	73,015	1,345	68,456
Boston, MA	112	1,273	16.3	7,694	77,992	5,455	77,907	3,218	77,747
Buffalo, NY	292	1,779	3.8	23,929	105,839	14,735	105,521	5,563	101,945
Burlington, VT	204	2,215	4.3	13,182	147,122	8,485	143,824	3,783	134,634
Cheyenne, WY	224	2,152	−15.8	20,061	126,714	11,931	125,635	3,782	120,915
Chicago, IL, O'Hare International Airport	124	1,854	3.8	8,501	116,663	5,402	112,763	2,252	100,427
Cleveland, OH	188	1,570	8.8	11,419	86,539	7,359	85,470	3,208	80,851
Colorado Springs, CO	159	1,925	−8.8	11,137	97,060	7,089	96,847	3,026	96,244
Columbus, OH, International Airport	92	1,429	12.8	4,581	71,037	2,972	68,002	1,367	62,038
Des Moines, IA	127	1,954	−1.8	10,884	128,140	6,796	125,931	2,654	116,545
Detroit, MI, Metro Airport	153	1,781	11.3	10,199	104,404	6,467	102,289	2,704	95,777
Duluth, MN	238	3,206	0.3	20,838	251,218	12,423	236,657	3,969	187,820
Ely, NV	153	2,445	13.3	7,421	141,288	5,268	139,242	3,098	136,920
Eugene, OR	18	481	15.8	841	17,018	634	16,997	429	16,992
Fairbanks, AK	288	4,258	−15.8	19,803	343,674	11,700	318,880	3,559	194,237
Baltimore, MD, BWI Airport	56	957	16.3	3,827	45,132	2,970	45,132	2,121	45,130
Great Falls, MT	233	1,907	−15.8	19,703	123,801	11,731	120,603	3,736	101,712
Indianapolis, IN	96	1,473	10.8	6,558	80,942	4,132	78,532	1,705	75,926
Lexington, KY	50	1,106	13.3	2,696	54,084	1,718	52,278	733	45,859
Madison, WI	161	2,308	5.3	11,404	149,363	7,279	147,112	3,094	140,108
Memphis, TN	13	473	12.8	1,010	21,756	691	21,518	373	21,102
Milwaukee, WI	161	1,960	7.3	11,678	127,230	7,564	123,960	3,431	119,945
Minneapolis-St. Paul, MN	199	2,513	0.3	16,532	183,980	10,325	178,495	4,097	166,921
New York, NY, JFK Airport	61	885	18.3	4,193	50,680	2,988	50,467	1,797	50,049
Oklahoma City, OK	35	686	6.8	2,955	40,957	1,850	39,725	741	38,308
Omaha, NE	94	1,981	−2.3	7,425	124,274	4,613	119,565	1,790	112,700
Peoria, IL	91	1,748	2.3	6,544	104,380	4,078	100,581	1,606	94,045
Philadelphia, PA, International Airport	56	992	18.3	3,758	50,494	2,669	50,412	1,588	50,203
Pittsburgh, PA, International Airport	168	1,514	9.3	10,029	79,312	6,350	77,750	2,626	72,361
Portland, ME	157	1,996	7.3	13,318	115,248	8,969	115,196	4,630	114,836
Portland, OR	15	329	21.8	623	13,399	464	13,194	310	12,918
Rapid City, SD	177	2,154	−4.8	16,889	137,523	9,738	135,024	2,535	106,102
Reno, NV	63	1,436	16.3	2,293	54,713	1,792	54,706	1,302	54,703
Salt Lake City, UT	142	1,578	16.3	5,263	70,254	4,271	69,927	3,286	69,927
Sault Ste. Marie, MI	425	2,731	−0.3	34,249	176,517	20,779	174,506	7,250	155,508
Seattle, WA	27	260	17.8	1,212	10,482	943	10,473	682	10,452
Spokane, WA	144	1,832	10.8	6,909	81,000	4,721	79,177	2,512	75,659
Springfield, MO	58	1,108	6.8	4,401	57,165	2,950	56,929	1,503	56,238
St. Louis, MO, International Airport	62	1,150	6.8	4,516	64,668	2,981	63,428	1,446	60,764
Topeka, KS	61	1,409	−1.8	4,507	75,598	2,821	74,028	1,126	68,402
Wichita, KS	60	1,223	0.3	4,961	69,187	3,106	67,828	1,229	60,991

Annual Energy Requirement per Unit Area at Steady-State Conditions,[a] Btu/ft^2 [b]

[a]Does not include back and edge heat losses.   [b]Multiply values listed by 2.93×10^{-4} to convert to kWh/ft^2.

Table 4 Required Aggregate Size and Air Content

Maximum Size Crushed Rock Aggregate, in.	Air Content, %
2.5	5 ± 1
1	6 ± 1
0.5	7.5 ± 1

Source: Potter (1967).   *Note*: Do not use river gravel or slag.

water used in the concrete mix, shrinkage during cure may be up to 0.75 in. per 100 ft. If the slab is more than 15 ft long, the concrete does not have sufficient strength to overcome friction between it and the ground while shrinking during the cure period.

If the slabs are poured in two separate layers, the top layer, which contains the snow-melting apparatus, usually does not contribute toward total slab strength; therefore, the lower layer must be designed to provide the total strength.

The concrete mix of the top layer should give maximum weatherability. The compressive strength should be 4000 to 5000 psi; recommended slump is 3 in. maximum, 2 in. minimum. Aggregate size and air content should be as follows:

The pipe or cable may be placed in contact with an existing sound slab (either concrete or asphalt) and then covered as described in the sections on Hydronic System Design and Electric System Design. If there are signs of cracking or heaving, the slab should be replaced. Pipe or cable should not be placed over existing expansion-contraction, control, or construction joints. The finest grade of asphalt is best for the top course; stone diameter should not exceed 0.38 in.

A moisture barrier should be placed between any insulation and the fill. If insulation is used, it should be nonhygroscopic. The joints in the barrier should be sealed and the fill made smooth enough to eliminate holes or gaps for moisture transfer. Also, the edges of the barrier should be flashed to the surface of the slab to seal the ends.

Snow-melting systems should have good surface drainage. When the ambient air temperature is 32°F or below, runoff from melting snow freezes immediately after leaving the heated area. Any water that gets under the slab also freezes when the system is shutdown, causing extreme frost heaving. Runoff should be piped away in drains that are heated or below the frost line. If the snow-melting surface is inclined (e.g., a ramp), surface runoff may collect at the lowest point. In addition to effective drainage down the ramp, the adjacent area may require heating to prevent freezing the accumulated runoff.

The area to be protected by the snow-melting system must first be measured and planned. For total snow removal, hydronic or electric heat must cover the entire area. In larger installations, it may be desirable to melt snow and ice from only the most frequently used areas, such as walkways and wheel tracks for trucks and autos. Planning for separate circuits should be considered so that areas within the system can be heated individually, as required.

Where snow-melting apparatus must be run around obstacles (e.g., a storm sewer grate), the pipe or cable spacing should be uniformly reduced. Because some drifting will occur adjacent to walls or vertical surfaces, extra heating capacity should be provided in these areas, and if possible also in the vertical surface. Drainage flowing through the area expected to be drifted tends to wash away some snow.

CONTROL

Manual Control

Manual operation is strictly by on-off control; an operator must activate and deactivate the system when snow falls. The system should be started at an idling load before snowfall, whenever possible, to minimize thermal stresses in the system and the slab. The system must continue to operate until all the snow on the slab melts and evaporates.

Automatic Control

If the snow-melting system is not turned on until snow starts falling, it may not melt snow effectively for several hours, giving additional snowfall a chance to accumulate and increasing the time needed to melt the area. Automatic controls provide satisfactory operation by activating the system when light snow starts, allowing adequate warm-up before heavy snowfall develops. Automatic deactivation reduces operating costs.

Snow Detectors. Snow detectors monitor precipitation and temperature. They allow operation only when snow is present and may incorporate a delay-off timer. Snow detectors located in the heated area activate the snow-melting system when precipitation (snow) occurs at a temperature below the preset slab temperature (usually 40°F). Another type of snow detector is mounted above ground, adjacent to the heated area, without cutting into the existing system; however, it does not detect tracked or drifting snow. Both types of sensors should be located so that they are not affected by overhangs, trees, blown snow, or other local conditions.

Slab Temperature Sensor. To limit energy waste during normal and light snow conditions, it is common to include a remote temperature sensor installed midway between two pipes or cables in the slab; the set point is adjusted between 40 and 60°F. Thus, during mild weather snow conditions, the system is automatically modulated or cycled on and off to keep the slab temperature (at the sensor) at set point.

Outdoor Thermostat. The control system may include an outdoor thermostat that deactivates the system when the outdoor ambient temperature rises above 35 to 40°F as automatic protection against accidental operation in summer or mild weather.

Control Selection

For optimum operating convenience and minimum operating cost, all of the previously mentioned controls should be incorporated in the snow-melting system.

Operating Cost

To evaluate operating cost during idling or melting, use the annual output data from Table 3. Idling and melting data are based on slab surface temperature control at 32°F during idling, which requires a slab temperature sensor. Without a slab temperature sensor, operating costs will be substantially higher.

HYDRONIC SYSTEM DESIGN

Hydronic system design includes selection of the following components: (1) heat transfer fluid, (2) piping, (3) fluid heater, (4) pump(s) to circulate the fluid, and (5) controls. With concrete slabs, thermal stress is also a design consideration.

Heat Transfer Fluid

Various fluids, including brine, oils, and glycol-water, are suitable for transferring heat from the fluid heater to the slab. Freeze protection is essential because most systems will not be operated continuously in subfreezing weather. Without freeze protection, power loss or pump failure could cause freeze damage to the piping and slab.

Brine is the least costly heat transfer fluid, but it has a lower specific heat than glycol. Using brine may be discouraged because of the cost of heating equipment that resists its corrosive potential.

Although **heat transfer oils** are not corrosive, they are more expensive than brine or glycol and have a lower specific heat and higher viscosity. Petroleum distillates used in snow-melting systems are classified as nonflammable but have fire points between 300 and 350°F. When oils are used as heat transfer fluids, any oil dripping from pump seals should be collected. It is good practice to place a barrier between the oil lines and the boiler so that a flashback from the boiler cannot ignite a possible oil leak. Other nonflammable fluids, such as those used in some transformers, can be used as antifreeze.

Glycols (ethylene glycol and propylene glycol) are the most popular in snow-melting systems because of their moderate cost, high specific heat, and low viscosity; ease of corrosion control is another advantage. Automotive glycols containing silicates are not recommended because they can cause fouling, pump seal wear, fluid gelation, and reduced heat transfer. The piping should be designed for periodic addition of an inhibitor. Glycols should be tested annually to determine any change in reserve alkalinity and freeze protection. Only inhibitors obtained from the manufacturer of the glycol should be added. Heat exchanger surfaces should be kept below 285°F, which corresponds to about 40 psig steam. Temperatures above 300°F accelerate deterioration of the inhibitors.

Because ethylene glycol and petroleum distillates are toxic, no permanent connection should be installed between the snow-melting system and the drinking water supply. Gordon (1950) discusses precautions concerning internal corrosion, flammability, toxicity, cleaning, joints, and hook-up that should be taken during installation of hydronic piping. The properties of brine and glycol are discussed in Chapter 31 of the 2009 *ASHRAE Handbook—Fundamentals*. The effect of glycol on system performance is detailed in Chapter 12 of the 2008 *ASHRAE Handbook—HVAC Systems and Equipment*.

Piping

Piping may be metal, plastic, or ethylene-propylene terpolymer (EPDM). Steel, iron, and copper pipes have long been used, but steel and iron may corrode rapidly if the pipe is not shielded by a coating and/or cathodic protection. Both the use of salts for deicing and elevated temperature accelerate corrosion of metallic components. NACE (1978) states that the corrosion rate roughly doubles for each 18°F rise in temperature.

Chapman (1952) derived the equation for the fluid temperature required to provide an output q_o. For construction as shown in Figure 3, the equation is

$$t_m = 0.5q_o + t_f \tag{16}$$

where t_m = average fluid (antifreeze solution) temperature, °F. Equation (16) applies to 1 in. as well as 3/4 in. IPS pipe (Figure 3).

Design information about heated slabs given in Chapter 6 of the 2008 *ASHRAE Handbook—HVAC Systems and Equipment* may be used for all other types of slab construction, pipe spacing, and pipe material. In using these design equations, snow cover or water film on the slab may be treated as surface covers.

For specific conditions or for cities other than those given in Table 1, Equations (1) and (16) are used. Table 5 gives solutions to these equations at a relative humidity of 80%. Splitting the surface heat flux among four components as in Equation (1) also affects the required water temperature; therefore, Equation (16) and Table 5 should be used with caution. Table 5 may also be used to determine successful systems operation conditions. For example, if a system as shown in Figure 3 is designed for 250 Btu/h·ft², it will satisfy eight severe snow conditions such as 10°F, 0.16 in./h, and 10 mph wind speed (Chapman 1999).

Satisfactory standard practices is to use 3/4 in. pipe or tube on 12 in. centers, unless the snow-melting surface heat flux is too high. If pumping loads require reduced friction, the pipe size can be

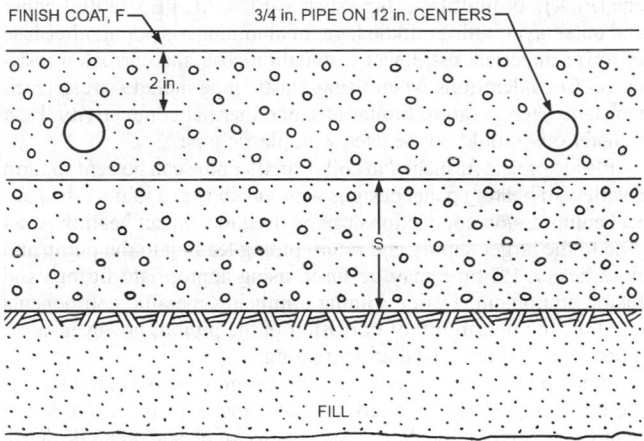

F = depth of finish coat—assumed to be 0.5 in. of concrete. Finish coat may be asphalt, but then cover slab should be reduced from 3 in. Depth of slab should always keep thermal resistance equal to 3 in. of concrete.

S = depth required by structural design (should be at least 2 in. of concrete).

Fig. 3   Detail of Typical Hydronic Snow-Melting System

Table 5   Steady-State Surface Heat Fluxes and Average Fluid Temperature for Hydronic Snow-Melting System in Figure 3

(Average fluid temperature based on 12 in. tube spacing)

s, Rate of Snowfall in/h	A_r		$t_a = 0°F$ Wind Speed V, mph 5	10	15	$t_a = 10°F$ Wind Speed V, mph 5	10	15	$t_a = 20°F$ Wind Speed V, mph 5	10	15	$t_a = 30°F$ Wind Speed V, mph 5	10	15
0.08	1.0	q_o	166	222	272	138	180	217	108	135	159	76	86	94
		t_m	116	144	169	102	123	142	87	100	112	71	76	80
	0.0	q_o	67	67	67	65	65	65	63	63	63	61	61	61
		t_m	66	66	66	65	65	65	64	64	64	63	63	63
0.16	1.0	q_o	233	289	339	203	244	282	171	197	221	136	146	155
		t_m	149	177	202	134	155	174	118	132	144	101	106	110
	0.0	q_o	133	133	133	129	129	129	125	125	125	121	121	121
		t_m	100	100	100	98	98	98	96	96	96	93	93	93
0.25	1.0	q_o	308	363	414	275	317	354	241	268	292	204	214	223
		t_m	187	215	240	171	192	210	154	167	179	135	140	144
	0.0	q_o	208	208	208	202	202	202	195	195	195	189	189	189
		t_m	137	137	137	134	134	134	131	131	131	127	127	127

Note: Table based on a characteristic slab length of 20 ft, standard air pressure, a water film temperature of 33°F, and relative humidity of 80%.

A_r = snow-free area ratio
q_o = slab heating flux, Btu/h·ft²

t_a = atmospheric dry-bulb temperature, °F
t_m = average fluid temperature based on construction shown in Figure 3, °F

Table 6 Typical Dependency of Maximum Heat Flux Deliverable by Plastic Pipes on Pipe Spacing and Concrete Overpour

(Average fluid temperature = 130°F)

Heat Flux,[a] Btu/h·ft²	Pipe Spacing on Centers,[b] in.
200	12
250	9
300	6
400	4

[a]Heat flux per unit area. Includes 30% back loss. Increase to 50% for bridges and structures with exposed back.

[b]Space pipes 1 in. closer for each 1 in. of concrete cover over 2 in. Space pipes 2 in. closer for each 1 in. of brick paver and mortar.

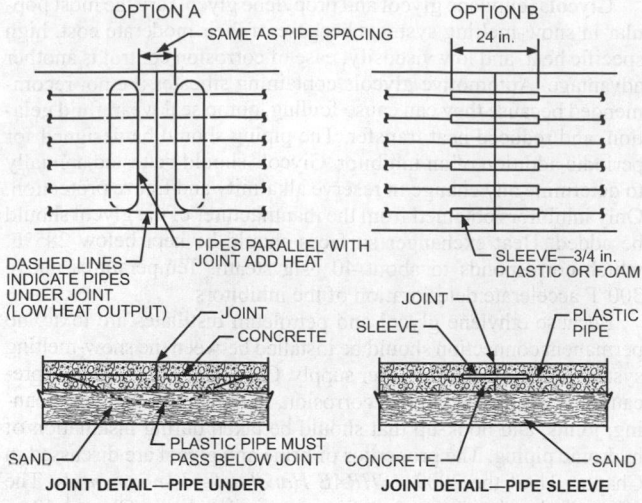

Fig. 4 Piping Details for Concrete Construction

increased to 1 in., but slab thickness must be increased accordingly. Piping should be supported by a minimum of 2 in. of concrete above and below. This requires a 5 in. slab for 3/4 in. pipe and a 5.4 in. slab for 1 in. pipe.

Plastic Pipe. Plastic [polyethylene (PE), cross-linked polyethylene (PEX)], or multilayer pipe such as PEX-AL-PEX (a PEX inner and outer layer with a middle layer of aluminum) is popular because of lower material cost, lower installation cost, and corrosion resistance. Considerations when using plastic pipe include stress crack resistance, temperature limitations, and thermal conductivity. Heat transfer oils should not be used with plastic pipe.

Plastic pipe is furnished in coils. Smaller pipe can be bent to form a variety of heating panel designs without elbows or joints. Mechanical compression connections can be used to connect heating panel pipe to the larger supply and return piping leading to the pump and fluid heater. PE pipe may be fused using appropriate fittings and fusion equipment. Fusion joining eliminates metallic components and thus the possibility of corrosion in the piping; however, it requires considerable installation training.

When plastic pipe is used, the system must be designed so that the fluid temperature required will not damage the pipe. If a design requires a temperature above the tolerance of plastic pipe, the delivered heat flux will never meet design requirements. The PE temperature limit is typically 140°F. For PEX, the temperature is 200°F (up to 80 psi) sustained fluid temperature, or 180°F (up to 100 psi). A solution to temperature limitations is to decrease the pipe spacing. Closer pipe spacing also helps eliminate striping of snow (unmelted portions between adjacent pipe projections on the surface). Adlam (1950) addresses the parameter of pipe size and the effect of pipe spacing on heat output. A typical solution is summarized in Table 6, which shows a way of designing pipe spacing according to flux requirements. This table also shows adjustments for the effect of more than 2 in. of concrete or paver over the pipe.

Oxygen permeation through plastic pipes may lead to corrosion on metal surfaces in the entire system unless plastic pipes are equipped with an oxygen barrier layer. Otherwise, either a heat exchanger must separate the plastic pipe circuitry from the rest of the system or corrosion-inhibiting additives must be used in the entire hydronic system.

Pipe Installation. It is good design practice to avoid passing any embedded piping through a concrete expansion joint; otherwise, the pipe may be stressed and possibly ruptured. Figure 4 shows a method of protecting piping that must pass through a concrete expansion joint from stress under normal conditions.

After pipe installation, but before slab installation, all piping should be air-tested to about 100 psig. This pressure should be maintained until all welds and connections have been checked for leaks. Isolate the air pressure test to manifold and piping, because boilers or other energy-converting, accumulating, or conditioning equipment may have lower pressure test limits. For example, boilers normally have an air test capability of 30 psi. Testing should not be done with water because (1) small leaks may not be observed during slab installation; (2) water leaks may damage the concrete during installation; (3) the system may freeze before antifreeze is added;

and (4) it is difficult to add antifreeze when the system is filled with water.

Air Control. Because introducing air causes deterioration of the antifreeze, the piping should not be vented to the atmosphere. It should be divided into smaller zones to facilitate filling and allow isolation when service is necessary.

Air can be eliminated from piping during initial filling by pumping the antifreeze from an open container into isolated zones of the piping. A properly sized pump and piping system that maintains adequate fluid velocity, together with an air separator and expansion tank, will keep air from entering the system during operation.

A strainer, sediment trap, or other means for cleaning the piping system may be provided. It should be placed in the return line ahead of the heat exchanger and must be cleaned frequently during initial system operation to remove scale and sludge. A strainer should be checked and cleaned, if necessary, at the start of and periodically during each snow-melting season.

An ASME safety relief valve of adequate capacity should be installed on a closed system.

Fluid Heater

The heat transfer fluid can be heated using any of a variety of energy sources, depending on availability. A fluid heater can use steam, hot water, gas, oil, or electricity. In some applications, heat may be available from secondary sources, such as engine generators, condensate, and other waste heat sources. Other low-temperature waste, or alternative energy resources may also be used with or without heat pumps or heat pipes. In a district heating system, the snow-melting system may be tied to the return piping of the district, which increases the overall temperature drop in the district heating system (Brown 1999).

The design capacity of the fluid heater can be established by evaluating the data in the section on Snow-Melting Heat Flux Requirement; it is usually 200 to 300 Btu/h·ft², which includes back and edge losses.

Design of the fluid heater should follow standard practice, with adjustments for the film coefficient. Consideration should be given to flue gas condensation and thermal shock in boilers because of low fluid temperatures. Bypass flow and temperature controls may be necessary to maintain recommended boiler temperatures. Boilers should be derated for high-altitude applications.

Pump Selection

The proper pump is selected based on the (1) fluid flow rate; (2) energy requirements of the piping system; (3) specific heat of the fluid; and (4) viscosity of the fluid, particularly during a cold start-up.

Pump Selection Example

Example 4. An area of 10,000 ft^2 is to be designed with a snow-melting system. Using the criteria of Equation (1), a heat flux (including back and edge losses) of 250 Btu/h·ft^2 is selected. Thus, the total fluid heater output must be 2,500,000 Btu/h, neglecting energy distribution losses. Size the pump for the heat transfer fluid used.

Solution: For a fluid with a specific heat of 0.85 Btu/lb·°F, the flow or temperature drop must be adjusted by 15%. From Equation (18) in Chapter 12 of the 2008 *ASHRAE Handbook—HVAC Systems and Equipment*, for a 23°F temperature drop, flow would be 250 gpm. If glycol is used at 130°F, the effect on pipe pressure loss and pump performance is negligible. If the system head is 40 ft and pump efficiency 60%, pump horsepower is 4.2. Centrifugal pump design criteria may be found in Chapter 43 of the 2008 *ASHRAE Handbook—HVAC Systems and Equipment*.

Controls

The controls discussed in the section on Control (Hydronic and Electric) provide convenience and operating economy but are not required for operation. Hydronic systems require fluid temperature control for safety and for component longevity. Slab stress and temperature limits of the heat transfer fluid, pipe components, and fluid heater need to be considered. Certain nonmetallic pipe materials should not be subjected to temperatures above 140°F. If the primary control fails, a secondary fluid temperature sensor should deactivate the snow-melting system and possibly activate an alarm.

Thermal Stress

Chapman (1955) discusses the problems of thermal stress in a concrete slab. In general, thermal stress will cause no problems if the following installation and operation rules are observed:

- Minimize the temperature difference between the fluid and the slab surface by maintaining (1) close pipe spacing (see Figure 3), (2) a low temperature differential in the fluid (less than 20°F), and (3) continuous operation (if economically feasible). According to Shirakawa et al. (1985), the temperature difference between the slab surface and the heating element skin should not exceed 70°F during operation.
- Install pipe within about 2 in. of the surface.
- Use reinforcing steel designed for thermal stress if high structural loads are expected (such as on highways).

Thermal shock to the slab may occur if heated fluid is introduced from a large source of residual heat such as a storage tank, a large piping system, or another snow-melting area. The slab should be brought up to temperature by maintaining the fluid temperature differential at less than 20°F.

ELECTRIC SYSTEM DESIGN

Snow-melting systems using electricity as an energy source have heating elements in the form of (1) mineral-insulated (MI) cable, (2) self-regulating cable, (3) constant-wattage cable, or (4) high-intensity infrared heaters.

Heat Flux

The basic load calculations for electric systems are the same as presented in the section on Snow-Melting Heat Flux Requirement. However, because electric system output is determined by the resistance installed and the voltage impressed, it cannot be altered by fluid flow rates or temperatures. Consequently, neither safety factors nor marginal capacity systems are design considerations.

Heat flux within a slab can be varied by altering the heating cable spacing to compensate for anticipated drift areas or other high-heat-loss areas. Power density should not exceed 120 W/ft^2 (NFPA *Standard* 70).

Electrical Equipment

Installation and design of electric snow-melting systems is governed by Article 426 of the *National Electrical Code®* (NEC, or NFPA *Standard* 70), which requires that each electric snow-melting circuit (except mineral-insulated, metal-sheathed cable embedded in a noncombustible medium) be provided with a ground fault protection device. An equipment protection device (EPD) with a trip level of 30 mA should be used to reduce the likelihood of nuisance tripping.

Double-pole, single-throw switches or tandem circuit breakers should be used to open both sides of the line. The switchgear may be in any protected, convenient location. It is also advisable to include a pilot lamp on the load side of each switch so that there is a visual indication when the system is energized.

Junction boxes located at grade level are susceptible to water ingress. Weatherproof junction boxes installed above grade should be used for terminations.

The power supply conduit is run underground, outside the slab, or in a prepared base. With concrete slab, this conduit should be installed before the reinforcing mesh.

Mineral-Insulated Cable

Mineral-insulated (MI) heating cable is a magnesium oxide (MgO)-filled, die-drawn cable with one or two copper or copper alloy conductors and a seamless copper or stainless steel alloy sheath. Copper sheath versions are usually protected from salts and other chemicals by a polyvinyl chloride (PVC) or high-density polyethylene jacket.

Cable Layout. To determine the characteristics of the MI heating cable needed for a specific area, the following must be known:

- Heated area size
- Power density required
- Voltage(s) available
- Approximate cable length needed

To find the approximate MI cable length, estimate 2 ft of cable per square foot of concrete. This corresponds to 6 in. on-center spacing. Actual cable spacing will vary between 3 and 9 in. for proper power density.

Cable spacing is dictated primarily by the heat-conducting ability of the material in which the cable is embedded. Concrete has a higher heat transmission coefficient than asphalt, permitting wider cable spacing. The following is a procedure to select the proper MI heating cable:

1. Determine total power required for each heated slab.

$$W = Aw \tag{17}$$

2. Determine total resistance.

$$R = E^2/W \tag{18}$$

3. Calculate cable resistance per foot.

$$r_1 = R/L_1 \tag{19}$$

where

W = total power needed, W
A = heated area of each heated slab, ft^2
w = required power density input, W/ft^2
R = total resistance of cable, Ω
E = voltage available, V
r_1 = calculated cable resistance, Ω per foot of cable
L_1 = estimated cable length, ft
L = actual cable length needed, ft
r = actual cable resistance, Ω/ft
M = cable on-center spacing, in.
I = total current per MI cable, A

Commercially available mineral-insulated heating cables have actual resistance values (if there are two conductors, the value is the total of the two resistances) ranging from 0.0016 to 0.6 Ω/ft. Manufacturing tolerances are ±10% on these values. MI cables are die-drawn, with the internal conductor drawn to size indirectly by pressures transmitted through the mineral insulation.

4. From manufacturers' literature, choose a cable with a resistance r closest to the calculated r_1. Note that r is generally listed at ambient room temperature. At the specific temperature, r may drift from the listed value. It may be necessary to make a correction as described in Chapter 6 of the 2008 *ASHRAE Handbook—HVAC Systems and Equipment*.

5. Determine the actual cable length needed to give the wattage desired.

$$L = R/r \qquad (20)$$

6. Determine cable spacing within the heated area.

$$M = 12A/L \qquad (21)$$

For optimum performance, heating cable spacing should be within the following limits: in concrete, 3 to 9 in.; in asphalt, 3 to 6 in.

Because the manufacturing tolerance on cable length is ±1%, and installation tolerances on cable spacing must be compatible with field conditions, it is usually necessary to adjust the installed cable as the end of the heating cable is rolled out. Cable spacing in the last several passes may have to be altered to give uniform heat distribution.

The installed cable within the heated areas follows a serpentine path originating from a corner of the heated area (Figure 5). As heat is conducted evenly from all sides of the heating cable, cables in a concrete slab can be run within half the spacing dimension of the perimeter of the heated area.

7. Determine the current required for the cable.

$$I = E/R, \text{ or } I = W/E \qquad (22)$$

8. Choose cold-lead cable as dictated by typical design guidelines and local electrical codes (see Table 7).

Cold-Lead Cable. Every MI heating cable is factory-fabricated with a non-heat-generating cold-lead cable attached. The cold-lead cable must be long enough to reach a dry location for termination and of sufficient wire gage to comply with local and *NEC* standards. The *NEC* requires a minimum cold-lead length of 6 in. within the junction box. MI cable junction boxes must be located such that the box remains dry and at least 3 ft of cold-lead cable is available at the end for any future service (Figure 5). Preferred junction box locations are indoors; on the side of a building, utility pole, or wall; or inside a manhole on the wall. Boxes should have a hole in the bottom to drain condensation. Outdoor boxes should be completely watertight except for the condensation drain hole. Where junction boxes are mounted below grade, the cable end seals must be coated with an epoxy to prevent moisture entry. Cable end seals should extend into the junction box far enough to allow the end seal to be removed if necessary.

Although MgO, the insulation in MI cable, is hygroscopic, the only vulnerable part of the cable is the end seal. However, should moisture penetrate the seal, it can easily be detected with a megohmmeter and driven out by applying a torch 2 to 3 ft from the end and working the flame toward the end.

Installation. When MI electric heating cable is installed in a concrete slab, the slab may be poured in one or two layers. In single-pour application, the cable is hooked on top of the reinforcing mesh before the pour is started. In two-layer application, the cable is laid on top of the bottom structural slab and embedded in the finish layer. For a proper bond between layers, the finish slab should be poured within 24 h of the bottom slab, and a bonding grout should be applied. The

Table 7 Mineral-Insulated Cold-Lead Cables (Maximum 600 V)

Single-Conductor Cable		Two-Conductor Cable	
Current Capacity, A	American Wire Gage	Current Capacity, A	American Wire Gage
35	14	25	14/2
40	12	30	12/2
55	10	40	10/2
80	8	55	8/2
105	6	75	6/2
140	4	95	4/2
165	3		
190	2		
220	1		

Source: *National Electrical Code®* (NFPA *Standard* 70).

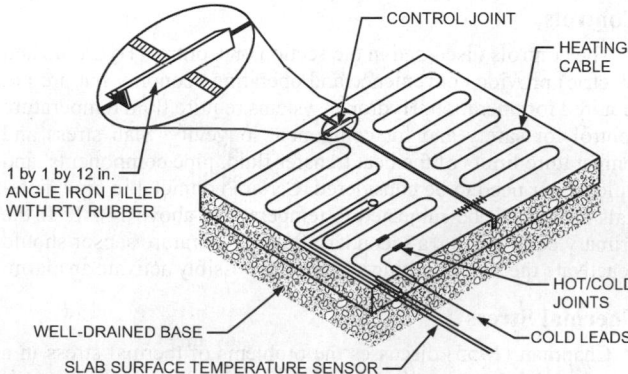

Fig. 5 Typical Mineral Insulated Heating Cable Installation in Concrete Slab

finish slab should be at least 2 in. thick. Cable should not run through expansion, control, or dummy joints (score or groove). If the cable must cross such a joint, it should cross the joint as few times as possible and be protected at the point of crossing with RTV rubber and a 1 by 1 by 12 in. angle iron as shown in Figure 5.

The cable is uncoiled from reels and laid as described in the section on Cable Layout. Prepunched copper or stainless steel spacing strips are often nailed to the lower slab for uniform spacing.

A high-density polyethylene (HDPE) or polyvinyl chloride (PVC) jacket is extruded by the manufacturer over the cable to protect from chemical damage and to protect the cable from physical damage without adding excessive thermal insulation.

If unjacketed MI cables are used, calcium chloride or other chloride additives should not be added to a concrete mix in winter because chlorides are destructive to copper. Cinder or slag fill under snow-melting slabs should also be avoided. The cold-lead cable should exit the slab underground in suitable conduits to prevent physical and chemical damage.

In asphalt slabs, the MI cable is fixed in place on top of the base pour with prepunched stainless steel strips or 6 in. by 6 in. wire mesh. A coat of bituminous binder is applied over the base and the cable to prevent them from floating when the top layer is applied. The layer of asphalt over the cable should be 1.5 to 3 in. thick (Figure 6).

Testing. Mineral-insulated heating cables should be thoroughly tested before, during, and after installation to ensure they have not been damaged either in transit or during installation.

Because MgO insulation is hygroscopic, damage to the cable sheath is easily detectable with a 500 V field megohmmeter. Cable insulation resistance should be measured on arrival of the cable. Cable with insulation resistance of less than 20 MΩ should not be used.

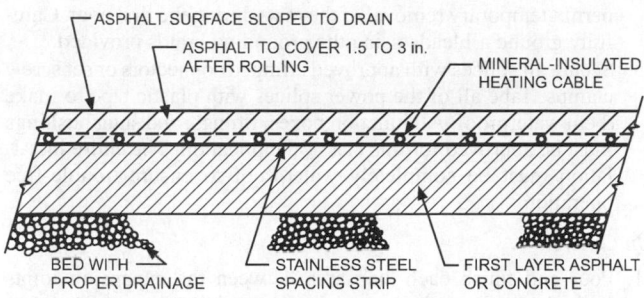

**Fig. 6 Typical Section, Mineral-Insulated
Heating Cable in Asphalt**
(Potter 1967)

Cable that shows a marked loss of insulation resistance after installation should be investigated for damage. Cable should also be checked for electrical continuity.

Self-Regulating Cable

Self-regulating heating cables consist of two parallel conductors embedded in a heating core made of conductive polymer. These cables automatically adjust their power output to compensate for local temperature changes. Heat is generated as electric current passes through the core between the conductors. As the slab temperature drops, the number of electrical paths increases, and more heat is produced. Conversely, as the slab temperature rises, the core has fewer electrical paths, and less heat is produced.

Power output of self-regulating cables may be specified as watts per unit length at a particular temperature or in terms of snow-melting performance at a given cable spacing. In typical slab-on-grade applications, adequate performance may be achieved with cables spaced up to 12 in. apart. Narrower cable spacings may be required to achieve the desired snow-melting performance. The parallel construction of the self-regulating cable allows it to be cut to length in the field without affecting the rated power output.

Layout. For uniform heating, the heating cable should be arranged in a serpentine pattern that covers the area with 12 in. on-center spacing (or alternative spacing determined for the design). The heating cable should not be routed closer than 4 in. to the edge of the slab, drains, anchors, or other material in the concrete.

Crossing expansion, control, or other slab joints should be avoided. Self-regulating heating cables may be crossed or overlapped as necessary. Because the cables limit power output locally, they will not burn out.

Both ends of the cable should terminate in an aboveground weatherproof junction box. Junction boxes installed at grade level are susceptible to water ingress. An allowance of heating cable should be provided at each end for termination.

The maximum circuit length published by the manufacturer for the cable type should be respected to prevent tripping of circuit breakers. Use ground fault circuit protection as required by national and local electrical codes.

Installation. Figure 7 shows a typical self-regulating cable installation. The procedure for installing a self-regulating system is as follows:

1. Hold a project coordination meeting to discuss the role of each trade and contractor. Good coordination helps ensure a successful installation.
2. Attach the heating cable to the concrete reinforcing steel or wire mesh using plastic cable ties at approximately 12 in. intervals. Reinforcing steel or wire mesh is necessary to ensure that the slab is structurally sound and that the heating cable is installed at the design depth.

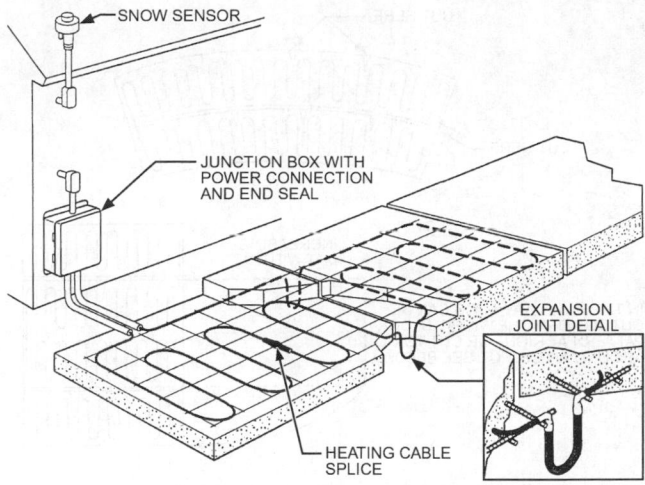

Fig. 7 Typical Self-Regulating Cable Installation

3. Test the insulation resistance of the heating cable using a 2500 V dc megohmmeter connected between the braid and the two bus wires. Readings of less than 20 MΩ indicate cable jacket damage. Replace or repair damaged cable sections before the slab is poured.
4. Pour the concrete, typically in one layer. Take precautions to protect the cable during the pour. Do not strike the heating cable with sharp tools or walk on it during the pour.
5. Terminate one end of the heating cable to the power wires, and seal the other end using connection components provided by the manufacturer.

Constant-Wattage Systems

In a constant-wattage system, the resistance elements may consist of a length of copper wire or alloy with a given amount of resistance. When energized, these elements produce the required amount of heat. Witsken (1965) describes this system in further detail.

Elements are either solid-strand conductors or conductors wrapped in a spiral around a nonconducting fibrous material. Both types are covered with a layer of insulation such as PVC or silicone rubber.

The heat-generating portion of an element is the conductive core. The resistance is specified in ohms per linear foot of core. Alternately, a manufacturer may specify the wire in terms of watts per foot of core, where the power is a function of the resistance of the core, the applied voltage, and the total length of core. As with MI cable, the power output of constant-wattage cable does not change with temperature.

Considerations in the selection of insulating materials for heating elements are power density, chemical inertness, application, and end use. Polyvinyl chloride is the least expensive insulation and is widely used because it is inert to oils, hydrocarbons, and alkalies. An outer covering of nylon is often added to increase its physical strength and to protect it from abrasion. The linear power density of embedded PVC is limited to 5 W/ft. Silicone rubber is not inert to oils or hydrocarbons. It requires an additional covering—metal braid, conduit, or fiberglass braid—for protection. This material can dissipate heat of up to 10 W/ft.

Lead can be used to encase resistance elements insulated with glass fiber. The lead sheath is then covered with a vinyl material. Output is limited to approximately 10 W/ft by the PVC jacket.

Teflon® has good physical and electrical properties and can be used at temperatures up to 500°F.

Low-power-density (less than 10 W/ft) resistance wires may be attached to plastic or fiber mesh to form a mat unit. Prefabricated

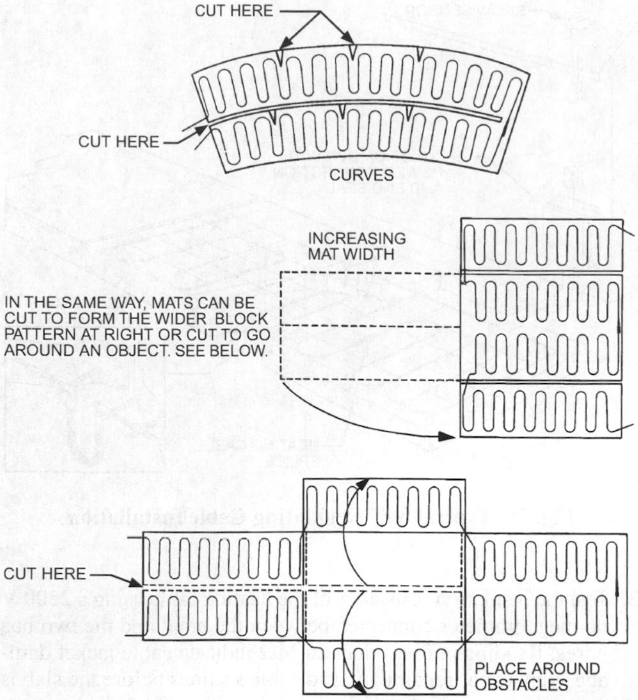

Fig. 8 Shaping Heating Mats Around Curves and Obstacles

factory-assembled mats are available in a variety of watt densities for embedding in specified paving materials to match desired snow-melting capacities. Mats of lengths up to 60 ft are available for installation in asphalt sidewalks and driveways.

Preassembled heating mats of appropriate widths are also available for **stair steps**. Heating mats are seldom made larger than 60 ft^2, because larger ones are more difficult to install, both mechanically and electrically. With a series of cuts, in the plastic or fiber mesh heating mats can be tailored to follow contours of curves and fit around objects, as shown in Figure 8. Extreme care should be exercised to prevent damage to the heater wire (or lead) insulation during this operation.

Mats should be installed 1.5 to 3.0 in. below the finished surface of asphalt or concrete. Installing mats deeper decreases the snow-melting efficiency. Only mats that can withstand hot-asphalt compaction should be used for asphalt paving.

Layout. Heating wires should be long enough to fit between the concrete slab dummy groove control or construction joints. Because concrete forms may be inaccurate, 2 to 4 in. of clearance should be allowed between the edge of the concrete and the heating wire. Approximately 4 in. should be allowed between adjacent heating wires at the control or construction joints.

For asphalt, the longest wire or largest heating mat that can be used on straight runs should be selected. The mats must be placed at least 12 in. in from the slab edge. Adjacent mats must not overlap. Junction boxes should be located so that each accommodates the maximum number of mats. Wiring must conform to requirements of the *NEC* (NFPA *Standard* 70). It is best to position junction boxes adjacent to or above the slab.

Installation

General

1. Check the wire or heating mats with an ohmmeter before, during, and after installation.
2. Temporarily lay the mats in position and install conduit feeders and junction boxes. Leave enough slack in the lead wires to

permit temporary removal of the mats during the first pour. Carefully ground all leads using the grounding braids provided.
3. Secure all splices with approved crimped connectors or set screw clamps. Tape all of the power splices with plastic tape to make them waterproof. All junction boxes, fittings, and snug bushings must be approved for this class of application. The entire installation must be completely waterproof to ensure trouble-free operation.

In Concrete

1. Pour and finish each slab area between the expansion joints individually. Pour the base slab and rough level to within 1.5 to 2 in. of the desired finish level. Place the mats in position and check for damage.
2. Pour the top slab over the mats while the rough slab is still wet, and cover the mats to a depth of at least 1.5 in., but not more than 2 in.
3. Do not walk on the mats or strike them with shovels or other tools.
4. Except for brief testing, do not energize the mats until the concrete is completely cured.

In Asphalt

1. Pour and level the base course. If units are to be installed on an existing asphalt surface, clean it thoroughly.
2. Apply a bituminous binder course to the lower base, install the mats, and apply a second binder coating over the mats. The finish topping over the mats should be applied in a continuous pour to a depth of 1.25 to 1.5 in. *Note*: Do not dump a large mass of hot asphalt on the mats because the heat could damage the insulation.
3. Check all circuits with an ohmmeter to be sure that no damage occurred during the installation.
4. Do not energize the system until the asphalt has completely hardened.

Infrared Snow-Melting Systems

Although overhead infrared systems can be designed specifically for snow-melting and freeze protection, they are usually installed for additional features they offer. Infrared systems provide comfort heating, which is particularly useful at entrances of plants, office buildings, and hospitals or on loading docks. Infrared lamps can improve a facility's security, safety, and appearance. These additional benefits may justify the somewhat higher cost of infrared systems.

Infrared fixtures can be installed under entrance canopies, along building facades, and on freestanding poles. Approved equipment is available for recessed, surface, and pendant mounting.

Infrared Fixture Layout. The same infrared fixtures used for comfort heating installations (as described in Chapter 15 of the 2008 *ASHRAE Handbook—HVAC Systems and Equipment*) can be used for snow-melting systems. The major differences are in the orientation of the target area: whereas in comfort applications, the *vertical* surfaces of the human body constitute the target of irradiation, in snow-melting applications, a *horizontal* surface is targeted. When snow melting is the primary design concern, fixtures with narrow beam patterns confine the radiant energy within the target area for more efficient operation. Asymmetric reflector fixtures, which aim the thermal radiation primarily to one side of the fixture centerline, are often used near the periphery of the target area.

Infrared fixtures usually have a longer energy pattern parallel to the long dimension of the fixture than at right angles to it (Frier 1965). Therefore, fixtures should be mounted in a row parallel to the longest dimension of the area. If the target area is 8 ft or more in width, it is best to locate the fixtures in two or more parallel rows. This arrangement also provides better comfort heating because radiation is directed across the target area from both sides at a more favorable incident angle.

Radiation Spill. An ideal energy distribution is uniform throughout the snow-melting target area at a density equal to the design requirement. The design of heating fixture reflectors determines the percentage of the total fixture radiant output scattered outside the target area design pattern.

Even the best-controlled beam fixtures do not produce a completely sharp cutoff at the beam edges. Therefore, if uniform distribution is maintained for the full width of the area, a considerable amount of radiant energy falls outside the target area. For this reason, infrared snow-melting systems are designed so that the power density on the slab begins to decrease near the edge of the area (Frier 1964). This design procedure minimizes stray radiant energy losses.

Figure 9 shows the power densities obtained in a sample snow-melting problem (Frier 1965). The sample design average is 45 W/ft². It is apparent that the incident power density is above the design average value at the center of the target area and below average at the periphery. Figure 9 shows how the power density and distribution in the snow-melting area depend on the number, wattage, beam pattern, and mounting height of the heaters, and on their position relative to the slab (Frier 1964).

With distributions similar to the one in Figure 9, snow begins to collect at the edges of the area as the energy requirements for snow melting approach or exceed system capacity. As snowfall lessens, the snow at the edges of the area and possibly beyond is then melted if the system continues to operate.

Target Area Power Density. Theoretical target area power densities for snow melting with infrared systems are the same as those for commercial applications of constant-wattage systems except that back and edge heat losses are smaller. However, note that theoretical density values are for radiation incident on the slab surface, not that emitted from the lamps. Merely multiplying the recommended snow-melting power density by the slab area to obtain the total power input for the system does not result in good performance. Experience has shown that multiplying this product by a correction factor of 1.6 gives a more realistic figure for the total required power input. The resulting wattage compensates not only for the radiant inefficiency involved, but also for the radiation falling outside the target area. For small areas, or when the fixture mounting height exceeds 16 ft, the multiplier can be as large as 2.0; large areas with sides of approximately equal length can have a multiplier of about 1.4.

The point-by-point method is the best way to calculate the fixture requirements for an installation. This method involves dividing the target area into 1 ft squares and adding the radiant energy from each infrared fixture incident on each square (Figure 9). The radiant energy distribution of a given infrared fixture can be obtained from the equipment manufacturer and should be followed for that fixture size and placement.

With infrared energy, the target area can be preheated to snow-melting temperatures in 20 to 30 min, unless the air temperature is well below 20°F or wind velocity is high (Frier 1965). This short warm-up time makes it unnecessary to turn on the system before snow begins to fall. The equipment can be turned on either manually or with a snow detector. A timer is sometimes used to turn the system off 4 to 6 h after snow stops falling, allowing time for the slab to dry completely.

If the snow is allowed to accumulate before the infrared system is turned on, there will be a delay in clearing the slab, as with embedded hydronic or electric systems. Because infrared energy is absorbed in the top layer of snow rather than by the slab surface, the time needed depends on snow depth and on atmospheric conditions. Generally, a system that maintains a clear slab by melting 1 in. of snow per hour as it falls requires 1 h to clear 1 in. of accumulated snow under the same conditions.

To ensure maximum efficiency, fixtures should be cleaned at least once a year, preferably at the beginning of the winter season. Other maintenance requirements are minimal.

Snow Melting in Gutters and Downspouts

Electrical heating cables are used to prevent heavy snow and ice accumulation on roof overhangs and to prevent ice dams from forming in gutters and downspouts (Lawrie 1966). Figure 10 shows a typical cable layout for protecting a roof edge and downspout. Cable for this purpose is generally rated at approximately 6 to 16 W/ft, and about 2.5 ft of wire is installed per linear foot of roof edge. One foot of heated wire per linear foot of gutter or downspout is usually adequate.

If the roof edge or gutters (or both) are heated, downspouts that carry away melted snow and ice must also be heated. A heated length of cable (weighted, if necessary) is dropped inside the downspout to the bottom, even if it is underground.

Lead wires should be spliced or plugged into the main power line in a waterproof junction box, and a ground wire should be installed from the downspout or gutter. Ground fault circuit protection is required per the *NEC* (NFPA *Standard* 70).

The system can be controlled with a moisture/temperature controller, ambient thermostat, or manual control. The moisture/temperature controller is the most energy efficient. If manual control is used, a protective thermostat should also be used to prevent system operation at ambient temperatures above 41°F.

FREEZE PROTECTION SYSTEMS

If the slab surface temperature is below 32°F, any water film present on that surface freezes. Water may be present because of accidental spillage, runoff from a nearby source, or premature shutoff of snow-melting operation after precipitation ends but before the surface dries (temperature usually drops rapidly after snow). Therefore, for cases with $A_r < 1$, if the system is shut off too soon, the remaining snow and fluid film on the surface may freeze (Adlam 1950). As previously discussed, idling keeps the slab surface from freezing by maintaining a surface temperature of at least 33.5°F and also reduces the required start-up surface heat flux for snow-melting.

To calculate surface heat flux during idling, the surface may be assumed to be free of snow, covered with a film of water. Unless there is a constant influx of water from the vicinity, the evaporation heat flux of that film may be ignored and the surface may be assumed to be uncovered because the insulation effect of the water film is negligible. In this case, surface heat flux can be calculated by Equation (5).

The surface is free of snow; therefore, Equation (5) approximates the surface heat flux. The mean radiant temperature that appears in

INTENSITY ON PAVEMENT FROM FOUR INFRARED FIXTURES, W/ft²														
14.7	16.0	19.75	23.7	25.5	27.5	28.2	28.0	28.2	27.5	25.5	23.7	19.75	16.0	14.7
23.7	24.8	28.7	31.7	35.7	38.2	38.5	39.4	38.5	38.2	35.7	31.7	28.7	24.8	23.7
25.7	31.7	37.5	42.7	46.4	49.4	52.2	53.0	52.2	49.4	46.4	42.7	37.5	31.7	25.7
28.2	34.3	42.8	46.7	51.2	55.7	58.5	63.0	58.5	55.7	51.2	46.7	42.8	34.3	28.2
28.2	34.3	42.8	46.7	51.2	55.7	58.5	63.0	58.5	55.7	51.2	46.7	42.8	34.3	28.2
25.7	31.7	37.5	42.7	46.4	49.4	52.2	53.0	52.2	49.4	46.4	42.7	37.5	31.7	25.7
23.7	24.8	28.7	31.7	35.7	38.2	38.5	39.4	38.5	38.2	35.7	31.7	28.7	24.8	23.7
14.7	16.0	19.75	23.7	25.5	27.5	28.2	28.0	28.2	27.5	25.5	23.7	19.75	16.0	14.7

Fig. 9   Typical Power Density Distribution for Infrared Snow-Melting System

8 by 15 ft target area, four single-element quartz lamps located 10 ft above floor (Potter 1967).

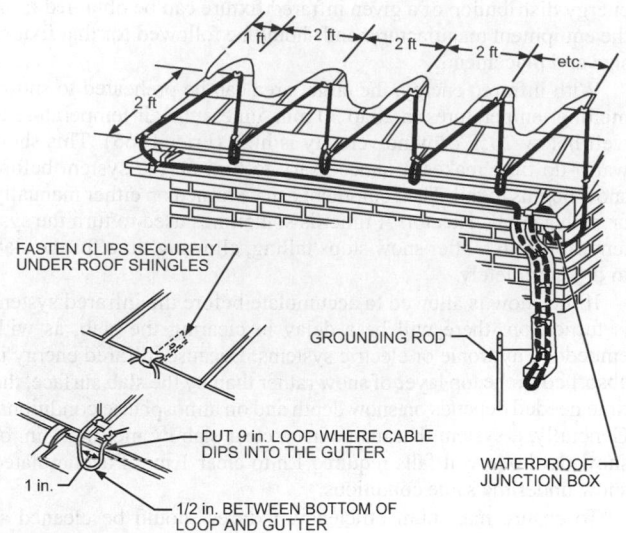

Fig. 10   Typical Insulated Wire Layout to Protect Roof Edge and Downspout

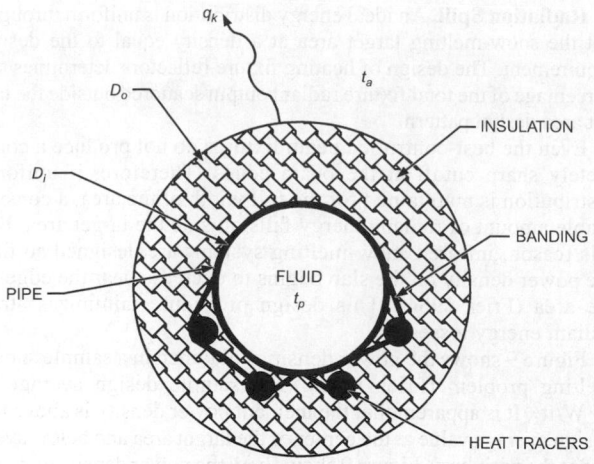

Fig. 11   Typical Heat Tracing Arrangement (Hydronic or Electric)

Equation (5) is evaluated using Equations (8), (9), and (10). The fraction F_{sc} of radiation between the surface and the clouds is equal to the cloud cover fraction in the meteorological data.

Chapman (1952) also proposed the following equation to determine the required surface heat flux for idling q_i in Btu/h·ft², if the mean ambient air temperature t_m during freezing is known:

$$q_i = (0.27V + 3.3)(32 - t_m) \tag{23}$$

A slab surface temperature monitor may control the freeze protection system. Whenever the slab surface temperature drops below 33°F, the system activates. However, idling the slab during the entire winter, as given in Table 3, may be too costly—and unnecessary if the main purpose is to reduce high snow-melting surface heat flux at start-up. For example, the annual energy requirement for idling is 45 times more than that for snow-melting in Chicago, $A_r = 0.5$. Therefore, a cost-effective operation may require starting the system to idle only before an anticipated snow. The lead time may be determined by the thermal mass of the slab, local meteorological conditions, idling and start-up snow-melting heat fluxes, and energy cost. Depending on local weather conditions, idling may also be started automatically when prevailing atmospheric conditions make snowfall likely.

Freeze protection systems may also be used in a variety of applications. For example, the foundation of a cold-storage warehouse may be protected from heaving by using a heated floor slab similar to a snow-melting system. The slab must be insulated at the top as well as the back and edges. Top insulation prevents the heated slab from interfering with the space-cooling process in the warehouse. Edge insulation must penetrate below the freezing line. Generally, design heat flux is taken to be between 5 and 10 Btu/(h·ft²) and the system is operated year-round.

Another freeze protection application is pipe tracing, where a pipe or conduit exposed to the atmosphere is protected against freezing of the fluid within. If a highly viscous fluid is transported, the desired pipe (fluid) temperature may need to be higher than the fluid-freezing temperature to maintain the viscosity required for fluid flow. Figure 11 shows a typical application in which small "tracer" pipes or electrical heating cables are banded along the lower surface of the pipe (Kenny 1999). In a hydronic tracing system, hot fluid or steam may be used. In electric systems, heating cable or mats may be used. Pipe and the tracing elements are covered with' thermal insulating material such as fiberglass, polyurethane,

calcium silicate, or cellular glass. Sometimes multiple insulation layers may be used. Insulating material must be protected from rain and other external conditions by a weather barrier. Pipe-tracing heat load per unit pipe length q_k is given by the following formula (IEEE 1983):

$$q_k = \frac{(t_p - t_a)}{\dfrac{1}{\pi D_i h_i} + \dfrac{\ln(D_o/D_i)}{2\pi k_1} + \dfrac{\ln(D_3/D_o)}{2\pi k_2} + \dfrac{1}{\pi D_3 h_{co}} + \dfrac{1}{\pi D_3 h_o}} \tag{24}$$

where

q_k = pipe tracing heat load per unit pipe length, Btu/h·ft
t_p = desired pipe temperature, °F
t_a = design ambient temperature, °F
D_i = inside diameter of inner insulation layer (and outer diameter of pipe), ft
D_o = outside diameter of inner insulation layer (and inside diameter of outer insulation layer, if present), ft
D_3 = outside diameter of outer insulation layer (if present), ft. Otherwise, the expression $\ln(D_3/D_o)/2\pi k_2$ in Equation (24) is dropped and D_3 in the last two terms in the denominator of the same equation is replaced by D_o.
k_1 = thermal conductance of inner insulation layer, Btu/h·ft·°F (evaluated at its average operating temperature)
k_2 = thermal conductance of outer insulation layer, if present, Btu/h·ft·°F (evaluated at its average operating temperature)
h_i = thermal convection coefficient of air film between pipe and inner insulation surface, Btu/h·ft·°F.
h_{co} = thermal convection coefficient of air between outer insulation surface and weather barrier (if present), Btu/h·ft·°F.
h_o = combined surface heat transfer coefficient for radiation and convection between weather barrier, if present (otherwise, the outer insulation layer), to ambient, Btu/h·ft·°F. Values for h_o may be calculated from information in Chapter 4 of the 2009 *ASHRAE Handbook—Fundamentals* and Chapter 15 of the 2008 *ASHRAE Handbook—HVAC Systems and Equipment*.

An appropriate pipe-tracing system is selected to satisfy q_k. For ease of product selection, some manufacturers offer design software or simple charts and graphs of heat losses for various pipe temperatures and insulation configurations. Safety factors are usually added by explicitly increasing the calculated heat load, by decreasing the design ambient temperature, or by conservative selection of k and h values. Pipe- or conduit-tracing heat loads can be more complex because of the heat sinks that penetrate the insulation surface(s); they often require a complex analysis to determine total heat loss, so a heat trace supplier should be consulted.

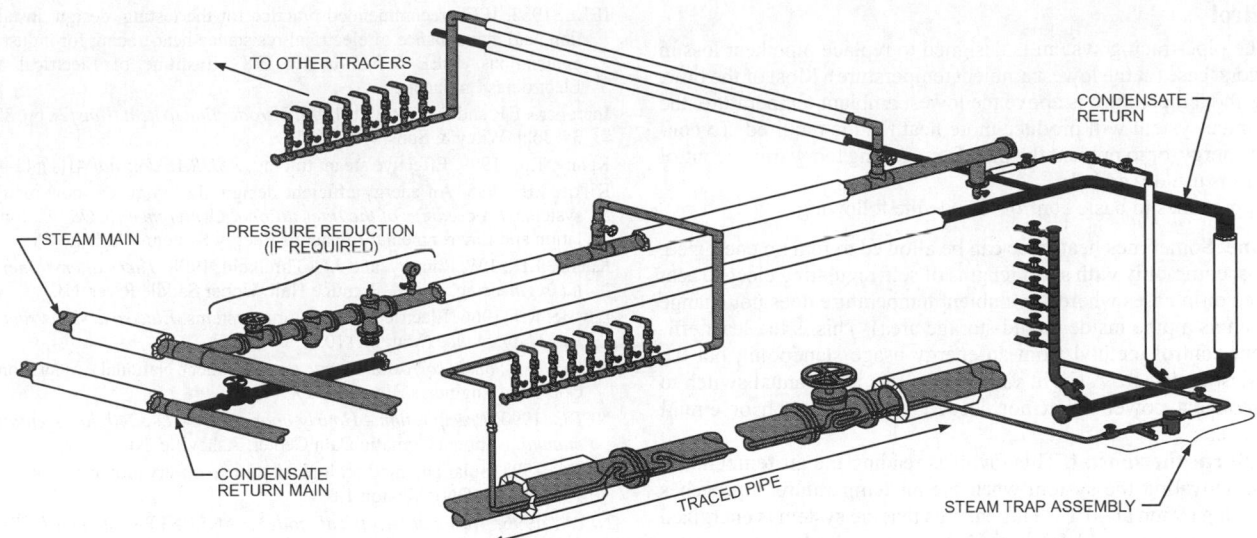

Fig. 12   Typical Pipe-Tracing System with Steam System

Steam Pipe-Tracing Systems

Steam tracing involves circulating steam in a pipe or tube that runs parallel to the pipe being traced. As the steam recondenses, it releases its latent heat and transfers it into the traced pipe. A typical steam pipe-tracing system is shown in Figure 12.

Steam systems have relatively high installed costs, particularly if an appropriately sized boiler and header system is not already in place. Steam is widely used for industrial applications, and the design is familiar to pipe installers. Steam is well suited for applications that require a high heat flux, but often is not as efficient for lower-heat-flux applications such as pipe freeze protection.

Electric Pipe-Tracing Systems

Electric pipe-tracing systems involve placing (tracing) an electrical resistance wire parallel to the pipe or tube being traced. This electric heater provides heat to the pipe to balance heat loss to the lower-ambient surrounding. A typical electric system is shown in Figure 13.

Types of electric heating cables include the following:

- **Self-regulating heating cables** consist of two parallel conductors embedded in a heating core made of conductive polymers. These cables automatically adjust their power output to compensate for local temperature changes. Heat is generated as electric current passes through the conductive core between the conductors. As the pipe temperature drops, the number of electrical paths increases, and more heat is produced. Power output of self-regulating heating cables is specified as watts per unit length at a particular temperature. The self-regulating feature makes the heating cables more energy efficient, because they change their power output based on the need at that point in the pipe. Because they are parallel, they can be cut to length and spliced in the field. Disadvantages are that they are often more expensive and have shorter maximum run lengths because of inrush current.

- **Series heating cables** are one or two copper, copper alloy, or nichrome elements surrounded by a polymer insulating jacket. Power outputs of series cables are specified in watts per unit length. These heating cables are usually inexpensive. Their main disadvantage is that the length cannot be adjusted without changing the power output.

- **Mineral-insulated (MI) heating cables** are series heating cables composed of a magnesium oxide (MgO)-filled, die-drawn cable

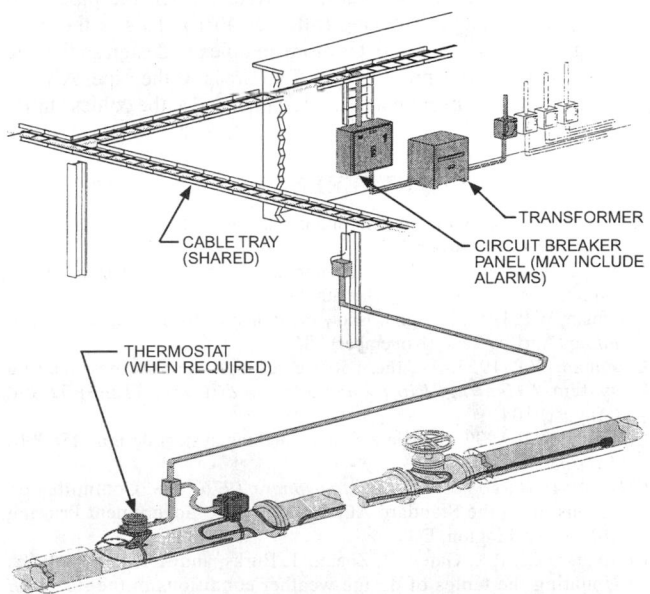

Fig. 13   Typical Pipe Tracing with Electric System

with one or two copper or copper alloy conductors and a seamless copper or stainless steel alloy sheath. Power output of an MI heating cable is specified in watts per unit length. MI heating cables are rugged and can withstand high temperatures, but are not very flexible. They generally must be ordered in the size needed; splicing in the field is craft-sensitive.

- **Zone heaters** consist of two insulated copper bus wires wrapped with a small-gage (38 to 41 AWG) nichrome heating wire, covered with polymer insulation. The heating wire is connected to alternate bus wires at nodes spaced 1 to 4 ft apart. Current flowing between the bus wires on the heating element generates heat. Power output of a zone heating cable is specified in watts per unit length. Zone heaters are parallel heaters, and thus can be cut to length and spliced in the field. Care must be used to prevent the thin heating wire from being damaged.

Control

The pipe-tracing system is designed to replace pipe heat loss in the worst case (at the lowest ambient temperature). Most of the time, when the temperature is above the lowest ambient temperature, the heat trace system will produce more heat than is required. To conserve energy, or to prevent the pipe from getting too warm, a control system is usually added.

Approaches to basic control include the following:

- **None.** Sometimes heat trace can be allowed to remain energized, most commonly with short lengths of self-regulating electric heat trace or in cases where the ambient temperature does not change (such as a pipe inside a cold-storage area). This is the least efficient control method from an energy usage standpoint, but the easiest to design. A slight variation on this is a manual switch to disconnect power when not needed, using a switch or circuit breaker.
- **Ambient thermostat.** This involves reading the air temperature and activating the system when the air temperature approaches freezing (often at 40°F). This ensures that the system is energized when the pipe could freeze and de-energized when it is warm. This system is often the best compromise between energy efficiency and ease of design.
- **Pipe-sensing thermostat.** This involves reading the pipe temperature and activating the heat trace system when the pipe temperature approaches freezing (often at 40°F). This is the most energy-efficient system, but is more complex to design so that the sensor reading is representative of all areas of the pipe. Always put the control sensor on the smallest pipe and at the coldest anticipated location.

REFERENCES

Adlam, T.N. 1950. *Snow melting.* Industrial Press, New York; University Microfilms, Ann Arbor, MI.

Brown, B. 1999. Klamath Falls geothermal district heating systems, GHC Bulletin, March 1999:5-9, Klamath, OR.

Chapman, W.P. 1952. Design of snow melting systems. *Heating and Ventilating* (April):95 and (November):88.

Chapman, W.P. 1955. Are thermal stresses a problem in snow melting systems? *Heating, Piping and Air Conditioning* (June):92 and (August):104.

Chapman, W.P. 1999. A review of snow melting system design. *ASHRAE Transactions* 105(1).

COESA. 1976. *U.S. Standard Atmosphere 1976.* U.S. Committee on Extension to the Standard Atmosphere. U.S. Government Printing Office, Washington, D.C.

Colliver, D.G., R.S. Gates, H. Zhang, T. Burks, and K.T. Priddy. 1998. Updating the tables of design weather conditions in the *ASHRAE Handbook—Fundamentals.* ASHRAE RP-890 *Final Report.*

Frier, J.P. 1964. Design requirements for infrared snow melting systems. *Illuminating Engineering* (October):686. Also discussion (December).

Frier, J.P. 1965. Snow melting with infrared lamps. *Plant Engineering* (October):150.

Gordon, P.B. 1950. Antifreeze protection for snow melting systems. *Heating, Piping and Air Conditioning Contractors National Association Official Bulletin* (February):21.

IEEE. 1983. IEEE recommended practice for the testing, design, installation, and maintenance of electrical resistance heat-tracing for industrial applications. IEEE *Standard* 515-1983. Institute of Electrical and Electronics Engineers.

Incropera, F.P. and D.P. DeWitt. 1996. *Introduction to heat transfer,* pp. 332-34. John Wiley & Sons, New York.

Kenny, T.M. 1999. Effective steam tracing, *ASHRAE Journal* 41(1):42-44.

Kilkis, I.B. 1995. An energy efficient design algorithm for snow-melting systems. *Proceedings of the International Conference ECOS '95.* Simulation and Environmental Impact of Energy Systems, ASME, NY.

Kuehn, T.H., J.W. Ramsey, and J.L. Threlkeld. 1998. *Thermal environmental engineering,* 3rd ed. Prentice Hall, Upper Saddle River, NJ.

Lawrie, R.J. 1966. Electric snow melting systems. *Electrical Construction and Maintenance* (March):110.

NACE. 1978. *Basic corrosion course text* (October). National Association of Corrosion Engineers, Houston, TX.

NCDC. 1990. *Precipitation—Hourly cooperative TD-3240 documentation manual.* National Climatic Data Center, Asheville, NC.

NCDC. 1993. Solar and meteorological surface observation network 1961-1990 (SAMSON), Version 1.0.

NFPA. 1996. *National electrical code®.* ANSI/NFPA *Standard* 70-96. National Fire Protection Association, Quincy, MA.

Potter, W.G. 1967. Electric snow melting systems. *ASHRAE Journal* 9(10):35-44.

Ramsey, J.W., H.D. Chiang, and R.J. Goldstein. 1982. A study of the incoming long-wave atmospheric radiation from a clear sky. *Journal of Applied Meteorology* 21:566-578.

Ramsey, J.W., M.J. Hewett, T.H. Kuehn, and S.D. Petersen. 1999. Updated design guidelines for snow melting system. *ASHRAE Transactions* 105(1):1055-1065.

Shirakawa, K., S. Kobayashi, S. Koyama, and M. Syuniji. 1985. Snow melting pavement with steel reinforced concrete. *Summimoto Research* 31.

Spitler, J.D., S.J. Rees, and X. Xia. 2002. Transient analysis of snow-melting system performance. *ASHRAE Transactions* 108(2):406-425.

Witsken, C.H. 1965. Snow melting with electric wire. *Plant Engineering* (September):129.

BIBLIOGRAPHY

Chapman, W.P. 1955. Snow melting system hydraulics. *Air Conditioning, Heating and Ventilating* (November).

Chapman, W.P. 1957. Calculating the heat requirements of a snow melting system. *Air Conditioning, Heating and Ventilating* (September through August).

Chapman, W.P. and S. Katunich. 1956. Heat requirements of snow melting systems. *ASHAE Transactions* 62:359.

Erickson, C.J. 1995. *Handbook of electrical heating for industry.* Institute of Electrical and Electronic Engineers.

Hydronics Institute. 1994. *Snow melting calculation and installation guide.* Berkeley Heights, NJ.

Kilkis, I.B. 1994. Design of embedded snow-melting systems: Part 1, Heat requirements—An overall assessment and recommendations. *ASHRAE Transactions* 100(1):423-433.

Kilkis, I.B. 1994. Design of embedded snow-melting systems: Part 2, Heat transfer in the slab—A simplified model. *ASHRAE Transactions* 100(1):434-441.

Mohinder, L.N. 2000. *Piping handbook,* 7th ed. McGraw-Hill, New York.

EVAPORATIVE COOLING

EVAPORATIVE cooling is energy-efficient, environmentally friendly, and cost-effective in many applications and all climates. Applications range from comfort cooling in residential, agricultural, commercial, and institutional buildings, to industrial applications for spot cooling in mills, foundries, power plants, and other hot environments. Several types of apparatus cool by evaporating water directly in the airstream, including (1) direct evaporative coolers, (2) spray-filled and wetted-surface air washers, (3) sprayed-coil units, and (4) humidifiers. Indirect evaporative cooling equipment combines the evaporative cooling effect in a secondary airstream with a heat exchanger to produce cooling without adding moisture to the primary airstream.

Direct evaporative cooling reduces the dry-bulb temperature and increases the relative humidity of the air. It is most commonly applied to dry climates or to applications requiring high air exchange rates. Innovative schemes combining evaporative cooling with other equipment have resulted in energy-efficient designs.

When temperature and/or humidity must be controlled within narrow limits, heat and mechanical refrigeration can be combined with evaporative cooling in stages. Evaporative cooling equipment, including unitary equipment and air washers, is covered in Chapter 40 of the 2008 *ASHRAE Handbook—HVAC Systems and Equipment*.

GENERAL APPLICATIONS

Cooling

Evaporative cooling is used in almost all climates. The wet-bulb temperature of the entering airstream limits direct evaporative cooling. The wet-bulb temperature of the secondary airstream limits indirect evaporative cooling.

Design wet-bulb temperatures are rarely higher than 78°F, making direct evaporative cooling economical for spot cooling, kitchens, laundries, agricultural, and industrial applications. At lower wet-bulb temperatures, evaporative cooling can be effectively used for comfort cooling, although some climates may require mechanical refrigeration for part of the year.

Indirect applications lower the air wet-bulb temperature and can produce leaving dry-bulb temperatures that approach the wet-bulb temperature of the secondary airstream. Using room exhaust as secondary air or incorporating precooled air in the secondary airstream lowers the wet-bulb temperature of the secondary air and further enhances the cooling capability of the indirect evaporative cooler.

Direct evaporative cooling is an adiabatic exchange of heat. Heat must be added to evaporate water. The air into which water is evaporated supplies the heat. The dry-bulb temperature is lowered, and sensible cooling results. The amount of heat removed from the air equals the amount of heat absorbed by the water evaporated as heat of

vaporization. If water is recirculated in the direct evaporative cooling apparatus, the water temperature in the reservoir approaches the wet-bulb temperature of the air entering the process. By definition, no heat is added to, or extracted from, an adiabatic process; the initial and final conditions fall on a line of constant wet-bulb temperature, which nearly coincides with a line of constant enthalpy.

The maximum reduction in dry-bulb temperature is the difference between the entering air dry- and wet-bulb temperatures. If air is cooled to the wet-bulb temperature, it becomes saturated and the process would be 100% effective. *Effectiveness* is the depression of the dry-bulb temperature of the air leaving the apparatus divided by the difference between the dry- and wet-bulb temperatures of the entering air. Theoretically, adiabatic direct evaporative cooling is less than 100% effective, although evaporative coolers are 85 to 95% (or more) effective.

When a direct evaporative cooling unit alone cannot provide desired conditions, several alternatives can satisfy application requirements and still be energy-effective and economical to operate. The recirculating water supplying the direct evaporative cooling unit can be increased in volume and chilled by mechanical refrigeration to provide lower leaving wet- and dry-bulb temperatures and lower humidity. Compared to the cost of using mechanical refrigeration only, this arrangement reduces operating costs by as much as 25 to 40%. Indirect evaporative cooling applied as a first stage, upstream from a second, direct evaporative stage, reduces both the entering dry- and wet-bulb temperatures before the air enters the direct evaporative cooler. Indirect evaporative cooling may save as much as 60 to 75% or more of the total cost of operating mechanical refrigeration to produce the same cooling effect. Systems may combine indirect evaporative cooling, direct evaporative cooling, heaters, and mechanical refrigeration, in any combination.

The psychrometric chart in Figure 1 illustrates what happens when air is passed through a direct evaporative cooler. In the example shown, assume an entering condition of 95°F db and 75°F wb. The initial difference is $95 - 75 = 20$°F. If the effectiveness is 80%, the depression is $0.80 \times 20 = 16$°F db. The dry-bulb temperature leaving the direct evaporative cooler is $95 - 16 = 79$°F. In the adiabatic evaporative cooler, only part of the water recirculated is assumed to evaporate and the water supply is recirculated. The recirculated water reaches an equilibrium temperature approximately the same as the wet-bulb temperature of the entering air.

The performance of an indirect evaporative cooler can also be shown on a psychrometric chart (Figure 1). Many manufacturers define effectiveness similarly for both direct and indirect evaporative cooling equipment. In indirect evaporative cooling, the cooling process in the primary airstream follows a line of constant moisture content (constant dew point). Indirect evaporative cooling effectiveness is the dry-bulb depression in the primary airstream divided by the difference between the entering dry-bulb temperature of the primary airstream and the entering wet-bulb temperature of the secondary air. Depending on heat exchanger design and relative

The preparation of this chapter is assigned to TC 5.7, Evaporative Cooling.

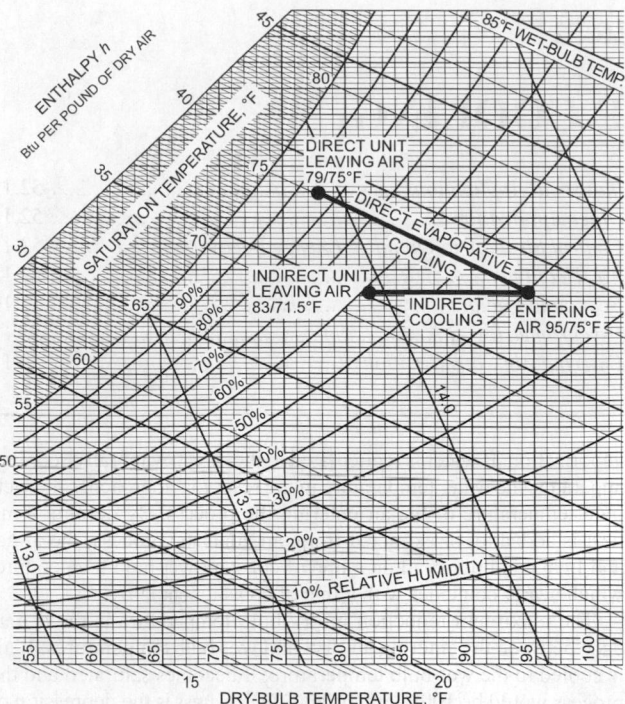

Fig. 1 Psychrometrics of Evaporative Cooling

quantities of primary and secondary air, effectiveness ratings may be as high as 85%.

Assuming 60% effectiveness, and assuming both primary and secondary air enter the apparatus at the outside condition of 95°F db and 75°F wb, the dry-bulb depression is 0.60(95 − 75) = 12°F. The dry-bulb temperature leaving the indirect evaporative cooling process is 95 − 12 = 83°F. Because the process cools without adding moisture, the wet-bulb temperature is also reduced. Plotting on the psychrometric chart shows that the final wet-bulb temperature is 71.5°F. Because both wet- and dry-bulb temperatures in the indirect evaporative cooling process are reduced, indirect evaporative cooling can substitute for part of the refrigeration load in many applications.

Humidification

Air can be humidified with a direct evaporative cooler by three methods: (1) using recirculated water without prior treatment of the air, (2) preheating the air and treating it with recirculated water, or (3) heating recirculated water. Air leaving an evaporative cooler used as either a humidifier or a dehumidifier is substantially saturated when in operation. Usually, the spread between leaving dry- and wet-bulb temperatures is less than 1°F. The temperature difference between leaving air and leaving water depends on the difference between entering dry- and wet-bulb temperatures and on certain physical features, such as the length and height of a spray chamber, cross-sectional area and depth of the wetted media used, quantity and velocity of air, quantity of water, and spray pattern. In any direct evaporative humidifier installation, air should not enter with a dry-bulb temperature of less than 39°F; otherwise, the water may freeze.

Recirculated Water. Except for the small amount of energy added by shaft work from the recirculating pump and the small amount of heat leakage into the apparatus through the unit enclosure, evaporative humidification is strictly adiabatic. As the recirculated liquid evaporates, its temperature approaches the thermodynamic wet-bulb temperature of the entering air.

The airstream cannot be brought to complete saturation, but its state point changes adiabatically along a line of constant wet-bulb

temperature. Typical saturation or humidifying effectiveness of various air washer spray arrangements is between 50 and 98%. The degree of saturation depends on the extent of contact between air and water. Other conditions being equal, low-velocity airflow is conducive to higher humidifying effectiveness.

Preheated Air. Preheating air increases both the dry- and wet-bulb temperatures and lowers the relative humidity; it does not, however, alter the humidity ratio (i.e., mass ratio of water vapor to dry air) or dew-point temperature of the air. At a higher wet-bulb temperature, but with the same humidity ratio, more water can be absorbed per unit mass of dry air in passing through the direct evaporative humidifier. Analysis of the process that occurs in the direct evaporative humidifier is the same as that for recirculated water. The desired conditions are achieved by heating to the desired wet-bulb temperature and evaporatively cooling at constant wet-bulb temperature to the desired dry-bulb temperature and relative humidity. Relative humidity of the leaving air may be controlled by (1) bypassing air around the direct evaporative humidifier or (2) reducing the number of operating spray nozzles or the area of media wetted.

Heated Recirculated Water. Heating humidifier water increases direct evaporative humidifier effectiveness. When heat is added to the recirculated water, mixing in the direct evaporative humidifier may still be modeled adiabatically. The state point of the mixture should move toward the specific enthalpy of the heated water. By raising the water temperature, the air temperature (both dry- and wet-bulb) may be raised above the dry-bulb temperature of entering air. The relative humidity of leaving air may be controlled by methods similar to those used with preheated air.

Dehumidification and Cooling

Direct evaporative coolers may also be used to cool and dehumidify air. If the entering water temperature is cooled below the entering wet-bulb temperature, both the dry- and wet-bulb temperatures of the leaving air are lowered. Dehumidification results if the leaving water temperature is maintained below the entering air dew point. Moreover, the final water temperature is determined by the sensible and latent heat absorbed from the air and the amount of circulated water, and it is 1 to 2°F below the final required dew-point temperature.

The air leaving a direct evaporative cooler being used as a dehumidifier is substantially saturated. Usually, the spread between dry- and wet-bulb temperatures is less than 1°F. The temperature difference between leaving air and leaving water depends on the difference between entering dry- and wet-bulb temperatures and on certain design features, such as the cross-sectional area and depth of the media or spray chamber, quantity and velocity of air, quantity of water, and the water distribution.

Air Cleaning

Direct evaporative coolers of all types perform some air cleaning. See the section on Air Cleaning later in this chapter for detailed information.

INDIRECT EVAPORATIVE COOLING SYSTEMS FOR COMFORT COOLING

Five types of indirect evaporative cooling systems are most commonly used for commercial, institutional, and industrial cooling applications. Figures 2 to 6 show schematics of these dry evaporative cooling systems.

Indirect evaporative cooling efficiency is measured by the approach of the outdoor air dry-bulb condition to either the room return air or scavenger outdoor air wet-bulb condition on the wet side of the air-to-air heat exchanger. The **indirect evaporative cooling percent effectiveness (IEE)** is expressed as follows:

$$IEE = \frac{t_1 - t_2}{t_1 - t_3} \times 100$$

where

t_1 = supply air inlet dry-bulb temperature, °F
t_2 = supply air outlet dry-bulb temperature, °F
t_3 = wet-side air inlet wet-bulb temperature, °F

The heat pipe air-to-air heat exchanger in Figure 2 uses a direct water spray from a recirculation sump on the wet side of the heat pipe tubes (Scofield and Taylor 1986). When either room return or scavenger outdoor air passes over the wet surface, outdoor air entering the building is dry-cooled and produces an approach to the wet-side wet-bulb temperature in the range of 60 to 80% IEE for equal mass flow rates on both sides of the heat exchanger. The IEE is a function of heat exchanger surface area, face velocity, and completeness of wetting achieved for the wet-side heat exchanger surface. Face velocities on the wet side are usually selected in the range of 400 to 450 fpm.

Figure 3 shows a cross-flow plate type air-to-air indirect evaporative cooling heat exchanger (Yellott and Gamero 1984). In this configuration, outdoor air is supplied for both wet- and dry-side flows through the heat exchanger. Recirculation water from the sump below the heat exchanger is sprayed downward through the vertical wet-side air path, counterflow to the air which moves vertically up through the heat exchanger. The horizontal airflow path is dry-cooled with a 60 to 80% IEE for equal mass flows on both sides of the heat exchanger. Again, the approach to the ambient wet bulb is a function of heat exchanger surface area and the effectiveness of the water spray system at completely wetting the wet-side surface of the air-to-air heat exchanger.

The heat wheel (Figure 4) and the run-around coil (Figure 5) both use a direct evaporative cooling component on the cold side to enhance the dry-cooling effect on the makeup air side. The heat wheel (sensible transfer), when sized for 500 fpm face velocity with equal mass flows on both sides, has an IEE around 60 to 70%. The run-around coil system at the same conditions produces an IEE of 35 to 50%. The adiabatic cooling component is usually selected for an effectiveness of 85 to 95%. Water coil freeze protection is required in cold climates for the run-around coil loop.

Air-to-air heat exchangers that are directly wetted produce a closer approach to the cold-side wet-bulb temperature, all things being equal. First cost, physical size, and parasitic losses are also reduced by direct wetting of the heat exchanger. In applications having extremely hard makeup water conditions, using a direct evaporative cooling device in lieu of directly wetting the air-to-air heat exchanger may reduce maintenance costs and extend the useful life of the system.

All of the air-to-air heat exchangers (Figures 2 to 5) produce beneficial winter heat recovery when using building return air with the sprays or adiabatic cooling component turned off.

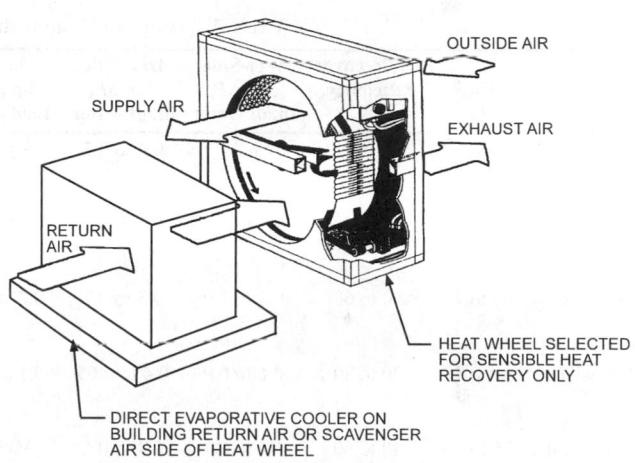

Fig. 4 Rotary Heat Exchanger with Direct Evaporative Cooling

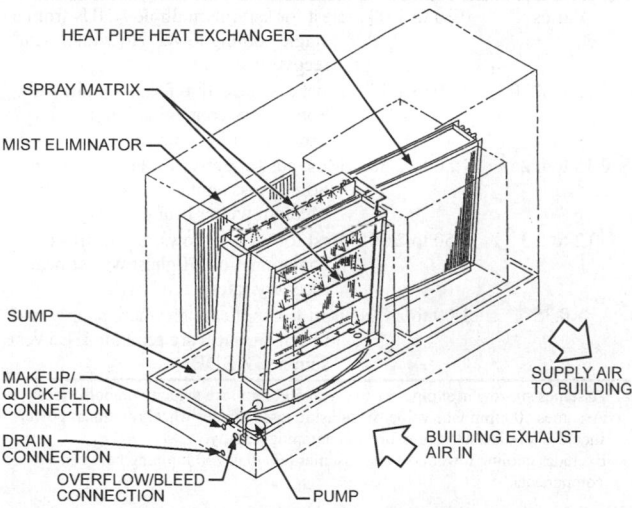

Fig. 2 Heat Pipe Air-to-Air Heat Exchanger with Sump Base

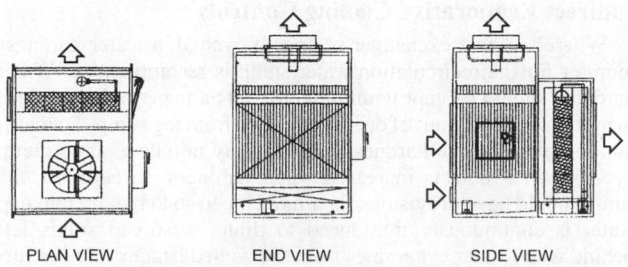

PLAN VIEW   END VIEW   SIDE VIEW

Fig. 3 Cross-Flow Plate Air-to-Air Indirect Evaporative Cooling Heat Exchanger with Direct Evaporative Cooling Second Stage

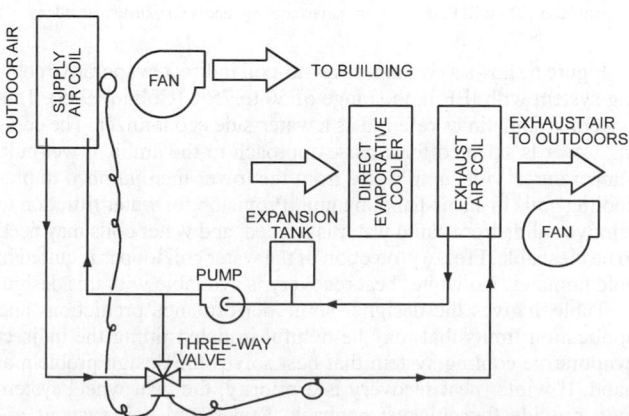

Fig. 5 Coil Energy Recovery Loop with Direct Evaporative Cooling

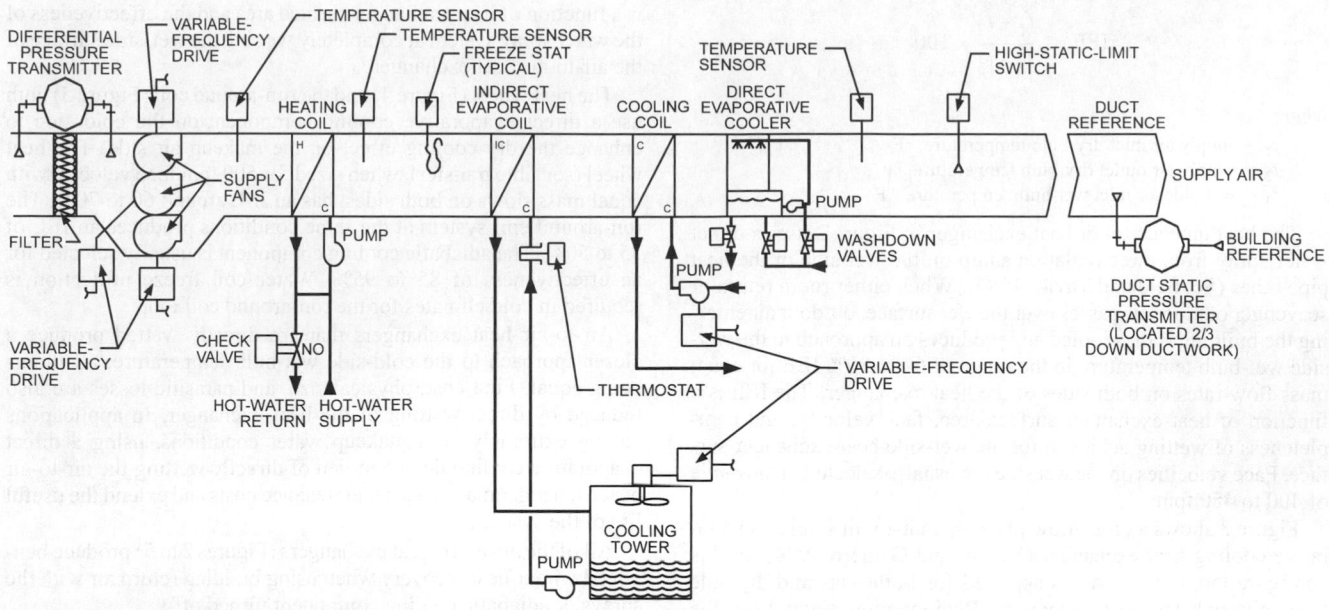

Fig. 6   Cooling-Tower-to-Coil Indirect Evaporative Cooling

Table 1   Indirect Evaporative Cooling Systems Comparison

System Type[a]	IEE,[b] %	Heat Recovery Efficiency, %	Wet-Side Air ΔP, in. of water	Dry-Side Air ΔP, in. of water	Pump hp per 10,000 cfm	Parasitic Loss Range,[e] kW/ton of Cooling	Equipment Cost Range,[f] $/Supply cfm	Notes
Cooling tower to coil	40 to 60	NA	NA	0.4 to 0.7	Varies	Varies	0.50 to 1.00	Best for serving multiple AHUs from a single cooling tower. No winter heat recovery.
Crossflow plate	60 to 85	40 to 50	0.7 to 1.0	0.4 to 0.7	0.1 to 0.2	0.12 to 0.20	1.20 to 1.70	Most cost-effective for lower airflows. Some cross contamination possible. Low winter heat recovery.
Heat pipe[c]	65 to 75	50 to 60	0.7 to 1.0	0.5 to 0.7	0.2 to 0.4	0.15 to 0.25	1.50 to 2.50	Most cost-effective for large airflows. Some cross contamination possible. Medium winter heat recovery.
Heat wheel[d]	60 to 70	70 to 80	0.6 to 0.9	0.4 to 0.65	0.1 to 0.2	0.2 to 0.3	1.50 to 2.50	Best for high airflows. Some cross contamination. Highest winter heat recovery rates.
Runaround coil[d]	35 to 50	40 to 60	0.6 to 0.8	0.4 to 0.6	Varies	> 0.35	1.00 to 2.00	Best for applications where supply and return air ducts are separated. Lowest summer WBDE.

IEE = indirect evaporative effectiveness
Notes:
[a]All air-to-air heat exchangers have equal mass flow on supply and exhaust sides.
[b]Plate and heat pipe are direct spray on exhaust side. Heat wheel and runaround coil systems use 90% WBDE direct evaporative cooling media on exhaust air side.

[c]Assumes six-row heat pipe, 11 fpi, with 500 fpm face velocity on both sides.
[d]Assumes 500 fpm face velocity. Parasitic loss includes wheel rotational power.
[e]Includes air-side static pressure and pumping penalty.
[f]Excludes cooling tower cost and assumes less than 200 ft piping between components.

Figure 6 shows a cooling-tower-to-coil indirect evaporative cooling system with IEE in the range of 50 to 75% (Colvin 1995). This system is sometimes referred as a water-side economizer. The cooling tower is selected for a close approach to the ambient wet-bulb temperature, with sump water from the tower then pumped to precooling coils in an air-handling unit. Provision for water filtration to remove solids from sump water is needed, and water coils may need to be cleanable. Freeze protection of the water coil loop is required in cold climates. No winter heat recovery is available with this design.

Table 1 gives the designer some performance predictions and application limits that may be helpful in determining the indirect evaporative cooling system that best solves the design problem at hand. If winter heat recovery is a priority, the heat wheel system may provide the quickest payback. Runaround coil systems are applied where supply air and exhaust air ducts are remote from each other. The heat pipe adapts well to high-volume air-handling systems where cooling energy reduction is the priority. The plate heat

exchanger fits smaller-volume systems with high cooling requirements but with lower winter heat recovery potential.

Indirect Evaporative Cooling Controls

Where the heat exchanger is directly wetted, a water hardness monitor for the recirculation water sump is recommended. Water hardness should be kept within 200 to 500 parts per million (ppm) to minimize plating out of dissolved solids from the sump. To maintain its set point, the hardness monitor may initiate a sump dump cycle when it detects increased water hardness. In addition, the sump should have provisions for a fixed bleed so that extra makeup water is continuously introduced to dilute dissolved solids left behind when water evaporates from the wetted heat exchanger surface. Sumps should always be drained at the end of a duty cycle and refilled the next day when the system is turned on. For rooftop applications, sumps should be drained for freeze protection during low ambient temperatures.

Air-side control for a cooling system with a 55°F supply air set point may be set up as follows. The heat exchanger's cold-side sprays or direct evaporative cooling component should be activated whenever ambient dry-bulb temperatures exceed 65°F, if room return air is used on the wet side of the air-to-air heat exchanger. Air-conditioned buildings have a stable return air wet-bulb condition in the range of 60 to 65°F. Valuable precooling of outdoor air may be achieved when ambient dry-bulb temperatures exceed the return air wet-bulb condition.

Where outdoor air is used on the cold-air side of the heat exchanger, cooling may begin at ambient temperatures above 55°F, because the wet-bulb condition of outdoor air is always lower than its dry-bulb condition.

Parasitic losses generated by the heat exchanger static pressure penalty to supply and return air fans and by the water pump need to be evaluated. These losses may be mitigated by opening bypass dampers around the heat exchanger for pressure relief and shutting off the pump in the ambient temperature range of 55 to 65°F db. Where scavenger outdoor air is used on the cold side of an air-to-air heat exchanger, this temperature range may be reduced somewhat. A comparison of the energy penalty to the precooling energy avoided determines the optimum range of ambient conditions for this control strategy.

For variable-air-volume (VAV) supply and return fan systems, the static penalty reduces by the square of the airflow reduction from full design flow at summer peak design condition. As airflow rates decrease across an air-to-air heat exchanger, the IEE increases, thereby providing better precooling. Where scavenger outdoor air is used for indirect evaporative cooling, the wet-side airflow rate is usually constant volume.

Winter heat recovery may be initiated at ambient temperatures below the 55°F supply air set point. Where building return air is used with an air-to-air heat exchanger, the 70 to 75°F return air condition is used to preheat makeup air for the building. For a VAV supply air system, Figure 7 shows the increased ventilation potential of a heat pipe air-to-air heat exchanger that uses face and bypass dampers on the supply air side to mix unheated outdoor air with preheated outdoor air to maintain the 55°F building supply air set point (Scofield and Bergman 1997). The heat pipe leaving air temperature may also be controlled with a tilt control (see Chapter 25 of the 2008 *ASHRAE Handbook—HVAC Systems and Equipment*). With a heat pipe economizer, a minimum outdoor air ventilation rate of 20% would not be breached until ambient temperatures dropped below –15°F.

Runaround coils control leaving supply air temperature with a three-way valve (see Figure 5). Because of their higher parasitic losses, these systems may require a wider range of ambient conditions where pressure-relief bypass dampers are open and the pump system shut down. Some projects limit activation of these recovery systems to ambient temperatures above 85°F to below 40°F.

Indirect/Direct Evaporative Cooling with VAV Delivery

Coupling indirect and direct evaporative cooling to a variable-air-volume (VAV) delivery system in arid climates can effectively eliminate requirements for mechanical refrigeration cooling in many applications. Many cities in the western United States have summer design conditions suitable to deliver 55°F or lower supply air to a building using a 70% IEE indirect and a 90% effective direct evaporative cooling system.

Figure 8 shows plan and elevation views of an air-handling unit using a sprayed heat pipe air-to-air heat exchanger and a wetted-media direct evaporative cooling section augmented by a final-stage chilled-water cooling coil (Scofield and Bergman 1997). The 70% indirect IEE is achieved with a direct-sprayed heat pipe using a sump and a recirculation water system on the building return air side of the heat exchanger. The 90% effective direct evaporative cooling medium is split into two sections for two-stage cooling capacity control of the 55°F leaving air temperature. The direct evaporative cooling system also uses a sump and water recirculation. Supply-side heat pipe face and bypass dampers control the final supply air temperature (55°F) in both summer (when indirect sprays are on) and winter, to control the heat pipe's heat recovery capacity. Heat pipe dampers on both sides of the heat exchanger are powered to full open to mitigate system parasitic losses during ambient temperature conditions when the value of energy recovered is exceeded by the fan energy penalty. The recirculation damper is used for morning warm-up of the building and for blending building return air with preheated outdoor air during extreme cold ambient conditions (see Figure 7).

Table 2 uses ASHRAE bin weather data for the semi-arid climate of Sacramento, California, to illustrate potential cooling energy savings for a 10,000 cfm VAV design that turns down to 5000 cfm at winter design (Scofield and Bergman 1997). Compared to a conventional-refrigeration cooling VAV design with a 25% minimum outdoor air economizer, the two-stage evaporative cooling system reduces peak cooling load by 49% while introducing 100% outdoor air. For a building duty cycle of 8760 h per year, the ton-hour savings is $34,037, or a 60% reduction compared to the conventional air-side economizer system with mechanical cooling only. Within ambient bin conditions of 62°F db/54°F wb to 57°F db/52°F wb, there are 2376 cooling hours per year (27% of the annual cooling hours) where a 90% wet-bulb depression efficiency (WBDE) direct evaporative cooling system may be used for the 55°F supply air requirement without refrigeration.

Figure 9 uses typical meteorological year (TMY) data for 14 cities in the western United States to illustrate the evaporative cooling annual refrigeration avoidance per 10,000 cfm of VAV supply air, compared to a 25% minimum outdoor air economizer (Scofield and Bergman 1997). For thermal energy storage (TES) applications, the two-stage evaporative cooling design may significantly reduce chiller plant storage capacity and refrigeration equipment first cost.

Benefits of this design in dry climates include the following:

• Indoor air quality is improved by using all outdoor air during cooling, and increased ventilation in winter through the heat pipe economizer (see Figure 7).
• Energy demand is in the range of 0.15 to 0.25 kW per ton of cooling, versus air-cooled refrigeration at 1.2 to 1.3 kW per ton.
• Peak building electrical cooling and gas heating demand requirements are reduced, especially for applications that require higher amounts of outdoor air.

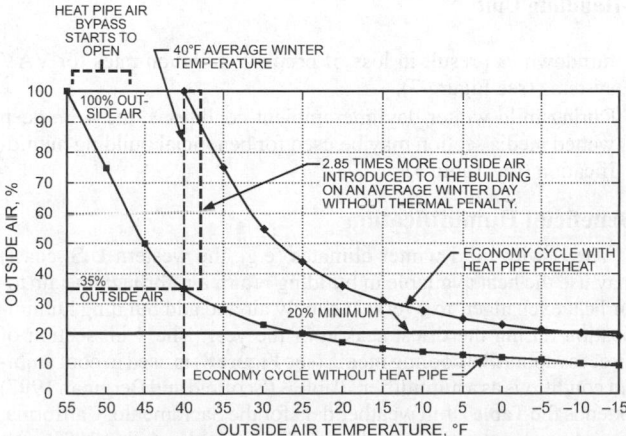

Fig. 7   Increased Winter Ventilation

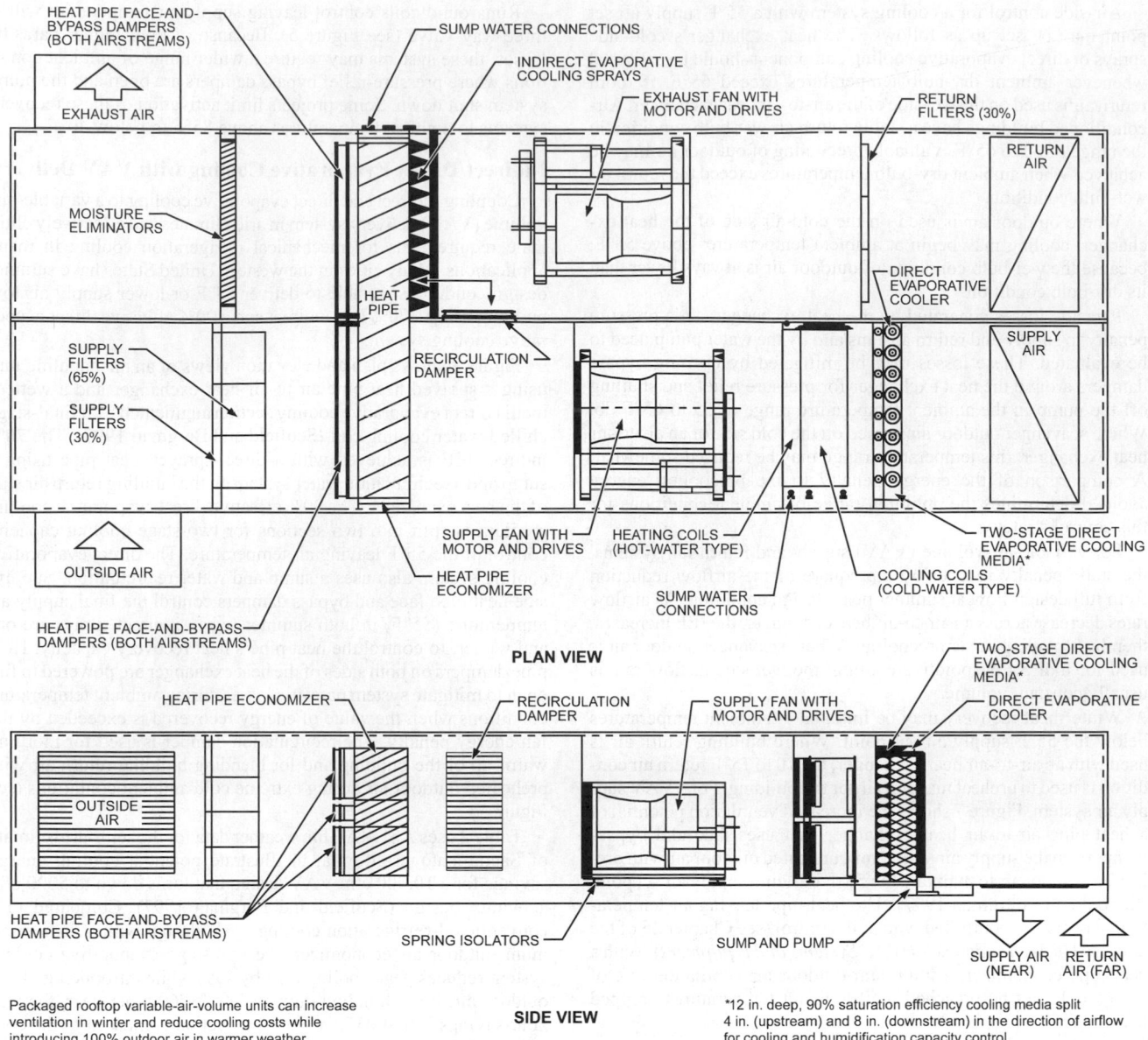

Packaged rooftop variable-air-volume units can increase ventilation in winter and reduce cooling costs while introducing 100% outdoor air in warmer weather.

SIDE VIEW

*12 in. deep, 90% saturation efficiency cooling media split 4 in. (upstream) and 8 in. (downstream) in the direction of airflow for cooling and humidification capacity control.

Fig. 8   Heat Pipe Air-Handling Unit

- Because VAV pinchdown terminals may reduce their minimum airflow settings and comply with ASHRAE *Standard* 62.1, supply and return fan energy savings are possible in cooler weather when using an all-outdoor-air design.
- VAV turndown of fans during cooler ambient conditions decreases fan parasitic energy losses because of the evaporative cooling system components.
- VAV turndown increases the IEE of both the air-to-air heat exchanger and direct evaporative cooling system.
- In semi-arid climates where a chilled-water final cooling stage is required, two-stage evaporative cooling allows central chilled-water plants to be turned off earlier in the fall and reactivated later in the spring. This results in significant maintenance and cooling energy cost savings.
- In cooler weather, resetting supply air down to 50°F and using only the direct evaporative cooler extends free cooling hours and reduces fan energy.
- When using building return air, winter heat recovery provides increased outdoor air quantities during the period when fan

turndown can result in loss of proper ventilation rates for VAV systems (see Figure 7).
- During mild winter daytime ambient conditions, the 4 in. deep wetted media section may be used for beneficial building humidification (see Table 3).

Beneficial Humidification

Areas with mild winter climates (e.g., the western U.S. coast) may use the heat available in building return air, through the air-to-air heat exchanger, to overheat supply air and add building humidification during the driest season of the year. The 4 in. section of direct evaporative cooling media (see Figure 8) is used in cool ambient conditions as a humidifier. Table 3 (Scofield and Bergman 1997) extends the Table 2 bin weather data for the Sacramento, California, site into winter ambient conditions. The table shows that 100% outdoor air may be introduced and humidity controlled between 54 and 32% for ambient conditions down to 37°F with a 60% heat pipe recovery effectiveness. ASHRAE *Standard* 62.1 recommends that humidities in occupied areas be maintained between 30 and 60% rh.

Table 2 Sacramento, California, Cooling Load Comparison

Outdoor Air db/wb, °F	VAV Supply, cfm	Hours Per Year[d]	100% Outdoor Air Indirect-Direct Evaporative Cooling				25% Outdoor Air Economizer		
			Indirect LAT db/wb, °F	Direct LAT db/wb, °F	Refrigeration,[a] tons	Refrigeration,[b] ton·h	Mixed Air db/ wb, °F	Refrigeration,[a] tons	Refrigeration,[b] ton·h
107/70	10,000	7	76.9/60	61.7/60	14.2	99.4	83/65.5	29.2	204.4
102/70	9,688	59	75.4/61.3	62.7/61.3	17.4	1026.6	81.8/65.5	28.3	1669.7
97/68	9,375	144	73.9/60.1	61.5/60.1	14	2016	80.5/65.0	25.9	3729.6
92/66	9,062	242	72.4/59.1	60.4/59.1	11.5	2783	79.2/64.5	23.7	5735.4
87/65	8,750	301	70.9/59.3	60.5/59.3	10.5	3160.5	78/64.3	22.3	6712.3
82/63	8,438	397	69.4/58.7	59.8/58.7	9.5	3771.5	76.8/63.9	20.3	8059.1
77/61	8,125	497	67.9/57.7	58.7/57.7	6.7	3329.9	77.5/63.3	18.3	9095.1
72/59	7,812	641	66.4/56.8	57.8/56.8	5.3	3397.3	72/59[b]	12.9	8268.9
67/57	7,500	821	64.9/56.0	56.9/56.0	3.9	3201.9	67/57[b]	9	7389
62/54	7,188	1086	62/54	54.8/54	0	0	62/54[b]	4.3	4669.8[c]
57/52	6,875	1290	57/52	52.5/52	0	0	57/52[b]	1	1290[c]
						Total ton·h = 22,786.1			Total ton·h = 56,823.3

LAT = leaving-air temperature

Notes:

[a]Amount of cooling required to reach 55°F db supply air requirements.

[b]Ambient conditions when dampers for air-side economizer introduce 100% outdoor air in arid climates.

[c]Ambient conditions when 90% saturation efficiency direct evaporative cooler may be used to eliminate refrigeration cooling.

Heat pipe bypass dampers should be open to minimize parasitic losses. Indirect water sprays should be off.

[d]Bin hours at each condition based on 24 h/day, 365 day/year duty cycle.

Table 3 Sacramento, California, Heat Recovery and Humidification

Outdoor Air db/wb, °F	VAV Supply,[a] cfm	Hours Per Year[b]	Heat Recovery Leaving Air db/wb, °F	Direct Evaporative Humidifier Leaving Air db/wb,[c] °F	Energy Savings,[d] Btu/h	Resultant Room rh
52/48	6562	1199	62.8/52.7	55/52.7	73,822	54%
47/44	6250	924	60.8/50.5	55/50.5	90,000	47%
42/40	5938	660	58.8/48.1	55/48.1	98,868	38%
37/36	5625	333	56.8/46.0	55/46.0	120,285	32%
32/31	5312	116	54.8/43.2	OFF	130,803	25%[e]
27/26	5000	30	52.8/41.0	OFF	250,776	21%[e]

Source: Scofield and Bergman (1997).

[a]VAV turndown airflow is assumed linear from summer design (10,000 cfm) to winter design (5000 cfm).

[b]Bin hours at each condition based on 24 h/day, 365 days/year duty cycle.

[c]Heat pipe overheats outdoor air to allow direct evaporative humidifier to add moisture.

[d]Recirculated building heat used for preheating 100% outdoor air and increasing humidity levels.

[e]Additional heat is required or recirculation damper must open during these bin conditions, to maintain both acceptable 30% indoor relative humidity and reach the 55°F supply air set point.

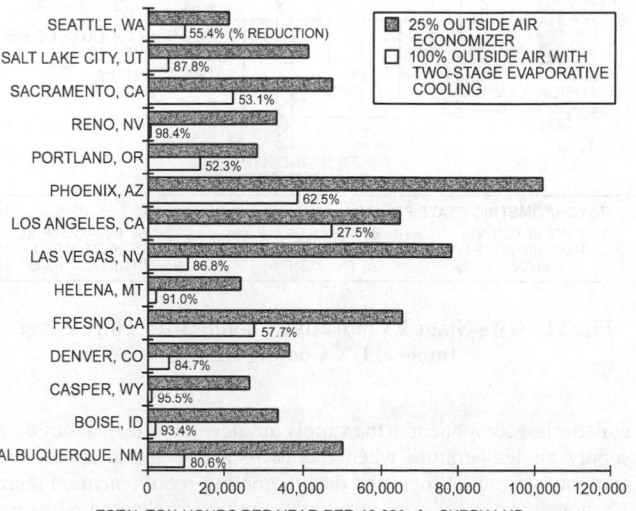

Note: Fourteen Western cities where indirect/direct evaporative cooling systems, using heat pipe (wet) indirect evaporative effectiveness of 70% and direct evaporative cooler saturation efficiency of 90%, can be used to introduce 100% outdoor air, with substantial reductions in ton-hour cooling requirements compared to conventional 24% outdoor air economizer damper design. Ton-hour totals for each system based on 24 h/day, 365 day/year duty cycle and 10,000 cfm VAV supply air. NREL hour by hour TMY data used to develop ton-hours listed. Fan heat not included.

Fig. 9 Refrigeration Reduction with Two-Stage Evaporative Cooling Design

There are only 146 bin hours below the 37°F ambient threshold during which the building recirculation air damper (see Figure 8) would have to open or additional heat be added with the hot-water coil to maintain the 55°F air delivery set point. The average winter temperature in Sacramento is 52.7°F, which is fairly typical for this region.

Indirect Evaporative Cooling With Heat Recovery

In indirect evaporative cooling, outside supply air passes through an air-to-air heat exchanger and is cooled by evaporatively cooled air exhausted from the building or application. The two airstreams never mix or come into contact, so no moisture is added to the supply airstream. Cooling the building's exhaust air results in a larger overall temperature difference across the heat exchanger and a greater cooling of the supply air. Indirect evaporative cooling requires only fan and water pumping power, so the coefficient of performance tends to be high. The principle of indirect evaporative cooling is effective in most air-conditioned buildings, because evaporative cooling is applied to exhaust air rather than to outside air.

Indirect evaporative cooling has been applied in a number of heat recovery applications (Mathur et al. 1993), such as plate heat exchangers (Scofield and DesChamps 1984; Wu and Yellot 1987), heat pipe exchangers (Mathur 1998; Scofield 1986), rotary regenerative heat exchangers, and two-phase thermosiphon loop heat exchangers (Mathur 1990). In residential air conditioning, the outside condensing unit can be evaporatively cooled to enhance performance (Mathur 1997; Mathur and Goswami 1995; Mathur et al. 1993). Indirect evaporative cooling with heat recovery is covered in

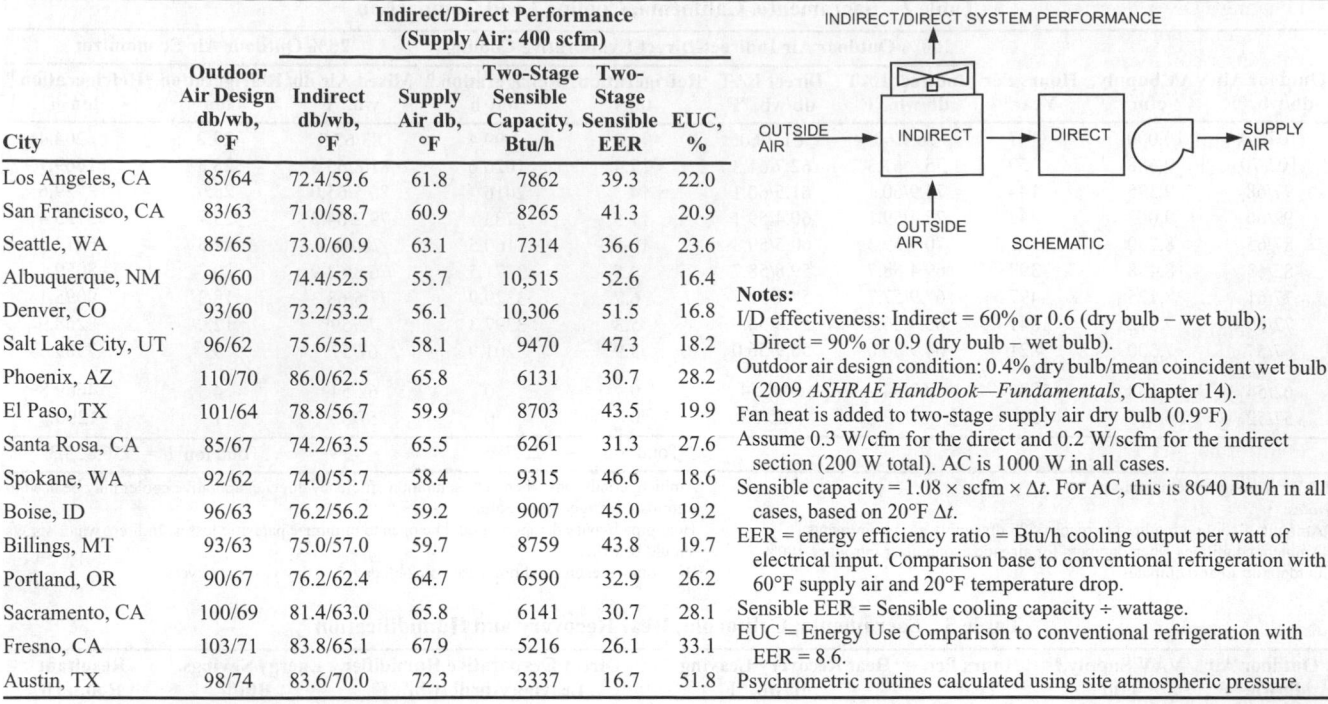

City	Outdoor Air Design db/wb, °F	Indirect db/wb, °F	Supply Air db, °F	Two-Stage Sensible Capacity, Btu/h	Two-Stage Sensible EER	EUC, %
Los Angeles, CA	85/64	72.4/59.6	61.8	7862	39.3	22.0
San Francisco, CA	83/63	71.0/58.7	60.9	8265	41.3	20.9
Seattle, WA	85/65	73.0/60.9	63.1	7314	36.6	23.6
Albuquerque, NM	96/60	74.4/52.5	55.7	10,515	52.6	16.4
Denver, CO	93/60	73.2/53.2	56.1	10,306	51.5	16.8
Salt Lake City, UT	96/62	75.6/55.1	58.1	9470	47.3	18.2
Phoenix, AZ	110/70	86.0/62.5	65.8	6131	30.7	28.2
El Paso, TX	101/64	78.8/56.7	59.9	8703	43.5	19.9
Santa Rosa, CA	85/67	74.2/63.5	65.5	6261	31.3	27.6
Spokane, WA	92/62	74.0/55.7	58.4	9315	46.6	18.6
Boise, ID	96/63	76.2/56.2	59.2	9007	45.0	19.2
Billings, MT	93/63	75.0/57.0	59.7	8759	43.8	19.7
Portland, OR	90/67	76.2/62.4	64.7	6590	32.9	26.2
Sacramento, CA	100/69	81.4/63.0	65.8	6141	30.7	28.1
Fresno, CA	103/71	83.8/65.1	67.9	5216	26.1	33.1
Austin, TX	98/74	83.6/70.0	72.3	3337	16.7	51.8

Notes:
I/D effectiveness: Indirect = 60% or 0.6 (dry bulb – wet bulb);
   Direct = 90% or 0.9 (dry bulb – wet bulb).
Outdoor air design condition: 0.4% dry bulb/mean coincident wet bulb
   (2009 *ASHRAE Handbook—Fundamentals*, Chapter 14).
Fan heat is added to two-stage supply air dry bulb (0.9°F)
Assume 0.3 W/cfm for the direct and 0.2 W/scfm for the indirect
   section (200 W total). AC is 1000 W in all cases.
Sensible capacity = $1.08 \times$ scfm $\times \Delta t$. For AC, this is 8640 Btu/h in all
   cases, based on 20°F Δt.
EER = energy efficiency ratio = Btu/h cooling output per watt of
   electrical input. Comparison base to conventional refrigeration with
   60°F supply air and 20°F temperature drop.
Sensible EER = Sensible cooling capacity ÷ wattage.
EUC = Energy Use Comparison to conventional refrigeration with
   EER = 8.6.
Psychrometric routines calculated using site atmospheric pressure.

Fig. 10   Indirect/Direct Two-Stage System Performance

detail in Chapter 25 of the 2008 *ASHRAE Handbook—HVAC Systems and Equipment*.

BOOSTER REFRIGERATION

Staged evaporative coolers can completely cool office buildings, schools, gymnasiums, sports facilities, department stores, restaurants, factory space, and other buildings. These coolers can control room dry-bulb temperature and relative humidity, even though one stage is a direct evaporative cooling stage. In many cases, booster refrigeration is not required. Supple (1982) showed that even in higher-humidity areas with a 1% mean wet-bulb design temperature of 75°F, 42% of the annual cooling load can be satisfied by two-stage evaporative cooling. Refrigerated cooling need supply only 58% of the load.

Figure 10 shows indirect/direct two-stage performance for 16 cities in the United States. Performance is based on 60% effectiveness of the indirect stage and 90% for the direct stage. Supply air temperatures (leaving the direct stage) at the 0.4% design dry-bulb mean coincident wet-bulb condition range from 56.1 to 72.3°F. Energy use ranges from 16.4 to 51.8%, compared to conventional refrigerated equipment.

Booster mechanical refrigeration provides inside design comfort conditions regardless of the outside wet-bulb temperature without having to size the mechanical refrigeration equipment for the total cooling load. If the inside humidity level becomes uncomfortable, the quantity of moisture introduced into the airstream must be limited to control room humidity. Where the upper relative humidity design level is critical, a life-cycle cost analysis favors a design with an indirect cooling stage and a mechanical refrigeration stage.

Figure 11 shows an air-handling unit design that uses building return air instead of outdoor air to develop the indirect (dry) evaporative cooling effect with a direct-sprayed, heat pipe, air-to-air, heat exchanger (Felver et al. 2001). The humid, cool air off the heat pipe is then used to reject the heat of refrigeration at a condenser coil downstream of the exhaust fan. The direct expansion (DX) cooling

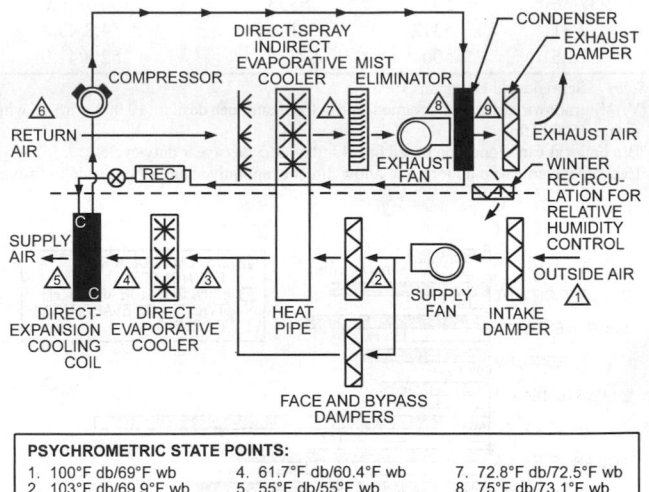

PSYCHROMETRIC STATE POINTS:

1. 100°F db/69°F wb	4. 61.7°F db/60.4°F wb	7. 72.8°F db/72.5°F wb
2. 103°F db/69.9°F wb	5. 55°F db/55°F wb	8. 75°F db/73.1°F wb
3. 74°F db/60.4°F wb	6. 75°F db/63°F wb	9. 95°F db/78.4°F wb

Fig. 11   Two-Stage Evaporative Cooling with Third-Stage Integral DX Cooling Design

coil, the last component in the supply air, develops the final building supply air temperature when the two-stage evaporative cooling components cannot meet the design cooling requirements. Figure 12 shows the process points for both supply and exhaust airstreams, using the Stockton, California, ASHRAE 0.4% summer dry-bulb design ambient condition. Several benefits accrue from this evaporative cooling design:

• Building return air has a more predictable and stable wet-bulb condition (60 to 65°F) than ambient air for use in generating the first stage of indirect (dry) evaporative cooling. Daytime absorption of moisture inside most buildings further enhances the first-stage cooling effect.

- Locating a DX condenser coil in sprayed exhaust off the heat pipe results in a more efficient rejection of refrigeration heat than a condenser coil located outdoors in the ambient air.
- Lower refrigeration condensing temperatures increase compressor capacity and compressor life, and reduce energy consumption.
- Central chilled-water plant or remote chiller installation and piping costs are eliminated.
- Evaporative cooling components provide back-up cooling capability in case of compressor failure. Figure 12 shows equilibrium conditions in the occupied area with indirect/direct evaporative cooling only.
- Peak refrigeration demand can be reduced 14 to 40% in California's semi-arid climate (Scofield 1994).
- Blow-through supply fan and draw-through exhaust fans provide:
  - Reduced supply fan heat addition for DX cooling system.
  - Reduced risk of cross contamination of supply air with exhaust air for hospital or laboratory applications.
  - Reduced fan noise breakout into building duct system.

There are several design considerations for the successful integration of DX refrigeration with two-stage evaporative cooling air-handling units, as shown in Figure 11.

For both constant-volume (CV) and variable-air-volume (VAV) units, the return air must closely match the supply airflow to ensure adequate heat rejection at the condenser coil. Buildings with large fixed-exhaust systems may not provide sufficient building return airflow for absorption of refrigeration heat at acceptable refrigerant condensing temperatures.

Face-and-bypass dampers are required around the condenser coil for control of the refrigerant condensing pressure and temperature.

Note that peak refrigeration requirements always occur during the highest ambient humidity (dew-point design) conditions. In semi-arid climates, this design condition occurs during reduced summer ambient dry-bulb temperatures (Ecodyne Corp. 1980). Review of site ASHRAE dew-point design conditions (Chapter 14 of the 2009 *ASHRAE Handbook—Fundamentals*) is required to determine the peak refrigeration cooling capacity needed to maintain the specified supply air temperature set point to the building.

RESIDENTIAL OR COMMERCIAL COOLING

In dry climates, evaporative cooling is effective at lower air velocities than those required in humid climates. Packaged direct evaporative coolers are used for residential and commercial application. Cooler capacity may be determined from standard heat gain calculations (see Chapters 17 and 18 of the 2009 *ASHRAE Handbook—Fundamentals*).

Detailed calculation of heat load, however, is usually not economically justified. Instead, one of several estimates gives satisfactory results. In one method, the difference between dry-bulb design temperature and coincident wet-bulb temperature divided by 10 is equal to the number of minutes needed for each air change. This or any other arbitrary method for equating cooling capacity with airflow depends on a direct evaporative cooler effectiveness of 70 to 80%. Obviously, the method must be modified for unusual conditions such as large unshaded glass areas, uninsulated roof exposure, or high internal heat gain. Also, such empirical methods make no attempt to predict air temperature at specific points; they merely establish an air quantity for use in sizing equipment.

Example 1. An indirect evaporative cooler is to be installed in a 50 by 80 ft one-story office building with a 10 ft ceiling and a flat roof. Outside design conditions are assumed to be 95°F db and 65°F wb. The following heat gains are to be used in the design:

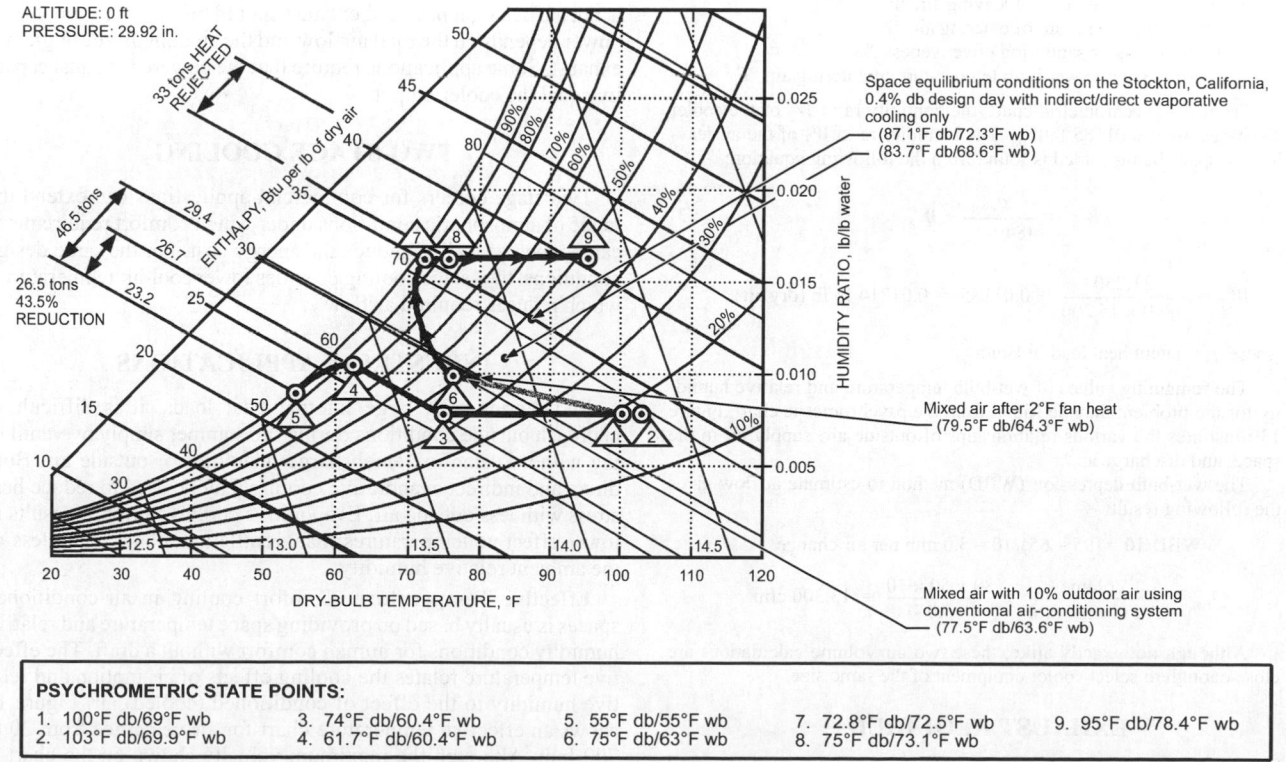

PSYCHROMETRIC STATE POINTS:

1. 100°F db/69°F wb	3. 74°F db/60.4°F wb	5. 55°F db/55°F wb	7. 72.8°F db/72.5°F wb	9. 95°F db/78.4°F wb
2. 103°F db/69.9°F wb	4. 61.7°F db/60.4°F wb	6. 75°F db/63°F wb	8. 75°F db/73.1°F wb	

Fig. 12   Psychrometrics of 100% OA, Two-Stage Evaporative Cooling Design (20,000 cfm Supply, 18,000 cfm Return) Compared with 10% OA Conventional System Operating at Stockton, California, ASHRAE 0.4% db Design Condition

	Heat Gains, Btu/h
All walls, doors, and roof	78,500
Glass area	5,960
Occupants (sensible load)	17,000
Lighting	62,700
Total sensible heat load	164,160
Total latent load (occupants)	21,250
Total heat load	185,410

Find the required air quantity, the temperature and humidity ratio of the air leaving the cooler (entering the office), and the temperature and humidity ratio of the air leaving the office.

Solution: A temperature rise of 10°F in the cooling air is assumed. The airflow rate that must be supplied by the indirect evaporative cooler may be found from the following equation:

$$Q_{ra} = \frac{q_s}{60 \rho c_p (t_1 - t_s)} = \frac{164,160}{60 \times 0.018 \times 10} = 15,200 \text{ cfm} \quad (1)$$

where

Q_{ra} = required airflow, cfm
q_s = instantaneous sensible heat load, Btu/h
t_1 = inside air dry-bulb temperature, °F
t_s = room supply air dry-bulb temperature, °F
ρc_p = density times specific heat of air ≈ 0.018 Btu/ft³·°F

This air volume represents a 2.6 min ($50 \times 80 \times 10/15,200$) air change for a building of this size. The indirect evaporative air cooler is assumed to have a saturation effectiveness of 80%. This is the ratio of the reduction of the dry-bulb temperature to the wet-bulb depression of the entering air. The dry-bulb temperature of the air leaving the indirect evaporative cooler is found from the following equation:

$$t_2 = t_1 - \frac{e_h}{100}(t_1 - t') = 95 - \frac{80}{100}(95 - 65) = 71°F \quad (2)$$

where

t_2 = dry-bulb temperature of leaving air, °F
t_1 = dry-bulb temperature of entering air, °F
e_h = humidifying or saturating effectiveness, %
t' = thermodynamic wet-bulb temperature of entering air, °F

From the psychrometric chart, the humidity ratio W_2 of the cooler discharge air is 0.01185 lb/lb$_{da}$. The humidity ratio W_3 of the air leaving the space being cooled is found from the following equation:

$$W_3 = \frac{q_e}{4840 Q_{ra}} + W_2 \quad (3)$$

$$W_3 = \frac{21,250}{4840 \times 15,200} + 0.01185 = 0.01214 \text{ lb/lb (dry air)}$$

where q_e = latent heat load in Btu/h.

The remaining values of wet-bulb temperature and relative humidity for the problem may be found from the psychrometric chart. Figure 13 illustrates the various relationships of outside air, supply air to the space, and discharge air.

The wet-bulb depression (WBD) method to estimate airflow gives the following result:

$$\text{WBD}/10 = (95 - 65)/10 = 3.0 \text{ min per air change}$$

$$Q_{ra} = \frac{\text{Volume}}{\text{Air change rate}} = \frac{80 \times 50 \times 10}{3.0} = 13,300 \text{ cfm}$$

Although not exactly alike, these two air volume calculations are close enough to select cooler equipment of the same size.

EXHAUST REQUIRED

If air is not exhausted freely, the increased static pressure will reduce airflow through the evaporative cooler. The result is a marked increase in the moisture and heat absorbed per unit mass of

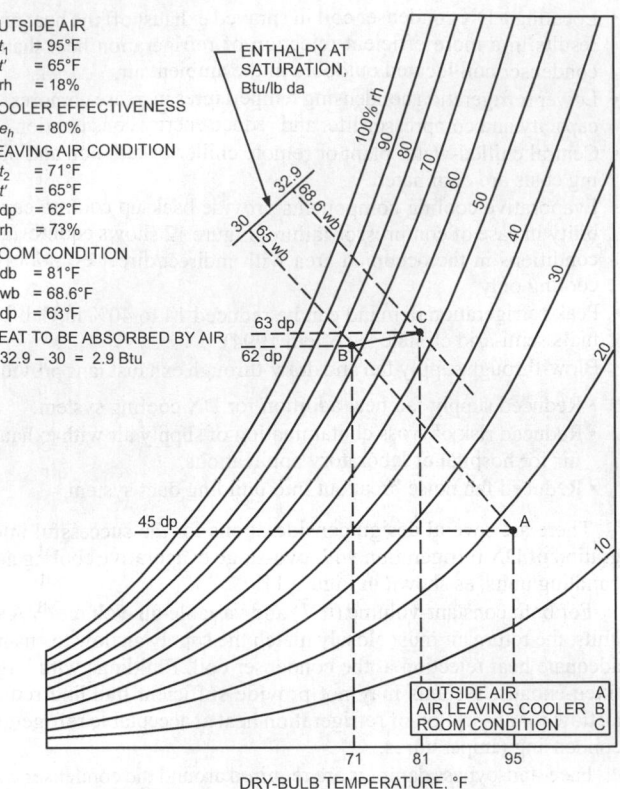

Fig. 13 Psychrometric Diagram for Example 1

air velocity in the room. The combination of these effects reduces the comfort level. Properly designed systems should have a minimum of 2 ft² of exhaust area for every 1000 cfm. If the exhaust area is not sufficient, a powered exhaust should be used. The amount of power depends on the total airflow and the amount of free or gravity exhaust. Some applications require that the powered exhaust capacity equal the cooler output.

TWO-STAGE COOLING

Two-stage coolers for commercial applications can extend the range of atmospheric conditions under which comfort requirements can be met, as well as reduce the energy cost. For the same design conditions, two-stage cooling provides lower cool-air temperatures, which reduces required airflow.

INDUSTRIAL APPLICATIONS

In factories with large internal heat loads, it is difficult to approach outside conditions during the summer simply by ventilating without using extremely large quantities of outside air. Both direct and indirect evaporative cooling may be used to reduce heat stress with less outside air. Evaporative cooling normally results in lower effective temperatures than ventilation alone, regardless of the ambient relative humidity.

Effective Temperature. Comfort cooling in air-conditioned spaces is usually based on providing space temperature and relative humidity conditions for human comfort without a draft. The effective temperature relates the cooling effects of air motion and relative humidity to the effect of conditioned (cooled) air. Figure 14 shows an effective temperature chart for air velocities from 20 to 700 fpm. Although the maximum velocity shown on the chart is 700 fpm, workers exposed to high-heat-producing operations may prefer air movement up to 4000 fpm to offset the radiant heat effect of equipment. Because the normal working range of the chart is approximately midway between the vertical dry- and wet-bulb scales, changes in either dry- or wet-bulb temperatures have similar effects on worker comfort. A reduction in either one decreases the effective temperature by about one-half of the reduction. Lines ED and CD on the chart illustrate this.

A condition of 95°F db and 75°F wb was chosen as the original state, because this condition is usually considered the summer design criterion in most areas. Reducing the temperature 15°F by evaporating water adiabatically provides an effective temperature reduction of 5.5°F for air moving at 20 fpm and a reduction of 9.5°F for air moving at 700 fpm, an improvement of 4°F.

The reduction in dry-bulb temperature through water evaporation increases the effectiveness of the cooling power of moving air in this example by 137%. On Line ED, the effective temperature varies from 83°F at 20 fpm to 79.5°F at 700 fpm with unconditioned air, whereas Line CD indicates an effective temperature of 77.5°F at 20 fpm and 70°F at 700 fpm with air cooled by a simple direct evaporative process. In the unconditioned case, increasing the air velocity from 20 to 700 fpm resulted in only a 3.5°F decrease in effective temperature. This contrasts with a 7.5°F decrease in effective temperature for the same range of air movement when the dry-bulb temperature was lowered by water evaporation. This demonstrates that direct evaporative cooling can provide a more comfortable environment regardless of geographical location.

Two methods are demonstrated to illustrate the environmental improvement that may be achieved with evaporative coolers. In one method, shown in Figure 15, temperature is plotted against time of day to illustrate effective temperature depression over time. Curve A shows ambient maximum dry-bulb temperature recordings. Curve B shows the corresponding wet-bulb temperatures. Curve C depicts the effective temperature when unconditioned air is moved over a person at 300 fpm. Curve D illustrates air conditioned in an 80% effective direct evaporative cooler before being projected over the person at 300 fpm. Curve E shows the additional decrease in effective temperature with air velocities of 700 fpm. Although a maximum suggested effective temperature of 80°F is briefly exceeded with unconditioned air at 300 fpm (Curve C), both the differential and total hours are substantially reduced from still-air conditions. Curves D and E illustrate that, in spite of the high wet-bulb temperatures, the in-plant environment can be continuously maintained below the suggested upper limit of 80°F effective temperature. This demonstration assumes that the combination of air velocity, duct length, and insulation between evaporative cooler and duct outlet is such that there is little heat transfer between air in ducts and warmer air under the roof.

Figure 16 illustrates another method of demonstrating the effect of using direct evaporative coolers by plotting effective comfort zones using ambient wet- and dry-bulb temperatures on a psychrometric chart (Crow 1972). The dashed lines show the expected improvement when using an 80% effective direct evaporative cooler.

Area Cooling

Both direct and indirect evaporative cooling may be used for area or spot cooling of industrial buildings. Both can be controlled either automatically or manually. In addition, evaporative coolers can supply tempered air during fall, winter, and spring. Gravity or power ventilators exhaust the air. Area cooling works well in buildings where personnel move about and workers are not subjected to concentrated, radiant heat sources. Area cooling may be used in either high- or low-bay industrial buildings, but may provide significant advantages in high-bay construction where cooling loads associated with roofs, lighting, and heat from equipment may be

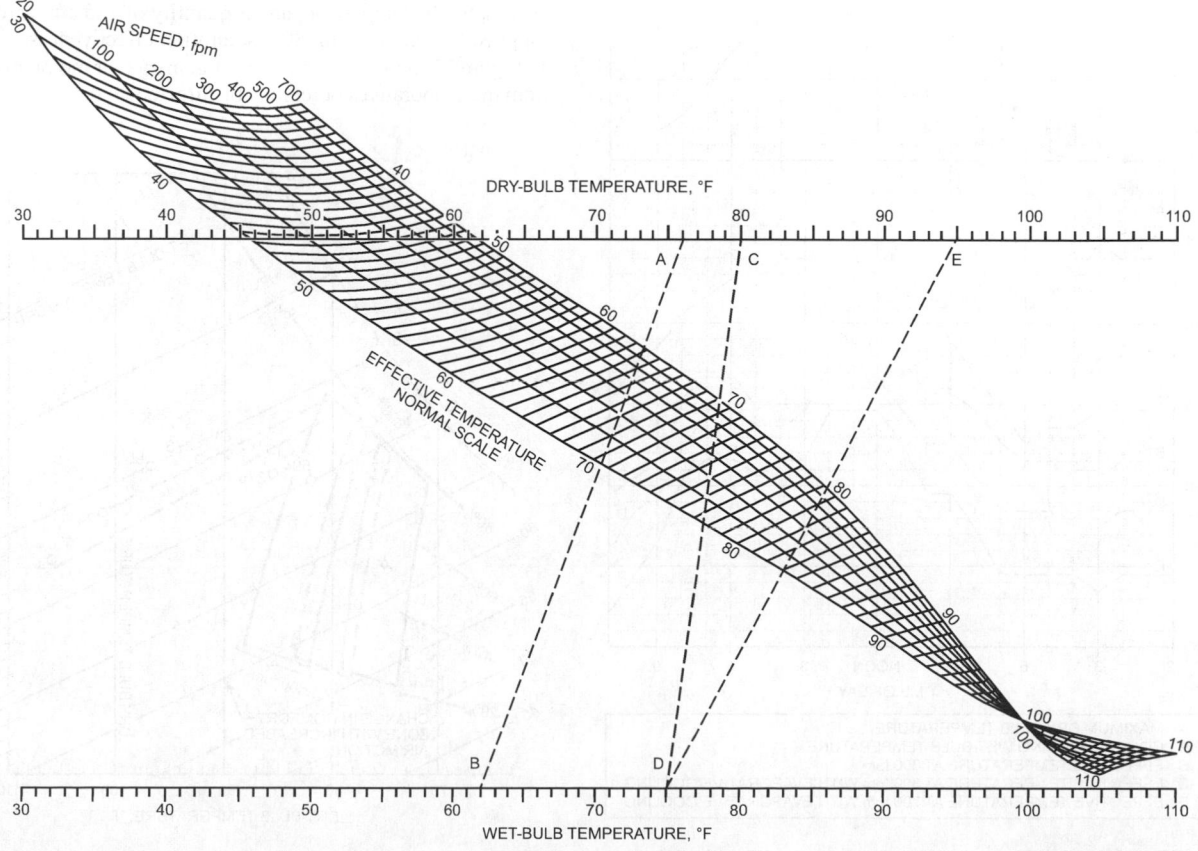

Fig. 14 Effective Temperature Chart

effectively eliminated by taking advantage of stratification. When cooling an area, ductwork should be designed to distribute air to the lower 10 ft of the space to ensure that cooler air is supplied to the workers.

Cooling requirements change from day to day and season to season, so, if discharge grilles are used, they should be adjustable to prevent drafts. The horizontal blades of an adjustable grille can be adjusted so that air is discharged above workers' heads rather than directly on them. In some cases, the air volume can be adjusted, either at each outlet or for the entire system, in which case the exhaust volume may need to be varied accordingly.

Spot Cooling

Spot cooling is a more efficient use of equipment when personnel work in one spot. Cool air is brought to the spot at levels below 10 ft, and may even be delivered from floor outlets. Duct height may depend on the location of other equipment in the area. For best results, air velocity should be kept low. Controls may be automatic or manual, with the fan often operating throughout the year. Workers are especially appreciative of spot cooling in hot environments, such as in chemical plants and die casting shops, and near glass-forming machines, billet furnaces, and pig and ingot casting.

When spot-cooling a worker, the air volume depends on the throw of the air jet, activity of the worker, and amount of heat that must be overcome. Air volumes can vary from 200 to 5000 cfm per worker, with target velocities ranging between 200 to 4000 fpm. Outlets should be between 4 to 10 ft from workstations to avoid entrainment of warm air and to effectively blanket workers with cooler air. Provisions should be made for workers to control the direction of air discharge, because air motion that is appropriate for hot weather may be

too great for cool weather or even cool mornings. Volume controls may be required to prevent overcooling the building and to minimize excessive grille blade adjustment.

Spot cooling is useful in rooms with elevated temperatures, regardless of climatic or geographical location. When the dry-bulb temperature of the air is below skin temperature, convection rather than evaporation cools workers. In these conditions, an 80°F airstream can provide comfort regardless of its relative humidity.

Cooling Large Motors

Electrical generators and motors are generally rated for a maximum ambient temperature of 104°F. When this temperature is exceeded, excessive temperatures develop in the electrical windings unless the load on the motor or generator is reduced. By providing evaporatively cooled air to the windings, this equipment may be safely operated without reducing the load. Likewise, transformer capacity can be increased using evaporative cooling.

The heat emitted by high-capacity electrical equipment may also be sufficient to raise the ambient condition to an uncomfortable level. With mill drive motors, an additional problem is often encountered with the commutator. If the air used to ventilate the motor is dry, the temperature rise through the motor results in a still lower relative humidity, at which the brush film can be destroyed, with unusual brush and commutator wear as well as the occurrence of dusting.

As a rule, a motor with a temperature rise of 25°F requires approximately 120 cfm of ventilating air per kilowatt hour of loss. If inlet air to the motor is 95°F, air leaving the motor would be 120°F. This average motor temperature of over 107°F is 3°F higher than it should be for the normal 104°F ambient. The same quantity of 95°F db inlet air at 75°F wb can be cooled by a direct evaporative cooler with a 97% saturation effectiveness. The resulting 88°F average motor temperature eliminates the need for special high-temperature insulation and improves the motor's ability to absorb temporary overloads. By comparison, an air quantity of 185 cfm is required if supplied by a cooler with 80% saturation effectiveness.

Figure 17 shows three basic arrangements for motor cooling. Air from the evaporative cooler may be directed on the motor windings,

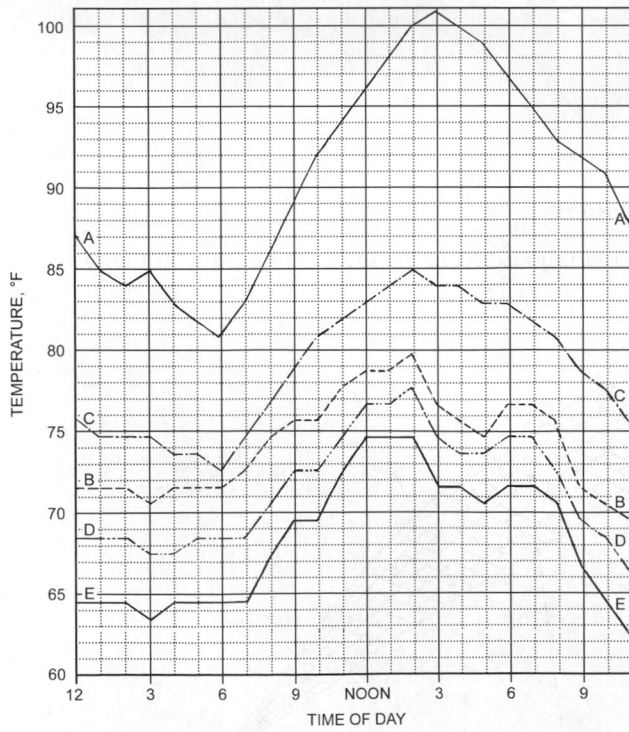

A   MAXIMUM DRY-BULB TEMPERATURE
B   CORRESPONDING WET-BULB TEMPERATURE
C   EFFECTIVE TEMPERATURE AT 300 fpm
D   EFFECTIVE TEMPERATURE AT 300 fpm WITH EVAPORATIVE COOLING
E   EFFECTIVE TEMPERATURE AT 700 fpm WITH EVAPORATIVE COOLING

Fig. 15   Effective Temperature for Summer Day in Kansas City, Missouri (Worst-Case Basis)

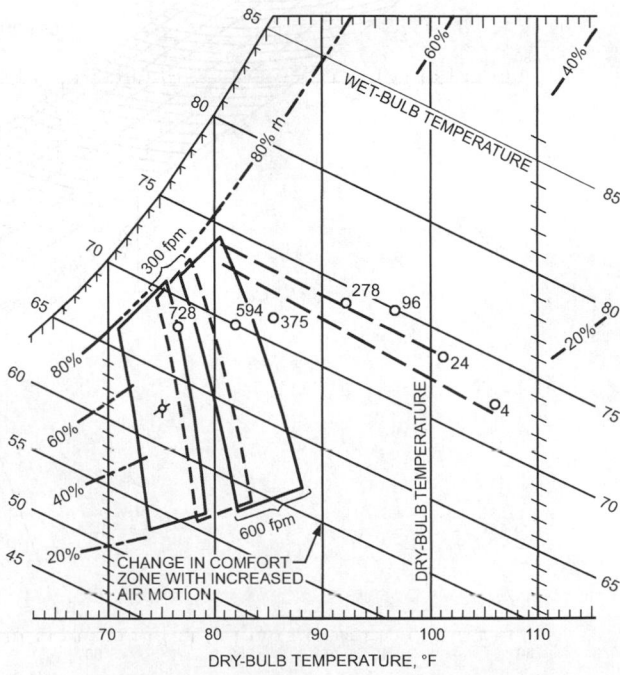

Fig. 16   Change in Human Comfort Zone as Air Movement Increases

or into the room, which requires an increased air volume to compensate for the building heat load. Operation of a direct evaporative cooler should be keyed to motor operation to ensure that (1) saturated or nearly saturated air is never introduced into a motor until it has had time to warm up, and (2) if more than one motor is served by a single system, air circulation through idle motors should be prevented.

Cooling Gas Turbine Engines and Generators

Combustion turbines used for electric power production are normally rated at 59°F. Their performance is greatly influenced by the compressor inlet air temperature because temperature affects air density and therefore mass flow. As ambient temperature increases, demand on electric utilities increases and the capacity of the combustion turbine decreases. Recovery of capacity because of inlet air cooling is approximately 0.4%/°F (cooling). Direct and indirect evaporative cooling is beneficial to gas turbine performance in almost all climates because when the air is the hottest, it generally has the lowest relative humidity. Expected increases in output using direct evaporative cooling range from 5.8% in Albany, New York, to 14% in Yuma, Arizona. In addition to increasing gas turbine output, direct evaporative cooling also improves heat rate and reduces NO_x emissions.

For an installation of this type, the following precautions must be taken: (1) mist eliminators must be provided to stop entrainment of free moisture droplets, (2) coolers must be turned off at a temperature below 45°F to prevent icing, and (3) water quality must be monitored closely (Stewart 1999).

Process Cooling

In the manufacture of textiles and tobacco and in processes such as spray coating, the required accurate relative humidity control can be provided by direct evaporative coolers. For example, textile manufacturing requires relatively high humidity and the machinery load is heavy, so a split system is customarily used to introduce free

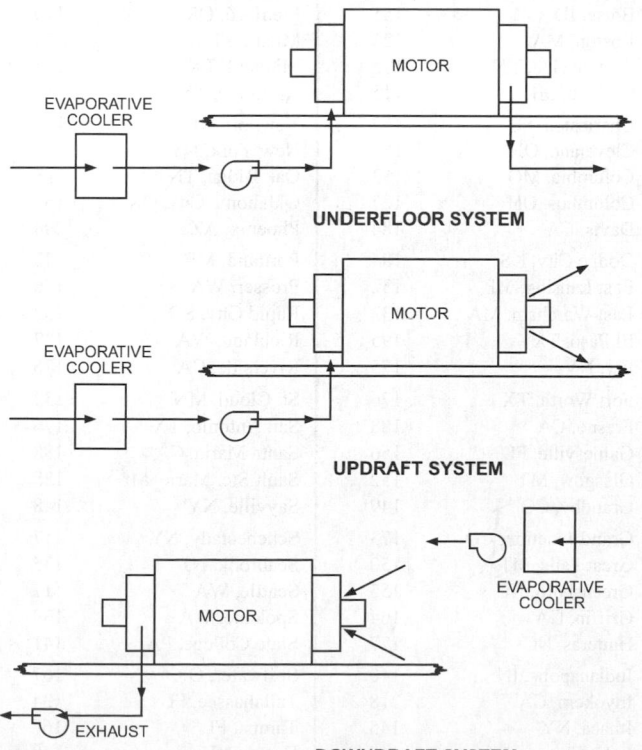

UNDERFLOOR SYSTEM

UPDRAFT SYSTEM

DOWNDRAFT SYSTEM

Fig. 17   Arrangements for Cooling Large Motors

moisture directly into the room. The air handled is reduced to approximately 60% of that normally required by an all-outside-air, direct evaporative cooler.

Cooling Laundries

Laundries have one of the most severe environments in which direct evaporative air cooling is applied because heat is produced not only by the processing equipment, but by steam and water vapor as well. A properly designed direct evaporative cooler reduces the temperature in a laundry 5 to 10°F below the outside temperature. With only fan ventilation, laundries usually exceed the outside temperature by at least 10°F. Air distribution should be designed for a maximum throw of not more than 30 ft. A minimum circulated velocity of 100 to 200 fpm should prevail in the occupied space. Ducts can be located to discharge the air directly onto workers in exceptionally hot areas, such as pressing and ironing departments. For these outlets, there should be some means of manual control to direct the air where it is desired, with at least 500 to 1000 cfm at a target velocity of 600 to 900 fpm for each workstation.

Cooling Wood and Paper Products Facilities

Wood-processing plants and paper mills are good applications for evaporative cooling because of the high temperatures and gases associated with wood-processing equipment. Wood dust should be kept out of the recirculation sumps of evaporative coolers, because the dust contains microorganisms and worm larvae that will grow in sumps.

Because of the types of gases and particulates present in most paper plants, water-cooled systems are preferred over air-cooled systems. The most prevalent contaminant is wood dust. Chlorine gas, caustic soda, sulfur, hydrogen sulfide, and other compounds are also serious problems, because they accelerate the corrosion of steel and yellow metals. With more efficient air scrubbing, ambient air quality in and about paper mills has become less corrosive, allowing use of equipment with well-analyzed and properly applied coatings on coils and housings. Phosphor-free brazed coil joints should be used in areas where sulfur compounds are present.

Heat is readily available from processing operations and should be used whenever possible. Most plants have good-quality hot water and steam, which can be readily geared to unit heater, central station, or reheat use. Newer plant air-conditioning methods, including evaporative cooling, that use energy-conservation techniques (such as temperature stratification) lend themselves to this type of large structure. Chapter 26 has further information on air-conditioning of paper facilities.

OTHER APPLICATIONS

Cooling Power-Generating Facilities

An appropriate air-cooling system can be selected once preliminary heating and cooling loads are determined and criteria are established for temperature, humidity, pressure, and airflow control. The same considerations for selection apply to power-generating facilities and industrial facilities.

Cooling Mines

Chapter 29 describes evaporative cooling methods for mines.

Cooling Animals

The design criteria for farm animal environments and the need for cooling animal shelters are discussed in Chapter 24. Direct evaporative cooling is ideally suited to farm animal shelters because 100% outside air is used. Fresh air removes odors and reduces the harmful effects of ammonia fumes. At night and in the spring and fall, direct evaporative cooling can also be used for ventilation.

Equipment should be sized to change the air in the shelter in 1 to 2 min, assuming the ceiling height does not exceed 10 ft. This flow rate usually keeps the shelter at or below 80°F. In addition, conditions can be improved with portable or packaged spot coolers.

For poultry housing, most applications require an air change every 0.75 to 1.5 min, with the majority at 1 min. Placing the fans at the ends or the center of the house, with the direct evaporative cooler located at the opposite end, creates a tunnel ventilation system with an air velocity of 300 to 500 fpm. Fans are generally selected for a total pressure drop of 0.125 in. of water, which means that the direct evaporative cooling media cannot have a pressure drop in excess of 0.075 in. of water. Thus, to prevent an inadequate volume of air being pulled through the poultry house, the designer must carefully size the media selected.

Using direct evaporative cooling for poultry broiler houses decreases bird mortality, improves feed conversion ratio, and increases the growth rate. Poultry breeder houses are evaporatively cooled to improve egg production and fertility during warm weather. Evaporative cooling of egg layers improves feed conversion, shell quality, and egg size. When the ambient outside temperature exceeds 100°F, evaporative cooling is often the only way to keep a flock alive. Direct evaporative cooling is also used to cool swine farrowing and gestation houses to improve production.

Produce Storage Cooling

Potatoes. Direct evaporative cooling for bulk potato storage should pass air directly through the pile. The ventilation and cooling system should provide 1.0 to 1.5 cfm/100 lb of potatoes. Average potato density is 45 lb/ft^3 in the pile. Pile depths range from 12 to 20 ft, which creates a static pressure of 0.15 to 0.25 in. of water. Ventilation consists of fresh air inlets, return air openings, exhaust air openings, main air ducts, and lateral ducts with holes or slots to distribute air uniformly through the pile. Distribution ducts should be placed no farther apart than 80% of the potato pile depth, and should extend to within 18 in. of the storage walls. Ducts, the direct evaporative cooling media, and any refrigeration coils cause a static pressure ranging from 0.5 to 1.0 in. of water. Typically the total static pressure ranges from 0.75 to 1.25 in. of water, depending on the equipment. Air speed through each of the openings in the ventilation/cooling system should be as listed in Table 4.

Direct evaporative cooling media should be 90 to 95% effective, depending on the climate. In arid regions, 95% effective media are recommended. In more humid climates, such as in the midwestern and eastern United States, 90% effective media are commonly used. Air speed through the media should be 500 to 550 fpm to ensure high pad efficiency with low static-pressure penalty.

For more information, see Chapter 37 of the 2010 *ASHRAE Handbook—Refrigeration.*

Apples. Direct evaporative cooling for apple storage without refrigeration should distribute cool air to all parts of the storage. The evaporative cooler may be floor-mounted or located near the ceiling in a fan room. Air should be discharged horizontally at ceiling level. Because the prevailing wet-bulb temperature limits the degree of cooling, a cooler with maximum reasonable size should be installed to reduce the storage temperature rapidly and as close to the wet-bulb temperature as possible. Generally, a cooler designed to exchange air every 3 min (20 air changes per hour) is the largest that can be

installed. This capacity provides a complete air change every 1 to 1.5 min (40 to 60 air changes per hour) when the storage is loaded.

For further information on apple storage, see Chapter 35 of the 2010 *ASHRAE Handbook—Refrigeration.*

Citrus. The chief purpose of evaporative-cooling fruits and vegetables is to provide an effective, inexpensive means of improving storage. However, it also serves a special function in the case of oranges, grapefruit, and lemons. Although mature and ready for harvest, citrus fruits are often still green. Color change (degreening) is achieved through a sweating process in rooms equipped with direct evaporative cooling. Air with a high relative humidity and a moderate temperature is circulated continuously during the operation. Ethylene gas, the concentration depending on the variety and intensity of green pigment in the rind, is discharged into the rooms. Ethylene destroys chlorophyll in the rind, allowing the yellow or orange color to become evident. During degreening, a temperature of 70°F and a relative humidity of 88 to 90% are maintained in the sweat room. (In the Gulf States, 82 to 85°F with 90 to 92% rh is used.) The evaporative cooler is designed to deliver 11 cfm per pound of fruit.

Direct and indirect evaporative cooling is also used as a supplement to refrigeration in the storage of citrus fruit. Citrus storage requires refrigeration in the summer, but the required conditions can often be obtained using evaporative cooling during the fall, winter, and spring when the outside wet-bulb temperature is low. For

Table 4 Air Speeds for Potato Storage Evaporative Cooler

Opening	Minimum Speed, fpm	Maximum Speed, fpm	Desired Speed, fpm
Fresh air inlet	1000	1400	1200
Return air opening	1000	1400	1200
Exhaust opening	1000	1200	1100
Main duct	500	900	700
Lateral duct	750	1100	900
Slot	900	1300	1050

Table 5 Three-Year Average Solar Radiation for Horizontal Surface During Peak Summer Month

City	Btu/h·ft^2	City	Btu/h·ft^2
Albuquerque, NM	198	Lemont, IL	142
Apalachicola, FL	170	Lexington, KY	170
Astoria, OR	132	Lincoln, NE	150
Atlanta, GA	158	Little Rock, AR	148
Bismarck, ND	140	Los Angeles, CA	162
Blue Hill, MA	128	Madison, WI	138
Boise, ID	155	Medford, OR	170
Boston, MA	125	Miami, FL	153
Brownsville, TX	175	Midland, TX	177
Caribou, ME	115	Nashville, TN	154
Charleston, SC	152	Newport, RI	138
Cleveland, OH	152	New York, NY	140
Columbia, MO	153	Oak Ridge, TN	148
Columbus, OH	127	Oklahoma City, OK	165
Davis, CA	184	Phoenix, AZ	200
Dodge City, KS	184	Portland, ME	133
East Lansing, MI	132	Prosser, WA	176
East Wareham, MA	132	Rapid City, SD	152
El Paso, TX	195	Richland, WA	137
Ely, NV	175	Riverside, CA	176
Fort Worth, TX	176	St. Cloud, MN	132
Fresno, CA	188	San Antonio, TX	176
Gainesville, FL	156	Santa Maria, CA	188
Glasgow, MT	152	Sault Ste. Marie, MI	138
Grandby, CO	149	Sayville, NY	148
Grand Junction, CO	173	Schenectady, NY	117
Great Falls, MT	150	Seabrook, NJ	135
Greensboro, NC	155	Seattle, WA	117
Griffin, GA	164	Spokane, WA	139
Hatteras, NC	177	State College, PA	141
Indianapolis, IN	140	Stillwater, OK	167
Inyokern, CA	218	Tallahassee, FL	134
Ithaca, NY	145	Tampa, FL	167
Lake Charles, LA	160	Upton, NY	148
Lander, WY	177	Washington, D.C.	142
Las Vegas, NV	195		

further information, see Chapter 36 of the 2010 *ASHRAE Handbook—Refrigeration*.

Cooling Greenhouses

Proper regulation of greenhouse temperatures during the summer is essential for developing high-quality crops. The principal load on a greenhouse is solar radiation, which at sea level at about noon in the temperate zone is approximately 200 Btu/h·ft². Smoke, dust, or heavy clouds reduce the radiation load. Table 5 gives solar radiation loads for representative cities in the United States. Note that the values cited are average solar heat gains, not peak loads. Temporary rises in temperature inside a greenhouse can be tolerated; an occasional rise above design conditions is not likely to cause damage.

Not all solar radiation that reaches the inside of the greenhouse becomes a cooling load. About 2% of the total solar radiation is used in photosynthesis. Transpiration of moisture varies by crop, but typically uses about 48% of the solar radiation. This leaves 50% to be removed by the cooler. Example 2 shows a method for calculating the size of a greenhouse evaporative cooling system.

Example 2. A direct evaporative cooler is to be installed in a 50 by 100 ft greenhouse. Design conditions are 92°F db and 73°F wb, and average solar radiation is 138 Btu/h·ft². An inside temperature of 90°F db must not be exceeded at design conditions.

Solution: The direct evaporative air cooler is assumed to have a saturation effectiveness of 80%. Equation (2) may be used to determine the dry-bulb temperature of the air leaving the direct evaporative cooler:

$$t_2 = 92 - \frac{80}{100}(92 - 73) = 77°F$$

The following equation, a modification of Equation (1), may be used to calculate the airflow rate that must be supplied by the direct evaporative cooler:

$$Q_{ra} = \frac{0.5 A I_t}{60 \rho c_p} \tag{4}$$

where

A = greenhouse floor area, ft²
I_t = total incident solar radiation, Btu/h·ft² of receiving surface
$60\rho c_p$ = density times specific heat of air times 60 min/h ≈ 1.0 Btu/ft³·°F at design conditions

For this problem

$$Q_{ra} = \frac{0.5 \times 50 \times 100 \times 138}{1.0(90 - 77)} = 26,500 \text{ cfm}$$

Horizontal illumination from the direct rays of noonday summer sun with clear sky can be as much as 10,000 footcandles (fc); under clear glass, this is approximately 8500 fc. Crops such as chrysanthemums and carnations grow best in full sun, but many foliage plants, such as gloxinias and orchids, do not need more than 1500 to 2000 fc. Solar radiation is nearly proportional to light intensity. Thus, the greater the amount of shade, the smaller the cooling capacity required. A value of 100 fc is approximately equivalent to 3 Btu/h·ft². Although atmospheric conditions such as clouds and haze affect the relationship, this is a safe conversion factor. This relationship should be used instead of Table 5 when illumination can be determined by design or measurement.

Direct evaporative cooling for greenhouses may be under either positive or negative pressure. Regardless of the type of system used, the length of air travel should not exceed 160 ft. The temperature rise of the cool air limits the throw to this value. Air movement must be kept low because of possible mechanical damage to the plants, but it should generally not be less than 100 fpm in areas occupied by workers.

CONTROL STRATEGY TO OPTIMIZE ENERGY RECOVERY

Figures 18A and 18B show a heat pipe air-to-air heat exchanger used in a hospital for winter heat recovery and summer indirect (dry) evaporative cooling. The heat pipe has a double-walled partition between the clean outdoor air (OA) flow and the contaminated building exhaust air (EA). With this partition, leakage from the EA side of the heat exchanger to the supply air (SA) side is eliminated and fans may be positioned as shown for blow-through exhaust and draw-through supply. The heat pipe has a direct spray manifold on the building return air side of the heat exchanger for indirect evaporative cooling. The spray pump is located in a water sump below the wet side of the heat pipe and uses recirculated water from the sump to wet the heat pipe. Potable makeup water is supplied to replace the water evaporated in the indirect cooling process along with wasted water required to maintain dissolved solids in the sump at acceptable levels. Using the building return air wet-bulb condition of 60 to 65°F, the summer cooling effect is greatly increased over dry-to-dry heat recovery.

The operation of both outdoor (OA) and exhaust air (EA) face and bypass dampers, working in concert with supply and return fan variable-frequency drives (VFD), allow parasitic fan static pressure losses to be minimized during favorable climatic conditions. Figure 18B shows a total bypass of the heat exchanger in the range of ambient temperatures of 55 to 65°F. The value of energy recovered is exceeded by the fan energy penalty during these outdoor air temperatures.

AIR CLEANING

Evaporative coolers are effective for improving IAQ in many ways. Their similarity to wet scrubbers means they can remove particulates and soluble gases. Direct evaporative coolers of all types perform some air cleaning. Rigid-media direct evaporative coolers are effective at removing particles down to about 1 μm. Air washers are effective down to about 10 μm.

The dust removal efficiency of direct evaporative coolers depends largely on the size, density, wettability, and solubility of the dust particles. Larger, more wettable particles are the easiest to remove. Separation is largely a result of the impingement of particles on the wetted surface of the eliminator plates or on the surface of the media. Because the force of impact increases with the size of the solid, the impact (together with the adhesive quality of the wetted surface) determines the cooler's usefulness as a dust remover.

The standard low-pressure spray is relatively ineffective in removing most atmospheric dusts. Direct evaporative coolers are of little use in removing soot particles because their greasy surface will not adhere to the wet plates or media. Direct evaporative coolers are also ineffective in removing smoke, because the small particles (less than 1 μm) do not impinge with sufficient impact to pierce the water film and be held on the media. Instead, the particles follow the air path between the media surfaces.

In the case of cross-corrugated media, the particles are removed from the media by the recirculated water. In locations with high particulate contamination, the sump and water distribution system should be flushed at least quarterly. If the particulate contains organic matter, it can contribute to biological growth on the media.

Control of Gaseous Contaminants

When used in a makeup air system comprised of a mixture of outside air and recirculated air, direct evaporative coolers function as scrubbers and reduce some gaseous contaminants found in outside air. These contaminants may concentrate in the recirculating water, so some water must be bled off. For more information regarding control of gaseous contaminants, see Chapter 46.

Evaporative coolers near sources of airborne nitric acid, chlorine, or ammonia absorb these chemicals, which can damage the cooler.

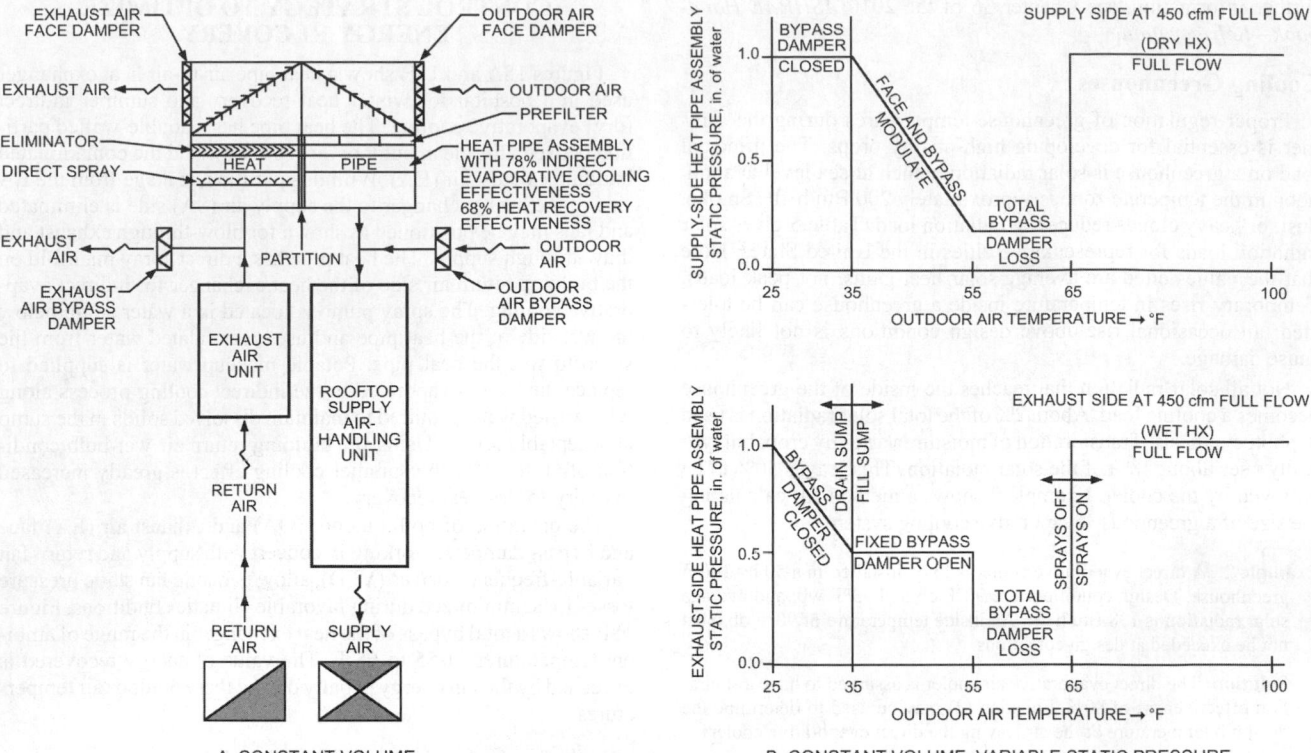

Fig. 18 Schematics for 100% Outdoor Air Used in Hospital: (A) Constant-Volume Control and (B) Constant-Volume, Variable-Static-Pressure Control

Table 6 Particulate Removal Efficiency of Rigid Media

Media Depth, in.	Particle Sizes, μm					
	0.3 to 0.5	0.5 to 0.7	0.7 to 1	1 to 5	5 to 10	>10
6,150	1.7%	21.3%	25.6%	43.6%	46.2%	61.3%
12,300	9.6%	31.8%	55.4%	87.2%	96.5%	97.3%

Source: Data courtesy of Munters Corporation.

Table 7 Insertion Loss for 12 in. Depth of Rigid Media, dB

Media Orientation	Octave Band Center Frequency, Hz							
	63	125	250	500	1000	2000	4000	8000
Dry, forward flow	2	1	2	5	4	5	10	14
Reverse flow	4	1	2	4	5	4	9	13
Wet, forward flow	1	0	3	3	3	4	6	9
Reverse flow	3	1	3	3	3	3	4	8

Source: Data courtesy of Munters Corporation.

The amount of soluble gases cleaned from the air depends on the air/water mixing, retention time, the water's pH, and the bleed rate. When exposed to soluble gases, evaporative coolers should be operated with a high bleed rate.

Ozone levels of the airstream can be reduced using evaporative coolers and air washers. Ozone is fairly unstable in a watery solution, decaying to ordinary diatomic oxygen (Ozone Information.com 2008). The stability of ozone absorbed in water depends on water temperature, ozone concentration, and length of holding time. Higher room humidity can vastly improve the rate at which it decays back to oxygen (Sterling et. al 1985).

Legionnaires' Disease. There have been no known cases of Legionnaires' disease with air washers or wetted-media evaporative air coolers. This is can be attributed to the low temperature of the recirculated water, which is not conducive to *Legionella* bacteria growth, as well as the absence of aerosolized water carryover that could transmit the bacteria to a host (ASHRAE *Guideline* 12-2000).

Evaporative cooler media has some sound attenuation properties. This insertion loss varies, depending on whether the media is wet or dry and whether the sound is travelling counter to (reverse flow) or with (forward flow) the airstream. Sound attenuation for 12 in. depth of rigid media can be found in Table 7.

ECONOMIC FACTORS

Design of direct and indirect evaporative cooling systems and sizing of equipment is based on the load requirements of the application and on the local dry- and wet-bulb design conditions, which may be found in Chapter 14 of the 2009 *ASHRAE Handbook—Fundamentals*. Total energy use for a specific application during a set period may be forecasted by using annual weather data. Dry-bulb and mean coincident wet-bulb temperatures, with the hours of occurrence, can be summarized and used in a modified bin procedure. The calculations must reflect the hours of use, conditions of load, and occupancy. Because of annual variations in dry- and wet-bulb temperatures and the effect of increasing cooling capacity with decreasing wet-bulb temperatures, bin calculations using mean coincident wet-bulb temperatures generally produce conservative results. When comparing various cooling systems, cost analysis should include annual energy reduction at the applicable electrical rate, plus anticipated energy cost escalation over the expected life.

Many areas have time-of-day electrical metering as an incentive to use energy during off-peak hours when rates are lowest. Reducing air-conditioning kilowatt demand is especially important in areas with ratcheted demand rates (Scofield and DesChamps 1980). Thermal storage using ice banks or chilled-water storage may be used as part of a multistage evaporatively refrigerated cooler to combine the energy-saving advantages of evaporative cooling and off-peak savings of thermal storage (Eskra 1980).

Direct Evaporation Energy Saving

Direct evaporative cooling may be used in all climates to save cooling and humidification energy. In humid climates, the benefits of direct evaporation are realized during periods when outside air is warm and dry, but cooling savings are unlikely to be realized during peak design conditions. In more arid areas, direct evaporative cooling may partially or fully offset mechanical cooling at peak load conditions. Humidification energy savings may be realized during the heating season when outside air is used to provide cooling and humidification. If properly controlled, direct evaporative cooling can use waste heat otherwise rejected from buildings when outside air is used for cooling.

Indirect Evaporation Energy Saving

Indirect evaporative cooling may be used in all climates to save cooling and, in some applications, heating energy. In humid climates, indirect evaporative cooling may be used throughout the cooling cycle to precool outside air. Indirect evaporative cooling can be used to extend the range of 100% outside air ventilation to both higher and lower temperatures, and to increase the percentage of outside air a system can support at any given temperature through heat recovery. In high-humidity areas, indirect evaporative cooling may be used to (1) partially offset mechanical cooling requirements at peak load conditions and (2) provide better control over low-load humidity conditions by permitting the use of smaller refrigeration equipment to provide ventilation over a wider range of outside air conditions. The cost of heating may be reduced when operating below temperatures at which minimum outside air quantities exceed the rates of ventilation required for free cooling by using heat recovered from building exhausts.

Water Cost for Evaporative Cooling

Typically, domestic service water is used for evaporative cooling to avoid excessive scaling and associated problems with poor water quality. In designing evaporative coolers, the cost of water treatment is included in the overall project cost. However, water cost is typically ignored for evaporative coolers because it is usually an insignificant part of the operational cost. Depending on the ambient dry-bulb temperature and wet-bulb depression for a specific location, the cost of water could become a significant part of the operational cost, because the greater the differential between dry- and wet-bulb temperatures, the greater the amount of water evaporated (Mathur 1997, 1998).

PSYCHROMETRICS

Figure 19 shows the two-stage (indirect/direct) process applied to nine cities in the western United States. The examples indicated are primarily shown for arid areas, but the principles also apply to moderately humid and humid areas when weather conditions permit. For each city indicated, the entering conditions to the first-stage indirect unit are at or near the 0.4% design dry- and wet-bulb temperatures in Chapter 14 of the 2009 *ASHRAE Handbook—Fundamentals*. Although higher effectiveness can be achieved for both the indirect and direct evaporative processes modeled, the effectiveness ratings are 60% for the first (indirect) stage and 90% for the second (direct) stage. Leaving air temperatures range from 52 to 70°F, with leaving conditions approaching saturation.

Figure 20 projects space conditions in each city at 78°F db for these second-stage supply temperatures based on a 95% room sensible heat factor (i.e., room sensible heat/room total heat). Except in Wichita, Los Angeles, and Seattle, room conditions can be maintained in the comfort zone without a refrigerated third stage. But even in these cities, third-stage refrigeration requirements are sharply reduced as compared to conventional mechanical cooling.

However, Figures 19 and 20 indicate the need to consider the following factors when deciding whether to include a third cooling stage:

- As the room sensible heat factor decreases, the supply air temperature required to maintain a given room condition decreases.
- As supply air temperature increases, the supply air quantity must increase to maintain space temperature, which results in higher air-side initial cost and increased supply air fan power.
- A decrease in the required room dry-bulb temperature requires an increase in the supply air quantity. For a given room sensible heat factor, a decrease in room dry-bulb temperature may cause the relative humidity to exceed the comfort zone.
- The suggested 0.4% entering design (dry-bulb/mean wet-bulb) conditions are only one concern. Partial-load conditions must also be considered, along with the effect (extent and duration) of spike wet-bulb temperatures. Mean wet-bulb temperatures can be used to determine energy use of the indirect/direct system. However, the higher wet-bulb temperature spikes should be considered to determine their effect on room temperatures.

An ideal condition for maximum use with minimum energy consumption of a two- and three-stage indirect/direct system is a room sensible heat factor of 90% and higher, a supply air temperature of 60°F, and a dry-bulb room design temperature of 78°F. In many cases, third-stage refrigeration is required to ensure satisfactory dry-

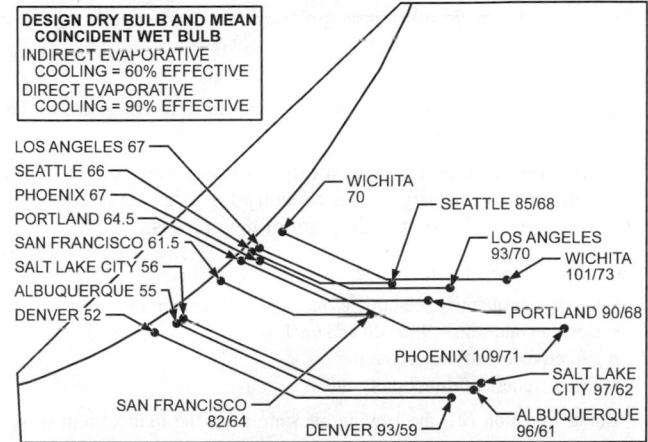

Fig. 19   Two-Stage Evaporative Cooling at 0.4% Design Condition in Various Cities in Western United States

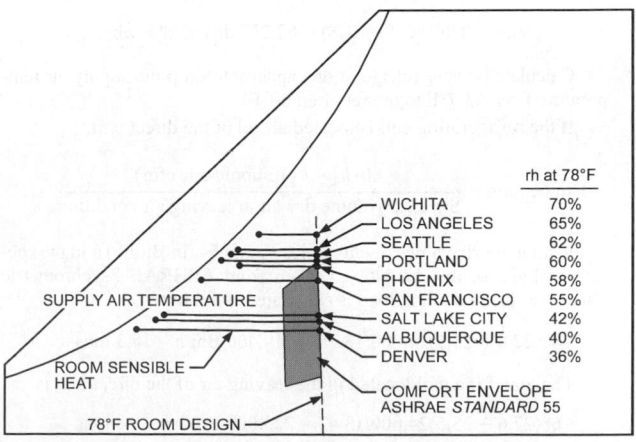

Fig. 20   Final Room Design Conditions After Two-Stage Evaporative Cooling

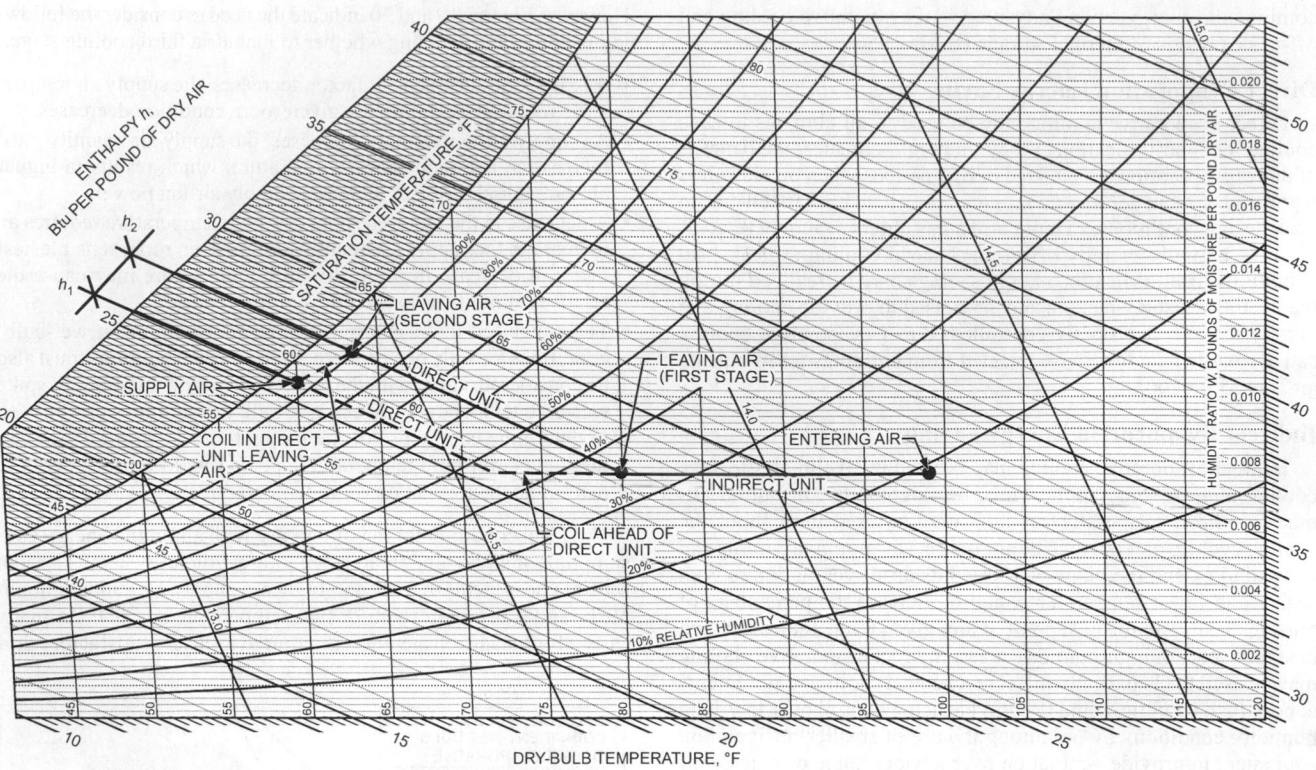

Fig. 21   Psychrometric Diagram of Three-Stage Evaporative Cooling Example 3

bulb temperature and relative humidity. Example 3 shows a method for determining the refrigeration capacity for three-stage cooling. Figure 21 is a psychrometric diagram of the process.

Example 3. Assume the following:

- Supply air quantity = 24,000 cfm; supply air temperature = 60°F
- Design condition = 99°F db and 68°F wb
- Effectiveness of indirect unit = 60%;
- Effectiveness of direct unit = 90%

Using Equation (2), the leaving air state from the indirect unit (first stage) is

$$99 - 0.60(99 - 68) = 80.4°F \text{ db } (61.8°F \text{ wb})$$

Using Equation (2), the leaving air state from the direct unit (second stage) is

$$80.4 - 0.90(80.4 - 61.8) = 63.7°F \text{ db } (61.8°F \text{ wb})$$

Calculate booster refrigeration capacity to drop the supply air temperature from 63.7°F to the required 60°F.

If the refrigerating coil is located ahead of the direct unit,

$$\text{Btu cooling} = \frac{60(h_1 - h_2)(\text{supply air, cfm})}{\text{Specific volume dry air at leaving air condition}}$$

With numeric values of enthalpies h_1 and h_2 (in Btu/lb) and the specific volume of air (in ft³/lb$_{da}$) taken from ASHRAE Psychrometric Chart No. 1, the cooling load is calculated as follows:

$$60(27.6 - 25.5)24,000/13.78 = 219,400 \text{ Btu/h} = 18.3 \text{ tons}$$

The load for a coil located in the leaving air of the direct unit is

$$60(27.6 - 25.5)24,000/13.43 = 225,000 \text{ Btu/h} = 18.8 \text{ tons}$$

Depending on the location of the booster coil, the preceding calculations can be used to determine third-stage refrigeration capacity and to select a cooling coil.

Using this example, refrigeration sizing can be compared to conventional refrigeration without staged evaporative cooling. Assuming mixed air conditions to the coil of 81°F db and 66.5°F wb, and the same 60°F db supply air as shown in Figure 20, the refrigerated capacity is

$$60(31.1 - 25.7)24,000/13.31 = 584,200 \text{ Btu/h} = 48.7 \text{ tons}$$

This represents an increase of 30.4 tons. The staged evaporative effect reduces the required refrigeration by 62.4%.

ENTERING AIR CONSIDERATIONS

The effectiveness of direct and indirect evaporative cooling depends on the entering air condition. Where outside air is used in a direct evaporative cooler, the design is affected by the prevailing outside dry- and wet-bulb temperatures as well as by the application. Where conditioned exhaust air is used as secondary air for indirect evaporative cooling, the design is less affected by local weather conditions, which makes evaporative cooling viable in hot and humid environments.

For example, in arid areas like Reno, Nevada, a simple, direct evaporative cooler with an effectiveness of 80% provides a leaving air temperature of 68°F when dry- and wet-bulb temperatures of the entering air are 96 and 61°F, respectively. In the same location, adding an indirect evaporative precooling stage with an effectiveness of 80% produces a leaving air condition of 53.6°F.

In a location such as Atlanta, Georgia, with design temperatures of 94 and 74°F, the same direct evaporative cooler could supply only 78°F. This could be reduced to 71.1°F by adding an 80% effective indirect evaporative precooling stage (Supple 1982). If exhaust air from the building served is provided at a stable 75°F db and 62.5°F wb, an indirect evaporative precooler could deliver air at 68.8°F, substantially reducing outside air cooling loads. Under these conditions, indirect evaporative precoolers can provide limited dehumidification capabilities.

Long-term benefits to owners of direct evaporative cooling systems include a 20 to 40% reduction of utility costs compared to mechanical refrigeration (Watt 1988). When used to control humidity, the reduction in cooling and humidification energy use ranges from 35 to 90% (Lentz 1991). Although direct evaporative cooling does not reduce peak cooling loads in other than arid areas, it can reduce both total cooling energy and humidification energy requirements in a wide range of environments, including hot and humid ones.

Indirect evaporative cooling lowers the temperature (both dry- and wet-bulb) of the air entering a direct evaporative cooling stage and, consequently, lowers the supply air temperature. When used with mechanical cooling on 100% outside air systems, with the secondary air taken from the conditioned space, the precooling effect may reduce peak cooling loads between 50 and 70%. Total cooling requirements may be reduced between 40 and 85% annually depending on location, system configuration, and load characteristics. Indirect evaporative coolers may also function as heat recovery systems, which expands the range of conditions over which the process is used. Indirect evaporative cooling, when used with building exhaust air, is especially effective in hot and humid climates.

REFERENCES

ASHRAE. 2000. Minimizing the risk of Legionellosis associated with building water systems. *Guideline* 12-2000.

ASHRAE. 2007. Ventilation for acceptable indoor air quality. ANSI/ASHRAE *Standard* 62.1-2007.

Colvin, T.D. 1995. Office tower reduces operating costs with two-stage evaporative cooling system. *ASHRAE Journal* 37(3):23-24.

Crow, L.W. 1972. Weather data related to evaporative cooling. Research Report 2223, *ASHRAE Transactions* 78(1):153-164.

Ecodyne Corp. 1980. *Weather data handbook*. McGraw-Hill, New York.

Eskra, N. 1980. Indirect/direct evaporative cooling systems. *ASHRAE Journal* 22(5):22.

Felver, T.G., et al. 2001. Cooling California's computer centers. *HPAC Magazine*, pp. 60-61.

Lentz, M.S. 1991. Adiabatic saturation and variable-air-volume: a prescription for economy in close environmental control. *ASHRAE Transactions* 97(1):477-485.

Mathur, G.D. 1990. Indirect evaporative cooling with two-phase thermosiphon coil loop heat exchangers. *ASHRAE Transactions* 96(1):1241-1249.

Mathur, G.D. 1997. Performance enhancement of existing air conditioning systems. *Intersociety Energy Conversion Engineering Conference, American Institute of Chemical Engineers* 3:1618-1623.

Mathur, G.D. 1998. Predicting yearly energy savings using bin weather data with heat pipe exchangers with indirect evaporative cooling. Intersociety Energy Conversation Engineering Conference, *Paper* 98-IECEC-049.

Mathur, G.D. and D.Y. Goswami. 1995. Indirect evaporative cooling retrofit as a demand side management strategy for residential air conditioning. *Intersociety Energy Conversion Engineering Conference, ASME* 2:317-322.

Mathur, G.D., D.Y. Goswami, and S.M. Kulkarni. 1993. Experimental investigation of a residential air conditioning system with an evaporatively cooled condenser. *Journal of Solar Energy Engineering* 115:206-211.

Scofield, C.M. 1986. The heat pipe used for dry evaporative cooling. *ASHRAE Transactions* 92(1B):371-381.

Scofield, C.M. 1994. California classroom VAV with IAQ and energy savings, too. *HPAC Magazine*, p. 89.

Scofield, M. and J. Bergman. 1997. ASHRAE *Standard* 62R: A simple method of compliance. *HPAC Magazine* (October):67

Scofield, M. and N. DesChamps. 1980. EBTR compliance and comfort too. *ASHRAE Journal* 22(6):61.

Scofield, C.M. and N.H. DesChamps. 1984. Indirect evaporative cooling using plate-type heat exchangers. *ASHRAE Transactions* 90(1B):148-153.

Scofield, M. and J. Taylor. 1986. A heat pipe economy cycle. *ASHRAE Journal* 28(10):35-40.

Sterling, E.M., A. Arundel, and T.D. Sterling. 1985. Criteria for human exposure to humidity in occupied buildings. *ASHRAE Transactions* 91(1B):611-622.

Stewart, W.E., Jr. 1999. *Design guide for combustion turbine inlet air cooling systems.* ASHRAE.

Supple, R.G. 1982. Evaporative cooling for comfort. *ASHRAE Journal* 24(8):42.

Watt, J.R. 1988. Power cost comparisons: Evaporative vs. refrigerative cooling. *ASHRAE Transactions* 94(2):1108-1115.

Wu, H. and J.L. Yellot. 1987. Investigation of a plate-type indirect evaporative cooling system for residences in hot and arid climates. *ASHRAE Transactions* 93(1):1252-1260.

Yellott, J.I. and J. Gamero. 1984. Indirect evaporative air coolers for hot, dry climates. *ASHRAE Transactions* 90(1B):139-147.

BIBLIOGRAPHY

ASHRAE. 2004. Thermal environmental conditions for human occupancy. ANSI/ASHRAE *Standard* 55-2004.

Peterson, J.L. and B.D. Hunn. 1992. Experimental performance of an indirect evaporative cooler. *ASHRAE Transactions* 98(2):15-23.

Stewart, W.E., Jr. and L.A. Stickler. 1999. Designing for combustion turbine inlet air cooling. *ASHRAE Transactions* 105(1).

Strock, C., ed. 1959. *Handbook of air conditioning, heating & ventilation.* Industrial Press, New York.

Watt, J.R. 1997. *Evaporative air conditioning handbook*, 3rd ed. Chapman & Hall, New York.

FIRE AND SMOKE MANAGEMENT

IN building fires, smoke often flows to locations remote from the fire, threatening life and damaging property. Stairwells and elevators frequently fill with smoke, thereby blocking or inhibiting evacuation. Smoke causes the most deaths in fires. **Smoke** includes airborne solid and liquid particles and gases produced when a material undergoes pyrolysis or combustion, together with air that is entrained or otherwise mixed into the mass.

The idea of using pressurization to prevent smoke infiltration of stairwells began to attract attention in the late 1960s. This concept was followed by the idea of the pressure sandwich (i.e., venting or exhausting the fire floor and pressurizing the surrounding floors). Frequently, a building's ventilation system is used for this purpose. **Smoke control** systems use fans to pressurize appropriate areas to limit smoke movement in fire situations. **Smoke management** systems include pressurization and all other methods that can be used singly or in combination to modify smoke movement.

This chapter discusses fire protection and smoke control systems in buildings as they relate to HVAC. For a more complete discussion, refer to *Principles of Smoke Management* (Klote and Milke 2002). National Fire Protection Association (NFPA) *Standard* 204 provides information about venting large industrial and storage buildings. For further information, refer to NFPA *Standards* 92A and 92B.

The objective of fire safety is to provide some degree of protection for a building's occupants, the building and property inside it, and neighboring buildings. Various forms of analysis have been used to quantify protection. Specific life safety objectives differ with occupancy; for example, nursing home requirements are different from those for office buildings.

Two basic approaches to fire protection are (1) to prevent fire ignition and (2) to manage fire effects. Figure 1 shows a decision tree for fire protection. Building occupants and managers have the primary role in preventing fire ignition. The building design team may incorporate features into the building to assist the occupants and managers in this effort. Because it is impossible to prevent fire ignition completely, managing fire effect has become significant in fire protection design. Examples include compartmentation, suppression, control of construction materials, exit systems, and smoke management. The *Fire Protection Handbook* (NFPA 2008) and *Smoke Movement and Control in High-Rise Buildings* (Tamura 1994) contain detailed fire safety information.

Historically, fire safety professionals have considered the HVAC system a potentially dangerous penetration of natural building membranes (walls, floors, etc.) that can readily transport smoke and fire. For this reason, HVAC has traditionally been shut down when fire is discovered; this prevents fans from forcing smoke flow, but does not prevent ducted smoke movement caused by buoyancy,

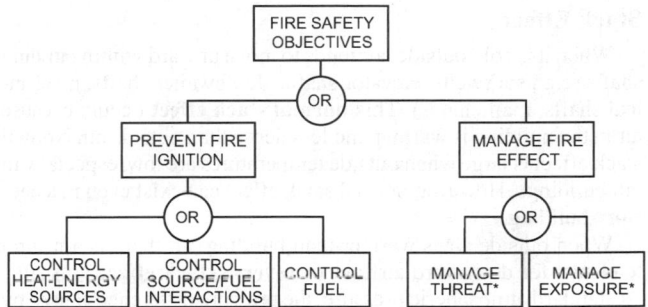

*Note: Smoke management is one of many fire protection tools that can be used to help manage the threat and exposure of fire.

Fig. 1 Simplified Fire Protection Decision Tree

stack effect, or wind. To solve the problem of smoke movement, methods of smoke control have been developed; however, smoke control should be viewed as only one part of the overall building fire protection system.

FIRE MANAGEMENT

Although most of this chapter discusses smoke management, fire management at HVAC penetrations is an additional concern for the HVAC engineer. The most efficient way to limit fire damage is through compartmentation. Fire-rated assemblies, such as the floor or walls, keep the fire in a given area for a specific period. However, fire can easily pass through openings for plumbing, HVAC ductwork, communication cables, or other services. Therefore, fire stop systems are installed to maintain the rating of the fire-rated assembly. The rating of a fire stop system depends on the number, size, and type of penetrations, and the construction assembly in which it is installed.

Performance of the entire fire stop system, which includes the construction assembly with its penetrations, is tested under fire conditions by recognized independent testing laboratories. ASTM *Standard* E814 and UL *Standard* 1479 describe ways to determine performance of **through-penetration fire stopping (TPFS)**.

TPFS is required by building codes under certain circumstances for specific construction types and occupancies. In the United States, model building codes require that most penetrations meet the ASTM E814 test standard. TPFS classifications are published by testing laboratories. Each classification is proprietary, and each applies to use with a specific set of conditions, so numerous types are usually required on any given project.

The construction manager and general contractor, not the architects and engineers, make work assignments. Sometimes they assign fire stopping to the discipline making the penetration; other times, they assign it to a specialty fire-stopping subcontractor. The

---

The preparation of this chapter is assigned to TC 5.6, Control of Fire and Smoke.

Construction Specifications Institute (CSI) assigns fire-stopping specifications to Division 7, which

- Encourages continuity of fire-stopping products on the project by consolidating their requirements (e.g., TPFS, expansion joint fire stopping, floor-to-wall joint fire stopping, etc.)
- Maintains flexibility of work assignments for the general contractor and construction manager
- Encourages prebid discussions between the contractor and subcontractors regarding appropriate work assignments

SMOKE MOVEMENT

A smoke control system must be designed so that it is not overpowered by the driving forces that cause smoke movement, which include stack effect, buoyancy, expansion, wind, and the HVAC system. During a fire, smoke is generally moved by a combination of these forces.

Stack Effect

When it is cold outside, air tends to move upward within building shafts (e.g., stairwells, elevator shafts, dumbwaiter shafts, mechanical shafts, mail chutes). This **normal stack effect** occurs because air in the building is warmer and less dense than outside air. Normal stack effect is large when outside temperatures are low, especially in tall buildings. However, normal stack effect can exist even in a one-story building.

When outside air is warmer than building air, there is a natural tendency for downward airflow, or **reverse stack effect**, in shafts. At standard atmospheric pressure, the pressure difference caused by either normal or reverse stack effect is expressed as

$$\Delta p = 7.64 \left(\frac{1}{T_o} - \frac{1}{T_i} \right) h \qquad (1)$$

where

- Δp = pressure difference, in. of water
- T_o = absolute temperature of outside air, °R
- T_i = absolute temperature of air inside shaft, °R
- h = distance above neutral plane, ft

For a building 200 ft tall with a neutral plane at midheight, an outside temperature of 0°F (460°R), and an inside temperature of 70°F (530°R), the maximum pressure difference from stack effect is 0.22 in. of water. This means that, at the top of the building, pressure inside a shaft is 0.22 in. of water greater than the outside pressure. At the base of the building, pressure inside a shaft is 0.22 in. of water lower than the outside pressure. Figure 2 diagrams the pressure

difference between a building shaft and the outside. A positive pressure difference indicates that shaft pressure is higher than the outside pressure, and a negative pressure difference indicates the opposite. Figure 3 illustrates air movement in buildings caused by both normal and reverse stack effect.

Figure 4 can be used to determine the pressure difference caused by stack effect. For normal stack effect, $\Delta p/h$ is positive, and the pressure difference is positive above the neutral plane and negative below it. For reverse stack effect, $\Delta p/h$ is negative, and the pressure difference is negative above the neutral plane and positive below it.

In unusually tight buildings with exterior stairwells, Klote (1980) observed reverse stack effect even with low outside air temperatures. In this situation, the exterior stairwell temperature is considerably lower than the building temperature. The stairwell represents the cold column of air, and other shafts within the building represent the warm columns of air.

If leakage paths are uniform with height, the neutral plane is near the midheight of the building. However, when the leakage paths are not uniform, the location of the neutral plane can vary considerably, as in the case of vented shafts. McGuire and Tamura (1975) provide methods for calculating the location of the neutral plane for some vented conditions.

Smoke movement from a building fire can be dominated by stack effect. In a building with normal stack effect, the existing air currents (as shown in Figure 3) can move smoke considerable distances from the fire origin. If the fire is below the neutral plane, smoke moves with building air into and up the shafts. This upward smoke

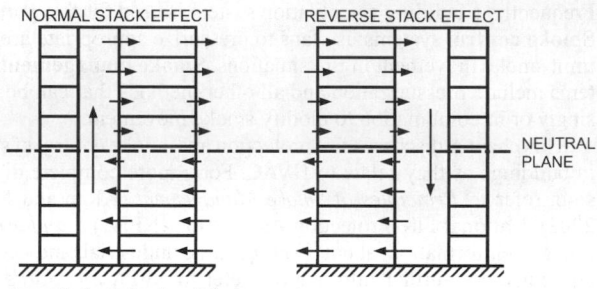

Note: Arrows indicate direction of air movement.

Fig. 3 Air Movement Caused by Normal and Reverse Stack Effect

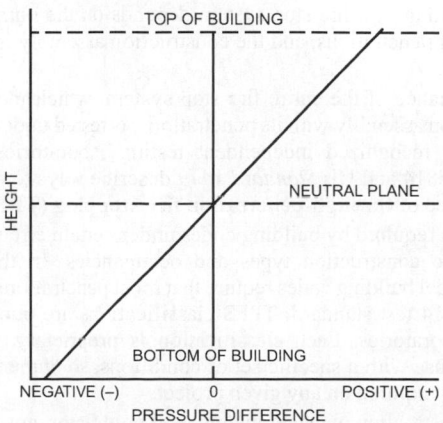

Fig. 2 Pressure Difference Between Building Shaft and Outdoors Caused by Normal Stack Effect

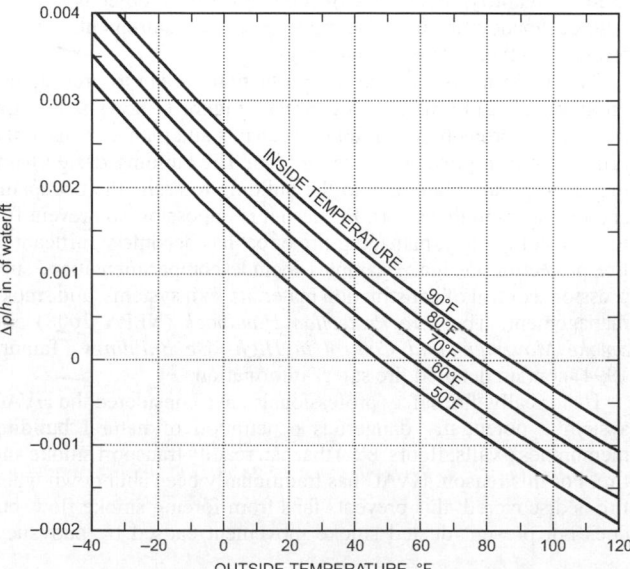

Fig. 4 Pressure Difference Caused by Stack Effect

flow is enhanced by buoyancy forces due to the temperature of the smoke. Once above the neutral plane, smoke flows from the shafts into the upper floors of the building. If leakage between floors is negligible, floors below the neutral plane (except the fire floor) remain relatively smoke-free until more smoke is produced than can be handled by stack effect flows.

Smoke from a fire located above the neutral plane is carried by building airflow to the outside through exterior openings in the building. If leakage between floors is negligible, all floors other than the fire floor remain relatively smoke-free until more smoke is produced than can be handled by stack effect flows. When leakage between floors is considerable, smoke flows to the floor above the fire floor.

Air currents caused by reverse stack effect (see Figure 3) tend to move relatively cool smoke down. In the case of hot smoke, buoyancy forces can cause smoke to flow upward, even during reverse stack effect conditions.

Buoyancy

High-temperature smoke has buoyancy because of its reduced density. At sea level, the pressure difference between a fire compartment and its surroundings can be expressed as follows:

$$\Delta p = 7.64 \left(\frac{1}{T_s} - \frac{1}{T_f} \right) h \tag{2}$$

where

Δp = pressure difference, in. of water
T_s = absolute temperature of surroundings, °R
T_f = average absolute temperature of fire compartment, °R
h = distance above neutral plane, ft

The pressure difference caused by buoyancy can be obtained from Figure 5 for surroundings at 68°F (528°R). The neutral plane is the plane of equal hydrostatic pressure between the fire compartment and its surroundings. For a fire with a fire compartment temperature at 1470°F (1930°R), the pressure difference 5 ft above the neutral plane is 0.052 in. of water. Fang (1980) studied pressures caused by room fires during a series of full-scale fire tests. During these tests, the maximum pressure difference reached was 0.064 in. of water across the burn room wall at the ceiling.

Much larger pressure differences are possible for tall fire compartments where the distance h from the neutral plane can be larger.

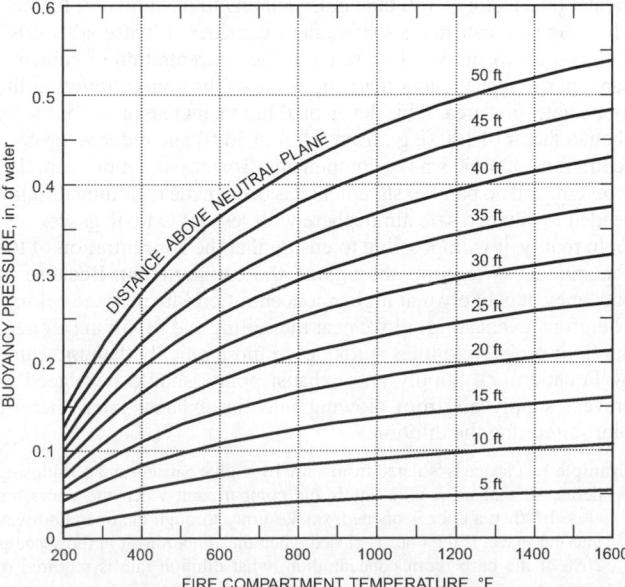

Fig. 5 Pressure Difference Caused by Buoyancy

If the fire compartment temperature is 1290°F (1750°R), the pressure difference 35 ft above the neutral plane is 0.35 in. of water. This is a large fire, and the pressures it produces are beyond present smoke control methods. However, the example illustrates the extent to which Equation (2) can be applied.

In sprinkler-controlled fires, the temperature in the fire room remains at that of the surroundings except for a short time before sprinkler activation. Sprinklers are activated by the **ceiling jet**, a thin (2 to 4 in.) layer of hot gas under the ceiling. The maximum temperature of the ceiling jet depends on the location of the fire, the activation temperature of the sprinkler, and the thermal lag of the sprinkler heat-responsive element. For most residential and commercial applications, the ceiling jet is between 180 and 300°F. In Equation (2), T_f is the average temperature of the fire compartment. For a sprinkler-controlled fire,

$$T_f = \frac{T_s(H - H_j) + T_j H_j}{H} \tag{3}$$

where

H = floor-to-ceiling height, ft
H_j = thickness of ceiling jet, ft
T_j = absolute temperature of ceiling jet, °R

For example, for H = 96 in., H_j = 4 in., T_s = 68 + 460 = 528°R, and T_j = 300 + 460 = 760°R,

$$T_f = [528(96 - 4) + 760 \times 4]/96 = 538°R \text{ or } 78°F$$

In Equation (2), this results in a pressure difference of 0.002 in. of water, which is insignificant for smoke control applications.

Expansion

Energy released by a fire can also move smoke by expansion. In a fire compartment with only one opening to the building, building air will flow in, and hot smoke will flow out. Neglecting the added mass of the fuel, which is small compared to airflow, the ratio of volumetric flows can be expressed as a ratio of absolute temperatures:

$$\frac{Q_{out}}{Q_{in}} = \frac{T_{out}}{T_{in}} \tag{4}$$

where

Q_{out} = volumetric flow rate of smoke out of fire compartment, cfm
Q_{in} = volumetric flow rate of air into fire compartment, cfm
T_{out} = absolute temperature of smoke leaving fire compartment, °R
T_{in} = absolute temperature of air into fire compartment, °R

For smoke at 1290°F (1750°R) and entering air at 67°F (527°R), the ratio of volumetric flows is 3.32. Note that absolute temperatures are used in the calculation. In such a case, if air enters the compartment at 3000 cfm, then smoke flows out at 9960 cfm, with the gas expanding to more than three times its original volume.

For a fire compartment with open doors or windows, the pressure difference across these openings caused by expansion is negligible. However, for a tightly sealed fire compartment, the pressure differences from expansion may be important.

Wind

Wind can have a pronounced effect on smoke movement within a building. The pressure wind exerts on a surface can be expressed as

$$p_w = 0.00643 C_w \rho_o V^2 \tag{5}$$

where

p_w = pressure exerted by wind, in. of water
C_w = pressure coefficient, dimensionless
ρ_o = outside air density, lb_m/ft^3
V = wind velocity, mph

The pressure coefficients C_w are in the range of −0.8 to 0.8, with positive values for windward walls and negative values for leeward walls. The pressure coefficient depends on building geometry and varies locally over the wall surface. In general, wind velocity increases with height from the surface of the earth. Houghton and Carruther (1976), MacDonald (1975), Sachs (1972), and Simiu and Scanlan (1978) give detailed information concerning wind velocity variations and pressure coefficients. Shaw and Tamura (1977) developed specific information about wind data with respect to air infiltration in buildings.

With a pressure coefficient of 0.8 and air density of 0.075 lb_m/ft^3, a 35 mph wind produces a pressure on a structure of 0.47 in. of water. The effect of wind on air movement within tightly constructed buildings with all exterior doors and windows closed is slight. However, wind effects can be important for loosely constructed buildings or for buildings with open doors or windows. Usually, the resulting airflows are complicated, and computer analysis is required.

Frequently, a window breaks in the fire compartment. If the window is on the leeward side of the building, the negative pressure caused by the wind vents the smoke from the fire compartment. This reduces smoke movement throughout the building. However, if the broken window is on the windward side, wind forces the smoke throughout the fire floor and to other floors, which endangers the lives of building occupants and hampers fire fighting. Wind-induced pressure in this situation can be large and can dominate air movement throughout the building.

HVAC Systems

Before methods of smoke control were developed, HVAC systems were shut down when fires were discovered because the systems frequently transported smoke during fires.

In the early stages of a fire, the HVAC system can aid in fire detection. When a fire starts in an unoccupied portion of a building, the system can transport the smoke to a space where people can smell it and be alerted to the fire. However, as the fire progresses, the system transports smoke to every area it serves, thus endangering life in all those spaces. The system also supplies air to the fire space, which aids combustion. Although shutting the system down prevents it from supplying air to the fire, it does not prevent smoke movement through the supply and return air ducts, air shafts, and other building openings because of stack effect, buoyancy, or wind.

SMOKE MANAGEMENT

In this chapter, smoke management includes all methods that can be used singly or in combination to modify smoke movement for the benefit of occupants or firefighters or for reducing property damage. Barriers, smoke vents, and smoke shafts are traditional methods of smoke management. The effectiveness of barriers is limited by the extent to which they are free of leakage paths. Smoke vents and smoke shafts are limited by the fact that smoke must be sufficiently buoyant to overcome any other driving forces that could be present. In the last few decades, fans have been used to overcome the limitations of traditional approaches. Compartmentation, dilution, pressurization, airflow, and buoyancy are used by themselves or in combination to manage smoke conditions in fire situations. These mechanisms are discussed in the following sections.

Compartmentation

Barriers with sufficient fire endurance to remain effective throughout a fire exposure have long been used to protect against fire spread. In this approach, walls, partitions, floors, doors, and other barriers provide some level of smoke protection to spaces remote from the fire. This section discusses passive compartmentation; using compartmentation with pressurization is discussed in the section on Pressurization (Smoke Control). Many codes, such as NFPA *Standard* 101 and the International Building Code® (ICC 2009), provide specific criteria for construction of smoke barriers (including doors) and their smoke dampers. The extent to which smoke leaks through such barriers depends on the size and shape of the leakage paths in the barriers and the pressure difference across the paths.

Dilution Remote from Fire

Smoke dilution is sometimes referred to as **smoke purging**, **smoke removal**, **smoke exhaust**, or **smoke extraction**. Dilution can be used to maintain acceptable gas and particulate concentrations in a compartment subject to smoke infiltration from an adjacent space. It can be effective if the rate of smoke leakage is small compared to either the total volume of the safeguarded space or the rate of purging air supplied to and removed from the space. Also, dilution can be beneficial to the fire service for removing smoke after a fire has been extinguished. Sometimes, when doors are opened, smoke flows into areas intended to be protected. Ideally, the doors are only open for short periods during evacuation. Smoke that has entered spaces remote from the fire can be purged by supplying outside air to dilute the smoke.

The following is a simple analysis of smoke dilution for spaces in which there is no fire. Assume that at time zero ($\theta = 0$), a compartment is contaminated with some concentration of smoke and that no more smoke flows into the compartment or is generated within it. Also, assume that the contaminant. is uniformly distributed throughout the space. The concentration of contaminant in the space can be expressed as

$$\frac{C}{C_o} = e^{-at} \tag{6}$$

The dilution rate can be determined from the following equation:

$$a = \frac{1}{t} \ln\left(\frac{C_o}{C}\right) \tag{7}$$

where

C_o = initial concentration of contaminant
C = concentration of contaminant at time θ
a = dilution rate, air changes per minute
t = time after smoke stops entering space or smoke production has stopped, min
e = base of natural logarithm (approximately 2.718)

Concentrations C_o and C must be expressed in the same units, but can be any units appropriate for the particular contaminant being considered.

McGuire et al. (1970) evaluated the maximum levels of smoke obscuration from a number of fire tests and a number of proposed criteria for tolerable levels of smoke obscuration. Based on this evaluation, they state that the maximum levels of smoke obscuration are greater by a factor of 100 than those relating to the limit of tolerance. Thus, they indicate that a space can be considered "reasonably safe" with respect to smoke obscuration if the concentration of contaminants in the space is less than about 1% of the concentration in the immediate fire area. This level of dilution increases visibility by about a factor of 100 (e.g., from 0.5 ft to 50 ft) and reduces the concentrations of toxic smoke components. Toxicity is a more complex problem, and no parallel statement has been made regarding dilution needed to obtain a safe atmosphere with respect to toxic gases.

In reality, it is impossible to ensure that the concentration of the contaminant is uniform throughout the compartment. Because of buoyancy, it is likely that higher concentrations are near the ceiling. Therefore, exhausting smoke near the ceiling and supplying air near the floor probably dilutes smoke even more quickly than indicated by Equation (7). Supply and exhaust points should be placed to prevent supply air from blowing into the exhaust inlet, thereby short-circuiting the dilution.

Example 1. A space is isolated from a fire by smoke barriers and self-closing doors, so that no smoke enters the compartment when the doors are closed. When a door is opened, smoke flows through the open doorway into the space. If the door is closed when the contaminant in the space is 20% of the burn room concentration, what dilution rate is required to reduce the concentration to 1% of that in the burn room in 6 min?

The time $t = 6$ min and $C_o/C = 20$. From Equation (7), the dilution rate is about 0.5 air changes per minute, or 30 air changes per hour.

Caution about Dilution near Fire. Many people have unrealistic expectations about what dilution can accomplish in the fire space. Neither theoretical nor experimental evidence indicates that using a building's HVAC system for smoke dilution will significantly improve tenable conditions in a fire space. The exception is an unusual space where the fuel is such that fire size cannot grow above a specific limit; this occurs in some tunnels and underground transit situations. Because HVAC systems promote a considerable degree of air mixing in the spaces they serve and because very large quantities of smoke can be produced by building fires, it is generally believed that smoke dilution by an HVAC system in the fire space does not improve tenable conditions in that space. Thus, any attempt to improve hazard conditions in the fire space, or in spaces connected to the fire space by large openings, with smoke purging will be ineffective.

Pressurization (Smoke Control)

Systems that pressurize an area using mechanical fans are referred to as smoke control in this chapter and in NFPA *Standard* 92A. A pressure difference across a barrier can control smoke movement, as illustrated in Figure 6. Within the barrier is a door. The high-pressure side of the door can be either a refuge area or an egress route. The low-pressure side is exposed to smoke from a fire. Airflow through gaps around the door and through construction cracks prevents smoke infiltration to the high-pressure side.

For smoke control analysis, the orifice equation can be used to estimate the flow through building flow paths:

$$Q = 776CA\sqrt{2\Delta p/\rho} \tag{8}$$

where

Q = volumetric airflow rate, cfm
C = flow coefficient
A = flow area (leakage area), ft^2
Δp = pressure difference across flow path, in. of water
ρ = density of air entering flow path, lb$_m$/ft^3

The flow coefficient depends on the geometry of the flow path, as well as on turbulence and friction. In the present context, the flow coefficient is generally 0.6 to 0.7. For $\rho = 0.075$ lb$_m$/ft^3 and $C = 0.65$, Equation (8) can be expressed as

$$Q = 2610A\sqrt{\Delta p} \tag{9}$$

The flow area is frequently the same as the cross-sectional area of the flow path. A closed door with a crack area of 0.11 ft^2 and a pressure difference of 0.01 in. of water has an air leakage rate of approximately 29 cfm. If the pressure difference across the door is increased to 0.30 in. of water, the flow is 157 cfm.

Frequently, in field tests of smoke control systems, pressure differences across partitions or closed doors have fluctuated by as

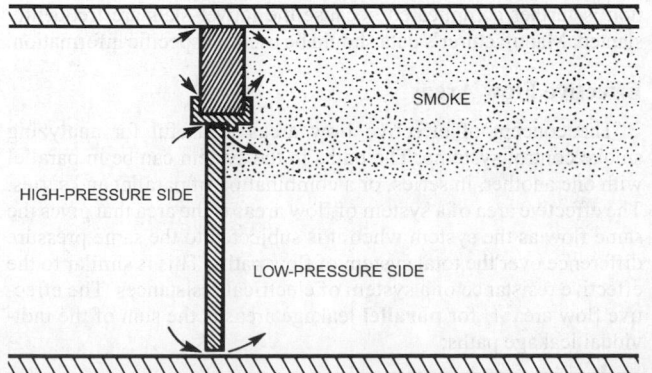

Fig. 6   Smoke Control System Preventing Smoke Infiltration to High-Pressure Side of Barrier

much as 0.02 in. of water. These fluctuations have generally been attributed to wind, although they could have been due to the HVAC system or some other source. To control smoke movement, the pressure difference produced by a smoke control system must be large enough to overcome pressure fluctuations, stack effect, smoke buoyancy, and wind pressure. However, the pressure difference should not be so large that the door is difficult to open.

Airflow

Airflow has been used extensively to manage smoke from fires in subway, railroad, and highway tunnels (see Chapter 15). Large airflow rates are needed to control smoke flow, and these flow rates can supply additional oxygen to the fire. Because of the need for complex controls, airflow is not used as extensively in buildings. The control problem consists of having very small flows when a door is closed and then significantly increased flows when that door is open. Furthermore, it is a major concern that the airflow supplies oxygen to the fire. This section presents the basics of smoke control by airflow and demonstrates why this technique is rarely recommended.

Thomas (1970) determined that in a corridor in which there is a fire, airflow can almost totally prevent smoke from flowing upstream of the fire. Molecular diffusion is believed to transfer trace amounts of smoke, which are not hazardous but which are detectable as the smell of smoke upstream. Based on work by Thomas, the critical air velocity for most applications can be approximated as

$$V_k = 5.68\left(\frac{q_c}{W}\right)^{1/3} \tag{10}$$

where

V_k = critical air velocity to prevent smoke backflow, fpm
q_c = heat release rate into corridor, Btu/h
W = corridor width, ft

This relation can be used when the fire is in the corridor or when smoke enters the corridor through an open doorway, air transfer grille, or other opening. Although critical velocities calculated from Equation (10) are general and approximate, they indicate the kind of air velocities required to prevent smoke backflow from fires of different sizes. For specific applications, other equations may be more appropriate: for tunnel applications, see Chapter 15; for smoke management in atriums and other large spaces, see Klote and Milke (2002) and NFPA *Standard* 92B.

Although Equation (10) can be used to estimate the airflow rate necessary to prevent smoke backflow through an open door, the oxygen supplied is a concern. Huggett (1980) evaluated the oxygen consumed in the combustion of numerous natural and synthetic solids. He found that, for most materials involved in building fires, the energy released is approximately 5630 Btu per pound of oxygen. Air is 23.3% oxygen by mass. Thus, if all the oxygen in a pound of air is consumed, 1300 Btu is liberated. If all the oxygen in 1 cfm of air with a density of 0.075 lb$_m$/ft^3 is consumed by fire, 5850 Btu/h is liberated.

Examples 2 and 3 demonstrate that the air needed to prevent smoke backflow can support an extremely large fire. Most commercial and residential buildings contain enough fuel (paper, cardboard, furniture, etc.) to support very large fires. Even when the amount of fuel is normally very small, short-term fuel loads (during building renovation, material delivery, etc.) can be significant. Therefore, using airflow for smoke control is not recommended, except when the fire is suppressed or in the rare cases when fuel can be restricted with confidence.

Example 2. What airflow at a doorway is needed to stop smoke backflow from a room fully involved in fire, and how large a fire can this airflow support?

A room fully involved in fire can have an energy release rate on the order of 8×10^6 Btu/h. Assume the door is 3 ft wide and 7 ft high. From

Equation (10), $V_k = 5.68(8 \times 10^6/3)^{1/3} = 790$ fpm. A flow through the doorway of $790 \times 3 \times 7 = 16,600$ cfm is needed to prevent smoke from backflowing into the area.

If all the oxygen in this airflow is consumed in the fire, the heat liberated is 16,600 cfm $\times$ 5850 Btu/h·cfm = 9.7×10^7 Btu/h. This is over 10 times more than the heat generated by the fully involved room fire and indicates why airflow is generally not recommended for smoke control in buildings.

Example 3. What airflow is needed to stop smoke backflow from a wastebasket fire, and how large a fire can this airflow support?

A wastebasket fire can have an energy release rate on the order of 5×10^5 Btu/h. As in Example 2, $V_k = 5.68(5 \times 10^5/3)^{1/3} = 310$ fpm. A flow through the doorway of $310 \times 3 \times 7 = 6500$ cfm is needed to prevent smoke backflow.

If all the oxygen in this airflow is consumed in the fire, the heat liberated is 6500 cfm $\times$ 5850 Btu/h·cfm = 3.8×10^7 Btu/h. This is still many times greater than the fully involved room fire and further indicates why airflow is generally not recommended for smoke control in buildings.

Buoyancy

The buoyancy of hot combustion gases is used in both fan-powered and non-fan-powered venting systems. Fan-powered venting for large spaces is commonly used for atriums and covered shopping malls, and non-fan-powered venting is commonly used for large industrial and storage buildings. There is a concern that sprinkler flow will cool the smoke, reducing buoyancy and thus the system effectiveness. Research is needed in this area. Refer to Klote and Milke (2002) and NFPA *Standards* 92B and 204 for detailed design information about these systems.

SMOKE CONTROL SYSTEM DESIGN

Door-Opening Forces

The door-opening forces resulting from the pressure differences produced by a smoke control system must be considered. Unreasonably high door-opening forces can make it difficult or impossible for occupants to open doors to refuge areas or escape routes.

The force required to open a door is the sum of the forces to overcome the pressure difference across the door and to overcome the door closer. This can be expressed as

$$F = F_{dc} + \frac{5.20 WA \Delta p}{2(W-d)} \tag{11}$$

where

F = total door-opening force, lb$_f$
F_{dc} = force to overcome door closer, lb$_f$
W = door width, ft
A = door area, ft^2
Δp = pressure difference across door, in. of water
d = distance from doorknob to edge of knob side of door, ft

This relation assumes that the door-opening force is applied at the knob. Door-opening force F_p caused by pressure difference can be determined from Figure 7 for a value of $d = 3$ in. The force to overcome the door closer is usually greater than 3 lb$_f$ and, in some cases, can be as great as 20 lb$_f$. For a door that is 7 ft high and 3 ft wide and subject to a pressure difference of 0.30 in. of water, the total door-opening force is 30 lb$_f$, if the force to overcome the door closer is 12 lb$_f$.

Flow Areas

In designing smoke control systems, airflow paths must be identified and evaluated. Some leakage paths are obvious, such as cracks around closed doors, open doors, elevator doors, windows, and air transfer grilles. Construction cracks in building walls are less obvious, but they are equally important.

The flow area of most large openings, such as open windows, can be calculated easily. However, flow areas of cracks are more difficult to evaluate. The area of these leakage paths depends on such features as workmanship, door fit, and weatherstripping. A 3 by 7 ft

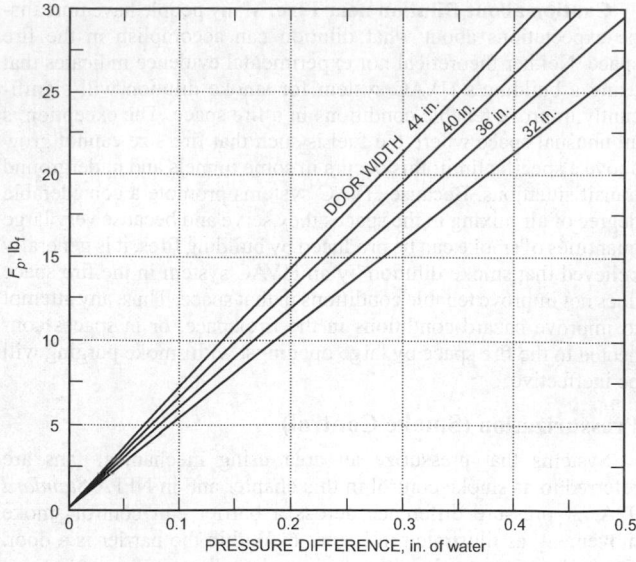

Fig. 7 Door-Opening Force Caused by Pressure Difference

door with an average crack width of 1/8 in. has a leakage area of 0.21 ft^2. However, if this door is installed with a 3/4 in. undercut, the leakage area is 0.36 ft^2, a significant difference. The leakage area of elevator doors is in the range of 0.55 to 0.70 ft^2 per door.

For open stairwell doorways, Cresci (1973) found complex flow patterns; the resulting flow through open doorways was considerably below that calculated using the doorway's geometric area as the flow area in Equation (8). Based on this research, it is recommended that the design flow area of an open stairwell doorway be *half* the geometric area (door height × width) of the doorway. An alternative for open stairwell doorways is to use the geometric area as the flow area and use a reduced flow coefficient. Because it does not allow the direct use of Equation (8), this approach is not used here.

Typical leakage areas for walls and floors of commercial buildings are tabulated as area ratios in Table 1. These data are based on a relatively small number of tests performed by the National Research Council of Canada (Shaw et al. 1993; Tamura and Shaw 1976a, 1976b, 1978; Tamura and Wilson 1966). Actual leakage areas depend primarily on workmanship rather than on construction materials, and in some cases, the flow areas in particular buildings may vary from the values listed. Data concerning air leakage through building components are also provided in Chapter 16 of the 2009 *ASHRAE Handbook—Fundamentals*.

Because a vent surface is usually covered by a louver and screen, a vent's flow area is less than its area (vent height × width). Calculation is further complicated because the louver slats are frequently slanted. Manufacturer's data should be used for specific information.

Effective Flow Areas

The concept of effective flow areas is useful for analyzing smoke control systems. The paths in the system can be in parallel with one another, in series, or a combination of parallel and series. The effective area of a system of flow areas is the area that gives the same flow as the system when it is subjected to the same pressure difference over the total system of flow paths. This is similar to the effective resistance of a system of electrical resistances. The effective flow area A_e for **parallel** leakage areas is the sum of the individual leakage paths:

$$A_e = \sum_{i=1}^{n} A_i \tag{12}$$

Table 1   Typical Leakage Areas for Walls and Floors of Commercial Buildings

Construction Element	Wall Tightness	Area Ratio
		A/A_w
Exterior building walls[a]	Tight	0.50×10^{-4}
(includes construction cracks and	Average	0.17×10^{-3}
cracks around windows and doors)	Loose	0.35×10^{-3}
	Very Loose	0.12×10^{-2}
Stairwell walls[a]	Tight	0.14×10^{-4}
(includes construction cracks but not	Average	0.11×10^{-3}
cracks around windows or doors)	Loose	0.35×10^{-3}
Elevator shaft walls[a]	Tight	0.18×10^{-3}
(includes construction cracks but	Average	0.84×10^{-3}
not cracks around doors)	Loose	0.18×10^{-2}
		A/A_f
Floors[b]	Tight	0.66×10^{-5}
(includes construction cracks and	Average	0.52×10^{-4}
gaps around penetrations)	Loose	0.17×10^{-3}

A = leakage area; A_w = wall area; A_f = floor area
Leakage areas evaluated at [a]0.3 in. of water; [b]0.1 in. of water.

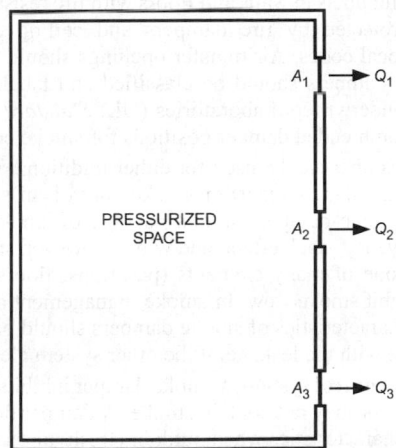

Fig. 8   Leakage Paths in Parallel

where n is the number of flow areas A_i in parallel.

For example, the effective area A_e for the three parallel leakage areas in Figure 8 is

$$A_e = A_1 + A_2 + A_3 \qquad (13)$$

If A_1 is 1.0 ft^2 and A_2 and A_3 are each 0.5 ft^2, then the effective flow area A_e is 2.0 ft^2.

The general rule for any number of leakage areas in **series** is

$$A_e = \left[\sum_{i=1}^{n} \frac{1}{A_i^2} \right]^{-0.5} \qquad (14)$$

where n is the number of leakage areas A_i in series.

Three leakage areas in series from a pressurized space are illustrated in Figure 9. The effective flow area of these paths is

$$A_e = \left(\frac{1}{A_1^2} + \frac{1}{A_2^2} + \frac{1}{A_3^2} \right)^{-0.5} \qquad (15)$$

In smoke control analysis, there are frequently only two paths in series, and the effective leakage area is

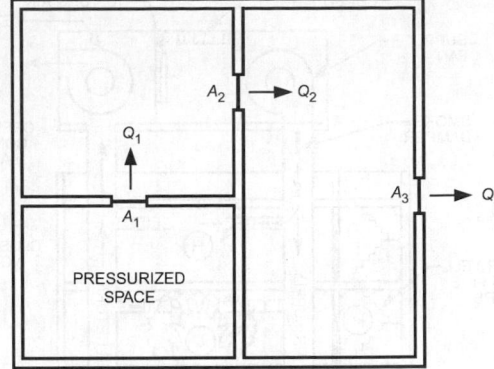

Fig. 9   Leakage Paths in Series

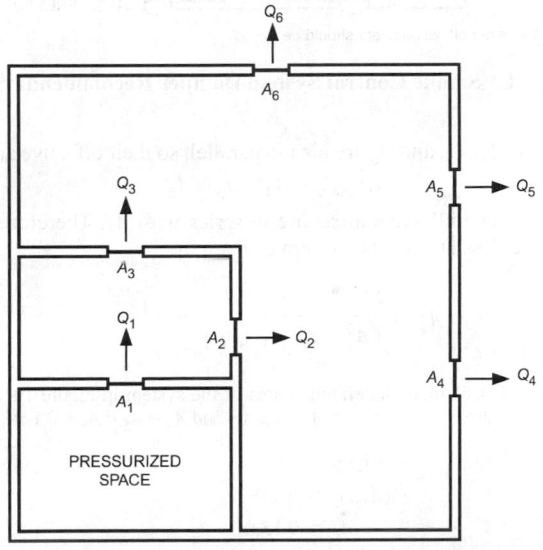

Fig. 10   Combination of Leakage Paths in Parallel and Series

$$A_e = \frac{A_1 A_2}{\sqrt{A_1^2 + A_2^2}} \qquad (16)$$

Example 4. Calculate the effective leakage area of two equal flow paths in series. Let $A = A_1 = A_2 = 0.20$ ft^2. From Equation (16),

$$A_e = \frac{A^2}{\sqrt{2A^2}} = 0.14 \text{ ft}^2$$

Example 5. Calculate the effective flow area of two flow paths in series, where $A_1 = 0.20$ ft^2 and $A_2 = 2.0$ ft^2. From Equation (16),

$$A_e = \frac{A_1 A_2}{\sqrt{A_1^2 + A_2^2}} = 0.199 \text{ ft}^2$$

Example 5 illustrates that when two paths are in series, and one is much larger than the other, the effective flow area is approximately equal to the smaller area.

Developing an effective area for a system of both parallel and series paths requires combining groups of parallel paths and series paths systematically. The system illustrated in Figure 10 is analyzed as an example. The figure shows that A_2 and A_3 are in parallel; therefore, their effective area is

$$(A_{23})_e = A_2 + A_3$$

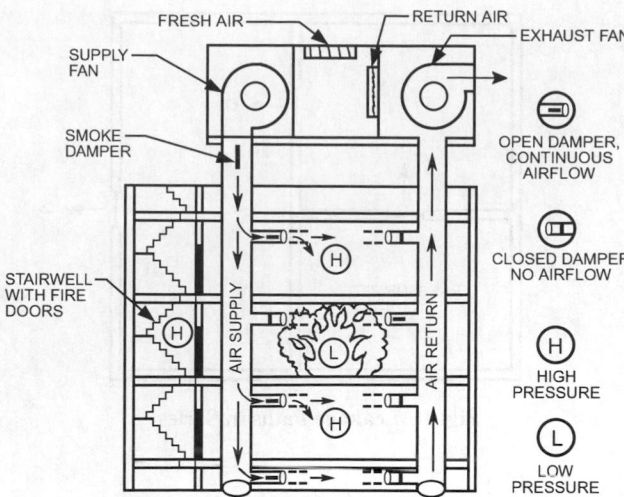

Fig. 11 Smoke Control System Damper Recommendation

Areas A_4, A_5, and A_6 are also in parallel, so their effective area is

$$(A_{456})_e = A_4 + A_5 + A_6$$

These two effective areas are in series with A_1. Therefore, the effective flow area of the system is given by

$$A_e = \left[\frac{1}{A_1^2} + \frac{1}{(A_{23})_e^2} + \frac{1}{(A_{456})_e^2}\right]^{-0.5}$$

Example 6. Calculate the effective area of the system in Figure 10, if the leakage areas are $A_1 = A_2 = A_3 = 0.2$ ft² and $A_4 = A_5 = A_6 = 0.1$ ft².

$$(A_{23})_e = 0.4 \text{ ft}^2$$
$$(A_{456})_e = 0.3 \text{ ft}^2$$
$$A_e = 0.15 \text{ ft}^2$$

Design Weather Data

Little weather information has been developed specifically for smoke control system design. Design temperatures for heating and cooling found in Chapter 14 of the 2009 *ASHRAE Handbook—Fundamentals* (on the CD-ROM accompanying that volume) may be used. Extreme temperatures can be considerably lower than the winter design temperatures. For example, the 99% design temperature for Tallahassee, Florida, is 28°F, but the lowest temperature observed there was –2°F (NOAA 1979).

Temperatures are generally below the design values for short periods, and because of the thermal lag of building materials, these short intervals of low temperature usually do not cause problems with heating. However, there is no time lag for a smoke control system; it is therefore subjected to all the extreme forces of stack effect that exist the moment it operates. If the outside temperature is below the system's winter design temperature, stack effect problems may result. A similar situation can occur with summer design temperatures and reverse stack effect.

Extreme wind data for smoke management design are listed in Chapter 14 of the 2009 *ASHRAE Handbook—Fundamentals*.

Design Pressure Differences

Both the maximum and minimum allowable pressure differences across the boundaries of smoke control should be considered. The maximum allowable pressure difference should not cause excessive door-opening forces.

The minimum allowable pressure difference across a boundary of a smoke control system might be the difference such that no smoke leakage occurs during building evacuation. In this case, the smoke control system must produce sufficient pressure differences to overcome forces of wind, stack effect, or buoyancy of hot smoke. Pressure differences caused by wind and stack effect can be large in the event of a broken window in the fire compartment. Evaluation of these pressure differences depends on evacuation time, rate of fire growth, building configuration, and the presence of a fire suppression system. NFPA *Standard* 92A suggests values of minimum and maximum design pressure difference.

Open Doors

Another design concern is the number of doors that could be opened simultaneously when the smoke control system is operating. A design that allows all doors to be open simultaneously may ensure that the system always works, but often adds to system cost.

The number of doors that may be open simultaneously depends largely on building occupancy. For example, in a densely populated building, it is likely that all doors will be open during evacuation. However, if a staged evacuation plan or refuge area concept is incorporated in the building fire emergency plan, or if the building is sparsely occupied, only a few of the doors may be open during a fire.

FIRE AND SMOKE DAMPERS

Openings for ducts in walls and floors with fire resistance ratings should be protected by fire dampers and ceiling dampers, as required by local codes. Air transfer openings should also be protected. These dampers should be classified and labeled in accordance with Underwriters Laboratories (UL) *Standard* 555. Figure 11 shows recommended damper positions for smoke control.

A smoke damper can be used for either traditional smoke management (smoke containment) or smoke control. In **smoke management**, a smoke damper inhibits passage of smoke under the forces of buoyancy, stack effect, and wind. However, smoke dampers are only one of many elements (partitions, floors, doors) intended to inhibit smoke flow. In smoke management applications, the leakage characteristics of smoke dampers should be selected to be appropriate with the leakage of the other system elements.

In a **smoke control system**, a smoke damper inhibits the passage of air that may or may not contain smoke. A damper does not need low leakage characteristics when outdoor (fresh) air is on the high-pressure side of the damper, as is the case for dampers that shut off supply air from a smoke zone or that shut off exhaust air from a non-smoke zone. In these cases, moderate leakage of smoke-free air through the damper does not adversely affect control of smoke movement. It is best to design smoke control systems so that only smoke-free air is on the high-pressure side of a closed smoke damper.

Smoke dampers should be classified and listed in accordance with UL *Standard* 555S for temperature, leakage, and operating velocity. The velocity rating of a smoke damper is the velocity at which the actuator will open and close the damper.

At locations requiring both smoke and fire dampers, combination dampers meeting the requirements of both UL *Standards* 555 and 555S can be used. The combination fire/smoke dampers must close when they reach their UL *Standard* 555S temperature rating to maintain the integrity of the firewall.

Fire, ceiling, and smoke dampers should be installed in accordance with the manufacturers' instructions. NFPA *Standard* 90A gives general guidelines on locations requiring these dampers.

The supply and return/smoke dampers should be a minimum of Class II leakage at 250°F. The return air damper should be a minimum of Class I leakage at 250°F to prevent recirculation of smoke exhaust. The operating velocity of the dampers should be evaluated when the dampers are in smoke control mode. To minimize velocity build-up, only zones adjacent to the fire need to be pressurized.

The exhaust ductwork and fan must be designed to handle the temperature of the exhaust smoke. This temperature can be lowered by making the smoke control zones large or by pressurizing only the zones adjacent to the fire zone and leaving all the other zones operating normally.

Fans Used to Exhaust Smoke

Understanding building code requirements for high-temperature fans in smoke control systems is important for both designers, who must select fans that can operate satisfactorily at elevated temperatures, and manufacturers, who can then design suitable off-the-shelf fans rather than customizing fans for each application. Only fans designed for use under elevated temperatures should be used in smoke management applications; other types may fail, or their performance may change because of component deformation or altered clearances among components. Also, some smoke exhaust applications (e.g., transit tunnels) require that smoke-handling fans reverse direction repeatedly on demand. Until recently, standards did not address reversibility or airflow performances of high-temperature fans at ambient and elevated temperatures. To allow manufacturers to provide suitable off-the-shelf products, a standard method of test (MOT) and ratings scale have been developed.

ANSI/ASHRAE *Standard* 149 provides testing laboratories with standard testing methods for fan characteristics specific to smoke exhaust functions, including (1) aerodynamic performance, (2) operation at specified elevated temperature, (3) reversal, and (4) damper performance (for dampers included with the fan).

AMCA *Publication* 212 establishes ratings to allow consistent comparison among catalog test data. Model code requirements for elevated temperature and duration of operation are charted on a graph, which is divided into several fan performance groups. Manufacturers can request that laboratories test fans according to ANSI/ASHRAE *Standard* 149; those data can then be incorporated into catalogs for off-the-shelf products according to AMCA *Publication* 212 ratings, allowing designers to select the most appropriate models and performances for their specific applications. This allows designers and code officials to compare different manufacturers' products more easily, and enhances confidence that products will perform as intended; it also allows manufacturers to provide more cost-efficient off-the-shelf products rather than custom-designing fans for each application.

PRESSURIZED STAIRWELLS

Many pressurized stairwells have been designed and built to provide a smoke-free escape route in the event of a building fire. They also provide a smoke-free staging area for firefighters. On the fire floor, a pressurized stairwell must maintain a positive pressure difference across a closed stairwell door to prevent smoke infiltration.

During building fires, some stairwell doors are opened intermittently during evacuation and fire fighting, and some doors may even be blocked open. Ideally, when the stairwell door is opened on the fire floor, airflow through the door should be sufficient to prevent smoke backflow. Designing a system to achieve this goal is difficult because of the many combinations of open stairwell doors and weather conditions affecting airflow.

Stairwell pressurization systems may be single- or multiple-injection systems. A **single-injection system** supplies pressurized air to the stairwell at one location, usually at the top. Associated with this system is the potential for smoke to enter the stairwell through the pressurization fan intake. Therefore, automatic shutdown during such an event should be considered.

For tall stairwells, single-injection systems can fail when a few doors are open near the air supply injection point, especially in bottom-injection systems when a ground-level stairwell door is open.

For tall stairwells, supply air can be supplied at a number of locations over the height of the stairwell. Figures 12 and 13 show two

examples of **multiple-injection systems** that can be used to overcome the limitations of single-injection systems. In these figures, the supply duct is shown in a separate shaft. However, systems have been built that eliminated the expense of a separate duct shaft by locating the supply duct in the stairwell itself. In such a case, care must be taken that the duct does not obstruct orderly building evacuation.

Stairwell Compartmentation

Compartmentation of the stairwell into a number of sections is one alternative to multiple injection (Figure 14). When the doors between compartments are open, the effect of compartmentation is lost. For this reason, compartmentation is inappropriate for densely populated buildings where total building evacuation by the stairwell is planned in the event of fire. However, when a staged evacuation plan is used and the system is designed to operate successfully with the maximum number of doors between compartments open, compartmentation can effectively pressurize tall stairwells.

Stairwell Analysis

This section presents an analysis for a pressurized stairwell in a building without vertical leakage. This method closely approximates the performance of pressurized stairwells in buildings with-

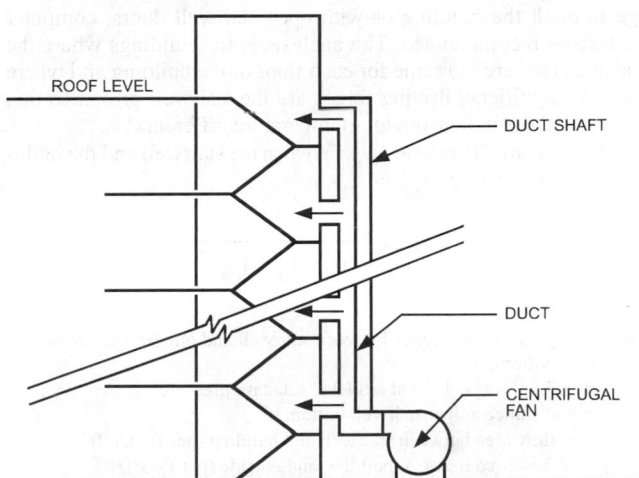

Fig. 12 Stairwell Pressurization by Multiple Injection with Fan Located at Ground Level

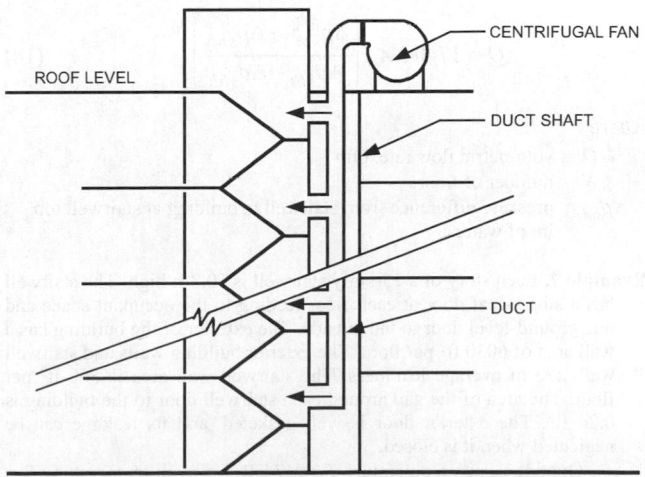

Fig. 13 Stairwell Pressurization by Multiple Injection with Roof-Mounted Fan

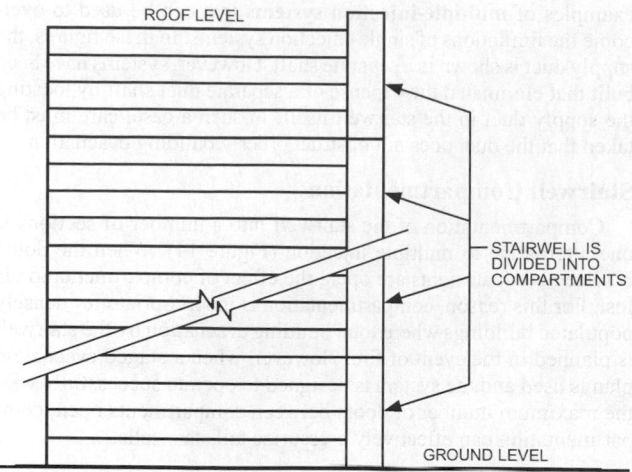

Note: Each four-floor compartment has a least one supply air injection point.

Fig. 14   Compartmentation of Pressurized Stairwell

out elevators. It is also useful for buildings with vertical leakage because it yields conservative results. For evaluating vertical leakage through the building or with open stairwell doors, computer analysis is recommended. The analysis is for buildings where the leakage areas are the same for each floor of the building and where the only significant driving forces are the stairwell pressurization system and the indoor-outdoor temperature difference.

The pressure difference Δp_{sb} between the stairwell and the building can be expressed as

$$\Delta p_{sb} = \Delta p_{sbb} + \frac{By}{1 + (A_{sb}/A_{bo})^2} \qquad (17)$$

where

Δp_{sbb} = pressure difference between stairwell and building at stairwell bottom, in. of water
B = 7.64($1/T_o$ – $1/T_s$) at sea level standard pressure
y = distance above stairwell bottom, ft
A_{sb} = flow area between stairwell and building (per floor), ft²
A_{bo} = flow area between building and outside (per floor), ft²
T_o = temperature of outside air, °R
T_s = temperature of stairwell air, °R

For a stairwell with no leakage directly to the outside, the flow rate of pressurization air is

$$Q = 1740 \, NA_{sb} \left(\frac{\Delta p_{sbt}^{3/2} - \Delta p_{sbb}^{3/2}}{\Delta p_{sbt} - \Delta p_{sbb}} \right) \qquad (18)$$

where

Q = volumetric flow rate, cfm
N = number of floors
Δp_{sbt} = pressure difference from stairwell to building at stairwell top, in. of water

Example 7. Each story of a 15-story stairwell is 10.8 ft high. The stairwell has a single-leaf door at each floor leading to the occupant space and one ground-level door to the outside. The exterior of the building has a wall area of 6030 ft² per floor. The exterior building walls and stairwell walls are of average leakiness. The stairwell wall area is 560 ft² per floor. The area of the gap around each stairwell door to the building is 0.26 ft². The exterior door is well gasketed, and its leakage can be neglected when it is closed.

Outside design temperature T_o = 474°R; stairwell temperature T_s = 530°R; maximum design pressure differences when all stairwell doors are closed is 0.35 in. of water; the minimum allowable pressure difference is 0.052 in. of water.

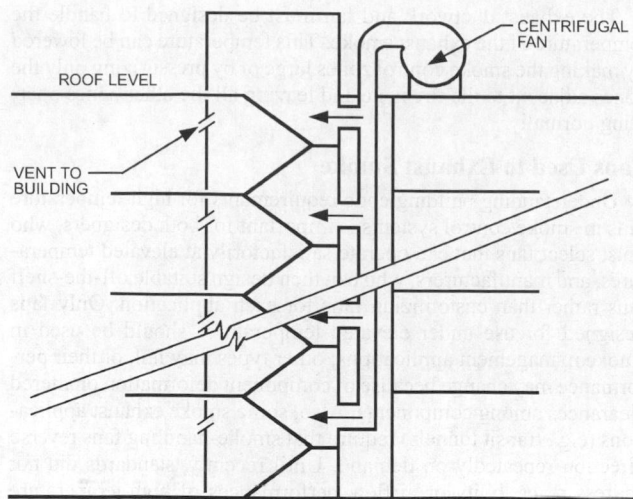

Notes: 1. Vents to building have barometric damper and fire damper in series.
2. Roof-mounted supply fan is shown; however, fan may be located at any level.
3. Manually operated damper may be located at stairwell top for smoke purging by fire department.

Fig. 15   Stairwell Pressurization with Vents to Building at Each Floor

Using the leakage ratio for an exterior building wall of average tightness from Table 1, A_{bo} = 6030(0.17 × 10⁻³) = 1.025 ft². Using the leakage ratio for a stairwell wall of average tightness from Table 1, the leakage area of the stairwell wall is 560(0.11 × 10⁻³) = 0.06 ft². The value of A_{sb} equals the leakage area of the stairwell wall plus the gaps around the closed doors: A_{sb} = 0.06 + 0.26 = 0.32 ft². The temperature factor B is calculated at 0.00170 in. of water/ft. The pressure difference at the stairwell bottom is selected as Δp_{sbb} = 0.080 in. of water to provide an extra degree of protection above the minimum allowable value of 0.052 in. of water. The pressure difference Δp_{sbt} is calculated from Equation (17) at 0.331 in. of water, using y = 15(10.8) = 162 ft. Thus, Δp_{sbt} does not exceed the maximum allowable pressure. The flow rate of pressurization air is calculated from Equation (18) at 5585 cfm.

The flow rate depends strongly on the leakage area around the closed doors and on the leakage area in the stairwell walls. In practice, these areas are difficult to evaluate and even more difficult to control. If flow area A_{sb} in Example 7 were 0.54 ft² rather than 0.32 ft², Equation (18) would give a flow rate of pressurization air of 9500 cfm. A fan with a sheave allows adjustment of supply air to offset for variations in actual leakage from the values used in design calculations.

Stairwell Pressurization and Open Doors

The simple pressurization system discussed in the previous section has two limitations regarding open doors. First, when a stairwell door to the outside and building doors are open, the simple system cannot provide enough airflow through building doorways to prevent smoke backflow. Second, when stairwell doors are open, pressure difference across the closed doors can drop to low levels. Two systems used to overcome these problems are overpressure relief (Tamura 1990) and supply fan bypass.

Overpressure Relief. The total airflow rate is selected to provide the minimum air velocity when a specific number of doors are open. When all the doors are closed, part of this air is relieved through a vent to prevent excessive pressure build-up, which could cause excessive door-opening forces. This excess air should be vented from the stairwell to the street-level floor. Fire and relief dampers should be the low-leakage type. Stairwell doors should have gasket seals at sides and top, leaving the bottom gap open for relief.

Barometric dampers that close when pressure drops below a specified value can minimize air loss through the vent when doors are open. Figure 15 illustrates a pressurized stairwell with overpressure relief vents to the building at each floor. In systems with

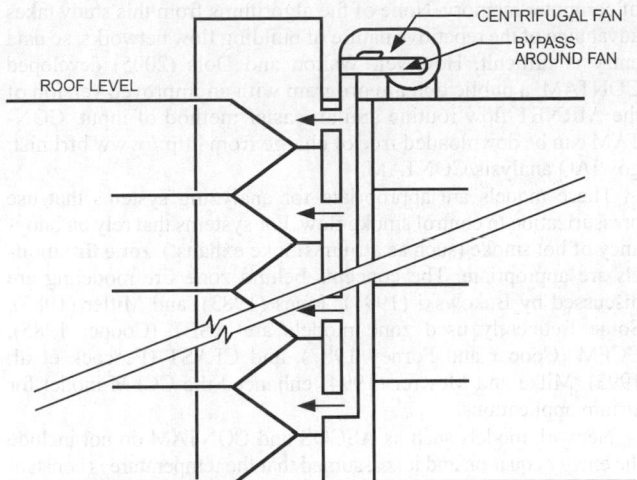

Notes: 1. Fan bypass controlled by one or more static-pressure sensors located between stairwell and building.
2. Roof-mounted supply fan is shown; however, fan may be located at any level.
3. Manually operated damper may be located at stairwell top for smoke purging by fire department.

Fig. 16 Stairwell Pressurization with Bypass Around Supply Fan

vents between stairwell and building, the vents typically have a fire damper in series with the barometric damper. To conserve energy, these fire dampers are normally closed, but they open when the pressurization system is activated. This arrangement also reduces the possibility of the annoying damper chatter that frequently occurs with barometric dampers.

An exhaust duct can provide overpressure relief in a pressurized stairwell. The system is designed so that the normal resistance of a nonpowered exhaust duct maintains pressure differences within the design limits.

Exhaust fans can also relieve excess pressure when all stairwell doors are closed. An exhaust fan should be controlled by a differential pressure sensor, so that it will not operate when the pressure difference between stairwell and building falls below a specified level. This control should prevent the fan from pulling smoke into the stairwell when a number of open doors have reduced stairwell pressurization. The exhaust fan should be specifically sized so that the pressurization system performs within design limits. A wind shield is recommended because an exhaust fan can be adversely affected by the wind.

An alternative method of venting a stairwell is through an automatically opening stairwell door to the outside at ground level. Under normal conditions, this door would be closed and, in most cases, locked for security reasons. Provisions are needed to prevent this lock from conflicting with the automatic operation of the system. Possible adverse wind effects are also a concern with a system that uses an open outside door as a vent. Occasionally, high local wind velocities develop near the exterior stairwell door; such winds are difficult to estimate without expensive modeling. Nearby obstructions can act as windbreaks or wind shields.

Supply Fan Bypass. In this system, the supply fan is sized to provide at least the minimum air velocity when the design number of doors are open. Figure 16 illustrates such a system. The flow rate of air into the stairwell is varied by modulating bypass dampers, which are controlled by one or more static pressure sensors that sense the pressure difference between the stairwell and the building. When all the stairwell doors are closed, the pressure difference increases and the bypass damper opens to increase the bypass air and decrease the flow of supply air to the stairwell. In this manner, excessive stairwell pressures and excessive pressure differences between the stairwell and the building are prevented.

ELEVATORS

Elevator smoke control systems intended for use by firefighters should keep elevator cars, elevator shafts, and elevator machinery rooms smoke-free. Small amounts of smoke in these spaces are acceptable, provided that the smoke is nontoxic and that operation of elevator equipment is not affected. Elevator smoke control systems intended for fire evacuation of people unable to self-rescue or other building occupants should also keep elevator lobbies smoke-free or nearly smoke-free. Obstacles to fire evacuation by elevators include

- Logistics of evacuation
- Reliability of electrical power
- Jamming of elevator doors
- Fire and smoke protection

All these obstacles, except smoke protection, can be addressed by existing technology (Klote 1984).

Klote and Tamura (1986) studied conceptual elevator smoke control systems for evacuation of people unable to self-rescue. The major problem was maintaining pressurization with open building doors, especially doors on the ground floor. Of the systems evaluated, only one with a supply fan bypass with feedback control maintained adequate pressurization with any combination of open or closed doors. There are probably other systems capable of providing adequate smoke control; the procedure used by Klote and Tamura can be viewed as an example of a method of evaluating the performance of a system to determine whether it suits the particular characteristics of a building under construction.

Transient pressures caused by **piston effect** when an elevator car moves in a shaft have been a concern in elevator smoke control. Piston effect is not a concern for slow-moving cars in multiple-car shafts, but can be considerable for fast cars in single-car shafts.

ZONE SMOKE CONTROL

Klote (1990) conducted a series of tests on full-scale fires that demonstrated that zone smoke control can restrict smoke movement to the zone where a fire starts.

Pressurized stairwells are intended to prevent smoke infiltration into stairwells. However, in a building with only stairwell pressurization, smoke can flow through cracks in floors and partitions and through shafts to damage property and threaten life at locations remote from the fire. Zone smoke control is intended to limit this smoke movement.

A building is divided into a number of smoke control zones, each separated from the others by partitions, floors, and doors that can be closed to inhibit smoke movement. In the event of a fire, pressure differences and airflows produced by mechanical fans limit spread of smoke from the zone in which the fire started. The concentration of smoke in this zone goes unchecked; thus, in zone smoke control systems, occupants should evacuate the smoke zone as soon as possible after fire detection.

A smoke control zone can consist of one floor, more than one floor, or part of a floor. Sprinkler zones and smoke control zones should be coordinated so that sprinkler water flow activates the zone's smoke control system. Some arrangements of smoke control zones are illustrated in Figure 17. All the nonsmoke zones in the building may be pressurized. The term **pressure sandwich** describes cases where only zones adjacent to the smoke zone are pressurized, as in Figures 17B and 17D.

Zone smoke control is intended to limit smoke movement to the smoke zone by the use of pressurization. Pressure differences in the desired direction across the barriers of a smoke zone can be achieved by supplying outside (fresh) air to nonsmoke zones, by venting the smoke zone, or by a combination of these methods.

Venting smoke from a smoke zone prevents significant overpressure from thermal expansion of gases caused by the fire. This venting can be accomplished by exterior wall vents, smoke shafts, and

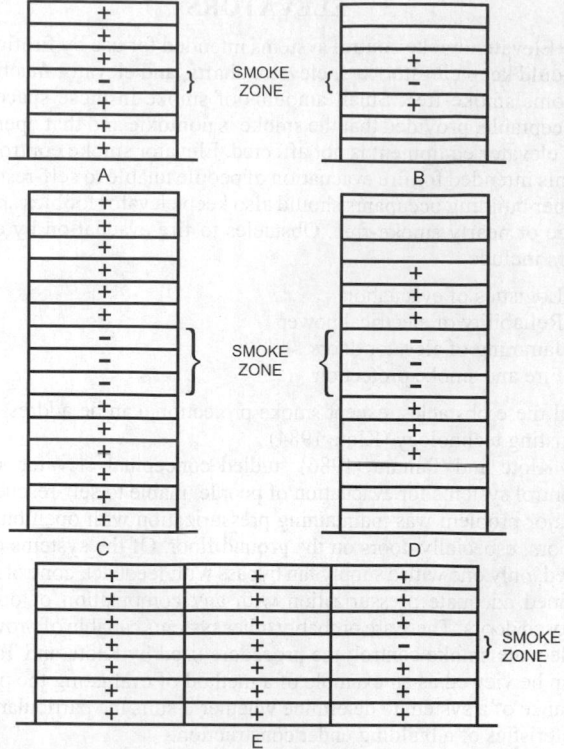

Note:
In the above figures, the smoke zone is indicated by a minus sign, and pressurized spaces are indicated by a plus sign. Each floor can be a smoke control zone as in A and B, or a smoke zone can consist of more than one floor as in C and D. All nonsmoke zones adjacent to the smoke zone may be pressurized, as in A and C, or only nonsmoke zones adjacent to the smoke zone may be pressurized, as in B and D. A smoke zone can also be limited to a part of a floor as in E.

Fig. 17   Some Arrangements of Smoke Control Zones

mechanical venting (exhausting). However, venting only slightly reduces smoke concentration in the smoke zone.

COMPUTER ANALYSIS FOR PRESSURIZATION SYSTEMS

Because of the complex airflow in buildings, network computer programs were developed to model the airflow with pressurization systems. These models represent rooms and shafts by nodes; airflow is from nodes of high pressure to nodes of lower pressure. Some programs calculate steady-state airflow and pressures throughout a building (Sander 1974; Sander and Tamura 1973). Other programs go beyond this to calculate the smoke concentrations that would be produced throughout a building in the event of a fire (Evers and Waterhouse 1978; Rilling 1978; Wakamatsu 1977; Yoshida et al. 1979).

The ASCOS program was developed specifically for analyzing pressurization smoke control systems (Klote 1982). ASCOS was the most widely used program for smoke control analysis (Said 1988), and has been validated against field data from flow experiments at an eight-story tower in Champs sur Marne, France (Klote and Bodart 1985). ASCOS and the other network models have been used extensively for design and for parametric analysis of the performance of smoke control systems. However, ASCOS was intended as a research tool for application to 10- and 20-story buildings. Not surprisingly, convergence failures have been encountered with applications to much larger buildings.

Wray and Yuill (1993) evaluated several flow algorithms to find the most appropriate one for analysis of smoke control systems. They selected the AIRNET flow routine developed by Walton (1989) as the best algorithm based on computational speed and use

of computer memory. None of the algorithms from this study takes advantage of the repetitive nature of building flow networks, so data entry is difficult. However, Walton and Dols (2005) developed CONTAM, a public domain program with an improved version of the AIRNET flow routine and an easier method of input. CONTAM can be downloaded free of charge from http://www.bfrl.nist.gov/IAQ analysis/CONTAM.

These models are appropriate for analyzing systems that use pressurization to control smoke flow. For systems that rely on buoyancy of hot smoke (such as atrium smoke exhaust), **zone fire models** are appropriate. The concepts behind zone fire modeling are discussed by Bukowski (1991), Jones (1983), and Mitler (1985). Some frequently used zone models are ASET (Cooper 1985), CCFM (Cooper and Forney 1987), and CFAST (Peacock et al. 1993). Milke and Mowrer (1994) enhanced the CCFM model for atrium applications.

Network models such as ASCOS and CONTAM do not include the energy equation and it is assumed that the temperature is constant for all nodes in the network. Therefore, these models cannot be used to simulate smoke movement near a fire. Conditions in the fire compartment and adjacent areas can be simulated using zone models. However, the number of rooms or compartments that can be simulated using a zone model is limited by the stability of the numerical solver and computational time (Fu et al. 2002). There have been simulations of smoke flow in a large, multicompartmented building using two separate models: (1) a zone model, such as CFAST (Jones et al. 2005), for conditions in compartments in the immediate fire area, and (2) a network model, such as CONTAM (Walton 1997), to determine smoke flow away from the fire source. Data from the zone model were used as input to the network model. This is an approximate method with only the mass flow determined over the entire building. Energy transfer is limited to the domain of the zone model.

ASHRAE research project RP-1328 used a zone model algorithm to simulate conditions in the fire region, and input those results directly into a network model that included the energy equation (Kashef and Hadjisophocleous 2010). This model allows a reasonable numerical simulation (time and accuracy) of the fire process, which determines both mass flow and energy transfer over an entire high-rise building using a standard personal computer.

SMOKE MANAGEMENT IN LARGE SPACES

In recent years, atrium buildings have become commonplace. Other large, open spaces include enclosed shopping malls, arcades, sports arenas, exhibition halls, and airplane hangars. For simplicity, the term **atrium** is used in this chapter in a generic sense to mean any of these large spaces. Traditional fire protection by compartmentation is not applicable to these large-volume spaces.

Most atrium smoke management systems are designed to prevent exposure of occupants to smoke during evacuation; this is the approach described in this section. An alternative goal is to maintain tenable conditions even when occupants have some contact with smoke, as discussed in the section on Tenability Systems.

The following approaches can be used to manage smoke in atriums:

- **Smoke filling.** This approach allows smoke to fill the atrium space while occupants evacuate the atrium. It applies only to spaces where the smoke-filling time is sufficient for both decision making and evacuation. Nelson and Mowrer (2008), Chapter 4 of Klote and Milke (2002), Proulx (2008), and Tubbs and Meacham (2007) have information on people movement during evacuation. The filling time can be estimated either by zone fire models or by filling equations [e.g., Equation (21)].

- **Unsteady clear height with upper layer exhaust.** This approach exhausts smoke from the top of the atrium at a rate such that occupants have sufficient time for decision making and evacuation. It requires analysis of people movement and fire model analysis of smoke filling.

- **Steady clear height with upper layer exhaust.** This approach exhausts smoke from the top of the atrium to achieve a steady clear height for a steady fire (Figure 18). A calculation method is presented in the section on Steady Clear Height with Upper Layer Exhaust.

Design Fires

The design fire has a major effect on the atrium smoke management system. Fire size is expressed in terms of rate of heat release. Fire growth is the rate of change of the heat release rate and is sometimes expressed as a growth constant that identifies the time required for the fire to attain a particular rate of heat release. Designs may be based on either steady fires or unsteady fires.

Fires are by nature unsteady, but the steady fire is a very useful idealization. **Steady fires** have a constant heat release rate. In many applications, using a steady design fire leads to straightforward and conservative design.

Morgan (1979) suggests 44 Btu/s·ft² as a typical rate of heat release per unit floor area for mercantile occupancies. Fang and Breese (1980) found about the same rate of heat release for residential occupancies. Law (1982) and Morgan and Hansell (1987) suggest a heat release rate per unit floor area for office buildings of 20 Btu/s·ft².

In many atriums, fuel loading is severely restricted with the intent of restricting fire size. Such atriums are characterized by interior finishes of metal, brick, stone, or gypsum board and furnished with objects made of similar materials plus plants. Even in such a **fuel-restricted atrium**, many combustible objects are present for short periods. Packing materials, holiday decorations, displays, construction materials, and furniture being moved into another part of the building are a few examples of **transient fuels**.

In this chapter, a heat release rate per floor area of 20 Btu/s·ft² is used for a fuel-restricted atrium, and 44 Btu/s·ft² is used for atriums containing furniture, wood, or other combustible materials.

Transient fuels must not be overlooked when selecting a design fire. Klote and Milke (2002) suggest incorporating transient fuels in a design fire by considering the fire occurring over 100 ft² of floor space. This results in a design fire of 2000 Btu/s for fuel-restricted atriums. In an atrium with combustibles, the design fire would be 4400 Btu/s. However, the area involved in fire may be much greater; flame spread considerations must be taken into account (Klote and Milke 2002; NFPA *Standard* 92B). A large atrium fire of 25,000 Btu/s would involve an area of 568 ft² at 44 Btu/s·ft². Table 2 lists some steady design fires.

Unsteady fires are often characterized by the following equation:

$$q = 1000 \left(\frac{t}{t_g} \right)^2 \tag{19}$$

where

q = heat release rate of fire, Btu/s
t = time, s
t_g = growth time, s

These unsteady fires are called *t*-**squared fires**; typical growth times are listed in Table 3.

Zone Fire Models

Atrium smoke management design is based on the zone fire model concept. This concept has been applied to several computer models used for atrium smoke management design analysis, including the Harvard Code (Mitler and Emmons 1981), ASET (Cooper 1985), the BRI Model (Tanaka 1983), CCFM (Cooper and Forney 1987), and CFAST (Peacock et al. 1993). The University of Maryland modified CCFM specifically for atrium smoke management design (Milke and Mowrer 1994). Although each of these models has unique features, they all share the same basic two-zone model concept.

For more information about zone models, see Mitler (1984), Mitler and Rockett (1986), and Quintiere (1989). The ASET-B model (Walton 1985) is a good starting point for learning about zone models.

Zone models were developed for room fires. In a room fire, hot gases rise above the fire, forming a **plume**. As the plume rises, it entrains air from the room so that the diameter and mass flow rate of the plume increase with elevation. Accordingly, plume temperature decreases with elevation. Fire gases from the plume flow up to the ceiling and form a hot stratified layer under the ceiling. Hot gases can flow through openings in walls to other spaces; this flow is referred to as a **door jet**, which is similar to a plume, except that it flows through an opening in a wall.

Figure 19A is a sketch of a room fire. Zone modeling is an idealization of the room fire conditions, as illustrated in Figure 19B. For this idealization, the temperatures of the hot upper and lower layers of the room are uniform. The height of the discontinuity between these layers is the same everywhere. The dynamic effects on pressure are considered negligible, so pressures are treated as hydrostatic. Other properties are considered uniform for each layer. Algebraic equations are used to calculate the mass flows caused by plumes and door jets.

Many computer zone models allow exhaust from the upper layer, which is essential for simulating atrium smoke exhaust systems. Heat transfer is estimated by methods ranging from a simple allowance as a fraction of the heat released by the fire to a complicated simulation including the effects of conduction, convection, and radiation.

Atrium Smoke Filling by a Steady Fire

The following experimental correlation of the accumulation of smoke in a space by a steady fire is the **steady-filling equation**:

Table 2 Steady Design Fire Sizes for Atriums

	Btu/s
Minimum fire for fuel-restricted atrium	2,000
Minimum fire for atrium with combustibles	5,000
Large fires	25,000

Table 3 Typical Fire Growth Times

t-Squared Fires	Growth Time t_g, s
Slow[a]	600
Medium[a]	300
Fast[a]	150
Ultrafast[b]	75

[a]Constants based on data from NFPA *Standards* 92B *and* 204.
[b]Constant based on data from Nelson (1987).

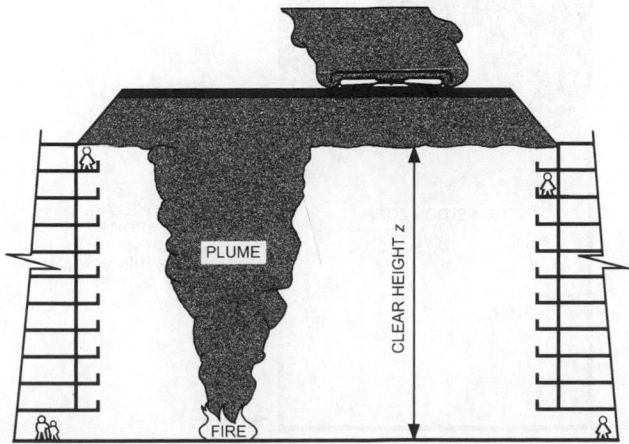

Fig. 18 Smoke Exhaust to Maintain Steady Clear Height

$$\frac{z}{H} = 0.67 - 0.28 \ln \left(\frac{tq^{1/3} H^{-4/3}}{A/H^2} \right) \qquad (20)$$

where

z = height of first indication of smoke above fire, ft
H = ceiling height above fire, ft
t = time, s
q = heat release rate from steady fire, Btu/
A = cross-sectional area of atrium, ft²

Equation (20) is conservative in that it estimates the height of the first indication of smoke above the fire rather than the smoke interface, as illustrated in Figure 20. In the idealized zone model, the smoke interface is considered to be a height where there is smoke above and none below. In actual fires, there is a gradual transition zone between the lower cool layer and upper hot layer. The first indication of smoke can be thought of as the bottom of the transition zone. Another factor making Equation (20) conservative is that it is based on a plume that has no contact with the walls, which would reduce entrainment of air.

Equation (20) is for a constant cross-sectional area with respect to height. For other atrium shapes, physical modeling or computational fluid dynamics (see Chapter 13 of the 2009 *ASHRAE Handbook—Fundamentals* for more information) can be used. Alternatively, a sensitivity analysis can be made using Equation (20) to set bounds on the filling time for an atrium of complex shape. The equation is appropriate for A/H^2 from 0.9 to 14 and for values of z greater than or equal to 20% of H. A value of z/H greater than 1 means that the smoke layer under the ceiling has not yet begun to descend. These conditions can be expressed as

$$A = \text{Constant with respect to } H$$

$$0.2 \leq \frac{z}{H} < 1.0$$

$$0.9 \leq \frac{A}{H^2} \leq 14$$

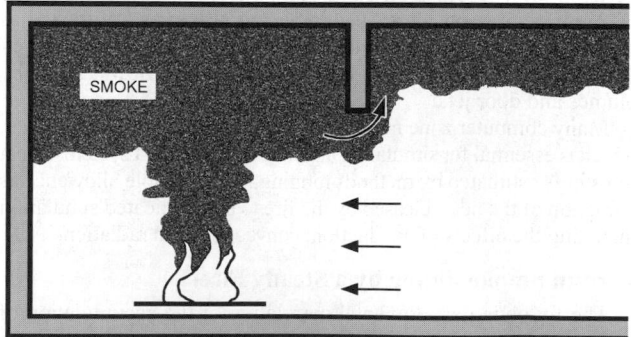

A. SKETCH OF ROOM FIRE

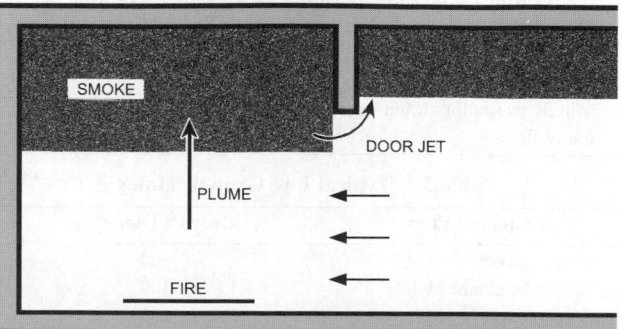

B. ZONE MODEL IDEALIZATION OF ROOM FIRE

Fig. 19   Room Fire and Zone Fire Model Idealization

When Equation (20) is solved for z/H, z/H is often outside the acceptable range. Equation (20) can be solved for time.

$$t = \frac{A}{H^2} \times \frac{H^{4/3}}{q^{1/3}} \exp \left[\frac{1}{0.28} \left(0.67 - \frac{z}{H} \right) \right] \qquad (21)$$

Atrium Smoke Filling by an Unsteady Fire

To analyze atrium smoke filling with an unsteady fire, use zone fire or CFD models. Klote and Milke (2002) provide an algebraic equation for the smoke-filling time for an atrium with a *t*-squared fire, but this equation has limited applicability because *t*-squared fires can be extremely large for the times considered in many smoke-filling applications.

Steady Clear Height with Upper Layer Exhaust

Figure 18 illustrates smoke exhaust from the hot smoke layer at the top of an atrium to maintain a steady clear height. Smoke flow into the upper layer from the fire plume depends on the fire's heat release rate, clear height, fuel type, and fuel orientation. The following is a generalized plume approximation that does not take into account the specifics of the material being burned.

$$\dot{m} = 0.022 q_c^{1/3} z^{5/3} + 0.0042 q_c \qquad (22)$$

where

$\dot{m}$ = mass flow of plume, lb$_m$/s
q_c = convective heat release rate of fire, Btu/s
z = clear height above top of fuel, ft

Clear height z is the distance from the top of the fuel to the interface between the "clear" space and the smoke layer. Because a smoke management system generally must protect against fire at any location, it is suggested that the top of the fuel be considered at the floor level.

Equation (22) is not applicable when the mean flame height is greater than the clear height. An approximate relationship for mean flame height is

$$z_f = 0.533 q_c^{2/5} \qquad (23)$$

where z_f = mean flame height, ft.

The convective portion q_c of the heat release rate can be expressed as

$$q_c = \xi q \qquad (24)$$

where ξ is the convective fraction of heat release. The convective fraction depends on the material being burned, heat conduction

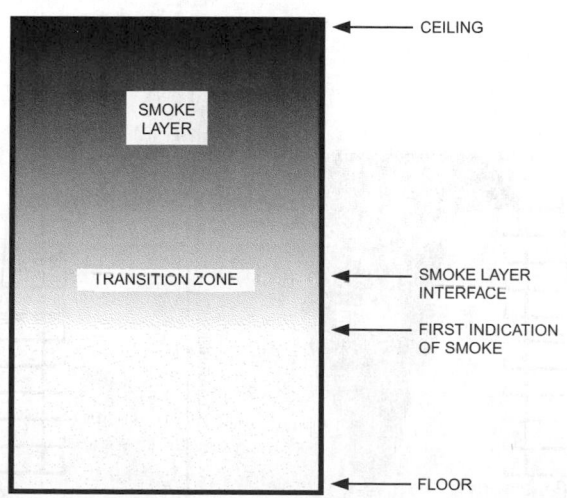

Fig. 20   Smoke Layer Interface

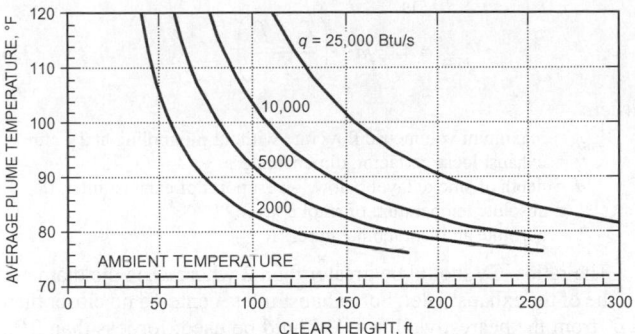

Note: Plume equations should not be used when plume temperature is less than 4°F above ambient.

Fig. 21   Average Plume Temperature

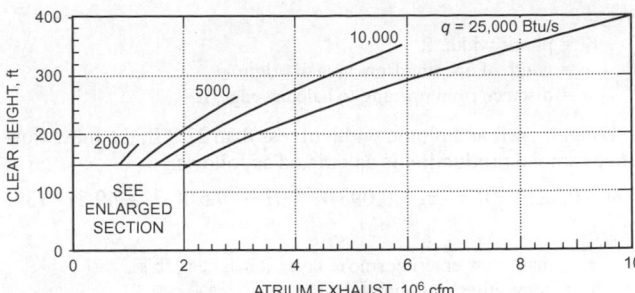

Fig. 22   Atrium Exhaust to Maintain Steady Clear Height

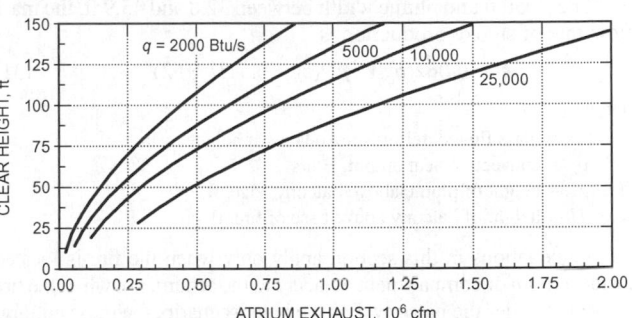

Fig. 23   Enlarged Scale for Figure 22

through the fuel, and the radiative heat transfer of the flames, but a value of 0.7 is often used.

The temperature of smoke entering the upper smoke layer is

$$T_p = T_a + \frac{q_c}{\dot{m}c_p} \tag{25}$$

where

T_p = plume temperature at clear height, °R
T_a = ambient temperature, °R
$\dot{m}$ = mass flow of plume, lb_m/s
q_c = convective heat release rate of fire, Btu/s
c_p = specific heat of plume gases, Btu/lb·°F

Figure 21 shows plume temperature as a function of height above the fuel as calculated from Equations (22) and (25). Smoke plumes consist primarily of air mixed with combustion products, and the specific heat of plume gases is generally taken to be the same as that of air (c_p = 0.24 Btu/lb·°F). Equation (22) was developed for strongly buoyant plumes. For small temperature differences between the plume and ambient, errors because of low buoyancy could be significant. This topic needs study, and, in the absence of better data, it is recommended that the plume equations not be used when this temperature difference is small (less than 4°F).

The density of smoke gases can be calculated from the perfect gas law:

$$\rho = \frac{p}{RT} \tag{26}$$

where

ρ = density, lb_m/ft^3
p = absolute pressure, lb_f/ft^2
R = gas constant, $ft·lb_f/lb_m·°R$
T = absolute temperature of smoke gases, °R

Volumetric flow is expressed as

$$Q = \frac{60\dot{m}}{\rho} \tag{27}$$

where

$\dot{m}$ = mass flow of plume or exhaust air, lb_m/s
Q = volumetric flow of exhaust gases, cfm
ρ = density of plume or exhaust gases, lb_m/ft^3

Atrium exhaust should equal the mass flow of the plume plus any leakage flow into the atrium above the clear height.

For an atrium with negligible heat loss from the smoke layer and negligible air leakage into the smoke layer from the outside, exhaust equals the plume's mass flow rate from Equation (22) at the same temperature as the plume from Equation (25). Figures 22 and 23 show the exhaust rate needed to maintain a constant clear height for

an atrium with negligible heat loss from the smoke layer and negligible air leakage into the smoke layer from the outside.

The major assumptions of the analysis plotted in Figures 22 and 23 are as follows:

- Plume has space to flow to top of atrium without obstructions
- Heat release rate of fire is constant
- Clear height is greater than mean flame height
- Smoke layer is adiabatic
- Plume flow and exhaust are the only significant mass flows into or out of smoke layer (i.e., outside airflow, either as leakage or as makeup air, into smoke layer is insignificant)

Balcony Spill Plumes

In addition to a fire on the floor of an atrium, another scenario that must be considered in the design of an atrium smoke management system is a balcony spill plume. In this scenario, the fire is located in an adjacent compartment and smoke enters the atrium through the compartment opening and flows under a balcony to form a plume in the atrium. Generalized plume approximations have been developed for this scenario. For scenarios in which the design height for the base of the smoke layer is <50 ft above the balcony edge, the mass entrainment into the plume is

$$m = 0.12(qW^2)^{1/3}(z_b + 0.25H) \tag{28}$$

where

m = mass flow rate in plume, lb/s
q = heat release rate of the fire, Btu/s
W = width of plume as it spills under balcony, ft
z_b = height above underside of balcony to smoke layer interface, ft
H = height of balcony above base of fire, ft

Physical barriers can be used to restrict the horizontal spread of smoke under the balcony. Draft curtains used for this application must extend at least 10% of the floor-to-ceiling height below the balcony. If the plume under the balcony is unrestricted, the width of the plume is determined as

$$W = w + b \tag{29}$$

where

    W = plume width, ft
    w = width of opening from area of origin, ft
    b = distance from opening to balcony edge, ft

For $z_b \geq 50$ ft and plume width of less than 32.8 ft, the mass flow rate of smoke production is calculated as follows:

$$m = 0.32q_c^{1/3}W^{1/5}(z_b + 0.098W^{7/15}H + 19.5W^{7/15} - 49.2) \quad (30)$$

where

    m = mass flow entering smoke layer at height z_b, lb/s
    q_c = convective heat output, Btu/s
    z_b = height of plume above balcony edge, ft
    H = height of balcony above base of fire, ft

For $z_b \geq 50$ ft and plume width between 32.8 and 45.9 ft, the mass flow rate of smoke production is

$$m = 0.062(q_c W^2)^{1/3}(z_b + 0.51H + 52) \quad (31)$$

where

    m = mass flow entering smoke layer at height z_b, lb/s
    q_c = convective heat output, Btu/s
    z_b = height of plume above balcony edge, ft
    H = height of balcony above base of fire, ft

The equations in this section apply only when the fire is located inside a room or compartment adjacent to the atrium, not when the fire is located under the balcony. For the latter scenario, the mass entrainment into the plume can only be determined using CFD modeling.

Makeup Air

Makeup air must be provided to ensure that exhaust fans are able to move the design air quantities and to ensure that door-opening force requirements are not exceeded. Makeup air can be provided using fans, openings to the outside, or a combination of fans and openings. The supply points for makeup air must be below the smoke layer interface.

It is recommended that the makeup air system be designed to provide 85 to 95% of the exhaust mass flow rate. The remaining 5 to 15% of makeup air will enter through cracks in the construction, including gaps around closed doors and windows.

Hadjisophocleous and Zhou (2008) and Zhou and Hadjisophocleous (2008) show that, for makeup air velocities exceeding 200 ft/min, the plume can be deflected, resulting in an increase in smoke production. For even higher velocities, the plume and smoke layer interface can be disrupted. The maximum air velocity must not exceed 200 ft/min where the makeup air could come into contact with the smoke plume, unless a higher velocity is supported by engineering analysis.

For systems using fans, the exhaust fans should operate before the makeup air system does.

Minimum Smoke Layer Depth

An atrium smoke management system must be designed with a smoke layer deep enough to accommodate a **ceiling jet**, a radial jet of smoke formed when a plume hits the ceiling. Usual estimates of ceiling jet depth are 10 to 20% of the distance between the base of the fuel and the ceiling (the ceiling jet itself is only about 10% of this distance, but at the walls the jet reverses and flows under itself). Generally, the smoke layer depth should be at least 20% of the distance between the base of the fuel and the ceiling.

Number of Exhaust Inlets

When the flow rate of a smoke exhaust inlet is relatively large, cold air from the lower layer can be pulled into the smoke exhaust. This phenomenon is called **plugholing**. A number of exhaust air inlets may be needed to prevent plugholing. The maximum volumetric flow rate that can be exhausted by a single exhaust inlet without plugholing is calculated by

$$V_{max} = 452\gamma d^{5/2}\left(\frac{T_s - T_o}{T_o}\right)^{1/2} \quad (32)$$

where

    V_{max} = maximum volumetric flow rate without plugholing at T_s, cfm
    γ = exhaust location factor, dimensionless
    d = depth of smoke layer below lowest point of exhaust inlet, ft
    T_s = absolute temperature of smoke layer, °R
    T_o = absolute ambient temperature, °R

The ratio d/D_i should be greater than 2 where D_i is the diameter of the of the exhaust inlet. For exhaust inlets centered no closer than $2D_i$ from the nearest wall, $\gamma = 1$ should be used; for less than $2D_i$, $\gamma = 0.5$ should be used. For exhaust inlets on a wall, use $\gamma = 0.5$.

For rectangular exhaust inlets, calculate D_i as

$$D_i = \frac{2ab}{a + b} \quad (33)$$

where

    a = length of inlet
    b = width of inlet

Where multiple inlets are needed to prevent plugholing, the minimum separation between inlets should be

$$S_{min} = 0.065V_e^{1/2} \quad (34)$$

where

    S_{min} = minimum edge-to-edge separation between inlets, ft
    V_e = volumetric flow rate of one exhaust inlet, cfm

This approach for calculating V_{max} and S_{min} is consistent with that of NFPA *Standard* 92B. A less conservative approach was in earlier versions of this *Standard* and Klote and Milke (2002); research is needed to evaluate these approaches.

Separation Between Inlets

When exhaust at an inlet is near the maximum flow rate Q_{max}, adequate separation between exhaust inlets must be maintained to minimize interaction between flows near the inlets. One criterion for separation between inlets is that it be at least the distance from a single inlet that would result in an arbitrarily small velocity based on sink flow. Using 40 fpm as the arbitrary velocity, the minimum separation distance for inlets located in a wall near the ceiling (or in the ceiling near the wall) is

$$S_{min} = 0.023\beta\sqrt{Q_e} \quad (35)$$

where

    S_{min} = minimum separation between inlets, ft
    Q_e = volumetric flow rate, cfm
    β = exhaust location factor, dimensionless

Prestratification and Detection

A layer of hot air often forms under the ceiling of an atrium because of solar radiation on the atrium roof. Although no studies have been made of this **prestratification layer**, building designers indicate that the temperature of such a layer can exceed 120°F. Temperatures below this layer are controlled by the building's heating and cooling system; the temperature can be considered to increase significantly over a small increase in elevation, as shown in Figure 24. The analysis of smoke stratification given in NFPA *Standard* 92B is not appropriate for the temperature profile addressed in this section because it is for a constant temperature increase per unit elevation.

When the average temperature of the plume is lower than that of the prestratification layer, the smoke will form a stratified layer beneath the prestratification layer, as shown in Figure 25. Average plume temperatures can be calculated from Equations (22) and (25); they are plotted in Figure 21, which shows that the average plume temperature is usually less than expected temperatures of the hot air layer. Thus, when there is a prestratified layer, smoke cannot be expected to reach the atrium ceiling, and smoke detectors mounted on that ceiling cannot be expected to go into alarm.

Beam smoke detectors can overcome this detection difficulty. The following approaches can provide prompt detection regardless of air temperature under the ceiling when a fire begins:

- **Upward-Angled Beam to Detect Smoke Layer.** One or more beams are aimed upward to intersect the smoke layer regardless of the level of smoke stratification. For redundancy, more than one beam smoke detector is recommended. Advantages include not needing to locate several horizontal beams, and the minimized risk of false activation by sunlight (a risk with some beam smoke detectors) because the receivers are angled downward. Figure 26 illustrates the upward-angled beam approach.
- **Horizontal Beams at Various Levels to Detect Smoke Layer.** One or more beam detectors are located at roof level, with

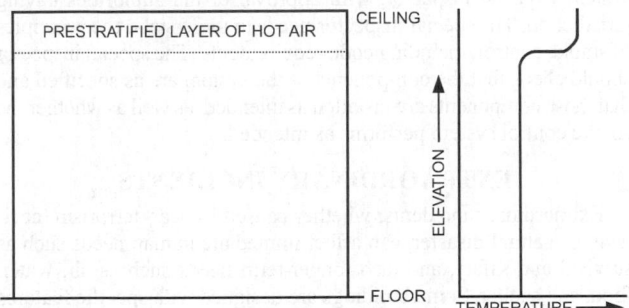

Note: Temperature below the hot layer is controlled by the building's heating and cooling system.

Fig. 24 Prestratified Layer of Hot Air under Atrium Ceiling and Resulting Temperature Profile

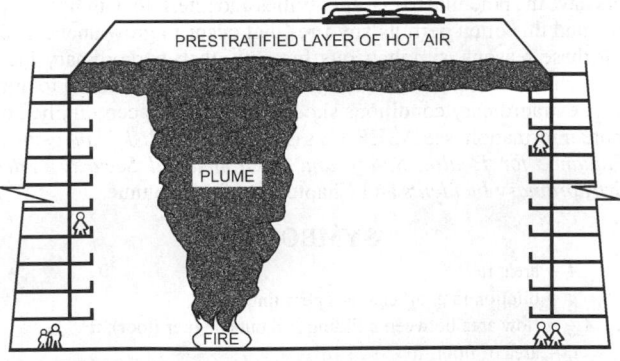

Fig. 25 Smoke Filling a Prestratified Atrium

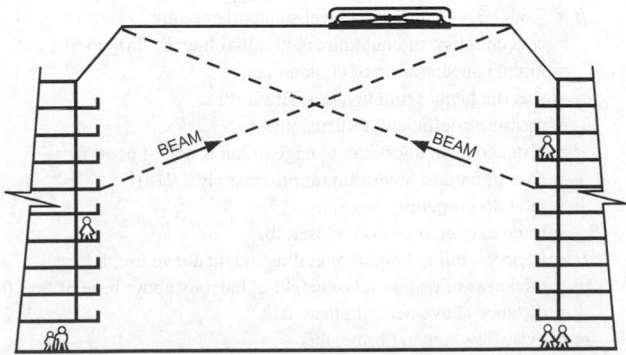

Fig. 26 Beam Detectors Used for Activation of Atrium Smoke Management System

additional detectors at lower levels. Exact beam positioning depends on the specific design, but should include beams at the bottom of identified unconditioned spaces and at or near the design smoke level, with several beams at intermediate positions.
- **Horizontal Beams to Detect Smoke Plume.** Beams are arranged at a level below the lowest expected stratification level. These beams must be close enough to each other to ensure intersection of the plume; spacing should be based on the width of the beam at the least elevation above a point of fire potential.

All components of a beam smoke detector must be accessible for maintenance. For the arrangement shown in Figure 26, a roof opening (not shown) could provide access for maintenance.

Loss of Buoyancy in Atriums

For some applications, loss of buoyancy can cause the smoke layer to descend and threaten occupants. There is little research on this loss of buoyancy, but the geometry of the large-volume space and the fire's heat release rate are major factors. Spaces that are unusually large or unusually long are of particular concern; for these cases, draft curtains can divide up the atrium into several smaller spaces. Theoretically, CFD modeling can predict loss of buoyancy in a large-volume space, but this has not been experimentally verified.

TENABILITY SYSTEMS

The intent of smoke control systems is to provide a tenable environment in the means of egress or other locations during building fires. A **tenable environment** is one in which combustion products, including heat, are limited to a level that is not life threatening. Analysis of a tenability system consists of a smoke transport analysis and a tenability evaluation. For most applications, smoke transport calculations are done by either a computational fluid dynamic (CFD) model or a network model.

Tenability Evaluation

Tenability evaluation considers the effects of exposure to toxic gas, heat, and thermal radiation, as well as reduced visibility. Toxic gas, heat, and thermal radiation exposure are direct threats to life. The severity of each threat depends on the intensity and length of exposure.

Reduced visibility does not directly threaten life, but it is an indirect hazard. Reduced visibility can reduce walking speed. When occupants and firefighters cannot see very much, they often become disoriented and cannot get away from the smoke, thus prolonging their exposure. Another concern is that a disoriented person can fall from an atrium balcony, which can be fatal.

Considerable research has been conducted regarding tenability, and methods are available for calculating exposures to combustion gases and reduced visibility (ISO *Standard* 13571; Jin 2008; Klote and Milke 2002; Purser 2008). There is no broad consensus on visibility criteria, but Jin (2008) suggests 43 ft for applications where occupants are unfamiliar with the building (e.g., museums, casinos), and 13 ft when occupants are familiar with the building (e.g., residential applications). The U.K. Chartered Institution of Building Service Engineers (CIBSE 2003) suggests 26 to 33 ft.

When combustion products from most materials are diluted enough to meet any of these visibility criteria, the hazards to life from toxic gases, heat, and thermal radiation are also eliminated for exposures up to 20 min. This means that, for most fires, tenability can be evaluated by calculating visibility, if the hazards of other exposures are checked.

Atria and Other Large Spaces

For atria and other large spaces, it is appropriate to use a CFD model (see Chapter 13 of the 2009 *ASHRAE Handbook—Fundamentals* for CFD information) to calculate tenability. CFD has been used for a wide range of applications, including aircraft design, automotive design, boiler design, and weather forecasting, and is extensively used for atrium smoke management systems. For examples of CFD simulations of smoke transport in large spaces, see Hadjisophocleous et al. (1999), Kashef et al. (2002), Klote (2005), and Lougheed and Hadjisophocleous (2000).

The idea of CFD is to divide the space of interest into a large number of cells and to solve the governing equations for each cell. For atrium applications, the number of cells typically ranges from 100,000 to 1,000,000 or more. Obstructions such as walls, balconies, and stairs need to be taken into account. Conditions at the boundaries need to be defined. Exhaust flow at or near the top of the atrium is specified, and makeup air conditions are also defined. This allows simulation of fluid flow in considerable detail.

Although CFD modeling has significant advantages in realistically simulating smoke flow, it is computationally intensive and requires a lot of computer memory and time; it is not uncommon for a CFD simulation to run for hours and sometimes days. Because CFD is computationally intensive, it produces so many numbers that the usual ways to evaluate computer output are inappropriate. Visualization methods have been developed so people can understand CFD results.

Several general-purpose CFD models are commercially available that can be used for atrium smoke control. The U.S. National Institute of Standards and Technology (NIST) developed the Fire Dynamics Simulator (FDS) model (McGrattan 2004; McGrattan and Forney 2004), and its visualization software called Smokeview (Forney and McGrattan 2004). The FDS model includes numerous fire-modeling applications, such as smoke plume flow, fires in enclosures, a burning townhouse, sprinklered fires, an oil tank fire, fires in aircraft hangars, rack storage commodity fire, and a brush fire advancing toward a house. FDS and Smokeview can be downloaded free of charge (www.fire.nist.gov/fds).

Large Multicompartmented Buildings

Because of the number of rooms and shafts in large multicompartmented buildings, it is often not feasible to use CFD to simulate smoke transport. Ideally, the model used for such simulations would be able to simulate a fire and mass and heat transfer throughout the building. No practical model can do this, but research is under way to develop such a model. Currently, the network model CONTAM is used to simulate smoke transport in large multicompartmented buildings. See the section on Computer Analysis for Pressurization Systems for more information about CONTAM. Ferreira (2002), Hadjisophocleous et al. (2002), Klote (2002, 2004), and Chapter 9 of Klote and Milke (2002) provide examples of tenability calculations using CONTAM. Because CONTAM requires that temperatures be supplied as data input by the user, the temperatures need to be calculated or estimated by the user.

ACCEPTANCE TESTING

Regardless of the care, skill, and attention to detail with which a smoke control system is designed, an acceptance test is needed as assurance that the system, as built, operates as intended.

An acceptance test should be composed of two levels of testing. The first is functional: an initial check of the system components. The importance of the initial check has become apparent because of problems encountered during tests of smoke control systems: fans operating backward, fans to which no electrical power was supplied, controls that did not work properly, etc.

The second level of testing is of performance, to determine whether the system performs adequately under all required modes of operation. This can consist of measuring pressure differences across barriers under various modes of smoke control system operation. If airflows through open doors are important, these should be measured. Chemical smoke from smoke candles (sometimes called smoke bombs) is not recommended for performance testing because it normally lacks the buoyancy of hot smoke from a real building fire. Smoke near a flaming fire has a temperature of 1000 to 2000°F.

Heating chemical smoke to such temperatures to emulate smoke from a real fire is not recommended unless precautions are taken to protect life and property. The same comments about buoyancy apply to tracer gases. Thus, pressure-difference testing is the most practical performance test. However, chemical smoke can be used to aid flow visualization.

ASHRAE *Guideline* 5 covers the commissioning of smoke management systems.

SPECIAL INSPECTOR

Some building codes require special inspection and tests of smoke control systems in addition to the ordinary inspection of and test requirements for buildings, structures, and parts of buildings. These special inspections and tests should verify the proper commissioning of the smoke control design in its final installed condition. Procedures for inspection and testing should be developed by the smoke control system's special inspector, with approval of the authorities having jurisdiction. The special inspector needs to understand the principles of smoke control, including code requirements. The special inspector should check that the components of the system are as specified and that those components are installed as intended, as well as whether the smoke control system performs as intended.

EXTRAORDINARY INCIDENTS

Extraordinary incidents, whether caused by war, terrorism, accident, or natural disaster, can affect immediate human needs such as survival and safety, and also longer-term needs such as air, water, food, and shelter. Some buildings are designed with specific features intended to make them less susceptible to extraordinary incidents. It is recommended that actuation of systems for fire and smoke protection be of higher priority than possibly conflicting automatic strategies designed to respond to other extraordinary conditions.

Some acts of terrorism use fire, and those using bombs often lead to fires. It is well known that war, terrorist attacks, and natural disasters have the potential to disrupt utilities and interfere with fire fighting, and this often permits any fires that occur to grow unchecked. For these reasons, simultaneous fire and other extraordinary incidents should be considered likely, and any features intended to mitigate extraordinary conditions should be designed accordingly. For more information, see ASHRAE's (2003) report, *Risk Management Guidance for Health, Safety and Environmental Security under Extraordinary Incidents* and Chapter 59 of this volume.

SYMBOLS

A = area, ft^2

a = dilution rate, air changes per minute

A_{bo} = flow area between building and outside (per floor), ft^2

A_f = area of floor, ft^2

A_{sb} = flow area between stairwell and building (per floor), ft^2

A_w = area of wall, ft^2

b = distance from opening to balcony edge, ft

B = $7.64(1/T_o - 1/T_s)$ at sea level standard pressure

C = concentration of contaminant at initial time θ; flow coefficient

C_o = initial concentration of contaminant

c_p = specific heat of plume gases, Btu/lb·°F

C_w = pressure coefficient, dimensionless

d = distance from doorknob to edge of knob side of door, ft

e = base of natural logarithm (approximately 2.718)

F = total door-opening force, lb$_f$

F_{dc} = force to overcome door closer, lb$_f$

H = floor-to-ceiling height, or ceiling height above fire, ft

H_j = thickness of ceiling jet or height of balcony above base of fire, ft

h = distance above neutral plane, ft

m = mass flow rate in plume, lb/s

$\dot{m}$ = mass flow of plume or exhaust air, lb$_m$/s

N = number of floors

p = absolute pressure, lb$_f$/ft^2

p_w = pressure exerted by wind, in. of water
Q = volumetric flow rate, cfm
q = heat release rate of fire, Btu/s
q_c = convective heat release rate, Btu/h or Btu/s
Q_e = volumetric flow rate, cfm
Q_{in} = volumetric flow rate of air into fire compartment, cfm
Q_{out} = volumetric flow rate of smoke out of fire compartment, cfm
R = gas constant, ft·lb$_f$·/lb$_m$·°R
S_{min} = minimum separation between inlets, ft
T = absolute temperature of smoke gases, °R
t = time, min or s
T_a = ambient temperature, °R
T_f = average absolute temperature of fire compartment, °R
t_g = growth time, s
T_i = absolute temperature of air inside shaft, °R
T_{in} = absolute temperature of air into fire compartment, °R
T_j = absolute temperature of ceiling jet, °R
T_o = absolute temperature of outside air, °R
T_{out} = absolute temperature of smoke leaving fire compartment, °R
T_p = plume temperature at clear height, °R
T_s = absolute temperature of stairwell air or surroundings, °R
V = wind velocity, mph
V_k = critical air velocity to prevent smoke backflow, fpm
w = width of opening from area of origin, ft
W = corridor, door, or plume spill width or length, ft
y = distance above stairwell bottom, ft
z = height of first indication of smoke above fire, or clear height above top of fuel, ft
z_b = height above underside of balcony to smoke layer interface, ft
z_f = mean flame height, ft

Greek

β = exhaust location factor, dimensionless
Δp = pressure difference, in. of water
Δp_{sb} = pressure difference between stairwell and building, in. of water
Δp_{sbb} = pressure difference between stairwell and building at stairwell bottom, in. of water
Δp_{sbt} = pressure difference from stairwell to building at stairwell top, in. of water
θ = time
ρ = density, lb$_m$/ft^3
ρ_o = outside air density, lb$_m$/ft^3
ξ = convective fraction of heat release

REFERENCES

AMCA. 2007. Certified ratings program—Product rating manual for smoke management fan performance. *Publication* 212-07. Air Movement and Control Association, Arlington Heights, IL.

ASHRAE. 2009. Laboratory methods of testing fans used to exhaust smoke in smoke management systems. ANSI/ASHRAE *Standard* 149-2000 (RA 2009).

ASHRAE. 2001. Commissioning smoke management systems. *Guideline* 5-1994 (RA 2001).

ASHRAE. 2003. Risk management guidance for health, safety and environmental security under extraordinary incidents. *Report*, Presidential Ad Hoc Committee for Building Health and Safety under Extraordinary Incidents.

ASTM. 2010. Test method for fire tests of through-penetration fire stops. *Standard* E814. American Society for Testing and Materials, West Conshohocken, PA.

Bukowski, R.W. 1991. Fire models, the future is now! *NFPA Journal* 85(2): 60-69.

CIBSE. 2003. Fire engineering. CIBSE *Guide* E, 2nd ed. Chartered Institution of Building Services Engineers, London.

Cooper, L.Y. 1985. ASET—A computer program for calculating available safe egress time. *Fire Safety Journal* 9:29-45.

Cooper, L.Y. and G.P. Forney. 1987. Fire in a room with a hole: A prototype application of the consolidated compartment fire model (CCFM) computer code. Presented at the 1987 Combined Meetings of Eastern Section of Combustion Institute and NBS Annual Conference on Fire Research.

Cresci, R.J. 1973. Smoke and fire control in high-rise office buildings—Part II, Analysis of stair pressurization systems. Symposium on Experience and Applications on Smoke and Fire Control, ASHRAE Annual Meeting, June.

Evers, E. and A. Waterhouse. 1978. *A computer model for analyzing smoke movement in buildings.* Building Research Establishment, Fire Research Station, Borehamwood, Hertsfordshire, U.K.

Fang, J.B. 1980. *Static pressures produced by room fires.* NBSIR 80-1984. National Bureau of Standards. Available from National Institute of Standards and Technology, Gaithersburg, MD.

Fang, J.B. and J.N. Breese. 1980. *Fire development in residential basement rooms.* NBSIR 80-2120. National Bureau of Standards. Available from National Institute of Standards and Technology, Gaithersburg, MD.

Ferreira, M.J. 2002. Use of multizone modeling for high-rise smoke control system design. *ASHRAE Transactions* 108(2):837-846.

Forney, G.P. and K. McGrattan. 2004. User's guide for Smokeview version 4—A tool for visualizing fire dynamics simulation data. NIST *Special Publication* 1017. National Institute of Standards and Technology, Gaithersburg, MD.

Fu, Z., A. Kashef, N. Benichou, and G.V. Hadjisophocleous. 2002. FIERA system theory documentation: Smoke movement model. *Internal Report* IR-835. National Research Council of Canada, Ottawa.

Hadjisophocleous, G.V., G.D. Lougheed, and S. Cao. 1999. Numerical study of the effectiveness of atrium smoke exhaust systems. *ASHRAE Transactions* 105(1):699-715.

Hadjisophocleous, G. V. and Zhou, J., Evaluation of atrium smoke exhaust make-up air velocity. *ASHRAE Transactions* 114(1):147-153.

Hadjisophocleous, G., F. Zhuman, and G. Lougheed. 2002. Computational and experimental study of smoke flow in the stair shaft of a 10-story tower. *ASHRAE Transactions* 108(1):724-730.

Houghton, E.L. and N.B. Carruther. 1976. *Wind forces on buildings and structures.* John Wiley & Sons, New York.

Huggett, C. 1980. Estimation of heat release by means of oxygen consumption measurements. *Fire and Materials* 4(2).

ICC. 2009. *International building code®.* International Code Council, Washington, D.C.

ISO. 2002. Life-threatening components of fire—Guidelines for the estimation of time available for escape using fire data. ISO/TS *Standard* 13571. International Organization for Standardization, Geneva.

Jin, T. 2008. *SFPE handbook of fire protection engineering*, Ch. 2-4, Visibility and human behavior in fire smoke. Society of Fire Protection Engineers, Bethesda, MD.

Jones, W.W. 1983. *A review of compartment fire models.* NBSIR 83-2684. National Bureau of Standards. Available from National Institute of Standards and Technology, Gaithersburg, MD.

Jones, W.W., R.D. Peacock, G.P. Forney, and P.A. Reneke. 2005. CFAST—Consolidated model of fire growth and smoke transport (Version 6). NIST *Special Publication* 1026. National Institute of Standards and Technology, Gaithersburg, MD.

Kashef, A., N. Benichou, G.D. Lougheed, and C. McCartney. 2002. A computational and experimental study of fire growth and smoke movement in large spaces. *10th Annual Conference of the Computational Fluid Dynamics Society of Canada*, Windsor, Ontario, pp. 1-6.

Kashef, A. and G.V. Hadjisophocleous. 2010. Algorithm for smoke modeling in large, multi-compartmented buildings (TRP-1328). ASHRAE Research Project, *Final Report.*

Klote, J.H. 1980. Stairwell pressurization. *ASHRAE Transactions* 86(1): 604-673.

Klote, J.H. 1982. *A computer program for analysis of smoke control systems.* NBSIR 82-2512. National Bureau of Standards. Available from National Institute of Standards and Technology, Gaithersburg, MD.

Klote, J.H. 1984. Smoke control for elevators. *ASHRAE Journal* 26(4): 23-33.

Klote, J.H. 1990. Fire experiments of zoned smoke control at the Plaza Hotel in Washington, D.C. *ASHRAE Transactions* 96(2):399-416.

Klote, J.H. 2002. Smoke management applications of CONTAM. *ASHRAE Transactions* 108(2):827-836.

Klote, J.H. 2004. Tenability and open doors in pressurized stairwells. *ASHRAE Transactions* 110(1):613-637.

Klote, J.H. 2005. CFD analysis of atrium smoke control at the Newseum. *ASHRAE Transactions* 111(2):567-574.

Klote, J.H. and X. Bodart. 1985. Validation of network models for smoke control analysis. *ASHRAE Transactions* 91(2B):1134-1145.

Klote, J.H. and J.A. Milke. 2002. *Principles of smoke management.* ASHRAE.

Klote, J.H. and G.T. Tamura. 1986. Smoke control and fire evacuation by elevators. *ASHRAE Transactions* 92(1).

Law, M. 1982. Air-supported structures: Fire and smoke hazards. *Fire Prevention* 148:24-28.

Lougheed, G.D. and G.V. Hadjisophocleous. 2000. The smoke hazard from a fire in high spaces. *ASHRAE Transactions* 107(1):720-729.

MacDonald, A.J. 1975. *Wind loading on buildings.* John Wiley & Sons, New York.

McGrattan, K. 2004. Fire Dynamics Simulator (version 4)—Technical reference guide. NIST *Special Publication* 1018. National Institute of Standards and Technology, Gaithersburg, MD.

McGrattan, K. and G.P. Forney. 2004. Fire Dynamics Simulator (version 4)—User's guide. NIST *Special Publication* 1019. National Institute of Standards and Technology, Gaithersburg, MD.

McGuire, J.H. and G.T. Tamura. 1975. Simple analysis of smoke flow problems in high buildings. *Fire Technology* 11(1):15-22.

McGuire, J.H., G.T. Tamura, and A.G. Wilson. 1970. Factors in controlling smoke in high buildings. Symposium on Fire Hazards in Buildings, ASHRAE Winter Meeting.

Milke, J.A. and F.W. Mowrer. 1994. Computer-aided design for smoke management in atria and covered malls. *ASHRAE Transactions* 100(2): 448-456.

Mitler, H.E. 1984. Zone modeling of forced ventilation fires. *Combustion Science and Technology* 39:83-106.

Mitler, H.E. 1985. *Comparison of several compartment fire models: An interim report.* NBSIR 85-3233. National Bureau of Standards. Available from National Institute of Standards and Technology, Gaithersburg, MD.

Mitler, H.E. and H.W. Emmons. 1981. Documentation for CFC V, the fifth Harvard computer code. Home Fire Project *Technical Report* 45. Harvard University, Cambridge, MA.

Mitler, H.E. and J.A. Rockett. 1986. *How accurate is mathematical fire modeling?* NBSIR 86-3459. National Bureau of Standards. Available from National Institute of Standards and Technology, Gaithersburg, MD.

Morgan, H.P. 1979. *Smoke control methods in enclosed shopping complexes of one or more stories: A design summary.* Fire Research Station, Borehamwood, Hertsfordshire, U.K.

Morgan, H.P. and G.O. Hansell. 1987. Atrium buildings: Calculating smoke flows in atria for smoke control design. *Fire Safety Journal* 12:9-12.

Nelson, H.E. 1987. *An engineering analysis of the early stages of fire development—The fire at the Dupont Plaza Hotel and Casino—Dec. 31, 1986.* NISTIR 87-3560. National Institute of Standards and Technology, Gaithersburg, MD.

Nelson, H.E. and F.W. Mowrer. 2002. *SFPE handbook of fire protection engineering,* Ch. 3-14, Emergency movement. Society of Fire Protection Engineers, Bethesda, MD.

NFPA. 2009. Installation of air-conditioning and ventilating systems. *Standard 90A-02.* National Fire Protection Association, Quincy, MA.

NFPA. 2009. Recommended practice for smoke-control systems. *Standard 92A.* National Fire Protection Association, Quincy, MA.

NFPA. 2009. Standard for smoke management systems in malls, atria, and large areas. *Standard 92B.* National Fire Protection Association, Quincy, MA.

NFPA. 2009. *Life safety code®. Standard 101.* National Fire Protection Association, Quincy, MA.

NFPA. 2007. Guide for smoke and heat venting. *Standard 204.* National Fire Protection Association, Quincy, MA.

NFPA. 2008. *Fire protection handbook,* 20th ed. National Fire Protection Association, Quincy, MA.

NOAA. 1979. *Temperature extremes in the United States.* National Oceanic and Atmospheric Administration (U.S.), National Climatic Center, Asheville, NC.

Peacock, R.D., G.P. Forney, P. Reneke, R. Portier, and W.W. Jones. 1993. CFAST, the consolidated model of fire growth and smoke transport. NIST *Technical Note* 1299. National Institute of Standards and Technology, Gaithersburg, MD.

Proulx, G. 2008. *SFPE handbook of fire protection engineering,* Ch. 3-12, Evacuation Time. National Fire Protection Association, Quincy, MA.

Purser, D.A. 2008. *SFPE handbook of fire protection engineering,* Ch. 2-6, Assessment of hazards to occupants from smoke, toxic gases, and heat. National Fire Protection Association, Quincy, MA.

Quintiere, J.G. 1989. Fundamentals of enclosure fire "zone" models. *Journal of Fire Protection Engineering* 1(3):99119.

Rilling, J. 1978. *Smoke study, 3rd phase: Method of calculating the smoke movement between building spaces.* Centre Scientifique et Technique du Bâtiment (CSTB), Champs sur Marne, France.

Sachs, P. 1972. *Wind forces in engineering.* Pergamon, New York.

Said, M.N.A. 1988. A review of smoke control models. *ASHRAE Journal* 30(4):36-40.

Sander, D.M. 1974. FORTRAN IV program to calculate air infiltration in buildings. DBR *Computer Program* 37. National Research Council, Ottawa, Canada.

Sander, D.M. and G.T. Tamura. 1973. FORTRAN IV program to simulate air movement in multi-story buildings. DBR *Computer Program* 35. National Research Council, Ottawa, Canada.

Shaw, C.Y., J.T. Reardon, and M.S. Cheung. 1993. Changes in air leakage levels of six Canadian office buildings. *ASHRAE Journal* 35(2):34-36.

Shaw, C.Y. and G.T. Tamura. 1977. The calculation of air infiltration rates caused by wind and stack action for tall buildings. *ASHRAE Transactions* 83(2):145-158.

Simiu, E. and R.H. Scanlan. 1978. *Wind effects on structures: An introduction to wind engineering.* John Wiley & Sons, New York.

Tamura, G.T. 1990. Field tests of stair pressurization systems with overpressure relief. *ASHRAE Transactions* 96(1):951-958.

Tamura, G.T. 1994. *Smoke movement and control in high-rise buildings.* National Fire Protection Association, Quincy, MA.

Tamura, G.T. and C.Y. Shaw. 1976a. Studies on exterior wall air tightness and air infiltration of tall buildings. *ASHRAE Transactions* 83(1):122-134.

Tamura, G.T. and C.Y. Shaw. 1976b. Air leakage data for the design of elevator and stair shaft pressurization systems. *ASHRAE Transactions* 83(2):179-190.

Tamura, G.T. and C.Y. Shaw. 1978. Experimental studies of mechanical venting for smoke control in tall office buildings. *ASHRAE Transactions* 86(1):54-71.

Tamura, G.T. and A.G. Wilson. 1966. Pressure differences for a 9-story building as a result of chimney effect and ventilation system operation. *ASHRAE Transactions* 72(1):180-189.

Tanaka, T. 1983. *A model of multiroom fire spread.* NBSIR 83-2718. National Bureau of Standards. Available from National Institute of Standards and Technology, Gaithersburg, MD.

Thomas, P.H. 1970. Movement of smoke in horizontal corridors against an airflow. *Institution of Fire Engineers Quarterly* 30(77):45-53.

Tubbs, J.S. and B.J. Meacham. 2007. *Egress design solutions: A guide to evacuation and crowd management.* John Wiley & Sons, Hoboken, NJ.

UL. 2010. Fire dampers. ANSI/UL *Standard* 555-10. Underwriters Laboratories, Northbrook, IL.

UL. 2010. Smoke dampers. ANSI/UL *Standard* 555S-10. Underwriters Laboratories, Northbrook, IL.

UL. 2010. Fire tests of through-penetration firestops. ANSI/UL *Standard* 1479-10. Underwriters Laboratories, Northbrook, IL.

Wakamatsu, T. 1977. Calculation methods for predicting smoke movement in building fires and designing smoke control systems. In *Fire Standards and Safety,* ASTM STP 614, pp. 168-193. American Society for Testing and Materials, West Conshohocken, PA.

Walton, G.N. 1989. *AIRNET—A computer program for building airflow network modeling.* National Institute of Standards and Technology, Gaithersburg, MD.

Walton, W.D. 1985. *ASETB: A room fire program for personal computers.* NBSIR 8531441. National Bureau of Standards. Available from National Institute of Standards and Technology, Gaithersburg, MD.

Walton, G.N. 1997. *CONTAM96 user manual.* NISTIR 6056. National Institute of Standards and Technology, Gaithersburg, MD.

Walton, G.N. and W.S. Dols. 2005. *CONTAM 2.4 user guide and program documentation.* NISTIR 7251. National Institute of Standards and Technology, Gaithersburg, MD.

Wray, C.P. and G.K. Yuill. 1993. An evaluation of algorithms for analyzing smoke control systems. *ASHRAE Transactions* 99(1):160-174.

Yoshida, H., C.Y. Shaw, and G.T. Tamura. 1979. A FORTRAN IV program to calculate smoke concentrations in a multi-story building. DBR *Computer Program* 45. National Research Council, Ottawa, Canada.

Zhou, J. and G.V. Hadjisophocleous. 2008. Parameters affecting fire plumes. *ASHRAE Transactions* 114(1):140-146.

BIBLIOGRAPHY

CIBSE. 1995. *Relationships for smoke control calculations.* TM19:1995. Chartered Institution of Building Service Engineers, London.

Lougheed, G.D. and G.V. Hadjisophocleous. 1997. Investigation of atrium smoke exhaust effectiveness. *ASHRAE Transactions* 103(2):519-533.

Sprat, D. and A.J.M. Heselden. 1974. Efficient extraction of smoke from a thin layer under a ceiling. *Fire Research Note* 1001.

RADIANT HEATING AND COOLING

R ADIANT heating and cooling applications are classified as panel heating or cooling if the panel surface temperature is below 300°F, and as low-, medium-, or high-intensity radiant heating if the surface or source temperature exceeds 300°F. In thermal radiation heat is transferred by electromagnetic waves that travel in straight lines, and can be reflected. Thermal radiation principally occurs between surfaces or between a source and a surface. In a conditioned space, air is not heated or cooled in this process. Because of these characteristics, radiant systems are effective for both spot heating and space heating or cooling requirements for an entire building.

Sensible heating loads may be reduced by 4 to 16% compared to ASHRAE standard design load. Percent reduction increases with the air change rate (Suryanarayana and Howell 1990).

LOW-, MEDIUM-, AND HIGH-INTENSITY INFRARED HEATING

Low-, medium-, and high-intensity infrared heaters are compact, self-contained direct-heating devices used in hangars, warehouses, factories, greenhouses, and gymnasiums, as well as in areas such as loading docks, racetrack stands, outdoor restaurants, animal breeding areas, swimming pool lounge areas, and areas under marquees. Infrared heating is also used for snow melting and freeze protection (e.g., on stairs and ramps) and process heating (e.g., paint baking and drying). An infrared heater may be electric, gas-fired, or oil-fired and is classified by the source temperature as follows:

- Low-intensity (source temperatures to 1200°F)
- Medium-intensity (source temperatures to 1800°F)
- High-intensity (source temperatures to 5000°F)

The source temperature is determined by such factors as the source of energy, the configuration, and the size. Reflectors can be used to direct the distribution of thermal radiation in specific patterns. Chapter 15 of the 2008 *ASHRAE Handbook—HVAC Systems and Equipment* covers radiant equipment in detail.

PANEL HEATING AND COOLING

Panel heating and cooling systems provide a comfortable environment by controlling surface temperatures and minimizing air motion within a space. They include the following designs:

- Ceiling panels
- Embedded hydronic tubing or attached piping in ceilings, walls, or floors
- Air-heated or cooled floors or ceilings
- Electric ceiling or wall panels
- Electric heating cable or wire mats in ceilings or floors
- Deep heat, a modified storage system using electric heating cable or embedded hydronic tubing in ceilings or floors

---

The preparation of this chapter is assigned to TC 6.5, Radiant Heating and Cooling.

In these systems, generally more than 50% of the heat transfer between the temperature-controlled surface and other surfaces is by thermal radiation. Panel heating and cooling systems are used in residences, office buildings, classrooms, hospital patient rooms, swimming pool areas, repair garages, and in industrial and warehouse applications. Additional information is available in Chapter 6 of the 2008 *ASHRAE Handbook—HVAC Systems and Equipment*.

Some radiant panel systems, referred to as hybrid HVAC systems, combine radiant heating and cooling with central air conditioning (Scheatzle 2003). They are used more for cooling than for heating (Wilkins and Kosonen 1992). The controlled-temperature surfaces may be in the floor, walls, or ceiling, with temperature maintained by electric resistance or circulation of water or air. The central station can be a basic, one-zone, constant-temperature, or constant-volume system, or it can incorporate some or all the features of dual-duct, reheat, multizone, or variable-volume systems. When used in combination with other water/air systems, radiant panels provide zone control of temperature and humidity.

Metal ceiling panels may be integrated into the central heating and cooling system to provide individual room or zone heating and cooling. These panels can be designed as small units to fit the building module, or they can be arranged as large continuous areas for economy. Room thermal conditions are maintained primarily by direct transfer of radiant energy, normally using four-pipe hot and chilled water. These systems have generally been used in hospital patient rooms. Metal ceiling panel systems are discussed in Chapter 6 of the 2008 *ASHRAE Handbook—HVAC Systems and Equipment*.

ELEMENTARY DESIGN RELATIONSHIPS

When considering radiant heating or cooling for human comfort, the following terms describe the temperature and energy characteristics of the total radiant environment:

- **Mean radiant temperature** (MRT) $\bar{t}_r$ is the temperature of an imaginary isothermal black enclosure in which an occupant would exchange the same amount of heat by radiation as in the actual nonuniform environment.
- **Ambient temperature** t_a is the temperature of the air surrounding the occupant.
- **Operative temperature** t_o is the temperature of a uniform isothermal black enclosure in which the occupant would exchange the same amount of heat by radiation and convection as in the actual nonuniform environment.

    For air velocities less than 80 fpm and mean radiant temperatures less than 120°F, the operative temperature is approximately equal to the adjusted dry-bulb temperature, which is the average of the air and mean radiant temperatures.

- **Adjusted dry-bulb temperature** is the average of the air temperature and the mean radiant temperature at a given location. The adjusted dry-bulb temperature is approximately equivalent to the operative temperature for air motions less than 80 fpm and mean radiant temperatures less than 120°F.

- **Effective radiant flux** (ERF) is defined as the net radiant heat exchanged at ambient temperature t_a between an occupant, whose surface is hypothetical, and all enclosing surfaces and directional heat sources and sinks. Thus, ERF is the net radiant energy received by the occupant from all surfaces and sources whose temperatures *differ* from t_a. ERF is particularly useful in high-intensity radiant heating applications.

The relationship between these terms can be shown for an occupant at surface temperature t_{sf}, exchanging sensible heat H_m in a room with ambient air temperature t_a and mean radiant temperature $\bar{t}_r$. The linear radiative and convective heat transfer coefficients are h_r and h_c, respectively; the latter is a function of the relative movement between the occupant and air movement V. The heat balance equation is

$$H_m = h_r(t_{sf} - \bar{t}_r) + h_c(t_{sf} - t_a) \tag{1}$$

During thermal equilibrium, H_m is equal to metabolic heat minus work and evaporative cooling by sweating. By definition of operative temperature,

$$H_m = (h_r + h_c)(t_{sf} - t_o) = h(t_{sf} - t_o) \tag{2}$$

The combined heat transfer coefficient is h, where $h = h_r + h_c$. Using Equations (1) and (2) to solve for t_o yields

$$t_o = \frac{h_r\bar{t}_r + h_c t_a}{h_r + h_c} = t_a + \left(\frac{h_r}{h}\right)(\bar{t}_r - t_a) \tag{3}$$

Thus, t_o is an average of $\bar{t}_r$ and t_a, weighted by their respective heat transfer coefficients; it represents how people sense the thermal level of their total environment as a single temperature.

Rearranging Equation (1) and substituting $h - h_r$ for h_c,

$$H_m + h_r(\bar{t}_r - t_a) = h(t_{sf} - t_a) \tag{4}$$

where $h_r(\bar{t}_r - t_a)$ is, by definition, the effective radiant flux (ERF) and represents the radiant energy absorbed by the occupant from all sources whose temperatures differ from t_a.

The principal relationships between $\bar{t}_r$, t_a, t_o, and ERF are as follows:

$$\text{ERF} = h_r(\bar{t}_r - t_a) \tag{5}$$

$$\text{ERF} = h(t_o - t_a) \tag{6}$$

$$\bar{t}_r = t_a + \text{ERF}/h_r \tag{7}$$

$$t_o = t_a + \text{ERF}/h \tag{8}$$

$$\bar{t}_r = t_a + (h/h_r)(t_o - t_a) \tag{9}$$

$$t_o = t_a + (h_r/h)(\bar{t}_r - t_a) \tag{10}$$

In Equations (1) to (10), the radiant environment is treated as a blackbody with temperature $\bar{t}_r$. The effect of the emittance of the source, radiating at absolute temperature in degrees Rankine, and the absorptance of skin and clothed surfaces is reflected in the effective values of $\bar{t}_r$ or ERF and not in h_r, which is generally given by

$$h_r = 4\sigma f_{eff}\left[\frac{(\bar{t}_r + t_a)}{2} + 460\right]^3 \tag{11}$$

where

h_r = linear radiative heat transfer coefficient, Btu/h·ft²·°F
f_{eff} = ratio of radiating surface of the human body to its total DuBois surface area $A_D = 0.71$
σ = Stefan-Boltzmann constant = 0.1712×10^{-8} Btu/h·ft²·°R⁴

The convective heat transfer coefficient for an occupant depends on the relative velocity between the occupant and the surrounding air, as well as the occupant's activity:

- If the occupant is walking in still air,

$$h_c = 0.092 V^{0.53} \qquad \{100 < V < 400 \text{ fpm}\} \tag{12a}$$

where V is the occupant's walking speed.
- If the occupant is sedentary with moving air,

$$h_c = 0.061 V_a^{0.6} \qquad \{40 < V_a < 800 \text{ fpm}\} \tag{12b}$$

$$h_c = 0.55 \qquad \{0 < V_a < 40 \text{ fpm}\} \tag{12c}$$

More information about h_c may be found in Chapter 9 of the 2009 *ASHRAE Handbook—Fundamentals*.

When $\bar{t}_r > t_a$, ERF adds heat to the body; when $t_a > \bar{t}_r$, heat is lost from the body because of thermal radiation. ERF is independent of the occupant's surface temperature and can be measured directly by a black globe thermometer or any blackbody radiometer or flux meter using the ambient air t_a as its heat sink.

In these definitions and for radiators below 1700°F (2160°R), the body clothing and skin surface are treated as blackbodies, exchanging radiation with an imaginary blackbody surface at temperature $\bar{t}_r$. The effectiveness of a radiating source on human occupants is governed by the absorptance α of the skin and clothing surface for the color temperature (in °R) of that radiating source. The relationship between α and temperature is illustrated in Figure 1. Values for α are those expected relative to the matte black surface normally found on globe thermometers or radiometers measuring radiant energy. A gas radiator usually operates at 1700°F (2160°R); a quartz lamp, for example, radiates at 4000°F (4460°R) with 240 V; and the sun's radiating temperature is 10,000°F (10,460°R). The use of α in estimating the ERF and t_o caused by sources radiating at temperatures above 1700°F (2160°R) is discussed in the section on Testing Instruments for Radiant Heating.

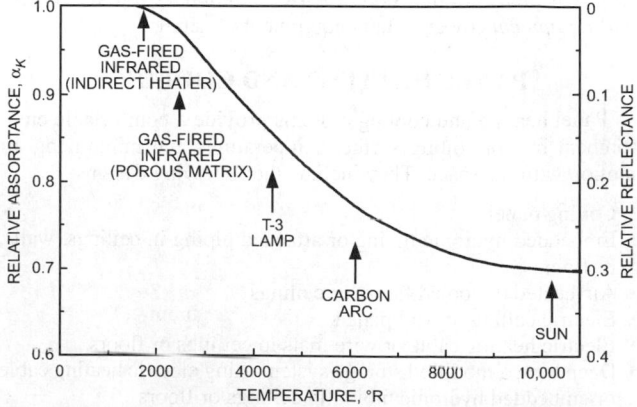

Fig. 1 Relative Absorptance and Reflectance of Skin and Typical Clothing Surfaces at Various Color Temperatures

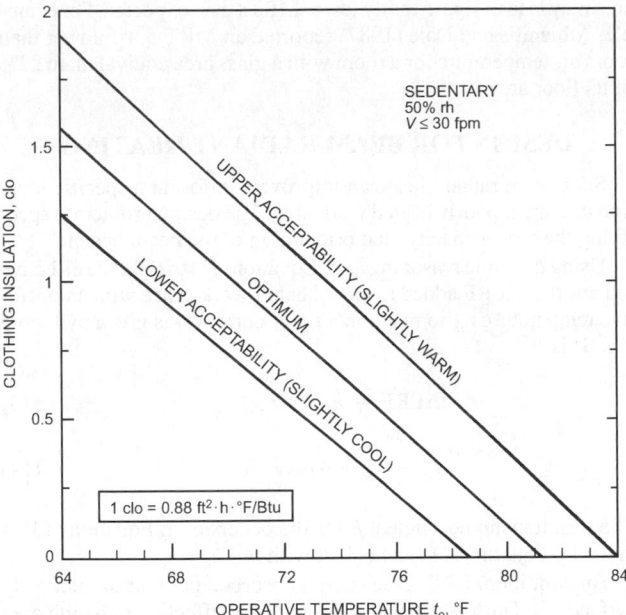

Fig. 2 Range of Thermal Acceptability for Sedentary People with Various Clothing Insulations and Operative Temperatures

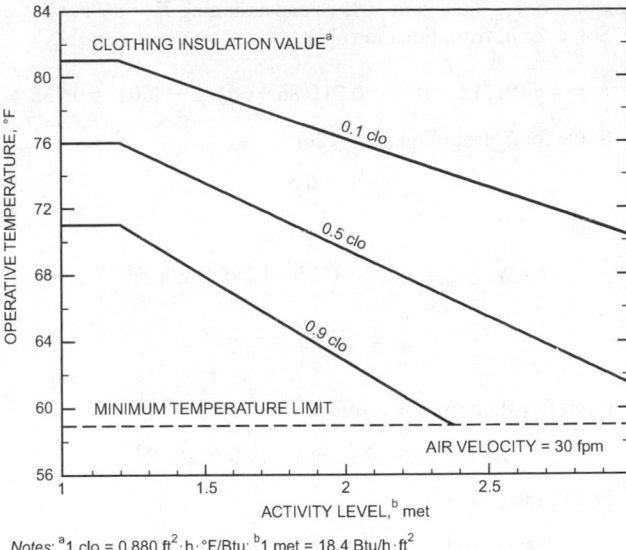

Notes: [a]1 clo = 0.880 ft^2·h·°F/Btu; [b]1 met = 18.4 Btu/h·ft^2

Fig. 3 Optimum Operative Temperatures for Active People in Low-Air-Movement Environments

DESIGN CRITERIA FOR ACCEPTABLE RADIANT HEATING

Perceptions of comfort, temperature, and thermal acceptability are related to activity, body heat transfer from the skin to the environment, and the resulting physiological adjustments and body temperature. The split between convective and radiant heat transfer from or to a body is also a matter of subjective human comfort. The optimum split is about 60% by thermal radiation and 40% by thermal convection. Heat transfer is affected by ambient air temperature, thermal radiation, air movement, humidity, and clothing worn. Thermal sensation is described as feelings of hot, warm, slightly warm, neutral, slightly cool, cool, and cold. An acceptable environment is defined as one in which at least 80% of the occupants perceive a thermal sensation between "slightly cool" and "slightly warm." Comfort is associated with a neutral thermal sensation during which the human body regulates its internal temperature with minimal physiological effort for the activity concerned. In contrast, warm discomfort is primarily related to the physiological strain necessary to maintain the body's thermal equilibrium rather than to the temperature sensation experienced. For a full discussion of the interrelation of physical, psychological, and physiological factors, refer to Chapter 9 of the 2009 *ASHRAE Handbook—Fundamentals*.

ANSI/ASHRAE *Standard* 55-1992 shows a linear relationship between clothing insulation worn and the operative temperature t_o for comfort (Figure 2). Figure 3 shows the effect of both activity and clothing on the t_o for comfort. Figure 4 shows the slight effect humidity has on the comfort of a sedentary person wearing average clothing. Figures 2, 3, and 4 are adapted from ANSI/ASHRAE *Standard* 55, Thermal Environmental Conditions for Human Occupancy.

A comfortable t_o at 50% rh is perceived as slightly warmer as humidity increases, or as slightly cooler as humidity decreases. Changes in humidity have a much greater effect on warm and hot discomfort. In contrast, cold discomfort is only slightly affected by humidity and is very closely related to a cold thermal sensation.

Determining the specifications for a radiant heating installation designed for human occupancy and acceptability involves the following steps:

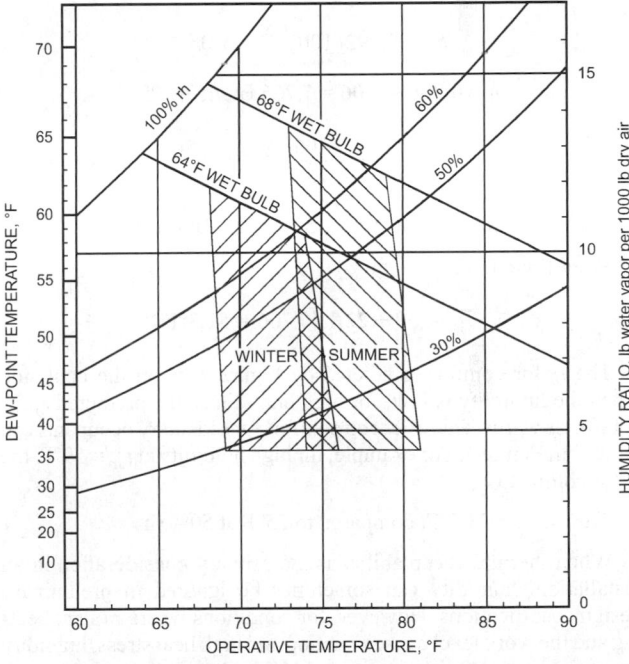

Fig. 4 ASHRAE Comfort Chart for Sedentary Occupants

1. Define the probable activity (metabolism) level of, and clothing worn by, the occupant and the relative air velocity. The following are two examples:

Case 1: Sedentary (1.1 met)
     Clothing insulation = 0.6 clo; V = 30 fpm
Case 2: Light work (walking) (2 met)
     Clothing insulation = 0.9 clo; V = 100 fpm

2. From Figure 2 (sedentary) or 3 (active), determine the optimum t_o for comfort and acceptability:

$$Case\ 1: t_o = 75°F; \quad Case\ 2: t_o = 62.5°F$$

3. For ambient air temperature t_a, calculate the mean radiant temperature $\bar{t}_r$ and/or ERF necessary for comfort and thermal acceptability.

Case 1: For $t_a = 60°F$ and 50% rh and assuming $\bar{t}_r = 86°F$, Solve for h_r from Equation (11):

$$h_r = 4 \times 0.1712 \times 10^{-8} \times 0.71[(86 + 60)/2 + 460]^3 = 0.736$$

Solve for h_c from Equation (12c):

$$h_c = 0.55$$

Then,

$$h = h_r + h_c = 0.736 + 0.55 = 1.286 \text{ Btu/h·ft}^2\text{·°F}$$

$$\frac{h_r}{h} = \frac{0.736}{1.286} = 0.57$$

From Equation (6), for comfort,

$$\text{ERF} = 1.286(75 - 60) = 19.3 \text{ Btu/h·ft}^2$$

From Equation (9),

$$\bar{t}_r = 60 + (1.286/0.736)(75 - 60) = 86.2°F$$

Case 2: For $t_a = 50°F$ and assuming $\bar{t}_r = 81°F$,

$$h_r = 4 \times 0.1712 \times 10^{-8} \times 0.71[(81 + 50)/2 + 460]^3 = 0.706$$

$$h_c = 0.092(100)^{0.53} = 1.06$$

$$h = 0.706 + 1.06 = 1.766 \text{ Btu/h·ft}^2\text{·°F}$$

$$\frac{h_r}{h} = \frac{0.706}{1.766} = 0.40$$

$$\text{ERF} = 1.766(62.5 - 50) = 22.1 \text{ Btu/h·ft}^2$$

From Equation (7),

$$\bar{t}_r = 50 + 22.1/0.706 = 81.3°F$$

The t_o for comfort, predicted by Figure 2, is on the cool side when the humidity is low; for high humidities, the predicted t_o for comfort is warm. This effect on comfort for sedentary occupants can be seen in Figure 4. For example, for high humidity at $t_{dp} = 59°F$, the t_o for comfort is

Case 1: $t_o = 73.5°F$, compared to 75°F at 50% rh

When thermal acceptability is the primary consideration in an installation, humidity can sometimes be ignored in preliminary design specifications. However, for conditions where radiant heating and the work level cause sweating and high heat stress, humidity is a major consideration and a hybrid HVAC system should be used.

Equations (3) to (12) can also be used to determine the ambient air temperature t_a required when the mean radiant temperature MRT is maintained by a specified radiant system.

When calculating heat loss, t_a must be determined. For a radiant system that is to maintain a MRT of $\bar{t}_r$, the operative temperature t_o can be determined from Figure 4. Then, t_a can be calculated by recalling that t_o is approximately equal to the average of t_a and t_r. For example, for a t_o of 73°F and a radiant system designed to maintain an MRT of 78°F, the t_a would be 68°F.

When the surface temperature of outside walls, particularly those with large areas of glass, deviates too much from room air temperature and from the temperature of other surfaces, simplified calculations for load and operative temperature may lead to errors in sizing and locating the panels. In such cases, more detailed radiant exchange calculations may be required, with separate estimation of heat exchange between the panels and each surface. A large window

area may lead to significantly lower MRTs than expected. For example, Athienitis and Dale (1987) reported an MRT 5.4°F lower than room air temperature for a room with a glass area equivalent to 22% of its floor area.

DESIGN FOR BEAM RADIANT HEATING

Spot beam radiant heat can improve comfort at a specific location in a large, poorly heated work area. The design problem is specifying the type, capacity, and orientation of the beam heater.

Using the same reasoning as in Equations (1) to (10), the effective radiant flux ΔERF added to an unheated work space with an operative temperature t_{uo} to result in a t_o for comfort (as given by Figure 2 or 3) is

$$\Delta\text{ERF} = h(t_o - t_{uo}) \tag{13}$$

or

$$t_o = t_{uo} + \Delta\text{ERF}/h \tag{14}$$

The heat transfer coefficient h for the occupant in Equation (13) is given by Equations (11) and (12), with $h = h_r + h_c$.

By definition, ERF is the energy absorbed per unit of total body surface A_D (DuBois area) and *not* the total effective radiating area A_{eff} of the body.

Geometry of Beam Heating

Figure 5 illustrates the following parameters that must be considered in specifying a beam radiant heater designed to produce the ERF, or mean radiant temperature $\bar{t}_r$, necessary for comfort at an occupant's workstation:

Ω = solid angle of heater beam, steradians (sr)
I_K = irradiance from beam heater, Btu/h·sr
K = subscript for absolute temperature of beam heater, °R
β = elevation angle of heater, degrees (at 0°, beam is horizontal)
ϕ = azimuth angle of heater, degrees (at 0°, beam is facing subject)
d = distance from beam heater to center of occupant, ft
A_p = projected area of occupant on a plane normal to direction of heater beam (ϕ, β), ft^2
α_K = absorptance of skin-clothing surface at emitter temperature (see Figure 1)

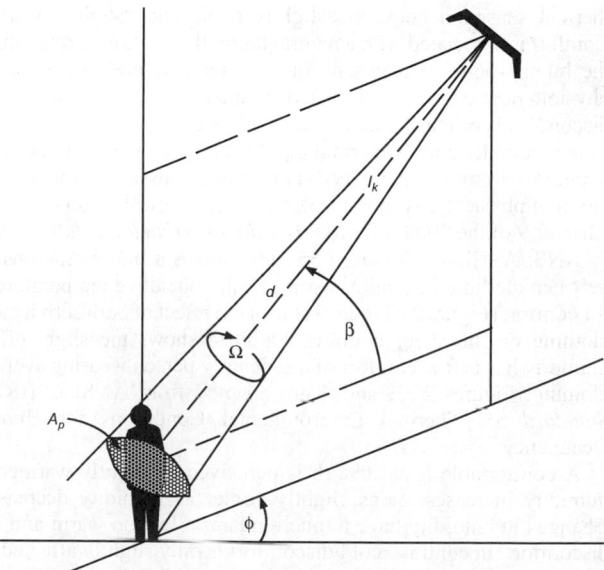

Fig. 5　Geometry and Symbols for Describing Beam Heaters

ERF may also be measured as the heat absorbed at the occupant's clothing and skin surface from a beam heater at absolute temperature:

$$ERF = \frac{\alpha_K I_K A_p}{d^2 A_D} \qquad (15)$$

where ERF is in Btu/h·ft² and (A_p/d^2) is the solid angle subtended by the projected area of the occupant from the radiating beam heater I_K, which is treated here as a point source. A_D is the DuBois area:

$$A_D = 0.0621 \, W^{0.425} H^{0.725}$$

where

   W = occupant weight, lb
   H = occupant height, ft

For additional information on radiant flux distribution patterns and sample calculations of radiation intensity I_K and ERF, refer to Chapter 15 of the 2008 *ASHRAE Handbook—HVAC Systems and Equipment.*

Floor Reradiation

In most low-, medium-, and high-intensity radiant heater installations, local floor areas are strongly irradiated. The floor absorbs most of this energy and warms to an equilibrium temperature t_f, which is higher than the ambient air temperature t_a and the unheated room enclosure surfaces. Part of the energy directly absorbed by the floor is transmitted by conduction to the colder layers beneath (or, for slabs-on-grade, to the ground), part is transferred by natural convection to room air, and the rest is reradiated. The warmer floor will raise ERF or $\bar{t}_r$ over that caused by the heater alone.

For a person standing on a large, flat floor that has a temperature raised by direct radiation t_f, the linearized $\bar{t}_r$ due to the floor and unheated walls is

$$\bar{t}_{rf} = F_{p-f} t_f + (1 - F_{p-f}) t_a \qquad (16)$$

where the unheated walls, ceiling, and ambient air are assumed to be at t_a, and F_{p-f} is the angle factor governing radiation exchange between the heated floor and the person.

The ERF_f from the floor affecting the occupant, which is due to the $(t_f - t_a)$ difference, is

$$ERF_f = h_r(\bar{t}_{rf} - t_a)$$
$$= h_r F_{p-f}(t_f - t_a) \qquad (17)$$

where h_r is the linear radiative heat transfer coefficient for a person as given by Equation (11). For a standing or sitting subject when the walls are farther than 16 ft away, F_{p-f} is 0.44 (Fanger 1973). For an average-sized 16 by 16 ft room, a value of 0.35 for F_{p-f} is suggested. For detailed information on floor reradiation, see Chapter 15 in the 2008 *ASHRAE Handbook—HVAC Systems and Equipment.*

In summary, when radiant heaters warm occupants in a selected area of a poorly heated space, the radiation heat necessary for comfort consists of two additive components: (1) ERF directly caused by the heater and (2) reradiation ERF_f from the floor. The effectiveness of floor reradiation can be improved by choosing flooring with a low specific conductivity. Flooring with high thermal inertia may be desirable during radiant transients, which may occur as heaters are cycled by a thermostat set to the desired operative temperature t_o.

Asymmetric Radiant Fields

In the past, comfort heating has required flux distribution in occupied areas to be uniform, which is not possible with beam radiant heaters. Asymmetric radiation fields, such as those experienced when lying in the sun on a cool day or when standing in front of a warm fire, can be pleasant. Therefore, a limited amount of asymmetry, which is allowable for comfort heating, is referred to as "reasonable uniform radiation distribution" and is used as a design requirement.

To develop criteria for judging the degree of asymmetry allowable for comfort heating, Fanger et al. (1980) proposed defining radiant temperature asymmetry as the difference in the plane radiant temperature between two opposing surfaces. Plane radiant temperature is the equivalent $\bar{t}_{r1}$ caused by radiation on one side of the subject, compared with the equivalent $\bar{t}_{r2}$ caused by radiation on the opposite side. Gagge et al. (1967) conducted a study of subjects (eight clothed and eight unclothed) seated in a chair and heated by two lamps. Unclothed subjects found a $(\bar{t}_r - t_a)$ asymmetry as high as 20°F to be comfortable, but clothed subjects were comfortable with an asymmetry as high as 31°F.

For an unclothed subject lying on an insulated bed under a horizontal bank of lamps, neutral temperature sensation occurred for a t_o of 72°F, which corresponds to a $(t_o - t_a)$ asymmetry of 20°F or a $(\bar{t}_r - t_a)$ asymmetry of 27°F, both averaged for eight subjects (Stevens et al. 1969). In studies of heated ceilings, 80% of eight male and eight female clothed subjects voted conditions as comfortable and acceptable for asymmetries as high as 20°F. The study compared the floor and heated ceilings. Asymmetry in the MRTs for direct radiation from three lamps and for floor reradiation is about 1°F, which is negligible.

In general, the human body has a great ability to sum sensations from many hot and cold sources. For example, a study of Australian aborigines sleeping unclothed next to open fires in the desert at night found a t_a of 43°F (Scholander 1958). The $\bar{t}_r$ caused directly by three fires was 171°F, and the sky $\bar{t}_r$ was 30°F; the resulting t_o was 82°F, which is acceptable for human comfort.

According to the limited field and laboratory data available, an allowable design radiant asymmetry of 22 ± 5°F should cause little discomfort over the comfortable t_o range used by ASHRAE *Standard* 55 and in Figures 2 and 3. Increased clothing insulation allows increases in the acceptable asymmetry, but increased air movement reduces it. Increased activity also reduces human sensitivity to changing $\bar{t}_r$ or t_o and, consequently, increases the allowable asymmetry. The design engineer should use caution with an asymmetry greater than 27°F, as measured by a direct beam radiometer or estimated by calculation.

RADIATION PATTERNS

Figure 6 indicates basic radiation patterns commonly used in design for radiation from point or line sources (Boyd 1962). A point source radiates over an area proportional to the square of the distance from the source. The area for a (short) line source also varies substantially as the square of the distance, with about the same area as the circle actually radiated at that distance. For line sources, the pattern width is determined by the reflector shape and position of the element within the reflector. The rectangular area used for installation purposes as the pattern of radiation from a line source assumes a length equal to the width plus the fixture length. This assumed length is satisfactory for design, but is often two or three times the pattern width.

Electric infrared fixtures are often identified by their beam pattern (Rapp and Gagge 1967), which is the radiation distribution normal to the line source element. The beam of a high-intensity infrared fixture may be defined as that area in which the intensity is at least 80% of the maximum intensity encountered anywhere within the beam. This intensity is measured in the plane in which maximum control of energy distribution is exercised.

The beam size is usually designated in angular degrees and may be symmetrical or asymmetrical in shape. For adaptation to their

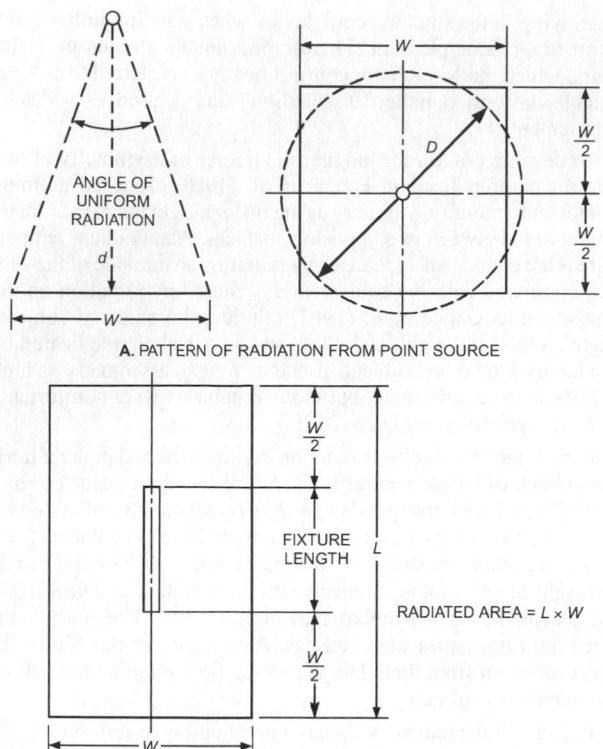

A. PATTERN OF RADIATION FROM POINT SOURCE

B. PATTERN OF RADIATION FROM LINE SOURCE

Note: The projected area W^2 normal to a beam that is Ω steradians wide at distance d is Ωd^2. The floor area irradiated by a beam heater at an angle elevation β is $W^2/\sin \beta$. Fixture length L increases the area irradiated by $(1 + L/W)$.

Fig. 6 Basic Radiation Patterns for System Design
(Boyd 1962)

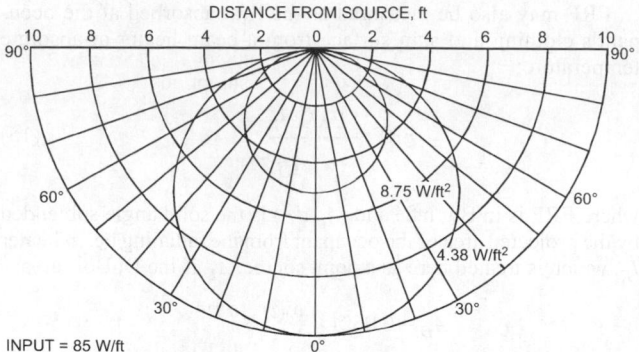

Fig. 7 Lines of Constant Radiant Flux for a Line Source

design specifications, some manufacturers indicate beam characteristics based on 50% maximum intensity.

The control used for an electric system affects the desirable maximum end-to-end fixture spacing. Actual pattern length is about three times the design pattern length, so control in three equal stages is achieved by placing every third fixture on the same circuit. If all fixtures are controlled by input controllers or variable voltage to electric units, end-to-end fixture spacing can be nearly three times the design pattern length. Side-to-side minimum spacing is determined by the fixture's distribution pattern and is not influenced by the control method.

Low-intensity equipment typically consists of a steel tube hung near the ceiling and parallel to the outside wall. Circulation of combustion products inside the tube elevates the tube temperature, and radiant energy is emitted. The tube is normally provided with a reflector to direct the radiant energy down into the space to be conditioned.

Radiant ceiling panels for heating only are installed with spacing between them to enhance downward convection, and are usually the primary heat source for the space. Usually the source is a surface. Details of radiant and convective heat transfer for panel heating and cooling are given in Chapter 6 of the 2008 *ASHRAE Handbook—HVAC Systems and Equipment*. Flux density is inversely proportional to distance from the source if the source is a recessed line, and is inversely proportional to the square of the distance from the source if the source is a point. In practice, actual dependence will be between these limits.

The rate of radiation exchange between a panel and a particular object depends on their temperatures, surface emittances, and geometrical orientation (i.e., shape factor). It also depends on the

temperatures and configurations of all the other objects and walls within the space.

The heat flux for an ideal line source is shown in Figure 7. All objects (except the radiant source) are at the same temperature, and the radiant source is suspended symmetrically in the room.

DESIGN FOR TOTAL SPACE HEATING

Radiant heating differs from conventional heating by a moderately elevated ERF, $\bar{t}_r$, or t_o over ambient temperature t_a. Standard methods of design are normally used, although informal studies indicate that radiant heating results in a lower heating load than convection heating (Zmeureanu et al. 1987). Buckley and Seel (1987) demonstrated that a combination of elevated floor temperature, higher mean radiant temperature, and reduced ambient temperature results in lower thermostat settings, reduced temperature differential across the building envelope, and thus lower heat loss and heating load for the structure. In addition, peak load may be decreased because of heat (cool) stored in the structure when embedded panels in the floor, walls, or ceiling are used (Kilkis 1992).

Most gas radiation systems for full-building heating concentrate the bulk of capacity at mounting heights of 10 to 16 ft at the perimeter, directed at the floor near the walls. Units can be mounted considerably higher. Successful application depends on supplying the proper amount of heat in the occupied area. Heaters should be located to take maximum advantage of the radiation pattern produced. Exceptions to perimeter placement include walls with high transmission losses and extreme height, as well as large roof areas where roof heat losses exceed perimeter and other heat losses.

Electric infrared systems installed indoors for complete building heating use layouts that uniformly distribute the radiation throughout the area used by people, as well as layouts that emphasize perimeter placement, such as in ice hockey rinks. Some electric radiant heaters emit significant visible radiation and provide both heating and illumination.

The orientation of equipment and people is less important for general area heating (large areas within larger areas) than for spot heating. With reasonably uniform radiation distribution in work or living areas, the exact orientation of the units is not important. Higher intensities of radiation may be desirable near walls with outside exposure. Radiation shields (reflective to infrared) fastened a few inches from the wall to allow free air circulation between the wall and shield are effective for frequently occupied work locations close to outside walls.

In full-building heating, units should be placed where their radiant and convective output best compensates for the structure's heat loss. The objective of a complete heating system is to provide a warm floor with low conductance to the heat sink beneath the floor. This thermal storage may permit cycling of units with standard controls.

TESTING INSTRUMENTS FOR RADIANT HEATING

In designing a radiant heating system, calculation of radiant heat exchange may involve some untested assumptions. During field installation, the designer must test and adjust the equipment to ensure that it provides acceptable comfort conditions. The black globe thermometer and directional radiometer can be used to evaluate the installation.

Black Globe Thermometer

The classic (Bedford) globe thermometer is a thin-walled, matte black, hollow sphere with a thermocouple, thermistor, or thermometer placed at the center. It can directly measure $\bar{t}_r$, ERF, and t_o. When a black globe is in thermal equilibrium with its environment, the gain in radiant heat from various sources is balanced by the convective loss to ambient air. Thus, in terms of the globe's linear radiative and convective heat transfer coefficients, h_{rg} and h_{cg}, respectively, the heat balance at equilibrium is

$$h_{rg}(\bar{t}_{rg} - t_g) = h_{cg}(t_g - t_a) \qquad (18)$$

where $\bar{t}_{rg}$ is the mean radiant temperature measured by the globe and t_g is the temperature in the globe.

In general, the $\bar{t}_{rg}$ of Equation (18) equals the $\bar{t}_r$ affecting a person when the globe is placed at the center of the occupied space and when the radiant sources are distant from the globe.

The effective radiant flux measured by a black globe is

$$\text{ERF}_g = h_{rg}(\bar{t}_{rg} - t_a) \qquad (19)$$

which is analogous to Equation (5) for occupants. From Equations (18) and (19), it follows that

$$\text{ERF}_g = (h_{rg} + h_{cg})(t_g - t_a) \qquad (20)$$

If the ERF_g of Equation (20) is modified by the skin-clothing absorptance α_K and the shape f_{eff} of an occupant relative to the black globe, the corresponding ERF affecting the occupant is

$$\text{ERF (for an occupant)} = f_{eff}\,\alpha_K\,\text{ERF}_g \qquad (21)$$

where α_K is defined in the section on Geometry of Beam Heating and f_{eff}, which is defined after Equation (11), is approximately 0.71 and equals the ratio h_r/h_{rg}. The t_o affecting an occupant, in terms of t_g and t_a, is given by

$$t_o = Kt_g + (1 - K)t_a \qquad (22)$$

where the coefficient K is

$$K = \alpha_K f_{eff}(h_{rg} + h_{cg})/(h_r + h_c) \qquad (23)$$

Ideally, when K is unity, the t_g of the globe would equal the t_o affecting an occupant.

For an average comfortable equilibrium temperature of 77°F and noting that f_{eff} for the globe is unity, Equation (11) yields

$$h_{rg} = 1.06 \text{ Btu/h·ft}^2\text{·°F} \qquad (24)$$

and Equation (12) yields

$$h_{cg} = 0.296 D^{-0.4} V_2^{0.53} \qquad \{100 \text{ fpm} < V_a < 400 \text{ fpm}\} \qquad (25)$$

where

D = globe diameter, in.
V_a = air velocity, fpm

Equation (25) is Bedford's convective heat transfer coefficient for a 6 in. globe's convective loss, modified for D. For any radiating source below 1700°F, the ideal diameter of a sphere that makes

Table 1 Value of K for Various Air Velocities and Globe Thermometer Diameters ($\alpha_g = 1$)

Air Velocity V_a, fpm	Approximate Globe Thermometer Diameter D, in.			
	2	4	6	8
50	1.35	1.15	1.05	0.99
100	1.43	1.18	1.06	0.99
200	1.49	1.21	1.07	1.00
400	1.54	1.23	1.08	1.00
800	1.59	1.26	1.09	1.00

$K = 1$ that is independent of air movement is 8 in. (see Table 1). Table 1 shows the value of K for various values of globe diameter D and ambient air movement V. The table shows that the uncorrected temperature of the traditional 6 in. globe would overestimate the true $(t_o - t_a)$ difference by 6% for velocities up to 200 fpm, and the probable error of overestimating t_o by t_g uncorrected would be less than 0.9°F. Globe diameters between 6 and 8 in. are optimum for using the uncorrected t_g measurement for t_o. The exact value for K may be used for smaller globes when estimating t_o from t_g and t_a measurements. The value of $\bar{t}_r$ may be found by substituting Equations (24) and (25) in Equation (18), because $\bar{t}_r$ (occupant) is equal to $\bar{t}_{rg}$. The smaller the globe, the greater the variation in K caused by air movement. Globes with D greater than 8 in. will overestimate the importance of radiation gain versus convection loss.

For sources radiating at high temperatures (1300 to 10,000°F), the ratio α_m/α_g may be set near unity by using a pink-colored globe surface, whose absorptance for the sun is 0.7, a value similar to that of human skin and normal clothing (Madsen 1976).

In summary, the black globe thermometer is simple and inexpensive and may be used to determine $\bar{t}_r$ [Equation (18)] and the ERF [Figure 1 and Equations (20) and (21)]. When the radiant heater temperature is less than 1700°F, the uncorrected t_g of a 6 to 8 in. black globe is a good estimate of the t_o affecting the occupants. A pink globe extends its usefulness to sun temperatures (10,000°F). A globe with a low mass and low thermal capacity is more useful because it reaches thermal equilibrium in less time.

Using the heat exchange principles described, many heated and unheated instruments of various shapes have been designed to measure acceptability in terms of t_o, $\bar{t}_r$, and ERF, as sensed by their own geometric shapes. Madsen (1976) developed an instrument that can determine the predicted mean vote (PMV) from the $(t_g - t_a)$ difference, as well as correct for clothing insulation, air movement, and activity (ISO 1994).

Directional Radiometer

The angle of acceptance (in steradians) in commercial radiometers allows the engineer to point the radiometer directly at a wall, floor, or high-temperature source and read the average temperature of that surface. Directional radiometers are calibrated to measure either the radiant flux accepted by the radiometer or the equivalent blackbody radiation temperature of the emitting surface. Many radiometers collimate (make parallel) input to sense small areas of body, clothing, wall, or floor surfaces. A directional radiometer allows rapid surveys and analyses of important radiant heating factors such as the temperature of skin, clothing surfaces, and walls and floors, as well as the radiation intensity I_K of heaters on the occupants. One radiometer for direct measurement of the equivalent radiant temperature has an angle of acceptance of 2.8° or 0.098 sr, so that at 1 ft, it measures the average temperature over a projected circle about 3/8 in. in diameter.

APPLICATIONS

When installing radiant heaters in specific applications, consider the following factors:

- Gas and electric high-temperature infrared heaters must not be placed where they could ignite flammable dust or vapors, or decompose vapors into toxic gases.
- Fixtures must be located with recommended clearances to ensure proper heat distribution. Stored materials must be kept far enough from fixtures to avoid hot spots. Manufacturers' recommendations must be followed.
- Unvented gas heaters inside tight, poorly insulated buildings may cause excessive humidity with condensation on cold surfaces. Proper insulation, vapor barriers, and ventilation prevent these problems.
- Combustion heaters in tight buildings may require makeup air to ensure proper venting of combustion gases. Some infrared heaters are equipped with induced-draft fans to relieve this problem.
- Some transparent materials may break because of uneven application of high-intensity infrared. Infrared energy is transmitted without loss from the radiator to the absorbing surfaces. The system must produce the proper temperature distribution at the absorbing surfaces. Problems are rarely encountered with glass 0.25 in. or less in thickness.
- Comfort heating with infrared heaters requires a reasonably uniform heat flux distribution in the occupied area. Although thermal discomfort can be relieved in warm areas with high air velocity, such as on loading docks, the full effectiveness of a radiant-heater installation is reduced by high air velocity.
- Radiant spot heating and zoning in large undivided areas with variable occupancy patterns provides localized heating just where and when people are working, which reduces the heating cost.

By building periodic microstructure into a thin silicon carbide wafer, infrared panel surfaces that are highly directional and coherent over many wavelengths can be obtained (Greffet et al. 2002).

Low-, Medium-, and High-Intensity Infrared Applications

Low-, medium-, and high-intensity infrared equipment is used extensively in industrial, commercial, and military applications. This equipment is particularly effective in large areas with high ceilings, such as in buildings with large air volumes and in areas with high infiltration rates, large access doors, or large ventilation requirements.

Factories. Low-intensity radiant equipment suspended near the ceiling around the perimeter of facilities with high ceilings enhances the comfort of employees because it warms floors and equipment in the work area. For older, uninsulated buildings, the energy cost for low-intensity radiant equipment is less than that of other heating systems. High-intensity infrared for spot heating and low-intensity infrared for zone temperature control effectively heat large unheated facilities.

Warehouses. Low- and high-intensity infrared are used for heating warehouses, which usually have a large volume of air, poor insulation, and high infiltration. Low-intensity infrared equipment is installed near the ceiling around the building's perimeter. High-level mounting near the ceiling leaves floor space available for product storage. Both low- and high-intensity infrared are arranged to control radiant intensity and provide uniform heating at the working level and frost protection areas, which is essential for perishable goods storage.

Garages. Low-intensity infrared provides comfort for mechanics working near or on the floor. With elevated MRT in the work area, comfort is provided at a lower ambient temperature.

In winter, opening the large overhead doors to admit equipment for service allows a substantial entry of cold outdoor air. On closing the doors, the combination of reradiation from the warm floor and radiant heat warming the occupants (not the air) provides rapid recovery of comfort. Radiant energy rapidly warms the cold vehicles. Radiant floor panel heating systems are also effective.

Low-intensity equipment is suspended near the ceiling around the perimeter, often with greater concentration near overhead doors. High-intensity equipment is also used to provide additional heat near doors.

Aircraft Hangars. Equipment suspended near the roofs of hangars, which have high ceilings and large access doors, provides uniform radiant intensity throughout the working area. A heated floor is particularly effective in restoring comfort after an aircraft has been admitted. As in garages, the combination of reradiation from the warm floor and radiation from the radiant heating system provides rapid regain of comfort. Radiant energy also heats aircraft moved into the work area.

Greenhouses. In greenhouse applications, a uniform flux density must be maintained throughout the facility to provide acceptable growing conditions. In a typical application, low-intensity units are suspended near, and run parallel to, the peak of the greenhouse.

Sometimes the soil is warmed by embedding pipes in the ground. The same system may also contribute to sensible heating of the greenhouse.

Outdoor Applications. Applications include loading docks, racetrack stands, outdoor restaurants, and under marquees. Low-, medium-, and high-intensity infrared are used in these facilities, depending on their layout and requirements.

Other Applications. Radiant heat may be used in a variety of large facilities with high ceilings, including religious compounds, day-care facilities, gymnasiums, swimming pools, enclosed stadiums, and facilities that are open to the outdoors. Radiant energy is also used to control condensation on surfaces such as large glass exposures. Careful sizing and design of direct gas-fired systems may improve thermal comfort in places of assembly by affecting air and wall surface temperatures. A numerical study with computational fluid design (CFD) models may be essential to achieve these benefits and reduce condensation risk (Xiang et al. 2001).

Low-, medium-, and high-intensity infrared are also used for other industrial applications, including process heating for component or paint drying ovens, humidity control for corrosive metal storage, and snow control and freeze-protection for parking or loading areas.

Panel Heating and Cooling

Residences. Embedded pipe systems, electric resistance panels, and forced warm-air panel systems have all been used in residences. The embedded pipe system is most common, using plastic or rubber tubing in the floor slab or copper tubing in older plaster ceilings. These systems are suitable for conventionally constructed residences with normal amounts of glass. Lightweight hydronic metal panel ceiling systems have also been applied to residences, and prefabricated electric panels are advantageous, particularly in rooms that have been added on.

Office Buildings. A panel system is usually applied as a perimeter heating system. Panels are typically piped to provide exposure control with one riser on each exposure and all horizontal piping incorporated in the panel piping. In these applications, the air system provides individual room control. Perimeter radiant panel systems have also been installed with individual zone controls. However, this type of installation is usually more expensive and, at best, provides minimal energy savings and limited additional occupant comfort. Radiant panels can be used for cooling as well as heating. Cooling installations are generally limited to retrofit or renovation jobs where ceiling space is insufficient for the required duct sizes. In these installations, the central air supply system provides ventilation air, dehumidification, and some sensible cooling. Two- and four-pipe water distribution systems may be used. Hot-water supply temperatures are commonly reset by outside temperature, with additional offset or flow control to compensate for solar load. Panel systems are readily adaptable to accommodate most changes in partitioning. Electric panels in lay-in ceilings have been used for full perimeter heating.

Schools. In all areas except gymnasiums and auditoriums, panels are usually selected for heating only, and may be used with any type of approved ventilation system. The panel system is usually sized to offset the transmission loads plus any reheating of the air. If the school is air conditioned by a central air system and has perimeter heating panels, single-zone piping may be used to control the panel heating output, and the room thermostat modulates the supply air temperature or volume. Heating and cooling panel applications are similar to those in office buildings. Panel heating and cooling for classroom areas has no mechanical equipment noise to interfere with instructional activities.

Hospitals. The principal application of heating and cooling radiant panels has been for hospital patient rooms. Perimeter radiant heating panels are typically applied in other areas of hospitals. Compared to conventional systems, radiant heating and cooling systems are well suited to hospital patient rooms because they (1) provide a draft-free, thermally stable environment, (2) have no mechanical equipment or bacteria and virus collectors, and (3) do not take up space in the room. Individual room control is usually achieved by throttling the water flow through the panel. The supply air system is often 100% outdoor air; minimum air quantities delivered to the room are those required for ventilation and exhaust of the toilet room and soiled linen closet. The piping system is typically a four-pipe design. Water control valves should be installed in corridors so that they can be adjusted or serviced without entering patient rooms. All piping connections above the ceiling should be soldered or welded and thoroughly tested. If cubicle tracks are applied to the ceiling surface, track installation should be coordinated with the radiant ceiling. In panel cooling, surface condensation may need a separate latent system (Isoardi and Brasselet 1995). Security panel ceilings are often used in areas occupied by mentally disturbed patients so that equipment cannot be damaged by a patient or used to inflict injury.

Swimming Pools. A partially clothed person emerging from a pool is very sensitive to the thermal environment. Panel heating systems are well suited to swimming pool areas. Floor panel temperatures must be controlled so they do not cause foot discomfort. Ceiling panels are generally located around the perimeter of the pool, not directly over the water. Panel surface temperatures are higher to compensate for the increased ceiling height and to produce a greater radiant effect on partially clothed bodies. Ceiling panels may also be placed over windows to reduce condensation.

Apartment Buildings. For heating, pipe coils are embedded in the masonry slab. The coils must be carefully positioned so as not to overheat one apartment while maintaining the desired temperature in another. The slow response of embedded pipe coils in buildings with large glass areas may be unsatisfactory. Installations for heating and cooling have been made with pipes embedded in hung plaster ceilings. A separate minimum-volume dehumidified air system provides the necessary dehumidification and ventilation for each apartment. The application of electric resistance elements embedded in floors or behind a skim coat of plaster at the ceiling has increased. Electric panels are easy to install and simplify individual room control.

Industrial Applications. Panel systems are widely used for general space conditioning of industrial buildings in Europe (Petráš 2001). For example, the walls and ceilings of an internal combustion engine test cell are cooled with chilled water. Although the ambient air temperature in the space reaches up to 95°F, the occupants work in relative comfort when 55°F water is circulated through the ceiling and wall panels.

Other Buildings. Metal panel ceiling systems can be operated as heating systems at elevated water temperatures and have been used in airport terminals, convention halls, lobbies, and museums, especially those with large glass areas. Cooling may also be applied. Because radiant energy travels through the air without warming it, ceilings can be installed at any height and remain effective. One par-

ticularly high ceiling installed for a comfort application is 50 ft above the floor, with a panel surface temperature of approximately 285°F for heating. The ceiling panels offset the heat loss from a single-glazed, all-glass wall.

The high lighting levels in television studios make them well suited to panels that are installed for cooling only and are placed above lighting to absorb the radiation and convection heat from the lights and normal heat gains from the space. The panel ceiling also improves the acoustical properties of the studio.

Metal panel ceiling systems are also installed in minimum- and medium-security jail cells and in facilities where disturbed occupants are housed. The ceiling is strengthened by increasing the gage of the ceiling panels, and security clips are installed so that the ceiling panels cannot be removed. Part of the perforated metal ceiling can be used for air distribution.

District Energy Systems. Panel heating and cooling increases the thermal efficiency and use rate of low-enthalpy energy resources like waste heat. Buildings equipped with panels may be cascaded with buildings with high-enthalpy HVAC systems, which increases the use rate of district energy systems (Kilkis 2002).

New Techniques. The introduction of thermoplastic and rubber tubing and new design techniques have improved radiant panel heating and cooling equipment. The systems are energy-efficient and use low water temperatures available from solar collectors and heat pumps (Kilkis 1993). Metal radiant panels can be integrated into ceiling design to provide a narrow band of radiant heating around the building perimeter. These new radiant systems are more attractive, provide more comfortable conditions, operate more efficiently, and have a longer life than some baseboard or overhead air systems.

SYMBOLS

A_D = total DuBois surface area of person, ft^2

A_{eff} = effective radiating area of person, ft^2

A_p = projected area of occupant normal to the beam, ft^2

clo = unit of clothing insulation equal to 0.880 ft$^2 \cdot$°F·h/Btu

D = diameter of globe thermometer, in.

d = distance of beam heater from occupant, ft

ERF = effective radiant flux (person), Btu/h·ft^2

ERF_f = radiant flux caused by heated floor on occupant, Btu/h·ft^2

ERF_g = effective radiant flux (globe), Btu/h·ft^2

F_{p-f} = angle factor between occupant and heated floor

f_{eff} = ratio of radiating surface (person) to its total area (DuBois)

H_m = net metabolic heat loss from body surface, Btu/h·ft^2

h = combined heat transfer coefficient (person), Btu/h·ft$^2 \cdot$°F

h_c = convective heat transfer coefficient for person, Btu/h·ft$^2 \cdot$°F

h_{cg} = convective heat transfer coefficient for globe, Btu/h·ft$^2 \cdot$°F

h_r = linear radiative heat transfer coefficient (person), Btu/h·ft$^2 \cdot$°F

h_{rg} = linear radiative heat transfer coefficient for globe, Btu/h·ft$^2 \cdot$°F

I_K = irradiance from beam heater, Btu/h·sr

K = coefficient that relates t_a and t_g to t_o [Equation (22)]

K = subscript indicating absolute irradiating temperature of beam heater, °R

L = fixture length, ft

met = unit of metabolic energy equal to 18.4 Btu/h·ft^2

t_a = ambient air temperature near occupant, °F

t_f = floor surface temperature, °F

t_g = globe temperature, °F

t_o = operative temperature, °F

$\overline{t_r}$ = mean radiant temperature affecting occupant, °F

$\overline{t_{rf}}$ = linearized $\overline{t_r}$ caused by floor and unheated walls on occupant, °F

t_{sf} = exposed surface temperature of occupant, °F

t_{uo} = operative temperature of unheated workspace, °F

V = relative velocity between occupant and air, fpm

V_a = air velocity, fpm

W = width of a square equivalent to the projected area of a beam of angle Ω steradians at a distance d, ft

α = relative absorptance of skin-clothing surface to that of matte black surface

α_g = absorptance of globe

α_K = absorptance of skin-clothing surface at emitter temperature

α_m = absorptance of skin-clothing surface at emitter temperatures above 1700°F
β = elevation angle of beam heater, degrees
σ = Stefan-Boltzmann constant = 0.1712×10^{-8} Btu/h·ft²·°R⁴
ϕ = azimuth angle of heater, degrees
Ω = radiant beam width, sr

REFERENCES

ASHRAE. 1992. Thermal environmental conditions for human occupancy. ANSI/ASHRAE *Standard* 55-1992.

Athienitis, A.K. and J.D. Dale. 1987. A study of the effects of window night insulation and low emissivity coating on heating load and comfort. *ASHRAE Transactions* 93(1A):279-294.

Boyd, R.L. 1962. Application and selection of electric infrared comfort heaters. *ASHRAE Journal* 4(10):57.

Buckley, N.A. and T.P. Seel. 1987. Engineering principles support an adjustment factor when sizing gas-fired low-intensity infrared equipment. *ASHRAE Transactions* 93(1):1179-1191.

Fanger, P.O. 1973. *Thermal comfort.* McGraw-Hill, New York.

Fanger, P.O., L. Banhidi, B.W. Olesen, and G. Langkilde. 1980. Comfort limits for heated ceiling. *ASHRAE Transactions* 86(2):141-156.

Gagge, A.P., G.M. Rapp, and J.D. Hardy. 1967. The effective radiant field and operative temperature necessary for comfort with radiant heating. *ASHRAE Transactions* 73(1):I.2.1-I.2.9; and *ASHRAE Journal* 9(5):63-66.

Greffet, J-J., R. Carminati, K. Joulain, J-P. Mulet, S. Mainguy, and Y. Chen. 2002. Coherent emission of light by thermal sources. *Nature* 416:61-64.

ISO. 1994. Moderate thermal environments—Determination of the PMV and PPD indices and specification of the conditions for thermal comfort. *Standard* 7730-1994. International Standard Organization, Geneva.

Isoardi, J.P. and J.P. Brasselet. 1995. Surface cooling in hospitals: Use of computer-aided design packages. *ASHRAE Transactions* 101(2):717-720.

Kilkis, B. 1992. Enhancement of heat pump performance using radiant floor heating systems. *ASME Winter Meeting: Advanced Energy Systems, Recent Research in Heat Pump Design, Analysis, and Application* 28:119-127.

Kilkis, I.B. 1993. Radiant ceiling cooling with solar energy: Fundamentals, modeling, and a case design. *ASHRAE Transactions* 99(2):521-533.

Kilkis, I.B. 2002. Rational use and management of geothermal energy resources. *International Journal of Global Energy Issues* 17(1/2):35-59.

Madsen, T.L. 1976. Thermal comfort measurements. *ASHRAE Transactions* 82(1):60-70.

Petráš, D. 2001. Hybrid heating and ventilating large industrial halls connected to district energy systems. *ASHRAE Transactions* 101(1):390-393.

Rapp, G.M. and A.P. Gagge. 1967. Configuration factors and comfort design in radiant beam heating of man by high temperature infrared sources. *ASHRAE Transactions* 73(3):1.1-1.8.

Scheatzle, D.G. 2003. Establishing a baseline data set for the evaluation of hybrid (radiant/convective) HVAC systems. ASHRAE *Research Project* RP-1140, Final Report.

Scholander, P.E. 1958. Cold adaptation in the Australian aborigines. *Journal of Applied Physiology* 13:211-218.

Stevens, J.C., L.E. Marks, and A.P. Gagge. 1969. The quantitative assessment of thermal comfort. *Environmental Research* 2:149-65.

Suryanarayana, S. and R.H. Howell. 1990. Sizing of radiant heating systems, Part II—Heated floors and infrared units. *ASHRAE Transactions* 96(1):666-675.

Wilkins, C.K. and R. Kosonen. 1992. Cool ceiling system: A European air-conditioning alternative. *ASHRAE Journal* 34(8):41-45.

Xiang, W., S.A. Tassou, and M. Kolokotroni. 2001. Heating of church buildings of historic importance with direct gas-fired heating systems. *ASHRAE Transactions* 107(1):357-364.

Zmeureanu, R., P.P. Fazio, and F. Haghighat. 1987. Thermal performance of radiant heating panels. *ASHRAE Transactions* 94(2):13-27.

SEISMIC- AND WIND-RESISTANT DESIGN

ALMOST all inhabited areas of the world are susceptible to the damaging effects of either earthquakes or wind. Restraints that are designed to resist one may not be adequate to resist the other. Consequently, when exposure to either earthquake or wind loading is a possibility, strength of equipment and attachments should be evaluated for all appropriate conditions.

Earthquake damage to inadequately restrained HVAC&R equipment can be extensive. Mechanical equipment that is blown off the support structure can become a projectile, threatening life and property. The cost of properly restraining the equipment is small compared to the high costs of replacing or repairing damaged equipment, liability for loss of life, or compared to the cost of building downtime due to damaged facilities.

Design and installation of seismic and wind restraints have the following primary objectives:

- To reduce the possibility of injury and the threat to life.
- To reduce long-term costs due to equipment damage and resultant downtime.

Note: The intent of building codes with respect to seismic design is not to prevent damage to property or the restrained equipment itself.

This chapter covers the design of restraints to limit equipment movement and to keep the equipment captive during an earthquake or extreme wind loading. Seismic restraints and isolators do not reduce the forces transmitted to the restrained equipment. Instead, properly designed and installed seismic restraints and isolators have the necessary strength to withstand the imposed forces. However, equipment that is to be restrained must also have the necessary strength to remain attached to the restraint.

The *International Building Code®* (IBC) (ICC 2009) provides a prescriptive approach for applying an equivalent static force representing the dynamic forces transmitted to the equipment by seismic or high-wind events. For mechanical systems, analysis of seismic and wind loading conditions can use static analysis from the prescriptive approach. Conservative safety factors are applied to reduce the complexity of earthquake and wind loading response analysis and evaluation. The following three aspects are considered in a properly designed restraint system:

- *Attachment of equipment to restraint.* The equipment must be positively attached to the restraint, and must have sufficient strength to withstand the imposed forces, and to transfer the forces to the restraint.

- *Restraint design.* The restraint also must be strong enough to withstand the imposed forces. This should be determined by the manufacturer by tests and/or analysis.
- *Attachment of restraint to substructure.* Attachment may be by bolts, lag bolts, welds, or concrete anchors. The substructure must be capable of surviving the imposed forces.

SEISMIC-RESISTANT DESIGN

Most seismic requirements adopted by local jurisdictions in North America are based on model codes developed by the International Code Council (ICC), such as the *International Building Code* (IBC). The *National Building Code of Canada* (NRC-IRC 2010) is Canada's equivalent version of the IBC. Local building officials must be contacted for specific requirements that may be more stringent than those presented in this chapter.

Other sources of seismic restraint information include

- *Seismic Restraint Manual: Guidelines for Mechanical Systems* (SMACNA 2008), includes seismic restraint information for mechanical equipment subjected to seismic forces of up to 1.0*g*.
- The most current National Fire Protection Association (NFPA) standards on restraint design are compliant with the IBC.
- U.S. Department of Energy DOE 430.1A and ASME AG-1 cover restraint design for nuclear facilities.
- DOD (2007) provides guidance for seismic design for U.S. Department of Defense (DoD) and Department of State (DoS) facilities, and DOD (2005) provides the seismic and wind design constants.
- *A Practical Guide to Seismic Restraint* (ASHRAE 2000) covers a broad range of seismic restraint design issues.
- *Federal Emergency Management Agency (FEMA)* installation manuals FEMA-412, FEMA-413, and FEMA-414 are available at the FEMA Web site. They provide a step-by-step process with details for installing seismic restraint devices.

In seismically active areas where governmental agencies regulate the earthquake-resistive design of buildings (e.g., California), the HVAC engineer usually does not prepare the code-required seismic restraint calculations. The HVAC engineer selects the heating and cooling equipment and, with the assistance of the acoustical engineer (if applicable to the project), selects the required vibration isolation devices. Seismic restraint calculations are performed for nonstructural components, and designs for piping, ductwork, and conduits are designed and detailed. For design-build projects, the design is reviewed by the registered design professional. Nonstructural restraint components are designed and constructed to resist the aftereffects of earthquake motions as required by the applicable

The preparation of this chapter is assigned to TC 2.7, Seismic and Wind Restraint Design.

code and in accordance with local building officials. Reviewed designs are submitted for approval by the authority having jurisdiction.

To ensure proper design factors are used, a designer should obtain information on the seismic design conditions (site class and occupancy category). The importance of the equipment and systems affected should be understood for code applications to include those items that must be functional after seismic events.

The owner or building officials maintain the code-required quality control over the design by requiring construction documents, special inspection requirements, and certification requirements prepared by the registered design professional and approved by the authority having jurisdiction. Upon completion of installation, the supplier of the seismic restraints, or a qualified representative, should inspect the installation and verify that all restraints and force-resisting systems are installed properly and comply with specifications.

TERMINOLOGY

Base plate thickness. Thickness of the equipment bracket fastened to the floor.

Effective shear force V_{eff}. Maximum shear force of one seismic restraint or tie-down bolt.

Effective tension force T_{eff}. Maximum tension force or pullout force on one seismic restraint or tie-down bolt.

Equipment. Any HVAC&R component that must be restrained from movement during an earthquake.

Resilient support. An active seismic device (such as a spring with a bumper) to prevent equipment from moving more than a specified amount.

Response spectra. Relationship between the acceleration response of the ground and the peak acceleration of the earthquake in a damped single degree of freedom at various frequencies. The ground motion response spectrum varies with soil conditions.

Rigid support. Passive seismic device used to restrict any movement.

Shear force V. Horizontal force generated at the plane of the seismic restraints, acting to cut the restraint at the base.

Seismic restraint. Device designed to withstand seismic forces and hold equipment in place during an earthquake.

Seismic force levels. The geographic location of a facility determines its seismic spectral response acceleration levels, as given in the *International Building Code*.

Snubber. Device made of steel-housed resilient bushings arranged to prevent equipment from moving beyond an established gap.

Tension force T. Force generated by overturning moments at the plane of the seismic restraints, acting to pull out the bolt.

CALCULATIONS

Sample calculations presented here assume that the equipment support is an integrated resilient support and restraint device. When the two functions of resilient support and motion restraint are separate or act separately, additional spring loads may need to be added to the anchor load calculation for the restraint device. Internal loads within integrated devices are not addressed in this chapter. These devices must be designed to withstand the full anchorage loads plus any internal spring loads.

Both static and dynamic analyses reduce the force generated by an earthquake to an equivalent statically applied force, which acts in a horizontal or vertical direction at the component's center of gravity. The resulting overturning moment is resisted by shear and tension (pullout) forces on the tie-down bolts. Static analysis is used for both rigidly mounted and resiliently mounted equipment.

Dynamic Analysis

Dynamic analysis of the isolation and snubber systems may be based on ground-level response spectra given in the IBC and reference standard ASCE 7 (ASCE 2005), which can be used as input for a dynamic analysis.

Response spectra applied to nonstructural components can be developed from ICC-ES acceptance criteria AC 156 (ICC-ES 2007). Site-specific ground response spectra developed by a geotechnical or soils engineer may be used, as well. The computer analysis used must be capable of analyzing nonlinear supports and site-specific ground motions. This dynamic analysis provides the maximum seismic input accelerations to the equipment components, allowing comparison to three-dimensional shock (drop) or shaker test fragility levels to determine equipment survivability. Actual drop or shaker test data for all HVAC equipment may not be available for the next several years.

Using the response spectra in the code for ground-floor inputs, or the spectra in ATC 29-2 for upper floors, a dynamic analysis can yield maximum input accelerations to equipment components. Comparing them to the allowable acceleration values in the table helps the engineer assess equipment survivability. Dynamic analysis can also provide maximum movement at all connections and, when added to the floor-to-ceiling code-mandated movements, allows the engineer to design these flexible connections and avoid pull-out or shear failures at these locations.

Under some conditions, Chapter 17 of IBC requires certificates of compliance for components and their attachments for a component importance factor I_p of 1.0 or 1.5. This is a life-safety issue as well as an essential equipment issue. Essential equipment with an $I_p = 1.5$ must have a certificate of compliance. Issuance of a certificate of compliance to the engineer of record and building official can be based on dynamic analysis. Most building officials require a stamp by a registered professional to be part of the calculations and certificate of compliance. Table 1 provides guidance on type of analysis (static or dynamic) and certificate of compliance documentation is required. Sample dynamic analysis is beyond the scope of this chapter and should be provided by experienced registered professionals. A common approach assumes an elastic response spectrum. The results of the dynamic analysis can then be scaled up or down as a percentage of the total lateral force obtained from the static analysis performed on the building.

Dynamic analysis of piping, ductwork, and equipment reflects the response of the equipment for all earthquake-generated frequencies. Especially for piping and equipment, when the earthquake forcing frequencies match the natural frequencies of the system, the resulting applied forces increase.

Static Analysis as Defined in the
International Building Code

The IBC specifies a design lateral force F_p for nonstructural components as

$$F_p = (0.4a_P S_{DS} W_P)\frac{I_p}{R_p}\left(1 + 2\frac{Z}{h}\right) \qquad (1)$$

but F_p need not be greater than

$$F_p = 1.6 S_{DS} I_p W_p \qquad (2)$$

nor less than

$$F_p = 0.3 S_{DS} I_p W_p \qquad (3)$$

where S_{DS} is determined by

$$S_{DS} = 2F_a S_S/3 \qquad (4)$$

where

a_p = component amplification factor in accordance with Table 2.
S_{DS} = design spectral response acceleration at short periods. S_S is the mapped spectral acceleration from Tables 4 and 5. (*Note*: More

Table 1   IBC Seismic Analysis Requirements

Component Operation Required for Life Safety	Building Seismic Design Category*	Required Analysis Type			
		Anchorage	Equipment Structural Capacity	Equipment Operational Capacity	Certificate of Compliance
No	A	Not required	Not required	Not required	Not required
No	B, C	Not required	Not required	Not required	Not required
No	D	Static	Dynamic or test	Not required	For mounting only
Yes	C, D	Static	Dynamic or test	Dynamic or test	For continued operation
No	E	Static	Dynamic or test	Dynamic or test	For continued operation
No	C, D	Static	Not required	Not required	Not required
Yes	C, D	Static	Dynamic or test	Dynamic or test	For continued operation
No	F	Static	Dynamic or test	Not required	For mounting only
Yes	F	Static	Dynamic or test	Dynamic or test	For continued operation

*If in question, reference structural documents.

Table 2   Coefficients for Mechanical Components

Mechanical and Electrical Component or Element	a_p	R_p
General Mechanical		
Boilers and furnaces	1.0	2.5
Piping		
High-deformability elements and attachments	1.0	3.5
Limited-deformability elements and attachments	1.0	2.5
Low-deformability elements or attachments	1.0	1.25
HVAC Equipment		
Vibration isolated	2.5	2.5
Non-vibration isolated	1.0	2.5
Mounted in-line with ductwork	1.0	2.5

Source: IBC (2006).

Table 3   Values of Site Coefficient F_a as Function of Site Class and Spectral Response Acceleration at Short Period (S_s)

Site Class	Soil Profile Name	Mapped Spectral Response Acceleration at Short Periods[a]				
		$S_s \leq 0.25$	$S_s = 0.50$	$S_s = 0.75$	$S_s = 1.00$	$S_s \geq 1.25$
A	Hard rock	0.8	0.8	0.8	0.8	0.8
B	Rock	1.0	1.0	1.0	1.0	1.0
C	Very dense soil and soft rock	1.2	1.2	1.1	1.0	1.0
D[c]	Stiff soil profile	**1.6**	**1.4**	**1.2**	**1.1**	**1.0**
E	Soft soil profile	2.5	1.7	1.2	0.9	b
F		See IBC for more information				

[a]Use straight-line interpolation for intermediate values of mapped spectral acceleration at short period S_s.
[b]Site-specific geotechnical investigation and dynamic site response analyses must be performed to determine appropriate values.
[c]D is the default Site Class unless otherwise stated in the approved geotechnical report.

detailed maps for the United States are available at the U.S. Geological Survey Web site: www.usgs.gov)

F_a = function of site soil characteristics and must be determined in consultation with either project geotechnical (soils) or structural engineer. Values for F_a for different soil types are given in Table 3. (*Note*: Without an approved geotechnical report, the default site soil classification is assumed to be site class D.)

R_p = component response modification factor in accordance with IBC.

I_p = component importance factor (see the IBC for explanation and determination of I_p).

$1 + 2z/h$ = height amplification factor where z is the height of attachment in the structure and h is the average height of the roof above grade. The value of $z \geq 0$ and z/h need not exceed 1.

$W_p(D)$ = weight of equipment, which includes all items attached or contained in the equipment.

The forces acting on the equipment are the lateral and vertical forces resulting from the earthquake, the force of gravity, and the forces of the restraint holding the equipment in place act on the center of gravity. The analysis assumes the equipment does not move during an earthquake; thus, the sum of the forces and moments must be zero. When calculating the overturning moment, including an uplift factor, the vertical component F_{pv} at the center of gravity is typically defined (for the IBC) to be

$$F_{pv} = 0.2S_{DS}D \quad (5)$$

If the equipment being analyzed is isolated, the final computed force must be doubled per section 1621.3.1 of the code.

Per section 1621.1.7 of the code, forces used when computing the loads for shallow (under 8 bolt diameter) embedment anchors are to be increased by a factor of $1.3R_p$.

Per section 1621.3.12.2 of the code, the only permitted expansion anchors for non-vibration-isolated equipment over 10 hp are undercut anchors.

Tables 4 and 5 contain brief listings of S_s factors that can be used to calculate the magnitude of the horizontal static seismic force acting at the equipment center of gravity. Values for IBC 2006 are available on the USGS Web site for U.S. locations or in Tables F-2 and G-2 of DOD (2005) for worldwide locations.

APPLYING STATIC ANALYSIS

The prescriptive method in the IBC allows that an equivalent static force can be calculated that represents the dynamic motions of an earthquake. The static forces acting on a piece of equipment are vertical and lateral forces resulting from the earthquake, the force of gravity, and forces at the restraints that hold the equipment in place. The analysis assumes that the equipment does not move during the earthquake and that the relative accelerations between its center of gravity and the ground generate forces that must be balanced by reactions at the restraints. Guidance from the code bodies indicates that equipment can be analyzed as though it were a rigid component; however, a factor a_p is applied in the computation to address flexibility issues on particular equipment types or flexible mounting arrangements. (*Note*: for dynamic analysis, it is common to use a 5% damping factor for equipment and a 1% damping factor for piping.) Although the basic force computation is different, the details of load distribution in the examples that follow apply independently of the code used.

The forces acting on the restraints include both shear and tensile components. The application direction of the lateral seismic acceleration can vary and is unknown. Depending on its direction, it is likely that not all of the restraints will be affected or share the load equally. It is important to determine the worst-case combination of forces at all restraint points for any possible direction that the lateral wave front can follow to ensure that the attachment is adequate.

Table 4 S_s Numbers* for Selected U.S. Locations (U.S. COE 1998)

State, City	ZIP	S_s	State, City	ZIP	S_s	State, City	ZIP	S_s	State, City	ZIP	S_s
Alabama			Ft. Wayne	46835	0.162	Butte	59701	0.599	**Rhode Island**		
Birmingham	35217	0.328	Gary	46402	0.173	Great Falls	59404	0.248	Providence	02907	0.267
Mobile	36610	0.124	Indianapolis	46260	0.182	**Nebraska**			**South Carolina**		
Montgomery	36104	0.170	South Bend	46637	0.121	Lincoln	68502	0.177	Charleston	29406	1.56
Arkansas			**Kansas**			Omaha	68144	0.127	Columbia	29203	0.578
Little Rock	72205	0.461	Kansas City	66103	0.122	**Nevada**			**South Dakota**		
Arizona			Topeka	66614	0.184	Las Vegas	89106	0.637	Rapid City	57703	0.153
Phoenix	85034	0.226	Wichita	67217	0.142	Reno	89509	1.29	Sioux Falls	57104	0.113
Tuscon	85739	0.325	**Kentucky**			**New York**			**Tennessee**		
California			Ashland	41101	0.221	Albany	12205	0.275	Chattanooga	37415	0.500
Fresno	93706	0.592	Covington	41011	0.186	Binghampton	13903	0.185	Knoxville	37920	0.589
Los Angeles	90026	1.50	Louisville	40202	0.247	Buffalo	14222	0.319	Memphis	38109	1.25
Oakland	94621	1.55	**Louisiana**			Elmira	14905	0.173	Nashville	37211	0.305
Sacramento	95823	0.568	Baton Rouge	70807	0.144	New York	10014	0.425	**Texas**		
San Diego	92101	1.54	New Orleans	70116	0.130	Niagara Falls	14303	0.311	Amarillo	79111	0.166
San Francisco	94114	1.50	Shreveport	71106	0.165	Rochester	14619	0.248	Austin	78703	0.088
San Jose	95139	2.05	**Massachusetts**			Schenectady	12304	0.278	Beaumont	77705	0.116
Colorado			Boston	02127	0.325	Syracuse	13219	0.192	Corpus Christi	78418	0.093
Colorado Springs	80913	0.178	Lawrence	01843	0.376	Utica	13501	0.250	Dallas	75233	0.117
Denver	80239	0.187	Lowell	01851	0.355	**North Carolina**			El Paso	79932	0.358
Connecticut			New Bedford	02740	0.261	Charlotte	28216	0.345	Ft. Worth	76119	0.110
Bridgeport	06606	0.332	Springfield	01107	0.260	Greensboro	27410	0.255	Houston	77044	0.107
Hartford	06120	0.274	Worchester	01602	0.271	Raleigh	27610	0.211	Lubbock	79424	0.099
New Haven	06511	0.285	**Maryland**			Winston-Salem	27106	0.281	San Antonio	78235	0.133
Waterbury	06702	0.287	Baltimore	21218	0.199	**North Dakota**			Waco	76704	0.095
Florida			**Maine**			Fargo	58103	0.073	**Utah**		
Ft. Lauderdale	33328	0.070	Augusta	04330	0.318	Grand Forks	58201	0.054	Salt Lake City	84111	1.79
Jacksonville	32222	0.142	Portland	04101	0.369	**Ohio**			**Virginia**		
Miami	33133	0.061	**Michigan**			Akron	44312	0.179	Norfolk	23504	0.132
St. Petersburg	33709	0.078	Detroit	48207	0.123	Canton	44702	0.316	Richmond	23233	0.300
Tampa	33635	0.083	Flint	48506	0.091	Cincinnati	45245	0.191	Roanoke	24017	0.290
Georgia			Grand Rapids	49503	0.087	Cleveland	44130	0.197	**Vermont**		
Atlanta	30314	0.258	Kalamazoo	49001	0.116	Columbus	43217	0.164	Burlington	05401	0.446
Augusta	30904	0.419	Lansing	48910	0.109	Dayton	45440	0.206	**Washington**		
Columbia	31907	0.169	**Minnesota**			Springfield	45502	0.216	Seattle	98108	1.51
Savannah	31404	0.402	Duluth	55803	0.056	Toledo	43608	0.171	Spokane	99201	0.315
Iowa			Minneapolis	55422	0.057	Youngstown	44515	0.163	Tacoma	98402	1.23
Council Bluffs	41011	0.186	Rochester	55901	0.055	**Oklahoma**			**Washington, D.C.**		
Davenport	52803	0.130	St. Paul	55111	0.056	Oklahoma City	73145	0.339	Washington	20002	0.178
Des Moines	50310	0.073	**Missouri**			Tulsa	74120	0.160	**Wisconsin**		
Idaho			Carthage	64836	0.149	**Oregon**			Green Bay	54302	0.066
Boise	83705	0.344	Columbia	65202	0.178	Portland	97222	1.04	Kenosha	53140	0.133
Pocatello	83201	0.553	Jefferson City	65109	0.207	Salem	97301	0.929	Madison	53714	0.114
Illinois			Joplin	64801	0.138	**Pennsylvania**			Milwaukee	53221	0.120
Chicago	60620	0.190	Kansas City	64108	0.122	Allentown	18104	0.289	Racine	53402	0.124
Moline	61265	0.135	Springfield	65801	0.120	Bethlehem	18015	0.304	Superior	54880	0.055
Peoria	61605	0.174	St. Joseph	64501	0.120	Erie	16511	0.164	**West Virginia**		
Rock Island	61201	0.131	St. Louis	63166	0.586	Harrisburg	17111	0.224	Charleston	25303	0.206
Rockford	61108	0.170	**Mississippi**			Philadelphia	19125	0.326	Huntington	25704	0.221
Springfield	62703	0.263	Jackson	39211	0.191	Pittsburgh	15235	0.129	**Wyoming**		
Indiana			**Montana**			Reading	19610	0.293	Casper	82601	0.341
Evansville	47712	0.754	Billings	59101	0.134	Scranton	18504	0.232	Cheyenne	82001	0.183

*Nominal values based on ZIP codes. See www.usgs.gov for calculator to check actual S_s using latitude and longitude for best results.

Once the overall seismic forces F_p and F_{pv} have been determined (as indicated in the previous section or per the local code requirement), the loads at the restraint points can be determined. There are many different valid methods that can be used to determine these loads, but this section suggests a couple of simple approaches.

Under some instances (particularly those relating to life-support issues in hospital settings), newer code requirements indicate that critical equipment must be seismically qualified to ensure its continued operation during and after a seismic event. Special care must be taken in these situations to ensure that equipment has been shaker

Table 5 S_s Numbers for Selected International Locations (U.S. COE 1998)

Country	City	S_s
Africa		
Algeria	Alger	1.24
	Oran	1.24
Angola	Luanda	0.06
Benin	Colonou	0.06
Botswana	Gaborone	0.06
Burkina Faso	Ougadougou	0.06
Burundi	Bujumbura	1.24
Cameroon	Douala	0.06
	Yaounde	0.06
Cape Verde	Praia	0.06
Central African Republic		
	Bangui	0.06
Chad	Ndjamena	0.06
Congo	Brazaville	0.06
Djibouti	Djibouti	1.24
Egypt	Alexandria	0.62
	Cairo	0.62
	Port Said	0.62
Equatorial Guinea		
	Malabo	0.06
Ethiopia	Addis Ababa	1.24
	Asmara	1.24
Gabon	Libreville	0.06
Gambia	Banjul	0.06
Ghana	Accra	1.24
Guinea	Bissau	0.31
	Conakry	0.06
Ivory Coast	Abidjan	0.06
Kenya	Nairobi	0.62
Lesotho	Maseru	0.62
Liberia	Monrovia	0.31
Libya	Tripoli	0.62
	Wheelus AFB	0.62
Madagascar	Tananarive	0.06
Malawi	Blantyre	1.24
	Lilongwe	1.24
	Zomba	1.24
Mali	Bamako	0.06
Mauritania	Nouakchott	0.06
Mauritius	Port Louis	0.06
Morocco	Casablanca	0.62
	Port Lyautey	0.31
	Rabat	0.62
	Tangier	1.24
Mozambique	Maputo	0.62
Niger	Niamey	0.06
Nigeria	Ibadan	0.06
	Kaduna	0.06
	Lagos	0.06
Rwanda	Kigali	1.24
Senegal	Dakar	0.06
Seychelles	Victoria	0.06
Sierra Leone	Freetown	0.06
Somalia	Mogadishu	0.06
South Africa	Cape Town	1.24
	Durban	0.62
	Johannesburg	0.62
	Natal	0.31
	Pretoria	0.62
Swaziland	Mbabane	0.62
Tanzania	Dar es Salaam	0.62
	Zanzibar	0.62
Togo	Lome	0.31
Tunisia	Tunis	1.24
Uganda	Kampaia	0.62
Zaire	Bukavu	1.24
	Kinshasa	0.06
	Lubumbashi	0.62
Zambia	Lusaka	0.62
Zimbabwe	Harare (Sallsbury)	1.24
Asia		
Afghanistan	Kabul	1.65
Bahrain	Manama	0.06
Bangladesh	Dacca	1.24
Brunei	Bandar Seri	
	Begawan	0.31
China	Canton	0.62
	Chengdu	1.24
	Hong Kong	0.62
	Nanking	0.62
	Peking	1.65
	Shanghai	0.62
	Shengyang	1.65
	Tibwa	1.65
	Tsingtao	1.24
	Wuhan	0.62
Cyprus	Nicosia	1.24
India	Bombay	1.24
	Calcutta	0.62
	Madras	0.31
	New Delhi	1.24
Indonesia	Bandung	1.65
	Jakarta	1.65
	Medan	1.24
	Surabaya	1.65
Iran	Isfahan	1.24
	Shiraz	1.24
	Tabriz	1.65
	Tehran	1.65
Iraq	Baghdad	1.24
	Basra	0.31
Israel	Haifa	1.24
	Jerusalem	1.24
	Tel Aviv	1.24
Japan	Fukuoka	1.24
	Itazuke AFB	1.24
	Misawa AFB	1.24
	Naha, Okinawa	1.65
	Osaka/Kobe	1.65
	Sapporo	1.24
	Tokyo	1.65
	Wakkanai	1.24
	Yokohama	1.65
	Yokota	1.65
Jordan	Amman	1.24
Korea	Kwangju	0.31
	Kimhae	0.31
	Pusan	0.31
	Seoul	0.06
Kuwait	Kuwait	0.31
Laos	Vientiane	0.31
Lebanon	Beirut	1.24
Malaysia	Kuala Lumpur	0.31
Myanmar	Mandalay	1.24
	Rangoon	1.24
Nepal	Kathmandu	1.65
Oman	Muscat	0.62
Pakistan	Islamabad	1.68
	Karachi	1.65
	Lahore	0.62
	Peshawar	1.65
Qatar	Doha	0.06
Saudi Arabia	Al Badi	0.31
	Dhahran	0.31
	Jiddah	1.24
	Khamis Mushayt	0.31
	Riyadh	0.06
Singapore	All	0.31
South Yemen	Aden City	1.24
Sri Lanka	Colombo	0.06
Syria	Aleppo	1.24
	Damascus	1.24
Taiwan	All	1.65
Thailand	Bangkok	0.31
	Chinmg Mai	0.62
	Songkhia	0.06
	Udom	0.31
Turkey	Adana	0.62
	Ankara	0.62
	Istanbul	1.65
	Izmir	1.65
	Karamursel	1.24
United Arab Emirates		
	Abu Dhabi	0.06
	Dubai	0.06
Viet Nam		
	Ho Chi Minh City (Saigon)	0.06
Yemen	Sanaa	1.24
Atlantic Ocean Area		
Azorea	All	0.62
Bermuda	All	0.31
Caribbean Sea		
Bahama Islands	All	0.31
Cuba	All	0.62
Dominican Republic		
	Santo Domingo	1.24
French West Indies		
	Martinique	1.24
Grenada	Saint Georges	1.24
Haiti	Port au Prince	1.24
Jamaica	Kingston	1.24
Leeward Islands	All	1.24
Puerto Rico	All	0.83
Trinidad & Tobago		
	All	1.24
Central America		
Belize	Belmopan	0.62
Canal Zone	All	0.62
Costa Rica	San Jose	1.24
El Salvador	San Salvador	1.65
Guatemala	Guatemala	1.65
Honduras	Tegucigalpa	1.24
Mexico	Ciudad Juarez	0.62
	Guadalajara	1.24
	Hermosillo	1.24
	Matamoros	0.06
	Mazatlan	0.60
	Merida	0.06
	Mexico City	1.24
	Monterrey	0.06
	Nuevo Laredo	0.06
	Tijuana	1.24
Nicaragua	Managua	1.65
Panama	Colon	1.24
	Galeta	0.83
Europe		
Albania	Tirana	1.24
Austria	Salzburg	0.62
	Vienna	0.62
Belgium	Antwerp	0.31
	Brussels	0.62
Bulgaria	Sofia	1.24
Croatia	Zagreb	1.24
Czech Republic	Bratislava	0.62
	Prague	0.31
Denmark	Copenhagen	0.31
Finland	Helsinki	0.31
France	Bordeaux	0.62
	Lyon	0.31
	Marseille	1.24
	Nice	1.24
	Strasbourg	0.62
Germany	Berlin	0.06
	Bonn	0.62
	Bremen	0.06
	Dusseldorf	0.31
	Frankfurt	0.62
	Hamburg	0.06
	Munich	0.31
	Stuttgart	0.62
	Vaihingen	0.62
Greece	Athens	1.24
	Kavalla	1.65
	Makri	1.65
	Rhodes	1.24
	Sauda Bay	1.65
	Thessaloniki	1.65
Hungary	Budapest	0.62
Iceland	Keflavik	1.24
	Reykjavik	1.65
Ireland	Dublin	0.06
Italy	Aviano AFB	1.24
	Brindisi	0.06
	Florence	1.24
	Genoa	1.24
	Milan	0.62
	Naples	1.24
	Palermo	1.24
	Rome	0.62
	Sicily	1.24
	Trieste	1.24
	Turin	0.62
Luxembourg	Luxembourg	0.31
Malta	Valletta	0.62
Netherlands	All	0.06
Norway	Oslo	0.62
Poland	Krakow	0.62
	Poznan	0.31
	Waraszawa	0.31
Portugal	Lisbon	1.65
	Oporto	1.24
Romania	Bucharest	1.24
Russia	Moscow	0.06
	St. Petersburg	0.06
Serbia	Belgrade	0.62
Spain	Barcelona	0.62
	Bilbao	0.62
	Madrid	0.06
	Rota	0.62
	Seville	0.62
Sweden	Goteborg	0.62
	Stockholm	0.31
Switzerland	Bern	0.62
	Geneva	0.31
	Zurich	0.62
Ukraine	Kiev	0.06
United Kingdom	Belfast	0.06
	Edinburgh	0.31
	Glasgow/Renfrew	0.31
	Hamilton	0.31
	Liverpool	0.31
	London	0.62
	Londonderry	0.31
	Thurso	0.31
North America		
Greenland	All	0.31
Canada	Argentia NAS	0.62
	Calgary, AB	0.31
	Churchill, MB	0.06
	Cold Lake, AB	0.31
	Edmonton, AB	0.31
	E. Harmon, AFB	0.62
	Fort Williams, ON	0.06
	Frobisher, NT	0.06
	Goose Airport	0.31
	Halifax, NS	0.31
	Montreal, QC	1.24
	Ottawa, ON	0.62
	St. John's, NL	1.24
	Toronto, ON	0.31
	Vancouver, BC	1.24
	Winnipeg, MB	0.31
South America		
Argentina	Buenos Aires	0.25
Brazil	Belem	0.06
	Belo Horizonte	0.06
	Brasilia	0.06
	Manaus	0.06
	Porto Alegre	0.06
	Recife	0.06
	Rio de Janeiro	0.06
	Salvador	0.06
	Sao Paulo	0.31
Bolivia	La Paz	1.24
	Santa Cruz	0.31
Chile	Santiago	1.65
	Valparaiso	1.65
Colombia	Bogota	1.24
Ecuador	Quito	1.65
	Guayaquil	1.24
Paraguay	Asuncion	0.06
Peru	Lima	1.65
	Piura	1.65
Uruguay	Montevideo	0.06
Venezuela	Maracaibo	0.62
	Caracas	1.65
Pacific Ocean Area		
Australia	Brisbane	0.31
	Canberra	0.31
	Melbourne	0.31
	Perth	0.31
	Sydney	0.31
Caroline Islands	Koror, Palau	0.62
	Ponape	0.06
Fiji	Suva	1.24
Johnson Island	All	0.31
Mariana Islands	Guam	1.24
	Saipan	1.24
	Tinian	1.24
Marshall Islands	All	0.31
New Zealand	Auckland	1.24
	Wellington	1.65
Papau New Guinea		
	Port Moresby	1.24
Phillippine Islands	Cebu	1.65
	Manila	1.65
	Bagulo	1.24
Samoa	All	1.24
Wake Island	All	0.06

Table 6 Load Combinations

(Equation Numbers as Referenced in IBC)

ASD	LRFD
5. $(1.0 + 0.14S_{DS})D + H + F + 0.7\rho Q_E$	5. $(1.2 + 0.2S_{DS})D + \rho Q_E + L + 0.2S$
8. $(0.6 + 0.14S_{DS})D + 0.7\rho Q_E + H$	7. $(0.9 - 0.2S_{DS})D + \rho Q_E + 1.6H$

tested or otherwise certified to meet the maximum anticipated seismic load. Table 6 illustrates some load combination calculations.

COMPUTATION OF LOADS AT BUILDING CONNECTION

ASCE *Standard* 7 is based on **load- and resistance-factor design (LRFD)**. In the past, building codes have been based on **allowable stress design (ASD)**. Both are allowed for seismic restraint design. Load factors and load combinations that must be considered in design are defined in Chapter 2 of ASCE *Standard* 7. If a component is anchored with post-installed anchors, the design is usually accomplished using provisions of LRFD.

The Load Combinations of Section 2.4.1 of ASCE *Standard* 7 must be considered in the design. Generally, for rigidly mounted components, Combination 8 is the critical combination to be considered.

The forces of the restraint holding the equipment in position include shear and tensile forces. It is important to determine the number of bolts that are affected by the earthquake forces. The direction of the lateral force should be evaluated in both horizontal directions, as shown in Figure 1. All bolts or as few as a single bolt may be affected.

Simple Case

Figure 1 shows a rigid floor-mount installation of a piece of equipment with the center of gravity at the approximate center of the restraint pattern. To calculate the shear force, the sum of the forces in the horizontal plane is

$$0 = F_p - V \qquad (6)$$

The equipment shown in Figure 1 has two bolts on each side, so that four bolts are in shear. Using a single-axis moment equation to calculate the tension force, the sum of the moments for overturning results in an overturning moment (OTM) and resisting moment (RM).

For Figure 1, two bolts are in tension. The effective tension force T_{eff}, where overturning affects only one side. See Example 1 for applications of the OTM and RM. See ASCE *Standard* 7 for load combinations that adjust the D (dead load) and E (earthquake load). Shear and tension forces V and T should be calculated independently for both axes, as shown in the front and side views. See the examples for complete analysis.

General Case

The classic method used to distribute seismic loads equally distributes lateral loads among the restraints and then modifies these loads as a function of the weight eccentricity. Worst-case weight, vertical seismic load, and overturning components are combined to determine a maximum vertical load component. This **polar method** is in common use and works well for most applications.

A second, **lump mass method**, proportions the restraint loads based on the equipment weight and distribution. When working with larger seismic forces or unstable equipment, this offers the option of more evenly distributing the seismic load, reducing anchor size and peak restraint requirements. Eccentric center of gravity (cg) loads are not required to be carried out to the corner restraints as in the polar method; this technique deemphasizes stresses in the equipment frame and is more suitable for nonrigid equipment types. Eccentric loads can be addressed with either the polar method or the lump mass method.

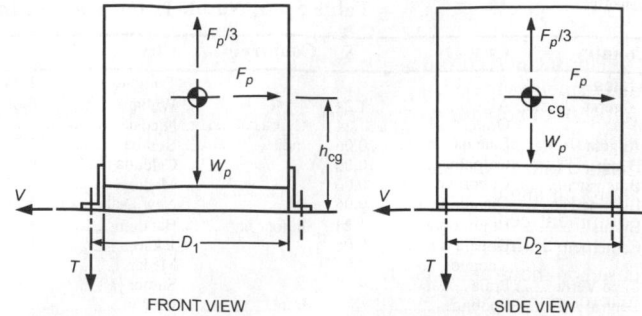

Fig. 1 Equipment with Rigidly Mounted Structural Bases

Note: Although only two methods of computing forces for more general equipment cases are illustrated here, there are many other valid methods that can be used to distribute the restraint forces. It is important that any method used include the ability to account for equipment weight, seismic uplift forces, overturning forces, and an offset center of gravity within the equipment.

Polar Method

Lateral forces are equally distributed among the restraints. If the equipment's center of gravity does not coincide with its geometric center, a rotational factor is added to account for the imbalance. This factor is determined in three steps. First, compute the true chord length in the horizontal plane between the equipment's center of gravity and the restraints' geometric center. Second, multiply the equipment total seismic lateral force by this length (to obtain a rotational moment). Third, divide this figure by the number of moment-resisting restraints times their distance from the geometric center. (The moment-resisting restraints are those farthest and equally spaced from the geometric center.) The resulting load can then be added to the original (balanced) figure. This method transfers all imbalance loads to the corner restraints and provides a valid method of restraint as long as the equipment acts as a rigid body. The assumption that a piece of equipment can transfer these loads out to the corners becomes less accurate as the equipment becomes less rigid.

Calculation of the tensile/compressive forces at the restraints is more complex than that for determining the shear loads, and must include weight, vertical seismic force, overturning forces, and (if isolated) the type of isolator/restraint system used. The total tensile and compressive forces are the worst-case summation of each of these components. For clarity, each component is addressed here as a separate entity.

The nominal weight component at each restraint is simply the total operating weight divided by the number of restraints. The vertical seismic force is simply the weight component at each location multiplied by the vertical seismic force factor in terms of the total F_{pv} load expressed in gs, the gravitational constant (F_{pv}/W_p, where F_{pv} is the vertical seismic load component as defined by the code and W_p is the total operating weight of the equipment). This can be directed either upward or downward when summing forces.

Lump Mass Method

In the lump mass method, the total equipment weight is distributed among the restraints in a manner that reflects the equipment's actual weight distribution. There are many methods of determining the distribution analytically or by testing, although they are not addressed in this section. Frequently, a weight distribution can be obtained from the equipment manufacturer.

Once the static point loads are obtained or computed for each restraint location, they can be multiplied by the lateral seismic acceleration factor (F_p/W_p) to determine lateral forces at each restraint point. Thus, if the weight at each restraint point is W_n, then

$$V_{eff} = (F_p/W_p)W_n \qquad (7)$$

This method considers the loads at all the restraints individually and computes the overturning forces for each in 1° increments for a full 360° of possible seismic wave front angle; it is only practical to perform using a spreadsheet. The total lateral seismic force F_p is divided into x- and y-axis components for each possible wave front approach angle. These forces are multiplied by the height of the equipment center of gravity above the point of restraint h_{cg}. The resulting moments are then resolved into forces at each restraint based on the x- and y-axis moment arms associated with the particular restraint location and the proportion of the load that it will bear.

Resilient Support Factors

If the equipment being restrained is isolated, the following three factors must be considered:

- For all forces that are not directed along the principal axes, only the corner restraints can be considered to be effective. Thus, for either distribution method, only the corner restraints can be considered capable of absorbing vertical loads.
- If the restraints are independent (separate entities) from the spring isolation elements and if, when exposed to uplift loads, vertical spring forces are not absorbed within the housing of an integral isolator/restraint assembly, the weight factor determined in the first step of the vertical load analysis should be ignored. (This is because any effect that a weight reduction has on the attachment hardware forces is replaced by an approximately equal vertical force component from the spring.)
- If the gap in the restraint element exceeds 1/4 in., the final computed forces must be doubled per the IBC.

Building Attachment

The common attachment arrangements are directly bolting with steel bolts and lag bolts, welding, or anchoring to concrete using post-installed anchors. To evaluate the combined effective tensile and shear forces that act simultaneously on these connections, a separate analysis is required.

If allowable stress design (ASD) data are used to size hardware for through-bolted connections for the IBC codes, which are strength based, the loads may be reduced by a factor of 1.4. If LRFD is used when selecting for the hardware, the 1.4 factor does not apply. All allowable capacities used for concrete post-installed anchor bolts selection should be drawn from ICC-ES test reports. These values reflect test data on a single anchor and should be derated for applications where embedment, edge distances, spacing, or location vary from the test conditions. Anchor manufacturers may have selection software to determine anchor bolt capacities that consider installation conditions. It should also be noted that the values published in ES reports may be either ASD or LRFD values and may need to be converted for compatibility with the (LRFD) IBC code being used.

ANSI STEEL BOLTS

For direct attachment with through bolts using ASD criteria, the design capacity of the attachment hardware should be based on criteria established in the American Institute of Steel Construction (AISC) manual. Based on the use of A307 bolts, the basic formula for computing allowable tensile stress when shear stresses are present is

$$T_{allow} = 26{,}000 - 1.8S_v \qquad (8)$$

where S_v is the shear stress in the bolt in psi. T_{allow}, the maximum allowable tensile stress, must not exceed 20,000 psi.

However, because these stresses are appropriate for dead- plus live-load combinations, they can be appropriately inflated by 1.33

when allowable stress design provisions are used and when they are used to resist wind and seismic loads as well. Peak bolt loads are based on the maximum permitted stress multiplied by the nominal bolt area.

LAG SCREWS INTO TIMBER

Acceptable loads for lag screws into timber can be obtained from the *National Design Specification®* (NDS®) *for Wood Construction* (AWC 2005). Selected fasteners must be secured to solid lumber, not to plywood or other similar material. Withdrawal force design values are a function of the screw size, penetration depth, and wood density and can be increased by a factor of 1.6 for short-term seismic or wind loads. Table 9.2A in the NDS identifies withdrawal forces on a force/embedment depth basis. Note that the values published in this table are capacities in both ASD and LRFD. In addition, NDS Table 9.4.2 introduces deration factors for reduced edge distance and bolt spacing.

In timber construction, the interaction formula given in Equation (8) does not apply. Instead, per Section 9.3.5 of the NDS, the equation is

$$Z_a' = (W'p)Z'/[(W'p)\cos^2\alpha + Z'\sin^2\alpha] \qquad (9)$$

where

Z' = shear capacity drawn from Table 9.3A
W' = side grain withdrawal force = $1800G^{3/2}D^{3/4}$
G = specific gravity of the timber
D = diameter
p = embedment depth of screw
α = angle of composite force measured flat with surface of timber

CONCRETE POST-INSTALLED ANCHOR BOLTS

Capacities are manufacturer/anchor-type specific. Capacity data should be obtained from the anchor's current ES report. Where failure of the steel does not govern the tensile load, strength-based design (LRFD) should be used. Obtain anchor information from the anchor ICC-ES (formerly ICBO-ESR) report based on anchor and installation factors. For groups of anchors, special factors are required and American Concrete Institute *Standard* 318-08 should be consulted.

ASD Applications

Interaction Formula. To evaluate the combined effective tension and shear forces that act simultaneously on the bolt, use the either of the following equations:

$$(T_{eff}/T_{allow\,ASD})^{5/3} + (V_{eff}/V_{allow\,ASD})^{5/3} \leq 1.0 \qquad (10)$$

or

$$(T_{eff}/T_{allow\,ASD}) + (V_{eff}/V_{allow\,ASD}) \leq 1.2 \qquad (11)$$

However, if $T_{eff} \leq 0.2$; $T_{allow\,ASD}$ the full T_{eff} can equal $T_{allow\,ASD}$, or if $V_{eff} \leq 0.2$; $V_{allow\,ASD}$ the full V_{eff} can equal $V_{allow\,ASD}$.

LRFD Applications

The engineer must select an anchor for use from a current evaluation report for anchors that satisfy provisions of ACI 318 Appendix D or ACI 355 (the provisions are the same). From ACI 318, the capacity of the anchor must be reduced in accordance with the following:

$$T = 0.75\phi N \qquad (12)$$

$$V = 0.75\phi V \qquad (13)$$

The interaction equation for LFRD is modified as follows:

$$(T_{eff}/T_{allow}) + (V_{eff}/V_{allow}) \leq 1.2 \qquad (14)$$

Types of Concrete Post-Installed Anchors

Several types of anchor bolts for insertion in concrete are manufactured. Wedge and undercut anchors perform better than self-drilling, sleeve, or drop-in types. Adhesive anchors are stronger than other anchors, but lose their strength at elevated temperatures (e.g., on rooftops and in areas damaged by fire).

Wedge anchors have a wedge on the end with a small clip around the wedge. After a hole is drilled, the bolt is inserted and the external nut tightened. The wedge expands the small clip, which bites into the concrete.

Undercut anchors expand to seat against a shoulder cut in the bottom of the anchor hole. Although these have the highest capacity of commonly available anchor types, the cost of the extra operation to cut the shoulder in the hole greatly limits the frequency of their use in the field.

A **self-drilling anchor** is basically a hollow drill bit. The anchor is used to drill the hole and is then removed. A wedge is then inserted on the end of the anchor, and the assembly is drilled back into place; the drill twists the assembly fully in place. The self-drilling anchor is heavily affected by the skill of the craft and usually not rated for seismic applications.

Drop-in expansion anchors are hollow cylinders with a tapered end. After they are inserted in a hole, a small rod is driven through the hollow portion, expanding the tapered end. These anchors are only for shallow installations because they have no reserve expansion capacity. These anchors are usually not rated for seismic applications.

A **sleeve anchor** is a bolt covered by a threaded, thin-wall, split tube. As the bolt is tightened, the thin wall expands. Additional load tends to further expand the thin wall. The bolt must be properly preloaded or friction force will not develop the required holding force. These anchors are typically not used in seismic applications because of the limited reserve capacity.

Large screw anchors are one-piece anchors that have a concrete cutting thread. These anchors were initially designed to be installed without a specified torque, but torque is used to ensure contact at the rated embedment

Adhesive anchors may be in glass capsules or installed with various tools. Pure epoxy, polyester, or vinyl ester resin adhesives are used with a threaded rod supplied by the contractor or the adhesive manufacturer. Some adhesives have a problem with shrinkage; others are degraded by heat. However, some adhesives have been tested without protection to 1100°F before they fail (all mechanical anchors will fail at this temperature). Where required, or if there is a concern, anchors should be protected with fire retardants similar to those applied to steel decks in high-rise buildings.

The manufacturer's instructions for installing the anchor bolts should be followed. ES reports have further information on allowable forces for design. Use a safety factor of 2 or as required by ES reports if the installation has not been inspected as required by the IBC Chapter 17 on special inspection.

Stainless steel anchors are required for use in outdoor applications noted in the latest versions of the IBC code.

WELD CAPACITIES

Weld capacities may be calculated to determine the size of welds needed to attach equipment to a steel plate or to evaluate raised support legs and attachments. A static analysis provides the effective tension and shear forces. The capacity of a weld is given per unit length of weld based on the shear strength of the weld material. For steel welds, the allowable shear strength capacity is 16,000 psi on the throat section of the weld. The section length is 0.707 times the specified weld size.

For a 1/16 in. weld, the length of shear in the weld is 0.707 × 1/16 = 0.0442 in. The allowable weld force $(F_w)_{allow}$ for a 1/16 in. weld is

$$(F_w)_{allow} = 0.0442 \times 16{,}000 = 700 \text{ lb per inch of weld} \qquad (15)$$

For a 1/8 in. weld, the capacity is 1400 lb/in.

The effective weld force is the sum of the vectors calculated in terms of effective shear and tension shall be reduced. Because the vectors are perpendicular, they are added by the method of the square root of the sum of the squares (SRSS), or

$$(F_w)_{allow} = \sqrt{(T_{eff})^2 + (V_{eff})^2} \qquad (16)$$

The length of weld required is given by the following equation:

$$\text{Weld length} = (F_w)_{eff}/(F_w)_{allow} \qquad (17)$$

SEISMIC SNUBBERS

Several types of snubbers are manufactured or field fabricated. All snubber assemblies should meet the following minimum requirements to avoid imparting excessive accelerations to HVAC&R equipment:

- Impact surface should have a high-quality elastomeric surface that is not cemented in place.
- Resilient material should be easy to inspect for damage and be replaceable.
- Snubber system must provide restraint in all directions.
- Snubber capacity should be verified either through test or by analysis and should be certified by an independent, registered engineer to avoid serious design flaws.

Typical snubbers are classified as Types A through J (Figure 2).

Type A. Snubber built into a resilient mounting. All-directional, molded bridge-bearing quality neoprene element is a minimum of 1/8 in. thick.

Type B. Isolator/restraint. Stable isolation spring bears on the base plate of the fixed restraining member. Earthquake motion of isolated equipment is restrained close to the base plate, minimizing pullout force to the base plate anchorage.

Type C. Spring isolator with built-in all-directional restraints. Restraints have molded neoprene elements with a minimum thickness of 1/8 in. A neoprene sound pad should be installed between the spring and base plate. Sound pads below the base plate are not recommended for seismic installations.

Type D. Integral all-directional snubber/restrained spring isolator with neoprene element.

Type E. Fully bonded neoprene mount capable of withstanding seismic loads in all directions with no metal-to-metal contact.

Type F. All-directional three-axis snubber with neoprene element. Neoprene element of bridge-bearing quality is a minimum of 3/16 in. thick. Snubber must have a minimum of two anchor bolt holes.

Type G. Lateral snubber. Neoprene element is a minimum of 1/4 in. thick. Upper bracket is welded to the equipment.

Type H. Restraint for floor-mounted equipment consisting of interlocking steel assemblies lined with resilient elastomer. Bolted to equipment and anchored to structure through slotted holes to allow field adjustment. After final adjustment, weld anchor to floor bracket and weld angle clip to equipment or, alternatively, fill slots with adhesive grout to prevent slip.

Type I. Single-axis, single-direction lateral snubber. Neoprene element is a minimum of 1/4 in. thick. Minimum floor mounting is with two anchor bolts. Must be used with a minimum of eight, two per corner.

Type J. A telescopic snubber for floor-mounted equipment, with a molded neoprene element, designed to distribute the seismic force to a larger surface area in a pipe section. This snubber is an

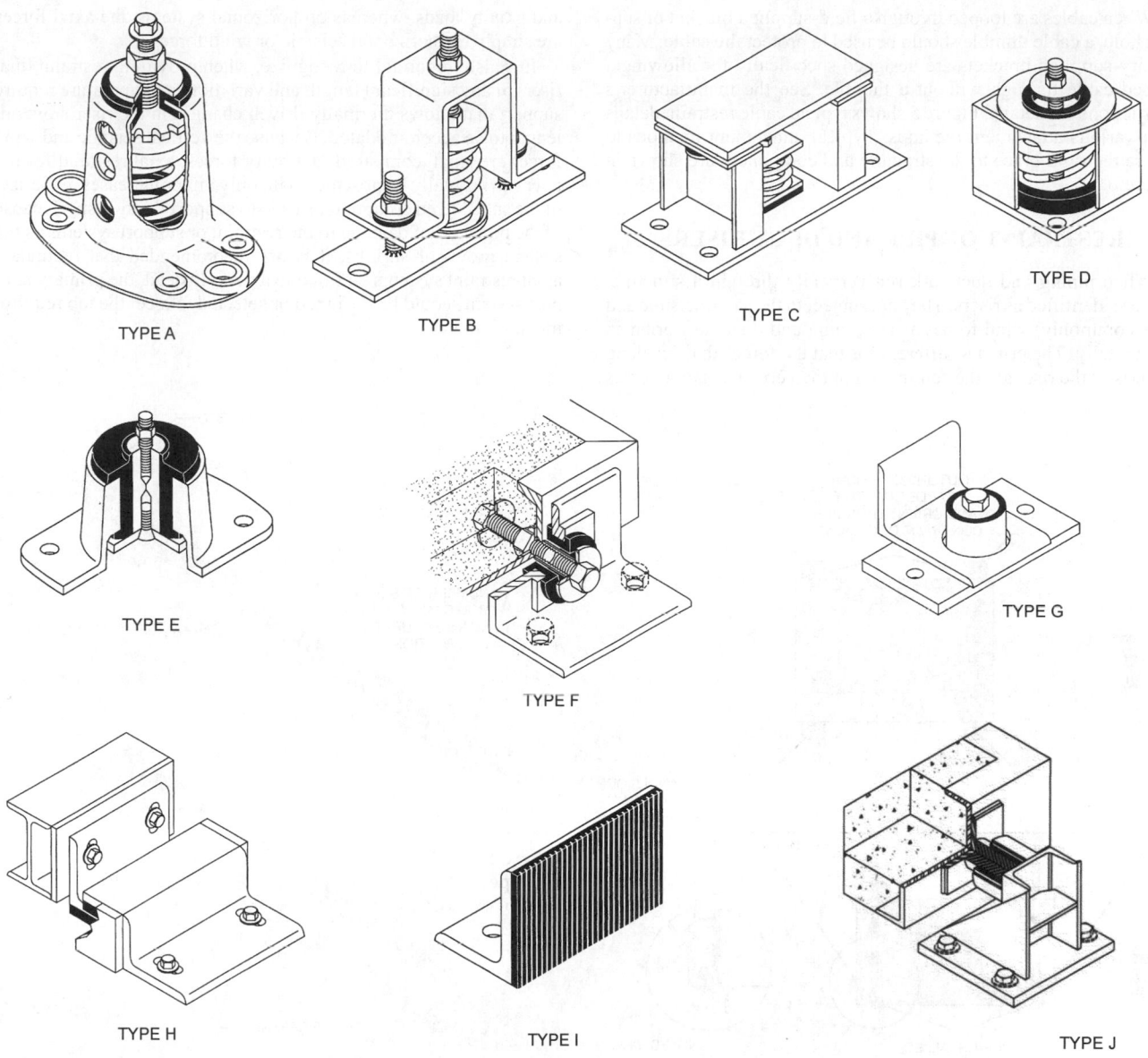

TYPE A

TYPE B

TYPE C

TYPE D

TYPE E

TYPE F

TYPE G

TYPE H

TYPE I

TYPE J

Fig. 2   Seismic Snubbers

all-directional restraint system when a minimum of four snubbers are installed. The molded neoprene element is a minimum of 1/4 in. thick and is installed with an air gap of 3/16 in. not to exceed 1/4 in.

SEISMIC RESTRAINTS

For suspended equipment, pipes, ducts, and raceways, it is necessary to restrain lateral movement resulting from seismic acceleration applied to the component. Unrestrained, suspended equipment and related systems will sway violently back and forth, impacting nearby building material and possibly overstressing the hanger rods. The overstressed rods will eventually break, and the equipment crash down. To prevent this swaying, suspended components are restrained by one of two methods: a wire rope system, or a rigid brace using steel struts, angles, or other steel elements.

Wire rope restraints are a restraint assembly for suspended equipment, piping, or ductwork consisting of high-strength, galvanized steel aircraft cable. A typical cable restraint system is shown in Figure 3. Cable should have a certified break strength. Some models are color-coded for easy field verification. Cable must be

manufactured to meet or exceed minimum materials and standard requirements. Break strengths must be per ASTM E-8 procedures. A safety factor of 2 may be used when prestretched cable is used with end connections designed to meet the cable break strength. Cables are installed to prevent excessive seismic motion and arranged so they do not engage during normal operation. Equipment suspended with vibration isolators must use cable restraints to avoid degrading the isolation. Rigid type bracing will short out the vibration isolators. To prevent buckling of the hanger rods, add a rod stiffener as shown in Figure 4.

Secure the cable to structure and to the braced component through a bracket or stake eye designed to meet the cable restraint rated capacity. Cables are typically secured using one of the following methods:

- Factory-installed permanent stake eye
- Field-looped through bracket and secured with cable grips
- Field-looped through bracket and secured with oval sleeve
- Factory brackets with integral cable clamps to secure cable

When cables are looped through a field-supplied bracket or support hole, a cable thimble should be used to protect the cable. Many factory-supplied brackets are designed specifically for allowing a looped cable through without a thimble. See the manufacturer's instructions for details. Figure 5 shows typical cable restraint details with various attachment methods. Typical attachment methods to secure the rigid brace to the structure and component are shown in Figure 6.

RESTRAINT OF PIPE AND DUCT RISERS

When piping and ductwork run vertically through a structure, they are identified as risers. They are subject to the same seismic and (less commonly) wind forces as are piping and ductwork oriented horizontally. The primary difference is that the forces that act along the axis of the riser are the summation of the vertical seismic forces and gravity loads, whereas on horizontal systems, the axial forces are simply the horizontal seismic or wind force.

It is also important to recognize, when providing restraint, that risers of any significant length and variation in temperature require support that allows thermally driven changes in the riser's overall length to be accommodated. Because the vertical seismic and wind forces are small compared to gravity forces, axial restraint for the riser can normally be provided with only minor increases in the size of the specialized components used to support the system. Because of the potential of damage to the restraint or support systems as the system grows or shrinks, it is not recommended that redundant axial restraint systems be fitted to a riser. Instead, the primary support system should be designed or selected to meet the job requirement.

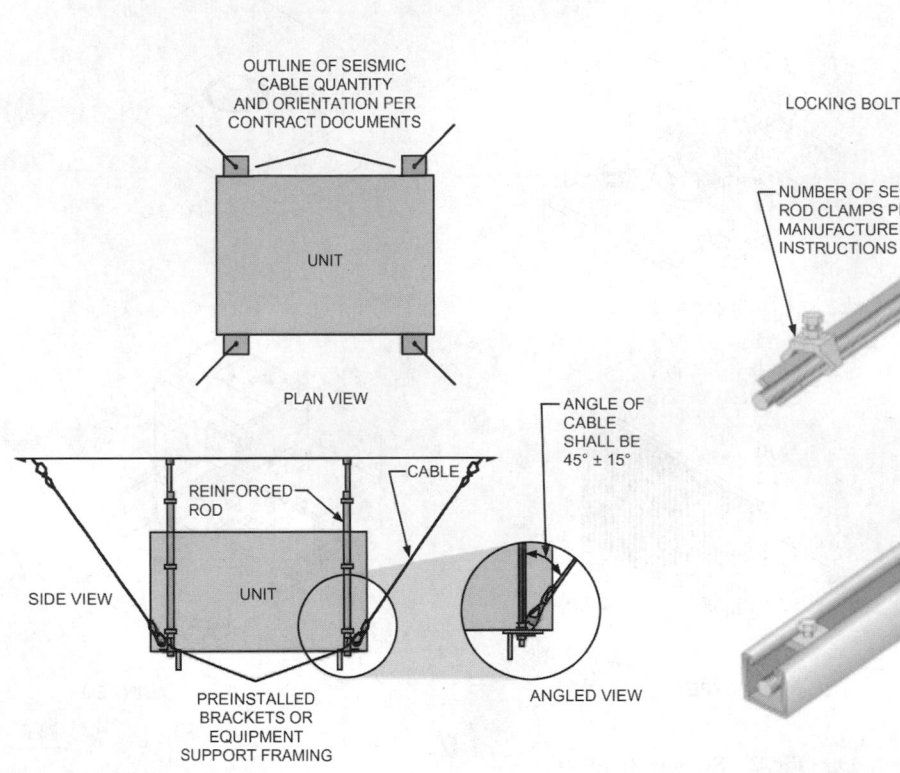

Fig. 3   Cable Restraint

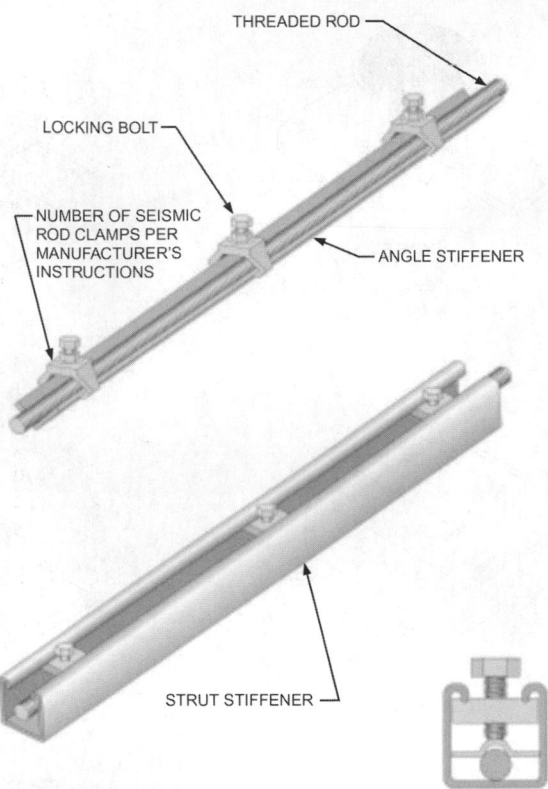

Fig. 4   Rod Stiffener

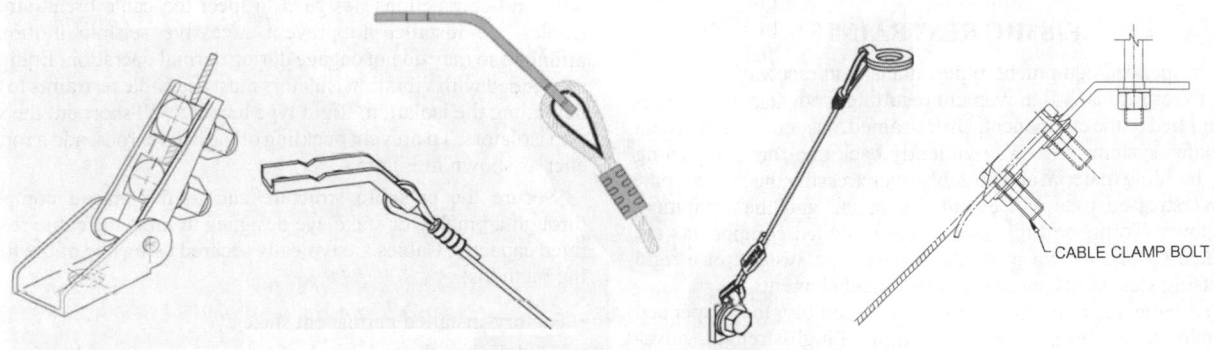

Fig. 5   Types of Cable Connections

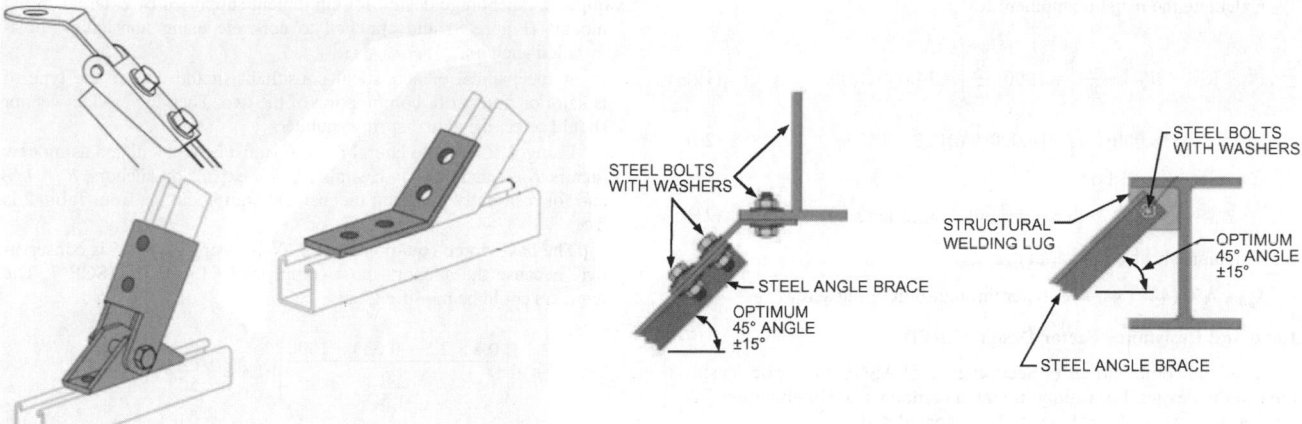

Fig. 6　Strut End Connections

Risers of significant length are also fitted with some type of stabilization devices. These can be as simple as snug-fitting holes in the floors that the risers penetrate, to specialized brackets or guiding devices that maintain the alignment of the piping or duct while still allowing it to expand or contract. As is the case with the vertical forces, the components used for guidance can frequently be used to provide resistance against seismic or wind events if they are sized and attached appropriately.

If, in the lateral load case, the components used to provide guidance are not adequate to resist the design seismic or wind load conditions, redundant, seismically qualified systems should be fitted to perform this task.

All axial and lateral restraints fitted to risers must be effective against forces that may act in any horizontal or vertical direction as applicable. In addition, the attachment hardware used must be seismically qualified components (e.g., anchors), installed in accordance with seismically qualified procedures.

EXAMPLES

The following examples are provided to assist in the design of equipment anchorage to resist seismic forces. For Examples 1 through 4, assume the provisions contained in ASCE 7-05 apply, $I_p = 1.5$, $S_s = 0.85$, site soil class is C, and the equipment is located at the top of a 50 ft building. Also include an uplift force component $F_{pv} = 0.2 S_{DS} D$ where D is the dead load for all examples. Examples 1 through 5 are solved using the polar method of analysis while Example 6 is solved by the lump mass method.

Note: These examples assume that $I_p = 1.5$. This assumes that the equipment being considered is essential to the continued function of the building following an earthquake, or contains hazardous materials. ASCE 7-05 Section 13.2.2 requires that this equipment be certified as being operable after the design earthquake.

Example 1. Anchorage design for equipment rigidly mounted to the structure (see Figure 7).

From Equations (1) to (4), calculate the lateral seismic force and its vertical component. Note that for post-installed, if the anchor satisfies the requirements of ASCE 7, Section 13.4.2, the value of R_p is the same as the component being considered. Post-installed anchors with current evaluation reports published by ICC or other agencies are deemed to be compliant with the provisions of ACI 355 or 318 Appendix D and thus satisfy Section 13.4.2. For rigidly mounted mechanical equipment (period < 0.06 s or > 16.7 Hz), a_p from Table 2 is 1.0, otherwise $a_p = 2.5$.

ASCE 7 is based on load and resistance factor design (LRFD). In the past, building codes have been based on allowable stress design (ASD). Load factors and load combinations defined in Chapter 2 of

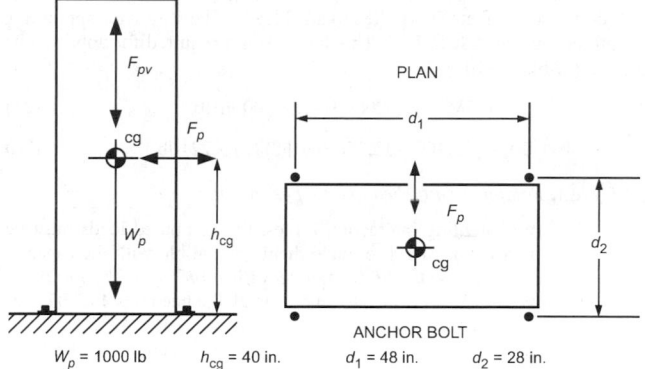

$W_p = 1000$ lb　　$h_{cg} = 40$ in.　　$d_1 = 48$ in.　　$d_2 = 28$ in.

Fig. 7　Equipment Rigidly Mounted to Structure (Example 1)

ASCE 7 must be considered in design. If a component is anchored with post-installed anchors, the design can only be accomplished using LRFD.

The first step in the load determination process is to determine S_{DS} using the following equation and $F_a = 1.1$ (from Table 3, site class C):

$$S_{DS} = 2 F_a S_s / 3 = 2 \times 1.1 \times 0.85/3 = 0.623$$

Using this value for S_{DS} Equation (1) gives

$$F_P = \left[\frac{0.4 \times 1.0 \times 0.623 \times 1000}{\frac{2.5}{1.5}} \right] \left(1 + 2 \times \frac{50}{50} \right) = 450 \text{ lb}$$

Equation (2) shows that F_p need not be greater than

$$1.6 \times 0.623 \times 1.5 \times 1000 = 1495 \text{ lb}$$

Equation (3) shows that F_p must not be less than

$$0.3 \times 0.623 \times 1.5 \times 1000 = 280 \text{ lb}$$

Therefore $F_p = 450$ lb.

When considering provisions of LRFD, a vertical acceleration component must be considered per ASCE 7, Section 12.4.2.2.

$$F_{PV} = 0.2 \times S_{DS} \times D = 0.2 \times 0.623 \times 1000 = 125 \text{ lb}$$

For Allowable Stress Design (ASD)

The load combinations of Section 2.4.1 of ASCE 7 must be considered in the design. For rigidly mounted components, Combination 8 is generally the critical combination to be considered.

Calculate the overturning moment OTM:

$$\text{OTM} = F_P h_{cg} = 450 \times 40 = 18,000 \text{ in} \cdot \text{lb} \tag{18}$$

Calculate the resisting moment RM:

$$RM = W_P\left(\frac{d_{min}}{2}\right) = 1000\left(\frac{28}{2}\right) = 14{,}000 \text{ in·lb} \tag{19}$$

$$T = [18{,}000(0.7) - 14{,}000(0.6)]/28 = 150 \text{ lb} \tag{20}$$

Calculate T_{eff} per bolt:

$$T_{eff} = 150/2 = 75 \text{ lb per through-bolt or lag screw} \tag{21}$$

Calculate shear force per bolt:

$$V_{eff} = 450/(4 \times 1.4) = 78 \text{ lb per through-bolt or lag screw} \tag{22}$$

Load and Resistance Factor Design (LRFD)

The load combinations of Section 2.3.2 of ASCE 7 must be considered in the design. For rigidly mounted components, Combination 7 is generally the critical combination to be considered.

$$RM = (W_p - F_{pv})d_{min}/2 = (1000 - 125)28/2 = 12{,}250 \text{ in·lb} \tag{23}$$

$$T_{eff} = [18{,}000 - 0.9(12{,}250)]/28(2) = 125 \text{ lb} \tag{24}$$

Per ASCE 7 Section 13.4.2, if post-installed anchors are used, an additional 1.3 factor is applied to all E loads. The 1.3 factor applies for projects using ASCE 7-05. This factor is not required for applications using ASCE 7-10.

$$OTM = 18{,}000(1.3) = 23{,}400 \text{ in·lb} \tag{25}$$

$$T_{eff} = [23{,}400 - 12{,}250(0.9)]/28(2) = 221 \text{ lb} \tag{26}$$

Case 1. *Equipment attached to a timber structure*

Before computing interaction forces, the computed loads must be reduced by a factor of 1.4 to make them compatible with the capacity data listed in the *National Design Specification® (NDS®) for Wood Construction* (AWC 1997). The lateral load V_{eff} becomes 112.5/1.4 or 80.4 lb per bolt and the pullout load T_{eff} becomes 103/1.4 = 73.5 lb per bolt. For the capacity of the connection, a resulting combined load and angle relative to the mounting surface must be computed. The combined load is

$$T\alpha_{eff} = \sqrt{(T_{eff})^2 + (V_{eff})^2} = \sqrt{(80.4)^2 + (73.5)^2} = 109 \text{ lb}$$

The angle $\alpha = \arcsin(T_{eff}/Z'_\alpha) = 42.5°$, where $Z\alpha$ is the allowable lag screw load multiplied by applicable factors and $Z'_\alpha\alpha$ is the factored allowable lag screw load at angle α from the mounting surface.

Selected fasteners must be secured to solid lumber, not to plywood or other similar material. The following calculations are made to determine whether a 1/2 in. diameter, 4 in. long lag screw in redwood will hold the required load. For this computation, it is assumed that bolt spacing, edge distance, temperature, and other factors do not reduce the bolt capacity (see NDS for further details) and that the load allowable factor for short-term wind or seismic loads is 1.6.

From Table 9.3A in the NDS, for redwood, $G = 0.37$, and Z perpendicular to the grain is 512 lb.

From Table 9.2A in the NDS, for $G = 0.37$ and 3.5 in. full thread, $W = 385 \times 3.5 = 1350$ lb.

Substituting into the combined load for lag bolts [Equation (9)] gives

$$Z'_\alpha = \frac{(385 \times 3.5)512}{(385 \times 3.5)42.5 + 512\sin^2 42.5} = 714 \text{ lb}$$

Therefore, a 1/2 in.) diameter, 4 in. long lag screw can be used at each corner of the equipment.

Case 2. *Equipment attached to steel*

For equipment attached directly to a steel member, analysis is the same as that shown in case 1. Capacities for the attaching bolts are given in the *Manual of Steel Construction* (AISC 1989). See Chapter J of the AISC Specification for design provisions.

For this example $T_{eff}/T_{ASD} = 125/4410 = 0.02 < 0.2$; therefore a combined tension shear check need not be performed on the connection. Therefore, 1/2 in. diameter bolts can be used.

Example 2. Anchorage design for equipment supported by external spring mounts (Figure 8) and attached to concrete using nonshallow post-installed anchors.

A mechanical or acoustical consultant should choose the type of isolator or snubber or combination of the two. Then the product vendor should select the actual spring snubber.

Using ASCE 7, the lateral force F_p must be recalculated using new factors. S_{DS} remains as in Example 1. For expansion anchors, $R_p = 1.5$, and for resiliently mounted mechanical equipment, a_p from Table 2 is 2.5.

The basic force equation is then (*Note*: using $R_p = 1.5$ is conservative, because the anchors must comply with 13.4.2 of ASCE 7. The numbers could be modified)

$$F_P = \left(\frac{0.4 \times 2.5 \times 0.623 \times 1000}{\frac{1.5}{1.5}}\right)\left(1 + 2 \times \frac{50}{50}\right) = 1869 \text{ lb}$$

Equation (2) indicates that F_p need not be greater than

$$1.6 \times 0.623 \times 1.5 \times 1000 = 1495 \text{ lb}$$

Equation (3) indicates that F_p must not be less than

$$0.3 \times 0.623 \times 1.5 \times 1000 = 280 \text{ lb}$$

The vertical force F_{pv} equals

$$F_{PV} = 0.2S_{DS}D = 0.2 \times 0.623 \times 1000 = 125 \text{ lb}$$

Because the equipment is resiliently supported, footnote b to Table 13.6-1 of ASCE 7 indicates that the computed forces may need to be doubled. Therefore, F_p is $1495 \times 2 = 2990$ lb and F_{pv} is $125 \times 2 = 250$ lb.

Assume that the center of gravity cg of the equipment coincides with the center of gravity of the isolator group.

If T = maximum tension on isolator and C = maximum compression on isolator, then

$$T, C = \frac{-W_P + F_{PV}}{4} \pm F_p h_{cg}\frac{\cos\theta}{2b} + F_p h_{cg}\frac{\sin\theta}{2a}$$

$$= \frac{-W_P + F_{PV}}{4} \pm \frac{F_p h_{cg}}{2}\left(\frac{\cos\theta}{2b} + \frac{\sin\theta}{2a}\right) \tag{27}$$

To find maximum T or C, set $dT/d\theta = 0$:

$$\frac{dT}{d\theta} = \frac{F_p h_{cg}}{2}\left(\frac{\cos\theta}{2b} + \frac{\sin\theta}{2a}\right) = 0 \tag{28}$$

$$\theta_{max} = \tan^{-1}(b/a) = \tan^{-1}(28/48) = 30.26° \tag{29}$$

$$T = \frac{-W_P + F_{PV}}{4} + \frac{F_p h_{cg}}{2}\left(\frac{\cos\theta_{max}}{2b} + \frac{\sin\theta_{max}}{2a}\right) \tag{30}$$

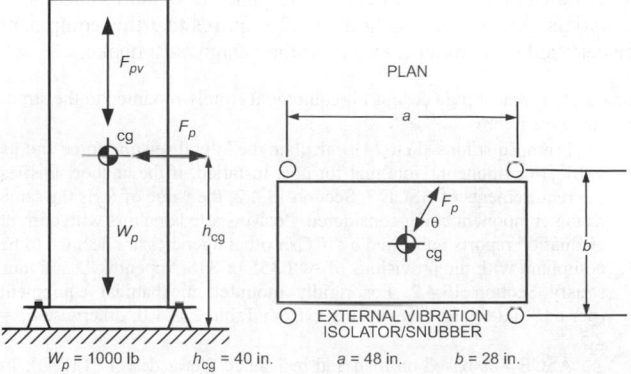

Fig. 8   **Equipment Supported by External Spring Mounts**

$$C = \frac{-W_P + F_{PV}}{4} + \frac{F_P h_{cg}}{2}\left(\frac{\cos\theta_{max}}{2b} + \frac{\sin\theta_{max}}{2a}\right) \quad (31)$$

$$T = \frac{-1000 + 250}{4} + \frac{2990 \times 40}{2}\left(\frac{\cos 30.26}{28} + \frac{\sin 30.26}{48}\right) = 2285 \text{ lb}$$

$$C = \frac{-1000 + 250}{4} + \frac{2990 \times 40}{2}\left(\frac{\cos 30.26}{28} + \frac{\sin 30.26}{48}\right) = -2660 \text{ lb}$$

Calculate the shear force per isolator:

$$V = (F_P/N_{iso}) = 2990/4 = 748 \text{ lb} \quad (32)$$

This shear force is applied at the operating height of the isolator. Uplift tension T on the vibration isolator is the worst condition for the design of the anchor bolts. The compression force C must be evaluated to check the adequacy of the structure to resist the loads.

$$(T_1)_{eff} \text{ per bolt} = T/2 = 2285/2 = 1143 \text{ lb} \quad (33)$$

The value of $(T_2)_{eff}$ per bolt due to overturning on the isolator is

$$(T_2)_{eff} = V \times \text{operating height}/dN_{bolt} \quad (34)$$

where d is the distance from edge of isolator base plate to center of bolt hole.

$$(T_2)_{eff} = (748 \times 8)/(3 \times 2) = 997 \text{ lb} \quad (35)$$

$$(T_{max})_{eff} = (T_1)_{eff} + (T_2)_{eff} = 1143 + 997 = 2140 \text{ lb} \quad (36)$$

$$V_{eff} = 748/2 = 374 \text{ lb} \quad (37)$$

See Example 1 for the design of the connections to the structural system.

Example 3. Anchorage design for equipment with a center of gravity different from that of the isolator group (Figure 10).

Anchor properties

$$I_x = 4B^2; \quad I_y = 4L^2 \quad (38)$$

Angles:

$$\theta = \tan^{-1}(B/L) \quad (39)$$

$$\alpha = \tan^{-1}(e_x/e_y) \quad (40)$$

$$\beta = 180 |\alpha - \theta| \quad (41)$$

$$\phi = \tan^{-1}(LI_x/BI_y) \quad (42)$$

Vertical reactions

$$(W_n)_{max/min} = W_P \pm F_{pv} \quad (43)$$

Vertical reaction caused by overturning moment

$$T_m = \pm F_P\left(\frac{Bh_{cg}}{I_x}\cos\theta + \frac{Lh_{cg}}{I_y}\sin\theta\right) \quad (44)$$

Vertical reaction caused by eccentricity

$$(T_e)_{max/min} = (W_n)_{max/min}\left(\frac{Be_y}{I_x} + \frac{Le_y}{I_y}\right) \quad (45)$$

Vertical reaction caused by W_p

$$(T_w)_{max/min} = (W_n)_{max/min}/4 \quad (46)$$

$$T_{eff} = T_m + (T_e)_{max} + (T_w)_{max} \text{ (always compression)} \quad (47)$$

$$T_{eff} = -T_m + (T_e)_{min} + (T_w)_{min} \text{ (tension if negative)} \quad (48)$$

Horizontal reactions

Horizontal reaction caused by rotation

$$V_{rot} = F_P\left(\frac{e_x^2 + e_y^2}{16(B^2 + L^2)}\right)^{0.5} \quad (49)$$

$$V_{dir} = F_p/4 \quad (50)$$

$$V_{max} = (V_{rot}^2 + V_{dir}^2 - 2V_{rot}V_{dir}\cos\beta)^{0.5} \quad (51)$$

See Example 1 for the design of the connections to the structural system.

The values of T_{min} and V_{max} are used to design the anchorage of the isolators and/or snubbers, and T_{max} is used to verify the structure's adequacy to resist the vertical loads.

Example 4. Anchorage design for equipment with supports and bracing for suspended equipment (Figure 11). Equipment weight $W_p = 500$ lb.

Because post-installed anchors may not withstand published allowable static loads when subjected to vibratory loads, vibration isolators should be used between the equipment and the structure to damp vibrations generated by the equipment.

Anchor properties

$$I_x = 4B^2; \quad I_y = 4L^2 \quad (52)$$

Angle

$$\phi = \tan^{-1}(LI_x/BI_y) = 36.86° \quad (53)$$

From Equation (43),

$$(W_n)_{max/min} = 500 \pm 124 = 624 \text{ lb or } 376 \text{ lb}$$

From Equation (44),

$$T_m = \pm 1122(0.132 + 0.075) = \pm 233 \text{ lb}$$

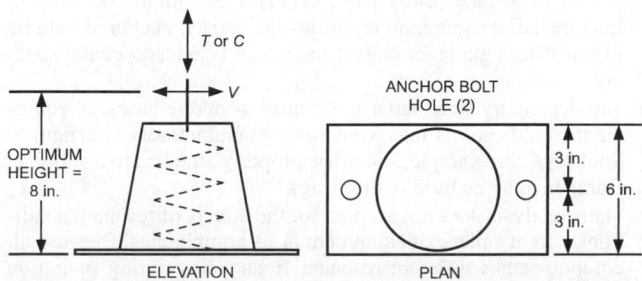

Fig. 9   Spring Mount Detail (Example 2)

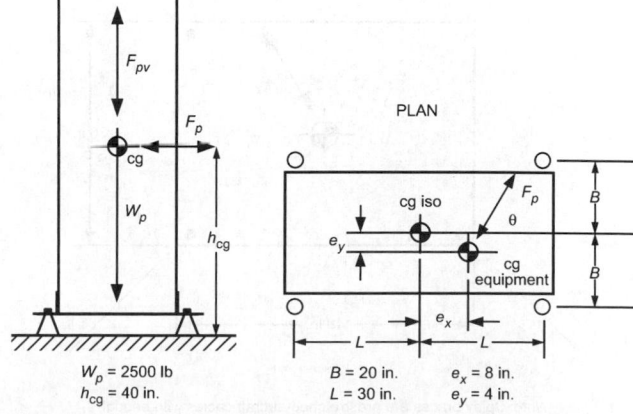

Fig. 10   Equipment with Center of Gravity Different from Isolator Group (in Plan View)

From Equation (45),

$$T_e = 0$$

From Equation (46),

$$(T_w)_{max/min} = 156 \text{ lb or } 94 \text{ lb}$$

From Equation (47),

$$(T_{eff})_{max} = 233 + 0 + 156 = 389 \text{ lb (downward)}$$

From Equation (48),

$$(T_{eff})_{max} = -233 + 0 + 94 = -139 \text{ lb (upward)}$$

Forces in the hanger rods:

Maximum tensile = 389 lb

Maximum compression = 139 lb

Force in the splay brace = $F_p \sqrt{2}$ = 1587 lb at a 1:1 slope

Because of the force being applied at the critical angle, as in Example 2, only one splay brace is effective in resisting the lateral load F_p.

Design of hanger rod/vibration isolator and connection to structure

When post-installed anchors are mounted to the underside of a concrete beam or slab, the allowable tension loads on the anchors must be reduced to account for cracking of the concrete. A general rule is to use half the allowable load. Some manufacturers have ICC reports that provide allowable values for anchors installed under the slab.

Determine whether a 1/2 in. wedge anchor with special inspection provisions will hold the required load.

$$T_{allow} = 600 \times 0.5 \times 2 = 600 \text{ lb} > T_{eff} = 389 \text{ lb}$$

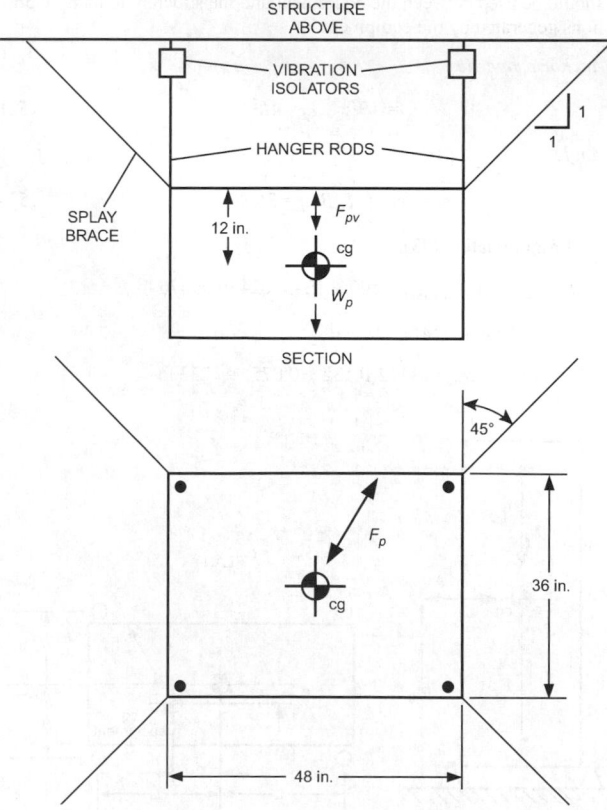

STRUCTURE ABOVE

VIBRATION ISOLATORS

HANGER RODS

SPLAY BRACE

12 in.

F_{pv}

cg

W_p

SECTION

45°

F_p

cg

36 in.

48 in.

Note: Splay braces are prestretched aircraft cables with enough slack so that isolators can fully function vertically.

Fig. 11   Supports and Bracing for Suspended Equipment

Therefore, a 1/2 in. rod and post drill-in anchor should be used at each corner of the unit.

For anchors installed without special inspection,

$$T_{allow} = 600 \times 0.5 = 300 \text{ lb} > T_{eff} = 389 \text{ lb}$$

Therefore, a larger anchor should be chosen.

Determine if the 1/2 in. hanger rod would require a stiffener if it is 36 in. long.

Design of splay brace and connection to structure

Force in the slack cable = 1587 lb

Because all of the load must be resisted by a single cable, the forces in the connection to the structure are

$$V_{max} = 1122 \text{ lb} \qquad T_{max} = F_p = 1122 \text{ lb}$$

Because the cable forces are relatively small, a 3/8 in. aircraft cable attached to clips with cable clamps should be used. The clips, in turn, may be attached to either the structure or the equipment.

The design of a post-installed anchor installation is similar to that shown in Example 1. Anchors installed through a metal deck will have lower capacities than anchors installed in a flat slab because of limited embedment depths. Take care to ensure that the design also satisfies the requirements contained in the evaluation report for the anchor specified.

Prescriptive provisions of ASCE 7 can be summarized as follows:

- Formulas for relative displacement of floor and ceiling can be conservatively estimated at 1% of the floor-to-ceiling height. This displacement must be used to determine the required horizontal flexibility of the pipe, duct, or electrical connections at the equipment interface.
- In ASCE 7, using all-directional snubbers with clearance of more than 1/4 in. increases F_p by a factor of 2.
- Component supports must be designed to accommodate component movement to prevent pounding on the structure or other components. This affects internal isolators and snubbers.
- Equipment components exposed to seismic impact forces and using nonductile housings must be designed using 25% of material yield stresses.
- Nonessential equipment, failure of which can cause essential equipment failure, must be designed as essential equipment.
- If the structure's site class is not provided in the contract documents, assume site class D, subject to change by the building official.
- For pipe or duct on any given run, if the distance from the bottom of the structure to the top of the support is 12 in. or less for all supports in that run, then that run does not need sway braces.
- Pipe and ducts may not be required to have sway braces, depending on size, material content, and importance factor. These conditions are defined in Chapter 13 of ASCE 7.

INSTALLATION PROBLEMS

The following should be considered when installing seismic restraints.

- Anchor location affects the required strengths. Concrete anchors should be located away from edges, stress joints, or existing fractures. The evaluation report for the chosen anchor should be followed as a guide for edge distances and center-to-center spacing.
- Supplementary steel bases and frames, concrete bases, or equipment modifications may void some manufacturers' warranties. Snubbers, for example, should be properly attached to a subbase. Bumpers may be used with springs.
- Static analysis does not account for the effects of resonant conditions within a piece of equipment or its components. Because all equipment has different resonant frequencies during operation and nonoperation, the equipment itself might fail even if the restraints do not. Equipment mounted inside a housing should be

seismically restrained to meet the same criteria as the exterior restraints.

- Snubbers used with spring mounts should withstand motion in all directions. Some snubbers are only designed for restraint in one direction; sets of snubbers or snubbers designed for multidirectional purposes should be used.
- Equipment must be strong enough to withstand the high deceleration forces developed by resilient restraints.
- Flexible connections should be provided between equipment that is braced and piping and ductwork that need not be braced.
- Flexible connections should be provided between isolated equipment and braced piping and ductwork.
- Bumpers installed to limit horizontal motion should be outfitted with resilient neoprene pads to soften the potential impact loads of the equipment.
- Anchor installations must be inspected (usually required for anchors resisting seismic forces); in many cases, damage occurs because bolts were not properly installed. To develop the rated restraint, bolts should be installed according to manufacturer's recommendations.
- Brackets in structural steel attachments should be matched to reduce bending and internal stresses at the joint.
- With the exception of heavy-duty clamps used to attach longitudinal restraints to piping systems, friction must not be relied on to resist any load. All connections should be positive and all holes should be tight-fitting or grouted to ensure minimal clearance at the attachment points.

WIND-RESISTANT DESIGN

Damage done to HVAC&R equipment by both sustained and gusting wind forces has increased concern about the adequacy of equipment protection defined in design documents. Two main areas of the HVAC&R system are exposed to wind events: the HVAC&R equipment and the exterior wall-mounted cladding components, such as intake and exhaust louvers. For HVAC&R equipment, the following calculative procedure generates the same type of total design lateral force used in static analysis of the seismic restraint. The value determined for the design wind force F_w can be substituted for the total design lateral seismic force F_p when evaluating and choosing restraint devices. For wall-mounted components, a design wind pressure P is determined, which can be used to specify

Table 7 Definition of Exposure Categories

Exposure B. Urban and suburban areas, wooded areas, or other terrain with numerous closely spaced obstructions the size of single-family dwellings or larger. Use of this exposure category is limited to those areas for which terrain representative of Exposure B prevails upwind for at least 2600 ft or 20 times the height of the building or structure, whichever is greater.

Exception: For buildings with mean roof height less than or equal to 30 ft, the upwind distance may be reduced to 1500 ft.

Exposure C. Open terrain with scattered obstructions having heights generally less than 30 ft. This category includes flat open country, grasslands, and all water surfaces in hurricane-prone areas. Exposure C applies for all cases where Exposure B or D do not apply.

Exposure D. Flat, unobstructed areas exposed to wind and flowing over open water outside of hurricane-prone regions for a distance of at least 1 mile. This exposure applies to structures exposed to the wind coming from over the water as well as smooth mud flats, salt flats, and unbroken ice.

Reprinted with permission from ASCE (2005).
Notes:
1. For a site located in a transition zone between exposure categories, the exposure resulting in the largest wind forces must be used.
2. Exposure Category D extends into downwind areas of Exposures B or C for a distance of 600 ft or 20 times the height of the building, whichever is greater.
3. The responsibility for determining the exposure category for a given new building project falls on the structural engineer of record. This value is documented in the structural notes drawing (the first of the structural drawings) for the project.

equipment performance levels and design anchors to adequately brace wall-mounted cladding components to the building structure.

The American Society of Civil Engineers' (ASCE) *Standard* 7-05 includes design guidelines for wind, snow, rain, and earthquake loads. Note that the equations, guidelines, and data presented here only cover nonstructural components. The current standard (2005) includes more comprehensive and rigorous procedures for evaluating wind forces and wind restraint. Refer to the latest version of ASCE *Standard* 7 adopted by the local jurisdiction.

TERMINOLOGY

Classification. Buildings and other structures are classified for wind load design exposure according to Table 7.

Basic wind speed. The fastest mile-per-hour wind speed at 33 ft above the ground of Terrain Exposure C (see Table 7) having an annual probability of occurrence of 0.02. Data in ASCE *Standard* 7 or regional climatic data may be used to determine basic wind speeds. ASCE data do not include all special wind regions (such as mountainous terrains, gorges, and ocean promontories) where records or experience indicate that the wind speeds are higher than what is shown in appropriate wind data tables. For these circumstances, regional climatic data may be used provided that both acceptable extreme-value statistical analysis procedures were used in reducing the data and that due regard was given to the length of record, averaging time, anemometer height, data quality, and terrain exposure. One final exclusion is that tornadoes were not considered in developing the basic wind speed distributions.

Components and Cladding. Elements of the building envelope that do not qualify as part of the main wind-force resisting system.

Corner Zone. Areas of building walls and roofs adjacent to building corners that experience increased external pressure from wind.

Design wind force. Equivalent static force that is assumed to act on a component in a direction parallel to the wind and not necessarily normal to the surface area of the component. This force varies with respect to height above ground level.

Importance factor *I*. A factor that accounts for the degree of hazard to human life and damage to HVAC components (Table 8). For hurricanes, the value of the importance factor can be linearly interpolated between the ocean line and 100 miles inland because wind effects are assumed negligible at this distance inland.

Gust response factor *G*. A factor that accounts for the fluctuating nature of wind and the corresponding additional loading effects on HVAC components.

Table 8 Wind Importance Factor *I* (Wind Loads)

Category	*I*
I	0.87
II	1
III	1.15
IV	1.15

Note: See Table 9 for categories.

Table 9 Exposure Category Constants

Exposure Category	α	Z_g, ft	Gust Factor *G*
B	7	1200	0.85
C	9.5	900	0.85
D	11.5	700	0.85

Reprinted with permission from ASCE (2005).
Note: See Table 7 for definitions of exposure categories.

Table 10   Force Coefficients for HVAC Components, Tanks, and Similar Structures

Shape	Type of Surface	C_f for h/D Values of		
		1	7	25
Square (wind normal to face)	All	1.3	1.4	2.0
Square (wind along diagonal)	All	1.0	1.1	1.5
Hexagonal or octagonal $D\sqrt{Q_z} > 2.5$	All	1.0	1.2	1.4
Round $D\sqrt{Q_z} > 2.5$	Moderately smooth	0.5	0.6	0.7
	Rough ($D'/D = 0.02$)	0.7	0.8	0.9
	Very rough ($D'/D = 0.08$)	0.8	1.0	1.2
Round $D\sqrt{Q_z} \le 2.5$	All	0.7	0.8	1.2

Reprinted with permission from ASCE (2005).
Notes:
1. Design wind force calculated based on area of structure projected on a plane normal to the wind direction. Force is assumed to act parallel to wind direction.
2. Linear interpolation may be used for h/D values other than shown.
3. Nomenclature:
    D = diameter or least horizontal dimension, ft
    D' = depth of protruding elements such as ribs and spoilers, ft
    h = structure (top of equipment) height (above ground), ft
    Q_z = velocity pressure evaluated at height z above ground level, lb/ft²

Minimum design wind load. The wind load may not be less than 10 lb/ft² multiplied by the area of the HVAC component projected on a vertical plane that is normal to the wind direction.

CALCULATIONS

Two procedures are used to determine the design wind load on HVAC components. The **analytical procedure**, described here, is the most common method for standard component shapes, based on the requirements in ASCE 7. The second method, the **wind-tunnel procedure**, is used in the analysis of complex and unusually shaped components or equipment located on sites that produce wind channeling or buffeting because of upwind obstructions. The analytical procedure produces design wind forces that are expected to act on HVAC components for durations of 1 to 10 s. The various factors, pressure, and force coefficients incorporated in this procedure are based on a mean wind speed that corresponds to the fastest wind speed.

Analytical Procedure

The design wind force is determined by the following equation:

$$F_w = Q_z GC_f A_F \quad (54)$$

where

   F_w = design wind force, lb
   Q_z = velocity pressure evaluated at height z above ground level, lb/ft²
   G = gust response factor for HVAC components evaluated at height z above ground level
   C_f = force coefficient (Table 10)
   A_f = area of HVAC component projected on a plane normal to wind direction, ft²

Certain of the preceding factors must be calculated from equations that incorporate site-specific conditions that are defined as follows:

Velocity Pressure. The design wind speed must be converted to a velocity pressure that is acting on an HVAC component at a height z above the ground. The equation is

$$Q_z = 0.00256 K_z K_{zt} K_d V^2 I \quad (55)$$

where

   K_z = velocity pressure exposure coefficient from Table 12
   K_{zt} = topographic factor = 1.0
   K_d = wind directionality factor = 1.0
   V = velocity from Figure 12, mph
   I = importance factor from Table 8

The force generated by the wind is calculated by

$$F_w = Q_z GC_f A_f \quad (56)$$

where

   F_w = design wind force, lb
   Q_z = velocity pressure evaluated at height z above ground level, lb/ft²
   G = gust response factor for HVAC components evaluated at height z above ground level
   C_f = force coefficient (Table 10)
   A_f = area of HVAC component projected on a plane normal to wind direction, ft²

The following example calculations are for a 400 ton cooling tower:

   Tower height h = 10 ft
   Tower width D = 10 ft
   Tower length l = 20 ft
   Tower operating weight W_p = 19,080 lb

   Tower diagonal dimension = $\sqrt{10^2 + 20^2}$ = 22.4 ft
   Area normal to wind direction A_f = 10 × 22.4 = 224 ft²

From Table 10, C_f = 1.0 for wind acting along diagonal with h/D = 10/10 = 1.

Example 5. Suburban hospital in Omaha, Nebraska. The top of the cooling tower is 100 ft above ground level. Building width normal to the wind B = 3000 ft, and building height H = 90 ft.

  Solution:
     From Figure 12, the design wind speed is found to be 90 mph.
     From Table 9, use Category IV.
     From Table 7, use Exposure B.
     From Table 8, I = 1.15.
     From Table 12, K_z = 0.99.
     From Figure 13, K_d = 0.9.
     From Table 9, G = 0.85.

Substitution into Equation (58) yields

Q_z = 0.00256 × 0.99 × 1.0 x 0.9 × (90)² × 1.15 = 21.25 lb/ft² = 2881 psi

Building height is greater than 60 ft; therefore, E_f = 1.0.
Substitution into Equation (54) yields the design wind force as

$$F_w = 21.25 \times 0.85 \times 1.0 \times 224 \times 1.0 = 4046 \text{ lb}$$

Example 6. Office building in New York City. Top of tower is 600 ft above ground level. Building wall normal to the wind B = 600 ft and building height H = 590 ft.

  Solution:
     From Figure 12, the design wind speed is 120 mph.
     From Table 9, use Category II.
     From Table 7, use Exposure B.
     From Table 8, I = 1.0.
     From Figure 13, K_d = 0.9.
     Because z > 500 ft, K_z must be determined from Note 2 of Table 12.
     From Table 9, α = 7.0, z_g = 1200, and G = 0.85.

Substituting into the first equation in Note 2 yields

$$K_z = 2.10(Z/Z_g)^{2/\alpha} = 1.72$$

Substituting into Equation (55) yields

$$Q_z = 0.00256 \times 1.72 \times 1.0 \times 0.9 \times (120)^2 \times 1.15 = 65.6 \text{ lb/ft}^2$$

Building height is greater than 60 ft, therefore E_f = 1.0.

Substituting into Equation (56) yields the design force wind as

$$F_W = 65.6 \times 0.85 \times 1.0 \times 224 \times 1.0 = 12,490 \text{ lb}$$

Example 7. Church in Key West, Florida. The top of the tower is 50 ft above ground level. Building wall normal to the wind $B = 300$ ft and building height $H = 40$ ft.

Solution:

From Figure 12, the design speed is found to be 150 mph.

From Table 9, use Category III.

From Table 5, use Exposure C (as this is a hurricane-prone region).

From Table 8, $I = 1.15$.

From Table 9, $G = 0.85$.

From Table 13, $K_d = 0.9$.

From Table 12, $K_z = 1.09$ (for exp category C).

From Equation (55):

$$Q_z = 0.00256 \times 1.09 \times 1.0 \times 0.9 \times (150)^2 \times 1.15 = 65 \text{ lb/ft}^2$$

Building height is less than 60 ft, $A_f /(B \times H) = 224/(300 \times 40) = 0.02$, therefore, $E_f = 1.9$

Substituting into Equation (56) gives the design wind force as

$$F_w = 65 \times 0.85 \times 1.0 \times 224 \times 1.9 = 23,514 \text{ lb}$$

WALL-MOUNTED HVAC&R COMPONENT CALCULATIONS (LOUVERS)

For many projects, the structural engineer of record will determine the components and cladding wind pressures provided on the structural notes drawing. If these wind pressures are not provided, the two following procedures (described previously) are used to determine the design wind load on HVAC cladding components.

Analytical Procedure

Velocity Pressure. The design wind speed must be converted to a velocity pressure that is acting on an HVAC component at height z above the ground. This is done using Equation (54). Once the velocity pressure has been determined, the design wind pressure can be calculated.

Low-Rise Buildings and Buildings with $h \leq 60\,ft$

The design wind pressure for cladding is determined by the following equation:

$$P_w = Q_h(GC_p - GC_{pi}) \tag{57}$$

where

P_w = design wind pressure, lb/ft^2

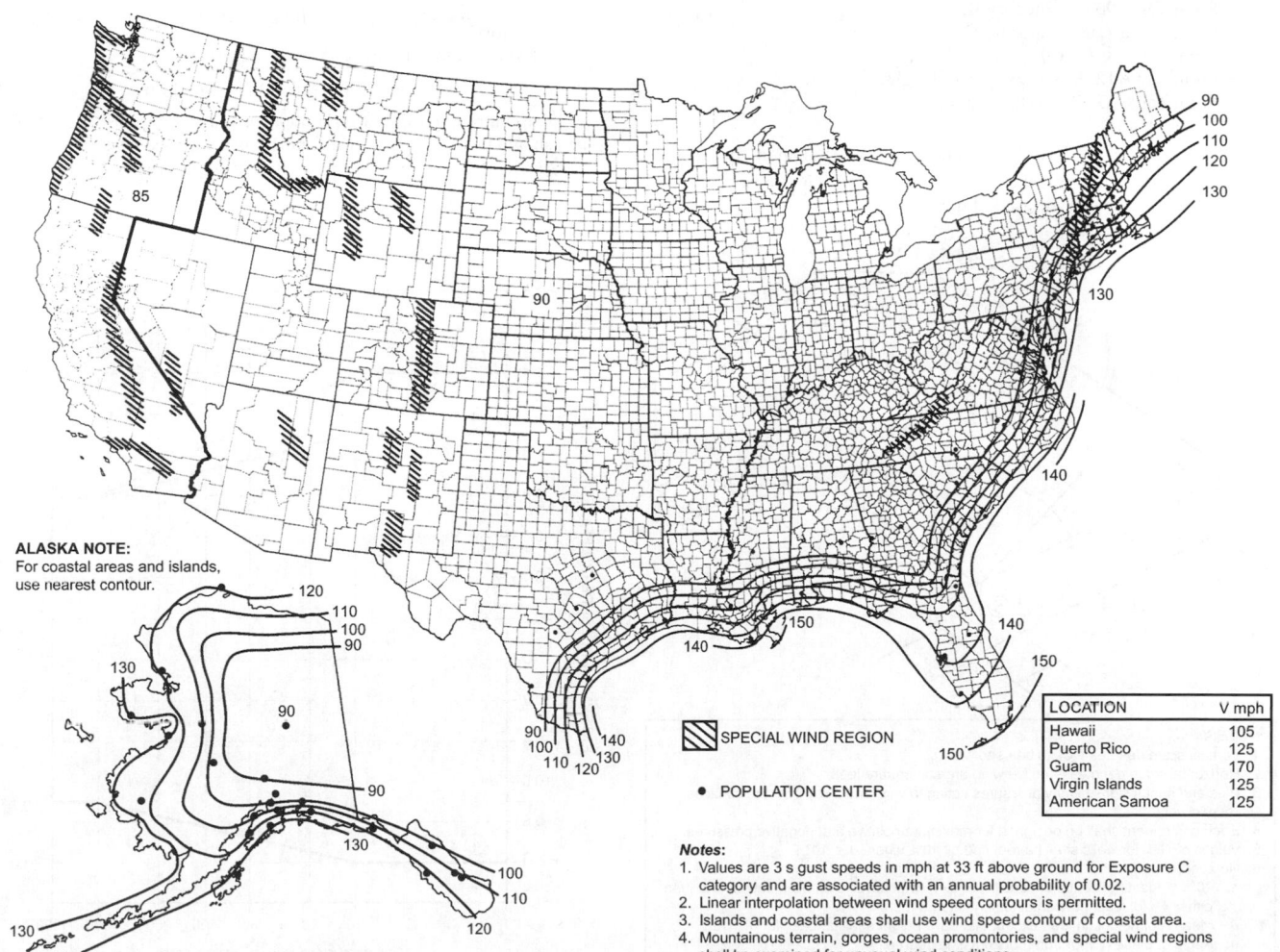

ALASKA NOTE:
For coastal areas and islands, use nearest contour.

▨ SPECIAL WIND REGION

● POPULATION CENTER

LOCATION	V mph
Hawaii	105
Puerto Rico	125
Guam	170
Virgin Islands	125
American Samoa	125

Notes:
1. Values are 3 s gust speeds in mph at 33 ft above ground for Exposure C category and are associated with an annual probability of 0.02.
2. Linear interpolation between wind speed contours is permitted.
3. Islands and coastal areas shall use wind speed contour of coastal area.
4. Mountainous terrain, gorges, ocean promontories, and special wind regions shall be examined for unusual wind conditions.

Fig. 12   Wind Speed Data
Reprinted with permission from ASCE (2005)

Q_h = velocity pressure evaluated at mean roof height h above ground level, lb/ft²
GC_p = external pressure coefficient given in Figure 13
GC_{pi} = internal pressure coefficient given in Table 14

Buildings with h > 60 ft

The design wind pressure is determined by the following equation:

$$P_w = Q(GCp) - q_i(GC_{pi}) \qquad (58)$$

where

P_w = design wind pressure, lb/ft²
Q_z = velocity pressure for windward walls calculated at height z above the ground of the component being examined
Q_h = velocity pressure for leeward walls, side walls and roofs, evaluated at height h of the roof
Q_i = velocity pressure for windward walls, side walls, leeward walls, and roofs, evaluated at height h of the roof
GC_p = external pressure coefficient given in Figure 14
GC_{pi} = internal pressure coefficient given in Table 14

Example 8. Office building in Houston, Texas. The top of the building is 30 ft above grade located in a newly developed suburban area. It is necessary to determine the wind pressures on louver 1 and louver 2 shown on the building elevation in Figure 15.

Solution:
From Figure 12, the design speed is found to be 120 mph.
From Table 9, use Category II.
From Table 7, use Exposure C.
From Table 8, $I = 1.0$.
From Table 12, $K_z = 0.98$, at roof height, $h = 30$ ft.
From Table 13, $K_d = 0.85$.
K_{zt} assumed to be 1.0.

Determine GC_p: Building height h is less than 60 ft; therefore, Equation (55) is used for the pressure evaluations. GC_p must be determined from Figure 15 for each of the louvers.

Louver 1: from the notes on Figure 15, it is necessary to determine the a dimension, which establishes the corner zone 5. The least horizontal dimension coming into the corner is 32 ft from the plan view. Ten percent of this value is 3.2 ft. The minimum value for the corner dimension is 3 ft. Louver 1 is located 1 ft 2 11/16 in. from the corner and is therefore in corner zone 5.

From Figure 15, $GC_p = +0.95$ or -1.3 for a 20 ft² wind area. A positive GC_p indicates a positive pressure on the windward side of the building. A negative GC_p indicates a suction pressure on the leeward side of the building. Both cases must be evaluated.

Louver 2: based on the corner calculation, louver 2 is in noncorner zone 4. From Figure 15, $GC_p = +0.9$ or -1.0 for a 30 ft² wind area.

Determine GC_{pi}: See Figure 14. Most buildings without significant wall openings are enclosed buildings. For the purposes of this example an enclosed building is assumed. $GC_{pi} = +0.18$ or -0.18. A positive sign indicates pressure outward on all structure walls. A negative sign indicates pressure inward on all structure walls.

Determine velocity pressure at roof elevation h from Equation (55):

$$Q_h = 0.00256 \times 0.98 \times 1.0 \times 0.85 \times (120)^2 \times 1.0 = 30.7 \text{ lb/ft}^2$$

Determine design wind pressure P from Equation (56):
Louver 1, case 1: positive external, positive internal

$$P = 30.7 \times (0.95 - 0.18) = 23.6 \text{ lb/ft}^2$$

Louver 1, case 2: positive external, negative internal

$$P = 30.7 \times [0.95 - (-0.18)] = 34.7 \text{ lb/ft}^2$$

Louver 1, case 3: negative external, positive internal

$$P = 30.7 \times [(-1.3) - 0.18] = -45.4 \text{ lb/ft}^2$$

Louver 1, case 4: negative external, negative internal

$$P = 30.7 \times [(-1.3) - (-0.18)] = -34.4 \text{ lb/ft}^2$$

The controlling values for P for louver 1 are 34.7 lb/ft², –45.4 lb/ft² and should be used to specify equipment performance levels.
Louver 2, case 1: positive external, positive internal

$$P = 30.7 \times (0.90 - 0.18) = 22.1 \text{ lb/ft}^2$$

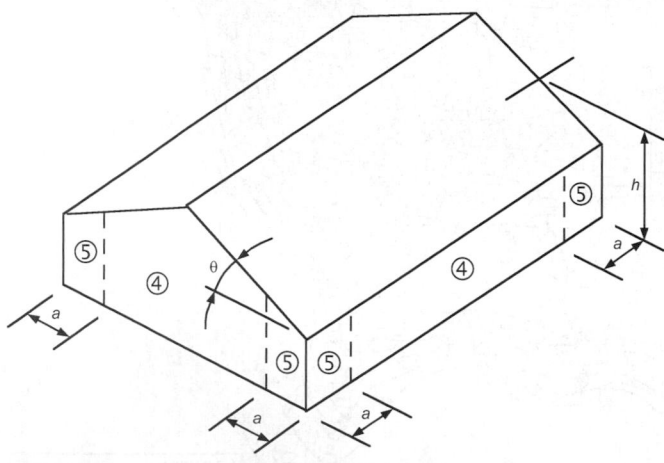

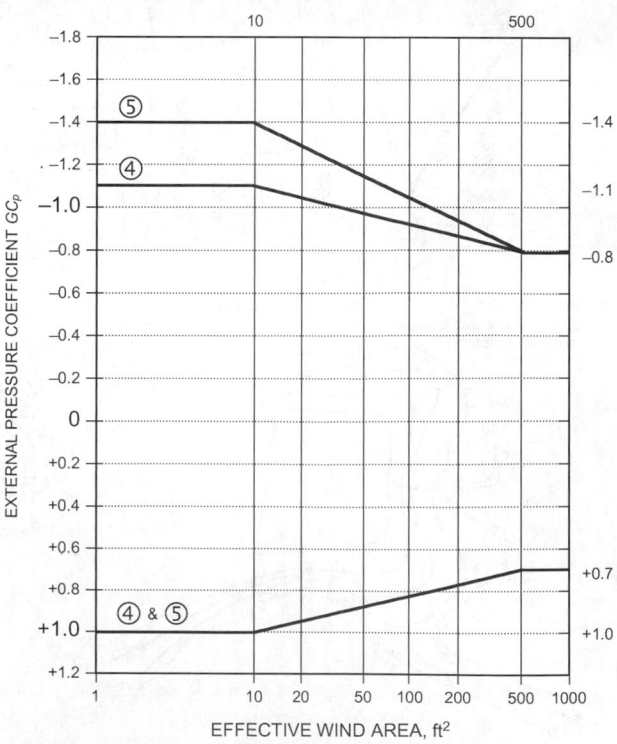

NOTES:
1. Vertical scale denotes GC_p to be used with q_h.
2. Horizontal scale denotes effective wind area, in square feet.
3. Plus and minus signs signify pressures acting toward and away from the surfaces, respectively.
4. Each component shall be designed for maximum positive and negative pressures.
5. Values of GC_p for walls shall be reduced by 10% when $\theta \leq 10°$.
6. Notation:
   a: 10% of least horizontal dimension or 0.4h, whichever is smaller, but not less than either 4% of least horizontal dimention or 3 ft.
   h: Mean roof height, in feet, except that eave height shall be used for $\theta \leq 10°$.
   θ: Angle of plane of roof from horizontal, in degrees.

Fig. 13 External Pressure Coefficient GC_p for Walls for $h \leq 60$ ft
Reprinted with permission from ASCE (2005)

Louver 2, case 2: positive external, negative internal

$$P = 30.7 \times [0.90 - (-0.18)] = 33.2 \ lb/ft^2$$

Louver 2, case 3: negative external, positive internal

$$P = 30.7 \times [(-1.0) - 0.18] = -36.2 \ lb/ft^2$$

Louver 2, case 4: negative external, negative internal

$$P = 30.7 \times [(-1.0) - (-0.18)] = -25.2 \ lb/ft^2$$

The controlling values for P for louver 2 are 33.2 lb/ft^2, $-$36.2 lb/ft^2 and should be used to specify equipment performance levels.

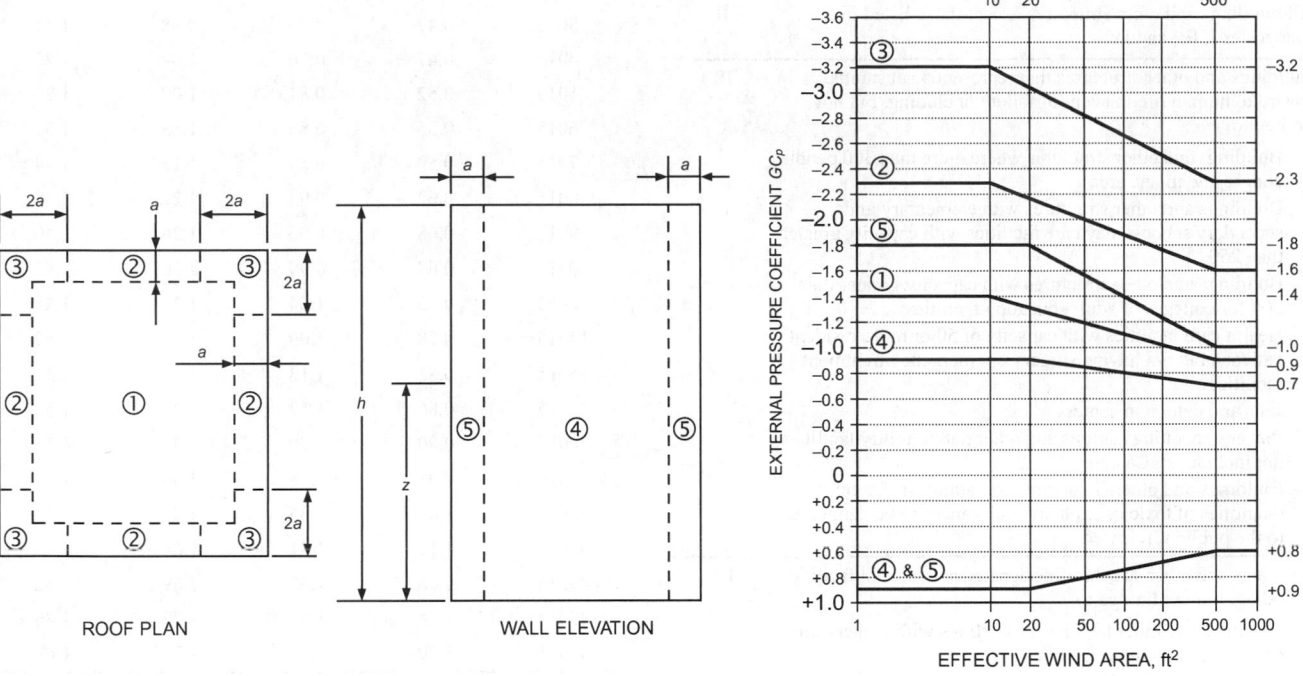

ROOF PLAN   WALL ELEVATION   EFFECTIVE WIND AREA, ft^2

NOTES:
1. Vertical scale denotes GC_p to be used with appropriate q_z or q_h.
2. Horizontal scale denotes effective wind area A, in square feet.
3. Plus and minus signs signify pressures acting toward and away from the surfaces, respectively.
4. Use q_z with positive values of GC_p and q_h with negative values of GC_p.
5. Each component shall be designed for maximum positive and negative pressures.
6. Coefficients are for roofs with angle $\theta \leq 10°$. For other roof angles and geometry, use GC_p values from Figure 6-11 and attendant q_h based on exposure defined in 6.5.6.

7. If a parapet equal to or higher than 3 ft is provided around the perimeter of the roof with $\theta \leq 10°$, Zone 3 shall be treated as Zone 2.
8. Notation:
   a: 10% of least horizontal dimension, but not less than 3 ft.
   h: Mean roof height, in feet, except that eave height shall be used for $\theta \leq 10°$.
   z: Height above ground, in feet.
   θ: Angle of plane of roof from horizontal, in degrees.

Fig. 14   External Pressure Coefficient GC_p for Walls for $h > 60$ ft
Reprinted with permission from ASCE (2005)

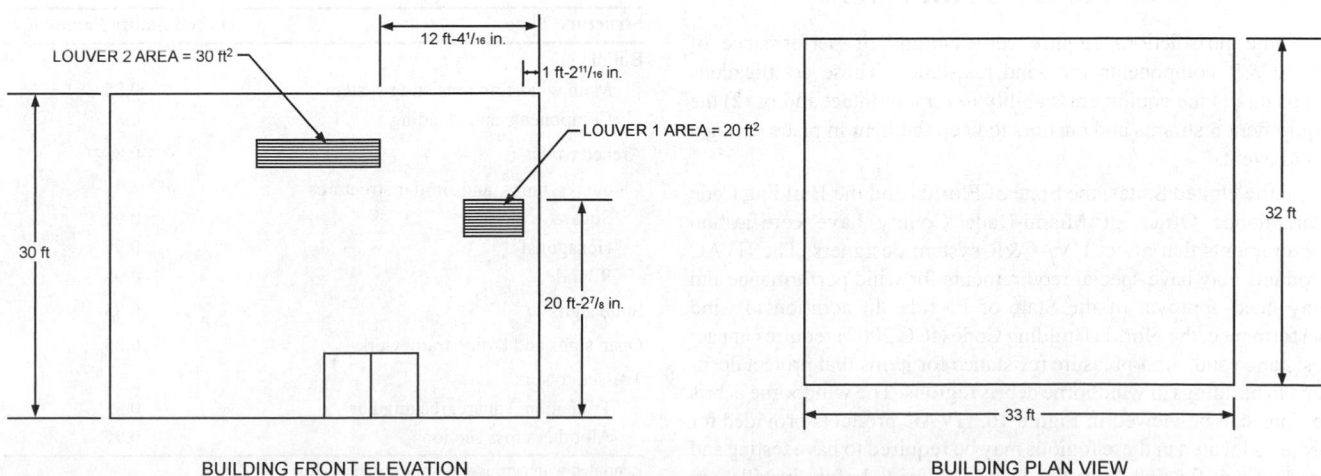

BUILDING FRONT ELEVATION   BUILDING PLAN VIEW

Fig. 15   Office Building, Example 8

Table 11 Classification of Buildings and Other Structures for Wind Loads

Nature of Occupancy	Category
Buildings and other structures that represent a low hazard to human life in event of failure, including, but not limited to, agricultural facilities, certain temporary facilities, and minor storage facilities	I
All buildings and other structures except those listed in Categories I, III, and IV	II
Buildings and other structures that represent a substantial hazard to human life in event of failure, including, but not limited to,	III
- Buildings and other structures where more than 300 people congregate in one area.	
- Buildings and other structures with elementary and secondary schools, day care facilities with capacity greater than 250	
- Buildings and other structures with capacity greater than 500 for colleges or adult education facilities	
- Health care facilities with capacity of 50 or more resident patients, but not having surgery or emergency treatment facilities	
- Jails and detention centers	
- Power generating stations and other public utility facilities not included in Category IV	
- Buildings and others structures containing sufficient quantities of toxic or explosive substances to be dangerous to the public if released	
Buildings and other structures designated as essential facilities including, but not limited to,	IV
- Hospitals and other health care facilities with surgery and emergency treatment facilities	
- Fire, rescue, and police stations and emergency vehicle garages	
- Designated earthquake, hurricane, or other emergency shelters	
- Communication center and other facilities required for emergency response	
- Power generating stations and other public utility facilities required in an emergency	
- Buildings and other structures with critical national defense functions	

Reprinted with permission from ASCE (2005).

CERTIFICATION OF HVAC&R COMPONENTS FOR WIND

Some jurisdictions require certifications of performance of HVAC&R components for wind resistance. These certifications focus on (1) the equipment's ability to remain intact and/or (2) the equipment restraints and anchors to keep the item in place during a wind event.

In the United States, the State of Florida and the Building Code Compliance Office of Miami-Dade County have certification requirements that affect HVAC&R system designers. The HVAC products may have special requirements for wind performance and may need approval of the State of Florida. In addition to wind performance, the Florida Building Code (ICC 2007) requires impact resistance and wind-pressure resistance for items that protect openings in buildings in windborne debris regions. The windborne debris regions can be viewed in Figure 16. HVAC products provided for projects located in these regions may be required to have testing and product certification from the State of Florida before installation. Other states, such as Texas, also have requirements for wind-pressure and impact testing. To ensure that the HVAC&R equipment

Table 12 Velocity Pressure Exposure Coefficient K_z

Height above ground level z, ft	Exposure			
	A	B	C	D
0 to 15	0.32	0.57	0.85	1.03
2015	0.36	0.62	0.90	1.08
2515	0.39	0.66	0.94	1.12
3015	0.42	0.70	0.98	1.16
4015	0.47	0.76	1.04	1.22
5015	0.52	0.81	1.09	1.27
6015	0.55	0.85	1.13	1.31
7015	0.59	0.89	1.17	1.34
8015	0.62	0.93	1.21	1.38
9015	0.68	0.96	1.24	1.40
10015	0.68	0.99	1.26	1.43
12015	0.73	1.04	1.31	1.48
14015	0.78	1.09	1.36	1.52
16015	0.82	1.13	1.39	1.55
18015	0.86	1.17	1.43	1.58
20015	0.90	1.20	1.46	1.61
25015	0.98	1.28	1.53	1.68
30015	1.05	1.35	1.59	1.73
35015	1.12	1.41	1.64	1.78
40015	1.18	1.47	1.69	1.82
45015	1.24	1.52	1.73	1.86
50015	1.29	1.56	1.77	1.89

Reprinted with permission from ASCE (2005).

Notes:
1. Linear interpolation for intermediate values of height z is acceptable.
2. For values of height z greater than 500 ft, K_z must be calculated using the following equations:

$$K_z = 2.01(z/z_g)^{2/\alpha} \quad \text{For} \quad 15 \text{ ft} \leq z \leq z_g$$

or

$$K_z = 2.01(15/z_g)^{2/\alpha} \quad \text{For} \quad z < 15 \text{ ft}$$

3. Exposure categories are defined in Table 7.
4. Values for alpha (α) and z_g are found in Table 9.

Table 13 Directionality Factor K_d

Structure Type	Directionality Factor K_d*
Buildings	
Main wind-force-resisting system	0.85
Components and cladding	0.85
Arched roofs	0.85
Chimneys, tanks, and similar structures	
Square	0.90
Hexagonal	0.95
Round	0.95
Solid signs	0.85
Open signs and lattice framework	0.85
Trussed towers	
Triangular, square, rectangular	0.85
All other cross sections	0.95

Reprinted with permission from ASCE (2005).
*Directionality factor K_d has been calibrated with combinations of load specified in Section 2. This factor shall only be applied when used in conjunction with load combinations specified in 2.3 and 2.4

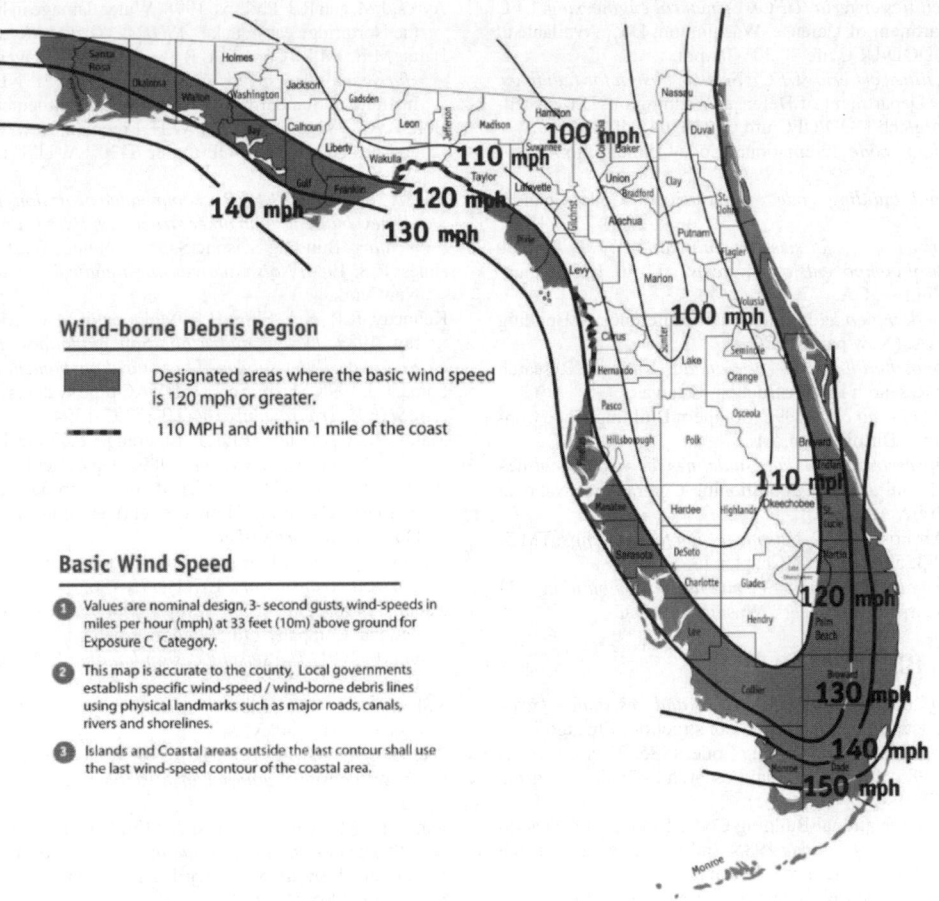

Fig. 16   State of Florida Windborne Debris Regions
ICC (2007)

Table 14   Internal Pressure Coefficient GC_{pi}

Enclosure Classification	GC_{pi}
Open buildings	0.00
Partially enclosed buildings	+0.55
	−0.55
Enclosed buildings	+0.18
	−0.18

Reprinted with permission from ASCE (2005).
Notes:
1. Plus and minus signs signify pressures acting toward and away from the internal surfaces, respectively.
2. Values of GC_{pi} shall be used with q_x or q_h as specified in 6.5.12.
3. Two cases shall be considered to determine the critical load requirements for the appropriate condition:
   (i) a positive value of GC_{pi} applied to all internal surfaces
   (ii) a negative value of GC_{pi} applied to all internal surfaces

supplied is compliant, designers should contact the local building code official in their project location.

Some of the testing protocols for the State of Florida are

- TAS 201-94: Impact Testing Procedures

- TAS 202-94: Criteria for Testing Impact and Non-Impact Resistant Building Envelope Components Using Uniform Static Air Pressure

- TAS 203-94: Criteria for Testing Products Subject to Cyclic Wind Pressure Loading

REFERENCES

ACI. 2008. Building code requirements for structural concrete (ACI 318-08) and commentary. *Standard* 318-08. American Concrete Institute, Farmington Hills, MI.

AISC. 1989. *Manual of steel construction—Allowable stress design*, 9th ed. American Institute of Steel Construction, Chicago.

ASCE. 2005. Minimum design loads for buildings and other structures. *Standard* ASCE 7-05. American Society of Civil Engineers, Reston, VA.

ASHRAE. 2000. A practical guide to seismic restraint. Research Project RP-812, *Final Report*.

ASME. 2003. Nuclear air and gas treatment. Code AG-1-2003. American Society of Mechanical Engineers, New York.

ASTM. 1996. Test methods for strength of anchors in concrete and masonry elements. *Standard* E488-96 (R2003). American Society for Testing and Materials, West Conshohocken, PA.

ATC. *Proceedings of seminar on seismic design, performance, and retrofit of nonstructural components on critical facilities*. ATC 29-2. Applied Technology Council, Washington, D.C.

AWC. 1997. *National design specification (NDS®) for wood construction*. American Wood Council, Washington, D.C.

BOCA. 1996. *The BOCA national building code*, 13th ed. Building Officials & Code Administrators International, Inc., Country Club Hills, IL.

Cover, L.E., et al. 1985. Handbook of nuclear power plant seismic fragilities. *Report* NUREG/CR-3558. Lawrence Livermore National Laboratory and U.S. Nuclear Regulatory Commission, Washington, D.C.

DOD. 1990. Structures to resist the effects of accidental explosions. *Technical Manual* TM 5-1300. U.S. Department of Defense, Washington, D.C.

DOD. 2002. Design and analysis of hardened structures to conventional weapons effects. *Technical Manual* TM 5-855-1. U.S. Department of Defense, Washington, D.C.

DOD. 2005. *Unified facilities criteria (UFC): Structural engineering*. UFC 3-310-01. U.S. Department of Defense, Washington, D.C. Available at www.wbdg.org/ccb/DOD/UFC/ufc_3_301_01.pdf.

DOD. 2007. *Unified facilities criteria (UFC): Seismic design for buildings*. UFC 3-310-04. U.S. Department of Defense, Washington, D.C. Available at www.wbdg.org/ccb/DOD/UFC/ufc_3_310_04.pdf.

ICC. 2007. *Florida building code*. International Code Council, Inc., Washington, D.C.

ICC. 2009. *International building code*®. International Code Council, Washington, D.C.

ICC-ES. 2007. *Acceptance criteria for seismic qualification by shake-table testing of nonstructural components and systems*. AC156. ICC Evaluation Service, Inc., Whittier, CA.

ICBO. 1997. *Uniform building code*. International Conference of Building Officials, Whittier, CA. (Now part of ICC.)

NRC-IRC. 2010. *National building code of Canada*. National Research Council Institute for Research in Construction, Ottawa.

SBCCI. 1994. *Standard building code* 1996. Southern Building Code Congress International, Inc., Birmingham, AL.

SMACNA. 2008. *Seismic restraint manual: Guidelines for mechanical systems*, 3rd ed. Sheet Metal and Air Conditioning Contractors' National Association, Chantilly, VA.

U.S. Army, Navy, and Air Force. 1992. *Seismic design for buildings*. TM 5-809-10, NAVFAC P-355, AFN 88-3, Chapter 13.

U.S. COE. 1998. *Technical instructions: Seismic design for buildings*. TI 809-04. U.S. Army Corps of Engineers, Washington, D.C.

BIBLIOGRAPHY

AISC. 1995. *Manual of steel construction—Load and resistance factor design*, 2nd ed. American Institute of Steel Construction, Chicago.

Associate Committee on the National Building Code. 1985. *National building code of Canada* 1985, 9th ed. National Research Council of Canada, Ottawa.

Associate Committee on the National Building Code. 1986. *Supplement to the National Building Code of Canada* 1985, 2nd ed. National Research Council of Canada, Ottawa. First errata, January.

ATC. *Proceedings of seminar and workshop on seismic design and performance of equipment and nonstructural elements in buildings and industrial structures*. ATC 29, NCEER (New York) & NSF (Washington D.C.).

ATC. *Seminar on seismic design, retrofit, and performance of nonstructural components*. ATC 29-1, NCEER (New York) & NSF (Washington D.C.).

AWS. 2000. *Structural welding code*. AWS D1.1-2000. Steel American Welding Society, Miami.

Ayres, J.M. and R.J. Phillips. 1998. Water damage in hospitals resulting from the Northridge earthquake. *ASHRAE Transactions* 104(1B):1286-1296.

Batts, M.E., M.R. Cordes, L.R Russell, J.R. Shaver, and E. Simiu. 1980. *Hurricane wind speeds in the United States*. NBS BSS 124. National Institute of Standards and Technology, Gaithersburg, MD.

Bolt, B.A. 1988. *Earthquakes*. W.H. Freeman, New York. DOE. 1989. General design criteria. DOE Order 6430.1A. U.S. Department of Energy, Washington, D.C.

FEMA 368 & 369. *NEHRP recommended provisions for seismic regulations for new buildings and other structures. Part 1, Provisions; Part 2, Commentary*. Building Seismic Safety Council, Washington, D.C.

Jones, R.S. 1984. *Noise and vibration control in buildings*. McGraw-Hill, New York.

Kennedy, R.P., S.A. Short, J.R. McDonald, M.W. McCann, and R.C. Murray. 1989. *Design and evaluation guidelines for the Department of Energy facilities subjected to natural phenomena hazards*.

Lama, P.J. 1998. Seismic codes, HVAC pipe systems and practical solutions. *ASHRAE Transactions* 104(1B):1297-1304.

Maley, R., A. Acosta, F. Ellis, E. Etheredge, L. Foote, D. Johnson, R. Porcella, M. Salsman, and J. Switzer. 1989. Department of the Interior, U.S. geological survey. U.S. geological survey strong-motion records from the Northern California (Loma Prieta) earthquake of October 17, 1989. Open-file *Report* 89-568.

Meisel, P.W. 2001. Static modeling of equipment acted on by seismic forces. *ASHRAE Transactions* 107(1):775-786.

Naeim, F. 1989. *The seismic design handbook*. Van Nostrand Reinhold International Company Ltd., London, England.

Naeim, F. 2001. *The seismic design handbook,* 2nd ed. Kluwer Academic, Boston.

NFPA. 2002. *Installation of sprinkler systems*. National Fire Protection Association, Quincy, MA.

Peterka, J.A., and J.E. Cermak. 1974. Wind pressures on buildings—Probability densities. *Journal of Structural Division*, ASCE 101(6):1255-1267.

Simiu, E., M.J. Changery, and J.J. Filliben. 1979. *Extreme wind speeds at 129 stations in the contiguous United States*. U.S. NBS BSS 118. National Institute of Standards and Technology, Gaithersburg, MD.

SMACNA. 2005. *HVAC duct construction standard—metal and flexible*, 3rd ed. Sheet Metal and Air Conditioning Contractors' National Association, Chantilly, VA.

Wasilewski, R.J. 1998. Seismic restraints for piping systems. *ASHRAE Transactions* 104(1B):1273-1295.

Weigels, R.L. 1970. *Earthquake engineering*, 10th ed. Prentice-Hall, Englewood Cliffs, NJ.

ELECTRICAL CONSIDERATIONS

PRODUCTION, delivery, and use of electricity involves countless decisions made along the way, by hundreds of people and companies. This chapter focuses on the decisions to be made about the building and equipment. Creating a building that works means including the best designs available, communicating needs and capabilities, and planning ahead.

For an owner-occupied building, the benefits of a properly designed building return to the owner throughout the building's life. For tenant-occupied spaces, good design means fewer problems with tenant and building system interference (e.g., lighting or appliances in one suite disrupting computers in a neighboring suite).

Because HVAC&R equipment can have a large effect on buildings, it is necessary to address electrical issues in buildings that specifically are caused by or have an effect on HVAC&R equipment.

TERMINOLOGY

Volt (V): practical unit of electric pressure; the pressure that will produce a current of 1 A against a resistance of 1 Ω; equal to 1 J/s. Also called the **electromotive force (emf)**.

Current (*I*): movement of electrons through a conductor; measured in amperes.

Ampere (A): practical unit of electric current flow. If a 1 Ω resistance is connected to a 1 V source, 1 A will flow.

Alternating current (ac): a current that reverses at regular, recurring intervals of time and that has alternately positive and negative values. The values vary over time in a sinusoidal manner.

Watt (W): unit of real electrical power, equal to the power developed in a circuit by a current of 1 A flowing through a potential difference of 1 V.

Volt-ampere (VA): amount of apparent power in an alternating current circuit equal to a current of 1 A at an emf of 1 V. It is dimensionally equivalent to watts. Volt-ampere is equal to watts when voltage and current are in phase.

Volt-ampere-reactive (VAR): unit for reactive power. The symbols Q and sometimes N are used for the quantity measured in VARs. VARs represent the power consumed by a reactive load (i.e., when there is a phase difference between applied voltage and current).

Power factor: for an ac electric power system, the ratio of the real power to the apparent power, or W/VA.

Three-phase power: supplied by three conductors, with the currents (or voltages) of any two 120° out of phase with each other.

Y (or "wye") connection: a configuration of wiring so that each winding of a polyphase transformer (or three single-phase transformers) is connected to a common point, the "neutral."

Delta-connected circuit: a three-phase circuit that is mesh-connected, so the windings of each phase of a three-phase transformer are connected in a series for a closed circuit (i.e., in a triangle or "delta" configuration).

The preparation of this chapter is assigned to TC 1.9, Electrical Systems.

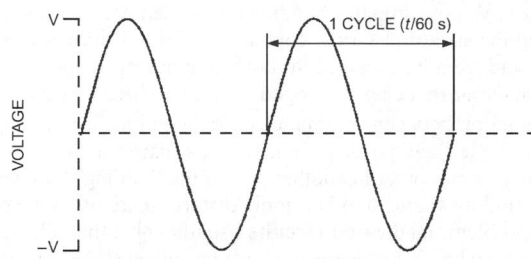

Fig. 1 Fundamental Voltage Wave

Fundamental voltage: produced by an electric ac generator and has a sinusoidal waveform with a frequency of 60 cycles per second, or 60 Hz (in the United States). Other countries may have a similar waveform but at 50 cycles per second of 50 Hz.

Cycle: the part of the fundamental waveform where the electrical potential goes from zero to a maximum to zero to a minimum, and back to zero again (i.e., one complete wave; see Figure 1). At 60 Hz, there are 60 cycles in 1 second.

RMS (root-mean-squared) voltage: an effective way to compare ac to dc value. For a pure sinusoidal waveform, RMS value is equal to 0.707 times the peak magnitude.

System voltage: the RMS phase-to-phase voltage of a portion of an ac electric utility system. Each system voltage pertains to a part of the system bounded by transformers or end-use equipment.

Service voltage: the voltage at the point where the electric systems of the supplier and the user are connected.

Utilization voltage: the voltage at the terminals of the utilization equipment.

Nominal system voltage: the rated system voltage level (i.e., 480 volts) at which the electrical system normally operates. To allow for operating contingencies, utility systems generally operate at voltage levels within –5% to +5% of nominal system voltage.

SAFETY

The greatest danger from electricity is that it is taken for granted and not taken seriously as a hazardous energy source. Electricity can produce bodily harm and property damage, and shut down entire operations. The type of damage from electricity ranges from a mild shock to the body to a major electrical fire. Electrical safety is important in all occupational settings. See information on safety codes in the Electrical Codes section.

PERFORMANCE

In the United States, the *National Electrical Code*® (NEC) is generally accepted as the minimum safety requirements for wiring and grounding in a structure. Other countries have similar requirements. The NEC ensures building design is safe, but may not provide the performance that a modern building requires. Rapid changes in electronic technologies have rendered many traditional electrical distribution practices obsolete and must be replaced with new designs. Electrical power distribution decisions made during design affect occupants' productivity for the life of the building. Many improvements over the minimum requirements are relatively inexpensive to implement as a building is constructed.

Power quality, like quality in other goods and services, is difficult to define. There are standards for voltage and waveshape, but the final measure of power quality is determined by the performance and productivity of equipment used by the building occupant. If the electric power is inadequate for those needs, then the quality is lacking.

Specifications for electric power are set down in recognized national standards. These are voltage levels and tolerances that should be met, on the average, over a long period of time. Electric utilities and building distribution systems generally meet such specifications. Voltage drop in a building is a fundamental reason for calculating the size of electrical conductors. Brief disturbances on the power line are not addressed in these time-averaged specifications; new standards are being developed to address these new concerns.

Interaction between tenants' electrical equipment is an ongoing problem. Often, a large load in one tenant's space can disrupt a small appliance or computer in another part of the building. Voltage drop along building wiring and harmonic distortion are often the causes of the problem. **Dedicated circuits** usually solve the voltage drop problem, but harmonic distortion must be solved at the contributing loads. By eliminating much of the wiring common to both pieces of equipment, the original performance of each is restored. With modern electronic loads, the interaction might easily involve a large load that interferes with smaller, more sensitive equipment. Disturbances might travel greater distances or through nondirect paths, so diagnostics are more difficult.

For tenants of a building with ordinary power distribution, lost productivity associated with power quality problems is an additional operating expense. The disturbance may last only milliseconds, but the disruption to business may require hours of recovery. This multiplication of lost time makes power quality a significant business problem.

Lost productivity may be the time it takes to restart a chiller, to repair a critical piece of equipment, or to retype a document. Another aspect of lost productivity is the stress on employees whose work is lost. The building owner may suffer loss, as well. Certainly the building equipment itself may suffer from the same damage or losses as tenant equipment. Sophisticated energy management systems, security systems, elevator controls, HVAC&R systems, and communications facilities are susceptible to disruption and vulnerable to damage.

ELECTRICAL SYSTEM COMPONENTS AND CONCEPTS

Voltage differential causes electrons to flow. In a direct current (dc) electrical system, electrons flow in only one direction. In an alternating current (ac) electrical system, electrons continually alternate or change direction at a prescribed number of times per second. The main disadvantage of dc voltage is its inability to be boosted or attenuated easily and efficiently. The alternating magnetic fields of ac make boosting or decreasing voltage with transformers feasible, which is why ac has been widely adopted. Electrons flow more efficiently at lower currents because I^2R losses are minimized. For the same load, raising the voltage level reduces the current while delivering the same power. When long distances are involved, electric utility companies step up voltages to very high levels for transmission. However, these voltages are extremely dangerous, so they must be stepped down to a safer, lower, usable voltage before use. Transformers offer an efficient way to change voltage levels (step-up or step-down) for an alternating current power source.

Transformers

Transformers are used to change one voltage to another voltage, typically to step up voltage levels from generators. Power can then be transmitted at a low current (with less loss). At the end of the transmission line, a step-down transformer reduces the voltage to a usable level.

A transformer consists of a ferromagnetic core wrapped with multiple coils, or windings, of wire. The input line is connected to the primary coil, and the output line is connected to the secondary coil. Alternating current I_1 in the primary coil induces an alternating magnetic flux ϕ that flows through the ferromagnetic core, changing direction during each electrical cycle. This flux in turn induces alternating current I_2 in the secondary coil. The voltage V_2 at the secondary coil is directly related to the primary voltage by the turns ratio (i.e., the number of turns N_1 in the primary coil divided by the number turns N_2 in the secondary coil).

An **ideal transformer** with two windings wrapped around a magnetized core is shown in Figure 2. The ideal model for a transformer assumes I^2R losses, core losses, leakage flux, and core reluctance are insignificant. A practical model includes these losses.

Transformer **losses** can be divided into core (or iron) losses, copper losses, and stray losses. Core losses include hysteresis losses and eddy current losses. All ferromagnetic materials tend to retain some degree of magnetization after exposure to an external magnetic field. This tendency to stay magnetized is called hysteresis, and it takes energy to overcome this opposition to change every time the magnetic field produced by the primary winding changes polarity. Eddy current losses result from induced currents circulating in the magnetic core perpendicular to the flux. Because iron conducts both electricity and magnetic flux, eddy currents are induced in the iron just as in the secondary windings from the alternating magnetic field.

Three identical single-phase, two-winding transformers may be connected to form a **three-phase bank**. The four possible connections are Y-Y, Y-Δ, Δ-Y, and Δ-Δ. The U.S. standard for marking three-phase transformers uses H_1, H_2, and H_3 on the high-voltage terminals and X_1, X_2, and X_3 on the low-voltage terminals; A, B, and C identify phases on the high-voltage side of the transformer, and a, b, and c identify phases on the low-voltage side. Typically, three-phase voltages present the higher voltage (phase-to-phase) first, followed by the lower voltage (phase-to-neutral). Single-phase voltages typically present the lower voltage (phase-to-neutral) first, followed by the higher voltage (phase-to-phase). For example, 208/120 V is three-phase and 120/240 V is single-phase.

Y-Y connections (Figure 3) are rarely used because of balancing and harmonics problems.

Y-Δ connections (Figure 4) are typically used for stepping down from high to medium voltage.

The Δ-Y transformer (Figure 5) is commonly used as a generator step-up transformer, where the Δ winding is connected to the generator terminals and the Y winding is connected to the transmission

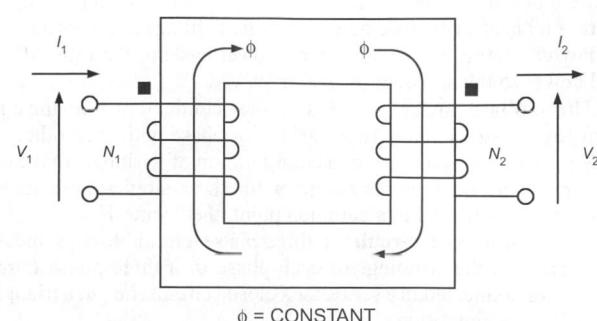

ϕ = CONSTANT

Fig. 2 Ideal Transformer

line. One advantage of the high-voltage Y winding is that a neutral point N is provided for grounding on the high-voltage side.

The Δ-Δ transformer (Figure 6) has the advantage that one phase can be removed for repair or maintenance while the remaining phases continue to operate as a three-phase bank. The open Δ connection allows balanced three-phase operation with the kVA rating reduced to 58% of the original bank. These Δ-Δ connections are typically used in distribution networks.

An **autotransformer** has two windings connected in series (Figure 7). Whereas a typical transformer's windings are only coupled magnetically via the mutual core flux, an autotransformer's windings are both electrically and magnetically coupled.

An autotransformer has smaller per-unit leakage impedances than a two-winding transformer; this results in both smaller series voltage drops and higher short-circuit currents. It also has lower per-unit losses, lower excitation current, and lower cost, if the turns ratio is not large. An autotransformer is not isolated as well as a typical two-winding transformer; transient overvoltages pass through the autotransformer more easily because the windings are connected electrically.

Transformer Coolants and Insulators. Because heat is created by the flow of electrical current through the windings, a liquid (e.g., **oil** or **silicone**) is often used as a coolant inside the transformer. Such liquids are also good electrical insulators for the wire windings and iron core. **Dry transformers** do not require a liquid for cooling, instead using ambient air for cooling as well as insulation. Dust, dirt, moisture, and other contaminants in the air can reduce its insulating capabilities and deteriorate exposed parts, and may cause premature failure of the transformer.

Emergency and Standby Power Systems

Emergency Power Supplies. Diesel **generators** are still the dominant source of emergency power, particularly for emergency loads such as fire pumps or elevators, but the constraints of emergency power systems must be understood to ensure proper operation of electrical equipment. This is especially true as standby systems are expanded to carry the full facility load, including HVAC&R systems, and as some diesel engines are replaced with

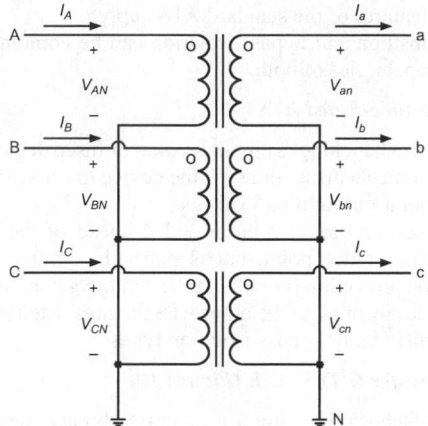

Fig. 3   Three-Phase Y-Y Transformer

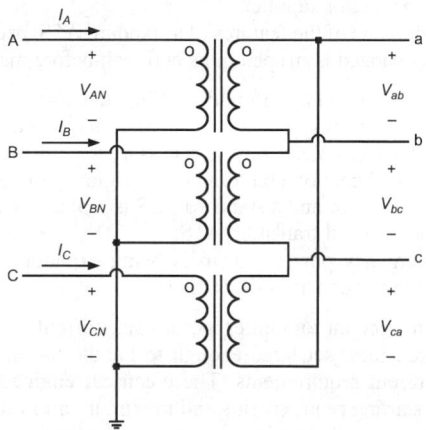

Fig. 4   Three-Phase Y-Δ Transformer

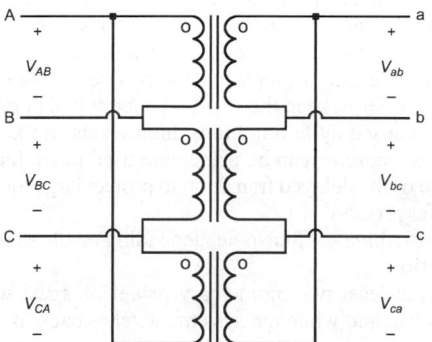

Fig. 6   Three-Phase Δ-Δ Transformer

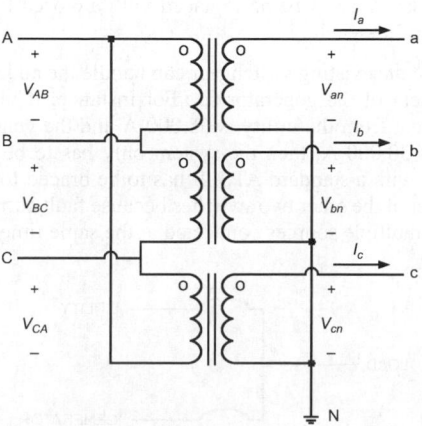

Fig. 5   Three-Phase Δ-Y Transformer

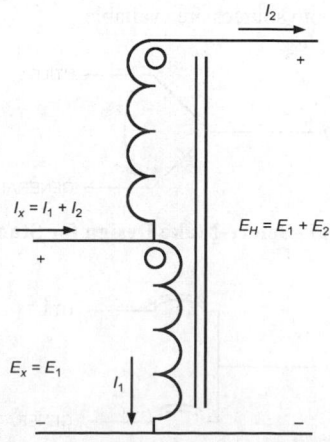

Fig. 7   Typical Autotransformer

natural gas engines, which have very different acceptable load step characteristics.

Most codes require diesel generators used for life safety loads to be online within 10 s. However, age, battery wear, and even improper exercising of the engine can cause delays and even failures to start. Even if the engine does start within 10 s, some type of ride-through device must support electronic loads (including computers and energy management systems) until the generator has started.

Uninterruptible power supply (UPS) units are most commonly used to keep computers and controls running during the transition. Batteries are sized to meet the load requirement for a specified period of time (usually about 15 min). The UPS provides power through the interruption and, if generators fail to start, provides enough time for orderly system shutdown.

Flywheel technology is gaining ground where larger (e.g., whole-building) loads are supported. The flywheels are also sized for time and load, but may be cost-prohibitive for orderly shutdown and even for engines that do not start until the second or third try (20 to 30 s). It is rare that one is installed to support a natural gas engine that may take between one to three minutes to start.

A key component of the emergency generator system is the automatic transfer switch (ATS) used to switch from the primary source (usually the utility) to the secondary source (usually the generator). The type of switch selected has important ramifications to the overall electrical system. The types of automatic transfer switches are as follows:

Standard ATS (Figure 8)

- *Break-* (from one source) *before-make* (to another source) design.
- Mechanically interlocked electrical contactors move loads from one power source to another.
- Most can optionally be equipped for delayed transition (switch delays in the neutral position for a preset number of seconds). This allows larger (>50 hp) motors' residual voltage to decay before connection to the other source. Other timers are usually available to provide ride-through for one or two recloser operations (used by utilities to keep their system operational during momentary faults caused by falling limbs, high winds, etc.).
- An *in-phase monitor* can be preset and used in conjunction with or in place of the delayed transition to protect large motors during the transfer process.
- Can be two-, three- or four-pole, depending on phase and grounding scenario.
- Results in at least two momentary power outages: when utility service is lost and when the system switches back to the utility.

Closed-Transition ATS (Figure 9)

- Provides momentary (less than six cycles) connection of load to both power sources within acceptable phase angle for seamless transfer when both sources are available.

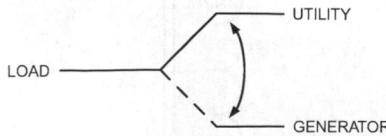

Fig. 8   Break-Before-Make Design for Standard ATS

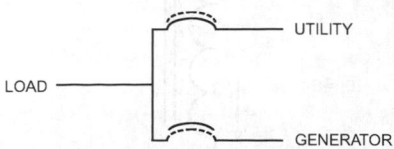

Fig. 9   Closed-Transition ATS

- Eliminates the second outage associated with a standard ATS and, in some cases, can be used to operate the generator as a peaking power source.

Bypass-Isolation ATS

- Same as a standard ATS except for its separate bypass mechanism to allow continuation of power to the load from either source while isolating the automatic switch device for maintenance or repair. Typically, the automatic portion of the switch can be mechanically isolated from the frame during this process.
- All other features of the standard ATS apply.
- Closed-transition and bypass-isolation can be combined to provide the capabilities of both.

Service-Entrance-Rated ATS

- Transfer switch includes a circuit breaker or fused device ahead of the transfer mechanism, allowing the device to act as the service-entrance point for building load.
- Fused protection can also be provided ahead of the secondary (generator) entrance point, but is normally omitted in lieu of locating the protection (circuit breaker) at the generator set.
- Can include any or all of the other ATS features listed in standard, closed-transition, or bypass-isolation types.

Parallel-Transfer (PT) Switch (Figure 10)

- Parallel switchgear in a simple one- or two-breaker design allows for parallel connection of the load to both power sources for an indefinite time period.
- Should include all protective relay devices as required by the local utility and generator supplier.
- Can include some of the features of a standard ATS, because most can also be operated as an open-transfer (break-before-make) switch.

Switch selection can profoundly affect the electrical system, especially if an existing back-up system is upgraded for an expanded role. For example, it may be necessary to completely replace the existing switchgear of a building if an original generator system with a small generator and a standard ATS is replaced with a larger generator and a closed-transition ATS.

If the emergency power system is being expanded to include large motor loads, such as chillers, then

- Set delay to account for chiller motor back current to dissipate.
- Size the generator set large enough to handle the motor loads' starting current requirements. The electrical engineer needs to know the starting requirements and maximum allowable voltage drop that the overall electrical system can handle, especially computer and control loads. Most major generator manufacturers can provide sizing software.

If a standard ATS is being replaced with a closed-transition or parallel-transfer switch,

- Verify that the existing switchgear can handle the added potential fault current of the generator set. For instance, if the available fault current from the utility is 65,000 A and the generator fault current is 30,000 A, then the system only has to be braced for 65,000 A with a standard ATS. It has to be braced for 95,000 A with either of the other two switches because fault current is additive with multiple sources connected at the same time.

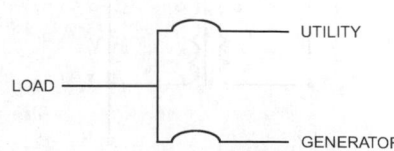

Fig. 10   Parallel-Transfer Switch

If the emergency power system is being converted from diesel to natural gas, then

- Check the acceptable load steps of the new engine: the diesel likely accepted 100% load, but the natural gas engine may be limited to as low as 20% load. Chillers and other large loads frequently exceed the acceptable load increase.
- Determine whether large-motor-load starting systems can be changed to reduce starting requirements and meet acceptable load steps.

If the emergency system is being modified for peaking or curtailment uses, then

- Obtain the specific utilities interconnection requirements; these tend to be utility-specific and may exceed normal NEC or local codes.
- Check the existing system for closed-transition or parallel-transfer switch issues and compatibility.
- Investigate emissions issues; standby emissions permit requirements are almost always less stringent than peaking requirements.

If the emergency system includes a UPS or flywheel system, then

- Size the UPS or flywheel for the maximum required load for a specific time; remember that capacity of these systems is x kW for y min.
- Avoid motor loads as much as possible.

Motors

Motor Control and Protection. Motor control must be effective without damaging the motor or its associated equipment. Control must be designed to prevent inadvertent motor starting caused by a fault in the control device. The control should be able to sense motor conditions to keep the motor windings from getting too warm.

Motor protection involves sensing motor current and line voltage, and can include bearing vibration, winding temperature, bearing temperature, etc. Motor temperature increase has two basic sources. Heating occurs when dirt or debris blocks airflow over or through the motor, or it comes from the motor current, commonly referred to as I^2R, where I is motor current and R is motor resistance. Because the current is squared, its contribution is exponential. R is the motor winding resistance, and is quite small. Therefore, motor resistance contributes a linear function to heating (and therefore temperature rise) in the motor. Motor windings can withstand temperature rise, depending on the motor winding temperature rating. The second source of motor temperature increase is lack of motor cooling. The primary source of cooling is moving air, usually from a shaft-driven fan. As a motor slows down, the fan runs more slowly; therefore, the less air movement, the less cooling. Because fan loads are also exponential, a small decrease in motor speed greatly reduces airflow on the motor, reducing cooling. To compound the issue, the slower the motor runs, the greater the slip, and the greater the motor current. This then becomes a vicious circle.

Motor Starters and Thermal Overloads. Motor starters start, stop, protect, and control the motor. They sense motor current based on a time curve: the shorter the time, the more current they let through. They may also limit the number of motor starts in a given period of time so the motor does not exceed its ANSI rating. Another cause for concern is the starter's ambient temperature: if it is different from the ambient temperature around the motor controller, the thermal overloads need to be sized accordingly.

Several motor starter types are available, and they can function several different ways. The **across-the-line** starter is the simplest but may disturb electric service because of high motor-starting current requirements. The **part-winding**, **reduced-voltage**, and **wye-delta** starters all act by reducing the starting voltage on the motor. Fan and pump loads can usually be started this way; this reduces starting current draw and demand, but sacrifices motor starting

speed. **Soft-start** starters also reduce the voltage and limit demand during starting, thereby easing stress on the electric system during motor starting. **Variable-frequency drives (VFDs)** are also used to soft-start a motor, and are designed to control acceleration during starting as well as optimize motor speed to its load. See the section Motor-Starting Methods for more information.

A VFD converts ac to dc and then back to variable-frequency, variable-voltage ac. Motor speed is varied by varying the frequency of the ac output voltage, typically using pulse-width modulation (PWM). This power conversion process results in distorted input current and may contribute to building power system voltage distortion. When motors and other loads are supplied from a distorted voltage source, their operating temperatures may rise.

Phase Loss Protection. Phase loss can be detected by sensing either voltage loss or current loss on one of the phases. Motor overloads often are set based on the nameplate data of the motor, but the real motor current draw is less because the motor may not be fully loaded. If one of three phases is lost, the current increases in the other two phases, but not by enough to trip the overloads. However, it will overheat and possibly damage the motor. Some phase loss detectors also check phase current to be sure that it is balanced within 10%. This protects the motor and makes the operator aware of potential motor or line problems.

Motor-Starting Effects. The following effects do not occur with all motors. For instance, brushless dc and inverter-driven motors, which are electronically controlled, do not have inrush currents that cause light dimming or sags.

Light Dimming or Voltage Sags. Light dimming or voltage sag associated with motor starts can be more than a nuisance. Motors have the undesirable effect of drawing several times their full load current while starting. This large current, by flowing through system impedances, may cause voltage sag that can dim lights, cause contactors to drop out, and disrupt sensitive equipment. The situation is worsened by an extremely poor starting displacement factor, usually in the range of 15 to 30%. If the motor-starting-induced voltage sag deepens, the time required for the motor to accelerate to rated speed increases. Excessive sag may prevent the motor from starting successfully. Motor-starting sags can persist for many seconds.

The Illuminating Engineering Society of North America (IESNA) is precise in describing lighting reactions. *Dimming* is an intentional technique to enhance the ambiance of surroundings by varying the perceived lighting levels. It is also used to reduce the electrical power used by lamps when adequate natural lighting is available; the controller electronically follows the natural variations in a way that will not be optically perceived. Similarly, *flicker* is deliberately selected in some lamps to resemble a candle's flame. In contrast, this chapter discusses possible causes of *undesirable, perceptible* reductions of a lamp's lumen output.

Momentary undesirable lighting reduction was quite common when building owners selected incandescent lamps, which compete with motors for available electrical current, to provide the desired lighting levels. More efficient lamps are now the norm for almost every application. For fluorescent lamps, electronic ballasts convert the 60 Hz electrical service to much higher frequencies, which eliminates perception of irritating flicker, and provides greater lumens per watt than the former magnetic ballasts. Similar improvements have occurred for mercury, metal halide, and sodium lamps, typically found in large stores, sports, and industrial applications. Including electronic circuitry in the lamps and lighting systems makes power quality even more important for building owners: electronic components are sensitive to both lower and upper threshold voltages, yet the building's voltage may experience momentary voltage sag as large motors are started. Specific product literature for lamps, lamp ballasts, and lighting controls should be reviewed to ensure client satisfaction with operation of the HVAC system's motors and controls.

Motor-Starting Methods. The following motor-starting methods can reduce voltage sag from motor starts.

An **across-the-line start**, energizing the motor in a single step (full-voltage starting), provides low cost and allows the most rapid acceleration. It is the preferred method unless the resulting voltage sag or mechanical stress is excessive.

Autotransformer starters have two autotransformers connected to open delta (similar to a delta connection using three single-phase transformers, but with one transformer removed; carries 57.7% of a full delta load). Taps provide a motor voltage of 80, 65, or 50% of system voltage during start-up. Starting torque varies with the square of the voltage applied to the motor, so the 50% tap delivers only 25% of the full-voltage starting torque. The lowest tap that will supply the required starting torque is selected. Motor current varies as the voltage applied to the motor, but line current varies with the square of the tap used, plus transformer losses of ~3%.

Resistance and **reactance starts** initially insert impedance in series with the motor. After a time delay, this impedance is shorted out. Starting resistors may be shorted out over several steps; starting reactors are shorted out in a single step. Line current and starting torque vary directly with the voltage applied to the motor, so for a given starting voltage, these starters draw more current than the line with autotransformer starts, but provide higher starting torque. Reactors are typically provided with 50, 45, and 37.5% taps.

Part-winding starters are attractive for use with dual-rated motors (220/440 V or 230/460 V). The stator of a dual-rated motor consists of two windings connected in parallel at the lower voltage rating, or in series at the higher voltage rating. When operated with a part-winding starter at the lower voltage rating, only one winding is energized initially, limiting starting current and torque to 50% of the values seen when both windings are energized simultaneously.

Delta-wye starters connect the stator in wye for starting, then after a time delay, reconnect the windings in delta. The wye connection reduces the starting voltage to 57% of the system line-line voltage, starting current and starting torque are reduced to 33% of their values for full voltage start.

Utilization Equipment Voltage Ratings

Utilization equipment is electrical equipment that converts electric power into some other form of energy, such as light, heat, or mechanical motion. Every item of utilization equipment should have a nameplate listing, which includes, among other things, the rated voltage for which the equipment is designed. In some cases the nameplate will also indicate the maximum and minimum voltage for proper operation. With one major exception, most utilization equipment carries a nameplate rating that is the same as the voltage system on which it is to be used: that is, equipment to be used on 120 V systems is rated 120 V. The major exception is motors and equipment containing motors, where performance peaks in the middle of the tolerance range of the equipment: better performance can be obtained over the tolerance range specified in ANSI *Standard* C84.1 by selecting a nameplate rating closer to the middle of this tolerance range. The difference between the nameplate rating of utilization equipment and the system nominal voltage is necessary because the performance guarantee for utilization equipment is based on the nameplate rating and not on the system nominal voltage.

The voltage tolerance limits in ANSI *Standard* C84.1 are based on ANSI/NEMA *Standard* MG 1, Motors and Generators edition, which establishes voltage tolerance limits of the standard low-voltage induction motor at ±10% of nameplate voltage ratings of 230 and 460 V. Because motors represent the major component of utilization equipment, they were given primary consideration in the establishment of this voltage standard. Figure 11 compares utilization voltages to nameplate ratings.

Voltage Level Variation Effects

Whenever voltage at the terminals of utilization equipment varies from its nameplate rating, equipment performance and life expectancy change. The effect may be minor or serious, depending

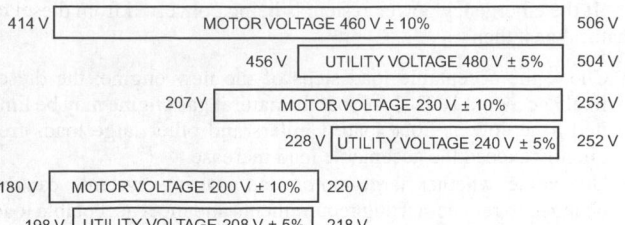

Fig. 11   Utilization Voltages Versus Nameplate Ratings

on the equipment characteristics and the amount of voltage deviation from the nameplate rating. NEMA standards provide tolerance limits within which performance will normally be acceptable. In precise operations, however, closer voltage control may be required. In general, a change in the applied voltage causes a proportional change in the current. Because the effect on the load equipment is proportional to the product of the voltage and the current, and because the current is proportional to the voltage, the total effect is approximately proportional to the square of the voltage.

However, the change is only approximately proportional and not exact: the change in the current affects the operation of the equipment, so the current continues to change until a new equilibrium position is established. For example, when the load is a resistance heater, the increase in current increases the heater temperature, which increases its resistance and, in turn, reduces the current. This effect will continue until a new equilibrium current and temperature are established. In the case of an induction motor, a voltage reduction reduces the current flowing to the motor, causing the motor to slow down. This reduces the impedance of the motor, increasing the current until a new equilibrium position is established between the current and motor speed.

Voltage Selection

Generally, the preferred utilization voltage for large commercial buildings is 480Y/277 V, three-phase. The three-phase power load is connected directly to the system at 480 V, and fluorescent ceiling lighting is connected phase-to-neutral at 277 V. Dry-type transformers rated 480 V/208Y/120 V are used to provide 120 V single-phase for convenience outlets and 208 V three-phase for other building equipment. Single-phase transformers with secondary ratings of 120/240 V may also be used to supply lighting and small office equipment. However, single-phase transformers should be connected in sequence on the primary phases to maintain balanced load on all phases of the primary system.

Where the supplying utility furnishes the distribution transformers, the choice of voltages will be limited to those the utility will provide. For tall buildings, space will be required on upper floors for transformer installations and the primary distribution cables supplying the transformers. Apartment buildings generally have the option of using either 208Y/120 V three-phase/four-wire systems, or 120/240 V single-phase systems, because the major load in residential occupancies consists of 120 V lighting fixtures and appliances. The 208Y/120 V systems are often more economical for large apartment buildings. Single-phase 120/240 V systems should be satisfactory for small apartment buildings and other small buildings.

However, large single-phase appliances, such as electric ranges and water heaters rated for use on 120/240 V single-phase systems, will not perform to the rated wattage on a 208Y/120 V systems, because the line-to-line voltage is appreciably below the rated voltage of the appliance.

POWER QUALITY VARIATIONS

Power quality refers to varied parameters that characterize the voltage and current for a given time and at a given point on the electric system. A power quality problem is usually any variation in the voltage or current that actually results in failure or misoperation of equipment in the facility. Therefore, power quality evaluations are a function of both the power system characteristics and the sensitivity of equipment connected to the power system.

This section defines the different kinds of power quality variations that may affect equipment operation. Important reasons for categorizing power anomalies include the following:

- Identifying the cause of the power anomalies. Understanding the characteristics of a power quality variation can often help identify the cause.
- Identifying the possible effects on equipment operation. A transient voltage can cause failure of equipment insulation; a sag in voltage may result in dropout of sensitive controls based on an undervoltage setting.
- Determining the requirements for measurement. Some power quality variations can be characterized with simple voltmeters, ammeters, or strip chart recorders. Other conditions require special-purpose disturbance monitors or harmonic analyzers.
- Identifying methods to improve the power quality. Solutions depend on the type of power quality variation. Transient disturbances can be controlled with surge arrestors, whereas momentary interruptions could require an uninterruptible power supply (UPS) system for equipment protection. Harmonic distortion may require special-purpose harmonic filters.

Power quality can be described in terms of *disturbances* and *steady-state variations*.

Disturbances. Disturbances are one-time, momentary events. Measurement equipment can characterize these events by using thresholds and triggering when disturbance characteristics exceed specified thresholds. Examples include transients, voltage sags and swells, and interruptions.

Steady-State Variations. Changes in the long-term or steady-state conditions can also result in equipment misoperation. High harmonic distortion levels can cause equipment heating and failure, as can long-term overvoltages or unbalanced voltages. These are variations best characterized by monitoring over a longer period of time with periodic sampling of the voltages and currents. Steady-state variations are best analyzed by plotting trends of the important quantities (e.g., RMS voltages, currents, distortion levels).

These two types of power quality are further defined in seven major categories and numerous subcategories. There are three primary attributes used to differentiate among subcategories within a power quality category: frequency components, magnitude, and duration. These attributes are not equally applicable to all categories. For instance, it is difficult to assign a time duration to a voltage flicker, and it is not useful to assign a spectral frequency content to variations in the fundamental frequency magnitude (sags, swells, overvoltages, undervoltages, interruptions).

Each category is defined by its most important attributes for that particular power quality condition. These attributes are useful for evaluating measurement equipment requirements, system characteristics affecting power quality variations, and possible measures to correct problems. The terminology has been selected to agree as much as possible with existing terminology used in technical papers and standards.

The following descriptions focus on causes of the power quality variations, important parameters describing the variation, and effects on equipment.

Transients

Transients are probably the most common disturbance on distribution systems in buildings and can be the most damaging. Transients can be classified as impulsive or oscillatory. These terms reflect the waveshape of a current or voltage transient.

Impulsive Transient. An impulsive transient (spikes or notches) is considered unidirectional; that is, the transient voltage or current wave is primarily of a single polarity (Figures 12 and 13). Impulsive transients are often characterized simply by magnitude and duration. Another important component that strongly influences the effect on many types of electronic equipment is the **rate of rise**, or rise time of the impulse. This rate of rise can be quite steep, and can be as fast as several nanoseconds. Repetitive subtractive transients (Figure 13), often caused by thyristors such as silicon controlled rectifiers (SCRs), are referred to as voltage notches.

The high-frequency components and high rate of rise are important considerations for monitoring impulses. Very fast sampling rates are required to characterize impulses with actual waveforms. In many power quality monitors, simple circuits are used to detect the transient's peak magnitude and duration (or volt-seconds). If impulse waveshapes are recorded, they usually do not include the fundamental frequency (60 Hz) component. When evaluating these disturbances, it is important to remember that stress on equipment is based on the impulse magnitude plus the magnitude of the fundamental component at the instant of the impulse. The voltage, current available, and pulse width determine the amount of energy available in a transient.

Oscillatory Transient. An oscillatory transient (Figure 14) is a voltage or current that changes polarity rapidly. Because the term "rapidly" is nebulous, the frequency content is used to divide

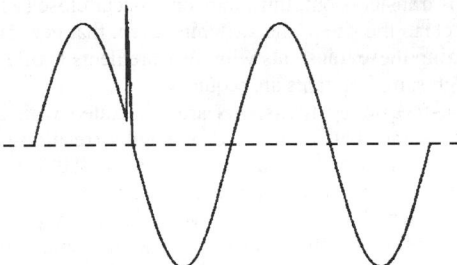

Fig. 12 Example of Spike

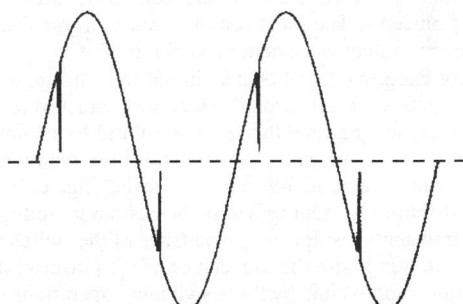

Fig. 13 Example of Notch

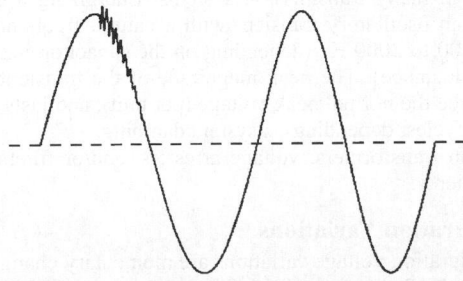

Fig. 14 Example of Oscillatory Transient

oscillatory transients into three subcategories: high, medium, and low frequency. Frequency ranges from these classifications are chosen to coincide with common types of power system oscillatory transient phenomena.

As with impulsive transients, oscillatory transients can be measured with or without including the fundamental frequency. One way to trigger on transients is to continually test for deviation in the waveform from one cycle to the next. This method will record any deviation exceeding the set threshold. When characterizing the transient, it is important to indicate the magnitude with and without the 60 Hz fundamental component.

Transients are generally caused by a switching event or by the response of the system to a lightning strike or fault. The oscillations result from interactions between system capacitances and inductances, and occur at the natural frequencies of the system excited by the switching event or fault.

High-frequency transients can occur at locations very close to the initiating switching event. Rise times created by closing a switch can be as fast as a few tens of nanoseconds. Short lengths of circuit have very high natural oscillation frequencies that can be excited by a step change in system conditions (e.g., operating a switch). Power electronic devices such as transistors and thyristors/SCRs can cause high-frequency transients many times during each cycle of the fundamental frequency. The transients can be in the tens or hundreds of kilohertz, and occasionally higher.

Because of the high frequencies involved, circuit resistance typically damps transients out; thus, they only occur close (within hundreds of feet) to the site of the switching event that generates them. Characterizing these transients with measurements is often difficult because high sampling rates are required.

Medium-frequency transients are associated with switching events with somewhat longer circuit lengths (resulting in lower natural frequencies). Switching events on most 480 V distribution systems in a facility cause transient oscillations within this frequency range, which can propagate over a significant portion of the low-voltage system. Motor interruption (definite interruption) is a good example of a common switching event that can excite transients in this frequency range.

Transients coupled from the primary power system (e.g., coupled through the step down transformer) can also cause medium-frequency transients. The most common cause of transients on the primary power system is capacitor switching.

Capacitor energizing results in an initial step change in the voltage, which gets coupled through stepdown transformers by the transformer capacitance and then excites natural frequencies of the low-voltage system (typically 2 to 10 kHz). **Low-frequency transients** are usually caused by capacitor switching, either on the primary distribution system or within the customer facility. Lower-frequency transients result from capacitance of the switched capacitor bank oscillating with the inductance of the power system. The natural frequencies excited by these switching operations are much lower than those of the low-voltage system without the capacitor bank, because of the large capacitance of the capacitor bank itself.

Capacitor switching operations are common on most distribution systems and many transmission systems. Energizing a capacitor results in an oscillatory transient with a natural frequency in the range of 300 to 2000 Hz (depending on the capacitor size and the system inductance). The peak magnitude of the transient can approach twice the normal peak voltage (per unit), and lasts between 0.5 and 3 cycles, depending on system damping.

Isolation transformers, voltage arresters, and/or filters can reduce transients.

Short-Duration Variations

Short-duration voltage variations are momentary changes in the fundamental voltage magnitude. Common causes are faults on the power system (short circuits between phases or from phase to ground). Depending on the fault location and system conditions, the fault can cause either momentary voltage rises (swells) or momentary voltage drops (sags). The fault condition can be close to or remote from the point of interest.

Sags. Sags (Figure 15) are often associated with system faults but can also be caused by switching heavy loads or starting large motors (usually a longer-duration variation). Figure 15 shows a typical voltage sag that can be associated with a remote fault condition. For instance, a fault on a parallel feeder circuit (on the primary distribution system) results in a voltage drop at the substation bus that affects all of the other feeders until the fault is cleared by opening a fuse or circuit breaker.

The percent drop in the RMS voltage magnitude and duration of the low-voltage condition are used to characterize sags. Voltage sags are influenced by system characteristics, system protection practices, fault location, and system grounding. The most common problem caused by voltage sags is tripping sensitive controls (e.g., adjustable-speed drives or process controllers), relays or contactors dropping out, and failure of power supplies to ride through the sag. Many types of voltage regulators are not fast enough to provide voltage support during sags, but ferroresonant transformers and some other line conditioners can provide some ride-through capability or can quickly compensate for deep sags.

In practice, sags are the type of power quality variation that most frequently causes problems. Fault conditions remote from a particular customer can still cause voltage sags that can cause equipment problems. Because there are no easy ways to eliminate faults on the power system, it is always necessary for customers to consider the effects of sags.

Swells. Swells or **surges** can also be associated with faults on the primary distribution system (Figure 16). They can occur on non-faulted phases when there is a single-line-to-ground fault.

Swells are characterized by their magnitude (RMS value) and duration. The severity of a voltage swell is a function of fault location, system impedance, and grounding. On a three-phase ungrounded system (delta), the line-to-ground voltages on the ground phases are 1.73 per unit (i.e., 1.73 times the normal line-to-ground voltage) during a single-line-to-ground fault condition. Close to the substation on a grounded system, there is no voltage rise on the ground phases because the substation transformer is usually delta-wye, providing a low-impedance path for the fault current.

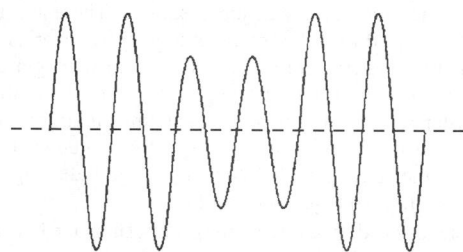

Fig. 15   Example of Sag

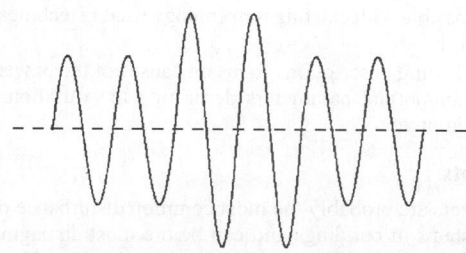

Fig. 16   Example of Swell (Surge)

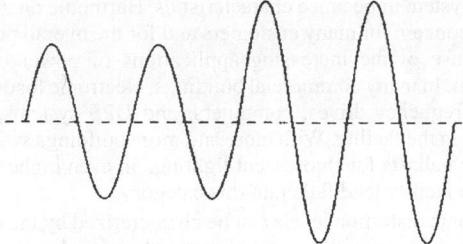

Fig. 17   Example of Overvoltage

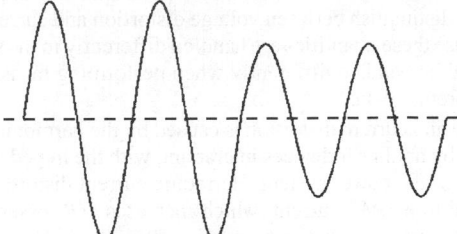

Fig. 18   Example of Undervoltage

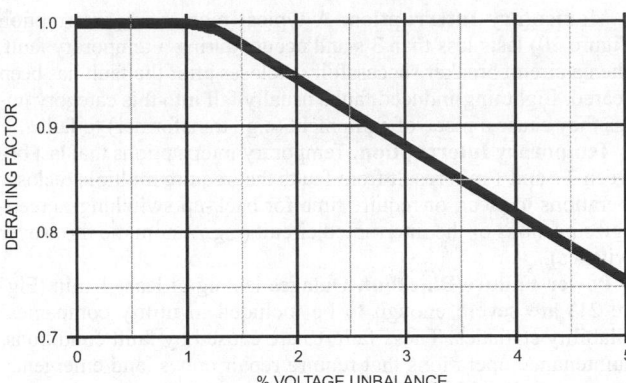

Fig. 19   Derating Factor Curve

Long-Duration Variations

Long-duration RMS voltage deviations generally do not result from system faults. They are caused by load changes on the system and system switching operations. The duration of these voltage variations depends on the operation of voltage regulators and other types of voltage control on the power system (e.g., capacitor controls, generator exciter controls). The time required for these voltage controllers to respond to system changes ranges from large fractions of a second to seconds. Long-duration variations can be overvoltages or undervoltages, depending on the cause of the variation. Voltage unbalance should be considered when evaluating steady-state or long-duration voltage variations. Unbalanced voltages can be one of the major causes of motor overheating and failure. With increasing emphasis on energy-efficient motors, requirements for voltage balance (i.e., limitations on negative sequence voltage magnitudes) may become even more important.

Overvoltages. Overvoltages (Figure 17) can result from load switching (e.g., switching off a large load), variations in system generation, or variations in reactive compensation on the system (e.g., switching a capacitor bank on). These voltages must be evaluated against the long-duration voltage capability of loads and equipment on the system. For instance, most equipment on the power system is only rated to withstand a voltage 10% above nominal for any length of time. Many sensitive loads can have even more stringent voltage requirements.

Long-duration overvoltages must also be evaluated with respect to the long-time overvoltage capability of surge arresters. Metal oxide variation (MOV) arresters in particular can overheat and fail due to high voltages for long durations (e.g., seconds).

Overvoltages can be controlled with voltage regulation equipment either on the power system or in a customer's facility. This can include various tap-changing regulators, ferroresonant regulators, line power conditioners, motor-generator sets, and uninterruptible power supplies.

Undervoltages. Undervoltages (Figure 18) have the opposite causes of overvoltages. Adding a load or removing a capacitor bank will cause an undervoltage until voltage regulation equipment on the system can bring the voltage back to within tolerances.

Motor starting is one of the most common causes of undervoltages. An induction motor draws 6 to 10 times its full load current during starting. This lagging current causes voltage drops in the system impedance. If the started motor is large enough relative to the system strength, these voltage drops can result in a significant system undervoltage. The magnitude of this starting current decreases over a period ranging from 1 s to minutes, depending on the inertia of the motor and the load, until the motor reaches full speed. This type of undervoltage can be mitigated by using various starting techniques to limit the starting current and is largely self-corrected when the starting is completed. For more information, see the section on Motor-Starting Effects.

Voltage Unbalance. Ideally, all phase-to-phase voltages to a three-phase motor should be equal or balanced. Unbalance between the individual phase voltages is caused by unbalanced loading on the system and by unbalances in the system impedances. Voltage unbalance is an important parameter for customers with motors because most three-phase motors have fairly stringent limitations on negative-sequence voltage (a measure of voltage unbalance), which is generated in the motor by unbalances in supplied voltages. Negative sequence currents heat the motor significantly. Voltage unbalance limitations (and steady-state voltage requirements in general) are discussed in ANSI *Standard* C84.1. The National Electrical Manufacturers Association (NEMA) has developed standards and methods for evaluating and calculating voltage unbalance. Unbalance, as defined by NEMA, is calculated by the following equation:

$$\% \text{ Voltage unbalance} = 100 \times \frac{\text{Maximum deviation from average voltage}}{\text{Average voltage}}$$

The motor derating factor caused by the unbalanced voltage curve from NEMA *Standard* MG 1 shows the nonlinear relationship between the percent of voltage unbalance and the associated derating factor for motors (Figure 19). A balanced-voltage three-phase power supply to the motor is essential for efficient system operation. For example, a voltage unbalance of 3.5% can increase motor losses by approximately 15%.

Interruptions and Outages

Interruptions can result from power system faults, equipment failures, generation shortages, control malfunctions, or scheduled maintenance. They are measured by their duration (because the voltage magnitude is always zero), which is affected by utility protection system design and the particular event causing the interruption.

Interruptions of any significant duration can potentially cause problems with a wide variety of different loads. Computers, controllers, relays, motors, and many other loads are sensitive to interruptions. The only protection for these loads during an interruption is a back-up power supply, a back-up generator (requires time to get started) or a UPS system (can constantly be online).

Momentary Interruption. A typical momentary interruption (Figure 20) lasts less than 3 s and occurs during a temporary fault, when a circuit breaker successfully recloses after the fault has been cleared. Lightning-induced faults usually fall into this category unless they cause a piece of equipment (e.g., transformer) to fail.

Temporary Interruption. Temporary interruptions that last between 3 s and 1 min result from faults that require multiple recloser operations to clear, or require time for back-up switching to reenergize portions of the interrupted circuit (e.g., automatic throwover switches).

Power Failure/Blackout. Outages lasting at least 1 min (Figure 21) are severe enough to be included in utility companies' reliability statistics. These failures are caused by fault conditions, maintenance operations that require repair crews, and emergency situations called blackouts.

Solutions involve using either UPS systems or back-up generators, depending on the critical nature of the load. UPS systems typically can provide uninterrupted supply for at least 15 min (based on battery capacity). This covers all momentary and temporary interruptions and provides sufficient time for an orderly shutdown. A UPS can be used in conjunction with a switching scheme involving multiple feeds from the utility to provide an even higher level of reliability. If back-up power is required beyond the capability of a UPS system, and multiple feeds are not realistic or adequate, then back-up generators are needed. On-site generators are typically used in these applications.

Brownout. A brownout is a long-term voltage reduction, usually of 3 to 5%. This is an intentional reduction to reduce load under emergency system conditions.

Harmonic Distortion

Harmonic distortion of the voltage waveform occurs because of the nonlinear characteristics of devices and loads on the power system. These nonlinear devices fall into one of three categories:

• Power electronics
• Ferromagnetic devices (e.g., transformers)
• Arcing devices

These devices usually generate harmonic currents, and voltage distortion on the system results from these harmonics interacting with the system impedance characteristics. Harmonic distortion is a growing concern for many customers and for the overall power system because of the increasing applications of power electronic equipment. In many commercial buildings, electronic loads, such as variable-frequency drives, computers, and UPS systems, may be dominant in the facility. With more and more buildings switching to electronic ballasts for fluorescent lighting, an even higher percentage of the facility load falls into this category.

Harmonic distortion levels can be characterized by the complete harmonic spectrum with magnitudes and phase angles of each individual harmonic component. However, it is more common to use a single quantity, the **total harmonic distortion (THD)**, to characterize harmonic distortion of a particular waveform. It is important in general to distinguish between voltage distortion and current distortion because these quantities are handled differently in the standards and should be handled differently when performing measurements and interpreting data.

Voltage and current distortion is caused by the harmonic currents generated by nonlinear devices interacting with the impedance characteristics of the power system. Harmonic current distortion results in elevated true RMS current, which increases I^2R losses and elevated peak current. Because harmonics flow at frequencies higher than the fundamental frequency (e.g., 180 Hz, 300 Hz, 420 Hz), additional losses are experienced because of the reduction of conductor effective cross-sectional area, a phenomenon known as **skin effect**. These losses are attributed to I^2X_L at each harmonic frequency.

A particular concern is when resonance conditions on the power system magnify harmonic currents and high-voltage distortion levels. The natural resonance of the power system varies based on system inductance and capacitance and should be evaluated when adding nonlinear devices (equipment) or power factor capacitors to the system. Capacitors offer a low-impedance path to harmonic frequencies and can therefore attract harmonics, often resulting in reduced life, fuse blowing, or capacitor failure. Figure 22 illustrates the voltage waveform with harmonic content. Figures 23 and 24 illustrate distorted current waveforms.

Harmonic distortion can be reduced by adding an ac line reactor or harmonic filter at the input of individual nonlinear loads (Figure 25). A 5% impedance line reactor typically reduces

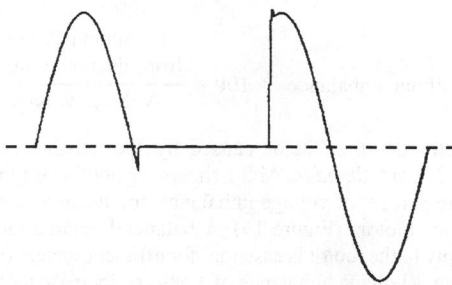

Fig. 20 Example of Momentary Interruption

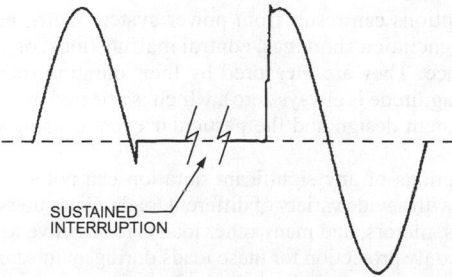

SUSTAINED
INTERRUPTION

Fig. 21 Example of Blackout or Power Failure Waveform

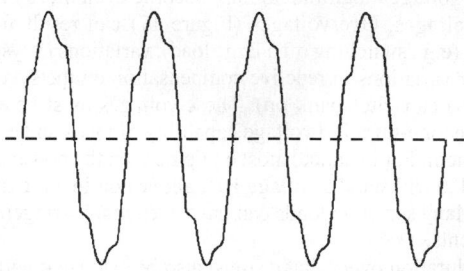

Fig. 22 Example of Harmonic Voltage Distortion

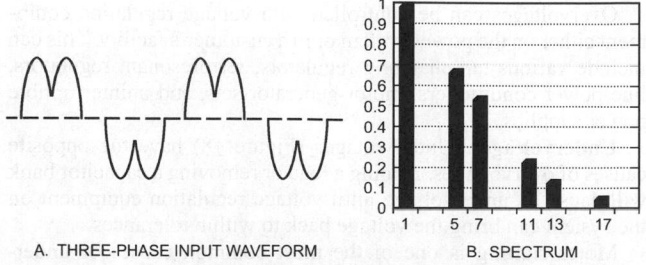

A. THREE-PHASE INPUT WAVEFORM   B. SPECTRUM

Fig. 23 Example of Harmonic Current Distortion for Six-Pulse Rectifier with 5% Impedance Reactor

harmonic distortion for three-phase nonlinear loads to about 35% of the fundamental current.

Harmonic current distortion can be reduced to levels of 5 to 8% total harmonic current distortion, at the individual load, using a typical low-pass harmonic filter (Figure 26). The benefit of reducing harmonics right at the contributing load is that the entire upstream power system benefits from reduced levels of current and voltage distortion.

Voltage Notches. A voltage waveform with notches (see Figure 13) caused by operating power electronics, especially where SCRs are involved [e.g., adjustable-speed drives (ASDs)], can be considered a special case that falls in between transients and harmonic distortion. Because notching occurs continuously (steady state), it can be characterized by the harmonic spectrum of the affected voltage. However, frequency components associated with the notching can be quite high, and it may not be possible to characterize them with measurement equipment normally used for harmonic analysis. It is usually easier to measure with an oscilloscope or transient disturbance monitor.

Three-phase SCR rectifiers (those typically in dc drives, UPS systems, ASDs, etc.) with continuous dc current are the most common cause of voltage notching. The notches occur when the current commutates from one phase to another on the ac side of the rectifier. During this period, there is a momentary short circuit between two phases. The severity of the notch at any point in the system is determined by the source of inductance and the inductance between the rectifier and the point being monitored.

Often, an isolation transformer or 3% impedance ac line reactor (inductor) can be used in the circuit to reduce the effect of notching on the source side. The additional inductance increases the severity of voltage notches at the rectifier terminals (commutation time, or width of the notch, increases with increased commutation reactance); however, most of the notching voltage appears across the ac inductor and notching is less severe on the source side, where other equipment shares a common voltage source.

Steep voltage changes caused by notching can also result in ringing (oscillation) because of capacitances and inductances in the supply circuit. This oscillation can disturb sensitive controls connected to the affected circuit. The high frequencies involved can also cause noise be capacitively coupled to adjacent electrical or communication circuits.

Voltage Flicker

Loads that vary with time, especially in the reactive component, can cause voltage flicker; the varying voltage magnitudes can affect lighting intensity. Arc furnaces are the most common cause of voltage flicker. The envelope of the 60 Hz variations is defined as the flicker signal V, and its RMS magnitude is expressed as a percentage of the fundamental. Voltage flicker is measured with respect to sensitivity of the human eye. A typical plot of the 60 Hz voltage envelope characterizing voltage flicker is shown in Figure 27.

The characteristics of voltage flicker are mainly determined by load characteristics and the system short-circuit capacity. For a critical load, it may be necessary to provide a dedicated feed so that it is not on the same circuit with a major load that causes voltage flicker. Using fast switching compensation, such as a static volt-ampere-reactive (VAR) system, or dynamic VAR compensation system, can mitigate the problem. Another method is to effectively increase the short-circuit capacity at the point of common coupling with other loads by using a series capacitor. Protecting the series capacitor during fault conditions requires careful design.

Voltage flicker appears as a modulation of the fundamental frequency (similar to amplitude modulation of an AM radio signal).

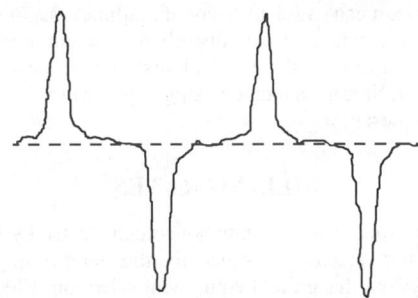

Fig. 24   Example of Harmonic Current Distortion for One-Phase Input Current for Single Personal Computer

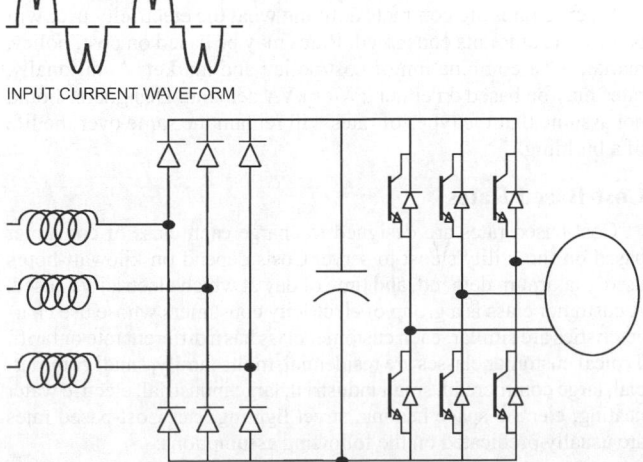

INPUT CURRENT WAVEFORM

Fig. 25   Example of VFD with ac Line Reactor

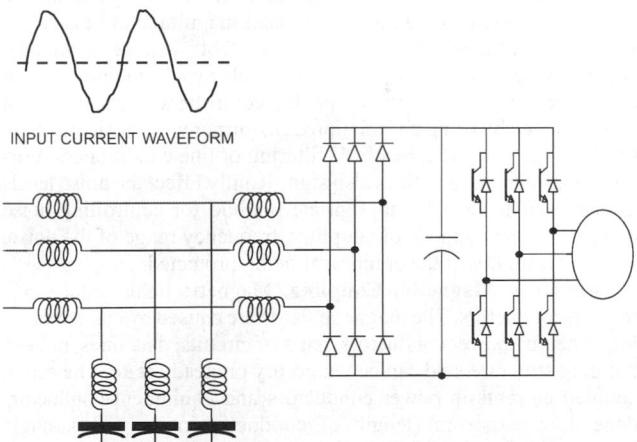

INPUT CURRENT WAVEFORM

Fig. 26   Example of VFD with Low-Pass Harmonic Filter

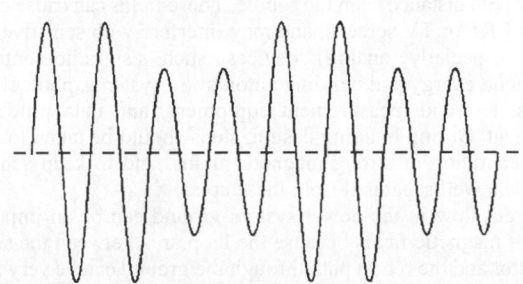

Fig. 27   Example of Flicker

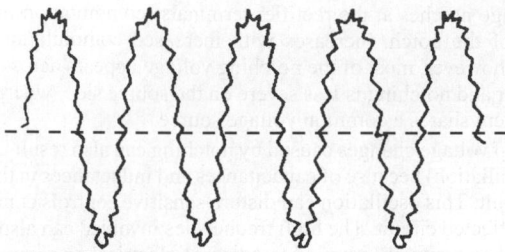

Fig. 28 Example of Electrical Noise

Therefore, it is easiest to define a magnitude for voltage flicker as the RMS magnitude of the modulation signal, which can be found by demodulating the waveform to remove the fundamental frequency and then measuring the magnitude of the modulation. Typically, magnitudes as low as 0.5% can result in perceptible light flicker if frequencies are in the range of 1 to 10 Hz. Flicker limitations are discussed in ANSI/IEEE *Standard* 146.

Light dimming, another type of light flicker, is caused by starting motors. Large single- or three-phase motors, as used in air conditioners, have the undesirable effect of drawing 6 to 10 times their full load current while starting. This large current, by flowing through system impedances, causes voltage sag that may dim lights.

Noise

Noise, a continuous, unwanted signal on the power circuits (Figure 28), can have a wide variety of different causes (e.g., switching, arcing, electric fields, magnetic fields, radio waves) and can be coupled onto the power circuit in a number of different ways. The noise source and susceptible circuit can be coupled by electric or magnetic fields or by electromagnetic interference (EMI).

The frequency range and magnitude of noise depend on the source that produces the noise. A typical magnitude of noise measured in the voltage is less than 1% of the RMS voltage magnitude. Noise having enough amplitude disturbs electronic equipment such as microcomputers and programmable controllers. Some noise can be eliminated by using an isolation transformer with an electrostatic shield; other noise requires EMI filtering or line conditioners. Wiring and grounding practices also significantly affect the noise levels at particular loads. The appropriate method for controlling noise depends on the methods of coupling, frequency range of the noise, and susceptibility of the equipment being protected.

Inductive (Magnetic) Coupling. Magnetic fields induce currents in conductors. The magnetic fields are caused by current flowing in nearby power conductors, parts of circuits, data lines, or even building structure, and can be temporary or steady state. The actual coupled currents in power conductors and equipment conductors depend on exposure (length of conductor in the field), angle between conductor and field, and magnetic field strength. Conductors carrying large currents and/or with large spacing between conductors can create strong 60 Hz magnetic fields that do not decay quickly with distance from the source. These fields can cause distortion on CRT or TV screens, and may interfere with sensitive electronic (especially analog) devices such as radio-controlled equipment, energy and building automation systems, photoelectric sensors, test and measurement equipment, and data processing equipment. During building design, steps should be taken to minimize generation of strong magnetic fields, and to keep sensitive equipment well separated from these areas.

Current flow in the power system ground can be an important cause of magnetic fields because the loop area between the supply conductor and the return path through the ground can be very large. Therefore, grounding techniques to minimize noise levels can help reduce magnetic field problems. To minimize interference, power conductors should always be physically separated from control circuit conductors. Magnetic shielding can also help.

Capacitive (Electrostatic) Coupling. Capacitive coupling between conductors results in coupling of transient voltage signals between circuits. Transient voltages with high-frequency components or high rates of rise are the most likely to be capacitively coupled between circuits (the coupling capability of a capacitor increases with frequency). Switch operations, arcing, lightning, or electrostatic discharge can cause these transients. Electric fields capacitively couple voltage between conductors. Strength is measured in volts per metre and may range from very slight to many kilovolts per metre.

High magnitudes of capacitively coupled voltages can affect normal operations of various types of electronic devices, or even cause discharges and damage. Possible solutions include applying shielding and coating, and improving design of power equipment to reduce the generation of high-transient-voltage conditions. To minimize interference, power conductors should always be physically separated from control circuit conductors.

Electromagnetic Interference (EMI). EMI refers to interference caused by electromagnetic waves over a wide range of frequencies. Many interference sources start out as either strongly magnetic or strongly electric, but within about half a wavelength the fields convert to a balanced ratio of electric and magnetic fields (an electromagnetic field). Transients such as electrostatic discharge, arcing, contacts, power electronic switching, fluorescent lighting, and lightning cause electromagnetic waves. Steady-state EMI can occur in the form of radio frequency interference (RFI) from microwave and radar transmissions, radio and TV broadcasts, corona of high-voltage transmission line, arc welding, and other sources that generate radio-frequency electromagnetic waves. Although RFI is not destructive, it can cause a variety of malfunctions of susceptible electronic equipment and can disturb microcomputers and programmable controllers; the level of disturbance depends on the amount of RFI. Solutions include using appropriate shielding or filtering techniques.

BILLING RATES

Equipment specifications state how much electricity is used, but the cost of that electricity is usually the determining factor in HVAC&R system design and equipment selection. Electricity tariffs or rates set prices for

- How much electricity is used; energy (kWh)
- Rate at which electricity is used; demand (kW)
- Quality of electricity used; power factor (VAR or kVAR)

Electric rates are contracts defining what the electricity user will pay for the amounts consumed. Rates may be based on cost, policy, market, or a combination of cost/policy and market. Additionally, rates may be based on either kW or kVA demand. Designers should not assume that the types of rates will remain the same over the life of a building.

Cost-Based Rates

Cost-based rates are designed to charge each class of consumer based on the utility's cost to serve. Costs depend on kilowatt-hours used, maximum demand, and time of day at which electricity is used. A customer class is a group of electricity consumers whose use characteristics are similar; each customer class has a different rate or tariff. Typical customer classes are residential, multi-family, small commercial, large commercial, small industrial, large industrial, electric water heating, electric space heating, street lighting, etc. Cost-based rates are usually predicated on the following assumptions:

- The more electricity a customer uses the less it costs, per kilowatt-hour, to serve that customer.

- The higher the demand of a customer, the more it costs to serve that customer.
- It costs less per kilowatt-hour to serve a customer with a higher load factor. Load factor LF is defined as the customer's average demand divided by the peak demand, or the energy consumed in the billing period divided by the peak demand times the number of hours in the billing period. [LF = kWh used in the billing period/(peak demand × hours in billing period)]
- It costs less for a utility to deliver electricity at times of low system load than at times of high system load.

Energy Charge. The consumer pays the utility a fixed amount for every kilowatt-hour used. Small customers, especially residential, often simply pay for energy used. Certain loads may have usage profiles that allow the utility to provide electricity at times of low production costs (e.g., street lighting) or cost less per kilowatt-hour because the customer uses significantly more than a typical customer (e.g., electric space heating). These loads are often metered separately and are charged a lower cost per kilowatt-hour than general-service usage.

Fuel Adjustment Clause (FAC). A significant part of the cost of electricity is the fuel needed to generate it. In the 1970s, the price of primary energy sources (especially oil) became extremely volatile, and the fuel adjustment clause was designed to accommodate this without requiring frequent rate adjustments. Energy charges with FACs consist of two parts: a fixed charge per kilowatt-hour and a variable charge per kilowatt-hour that depends on the average price of fuel purchased by the power generator. During periods of low fossil fuel prices or high hydro runoff, for example, the FAC results in lower prices to the consumer.

Demand Charge. "Demand" is the maximum rate of use of electricity. It is expressed in kilowatts (kW) or kilovolt-amps (kVA) and is typically measured over 15, 30, or 60 min periods. For example, 1 kWh used in 15 min is equal to 4 kW demand (1 kWh/0.25 hours). Demand charges are designed to cover the system capacity cost to deliver energy to a customer and/or the marginal generation cost to produce electricity at time of highest usage. To deliver electricity to a consumer, a utility must install wires, transformers, and meters; the higher the customer's projected demand, the larger the capacity of wires and transformers serving the customer must be. From a system-wide perspective, the utility must also build enough generation and transmission capacity to serve the system load at its highest peak level. A **noncoincident demand (NCD) charge** is a charge per kilowatt for the customer's maximum demand for electricity (in any 15, 30, or 60 min period) during the billing cycle. Noncoincident demand charges are generally imposed to cover the cost to transform and deliver energy to the consumer. A **coincident demand (CD) charge** (also known as a **peak** or **on-peak demand charge**) is a charge per kilowatt for the customer's maximum demand for electricity in any 15, 30, or 60 min period occurring during times of high system load. For example, in a summer-peaking utility, the demand charge may be applied only to the maximum demand occurring between 11:30 AM and 6:00 PM on weekdays from May to September. CD charges are designed to pay for additional generation and other system reinforcement costs needed to meet peak demands.

Ratcheted Demand Charge. Demand charges may be calculated based on the customer's maximum demand during each billing cycle, or the maximum demand during the 12 mo preceding the electricity bill. A high demand in one month "ratchets" the demand charge up for some or all of the following 11 months.

Seasonal Rate. Some utilities' generation costs vary significantly from one season to another. For example, spring runoff may yield more low-cost hydroelectric energy, or high summer air-conditioning loads may result in more power generation by less efficient peaking plants. A seasonally adjusted rate reflects this by setting different kilowatt and/or kilowatt-hour charges for different seasons.

Time of Use (TOU) Rate. A utility's average production cost for electricity usually varies with the total system load. As demand increases, less efficient (i.e., more costly) generators are used. Increasing peak loads also require that a utility invest in greater generation, transmission, and distribution capacity to meet the peak. A TOU rate is designed to recover the increased production or capacity costs during times of system peak. Electricity use is recorded by multiregister meters and priced at different levels, depending on whether the peak/off-peak or peak/shoulder/off-peak model is used. TOU rates are usually designed not only to "recover" time-differentiated production costs, but also to induce consumers to shift their electricity usage from peak periods to times of lower system load. In this way, TOU rates are both "cost-based" and "policy-based." Thermal energy storage (TES) systems are one method for consumers to reduce their on-peak demand or energy charges and to consume electricity during lower-cost, off-peak times. With TES, a consumer may use the same or more total electricity but will pay less for it.

Declining Block Rate. In designing cost-based rates, the cost to serve is usually inversely proportional to the amount consumed. To reflect this, energy may be priced according to a declining block rate, where the cost per kilowatt-hour decreases as usage increases. For example, a residential customer may pay $0.12/kWh for the first 800 kWh used in a month, $0.10/kWh for the next 550 kWh, and $0.08/kWh for all electricity above 1350 kWh.

Demand-Dependent Block Rate. For larger customers, the size of the blocks of a block rate may depend on the measured demand. For example, a commercial consumer may pay $0.12/kWh for the first 75 kWh per kilowatt of billing demand, $0.10/kWh for the next 150 kWh per kilowatt of billing demand, and $0.08/kWh for every kWh above 225 kWh per kilowatt of billing demand.

Load Factor Penalty. Utilities recover their fixed costs (e.g., capital cost of a transformer) as well as production costs through energy (kWh) charges as well as demand charges. If a consumer's load factor is less than expected, then the utility's kilowatt-hour revenues may not be sufficient to recover its fixed costs. This may occur for a customer with self-owned on-site generation who relies on the utility mainly when the on-site generator is being maintained. In this case, the utility may impose a load factor penalty or surcharge.

Power Factor Penalty. A power system is more efficient and stable when all three phases are equally loaded (balanced) and serving pure resistive load (100% power factor). Some inductive loads, such as most lighting and motors, require reactive power, which may be more difficult and expensive for the utility to supply. Moreover, reactive power (kVAR) is not always measured, and therefore not billed, by typical kilowatt-hour meters. Therefore, utilities often impose a power factor penalty or surcharge on customers with very reactive or inductive loads (power factor not close to 100%) to pay for the VARs that it must supply.

Customer Charge. This is a monthly, usually fixed charge that a consumer must pay, regardless of whether any electricity is used, for being a customer of a utility. This charge typically covers customer services such as the costs of billing, metering, and customer support services, such as call centers.

Connection Fee. When electric service is initiated, especially when construction (additional distribution lines, substations, distribution transformers, etc.) is required, the utility may charge a connection fee. For example, a utility may charge a residential customer a fixed cost per foot of distribution line that must be constructed beyond an initial 500 ft of line. Some customers may require redundant facilities to ensure reliable service (e.g., a second feeder and load transfer switch for a hospital). The utility would probably include such costs in its connection fee.

Policy-Based Rates

Policy-based rates are designed to encourage consumers to modify their energy use to better conform with the objectives of the utility or legislative or regulatory body (e.g., using nonpolluting or renewable energy sources, deferring grid expansion or generator construction by shifting electricity demand from peak periods, better using waste heat from industrial processes, etc.). It can be argued that some policy-based rates are in fact cost-based, but they incorporate "externalities," or external costs that cannot be directly allocated to a consumer's electricity use. The time of use rate could fall under either category.

Inverted Block Rate. The marginal cost to provide an existing customer with additional energy decreases as energy use per kilowatt of connected load increases. However, additional energy use often hastens the need to construct new generation facilities. An inverted block rate motivates the consumer to reduce energy use by charging less for the first kilowatt-hour used. For example, a residential consumer may pay $0.08 for the first 850 kWh per month, $0.10 for the next 500 kWh, and $0.12 for all usage above 1350 kWh. As with the declining block rate, the break points between "blocks" may be based on demand for that billing cycle.

Lifeline Rate. Lifeline rates are designed so poor consumers will still be able to afford necessary electricity (e.g., enough for refrigeration, lights, adequate heat in winter, etc.). For example, a consumer pays a subsidized price per kilowatt-hour for a minimal amount of electricity and the market rate for any usage above the minimum (e.g., $0.04/kWh for the first 675 kWh and $0.12/kWh for usage above 675 kWh/month).

Net Metering. This is applicable for consumers who own their own on-site generation, still buy electricity from the grid, but sometimes can generate more electricity than is needed in their facility. Net metering is a contract in which the customer pays the utility for the net electricity purchased (i.e., excess on-site generation is sold to the grid and offsets what the customer owes the utility for purchases from the grid).

Green Power Rate. Customers may sign up for blocks of electricity produced by renewable sources, such as wind turbines. Because of electricity from such sources usually costs more than electricity produced by conventional generation, the green energy is sold at a premium.

Surcharges. Government agencies may add special surcharges to electric bills to provide funding for specific energy-related or general-purpose programs. For example, in the United States, many state governments collect public benefit funds through electric bills to provide money for energy efficiency, renewable energy, low-income weatherization, and research programs. Other surcharges may be for power plant upgrades or putting electric distribution lines underground. Surcharges can be fixed or charged on a per-kWh basis. Some surcharges may be capped at a specific dollar amount per type of customer.

Taxes. Government agencies may collect sales and other specific taxes through utility bills. The taxes may be calculated as a percentage total of the entire bill, or on a per-kWh basis (such as $0.001 per kWh). For some customers, taxes and surcharges may comprise a significant portion of the total bill.

Market-Based Rates

Electric utilities are restructuring to disaggregate electricity production, transmission, and distribution and open the market to competition. The theory is that a competitive electricity sector, governed by market rules, is more efficient, lower-cost, and more congruent with consumer needs than regulated electric utilities. Market-based rates are not new, but they are becoming more prevalent. Rates tend to be volatile, and are structured as a contract between the consumer and energy supplier, rather than as a traditional tariff. As a result, they tend to be more customer-specific than uniform over a customer class.

Real-Time Pricing (RTP). Under this scheme, the cost of electricity varies with each hour. The supplier sets a price for electricity based on its forecasted cost to produce or provide the electricity. Hour-by-hour prices are communicated to the customer from 1 to 24 h in advance, and the consumer decides what, if any, action to take in response to the forecasted prices. The most common RTP programs send prices to consumers each evening, to cover the next day. Some programs send prices 4 h in advance, and several also allow 1 h alerts for "surprise" prices during system emergencies or forced outages. Consumers who were on a demand (kilowatt) and energy (kilowatt-hour) rate often are billed only for energy use (kilowatt-hour) on RTP, because the hourly energy cost incorporates the demand charge. The RTP may only apply to the generation portion of the electric bill if transmission and distribution charges are still regulated.

Fixed Pricing. Some consumers in a deregulated market may opt for fixed prices for their electric supply. In these situations, the consumer may pay a price that is higher or lower than the RTP or spot prices. With this type of pricing, the burden of price risk goes to the supplier, who may charge a risk premium to the customer.

Spot Pricing. Consumers in some regions may purchase some or all of their electricity on the spot market, based on the current marginal cost of electricity. This is done through a power exchange, with the consumer either contracting directly with the exchange or going through a third-party electricity broker. Consumers may also purchase electricity through a combination of long-term contracts and spot market purchases.

Interruptible Rates and Responsive Loads. At times of high marginal electricity costs, it may be more cost-effective for a utility to pay its customers to reduce electricity consumption than to contract for additional electricity supplies. With interruptible rates, a consumer agrees to reduce power consumption to or below an agreed-upon level (or go off-line) when requested to do so by the utility, in return for a lower price. The utility's requests may be limited in number of times per year (or month) the consumer can be asked to reduce load, maximum duration of the load reduction, and minimum notice required (typically 1 to 4 h) before electric load is reduced. In other cases, there is no limit to the number or length of utility requests.

Direct Load Control. This is similar to interruptible load, but instead of the consumer's complying with the utility's request, the utility can exercise direct control over the consumer's appliances. Appliances commonly contracted for load control programs are water heaters, swimming pool pumps, air conditioners, heat pumps, resistance space heat, and controllable thermostats for HVAC&R systems. The tariff usually is in the form of a monthly rebate or fixed bill credit for each controllable appliance.

Performance-Based Rates. These are designed to ensure that the consumer receives acceptable quality and reliability of electric supply. The utility and consumer agree on a minimum standard for service quality in terms of number and duration of outages, voltage sags and swells, harmonic levels, or other transient phenomena. If these minimum service quality levels are not met, the utility must rebate money to the consumer; the amount rapidly increases as performance or service quality declines. In some cases, the rate, or type of rate charged, may not change, but customers receive bill credits or rebates based on a reduction in the electric company's rate of return.

Performance Contracting. In performance contracting, a consumer contracts with a third party to pay for the end-use applications of electricity. This usually involves an agreement where the performance contractor [often called an energy service company (ESCO)] installs and sometimes operates and maintains improved equipment for HVAC&R systems, lighting, building or process energy management, etc. The consumer's payments are indexed to successful equipment performance. That performance is often evaluated in terms of the facility's utility bills and calculated cost savings comparing the actual energy costs with estimates of what the costs would have been without the ESCO's intervention. A more detailed explanation of performance contracting is presented in ASHRAE *Guideline* 14-2002.

CODES AND STANDARDS

NEC®

The *National Electrical Code*® (NEC) is devised and published by the National Fire Protection Association, a consensus standards writing industry group. It is revised every three years. The code exists in several versions: the full text, an abridged edition, and the NEC Handbook (which contains the authorized commentary on the code, as well as the full text). It sets minimum electrical safety standards, and is widely adopted.

UL Listing

Underwriters Laboratories (UL), formerly an insurance industry organization, is now independent and nonprofit. It tests electrical components and equipment for potential hazards. When a device is UL-listed, UL has tested the device, and it meets their requirements for safety (i.e., fire or shock hazard). It does not necessarily mean that the device actually does what it is supposed to do. The UL does not have power of law in the United States; non-UL-listed devices are legal to install. However, insurance policies may have clauses that limit their liability in a claim related to failure of a non-UL-listed device. The NEC requires that a wiring component used for a specific purpose is UL-listed for that purpose. Thus, certain components must be UL-listed before inspector approval and/or issuance of occupancy permits.

CSA Approved

The Canadian Standards Association (CSA) is made up of various government agencies, power utilities, insurance companies, electrical manufacturers, and other organizations. They update CSA *Standard* C22.1, the *Canadian Electrical Code* (CEC), every two or three years.

The Canadian Standards Association (or recognized equivalent) must certify every electrical device or component before it can be sold in Canada. Implicit in this is that all wiring must be done with CSA-approved materials. Testing is similar to UL testing (a bit more stringent), except that CSA approval is required by law. Like the UL, if a fire is caused by non-CSA-approved equipment, the insurance company may not pay the claim.

ULC

Underwriters Laboratory of Canada (ULC) is an independent organization that undertakes the quarterly inspection of manufacturers to ensure continued compliance of UL listed/recognized products to agency reports and safety standards. This work is done under contract to UL, Inc.; they are not a branch or subsidiary of UL.

NAFTA Wiring Standards

Since the North America Free Trade Agreement (NAFTA) came into effect on January 1, 1994, CSA approval of a device is legally considered equivalent to UL approval in the United States, and UL listing is accepted as equivalent to CSA approval in Canada. Devices marked only with UL approval are acceptable in the CEC, and CSA approval by itself of a device is accepted by the NEC. This allows much freer trade in electrical materials between the two countries. This does not affect the electrical codes themselves, so differences in practice between the NEC and CEC remain.

IEEE

The Institute of Electrical and Electronic Engineers' *Standard* 519 suggests limits for both harmonic current and voltage distortion, based on electrical system conditions.

BIBLIOGRAPHY

ANSI. 2006. Electric power systems and equipment—Voltage ratings (60 Hz). *Standard* C84.1-2006. American National Standards Institute, Washington, D.C.

ASHRAE. 2002. Measurement of energy and demand savings. *Guideline* 14-2002.

CSA. 2006. Canadian electrical code, part I, 20th ed.: Safety standard for electrical installations. *Standard* C22.1-06. Canadian Standards Association, Toronto.

IEEE. 1980. Definitions of fundamental waveguide terms. ANSI/IEEE *Standard* 146-1980. Institute of Electrical and Electronics Engineers, Piscataway, NJ.

IEEE. 1992. Recommended practices and requirements for harmonic control in electrical power systems. ANSI/IEEE *Standard* 519-1992. Institute of Electrical and Electronics Engineers, Piscataway, NJ.

NEMA. 2007. Condensed information guide for general purpose industrial ac small and medium squirrel-cage induction motor standards. ANSI/NEMA *Standard* MG 1-2007. National Electrical Manufacturers Association, Rosslyn, VA.

NFPA. 2011. National electrical code®. *Standard* 70-2011. National Fire Protection Association, Quincy, MA.

ROOM AIR DISTRIBUTION

ROOM air distribution systems, like other HVAC systems, are intended to achieve required thermal comfort and ventilation for space occupants and processes. Although air terminals (inlets and outlets), terminal units, local ducts, and the rooms themselves may affect room air distribution, this chapter addresses only air terminals and their effect on occupant comfort. This chapter is intended to help HVAC designers apply air distribution systems to occupied spaces, providing information on characteristics of various air distribution strategies, and tools and guidelines for applications and system design. Naturally ventilated spaces are not addressed; see Chapter 16 of the 2009 *ASHRAE Handbook—Fundamentals* for details. Also see Chapter 20 of the 2009 *ASHRAE Handbook—Fundamentals* for more information on space air diffusion; Chapter 19 of the 2008 *ASHRAE Handbook—HVAC Systems and Equipment* for information on room air distribution equipment; and Chapter 48 of this volume for sound and vibration control guidance.

Room air distribution systems can be classified by (1) their primary objective and (2) the method by which they attempt to accomplish that objective. The objective of any air distribution system can be classified as one of the following:

- Conditioning and/or ventilation of the space for occupant thermal comfort
- Conditioning and/or ventilation to support processes within the space
- A combination of these

As a general guideline, the **occupied zone** in a space is any location where occupants normally reside, and may differ from project to project; it is application-specific, and should be carefully defined by the designer. The occupied zone is generally considered to be the room volume between the floor level and 6 ft above the floor. Standards and guidelines, such as ANSI/ASHRAE *Standards* 55 and 62.1, further define the occupied zone (e.g., *Standard* 55 exempts areas near walls).

Occupant comfort is defined in detail in ANSI/ASHRAE *Standard* 55-2004. Figure 5.2.1.1 of the standard shows acceptable ranges of temperature and humidity for spaces. As a general guide, 80% of occupants in typical office spaces can be satisfied with thermal environments over a wide range of temperatures and relative humidities. Designers often target indoor dry-bulb temperatures between 73 and 77°F, relative humidities between 25 and 60%, and occupied zone air velocities below 50 fpm. ANSI/ASHRAE *Standard* 113 describes a method for evaluating effectiveness of various room air distribution systems in achieving thermal comfort.

Room air distribution methods can be classified as one of the following:

- **Mixed systems** (e.g., overhead distribution) have little or no thermal stratification of air in the occupied and/or process space.
- **Full thermal stratification systems** (e.g., thermal displacement ventilation) have little or no air mixing in the occupied and/or process space.
- **Partially mixed systems** (e.g., most underfloor air distribution designs) provide limited air mixing in the occupied and/or process space.
- **Task/ambient air distribution** (e.g., personally controlled desk outlets, spot conditioning systems) focuses on conditioning only part of the space for thermal comfort and/or process control.

Because task/ambient design requires a high degree of individual control, it is not covered in this chapter; see Chapter 20 of the 2009 *ASHRAE Handbook—Fundamentals* for details. Limited design guidance is also provided by Bauman and Daly (2003).

Figure 1 illustrates the spectrum between the two extremes (full mixing and full stratification) of room air distribution strategies.

INDOOR AIR QUALITY AND SUSTAINABILITY

Air distribution systems affect not only indoor air quality (IAQ) and thermal comfort, but also energy consumption over the entire life of the project. Choices made early in the design process are important. ANSI/ASHRAE/IESNA *Standard* 90.1 provides energy efficiency requirements that affect supply air characteristics.

The U.S. Green Building Council's (USGBC) Leadership in Energy and Environmental Design (LEED®) Green Building Rating System™ was originally created in response to indoor air quality concerns, and has evolved to include prerequisites and credits for increasing ventilation effectiveness and improving thermal comfort (USGBC 2009). These requirements and optional points are relatively easy to achieve if good room air distribution design principles, methods, and standards are followed.

Environmental tobacco smoke (ETS) control is a LEED prerequisite. Banning indoor smoking is a common approach, but if indoor smoking is to be allowed, ANSI/ASHRAE *Standard* 62.1-2010 requires that more than the base non-ETS ventilation air be provided where ETS is present in all or part of a building. Rock (2006) provides additional advice on dealing with ETS.

Ventilation effectiveness is affected directly by the room air distribution system's design, construction, and operation, but is very difficult to predict. Many attempts have been made to quantify ventilation effectiveness, including ASHRAE *Standard* 129. However, this standard is only for experimental tests in well-controlled laboratories and should not be applied directly to real buildings.

Because of the difficulty in predicting ventilation effectiveness, ASHRAE *Standard* 62.1 provides a table of typical values that were determined through the experiences of its Standard Project Committee and reviewers or extracted from research literature; for example, well-designed ceiling-based air diffusion systems produce near-perfect air mixing in cooling mode, and yield an air change effectiveness of almost 1.0. More information on ASHRAE *Standard* 62.1 is available in its user's manual (ASHRAE 2011).

Displacement and underfloor air distribution (UFAD) systems have the potential for values greater than 1.0. More information on ceiling- and wall-mounted air inlets and outlets can be found in Rock and Zhu (2002). Performance of displacement systems is described by Chen and Glicksman (2003), and UFAD is discussed in detail by Bauman and Daly (2003).

The preparation of this chapter is assigned to TC 5.3, Room Air Distribution.

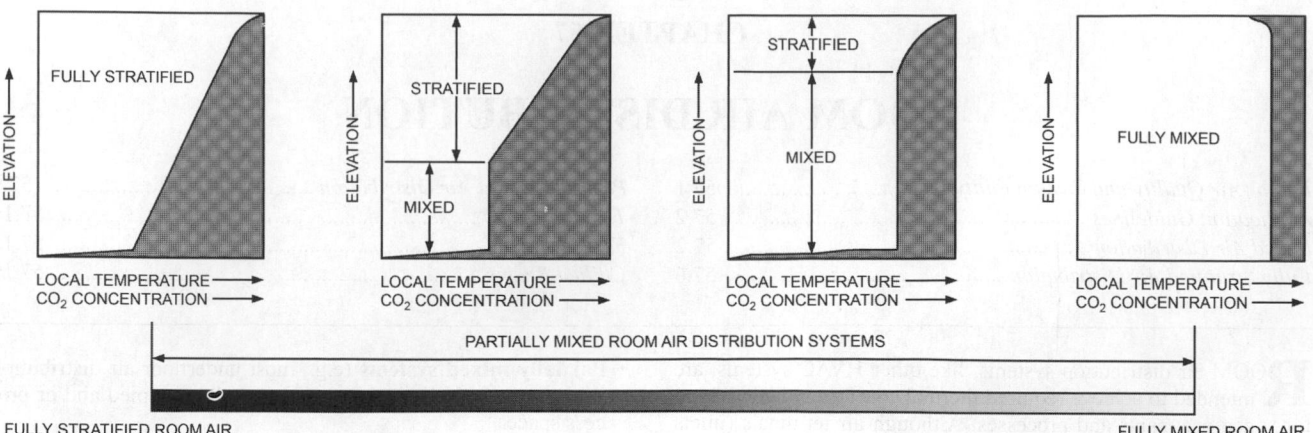

EXAMPLES:
• Thermal displacement using low-velocity cool air
• Natural ventilation

EXAMPLES:
• Underfloor air distribution (using room air induction) in cooling mode operation
• Underseat air distribution (using room air induction) in a cooling mode operation
• Task/ambient cooling (using furniture-based outlets)
• Task/ambient (spot) cooling or heating (industrial applications)

EXAMPLES:
• Overhead mixed air supply in cooling operation
• Fan-coil units and unit ventilators
• High-velocity floor-based supply in heating operation

Fig. 1   Classification of Air Distribution Strategies

Air terminals, such as diffusers or grilles, may become unsightly over time because of accumulation of dirt on their faces (smudging). Instead of replacing air terminals, and thus requiring new materials and energy for manufacturing, they can often be cleaned in place to restore their appearance. Those that cannot be cleaned and must be replaced should be recycled, not discarded, to recover the various metals and other desirable materials of construction.

APPLICATION GUIDELINES

Design Constraints

Space design constraints affect room air distribution system choices and how air inlets and outlets are used. Space constraints may include the following:

• Dimensions
• Heat gain and loss characteristics
• Use
• Acoustical requirements
• Available locations for air inlets and outlets

Inlet and outlet characteristics are discussed in Chapter 19 of the 2008 *ASHRAE Handbook—HVAC Systems and Equipment*. This chapter discusses more specific application considerations for air inlets and outlets.

Sound

Sound emitted from inlets and outlets is directly related to the airflow quantity and free area velocity. The airflow sound intensity in a space also depends on the room's acoustical absorption and the observer's distance from air distribution devices. For more information, see Chapter 48 of this volume and Chapter 8 in the 2009 *ASHRAE Handbook—Fundamentals*.

Inlet Conditions to Air Outlets

The way an airstream approaches an outlet is important. For good air diffusion, the inlet configuration should create a uniform discharge velocity profile from the outlet, or the outlet may not perform as intended.

The outlet usually cannot correct effects of improper duct approach. Many sidewall outlets are installed either at the end of vertical ducts or in the side of horizontal ducts, and most ceiling outlets are attached either directly to the bottom of horizontal ducts or

Table 1   Recommended Return Inlet Face Velocities

Inlet Location	Velocity Across Gross Area, fpm
Above occupied zone	>800
In occupied zone, not near seats	600 to 800
In occupied zone, near seats	400 to 600
Door or wall louvers	200 to 300
Through undercut area of doors	200 to 300

to special vertical takeoff ducts that connect the outlet with the horizontal duct. In all these cases, devices for directing and equalizing the airflow may be necessary for proper direction and diffusion of the air.

Return Air Inlets

The success of a mixed air distribution system depends primarily on supply diffuser location. Return grille location is far less critical than with outlets. In fact, the return air intake affects room air motion only immediately around the grille. Measurements of velocity near a return air grille show a rapid decrease in magnitude as the measuring device is moved away from the grille face. Table 1 shows recommended maximum return air grille velocities as a function of grille location. Every enclosed space should have return/transfer inlets of adequate size per this table.

For stratified and partially mixed air distribution systems, there are advantageous locations for return air inlets. For example, an intake can be located to return the warmest air in cooling season.

If the outlet is selected to provide adequate throw and directed away from returns or exhausts, supply short-circuiting is normally not a problem. The success of this practice is confirmed by the availability and use of combination supply and return diffusers.

MIXED AIR DISTRIBUTION

In mixed air systems, high-velocity supply jets from air outlets maintain comfort by mixing room air with supply air. This air mixing, heat transfer, and resultant velocity reduction should occur outside the occupied zone. Occupant comfort is maintained not directly by motion of air from outlets, but from secondary air motion from mixing in the unoccupied zone. Comfort is maximized when uniform temperature distribution and room air velocities of less than 50 fpm are maintained in the occupied zone.

Maintaining velocities less than 50 fpm in the occupied zone is often overlooked by designers, but is critical to maintaining comfort. The outlet's selection, location, supply air volume, discharge velocity, and air temperature differential determine the resulting air motion in the occupied zone.

Principles of Operation

Mixed systems generally provide comfort by entraining room air into discharge jets located outside occupied zones, mixing supply and room air. Ideally, these systems generate low-velocity air motion (less than 50 fpm) throughout the occupied zone to provide uniform temperature gradients and velocities. Proper selection of an air outlet is critical for proper air distribution; improper selection can result in room air stagnation, unacceptable temperature gradients, and unacceptable velocities in the occupied zone that may lead to occupant discomfort.

The location of a discharge jet relative to surrounding surfaces is important. Discharge jets attach to parallel surfaces, given sufficient velocity and proximity. When a jet is attached, the throw increases by about 40% over a jet discharged in an open area. This difference is important when selecting an air outlet. For detailed discussion of the surface effect on discharge jets, see Chapter 20 of the 2009 *ASHRAE Handbook—Fundamentals*.

Space Ventilation and Contaminant Removal

These systems are intended to maintain acceptable indoor air quality by mixing supply and room air (dilution ventilation). Supply air is typically a conditioned mixture of ventilation and recirculated air. Outlet type and discharge velocity determine the mixing rate of the space and should be a design consideration. The room's return or exhaust air carries away diluted air contaminants. Space air ventilation rates are mandated under ASHRAE *Standard* 62.1-2010, but supply airflow rates are often higher because of thermal loads.

Benefits and Limitations

Benefits of fully mixed systems include the following:

- Most office applications can use lower supply dry-bulb temperatures, for smaller ductwork and lower supply air quantities.
- Air can be supplied at a lower moisture content, possibly eliminating the need for a more complex humidity control system.
- Vertical temperature gradients are lower for cooling applications with high internal heat gains, which may improve thermal comfort.
- Mixed systems are the most common design for distribution systems, because designers and installers are familiar with the required system components and installation.

Limitations of mixed systems include the following:

- Partial-load operation in variable-air-volume (VAV) systems may reduce outlet velocities, reducing room air mixing and compromising thermal comfort. Designers should consider this when selecting outlets.
- Cooling and heating with the same ceiling or high-sidewall diffuser may cause inadequate performance in heating mode and/or excessive velocity in cooling mode.
- Ceilings more than 12 ft high may require special design considerations to provide acceptable comfort in the occupied zone. Care should be taken to select the proper outlet for these applications.
- Because mixed systems typically use high-velocity jets of air, any obstructions in the space (e.g., bookshelves, wall partitions, furniture) can reduce comfort.
- Lighter-than-air contaminants are uniformly mixed in the space and typically result in higher contaminant concentrations, which may compromise indoor air quality.

Mixed air systems typically use either ceiling or sidewall outlets discharging air horizontally, or floor- or sill-mounted outlets dis-

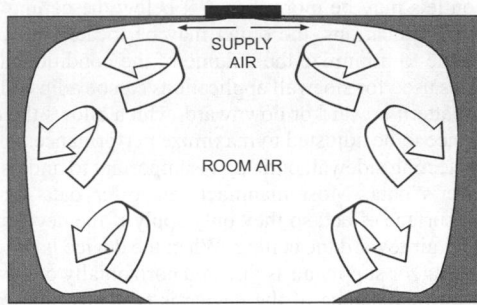

Fig. 2 Air Supplied at Ceiling Induces Room Air into Supply Jet

charging air vertically. They are the most common method of air distribution in North America.

Horizontal Discharge Cooling with Ceiling-Mounted Outlets

Ceiling-mounted outlets typically use the surface effect to transport supply air in the unoccupied zone. The supply air projects across the ceiling and, with sufficient velocity, can continue down wall surfaces and across floors, as shown in Figure 2. In this application, supply air should remain outside the occupied zone until it is adequately mixed and tempered with room air. Air motion in the occupied zone is generated by room air entrainment into the supply air (Nevins 1976).

Overhead outlets may also be installed on exposed ducts, in which case the surface effect does not apply. Typically, if the outlet is mounted 1 ft or more below a ceiling surface, discharge air will not attach to the surface. The unattached supply air has a shorter throw and can project downward, resulting in high air velocities in the occupied zone. Some outlets are designed for use in exposed duct applications. Typical outlet performance data presented by manufacturers are for outlets with surface effect; consult manufacturers for information on exposed duct applications.

Vertical-Discharge Cooling or Heating with Ceiling-Mounted Outlets

Vertically projected outlets are typically selected for high-ceiling applications that require forcing supply air down to the occupied zone. It is important to keep cooling supply air velocity below 50 fpm in the occupied zone. For heating, supply air should reach the floor.

There are outlets specifically designed for vertical projection and it is important to review the manufacturer's performance data notes to understand how to apply catalog data. Throws for heating and cooling differ and also vary depending on the difference between supply and room air temperatures.

Cooling with Sidewall Outlets

Sidewall outlets are usually selected when access to the ceiling plenum is restricted. Sidewall outlets within 1 ft of a ceiling and set for horizontal or a slightly upward projection the sidewall outlet provide a discharge pattern that attaches to the ceiling and travels in the unoccupied zone. This pattern entrains air from the occupied zone to provide mixing.

In some applications, the outlet must be located 2 to 4 ft below the ceiling. When set for horizontal projection, the discharge at some distance from the outlet may drop into the occupied zone. Most devices used for sidewall application can be adjusted to project the air pattern upwards toward the ceiling. This allows the discharge air to attach to the ceiling, increasing throw distance and minimizing drop. This application provides occupant comfort by inducing air from the occupied zone into the supply air.

Some outlets may be more than 4 ft below the ceiling (e.g., in high-ceiling applications, the outlet may be located closer to the occupied zone to minimize the volume of the conditioned space). Most devices used for sidewall applications can be adjusted to project the air pattern upward or downward, which allows the device's throw distance to be adjusted to maximize performance.

When selecting sidewall outlets, it is important to understand the manufacturer's data. Most manufacturers offer data for outlets tested with surface effect, so they only apply if the device is set to direct supply air toward the ceiling. When the device is 4 ft or more below a ceiling, or supply air is directed horizontally or downward, the actual throw distance of the device is typically shorter. Many sidewall outlets can be adjusted to change the spread of supply air, which can significantly change throw distance. Manufacturers usually publish throw distances based on specific spread angles.

Cooling with Floor-Mounted Air Outlets

Although not typically selected for nonresidential buildings, floor-mounted outlets can be used for mixed system cooling applications. In this configuration, room air from the occupied zone is induced into the supply air, providing mixing. When cooling, the device should be selected to discharge vertically along windows, walls, or other vertical surfaces. Typical nonresidential applications include lobbies, long corridors, and houses of worship.

It is important to select a device that is specially designed for floor applications. It must be able to withstand both the required dynamic and static structural loads (e.g., people walking on them, loaded carts rolling across them). Also, many manufacturers offer devices designed to reduce the possibility of objects falling into the device. It is strongly recommended that obstructions are not located above these in-floor air terminals, to avoid restricting their air jets.

Long floor-mounted grilles generally have both functioning and nonfunctioning segments. When selecting air outlets for floor mounting, it is important to note that the throw distance and sound generated depend on the length of the active section. Most manufacturers' catalog data include correction factors for length's effects on both throw and sound. These corrections can be significant and should be evaluated. Understanding manufacturers' performance data and corresponding notes is imperative.

Cooling with Sill-Mounted Air Outlets

Sill-mounted air outlets are commonly used in applications that include unit ventilators and fan coil units. The outlet should be selected to discharge vertically along windows, walls, or other vertical surfaces, and project supply air above the occupied zone.

As with floor-mounted grilles, when selecting and locating sill grilles, consider selecting devices designed to reduce the nuisance of objects falling inside them. It is also recommended that sills be designed to prevent them from being used as shelves.

Heating and Cooling with Perimeter Ceiling-Mounted Outlets

When air outlets are used at the perimeter with vertical projection for heating and/or cooling, they should be located near the perimeter surface, and selected so that the published 150 fpm isothermal throw extends at least halfway down the surface or 5 ft above the floor, whichever is lower. In this manner, during heating, warm air mixes with the cool downdraft on the perimeter surface, to reduce or even eliminate drafts in the occupied space.

If a ceiling-mounted air outlet is located away from the perimeter wall, in cooling mode, the high-velocity cool air reduces or overcomes the thermal updrafts on the perimeter surface. To accomplish this, the outlet should be selected for horizontal discharge toward the wall. Outlet selection should be such that isothermal throw to the terminal velocity of 150 fpm should include the distance from the outlet to the perimeter surface. For heating, the supply air temperature should not exceed 15°F above the room air temperature.

Space Temperature Gradients and Airflow Rates

A fully mixed system creates homogeneous thermal conditions throughout the space. As such, thermal gradients should not be expected to exist in the occupied zone. Improper selection, sizing, or placement may prevent full mixing and can result in stagnant areas, or having high-velocity air entering the occupied zone.

Supply airflow requirements to satisfy space sensible heat gains or losses are inversely proportional to the temperature difference between supply and return air. The following equation can be used to calculate space airflow requirements (at standard conditions):

$$Q = \frac{q_s}{1.08(t_r - t_s)} \tag{1}$$

where

Q = required supply airflow rate to meet sensible load, cfm
q_s = net sensible heat gain in the space, Btu/h
t_r = return or exhaust air temperature, °F
t_s = supply air temperature, °F

For fully mixed systems with conventional ceiling heights, the return (or exhaust) and room air temperatures are the same; for example, a room with a set-point temperature of 75°F has, on average, a 75°F return or exhaust air temperature.

Methods for Evaluation

The objective of air diffusion is to create the proper combination of room air temperature, humidity, and air motion in the occupied zone to provide thermal comfort and acceptable indoor environmental quality. There are three recommended methods of selecting outlets for mixed air systems using manufacturers' data:

- By appearance, flow rate, and sound data
- By isovels (lines of constant velocity) and mapping
- By comfort criteria

These selection methods are not meant to be independent. It is the designer's choice as to which to start with, but it is recommended that at least two methods be used for any design.

Variation from accepted thermal limits (ASHRAE *Standard* 55), lack of uniform thermal conditions in the space, or excessive fluctuation of conditions in one part of the space may produce discomfort. Thermal discomfort also can arise from any of the following conditions:

- Excessive air motion (draft)
- Excessive room air temperature stratification (horizontal, vertical, or both)
- Failure to deliver or distribute air according to load requirements at different locations
- Rapid fluctuation of room temperature

Selection

By Appearance, Flow Rate, and Sound Data. For a given appearance, flow rate, pressure drop, and sound level criteria, designers can select outlets from manufacturers' catalogs, using the following steps:

1. Determine air volumetric flow requirements based on load and room size. For VAV systems, evaluation should include the range of flow rates from minimum occupied to design load.
2. Determine acceptable outlet noise criterion (NC); consult Chapter 48 of this volume, or Chapter 8 in the 2009 *ASHRAE Handbook—Fundamentals*.
3. Locate a range of products from manufacturers' catalogs that meet the airflow and NC requirements. Multiple outlets in a space at the same cataloged NC, and other design considerations, may result in actual sound levels greater than cataloged values.

Table 2 Effect of Neck-Mounted Damper on Air Outlet NC

Total Pressure Ratio*	100%	150%	200%	400%
dB Increase	0	4.5	8	16

*Ratio of air pressure before and after damper.

Manufacturers' data are obtained using ideal inlet conditions, and may vary from field installations. From experience,

- For identical outlets 10 ft or more apart, the cataloged NC rating applies.
- Identical outlets within 10 ft of each other add no more than 3 dB to the sound pressure level.
- For continuous linear outlets, only the sound produced by the closest 10 ft need be considered.
- A wide-open damper installed in the neck of a diffuser can add 4 to 5 NC to the cataloged NC value.
- Significantly closed balancing dampers can add more than 10 NC, depending on duct pressure and how far upstream it is installed. Table 2 gives an example.

4. Select air terminals from manufacturers' catalogs that meet aesthetic and physical needs.

Although these selections may meet the sound requirements for a project, the results do not fully address occupant comfort. Without evaluating the throw of the outlets or room air mixing, this selection method may result in excessive air velocities in the occupied zone, or limited mixing and resultant stagnation. It is recommended that the designer consider selection by isovel mapping or by comfort criteria in addition to selection by appearance, flow rate, and sound data. Either of these methods addresses resulting air motion in the occupied zone and occupant comfort.

By Isovels and Mapping. Using manufacturers' catalog throw data, a designer can predict the path of an outlet's discharge jet. Most manufacturers' catalogs list the distance a jet travels to reach a terminal velocity of 150 to 50 fpm. With this information, the designer can map the path of the discharge jet for a given outlet. This evaluation can prevent problems such as excessively high air velocities in the occupied zone, or stagnation in a given area. Note that most manufacturers' throw data are based on isothermal supply air; the supply jet temperature is equal to the room air temperature. When using this mapping method, consider the positive or negative buoyancy of nonisothermal (heated or cooled) supply air. In both heating and cooling, a discharge jet should travel the distance shown in the catalog to a terminal velocity of 150 fpm without much influence from buoyancy. When evaluating a jet at lower terminal velocities (e.g., 100 or 50 fpm), consider buoyancy's effect on the distance the jet will travel.

A cool air jet travels less far along a horizontal surface than an isothermal jet does. If an outlet is selected so that the horizontal jet does not have enough velocity to reach a vertical surface, the jet can separate from the horizontal surface and project down into the occupied zone, causing drafts and discomfort. Manufacturers' tables show the drop of a cool air jet so the designer can predict the resulting path.

When evaluating heated air, the designer should consider the positive buoyancy of the discharge jet. A heated jet projecting along a horizontal surface or in an upward vertical pattern travels farther than an isothermal jet. In downward vertical discharge, a heated jet travels a shorter distance than an isothermal jet.

Combining selection by isovels and mapping with acoustical selection allows discharge jet location and intensity in a space to be predicted. Outlet selection should be evaluated at the space's typical operating points (i.e., maximum heating and cooling, and minimum heating and cooling). The following steps may be used:

1. Identify the occupied zone for the space.

2. Select outlet(s) that meet design NC, pressure drop, and flow rate requirements. Identify the supply jet location using cataloged throw data.
3. Evaluate air jet mapping to ensure terminal velocities in the occupied zone do not exceed 50 fpm.
4. For overhead heating applications, $\Delta t < 15°F$ (see Chapter 20 of the 2009 *ASHRAE Handbook—Fundamentals*), evaluate the diagram to ensure that jet velocities 5 ft from the floor are at least 150 fpm.

Other design considerations include the following:

- In multiple-outlet applications, jets should not collide to cause a downward projection of air resulting in velocities greater than 50 fpm in the occupied zone.
- For VAV applications, consider both minimum and maximum flow conditions.

Selection by Comfort Criteria T_{50}/L. Selection by isovels and mapping is effective at predicting the path of the discharge jet from an outlet and evaluating resultant occupant comfort. However, there is an established method to quantify occupant comfort during cooling conditions, based on space dimensions and isothermal catalog throw data. This method can be used to predict a space's resulting air diffusion performance index (ADPI).

The comfort criteria T_{50}/L method was developed to predict occupant comfort during cooling conditions, using manufacturers' isothermal catalog throw data (T, usually for 50 fpm terminal velocity) and the dimensions available for throw (L) on the plan view of a mechanical drawing. By using the ratio of T_{50}/L, the designer can predict the level of comfort with a single rating number: ADPI. ADPI can provide further information about the comfort level in a space for results obtained from the NC and mapping selection methods.

Air Distribution Performance Index (ADPI). The air distribution performance index was developed as a way to quantify the comfort level for a space conditioned by a mixed air system in cooling. ADPI uses the effective draft temperature collected at an array of points taken within the occupied zone to predict comfort. ADPI is the percentage of points in a space where the effective draft temperature is between –3 and +2°F and the air velocity is less than 70 fpm. A high percentage of people have been found to be comfortable in cooling applications for office-type occupations where these conditions are met. High ADPI values generally correlate to high space thermal comfort levels with the maximum obtainable value of 100. Select outlets to provide a minimum ADPI value of 80.

The effective draft temperature provides a quantifiable indication of comfort at a discrete point in a space by combining the physiological effects of air temperature and air motion on a human body. The effective draft temperature t_{ed} (the difference in temperature between any point in the occupied zone and the control condition) can be calculated using the following equation proposed by Rydberg and Norback (1949) and modified by Straub (Straub and Chen 1957; Straub et al. 1956) in discussion of a paper by Koestel and Tuve (1955):

$$t_{ed} = (t_x - t_c) - 0.07(V_x - 30) \qquad (2)$$

where

t_{ed} = effective draft temperature, °F
t_x = local airstream dry-bulb temperature, °F
t_c = average (control) room dry-bulb temperature, °F
V_x = local airstream centerline velocity, fpm

T_{50}/L *Selection Method.* This method uses the ratio of cataloged isothermal throw data at 50 fpm to the characteristic length for a given device (Table 3).

Each type of diffuser has different performance characteristics and therefore may provide a different ADPI value for the same cool-

Table 3 Characteristic Room Length for Several Diffusers
(Measured from Center of Air Outlet)

Diffuser Type	Characteristic Length L
High sidewall grille	Distance to wall perpendicular to jet
Circular ceiling diffuser	Distance to closest wall or intersecting air jet
Sill grille	Length of room in direction of jet flow
Ceiling slot diffuser	Distance to wall or midplane between outlets
Light troffer diffusers	Distance to midplane between outlets plus distance from ceiling to top of occupied zone
Perforated, louvered ceiling diffusers	Distance to wall or midplane between outlets

Table 4 Air Diffusion Performance Index (ADPI)
Selection Guide

Terminal Device	Room Load, Btu/h·ft²	T_{50}/L for Maximum ADPI	Maximum ADPI	For ADPI Greater Than	Range of T_{50}/L
High sidewall grilles	80	1.8	68	—	—
	60	1.8	72	70	1.5 to 2.2
	40	1.6	78	70	1.2 to 2.3
	20	1.5	85	80	1.0 to 1.9
Circular ceiling diffusers	80	0.8	76	70	0.7 to 1.3
	60	0.8	83	80	0.7 to 1.2
	40	0.8	88	80	0.5 to 1.5
	20	0.8	93	90	0.7 to 1.3
Sill grille, straight vanes	80	1.7	61	60	1.5 to 1.7
	60	1.7	72	70	1.4 to 1.7
	40	1.3	86	80	1.2 to 1.8
	20	0.9	95	90	0.8 to 1.3
Sill grille, spread vanes	80	0.7	94	90	0.6 to 1.5
	60	0.7	94	80	0.6 to 1.7
	40	0.7	94	—	—
	20	0.7	94	—	—
Ceiling slot diffusers (for T_{100}/L)	80	0.3	85	80	0.3 to 0.7
	60	0.3	88	80	0.3 to 0.8
	40	0.3	91	80	0.3 to 1.1
	20	0.3	92	80	0.3 to 1.5
Light troffer diffusers	60	2.5	86	80	<3.8
	40	1.0	92	90	<3.0
	20	1.0	95	90	<4.5
Perforated, louvered ceiling diffusers	11 to 50	2.0	96	90	1.4 to 2.7
				80	1.0 to 3.4

ing application at the same conditions. Calculating T_{50}/L for a given outlet can predict the level of cooling comfort for a space. Using Table 4, the designer can optimize not only the type of diffuser to select but also the size and capacity.

Using T_{50}/L helps designers maximize space cooling comfort; however, this method is not meant to, nor may it be practical to, evaluate T_{50}/L values for each outlet on a project.

Design Procedures. T_{50}/L can be used as a general tool to evaluate cooling comfort levels in a space, at the beginning of design to optimize outlet selection (as shown in the following steps), or at the end of the process to predict cooling comfort levels in spaces designed using NC and mapping methods:

1. Determine air volumetric flow requirements based on load and room size. For VAV systems, evaluation should include both minimum occupied and maximum design flow rates.
2. Select tentative diffuser type and location in room.
3. Determine room's characteristic length L (Table 3).
4. Select recommended T_{50}/L (or T_{100}/L) ratio from Table 4.
5. Calculate throw distance T_{50} by multiplying recommended T_{50}/L (T_{100}/L for linear slots) ratio from Table 4 by available length L.

6. Locate appropriate outlet size from manufacturer's catalog.
7. Ensure that this outlet meets other imposed specifications (e.g., noise, static pressure loss).

Example 1. For a 20 by 12 ft room, with 9 ft ceiling, with uniform loading of 10 Btu/h·ft² or 2400 Btu/h, and air volumetric flow of 1 cfm/ft² or 240 cfm for one outlet, find the size for a high sidewall grille located at the center of 12 ft end wall, 9 in. from ceiling.

Solution:
Characteristic length L = 20 ft (length of room: Table 3)
Recommended T_{50}/L = 1.5 (Table 4)
Throw to 50 fpm T_{50} = 1.5 × 20 = 30 ft
Refer to the manufacturer's catalog for a size that gives this isothermal throw to 50 fpm. One manufacturer recommends the following sizes, when vanes are straight, discharging at 240 cfm: 16 by 4 in., 12 by 5 in., or 10 by 6 in.

More information on conventional mixing systems, and many more design examples, can be found in Rock and Zhu (2002).

FULLY STRATIFIED AIR DISTRIBUTION

Systems that discharge cool air at low sidewall or floor locations with very little entrainment of (and thus mixing with) room air create (vertical) thermal stratification throughout the space. These **displacement ventilation** systems have been popular in northern Europe for some time. Floor-based outlets in underfloor applications may also be used to provide fully stratified air distribution.

Principles of Operation

Thermal displacement ventilation (TDV) systems (Figure 3) use very low discharge velocities, typically 50 to 70 fpm, to deliver cool supply air to the space. The discharge temperature of the supply air is generally above 60°F, although lower temperatures may be used in industrial applications, exercise or sports facilities, and transient areas. The cool air is negatively buoyant compared to ambient air and drops to the floor after discharge. It then spreads across the lower level of the space.

As convective heat sources (Figure 3) in the space transfer heat to the cooler air around them, natural convection currents form and rise along the heat transfer boundary. Without significant room air movement, these currents rise to form a convective heat plume around and above the heat source. As the plume rises, it expands by entraining surrounding air. Its growth and ascent are proportional to the heat source's size and intensity and temperature of ambient air above it. Ambient air from below and around the heat source fills the void created by the rising plume. If the heat source is near the floor (e.g., an occupant), the plume entrains cool, conditioned air from the floor level, which is drawn to the respiration level, and serves as the source of inhaled air. Exhaled air rises with the escaping heat plume, because it is warmer and more humid than the ambient air. Convective heat from sources located above the occupied zone has little effect on occupied-zone air temperature.

At a certain height, where plume temperature equals ambient temperature, the plume disintegrates and spills horizontally. Two distinct zones are thus formed in the room: a lower occupied zone with little or no recirculation flow (close to displacement flow), and an upper zone with recirculation flow. The boundary between these two zones is called **shift zone**. The shift zone height is calculated as the height above the floor where the total amount of air carried in convective plumes above heat sources equals the supply airflow distributed through displacement diffusers. Actual and simplified representations of the temperature gradient in the space are shown in Figure 4.

Thermal displacement ventilation systems can be modeled as shown in Figure 4. A thin layer of conditioned supply air lies adjacent to the floor. Next is a lower zone in which both ambient air temperature and contaminant concentration levels increase with height. Finally, a pool of warm used or contaminated air (the upper zone) may form next to the ceiling, depending on the supplied airflow rate in proportion to the volume of thermal plumes rising through the space.

Space Ventilation and Contaminant Removal

Thermal displacement ventilation is very effective at removing airborne contaminants that are equal to or lighter than the ambient air (e.g., respiratory-produced contaminants, tobacco smoke). According to ASHRAE *Standard* 62.1-2010, these systems have zone air distribution effectiveness E_Z values of 1.2, compared to maximum values of 1.0 for mixed air systems.

Typical Applications

Thermal displacement ventilation systems typically have higher return air temperatures than mixed systems. Thus, they may allow extended periods of air- or water-side economizer operation, especially in mild, relatively dry climates.

Thermal displacement ventilation systems are commonly used in applications such as

- Restaurants
- Casinos

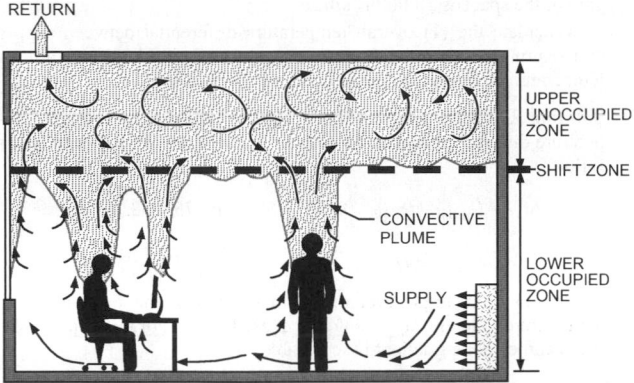

Fig. 3 Displacement Ventilation System Characteristics

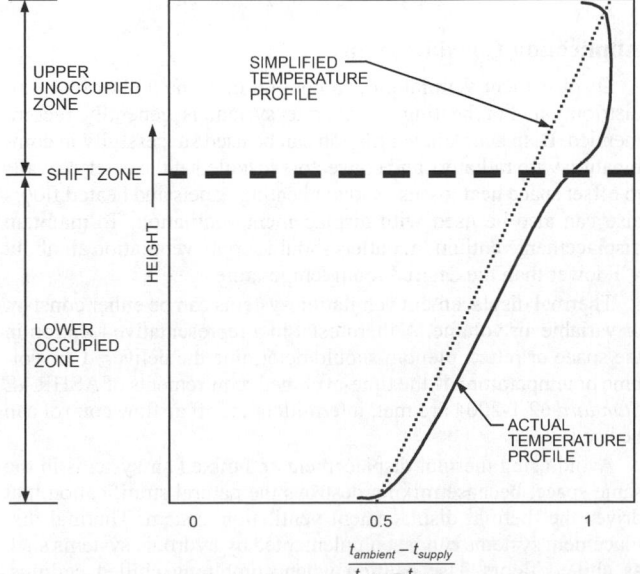

Fig. 4   Temperature Profile of Displacement Ventilation System

- Large open-plan offices, classrooms, lecture halls, and meeting rooms
- Theaters and auditoriums
- Hospitals and clean rooms
- Other spaces with high ceilings

Benefits and Limitations

Benefits of displacement ventilation systems include the following:

- Removal of airborne contaminants is more effective.
- In mild climates, significantly less energy may be used to maintain the same space occupied zone air temperature in cooling mode.
- Air distribution effectiveness is high: less outside air is required to meet ASHRAE *Standard* 62.1-2010 requirements.
- Diffuser noise level is lower.
- Lower turbulence intensity can reduce draft-related complaints.

Some applications do not favor use of thermal displacement ventilation. Small offices, especially with perimeter exposures, often do not have room for the large outlets that may be required. The following types of areas may be better served by a mixed system:

- Spaces with ceiling heights less than 9 ft
- Spaces with exceptionally high occupied zone heat loads (Bauman and Daly 2003)
- Spaces with ceiling heights below 10 ft that are subjected to significant room air disturbances
- Applications where contaminants are heavier and/or colder than ambient air

When thermal displacement systems are used in humid climates, it may be necessary to dehumidify and possibly reheat supply air to maintain desired space conditions. As with all HVAC air systems' design, a psychrometric analysis is advised.

Outlet Characteristics

Displacement outlets are designed for average face velocities between 50 and 70 fpm, and are typically in a low sidewall or floor location. Return or exhaust air intakes should always be located above the occupied zone for human thermal comfort applications.

Displacement outlets are available in a number of configurations and sizes. Some models are designed to fit in corners or along sidewalls, or stand freely as columns. It is important to consider the degree of flow equalization the outlet achieves, because use of the entire outlet surface for air discharge is paramount to minimizing clear zones and maintaining acceptable temperatures at the lower levels of the space.

Stationary occupants should not be subjected to discharge velocities exceeding about 40 fpm because air at the ankle level within this velocity envelope tends to be quite cool. As such, most outlet manufacturers define a **clear zone** in which location of stationary, low-activity occupants is strongly discouraged, but transient occupancy, such as in corridors or aisles, is possible. Occupants with high activity levels may also find the clear zone acceptable.

Space Temperature Gradients and Airflow Rates

Figure 5 illustrates thermal temperature gradients that might be expected for a classroom with a 10 ft ceiling, served by thermal displacement ventilation. If loads are typical to the application and proper space airflow is supplied, Skistad et al. (2002) indicate that approximately 50% of the total temperature difference between supply air and return or exhaust air is dissipated in clear zone(s) next to the outlet(s). The other half of the temperature gradient is the **space temperature gradient (STG)**, assumed to be linear with air temperature, increasing gradually from floor to ceiling.

For stationary, low-activity occupants, keep supply air temperatures above 60°F. When occupants are very near outlets (e.g., in underseat delivery), keep supply air temperatures at or above 64°F.

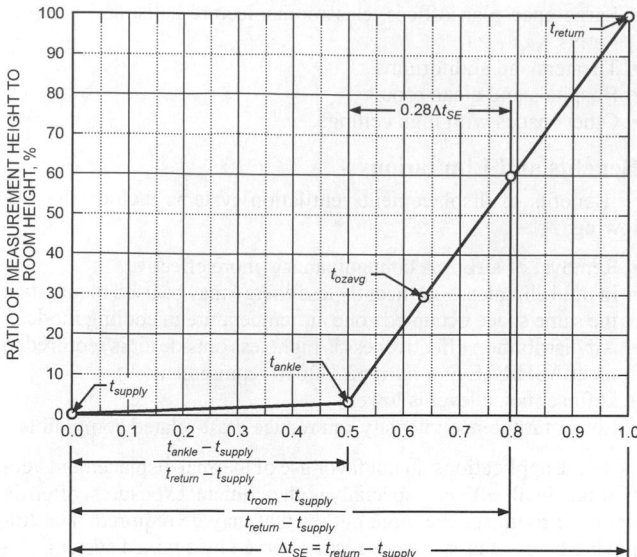

Fig. 5   Temperature Gradient Relationships for Thermal Displacement Ventilation System in Typical Classroom or Office with 10 ft Ceiling

Methods of Evaluation

Unlike mixed systems, outlets in thermal displacement systems discharge air at very low velocities, resulting in very little mixing. As such, design of these systems primarily involves determining a supply airflow rate to manage the thermal gradients in the space in accordance with ASHRAE comfort guidelines. ASHRAE *Standard* 55 recommends that the vertical temperature difference between the ankle and head levels of space occupants be limited to no more than 5.4°F to maintain a high degree (>95%) of occupant satisfaction.

Design Procedures

Displacement ventilation system design is somewhat different than for mixing ventilation. For mixing ventilation systems, where air is mixed relatively evenly throughout the space, the return/exhaust air temperature is assumed to equal the space temperature. In displacement ventilation systems, the space is divided into two vertical zones. The desired space air temperature is maintained only in the lower zone and is always higher in the upper zone because of the temperature stratification created by natural convection.

Depending on space requirements, two types of design methods are used. The most common is temperature-based, and is used when heat removal is the main objective of the air-conditioning system design (e.g., in schools, offices, auditoriums, sport facilities). The other, the shift-zone method, is used when contaminant removal should be considered (e.g., smoking rooms and other facilities with emissions of gaseous, equal- or lighter-than-air contaminants). The objective of the temperature-based method is satisfying thermal comfort in the occupied zone; the shift-zone design addresses this concern and also ensures that contaminants rise above the occupants' breathing level.

Space temperature gradient (STG) is affected by the strength and location of heat sources in the space, heat exchange by radiation between surfaces in the space, and supply airflow. The design procedure presented in this section is based on Skistad et al.'s (2002) simplified method of estimating temperature gradient (Figure 5). This method is applicable for typical spaces with a ceiling height up to 12 ft, such as classrooms, office spaces, and meeting rooms. When designing more complex spaces, computational fluid dynamics (CFD) software programs may be used (see Chapter 13 of the 2009 *ASHRAE Handbook—Fundamentals* for more information).

The thermal gradient relationships illustrated in Figure 5 can be used to establish an acceptable supply-to-return air temperature differential Δt_{SR} from which the supply airflow rate is calculated. Because the space temperature gradient is assumed to be linear, the occupied gradient in the occupied zone is proportional to the volume of the space it represents. For example, if return height is 10 ft and the occupied zone is 5 ft high, its gradient comprises 50% of the space temperature gradient, or 25% of Δt_{SR}. The temperature difference between room air at the top of the occupied zone and the supply air is therefore 75% of Δt_{SR}.

Determining an acceptable Δt_{SR} should consider both the room-to-supply temperature differential and the occupied zone temperature gradient (as limited to 5.4°F by ASHRAE *Standard* 55).

In general, high-ceiling applications allow larger supply-to-return air temperature differentials, because the occupied zone is a smaller percentage of total room air volume. However, the differential may be reduced by limitations on the supply air temperature, as shown in Example 2.

The supply airflow rate Q to achieve Δt_{SR} is calculated from Equation (1).

Example 2. A classroom with a 10 ft ceiling is to be cooled by thermal displacement ventilation. The supply air temperature is 62°F and room temperature is maintained at 76°F at 5 ft level. The total sensible heat gain of the space is 28,000 Btu/h.

Calculate the (1) overall temperature differential between supply and return airflow and (2) required space airflow. Identify return air temperature and temperature at occupants' ankle level.

Solution: Using the relationships in Figure 5, the supply-to-return temperature differential Δt_{SR} and return air temperature can be predicted as follows:

$$\Delta t_{SR} = (t_{room} - t_{supply})/0.75 = (76 - 62)/0.75 = 18.7°F$$

$$t_{return} = t_{supply} + \Delta t_{SR} = 62 + 18.7 = 80.7°F$$

To ensure a high level of thermal comfort, the occupied-zone temperature gradient Δt_{oz} should not exceed 5.4°F. For this application, the occupied zone gradient is acceptable:

$$\Delta t_{oz} = \Delta t_{SR} \times 0.25 = 18.7 \times 0.25 = 4.7°F$$

From Equation (1), the airflow required to maintain this gradient is

$$Q = 28,000/(1.08 \times 18.7) = 1386 \text{ cfm}$$

Application Considerations

Displacement ventilation is a cooling-only method of room air distribution. For heating, a separate system is generally recommended. Displacement ventilation can be used successfully in combination with radiators and convectors installed at the exterior walls to offset space heat losses. Radiant heating panels and heated floors also can also be used with displacement ventilation. To maintain displacement ventilation, outlets should supply ventilation air about 4°F lower than the desired room temperature.

Thermal displacement ventilation systems can be either constant or variable air volume. A thermostat in a representative location in the space or return plenum should determine the delivered air volume or temperature. If the time-averaged requirements of ASHRAE *Standard* 62.1-2004 are met, intermittent on/off airflow control can be used.

Avoid using thermal displacement and mixed air systems in the same space, because mixing destroys the natural stratification that drives the thermal displacement ventilation system. Thermal displacement systems can be complemented by hydronic systems such as chilled floors. Use caution when combining chilled ceilings, beams, or panels with fully stratified systems, because cold surfaces in the upper zone of the space may recirculate contaminants stratified in the upper zone back into the occupied zone.

Chen and Glicksman (2003) provide additional information on fully stratified air distribution systems.

PARTIALLY MIXED AIR DISTRIBUTION

A partially mixed system's characteristics fall between a fully mixed system and a fully stratified system. It includes both a high-velocity mixed air zone and a low-velocity stratified zone where room air motion is caused by thermal forces. For example, floor-based outlets, when operating in a cooling mode with relatively high discharge velocities (>150 fpm), create mixing, thus affecting the amount of stratification in the lower portions of the room. In the upper portions of the room, away from the influence of floor outlets, room air often remains thermally stratified in much the same way as displacement ventilation systems.

Principles of Operation

Supply air is discharged, usually vertically, at relatively high velocities and entrains room air in a similar fashion to outlets used in mixed air systems. This entrainment, as shown in Figure 6, reduces the temperature and velocity differentials between supply and ambient room air. This discharge results in a vertical plume that rises until its velocity is reduced to about 50 fpm. At this point, its kinetic energy is insufficient to entrain much more room air, so mixing stops. Because air in the plume is still cooler than the surrounding air, the supply air spreads horizontally across the space, where it is entrained by rising thermal plumes generated by nearby heat sources.

Research and experience have shown that the amount of room air stratification varies depending on design, commissioning, and operation. Control of stratification includes the following considerations:

- By reducing airflow and mixing in the occupied zone, fan energy can be reduced and stratification can be increased, approaching a reasonable target at 3 to 4°F temperature difference from head to ankle height, which satisfies ASHRAE *Standard* 55-2010.
- By increasing airflow and mixing in the occupied zone, excessive stratification can be avoided, thereby improving thermal comfort.

In practice, successful installation requires an optimal balance of these issues (Webster and Bauman 2006).

Figure 6 shows one example of the resulting room air distribution in which the room air is mixed in the **lower mixed zone**, which is bounded by the floor and the elevation (**throw height**) at which the 50 fpm terminal velocity occurs. At this elevation, stratification begins to occur and a linear temperature gradient, similar to that found in thermal displacement systems, forms and extends through the **stratified zone**. As with thermal displacement ventilation, convective heat plumes from space heat sources draw conditioned air from the lower (mixed) level through the stratified zone and to the overhead return location. A third zone, referred to as the **upper mixed zone**, may exist where the volume of rising heat plumes

terminate. Although velocities in this area are quite low, the air tends to be mixed.

Space Ventilation and Contaminant Removal

Partially mixed systems' ventilation and contaminant removal efficiencies vary considerably. Restricting mixed conditions to below the breathing level results in most respiratory-associated contaminants being conveyed directly to the overhead return by heat plumes rising from occupants. If the lower mixed zone extends above the breathing level, contaminants are entrained and horizontally transmitted across occupied levels of the space, as occurs in mixed air (dilution ventilation) systems.

According to ASHRAE *Standard* 62.1, these systems may have zone air distribution effectiveness E_Z values that exceed those of fully mixed systems.

Typical Applications

Partially mixed systems are commonly used in applications such as the following:

- Office buildings with raised floors
- Call centers
- Libraries
- Casinos
- Other spaces with open or high ceilings

Many underfloor air distribution (UFAD) systems can be classified as partially mixed systems. These systems are popular because of their relocation flexibility when used in conjunction with raised-access flooring systems. Outlet accessibility also allows easy occupant adjustment of space airflow delivery. The cavity beneath the access floor tiles is generally pressurized and used as a supply air plenum. Supply outlets placed in access floor tiles are commonly tapped directly into the pressurized plenum, but may be ducted from a fan-assisted terminal unit mounted beneath the floor.

Benefits and Limitations

Benefits of UFAD systems include the following:

- Using a raised floor system may substantially reduce air distribution ductwork and terminal requirements.
- Central fan energy consumption may be lower.
- The space service flexibility of the access floor platform is extended to include HVAC services as well. Nonducted outlets can be easily added or relocated.
- Because most outlets are sized to handle loads typical to an interior single-occupant office or workstation, they can be placed within the workstation to give occupants thermal control over their individual work environment. This makes higher individual occupant comfort levels possible.
- Air- and water-side economizer opportunities are extended, especially in mild and relatively dry climates.

Applications where contaminants are heavier and/or colder than ambient air may be better served by a mixed air system. As with thermal displacement systems, partially stratified systems in humid climates require that the outside air be sufficiently dehumidified to satisfy space latent requirements. The temperature of dehumidified air must often be increased before introduction to the occupied space.

Outlet Characteristics

One outlet type is a swirl diffuser with a high-induction core, which induces large amounts of room air to quickly reduce supply to ambient air velocity and temperature differentials. Supply air is injected into the room as a swirling vertical plume close to the outlet. Properly selected, these outlets produce a limited vertical projection of the supply air plume, restricting mixing to the lower portions of the space. Most of these outlets allow occupants to adjust

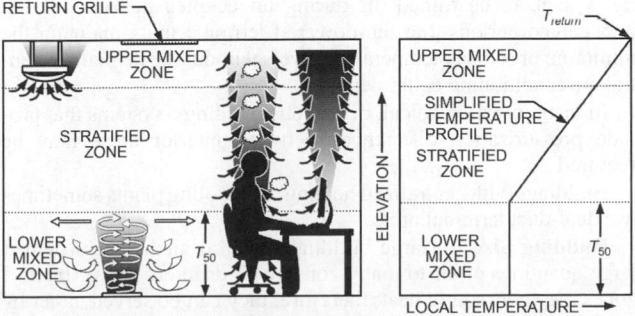

Fig. 6   UFAD System in Partially Stratified Application

the outlet airflow rate easily. Other versions incorporate automatically controlled dampers that are repositioned by a signal from the space thermostat and/or central control system.

Another category includes more conventional floor grilles designed for directional discharge of supplied airflow. These grilles may be either linear or modular in design, and may allow occupants to adjust the discharge air pattern by repositioning the core of the outlet. Most floor grilles include an integral actuated damper, or other means, that automatically throttles the volume of air in response to the zone conditioning requirements.

Room air induction allows UFAD diffusers to comfortably deliver supply air a few degrees cooler than possible with outlets used for thermal displacement ventilation outlets. The observance of clear, or adjacent, zones above and around the diffusers, where stationary occupants should not reside, is recommended. Outlet manufacturers typically identify such restrictive areas in their product literature.

Space Temperature Gradients and Airflow Rates

The objective of partially mixed systems is to condition the air in the occupied zone while allowing stratification to naturally occur. By allowing this stratification, some of the space heat gain can be removed by return or exhaust instead of by supply air delivery to the space. If the supply airflow rate and sensible heat gains affecting the lower zone are balanced, an acceptable temperature gradient (<5°F) can be achieved in the occupied zone. Supply airflow beyond that required by these heat gains reduces the degree of stratification shown in Figure 6. If the supply airflow rate is insufficient, excessive vertical space temperature gradients may occur.

Accurate calculation of the space design supply airflow rate requires analysis of all space sensible heat gains to determine their contribution to the lower zone. Although there is not yet a single recognized procedure for calculating these airflow rates, most UFAD equipment manufacturers offer guidance.

Methods of Evaluation

As for thermal displacement systems, design involves determining a supply airflow rate that limits thermal gradients in the occupied zone in accordance with ASHRAE *Standard* 55 guidelines. ASHRAE *Standard* 55 recommends that the vertical temperature difference between the ankle and head level of space occupants be limited to no more than 5.4°F if a high degree (>95%) of occupant comfort is to be maintained.

Design Procedures

The design of partially mixed air distribution systems requires identifying both thermal and contaminant removal objectives:

- The desired space temperature, the elevation to which it applies, and an appropriate supply air temperature must be identified.
- The supply air temperature for UFAD systems served by a pressurized or neutral pressure floor plenum should be limited to that which results in a relative humidity level below 80% in the floor cavity, to minimize the threat of mold or fungus growth.
- Supply air temperatures tend to rise as air moves through the floor cavity; therefore, the supply air temperature varies with its distance traveled. When determining space airflow requirements, supply temperatures should be modified accordingly to avoid undercooling the occupied space. Bauman and Daly (2003) discuss this subject further.
- If the objective is to provide displacement ventilation of respiratory contaminants in the stratified zone, mixing must be limited to below the breathing level of most space occupants.
- Outlets should be located far enough from stationary occupants to ensure that they are not subjected to drafts that might cause thermal discomfort. Outlet manufacturers generally prescribe clear zones that quantify this separation distance.

Application Considerations

ASHRAE's *Underfloor Air Distribution Design Guide* (Bauman and Daly 2003) includes a thorough discussion of issues involved in the design, application, and commissioning of UFAD systems. Some considerations include the following:

- Supply temperatures in the access floor cavity should be kept at 60°F or above, to minimize the risk of condensation and subsequent mold growth.
- Most UFAD outlets can be adjusted automatically by a space thermostat or other control system, or manually by the occupant. In the latter case, outlets should be located within the workstation they serve.
- Use of manually adjusted outlets should be restricted to open office areas where cooling loads do not tend to vary considerably or frequently. Perimeter areas and conference rooms require automatic control of supply air temperatures and/or flow rates because their thermal loads are highly transient.
- Heat transfer to and from the floor slab affects discharge air temperature and should be considered when calculating space airflow requirements. Floor plenums should be well sealed to minimize air leakage, and exterior walls should be well insulated and have good vapor retarders. Night and holiday temperature setbacks should likely be avoided, or at least reduced, to minimize plenum condensation and thermal mass effect problems. With air-side economizers, using enthalpy control rather than dry-bulb control can help reduce hours of admitting high moisture-content air, thus also reducing the potential for condensation in the floor plenums.
- Avoid using stratified and mixed air systems in the same space, because mixing destroys the natural stratification that drives the stratified system.
- Return static pressure drop should be relatively equal throughout the spaces being served by a common UFAD plenum. This reduces the chance of unequal pressurization in the UFAD plenum.

TERMINAL UNITS

System Selection

Designers have various systems (terminal units and their associated controls) to choose from when designing a building. The owner's needs must be met for installation, application, and cost of operation. The designer must consider performance, capacity, reliability, energy consumption, sustainability, and spatial requirements and restrictions. The following guidelines describe different types of equipment and their general uses, restrictions, and limitations. Table 5 summarizes the different types of terminal units and their suitability for particular commercial building applications.

Building Use. Before specifying equipment types, the designer must consider the building's intended use. Office buildings with daily operational schedules frequently use fan-powered terminal units. Usually, fan-powered terminals with auxiliary heaters (supplementary heat) are used in perimeter zones; these units allow the greatest flexibility for individual zones while also allowing the central system to be turned off during unoccupied periods. During unoccupied periods, the fan-powered terminal units maintain the minimum or setback temperature levels without the help of the central air-conditioning equipment.

In institutional, medical, or campus buildings, systems that provide pressurization differences between interior areas may be required.

Buildings with centralized heating and cooling plants sometimes use dual-duct terminal units.

Building Size. In large buildings, central air handlers deliver large quantities of air to many zones with different needs. Interior zones may not require heat; therefore, they can be served either by single duct or fan-powered units with no supplemental heat. Unless the building is located in a tropical climate, the perimeter zones

Table 5 Suitability of Terminal Units for Various Applications

| | Office Space, Educational, and Institutional Buildings | | | | | Hospitals, Cleanrooms, and Laboratories[a] | | | Noise-Sensitive Applications[b] | | | Other Facilities | | |
| | Large | | | Small | | | | | | | | | | |
Terminal Types	Interior Zone	Exterior Zone	Low Temperature	Interior Zone	Exterior Zone	Patient Areas	Operating Areas	Laboratory Space	Broadcast Studios	Theaters	Libraries	Public Use	Shopping Centers	Mixed Use
Single-Duct														
VAV without reheat	•	•	•	•	•	•	P	•	•	•	•	•	•	•
VAV with reheat	•	•	•	•	•	•	N	P	•	•	•	•	•	•
Dual-Duct														
VAV no mixing	•	•	N	N	N	N	N	N	•	•	•	•	•	•
VAV with mixing	•	•	N	N	N	•	•	N	•	•	•	•	•	•
Constant volume	•	•	N	N	N	P	P	•	P	•	•	•	N	•
Exhaust Terminal	•	•	N	N	N	P	P	P	•	•	•	•	•	•
Induction Terminal														
VAV with heat	•	•	•	•	•	•	N	•	•	•	•	•	•	•
VAV without heat	•	•	•	•	•	N	N	N	•	•	•	•	•	•
Fan-Powered														
Parallel with heat	N	•	•	N	•	N	N	N	N	•	•	P	•	•
Series without heat	P	•	•	•	•	N	N	N	P	P	P	P	P	P
Series with heat	•	P	•	•	•	•	N	•	P	P	P	P	P	P
Low-temperature	N	N	P	N	N	N	N	N	N	N	N	N	N	N
Bypass	N	N	N	•	•	N	N	N	N	N	N	N	N	N

P = Preferred for this application.
• = Used for this application.
N = Not recommended for this application.
[a]Sealed lining is recommended to minimize entrainment of airborne fibers from liner to occupied spaces.
[b]Special consideration should be given to selecting very quiet operating equipment and use of attenuators or silencers.

require heat, typically electric or hot water. These are usually included with the terminal units, but sometimes separate heating systems are used, such as dual-duct or baseboard. In buildings where the owner desires low operating costs, the static pressure in the ducts should be lowered in accordance with ASHRAE *Standard* 90.1 to the minimum pressure, which sets at least one VAV damper to near-full open. Interior zones in these buildings can use fan-powered terminal units to keep the static pressure low. Buildings with parallel-type fan-powered terminal units usually use single-duct units in the interior zones and require higher system static pressures.

In small building, such as shopping malls and other low rise buildings where each tenant area is small, it is common to use small packaged air conditioners. If terminal units are used on these systems, usually single duct or bypass units are selected. A variation of this system, variable-volume variable-temperature (VVT), uses pressure-dependent single-duct units with a main bypass damper in the supply duct. The bypass damper is regulated by static pressure in the supply duct. A nearly constant pressure can be maintained, allowing the packaged units to operate at constant volume and the individual zones to be pressure-dependent VAV.

Building Controls. The type of control system to be used is somewhat dependent on the size and type of building. Controls can be electric, pneumatic, analog electronic, or digital electronic.

- Electric controls are pressure dependent where the damper responds to a single control input. For example, the thermostat sends a signal to the damper to open or close based solely on room sensible temperature.
- Pneumatic controls are usually used for building renovation or buildout of existing buildings where the base building already has a pneumatic system installed. They can be pressure dependent or pressure independent. Pneumatic controls require regular system maintenance and may need to be periodically rebalanced.

- Analog controls are often applied to smaller buildings that do not have a building operation staff. Typically, these controls do not communicate with other zones or other equipment in the building.
- Digital controls are typically used on buildings that have a building operation staff. These controls provide individual zone control and communication to the building management system.

Acoustical Considerations. Terminal units and room air distribution devices are typical equipment sound sources for the room. However, they are not the only sources that affect room acoustics. Refer to Chapter 48 as well as AHRI *Standard* 885 and other standards for guidance on space acoustics. Broadcast studios, theaters, and libraries require very low noise levels. Equipment selection and location is important here, and careful examination of the equipment sound performance is imperative. Radio-frequency interference (RFI) and electromagnetic interference (EMI) should also be considered when designing television studios.

Environmental Factors. Environmental factors play an important role in system selection. They include the climate and air conditions inside as well as outside the building. They also include legislative requirements such as outside air ventilation rates and local building codes. When high ventilation rates are required, reheat is often required to maintain human comfort. Fan-powered terminal units are usually used where the thermal load changes significantly and heating is required. Single-duct terminal units are usually applied in the interior where the thermal load is normally stable.

Contamination Considerations. Hospitals, cleanrooms, and laboratories pose special challenges. Protective isolation spaces such as operating rooms, bone marrow transplant patient rooms, AIDS patient areas, and cleanrooms require positively pressurized environments. Infectious isolation spaces such as tuberculosis patient rooms require negatively pressurized environments. See ASHRAE *Standard* 170 for details on room pressurization, Chapter

8 for specifics on health care requirements, and Chapter 18 for details on clean spaces. Hospital rooms and cleanrooms frequently also require constant high ventilation rates, which tend to favor single- or dual-duct terminal units. Pressure-independent, variable-speed motor technology has lead to the development of fan-powered pressurization units.

To minimize entrainment of fibers into the airstream, either do not use internal insulation or use special liners in the terminal units and duct systems. Insulations can be isolated from the airstream by metal, foil, or polymer liners inside silencers and terminal units. All of these liners have different thermal, acoustic, and other physical properties and should be evaluated for each job.

Maintenance and Accessibility.

Typical Applications. Terminal units are typically not easily accessible after building occupation; they should be selected and located with consideration for required maintenance. Review the applicable building codes [e.g., ICC (2009)] for required access.

Critical Environments. Some applications, such as cleanrooms and operating theaters, require high levels of reliability from terminal units because of the difficulty and cost associated with servicing or maintaining the equipment. In a cleanroom, for example, if the ceiling must be opened, the space may require disinfection before it can be used again. Associated costs might include lost production time as well as the cost for wipedown and/or disinfecting the room and/or equipment. In cases like these, the designer should consider locating the equipment outside of the clean space or consider highly reliable, very-low-maintenance, very basic equipment and controls. Terminal units may require access to internal components for cleaning in the case of contamination.

Cost Factors. Costs should be considered before system selection is finalized. Installation, operation, and maintenance all contribute to total cost. Often, one of these costs overrides the others. Electric heaters usually have a lower installed cost than hot water coils, but they may have a higher operating cost. Local utility rates and building codes should be researched to arrive at the correct decision before making the final selection.

Applications

Single Duct. The basic single-duct unit consists of an airflow regulator and may also include an actuator, an airflow-measuring device, and selected controls. Accessory discharge attenuators and multiple outlet attenuators are also frequently used.

Typical applications include

- Where the supply air system is not tasked with the space heating requirements
- With VVT or other auto-changeover controls
- Constant air volume
- Constant pressure control

A **single duct with reheat** has an added heating coil (hot water or electric). They are typically applied in zones where heat losses create a need for heating. The terminal usually reheats at the minimum airflow setting. An auxiliary higher heating setting may be available as an option with additional controls.

Interior zones, where ventilation requirements may be larger than the desired heating airflow, may require additional reheat.

A **single-duct exhaust** consists of an airflow regulator and may also include an actuator, an airflow-measuring device, and selected controls. Accessory inlet attenuators are sometimes used. Specialty materials may need to be considered for corrosion resistance. The designer should consider pressure drop and inlet versus outlet designs.

Single-duct exhaust units are typically used in isolation wards, operating theaters, laboratories, cleanrooms, and systems that require supply and exhaust tracking.

Dual Duct. A **nonmixing dual duct** is effectively two single-duct terminal units side by side. The basic unit incorporates separate cold and hot (sometimes neutral) air inlets. They are usually applied in exterior zones in buildings where overhead heating and cooling are desired but not auxiliary heat, and zero minimum flow is acceptable during changeover between heating and cooling.

A **mixing dual duct** is the same, with an integral mixing/attenuator section on the downstream end of the terminal unit to minimize temperature stratification in the discharge airstream. Interior and exterior zones in buildings (e.g., hospitals) where overhead heating and cooling are desired but an auxiliary heating coil is not, and zero minimum flow is unacceptable during changeover between heating and cooling. Mixing performance should be evaluated to fit the application.

Fan-Powered Parallel Flow. Sometimes called **variable-volume** or **intermittent-fan units**, these consist of a single duct unit, blower/motor, backdraft damper, and selected controls where the motor and primary damper are arranged such that mixing occurs downstream of the blower. Supplemental heating coils (either hot water or electric) are generally required. Electric heaters are typically located on the discharge of the unit. Water coils may be on the discharge or the induction port, although location on the discharge adds to the supply air system's static pressure requirements and increases leakage through the backdraft damper, as shown in ASHRAE research project RP-1292 (Davis et al. 2007). Heating coils on the induction port increase ambient temperature at the motor and decrease motor life.

Fan-powered parallel-flow units are used in exterior zones where heating and cooling loads may vary considerably, and in buildings where heating is needed when the central system is shut down during unoccupied hours.

Fan-Powered Series Flow. The basic unit consists of a single-duct unit, blower/motor, and selected controls where the motor and primary damper are arranged such that mixing occurs upstream of the blower. These units are also called **constant-volume** or **constant-fan**. Supplemental heating coils (either hot water or electric) are generally required. Electric heaters are typically located on the discharge of the unit; water coils may be on the discharge or the induction port. Heating coils on the induction port increase ambient temperature at the motor and decrease motor life.

Fan-powered series-flow units generally are used in the following situations:

- Exterior zones where heating and cooling loads may vary considerably
- Buildings where heating is desired when the central system is shut down during unoccupied hours
- To allow lower central system static pressure
- Where occupant comfort can be optimized, because the high- (sometimes constant-) volume variable-temperature air delivery produces consistent air distribution, acoustics, and ventilation

Low-Profile Fan-Powered Series or Parallel Flow. Similar in construction to the standard series or parallel flow terminal described earlier, these units are typically less than 12 in. (300 mm) high for all sizes, to minimize the depth of ceiling space required. Unlike standard fan-powered terminals, the fan/motor assembly is installed flat on its side with the wheel rotating in a horizontal plane.

Typical applications are the same as for regular series or parallel units, but low-profile versions are commonly used where zoning requirements limit building height and the architect wishes to maximize the number of floors, because these units fit in a shallow ceiling plenum. Designers should pay special attention to available space and unit heights.

Ventilation Air Inlet Fan-Powered Series Flow. These units are similar in construction to the standard series-flow terminal, but have an added secondary air inlet that provides a direct connection to the terminal unit for ventilation air. They are commonly used in buildings where ventilation air is piped in a dedicated ventilation duct system to each terminal unit; this is generally done where it is desirable to monitor ventilation air quantities to each zone.

Low-Temperature Fan-Powered Series Flow. These units are the same as fan-powered series flow, but have special construction to minimize the potential for condensation. They can be used with cold-water/ice storage systems that provide low-temperature central system air distribution to the zone terminals when there is potential for condensation, or where standard terminals may be exposed to high humidity.

Underfloor Fan-Powered Series Flow. This unit is a fan-powered series flow terminal designed to fit between the pedestal support grids of a raised- or access-floor HVAC system, without modification to the floor. They are available in several unit sizes, but with limited height and width.

Primary and induction ports, if any, to the unit may or may not be ducted. Typically, air under the raised floor is cool air supplied directly to the space, although heated air may also be ducted to the unit. In these cases, a control system is required to select the proper damper sequence to control room air distribution to maintain the proper ambient conditions in the occupied space.

Bypass Terminals. The basic bypass terminal consists of a diverter-type damper, actuator, bypass port, and selected pressure-dependent controls. A balancing damper is recommended ahead of the inlet. Use of reheat coils is discouraged, and electric reheat should be prohibited because of the potential fire hazard.

Bypass terminals are used primarily with packaged rooftop air conditioning (PTAC) equipment with a direct-expansion (DX) coil where zoning is desired, but relatively constant airflows across the system components (e.g., coils, fans) are required to minimize the potential for freeze-up. The system offers an economical VAV supply design with low first cost. It does not provide the energy-saving advantages of variable fan volume, but avoids the expense of a more sophisticated system.

Comparison of Series- and Parallel-Flow Fan-Powered Terminal Units

Series (constant-volume) units have a continuously operating fan during occupied mode, and are typically installed in the ceiling plenum. Primary air is ducted from a central air handler. Induction air is either from the ceiling plenum or occasionally ducted from the conditioned space. Fan air is a combination of plenum and primary air supplied to the zone.

Parallel (variable-volume) units should have an intermittently operating fan during the occupied mode and are typically installed in the ceiling plenum. Primary air is ducted from the central air handler and should flow directly to the zone in a variable volume cooling mode. Fan air is induced from the ceiling plenum or may be ducted from the conditioned space. The fan should run in the dead band and heating modes. When the fan runs, the zone air is induced air or a combination of primary and induced air:

- **Primary air** is that delivered to the space for the purposes of satisfying ventilation, latent, and all or part of the sensible load.
- **Secondary air** is that circulated from the return air plenum.
- **Supply air** is a mixture of primary and secondary air delivered to the space.

Configuration.
Series. The fan and VAV damper are aligned so that all conditioned air and all induced air independently enter the mixing section and must go through the fan to exit the unit. The mixing section is between the VAV damper and the fan.

Parallel. The fan and VAV damper are arranged so that all induced air enters the fan, and conditioned air bypasses the fan. Any mixing of conditioned air with the induced plenum air occurs on the discharge side of the fan. A backdraft damper inhibits air from exiting the unit through the fan when not running.

Terminal Fan Selection.
Series. In occupied mode, the fan runs continuously, supplying either a constant or modulating volume to the space. Some direct digital controls (DDCs) provide an optional output that may be used for controlling fan airflow using the building management system (BMS). This allows dynamic fan volume control, which may be either modulating or multiple-speed operation from a single speed motor. The fan must be sized to match the maximum airflow to be supplied to the zone. These terminal fan airflows are usually larger than those with parallel units for similar zones. When modeling energy consumption for this unit, it is important to model the energy consumption and heat generated by the fan motor for both unoccupied and occupied periods. The motor heat should also be included in the heating mode calculation (Davis et al. 2007).

Parallel. Typically, the fan runs only in heating or deadband modes, supplying a relatively constant induced air volume to the space. The fan must be sized to supply the required heating airflow to the zone, which requires overcoming the pressure created in the mixing chamber caused by the inclusion of primary air. These units usually have smaller terminal fan airflows than series units for similar zones. When modeling energy consumption for this unit, it is important to model casing and backdraft damper leakage as well as fan energy consumption for both the terminal unit and the air handler during operation. Motor heat should also be included in the heating mode calculation (Davis et al. 2007).

VAV Cooling and Inlet Static Pressure Requirements.
Series. Additional savings above that for the single- or dual-duct units can be realized because of the low inlet static pressure requirement of the series fan powered terminal unit. The pressure at the air handler can be reduced to that required to push the conditioned air through the ducts to the unit and across the VAV damper into the mixing section.

Parallel. Like the single or dual duct unit, the air handler must push the conditioned air through the ducts to the parallel unit, across the VAV damper, into the mixing section, through the discharge duct from the unit and across the diffuser(s) into the room.

Control Sequence.
Series. The fan runs constantly during occupied periods. On a call for cooling, the controls modulate the VAV damper toward maximum airflow, delivering primary air to the mixing chamber. If the fan is set at the same airflow as the primary air, no air is induced from the plenum. If the fan is set at a higher airflow than the primary air (e.g., as in a low-temperature application), air is induced from the plenum to meet the fan's set point. The primary and induced air are blended before they enter the fan. Constant-volume, variable-temperature air is then discharged into the downstream duct and into the conditioned space.

As cooling demand decreases, the VAV damper modulates toward minimum airflow, reducing the primary air into the mixing chamber. This increases the volume of warmer induced air into the mixing chamber. Fan air can also be reduced [typically with an electronically commutated motor (ECM)] as load changes. The increased percentage of plenum air causes the discharge temperature to rise to approach the plenum temperature, taking advantage of recaptured heat from lights, people, and machinery.

After a further decrease in zone temperature, the controls automatically energize the supplemental heat (optional equipment), which can be either electric or hot water coils. The discharge temperature increases as heat is applied. As the temperature increases in the zone, the sequence reverses.

Parallel. On a call for cooling, the controls modulate the VAV damper toward maximum airflow while the fan is off. Variable-volume, constant-temperature air is then discharged into the downstream ducts and into the conditioned space.

On a decreasing call for cooling, the VAV damper modulates toward minimum airflow. The unit delivers variable-volume, constant-temperature air to the zone.

In deadband, the controls energize the fan. Fan air and primary air are blended in the mixing chamber on the fan's discharge side. The increased plenum air causes the discharge temperature to rise. Constant-volume, constant-temperature air is delivered to the zone.

On a call for heating, the heat is energized. The heater may be staged or modulating. Constant-volume, constant- or variable-temperature air can be delivered to the zone.

Because the largest energy-consuming characteristic of parallel boxes is leakage and not motor energy, and because modulating fan air in the heating mode increases reheat, ECMs are not recommended for parallel boxes.

Fan Interlocks.

Series. Typically, series units are designed to run continuously. Usually, they are energized only during occupied periods or when needed for emergency heating during unoccupied periods. Care should be taken to interlock the unit fan with air handlers in the building to ensure that the terminal unit fans start during occupied periods. Series-unit fans should be started ahead of the air handler to prevent backflow into the plenum and backward rotation of the fan.

Parallel. Fans in parallel units are designed to be energized in the deadband and heating modes throughout the day. Primary air enters the mixing chamber at the fan discharge. When the fan is not energized, there is a positive pressure at the fan's discharge. Typically, this would cause the blower and motor to rotate backwards; however, parallel units are equipped with backdraft dampers at the fan, which inhibit backward airflow through the fan into the plenum.

Acoustics.

Series. Series fans are sized to match the maximum airflow required in the zone. The fan runs constantly during occupied periods. There are two sound sources in the unit: the fan motor and the VAV damper. Although both contribute to the overall discharge and radiated sound emitted from the unit, the fan is primarily responsible for discharge noise, and both the damper and the fan are responsible for radiated sound. Usually, sound radiated into the occupied space is the greater and usually more difficult to attenuate.

Comparing the sound level between a series and a parallel unit in similar zones, the fan in the series unit might generate a slightly higher sound power level. In the series unit, the fan and damper would be at their peak when the unit operates at full cooling capacity, the loudest point in the sequence of operation for sound generation. As the primary air decreases, the sound generated by the fan typically masks the sound from the damper. Sequences that also command the fan to modulate in this condition create lower ambient sound levels.

Series fan power sound levels are more consistent compared to ambient background sound levels than the rapid cyclic nature of the parallel unit.

Parallel. Parallel fans are sized to match the heating airflow required in the zone. The fan runs intermittently when the mode changes from cooling to deadband. There are two sound sources in the unit: the fan/blower and the VAV damper. Both damper and fan are responsible for radiated and discharge sound. Usually, sound radiated into the occupied space is the greater and usually more difficult to attenuate.

Comparing the sound level between a parallel and a series unit in similar zones, the parallel unit may generate slightly less sound. In the parallel unit the fan and damper would normally not peak simultaneously. When the unit operates at full cooling capacity, the damper is at its peak sound generation and the fan is off. In heating mode, the fan peaks while the damper is at minimum sound generation.

Damper sound must be considered as the sound increases with increasing inlet static pressure. Parallel units require significantly higher inlet static pressures at the unit. Fan sound is consistent into the zone when the fan is running; however, the fan is intermittent during much of the day. This rapid fan cycling can create a variation in sound levels and subsequent noise in the space, which can be more annoying than a higher consistent sound level.

Energy Consumption.

Fan-powered VAV terminal units take advantage of typical VAV savings at the air handler and chiller during cooling periods, and even more savings are realized when heating is required. Fan-powered terminals induce warm plenum air from the ceiling and blend it with the primary air at minimum ventilation requirements as required by ASHRAE *Standard* 90.1 during the heating sequence. This recaptures much of the heat created in the zone and plenum by lights, occupants, solar loading, and machinery or equipment such as computers, coffee machines, copiers, etc. The unit returns this heat as free heating rather than losing it at the air handler. If additional heating is required, supplemental heat is added to the sequence, but the unit still saves energy by warming blended air, for example, at 72°F rather than reheating primary cooled air at 55°F, saving the cost of 17°F at the heating airflow. According to ASHRAE research project RP-1292 (Davis et al. 2007), there is very little difference in total building energy use between series and parallel units when equipped with permanent split-capacitor (PSC) motors.

Series. Series fans run constantly during occupied periods, and the fan is sized for the full airflow to the zone. This causes the energy consumption from the fan to be higher than that of a parallel fan in a similar zone. ASHRAE RP-1292 identified the motor as the biggest energy user in a series unit (Davis et al. 2007). The fan energy raises the air temperature across the motor by 1 to 3°F. This means that total energy use can be reduced if the fan energy is reduced. Using electronically commutated motors (ECMs) can significantly increase motor lifetimes and provide significant energy savings in excess of 50%.

Series units are designed for very low inlet static pressures. This saves energy at the air handler compared to a parallel unit for a similar zone.

Casing leakage does not affect energy consumption of series units. All the primary air injected into the unit gets delivered to the occupied zone. This causes ceiling plenum temperatures to be warmer for installations with series units than those with parallel units, and more heat can be reclaimed from the ceiling plenum when the unit is in heating mode.

Because the motor heat is in the airstream, more heat is introduced into the occupied zone during cooling mode when compared to a parallel unit.

Parallel. Parallel unit fans operate only when required during heating and deadband modes. The fan is sized for the induced airflow required in heating mode, which may be much less than the zone's total airflow requirement.

Because the minimum primary air and induced air are mixed downstream of the terminal fan, the terminal inlet static pressure requirement is greater than for a series terminal. This adds operating cost at the air handler. When just one duct path is designated as the critical path for static pressure set point, this can cause the system pressure to increase in excess of 1 in. of water compared to a building with series units. This can greatly increase energy usage. However, ASHRAE *Standard* 90.1 procedures reset the system static pressure to always drive at least one VAV damper to its nearly full-open position, making the increased static pressure required for parallel units not as great.

According to ASHRAE RP-1292, the largest single energy issue in both types of terminal unit, other than operating schedule, was leakage in the parallel unit (Davis et al. 2007). Leakage is typically between 10 and 15% of the total airflow through the parallel unit, and is highest at full cooling. This means that parallel units may need to be oversized to cover the total load to the occupied zone. It also means that the ceiling plenums are cooler with parallel units than those with series units, reducing the amount of free heat that can be reclaimed in heating mode.

Because the motor runs in deadband and heating modes, any heat generated by the motor is delivered to the occupied space. Consequently, the motor energy does not significantly affect the total energy consumed by the parallel unit. Because of this, using more efficient motors will not measurably improve the unit's energy consumption.

FAN SELECTION

When selecting a unit for a particular set of conditions, care should be taken so that the air delivery is designed to meet the room's sound criteria and system's static requirements. Specific sound data can be found in manufacturers' catalogs for various airflow deliveries for each unit. This should be the guiding factor in selecting unit sizes. A simple rule of thumb is that, when considering a unit selection for a typical office space, the fan should be selected for performance down from the high end performance by 20 to 25% of the distance to the low end of the fan curve at the specified external static requirement. This allows for low room sound levels while maintaining some flexibility for future changes in the zone. When selecting equipment for large, open areas where sound criteria may not be as critical, select equipment closer to the high end of the fan curve. For a meeting room or an executive office, select equipment slightly below the halfway point on the fan curve. For an auditorium, chief executive offices, conference room, or some similarly sensitive area, select operation nearer the low end of the fan curve.

Avoid selecting equipment right on the maximum or minimum curves. This leaves no flexibility, either in the equipment for future changes or for variations that may occur due to power variations or duct fittings.

Fan Airflow Control on Fan-Powered Terminals

When designing air systems and using fan-powered VAV terminal units, it is important to match the fan air and primary air capacities to the space requirements. Series units require precise adjustment of fan airflow in relation to the primary air. The parallel unit's fan airflow requires less critical adjustment. Fan-powered terminals nearly always use single-phase motors, most commonly with electronic fan speed control [sometimes called **wave choppers**, **thyristor controllers**, or **silicon-controlled rectifiers (SCRs)**] or with electronically commutated motors.

Fan Shift in Fan-Powered Terminal Units. Before adjusting the fan, the possibility of fan shift must be considered. This occurs when the blower is subjected to variations in pressure or airflow patterns. As the primary airflow changes, the pressure drop and changes in local jets may cause the fan to shift its performance as it rides the fan curve. Consequences from the phenomenon vary from building to building and zone to zone. Noise levels may change greatly as the volume changes, and this may be annoying. Design ventilation rates can also vary, sometimes by more than 20%. This can be aggravated by undersizing the terminal unit.

Electronic Fan Speed Control (PSC Motors). Electronic fan speed controls use a thyristor to adjust the fan's electrical input ac voltage. This is called **phase proportioning** or **wave chopping**. When the current sine wave crosses the zero point, the thyristor acts as a timing device, holding the voltage off the motor for some preset period of time. When the thyristor is turned on, the voltage seeks out the sine wave and then follows the curve to the next current zero crossing, where the process begins again on the opposite side of the sine wave. Basically, this reduces the root mean square (RMS) value of the voltage supplied to the motor. This in turn reduces the torque

available to turn the rotor and lowers the motor speed. Amp draw is slightly affected during this process if the motors and blowers are sized properly. Some units may suffer from large changes in amp draw that significantly affect the motor's efficiency and operating characteristics. Reducing the voltage while keeping the amperage draw constant reduces the motor's power consumption.

Nameplate Ratings. The standard that covers fan-powered terminal unit nameplate ratings is ANSI/UL *Standard* 1995. This standard relates to equipment manufacturers and not field issues covered in international and local codes. Nameplate ratings on the unit usually do not match the nameplate ratings on the motor. Amp draw can be above or below the motor nameplate. Even voltage can vary. Differences between the motor label and the unit label may be significant in some cases, but these different ratings do not generally affect the performance or lifetime of the motor or unit. Be sure to refer to the unit nameplate ratings and not the motor nameplate ratings when sizing supply circuit requirements. These ratings are set at the safest possible condition. Because static and set points vary on each unit, performance may not be what is on the unit nameplate.

Electronically Controlled Motor (ECM) Technology. ECMs may provide significant energy savings and superior controllability. They can provide significant power savings on series units; however, they may not provide any savings on parallel units for reasons discussed in Davis et al. (2007).

Sizing Fan-Powered Terminals

Selection of fan-powered terminal units involves four elements: primary air valve, fan size, heating coil, and acoustics. How these elements are selected and their interactive effects determine the overall performance of the units.

Primary Air Valve. Identify the type of controller that is desired and select an inlet size that meets the minimum and maximum airflow desired from the recommended primary air airflow range table in the Performance Data section of the manufacturer's catalog. Selecting terminals near the top of their range may reduce cost, but increases velocity and noise. Selecting terminals toward the bottom may reduce noise, but may reduce controllability of the minimum airflow. Selecting the maximum airflow setting at between 70 and 85% of full capacity (approximately 2000 fpm inlet velocity) is a good tradeoff to avoid possible low-velocity control problems and sound problems at higher velocities.

Fan Size. Parallel fan size is determined by calculating the difference between the unit design heating airflow and minimum primary airflow. If minimum airflow is zero, then fan airflow is the heating airflow. In most cases, the fan can be downsized compared to a series terminal, reducing both first cost and operating cost because the fan only requires the capacity to handle the secondary airflow at reduced downstream static pressure compared to the maximum design airflow. In most applications of a parallel terminal, a minimum primary airflow is required to meet ventilation requirements. This primary airflow contributes to the total resistance experienced by the fan and should be accounted for along with all components downstream of the fan, such as heaters, ductwork, and diffusers. Hot-water coils may be positioned out of the primary airflow (i.e., on the inlet side of the fan, where they would not affect the primary airflow static pressure). In this configuration, heat generated by the water coil shortens the motor life.

Series fan terminal units require the fan to be sized to handle the maximum design airflow. The fan airflow must be at least equal to the primary airflow to ensure the mixing chamber in the terminal does not become pressurized, resulting in primary air spilling out into the ceiling plenum through the induction ports. The external static pressure requirements are the sum of the ductwork and diffusers downstream at design airflow plus an applicable hot-water coil or electric heater, if required. When fan airflow and downstream static pressure have been determined, select the fan size from the fan curves in the Performance Data section of the manufacturer's catalog. Selecting toward the upper end of the range reduces first cost and optimizes fan operating efficiency. Upsizing the fan and operating it at a reduced speed can result in quieter operation. When electric or hot-water coils are required, be sure to include the static pressure required for those items when referring to the fan curves.

Heating Coil. First, determine the heating supply air temperature to the space by calculation using the heat transfer equation:

$$q = 1.085 \times \text{cfm} \times \Delta t$$

where

  q = design heat loss in space, Btu
  Δt = supply air temperature (SAT) – room design temperature.

The supply air temperature (SAT) to the space equals the leaving air temperature (LAT) for the terminal unit.

For a series unit, once the terminal LAT is determined, the heating requirements for the coil can be calculated. The leaving air temperature for the coil varies based on the coil design and terminal unit model. It is generally a good idea to limit leaving air temperature to 15°F above room temperature. The LAT is the temperature leaving heating coil for a series unit, but requires calculations using a mixed-air equation for the primary and fan air volumes and temperatures for a parallel unit with the coil located out of the primary airstream. These leaving air temperatures can be effectively used to warm the room, because their airstreams are not so buoyant that they cannot be driven to the floor in overhead applications, and are warm enough to not produce chills from drafts.

Once both coil entering air temperature (EAT) and LAT are calculated, the heat transfer q for the coil must be calculated, using the heat transfer equation. For electric heat, the capacity must be converted from Btu/h to kilowatts for selection. The required kilowatts and number of steps desired should be checked with availability from the charts in the Performance Data section of the manufacturer's catalog. For hot-water coils, reference the capacity charts in the Performance Data to select the appropriate coil.

$$\text{EAT of coil} = (T_1 Q_1 + T_2 Q_2)/QT$$

where

  T_1 = plenum air temperature, °F
  T_2 = primary air temperature, °F
  Q_1 = plenum air quantity, cfm
  Q_2 = primary air quantity, cfm
  QT = total air moved by terminal fan, cfm

These processes are for calculating the total heat requirement based on the room load calculations. At part-load conditions, it may be desirable to modulate the airflow through the terminal unit as well as the heat output to maintain an acceptable discharge air temperature. This can be done with modulating valves on coils or SCRs on electric heaters. Staging the electric heaters can create similar results at a lower equipment cost. Modulating the heat causes the heaters to run longer, but at lower energy consumption. This can make the room more comfortable without increasing energy costs.

Acoustics. Sound levels are affected by air-regulator-generated noise and fan-generated noise. The maximum noise generated by a given air regulator size is determined by the difference between the highest inlet static pressure and external static pressure at the design cooling airflow for a parallel unit. For a series unit, it is the highest inlet static pressure. This represents the most extreme operating condition. To determine fan noise levels, fan airflow (adjusted within its range by the speed controller) and external static pressure conditions are required.

The acoustical performance data are presented in two formats for the parallel and series types, because their sequence of operation differs. With a parallel unit, air regulator and fan operation are evaluated separately because their operations are not simultaneous under most conditions. With a series unit, air regulator and fan are

evaluated together, because they operate simultaneously, and fan only for heating with no minimum primary airflow.

From the performance data, determine the sound power levels and predicted room noise criteria for both discharge and radiated path under the appropriate operating conditions. Radiated noise from the unit casing typically dictates the noise level when the terminal unit is installed above the occupied space.

Care should be taken because some published room noise criteria are based on certain path attenuation assumptions that may not be indicative of a specific design. The size of the terminal may be increased to reduce noise, but it is also preferable to evaluate the room noise criteria to ensure the necessary reductions are achieved and finished levels do not exceed the design goal in the occupied space. To do this properly, the engineer must specify all factors in the building specifications that affect sound attenuation. An ideal specification specifies maximum allowed discharge and radiated sound power by octave band rather than just catalog-based NC values.

Example 3. Parallel Terminal with Hot-Water Heat. Select a unit inlet for a maximum/minimum primary airflow at 1000/250 cfm with 1 in. of water inlet static pressure.

The heating airflow required is 600 cfm. Downstream resistance at 1000 cfm is 0.4 in. of water. Zone design heat loss is 20,000 Btu/h, design room temperature is 72°F, plenum air temperature is 75°F, and primary air temperature is 55°F.

Solution:

Air Valve Selection. Based on a good design inlet velocity of 2000 fpm, choose a 10 in. inlet.

Fan Selection. Fan heating airflow = Heating airflow (600 cfm) – Primary airflow (250 cfm) = 350 cfm. The downstream static pressure the fan must overcome is the fan airflow plus primary airflow (600 cfm), and because this is less than maximum design airflow (1000 cfm), fan downstream static pressure = $(600/1000)^2 \times 0.4 = 0.144$ in. of water. Refer to fan curves to select the proper unit. The correct unit will handle 350 cfm at 0.144 in. of water static pressure with correct setting of the speed controller, and allows for the selection of a one- or two-row hot water coil.

Heating Coil Selection. For heating, the temperature difference (Δt) is the zone supply air temperature (SAT) minus the design set point temperature. Using the heat transfer equation,

$$20{,}000 \text{ Btu/h} = 1.085 \times 600 \times (\text{SAT} - 72)$$

$$\text{SAT} = 103°F$$

As the heating coil is on the unit discharge, the unit supply temperature equals the coil LAT. Coil entering air temperature (EAT) is a mixture of plenum and minimum primary air.

Design heating flow × Coil EAT =
(Primary airflow × Primary air temperature)
+ [(Design heating airflow – Primary airflow) × Plenum temperature]

$$600 \times \text{Coil EAT} = 250 \times 55 + (600 - 250) \times 75$$

$$\text{Coil EAT} = 67°F$$

For the heating coil, the temperature difference is the coil LAT minus the coil EAT.

Coil heat pickup $q = 1.085 \times$ Design airflow (cfm)
× (Coil LAT – Coil EAT)

Coil $q = 1.085 \times 600 \times (103 - 67) = 21{,}600$ Btu/h

From the hot-water coil data, select a two-row coil at 600 cfm to provide 21,600 Btu/h at about 0.8 gpm (based on a Δt of 110°F between entering air and entering water).

Note 1: The coil selection in this example produces a discharge air temperature that is too high for normal applications. A discharge air temperature limit of 87°F should be used. If additional heat is required, airflow should be increased.

Note 2: The mixed-air condition does not bring the EAT to room temperature. Additional induction or plenum air should be added to increase the mixed-air temperature to near room temperature to avoid reheat.

Note 3: When using a PSC motor, the motor will add 1 to 3°F to the airstream.

Example 4. Series Terminal with Electric Heat. Select a unit to supply a constant 1500 cfm with 0.5 in. of water inlet static pressure. Minimum primary airflow is 375 cfm and downstream resistance caused by duct-work and diffusers is 0.4 in. of water. Zone design heat loss is 45,000 Btu/h, design room temperature is 72°F, plenum air temperature is 75°F, and primary air temperature is 55°F.

Solution:

Air Valve Selection. Based on a good design inlet velocity of 2000 fpm, choose a 12 in. inlet

Fan Selection. Fan airflow equals design airflow with a series unit. Fan external static pressure equals downstream static pressure (duct-work and diffusers). The resistance of electric and hot-water heating coils and their associated additional pressure drop may or may not be taken into account on the fan curves. Be sure it is included in the final static needs. From the fan curves, select a unit that will handle 1500 cfm at 0.4 in. of water and falls in the middle of the fan range as recommended in the section on Fan Size.

Heating Coil Selection. For heating, the temperature difference (Δt) is the zone supply air temperature (SAT) minus the design set point temperature.

$$45,000 \text{ Btu/h} = 1.085 \times 1500 \times (\text{SAT} - 72)$$

$$\text{SAT} = 100°F$$

Because the heating coil is on the unit discharge, the unit supply temperature equals the coil LAT. Coil entering air temperature (EAT) is a mixture of plenum and minimum primary air.

Design heating flow × Coil EAT =
(Primary airflow × Primary air temperature) +
[(Design heating airflow − Primary airflow) × Plenum temperature]

$$1500 \times \text{Coil EAT} = 375 \times 55 + (1500 - 375) \times 75$$

$$\text{Coil EAT} = 70°F$$

For the heating coil, the temperature difference is the coil LAT minus the coil EAT.

Coil heat loss $q = 1.085 \times$ Design airflow (cfm)
$\times$ (Coil LAT − Coil EAT)

$$\text{Coil } q = 1.085 \times 1500 \times (100 - 70) = 48,825 \text{ Btu/h}$$

To convert to kWh,

$$48,825/3413 = 14.3 \text{ kWh}$$

From the manufacturer's catalog, select an electric heater with the proper input voltage (208, 240 or 480 volt/three-phase electric heat coil) that could be available with up to three stages with pneumatic or digital control or two stages with electronic control.

Note 1: Although there are air-side pressure drop data in the catalog, it is only necessary to calculate the drop if it is not included in the fan curves.

Note 2: The coil selection in this example produces a discharge air temperature that is too high for normal applications. A discharge air temperature limit of 87°F should be used. If additional heat is required, airflow should be increased.

Note 3: The mixed-air condition did not bring the EAT to room temperature. Additional induction or plenum air should be added to increase the mixed-air temperature to near room temperature to avoid reheat.

Note 4: Using a PSC motor adds 1 to 3°F to the airstream.

Installation and Application Precautions: Avoiding Common Errors and Problems

Sizing Terminals.

- Select terminals based on recommended air volume ranges. The pressure-independent terminal's main feature is its ability to accept factory-recommended minimum and maximum airflow limits that correspond to the designer's space load and ventilation requirements for a given zone. A common misconception is that oversizing a terminal makes the unit's operation quieter. In reality, the oversized terminal damper must operate in a pinched-down condition most of the time, which may actually increase noise levels to the space. Control accuracy may suffer because the terminal is only using a fraction of its total damper travel or stroke. In addition, the low inlet velocities may be insufficient to produce a readable signal for the velocity pressure measuring device and reset controller. This means minimum settings may not hold with a resultant loss of control accuracy and undesirable hunting.

- To maximize performance, size the terminal's maximum airflow limit for 70 to 85% of its rated capacity (approximately 2000 fpm) in accordance with the catalog recommendations. For accurate control, the minimum setting guideline should not be lower than 400 fpm inlet neck velocity for units using inlet velocity sensors. Other minimum guidelines may apply for unit with specialty controls.

- Oversizing the discharge duct may create low static conditions, requiring the fan to operate outside its recommended operating range.

- A problem associated with oversizing terminals with electric heat is insufficient total pressure, which can occasionally trip the air-flow safety switch.

Space Restrictions. During design, try to ensure that terminals are located for ease of installation, optimum performance, and maintenance accessibility.

Optimizing Inlet Conditions.

- The type of duct and its approach may have a large and adverse impact on both pressure drop and control accuracy. Although multipoint velocity pressure measuring devices can compensate to a large degree, good design practice should always prevail. Wherever possible, a straight duct inlet connection with a minimum length of three duct diameters and the same diameter as the inlet should be provided.

- Terminal collars are undersized to suit nominal ductwork dimensions. The inlet duct slips over the terminal inlet collar and is fastened and sealed in accordance with job specifications. Never insert a duct inside the inlet collar, or control calibration will be adversely affected.

- Sometimes space restrictions make it impossible to provide an ideal inlet condition. In this case, field adjustment of the airflow settings may be required to compensate for error in the flow measurement. The use of flow-straightening devices (equalizing grids) is recommended after short-radius elbows that are immediately ahead of the terminal and where terminals are unavoidably tapped directly off the main duct. Use of these devices typically increases sound levels.

- The balancing contractor should validate flow rates. See ASHRAE *Standard* 111.

Zoning Requirements. Correctly sizing terminals with regard to the physical conditions of the occupied space is vital to ensure acceptable performance. One large terminal serving a space with divided work areas may result in the single thermostat only providing acceptable temperature control for the area where the thermostat is located. The other area(s) served may be too cold or too hot if they have differing space load requirements.

Optimizing Discharge Conditions. Poor discharge duct connections may have an adverse affect on pressure drop. Try to avoid installing tees, transitions, and elbows close to the unit discharge. Avoid long runs of flex, and keep short flex runs as straight as possible. Make curves as shallow as possible, and ensure that the entrance condition to diffuser outlet is straight. Discharge ducts should be designed for a maximum velocity of 1000 fpm.

Noncompliance with Local Electrical Codes. Some local jurisdictions have more exacting codes than the minimum requirements of national codes and standards such as the International Code Council's (ICC) *International Building Code®* (IBC). One example is the primary fusing required of the power circuit in some areas.

Power Source Compatibility. Terminals with an electrical power supply, such as fan-powered terminals and single-duct terminals with electric heat, should be checked for compatibility with source. Voltage, phase and frequency must match. Where motor voltage differs, the single-phase voltage requirement may have to be tapped from a three-phase (four-wire wye) power source.

Avoiding Excessive Air Temperature Rise. Terminals with electric or hot-water reheat coils should be designed to satisfy load conditions, but attention should be paid to the temperature differential (Δt) between the entering room air and ambient temperature. Chapter 19 of the 2009 *ASHRAE Handbook—Fundamentals* recommends a maximum Δt of 15°F to avoid possible stratification when heating from overhead caused by the excessive buoyancy of the warm air. This ensures good room mixing and temperature equalization. Exceeding a Δt of 15°F requires an increase of 25% in the ventilation air per ASHRAE *Standard* 62.1. Absolute maximum discharge air temperature is 120°F. Although this temperature will probably keep the equipment on line, it will not provide comfortable temperatures in the space.

Correctly Supporting Terminals. Although the basic single-duct terminal is light enough that it usually can be supported by the ductwork in which it is installed, these units should be independently supported. When accessory modules such as heating coils, attenuators, or multiple-outlet plenums are included, the assembly must be supported independently. Larger terminals such as fan-powered terminal units should always be independently supported, secured to building structure, and may require isolation mounting. Be careful not to block access panels with straps, thread rods, or trapeze supports. Be sure to comply with all building and local codes regarding seismic restraints (see Chapter 55).

Minimizing Duct Leakage. To prevent excess air leakage and minimize energy waste, all joints should be sealed with a UL-approved duct sealer. Most leakage can be avoided by practicing good fabrication and installation techniques, particularly upstream of the terminal, which may be required to hold significantly higher pressures than downstream of the terminal.

Acoustic Design and Installation. To help ensure an acceptable room noise level in the occupied space, engineers can minimize the sound contribution of air terminals by taking into account several design considerations and by using the following guidelines.

- Design systems to operate at low (minimum) supply static pressure at the primary air inlet. This reduces the generated sound level, provides more energy-efficient operation, and allows the central fan to be downsized. Excessive static pressure generates noise.
- Use of metal ducts before the inlet can reduce breakout noise from the damper. Between the terminal unit and the air outlet, flexible duct can be more effective than lined duct at reducing terminal unit noise. Flexible duct can also generate sound if bends or sagging is present. Sometimes, flexible couplers can reduce vibration passed from the terminal unit to the duct connections.
- Select terminals to operate toward the middle area of their operating range. Larger inlets reduce velocity and hence noise in low-pressure applications, but may increase noise in higher-pressure applications. For fan-powered terminals, lower fan speeds generally produce lower sound levels. Sound emissions are lower when fan-speed controllers are used to reduce fan rotational speed rather than using mechanical dampers to restrict airflow.
- Whenever possible, locate terminals above noncritical areas that are less sensitive to noise, such as corridors, copy rooms, or

storage/file rooms. This isolates critical areas from potential radiated noise.
- Locate terminals in the largest ceiling plenum space available to maximize radiated noise reduction. Install terminals at the highest practical point above ceiling to optimize radiated sound dissipation.
- Avoid locating terminals near return air openings or light fixtures. This decreases the potential for direct paths for radiated sound to enter the space without the benefit of ceiling attenuation.
- Locate terminals to allow use of lined discharge ductwork to help attenuate discharge sound.
- To avoid possible aerodynamic noise, keep airflow velocities below 1000 fpm in branch ducts and below 800 fpm in runouts to air outlet devices.
- In large spaces, consider using a larger number of smaller air outlets to minimize outlet-generated sound.
- Insulated flexible duct on diffuser runouts reduces room noise levels
- Using ceilings with a high sound transmission loss classification helps reduce radiated sound.

See Chapter 8 in the 2009 *ASHRAE Handbook—Fundamentals* and Chapter 48 in this volume for more information on sound and vibration control.

CHILLED BEAMS

An **active chilled beam** is an air diffusion device that introduces conditioned air to the space for the purposes of temperature and latent control. Primary air (i.e., air delivered to meet ventilation, latent, and all or part of the sensible load) is delivered through a series of nozzles, creating induction of room air through a unit-mounted chilled-water coil, which conditions air before reintroducing it to the space. Depending on the nozzle size and configuration, active beams typically induce two to five parts of room air for every part of primary air they deliver to the space. Sensible heat removal by the beam's integral cooling coil complements the cooling effect of the primary air supply.

Passive chilled beams rely on the natural buoyancy of air currents associated with convective heat sources to transport warm air to the upper portion of the space. Upon contact with the beam's integral heat transfer coil, this air is cooled and falls back into the space. Primary air is delivered to the space for the purposes of ventilation and latent control via a separate system. Air circulated through the chilled-beam coil is called secondary air; supply air is comprised of mixed primary and secondary air.

Codes and Standards

Chilled-beam system designs should conform to the following building codes and standards, as well as any applicable local code requirements:

- ISO *Standard* 7730 and ASHRAE *Standard* 55
- ASHRAE *Standard* 62.1
- ASHRAE *Standard* 90.1
- IBC
- NFPA *Standard* 90A and 90B
- ANSI *Standard* S12.65

Application Considerations

Chilled-beam systems must be designed to treat sensible and latent space heat gains, provide adequate space ventilation, and maintain occupant comfort in conformance with ASHRAE *Standard* 55 and other applicable codes.

In general, chilled beams offer the opportunity to capitalize on the benefits of decoupled ventilation systems. They also offer the designer the opportunity to manage sensible loads in the space separately from ventilation and latent needs. Chilled beams work well with dedicated outdoor air and demand control ventilation systems.

Benefits. Heat extraction or addition by the coil allows for significant reduction in primary airflow requirements over all-air ducted systems. Energy to transport cooling and/or heating media is reduced because of water's high specific heat and density. As a result, chilled-beam systems require less space for the mechanical services, because of smaller ductwork and air-handling unit sizes. This reduction in mechanical service space requirements means it is possible to reduce the floor-to-floor height of a multistory building.

Effectively, a 1 in. diameter water pipe delivering 57°F water can transport as much energy as a 14×14 in. duct delivering air at 55°F. The transport energy required to deliver cooling by water is only 15 to 20% of that which would be required by air. Water provides 3500 times the heat-carrying capability of air.

Free cooling opportunities may be extended as a result of the lower secondary chilled-water temperatures, and provide an improved selection of available system options (e.g., geothermal, dry coolers, closed-circuit fluid coolers). Chilled-beam systems offer opportunities to enhance chiller efficiencies and provide broader evaporator ranges, because of higher chilled-water temperatures, and cascaded evaporator flows between primary and secondary chilled-water loops. Higher inlet water temperatures also make chilled-beam systems as an excellent choice for geothermal applications, where suitable electrical costs allow this as an option.

Chilled-beam systems are typically operated with a constant-(minimal) volume supply airflow to the space. Constant-volume systems offer enhanced thermal comfort because of their consistent room air movement, and an improved acoustic environment.

Maintenance of chilled beams is virtually nonexistent. Vacuuming the coils is occasionally required, and is typically guided by the needs of the space. Often, it is expected that service intervals could extend to 3 to 5 years. The lack of moving parts in chilled beams inherently produces a highly reliable system, and because most beams do not contain filters, servicing costs are minimal.

Passive chilled beams can provide a very efficient means of perimeter-area temperature control when coupled with underfloor air distribution (UFAD) systems. For more information on UFAD systems, see Chapter 20 of the 2009 *ASHRAE Handbook—Fundamentals*.

Limitations. Space humidity levels must be managed closely because of the limited dehumidification capability of the primary air. This is particularly important where operable windows are used or high infiltration levels are encountered in a humid climate.

Chilled beams may be considered beneficial for the following applications/spaces:

- Environments with moderate to high sensible heat ratios, such as offices, apartments, high rise condos, or classrooms
- Laboratories or other spaces with significant imbalances between sensible loads and ventilation requirements
- Retrofits, because of minimal mechanical space requirements, and in cases of suitable envelope construction

Chilled beams may be inappropriate for the following spaces:

- Environments with high latent gains, such as kitchens, bathrooms, or locker rooms
- Swimming pools, natatoriums, and sauna areas
- Poor building envelopes or buildings with unmanageable latent loads
- Mixed-mode ventilation (operable windows) without proper condensation safeguards

Chilled beams must be independently supported from ceiling grid systems, and subsequently positioned into the grid. Appropriate care must be used when beams are installed in seismic zones, to ensure compliance to all building and safety codes.

Passive beams are restricted to cooling-only applications, and require a separate heating system. Careful consideration is needed in the placement and control of passive chilled beams to ensure thermal comfort.

Cooling

The objective of chilled-beam design is to minimize primary airflow rates, ideally reducing them to the space minimum ventilation requirement. However, where zone cooling requirements cannot be achieved with the minimum primary airflow rate, chilled beams may be used with air-handling units that mix return and outdoor air volumes.

Active and passive chilled beams rely on the primary air supply for dew-point control. As such, the design conditions must be evaluated to confirm that the primary air is appropriately treated to manage the latent space loads. Chilled beams are intended to operate without condensation. Consequently, active chilled-beam supply water temperatures should be maintained at or above the room dew-point temperature to prevent condensation on the coil and its supply water piping. Passive chilled-beam water supply temperatures should be maintained slightly (1 to 2°F) above the room dew-point temperature. In both cases, the chilled-water supply piping must be adequately insulated to prevent condensation on the pipe itself. Where adequate control of space humidity levels cannot be ensured, higher supply water temperatures and/or condensation controls should be considered. This is discussed in the following sections.

Terminal filtration and condensate pans are not required with a properly designed primary air system and chilled-water temperatures maintained above the room dew point. Heating coils provide sensible heat only, and thus filtration and condensate capture devices are not necessary. Chilled-beam systems designed with noncondensing (dry) coils should be treated similarly.

Heating

Heating is limited to active chilled-beam systems; heating with overhead passive chilled beams is not effective.

The hot water serving the active beam's coil must be chosen to limit the discharge air temperature to less than 15°F above the room design set point. Additionally, to ensure proper room air distribution, the discharge velocity should be selected in accordance with guidance presented previously.

Resetting the primary air temperature with a duct-mounted booster coil allows the primary air serving the interior spaces to continue to provide cooling, while the perimeter duct adds heating capability through this reset. Assuming the active chilled beams are the primary heating system, caution is recommended in placing the equipment at ceiling level with respect to the curtain wall's orientation, to ensure air movement across these surfaces promotes a comfortable environment.

Thermal Comfort

Chilled-beams systems are designed to optimize delivery of cooling to the space, but the paramount consideration in sizing and locating beams in the room should focus on occupant thermal comfort. ASHRAE *Standard* 55 defines limits on local air temperatures and velocities that maintain acceptable levels of occupant thermal comfort. The standard defines the **occupied zone** as the one in which stationary occupants reside; the height of this zone is generally considered 67 in. for standing occupants or 43 in. for predominantly seated occupants. Velocities in the occupied zone should not exceed 50 fpm, and occupied-zone vertical temperature gradients should be maintained at 5.4°F or less.

Properly applied passive chilled beams have a limited effect on occupant thermal comfort; however, their complementary primary air supply system often does. Stratified or partially mixed air diffusion strategies are commonly used with passive beams because of their minimal influence on the natural buoyancy driven air patterns associated with the chilled-beam operation. The secondary air circulation through the passive beam transports upper-level air back to the occupied zone, possibly altering the level of stratification in the space.

Active chilled beams directly supply a mixture of primary and secondary air to the space and should therefore be treated like the other air distribution devices used in fully mixed air distribution systems. Because the temperature of the chilled water supplying the coil must be at (or above) the space dew-point temperature, it is typically 56 to 60°F; thus, reconditioned air leaving the coil is several degrees warmer than the primary air with which it is subsequently mixed. This results in beam design discharge air temperatures ranging from 58 to 60°F, warmer than those normally used by conventional all-air systems. Because the required supply airflow is inversely proportional to the room to supply air differential, active beams must discharge 15 to 25% more air to the space to satisfy its sensible heat gains.

Space Temperature Control and Zoning

Chilled-beam systems' primary airflow rates are much closer to the space ventilation rates than those of all-air systems, so primary control of the space temperature is normally accomplished by throttling the chilled-water flow. Simple on/off operation of two-position water valves provides adequate control of active chilled beams. Proportional valves are recommended for passive beams, and for active beams in applications where more precise space temperature control is required.

In applications where primary air is supplied at conventional temperatures (55 to 57°F) to spaces with significant sensible load variations, it may also be necessary to reset the primary airflow rate or temperature during low-load conditions. One approach is to vary the primary airflow rate in reaction to thermal demands and/or occupancy of the space.

Supply of primary air at or close to room temperature overcomes the potential for overcooling. This, however, results in reduced beam cooling capacities and necessitates the use of more or larger beams. It is also likely to require desiccants to provide adequate dehumidification of the primary air. Spaces with high ventilation requirements and significant cooling turndown rates, such as large conference rooms, may be better served by a VAV solution.

Thermal zoning of chilled-beam systems should be performed in a manner generally consistent with other HVAC systems. Each thermal zone consists of a space thermostat, a chilled- (and, where applicable, hot-) water control valve, and multiple chilled beams.

Selection and Location

Chilled beams may be exposed or integrated with an acoustical ceiling system. Active chilled beams may be of either open or closed design. Closed beams induce secondary air from below, whereas open beams induce through their top or sides. When passive or open active beams are applied, an adequate air path must be provided for secondary air to enter the beam.

For sizing and selection purposes, secondary air entering a chilled beam should generally be considered equal to that maintained within the occupied zone unless solid evidence indicates otherwise.

Most active beams suppliers offer various nozzle sizes and configurations. Nozzle configuration affects the beam's pressure requirement and acoustical performance as well as its induction function. Active beams with adjustable discharge or nozzle patterns may also allow for field alteration of the beam's air distribution characteristics. This may also affect the beam's cooling capacity, so changes should be made with caution.

Beam sizing and location must consider cooling capacity, acoustics, thermal comfort, and integration with other equipment and services. Active beams use a horizontal discharge of their supply air mixture through linear openings along their perimeter, and thus

display room air diffusion characteristics similar to those of linear slot diffusers. As such, active beams should be selected and located such that velocities within the occupied zone are limited to 50 fpm or less if compliance with ASHRAE *Standard* 55 is the design intent. The use of mapping techniques (see the section on Mixed Air Distribution) and/or selection of active beam throw values within the ranges of Table 4 may be used to estimate compliance with these comfort recommendations.

Locating stationary occupants directly below passive beams can result in thermal discomfort. Care must be taken to ensure that the velocity and temperature of the descending airstream entering the occupied zone comply with the thermal comfort requirements of ASHRAE *Standard* 55.

Operational Considerations

Water supply service to active and passive beams should not be activated until space dew-point temperatures are at or below the chilled water's supply temperature.

Where maintenance of adequate space dew-point temperatures cannot be ensured, some type of condensation detection and mitigation strategy should be used. Designers have various methods of accomplishing this, including the following:

- Sensors, attached to the supply water pipe, that sense formation of surface moisture and discontinue chilled-water flow until the moisture has evaporated. This method is relatively inexpensive but also reactive, and results in termination of secondary air cooling while condensation exists.
- Dew-point calculation and reset of the chilled-water supply temperature is a proactive strategy that does not involve full suspension of secondary cooling. This method can be applied on a room-by-room basis, but calculation on a floor-by-floor basis is usually sufficient and less costly.
- For applications in spaces with operable windows or doors, occupants and staff should be educated on the effect this has on their thermal environment.
- In certain applications, condensate trays may be used to collect temporary and infrequent condensation. Where applied, adequate condensate removal means must be ensured; evaporation of the condensate should not be assumed. This may involve using condensate pumps when gravity drainage cannot be accomplished.

REFERENCES

AHRI. 2008. Procedure for estimating occupied space sound levels in the application of air terminals and air outlets. *Standard* 885. Air-Conditioning, Heating, and Refrigeration Institute, Arlington, VA.

ANSI. 2011. Rating noise with respect to speech interference. ANSI/ASA *Standard* S12.65-2006 (R2011). Acoustical Society of America, Melville, NY.

ASHRAE. 2010. Thermal environmental conditions for human occupancy. ANSI/ASHRAE *Standard* 55-2010.

ASHRAE. 2010. Ventilation for acceptable indoor air quality. ANSI/ASHRAE *Standard* 62.1-2010.

ASHRAE. 2010. Energy standard for buildings except low-rise residential buildings. ANSI/ASHRAE/IESNA *Standard* 90.1-2010.

ASHRAE. 2008. Measurement, testing, adjusting, and balancing of building HVAC systems. ANSI/ASHRAE *Standard* 111-2008.

ASHRAE. 2009. Method of testing for room air diffusion. ANSI/ASHRAE *Standard* 113-2009.

ASHRAE. 2002. Measuring air change effectiveness. ANSI/ASHRAE *Standard* 129-1997 (RA 02).

ASHRAE. 2008. Ventilation of health care facilities. ANSI/ASHRAE *Standard* 170-2008.

ASHRAE. 2011. Standard *62.1-2010 user's manual.*

Bauman, F.S. and A. Daly. 2003. *Underfloor air distribution design guide.* ASHRAE.

Chen, Q.Y. and L. Glicksman. 2003. *System performance evaluation and design guidelines for displacement ventilation.* ASHRAE.

Davis, M., J.A. Bryant, D.L. O'Neal, A. Hervey, and A. Cramlet. 2007. Comparison of the total energy consumption of series versus parallel fan powered VAV terminal units, phases I and II. ASHRAE Research Project RP-1292, *Final Report*.

ISO. 2005. Ergonomics of the thermal environment—Analytical determination and interpretation of thermal comfort using calculation of the PMB and PPD indices and local thermal comfort criteria. *Standard* 7730-2005.

ICC. 2009. *2009 international building code®*. International Code Council, Washington, D.C.

Koestel, A. and G.L. Tuve. 1955. Performance and evaluation of room air distribution systems. *ASHAE Transactions* 61:533.

Nevins, R.G. 1976. *Air diffusion dynamics*. Business News Publishing, Birmingham, MI.

NFPA. 2009. Standard for the installation of air-conditioning and ventilating systems. *Standard* 90A. National Fire Protection Association, Quincy, MA.

NFPA. 2009. Standard for the installation of warm air heating and air-conditioning systems. *Standard* 90B. National Fire Protection Association, Quincy, MA.

Rock, B.A. 2006. *Ventilation for environmental tobacco smoke*. Elsevier Science, New York.

Rock, B.A. and D. Zhu. 2002. *Designer's guide to ceiling-based air diffusion*. ASHRAE.

Rydberg, J. and P. Norback. 1949. Air distribution and draft. *ASHVE Transactions* 55:225.

Skistad, H., E. Mundt, P. Nielsen, K. Hagström, and J. Railio. 2002. Displacement ventilation in non-industrial premises. REHVA *Guidebook* 1. Federation of European Heating and Air-Conditioning Associations, Brussels.

Straub, H.E. and M.M. Chen. 1957. Distribution of air within a room for year-round air conditioning—Part II. University of Illinois Engineering Experiment Station *Bulletin* 442.

Straub, H.E., S.F. Gilman, and S. Konzo. 1956. Distribution of air within a room for year-round air conditioning—Part I. University of Illinois Engineering Experiment Station *Bulletin* 435.

UL. 2005. Heating and cooling equipment. ANSI/UL *Standard* 1995. Underwriters Laboratories, Northbrook, IL.

Webster, T. and F. Bauman. 2006. Design guidelines for stratification in UFAD systems. *HPAC Engineering* 78(6):16.

INTEGRATED BUILDING DESIGN

INTEGRATED building design (IBD) is a collaborative process of preparing design and construction documents that result in optimized project system solutions. For IBD to succeed and be beneficial to the project, the entire project delivery team must be committed to, understand, and remain engaged and involved in the process from project inception through operation and maintenance.

This chapter provides a working knowledge of IBD, highlights activities that support collaboration, and helps the HVAC design professional develop a structured and integrated approach to project delivery. The basic process framework is outlined, and major milestones are identified.

The resources in the References and Bibliography, as well as other Handbook chapters and ASHRAE guidelines and standards offer in-depth guidance on various IBD application requirements and should be referred to for more information.

PROJECT DELIVERY

Delivery of solutions in the built world is accomplished in many ways and through various delivery techniques. Whether it is design-bid-build, design-build, design-construction manager (design-CM), etc., each delivery method requires interaction between design professionals representing inclusive elements of the project.

Sequential Design Process

In a sequential design process (SDP), elements of the built solution are defined and developed in a systematic, sequential, and somewhat isolated process. As each element is resolved, a new element is added to build on the overall project solution. A typical SDP can be briefly outlined as follows:

1. The architect develops the building program and proposes a facility layout to support the program.
2. Trade budgets are set to comply with the overall construction budget.
3. The structural engineer defines the skeleton to support the proposed facility.
4. Facility services engineers are given the proposed layout, structural system, and trade budgets, then directed to apply mechanical, electrical, and plumbing (MEP) solutions.
5. Discipline design evolves on individual paths to support the allocated trade budgets.
6. Coordination impacts are shared with the project team.
7. The owner's project team establishes the project program, including project scope and overall site plans.

The preparation of this chapter is assigned to TC 7.1, Integrated Building Design.

Integrated Design Process

An integrated design process (IDP) discourages sequential philosophy and promotes holistic collaboration of the project team members during all phases of project delivery. Emphasis is placed on optimizing system solutions that are responsive to the objectives defined for the project. Optimizing system solutions requires the participation of all team members.

Effort Shift

Execution of the IDP requires a departure from conventional SDP methodology. This change includes a true shift in effort during the classical design phases (schematic design, design development, and construction documents), but also includes the addition of enhanced efforts outside the traditional design phases.

The classical design phases promote a certain level of SDP based on their inherent percentage development model. SDP typically unfolds along the following development path:

1. During the schematic design phase, the basic form and function of the facility are defined to meet the program. Facility services are described in concept to support the developed facility construction.
2. During the design development phase, systems proceed in parallel development paths to find best-fit solutions that meet the project budget and provide compliance with prescribed regulatory requirements such as building, life safety, energy, and ventilation codes. Life-cycle cost activities (LCCA) may be used to optimize individual discipline solutions in response to the schematic design development package.
3. During the construction document phase, all building components are detailed into work results suitable for procurement and construction.

Integrated design alters this traditional delivery model by front-loading collaboration efforts to optimize building system solutions in response to the defined project objectives. Integrated design is most effective when key issues are addressed early in design and planning (see Figure 1). An important point to remember is that implementation of the IDP does not necessarily mean getting more things done early; it means getting the right things done early. To accomplish this, traditional team roles that lagged early design activities must participate on equal footing earlier in the process, so that holistic issues are considered before it becomes too late to effect responsible inclusion.

OBJECTIVES

IDP is accomplished by responding to project objectives. These may be defined by the owner before team selection, or developed by the project team during any project phase. The key is to define substantive objectives that can materialize into practical, constructable results. Prominent objectives are outlined in this section.

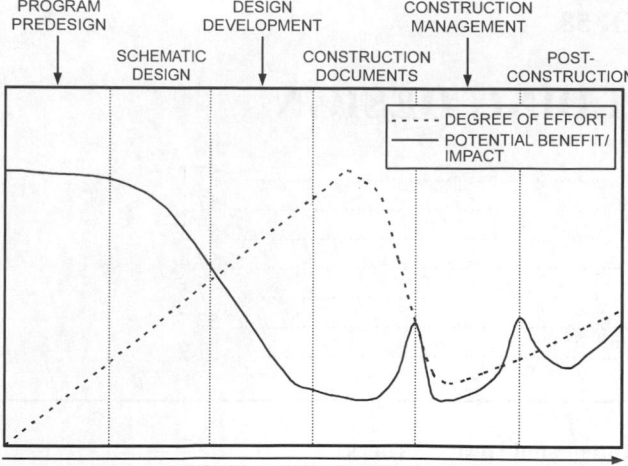

Fig. 1   Benefits of Early Design Collaboration
(Lewis 2004)

Definitions of objectives in the IDP are only highlighted here; consult application-specific material when considering the effects of each objective.

Various organizations and owner entities have developed sustainable, green, and high-performance prescriptive point systems to promote select project delivery objectives. Typically, these point structures address sustainable sites, water efficiency, energy and atmosphere, materials and resources, and indoor environmental quality. Although incorporation of these objectives is a highly desirable pursuit and promotes responsible environmental stewardship, the team must realize that desired accumulation of points in a prescriptive rating system should not shift focus from meeting the defined project objectives.

Energy Use

Energy performance objectives can be as simple as providing minimum prescriptive energy code compliance, or as detailed as providing a net-zero-energy performance facility. The extent and complexity of the objectives must be tailored to each individual project. Sample objectives that may be encountered include the following:

- Provide minimum prescriptive compliance per applicable energy code requirements
- Improve energy performance by an owner-defined percentage beyond applicable energy code benchmark(s)
- Provide a facility site energy density of less than owner-defined consumption per unit area
- Provide a facility source energy density of less than owner-defined consumption per unit area
- Provide owner-defined percentage of facility's source energy from renewable resources
- Limit owner-defined percentage of facility's source energy to nonrenewable or consumable resources

Typically, energy-related objectives address consumption, efficiency, and generation (site and source) issues. There are a myriad of variations, combinations, and themes that objectives can take regarding these factors. These combinations serve as an excellent example of why the underlying objectives for the project should be fulfilled before performance-rating-system point tabulation becomes the primary focus.

Indoor Environmental Quality (IEQ)

IEQ objectives vary with the programmed use for the building. Each aspect of IEQ must be considered.

Acoustical comfort may require specific focus because of the nature of activities inside and around the facility. Theaters, for example, have certain prescribed noise criteria that must be provided to allow for intended operation. Achieving prescribed noise criteria for specific building operations requires knowledgeable collaboration by all parties that control the source noise, transmission paths, and measured point of sound pressure.

Depending on the facility, **thermal comfort** may or may not be a critical objective. The project team needs to clearly understand the individual facility thermal conditions and the range of acceptable variation. This criterion has a significant effect on the size, type, and complexity of potential infrastructure solutions.

Depending on the climate and operational needs, **humidity** or **moisture control** may be an appropriate objective. This objective can be further expanded to address building protection, occupant comfort, or process needs.

Ventilation effectiveness deals with the practical and reliable means of providing ventilation air into the breathing zone of the facility occupants. Table 6-2 in ASHRAE *Standard* 62.1-2007 identifies zone air distribution effectivenesses E_z ranging from 0.5 to 1.2 for various air distribution configurations. An objective that may be defined is to limit HVAC solution configuration to systems that provide an E_z value of 1.0 or greater.

Visual quality can be a concern for certain operations. The quality of ambient light in a space can have direct effect on occupants' productivity. Properly applied and controlled, daylighting can improve the visual quality of the occupied space and reduce the capacity of the HVAC systems by decreasing the need for artificial indoor lighting systems.

Water Usage

Potable water is essential to life and faces continuous pressures relative to availability and quality. IBD objectives for water usage typically focus on conservation and reclamation efforts. Water has a cost associated with its use. This use should be modeled in the total ownership cost of a facility.

Conservation and reclamation of water do not apply only to plumbing systems. HVAC systems can consume significant amounts of water and are prime candidates for environmentally responsible project objectives. Sample objectives that have an HVAC influence include the following:

- Reclaim all cooling condensate discharge for use in gray-water systems. Note that reclaimed gray water can be applied to a host of facility service applications, such as cooling tower makeup, landscape irrigation, urinal flushing, etc.
- Capture all facility storm water drainage for use as gray-water makeup for HVAC, plumbing, and landscaping needs.
- Increase concentration limits and/or decrease cycles on cooling tower blowdown to conserve water consumption. This, of course, must be balanced against the suitability of an integrated maintenance program and limited to local water quality characteristics that contribute to scale, corrosion, fouling, and microbial growth.

Vulnerability

Global events and operational needs may dictate that built solutions address vulnerability objectives. The facility infrastructure may require protection from outside seismic incidents, blast events, or chemical and biological contamination. Inside operations that create explosion, chemical, biological, or radiological hazards may require focus. Additionally, protecting occupants in the facility may be an inclusive or standalone priority. In any case, vulnerability objectives create some challenging opportunities for collaboration and demand that the project team have an effective prioritization system in force on the project. See Chapter 59 for more information on the topic.

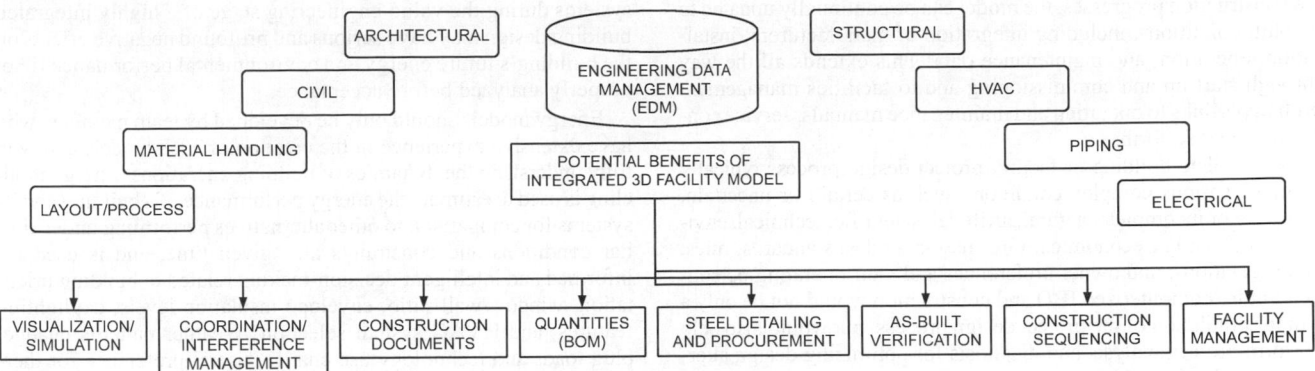

Fig. 2   Overview of BIM Benefits

Environmental Stewardship

Waste reduction is a pressing need in the built world. The capacity of landfills to absorb construction debris is not limitless. Reuse and recycling of construction debris can help mitigate landfill overuse. Also, when materials cannot be harvested or obtained from the project site, using new construction materials that include recycled content is a proactive consideration.

As concerns with global warming and greenhouse gases increase, minimizing the carbon footprint of the facility may become a critical objective. This will require a unique collaborative effort to minimize the sum of the embodied energy and carbon emissions of all processes and components required to construct, own, operate, and maintain a facility.

Critical Operations

Certain objectives are critical to operations for data centers, emergency response, law enforcement, government, health care, shelters, manufacturing, and pharmaceutical facilities. For example,

- Facilities that require high reliability must focus on ensuring that systems and components meet the specified probability they will operate for the duration of use. As the required reliability increases, infrastructure design must respond in kind with system redundancy and diversity.
- Facilities that require high availability must focus on ensuring that systems and components meet the specified probability they will operate and be accessible when required for use.
- Scalability may dictate that infrastructure have provisions for expansion and growth relative to dynamic business factors and technology development.

General Operations

Accessibility priorities may dictate that some infrastructure have unique requirements to ensure proper performance and serviceable attention during the operational life. Accessibility has an infrastructure cost effect that must be factored into the total ownership cost.

Replaceability objectives may define where facility infrastructure can be located so that replacements can be made when the useful life has expired. Total ownership solutions should plan for the costs to replace equipment and not leave this as a hidden burden for the facility owner to bear later.

Many owners face a dynamic known as churn (reconfiguring a space or changing its use). Objectives that plan for churn can help mitigate complete replacement of facility services if changes need to be made.

TOOLS

IDP requires detailed simulation and evaluation of system solutions across multiple design responsibilities. Performing these simulations by hand can be onerous. Tools are readily available in the industry to assist the project team in maximizing their collaboration efforts.

This section discusses three critical, fundamental types of tools suggested for use in executing IDP. These are by no means the only tools available; an abundance of tools are available from government, utility, commercial, manufacturing, and technical society sources. Comprehensive listings of potential resources are available on the National Institute of Building Sciences (NIBS) Whole Building Design Guide Web site (http://www.wbdg.org/index.php).

Building Information Modeling

Building information modeling (BIM) is the process of using intelligent graphic and data modeling software to create optimized and integrated building design solutions. As such, it is an enabling tool for IBD (see Figure 2). The ultimate goal of BIM is to assemble a single database of fully integrated and interoperable information that can be used seamlessly and holistically by all members of the design and construction team, and ultimately by owners/operators throughout a facility's life cycle. The desired result is a BIM model where three-dimensional (3D) graphical imaging carries real-time (i.e., immediate and dynamic access) data, and where every line and every object carries real-life intelligent physical and performance data. That model includes the aesthetic, physical, and thermal properties of each component, as well as specification and cost data. The design team interfaces with the model to seamlessly generate comprehensive simulation evaluations, including natural daylight modeling, energy modeling, and life-cycle cost analysis of the building.

The modeling technology can start with direct data transfer from the design calculation software into graphic layouts (for systems such as structural steel, fire protection, or other modular elements). Alternatively, it can use the graphic layouts as direct input to load calculations (e.g., energy simulation, pipe sizing, duct sizing). Modeling programs can also link to specifications and to manufacturers' Web sites for data input. Either way, building information modeling technology already extends into fully integrated 4D modeling (adding the fourth dimension of time for scheduling software) and 5D modeling (adding the fifth dimension of cost for estimating and budget control).

After design optimization is complete, the original modeling software can compile data from each discipline and generate a set of digital 2D or 3D construction documents for use in procuring construction bids. The model can interface with a contractor's cost estimating, scheduling, and project management software and manufacturers' material, fabrication, and cost databases to generate optimized cost estimates and construction schedules. The development can continue with the provision of automatic bills of material (BOM) and generation of automatic shop drawings for everything from structural steel to sheet metal duct fabrication, fire protection and piping fabrication, electrical cabling and bus duct layouts, etc.

As construction progresses, the model can be continually updated to as-built conditions, including integration of manufacturers' installation, operation, and maintenance data. This extends all the way through start-up and commissioning and to facilities management with hyperlinks to operating and maintenance manuals, service contractors, and so forth.

A complete building or facility project design process typically involves various complex conditions such as certain or uncertain, complete or incomplete, natural, artificial, scientific, technical, environmental, and/or economical information; codes, standards, rules, and regulations; and owner preferences and their operating personnel. Future computerized IBD and construction would not be only a simple simulation of the project design process, but would also help perform project analysis and diagnosis for optimizing design alternatives and project decisions.

There is a significant amount of work being done worldwide on software tool and protocol developments by governmental agencies, nonprofit and research organizations, as well as commercial entities, to facilitate and promote BIM technology. There is also significant new interoperable application software under development using BIM.

A major key to the success of these efforts is establishing common software protocols.

The International Alliance for Interoperability (IAI) was formed in 1995 to define and develop standards or protocols as a framework for data exchange, creating the industry foundation classes (IFCs) and gaining ISO recognition. IFCs are maintained by buildingSMART International, the current name for IAI. The buildingSMART Web site (http://www.buildingsmart.com) provides the IFC specifications, definitions, and model documentation that can be used to develop BIM interoperable software applications. The latest IFC model is called the buildingSMART data model, codified as ISO/PAS *Standard* 16739.

The National Institute of Building Sciences (NIBS) established the buildingSMART alliance™ to advance open interoperability and full-life-cycle implementation of BIM, and their National BIM Standard (NBIMS 2007) committee establishes standard definitions for building information exchanges to support critical business contexts. NBIMS addresses the need for a life-cycle view of building supply chain processes, the scope of work necessary to define and standardize information exchange between trading partners, suggestions for a methodology to address this work, and examples of work in progress.

Energy Modeling

Energy modeling uses scientific methods and analytical tools to estimate the energy consumption patterns of a given facility, constructed of given materials, located in a given climate zone, and operated according to given schedules. These tools and methods range from simple hand calculations and spreadsheets to the most sophisticated software tools designed to consider numerous building configurations, various zoning options, and multiple systems. Some of the more common software tools include programs free for download such as the U.S. Department of Energy's (DOE) DOE2, Energy10, EnergyPlus, and eQuest. Commercial entities and equipment manufacturers also have products available to support building load calculation and detailed energy performance modeling.

Energy modeling should be used to help integrate and optimize a building's energy performance over the facility's expected life cycle. Successful application of this tool comes from evaluating system solutions as early as possible to develop best-fit solutions for the developing design, thus minimizing radical design changes late in the design phases.

Energy modeling may also be used if it becomes necessary to value-engineer a project after the design phase is complete. Simple substitutions of less costly materials, products, equipment, or

systems during the value-engineering stage of a highly integrated building design may have serious and profound negative effects on the building's future energy and environmental performance if not properly analyzed before acceptance.

Energy models should only be developed by team members who have extensive experience in the creation of such models and who truly understand the dynamics of building operations. Energy modeling is used to estimate the energy performance of a building and its systems for comparison to other alternatives performing under similar conditions and constraints at a given time, and is used for informed and intelligent decision-making related to building orientation, window/wall ratio, envelope insulation levels, daylighting features, and HVAC system selection. Weather patterns change; plug loads and technology use change; users' preference for thermostat set points often differ from those modeled; material properties change and degrade over time; system and equipment maintenance may be kept current or deferred after owner occupancy; and hours of usage and operation change; these are just a few reasons why modeled energy use rarely tracks favorably with actual energy use. The following points should always be kept in mind when using energy models for system evaluation:

- Model results are not a guarantee of actual or future performance
- Model results are not a guarantee of actual or future energy costs

Refer to Chapter 19 in the 2009 *ASHRAE Handbook—Fundamentals* for an in-depth discussion on modeling methods for systems design and design optimization.

Life-Cycle Analysis Tools

All system evaluations share a common need to demonstrate what the financial effects are relative to total ownership cost. This requires a comprehensive comparison of capital, utility, energy, maintenance, replacement, disposal, and occupant costs for the facility's projected life. Life-cycle cost analysis (LCCA) provides a means of examining how each of these factors impact the owner's cost obligations.

A comprehensive methodology for facilitating life-cycle comparisons can be found in the National Institute of Standards and Technology's *Handbook* 135, Life-Cycle Costing Manual for the Federal Energy Management Program (NIST 1996). NIST provides a number of supplemental publications and tools that should be used in conjunction with this source, including the following:

- Annual supplements to *Handbook* 135, providing annually updated energy price indices and discount factor multipliers
- The DOE's Building Life-Cycle Cost (BLCC) Computer Program, which provides an electronic means of applying the methodology of *Handbook* 135

All of NIST's life-cycle publications, tools, and annual updates may be downloaded from the U.S. Department of Energy's Federal Energy Management Program Web site.

Chapter 36 in the 2003 *ASHRAE Handbook—HVAC Applications* contains expanded LCCA reference material that focuses on direct HVAC system and component effects, as well as valuable perspectives and data on owning, operating, and maintenance cost factors.

OWNER PROCESS

Project inception begins with the owner and a need to begin construction of a facility. Successful incorporation of IBD objectives is set at this point, before design begins.

Programming

When a need for facility space is created, the owner must first evaluate options available to meet the required facility need. The scenarios of construction, reduced construction, or no construction should be debated to determine which option provides the best-fit alternative. Questions to consider include the following:

- Are adaptive reuse alternatives suitable in the project and available?
- Does the program need have redundancies that contribute to wasteful infrastructure construction?
- Can operating schedules be adjusted to minimize built space?
- Do consolidation opportunities exist within and outside the organization that could foster more environmentally responsive built solutions?
- Do multiple use opportunities exist that can support additional program uses, expanded use potential, and operational scalability?
- What are the real and artificially generated work sequence needs? If an accelerated project delivery is required, what are the consequences for not allowing for optimized solutions to be thoroughly developed and evaluated?
- What are the project objectives as seen from the owner's perspective? Early definition of objectives is instrumental in assembling the correct project team members.
- Based on definition of the project objectives, who will best serve the role as IBD champion for the owner's interest?

Siting

The location of the proposed built facility directly influences the course the IDP takes. Site alternatives need to be researched to determine whether site contamination may influence infrastructure solutions, or whether wetland offset measures need to be implemented at other sites.

The location and suitability of utility resources must be identified. Sites should be selected that allow options and flexibility when system solutions are being considered. Lack of available utilities affects optimization models. Basic features to consider include the following:

- Is potable water available from a municipal source, or will well water be required? If provided from a municipal source, what line pressures are available?
- Is a municipal sanitary sewer available, or will an on-site septic system be required?
- Is a municipal stormwater system available; if not, what are the alternatives for handling stormwater runoff?
- What is the proximity of electricity, gas, and district energy systems?
- Is the site conducive to implementation of renewable site/source energy?

Local and regional transportation conditions must also be identified, such as whether local and regional road systems are

- Suitable for moving in process raw materials and transporting processed products out
- Sufficient for transporting materials and products consistently

Neighbor relations present another decision matrix for the owner to consider. The proposed location of a built facility and the specific operational dynamics may contribute negatively to the surrounding community. Questions to ask that may lead to unique project mitigation objectives include the following:

- What will the effects on traffic be for building occupants and neighboring communities?
- Do operational usage schedules complement neighboring communities?
- Will light-trespass mitigation and ambient-noise abatement be required?
- What are the effects of stormwater runoff to adjacent properties?

Budgeting

Once the owner's program is set, siting options have been evaluated, and mitigation needs have been defined, the owner needs to appropriate funding to support design, construction, and operation of the facility. Funding sources should be planned to cover anticipated professional services fees, capital construction costs, contingencies, escalation costs, maintenance costs, utility/energy costs, and occupant costs. Allocations should be made with an eye towards encouraging collaborative optimization efforts and leveraging noncapital costs to minimize overall total ownership cost.

Budget shortfalls during some point of the project are not uncommon. Early in the budgeting process, a reasonable contingency (~10 to 20%) should be incorporated into the budget to allow for unknowns and minor changes as the project evolves.

Team Selection

Design team members should have a high degree of competence and knowledge gained though education and experience on similar projects. Team selection can vary, depending on the client's in-house capabilities. Typically, the design team needs to assist the owner during initial programming and siting activities. Regardless of the timing, it is imperative that the team be selected based on qualifications that support the project objectives.

COLLABORATION

Collaborative design requires that all members of the design team possess demonstrated expertise, an ability to work collectively in a nonisolated setting, and a drive of stewardship to support the IDP vision. Team members should share similar corporate philosophies, have compatible operating procedures, use common optimization tools, and be committed to adhering to consistent interdisciplinary quality assurance/quality control (QA/QC) procedures.

Teamwork

Working with a team requires that participants engage in joint decision making. Individual thinking and processes must give way to support the team and a decision-making mentality that supports the direction of the team. Individuals must keep in mind that their actions and reactions affect integrated system solutions. Design in isolation does not support team collaboration.

Team members must foster a professional level of respect for each other. When individuals suggest new strategies to improve the whole, dissident views will occur. Emotions must be removed from these events. Evaluations must be made on objective application and support of meeting project objectives.

The project team leader should be trained to handle conflict management and dissident views in a professional manner. Consensus agreement will not always be apparent, and the project team must avoid fracture of the collaborative effort when differences occur.

Effective, concise, and complete communication must be adopted to keep the team informed of all decisions across all design disciplines. Communications within the project team should be standardized as much as possible. Each form of communication should contain the date the communication originated, any revision dates, project name, project number, and the originator's contact information. In addition, a clear and concise subject line should be included to focus recipients to the subject matter at hand. For collaboration to work, all team members must be kept in the communication loop so that each understands where the collaboration process stands.

Team Formation

The importance of experience in an integrated design team cannot be stressed enough. Systems thinking requires input from individuals who have extensive design experience and understand how systems and components interact. IDP cannot succeed when key representatives are cast into on-the-job training. On the other hand, IDP provides an excellent opportunity to mentor and train supporting staff in system integration, and the opportunity should be fully exploited.

Participation in IDP requires individuals to have a proactive attitude that supports the ups and downs of iterative system evaluation. Individuals who can see the big picture and appreciate that the whole will be better than the sum of the parts enhance the team efforts.

Participants should have experience with optimization techniques. This equates to more than being able to run a load calculation. True optimization expertise requires understanding how building systems interact, what elements can be examined for the benefit of the whole, and how to evaluate results in detailed financial models that consider all ownership costs.

Some projects may require adding specialty consultants to the project team to support activities such as smoke control, acoustics, seismic restraint, or food service. Rarely do all expertise needs reside within the same firm on complex projects. Managing outside specialty consultants is an added responsibility that must be factored into the collaborative process.

As with the owner, the design team requires an IBD champion to keep focus on the project objectives.

Decision-Making Criteria

As system evaluations are made, benchmarks or baselines must be established to measure the suitability of system evolution. These decision-making criteria must be agreed to with the owner before any design work begins. Criteria need to be realistic, allow for maintaining pace with the budget limitations, accommodate strategic flexibility, and address budget correction possibilities when the need arises. The basis of the criteria depends totally on the owner's resource capabilities and financial position, but in all cases should be based on meaningful life-cycle analysis.

Scenarios may arise that challenge the design criteria and defined objectives for the project. It is not uncommon for project scopes or financial factors to change during project design, procurement, or construction. This deviation from the original path may require a reprioritization of objectives and a change in the decision-making criteria. The team should be prepared to adapt to such dynamics and be able to refocus the IDP in a responsive and efficient manner.

Strategy Development

To proceed with the IDP, proposed strategies should lead the team toward a built facility of integrated systems. Development of these strategies is influenced by the prioritization of project objectives and by the direction of the developing building solution. Using the simple, but not quantified, objective of reduced energy use as an example, the following broad strategies and substrategies reflect the progressive optimization for a commercial office building in the southeastern United States. Note that the example does not provide an all-inclusive strategy roadmap, or represent a fixed recommendation.

1. Minimize building envelope load
   - Optimize siting and footprint aspect ratio to maximize east/west orientation
   - Incorporate effects of major sky obstructions such as buildings, trees, and geological features
   - Minimize east/west-facing glass
   - Incorporate overhangs to reduce solar radiation component during warm months
   - Optimize glass thermal quality and solar transmittance capacity per orientation
   - Optimize roof thermal and solar reflectance performance
   - Optimize wall and floor thermal performance
2. Minimize building performance load effects
   - Incorporate daylighting to improve visual comfort and to reduce connected ambient light load
     - Modify glass to control solar radiation, but allow visible light transmittance

- Adjust initial space layout to provide open areas on building perimeter
- Incorporate passive solar reclaim on south exposure to offset winter heating load
- Incorporate green roof systems to stabilize year-round plenum temperature

3. Minimize connected internal load
   - Identify a realistic occupant use schedule and integrate with internal load components
   - Optimize applied lighting efficiency and staging controls
   - Maximize efficiency of office equipment
4. Develop HVAC infrastructure based on optimized load profile (for simplicity, other MEP infrastructure system dependencies are excluded from the primary demonstration)
   - Select best-fit strategy for HVAC systems based on minimized building load, use profile, and ventilation requirements. Note that this exercise requires extensive modeling by the HVAC engineer to find the best-fit solution. (For this example, assume that a water-cooled chilled-water plant, hot-water gas boiler plant, gas-fired desiccant dedicated outside air unit, and four-pipe fan-coil terminal units have been shown to provide the best-fit energy performance solution that maintains budget control. There is no basis to this selection other than to demonstrate subsequent evaluation strategies.)
5. Optimize HVAC systems
   - *Chiller plant*
     - Analyze terminal unit cooling coil selections to determine water temperature differential needed to provide cooling performance, minimize distribution pipe size, and minimize pumping capacity
     - Optimize condenser water temperature differential to find best fit for chiller, cooling tower, and condenser water pumps
     - Evaluate effect of water-side economizers
     - Evaluate application of condenser water heat recovery as a waste heat source
   - *Boiler plant*
     - Analyze terminal unit heating coil selections to determine water temperature differential needed to provide heating performance, minimize distribution pipe size, and minimize pumping capacity
   - *Ventilation system*
     - Analyze effect of energy recovery on dedicated outside air unit and central plant capacities
     - Evaluate distribution options and effect on air horsepower for delivery of ventilation air to the occupied space
6. Optimize HVAC components
   - *Chiller plant*
     - Optimize chiller for best-fit performance compared to projected load profile and condenser water relief opportunities
     - Optimize cooling-tower performance using air or water modulation compared to projected load profile
     - Examine effect of motor efficiency improvements on pumping systems
     - Optimize sizing of noncritical path piping mains and branches
   - *Boiler plant*
     - Optimize boiler performance compared to projected load profile
     - Examine effect of motor efficiency improvements on pumping systems
     - Optimize sizing of noncritical path piping mains and branches
   - *Ventilation system*
     - Examine effect of motor efficiency improvements on fan systems
     - Optimize sizing of noncritical path duct mains and branches

This simple example demonstrates that there are many possible strategies for a project. The magnitude is greatly expanded when multiple objectives are pursued. In the example, note that HVAC system design is not even a factor until development of the least-load-impact building is assembled. Only then do applied MEP system solutions come into focus. The example strategy development supports the object: reduce energy use by minimizing load, right-sizing systems to meet the load, and then maximizing component efficiency to match the use profile.

Interdisciplinary Integration

Coordinate with other disciplines to optimize the design arrangement inside the building to comprehensively use project conditions, shorten construction duration, reduce overall project cost, and enhance convenience for interdisciplinary operation and maintenance.

The entire building project could be considered as an integrated system with a unified overall scope of work and unified timeline schedule to be performed and achieved by an integrated design and construction team. In this integrated system, all individual systems and their components could be considered as subsystems of the overall integrated building project. The subsystems in a complete project include the following:

- Overall site plan, entrances and gates, roads and transportation, landscape, electrical substations or electrical main connection, gas or other fuel main intake pressure regulation station, water, sewer, sanitary and storm drain piping and main connections, telephone, network, security and fire protection system main connections, outdoor lighting, etc.
- Building foundations, structure system, walls, roofs, ceilings, floors, elevators, electrical, gas, fuel, water, sewer, plumbing, sanitary, mechanical, HVAC, chiller, boiler, noise control, lighting, process systems, energy and process material recovery systems, exhaust air and wastewater treatment systems, hazard control systems, explosionproof and corrosionproof issues, instrumentation and control systems, fire protection systems, etc.
- Mini environmental booths; supply, recirculation, makeup, and exhaust air systems; lighting, process mechanical, chemical, electrical, and control systems; production lines; process conveyers; special gas supply systems; acoustics; operating personnel; material and products access doors, windows, or openings; air showers; room temperature, humidity, static electricity, CO_2, and pressurization control systems; fire protection and after-fire recovery systems; seismic design and emergency response facilities, etc.
- Construction service and construction; installation work; system start-up; test, adjust, and balance; commissioning; and handover to operation
- Building management and system operation and maintenance

Iterative Evaluation and Analysis

As the example strategy for minimizing energy use demonstrates, there are multiple development steps that must be followed in pursuit of IBD objectives. Design of the final product depends on looking at each system-level component contribution and determining whether incorporation of the proposed strategy improves the project whole within the guidelines of the decision criteria.

As each strategy is explored, numerous *what-if* questions must be evaluated and accepted or discarded. Acceptance allows the team to move forward, but does not preclude returning for reevaluation later. Nonacceptance may progress into alternative solutions for the target strategy, or it may lead to totally discarding the strategy and moving forward. Discarding a strategy does not preclude returning to it later.

Obviously, IDP requires significant effort and creativity to (1) develop applicable strategies to meet project objectives, (2) refine strategy flow as the building evolves, (3) define component variables as each strategy is evaluated, (4) analyze the financial effects of each strategy, and (5) repeat the process over many iterations.

DESIGN ACTIVITIES AND DELIVERABLES

Some of the more common design activities and deliverables are affected by IDP. Some schools of thought suggest that these items be decoupled from integrated practice, but this perspective unfortunately misses the underlying premise that all activities in the project delivery process have a certain level of interdependency. When collaborative design strategies are used, these interdependencies require increased stewardship.

Drawings

The drawings are graphic representations of the work on a project, and include plans, elevations, sections, details, legends, notes, abbreviations, and schedules. They are often diagrammatic, and rarely show every detail required to construct a facility. Drawings show quantities, extents, and spatial relationships of the elements of construction to each other and existing conditions and surroundings. They may identify a particular product, material, finish, or process many times. However, the particular product, material, or process should be specified only one time in the specifications. Descriptions and identifiers on the drawings should be simple, concise, and generic. IBD does not change this basic definition.

IBD does have an effect when it comes time to communicate the system solutions onto drawings that will be used for construction. Coordination now becomes an appropriate and critical IDP tool. The project team must take time to ensure that the integrated work results are correctly identified throughout the drawing set.

The project team should avoid issuing drawings in decoupled groups or individual sheets during the procurement phase. Bidding in isolation is just as detrimental as design in isolation when it comes to achieving integrated solutions.

Specifications

The project manual is the textual description of the work and other requirements for a project; it includes procurement and contracting requirements, general requirements, and technical specifications for the work of the project.

Specifications describe the administration, quality, products, materials, workmanship, warranty, testing, and start-up requirements of the work of a project (CSI 2005). For uniformity in structure, location of information, consistency, and quality control, it is best if the specifications are organized into divisions and subdivisions (sections) that correspond to the major divisions of work required to complete the project as defined in *MasterFormat* (CSI 2004).

The 2004 edition of *MasterFormat* includes some very powerful new sections to assist in IBD delivery, including the following level 2 and 3 additions to division 01, General Requirements:

- 01 33 00, Submittal Procedures
  59. 01 33 29, Sustainable Design Reporting
- 01 78 00, Closeout Submittals
  60. 01 78 53, Sustainable Design Closeout Documentation
- 01 81 00, Facility Performance Requirements
  61. 01 81 13, Sustainable Design Requirements
  62. 01 81 16, Facility Environmental Requirements
  63. 01 81 19, Indoor Air Quality Requirements
- 01 86 00, Facility Services Performance Requirements
  64. 01 86 19, HVAC Performance Requirements
  65. 01 81 23, Integrated Automation Requirements
- 01 91 00, Commissioning
  66. 01 91 13, General Commissioning Requirements
- 01 92 00, Facility Operation
  67. 01 92 13, Facility Operation Procedures
- 01 93 00, Facility Maintenance
  68. 01 91 13, Facility Maintenance Procedures

Tools are in place in the industry to support communication of integrated system design into work results that can be consistently located. Further study of the *MasterFormat* structure will demonstrate that individual facility services, such as HVAC, have defined specification structures to support effective communication of system solutions.

Value Engineering

Value engineering (VE) is similar to life-cycle cost analysis, and may be performed at any phase of design. It is most often performed when potential construction cost overruns have been identified, or alternative systems or substitute equipment is being considered. Applying the same methods described for life-cycle cost analysis, the alternative system or component being value engineered should be analyzed for its total ownership cost effect. The results should then be compared to the project objectives to see if incorporation has merit.

When VE is complete, recommendations should be made to the owner describing the process and methods used to determine the total cost of the system(s) analyzed. The system that best suits the owner's needs and expectations should be indicated. Recommendation of proposed VE alternatives should indicate how the defined project objectives are affected.

Risk Management

Risk management includes the following:

- Systematic, consistent application of written standard office procedures
- Judicious implementation of QA/QC procedures
- Comprehensive record keeping
- Timely and accurate communications
- Written contracts that include certain basic terms and conditions for all services rendered

Because IDP involves significant collaboration, team members need to practice a policy of keeping good, complete, contemporaneous records of the facts discussed and decisions made (and by whom) in meetings, during site visits, in e-mails, and during telephone conversations. Most errors and omissions (E&O) and liability insurance carriers and their legal counsels offer guidance, and customarily provide publications on risk management as part of their service to their insured. Team members should be well versed in how to practice proactive risk management so that liability paralysis does not reduce collaborative participation.

Budget Control

Traditionally, there are two types of budgets the design team must manage during the design phase: design cost and construction cost. Note that this is not an absolute, because there could be a deviation if progressive-thinking owners incorporate all or portions of operating-related budgets into the equation.

Design cost control begins with design team resource allocation, budgeting, and scheduling while preparing the fee proposal. Once a complete scope of work has been defined, a project budget analysis should be prepared and submitted with the fee proposal to the client. In the SDP model, regular monitoring of actual design cost as compared to the original project budget analysis and scope of work should help avoid scope creep and ensure that projects are delivered within the design fee budget. The IDP model requires that design fee budget control include an additional oversight element. Although infinite evaluations may lead to the absolute best built solution, design fee structures have a practical limit on how many evaluations are affordable. It is therefore financially critical for the design professional to develop a clear strategy at the time of fee negotiation so that all parties agree on the extent and quantity of strategy evaluations, how the fee is structured to reflect the applied effort at the time of service, and how additional services are accommodated if additional evaluations are required.

Responsible control of construction cost budgets can vary depending on the project delivery model. Design-bid-build models place the design team in an oversight role. Design-build allows the contracting entity to control cost of the delivered solution. Design-CM brings in a third-party construction manager, who is responsible for delivery of the project within the defined construction budget. IBD is achievable under any of these delivery models. Each requires, however, accurate cost projections to support realistic system evaluation. Likewise, the construction budget needs to represent a level of funding that supports construction of the final system solutions. Cost projection and cost control play hand-in-hand throughout the iterative evaluation process.

Constructability Review

Constructability is a measure of how well construction documents provide the construction team with the information necessary to complete and deliver a project that meets the owner's expectations and documented project requirements. A constructability review is an organized process of reviewing construction documents during the design phases to make recommendations to the owner and design team about how the design may better define expected construction work results. The IDP is well served when knowledgeable construction representatives provide objective feedback on the constructability of developing system solutions.

Operational Review

Operational reviews should be conducted during the design development and construction document phases of design. Depending on the owner, this type of review may be increased to correspond with evaluation scenarios. Operational review can also be one of the decision-making criteria used on a project.

Reviewers should be knowledgeable about systems, equipment, controls, operation, and maintenance. Ideally, the review should include representation from the group that will be ultimately responsible for operating the facility. During the review, sequences of operation should be thoroughly reviewed to ensure that integrated solutions are truly integrated. Equipment location should be reviewed to verify that required maintenance clearance and accessibility are provided. Drawings and specifications should be checked to ensure that (1) the appropriate level of system and component commissioning has been prescribed, (2) adequate and usable closeout documentation has been itemized, and (3) sufficient training has been scheduled for operational staff.

Operational constraints must be considered when system solutions are developed. Nonconventional systems and equipment can be somewhat intimidating for building operators. The issues of perceived complexity and risk must be mitigated. Solutions must be kept in perspective with the client's ability to operate and maintain the facility. Operational review is an excellent process to address these concerns.

Commissioning

Commissioning is a systematic process of applying QA/QC procedures to the design and construction of a building, to verify that key elements of the design are, in fact, constructed as designed, and started, tested, operated, and maintained so that the building meets the designer's intent and owner's expectations.

ASHRAE defines commissioning of HVAC systems as "the process of ensuring that systems are designed, installed, functionally tested, and capable of being operated and maintained to perform in conformity with the design intent. Commissioning begins with planning and includes design, construction, start-up, acceptance and training, and can be applied throughout the life of the building" (ASHRAE *Guideline* 1).

Several types of HVAC commissioning processes are available: overall, construction, and existing building commissioning (or retrocommissioning). The commissioning process described here applies to new construction and major renovations.

In new construction projects, the overall HVAC project commissioning approach is recommended. It starts at the inception of a building project during predesign and continues through the design, construction, acceptance, training, operation, maintenance, and postacceptance phases, integrated as part of the entire project.

The owner selects and contracts with an HVAC commissioning authority (CA) at very beginning of the predesign phase. The commissioning authority develops the scope of commissioning and reviews design intent during predesign; during the design phase, the CA reviews the design to ensure the HVAC project accommodates the commissioning process. The CA coordinates with the owner, design engineer, and HVAC contractor, and issues commissioning specifications to address owner requirements, define contractors' responsibilities, and review contractors' submittals. This leads to HVAC construction commissioning, and completion of the rest of the commissioning process.

Commissioning is a rigorous and intense process that should be used when integrated-system-based design solutions are provided. See Chapter 43 for more information.

PROJECT DELIVERY SEQUENCE FOCUS

IDP applies to all phases of project delivery, not just the standard design phases. Successful delivery requires that the entire project team contribute to their role at the appropriate time. The project team should be cognizant of the unique effort focus required within each phase. The following outline identifies key focus elements that IDP should consider:

- Owner planning
  69. Determine best-fit construction need
  70. Identify least-impact, best-fit siting option
  71. Define project objectives
  72. Structure project delivery budget to support IDP delivery
  73. Select team to support project objectives
- Predesign
  74. Refine project objectives
  75. Define decision-making criteria
  76. Develop strategies to support project objectives
  77. Develop initial program solution
- Schematic design
  78. Develop facility solution assembly
  79. Optimize facility systems
  80. Apply facility services solutions to support optimized facility
- Design development
  81. Optimize all facility and services systems
- Construction documents
  82. Optimize all facility and services components
  83. Communicate system solution accomplishments into constructable work results
- Procurement
  84. Provide oversight to ensure that procurement document revisions do not negatively affect developed system solutions
  85. Adjust project objectives and conduct new strategy evaluations as required to respond to dynamic scope modifications
- Construction
  86. Adjust project objectives and conduct new strategy evaluations as required to respond to dynamic scope modifications
  87. Provide oversight to ensure that product substitutions and contract document revisions do not negatively affect developed system solutions
  88. Provide oversight of facility and services construction to ensure that work results comply with design intent

- Operation
  89. Verify that systems and components operate as intended across all seasonal and usage conditions
  90. Verify that operational staff fully understand how systems work holistically
  91. Verify that applicable maintenance procedures and plans are understood to support systems and components for the facility life cycle.

REFERENCES

ASHRAE. 1996. The HVAC commissioning process. *Guideline* 1-1996.

ASHRAE. 2004. Ventilation for acceptable indoor air quality. ANSI/ASHRAE *Standard* 62.1-2004.

buildingSMART. 2009. *Building SMART: International Alliance for Interoperability.* http://www.buildingsmart.com.

CSI. 2004. *MasterFormat.* Construction Specification Institute, Alexandria, VA.

CSI. 2005. *Project resource manual*, 5th ed. Construction Specification Institute, Alexandria, VA.

ISO. 1994. Industrial automation and systems integration. *Standard* 10303-11. International Organization for Standardization, Geneva.

ISO/PAS. 2005. Industry foundation classes, platform specification. *Standard* 16739. International Organization for Standardization, Geneva.

Lewis, M. 2004. Integrated design for sustainable buildings. *ASHRAE Journal* 46(9):S22-S30.

NIBS. 2007. *National building information model standard*, vol. 1—Part 1: *Overviews, principles, and methodologies.* National Institute of Building Sciences, Washington, D.C.

NIBS. 2007. *Whole building design guide online.* National Institute of Building Sciences, Washington, D.C.

BIBLIOGRAPHY

AIA. 1997. Standard form of agreement between owner and architect. *Document* B141-1997. American Institute of Architects, Washington, D.C.

AIA. 1995. Project checklist. *Document* D200-1995. American Institute of Architects, Washington, D.C.

AIA. 2001. *Architect's handbook of professional practice*, 13th ed. John Wiley & Sons, New York.

ASHRAE. 2004. *Advanced energy design guide for small office buildings.* ASHRAE Special Project SP-102.

ASHRAE. 2006. *ASHRAE GreenGuide: The design, construction, and operation of sustainable buildings*, 2nd ed. David L. Grumman, ed.

CII. 2006. *Constructability implementation guide*, 2nd ed. SP34-1. Construction Industry Institute, Austin.

EJCDC. 2002. Standard Form of Agreement Between Owner and Engineer for Professional Services (E-500). Engineers Joint Contract Document Committee, National Society of Professional Engineers, Alexandria, VA.

Holness, G.V.R. 2006. Building information modeling: Future direction of the design and construction industry. *ASHRAE Journal* 48(8):38-46.

NIST. 1996. *Life-cycle costing manual for the Federal Energy Management Program*, 1995 ed. *Handbook* 135, S.K. Fuller and S.R. Petersen, eds. National Institute of Standards and Technology, Gaithersburg, MD, and U.S. Department of Energy, Washington, D.C. Available from http://fire.nist.gov/bfrlpubs/build96/art121.html.

PECI. 2006. *Model commissioning plan and guide specifications online*, v2.05. Portland Energy Conservation, Inc., OR. Available from http://www.peci.org/large-commercial/mcpgs.html.

USGBC. 2006. *Leadership in Energy and Environmental Design online.* U.S. Green Building Council, Washington, D.C. http://www.usgbc.org

RESOURCES

BLCC	www1.eere.energy.gov/femp/information/download_blcc.html
DOE2 and Energy10	www1.eere.energy.gov/buildings/commercial_initiative_modeling_software.html
EnergyPlus	www.eere.energy.gov/buildings/energyplus/
eQuest	www.doe2.com/equest

CHAPTER 59

HVAC SECURITY

THIS chapter is intended to be an overview of HVAC security considerations relative to chemical, biological, radiological, and explosive (CBRE) incidents that do not cause major structural damage to a building or its infrastructure. The focus is on CBRE incidents whether occurring by accident (e.g., an industrial spill) or a premeditated occurrence.

Because of the nature of security, there is not a lot of documentation available pertaining to designing, constructing, operating, or maintaining HVAC equipment and systems from a security standpoint. Organizations such as the Department of Defense have guidelines that are considered highly confidential and are only shared with others on an as-needed basis. In other situations, special security organizations follow behind the design and/or construction teams with security measures that are not shared with these design/construction organizations.

In general, HVAC security applies to all building applications based on a broad range of reasons, needs, and requests. HVAC security plays a particularly important role for businesses such as pharmaceutical companies, property managers of high-profile commercial buildings where workers and visitors come and go on a regular basis throughout the day and night, and convention centers and sport stadiums entertainment venues where thousands of people are present for a few hours.

This chapter is not intended to be used for design or development of life safety systems or procedures, or for protection of personnel during an incident; rather, it offers an approach to HVAC security, details the need to provide proactive maintenance of these components and systems, and provides descriptions of some CBRE incidents and their associated effects on buildings, building equipment, and occupants, along with general guidelines for how to deal with their effects on building infrastructure.

Since September 11, 2001, more published information has been available about procedures for preventing, mitigating, and remediating terrorist or other CBRE incidents. ASHRAE's (2003a) *Report of Presidential Ad Hoc Committee for Building Health and Safety under Extraordinary Incidents* discusses many aspects of buildings, building infrastructure, and measures that can both reduce the threat and/or damage from such incidents. Several departments of the U.S. federal government, including the Federal Emergency Management Agency (FEMA), Department of Homeland Security (DHS), National Institute for Occupational Safety and Health (NIOSH), Centers for Disease Control (CDC), and Department of Defense (DOD), have produced reports and guidelines for dealing with terrorist threats to buildings (see the Bibliography). Emphasis is generally on actions to reduce the potential

harm to building occupants, both by reducing the threat and by instituting procedures that reduce the hazard during an incident.

In almost any case of a terrorist event affecting a building, its infrastructure, or its occupants, the affected building and its immediate surroundings are likely to be in police or military control for several days (or longer) after the event. During this period, the role of the building(s) owner or physical plant staff is to assist in controlling or remediating the affected areas through their knowledge of the building and its infrastructure systems. Assessment of damage or remaining danger to the building or personnel is difficult, particularly with chemical, biological, and radiological events, in which the contaminating agent often is invisible and is only revealed through adverse health effects. As such, there are no specific guidelines for how or when a building can be brought back online and readied for occupancy; each event is unique. Any preparation or response protocol for CBRE incidents internal or external to the facility should be designed to consider the specifics of the building and its occupants. It is impossible to provide general guidelines for incidents that are so unpredictable and potentially so devastating. This chapter attempts to shed light on design intent, construction administration, commissioning, and recommissioning, and some of the possible effects on buildings, their systems, and their occupants, which may aid in the development of a more specific protocol in line with a particular facility's needs.

REQUIREMENTS ANALYSIS

The initial process of any HVAC system design begins with the owner's project requirements (OPR). The OPR covers a wide range of categories to document the owner's intent in investing in this new construction, renovation, or infrastructure project. The OPR identifies the drivers that will shape the design, how it will be constructed, and how it will be operated and maintained over the building's life. It also sets a construction budget and project timeline. Confidential security requirements are addressed, but may remain confidential, with limited documentation between the owner's security professionals and the designer. It is also possible that a separate design team and construction team may be brought into the project after design is completed and before the owner begins occupancy, to fulfill the basis of design (BOD).

When drafting an OPR, the design team should consider the following:

- Who are the main occupants?
- What is the intended use of the building?
- What is the total planned population of the building, including visitors and service staff?
- What type of operations will the facility and/or occupants conduct?

The preparation of this chapter is assigned to TG2.HVAC, Heating, Ventilation, and Air-Conditioning Security.

- What is the planned response to an incident? Will occupants evacuate, shelter in place, or carry on normal activities uninterrupted?
- How will the building staff become aware of a threat, and what is the likely notification time?
- What level of protection is required against threats?
- Will some occupants have planned responses that differ significantly from the general building plan?
- What level of access will the general public be allowed in the building?
- Does the owner have a dedicated security team and/or consultant?
- What life safety measures are planned for the building?
- Will occupants be required to remain in the building after an incident (e.g., in a high-security prison)?
- Are there any unique environmental health concerns (e.g., explosive atmosphere, laboratory)?

The OPR should be complete and exhaustive, and should adequately cover the owner's overall goals for the building's HVAC security. Many HVAC security measures are relatively low in cost and effort during new construction or renovation planning, but may require significant cost and effort if implemented after construction is completed. Therefore, it is critical to capture these requirements early in any design process.

All projects should include some minimal level of HVAC security design and planning. These measures are typically included in the life safety requirements, specific designs, or best practices typically applied to building construction. These **baseline measures** include equipment or design features that can be applied to all buildings at a minimal cost and effort to provide basic protection against internal and external threats. Baseline measures support the safe sheltering in place and/or evacuation of occupants during an incident. Many baseline measures can be implemented in an existing facility with little or no additional engineering design or cost, and with minor alteration to the facility operations. **Enhanced measures** include equipment or design features beyond the baseline level, and are intended for facilities with identified risks or critical operations. Costs for design, construction, and sustainment can be significant, depending on the measures selected; however, the protection afforded by these systems typically allows longer-term sheltering in place or continuous uninterrupted operations for the duration of the incident. Specific design features and equipment to provide these measures of protection are discussed later in this chapter. A building's particular HVAC security design uses its own unique collection of HVAC security measures, depending on risk and requirements.

RISK EVALUATION

In parallel with development of the OPR, a risk evaluation should be conducted for the building and its planned design. FEMA and other industry organizations have developed various guidance documents and software to assess the risk and appropriate response from both external and internal events. Significant detail regarding risk management for catastrophic events is included in ASHRAE *Guideline* 29-2009.

Risk is a function of the **probability** of an event occurring and the **consequence** of this event. For HVAC security planning, the probability of a catastrophic event occurring is typically nearly zero. This probability is shaped by multiple factors, including the facility's occupants, its location, and nearby objects that may pose a threat. The consequence of an event, however, is usually considered extremely high. Factors to be considered include potential loss of life, failure of critical infrastructure, and remediation time and effort.

Figure 1 presents a generalized framework for managing building security risks. The key considerations for a risk analysis include the following:

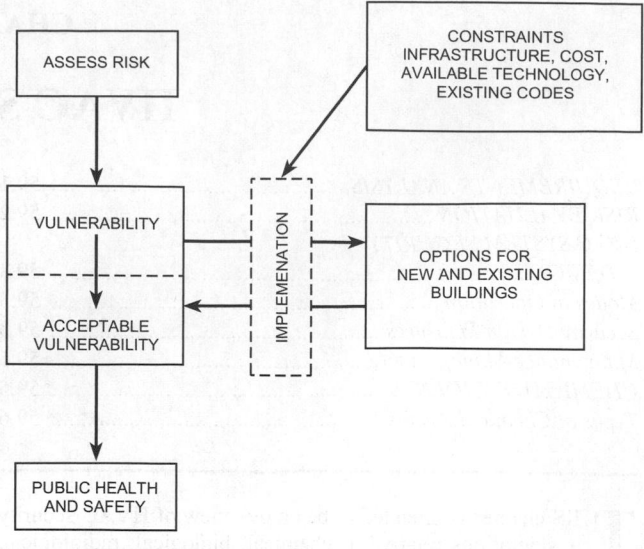

Fig. 1 Risk Management Framework
(Adapted from ASHRAE *Guideline* 29-2009)

- Vulnerabilities: what elements of the building design, construction, location, or operations present opportunities for catastrophic events?
- Acceptable vulnerabilities: what identified vulnerabilities cannot or should not be addressed, and thus must be accepted as operational risks?
- Impact: what are the consequences of an adverse event, including remediation, reconstruction, and lost business, and how does this compare to the cost of implementing HVAC security measures?
- Constraints: what limitations exist that would shape the HVAC security design of a building?

Successful risk evaluation should include a review of all facets of the planned building design and operations to determine the risk. This evaluation may include the following areas of assessment:

Building and occupants
- Identification of potential high-value targets
- Identification of specific vulnerabilities
- Classification of occupants and operations
- Assessment of benefits of containment versus evacuation

Potential threats and vulnerabilities
- Identification of potential aggressors
- Identification of potential delivery systems

Likely support mechanisms
- Identification of likely first-responder units
- Identification of infrastructural support

Postevent remediation
- Consideration for potential consequences if building is unoccupied or unusable for extended periods
- Estimation of relative magnitude of remediation measures

Examples of considerations that may increase the overall risk assessment of the facility include the following:

- Potential effect of building remaining unoccupied for extended periods
- Military and government command centers
- Significant landmarks
- Critical infrastructure elements

- Single-failure point operations or equipment
- Corporate headquarters or critical operations centers
- Transportation hubs
- Communications nodes
- Popular tourist destinations
- Approach and takeoff areas for major airports
- Sites open to the public
- Sites frequently targeted by protests and demonstrations
- Locations near significant potential hazards such as nuclear power plants or chemical manufacturing facilities
- Locations adjacent to major shipping and transportation routes
- Sites frequently subjected to severe natural or weather events, such as tornadoes, hurricanes, or earthquakes

Obviously, this list is not exhaustive, and many buildings may have unique circumstances or characteristics that warrant specialized HVAC security measures. Generally, it is difficult or impossible to completely mitigate against all risks; thus, the overall goal of a risk evaluation and implementation of security measures is to move subsequent evaluated risks to lower levels. That is, if a building's overall risk is assessed as high, measures should be implemented such that subsequent risk assessments for the same building would be medium or low.

HVAC SYSTEM SECURITY DESIGN

Building design and operations during a CBRE event should leverage strongly from the OPR and risk evaluation documents. This section presents generalized recommendations and practices for HVAC system security design, and detailed information on specific threats is provided in subsequent sections.

MODES OF OPERATION

Three main modes of operation can be considered in HVAC security design: evacuation, sheltering in place, and uninterrupted. Each mode presents unique challenges, costs, and benefits, and each should be reviewed and compared to the risk assessment and OPR to determine which best meets the building owner's needs.

Evacuation

Evacuation is the immediate, rapid, and controlled egress of occupants from a facility in the event of an emergency. This mode is commonly used in fire protection engineering. Planning and design for evacuation includes measures to prevent catastrophic failure of the facility for a short duration, and egress direction support, including emergency lighting, signage, and doors.

This mode is effective in many cases and generally is the easiest to implement. However, this mode may not be effective against external threats where personnel may evacuate directly into the path of a threat, and can present difficulty for triage and containment. Typically, minimal cost and design efforts are required to implement an evacuation mode during an event, because most buildings are required to include similar measures and equipment for fire protection. Also, typically limited or no additional training is required for building occupants, most of whom are familiar with normal evacuation procedures.

Shelter-in-Place

This is short-duration occupancy of a facility or section thereof to avoid immediate threats. This mode requires occupants to remain in the building during the event and seal the building against intake or further dissemination of threats until the immediate danger has passed. Although some facilities provide full-building shelter-in-place coverage, many designate discrete rooms or areas for occupants to remain in. The normal expected duration is on the order of several hours or less.

Sheltering in place is an effective protective measure when implemented properly and quickly after identification of a threat. Many government and military facilities implement some level of sheltering in place. Effective application requires immediate identification of a threat and subsequent shutdown of HVAC systems to prevent the spread of contamination in the building. Additionally, designated shelter in place locations may include some food and water supply for occupant consumption during an event, as well as means of communications with emergency responder personnel. Generally during a shelter in place event, occupants do not continue normal work because of the anxiety of the event and the potential requirement for occupants to move into a common area (e.g., conference room, break room) without normal work equipment. Because durations are typically expected to be several hours or less, no bedding is required, and the relative population density can be high. Implementation of a shelter-in-place strategy requires early coordination and training of building occupants to avoid confusion during an event.

Uninterrupted Operation

Uninterrupted operation is the continuous occupancy and use of a building, or some portions therein, during an event without contamination of personnel and equipment. This mode allows for occupants to continue doing work without evacuating or sheltering in place. Although some buildings may complete this mode by providing personal protective equipment to occupants, many have installed collective protection systems in the building HVAC systems, including advanced filtration, airflow balancing, controls, and architectural modifications. Collective protection is achieved by filtering all incoming air to a building and providing this air at an overpressure to spaces, thus creating a protective zone where personnel can continue to operate during an event.

This mode can be extremely effective against both internal and external threats, depending upon configuration and design. Typically, buildings can operate in these modes for hours or days, depending upon the threat; however, with this extended duration, bedding, food, and water, along with lower population densities, should be considered. The relative cost (both capital and sustainment) for this mode can be extremely high, so it is generally used only for critical facilities such as command centers and vital infrastructure elements.

SECURITY DESIGN MEASURES

Multiple measures can be implemented to provide HVAC security and protections, some of which are discussed in this section. It is important to consider that these measures are typically combined with other design elements to enhance building protection, and that not all measures may be applicable to all buildings or locations. The BOD is the documentation of design criteria, set points, parameters, and narrative that outlines the HVAC system security. Again, because of the nature of the security OPR, BOD documentation may not be published. BOD considerations may include some or all of the following measures.

Emergency Power

HVAC systems that are designed to respond to a CBRE incident by continuing to operate should be powered from an emergency electrical power distribution system. Some considerations may include redundant external feed from two sources, generator sets, and/or uninterruptible power supplies (UPS). Any of these options may require significant first cost as well as maintenance and sustainment costs. In designing emergency power for the HVAC system, consideration should be given to the location of emergency generator(s) and associated switchgear and motor control center so that they are well above flood water levels. Other considerations are to ensure electrical power is provided to the building automation system

(BAS) so that the HVAC security system functions with the building automation computer under emergency power.

Redundant Design

Similar to power loss, failure of a critical component can place overall building HVAC security in jeopardy. Building designers should consider including redundant equipment in systems, such as air-handling units, blowers, or motors. Robust systems should include automatic control of these components, thus allowing for switchover from a failed component to the back-up immediately and automatically. If redundant systems cannot be installed, the owner should consider stocking critical or hard-to-find components in the facility so failed components can be replaced quickly in case of a failure.

System Shutdown and/or Isolation

Rapid shutdown and/or isolation of air-handling units, including outside air intakes, can prevent or limit intake of contaminants into the air distribution system and thus decrease the potential spread of these agents. Many facilities use digital controls networks that readily allow the addition of a shutoff actuator button. For manual initiation, these buttons should be located in one or more areas regularly accessible by building occupants or normally staffed locations (e.g., security guard stations, central lobbies, reception desks) and treated in a manner similar to fire alarm activation stations. Automatic initiation methods using external detectors may be used, but the rate of false positives and negatives, capabilities of existing detection technology, and overall reaction time should be considered. On activation, the system should initiate rapid closure of air distribution dampers and rampdown of equipment to prevent movement of air through areas of the building. In these cases, consider using a spring-shut damper, although precautions such as bypass or relief duct systems should be taken in case of potential system damage by these closures.

Protective Equipment

Many facilities have begun to distribute **personal protective equipment (PPE)** to facility occupants. This equipment, including escape hoods and respirators, may be issued to building staff as well as being placed in centrally available locations for occupant and visitor use. Equipment is generally intended for single-time use to allow occupants to safely evacuate the building during an event. Several manufacturers can provide this equipment, with varying levels of protection, shelf life, and recertification requirements. The designer should consider the overall burden to the building when providing PPE to occupants, including capital costs, training, shelf life, and life-cycle costs for the equipment.

100% Outdoor Air Operation

Normal air-handling systems using returns present an issue for internal release scenarios, because these returns can carry contaminants from the release point in the zone and redistribute throughout the HVAC systems, possibly contaminating the entire building. Designing air-handling systems that use 100% outdoor air is a good alternative to prevent this type of distribution, and limits the spread of contaminants in a space. Although 100% outdoor air systems can present a significant cost and energy burden, this may be offset by the added capability for occupant protection. Typically, this approach is suitable for small buildings or sections of a building. Alternatively, the designer may consider using local terminal units with integrated fans to provide local recirculation and meet space heating or cooling demands, thus reducing the outdoor air required to each unit.

HVAC Zoning

Using multiple HVAC zones in a building allows localized control of the air movement equipment, and can limit the transport mechanisms for contaminant spread. Each zone, especially when enclosed with walls or partitions, can contain airborne contaminants without widespread movement to adjacent spaces. HVAC zoning can provide occupants with enhanced control over the systems in their spaces and can also help limit the spread of airborne diseases such as influenza.

Increased Standoff Distances

Close proximity to publicly accessible areas increases the risk of an external event having catastrophic consequences for a facility. The presence of a buffer with controlled or limited access can significantly lessen the effect of an airborne contaminant release or blast. This standoff area must have limited access to the general public and should limit vehicular traffic to emergency access, deliveries, and facility maintenance. Increased standoff distances also provide additional area for emergency first responders during or after an event. Although these buffer areas present an additional cost associated with capital investment, they also provide occupants with aesthetic benefits, including additional green space.

Occupant Notification Systems

The moments immediately before, during, and after an event can be confusing for building staff and occupants, especially if some occupants panic and not fully understand what actions should be taken. Most buildings include some type of notification system that can be used to communicate to occupants in critical situations. Systems include loudspeakers, alarm horns and strobes, automated telephone alerts, or computer notification systems; at minimum, many buildings can implement mass e-mail notification or designation of certain personnel to serve as runners with little or no cost. Building managers may consider providing emergency action information cards for all occupants to keep in their work spaces, to refer to during an emergency.

Air Intake Protection

For new buildings and HVAC systems, fresh-air intakes should be elevated to help prevent malicious acts (e.g., inserting a hazardous material directly into the intake) and minimize the concentration of hazardous materials during a ground-level release. Intakes should be placed at the highest practical level on the building, at least 10 ft above grade. Most ground-level releases near the building will remain close to ground level, and the concentration of hazardous material in the air decreases with increasing height. Existing fresh-air intakes close to ground level can be modified to prevent physical tampering by placing fencing or barriers or building a plenum around the intake to limit potential intake of contaminants. Physical access to system intakes should be limited, and security cameras focused on intake areas may be considered. To prevent direct tampering of intakes, a sloped screen should be installed at the top of the intake to prevent direct insertion of any hazardous substance or container.

Increased Prefiltration Efficiency

A relatively simple and cost-effective protective measure that can be undertaken in most every facility is upgrading existing prefilters to a higher-efficiency model. An increase in prefiltration efficiency can prevent the intake of a significant fraction of external airborne material, including biological and radiological particles; a related benefit includes the reduction of airborne allergens entering a building, thus resulting in a potential decrease in worker health issues and absenteeism. Typically, increased efficiency prefilters present a relatively minimal cost increase as compared to standard prefilters, and require no system modifications or additional maintenance.

Additional Filtration

As with increased prefiltration, adding additional filtration can reduce or eliminate airborne threats entering a building. Multiple options and levels of efficiency are available, ranging from cost-efficient low-efficiency models to military-grade filtration. Full-time filtration provides occupants with protection without the requirement for advance notification; however, part-time or standby systems can be effective if advance warning or detection is available. The current standard used in multiple government, military, and private-sector buildings includes high-efficiency air filtration (HEPA) filters for biological and radiological threats, and activated and impregnated carbon filters for chemical threats. These filtration measures present significant capital and sustainment costs, which may make these systems unaffordable for lower-cost, noncritical buildings.

Location of Mechanical Equipment

When designers develop plans for building mechanical systems, one of the main considerations is system accessibility for regular maintenance and replacement. However, in some cases, the placement of mechanical systems may present security risks. Mechanical and electrical rooms should be placed in secure areas of the building that are not accessible to the general building population, and should be located away from any potential hazards such as flood areas, hazardous materials storage, loading docks, central lobbies, and areas that may be vulnerable to vehicle impact. Where possible, mechanical spaces should be accessible by maintenance personnel from within the facility to allow repair during an event.

Physical Security Measures

Many physical security measures can be applied to HVAC systems and overall building protection that may prevent the release of a hazardous material or contaminant. Security screening at entry points can help detect containers that may contain hazardous materials. This screening may include x-ray scanning, metal detectors, or manual searching of personal belongings such as briefcases and handbags. Rooftop access should be restricted to authorized personnel, because mechanical equipment, exhaust stacks, and ducting may allow introduction of contaminants. Rooftop entries and exits should be monitored and controlled by the building security system.

Air Supply Quantities and Pressure Gradients

Many contaminant releases depend on air movement to move contaminants throughout the building. This transport can be influenced by small differential pressures between spaces, often less than 0.10 in. of water. These gradients may be effectively used to limit the spread of airborne contaminants between offices, corridors, and common areas. HVAC designers may consider providing a small excess of air to selected areas to effectively overpressurize these spaces with respect to adjacent spaces. This is of particular importance in systems where the HVAC system includes filtration equipment, allowing the protected space to be maintained at an overpressure with clean filtered air.

Sensors

Detection and early warning of a threat are extremely important to building protection. With rapid notification, building staff can implement measures to protect against the threat, including initiating sheltering in place or evacuation. However, implementing a robust detection system can pose many challenges. Technology is being developed in both the government/military and commercial sectors, but these new devices still have limitations. Currently, although many products exist for point and standoff detection, some of these methods still place a significant burden on the building staff, such as laboratory-scale analysis for confirmation and specially trained personnel. In all cases, designers should consider the sustainment cost and relative frequency of false positives/negatives when specifying detection equipment.

Mailroom and Lobby Measures

The mailroom and central lobbies of buildings are highly vulnerable areas: they act as building interfaces with the general public. In most buildings, these are areas where uncleared personnel or packages come in proximate contact with the facility and where threats can cause the greatest harm. These areas should be given special consideration and additional protective measures to ensure all threats are minimized.

In entry areas, many buildings have mandatory security access procedures for regular building occupants as well as visitors. Security measures such as magnetometers, x-ray scanners, and personnel screening may be used to limit potential hazards from entering the building. Designers should consider using segregated HVAC systems in lobby areas with dedicated air-handling units (AHUs) for the lobby area, and maintaining the lobby spaces at a negative differential pressure with respect to interior spaces. Lobby windows and doors should include blast-resistant glazing and construction, and walls between the lobby and general building interior may include enhanced blast resistance ratings. Security or reception personnel should have controls in their work areas to allow rapid lockdown of all building entries and exits.

The anthrax mailings of 2001 highlighted the vulnerability of buildings to attack by mailborne threats. The U.S. Postal Service and package delivery companies have since implemented enhanced security, but building owners may consider additional measures. Mailrooms and package-receiving areas should include the measures described for lobbies, especially segregated HVAC systems that maintain these areas at a negative pressure compared to the rest of the building. Some facilities have enhanced mail- and package-screening procedures, including separate mail-handling facilities, x-ray or metal detection scans, individual parcel opening and screening, and laboratory analysis of packages. At minimum, mailroom staff should review incoming mail and immediately notify law enforcement of any suspicious letters and packages, such as those with exposed wires, irregular shapes or weights, misaddressed labels, or unexpected senders or locations.

MAINTENANCE MANAGEMENT

Because of the nature of HVAC security, operation is a critical requirement as it pertains to reliability and repeatability. Similar to how emergency generators are operated once a week, once a month, and fully loaded annually, HVAC security systems need to be operated on a scheduled basis to assure the equipment will respond in an emergency situation. Operation of the building automation system is an integral part of this routine exercising of the HVAC systems.

Proactive maintenance management also contributes to system performance and reliability. Modern facilities will most likely have computerized maintenance management software (CMMS) system to manage the maintenance process and documentation of this process. It is important to emphasize the value of documentation management; the design engineer should account for this in the design phase, specifying in the operation and maintenance requirements that the contract will be required before project closeout. These requirements should complement the CMMS database criteria so that project closeout documents are electronic and compatible with the CMMS system.

CHEMICAL INCIDENTS

A chemical incident is defined as the accidental or intentional release of a gaseous or vaporous compound into breathable air. Releases of toxic liquids, solids, or powders are not addressed in this chapter. A release may occur inside or outside a building, and may be of short duration (e.g., from a broken container, an accidental valve opening, or a terrorist incident) or sustained (e.g., from a leaking storage tank or broken supply line). Descriptions of classes of and individual air contaminants, including chemicals, are found in Chapter 11 of the 2009 *ASHRAE Handbook—Fundamentals*, and removal techniques are covered in Chapter 29 of the 2008 *ASHRAE Handbook—HVAC Systems and Equipment* and Chapter 46 of this volume. Discussions in this chapter are limited to chemicals that are considered acutely toxic or corrosive, and present immediate danger to building occupants or systems.

Industrial buildings, where harmful chemicals may be used routinely, are likely at higher risk for internal chemical incidents than a typical commercial building, but, because of training, established procedures, and experienced personnel, they are also likely to be more prepared to handle an incident. Most commercial buildings, except for some government and high-profile buildings, do not have procedures in place for handling a chemical incident. A terrorist chemical event in a typical commercial building adds new difficulties, because details of the release are not known until long after the incident, and affected buildings and occupants are generally caught off guard, with little or no procedure in place for handling the event.

Chemical substances that can cause physical distress when introduced into breathable air are numerous; this chapter addresses only gaseous or vaporous compounds, and of those, addresses only two groups (1) those specifically known as chemical agents (in terms of warfare/terrorist activities) that might be intentionally introduced into a building's environment, and (2) a few common industrial gaseous substances that might accidentally be introduced into a building HVAC system through external or internal release, thus requiring HVAC or facility remediation of some kind. The purpose of this section is to address buildings that have no expectation of an accidental chemical release, based on the activities performed within the facility, as opposed to those of surrounding, related facilities (e.g., industrial facilities that have their own response plans). For control of airborne gases and vapors that are used as part of the building's normal operation, such as in laboratories or industrial processes, see Chapter 29 of the 2008 *ASHRAE Handbook—HVAC Systems and Equipment* and any of the application-specific chapters in this volume.

TYPES OF CHEMICAL AGENTS

Intentional contamination of facilities and their HVAC systems (and thus very ready dispersion to occupants) with gaseous or vaporous chemical substances has become a real concern. Chemical agents are classified by the U.S. Army (2005) either as toxic or incapacitating. Toxic chemical agents include nerve, blister, lung-damaging, and blood agents. Any of these agents may be introduced in sufficient quantity so as to injure building occupants and, in the process, compromise the building's HVAC system. Irritating agents (e.g., tear gas), which cause temporary trauma through reflexive action but are not generally lethal, are not considered by the U.S. military to be chemical agents.

Incapacitating Agents

Incapacitating agents are defined by the U.S. DOD as chemical agents that produce temporary physiological or mental effects, or both, that make individuals unable to make a concerted effort to perform their assigned duties. In occupational medicine, *incapacitation* generally means *disability*, and denotes the inability to perform a task because of a quantifiable physical or mental impairment. Thus, by definition, any of the chemical warfare agents may incapacitate a victim; however, by the military definition, incapacitation refers to impairments that are temporary and nonlethal, and does not include low-dose "lethal" agents. Incapacitating agents may cause symptoms that persist for hours to days, but are temporary and

recoverable even without treatment. Incapacitating agents can be classified as either central nervous system (CNS) depressants or stimulants.

CNS depressants are compounds that depress or block activity of the CNS by inhibiting the transfer of information across synapses. Common CNS depressants include

- 3-quinuclidinyl benzilate (BZ)
- Cannabinols
- Phenothiazines
- Fentanyls
- Hypnotics

CNS stimulants cause excessive nervous activity by facilitating transmission of impulses across certain synapses that may otherwise be insufficient pathways. The brain becomes flooded with information, making concentration and decision making difficult. The most common CNS stimulant is d-lysergic acid diethyl amide (LSD).

Symptoms of poisoning by these agents include confusion, disorientation, restlessness, dizziness, staggering, or vomiting. Some may cause dryness of mouth, elevated temperature, pupil dilation, slurred/nonsensical speech, inappropriate behavior, and hallucinations. If several personnel exhibit any such behavior, it is prudent to move outside the building, because these agents are usually delivered by smoke-producing munitions or aerosols and are introduced through the respiratory system.

Irritants

Irritants can be classified as either tear-producing or vomiting-producing agents. The sole purpose of irritants, which include tear gas, riot control agents, and lachrimators, is to produce immediate discomfort and eye closure, thus rendering the victim incapable of fighting or resisting. Irritants cause eye discomfort, and some may cause vomiting; all are usually introduced to an environment as a gas. Police forces use irritants for crowd control. Irritants were used before World War I, and, during the war, they were the first chemical agents used, well before better-known agents such as chlorine, phosgene, and mustard gas.

Tear gas (CS) and chloroacetophenone (CN; sold in diluted form as a protective spray) are by far the most important pulmonary irritants. Capsaicin (methyl vanillyl nonenamide) is the active ingredient in pepper spray, also called OC (oleoresin capsicum). Pepper spray has, to some extent, replaced CN as a personal protective agent, with less dangerous effects. As its common name implies, the active ingredient is the burning agent in pepper plant fruits.

Although CS and CN are the most important agents in this class, several others require mention. Chloropicrin (PS) and bromobenzenecyanide (CA) were developed before World War I. Both largely have been replaced, because they were too lethal for their intended effects but not lethal enough to compete with the more effective blistering and nerve agents. PS still is used occasionally as a soil sterilant or grain disinfectant.

Toxic Chemical Agents

Nerve Agents. Nerve agents are organophosphate ester derivatives of phosphoric acid, and are among the deadliest of rapid-onset chemical agents. Nerve agents can be divided into G and V agents. **G agents** are fluorine- or cyanide-containing organophosphates. These agents are colorless and have an odor that ranges from weakly "fruity" to odorless. In an unmodified state, G agents are highly volatile, resulting in low persistency. However, they can be combined with various thickening substances, increasing persistency and penetration of intact skin. The primary hazard of G agents is vapor contact because of their high volatilities.

V agents are sulfur-containing organophosphates. These agents are low-volatile oily liquids, resulting in increased persistency. The increased persistency makes V agents a primarily contact hazard.

Common nerve agents include

- VX
- Tabun (GA)
- Sarin (GB)
- Soman (GD)
- Cyclosarin (GF)

Both G and V agents are potent inhibitors of the enzyme acetylcholinesterase (AChE) and present the same symptoms after exposure. Inhibiting AChE allows acetylcholine to accumulate, which mimics a massive release of acetylcholine in the nervous system. Nerve agents may be absorbed through any body surface (skin, eyes, respiratory) or ingested. Symptoms of nerve agent poisoning include

- Sweating and/or muscular twitching
- Pupil contraction, eye pain, or blurred vision
- Headache, pain
- Weakness
- Nausea, vomiting (particularly in ingestion)
- Mucous secretions in respiratory pathways, nose, or throat
- Wheezing, coughing
- Severe exposure: convulsions; vomiting; red, pinpoint eyes; unconsciousness; or respiratory failure

Mild exposure to nerve agents may cause anxiety, restlessness, and giddiness. Further exposure results in the listed symptoms and/or memory impairment, slowed reactions, or difficulty in concentration. Moderate exposure, if diagnosed and monitored, shows abnormalities in electroencephalograms (EEGs) as well as the symptoms listed. Reactions to nerve agents are immediate (i.e., within minutes of exposure). Recovery from nerve agent exposure is slow, usually days, and susceptibility to the agent is increased for months afterward.

Nerve agents are liquid at room temperature, but their volatilities can vary. Highly volatile agents (G agents) can be easily introduced as vapors into HVAC systems, whereas low-volatility agents (V and thickened G agents) can be introduced as droplets or vapors by mechanical means. Highly volatile agents are less persistent and require less intense cleanup than naturally persistent, highly volatile agents, which require intense cleanup if introduced into a building. For the most part, these agents are moderately soluble in water and highly soluble in lipids. They are rapidly inactivated by strong alkalis and chlorinating compounds, which are used in the decontamination/neutralization of these agents.

If nerve agents are suspected, evacuate the facility immediately. Because many nerve agents are (or can be made) persistent and dose is accumulative, evacuation is necessary. A facility must be decontaminated if exposed to nerve agents.

Blister Agents. Blister agents (vesicants) can be classified as mustards, arsenicals, and urticants. These agents are generally used as warfare agents meant to degrade fighting efficiency rather than to kill. They are usually thickened to make them persistent and contaminate surfaces, but may be introduced as a gas or vapor. Vesicants result in burns and blisters to the skin, eyes, and/or respiratory tract.

Mustard agents contain either sulfur or nitrogen and are persistent in cold and temperate conditions. They can be combined with other substances to thicken the agent, increasing their persistency. Warmer temperatures decrease persistency, but concentrations in air can be high because of the greater evaporation rate. Common mustard agents include

- Sulfur mustard (H and HD)
- Nitrogen mustards (HN)

Arsenical agents contain a central arsenic atom. These agents hydrolyze rapidly with water and lose most of their vesicant properties.

Arsenicals are more volatile than mustards and are less toxic than other blister agents. Common arsenical agents include

- Lewisite (L)
- Mustard-lewisite (HL)
- Phenyldichloroarsine (PD)

Urticants are halogenated oximes and have a disagreeable, penetrating odor. The most recognized urticant is phosgene oxime (CX), which is one of the most irritating substances known.

The most likely routes of exposure are inhalation, dermal contact, and ocular contact. Depending on the particular vesicant, clinical effects may occur immediately (as with phosgene oxime or lewisite) or may be delayed for 2 to 24 h (as with mustards). Blister agents must be cleaned from the skin and membranes immediately to lessen their effects. Persons exposed to blister agents must be handled so as not contaminate those helping them. Evacuation is necessary, and contaminated people should be kept outdoors to prevent accumulation of the vesicant in a confined space. Effects of exposure include

- Mild to severe conjunctivitis, possibly progressing to ulceration
- Lesions on skin; burns
- Itching
- Pain (immediate with exposure to lewisite)
- Respiratory damage (in small doses may take time to appear as bronchitis, etc.)

Vesicants are, for the most part, soluble in nonaqueous solvents, not in water. They are more dangerous as liquids, because the degree to which they cause health problems is related to their concentration on body surfaces. In general, they have high vapor pressures and thus are easily vaporized in a confined space. Decontamination of exposed surfaces is needed.

Lung-Damaging Agents. Lung-damaging (choking) agents are those that primarily attack lung tissue, causing pulmonary edema. Examples include

- Phosgene (CG)
- Diphosgene (DP)
- Chlorine
- Chloropicrin (PS)

As choking agents, most of these agents (except for diphosgene) exist as gases at room temperature and pressure, and are thus easily spread through ventilation systems. Exposure symptoms include

- Choking sensations, coughing
- Tightness in chest
- Nausea, vomiting
- Headache

Because these agents are gaseous, they will disperse. They all have specific odors; for example, CG smells like fresh-mown hay. Thorough ventilation of contaminated areas is necessary. Choking agent gases typically are heavier than air, and thus tend to accumulate in low-lying areas.

Blood Agents. Blood agents, also known as cyanogens, interfere with the absorption and use of oxygen at the cellular level, and thus are usually introduced through the respiratory system. Examples include hydrogen cyanide (AC) and cyanogen chloride (CK). These agents are highly volatile and gaseous at temperatures over 70°F and are nonpersistent even at low temperatures. They therefore dissipate quickly in air, especially hydrogen cyanide, which is light; cyanogen chloride is heavier than air and tends to collect in low places. These two blood agents have different symptoms. Symptoms of exposure to AC include

- Faint odor of almonds
- Internal hemorrhaging
- Pink skin color
- Highly toxic, high concentrations can cause immediate death

CK symptoms include

- Intense irritation to the lungs and eyes
- Coughing
- Tightness in chest
- Dizziness
- Unconsciousness
- Respiratory failure

Because these agents are not persistent, thorough ventilation should dissipate the gases.

Other HVAC-Compromising Gases and Vapors

Accidental contamination of facility HVAC systems by gaseous or vaporous chemical substances has been a real concern for years, mainly because of the extensive production, use, and transport of large quantities of hazardous materials for manufacturing purposes. Intentional contamination of a facility could be accomplished with chemicals other than the specific chemical agents discussed previously. Contamination inside a building should result in immediate evacuation. However, external contamination might entail evacuation to a more distant location or shelter in place (i.e., not evacuating). Contamination from an incident in the immediate vicinity of (but external to) a facility might require shutdown of the facility's HVAC system for a short period of time. For instance, a large corrosive spill nearby might necessitate staying in a building for protection while transportation is arranged (if not available immediately). This might occur at a school where children would be more susceptible to injury upon exiting the building, having no way to evacuate a safe distance. Because the situations and possibilities are so varied, this discussion is limited to more typical scenarios.

Toxic Gases. The most common toxic gas that might threaten a facility and its personnel is carbon monoxide (CO), which is colorless, odorless, and tasteless. Carbon monoxide is produced by incomplete combustion of fossil fuels (gas, oil, coal, wood) used in boilers, engines, oil burners, gas fires, water heaters, solid-fuel appliances, and open fires. Dangerous amounts of CO can accumulate when, as a result of poor installation, poor maintenance, or failure, an appliance's fuel is not burned properly, or when rooms are poorly ventilated and the carbon monoxide is unable to escape. Because CO has no smell, taste, or color, it is important to have good ventilation, maintain all appliances regularly, and have reliable detector alarms installed to give both a visual and audible warning in case of a dangerous build-up of CO. Scenarios involving toxic gases usually entail evacuation to a safe distance. HVAC systems normally require cleaning using clean purge air the through the ventilation distribution system.

Corrosive Substances. Corrosive gases and vapors encompass a large class of materials. A few are purely gaseous in nature at room conditions, but some vapors result from the vapor pressure created by a liquid (or solid) presence. Some examples of corrosive gases and vapors are given in Table 1.

Corrosive gases and vapors are hazardous to all parts of the body, although some organs (e.g., eyes, respiratory tract) are particularly sensitive. The magnitude of the effect is related to the solubility of the material in body fluids. Highly soluble gases (e.g., ammonia, hydrogen chloride) cause severe nose and throat irritation, whereas lower-solubility substances (e.g., nitrogen dioxide, phosgene, sulfur dioxide) can penetrate deep into the lungs. Exposed skin may also be at risk for irritation or burns at higher concentrations or longer-term exposures. For some substances, warnings such as odor or eye, nose, or respiratory tract irritation may be inadequate. Accidents involving corrosive substances inside or outside a building require cleanup and decontamination of the facility's HVAC system and other equipment, because of the substances' persistence. In some cases, physical damage to a building's infrastructure may result from exposure to corrosives (e.g., etching of metal surfaces, which can lead to holes in ducting and compromised wiring). Building

Table 1 Corrosive Gases and Vapors

Corrosive Gases	Corrosive Acidic Vapors	Corrosive Basic Vapors
Hydrogen cyanide	Hydrochloric acid	Sodium hydroxide
Ammonia	Sulfuric acid	Ammonium hydroxide
Sulfur dioxide	Nitric acid	Caustic soda
Chlorine	Hydrofluoric acid	Potassium hydroxide
Hydrogen bromide	Acetic acid	Other hydroxides
Boron trichloride	Other acids	
Monomethylamine		
Phosphorus pentafluoride		

codes outline methods of design and installation for mechanical and electrical systems in corrosive environments (NFPA 2005), but in buildings that are not classified as such, and thus are not constructed accordingly, systems may be damaged when exposed to corrosive chemicals. In such a case, the building and its systems should be thoroughly inspected and tested before reoccupation.

BIOLOGICAL INCIDENTS

Biological incidents involve the intentional or accidental release of unwanted bioaerosols and/or biocontaminants in or around a building, such that the building's integrity or usefulness is compromised. Bioaerosols are airborne particulates derived from living organisms and include living microorganisms, viruses, spores, and toxins derived from remnants or fragments of living tissue. Bioaerosols are in the air, both indoors and outdoors, and their presence mostly goes unnoticed except for seasonal allergies or an occasional cold. There is an evolved balance between the types and levels of bioaerosols in the ambient air and the animals who breathe that air. That balance can be disturbed locally by the purposeful or accidental release of a bioaerosol in or around a building. Unfortunately, bioaerosols are difficult to detect and identify in real time, because identification generally involves DNA analysis or other skilled analytical techniques. As a consequence, bioaerosols may be fully distributed in a building hours or days before anything is detected, much less identified, and the first sign of an incident may be symptoms of personnel.

There are hundreds of known bioaerosols that are pathogenic to humans to varying degrees. These include the spore-forming bacteria *Bacillus anthracis* (commonly known as anthrax), *Variola* spp. (the virus that causes smallpox), the bacteria *Yersinia pestis* (cause of bubonic plague), and many others. Human susceptibility varies by microorganism, and is gaged by several dose measures:

- **ID_{50}, mean infectious dose**, is the number of microorganisms or bioaerosol particles that causes 50% of an exposed population to be infected.
- **LD_{50}, mean lethal dose**, is the number of microorganisms or bioaerosol particles that causes death in 50% of an exposed population.

One of the greatest threats in a biological incident or attack is toxins, which are poisonous chemicals produced by living organisms and that may have effects resembling those of chemical agents. Toxicity and lethality of toxins vary, but highly toxic, stable toxins pose a risk for weaponization. Two classifications of potentially threatening toxins are neuro- and cytotoxins. **Neurotoxins** interfere with nerve impulse transmission and have significant effects on the nervous system. However, they can work in different manners, inhibiting or stimulating various enzymes and blocking various receptors. Effects are similar to those of chemical nerve agents, and include convulsions or paralysis, blurred vision, seizures, and muscle fatigue. **Cytotoxins** disrupt or destroy cells and cellular processes such as protein synthesis and other biochemical process. Symptoms may be similar to those of chemical blister, choking, and

Table 2 Limited List of Human Pathogenic Microorganisms

Bioaerosol	Incubation Period, Days	ID_{50}, Organisms	LD_{50}, Organisms
Bacillus anthracis (anthrax)	2 to 3	10,000	28,000
Ebolavirus spp. (Ebola)	14 to 21	10	Low
Francisella tularensis (tularemia)	1 to 14	10	Low
Hantavirus (Hanta)	14 to 30	N/A	N/A
Variola spp. (smallpox)	12	N/A	N/A
Yersinia pestis (bubonic plague)	2 to 6	N/A	N/A

vomiting agents, as well as nausea, diarrhea, rashes, inflammation, and necrosis. Toxins can be produced by a variety of organisms such as bacteria, fungi, mold, algae, plants, and animals.

A summary of potential bioaerosol weapon agents is given in Table 2 along with ID_{50} and LD_{50} values. A more comprehensive list can be found in Kowalski (2003). More detailed descriptions of bioaerosols, their health effects, and methods of measurement can be found in most epidemiology texts. *Bioaerosols: Assessment and Control*, from the American Conference of Governmental Industrial Hygienists (ACGIH 1999), is a recommended starting point for quantitative determination of bioaerosol levels. General information on bioaerosols, their health effects, and their removal from building airstreams is found in Chapters 10 and 11 of the 2009 *ASHRAE Handbook—Fundamentals*, as well as Chapter 28 of the 2008 *ASHRAE Handbook—HVAC Systems and Equipment*.

The primary threat to buildings from airborne biological incidents is adverse health effects to building occupants. Once an event happens, there is an immediate danger to building occupants from the initial dose, but there is also the risk of prolonged exposure from contaminated surfaces and reaerosolization of the agent. Depending on the agent, the prolonged exposure risk may dissipate quickly if the pathogenic organism has a short life outside a host, or it may remain indefinitely until the contaminating agent is fully removed. Anthrax, which is a spore-forming bacterium, falls into this latter category because it can lie dormant in many environments for long periods of time, only to come out of dormancy when exposed to a proper host. Excluding incidents of extreme mold growth [which is not covered in this chapter; see, e.g., ASHRAE (2003b) for information], biological incidents present no real threat to the integrity of building equipment; however, building equipment may play an important role in both distribution and possible removal of air contaminants. Remediation after an incident is likely to involve comprehensive cleaning of building equipment (particularly air-handling equipment), and may require removal and replacement of contaminated systems. Techniques for remediating contaminated equipment include surface cleaning with bleach or alcohol solutions, treatment with ultraviolet (UV) light, and volume gaseous treatments with ozone, hydrogen peroxide, or gas plasma. New technologies and procedures are under development, particularly since the anthrax events of 2001 in the United States. See the Bibliography and Online Resources for sources of information on the latest developments in remediation technology.

It is important to determine as much as possible about a release, whether purposeful or accidental, as rapidly as possible. It may be more difficult to completely assess the nature of a purposeful release, because it may contain more than one pathogenic agent, with different incubation periods, and the release may have taken place in several locations. Accidental releases are more likely to be a single pathogen at a single location, and the release is more likely to have been known to occur.

Biological pathogens have been weaponized to enable delivery in a variety of forms. Effective delivery of bioagents to a large population is difficult because of the need to get relatively large doses to large numbers of people. Dilution of contaminants in ambient air is rapid, and very large numbers of organisms are required to produce lethal concentrations. The confines of a building and controlled air exchanges rates can help maintain concentrations of agents for longer periods of time than would occur in outdoor air. However, filtration and real-time killing mechanisms in building air-handling systems can remove or render ineffective airborne bioaerosols. Engineering requirements for design of filtration or other techniques for treating indoor air are addressed more fully in other publications [e.g., NIOSH (2003)]. Information is rapidly evolving; for the latest, consult the most recent versions of publications on building protection.

RADIOLOGICAL INCIDENTS

The occurrence of a significant accidental or intentional radiological release to the environment is of low probability because there are limited locations where considerable amounts of radiological material reside. These sources include spent fuel or low-level radioactive waste (radwaste) storage facilities, nuclear generating stations, and weapons fabrication and storage facilities. These facilities are usually analyzed beforehand, as part of the construction licensing process, for postulated accidental releases of radiological material and the consequences to both on- and off-site personnel.

An intentional release of radiological material is most likely to be in the form of the deployment of a nuclear weapon or a radiological dispersal device (RDD), sometimes called a dirty bomb. It is normally assumed that a terrorist group is highly unlikely to possess and use conventional, sophisticated nuclear weapons because of the difficulties of obtaining or independently developing the necessary materials and technology. Development and deployment of an RDD, however, is considered viable because of its simplicity of design. RDDs combine conventional explosives and radioactive material, and are designed to scatter dangerous amounts of radioactive material over a general area. Terrorist use of RDDs also seems more likely because radiological materials used in medicine, agriculture, industry, and research are comparatively more obtainable than weapons-grade uranium or plutonium. A significant amount of the damage from an RDD would be from the initial blast. See the section on Explosive Incidents for design of HVAC system protection against blast effects.

RADIOACTIVE MATERIALS' EFFECTS AND SOURCES

Decay of radioactive materials produces energetic emissions (ionizing radiation) that can effect changes in human tissue cells. These energetic emissions are divided into **alpha** particles, **beta** particles, and **gamma/x-rays**. Alpha and beta radiation can only travel very short distances (several feet, maximum) and do not have enough energy to penetrate the outer layers of human skin; they are, however, a hazard if directly inhaled or ingested. Gamma and x-rays can travel long distances in air and can pass through the body, potentially exposing internal organs to significant damage, depending on the amount absorbed. The radiation effect on humans is usually measured in röntgen equivalent man (rem), the product of the absorbed dose and the biological efficiency of the radiation.

In developing an RDD, a significant quantity of radioactive material must first be collected. Some common radioactive materials currently used in industry include

- Colbalt-60 (Co-60), cesium-137 (Cs-137), and iridium-192 (Ir-192) are used in cancer therapy, industrial radiography and gages, food irradiation, oil well production, and medical implants. These are all considered gamma emitters.
- Strontium-90 (Sr-90) is used in the production of radioisotope thermoelectric generators (RTGs), which produce electricity for remote devices such as spacecraft. This is considered a beta emitter.

- Plutonium-238 (Pu-238) and americium-241 (Am-241) are used in oil well production, RTGs, and industrial gauges. These are considered alpha emitters.

RADIOLOGICAL DISPERSION

Dispersion may be by conventional explosives, using aircraft to disperse the material in the form of an aerosol or particulate, or simply placing a container of radioactive material within a confined area or facility. In most cases, a dirty bomb or other RDD would have localized effects (based on the strength of material used) ranging from less than a city block to several square miles. The area affected by the dispersion of the material is a function of various factors, including

- Meteorological conditions, including atmospheric stability and wind speed
- Local topography, location of buildings, and other landscape characteristics
- Amount and type of radioactive material dispersed
- Dispersal mechanism (e.g., particulate, aerosol)
- Physical and chemical form of the radioactive material (e.g., dispersal as fine particles versus heavier droplets or particulate)

Radioactive material released as either an aerosol or fine particulate in a plume spreads roughly at the speed and direction of the prevailing wind velocity. The conditions of atmospheric stability (sometimes referred to in terms of the Pasquill stability classification) also determine the fallout's overall spread and concentration. Atmospheric dispersion computer models are sometimes used to predict the spread, location, and concentration of a postulated radioactive plume. These analytical tools can be useful in providing early warning to residences and facilities projected to be in the affected fallout path.

RADIATION MONITORING

Radioactivity cannot be seen, smelled, or tasted by humans. However, in the United States, there are many radiation-monitoring programs available at the federal, state, and local level that can measure radiation levels and/or track the released radiation plume. There may be a local facility (e.g., a nuclear power plant) that can track radiological fallout in the affected area. State-level officials have access to various monitoring programs for their areas. These programs use current weather patterns and wind velocities to track radiological plumes and provide public warning. Depending on the severity of the release, the radioactive plume may travel several hundred miles or, more typically, be localized.

FACILITY RESPONSE

Physical safety of personnel should be of primary concern in responding to a radiological event. Time, distance, and shielding are the three most important aspects to minimizing the effects of human exposure to ionizing radiation. Shielding with stone, concrete, or other dense materials is usually not considered in the initial design of most commercial buildings. However, most facilities use these materials for structural strength (foundations and basements) or for fire protection in protective corridors and stairways. Procedures should be considered to instruct building occupants to immediately move to identified safe locations in the facility in the event of a known or suspected release. Limiting the time of exposure to a radiation field also helps reduce the total amount of exposure and subsequent health effects. Distance from the radiation source is the greatest factor in reducing the amount of direct (deep-dose) exposure, especially if the hazard is present for a considerable time period. The distance required to minimize this dose is usually small (several feet).

Perhaps the greatest potential hazard from the release of low-level radiological debris from an RDD is inhalation or ingestion.

Health effects from ingested contamination particles that enter the body from breathing, cuts/abrasions, eating, or drinking can be more severe than external exposure, depending on the amount of material consumed. Individuals should be instructed to move to more isolated rooms, or areas in the facility that may be isolated or filtered from significant inleakage of contaminated air. Distribution of personal protection equipment (PPE) should also be considered, depending on the risk assessment of a potential threat. If the release is determined to be internal, an organized evacuation should occur (consider developing evacuation procedures). Because evacuees may have become externally contaminated, an egress plan should be considered that includes a radiation detection monitoring procedure implemented either by internal personnel or by local emergency authorities. This procedure may include instruction to dispose of outer clothing and the use of showers to remove radioactive particles from body surfaces.

EXPLOSIVE INCIDENTS

Detonation of high explosives near a building generates pressures that act on all exposed surfaces. The magnitude of pressure depends on the size and shape of the charge, distance from the charge, and any intervening barriers. In addition to increased pressures, blasts may generate projectiles from either loose materials or fragments from damaged components. This section considers only loads that do not generate significant structural damage; it is assumed that, if the structure is severely damaged, continued operation of the HVAC systems is not crucial. Also, it is important to remember that life safety is the primary goal of all protective systems. This section deals strictly with HVAC equipment and systems, but any solutions must not compromise the safety of building occupants.

LOADING DESCRIPTION

Detonation of high explosives generates a pressure wave (blast wave) that propagates out from the explosion with decreasing velocity. The free-field blast wave, far from any surfaces, is characterized by a rapid, almost instantaneous increase in pressure, followed by a gradual decay in pressure and a negative-pressure phase. A typical free-field blast overpressure P_{so}, reaching the target at time t_A, is shown in Figure 2. When the blast wave impinges on a rigid surface, the pressure is reflected (P_r) and magnified over the free-field values. Peak pressure varies inversely with the cube of the distance from the explosion. Intervening barriers may reduce the blast load, but quantifying the effect is difficult, and great care must be taken when determining the resulting loads.

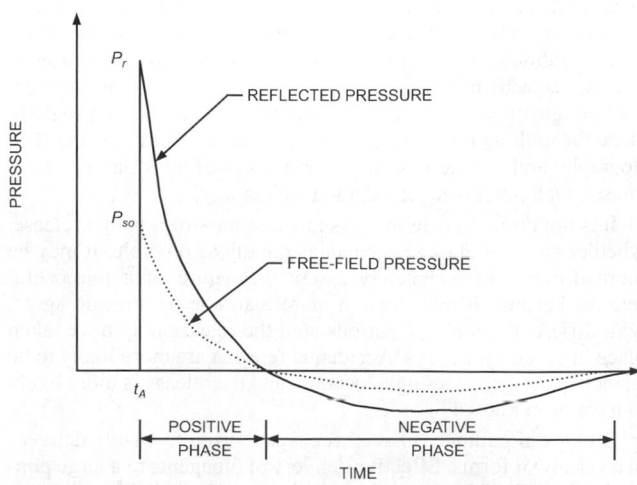

Fig. 2   Free-Field and Reflected Pressure Wave Pulses

Internal explosions can generate extremely large loads on HVAC equipment. In addition to short-duration reflected pressures generated by the explosion, a quasi-static pressure may develop, greatly increasing the impulse to which the equipment is exposed. The magnitude and duration of quasi-static loading is a function of room volume and vented area.

In addition to direct air blast, HVAC equipment in buildings subjected to explosions can experience large accelerations and relative displacements caused by the resulting structure motion. The structure responds to loads imposed both through the foundation (ground shock) and from the air blast (air shock). In general, accurately predicting the time history of motion in the building is extremely difficult. However, because structural design of equipment is based on peak loads, there is usually no need to determine the full time history. Methods are available for predicting the peak motion caused by a given event, including the frequency dependence of the response.

Blast loads can also enter the interior space of buildings through utility openings, even if the building shell is undamaged. The blast can damage the passageway itself, and can also build up pressure in the building and subsequently damage other equipment. Pressure build-up depends upon the opening area and volume of interior space, as well as the pressure differential.

Explosions can produce primary and secondary fragments that may damage equipment and piping. Primary fragments are generated from the explosive casing, whereas secondary fragments result from damage to the structure and nonstructural components (e.g., concrete spalling, glass breakage). Predictions of fragment size and velocity can be used to estimate damage to equipment, ducts, and piping.

DESIGN CONSIDERATIONS

The best protection from any explosive effects is to locate equipment in a nonvulnerable area. Assuming the exterior building shell is adequately designed, any interior room may be considered protected. Barrier walls can protect equipment that must be externally located, or equipment can be positioned far enough away from any possible blast location that pressures are reduced below damaging levels. In general, correctly locating equipment is the least expensive option for handling blast loads. Hardening equipment, anchorage, and connection should be considered only after relocation has been eliminated as a possibility.

In addition to the direct air blast, equipment in a building subject to a blast experiences a support shock loading. Guidelines for the maximum shock that can be withstood by various equipment types are given in Chapter 55. Although general information can be obtained from these tables, it is important to realize that the data are several decades old and may not apply to modern HVAC equipment. If the shock is greater than the equipment's capacity, it may be possible to provide a shock isolation system to lower the demand. The isolation system works by decoupling horizontal motion of the support structure from that of the equipment. Note that this is different from typical equipment isolation applications, which are generally concerned with vertical vibrations of the equipment itself.

Although it has been stated that proper seismic design also protect against blast loads, this is not generally true. The effect of blast loading on equipment has some similarities to that of seismic loads, but there are some key differences. Both loads generate horizontal and vertical forces that act on equipment. However, in both magnitude and distribution, seismic forces are proportional to equipment mass, whereas blast loads are proportional to the equipment surface area. The effect is essentially the same for some types of equipment, such as pumps, where the mass and surface area distributions are approximately identical. However, equipment covered by a sheet metal shell is loaded very differently. Seismic loads are applied directly to the heavy components in the shell, whereas blast loads are applied to the shell itself, with little or no load acting directly on the interior components. Thus, even equipment that has been seismically rated or certified needs additional investigation for blast resistance.

In contrast, anchorage design is identical for blast and seismic loads. Properly designed seismic anchorage for most equipment in moderate to high seismic zones is adequate for reasonable levels of blast loading. In either case, loads applied to equipment are used to determine shear and uplift loads on the anchorage that are checked against the allowable load for the specific attachment hardware. Although in reality the dynamic reactions from the resulting equipment motion are the actual anchorage force, for nonisolated equipment it is usually conservative to assume a static distribution of the maximum applied loads. This is not always the case for isolated equipment, because resonance of the load, equipment, and isolation system may produce dynamic forces well above those predicted from a static analysis.

When designing HVAC systems for blast load, it is important to remember that there are two types of failure. The first is a temporary loss of service, such as might be caused by tripping a breaker. The second, more serious case involves actual damage to the equipment or system. It is important to determine which scenario is important for the system, and design accordingly. Preventing temporary outages is, in general, much more expensive than preventing a catastrophic failure.

Exposed piping and ductwork are also subject to both pressure and fragment loading. Pressure loading can be carried through proper selection and spacing of supports. The flexure and shear capacities of the pipe or duct, and the capacity of the support, determine the spacing. Fragment effects are more difficult to analyze, because the exact size and velocity of any fragments are impossible to predict. The only way to fully protect against fragments is to locate the pipe or duct where fragment impact is not possible.

Openings in the building for HVAC or other purposes must also be designed for blast effects. HVAC systems can be damaged by pressure propagated through the opening or blockage of the opening. Grilles or louvers can be analytically designed to resist the blast load, preventing blockage and allowing continued operation. Additionally, the pressure increase in ducts, and subsequently in interior rooms, can be calculated. Properly designed silencers may reduce pressure in ducts. Design of openings for blast resistance must also be closely coordinated with protection from chemical and biological agents.

REFERENCES

ASHRAE. 2003a. *Risk management guidance for health, safety and environmental security under extraordinary incidents.* Report of Presidential Ad Hoc Committee for Building Health and Safety under Extraordinary Incidents.

ASHRAE. 2003b. *Mold and moisture management in buildings.*

ASHRAE. 2009. Guideline for the risk management of public health and safety in buildings. *Guideline* 29-2009.

NFPA. 2011. National electric code®. *Standard* 70. National Fire Protection Association, Quincy, MA.

NIOSH. 2003. *Guidance for filtration and air-cleaning systems to protect building environments from airborne chemical, biological, or radiological attacks.* U.S. Department of Health and Human Services, National Institute for Occupational Safety and Health, Washington, D.C. Available at http://www.cdc.gov/niosh/docs/2003-136/pdfs/2003-136.pdf.

U.S. Departments of the Army, Navy, Air Force, and Marine Corps. 2005. Potential military chemical/biological agents and compounds. *Field Manual* FM 3-11.9. Available from http://www.train.army.mil.

BIBLIOGRAPHY

ACGIH. 1999. *Bioaerosols: Assessment and control.* J. Macher, ed. American Conference of Governmental Industrial Hygienists, Cincinnati.

Archibald, R.W., J.J. Medby, B. Rosen, and J. Schachter. 2002. *Security and safety in Los Angeles high-rise buildings after 9/11.* RAND, Santa Monica, CA.

ASCE. 1999. *Structural design for physical security: State of the practice.* American Society of Civil Engineers, Reston, VA.

DHS. 2003. *Radiological dispersion devices fact sheet.* U.S. Department of Homeland Security, Washington, D.C. Available at http://www.dhs.gov/xnews/releases/press_release_0085.shtm.

DHS. 2009. *National infrastructure protection plan.* U.S. Department of Homeland Security, Washington, D.C. Available from http://www.dhs.gov/xlibrary/assets/NIPP_Plan.pdf.

FEMA. 2003. Reference manual to mitigate potential terrorist attacks against buildings. *Report* 426. U.S. Federal Emergency Management Agency, Washington, D.C. Available from http://www.fema.gov/plan/prevent/rms/rmsp426.

FEMA. 2004. Primer for design of commercial buildings to mitigate terrorist attacks. *Report* 427. U.S. Federal Emergency Management Agency, Washington, D.C. Available from http://www.fema.gov/library/viewRecord.do?id=1560.

FEMA. 2003. Primer to design safe school projects in case of terrorist attacks. *Report* 428. U.S. Federal Emergency Management Agency, Washington, D.C. Available from http://www.fema.gov/library/viewRecord.do?id=1561.

FEMA. 2005. Risk assessment: A how-to guide to mitigate potential terrorist attacks against buildings. *Report* 452. U.S. Federal Emergency Management Agency, Washington, D.C. Available from http://www.fema.gov/library/viewRecord.do?id=1938.

Hunter, P. and S.T. Oyama. 2000. *Control of volatile organic compound emissions: Conventional and emerging technologies.* John Wiley & Sons, Hoboken, NJ.

Ketchum, J.S. and F.R. Sidell. 1997. Incapacitating agents. Ch. 11 in *Medical aspects of chemical & biological warfare.* U.S. Office of the Surgeon General, Medical Nuclear, Biological, and Chemical (NBC). Available from http://www.bordeninstitute.army.mil/published_volumes/chembio/chembio.html.

Kowalski, W.J. 2003. *Immune building systems technology.* McGraw-Hill, New York.

LBL. 2005. *Advice for safeguarding buildings against chemical or biological attack.* Lawrence Berkeley National Laboratory, Berkley. Available at http:// securebuildings.lbl.gov.

MFMER. 2005. *Biological and chemical weapons: Arm yourself with information.* MH00027. Mayo Foundation for Medical Education and Research, Rochester, MN.

National Academies. 2004. Radiological attack—Dirty bombs and other devices. *News & Terrorism: Communicating in a Crisis.* National Academies and U.S. Department of Homeland Security, Washington, D.C. Available at http://www.nae.edu/File.aspx?id=11317.

NFPA. 2010. Disaster/emergency management and business continuity programs. *Standard* 1600. National Fire Protection Association, Quincy, MA.

NIOSH. 2002. *Guidance for protecting building environments from airborne chemical, biological or radiological attacks.* U.S. Department of Health and Human Services, National Institute for Occupational Safety and Health, Washington, D.C. Available from http://www.cdc.gov/niosh/docs/2002-139/.

ORNL. 2005. *Mitigation of CBRN incidents for HVAC systems in federal facilities.* ORNL/TM-2004/260. Oak Ridge National Laboratory, Oak Ridge, TN. Available at http://www.ornl.gov/~webworks/cppr/y2001/rpt/121921.pdf.

Price, P.N., M.D. Sohn, A.J. Gadgil, W.W. Delp, D.M. Lorenzetti, E.U. Finlayson, T.L. Thatcher, R.G. Sextro, E.A. Derby, and S.A. Jarvis. 2003. *Protecting buildings from a biological or chemical attack: Actions to take before or during a release.* LBNL-51959. Lawrence Berkeley National Laboratory, Berkeley. Available at http://securebuildings.lbl.gov/images/BldgAdvice.pdf.

Sidell, F.R., W.C. Patrick IIII, and T.R. Dashiell. 1998. *Jane's chem-bio handbook,* 3rd ed. Jane's Information Group, Alexandria, VA.

U.S. Department of the Army. 1986. Fundamentals of protective design for conventional weapons. *Technical Manual* TM 5-855-1. Washington, D.C.

U.S. Departments of the Army, Navy, and Air Force. 1990. Structures to resist the effects of accidental explosions, rev. 1. Department of the Army *Technical Manual* TM 5-1300, Department of the Navy *Publication* NAVFAC P-397, Department of the Air Force *Manual* AFM 88-22. Available from http://www.ddesb.pentagon.mil/tm51300.htm.

U.S. Departments of the Army, Navy, and Air Force, and Commandant, Marine Corps. Treatment of chemical agent casualties and conventional military chemical injuries. *Field Manual* FM8-285. Falls Church, VA. Available from http://www.globalsecurity.org/wmd/library/policy/army/fm/8-285/.

U.S. Department of Defense. 2008. Security engineering: Procedures for designing airborne chemical, biological, and radiological protection for buildings. *Unified Facilities Criteria* UFC 4-024-01. Available from http://www.wbdg.org.

U.S. NRC. 2007. Meteorological programs in support of nuclear power plants. *Regulatory Guide* 1.23. U.S. Nuclear Regulatory Commission. Available at http://www.nrc.gov/reading-rm/doc-collections/reg-guides/power-reactors/rg/01-023/01-023r1.pdf.

ONLINE RESOURCES

U.S. Centers for Disease Control and Prevention: www.bt.cdc.gov/ Emergency Preparedness and Response

U.S. Department of Homeland Security, Ready Web site: www.ready.gov/

ULTRAVIOLET AIR AND SURFACE TREATMENT

ULTRAVIOLET germicidal irradiation (UVGI) uses short-wave ultraviolet (UVC) energy to inactivate viral, bacterial, and fungal organisms so they are unable to replicate and potentially cause disease. UVC energy disrupts the DNA of a wide range of microorganisms, rendering them harmless (Brickner et al. 2003; CIE 2003). Early work established that the most effective UV wavelength range for inactivation of microorganisms was between 220 and 280 nm, with peak effectiveness near 265 nm. The standard source of UVC in commercial systems is low-pressure mercury vapor lamps, which emit mainly near-optimal 253.7 nm UVC. Use of germicidal ultraviolet (UV) lamps and lamp systems to disinfect room air and air streams dates to about 1900 (Reed 2010). Riley (1988) and Shechmeister (1991) have written extensive reviews of UVGI disinfection. Application of UVGI is becoming increasingly frequent as concerns about indoor air quality increase. UVGI is now used as an engineering control to interrupt the transmission of pathogenic organisms, such as *Mycobacterium tuberculosis* (TB), influenza viruses, mold, and potential bioterrorism agents (Brickner et al. 2003; CDC 2002, 2005; GSA 2010; McDeVitt et al. 2008; Rudnick et al. 2009).

UVC lamps and lamp systems are placed in air-handling systems and in room settings for the purpose of air and surface disinfection (Figure 1). The control of bioaerosols using UVGI can improve indoor air quality (IAQ) and thus enhance occupant health, comfort, and productivity (ASHRAE 2009; Menzies et al. 2003). Detailed descriptions of UVGI components and systems are given in Chapter 16 of the 2008 *ASHRAE Handbook—HVAC Systems and Equipment*. Upper-air (or upper-room) systems are installed in occupied spaces to control bioaerosols (suspended viruses, bacteria, and fungi contained in droplet nuclei and other carrier particles) in the space. In-duct systems are installed in air-handling units to control bioaerosols in recirculated return air that may be collected from many spaces, and to control microbial growth on cooling coils and other surfaces. Keeping the coils free of biofilm buildup can help reduce pressure drop across the coils (and therefore the energy required to move the air), and eliminates one potential air contamination source that could degrade indoor air quality. UVGI is typically combined with conventional air quality control methods, including dilution ventilation and particulate filtration, to optimize their cost and energy use (Ko et al. 2001).

This chapter discusses these common approaches to the application of UVGI systems. It also surveys the most recent UVGI design guidelines, standards, and practices and discusses energy use and economic considerations that arise when applying UVGI systems.

FUNDAMENTALS

Ultraviolet energy is electromagnetic radiation with a wavelength shorter than that of visible light and longer than x-rays

(Figure 2). The International Commission on Illumination (CIE 2003) defines the UV portion of the electromagnetic spectrum as radiation having wavelengths between 100 and 400 nm. The UV spectrum is further divided into UV-A (wavelengths of 400 to 315 nm), UV-B (315 to 280 nm), and UV-C (280 to 100 nm) (IESNA 2000). UVGI applications discussed in this chapter typically use UVC from 200 to 280 nm. The optimal wavelength for inactivating microorganisms is 265 nm (Figure 3). As shown in Figure 3, the germicidal effect decreases rapidly if the wavelength is not optimal.

UV Dose and Microbial Response

This section is based on Martin et al. (2008).

UVGI inactivates microorganisms by damaging the structure of nucleic acids and proteins at the molecular level, making them incapable of reproducing. The most important of these is deoxyribonucleic acid (DNA), which is responsible for cell replication (Harm 1980). The nucleotide bases (pyrimidine derivatives thymine and cytosine, and purine derivatives guanine and adenine) absorb most of the UV energy responsible for cell inactivation (Diffey 1991; Setlow 1966). Absorbed UV photons can damage DNA in a variety of ways, but the most significant damage event is the creation of pyrimidine dimers, where two adjacent thymine or cytosine bases bond with each other, instead of across the double helix as usual (Diffey 1991). In general, the DNA molecule with pyrimidine dimers is unable to function properly, resulting in the organism's inability to replicate or even its death (Diffey 1991; Miller et al. 1999; Setlow 1997; Setlow and Setlow 1962). An organism that cannot reproduce is no longer capable of causing disease.

UVGI effectiveness depends primarily on the UV dose (D_{UV}, $\mu J/cm^2$) delivered to the microorganisms:

$$D_{UV} = It \tag{1}$$

where I is the average irradiance in $\mu W/cm^2$, and t is the exposure time in seconds (note that 1 J = 1 W/s). Although Equation (1) appears quite simple, its application can be complex (e.g., when calculating the dose received by a particle following a tortuous path through a device in which the irradiance varies spatially). The dose is generally interpreted as that occurring on a single pass through the device. Although the effects of repeated exposure of microorganisms entrained in recirculated air may be cumulative, this effect has not been quantified and it is conservative to neglect it.

The survival fraction S of a microbial population exposed to UVGI is an exponential function of dose:

$$S = e^{-kD_{UV}} \tag{2}$$

where k is a species-dependent deactivation rate constant, in $cm^2/\mu J$. The resulting single-pass inactivation rate η is the complement of S:

$$\eta = 1 - S \tag{3}$$

The preparation of this chapter is assigned to TC 2.9, Ultraviolet Air and Surface Treatment.

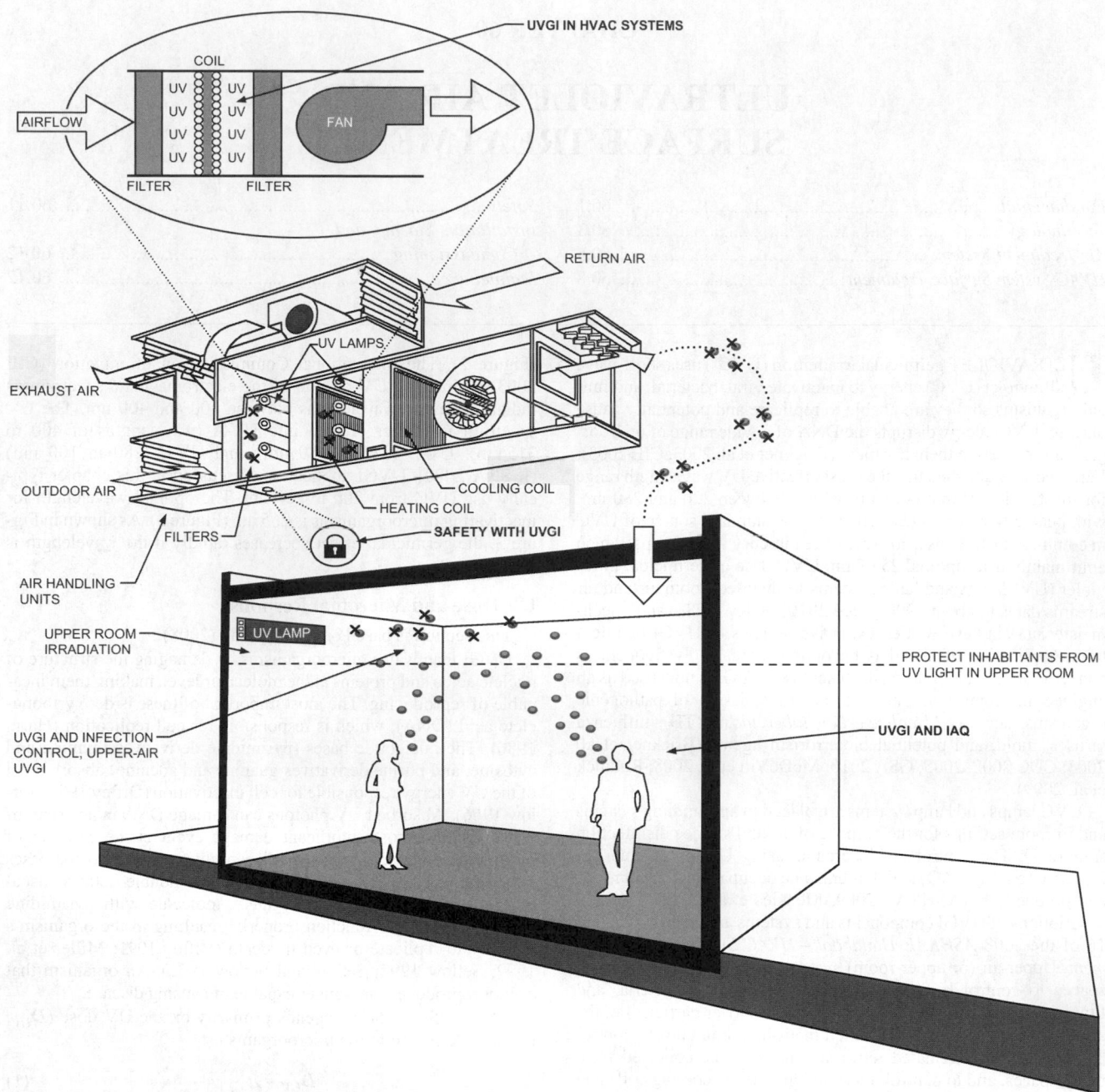

Fig. 1   Potential Applications of UVGI to Control Microorganisms in Air and on Surfaces
(ASHRAE 2009)

and is a commonly used indicator of overall UVGI effectiveness, representing the percentage of the microbial population inactivated after one pass through the irradiance field(s).

Deactivation rate constants (k-values) are species-dependent and relate the susceptibility of a given microorganism population to UV radiation (Hollaender 1943; Jensen 1964; Sharp 1939, 1940). Measured k-values for many species of viruses, bacteria, and fungi have been published in the scientific literature and previously summarized (Brickner et al. 2003; Kowalski 2009; Philips 2006). As shown in Figure 3, bacteria are generally more susceptible to UVGI than fungi, but this is not always the case (see Chapter 16 of the 2008 *ASHRAE Handbook—HVAC Systems and Equipment*). It is more difficult to generalize when it comes to viruses. Reported k-values

for different species of microorganisms vary over several orders of magnitude. Consequently, choosing which k-value to use for UVGI system design is often difficult and confusing. The variation in reported k-values makes generalizing the use of Equation (2) particularly complicated for heterogeneous microbial populations. Even accurately determining S for one specific microorganism can be difficult, because the reported k-values for the same species sometimes differ significantly.

Variations in published k-values may relate to differences in conditions under which the UV irradiance was conducted (in air, in water, or on surfaces), the methods used to measure the irradiance level, and errors related to the microbiological culture-based measurements of microbial survival (Martin et al. 2008). Because no

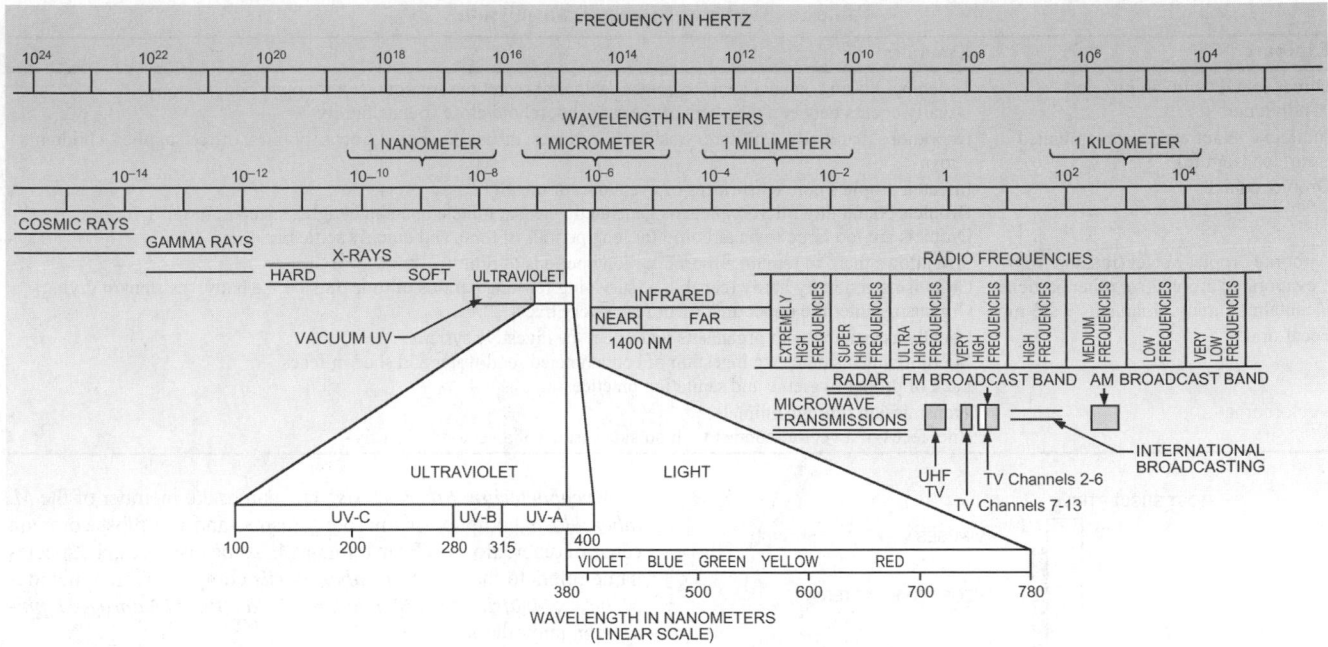

Fig. 2 Electromagnetic Spectrum
(IESNA 2000)

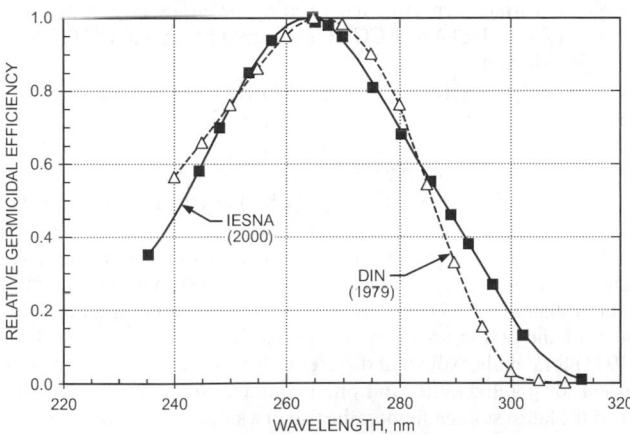

Fig. 3 Standardized Germicidal Response Functions

standard methods are currently available for the determination of deactivation rate constants, care needs to be taken when applying values reported in the literature to applications under different environmental conditions.

UV Inactivation of Biological Contaminants

The focus of this chapter is application of UVGI to deactivate microorganisms, specifically bacteria, fungi, and viruses on surfaces and in air streams. The application of UVGI for upper room treatment generally applies to pathogenic bacteria and viruses. Under some circumstances, these pathogens have the potential to be transmitted throughout the HVAC system.

As shown in Table 1, infectious diseases can be transmitted by a variety of means. UVC is effective against microorganisms in the air that flows through the UVGI field and on irradiated surfaces.

As shown in Table 2 and Figure 4, viruses and vegetative bacteria are the generally most susceptible, followed by the Mycobacteria, then bacterial spores and finally the fungal spores, the most

resistant. Within each group, an individual species may be significantly more resistant or susceptible, so care should be taken to use this ranking only as a guideline. It should be noted that the spore-forming bacteria and fungi also have vegetative forms, which are markedly more susceptible to inactivation than are the spore forms. Viruses are a separate case. As a group, their susceptibility to inactivation is even broader than for the bacteria or fungi.

TERMINOLOGY

Just as it is customary to express the size of aerosols in micrometers and electrical equipment's power consumption in watts, regardless of the prevailing unit system, it is also customary to express total lamp UVC output, UVC fluence, and UVC dose using SI units.

Multiply I-P	By	To Obtain SI
Btu/ft^2 (International Table)	1135.65	μJ/cm^2
Btu/h·ft^2	315.46	μW/cm^2
To Obtain I-P	**By**	**Divide SI**

Burn-in time. Period of time that UV lamps are powered on before being put into service, typically 100 h.

Cutaneous damage. Any damage to the skin, particularly that caused by exposure to UVC energy.

Disinfection. Compared to sterilization, a less lethal process of inactivating microorganisms.

Droplet nuclei. Residual viable microorganisms in air, following evaporation of surrounding moisture. These microscopic particles are produced when an infected person coughs, sneezes, shouts, or sings. The particles can remain suspended for prolonged periods and can be carried on normal air currents in a room and beyond to adjacent spaces or areas receiving exhaust air.

Erythema (actinic). Reddening of the skin, with or without inflammation, caused by the actinic effect of solar radiation or artificial optical radiation. See CIE (1987) for details. (Nonactinic erythema can be caused by various chemical or physical agents.)

Exposure. Being subjected to infectious agents, irradiation, particulates, or chemicals that could have harmful effects.

Table 1 Modes of Disease Transmission

Exposure	Examples
Direct contact with an infected individual	Touching, kissing, sexual contact, contact with oral secretions, or contact with open body lesions
	Usually occurs between members of the same household/close friends/family
Indirect contact with a contaminated surface (fomite)	Doorknobs, handrails, furniture, washroom surfaces, dishes, keyboards, pens, phones, office supplies, children's toys
Droplet contact	Infected droplets contact surfaces of eye, nose, or mouth
	Droplets containing microorganisms generated when an infected person coughs, sneezes, or talks
	Droplets are too large to be airborne for long periods of time, and quickly settle out of air
Airborne droplet nuclei (residue from evaporated droplets) or other particles containing microorganisms ~ ≤ 5 μm	Size allows them to remain airborne for long periods of time
	Organisms generally hardy (capable of surviving for long periods of time outside the body, resistant to drying)
	Organisms enter the upper and lower respiratory tracts
Fecal-oral	Usually associated with organisms that infect the digestive system
	Microorganisms enter via ingestion of contaminated food/water and shed in feces
	Lack of proper hygienic and sanitation practices
Vectorborne	Transmission through animals
	Bite, feces of a vector, contact with outside surface of a vector (e.g., a fly)

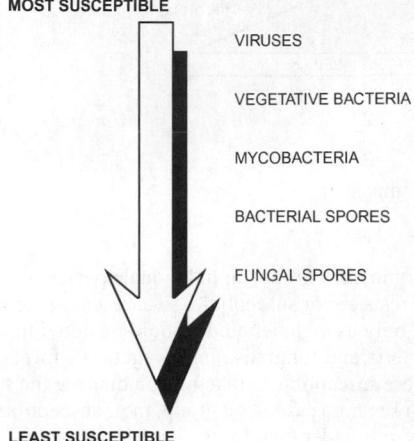

Fig. 4 General Ranking of Susceptibility to UVC Inactivation of Microorganisms by Group

Table 2 Representative Members of Organism Groups

Organism Group	Member of Group
Vegetative Bacteria	Staphylococcus aureus
	Streptococcus pyogenes
	Escherichia coli
	Pseudomonas aeruginosa
	Serratia marcescens
Mycobacteria	Mycobacterium tuberculosis
	Mycobacterium bovis
	Mycobacterium leprae
Bacterial Spore	Bacillus anthracis
	Bacillus cereus
	Bacillus subtilis
Fungal Spores	Aspergillus versicolor
	Penicillium chrysogenum
	Stachybotrys chartarum
Viruses	Influenza viruses
	Measles
	SARS
	Smallpox

Fluence. Radiant flux passing from all directions through a unit area, often expressed as J/m^2, J/cm^2, or $\mu W \cdot s/cm^2$.

Irradiance. Power of electromagnetic radiation incident on a surface per unit surface area, typically reported in microwatts per square centimeter ($\mu W/cm^2$). See CIE (1987) for details.

Mycobacterium tuberculosis. The namesake member of the *M. tuberculosis* complex of microorganisms, and the most common cause of tuberculosis (TB) in humans. In some instances, the species name refers to the entire *M. tuberculosis* complex, which includes *M. bovis, M. africanum, M. microti, M. canettii, M. caprae, M. pinnipedii,* and others.

Ocular damage. Any damage to the eye, particularly that caused by exposure to UV energy.

Permissible exposure time (PET). Calculated time period that humans, with unprotected eyes and skin, can be exposed to a given level of UV irradiance without exceeding the NIOSH recommended exposure limit (REL) or ACGIH Threshold Limit Value® (TLV®) for UV radiation.

Personal protective equipment (PPE). Protective clothing, helmets, goggles, respirators, or other gear designed to protect the wearer from injury from a given hazard, typically used for occupational safety and health purposes.

Photokeratitis. Defined by CIE (1993) as corneal inflammation after overexposure to ultraviolet radiation.

Photokeratoconjunctivitis. Inflammation of cornea and conjunctiva after exposure to UV radiation. Exposure to wavelengths shorter than 320 nm is most effective in causing this condition. The peak of the action spectrum is approximately 270 nm. See CIE (1993) for details. Note that different action spectra have been published for photokeratitis and photoconjuctivitis (CIE 1993); however, the latest studies support the use of a single action spectrum for both ocular effects.

Radiometer. An instrument used to measure radiometric quantities, particularly UV irradiance or fluence.

Threshold Limit Value® (TLV®). An exposure level under which most people can work consistently for 8 h a day, day after day, without adverse effects. Used by the ACGIH to designate degree of exposure to contaminants. TLVs can be expressed as approximate milligrams of particulate per cubic meter of air (mg/m^3). TLVs are listed either for 8 h as a time-weighted average (TWA) or for 15 min as a short-term exposure limit (STEL).

Ultraviolet radiation. Optical radiation with a wavelength shorter than that of visible radiation. [See CIE (1987) for details.] The range between 100 and 400 nm is commonly subdivided into

> UVA: 315 to 400 nm
> UVB: 280 to 315 nm
> UVC: 200 to 280 nm
> Vacuum UV 100 to 200 nm

Ultraviolet germicidal irradiation (UVGI). Ultraviolet radiation that inactivates microorganisms. UVGI is generated by germicidal lamps that kill or inactivate microorganisms by emitting radiation predominantly at a wavelength of 253.7 nm.

UV dose. Product of UV irradiance and specific exposure time on a given microorganism or surface, typically reported in millijoules per square centimeter ($\mu J/cm^2$).

Wavelength. Distance between repeating units of a wave pattern, commonly designated by the Greek letter lambda (λ).

AIR TREATMENT SYSTEMS

Design Guidance

Early guidelines published by General Electric (1950), Philips (1985), and Westinghouse (1982) are still used by many system designers today. First et al. (1999), Kowalski (2003, 2006, 2009), NIOSH (2010), and Riley et al. (1976) have made meaningful advances in the analysis and modeling of UVGI systems that have improved guidance for system design, yet no consensus guidelines exist that comprehensively address all aspects of UVGI system design required to ensure desired performance.

Likewise, application support for UVGI technologies is growing and many successful systems have been installed; however, there are no industry standards for rating the effectiveness of UVGI devices and systems. The EPA (2007) stated, "The most important needs in the area of UVGI are industry standards to rate devices and installations, as well as guidance for installation and maintenance."

UVGI system design today relies on performance data from lamp, ballast, and fixture manufacturers and the experience of system designers. Many equipment manufacturers have methods for estimating the UV dose delivered, which may include using tabulated data charts, mathematical modeling, and complex formulas. Like most HVAC components, UVGI systems are often oversized to ensure performance. This oversizing, though conservative, can potentially increase equipment and utility costs, and may result in less energy-efficient systems.

ASHRAE Technical Committee 2.9, Ultraviolet Air and Surface Treatment, was created in 2003 (initially as a Task Group, converted to a standing Technical Committee in 2007) in part to address these deficiencies by initiating research programs, preparing *ASHRAE Handbook* chapters, and by serving as cognizant committee for development of needed standards. Proposed standards for UVGI air (185.1) and surface (185.2) disinfection systems are under development. These test standards will provide ratings of equipment performance that will aid in selection of appropriate components by UVGI system designers.

Upper-Room Air Disinfection

The primary objective of upper-room air UVGI placement is to interrupt the transmission of airborne infectious pathogens within the indoor environment. The source of these infectious organisms may be infected humans, animals, or bioaerosols introduced for terrorism purposes. Humans are the most important sources of airborne agents that infect people (ACGIH 1999). The measles and influenza viruses and the tuberculosis bacterium are three important infectious organisms known to be transmitted indoors by means of air shared between infected and susceptible persons. Studies of person-to-person outbreaks indicate two transmission patterns: within-room exposure such as in a congregate space, and transmissions beyond a room through corridors and by entrainment in ventilation ductwork, where air is then recirculated throughout the building. ASHRAE provides guidance on protecting buildings from extraordinary incidents in which a bioterror agent is aerosolized into a building (ASHRAE 2003).

UVGI is used, in combination with other environmental controls, to protect building occupants in all of these areas of concern (Brickner et al. 2003; Kowalski and Bahnfleth 2003). Since the 1930s (Riley and O'Grady 1961; Wells 1955) and continuing to the present day (First et al. 2007a, 2007b; Miller et al. 2002; Xu et al. 2003), numerous experimental studies have demonstrated the efficacy of upper-room UVGI. Additionally, evidence of effectiveness

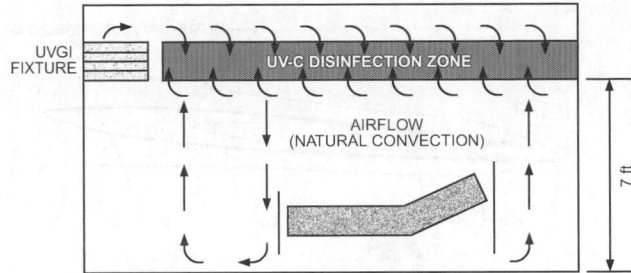

Fig. 5 Typical Elevation View of Upper-Room UV Applied in Hospital Patient Room

has been shown for inactivation of tuberculosis (Escombe et al. 2009); in reducing measles transmission in a school, and interruption of influenza transmission within a hospital (McLean 1961). For specific applications, such as hospital surgical rooms where occupants wear protective equipment, and unoccupied rooms, whole-room UVGI is also used.

For any application, the ability of UVC to inactivate microorganism is a function of dose. **Dose** is the length of time of exposure multiplied by the irradiance measured in $\mu W/cm^2$ (see Chapter 16 in the 2008 *ASHRAE Handbook—HVAC Systems and Equipment* for more details). A key difference between surface decontamination and airborne inactivation of organisms is exposure time. In a duct system, exposure time is on the order of seconds or fractions of seconds because of the rapid movement of air through the duct. Therefore, the irradiance must be sufficiently high to provide the dose necessary to inactivate the pathogen in seconds or a fraction of a second, depending upon the configuration and characteristics of the UVC system.

Organisms differ in their susceptibility to UVC inactivation, as discussed in the section on Fundamentals. Depending on the application, a public health or medical professional, microbiologist, or other individual with knowledge of the threat or organisms of concern should be consulted.

Various upper-room UVGI systems are designed to generate a controlled UVC field above the heads of occupants and to minimize UVC in the lower room. Settings appropriate to upper-room UVGI placement include congregate spaces, where unknown and potentially infected persons may share the same space with uninfected persons (e.g., a medical waiting room or homeless shelter). Additionally, circulation paths where unknown infected persons may go for treatment in a medical facility would be covered, such as corridors. Upper-room UVGI covers situations where untreated recirculated air might enter an occupied space (see Figures 5 and 6 for illustrations of pathogen control using UVGI). UVGI is most effective in areas with minimal ventilation, 2 air changes per hour (ach) up to normal recommended levels of 6 ach. Ventilation patterns (natural and mechanical) should be designed to promote good air mixing within the space equipped with UVGI so that infectious microorganisms encounter the UVC beam and are irradiated, thus reducing the risk of exposure of occupants to airborne infectious agents. Recent studies that have employed the use of natural ventilation and UVGI have shown that upper-room UVGI is an effective, low-cost intervention for use in TB infection control (Escombe et al. 2009). In specialized applications such as a hospital operating room, the whole room is flooded with UV. In such examples, the occupants must be fully protected from exposure, or UVC can be active while the room is unoccupied (Hart 1960).

Upper-Air UVGI Systems

Upper-air UVGI systems are designed to irradiate only air in the upper part of the room (Figures 7 and 8). Parameters for UVGI effectiveness include room configuration, UV fixture placement,

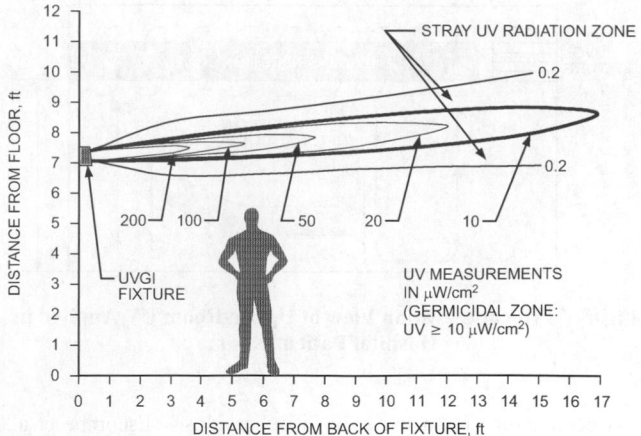

Fig. 6 Typical Elevation View Showing UVGI Energy Place above Heads of Room Occupants, Maintaining Safety

Fig. 8 Upper-Room UVGI Fixtures and Natural Ventilation in Corridor of TB Facility in Brazil
(Centers for Disease Control and Prevention)

Fig. 7 Upper-Room UVGI Treating Congregate Setting
(TUSS Project, St. Vincent's Hospital, New York City)

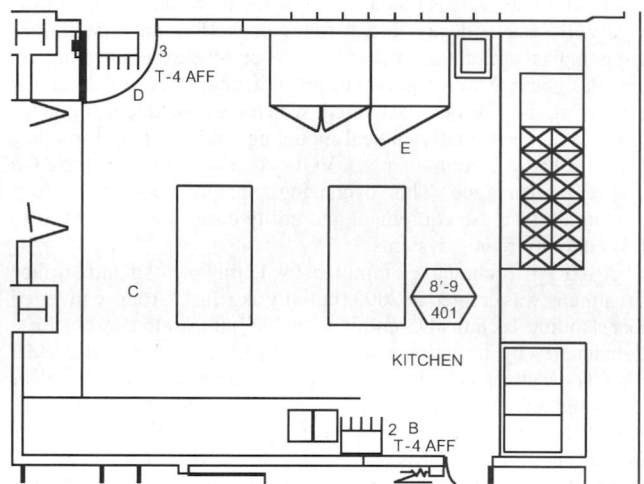

Fig. 9 Typical Layout of UVGI Fixtures for Patient Isolation Room
(First et al. 1999)

and the adequacy of airflow in bringing contaminated air into the irradiated upper room. UVGI fixtures should be placed appropriately, spaced to accommodate the area, shape, and height of the space in which air is to be disinfected. Figures 9 to 11 show some examples of upper-air fixture placement. UVGI fixtures should be selected based on the floor-to-ceiling height. For spaces at least 8 ft high, UVGI fixtures that produce a narrow, focused beam should be used. The fixture should be mounted so that its UV emission is distributed parallel to the plane of the ceiling. This placement prevents stray ultraviolet rays from impinging on occupants below. A new computer-aided lighting software program is being modified to help automate the placement of fixtures, and to calculate the uniformity and average UV provided (Brickner et al. 2009).

Upper-air systems rely on air convection and mixing to move air from the lower to the upper portion of the room, where it can be irradiated and airborne microorganisms inactivated (Kethley and Branc 1972). Many fixtures incorporate a safety switch that breaks the circuit when fixtures are opened for servicing, and should contain baffles or louvers appropriately positioned to direct UV irradiation to the upper air space. Baffles and louvers must never be bent or deformed.

Upper-room UVGI fixtures typically use low-pressure UVC lamps in tubular and compact shapes, and require a variety of elec-

trical wattages. Beyond lamp size, shape, and ballast, fixtures are designed to be open or restricted in distribution, depending on the physical space to be treated. Ceiling heights above 10 ft allow for more open fixtures, which are more efficient. For occupied spaces with lower ceilings (less than 10 ft), various louvered upper-room UVGI fixtures (wall, pendant, and corner) are available to be mounted in combinations at least 7 ft from the floor to the bottom of the fixture. Figure 5 shows a typical elevation and corresponding UV level, and Figure 6 illustrates distribution in a room.

Application guidance with placement criteria for UV equipment is provided by Boyce (2003), CDC (2005), CIE (2003), Coker et al. (2001), First et al. (1999), IESNA (2000), and NIOSH (2010). An example of the guidance provided by Coker et al. is shown in Table 3. Additionally, manufacturer-specific advice on product operations should be followed.

A basic standard for upper-room installations has been one 30 W (electrical input) fixture for every 200 ft^2 (Riley et al. 1976). A UVGI installation that produces a maintained, uniform distribution of UV averaging between 30 and 50 μW/cm^2 should be effective in inactivating most airborne droplet nuclei containing

Table 3   UVGI Fixture Mounting Height Guidance
(Coker et al. 2001)

	Wall-Mounted Fixtures*		Ceiling-Mounted Figures*	
	Corner Mount	Wall Mount	Pendant	Pendant with Fan
Beam pattern	90°	180°	360°	360°
Minimum ceiling height	2.44 m	2.44 m	2.89 m	2.89 m
Fixture mounted height	2.1 m	2.1 m	2.4 m	2.4 m
Ideal UV-C intensity for effective disinfection	> 10 µW/cm²	> 10 µW/cm²	> 10 µW/cm²	> 10 µW/cm²

*Appropriately designed UV fixtures are available for all locations. Only the most commonly used have been included in the table.

mycobacteria, and is presumably effective against viruses, as well (First et al. 2007a, 2007b; Miller et al. 2002; Xu et al. 2003). Beyond UVC emission strength, effectiveness of upper-air UVGI is related to air mixing, relative humidity, and the inherent characteristics of the pathogenic organisms being addressed (Ka et al. 2004; Ko et al. 2000; Rudnick 2007). Effectiveness improves greatly with well-mixed air (First et al. 2007a, 2007b; Miller et al. 2002; Riley et al. 1971a, 1971b). Ventilation systems should maximize air mixing to receive the greatest benefit from upper-room UVGI. Relative humidity should be less than 60%; levels over 80% RH may reduce effectiveness (Kujundzic et al. 2007; Xu et al. 2003). To maintain efficient output from low-pressure UVC lamps in the upper room, room temperature should be within recommended ASHRAE *Standard* 55 guidelines for occupant comfort. For example, in high-risk areas such as corridors of infectious disease wards, a minimum of 0.4 µW/cm² at eye level is a good engineering guide (Coker et al. 2001). No long-term health effects of UVC exposure at levels found in the lower occupied part of rooms are known.

The overall effectiveness of upper-air UVGI systems improves significantly when the air in the space is well mixed. Although convection air currents created by occupants and equipment can provide adequate air circulation in some settings, mechanical ventilation systems that maximize air mixing are preferable. If air mixing with mechanical ventilation is not possible, fans can be placed in the room to enhance mixing.

Depending on the disinfection goals, upper-air systems should be operated in a fashion similar to corresponding in-duct systems. Those systems designed to reduce or eliminate the spread of airborne infectious diseases in buildings with continuous occupancy and/or with immunocompromised populations, should be operated 24 h per day, 7 days per week. Upper-air systems designed for improved indoor air quality installed in more traditional commercial buildings can be operated intermittently, or powered on during hours of normal building occupancy and powered off when the facility is empty. This may provide acceptable indoor air quality during periods of building occupancy while saving energy and requiring less frequent lamp replacements. However, intermittent operation must be factored into the initial system design because of the effect that cycling UV lamps on and off may have on lamp performance and lamp life.

In-Duct Systems: Airstream Disinfection

The principal design objective for an in-duct system is to distribute UV energy uniformly in all directions throughout the length of the duct or air-handling unit (AHU) to deliver the dose to air moving through the irradiated zone with the minimum energy system power. Enhancing the overall reflectivity of the inside of the air handler can improve UVGI system performance by reflecting UVC energy back into the irradiated zone, thus increasing the effective UV dose. Using materials such as aluminum or other highly reflective materials can increase reflectivity. Properly designed in-duct UV air disinfection systems also maintain the cleanliness of cooling coil surfaces and condensate pans because of the high power required for this ap-

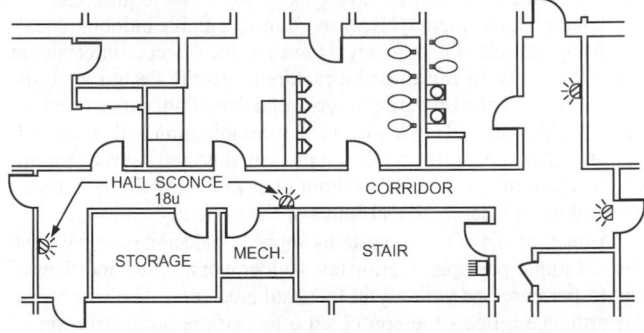

**Fig. 10   Upper-Room UVGI Fixtures with 180° Emission
Profile Covering Corridors**
Originally published in *Public Health Reports* (Brickner et al. 2003)

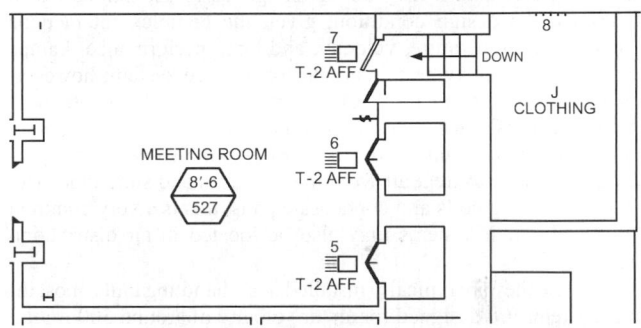

SPACE USAGE	ROOM DIMENSIONS			UVGI FIXTURE COVERAGE			
CONFERENCE/OPEN OFFICE ROOM	*L* ft	*W* ft	*H* ft	*MH* ft	*A* ft²	*V* ft³	(*N*) WM
	8.9	5.5	2.6	2.2	49	117.4	(3) 8.5 W

Design Concept: Congregate setting, high occupancy, shared air with adjacent auditorium (multipurpose room). Dropped ceiling for an open office plan. Look for long path lengths, evenly space fixtures along one wall.

L = length, *W* = width, *H* = floor-to-ceiling height, *A* = area covered, *V* = room volume, *MH* = mounting height above finished floor, *N* = number of UVGI fixtures, *W* = nominal wattage, WM = wall mounted, CM = ceiling mounted

Fig. 11   Example Upper-Air UVGI Layout for Meeting Room

plication. Systems designed specifically for coil and condensate pan applications are likely not adequate for proper air disinfection.

Design dose is a function of the design-basis microbe (i.e., the targeted microorganism with the smallest *k*-value) and the desired level of disinfection. Generally, a single-pass inactivation efficiency is specified, analogous to the specification of a particulate filter MERV rating. In some cases, the design disinfection level may be a true performance specification based on the exposure in an occupied space. Determining this value requires analysis of the entire system that is used to determine the single-pass performance based. Which approach is selected depends on the type of application. Laboratory/hospital installations are more likely to have specific, identified targets than, for example, school or office installations. The required average irradiance for a typical in-duct system is on the order of 1000 to 10,000 $\mu W/cm^2$, but it could be higher or lower depending on the application requirements.

In-duct air disinfection systems designed to reduce the spread of airborne infectious diseases (e.g., tuberculosis, influenza) in buildings with continuous occupancy and/or with immunocompromised populations (e.g., hospitals, prisons, homeless shelters) should be operated on a continuous basis. However, properly designed systems installed in more traditional commercial buildings (e.g., offices, retail) can be operated intermittently, or powered on during hours of normal building occupancy and powered off when the facility is empty. This may save energy costs and require less frequent lamp replacement while providing acceptable indoor air quality during periods of occupancy. However, the effect of intermittent operation on lamp life must be factored into the design analysis: cycling reduces the operating hours to failure of hot cathode lamps. In-duct UVGI should always be used in combination with proper filtration. Filters help protect UV lamps from dust and debris accumulation which may reduce UV output over time, and filters enhance the overall air cleaning capabilities of the system.

In-duct air disinfection systems should be designed to have the desired single-pass inactivation level under worst-case conditions of air temperature and velocity in the irradiated zone. The worst-case performance reflects the combined effect of the number/power of UVC generators; air residence time, which is inversely proportional to air velocity; and lamp/ballast characteristics, including wind chill effect and depreciation (as discussed in Chapter 16 of the 2008 *ASHRAE Handbook—HVAC Systems and Equipment*). Lee et al. (2009) showed that it may be advantageous to use simulation to determine the design condition, given the complex interactions between air temperature, velocity, and lamp performance. Lamps may be located anywhere in an air conveyance system; however, some locations provide more efficiency and potentially greater benefit. In most cases, the lowest maximum velocity in a system occurs inside air-handling units. For this reason, and also because it provides the ability to treat air from many spaces and simultaneously irradiate cooling coils and condensate pans, this is a very common choice, although systems may also be located in air distribution ducts.

Because they are typically installed in air handling units, most in-duct systems are designed for an air velocity of around 500 ft/min. At this velocity, an irradiance zone 8 ft in length achieves a 1 s exposure. As a rule of thumb, a minimum 0.25 s exposure should be provided; otherwise, system cost and power consumption will be excessive. UVGI fixtures are most often located downstream of the heating/cooling coils. However, in some cases, mounting fixtures upstream of the coil may result in lower air velocity and in-duct temperatures, resulting in a more optimum lamp performance temperature and more cost-effective disinfection. The tradeoff is reducing the effectiveness of disinfection of the cooling coil and forgoing irradiation of the drain pan that lamps mounted downstream of the coil also provide.

Studies of Airstream Disinfection Effectiveness

Recent laboratory studies [e.g., RTI (2005); VanOsdell and Foarde (2002)] conclusively demonstrate the ability of commercially available equipment to achieve a high level of disinfection of moving airstreams. These studies have generally involved tests with surrogates rather than infectious disease agents, but it can be assumed that an infectious agent with a *k*-value similar to an experimental surrogate will be similarly inactivated. Previous field studies showed clinical effectiveness (i.e., reduced incidence of infection) (Nagy et al. 1954; Rentschler and Nagy 1940), but similar recent studies are lacking. Although pilot studies have begun (Bierman and Brons 2007; Rudnick et al. 2009), further recorded field studies are needed to benchmark installed system performance. Many UV airstream disinfection systems have been installed in hospital environments to help reduce pathogens to complement conventional dilution/filtration systems.

HVAC SYSTEM SURFACE TREATMENT

Coil and Drain Pan Irradiation

Conditions in HVAC systems can promote the growth of bacteria and mold-containing biofilms on damp or wet surfaces such as cooling coils, drain pans (Levetin et al. 2001), plenum walls, humidifiers, fans, energy recovery wheels, and filters. Locations in and downstream of the cooling coil section are particularly susceptible because of condensation and carryover of moisture from coil fins. Cooling coil fouling by biofilms may increase coil pressure drop, and reduce airflow and heat exchange efficiency (Montgomery and Baker 2006). Filters capture bacteria, mold, and dust, which may lead to microbial growth in damp filter media. As the growth proliferates, a filter's resistance to airflow can increase. This can result in more frequent filter changeouts and increased exposure to microbes for maintenance workers and building occupants. As performance degrades, so does the air quality in occupied spaces (Kowalski 2006).

Conventional methods for maintaining air-handling system components include chemical and mechanical cleaning, which can be costly, difficult to perform, and dangerous to maintenance staff and building occupants. Vapors from cleaning agents can contribute to poor air quality, chemical runoff contributes to groundwater contamination, and mechanical cleaning can reduce component life. Furthermore, system performance can begin to degrade again shortly after cleaning as microbial deposits reappear or reactivate.

UVGI can be applied to HVAC systems to complement conventional system maintenance procedures. A large dose can be delivered to a stationary surface with a low UVC irradiance because of the essentially infinite exposure time, making it relatively easy to cost-effectively prevent the growth of bacteria and mold on system components. In contrast to air disinfection irradiance levels, which may exceed 1000 $\mu W/cm^2$, coil surface irradiance levels on the order of 1 $\mu W/cm^2$ can be effective (Kowalski 2009), although 50 to 100 $\mu W/cm^2$ is more typical. Using reflectors to focus lamp output on surfaces can reduce the power required for surface treatment, but at the expense of reducing air treatment effectiveness. Potential advantages of UVGI surface treatment include keeping surfaces clean continuously rather than periodically restoring fouled surfaces, no use of chemicals, lower maintenance cost, and potentially better HVAC system performance.

Fixtures can be installed to target problematic components such as cooling coils, condensate pans, or filters (Figure 12), or applied to give broad distribution of UVC energy over an entire enclosure (e.g., mixing box/plenum) that might have microbial activity. Like in-duct air-treatment equipment, systems for surface treatment in air-handling units should be designed to withstand moisture and condensate and selected to operate over a full range of system operating conditions.

Room Surface Treatment

UVGI for surface and air disinfection, particularly in health care settings, has been applied to reduce the number of pathogenic organisms on surfaces and in air. Environmental contamination in health care settings and transmission of health-care-associated pathogens to patients occurs most frequently via contaminated hands of health care workers and transmission of pathogens to patient (Boyce 2010). A primary concern in health care settings has been reducing nosocomial infections and finding new approaches for these environments to help eliminate infections from hospital settings. Hospital-acquired infections generate a high financial burden for the health care industry and the consumer. In the United States, an estimated 1.7 million hospital-acquired infections occur annually in hospitals, leading to about 100,000 deaths (U.S. DHHS 2009).

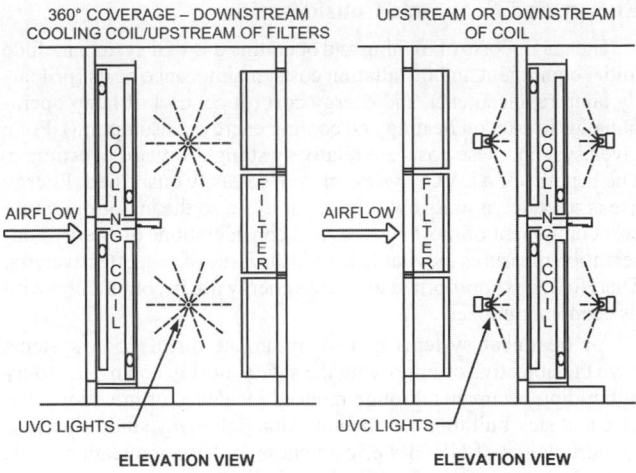

Fig. 12 Section View of Typical HVAC Surface Treatment Installations

UVGI is one of the tools that can be used to reduce pathogens in health care environments. Various portable UVGI devices are available, which can be easily moved into patient rooms, surgical suites, ICUs, and other critical areas that need surface and air disinfection during a terminal cleaning process or when a patient is diagnosed with a disease transferred by pathogens. Some of the pathogens of interest and their reduction in health care settings are multidrug-resistant, such as methicillin-resistant *Staphylococcus aureus* (MRSA), *Clostridium difficile*, *Acinetobacter baumannii*, and vancomycin-resistant *Enterococci* (VRE). These pathogens can be deactivated by proper application of UVGI. A study by Rastogi et al. (2007) investigated the efficacy of UVGI disinfection of *Acinetobacter baumannii* on contaminated surfaces relevant to medical treatment facilities. The UVGI exposure to surfaces resulted in ≥4-log (CFU) reduction in viable cells of *A. baumannii*.

UVGI fixtures can also be installed in surgical suites to disinfect surfaces and air between or during procedures. A 19-year study on UVGI during orthopedic surgery showed that 47 infections occurred following 5980 joint replacements. The infection rates for total hip replacements decreased from 1.03% to 0.72% ($p = 0.5407$), and for total knee replacements from 2.20% to 0.5% ($p < 0.0001$). The study concluded that UVGI appears to be an effective way to lower the risk of infection in the operating room during total joint replacement (Ritter et al. 2007). Safety precautions must be followed when applying UVGI during surgery to protect workers from accidental exposure (see the following discussion of intensity of source).

Application of UVC to other surfaces is based on time, intensity, distance, and desired target, all of which determine the effectiveness of the application. Surfaces away from the HVAC systems may be irradiated with UVC to achieve the same desired effects as with cooling coil or air disinfection applications inside an air handler. Locations such as schools, hospitals, morgues, homeless shelters, and health care settings can be irradiated with fixed or portable in-room UVC fixtures that serve as part of the room's disinfection methodology. UVC irradiates all line-of-site objects and shadow areas contained in these facilities, such as tables, chairs, surgical equipment, and objects. Tools used in microbiological applications can be irradiated with UVC for simple surface disinfection. Different materials absorb and reflect UVC energy at different rates; depending on the overall reflectivity of the materials, irradiation time, intensity of the UVC fixture, and corresponding microbial k-value, various levels of disinfection can be achieved. Furthermore, humidity levels in the space may also affect UVC transmission, so the local environment plays a role, as well. Consideration must be given to all room materials exposed. No living organisms, including

animals and plants, should be in the room when UVC is used. Some materials degrade when exposed to UVC energy.

The dose ($\mu J/cm^2$) of UVC needed to eliminate a desired microbe for in-room applications or tool disinfection depends on the selected target and desired deactivation rate. Different microorganisms require various levels of UVC energy for deactivation, as discussed previously. Vegetative forms of bacteria tend to be more susceptible to UVC energy than spore-forming microorganisms are.

The same principles as for in-duct applications apply here. There are two primary methods of UVC delivery: direct (line of sight) and indirect (reflection). Most surface applications use a direct source, where the source (typically a mercury vapor lamp) is contained in an assembly designed to direct the UVC energy at a particular surface or in a particular direction with no impedance to the energy beam. In an indirect application, the energy is either attenuated or reflected onto a surface using a quartz or plastic filtering material or a reflective material. In the latter case, additional calculations are needed to determine the rate of attenuation caused by the filter or reflector. The reflected UVC energy can also be measured to determine accurately when a given amount of the UVC dose has been delivered to the desired target.

The basics of determining the radiant energy levels to a surface are as follows:

Length of exposure. When disinfecting surfaces, it must be first determined if the target is moving or stationary. This helps to determine if there are any limiting factors associated with the length of exposure time. In most surface disinfection applications, time is relative to intensity, meaning that increasing the intensity of the source can decrease the exposure time necessary. It is important to remember that microorganisms require a higher or lower intensity for deactivation, depending on their structure (Brickner et al. 2003).

Intensity of source. UVC lamp and equipment manufacturers normally provide the intensity of a given source (lamp or fixture) at a given distance. A distance correction factor may be needed when calculating a desired dose or intensity for a surface. UVC energy follows the same inverse square rule for intensity as visible energy and other electromagnetic sources: the amount of energy at the surface is measured in proportion to the square of the distance from the energy's source (UVC lamp), assuming no loss due to scattering or absorption. Temperature and airflow corrections may also be necessary, depending on the location of the application. The intensity of a source is given in power per unit area (i.e., $\mu W/cm^2$).

Distance from source to substrate. In a point irradiation application, the distance is relatively easy to calculate. Calculating time requirements and intensity levels for a 3D object or space is more complex. The varying distances from the source are the first challenge, because the object itself creates a shadowing effect, and any shadows from the local environment must be taken into consideration. However, portable devices are available that effectively measure the reflected dose from shadow areas and offer quantifiable results.

Recent studies on in-room UVGI disinfection devices have shown that UVGI can be successfully applied to reduce microbiological loads of surfaces located in shadow areas in addition to line of sight. The reductions were up to 4-log for organisms such as MRSA, VRE, *Acinetobacter*, and *C. difficile*. Furthermore, it was concluded that UV room decontamination reduced colony counts of pathogens by greater than 99.9% within 20 min with the device tested (Rutala 2009). Note that, depending on the portable or stationary UVGI device, performance could greatly differ with respect to irradiation time, because overall dose delivered to surfaces is the critical measure of portable device performance.

Alternative and Complementary Systems

ASHRAE (2009) identifies the following demonstrated ways of reducing airborne infectious disease transmission:

- Dilution, personalized, and source capture ventilation
- In-room airflow control
- Room pressure differentials
- Filtration
- UVGI

From one perspective, these may be viewed as distinct, mutually exclusive alternatives for bioaerosol control. In principle, ventilation alone, filtration alone, or UVGI alone can yield the same level of control of a given contaminant source. However, in most cases, multiple modes of air quality control are used in the same system, often as a result of code requirements. For example, air quality codes for commercial buildings based on ASHRAE *Standard* 62.1 minimally require both dilution ventilation and particulate filtration at prescribed levels.

When used in combination with other mandatory air treatment modes, UVGI provides an incremental benefit. For example, if a particulate filter removes 85% of a given agent in an incoming airstream and a UVGI system with a single-pass efficiency of 85% for the same contaminant is installed in series with it, the combined filter/UVGI system would have a combined single-pass capture and inactivation efficiency of approximately 98% (i.e., the incremental benefit of adding an 85% efficient device is only 13%). Situations involving ventilation, filtration, and UVGI can be evaluated quantitatively by analyzing the entire system.

An example of this type of analysis was given by Nazaroff and Wechsler (2009) for several common arrangements of air cleaners in combination with ventilation. The performance of an air cleaner added to a system with ventilation is defined in terms of an effectiveness ε, which is the difference in contaminant concentration in a space of interest caused by adding an air cleaner and the concentration that would exist without the air cleaner:

$$\varepsilon = \frac{C_{baseline} - C_{control}}{C_{baseline}} \qquad (4)$$

where $C_{baseline}$ is the concentration without the air cleaner and $C_{control}$ is the concentration after addition of the air cleaner. This performance measure would show, for example, that adding UVGI to a system with a low ventilation rate would have a higher effectiveness (i.e., greater impact) than adding the same device to the same system with a higher ventilation rate. The extension of this concept to multiple-space systems and multiple air cleaners and air cleaner types is straightforward. System designers can use such methods to obtain more accurate cost/benefit estimates and also to optimize the characteristics and placement of air cleaners.

Even in the absence of the constraints imposed by building codes, the system designer should consider the potential benefits of combining air treatment methods. For example, the cost of particulate filters and their negative impact on fan energy use increase in inverse relation to the sizes of particles to be controlled (i.e., filters for smaller particles tend to be more expensive and have higher pressure drop than filters for larger particles). On the other hand, many larger microorganisms that are relatively resistant to UVGI, such as fungal spores, can be captured effectively by filters of moderately high efficiency and cost (Kowalski 2009). In addition, using UVGI to suppress microbial growth on filters that capture but do not kill is a potential complementary use of these two technologies. Ultimately, the decision to use or not use one of the available, effective microbial control methods should be based on a complete analysis that considers overall performance goals for air quality, impact on energy use, and economic factors. Such an analysis is illustrated for a typical air disinfection system by Lee et al. (2009), as discussed in the following section.

Energy and Economic Considerations

The major costs of owning and operating a UVGI system include initial equipment and installation costs, maintenance costs (primarily lamp replacement), and energy cost (direct cost of lamp operation plus impact on heating and cooling energy consumption). For a given system, these costs are relatively straightforward to estimate. The benefits of a UVGI system are not so easily quantified. Energy use is of concern in its own right, but it is also the major operating cost component of most systems and considerations of energy conservation measures inevitably lead to the issue of cost effectiveness. Therefore, it is appropriate to discuss energy use in conjunction with its economic impact.

Air treatment systems and room surface disinfection systems have the objective of improving the safety, health, and productivity of building occupants through reduced incidence of infectious disease and sick building complaints. Although many studies exist to support claims of UVGI's effectiveness in these applications, it is difficult to express the resulting benefits in economic terms. A conservative approach to economic evaluation is to compare the costs of alternative approaches such as dilution ventilation and particulate filtration that have the same effectiveness.

When alternative systems are compared with UVGI, all associated costs must be carefully estimated. Increased ventilation adds to heating- and cooling-coil loads and may also affect fan energy use. Particle filtration systems have their own associated installation and maintenance costs, and may significantly increase air-side pressure drop and, therefore, fan energy consumption. It must also be kept in mind that dilution potentially affects all types of contaminants, and particulate filtration removes potentially harmful matter from the air, whereas UVGI only treats microorganisms and does not remove them from the air.

Cooling-coil treatment systems have the two-fold objectives of maintaining coil performance and minimizing energy use by reducing air-side flow resistance and increasing the overall heat transfer coefficient relative to a conventionally maintained, mechanically and chemically cleaned coil. As in the case of air disinfection systems, the costs to install and operate such systems are easily estimated, but although there are many anecdotal reports of significant improvement in performance, there is little peer-reviewed literature documenting its effectiveness.

Upper-Air UVGI

The effectiveness of upper-air UVGI performance has often been described in terms of equivalent air changes per hour (ach): that is, by the rate of outside airflow measured in room volumes per hour that would achieve the same reduction of microbial air contamination in a well-mixed space. Riley's (1976) study of UVGI efficacy found that one 17 W UVC lamp covering 200 ft^2 produced 10 equivalent ach versus a natural die-off of 2 ach when a surrogate for tuberculosis was released in the room. The UVC lamp took less than 20 min to inactivate the bioaerosol vs. over 30 minutes for a natural die-off. In a bioaerosol room study, McDevitt et al. (2008) showed seasonal variations of between 20 to 1000 equivalent ach for a surrogate for smallpox. Ko et al. (2001) modeled the cost of using three air cleansing strategies to control transmission of tuberculosis in a medical waiting room. They calculated a present value per avoided tuberculin skin test conversion (evidence of infection) of $1708 for increased ventilation, $420 for HEPA filtration, and $133 for upper-air UVGI: that is, UVGI was less expensive by a factor of 3 to 13. Another metric is cost to provide a typical level of treatment per unit of floor area. The estimated health care benefit, typical of such analyses, was much larger than the cost: roughly $40/ft^2 per year. The drawback to the more economically advantageous upstream UVGI location is that it is considered a less favorable location for cooling coil irradiation, which many air treatment systems are designed to do as a secondary benefit.

In-Duct Air Treatment

Bahnfleth et al. (2009) and Lee et al. (2009) used simulation to investigate the energy use and operating cost of in-duct UVGI air treatment applied upstream or downstream of the cooling coil in a cooling-only variable-air-volume system located in New York and compared it with equivalent added particulate filter. A representative MERV 12 filter was estimated to provide the same performance as UVGI designed for 85% single-pass inactivation under design conditions. They computed not only the costs associated with the alternatives considered, but also estimated the health benefit using a method based on the Wells-Riley equation as applied by Fisk et al. (2005). They found that locating the UVGI system upstream of the cooling coil in the normally warmer mixed-air section of the air-handling unit reduced its required size by roughly 50% relative to a downstream location using typical in-duct lamp characteristics. Annual energy cost at an average electric rate of $0.10/kW·h ($0.03/1000 Btu) was approximately $0.02/ft^2 for the downstream location and $0.01/ft^2 for the upstream location, whereas the additional MERV 12 filter cost $0.10/ft^2. Annualized life-cycle cost, including installation and maintenance, was $0.74/ft^2 for the downstream location, $0.38/ft^2 for the upstream location, and $1.79/ft^2 for MERV 12 filtration.

Upper Air Versus In-Duct

Economic factors clearly favor an upper-air system when the building being treated with UVGI has no air distribution system. When a recirculating central air distribution system is present, a choice becomes possible between upper air systems, which must be distributed throughout occupied spaces, and in-duct systems, which can be centralized. As noted in the preceding discussion of in-duct systems, an annual operating cost of $0.01 to 0.02/ft^2 is possible at an electric rate of $0.10/kW·h ($0.03/1000 Btu). The same study (Lee et al. 2009) estimated an installed cost for equipment of $0.13 to 0.25/ft^2. By comparison, a typical upper air system might cost more than $2/ft^2 to install and more than $0.10/ft^2 to operate, based on typical sizing procedures and current equipment costs. This comparison seems to strongly favor in-duct systems where they are applicable, but is based on an assumption of equal performance that may not be valid.

Cooling Coil Surface Treatment

Cooling coil surface treatment, as described elsewhere, is done as an alternative to periodic mechanical and chemical cleaning of coils. By suppressing the formation of biofilms (and in the worst cases, extensive mold growth) on coils, coil irradiation should reduce air-side pressure drop, increase heat transfer coefficient, and reduce both fan and refrigeration system energy consumption. Several studies have documented the ability of coil irradiation to reduce microbial growth (Levetin et al. 2001; Shaughnessy et al. 1998). No peer-reviewed studies have yet been published to document the impact of coil irradiation on energy consumption, but there are many strong anecdotal reports of its effectiveness. As noted previously, the U.S. General Services Administration has sufficient confidence in this application to include it in its mechanical requirements (GSA 2010).

SAFETY

Hazards of Ultraviolet Radiation to Humans

UVC is a low-penetrating form of UV compared to UVA or UVB. Measurements of human tissue show that 4 to 7% of UVC (along with a wide range of wavelengths, 250 to 400 nm) is reflected (Diffey 1983) and absorbed in the first 2 μm of the stratum corneum (outer dead layer of human skin), thus minimizing the amount of UVC transmitted through the epidermis (Bruls 1984).

Although UV is more energetic than the visible portion of the electromagnetic spectrum, it is invisible to humans. Therefore, exposure to ultraviolet energy may result in transient corneal damage, which may initially go unnoticed.

Ocular damage generally begins with **photokeratitis** (inflammation of the cornea), but can also result in **photokeratoconjunctivitis** [inflammation of the conjunctiva (ocular lining)]. Symptoms, which may not be evident until several hours after exposure, may include an abrupt sensation of sand in the eyes, tearing, and eye pain, possibly severe. These symptoms usually appear within 6 to 12 h after UV exposure, and resolve fully within 24 to 48 h. Acute overexposure to UVC band radiation may cause incapacity due to eye discomfort, but this generally abates after several days, leaving no permanent damage.

Cutaneous damage consists of erythema, a reddening of the skin akin to sunburn but without tanning. The maximum effect of erythema occurs at a wavelength of 296.7 nm in the UVB band. UVC radiation at a wavelength of 253.7 nm is less effective in causing erythema. Because ultraviolet radiation is carcinogenic, questions have been raised concerning open-air UVGI systems. The International Commission on Illumination (CIE) completed a review of UVC photocarcinogenesis risks from germicidal lamps using basic biophysical principles: because of the attenuation provided by the stratum corneum and epithelial tissues of the skin, upper-air disinfection can be safely used without significant risk for long-term delayed effects such as skin cancer (CIE 2010).

Sources of UV Exposure

UVC energy does not normally penetrate through solid substances, and is attenuated by most materials. Quartz glass, soda barium glass, and TFPE plastic have high transmissions for UVC radiation.

UVC energy can reflect from polished metals and several types of painted and nonpainted surfaces; however, a surface's ability to reflect visible light cannot be used to indicate its UV reflectance. The fact that a blue glow can be observed on the metal surface from an operating low-pressure UV fixture lamp could indicate the presence of UV, and a measurement should be performed to ensure there is no exposure risk. The lack of reflected blue light clearly indicates the absence of UV energy. Note that ultraviolet energy is invisible to the normal human eye; however, it follows the same optical path as the visible blue light spectrum generated by the UVC lamp.

Well-designed and commissioned UVGI installations, education of maintenance personnel, signage, and use of safety switches can help to avoid overexposure. During commissioning and before operation of the UVGI installation, hand-held radiometers with sensors tuned to read the specific 254 nm wavelength should be used to measure stray UVC energy (primarily in upper-air systems).

Exposure Limits

In 1972, the Centers for Disease Control and Prevention (CDC) and National Institute for Occupational Safety and Health (NIOSH) published a **recommended exposure limit (REL)** for occupational exposure to UV radiation. The REL is intended to protect workers from the acute effects of UV exposure, although photosensitive persons and those exposed concomitantly to photoactive chemicals might not be protected by the recommended standard.

Exposures exceeding CDC/NIOSH REL levels require that workers use personal protective equipment (PPE), which consists of eyewear and clothing known to be nontransparent to UVC penetration and which covers exposed eyes and skin.

UV inspection, maintenance, and repair workers typically do not remain in one location during the course of their workday, and therefore are not exposed to UV irradiance levels for 8 h. Threshold Limit Value® (TLV®) consideration should be based on real-time occupancy of spaces treated by UVGI (ACGIH 2007). This recommendation is supported by UV monitoring data from First et al. (2005), which found that peak meter readings poorly predict actual exposure of room occupants.

Some indoor plants do not tolerate prolonged UVC exposure and should not be hung in the upper room.

Evidence of Safety

During the height of the tuberculosis resurgence in the United States in the 1990s, the Tuberculosis Ultraviolet Shelter Study (TUSS), a double-blind, placebo-controlled field trial of upper-room UVGI, was conducted at 14 homeless shelters in six U.S. cities from 1997 to 2004 (Brickner et al. 2000). Following available recommended placement, installation, and maintenance guidelines, each building in the study was evaluated for treatment with upper-room UVGI fixtures. At the conclusion of the study, the safety of room occupants was evaluated using data from a total of 3,611 staff and homeless study subjects regarding eye and skin irritation. Analysis showed no statistically significant difference in the number of reports of symptoms between the active and placebo periods. There was one definite instance of UV-related photokeratoconjunctivitis (from eye overexposure). This occurred from a placement of a bunk bed in a dormitory where a single bed had been used when the UV fixtures were first installed. By moving the UV fixture, this incident was resolved. This study demonstrated that, with careful application, side effects of UV overexposure can be avoided. Because of the enclosed nature of in-duct UVGI systems, with careful adherence to safety guidelines, these systems should not result in UV exposure.

Safety Design Guidance

In-duct systems should be fully enclosed to prevent leakage of UV radiation to unprotected persons or materials outside of the HVAC equipment.

All access panels or doors to the lamp chamber and panels or doors to adjacent chambers where UV radiation may penetrate or be reflected should have warning labels posted in appropriate languages. Labels should be placed on the outside of each panel or door, in a prominent location visible to people accessing the system.

Lamp chambers should have electrical disconnect devices. Positive disconnection devices are preferred over switches. Disconnection devices must be able to be locked or tagged out, and should be located outside the lamp chamber, next to the chamber's primary access panel or door. Switches should be wired in series so that opening any access deenergizes the system. On/off switches for UV lamps must not be located in the same location as general room lighting; instead, they must be in a location that only authorized persons can access, and should be locked or password protected to ensure that they are not accidentally turned on or off.

The lamp chamber should have one or more viewports of UVC-absorbing materials. Viewports should be sized and located to allow an operating UV system to be viewed from outside of the HVAC equipment.

Upper-air systems should have on/off switches and an electrical disconnect device on the louvers. If UV radiation measurements at the time of initial installation exceed the recommended exposure limit, all highly UV-reflecting materials should be removed, replaced, or covered. UV-absorbing paints containing titanium oxide can be used on ceilings and walls to minimize reflectance in the occupied space.

Warning labels must be posted on all upper-air UV fixtures to alert personnel to potential eye and skin hazards. Damaged or illegible labels must be replaced as a high priority. Warning labels must contain the following information:

- Wall sign for upper-air UVGI
  Caution: Ultraviolet energy. Switch off lamps before entering upper room.
- General warning posted near UVGI lamps
  Caution: Ultraviolet energy. Protect eyes and skin.
- Multi-lingual warning posted on the door of air handlers where UVGI is present in ductwork

Caution: Ultraviolet energy in duct. Do not switch off safety button or activate lamps with door open.

UVGI Fixtures

Upper Air Fixtures. Wall-mounted fixtures with louvers are designed to keep ultraviolet rays above eye and head level. Specially designed louvers keep the rays from bouncing off the ceiling and in a vertical path above 7 ft. These fixtures reduce airborne microorganisms as normal air convection moves them into the path of the ultraviolet rays, where they are inactivated. Some fixtures also use small fans to help circulate the air by the UV fixture.

Ceiling-suspended fixtures are designed in the same way as wall mounted fixtures, incorporating louvers and/or UV traps to keep the rays out of eye and head level.

In-Duct Units. In-duct UVGI systems are designed to mount inside an air-conditioning duct; the UV energy is confined to the inside of the duct. They are normally installed with a safety interlock, so if the fixture is removed from the duct or an access door is opened, the lamps turn off to avoid accidental human exposure to UV energy.

In-duct UVGI systems can be used to irradiate cooling coils, drain pans, and other HVAC components, or to disinfect moving air.

INSTALLATION, STARTUP, AND COMMISSIONING

UVC Radiation Measurements

Those responsible for the commissioning process should inspect fixture placement and eye level irradiance measurements using a 254 nm selective radiometer.

UVC levels can be measured with a UV radiometer directly facing the device at eye height at various locations in a room, and must be taken in the same location each time. If the readings indicate a dosage exceeding 6 $\mu J/cm^2$, the UV systems must be deactivated until adjustments can be made or the manufacturer can be contacted.

Fixtures must be adjusted if eye-level exposure exceeds the 8 h TLV for UVC 254 nm wavelength.

UVC measurements should be taken at eye level (between 5.5 and 6.0 ft) at compass points from each figure. Check reflective surfaces (e.g., TVs, monitors).

Incorporate readings into final commissioned drawings.

Measurements should be made at initial installation, whenever new UV lamps are installed (newer lamp designs may provide increased irradiance), and whenever modifications are made to the UVGI system or room (e.g., adjustment of fixture height, relocation or reposition of louvers, addition of UV-absorbing or -reflecting materials, changes in room dimension or modular partition height).

The operating instructions and advice of UVGI system designers and lamp manufacturers should always be followed to ensure the proper operation of any UVGI system. It is important to operate any such system within the temperature and relative humidity ranges considered during the system design process. The following section presents some general guidelines for maintaining adequate system performance.

MAINTENANCE

All UVGI systems require periodic inspection, maintenance, and lamp replacement to ensure proper system performance. Whenever maintenance is performed on UVGI systems, the appropriate safety guidelines outlined elsewhere in this chapter should be carefully followed.

Material Degradation

UVC energy can be detrimental to some organic materials. If the UVC is not applied properly and UVC-sensitive materials are not shielded or substituted, degradation can occur. However, the degradation may not be enough to cause failure of the material if UVC only penetrates on the micrometer scale into the material before the degradation plateaus off, leaving a still fully functional material, as found by ASHRAE research project RP-1509 sponsored by TC 2.9. For details, see Kaufman (2010).

Visual Inspection

Maintenance personnel should routinely perform periodic visual inspection of the UVC lamp assembly. Typically, a viewing port or an access door window is sufficient for in-duct applications. Closer visual inspection may be required for upper-air systems because a single burned-out lamp in a multilamp fixture may not be apparent from the lower room. Personal protective measures are required for this close-up inspection.

Any burned-out or failing lamps should be replaced immediately. If lamps become dirty because of inadequate filtration used with in-duct UVGI systems or upper-air systems installed in dusty environments, they should be cleaned with a lint-free cloth and isopropyl alcohol. Care should be taken to ensure no film remains on the surface of the lamps after cleaning. This film could reduce UV output from the lamp. Complete lamp fixtures should be replaced whenever they are visibly damaged or in accordance with manufacture warranty guidelines.

Radiometer

Another means of monitoring UVGI lamps is through the use of a stationary or portable radiometer. These are generally used to monitor the "relative" output of the UVGI system by measuring the UV intensity produced by the lamps. Caution is needed when using a radiometer in critical applications, because these devices are intended only to give a relative indication of the lamp output. Radiometer sensors can degrade over time with constant exposure to UV. If an accurate measurement of UV intensity is required, a calibrated laboratory radiometer should be used.

Lamp Replacement

UVC lamps should be replaced at the end of their useful life, based on radiometer measurements or equipment manufacturer recommendations. Where applicable, it may be prudent simply to change lamps annually (8760 h when lamps are run continuously) to ensure that adequate UV energy is supplied by a given system. Lamps can operate long after their useful life, but at greatly reduced performance, and require regular measurement to ensure that a maintained level of UVC is being generated. A blue visible light emitted from the lamp does *not* indicate that UVC is present. The typical rated life of UVC lamps is in the range of 6000 to 10,000 h of operation. Switching lamps on and off too often may lead to early lamp failure, depending on the ballast type used. Consult the lamp manufacturer for specific information on expected lamp life and effects of switching.

Lamp and Ballast Disposal

UVC lamps should be treated in the same manner as other mercury-containing devices, such as fluorescent bulbs. Most lamps must be treated as hazardous waste and cannot be discarded with regular waste, although low mercury bulbs often can be discarded as regular waste; however, some state and local jurisdictions classify these lamps as hazardous waste. The U.S. EPA's universal waste regulations allow users to treat mercury lamps as regular waste for the purpose of transport to a recycling facility (EPA 2008). This simplified process was developed to promote recycling. The National Electrical Manufacturers Association maintains an online list of companies claiming to recycle or handle used mercury lamps (NEMA 2009). The most stringent of local, state, or federal regulations for disposal should be followed.

UVGI systems currently depend on the use of an electronic ballast to provide the UV lamp with power; however, many older systems used magnetic ballasts instead. Magnetic ballasts manufactured before 1979 contain polychlorinated biphenols (PCB) in the dielectric of their capacitors. (EPA 2007). Recycling is the best way to dispose of a magnetic ballast. The process allows the reuse of copper and aluminum wire, steel laminations, and steel cases, and disposes of capacitors and potting compound as hazardous waste in high-temperature incinerators.

As electronic ballasts fail, they should be treated as electronic waste. Many lamp and ballast recyclers are expanding their businesses and becoming certified to accept electronic waste. Some recyclers now accept both lamps and electronic ballasts.

Personnel Safety Training

Workers should be provided with as much training as necessary, including health and safety training, and some degree of training in handling lamps and materials. Workers should be made aware of hazards in the work area and trained in precautions to protect themselves. Training topics include the following:

- UVC exposure hazards
- Electrical safety
- Lock-out/tag-out (for induct units)
- Health hazards of mercury
- Rotating machinery (for induct units)
- Slippery condensate pans (for induct units)
- Sharp unfinished edges (for induct units)
- Confined-space entry (if applicable) (for induct units)
- Emergency procedures

Workers expected to clean up broken lamps should be trained in proper protection, cleanup, and disposal.

No personnel should be subject to direct UV exposure, but if exposure is unavoidable, personnel should wear protective clothing (no exposed skin), protective eyewear, and gloves. Most types of eyewear, including prescription glasses, are sufficient to protect eyes from UV, but not all offer complete coverage. Standard-issue protective goggles may be the best alternative.

If individual lamp operating conditions must be observed, this should preferably be done using the viewing window(s).

During maintenance, renovation, or repair work in rooms where upper-room UV systems are present, all UVGI systems must be deactivated before personnel enter the upper part of the room.

For in-duct systems, access to lamps should be allowed only when lamps are de-energized. The lamps should be turned off before air-handling unit (AHU) or fan shutdown to allow the lamps to cool and to purge any ozone in the lamp chamber (if ozone-producing lamps are used). If AHUs or fans are de-energized first, the lamp chamber should be opened and allowed to ventilate for several minutes. Workers should always wear protective eyewear and puncture-resistant gloves for protection in case a lamp breaks.

Access to the lamp chamber should follow a site-specific lockout/tag-out procedure. Do not rely on panel and door safety switches as the sole method to ensure lamp deenergizing. Doors may be inadvertently closed or switches may be inadvertently contacted, resulting in unexpected lamp activation.

If workers enter the condensate area of equipment, the condensate pan should be drained and any residual water removed.

In general, avoid performing readings with the fan running and workers inside an AHU (e.g., to test for output reduction caused by air cooling). Tests of this nature should be instrumented and monitored from outside the equipment.

Lamp Breakage

If a lamp breaks, all workers must exit the HVAC equipment area. Panels or doors should be left open and any additional lamp chamber access points should also be opened. Do not turn air-handling unit fans back on. After a period of 15 minutes, workers may reenter the HVAC equipment to begin bulb clean-up.

If a lamp breaks in a worker's hand, the worker should not exit the HVAC equipment with the broken bulb. Carefully set the broken bulb down, and then exit the space. When possible, try not to set the broken lamp in any standing condensate water. Follow standard ventilation and reentry procedures.

Cleanup requires special care because of mercury drop proliferation, and should be performed by trained workers. As a minimum, workers should wear cut-resistant gloves, as well as safety glasses to protect eyes from glass fragments. Large bulb pieces should be carefully picked up and placed in an impervious bag. HEPA-vacuum the remaining particles, or use other means to avoid dust generation.

REFERENCES

ACGIH. 1999. *Bioaerosols: Assessment and control*, Ch. 9: Respiratory infections—Transmission and environmental control, by E.A. Nardell and J.M. Macher. American Conference on Governmental Industrial Hygienists, Cincinnati, OH.

ACGIH. 2007. *TLVs® and BEIs®*. American Conference of Governmental Industrial Hygienists, Cincinnati, OH.

ASHRAE. 2004. Thermal environmental conditions for human occupancy. ANSI/ASHRAE *Standard* 55-2004.

ASHRAE. 2010. Ventilation for acceptable indoor air quality. ANSI/ASHRAE *Standard* 62.1-2010.

ASHRAE. 2003. Risk management guidance for health, safety, and environmental security under extraordinary incidents. *Report*, Presidential Ad Hoc Committee for Building Health and Safety under Extraordinary Incidents.

ASHRAE. 2009. *Indoor air quality guide: Best practices for design, construction, and commissioning.*

Bahnfleth, W., B. Lee, J. Lau, and J. Freihaut. 2009. Annual simulation of in-duct ultraviolet germicidal irradiation system performance. Proceedings of Building Simulation 2009, The 11th International Building Performance Simulation Association Conference and Exhibition, July 2009, Glasgow, Scotland.

Bierman, A. and J. Brons. 2007. *Field evaluation of ultraviolet germicidal irradiation (UVGI) in an air duct system*. Lighting Research Center, RPI, Troy, NY. http://www.lrc.rpi.edu/researchAreas/pdf/FieldEvaluationUVGIReport.pdf

Boyce, P. 2003. *Controlling tuberculosis transmission with ultraviolet irradiation*. Rensselaer Polytechnic Institute, Troy, NY.

Boyce, J. 2010. When the patient is discharged: Terminal disinfection of hospital rooms. *Medscape.com*. http://www.medscape.com/viewarticle/723217 (requires free registered account).

Brickner, P.W., R.L. Vincent, E.A. Nardell, C. Pilek, W.T. Chaisson, M. First et al. 2000. Ultraviolet upper room air disinfection for tuberculosis control: An epidemiological trial. *Journal of Healthcare Safety Compliance & Infection Control* 4(3):123-131.

Brickner, P.W., R.L. Vincent, M. First, E. Nardell, M. Murray, and W. Kaufman. 2003. The application of ultraviolet germicidal irradiation to control transmission of airborne disease: Bioterrorism countermeasure. *Public Health Report* 118(2):99-114.

Brickner, P.W., et al. 2009. Computer aided design for UVGI. NYSERDA Project 9425. St. Vincent's Hospital, New York.

Bruls, W. 1984. Transmission of human epidermis and stratum corneum as a function of thickness in the ultraviolet and visible wavelengths. *Journal of Photochemistry and Photobiology* 40:485-494.

CDC. 2002. *Comprehensive procedures for collecting environmental samples for culturing* Bacillus anthracis. Centers for Disease Control and Prevention, Atlanta, GA. http://www.bt.cdc.gov/agent/anthrax/environmental-sampling-apr2002.asp.

CDC. 2005. Guidelines for preventing the transmission of *Mycobacterium tuberculosis* in health-care settings. *Morbidity and Mortality Weekly Report (MMWR)* 37-38, 70-75.

CIE. 1987. *International lighting vocabulary*, 4th ed. Commission Internationale de L'Eclairage, Vienna.

CIE. 1993. CIE collection in photobiology and photochemistry. *Publications* 106/1 (Determining ultraviolet action spectra), 106/2 (Photokeratitis), and 106/3 (Photoconjuctivitis). Commission Internationale de L'Eclairage, Vienna.

CIE. 2003. *Ultraviolet air disinfection*. Commission Internationale de L'Eclairage, Vienna.

CIE. 2010. *UV-C photocarcinogenesis risks from germicidal lamps*. CIE 187:2010, Commission Internationale de L'Eclairage, Vienna.

Coker, A., E. Nardell, P. Fourie, W. Brickner, S. Parsons, N. Bhagwandin, and P. Onyebujoh. 2001. *Guidelines for the utilization of ultraviolet germicidal irradiation (UVGI) technology in controlling the transmission of tuberculosis in health care facilities in South Africa*. South African Centre for Essential Community Services and National Tuberculosis Research Programme, Medical Research Council, Pretoria.

Diffey, B.L. 1983. A mathematical model for ultraviolet optics in skin. *Physics in Medicine and Biology* 28:657-747.

DIN. 1979. Optical radiation physics and illumination engineering. *Standard* 5031. German Institute for Standardization, Berlin.

EPA. 2007. *Polychlorinated biphenyls (PCBs), storage and disposal: Ballasts*. Available at http://www.epa.gov/region09/toxic/pcb/ballast.html.

EPA. 2008. *Universal waste*. Available at http://www.epa.gov/epawaste/index.htm.

Escombe, A.R., R.H. Gilman, M. Navincopa, E. Ticona, B. Mitchell, C. Noakes, C. Martínez, P. Sheen, R. Ramirez, W. Quino, A. Gonzalez, J.S. Friedland, and C.A. Evans. 2009. Upper-room ultraviolet light and negative air ionization to prevent tuberculosis transmission. *PLoS Med* 17(6).

First, M.W., E.A. Nardell, W.T. Chaisson, and R.L. Riley. 1999. Guidelines for the application of upper-room ultraviolet irradiation for preventing transmission of airborne contagion—Part 1: Basic principles. *ASHRAE Transactions* 105(1):869-876.

First, M.W., R.A. Weker, S. Yasui, and E.A. Nardell. 2005. Monitoring human exposures to upper-room germicidal ultraviolet irradiation. *Journal of Occupational and Environmental Hygiene* 2:285-292.

First, M.W., F.M. Rudnick, K. Banahan, R.L. Vincent, and P.W. Brickner. 2007a. Fundamental factors affecting upper-room ultraviolet germicidal irradiation—Part 1: Experimental. *Journal of Environmental Health* 4:1-11.

First, M.W., K. Banahan, and T.S. Dumyahn. 2007b. Performance of ultraviolet light germicidal irradiation lamps and luminaires in long-term service. *LEUKOS* 3:181-188.

Fisk, W.J., O. Seppanen, D. Faulkner, and J. Huang. 2005. Economic benefits of an economizer system: Energy savings and reduced sick leave. *ASHRAE Transactions* 111(2).

GSA. 2010. *The facilities standards for the Public Buildings Service*. Public Buildings Service of the General Services Administration, Washington, D.C.

Harm, W. 1980. *Biological effects of ultraviolet radiation*. Cambridge University Press, New York.

Hart, D. 1960. Bactericidal ultraviolet radiation in the operating room. *Journal of the American Medical Association* 172:1019-1028.

Hollaender, A. 1943. Effect of long ultraviolet and short visible radiation (3500 to 4900 Å) on *Escherichia coli. Journal of Bacteriology* 46(6):531-541.

IESNA. 2000. *The IESNA lighting handbook*, 9th ed., Ch. 5: Nonvisual effects of optical radiation. M.S. Rea ed. Illuminating Engineering Society of North America, New York, NY.

Jensen, M.M. 1964. Inactivation of airborne viruses by ultraviolet irradiation. *Applied Microbiology* 12(5):418-420.

Ka, M., H.A.B. Lai, and M.W. First. 2004. Size and UV germicidal irradiation susceptibility of *Serratia marcescens* when aerosolized from different suspending media. *Applied and Environmental Microbiology* (April):2021-2027.

Kaufman, R.E. 2010. Study the degradation of typical HVAC materials, filters and components irradiated by UVC energy. ASHRAE Research Project RP-1509, *Final Report*.

Kethley, T.W. and K. Branc. 1972. Ultraviolet lamps for room air disinfection: Effect of sampling location and particle size of bacterial aerosol. *Archives of Environmental Health* 25(3):205-214.

Ko, G., M.W. First, and H.A. Burge. 2000. Influence of relative humidity on particle size and UV sensitivity of *Serratia marcescens* and *Mycobacterium bovis* BCG aerosols. *Tubercle and Lung Disease* 80(4-5):217-228.

Ko, G., H. Burge, E. Nardell, and K. Thompson. 2001. Estimation of tuberculosis risk and incidence under upper room ultraviolet germicidal irradiation in a waiting room in a hypothetical scenario. *Risk Analysis* 21(4):657-673.

Kowalski, W.J. 2003. *Immune building systems technology*. McGraw-Hill, New York.

Kowalski, W.J. 2006. *Aerobiological engineering handbook.* McGraw-Hill, New York.

Kowalski, W. 2009. *Ultraviolet germicidal irradiation handbook.* Springer-Verlag, Berlin.

Kowalski, W. and W. Bahnfleth. 2003. Immune building technology and bioterrorism defense. *HPAC Engineering* 75(1):57-62. http://www.engr.psu.edu/ae/faculty/bahnfleth/publications/immune_building_technology.pdf.

Kujundzic, E., M. Hernandez, and S.L. Miller. 2007. Ultraviolet germicidal irradiation inactivation of airborne fungal spores and bacteria in upper-room air and HVAC in-duct configurations. *Journal of Environmental Engineering Science* 6:1-9.

Lee, B., W. Bahnfleth, and K. Auer. 2009. Life-cycle cost simulation of in-duct ultraviolet germicidal irradiation systems. *Proceedings of Building Simulation 2009, 11th International Building Performance Simulation Association Conference and Exhibition*, July, Glasgow.

Levetin, E., R. Shaughnessy, C. Rogers, and R. Scheir. 2001. Effectiveness of germicidal UV radiation for reducing fungal contamination within air-handling units. *Applied and Environmental Microbiology* 67(8):3712-3715.

Martin, S.B., C. Dunn, J.D. Freihaut, W.P. Bahnfleth, J. Lau, and A. Nedeljkovic-Davidovic. 2008. Ultraviolet germicidal irradiation current best practices. *ASHRAE Journal* (August):28-36.

McDevitt, J.J., D.K. Milton, S.N. Rudnick, and M.W. First. 2008. Inactivation of poxviruses by upper-room UVC light in a simulated hospital room environment. *PLoS ONE* 3(9):e3186. http://www.plosone.org/article/info:doi/10.1371/journal.pone.0003186.

McLean, R.L. 1961. General discussion: The mechanism of spread of Asian influenza. Presented at the International Conference of Asian Influenza, Bethesda, MD; *American Review of Respiratory Diseases* 83(2 Part 2):36-38.

Menzies, D., J. Popa, J. Hanley, T. Rand, and D. Milton. 2003. Effect of ultraviolet germicidal lights installed in office ventilation systems on workers' health and well being: Double-blind multiple cross over trial. *Lancet* 363:1785-1792.

Miller, R.V., W. Jeffrey, D. Mitchell and M. Elasri. 1999. Bacterial responses to ultraviolet light. *American Society for Microbiology (ASM) News* 65(8):535-541.

Miller, S.L., M. Fennelly, M. Kernandez, K. Fennelly, J. Martyny, J. Mache, E. Kujundzic, P. Xu, P. Fabian, J. Peccia, and C. Howard. 2002. Efficacy of ultraviolet irradiation in controlling the spread of tuberculosis. *Final Report*, Centers for Disease Control, Atlanta, GA, and National Institute for Occupational Safety and Health, Washington, D.C.

Montgomery, R. and R. Bakcr. 2006. Study verifies coil cleaning saves energy. *ASHRAE Journal* 48(11):34-36.

Nagy, R., G. Mouromseff, and F.H. Rixton. 1954. Disinfecting air with sterilizing lamps. *Heating, Piping, and Air Conditioning* 26(April):82-87.

Nazaroff, W. and C. Weschler. 2009. Air cleaning effectiveness for improving indoor air quality: Open-path and closed-path configurations. *Proceedings of Healthy Buildings 2009*, Syracuse, NY, Paper 376.

NEMA. 2009. *Lamprecycle.org: Environmental responsibility starts here.* National Electrical Manufacturers Association. http://www.lamprecycle.org/.

NIOSH. 1972. Criteria for a recommended standard: Occupational exposure to ultraviolet radiation. *Publication* 73-11009. National Institute for Occupational Safety and Health, Washington, D.C.

NIOSH. 2010. Environmental control for tuberculosis: Basic upper-room ultraviolet germicidal irradiation guidelines for healthcare settings. NIOSH *Publication* 2009-105. http://www.cdc.gov/niosh/docs/2009-105/.

Philips. 1985. *Germicidal lamps and applications.* Philips Lighting Division, Roosendaal, the Netherlands.

Philips. 2006. *Ultraviolet purification application information.* Philips Lighting B.V., Roosendaal, the Netherlands.

Rastogi, V.K., L. Wallace, and L.S. Smith. 2007. Disinfection of *Acinetobacter baumannii*-contaminated surfaces relevant to medical treatment facilities with ultraviolet C light. Military Medicine 172(11):1166.

Reed, N.G. 2010. The history of ultraviolet germicidal irradiation for air disinfection. *Public Health Reports* 125:15-27.

Riley, R.L. 1988. Ultraviolet air disinfection for control of respiratory contagion. In *Architectural design and indoor microbial pollution*, pp. 179-197. Oxford University Press, New York.

Riley, R.L. and F. O'Grady. 1961. *Airborne infection—Transmission and Control.* Macmillan, New York.

Riley, R.L. and S. Permutt. 1971a. Room air disinfection by ultraviolet irradiation of upper air: Air mixing and germicidal effectiveness. *Archives of Environmental Health* 22(2):208-219.

Riley, R.L., S. Permutt, and J.E. Kaufman. 1971b. Convection, air mixing, and ultraviolet air disinfection in rooms. *Archives of Environmental Health* 22(2):200-207.

Riley, R.L., M. Knight, and G. Middlebrook. 1976. Ultraviolet susceptibility of BCG and virulent tubercle bacilli. *American Review of Respiratory Disease* 113:413-418.

Ritter, M.A., E.M. Olberding, and R.A. Malinzak. 2007. Ultraviolet lighting during orthopaedic surgery and the rate of infection. The Journal of Bone & Joint Surgery 89:1935-1940.

RTI. 2005. *Test/QA plan for biological inactivation efficiency by HVAC in-duct ultraviolet light air cleaners.* Research Triangle Institute, Research Triangle Park, NC.

Rudnick, S. 2007. Fundamental factors affecting upper-room germicidal irradiation—Part 2: Predicting effectiveness. *Journal of Occupational and Environmental Hygiene* 4(5):352-362.

Rudnick, S.N., M.W. First, R.L. Vincent, and P.W. Brickner. 2009. In-place testing of in-duct ultraviolet germicidal irradiation. *HVAC&R Research* 15(3).

Rutala, W. 2009. Disinfection and sterilization: Successes and failures. Presented at APIC Convention, June, Ft. Lauderdale, FL.

Setlow, J.K. 1966. The molecular basis of biological effects of ultraviolet radiation and photoreactivation. *Current Topics in Radiation Research* 2:195-248.

Setlow, R.B. 1997. DNA damage and repair: A photobiological odyssey. *Photochemistry and Photobiology* 65S:119S-122S.

Setlow, R.B. and J.K. Setlow. 1962. Evidence that ultraviolet-induced thymine dimers in DNA Cause Biological Damage. *Proceedings of the National Academy of Sciences* 48(7):1250-1257.

Sharp, D.G. 1939. The lethal action of short ultraviolet rays on several common pathogenic bacteria. *Journal of Bacteriology* 37(4):447-460.

Sharp, D.G. 1940. The effects of ultraviolet light on bacteria suspended in air. *Journal of Bacteriology* 39(5):535-547.

Shaughnessy, R., E. Levetin, and C. Rogers. 1998. The effects of UV-C on biological contamination of AHUs in a commercial office building: preliminary results. *Proceedings of IAQ and Energy '98*, pp. 229-236.

Shechmeister, I.L. 1991. Sterilization by ultraviolet radiation. In *Disinfection, sterilization and preservation*, pp. 535-565. Lea and Febiger, Philadelphia.

U.S. DHHS. 2009. National healthcare quality report. U.S. Department of Health and Human Services. Agency for Healthcare Research and Quality (AHRQ) *Publication* 10-0003. March 2010. http://www.ahrq.gov/qual/qrdr09.htm.

U.S. EPA. 2006. *Biological inactivation efficiency of HVAC in-duct ultraviolet light devices.* EPA/600/S-11/002. U.S. Environmental Protection Agency, Washington, D.C.

VanOsdell, D. and K. Foarde. 2002. Defining the effectiveness of UV lamps installed in circulating air ductwork. *Final Report*, Air-Conditioning and Refrigeration Technology Institute 21-CR Project 61040030.

Wells, W.F. 1955. *Airborne contagion and air hygiene; an ecological study of droplet infections.* Cambridge: Published for the Commonwealth Fund by Harvard University Press.

Westinghouse. 1982. Westinghouse sterilamp germicidal ultraviolet tubes. Westinghouse *Engineering Notes* A-8968.

Xu, P., J. Peccia, P. Fabian, J.W. Martyny, K.P. Fennelly, M. Hernandez, and S.L. Miller. 2003. Efficacy of ultraviolet germicidal irradiation of upper-room air in inactivating airborne bacterial spores and mycobacteria in full-scale studies. *Atmospheric Environment* 37(3):405-419.

BIBLIOGRAPHY

Abshire, R.L. and H. Dunton. 1981. Resistance of selected strains of *Pseudomonas aeruginosa* to low-intensity ultraviolet radiation. *Applied Environmental Microbiology* 41(6):1419-1423.

Bahnfleth, W.P. and W.J. Kowalski. 2004. Clearing the air on UVGI systems. *RSES Journal*, pp. 22-24.

Bernstein, J.A., R.C. Bobbitt, L. Levin, R. Floyd, M.S. Crandall, R.A. Shalwitz, A. Seth, and M. Glazman. 2006. Health effects of ultraviolet irradiation in asthmatic children's homes. *Journal of Asthma* 43(4):255-262.

Blatt, M.S., T. Okura, and B. Meister. 2006. Ultraviolet light for coil cleaning in schools. *Engineered Systems* (March):50-61.

Bolton, J.R. 2001. *Ultraviolet applications handbook.* Photosciences, Ontario.

Department of General Services. 2001. *Working with ultraviolet germicidal irradiation (UVGI) lighting systems: Code of safe practice.* County of Sacramento, CA.

Dumyahn, T. and M.W. First. 1999. Characterization of ultraviolet upper room air disinfection devices. *American Industrial Hygiene Association Journal* 60:219-227.

Luckiesh, M. 1946. *Applications of germicidal, erythemal and infrared energy.* D. Van Nostrand, New York.

Masschelein, W.J. 2002. *Ultraviolet light in water and wastewater sanitation,* R.G. Rice, ed. Lewis Publishers, New York.

Nardell, E.A., S.J. Bucher, P.W. Brickner, C. Wang, R.L. Vincent, K. Becan-McBride et al. 2008. Safety of upper-room ultraviolet germicidal air disinfection for room occupants: Results from the tuberculosis ultraviolet shelter study. *Public Health Report* 123(1):52-60.

NEHC. 1992. *Ultraviolet radiation guide.* Navy Environmental Health Center, Bureau of Medicine and Surgery, Norfolk, VA.

NEMA. 2004. Performance testing for lighting controls and switching devices with electronic fluorescent ballasts. *Standard* 410-2004. National Electrical Manufacturers Association, Rosslyn, VA.

Philips Lighting. 1992. *Disinfection by UV-radiation.* Eindhoven, the Netherlands.

RLW Analytics. 2006. Improving indoor environment quality and energy performance of California K-12 schools, project 3: Effectiveness of UVC light for improving school performance. *Final Report,* California Energy Commission Contract 59903-300.

Scheir, R. and F.B. Fencl. 1996. Using UVGI technology to enhance IAQ. *Heating, Piping and Air Conditioning* 68:109-124.

Siegel, J., I. Walker, and M. Sherman. 2002. Dirty air conditioners: Energy implications of coil fouling. *Proceedings of the ACEEE Summer Study on Energy Efficiency in Buildings.* pp. 287-299.

Sylvania. 1982. Germicidal and short-wave ultraviolet radiation. Sylvania *Engineering Bulletin* 0-342.

Vincent, R. and P. Brickner. 2008. Safety and UV exposure. *IAQ Applications* 9(3)

WHO. 2006. Solar ultraviolet radiation: Global burden of disease from solar ultraviolet radiation. *Environmental Burden of Disease Series* 13. World Health Organization, Geneva. Available from http://www.who.int/quantifying_ehimpacts/publications/ebd13/en/index.html.

Witham, D. 2005. Ultraviolet—A superior tool for HVAC maintenance. *IUVA Congress, Tokyo.*

CHAPTER 61

CODES AND STANDARDS

THE Codes and Standards listed here represent practices, methods, or standards published by the organizations indicated. They are useful guides for the practicing engineer in determining test methods, ratings, performance requirements, and limits of HVAC&R equipment. Copies of the standards can be obtained from most of the organizations listed in the Publisher column, from Global Engineering Documents at **global.ihs.com**, or from CSSINFO at **cssinfo.com**. Addresses of the organizations are given at the end of the chapter. A comprehensive database with over 250,000 industry, government, and international standards is at **www.nssn.org**.

Selected Codes and Standards Published by Various Societies and Associations

Subject	Title	Publisher	Reference
Air	Commercial Application, Systems, and Equipment, 1st ed.	ACCA	ACCA Manual CS
Conditioners	Residential Equipment Selection, 2nd ed.	ACCA	ANSI/ACCA Manual S
	Methods of Testing Air Terminal Units	ASHRAE	ANSI/ASHRAE 130-2008
	Non-Ducted Air Conditioners and Heat Pumps—Testing and Rating for Performance	ISO	ISO 5151:1994
	Ducted Air-Conditioners and Air-to-Air Heat Pumps—Testing and Rating for Performance	ISO	ISO 13253:1995
	Guidelines for Roof Mounted Outdoor Air-Conditioner Installations	SMACNA	SMACNA 1998
	Heating and Cooling Equipment (2005)	UL/CSA	ANSI/UL 1995/C22.2 No. 236-05
Central	Performance Standard for Single Package Central Air-Conditioners and Heat Pumps	CSA	CAN/CSA-C656-05
	Performance Standard for Rating Large and Single Packaged Air Conditioners and Heat Pumps	CSA	CAN/CSA-C746-06
	Performance Standard for Split-System and Single-Package Central Air Conditioners and Heat Pumps	CSA	CAN/CSA-C656-05
	Heating and Cooling Equipment (2005)	UL/CSA	ANSI/UL 1995/C22.2 No. 236-05
Gas-Fired	Gas-Fired, Heat Activated Air Conditioning and Heat Pump Appliances	CSA	ANSI Z21.40.1-1996 (R2002)/CGA 2.91-M96
	Gas-Fired Work Activated Air Conditioning and Heat Pump Appliances (Internal Combustion)	CSA	ANSI Z21.40.2-1996 (R2002)/CGA 2.92-M96
	Performance Testing and Rating of Gas-Fired Air Conditioning and Heat Pump Appliances	CSA	ANSI Z21.40.4-1996 (R2002)/CGA 2.94-M96
Packaged Terminal	Packaged Terminal Air-Conditioners and Heat Pumps	AHRI/CSA	AHRI 310/380-04/CSA C744-04
Room	Room Air Conditioners	AHAM	ANSI/AHAM RAC-1-2008
	Method of Testing for Rating Room Air Conditioners and Packaged Terminal Air Conditioners	ASHRAE	ANSI/ASHRAE 16-1983 (RA09)
	Method of Testing for Rating Room Air Conditioner and Packaged Terminal Air Conditioner Heating Capacity	ASHRAE	ANSI/ASHRAE 58-1986 (RA09)
	Method of Testing for Rating Fan-Coil Conditioners	ASHRAE	ANSI/ASHRAE 79-2002 (RA06)
	Performance Standard for Room Air Conditioners	CSA	CAN/CSA-C368.1-M90 (R2007)
	Room Air Conditioners	CSA	C22.2 No. 117-1970 (R2007)
	Room Air Conditioners (2007)	UL	ANSI/UL 484
Unitary	Unitary Air-Conditioning and Air-Source Heat Pump Equipment	AHRI	ANSI/AHRI 210/240-2006
	Sound Rating of Outdoor Unitary Equipment	AHRI	AHRI 270-95
	Application of Sound Rating Levels of Outdoor Unitary Equipment	AHRI	AHRI 275-97
	Commercial and Industrial Unitary Air-Conditioning and Heat Pump Equipment	AHRI	AHRI 340/360-2007
	Methods of Testing for Rating Electrically Driven Unitary Air-Conditioning and Heat Pump Equipment	ASHRAE	ANSI/ASHRAE 37-2009
	Methods of Testing for Rating Heat-Operated Unitary Air-Conditioning and Heat Pump Equipment	ASHRAE	ANSI/ASHRAE 40-2002 (RA06)
	Methods of Testing for Rating Seasonal Efficiency of Unitary Air Conditioners and Heat Pumps	ASHRAE	ANSI/ASHRAE 116-2010
	Method of Testing for Rating Computer and Data Processing Room Unitary Air Conditioners	ASHRAE	ANSI/ASHRAE 127-2007
	Method of Rating Unitary Spot Air Conditioners	ASHRAE	ANSI/ASHRAE 128-2001
Ships	Specification for Mechanically Refrigerated Shipboard Air Conditioner	ASTM	ASTM F1433-97 (2004)
Accessories	Flashing and Stand Combination for Air Conditioning Units (Unit Curb)	IAPMO	IAPMO PS 120-2004
Air	Commercial Application, Systems, and Equipment, 1st ed.	ACCA	ACCA Manual CS
Conditioning	Heat Pump Systems: Principles and Applications, 2nd ed.	ACCA	ACCA Manual H
	Residential Load Calculation, 8th ed.	ACCA	ANSI/ACCA Manual J
	Commercial Load Calculation, 4th ed.	ACCA	ACCA Manual N
	Comfort, Air Quality, and Efficiency by Design	ACCA	ACCA Manual RS
	Environmental Systems Technology, 2nd ed. (1999)	NEBB	NEBB

Selected Codes and Standards Published by Various Societies and Associations (*Continued*)

Subject	Title	Publisher	Reference
	Installation of Air Conditioning and Ventilating Systems	NFPA	NFPA 90A-02
	Standard of Purity for Use in Mobile Air-Conditioning Systems	SAE	SAE J1991-1999
	HVAC Systems—Applications, 1st ed.	SMACNA	SMACNA 1987
	HVAC Systems—Duct Design, 4th ed.	SMACNA	SMACNA 2006
	Heating and Cooling Equipment (2005)	UL/CSA	ANSI/UL 1995/C22.2 No. 236-05
Aircraft	Air Conditioning of Aircraft Cargo	SAE	SAE AIR806B-1997
	Aircraft Fuel Weight Penalty Due to Air Conditioning	SAE	SAE AIR1168/8-1989
	Air Conditioning Systems for Subsonic Airplanes	SAE	SAE ARP85E-1991
	Environmental Control Systems Terminology	SAE	SAE ARP147E-2001
	Testing of Airplane Installed Environmental Control Systems (ECS)	SAE	SAE ARP217D-1999
	Guide for Qualification Testing of Aircraft Air Valves	SAE	SAE ARP986C-1997
	Control of Excess Humidity in Avionics Cooling	SAE	SAE ARP987A-1997
	Engine Bleed Air Systems for Aircraft	SAE	SAE ARP1796-2007
	Aircraft Ground Air Conditioning Service Connection	SAE	SAE AS4262A-1997
	Air Cycle Air Conditioning Systems for Military Air Vehicles	SAE	SAE AS4073-2000
Automotive	Refrigerant 12 Automotive Air-Conditioning Hose	SAE	SAE J51-2004
	Design Guidelines for Air Conditioning Systems for Off-Road Operator Enclosures	SAE	SAE J169-1985
	Test Method for Measuring Power Consumption of Air Conditioning and Brake Compressors for Trucks and Buses	SAE	SAE J1340-2003
	Information Relating to Duty Cycles and Average Power Requirements of Truck and Bus Engine Accessories	SAE	SAE J1343-2000
	Rating Air-Conditioner Evaporator Air Delivery and Cooling Capacities	SAE	SAE J1487-2004
	Recovery and Recycle Equipment for Mobile Automotive Air-Conditioning Systems	SAE	SAE J1990-1999
	R134a Refrigerant Automotive Air-Conditioning Hose	SAE	SAE J2064-2005
	Service Hose for Automotive Air Conditioning	SAE	SAE J2196-1997
Ships	Mechanical Refrigeration and Air-Conditioning Installations Aboard Ship	ASHRAE	ANSI/ASHRAE 26-2010
	Practice for Mechanical Symbols, Shipboard Heating, Ventilation, and Air Conditioning (HVAC)	ASTM	ASTM F856-97 (2004)
Air Curtains	Laboratory Methods of Testing Air Curtains for Aerodynamic Performance	AMCA	AMCA 220-05
	Air Terminals	AHRI	AHRI 880-98
	Standard Methods for Laboratory Airflow Measurement	ASHRAE	ANSI/ASHRAE 41.2-1987 (RA92)
	Method of Testing the Performance of Air Outlets and Inlets	ASHRAE	ANSI/ASHRAE 70-2006
	Rating the Performance of Residential Mechanical Ventilating Equipment	CSA	CAN/CSA C260-M90 (R2007)
	Air Curtains for Entranceways in Food and Food Service Establishments	NSF	NSF/ANSI 37-2007
Air Diffusion	Air Distribution Basics for Residential and Small Commercial Buildings, 1st ed.	ACCA	ACCA Manual T
	Method of Testing the Performance of Air Outlets and Inlets	ASHRAE	ANSI/ASHRAE 70-2006
	Method of Testing for Room Air Diffusion	ASHRAE	ANSI/ASHRAE 113-2009
Air Filters	Comfort, Air Quality, and Efficiency by Design	ACCA	ACCA Manual RS
	Industrial Ventilation: A Manual of Recommended Practice, 26th ed. (2007)	ACGIH	ACGIH
	Air Cleaners	AHAM	ANSI/AHAM AC-1-2006
	Residential Air Filter Equipment	AHRI	AHRI 680-2004
	Commercial and Industrial Air Filter Equipment	AHRI	AHRI 850-2004
	Agricultural Cabs—Engineering Control of Environmental Air Quality—Part 1: Definitions, Test Methods, and Safety Procedures	ASABE	ANSI/ASAE S525-1.2-2003
	Part 2: Pesticide Vapor Filters—Test Procedure and Performance Criteria	ASABE	ANSI/ASAE S525-2-2003
	Method of Testing General Ventilation Air-Cleaning Devices for Removal Efficiency by Particle Size	ASHRAE	ANSI/ASHRAE 52.2-2007
	Code on Nuclear Air and Gas Treatment	ASME	ASME AG-1-2003
	Nuclear Power Plant Air-Cleaning Units and Components	ASME	ASME N509-2002
	Testing of Nuclear Air-Treatment Systems	ASME	ASME N510-2007
	Specification for Filter Units, Air Conditioning: Viscous-Impingement and Dry Types, Replaceable	ASTM	ASTM F1040-87 (2007)
	Test Method for Air Cleaning Performance of a High-Efficiency Particulate Air Filter System	ASTM	ASTM F1471-93 (2001)
	Specification for Filters Used in Air or Nitrogen Systems	ASTM	ASTM F1791-00 (2006)
	Method for Sodium Flame Test for Air Filters	BSI	BS 3928:1969
	Particulate Air Filters for General Ventilation: Determination of Filtration Performance	BSI	BS EN 779:2002
	Electrostatic Air Cleaners (2000)	UL	ANSI/UL 867
	High-Efficiency, Particulate, Air Filter Units (1996)	UL	ANSI/UL 586
	Air Filter Units (2004)	UL	ANSI/UL 900
	Exhaust Hoods for Commercial Cooking Equipment (1995)	UL	UL 710
	Grease Filters for Exhaust Ducts (2000)	UL	UL 1046

Selected Codes and Standards Published by Various Societies and Associations (*Continued*)

Subject	Title	Publisher	Reference
Air-Handling Units	Commercial Application, Systems, and Equipment, 1st ed.	ACCA	ACCA Manual CS
	Central Station Air-Handling Units	AHRI	ANSI/AHRI 430-99
	Non-Recirculating Direct Gas-Fired Industrial Air Heaters	CSA	ANSI Z83.4-2003/CSA 3.7-2003
Air Leakage	Residential Duct Diagnostics and Repair (2003)	ACCA	ACCA
	Air Leakage Performance for Detached Single-Family Residential Buildings	ASHRAE	ANSI/ASHRAE 119-1988 (RA04)
	Method of Determining Air Change Rates in Detached Dwellings	ASHRAE	ANSI/ASHRAE 136-1993 (RA06)
	Test Method for Determining Air Change in a Single Zone by Means of a Tracer Gas Dilution	ASTM	ASTM E741-00 (2006)
	Test Method for Determining Air Leakage Rate by Fan Pressurization	ASTM	ASTM E779-03
	Test Method for Field Measurement of Air Leakage Through Installed Exterior Window and Doors	ASTM	ASTM E783-02
	Practices for Air Leakage Site Detection in Building Envelopes and Air Retarder Systems	ASTM	ASTM E1186-03
	Test Method for Determining the Rate of Air Leakage Through Exterior Windows, Curtain Walls, and Doors Under Specified Pressure and Temperature Differences Across the Specimen	ASTM	ASTM E1424-91 (2000)
	Test Methods for Determining Airtightness of Buildings Using an Orifice Blower Door	ASTM	ASTM E1827-96 (2007)
	Practice for Determining the Effects of Temperature Cycling on Fenestration Products	ASTM	ASTM E2264-05
	Test Method for Determining Air Flow Through the Face and Sides of Exterior Windows, Curtain Walls, and Doors Under Specified Pressure Differences Across the Specimen	ASTM	ASTM E2319-04
	Test Method for Determining Air Leakage of Air Barrier Assemblies	ASTM	ASTM E2357-05
Boilers	Packaged Boiler Engineering Manual (1999)	ABMA	ABMA 100
	Selected Codes and Standards of the Boiler Industry (2001)	ABMA	ABMA 103
	Operation and Maintenance Safety Manual (1995)	ABMA	ABMA 106
	Fluidized Bed Combustion Guidelines (1995)	ABMA	ABMA 200
	Guide to Clean and Efficient Operation of Coal Stoker-Fired Boilers (2002)	ABMA	ABMA 203
	Guideline for Performance Evaluation of Heat Recovery Steam Generating Equipment (1995)	ABMA	ABMA 300
	Guidelines for Industrial Boiler Performance Improvement (1999)	ABMA	ABMA 302
	Measurement of Sound from Steam Generators (1995)	ABMA	ABMA 304
	Guideline for Gas and Oil Emission Factors for Industrial, Commercial, and Institutional Boilers (1997)	ABMA	ABMA 305
	Combustion Control Guidelines for Single Burner Firetube and Watertube Industrial/Commercial/Institutional Boilers (1999)	ABMA	ABMA 307
	Combustion Control Guidelines for Multiple-Burner Boilers (2001)	ABMA	ABMA 308
	Boiler Water Quality Requirements and Associated Steam Quality for Industrial/Commercial and Institutional Boilers (2005)	ABMA	ABMA 402
	Commercial Application, Systems, and Equipment, 1st ed.	ACCA	ACCA Manual CS
	Method of Testing for Annual Fuel Utilization Efficiency of Residential Central Furnaces and Boilers	ASHRAE	ANSI/ASHRAE 103-2007
	Boiler and Pressure Vessel Code—Section I: Power Boilers; Section IV: Heating Boilers	ASME	BPVC-2007
	Fired Steam Generators	ASME	ASME PTC 4-1998
	Boiler, Pressure Vessel, and Pressure Piping Code	CSA	CSA B51-2003 (R2007)
	Testing Standard for Commercial Boilers, 2nd ed. (2007)	HYDI	HYDI BTS-2007
	Rating Procedure for Heating Boilers, 6th ed. (2005)	HYDI	IBR
	Prevention of Furnace Explosions/Implosions in Multiple Burner Boilers	NFPA	ANSI /NFPA 8502-99
	Heating, Water Supply, and Power Boilers—Electric (2004)	UL	ANSI/UL 834
	Boiler and Combustion Systems Hazards Code	NFPA	NFPA 85-07
Gas or Oil	Gas-Fired Low-Pressure Steam and Hot Water Boilers	CSA	ANSI Z21.13-2004/CSA 4.9-2004
	Controls and Safety Devices for Automatically Fired Boilers	ASME	ASME CSD-1-2006
	Industrial and Commercial Gas-Fired Package Boilers	CSA	CAN 1-3.1-77 (R2006)
	Oil-Burning Equipment: Steam and Hot-Water Boilers	CSA	B140.7-2005
	Single Burner Boiler Operations	NFPA	ANSI/NFPA 8501-01
	Prevention of Furnace Explosions/Implosions in Multiple Burner Boilers	NFPA	ANSI/NFPA 8502-99
	Oil-Fired Boiler Assemblies (1995)	UL	UL 726
	Commercial-Industrial Gas Heating Equipment (2006)	UL	UL 795
	Standards and Typical Specifications for Tray Type Deaerators, 7th ed. (2003)	HEI	HEI 2954
Terminology	Ultimate Boiler Industry Lexicon: Handbook of Power Utility and Boiler Terms and Phrases, 6th ed. (2001)	ABMA	ABMA 101
Building Codes	ASTM Standards Used in Building Codes	ASTM	ASTM
	Practice for Conducting Visual Assessments for Lead Hazards in Buildings	ASTM	ASTM E2255-04
	Standard Practice for Periodic Inspection of Building Facades for Unsafe Conditions	ASTM	ASTM E2270-05
	Structural Welding Code—Steel	AWS	AWS D1.1M/D1.1:2008
	BOCA National Building Code, 14th ed. (1999)	BOCA	BNBC
	Uniform Building Code, vol. 1, 2, and 3 (1997)	ICBO	UBC V1, V2, V3

Selected Codes and Standards Published by Various Societies and Associations (*Continued*)

Subject	Title	Publisher	Reference
	International Building Code (2009)	ICC	IBC
	International Code Council Performance Code (2009)	ICC	ICC PC
	International Existing Building Code (2009)	ICC	IEBC
	International Energy Conservation Code (2009)	ICC	IECC
	International Property Maintenance Code (2009)	ICC	IPMC
	International Residential Code (2009)	ICC	IRC
	Directory of Building Codes and Regulations, State and City Volumes (annual)	NCSBCS	NCSBCS (electronic only)
	Building Construction and Safety Code	NFPA	ANSI/NFPA 5000-2006
	National Building Code of Canada (2005)	NRCC	NRCC
	Standard Building Code (1999)	SBCCI	SBC
Mechanical	Safety Code for Elevators and Escalators	ASME	ASME A17.1-2004
	Natural Gas and Propane Installation Code	CSA	CAN/CSA-B149.1-05
	Propane Storage and Handling Code	CSA	CAN/CSA-B149.2-05
	Uniform Mechanical Code (2006)	IAPMO	IAPMO
	International Mechanical Code (2009)	ICC	IMC
	International Fuel Gas Code (2009)	ICC	IFGC
	Standard Gas Code (1999)	SBCCI	SBC
Burners	Guidelines for Burner Adjustments of Commercial Oil-Fired Boilers (1996)	ABMA	ABMA 303
	Domestic Gas Conversion Burners	CSA	ANSI Z21.17-1998 (R2004)/ CSA 2.7-M98
	Installation of Domestic Gas Conversion Burners	CSA	ANSI Z21.8-1994 (R2002)
	Installation Code for Oil Burning Equipment	CSA	CAN/CSA-B139-06
	Oil-Burning Equipment: General Requirements	CSA	CAN/CSA-B140.0-03
	Vapourizing-Type Oil Burners	CSA	B140.1-1966 (R2006)
	Oil Burners: Atomizing-Type	CSA	CAN/CSA-B140.2.1-M90 (R2005)
	Pressure Atomizing Oil Burner Nozzles	CSA	B140.2.2-1971 (R2006)
	Oil Burners (2003)	UL	ANSI/UL 296
	Waste Oil-Burning Air-Heating Appliances (1995)	UL	ANSI/UL 296A
	Commercial-Industrial Gas Heating Equipment (2006)	UL	UL 795
	Commercial/Industrial Gas and/or Oil-Burning Assemblies with Emission Reduction Equipment (2006)	UL	UL 2096
Chillers	Commercial Application, Systems, and Equipment, 1st ed.	ACCA	ACCA Manual CS
	Absorption Water Chilling and Water Heating Packages	AHRI	AHRI 560-2000
	Water Chilling Packages Using the Vapor Compression Cycle	AHRI	AHRI 550/590-2003
	Method of Testing Liquid-Chilling Packages	ASHRAE	ANSI/ASHRAE 30-1995
	Performance Standard for Rating Packaged Water Chillers	CSA	CAN/CSA C743-02 (R2007)
Chimneys	Specification for Clay Flue Liners	ASTM	ASTM C315-07
	Specification for Industrial Chimney Lining Brick	ASTM	ASTM C980-88 (2007)
	Practice for Installing Clay Flue Lining	ASTM	ASTM C1283-07a
	Guide for Design and Construction of Brick Liners for Industrial Chimneys	ASTM	ASTM C1298-95 (2007)
	Guide for Design, Fabrication, and Erection of Fiberglass Reinforced Plastic Chimney Liners with Coal-Fired Units	ASTM	ASTM D5364-93 (2002)
	Chimneys, Fireplaces, Vents, and Solid Fuel-Burning Appliances	NFPA	ANSI/NFPA 211-06
	Medium Heat Appliance Factory-Built Chimneys (2001)	UL	ANSI/UL 959
	Factory-Built Chimneys for Residential Type and Building Heating Appliance (2001)	UL	ANSI/UL 103
Cleanrooms	Practice for Cleaning and Maintaining Controlled Areas and Clean Rooms	ASTM	ASTM E2042-04
	Practice for Design and Construction of Aerospace Cleanrooms and Contamination Controlled Areas	ASTM	ASTM E2217-02 (2007)
	Practice for Tests of Cleanroom Materials	ASTM	ASTM E2312-04
	Practice for Aerospace Cleanrooms and Associated Controlled Environments— Cleanroom Operations	ASTM	ASTM E2352-04
	Test Method for Sizing and Counting Airborne Particulate Contamination in Clean Rooms and Other Dust-Controlled Areas Designed for Electronic and Similar Applications	ASTM	ASTM F25-04
	Practice for Continuous Sizing and Counting of Airborne Particles in Dust-Controlled Areas and Clean Rooms Using Instruments Capable of Detecting Single Sub-Micrometre and Larger Particles	ASTM	ASTM F50-07
	Procedural Standards for Certified Testing of Cleanrooms, 2nd ed. (1996)	NEBB	NEBB
Coils	Forced-Circulation Air-Cooling and Air-Heating Coils	AHRI	AHRI 410-2001
	Methods of Testing Forced Circulation Air Cooling and Air Heating Coils	ASHRAE	ANSI/ASHRAE 33-2000
Comfort Conditions	Threshold Limit Values for Physical Agents (updated annually)	ACGIH	ACGIH
	Good HVAC Practices for Residential and Commercial Buildings (2003)	ACCA	ACCA
	Comfort, Air Quality, and Efficiency by Design (1997)	ACCA	ACCA Manual RS
	Thermal Environmental Conditions for Human Occupancy	ASHRAE	ANSI/ASHRAE 55-2010

Selected Codes and Standards Published by Various Societies and Associations (*Continued*)

Subject	Title	Publisher	Reference
	Classification for Serviceability of an Office Facility for Thermal Environment and Indoor Air Conditions	ASTM	ASTM E2320-04
	Hot Environments—Estimation of the Heat Stress on Working Man, Based on the WBGT Index (Wet Bulb Globe Temperature)	ISO	ISO 7243:1989
	Ergonomics of the Thermal Environment—Analytical Determination and Interpretation of Thermal Comfort Using Calculation of the PMV and PPD Indices and Local Thermal Comfort Criteria	ISO	ISO 7730:2005
	Ergonomics of the Thermal Environment—Determination of Metabolic Rate	ISO	ISO 8996:2004
	Ergonomics of the Thermal Environment—Estimation of the Thermal Insulation and Water Vapour Resistance of a Clothing Ensemble	ISO	ISO 9920:2007
Compressors	Displacement Compressors, Vacuum Pumps and Blowers	ASME	ASME PTC 9-1970 (RA97)
	Performance Test Code on Compressors and Exhausters	ASME	ASME PTC 10-1997 (RA03)
	Compressed Air and Gas Handbook, 6th ed. (2003)	CAGI	CAGI
Refrigerant	Positive Displacement Condensing Units	AHRI	AHRI 520-2004
	Positive Displacement Refrigerant Compressors and Compressor Units	AHRI	AHRI 540-2004
	Safety Standard for Refrigeration Systems	ASHRAE	ANSI/ASHRAE 15-2010
	Methods of Testing for Rating Positive Displacement Refrigerant Compressors and Condensing Units	ASHRAE	ANSI/ASHRAE 23.1-2010
	Testing of Refrigerant Compressors	ISO	ISO 917:1989
	Refrigerant Compressors—Presentation of Performance Data	ISO	ISO 9309:1989
	Hermetic Refrigerant Motor-Compressors (1996)	UL/CSA	UL 984/C22.2 No.140.2-96 (R2001)
Computers	Method of Testing for Rating Computer and Data Processing Room Unitary Air Conditioners	ASHRAE	ANSI/ASHRAE 127-2007
	Method of Test for the Evaluation of Building Energy Analysis Computer Programs	ASHRAE	ANSI/ASHRAE 140-2007
	Protection of Electronic Computer/Data Processing Equipment	NFPA	NFPA 75-03
Condensers	Commercial Application, Systems, and Equipment, 1st ed.	ACCA	ACCA Manual CS
	Water-Cooled Refrigerant Condensers, Remote Type	AHRI	AHRI 450-2007
	Remote Mechanical-Draft Air-Cooled Refrigerant Condensers	AHRI	AHRI 460-2005
	Remote Mechanical Draft Evaporative Refrigerant Condensers	AHRI	AHRI 490-2003
	Safety Standard for Refrigeration Systems	ASHRAE	ANSI/ASHRAE 15-2010
	Method of Testing for Rating Remote Mechanical-Draft Air-Cooled Refrigerant Condensers	ASHRAE	ANSI/ASHRAE 20-1997 (RA06)
	Methods of Testing for Rating Water-Cooled Refrigerant Condensers	ASHRAE	ANSI/ASHRAE 22-2007
	Methods of Laboratory Testing Remote Mechanical-Draft Evaporative Refrigerant Condensers	ASHRAE	ANSI/ASHRAE 64-2005
	Steam Surface Condensers	ASME	ASME PTC 12.2-1998
	Standards for Steam Surface Condensers, 10th ed.	HEI	HEI 2629
	Standards for Direct Contact Barometric and Low Level Condensers, 7th ed. (1995)	HEI	HEI 2634
	Refrigerant-Containing Components and Accessories, Nonelectrical (2001)	UL	ANSI/UL 207
Condensing Units	Commercial Application, Systems, and Equipment, 1st ed.	ACCA	ACCA Manual CS
	Commercial and Industrial Unitary Air-Conditioning Condensing Units	AHRI	AHRI 365-2002
	Methods of Testing for Rating Positive Displacement Refrigerant Compressors and Condensing Units	ASHRAE	ANSI/ASHRAE 23.1-2010
	Heating and Cooling Equipment (2005)	UL/CSA	ANSI/UL 1995/C22.2 No. 236-95
Containers	Series 1 Freight Containers—Classifications, Dimensions, and Ratings	ISO	ISO 668:1995
	Series 1 Freight Containers—Specifications and Testing; Part 2: Thermal Containers	ISO	ISO 1496-2:1996
	Animal Environment in Cargo Compartments	SAE	SAE AIR1600A-1997
Controls	Temperature Control Systems (2002)	AABC	National Standards, Ch. 12
	BACnet®—A Data Communication Protocol for Building Automation and Control Networks	ASHRAE	ANSI/ASHRAE 135-2008
	Method of Test for Conformance to BACnet®	ASHRAE	ANSI/ASHRAE 135.1-2009
	Temperature Indicating and Regulating Equipment	CSA	C22.2 No. 24-93 (R2003)
	Performance Requirements for Electric Heating Line-Voltage Wall Thermostats	CSA	C273.4-M1978 (R2003)
	Performance Requirements for Thermostats Used with Individual Room Electric Space Heating Devices	CSA	CAN/CSA C828-06
	Solid-State Controls for Appliances (2003)	UL	UL 244A
	Limit Controls (1994)	UL	ANSI/UL 353
	Primary Safety Controls for Gas- and Oil-Fired Appliances (1994)	UL	ANSI/UL 372
	Temperature-Indicating and -Regulating Equipment (2007)	UL	UL 873
	Tests for Safety-Related Controls Employing Solid-State Devices (2004)	UL	UL 991
	Control Centers for Changing Message Type Electric Signals (2003)	UL	UL 1433
	Automatic Electrical Controls for Household and Similar Use; Part 1: General Requirements (2002)	UL	UL 60730-1A
	Process Control Equipment (2002)	UL	UL 61010C-1

Selected Codes and Standards Published by Various Societies and Associations (*Continued*)

Subject	Title	Publisher	Reference
Commercial and Industrial	Guidelines for Boiler Control Systems (Gas/Oil Fired Boilers) (1998)	ABMA	ABMA 301
	Guideline for the Integration of Boilers and Automated Control Systems in Heating Applications (1998)	ABMA	ABMA 306
	Industrial Control and Systems: General Requirements	NEMA	NEMA ICS 1-2000 (R2005)
	Preventive Maintenance of Industrial Control and Systems Equipment	NEMA	NEMA ICS 1.3-1986 (R2001)
	Industrial Control and Systems, Controllers, Contactors, and Overload Relays Rated Not More than 2000 Volts AC or 750 Volts DC	NEMA	NEMA ICS 2-2000 (R2004)
	Industrial Control and Systems: Instructions for the Handling, Installation, Operation and Maintenance of Motor Control Centers Rated Not More than 600 Volts	NEMA	NEMA ICS 2.3-1995 (R2002)
	Industrial Control Equipment (1999)	UL	ANSI/UL 508
Residential	Manually Operated Gas Valves for Appliances, Appliance Connector Valves and Hose End Valves	CSA	ANSI Z21.15-1997 (R03)/CGA 9.1-1997
	Gas Appliance Pressure Regulators	CSA	ANSI Z21.18-2007/CSA 6.3-2007
	Automatic Gas Ignition Systems and Components	CSA	ANSI Z21.20-2007/C22.2 No. 199-2007
	Gas Appliance Thermostats	CSA	ANSI Z21.23-2000 (R2005)
	Manually-Operated Piezo-Electric Spark Gas Ignition Systems and Components	CSA	ANSI Z21.77-2005/CGA 6.23-2005
	Manually Operated Electric Gas Ignition Systems and Components	CSA	ANSI Z21.92-2005/CSA 6.29-2005 (R2007)
	Residential Controls—Electrical Wall-Mounted Room Thermostats	NEMA	NEMA DC 3-2003
	Residential Controls—Surface Type Controls for Electric Storage Water Heaters	NEMA	NEMA DC 5-2002
	Residential Controls—Temperature Limit Controls for Electric Baseboard Heaters	NEMA	NEMA DC 10-1983 (R2003)
	Hot-Water Immersion Controls	NEMA	NEMA DC 12-1985 (R2002))
	Line-Voltage Integrally Mounted Thermostats for Electric Heaters	NEMA	NEMA DC 13-1979 (R2002)
	Residential Controls—Class 2 Transformers	NEMA	NEMA DC 20-1992 (R2003)
	Safety Guidelines for the Application, Installation, and Maintenance of Solid State Controls	NEMA	NEMA ICS 1.1-1984 (R2003)
	Electrical Quick-Connect Terminals (2003)	UL	ANSI/UL 310
Coolers	Refrigeration Equipment	CSA	CAN/CSA-C22.2 No. 120-M91 (R2004)
	Unit Coolers for Refrigeration	AHRI	AHRI 420-2000
	Refrigeration Unit Coolers (2004)	UL	ANSI/UL 412
Air	Methods of Testing Forced Convection and Natural Convection Air Coolers for Refrigeration	ASHRAE	ANSI/ASHRAE 25-2001 (RA06)
Drinking Water	Methods of Testing for Rating Drinking-Water Coolers with Self-Contained Mechanical Refrigeration	ASHRAE	ANSI/ASHRAE 18-2008
	Drinking-Water Coolers (1993)	UL	ANSI/UL 399
	Drinking Water System Components—Health Effects	NSF	NSF/ANSI 61-2007a
Evaporative	Method of Testing Direct Evaporative Air Coolers	ASHRAE	ANSI/ASHRAE 133-2008
	Method of Test for Rating Indirect Evaporative Coolers	ASHRAE	ANSI/ASHRAE 143-2007
Food and Beverage	Terminology for Milking Machines, Milk Cooling, and Bulk Milk Handling Equipment	ASABE	ASAE S300.3-2003
	Methods of Testing for Rating Vending Machines for Bottled, Canned, and Other Sealed Beverages	ASHRAE	ANSI/ASHRAE 32.1-2010
	Methods of Testing for Rating Pre-Mix and Post-Mix Beverage Dispensing Equipment	ASHRAE	ANSI/ASHRAE 32.2-2003 (RA07)
	Manual Food and Beverage Dispensing Equipment	NSF	NSF/ANSI 18-2005
	Commercial Bulk Milk Dispensing Equipment	NSF	NSF/ANSI 20-2007
	Refrigerated Vending Machines (1995)	UL	ANSI/UL 541
Liquid	Refrigerant-Cooled Liquid Coolers, Remote Type	AHRI	AHRI 480-2007
	Methods of Testing for Rating Liquid Coolers	ASHRAE	ANSI/ASHRAE 24-2009
	Liquid Cooling Systems	SAE	SAE AIR1811A-1997
Cooling Towers	Cooling Tower Testing (2002)	AABC	National Standards, Ch 13
	Commercial Application, Systems, and Equipment, 1st ed.	ACCA	ACCA Manual CS
	Bioaerosols: Assessment and Control (1999)	ACGIH	ACGIH
	Atmospheric Water Cooling Equipment	ASME	ASME PTC 23-2003
	Water-Cooling Towers	NFPA	NFPA 214-05
	Acceptance Test Code for Water Cooling Towers	CTI	CTI ATC-105 (00)
	Code for Measurement of Sound from Water Cooling Towers (2005)	CTI	CTI ATC-128 (05)
	Acceptance Test Code for Spray Cooling Systems (1985)	CTI	CTI ATC-133 (85)
	Nomenclature for Industrial Water Cooling Towers (1997)	CTI	CTI NCL-109 (97)
	Recommended Practice for Airflow Testing of Cooling Towers (1994)	CTI	CTI PFM-143 (94)
	Fiberglass-Reinforced Plastic Panels (2002)	CTI	CTI STD-131 (02)
	Certification of Water Cooling Tower Thermal Performance (R2004)	CTI	CTI STD-201 (04)
Crop Drying	Density, Specific Gravity, and Mass-Moisture Relationships of Grain for Storage	ASABE	ANSI/ASAE D241.4-2003
	Dielectric Properties of Grain and Seed	ASABE	ASAE D293.2-1989 (R2005)
	Thermal Properties of Grain and Grain Products	ASABE	ASAE D243.4-2003
	Moisture Relationships of Plant-Based Agricultural Products	ASABE	ASAE D245.5-19995 (R2001)
	Construction and Rating of Equipment for Drying Farm Crops	ASABE	ASAE S248.3-1976 (R2005)

Selected Codes and Standards Published by Various Societies and Associations (*Continued*)

Subject	Title	Publisher	Reference
	Cubes, Pellets, and Crumbles—Definitions and Methods for Determining Density, Durability, and Moisture Content	ASABE	ASAE S269.4-1991
	Resistance to Airflow of Grains, Seeds, Other Agricultural Products, and Perforated Metal Sheets	ASABE	ASAE D272.3-1996
	Shelled Corn Storage Time for 0.5% Dry Matter Loss	ASABE	ASAE D535-2005
	Moisture Measurement—Unground Grain and Seeds	ASABE	ASAE S352.2-2003
	Moisture Measurement—Meat and Meat Products	ASABE	ASAE S353-2003
	Moisture Measurement—Forages	ASABE	ASAE S358.2-2003
	Moisture Measurement—Peanuts	ASABE	ASAE S410.1-2003
	Energy Efficiency Test Procedure for Tobacco Curing Structures	ASABE	ASAE S416-2003
	Thin-Layer Drying of Agricultural Crops	ASABE	ANSI/ASAE S448.1-2001 (R2006)
	Moisture Measurement—Tobacco	ASABE	ASAE S487-2003
	Thin-Layer Drying of Agricultural Crops	ASABE	ASAE S488-1990 (R2005)
	Temperature Sensor Locations for Seed-Cotton Drying Systems	ASABE	ASAE 530.1-2007
Dehumidifiers	Commercial Application, Systems, and Equipment, 1st ed.	ACCA	ACCA Manual CS
	Bioaerosols: Assessment and Control (1999)	ACGIH	ACGIH
	Dehumidifiers	AHAM	ANSI/AHAM DH-1-2008
	Method of Testing for Rating Desiccant Dehumidifiers Utilizing Heat for the Regeneration Process	ASHRAE	ANSI/ASHRAE 139-2007
	Moisture Separator Reheaters	ASME	PTC 12.4-1992 (RA04)
	Dehumidifiers	CSA	C22.2 No. 92-1971 (R2004)
	Performance of Dehumidifiers	CSA	CAN/CSA C749-07
	Dehumidifiers (2004)	UL	ANSI/UL 474
Desiccants	Method of Testing Desiccants for Refrigerant Drying	ASHRAE	ANSI/ASHRAE 35-2010
Documentation	Preparation of Operating and Maintenance Documentation for Building Systems	ASHRAE	ASHRAE *Guideline* 4-2008
Driers	Liquid-Line Driers	AHRI	ANSI/AHRI 710-2004
	Method of Testing Liquid Line Refrigerant Driers	ASHRAE	ANSI/ASHRAE 63.1-1995 (RA01)
	Refrigerant-Containing Components and Accessories, Nonelectrical (2001)	UL	ANSI/UL 207
Ducts and Fittings	Hose, Air Duct, Flexible Nonmetallic, Aircraft	SAE	SAE AS1501C-1994
	Ducted Electric Heat Guide for Air Handling Systems, 2nd ed.	SMACNA	SMACNA 1994
	Factory-Made Air Ducts and Air Connectors (2005)	UL	ANSI/UL 181
Construction	Industrial Ventilation: A Manual of Recommended Practice, 26th ed. (2007)	ACGIH	ACGIH
	Preferred Metric Sizes for Flat, Round, Square, Rectangular, and Hexagonal Metal Products	ASME	ASME B32.100-2005
	Sheet Metal Welding Code	AWS	AWS D9.1M/D9.1:2006
	Fibrous Glass Duct Construction Standards, 5th ed.	NAIMA	NAIMA AH116
	Residential Fibrous Glass Duct Construction Standards, 3rd ed.	NAIMA	NAIMA AH119
	Thermoplastic Duct (PVC) Construction Manual, 2nd ed.	SMACNA	SMACNA 1995
	Accepted Industry Practices for Sheet Metal Lagging, 1st ed.	SMACNA	SMACNA 2002
	Fibrous Glass Duct Construction Standards, 7th ed.	SMACNA	SMACNA 2003
	HVAC Duct Construction Standards, Metal and Flexible, 3rd ed.	SMACNA	SMACNA 2005
	Rectangular Industrial Duct Construction Standards, 2nd ed.	SMACNA	SMACNA 2004
Industrial	Round Industrial Duct Construction Standards, 2nd ed.	SMACNA	SMACNA 1999
	Rectangular Industrial Duct Construction Standards, 2nd ed.	SMACNA	SMACNA 2004
Installation	Flexible Duct Performance and Installation Standards, 4th ed.	ADC	ADC-91
	Installation of Air Conditioning and Ventilating Systems	NFPA	NFPA 90A-06
	Installation of Warm Air Heating and Air-Conditioning Systems	NFPA	NFPA 90B-06
Material Specifications	Specification for General Requirements for Flat-Rolled Stainless and Heat-Resisting Steel Plate, Sheet and Strip	ASTM	ASTM A480/A480M-06b
	Specification for General Requirements for Steel, Sheet, Carbon, and High-Strength, Low-Alloy, Hot-Rolled and Cold-Rolled	ASTM	ASTM A568/A568M-07a
	Specification for Steel Sheet, Zinc-Coated (Galvanized) or Zinc-Iron Alloy-Coated (Galvannealed) by the Hot-Dipped Process	ASTM	ASTM A653/A653M-07
	Specification for General Requirements for Steel Sheet, Metallic-Coated by the Hot-Dip Process	ASTM	ASTM A924/A924M-07
	Specification for Steel, Sheet and Strip, Cold-Rolled, Carbon, Structural, High-Strength Low-Alloy and High-Strength Low-Alloy with Improved Formability	ASTM	ASTM A1008/A1008M-07a
	Specification for Steel, Sheet and Strip, Hot-Rolled, Carbon, Structural, High-Strength Low-Alloy and High-Strength Low-Alloy with Improved Formability	ASTM	ASTM A1011/A1011M-07
	Practice for Measuring Flatness Characteristics of Coated Sheet Products	ASTM	ASTM A1030/A1030M-05
System Design	Installation Techniques for Perimeter Heating and Cooling, 11th ed.	ACCA	ACCA Manual 4
	Residential Duct Systems	ACCA	ANSI/ACCA Manual D
	Commercial Low Pressure, Low Velocity Duct System Design, 1st ed.	ACCA	ACCA Manual Q
	Air Distribution Basics for Residential and Small Commercial Buildings, 1st ed.	ACCA	ACCA Manual T
	Method of Test for Determining the Design and Seasonal Efficiencies of Residential Thermal Distribution Systems	ASHRAE	ANSI/ASHRAE 152-2004

Selected Codes and Standards Published by Various Societies and Associations (*Continued*)

Subject	Title	Publisher	Reference
Testing	Closure Systems for Use with Rigid Air Ducts (2005)	UL	ANSI/UL 181A
	Closure Systems for Use with Flexible Air Ducts and Air Connectors (2005)	UL	ANSI/UL 181B
	Duct Leakage Testing (2002)	AABC	National Standards, Ch 5
	Residential Duct Diagnostics and Repair (2003)	ACCA	ACCA
	Flexible Air Duct Test Code	ADC	ADC FD-72 (R1979)
	Test Method for Measuring Acoustical and Airflow Performance of Duct Liner Materials and Prefabricated Silencers	ASTM	ASTM E477-06a
	Method of Testing to Determine Flow Resistance of HVAC Ducts and Fittings	ASHRAE	ANSI/ASHRAE 120-2008
	Method of Testing HVAC Air Ducts and Fittings	ASHRAE	ANSI/ASHRAE/SMACNA 126-2008
	HVAC Air Duct Leakage Test Manual, 1st ed.	SMACNA	SMACNA 1985
	HVAC Duct Systems Inspection Guide, 3rd ed.	SMACNA	SMACNA 2006
Electrical	Electrical Power Systems and Equipment—Voltage Ratings	ANSI	ANSI C84.1-2006
	Test Method for Bond Strength of Electrical Insulating Varnishes by the Helical Coil Test	ASTM	ASTM D2519-07
	Standard Specification for Shelter, Electrical Equipment, Lightweight	ASTM	ASTM E2377-04
	Canadian Electrical Code, Part I (20th ed.)	CSA	C22.1-06
	Part II—General Requirements	CSA	CAN/CSA-C22.2 No. 0-M91 (R2006)
	ICC Electrical Code, Administrative Provisions (2006)	ICC	ICCEC
	Enclosures for Electrical Equipment (1000 Volts Maximum)	NEMA	ANSI/NEMA 250-2003
	Low Voltage Cartridge Fuses	NEMA	NEMA FU 1-2002 (R2007)
	Industrial Control and Systems: Terminal Blocks	NEMA	NEMA ICS 4-2005
	Industrial Control and Systems: Enclosures	NEMA	ANSI/NEMA ICS 6-1993 (R2006)
	Application Guide for Ground Fault Protective Devices for Equipment	NEMA	ANSI/NEMA PB 2.2-2004
	General Color Requirements for Wiring Devices	NEMA	NEMA WD 1-1999 (R2005)
	Wiring Devices—Dimensional Requirements	NEMA	ANSI/NEMA WD 6-2002
	National Electrical Code	NFPA	NFPA 70-08
	National Fire Alarm Code	NFPA	NFPA 72-07
	Compatibility of Electrical Connectors and Wiring	SAE	SAE AIR1329A-1988
	Molded-Case Circuit Breakers, Molded-Case Switches, and Circuit-Breaker Enclosures	UL	ANSI/UL489
Energy	Air-Conditioning and Refrigerating Equipment Nameplate Voltages	AHRI	AHRI 110-2002
	Comfort, Air Quality, and Efficiency by Design	ACCA	ACCA Manual RS
	Energy Standard for Buildings Except Low-Rise Residential Buildings	ASHRAE	ANSI/ASHRAE/IES 90.1-2010
	Energy-Efficient Design of Low-Rise Residential Buildings	ASHRAE	ANSI/ASHRAE/IES 90.2-2007
	Energy Conservation in Existing Buildings	ASHRAE	ANSI/ASHRAE/IES 100-2006
	Methods of Measuring, Expressing, and Comparing Building Energy Performance	ASHRAE	ANSI/ASHRAE 105-2007
	Method of Test for the Evaluation of Building Energy Analysis Computer Programs	ASHRAE	ANSI/ASHRAE 140-2007
	Method of Test for Determining the Design and Seasonal Efficiencies of Residential Thermal Distribution Systems	ASHRAE	ANSI/ASHRAE 152-2004
	Standard for the Design of High-Performance, Green Buildings Except Low-Rise Residential Buildings	ASHRAE/ USGBC	ANSI/ASHRAE/USGBC/IES 189.1-2009
	Fuel Cell Power Systems Performance	ASME	PTC 50-2002
	International Energy Conservation Code (2009)	ICC	IECC
	International Green Construction Code™	ICC	IGCC
	Uniform Solar Energy Code (2000)	IAPMO	IAPMO
	Energy Management Guide for Selection and Use of Fixed Frequency Medium AC Squirrel-Cage Polyphase Induction Motors	NEMA	NEMA MG 10-2001 (R2007)
	Energy Management Guide for Selection and Use of Single-Phase Motors	NEMA	NEMA MG 11-1977 (R2007)
	HVAC Systems—Commissioning Manual, 1st ed.	SMACNA	SMACNA 1994
	Building Systems Analysis and Retrofit Manual, 1st ed.	SMACNA	SMACNA 1995
	Energy Systems Analysis and Management, 1st ed.	SMACNA	SMACNA 1997
	Energy Management Equipment (2007)	UL	UL 916
Exhaust Systems	Fan Systems: Supply/Return/Relief/Exhaust (2002)	AABC	National Standards, Ch 10
	Commercial Application, Systems, and Equipment, 1st ed.	ACCA	ACCA Manual CS
	Industrial Ventilation: A Manual of Recommended Practice, 26th ed. (2007)	ACGIH	ACGIH
	Fundamentals Governing the Design and Operation of Local Exhaust Ventilation Systems	AIHA	ANSI/AIHA Z9.2-2006
	Safety Code for Design, Construction, and Ventilation of Spray Finishing Operations	AIHA	ANSI/AIHA Z9.3-2007
	Laboratory Ventilation	AIHA	ANSI/AIHA Z9.5-2003
	Recirculation of Air from Industrial Process Exhaust Systems	AIHA	ANSI/AIHA Z9.7-2007
	Method of Testing Performance of Laboratory Fume Hoods	ASHRAE	ANSI/ASHRAE 110-1995
	Ventilation for Commercial Cooking Operations	ASHRAE	ANSI/ASHRAE 154-2003
	Performance Test Code on Compressors and Exhausters	ASME	PTC 10-1997 (RA03)
	Flue and Exhaust Gas Analyses	ASME	PTC 19.10-1981
	Mechanical Flue-Gas Exhausters	CSA	CAN B255-M81 (R2005)

Selected Codes and Standards Published by Various Societies and Associations (*Continued*)

Subject	Title	Publisher	Reference
	Exhaust Systems for Air Conveying of Vapors, Gases, Mists, and Noncombustible Particulate Solids	NFPA	ANSI/NFPA 91-04
	Draft Equipment (2006)	UL	UL 378
Expansion Valves	Thermostatic Refrigerant Expansion Valves	AHRI	ANSI/AHRI 750-2007
	Method of Testing Capacity of Thermostatic Refrigerant Expansion Valves	ASHRAE	ANSI/ASHRAE 17-2008
Fan-Coil Units	Industrial Ventilation: A Manual of Recommended Practice, 26th ed. (2007)	ACGIH	ACGIH
	Room Fan-Coils	AHRI	AHRI 440-2005
	Methods of Testing for Rating Fan-Coil Conditioners	ASHRAE	ANSI/ASHRAE 79-2002 (RA06)
	Heating and Cooling Equipment (2005)	UL/CSA	ANSI/UL 1995/C22.2 No. 236-95
Fans	Residential Duct Systems	ACCA	ANSI/ACCA Manual D
	Commercial Low Pressure, Low Velocity Duct System Design, 1st ed.	ACCA	ACCA Manual Q
	Industrial Ventilation: A Manual of Recommended Practice, 26th ed. (2007)	ACGIH	ACGIH
	Standards Handbook	AMCA	AMCA 99-03
	Drive Arrangements for Centrifugal Fans	AMCA	ANSI/AMCA 99-2404-03
	Inlet Box Positions for Centrifugal Fans	AMCA	ANSI/AMCA 99-2405-03
	Designation for Rotation and Discharge of Centrifugal Fans	AMCA	ANSI/AMCA 99-2406-03
	Motor Positions for Belt or Chain Drive Centrifugal Fans	AMCA	ANSI/AMCA 99-2407-03
	Operating Limits for Centrifugal Fans	AMCA	AMCA 99-2408-69
	Drive Arrangements for Tubular Centrifugal Fans	AMCA	ANSI/AMCA 99-2410-03
	Impeller Diameters and Outlet Areas for Centrifugal Fans	AMCA	ANSI/AMCA 99-2412-03
	Impeller Diameters and Outlet Areas for Industrial Centrifugal Fans	AMCA	ANSI/AMCA 99-2413-03
	Impeller Diameters and Outlet Areas for Tubular Centrifugal Fans	AMCA	ANSI/AMCA 99-2414-03
	Dimensions for Axial Fans	AMCA	ANSI/AMCA 99-3001-03
	Drive Arrangements for Axial Fans	AMCA	ANSI/AMCA 99-3404-03
	Air Systems	AMCA	AMCA 200-95 (R2007)
	Fans and Systems	AMCA	AMCA 201-02 (R2007)
	Troubleshooting	AMCA	AMCA 202-98 (R2007)
	Field Performance Measurement of Fan Systems	AMCA	AMCA 203-90 (R2007)
	Balance Quality and Vibration Levels for Fans	AMCA	ANSI/AMCA 204-05
	Laboratory Methods of Testing Air Circulator Fans for Rating	AMCA	ANSI/AMCA 230-07
	Laboratory Method of Testing Positive Pressure Ventilators for Rating	AMCA	ANSI/AMCA 240-06
	Reverberant Room Method for Sound Testing of Fans	AMCA	AMCA 300-05
	Methods for Calculating Fan Sound Ratings from Laboratory Test Data	AMCA	AMCA 301-06
	Application of Sone Ratings for Non-Ducted Air Moving Devices	AMCA	AMCA 302-73 (R2008)
	Application of Sound Power Level Ratings for Fans	AMCA	AMCA 303-79 (R2008)
	Recommended Safety Practices for Users and Installers of Industrial and Commercial Fans	AMCA	AMCA 410-96
	Industrial Process/Power Generation Fans: Site Performance Test Standard	AMCA	AMCA 803-02
	Mechanical Balance of Fans and Blowers	AHRI	AHRI *Guideline* G-2002
	Acoustics—Measurement of Noise and Vibration of Small Air-Moving Devices—Part 1: Airborne Noise Emission	ASA	ANSI S12.11-2003/Part 1/ISO 10302:1996 (MOD)
	Part 2: Structure-Borne Vibration	ASA	ANSI S12.11-2003/Part 2
	Laboratory Methods of Testing Fans for Certified Aerodynamic Performance Rating	ASHRAE/ AMCA	ANSI/ASHRAE 51-2007 ANSI/AMCA 210-07
	Laboratory Method of Testing to Determine the Sound Power in a Duct	ASHRAE/ AMCA	ANSI/ASHRAE 68-1997 ANSI/AMCA 330-97
	Laboratory Methods of Testing Fans Used to Exhaust Smoke in Smoke Management Systems	ASHRAE	ANSI/ASHRAE 149-2000 (RA09)
	Ventilation for Commercial Cooking Operations	ASHRAE	ANSI/ASHRAE 154-2003
	Fans	ASME	ANSI/ASME PTC 11-1984 (RA03)
	Fans and Ventilators	CSA	C22.2 No. 113-M1984 (R2004)
	Rating the Performance of Residential Mechanical Ventilating Equipment	CSA	CAN/CSA C260-M90 (R2007)
	Energy Performance of Ceiling Fans	CSA	CAN/CSA C814-96 (R2007)
	Electric Fans (1999)	UL	ANSI/UL 507
	Power Ventilators (2004)	UL	ANSI/UL 705
Fenestration	Test Method for Accelerated Weathering of Sealed Insulating Glass Units	ASTM	ASTM E773-01
	Practice for Calculation of Photometric Transmittance and Reflectance of Materials to Solar Radiation	ASTM	ASTM E971-88 (2003)
	Test Method for Solar Photometric Transmittance of Sheet Materials Using Sunlight	ASTM	ASTM E972-96 (2007)
	Test Method for Solar Transmittance (Terrestrial) of Sheet Materials Using Sunlight	ASTM	ASTM E1084-86 (2003)
	Practice for Determining the Load Resistance of Glass in Buildings	ASTM	ASTM E1300-07e1
	Practice for Installation of Exterior Windows, Doors and Skylights	ASTM	ASTM E2112-07
	Test Method for Insulating Glass Unit Performance	ASTM	ASTM E2188-02
	Test Method for Testing Resistance to Fogging Insulating Glass Units	ASTM	ASTM E2189-02
	Specification for Insulating Glass Unit Performance and Evaluation	ASTM	ASTM E2190-02

Selected Codes and Standards Published by Various Societies and Associations (*Continued*)

Subject	Title	Publisher	Reference
	Guide for Assessing the Durability of Absorptive Electrochemical Coatings within Sealed Insulating Glass Units	ASTM	ASTM E2354-04
	Tables for Reference Solar Spectral Irradiance: Direct Normal and Hemispherical on 37° Tilted Surface	ASTM	ASTM G173-03e1
	Windows	CSA	A440-08
	Energy Performance of Windows and Other Fenestration Systems	CSA	A440.3-04
	Window, Door, and Skylight Installation	CSA	A440.4-98
	Energy Performance Evaluation of Swinging Doors	CSA	A453-95 (R2000)
Filter-Driers	Flow-Capacity Rating of Suction-Line Filters and Suction-Line Filter-Driers	AHRI	AHRI 730-2005
	Method of Testing Liquid Line Filter-Drier Filtration Capability	ASHRAE	ANSI/ASHRAE 63.2-1996 (RA2010)
	Method of Testing Flow Capacity of Suction Line Filters and Filter-Driers	ASHRAE	ANSI/ASHRAE 78-1985 (RA07)
Fireplaces	Factory-Built Fireplaces (1996)	UL	ANSI/UL 127
	Fireplace Stoves (2007)	UL	ANSI/UL 737
Fire Protection	Test Method for Surface Burning Characteristics of Building Materials	ASTM/NFPA	ASTM E84-08
	Test Methods for Fire Test of Building Construction and Materials	ASTM	ASTM E119-08
	Test Method for Room Fire Test of Wall and Ceiling Materials and Assemblies	ASTM	ASTM E2257-03
	Test Method for Determining Fire Resistance of Perimeter Fire Barriers Using Intermediate-Scale Multi-Story Test Apparatus	ASTM	ASTM E2307-04e1
	Guide for Laboratory Monitors	ASTM	ASTM E2335-04
	Test Method for Fire Resistance Grease Duct Enclosure Systems	ASTM	ASTM E2336-04
	Practice for Specimen Preparation and Mounting of Paper or Vinyl Wall Coverings to Assess Surface Burning Characteristics	ASTM	ASTM E2404-07a
	BOCA National Fire Prevention Code, 11th ed. (1999)	BOCA	BNFPC
	Uniform Fire Code	IFCI	UPC 1997
	International Fire Code (2009)	ICC	IFC
	International Mechanical Code (2009)	ICC	IMC
	International Urban-Wildland Interface Code (2009)	ICC	IUWIC
	Fire-Resistance Tests—Elements of Building Construction; Part 1: Gen. Requirements	ISO	ISO 834-1:1999
	Fire-Resistance Tests—Door and Shutter Assemblies	ISO	ISO 3008:2007
	Reaction to Fire Tests—Ignitability of Building Products Using a Radiant Heat Source	ISO	ISO 5657:1997
	Fire-Resistance Tests—Ventilating Ducts	ISO	ISO 6944:1985
	Fire Service Annunciator and Interface	NEMA	NEMA SB 30-2005
	Fire Protection Handbook (2008)	NFPA	NFPA
	National Fire Codes (issued annually)	NFPA	NFPA
	Fire Protection Guide to Hazardous Materials	NFPA	NFPA HAZ-01
	Uniform Fire Code	NFPA	NFPA 1-06
	Installation of Sprinkler Systems	NFPA	NFPA 13-2007
	Flammable and Combustible Liquids Code	NFPA	NFPA 30-08
	Fire Protection for Laboratories Using Chemicals	NFPA	NFPA 45-04
	National Fire Alarm Code	NFPA	NFPA 72-07
	Fire Doors and Fire Windows	NFPA	NFPA 80-07
	Health Care Facilities	NFPA	NFPA 99-05
	Life Safety Code	NFPA	NFPA 101-06
	Methods of Fire Tests of Door Assemblies	NFPA	NFPA 252-08
	Standard Fire Code (1999)	SBCCI	SFPC
	Fire, Smoke and Radiation Damper Installation Guide for HVAC Systems, 5th ed.	SMACNA	SMACNA 2002
	Fire Tests of Door Assemblies (2008)	UL	ANSI/UL 10B
	Heat Responsive Links for Fire-Protection Service (2003)	UL	ANSI/UL 33
	Fire Tests of Building Construction and Materials (2003)	UL	ANSI/UL 263
	Fire Dampers (2006)	UL	ANSI/UL 555
	Fire Tests of Through-Penetration Firestops (2003)	UL	ANSI/UL 1479
Smoke Management	Commissioning Smoke Management Systems	ASHRAE	ASHRAE *Guideline* 5-1994 (RA01)
	Laboratory Methods of Testing Fans Used to Exhaust Smoke in Smoke Management Systems	ASHRAE	ANSI/ASHRAE 149-2000 (RA09)
	Recommended Practice for Smoke-Control Systems	NFPA	NFPA 92A-06
	Smoke Management Systems in Malls, Atria, and Large Areas	NFPA	NFPA 92B-05
	Ceiling Dampers (2006)	UL	ANSI/UL 555C
	Smoke Dampers (1999)	UL	ANSI/UL 555S
Freezers	Energy Performance and Capacity of Household Refrigerators, Refrigerator-Freezers, and Freezers	CSA	C300-00 (R2005)
	Energy Performance Standard for Food Service Refrigerators and Freezers	CSA	C827-98 (R2003)
	Refrigeration Equipment	CSA	CAN/CSA-C22.2 No. 120-M91 (R2004)
Commercial	Dispensing Freezers	NSF	NSF/ANSI 6-2007
	Commercial Refrigerators and Freezers	NSF	NSF/ANSI 7-2007

Selected Codes and Standards Published by Various Societies and Associations (*Continued*)

Subject	Title	Publisher	Reference
Household	Commercial Refrigerators and Freezers (2006)	UL	ANSI/UL 471
	Ice Makers (1995)	UL	ANSI/UL 563
	Ice Cream Makers (2005)	UL	ANSI/UL 621
	Household Refrigerators, Refrigerator-Freezers and Freezers	AHAM	ANSI/AHAM HRF-1-2007
	Household Refrigerators and Freezers (1993)	UL/CSA	ANSI/UL 250/C22.2 No. 63-93 (R1999)
Fuels	Threshold Limit Values for Chemical Substances (updated annually)	ACGIH	ACGIH
	International Gas Fuel Code (2006)	AGA/NFPA	ANSI Z223.1/NPFA 54-2006
	Reporting of Fuel Properties when Testing Diesel Engines with Alternative Fuels Derived from Biological Materials	ASABE	ASAE EP552-1996
	Coal Pulverizers	ASME	PTC 4.2 1969 (RA03)
	Classification of Coals by Rank	ASTM	ASTM D388-05
	Specification for Fuel Oils	ASTM	ASTM D396-08
	Test Method for Determination of Homogeneity and Miscibility in Automotive Engine Oils	ASTM	ASTM D922-00a (2006)
	Specification for Diesel Fuel Oils	ASTM	ASTM D975-07b
	Specification for Gas Turbine Fuel Oils	ASTM	ASTM D2880-03
	Specification for Kerosene	ASTM	ASTM D3699-07
	Practice for Receipt, Storage and Handling of Fuels	ASTM	ASTM D4418-00 (2006)
	Test Method for Determination of Yield Stress and Apparent Viscosity of Used Engine Oils at Low Temperature	ASTM	ASTM D6896-03 (2007)
	Test Method for Total Sulfur in Naphthas, Distillates, Reformulated Gasolines, Diesels, Biodiesels, and Motor Fuels by Oxidative Combustion and Electrochemical Detection	ASTM	ASTM D6920-07
	Test Method for Measurement of Hindered Phenolic and Aromatic Amine Antioxidant Content in Non-Zinc Turbine Oils by Linear Sweep Voltammetry	ASTM	ASTM D6971-04
	Practice for Enumeration of Viable Bacteria and Fungi in Liquid Fuels—Filtration and Culture Procedures	ASTM	ASTM D6974-04a
	Test Method for Evaluation of Aeration Resistance of Engine Oils in Direct-Injected Turbocharged Automotive Diesel Engine	ASTM	ASTM D6984-07a
	Specification for Middle Distillate Fuel Oil-Military Marine Applications	ASTM	ASTM D6985-04a
	Test Method for Determination of Ignition Delay and Derived Cetane Number DCN of Diesel Fuel Oils by Combustion in a Constant Volume Chamber	ASTM	ASTM D6890-07b
	Test Method for Determination of Total Sulfur in Light Hydrocarbon, Motor Fuels, and Oils by Online Gas Chromatography with Flame Photometric Detection	ASTM	ASTM D7041-04
	Test Method for Sulfur in Gasoline and Diesel Fuel by Monochromatic Wavelength Dispersive X-Ray Fluorescence Spectrometry	ASTM	ASTM D7044-04a
	New Draft Standard Test Method for Flash Point by Modified Continuously Closed Cup Flash Point Tester	ASTM	ASTM D7094-04
	Test Method for Determining the Viscosity-Temperature Relationship of Used and Soot-Containing Engine Oils at Low Temperatures	ASTM	ASTM D7110-05a
	Test Method for Determination of Trace Elements in Middle Distillate Fuels by Inductively Coupled Plasma Atomic Emission Spectrometry (ICPAES)	ASTM	ASTM D7111-05
	Test Method for Determining Stability and Compatibility of Heavy Fuel Oils and Crude Oils by Heavy Fuel Oil Stability Analyzer (Optical Detection)	ASTM	ASTM D7112-05a
	Test Method for Determination of Intrinsic Stability of Asphaltene-Containing Residues, Heavy Fuel Oils, and Crude Oils	ASTM	ASTM D7157-05
	Test Method for Hydrogen Content of Middle Distillate Petroleum Products by Low-Resolution Pulsed Nuclear Magnetic Resonance Spectroscopy	ASTM	ASTM D7171-05
	Gas-Fired Central Furnaces	CSA	ANSI Z21.47-2006/CSA 2.3-2006
	Gas Unit Heaters and Gas-Fired Duct Furnaces	CSA	ANSI Z83.8-2006/CSA-2.6-2006
	Industrial and Commercial Gas-Fired Package Furnaces	CSA	CGA 3.2-1976 (R2003)
	Uniform Mechanical Code (2006)	IAPMO	Chapter 13
	Uniform Plumbing Code (2006)	IAPMO	Chapter 12
	International Fuel Gas Code (2009)	ICC	IFGC
	Standard Gas Code (1999)	SBCCI	SGC
	Commercial-Industrial Gas Heating Equipment (2006)	UL	UL 795
Furnaces	Commercial Application, Systems, and Equipment, 1st ed.	ACCA	ACCA Manual CS
	Residential Equipment Selection, 2nd ed.	ACCA	ANSI/ACCA Manual S
	Method of Testing for Annual Fuel Utilization Efficiency of Residential Central Furnaces and Boilers	ASHRAE	ANSI/ASHRAE 103-2007
	Prevention of Furnace Explosions/Implosions in Multiple Burner Boilers	NFPA	NFPA 8502-99
	Residential Gas Detectors (2000)	UL	ANSI/UL 1484
	Heating and Cooling Equipment (2005)	UL/CSA	ANSI/UL 1995/C22.2 No. 236-95
	Single and Multiple Station Carbon Monoxide Alarms (2008)	UL	ANSI/UL 2034
Gas	International Gas Fuel Code (2006)	AGA/NFPA	ANSI Z223.1/NFPA 54-2006
	Gas-Fired Central Furnaces	CSA	ANSI Z21.47-2006/CSA 2.3-2006

Selected Codes and Standards Published by Various Societies and Associations (*Continued*)

Subject	Title	Publisher	Reference
	Gas Unit Heaters and Gas-Fired Duct Furnaces	CSA	ANSI Z83.8-2006/CSA-2.6-2006
	Industrial and Commercial Gas-Fired Package Furnaces	CSA	CGA 3.2-1976 (R2003)
	International Fuel Gas Code (2009)	ICC	IFGC
	Standard Gas Code (1999)	SBCCI	SGC
	Commercial-Industrial Gas Heating Equipment (2006)	UL	UL 795
Oil	Specification for Fuel Oils	ASTM	ASTM D396-08
	Specification for Diesel Fuel Oils	ASTM	ASTM D975-07b
	Test Method for Smoke Density in Flue Gases from Burning Distillate Fuels	ASTM	ASTM D2156-94 (2003)
	Standard Test Method for Vapor Pressure of Liquefied Petroleum Gases (LPG) (Expansion Method)	ASTM	ASTM D6897-2003a
	Oil Burning Stoves and Water Heaters	CSA	B140.3-1962 (R2006)
	Oil-Fired Warm Air Furnaces	CSA	B140.4-04
	Installation of Oil-Burning Equipment	NFPA	NFPA 31-06
	Oil-Fired Central Furnaces (2006)	UL	UL 727
	Oil-Fired Floor Furnaces (2003)	UL	ANSI/UL 729
	Oil-Fired Wall Furnaces (2003)	UL	ANSI/UL 730
Solid Fuel	Installation Code for Solid-Fuel-Burning Appliances and Equipment	CSA	B365-01 (R2006)
	Solid-Fuel-Fired Central Heating Appliances	CSA	CAN/CSA-B366.1-M91 (R2007)
	Solid-Fuel and Combination-Fuel Central and Supplementary Furnaces (2006)	UL	ANSI/UL 391
Green Buildings	Standard for the Design of High-Performance, Green Buildings Except Low-Rise Residential Buildings	ASHRAE/ USGBC	ANSI/ASHRAE/USGBC/IES 189.1-2009
	International Green Construction Code™	ICC	IGCC
Heaters	Gas-Fired High-Intensity Infrared Heaters	CSA	ANSI Z83.19-2001/CSA 2.35-2001 (R2005)
	Gas-Fired Low-Intensity Infrared Heaters	CSA	ANSI Z83.20-2008/CSA 2.34-2008
	Threshold Limit Values for Chemical Substances (updated annually)	ACGIH	ACGIH
	Industrial Ventilation: A Manual of Recommended Practice, 26th ed. (2007)	ACGIH	ACGIH
	Thermal Performance Testing of Solar Ambient Air Heaters	ASABE	ANSI/ASAE S423-1991
	Air Heaters	ASME	ASME PTC 4.3-1968 (RA91)
	Guide for Construction of Solid Fuel Burning Masonry Heaters	ASTM	ASTM E1602-03
	Non-Recirculating Direct Gas-Fired Industrial Air Heaters	CSA	ANSI Z83.4-2003/CSA 3.7-2003
	Electric Duct Heaters	CSA	C22.2 No. 155-M1986 (R2004)
	Portable Kerosene-Fired Heaters	CSA	CAN3-B140.9.3 M86 (R2006)
	Standards for Closed Feedwater Heaters, 7th ed. (2004)	HEI	HEI 2622
	Electric Heating Appliances (2005)	UL	ANSI/UL 499
	Electric Oil Heaters (2003)	UL	ANSI/UL 574
	Oil-Fired Air Heaters and Direct-Fired Heaters (1993)	UL	UL 733
	Electric Dry Bath Heaters (2004)	UL	ANSI/UL 875
	Oil-Burning Stoves (1993)	UL	ANSI/UL 896
Engine	Electric Engine Preheaters and Battery Warmers for Diesel Engines	SAE	SAE J1310-1993
	Selection and Application Guidelines for Diesel, Gasoline, and Propane Fired Liquid Cooled Engine Pre-Heaters	SAE	SAE J1350-1988
	Fuel Warmer—Diesel Engines	SAE	SAE J1422-1996
Nonresidential	Installation of Electric Infrared Brooding Equipment	ASABE	ASAE EP258.3-2004
	Gas-Fired Construction Heaters	CSA	ANSI Z83.7-00 (R2005)/CSA 2.14-00 (R2006)
	Recirculating Direct Gas-Fired Industrial Air Heaters	CSA	ANSI Z83.18-2004
	Portable Industrial Oil-Fired Heaters	CSA	B140.8-1967 (R2006)
	Fuel-Fired Heaters—Air Heating—for Construction and Industrial Machinery	SAE	SAE J1024-1989
	Commercial-Industrial Gas Heating Equipment (2006)	UL	UL 795
	Electric Heaters for Use in Hazardous (Classified) Locations (2006)	UL	ANSI/UL 823
Pool	Methods of Testing and Rating Pool Heaters	ASHRAE	ANSI/ASHRAE 146-2006
	Gas-Fired Pool Heaters	CSA	ANSI Z21.56-2006/CSA 4.7-2006
	Oil-Fired Service Water Heaters and Swimming Pool Heaters	CSA	B140.12-03
Room	Specification for Room Heaters, Pellet Fuel Burning Type	ASTM	ASTM E1509-04
	Gas-Fired Room Heaters, Vol. II, Unvented Room Heaters	CSA	ANSI Z21.11.2-2007
	Gas-Fired Unvented Catalytic Room Heaters for Use with Liquefied Petroleum (LP) Gases	CSA	ANSI Z21.76-1994 (R2006)
	Vented Gas-Fired Space Heating Appliances	CSA	ANSI Z21.86-2004/CSA 2.32-2004
	Vented Gas Fireplace Heaters	CSA	ANSI Z21.88-2005/CSA 2.33-2005
	Unvented Kerosene-Fired Room Heaters and Portable Heaters (1993)	UL	UL 647
	Movable and Wall- or Ceiling-Hung Electric Room Heaters (2000)	UL	UL 1278
	Fixed and Location-Dedicated Electric Room Heaters (1997)	UL	UL 2021
	Solid Fuel-Type Room Heaters (1996)	UL	ANSI/UL 1482

Selected Codes and Standards Published by Various Societies and Associations (*Continued*)

Subject	Title	Publisher	Reference
Transport	Heater, Airplane, Engine Exhaust Gas to Air Heat Exchanger Type	SAE	SAE ARP86-1996
	Installation, Heaters, Airplane, Internal Combustion Heater Exchange Type	SAE	SAE ARP266-2001
	Heater, Aircraft, Internal Combustion Heat Exchanger Type	SAE	SAE AS8040-1996
	Motor Vehicle Heater Test Procedure	SAE	SAE J638-1998
	Heater, Aircraft, Internal Combustion Heat Exchanger Type	SAE	SAE AS8040-1996
Unit	Gas Unit Heaters and Gas-Fired Duct Furnaces	CSA	ANSI Z83.8-2006/CSA-2.6-2006
	Oil-Fired Unit Heaters (1995)	UL	ANSI/UL 731
Heat Exchangers	Remote Mechanical-Draft Evaporative Refrigerant Condensers	AHRI	AHRI 490-2003
	Method of Testing Air-to-Air Heat/Energy Exchangers	ASHRAE	ANSI/ASHRAE 84-2008
	Boiler and Pressure Vessel Code—Section VIII, Division 1: Pressure Vessels	ASME	ASME BPVC-2007
	Single Phase Heat Exchangers	ASME	ASME PTC 12.5-2000 (RA05)
	Air Cooled Heat Exchangers	ASME	ASME PTC 30-1991 (RA05)
	Standard Methods of Test for Rating the Performance of Heat-Recovery Ventilators	CSA	C439-00 (R2005)
	Standards for Power Plant Heat Exchangers, 4th ed. (2004)	HEI	HEI 2623
	Standards of Tubular Exchanger Manufacturers Association, 9th ed. (2007)	TEMA	TEMA
	Refrigerant-Containing Components and Accessories, Nonelectrical (2001)	UL	ANSI/UL 207
Heating	Commercial Application, Systems, and Equipment, 1st ed.	ACCA	ACCA Manual CS
	Comfort, Air Quality, and Efficiency by Design	ACCA	ACCA Manual RS
	Residential Equipment Selection, 2nd ed.	ACCA	ANSI/ACCA Manual S
	Heating, Ventilating and Cooling Greenhouses	ASABE	ANSI/ASAE EP406.4-2003
	Heater Elements	CSA	C22.2 No. 72-M1984 (R2004)
	Determining the Required Capacity of Residential Space Heating and Cooling Appliances	CSA	CAN/CSA-F280-M90 (R2004)
	Heat Loss Calculation Guide (2001)	HYDI	HYDI H-22
	Residential Hydronic Heating Installation Design Guide	HYDI	IBR Guide
	Radiant Floor Heating (1995)	HYDI	HYDI 004
	Advanced Installation Guide (Commercial) for Hot Water Heating Systems (2001)	HYDI	HYDI 250
	Environmental Systems Technology, 2nd ed. (1999)	NEBB	NEBB
	Pulverized Fuel Systems	NFPA	NFPA 8503-97
	Aircraft Electrical Heating Systems	SAE	SAE AIR860-2000
	Heating Value of Fuels	SAE	SAE J1498-2005
	Performance Test for Air-Conditioned, Heated, and Ventilated Off-Road Self-Propelled Work Machines	SAE	SAE J1503-2004
	HVAC Systems—Applications, 1st ed.	SMACNA	SMACNA 1987
	Electric Baseboard Heating Equipment (1994)	UL	ANSI/UL 1042
	Electric Duct Heaters (2004)	UL	ANSI/UL 1996
	Heating and Cooling Equipment (2005)	UL/CSA	ANSI/UL 1995/C22.2 No. 236-95
Heat Pumps	Commercial Application, Systems, and Equipment, 1st ed.	ACCA	ACCA Manual CS
	Geothermal Heat Pump Training Certification Program	ACCA	ACCA Training Manual
	Heat Pumps Systems, Principles and Applications, 2nd ed.	ACCA	ACCA Manual H
	Residential Equipment Selection, 2nd ed.	ACCA	ANSI/ACCA Manual S
	Industrial Ventilation: A Manual of Recommended Practice, 26th ed. (2007)	ACGIH	ACGIH
	Water-Source Heat Pumps	AHRI	AHRI 320-98
	Ground Water-Source Heat Pumps	AHRI	AHRI 325-98
	Ground Source Closed-Loop Heat Pumps	AHRI	AHRI 330-98
	Commercial and Industrial Unitary Air-Conditioning and Heat Pump Equipment	AHRI	AHRI 340/360-2007
	Methods of Testing for Rating Electrically Driven Unitary Air-Conditioning and Heat Pump Equipment	ASHRAE	ANSI/ASHRAE 37-2009
	Methods of Testing for Rating Seasonal Efficiency of Unitary Air-Conditioners and Heat Pumps	ASHRAE	ANSI/ASHRAE 116-2010
	Performance Standard for Split-System and Single-Package Central Air Conditioners and Heat Pumps	CSA	CAN/CSA-C656-05
	Installation Requirements for Air-to-Air Heat Pumps	CSA	C273.5-1980 (R2002)
	Performance of Direct-Expansion (DX) Ground-Source Heat Pumps	CSA	C748-94 (R2005)
	Water-Source Heat Pumps—Testing and Rating for Performance, Part 1: Water-to-Air and Brine-to-Air Heat Pumps	CSA	CAN/CSA C13256-1-01
	Part 2: Water-to-Water and Brine-to-Water Heat Pumps	CSA	CAN/CSA C13256-2-01 (R2005)
	Heating and Cooling Equipment (2005)	UL/CSA	ANSI/UL 1995/C22.2 No. 236-95
Gas-Fired	Gas-Fired, Heat Activated Air Conditioning and Heat Pump Appliances	CSA	ANSI Z21.40.1-1996 (R2002)/CGA 2.91-M96
	Gas-Fired, Work Activated Air Conditioning and Heat Pump Appliances (Internal Combustion)	CSA	ANSI Z21.40.2-1996 (R2002)/CGA 2.92-M96
	Performance Testing and Rating of Gas-Fired Air Conditioning and Heat Pump Appliances	CSA	ANSI Z21.40.4-1996 (R2002)/CGA 2.94-M96

Selected Codes and Standards Published by Various Societies and Associations (*Continued*)

Subject	Title	Publisher	Reference
Heat Recovery	Gas Turbine Heat Recovery Steam Generators	ASME	ANSI/ASME PTC 4.4-1981 (RA03)
	Water Heaters, Hot Water Supply Boilers, and Heat Recovery Equipment	NSF	NSF/ANSI 5-2007
High-Performance Buildings	Standard for the Design of High-Performance, Green Buildings Except Low-Rise Residential Buildings	ASHRAE/ USGBC	ANSI/ASHRAE/USGBC/IES 189.1-2009
	International Green Construction Code™	ICC	IGCC
Humidifiers	Commercial Application, Systems, and Equipment, 1st ed.	ACCA	ACCA Manual CS
	Comfort, Air Quality, and Efficiency by Design	ACCA	ACCA Manual RS
	Bioaerosols: Assessment and Control (1999)	ACGIH	ACGIH
	Humidifiers	AHAM	ANSI/AHAM HU-1-2006
	Central System Humidifiers for Residential Applications	AHRI	AHRI 610-2004
	Self-Contained Humidifiers for Residential Applications	AHRI	AHRI 620-2004
	Commercial and Industrial Humidifiers	AHRI	ANSI/AHRI 640-2005
	Humidifiers (2001)	UL/CSA	ANSI/UL 998/C22.2 No. 104-93
Ice Makers	Performance Rating of Automatic Commercial Ice Makers	AHRI	AHRI 810-2007
	Ice Storage Bins	AHRI	AHRI 820-2000
	Methods of Testing Automatic Ice Makers	ASHRAE	ANSI/ASHRAE 29-2009
	Refrigeration Equipment	CSA	CAN/CSA-C22.2 No. 120-M91 (R2004)
	Performance of Automatic Ice-Makers and Ice Storage Bins	CSA	C742-98 (R2003)
	Automatic Ice Making Equipment	NSF	NSF/ANSI 12-2007
	Ice Makers (1995)	UL	ANSI/UL 563
Incinerators	Incinerators and Waste and Linen Handling Systems and Equipment	NFPA	NFPA 82-04
	Residential Incinerators (2006)	UL	UL 791
Indoor Air Quality	Good HVAC Practices for Residential and Commercial Buildings (2003)	ACCA	ACCA
	Comfort, Air Quality, and Efficiency by Design (Residential) (1997)	ACCA	ACCA Manual RS
	Bioaerosols: Assessment and Control (1999)	ACGIH	ACGIH
	Ventilation for Acceptable Indoor Air Quality	ASHRAE	ANSI/ASHRAE 62.1-2010
	Ventilation and Acceptable Indoor Air Quality in Low-Rise Residential Buildings	ASHRAE	ANSI/ASHRAE 62.2-2010
	Test Method for Determination of Volatile Organic Chemicals in Atmospheres (Canister Sampling Methodology)	ASTM	ASTM D5466-01 (2007)
	Guide for Using Probability Sampling Methods in Studies of Indoor Air Quality in Buildings	ASTM	ASTM D5791-95 (2006)
	Guide for Using Indoor Carbon Dioxide Concentrations to Evaluate Indoor Air Quality and Ventilation	ASTM	ASTM D6245-07
	Guide for Placement and Use of Diffusion Controlled Passive Monitors for Gaseous Pollutants in Indoor Air	ASTM	ASTM D6306-98 (2003)
	Test Method for Determination of Metals and Metalloids Airborne Particulate Matter by Inductively Coupled Plasma Atomic Emissions Spectrometry (ICP-AES)	ASTM	ASTM D7035-04
	Test Method for Metal Removal Fluid Aerosol in Workplace Atmospheres	ASTM	ASTM D7049-04
	Practice for Emission Cells for the Determination of Volatile Organic Emissions from Materials/Products	ASTM	ASTM D7143-05
	Practice for Collection of Surface Dust by Micro-Vacuum Sampling for Subsequent Metals Determination	ASTM	ASTM D7144-05a
	Test Method for Determination of Beryllium in the Workplace Using Field-Based Extraction and Fluorescence Detection	ASTM	ASTM D7202-06
	Practice for Referencing Suprathreshold Odor Intensity	ASTM	ASTM E544-99 (2004)
	Guide for Specifying and Evaluating Performance of a Single Family Attached and Detached Dwelling—Indoor Air Quality	ASTM	ASTM E2267-04
	Classification for Serviceability of an Office Facility for Thermal Environment and Indoor Air Conditions	ASTM	ASTM E2320-04
	Practice for Continuous Sizing and Counting of Airborne Particles in Dust-Controlled Areas and Clean Rooms Using Instruments Capable of Detecting Single Sub-Micrometre and Larger Particles	ASTM	ASTM F50-07
	Ambient Air—Determination of Mass Concentration of Nitrogen Dioxide—Modified Griess-Saltzman Method	ISO	ISO 6768:1998
	Air Quality—Exchange of Data	ISO	ISO 7168:1999
	Environmental Tobacco Smoke—Estimation of Its Contribution to Respirable Suspended Particles—Determination of Particulate Matter by Ultraviolet Absorptance and by Fluorescence	ISO	ISO 15593:2001
	Indoor Air—Part 3: Determination of Formaldehyde and Other Carbonyl Compounds—Active Sampling Method	ISO	ISO 16000-3:2001
	Workplace Air Quality—Sampling and Analysis of Volatile Organic Compounds by Solvent Desorption/Gas Chromatography—Part 1: Pumped Sampling Method	ISO	ISO 16200-1:2001
	Part 2: Diffusive Sampling Method	ISO	ISO 16200-2:2000
	Workplace Air Quality—Determination of Total Organic Isocyanate Groups in Air Using 1-(2-Methoxyphenyl) Piperazine and Liquid Chromatography	ISO	ISO 16702:2007

Selected Codes and Standards Published by Various Societies and Associations (*Continued*)

Subject	Title	Publisher	Reference
Aircraft	Installation of Household Carbon Monoxide (CO) Warning Equipment	NFPA	NFPA 720-2005
	Indoor Air Quality—A Systems Approach, 3rd ed.	SMACNA	SMACNA 1998
	IAQ Guidelines for Occupied Buildings Under Construction, 2nd ed.	SMACNA	ANSI/SMACNA 008-2008
	Single and Multiple Station Carbon Monoxide Alarms (1996)	UL	ANSI/UL 2034
	Guide for Selecting Instruments and Methods for Measuring Air Quality in Aircraft Cabins	ASTM	ASTM D6399-04
	Guide for Deriving Acceptable Levels of Airborne Chemical Contaminants in Aircraft Cabins Based on Health and Comfort Considerations	ASTM	ASTM D7034-05
Insulation	Guidelines for Use of Thermal Insulation in Agricultural Buildings	ASABE	ANSI/ASAE S401.2-2003
	Terminology Relating to Thermal Insulating Materials	ASTM	ASTM C168-05a
	Test Method for Steady-State Heat Flux Measurements and Thermal Transmission Properties by Means of the Guarded-Hot-Plate Apparatus	ASTM	ASTM C177-04
	Test Method for Steady-State Heat Transfer Properties of Horizontal Pipe Insulations	ASTM	ASTM C335-05ae1
	Practice for Prefabrication and Field Fabrication of Thermal Insulating Fitting Covers for NPS Piping, Vessel Lagging, and Dished Head Segments	ASTM	ASTM C450-02
	Test Method for Steady-State and Thermal Transmission Properties by Means of the Heat Flow Meter Apparatus	ASTM	ASTM C518-04
	Specification for Preformed Flexible Elastometric Cellular Thermal Insulation in Sheet and Tubular Form	ASTM	ASTM C534-07a
	Specification for Cellular Glass Thermal Insulation	ASTM	ASTM C552-07
	Specification for Rigid, Cellular Polystyrene Thermal Insulation	ASTM	ASTM C578-07
	Practice for Inner and Outer Diameters of Rigid Thermal Insulation for Nominal Sizes of Pipe and Tubing (NPS System)	ASTM	ASTM C585-90 (2004)
	Specification for Unfaced Preformed Rigid Cellular Polyisocyanurate Thermal Insulation	ASTM	ASTM C591-07
	Practice for Determination of Heat Gain or Loss and the Surface Temperature of Insulated Pipe and Equipment Systems by the Use of a Computer Program	ASTM	ASTM C680-04e4
	Specification for Adhesives for Duct Thermal Insulation	ASTM	ASTM C916-85 (2007)
	Classification of Potential Health and Safety Concerns Associated with Thermal Insulation Materials and Accessories	ASTM	ASTM C930-05
	Practice for Thermographic Inspection of Insulation Installations in Envelope Cavities of Frame Buildings	ASTM	ASTM C1060-90 (2003)
	Specification for Fibrous Glass Duct Lining Insulation (Thermal and Sound Absorbing Material)	ASTM	ASTM C1071-05
	Specification for Faced or Unfaced Rigid Cellular Phenolic Thermal Insulation	ASTM	ASTM C1126-04
	Practice for Installation and Use of Radiant Barrier Systems (RBS) in Building Construction	ASTM	ASTM C1158-05
	Test Method for Steady-State and Thermal Performance of Building Assemblies by Means of a Hot Box Apparatus	ASTM	ASTM C1363-05
	Specification for Perpendicularly Oriented Mineral Fiber Roll and Sheet Thermal Insulation for Pipes and Tanks	ASTM	ASTM C1393-00a (2006)
	Guide for Measuring and Estimating Quantities of Insulated Piping and Components	ASTM	ASTM C1409-98 (2003)
	Specification for Cellular Melamine Thermal and Sound Absorbing Insulation	ASTM	ASTM C1410-05a
	Guide for Selecting Jacketing Materials for Thermal Insulation	ASTM	ASTM C1423-98 (2003)
	Specification for Preformed Flexible Cellular Polyolefin Thermal Insulation in Sheet and Tubular Form	ASTM	ASTM C1427-07
	Specification for Polyimide Flexible Cellular Thermal and Sound Absorbing Insulation	ASTM	ASTM C1482-04
	Specification for Cellulosic Fiber Stabilized Thermal Insulation	ASTM	ASTM C1497-04
	Test Method for Characterizing the Effect of Exposure to Environmental Cycling on Thermal Performance of Insulation Products	ASTM	ASTM C1512-07
	Specification for Flexible Polymeric Foam Sheet Insulation Used as a Thermal and Sound Absorbing Liner for Duct Systems	ASTM	ASTM C1534-07
	Standard Guide for Development of Standard Data Records for Computerization of Thermal Transmission Test Data for Thermal Insulation	ASTM	ASTM C1558-03 (2007)
	Guide for Determining Blown Density of Pneumatically Applied Loose Fill Mineral Fiber Thermal Insulation	ASTM	ASTM C1574-04
	Test Method for Determining the Moisture Content of Inorganic Insulation Materials by Weight	ASTM	ASTM C1616-07e1
	Specification for Cellular Polypropylene Thermal Insulation	ASTM	ASTM C1631-05
	Classification for Rating Sound Insulation	ASTM	ASTM E413-04
	Test Method for Determining the Drainage Efficiency of Exterior Insulation and Finish Systems (EIFS) Clad Wall Assemblies	ASTM	ASTM E2273-03
	Practice for Use of Test Methods E96 for Determining the Water Vapor Transmission (WVT) of Exterior Insulation and Finish Systems	ASTM	ASTM E2321-03
	Thermal Insulation—Vocabulary	ISO	ISO 9229:2007
	National Commercial and Industrial Insulation Standards, 6th ed.	MICA	MICA
	Accepted Industry Practices for Sheet Metal Lagging, 1st ed.	SMACNA	SMACNA 2002

Selected Codes and Standards Published by Various Societies and Associations (*Continued*)

Subject	Title	Publisher	Reference
Louvers	Laboratory Methods of Testing Dampers for Rating	AMCA	AMCA 500-D-07
	Laboratory Methods of Testing Louvers for Rating	AMCA	AMCA 500-L-07
Lubricants	Methods of Testing the Floc Point of Refrigeration Grade Oils	ASHRAE	ANSI/ASHRAE 86-1994 (RA06)
	Test Method for Pour Point of Petroleum Products	ASTM	ASTM D97-07
	Classification of Industrial Fluid Lubricants by Viscosity System	ASTM	ASTM D2422-97 (2007)
	Test Method for Relative Molecular Weight (Relative Molecular Mass) of Hydrocarbons by Thermoelectric Measurement of Vapor Pressure	ASTM	ASTM D2503-92 (2007)
	Test Method for Determination of Moderately High Temperature Piston Deposits by Thermo-Oxidation Engine Oil Simulation Test	ASTM	ASTM D7097-06a
	Petroleum Products—Corrosiveness to Copper—Copper Strip Test	ISO	ISO 2160:1998
Measurement	Industrial Ventilation: A Manual of Recommended Practice, 26th ed. (2007)	ACGIH	ACGIH
	Engineering Analysis of Experimental Data	ASHRAE	ASHRAE *Guideline* 2-2010
	Standard Method for Measurement of Proportion of Lubricant in Liquid Refrigerant	ASHRAE	ANSI/ASHRAE 41.4-1996 (RA06)
	Standard Method for Measurement of Moist Air Properties	ASHRAE	ANSI/ASHRAE 41.6-1994 (RA06)
	Method of Measuring Solar-Optical Properties of Materials	ASHRAE	ANSI/ASHRAE 74-1988
	Methods of Measuring, Expressing, and Comparing Building Energy Performance	ASHRAE	ANSI/ASHRAE 105-2007
	Method for Establishing Installation Effects on Flowmeters	ASME	ASME MFC-10M-2000
	Test Uncertainty	ASME	ASME PTC 19.1-2005
	Measurement of Industrial Sound	ASME	ANSI/ASME PTC 36-2004
	Test Methods for Water Vapor Transmission of Materials	ASTM	ASTM E96/E96M-05
	Specification for Temperature-Electromotive Force (EMF) Tables for Standardized Thermocouples	ASTM	ASTM E230-03
	Practice for Continuous Sizing and Counting of Airborne Particles in Dust-Controlled Areas and Clean Rooms Using Instruments Capable of Detecting Single Sub-Micrometre and Larger Particles	ASTM	ASTM F50-07
	Use of the International System of Units (SI): The Modern Metric System	IEEE/ASTM	IEEE/ASTM-SI10-2002
	Ergonomics of the Thermal Environment—Instruments for Measuring Physical Quantities	ISO	ISO 7726:1998
	Ergonomics of the Thermal Environment—Determination of Metabolic Rate	ISO	ISO 8996:2004
	Ergonomics of the Thermal Environment—Estimation of the Thermal Insulation and Water Vapour Resistance of a Clothing Ensemble	ISO	ISO 9920:2007
Fluid Flow	Standard Methods of Measurement of Flow of Liquids in Pipes Using Orifice Flowmeters	ASHRAE	ANSI/ASHRAE 41.8-1989
	Calorimeter Test Methods for Mass Flow Measurements of Volatile Refrigerants	ASHRAE	ANSI/ASHRAE 41.9-2000 (RA06)
	Flow Measurement	ASME	ASME PTC 19.5-2004
	Glossary of Terms Used in the Measurement of Fluid Flow in Pipes	ASME	ASME MFC-1M-2003
	Measurement Uncertainty for Fluid Flow in Closed Conduits	ASME	ANSI/ASME MFC-2M-1983 (RA01)
	Measurement of Fluid Flow in Pipes Using Orifice, Nozzle, and Venturi	ASME	ASME MFC-3M-2004
	Measurement of Liquid Flow in Closed Conduits Using Transit-Time Ultrasonic Flowmeters	ASME	ASME MFC-5M-1985 (RA01)
	Measurement of Fluid Flow in Pipes Using Vortex Flowmeters	ASME	ASME MFC-6M-1998 (RA05)
	Fluid Flow in Closed Conduits: Connections for Pressure Signal Transmissions Between Primary and Secondary Devices	ASME	ASME MFC-8M-2001
	Measurement of Liquid Flow in Closed Conduits by Weighing Method	ASME	ASME MFC-9M-1988 (RA01)
	Measurement of Fluid Flow by Means of Coriolis Mass Flowmeters	ASME	ASME MFC-11M-2006
	Measurement of Fluid Flow Using Small Bore Precision Orifice Meters	ASME	ASME MFC-14M-2003
	Measurement of Fluid Flow in Closed Conduits by Means of Electromagnetic Flowmeters	ASME	ASME MFC-16M-1995 (R01)
	Measurement of Fluid Flow Using Variable Area Meters	ASME	ASME MFC-18M-2001
	Test Method for Determining the Moisture Content of Inorganic Insulation Materials by Weight	ASTM	ASTM C1616-07e1
	Test Method for Indicating Wear Characteristics of Petroleum Hydraulic Fluids in a High Pressure Constant Volume Vane Pump	ASTM	ASTM D6973-05
	Test Method for Dynamic Viscosity and Density of Liquids by Stabinger Viscometer and the Calculation of Kinematic Viscosity	ASTM	ASTM D7042-04
	Test Method for Indicating Wear Characteristics of Petroleum and Non-Petroleum Hydraulic Fluids in a Constant Volume Vane Pump	ASTM	ASTM D7043-04a
	Practice for Calculating Viscosity of a Blend of Petroleum Products	ASTM	ASTM D7152-05e1
	Test Method for Same-Different Test	ASTM	ASTM E2139-05
	Practice for Field Use of Pyranometers, Pyrheliometers, and UV Radiometers	ASTM	ASTM G183-05
Gas Flow	Standard Methods for Laboratory Airflow Measurement	ASHRAE	ANSI/ASHRAE 41.2-1987 (RA92)
	Method of Test for Measurement of Flow of Gas	ASHRAE	ANSI/ASHRAE 41.7-1984 (RA06)
	Measurement of Gas Flow by Turbine Meters	ASME	ANSI/ASME MFC-4M-1986 (RA03)
	Measurement of Gas Flow by Means of Critical Flow Venturi Nozzles	ASME	ANSI/ASME MFC-7M-1987 (RA01)
Pressure	Standard Method for Pressure Measurement	ASHRAE	ANSI/ASHRAE 41.3-1989
	Pressure Gauges and Gauge Attachments	ASME	ASME B40.100-2005

Selected Codes and Standards Published by Various Societies and Associations (*Continued*)

Subject	Title	Publisher	Reference
	Pressure Measurement	ASME	ANSI/ASME PTC 19.2-1987 (RA04)
Temperature	Standard Method for Temperature Measurement	ASHRAE	ANSI/ASHRAE 41.1-1986 (RA06)
	Thermometers, Direct Reading and Remote Reading	ASME	ASME B40.200-2001
	Temperature Measurement	ASME	ASME PTC 19.3-1974 (RA04)
	Total Temperature Measuring Instruments (Turbine Powered Subsonic Aircraft)	SAE	SAE AS793-2001
Thermal	Method of Testing Thermal Energy Meters for Liquid Streams in HVAC Systems	ASHRAE	ANSI/ASHRAE 125-1992 (RA06)
	Test Method for Steady-State Heat Flux Measurements and Thermal Transmission Properties by Means of the Guarded-Hot-Plate Apparatus	ASTM	ASTM C177-04
	Test Method for Steady-State Heat Flux Measurements and Thermal Transmission Properties by Means of the Heat Flow Meter Apparatus	ASTM	ASTM C518-04
	Practice for In-Situ Measurement of Heat Flux and Temperature on Building Envelope Components	ASTM	ASTM C1046-95 (2007)
	Practice for Determining Thermal Resistance of Building Envelope Components from In-Situ Data	ASTM	ASTM C1155-95 (2007)
	Test Method for Thermal Performance of Building Materials and Envelope Assemblies by Means of a Hot Box Apparatus	ASTM	ASTM C1363-05
Mobile Homes and Recreational Vehicles	Residential Load Calculation, 8th ed.	ACCA	ANSI/ACCA Manual J
	Recreational Vehicle Cooking Gas Appliances	CSA	ANSI Z21.57-2007
	Oil-Fired Warm Air Heating Appliances for Mobile Housing and Recreational Vehicles	CSA	B140.10-06
	Mobile Homes	CSA	CAN/CSA-Z240 MH Series-92 (R2005)
	Recreational Vehicles	CSA	CAN/CSA-Z240 RV Series-08
	Gas Supply Connectors for Manufactured Homes	IAPMO	IAPMO TS 9-2003
	Fuel Supply: Manufactured/Mobile Home Parks & Recreational Vehicle Parks	IAPMO	Chapter 13, Part II
	Manufactured Housing Construction and Safety Standards	ICC/ANSI	ICC/ANSI 2.0-1998
	Manufactured Housing	NFPA	NFPA 501-05
	Recreational Vehicles	NFPA	NFPA 1192-08
	Plumbing System Components for Recreational Vehicles	NSF	NSF/ANSI 24-2006
	Low Voltage Lighting Fixtures for Use in Recreational Vehicles (2005)	UL	ANSI/UL 234
	Liquid Fuel-Burning Heating Appliances for Manufactured Homes and Recreational Vehicles (1995)	UL	ANSI/UL 307A
	Gas-Burning Heating Appliances for Manufactured Homes and Recreational Vehicles (2006)	UL	UL 307B
	Gas-Fired Cooking Appliances for Recreational Vehicles (2006)	UL	UL 1075
Motors and Generators	Installation and Maintenance of Farm Standby Electric Power	ASABE	ANSI/ASAE EP364.3-2006
	Nuclear Power Plant Air-Cleaning Units and Components	ASME	ASME N509-2002
	Testing of Nuclear Air Treatment Systems	ASME	ASME N510-2007
	Fired Steam Generators	ASME	ASME PTC 4-1998
	Gas Turbine Heat Recovery Steam Generators	ASME	ASME PTC 4.4-1981 (RA03)
	Test Methods for Film-Insulated Magnet Wire	ASTM	ASTM D1676-03
	Test Method for Evaluation of Engine Oils in a High Speed, Single-Cylinder Diesel Engine—Caterpillar 1R Test Procedure	ASTM	ASTM D6923-05
	Test Method for Evaluation of Diesel Engine Oils in the T-11 Exhaust Gas Recirculation Diesel Engine	ASTM	ASTM D7156-07a
	Energy Efficiency Test Methods for Three-Phase Induction Motors	CSA	C390-98 (R2005)
	Motors and Generators	CSA	C22.2 No. 100-04
	Emergency Electrical Power Supply for Buildings	CSA	CSA C282-05
	Energy Efficiency Test Methods for Single- and Three-Phase Small Motors	CSA	CAN/CSA C747-94 (R2005)
	Standard Test Procedure for Polyphase Induction Motors and Generators	IEEE	IEEE 112-1996
	Motors and Generators	NEMA	NEMA MG 1-2006
	Energy Management Guide for Selection and Use of Fixed Frequency Medium AC Squirrel-Cage Polyphase Industrial Motors	NEMA	NEMA MG 10-2001 (R2007)
	Energy Management Guide for Selection and Use of Single-Phase Motors	NEMA	NEMA MG 11-1977 (R2007)
	Magnet Wire	NEMA	NEMA MW 1000-2003
	Motion/Position Control Motors, Controls, and Feedback Devices	NEMA	NEMA ICS 16-2001
	Electric Motors (1994)	UL	UL 1004
	Electric Motors and Generators for Use in Division 1 Hazardous (Classified) Locations (2003)	UL	ANSI/UL 674
	Overheating Protection for Motors (1997)	UL	ANSI/UL 2111
Pipe, Tubing, and Fittings	Scheme for the Identification of Piping Systems	ASME	ASME A13.1-2007
	Pipe Threads, General Purpose (Inch)	ASME	ANSI/ASME B1.20.1-1983 (RA01)
	Wrought Copper and Copper Alloy Braze-Joint Pressure Fittings	ASME	ASME B16.50-2001
	Power Piping	ASME	ASME B31.1-2007
	Fuel Gas Piping	ASME	ASME B31.2-1968
	Process Piping	ASME	ASME B31.3-2006

Selected Codes and Standards Published by Various Societies and Associations (*Continued*)

Subject	Title	Publisher	Reference
	Refrigeration Piping and Heat Transfer Components	ASME	ASME B31.5-2006
	Building Services Piping	ASME	ASME B31.9-2004
	Practice for Obtaining Hydrostatic or Pressure Design Basis for "Fiberglass" (Glass-Fiber-Reinforced Thermosetting-Resin) Pipe and Fittings	ASTM	ASTM D2992-06
	Specification for Welding of Austenitic Stainless Steel Tube and Piping Systems in Sanitary Applications	AWS	AWS D18.1:1999
	Standards of the Expansion Joint Manufacturers Association, 8th ed. (2003)	EJMA	EJMA
	Pipe Hangers and Supports—Materials, Design and Manufacture	MSS	MSS SP-58-2002
	Pipe Hangers and Supports—Selection and Application	MSS	ANSI/MSS SP-69-2003
	General Welding Guidelines (2002)	NCPWB	NCPWB
	International Fuel Gas Code	AGA/NFPA	ANSI Z223.1/NFPA 54-2006
	Refrigeration Tube Fittings—General Specifications	SAE	SAE J513-1999
	Seismic Restraint Manual—Guidelines for Mechanical Systems, 3rd ed.	SMACNA	ANSI/SMACNA 001-2008
	Tube Fittings for Flammable and Combustible Fluids, Refrigeration Service, and Marine Use (1997)	UL	ANSI/UL 109
Plastic	Specification for Acrylonitrile-Butadiene-Styrene (ABS) Plastic Pipe, Schedules 40 and 80	ASTM	ASTM D1527-99 (2005)
	Specification for Poly (Vinyl Chloride) (PVC) Plastic Pipe, Schedules 40, 80, and 120	ASTM	ASTM D1785-06
	Specification for Polyethylene (PE) Plastic Pipe, Schedule 40	ASTM	ASTM D2104-03
	Test Method for Obtaining Hydrostatic or Pressure Design Basis for Thermoplastic Pipe Products	ASTM	ASTM D2837-04e1
	Specification for Polybutylene (PB) Plastic Hot- and Cold-Water Distribution Systems	ASTM	ASTM D3309-96a (2002)
	Specification for Perfluoroalkoxy (PFA)-Fluoropolymer Tubing	ASTM	ASTM D6867-03
	Specification for Polyethylene Stay in Place Form System for End Walls for Drainage Pipe	ASTM	ASTM D7082-04
	Specification for Chlorinated Poly (Vinyl Chloride) (CPVC) Plastic Pipe, Schedules 40 and 80	ASTM	ASTM F441/F441M-02
	Specification for Crosslinked Polyethylene/Aluminum/Crosslinked Polyethylene Tubing OD Controlled SDR9	ASTM	ASTM F2262-05
	Test Method for Evaluating the Oxidative Resistance of Polyethylene (PE) Pipe to Chlorinated Water	ASTM	ASTM F2263-07e1
	Specification for 12 to 60 in. Annular Corrugated Profile-Wall Polyethylene (PE) Pipe and Fittings for Gravity-Flow Storm Sewer and Subsurface Drainage Applications	ASTM	ASTM F2306/F2306M-07
	Specification for Series 10 Poly (Vinyl Chloride) (PVC) Closed Profile Gravity Pipe and Fittings Based on Controlled Inside Diameter	ASTM	ASTM F2307-03
	Standard Test Method for Evaluating the Oxidative Resistance of Multilayer Polyolefin Tubing to Hot Chlorinated Water	ASTM	ASTM F2330-04
	Test Method for Determining Chemical Compatibility of Thread Sealants with Thermoplastic Threaded Pipe and Fittings Materials	ASTM	ASTM F2331-04e1
	Test Method for Determining Thermoplastic Pipe Wall Stiffness	ASTM	ASTM F2433-05
	Specification for Steel Reinforced Polyethylene (PE) Corrugated Pipe	ASTM	ASTM F2435-07
	Electrical Polyvinyl Chloride (PVC) Tubing and Conduit	NEMA	NEMA TC 2-2003
	PVC Plastic Utilities Duct for Underground Installation	NEMA	NEMA TC 6 and 8-2003
	Smooth Wall Coilable Polyethylene Electrical Plastic Duct	NEMA	NEMA TC 7-2005
	Fittings for PVC Plastic Utilities Duct for Underground Installation	NEMA	NEMA TC 9-2004
	Electrical Nonmetallic Tubing (ENT)	NEMA	NEMA TC 13-2005
	Plastics Piping System Components and Related Materials	NSF	NSF/ANSI 14-2007
	Rubber Gasketed Fittings for Fire-Protection Service (2004)	UL	ANSI/UL 213
Metal	Welded and Seamless Wrought Steel Pipe	ASME	ASME B36.10M-2004
	Stainless Steel Pipe	ASME	ASME B36.19M-2004
	Specification for Pipe, Steel, Black and Hot-Dipped, Zinc-Coated, Welded and Seamless	ASTM	ASTM A53/53M-07
	Specification for Seamless Carbon Steel Pipe for High-Temperature Service	ASTM	ASTM A106/A106M-06a
	Specification for Pipe, Steel, Electric-Fusion Arc-Welded Sizes NPS 16 and Over	ASTM	ASTM A1034-05b
	Specification for Steel Line Pipe, Black, Furnace-Butt-Welded	ASTM	ASTM A1037/A1037M-05
	Specification for Composite Corrugated Steel Pipe for Sewers and Drains	ASTM	ASTM A1042/A1042M-04
	Specification for Seamless Copper Pipe, Standard Sizes	ASTM	ASTM B42-02e1
	Specification for Seamless Copper Tube	ASTM	ASTM B75-02
	Specification for Seamless Copper Water Tube	ASTM	ASTM B88-03
	Specification for Seamless Copper Tube for Air Conditioning and Refrigeration Field Service	ASTM	ASTM B280-03
	Specification for Hand-Drawn Copper Capillary Tube for Restrictor Applications	ASTM	ASTM B360-01
	Specification for Welded Copper Tube for Air Conditioning and Refrigeration Service	ASTM	ASTM B640-07
	Specification for Copper-Beryllium Seamless Tube UNS Nos. C17500 and C17510	ASTM	ASTM B937-04
	Test Method for Rapid Determination of Corrosiveness to Copper from Petroleum Products Using a Disposable Copper Foil Strip	ASTM	ASTM D7095-04
	Thickness Design of Ductile-Iron Pipe	AWWA	ANSI/AWWA C150/A21.50-02
	Fittings, Cast Metal Boxes, and Conduit Bodies for Conduit and Cable Assemblies	NEMA	NEMA FB 1-2007

Selected Codes and Standards Published by Various Societies and Associations (*Continued*)

Subject	Title	Publisher	Reference
	Polyvinyl-Chloride (PVC) Externally Coated Galvanized Rigid Steel Conduit and Intermediate Metal Conduit	NEMA	NEMA RN 1-2005
Plumbing	Backwater Valves	ASME	ASME A112.14.1-2003
	Plumbing Supply Fittings	ASME	ASME A112.18.1-2005
	Plumbing Waste Fittings	ASME	ASME A112.18.2-2005
	Performance Requirements for Backflow Protection Devices and Systems in Plumbing Fixture Fittings	ASME	ASME A112.18.3-2002
	Uniform Plumbing Code (2006) (with IAPMO Installation Standards)	IAPMO	IAPMO
	International Plumbing Code (2009)	ICC	IPC
	International Private Sewage Disposal Code (2009)	ICC	IPSDC
	2006 National Standard Plumbing Code (NSPC)	PHCC	NSPC 2003
	2006 National Standard Plumbing Code—Illustrated	PHCC	PHCC 2003
	Standard Plumbing Code (1997)	SBCCI	SPC
Pumps	Centrifugal Pumps	ASME	ASME PTC 8.2-1990
	Specification for Horizontal End Suction Centrifugal Pumps for Chemical Process	ASME	ASME B73.1-2001
	Specification for Vertical-in-Line Centrifugal Pumps for Chemical Process	ASME	ASME B73.2-2003
	Specification for Sealless Horizontal End Suction Metallic Centrifugal Pumps for Chemical Process	ASME	ASME B73.3-2003
	Specification for Thermoplastic and Thermoset Polymer Material Horizontal End Suction Centrifugal Pumps for Chemical Process	ASME	ASME B73.5M-1995 (RA01)
	Liquid Pumps	CSA	CAN/CSA-C22.2 No. 108-01
	Energy Efficiency Test Methods for Small Pumps	CSA	CAN/CSA C820-02 (R2007)
	Performance Standard for Liquid Ring Vacuum Pumps, 3rd ed. (2005)	HEI	HEI 2854
	Centrifugal Pumps for Nomenclature and Definitions	HI	ANSI/HI 1.1-1.2 (2000)
	Centrifugal Pumps for Design and Applications	HI	ANSI/HI 1.3 (2000)
	Centrifugal Pumps for Installation, Operation, and Maintenance	HI	ANSI/HI 1.4 (2000)
	Vertical Pumps for Nomenclature and Definitions	HI	ANSI/HI 2.1-2.2 (2000)
	Vertical Pumps for Design and Application	HI	ANSI/HI 2.3 (2000)
	Vertical Pumps for Installation, Operation, and Maintenance	HI	ANSI/HI 2.4 (2000)
	Rotary Pumps for Nomenclature, Definitions, Application, and Operation	HI	ANSI/HI 3.1-3.5 (2000)
	Sealless Rotary Pumps for Nomenclature, Definitions, Application, Operation, and Test	HI	ANSI/HI 4.1-4.6 (2000)
	Sealless Centrifugal Pumps for Nomenclature, Definitions, Application, Operation, and Test	HI	ANSI/HI 5.1-5.6 (2000)
	Reciprocating Pumps for Nomenclature, Definitions, Application, and Operation	HI	ANSI/HI 6.1-6.5 (2000)
	Direct Acting (Steam) Pumps for Nomenclature, Definitions, Application, and Operation	HI	ANSI/HI 8.1-8.5 (2000)
	Pumps—General Guidelines for Types, Definitions, Application, Sound Measurement and Decontamination	HI	ANSI/HI 9.1-9.5 (2000)
	Centrifugal and Vertical Pumps for Allowable Nozzle Loads	HI	ANSI/HI 9.6.2 (2001)
	Centrifugal and Vertical Pumps for Allowable Operating Region	HI	ANSI/HI 9.6.3 (1997)
	Centrifugal and Vertical Pumps for Vibration Measurements and Allowable Values	HI	ANSI/HI 9.6.4 (2001)
	Centrifugal and Vertical Pumps for Condition Monitoring	HI	ANSI/HI 9.6.5 (2000)
	Pump Intake Design	HI	ANSI/HI 9.8 (1998)
	Engineering Data Book, 2nd ed.	HI	HI (1990)
	Circulation System Components and Related Materials for Swimming Pools, Spas/Hot Tubs	NSF	NSF/ANSI 50-2007
	Pumps for Oil-Burning Appliances (1997)	UL	UL 343
	Motor-Operated Water Pumps (2002)	UL	ANSI/UL 778
	Swimming Pool Pumps, Filters, and Chlorinators (2008)	UL	ANSI/UL 1081
Radiators	Testing and Rating Standard for Baseboard Radiation, 8th ed. (2005)	HYDI	IBR
	Testing and Rating Standard for Finned Tube (Commercial) Radiation, 6th ed. (2005)	HYDI	IBR
Receivers	Refrigerant Liquid Receivers	AHRI	AHRI 495-2005
	Refrigerant-Containing Components and Accessories, Nonelectrical (2001)	UL	ANSI/UL 207
Refrigerants	Threshold Limit Values for Chemical Substances (updated annually)	ACGIH	ACGIH
	Specifications for Fluorocarbon Refrigerants	AHRI	AHRI 700-2006
	Refrigerant Recovery/Recycling Equipment	AHRI	AHRI 740-98
	Refrigerant Information Recommended for Product Development and Standards	ASHRAE	ASHRAE *Guideline* 6-2008
	Method of Testing Flow Capacity of Refrigerant Capillary Tubes	ASHRAE	ANSI/ASHRAE 28-1996 (RA2010)
	Designation and Safety Classification of Refrigerants	ASHRAE	ANSI/ASHRAE 34-2010
	Sealed Glass Tube Method to Test the Chemical Stability of Materials for Use Within Refrigerant Systems	ASHRAE	ANSI/ASHRAE 97-2007
	Refrigeration Oil Description	ASHRAE	ANSI/ASHRAE 99-2006
	Reducing the Release of Halogenated Refrigerants from Refrigerating and Air-Conditioning Equipment and Systems	ASHRAE	ANSI/ASHRAE 147-2002
	Test Method for Acid Number of Petroleum Products by Potentiometric Titration	ASTM	ASTM D664-07
	Test Method for Concentration Limits of Flammability of Chemical (Vapors and Gases)	ASTM	ASTM E681-04
	Refrigerant-Containing Components for Use in Electrical Equipment	CSA	C22.2 No. 140.3-M1987 (R2004)

Selected Codes and Standards Published by Various Societies and Associations (*Continued*)

Subject	Title	Publisher	Reference
	Refrigerants—Designation System	ISO	ISO 817:2005
	Procedure Retrofitting CFC-12 (R-12) Mobile Air-Conditioning Systems to HFC-134a (R-134a)	SAE	SAE J1661-1998
	Recommended Service Procedure for the Containment of CFC-12 (R-12)	SAE	SAE J1989-1998
	Standard of Purity for Recycled HFC-134a for Use in Mobile Air-Conditioning Systems	SAE	SAE J2099-1999
	HFC-134a (R-134a) Service Hose Fittings for Automotive Air-Conditioning Service Equipment	SAE	SAE J2197-1997
	Recommended Service Procedure for the Containment of HFC-134a	SAE	SAE J2211-1998
	HFC-134a (R-134a) Recovery/Recycling Equipment for Mobile Air-Conditioning Systems	SAE	SAE J2210-1999
	CFC-12 (R-12) Refrigerant Recovery Equipment for Mobile Automotive Air-Conditioning Systems	SAE	SAE J2209-1999
	Refrigerant-Containing Components and Accessories, Nonelectrical (2001)	UL	ANSI/UL 207
	Refrigerant Recovery/Recycling Equipment (2005)	UL	ANSI/UL 1963
	Field Conversion/Retrofit of Products to Change to an Alternative Refrigerant—Construction and Operation (1993)	UL	ANSI/UL 2170
	Field Conversion/Retrofit of Products to Change to an Alternative Refrigerant—Insulating Material and Refrigerant Compatibility (1993)	UL	ANSI/UL 2171
	Field Conversion/Retrofit of Products to Change to an Alternative Refrigerant—Procedures and Methods (1993)	UL	ANSI/UL 2172
	Refrigerants (2006)	UL	ANSI/UL 2182
Refrigeration	Safety Standard for Refrigeration Systems	ASHRAE	ANSI/ASHRAE 15-2010
	Mechanical Refrigeration Code	CSA	B52-05
	Refrigeration Equipment	CSA	CAN/CSA-C22.2 No. 120-M91 (R2004)
	Equipment, Design and Installation of Ammonia Mechanical Refrigerating Systems	IIAR	ANSI/IIAR 2-1999
	Refrigerated Medical Equipment (1993)	UL	ANSI/UL 416
Refrigeration Systems	Ejectors	ASME	ASME PTC 24-1976 (RA82)
	Safety Standard for Refrigeration Systems	ASHRAE	ANSI/ASHRAE 15-2010
	Designation and Safety Classification of Refrigerants	ASHRAE	ANSI/ASHRAE 34-2010
	Reducing the Release of Halogenated Refrigerants from Refrigerating and Air-Conditioning Equipment and Systems	ASHRAE	ANSI/ASHRAE 147-2002
	Testing of Refrigerating Systems	ISO	ISO 916-1968
	Standards for Steam Jet Vacuum Systems, 6th ed.	HEI	HEI 2866-1
Transport	Mechanical Transport Refrigeration Units	AHRI	AHRI 1110-2006
	Mechanical Refrigeration and Air-Conditioning Installations Aboard Ship	ASHRAE	ANSI/ASHRAE 26-2010
	General Requirements for Application of Vapor Cycle Refrigeration Systems for Aircraft	SAE	SAE ARP731-2003
	Safety Standard for Motor Vehicle Refrigerant Vapor Compression Systems	SAE	SAE J639-2005
Refrigerators	Method of Testing Commercial Refrigerators and Freezers	ASHRAE	ANSI/ASHRAE 72-2005
Commercial	Energy Performance Standard for Commercial Refrigerated Display Cabinets and Merchandise	CSA	C657-04
	Energy Performance Standard for Food Service Refrigerators and Freezers	CSA	C827-98 (R2003)
	Gas Food Service Equipment	CSA	ANSI Z83.11-2006/CSA 1.8A-2006
	Mobile Food Carts	NSF	NSF/ANSI 59-2002e
	Food Equipment	NSF	NSF/ANSI 2-2007
	Commercial Refrigerators and Freezers	NSF	NSF/ANSI 7-2007
	Refrigeration Unit Coolers (2004)	UL	ANSI/UL 412
	Refrigerating Units (2006)	UL	ANSI/UL 427
	Commercial Refrigerators and Freezers (2006)	UL	ANSI/UL 471
Household	Household Refrigerators, Refrigerator-Freezers and Freezers	AHAM	ANSI/AHAM HRF-1-2007
	Refrigerators Using Gas Fuel	CSA	ANSI Z21.19-2002/CSA1.4-2002
	Energy Performance and Capacity of Household Refrigerators, Refrigerator-Freezers, and Freezers	CSA	CAN/CSA C300-00 (R2005)
	Household Refrigerators and Freezers (1993)	UL/CSA	ANSI/UL 250-1997/C22.2 No. 63-93
Retrofitting			
Building	Residential Duct Diagnostics and Repair (2003)	ACCA	ACCA
	Good HVAC Practices for Residential and Commercial Buildings (2003)	ACCA	ACCA
	Building Systems Analysis and Retrofit Manual, 1st ed.	SMACNA	SMACNA 1995
Refrigerant	Procedure for Retrofitting CFC-12 (R-12) Mobile Air Conditioning Systems to HFC-134a (R-134a)	SAE	SAE J1661-1998
	Field Conversion/Retrofit of Products to Change to an Alternative Refrigerant—Construction and Operation (1993)	UL	ANSI/UL 2170
	Field Conversion/Retrofit of Products to Change to an Alternative Refrigerant—Insulating Material and Refrigerant Compatibility (1993)	UL	ANSI/UL 2171
	Field Conversion/Retrofit of Products to Change to an Alternative Refrigerant—Procedures and Methods (1993)	UL	ANSI/UL 2172

Selected Codes and Standards Published by Various Societies and Associations (*Continued*)

Subject	Title	Publisher	Reference
Roof	Commercial Low Pressure, Low Velocity Duct System Design, 1st ed.	ACCA	ACCA Manual Q
Ventilators	Power Ventilators (2004)	UL	ANSI/UL 705
Solar	Thermal Performance Testing of Solar Ambient Air Heaters	ASABE	ANSI/ASAE S423-1991
Equipment	Testing and Reporting Solar Cooker Performance	ASABE	ASAE S580-2003
	Method of Measuring Solar-Optical Properties of Materials	ASHRAE	ASHRAE 74-1988
	Methods of Testing to Determine the Thermal Performance of Solar Collectors	ASHRAE	ANSI/ASHRAE 93-2010
	Methods of Testing to Determine the Thermal Performance of Solar Domestic Water Heating Systems	ASHRAE	ANSI/ASHRAE 95-1987
	Methods of Testing to Determine the Thermal Performance of Unglazed Flat-Plate Liquid-Type Solar Collectors	ASHRAE	ANSI/ASHRAE 96-1980 (RA89)
	Practice for Installation and Service of Solar Space Heating Systems for One and Two Family Dwellings	ASTM	ASTM E683-91 (2007)
	Practice for Evaluating Thermal Insulation Materials for Use in Solar Collectors	ASTM	ASTM E861-94 (2007)
	Practice for Installation and Service of Solar Domestic Water Heating Systems for One and Two Family Dwellings	ASTM	ASTM E1056-85 (2007)
	Reference Solar Spectral Irradiance at the Ground at Different Receiving Conditions—Part 1: Direct Normal and Hemispherical Solar Irradiance for Air Mass 1.5	ISO	ISO 9845-1:1992
	Solar Collectors	CSA	CAN/CSA F378-87 (R2004)
	Solar Domestic Hot Water Systems (Liquid to Liquid Heat Transfer)	CSA	CAN/CSA F379.1-88 (R2006)
	Seasonal Use Solar Domestic Hot Water Systems	CSA	CAN/CSA F379.2-M89 (R2006)
	Installation Code for Solar Domestic Hot Water Systems	CSA	CAN/CSA F383-87 (R2005)
	Solar Heating—Domestic Water Heating Systems—Part 2: Outdoor Test Methods for System Performance Characterization and Yearly Performance Prediction of Solar-Only Systems	ISO	ISO 9459-2:1995
	Test Methods for Solar Collectors—Part 1: Thermal Performance of Glazed Liquid Heating Collectors Including Pressure Drop	ISO	ISO 9806-1:1994
	Part 2: Qualification Test Procedures	ISO	ISO 9806-2:1995
	Part 3: Thermal Performance of Unglazed Liquid Heating Collectors (Sensible Heat Transfer Only) Including Pressure Drop	ISO	ISO 9806-3:1995
	Solar Water Heaters—Elastomeric Materials for Absorbers, Connecting Pipes and Fittings—Method of Assessment	ISO	ISO 9808:1990
	Solar Energy—Calibration of a Pyranometer Using a Pyrheliometer	ISO	ISO 9846:1993
Solenoid Valves	Solenoid Valves for Use with Volatile Refrigerants	AHRI	AHRI 760-2007
	Methods of Testing Capacity of Refrigerant Solenoid Valves	ASHRAE	ANSI/ASHRAE 158.1-2004
	Electrically Operated Valves (1999)	UL	UL 429
Sound	Threshold Limit Values for Physical Agents (updated annually)	ACGIH	ACGIH
Measurement	Specification for Sound Level Meters	ASA	ANSI S1.4-1983 (R2006)
	Specification for Octave-Band and Fractional-Octave-Band Analog and Digital Filters	ASA	ANSI S1.11-2004
	Microphones, Part 1: Specifications for Laboratory Standard Microphones	ASA	ANSI S1.15-1997/Part 1 (R2006)
	Part 2: Primary Method for Pressure Calibration of Laboratory Standard Microphones by the Reciprocity Technique	ASA	ANSI S1.15-2005/Part 2
	Specification for Acoustical Calibrators	ASA	ANSI S1.40-2006
	Measurement of Industrial Sound	ASME	ASME PTC 36-2004
	Test Method for Measuring Acoustical and Airflow Performance of Duct Liner Materials and Prefabricated Silencers	ASTM	ASTM E477-06a
	Test Method for Determination of Decay Rates for Use in Sound Insulation Test Methods	ASTM	ASTM E2235-04e1
	Sound and Vibration Design and Analysis (1994)	NEBB	NEBB
Fans	Reverberant Room Method for Sound Testing of Fans	AMCA	AMCA 300-05
	Methods for Calculating Fan Sound Ratings from Laboratory Test Data	AMCA	AMCA 301-06
	Application of Sone Ratings for Non-Ducted Air Moving Devices	AMCA	AMCA 302-73 (R2008)
	Application of Sound Power Level Ratings for Fans	AMCA	AMCA 303-79 (R2008)
	Acoustics—Measurement of Noise and Vibration of Small Air-Moving Devices—Part 1: Airborne Noise Emission	ASA	ANSI S12.11/1-2003/ISO 10302:1996 (MOD-2003)
	Part 2: Structure-Borne Vibration	ASA	ANSI S12.11/2-2003
	Laboratory Method of Testing to Determine the Sound Power in a Duct	ASHRAE/ AMCA	ANSI/ASHRAE 68-1997/ AMCA 330-97
Other	Sound Rating of Outdoor Unitary Equipment	AHRI	AHRI 270-95
Equipment	Application of Sound Rating Levels of Outdoor Unitary Equipment	AHRI	AHRI 275-97
	Sound Rating and Sound Transmission Loss of Packaged Terminal Equipment	AHRI	AHRI 300-2000
	Sound Rating of Non-Ducted Indoor Air-Conditioning Equipment	AHRI	AHRI 350-2000
	Sound Rating of Large Outdoor Refrigerating and Air-Conditioning Equipment	AHRI	AHRI 370-2001
	Method of Rating Sound and Vibration of Refrigerant Compressors	AHRI	AHRI 530-2005
	Method of Measuring Machinery Sound Within an Equipment Space	AHRI	AHRI 575-94
	Statistical Methods for Determining and Verifying Stated Noise Emission Values of Machinery and Equipment	ASA	ANSI S12.3-1985 (R2006)

Selected Codes and Standards Published by Various Societies and Associations (*Continued*)

Subject	Title	Publisher	Reference
Techniques	Sound Level Prediction for Installed Rotating Electrical Machines	NEMA	NEMA MG 3-1974 (R2006)
	Preferred Frequencies, Frequency Levels, and Band Numbers for Acoustical Measurements	ASA	ANSI S1.6-1984 (R2006)
	Reference Quantities for Acoustical Levels	ASA	ANSI S1.8-1989 (R2006)
	Measurement of Sound Pressure Levels in Air	ASA	ANSI S1.13-2005
	Procedure for the Computation of Loudness of Steady Sound	ASA	ANSI S3.4-2007
	Criteria for Evaluating Room Noise	ASA	ANSI S12.2-1995 (R1999)
	Methods for Determining the Insertion Loss of Outdoor Noise Barriers	ASA	ANSI S12.8-1998 (R2003)
	Engineering Method for the Determination of Sound Power Levels of Noise Sources Using Sound Intensity	ASA	ANSI S12.12-1992 (R2007)
	Procedures for Outdoor Measurement of Sound Pressure Level	ASA	ANSI S12.18-1994 (R2004)
	Methods for Measurement of Sound Emitted by Machinery and Equipment at Workstations and Other Specified Positions	ASA	ANSI S12.43-1997 (R2007)
	Methods for Calculation of Sound Emitted by Machinery and Equipment at Workstations and Other Specified Positions from Sound Power Level	ASA	ANSI S12.44-1997 (R2007)
	Acoustics—Determination of Sound Power Levels of Noise Sources Using Sound Pressure—Precision Method for Reverberation Rooms	ASA	ANSI S12.51-2002 (R2007)/ISO 3741:1999
	Acoustics—Determination of Sound Power Levels of Noise Sources—Engineering Methods for Small, Movable Sources in Reverberant Fields—Part 1: Comparison Method for Hard-Walled Test Rooms	ASA	ANSI S12.53/Part 1-1999 (R2004)/ ISO 3743-1:1994
	Part 2: Methods for Special Reverberation Test Rooms	ASA	ANSI S12.53/Part 2-1999 (R2004)/ ISO 3743-2:1994
	Acoustics—Determination of Sound Power Levels of Noise Sources Using Sound Pressure—Engineering Method in an Essentially Free Field over a Reflecting Plane	ASA	ANSI S12.54-1999 (R2004)/ISO 3744:1994
	Acoustics—Determination of Sound Power Levels of Noise Sources Using Sound Pressure—Survey Method Using an Enveloping Measurement Surface over a Reflecting Plane	ASA	ANSI S12.56-1999 (R2004)/ISO 3746:1995
	Test Method for Impedance and Absorption of Acoustical Materials by the Impedance Tube Method	ASTM	ASTM C384-04
	Test Method for Sound Absorption and Sound Absorption Coefficients by the Reverberation Room Method	ASTM	ASTM C423-07a
	Test Method for Measurement of Airborne Sound Insulation in Buildings	ASTM	ASTM E336-07
	Test Method for Impedance and Absorption of Acoustical Materials Using a Tube, Two Microphones and a Digital Frequency Analysis System	ASTM	ASTM E1050-08
	Test Method for Evaluating Masking Sound in Open Offices Using A-Weighted and One-Third Octave Band Sound Pressure Levels	ASTM	ASTM E1573-02
	Test Method for Measurement of Sound in Residential Spaces	ASTM	ASTM E1574-98 (2006)
	Acoustics–Measurement of Sound Insulation in Buildings and of Building Elements; Part 1: Requirements for Laboratory Test Facilities with Suppressed Flanking Transmission	ISO	ISO 140-1:1997
	Part 4: Field Measurements of Airborne Sound Insulation Between Rooms	ISO	ISO 140-4:1998
	Part 5: Field Measurements of Airborne Sound Insulation of Facade Elements and Facades	ISO	ISO 140-5:1998
	Part 6: Laboratory Measurements of Impact Sound Insulation of Floors	ISO	ISO 140-6:1998
	Part 7: Field Measurements of Impact Sound Insulation of Floors	ISO	ISO 140-7:1998
	Part 8: Laboratory Measurements of the Reduction of Transmitted Impact Noise by Floor Coverings on a Heavyweight Standard Floor	ISO	ISO 140-8:1997
	Acoustics—Method for Calculating Loudness Level	ISO	ISO 532:1975
	Acoustics—Determination of Sound Power Levels of Noise Sources Using Sound Intensity; Part 1: Measurement at Discrete Points	ISO	ISO 9614-1:1993
	Part 2: Measurement by Scanning	ISO	ISO 9614-2:1996
	Procedural Standards for Measurement and Assessment of Sound and Vibration, 2nd ed. (2006)	NEBB	NEBB
Terminology	Acoustical Terminology	ASA	ANSI S1.1-1994 (R2004)
	Terminology Relating to Environmental Acoustics	ASTM	ASTM C634-02e1
Space Heaters	Methods of Testing for Rating Combination Space-Heating and Water-Heating Appliances	ASHRAE	ANSI/ASHRAE 124-2007
	Gas-Fired Room Heaters, Vol. II, Unvented Room Heaters	CSA	ANSI Z21.11.2-2007
	Vented Gas-Fired Space Heating Appliances	CSA	ANSI Z21.86-2004/CSA 2.32-2004
	Movable and Wall- or Ceiling-Hung Electric Room Heaters (2000)	UL	UL 1278
	Fixed and Location-Dedicated Electric Room Heaters (1997)	UL	UL 2021
Sustainability	Standard for the Design of High-Performance, Green Buildings Except Low-Rise Residential Buildings	ASHRAE/ USGBC	ANSI/ASHRAE/USGBC/IES 189.1-2009
	International Green Construction Code™	ICC	IGCC
Symbols	Graphic Electrical/Electronic Symbols for Air-Conditioning and Refrigerating Equipment	AHRI	AHRI 130-88
	Graphic Symbols for Heating, Ventilating, Air-Conditioning, and Refrigerating Systems	ASHRAE	ANSI/ASHRAE 134-2005

Selected Codes and Standards Published by Various Societies and Associations (*Continued*)

Subject	Title	Publisher	Reference
	Graphical Symbols for Plumbing Fixtures for Diagrams Used in Architecture and Building Construction	ASME	ANSI/ASME Y32.4-1977 (RA04)
	Symbols for Mechanical and Acoustical Elements as Used in Schematic Diagrams	ASME	ANSI/ASME Y32.18-1972 (RA03)
	Practice for Mechanical Symbols, Shipboard Heating, Ventilation, and Air Conditioning (HVAC)	ASTM	ASTM F856-97 (2004)
	Standard Symbols for Welding, Brazing, and Nondestructive Examination	AWS	AWS A2.4:2007
	Standard Letter Symbols for Quantities Used in Electrical Science and Electrical Engineering	IEEE	IEEE 280-1982 (R2003)
	Graphic Symbols for Electrical and Electronics Diagrams	IEEE	ANSI 315-1975 (R1986)/IEEE 315A-1986
	Standard for Logic Circuit Diagrams	IEEE	IEEE 991-1986 (R1994)
	Use of the International System of Units (SI): The Modern Metric System	IEEE/ASTM	IEEE/ASTM-SI10-2002
	Abbreviations and Acronyms	ASME	ASME Y14.38-1999
	Engineering Drawing Practices	ASME	ASME Y14.100-2004
	Safety Color Code	NEMA	ANSI/NEMA Z535-2002
Terminals, Wiring	Electrical Quick-Connect Terminals (2003)	UL	ANSI/UL 310
	Wire Connectors (2003)	UL	ANSI/UL 486A-486B
	Splicing Wire Connectors (2004)	UL	ANSI/UL 486C
	Equipment Wiring Terminals for Use with Aluminum and/or Copper Conductors (1994)	UL	ANSI/UL 486E
Testing and Balancing	AABC National Standards for Total System Balance (2002)	AABC	AABC
	Industrial Process/Power Generation Fans: Site Performance Test Standard	AMCA	AMCA 803-02
	Guidelines for Measuring and Reporting Environmental Parameters for Plant Experiments in Growth Chambers	ASABE	ANSI/ASAE EP411.4-2002
	HVAC&R Technical Requirements for the Commissioning Process	ASHRAE	ASHRAE *Guideline* 1.1-2007
	Measurement, Testing, Adjusting, and Balancing of Building HVAC Systems	ASHRAE	ANSI/ASHRAE 111-2008
	Practices for Measuring, Testing, Adjusting, and Balancing Shipboard HVAC&R Systems	ASHRAE	ANSI/ASHRAE 151 2010
	Centrifugal Pump Tests	HI	ANSI/HI 1.6 (M104) (2000)
	Vertical Pump Tests	HI	ANSI/HI 2.6 (M108) (2000)
	Rotary Pump Tests	HI	ANSI/HI 3.6 (M110) (2000)
	Reciprocating Pump Tests	HI	ANSI/HI 6.6 (M114) (2000)
	Pumps—General Guidelines for Types, Definitions, Application, Sound Measurement and Decontamination	HI	HI 9.1-9.5 (M117) (2000)
	Submersible Pump Tests	HI	ANSI/HI 11.6 (M126) (2001)
	Procedural Standards for Certified Testing of Cleanrooms, 2nd ed. (1996)	NEBB	NEBB
	Procedural Standards for Testing, Adjusting, Balancing of Environmental Systems, 7th ed. (2005)	NEBB	NEBB
	HVAC Systems Testing, Adjusting and Balancing, 3rd ed.	SMACNA	SMACNA 2002
Thermal Storage	Thermal Energy Storage: A Guide for Commercial HVAC Contractors	ACCA	ACCA
	Method of Testing Active Latent-Heat Storage Devices Based on Thermal Performance	ASHRAE	ANSI/ASHRAE 94.1-2010
	Method of Testing Thermal Storage Devices with Electrical Input and Thermal Output Based on Thermal Performance	ASHRAE	ANSI/ASHRAE 94.2-2010
	Method of Testing Active Sensible Thermal Energy Devices Based on Thermal Performance	ASHRAE	ANSI/ASHRAE 94.3-2010
	Measurement, Testing, Adjusting, and Balancing of Building HVAC Systems	ASHRAE	ANSI/ASHRAE 111-2008
	Method of Testing the Performance of Cool Storage Systems	ASHRAE	ANSI/ASHRAE 150-2000 (RA04)
Transformers	Minimum Efficiency Values for Liquid-Filled Distribution Transformers	CSA	CAN/CSA C802.1-00 (R2005)
	Minimum Efficiency Values for Dry-Type Transformers	CSA	CAN/CSA C802.2-06
	Maximum Losses For Power Transformers	CSA	CAN/CSA C802.3-01 (R2007)
	Guide for Determining Energy Efficiency of Distribution Transformers	NEMA	NEMA TP-1-2002
Turbines	Steam Turbines	ASME	ASME PTC 6-2004
	Steam Turbines for Combined Cycle	ASME	ASME PTC 6.2-2004
	Hydraulic Turbines and Pump-Turbines	ASME	ASME PTC 18-2002
	Gas Turbines	ASME	ASME PTC 22-2005
	Wind Turbines	ASME	ASME PTC 42-1988 (RA04)
	Specification for Stainless Steel Bars for Compressor and Turbine Airfoils	ASTM	ASTM A1028-03
	Specification for Gas Turbine Fuel Oils	ASTM	ASTM D2880-03
	Land Based Steam Turbine Generator Sets, 0 to 33,000 kW	NEMA	NEMA SM 24-1991 (R2002)
	Steam Turbines for Mechanical Drive Service	NEMA	NEMA SM 23-1991 (R2002)
Valves	Face-to-Face and End-to-End Dimensions of Valves	ASME	ASME B16.10-2000 (R03)
	Valves—Flanged, Threaded, and Welding End	ASME	ASME B16.34-2004
	Manually Operated Metallic Gas Valves for Use in Aboveground Piping Systems up to 5 psi	ASME	ASME B16.44-2002
	Pressure Relief Devices	ASME	ASME PTC 25-2001
	Methods of Testing Capacity of Refrigerant Solenoid Valves	ASHRAE	ANSI/ASHRAE 158.1-2004

Selected Codes and Standards Published by Various Societies and Associations (*Continued*)

Subject	Title	Publisher	Reference
	Relief Valves for Hot Water Supply	CSA	ANSI Z21.22-1999 (R2003)/ CSA 4.4-M99 (R2004)
	Control Valve Capacity Test Procedures	ISA	ANSI/ISA-S75.02-1996
	Flow Equations for Sizing Control Valves	ISA	ANSI/ISA-S75.01.01-2002
	Industrial Valves—Part-Turn Actuator Attachments	ISO	ISO 5211-1:2001
	Metal Valves for Use in Flanged Pipe Systems—Face-to-Face and Centre-to-Face Dimensions	ISO	ISO 5752:1982
	Safety Valves for Protection Against Excessive Pressure, Part 1: Safety Valves	ISO	ISO 4126-1:2004
	Oxygen System Fill/Check Valve	SAE	SAE AS1225A-1997
	Valves for Anhydrous Ammonia and LP-Gas (Other Than Safety Relief) (2007)	UL	ANSI/UL 125
	Safety Relief Valves for Anhydrous Ammonia and LP-Gas (2007)	UL	ANSI/UL 132
	LP-Gas Regulators (1999)	UL	ANSI/UL 144
	Electrically Operated Valves (1999)	UL	UL 429
	Valves for Flammable Fluids (2007)	UL	ANSI/UL 842
Gas	Manually Operated Metallic Gas Valves for Use in Gas Piping Systems up to 125 psig (Sizes NPS 1/2 through 2)	ASME	ASME B16.33-2002
	Large Metallic Valves for Gas Distribution (Manually Operated, NPS-2 1/2 to 12, 125 psig Maximum)	ASME	ANSI/ASME B16.38-2007
	Manually Operated Thermoplastic Gas Shutoffs and Valves in Gas Distribution Systems	ASME	ASME B16.40-2002
	Manually Operated Gas Valves for Appliances, Appliance Connection Valves, and Hose End Valves	CSA	ANSI Z21.15-1997 (R2003)/CGA 9.1-M97
	Automatic Valves for Gas Appliances	CSA	ANSI Z21.21-2005/CGA 6.5-2005
	Combination Gas Controls for Gas Appliances	CSA	ANSI Z21.78-2005/CGA 6.20-2005
	Convenience Gas Outlets and Optional Enclosures	CSA	ANSI Z21.90-2001/CSA 6.24-2001 (R2005)
Refrigerant	Thermostatic Refrigerant Expansion Valves	AHRI	AHRI 750-2007
	Solenoid Valves for Use with Volatile Refrigerants	AHRI	AHRI 760-2007
	Refrigerant Pressure Regulating Valves	AHRI	AHRI 770-2001
Vapor Retarders	Practice for Selection of Vapor Retarders for Thermal Insulation	ASTM	ASTM C755-03
	Practice for Determining the Properties of Jacketing Materials for Thermal Insulation	ASTM	ASTM C921-03a
	Specification for Flexible, Low Permeance Vapor Retarders for Thermal Insulation	ASTM	ASTM C1136-06
	Test Method for Water Vapor Transmission Rate of Flexible Barrier Materials Using an Infrared Detection Technique	ASTM	ASTM F372-99 (2003)
Vending Machines	Methods of Testing for Rating Vending Machines for Bottled, Canned, and Other Sealed Beverages	ASHRAE	ANSI/ASHRAE 32.1-2010
	Methods of Testing for Rating Pre-Mix and Post-Mix Beverage Dispensing Equipment	ASHRAE	ANSI/ASHRAE 32.2-2003 (RA07)
	Vending Machines	CSA	C22.2 No. 128-95 (R2004)
	Energy Performance of Vending Machines	CSA	CAN/CSA C804-96 (R2007)
	Vending Machines for Food and Beverages	NSF	NSF/ANSI 25-2007
	Refrigerated Vending Machines (1995)	UL	ANSI/UL 541
	Vending Machines (1995)	UL	ANSI/UL 751
Vent Dampers	Automatic Vent Damper Devices for Use with Gas-Fired Appliances	CSA	ANSI Z21.66-1996 (R2001)/CSA 6.14-M96
	Vent or Chimney Connector Dampers for Oil-Fired Appliances (1994)	UL	ANSI/UL 17
Ventilation	Commercial Application, Systems, and Equipment, 1st ed.	ACCA	ACCA Manual CS
	Commercial Low Pressure, Low Velocity Duct System Design, 1st ed.	ACCA	ACCA Manual Q
	Comfort, Air Quality, and Efficiency by Design	ACCA	ACCA Manual RS
	Guide for Testing Ventilation Systems (1991)	ACGIH	ACGIH
	Industrial Ventilation: A Manual of Recommended Practice, 26th ed. (2007)	ACGIH	ACGIH
	Design of Ventilation Systems for Poultry and Livestock Shelters	ASABE	ASAE EP270.5-2003
	Design Values for Emergency Ventilation and Care of Livestock and Poultry	ASABE	ANSI/ASAE EP282.2-2004
	Heating, Ventilating and Cooling Greenhouses	ASABE	ANSI/ASAE EP406.4-2003
	Guidelines for Selection of Energy Efficient Agricultural Ventilation Fans	ASABE	ASAE EP566-1997
	Uniform Terminology for Livestock Production Facilities	ASABE	ASAE S501-1990
	Agricultural Ventilation Constant Speed Fan Test Standard	ASABE	ASABE S565-2005
	Ventilation for Acceptable Indoor Air Quality	ASHRAE	ANSI/ASHRAE 62.1-2010
	Ventilation and Acceptable Indoor Air Quality in Low-Rise Residential Buildings	ASHRAE	ANSI/ASHRAE 62.2-2010
	Method of Testing for Room Air Diffusion	ASHRAE	ANSI/ASHRAE 113-2009
	Measuring Air Change Effectiveness	ASHRAE	ANSI/ASHRAE 129-1997 (RA02)
	Method of Determining Air Change Rates in Detached Dwellings	ASHRAE	ANSI/ASHRAE 136-1993 (RA06)
	Ventilation for Commercial Cooking Operations	ASHRAE	ANSI/ASHRAE 154-2003
	Residential Mechanical Ventilation Systems	CSA	CAN/CSA F326-M91 (R2005)
	Parking Structures	NFPA	NFPA 88A-07
	Installation of Air Conditioning and Ventilating Systems	NFPA	NFPA 90A-02

Selected Codes and Standards Published by Various Societies and Associations (*Continued*)

Subject	Title	Publisher	Reference
	Ventilation Control and Fire Protection of Commercial Cooking Operations	NFPA	NFPA 96-08
	Food Equipment	NSF	NSF/ANSI 2-2007
	Class II (Laminar Flow) Biosafety Cabinetry	NSF	NSF/ANSI 49-2007
	Aerothermodynamic Systems Engineering and Design	SAE	SAE AIR1168/3-1989
	Heater, Airplane, Engine Exhaust Gas to Air Heat Exchanger Type	SAE	SAE ARP86-1996
	Test Procedure for Battery Flame Retardant Venting Systems	SAE	SAE J1495-2005
Venting	Commercial Application, Systems, and Equipment, 1st ed.	ACCA	ACCA Manual CS
	Draft Hoods	CSA	ANSI Z21.12-1990 (R2000)
	National Fuel Gas Code	AGA/NFPA	ANSI Z223.1/NFPA 54-2006
	Explosion Prevention Systems	NFPA	NFPA 69-08
	Smoke and Heat Venting	NFPA	NFPA 204-07
	Chimneys, Fireplaces, Vents and Solid Fuel-Burning Appliances	NFPA	NFPA 211-06
	Guide for Steel Stack Construction, 2nd ed.	SMACNA	SMACNA 1996
	Draft Equipment (2006)	UL	UL 378
	Gas Vents (1996)	UL	ANSI/UL 441
	Type L Low-Temperature Venting Systems (1995)	UL	ANSI/UL 641
Vibration	Balance Quality and Vibration Levels for Fans	AMCA	ANSI/AMCA 204-05
	Techniques of Machinery Vibration Measurement	ASA	ANSI S2.17-1980 (R2004)
	Mechanical Vibration and Shock—Resilient Mounting Systems—Part 1: Technical Information to Be Exchanged for the Application of Isolation Systems	ASA	ISO 2017-1:2005
	Evaluation of Human Exposure to Whole-Body Vibration—Part 2: Vibration in Buildings (1 Hz to 80 Hz)	ISO	ISO 2631-2:2003
	Guidelines for the Evaluation of the Response of Occupants of Fixed Structures, Especially Buildings and Off-Shore Structures, to Low-Frequency Horizontal Motion (0.063 to 1 Hz)	ISO	ISO 6897:1984
	Procedural Standards for Measurement and Assessment of Sound and Vibration, 2nd ed. (2006)	NEBB	NEBB
	Sound and Vibration Design and Analysis (1994)	NEBB	NEBB
Water Heaters	Desuperheater/Water Heaters	AHRI	AHRI 470-2006
	Safety for Electrically Heated Livestock Waterers	ASABE	ASAE EP342.2-1995 (R2005)
	Methods of Testing to Determine the Thermal Performance of Solar Domestic Water Heating Systems	ASHRAE	ANSI/ASHRAE 95-1987
	Method of Testing for Rating Commercial Gas, Electric, and Oil Service Water Heating Equipment	ASHRAE	ANSI/ASHRAE 118.1-2008
	Method of Testing for Rating Residential Water Heaters	ASHRAE	ANSI/ASHRAE 118.2-2006
	Methods of Testing for Rating Combination Space-Heating and Water-Heating Appliances	ASHRAE	ANSI/ASHRAE 124-2007
	Methods of Testing for Efficiency of Space-Conditioning/Water-Heating Appliances That Include a Desuperheater Water Heater	ASHRAE	ANSI/ASHRAE 137-2009
	Boiler and Pressure Vessel Code—Section IV: Heating Boilers	ASME	BPVC-2007
	Section VI: Recommended Rules for the Care and Operation of Heating Boilers	ASME	BPVC-2007
	Gas Water Heaters—Vol. I: Storage Water Heaters with Input Ratings of 75,000 Btu per Hour or Less	CSA	ANSI Z21.10.1-2004/CSA 4.1-2004
	Vol. III: Storage, with Input Ratings Above 75,000 Btu per Hour, Circulating and Instantaneous Water Heaters	CSA	ANSI Z21.10.3-2004/CSA 4.3-2004
	Oil Burning Stoves and Water Heaters	CSA	B140.3-1962 (R2006)
	Oil-Fired Service Water Heaters and Swimming Pool Heaters	CSA	B140.12-03
	Construction and Test of Electric Storage-Tank Water Heaters	CSA	CAN/CSA-C22.2 No. 110-94 (R2004)
	Performance of Electric Storage Tank Water Heaters for Household Service	CSA	C191-04
	Energy Efficiency of Electric Storage Tank Water Heaters and Heat Pump Water Heaters	CSA	CSA C745-03
	One Time Use Water Heater Emergency Shut-Off	IAPMO	IGC 175-2003
	Water Heaters, Hot Water Supply Boilers, and Heat Recovery Equipment	NSF	NSF/ANSI 5-2007
	Household Electric Storage Tank Water Heaters (2004)	UL	ANSI/UL 174
	Oil-Fired Storage Tank Water Heaters (1995)	UL	ANSI/UL 732
	Commercial-Industrial Gas Heating Equipment (2006)	UL	UL 795
	Electric Booster and Commercial Storage Tank Water Heaters (2004)	UL	ANSI/UL 1453
Welding and Brazing	Boiler and Pressure Vessel Code—Section IX: Welding and Brazing Qualifications	ASME	BPVC-2007
	Structural Welding Code—Steel	AWS	AWS D1.1M/D1.1:2008
	Specification for Welding of Austenitic Stainless Steel Tube and Piping Systems in Sanitary Applications	AWS	AWS D18.1:1999
Wood-Burning Appliances	Threshold Limit Values for Chemical Substances (updated annually)	ACGIH	ACGIH
	Specification for Room Heaters, Pellet Fuel Burning Type	ASTM	ASTM E1509-04
	Guide for Construction of Solid Fuel Burning Masonry Heaters	ASTM	ASTM E1602-03
	Installation Code for Solid-Fuel-Burning Appliances and Equipment	CSA	CAN/CSA-B365-01 (R2006)

Selected Codes and Standards Published by Various Societies and Associations (*Continued*)

Subject	Title	Publisher	Reference
	Solid-Fuel-Fired Central Heating Appliances	CSA	CAN/CSA-B366.1-M91 (R2007)
	Chimneys, Fireplaces, Vents, and Solid Fuel-Burning Appliances	NFPA	ANSI/NFPA 211-06
	Commercial Cooking, Rethermalization and Powered Hot Food Holding and Transport Equipment	NSF	NSF/ANSI 4-2007e

ORGANIZATIONS

Abbrev.	Organization	Address	Telephone	http://www.
AABC	Associated Air Balance Council	1518 K Street NW, Suite 503 Washington, D.C. 20005	(202) 737-0202	aabchq.com
ABMA	American Boiler Manufacturers Association	8221 Old Courthouse Road, Suite 207 Vienna, VA 22182	(703) 356-7171	abma.com
ACCA	Air Conditioning Contractors of America	2800 Shirlington Road, Suite 300 Arlington, VA 22206	(703) 575-4477	acca.org
ACGIH	American Conference of Governmental Industrial Hygienists	1330 Kemper Meadow Drive Cincinnati, OH 45240	(513) 742-2020	acgih.org
ADC	Air Diffusion Council	1901 N Roselle Road, Suite 800 Schaumburg, IL 60195	(847) 706-6750	flexibleduct.org
AGA	American Gas Association	400 N. Capitol Street NW, Suite 450 Washington, D.C. 20001	(202) 824-7000	aga.org
AHAM	Association of Home Appliance Manufacturers	1111 19th Street NW, Suite 402 Washington, D.C. 20036	(202) 872-5955	aham.org
AIHA	American Industrial Hygiene Association	2700 Prosperity Avenue, Suite 250 Fairfax, VA 22031	(703) 849-8888	aiha.org
AMCA	Air Movement and Control Association International	30 West University Drive Arlington Heights, IL 60004-1893	(847) 394-0150	amca.org
ANSI	American National Standards Institute	1819 L Street NW, 6th Floor Washington, D.C. 20036	(202) 293-8020	ansi.org
AHRI	Air-Conditioning, Heating, and Refrigeration Institute	2111 Wilson Boulevard, Suite 500 Arlington, VA 22201	(703) 524-8800	ari.org
ASA	Acoustical Society of America	2 Huntington Quadrangle, Suite 1NO1 Melville, NY 14747-4502	(516) 576-2360	asa.aip.org
ASABE	American Society of Agricultural and Biological Engineers	2950 Niles Road St. Joseph, MI 49085-9659	(269) 429-0300	asabe.org
ASHRAE	American Society of Heating, Refrigerating and Air-Conditioning Engineers	1791 Tullie Circle, NE Atlanta, GA 30329	(404) 636-8400	ashrae.org
ASME	ASME	3 Park Avenue New York, NY 10016-5990	(973) 882-1170	asme.org
ASTM	ASTM International	100 Barr Harbor Drive West Conshohocken, PA 19428-2959	(610) 832-9500	astm.org
AWS	American Welding Society	550 NW LeJeune Road Miami, FL 33126	(305) 443-9353	aws.org
AWWA	American Water Works Association	6666 W. Quincy Avenue Denver, CO 80235	(303) 794-7711	awwa.org
BOCA	Building Officials and Code Administrators International	(see ICC)		
BSI	British Standards Institution	389 Chiswick High Road London W4 4AL, UK	44(0)20-8996-9001	bsi-global.com
CAGI	Compressed Air and Gas Institute	1300 Sumner Avenue Cleveland, OH 44115-2851	(216) 241-7333	cagi.org
CSA	Canadian Standards Association International	5060 Spectrum Way, Suite 100 Mississauga, ON L4W 5N6, Canada	(416) 747-4000	csa.ca
	Also available from CSA America	8501 East Pleasant Valley Road Cleveland, OH 44131-5575	(216) 524-4990	csa-america.org
CTI	Cooling Technology Institute	P.O. Box 73383 Houston, TX 77273-3383	(281) 583-4087	cti.org
EJMA	Expansion Joint Manufacturers Association	25 North Broadway Tarrytown, NY 10591	(914) 332-0040	ejma.org
HEI	Heat Exchange Institute	1300 Sumner Avenue Cleveland, OH 44115-2815	(216) 241-7333	heatexchange.org
HI	Hydraulic Institute	6 Campus Drive, First Floor North Parsippany, NJ 07054-4406	(973) 267-9700	pumps.org
HYDI	Hydronics Institute Division of GAMA	(see AHRI)		
IAPMO	International Association of Plumbing and Mechanical Officials	5001 E. Philadelphia Street Ontario, CA 91761-2816	(909) 472-4100	iapmo.org

ORGANIZATIONS (*Continued*)

Abbrev.	Organization	Address	Telephone	http://www.
ICBO	International Conference of Building Officials	(*see* ICC)		
ICC	International Code Council	500 New Jersey Ave NW, 6th Floor Washington, D.C. 20001	(888) 422-7233	iccsafe.org
IEEE	Institute of Electrical and Electronics Engineers	45 Hoes Lane Piscataway, NJ 08854-4141	(732) 981-0060	ieee.org
IES	Illuminating Engineering Society	120 Wall Street, Floor 17 New York, NY 10005-4001	(212) 248-5000	iesna.org
IFCI	International Fire Code Institute	(*see* ICC)		
IIAR	International Institute of Ammonia Refrigeration	1001 N. Fairfax St, Suite 503 Alexandria, VA 22314	(703) 312-4200	iiar.org
ISA	The Instrumentation, Systems, and Automation Society	67 Alexander Drive, P.O. Box 12777 Research Triangle Park, NC 27709	(919) 549-8411	isa.org
ISO	International Organization for Standardization	1, ch. de la Voie-Creuse, Case postale 56 CH-1211 Geneva 20, Switzerland	41-22-749-01 11	iso.org
MCAA	Mechanical Contractors Association of America	1385 Piccard Drive Rockville, MD 20850	(301) 869-5800	mcaa.org
MICA	Midwest Insulation Contractors Association	16712 Elm Circle Omaha, NE 68130	(800) 747-6422	micainsulation.org
MSS	Manufacturers Standardization Society of the Valve and Fittings Industry	127 Park Street NE Vienna, VA 22180-4602	(703) 281-6613	mss-hq.com
NAIMA	North American Insulation Manufacturers Association	44 Canal Center Plaza, Suite 310 Alexandria, VA 22314	(703) 684-0084	naima.org
NCPWB	National Certified Pipe Welding Bureau	1385 Piccard Drive Rockville, MD 20850-4340	(301) 869-5800	mcaa.org/ncpwb
NCSBCS	National Conference of States on Building Codes and Standards	505 Huntmar Park Drive, Suite 210 Herndon, VA 20170	(703) 437-0100	ncsbcs.org
NEBB	National Environmental Balancing Bureau	8575 Grovemont Circle Gaithersburg, MD 20877	(301) 977-3698	nebb.org
NEMA	Association of Electrical and Medical Imaging Equipment Manufacturers	1300 North 17th Street, Suite 1752 Rosslyn, VA 22209	(703) 841-3200	nema.org
NFPA	National Fire Protection Association	1 Batterymarch Park Quincy, MA 02169-7471	(617) 770-3000	nfpa.org
NRCC	National Research Council of Canada, Institute for Research in Construction	1200 Montreal Road, Bldg M-58 Ottawa, ON K1A 0R6, Canada	(877) 672-2672	nrc-cnrc.ca
NSF	NSF International	P.O. Box 130140, 789 N. Dixboro Road Ann Arbor, MI 48113-0140	(734) 769-8010	nsf.org
PHCC	Plumbing-Heating-Cooling Contractors Association	180 S. Washington Street, P.O. Box 6808 Falls Church, VA 22046	(703) 237-8100	phccweb.org
SAE	Society of Automotive Engineers International	400 Commonwealth Drive Warrendale, PA 15096-0001	(724) 776-4841	sae.org
SBCCI	Southern Building Code Congress International	(*see* ICC)		
SMACNA	Sheet Metal and Air Conditioning Contractors' National Association	4201 Lafayette Center Drive Chantilly, VA 20151-1209	(703) 803-2980	smacna.org
TEMA	Tubular Exchanger Manufacturers Association	25 North Broadway Tarrytown, NY 10591	(914) 332-0040	tema.org
UL	Underwriters Laboratories	333 Pfingsten Road Northbrook, IL 60062-2096	(847) 272-8800	ul.com

Additions and Corrections

This report includes additional information, and technical errors found between June 15, 2008, and April 15, 2011, in the inch-pound (I-P) editions of the 2008, 2009, and 2010 *ASHRAE Handbook* volumes. Occasional typographical errors and nonstandard symbol labels will be corrected in future volumes. The most current list of Handbook additions and corrections is on the ASHRAE Web site (www.ashrae.org).

The authors and editor encourage you to notify them if you find other technical errors. Please send corrections to: Handbook Editor, ASHRAE, 1791 Tullie Circle NE, Atlanta, GA 30329, or e-mail mowen@ashrae.org.

2008 HVAC Systems and Equipment

p. 6.4, Eqs. (9a) and (10). The equations should read as follows:

$$q_c = 0.02(t_p - t_a)^{0.25}(t_p - t_a) \tag{9a}$$

$$q_c = 0.31|t_p - t_a|^{0.31}(t_p - t_a) \tag{10}$$

p. 12.11, Eq. (18). Change the "D*h*" to "Δh."

p. 12.21, 1st col., 2nd paragraph. The reference to Equation (23) should be to Equation (20).

p. 12.24, 2nd col., 1st full paragraph. The reference to Equation (23) should be to Equation (20).

p. 21.1, Fig. 1. Replace the figure with the one shown below.

p. 22.1, Fig. 1. Replace the figure with the one shown at right.

p. 22.8, Example 1. In the third equation in the solution, delete "1000 × 2.5 × 4.18."

p. 23.4, Fig. 7. Replace the figure with the one shown at right, bottom.

p. 23.10, Dry Air-Conditioning Systems, 4th paragraph. After the second sentence, add, "Dehumidified ventilation air that positively pressurizes the building may help counter moist air infiltration."

p. 25.8, Paragraphs 7 to 11 repeat text earlier on the page; delete them. In Eq. (24), in both denominators, change $\dot{m}_e$ to $\dot{m}_s$.

p. 31.7, 2nd col., 7th line. Change "sue" to "use."

p. 34.36, References. The Rutz (1992) resource is no longer available from the ASHRAE Handbook editor; instead, it is available from NTIS at http://www.ntis.gov/search/product.aspx?ABBR= PB92217504.

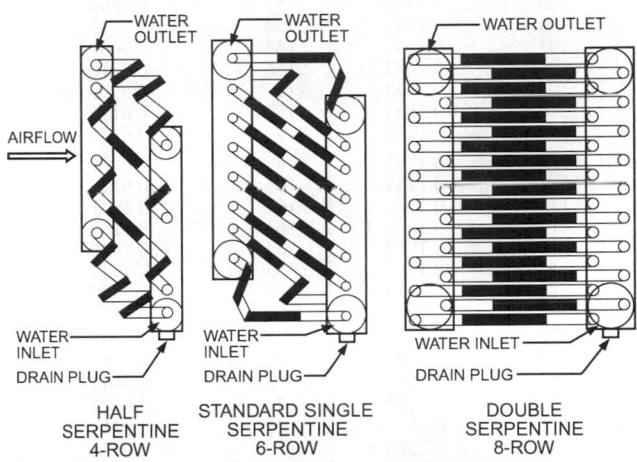

Fig. 1 Typical Water Circuit Arrangements
(2008 HVAC Systems and Equipment, Chapter 22, p. 1)

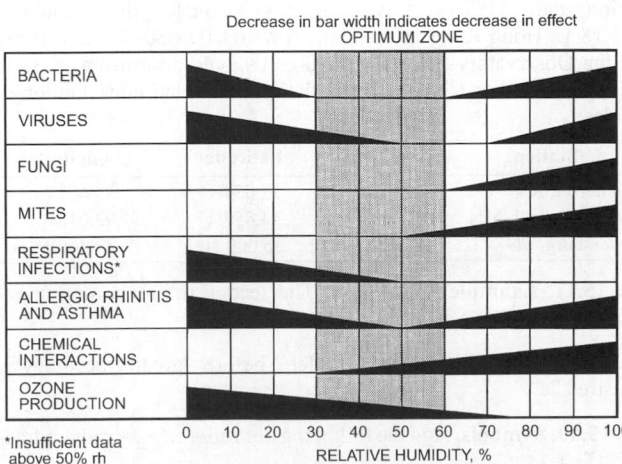

Fig. 1 Optimum Humidity Range for Human Comfort and Health
(Adapted from Sterling et al. 1985)
(2008 HVAC Systems and Equipment, Chapter 21, p. 1)

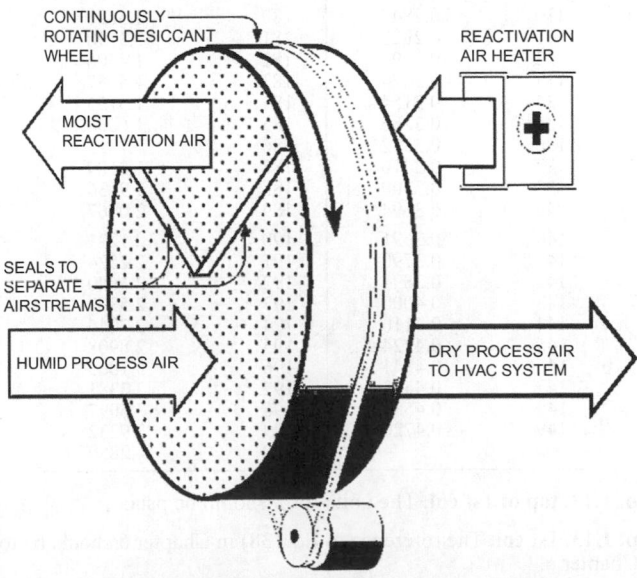

Fig. 7 Typical Rotary Dehumidification Wheel
(2008 HVAC Systems and Equipment, Chapter 23, p. 4)

p. 45.12, Table 11, Notes. L is determined from Equation (5), not Equation (4).

p. 49.3, 1st col., 3rd line. The Energy Policy and Conservation Act of 2005 should be Public Law 109-58.

2009 Fundamentals

Contributors List. For Ch. 15, add the following contributors: Hakim Elmahdy, National Research Council of Canada; Brian Crooks; John L. Wright, University of Waterloo; and William R. McCluney, SunPine Consulting.

p. 1.6, Table 2. Correct values for s_s for temperatures from 100 to 200°F are given below.

Table 2   Thermodynamic Properties of Moist Air at Standard Atmospheric Pressure, 14.696 psia

(2009 Fundamentals, Chapter 1, p. 6; partial)

Temp., °F t	s_s	Temp., °F t	s_s
100	0.1376	150	0.4855
101	0.1408	151	0.4995
102	0.1442	152	0.5140
103	0.1476	153	0.5291
104	0.1511	154	0.5447
105	0.1547	155	0.5609
106	0.1584	156	0.5778
107	0.1622	157	0.5953
108	0.1661	158	0.6136
109	0.1701	159	0.6325
110	0.1742	160	0.6523
111	0.1784	161	0.6728
112	0.1828	162	0.6943
113	0.1872	163	0.7167
114	0.1918	164	0.7400
115	0.1965	165	0.7644
116	0.2013	166	0.7900
117	0.2062	167	0.8167
118	0.2113	168	0.8447
119	0.2165	169	0.8741
120	0.2219	170	0.9049
121	0.2274	171	0.9372
122	0.2331	172	0.9713
123	0.2389	173	1.0072
124	0.2449	174	1.0450
125	0.2510	175	1.0849
126	0.2574	176	1.1271
127	0.2639	177	1.1717
128	0.2706	178	1.2190
129	0.2776	179	1.2693
130	0.2847	180	1.3228
131	0.2920	181	1.3798
132	0.2996	182	1.4406
133	0.3074	183	1.5057
134	0.3154	184	1.5755
135	0.3236	185	1.6506
136	0.3322	186	1.7315
137	0.3410	187	1.8189
138	0.3500	188	1.9136
139	0.3594	189	2.0167
140	0.3691	190	2.1291
141	0.3790	191	2.2524
142	0.3894	192	2.3880
143	0.4000	193	2.5379
144	0.4110	194	2.7046
145	0.4224	195	2.8908
146	0.4341	196	3.1005
147	0.4463	197	3.3381
148	0.4589	198	3.6097
149	0.4720	199	3.9232
		200	4.2889

p. 1.13, top of 1st col. The units for p should be psia.

p. 1.13, 1st col. The reference to Eq. (38) in Chapter 5 should be to Chapter 6.

p. 1.16, Example 3. The reference to Table 2 should be to Table 3.

p. 3.2, 2nd col., last paragraph. Delete the second instance of the following sentence: "The terms E_M and $H_M (= E_M/g)$ are defined as

positive, and represent energy added to the fluid by pumps or blowers."

p. 4.19, Table 8, Eq. (T8.12). Change 0.37 to 0.037.

p. 4.22, Table 10. Correct Eqs. (T10.1) and (T10.7) as follows:

$$\frac{1 - \exp[-N(1 + c_r)]}{1 + c_r} \tag{T10.1}$$

$$\frac{1 - \exp(-c_r \gamma)}{c_r} \tag{T10.7}$$

p. 4.24, Table 11. In the Muley and Manglik (1999) equation for Nu, the last line should read "$\times \text{Re}^{0.728-0.0543\sin\{[\pi(90-\beta)/45+3.7]\}}$."

p. 9.19, Eq. (79). The equation should be $t_{oc} = 54.1 + 0.31 t_{out.}$

p. 9.17, between Eqs. (64) and (65). The reference to Equation (58) should be to Equation (60).

p. 13.14, 2nd col., 3rd full paragraph. The equation should read $\rho = P/(R_{air}T)$.

p. 14.1, 1st paragraph. The StationFinder utility is included only with the HandbookCD+ version of this chapter, not with the CD-ROM included with the print edition.

p. 14.7, Eq. (5). Add a closing parenthesis at the end of the equation.

p. 14.8, bottom of page. The first column should end with "the following equation provides sufficient accuracy." The line beginning "**Solution:**" should be the next-to-last line in the second column.

p. 14.9, Eq. (20). The first operator should be +, not –.

Ch. 14, Appendix: Design Conditions for Selected Locations. The 99% and 99.6% heating design dry-bulb temperatures for the following Alaska stations should be

	99.6% DB	**99% DB**
Anchorage/Elmendorf	−14.8	−9.3
Lake Hood Seaplane	−8.7	−4.1
Anchorage Intl AP	−8.9	−4.4
Anchorage Merrill Field	−11.0	−6.9

Ch. 14, Appendix: Design Conditions for Selected Locations, and Climatic Design Conditions Tables (on the CD-ROM). The main entry for Taiwan should read "Taiwan." In the Russian Federation, station UST-ISIM, WMO ID 283820, the longitude should be 71.18 E. Hong Kong International (WMO ID 450070) and Hong Kong Observatory (WMO ID 450050) should appear under Hong Kong rather than China. Correct the following latitudes and longitudes:

Station	**WMO ID**	**Latitude**	**Longitude**
Somerset, KY	724354	37.054N	84.601W
South St. Paul, MN	726603	44.857N	93.018W
Dyersburg, TN	723347	36.000 N	89.400W

p. 15.11, Example 2, Solution. The second reference to Table 4 should be to Table 1.

p. 15.16, 1st col., top. Add "when" before "greater accuracy is desired."

p. 15.60, Symbols. Add the following definition: "F_R = radiant fraction."

p. 15.62, References. Add the following source before LBL 2003:

Laoudi, A., A.D. Galasiu, M.R. Atif, and A. Haqqani. 2003. SkyVision: A software tool to calculate the optical characteristics and daylighting performance of skylights. *Building Simulation, 8th IBPSA Conference,* Eindhoven, Netherlands, pp. 705-712.

p. 16.7, Eq. (26). On the third line, change "$[C_p(3) - C_p(4)]$" to "$[C_p(3) + C_p(4)]$."

p. 16.23, Eq. (48). The equation should use the absolute value of Δt.

p. 17.7, Eq. (11). The correct equation is as follows:

$$Q_v = 0.01 A_{cf} + 7.5(N_{br} + 1)$$

p. 17.8, Eq. (21). The closing parenthesis should be at the end of the equation.

p. 18.2, 2nd col., Cooling Load Calculations in Practice section. On the third line, change "filtration" to "infiltration."

pp. 18.8-18.11, Tables 5A to 5E. Items with an asterisk appear only in Swierzycna (2009); all others appear in both Swierzycna (2008) and (2009).

p. 18.31, Eq. (39). The denominator of the first fraction should be $\pi(z_2 - z_1)$.

p. 18.31, Eq. (40). The correct equation is as follows:

$$U_{avg,bf} = \frac{2k_{soil}}{\pi w_b}$$
$$\times \left[\ln\left(\frac{w_b}{2} + \frac{z_f}{2} + \frac{k_{soil}R_{other}}{\pi} \right) - \ln\left(\frac{z_f}{2} + \frac{k_{soil}R_{other}}{\pi} \right) \right]$$

(40)

p. 18.33, 2nd col. In the list of variables under item 2, T_r = room or return air temperature, °F.

p. 18.41, Table 29. Correct the diffuse solar heat gain values for the following rows:

Local Std. Hour	Diff. Solar Heat Gain, Btu/h	Local Std. Hour	Diff. Solar Heat Gain, Btu/h
6	106	13	2436
7	569	14	2614
8	1002	15	2648
9	1371	16	2479
10	1665	17	2072
11	1887	18	1429
12	2177	19	599

p. 18.43, Table 31. Use these values for the following two columns:

Convective 54%, Btu/h	Radiant 46%, Btu/h	Convective 54%, Btu/h	Radiant 46%, Btu/h
−29	−25	1615	1376
−48	−41	2293	1953
−65	−56	2853	2431
−80	−68	3093	2635
−90	−76	2847	2425
−35	−29	2102	1791
199	169	867	739
469	399	162	138
720	614	116	99
930	793	70	60
1100	937	31	27
1275	1086	0	0

p. 20.13, 2nd col., Air Diffusion Performance Index (ADPI), 1st paragraph. The reference should be to Eq. (15).

p. 20.14, 2nd col., Convective Flows Associated with Space Heat Sources, last paragraph. The reference should be to Eq. (12).

p. 21.6, Eq. (19). The correct equation is as follows:

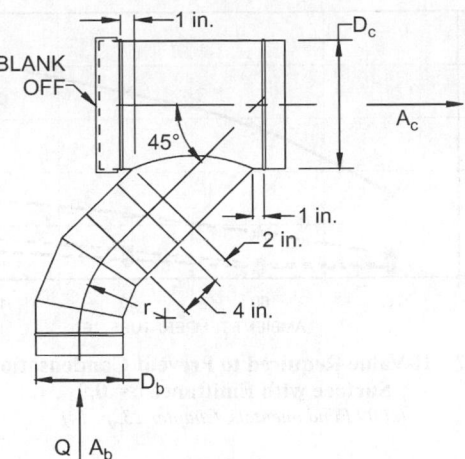

Figure for ED5-6
(2009 Fundamentals, Chapter 21, p. 44)

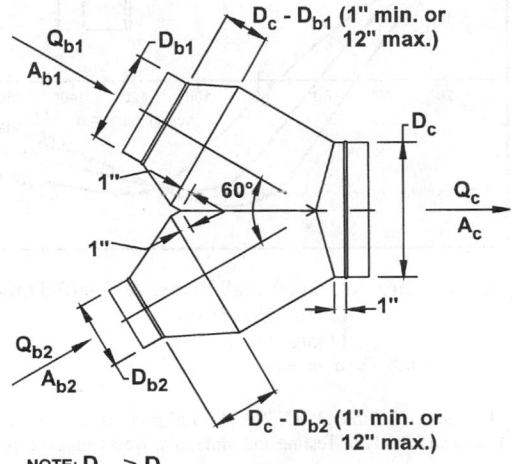

NOTE: $D_{b1} \geq D_{b2}$

Figure for ED5-9
(2009 Fundamentals, Chapter 21, p. 44)

$$\frac{1}{\sqrt{f}} = -2 \log\left(\frac{12\varepsilon}{3.7 D_h} + \frac{2.51}{Re\sqrt{f}} \right)$$

(19)

p. 21.33, Table for ED5-1, Wye, 30°, Converging. In the expression centered over the table columns, change Q_b to Q_s.

pp. 21.42-44, Tables for ED5-3, D_c > 10 in., Converging. In the expressions centered over the table columns, change Q_b to Q_s (three places).

p. 21.44, Figures for ED5-6 and ED5-9. The correct figures are shown above.

pp. 21.45 (bottom table) and 21.46 (top table). Change "C_{b1} Values" to "C_{b2} Values."

p. 23.14, Fig. 7. The correct figure is shown on p. A.4, top left.

p. 23.15, Eq. (7). Replace § with /.

p. 24.5, Fig. 7. The curves for L/W = 1/4 and L/W = 4 should be swapped. The correct figure is supplied on p. A.4, top right.

pp. 26.5-9, Table 4. Please replace the table with the one that begins on page A.5. Cells with updated values are highlighted.

p. 26.19, References. Please add the following entry:

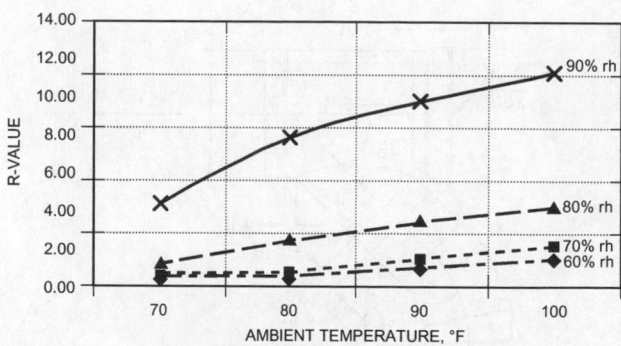

Fig. 7 R-Value Required to Prevent Condensation on Surface with Emittance ε = 0.9
(2009 Fundamentals, Chapter 23, p. 14)

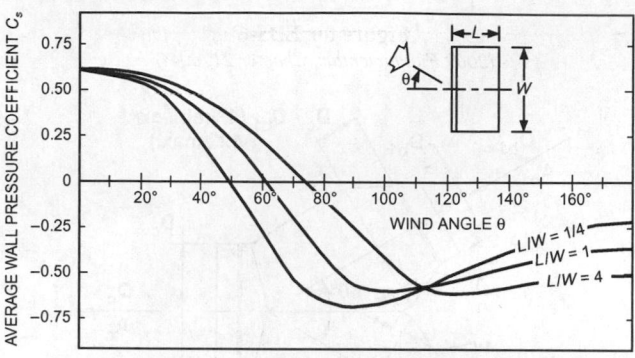

Fig. 7 Surface-Averaged Wall Pressure Coefficients for Tall Buildings
(Akins et al. 1979)
(2009 Fundamentals, Chapter 24, p. 5)

Cardenes, T.J. and G.T. Bible. 1987. *The thermal properties of wood—Data base.* American Society of Testing and Materials, West Conshohocken, PA.

p. 27.9, Step 4, last line. The reference should be to Table 3 in Chapter 1. To convert table values from psia to in. Hg, multiply them by 2.036. Thus, 0.36334 psia × 2.036 = 0.7398 in. Hg.

p. 31.6, 8, 10, and 11. Units for viscosity in Figures 12 and 16 and Tables 9 and 13 should be lb/ft·h.

p. 36.2, 1st col., Hydraulic diameter D_h. The second sentence should read "For a rectangular duct with dimensions $W \times H$, the hydraulic diameter $D_h = 2WH/(W + H)$."

p. 39.2, Table 2, Viscosity (absolute). In the fourth data row, change 0.0671955 to 0.671955.

2010 Refrigeration

Ch. 1. All existing references to Table 19 should be to Table 20. All existing references to Table 20 should be to Table 19.

pp. 1, 1st col., Hydraulic diameter D_h. The second sentence should read "For a rectangular duct with dimensions $W \times H$, the hydraulic diameter $D_h = 2WH/(W + H)$."

p. 11.23, Table 2. Amend the table values as follows:

Table 2 Values of f for Discharge Capacity of Pressure Relief Devices
(2010 Refrigeration, Chapter 11, p. 23)

Refrigerant	f
On the low side of a limited-charge cascade system:	
R-13, R-13B1, R-503	2.0 (0.163)
R-14	2.5 (0.203)
R-23, R-170, R-508A, R-508B, R-744, R-1150	1.0 (0.082)
Other applications:	
R-11, R-32, R-113, R-123, R-142b, R-152a, R-290, R-600, R-600a, R-764	1.0 (0.082)
R-12, R-22, R-114, R-124, R-134a, R-401A, R-401B, R-401C, R-405A, R-406A, R-407C, R-407D, R-407E, R-409A, R-409B, R-411A, R-411B, R-411C, R-412A, R-414A, R-414B, R-500, R-1270	(0.131)
R-115, R-402A, R-403B, R-404A, R-407B, R-410A, R-410B, R-502, R-507A, R-509A	2.5 (0.203)
R-143a, R-402B, R-403A, R-407A, R-408A, R-413A	2.0 (0.163)
R-717	0.5 (0.041)
R-718	0.2 (0.016)

Notes:
1. Listed values of f do not apply if fuels are used within 20 ft of pressure vessel. In this case, use methods in API (2000, 2003) to size pressure-relief device.
2. When one pressure-relief device or fusible plug is used to protect more than one pressure vessel, required capacity is the sum of capacities required for each pressure vessel.
3. For refrigerants not listed, consult ASHRAE *Standard* 15.

Table 4   Typical Thermal Properties of Common Building and Insulating Materials: Design Values[a]

(2009 Fundamentals, Chapter 26, pp. 5-9)

Description	Density, lb/ft³	Conductivity[b] k, Btu·in/h·ft²·°F	Resistance R, h·ft²·°F/Btu	Specific Heat, Btu/lb·°F	Reference[n]
Building Board and Siding					
Board					
Asbestos/cement board	120	4	—	0.24	Nottage (1947)
Cement board	71	1.7	—	0.2	Kumaran (2002)
Fiber/cement board	86	1.7	—	0.2	Kumaran (2002)
	61	1.3	—	0.2	Kumaran (1996)
	26	0.5	—	0.45	Kumaran (1996)
	20	0.4	—	0.45	Kumaran (1996)
Gypsum or plaster board	40	1.1	—	0.21	Kumaran (2002)
Oriented strand board (OSB) 7/16 in.	41	—	0.62	0.45	Kumaran (2002)
1/2 in.	41	—	0.68	0.45	Kumaran (2002)
Plywood (douglas fir) 1/2 in.	29	—	0.79	0.45	Kumaran (2002)
5/8 in.	34	—	0.85	0.45	Kumaran (2002)
Plywood/wood panels 3/4 in.	28	—	1.08	0.45	Kumaran (2002)
Vegetable fiber board				—	
Sheathing, regular density[e] 1/2 in.	18	—	1.32	0.31	Lewis (1967)
intermediate density[e] 1/2 in.	22	—	1.09	0.31	Lewis (1967)
Nail-base sheathing[e] 1/2 in.	25	—	1.06	0.31	
Shingle backer 3/8 in.	18	—	0.94	0.3	
Sound-deadening board 1/2 in.	15	—	1.35	0.3	
Tile and lay-in panels, plain or acoustic	18	0.4	—	0.14	
Laminated paperboard	30	0.5	—	0.33	Lewis (1967)
Homogeneous board from repulped paper	30	0.5	—	0.28	
Hardboard[e]					
medium density	50	0.73	—	0.31	Lewis (1967)
high density, service-tempered and service grades	55	0.82	—	0.32	Lewis (1967)
high density, standard-tempered grade	63	1	—	0.32	Lewis (1967)
Particleboard[e]					
low density	37	0.71	—	0.31	Lewis (1967)
medium density	50	0.94	—	0.31	Lewis (1967)
high density	62	1.18	0.85	—	Lewis (1967)
underlayment 5/8 in.	40	—	0.82	0.29	Lewis (1967)
Waferboard	44	0.73	—	0.45	Kumaran (1996)
Shingles					
Asbestos/cement	120	—	0.21	—	
Wood, 16 in., 7 1/2 in. exposure	—	—	0.87	0.31	
Wood, double, 16 in., 12 in. exposure	—	—	1.19	0.28	
Wood, plus ins. backer board 5/16 in.	—	—	1.4	0.31	
Siding				—	
Asbestos/cement, lapped 1/4 in.	—	—	0.21	0.24	
Asphalt roll siding	—	—	0.15	0.35	
Siding					
Asbestos/cement, lapped 1/4 in.	—	—	0.21	0.24	
Asphalt roll	—	—	0.15	0.35	
Asphalt insulating siding (1/2 in. bed)	—	—	1.46	0.35	
Hardboard siding 7/16 in.	—	—	0.67	0.28	
Wood, drop, 8 in. 1 in.	—	—	0.79	0.28	
Wood, bevel					
8 in., lapped 1/2 in.	—	—	0.81	0.28	
10 in., lapped 3/4 in.	—	—	1.05	0.28	
Wood, plywood, 3/8 in., lapped	—	—	0.59	0.29	
Aluminum, steel, or vinyl,[j, k] over sheathing				—	
hollow-backed	—	—	0.62	0.29[k]	
insulating-board-backed				—	
3/8 in.	—	—	1.82	0.32	
foil-backed 3/8 in.	—	—	2.96	—	
Architectural (soda-lime float) glass	158	6.9	—	0.21	
Building Membrane					
Vapor-permeable felt	—	—	0.06	—	
Vapor: seal, 2 layers of mopped 15 lb felt	—	—	0.12	—	

Table 4 Typical Thermal Properties of Common Building and Insulating Materials: Design Values[a] (Continued)

(2009 Fundamentals, Chapter 26, pp. 5-9)

Description	Density, lb/ft³	Conductivity[b] k, Btu·in/h·ft²·°F	Resistance R, h·ft²·°F/Btu	Specific Heat, Btu/lb·°F	Reference[n]
Vapor: seal, plastic film	—	—	Negligible	—	
Finish Flooring Materials					
Carpet and rebounded urethane pad ... 3/4 in.	7	—	2.38	—	NIST (2000)
Carpet and rubber pad (one-piece) ... 3/8 in.	20	—	0.68	—	NIST (2000)
Pile carpet with rubber pad ... 3/8 to 1/2 in.	18	—	1.59	—	NIST (2000)
Linoleum/cork tile ... 1/4 in.	29	—	0.51	—	NIST (2000)
PVC/Rubber floor covering	—	2.8	—	—	CIBSE (2006)
Rubber tile ... 1.0 in.	119	—	0.34	—	NIST (2000)
Terrazzo ... 1.0 in.	—	—	0.08	0.19	
Insulating Materials					
Blanket and batt[c,d]					
Glass-fiber batts ... 3 to 3 1/2 in.	0.6 to 0.9	0.30	—	0.2	Kumaran (2002)
... 6 in.	0.5 to 0.8	0.31 to 0.33	—	0.2	Kumaran (2002)
Mineral fiber ... 5 1/2 in.	2	0.25	—	0.2	Kumaran (1996)
Mineral wool, felted	1 to 3	0.28	—	—	CIBSE (2006), NIST (2000)
	4 to 8	0.24	—	—	NIST (2000)
Slag wool	3 to 12	0.26	—	—	Raznjevic (1976)
	16	0.28	—	—	Raznjevic (1976)
	19	0.30	—	—	Raznjevic (1976)
	22	0.33	—	—	Raznjevic (1976)
	25	0.35	—	—	Raznjevic (1976)
Board and slabs					
Cellular glass	8	0.33	—	0.18	(Manufacturer)
Cement fiber slabs, shredded wood with Portland cement binder	25 to 27	0.50 to 0.53	—	—	
with magnesia oxysulfide binder	22	0.57	—	0.31	
Glass fiber board	10	0.22 to 0.28	—	0.2	Kumaran (1996)
Expanded rubber (rigid)	4	0.2	—	0.4	Nottage (1947)
Expanded polystyrene extruded (smooth skin)	1.6 to 2.4	0.15 to 0.21	—	0.35	Kumaran (1996)
Expanded polystyrene, molded beads	0.9 to 1.6	0.22 to 0.27	—	0.35	Kumaran (1996)
Mineral fiberboard, wet felted	10	0.26	—	0.2	Kumaran (1996)
core or roof insulation	16 to 17	0.34	—	—	
acoustical tile[g]	18	—	—	0.19	
	21	0.37	—	—	
wet-molded, acoustical tile[g]	23	0.42	—	0.14	
Perlite board	10	0.36	—	—	Kumaran (1996)
Polyisocyanurate, aged					
unfaced	1.6 to 2.3	0.14 to 0.19	—	—	Kumaran (2002)
with facers	4	0.13	—	0.35	Kumaran (1996)
Phenolic foam board with facers, aged	4	0.13	—	—	Kumaran (1996)
Loose fill					
Cellulosic (milled paper or wood pulp)	2 to 3.5	0.26 to 0.31	—	0.45	NIST (2000), Kumaran (1996)
fiberized	1.2 to 2.0				
Perlite, expanded	2 to 4	0.27 to 0.31	—	0.26	(Manufacturer)
	4 to 7.5	0.31 to 0.36	—	—	(Manufacturer)
	7.5 to 11	0.36 to 0.42	—	—	(Manufacturer)
Mineral fiber (rock, slag, or glass)[d]					
approx. 3 3/4 to 5 in.	0.6 to 2.0	—	11.0	0.17	
approx. 6 1/2 to 8 3/4 in.	0.6 to 2.0	—	19.0	—	
approx. 7 1/2 to 10 in.	0.6 to 2.0	—	22.0	—	
approx. 10 1/4 to 13 3/4 in.	0.6 to 2.0	—	30.0	—	
approx. 3 1/2 in. (closed sidewall application)	2.0 to 4.0	—	12.0 to 14.0	—	
Vermiculite, exfoliated	7.0 to 8.2	0.47	—	0.32	Sabine et al. (1975)
	4.0 to 6.0	0.44	—	—	(Manufacturer)
Spray-applied					
Cellulosic fiber	3.5 to 6.0	0.29 to 0.34	—	—	Yarbrough et al. (1987)
Glass fiber	3.5 to 4.5	0.26 to 0.27	—	—	Yarbrough et al. (1987)
Polyurethane foam (low density)	0.4 to 0.5	0.29	—	0.35	Kumaran (2002)
	2.4	0.18	—	0.35	Kumaran (2002)
aged and dry ... 1 1/2 in.	2.0	—	9.09	0.35	Kumaran (1996)
... 2 in.	3.5	—	10.9	0.35	Kumaran (1996)

Table 4 Typical Thermal Properties of Common Building and Insulating Materials: Design Values[a] *(Continued)*

(2009 Fundamentals, Chapter 26, pp. 5-9)

Description	Density, lb/ft³	Conductivity[b] k, Btu·in/h·ft²·°F	Resistance R, h·ft²·°F/Btu	Specific Heat, Btu/lb·°F	Reference[n]
.... 4 1/2 in.	2.0	—	20.95	—	Kumaran (1996)
Ureaformaldehyde foam, dry	0.5 to 1.2	0.21 to 0.22	—	—	CIBSE (2006)
Metals					
(See Chapter 33, Table 3)					
Roofing					
Asbestos/cement shingles	120	—	0.21	0.24	
Asphalt (bitumen with inert fill)	100	2.98	—	—	CIBSE (2006)
....	119	4.0	—	—	CIBSE (2006)
....	144	7.97	—	—	CIBSE (2006)
Asphalt roll roofing...........................	70	—	0.15	0.36	
Asphalt shingles...........................	70	—	0.44	0.3	
Built-up roofing .... 3/8 in.	70	—	0.33	0.35	
Mastic asphalt (heavy, 20% grit)...........................	59	1.32	—	—	CIBSE (2006)
Reed thatch...........................	17	0.62	—	—	CIBSE (2006)
Roofing felt...........................	141	8.32	—	—	CIBSE (2006)
Slate........................... 1/2 in.	—	—	0.05	0.3	
Straw thatch	15	0.49	—	—	CIBSE (2006)
Wood shingles, plain and plastic-film-faced...............	—	—	0.94	0.31	
Plastering Materials					
Cement plaster, sand aggregate...........................	116	5.0	—	0.2	
Sand aggregate					
.... 3/8 in.	—	—	0.08	0.2	
.... 3/4 in.	—	—	0.15	0.2	
Gypsum plaster	70	2.63	—	—	CIBSE (2006)
	80	3.19	—	—	CIBSE (2006)
Lightweight aggregate					
.... 1/2 in.	45	—	0.32	—	
.... 5/8 in.	45	—	0.39	—	
on metal lath........................... 3/4 in.	—	—	0.47	—	
Perlite aggregate...........................	45	1.5	—	0.32	
Sand aggregate	105	5.6	—	0.2	
on metal lath........................... 3/4 in.	—	—	0.13	—	
Vermiculite aggregate	30	1	—	—	CIBSE (2006)
....	40	1.39	—	—	CIBSE (2006)
....	45	1.7	—	—	CIBSE (2006)
....	50	1.8	—	—	CIBSE (2006)
....	60	2.08	—	—	CIBSE (2006)
Perlite plaster	25	0.55	—	—	CIBSE (2006)
....	38	1.32	—	—	CIBSE (2006)
Pulpboard or paper plaster	38	0.48	—	—	CIBSE (2006)
Sand/cement plaster, conditioned	98	4.4	—	—	CIBSE (2006)
Sand/cement/lime plaster, conditioned	90	3.33	—	—	CIBSE (2006)
Sand/gypsum (3:1) plaster, conditioned...........................	97	4.5	—	—	CIBSE (2006)
Masonry Materials					
Masonry units					
Brick, fired clay...........................	150	8.4 to 10.2	—	—	Valore (1988)
....	140	7.4 to 9.0	—	—	Valore (1988)
....	130	6.4 to 7.8	—	—	Valore (1988)
....	120	5.6 to 6.8	—	0.19	Valore (1988)
....	110	4.9 to 5.9	—	—	Valore (1988)
....	100	4.2 to 5.1	—	—	Valore (1988)
....	90	3.6 to 4.3	—	—	Valore (1988)
....	80	3.0 to 3.7	—	—	Valore (1988)
....	70	2.5 to 3.1	—	—	Valore (1988)
Clay tile, hollow...........................					
1 cell deep........................... 3 in.	—	—	0.80	0.21	Rowley (1937)
.... 4 in.	—	—	1.11	—	Rowley (1937)
2 cells deep........................... 6 in.	—	—	1.52	—	Rowley (1937)
.... 8 in.	—	—	1.85	—	Rowley (1937)
.... 10 in.	—	—	2.22	—	Rowley (1937)
3 cells deep........................... 12 in.	—	—	2.50	—	Rowley (1937)
Lightweight brick...........................	50	1.39	—	—	Kumaran (1996)
....	48	1.51	—	—	Kumaran (1996)

Table 4　Typical Thermal Properties of Common Building and Insulating Materials: Design Values[a] (Continued)

(2009 Fundamentals, Chapter 26, pp. 5-9)

Description	Density, lb/ft³	Conductivity[b] k, Btu·in/h·ft²·°F	Resistance R, h·ft²·°F/Btu	Specific Heat, Btu/lb·°F	Reference[n]
Concrete blocks[h, i]					
Limestone aggregate					
8 in., 36 lb, 138 lb/ft³ concrete, 2 cores	—	—	—	—	
with perlite-filled cores	—	—	2.1	—	Valore (1988)
12 in., 55 lb, 138 lb/ft³ concrete, 2 cores	—	—	—	—	
with perlite-filled cores	—	—	3.7	—	Valore (1988)
Normal-weight aggregate (sand and gravel)					
8 in., 33 to 36 lb, 126 to 136 lb/ft³ concrete, 2 or 3 cores	—	—	1.11 to 0.97	0.22	Van Geem (1985)
with perlite-filled cores	—	—	2.0	—	Van Geem (1985)
with vermiculite-filled cores	—	—	1.92 to 1.37	—	Valore (1988)
12 in., 50 lb, 125 lb/ft³ concrete, 2 cores	—	—	1.23	0.22	Valore (1988)
Medium-weight aggregate (combinations of normal and lightweight aggregate)					
8 in., 26 to 29 lb, 97 to 112 lb/ft³					
concrete, 2 or 3 cores	—	—	1.71 to 1.28	—	Van Geem (1985)
with perlite-filled cores	—	—	3.7 to 2.3	—	Van Geem (1985)
with vermiculite-filled cores	—	—	3.3	—	Van Geem (1985)
with molded-EPS-filled (beads) cores	—	—	3.2	—	Van Geem (1985)
with molded EPS inserts in cores	—	—	2.7	—	Van Geem (1985)
Lightweight aggregate (expanded shale, clay, slate or slag, pumice)					
6 in., 16 to 17 lb, 85 to 87 lb/ft³					
concrete, 2 or 3 cores	—	—	1.93 to 1.65	—	Van Geem (1985)
with perlite-filled cores	—	—	4.2	—	Van Geem (1985)
with vermiculite-filled cores	—	—	3.0	—	Van Geem (1985)
8 in., 19 to 22 lb, 72 to 86 lb/ft³ concrete	—	—	3.2 to 1.90	0.21	Van Geem (1985)
with perlite-filled cores	—	—	6.8 to 4.4	—	Van Geem (1985)
with vermiculite-filled cores	—	—	5.3 to 3.9	—	Shu et al. (1979)
with molded-EPS-filled (beads) cores	—	—	4.8	—	Shu et al. (1979)
with UF foam-filled cores	—	—	4.5	—	Shu et al. (1979)
with molded EPS inserts in cores	—	—	3.5	—	Shu et al. (1979)
12 in., 32 to 36 lb, 80 to 90 lb/ft³,					
concrete, 2 or 3 cores	—	—	2.6 to 2.3	—	Van Geem (1985)
with perlite-filled cores	—	—	9.2 to 6.3	—	Van Geem (1985)
with vermiculite-filled cores	—	—	5.8	—	Valore (1988)
Stone, lime, or sand	180	72	—	—	Valore (1988)
Quartzitic and sandstone	160	43	—	—	Valore (1988)
	140	24	—	—	Valore (1988)
	120	13	—	0.19	Valore (1988)
Calcitic, dolomitic, limestone, marble, and granite	180	30	—	—	Valore (1988)
	160	22	—	—	Valore (1988)
	140	16	—	—	Valore (1988)
	120	11	—	0.19	Valore (1988)
	100	8	—	—	Valore (1988)
Gypsum partition tile					
3 by 12 by 30 in., solid	—	—	1.26	0.19	Rowley (1937)
4 cells	—	—	1.35	—	Rowley (1937)
4 by 12 by 30 in., 3 cells	—	—	1.67	—	Rowley (1937)
Limestone	150	3.95	—	0.2	Kumaran (2002)
	163	6.45	—	0.2	Kumaran (2002)
Concretes[i]					
Sand and gravel or stone aggregate concretes (concretes with >50% quartz or quartzite sand have conductivities in higher end of range)	150	10.0 to 20.0	—	—	Valore (1988)
	140	9.0 to 18.0	—	0.19 to 0.24	Valore (1988)
	130	7.0 to 13.0	—	—	Valore (1988)
Lightweight aggregate or limestone concretes	120	6.4 to 9.1	—	—	Valore (1988)
Expanded shale, clay, or slate; expanded slags; cinders; pumice (with density up to 100 lb/ft³); scoria (sanded concretes have conductivities in higher end of range)	100	4.7 to 6.2	—	0.2	Valore (1988)
	80	3.3 to 4.1	—	0.2	Valore (1988)
	60	2.1 to 2.5	—	—	Valore (1988)
	40	1.3	—	—	Valore (1988)

Table 4 Typical Thermal Properties of Common Building and Insulating Materials: Design Values[a] *(Continued)*
(2009 Fundamentals, Chapter 26, pp. 5-9)

Description	Density, lb/ft^3	Conductivity[b] k, Btu·in/h·ft^2·°F	Resistance R, h·ft^2·°F/Btu	Specific Heat, Btu/lb·°F	Reference[n]
Gypsum/fiber concrete					
(87.5% gypsum, 12.5% wood chips)	51	1.66	—	0.2	Rowley (1937)
Cement/lime, mortar, and stucco	120	9.7	—	—	Valore (1988)
	100	6.7	—	—	Valore (1988)
	80	4.5	—	—	Valore (1988)
Perlite, vermiculite, and polystyrene beads	50	1.8 to 1.9	—	—	Valore (1988)
	40	1.4 to 1.5	—	0.15 to 0.23	Valore (1988)
	30	1.1	—	—	Valore (1988)
	20	0.8	—	—	Valore (1988)
Foam concretes	120	5.4	—	—	Valore (1988)
	100	4.1	—	—	Valore (1988)
	80	3.0	—	—	Valore (1988)
	70	2.5	—	—	Valore (1988)
Foam concretes and cellular concretes	60	2.1	—	—	Valore (1988)
	40	1.4	—	—	Valore (1988)
	20	0.8	—	—	Valore (1988)
Aerated concrete (oven-dried)	27 to 50	1.4	—	0.2	Kumaran (1996)
Polystyrene concrete (oven-dried)	16 to 50	2.54	—	0.2	Kumaran (1996)
Polymer concrete	122	11.4	—	—	Kumaran (1996)
	138	7.14	—	—	Kumaran (1996)
Polymer cement	117	5.39	—	—	Kumaran (1996)
Slag concrete	60	1.5	—	—	Touloukian et al (1970)
	80	2.25	—	—	Touloukian et al. (1970)
	100	3	—	—	Touloukian et al. (1970)
	125	8.53	—	—	Touloukian et al. (1970)
Woods (12% moisture content)[l]					
Hardwoods	—	—	—	0.39[m]	Wilkes (1979)
Oak	41 to 47	1.12 to 1.25	—	—	Cardenas and Bible (1987)
Birch	43 to 45	1.16 to 1.22	—	—	Cardenas and Bible (1987)
Maple	40 to 44	1.09 to 1.19	—	—	Cardenas and Bible (1987)
Ash	38 to 42	1.06 to 1.14	—	—	Cardenas and Bible (1987)
Softwoods	—	—	—	0.39[m]	Wilkes (1979)
Southern pine	36 to 41	1.00 to 1.12	—	—	Cardenas and Bible (1987)
Southern yellow pine	31	1.06 to 1.16	—	—	Kumaran (2002)
Eastern white pine	25	0.85 to 0.94	—	—	Kumaran (2002)
Douglas fir/larch	34 to 36	0.95 to 1.01	—	—	Cardenas and Bible (1987)
Southern cypress	31 to 32	0.90 to 0.92	—	—	Cardenas and Bible (1987)
Hem/fir, spruce/pine/fir	24 to 31	0.74 to 0.90	—	—	Cardenas and Bible (1987)
Spruce	25	0.74 to 0.85	—	—	Kumaran (2002)
Western red cedar	22	0.83 to 0.86	—	—	Kumaran (2002)
West coast woods, cedars	22 to 31	0.68 to 0.90	—	—	Cardenas and Bible (1987)
Eastern white cedar	22	0.82 to 0.89	—	—	Kumaran (2002)
California redwood	24 to 28	0.74 to 0.82	—	—	Cardenas and Bible (1987)
Pine (oven-dried)	23	0.64	—	0.45	Kumaran (1996)
Spruce (oven-dried)	25	0.69	—	0.45	Kumaran (1996)

COMPOSITE INDEX
ASHRAE HANDBOOK SERIES

This index covers the current Handbook series published by ASHRAE. The four volumes in the series are identified as follows:

S = 2008 HVAC Systems and Equipment

F = 2009 Fundamentals

R = 2010 Refrigeration

A = 2011 HVAC Applications

Alphabetization of the index is letter by letter; for example, **Heaters** precedes **Heat exchangers**, and **Floors** precedes **Floor slabs**.

The page reference for an index entry includes the book letter and the chapter number, which may be followed by a decimal point and the beginning page in the chapter. For example, the page number S31.4 means the information may be found in the 2008 HVAC Systems and Equipment volume, Chapter 31, beginning on page 4.

Each Handbook volume is revised and updated on a four-year cycle. Because technology and the interests of ASHRAE members change, some topics are not included in the current Handbook series but may be found in the earlier Handbook editions cited in the index.

Pulldown load, R15.5
Pumps
  centrifugal, S43
    affinity laws, S43.7
    antifreeze effect on, S12.24
    arrangement, S43.10
      compound, S12.8
      pumping
        distributed, S43.12
        parallel, S12.7; S43.10
        primary-secondary, S12.8; S43.11
        series, S12.7; S43.10
        variable-speed, S43.12
      standby pump, S12.8; S43.11
      two-speed motors, S43.11
    casing, S43.2
    cavitation, S43.8
    commissioning, S43.13
    construction, S43.1
    efficiency, best efficiency point (BEP),
        S43.6
    energy conservation, S43.13
    impellers
      operation, S43.2
      trimming, S43.6, 7, 8
    installation, S43.13
    mixing, S12.8
    motors, S43.12
    operation, S43.2, 13
    performance
      net positive suction, S43.8
      operating point, S43.5
      pump curves, S12.6; S43.3, 5
    power, S43.6
    radial thrust, S43.8
    selection, S43.9
    types, S43.2
    variable-speed, S12.8
  chilled-water, A42.12, 25, 27
    sequencing, A42.25, 29
  condenser water, A42.12
  as fluid flow indicators, A38.13
  geothermal wells, A34.28
    lineshaft, A34.6
    submersible, A34.6
  hydronic snow melting, A51.12
  liquid overfeed systems, R4.4
  solar energy systems, A35.12
  systems, water, S12.6; S14.6
  variable-speed, A42.14, 27
Purge units, centrifugal chillers, S42.11
Radiant heating and cooling, A55; S6.1; S15;
    S33.4. (*See also* **Panel heating and cooling**)
  applications, A54.8
  asymmetry, A54.5
  beam heating design, A54.4; S15.5
  control, A47.4
  design, A54.2, 3
  direct infrared, A54.1, 4, 8
  equations, A54.2
  floor reradiation, A54.5
  infrared, A54.1, 4, 8; S15
    beam heater design, S15.5
    control, S15.4
    efficiency, S15.3
    electric, S15.2
    energy conservation, S15.1
    gas-fired, S15.1

  indirect, S15.2
  maintenance, S15.5
  oil-fired, S15.3
  precautions, S15.4
  reflectors, S15.4
  installation, A54.7
  intensity, S15.1
  panels, A54.1, 8; S33.4; S35.6
    applications, A54.8
    control, A47.4
    heating, S33.4
    hydronic systems, S35.6
  radiation patterns, A54.5
  snow-melting systems, A51.16
  terminology
    adjusted dry-bulb temperature, A54.1
    ambient temperature, A54.1
    angle factor, S15.5
    effective radiant flux (ERF), A54.2; S15.5
    fixture efficiency, S15.4
    mean radiant temperature (MRT),
        A54.1; S6.1
    operative temperature, A54.1
    pattern efficiency, S15.4
    radiant flux distribution, S15.6
    radiation-generating ratio, S15.4
  test instruments, A54.7
  total space heating, A54.6
Radiant time series (RTS) method, F18.2, 20
  factors, F18.21
  load calculations, nonresidential, F18.1
Radiation
  atmospheric, A35.5
  diffuse, F15.15, 18
  electromagnetic, F10.16
  ground-reflected, F15.15
  optical waves, F10.18
  radiant balance, F4.15
  radio waves, F10.18
  solar, A35.3
  thermal, F4.2, 11; S6.1
    angle factors, F4.13
    blackbody, F4.11
      spectral emissive power, F4.12
    black surface, F4.2
    display cases, R15.4
    energy transfer, F4.11
    exchange between surfaces, F4.14
    in gases, F4.16
    gray, F4.2, 12
    heat transfer, F4.2
    infrared, F15.16
    Kirchoff's law, F4.12
    monochromatic emissive power,
        F4.12
    nonblack, F4.12
    transient, F4.8
Radiators, S35.1, 5
  design, S35.3
  nonstandard condition corrections, S35.3
  types, S35.1
Radioactive gases, contaminants, F11.19
Radiometers, A54.7
Radon, F10.10, 12, 17
  control, A46.13; F16.20
  indoor concentrations, F11.17
Rail cars
  air conditioning, A11.5

  air distribution, A11.7
  heaters, A11.7
  vehicle types, A11.5
Railroad tunnels, ventilation
  design, A15.17
  diesel locomotive facilities, A15.27
  equipment, A15.33
  locomotive cooling requirements, A15.17
  tunnel aerodynamics, A15.17
  tunnel purge, A15.18
Rain, and building envelopes, F25.3
RANS. *See* **Reynolds-Averaged Navier-Stokes
    (RANS) equation**
Rapid-transit systems. *See* **Mass-transit
    systems**
Rayleigh number, F4.18
RC curves. *See* **Room criterion (RC) curves**
Receivers
  ammonia refrigeration systems
    high-pressure, R2.3
    piping, R2.16
    through-type, R2.16
  halocarbon refrigerant, R1.21
  liquid overfeed systems, R4.7
Recycling refrigerants, R9.3
Refrigerant/absorbent pairs, F2.15
Refrigerant-control devices, R11
  air conditioners, S48.6; S49.2
  automobile air conditioning, A10.8
  capillary tubes, R11.24
  coolers, liquid, S41.5
  heat pumps
    system, S8.7
    unitary, S48.10
  lubricant separators, R11.23
  pressure transducers, R11.4
  sensors, R11.4
  short-tube restrictors, R11.31
  switches
    differential control, R11.2
    float, R11.3
    pressure control, R11.1
  valves, control
    check, R11.21
    condenser pressure regulators,
        R11.15
    condensing water regulators, R11.20
    expansion,
      constant pressure, R11.11, 14
      electric, R11.10
      thermostatic, R11.5
    float, R11.17
    pressure relief devices, R11.22
    solenoid, R11.17
    suction pressure regulators, R11.14
Refrigerants, F29.1
  absorption solutions, F30.1, 69
  ammonia, F30.1, 34–35; R6.4
    chemical reactions, R6.5
    refrigeration system practices, R2.1
    refrigeration systems, R3.1
  ammonia/water, F30.1, 69
  analysis, R6.4
  automobile air conditioning, A10.11
  azeotropic, F2.6
  bakeries, R41.7
  carbon dioxide, F30.1, 38–39
    refrigeration systems, R3.1

COMMENT PAGE

ASHRAE publications strive to present the most current and useful information possible. If you would like to comment on chapters in this or any volume of the *ASHRAE Handbook*, please use one of the following methods:

- Fill out the comment form on the ASHRAE Web site (www.ashrae.org)
- E-mail the editor at mowen@ashrae.org

- Cut out this page and fax it to the editor at 678-539-2187, or mail it to

  Handbook Editor
  ASHRAE
  1791 Tullie Circle
  Atlanta, GA 30329 USA

Please provide your contact information if you would like a response. (Personal identification information will not be used for any purpose beyond responding to your comments.)

Name: _____

Phone: _____

E-mail: _____

Fax: _____

Address: _____

Preferred Contact Method(s): _____

COMMENT PAGE

ASHRAE publications strive to present the most current and useful information possible. If you would like to comment on chapters in this or any volume of the *ASHRAE Handbook*, please use one of the following methods:

- Fill out the comment form on the ASHRAE Web site (www.ashrae.org)
- E-mail the editor at mowen@ashrae.org

- Cut out this page and fax it to the editor at 678-539-2187, or mail it to

  Handbook Editor
  ASHRAE
  1791 Tullie Circle
  Atlanta, GA 30329 USA

Please provide your contact information if you would like a response. (Personal identification information will not be used for any purpose beyond responding to your comments.)

Name: _____

E-mail: _____

Address: _____
_____

Phone: _____

Fax: _____

Preferred Contact Method(s): _____
_____

COMMENT PAGE

ASHRAE publications strive to present the most current and useful information possible. If you would like to comment on chapters in this or any volume of the *ASHRAE Handbook*, please use one of the following methods:

- Fill out the comment form on the ASHRAE Web site (www.ashrae.org)
- E-mail the editor at mowen@ashrae.org

- Cut out this page and fax it to the editor at 678-539-2187, or mail it to

  Handbook Editor
  ASHRAE
  1791 Tullie Circle
  Atlanta, GA 30329 USA

Please provide your contact information if you would like a response. (Personal identification information will not be used for any purpose beyond responding to your comments.)

Name: _____   Phone: _____

E-mail: _____   Fax: _____

Address: _____   Preferred Contact Method(s): _____

COMMENT PAGE

ASHRAE publications strive to present the most current and useful information possible. If you would like to comment on chapters in this or any volume of the *ASHRAE Handbook*, please use one of the following methods:

- Fill out the comment form on the ASHRAE Web site (www.ashrae.org)
- E-mail the editor at mowen@ashrae.org

- Cut out this page and fax it to the editor at 678-539-2187, or mail it to

Handbook Editor
ASHRAE
1791 Tullie Circle
Atlanta, GA 30329 USA

_____

_____

_____

_____

_____

_____

_____

_____

_____

_____

_____

_____

_____

_____

_____

_____

_____

_____

_____

_____

_____

_____

Please provide your contact information if you would like a response. (Personal identification information will not be used for any purpose beyond responding to your comments.)

Name: _____   Phone: _____

E-mail: _____   Fax: _____

Address: _____   Preferred Contact Method(s): _____

_____   _____

COMMENT PAGE

ASHRAE publications strive to present the most current and useful information possible. If you would like to comment on chapters in this or any volume of the *ASHRAE Handbook*, please use one of the following methods:

- Fill out the comment form on the ASHRAE Web site (www.ashrae.org)
- E-mail the editor at mowen@ashrae.org

- Cut out this page and fax it to the editor at 678-539-2187, or mail it to

  Handbook Editor
  ASHRAE
  1791 Tullie Circle
  Atlanta, GA 30329 USA

Please provide your contact information if you would like a response. (Personal identification information will not be used for any purpose beyond responding to your comments.)

Name: _____   Phone: _____

E-mail: _____   Fax: _____

Address: _____   Preferred Contact Method(s): _____

_____   _____